SOLAR SYSTEM DATA

EARTH

Mass	$M_E = 5.98 \times 10^{24}$ kg
Equatorial radius	$R_E = 6.378 \times 10^6$ m
Polar radius	$R'_E = 6.357 \times 10^6$ m
Mean density	5520 kg/m^3
Surface gravity	$g = 9.81$ m/s^2 $= 32.2$ ft/s^2
Period of rotation	1 sidereal day = 23 h 56 min 4 s = 8.616×10^4 s
Moment of inertia:	
about polar axis	$I = 0.331 M_E R_E^2$
about equatorial axis	$I' = 0.329 M_E R_E^2$
Mean distance from Sun	1.50×10^{11} m
Period of revolution (period of orbit)	1 year = 365 days 6 h = 3.16×10^7s
Orbital speed	29.8 km/s

MOON

Mass	7.35×10^{22} kg
Radius	1.74×10^6 m
Mean density	3340 kg/m^3
Surface gravity	1.62 m/s^2
Period of rotation	27.3 days
Mean distance from Earth	3.84×10^8 m
Period of revolution	1 sidereal month = 27.3 days

SUN

Mass	$M_S = 1.99 \times 10^{30}$ kg
Radius	6.96×10^8 m
Mean density	1410 kg/m^3
Surface gravity	274 m/s^2
Period of rotation	~ 26 days
Luminosity	3.9×10^{26} W

PLANET	MEAN DISTANCE FROM SUN	PERIOD OF REVOLUTION	MASS	EQUATORIAL RADIUS	SURFACE GRAVITY	PERIOD OF ROTATION
Mercury	57.9×10^6 km	0.241 year	3.30×10^{23} kg	2 439 km	0.38g	58.6 days
Venus	108	0.615	4.87×10^{24}	6 052	0.91	243
Earth	150	1.00	5.98×10^{24}	6 378	1.00	0.997
Mars	228	1.88	6.42×10^{23}	3 393	0.38	1.026
Jupiter	778	11.9	1.90×10^{27}	71 398	2.53	0.41
Saturn	1430	29.5	5.67×10^{26}	60 000	1.07	0.43
Uranus	2870	84.0	8.70×10^{25}	25 400	0.92	0.65
Neptune	4500	165	1.03×10^{26}	24 300	1.19	0.77
Pluto	5890	248	1.5×10^{22}	1 500	0.045	6.39

Physics for Engineers and Scientists

Third Edition

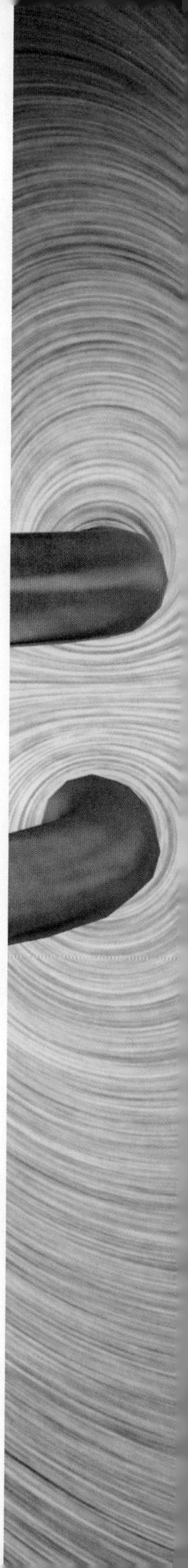

Extended Edition (Chapters 1–41)

MOTION, FORCE, AND ENERGY

OSCILLATIONS, WAVES, AND FLUIDS

TEMPERATURE, HEAT, AND THERMODYNAMICS

ELECTRICITY AND MAGNETISM

WAVES AND OPTICS

RELATIVITY, QUANTA, AND PARTICLES

W • W • NORTON & COMPANY   NEW YORK • LONDON

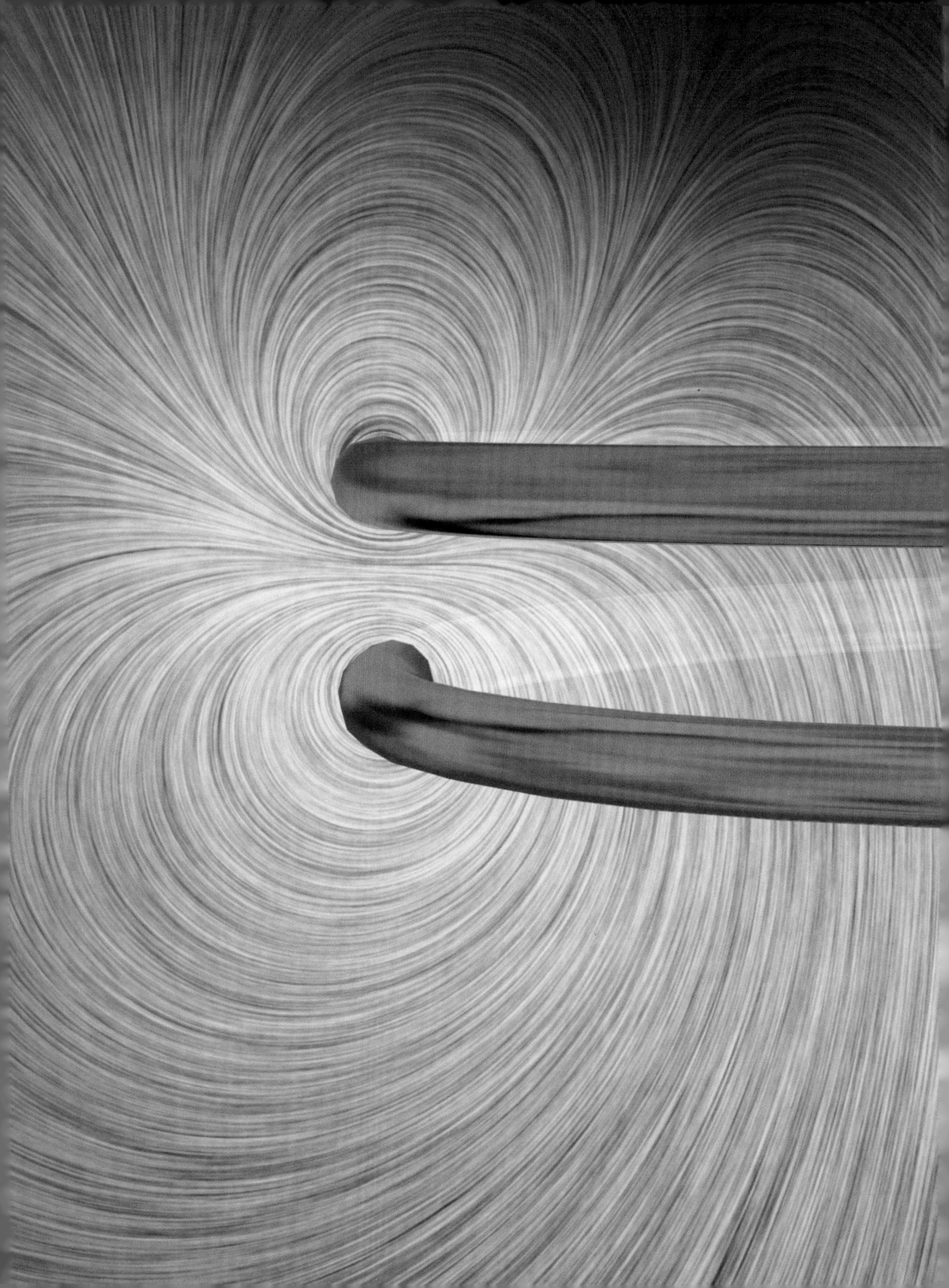

Physics for Engineers and Scientists Third Edition

Extended Edition (Chapters 1–41)

MOTION, FORCE, AND ENERGY
OSCILLATIONS, WAVES, AND FLUIDS
TEMPERATURE, HEAT, AND THERMODYNAMICS
ELECTRICITY AND MAGNETISM
WAVES AND OPTICS
RELATIVITY, QUANTA, AND PARTICLES

HANS C. OHANIAN,
UNIVERSITY OF VERMONT

JOHN T. MARKERT,
UNIVERSITY OF TEXAS AT AUSTIN

W • W • NORTON & COMPANY   NEW YORK • LONDON

To Susan Ohanian, writer, who gently tried to teach me some of her craft.—H.C.O.

To Frank D. Markert, a printer by trade; to Christiana Park, for her thirst for new knowledge; and to Erin, Ryan, Sean, and Gwen, for their wonder and clarity.—J.T.M.

Printed in the United States of America
Third Edition

Composition: Techbooks
Manufacturing: RR Donnelley & Sons Company
Editor: Leo A. W. Wiegman
Media Editor: April E. Lange
Director of Manufacturing—College: Roy Tedoff
Senior Project Editor: Christopher Granville
Photo Researcher: Kelly Mitchell
Editorial Assistant: Lisa Rand, Sarah L. Mann
Copy Editor: Richard K. Mickey
Book designer: Sandy Watanabe
Layout artist: Paul Lacy
Illustration Studio: J. B. Woolsey and Penumbra Design, Inc.
Cover Illustration: John Belcher, inter alia.
Cover Design: Joan Greenfield

Markert. -- 3rd ed., extended
p. cm
Previously published: Physics, 2nd ed. New York : Norton, c1989.
"Extended ed., chapters 1-41."
Includes index.
ISBN-13: 978-0-393-92631-6
ISBN-10: 0-393-92631-1
I. Physics. I. Markert, John T. II. Title.
QC21.3.053 2007b
530--dc22

2007061157

W. W. Norton & Company, Inc., 500 Fifth Avenue, New York, N.Y. 10110
www.wwnorton.com

W. W. Norton & Company Ltd., Castle House, 75/76 Wells Street, London W1T 3QT

1234567890

W. W. Norton & Company has been independent since its founding in 1923, when William Warder Norton and Mary D. Herter Norton first published lectures delivered at the People's Institute, the adult education division of New York City's Cooper Union. The Nortons soon expanded their program beyond the Institute, publishing books by celebrated academics from America and abroad. By mid-century, the two major pillars of Norton's publishing program—trade books and college texts— were firmly established. In the 1950s, the Norton family transferred control of the company to its employees, and today—with a staff of four hundred and a comparable number of trade, college, and professional titles published each year—W. W. Norton & Company stands as the largest and oldest publishing house owned wholly by its employees.

Brief Contents

Chapters 1–21 appear in Volume 1; Chapters 22–36 appear in Volume 2; Chapters 36–41 appear in Volume 3.

Table of Contents

Chapters 1–21 appear in Volume 1; Chapters 22–36 appear in Volume 2; and Chapters 36–41 appear in Volume 3.

APPENDICES

Preface

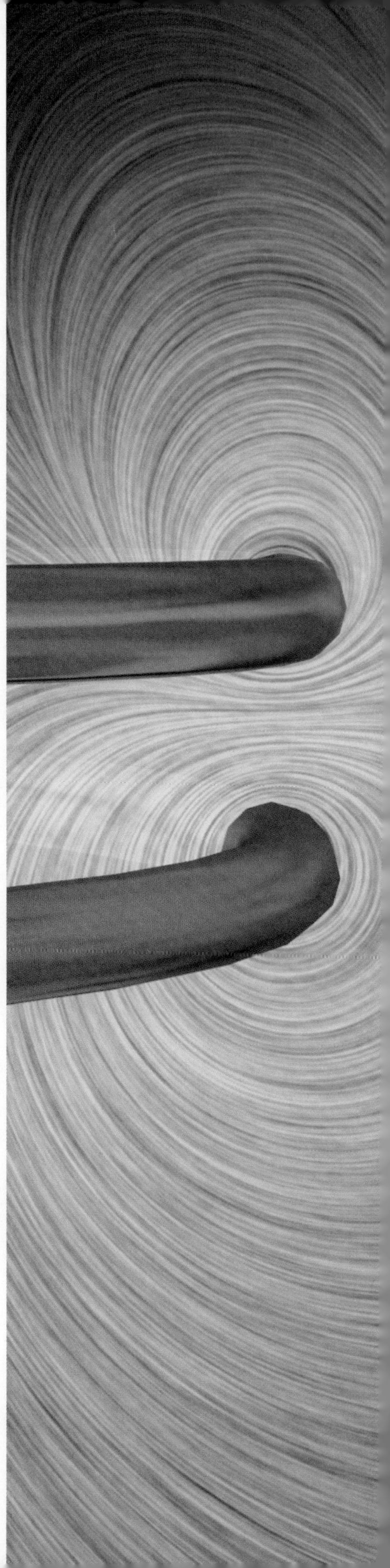

Our aim in *Physics for Engineers and Scientists,* Third Edition, is to present a modern view of classical mechanics and electromagnetism, including some optics and quantum physics. We also want to offer students a glimpse of the practical applications of physics in science, engineering, and everyday life.

The book and its learning package emerged from a collaborative effort that began more than six years ago. We adapted the core of Ohanian's earlier *Physics* (Second Edition, 1989) and combined it with relevant findings from recent physics education research on how students learn most effectively. The result is a text that presents a clear, uncluttered explication of the core concepts in physics, well suited to the needs of undergraduate engineering and science students.

Organization of Topics

The 41 chapters of the book cover the essential topics of introductory physics: mechanics of particles, rigid bodies, and fluids; oscillations, wave motion, heat and thermodynamics; electricity and magnetism; optics; special relativity; and atomic and subatomic physics.

Our arrangement and treatment of topics are fairly traditional with a few deliberate distinctions. We introduce the principle of superposition of forces early in Chapter 5 on Newton's laws of motion, and we give the students considerable exposure to the vector superposition of gravitational forces in Chapter 9. This leaves the students well prepared for the later application of vector superposition of electric and magnetic forces generated by charge or current distributions. We place gravitation in Chapter 9 immediately after the chapters on work and energy, because we regard gravitation as a direct application of these concepts (instructors who prefer to postpone gravitation can, of course, do so). We introduce forces on stationary electric charges in a detailed, complete exposition in Chapter 22, before proceeding to the less obvious concept of the electric field in Chapter 23. We start the study of magnetism in Chapter 29 with the force on a moving charged particle near a current, instead of the more common practice of starting with a postulate about the magnetic field in the abstract. With our approach, the observed magnetic forces on moving charges lead naturally to the magnetic field, and this progression from magnetic force to magnetic field will remind students of the closely parallel progression from electric force to electric field. For efficiency and brevity, we sometimes combine in one chapter closely related topics that other authors elect to spread over more than one chapter. Thus, we cover induction and inductance together in Chapter 31 and interference and diffraction together in Chapter 35.

Concise Writing with Sharp Focus on Core Concepts

Our goal is concise exposition with a sharp focus on core concepts. Brevity is desirable because long chapters with a large number of topics and excessive verbiage are confusing and tedious for the student. In our writing, we obey the admonitions of Strunk and White's *Elements of Style*: use the active voice; make statements in positive form; use definite, specific, concrete language; omit needless words.

We strove for simplicity in organizing the content. Each chapter covers a small set of core topics—rarely more than five or six—and we usually place each core topic in a section of its own. This divides the content into manageable segments and gives the chapter a clear and clean outline. Transitional sentences at the beginning or end of sections spell out the logical connections between each section and the next. Within each section, we strove for a seamless narrative leading from the discussions of concepts to their applications in Example problems. We sought to avoid the patchy, cobbled structure of many texts in which the discussions appear to serve as filler between one equation and the next.

Emphasis on the Atomic Structure of Matter

Throughout the book, we encourage students to keep in mind the atomic structure of matter and to think of the material world as a multitude of restless electrons, protons, and neutrons. For instance, in the mechanics chapters, we emphasize that all macroscopic bodies are systems of particles and that the equations of motion for macroscopic bodies emerge from the equations of motion of the individual particles. We emphasize that macroscopic forces are the result of a superposition of the forces among the particles of the system, and we consider atoms and their bonds in the qualitative discussions of elasticity, thermal expansion, and changes of state. By exposing students to the atomic structure of matter in the first semester, we help them to grasp the nature of the charged particles that play a central role in the treatment of electricity and magnetism in the second semester. Thus, in the electricity chapters, we introduce the concepts of positive and negative charge by referring to protons and electrons, not by referring to the antiquated procedure of rubbing glass rods with silk rags.

We try to make sure that students are always aware of the limitations of the nineteenth-century fiction that matter and electric charge are continua. Blind reliance on this old fiction has often been justified by the claim that, although engineering students need physics as a problem-solving tool, the atomic structure of matter is of little concern to them. This supposition may be adequate for a superficial treatment of mechanical engineering. Yet much of modern engineering—from materials science to electronics—hinges on understanding the atomic structure of matter. For this purpose, engineers need a physicist's view of physics.

Real-World Examples Begin Each Chapter

Each chapter opens with a "Concepts in Context" photograph illustrating a practical application of physics. The caption for this photo explores various core concepts in a concrete real-world context. The questions included in the caption are linked to several solved Examples or discussions later in the chapter. Such revisiting of the chapter-opening application provides layers of learning, as new concepts are carefully built upon a foundation firmly planted in the real world. The emphasis on real-world data is also evident throughout other Examples and in the end-of-chapter problems. By exposing students to realistic data, we give them confidence to apply physics in their later science or engineering courses.

Conceptual Discussions Precede and Motivate the Math

Only after a careful exposition of the conceptual foundations in a qualitative physical context does each section proceed to the mathematical treatment. Thus, we ensure that the mathematical formulas and their consequences and variations are rooted in a firm conceptual foundation. We were very careful to provide clear, thorough, and accurate explanations and derivations of all mathematical statements, to ensure that students acquire a good intuition about why particular equations are applied. Immediately after such derivations, we provide solved Examples to establish a firm connection between theory and concrete practical applications.

Examples Enliven the Text

We devote significant portions of each chapter to carefully selected Examples of solved problems—about 390 altogether or 9 on average per chapter. These Examples are concrete illustrations of the preceding conceptual discussions. They build cumulatively upon each other, from simple to more complicated as the chapter progresses. To enliven the text, we employ realistic data in the Examples, such as students would actually encounter outside the classroom. The solved Examples are designed to cover most variations of possible problems, with solutions that include both general approaches and specific details on how to extract the important information for the given problem. For instance, when such keywords as *initially* or *at rest* occur in a solved Example, we are careful to point out their importance in the problem-solving process. Comments appended to some Examples draw attention to limitations in the solution or to wider implications.

Checkup Questions Implement Active Learning

We conclude each section of a chapter with a series of brief *Checkup* questions. These permit students to test their mastery of core concepts, and they can be of great help in clearing up common misconceptions. Checkup questions include variations and "flip sides" of simple concepts that often occur to students but are rarely addressed. We give detailed answers to each Checkup question at the back of the chapter. The entire book contains roughly 5 Checkup questions per section—comprising a total of about 800 Checkup questions.

The final Checkup question of each section is always in multiple-choice forrmat—specifically designed for interactive teaching. At the University of Texas, instructors use such multiple-choice questions as classroom concept quizzes for welcome breaks in conventional lecturing. When more than one answer is popular, the instructor and class immediately know that more discussion or more examples are needed. Such occasions lend themselves well to peer instruction, in which the students explain to one another their reasoning before responding. This pedagogy implements an active, participatory alternative to the traditional lecture format. In addition, several supplements to the textbook, including the Student Activity Workbook, Online Concept Tutorials, Smartwork online homework, and PhysiQuizzes also implement active learning and a mastery-based approach.

Problem-Solving Techniques

Many chapters have inserts in the form of boxes devoted to Problem-Solving Techniques. These 39 skill boxes summarize the main steps or approaches for the solution of common classes of problems. Often deployed after several seemingly disparate Examples, the Problem-Solving Techniques boxes underscore the unity and generality of the techniques used in the Examples. The boxes list the steps or approaches to be taken, providing a handy reference and review.

Math Help

We have placed a Math Help box wherever students encounter a mathematical concept or technique that may be difficult or unfamiliar. These 6 skill boxes briefly review and summarize such topics as trigonometry, derivatives, integrals, and ellipses. Students can find more detailed help in Appendix 2 on basic algebra, 3 on trigonometry and geometry, 4 on calculus, and 5 on propagation of uncertainties.

Physics in Practice

Many chapters have a short essay on Physics in Practice that illustrates an application of physics in engineering and everyday life. These 27 essay boxes discuss practical topics, such as ultracentrifuges, communication and weather satellites, magnetic levitation, etc. Each of these essays provides a wealth of interesting detail and offers a practical supplement to some of the chapter topics. They have been designed to be engaging, yet sufficiently qualitative to provide some respite from the more analytical discussions, Examples, and Questions.

Figures and Balloon Captions

Over 1,500 figures illustrate the text. We made every effort to assemble a visual narrative as clear as the verbal narrative. Each figure in a sequence carefully builds upon the visual information in the figure that precedes it. Many figures in the text contain a caption in "balloon" that points to important features within the figure. The balloon caption is a concise and informative supplement to the conventional figure caption. The balloons make immediately obvious some details that would require a long, wordy explanation in the conventional caption. Often the balloon captions are arranged so that some cause-effect or other sequential thought process becomes immediately evident. All drawn figures are available to instructors in digital form for use in the course.

End–of–Chapter Summary

Each chapter narrative closes with several support elements, starting with a brief Summary. The Summary contains the essential physical laws, quantities, definitions, and key equations introduced in the chapter. A page reference, key equation number, and often a thumbnail figure accompany these laws, definitions, and equations. The Summary does not include repetition of the detailed explanations of the chapter. The Summary is followed by Questions for Discussion, Problems, Review Problems, and Answers to Checkups.

Questions for Discussion

After the chapter's Summary, we include a large selection of qualitative Questions for Discussion — about 700 in the entire book or roughly 17 per chapter. We intend these qualitative end-of-chapter Questions to stimulate student thinking. Some of these questions are deliberately formulated so as to have no unique answer, which is intended to promote class discussion.

Problems

After the chapter's qualitative Questions, we include computational Problems grouped by chapter section — about 3000 in the entire book, or roughly 73 per chapter. Each problem's level of difficulty is indicated by no asterisk, one asterisk (*), or two asterisks (**). Most no-asterisk Problems are easy and straightforward, only requiring students to "plug in" the correct values to compute answers or to retrace the steps of an Example. One-asterisk Problems are of medium difficulty. They contain a few complications requiring the combination of several concepts or the manipulation of several formulas. Two-asterisk Problems are difficult and challenging. They demand considerable thought and perhaps some insight, and occasionally require appreciable mathematical skill. When an Online Concept Tutorial (see below) is available for help in mastering the concepts in a given section, a dagger footnote (†) tells students where to find the tutorial.

We tried to make the Problems interesting for students by drawing on realistic examples from technology, science, sports, and everyday life. Many of the Problems are based on data extracted from engineering handbooks, car repair manuals, *Jane's Book of Aircraft*, *The Guinness Book of World Records*, newspaper reports, research and industrial instrumentation manuals, etc. Many other Problems deal with atoms and subatomic particles. These Problems are intended to reinforce the atomistic view of the material world. In some cases, experts will perhaps consider the use of classical physics somewhat objectionable in a problem that really ought to be handled by quantum mechanics. But we believe that the advantages of familiarization with atomic quantities and magnitudes outweigh the disadvantages of an occasional naive use of classical mechanics.

Among the Problems are a smaller number of somewhat contrived, artificial Problems that make no pretense of realism (for example, "A block slides on an inclined plane tied by string..."). Such unrealistic Problems are sometimes the best way to bring an important concept into sharp focus. Some Problems are formulated as guided problems, with a series of questions that take the student through an important problem-solving procedure, step by step.

Review Problems

After the Problems section of each chapter, we offer an extra selection of Review Problems — about 600 in the entire book or roughly 15 per chapter. We wrote these Review Problems specifically to help students prepare for examinations. Hence, Review Problems often test comprehension by requiring students to apply concepts from more than one section of the chapter and occasionally from prior, related chapters. Answers to all odd-numbered Problems and Review Problems are given in Appendix 11.

Units and Significant Figures

We use the SI system of units exclusively throughout the text. In the abbreviations for the units we follow the recommendations of the International Committee for Weights and Measures (CIPM), although we retain some traditional units, such as revolution and calorie that have been discontinued by the CIPM. In addition, for the sake of clarity we spell out the name of the unit in full whenever the abbreviation is likely to lead to ambiguity and confusion (for instance, in the case of V for volt, which is easily confused with V for potential; or in the case of C for coulomb, which might be confused with C for capacitance). We try to use realistic numbers of significant figures, with most Examples and Problems using two or three. In cases where it is natural to employ some data with two significant figures and some with three, we have been careful to propagate the appropriate number of significant figures to the result.

For reference purposes, we give the definitions of the British units. Currently only the United States, Bangladesh, and Liberia still adhere to these units. In the United States, automobile manufacturers have already switched to metric units for design and construction. The U. S. Army has also switched to metric units, so soldiers give distances in meters and kilometers (in army slang, the kilometer is called a "klick," a usage that is commendable itself for its brevity). British units are not used in examples or in problems, with the exception of a handful of problems in the early chapters. In the definitions of the British units, the pound (lb) is taken to be the unit of mass, and the pound force (lbf) is taken to be the unit of force. This is in accord with the practice approved by the American National Standards Institute (ANSI), the Institute of Electrical and Electronics Engineers (IEEE), and the U. S. Department of Defense.

Optional Sections and Chapters

We recognize course content varies from institution to institution. Some sections and some chapters can be regarded as optional and can be omitted without loss of continuity. These optional sections are marked by asterisks in the Table of Contents.

Mathematical Prerequisites

In order to accommodate students who are taking an introductory calculus course concurrently, derivatives are used slowly at first (Chapter 2), and routinely later on. Likewise, the use of integrals is postponed as long as possible (Chapter 7), and they come into heavy use only in the second volume (after Chapter 21). For students who need a review of calculus, Appendix 4 contains a concise primer on derivatives and integrals.

Acknowledgments

We have had the benefit of a talented author team for our support resources. In addition to their primary role in the assembly of the learning package, they all have also made substantial contributions to the accuracy and clarity of the text.

Stiliana Antonova, Barnard College
Charles Chiu, University of Texas-Austin
William J. Ellis, University of California-Davis
Mirela Fetea, University of Richmond
Rebecca Grossman, Columbia University
David Harrison, University of Toronto
Prabha Ramakrishnan, North Carolina State University
Hang Deng-Luzader, Frostburg State University
Stephen Luzader, Frostburg State University
Kevin Martus, William Paterson University
David Marx, Illinois State University
Jason Stevens, Deerfield Academy
Brian Woodahl, Indiana University–Purdue University-Indianapolis
Raymond Zich, Illinois State University

And at Sapling Systems and Science Technologies in Austin, Texas, for content, James Caras, Ph.D.; Jon Harmon, B.S.; Kevin Nelson, Ph.D.; John A. Underwood, Ph.D.; and Jason Vestuto, M.S. and for animation and programming, Jeff Sims and Nathan Wheeler.

Our manuscript was subjected to many rounds of peer review. The reviewers were instrumental in identifying myriad improvements, for which we are grateful:

Yildirim Aktas	University of North Carolina–Charlotte
Patricia E. Allen	Appalachian State University
Steven M. Anlage	University of Maryland
B. Antanaitis	Lafayette College
Laszlo Baksay	Florida Institute of Technology
Marco Battaglia	University of California-Berkeley
Lowell Boone	University of Evansville
Marc Borowczak	Walsh University
Amit Chakrabarti	Kansas State University
D. Cornelison	Northern Arizona University
Corbin Covault	Case Western Reserve University
Kaushik De	University of Texas at Arlington
William E. Dieterle	California University of Pennsylvania
James Dunne	Mississippi State University
R. Eagleton	California Polytechnic University-Pomona
Gregory Earle	University of Texas-Dallas
William Ellis	University of California-Davis
Mark Eriksson	University of Wisconsin-Madison
Morten Eskildsen	University of Notre Dame
Bernard Feldman	University of Missouri–St. Louis
Mirela Fetea	University of Richmond
J. D. Garcia	University of Arizona
U. Garg	University of Notre Dame
Michael Gurvitch	State University of New York at Stony Brook
David Harrison	University of Toronto
John Hernandez	University of North Carolina–Chapel Hill
L. Hodges	Iowa State University
Jean-Pierre Jouas	United Nations International School
Kevin Kimberlin	Bradley University
Sebastian Kuhn	Old Dominion University
Tiffany Landry	Folsom Lake College
Dean Lee	North Carolina State University
Frank Lee	George Washington University
Stephen Luzader	Frostburg State University
Kevin Martus	William Paterson University
M. Matkovich	Oakton Community College
David McIntyre	Oregon State University
Rahul Mehta	University of Central Arkansas
Kenneth Mendelson	Marquette University
Laszlo Mihaly	State University of New York at Stony Brook
Richard Mistrick	Pennsylvania State University
Rabindra Mohapatra	University of Maryland
Philip P. J. Morrison	University of Texas at Austin
Greg Mowry	University of Saint Thomas
David Murdock	Tennessee Technological University
Anthony J. Nicastro	West Chester University
Scott Nutter	Northern Kentucky University
Robert Oerter	George Mason University
Ray H. O'Neal, Jr.	Florida A & M University
Frederick Oho,	Winona State University
Paul Parris	University of Missouri–Rolla
Ashok Puri	University of New Orleans
Michael Richmond	Rochester Institute of Technology
John Rollino	Rutgers University–Newark
David Schaefer	Towson State University
Joseph Serene	Georgetown University
H. Shenton	University of Delaware
Jason Stevens	Deerfield Academy

Jay Strieb	Villanova University
John Swez	Indiana State University
Devki N. Talwar	Indiana University of Pennsylvania
Chin-Che Tin	Auburn University
Tim Usher	California State University-San Bernardino
Andrew Wallace	Angelo State University
Barrett Wells	University of Connecticut
Edward A.P. Whittaker	Stevens Institute of Technology
David Wick	Clarkson University
Don Wieber	Contra Costa College
J. William Gary	University of California-Riverside
Suzanne Willis	Northern Illinois University
Thomas Wilson	Marshall University
William. J. F. Wilson	University of Calgary
Brian Woodahl	Indiana University–Purdue University-Indianapolis
Hai-Sheng Wu	Mankato State University

We thank John Belcher, Michael Danziger, and Mark Bessette of the Massachusetts Institute of Technology for creating the cover image. It illustrates the magnetic field generated by two currents in two copper rings. This is one frame of a continuous animation; at the instant shown, the current in the upper ring is opposite to that in the lower ring and is of smaller magnitude. The magnetic field structure shown in this picture was calculated using a modified intregration technique. This image was created as part of the Technology Enabled Active Learning (TEAL) program in introductory physics at MIT, which teaches physics interactively, combining desktop experiments with visualizations of those experiments to "make the unseen seen."

We thank the several editors that supervised this project: first Stephen Mosberg, then Richard Mixter, John Byram, and finally Leo Wiegman, who had the largest share in the development of the text, and also gave us the benefit of his incisive line-by-line editing of the proofs, catching many slips and suggesting many improvements. We also thank the editorial staff at W. W. Norton & Co., including Chris Granville, April Lange, Roy Tedoff, Rubina Yeh, Rob Bellinger, Kelly Mitchell, Neil Hoos, Lisa Rand, and Sarah Mann, as well as the publishing professionals whom Norton engaged, such as Paul Lacy, Richard K. Mickey, Susan McLaughlin, and John B. Woolsey for their enthusiasm and their patience in dealing with the interminable revisions and corrections of the text and its support package. In addition, JTM is grateful to Robert W. Christy of Dartmouth University for various pointers on textbook writing.

HANS C. OHANIAN
Burlington, Vermont
hohanian@uvm.edu

JOHN T. MARKERT
Austin, Texas
jmarkert@physics.utexas.edu

Publication Formats

Physics for Engineers and Scientists comprises six parts. The text is published in two hardcover versions and several paperback versions.

Hardcover Versions

Third Extended Edition, Parts I–VI, 1450 pages, ISBN 0-393-92631-1
    (Chapters 1–41 including Relativity, Quanta and Particles)
Third Edition, Parts I–V, 1282 pages, ISBN 0-393-97422-7
    (Chapters 1–36, including Special Relativity)

Paperback Versions

Volume 1, (Chapters 1–21) 778 pages, ISBN 0-393-93003-3
    Part I Motion, Force, and Energy (Chapters 1–14)
    Part II Oscillations, Waves, and Fluids (Chapters 15–18)
    Part III Temperature, Heat, and Thermodynamics (Chapter 19–21)
Volume 2, (Chapters 22–36) 568 pages, ISBN 0-393-93004-1
    Part IV Electricity and Magnetism (Chapters 22–32)
    Part V Waves and Optics (Chapters 33–35 and Chapter 36 on Special Relativity)
Volume 3, (Chapters 36–41) 250 pages, ISBN 0-393-92969-8
    Part VI Relativity, Quanta, and Particles

In addition, to explore customized versions, please contact your Norton representative.

Two Norton ebook Options

Physics for Engineers and Scientists is available in a Norton ebook format that retains the content of the print book. The ebook offers a variety of tools for study and review, including sticky notes, highlighters, zoomable images, links to Online Concept Tutorials, and a search function. Purchased together, the SmartWork with integrated ebook bundle makes it easy for students to check text references when completing online homework assignments.

The ebook may also be purchased as a standalone item. The downloadable PDF version is available for purchase from Powells.com.

nortonebooks.com

Package Options

Each version of the text purchased from Norton—with or without SmartWork—will come with free access to our website at Norton's StudySpace that includes the valuable Online Concept Tutorials. Each version of the text may be purchased as a stand-alone book or as a package that includes—each for a fee—Norton's new SmartWork online homework system or the Student Activity Workbook by David Harrison and William Ellis. Hence, several optinal packages are available to instructors:

- Textbook–StudySpace–Online Concept Tutorials + Student Activity Workbook
- Textbook–StudySpace–Online Concept Tutorials + SmartWork/ebook
- Textbook–StudySpace–Online Concept Tutorials + SmartWork/ebook + Student Activity Workbook

The Support Program

To enhance individual learning and also peer instruction, a carefully integrated support program accompanies the text. Each element of the support program has two goals. First, each support resource mirrors the text's emphasis on sharply focused core concepts. Second, treatment of a core concept in a support resource offers a perspective that is different from but compatible with that of the text. If a student needs help beyond the text, he or she would more likely benefit from a fresh presentation on the same concept rather than from one that simply repeats the text presentation.

Hence, the text and its support package offers three or more different approaches to the core concepts. For example, Newton's First and Second Laws are rendered with interactive animations in the Online Concept Tutorial "Forces," with pencil-and- paper exercises in Chapter 5 of the Student Activity Workbook crafted by David Harrison and William Ellis, and with concept test inquiries in PhysiQuiz questions written by Charles Chiu and edited by Jason Stevens.

Both printed and digital resources are offered within the support program. Outstanding web-based resources for both instructors and students include tutorials and a homework system.

SmartWork Online Homework System

smartwork
www.wwnorton.com/physics

SmartWork—Norton's online homework management system—provides ready-made automatically graded assignments, including guided problems, simple feedback questions, and animated tutorials—all specifically designed to extend the text's emphasis on core concepts and problem-solving skills.

Developed in collaboration with Sapling Systems, SmartWork features an intuitive, easy-to-use interface that offers instructors flexible tools to manage assignments, while making it easy for students to compose mathematical expressions, draw vectors and graphs, and receive helpful and immediate feedback. Two different types of questions expand upon the exposition of concepts in the text:

Simple Feedback Problems present students with problems that anticipate common misconceptions and offer prompts at just the right moment to help them discover the correct solution.

Guided Tutorial Problems addresses more challenging topics. If a student answers a problem incorrectly, SmartWork guides the student through a series of discrete tutorial steps that lead to a general solution. Each step is a simple feedback question that the student answers, with hints if necessary. After completing all of the tutorial steps, the student returns to the original problem ready to apply this newly-obtained knowledge.

SmartWork problems use algorithmic variables so two students are unlikely to see exactly the same problem. Instructors can use the problem sets provided, or can customize these ready-made questions and assignments, or use SmartWork to create their own.

SmartWork is available bundled with the Norton ebook of *Physics for Engineers and Scientists*. Where appropriate, SmartWork prompts students to review relevant sections in the textbook. Links to the **ebook** make it easy for students to consult the text while working through problems online.

Online Concept Tutorials

Online Concept Tutorial
www.wwnorton.com/physics

Developed in collaboration with Science Technologies specifically for this course, these 45 tutorials feature interactive animations that reinforce conceptual understanding and develop students' quantitative skills. In-text icons alert students to the availability of a tutorial. All Online Concept Tutorials are available on the free StudySpace web site and are integrated into SmartWork. Tutorials can also be accessed from a CD-ROM that requires no installation, browser tune-ups, or plug-ins.

StudySpace Website

STUDYSPACE www.wwnorton.com/physics. This free and open website is the portal for both public and premium content. Free content at StudySpace includes the Online Concept Tutorials and a Study Plan for each chapter in *Physics for Engineers and Scientists*. Premium content at StudySpace includes links to the online ebook and to SmartWork.

www.wwnorton.com/physics

WebAssign

Selected end-of-chapter problems from Physics for Engineers and Scientists are available in WebAssign, with additional problems available to adopting instructors by request to WebAssign.

Additional Instructor Resources

TEST BANK by Mirela Fetea, University of Richmond; Kevin Martus, William Paterson University; and Brian Woodahl, Indiana University-Purdue University-Indianapolis. The Test Bank offers approximately 2000 multiple-choice questions, available in ExamView, WebCT, BlackBoard, rich-text, and printed format.

INSTRUCTOR SOLUTIONS MANUAL by Stephen Luzader and Hang-Deng Luzader, both of Frostburg State University, and David Marx of Illinois State University. The Instructor Solution Manual offers solutions to all end-of-chapter Problems and Review Problems, checked for accuracy and clarity.

PHYSIQUIZ "CLICKER" QUESTIONS by Charles Chiu, University of Texas at Austin, with Jason Stevens, Deerfield Academy. The PhysiQuiz multiple-choice questions are designed for use with classroom response, or "clicker", systems. The 300 PhysiQuiz questions are available as PowerPoint slides, in printed format, and as transparency masters.

NORTON MEDIA LIBRARY INSTRUCTOR CD-ROM The Media Library for instrutors includes selected figures, tables, and equations from the text in JPEG and PowerPoint formats, PhysiQuiz "clicker" questions, and PowerPoint-ready offline versions of the Online Concept Tutorials.

INSTRUCTOR RESOURCE MANUAL offers a guide to the support package with descriptions of the Online Concept Tutorials, information about the SmartWork homework problems available for each chapter, printed PhysiQuiz "clicker" questions, and instructor notes for the workshop activities in the Student Activity Workbook.

TRANSPARENCY ACETATES Approximately 200 printed color acetates of key figures from the text.

BLACKBOARD AND WEBCT COURSE CARTRIDGES Course Cartridges for BlackBoard and WebCT include access to the Online Concept Tutorials, a Study Plan for each chapter, multiple-choice tests, plus links to the premium, password-protected contents of the Norton ebook and SmartWork.

Additional Student Resources

STUDENT ACTIVITY WORKBOOK by David Harrison, University of Toronto, and William Ellis, University of California Davis. The *Student Activity Workbook* is an important part of the learning package. For each chapter of *Physics for Engineers and Scientists*, the Workbook's Activities break down a physical condition into constituent parts. The Activities are pencil and paper exercises well suited to either individual or small group collaboration. The Activities include both conceptual and quantitative exercises. Some Activities are guided problems that pose a question and present a solution scheme via follow up questions. The Workbook is available in two paperback volumes: Volume 1 comprises Chapters 1–21 and Volume 2 comprises Chapters 22–41.

STUDENT SOLUTIONS MANUAL by Stephen Luzader and Hang-Deng Luzader, both of Frostburg State University, and David Marx of Illinois State University. The Student Solutions Manual contains detailed solutions to approximately 25% of the problems in the book, chosen from the odd-numbered problems whose answers appear in the back of the book. The Manual is available in two paperback volumes: Volume 1 comprises Chapters 1–21 and Volume 2 comprises Chapters 22–41.

ONLINE CONCEPT TUTORIALS CD-ROM The 45 Online Concept Tutorials (see above) can also be accessed from an optional CD-ROM that requires no installation, browser tune-up, or plug-in.

About the Authors

Hans C. Ohanian received his B.S. from the University of California, Berkeley, and his Ph.D from Princeton University, where he worked with John A. Wheeler. He has taught at Rensselaer Polytechnic Institute, Union College, and the University of Vermont. He is the author of several textbooks spanning all undergraduate levels: *Physics*, *Principles of Physics*, *Relativity: A Modern Introduction*, *Modern Physics*, *Principles of Quantum Mechanics*, *Classical Electrodynamics*, and, with Remo Ruffini, *Gravitation and Spacetime*. He is also the author of dozens of articles dealing with gravitation, relativity, and quantum theory, including many articles on fundamental physics published in the *American Journal of Physics*, where he served as associate editor for some years. He lives in Vermont. hohanian@uvm.edu

John T. Markert received his B.A. in physics and mathematics from Bowdoin College (1979), and his M.S. (1984) and Ph.D. (1987) in physics from Cornell University, where he was recipient of the *Clark Award for Excellence in Teaching*. After postdoctoral research at the University of California, San Diego, he joined the faculty at the University of Texas at Austin in 1990, where he has received the *College of Natural Sciences Teaching Excellence Award* and is currently Professor of Physics and Department Chair. His introductory physics teaching methods emphasize context-based approaches, interactive techniques, and peer instruction. He is author or coauthor of over 120 journal articles, including experimental condensed-matter physics research in superconductivity, magnetism, and nanoscience. He lives in Austin, Texas, with his spouse and four children. jmarkert@physics.utexas.edu

Owner's Manual for *Physics for Engineers and Scientists*

These pages give a brief tour of the features of *Physics for Engineers and Scientists* and its study resources. Some resources are found within the book. Others are located in accompaning paperback publications or at the StudySpace web portal. Features on the text pages shown here come chiefly from the discussion of friction in Chapter 6, but are common in other chapters.

The learning resources listed below help students study by offering alternative explanations of the core concepts found in the text. These student resources are briefly described at the end of this owner's manual:

- Online Concept Tutorials
- SmartWork Online Homework
- StudySpace
- Student Activity Workbook
- Student Solutions Manual

Each chapter of the textbook starts with a real-world example of a core concept. Chapter 6 opens with the concept of friction and uses automobile tires as an example of friction that is revisited in several different conditions. The opening photograph, it's caption and the caption's closing questions all discuss this example.

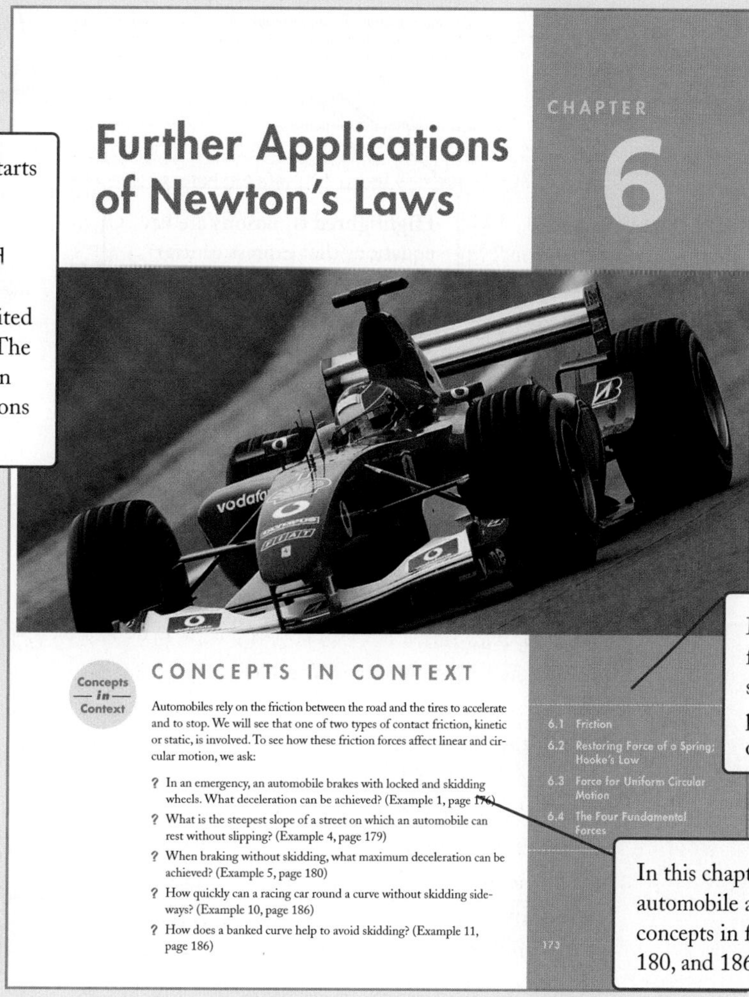

CHAPTER 6

Further Applications of Newton's Laws

Concepts in Context

CONCEPTS IN CONTEXT

Automobiles rely on the friction between the road and the tires to accelerate and to stop. We will see that one of two types of contact friction, kinetic or static, is involved. To see how these friction forces affect linear and circular motion, we ask:

? In an emergency, an automobile brakes with locked and skidding wheels. What deceleration can be achieved? (Example 1, page 176)

? What is the steepest slope of a street on which an automobile can rest without slipping? (Example 4, page 179)

? When braking without skidding, what maximum deceleration can be achieved? (Example 5, page 180)

? How quickly can a racing car round a curve without skidding sideways? (Example 10, page 186)

? How does a banked curve help to avoid skidding? (Example 11, page 186)

173

Most chapters have six or fewer sections. Most sections are four or five pages in length and cover one major topic.

In this chapter, the rubber tires of an automobile are revisited to explore concepts in friction on pages 176, 179, 180, and 186, as indicated.

To find a solution of the equation of motion means to find a force **F** and a corresponding acceleration **a** such that Newton's equation $m\mathbf{a} = \mathbf{F}$ is satisfied. For a physicist, the typical problem involves a known force and an unknown motion; for example, the physicist knows the forces between the planets and the Sun, and she seeks …

In mathematical expressions, such as $m\mathbf{a}$=**F**, the **bold type** indicates a **vector** and *italic* indicates variables that are not vectors.

The icon indicates an **Online Concept Tutorial** is available for a key concept. Each such icon includes the identification number of the tutorial—8, in this case. These tutorials offer a visual guide and self-quiz for the concept at hand. Find all the Tutorials at www.wwnorton.com/physics.

Text in *italic type* indicates major definitions of laws or statements of general principles.

Text in **bold type** highlights the first use of a key term and is generally accompanied by an explanation.

Online Concept Tutorial 8

6.1 FRICTION

Friction forces, which we have ignored up to now, play an important role in our environment and provide us with many interesting examples of motion with constant force. For instance, if the driver of …

LEONARDO da VINCI (1452–1519) *Italian artist, engineer, and scientist. Famous for his brilliant achievements in painting, sculpture, and architecture. Leonardo also made pioneering contributions to science. But Leonardo's investigations of friction were forgotten, and the laws of friction were rediscovered 200 years later by Guillaume Amontons, a French physicist.*

Short **biographical sketches** appear in the margins of this text. Each offer a brief glimpse into the life of some major contributor to our knowledge about the physical world—in this case, Italian artist and engineer Leonardo da Vinci.

Key concepts or important variants of these concepts have a key-term label in the margin.

The magnitude of the friction force between unlubricated, dry surfaces sliding one over the other is proportional to the magnitude of the normal force acting on the surfaces and is independent of the area of contact and of the relative speed.

Friction involving surfaces in relative motion is called **sliding friction,** or **kinetic friction**. According to the above law, the magnitude of the force of kinetic friction can be written mathematically as

force of kinetic friction

$$f_k = \mu_k N \tag{6.1}$$

where μ_k is the **coefficient of kinetic friction,** a constant characteristic of the material involved. Table 6.1 lists typical friction coefficients for various materials.

Note that Eq. (6.1) states that the magnitudes of the friction force and the normal force are proportional. The *directions* of these forces are, however, quite different: the normal fo… …rce $\mathbf{f}_k$ is parallel …

Highlighted equations are key equations that express central physics concepts mathematically.

… ations in many everyday engineering problems in which the speeds are not extreme. The simple friction law is then a reasonably good approximation for a wide range of materials, and it is at its best for metals sliding on metals.

The fact that the friction force is independent of the area of contact means that the friction force of the block sliding on the tabletop is the same whether the block slides on a large face or on one of the small faces (see Fig. 6.2). This may seem surprising at first—we might expect the friction force to be larger when the block slides on the larger face, with more area in contact with the tabletop. However, the normal force is then distributed over a larger area, and is therefore less effective in pressing the atoms together; and the net result is that the friction force is independent of the area of contact.

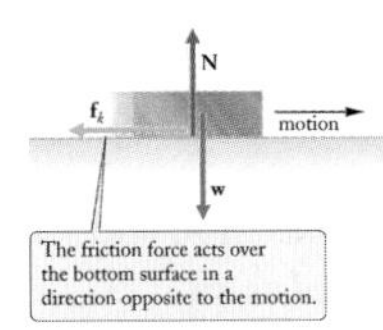

FIGURE 6.1 Forces on a block sliding on a plate.

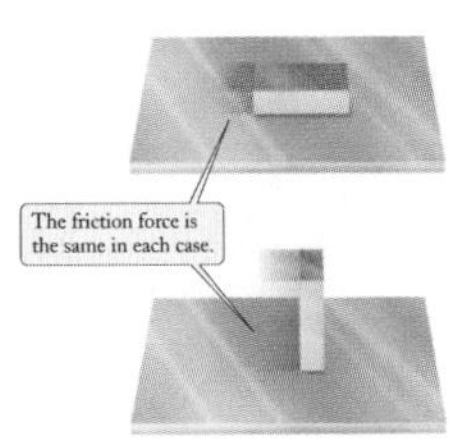

FIGURE 6.2 Steel block on a steel plate, sliding on a large face or on a small face.

TABLE 6.1 KINETIC AND STATIC FRICTION COEFFICIENTS[a]

MATERIALS	μ_k	μ_s
Steel on steel	0.6	0.7
Steel on lead	0.9	0.9
Steel on copper	0.4	0.5
Copper on cast iron	0.3	1.1
Copper on glass	0.5	0.7
Waxed ski on snow		
at −10°C	0.2	—
at 0°C	0.05	—
Rubber on concrete	≈1	≈1

[a] The friction coefficient depends on the condition of the surfaces. The values in this table are typical for dry surfaces but not entirely reliable.

Examples are a critical part of each chapter.
- Examples provide concrete illustrations of the concepts being discussed.
- As the chapter unfolds, Examples progress from simple to more complex.

Throughout the text, **figures** often build on each other with a new layer of information.
- **Balloon comments** often point out components of special note in the figure.

The **Concept in Context** icon here indicates the chapter-opening example —automobile tires—is being revisited. In this Example, we explore the slowing down of a skidding automobile with a specific coefficient of kinetic friction for a rubber tire.

EXAMPLE 3 A man pushes a heavy crate over a floor. The man pushes downward and forward, so his push makes an angle of 30° with the horizontal (Fig. 6.5a). The mass of the crate is 60 kg, and the coefficient of sliding friction is $\mu_k = 0.50$. What force must the man exert to keep the crate moving at uniform velocity?

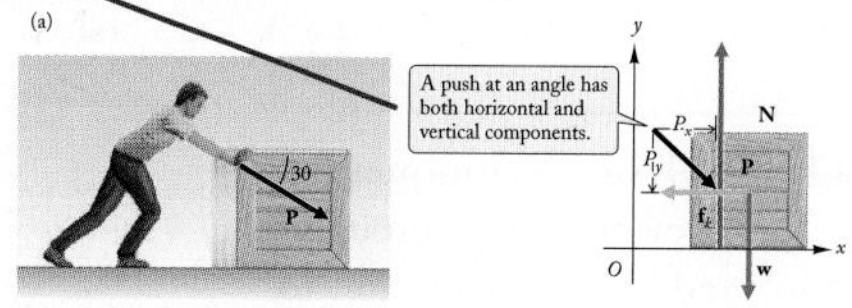

FIGURE 6.5 (a) Man pushing a crate. (b) "Free-body" diagram for the crate.

SOLUTION: Figure 6.5b is a "free-body" diagram for the crate. The forces on the crate are the push **P** of the man, the weight **w**, the normal force **N**, and the friction force $\mathbf{f}_k$. Note that because the man pushes the crate down against the floor, the magnitude of the normal force is not equal to *mg;* we will have to treat the magnitude of the normal force as unknown. Taking the *x* axis horizontal and the *y* axis vertical, we see from Fig. 6.5b that the *x* and *y* components of the forces are (see also Fig. 5.37)

$$P_x = P\cos 30^\circ \qquad P_y = -P\sin 30^\circ$$
$$w_x = 0 \qquad w_y = -mg$$
$$N_x = 0 \qquad N_y = N$$
$$f_{k,x} = -\mu_k N \qquad f_{k,y} = 0$$

…cceleration of the crate is zero in both the *x* and the *y* directions, the …n each of these directions must be zero:

$$P\cos 30^\circ + 0 + 0 - \mu_k N = 0$$
$$-P\sin 30^\circ - mg + N + 0 = 0$$

…wo equations for the two unknowns *P* and *N*. By multiplying the second …y μ_k and then adding the resulting equation to the first, we can elimi-…d we find an equation for *P*:

$$P\cos 30^\circ - \mu_k P \sin 30^\circ - \mu_k mg = 0$$

…is for *P*, we find

$$P = \frac{\mu_k mg}{\cos 30^\circ - \mu_k \sin 30^\circ} = \frac{0.50 \times 60\ \text{kg} \times 9.81\ \text{m/s}^2}{\cos 30^\circ - 0.50 \times \sin 30^\circ} \qquad (6.4)$$
$$= 4.8 \times 10^2\ \text{N}$$

Solutions in Examples may cover both general approaches and specific details on how to extract the information from the problem statement.

Concepts in Context

EXAMPLE 1 Suppose that the coefficient of kinetic friction of the hard rubber of an automobile tire sliding on the pavement of a street is $\mu_k = 0.8$. What is the deceleration of an automobile on a flat street if the driver brakes sharply, so all the wheels are locked and skidding? (Assume the vehicle is an economy model without an antilock braking system.)

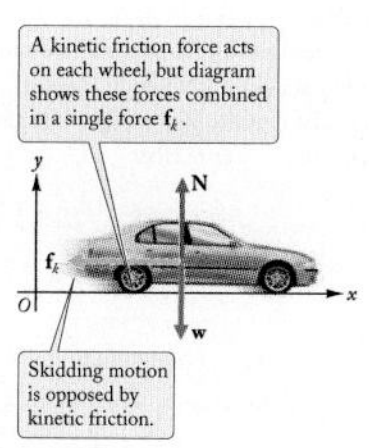

FIGURE 6.3 "Free-body" diagram for an automobile skidding with locked wheels.

SOLUTION: Figure 6.3 shows the "free-body" diagram with all the forces on the automobile. These forces are the weight **w**, the normal force **N** exerted by the street, and the friction force $\mathbf{f}_k$. The normal force must balance the weight; hence the magnitude of the normal force is the same as the magnitude of the weight, or $N = w = mg$. According to Eq. (6.1), the magnitude of the friction force is then

$$f_k = \mu_k N = 0.8 \times mg$$

Since this friction force is the only horizontal force on the automobile, the deceleration of the automobile along the street is

$$a_x = -\frac{f_k}{m} = -\frac{0.8 \times mg}{m} = -0.8 \times g = -0.8 \times 9.8\ \text{m/s}^2$$
$$= -8\ \text{m/s}^2$$

COMMENT: The normal forces and the friction forces act on all the four wheels of the automobile; but in Fig. 6.3 (and in other "free-body" diagrams in this chapter) these forces have been combined into a net force **N** and a net friction force $\mathbf{f}_k$, which, for convenience, are shown as though acting at the center of the automobile. To the extent that the motion is treated as purely translational motion (that is, particle motion), it makes no difference at what point of the automobile the forces act. Later, in Chapter 13, we will study how forces affect the rotational motion of bodies, and it will then become important to keep track of the exact point at which each force acts.

Comments occasionally close an Example to point out the particular limitations and broader implications of a Solution.

EXAMPLE 2 A ship is launched toward the water on a slipway making an angle of 5° with the horizontal direction (see Fig. 6.4). The coefficient of kinetic friction between the bottom of the ship and the slipway is $\mu_k = 0.08$. What is the acceleration of the ship along the slipway? What is the speed of the ship after accelerating from rest through a distance of 120 m down the slipway to the water?

SOLUTION: Figure 6.4b is the "free-body" diagram for the ship. The forces shown are the weight **w**, the normal force exerted by the slipway **N**, and the friction force $\mathbf{f}_k$. The magnitude of the weight is $w = mg$.

Since there is no motion in the direction perpendicular to the slipway, we find, as in Eq. (5.36), that the normal force is

$$N = mg\cos\theta$$

and the magnitude of the friction force is

$$f_k = \mu_k N = \mu_k mg\cos\theta \qquad (6.2)$$

With the *x* axis parallel to the slipway, the *x* component of the weight is (see Fig. 6.4c)

$$w_x = mg\sin\theta$$

A **Checkup** appears at the end of each section within a chapter.
- Each Checkup is a self-quiz to test the reader's mastery of the concepts in the preceding section.
- Each Checkup has an answer (see below).

Problem-Solving Techniques boxes appear in relevant places throughout the book and offers tips on how to approach problems of a particular kind—in this case, problems involving the use of friction or centripetal force.

Answers to Checkups appear at the very back of each chapter, after the Review Problems.

190 CHAPTER 6 Further Applications of Newton's Laws

Checkup 6.3

QUESTION 1: A stone is being whirled around a circle at the end of a string when the string suddenly breaks. Describe the motion of the stone after the string breaks; ignore gravity.

QUESTION 2: At an intersection, a motorcycle makes a right turn at constant speed. During this turn the motorcycle travels along a 90° arc of a circle. What is the direction of the acceleration of the motorcycle during this turn?

QUESTION 3: A car moves at constant speed along a road leading over a small hill with a spherical top. What is the direction of the acceleration of the car when at the top of this hill?

QUESTION 4: In Example 12, for the aircraft looping the loop, does the chair exert a centripetal or a centrifugal force on the pilot? Does the pilot exert a centripetal or a centrifugal force on the chair? What is the direction of the pilot's apparent, increased weight at the instant the aircraft passes through the bottom of the loop? Does the direction of the apparent weight change as the aircraft climbs up the loop?

QUESTION 5: Two cars travel around a traffic circle in adjacent (outer and inner) lanes. If the two cars travel at the same constant speed, which completes the circle first? Which has the larger acceleration?

(A) Outer; outer. (B) Inner; outer.
(C) Outer; inner. (D) Inner; inner.

PROBLEM-SOLVING TECHNIQUES — FRICTION FORCES AND CENTRIPETAL FORCES

The problems involving applications of Newton's laws in this chapter can be solved by the techniques discussed in the preceding chapter. In dealing with friction forces and with the centripetal force for uniform circular motion, pay special attention to the directions of the forces.

1. The magnitude of the sliding friction force is proportional to the magnitude of the normal force, but the direction is not the direction of the normal force. Instead, the sliding friction force is always parallel to the sliding surfaces, opposite to the direction of motion.
2. The static friction force is also always parallel to the sliding surfaces, opposite to the direction in which the body tends to move. If you have any doubts about the direction of the static friction force, pretend that the friction is absent, and ask yourself in what direction the body would then move; the static friction force is in the opposite direction.
3. Uniform circular motion requires a force toward the center of the circle, that is, a centripetal force. When preparing a "free-body" diagram for a body in uniform circular motion, include all the pushes and pulls acting on the moving body, but do *not* include a "centripetal mv^2/r force." This would be a mistake, like including an "ma force" in the "free-body" diagram for a body with some kind of translational motion. The quantity mv^2/r is not a force; it is merely the product of mass and centripetal acceleration. This acceleration is caused by one force or by the resultant of several forces already included among the pushes and pulls displayed in the "free-body" diagram. For instance, in Example 11 the resultant force is $w \tan\theta$, in Example 12 the resultant force is $N - mg$, and these resultants equal mv^2/r by Newton's Second Law [see Eqs. (6.17) and (6.18)]. To prevent confusion, do not include the resultant in the "free-body" diagram for a body in uniform circular motion. Instead, draw the resultant on a separate diagram (see Fig. 6.21b).

202 CHAPTER

*85. Two springs of constants 2.0×10^3 N/m and $3.0 \times$ are connected in tandem, and a mass of 5.0 kg hang from this spring. By what amount does the mass str combined spring? Each individual spring?

*86. A block of mass 1.5 kg is placed on a flat surface, a being pulled horizontally by a spring with a spring 1.2×10^3 N/m (see Fig. 6.48). The coefficient of st between the block and the table is $\mu_s = 0.60$, and th cient of sliding friction is $\mu_k = 0.40$.
(a) By what amount must the spring be stretched t block moving?
(b) What is the acceleration of the block if the stre spring is maintained at a constant value equal t required to start the motion?
(c) By what amount must the spring be stretched to keep the mass moving at constant speed?

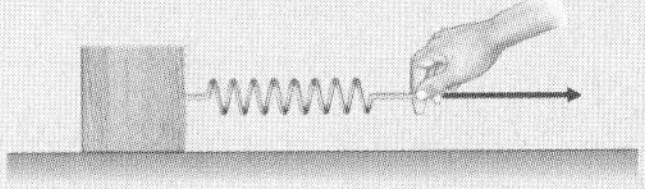

FIGURE 6.48 Mass pulled by spring.

*87. A block of mass 1.5 kg is placed on a plane inclined at 30°, and it is being pulled upward by a spring with a spring constant 1.2×10^3 N/m (see Fig. 6.49). The direction of pull of the spring is parallel to the inclined plane. The coefficient of static ...lined plane is $\mu_s = 0.60$, ... $\mu_k = 0.40$.
...e stretched to start the
...ck if the stretch of the ...value equal to that
(c) By what amount must the spring be stretched to keep the mass moving at constant speed?

*88. A mass m_1 slides on a smooth, frictionless table. The mass is constrained to move in a circle by a string that passes through a hole in the center of the table and is attached to a second mass m_2 hanging vertically below the table (Fig. 6.50). If the radius of the circular motion of the first mass is r, what must be its speed?

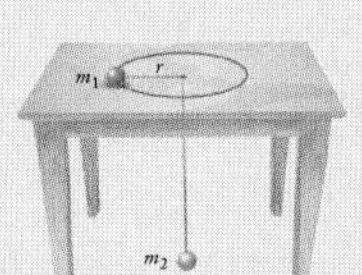

FIGURE 6.50 Mass in circular motion and hanging mass.

89. An automobile enters a curve of radius 45 m at 70 km/h. Will the automobile skid? The curve is not banked, and the coefficient of static friction between the wheels and the road is 0.80.

*90. A stone of 0.90 kg attached to a rod is being whirled around a vertical circle of radius 0.92 m. Assume that during this motion the speed of the stone is constant. If at the top of the circle the tension in the rod is (just about) zero, what is the tension in the rod at the bottom of the circle?

Answers to Checkups

Checkup 6.1

1. The weight of the second book results in a normal force between the first book and the table that is twice as large, so the friction force, and thus the horizontal push to overcome it, will be twice as large, or 20 N. If the first book pushes the second, then the friction force of the second book on the first adds to the friction force of the first to require a push also twice as large as the original, or 20 N.
2. While the block coasts up the incline, the friction, which always opposes the *motion*, is directed down the plane (the corresponding "free-body" diagram would have the weight

MATH HELP ELLIPSES

An ellipse is defined geometrically by the condition that the sum of the distance from one focus of the ellipse and the distance from the other focus is the same for all points on the ellipse. This geometrical condition leads to a simple method for the construction of an ellipse: Stick pins into the two foci and tie a length of string to these points. Stretch the string taut to the tip of a pencil, and move this pencil around the foci while keeping the string taut (see Fig. 1a).

An ellipse can also be constructed by slicing a cone obliquely (see Fig. 1b). Because of this, an ellipse is said to be a conic section.

The largest diameter of the ellipse is called the major axis, and the smallest diameter is called the minor axis. The semimajor axis and the semiminor axis are one-half of these diameters, respectively (see Fig. 1c).

If the semimajor axis of length a is along the x axis and the semiminor axis of length b is along the y axis, then the x and y coordinates of an ellipse centered on the origin satisfy

$$\frac{x^2}{a^2} + \frac{y^2}{b^2} = 1$$

The foci are on the major axis at a distance f from the origin given by

$$f = \sqrt{a^2 - b^2}$$

The separation between a planet and the Sun is $a - f$ at perihelion and is $a + f$ at aphelion.

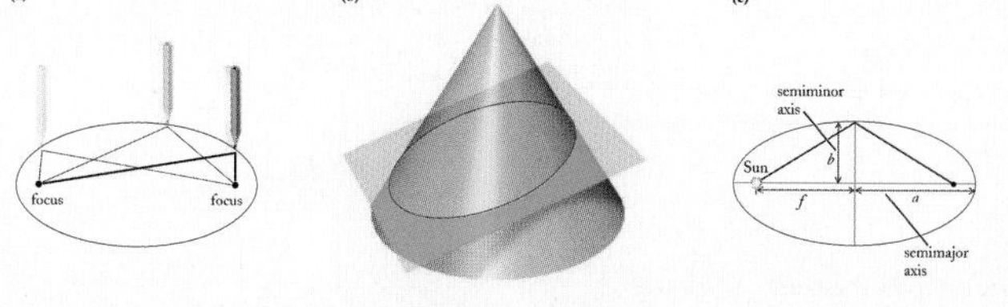

FIGURE 1 (a) Constructing an ellipse. (b) Ellipse as a conic section. (c) Focal distance f, semimajor axis a, and semiminor axis b of an ellipse.

Math Help boxes offer specific mathematical guidance at the initial location in the text where that technique is most relevant.
- In this case in Chapter 9, ellipses are important in studying orbits.
- Additional math help is available in Appendices 2, 3, 4, and 5 at the back of the textbook.

Throughout the text, **Physics in Practice** boxes offer specific details on a real-world application of the concept under discussion—in this case, forces at work in automobile collisions in Chapter 11.

Figure 9.10 illustrates this law. The two colored areas are equal, and the planet takes equal times to move from P to P' and from Q to Q'. According to Fig. 9.10, the speed of the planet is larger when it is near the Sun (at Q) than when it is far from the Sun (at P).

Kepler's Second Law, also called the law of areas, is a direct consequence of the central direction of the gravitational force. We can prove this law by a simple geometrical argument. Consider three successive positions P, P', P'' on the orbit, separated by a relatively small distance. Suppose that the time intervals between P, P' and between P', P'' are equal—say, each of the two intervals is one second. Figure 9.11 shows the positions P, P', P''. Between these positions the curved orbit can be approximated by straight line segments PP' and $P'P''$. Since the time intervals are one unit of time (1 second), the lengths of the segments PP' and $P'P''$ are in proportion to the

PHYSICS IN PRACTICE AUTOMOBILE COLLISIONS

Concepts — in — Context

We can fully appreciate the effects of the secondary impact on the human body if we compare the impact speeds of a human body on the dashboard or the windshield with the speed attained by a body in free fall from some height. The impact of the head on the windshield at 15 m/s is equivalent to falling four floors down from an apartment building and landing headfirst on a hard surface. Our intuition tells us that this is likely to be fatal. Since our intuition about the dangers of heights is much better than our intuition about the dangers of speeds, it is often instructive to compare impact speeds with equivalent heights of fall. The table lists impact speeds and equivalent heights, expressed as the number of floors the body has to fall down to acquire the same speed.

The number of fatalities in automobile collisions has been reduced by the use of air bags. The air bag helps by cushioning the impact over a longer time, reducing the time-average force. To be effective, the air bag must inflate quickly, before the passenger reaches it, typically in about 10 milliseconds. Because of this, a passenger, especially a child, too near an air bag prior to inflation can be injured or killed by the impulse from the inflation. But for a properly seated adult passenger, the inflated air bag cushions the passenger, reducing the severity of injuries.

However, the impact can still be fatal—you wouldn't expect to survive a jump from an 11-floor building onto an air mattress.

For maximum protection, a seat belt should always be worn even in vehicles equipped with air bags. In lateral collisions, in repeated collisions (such as in car pileups), and in rollovers, an air bag is of little help, and a seat belt is essential. The effectiveness of seat belts is well demonstrated by the experiences of race car drivers. Race car drivers wear lap belts and crossed shoulder belts. Even in spectacular crashes at very high speeds (see the figure), the drivers rarely suffer severe injuries.

In a race at the California Speedway in October 2000, a car flips over and breaks in half after a crash, but the driver, Luis Diaz, walks away from the wreck.

COMPARISON OF IMPACT SPEEDS AND HEIGHTS OF FALL

SPEED	SPEED	EQUIVALENT HEIGHT (NUMBER OF FLOORS)[a]
15 km/h	9 mi/h	$\frac{1}{3}$
30	19	1
45	28	3
60	37	5
75	47	8
90	56	11
105	65	15

[a]Each floor is 2.9 m.

SOLUTION: The only horizontal force on the ball is the normal force exerted by the wall; this force reverses the motion of the ball (see Fig. 11.3). Since the wall is very massive, the reaction force of the ball on the wall will not give the wall any appreciable velocity. Hence the kinetic energy of the system, both before and after the collision, is merely the kinetic energy of the ball. Conservation of this

The text frequently offers **tables of typical values** of physical quantities.
- Such tables usually are labeled "**Some ...**," as in this case, from Chapter 5.
- These tables give some impression of the magnitudes encountered in the real world.

TABLE 5.1 SOME FORCES

Gravitational pull of Sun on Earth	3.5×10^{22} N
Thrust of Saturn V rocket engines (a)	3.3×10^{7} N
Pull of large tugboat	1×10^{6} N
Thrust of jet engines (Boeing 747)	7.7×10^{5} N
Pull of large locomotive	5×10^{5} N
Decelerating force on automobile during braking	1×10^{4} N
Force between two protons in a nucleus	$\approx 10^{4}$ N
Accelerating force on automobile	7×10^{3} N
Gravitational pull of Earth on man	7.3×10^{2} N
Maximum upward force exerted by forearm (isometric)	2.7×10^{2} N
Gravitational pull of Earth on apple (b)	2 N
Gravitational pull of Earth on 5¢ coin	5.1×10^{-2} N
Force between electron and nucleus of atom (hydrogen)	8×10^{-8} N
Force on atomic-force microscope tip	10^{-12} N
Smallest force detected (mechanical oscillator)	10^{-19} N

(a)

(b)

Each chapter closes with a Summary followed by Questions for Discussion, Problems, Review Problems, and Answers to Checkups.

A **Summary** lists the subjects and page references for any special content in this chapter—such as Math Help, Problem-Solving Techniques, or Physics in Practice boxes.
- Next the Summary lists the chapter's core concepts in the order they are treated. The concept appears on the left in bold.
- The mathematical expression for the concept appears in the middle column with an equation number on the far right.

SUMMARY

PHYSICS IN PRACTICE Ultracentrifuges (page 188)

PROBLEM-SOLVING TECHNIQUES Friction Forces and Centripetal Forces (page 190)

KINETIC FRICTION FORCE (Direction *opposes* motion.) $f_k = \mu_k N$ (6.1)

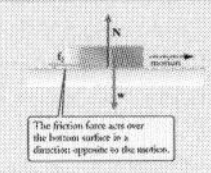

STATIC FRICTION FORCE (Direction opposes force which tries to move body; magnitude varies in response to applied force.) $f_{s,\max} = \mu_s N$ (6.5)

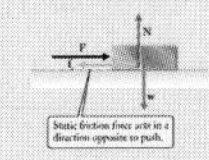

RESTORING FORCE OF A SPRING (HOOKE'S LAW) (Direction is toward relaxed position; x is measured from relaxed position.) $F = -kx$ (6.11)

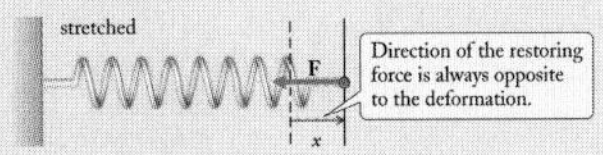

FORCE DUE TO AIR RESISTANCE At high speed v, where C is a dimensionless aerodynamic constant, rho is the density of air, and A is the cross-sectional area. $f_{\text{air}} = \frac{1}{2}C\rho A v^2$

FORCE REQUIRED FOR UNIFORM CIRCULAR MOTION (Direction is centripetal.) $F = \dfrac{mv^2}{r}$ (6.13)

About 15 or more **Questions for Discussion** follow the Summary in each chapter.
- These questions require thought, but not calculation; e.g. "Why are wet streets slippery?"
- Some of these questions are intended as brain teasers that have no unique answer, but will lead to provocative discussions.

QUESTIONS FOR DISCUSSION

1. According to the adherents of parapsychology, some people are endowed with the supernormal power of psychokinesis, e.g., spoon-bending-at-a-distance via mysterious psychic forces emanating from the brain. Physicists are confident that the only forces acting between pieces of matter are those listed in Section 6.4, none of which are implicated in psychokinesis. Given that the brain is nothing but a (very complicated) piece of matter, what conclusions can a physicist draw about psychokinesis?
2. If you carry a spring balance from London to Hong Kong, do you have to recalibrate it? If you carry a beam balance?
3. When you stretch a rope horizontally between two fixed points, it always sags a little, no matter how great the tension. Why?
4. What are the forces on a soaring bird? How can the bird gain altitude without flapping its wings?
5. How could you use a pendulum suspended from the roof of your automobile to measure its acceleration?
6. When an airplane flies along a parabolic path similar to that of a projectile, the passengers experience a sensation of weightlessness. How would the airplane have to fly to give the passengers a sensation of enhanced weight?
7. A frictionless chain hangs over two adjoining inclined planes (Fig. 6.24a). Prove the chain is in equilibrium, i.e., the chain will not slip to the left or to the right. [Hint: One method of proof, due to the seventeenth-century engineer and mathematician Simon Stevin, asks you to pretend that an extra piece of chain is hung from the ends of the original chain (Fig. 6.24b). This makes it possible to conclude that the original chain cannot slip.]

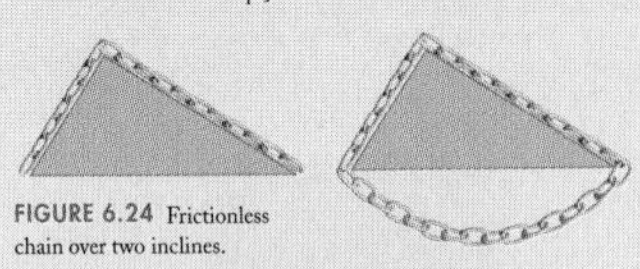

FIGURE 6.24 Frictionless chain over two inclines.

8. Seen from a reference frame moving with the wave, the motion of a surfer is analogous to the motion of a skier down a mountain.[2] If the wave were to last forever, could the surfer ride it forever? In order to stay on the wave as long as possible, in what direction should the surfer ski the wave?
9. Excessive polishing of the surfaces of a block of metal increases its friction. Explain.
10. Some drivers like to spin the wheels of their automobiles for a quick start. Does this give them greater acceleration? (Hint: $\mu_s > \mu_k$.)
11. Cross-country skiers like to use a ski wax that gives their skis a large coefficient of static friction, but a low coefficient of kinetic friction. Why is this useful? How do "waxless" skis achieve the same effect?
12. Designers of locomotives usually reckon that the maximum force available for moving the train ("tractive force") is one-fourth or one-fifth of the weight resting on the drive wheels of the locomotive. What value of the friction coefficient between the wheels and the track does this implicitly assume?
13. When an automobile with rear-wheel drive accelerates from rest, the maximum acceleration that it can attain is less than the maximum deceleration that it can attain while braking. Why? (Hint: Which wheels of the automobile are involved in acceleration? In braking?)
14. Can you think of some materials with $\mu_k > 1$?
15. For a given initial speed, the stopping distance of a train is much longer than that of a truck. Why?
16. Why does the traction on snow or ice of an automobile with rear-wheel drive improve when you place extra weight over the rear wheels?
17. Why are wet streets slippery?
18. In order to stop an automobile on a slippery street in the shortest distance, it is best to brake as hard as possible without initiating a skid. Why does skidding lengthen the stopping distance? (Hint: $\mu_s > \mu_k$.)
19. Suppose that in a panic stop, a driver locks the wheels of his automobile and leaves skid marks on the pavement. How can you deduce his initial speed from the length of the skid marks?
20. Hot-rod drivers in drag races find it advantageous to spin their wheels very fast at the start so as to burn and melt the rubber on their tires (Fig. 6.25). How does this help them to attain a larger acceleration than expected from the static coefficient of friction?

FIGURE 6.25 Drag racer at the start of the race.

[2] There is, however, one complication: surf waves grow higher as they approach the beach. Ignore this complication.

About 70 **Problems** and 15 **Review Problems** follow each chapter's Questions for Discussion.
- The Problem's statement contains data and conditions upon which a solution will hinge.
- Problems are grouped by chapter section and proceed from simple to more complex within each section.
- Many Problems employ real-world data and occasionally may introduce applications beyond those treated in the chapter.

194 CHAPTER 6 Further Applications of Newton's Laws

21. A curve on a highway consists of a quarter circle connecting two straight segments. If this curve is banked perfectly for motion at some given speed, can it be joined to the straight segments without a bump? How could you design a curve that is banked perfectly along its entire length and merges smoothly into straight segments without any bump?
22. Automobiles with rear engines (such as the old VW "Beetle") tend to oversteer; that is, in a curve the rear end tends to swing toward the outside of the curve, turning the car excessively into the curve. Explain.
23. When rounding a curve in your automobile, you get the impression that a force tries to pull you toward the outside of the curve. Is there such a force?
24. If the Earth were to stop spinning (other things remaining equal), the value of g at all points of the surface except the poles would become slightly larger. Why?
25. (a) If a pilot in a fast aircraft very suddenly pulls out of a dive (Fig. 6.26a), he will suffer blackout caused by loss of blood pressure in the brain. If he suddenly begins a dive while climbing (Fig. 6.26b), he will suffer *redout* caused by excessive blood pressure in the brain. Explain.
(b) A pilot wearing a G suit—a tightly fitting garment that squeezes the tissues of the legs and abdomen—can tolerate $8g$ while pulling out of a dive (Fig. 6.26c). How does this G suit prevent blackout? A pilot can tolerate no more than $-2g$ while beginning a dive. Why does the G suit not help against redout?
26. While rounding a curve at high speed, a motorcycle rider leans the motorcycle toward the center of the curve. Why?

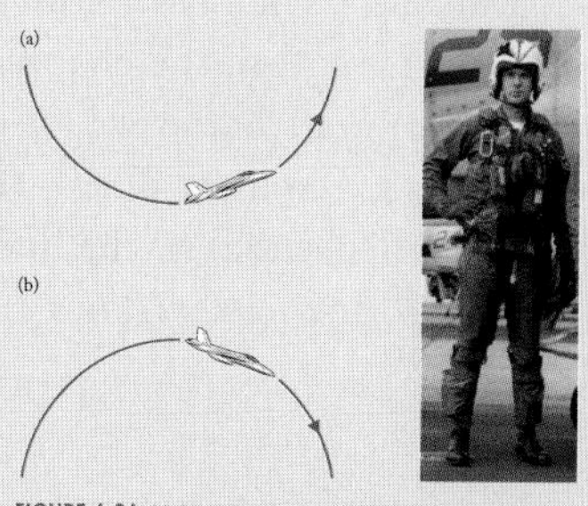

FIGURE 6.26 (a) Aircraft pulling out of a dive. (b) Aircraft beginning a dive. (c) Pilot wearing a G suit.

Review Problems are specifically designed to help students prepare for examinations.
- Review Problems often test comprehension of concepts from more than one section within the chapter.
- Review Problems often take a guided approach by posing series of questions that build on each other.

PROBLEMS

6.1 Friction†

1. The ancient Egyptians moved large stones by dragging them across the sand in sleds. How many Egyptians were needed to drag an obelisk of 700 metric tons? Assume that $\mu_k = 0.30$ for the sled on sand and that each Egyptian exerted a horizontal force of 360 N.
2. The base of a winch is bolted to a mounting plate with four bolts. The base and the mounting plate are flat surfaces made of steel; the friction coefficient of these surfaces in contact is $\mu_s = 0.40$. The bolts provide a normal force of 2700 N each. What maximum static friction force will act between the steel surfaces and help oppose lateral slippage of the winch on its base?
3. According to tests performed by the manufacturer, an automobile with an initial speed of 65 km/h has a stopping distance of 20 m on a level road. Assuming that no skidding occurs during braking, what is the value of μ_s between the wheels and the road required to achieve this stopping distance?
4. A crate sits on the load platform of a truck. The coefficient of friction between the crate and the platform is $\mu_s = 0.40$. If the truck stops suddenly, the crate will slide forward and crash into the cab of the truck. What is the maximum braking deceleration that the truck may have if the crate is to stay put?
5. When braking (without skidding) on a dry road, the stopping distance of a sports car with a high initial speed is 38 m. What would have been the stopping distance of the same car with the same initial speed on an icy road? Assume that $\mu_s = 0.85$ for the dry road and $\mu_s = 0.20$ for the icy road.
6. In a remarkable accident on motorway M1 (in England), a Jaguar car initially speeding "in excess of 100 mph" skidded 290 m before coming to a rest. Assuming that the wheels were completely locked during the skid and that the coefficient of kinetic friction between the wheels and the road was 0.80, find the initial speed.

† For help, see Online Concept Tutorial 8 at www.wwnorton.com/physics

Review Problems

REVIEW PROBLEMS

76. At liftoff, the Saturn V rocket used for the Apollo missions has a mass of 2.45×10^6 kg.
(a) What is the minimum thrust that the rocket engines must develop to achieve liftoff?
(b) The actual thrust that the engines develop is 3.3×10^7 N. What is the vertical acceleration of the rocket at liftoff?
(c) At burnout, the rocket has spent its fuel, and its remaining mass is 0.75×10^6 kg. What is the acceleration just before burnout? Assume that the motion is still vertical and that the strength of gravity is the same as when the rocket is on the ground.
77. If the coefficient of static friction between the tires of an automobile and the road is $\mu_s = 0.80$, what is the minimum distance the automobile needs in order to stop without skidding from an initial speed of 90 km/h? How long does it take to stop?
78. Suppose that the last car of a train becomes uncoupled while the train is moving upward on a slope of 1:6 at a speed of 48 km/h.
(a) What is the deceleration of the car? Ignore friction.
(b) How far does the car coast up the slope before it stops?
79. A 40-kg crate falls off a truck traveling at 80 km/h on a level road. The crate slides along the road and gradually comes to a halt. The coefficient of kinetic friction between the crate and the road is 0.80.
(a) Draw a "free-body" diagram for the crate sliding on the road.
(b) What is the normal force the road exerts on the crate?
(c) What is the friction force the road exerts on the crate?
(d) What is the weight force on the crate? What is the net force on the crate?
(e) What is the deceleration of the crate? How far does the crate slide before coming to a halt?
80. A 2.0-kg box rests on an inclined plane which makes an angle of 30° with the horizontal. The coefficient of static friction between the box and the plane is 0.90.
(a) Draw a "free-body" diagram for the box.
(b) What is the normal force the inclined plane exerts on the box?
(c) What is the friction force the inclined plane exerts on the box?
(d) What is the net force the inclined plane exerts on the box? What is the direction of this force?
81. The body of an automobile is held above the axles of the wheels by means of four springs, one near each wheel. Assume that the springs are vert… springs are the same. T… is 1200 kg, and the spri… N/m. When the autom… far are the springs comp…

*82. A block of wood rests o… The coefficient of static… paper is $\mu_s = 0.70$, and… $\mu_k = 0.50$. If you tilt the… begin to move?

*83. Two blocks of masses m_1 and m_2 are connected by a string. One block slides on a table, and the other hangs from the string, which passes over a pulley (see Fig. 6.46). The coefficient of sliding friction between the first block and the table is $\mu_k = 0.20$. What is the acceleration of the blocks?

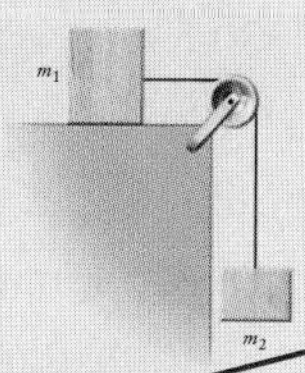

FIGURE 6.46 Mass on table, pulley, and hanging mass.

*84. A man of mass 75 kg is pushing a heavy box on a flat floor. The coefficient of sliding friction between the floor and the box is 0.20, and the coefficient of static friction between the man's shoes and the floor is 0.80. If the man pushes horizontally (see Fig. 6.47), what is the maximum mass of the box he can move?

FIGURE 6.47 Pushing a box.

The **dagger footnote** (†) that accompanies a Problem heading—in this case, "6.1 Friction"—indicates the availability of an Online Concept Tutorial on this specific topic and states its web address.

Problems and Review Problems are marked by **level of difficulty**:
- Those without an asterisk are the most common and require very little manipulation of existing equations; or they may merely require retracing the steps of a worked Example.
- Problems marked with one asterisk (*) are of medium difficulty and may require use of several concepts and manipulation more than one equation to isolate and solve for the unknown variable.
- Problems marked with two asterisks (**) are challenging, demand considerable thought, may require significant mathematical skill, and are the least common.

Online Concept Tutorials

www.wwnorton.com/physics

An **Online Concept Tutorial** accompanies many central topics in this textbook. When a Tutorial is available, its numbered icon appears at section heading within the chapter and a dagger footnote appears in the end-of-chapter Problems section as reminder. These Tutorials are digitally delivered, either via the Internet or via a CD-ROM for those without Internet access.

Online Concept Tutorial 8

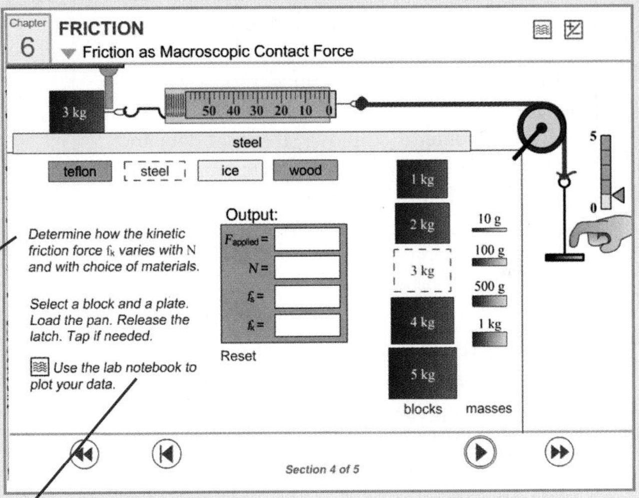

Many Tutorials contain **online experiments**—in this case, determining how the kinetic friction force varies with the normal force and with the choice of materials.

The online experiments allow students to change independent variables—in this case, mass and material.

- Students may collect and display data in a built-in **lab notebook**.
- Each Tutorial includes an interactive **self-quiz**.

The Online Concept Tutorials listed here indicate each textbook section supported by the tutorial (in paratheses).

smartwork
www.wwnorton.com/physics

SmartWork is a subscription-based online homework-management system that makes it easy for instructors to assign, collect and grade end-of-chapter problems from *Physics for Engineers and Scientists*. Built-in hinting and feedback address common misperceptions and help students get the maximum benefit from these assignments.

Simple Feedback Problems anticipate common misconceptions and offer prompts at just the right moment to help students reach the correct solution.

Guided Tutorial Problems address challenging topics.
• If a student solves one of these problems incorrectly, she is presented with a series of discrete tutorial steps that lead to a general solution.
• Each step includes hinting and feedback. After working through these remedial steps, the student returns to a restatement of the original problem, ready to apply this newly obtained knowledge.

SmartWork is available as a stand-alone purchase, or with an integrated ebook version of *Physics for Engineers and Scientists*.
• Where appropriate, SmartWork prompts students to review relevant sections of the text.
• Links to the ebook make it easy for students consult the text while working through problems.

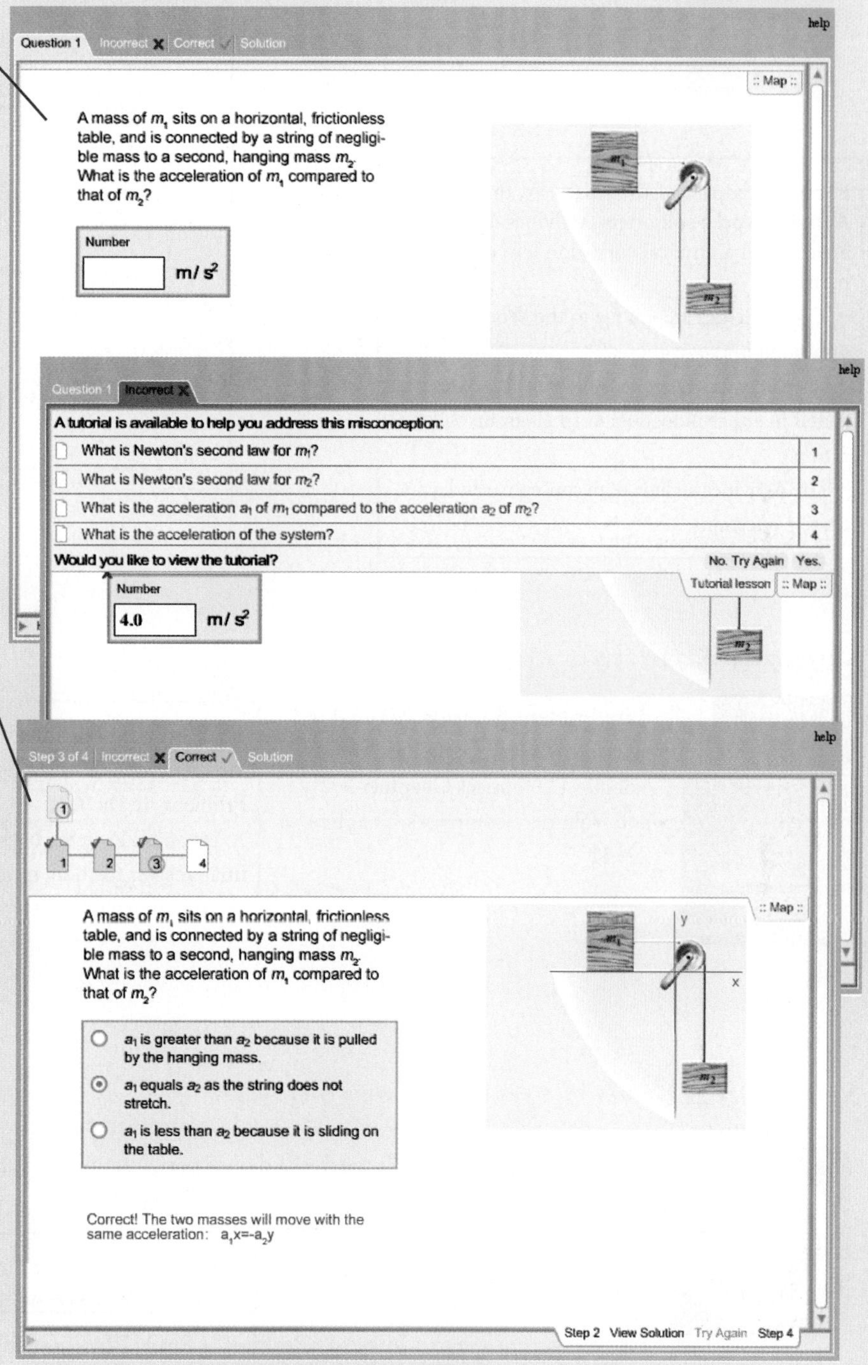

Student Activity Workbook

The Student Activity Workbook is available in two paperback volumes: Volume 1 comprises Chapters 1–21 and Volume 2 comprises Chapters 22–41.

For each chapter of the textbook, the **Student Activity Workbook** offers Activities designed to break down a physical condition into constituent parts.

- The Activities are unique to the Workbook and not found in the textbook.
- The Activities are pencil and paper exercises well suited to either individual or small group collaboration.
- The Activities include both conceptual and quantitative questions.

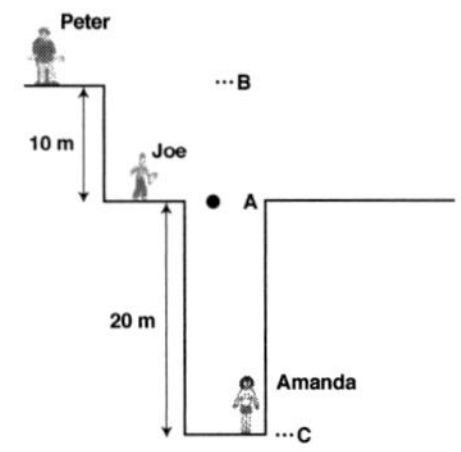

ACTIVITY 7

Joe is standing on the ground, Pete is standing on a 10m high cliff, and Amanda is at the bottom of a 20m deep pit, as shown. All three are using coordinate systems with the vertical axis directed up.

Joe's coordinate system has the zero of the vertical axis and zero of gravitational potential at ground level.

Peter's coordinate system has the zero of the vertical axis and zero of gravitational potential at the height of the cliff.

Amanda's coordinate system has the zero of the vertical axis and zero of gravitational potential at the bottom of the pit.

A ball of mass m is initially at rest at ground level, Position A.

a) What is the gravitational potential energy of the ball as measured by Joe, by Peter, and by Amanda?

b) The ball is then raised to the height of the cliff, Position B, and is held at rest. What is the gravitational potential energy of the ball at Position B as measured by Joe, by Peter, and by Amanda?

c) The ball is then released from rest and strikes the ground at the bottom of the pit, Position C. What is the gravitational potential energy of the ball at Position C as measured by Joe, by Peter, and by Amanda?

d) What is this change in gravitational potential energy measured by J, P, and A between positions A and C?

Student Solution Manual

The Student Solution Manual is available in two paperback volumes: Volume 1 comprises Chapters 1–21 and Volume 2 comprises Chapters 22–41.

The **Student Solutions Manual** contains worked solutions for about 50% of the odd-numbered Problems and Review Problems in the text.

- Appendix 11 in the back of the textbook contains only the final answer for odd-numbered problems in the chapters, not the intermediate steps of the solutions.

StudySpace www.wwnorton.com/physics

The StudySpace website is the free and open portal through which students access the resources that accompany this text.

Online Concept Tutorial

- 45 Online Concept Tutorials—at no additional cost.

STUDY PLANS

- 41 Study Plans, one for each chapter—at no additional cost.

smartwork

- Smartwork online homework system—a subscription service.

- ebook links to textbook chapters—as part of subscription service.

Prelude

The World of Physics

Physics is the study of matter. In a quite literal sense, physics is the greatest of all natural sciences: it encompasses the smallest particles, such as electrons and quarks; and it also encompasses the largest bodies, such as galaxies and the entire Universe. The smallest particles and the largest bodies differ in size by a factor of more than ten thousand billion billion billion billion! In the pictures on the following pages we will survey the world of physics and attempt to develop some rough feeling for the sizes of things in this world. This preliminary survey sets the stage for our explanations of the mechanisms that make things behave in the way they do. Such explanations are at the heart of physics, and they are the concern of the later chapters of this book.

Since the numbers we will be dealing with in this prelude and in the later chapters are often very large or very small, we will find it convenient to employ the **scientific notation** for these numbers. In this notation, numbers are written with powers of 10; thus, hundred is written as 10^2, thousand is written as 10^3, ten thousand is written as 10^4, and so on. A tenth is written as 10^{-1}, a hundredth is written as 10^{-2}, a thousandth is written as 10^{-3}, and so on. The following table lists some powers of ten:

$10 = 10^1$	$0.1 = 1/10 = 10^{-1}$
$100 = 10^2$	$0.01 = 1/100 = 10^{-2}$
$1000 = 10^3$	$0.001 = 1/1000 = 10^{-3}$
$10000 = 10^4$	$0.0001 = 1/10000 = 10^{-4}$
$100000 = 10^5$	$0.00001 = 1/100000 = 10^{-5}$
$1000000 = 10^6$	$0.000001 = 1/1000000 = 10^{-6}$ etc.

Note that the power of 10, or the exponent on the 10, simply tells us how many zeros follow the 1 in the number (if the power of 10 is positive) or how many zeros follow the 1 in the denominator of the fraction (if the power of 10 is negative).

In scientific notation, a number that does not coincide with one of the powers of 10 is written as a product of a decimal number and a power of 10. For example, in this notation, 1500000000 is written as 1.5×10^9. Alternatively, this number could be written as 15×10^8 or as 0.15×10^{10}; but in scientific notation it is customary to place the decimal point immediately after the first nonzero digit. The same rule applies to numbers smaller than 1; thus, 0.000015 is written as 1.5×10^{-5}.

The pictures on the following pages fall into two sequences. In the first sequence we zoom out: we begin with a picture of a woman's face and proceed step by step to pictures of the entire Earth, the Solar System, the Galaxy, and the Universe. This ascending sequence contains 27 pictures, with the scale decreasing in steps of factors of 10.

Most of our pictures are photographs. Many of these have become available only in recent years; they were taken by high-flying aircraft, Landsat satellites, astronauts, or sophisticated electron microscopes. For some of our pictures no photographs are available and we have to rely, instead, on drawings.

PART I: THE LARGE-SCALE WORLD

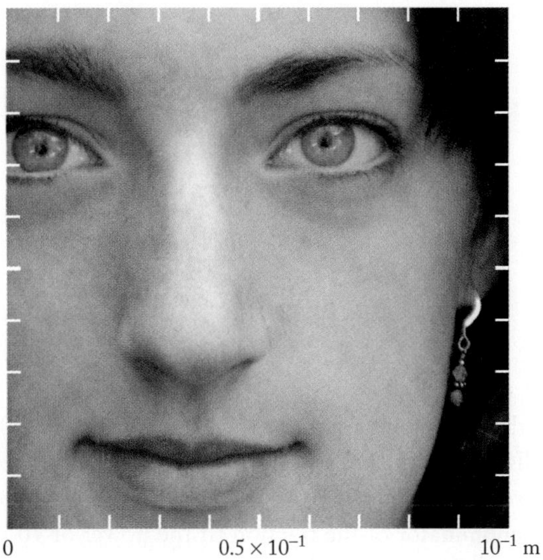

Fig. P1 SCALE 1:1.5 This is Erin, an intelligent biped of the planet Earth, Solar System, Orion Spiral Arm, Milky Way Galaxy, Local Group, Local Supercluster. Erin belongs to the phylum Chordata, class Mammalia, order Primates, family Hominidae, genus *Homo*, species *sapiens*. She is made of 5.4×10^{27} atoms, with 1.9×10^{28} electrons, the same number of protons, and 1.5×10^{28} neutrons.

Fig. P2 SCALE 1:1.5 × 10 Erin has a height of 1.7 meters and a mass of 57 kilograms. Her chemical composition (by mass) is 65% oxygen, 18.5% carbon, 9.5% hydrogen, 3.3% nitrogen, 1.5% calcium, 1% phosphorus, and 1.2% other elements.

The matter in Erin's body and the matter in her immediate environment occur in three states of aggregation: **solid**, **liquid**, and **gas**. All these forms of matter are made of atoms and molecules, but solid, liquid, and gas are qualitatively different because the arrangements of the atomic and molecular building blocks are different.

In a solid, each building block occupies a definite place. When a solid is assembled out of molecular or atomic building blocks, these blocks are locked in place once and for all, and they cannot move or drift about except with great difficulty. This rigidity of the arrangement is what makes the aggregate hard—it makes the solid "solid." In a liquid, the molecular or atomic building blocks are not rigidly connected. They are thrown together at random and they move about fairly freely, but there is enough adhesion between neighboring blocks to prevent the liquid from dispersing. Finally, in a gas, the molecules or atoms are almost completely independent of one another. They are distributed at random over the volume of the gas and are separated by appreciable distances, coming in touch only occasionally during collisions. A gas will disperse spontaneously if it is not held in confinement by a container or by some restraining force.

0 0.5×10^1 10^1 m

Fig. P3 SCALE 1:1.5 × 10² The building behind Erin is the New York Public Library, one of the largest libraries on Earth. This library holds 1.4×10^{10} volumes, containing roughly 10% of the total accumulated knowledge of our terrestrial civilization.

0 0.5×10^2 10^2 m

Fig. P4 SCALE 1:1.5 × 10³ The New York Public Library is located at the corner of Fifth Avenue and 42nd Street, in the middle of New York City, with Bryant Park immediately behind it.

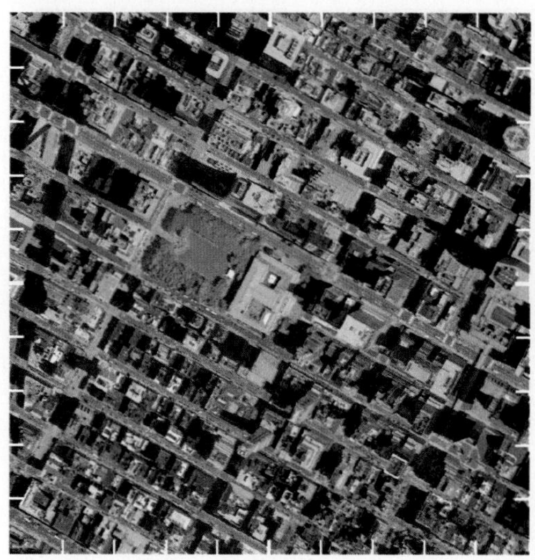

0 0.5×10^3 10^3 m

Fig. P5 SCALE 1:1.5 × 10⁴ This aerial photograph shows an area of 1 kilometer × 1 kilometer in the vicinity of the New York Public Library. The streets in this part of the city are laid out in a regular rectangular pattern. The library is the building in the park in the middle of the picture. The photograph was taken early in the morning, and the high buildings typical of New York cast long shadows.

The photograph was taken from an airplane flying at an altitude of a few thousand meters. North is at the top of the photograph.

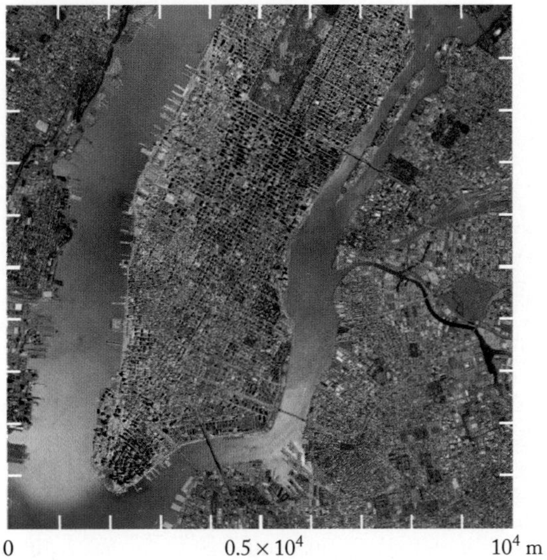

Fig. P6 SCALE 1:1.5 × 10^5 This photograph shows a large portion of New York City. We can barely recognize the library and its park as a small rectangular patch slightly above the center of the picture. The central mass of land is the island of Manhattan, with the Hudson River on the left and the East River on the right.

This photograph and the next two were taken by satellites orbiting the Earth at an altitude of about 700 kilometers.

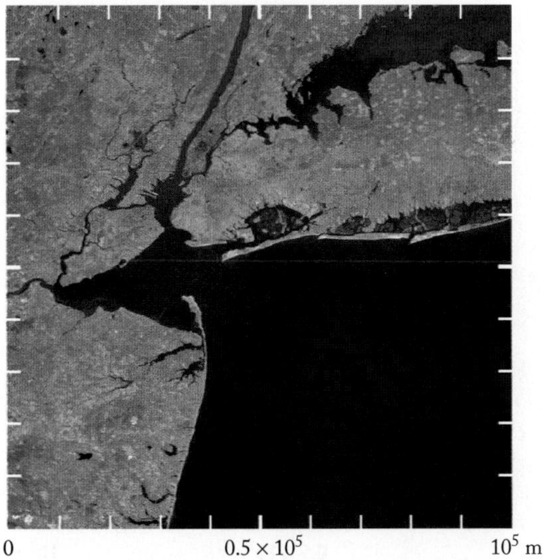

Fig. P7 SCALE 1:1.5 × 10^6 In this photograph, Manhattan is in the upper middle. On this scale, we can no longer distinguish the pattern of streets in the city. The vast expanse of water in the lower right of the picture is part of the Atlantic Ocean. The mass of land in the upper right is Long Island. Parallel to the south shore of Long Island we can see a string of very narrow islands; they almost look man-made. These are barrier islands; they are heaps of sand piled up by ocean waves in the course of thousands of years.

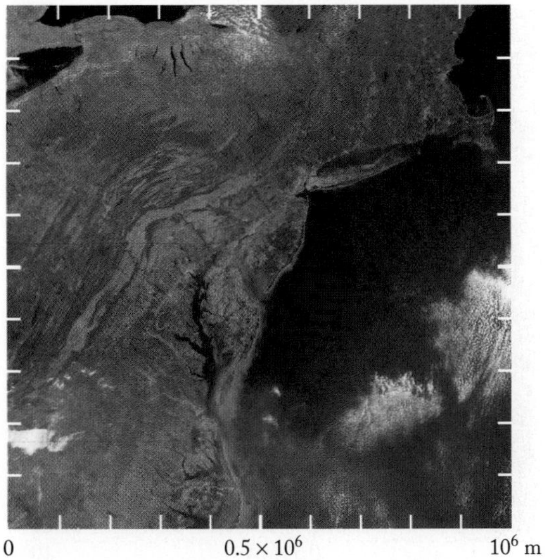

Fig. P8 SCALE 1:1.5 × 10^7 Here we see the eastern coast of the United States, from Cape Cod to Cape Fear. Cape Cod is the hook near the northern end of the coastline, and Cape Fear is the promontory near the southern end of the coastine. Note that on this scale no signs of human habitation are visible. However, at night the lights of large cities would stand out clearly.

This photograph was taken in the fall, when leaves had brilliant colors. Streaks of orange trace out the spine of the Appalachian mountains.

Fig. P9 SCALE 1:1.5 × 10^8 In this photograph, taken by the Apollo 16 astronauts during their trip to the Moon, we see a large part of the Earth. Through the gap in the clouds in the lower middle of the picture, we can see the coast of California and Mexico. We can recognize the peninsula of Baja California and the Gulf of California. Erin's location, the East Coast of the United States, is covered by a big system of swirling clouds on the right of the photograph.

Note that a large part of the area visible in this photograph is ocean. About 71% of the surface of the Earth is ocean; only 29% is land. The atmosphere covering this surface is about 100 kilometers thick; on the scale of this photograph, its thickness is about 0.7 millimeter. Seen from a large distance, the predominant colors of the planet Earth are blue (oceans) and white (clouds).

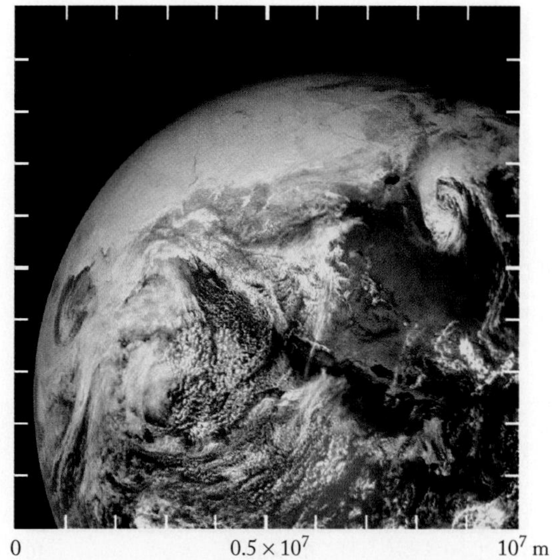

Fig. P10 SCALE 1:1.5 × 10^9 This photograph of the Earth was taken by the Apollo 16 astronauts standing on the surface of the Moon. Sunlight is striking the Earth from the top of the picture.

As is obvious from this and from the preceding photograph, the Earth is a sphere. Its radius is 6.37 × 10^6 meters and its mass is 5.98 × 10^{24} kilograms.

Fig. P11 SCALE 1:1.5 × 10^{10} In this drawing, the dot at the center represents the Earth, and the solid line indicates the orbit of the Moon around the Earth (many of the pictures on the following pages are also drawings). As in the preceding picture, the Sun is far below the bottom of the picture. The position of the Moon is that of January 1, 2000.

The orbit of the Moon around the Earth is an **ellipse**, but an ellipse that is very close to a circle. The solid red curve in the drawing is the orbit of the Moon, and the dashed green curve is a circle; by comparing these two curves we can see how little the ellipse deviates from a circle centered on the Earth. The point on the ellipse closest to the Earth is called the **perigee**, and the point farthest from the Earth is called the **apogee**. The distance between the Moon and the Earth is roughly 30 times the diameter of the Earth. The Moon takes 27.3 days to travel once around the Earth.

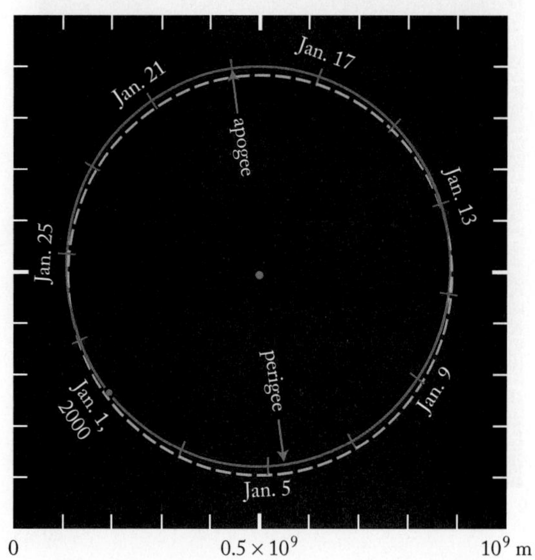

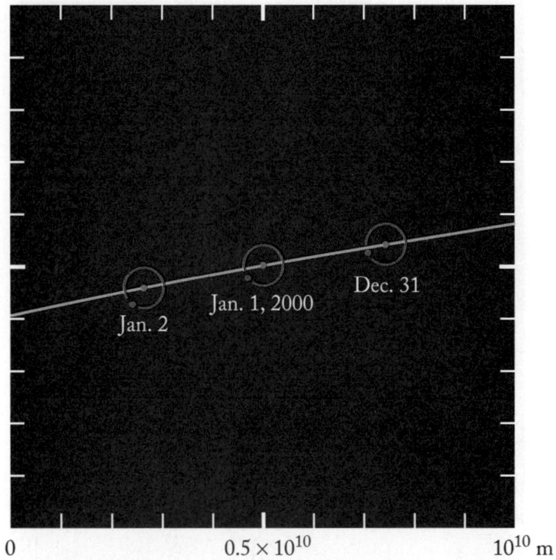

Fig. P12 SCALE 1:1.5 × 10^{11} This picture shows the Earth, the Moon, and portions of their orbits around the Sun. On this scale, both the Earth and the Moon look like small dots. Again, the Sun is far below the bottom of the picture. In the middle, we see the Earth and the Moon in their positions for January 1, 2000. On the right and on the left we see, respectively, their positions for 1 day before and 1 day after this date.

Note that the net motion of the Moon consists of the combination of two simultaneous motions: the Moon orbits around the Earth, which in turn orbits around the Sun.

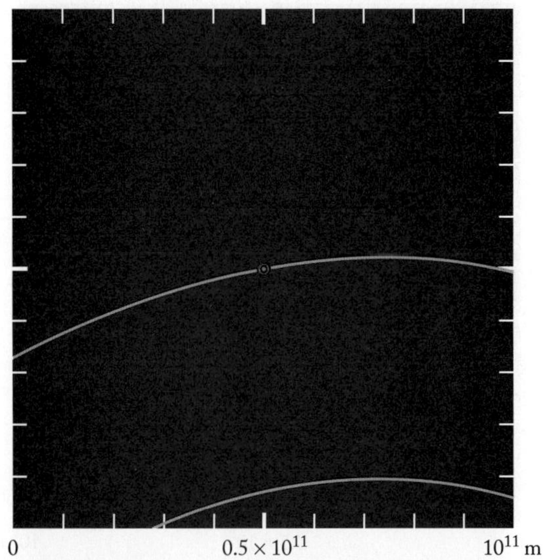

Fig. P13 SCALE 1:1.5 × 10^{12} Here we see the orbits of the Earth and of Venus. However, Venus itself is beyond the edge of the picture. The small circle is the orbit of the Moon. The dot representing the Earth is much larger than what it should be, although the artist has drawn it as minuscule as possible. On this scale, even the Sun is quite small; if it were included in this picture, it would be only 1 millimeter across.

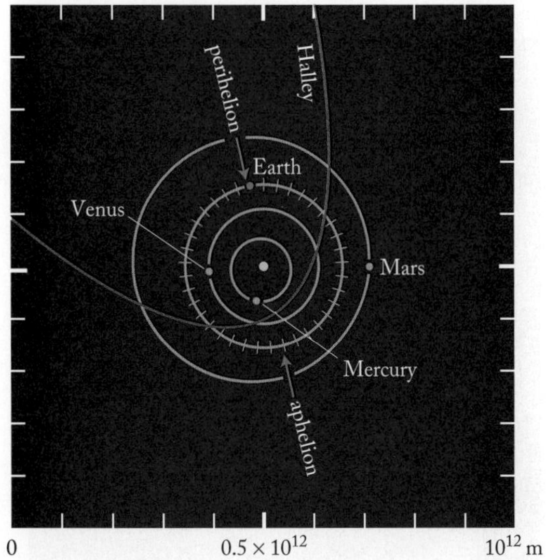

Fig. P14 SCALE 1:1.5 × 10^{13} This drawing shows the positions of the Sun and the inner planets: Mercury, Venus, Earth, and Mars. The positions of the planets are those of January 1, 2000. The orbits of all these planets are ellipses, but they are close to circles. The point of the orbit nearest to the Sun is called the **perihelion** and the point farthest from the Sun is called the **aphelion**. The Earth reaches perihelion about January 3 and aphelion about July 6 of each year.

All the planets travel around their orbits in the same direction: counterclockwise in our picture. The marks along the orbit of the Earth indicate the successive positions at intervals of 10 days.

Beyond the orbit of Mars, a large number of asteroids orbit around the Sun; these have been omitted to prevent excessive clutter. Furthermore, a large number of comets orbit around the Sun. Most of these have pronounced elliptical orbits. The comet Halley has been included in our drawing.

The Sun is a sphere of radius 6.96×10^8 meters. On the scale of the picture, the Sun looks like a very small dot, even smaller than the dot drawn here. The mass of the Sun is 1.99×10^{30} kilograms.

The matter in the Sun is in the **plasma** state, sometimes called the fourth state of matter. Plasma is a very hot gas in which violent collisions between the atoms in their random thermal motion have fragmented the atoms, ripping electrons off them. An atom that has lost one or more electrons is called an **ion**. Thus, plasma consists of a mixture of electrons and ions engaging in frequent collisions. These collisions are accompanied by the emission of light, making the plasma luminous.

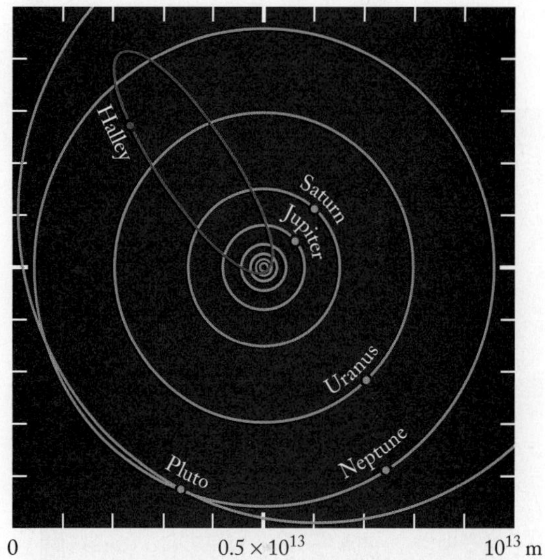

Fig. P15 SCALE 1:1.5 × 10^{14} This picture shows the positions of the outer planets of the Solar System: Jupiter, Saturn, Uranus, Neptune, and Pluto. On this scale, the orbits of the inner planets are barely visible. As in our other pictures, the positions of the planets are those of January 1, 2000.

The outer planets move slowly and their orbits are very large; thus they take a long time to go once around their orbit. The extreme case is that of Pluto, which takes 248 years to complete one orbit.

Uranus, Neptune, and Pluto are so far away and so faint that their discovery became possible only through the use of telescopes. Uranus was discovered in 1781, Neptune in 1846, and the tiny Pluto in 1930. Pluto is now known as one of several dwarf planets.

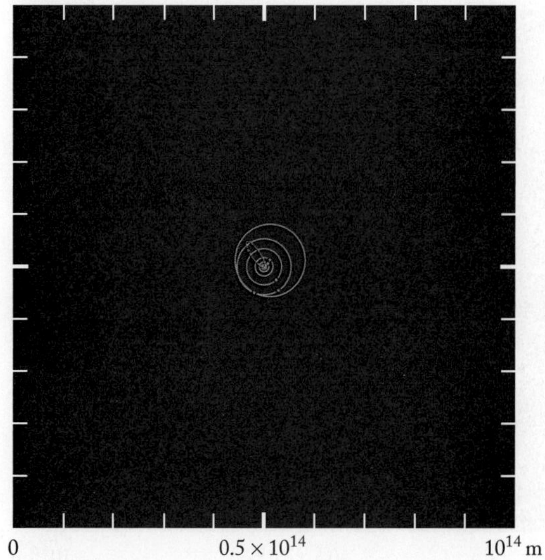

Fig. P16 SCALE 1:1.5 × 10^{15} We now see that the Solar System is surrounded by a vast expanse of space. Although this space is shown empty in the picture, the Solar System is encircled by a large cloud of millions of comets whose orbits crisscross the sky in all directions. Furthermore, the interstellar space in this picture and in the succeeding pictures contains traces of gas and of dust. The interstellar gas is mainly hydrogen; its density is typically 1 atom per cubic centimeter.

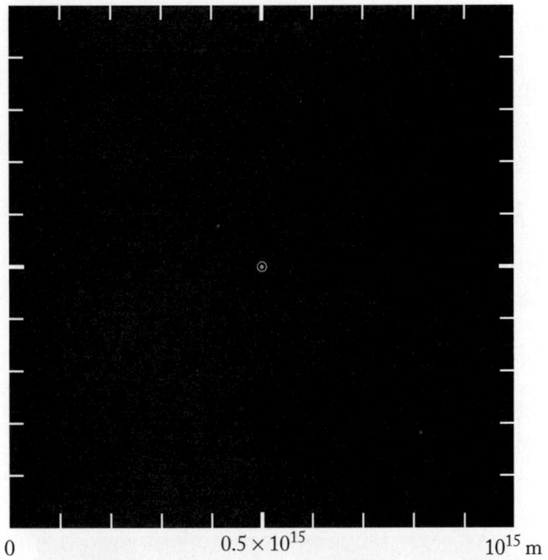

Fig. P17 SCALE 1:1.5 × 10^16 More interstellar space. The small circle is the orbit of Pluto.

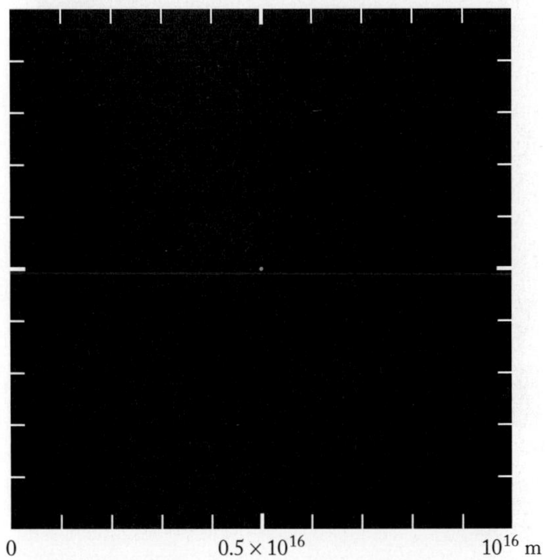

Fig. P18 SCALE 1:1.5 × 10^17 And more interstellar space. On this scale, the Solar System looks like a minuscule dot, 0.1 millimeter across.

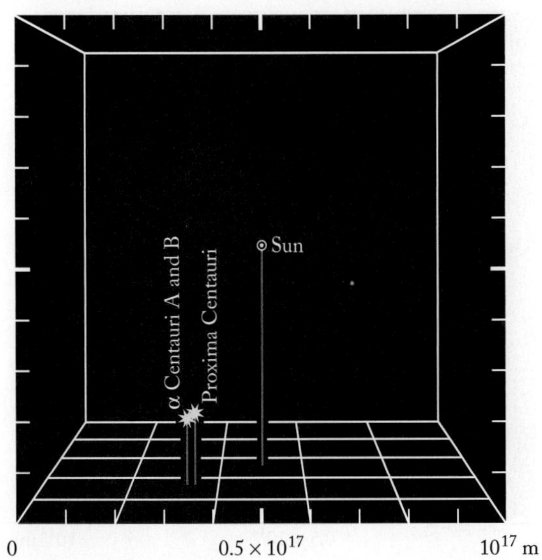

Fig. P19 SCALE 1:1.5 × 10^18 Here, at last, we see the stars nearest to the Sun. The picture shows all the stars within a cubical box 10^{17} meters × 10^{17} meters × 10^{17} meters centered on the Sun: Alpha Centauri A, Alpha Centauri B, and Proxima Centauri. All three are in the constellation Centaurus, in the southern sky.

The star closest to the Sun is Proxima Centauri. This is a very faint, reddish star (a "red dwarf"), at a distance of 4.0×10^{16} meters from the Sun. Astronomers like to express stellar distances in light-years: Proxima Centauri is 4.2 light-years from the Sun, which means light takes 4.2 years to travel from this star to the Sun.

Fig. P20 SCALE 1:1.5 × 10^19 This picture displays the brightest stars within a cubical box 10^{18} meters × 10^{18} meters × 10^{18} meters centered on the Sun. There are many more stars in this box besides those shown—the total number of stars in this box is close to 2000.

Sirius is the brightest of all the stars in the night sky. If it were at the same distance from the Earth as the Sun, it would be 28 times brighter than the Sun.

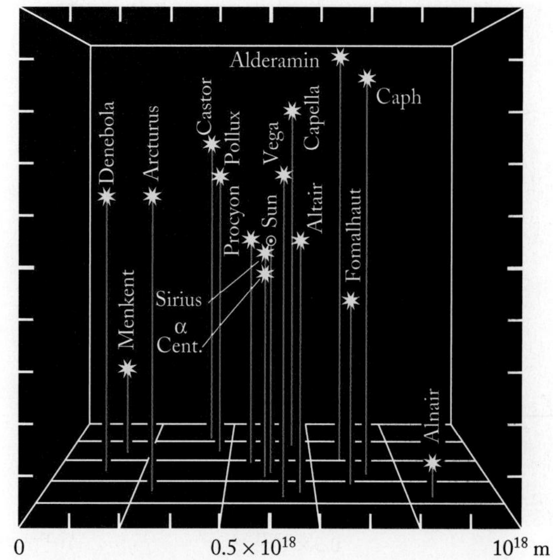

Fig. P21 SCALE 1:1.5 × 10^20 Here we expand our box to 10^{19} meters × 10^{19} meters × 10^{19} meters, again showing only the brightest stars and omitting many others. The total number of stars within this box is about 2 million. We recognize several clusters of stars in this picture: the Pleiades Cluster, the Hyades Cluster, the Coma Berenices Cluster, and the Perseus Cluster. Each of these has hundreds of stars crowded into a fairly small patch of sky. In this diagram, Starbursts signify single stars, circles with starbursts indicate star clusters, and a circle with a single star indicate a star cluster with its brightest star.

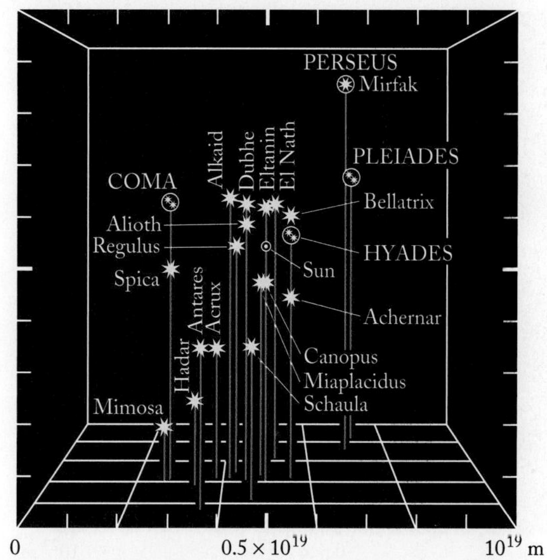

Fig. P22 SCALE 1:1.5 × 10^21 This photograph shows a view of the Milky Way in the direction of the constellation Sagittarius. Now there are so many stars in our field of view that they appear to form clouds of stars. There are about a million stars in this photograph, and there are many more stars too faint to show up distinctly. Although this photograph is not centered on the Sun, it is similar to what we would see if we could look toward the Solar System from very far away.

Fig. P23 SCALE 1:1.5 × 10^{22} This is the spiral galaxy NGC 2997. Its clouds of stars are arranged in spiral arms wound around a central bulge. The bright central bulge is the nucleus of the galaxy; it has a more or less spherical shape. The surrounding region, with the spiral arms, is the disk of the galaxy. This disk is quite thin; it has a thickness of only about 3% of its diameter. The stars making up the disk circle around the galactic center in a clockwise direction.

Our Sun is in a spiral galaxy of roughly similar shape and size: the **Milky Way Galaxy**. The total number of stars in this galaxy is about 10^{11}. The Sun is in one of the spiral arms, roughly one-third inward from the edge of the disk toward the center.

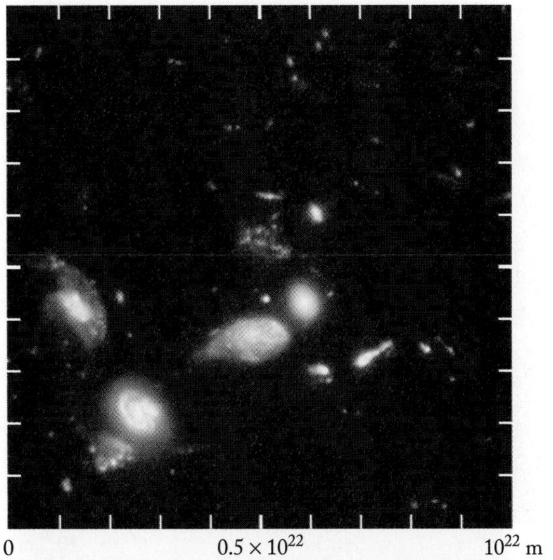

Fig. P24 SCALE 1:1.5 × 10^{23} Galaxies are often found in clusters of several galaxies. Some of these clusters consist of just a few galaxies, others of hundreds or even thousands. The photograph shows a cluster, or group, of galaxies beyond the constellation Fornax. The group contains an elliptical galaxy like a luminous yellow egg (center), three large spiral galaxies (left), and a spiral with a bar (bottom left).

Our Galaxy is part of a modest cluster, the **Local Group**, consisting of our own Galaxy, the great Andromeda Galaxy, the Triangulum Galaxy, the Large Magellanic Cloud, plus 16 other small galaxies.

According to recent investigations, the dark, apparently empty, space near galaxies contains some form of distributed matter, with a total mass 20 or 30 times as large as the mass in the luminous, visible galaxies. But the composition of this invisible, extragalactic **dark matter** is not known.

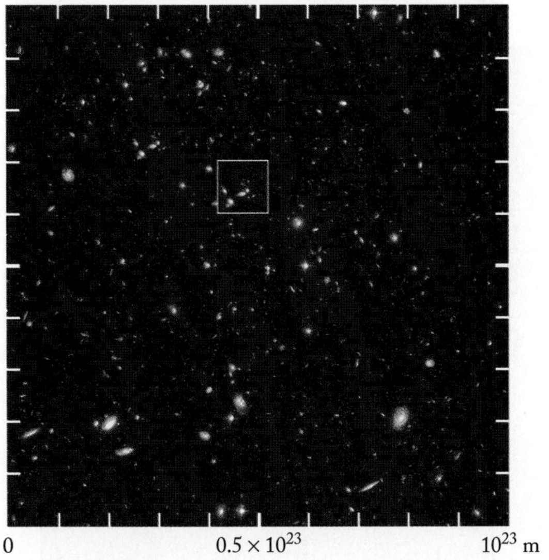

Fig. P25 SCALE 1:1.5 × 10^{24} The Local Group lies on the fringes of a very large cluster of galaxies, called the **Local Supercluster**. This is a cluster of clusters of galaxies. At the center of the Local Supercluster is the **Virgo Cluster** with several thousand galaxies. Seen from a large distance, our supercluster would present a view comparable with this photograph, which shows a multitude of galaxies beyond the constellation Fornax, all at a very large distance from us. The photograph was taken with the Hubble Space Telescope coupled to two very sensitive cameras using an exposure time of almost 300 hours.

All these distant galaxies are moving away from us and away from each other. The very distant galaxies in the photo are moving away from us at speeds almost equal to the speed of light. This motion of recession of the galaxies is analogous to the outward motion of, say, the fragments of a grenade after its explosion. The motion of the galaxies suggests that the Universe began with a big explosion, the **Big Bang**, that launched the galaxies away from each other.

Fig. P26 SCALE 1:1.5 × 10^{25} On this scale a galaxy equal in size to our own Galaxy would look like a fuzzy dot, 0.1 millimeter across. Thus, the galaxies are too small to show up clearly on a photograph. Instead we must rely on a plot of the positions of the galaxies. The plot shows the positions of about 200 galaxies. The dense cluster of galaxies in the lower half of the plot is the Virgo Cluster.

Since we are looking into a volume of space, some of the galaxies are in the foreground, some are in the background; but our plot takes no account of perspective.

The luminous stars in the galaxies constitute only a small fraction of the total mass of the Universe. The space around the galaxies and the clusters of galaxies contains dark matter, and the space between the clusters contains **dark energy**, a strange form of matter that causes an acceleration of the expansion of the Universe.

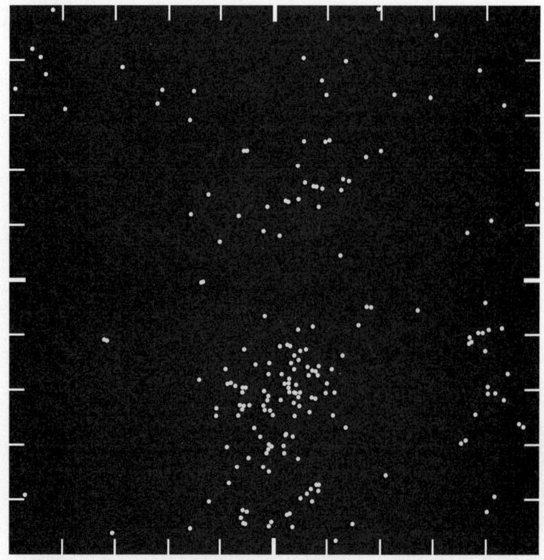

Fig. P27 SCALE 1:1.5 × 10^{26} This plot shows the positions of about 100,000 galaxies in a patch of the sky at distances of up to 1 × 10^9 light years from the Earth. The false color in this image indicates the distance–red for shorter distances, blue for larger distances.

The visible galaxies plotted here contribute only about 5% of the total mass in the universe. The dark matter near the galaxies contribute another 25%. The remaining 70% of the total mass in the universe is in the form of dark energy, which is uniformly distributed over the vast reaches of intergalactic space.

This is the last of our pictures in the ascending series. We have reached the limits of our zoom out. If we wanted to draw another picture, 10 times larger than this, we would need to know the shape and size of the entire Universe. We do not yet know that.

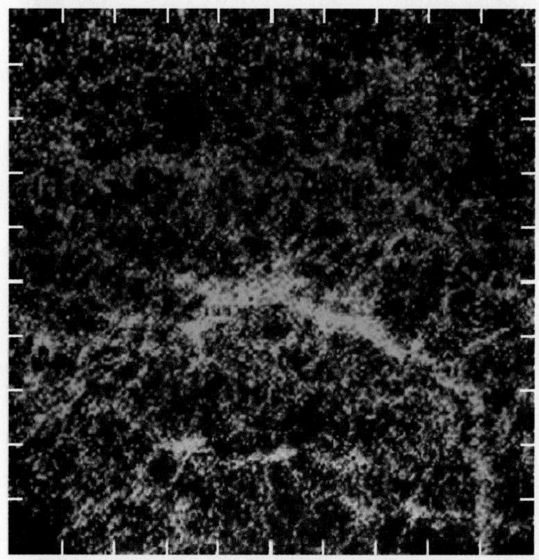

PART II: THE SMALL-SCALE WORLD

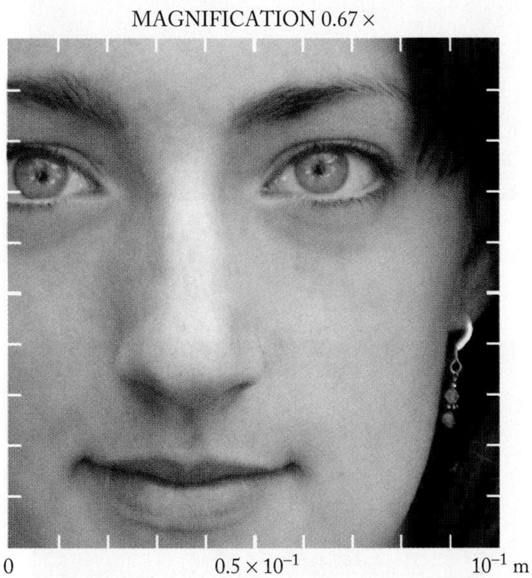

Fig. P28 SCALE 1:1.5 We now return to Erin and zoom in on her eye. The surface of her skin appears smooth and firm. But this is an illusion. Matter appears continuous because the number of atoms in each cubic centimeter is extremely large. In a cubic centimeter of human tissue there are about 10^{23} atoms. This large number creates the illusion that matter is continuously distributed—we see only the forest and not the individual trees. The solidity of matter is also an illusion. The atoms in our bodies are mostly vacuum. As we will discover in the following pictures, within each atom the volume actually occupied by subatomic particles is only a very small fraction of the total volume.

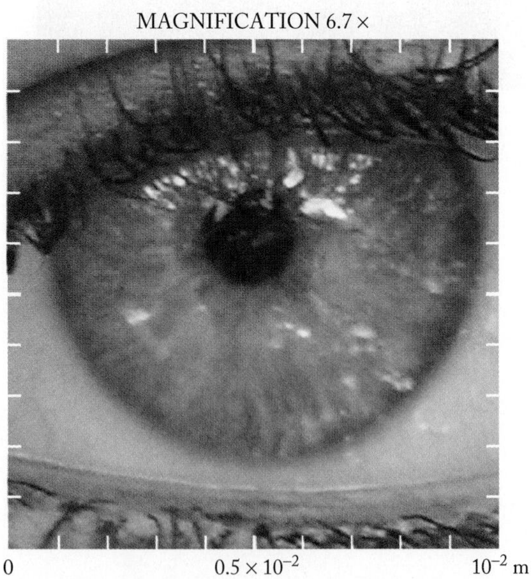

Fig. P29 SCALE 1:1.5 × 10^{-1} Our eyes are very sophisticated sense organs; they collect more information than all our other sense organs taken together. The photograph shows the pupil and the iris of Erin's eye. Annular muscles in the iris change the size of the pupil and thereby control the amount of light that enters the eye. In strong light the pupil automatically shrinks to about 2 millimeters; in very weak light it expands to as much as 7 millimeters.

Fig. P30 SCALE 1:1.5 × 10^{-2} This false-color photograph shows the delicate network of blood vessels on the front surface of the retina, the light-sensitive membrane lining the interior of the eyeball. The rear surface of the retina is densely packed with two kinds of cells that sense light: cone cells and rod cells. In a human retina there are about 6 million cone cells and 120 million rod cells. The cone cells distinguish colors; the rod cells distinguish only brightness and darkness, but they are more sensitive than the cone cells and therefore give us vision in faint light ("night vision").

This and the following photographs were made with various kinds of **electron microscopes**. An ordinary microscope uses a beam of light to illuminate the object; an electron microscope uses a beam of electrons. Electron microscopes can achieve much sharper contrast and much higher magnification than ordinary microscopes.

MAGNIFICATION $6.7 \times 10\times$

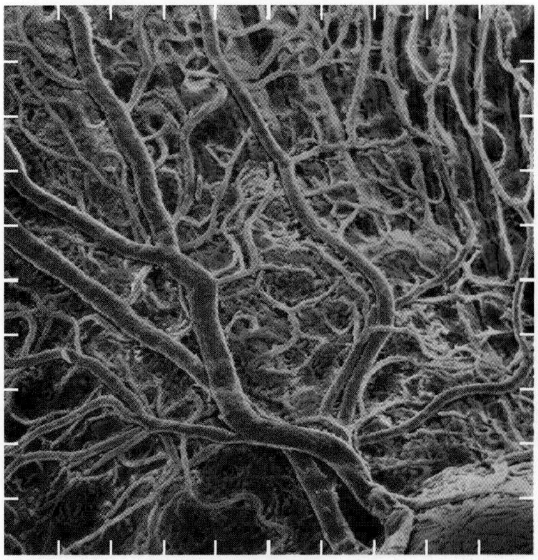

0 0.5×10^{-3} 10^{-3} m

Fig. P31 SCALE 1:1.5 × 10^{-3} Here we have a false-color photograph of rod cells prepared with a scanning electron microscope (SEM). For this photograph, the retina was cut apart and the microscope was aimed at the edge of the cut. In the top half of the picture we see tightly packed rods. Each rod is connected to the main body of a cell containing the nucleus. In the bottom part of the picture we can distinguish tightly packed cell bodies of the cell.

MAGNIFICATION $6.7 \times 10^{2}\times$

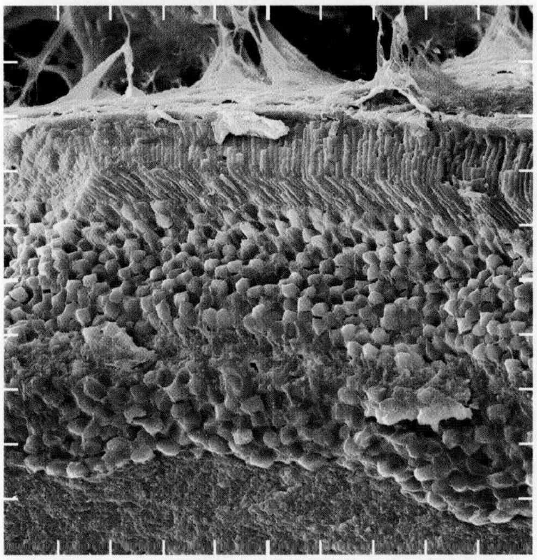

0 0.5×10^{-4} 10^{-4} m

Fig. P32 SCALE 1:1.5 × 10^{-4} This is a close-up view of a few rods cells. The upper portions of the rods contain a special pigment—visual purple—which is very sensitive to light. The absorption of light by this pigment initiates a chain of chemical reactions that finally trigger nerve pulses from the eye to the brain.

MAGNIFICATION $6.7 \times 10^{3}\times$

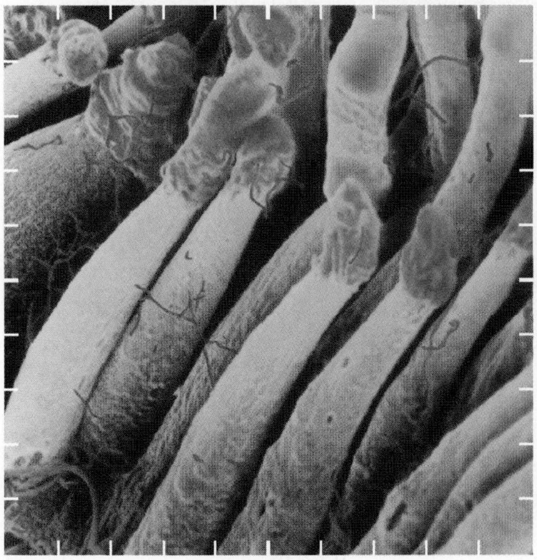

0 0.5×10^{-5} 10^{-5} m

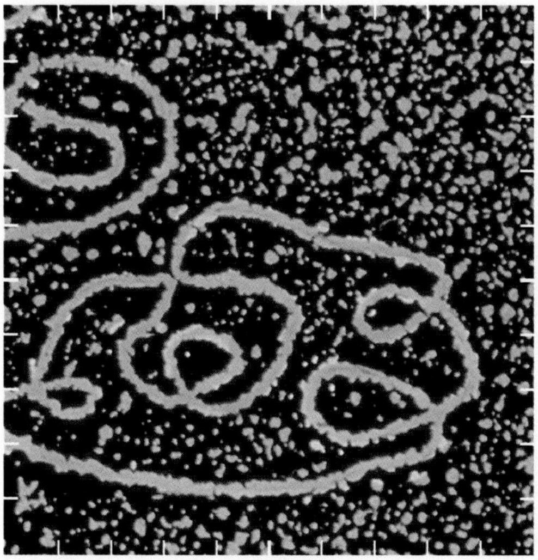

Fig. P33 SCALE 1:1.5 × 10^{-5} These are strands of DNA, or deoxyribonucleic acid, as seen with a transmission electron microscope (TEM) at very high magnification. DNA is found in the nuclei of cells. It is a long molecule made by stringing together a large number of nitrogenous base molecules on a backbone of sugar and phosphate molecules. The base molecules are of four kinds, the same in all living organisms. But the sequence in which they are strung together varies from one organism to another. This sequence spells out a message—the base molecules are the "letters" in this message. The message contains all the genetic instructions governing the metabolism, growth, and reproduction of the cell.

The strands of DNA in the photograph are encrusted with a variety of small protein molecules. At intervals, the strands of DNA are wrapped around larger protein molecules that form lumps looking like the beads of a necklace.

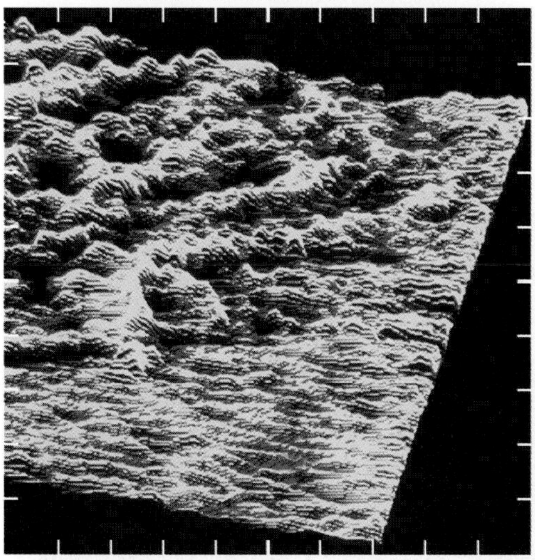

Fig. P34 SCALE 1:1.5 × 10^{-6} The highest magnifications are attained by a newer kind of electron microscope, the scanning tunneling microscope (STM). This picture was prepared with such a microscope. The picture shows strands of DNA deposited on a substrate of graphite. In contrast to the strands of the preceding picture, these strands are uncoated; that is, they are without protein encrustations.

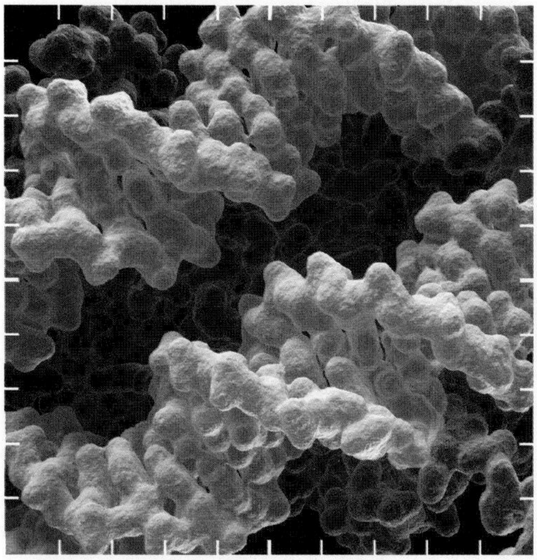

Fig. P35 SCALE 1:1.5 × 10^{-7} This close-up picture of strands of DNA reveals the helical structure of this molecule. The strand consists of a pair of helical coils wrapped around each other. This picture was generated by a computer from data obtained by illuminating DNA samples with X rays (X-ray scattering).

Fig. P36 SCALE 1:1.5 × 10^{-8} This picture shows a layer of palladium atoms on surface of graphite as seen with an STM. Here we have visual evidence of the atomic structure of matter. The palladium atoms are arranged in a symmetric, repetitive hexagonal pattern. Materials with such regular arrangements of atoms are called **crystals**.

Each of the palladium atoms is approximately a sphere, about 3×10^{-10} meter across. However, the atom does not have a sharply defined boundary; its surface is somewhat fuzzy. Atoms of other elements are also approximately spheres, with sizes that range from 2×10^{-10} to 4×10^{-10} meter across.

At present we know of more than 100 kinds of atoms or chemical elements. The lightest atom is hydrogen, with a mass of 1.67×10^{-27} kilogram; the heaviest is element 114, ununquadium, with a mass about 289 times as large.

MAGNIFICATION $6.7 \times 10^7 \times$

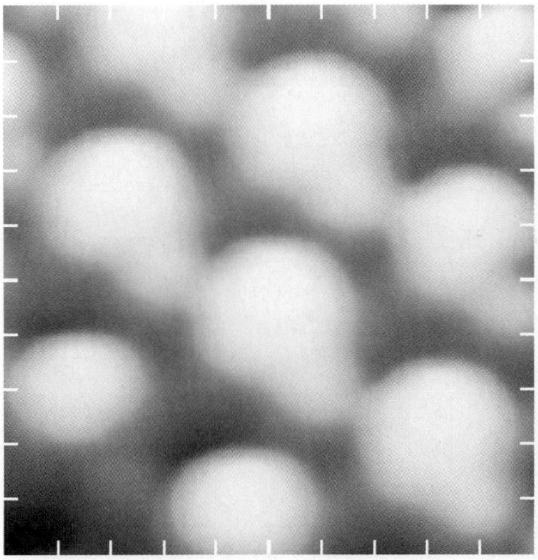

Fig. P37 SCALE 1:1.5 × 10^{-9} The drawing shows the interior of an atom of neon. This atom consists of 10 electrons orbiting around a nucleus. In the drawing, the electrons have been indicated by small dots, and the nucleus by a slightly larger dot at the center of the picture. These dots have been drawn as small as possible, but even so the size of these dots does not give a correct impression of the actual sizes of the electrons and of the nucleus. The electron is smaller than any other particle we know; maybe the electron is truly pointlike and has no size at all. The nucleus has a finite size, but this size is much too small to show up on the drawing. Note that the electrons tend to cluster near the center of the atom. However, the overall size of the atom depends on the distance to the outermost electron; this electron defines the outer edge of the atom.

The electrons move around the nucleus in a very complicated motion, and so the resulting electron distribution resembles a fuzzy cloud, similar to the STM image of the previous picture. This drawing, however, shows the electrons as they would be seen at one instant of time with a hypothetical microscope that employs gamma rays instead of light rays to illuminate an object; no such microscope has yet been built.

The mass of each electron is 9.11×10^{-31} kilogram, but most of the mass of the atom is in the nucleus; the 10 electrons of the neon atom have only 0.03% of the total mass of the atom.

MAGNIFICATION $6.7 \times 10^8 \times$

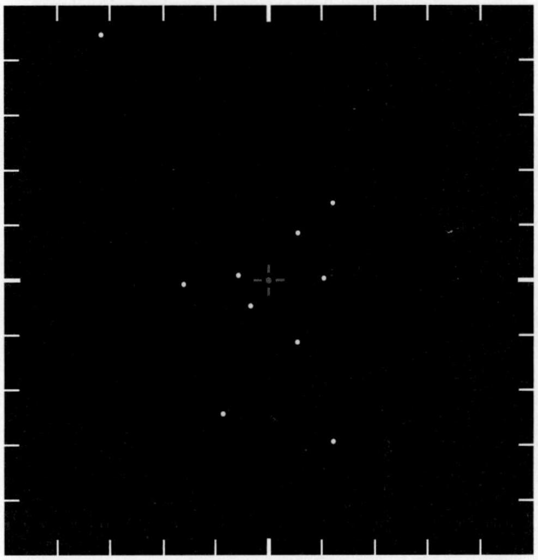

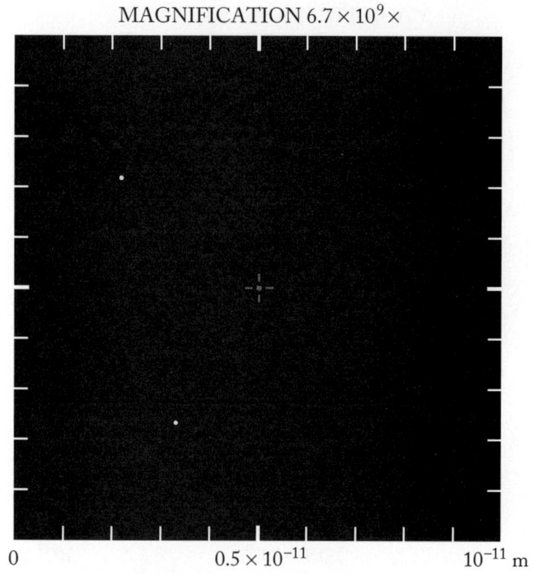

Fig. P38 SCALE 1:1.5 × 10^{-10} Here we are closing in on the nucleus. We are seeing the central part of the atom. Only two electrons are in our field of view; the others are beyond the margin of the drawing. The size of the nucleus is still much smaller than the size of the dot at the center of the drawing.

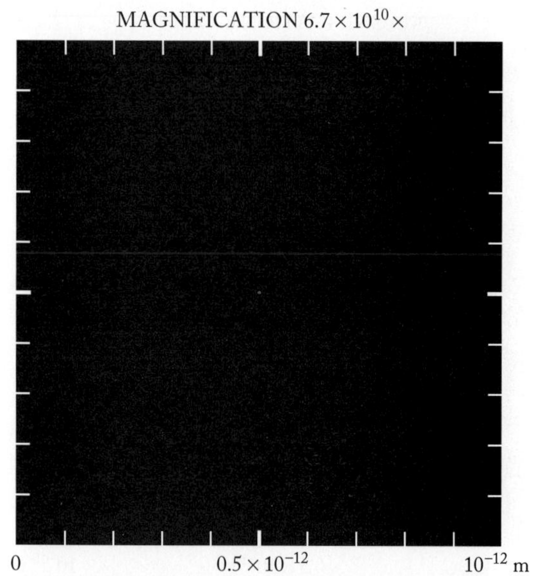

Fig. P39 SCALE 1:1.5 × 10^{-11} In this drawing we finally see the nucleus in its true size. At this magnification, the nucleus of the neon atom looks like a small dot, 0.5 millimeter in diameter. Since the nucleus is extremely small and yet contains most of the mass of the atom, the density of the nuclear material is enormous. If we could assemble a drop of pure nuclear material of a volume of 1 cubic centimeter, it would have a mass of 2.3×10^{11} kilograms, or 230 million metric tons!

Our drawings show clearly that most of the volume within the atom is empty space. The nucleus occupies only a very small fraction of this volume.

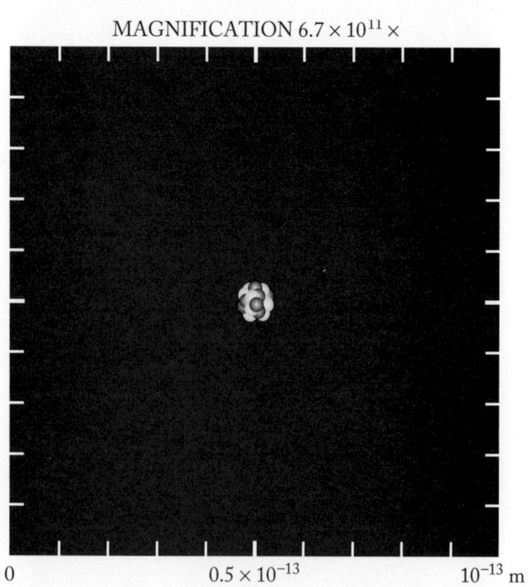

Fig. P40 SCALE 1:1.5 × 10^{-12} We can now begin to distinguish the nuclear structure. The nucleus has a nearly spherical shape, but its surface is slightly fuzzy.

Fig. P41 SCALE 1:1.5 × 10^{-13} At this extreme magnification we can see the details of the nuclear structure. The nucleus of the neon atom is made up of 10 protons (white balls) and 10 neutrons (red balls). Each proton and each neutron is a sphere with a diameter of about 2×10^{-15} meter, and a mass of 1.67×10^{-27} kilogram. In the nucleus, these protons and neutrons are tightly packed together, so tightly that they almost touch. The protons and neutrons move around the volume of the nucleus at high speed in a complicated motion.

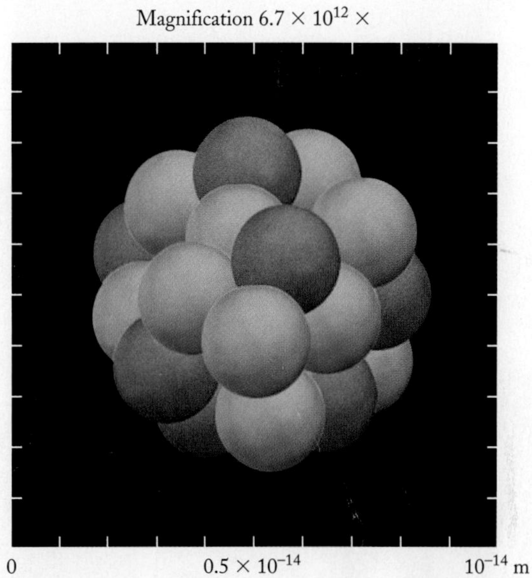

Fig. P42 SCALE 1:1.5 × 10^{-14} This final picture shows three pointlike bodies within a proton. These pointlike bodies are **quarks**—each proton and each neutron is made of three quarks. Recent experiments have told us that the quarks are much smaller than protons, but we do not yet know their precise size. Hence the dots in the drawing probably do not give a fair description of the size of the quarks. The quarks within protons and neutrons are of two kinds, called **up** and **down**. The proton consists of two *up* quarks and one *down* quark joined together; the neutron consists of one *up* quark and two *down* quarks joined together.

This final picture takes us to the limits of our knowledge of the subatomic world. As a next step we would like to zoom in on the quarks and show what they are made of. According to a speculative theory, they are made of small snippets or loops of strings, 10^{-35}m long. But we do not yet have any evidence for this theory.

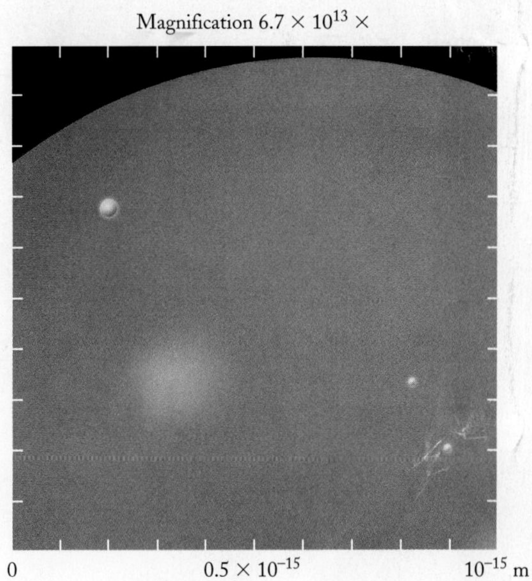

NASA

PART

1

Motion, Force, and Energy

CONTENTS

At launch, the space shuttle assembly, including the external fuel tank and auxiliary boosters, has a mass of 2×10^6 kg. The thrust of the powerful rocket engines, including the shuttle orbiter's main engines shown here, accelerates the entire launch vehicle to the speed of sound in just 45 seconds.

CHAPTER

1 Space, Time, and Mass

CONCEPTS IN CONTEXT

Concepts in Context

A laser-ranging device is used by geologists and surveyors to measure distance. It emits a pulse of laser light toward a mirror placed at some unknown distance and measures the time taken by the pulse to travel to the mirror and back. From this round-trip time and the known speed of light, it then calculates the distance. Since the speed of light is very large, the round-trip time is very small.

With the definitions of the units of distance and time given in this chapter, we can consider the following questions:

? How far does light travel in a small fraction of a second? (Example 1, page 10)

? How is the precision of the distance determination limited by the precision of the time measurement? (Example 3, page 15)

The investigator of any phenomenon—an earthquake, a flash of lightning, a collision between two ships—must begin by asking, Where and when did it happen? Phenomena happen at points in space and at points in time. A complicated phenomenon, such as a collision between two ships, is spread out over many points of space and time. But no matter how complicated, any phenomenon can be fully described by stating what happened at diverse points of space at successive instants of time. Measurements of positions and times require the use of coordinate grids and reference frames, which we will discuss in the first section of this chapter.

Ships and other macroscopic bodies are made of atoms. Since the sizes of the atoms are extremely small compared with the sizes of macroscopic bodies, we can regard atoms as almost pointlike masses for most practical purposes. *A pointlike mass of no discernible size or internal structure is called an* **ideal particle**. At any given instant of time, the ideal particle occupies a single point of space. Furthermore, the particle has a mass. And that is all: if we know the position of an ideal particle at each instant of time and we know its mass, then we know everything that can be known about the particle. *Position, time, and mass give a complete description of the behavior and the attributes of an ideal particle.*[1] Since every macroscopic body consists of particles, we can describe the behavior and the attributes of such a body by describing the particles within the body. Thus, measurements of position, time, and mass are of fundamental significance in physics. We will discuss the units for these measurements in later sections of this chapter.

1.1 COORDINATES AND REFERENCE FRAMES

If you are lost somewhere on the highways in Canada and you stop at a service station to ask for directions to Moose Jaw, the attendant might instruct you to go 90 kilometers north on Route 6 and then 70 kilometers west on Route 1 (see Fig. 1.1). In giving these instructions, the attendant is taking the service station as **origin**, and he is specifying the position of Moose Jaw relative to this origin. To achieve a precise, quantitative description of the position of a particle, physicists use much the same procedure. They first take some convenient point of space as origin and then specify the position of the particle relative to this origin. For this purpose, they imagine a grid of lines around the origin and give the location of the particle within this grid; that is, they imagine that the ground is covered with graph paper, and they specify the position of the particle by means of coordinates read off this graph paper.

The most common coordinates are **rectangular coordinates** x and y, which rely on a rectangular grid. Figure 1.2 shows such a rectangular grid. The mutually perpendicular lines through the origin O are called the x and y axes. The coordinates of the grid point P, where the particle is located, simply tell us how far we must move parallel to the corresponding axis in order to go from the origin O to the point P. For example, the point P shown in Fig. 1.2a has coordinates $x = 3$ units and $y = 5$ units. If we move from the origin in a direction opposite to that indicated by the arrow on the axis, then the coordinate is negative; thus, the point P shown in Fig. 1.2b has a negative x coordinate, $x = -3$ units.

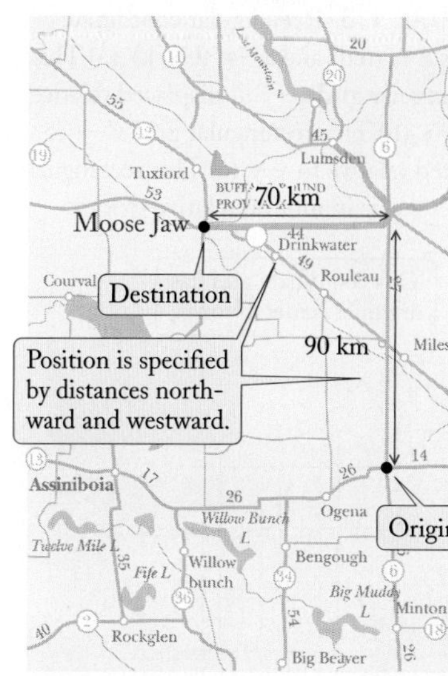

FIGURE 1.1 To reach Moose Jaw, Canada, the automobile has to travel 90 km north and then 70 km west.

[1] We will disregard for now the possibility that the particle also has an electric charge. Electricity is the subject of Chapters 22–33.

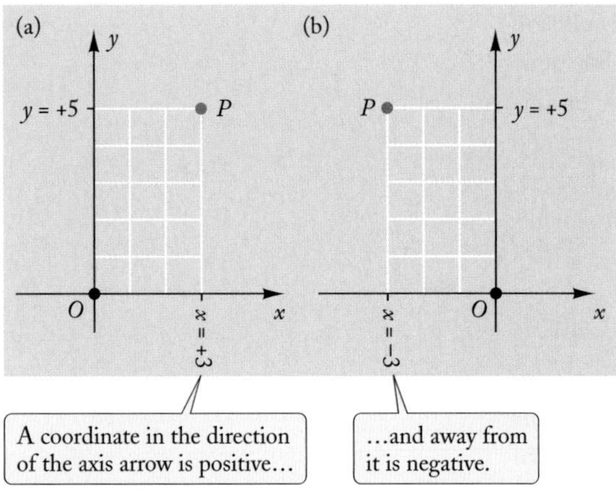

FIGURE 1.2 Rectangular coordinates x and y of a point P. (a) Both coordinates are positive; (b) the x coordinate is negative.

The two-dimensional grid shown in Fig. 1.2 is adequate when we want to describe the two-dimensional (east–west and north–south) motion of an automobile traveling on flat ground or the motion of a ship on the (nearly) flat surface of the water of a harbor. However if we want to describe the three-dimensional (east–west, north–south, and up–down) motion of an aircraft flying through the air or a submarine diving through the ocean, then we need a three-dimensional grid, with x, y, and z axes. And if we want to describe the motion of an automobile along a straight road, then we need only a one-dimensional grid; that is, we need only the x axis, which we imagine placed along the road.

When we determine the position of a particle by means of a coordinate grid erected around some origin, we perform a *relative* measurement—the coordinates of the point at which the particle is located depend on the choice of origin and on the choice of coordinate grid. The choice of origin of coordinates and the choice of coordinate grid are matters of convenience. For instance, a harbormaster might use a coordinate grid with the origin at the harbor; but a municipal engineer might prefer a displaced coordinate grid with its origin at the center of town (see Fig. 1.3a) or a rotated coordinate grid oriented along the streets of the town (see Fig. 1.3b). The navigator of a ship might find it convenient to place the origin at the midpoint of her ship and to use a coordinate grid erected around this origin; the grid then moves with the ship (see Fig. 1.3c). If the navigator plots the track of a second ship on this grid, she can tell at a glance what the distance of closest approach will be, and whether the other ship is on a collision course (whether it will cross the origin).

For the description of the motion of a particle, we must specify both its position and the time at which it has this position. To determine the time, we use a set of synchronized clocks which we imagine arranged at regular intervals along the coordinate grid. When a particle passes through a grid point P, the coordinates give us the position of the particle in space, and the time registered by the nearby clock gives us the time t. *Such a coordinate grid with an array of synchronized clocks is called a* **reference frame**. Like the choice of origin and the choice of coordinate grid, the choice of reference frame is a matter of convenience. For instance, Fig. 1.4a shows a reference frame erected around the harbor, and Fig. 1.4b shows a reference frame erected around the ship. Reference frames are usually named after the body or point around which they are erected. Thus, we speak of the reference frame of the harbor, the reference frame of the ship, the reference frame of the laboratory, the reference frame of the Earth, etc.

FIGURE 1.3 Rectangular coordinate grids x–y (red) and x'–y' (black). (a) This rectangular grid x'–y' is displaced relative to x–y. (b) This rectangular grid x'–y' is rotated relative to x–y. (c) This rectangular grid x'–y' is in motion relative to x–y.

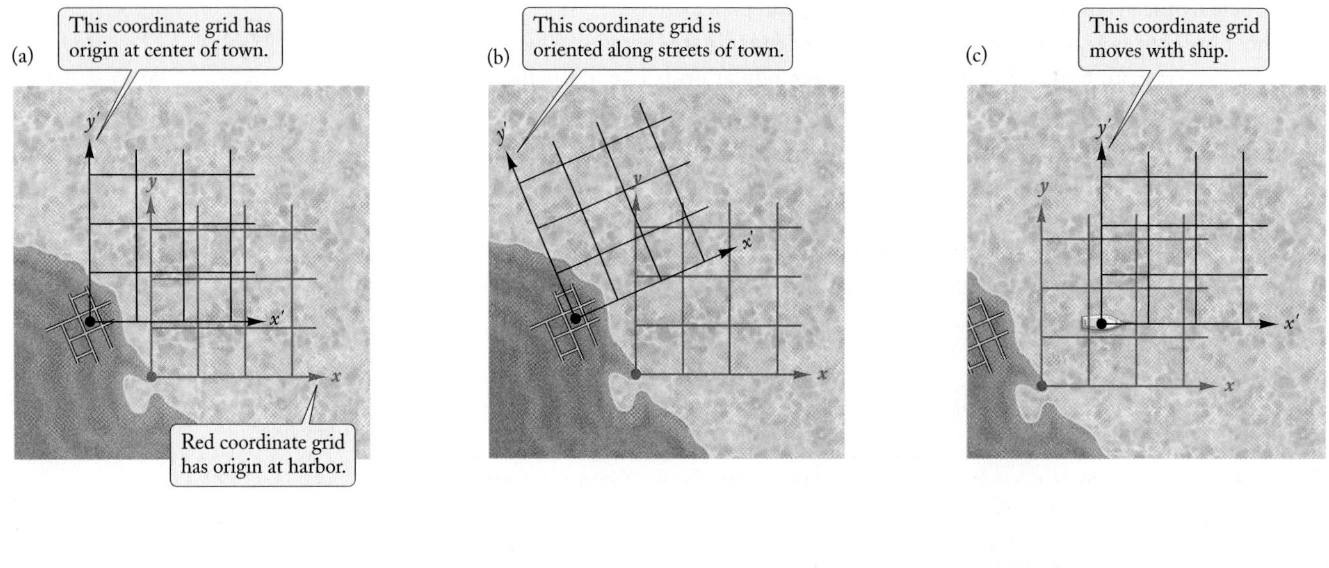

Checkup 1.1

QUESTION 1: For an ideal particle, position and mass are the only two measurable quantities (at any given instant). Consider an extended body, such as a bowling ball. What quantities can you measure about the bowling ball, besides position and mass? Do you know the units of any of these quantities?

QUESTION 2: Consider a coordinate grid x'–y' shifted relative to the grid x–y by a fixed amount, as in Fig. 1.3a. Mark a point P on this grid. Is the x' value for the given point P larger or smaller than the x value? What about the y' and y values?

QUESTION 3: What is the distinction between a coordinate grid and a reference frame?

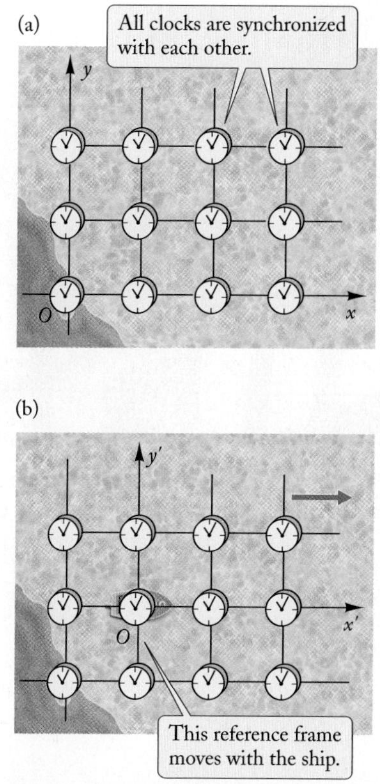

FIGURE 1.4 (a) A reference frame consists of a coordinate grid and a set of synchronized clocks. (b) A reference frame erected around a ship.

1.2 THE UNIT OF LENGTH

In order to make numerical records of our measurements of position, time, and mass we need to adopt a unit of length, a unit of time, and a unit of mass, so we can express our measurements as numerical multiples or fractions of these units. In this book we will use the **metric system of units**, *which is based on the meter as the unit of length, the second as the unit of time, and the kilogram as the unit of mass.* These units of length, time, and mass, in conjunction with the unit of temperature and the unit of electric charge (to be introduced in later chapters), are sufficient for the measurement of any physical quantity. Scientists and engineers refer to this set of units as the **International System of Units**, or **SI units** (from the French, *Système International*).[2]

Originally, the standard of length that specified the size of one meter was the standard meter bar kept at the International Bureau of Weights and Measures at Sèvres, France. This is a bar made of platinum–iridium alloy with a fine scratch mark near each end (see Fig. 1.5). By definition, the distance between these scratch marks was taken to be exactly one meter. The length of the meter was originally chosen so as to make the polar circumference of the Earth exactly 40 million meters (see Fig. 1.6); however, modern determinations of this circumference show it to be about 0.02% more than 40 million meters.

FIGURE 1.5 International standard meter bar.

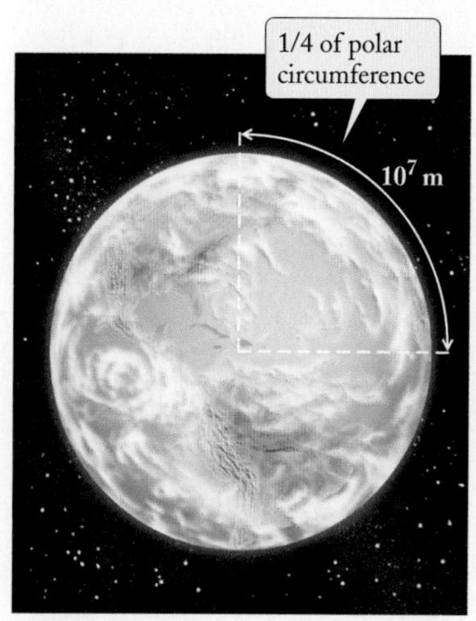

FIGURE 1.6 One-quarter of the polar circumference of the Earth equals approximately 10^7 m.

[2] SI units are also discussed in Appendix 5; more information may be found at the National Institute of Standards and Technology website, http://www.physics.nist.gov/cuu/Units.

FIGURE 1.7 Some gauge blocks, commonly used by machinists as length standards. The height or thickness of each block serves as a length standard.

FIGURE 1.8 Stabilized laser at the National Institute of Standards and Technology.

Copies of the prototype standard meter were manufactured in France and distributed to other countries to serve as secondary standards. The length standards used in industry and engineering have been derived from these secondary standards. For instance, Fig. 1.7 shows a set of gauge blocks commonly used as length standards in machine shops.

The precision of the standard meter is limited by the coarseness of the scratch marks at the ends. For the sake of higher precision, physicists developed improved definitions of the standard of length. The most recent improvement emerged from the development of stabilized lasers (see Fig. 1.8). These lasers emit light waves of extreme uniformity which make it possible to determine the speed of light with extreme precision. This led to the adoption of a new definition for the length of the meter in terms of the speed of light: the **meter** (1 m) is the length traveled by a light wave in vacuum in a time interval of 1/299 792 458 second. Note that, since the meter is adjusted so that light travels exactly one meter in 1/299 792 458 second, the speed of light is exactly

$$[\text{speed of light}] = 299\,792\,458 \text{ meters per second} \qquad (1.1)$$

Thus, the new definition of the meter amounts to the adoption of the speed of light as a standard of speed.

Table 1.1 lists a few distances and sizes, from the largest to the smallest. Many of the quantities listed in the table have already been mentioned in the Prelude. Quantities indicated with an $\approx$ (approximately equals) sign are not precisely defined; these quantities are rough approximations.

Table 1.2 lists some multiples and submultiples of the meter and their abbreviations. The **prefixes** used in Table 1.2, and other standard prefixes, are abbreviations for particular powers of ten. Prefixes representing powers of ten that differ by factors of 10^3 are often used with any unit, for convenience or conciseness. These standard prefixes and their abbreviations are listed in Table 1.3.

In the **British system of units**—abandoned by Britain and almost all other countries, but regrettably still in use in the United States—the unit of length is the **foot** (ft), which is exactly 0.3048 m. (For quick mental conversion from feet to meters, multiply by 0.3.) Table 1.4 gives the multiples and submultiples of the foot, but we will have little need for this table because we will hardly ever use British units in this book. Sporadic efforts to adopt metric units in the United States have failed, although most American automobile manufacturers now use metric units, and so does the U.S. Army (note that in U.S. Army slang, a kilometer is called a "klick," a usage commendable for its brevity).

TABLE 1.1 SOME DISTANCES AND SIZES

Distance to boundary of observable Universe	$\approx 1 \times 10^{26}$ m
Distance to Andromeda galaxy (a)	2.1×10^{22} m
Diameter of our Galaxy	7.6×10^{20} m
Distance to nearest star (Proxima Centauri)	4.0×10^{16} m
Earth–Sun distance	1.5×10^{11} m
Radius of Earth	6.4×10^{6} m
Wavelength of radio wave (AM band)	$\approx 3 \times 10^{2}$ m
Length of ship *Queen Elizabeth* (b)	3.1×10^{2} m
Height of average human male	1.8 m
Diameter of 5¢ coin (c)	2.1×10^{-2} m
Diameter of red blood cell (human)	7.5×10^{-6} m
Wavelength of visible light	$\approx 5 \times 10^{-7}$ m
Diameter of smallest virus (potato spindle) (d)	2×10^{-8} m
Diameter of atom	$\approx 1 \times 10^{-10}$ m
Diameter of atomic nucleus (iron)	$\approx 8 \times 10^{-15}$ m
Diameter of proton	$\approx 2 \times 10^{-15}$ m

(a)

(b)

(c)

(d)
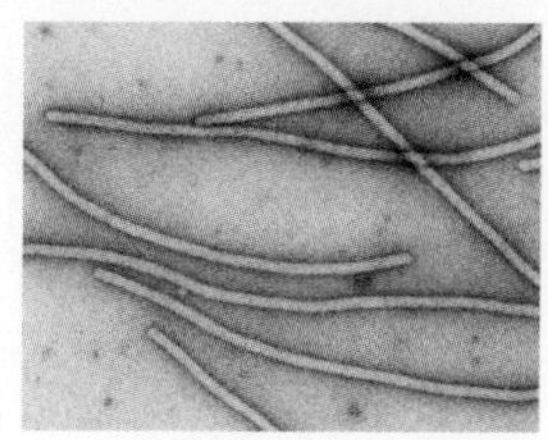

TABLE 1.2 MULTIPLES AND SUBMULTIPLES OF THE METER

kilometer (klick)	1 km = 10^{3} m
meter	1 m
centimeter	1 cm = 10^{-2} m
millimeter	1 mm = 10^{-3} m
micrometer (micron)	1 μm = 10^{-6} m
nanometer	1 nm = 10^{-9} m
angstrom	1 Å = 10^{-10} m
picometer	1 pm = 10^{-12} m
femtometer (fermi)	1 fm = 10^{-15} m

TABLE 1.3 PREFIXES FOR UNITS

MULTIPLICATION FACTOR	PREFIX	SYMBOL
10^{21}	zetta	Z
10^{18}	exa	E
10^{15}	peta	P
10^{12}	tera	T
10^{9}	giga	G
10^{6}	mega	M
10^{3}	kilo	k
10^{-3}	milli	m
10^{-6}	micro	μ
10^{-9}	nano	n
10^{-12}	pico	p
10^{-15}	femto	f
10^{-18}	atto	a
10^{-21}	zepto	z

TABLE 1.4 MULTIPLES AND SUBMULTIPLES OF THE FOOT

mile	1 mi = 5280 ft = 1609.38 m
yard	1 yd = 3 ft = 0.9144 m
foot	1 ft = 0.3048 m
inch	1 in. = $\frac{1}{12}$ ft = 2.540 cm
mil	1 mil = 0.001 in.

✔ Checkup 1.2

QUESTION 1: How many centimeters are there in one kilometer? How many millimeters in a kilometer?

QUESTION 2: How many microns are there in a fermi?

QUESTION 3: How many microns are there in an angstrom?

(A) 10^6 (B) 10^4 (C) 10^{-4} (D) 10^{-6}

1.3 THE UNIT OF TIME

The unit of time is the second. Originally one second was defined as $1/(60 \times 60 \times 24)$, or 1/86 400, of a mean solar day. The solar day is the time interval required for the Earth to complete one rotation relative to the Sun. The length of the solar day depends on the rate of rotation of the Earth, which is subject to a host of minor variations, both seasonal and long-term, which make the rotation of the Earth an imperfect timekeeper.

To avoid any variation in the unit of time, we now use an atomic standard of time. This standard is the period of one vibration of microwaves emitted by an atom of cesium. The **second** (1 s) is defined as the time needed for 9 192 631 770 vibrations of a cesium atom. Figure 1.9 shows one of the atomic clocks at the National Institute of Standards and Technology in Boulder, Colorado. In this clock, the feeble vibrations of the cesium atoms are amplified to a level that permits them to control the dial of the clock. Good cesium clocks are very, very good—they lose or gain no more than 1 second in 20 million years.

Precise time signals keyed to the cesium atomic clocks of the National Institute of Standards and Technology are continuously transmitted by radio station WWV, Fort Collins, Colorado. These time signals can be picked up worldwide on shortwave receivers tuned to 2.5, 5, 10, 15, or 20 megahertz. Precise time signals are also announced continuously by telephone [in the United States, the telephone number is (303) 499-7111; the precise time is also available online at www.time.gov]. The time announced on the radio and on the telephone is Coordinated Universal Time, or Greenwich time, which is exactly 5 hours ahead of Eastern Standard Time.

Table 1.5 lists some typical time intervals, and Table 1.6 gives multiples and submultiples of the second.

FIGURE 1.9 Cesium atomic clock at the National Institute of Standards and Technology.

TABLE 1.5 SOME TIME INTERVALS

Age of the Universe	$\approx 4 \times 10^{17}$ s
Age of the Solar System	1.4×10^{17} s
Age of oldest known fossils	1.1×10^{17} s
Age of human species	7.9×10^{12} s
Age of the oldest written records (Sumerian)	1.6×10^{11} s
Life span of man (average)	2.2×10^{9} s
Travel time for light from nearest star	1.4×10^{8} s
Revolution of Earth (1 year)	3.2×10^{7} s
Rotation of Earth (1 day)	8.6×10^{4} s
Life span of free neutron (average)	9.2×10^{2} s
Travel time for light from Sun	5×10^{2} s
Travel time for light from Moon	1.3 s
Period of heartbeat (human)	≈ 0.9 s
Period of sound wave (middle C)	3.8×10^{-3} s
Period of radio wave (AM band)	$\approx 1 \times 10^{-6}$ s
Period of light wave	$\approx 2 \times 10^{-15}$ s
Life span of shortest-lived unstable particle	$\approx 10^{-24}$ s

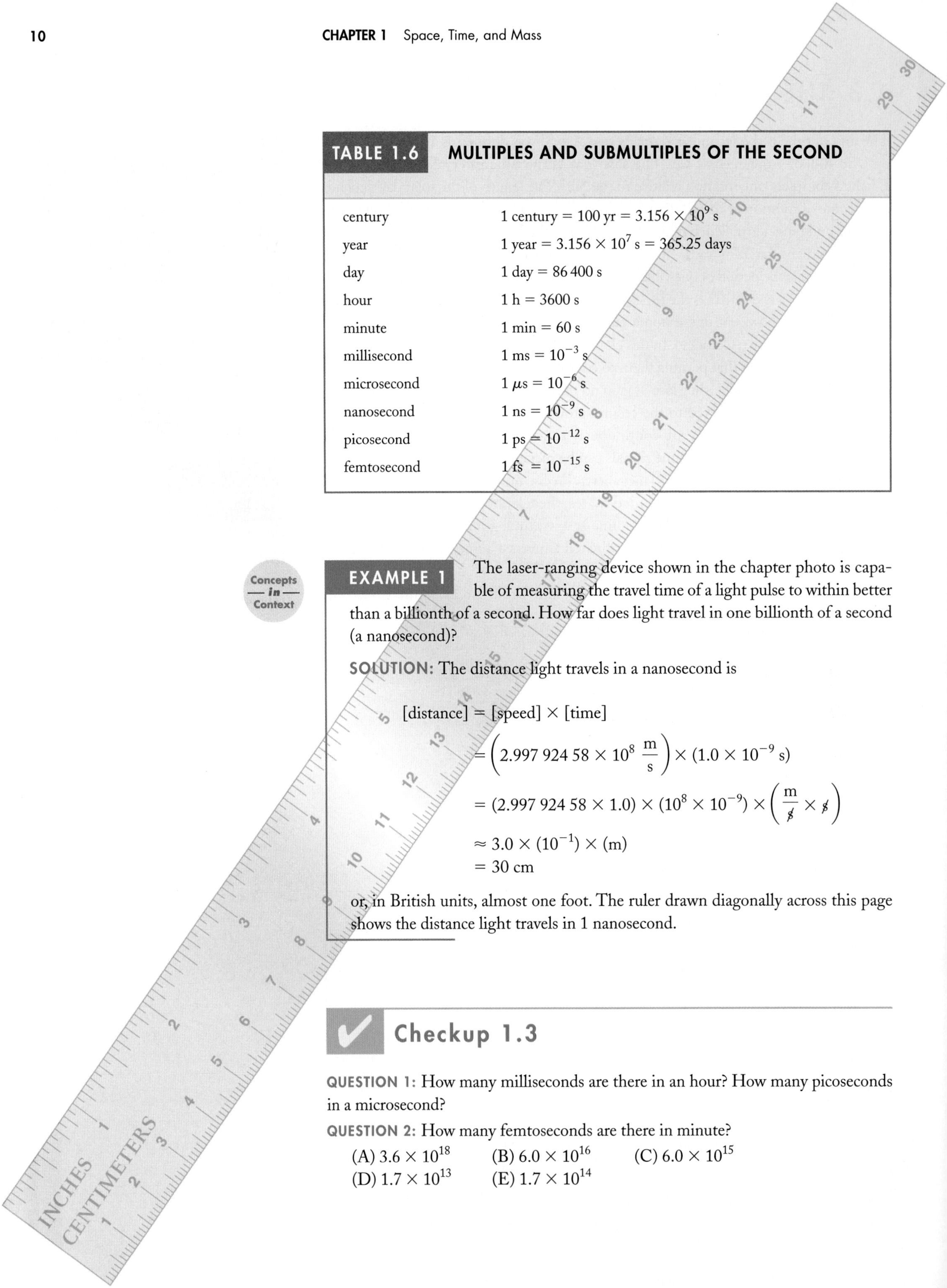

TABLE 1.6 MULTIPLES AND SUBMULTIPLES OF THE SECOND

century	1 century = 100 yr = 3.156×10^9 s
year	1 year = 3.156×10^7 s = 365.25 days
day	1 day = 86 400 s
hour	1 h = 3600 s
minute	1 min = 60 s
millisecond	1 ms = 10^{-3} s
microsecond	1 μs = 10^{-6} s
nanosecond	1 ns = 10^{-9} s
picosecond	1 ps = 10^{-12} s
femtosecond	1 fs = 10^{-15} s

Concepts — in — Context

EXAMPLE 1 The laser-ranging device shown in the chapter photo is capable of measuring the travel time of a light pulse to within better than a billionth of a second. How far does light travel in one billionth of a second (a nanosecond)?

SOLUTION: The distance light travels in a nanosecond is

$$\begin{aligned}[\text{distance}] &= [\text{speed}] \times [\text{time}] \\ &= \left(2.997\,924\,58 \times 10^8\ \frac{\text{m}}{\text{s}}\right) \times (1.0 \times 10^{-9}\ \text{s}) \\ &= (2.997\,924\,58 \times 1.0) \times (10^8 \times 10^{-9}) \times \left(\frac{\text{m}}{\not{\text{s}}} \times \not{\text{s}}\right) \\ &\approx 3.0 \times (10^{-1}) \times (\text{m}) \\ &= 30\ \text{cm}\end{aligned}$$

or, in British units, almost one foot. The ruler drawn diagonally across this page shows the distance light travels in 1 nanosecond.

✔ Checkup 1.3

QUESTION 1: How many milliseconds are there in an hour? How many picoseconds in a microsecond?

QUESTION 2: How many femtoseconds are there in minute?

(A) 3.6×10^{18} (B) 6.0×10^{16} (C) 6.0×10^{15}
(D) 1.7×10^{13} (E) 1.7×10^{14}

1.4 THE UNIT OF MASS

The unit of mass is the kilogram. The standard of mass is a cylinder of platinum–iridium alloy kept at the International Bureau of Weights and Measures (see Fig. 1.10). The **kilogram** (1 kg) is defined as exactly equal to the mass of this cylinder. Mass is the only fundamental unit for which we do not, as yet, have an atomic standard.

Mass is measured with a balance, an instrument that compares the *weight* of an unknown mass with a known force, such as the weight of the standard mass or the pull of a calibrated spring. Weight is directly proportional to mass, and hence equal weights imply equal masses (the precise distinction between mass and weight will be spelled out in Chapter 5). Figure 1.11 shows a watt balance, an extremely accurate balance especially designed by the National Institute of Standards and Technology. The watt balance is effectively a spring balance, but instead of a mechanical spring suspension it uses a magnetic suspension with calibrated magnetic forces.

FIGURE 1.10 International standard kilogram.

To relate the mass of an atom to the kilogram mass we need to know **Avogadro's number** N_A, or the number of atoms per mole. One **mole** *of any chemical element (or any chemical compound) is that amount of matter containing as many atoms (or molecules) as there are atoms in exactly* 12 *grams of carbon*-12. The "atomic mass" of a chemical element (or the "molecular mass" of a compound) is the mass of one mole expressed in grams. Thus, according to the table of atomic masses in Appendix 8, one mole of carbon atoms (C) has a mass of 12.0 grams, one mole of hydrogen atoms (H) has a mass of 1.0 gram, one mole of oxygen atoms (O) has a mass of 16.0 grams, one mole of oxygen molecules (O_2) has a mass of 32.0 grams, one mole of water molecules (H_2O) has a mass of 18.0 grams, and so on.

The available experimental data yield the following value for N_A:

$$N_A = 6.022\,14 \times 10^{23} \text{ atoms or molecules per mole} \tag{1.2}$$

Since there are N_A atoms in one mole, the mass of one atom is the mass of one mole, or the "atomic mass," divided by N_A:

$$[\text{mass of atom}] = \frac{[\text{"atomic mass"}]}{N_A} \tag{1.3}$$

FIGURE 1.11 A high-precision balance.

Hence, the mass of, say, a carbon-12 atom is

$$[\text{mass of carbon-12 atom}] = \frac{12 \text{ grams}}{6.022\,14 \times 10^{23}} = 1.992\,65 \times 10^{-23} \text{ gram}$$
$$= 1.992\,65 \times 10^{-26} \text{ kg} \tag{1.4}$$

Masses of atoms are often measured in terms of the **atomic mass unit** (1 u), which is exactly $\frac{1}{12}$ the mass of a carbon-12 atom:

$$1 \text{ atomic mass unit} = 1 \text{ u} = \frac{1.992\,65 \times 10^{-26} \text{ kg}}{12} \tag{1.5}$$

That is,

$$1 \text{ u} = 1.660\,54 \times 10^{-27} \text{ kg} \tag{1.6}$$

Note that with this definition of the atomic mass unit, the "atomic mass," or the number of grams in one mole, necessarily has the same numerical value as the mass of one atom expressed in u. Thus, the carbon atom has a mass of 12 u, the hydrogen atom has a mass of 1.0 u, the oxygen atom has a mass of 16.0 u, and so on.

EXAMPLE 2 How many atoms are there in a 5-cent coin? Assume that the coin is made of nickel and has a mass of 5.2×10^{-3} kg, or 5.2 grams. Atomic masses appear in Appendix 8.

SOLUTION: We recall that the atomic mass is the mass of one atom expressed in u. According to the periodic table of chemical elements in Appendix 8, the atomic mass of nickel is 58.69. Thus, the mass of one nickel atom is 58.69 u, or, $58.69 \times 1.66 \times 10^{-27}$ kg $= 9.74 \times 10^{-26}$ kg. The number of atoms in our 5.2×10^{-3} kg is then

$$\frac{5.2 \times 10^{-3}\ \cancel{\text{kg}}}{9.74 \times 10^{-26}\ \cancel{\text{kg}}/\text{atom}} = 5.3 \times 10^{22} \text{ atoms}$$

Table 1.7 gives some examples of masses expressed in kilograms, and Table 1.8 lists multiples and submultiples of the kilogram. In the British system of units, the unit of mass is the **pound**, which is exactly 0.453 592 37 kg.

For a quick grasp of the rough relationship between British and metric units, remember the following approximate equalities, good to within ±10% (more exact conversion factors are tabulated in Appendix 7):

$$1 \text{ yard} \approx 1 \text{ m}$$
$$1 \text{ mile} \approx 1.6 \text{ km}$$
$$1 \text{ pound} \approx \tfrac{1}{2} \text{ kg}$$
$$1 \text{ quart} \approx 1 \text{ liter}$$
$$1 \text{ gallon} \approx 4 \text{ liters}$$

TABLE 1.7 SOME MASSES

Observable Universe	$\approx 10^{53}$ kg
Galaxy	4×10^{41} kg
Sun	2.0×10^{30} kg
Earth (a)	6.0×10^{24} kg
Ship *Queen Elizabeth*	7.6×10^{7} kg
Jet airliner (Boeing 747, empty)	1.6×10^{5} kg
Automobile (b)	1.5×10^{3} kg
Man (average male)	73 kg
Apple (c)	0.2 kg
5¢ coin	5.2×10^{-3} kg
Raindrop	2×10^{-6} kg
Red blood cell (d)	9×10^{-14} kg
Smallest virus (potato spindle)	4×10^{-21} kg
Atom (iron)	9.5×10^{-26} kg
Proton	1.7×10^{-27} kg
Electron	9.1×10^{-31} kg

(a)

(b)

(c)

(d)
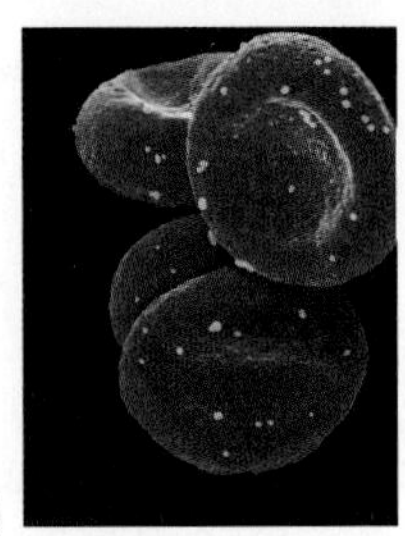

TABLE 1.8 MULTIPLES AND SUBMULTIPLES OF THE KILOGRAM	
metric ton (tonne)	$1\ \text{t} = 10^3\ \text{kg}$
kilogram	$1\ \text{kg}$
gram	$1\ \text{g} = 10^{-3}\ \text{kg}$
milligram	$1\ \text{mg} = 10^{-6}\ \text{kg}$
microgram	$1\ \mu\text{g} = 10^{-9}\ \text{kg}$
atomic mass unit	$1\ \text{u} = 1.66 \times 10^{-27}\ \text{kg}$
pound	$1\ \text{lb} = 0.454\ \text{kg}$
ounce ($\frac{1}{16}$ lb)	$1\ \text{oz} = 28.3\ \text{g}$
ton (2000 lb)	$1\ \text{ton} = 907\ \text{kg}$

To obtain a better feel for metric units, it helps to know that:

- The height of a person is typically 1.6 to 1.8 m.
- The mass of a person is typically 60 to 75 kg.
- The distance from the centerline of the body to the end of the outstretched arm is about 1 m.

✔ Checkup 1.4

QUESTION 1: How many grams are there in 1 metric ton? How many metric tons in one milligram?

QUESTION 2: How many atomic mass units (u) are there in 1 kilogram?

(A) 1.66×10^{-26} (B) 1.66×10^{-23} (C) 6.02×10^{23} (D) 6.02×10^{26}

1.5 DERIVED UNITS

Online Concept Tutorial

The meter, the second, and the kilogram are the fundamental units, or **base units**, of the metric system of units. Any other physical quantity can be measured by introducing a **derived unit** *constructed by some combination of the base units.* For example, **area** can be measured with a derived unit that is the square of the unit of length; thus, in the metric system, the unit of area is the square meter ($1\ \text{m} \times 1\ \text{m} = 1\ \text{m}^2$), which is the area of a square, one meter on a side (Fig. 1.12a). And **volume** can be measured with a derived unit that is the cube of the unit of length; in the metric system, the unit of volume is the cubic meter ($1\ \text{m} \times 1\ \text{m} \times 1\ \text{m} = 1\ \text{m}^3$), which is the volume of a cube, one meter on a side (Fig. 1.12b). Tables 1.9 and 1.10 give multiples and submultiples of these units.

Similarly, **density**, or mass per unit volume, can be measured with a derived unit that is the ratio of the unit of mass and the unit of volume. In the metric system, the unit of density is the kilogram per cubic meter ($1\ \text{kg/m}^3$). For example, the density of water is $1000\ \text{kg/m}^3$, which means that one cubic meter of water has a mass of 1000 kilograms. We will see in later chapters that other physical quantities, such as speed, acceleration, force, etc., are also measured with derived units.

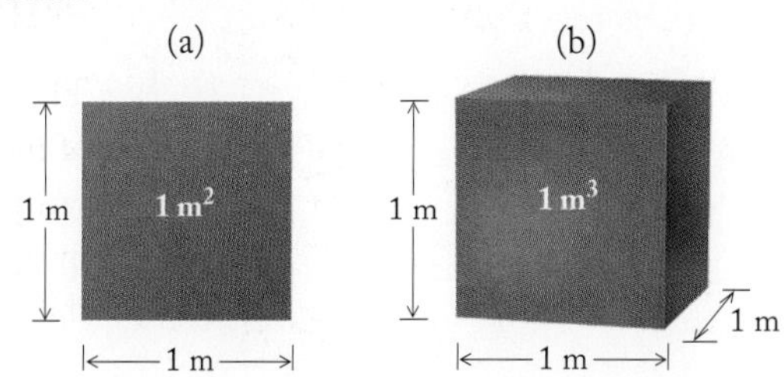

FIGURE 1.12 (a) One square meter. (b) One cubic meter.

TABLE 1.9 MULTIPLES AND SUBMULTIPLES OF THE SQUARE METER

square meter	$1\ m^2$
square kilometer	$1\ km^2 = 10^6\ m^2$
square centimeter	$1\ cm^2 = 10^{-4}\ m^2$
square millimeter	$1\ mm^2 = 10^{-6}\ m^2$

TABLE 1.10 MULTIPLES AND SUBMULTIPLES OF THE CUBIC METER

cubic meter	$1\ m^3$
cubic kilometer	$1\ km^3 = 10^9\ m^3$
liter	$1\ \text{liter} = 10^{-3}\ m^3 = 10^3\ cm^3$
cubic centimeter	$1\ cm^3 = 10^{-6}\ m^3$
cubic millimeter	$1\ mm^3 = 10^{-9}\ m^3$

The International System of Units, or SI, used in this book is the most widely accepted system of units in science and engineering. It is based on the meter, the second, and the kilogram, plus a special unit for temperature and a special unit for electric current.

✔ Checkup 1.5

QUESTION 1: How many square centimeters are there in a square meter? How many cubic centimeters are there in a cubic meter?

QUESTION 2: How many square millimeters are there in a square kilometer?

QUESTION 3: How many cubic millimeters are in a cubic kilometer?

(A) 10^{-18} (B) 10^{-6} (C) 10^6 (D) 10^{18}

Online Concept 2 Tutorial

1.6 SIGNIFICANT FIGURES; CONSISTENCY OF UNITS AND CONVERSION OF UNITS

Significant Figures

The numbers in Tables 1.1, 1.5, and 1.7 are written in scientific notation, with powers of ten. This not only has the advantage that very large or very small numbers can be written compactly, but it also serves to indicate the precision of the numbers. For instance, a scientist observing the 1998 Berlin marathon at which Ronaldo da Costa set the world record of 2 h 6 min 5.0 s would have reported the running time as 7.5650 ×

10^3 s, or 7.565×10^3 s, or 7.57×10^3 s, or 7.6×10^3 s, depending on whether the measurement of time was made with a stopwatch, or a wristwatch with a second hand (but no stop button), or a wristwatch without a second hand, or a "designer" watch with one of those daft blank faces without any numbers at all. The first of these watches permits measurements to within $\frac{1}{10}$ s, the second to within about 1 s, the third to within 10 or 20 s, and the fourth to within 1 or 2 minutes (if the scientist is good at guessing the position of the hand on the blank face). We will adopt the rule that only as many digits, or **significant figures**, are to be written down as are known to be fairly reliable. In accordance with this rule, the number 7.5650×10^3 s comprises five significant figures, of which the last (0) represents tenths of a second; the number 7.565 $\times 10^3$ s comprises four significant figures, of which the last (5) represents seconds; and so on. Thus, the scientific notation gives us an immediate indication of the precision to within which the number has been measured.

When numbers in scientific notation are multiplied or divided in calculations, the final result should always be rounded off so that it has no more significant figures than the original numbers, because the final result can be no more accurate than the original numbers on which it is based. Thus, the result of multiplying 7.57×10^3 s by 7.57×10^3 s is $5.73049 \times 10^4\ s^2$, which should be rounded off to $5.73 \times 10^4\ s^2$, because we were given only three significant figures in the original numbers. When numbers are added or subtracted, the result should be rounded to the largest decimal place among the last digits of the original numbers. Thus, $89.23 + 5.7 = 94.93$ should be rounded to 94.9, because one of the original numbers is known only to the tenths place.

EXAMPLE 3 A surveyor's laser-ranging device measures a time interval of 1.176×10^{-6} s for a laser light pulse to make a round trip from a marker. What is the round-trip distance, expressed with the correct number of significant figures?

Concepts — in — Context

SOLUTION: The distance light travels in a time interval of 1.176×10^{-6} s is

$$\begin{aligned}[\text{distance}] &= [\text{speed}] \times [\text{time}] \\ &= \left(2.997\,924\,58 \times 10^8\ \frac{\text{m}}{\not{\text{s}}}\right) \times (1.176 \times 10^{-6}\ \not{\text{s}}) \qquad (1.7) \\ &= 352.6\ \text{m}\end{aligned}$$

where the result which appears on a calculator, 352.555 9306 m, has been rounded in the last step to 352.6 m, that is, to the same number of digits as the least accurately known factor from which it was calculated. This fraction-of-a-meter accuracy agrees with the distance calculated for a nanosecond time interval in Example 1.

Sometimes even a number known to many significant figures is rounded off to fewer significant figures for the sake of convenience, when high accuracy is not required. For instance, the exact value of the speed of light is $2.997\,924\,58 \times 10^8$ m/s, but for most purposes, it is adequate to round this off to 3.00×10^8 m/s, and we will often employ this approximate value of the speed of light in our calculations.[3]

[3] When rounding a number, the following rule is employed (for example, in your handheld calculator): if the first of the digits to be rounded off is from 5 to 9, the prior digit is increased by one ("rounded up"); if the first digit of those to be rounded is 0 to 4, the prior digit is unchanged ("rounded down").

Consistency of Units

In all the equations of physics, the units on the left and the right sides of the equation must be consistent. This consistency is illustrated by the calculations in Example 3, where we see that on the right side of Eq. (1.7) the units of time cancel, and the final result for the right side then has units of length, in agreement with the units for the left side, which is a distance and requires units of length. It is a general rule that in any calculation with the equations of physics, the units can be multiplied and divided as though they were algebraic quantities, and this automatically yields the correct units for the final result. This requirement of consistency of units in the equations of physics can be reformulated in a more general way as a requirement of consistency of dimensions. In this context, the **dimensions** of a physical quantity are said to be length, time, mass, or some product or ratio of these if the units of this physical quantity are those of length, time, mass, or some product or ratio of these. Thus, volume has the dimensions of $[\text{length}]^3$, density has the dimensions of $[\text{mass}]/[\text{length}]^3$, speed has the dimensions of [length]/[time], and so on. In any equation of physics, *the dimensions of the two sides of the equation must be the same.* For instance, we can test the consistency of Eq. (1.7) by examining the dimensions of the quantities appearing in this equation:

$$[\text{length}] = \frac{[\text{length}]}{[\text{time}]} \times [\text{time}] \tag{1.8}$$

Dimensions are often used in preliminary tests of the consistency of equations, when there is some suspicion of a mistake in the equation. A test of the consistency of dimensions tells us no more than a test of the consistency of units, but has the advantage that we need not commit ourselves to a particular choice of units, and we need not worry about conversions among multiples and submultiples of the units. Bear in mind that if an equation fails this consistency test, it is proved wrong; but if it passes, it is not proved right.

Dimensions are sometimes used to find relationships between physical quantities. Such a determination of the appropriate proportionality between powers of relevant quantities is called **dimensional analysis**. Such analysis is performed by requiring the consistency of dimensions of units on each side of an equation. Dimensional analysis will prove useful when we have become familiar with more physical quantities and their dimensions.

The disastrous end of the space mission *Mars Climate Orbiter* on December 3, 1999 (see Fig. 1.13), teaches us a lesson on the importance of always attaching units to a number. Engineers at Lockheed Martin provided spacecraft operating data needed for navigation in British units rather than metric units. Flight controllers assumed the data were in metric units, and thus the probe did not behave as intended when the relevant thrusters were fired near Mars. The \$155 000 000 mission became a total loss when the spacecraft entered the atmosphere and crashed instead of going into orbit around Mars.

FIGURE 1.13 An artist's conception of the *Mars Climate Orbiter* spacecraft.

Conversion of Units

In many calculations, it is necessary to convert quantities expressed in one set of units to another set of units. Such conversions involve no more than simple substitutions of the equivalent amounts in the two sets of units (a comprehensive list of equivalent amounts in different units can be found in Appendix 7). For example, the density of water is 1.000×10^3 kg/m^3. To express this in g/cm^3, we substitute 1 kg = 1000 g and 1 m = 100 cm, and we find

$$1.000 \times 10^3 \, \frac{\text{kg}}{\text{m}^3} = 1.000 \times 10^3 \times \frac{1000 \text{ g}}{(100 \text{ cm})^3} = 1.000 \times 10^3 \times \frac{10^3 \text{ g}}{10^6 \text{ cm}^3}$$

$$= 1.000 \, \frac{\text{g}}{\text{cm}^3}$$

An alternative method for the conversion of units from one set of units to another takes advantage of multiplication by factors that are identically equal to 1. Since 1 kg = 1000 g, we have the identity

$$1 = \frac{1000 \text{ g}}{1 \text{ kg}}$$

and, similarly,

$$1 = \frac{1 \text{ m}}{100 \text{ cm}}$$

This means that any quantity can be multiplied by (1000 g)/(1 kg) or (1 m)/(100 cm) without changing its value. Thus, starting with 1.000×10^3 kg/m^3, we obtain

$$1.000 \times 10^3 \, \frac{\text{kg}}{\text{m}^3} = 1.000 \times 10^3 \, \frac{\text{kg}}{\text{m}^3} \times \frac{1000 \text{ g}}{1 \text{ kg}} \times \frac{1 \text{ m}}{100 \text{ cm}} \times \frac{1 \text{ m}}{100 \text{ cm}} \times \frac{1 \text{ m}}{100 \text{ cm}}$$

$$= 1.000 \times 10^3 \times 1000 \times \frac{1}{100} \times \frac{1}{100} \times \frac{1}{100} \times \frac{\cancel{\text{kg}}}{\cancel{\text{m}^3}} \times \frac{\text{g}}{\cancel{\text{kg}}} \times \frac{\cancel{\text{m}^3}}{\text{cm}^3}$$

Multiplying this out, and canceling the kg and m^3, we find

$$1.000 \times 10^3 \, \frac{\text{kg}}{\text{m}^3} = 1.000 \, \frac{\text{g}}{\text{cm}^3} \tag{1.9}$$

Ratios such as (1000 g)/(1 kg) or (1 m)/(100 cm) that are identically equal to 1 are called **conversion factors**. To change the units of a quantity, simply multiply the quantity by one or several conversion factors that will bring about the desired cancellation of the old units.

The ratio of two quantities with identical dimensions or units will have no dimensions at all. For example, the **slope** of a path relative to the horizontal direction is defined as the ratio of the increment of height to the increment of horizontal distance. Since this is the ratio of two lengths, it is a dimensionless quantity. Likewise, the sine of an angle is defined as the ratio of two lengths; in a right triangle, the sine of one of the acute angles is equal to the length of the opposite side divided by the length of the hypotenuse (see Math Help: Trigonometry of the Right Triangle for a review). Thus the slope and the sine, cosine, and tangent of an angle are examples of **dimensionless quantities**.

EXAMPLE 4 Immediately after takeoff, a jet airliner climbs away from the runway at an upward angle of 12° (see Fig. 1.14). What is the slope of the path of ascent of the airliner? What altitude does it reach at a horizontal distance of 2000 m, or 2.0 km, from the point of takeoff?

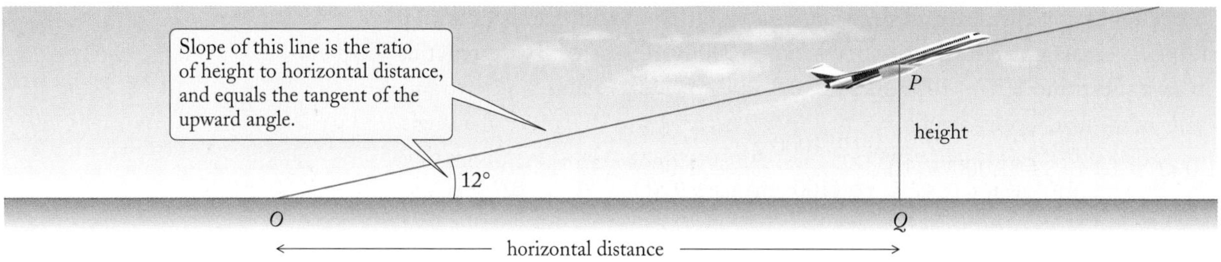

FIGURE 1.14 A jet airliner climbs after takeoff.

SOLUTION: If P is some point on the path at a horizontal distance OQ and a height PQ (see Fig. 1.13), then the slope of the path is

$$[\text{slope}] = \frac{[\text{height}]}{[\text{horizontal distance}]} = \frac{PQ}{OQ} \tag{1.10}$$

Trigonometry tells us that in the right triangle OPQ, the ratio PQ/OQ is the tangent of the angle θ (see the Math Help box). Therefore

$$[\text{slope}] = \tan\theta$$

With our calculator, we find that the tangent of 12° is 0.21. Hence,

$$[\text{slope}] = 0.21$$

This dimensionless number means that the airliner climbs 0.21 m for every 1 m it advances in the horizontal direction. Slopes are often quoted as ratios; thus a slope of 0.21 can be expressed in the alternative form 21:100.

By proportions, the height reached for 2000 m of horizontal advance must be 2000 times as large as the height for 1 m of horizontal advance; that is,

$$[\text{height}] = 2000\text{ m} \times 0.21 = 4.2 \times 10^2\text{ m}$$

PROBLEM-SOLVING TECHNIQUES — UNITS AND SIGNIFICANT FIGURES

In all calculations with the equations of physics, always include the units in your calculations, and multiply and divide them as though they were algebraic quantities. This will automatically yield the correct units for the final result. If it does not, you have made some mistake in the calculation. Thus, it is always worthwhile to keep track of the units in calculations, because this provides some extra protection against costly mistakes. A failure of the expected cancellations is a sure sign of trouble!

If it is necessary to change the units of a quantity, either substitute the old units for equal amounts of new units, or else multiply the old units by whatever conversion factors will bring about the cancellation of the old units.

Always round off your final result to as many significant figures as specified in the given data. For instance, in Example 3, we rounded off the final result to four significant figures, since four significant figures were specified in the time interval measured by the device. Any additional significant figures in the final result would be unreliable and misleading. In fact, even the fourth significant figure in the answer [Eq. (1.7)] is not quite reliable—the calculation from a time specified to 1 nanosecond accuracy would give a distance accurate to only 0.3 m, as in Example 1. It is always wise to doubt the accuracy of the last significant figure, in the final result and (sometimes) also in the initial data.

EXAMPLE 5 We can obtain a rough estimate of the size of a molecule by means of the following simple experiment. Take a droplet of oil and let it spread out on a smooth surface of water. When the oil slick attains its maximum area, it consists of a monomolecular layer; that is, it consists of a single layer of oil molecules which stand on the water surface side by side. Given that an oil droplet of mass 8.4×10^{-7} kg and of density 920 kg/m^3 spreads out into an oil slick of maximum area 0.55 m^2, calculate the length of an oil molecule.

SOLUTION: The volume of the oil droplet is

$$[\text{volume}] = \frac{[\text{mass}]}{[\text{density}]}$$

$$= \frac{8.4 \times 10^{-7}\ \text{kg}}{920\ \text{kg/m}^3} = 9.1 \times 10^{-10}\ \text{m}^3 \qquad (1.11)$$

The volume of the oil slick must be exactly the same. This latter volume can be expressed in terms of the thickness and the area of the oil slick:

$$[\text{volume}] = [\text{thickness}] \times [\text{area}]$$

Consequently,

$$[\text{thickness}] = \frac{[\text{volume}]}{[\text{area}]}$$

$$= \frac{9.1 \times 10^{-10}\ \text{m}^3}{0.55\ \text{m}^2} = 1.7 \times 10^{-9}\ \text{m} \qquad (1.12)$$

Since we are told that the oil slick consists of a single layer of molecules standing side by side, the length of a molecule is the same as the calculated thickness, 1.7×10^{-9} m.

MATH HELP TRIGONOMETRY OF THE RIGHT TRIANGLE

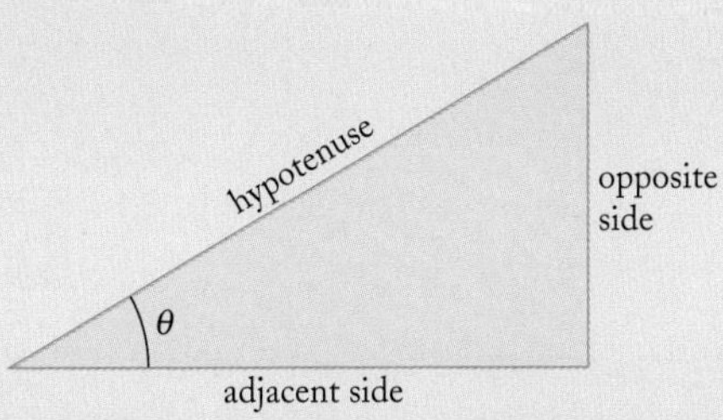

If θ is one of the acute angles of a right triangle, the side facing this angle is the *opposite side,* the side next to the angle is the *adjacent side,* and the side facing the right angle is the *hypotenuse.*

The figure shows a right triangle with an angle θ, its opposite and adjacent sides, and the hypotenuse. The sine, cosine, and tangent of θ are defined as follows:

$$\sin\theta = \frac{[\text{opposite side}]}{[\text{hypotenuse}]}$$

$$\cos\theta = \frac{[\text{adjacent side}]}{[\text{hypotenuse}]}$$

$$\tan\theta = \frac{[\text{opposite side}]}{[\text{adjacent side}]}$$

The Pythagorean theorem states:

$$[\text{hypotenuse}]^2 = [\text{opposite side}]^2 + [\text{adjacent side}]^2$$

This theorem implies that $1 = \sin^2\theta + \cos^2\theta$.

In principle, the numerical value of the sine, cosine, or tangent of any angle can be found by laying out a right triangle with this angle, measuring its sides, and evaluating the ratios given in the definitions. In practice, numerical values of tangents, cosines, and sines are obtained from handheld electronic calculators.

Appendix 3 gives a further review of trigonometry.

Checkup 1.6

QUESTION 1: If you use an ordinary ruler, marked in millimeters, to how many significant figures can you measure the length, width, and thickness of your textbook? Accordingly, to how many significant figures can you calculate the volume?

QUESTION 2: How many significant figures are there in the following numbers: 7.3, 1.24, 12.4, 4.85×10^6?

QUESTION 3: You multiply 7.3 and 1.24. How many significant figures are there in the result?

QUESTION 4: You add 73 and 1.2×10^2. What is the result? How many significant figures are there in the result?

QUESTION 5: What is the conversion factor for converting m to km? km to m? cm to km? m^2 to km^2? m^3 to km^3? m/s to mi/h?

QUESTION 6: What is the conversion factor for converting m/s to km/h?

(A) $(1\ \text{km}/10^3\ \text{m}) \times (3600\ \text{s}/1\ \text{h})$
(B) $(10^3\ \text{km}/1\ \text{m}) \times (3600\ \text{s}/1\ \text{h})$
(C) $(10^3\ \text{m}/1\ \text{km}) \times (3600\ \text{s}/1\ \text{h})$
(D) $(10^3\ \text{m}/1\ \text{km}) \times (1\ \text{h}/3600\ \text{s})$
(E) $(10^3\ \text{m}/1\ \text{km}) \times (3600\ \text{h}/1\ \text{s})$

SUMMARY

PROBLEM-SOLVING TECHNIQUES Units and Significant Figures **(page 18)**

MATH HELP Trigonometry of the Right Triangle **(page 19)**

IDEAL PARTICLE A pointlike mass, whose motion can be described completely by giving its position as a function of time

REFERENCE FRAME A coordinate grid with a set of synchronized clocks

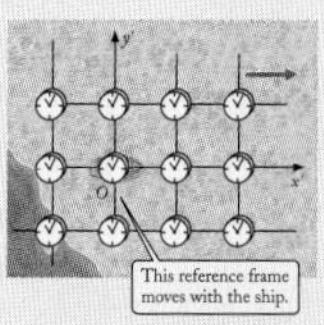

SI UNITS OF LENGTH, TIME, AND MASS Meter, second, kilogram

STANDARDS OF LENGTH, TIME, AND MASS Speed of light, cesium atomic clock, and standard cylinder of platinum–iridium

AVOGADRO'S NUMBER $N_A = 6.022 \times 10^{23}$ atoms or molecules per mole **(1.2)**

ATOMIC MASS UNIT $1\ \text{u} = 1.66 \times 10^{-27}\ \text{kg}$ **(1.5)**

MOLE That amount of matter containing as many atoms (or molecules) as in exactly 12 g of carbon-12

DERIVED UNIT A unit constructed from some combination of the base units of length, time, and mass

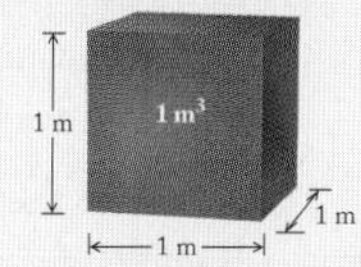

SIGNIFICANT FIGURES The digits in a number that are known with certainty (the last such digit is often not entirely reliable). When multiplying or dividing two or more quantities, the result has the same number of significant figures as the least number in the original quantities. When adding or subtracting two or more quantities, the number of significant figures in the result is determined by the largest decimal place among the last digits in the original quantities.

CONSISTENCY OF UNITS In any equation, the dimensions (the powers of length, time, and mass) on each side of the equation must be the same.

CONVERSION FACTORS Ratios that are identically equal to 1, used as factors to change the units of a quantity

QUESTIONS FOR DISCUSSION

1. Try to estimate by eye the lengths, in centimeters or meters, of a few objects in your immediate environment. Then measure them with a ruler or meterstick. How good were your estimates?
2. How close is your watch to standard time right now? Roughly how many minutes does your watch gain or lose per month?
3. What is meant by the phrase *a point in time*?
4. Mechanical clocks (with pendulums) were not invented until the tenth century A.D. What clocks were used by the ancient Greeks and Romans?
5. By counting aloud "One Mississippi, two Mississippi, three Mississippi," etc., at a fairly fast rate, you can measure seconds reasonably accurately. Try to measure 30 seconds in this way. How good a timekeeper are you?
6. Pendulum clocks are affected by the temperature and pressure of air. Why?
7. In 1761 an accurate chronometer built by John Harrison (see Fig. 1.15) was tested aboard HMS *Deptford* during a voyage at sea for 5 months. During this voyage, the chronometer accumulated an error of less than 2 minutes. For this achievement, Harrison was ultimately awarded a prize of £20 000 that the British government had offered for the discovery of an accurate method for the determination of geographical longitude at sea. Explain how the navigator of a ship uses a chronometer and observation of the position of the Sun in the sky to find longitude.

FIGURE 1.15 Harrison's chronometer H.4.

8. Captain Lecky's *Wrinkles in Practical Navigation,* a famous nineteenth-century textbook of celestial navigation, recommends that each ship carry three chronometers for accurate timekeeping. What can the navigator do with three chronometers that cannot be done with two?
9. Suppose that by an "act of God" (or by the act of a thief) the standard kilogram at Sèvres were destroyed. Would this destroy the metric system?

10. Estimate the masses, in grams or kilograms, of a few bodies in your environment. Check the masses with a balance if you have one available.

11. Consider the piece of paper on which this sentence is printed. If you had available suitable instruments, what physical quantities could you measure about this piece of paper? Make the longest list you can and give the units. Are all these units derived from the meter, second, and kilogram?

12. Could we take length, time, and density as the three fundamental units? What could we use as a standard of density?

13. Could we take length, mass, and density as the three fundamental units? Length, mass, and speed?

PROBLEMS

1.2 The Unit of Length

1. What is your height in feet? In meters?

2. With a ruler, measure the thickness of this book, excluding the cover. Deduce the thickness of each of the sheets of paper making up the book.

3. A football field measures 100 yd × $53\frac{1}{3}$ yd. Express each of these lengths in meters.

4. If each step you take is 0.60 m, how many steps do you need to cover 1.0 km?

5. The pica is a unit of length used by printers and book designers; 1 pica = $\frac{1}{6}$ in., which is the standard distance between one line of typing produced by a typewriter and the next (single-spaced). How many picas long and wide is a standard sheet of paper, 11 in. long and $8\frac{1}{2}$ in. wide?

6. Express the last four entries in Table 1.1 in inches.

7. Express the following fractions of an inch in millimeters: $\frac{1}{2}$, $\frac{1}{4}$, $\frac{1}{8}$, $\frac{1}{16}$, $\frac{1}{32}$, and $\frac{1}{64}$ in.

8. Express one mil (one thousandth of an inch) in micrometers (microns). Express one millimeter in mils.

9. Analogies can often help us to imagine the very large or very small distances that occur in astronomy or in atomic physics.
   (a) If the Sun were the size of a grapefruit, how large would the Earth be? How far away would the nearest star be?
   (b) If your head were the size of the Earth, how large would an atom be? How large would a red blood cell be?

10. One of the most distant objects observed by astronomers is the quasar Q1208+1011, at a distance of 12.4 billion light-years from the Earth. If you wanted to plot the position of this quasar on the same scale as the diagram at the top of page xliii of the Prelude, how far from the center of the diagram would you have to place this quasar?

11. On the scale of the second diagram on page xl of the Prelude, what would have to be the size of the central dot if it were to represent the size of the Sun faithfully?

12. An *interferometer* uses the pattern created by mixing laser light waves in order to measure distances extremely accurately. The wavelength of the laser light waves used is 633 nanometers. A fiber-optic interferometer can measure a distance 10^{-6} times the size of a wavelength. How does such precision compare with the diameter of an atom?

*13. The *thread* of a screw is often described either in terms of the number of complete turns required for the screw to advance one inch (English units) or in terms of the number of millimeters advanced in one complete turn (metric units). For delicate adjustments, scientists often use screws either with a thread of 80 turns per inch or a thread of 0.5 mm per turn. Express each of these in terms of the number of micrometers the screw advances for a partial turn through an angle of 5°.

*14. A nautical mile (nmi) equals 1.151 mi, or 1852 m. Show that the distance of 1 nmi along a meridian of the Earth corresponds to a change in latitude of 1 minute of arc.

**15. A physicist plants a vertical pole at the waterline on the shore of a calm lake. When she stands next to the pole, its top is at eye level, 175 cm above the waterline. She then rows across the lake and walks along the waterline on the opposite shore until she is so far away from the pole that the entire view of it is blocked by the curvature of the surface of the lake; that is, the entire pole is below the horizon (Fig. 1.16). She finds that this happens when her distance from the pole is 9.4 km. From this information, deduce the radius of the Earth.

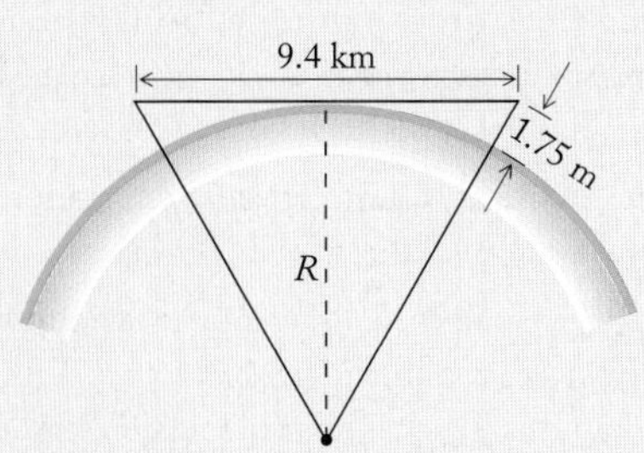

FIGURE 1.16 The distance between the physicist and the pole is 9.4 km.

1.3 The Unit of Time

16. What is your age in days? In seconds?

17. The age of the Earth is 4.5×10^9 years. Express this in seconds.

18. A computer can perform a single calculational step each nanosecond (10^{-9} s). How many steps can be performed in one hour?

19. An Olympic marathon record of 2 h 9 min 21 s was set by Carlos Lopes of Portugal in 1984. Express this time in seconds.

20. Joan Benoit of the United States set the women's Olympic marathon record in 1984 with a time of 2 h 24 min 52 s. Express this time in seconds.

21. The solar day is the interval for the Earth to complete one rotation in relation to the Sun, and the sidereal day is the interval for the Earth to complete one rotation in relation to distant stars. The solar day has exactly 24 hours. How many hours and minutes are there in one sidereal day? (Hint: 1 year is 365.24 solar days, but 366.24 sidereal days. Why?)

22. A mechanical wristwatch ticks 4 times per second. Suppose this watch runs for 10 years. How often does it tick in this time interval?

23. How many days is a million seconds?

24. How many hours are there in a week? How many seconds?

25. Your heart beats 71 times per minute. How often does it beat in a year?

*26. Each day at noon a mechanical wristwatch was compared with WWV time signals. The watch was not reset. It consistently ran late, as follows: June 24, late 4 s; June 25, late 20 s; June 26, late 34 s; June 27, late 51 s.

   (a) For each of the three 24-hour intervals, calculate the rate at which the wristwatch lost time. Express your answer in seconds lost per hour.

   (b) What is the average of the rates of loss found in part (a)?

   (c) When the wristwatch shows $10^{h}30^{m}$ on June 30, what is the correct WWV time? Do this calculation with the average rate of loss of part (b) and also with the largest rates of loss found in part (a). Estimate to within how many seconds the wristwatch can be trusted on June 30 after the correction for rate of loss has been made.

*27. The navigator of a sailing ship seeks to determine his longitude by observing at what time (Coordinated Universal Time) the Sun reaches the zenith at his position (local noon). Suppose that the navigator's chronometer is in error and is late by 1.0 s compared with Coordinated Universal Time. What will be the consequent error of longitude (in minutes of arc)? What will be the error in position (in kilometers) if the ship is on the equator?

1.4 The Unit of Mass

28. What is your mass in pounds? In kilograms? In atomic mass units?

29. What percentage of the mass of the Solar System is in the planets? What percentage is in the Sun? Use the data given in the table printed inside the cover of this book.

30. What is the ratio of the largest to the smallest length listed in Table 1.1? The ratio of the longest to the shortest time in Table 1.5? The ratio of the largest to the smallest mass in Table 1.7? Do you see any coincidences (or near-coincidences) between these numbers? Some physicists have proposed that coincidences between these large numbers must be explained by new cosmological theories.

31. The atom of uranium consists of 92 electrons, each of mass 9.1×10^{-31} kg, and a nucleus. What percentage of the total mass is in the electrons and what percentage is in the nucleus of the atom?

32. A laboratory microbalance can measure a mass of one-tenth of a microgram, a very small speck of matter. How many atoms are there in such a speck of gold, which has 197 grams in one mole?

33. English units use the ordinary pound, also called the *avoirdupois* pound, to specify the mass of most types of things. However, the *troy* pound is often used to measure precious stones, precious metals, and drugs, where 1 troy pound = 0.822 86 avoirdupois pound. If we adopt these different pounds, how many grams are there in a troy pound of gold? In an avoirdupois pound of feathers?

34. Mechanical nano-oscillators can detect a mass change as small as 10^{-21} kg. How many atoms of iron (55.85 g/mole) must be deposited on such an oscillator to produce a measurable mass change?

35. (a) How many molecules of water are there in one cup of water? A cup is about 250 cm^3.

   (b) How many molecules of water are there in the ocean? The total volume of the ocean is 1.3×10^{18} m^3.

   (c) Suppose you pour a cup of water into the ocean, allow it to become thoroughly mixed, and then take a cup of water out of the ocean. On the average, how many molecules originally in the cup will again be in the cup?

*36. How many atoms are there in the Sun? The mass of the Sun is 1.99×10^{30} kg, and its chemical composition (by mass) is approximately 70% hydrogen and 30% helium.

*37. The chemical composition of air is (by mass) 75.5% N_2, 23.2% O_2, and 1.3% Ar. What is the average "molecular mass" of air; that is, what is the mass of 6.02×10^{23} molecules of air?

*38. How many atoms are there in a human body of 73 kg? The chemical composition (by mass) of a human body is 65% oxygen, 18.5% carbon, 9.5% hydrogen, 3.3% nitrogen, 1.5% calcium, 1% phosphorus, and 0.35% other elements (ignore the "other elements" in your calculation).

1.5 Derived Units†

1.6 Significant Figures; Conversion of Units†

39. As seen from Earth, the Sun has an angular diameter of 0.53°. The distance between the Earth and Sun is 1.5×10^{11} m. From this, calculate the radius of the Sun.

40. The light-year is the distance that light travels in one year. Express the light-year in meters.

41. The distance from our Galaxy to the Andromeda galaxy is 2.2×10^6 light-years. Express this distance in meters.

42. In analogy with the light-year, we can define the light-second as the distance light travels in one second and the light-minute as the distance light travels in one minute. Express the Earth–Sun distance in light-minutes. Express the Earth–Moon distance in light-seconds.

43. Astronomers often use the astronomical unit (AU), the parsec (pc), and the light-year. The AU is the distance from the Earth to the Sun;[4] 1 AU = 1.496×10^{11} m. The pc is the distance at which 1 AU subtends an angle of exactly one second of arc (Fig. 1.17). The light-year is the distance that light travels in one year.

    (a) Express the pc in AU.

    (b) Express the pc in light-years.

    (c) Express the pc in meters.

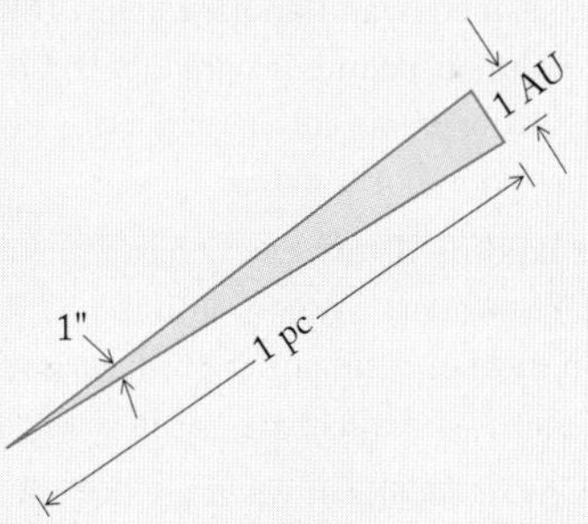

FIGURE 1.17 Geometry relating the astronomical unit (AU) to the parsec (pc).

44. How many square feet are there in a square meter?

45. How many cubic feet are there in a cubic meter?

46. A tennis court measures 78 ft × 27 ft. Calculate the area of this court. Express your result in square meters.

47. The tallest man was Robert Wadlow, who continued to grow throughout his life and attained 8 ft 11.1 in. a few days before his death in 1940. Express his height in meters. How many significant figures are there in your result?

48. A football field measures 100 yd × $53\frac{1}{3}$ yd. Calculate the area of this field; express your result in square meters.

49. The density of copper is 8.9 g/cm³. Express this in kg/m³, lb/ft³, and lb/in.³.

[4] Strictly, it is the semimajor axis of the Earth's orbit.

† For help, see Online Concept Tutorial 1 and 2 at www.wwnorton.com/physics

50. What is the volume of an average human body? (Hint: The density of the body is about the same as that of water.)

51. Your heart pumps 92 cm³ of blood per second (when you are resting). How much blood does it pump per day? Express the answer in m³.

52. As stated in the preceding problem, your heart pumps 97 cubic centimeters of blood per second. If your total volume of blood is 5.2 liters, what is the average travel time for your blood to complete one trip around your circulatory system?

53. Computers typically use many millions of transistors; each transistor occupies an area of approximately 10^{-6} m × 10^{-6} m = 10^{-12} m². How many such transistors can fit on a 1 cm² silicon chip? Future-generation computers may exploit a three-dimensional arrangement of transistors. If each layer of transistors is 10^{-7} m thick, how many transistors could fit in a 1-cm³ silicon cube?

54. Water has a density of 1.00 g/cm³. Express this in pounds per gallon.

55. Express the results of the following calculations in scientific notation with an appropriate number of significant figures:

    (a) $3.6 \times 10^4 \times 2.049 \times 10^{-2}$.

    (b) $2.581 \times 10^2 - 7.264 \times 10^1$.

    (c) $0.079832 \div 9.43$.

56. Our Sun has a radius of 7.0×10^8 m and a mass of 2.0×10^{30} kg. What is its average density? Express your answer in grams per cubic centimeter.

57. Pulsars, or neutron stars, typically have a radius of 20 km and a mass equal to that of the Sun (2.0×10^{30} kg). What is the average density of such a pulsar? Express your answer in metric tons per cubic centimeter.

58. The total volume of the oceans of the Earth is 1.3×10^{18} m³. What percentage of the mass of the Earth is in the oceans?

59. A fire hose delivers 300 liters of water per minute. Express this in m³/s. How many kilograms of water per second does this amount to?

60. Meteorologists usually report the amount of rain in terms of the depth in inches to which the water would accumulate on a flat surface if it did not run off. Suppose that 1 in. of rain falls during a storm. Express this in cubic meters of water per square meter of surface. How many kilograms of water per square meter of surface does this amount to?

61. The nuclei of all atoms have approximately the same density of mass. The nucleus of a copper atom has a mass of 1.06×10^{-25} kg and a radius of 4.8×10^{-15} m. The nucleus of a lead atom has a mass of 3.5×10^{-25} kg; what is its radius? The nucleus of an oxygen atom has a mass of 2.7×10^{-26} kg; what is its radius? Assume that the nuclei are spherical.

62. The table printed inside the book cover gives the masses and radii of the major planets. Calculate the average density of each planet and make a list of the planets in order of decreasing densities. Is there a correlation between the density of a planet and its distance from the Sun?

63. The roof of a house has a slope, or pitch, of 1:1 (that is, 45°). The roof has a complex shape, with several gables and dormers (see Fig. 1.18), but all roof surfaces have the same pitch. The area of the ground floor is 250 m^2. What is the area of the roof surface?

FIGURE 1.18 Roof of a house.

64. The Global Positioning System (GPS) used by navigators of ships and aircraft exploits radio signals from artificial satellites to determine the position of the ship or the aircraft. Portable GPS units for use on yachts (see Fig. 1.19) incorporate a radio receiver and a computer; they give the position to within ±15 m. What error in latitude angle corresponds to a north–south error of 15 m along the surface of the Earth?

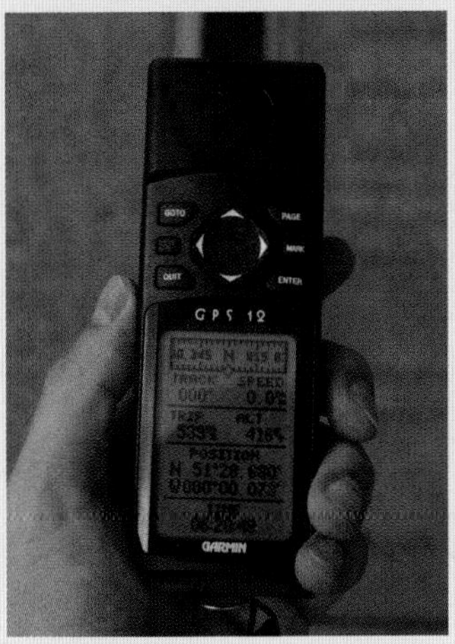

FIGURE 1.19 Global Positioning System (GPS) receiver.

65. Some crystals can be polished at an angle to produce sharp "atomic steps" (see Fig. 1.20.). What angle θ should be chosen to produce one step every five atoms? One step every ten atoms? How many atoms will there be along a step if a crystal is polished as flat as is normally practical, usually around 0.10°?

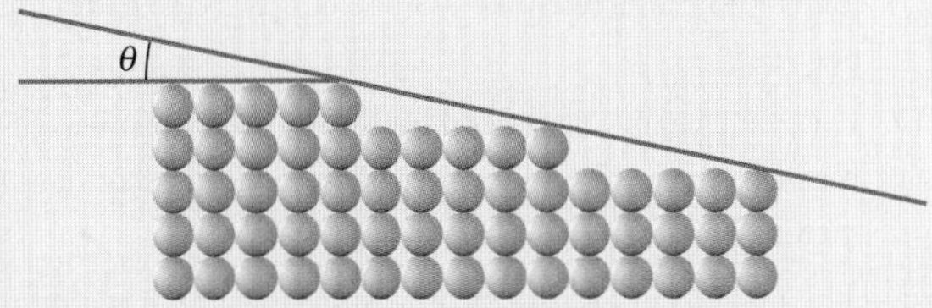

FIGURE 1.20 Atomic steps of a crystal.

66. You want to estimate the height of a skyscraper from the ground. To do so, you walk 150 steps (approximately 75 m) away from a vertical wall and, using a protractor, measure an angle of 78° that the line of sight to the top of the skyscraper makes with the horizontal. How tall is the skyscraper? How many significant figures are in your result?

67. In 1 astronomical year, or "tropical year," of 365.24 days, the Earth moves once around the Sun; that is, it moves 360° along its orbit and returns to the same point of the orbit. How far around the Sun, in degrees, does the Earth move in 1 calendar year of 365 days? How far, in degrees, does the Earth move in 4 consecutive calendar years (one of which is a leap year of 366 days)?

*68. In the Galápagos (on the equator) the small island of Marchena is 60 km west of the small island of Genovesa. If the sun sets at 8:00 P.M. at Genovesa, when will it set at Marchena?

*69. For tall trees, the diameter at the base (or the diameter at any given point of the trunk, such as the midpoint) is roughly proportional to the $\frac{3}{2}$ power of the length. The tallest sequoia in Sequoia National Park in California has a length of 81 m, a diameter of 7.6 m at the base, and a mass of 6100 metric tons. A petrified sequoia found in Nevada has a length of 90 m. Estimate its diameter at the base, and estimate the mass it had when it was still alive.

REVIEW PROBLEMS

70. The Earth is approximately a sphere of radius 6.37×10^6 m. Calculate the distance from the pole to the equator, measured along the surface of the Earth. Calculate the distance from the pole to the equator, measured along a straight line passing through the Earth.

71. The "atomic mass" of fissionable uranium is 235.0 g. What is the mass of a single uranium atom? Express your answer in kilograms and in atomic mass units.

72. How many water molecules are there in 1.0 liter of water? How many oxygen atoms? Hydrogen atoms?

73. How many molecules are there in one cubic centimeter of air? Assume that the density of air is 1.3 kg/m^3 and that it consists entirely of nitrogen molecules (N_2). The atomic mass of nitrogen is 14.0 g.

74. Normal human blood contains 5.1×10^6 red blood cells per cubic millimeter. The total volume of blood in a man of 70 kg is 5.2 liters. How may red blood cells does this man have?

75. An epoxy paint used for painting the hull of a ship is supposed to be applied at the rate of 1 liter of paint per 8 square meters of hull surface. How thick will the film of freshly applied paint be?

76. Smokestacks in the United States spew out about 8×10^6 metric tons of fly ash per year. If this stuff settles uniformly over all of the area of the United States (9.4×10^6 km^2), how many kilograms of fly ash will be deposited per square meter per year?

77. For many years, the federal highway speed limit was 55 mi/h. Express this in kilometers per hour, feet per second, and meters per second.

78. The nucleus of an iron atom is spherical and has a radius of 4.6×10^{-15} m; the mass of the nucleus is 9.5×10^{-26} kg. What is the density of the nuclear material? Express your answer in metric tons per cubic centimeter.

79. The longest officially verified human life span was attained by the Japanese man Shigechiyo Izumi, who died in 1986 at an age of 120 years and 237 days. Express this age in seconds. How many significant figures are there in your result?

80. A driveway up a hill has a slope of 1:9. How high does it ascend in 300 m of horizontal distance? What is the corresponding distance measured along the driveway?

81. A small single-engine plane is flying at a height of 5000 m at a (horizontal) distance of 18 km from the San Francisco airport when the engine quits. The pilot knows that, without the engine, the plane will glide downward at an angle of 15°. Can she reach San Francisco?

*82. You are crossing the Atlantic in a sailboat and hoping to make landfall in the Azores. The highest peak in the Azores has a height of 2300 m. From what distance can you see this peak just emerging over the horizon? Assume that your eye is (almost) at the level of the water.

*83. Some engineers have proposed that for long-distance travel between cities we should dig perfectly straight connecting tunnels through the Earth (see Fig. 1.21). A train running along such a tunnel would initially pick up speed in the first half of the tunnel as if running downhill; it would reach maximum speed at the midpoint of the tunnel; and it would gradually slow down in the second half of the tunnel, as if running uphill. Suppose that such a tunnel were dug between San Francisco and Washington, D.C. The distance between these cities, measured along the Earth's surface, is 3900 km.

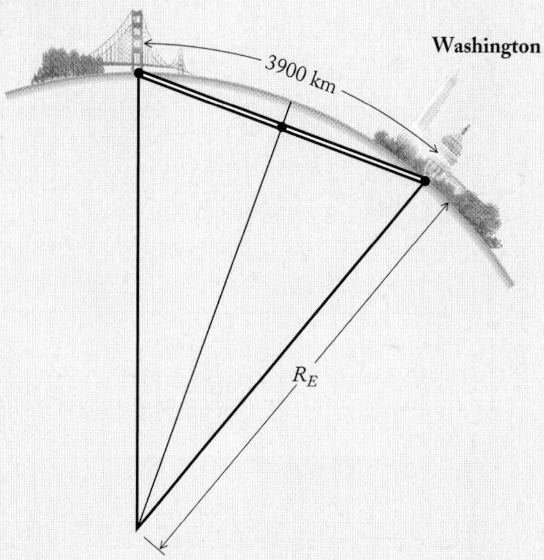

FIGURE 1.21 A proposed tunnel through the Earth.

(a) What is the distance along the straight tunnel?

(b) What is the depth of the tunnel at its midpoint, somewhere below Kansas?

(c) What is the downward slope of the tunnel relative to the horizontal direction at San Francisco?

Answers to Checkups

Checkup 1.1

1. Extended bodies have many characteristics that can be measured; for example, size and shape may be specified by measuring the bowling ball's diameter, surface area, or volume, with units of length, length-squared, and length-cubed, respectively. Other measurable quantities include density, hardness, temperature, color, and chemical composition; we will examine the units of such quantities in later chapters.

2. Since the origin of the x'–y' coordinate grid is shifted from the x–y origin by negative amount in the x direction, and a positive amount in the y direction any point will have larger values of x' compared to x and smaller values of y' compared to y, as in Fig. 1.3a.

3. A coordinate grid is used to specify positions in space. A reference frame includes clocks which specify the time when something occurs at a certain position.

Checkup 1.2

1. There are 100 centimeters in a meter, and there are 1000 meters in a kilometer, so there are $100 \times 1000 = 10^5$ centimeters in a kilometer. Similarly, with 10^3 millimeters in a meter, there are $10^3 \times 10^3 = 10^6$ millimeters in a kilometer.

2. Since there are 10^6 microns in a meter and there are 10^{-15} m in a fermi; there are $10^{-15} \times 10^6$ microns $= 10^{-9}$ microns in a fermi.

3. (C) 10^{-4}. There are 10^6 microns in a meter, and 10^{-10} m in an angstrom, so there are $10^{-10} \times 10^6$ microns $= 10^{-4}$ micron in an angstrom.

Checkup 1.3

1. There are $60 \times 60 = 3600$ seconds in an hour, and there are 1000 milliseconds in a second, so there are $1000 \times 3600 = 3.6 \times 10^6$ milliseconds in an hour. There are 10^{12} picoseconds in a second and 10^6 microseconds in a second, so there are 10^6 picoseconds in a microsecond.

2. (B) 6.0×10^{16}. There are 10^{15} femtoseconds in a second, and 60 seconds in a minute, so there are $60 \times 10^{15} = 6.0 \times 10^{16}$ femtoseconds in a minute.

Checkup 1.4

1. From Table 1.8, there are 10^3 kg in a metric ton. Since there are 10^3 grams in a kilogram, there are $10^3 \times 10^3 = 10^6$ grams in a metric ton. Furthermore, there are 10^9 milligrams in a metric ton, and therefore 10^{-9} ton in one milligram.

2. (D) 6.02×10^{26}. There are 1.66×10^{-27} kilograms in one u, so the number of u per kilogram is the inverse, 1 u/(1.66×10^{-27} kg) $= 6.02 \times 10^{26}$ u/kg.

Checkup 1.5

1. Since 1 m $= 10^2$ cm, then one square meter is $(1\text{ m})^2 = (10^2\text{ cm})^2 = 10^4\text{ cm}^2$. Similarly, a cubic meter is $(1\text{ m})^3 = (10^2\text{ cm})^3 = 10^6\text{ cm}^3$. These values can also be obtained directly from Tables 1.9 and 1.10.

2. Using Table 1.9, $1\text{ km}^2 = 10^6\text{ m}^2 \times (1\text{ mm}^2)/(10^{-6}\text{ m}^2) = 10^{12}\text{ mm}^2$.

3. (D) 10^{18}. From Table 1.10, $1\text{ km}^3 = 10^9\text{ m}^3 \times (1\text{ mm}^3)/(10^{-9}\text{ m}^3) = 10^{18}\text{ mm}^3$.

Checkup 1.6

1. Since the length and width are around 20 to 30 cm, a measurement of either to the nearest millimeter (1 mm = 0.1 cm) will provide three significant figures. The thickness is only a few centimeters, so its measurement to the nearest 0.1 cm will have two significant figures. The volume is the product of these three lengths, and will have only as many significant figures as the least number in the multiplied quantities; the volume thus has two significant figures.

2. One merely reads off the number of digits; the first number given has two significant figures, and the others have three.

3. The product has only as many significant figures as the least number of the multiplied quantities; here, the product has two significant figures.

4. When we add, only the largest decimal place among the last significant figures of the quantities added is significant in the result. Here, we must round to the nearest 10, since only the 10s digit was known in the second number given. Thus we write 1.9×10^2 for the sum.

5. For the conversion factors, we merely write the ratio of equal quantities that provides the desired units. For the given conversions, the factors are, respectively: 1 km/10^3 m; 10^3 m/1 km; 1 km/10^5 cm; 1 km^2/(10^3 m)^2; 1 km^3/(10^3 m)^3; (1 km/10^3 m) $\times$ (3600 s/1 h); (1 mi/1609 m) $\times$ (3600 s/1 h).

6. (A) (1 km/10^3 m) $\times$ (3600 s/1 h). When mulitplied by m/s, this both provides the desired units of km/h and contains correct conversion factors.

CHAPTER

2 Motion along a Straight Line

CONCEPTS IN CONTEXT

Concepts — *in* — Context

Aircraft are launched from the deck of an aircraft carrier by a catapult. In conjunction with the jet engines of the aircraft, the catapult quickly accelerates the aircraft to the speed required for takeoff. Typically, during such a "cat shot" the speed of the aircraft increases from zero to 260 km/h (160 mi/h) in just under two seconds. Some pilots find such extreme accelerations exhilarating.

The concepts of velocity and acceleration introduced in this chapter will permit us to answer the following questions:

? What is the magnitude of the acceleration during the launch? (Example 3, page 40)

? How is the distance traveled by the aircraft related to the acceleration and the time? (Example 5, page 44)

? How does the acceleration during launch compare with the acceleration of free fall? (Example 11, page 53)

The branch of physics that studies the motion of bodies is called **mechanics**. In antiquity, the science of mechanics (from the Greek *mechane,* machine) was the study of machines, and this is still what we have in mind when we call an automobile repairperson a mechanic. But physicists soon recognized that the essential aspect of the study of machines is motion, and mechanics thus became the study of motion. Broadly speaking, mechanics is divided into kinematics and dynamics. **Kinematics** deals with the mathematical description of motion in terms of position, velocity, and acceleration; **dynamics** deals with force as the cause of changes in motion.

In this chapter, and in the next nine chapters, we will be concerned only with the **translational motion** of a particle, that is, a change of position of the particle with time. In the case of an ideal particle—a pointlike mass of infinitesimal size—the dependence of position on time provides a complete description of the motion of the particle. In the case of a more complicated body—an automobile, a ship, or a planet—the dependence of position on time does not provide a complete description. Such a complicated body can rotate, or change its orientation in space; furthermore, the body has many internal parts which can move in relation to one another. We will examine such nontranslational motions later in the book. But insofar as we are not interested in the size, shape, orientation, and internal structure of a complicated body, we may find it useful to concentrate on its translational motion and ignore all rotational or internal motions. Under these circumstances, we may pretend that the motion of the complicated body is particle motion.

2.1 AVERAGE SPEED

If your automobile travels 160 km along a highway in 2 h, you would say that your average speed for the trip is 80 kilometers per hour. But this statement relies on the implicit assumption that the motion of the automobile can be regarded as particle motion, and that we ignore the size of the automobile and the complicated behavior of its internal machinery. To make the definition and the measurement of the average speed of the automobile precise and unambiguous, we must select some marker point that we imagine painted on the automobile. For instance, we might select the midpoint of the front bumper as our marker point. We reckon departure and arrival according to when the marker point crosses the starting line and the finishing line, and we reckon distance according to the path traced out by the marker point (see Fig. 2.1). We can then treat the motion of the automobile as particle motion.

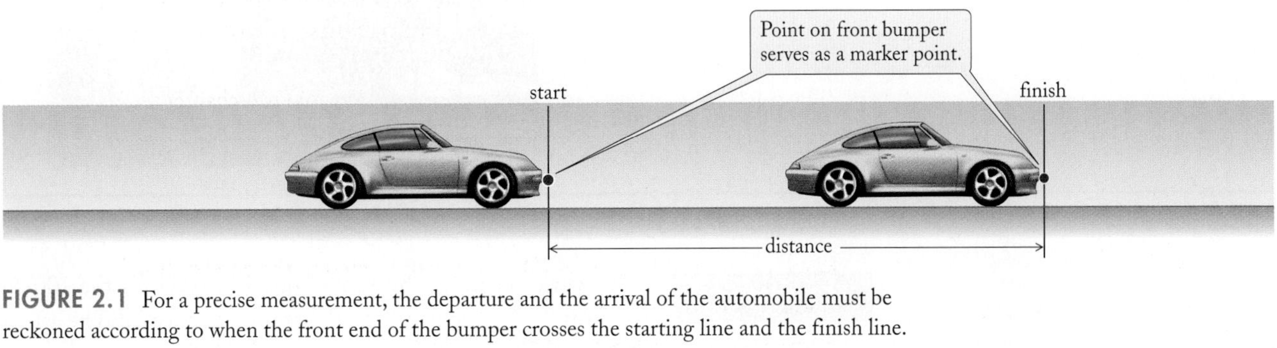

FIGURE 2.1 For a precise measurement, the departure and the arrival of the automobile must be reckoned according to when the front end of the bumper crosses the starting line and the finish line.

For a particle that travels along some path, straight or curved, the total distance traveled and the time taken to travel this distance have a precise, unambiguous meaning. The **average speed** *of the particle is defined as the ratio of this distance and the time taken*. Expressing this as a formula, we can write

average speed

$$[\text{average speed}] = \frac{[\text{total distance traveled}]}{[\text{time taken}]} \tag{2.1}$$

Thus, average speed is the rate of change of the distance, or the change of distance per unit time. We see from Eq. (2.1) that the unit of speed is the unit of length divided by the unit of time. In the SI system, *the unit of speed is the* **meter per second** (m/s). In practice, the speed of automobiles, aircraft, and other everyday objects is often measured in kilometers per hour (km/h):

$$1 \text{ km/h} = \frac{1000 \text{ m}}{3600 \text{ s}} = 0.278 \text{ m/s} \tag{2.2}$$

Table 2.1 gives some examples of typical speeds.

TABLE 2.1 SOME SPEEDS

Light	3.0×10^8 m/s
Recession of fastest known quasar	2.8×10^8 m/s
Electron around nucleus (hydrogen)	2.2×10^6 m/s
Earth around Sun	3.0×10^4 m/s
Rifle bullet (muzzle velocity) (a)	$\approx 7 \times 10^2$ m/s
Random motion of molecules in air (average)	4.5×10^2 m/s
Sound	3.3×10^2 m/s
Jet airliner (Boeing 747, maximum airspeed)	2.7×10^2 m/s
Cheetah (maximum) (b)	28 m/s
Typical highway speed limit (55 mi/h)	25 m/s
Human (maximum)	12 m/s
Human (walking briskly)	1.3 m/s
Snail	$\approx 10^{-3}$ m/s
Glacier (c)	$\approx 10^{-6}$ m/s
Rate of growth of hair (human)	3×10^{-9} m/s
Continental drift	$\approx 10^{-9}$ m/s

(a)

(b)

(c)

EXAMPLE 1 The world record set by Asafa Powell in 2005 for the 100-m run was 9.77 s (see Fig. 2.2). What average speed did he attain while setting this record?

SOLUTION: According to Eq. (2.1),

$$[\text{average speed}] = \frac{[\text{total distance traveled}]}{[\text{time taken}]} = \frac{100\text{ m}}{9.77\text{ s}} = 10.2\text{ m/s} \qquad (2.3)$$

Here the final result has been rounded to three significant figures, since the data for the problem have three significant figures.

FIGURE 2.2 Asafa Powell in 2005.

Motion and speed are relative; the value of the speed depends on the frame of reference with respect to which it is calculated. Example 1 gives the speed of the runner relative to a reference frame attached to the surface of the Earth. However, relative to the reference frame of a bicyclist riding in the same direction as the runner, the speed will be different (Fig. 2.3). For instance, if the ground speed of the bicyclist is also 10.2 m/s, then relative to her reference frame, the runner will have zero speed. Thus, questions regarding speed are meaningless unless the frame of reference is first specified. In everyday language, "speed" often means speed relative the the Earth's surface. If the speed is reckoned relative to some other body, this will usually be clear from the context. For example, in Table 2.1 the speed of the jet airliner is reckoned relative to the air (which may be in motion relative to the Earth). We will be careful to specify the frame of reference whenever it is not clear from the context.

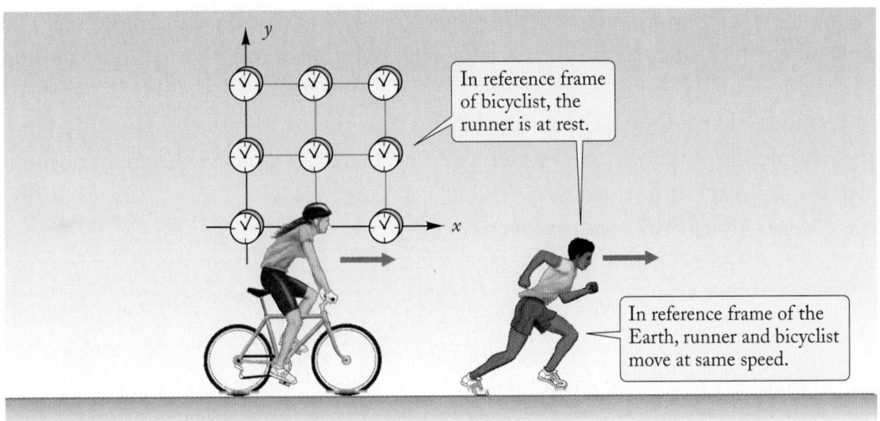

FIGURE 2.3 A bicyclist and her reference frame. If both the bicyclist and the runner are moving toward the right at the same speed, then the runner is at rest relative to the reference frame of the bicyclist.

Checkup 2.1

QUESTION 1: A man takes 100 s to walk 50 m along a straight road. What is his average speed? A woman jogs the same distance in 50 s, and then stands at the endpoint for 50 s. What is her average speed over the 100-s interval?

QUESTION 2: Consider a runner running at 10.1 m/s relative to the surface of the Earth and consider a spectator standing at rest on the surface of the Earth. A bicyclist is riding in the same direction as the runner, but at a different speed. Is it possible that in the reference frame of the bicyclist the spectator has a larger speed than the runner? That the spectator and the runner have equal speeds?

QUESTION 3: A car starts at rest ($v = 0$) and increases its speed to 30 m/s in 5.0 s; during this time, the car travels a distance of 60 m. The average speed of the car during this time interval is:

(A) 150 m/s (B) 60 m/s (C) 45 m/s (D) 15 m/s (E) 12 m/s

Online Concept Tutorial

2.2 AVERAGE VELOCITY FOR MOTION ALONG A STRAIGHT LINE

For the rest of this chapter, we will consider the special case of motion along a straight line, that is, motion in one dimension. For convenience, we will assume that the straight line coincides with the x axis (Fig. 2.4). We can then give a complete description of the motion of the particle by specifying the x coordinate at each instant of time. In mathematical language, this means that *the x coordinate is a function of time.* Graphically, we can represent the motion by means of a plot of the x coordinate vs. the time coordinate. For example, Fig. 2.5 is such a plot of the position coordinate x vs. the time coordinate t for an automobile that starts from rest, accelerates along a straight road for 10 seconds, and then brakes and comes to a full stop 4.3 seconds later (the plot is based on data from an acceleration test of a Maserati sports car). The position is measured from the starting point on the road to a marker point marked on the automobile. The plot shown in this figure gives us a complete description of the (translational) motion of the automobile.

FIGURE 2.4 An automobile moving along a straight line. The x axis coincides with this straight line.

Suppose that at time t_1 the automobile is at position x_1, and at a subsequent time t_2 the automobile is at position x_2 (see Fig. 2.6). Then $x_2 - x_1$ is the change of position that occurs in the time interval $t_2 - t_1$. The **average velocity** *is defined as the ratio of this change of position and the time interval*:

average velocity

$$\bar{v} = \frac{x_2 - x_1}{t_2 - t_1} \tag{2.4}$$

Equation (2.4) can also be written as

$$\bar{v} = \frac{\Delta x}{\Delta t} \tag{2.5}$$

with $\Delta x = x_2 - x_1$ and $\Delta t = t_2 - t_1$ (here, the overbar on the symbol for velocity is a standard notation used in science to indicate an *average* quantity; and Δ, the Greek capital letter *delta,* is a standard notation used to indicate a *change* in a quantity). Thus, the average velocity is the average rate of change of the position, or the average change of position per unit time.

Graphically, *in the plot of position vs. time, the average velocity is the ratio of the vertical separation between the points P_1 and P_2 and the horizontal separation.* If we draw a straight line connecting the points P_1 and P_2 in the plot, the ratio of the vertical separation and the horizontal separation between any two points on this line is the **slope** of the line. Note that this mathematical definition of slope agrees with the everyday notion of slope: a steep line, with a large vertical separation between the points P_1 and P_2, has a large slope; and a nearly horizontal line, with a small vertical separation between the points P_1 and P_2, has a small slope. With this mathematical definition of slope, we can say that the average velocity equals the slope of the straight line connecting the points P_1 and P_2 (see Fig. 2.6). For instance, if $t_1 = 8.0$ s and $t_2 = 14.3$ s,

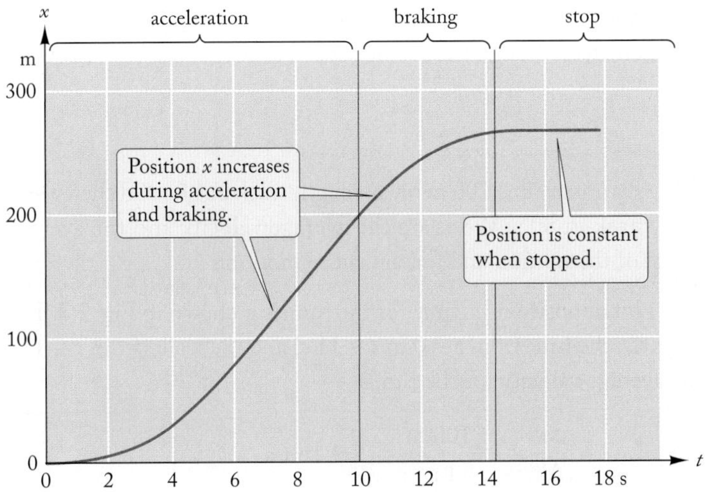

FIGURE 2.5 Plot of position vs. time for an automobile that accelerates ($0 \text{ s} < t < 10 \text{ s}$), then brakes ($10 \text{ s} < t < 14.3 \text{ s}$), and then stops ($t > 14.3 \text{ s}$) (based on data from a road test of a Maserati Bora by *Road & Track* magazine).

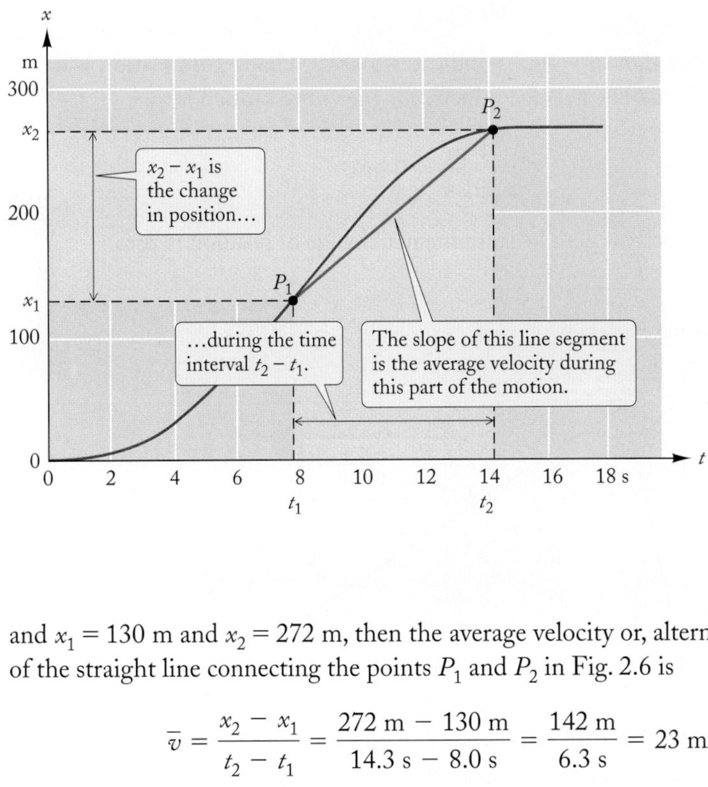

FIGURE 2.6 The average velocity for the interval from t_1 = 8.0 s to t_2 = 14.3 s is the slope of the straight line P_1P_2.

and $x_1 = 130$ m and $x_2 = 272$ m, then the average velocity or, alternatively, the slope of the straight line connecting the points P_1 and P_2 in Fig. 2.6 is

$$\bar{v} = \frac{x_2 - x_1}{t_2 - t_1} = \frac{272 \text{ m} - 130 \text{ m}}{14.3 \text{ s} - 8.0 \text{ s}} = \frac{142 \text{ m}}{6.3 \text{ s}} = 23 \text{ m/s} \tag{2.6}$$

A positive or negative slope of the position vs. time plot corresponds, respectively, to a positive or negative sign of the velocity. According to the general formula (2.4), the velocity is positive or negative depending on whether x_2 is larger or smaller than x_1, that is, depending on whether the x coordinate increases or decreases in the time interval $t_2 - t_1$. This means that the sign of the velocity depends on the direction of motion. If the motion is in the positive x direction—as in the example plotted in Fig. 2.5—the velocity is positive; if the motion is in the negative x direction, the velocity is negative. Thus, according to the precise terminology used in physics, speed [defined by Eq. (2.1)] and velocity [defined by Eq. (2.4)] are not the same thing, because *speed is always positive, whereas velocity is positive or negative depending on the direction of motion.* Furthermore, if the motion has one portion in the positive x

direction and another portion in the negative x direction, then it is possible for the average velocity to be zero even though the average speed is not zero, as the following example illustrates.

EXAMPLE 2 A runner runs 100 m on a straight track in 11 s and then walks back in 80 s. What are the average velocity and the average speed for each part of the motion and for the entire motion?

SOLUTION: The plot of position vs. time for the runner is shown in Fig. 2.7. The motion has two parts: the run (from $t = 0$ to $t = 11$ s) and the walk (from $t = 11$ s to $t = 91$ s). The average velocity for the run is

$$\bar{v} = \frac{\Delta x}{\Delta t} = \frac{+100\text{ m}}{11\text{ s}} = +9.1\text{ m/s}$$

The average velocity for the walk is

$$\bar{v} = \frac{-100\text{ m}}{80\text{ s}} = -1.3\text{ m/s}$$

(Here the minus sign in -100 m indicates that the change of position is in the negative direction.) The average velocity for the entire motion is

$$\bar{v} = \frac{0\text{ m}}{91\text{ s}} = 0\text{ m/s}$$

This average velocity is zero because the net change of position is zero.

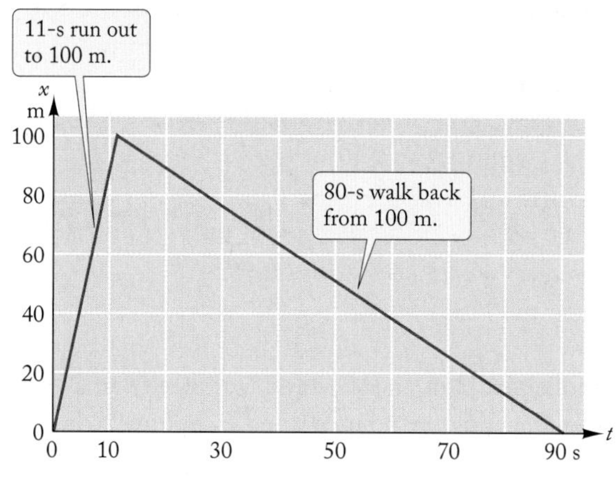

FIGURE 2.7 Plot of position vs. time for a runner.

The average speeds for the run and the walk are, respectively, 9.1 m/s and 1.3 m/s (here, there is no minus sign, since distance is always reckoned as positive). The average speed for the entire motion is the total distance traveled divided by the time taken,

$$[\text{average speed}] = \frac{200\text{ m}}{91\text{ s}} = 2.2\text{ m/s}$$

The average *speed* differs from the average *velocity* because the distance traveled (200 m) differs from the net change of position (zero).

Checkup 2.2

QUESTION 1: You lean out of a window and throw a ball straight down, so it bounces off the sidewalk and returns to your hand 2.0 s later. What is the average velocity of the ball for this motion? Can you calculate the average speed from the information given here?

QUESTION 2: Is it possible that for the trip of an automobile the speed is positive and the average velocity negative? That the speed is positive and the average velocity zero?

(A) No; no (B) Yes; no (C) No; yes (D) Yes; yes

2.3 INSTANTANEOUS VELOCITY

If, during the road test of the Maserati, the driver was keeping an eye on the speedometer, he would have seen the needle gradually climb from zero to a value well above the legal limit while he was stepping on the accelerator, and then quickly fall to zero when he slammed on the brakes. The speedometer of an automobile displays the instantaneous velocity, that is, the velocity at one instant of time. But although we are all familiar with the speedometers of automobiles, this familiarity does not fully enable us to grasp the mathematical definition of instantaneous velocity, which has to be based on purely kinematic concepts.

Since a moving automobile or a moving particle does not cover any distance in one instant, it is not immediately obvious how we should define the instantaneous velocity. Only in the exceptional case of a particle moving uniformly, with constant velocity, is the instantaneous velocity obvious—it then coincides with the average velocity. We can see this from an examination of the plot of position vs. time. If a particle moves at constant velocity, the plot of position vs. time is a straight line, with a slope equal to the velocity. For example, Fig. 2.8 shows a plot of position vs. time for an automobile moving along a straight road at a constant velocity of 25 m/s. This plot is a straight line of constant slope—the slope in any time interval is equal to the slope in any other time interval. Thus, the average velocity is the same for all time intervals—it is always 25 m/s. Since the velocity is always the same, we may regard the instantaneous velocity for this motion as identical to the average velocity.

If a particle moves with a varying velocity (accelerated motion), the plot of position vs. time is a curve. The plot of position vs. time for the accelerating automobile shown in Fig. 2.5 gives us an example of this: the automobile first accelerates and then decelerates, and the plot of position vs. time is a curve of varying slope. How can we construct a definition of the instantaneous velocity of the automobile on the basis of this plot?

To formulate a definition of the instantaneous velocity, consider the instant $t = 4$ s. We can find an *approximate* value for the velocity at this instant by taking a small time interval of, say, 0.10 s centered on 4 s, that is, a time interval from 3.95 s to 4.05 s. In this time interval the automobile covers some small distance Δx, and we can approximate the actual (curved) plot of position vs. time by a straight line segment connecting the endpoints of the interval (see Fig. 2.9a). According to the discussion at the beginning of this section, the instantaneous velocity associated with a straight plot of position vs. time is simply the slope of the plot; hence the instantaneous velocity at $t = 4$ s can be evaluated

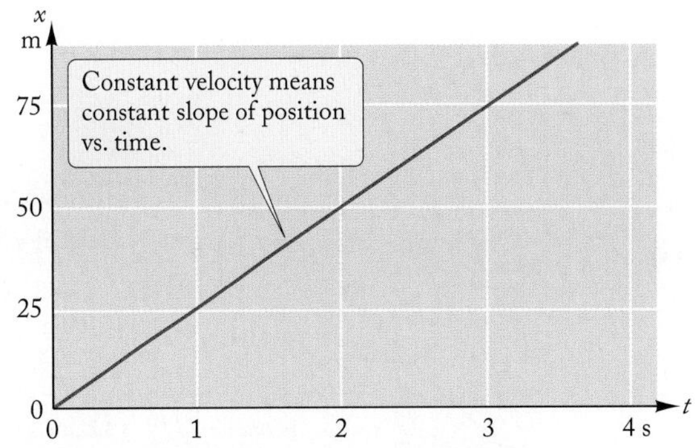

FIGURE 2.8 Plot of position vs. time for an automobile moving at constant velocity.

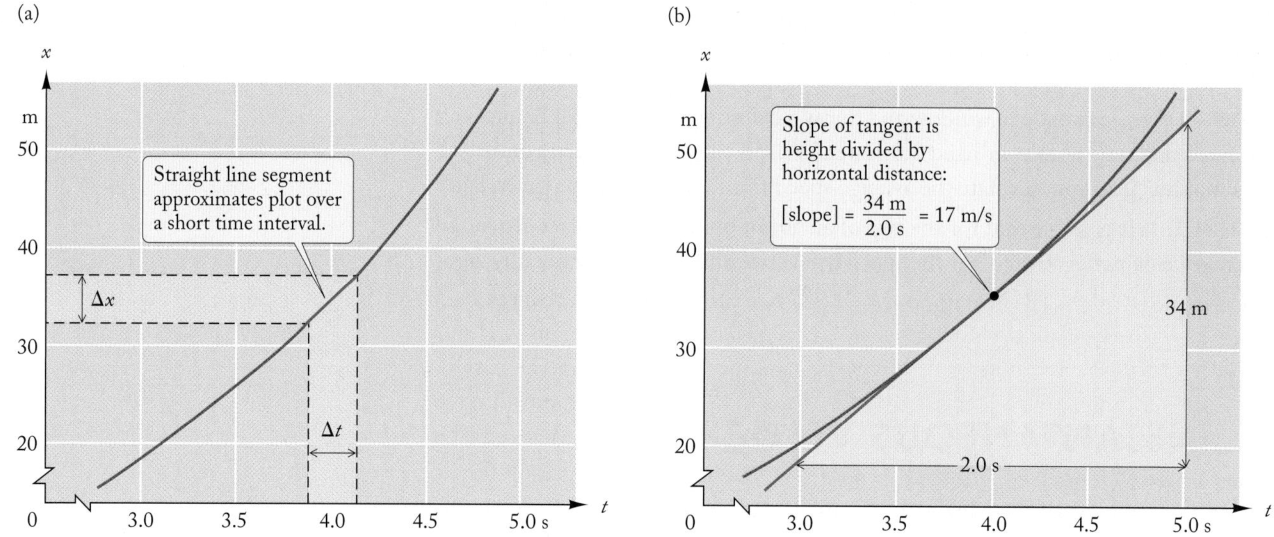

FIGURE 2.9 (a) Plot of position vs. time for an automobile moving with varying velocity. Over a small time interval, we can approximate the plot by a short straight line (blue). (b) The straight line tangent to the plot at 4.0 s rises by 34 m in 2.0 s; thus, the line has a slope of (34 m)/(2.0 s) = 17 m/s.

approximately as the slope of the short line segment shown in Fig. 2.9a. Whether this is a good approximation depends on how closely the straight line segment coincides with the actual curved plot. Obviously, the approximation can be improved by taking a shorter time interval, 0.0010 s or even less. In the limiting case of extremely small time intervals (infinitesimal time intervals), the line segment has the direction of the tangent that touches the plot at the point $t = 4$ s. Hence the **instantaneous velocity** *at any given time equals the slope of the tangent that touches the plot at that time.* For example, drawing the tangent that touches the plot at $t = 4$ s (Fig. 2.9b) and measuring its slope on the graph, we readily find that this slope is 17 m/s; hence the instantaneous velocity at $t = 4$ s is 17 m/s.

By drawing tangents at other points of the plot and measuring their slopes, as illustrated in Fig. 2.10, we can obtain a complete table of values of instantaneous velocities at different times. Figure 2.11 is a plot of the results of such a determination of the instantaneous velocities. The velocity is initially zero (zero slope in Fig. 2.10), then increases to a maximum of 34.9 m/s at $t = 10.0$ s (maximum slope in Fig. 2.10), and finally decreases to zero at $t = 14.3$ s (zero slope in Fig. 2.10).

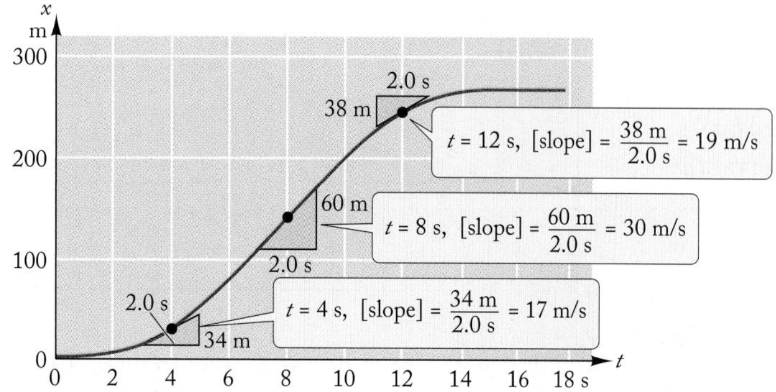

FIGURE 2.10 To find the instantaneous velocities at different times, we draw tangents to the plot at these times and measure their slopes.

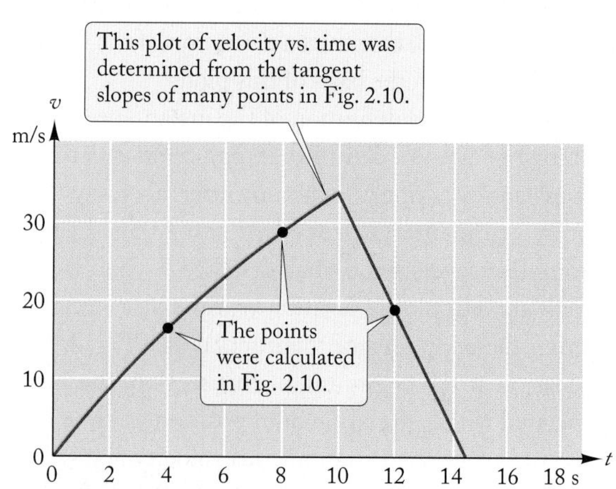

FIGURE 2.11 Instantaneous velocity as a function of time.

The graphical method for determining the instantaneous velocity has the disadvantage that we must first prepare a careful plot of position vs. time. If we have available a complete table of values of the position at different times or an exact mathematical formula for the position as a function of time, then we can calculate the instantaneous velocity by a numerical method, without bothering with a plot. We recall that in the graphical method we began with some small time interval centered on the relevant time, and then took an even smaller time interval, and ultimately considered what happens to the slope in the limiting case of extremely small time intervals. Likewise, in the numerical method, we begin with some small time interval Δt, obtain the change of position Δx, and evaluate the ratio $\Delta x/\Delta t$. Then we take a smaller time interval and repeat the calculation. And then we take an even smaller time interval and repeat the calculation again, and so on. Ultimately, we find that in the limiting case of extremely small time intervals (infinitesimal time intervals), the result of the calculation does not depend on the size of the time interval; that is, we obtain the same result for an extremely small time interval and for an even smaller time interval. This limiting result is the correct value of the instantaneous velocity. Thus, we can write the following formula for the instantaneous velocity:

$$v = \lim_{\Delta t \to 0} \frac{\Delta x}{\Delta t} \tag{2.7}$$

instantaneous velocity

Here, the notation lim $\Delta t \to 0$ indicates that $\Delta x/\Delta t$ is to be evaluated in the limiting case of an infinitesimal time interval Δt.

As an example of the application of this equation for the instantaneous velocity, consider the case of the accelerating Maserati whose motion is described by the graph plotted in Fig. 2.5. For a numerical calculation based on Eq. (2.7), we need a formula that gives us x at each instant t, that is, a formula that gives us x as a function of t. The initial, accelerated part of the motion is described by the formula

$$x = 2.376t^2 - 0.042t^3 \qquad \text{(for } t \text{ between 0 s and 10 s)} \tag{2.8}$$

where x is measured in meters and t in seconds. This formula is merely an alternative way of presenting the information contained in Fig. 2.5 for the interval from 0 s to 10 s. By making a table of values of x at different times t, it is possible to check that the graph plotted in Fig. 2.5 and the formula (2.8) agree—they give the same value of x for any specified time t [in fact, the formula (2.8) was constructed from the data in the graph so as to ensure this agreement].

If we want the instantaneous velocity at $t = 4$ s, we can take $t_1 = 3.9995$ s and $t_2 = 4.0005$ s, for which the formula (2.8) yields

$$x_1 = 2.376 \times (3.9995)^2 - 0.042 \times (3.9995)^3 = 35.3195 \text{ m}$$

and

$$x_2 = 2.376 \times (4.0005)^2 - 0.042 \times (4.0005)^3 = 35.3365 \text{ m}$$

so the instantaneous velocity at $t = 4$ s is approximately

$$v = \frac{\Delta x}{\Delta t} = \frac{x_2 - x_1}{0.001 \text{ s}} = \frac{35.3365 \text{ m} - 35.3195 \text{ m}}{0.001 \text{ s}} = \frac{0.017 \text{ m}}{0.001 \text{ s}} = 17 \text{ m/s}$$

To check the accuracy of this value of the instantaneous velocity, we take a smaller time interval, say, $t_1 = 3.9998$ s and $t_2 = 4.0002$ s. This yields $x_1 = 35.3246$ m and $x_2 = 35.3314$ m, so the instantaneous velocity at $t = 4$ s is approximately

$$v = \frac{x_2 - x_1}{0.0004 \text{ s}} = \frac{35.3314 \text{ m} - 35.3246 \text{ m}}{0.0004 \text{ s}} = 17 \text{ m/s} \tag{2.9}$$

MATH HELP — DIFFERENTIAL CALCULUS; RULES FOR DERIVATIVES

The derivative of the function $f = f(t) = t^n$:

$$\frac{d}{dt}f = \frac{d}{dt}t^n = nt^{n-1}$$

The derivative of a function f multiplied by a constant c:

$$\frac{d}{dt}(cf) = c\frac{df}{dt}$$

The derivative of the sum of two functions f and g:

$$\frac{d}{dt}(f+g) = \frac{df}{dt} + \frac{dg}{dt}$$

(For more on calculus, see Appendix 4.)

This agrees with the first calculation and confirms that our first choice of time interval was already small enough for an accurate calculation of the instantaneous velocity. Note that the result obtained by the numerical method also agrees with the result obtained by the graphical method, as it should.[1]

In the language of differential calculus, the limit $\Delta x/\Delta t$ for $\Delta t \to 0$ is written as dx/dt, and Eq. (2.7) becomes

instantaneous velocity as derivative of x with respect to t

$$v = \frac{dx}{dt} \tag{2.10}$$

The expression dx/dt is called the **derivative** of x with respect to t. Thus, the *instantaneous velocity is the derivative of the position with respect to time.*

The derivative of a function of t can be evaluated according to the rules of differential calculus. For example, consider a common function, namely, the time t taken to a power n. According to the rules of differential calculus, the derivative of the function t^n with respect to t equals nt^{n-1}. More generally, the derivative of ct^n (where c is some arbitrary constant) is cnt^{n-1}, and the derivative of the sum or difference of several such functions is the sum or difference of their derivatives.

For example, the derivative of $2.376t^2 - 0.042t^3$ with respect to time is $2.376 \times 2t - 0.042 \times 3t^2$, or $4.752t - 0.126t^2$. Accordingly, the instantaneous velocity implied by the formula (2.8) for x is

$$v = \frac{dx}{dt} = 4.752t - 0.126t^2 \quad \text{(for } t \text{ between 0 s and 10 s)} \tag{2.11}$$

where v is measured in m/s and t in s. With this formula we can readily evaluate the instantaneous velocity at any time. Thus, at $t = 4.0$ s, we obtain

$$v = 4.752 \times 4.0 - 0.126 \times (4.0)^2 = 17 \text{ m/s}$$

which agrees with our previous calculation.

Note that we now have available three methods for the calculation of the instantaneous velocity: the graphical method (based on a determination of the slope in the plot of position vs. time), the numerical method based on small time intervals, and the method of derivatives. These methods are equivalent—they all give the same result. But the method of derivatives is the most convenient and most accurate, and we will hereafter use it whenever we have available an explicit formula for the position as a function of time.

[1] It might seem more reasonable to calculate the desired instantaneous velocity by taking a much smaller time interval, say, 10^{-6} s or so. But if we use an ordinary calculator to evaluate x, we reach a point of diminishing returns. The calculator can handle only ten digits, and if the difference between x_2 and x_1 is smaller than the last digit, the calculator will reckon the difference as zero. Thus, we must take care not to use excessively small time intervals when evaluating Eq. (2.7) on such a calculator.

Checkup 2.3

QUESTION 1: What is the meaning of a negative velocity v? Can an aircraft have a negative velocity v of, say, -400 km/h?

QUESTION 2: You are driving an automobile along a straight road, and the speedometer registers 60 km/h (17 m/s). Is the velocity positive or negative? What information do you need to decide this question?

QUESTION 3: Suppose that your average velocity for a 1-hour bike trip on a straight road was 8 m/s. Is it possible that your instantaneous velocity during the entire trip was always larger than 8 m/s?

QUESTION 4: The position x in meters of a body as a function of the time t in seconds is given by $x = \frac{1}{4}t^4$. What is the instantaneous velocity at $t = 2.0$ s?

(A) 16 m/s. (B) 8.0 m/s. (C) 6.0 m/s.
(D) 4.0 m/s. (E) 2.0 m/s.

2.4 ACCELERATION

Online Concept Tutorial

Any motion with a change of velocity is accelerated motion. Thus, the motion of an automobile that speeds up is accelerated motion, but so is the motion of an automobile that slows down while braking—in both cases there is a *change* of velocity. If a particle has velocity v_1 at time t_1 and velocity v_2 at time t_2, then the **average acceleration** *for this time interval is defined as the change of velocity divided by the change of time*,

$$\bar{a} = \frac{v_2 - v_1}{t_2 - t_1} \tag{2.12}$$

average acceleration

or

$$\bar{a} = \frac{\Delta v}{\Delta t} \tag{2.13}$$

where $\Delta v = v_2 - v_1$ and $\Delta t = t_2 - t_1$. Accordingly, the average acceleration is the average rate of change of the velocity, or the average change of velocity per unit time. The unit of acceleration is the unit of velocity divided by the unit of time. Consequently, in the SI system, *the unit of acceleration is the meter per second per second*, or **meter per second squared** [(m/s)/s, or m/s^2]. Table 2.2 gives some typical values of accelerations.

The acceleration can be positive or negative, depending on the sign of the velocity change $v_2 - v_1$. If the velocity is positive and increasing in magnitude, the acceleration is positive; if the velocity is positive and decreasing in magnitude, the acceleration is negative. However, note that if the velocity is *negative* (motion in the negative x direction) and *increasing* in magnitude, that is, becoming *more* negative, the acceleration is *negative*. Thus, an automobile speeding up while moving in the negative x direction has negative acceleration;

TABLE 2.2 SOME ACCELERATIONS

Baseball being struck by bat	3×10^4 m/s^2
Soccer ball being struck by foot	3×10^3 m/s^2
Rat flea, starting a jump	2×10^3 m/s^2
Automobile crash (60 mi/h into fixed barrier)	1×10^3 m/s^2
Parachute opening (extreme)	3.2×10^2 m/s^2
Free fall on surface of Sun	2.7×10^2 m/s^2
Explosive seat ejection from aircraft (extreme)	1.5×10^2 m/s^2
Loss of consciousness of human ("blackout")	70 m/s^2
Free fall on surface of Earth	9.8 m/s^2
Braking of automobile	$\approx$8 m/s^2
Free fall on surface of Moon	1.7 m/s^2

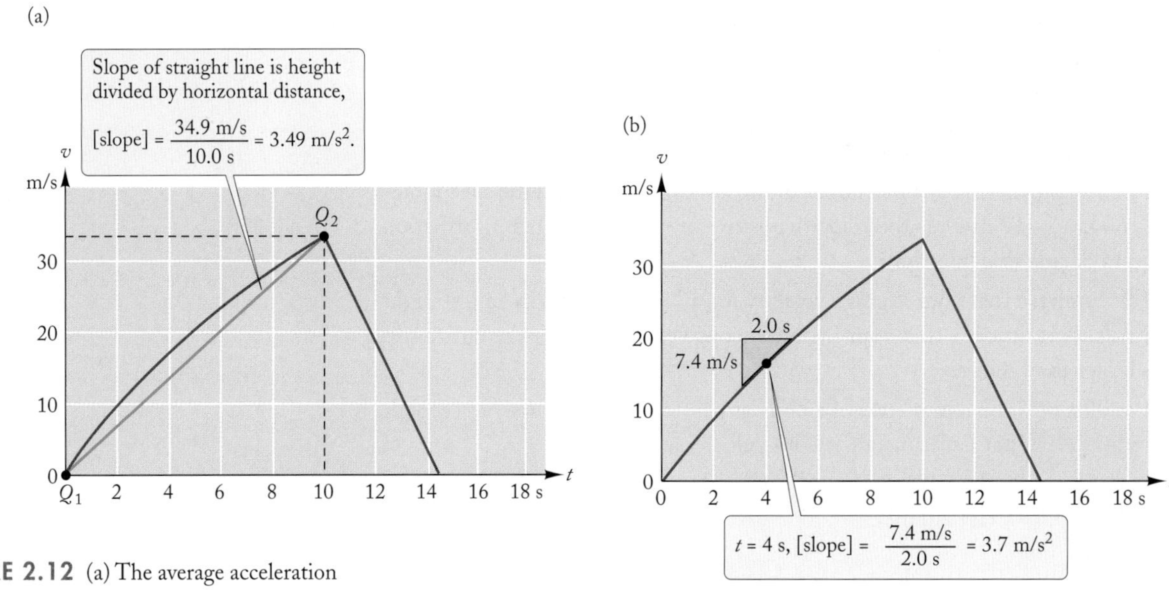

FIGURE 2.12 (a) The average acceleration for the interval from $t_1 = 0$ to $t_2 = 10.0$ s is the slope of the straight line Q_1Q_2. (b) The instantaneous acceleration at $t = 4$ s is the slope of the tangent at that point.

conversely, an automobile slowing down or "**decelerating**" while moving in the negative x direction has positive acceleration! These quirks of the formula (2.12) for the acceleration must be kept in mind.

On a plot of velocity vs. time, the average acceleration is the slope of the straight line connecting the points corresponding to t_1 and t_2 on the plot. For instance, in the case of the accelerating Maserati discussed above, the plot of velocity as a function of time is shown in Fig. 2.11, and the average acceleration for the time interval from $t_1 = 0$ to $t_2 = 10.0$ s is the ratio of the vertical separation to the horizontal separation between the two points Q_1 and Q_2 on the plot (see Fig. 2.12a):

$$\bar{a} = \frac{v_2 - v_1}{t_2 - t_1} = \frac{34.9\text{ m/s} - 0\text{ m/s}}{10.0\text{ s} - 0\text{ s}} = 3.49\text{ m/s}^2$$

Concepts in Context

EXAMPLE 3 The chapter photo shows a fighter jet being launched by a catapult from the deck of an aircraft carrier. During this launch, the fighter jet attains a speed of 260 km/h in only 1.8 s. What is the average acceleration of the jet during this time interval?

SOLUTION: We first convert $\Delta v = 260$ km/h to m/s:

$$\Delta v = 260\text{ km/h} \times \frac{1000\text{ m}}{1\text{ km}} \times \frac{1\text{ h}}{3600\text{ s}} = 72\text{ m/s}$$

With $\Delta v = 72$ m/s and $\Delta t = 1.8$ s, the average acceleration is then

$$\bar{a} = \frac{\Delta v}{\Delta t} = \frac{72\text{ m/s}}{1.8\text{ s}} = 4.0 \times 10^1\text{ m/s}^2 = 40\text{ m/s}^2$$

The **instantaneous acceleration** *at some instant of time is the slope of the tangent drawn on the plot of velocity vs. time.* For example, at $t = 4$ s, we can draw the tangent to the velocity curve in Fig. 2.11 and find that the slope, or the instantaneous acceleration, is 3.7 m/s^2 (see Fig. 2.12b). By drawing tangents for different times and measuring their slopes, we can prepare a complete table of values of the instantaneous acceleration at different times. Figure 2.13 shows a plot of the values of the instantaneous acceleration obtained from our road test of the Maserati. At the initial instant $t = 0$, the acceleration is large (large slope in Fig. 2.11); as the automobile gains velocity, the acceleration gradually drops (decreasing slope in Fig. 2.11); at $t = 10$ s, the brakes are applied, leading to a large negative acceleration of -8.1 m/s^2 (negative slope in Fig. 2.11); this negative acceleration is maintained until the Maserati comes to a halt. Note that when the driver steps on the brakes at time $t = 10$ s, the acceleration suddenly switches from positive to negative; thus, the plot in Fig. 2.13 has a jump at this point.

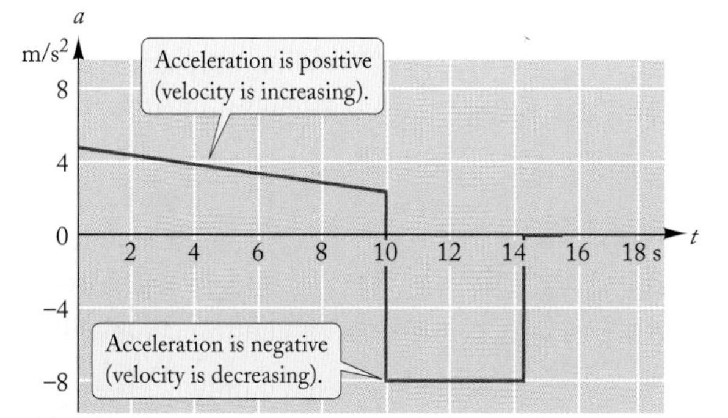

FIGURE 2.13 Instantaneous acceleration as a function of time.

As in the case of the instantaneous velocity, the instantaneous acceleration can also be calculated as the limit of the ratio of small increments:

$$a = \lim_{\Delta t \to 0} \frac{\Delta v}{\Delta t}$$

or

$$a = \frac{dv}{dt} \tag{2.14}$$

instantaneous acceleration as derivative of v with respect to t

This says that the acceleration is the derivative of the velocity with respect to time. Equivalently, we can say that the acceleration is the second derivative of the position x with respect to time, that is

$$a = \frac{d}{dt}\left(\frac{dx}{dt}\right) \quad \text{or} \quad a = \frac{d^2x}{dt^2} \tag{2.15}$$

For example, the acceleration calculated by differentiation of the velocity formula (2.11) is

$$\begin{aligned} a &= \frac{d}{dt}(4.752t - 0.126t^2) = 4.752 - 0.126 \times 2t \\ &= 4.752 - 0.252t \quad \text{(for } t \text{ between 0 s and 10 s)} \end{aligned} \tag{2.16}$$

where the acceleration is measured in m/s^2. For example, at $t = 4$ s, this relation gives $a = 3.74$ m/s^2, in agreement with the slope of the tangent to the velocity curve, $a = 3.7$ m/s^2, mentioned above.

✔ Checkup 2.4

QUESTION 1: What is the meaning of a negative acceleration? Can an "accelerating" automobile starting from rest have a negative acceleration a? Can a braking automobile have a positive acceleration a?

QUESTION 2: Give an example of motion with positive instantaneous velocity and a simultaneous negative acceleration. Give an example of motion with negative instantaneous velocity and a simultaneous positive acceleration.

QUESTION 3: Suppose that at one instant of time the velocity of a body is zero. Can this body have a nonzero acceleration at this instant? Give an example.

QUESTION 4: You drop a tennis ball on a hard floor, and it bounces upward. If the x axis is directed upward, what is the sign of the velocity before the ball hits the floor? After? What is the sign of the acceleration before the ball hits the floor? After? What is the sign of the acceleration while the ball is in contact with the floor?

QUESTION 5: A train strikes an automobile abandoned on a railroad crossing. The train drags the automobile along, and then stops at some distance beyond the crossing. Assume the x axis is along the railroad track, in the direction of travel of the train. What is the sign of the average acceleration of the automobile for the complete motion? What is the sign of the average acceleration of the train? What is the sign of the instantaneous acceleration of the automobile during the impact? After the impact?

QUESTION 6: Two automobiles collide head-on. Compare the signs of the velocities of the automobiles just before the collision. Compare the signs of their accelerations during the collision.

(A) Same; same
(B) Same; opposite
(C) Opposite; same
(D) Opposite; opposite

Online Concept 3 Tutorial

2.5 MOTION WITH CONSTANT ACCELERATION

The acceleration of a body may vary as a function of position or time, as in the instantaneous acceleration plot of Fig. 2.13; later, in Section 2.7, we will learn how to use integral calculus to determine the position of a body as a function of time for arbitrary, varying acceleration. However, it is very common for a body to experience a *constant acceleration*, at least for some interval of time; this permits a simpler analysis. Constant acceleration implies a constant slope in the plot of velocity vs. time; thus the plot is a straight line. In this case, the velocity simply increases (or decreases) by equal amounts in each 1-second time interval. For example, in the interval between 10.0 s and 14.3 s, the velocity plotted in Fig. 2.11 decreases by 8.1 m/s in each second while the automobile brakes.

In the case of constant acceleration, there are some simple relations between acceleration, velocity, position, and time that permit us to calculate one of these quantities from the others. Suppose that the initial velocity at time zero is v_0 and that the velocity increases at a constant rate given by the constant acceleration a. After a time t has elapsed, the velocity will have increased by an amount at, and it will have attained the value

$$v = v_0 + at \tag{2.17}$$

Suppose that the initial position is x_0 at time zero. After a time t has elapsed, the position will have changed by an amount equal to the product of the average velocity multiplied by the time; that is, the position will have changed from the initial value x_0 to

$$x = x_0 + \overline{v}t \tag{2.18}$$

Since the velocity increases uniformly with time, the average value of the velocity is simply the average of the initial value and the final values, or

$$\overline{v} = \tfrac{1}{2}(v_0 + v) \tag{2.19}$$

and therefore

$$x = x_0 + \tfrac{1}{2}(v_0 + v)t \tag{2.20}$$

Thus, the change in the position is

$$x - x_0 = \tfrac{1}{2}(v_0 + v)t \tag{2.21}$$

To express this in terms of the acceleration, we substitute Eq. (2.17) into (2.21), and we find

$$x - x_0 = \tfrac{1}{2}(v_0 + v_0 + at)t$$

or

$$x - x_0 = v_0 t + \tfrac{1}{2}at^2 \tag{2.22}$$

The right side of this equation consists of two terms: the term $v_0 t$ represents the change in position that the particle would suffer if moving at constant velocity v_0, and the term $\frac{1}{2}at^2$ represents the effect of the acceleration.

Equations (2.17) and (2.22) express velocity and position in terms of time. By eliminating the time t between these two equations, we obtain a direct relation between position and velocity, which is sometimes useful. According to Eq. (2.17),

$$t = \frac{v - v_0}{a} \tag{2.23}$$

and if we substitute this into Eq. (2.22), we obtain

$$\begin{aligned} x - x_0 &= v_0\left(\frac{v - v_0}{a}\right) + \tfrac{1}{2}a\left(\frac{v - v_0}{a}\right)^2 \\ &= \frac{v_0 v - v_0^2}{a} + \tfrac{1}{2}\frac{v^2 - 2vv_0 + v_0^2}{a} \\ &= \tfrac{1}{2}\frac{v^2 - v_0^2}{a} \end{aligned} \tag{2.24}$$

which we can rearrange as follows:

$$a(x - x_0) = \tfrac{1}{2}(v^2 - v_0^2) \tag{2.25}$$

Figure 2.14 shows graphs of acceleration, velocity, and position for motion with constant acceleration. For the sake of simplicity, the initial velocity and position have been taken as zero ($v_0 = 0$, $x_0 = 0$). In these graphs, the motion for negative values of t (instants before zero) has also been included with the assumption that a always has the same constant value (this corresponds, for instance, to a subway car that slows down as it reaches a terminal station at the end of the track, stops instantaneously, and then speeds up as it travels away from the station, back in the same direction it came from). Note that the plot in Fig. 2.14b is a straight line; this merely means that the velocity increases in direct proportion to the time, in accord with Eq. (2.17). The plot in Fig. 2.14c is a parabola; this parabola results because the change of position is proportional to the square of the time, as indicated by Eq. (2.22). The parabolic shape of the plot is a distinctive characteristic of motion with constant acceleration.

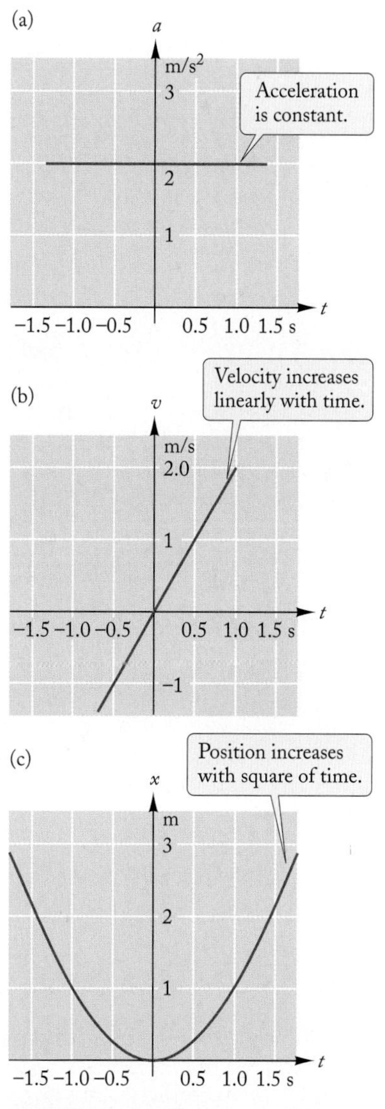

FIGURE 2.14 (a) Acceleration vs. time for motion with constant acceleration; this plot shows a constant value of 2.0 m/s². (b) Velocity vs. time; this plot is a straight line of slope 2.0 m/s². (c) Plot of position vs. time; the plot is a parabola.

EXAMPLE 4 Use the rules of differential calculus to evaluate the second derivative of x with respect to t from Eq. (2.22), and verify that this second derivative equals the acceleration a.

SOLUTION: From Eq. (2.22),

$$x = x_0 + v_0 t + \tfrac{1}{2}at^2$$

In the differentiation, x_0, v_0, and a are constants. Hence the first derivative of x is

$$\frac{dx}{dt} = \frac{d}{dt}\left(x_0 + v_0 t + \tfrac{1}{2}at^2\right) = 0 + v_0 + \tfrac{1}{2}a \times 2t = v_0 + at$$

and the second derivative is

$$\frac{d^2x}{dt^2} = \frac{d}{dt}(v_0 + at) = a$$

Equations (2.17), (2.21), (2.22), and (2.25) contain the acceleration a, the time t, and the instantaneous and initial positions and velocities, x, x_0, v, and v_0. In a typical problem of motion with constant acceleration, some of these quantities will be known, and the others will be unknown, to be calculated from the equations. Which of these equations are the most useful will depend on the problem.

Concepts — in — Context

EXAMPLE 5 In Example 3 we found that during a catapult launch from the deck of an aircraft carrier (see the chapter photo), the average acceleration of a fighter jet is 40 m/s^2 during a time interval of 1.8 s. Assuming the motion proceeds with constant acceleration, how far does the fighter jet travel along the deck during this time interval?

SOLUTION: For motion with constant acceleration, the relevant equations are Eqs. (2.17), (2.21), (2.22), and (2.25). To solve this problem, we must decide which of these equations we need. The unknown quantity is the distance, and the known quantities are the acceleration, the final and initial speeds, and the time. For convenience, we assume that the origin of our coordinates is at the initial position of the aircraft, so $x_0 = 0$ (see Fig. 2.15). Then the known and the unknown quantities are as follows:

UNKNOWN	KNOWN
x	$a = 40\ \text{m/s}^2$ $v_0 = 0$ $x_0 = 0$ $t = 1.8$ s

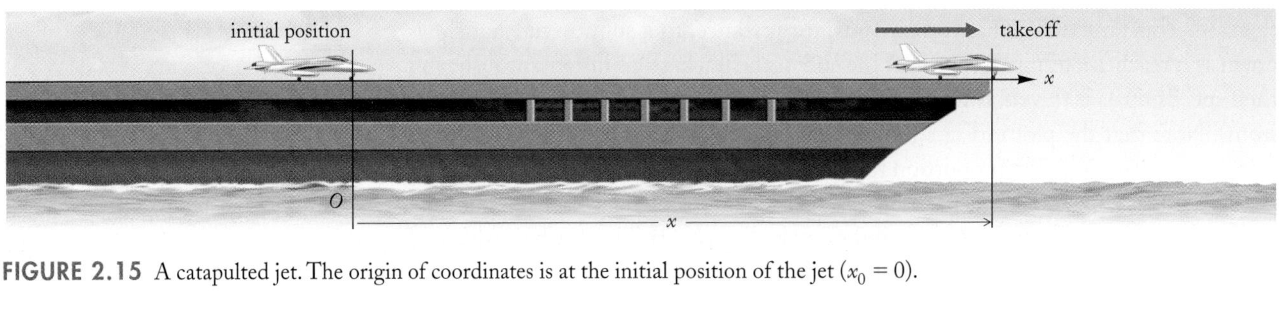

FIGURE 2.15 A catapulted jet. The origin of coordinates is at the initial position of the jet ($x_0 = 0$).

Equations (2.21), (2.22), and (2.25) contain the unknown quantity x, and they contain other quantities, all of which are known. We can use any one of these equations to find x. A simple choice is Eq. (2.22), since it is already solved for x:

$$x = x_0 + v_0 + \tfrac{1}{2}at^2 = 0 + 0 + \tfrac{1}{2} \times (40\ \text{m/s}^2) \times (1.8\ \text{s})^2$$

$$= 65\ \text{m}$$

EXAMPLE 6 An automobile initially traveling at 50 km/h crashes into a stationary, rigid barrier. The front end of the automobile crumples, and the passenger compartment comes to rest after advancing by 0.40 m (see Fig. 2.16). Assuming constant deceleration during the crash, what is the value of the deceleration? How long does it take the passenger compartment to stop?

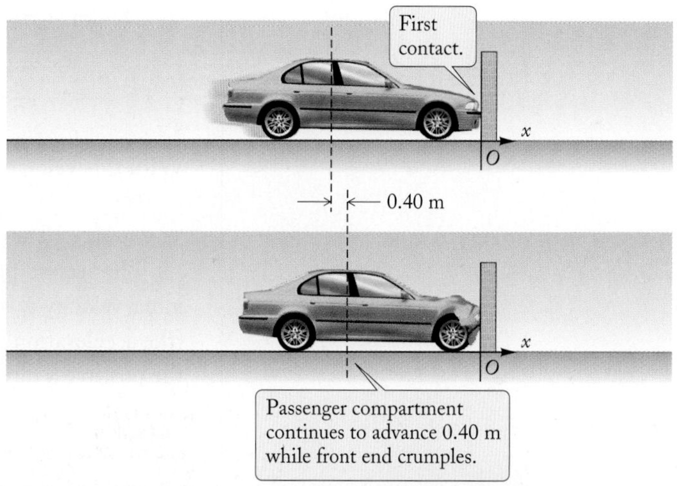

FIGURE 2.16 Deformation of the front end of an automobile crashing into a barrier. The origin of coordinates is at the barrier.

SOLUTION: The known quantities are the initial velocity ($v_0 = 50$ km/h just before the automobile comes in contact with the barrier), the final velocity ($v = 0$ when the passenger compartment comes to rest), and the change of position of the passenger compartment ($x - x_0 = 0.40$ m; see Fig. 2.16):

UNKNOWN	KNOWN
a	$x - x_0 = 0.40$ m
t	$v_0 = 50$ km/h
	$v = 0$

To find the unknown acceleration, it will be best to use Eq. (2.25), since there the acceleration appears as the *only* unknown quantity.

Solving Eq. (2.25) for a, we find

$$a = \frac{v^2 - v_0^2}{2(x - x_0)}$$

Substitution of the known quantities, with $v_0 = 50\ \text{km/h} = 50 \times (1000\ \text{m})/(3600\ \text{s}) = 13.9\ \text{m/s}$, yields

$$a = \frac{v^2 - v_0^2}{2(x - x_0)} = \frac{-(13.9\ \text{m/s})^2}{2 \times 0.40\ \text{m}} = -240\ \text{m/s}^2$$

This is a large deceleration. A passenger involved in such a crash would suffer severe injuries, unless well restrained by a snug seat belt or an air bag.

We can next calculate, from Eq. (2.17), the time the passenger compartment takes to stop:

$$t = \frac{v - v_0}{a} = \frac{-13.9\ \text{m/s}}{-240\ \text{m/s}^2} = 0.058\ \text{s}$$

EXAMPLE 7 An automobile is traveling at 86 km/h on a straight road when the driver spots a wreck ahead and slams on the brakes. The **reaction time** of the driver, that is, the time interval between seeing the wreck and stepping on the brakes, is 0.75 s. Once the brakes are applied, the automobile decelerates at 8.0 m/s^2. What is the total stopping distance (see Fig. 2.17)?

SOLUTION: The motion has two parts. The first part, before the brakes are applied, is motion at constant velocity; the second part, after the brakes are applied, is motion with constant (negative) acceleration.

The first part of the motion lasts for a time $\Delta t = 0.75$ s, with a constant velocity $v_0 = 86$ km/h, that is,

$$v_0 = 86\frac{\text{km}}{\text{h}} \times \frac{10^3\ \text{m}}{1\ \text{km}} \times \frac{1\ \text{h}}{3600\ \text{s}} = 24\ \text{m/s}$$

With this velocity, the automobile travels a distance

$$v_0\,\Delta t = 24\ \text{m/s} \times 0.75\ \text{s} = 18\ \text{m}$$

The second part of the motion therefore has an initial position $x_0 = 18$ m, an initial velocity $v_0 = 24$ m/s, a final velocity $v = 0$, and an acceleration $a = -8.0$ m/s^2 (the acceleration is negative since the automobile is decelerating while moving in the positive x direction). The final distance is the unknown:

UNKNOWN	KNOWN
x	$a = -8.0\ \text{m/s}^2$
	$v = 0$
	$v_0 = 24\ \text{m/s}$
	$x_0 = 18\ \text{m}$

The most suitable equation for the solution of this problem is Eq. (2.25), since it contains the unknown quantity and all the other quantities in it are known. Solving this equation for x, we find that the total stopping distance is

$$x = x_0 + \frac{v^2 - v_0^2}{2a}$$

$$= 18\ \text{m} + \frac{0 - (24\ \text{m/s})^2}{2 \times (-8.0\ \text{m/s}^2)} = 18\ \text{m} + 36\ \text{m} = 54\ \text{m}$$

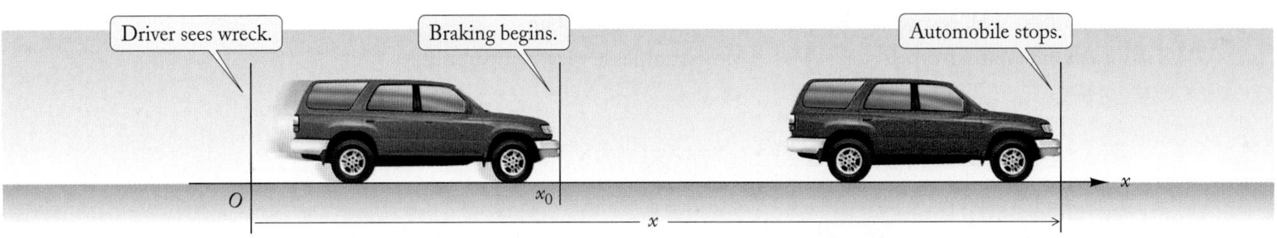

FIGURE 2.17 A braking automobile. The origin of coordinates is at the point where the driver spots wreck.

PHYSICS IN PRACTICE STOPPING DISTANCES

The accompanying table lists total stopping distances of an automobile for several initial speeds. These stopping distances were calculated as in Example 7. The reaction time of the driver was assumed to be 0.75 s, and the braking deceleration was assumed to be 8.0 m/s^2. At all speeds, the reaction time makes a significant contribution to the stopping distance. For an alert driver, the average reaction time is 0.75 s, but for a nonalert driver (such as a driver chatting on a cell phone), the reaction time can be much longer. The braking deceleration of 8 m/s^2 is the typical deceleration that an automobile with good tires can attain during sharp braking on a dry pavement. If the tires are worn or if the pavement is wet, the attainable deceleration is less than 8 m/s^2.

The deceleration also depends on the characteristics of the automobile. For instance, a high-performance sports car with a rear engine, such as the Porsche Carrera, can attain decelerations of nearly 11 m/s^2 (during sharp braking, the nose of a car tends to dive down, placing extra pressure on the front wheels; the large mass of the rear engine distributes the braking effort more equally over the front and rear wheels).

If the driver pushes the brakes too hard, the wheels will lock, and the automobile will skid, which results in a reduced deceleration and a substantial increase in stopping distance. Skidding often leads to loss of directional control, and the automobile might spin around several times and then crash into an obstacle or roll over. Cars equipped with an antilock braking system (ABS) avoid skidding by automatic, rapid, repeated application of the brakes.

AUTOMOBILE STOPPING DISTANCES

v_0	$v_0 \Delta t$	$-\dfrac{v_0^2}{2a}$	TOTAL STOPPING DISTANCE
15 km/h	3.1 m	1.1 m	4.2 m
30	6.3	4.3	10.6
45	9.4	10	19
60	12	18	30
75	16	27	43
90	19	39	58

EXAMPLE 8 On a foggy day, a minivan is traveling at 80 km/h along a straight road when the driver notices a truck ahead traveling at 25 km/h in the same direction. The driver begins to brake when the truck is 12 m ahead, decelerating the minivan at 8.0 m/s^2, while the truck continues at a steady 25 km/h. How long after this instant does the minivan collide with the truck? What is the speed of the minivan at the instant of collision?

SOLUTION: We designate the position, velocity, and acceleration of the minivan by x, v, and a and the position, velocity, and acceleration of the truck by x', v', and a'. We reckon the x and x' coordinates from the point at which braking begins (see Fig. 2.18).

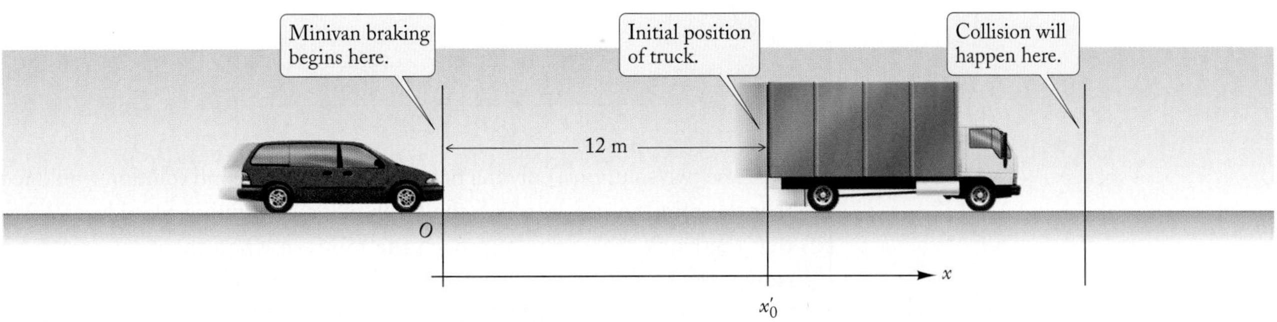

FIGURE 2.18 A braking minivan and a truck traveling at constant velocity. The origin of coordinates is at the point where braking begins. The initial position of the minivan is $x_0 = 0$, and the initial position of the truck (measured to the rear of the truck) is $x'_0 = 12$ m.

The time t of the collision and the positions x and x' at that time are unknown. The initial positions x_0 and x'_0, velocities v_0 and v'_0, and accelerations a and a' are known:

UNKNOWN	KNOWN
t	$x_0 = 0$
x	$x'_0 = 12$ m
x'	$v_0 = 80$ km/h
	$v'_0 = 25$ km/h
	$a = -8.0$ m/s^2
	$a' = 0$

To relate the unknowns x and x' to the known quantities, we use Eq. (2.22) for the minivan:

$$x = x_0 + v_0 t + \tfrac{1}{2}at^2 = v_0 t + \tfrac{1}{2}at^2$$

and for the truck:

$$x' = x'_0 + v'_0 t + \tfrac{1}{2}a't^2 = x'_0 + v'_0 t$$

Here we have two equations but three unknowns (t, x, x'). We can extract the unknown time t from these equations by taking into account that when the vehicles collide, $x = x'$. This condition tells us that

$$v_0 t + \tfrac{1}{2}at^2 = x'_0 + v'_0 t$$

This is a quadratic equation. Before proceeding with the solution, it is convenient to substitute the known numbers $a = -8.0\ \text{m/s}^2$, $v_0 = 80\ \text{km/h} = 80 \times 1000\ \text{m}/3600\ \text{s} = 22.2\ \text{m/s}$, $x'_0 = 12$ m, and $v'_0 = 25\ \text{km/h} = 25 \times 1000\ \text{m}/3600\ \text{s} = 6.9\ \text{m/s}$. Since the acceleration values are in m/s^2 and the velocity values in m/s, the time will be in seconds. Omitting the units, we obtain

$$22.2t - 4.0t^2 = 12 + 6.9t$$

and if we shift all the terms to the left side, we obtain

$$-4.0t^2 + 15.3t - 12 = 0$$

This has the standard form for a quadratic equation

$$at^2 + bt + c = 0$$

with the two solutions (see Appendix 2)

$$t = \frac{-b \pm \sqrt{b^2 - 4ac}}{2a}$$

$$= \frac{-15.3 \pm \sqrt{(15.3)^2 - 4 \times (-4.0) \times (-12)}}{2 \times (-4.0)} = 1.1\ \text{s or } 2.7\ \text{s}$$

Of these two solutions, only the first is relevant (the second solution would require that the minivan pass through the truck while continuing to brake at 1.1 s and that the truck then again approach the minivan when the minivan has nearly stopped at 2.7 s). Thus, the collision occurs at a time 1.1 s.

The speed of the minivan at this time is

$$v = v_0 + at = 22.2\ \text{m/s} - 8.0\ \text{m/s}^2 \times 1.1\ \text{s} = 13\ \text{m/s}$$

Checkup 2.5

QUESTION 1: According to Example 7, the braking distance for an automobile with an initial speed of 86 km/h is 36 m. What is the braking distance for the same automobile if the initial speed is twice as large?

QUESTION 2: According to Example 6, the deceleration of the passenger compartment of an automobile crashing into a barrier is 240 m/s^2 if its front end crumples 0.40 m. What would be the deceleration of an automobile with the same speed but a stiffer front end, which crumples only 0.20 m?

QUESTION 3: An automobile with initial velocity v_0 brakes to a stop with constant deceleration in a time t. If the initial velocity were twice as large but the constant deceleration were half as large, the time to stop would be:

(A) $8t$ (B) $4t$ (C) $2t$ (D) t (E) $\frac{1}{2}t$

2.6 THE ACCELERATION OF FREE FALL

Online Concept Tutorial 3

A body released near the surface of the Earth will accelerate downward under the influence of the pull of gravity exerted by the Earth. If the frictional resistance of the air has been eliminated (by placing the body in an evacuated container), then the body is in **free fall**, and the downward motion proceeds with constant acceleration. It is a remarkable fact that the value of this acceleration of free fall is exactly the same for all bodies released at the same location—the value of the acceleration is completely independent of the speeds, masses, sizes, shapes, chemical compositions, etc., of the bodies. Figure 2.19 shows a simple experiment that verifies this equality of the rates of free fall for two bodies of different masses. The universality of the rates of free fall is one of the most precisely and rigorously tested laws of nature; a long series of careful experiments have tested the equality of the rates of free fall of different bodies to within 1 part in 10^{10}, and in some special cases even to within 1 part in 10^{12}.

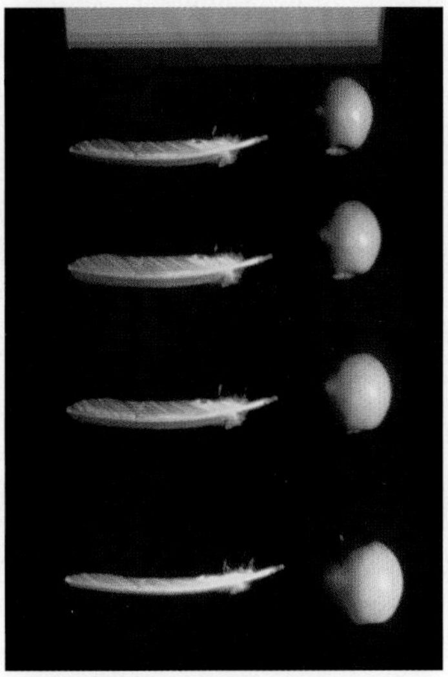

FIGURE 2.19 Stroboscopic photograph of an apple and a feather in free fall in a partially evacuated chamber. The apple and feather were released simultaneously from the trapdoor at the top. The photograph was made by leaving the shutter of the camera open and triggering a flash of light at regular intervals.

The downward acceleration of a freely falling body near the surface of the Earth is usually denoted by g. The numerical value of g is approximately

$$g \approx 9.81 \text{ m/s}^2 \tag{2.26}$$

acceleration of free fall

The exact value of the acceleration of free fall varies somewhat from location to location on the Earth, and it varies with altitude. However, at or near the surface of the Earth this variation amounts to a mere fraction of a percent, and we will neglect it.

For the description of free-fall motion, we can use the formulas for motion with constant acceleration developed in the preceding section. To apply these formulas, we must make a choice for the direction of the x axis. We can take the x axis positive in the upward direction or positive in the downward direction; but once we make one of these choices at the beginning of a problem, we must adhere to it to the end.

For the sake of uniformity, in all the examples in this section, we will take the x axis positive in the upward direction. With this choice of x axis, the acceleration for a freely falling particle is negative, that is, $a = -g$; and Eqs. (2.17), (2.22), and (2.25) become

$$v = v_0 - gt \tag{2.27}$$

free-fall motion

$$x = x_0 + v_0 t - \tfrac{1}{2}gt^2 \tag{2.28}$$

$$-g(x - x_0) = \tfrac{1}{2}(v^2 - v_0^2) \tag{2.29}$$

PROBLEM-SOLVING TECHNIQUES GENERAL GUIDELINES

The solving of problems is an art; there is no simple recipe for obtaining the solutions. Most of the problems in this and the following chapters are applications of the concepts and principles developed in the text. The examples scattered throughout each chapter illustrate typical cases of problem solving. Sometimes you will be able to solve a problem by imitating one of these examples. But if you can't see how to begin the solution, try the following steps:

1. Draw a sketch of the situation described in the problem and label all relevant quantities.
2. If the problem deals with some kind of motion, try to visualize the progress of the motion in time, as though you were watching a movie.
3. For problems that deal with motion, you need to introduce a coordinate axis to describe the motion. Velocities and accelerations must be reckoned as positive when they coincide with the direction of the axis, and negative if they are opposite to the direction of the axis. The origin can be chosen at any convenient location; once chosen, you must use it throughout the problem.
4. Prepare a complete list of the given (known) and sought (unknown) quantities.
5. Ask yourself what physical conditions and principles are applicable to the situation. For instance, does the motion proceed with constant velocity? With constant acceleration? Does the principle of universality of the rate of free fall ($a = -g$) apply?
6. Examine the formulas that are valid under the identified conditions. Then try to spot a formula that permits you to express the unknowns in terms of the known quantities (see Examples 6 and 7). Be discriminating in your selection of formulas—sometimes a formula will tempt you because it displays all the desired quantities, but it will be an invalid formula if the assumptions that went into its derivation are not satisfied in your problem.
7. You will sometimes find that you seem to have too many unknowns and too few equations. Then ask yourself, Are there any special conditions that relate the unknowns (see Example 8)? Are there any quantities that you can calculate from the known quantities? Do these calculated quantities bring you nearer to the answer?
8. It is good technique to solve all the equations by algebraic manipulations and substitute numbers only at the very end; this makes it easier to spot and correct mistakes.
9. When you substitute numbers, also include the units of these numbers. The units in your equations should then combine or cancel in such a way as to give the correct units for the final result. If the units do not combine or cancel in the expected way, something has gone wrong with your algebra.
10. After you have finished your calculations, always check whether the answer is plausible. For instance, if your calculation yields the result that a diver jumping off a cliff hits the water at 3000 km/h, then somebody has made a mistake somewhere!
11. Last, remember to round your final answer to the number of significant figures appropriate for the data given in the problem.

Strictly, these equations are valid only for bodies falling in a vacuum, where there is no frictional air resistance. But they are also good approximations for dense bodies, such as chunks of metal or stone, released in air. For such bodies the frictional resistance offered by the air is unimportant as long as the speed is low (the exact restriction to be imposed on the speed depends on the mass, size, and shape of the body, and on the desired accuracy of the calculation). Unless specifically mentioned, we will ignore the resistance of air, even when the speeds are not all that low.

GALILEO GALILEI (1564–1642)
Italian mathematician, astronomer, and physicist. Galileo demonstrated experimentally that all bodies fall with the same acceleration, and he deduced that the trajectory of a projectile is a parabola. He initiated a new approach to mechanics by recognizing that the natural state of motion of a body, in the absence of forces, is motion with uniform velocity. With a telescope of his own design, he discovered the satellites of Jupiter and sunspots. He vociferously defended the heliocentric system of Copernicus, for which he was condemned by the Inquisition.

EXAMPLE 9 At Acapulco, professional divers amuse tourists by jumping from a 36-m-high cliff into the sea (Fig. 2.20). How long do they fall? What is their impact velocity on the water?

SOLUTION: For this problem, the relevant equations are Eqs. (2.17), (2.22), and (2.25) [or their equivalents with $a = -g$, Eqs. (2.27), (2.28), and (2.29)]. The known quantities are the acceleration of free fall, the change of position, and the initial velocity; the unknown quantities are the time of fall and the final velocity. The initial velocity is $v_0 = 0$ (we assume that the diver merely drops from the initial position, without pushing up or down when jumping off the cliff). The change of position is $x - x_0 = -36$ m. This is negative because the final position is below the initial position; that is, the motion is in the negative x direction (recall that we choose $+x$ upward); see Fig. 2.20. Whether we choose the origin at the top of the cliff ($x_0 = 0$ and $x = -36$ m) or at the bottom ($x = 0$ and $x_0 = 36$ m), the change of position is the same ($x - x_0 = -36$ m).

UNKNOWN	KNOWN
t	$x - x_0 = -36$ m
v	$v_0 = 0$
	$g = 9.81\ \text{m/s}^2$

To calculate the time from the known quantities, we will use Eq. (2.28), in which the time is the only unknown. With $v_0 = 0$, Eq. (2.28) yields

$$x - x_0 = -\tfrac{1}{2}gt^2$$

which we can solve for t by dividing both sides by $-\frac{1}{2}g$ and taking the square root of both sides:

$$t = \sqrt{-\frac{2(x - x_0)}{g}} \tag{2.30}$$

$$= \sqrt{-\frac{2 \times (-36\ \text{m})}{9.81\ \text{m/s}^2}} = 2.7\ \text{s}$$

From Eq. (2.27), the impact velocity is then

$$v = 0 - gt = -9.81\ \text{m/s}^2 \times 2.7\ \text{s}$$
$$= -26\ \text{m/s}$$

This is about 94 km/h!

COMMENT: Remember that the sign of the velocity or acceleration tells you the direction of the velocity or the acceleration. For instance, the result $v = -26$ m/s means that the motion of the diver is opposite to the direction of the x axis; the x axis is directed upward, and the motion of the diver is directed downward.

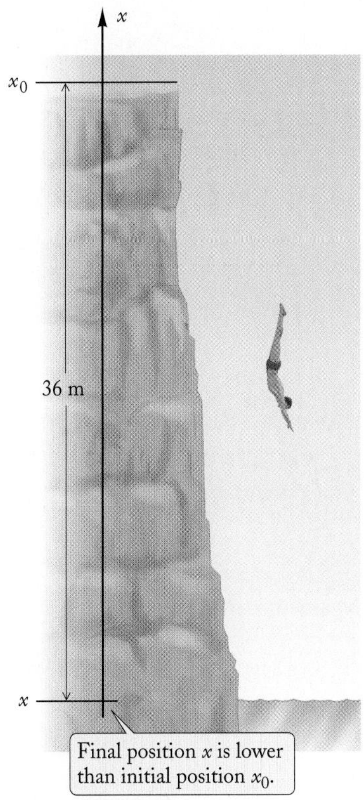

FIGURE 2.20 Jump of a diver. The change of position is negative ($x - x_0 < 0$).

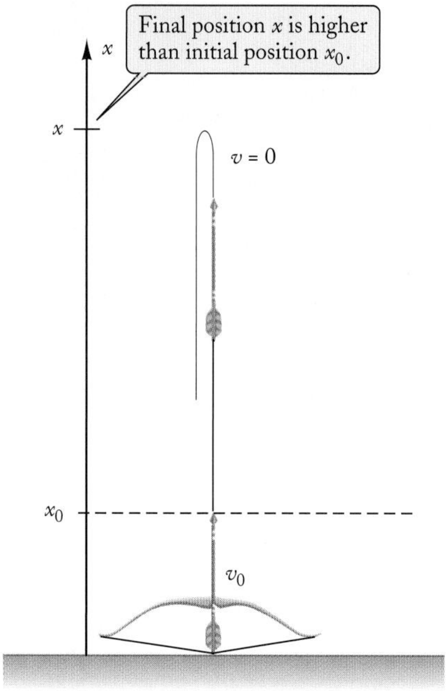

FIGURE 2.21 An ascending arrow. The change of position is positive ($x - x_0 > 0$).

EXAMPLE 10 A powerful bow, like one of those used to establish world records in archery, can launch an arrow at a velocity of 90 m/s. How high will such an arrow rise if aimed vertically upward? How long will it take to return to the ground? What will be its velocity when it hits the ground? For simplicity, ignore air friction and treat the arrow as an ideal particle.

SOLUTION: At the ground, the initial velocity is positive, $v_0 = 90$ m/s (see Fig. 2.21). The arrow moves upward while its velocity decreases at the rate of 9.81 m/s^2. At the highest point of the motion, the arrow ceases to move upward and is momentarily at rest; at this point the instantaneous velocity is zero, $v = 0$. For the upward motion, we can therefore regard the initial and final velocities as known. The height reached and the time are unknown:

UNKNOWN	KNOWN
$x - x_0$	$v_0 = 90$ m/s
t	$v = 0$
	$g = 9.81$ m/s^2

Equation (2.29) relates the height to the known quantities. Dividing this equation by $-g$, we find

$$x - x_0 = \frac{-(v^2 - v_0^2)}{2g} = \frac{-0 + (90\ \text{m/s})^2}{2 \times 9.81\ \text{m/s}^2} = 4.1 \times 10^2\ \text{m}$$

Equation (2.27) relates the time for the upward motion to the known quantities. Solving this equation for t, we find the time for the upward motion:

$$t = \frac{v_0 - v}{g} = \frac{90\ \text{m/s} - 0}{9.81\ \text{m/s}^2} = 9.2\ \text{s} \tag{2.31}$$

The downward motion is simply the reverse of the upward motion—during the downward motion, the arrow accelerates at the rate of 9.81 m/s^2, just as it decelerated at this same rate during the upward motion. The downward motion therefore takes exactly as long as the upward motion, and the total time required for the arrow to complete the up and down motion is twice the time required for the upward motion, that is, 2×9.2 s $= 18.4$ s.

The velocity of the arrow when it hits the ground is simply the reverse of the initial velocity; thus, it is -90 m/s.

COMMENT: Keep in mind that although the instantaneous velocity of the arrow is zero at the highest point of the motion, the acceleration is still the same as that at any other point, $a = -g$. The arrow is momentarily at rest, but it is still accelerating!

Acceleration is sometimes measured in multiples of the "standard" acceleration of gravity; we will call this the **standard** g, where

standard g

$$1\ \text{standard}\ g = 9.80665\ \text{m/s}^2 \approx 9.81\ \text{m/s}^2 \tag{2.32}$$

Note the distinction between g and standard g: g is the acceleration of gravity at or near the surface of the Earth—its value is approximately 9.81 m/s^2, but its exact value depends on location. The standard g is a unit of acceleration—to three significant figures, its value is 9.81 m/s^2 by definition, and it does not depend on location. Thus, the acceleration of gravity at the Earth's surface is approximately 1 standard g, the acceleration of gravity on the Moon's surface (see Table 2.2) is 0.17 standard g, and so on.

EXAMPLE 11 In Example 3, we saw that a catapult launching involved an acceleration of 40 m/s^2. Express this in standard g's.

Concepts — in — Context

SOLUTION: To express this acceleration in standard g's, we multiply by the conversion factor provided by the definition (2.32):

$$a = 40\ \text{m/s}^2 = 40\ \text{m/s}^2 \times \frac{1\ \text{standard}\ g}{9.81\ \text{m/s}^2} = 4.1\ \text{standard}\ g\text{'s}$$

It is sometimes said that the pilot is then "pulling 4.1 g's."

Finally, we make some brief comments on the effects of the frictional resistance of air on bodies falling at high speeds. By holding your hand out of the window of a speeding automobile, you can readily feel that air offers a substantial frictional resistance to motion at speeds in excess of a few tens of kilometers per hour. This frictional resistance increases with speed (roughly in proportion to the square of the speed). Hence a falling, accelerating body will experience a larger and larger frictional resistance as its speed increases. Ultimately, this resistance becomes so large that it counterbalances the pull of gravity—the body ceases to accelerate and attains a constant speed. This ultimate speed is called the **terminal speed**, or **terminal velocity**. The precise value of the terminal speed depends on the mass of the body, its size, and its shape; for instance, a skydiver falling with a closed parachute (see Fig. 2.22) attains a terminal speed of about 200 km/h, whereas a skydiver with an open parachute attains a terminal speed of only about 18 km/h. We will reconsider effects of air friction in Section 6.1, after we become familiar with the concept of force.

FIGURE 2.22 Skydivers falling with closed parachutes.

✓ Checkup 2.6

QUESTION 1: You drop an empty wine bottle and a full wine bottle out of a second-floor window. Does the full wine bottle have a larger acceleration? A larger impact speed on the ground?

QUESTION 2: You drop a full cup of water out of a second-floor window. Will the water spill out of the cup while the cup falls?

QUESTION 3: You throw a stone straight up, so it reaches a maximum height and then falls back down. At the instant the stone reaches maximum height, is its velocity positive, negative, or zero? Is its acceleration positive, negative, or zero? Assume the x axis is directed upward.

QUESTION 4: What is the acceleration of a falling skydiver who has attained his terminal velocity?

QUESTION 5: How would Eqs. (2.27) to (2.29) be different if, instead of taking the x axis positive in the upward direction, we took it positive in the downward direction?

QUESTION 6: According to Example 10, an arrow launched upward at 90 m/s reaches a maximum height of 410 m. What maximum height will the arrow reach when launched with a speed half as large?

(A) 820 m (B) 410 m (C) 205 m (D) 103 m (E) 51 m

2.7 INTEGRATION OF THE EQUATIONS OF MOTION[2]

In the sections above, we dealt with the special case of motion with constant acceleration. However, often the acceleration is not constant, as in Fig. 2.13. In this section we will see how to obtain the position as a function of time when the acceleration is time-dependent. Suppose that the instantaneous acceleration is $a = dv/dt$, where a is a function of time. We rearrange this relation and obtain

$$dv = a\,dt$$

We can integrate this expression directly, for example, from the initial value of the velocity v_0 at time $t = 0$ to some final value v at time t (in the following equations, the integration variables are indicated by primes to distinguish them from the upper limits of integration):

$$\int_{v_0}^{v} dv' = \int_0^t a\,dt'$$

$$v - v_0 = \int_0^t a\,dt' \tag{2.33}$$

This gives the velocity as a function of time:

$$v = v_0 + \int_0^t a\,dt' \tag{2.34}$$

[2] Students not familiar with integral calculus should read Section 7.2 before reading this section.

If the acceleration a were constant, we could bring it outside the integral and immediately obtain our previous result, Eq. (2.17). Equation (2.34) enables us to calculate the velocity as a function of time for any acceleration which is a known function of time.

If the velocity $v = dx/dt$ is a known function of time, then the position x can be obtained in a similar manner:

$$dx = v\,dt$$

$$\int_{x_0}^{x} dx' = \int_0^t v\,dt'$$

$$x - x_0 = \int_0^t v\,dt' \tag{2.35}$$

In the special case of constant acceleration a, we could insert Eq. (2.17), $v = v_0 + at$, into the last integral: we then would again obtain Eq. (2.22). When a is not constant, we must first use Eq. (2.34) to find v as a function of time and then insert this function in Eq. (2.35) to find the position as a function of time, as in the following example.

EXAMPLE 12 A motorcycle accelerates quickly from rest, with an acceleration that has an initial value $a_0 = 4.0\ \text{m/s}^2$ at $t = 0$ and decreases to zero during the interval $0 \le t \le 5.0$ s according to

$$a = a_0\left(1 - \frac{t}{5.0\ \text{s}}\right)$$

After $t = 5.0$ s, the motorcycle maintains a constant velocity. What is this final velocity? In the process of "getting up to speed," how far does the motorcycle travel?

SOLUTION: The acceleration a is given as an explicit function of time. Since we are beginning from rest, the initial velocity is $v_0 = 0$, so Eq. (2.34) gives v as a function of t:

$$v = v_0 + \int_0^t a\,dt' = 0 + \int_0^t a_0\left(1 - \frac{t'}{5.0\ \text{s}}\right)dt'$$

$$= a_0\left(\int_0^t dt' - \frac{1}{5.0\ \text{s}}\int_0^t t'\,dt'\right) = a_0\left(t'\Big|_0^t - \frac{1}{5.0\ \text{s}}\,\frac{t'^2}{2}\Big|_0^t\right)$$

$$= a_0\left(t - \frac{t^2}{10\ \text{s}}\right)$$

where we have used the properties that the integral of the sum is the sum of the integrals and that $\int t^n\,dt = t^{n+1}/(n+1)$. At $t = 5.0$ s, this velocity reaches its final value of

$$v = 4.0\ \text{m/s}^2 \times \left[5.0\ \text{s} - \frac{(5.0\ \text{s})^2}{10\ \text{s}}\right] = 10\ \text{m/s}$$

To obtain the distance traveled during the acceleration, we must insert the time-dependent velocity into Eq. (2.35):

$$x - x_0 = \int_0^t v\,dt' = \int_0^t a_0\left(t' - \frac{t'^2}{10\text{ s}}\right)dt'$$

$$= a_0\left(\int_0^t t'\,dt' - \frac{1}{10\text{ s}}\int_0^t t'^2\,dt'\right) = a_0\left(\frac{t'^2}{2}\Big|_0^t - \frac{1}{10\text{ s}}\frac{t'^3}{3}\Big|_0^t\right)$$

$$= a_0\left(\frac{t^2}{2} - \frac{t^3}{30\text{ s}}\right)$$

Evaluating this expression at $t = 5.0$ s, we find

$$x - x_0 = 4.0\text{ m/s}^2 \times \left[\frac{(5.0\text{ s})^2}{2} - \frac{(5.0\text{ s})^3}{30\text{ s}}\right] = 33\text{ m}$$

Sometimes the acceleration is a known function of the velocity, instead of a known function of time. This is true, for example, for the case of frictional air resistance discussed in Section 2.6 (we will examine this in detail in Chapter 6). Consider one-dimensional motion with an acceleration that is a function of the velocity. In this case, it is possible to integrate the relation $a = dv/dt$ by first rearranging it:

$$dt = \frac{dv}{a}$$

and integrating from time $t = 0$, when $v = v_0$, to some time t:

$$\int_0^t dt' = \int_{v_0}^{v} \frac{dv'}{a}$$

or simply

$$t = \int_{v_0}^{v} \frac{dv'}{a} \tag{2.36}$$

This provides t as a function of v (and v_0), which can sometimes be easily inverted to find v as a function of t.

We have now seen that direct integration of the equations of motion can be applied when the acceleration is known as a function of time or velocity.

✔ Checkup 2.7

QUESTION 1: Beginning at $t = 0$, a particle accelerates from rest and then moves in one dimension. The acceleration increases from zero in proportion to the time t. By what factor is the particle's speed at $t = 2$ s greater than it was at $t = 1$ s?

QUESTION 2: A water rocket accelerates from rest beginning at $t = 0$ so that its acceleration increases in proportion to the time t. The distance traveled increases with time in proportion to what function of time?

(A) $\ln t$ (B) t^2 (C) t^3 (D) t^4

SUMMARY

MATH HELP Differential Calculus; Rules for Derivatives **(page 38)**

PHYSICS IN PRACTICE Stopping Distances **(page 47)**

PROBLEM-SOLVING TECHNIQUES General Guidelines **(page 50)**

AVERAGE SPEED $\dfrac{[\text{total distance traveled}]}{[\text{time taken}]}$ **(2.1)**

AVERAGE VELOCITY $\bar{v} = \dfrac{\Delta x}{\Delta t}$ **(2.5)**

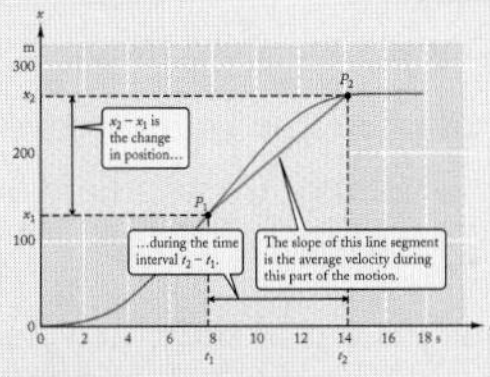

INSTANTANEOUS VELOCITY $v = \dfrac{dx}{dt}$ **(2.10)**

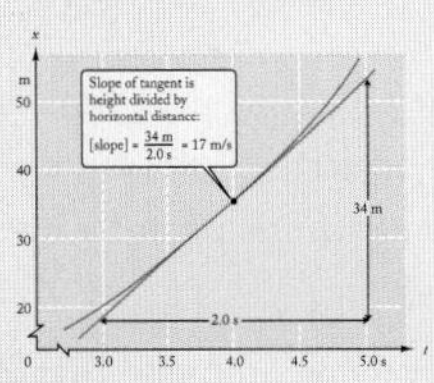

AVERAGE ACCELERATION $\bar{a} = \dfrac{\Delta v}{\Delta t}$ **(2.13)**

INSTANTANEOUS ACCELERATION $a = \dfrac{dv}{dt}$ **(2.14)**

MOTION WITH CONSTANT ACCELERATION

$$v = v_0 + at \qquad \textbf{(2.17)}$$

$$x - x_0 = \tfrac{1}{2}(v_0 + v)t \qquad \textbf{(2.21)}$$

$$x - x_0 = v_0 t + \tfrac{1}{2}at^2 \qquad \textbf{(2.22)}$$

$$a(x - x_0) = \tfrac{1}{2}(v^2 - v_0^2) \qquad \textbf{(2.23)}$$

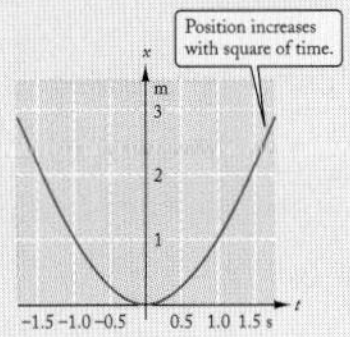

ACCELERATION OF FREE FALL $g \approx 9.81\ \text{m/s}^2$ **(2.26)**

STANDARD g 1 standard $g = 9.81\ \text{m/s}^2$ **(2.32)**

MOTION OF FREE FALL (x axis is upward)

$$v = v_0 - gt \qquad \textbf{(2.27)}$$

$$x - x_0 = v_0 t - \tfrac{1}{2}gt^2 \qquad \textbf{(2.28)}$$

$$-g(x - x_0) = \tfrac{1}{2}(v^2 - v_0^2) \qquad \textbf{(2.29)}$$

QUESTIONS FOR DISCUSSION

1. The motion of a runner can be regarded as particle motion, but the motion of a gymnast cannot. Explain.
2. A newspaper reports that the world record for speed skiing is 203.160 km/h. This speed was measured on a 100-m "speed trap." The skier took about 1.7 s to cross this trap. In order to calculate speed to six significant figures, we need to measure distance and time to six significant figures. What accuracy in distance and time does this require?
3. Do our sense organs permit us to feel velocity? Acceleration?
4. What is your velocity at this instant? Is this a well-defined question? What is your acceleration at this instant?
5. Does the speedometer on your car give you speed or velocity? Does the speedometer care whether you drive eastward or westward along a road?
6. Suppose at one instant in time the velocity of a body is zero. Can this body have a nonzero acceleration at this instant? Give an example.
7. Give an example of a body in motion with instantaneous velocity and acceleration of the same sign. Give an example of a body in motion with instantaneous velocity and acceleration of opposite signs.
8. Experienced drivers recommend that when driving in traffic you should stay at least 2 s behind the car in front of you. This is equivalent to a distance of about two car lengths for every 15 km/h. Why is it necessary to leave a larger distance between the cars when the speed is larger?
9. Is the average speed equal to the average magnitude of the velocity? Is the average speed equal to the magnitude of the average velocity?
10. In the seventeenth century Galileo Galilei investigated the acceleration of gravity by rolling balls down an inclined plane. Why did he not investigate the acceleration directly by dropping a stone from a tower?
11. Why did astronauts find it easy to jump on the Moon? If an astronaut can jump to a height of 20 cm on the Earth (with his space suit), how high can he jump on the Moon?
12. An elevator is moving upward with a constant velocity of 5 m/s. If a passenger standing in this elevator drops an apple, what will be the acceleration of the apple relative to the elevator?
13. Some people are fond of firing guns into the air when under the influence of drink or patriotic fervor. What happens to the bullets? Is this practice dangerous?
14. A woman riding upward in an elevator drops a penny in the elevator shaft when she is passing the third floor. At the same instant, a man standing at the elevator door at the third floor also drops a penny in the elevator shaft. Which coin hits the bottom first? Which coin hits with the higher speed? Neglect friction.
15. If you take the frictional resistance of air into account, how does this change your answers to Question 14?
16. Suppose that you drop a $\frac{1}{2}$-kg packet of sugar and a $\frac{1}{2}$-kg ball of lead from the top of a building. Taking air friction into account, which will take the shorter time to reach the ground? Suppose you place the sugar and the lead in identical sealed glass jars before dropping them. Which will now take the shorter time?
17. Is air friction important in the falling motion of a raindrop? If a raindrop were to fall without friction from a height of 300 m and hit you, what would it do to you?
18. An archer shoots an arrow straight up. If you consider the effects of the frictional resistance of air, would you expect the arrow to take a longer time to rise or to fall?

PROBLEMS

2.1 Average Speed

1. The speed of nerve pulses in mammals is typically 10^2 m/s. If a shark bites the tail of a 30-m-long whale, roughly how long will it take before the whale knows of this?
2. The world record for the 100-yard run is 9.0 s. What is the corresponding average speed in miles per hour?
3. Glaciers sometimes advance 20 m per year. Express this speed in m/s and in cm/day.
4. In the fall, the monarch butterfly migrates some 3500 km from the Northeastern United States to Mexico. Guess the speed of the butterfly, and estimate how long it takes to make this trip.
5. In 1958 the nuclear-powered submarine *Nautilus* took 6 days and 12 hours to travel submerged 5068 km across the Atlantic from Portland, England, to New York City. What was the average speed (in km/h) for this trip?

6. A galaxy beyond the constellation Corona Borealis is moving directly away from our Galaxy at the rate of 21 600 km/s. This galaxy is now at a distance of 1.4×10^9 light-years from our Galaxy. Assuming that the galaxy has always been moving at a constant speed, how many years ago was it right on top of our Galaxy?

7. On one occasion a tidal wave (tsunami) originating near Java was detected in the English Channel 32 h later. Roughly measure the distance from Java to England by sea (around the Cape of Good Hope) on a map of the world, and calculate the average speed of the tidal wave.

8. In 1971, Francis Chichester, in the yacht *Gypsy Moth V*, attempted to sail the 4000 nautical miles (nmi) from Portuguese Guinea to Nicaragua in no more than 20 days.
   (a) What minimum average speed (in nautical miles per hour) does this require?
   (b) After sailing 13 days, he still had 1720 nmi to go. What minimum average speed did he require to reach his goal in the remaining 7 days? Knowing that his yacht could at best achieve a maximum speed of 10 nmi/h, what could he conclude at this point?

9. The first "marathon" was run in 490 B.C. by the Greek Pheidippides, who ran the 35 km (22 mi) from the battlefield of Marathon to Athens in 2.5 h. What was his average speed?

10. In the open ocean, dolphins sometimes travel 110 km per day, and in short bursts, they can attain speeds of 32 km/h. Express the average travel speed and the burst speed in m/s.

11. The surface of the Earth consists of several large plates which move relative to each other. The Pacific plate (which includes a slice of the coastal part of California) is sliding northward along the continental North American plate at a speed of 4.0 cm per year. How long does it take for this plate to move 1.0 km? 1.0×10^3 km?

12. In setting a world record for speed, the horse Big Racket ran 402 m ($\frac{1}{4}$ mile) in 23.8 s. What was its average speed?

13. The speed-skating record for women is held by Bonnie Blair, who completed a 500-m race in 39.10 s. What is the corresponding average speed? Give the answer in m/s and in km/h.

14. The fastest speed ever measured for a tennis ball served by a player was 263 km/h. At this speed, how long does the tennis ball take to go from one end of the court to the other, a distance of 23.8 m?

15. The distance from New York to Belem, Brazil, is 5280 km. How long does it take you to travel this distance by airliner, at 900 km/h? How long does it take you by ship, at 35 km/h?

16. At the 1988 Olympics, Florence Griffiths-Joyner (Fig. 2.23) ran 100 m in 10.54 s, and she ran 200 m in 21.34 s. What was her average speed in each case?

17. In the case of the closest verdict for a championship ski race, the winner of a cross-country race reached the finish line one hundredth of a second ahead of his closest competitor. If both were moving at a speed of 6 m/s, what was the distance between them at the finish?

FIGURE 2.23 Florence Griffiths-Joyner wins the race.

*18. A hunter shoots an arrow at a deer running directly away from him. When the arrow leaves the bow, the deer is at a distance of 40 m. When the arrow strikes, the deer is at a distance of 50 m. The speed of the arrow is 65 m/s. What must have been the speed of the deer? How long did the arrow take to travel to the deer?

*19. The fastest land animal is the cheetah, which runs at a speed of up to 101 km/h. The second-fastest is the antelope, which runs at a speed of up to 88 km/h.
   (a) Suppose that a cheetah begins to chase an antelope. If the antelope has a head start of 50 m, how long does it take the cheetah to catch the antelope? How far will the cheetah have traveled by this time?
   (b) The cheetah can maintain its top speed for only about 20 s (and then has to rest), whereas the antelope can continue at top speed for a considerably longer time. What is the maximum head start the cheetah can allow the antelope?

20. Carl Lewis set an Olympic record for the 100-m run in 1991 with a time of 9.86 s. What was his average speed?

21. The distance for a marathon is 26 mi 385 yd. In 1984, Joan Benoit set an Olympic record with a time of 2 h 24 min 52 s. What was her average speed in m/s?

*22. On a particular wristwatch, the tips of the minute and second hands are each 0.90 cm from the center, whereas the tip of the hour hand is 0.50 cm from the center. What is the average speed of each of these three tips?

*23. The position of a runner is given by $x = 4.0t - 0.50t^2$, where x is in meters and t is in seconds. What is the average speed between $t = 0$ and $t = 8.0$ s? (Hint: Find the maximum value of x to determine each of the outward and backward distances.)

*24. You make a 100-km trip in an automobile, traveling the first 50 km at 60 km/h and the second 50 km at 80 km/h. What is your average speed for this trip? Explain why the average speed is not 70 km/h.

*25. The table printed inside the book cover gives the radii of the orbits of the planets (mean distance from the Sun) and the times required for moving around the orbit (period of revolution).

(a) Calculate the speed of motion of each of the nine planets in its orbit around the Sun. Assume that the orbits are circular.

(b) In a logarithmic graph of speed vs. radius, plot the logarithm of the speed of each planet and the logarithm of its radial distance from the Sun as a point. Draw a curve through the nine points. Can you represent this curve by a simple equation?

2.2 Average Velocity for Motion along a Straight Line†

26. Suppose you throw a baseball straight up so that it reaches a maximum height of 8.00 m and returns 2.55 s after you throw it. What is the average speed for this motion of the ball? What is the average velocity?

27. Consider the automobile with the position as a function of time plotted in Fig. 2.5. What is the average velocity for the interval from $t = 0$ to $t = 10.0$ s? From $t = 10.0$ s to $t = 14.3$ s?

28. An elevator travels up from the first floor to the ninth floor in 20 s, down from the ninth floor to the fifth floor in 12 s, and finally up from the fifth floor to the twelfth floor in 18 s. The spacing between any two adjacent floors is 4.0 m. For the entire period of 50 s, what is the average speed? What is the average velocity?

29. A messenger carries a package twelve city blocks north in 14 min 5 s; there, she receives a second package, which she brings six blocks south in 6 min 28 s; finally, she gets a third package, which she brings 3 blocks north in 3 min 40 s. If the city block spacing is 81 m, what is her average speed? What is her average velocity?

30. The position of a person is given by $x = 4.0t - 0.50t^2$, where x is in meters and t is in seconds. What is the average velocity between $t = 0$ and $t = 8.0$ s? Between $t = 8.0$ s and $t = 10.0$ s?

31. A horse trots around a quarter-mile track three complete times in 1 min 40 s. What is the average speed? The average velocity?

32. A dog runs 35 m to catch a Frisbee in 4.5 s, turns quickly and trots 22 m back in 3.6 s, but then stops to guard the Frisbee. What is the dog's average speed? The dog's average velocity?

33. A squirrel walks along a telephone cable, occasionally stopping, even turning back once before proceeding. The squirrel's position as a function of time is shown in Fig. 2.24. What is the average speed for the entire time shown, $0 \leq t \leq 30$ s? The average velocity for the entire time?

34. The position of an automobile as a function of time is plotted in Fig. 2.25. What is the average velocity for $0 \leq t \leq 2.0$ s? For $2.0 \leq t \leq 4.0$ s? Estimate the instantaneous velocity at $t = 1.0$ s and at $t = 3.0$ s.

† For help, see Online Concept Tutorial 3 at www.wwnorton.com/physics

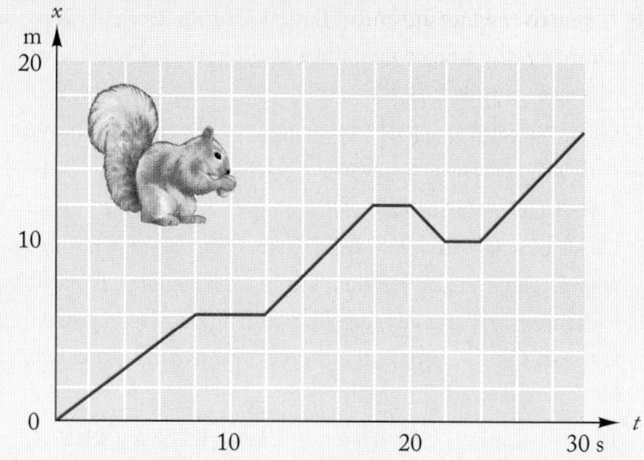

FIGURE 2.24 Position of a squirrel as a function of time.

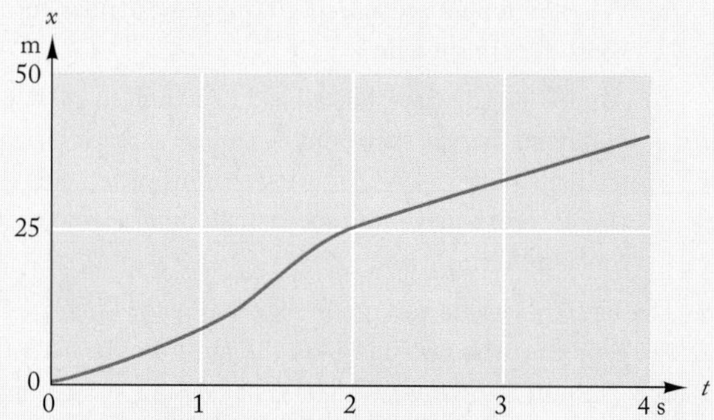

FIGURE 2.25 Automobile position as a function of time.

2.3 Instantaneous Velocity
2.4 Acceleration†

35. In an experiment with a water-braked rocket sled, an Air Force volunteer was subjected to an acceleration of 810 m/s^2 for 0.040 s. What was his change of speed in this time interval?

36. A Porsche racing car takes 2.2 s to accelerate from 0 to 96 km/h (60 mi/h). What is the average acceleration?

37. A football player kicks a stationary ball (Fig. 2.26) and sends it flying. Slow-motion photography shows that the ball is in contact with the foot for 8.0×10^{-3} s and leaves with a speed of 27 m/s. What is the average acceleration of the ball while in contact with the foot?

FIGURE 2.26 Kicking a football.

38. The driver of an automobile traveling at 80 km/h suddenly slams on the brakes and stops in 2.8 s. What is the average deceleration during braking?

†39. (a) The blue curve in Fig. 2.27 is a plot of velocity vs. time for a Triumph sports car undergoing an acceleration test. By drawing tangents to the velocity curve, find the accelerations at time $t = 0, 10, 20, 30$, and 40 s.

(b) The red curve in Fig. 2.27 is a plot of velocity vs. time for the same car when coasting with its gears in neutral. Find the accelerations at time $t = 0, 10, 20, 30$, and 40 s.

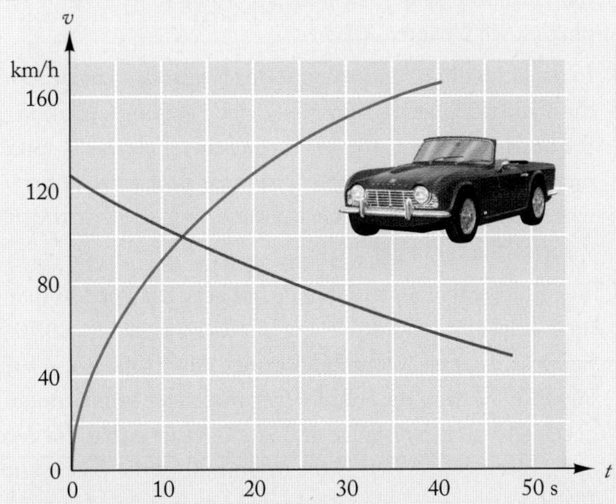

FIGURE 2.27 Instantaneous velocity as a function of time for a Triumph sports car in an acceleration test.

40. The position of a particle in meters is given by $x = 2.5t + 3.1t^2 - 4.5t^3$, where t is the time in seconds. What are the instantaneous velocity and instantaneous acceleration at $t = 0.0$ s? At $t = 2.0$ s? What are the average velocity and average acceleration for the time interval $0 \leq t \leq 2.0$ s?

41. The position of a particle in meters is given by $x = 3.6t^2 - 2.4t^3$, where t is in seconds. Find the two times when the velocity is zero. Calculate the position for each of those times. Sketch x as a function of t for $0 \leq t \leq 2$ s.

42. The velocity of an automobile as a function of time is given by $v = Bt - Ct^2$, where $B = 6.0\ \text{m/s}^2$ and $C = 2.0\ \text{m/s}^3$. At what times is the velocity zero? At what time is the acceleration zero? Sketch v as a function of t for $0 \leq t \leq 4.0$ s.

43. The velocity of a bicycle as a function of time is shown in Fig. 2.28. What is the average acceleration for $0 \leq t \leq 5.0$ s? For $5.0\ \text{s} \leq t \leq 10$ s? Estimate the instantaneous acceleration at $t = 3.0$ s.

*44. The velocity of a parachutist as a function of time is given by $v = v_f + (v_0 - v_f)e^{[-t/(2.5\ \text{s})]}$, where $t = 0$ corresponds to the instant the parachute is opened, $v_0 = 200$ km/h is the velocity before opening of the parachute, and $v_f = 18$ km/h is the final (terminal) velocity. What acceleration does the parachutist experience just after opening the parachute?

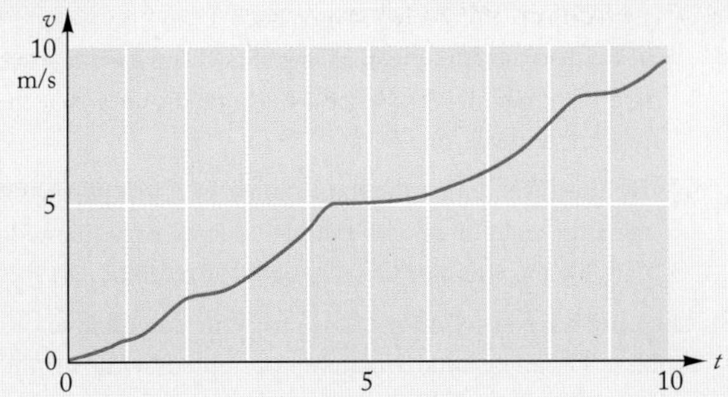

FIGURE 2.28 Bicycle velocity as a function of time.

*45. The velocity of a particle is given by $v = v_0/(1 + At^2)$, where $v_0 = 25$ m/s and $A = 2.0\ \text{s}^{-2}$. What is the acceleration at $t = 0$? At $t = 2.0$ s? At $t \to \infty$?

*46. Consider the red curve in Fig. 2.27 showing velocity vs. time for a decelerating sports car.

(a) From this curve, estimate the average velocity for the interval $t = 0$ s to $t = 5$ s. (Hint: The average velocity for a small time interval is roughly the average of the initial and final velocities; alternatively, it is roughly the velocity at the midpoint of the time interval.) Estimate how far the car travels in this time interval.

(b) Repeat the calculation of part (a) for every 5-s interval between $t = 5$ s and $t = 45$ s.

(c) What is the *total* distance that the car will have traveled at the end of 45 s?

*47. Table 2.3 gives the horizontal velocity as a function of time for a projectile of 100 lb fired horizontally from a 6-inch naval gun. The velocity decreases with time because of the frictional drag of the air.

TABLE 2.3 EFFECT OF AIR RESISTANCE ON A PROJECTILE

TIME	VELOCITY	TIME	VELOCITY
0 s	657 m/s	1.80 s	557 m/s
0.30	638	2.10	542
0.60	619	2.40	528
0.90	604	2.70	514
1.20	588	3.00	502
1.50	571		

(a) On a piece of graph paper, make a plot of velocity vs. time and draw a smooth curve through the points of the plot.

(b) Estimate the average velocity for each time interval [the estimated average velocity for the time interval $t = 0$ s to

† Due to the difficulty of accurately drawing tangents, answers for this problem that differ by up to 10% are acceptable.

$t = 0.30$ s is $\frac{1}{2}(657 + 638)$ m/s, etc.]. From the average velocities calculate the distances that the projectile travels in each time interval. What is the total distance that the projectile travels in 3.00 s?

(c) By directly counting the squares on your graph paper, estimate the area (in units of s·m/s) under the curve plotted in part (a), and compare this area with the result of part (b).

48. The instantaneous velocity of the projectile described in Problem 47 can be approximately represented by the formula (valid for $0 \text{ s} \leq t \leq 3.00$ s)

$$v = 655.9 - 61.14t + 3.26t^2$$

where v is measured in meters per second and t is in seconds. Calculate the instantaneous acceleration of the projectile at $t = 0$ s, $t = 1.50$ s, and $t = 3.00$ s.

*49. The position of a particle as a function of time is given by

$$x = A \cos bt$$

where A and b are constants. Assume that $A = 2.0$ m and $b = 1.0$ radian/s.

(a) Roughly plot the position of this particle for the time interval $0 \text{ s} \leq t \leq 7.0$ s.

(b) At what time does the particle pass the origin ($x = 0$)? What are its velocity and acceleration at this instant?

(c) At what time does the particle reach maximum distance from the origin? What are its velocity and acceleration at this instant?

*50. The motion of a rocket burning its fuel at a constant rate while moving through empty interstellar space can be described by

$$x = u_{ex}\,t + u_{ex}\,(1/b - t)\ln(1 - bt)$$

where u_{ex} and b are constants (u_{ex} is the exhaust velocity of the gases at the tail of the rocket, and b is proportional to the rate of fuel consumption).

(a) Find a formula for the instantaneous velocity of the rocket.

(b) Find a formula for the instantaneous acceleration.

(c) Suppose that a rocket with $u_{ex} = 3.0 \times 10^3$ m/s and $b = 7.5 \times 10^{-3}$/s takes 120 s to burn all its fuel. What is the instantaneous velocity at $t = 0$ s? At $t = 120$ s?

(d) What is the instantaneous acceleration at $t = 0$ s? At $t = 120$ s?

2.5 Motion with Constant Acceleration†

51. The takeoff speed of a jetliner is 360 km/h. If the jetliner is to take off from a runway of length 2100 m, what must be its acceleration along the runway (assumed constant)?

52. A British 6-inch naval gun has a barrel 6.63 m long. The muzzle speed of a projectile fired from the gun is 657 m/s. Assuming that upon detonation of the explosive charge the projectile moves along the barrel with a constant acceleration, what is the magnitude of this acceleration? How long does it take the projectile to travel the full length of the barrel?

53. The nearest star is Proxima Centauri, at a distance of 4.2 light-years from the Sun. Suppose we wanted to send a spaceship to explore this star. To keep the astronauts comfortable, we want the spaceship to travel with a constant acceleration of 9.81 m/s^2 at all times (this will simulate ordinary gravity within the spaceship). If the spaceship accelerates at 9.81 m/s^2 until it reaches the midpoint of its trip and then decelerates at 9.81 m/s^2 until it reaches Proxima Centauri, how long will the one-way trip take? What will be the speed of the spaceship at the midpoint? Do your calculations according to Newtonian physics (actually, the speed is so large that the calculation should be done according to relativistic physics; see Chapter 36).

54. In an accident on motorway M1 in England, a Jaguar sports car made skid marks 290 m long while braking. Assuming that the deceleration was 10 m/s^2 during this skid (this is approximately the maximum deceleration that a car with rubber wheels can attain on ordinary pavements), calculate the initial speed of the car before braking.

55. The front end of an automobile has been designed so that upon impact it progressively crumples by as much as 0.70 m. Suppose that the automobile crashes into a solid brick wall at 80 km/h. During the collision the passenger compartment decelerates over a distance of 0.70 m. Assume that the deceleration is constant. What is the magnitude of the deceleration? If the passenger is held by a safety harness, is he likely to survive? (Hint: Compare the deceleration with the acceleration listed for a parachutist in Table 2.2.)

56. A jet-powered car racing on the Salt Flats in Utah went out of control and made skid marks 9.6 km long. Assuming that the deceleration during the skid was about 5.0 m/s^2, what must have been the initial speed of the car? How long did the car take to come to a stop?

57. The operation manual of a passenger automobile states that the stopping distance is 50 m when the brakes are fully applied at 96 km/h. What is the deceleration? What is the stopping time?

58. With an initial speed of 260 km/h, the French TGV (*train à grande vitesse*) takes 1500 m to stop on a level track. Assume that the deceleration is constant. What is the magnitude of the deceleration? What is the time taken for stopping?

59. An automobile accelerates from rest for 20 s with a constant acceleration of 1.5 m/s^2. What is its final velocity? How far does it travel in that time?

60. At one instant, a body in motion with constant acceleration has velocity 3.0 m/s when its position is $x = 5.0$ m. At a time 4.0 s later, its position is $x = 1.0$ m. What is the acceleration? What is the body's velocity at the later instant?

61. A small airplane accelerates from rest at a constant acceleration of 1.2 m/s^2 along the runway. The plane rises from the runway after traveling 150 m from its starting point. How long did this take?

62. A Concorde jet traveling at 550 km/h accelerates uniformly at 0.60 m/s^2 for 90 s. What is its final velocity?

† For help, see Online Concept Tutorial 3 at www.wwnorton.com/physics

63. A type-A driver guns the engine when a light turns green, and then brakes somewhat quickly for the next red light; the acceleration as a function of time is shown in Fig. 2.29 for $0 \leq t \leq 10$ s. If the driver started from rest at the origin at $t = 0$, accurately sketch the velocity and position as functions of time.

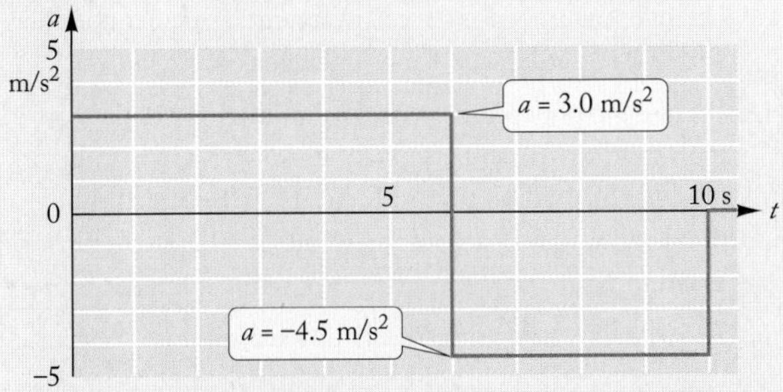

FIGURE 2.29 Automobile acceleration as a function of time.

*64. A truck is initially traveling on a highway at constant speed. The driver wishes to pass an automobile and so begins a 30-s period of constant acceleration at 0.050 m/s^2. During this period, the truck travels 700 m. What was the truck's initial speed? The truck's final speed?

*65. A train with velocity v_0 accelerates uniformly at 0.50 m/s^2 for 15 s. It travels 550 m during this time. What is its final velocity? What is v_0?

*66. In a drag race a car starts at rest and attempts to cover 440 yd in the shortest possible time. The world record for a piston-engined car is 5.637 s; while setting this record, the car reached a final speed of 250.69 mi/h at the 440-yd mark.

(a) What was the average acceleration for the run?

(b) Prove that the car did not move with constant acceleration.

(c) What would have been the final speed if the car had moved with constant acceleration so as to reach 440 yd in 5.637 s?

*67. In a large hotel, a fast elevator takes you from the ground floor to the 21st floor. The elevator takes 17 s for this trip: 5 s at constant acceleration, 7 s at constant velocity, and 5 s at constant deceleration. Each floor in the hotel has a height of 2.5 m. Calculate the values of the acceleration and deceleration (assume they are equal). Calculate the maximum speed of the elevator.

*68. (a) In a skyscraper, an elevator takes 55 s to descend from the top floor to ground level, a distance of 400 m. What is the average speed of the elevator for this trip?

(b) The elevator is at rest at the beginning and at the end of the trip. If you wanted to program the elevator so that it completes the trip in the specified time with a minimum acceleration and a minimum deceleration, how would you have to accelerate and decelerate the elevator? What would be this minimum value of the acceleration and deceleration? What would be the maximum speed during the trip?

*69. For a sleepy or drunk driver, the reaction time is much longer than for an alert driver. Recalculate the table of stopping distances for an automobile in Section 2.5 if the reaction time is 2.0 s instead of 0.75 s.

*70. As discussed in Section 2.5, the total stopping distance for an automobile has two contributions: the reaction-time contribution and the braking contribution. On the basis of the data given in the table on page 47, for what initial speed are these two contributions equal?

*71. In a collision, an automobile initially traveling at 50 km/h decelerates at a constant rate of 200 m/s^2. A passenger not wearing a seat belt crashes against the dashboard. Before the collision, the distance between the passenger and the dashboard was 0.60 m. With what speed, relative to the automobile, does the passenger crash into the dashboard? Assume that the passenger has no deceleration before contact with the dashboard.

**72. Figure 2.30 (adapted from a diagram in the operation manual of an automobile) describes the automobile's passing ability at low speed. From the data supplied in this figure, calculate the acceleration of the automobile during the pass and the time required for the pass. Assume constant acceleration.

*73. The speed of a body released from rest falling through a viscous medium (for instance, an iron pellet falling in a jar full of oil) is given by the formula

$$v = -g\tau + g\tau e^{-t/\tau}$$

where τ is a constant that depends on the size and shape of the body and on the viscosity of the medium and $e = 2.718\ldots$ is the basis of the natural logarithms.

(a) Find the acceleration as a function of time.

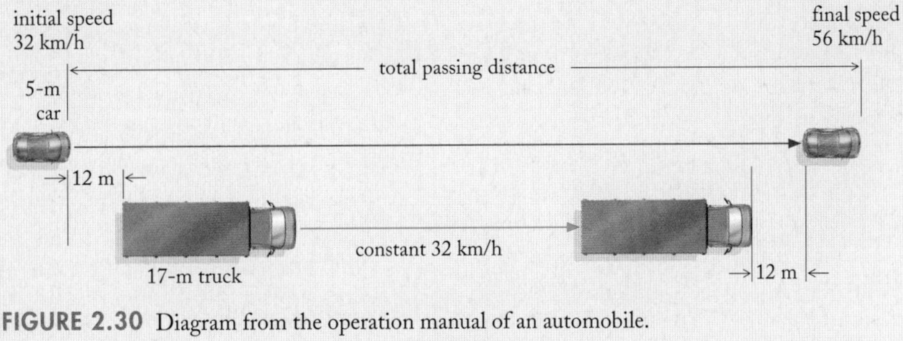

FIGURE 2.30 Diagram from the operation manual of an automobile.

(b) Show that for $t \to \infty$, the speed approaches the terminal value $-g\tau$.

(c) By differentiation, verify that the equation for the position as a function of time consistent with the above expression for the speed is

$$x = -g\tau t - g\tau^2 e^{-t/\tau} + g\tau^2 + x_0$$

(d) Show that for small values of t ($t \ll \tau$), the equation for x is approximately $x \approx -\frac{1}{2}gt^2 + x_0$.

2.6 The Acceleration of Free Fall†

74. An apple drops from the top of the Empire State Building, 380 m above street level. How long does the apple take to fall? What is its impact velocity on the street? Ignore air resistance.

75. Peregrine falcons dive on their prey with speeds of up to 130 km/h. From what height must a falcon fall freely to achieve this speed? Ignore air resistance.

76. Cats are known for their ability to survive falls from buildings several floors high. If a cat falls three floors down, a distance of 8.7 m, what is its speed when it hits the ground?

77. The world record for a high jump from a standing position is 1.90 m. With what speed must a jumper leave the ground to attain this height? (Treat this as particle motion, although it really is not.)

78. The muzzle speed of a .22-caliber bullet fired from a rifle is 366 m/s. If there were no air resistance, how high would this bullet rise when fired straight up?

79. An engineer standing on a bridge drops a penny toward the water and sees the penny splashing into the water 3.0 s later. How high is the bridge?

80. The volcano Loki on Io, one of the moons of Jupiter, ejects debris to a height of 200 km (Fig. 2.31). What must be the initial ejection velocity of the debris? The acceleration of gravity on Io is 1.80 m/s^2. There is no atmosphere on Io, hence no air resistance.

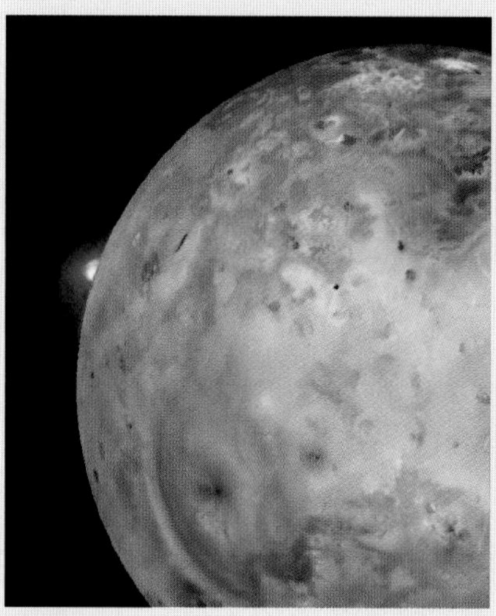

FIGURE 2.31 The volcano Loki on Io.

† For help, see Online Concept Tutorial 3 at www.wwnorton.com/physics

81. The crash of an automobile into a solid barrier can be simulated by dropping the automobile, nose down, from some height onto a hard platform (some of the early crash tests were performed in this manner). From what height must you drop the automobile to simulate a crash at 45 km/h? 75 km/h? 105 km/h? To gain a better appreciation of what these heights mean, express each height as a number of floors up a typical apartment building, with 2.9 m per floor (2.9 m is one floor up, etc.).

82. In diving competitions, a diver jumping from a platform 10 m high performs $1\frac{1}{2}$ somersaults (head up, to down, to up, to down) while falling toward the water. How much time does the diver have for each $\frac{1}{2}$ somersault (head up to down, or down to up)?

83. A hardball is thrown vertically upward, attains a height of 9.5 m above the point of release, falls downward, and is caught by the thrower. How long was it in the air? What was its initial velocity?

84. A Super Ball is thrown to the ground with an initial downward velocity of 1.0 m/s and pops back up to the hand of the thrower magnitude as the initial velocity; the total time elapsed is 0.75 s. What height was the Super Ball thrown from?

85. Standing at the edge of a building of height 9.2 m, a woman reaches out over the edge and throws a penny vertically upward. It rises, then falls past her, hitting the ground below 2.5 s after release. What height did the penny attain? What was its initial velocity?

86. A stone is released from rest and falls through a height h. What is the average speed of the stone during this fall?

*87. A rubber ball is thrown vertically downward from a height of 1.5 m and bounces back to the same height with a reversed velocity of the same in a total time of 0.90 s. What was the initial downward velocity? What was the velocity just before hitting the floor?

*88. A particle is initially at rest at some height. If the particle is allowed to fall freely, what distance does it cover in the time from $t = 0$ s to $t = 1$ s? From $t = 1$ s to $t = 2$ s? From $t = 2$ s to $t = 3$ s? Show that these successive distances are in the ratios 1:3:5:7 . . . .

*89. Galileo claimed in his *Dialogues* that "the variation of speed in air between balls of gold, lead, copper, porphyry, and other heavy materials is so slight that in a fall of 100 cubits a ball of gold would surely not outstrip one of copper by as much as four fingers. Having observed this, I came to the conclusion that in a medium totally void of resistance all bodies would fall with the same speed." One cubit is about 46 cm, and four fingers are about 10 cm. According to Galileo's data, what is the maximum percent difference between the accelerations of the balls of gold and of copper?

*90. Quechee gorge in Vermont has a depth of 45 m. If you want to measure this depth within 10% by timing the fall of a stone dropped from the bridge across the gorge, how accurately must you measure the time? Is an ordinary watch with a second hand adequate for this task, or do you need a stopwatch?

*91. A golf ball released from a height of 1.5 m above a concrete floor bounces back to a height of 1.1 m. If the ball is in contact with the floor for 6.2×10^{-4} s, what is the average acceleration of the ball while in contact with the floor?

*92. In 1978 the stuntman A. J. Bakunas died when he jumped from the 23rd floor of a skyscraper and hit the pavement. The air bag that was supposed to cushion his impact ripped.

(a) The height of his jump was 96 m. What was his impact speed?

(b) The air bag was 3.7 m thick. What would have been the man's deceleration had the air bag not ripped? Assume that his deceleration would have been uniform over the 3.7-m interval.

*93. The HARP (High-Altitude Research Project) gun can fire an 84-kg projectile containing scientific instruments straight up to an altitude of 180 km. If we pretend there is no air resistance, what muzzle speed is required to attain this altitude? How long does the projectile remain at a height in excess of 100 km, the height of interest for high-altitude research?

*94. The International Geodesy Association has adopted the following formula from the acceleration of gravity as a function of the latitude Θ (at sea level):

$$g = 978.0318\ \text{cm/s}^2 \times (1 + 53.024 \times 10^{-4} \sin^2 \Theta - 5.9 \times 10^{-6} \sin^2 2\Theta)$$

(a) According to this formula, what is the acceleration of gravity at the equator? At a latitude of 45°? At the pole?

(b) Show that according to this formula, g has a minimum at the equator, a maximum at the pole, and no minima or maxima at intermediate latitudes.

*95. At a height of 1500 m, a dive bomber in a vertical dive at 300 km/h shoots a cannon at a target on the ground. Relative to the bomber the initial speed of the projectile is 700 m/s. What will be the impact speed of the projectile on the ground? How long will it take to get there? Ignore air friction in your calculation.

**96. The elevators in the CN tower in Toronto travel at 370 m/min from ground level to the Skypod, 335 m up. Suppose that when an elevator begins to rise from ground level, you drop a penny from the Skypod down the elevator shaft. At what height does the elevator meet the penny?

**97. You throw a baseball straight up, so it bounces off a ceiling 10 m above your hand and returns to your hand 2.0 s later. What initial speed is required? What is the speed at impact? Assume that the impact on the ceiling reverses the velocity of the baseball but does not change its magnitude.

**98. Suppose you throw a stone straight up with an initial speed of 15.0 m/s.

(a) If you throw a second stone straight up 1.00 s after the first, with what speed must you throw the second stone if it is to hit the first at a height of 11.0 m? (There are two answers. Are both plausible?)

(b) If you throw the second stone 1.30 s after the first, with what speed must you throw the second stone if it is to hit the first at a height of 11.0 m?

***99. Raindrops drip from a spout at the edge of a roof and fall to the ground. Assume that the drops drip at a steady rate of n drops per second (where n is large) and that the height of the roof is h.

(a) How many drops are in the air at one instant?

(b) What is the median height of the drops (i.e., the height above and below which an equal number of drops are found)?

(c) What is the average of the heights of these drops?

2.7 Integration of the Equations of Motion

100. An automobile has an initial velocity $v_0 = 8.0$ m/s. During $0 \le t \le 3.0$ s, it experiences a time-dependent acceleration given by $a = Ct^2$, where $C = 0.25\ \text{m/s}^4$. What is the instantaneous velocity at $t = 3.0$ s? What is the change in position between $t = 0$ and $t = 1.0$ s?

101. A particle is initially at rest. Beginning at $t = 0$, it begins moving, with an acceleration given by $a = a_0\{1 - [t^2/(4.0\ \text{s}^2)]\}$ for $0 \le t \le 2.0$ s and $a = 0$ thereafter. The initial value is $a_0 = 20\ \text{m/s}^2$. What is the particle's velocity after 1.0 s? After a long time? How far has the particle traveled after 2.0 s?

102. A particle moves in one dimension and experiences an acceleration which varies with time, given by $a = At + Bt^2$, where $A = 15\ \text{m/s}^3$ and $B = 25\ \text{m/s}^4$. The particle is at rest at $t = 0$. What is its speed at $t = 2.0$ s? How far does it travel between $t = 1.0$ s and $t = 2.0$ s?

*103. A parachutist in free fall has a downward speed v_0 when she opens her parachute. Thereafter, she experiences an acceleration proportional to her speed given by $a = g - Av$, where A is a constant (which depends on the size and shape of the parachute, the air density, and the mass of the parachutist). Show that her speed as a function of time is given by

$$v = \frac{g}{A}(1 - e^{-At}) + v_0 e^{-At}$$

After a time $t \gg 1/A$, v approaches the constant value $v = g/A$; this is the terminal velocity.

*104. At moderate to high speeds, an automobile coasting in neutral experiences a deceleration (due to air resistance) of the form $a = -Bv^2$, where $B = 6.1 \times 10^{-4}\ \text{m}^{-1}$ and v is the speed in m/s. The driver of a car traveling horizontally at 120 km/h on a level road shifts into neutral. Calculate the amount of time required for the car to slow to 90 km/h.

REVIEW PROBLEMS

105. In 1993, Noureddine Morcelli set a new record of 3 min 44.39 s for a run of 1 mile, which superseded the previous record of 3 min 46.32 s by Steve Cram. If both runners had run together, at constant speed, how far behind would Cram have been at the finish line?

106. The world record for speed skating set in Salt Lake City in 2002 by Derek Parra was 1 min 43.95 s for 1500 m. What was his average speed?

107. A runner runs 100 m in 10 s, rests 60 s, and returns at a walk in 80 s. What is the average speed for the complete motion? What is the average velocity?

108. You travel in your car for 30 min at 35 km/h, and then for another 30 min at 85 km/h.

(a) What distance do you cover in the first 30 min? In the second 30 min?

(b) What is your average speed for the entire trip?

109. The fastest predator in the ocean is the sailfish. When pursuing prey, it swims at 109 km/h. Suppose a sailfish spots a mackerel at a distance of 20 m. The mackerel attempts to escape at 33 km/h, on a straight path. How long does the sailfish take to catch it? How far does the sailfish travel during the pursuit?

*110. The distance from San Francisco to Vancouver is 1286 km by air. A plane leaves San Francisco at 10:00 A.M. heading north toward Vancouver, and another plane leaves Vancouver at 11:00 A.M. heading south toward San Francisco. The first plane travels at 720 km/h, and the second plane, slowed by a headwind, travels at 640 km/h. Where do the planes meet? At what time?

*111. You are traveling on a highway at 80 km/h and you are overtaking a car traveling in the same direction at 50 km/h. How long does it take you to go from 10 m behind this car to 10 m ahead of this car? The length of the car is 4 m.

*112. The highest speed attained by a cyclist on level ground is 105 km/h. To attain this speed, the cyclist used a streamlined recumbent bike. Starting from rest, he gradually built up his speed by pedaling furiously over a distance of 3.2 km. If his acceleration was uniform over this distance, what was the acceleration and how long did the cyclist take to build up his final speed?

*113. A particle moves along the x axis, with the following equation for the position as a function of time:

$$x = 2.0 + 6.0t - 3.0t^2$$

where x is measured in meters and t is measured in seconds.

(a) What is the position of the particle at $t = 0.50$ s?

(b) What is the instantaneous velocity at this time?

(c) What is the instantaneous acceleration at this time?

*114. An automobile speeding at 100 km/h passes a stationary police cruiser. The police officer starts to move her cruiser in pursuit 8.0 s after the automobile passes. She accelerates uniformly to 120 km/h in 10 s, and then continues at uniform speed until she catches the speeder.

(a) How far ahead is the speeder when the police cruiser starts?

(b) How far ahead, relative to the cruiser, is the speeder when the cruiser reaches the uniform speed of 120 km/h?

(c) How long does the police cruiser take to catch the speeder? How far from the initial position?

*115. An automobile is traveling at 90 km/h on a country road when the driver suddenly notices a cow in the road 30 m ahead. The driver attempts to brake the automobile, but the distance is too short. With what velocity does the automobile hit the cow? Assume that, as in the table on page 47, the reaction time of the driver is 0.75 s and that the deceleration of the automobile is 8.0 m/s^2 when the brakes are applied.

*116. The nozzle of a fire hose discharges water at the rate of 280 liters/min at a speed of 26 m/s. How high will the stream of water rise if the nozzle is aimed straight up? How many liters of water will be in the air at any given instant?

*117. According to an estimate, a man who survived a fall from a 56-m cliff took 0.015 s to stop upon impact on the ground. What was his speed just before impact? What was his average deceleration during impact?

*118. A skydiver jumps out of an airplane at a height of 1000 m. Calculate how long she takes to fall to the ground. Assume that she falls freely with a downward acceleration g until she reaches the terminal speed of 200 km/h, and that she then continues to fall without acceleration until very near the ground (where she opens her parachute a few moments before reaching the ground).

**119. From a window on the fifth floor of a building, 13 m up, you drop two tennis balls with an interval of 1.0 s between the first and the second.

(a) Where is the first ball when you release the second? Where is the second ball when the first hits the ground?

(b) What is the instantaneous velocity of the first ball relative to the second just before the first hits the ground?

(c) What is the instantaneous acceleration of the first ball relative to the second?

Answers to Checkups

Checkup 2.1

1. The average speed is the total distance traveled divided by the time; for the man, this is (50 m)/(100 s), or 0.50 m/s. For the entire 100-s interval, the woman has the same average speed, 0.50 m/s (although her average speed for the first 50 s is twice as large, 1.0 m/s).
2. It is possible that, to the bicyclist, the spectator has a larger speed than the runner; for example, if the bicyclist travels at nearly the same speed as the runner (relative to the Earth), then the runner has a relative speed near zero, and the spectator has a relative speed near 10.1 m/s (but going in the backward direction). It is also possible that both the runner and the spectator have the same speed; this occurs in the special case when the bicyclist travels at half the runner's speed (5.05 m/s) relative to the Earth. In that case, the runner is traveling at 5.05 m/s relative to the bicyclist (in the forward direction), and the spectator is traveling at 5.05 m/s relative to the bicyclist (in the backward direction).
3. (E) 12 m/s. The average speed is the total distance divided by the time taken, (60 m)/(5.0 s) = 12 m/s.

Checkup 2.2

1. Since there is zero net change of position for the round-trip motion of the ball, the average velocity will be zero (the negative and positive parts cancel). We cannot calculate the average speed, since we would need to use the total distance, i.e., twice the distance from the hand to the sidewalk; this was not given (and cannot be calculated from the information given).
2. (D) Yes; yes. As in Example 2, the speed is always positive, but the average velocity can be positive or negative, depending on the relative locations of the starting and ending points for the coordinate system chosen. Also, for any round trip, the net change in position, and thus the average velocity, is zero, whereas, unless there was no motion at any time, the average speed will always be positive.

Checkup 2.3

1. A negative velocity refers to motion in the negative direction for the coordinate axis chosen; in one dimension, this would imply in the negative x direction. An aircraft could have a negative velocity if, for example, the x axis points east but the aircraft flies west.
2. We cannot tell if the velocity is positive or negative until the direction of the x axis is specified.
3. No. If the average velocity was 8 m/s, then the instantaneous velocity must have been sometimes larger than and sometimes smaller than (or always equal to) 8 m/s.
4. (B) 8.0 m/s. The instantaneous velocity is given by the derivative $v = dx/dt = \frac{1}{4} \times 4t^3 = t^3$. Evaluated at $t = 2.0$ s, this is $v = (2.0)^3 = 8.0$ m/s.

Checkup 2.4

1. A negative acceleration means that the velocity in the x direction is decreasing (by becoming either less positive or more negative). An automobile starting from rest can have a negative acceleration by beginning to move in the negative x direction (for example, if the x axis points east and the automobile moves west). A braking automobile can have a positive acceleration (same example: if the x axis points east and an automobile moving west slows down, the acceleration is in the positive x direction).
2. Let the x axis be pointing east; in that case, an automobile moving east and slowing down has a positive velocity and a negative acceleration. Let the x axis point west; in that case, the same automobile (moving east and slowing down) has a negative velocity and a positive acceleration.
3. Yes—for example, a ball bouncing off a wall is instantaneously at rest, but has a nonzero acceleration (since its velocity is changing from one direction to the other).
4. As the ball drops, the velocity is opposite to the chosen axis direction, and so it is negative. After it hits the floor, the velocity is positive (the x axis is upward) until the ball reaches its peak and starts to drop again. Before the ball hits the floor, the acceleration is negative (objects fall increasingly quickly because of gravity); after it hits the floor, the acceleration is still negative (it slows down as it goes back up). During the collision, the acceleration is positive; this acceleration changes the velocity from negative to positive.
5. Since the automobile begins at rest and ends at rest, its average acceleration for the complete motion is zero ($\bar{a} = \Delta v/\Delta t$). The average acceleration of the train is negative, since the x axis is in the direction of travel and the train slows to a stop ($v_2 < v_1$). During the impact, the automobile velocity increases from zero to a positive value, and so it is instantaneously positive. After the impact, the automobile slows with the train, and so it has negative instantaneous acceleration.
6. (D) Opposite; opposite. Just before the collision, the signs of the velocities will be opposite, one positive and one negative, since the velocities are in opposite directions ("head-on"). During the collision, the accelerations will also be opposite; the car with positive velocity will undergo a negative acceleration, and the car with a negative velocity will have a positive acceleration.

Checkup 2.5

1. In Example 7, we saw that the braking distance varies as the square of the velocity ($x - x_0 = -v_0^2/2a$). Thus, twice the initial speed will result in four times the braking distance, or 144 m.
2. In Example 6, we found that the deceleration varies inversely with the crumpling distance [$a = (v^2 - v_0^2)/2(x - x_0)$]. Thus, halving the crumpling distance doubles the deceleration to 480 m/s^2.
3. (B) $4t$. From $v = v_0 + at$, with final velocity $v = 0$ (stopped), we see the time to stop is $t = -v_0/a$. Thus if v_0 is twice as large and a is half as large, t is four times as long.

Checkup 2.6

1. No to both questions. The two bottles have the same acceleration and the same impact speed; the acceleration due to gravity is the same for different masses (neglecting air friction).
2. No. The water accelerates at the same rate as the cup (as usual, we neglect air friction effects).
3. The stone's velocity is zero at the maximum height, since it momentarily stops there. The velocity changes from positive on the way up to negative on the way down, and so it is zero at the top. The acceleration remains negative (and constant, $a = -g$) throughout the motion.
4. Zero. Terminal speed refers to the constant velocity attained under the combined influences of gravity and air resistance. Constant velocity means zero rate of change of velocity, that is, zero acceleration.
5. With the x axis positive down, the acceleration due to gravity becomes positive; that is, the quantity $-g$ in each of Eqs. (2.27) to (2.29) becomes $+g$.
6. (D) 103 m. Under constant acceleration, the change in position is proportional to the change in the square of the velocity [$a(x - x_0) = \frac{1}{2}(v^2 - v_0^2)$]. So if the velocity is half as large, the height will be one-fourth as high, or (410 m)/4 $\approx$ 103 m.

Checkup 2.7

1. The acceleration is given as proportional to t. Since the velocity, by Eq. (2.34), is the integral of the acceleration over time, the speed will be proportional to t^2. Thus the velocity will be 4 times as large at $t = 2$ s as it was at $t = 1$ s.
2. (C) t^3. If the acceleration a increases in proportion to the time t, then Eq. (2.34) indicates that the velocity $v = \int a\,dt$ increases in proportion to t^2 (for zero initial velocity). The distance then increases with time according to Eq. (2.35), $x - x_0 = \int v\,dt$, which is proportional to t^3, because $\int t^2\,dt = t^3/3$.

CHAPTER

3

Vectors

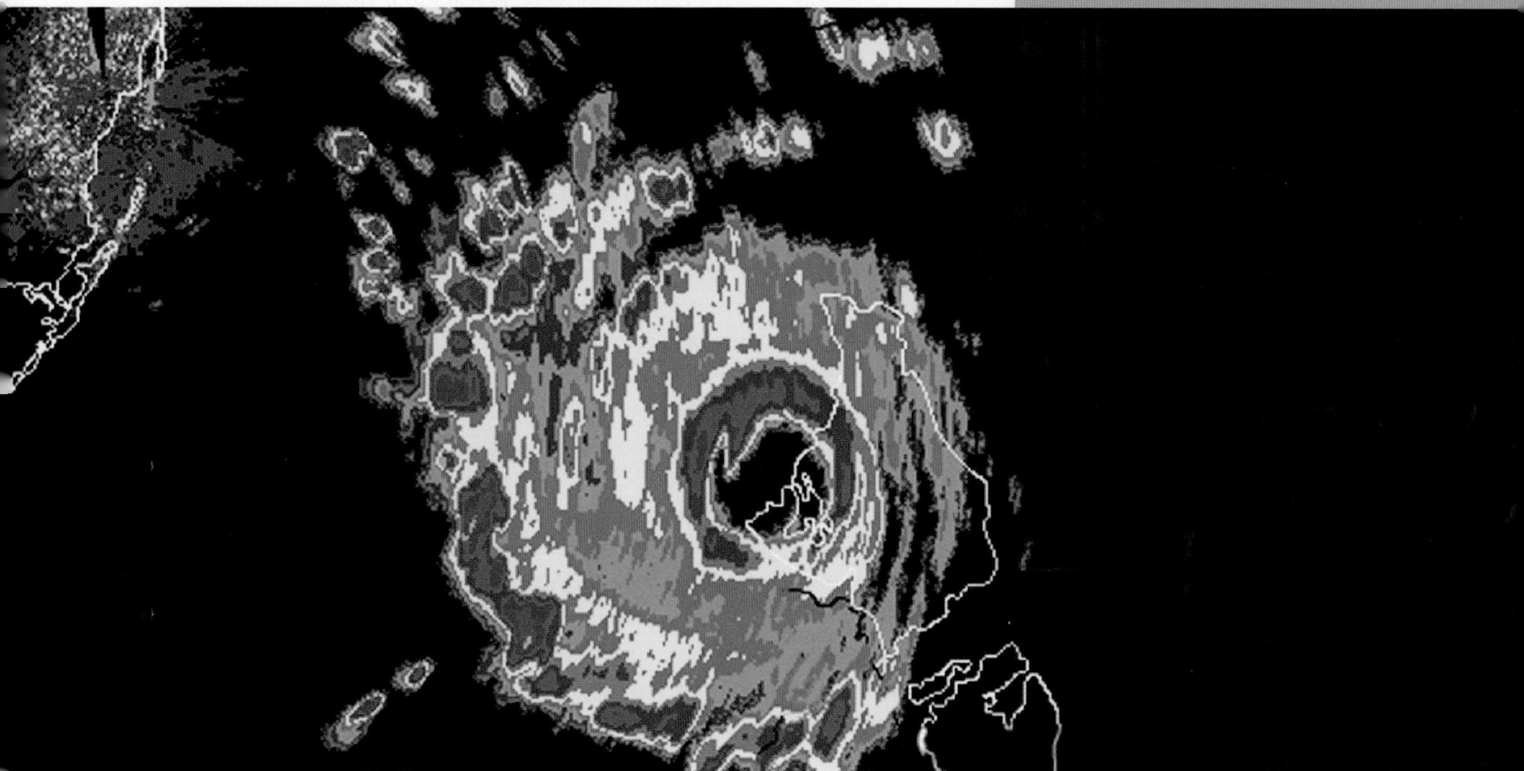

Concepts — *in* — Context

CONCEPTS IN CONTEXT

The colorful circular feature on this radar display is a hurricane off Miami, Florida. The distance and direction from Miami to the hurricane can be represented by a *displacement vector*. Such displacement vectors are commonly used in navigation. Many other types of vectors are used in science and engineering.

While we learn about vectors, we will examine such questions as:

? How are vectors used in navigation? (Physics in Practice: Vectors in Navigation, page 71)

? How can two positions be used to determine relative position? (Example 2, page 78)

? How can vector algebra be used to determine direction? (Example 5, page 83)

The mathematical concept of vector turns out to be very useful for the description of position, velocity, and acceleration in two- or three-dimensional motion. We will see that a vector description of the motion gives precise meaning to the intuitive notion that velocity and acceleration have a *direction* as well as a magnitude: the velocity and the acceleration of a particle can point north, or east, or up, or down, or in any direction in between. And we will see in later chapters that the vector concept is also useful for the description of many other physical quantities, such as force and momentum, which have both a magnitude and a direction. The present chapter is an introduction to vectors and their addition, subtraction, and multiplication. After developing these mathematical tools in this chapter, we will apply vectors to some aspects of motion in two and three dimensions in the next and subsequent chapters.

Online Concept Tutorial 4

3.1 THE DISPLACEMENT VECTOR AND OTHER VECTORS

We begin with the concepts of displacement and displacement vector. The displacement of a particle is simply a change of its position. If a particle moves from a point P_1 to a point P_2, we can represent the change of position graphically by an arrow, or directed line segment, from P_1 to P_2. *The directed line segment is the* **displacement vector** of the particle. For example, if a ship moves from Liberty Island to the Battery in New York harbor, then the displacement vector is as shown in Fig. 3.1. Note that the displacement vector tells us only where the final position (P_2) is in relation to the initial position (P_1); it does not tell us what path the ship followed between the two positions. Thus, any of the paths shown by the red lines in Fig. 3.2 results in the same final displacement vector.

A displacement vector has a length and a *direction,* which are graphically represented by the length of the arrow and the direction of its tip. Although every vector can be represented by drawing an arrow, not every drawing of an arrow you encounter in everyday life is a vector. For instance, the arrows commonly used in traffic signs are *not* vectors—they tell you only the direction in which you are required to travel, but not the distance (see Fig. 3.3).

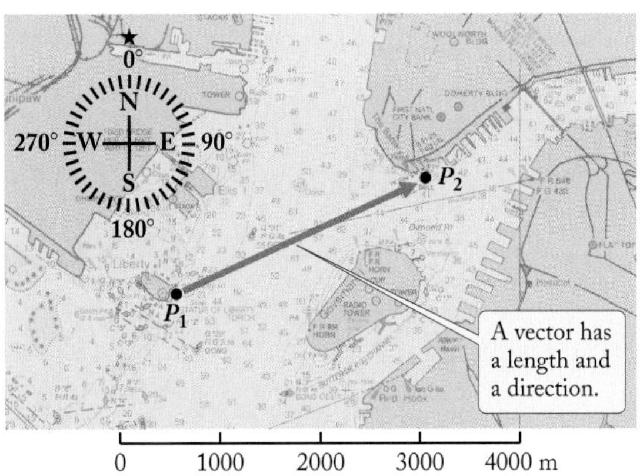

FIGURE 3.1 Displacement vector for a ship moving from Liberty Island to the Battery in New York harbor. The length of this vector is 2790 m; its direction is 65° east of north. (Adapted from National Ocean Survey Chart 12328.)

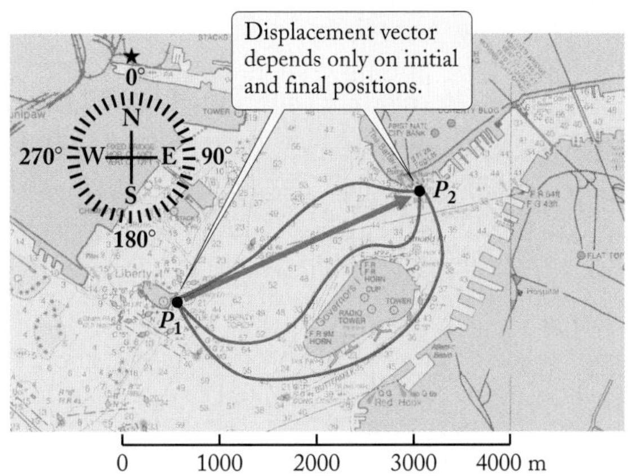

FIGURE 3.2 Several alternative paths from Liberty Island to the Battery (red lines). All of these result in the same displacement vector.

PHYSICS IN PRACTICE VECTORS IN NAVIGATION

Concepts — in — Context

Vectors are widely used by pilots and navigators of ships and aircraft in calculations of positions and directions of travel. For instance, the pilot of an aircraft approaching an airport might ask the airport controller, "Give me a vector," which means that the pilot wants to be told what the displacement vector is from her aircraft to the airport. The controller will consult the radar display showing aircraft positions in relation to the airport, and he might answer, "Heading one five zero, thirty miles," to indicate to the pilot the direction and the distance the aircraft needs to travel to reach the airport. In navigational practice, angles are reckoned clockwise from (magnetic) north; thus, a heading of "one five zero," or 150°, corresponds to 60° south of east.

Incidentally: It is common for the heading of an airport runway to be abbreviated to its first two digits; thus, if the airport controller orders the pilot to land on "runway 24" he means the runway that has a direction 240° clockwise from north, that is, a direction 30° south of west. Runways are used in one of two opposite directions depending on the wind; although only one number is used to specify the active direction, both directions (which always differ by 18, or 180°) appear on airport runway signs (for example, "6/24").

Instead of describing the vector graphically by drawing an arrow, we can describe it numerically by giving the numerical value of its length (in, say, meters) and the numerical value of the angle (in, say, degrees) it makes with some reference direction. For example, we can specify the displacement vector in Fig. 3.1 by stating that it is 2790 m long and points at an angle of 65° east of north. Note that the length, or magnitude, of a vector is always a positive quantity. If we want to construct a vector opposite to the vector shown in Fig. 3.1, we must reverse its direction, but we keep its length positive.

FIGURE 3.3 Arrows painted on a traffic sign.

Since the displacement vector describes a *change* in position, any two line segments of identical length and direction represent equal vectors, regardless of whether the initial points of the line segments are the same. Thus, the parallel directed line segments shown in Fig. 3.4 do not represent different vectors; they both involve the same *change* of position (same distance and same direction), and they both represent equal displacement vectors.

In printed books, vectors are usually indicated by boldface letters, such as **A**, and we will follow this most common convention. In handwritten calculations, an alternative notation consisting of either a small arrow over the letter, such as $\vec{A}$, or a wavy underline, such as $\underset{\sim}{A}$, is usually more convenient. Each means the same as our **A**. We will use an ordinary italic letter, such as *A*, to denote the (positive) length or magnitude of a displacement vector. The magnitude of a vector is also often expressed by placing the vector between vertical lines; that is, $A = |\mathbf{A}|$.

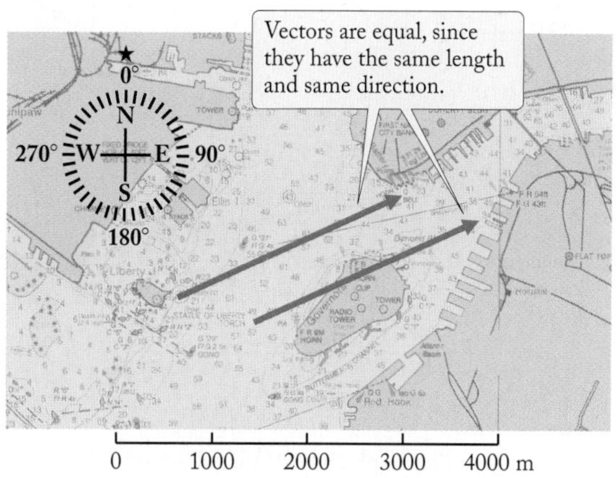

FIGURE 3.4 These two displacement vectors are equal.

The displacement vector serves as prototype for all other vectors. To decide whether some quantity endowed with both magnitude and direction is a vector, we compare its mathematical properties with those of the displacement vector. *Any quantity that has magnitude and direction and that behaves mathematically like the displacement vector is a* **vector**. For example, velocity, acceleration, and force are vectors; they can be represented graphically by directed line segments of a length equal to the magnitude of the velocity, acceleration, or force (in some suitable units) and a corresponding direction.

By contrast, any quantity that has a magnitude but *no* direction is called a **scalar**. For example, length, time, mass, area, volume, density, temperature, and energy are scalars; they can be completely specified by their numerical magnitude and units. Note that the *length* of a displacement vector is a quantity that has a magnitude but no direction—that is, the length is a scalar.

✔ Checkup 3.1

QUESTION 1: An airliner flies nonstop from San Francisco to New York. Another airliner flies from San Francisco to New Orleans, and from there to New York. Are the displacement vectors for the two airliners equal? Are the distances traveled equal?

QUESTION 2: Can the magnitude of a vector be negative? Zero?

QUESTION 3: Two aircraft are flying in formation on parallel paths separated by 200 m. The aircraft fly 3000 m due west. Are their displacement vectors equal?

QUESTION 4: The hurricane in the chapter photo is 200 km from Miami, in a direction 30° south of east. Consider a point due south of Miami and due west of the hurricane. How far is this point from Miami?

(A) 71 km (B) 87 km (C) 100 km
(D) 141 km (E) 173 km

Online Concept Tutorial 4

3.2 VECTOR ADDITION AND SUBTRACTION

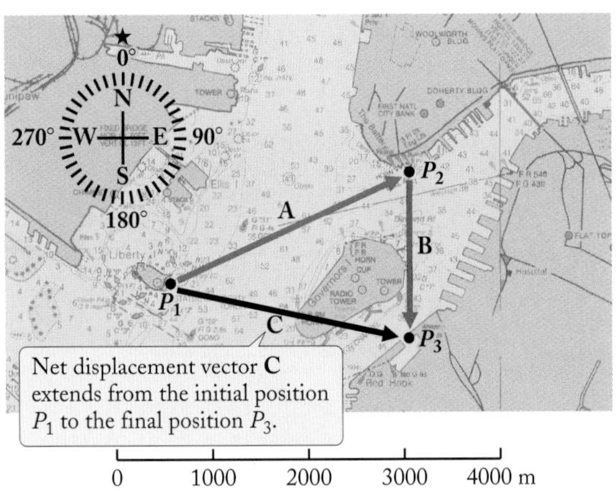

FIGURE 3.5 The displacement **A** is followed by the displacement **B**. The net displacement is **C**.

Since by definition all vectors have the mathematical properties of displacement vectors, we can investigate all the mathematical operations with vectors by looking at displacement vectors. The most important of these mathematical operations is **vector addition**.

Two displacements carried out in succession result in a net displacement, which is regarded as the vector sum of the individual displacements. For example, Fig. 3.5 shows a displacement vector **A** (from P_1 to P_2) and a displacement vector **B** (from P_2 to P_3). The net displacement vector is the directed line segment from the initial position P_1 to the final position P_3. This net displacement vector is denoted by **C** in Fig. 3.5. This vector **C** can be regarded as the sum of the individual displacements:

$$\mathbf{C} = \mathbf{A} + \mathbf{B} \tag{3.1}$$

The sum of two vectors is usually called the **resultant** of these vectors. Thus, **C** is called the resultant of **A** and **B**.

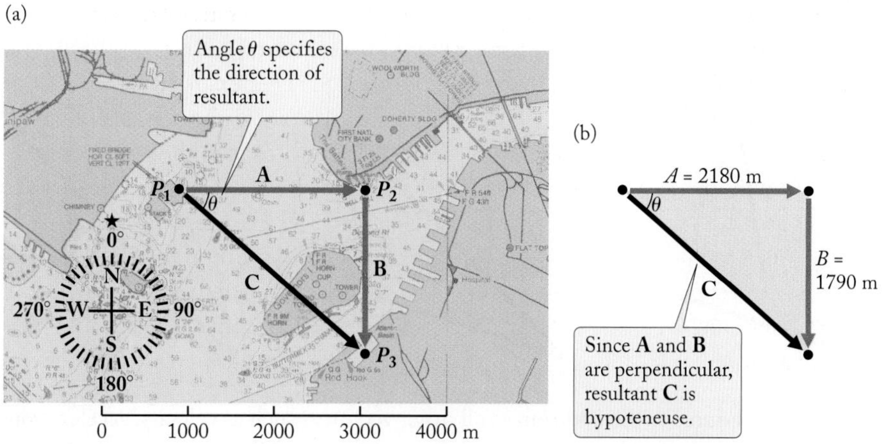

FIGURE 3.6 (a) The vectors **A** and **B** represent the displacements of the motorboat moving from Ellis Island (P_1) to the Battery (P_2) and from the Battery to the Atlantic Basin (P_3). The vector **C** is the sum of these two vectors. (b) The vectors **A**, **B**, and **C** form a right triangle. Two of the sides are known, and the hypotenuse is unknown.

EXAMPLE 1 A motorboat moves from Ellis Island in New York harbor to the Battery and from there to the Atlantic Basin (see Fig. 3.6). The first displacement vector is 2180 m due east, and the second is 1790 m due south. What is the resultant?

SOLUTION: The resultant of the two displacement vectors **A** and **B** is the vector **C**, from the tail of **A** to the head of **B**. For a graphical determination of **C**, we can measure the length of **C** directly on the chart using the scale of length marked on the chart, and we can measure the direction of **C** with a protractor (the way the navigator of the ship would solve the problem). This yields a length of about 2800 m and an angle of about 39° for the resultant vector **C**.

For a more precise (numerical) determination of **C**, we note that **A**, **B**, and **C** form a right triangle. We can therefore find **C** by using the standard trigonometric methods for the solution of triangles.[1] The lengths of the two known sides of this triangle are $A = 2180$ m and $B = 1790$ m (see Fig. 3.6). The unknown side is the the hypotenuse of the triangle. By the Pythagorean theorem, the length C of the hypotenuse is

$$C = \sqrt{A^2 + B^2} \tag{3.2}$$

from which

$$C = \sqrt{(2180 \text{ m})^2 + (1790 \text{ m})^2} = 2820 \text{ m}$$

The tangent of the angle θ is the ratio of the opposite side B to the adjacent side A:

$$\tan\theta = \frac{B}{A} = \frac{1790 \text{ m}}{2180 \text{ m}} = 0.821$$

With our calculator, we find that the angle whose tangent is 0.821 is

$$\theta = 39.4°$$

Thus, the resultant vector **C** has a length of 2820 m at an angle 39.4° south of east, which is consistent with values obtained by the graphical method.

[1] Appendix 3 gives a review of trigonometry.

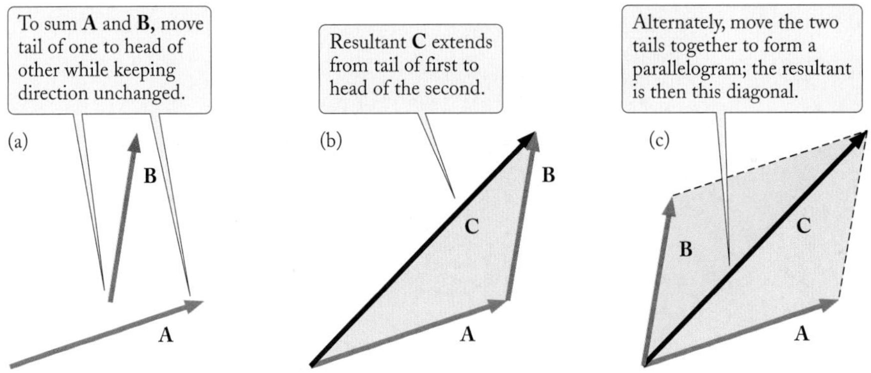

FIGURE 3.7 The vector sum **A** + **B**; the resultant is **C**. (a) The two vectors **A** and **B**. (b) Addition of **A** and **B** by the tail-to-head method. (c) Addition of **A** and **B** by the parallelogram method.

The procedure for the addition of all other vectors—such as velocity, acceleration, and force vectors—mimics that for displacement vectors. All such vectors can be represented by arrows. If **A** and **B** are two arbitrary vectors (see Fig. 3.7a), then their resultant can be obtained by placing the tail of **B** on the head of **A** while keeping the magnitude and direction of **B** unchanged; the directed line segment connecting the tail of **A** to the head of **B** is the resultant (see Fig. 3.7b). Alternatively, the resultant can be obtained by placing the tail of **B** on the tail of **A** and drawing a parallelogram with **A** and **B** as two of the sides; the diagonal of the parallelogram is then the resultant (see Fig. 3.7c).

Note that the order in which the two vectors are added makes no difference to the final result. Whether we place the tail of **B** on the head of **A** or the tail of **A** on the head of **B**, the resultant is the same (see Fig. 3.8). Hence

commutative law for vector addition

$$\mathbf{A} + \mathbf{B} = \mathbf{B} + \mathbf{A} \tag{3.3}$$

This identity is called the **commutative law** for vector addition; it indicates that, just as in ordinary addition of numbers, the order of the terms is irrelevant.

The magnitude of the resultant of two vectors is usually less than the sum of the magnitudes of the vectors. Thus, if

$$\mathbf{C} = \mathbf{A} + \mathbf{B} \tag{3.4}$$

then

$$C \leq A + B \tag{3.5}$$

This inequality simply expresses the fact that in a triangle (see Fig. 3.7b) the length of any side is less than the sum of the lengths of the other two sides. Only in the special case where **A** and **B** are parallel (see Fig. 3.9) will the magnitude of **C** equal the sum of the magnitudes of **A** and **B**; it can never exceed this sum.

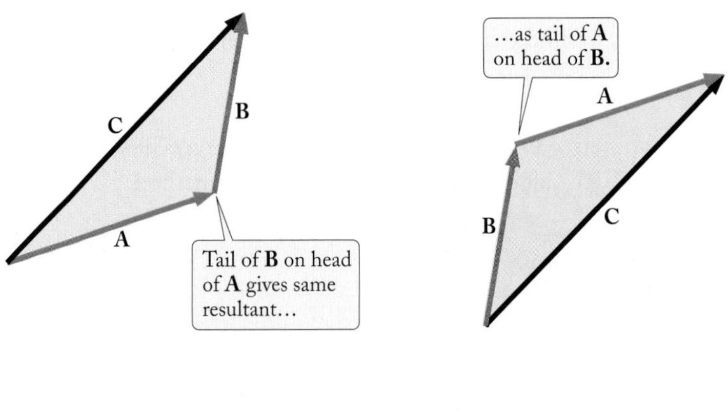

FIGURE 3.8 The resultant for the vector sum **A** + **B** is the same as for the vector sum **B** + **A** (compare Fig. 3.7).

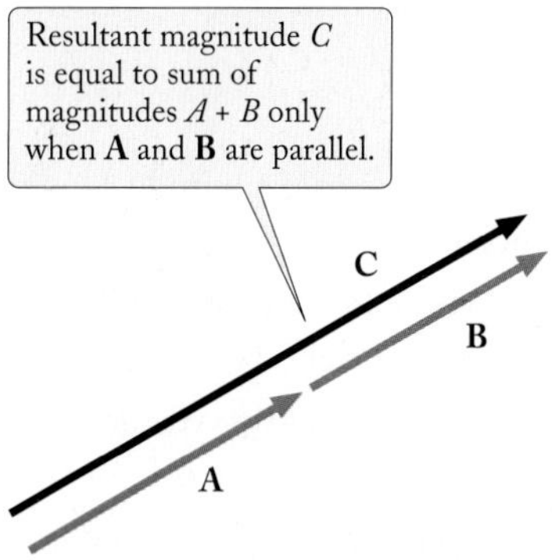

FIGURE 3.9 Parallel vectors **A** and **B**, and their resultant **C**.

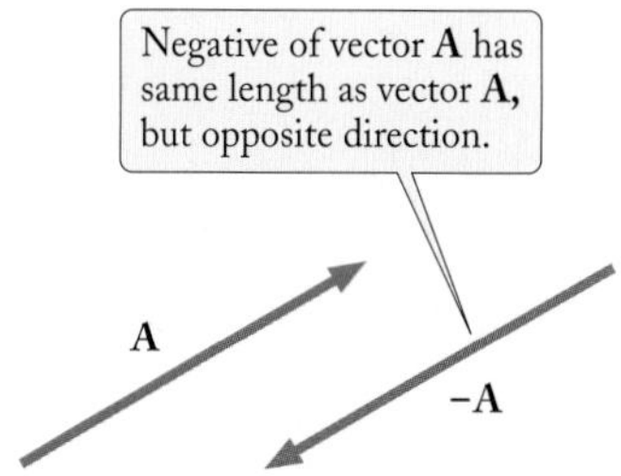

FIGURE 3.10 The vector **A** and its negative −**A**.

The negative of a given vector **A** is a vector of the same magnitude, but opposite direction; this new vector is denoted by −**A** (see Fig. 3.10). Obviously, the sum of a vector and its negative gives a vector of zero magnitude:

sum of vector and its negative

$$\mathbf{A} + (-\mathbf{A}) = 0 \tag{3.6}$$

The subtraction of two vectors **A** and **B** is defined as the sum of **A** and −**B**; that is,

$$\mathbf{A} - \mathbf{B} = \mathbf{A} + (-\mathbf{B}) \tag{3.7}$$

Figure 3.11a shows the vector difference **A** − **B** obtained by constructing the vector sum of **A** and −**B** by the parallelogram method. Alternatively, Fig. 3.11b shows how to obtain the vector difference **A** − **B** by drawing the directed line segment from the head of **B** to the head of **A**. Comparison of these two diagrams establishes that this directed line segment from the head of **B** to the head of **A** is equal to the vector difference **A** − **B**.

A vector can be multiplied by any positive or negative number. For instance, if **A** is a given vector, then 3**A** is a vector of the same direction and of a magnitude three times as large (see Fig. 3.12a), whereas −3**A** is a vector of the opposite direction and, again, of a magnitude three times as large (see Fig. 3.12b). In particular, if we multiply a vector by −1, we obtain the negative of that vector:

$$(-1)\mathbf{A} = -\mathbf{A} \tag{3.8}$$

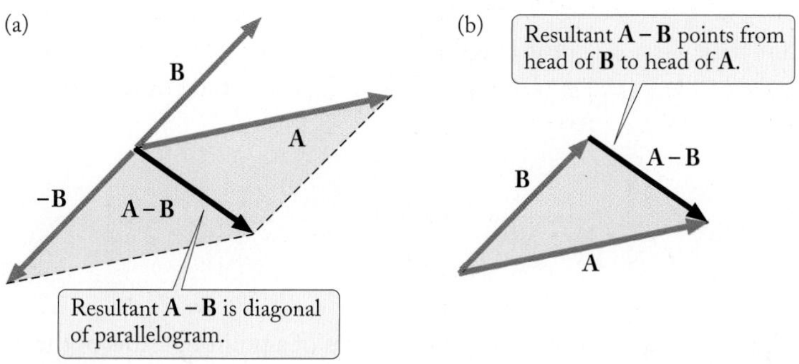

FIGURE 3.11 (a) To obtain the vector difference **A** − **B**, we first draw the vector −**B**, and then construct the vector sum **A** + (−**B**) by the parallelogram method. (b) Alternatively, we can draw the directed line segment from the head of **B** to the head of **A**; this gives the same result as the parallelogram method, and establishes that this directed line segment is equal to the vector difference **A** − **B**.

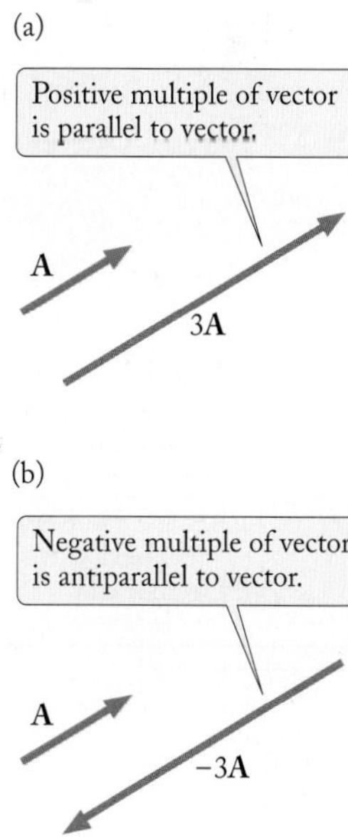

FIGURE 3.12 (a) The vector 3**A** has the same direction as the vector **A** and is three times as long. (b) The vector −3**A** has a direction opposite to that of **A** and is three times as long.

Checkup 3.2

QUESTION 1: Two vectors have nonzero magnitude. Under what conditions will their sum be zero? Their difference?

QUESTION 2: Suppose that two vectors are perpendicular to each other. Is the magnitude of their sum larger or smaller than the magnitude of each?

QUESTION 3: Is it possible for the sum of two vectors to have the same magnitude as the difference of the two vectors?

QUESTION 4: Three vectors have the same magnitude. Under what conditions will their sum be zero?

QUESTION 5: An automobile drives 3 km south and then drives 4 km west. What is the magnitude of the resultant displacement vector?

(A) 1 km (B) 5 km (C) 7 km (D) 16 km (E) 25 km

Online Concept Tutorial 4

3.3 THE POSITION VECTOR; COMPONENTS OF A VECTOR

In Section 1.1 we saw that to describe the position of a point in space, we must choose an origin and construct a coordinate grid. If the grid is rectangular, then the position of a point will be specified by the three rectangular coordinates x, y, and z. Alternatively, we can describe the position by means of the displacement vector from the origin to the point. For the sake of simplicity, we will usually restrict our discussion to points in a plane, so the position of a point is specified by two rectangular coordinates x and y. Figure 3.13 shows the point, its coordinates x and y, and the displacement vector from the origin to the point. This displacement vector is called the **position vector**, usually denoted by **r**.

As shown in Fig. 3.13, the position vector of a point has a length r, and it makes an angle θ with the x axis. Consider the right triangle shown in Fig. 3.13. The length of the position vector is the hypotenuse of this triangle, and the x and y coordinates form the sides. From the definitions of the cosine and the sine of the angle θ, we find

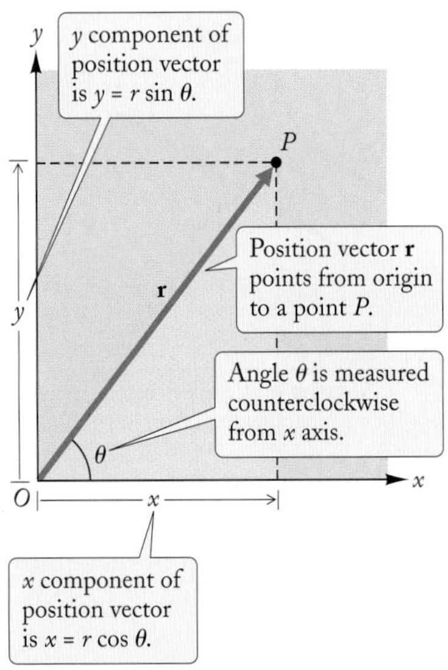

FIGURE 3.13 The position vector **r** in two dimensions and its components.

$$\cos\theta = \frac{x}{r}$$

$$\sin\theta = \frac{y}{r}$$

Hence

$$x = r\cos\theta \tag{3.9}$$

$$y = r\sin\theta \tag{3.10}$$

These two equations express the x and y coordinates of a point in terms of the length r and direction θ of the position vector. Conversely, we can express r and θ in terms of the coordinates x and y. To do this, we apply the Pythagorean theorem to the right triangle in Fig. 3.13:

$$r = \sqrt{x^2 + y^2} \quad (3.11)$$

Furthermore, from the definition of the tangent of the angle θ, we find

$$\tan\theta = \frac{y}{x} \quad (3.12)$$

These two equations make it easy to calculate r and θ from x and y.

According to the terminology introduced in Chapter 1, x and y are the components of the position. We will now adopt a vectorial terminology according to which *x and y are the* **components** *of the position vector.* Note that graphically the components of the position vector are determined by dropping perpendiculars from the head of the vector to the x and y axes (see Fig. 3.13).

In general, for an arbitrary vector **A**, the definition of the components is analogous to the definition of the components of the position vector. We place the tail of the vector at the origin, and we drop perpendiculars from the head of the vector to the x and y axes. The intercepts of these perpendiculars with the axes (which may be positive or negative) give us the x and y components of the vector **A** (see Fig. 3.14). Designating these components by A_x and A_y, we see from Fig. 3.14 that

components of a vector

$$A_x = A\cos\theta \quad (3.13)$$

$$A_y = A\sin\theta \quad (3.14)$$

These equations are analogous to Eqs. (3.9) and (3.10). Furthermore, we see from Fig. 3.14 that

magnitude in terms of components

$$A = \sqrt{A_x^2 + A_y^2} \quad (3.15)$$

and that

$$\tan\theta = \frac{A_y}{A_x} \quad (3.16)$$

These two equations are analogous to Eqs. (3.11) and (3.12).

Note that the components uniquely specify the vector—if we know the components, we can find the magnitude and direction of the vector from Eqs. (3.15) and (3.16).

The addition or subtraction of two vectors can be performed by adding or subtracting their components. Thus, if A_x, A_y and B_x, B_y are the components of the vectors **A** and **B**, then the components of the resultant $\mathbf{C} = \mathbf{A} + \mathbf{B}$ of these two vectors are

$$C_x = A_x + B_x$$
$$C_y = A_y + B_y \quad (3.17)$$

This is sometimes the most convenient method for evaluating sums or differences of vectors.

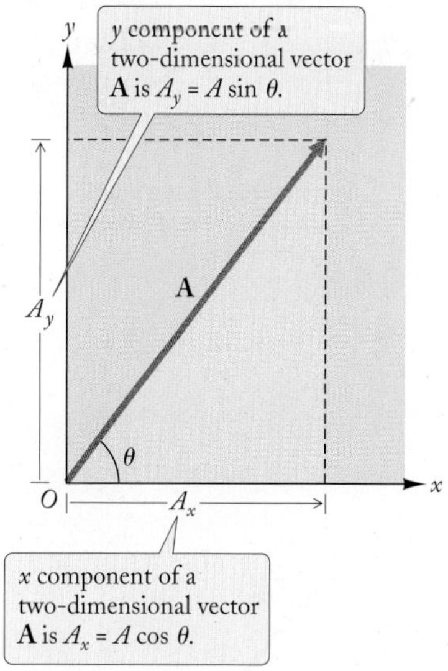

FIGURE 3.14 A vector **A** in two dimensions and its components A_x and A_y.

Concepts in Context

EXAMPLE 2 The eye of the hurricane in the chapter photo is 200 km from Miami on a bearing of 30° south of east. Suppose a reconnaissance aircraft is initially 100 km due north of Miami. What displacement vector will bring the aircraft to the eye of the hurricane?

SOLUTION: In Example 1 we saw how to use a graphical method and a trigonometric method for finding an unknown vector. Here we will see how to use the component method.

For this calculation, we need to make a choice of coordinate system, that is, a choice of origin and of axes. In Fig. 3.15, we have placed the origin on Miami, with the x axis eastward and the y axis northward. In this coordinate system, the airplane has a position vector **A** with components

$$A_x = 0 \text{ km} \qquad \text{and} \qquad A_y = 100 \text{ km}$$

The hurricane has a position vector **B** with components [see Eqs. (3.13) and (3.14)]

$$B_x = 200 \text{ km} \times \cos(-30°) = 173 \text{ km}$$
$$B_y = 200 \text{ km} \times \sin(-30°) = -100 \text{ km}$$

The displacement vector **C** from the airplane to the hurricane is the directed line segment from the head of **A** to the head of **B**. According to our discussion of vector subtraction, this directed line segment is equal to the vector difference **B** − **A**, that is, **C** = **B** − **A** (see Fig. 3.15; you may find it easier to recognize that **A** + **C** = **B**). The vector **C** therefore has components

$$C_x = B_x - A_x = 173 \text{ km} - 0 \text{ km} = 173 \text{ km}$$
$$C_y = B_y - A_y = -100 \text{ km} - 100 \text{ km} = -200 \text{ km}$$

The magnitude of the vector **C** is then, according to Eq. (3.15),

$$C = \sqrt{C_x^2 + C_y^2} = \sqrt{(173 \text{ km})^2 + (-200 \text{ km})^2} = 264 \text{ km}$$

and the angle between **C** and the x axis is given by [see Eq. (3.13)]

$$\cos\theta = \frac{C_x}{C} = \frac{173}{264} = 0.655$$

Our calculator then tells us

$$\theta = 49°$$

Hence a displacement of 264 km at 49° south of east will bring the airplane to the hurricane.

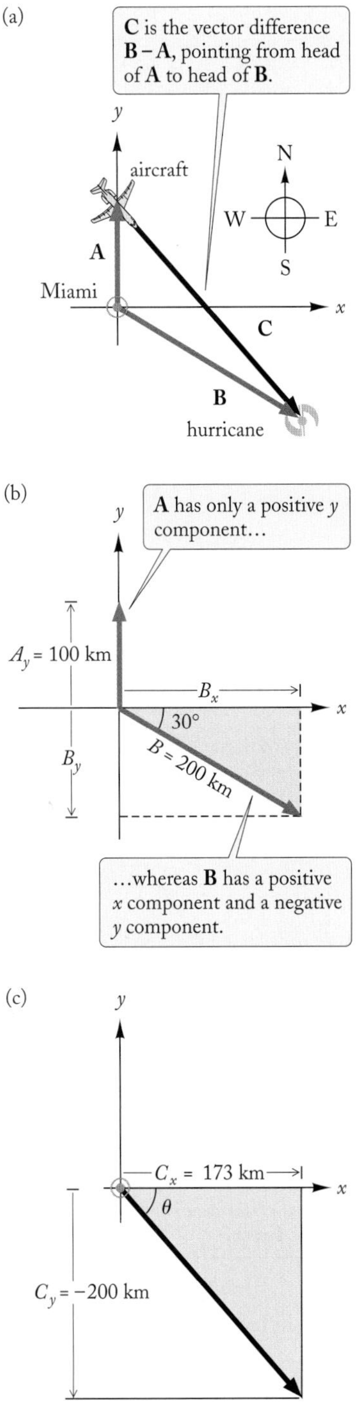

FIGURE 3.15 (a) Position vectors of the aircraft (**A**) and of the hurricane (**B**). The displacement vector from the aircraft to the hurricane is the difference between these vectors, **C** = **B** − **A**. (b) The x and y components of **A** and **B**. (c) The x and y components of **C**. To display these components, the tail of **C** has been shifted to the origin.

PROBLEM-SOLVING TECHNIQUES — VECTOR ADDITION AND SUBTRACTION

When dealing with the addition or subtraction of two or more vectors, it is usually convenient to use a method based on components, as in Example 2.

1. Begin by making some convenient choice of coordinate axes x and y. Selecting a coordinate system such that x or y is parallel to one of the vectors may help.
2. Draw the vectors. If possible, place the tails of the vectors at the origin, since this makes it easier to calculate their components. Try to draw the lengths and angles of the vectors fairly accurately, since this will help you to see roughly what answer to expect.
3. Drop perpendiculars from the heads (and, if necessary, the tails) of each vector, and mark the x and y components of each vector along the axes. The signs of the components will be obvious by inspection: a component is positive if it extends from the origin along the positive part of the axis, and a component is negative if it extends from the origin to the negative part of the axis.
4. Add or subtract the components as required to obtain the components of the resultant.
5. If the problem asks for the magnitude and the direction of the resultant (instead of components), then calculate these by means of Eq. (3.15), $A = \sqrt{A_x^2 + A_y^2}$, and Eq. (3.16), $\tan\theta = A_y/A_x$, using the components found in step 4.
6. Keep in mind that in any problem that has a vector as answer, you must either state all the components of the vector or else the magnitude *and* the direction of the vector. It is a common mistake to state merely the magnitude, without the direction—this is an incomplete answer.

In the preceding, we have dealt with vectors in only two dimensions, with only x and y components. More generally, a vector in three dimensions has x, y, and z components. To obtain these components, we place the tail of the vector at the origin and drop perpendiculars from the head of the vector to the x–y, x–z, and y–z planes, so as to form a "box" of which the vector is the diagonal (see Fig. 3.16). The sides of this box give us the components A_x, A_y, and A_z of the vector. The generalization of our two-dimensional case to three dimensions gives for the length of the vector

$$A = \sqrt{A_x^2 + A_y^2 + A_z^2}$$

magnitude of a 3D vector

The position vector and other vectors can be expressed in terms of **unit vectors** along the coordinate axes. The unit vectors along the x, y, and z axes are designated by **i**, **j**, and **k**, respectively (see Fig. 3.17). The magnitude of each of these vectors is 1; that is, $|\mathbf{i}| = 1$, $|\mathbf{j}| = 1$, and $|\mathbf{k}| = 1$. Thus, the magnitudes of these *unit vectors have no*

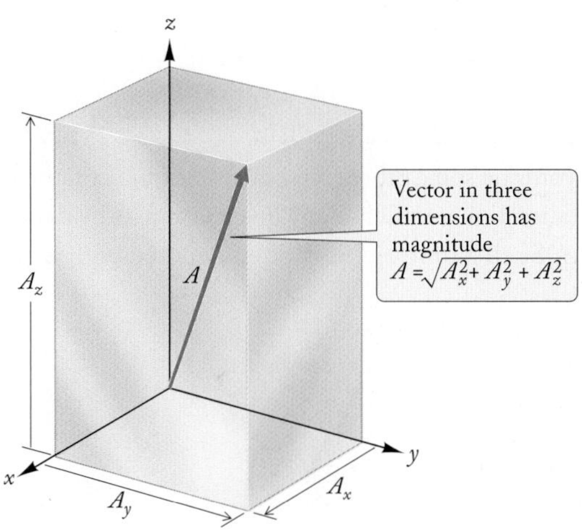

FIGURE 3.16 A vector **A** in three dimensions and its components A_x, A_y, and A_z. The components A_x, A_y, and A_z are represented by the sides of the rectangular box constructed by dropping perpendiculars from the head of the vector to the x–y, x–z, and y–z planes.

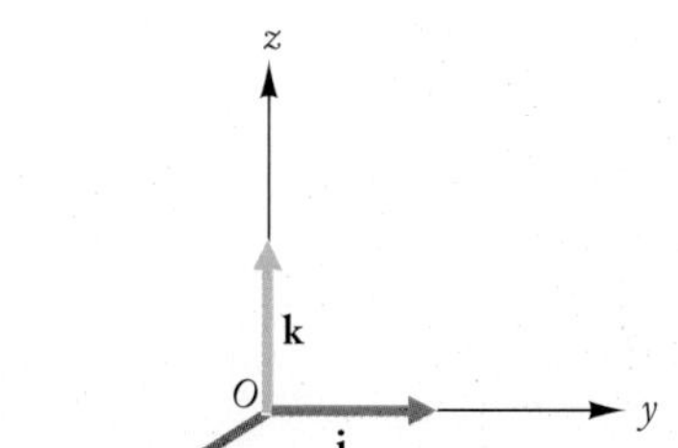

FIGURE 3.17 The unit vectors **i**, **j**, and **k**.

dimensions of length, time, or mass—they are pure numbers. Now consider the vector sum $x\mathbf{i} + y\mathbf{j} + z\mathbf{k}$. This vector sum consists of a displacement of magnitude x in the x direction, followed by a displacement of magnitude y in the y direction, followed by a displacement of magnitude z in the z direction. Thus, this vector sum brings us from the origin to the point with coordinates x, y, z. Hence, this vector sum coincides with the position vector:

$$\mathbf{r} = x\mathbf{i} + y\mathbf{j} + z\mathbf{k} \tag{3.18}$$

More generally, for any arbitrary vector with components A_x, A_y, and A_z, we can write

$$\mathbf{A} = A_x\mathbf{i} + A_y\mathbf{j} + A_z\mathbf{k} \tag{3.19}$$

The advantage of expressing vectors in terms of unit vectors is that we can then manipulate them algebraically, as illustrated in the following example and in the calculations of the next section.

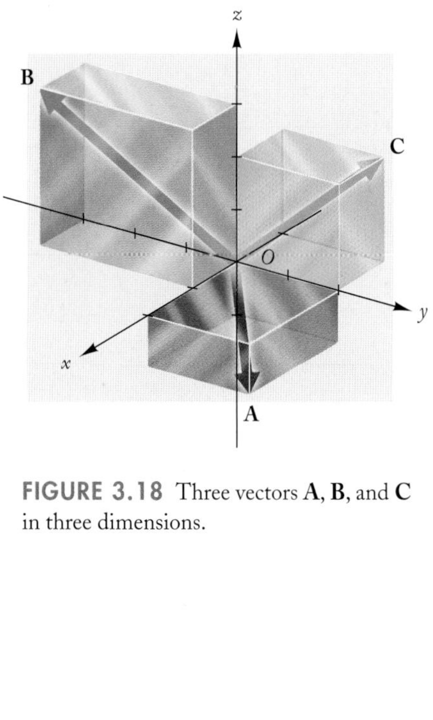

FIGURE 3.18 Three vectors **A**, **B**, and **C** in three dimensions.

EXAMPLE 3 Consider the following three vectors (see Fig. 3.18):

$$\mathbf{A} = 2\mathbf{i} + 2\mathbf{j} - \mathbf{k}$$
$$\mathbf{B} = \mathbf{i} - 3\mathbf{j} + 3\mathbf{k}$$
$$\mathbf{C} = -\mathbf{i} + 2\mathbf{j} + 2\mathbf{k}$$

What is the resultant $\mathbf{D} = \mathbf{A} + \mathbf{B} + \mathbf{C}$ of these three vectors? What is the magnitude of the resultant?

SOLUTION: When the vectors in a sum are expressed in terms of unit vectors, we can manipulate them algebraically, by collecting terms involving the same unit vectors:

$$\begin{aligned}\mathbf{D} = \mathbf{A} + \mathbf{B} + \mathbf{C} &= (2\mathbf{i} + 2\mathbf{j} - \mathbf{k}) + (\mathbf{i} - 3\mathbf{j} + 3\mathbf{k}) + (-\mathbf{i} + 2\mathbf{j} + 2\mathbf{k}) \\ &= (2\mathbf{i} + \mathbf{i} - \mathbf{i}) + (2\mathbf{j} - 3\mathbf{j} + 2\mathbf{j}) + (-\mathbf{k} + 3\mathbf{k} + 2\mathbf{k})\end{aligned}$$

Addition of the terms involving the same unit vectors is mathematically equivalent to addition of the components, as in Eq. (3.17). When we perform these additions, we find the resultant

$$\mathbf{D} = 2\mathbf{i} + \mathbf{j} + 4\mathbf{k}$$

From this addition we see that the components of the resultant are $D_x = 2$, $D_y = 1$, and $D_z = 4$. The magnitude D of the resultant is given by our three-dimensional generalization of the Pythagorean theorem:

$$D = \sqrt{D_x^2 + D_y^2 + D_z^2} = \sqrt{2^2 + 1^2 + 4^2} = \sqrt{21} = 4.6$$

We finish the section with a few words on some other vector quantities and notations.

We can obtain a unit vector in a direction parallel to any given vector **A** by dividing the vector **A** by its magnitude A. Such a unit vector is often denoted with a hat above the original vector:

$$\hat{\mathbf{A}} = \frac{\mathbf{A}}{A}$$

This can be calculated from our usual explicit forms $\mathbf{A} = A_x\mathbf{i} + A_y\mathbf{j} + A_z\mathbf{k}$ and $A = \sqrt{A_x^2 + A_y^2 + A_z^2}$. From these relations, we see that this form of a unit vector indeed has a magnitude identically equal to 1.

Many books in science and engineering use a compact notation for a vector: the x, y, and z components are placed in order between parentheses. Thus we can write

$$\mathbf{A} = (A_x, A_y, A_z)$$

For example, the resultant of Example 3 would be written $\mathbf{D} = (2, 1, 4)$. Which of the several notations is most convenient depends on the calculation; we will most often use the unit-vector form.

✔ Checkup 3.3

QUESTION 1: The magnitude of a vector is never smaller than the magnitude of any one component of the vector. Explain.

QUESTION 2: A vector is parallel to the y axis. What is its x component?

QUESTION 3: A vector has equal, positive x and y components and zero z component. What is the angle between this vector and the x axis?

QUESTION 4: A vector $\mathbf{A}$ has an x component $A_x = 3$ and a y component $A_y = -1$. What are the x and y components of the vector $2\mathbf{A}$? The vector $-4\mathbf{A}$?

QUESTION 5: What is the magnitude of the vector $\mathbf{i} + \mathbf{j}$?

(A) 0 (B) 1 (C) $\sqrt{2}/2$ (D) $\sqrt{2}$ (E) 2

3.4 VECTOR MULTIPLICATION[2]

There are several ways of multiplying vectors. The reason for this diversity is that in forming the "product" of two vectors, we must take into account both their magnitudes and their directions. Depending on how we combine these quantities, we obtain different kinds of products. The two most important kinds of products are the dot product and the cross product.

Dot Product

The **dot product** (also called the **scalar product**) of two vectors $\mathbf{A}$ and $\mathbf{B}$ is denoted by $\mathbf{A} \cdot \mathbf{B}$. This quantity is simply *the product of the magnitudes of the two vectors and the cosine of the angle between them* (Fig. 3.19):

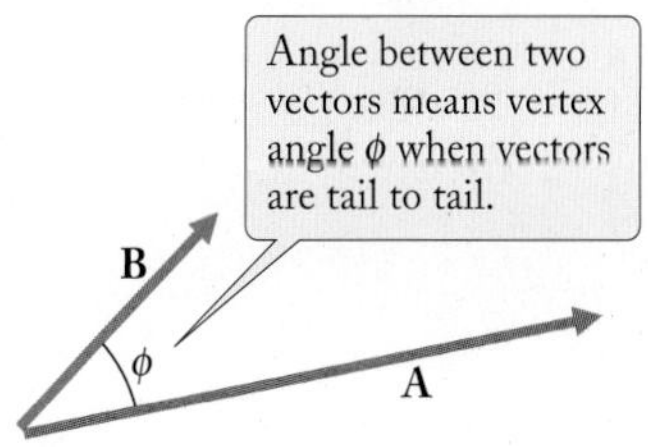

FIGURE 3.19 Two vectors **A** and **B**. The angle between them is ϕ.

$$\mathbf{A} \cdot \mathbf{B} = AB \cos \phi \tag{3.20}$$

dot product in terms of angle

Thus, the dot product of two vectors simply gives a number, that is, a scalar rather than a vector. The number will be positive if $\phi < 90°$, negative if $\phi > 90°$, and zero if $\phi = 90°$. If the two vectors are perpendicular, then their dot product is zero. Note that the dot product is commutative; as in ordinary multiplication, the order of the factors is irrelevant, that is, $\mathbf{A} \cdot \mathbf{B} = \mathbf{B} \cdot \mathbf{A}$.

[2] The dot product will first be used in Section 7.1 and the cross product will first be used in Section 13.4. Those sections contain brief, self-contained expositions of the dot product and cross products.

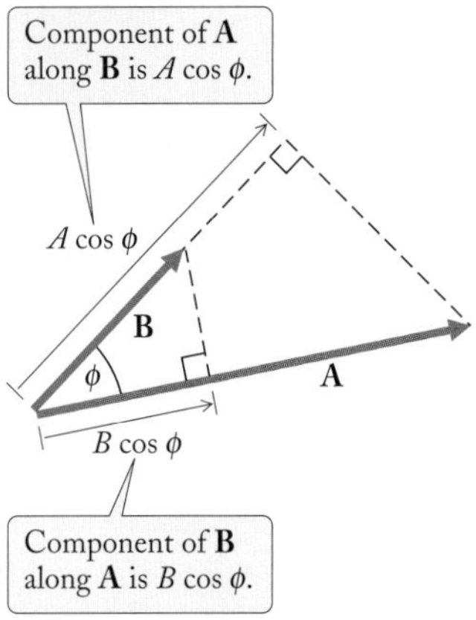

FIGURE 3.20 The component of **B** along **A** (obtained by dropping a perpendicular from the head of **B** toward **A**) equals $B \cos \phi$. Hence the scalar product $\mathbf{A} \cdot \mathbf{B} = AB \cos \phi$ equals the magnitude of **A** multiplied by the component of **B** along **A**.

We see from Fig. 3.20 that the dot product $\mathbf{A} \cdot \mathbf{B}$ can be regarded as B times the component of **A** along **B**, or as A times the component of **B** along **A**. The special case of the dot product of a vector with itself gives the square of the magnitude of the vector:

$$\mathbf{A} \cdot \mathbf{A} = AA \cos 0° = A^2 \tag{3.21}$$

By means of the the usual rules for the multiplication of a sum of terms, we can derive a useful expression for the dot product in terms of components. Suppose that $\mathbf{A} = A_x\mathbf{i} + A_y\mathbf{j}$ and $\mathbf{B} = B_x\mathbf{i} + B_y\mathbf{j}$; then

$$\mathbf{A} \cdot \mathbf{B} = (A_x\mathbf{i} + A_y\mathbf{j}) \cdot (B_x\mathbf{i} + B_y\mathbf{j})$$

and if we multiply this out term by term, we find

$$\mathbf{A} \cdot \mathbf{B} = A_xB_x\,\mathbf{i}\cdot\mathbf{i} + A_xB_y\,\mathbf{i}\cdot\mathbf{j} + A_yB_x\,\mathbf{j}\cdot\mathbf{i} + A_yB_y\,\mathbf{j}\cdot\mathbf{j} \tag{3.22}$$

Here, $\mathbf{i}\cdot\mathbf{i}$ is the product of a vector with itself, which is the square of the magnitude of the vector:

$$\mathbf{i}\cdot\mathbf{i} = 1^2 = 1$$

Likewise,

$$\mathbf{j}\cdot\mathbf{j} = 1^2 = 1$$

The middle two terms in Eq. (3.22) involve $\mathbf{i}\cdot\mathbf{j}$ and $\mathbf{j}\cdot\mathbf{i}$, both of which are zero, since **i** and **j** are perpendicular to each other (compare Fig. 3.17). Hence Eq. (3.22) reduces to

$$\mathbf{A} \cdot \mathbf{B} = A_xB_x + A_yB_y \tag{3.23}$$

More generally, we can show that in the three-dimensional case, the dot product can be expressed as

dot product in terms of components

$$\mathbf{A} \cdot \mathbf{B} = A_xB_x + A_yB_y + A_zB_z \tag{3.24}$$

Thus the dot product is simply the sum of the products of the x, y, and z components of the two vectors.

Finally, note that the components of a vector are equal to the dot product of the vector and the corresponding unit vector. For instance, if $\mathbf{A} = A_x\mathbf{i} + A_y\mathbf{j}$, then

$$\begin{aligned}\mathbf{i}\cdot\mathbf{A} &= \mathbf{i}\cdot(A_x\mathbf{i} + A_y\mathbf{j}) \\ &= A_x\mathbf{i}\cdot\mathbf{i} + A_y\mathbf{i}\cdot\mathbf{j} = A_x \times 1 + A_y \times 0 = A_x\end{aligned}$$

and

$$\begin{aligned}\mathbf{j}\cdot\mathbf{A} &= \mathbf{j}\cdot(A_x\mathbf{i} + A_y\mathbf{j}) \\ &= A_x\mathbf{j}\cdot\mathbf{i} + A_y\mathbf{j}\cdot\mathbf{j} = A_x \times 0 + A_y \times 1 = A_y\end{aligned}$$

EXAMPLE 4 Find the dot product of the vectors **A** and **B** of Example 2.

SOLUTION: The vector **A** has a magnitude $A = 100$ km and the vector **B** has a magnitude $B = 200$ km; the angle between the vectors is $\phi = 120°$. Hence

$$\begin{aligned}\mathbf{A} \cdot \mathbf{B} = AB \cos \phi &= 100 \text{ km} \times 200 \text{ km} \times \cos 120° \\ &= -10\,000 \text{ km}^2\end{aligned} \tag{3.25}$$

Alternatively, this calculation can be done using the components from Example 2:

$$\mathbf{A}\cdot\mathbf{B} = A_xB_x + A_yB_y$$
$$= 0\text{ km}\times 173\text{ km} + 100\text{ km}\times(-100\text{ km}) = -10\,000\text{ km}^2 \qquad (3.26)$$

This agrees with Eq. (3.25).

EXAMPLE 5 The displacement from Miami to the hurricane in Example 2 is $\mathbf{B} = (173\text{ km})\mathbf{i} + (-100\text{ km})\mathbf{j}$; the displacement from the initial position of the reconnaissance aircraft to the hurricane is $\mathbf{C} = (173\text{ km})\mathbf{i} + (-200\text{ km})\mathbf{j}$. Use the dot product to determine the angle between these two vectors (the angle ϕ in Fig. 3.21).

Concepts in Context

SOLUTION: Since

$$\mathbf{B}\cdot\mathbf{C} = BC\cos\phi$$

the cosine of the angle ϕ between the vectors is

$$\cos\phi = \frac{\mathbf{B}\cdot\mathbf{C}}{BC} \qquad (3.27)$$

Here, we can substitute for $\mathbf{B}\cdot\mathbf{C}$ the component expression (3.23), and for B and C the usual expressions (3.15) for the magnitudes:

$$\cos\phi = \frac{\mathbf{B}\cdot\mathbf{C}}{BC} = \frac{B_xC_x + B_yC_y}{\sqrt{B_x^2+B_y^2}\sqrt{C_x^2+C_y^2}}$$

$$= \frac{(173\text{ km})(173\text{ km}) + (-100\text{ km})(-200\text{ km})}{\sqrt{(173\text{ km})^2+(-100\text{ km})^2}\sqrt{(173\text{ km})^2+(-200\text{ km})^2}}$$

$$= \frac{49\,900}{200\times 264} = 0.945$$

With our calculator, we take the inverse cosine of 0.945 and find that the angle is

$$\phi = 19.1°$$

COMMENT: This trick for calculating the unknown angle between two vectors is also useful for calculating the angle between two lines in space; simply take any two vectors pointing along the lines and calculate the angle between them.

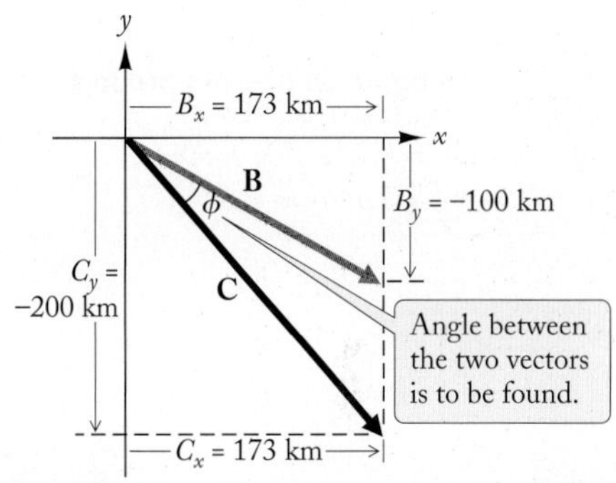

FIGURE 3.21 The two vectors **B** and **C** from Fig. 3.15.

Cross Product

In contrast to the dot product of two vectors, which is a scalar, the **cross product** (also called the **vector product**) of two vectors is a vector. The cross product of two vectors **A** and **B** is denoted by **A** × **B**. The magnitude of this vector is equal to *the product of the magnitudes of the two vectors and the sine of the angle between them.* Thus if we write the vector resulting from the cross product as

$$\mathbf{C} = \mathbf{A}\times\mathbf{B} \qquad (3.28)$$

then the magnitude of this vector is

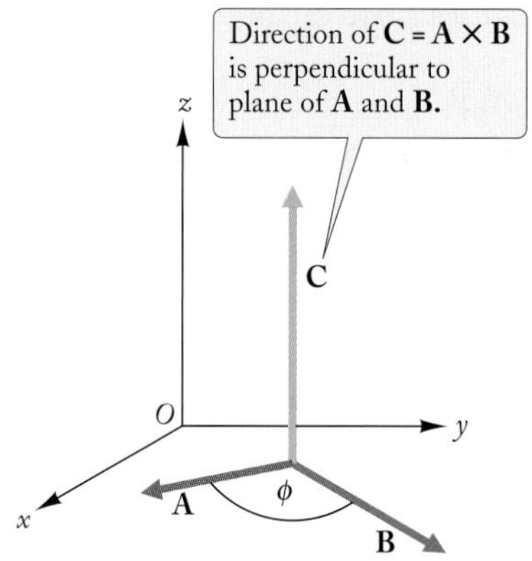

FIGURE 3.22 Two vectors **A** and **B** and their cross product $\mathbf{C} = \mathbf{A} \times \mathbf{B}$.

magnitude of cross product

$$C = AB \sin \phi \tag{3.29}$$

The direction of the vector **C** is defined to be along the perpendicular to the plane formed by **A** and **B** (Fig. 3.22). The direction of **C** along this perpendicular is given by the **right-hand rule**: put the fingers of your right hand along **A** (Fig. 3.23a), and curl them toward **B** in the direction of the smaller angle from **A** to **B** (Fig. 3.23b); the thumb then points along **C**. Note that the fingers must be curled from the first vector in the product toward the second. Thus, $\mathbf{A} \times \mathbf{B}$ is not the same as $\mathbf{B} \times \mathbf{A}$. For the latter product, the fingers must be curled from **B** toward **A** (rather than vice versa); hence, the direction of the vector $\mathbf{B} \times \mathbf{A}$ is opposite to that of $\mathbf{A} \times \mathbf{B}$:

cross product is anticommutative

$$\mathbf{B} \times \mathbf{A} = -\mathbf{A} \times \mathbf{B} \tag{3.30}$$

Accordingly, the cross product of two vectors is anticommutative; in contrast to ordinary multiplication, the result depends on the order of the factors and changes sign when the order of factors is reversed.

As we see from Fig. 3.24, the magnitude of $\mathbf{A} \times \mathbf{B}$ is equal to the magnitude of **B** times the component of **A** perpendicular to **B** (or the magnitude of **A** times the component of **B** perpendicular to **A**). Furthermore, from Fig. 3.24, we see that the magnitude of $\mathbf{A} \times \mathbf{B}$ is equal to the area of the parallelogram formed out of the vectors **A** and **B**.

If the vectors **A** and **B** are parallel, then their cross product is zero, since for two parallel vectors, $\phi = 0$ in Eq. (3.29). In particular, the cross product of any vector with

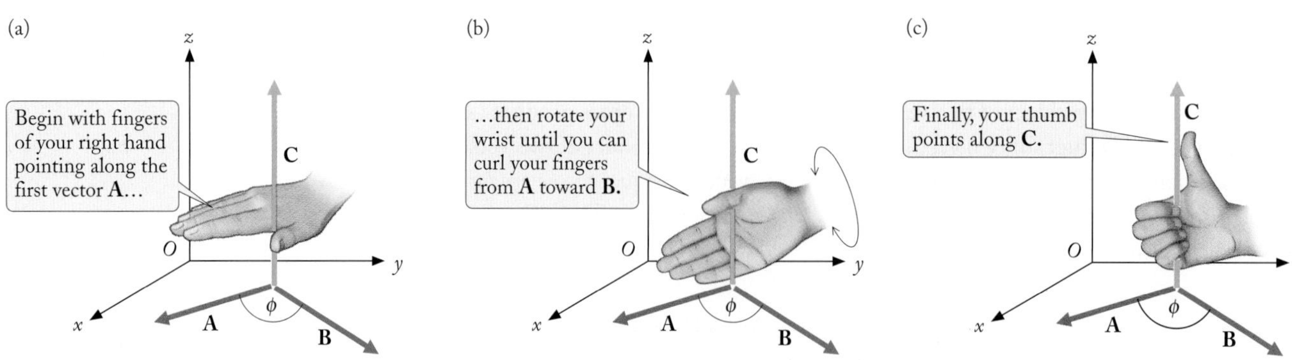

FIGURE 3.23 The right-hand rule for the cross product. If the fingers of the right hand curl from **A** toward **B**, the thumb points along **C**.

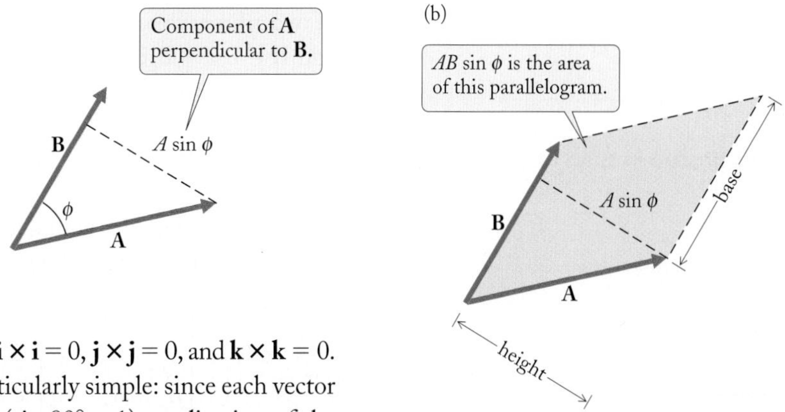

FIGURE 3.24 (a) The component of **A** perpendicular to **B** is $A \sin \phi$. (b) The area of the parallelogram that has **A** and **B** as sides equals the length of the base (B) multiplied by the height $A \sin \phi$.

itself is zero. In the case of the unit vectors, this tells us $\mathbf{i} \times \mathbf{i} = 0$, $\mathbf{j} \times \mathbf{j} = 0$, and $\mathbf{k} \times \mathbf{k} = 0$. The cross product of two different unit vectors is particularly simple: since each vector has unit magnitude and each pair is at right angles ($\sin 90° = 1$), application of the right-hand rule gives (see Fig. 3.17)

$$\mathbf{i} \times \mathbf{j} = \mathbf{k} \qquad \mathbf{j} \times \mathbf{k} = \mathbf{i} \qquad \mathbf{k} \times \mathbf{i} = \mathbf{j} \tag{3.31}$$

As in Eq. (3.30), reversing the order of multiplication changes the sign of the cross products:

$$\mathbf{j} \times \mathbf{i} = -\mathbf{k} \qquad \mathbf{k} \times \mathbf{j} = -\mathbf{i} \qquad \mathbf{i} \times \mathbf{k} = -\mathbf{j} \tag{3.32}$$

Notice that the unit vectors in these equations are in cyclic order (**ijkijk**) for the positive products, and in reverse order (**kjikji**) for the negative products.

EXAMPLE 6 What is the cross product of the two vectors **A** and **B** of Example 2?

SOLUTION: Figure 3.25 shows the vectors **A** and **B**. The magnitudes of these vectors are 100 km and 200 km, respectively; the angle between them is 120°.

Hence the magnitude of the cross product is

$$\begin{aligned} C &= AB \sin \phi \\ &= 100 \text{ km} \times 200 \text{ km} \times \sin 120° = 17\,300 \text{ km}^2 \end{aligned}$$

The direction of the vector **C** is perpendicular to the plane of **A** and **B**; according to the right-hand rule, **C** is along the negative z axis (Fig. 3.25).

Alternatively, we could have used the component notation throughout. From Example 2, we know

$$\begin{aligned} \mathbf{A} &= (100 \text{ km})\mathbf{j} \\ \mathbf{B} &= (173 \text{ km})\mathbf{i} + (-100 \text{ km})\mathbf{j} \end{aligned}$$

To find the cross product, we multiply out the expressions, taking care to maintain the order of the products:

$$\begin{aligned} \mathbf{C} = \mathbf{A} \times \mathbf{B} &= (100 \text{ km})\mathbf{j} \times [(173 \text{ km})\mathbf{i} + (-100 \text{ km})\mathbf{j}] \\ &= (100 \text{ km})(173 \text{ km})\mathbf{j} \times \mathbf{i} + (100 \text{ km})(-100 \text{ km})\mathbf{j} \times \mathbf{j} \\ &= (17\,300 \text{ km}^2)(-\mathbf{k}) + 0 = (-17\,300 \text{ km}^2)\mathbf{k} \end{aligned}$$

The second term was set to zero, since $\mathbf{j} \times \mathbf{j} = 0$.

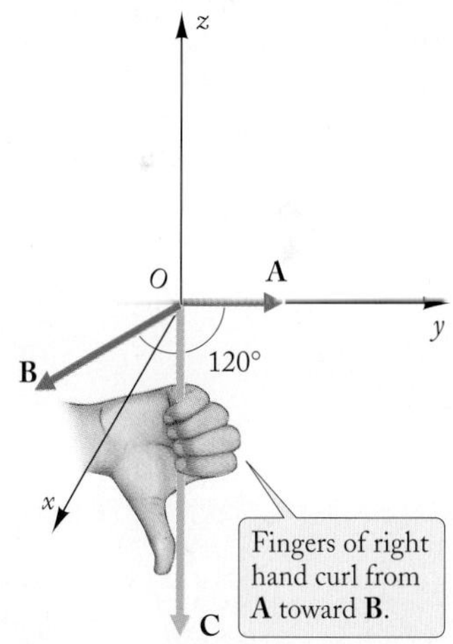

FIGURE 3.25 The two vectors **A** and **B**, and their cross product.

In most situations, one of the techniques of Example 6 will provide the most direct means of evaluating the cross product. The second technique, multiplying out the components, can also be used to write down a general expression for the cross product of two arbitrary vectors. By multiplying out the cross product of $\mathbf{A} = A_x\mathbf{i} + A_y\mathbf{j} + A_z\mathbf{k}$ and $\mathbf{B} = B_x\mathbf{i} + B_y\mathbf{j} + B_z\mathbf{k}$ and evaluating the nine resulting unit-vector cross products, we can obtain the general result (see Problem 56)

cross product in terms of components

$$\mathbf{A} \times \mathbf{B} = (A_yB_z - A_zB_y)\mathbf{i} + (A_zB_x - A_xB_z)\mathbf{j} + (A_xB_y - A_yB_x)\mathbf{k} \tag{3.33}$$

Those readers familiar with determinants can verify that the right side of Eq. (3.33) can be obtained by expanding a 3 × 3 determinant:

$$\mathbf{A} \times \mathbf{B} = \begin{vmatrix} \mathbf{i} & \mathbf{j} & \mathbf{k} \\ A_x & A_y & A_z \\ B_x & B_y & B_z \end{vmatrix} = \mathbf{i}\begin{vmatrix} A_y & A_z \\ B_y & B_z \end{vmatrix} - \mathbf{j}\begin{vmatrix} A_x & A_z \\ B_x & B_z \end{vmatrix} + \mathbf{k}\begin{vmatrix} A_x & A_y \\ B_x & B_y \end{vmatrix} \tag{3.34}$$

where a 2 × 2 determinant is given by $\begin{vmatrix} a & b \\ c & d \end{vmatrix} = ad - bc$.

As a final note, we point out that since the cross product is a vector, its calculation, like any relation equating vectors, really involves three separate equations, one for each component.

✓ Checkup 3.4

QUESTION 1: Two vectors have nonzero magnitude. Under what conditions will their dot product be zero? Their cross product?

QUESTION 2: Suppose that $\mathbf{A} \cdot \mathbf{i} = 0$ and $\mathbf{A} \cdot \mathbf{k} = 0$. What can you conclude about the direction of a nonzero vector **A**?

QUESTION 3: The dot product of two vectors **A** and **B** is −2. What is the dot product of the two vectors 2**A** and 5**B**?

QUESTION 4: Suppose that $\mathbf{A} \times \mathbf{j} = \mathbf{0}$. What can you conclude about the direction of a nonzero vector **A**?

QUESTION 5: Suppose that the vector **A** points west and the vector **B** points vertically up. What is the direction of $\mathbf{A} \times \mathbf{B}$? The direction of $\mathbf{B} \times \mathbf{A}$?

QUESTION 6: Suppose that $\mathbf{A} \cdot \mathbf{B} > 0$. What can you conclude about the angle ϕ between **A** and **B**?

(A) $0 \le \phi < 90°$ (B) $\phi = 90°$ (C) $90° < \phi < 180°$ (D) $\phi = 180°$

PROBLEM-SOLVING TECHNIQUES — DOT PRODUCT AND CROSS PRODUCT OF VECTORS

- We have available two alternative formulas for evaluating the dot product of two vectors, Eq. (3.20) and Eq. (3.24). Which is best depends on how the vectors are specified. If the magnitudes and the directions of the vectors are given, then Eq. (3.20) is best; if the components are given, then Eq. (3.24) is best.
- We also have available two formulas for the cross product, Eq. (3.29) and Eq. (3.33). If we know the magnitudes and the directions of the vectors; then Eq. (3.29) is best. If we know the components, Eq. (3.33) provides a general expression for the cross product, although it is rather messy. In many cases, it is easier to merely multiply out the components and evaluate the simpler unit-vector cross products of Eqs. (3.31) and (3.32), as we did in Example 6.
- Because the right hand is often busy with the pencil, it is a common mistake to use the left hand while trying to determine the direction of a cross product. Do not make this mistake—it gives the opposite direction.

SUMMARY

PHYSICS IN PRACTICE Vectors in Navigation **(page 71)**

PROBLEM-SOLVING TECHNIQUES Vector Addition and Subtraction **(page 79)**

PROBLEM-SOLVING TECHNIQUES Dot Product and Cross Product Vectors **(page 86)**

VECTOR Quantity with magnitude and direction; it behaves like a displacement.

ADDITION OF VECTORS Use the parallelogram method or the tail-to-head method; alternatively, add components. **(3.3)**

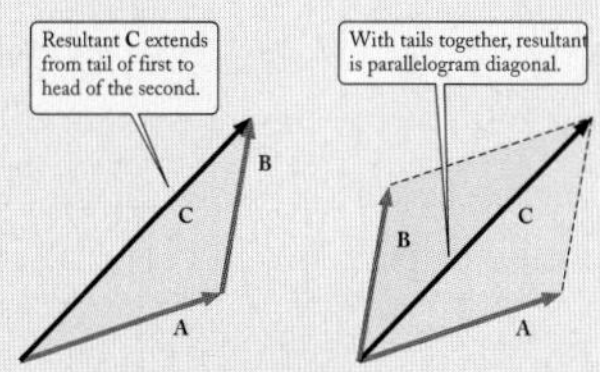

COMPONENTS OF A 2D VECTOR (θ is measured counterclockwise from the x axis.)

$$A_x = A\cos\theta \qquad (3.13)$$

$$A_y = A\sin\theta \qquad (3.14)$$

DIRECTION OF A 2D VECTOR

$$\tan\theta = \frac{A_y}{A_x} \qquad (3.15)$$

MAGNITUDE OF A 2D VECTOR

$$A = \sqrt{A_x^2 + A_y^2} \qquad (3.16)$$

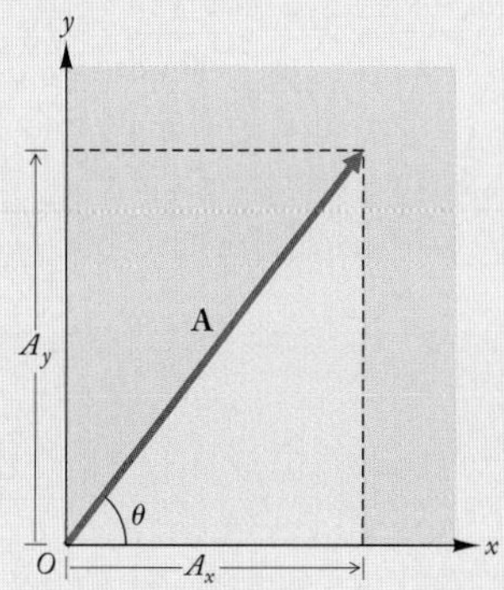

MAGNITUDE OF A 3D VECTOR

$$A = \sqrt{A_x^2 + A_y^2 + A_z^2} \qquad (3.18)$$

UNIT VECTORS **i**, **j**, and **k** along the x, y, and z axes, respectively

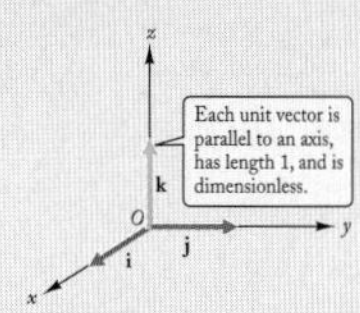

DOT PRODUCT $\mathbf{A}\cdot\mathbf{B} = AB\cos\phi = A_xB_x + A_yB_y + A_zB_z$ **(3.21)**

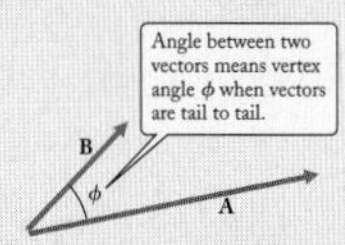

CROSS PRODUCT $\mathbf{C} = \mathbf{A}\times\mathbf{B}$ (direction of **C** is perpendicular to both **A** and **B**, given by right-hand rule: curl the fingers from **A** to **B** and the thumb gives the direction of **C**)

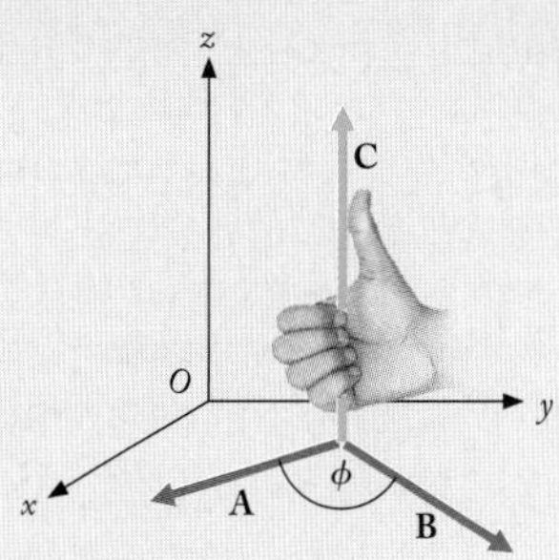

MAGNITUDE OF CROSS PRODUCT	$C = AB \sin \phi$	(3.30)
CROSS PRODUCTS OF UNIT VECTORS		
In cyclic order, positive:	$\mathbf{i} \times \mathbf{j} = \mathbf{k}$ $\mathbf{j} \times \mathbf{k} = \mathbf{i}$ $\mathbf{k} \times \mathbf{i} = \mathbf{j}$	(3.31)
In reverse order, negative:	$\mathbf{j} \times \mathbf{i} = -\mathbf{k}$ $\mathbf{k} \times \mathbf{j} = -\mathbf{i}$ $\mathbf{i} \times \mathbf{k} = -\mathbf{j}$	(3.32)
CROSS PRODUCT IN TERMS OF COMPONENTS	$\mathbf{A} \times \mathbf{B} = (A_yB_z - A_zB_y)\mathbf{i} + (A_zB_x - A_xB_z)\mathbf{j} + (A_xB_y - A_yB_x)\mathbf{k}$	(3.33)

QUESTIONS FOR DISCUSSION

1. A large oil tanker proceeds from Kharg Island (Persian Gulf) to Rotterdam via the Cape of Good Hope. A small oil tanker proceeds from Kharg Island to Rotterdam via the Suez Canal. Are the displacement vectors of the two tankers equal? Are the distances covered equal?
2. An airplane flies from Boston to Houston and back to Boston. Is the displacement zero in the reference frame of the Earth? In the reference frame of the Sun?
3. Does a vector of zero magnitude have a direction? Does it matter?
4. If **A** and **B** are any arbitrary vectors, then $\mathbf{A} \cdot (\mathbf{B} \times \mathbf{A}) = 0$. Explain.
5. Why is there no vector division? (Hint: If **A** and **B** are given and if $A = \mathbf{B} \cdot \mathbf{C}$, then there exist several vectors **C** that satisfy this equation, and similarly for $\mathbf{A} = \mathbf{B} \times \mathbf{C}$.)
6. Assume that **A** is some nonzero vector. If $\mathbf{A} \cdot \mathbf{B} = \mathbf{A} \cdot \mathbf{C}$, can we conclude that $\mathbf{B} = \mathbf{C}$? If $\mathbf{A} \times \mathbf{B} = \mathbf{A} \times \mathbf{C}$, can we conclude that $\mathbf{B} = \mathbf{C}$? What if *both* $\mathbf{A} \cdot \mathbf{B} = \mathbf{A} \cdot \mathbf{C}$ and $\mathbf{A} \times \mathbf{B} = \mathbf{A} \times \mathbf{C}$?

PROBLEMS

3.2 Vector Addition and Subtraction†

1. A ship moves from the Golden Gate Bridge in San Francisco Bay to Alcatraz Island and from there to Point Blunt. The first displacement vector is 10.2 km due east, and the second is 5.9 km due north. What is the resultant displacement vector?
2. In midtown Manhattan, the street blocks have a uniform size of 80 m × 280 m, with the shortest side oriented at 29° east of north ("uptown") and the long side oriented at 29° north of west. Suppose you walk three blocks uptown and then two blocks to the left. What is the magnitude and direction of your displacement vector?
3. Figure 3.26 shows the successive displacements of an aircraft flying a search pattern. The initial position of the aircraft is P and the final position is P'. What is the net displacement (magnitude and direction) between P and P'? Find the answer both graphically (by carefully drawing a page-size diagram with protractor and ruler and measuring the resultant) and trigonometrically (by solving triangles).

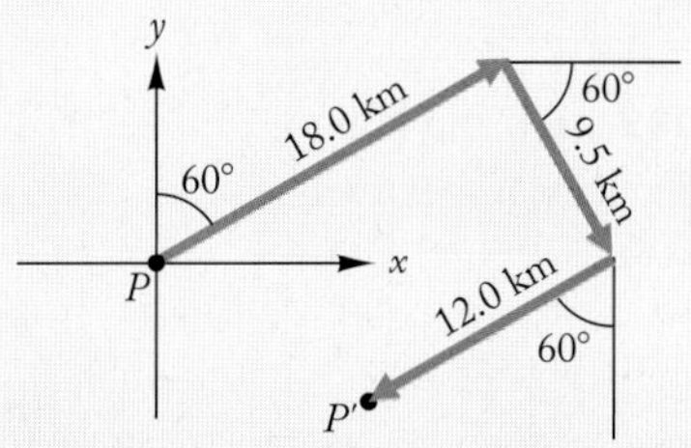

FIGURE 3.26 Successive displacement vectors of an aircraft.

4. A sailboat tacking against the wind moves as follows: 3.2 km at 45° east of north, 4.5 km at 50° west of north, 2.6 km at 45° east of north. What is the net displacement for the entire motion?
5. The recommended route for yachts entering the small harbor of South Bay at Mosquito Island (British Virgin Islands) is as follows: from the northeast side of Mosquito Rock proceed on a course of 135° (magnetic) for 1450 m; then alter course to 180° and proceed for 620 m; then alter course to 285° and proceed for 1190 m; finally alter course to 0° and proceed for

† For help, see Online Concept Tutorial 4 at www.wwnorton.com/physics

730 m to reach South Bay harbor. What is the displacement vector from Mosquito Rock to the harbor? Solve this problem graphically, by making an accurate, large diagram showing the successive displacement vectors.

6. A man runs 2.5 km north and then 1.5 km in a direction 30° east of north. A woman walks directly between the same initial and final points. What distance does the woman walk? In what direction?

7. An airplane flies 480 km in a direction 40° east of north to a hub city, and then 370 km in a direction 10° north of west to a final destination. By plotting these individual displacements on graph paper, carefully determine the magnitude and direction of the net displacement.

8. One speedboat travels 14 km in a direction 60° south of east. A second boat has the same displacement, but instead travels due east and then due south. What distance did the second boat travel east? What distance south?

9. The resultant of two displacement vectors has a length of 5.0 m and a direction due north. One of the displacement vectors has a length of 2.2 m and a direction 35° east of north. What is the other displacement vector?

10. Three displacement vectors **A**, **B**, and **C** are, respectively, 4.0 cm at 30° west of north, 8.0 cm at 30° east of north, and 3.0 cm due north. Carefully draw these vectors on a sheet of paper. Find $\mathbf{A} + \mathbf{B} + \mathbf{C}$ graphically. Find $\mathbf{A} + \mathbf{B} - \mathbf{C}$ graphically.

11. During the maneuvers preceding the Battle of Jutland, the British battle cruiser *Lion* moved as follows (distances are in nautical miles): 1.2 nmi due north, 6.1 nmi at 38° east of south, 2.9 nmi at 59° east of south, 4.0 nmi at 89° east of north, and 6.5 nmi at 31° east of north.

    (a) Draw each of these displacement vectors and draw the net displacement vector.

    (b) Graphically or algebraically find the distance between the initial position and the final position.

*12. The Earth moves around the Sun in a circle of radius 1.50×10^{11} m at (approximately) constant speed.

    (a) Taking today's position of the Earth as origin, draw a diagram showing the position vector 3 months, 6 months, 9 months, and 12 months later.

    (b) Draw the displacement vector between the 0-month and the 3-month positions, the 3-month and the 6-month positions, etc. Calculate the magnitude of the displacement vector for one of these 3-month intervals.

*13. Both Singapore and Quito are (nearly) on the Earth's equator; the longitude of Singapore is 104° east and that of Quito is 78° west. What is the magnitude of the displacement vector between these cities? What is the distance between them measured along the equator?

*14. By a method known as "doubling the angle on the bow," the navigator of a ship can determine his position relative to a fixed point, such as a lighthouse. Figure 3.27 shows the (straight) track of a ship passing by a lighthouse. At the point P, the navigator measures the angle α between the line of sight to the lighthouse and the direction of motion of the ship. He then measures how far the ship advances through the water until the angle between the line of sight and the direction of motion is twice as large as it was initially. Prove that the magnitude of the displacement vector PP' equals the magnitude of the position vector AP' of the ship relative to the lighthouse.

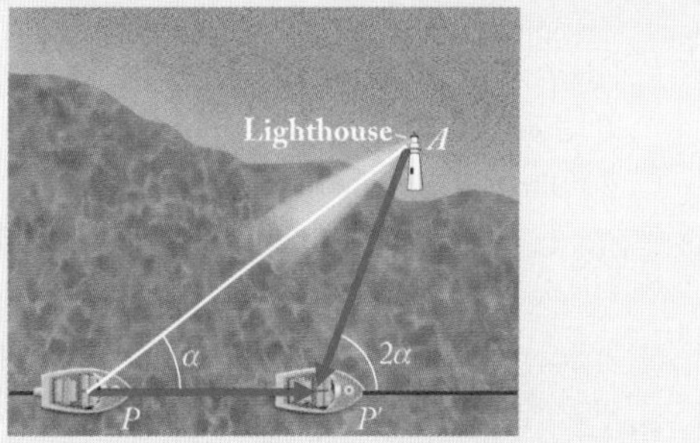

FIGURE 3.27 A lighthouse and two positions of a ship.

*15. The radar operator of a stationary Coast Guard cutter observes that at 10^h30^m an unidentified ship is at a distance of 9.5 km on a bearing of 60° east of north and at 11^h10^m the unidentified ship is at a distance of 4.2 km on a bearing of 33° east of north. Measured from its position at 10^h30^m, what is the displacement vector of the unidentified ship at 11^h10^m? Assuming that the unidentified ship continues on the same course at the same speed, what will be its displacement vector at 11^h30^m? What will be its distance and bearing from the cutter?

*16. Suppose that two ships proceeding at constant speeds are on converging straight tracks. Prove that the ships will collide if and only if the bearing of each remains constant as seen from the other. This constant-bearing rule is routinely used by mariners to check whether there is danger of collision. (Hint: A convenient method of proof is to draw the displacement vector from one ship to the other at several successive times.)

3.3 The Position Vector; Components of a Vector†

17. A displacement vector has a magnitude of 12.0 km in the direction 40.0° west of north. What is the north component of this vector? The west component?

18. A vector length of 5.0 m is in the x–y plane at an angle of 30° with the x axis. What is the x component of this vector? The y component?

19. A displacement vector has a magnitude of 4.0 m and a vertically downward direction. What is the component of this vector along a line sloping upward at 25°? (Hint: Place your x axis along the sloping line.)

20. Air traffic controllers usually describe the position of an aircraft relative to the airport by altitude, horizontal distance, and bearing. Suppose an aircraft is at altitude 500 m, distance 15 km, and bearing 35° east of north. What are the x, y, and z components (in meters) of the position vector? The x axis is east, the y axis is north, and the z axis is vertical.

† For help, see Online Concept Tutorial 4 at www.wwnorton.com/physics

21. The displacement vectors **A** and **B** are in the x–y plane. Their components are $A_x = 3.0$ cm, $A_y = 2.0$ cm, $B_x = -1.0$ cm, $B_y = 3.0$ cm.
    (a) Draw a diagram showing these vectors.
    (b) Calculate the resultant of **A** and **B**. Draw the resultant on your diagram.

22. A vector in the x–y plane has a magnitude of 8.0 units. The angle between the vector and the x axis is 52°. What are the x, y, and z components of this vector?

23. Given that a vector has a magnitude of 6.0 units and makes an angle of 45° and 85° with the x and y axes, respectively, find the x and y components of this vector. Does the given information determine the z component? What can you say about the z component?

24. What is the magnitude of the vector $3\mathbf{i} + 2\mathbf{j} - \mathbf{k}$?

25. Suppose that $\mathbf{A} = -5\mathbf{i} - 3\mathbf{j} + \mathbf{k}$ and $\mathbf{B} = 2\mathbf{i} + \mathbf{j} - 3\mathbf{k}$. Calculate the following:
    (a) $\mathbf{A} + \mathbf{B}$
    (b) $\mathbf{A} - \mathbf{B}$
    (c) $2\mathbf{A} - 3\mathbf{B}$

26. A position vector has x component 6.0 m, y component -8.0 m, and zero z component. What is the magnitude of this vector? What angle does it make with the x axis?

27. A vector of length 14 m points in a direction 135° counterclockwise from the x axis. What is its x component? Its y component?

28. Four vectors are given by $\mathbf{A} = 2.5\mathbf{i} + 3.5\mathbf{j}$, $\mathbf{B} = 1.0\mathbf{i} + 4.5\mathbf{j} + 2.5\mathbf{k}$, $\mathbf{C} = 1.5\mathbf{i} + 2.0\mathbf{j} + 3.0\mathbf{k}$, and $\mathbf{D} = 3.0\mathbf{j} + 1.5\mathbf{k}$. What is the vector $\mathbf{E} = \mathbf{A} + \mathbf{B} + \mathbf{C} + \mathbf{D}$? What is the vector $\mathbf{F} = \mathbf{A} - \mathbf{B} + \mathbf{C} - \mathbf{D}$? What is the magnitude of **E**? The magnitude of **F**?

29. What is the unit vector parallel to the vector $\mathbf{A} = 2.0\mathbf{i} + 4.0\mathbf{j} + 4.0\mathbf{k}$?

*30. Two vectors are given by $\mathbf{A} = 2.0\mathbf{i} + 3.0\mathbf{j} + 1.0\mathbf{k}$ and $\mathbf{B} = -1.0\mathbf{i} + 2.0\mathbf{j} + B_z\mathbf{k}$. The magnitude of the resultant $\mathbf{A} + \mathbf{B}$ is 6.0. What are the two possible values of B_z?

*31. Three vectors are given by $\mathbf{A} = 2.0\mathbf{i} + 3.0\mathbf{j}$, $\mathbf{B} = 1.0\mathbf{i} + 5.0\mathbf{j}$, and $\mathbf{C} = -1.0\mathbf{i} + 3.0\mathbf{j}$. Find constants c_1 and c_2 such that $c_1\mathbf{A} + c_2\mathbf{B} = \mathbf{C}$.

*32. An air traffic controller notices that one aircraft approaching the airport is at an altitude of 2500 m, (horizontal) distance 120 km, and bearing 20° south of east. A second aircraft is at altitude 3500 m, distance 110 km, and bearing 25° south of east. What is the displacement vector from the first aircraft to the second? Express your answer in terms of altitude, (horizontal) distance, and bearing.

*33. A remarkably fast crossing of the Atlantic by sail was achieved in 1916 by the four-masted ship *Lancing,* which sailed from New York (latitude 40°48′ north, longitude 73°58′ west) to Cape Wrath, Scotland (latitude 58°36′ north, longitude 5°1′ west), in $6\frac{3}{4}$ days. What was the magnitude of the displacement vector for this trip?

*34. A vector has components $A_x = 5.0$, $A_y = -3.0$, $A_z = 1.0$. What is the magnitude of this vector? What is the angle between this vector and the x axis? The y axis? The z axis?

*35. Find a vector that has the same direction as $3\mathbf{i} - 6\mathbf{j} + 2\mathbf{k}$ but a magnitude of 2 units.

*36. Given that $\mathbf{A} = 6\mathbf{i} - 2\mathbf{j}$ and $\mathbf{B} = -4\mathbf{i} - 3\mathbf{j} + 8\mathbf{k}$, find a vector **C** such that $3\mathbf{A} - 2\mathbf{C} = 4\mathbf{B}$.

3.4 Vector Multiplication

37. Calculate the dot product of the vectors $5\mathbf{i} - 2\mathbf{j} + \mathbf{k}$ and $2\mathbf{i} - \mathbf{k}$.

38. Calculate the dot product of the vectors **A** and **B** described in Example 1.

39. Find the magnitude of the vector $-2\mathbf{i} + \mathbf{j} + 2\mathbf{k}$. Find the magnitude of the vector $3\mathbf{i} - 6\mathbf{j} + 2\mathbf{k}$. Find the angle between these two vectors.

40. Show that, in three dimensions, the dot product of two vectors can be expressed as [see Eq. (3.24)]
$$\mathbf{A} \cdot \mathbf{B} = A_xB_x + A_yB_y + A_zB_z$$

41. Find the angle between the vector $\mathbf{A} = 3.0\mathbf{i} + 4.0\mathbf{j} + 2.0\mathbf{k}$ and the x axis.

42. For the two vectors $\mathbf{A} = 4.0\mathbf{i} + 3.0\mathbf{j} + 2.0\mathbf{k}$ and $\mathbf{B} = -1.0\mathbf{i} + 2.0\mathbf{j} + 1.0\mathbf{k}$, calculate $\mathbf{A} \cdot \mathbf{B}$ and $\mathbf{A} \times \mathbf{B}$.

43. The dot product of two vectors and the magnitude of the cross product of the same vectors are equal. What is the angle between the two vectors?

44. The displacement vector **A** has a length of 50 m and a direction of 30° east of north; the displacement vector **B** has a length of 35 m and a direction 70° west of north. What are the magnitude and direction of the cross product $\mathbf{A} \times \mathbf{B}$? The cross product $\mathbf{B} \times \mathbf{A}$?

45. Calculate the cross product of the vectors **A** and **B** described in Example 1.

46. Calculate the cross product of the vectors $2\mathbf{i} - 5\mathbf{j} + 3\mathbf{k}$ and $\mathbf{i} - 2\mathbf{k}$.

*47. Suppose that
$$\mathbf{A} = \mathbf{i}\cos\omega t + \mathbf{j}\sin\omega t$$
where ω is a constant. Find $d\mathbf{A}/dt$ (note that **i** and **j** behave as constants in differentiation). Show that $d\mathbf{A}/dt$ is perpendicular to **A**.

*48. The displacement vector **A** has a length of 30 m and a direction 20° south of east. The displacement vector **B** has a length of 40 m and a direction 20° west of north. Find the component of **A** along **B**. Find the component of **B** along **A**.

*49. The two vectors $\mathbf{A} = 5.0\mathbf{i} - 2.0\mathbf{j} + 3.0\mathbf{k}$ and $\mathbf{B} = B_x\mathbf{i} + 3.0\mathbf{j} + B_z\mathbf{k}$ have cross product $\mathbf{C} = \mathbf{A} \times \mathbf{B} = 2.0\mathbf{j} + C_z\mathbf{k}$. Find the values of B_x, B_z, and C_z.

*50. Two vectors **A** and **B** lie in the x–y plane. Show that the tangent of the angle θ between them is given by

$$\tan\theta = \frac{A_xB_y - A_yB_x}{A_xB_x + A_yB_y}$$

*51. A vector **A** has components $A_x = 2, A_y = -1, A_z = -4$. Find a vector (give its components) that has the same direction as **A** but a magnitude of 1 unit.

*52. Find a unit vector that bisects the angle between the vectors $\mathbf{j} + 2\mathbf{k}$ and $3\mathbf{i} - \mathbf{j} + \mathbf{k}$.

*53. Find a unit vector that points toward a position halfway between the two position vectors $4\mathbf{i} + 2\mathbf{j}$ and $-\mathbf{i} + 3\mathbf{j} + 2\mathbf{k}$.

*54. Find the angle between the diagonal of a cube and one of its edges. (Hint: Suppose that the edges of the cube are parallel to the vectors **i**, **j**, and **k**. What vector is then parallel to the diagonal?)

*55. The dot product of two vectors **A** and **B** is zero. The magnitudes of the two vectors are, respectively, $A = 4$ and $B = 6$. What can you say about the cross product of these two vectors?

*56. (a) Figure 3.17 displays the unit vectors **i**, **j**, and **k** along the *x*, *y*, and *z* axes. The positive directions for these axes, and the directions for the unit vectors, have been chosen according to the standard convention for a "right-handed" coordinate system (if, say, the positive direction of the *x* axis and the direction of **i** were reversed, the coordinate system would become "left-handed"). Use the right-hand rule to establish the following multiplication table for the unit vectors displayed in Fig. 3.17 [see Eqs. (3.31) and (3.32)]:

$\mathbf{i}\times\mathbf{i} = 0$	$\mathbf{j}\times\mathbf{i} = -\mathbf{k}$	$\mathbf{k}\times\mathbf{i} = \mathbf{j}$
$\mathbf{i}\times\mathbf{j} = \mathbf{k}$	$\mathbf{j}\times\mathbf{j} = 0$	$\mathbf{k}\times\mathbf{j} = -\mathbf{i}$
$\mathbf{i}\times\mathbf{k} = -\mathbf{j}$	$\mathbf{j}\times\mathbf{k} = \mathbf{i}$	$\mathbf{k}\times\mathbf{k} = 0$

(b) With the above multiplication table, derive Eq. (3.33) for the cross product of two vectors in terms of their components:

$$\mathbf{A}\times\mathbf{B} = (A_yB_z - A_zB_y)\mathbf{i} + (A_zB_x - A_xB_z)\mathbf{j} + (A_xB_y - A_yB_x)\mathbf{k}$$

(c) From this formula, verify that if **A** and **B** are parallel, then $\mathbf{A}\times\mathbf{B} = \mathbf{0}$.

*57. Verify that the evaluation of the determinant in Eq. (3.34) (according to the usual rules for determinants) yields the formula for the cross product given in Eq. (3.33):

$$\mathbf{A}\times\mathbf{B} = \begin{vmatrix} \mathbf{i} & \mathbf{j} & \mathbf{k} \\ A_x & A_y & A_z \\ B_x & B_y & B_z \end{vmatrix}$$

*58. Evaluate $((\mathbf{i}\times\mathbf{j})\times\mathbf{i})\times\mathbf{i}$. Evaluate $(((\mathbf{i}\times\mathbf{j})\times\mathbf{j})\times\mathbf{j})$.

*59. Given that $\mathbf{A} = 2\mathbf{i} - 3\mathbf{j} + 2\mathbf{k}$ and $\mathbf{B} = -3\mathbf{i} + 4\mathbf{k}$, calculate the cross product $\mathbf{A}\times\mathbf{B}$.

*60. The vectors **A**, **B**, and **C** have components $A_x = 3, A_y = -2, A_z = 2, B_x = 0, B_y = 0, B_z = 4, C_x = 2, C_y = -3, C_z = 0$. Calculate the following:

(a) $\mathbf{A}\cdot(\mathbf{B}+\mathbf{C})$

(b) $\mathbf{A}\times(\mathbf{B}+\mathbf{C})$

(c) $\mathbf{A}\cdot(\mathbf{B}\times\mathbf{C})$

(d) $\mathbf{A}\times(\mathbf{B}\times\mathbf{C})$

*61. Find a unit vector perpendicular to both $4\mathbf{i} + 3\mathbf{j}$ and $-\mathbf{i} - 3\mathbf{j} + 2\mathbf{k}$.

62. Show that the magnitude of $\mathbf{A}\cdot(\mathbf{B}\times\mathbf{C})$ is the volume of the parallelepiped determined by **A, **B**, and **C**.

63. Show that $\mathbf{A}\times(\mathbf{B}\times\mathbf{C}) = \mathbf{B}(\mathbf{A}\cdot\mathbf{C}) - \mathbf{C}(\mathbf{A}\cdot\mathbf{B})$. (Hint: Choose the orientation of your coordinate axes in such a way that **B is along the *x* axis and **C** is in the *x*–*y* plane.)

*64. In the vicinity of New York City, the direction of magnetic north is 11°55′ west of true north (that is, a magnetic compass needle points 11°55′ west of north). Suppose that an aircraft flies 5.0 km on a bearing of 56° east of magnetic north.

(a) What are the north and east components of this displacement in a coordinate system based on the direction of magnetic north?

(b) What are the north and east components of the displacement in a coordinate system based on the direction of true north?

*65. A vector has components $A_x = 6, A_y = -3, A_z = 0$ in a given rectangular coordinate system. Find a new coordinate system, with new directions of the *x* and *y* axes, such that the only nonzero component of the vector is A'_x.

REVIEW PROBLEMS

66. To reach Moose Jaw, Canada, you drive your automobile due north 90 km, and then due west 70 km. What are the magnitude and the direction of your displacement vector? What is the distance you traveled?

67. The displacement vector **A** has a length of 350 m in the direction 45° west of north; the displacement vector **B** has a length of 120 m in the direction 20° east of north. Find the magnitude and direction of the resultant of these vectors.

68. In Chapter 10 we will become acquainted with the center of mass, which for a collection of equal particles is simply the average position of the particles. Suppose that three particles have position vectors $5\mathbf{i} + 3\mathbf{k}$, $-2\mathbf{i} + \mathbf{j} - 3\mathbf{k}$, and $4\mathbf{i} + 2\mathbf{j} + \mathbf{k}$, respectively. What is the average of these position vectors? What is the length of this average position vector?

69. A vector drawn on a wall has a magnitude of 2.0 and makes an angle of 30° with the vertical direction. What are the vertical and horizontal components of this vector?

70. An aircraft flies 250 km in a direction 30° east of south, and then 250 km in a direction 30° west of south. What are the magnitude and the direction of the resultant displacement vector?

71. The vector **A** has a length of 6.2 units in a direction 30° south of east. The vector **B** has a length of 9.6 units in a direction due south. What is the sum **A** + **B** of these vectors? What is the difference **A** − **B**?

*72. A room measures 4 m in the x direction, 5 m in the y direction, and 3 m in the z direction. A lizard crawls along the walls from one corner of the room to the diametrically opposite corner. If the starting point is the origin of coordinates, what is the displacement vector? What is the length of the displacement vector? If the lizard chooses the shortest path along the walls, what is the length of its path?

73. What is the magnitude of the vector $2\mathbf{i} + \mathbf{j} - 4\mathbf{k}$?

74. Suppose that $\mathbf{A} = 4\mathbf{i} - 2\mathbf{j}$ and $\mathbf{B} = -3\mathbf{i} - 4\mathbf{j}$. Calculate the following:

(a) $\mathbf{A} + \mathbf{B}$

(b) $\mathbf{A} - \mathbf{B}$

(c) $3\mathbf{A} - \mathbf{B}$

75. The displacement vector **A** has a length of 50 m and direction of 30° east of north; the displacement vector **B** has a length of 35 m and a direction 70° west of north. What is the dot product of these vectors?

76. The displacement vector **A** has a length of 6.0 m in the direction 30° east of north; the displacement vector **B** has a length of 8.0 m in the direction 40° south of east. Find the magnitude and the direction of $\mathbf{A} \times \mathbf{B}$.

*77. Suppose that $\mathbf{A} = 3\mathbf{i} + 4\mathbf{j}$ and $\mathbf{B} = \mathbf{i} + 3\mathbf{j} - 2\mathbf{k}$. Find the component of **A** along the direction of **B**. Find the component of **B** along the direction of **A**.

*78. Given the vectors $\mathbf{A} = 2\mathbf{i} + 2\mathbf{j} - \mathbf{k}$ and $\mathbf{B} = 3\mathbf{i} - \mathbf{j}$, calculate

(a) The sum $\mathbf{A} + \mathbf{B}$

(b) The difference $\mathbf{A} - \mathbf{B}$

(c) The dot product $\mathbf{A} \cdot \mathbf{B}$

(d) The cross product $\mathbf{A} \times \mathbf{B}$

Answers to Checkups

Checkup 3.1

1. A displacement vector depends only on the net change in position, that is, on the beginning and ending points, and so the displacement vectors for the two airliners are equal. The distances traveled refer to the total distances covered, and so are unequal for the two different routes.

2. The magnitude of a vector is its length, which is defined as a positive quantity, and cannot be negative. Only the zero vector has zero length (and it is the only vector with no direction).

3. Yes, the two displacement vectors are equal. Both have the same direction (due west) and the same magnitude (3000 m); only the *change* in position matters, not any particular positions.

4. (C) 100 km. The vector from that point to Miami is one side of a 30°–60°–90° right triangle (the side opposite the 30° angle), a triangle with a 200-km hypotenuse. Thus the distance to Miami is (200 km) × (sin 30°) = 100 km.

Checkup 3.2

1. Their sum will be zero if the two vectors have the same magnitude but point in opposite directions (antiparallel). Their difference will be zero if they are identical vectors, that is, if they have the same magnitude and direction (parallel).

2. For two perpendicular vectors, the magnitude of the sum must be larger than the magnitude of each. This is so because two such vectors form the sides of a right triangle, with their sum forming the hypotenuse.

3. Yes; any two perpendicular vectors have the same magnitude for their sum and for their difference. (Picture adding two perpendicular vectors tail to head for the sum, and then reverse one of the vectors for the difference.)

4. By picturing adding the three vectors tail to head, a zero resultant requires that the head of the third vector close back on the tail of the first vector. Thus one can see that the only way three equal-magnitude vectors can sum to zero is if they form the sides of an equilateral triangle.

5. (B) 5 km. Since the two displacements are in perpendicular directions, the net displacement is given by the hypotenuse $C = \sqrt{A^2 + B^2} = \sqrt{(3\text{ km})^2 + (4\text{ km})^2} = 5\text{ km}$.

Checkup 3.3

1. The magnitude is obtained by squaring the components, adding them, and taking the square root of the result. Since the squares are positive, the sum of the squares will always be greater than or equal to the square of any given component (only equal when the other components are zero). See Eq. (3.15).

2. A vector parallel to the y axis has only a y component, and so its x component must be zero.

3. With equal x and y components, it must point in a direction halfway between the x and y directions, that is, at 45° to the x axis. One can also calculate this from $\tan\theta = y/x = 1$.

4. A constant multiplying a vector multiplies each component; thus 2**A** has x component $2A_x = 6$ and y component $2A_y = -2$. Similarly, -4**A** has x component $-4A_x = -12$ and y component $-4A_y = 4$.

5. (D) $\sqrt{2}$. Each of the two (perpendicular) components has unit magnitude, so the vector has magnitude $\sqrt{1^2 + 1^2} = \sqrt{2}$.

Checkup 3.4

1. The dot product will be zero if the vectors are perpendicular (since $\cos\phi = 0$ when $\phi = 90°$). The cross product will be zero when the vectors are parallel or antiparallel (since $\sin\phi = 0$ when $\phi = 0°$ or 180°).

2. The vector must be perpendicular to both the **i** and **k** directions; thus it has only a **j** component (along the $\pm y$ direction).

3. Multiplying each vector by a constant mutiplies the product by that constant, so the dot product is $2 \times 5 = 10$ times as large, or equals -20.

4. If the cross product is zero, then **A** must have only a **j** component, that is, **A** is along the $\pm y$ axis. Alternatively, the sine of the angle between **A** and **j** must be zero ($\phi = 0°$ or 180°), so that **A** must be parallel or antiparallel to **j**.

5. Pointing the fingers of the right hand along **A** (west) and orienting the hand so that the fingers can curl toward **B** (upward), we find that the thumb points north (so this is the direction of **A** × **B**). Similarly, pointing the fingers up and curling them toward the west, the thumb points south, the direction of **B** × **A**. Also, if we know **A** × **B** is north, then **B** × **A** = −**A** × **B** tells us that **B** × **A** is south.

6. (A) $0 \leq \phi < 90°$. If the dot product is greater than zero, then the cosine of the angle between the vectors is positive, that is, $\phi < 90°$.

CHAPTER

4 Motion in Two and Three Dimensions

CONCEPTS IN CONTEXT

Concepts in Context

This time-exposure photograph shows the curved trajectories of incandescent chunks of lava ejected during an eruption of the Stromboli volcano. Such chunks of lava, called "volcanic bombs," are often ejected at speeds of 600 km/h or more, and they can land at distances of several km from the volcano. They sometimes start fires at the point of impact.

The motion of volcanic bombs is an instance of projectile motion, and with the concepts developed in this chapter we can address the following questions:

? Given the initial speed and direction of motion of the projectile, what are the horizontal and vertical components of the velocity and what are the horizontal and vertical motions? (Example 2, page 100)

? How high does the projectile rise vertically? (Example 4, page 105)

? At what time does the projectile reach its maximum height? (Example 4, page 105)

? What is the shape of the trajectory? (Section 4.4, page 108)

In this chapter we will deal with translational motion in a plane, such as the motion of an automobile on the crisscrossing and curving streets of a flat city or the motion of a boat on the surface of a lake. This is two-dimensional motion, and it is a simple generalization of the one-dimensional motion we studied in Chapter 2. In essence, two-dimensional motion consists of two one-dimensional motions occurring simultaneously. Thus, we will have to apply the formulas of Chapter 2 separately to each of these one-dimensional motions. We can further generalize to the case of three-dimensional motion, such as the motion of an aircraft or the motion of an automobile on a mountain road, consisting of three one-dimensional motions. But we will rarely have to deal with three-dimensional motion, because one- or two-dimensional motion adequately describes most of the physical problems to be discussed in later chapters.

We will examine in detail some cases of two-dimensional motion: projectile motion and uniform circular motion. A stone or a ball thrown by hand or hit by a racket, a bomb released from an aircraft, and a volcanic bomb ejected from a volcano (see the chapter photo) are examples of projectile motion. The projectile always remains in the fixed vertical plane defined by its initial vertical and horizontal velocities, and it traces out a two-dimensional curved trajectory in this plane. An automobile traveling at constant speed around a traffic circle, a child riding a merry-go-round, and a communications satellite circling the Earth are examples of uniform circular motion. The moving particle remains in a fixed plane, and it traces out a circular path in this plane.

4.1 COMPONENTS OF VELOCITY AND ACCELERATION

Online Concept Tutorial 5

To describe the translational motion of a particle in a plane, we need two coordinates, say, an x coordinate and a y coordinate. For instance, if the particle is an automobile traveling on the streets of a (flat) city, we can describe the position by choosing an origin at, say, the library, and laying out an x axis in the eastward direction and a y axis in the northward direction (see Fig. 4.1). The x and y coordinates measured with respect to these axes then provide a complete description of the translational motion of the automobile.

For a moving particle, both the x and the y components of the position change with time. Correspondingly, there are x and y components of the velocity. In the case of the average velocity, we can define these x and y components by analogy with Eq. (2.5):

$$\bar{v}_x = \frac{\Delta x}{\Delta t} \tag{4.1}$$

and

$$\bar{v}_y = \frac{\Delta y}{\Delta t} \tag{4.2}$$

where $\Delta x = x_2 - x_1$ and $\Delta y = y_2 - y_1$ are the changes in the x and y components of the position in the time interval $\Delta t = t_2 - t_1$ (see Fig. 4.1). Stated in words: the x component of the velocity is the rate of change of the x coordinate, and the y component of velocity is the rate of change of the y coordinate.

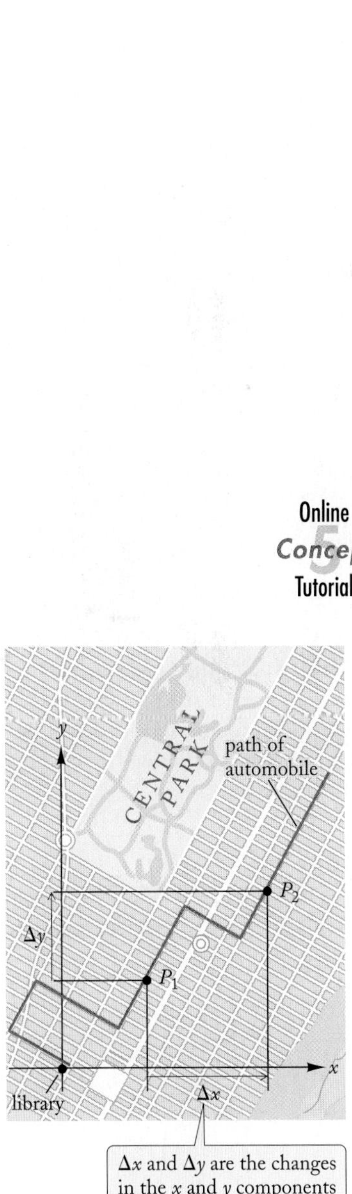

FIGURE 4.1 Path of an automobile along the streets of New York City.

Likewise, in the case of the instantaneous velocity, we can define the x and y components by analogy with Eq. (2.10):

components of the instantaneous velocity

$$v_x = \frac{dx}{dt} \tag{4.3}$$

and

$$v_y = \frac{dy}{dt} \tag{4.4}$$

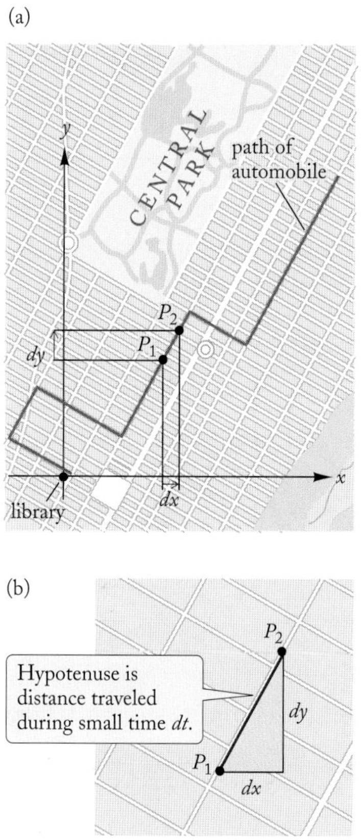

FIGURE 4.2 (a) In a small time interval dt, the displacement is P_1P_2, and the changes in the x and y coordinates are dx and dy. (b) the displacements dx and dy are the sides of a right triangle.

As we saw in Chapter 2, for one-dimensional motion the instantaneous speed (such as the speed indicated by the speedometer of an automobile) equals the magnitude of the instantaneous velocity. For two-dimensional motion, the velocity has two components and the relationship between speed and velocity is not that obvious. To discover what the relationship is, consider the motion of a particle in a small time interval dt. In this time interval the particle travels a distance P_1P_2 from the point P_1 to the point P_2, and the changes in its x and y coordinates are dx and dy. As indicated in Fig. 4.2, dx and dy form the sides of a small right triangle, and P_1P_2 is the hypotenuse of this triangle. According to the Pythagorean theorem,

$$[\text{distance traveled}] = P_1P_2 = \sqrt{(dx)^2 + (dy)^2} \tag{4.5}$$

Hence the instantaneous speed is

$$[\text{instantaneous speed}] = \frac{[\text{distance traveled}]}{[\text{time taken}]} = \frac{P_1P_2}{dt} \tag{4.6}$$

$$= \frac{\sqrt{(dx)^2 + (dy)^2}}{dt} = \sqrt{\left(\frac{dx}{dt}\right)^2 + \left(\frac{dy}{dt}\right)^2}$$

$$= \sqrt{v_x^2 + v_y^2} \tag{4.7}$$

Thus, the instantaneous speed is the square root of the sum of the squares of the components of the instantaneous velocity. If we use the letter v (without any subscript) to represent the instantaneous speed in two-dimensional motion, we can write Eq. (4.7) as

magnitude of velocity in terms of components

$$v = \sqrt{v_x^2 + v_y^2} \tag{4.8}$$

The definitions of the components of the acceleration in two dimensions can be formulated in much the same way. For the average acceleration we have

$$\bar{a}_x = \frac{\Delta v_x}{\Delta t} \tag{4.9}$$

and

$$\bar{a}_y = \frac{\Delta v_y}{\Delta t} \tag{4.10}$$

where Δv_x and Δv_y are the changes in the x and the y components of the velocity. For the instantaneous acceleration we have

$$a_x = \frac{dv_x}{dt} \qquad (4.11)$$

components of the instantaneous acceleration

and

$$a_y = \frac{dv_y}{dt} \qquad (4.12)$$

It is an important consequence of these definitions that *there is an acceleration whenever any of the components of the velocity change.* This means that not only is there an acceleration when an automobile increases or decreases its speed, but there is also an acceleration when the automobile travels around a curve at constant speed, as we will see in the following example.

EXAMPLE 1 An automobile, traveling at a constant speed of 25 m/s, enters a 90° curve and emerges from this curve 6.0 s later. What are the components of the average acceleration for this time interval?

SOLUTION: Figure 4.3 shows the path of the automobile and the orientation of the axes. The initial velocity has an x component ($v_x = 25$ m/s) and no y component. The final velocity has a y component ($v_y = -25$ m/s) and no x component. Hence the changes in the velocity components are

$$\Delta v_x = [\text{final } x \text{ velocity}] - [\text{initial } x \text{ velocity}] = 0 - 25 \text{ m/s} = -25 \text{ m/s}$$

and

$$\Delta v_y = [\text{final } y \text{ velocity}] - [\text{initial } y \text{ velocity}] = -25 - 0 \text{ m/s} = -25 \text{ m/s}$$

The components of the average acceleration are then

$$\bar{a}_x = \frac{\Delta v_x}{\Delta t} = \frac{-25 \text{ m/s}}{6.0 \text{ s}} = -4.2 \text{ m/s}^2$$

and

$$\bar{a}_y = \frac{\Delta v_y}{\Delta t} = \frac{-25 \text{ m/s}}{6.0 \text{ s}} = -4.2 \text{ m/s}^2$$

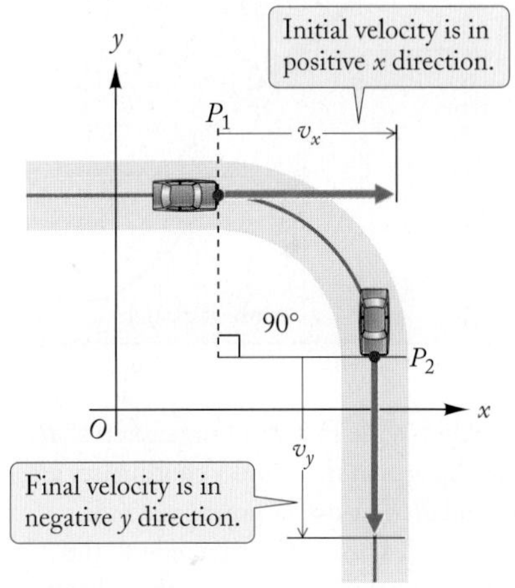

FIGURE 4.3 Automobile rounding a curve. The arrows indicate the directions of the motion.

✔ Checkup 4.1

QUESTION 1: Consider a ship moving on the sea. The x axis is eastward and the y axis is northward. For each of the following cases, state whether the x and y components of the velocity are positive, negative, or zero (+, −, 0): (a) ship is moving northwest; (b) ship is moving south; (c) ship is moving southeast.

QUESTION 2: The speedometer of your automobile shows that you are proceeding at a steady speed of 80 km/h. Is it nevertheless possible that your automobile is in accelerated motion?

QUESTION 3: A particle travels once around a circle at uniform speed. What are the average velocity and the average acceleration for this motion?

QUESTION 4: A motorcycle is traveling clockwise around a traffic circle at a steady speed of 50 km/h. The x axis is eastward and the y axis is northward. What are the x and y components of the velocity at the instant the motorcycle is at the eastern point of the traffic circle?

(A) $v_x = 50$ km/h; $v_y = 0$ (B) $v_x = -50$ km/h; $v_y = 0$
(C) $v_x = 0$; $v_y = 50$ km/h (D) $v_x = 0$; $v_y = -50$ km/h

Online Concept Tutorial 5

4.2 THE VELOCITY AND ACCELERATION VECTORS

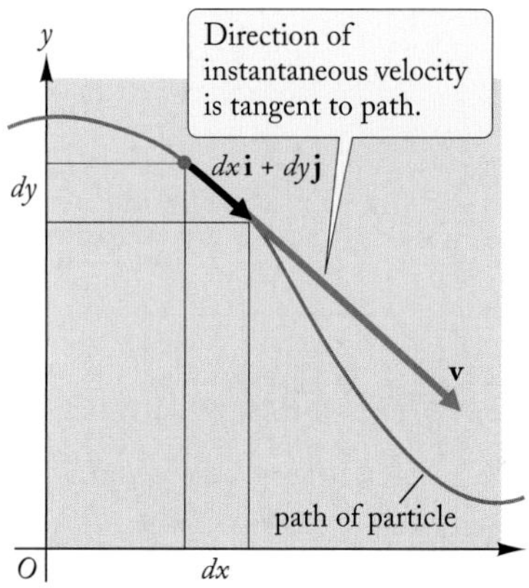

FIGURE 4.4 In a small time interval dt, the changes in the x and y coordinates are dx and dy. The displacement vector is $dx\mathbf{i} + dy\mathbf{j}$. This vector is tangent to the path of the particle, and so is the velocity vector $\mathbf{v} = (dx\mathbf{i} + dy\mathbf{j})/dt$.

In the preceding section we described two-dimensional velocity and acceleration by components. Now we will see how to describe these quantities more concisely by vectors.

The x coordinate and the y coordinate of a particle can be regarded as the x component and the y component of the position vector:

$$\mathbf{r} = x\mathbf{i} + y\mathbf{j} \tag{4.13}$$

Likewise, the x and y components of the velocity introduced in Eqs. (4.3) and (4.4) can be regarded as the x and y components of the velocity vector:

$$\mathbf{v} = v_x\mathbf{i} + v_y\mathbf{j} \tag{4.14}$$

or

$$\mathbf{v} = \frac{dx}{dt}\mathbf{i} + \frac{dy}{dt}\mathbf{j} \tag{4.15}$$

The velocity vector is the rate of change of the position vector. Note that the velocity vector equals the small displacement $dx\mathbf{i} + dy\mathbf{j}$ divided by dt. Thus, the direction of the velocity vector is the direction of the small displacement $dx\mathbf{i} + dy\mathbf{j}$, that is, the direction of the instantaneous velocity vector is the direction of the instantaneous motion, which is tangent to the path of the particle (see Fig. 4.4). For example, Fig. 4.5 shows the velocity vectors at different instants for a projectile launched upward at some angle; at each instant, the velocity is tangent to the path. The lengths of the velocity

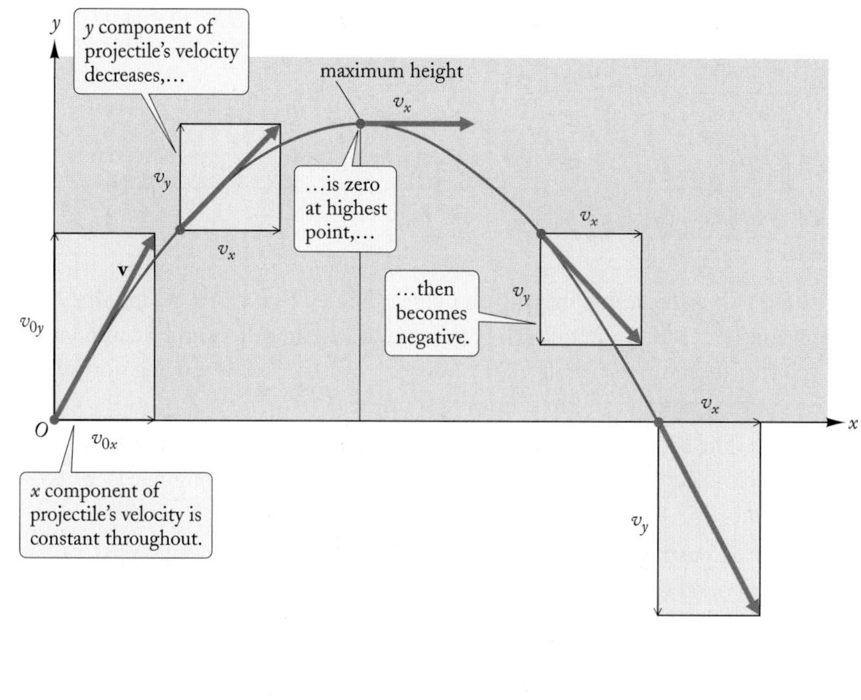

FIGURE 4.5 Velocity vectors of a projectile at different instants.

vectors indicate the magnitude of the velocity. In Fig. 4.5, we can see that the magnitude of the velocity is largest at the start of the motion, and it is smallest at the apex. In order to prepare this drawing of velocity vectors, the artist had to make a choice of scale for these vectors; for instance, a centimeter of length for a velocity vector in Fig. 4.5 might represent 10 m/s. Often, we will be mainly interested in how the relative magnitudes of various vectors look at different points, and for such comparisons we do not need to know the scale.

According to the usual equations for the components of a vector [Eqs. (3.13) and (3.14)],

$$v_x = v\cos\theta \tag{4.16}$$

$$v_y = v\sin\theta \tag{4.17}$$

where θ is the angle between the velocity vector and the x axis (see Fig. 4.6) and v is the magnitude of the velocity vector; that is, v is the speed. These relations permit us to calculate the x and y components of the velocity if we know the speed and the direction of motion.

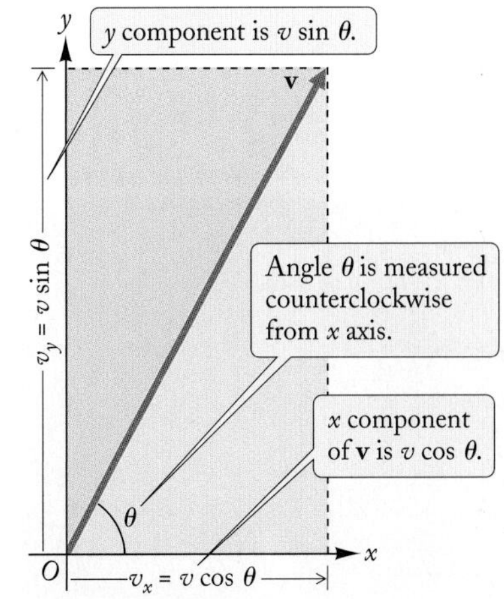

FIGURE 4.6 The x and y components of the velocity vector **v**.

PHYSICS IN PRACTICE VELOCITY VECTORS

Charts of velocity vectors are used to visualize the flow of bodies of water, such as the tidal flow in harbors, and the flow of air masses, such as the updrafts and downdrafts in thunderstorm cells or the flow of air around obstacles. For instance, Figure 1 displays the tidal flow in Tampa Bay. The velocity vectors indicate the direction and the magnitude of the flow of water at one particular time. Information on tidal flow is important to navigators of ships and also to engineers concerned with the construction of harbors and bridges or the dispersion of pollutants. Figure 2 shows the velocity vectors associated with the flow of air in one of the diffuser channels in the centrifugal compressor of a jet engine. The complex velocity vector pattern indicates a pressure surge and a compressor stall, with a reversal of the normal direction of flow. This results in loss of power and engine flameout. By understanding the details of the flow, engineers can optimize compressor design and provide protection against potentially catastrophic engine failures.

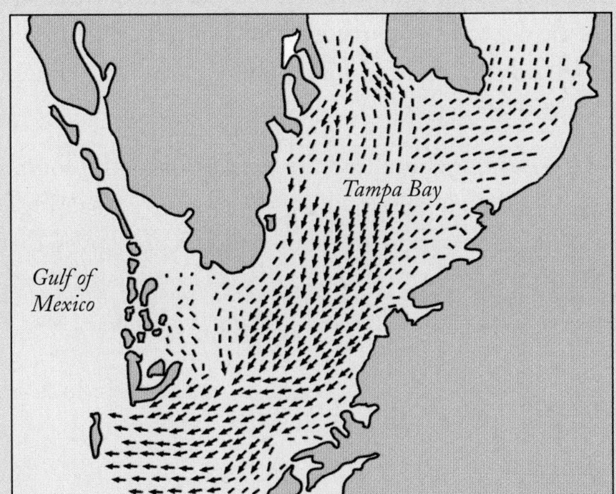

FIGURE 1 Tidal flow in Tampa Bay.

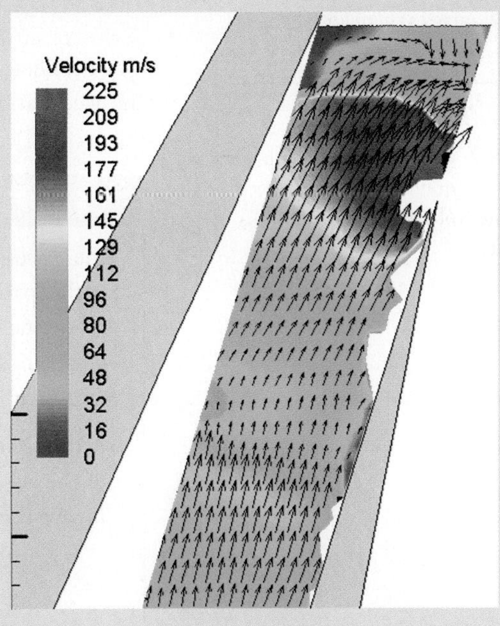

FIGURE 2 The airflow in one of the diffuser channels of a centrifugal compressor. The velocity vectors were measured by Digital Particle Imaging Velocimetry (DPIV). The airflow is seeded with a sprinkling of shiny particles illuminated with quick bursts of laser light. A high-speed digital camera records the positions of the particles, and velocities are calculated from changes in the positions.

Concepts — in — Context

(a)

(b)

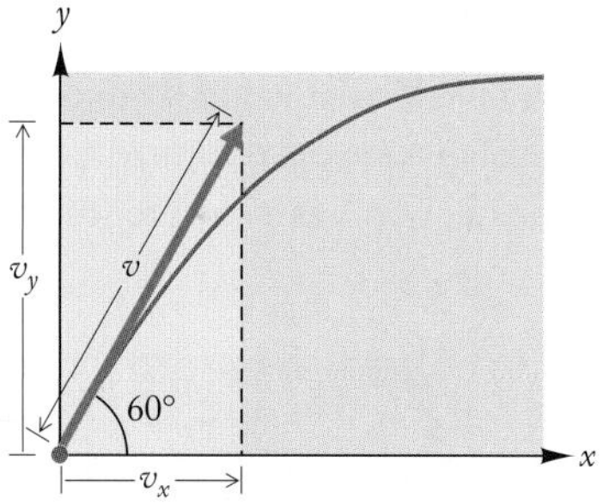

FIGURE 4.7 (a) A "volcanic bomb" after impact on the ground. (b) Initial velocity vector of the "volcanic bomb" and its components.

EXAMPLE 2 Volcanos sometime eject large "volcanic bombs," with masses of several tons (see the chapter photo and Figure 4.7a). Suppose that such a bomb is ejected eastward by a volcano with an initial speed of 330 km/h at an upward angle of 60.0° with the horizontal (see Fig. 4.7b). What are the components of the initial velocity of the bomb in the horizontal and vertical directions? Assume the x axis is horizontal and eastward and the y axis is vertical and upward.

SOLUTION: A speed of 330 km/h corresponds to

$$330\frac{\text{km}}{\text{h}} \times \frac{1000\text{ m}}{1\text{ km}} \times \frac{1\text{ h}}{3600\text{ s}} = 91.7\text{ m/s}$$

With $v = 91.7$ m/s and $\theta = 60.0°$, Eqs. (4.16) and (4.17) immediately give the components of the velocity:

$$v_x = v\cos\theta = 91.7\text{ m/s} \times \cos 60.0° = 91.7\text{ m/s} \times 0.500$$
$$= 45.8\text{ m/s}$$
$$v_y = v\sin\theta = 91.7\text{ m/s} \times \sin 60.0° = 91.7\text{ m/s} \times 0.866$$
$$= 79.4\text{ m/s}$$

The x and y components of the acceleration can be regarded as the components of the acceleration vector:

$$\mathbf{a} = a_x\mathbf{i} + a_y\mathbf{j} \tag{4.18}$$

or

$$\mathbf{a} = \frac{dv_x}{dt}\mathbf{i} + \frac{dv_y}{dt}\mathbf{j} \tag{4.19}$$

The acceleration vector is the rate of change of the velocity vector. The direction of the acceleration vector is the direction of the change of the velocity vector.

EXAMPLE 3 An aircraft in level flight releases a rocket spaceship at time $t = 0$. The engine of the spaceship ignites 2.0 s after release and gives the spaceship a horizontal acceleration of 6.0 m/s^2. Assume that the vertical motion of the spaceship is free-fall motion during the first few seconds after its release. Ignore air resistance. (a) What is the direction of the acceleration of the spaceship when the engine has ignited? (b) What is the direction of its velocity relative to the moving aircraft at $t = 2.0$ s? At $t = 3.0$ s?

SOLUTION: (a) With the x axis horizontal and the y axis vertical, the components of the acceleration are $a_x = 6.0$ m/s^2 and $a_y = -g = -9.8$ m/s^2 (see Fig. 4.8a). Hence the angle between the acceleration vector and the x axis is given by

$$\tan\theta = \frac{a_y}{a_x} = \frac{-9.8\text{ m/s}^2}{6.0\text{ m/s}^2} = -1.63$$

With our calculator, the inverse tangent of -1.63 gives

$$\theta = -58°$$

(b) At $t = 2.0$ s, the spaceship has a vertical component of velocity relative to the aircraft, but no horizontal component of velocity (until its engine ignites, the spaceship

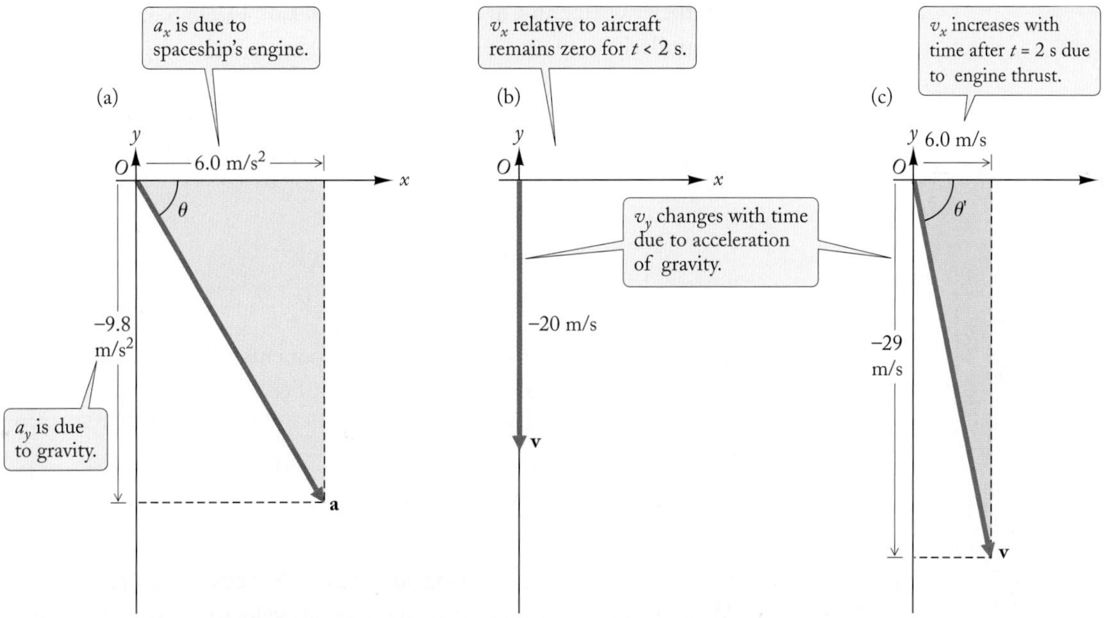

FIGURE 4.8 (a) The x and y components of the acceleration of the spaceship. (b) The components of the velocity at $t = 2.0$ s. (c) At $t = 3.0$ s.

moves forward with the same horizontal motion as the aircraft; that is, it has no horizontal velocity relative to the aircraft). The vertical component of velocity is simply that acquired by free fall in the interval 2.0 s:

$$v_y = a_y t = -gt = -9.8 \text{ m/s}^2 \times 2.0 \text{ s} = -20 \text{ m/s}$$

The direction of the velocity (relative to the aircraft) at this time is vertically down (see Fig. 4.8b).

At $t = 3.0$ s, the vertical component of the velocity is

$$v_y = -9.8 \text{ m/s}^2 \times 3.0 \text{ s} = -29 \text{ m/s}$$

The horizontal component of velocity of the spaceship (relative to the horizontal velocity of the aircraft) arises from the horizontal acceleration of 6.0 m/s² provided by the engine. This acceleration begins at $t = 2.0$ s, and in the time interval Δt from $t = 2.0$ s to 3.0 s it contributes a horizontal component

$$v_x = a_x \Delta t = 6.0 \text{ m/s}^2 \times 1.0 \text{ s} = 6.0 \text{ m/s}$$

The angle between the velocity vector and the x axis is then given by

$$\tan\theta' = \frac{v_y}{v_x} = \frac{-29 \text{ m/s}}{6.0 \text{ m/s}} = -4.8$$

from which our calculator finds

$$\theta' = -78°$$

This means that the direction of the velocity vector is 78° below the horizontal (see Fig. 4.8c).

Our various definitions and results are easily extended to three dimensions. The average and instantaneous velocities and accelerations are defined the same way for a z

component as they were for an x or a y component, in Eqs. (4.1)–(4.4) and (4.9)–(4.12). The speed is similarly

$$v = \sqrt{v_x^2 + v_y^2 + v_z^2}$$

and the position, velocity, and acceleration vectors contain a z component:

$$\mathbf{r} = x\mathbf{i} + y\mathbf{j} + z\mathbf{k}$$

$$\mathbf{v} = v_x\mathbf{i} + v_y\mathbf{j} + v_z\mathbf{k}$$

$$\mathbf{a} = a_x\mathbf{i} + a_y\mathbf{j} + a_z\mathbf{k}$$

However, we will rarely have to use all three components to examine any motion of interest; usually, a two-dimensional analysis will suffice.

✔ Checkup 4.2

QUESTION 1: An automobile is traveling around a traffic circle in a counterclockwise direction. What is the direction of its instantaneous velocity vector when the automobile is at the east, north, west, and south extremes of the circle?

QUESTION 2: The x component of the velocity of an aircraft is 150 km/h, and the y component of the velocity is also 150 km/h. What is the direction of the velocity vector relative to the x and y axes?

QUESTION 3: A tennis ball, initially traveling horizontally, collides with a wall and bounces back horizontally. What is the direction of the acceleration vector during the collision?

QUESTION 4: A skier accelerates along a slope at a rate of 6.0 m/s^2. The horizontal component of the skier's acceleration is 3.0 m/s^2. What is the angle of the slope with respect to horizontal?

(A) 27° (B) 30° (C) 45° (D) 60° (E) 63°

4.3 MOTION WITH CONSTANT ACCELERATION

For a particle moving in two dimensions with constant acceleration, we can derive equations relating acceleration, velocity, position, and time analogous to the equations that hold in one dimension (see Section 2.5). If the x component of the acceleration is a_x and the y component a_y, then, by analogy with Eq. (2.17), we see that the components of the velocity at time t will be

$$v_x = v_{0x} + a_x t \tag{4.20}$$

$$v_y = v_{0y} + a_y t \tag{4.21}$$

These two equations can be regarded as components of the following vector equation:

$$\mathbf{v} = \mathbf{v}_0 + \mathbf{a}t \tag{4.22}$$

where, as always, the subscript 0 (zero) indicates the values at the initial time zero. Furthermore, a mathematical argument analogous to that used in the one-dimensional case [see Eqs. (2.18)–(2.22)] leads us to the following expressions for the change in the position:

$$x - x_0 = v_{0x}t + \tfrac{1}{2}a_x t^2 \tag{4.23}$$

$$y - y_0 = v_{0y}t + \tfrac{1}{2}a_y t^2 \tag{4.24}$$

These equations can be regarded as components of the following vector equation:

$$\mathbf{r} - \mathbf{r}_0 = \mathbf{v}_0 t + \tfrac{1}{2}\mathbf{a}t^2 \tag{4.25}$$

Equations (4.20)–(4.25) state that *the x and y components of the motion evolve completely independently of one another.* Thus, the x acceleration affects only the x velocity, and the change in the x position is completely determined by the x acceleration and the initial x velocity. Figure 4.9 shows an experimental demonstration of this independence between the x and y components of the motion. Two balls were released simultaneously from a platform; one was merely dropped from rest, the other was launched with an initial horizontal velocity. According to our discussion of the acceleration of gravity in Chapter 2, the vertical (downward) free-fall accelerations of the two balls are the same. According to Eq. (4.24), the vertical motions (y motions) of the two balls should then be the same, even though their horizontal motions (x motions) differ. Furthermore, according to Eq. (4.23), the horizontal motion of the second ball should simply proceed with uniform horizontal velocity, even though this ball has a vertical acceleration.

The stroboscopic images of the balls recorded on the photograph at equal intervals of time confirm these predictions. The balls indeed fall downward in unison, reaching equal heights at the same instants of time. The red grid lines drawn in Fig. 4.9b permit us to verify that the vertical components of the positions of the two balls are always exactly the same, even though the horizontal components differ. Furthermore, the blue grid lines permit us to verify that the horizontal component of the velocity of the second ball is constant.

(a) (b)

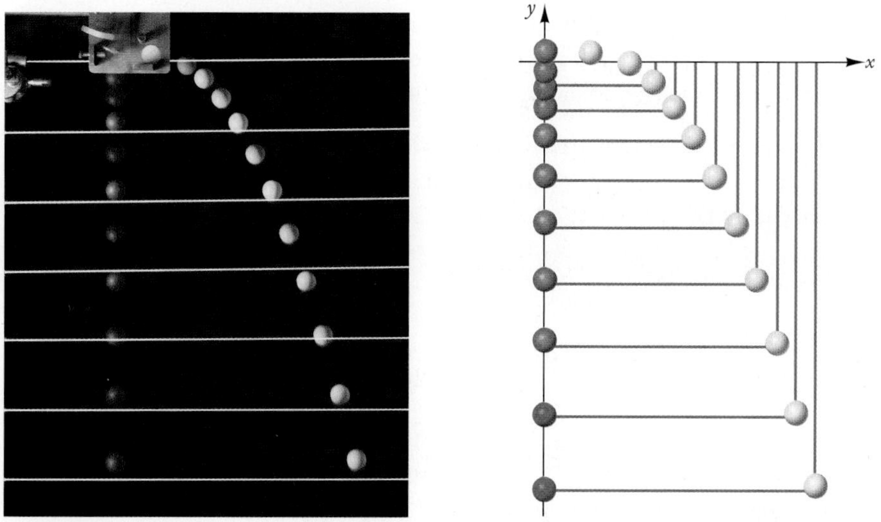

FIGURE 4.9 (a) Stroboscopic photograph showing multiple exposures of two balls that have been released simultaneously from a platform; one ball has a horizontal velocity, the other does not. The time interval between the exposures is 1/40 s. (b) For the analysis of the photograph, we draw horizontal (red) and vertical (blue) grid lines through the positions of the balls. The red grid lines verify that the vertical components of the positions of the two balls always coincide. The blue grid lines, by their uniform spacing, verify that the horizontal component of the velocity of the second ball remains constant.

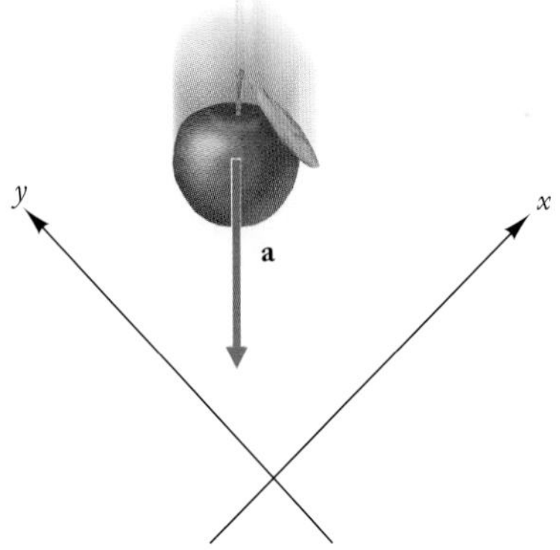

FIGURE 4.10 Apple falling in tilted coordinate system.

Checkup 4.3

QUESTION 1: Consider a motorboat maneuvering on the surface of a lake, with x and y axes along the surface. Is it possible for the boat to have accelerated motion in the x direction, and unaccelerated motion in the y direction? Is it possible for the boat to have accelerated motion in both the x and the y directions?

QUESTION 2: Suppose the direction of motion of a particle lies between the x and y axes. You know that the x motion is accelerated, and you see that the particle moves along a straight line. Is the y motion accelerated or unaccelerated?

QUESTION 3: Consider the motion of a falling apple in a coordinate system whose axes are arranged so the upward direction is at 45° between the x and the y axes (see Fig. 4.10). What are the accelerations a_x and a_y of the apple?

(A) $a_x = 0$; $a_y = 9.8\ \text{m/s}^2$
(B) $a_x = 6.9\ \text{m/s}^2$; $a_y = 6.9\ \text{m/s}^2$
(C) $a_x = 6.9\ \text{m/s}^2$; $a_y = -6.9\ \text{m/s}^2$
(D) $a_x = -6.9\ \text{m/s}^2$; $a_y = 6.9\ \text{m/s}^2$
(E) $a_x = -6.9\ \text{m/s}^2$; $a_y = -6.9\ \text{m/s}^2$

Online Concept Tutorial 5

4.4 THE MOTION OF PROJECTILES

We know that near the surface of the Earth, the pull of gravity gives a freely falling body a downward acceleration of about $9.81\ \text{m/s}^2$. If we ignore air resistance, this is the only acceleration that the body experiences when launched from some initial position with some initial velocity. Thus, the motion of a baseball thrown by hand (or hit by a bat) is motion with constant vertical acceleration and zero horizontal acceleration. This kind of motion is called **projectile motion**, or ballistic motion. In most cases of projectile motion, we deliberately launch the body with an initial upward component of velocity, so that it flies farther before striking the ground. But we can also launch the projectile horizontally, without any initial upward component of velocity (as in the case of the ball whose motion we examined in Fig. 4.9), and we can even launch the projectile with an initial downward component of velocity.

In any case, the initial velocity of the projectile can be characterized by its vertical and horizontal components. If we take the y axis in the upward direction[1] and the x axis in the direction of the initial horizontal velocity, we have $a_x = 0$, $a_y = -g = -9.81\ \text{m/s}^2$. Furthermore, let us assume that the origin of coordinates coincides with the initial position of the projectile, so $x_0 = 0$ and $y_0 = 0$. The components of the velocity and position at the time t will then be, according to Eqs. (4.20), (4.21), (4.23), and (4.24):

horizontal motion of projectile

$$v_x = v_{0x} \tag{4.26}$$

$$x = v_{0x}t \qquad (x_0 = 0) \tag{4.27}$$

vertical motion of projectile

$$v_y = v_{0y} - gt \tag{4.28}$$

$$y = v_{0y}t - \tfrac{1}{2}gt^2 \qquad (y_0 = 0) \tag{4.29}$$

[1] Note that in Section 2.6 we took the x axis in the upward direction, whereas now we are taking the y axis in the upward direction.

These equations represent a motion with constant velocity in the x direction and a simultaneous motion with constant (downward) acceleration in the y direction. According to Eq. (2.29), the vertical position and velocity also obey the relation

$$-gy = \tfrac{1}{2}(v_y^2 - v_{0y}^2) \qquad (4.30)$$

If the motion is initially upward, the projectile will ascend to some maximum height and then begin its descent. At the instant of maximum height, the vertical component of the velocity is zero, since at this one instant the projectile has ceased to move upward and has not yet begun to move downward. We can find the instant of maximum height and the value of the maximum height by inserting the condition $v_y = 0$ into Eqs. (4.28) and (4.30), respectively.

EXAMPLE 4 Consider the "volcanic bomb" described in Example 2. At what time does this projectile reach its maximum height? What is this maximum height?

Concepts — in — Context

SOLUTION: The known quantities are the components of the initial velocity; according to Example 2, these components are $v_{0x} = 45.8$ m/s and $v_{0y} = 79.4$ m/s. Furthermore, we know that when the projectile reaches its maximum height, $v_y = 0$ (see Fig. 4.11). Since the problem asks no questions about the horizontal motion, we can ignore this motion altogether, and we can concentrate on the vertical motion. For this vertical motion, Eqs. (4.28) and (4.29) are applicable. The known and unknown quantities are

UNKNOWN	KNOWN
t	$v_{0y} = 79.4$ m/s
y	$v_y = 0$
	$g = 9.81$ m/s^2

From these known values of v_{0y} and v_y, Eq. (4.28) determines the instant of maximum height:

$$0 = v_{0y} - gt$$

or, solving this for t,

$$t = \frac{v_{0y}}{g}$$

$$= \frac{79.4 \text{ m/s}}{9.81 \text{ m/s}^2} = 8.09 \text{ s}$$

With this value of the time, we can then calculate the maximum height from Eq. (4.29):

$$y = v_{0y}t - \tfrac{1}{2}gt^2$$

$$= 79.4 \text{ m/s} \times 8.09 \text{ s} - \tfrac{1}{2} \times 9.81 \text{ m/s}^2 \times (8.09 \text{ s})^2$$

$$= 321 \text{ m}$$

COMMENT: Note that we could instead have used Eq. (4.30), with $v_y = 0$, to calculate the same maximum height directly. Solving Eq. (4.30) for y we have

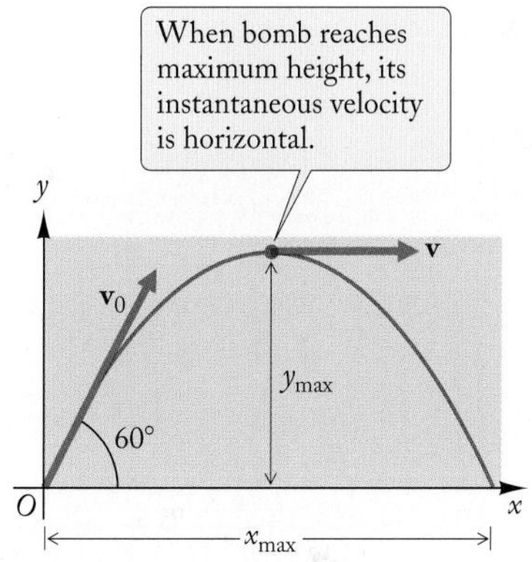

FIGURE 4.11 Trajectory of a volcanic bomb launched with an initial velocity $\mathbf{v}_0$.

$$y = \frac{\frac{1}{2}(-v_{0y}^2)}{-g} = \frac{v_{0y}^2}{2g}$$

$$= \frac{(79.4\ \text{m/s})^2}{2 \times 9.81\ \text{m/s}^2} = 321\ \text{m}$$

In this case, we did not need to determine the time t, because Eq. (4.30) was obtained by eliminating the variable t from Eqs. (4.28) and (4.29) (see Section 2.5).

In our calculation we have ignored the effects of air resistance. For a dense projectile of large mass, such as a large volcanic bomb, this is a good approximation.

EXAMPLE 5 You throw a baseball toward a wall, 10 m away. The initial direction of motion of the ball is horizontal, and the initial speed is 20 m/s. How far below its initial height does this projectile hit the wall?

SOLUTION: It is convenient to place the origin of coordinates at the initial height of the ball (see Fig. 4.12). We first consider the horizontal motion. This horizontal motion proceeds at uniform velocity, with $v_{0x} = 20$ m/s. Hence the time the ball takes to reach the wall is

$$t = \frac{x}{v_{0x}} = \frac{10\ \text{m}}{20\ \text{m/s}} = 0.50\ \text{s}$$

Next, we consider the vertical motion, which proceeds with uniform acceleration, according to Eqs. (4.28)–(4.30). The initial velocity is $v_{0y} = 0$, since the ball was thrown horizontally. The final time is now known, and the final value of y is unknown:

UNKNOWN	KNOWN
y	$v_{0y} = 0$
	$t = 0.50$ s
	$g = 9.81\ \text{m/s}^2$

From these known values, Eq. (4.29) immediately gives us y:

$$y = v_{0y}t - \tfrac{1}{2}gt^2 = 0 - \tfrac{1}{2} \times 9.81\ \text{m/s}^2 \times (0.50\ \text{s})^2$$

$$= -1.2\ \text{m}$$

The negative sign means that the ball has descended 1.2 m below its initial height by the time it reaches the wall.

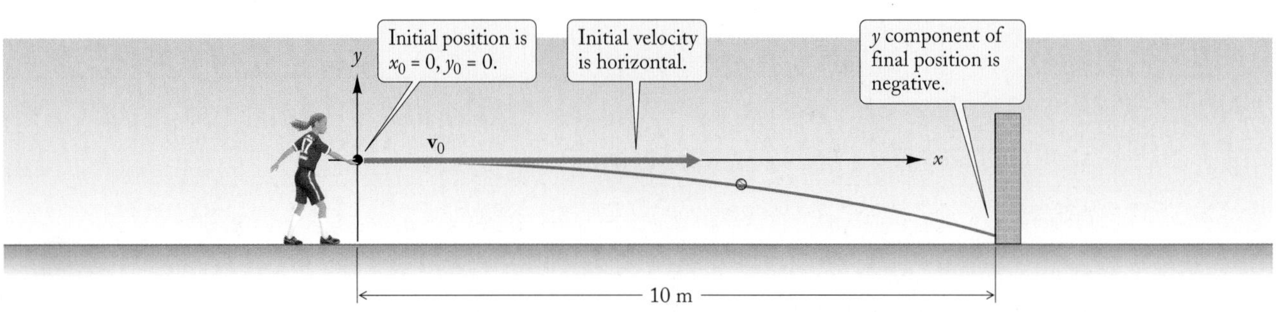

FIGURE 4.12 Trajectory of a baseball, thrown with an initial horizontal velocity.

EXAMPLE 6 In low-level bombing (at "smokestack level"), a bomber releases a bomb at a height of 50 m above the surface of the sea while in horizontal flight at a constant speed of 300 km/h. How long does the bomb take to fall to the surface of the sea? How far ahead (horizontally) of the point of release is the point of impact?

SOLUTION: It is convenient to place the origin of coordinates at the point of release, 50 m above the level of the sea, with the x axis along the horizontal path of the bomber (see Fig. 4.13). The initial velocity of the bomb is the same as that of the bomber:

$$v_{0x} = 300\ \text{km/h} = 300\,\frac{\text{km}}{\text{h}} \times \frac{1000\ \text{m}}{1\ \text{km}} \times \frac{1\ \text{h}}{3600\ \text{s}} = 83.3\ \text{m/s}$$

$$v_{0y} = 0$$

When the bomb reaches the level of the sea, its vertical position is $y = -50$ m (i.e., the bomb is 50 m below our origin of coordinates).

We begin with the vertical motion, described by Eqs. (4.28) and (4.29). The initial value of the velocity and the final value of y are known:

UNKNOWN	**KNOWN**
t	$y = -50$ m
	$v_{0y} = 0$
	$g = 9.81\ \text{m/s}^2$

With $v_{0y} = 0$, Eq. (4.29) then determines the time of impact:

$$y = -\tfrac{1}{2}gt^2$$

To solve for t, we divide both sides of this equation by $-\frac{1}{2}g$ and take the square root of both sides:

$$t = \sqrt{-\frac{2y}{g}}$$

$$= \sqrt{\frac{2 \times 50\ \text{m}}{g}} = \sqrt{\frac{2 \times 50\ \text{m}}{9.81\ \text{m/s}^2}} = 3.2\ \text{s}$$

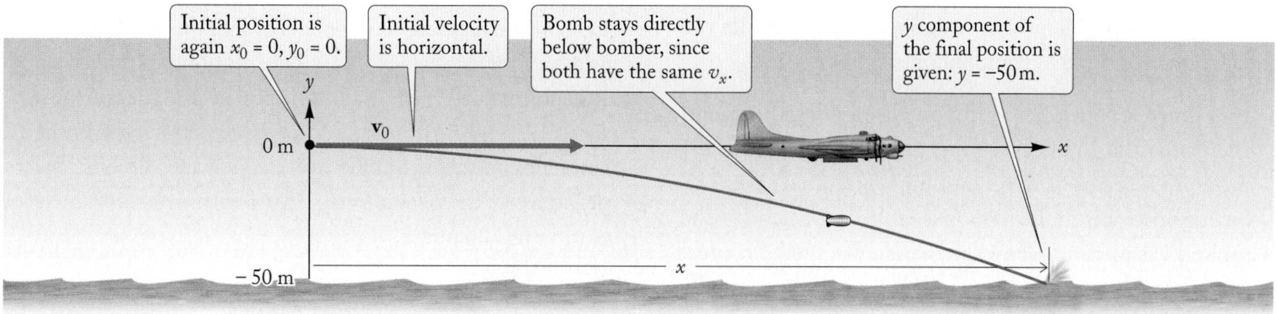

FIGURE 4.13 Trajectory of a bomb dropped by a bomber. The initial vertical component of the velocity is zero, and the initial horizontal component is the same as that of the bomber.

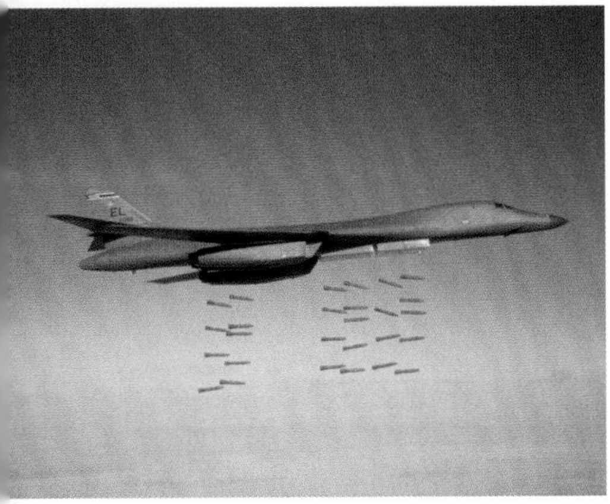

FIGURE 4.14 "Strings" of bombs released from a bomber. The bombs continue to move forward with the same horizontal velocity as that of the bomber.

Next, we consider the horizontal motion. The bomb moves with a constant horizontal velocity $v_{0x} = 83.3$ m/s. Hence at the time 3.2 s, the horizontal position of the bomb is

$$x = v_{0x}t = 83.3 \text{ m/s} \times 3.2 \text{ s} = 270 \text{ m}$$

Note that the bomber moves exactly the same *horizontal* distance in this time; that is, the bomb always remains directly below the bomber because both have exactly the same horizontal velocity $v_{0x} = 83.3$ m/s. Figure 4.14 shows bombs released by a bomber at successive instants of time, and demonstrates that they remain directly below the bomber.

The path of a baseball, a golf ball, a bomb, or any other projectile is a *parabola*. The mathematical proof of this statement rests on Eqs. (4.27) and (4.29). Suppose that time is reckoned from the instant the projectile reaches its maximum height (so t is negative before the instant of maximum height, and positive after). Then $v_{0y} = 0$ and Eqs. (4.27) and (4.29) become

$$x = v_{0x}t$$

$$y = -\tfrac{1}{2}gt^2$$

If we square both sides of the first equation and divide it into the second equation, we obtain

$$\frac{y}{x^2} = -\frac{1}{2}\frac{gt^2}{v_{0x}^2 t^2} \tag{4.31}$$

and, canceling the factors of t^2 and multiplying by x^2, we obtain

$$y = -\left(\frac{1}{2}\frac{g}{v_{0x}^2}\right)x^2 \tag{4.32}$$

Concepts — in — Context

This equation says that y is proportional to x^2, which is the equation for a parabola with apex at $y = 0$. The parabolic shape of the trajectory can be seen clearly in a multiple-exposure photograph, such as Fig. 4.15. It can also be seen clearly in the time-exposure photograph of incandescent volcanic bombs in the chapter photo.

PROBLEM-SOLVING TECHNIQUES — PROJECTILE MOTION

The solution of problems of projectile motion exploits the independence of the two components of the motion.

1 First solve for one of the components of the motion, and then the other. Whether you should begin with the y motion (vertical) or the x motion (horizontal) depends on circumstances. If the motion is limited vertically (by the ground), as in Examples 4 and 6, then you must first deal with the y motion. If the motion is limited horizontally (by a wall), as in Example 5, then you must first deal with the x motion.

2 Each of the separate components of motion can be treated as one-dimensional motion. The x motion proceeds with constant velocity; the y motion proceeds with constant acceleration. The solution of these two parts of the problem relies on the techniques we used for one-dimensional problems with constant velocity or with constant acceleration in Chapter 2.

If air resistance is negligible, the path of a projectile launched with some horizontal and vertical velocity is always some portion of a parabola. In the case of a bomb dropped from an airplane, the relevant portion of the parabola begins at the apex and descends to the level of the target. In the case of a ball or stone launched by hand, a bomb launched by a volcano, or a shot or bullet fired from a gun, the relevant portion begins at the level of the launch point or muzzle, rises to the apex, and then descends to the target. In the latter case, it is often important to calculate the **maximum height** reached, the **time of flight** (time between the instants of launch and impact), and the **range** (the horizontal distance between the points of launch and impact). This kind of calculation is illustrated in the following example.

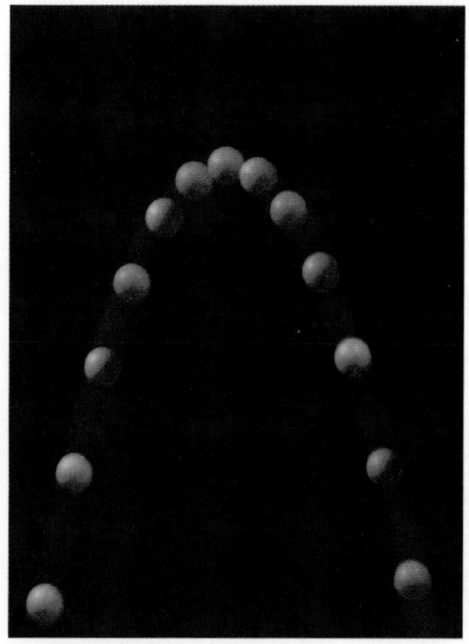

FIGURE 4.15 Stroboscopic photograph showing the path of a projectile (a golf ball) launched upward at an angle. The path is a parabola. At the apex of the parabola (the point of maximum height), the vertical component of the velocity is instantaneously zero.

EXAMPLE 7 A champion discus thrower (Fig. 4.16) throws a discus with an initial speed of 26 m/s. If the discus is thrown upward at an angle of 45°, what maximum height and what range will the discus attain? What is its time of flight? For the sake of simplicity, ignore the height of the hand over the ground, so the launch point and the impact point are at the same height.

SOLUTION: We take the origin of coordinates at the launch point. Equations (4.26)–(4.29) are then applicable to this problem. However, to extract the answers we want, we must manipulate these equations somewhat.

Since Eqs. (4.26)–(4.29) depend on the components v_{0x} and v_{0y} of the initial velocity, we begin by calculating these components. From Fig. 4.17 we see that

$$\begin{aligned} v_{0x} &= v_0 \cos\theta \\ &= v_0 \cos 45° = 26\text{ m/s} \times \cos 45° = 26\text{ m/s} \times 0.707 \\ &= 18.4\text{ m/s} \end{aligned} \tag{4.33}$$

and

$$\begin{aligned} v_{0y} &= v_0 \sin\theta \\ &= v_0 \sin 45° = 26\text{ m/s} \times \sin 45° = 26\text{ m/s} \times 0.707 \\ &= 18.4\text{ m/s} \end{aligned} \tag{4.34}$$

FIGURE 4.16 A champion discus thrower.

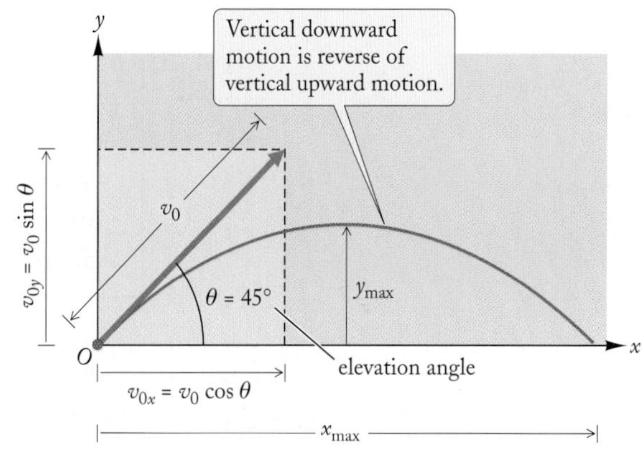

FIGURE 4.17 Trajectory of a discus thrown upward. The arrow indicates the initial velocity vector of the discus. The angle θ between this direction and the horizontal is the elevation angle.

As a next step, we calculate the time at which the discus reaches maximum height. From this we can immediately find the time of impact, or the time of flight, because this time is twice the time required to reach maximum height; we can recognize this if we concentrate on the y coordinate, for which the downward motion, after maximum height, is merely the reverse of the upward motion. Thus, the time of flight (which includes the upward and the downward motions) must be twice the time for the upward motion:

$$t_{\text{flight}} = 2t_{\text{height}} \tag{4.35}$$

The discus reaches maximum height when $v_y = 0$. Equation (4.28) then implies

$$0 = v_{0y} - gt_{\text{height}} \tag{4.36}$$

and, solving this for t_{height}, following the method we used in Example 4, we find

$$\begin{aligned} t_{\text{height}} &= \frac{v_{0y}}{g} \\ &= \frac{18.4\ \text{m/s}}{9.81\ \text{m/s}^2} = 1.87\ \text{s} \end{aligned} \tag{4.37}$$

Accordingly,

$$\begin{aligned} t_{\text{flight}} &= 2t_{\text{height}} = 2\frac{v_{0y}}{g} \\ &= 2 \times 1.87\ \text{s} = 3.7\ \text{s} \end{aligned} \tag{4.38}$$

At the time t_{height}, the height reached by the discus is, by Eq. (4.29),

$$\begin{aligned} y_{\max} &= v_{0y}t_{\text{height}} - \tfrac{1}{2}g(t_{\text{height}})^2 = v_{0y}\left(\frac{v_{0y}}{g}\right) - \tfrac{1}{2}g\left(\frac{v_{0y}}{g}\right)^2 = \frac{1}{2}\frac{v_{0y}^2}{g} \\ &= \frac{1}{2}\frac{(18.4\ \text{m/s})^2}{9.81\ \text{m/s}^2} = 17\ \text{m} \end{aligned} \tag{4.39}$$

The horizontal distance from the launch point attained at the time $t = t_{\text{flight}}$ is equal to the range:

$$\begin{aligned} x_{\max} &= v_{0x}t_{\text{flight}} = v_{0x} \times 2t_{\text{height}} = v_{0x} \times 2\frac{v_{0y}}{g} = \frac{2v_{0x}v_{0y}}{g} \\ &= \frac{2 \times 18.4\ \text{m/s} \times 18.4\ \text{m/s}}{9.81\ \text{m/s}^2} = 69\ \text{m} \end{aligned} \tag{4.40}$$

COMMENT: Note that in the intermediate stages of this calculation, we kept three significant figures, even though the data contain only two significant figures. It is often necessary to carry extra significant figures in intermediate stages of a calculation to prevent accumulation of errors from round-off.

An alternative method of solution is to substitute $v_{0x} = v_0 \cos\theta$ and $v_{0y} = v_0 \sin\theta$ into Eqs. (4.38), (4.39), and (4.40). These substitutions yield

time of flight

$$t_{\text{flight}} = \frac{2v_0 \sin\theta}{g} \tag{4.41}$$

$$y_{max} = \frac{v_0^2 \sin^2\theta}{2g} \tag{4.42}$$

maximum height

and, for the horizontal range,

$$x_{max} = \frac{2v_0^2 \sin\theta \cos\theta}{g} \tag{4.43}$$

range

These equations express parameters of interest explicitly in terms of quantities that are usually known: the initial speed v_0 (the launch speed) and the angle θ between the horizontal and the direction of the initial motion (the elevation angle). Thus, these equations permit a direct calculation of the answers to our problem, without further intermediate steps.

Equations (4.41)–(4.43) are often handy for the solution of other problems of projectile motion; for instance, these equations permit us to calculate the launch speed and the elevation angle required to hit a target at a known range. However, keep in mind that these equations do not give us a complete description of the motion. The complete description of the motion is contained in Eqs. (4.26)–(4.29), and we can always extract anything we want to know about the motion from these equations.

Figure 4.18 shows the trajectories of projectiles of the same launch speed as a function of the elevation angle. Note that *the range is maximum for an elevation angle of* 45°. We can see this by recalling that at a maximum, the derivative of a function is zero. If we take the derivative (as a function of the angle θ) of the angular part of the range function, Eq. (4.43), we have, using the product rule,

$$\begin{aligned}\frac{d}{d\theta}(\sin\theta\cos\theta) &= \left(\frac{d}{d\theta}\sin\theta\right) \times \cos\theta + \sin\theta \times \left(\frac{d}{d\theta}\cos\theta\right)\\ &= \cos\theta \times \cos\theta + \sin\theta \times (-\sin\theta) = \cos^2\theta - \sin^2\theta\end{aligned}$$

This is zero for $\cos^2\theta = \sin^2\theta$, or $\cos\theta = \sin\theta$, which indeed corresponds to an angle of 45°. This angle of 45° represents a compromise: an angle smaller than 45° gives the projectile a larger x velocity but reduces the time it spends in flight, whereas an angle larger than 45° increases the time it spends in flight but gives it a smaller x velocity.

By inspection of Fig. 4.18 we see that any two angles that are equal amounts above or below 45° yield equal ranges; for instance, 60° (15° above 45°) and 30° (15° below 45°) yield equal ranges. This is evident from Eq. (4.43), since $\sin\theta \cos\theta = \sin\theta \sin(90° - \theta)$, which is the same product for any two complementary angles, which differ from 45° by the same amount.

In Fig. 4.18, as in all calculations of this section, air resistance has been neglected. For a high-speed projectile, such as a rifle bullet, air resistance is quite important, and the parabolic trajectory is distorted into a trajectory of more complicated shape, called a **ballistic curve**. If we attempt to use the simple formulas of this section in the calculation of the motion of a high-speed projectile, our results will bear only a vague resemblance to reality.

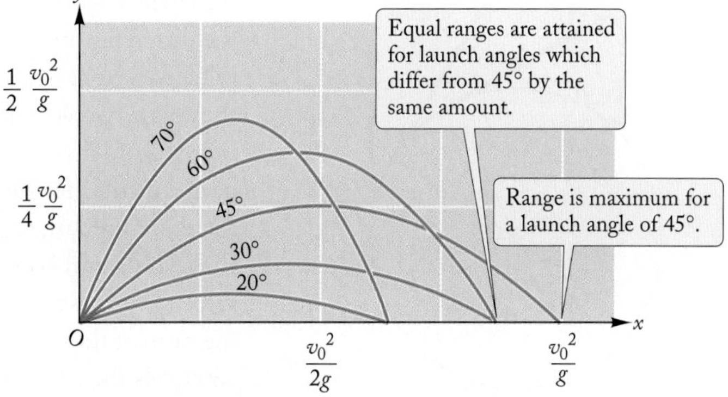

FIGURE 4.18 Trajectories of projectiles of the same launch speed, but different elevation angles. In this plot, the horizontal and the vertical coordinates are measured in fractions of v_0^2/g, which is the maximum range the projectile can achieve, when it is launched with an elevation angle of 45°.

Checkup 4.4

QUESTION 1: What is the instantaneous acceleration of a projectile when it reaches the top of its trajectory?

QUESTION 2: For a given, fixed launch speed, how should you launch a projectile to achieve the longest range? The longest time of flight? The largest height?

QUESTION 3: Consider a projectile launched with some elevation angle larger than 0° and smaller than 90°. Are the velocity and the acceleration of this projectile ever parallel? Perpendicular?

QUESTION 4: You launch two projectiles with the same elevation angle, but different launch speeds. If one projectile has twice the launch speed as the other, how much farther will it go?

QUESTION 5: Consider the trajectories for several projectiles with the same launch speed, but different elevation angles. Which of these projectiles will return to the ground in the shortest time?

QUESTION 6: You launch six projectiles with the same launch speed, but different elevation angles. The first projectile has an elevation angle of 20°. Which of the other projectiles has a shorter range than the first? The other projectiles have elevation angles of

(A) 40° (B) 50° (C) 60° (D) 70° (E) 80°

4.5 UNIFORM CIRCULAR MOTION

Uniform circular motion is motion with constant speed along a circular path, such as the motion of an automobile traveling around a traffic circle. Figure 4.19 shows the positions at different times for a particle in uniform circular motion. The velocity vector at any instant is tangent to the path, or tangent to the circle. All the velocity vectors shown have the same magnitude (same speed), but they differ in direction. Because of this change of direction, *uniform circular motion is accelerated motion.*

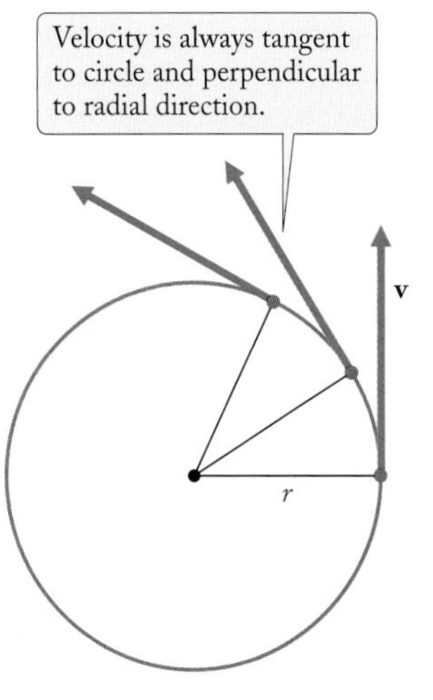

FIGURE 4.19 Instantaneous velocity vectors for a particle in uniform circular motion.

Suppose that the constant speed of the particle is v and the radius of the circle is r. To find the value of the instantaneous acceleration, we must look at the velocity change in a very short time interval Δt. We choose a convenient origin at the center of the circle. Figure 4.20a shows the particle at two positions $\mathbf{r}_1$ and $\mathbf{r}_2$ a short time apart; the difference between these vectors is $\Delta\mathbf{r} = \mathbf{r}_2 - \mathbf{r}_1$. The figure also shows the two velocity vectors $\mathbf{v}_1$ and $\mathbf{v}_2$; both velocity vectors have the same magnitude v; Fig. 4.20b shows the two velocity vectors tail to tail and their difference $\Delta\mathbf{v} = \mathbf{v}_2 - \mathbf{v}_1$. The position vectors make a small angle $\Delta\theta$ with each other; the velocity vectors are always perpendicular to the position vectors, and the angle between the velocity vectors is therefore the same as the angle between the position vectors. Since the angles $\Delta\theta$ in Figs. 4.20a and b are equal, the triangles formed by the position vectors and by the velocity vectors are similar—hence the ratio of the short sides of these triangles equals the ratio of their long sides; that is, the ratio of the magnitudes of the vectors $\Delta\mathbf{v}$ and $\Delta\mathbf{r}$ equals the ratio of v and r:

$$\frac{[\text{magnitude of } \Delta\mathbf{v}]}{[\text{magnitude of } \Delta\mathbf{r}]} = \frac{v}{r} \tag{4.44}$$

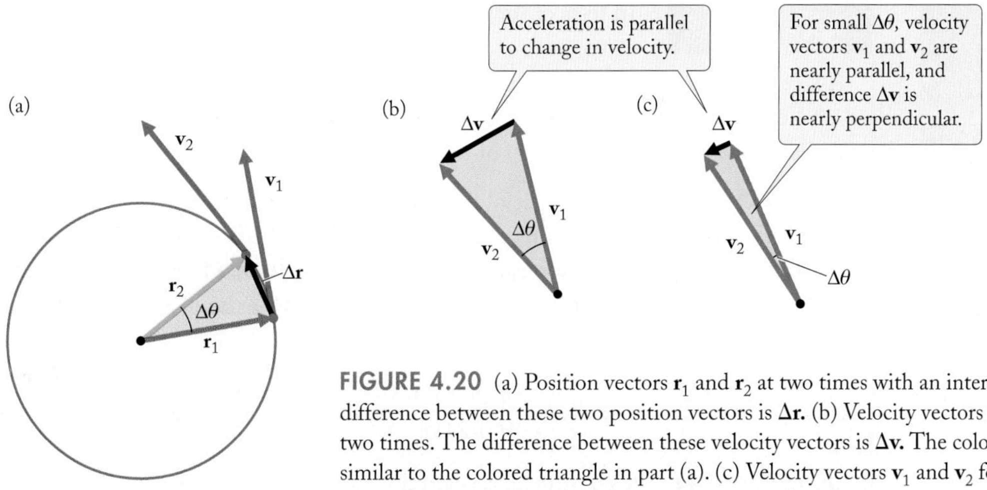

FIGURE 4.20 (a) Position vectors $\mathbf{r}_1$ and $\mathbf{r}_2$ at two times with an interval Δt. The difference between these two position vectors is $\Delta\mathbf{r}$. (b) Velocity vectors $\mathbf{v}_1$ and $\mathbf{v}_2$ at the two times. The difference between these velocity vectors is $\Delta\mathbf{v}$. The colored triangle is similar to the colored triangle in part (a). (c) Velocity vectors $\mathbf{v}_1$ and $\mathbf{v}_2$ for two times with a very small time interval Δt and a very small angle $\Delta\theta$.

We can rewrite this as

$$[\text{magnitude of } \Delta\mathbf{v}] = \frac{v}{r} \times [\text{magnitude of } \Delta\mathbf{r}] \tag{4.45}$$

If Δt is very small, then $\Delta\theta$ will also be very small, and the straight line segment $\Delta\mathbf{r}$ will approximately coincide with the circular arc from the tip of $\mathbf{r}_1$ to the tip of $\mathbf{r}_2$. The latter length is simply the distance traveled by the particle in the time Δt:

$$[\text{magnitude of } \Delta\mathbf{v}] \approx \frac{v}{r} \times [\text{distance traveled in time } \Delta t] \tag{4.46}$$

The magnitude of the acceleration is the magnitude of $\Delta\mathbf{v}$ divided by the time Δt:

$$a = \frac{[\text{magnitude of } \Delta\mathbf{v}]}{\Delta t} \approx \frac{v}{r} \times \frac{[\text{distance traveled in time } \Delta t]}{\Delta t} \tag{4.47}$$

But the distance traveled divided by the time Δt is the speed v, so

$$a \approx \frac{v}{r} \times v \tag{4.48}$$

This equation becomes exact if the time Δt is extremely small, giving us the result

$$a = \frac{v^2}{r} \tag{4.49}$$

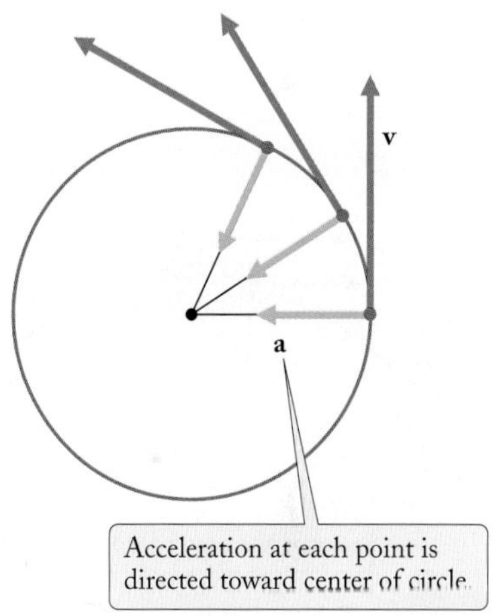

FIGURE 4.21 Instantaneous acceleration vectors for a particle in uniform circular motion.

centripetal acceleration

The *direction* of this acceleration remains to be determined. From Fig. 4.20c it is clear that, for the case of very small Δt, the direction of $\Delta\mathbf{v}$ will be perpendicular to the velocity vectors $\mathbf{v}_1$ and $\mathbf{v}_2$ (which will be nearly parallel in this limiting case). Hence the instantaneous acceleration is perpendicular to the instantaneous velocity. Since the velocity vector corresponding to circular motion is tangential to the circle, *the acceleration vector points inward along the radius, toward the center of the circle.* Figure 4.21 shows the velocity and acceleration vectors at several positions along the circular path. The acceleration of uniform circular motion, with a magnitude given by Eq. (4.49), is called **centripetal acceleration**, because it is directed toward the center of the circle.

FIGURE 4.22 Centrifuge at the NASA Spacecraft Center in Houston.

EXAMPLE 8 In tests of the effects of high acceleration on the human body, astronauts at the NASA Spacecraft Center in Houston are placed in a gondola that is whirled around a circular path of radius 15 m at the end of a revolving girder (see Fig. 4.22). If the girder makes 24 revolutions per minute, what is the acceleration of the gondola?

SOLUTION: The circumference of the circular path is $2\pi \times [\text{radius}] = 2\pi \times 15$ m. Since the gondola makes 24 revolutions per minute, or 24 revolutions per 60 seconds, the time per revolution is (60 s)/(24 revolutions); that is, the gondola takes (60/24) s to go around this circumference. The speed is the distance divided by the time; therefore

$$v = \frac{2\pi \times 15 \text{ m}}{(60/24) \text{ s}} = 38 \text{ m/s}$$

From Eq. (4.49), the centripetal acceleration is then

$$a = \frac{v^2}{r} = \frac{(38 \text{ m/s})^2}{15 \text{ m}} = 95 \text{ m/s}^2$$

This is almost 10 times the acceleration due to gravity (almost 10 standard g's); this is near the limit of human tolerance.

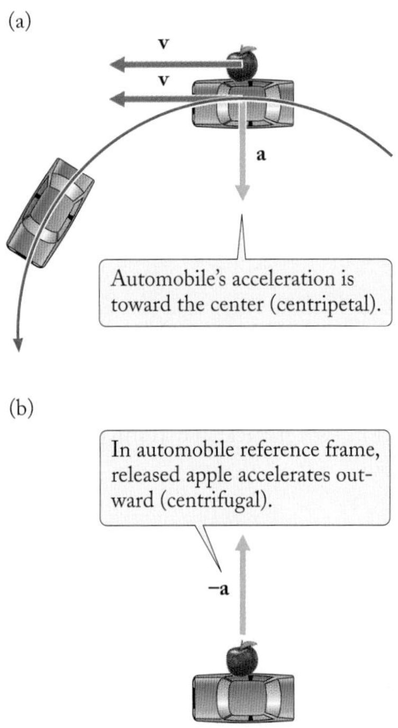

FIGURE 4.23 (a) Automobile in uniform circular motion. The driver has just released an apple, which now moves with constant velocity **v** relative to the ground. The automobile accelerates away from the apple with the centripetal acceleration **a**. (b) In the reference frame of the automobile, the apple has a centrifugal acceleration −**a**.

EXAMPLE 9 An automobile drives around a traffic circle of radius 30 m. If the wheels of the automobile can withstand a maximum transverse acceleration of 8.0 m/s^2 without skidding, what is the maximum permissible speed?

SOLUTION: With a radius of 30 m and a centripetal acceleration of 8.0 m/s^2, Eq. (4.49) gives

$$v^2 = ar = 8.0 \text{ m/s}^2 \times 30 \text{ m} = 240 \text{ (m/s)}^2$$

Hence

$$v = \sqrt{240 \text{ (m/s)}^2} = 15 \text{ m/s}$$

This is the same as 15 m/s × (1 km)/(1000 m) × (3600 s)/(1 h) = 56 km/h. If the driver tries to go around the circle faster than this, the automobile will skid out of the circle.

Note that the acceleration of a particle, or an automobile, in uniform circular motion is always centripetal (toward the center), never centrifugal (away from the center). However, with respect to the reference frame of the accelerating automobile, any body free to move in a horizontal direction—such as an apple that the driver of the automobile has placed on the (frictionless) dashboard—slides away and accelerates toward the outer side of the automobile with a centrifugal acceleration of magnitude equal to the centripetal acceleration of the automobile with respect to the ground (see Fig. 4.23). This centrifugal acceleration of the apple exists only in the reference frame of the automobile; it does not exist in the reference frame of the ground, where the apple merely continues its horizontal motion with constant velocity while the dashboard of the automobile accelerates sideways away from it.

Checkup 4.5

QUESTION 1: A truck drives along a curved road with an S-shaped curve. Describe the direction of the acceleration vector at different points of the S.

QUESTION 2: A particle travels once around a circle with uniform circular motion. What are the average velocity and the average acceleration for this motion?

QUESTION 3: A girl sits in a swing, which is swinging back and forth along an arc of a circle. What are the directions of the velocity vector and the acceleration vector when the girl passes through the lowest point of the circle?

QUESTION 4: A road leads over the crest of a hill. Near the crest, the vertical cross section of the hill is an arc of circle. If an automobile drives along the road at constant speed, what are the directions of the velocity vector and the acceleration vector at the crest of the hill?

(A) **v** forward, **a** upward
(B) **v** forward, **a** downward
(C) **v** upward, **a** upward
(D) **v** upward, **a** downward

4.6 THE RELATIVITY OF MOTION AND THE ADDITION OF VELOCITIES

Motion is relative—the values of the position, velocity, and acceleration of a particle depend on the frame of reference in which these quantities are measured. For example, consider one reference frame attached to the shore and a second reference frame attached to a ship moving away from the shore due east at a constant velocity of 5 m/s. Suppose that observers in both reference frames measure and plot the position vector of a fast motorboat passing by. The observers will then find different results for the position and the velocity of the motorboat. If the velocity of the motorboat is 12 m/s due east in the reference frame of the shore, it will be 7 m/s due east in the reference frame of the ship. The velocities of the motorboat in the two reference frames are related by a simple addition rule: the velocity relative to the shore is the velocity relative to the ship plus the velocity of the ship relative to the shore, that is, 12 m/s = 7 m/s + 5 m/s. This simple addition rule seems intuitively obvious, but to see where it comes from, we need to examine the position vectors of the motorboat in the two reference frames.

The position vector measured in the first reference frame (shore) will be denoted by **r**; that measured in the second reference frame (ship) will be denoted by **r**′. These two position vectors are different (see Fig. 4.24, where the position vector **r** is green and the position vector **r**′ is blue). The velocity of the ship relative to the shore will be denoted by $\mathbf{V}_O$. For the sake of simplicity, let us assume that the velocity of the ship relative to the shore is constant,

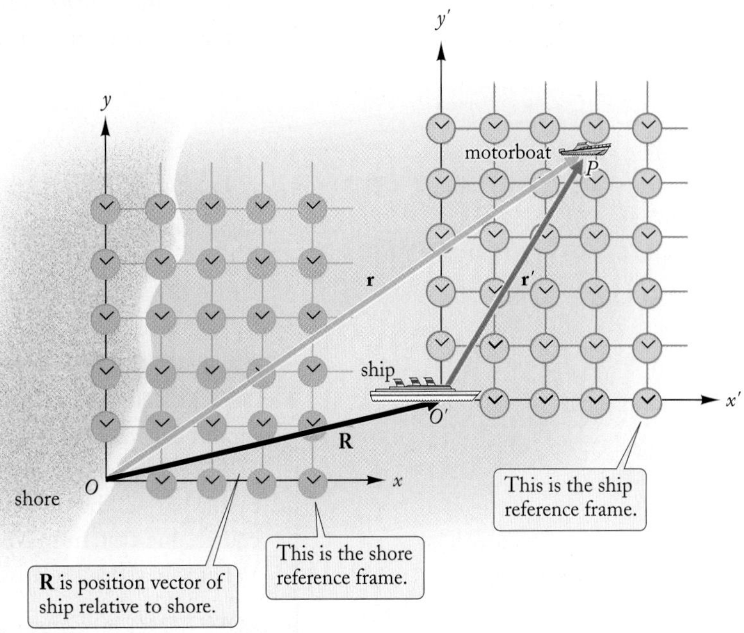

FIGURE 4.24 The coordinate grid x'–y' (blue) of the ship moves relative to the coordinate grid x–y (green) of the shore.

and that the ship started at the shore at time $t = 0$. The position vector of the ship relative to the shore is then $\mathbf{R} = \mathbf{V}_O t$. The vectors $\mathbf{r}$, $\mathbf{r}'$, and $\mathbf{R}$ form a vector triangle (see Fig. 4.24), and by inspection of this triangle we recognize that

$$\mathbf{r} = \mathbf{r}' + \mathbf{R}$$

or

$$\mathbf{r} = \mathbf{r}' + \mathbf{V}_O t \tag{4.50}$$

This equation merely says that the position vector of the motorboat relative to the shore is the position vector relative to the ship plus the position vector of the ship relative to the shore.

To extract the addition rule for velocities from Eq. (4.50), we contemplate what happens in a small time interval Δt. In such a time interval,the position vectors $\mathbf{r}$ and $\mathbf{r}'$ change by $\Delta\mathbf{r}$ and $\Delta\mathbf{r}'$. According to Eq. (4.50), these small changes are related by

$$\Delta\mathbf{r} = \Delta\mathbf{r}' + \mathbf{V}_O \Delta t \tag{4.51}$$

Dividing this by Δt, we find

$$\frac{\Delta\mathbf{r}}{\Delta t} = \frac{\Delta\mathbf{r}'}{\Delta t} + \mathbf{V}_O \tag{4.52}$$

In the limit $\Delta t \to 0$, $\Delta\mathbf{r}/\Delta t$ is the instantaneous velocity of the motorboat relative to the shore, and $\Delta\mathbf{r}'/\Delta t$ is the instantaneous velocity of the motorboat relative to the ship. Designating these velocites by $\mathbf{v}$ and $\mathbf{v}'$, respectively, we obtain

addition rule for velocities

$$\mathbf{v} = \mathbf{v}' + \mathbf{V}_O \tag{4.53}$$

This is the **addition rule for velocities**, also called the *Galilean velocity transformation,* because Galileo was the first to investigate the relativity of motion.

Note that since the velocities in the two reference frames differ by only a constant quantity $\mathbf{V}_O$, the *changes* in these velocities are the same in the two reference frames. Thus, the accelerations in the two reference frames are the same,

$$\mathbf{a}' = \mathbf{a} \tag{4.54}$$

If the motorboat is accelerating at, say, 3 m/s^2 relative to the shore, it will be accelerating at the same rate relative to the ship. This means that *for reference frames in uniform motion relative to one another, acceleration is an absolute quantity.*

EXAMPLE 10 Off the coast of Miami, the Gulf Stream current has a velocity of 4.8 km/h in a direction due north. The captain of a motorboat wants to travel from Miami to North Bimini island, due east of Miami. His boat has a speed of 18 km/h relative to the water. (a) If he heads his boat due east, what will be his actual course relative to the shore? (b) To attain a straight course, due east, from Miami to Bimini, in what direction must he head his boat? What is his speed relative to the shore?

SOLUTION: (a) Figure 4.25a shows the boat heading due east. The Gulf Stream current carries the boat northward, resulting in a northeast course. Figure 4.25b shows the velocity vector $\mathbf{V}_O$ of the water relative to the shore and the velocity

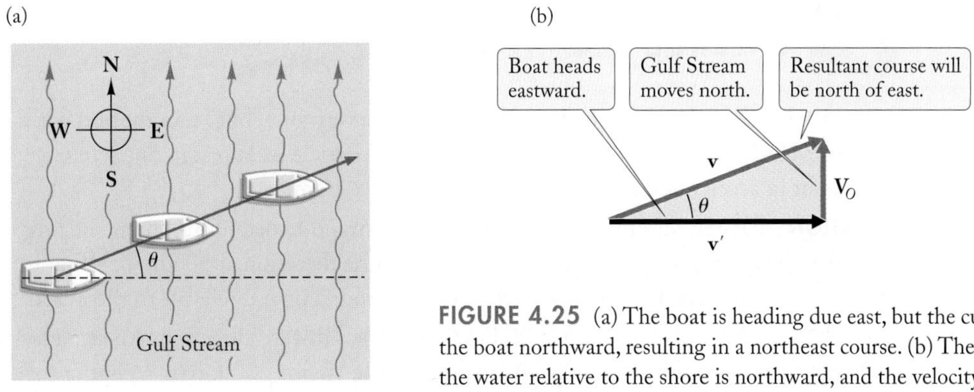

FIGURE 4.25 (a) The boat is heading due east, but the current carries the boat northward, resulting in a northeast course. (b) The velocity $\mathbf{V}_O$ of the water relative to the shore is northward, and the velocity $\mathbf{v}'$ of the boat relative to the water is eastward. The vector sum $\mathbf{v}' + \mathbf{V}_O$ is northeastward.

vector $\mathbf{v}'$ of the boat relative to the water. According to Eq. (4.53), the velocity vector $\mathbf{v}$ of the boat relative to the shore is the sum of $\mathbf{v}'$ and $\mathbf{V}_O$; that is,

$$\mathbf{v} = \mathbf{v}' + \mathbf{V}_O \tag{4.55}$$

From Fig. 4.25b we see that the tangent of the angle θ between $\mathbf{v}$ and the eastward direction is

$$\tan\theta = \frac{V_O}{v'} = \frac{4.8 \text{ km/h}}{18 \text{ km/h}} = 0.27$$

With a calculator, we find that the inverse tangent of 0.27, and thus the direction of the actual course of the motorboat, is

$$\theta = 15^\circ$$

(b) Figure 4.26a shows the boat on a due east course. To achieve this course, the captain must head his boat southeastward at an angle ϕ, to compensate for the Gulf Stream current. Figure 4.26b shows the velocity vector $\mathbf{V}_O$ of the water relative to the shore and the velocity vector $\mathbf{v}'$ of the boat relative to the water. By hypothesis, the velocity vector $\mathbf{v}$ relative to the shore points due east. Since the vector triangle is a right triangle, the sine of the angle ϕ between $\mathbf{v}'$ and the eastward direction is

$$\sin\phi = \frac{V_O}{v'} = \frac{4.8 \text{ km/h}}{18 \text{ km/h}} = 0.27 \tag{4.56}$$

With our calculator, we find that if the sine is 0.27, the angle is

$$\phi = 16^\circ$$

Thus, the boat must head 16° south of east.

The speed of the boat relative to the shore can be calculated from the right triangle of Fig. 4.26b. It is

$$v = v' \cos\phi = 18 \text{ km/h} \times \cos 16^\circ = 17 \text{ km/h}$$

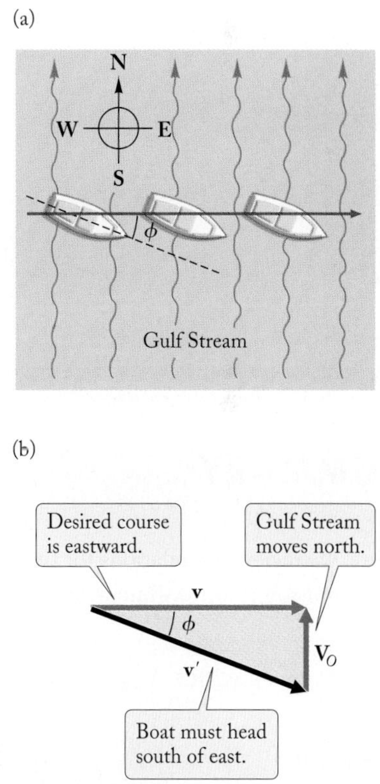

FIGURE 4.26 (a) The course of the boat is due east, but the heading of the boat is south of east. (b) The velocity $\mathbf{V}_O$ of the water relative to the shore is northward, and the velocity $\mathbf{v}'$ of the boat relative to the water is southeastward. The angle ϕ has been selected so that the vector sum $\mathbf{v}' + \mathbf{V}_O$ is due east.

Checkup 4.6

QUESTION 1: Can a particle be at rest in one reference frame, but be in motion with uniform velocity in another reference frame? Can a particle be at rest in one reference frame, but in accelerated motion in another reference frame?

QUESTION 2: A red automobile traveling along a straight road at 80 km/h is being overtaken by a black automobile traveling in the same direction at 90 km/h. What is the velocity of the black automobile relative to the red? The red relative to the black?

QUESTION 3: A wind is blowing at 30 km/h in the southward direction. An airplane is flying northward with an airspeed (relative to the air) of 200 km/h. What is the ground speed (relative to the ground) of the airplane?

QUESTION 4: To swim across a flowing river in the shortest possible time, in what direction should you swim? To reach a point on the shore directly opposite your starting point, how should you swim?

QUESTION 5: Relative to particle A, particle B is in uniform motion. Relative to particle B, particle C is in accelerated motion. Relative to particle A, is the motion of particle C accelerated or unaccelerated?

QUESTION 6: The wind on Lake Champlain is blowing from the north. A motorboat heads across the lake, in the west direction. Relative to the motorboat, the wind seems to blow from:

(A) East of north (B) West of north (C) East of south
(D) West of south (E) North

SUMMARY

Here, we review the relations for two dimensions. As discussed at the end of Section 4.2, the extension to three dimensions is straightforward, but rarely needed.

PHYSICS IN PRACTICE Velocity Vectors **(page 99)**

PROBLEM-SOLVING TECHNIQUES Projectile Motion **(page 108)**

AVERAGE VELOCITY COMPONENTS

$$\bar{v}_x = \frac{\Delta x}{\Delta t} \qquad \bar{v}_y = \frac{\Delta y}{\Delta t} \qquad (4.1; 4.2)$$

INSTANTANEOUS VELOCITY COMPONENTS

$$v_x = \frac{dx}{dt} \qquad v_y = \frac{dy}{dt} \qquad (4.3; 4.4)$$

AVERAGE ACCELERATION COMPONENTS

$$\bar{a}_x = \frac{\Delta v_x}{\Delta t} \qquad \bar{a}_y = \frac{\Delta v_y}{\Delta t} \qquad (4.9; 4.10)$$

INSTANTANEOUS ACCELERATION COMPONENTS

$$a_x = \frac{dv_x}{dt} \qquad a_y = \frac{dv_y}{dt} \qquad (4.11; 4.12)$$

POSITION VECTOR

$$\mathbf{r} = x\mathbf{i} + y\mathbf{j} \qquad (4.13)$$

VELOCITY VECTOR

$$\mathbf{v} = v_x\mathbf{i} + v_y\mathbf{j} = \frac{dx}{dt}\mathbf{i} + \frac{dy}{dt}\mathbf{j} \qquad (4.14; 4.15)$$

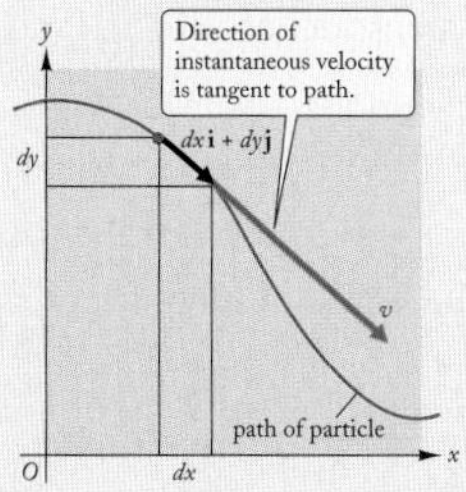

ACCELERATION VECTOR

$$\mathbf{a} = a_x\mathbf{i} + a_y\mathbf{j} = \frac{dv_x}{dt}\mathbf{i} + \frac{dv_y}{dt}\mathbf{j} \qquad (4.18; 4.19)$$

MOTION WITH CONSTANT ACCELERATION

$$v_x = v_{0x} + a_x t \qquad (4.20)$$

$$v_y = v_{0y} + a_y t \qquad (4.21)$$

$$x - x_0 = v_{0x}t + \tfrac{1}{2}a_x t^2 \qquad (4.23)$$

$$y - y_0 = v_{0y}t + \tfrac{1}{2}a_y t^2 \qquad (4.24)$$

PROJECTILE MOTION (With initial position $x_0 = 0, y_0 = 0$; $g \approx 9.81\ \text{m/s}^2$, and elevation angle θ measured from the horizontal.)

$$v_x = v_{0x} = v_0 \cos\theta \qquad (4.26)$$

$$v_y = v_{0y} - gt = v_0 \sin\theta - gt \qquad (4.28)$$

$$x = v_{0x}t \qquad (4.27)$$

$$y = v_{0y}t - \tfrac{1}{2}gt^2 \qquad (4.29)$$

RANGE x_{max}, MAXIMUM HEIGHT y_{max}, AND TIME OF FLIGHT t_{flight} (Over flat ground)

$$x_{max} = \frac{2v_0^2 \sin\theta\cos\theta}{g} \qquad (4.43)$$

$$y_{max} = \frac{v_0^2 \sin^2\theta}{2g} \qquad (4.42)$$

$$t_{flight} = \frac{2v_0 \sin\theta}{g} \qquad (4.41)$$

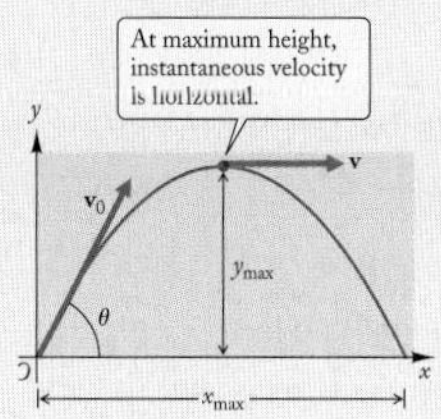

CENTRIPETAL ACCELERATION IN UNIFORM CIRCULAR MOTION (The direction of **a** is toward the center of the circle.)

$$a = \frac{v^2}{r} \qquad (4.49)$$

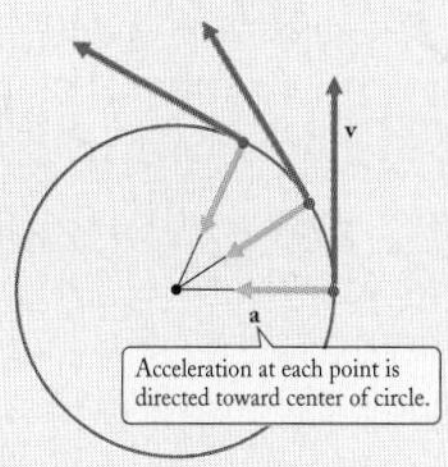

ADDITION RULE FOR VELOCITIES (Galilean velocity transformation)

$$\mathbf{v} = \mathbf{v}' + \mathbf{V}_O \qquad (4.53)$$

ACCELERATION IS ABSOLUTE (For uniform relative motion of reference frames.)

$$\mathbf{a} = \mathbf{a}' \qquad (4.54)$$

QUESTIONS FOR DISCUSSION

1. Can an automobile have eastward instantaneous velocity and northward instantaneous acceleration? Give an example.
2. Consider an automobile that is rounding a curve and braking at the same time. Draw a diagram showing the relative directions of the instantaneous velocity and acceleration.
3. A projectile is launched over level ground. Its initial velocity has a horizontal component v_{0x} and a vertical component v_{0y}. What is the average velocity of the projectile between the instants of launch and of impact?
4. If you throw a crumpled piece of paper, its trajectory is not a parabola. How does it differ from a parabola and why?
5. If a projectile is subject to air resistance, then the elevation angle for maximum range is not 45°. Do you expect the angle to be larger or smaller than 45°?
6. Baseball pitchers are fond of throwing curveballs. How does the trajectory of such a ball differ from the simple parabolic trajectory we studied in this chapter? What accounts for the difference?
7. A pendulum is swinging back and forth. Is this uniform circular motion? Draw a diagram showing the directions of the velocity and the acceleration at the top of the swing. Draw a similar diagram at the bottom of the swing.
8. Why do raindrops fall down at a pronounced angle with the vertical when seen from the window of a speeding train? Is this angle necessarily the same as that of the path of a water drop sliding down the outside surface of the window?
9. When a sailboat is sailing to windward ("beating"; see Fig. 4.27), the wind feels much stronger than when the sailboat is sailing downwind ("running"). Why?
10. Rain is falling vertically. If you run through the rain, at what angle should you hold your umbrella? If you don't have an umbrella, should you bend forward while running?
11. In the reference frame of the ground, the path of a sailboat beating to windward makes an angle of 45° with the direction of the wind. In the reference frame of the sailboat, the angle is somewhat smaller. Explain.
12. According to a theory proposed by Galileo, the tides on the oceans are caused by the Earth's rotational motion about its axis combined with its translational motion around the Sun. At midnight these motions are in the same direction; at noon they are in opposite directions (Fig. 4.28). Thus, any point on the Earth alternately speeds up and slows down. Galileo was of the opinion that the speeding up and slowing down of the ocean basins would make the water slosh back and forth, thus giving rise to tides. What is wrong with this theory? (Hint: What is the acceleration of a point on the Earth? Does this acceleration depend on the translational motion?)

FIGURE 4.27 Sailboat beating to windward.

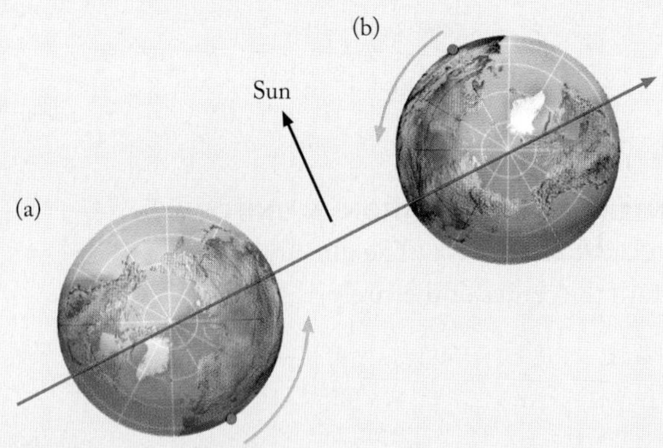

FIGURE 4.28 Rotational and translational motions of the Earth (a) at midnight and (b) at noon.

PROBLEMS

4.1 Components of Velocity and Acceleration†
4.2 The Velocity and Acceleration Vectors†

1. A sailboat tacking against the wind moves as follows: 3.2 km at 45° east of north, 4.5 km at 50° west of north, and 2.6 km at 45° east of north. The entire motion takes 1 h 15 min.

   (a) What is the total displacement for this motion?

   (b) What is the average velocity for this motion?

   (c) What is the speed if it is assumed to be constant?

2. In one-half year, the Earth moves halfway around its orbit, a circle of radius 1.50×10^{11} m centered on the Sun. What is the average speed, and what is the magnitude of the average velocity for this time interval?

3. The fastest bird is the spine-tailed swift, which reaches speeds of 171 km/h. Suppose that you wish to shoot such a bird with a .22-caliber rifle that fires a bullet with a speed of 366 m/s. If you fire at the instant when the bird is 30 m directly overhead, how many meters ahead of the bird must you aim the rifle? Ignore gravity in this problem.

4. An automobile with a drunken driver at the wheel travels round and round a traffic circle at 30 km/h. The automobile takes 80 s to go once around the circle. At $t = 0$, the automobile is at the east of the traffic circle; at $t = 20$ s it is at the north; at $t = 40$ s it is at the west; etc. What are the components of the velocity of the automobile at $t = 0$, $t = 10$ s, $t = 20$ s, $t = 30$ s, and $t = 40$ s? The x axis points eastward and the y axis points northward.

5. Suppose that a particle moving in three dimensions has a position vector

$$\mathbf{r} = (4 + 2t)\mathbf{i} + (3 + 5t + 4t^2)\mathbf{j} + (2 - 2t - 3t^2)\mathbf{k}$$

   where distance is measured in meters and time in seconds.

   (a) Find the instantaneous velocity vector.

   (b) Find the instantaneous acceleration vector. What are the magnitude and the direction of the acceleration?

6. A particle is moving in the x–y plane; the components of its position are

$$x = A\cos bt \qquad y = A\sin bt$$

   where A and b are constants.

   (a) What are the components of the instantaneous velocity vector? The instantaneous acceleration vector?

   (b) What is the magnitude of the instantaneous velocity? The instantaneous acceleration?

7. For the motion of the cruise missile described in Example 3, calculate the displacement of the missile relative to the aircraft at $t = 2.0$ s and at $t = 3.0$ s. What are the magnitude and the direction of the displacement vector at each of these times?

† For help, see Online Concept Tutorial 5 at www.wwnorton.com/physics

8. The components of the position of a body as a function of time are given by:

$$x = 5t + 4t^2 \qquad y = 3t^2 + 2t^3 \qquad z = 0$$

   where x and y are in meters and t is in seconds. What is the velocity vector as a function of time? What is the acceleration vector as a function of time? What is the speed at $t = 2.0$ s?

9. An airplane traveling at a constant speed of 300 km/h flies 30° north of east for 0.50 h and then flies 30° west of south for 1.00 h. What is the average velocity vector for the entire flight? What is the average acceleration vector for the entire flight?

10. The components of the position vector of a particle moving in the x–y plane are

$$x = A\cos bt \qquad y = Bt$$

   where A, b, and B are constants. What are the components of the instantaneous velocity vector? The instantaneous acceleration vector? What is the speed of the particle?

11. As an aircraft approaches landing, the components of its position are given by

$$x = 90t \qquad y = 500 - 15t$$

   where x and y are in meters and t is in seconds. What is the velocity vector of the aircraft during this descent? What is the value of its speed during the descent? What angle does the velocity vector make with the horizontal?

*12. Two football players are initially 20 m apart. The first player (a receiver) runs perpendicularly to the initial line joining the two players at a constant speed of 7.0 m/s. After two seconds, the second player (the quarterback) throws the ball at a horizontal speed of 15 m/s (ignore any vertical motion). In what horizontal direction should the quarterback aim so that the ball reaches the same spot the receiver will be? At what time will the ball be caught?

4.3 Motion with Constant Acceleration
4.4 The Motion of Projectiles†

13. Suppose that the acceleration vector of a particle moving in the x–y plane is

$$\mathbf{a} = 3\mathbf{i} + 2\mathbf{j}$$

   where the acceleration is measured in m/s^2. The velocity vector and the position vector are zero at $t = 0$.

   (a) What is the velocity vector of this particle as a function of time?

   (b) What is the position vector as a function of time?

14. The fastest recorded speed of a baseball thrown by a pitcher is 162.3 km/h (100.9 mi/h), achieved by Nolan Ryan in 1974 at

Anaheim Stadium. If the baseball leaves the pitcher's hand with a horizontal velocity of this magnitude, how far will the ball have fallen vertically by the time it has traveled 20 m horizontally?

15. At Acapulco, professional divers jump from a 36-m-high cliff into the sea (compare Example 9 in Chapter 2). At the base of the cliff, a rocky ledge sticks out for a horizontal distance of 6.4 m. With what minimum horizontal velocity must the divers jump off if they are to clear this ledge?

16. Consider the bomb dropped from the bomber described in Example 6.

    (a) What are the final horizontal and vertical components of the velocity of the bomb when it strikes the surface of the sea?

    (b) What is the final speed of the bomb? Compare this with the initial speed of the bomb.

17. A stunt driver wants to make his car jump over 10 cars parked side by side below a horizontal ramp (Fig. 4.29). With what minimum speed must he drive off the ramp? The vertical height of the ramp is 2.0 m, and the horizontal distance he must clear is 24 m.

18. A particle has an initial position vector $\mathbf{r} = 0$ and an initial velocity $\mathbf{v}_0 = 3\mathbf{i} + 2\mathbf{j}$ (where distance is measured in meters and velocity in meters per second). The particle moves with a constant acceleration $\mathbf{a} = \mathbf{i} - 4\mathbf{j}$ (measured in m/s^2). At what time does the particle reach a maximum y coordinate? What is the position vector of the particle at that time?

19. According to a reliable report, in 1795 a member of the Turkish embassy in England shot an arrow to a distance of 441 m. According to a less reliable report, a few years later the Turkish Sultan Selim shot an arrow to 889 m. In each of these cases calculate what must have been the minimum initial speed of the arrow.

20. A golfer claims that a golf ball launched with an elevation angle of 12° can reach a horizontal range of 250 m. Ignoring air friction, what would the initial speed of such a golf ball have to be? What maximum height would it reach?

21. A gunner wants to fire a gun at a target at a horizontal distance of 12 500 m from his position.

    (a) If his gun fires with a muzzle speed of 700 m/s and if $g = 9.81$ m/s^2, what is the correct elevation angle? Pretend that there is no air resistance.

    (b) If the gunner mistakenly assumes $g = 9.80$ m/s^2, by how many meters will he miss the target?

22. According to the *Guinness Book of World Records,* during a catastrophic explosion in Halifax on December 6, 1917, William Becker was thrown through the air for some 1500 m and was found, still alive, in a tree. Assume that Becker left the ground and returned to the ground (ignore the height of the tree) at an angle of 45°. With what speed did he leave the ground? How high did he rise? How long did he stay in flight?

23. In a circus act at the Ringling Bros. and Barnum & Bailey Circus, a "human cannonball" was fired from a large cannon with a muzzle speed of 87 km/h. Assume that the firing angle was 45° from the horizontal. How many seconds did the human cannonball take to reach maximum height? How high did he rise? How far from the cannon did he land?

24. The world record for the javelin throw by a woman established in 1976 by Ruth Fuchs in Berlin was 69.11 m (226 ft 9 in.). If Fuchs had thrown her javelin with the same initial velocity in Buenos Aires rather than in Berlin, how much farther would it have gone? The acceleration of gravity is 9.8128 m/s^2 in Berlin and 9.7967 m/s^2 in Buenos Aires. Pretend that air resistance plays no role in this problem.

25. The motion of an ICBM can be regarded as the motion of a projectile, because along the greatest part of its trajectory the missile is in free fall, outside of the atmosphere. Suppose that the missile is to strike a target 1000 km away. What minimum speed must the missile have at the beginning of its trajectory? What maximum height does it reach when launched with this minimum speed? How long does it take to reach its target? For these calculations assume that $g = 9.8$ m/s^2 everywhere along the trajectory and ignore the (short) portions of the trajectory inside the atmosphere.

26. The natives of the South American Andes throw stones by means of slings which they whirl (see Fig. 4.30). They can accurately throw a 0.20-kg stone to a distance of 50 m.

    (a) What is the minimum speed with which the stone must leave the sling to reach this distance?

    (b) Just before the release, the stone is being whirled around a circle of radius 1.0 m with the speed calculated in part (a). How many revolutions per second does the stone make?

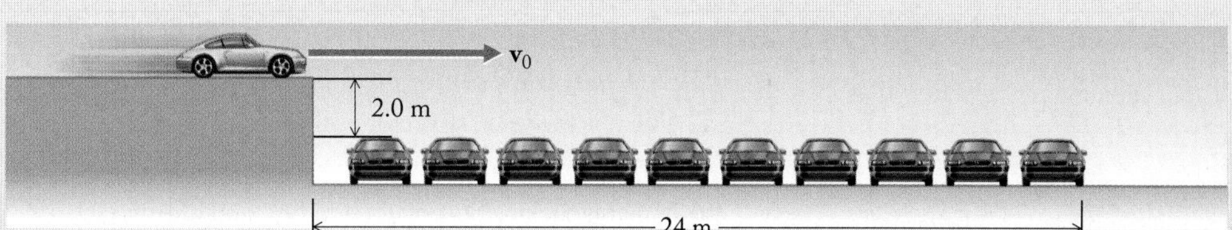

FIGURE 4.29 A stunt.

FIGURE 4.30 Whirling a stone before slinging.

27. The nozzle of a fire hose ejects 280 liters of water per minute at a speed of 26 m/s. How far away will the stream of water land if the nozzle is aimed at an angle of 35° with the horizontal? How many liters of water are in the air at any given instant?

28. According to an ancient Greek source, a stone-throwing machine on one occasion achieved a range of 730 m. If this is true, what must have been the minimum initial speed of the stone as it was ejected from the engine? When thrown with this speed, how long would the stone have taken to reach its target?

29. For what launch angle will the height and range of a projectile be equal?

30. A juggler tosses and catches balls at waist level; the balls are tossed at launch angles of 60°. If a ball attains a height 60 cm above waist level, how long is a ball in the air?

31. At $t = 0$, a small particle begins at the origin with initial velocity components $v_{0x} = -10$ m/s and $v_{0y} = 25$ m/s. Throughout its motion, the particle experiences an acceleration $\mathbf{a} = (2.0\mathbf{i} - 4.5\mathbf{j})$ m/s^2. Find the speed of the particle at $t = 3.0$ s. Find the position vector of the particle at $t = 3.0$ s.

32. A baseball is popped up, remaining aloft for 6.0 s before being caught at a horizontal distance of 75 m from the starting point. What was the launch angle?

33. An errant speeding bus launches from an unfinished highway ramp angled 10° upward. To complete the jump across a horizontal roadway gap of 15 m, what minimum initial speed must the bus have?

34. A child rolls a ball horizontally off the edge of a table. For what initial speed will the ball strike the floor a horizontal distance away from the table edge equal to the table height? In that case, what is the velocity of the ball just before it hits the floor?

*35. A boy stands at the edge of a cliff and launches a rock upward at an angle of 45.0°. The rock comes back down to the elevation where it was released 2.25 s later, then continues until it is seen to splash into the lake below 4.00 s after release. How far below the point of release is the lake surface? What horizontal distance from the point of release is the splash?

*36. A rock is thrown from a bridge at an upward launch angle of 30° with an initial speed of 25 m/s. The bridge is 30 m above the river. How much time elapses before the rock hits the water?

*37. A hockey player 25 m from the goal hits the hockey puck toward the goal, imparting a launch speed of 65 m/s at a launch angle of 10°. If the goal is 1.5 m high, does the shot score? At what vertical height does the puck pass the goal? How long does the puck take to reach the goal?

*38. (a) A golfer wants to drive a ball to a distance of 240 m. If he launches the ball with an elevation angle of 14.0°, what is the appropriate initial speed? Ignore air resistance.

(b) If the speed is too great by 0.6 m/s, how much farther will the ball travel when launched at the same angle?

(c) If the elevation angle is 0.5° larger than 14.0°, how much farther will the ball travel if launched with the speed calculated in part (a)?

*39. Show that for a projectile launched with an elevation angle of 45°, the maximum height reached is one-quarter of the range.

*40. During a famous jump in Richmond, Virginia, in 1903, the horse Heatherbloom with its rider jumped over an obstacle 8 ft 8 in. high while covering a horizontal distance of 37 ft. At what angle and with what speed did the horse leave the ground? Make the (somewhat doubtful) assumption that the motion of the horse is particle motion.

*41. With what elevation angle must you launch a projectile if its range is to equal twice its maximum height?

*42. In a baseball game, the batter hits the ball and launches it upward at an angle of 52° with a speed of 38 m/s. At the same instant, the center fielder starts to run toward the (expected) point of impact of the ball from a distance of 45 m. If he runs at 8.0 m/s, can he reach the point of impact before the ball?

*43. The gun of a coastal battery is emplaced on a hill 50 m above the water level. It fires a shot with a muzzle speed of 600 m/s at a ship at a horizontal distance of 12000 m. What elevation angle must the gun have if the shot is to hit the ship? Pretend there is no air resistance.

*44. In a flying ski jump, the skier acquires a speed of 110 km/h by racing down a steep hill and then lifts off into the air from a horizontal ramp. Beyond this ramp, the ground slopes downward at an angle of 45°.

(a) Assuming that the skier is in a free-fall motion after he leaves the ramp, at what distance down the slope will he land?

(b) In actual jumps, skiers reach distances of up to 165 m. Why does this not agree with the result you obtained in part (a)?

*45. Olympic target archers shoot arrows at a bull's-eye 12 cm across from a distance of 90.00 m. If the initial speed of the arrow is 70.00 m/s, what must be the elevation angle? If the archer misaims the arrow by 0.03° in the vertical direction, will it hit the bull's-eye? If the archer misaims the arrow by 0.03° in the horizontal direction, will it hit the bull's-eye? Assume that the height of the bull's-eye above the ground is the same as the initial arrow height of the bow and ignore air resistance.

*46. The muzzle speed for a Lee–Enfield rifle is 630 m/s. Suppose you fire this rifle at a target 700 m away and at the same level as the rifle.

(a) In order to hit the target, you must aim the barrel at a point above the target. How many meters above the target must you aim? Pretend there is no air resistance.

(b) What will be the maximum height that the bullet reaches along its trajectory?

(c) How long does the bullet take to reach the target?

*47. In artillery, it is standard practice to fire a sequence of trial shots at a target before commencing to fire "for effect." The artillerist first fires a shot short of the target, then a shot beyond the target, and then makes the necessary adjustment in elevation so that the third shot is exactly on target. Suppose that the first shot fired from a gun aimed with an elevation angle of 7°20′ lands 180 m short of the target; the second shot fired with an elevation of 7°35′ lands 120 m beyond the target. What is the correct elevation angle to hit the target?

*48. A hay-baling machine throws each finished bundle of hay 2.5 m up in the air so it can land on a trailer waiting 5.0 m behind the machine. What must be the speed with which the bundles are launched? What must be the angle of launch?

*49. Consider the trajectories for projectiles with the same launch speed, but different elevation angles. If you launch a large number of such projectiles simultaneously, will any of them ever collide while in flight? Explain carefully.

*50. Suppose that at the top of its parabolic trajectory a projectile has a horizontal speed v_{0x}. The segment at the top of the parabola can be approximated by a circle, called the osculating circle (Fig. 4.31). What is the radius of this circle? (Hint: The projectile is instantaneously in uniform circular motion at the top of the parabola.)

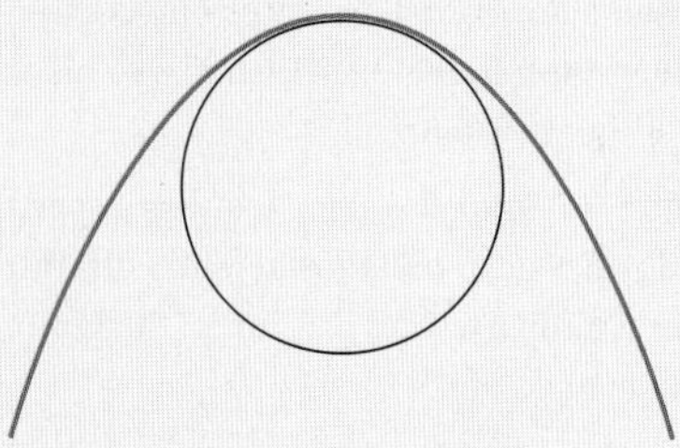

FIGURE 4.31 The osculating circle.

*51. A battleship steaming at 45 km/h fires a gun at right angles to the longitudinal axis of the ship. The elevation angle of the gun is 30°, and the muzzle velocity of the shot is 720 m/s; the gravitational acceleration is 9.8 m/s^2. What is the range of this shot in the reference frame of the ground? Pretend that there is no air resistance.

**52. The maximum speed with which you can throw a stone is about 25 m/s (a professional baseball pitcher can do much better than this). Can you hit a window 50 m away and 13 m up from the point where the stone leaves your hand? What is the maximum height of a window you can hit at this distance?

**53. A gun standing on sloping ground (see Fig. 4.32) fires up the slope. Show that the slant range of the gun (measured along the slope) is

$$l = \frac{2v_0^2 \cos^2\theta}{g\cos\alpha}(\tan\theta - \tan\alpha)$$

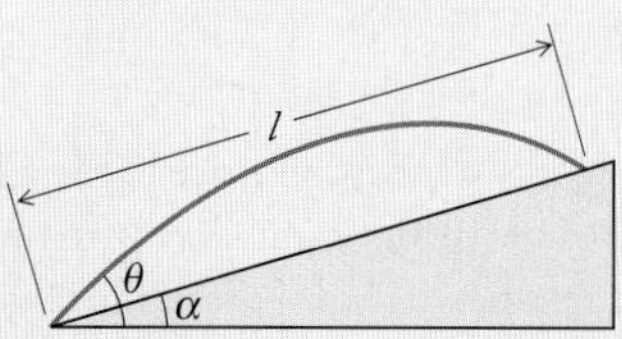

FIGURE 4.32 Projectile motion up a slope.

where α is the angle of the slope and the other symbols have their usual meaning. For what value of θ is this range a maximum?

**54. Two football players are initially 15 m apart. The first player (a receiver) runs perpendicular to the line joining the two players at a constant speed of 8.0 m/s. After two seconds, the second player (the quarterback) throws the ball with a horizontal component velocity of 20 m/s. In what horizontal direction and with what vertical launch angle should the quarterback throw so that the ball reaches the same spot the receiver will be? At what time will the ball be caught?

**55. When a tractor leaves a muddy field and drives on the highway, clumps of mud will sometimes come off the rear wheels and be launched into the air (see Fig. 4.33). In terms of the speed u of the tractor and the radius R of the wheel, find the maximum possible height that a clump of dirt can reach. In your calculation be careful to take into account both the initial velocity of the clump and the initial height at which it comes off the wheel. Evaluate numerically for $u = 30$ km/h and $R = 0.80$ m. (Hint: Solve this problem in the reference frame of the tractor.)

FIGURE 4.33 Tractor wheel flinging mud.

**56. A gun on the shore (at sea level) fires a shot at a ship which is heading directly toward the gun at a speed of 40 km/h. At the instant of firing, the distance to the ship is 15000 m. The muzzle speed of the shot is 700 m/s. Pretend that there is no air resistance.

(a) What is the required elevation angle for the gun? Assume $g = 9.8$ m/s^2.

(b) What is the time interval between firing and impact?

**57. A ship is steaming at 30 km/h on a course parallel to a straight shore at a distance of 17000 m. A gun emplaced on the shore (at sea level) fires a shot with a muzzle speed of 700 m/s when the ship is at the point of closest approach. If the shot is to hit the ship, what must be the elevation angle of the gun? How far ahead of the ship must the gun be aimed? Give the answer to the latter question both in meters and in minutes of arc. Pretend that there is no air resistance. (Hint: Solve this problem by the following method of successive approximations. First calculate the time of flight of the shot, neglecting the motion of the ship; then calculate how far the ship moves in this time; and then calculate the elevation angle and the aiming angle required to hit the ship at this new position.)

4.5 Uniform Circular Motion

58. An audio compact disk (CD) player is rotating at an angular velocity of 32.5 radians per second when playing a track at a radius of 4.0 cm. What is the linear speed at that radius? What is the rotation rate in revolutions per minute?

59. In science fiction movies, large, ring-shaped space stations rotate so that astronauts experience an acceleration, which feels the same as gravity. If the station is 200 m in radius, how many revolutions per minute are required to provide an acceleration of 9.81 m/s^2?

60. When drilling metals, excess heat is avoided by staying below a recommended linear cutting speed. A 3.0-mm-diameter hole and a 25-mm-diameter hole need to be drilled. At what maximum number of revolutions per minute can the drill bit rotate so that a point on its perimeter does not exceed the material's linear cutting speed limit of 3.0 m/s?

61. The Space Shuttle orbits the Earth on a circle of radius 6500 km every 87 minutes. What is the centripetal acceleration of the Space Shuttle in this orbit?

62. A mechanical pitcher hurls baseballs for batting practice. The arm of the pitcher is 0.80 m long and is rotating at 45 radians/second at the instant of release. What is the speed of the pitched ball?

63. An ultracentrifuge spins a small test tube in a circle of radius 10 cm at 1000 revolutions per second. What is the centripetal acceleration of the test tube? How many standard *g*'s does this amount to?

64. The blade of a circular saw has a diameter of 20 cm. If this blade rotates at 7000 revolutions per minute (its maximum safe speed), what are the speed and the centripetal acceleration of a point on the rim?

65. At the Fermilab accelerator (one of the world's largest atom smashers), protons are forced to travel in an evacuated tube in a circular orbit of diameter 2.0 km (Fig. 4.34). The protons have a speed nearly equal to the speed of light (99.999 95% of the speed of light). What is the centripetal acceleration of these protons? Express your answer in m/s^2 and in standard *g*'s.

66. A phonograph record rotates at $33\frac{1}{3}$ revolutions per minute. The radius of the record is 15 cm. What is the speed of a point at its rim?

67. The Earth moves around the Sun in a circular path of radius 1.50×10^{11} m at uniform speed. What is the magnitude of the centripetal acceleration of the Earth toward the Sun?

68. An automobile has wheels of diameter 64 cm. What is the centripetal acceleration of a point on the rim of this wheel when the automobile is traveling at 95 km/h?

*69. The Earth rotates about its axis once in one sidereal day of 23 h 56 min. Calculate the centripetal acceleration of a point located on the equator. Calculate the centripetal acceleration of a point located at a latitude of 45°.

*70. When looping the loop, the Blue Angels stunt pilots of the U.S. Navy fly their jet aircraft along a vertical circle of diameter 1000 m (Fig. 4.35). At the top of the circle, the speed is 350 km/h; at the bottom of the circle, the speed is 620 km/h. What is the centripetal acceleration at the top? At the bottom? In the reference frame of one of these aircraft, what is the acceleration that the pilot feels at the top and at the bottom; i.e., what is the acceleration relative to the aircraft of a small body, such as a coin, released by the pilot?

71. The table inside the book cover lists the radii of the orbits of the planets around the Sun and the time taken to complete an orbit ("period of revolution"). Assume that the planets move along circles at constant speed. Calculate the centripetal acceleration for each of the first three planets (Mercury, Venus, Earth). Verify that the centripetal accelerations are in proportion to the inverses of the squares of the orbital radii.

FIGURE 4.34 The main accelerator ring at Fermilab.

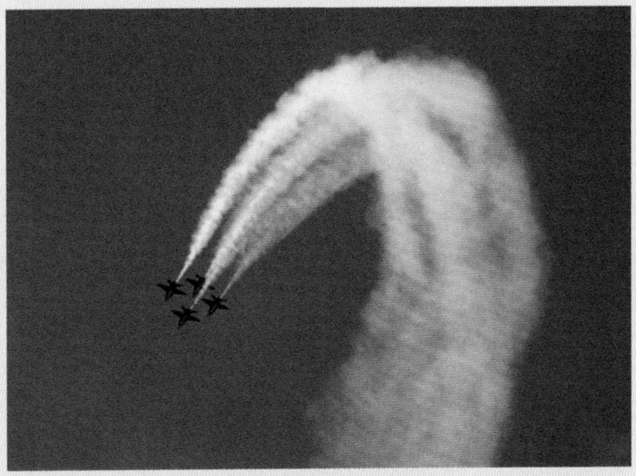

FIGURE 4.35 Blue Angels looping the loop.

4.6 The Relativity of Motion and the Addition of Velocities

72. On a rainy day, a steady wind is blowing at 30 km/h. In the reference frame of the *air*, the raindrops are falling vertically with a speed of 10 m/s. What are the magnitude and the direction of the velocity of the raindrops in the reference frame of the ground?

73. In an airport, a moving walkway has a speed of 1.5 m/s relative to the ground. What is the speed, relative to the ground, of a passenger running forward on this walkway at 4.0 m/s? What is the speed, relative to the ground, of a passenger running backward on this walkway at 4.0 m/s?

74. On a rainy day, raindrops are falling with a vertical velocity of 10 m/s. If an automobile drives through the rain at 25 m/s, what is the velocity (magnitude and direction) of the raindrops relative to the automobile?

75. A battleship steaming at 13 m/s toward the shore fires a shot in the forward direction. The elevation angle of the gun is 20°, and the muzzle speed of the shot is 660 m/s. What is the velocity vector of the shot relative to the shore?

76. A wind of 30 m/s is blowing from the west. What will be the speed, relative to the ground, of a sound signal traveling due north? The speed of sound, relative to air, is 330 m/s.

77. On a windy day, a hot-air balloon is ascending at a rate of 1.5 m/s relative to the air. Simultaneously, the air is moving with a horizontal velocity of 12.0 m/s. What is the velocity (magnitude and direction) of the balloon relative to the ground?

78. You can paddle your kayak at a speed of 3.5 km/h relative to the water. If a river is flowing at 2.5 km/h, how far can you paddle downstream in 40 minutes? How long will it take you to paddle back upstream from there?

79. As a train rolls by at 5.00 m/s, you see a cat on one of the flatcars. The cat is walking toward the back of the train at a speed of 0.50 m/s relative to the car. On the cat is a flea which is walking from the cat's neck to its tail at a speed of 0.10 m/s relative to the cat. How fast is the flea moving relative to you?

80. A boat with maximum speed v (relative to the water) is on one shore of a river of width d. The river is flowing at speed V. Traveling in a straight line, how long does it take to get to a point directly opposite? What is the fastest crossing time to any point?

81. Each step on an up escalator is 20 cm high and 30 cm deep. The escalator advances 1.5 step per second. If you also walk up the escalator stairs at a rate of 1.0 step per second, what is your velocity (magnitude and direction) relative to a fixed observer?

*82. A villain in a car traveling at 30 m/s fires a projectile along the direction of motion toward the front of the car with a launch speed of 50 m/s relative to the car. A hero standing nearby observes the projectile to travel straight up. What was the launch angle as viewed by the villain? What height does the projectile attain?

*83. A blimp is motoring at constant altitude. The airspeed indicator on the blimp shows that its speed relative to the air is 20 km/h, and the compass shows that the heading of the blimp is 10° east of north. If the air is moving over the ground with a velocity of 15 km/h due east, what is the velocity (magnitude and direction) of the blimp relative to the ground? For an observer on the ground, what is the angle between the longitudinal axis of the blimp and the direction of motion?

*84. A sailboat is moving in a direction 50° east of north at a speed of 14 km/h. The wind measured by an instrument aboard the sailboat has an apparent (relative to the sailboat) speed of 32 km/h coming from an apparent direction of 10° east of north. Find the true (relative to ground) speed and direction of the wind.

*85. (a) In still air, a high-performance sailplane has a rate of descent (or sinking rate) of 0.50 m/s at a forward speed (or airspeed) of 60 km/h. Suppose the plane is at an initial altitude of 1500 m. How far can it travel horizontally in still air before it reaches the ground?

(b) Suppose the plane is in a (horizontal) wind of 20 km/h. With the same initial conditions, how far can it travel in the downwind direction? In the upwind direction?

*86. A wind is blowing at 50 km/h from a direction 45° west of north. The pilot of an airplane wishes to fly on a route due north from an airport. The airspeed of the airplane is 250 km/h.

(a) In what direction must the pilot point the nose of the airplane?

(b) What will be the airplane's speed relative to the ground?

*87. At the entrance of Ambrose Channel at New York harbor, the tidal current at one time of the day has a velocity of 4.2 km/h in a direction 20° south of east. Consider a ship in this current; suppose that the ship has a speed of 16 km/h relative to the water. If the helmsman keeps the bow of the ship aimed due north, what will be the actual velocity (magnitude and direction) of the ship relative to the ground?

*88. A white automobile is traveling at a constant speed of 90 km/h on a highway. The driver notices a red automobile 1.0 km behind, traveling in the same direction. Two minutes later, the red automobile passes the white automobile.

(a) What is the average speed of the red automobile relative to the white?

(b) What is the speed of the red automobile relative to the ground?

*89. Two automobiles travel at equal speeds in opposite directions on two separate lanes of a highway. The automobiles move at constant speed v_0 on straight parallel tracks separated by a distance h. Find a formula for the rate of change of the distance between the automobiles as a function of time; take the instant of closest approach as $t = 0$. Plot v vs. t for $v_0 =$ 60 km/h, $h = 50$ m.

*90. A ferryboat on a river has a speed v relative to the water. The water of the river flows with speed V relative to the ground. The width of the river is d.

(a) Show that the ferryboat takes a time $2d/\sqrt{v^2 - V^2}$ to travel across the river and back.

(b) Show that the ferryboat takes a time $2dv/(v^2 - V^2)$ to travel a distance d up the river and back. Which trip takes a shorter time?

**91. An AWACS aircraft is flying at high altitude in a wind of 150 km/h from due west. Relative to the air, the heading of the aircraft is due north and its speed is 750 km/h. A radar operator on the aircraft spots an unidentified target approaching from northeast; relative to the AWACS aircraft, the bearing of the target is 45° east of north, and its speed is 950 km/h. What is the speed of the unidentified target relative to the ground?

REVIEW PROBLEMS

92. In the fastest-ever descent from Mt. Everest, two climbers slid down the side of the mountain on the seats of their pants, using their ice picks as brakes. They descended a height of 2340 m in 3.5 h. From these data, can you calculate the average of the vertical component of the velocity? The average of the horizontal component of the velocity? The average speed? What extra information do you need to calculate these quantities?

93. With its engine cut off, a small airplane glides downward at an angle of 15° below the horizontal at a speed of 240 km/h.

(a) What are the horizontal and the vertical components of its velocity?

(b) If the airplane is initially at a height of 2000 m above the ground, how long does it take to crash into the ground?

94. An automobile enters a 180° curve at a constant speed of 25 m/s and emerges from this curve 12 s later. What are the components of the average acceleration for this time interval?

95. At the entrance to Ambrose Channel at New York harbor, the maximum tidal current has a velocity of 1.2 km/h in a direction 20° south of east. What is the component of this velocity in the east direction? In the north direction?

96. A blimp motoring at a constant altitude has a velocity component of 15 km/h in the north direction and a velocity component of 15 km/h in the east direction. What is the speed of the blimp? What is the direction of motion of the blimp?

97. Suppose that the position vector of a particle is given by the following function of time:

$$\mathbf{r} = (6.0 + 2.0t^2)\mathbf{i} + (3.0 - 2.0t + 3.0t^2)\mathbf{j}$$

where distance is measured in meters and time in seconds.

(a) What is the instantaneous velocity vector at $t = 2.0$ s? What is the magnitude of this vector?

(b) What is the instantaneous acceleration vector? What are the magnitude and direction of this vector?

98. An archer shoots an arrow over level ground. The arrow leaves the bow at a height of 1.5 m with an initial velocity of 60 m/s in a horizontal direction.

(a) How long does this arrow take to fall to the ground?

(b) At what horizontal distance does this arrow strike the ground?

99. Volcanos on the Earth eject rocks at speeds of up to 700 m/s. Assume that the rocks are ejected in all directions; ignore the height of the volcano and ignore air friction.

(a) What is the maximum height reached by the rocks?

(b) What is the maximum horizontal distance reached by the rocks?

(c) Is it reasonable to ignore air friction in these calculations?

100. A large stone-throwing engine designed by Archimedes could throw a 77-kg stone over a range of 180 m. What must have been the initial speed of the stone if thrown at an initial angle of 45° with the horizontal?

101. The world record for the discus throw set by M. Wilkins in 1976 was 70.87 m. What is the initial speed of the discus required to achieve this range? Assume that the discus was launched with an elevation angle of 45°, and that the height of the hand over the ground was 2.0 m at the instant of launch.

*102. When you hold the nozzle of a water hose horizontally, at a height of 1.0 m above the ground, the stream of water lands 4.0 m from you. If you now aim the nozzle straight up, how high (above the nozzle) will the stream of water rise?

*103. An automobile travels at a steady 90 km/h along a road leading over a small hill. The top of the hill is rounded so, in the vertical plane, the road approximately follows an arc of a circle of radius 70 m. What is the centripetal acceleration of the automobile at the top of the hill?

*104. A lump of concrete falls off a crumbling overpass and strikes an automobile traveling on a highway below. The lump of concrete falls 5.0 m before impact, and the automobile has a speed of 90 km/h.

(a) What is the speed of impact of the lump in the reference frame of the automobile?

(b) What is the angle of impact?

*105. You are driving an automobile at a steady speed of 90 km/h along a straight highway. Ahead of you is a 10-m-long truck traveling at a steady speed of 60 km/h. You decide to pass this truck, and you switch into the passing lane when 40 m behind the truck.

(a) What is your speed in the reference frame of the truck?

(b) How long do you take to pass the truck, starting 40 m behind the truck and ending 40 m ahead of the truck? (Hint: Calculate this time in the reference frame of the truck.)

Answers to Checkups

Checkup 4.1

1. With the x axis eastward and the y axis northward, we must have: (a) for motion northwest, the x component is $-$ and the y component is $+$; (b) for southward motion, the x component is 0 and the y component is $-$; and (c) for motion southeast, the x component is $+$ and the y component is $-$.

2. Yes—if the direction of the velocity is changing, then the motion is accelerated. When traveling at constant speed, you experience acceleration when the path is curved.

3. Since the initial and final positions are the same ($\Delta x = 0$ and $\Delta y = 0$), the average velocity is zero. Since the initial and final velocities are the same ($\Delta v_x = 0$ and $\Delta v_y = 0$), the average acceleration is zero.

4. (C) $v_x = 0$; $v_y = 50$ km/h. We assume the motorcycle travels counterclockwise (looking from above) around the traffic circle. In that case, at the eastern point, the motorcycle is traveling northward, and so has $v_x = 0$ and $v_y = 50$ km/h.

Checkup 4.2

1. In general, the instantaneous velocity vector will be tangent to the circle wherever the automobile is heading on the circle, and so will be in the north, west, south, and east directions, respectively.

2. Since the x and y components of the velocity are equal, the direction of the aircraft is halfway between the x and y axes, that is, at an angle of 45° with respect to either axis. We can also see that $\tan\theta = v_y/v_x = 1$ implies $\theta = 45°$.

3. Since the change in velocity ($v_2 - v_1$) is opposite to the original motion, the acceleration vector during the collision is also opposite to the initial motion.

4. (D) 60°. The magnitude of the acceleration is $a = 6.0\ \text{m/s}^2$; the x component is $a_x = 3.0\ \text{m/s}^2$. But $a_x = a\cos\theta$, where θ is the angle with respect to the horizontal, so $\cos\theta = a_x/a = 3/6 = 1/2$. The inverse cosine of 1/2 is 60°.

Checkup 4.3

1. It is possible to have accelerated motion in the x direction only; for example, if the boat moves in the x direction while speeding up. It is possible to have accelerated motion in both the x and y directions; for example, if the boat moves in a direction between the x and y directions while speeding up.

2. If the y motion were unaccelerated, the particle would move equal distance in equal times in the y direction, while moving greater and greater distances in equal times in the (accelerated) x direction. Its path would thus not be a straight line. Since the particle moves along a straight line, and we know the x motion is accelerated, then the y motion must be accelerated also.

3. (E) $a_x = -6.9\ \text{m/s}^2$; $a_y = -6.9\ \text{m/s}^2$. Since the coordinate axes are tilted by 45° with respect to vertical, both the x and y directions will have equal accelerations; these will both be equal to the component of the acceleration due to gravity along the axes. For the x direction, we will have $a_x = -g \times \sin 45° = -9.8\ \text{m/s}^2 \times \sin 45° = -6.9\ \text{m/s}^2$; for the y direction, we will have $a_y = -g \times \cos 45° = -9.8\ \text{m/s}^2 \times \cos 45° = -6.9\ \text{m/s}^2$.

Checkup 4.4

1. At all points of its trajectory, the acceleration of a projectile is the same; it is always $a_y = -g = -9.81\ \text{m/s}^2$.

2. As we saw graphically and by maximizing the range function, the maximum range is achieved by launching at an angle of 45° (for a given fixed launch speed). From Eqs. (4.41) and (4.42), we see that both the maximum time of flight and the maximum height are achieved by launching the projectile vertically ($\sin\theta = 1$ when $\theta = 90°$).

3. The acceleration is constant and always downward ($a_y = -g$), but the velocity is never downward (since we are told that the projectile was launched at an angle *smaller* than 90°), and so is never parallel to the acceleration. At the top of its trajectory (y_{max}), the velocity is horizontal, and so is perpendicular to the acceleration there.

4. Since the range varies as the square of the launch speed [Eq. (4.43)], the projectile launched at twice the launch speed will go four times as far (for the same elevation angle).

5. Since the time of flight varies proportionally to $\sin\theta$, the sine of the launch angle, the projectile with the smallest launch angle will return to the ground in the shortest time.

6. (E) 80°. The range is largest at 45°, and drops off symmetrically above and below 45°. Thus, compared with 20° (which differs from 45° by 25°), elevation angles of 30°, 40°, 50°, and 60° will result in larger ranges (since they differ from 45° by smaller angles); the projectile launched at 70° will have the same range as the 20° one. However, the projectile launched at 80° will have a shorter range (since that elevation angle differs from 45° by the larger angle of 35°).

Checkup 4.5

1. By an "S-shaped curve" we usually mean one that turns first one way and then the other way before returning parallel to the original direction. For example, if one enters the S from below, the acceleration is first toward the left and then briefly zero as the curvature changes to the opposite direction, and then the acceleration is toward the right at the top of the S.
2. Since the velocity traces out all directions uniformly as the particle travels once around the circle, the average velocity is zero. Similarly, the acceleration points toward the center of the circle, and rotates around including all directions equally as the particle travels once around the circle, so the average acceleration is also zero.
3. At the lowest point, the instantaneous velocity is in the direction the girl is moving: straight ahead (or straight back) horizontally. For the acceleration, we note that although the motion is not uniform circular motion, at the bottom the speed goes through a maximum: she switches from speeding up on the way down to slowing down on the way back up. Since there is no change in speed at the bottom point, the acceleration is purely centripetal there, and thus is toward the center of the circle, or vertically up.
4. (B) **v** forward, **a** downward. At the moment the automobile crests the hill, its velocity is horizontal, straight ahead. At constant speed along the arc of a circle, the acceleration is purely centripetal, and so is toward the center of the circle, or vertical and downward.

Checkup 4.6

1. Yes, a particle at rest in one reference frame is in motion with uniform velocity in any reference frame that moves with uniform velocity with respect to the reference frame where the particle was at rest. Yes, a particle at rest in one reference frame is in accelerated motion in any reference frame that is in accelerated motion relative to the reference frame where the particle was at rest.
2. The black automobile is moving forward at 10 km/h relative to the red [see Eq. (4.53)]. The red automobile is moving backward relative to the black automobile, and so the red is moving at −10 km/h relative to the black.
3. Since the wind opposes the motion, the speed relative to the ground is $v = v' - V_O = 200\ \text{km/h} - 30\ \text{km/h} = 170\ \text{km/h}$.
4. To swim across in the minimum time, you would swim perpendicular to the river, so that all your swimming velocity was directed at crossing the river; however, you would wind up somewhat downstream. To reach a point on the shore directly opposite your starting point, you would have to swim somewhat upstream, as with the motorboat in Example 10.
5. For reference frames in uniform motion with respect to one another, acceleration is an absolute quantity; thus, relative to particle A, the motion of particle C is accelerated, and it is exactly the same acceleration as the motion of particle C relative to particle B.
6. (B) West of north. The westward motion of the motorboat makes the wind seem to have a component coming from the west (looking forward, you feel a breeze on your face); added to the actual wind from the north, the wind seems to come from west of north.

CHAPTER

5 Newton's Laws of Motion

CONCEPTS IN CONTEXT

Concepts in Context

Elevators remind us of familiar sensations triggered by vertical accelerated motion: When accelerating upward we feel heavy; the floor seems to push harder on our feet. As the upward motion slows toward a stop, we may feel the "butterflies in the stomach" sensation associated with free fall.

With the concepts of this chapter we will describe forces and predict accelerations, and we can ask:

? What force must the floor of an elevator apply to your feet to accelerate you upward? (Checkup 5.4, question 6, page 143; and Example 6, page 147)

? How do an elevator and its counterweight accelerate if the cable connecting them is permitted to run freely? (Example 10, page 154)

? What devices are installed on elevators to ensure their safety? (Physics in Practice: Elevators, page 157)

So far we have dealt only with the mathematical description of motion—the definitions of position, velocity, and acceleration and the relationships between these quantities. We did not inquire what causes a body to accelerate. In this chapter we will see that *the cause of acceleration is a force exerted on the body by some external agent.* The fundamental properties of force and the relationship between force and acceleration are contained in Newton's three laws of motion. The first of these laws describes the natural state of motion of a free body on which no net external force is acting, whereas the other two laws deal with the behavior of bodies under the influence of external forces.

SIR ISAAC NEWTON (1642–1727)
English mathematician and physicist, widely regarded as the greatest scientist of all time. His brilliant discoveries in mechanics were published in 1687 in his book Principia Mathematica, *one of the glories of the Age of Reason. In this book, Newton laid down the laws of motion and the Law of Universal Gravitation, and he demonstrated that planets in the sky as well as bodies on the Earth obey the same mathematical equations. For over 200 years, Newton's laws stood as the unchallenged basis of all our attempts at a scientific explanation of the physical world.*

The first law was actually discovered by Galileo Galilei early in the seventeenth century, but it remained for Isaac Newton, in the second half of the seventeenth century, to formulate a coherent theory of forces and to lay down a complete set of equations from which the motion of bodies under the influence of arbitrary forces can be calculated. The study of forces and their effects on the motion of bodies is called **dynamics**, and Newton's laws of motion are sometimes called the laws of dynamics.

5.1 NEWTON'S FIRST LAW

Everyday experience seems to suggest that a force—a push or a pull—is needed to keep a body moving at constant velocity. For example, if the wind pushing a sailboat suddenly ceases, the boat will coast along for some distance, but it will gradually slow down, stop, and remain stopped until a new gust of wind comes along. However, everyday experience misleads us: what actually slows down the sailboat is not the *absence* of a propulsive force but, rather, the *presence* of friction forces exerted by the water and the air, which oppose the motion. *Under ideal frictionless conditions, a body in motion would continue to move forever.* Experiments with pucks or gliders riding on a cushion of air on a low-friction air table or air track give a clear indication of the persistence of motion (see Fig. 5.1); but in order to eliminate friction entirely, it is best to use bodies moving in a vacuum, without even air against which to rub. Experiments with particles moving in vacuum show that a body left to itself, on which no net external force is acting, persists indefinitely in its state of uniform motion.

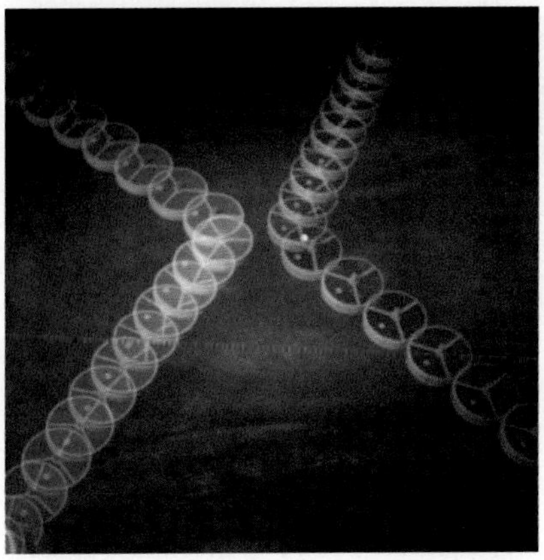

FIGURE 5.1 Multiple-exposure photograph of two pucks moving on an air table.

In this context, an external force is any force exerted on the body by some *other* body. By contrast, internal forces are those exerted by some part of a body on another part of the *same* body. For instance, the forces that the screws or bolts in the sailboat exert on its planks are internal forces; such internal forces do not affect the motion of the boat.

Newton's First Law summarizes experiments and observations on the motion of bodies on which no net external force is acting:

In the absence of a net external force, a body at rest remains at rest, and a body in motion continues to move at constant velocity.

Newton's First Law

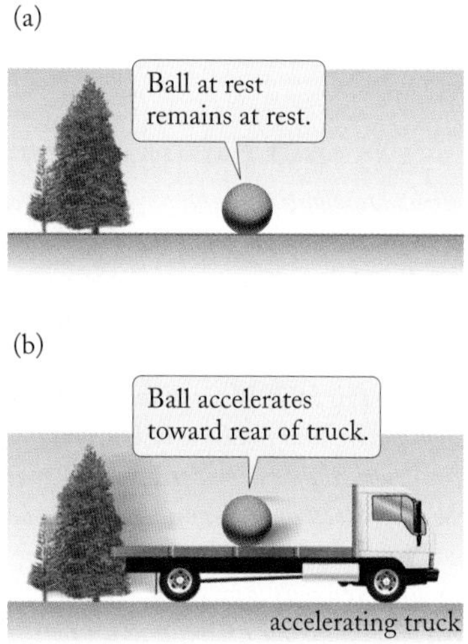

FIGURE 5.2 (a) In the absence of a net external force, a ball at rest on a level street remains at rest. (b) But a ball at rest on the platform of an accelerating truck acquires a "spontaneous" acceleration toward the rear of the truck (in the reference frame of the truck).

The tendency of a body to continue in its initial state of motion (a state of rest or a state of uniform velocity) is called its **inertia**. Accordingly, the First Law is often called the Law of Inertia.

A crucial restriction on Newton's First Law concerns the choice of reference frame: the law is not valid in all reference frames but only in certain special frames. If this law is valid in one given reference frame, then it cannot be valid in a second reference frame that has an accelerated motion relative to the first. For example, in the reference frame of the ground, a bowling ball at rest on a level street remains at rest, but in the reference frame of an accelerating truck, a bowling ball initially at rest on the platform of the truck acquires a "spontaneous" acceleration toward the rear of the truck, in contradiction to Newton's First Law (see Fig. 5.2). Those special reference frames in which the law is valid are called **inertial reference frames**. Thus, the reference frame of the ground is an inertial reference frame, but that of the accelerating truck is not.

Note that *if some first reference frame is inertial, any other reference frame in uniform translational motion relative to the first will also be inertial, and any other reference frame in accelerated motion relative to the first will not be inertial.* Thus, any two inertial reference frames can differ only by some constant relative velocity; they cannot differ by an acceleration. This implies that, as measured with respect to inertial reference frames, *acceleration is absolute:* when a particle has some acceleration in one inertial reference frame, then the particle will have exactly the same acceleration in any other [see Eq. (4.54)]. By contrast, the velocity of the particle is relative; the velocities are related by the addition rule for velocities [see Eq. (4.53)].

Finally we must address an important question: Which of the reference frames in practical use for everyday measurements are inertial? For the description of everyday phenomena, the most commonly used reference frame is one attached to the ground, with the origin of coordinates fixed at some point on the surface of the Earth (see Fig. 5.3). Although crude experiments indicate that this reference frame is inertial (for example, a ball placed on a level street remains at rest), more precise experiments show that this reference frame is not inertial. The Earth rotates about its axis, and this rotational motion gives points on the ground a centripetal acceleration; thus, a reference frame attached to the ground is an accelerated, noninertial reference frame. However, the numerical value of the centripetal acceleration of points on the surface of the Earth is fairly small—about $0.034\ \text{m/s}^2$ at the equator—and it can be neglected for most purposes. Our additional centripetal acceleration due to the motion of the Earth around the Sun is even smaller, about $0.002\ \text{m/s}^2$. Hereafter, unless otherwise stated, we will take it for granted that the reference frames in which we express the laws of physics are inertial reference frames, either exactly inertial or at least so nearly inertial that no appreciable deviation from Newton's First Law occurs within the region of space and time in which we are interested.

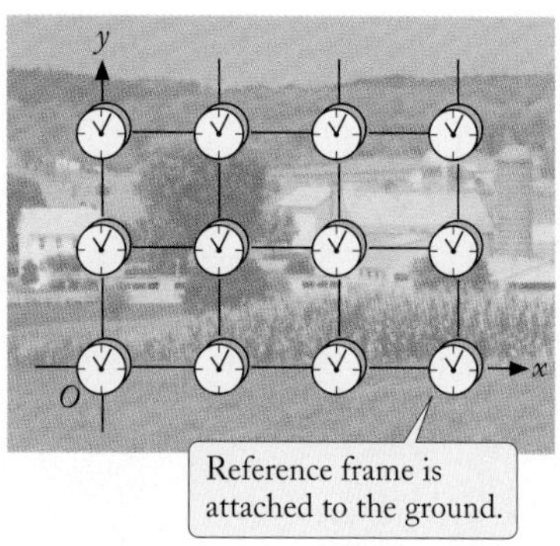

FIGURE 5.3 A reference frame with origin fixed at a point on the surface of the Earth.

✔ Checkup 5.1

QUESTION 1: To keep a stalled car moving steadily along a level street you have to keep pushing it. Does this contradict Newton's First Law?

QUESTION 2: When you roll a bowling ball on a level surface, you find it gradually slows down. Does this contradict Newton's First Law?

QUESTION 3: A car is traveling at constant speed along a straight, level road. Is the reference frame of this car an inertial reference frame? What if the car rounds a curve at constant speed? What if the car brakes?

QUESTION 4: An elevator is descending at constant speed. Is the reference frame of this elevator an inertial reference frame?

QUESTION 5: A diver is in free fall after jumping off a diving board. Is the reference frame of this diver an inertial reference frame?

QUESTION 6: Which of the following represents an inertial reference frame?

(A) The reference frame of an elevator in free fall (constant acceleration).
(B) The reference frame of a bird descending at constant velocity.
(C) The reference frame of a particle in uniform circular motion.
(D) The reference frame of a car slowing down while coasting uphill.

5.2 NEWTON'S SECOND LAW

Online Concept Tutorial 6

Newton's Second Law of motion establishes the relationship between the force acting on a body and the acceleration caused by this force. This law summarizes experiments and observations on bodies moving under the action of external forces. Qualitatively, a **force** is any push or pull exerted on a body, such as the push of the wind on a sailboat, or the pull of your hand on a doorknob. It is intuitively obvious that such a push or pull has a direction as well as a magnitude—in fact, force is a vector quantity, and it can be represented graphically by an arrow (see Fig. 5.4). For the sake of simplicity, we assume for now that only one force is acting on the body, but we will eliminate this assumption in the next section.

FIGURE 5.4 Man pushing an automobile. The force has a magnitude and a direction.

Newton's Second Law states:

An external force acting on a body gives it an acceleration that is in the direction of the force and has a magnitude directly proportional to the magnitude of the force and inversely proportional to the mass of the body:

$$\mathbf{a} = \frac{\mathbf{F}}{m} \tag{5.1}$$

or

$$m\mathbf{a} = \mathbf{F} \tag{5.2}$$

Newton's Second Law, for single force

According to Eq. (5.1) or (5.2), the acceleration vector is equal to the force vector divided by the mass; thus, this equation specifies both the magnitude and the direction of the acceleration, as asserted by the verbal statement of the law.

The Second Law is subject to the same restrictions as the First Law: it is valid only in inertial reference frames.[1]

[1] The validity of the Second Law requires that the clocks of the inertial reference frame be correctly synchronized. Such a correct synchronization can be achieved by slowly transporting a calibrating clock from place to place in the reference frame or by using light signals and making an allowance for light travel time (this will be discussed in Chapter 36).

Before we deal with applications of the Second Law, we must give the precise definitions of mass and of force. These definitions are contained in the Second Law itself; that is, the Second Law plays a dual role as a law of physics and as a definition of mass and force.

The definition of mass hinges on comparing the unknown mass with a standard mass, which is assumed known. To compare the two masses, we exert forces of identical magnitudes on each, and we measure the accelerations that these forces produce on each. For instance, we might attach identical rubber bands or springs to the unknown mass and to the standard mass, and stretch these rubber bands or springs by identical amounts, thereby producing forces of identical magnitudes. According to the Second Law, *if two bodies of different masses are subjected to forces of identical magnitudes, the accelerations will be in the inverse ratio of the masses.* If we designate the acceleration of the unknown mass by a and that of the standard mass by a_s and their masses by m and m_s, we can express this inverse ratio of masses and accelerations as

definition of mass

$$\frac{m}{m_s} = \frac{a_s}{a} \tag{5.3}$$

This relation serves to define the unknown mass m in terms of the standard mass m_s. The relation says that the unknown mass is large if its acceleration is small. This is of course quite reasonable. A large mass is hard to accelerate—it has a large inertia. If we pull a baseball bat with our rubber band, it will accelerate readily; but if we pull a supertanker, it will hardly accelerate at all. The precise definition given by Eq. (5.3) expresses the intuitive notion that mass is a measure of the resistance that the body offers to changes in its velocity.

kilogram (kg)

pound (lb)

As already mentioned in Section 1.4, the unit of mass in the SI system of units is the **kilogram** (1 kg), and the standard of mass is the standard kilogram, a cylinder of platinum–iridium alloy kept at the International Bureau of Weights and Measures. Table 1.7 gives some examples of masses expressed in kilograms, and Table 1.8 lists multiples and submultiples of the kilogram. Among these submultiples is the **pound** (1 lb = 0.4536 kg), which is a unit of mass in the British system of units.

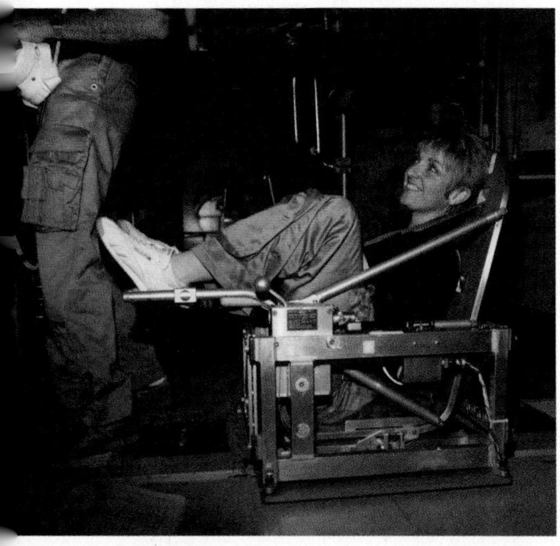

FIGURE 5.5 An astronaut trains on the body-mass measurement device (BMMD).

EXAMPLE 1 During the Skylab mission, three astronauts were kept for about 2 months in weightless conditions. Since an ordinary balance will not work under these conditions, scientists who wanted a daily record of the astronauts' masses had to invent a new mass measurement device. Figure 5.5 shows the device used aboard Skylab. It consisted of a small chair that could be accelerated back and forth by a spring attached to it. Instruments connected to the chair measured the acceleration. With a known standard mass of 66.9 kg placed in the chair, the bent spring produced an acceleration a_s. With the standard mass removed and with astronaut J. R. Lousma sitting in the chair, the bent spring (with the same amount of bending) produced an acceleration a, which was smaller by a factor of 0.779. Deduce the mass of Lousma. Ignore the mass of the chair.

SOLUTION: The bent spring provides the same force F when the standard mass is placed in the chair and when the astronaut is placed in the chair. Consequently, the accelerations must be in the inverse ratio of the masses, as in Eq. (5.3):

$$\frac{m}{m_s} = \frac{a_s}{a}$$

from which

$$m = \frac{a_s}{a} m_s$$

Since the ratio of the measured acceleration is $a_s / a = 1 / 0.779$, we then find

$$m = \frac{a_s}{a} m_s = \frac{1}{0.779} \times 66.9 \text{ kg} = 85.9 \text{ kg} \qquad (5.4)$$

for the mass of Lousma.

The quantitative definition of force also relies on the Second Law. To measure a given force—say, the force generated by a spring that has been stretched a certain amount—we apply this force to the standard kilogram. If the resulting acceleration of the standard kilogram is a_s, then the force has a magnitude

$$F = m_s a_s = 1 \text{ kg} \times a_s \qquad (5.5)$$

After the standard mass has been used to measure the force, any other masses to which this same force is applied will be found to obey the Second Law. In regard to these other masses, the Second Law is an assertion about the physical world that can be verified by experiments—it is a law of physics.

In the SI system of units, the unit of force is the **newton** (N); this is the force that will give a mass of 1 kg an acceleration of 1 m/s^2:

$$1 \text{ newton} = 1 \text{ N} = 1 \text{ kg} \cdot \text{m/s}^2 \qquad (5.6)$$

Table 5.1 lists the magnitudes of some typical forces.

TABLE 5.1 SOME FORCES

Gravitational pull of Sun on Earth	3.5×10^{22} N
Thrust of Saturn V rocket engines (a)	3.3×10^{7} N
Pull of large tugboat	1×10^{6} N
Thrust of jet engines (Boeing 747)	7.7×10^{5} N
Pull of large locomotive	5×10^{5} N
Decelerating force on automobile during braking	1×10^{4} N
Force between two protons in a nucleus	$\approx 10^{4}$ N
Accelerating force on automobile	7×10^{3} N
Gravitational pull of Earth on man	7.3×10^{2} N
Maximum upward force exerted by forearm (isometric)	2.7×10^{2} N
Gravitational pull of Earth on apple (b)	2 N
Gravitational pull of Earth on 5¢ coin	5.1×10^{-2} N
Force between electron and nucleus of atom (hydrogen)	8×10^{-8} N
Force on atomic-force microscope tip	10^{-12} N
Smallest force detected (mechanical oscillator)	10^{-19} N

(a)

(b)

FIGURE 5.6 *Spirit of America* on the Salt Flats of Utah.

EXAMPLE 2 The racing car *Spirit of America* (see Fig. 5.6), which set a world record for speed on the Salt Flats of Utah, had a mass of 4100 kg, and its jet engine could develop up to 68 000 N of thrust. What acceleration could this car achieve?

SOLUTION: According to Newton's Second Law, a horizontal force of magnitude 68 000 N produces an acceleration

$$a = \frac{F}{m} = \frac{68\,000 \text{ N}}{4100 \text{ kg}} = 17 \text{ m/s}^2 \tag{5.7}$$

EXAMPLE 3 Some small animals—locusts, beetles, and fleas—attain very large accelerations while starting a jump. The rat flea attains an acceleration of about 2.0×10^3 m/s^2. Calculate what force the hind legs of the flea must exert on the body while pushing it off with this acceleration. The mass of the flea is about 6.0×10^{-11} kg; neglect the mass of the legs.

SOLUTION: According to Newton's Second Law, the magnitude of the force is

$$F = ma = 6.0 \times 10^{-11} \text{ kg} \times 2.0 \times 10^3 \text{ m/s}^2 = 1.2 \times 10^{-7} \text{ N}$$

In the British system of units, the unit of force is the **pound-force** (lbf), which equals 4.4482 N. In everyday usage, the pound-force is often simply called pound, but we must be careful not to confuse the pound-force (a unit of force) with the pound (a unit of mass). The widespread confusion between the two kinds of pounds stems from their close relationship—the pound-force is the weight of a pound-mass.[2] Confusion is displayed on labels on grocery packages, which typically state "weight 1 lb" when they should state "mass 1 lb." Also, labels on tire gauges state "pressure lb/in^2" when they should state "pressure lbf/in^2."

We now turn to the question of the practical measurement of force and mass. Measurements of force can be conveniently performed with a spring balance (see Fig. 5.7), by matching the unknown force with a known force supplied by a stretched, calibrated spring. Alternatively, measurements of force can be performed by comparing the unknown force with a known weight. Weight is the downward pull that the gravity of the Earth exerts on a body. The weight of a body is proportional to its mass, and standard sets of weights are usually constructed by taking multiples and submultiples of the standard of mass (Fig. 5.8).

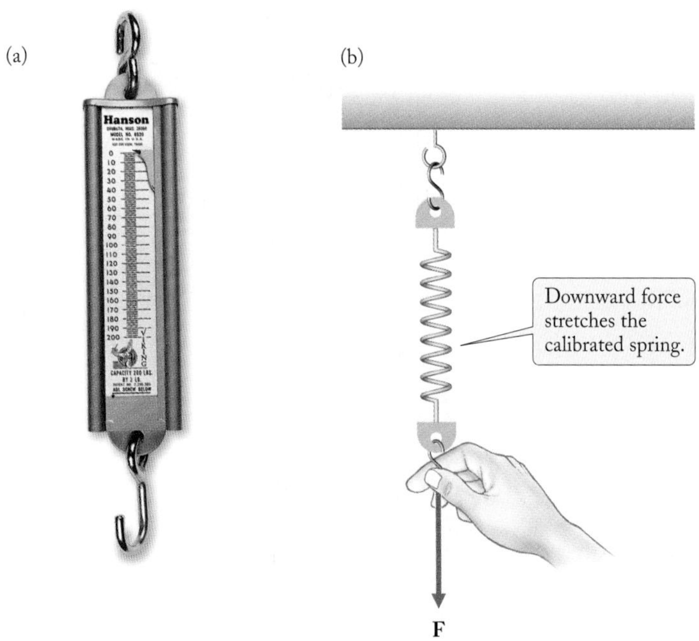

FIGURE 5.7 (a) Spring balance. (b) A spring balance, used to measure an unknown force **F**.

[2] Note that 1 lbf is *not* equal to 1 lb $\times$ 1 ft/s^2; instead, 1 lbf = 1 lb $\times\, g$ = 1 lb $\times$ 32.2 ft/s^2. If you use British units in Newton's Second Law, you must take into account this extra conversion between lb·ft/s^2 and lbf. An alternative British unit of mass is the **slug**; 1 slug = (1 lbf)/(1 ft/s^2) = 32.2 lb. With this unit of mass, Newton's Second Law automatically delivers the correct units, without any need for extra conversions. But the slug is hardly ever used by practicing engineers.

FIGURE 5.8
A set of standard weights.

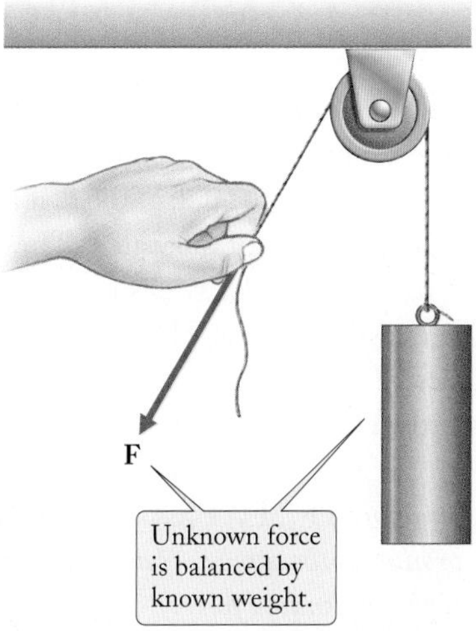

FIGURE 5.9 The unknown force **F** pulling on one end of the string is measured by balancing it with a known weight acting on the other end.

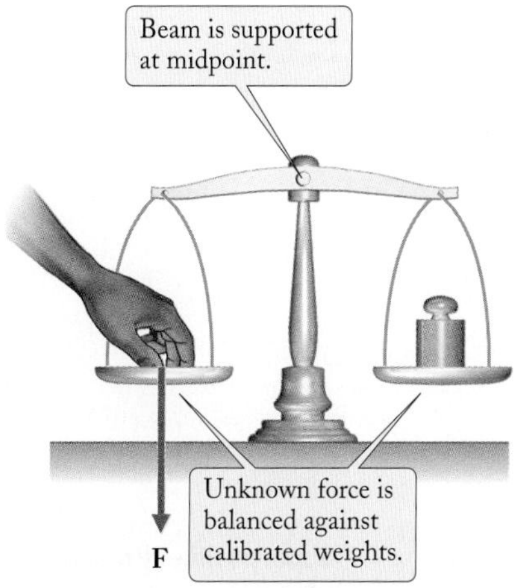

FIGURE 5.10 In a beam balance, the unknown force **F** pushing down on one pan is measured by balancing it with known weights placed in the other pan.

We will discuss the meaning of "weight" in more detail in Section 5.4. A simple comparison of an unknown force and a known weight can be performed by letting the force act on one end of a string while the weight acts on the other end (see Fig. 5.9). A more precise comparison can be achieved with a beam balance (see Fig. 5.10), by letting the force push down on one of the balance pans while a known weight is placed in the other balance pan.

Measurements of mass are commonly carried out with beam balances that compare the weights of the masses. Since the weight of a body is proportional to its mass, measurements of mass via weight give results consistent with those obtained by the primary procedure based on Eq. (5.3).

The masses of electrons and protons and the masses of ions (atoms with missing electrons or added electrons) are too small to be measured by their weight. Instead, they are measured with a procedure based on Eq. (5.2), by applying a known force to the particle, measuring the resulting acceleration, and then calculating the mass. Table 5.2 lists the masses of the electron, the proton, and the neutron.

TABLE 5.2

THE MASSES OF ELECTRONS, PROTONS, AND NEUTRONS

PARTICLE	MASS
Electron	9.11×10^{-31} kg
Proton	1.673×10^{-27} kg
Neutron	1.675×10^{-27} kg

✔ Checkup 5.2

QUESTION 1: Two cardboard boxes rest on a smooth, frictionless table. How can you determine which box has more mass without lifting them off the table?

QUESTION 2: To get your stalled automobile moving, you can either push against the rear end, or pull on the front end. What is the direction of the force in each case?

QUESTION 3: Tired of waiting for the wind, a sailor decides to stand up in his sailboat and push on the mast. Will this push accelerate the sailboat?

QUESTION 4: A bobsled slides on flat ice, without friction. A man pushing the empty bobsled as hard as possible gives it an acceleration of 4 m/s^2. What will be the acceleration of the sled if two men push on it equally? What will be the acceleration of a loaded bobsled, of twice the mass of the empty sled, if one man pushes? If two men push? Choose among the following respective quantities in units of m/s^2:

(A) 8, 1, 2 (B) 8, 1, 4 (C) 8, 2, 4 (D) 16, 1, 2 (E) 16, 1, 4

Online Concept Tutorial 7

5.3 THE COMBINATION OF FORCES

More often than not, a body will be subjected to the simultaneous action of several forces. For example, Fig. 5.11 shows a barge under tow by two tugboats. The forces acting on the barge are the pull of the first towrope, the pull of the second towrope, and the frictional resistance of the water.[3] These forces are indicated by the arrows in Fig. 5.11. Newton's Second Law tells us what each of these forces would do if acting by itself. The question now is, How can we calculate the simultaneous effect of two or more forces? The answer is supplied by an addition principle for forces, called the **Superposition Principle** for forces:

If several forces $\mathbf{F}_1, \mathbf{F}_2, \mathbf{F}_3, \ldots$ *act simultaneously on a body, then the acceleration they produce is the same as that produced by a single force* $\mathbf{F}_{\text{net}}$ *given by the vector sum of the individual forces,*

Superposition Principle

$$\mathbf{F}_{\text{net}} = \mathbf{F}_1 + \mathbf{F}_2 + \mathbf{F}_3 + \cdots \quad (5.8)$$

The single force $\mathbf{F}_{\text{net}}$ that has the same effect as the combination of the individual forces is called the **net force**, or the **resultant force**. The net force then determines the acceleration, and Newton's Second Law takes the form

Newton's Second Law, for net force

$$m\mathbf{a} = \mathbf{F}_{\text{net}} \quad (5.9)$$

We must emphasize that this Superposition Principle is a law of physics, which has the same status as Newton's laws. Crude tests of this principle can be performed in laboratory experiments by pulling on a body with known forces in known directions. But the most precise empirical test of this principle emerges from the study of planetary motion; there it is found that the net force on a planet is indeed the vector sum of all the gravitational pulls exerted by the Sun and by the other planets.

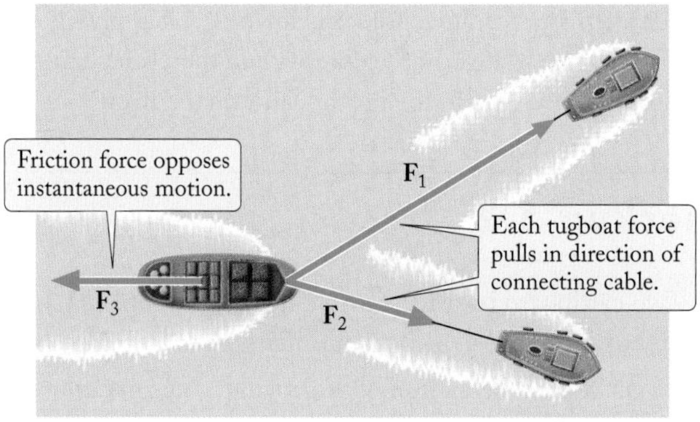

FIGURE 5.11 A barge under tow by two tugboats. $\mathbf{F}_1$ and $\mathbf{F}_2$ are the forces exerted by the tugboats; $\mathbf{F}_3$ is the frictional resistance of the water.

[3] These are the horizontal forces. There are also vertical forces: the downward pull of gravity (the weight) and the upward pressure of the water (the buoyancy). The vertical forces can be ignored, since they cancel each other, and do not contribute to the net force.

EXAMPLE 4 Suppose that the two towropes in Fig. 5.11 pull with horizontal forces of 2.5×10^5 N and 1.0×10^5 N, respectively, and that these forces make angles of 30° and 15° with the long axis of the barge (see Fig. 5.12). Suppose that the friction force is zero. What are the magnitude and direction of the net horizontal force the towropes exert on the barge?

SOLUTION: The net force is the vector sum

$$\mathbf{F}_{\text{net}} = \mathbf{F}_1 + \mathbf{F}_2 \tag{5.10}$$

where $\mathbf{F}_1$ is the force of the first towrope and $\mathbf{F}_2$ that of the second. The net force is shown in Fig. 5.12a. With the x and y axes arranged as in Fig. 5.12a, the forces can be resolved into x and y components. The x component of the net force is the sum of the x components of the individual forces (see Fig. 5.12b),

$$\begin{aligned} F_{\text{net},x} &= F_{1,x} + F_{2,x} \\ &= 2.5 \times 10^5\ \text{N} \times \cos 30^\circ + 1.0 \times 10^5\ \text{N} \times \cos 15^\circ \\ &= 2.5 \times 10^5\ \text{N} \times 0.866 + 1.0 \times 10^5\ \text{N} \times 0.966 \\ &= 3.1 \times 10^5\ \text{N} \end{aligned} \tag{5.11}$$

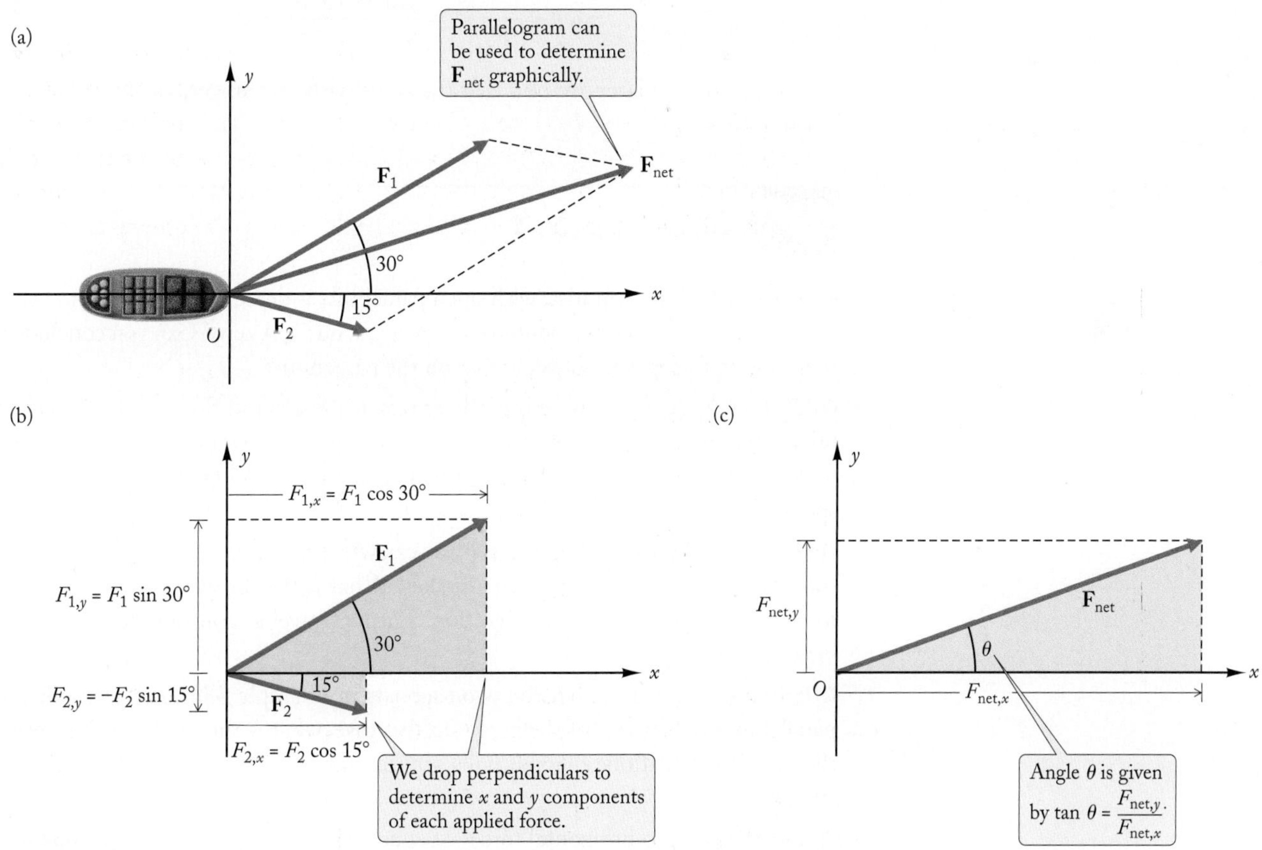

FIGURE 5.12 (a) One tugboat pulls with a force $\mathbf{F}_1$, and the other pulls with a force $\mathbf{F}_2$. The magnitudes of these forces are $F_1 = 2.5 \times 10^5$ N and $F_2 = 1.0 \times 10^5$ N, respectively. The net force $\mathbf{F}_{\text{net}}$ is the vector sum of the two forces $\mathbf{F}_1$ and $\mathbf{F}_2$. (b) The forces $\mathbf{F}_1$ and $\mathbf{F}_2$ and their x and y components. (c) The net force $\mathbf{F}_{\text{net}}$ and its x and y components.

Likewise, the y component of the net force is the sum of the y components of the individual forces,

$$\begin{aligned} F_{\text{net},y} &= F_{1,y} + F_{2,y} \\ &= 2.5 \times 10^5 \text{ N} \times \sin 30^\circ - 1.0 \times 10^5 \text{ N} \times \sin 15^\circ \\ &= 2.5 \times 10^5 \text{ N} \times 0.500 - 1.0 \times 10^5 \text{ N} \times 0.259 \\ &= 1.0 \times 10^5 \text{ N} \end{aligned} \tag{5.12}$$

The y components of the individual forces are of opposite sign because one tugboat pulls the barge to the left (up in Fig. 5.12) and the other to the right (down in Fig. 5.12).

The components $F_{\text{net},x}$ and $F_{\text{net},y}$ uniquely specify the net force, and we could end our calculation of the net force with these components. However, the problem asks for the magnitude and the direction of the net force, and we therefore have to take our calculation a step further. According to Eq. (3.15), the magnitude of the net force is the square root of the sum of squares of the components:

$$\begin{aligned} F_{\text{net}} &= \sqrt{(F_{\text{net},x})^2 + (F_{\text{net},y})^2} \\ &= \sqrt{(3.1 \times 10^5 \text{ N})^2 + (1.0 \times 10^5 \text{ N})^2} = 3.3 \times 10^5 \text{ N} \end{aligned} \tag{5.13}$$

The direction of the net force makes an angle θ with the x axis (see Fig. 5.12c). According to Eq. (3.16), this angle is given by

$$\tan \theta = \frac{F_{\text{net},y}}{F_{\text{net},x}} = \frac{1.0 \times 10^5 \text{ N}}{3.1 \times 10^5 \text{ N}} = 0.32 \tag{5.14}$$

With our calculator, we find that the angle with this tangent is 18°.

✔ Checkup 5.3

QUESTION 1: A parachutist, with open parachute, is descending at uniform velocity. Can you conclude that the net force on the parachutist is zero? Can you conclude that there are no forces whatsoever acting on the parachutist?

QUESTION 2: An elevator is initially at rest at the ground floor. It then accelerates briefly, and then continues to ascend at constant speed. What is the direction of the net force on the elevator when at rest? When accelerating? When ascending at constant speed?

QUESTION 3: You are riding in a subway car, which accelerates, then proceeds at constant velocity for a while, and then brakes. What is the direction of the net external force on your body during acceleration? During travel at constant velocity? During braking?

QUESTION 4: Suppose that the two tugboats in Example 4 both pull in a direction parallel to the long axis of the barge. In this case, what is the net force they exert on the barge? What if one tugboat pulls at the front of the barge and the other pushes from behind?

QUESTION 5: Two horizontal forces of equal magnitudes are acting on a box sliding on a smooth table. The direction of one force is 30° west of north, the other is in the west direction. What is the direction of the acceleration of the box?

(A) 15° north of west (B) 30° north of west (C) Directly northwest
(D) 30° west of north (E) 15° west of north

5.4 WEIGHT; CONTACT FORCE AND NORMAL FORCE

Online Concept Tutorial 6

The gravity of the Earth is the most familiar of all forces. When you hold a body, say, an apple, in your hands, you can feel the downward pull of gravity on the apple; and if you release the apple, you can see it accelerating under the influence of this pull. In the terminology of physics, the pull of gravity on a body is called the **weight** of the body. Thus, weight is a force; it is a vector quantity—it has a direction (downward) as well as a magnitude. The unit of weight is the unit of force, that is, the newton (N).

The magnitude of the weight force is directly proportional to the mass of the body. To understand this, consider a body of mass m in free fall near the surface of the Earth, say, an apple you have released from your hand (see Fig. 5.13). The body has a downward acceleration g. Since we attribute this acceleration to the weight force, Newton's Second Law tells us that the magnitude of the weight force acting on the body must be

$$F = ma = mg \tag{5.15}$$

We will denote the weight by the vector symbol **w**. According to Eq. (5.15), the magnitude of the weight is

$$w = mg \tag{5.16}$$

weight

If the body is not in free fall but is held in a stationary position by some support, then the weight is of course still the same as that given by Eq. (5.16); however, the support balances the downward weight force and prevents it from producing a downward motion.

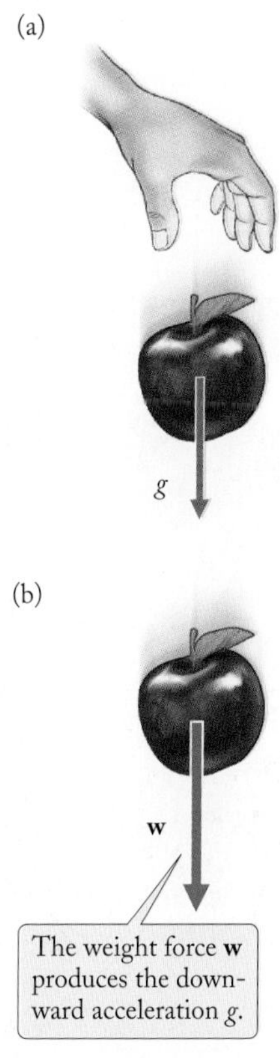

FIGURE 5.13 (a) An apple in free fall has a downward acceleration g. (b) The force on the apple is also downward, and it has a magnitude $w = mg$.

EXAMPLE 5 What is the weight of a 54-kg woman? Assume that $g = 9.81\ \text{m/s}^2$.

SOLUTION: By Eq. (5.16), the magnitude of the weight is

$$w = mg = 54\ \text{kg} \times 9.81\ \text{m/s}^2 = 530\ \text{N}$$

and its direction is downward.

COMMENT: Since the value of g depends on location, the weight of a body also depends on its location. For example, if the 54-kg woman travels from London ($g = 9.81\ \text{m/s}^2$) to Hong Kong ($g = 9.79\ \text{m/s}^2$), her weight will decrease from 530 N to 529 N, a difference of 1 N. And if this woman were to travel to the Moon ($g = 1.62\ \text{m/s}^2$), her weight would decrease to 87 N!

The preceding example illustrates an essential distinction between mass and weight. *Mass is an intrinsic property of a body,* measuring the inertial resistance with which the body opposes changes in its motion. The definition of mass is formulated in such a way that a given body has the same mass regardless of its location in the universe. *Weight is an extrinsic property of a body,* measuring the pull of gravity on the body. It depends on the (gravitational) environment in which the body is located, and it therefore depends on location.

A body deep in intergalactic space, far from the gravitational pull of any star or planet, will experience hardly any gravitational pull—the weight of the body will be nearly zero; that is, the body will be weightless. Although such a condition of true

FIGURE 5.14 This soccer player has jumped into the air. Both he and the ball are in free fall.

FIGURE 5.15 Astronauts training in an airplane.

weightlessness is impossible at any location on or near the Earth, *a condition of apparent weightlessness can be simulated on or near the Earth by means of a freely falling reference frame.* Consider an observer in free fall, such as the soccer player in Fig. 5.14, who has jumped into the air to kick a ball. The player and the ball both accelerate downward at the same rate; thus, the ball does not accelerate relative to the player. In the reference frame of the player (a freely falling reference frame accelerating downward with the acceleration g), the freely falling ball, or any other freely falling body, continues to move with constant velocity, as though there were no force acting on it. This means that in such a reference frame, the gravitational pull is *apparently* zero; the weight is *apparently* zero. Of course, this simulated weightlessness arises from the accelerated motion of the reference frame—in the unaccelerated, inertial reference frame of the ground, the weight of the ball is certainly not zero. Nevertheless, if the player insists on looking at things from his own reference frame, he will judge the weight of the ball, and the weight of his own body, as zero. This condition of weightlessness is also simulated within an airplane flying along a parabola, imitating the motion of a (frictionless) projectile (see Fig. 5.15); and it is also simulated in a spacecraft orbiting the Earth (see Fig. 5.16). Both of these motions are free-fall motions.

FIGURE 5.16 Astronaut floating in the cargo bay of the Space Shuttle.

The gravity of the Earth reaches from the Earth to any other body, even a body placed high above the surface. Gravity bridges empty space and requires no perceptible medium for its transmission. In contrast, most of the other forces familiar from everyday experience require direct contact between the bodies. You cannot exert a push on a box unless your hand is in contact with the box; and the box cannot exert a push on the floor unless it is in contact with the floor. The push that the surface of a body exerts on the adjacent surface of another body is called a **contact force**. If the two bodies are solid, the contact forces between their adjacent surfaces are of two kinds: the compressional force, or normal force, that arises when the surfaces are pressed together perpendicularly, and the friction force that arises when the surfaces are sliding, or trying to slide, past each other. We will deal with frictional forces in the next chapter.

The **normal force** represents the resistance that solid bodies offer to interpenetration. When you try to push two bodies together, their surfaces begin to repel as soon as they come into contact. You can feel this kind of repulsive contact force when you push with your hand or your foot against any solid surface. For instance, if you push with your hand horizontally against a wall, you can feel the wall pushing against your hand, stopping your hand from penetrating the wall (see Fig. 5.17). This push of the wall is called a normal force, because it is "normal," meaning perpendicular, to the wall. This normal force arises from the contact between the atoms of your hand and the atoms of the wall; the atoms of your hand and the atoms on the surface of the wall exert repulsive forces on each other, which oppose their interpenetration.

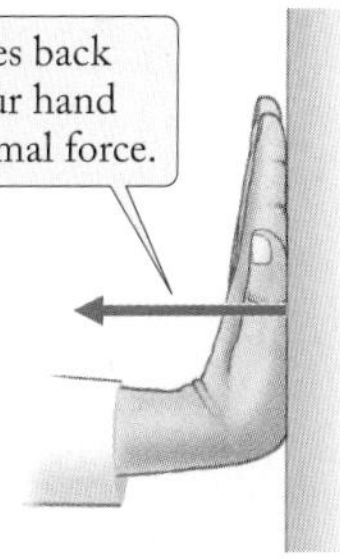

FIGURE 5.17 When you push against a wall with your hand, the wall pushes back against your hand and resists penetration.

How does the wall succeed in preventing your hand from penetrating the wall, regardless of how hard you push? The resistance offered by the wall results from a slight compression of the material of the wall. The atoms in the material in the wall behave like an array of miniature springs; these atomic springs compress slightly when you push your hand against them, and the force that these springs exert on your hand increases with the amount of compression. Hence your hand compresses the wall until the increasing force of the atomic springs stops your hand. If the material of the wall is hard—for example, concrete—the amount of compression is so slight as to be unnoticeable, and the wall seems impenetrable.

✔ Checkup 5.4

QUESTION 1: You throw a 1.0-kg stone straight up. What is the force of gravity on the stone while it is traveling upward? When it is instantaneously at rest at the top of its trajectory? When it is traveling back down?

QUESTION 2: A star deep in intergalactic space, far from the gravitational pull of any other star or planet, is weightless. Is it also massless?

QUESTION 3: The accelerations due to gravity on the surface of the Earth, the Moon, and Jupiter are 9.81 m/s^2, 1.62 m/s^2, and 24.8 m/s^2, respectively. Where would your weight be largest? Smallest?

QUESTION 4: An astronaut and her spacecraft are initially at rest on the launchpad. Then the rocket engines fire, and the spacecraft lifts off and ascends. After some minutes, the rocket engines cut off, and the spacecraft coasts through empty space. At what point will the astronaut begin to experience (apparent) weightlessness?

QUESTION 5: A book with a weight of 50 N lies on a table. What is the normal force that the table exerts on the book? If we place a second, identical book on top of the first, what is the normal force that the table exerts on the first book? What is the normal force that the first book exerts on the second?

QUESTION 6: An elevator traveling upward decelerates to stop at a floor. Is the normal force on the feet of a passenger during the deceleration larger or smaller than her weight? Another elevator traveling downward decelerates to stop at a floor. Is the normal force on the feet of a passenger during the deceleration larger or smaller than his weight?

Concepts in Context

(A) Larger; larger
(B) Larger; smaller
(C) Smaller; smaller
(D) Smaller; larger

Online 7 Concept Tutorial

5.5 NEWTON'S THIRD LAW

When you push with your hand against a body, such as a wall, the body pushes back at you. Thus, the mutual interaction of your hand and the wall involves two normal forces: the "action" force of the hand on the wall and the "reaction" force of the wall on the hand (see Fig. 5.18). These forces are said to form an **action–reaction pair.** Which of the forces is regarded as "action" and which as "reaction" is irrelevant. It may seem reasonable to regard the push of the hand as an action; then the push of the wall is a reaction. However, it is equally valid to regard the push of the wall on the hand as an action, and then the push of the hand on the wall is a reaction. At the microscopic level, both the hand and the wall consist of atoms, and when two atoms exert forces on each other, it is equally valid to regard the first atom as "acting" and the second as "reacting" or vice versa. The important point is that forces always occur in pairs; each of them cannot exist without the other. This is true not only for normal forces, as in the example of the hand and the wall, but for all forces.

Newton's Third Law gives the quantitative relationship between the action force and the reaction force:

Newton's Third Law

Whenever a body exerts a force on another body, the latter exerts a force of equal magnitude and opposite direction on the former.

For instance, if the push of your hand on a wall has a magnitude of 60 N and is perpendicular to the wall, then the push of the wall on your hand also has a magnitude of 60 N and is also perpendicular to the wall, but in the opposite direction (both your push and the push of the wall are normal forces). But if the push of your stationary hand on the wall makes an angle with the wall (see Fig. 5.19), then the push of the wall on your hand makes a corresponding angle (both your push and the push of the wall are then a combination of normal and friction forces).

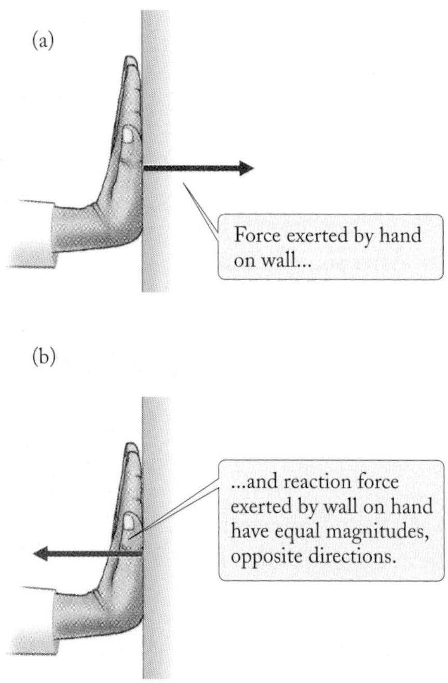

FIGURE 5.18 (a) Hand pushes on wall; (b) wall pushes on hand.

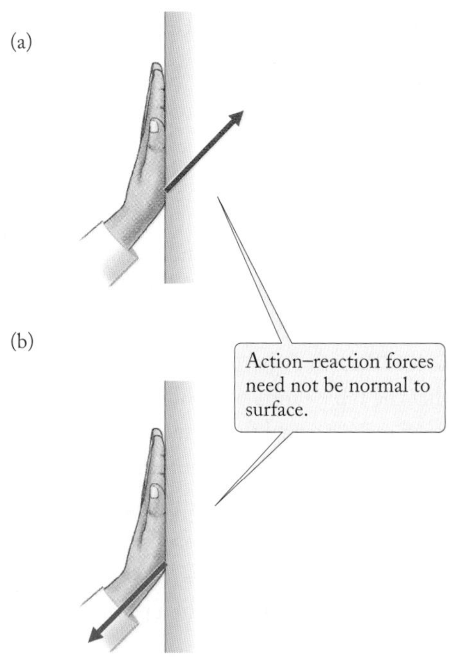

FIGURE 5.19 (a) Hand pushes on wall at an angle; (b) wall pushes on hand at an angle.

The equality of the magnitudes of action and reaction is valid even if the body you push against is not held in a fixed position (like a wall) but is free to move. For example, if you push on a cart with a force of 60 N, the cart will push back on you with a force of 60 N (see Fig. 5.20), even while the cart accelerates away from you. Note that although these action and reaction forces are of equal magnitudes, they act on different bodies and their effects are quite different: the first force gives an acceleration to the cart (if there is no other force acting on the cart), whereas the second force merely slows your hand and prevents it from accelerating as much as it would if the cart were not there. *Thus, although action and reaction are forces of equal magnitudes and of opposite directions, their effects do not cancel because they act on different bodies.*

FIGURE 5.20 Woman pushes on cart; cart pushes on woman.

We can express Newton's Third Law mathematically simply by equating the force exerted by a first body on a second body with the negative of the force exerted by the second body on the first body:

action and reaction forces

$$\mathbf{F}_{1\text{ on }2} = -\mathbf{F}_{2\text{ on }1} \qquad (5.17)$$

Reaction forces play a crucial role in all animals and machines that produce locomotion by pushing against the ground, water, or air. For example, a man walks by pushing backward on the ground; the reaction of the ground then pushes the man forward (see Fig. 5.21). An automobile moves by pushing backward on the ground with its wheels; the reaction of the ground then pushes the automobile forward (see Fig. 5.22). A tugboat moves by pushing backward against the water with its propeller; the reaction of the water on the propeller then pushes the tugboat forward (see Fig. 5.23). Even the propulsion of a jet aircraft or a rocket relies on reaction forces. The rocket engine expels exhaust gases; the reaction of the exhaust gases then pushes the engine and the rocket forward (see Fig. 5.24). The atmosphere is of no help in rocket propulsion; rather it is a hindrance, since it exerts a frictional resistance on the rocket.

FIGURE 5.21 Man pushes against ground; ground pushes against man.

FIGURE 5.22 Automobile pushes against ground; ground pushes against automobile.

FIGURE 5.23 Propeller pushes against water; water pushes against propeller.

Reaction forces exist even if the two interacting bodies are not in direct contact, so the forces between them must bridge the intervening empty space. For instance, consider an apple in free fall at some height above the ground. The Earth pulls on the apple by means of gravity. If this pull has a magnitude of, say, 2 N, then Newton's Third Law requires that the apple pull on the Earth with an opposite force of 2 N (see Fig. 5.25). This reaction force is also a form of gravity—it is the gravity that the apple exerts on the Earth. However, the effect of the apple on the motion of the Earth is insignificant because the mass of the Earth is so large that a force of only 2 N produces only a negligible acceleration of the Earth.

Keep in mind that although the two forces in an action–reaction pair are always of equal magnitudes and opposite directions, two forces of equal magnitudes and opposite directions are not always an action–reaction pair. For instance, consider a box of mass m sitting on the floor (see Fig. 5.26). There are two forces acting on the box: the weight $\mathbf{w}$ of the box pointing downward in the vertical direction, and the normal force of the floor $\mathbf{N}$ pointing upward in the vertical direction. Figure 5.27 shows the box and these two forces $\mathbf{w}$ and $\mathbf{N}$ acting on it. Since the box is supposed to remain at rest, the net force on the box must be zero, which requires that the two forces $\mathbf{w}$ and $\mathbf{N}$ have equal magnitudes. The magnitude of the weight is mg, and therefore the magnitude of the normal force must also be mg. However, although $\mathbf{w}$ and $\mathbf{N}$ are of equal magnitudes, they are *not* an action–reaction pair. Instead, the normal force $\mathbf{N}$ exerted by the floor upward on the box forms an action–reaction pair with the normal force $\mathbf{N}'$ exerted by the box downward on the floor; and the weight $\mathbf{w}$, or the gravitational pull exerted by the Earth on the box, forms an action–reaction pair with the gravitational pull $\mathbf{w}'$ exerted by the box on the Earth.

A diagram such as Fig. 5.27 that shows a body and all the external forces acting on the body, but not the reaction forces that the body exerts on its environment, is called a **"free-body" diagram**. (In this context *free* does not mean free of forces; it means that the body is shown free of its environment, and this environment is represented by the forces it exerts.) Thus, the floor on which the box rests is not shown in Fig. 5.27—the effects of the floor are entirely contained in the normal force $\mathbf{N}$. The "free-body" diagram eliminates clutter and helps us to focus on the body and on the forces that we need to formulate the equation of motion of the body.

Note that Figs. 5.20–5.25 are not "free-body" diagrams; for example, Figs. 5.21, 5.22, and 5.25 show the forces on the body (man, or car, or apple) and also the reaction forces on the ground in the same diagram.

FIGURE 5.24 Rocket pushes against exhaust gases; exhaust gases push against rocket.

FIGURE 5.25 Earth pulls on apple; apple pulls on Earth.

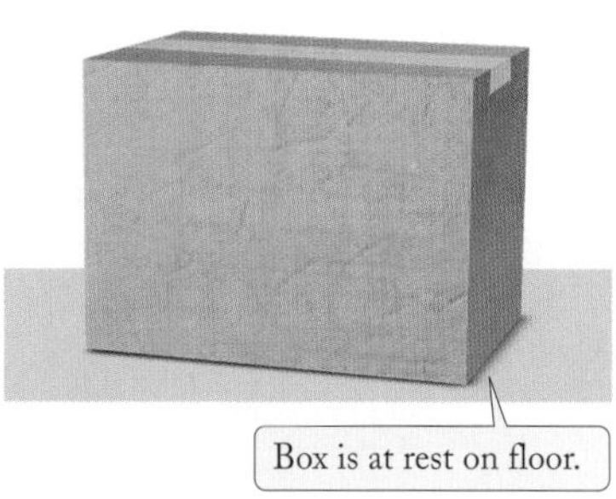

FIGURE 5.26 A box resting on a floor.

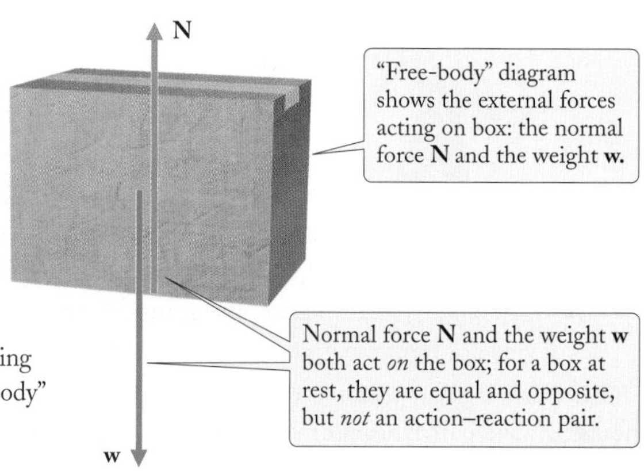

FIGURE 5.27 Forces acting on the box. This is a "free-body" diagram for the box.

EXAMPLE 6 A man of mass 75.0 kg is standing in an elevator which is accelerating upward at 2.00 m/s^2 (see Fig. 5.28). What is the normal force that the floor of the elevator exerts on the man? What is the normal force that the man exerts on the floor?

Concepts — in — Context

FIGURE 5.28 A man standing in an elevator accelerating upward.

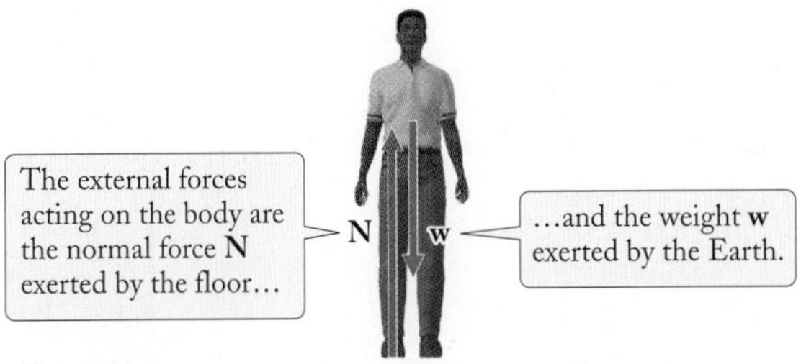

FIGURE 5.29 "Free-body" diagram for the man.

SOLUTION: The two forces on the man are his weight and the normal force; thus we need only consider vertical forces. These forces are shown in the "free-body" diagram of Fig. 5.29. The net force on the man is

$$F_{net} = N - mg$$

where the forces are regarded as positive when directed upward. Since the net force F_{net} gives the man an acceleration a, Newton's Second Law tells us that

$$ma = N - mg$$

or

$$N = ma + mg$$

Hence

$$N = 75.0\text{ kg} \times 2.00\text{ m/s}^2 + 75.0\text{ kg} \times 9.81\text{ m/s}^2 = 150\text{ N} + 736\text{ N} = 886\text{ N}$$

Thus, the normal force on the man is *larger* than his weight by 150 N.

According to Newton's Third Law, the normal reaction force that the man exerts downward on the floor of the elevator has the same magnitude, also 886 N, since the normal forces on the elevator and on the man form an action–reaction pair. If the man were standing on a spring balance (bathroom scale), the balance would register this larger "weight" of 886 N, as though gravity had increased, and the spring balance would indicate a reading of (886 N)/(9.81 m/s^2) = 90 kg, instead of 75 kg. Note that the direction of the velocity of the elevator is irrelevant; only the direction of the acceleration matters. If the elevator were descending and braking (again, a positive acceleration!), the result would be the same.

EXAMPLE 7 A tugboat tows a barge of mass 50 000 kg by means of a cable (see Fig. 5.30a). If the tugboat exerts a horizontal pull of 6000 N on the cable, what is the acceleration of the cable and the barge? What is the magnitude of the pull that the cable exerts on the barge? Assume that the mass of the cable can be neglected (i.e., assume the cable is practically massless), and ignore the friction of the water on the barge.

SOLUTION: Before drawing the "free-body" diagram, we must decide what is our "body." We could take the barge as our body, or the cable, or both jointly. Since the barge and the cable accelerate jointly, it will be best to take the barge and cable jointly as our body. The "free-body" diagram for this body is shown in Fig. 5.30b (only horizontal forces have been included in this diagram). The force of 6000 N exerted by the tugboat accelerates both the cable and the barge; that is, it accelerates a total mass of 50 000 kg. Hence, the resulting acceleration is

$$a = \frac{F}{m} = \frac{6000\ \text{N}}{50\,000\ \text{kg}} = 0.12\ \text{m/s}^2$$

By Newton's Third Law, the pull of the cable on the barge has the same magnitude as the pull of the barge on the cable. To find this magnitude we can examine the "free-body" diagram either for the barge or for the cable. Let us choose the cable; the "free-body" diagram for this body is shown in Fig. 5.31. The tugboat pulls at the forward end of the cable with a force of 6000 N, and the barge pulls at the rearward end. For a cable of zero mass, the net force on the cable must be zero (for a body of zero mass, $F = ma = 0 \times a = 0$). Hence the pull of the barge on the rearward end of the cable must match the pull of the tugboat at the forward end—both pulls must be 6000 N. Newton's Third Law then requires that the cable pull on the barge with a force of 6000 N.

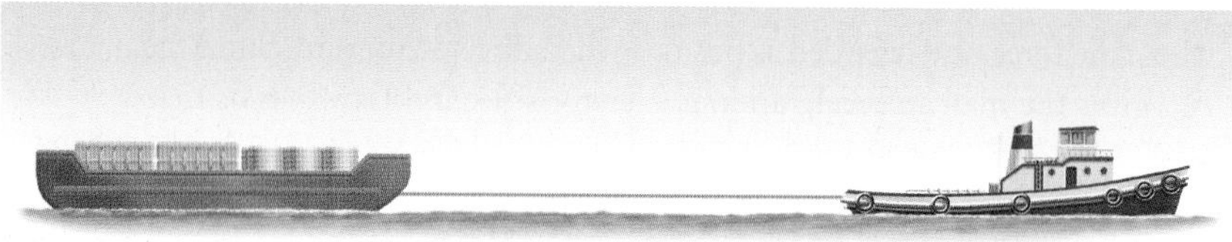

FIGURE 5.30(a) A tugboat tows a barge.

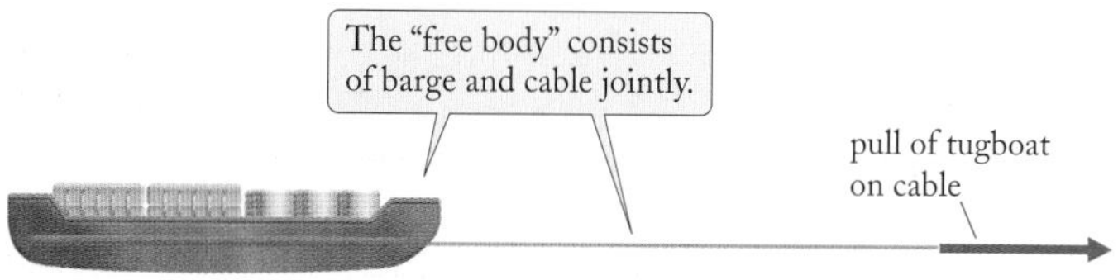

FIGURE 5.30(b) "Free-body" diagram for barge and cable.

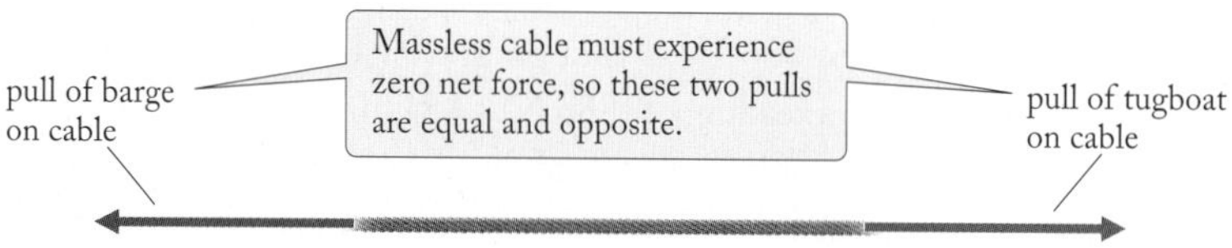

FIGURE 5.31 "Free body" diagram for the forces acting on the cable, showing both the external pulls exerted by the tugboat and the barge.

COMMENT: Note that although the pulls at the forward and rearward ends of the cable are two forces of equal magnitudes and opposite directions, these two forces are *not* an action–reaction pair. The pull at the forward end forms an action–reaction pair with the pull exerted by the cable on the tugboat, and the pull at the backward end forms an action–reaction pair with the pull exerted by the cable on the barge. These pulls of the cable are shown in red in Fig. 5.32.

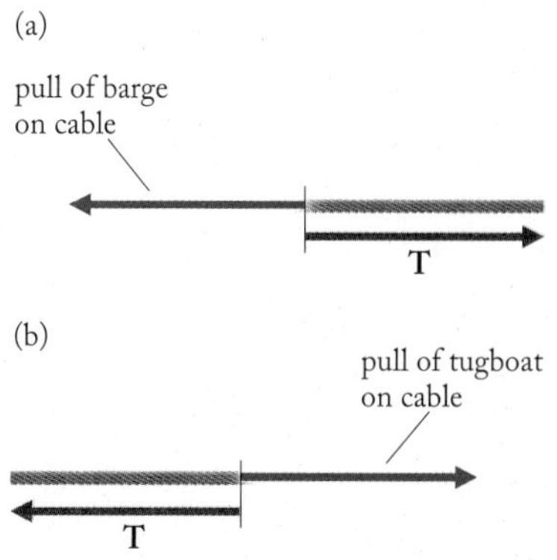

FIGURE 5.32 Action-reaction pairs. (a) Barge pulls on cable; cable pulls on barge with tension **T**. (b) Tugboat pulls on cable; cable pulls on tugboat with tension **T**.

The force with which a cable pulls on what is attached to it is called the **tension T**. *The direction of the tension is along the cable.* Thus, in Fig. 5.32, the magnitude of the tension of the cable is $T = 6000$ N at its forward end and also $T = 6000$ N at its rearward end. This equality of the magnitudes of the tensions at the forward and the rearward ends of the cable is a consequence of neglecting the mass of the cable. If we were to take into account the mass of the cable, then a net force would be needed to accelerate the cable. This means that the force pulling on the forward end of the cable would have to be *larger* than the force pulling on the rearward end—by Newton's Third Law, the tensions would be of unequal magnitudes.

For practical purposes, the mass of a cable, rope, string, wire, or chain can often be neglected compared with the mass of the body to which it is attached. Under these conditions, *the cable transmits the magnitude of the tension without change.* In subsequent problems we will always neglect the mass of the cable unless we explicitly state otherwise. The transmission of the magnitude of the tension without change occurs even if the cable is led around (frictionless and massless) pulleys, which change only the direction of the pull, as illustrated in the following example. We examine a frictional, locked pulley later in Example 10. We will examine pulleys that are not massless much later, in Chapter 13.

EXAMPLE 8 Figure 5.33 shows a traction apparatus used in hospitals to exert a steady pull on a broken leg, in order to keep the bones aligned. The middle pulley is attached to the cast, and the other two pulleys are attached to the bed or the wall. A flexible wire passes over these pulleys, and a brick hang ing from this wire provides a tension. The upper and the lower portions of the wire are oriented, respectively, upward and downward from the middle pulley at angles of 35° with respect to the horizontal. If the horizontal pull on the leg is to be 50 N, what tension must the brick provide at the end of the wire?

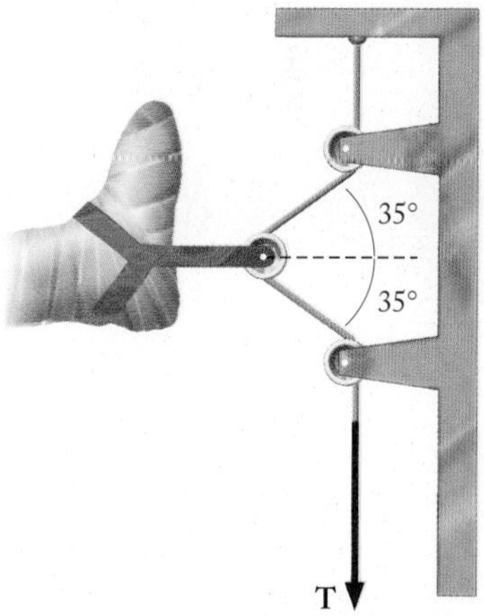

FIGURE 5.33 Traction apparatus.

SOLUTION: As discussed above, the tension is constant along the entire wire. If the magnitude of the tension at the lower end of the wire is T, the magnitude of the tension at all other points of the wire must also be T. Under static conditions, the upper and the lower portions of the wire may be regarded as attached to the middle pulley at the points of first contact. Thus the upper portion of the wire pulls upward at an angle of 35° with a force $\mathbf{T}_1$ of magnitude T, and the lower portion pulls downward at an angle of 35° with a force $\mathbf{T}_2$ of the same magnitude.

Figure 5.34a shows these forces $\mathbf{T}_1$ and $\mathbf{T}_2$ that the wire exerts on the middle pulley. The x axis is horizontal, and the y axis vertical. The vertical components of these forces cancel, since they have opposite signs and equal magnitudes. The horizontal components add, since they both have positive signs. The resultant force exerted by the wire on the middle pulley is therefore in the horizontal direction, that is, the x direction.

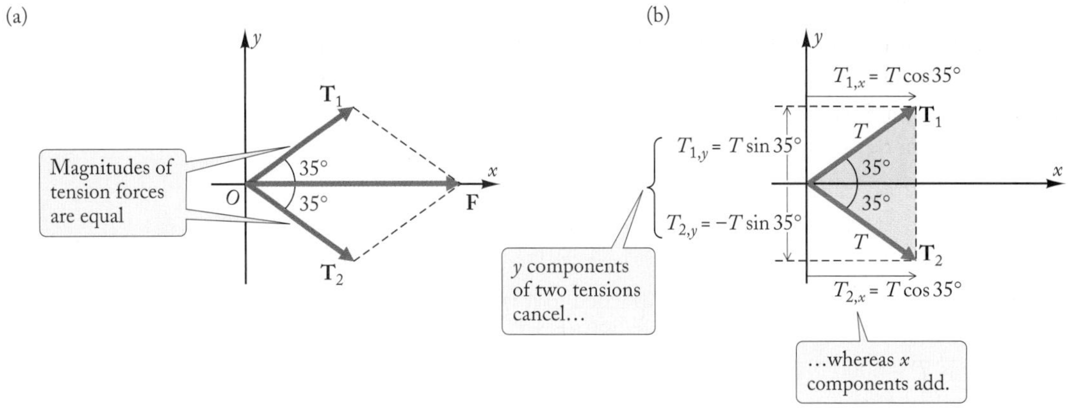

FIGURE 5.34 (a) Forces exerted by the wire on the middle pulley. The upper portion of the wire pulls with a tension $\mathbf{T}_1$ and the lower portion pulls with a tension $\mathbf{T}_2$. The magnitudes of these forces are equal: $T_1 = T_2 = T$. The resultant $\mathbf{F}$ is in the x direction. (b) Components of these forces.

The components of $\mathbf{T}_1$ and $\mathbf{T}_2$ in the x direction are $T \cos 35°$ (see Fig. 5.34b). Hence the x component of the resultant force is

$$F_x = T_{1,x} + T_{2,x} = T\cos 35° + T\cos 35° = 2T\cos 35° \tag{5.18}$$

Since the magnitude of F_x is supposed to be 50 N, we obtain

$$T = \frac{F_x}{2\cos 35°} = \frac{50\text{ N}}{2 \times 0.819} = 31\text{ N} \tag{5.19}$$

Therefore the brick must provide a tension of 31 N.

✔ Checkup 5.5

QUESTION 1: Draw a "free-body" diagram for an apple in free fall. Draw a "free-body" diagram for an apple at rest on a table.

QUESTION 2: An apple of weight 2 N hangs from the branch of a tree. The two forces on the apple are the weight and the upward pull exerted by the branch. If these two forces are regarded as actions, what are the reactions?

QUESTION 3: While driving a car, you accelerate, then proceed at constant velocity for a while, and then brake. What is the direction of the net external force on the car during acceleration? During travel at constant velocity? During braking? What external body exerts the force on the car during acceleration and during braking?

QUESTION 4: While sitting at the edge of a dock, you push with your feet against a supertanker with a force of 400 N. What is the force with which the supertanker pushes against your feet? If, instead, you push against a rowboat with a force of 400 N, what is the force with which the rowboat pushes against your feet? Is there any difference in the behaviors of the supertanker and the rowboat?

QUESTION 5: A book of weight 50 N lies on a table, and a second book lies on top of the first. Draw the "free-body" diagram for each book. List all the forces acting on the books. Which of these forces are action–reaction pairs?

QUESTION 6: A man pulls with a force of 150 N on one end of a rope, and a woman with a force of 150 N on the other end. What is the tension in the rope? If the woman now ties her end of the rope to a tree and walks away, while the man continues pulling, what will be the tension in the rope?

QUESTION 7: A mass of 10 kg hangs on a rope attached to a spring scale which hangs from the ceiling by a second rope (Fig. 5.35). Assume that the masses of the ropes and of the spring scale can be neglected. What is the tension in the first rope? What is the tension in the second rope? What is the weight (in N) on the spring scale? Choose among the following respective quantities:

(A) 0 N; 98 N; 98 N (B) 98 N; 0 N; 98 N
(C) 98 N; 98 N; 0 N (D) 98 N; 98 N; 98 N

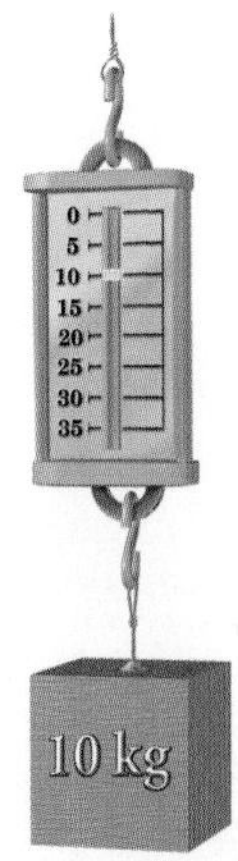

FIGURE 5.35 A 10-kg mass hanging on a spring balance.

Online Concept Tutorial 7

5.6 MOTION WITH A CONSTANT FORCE

equation of motion

Newton's Second Law is often called the **equation of motion**. If the force on a particle is known, then *the Second Law determines the acceleration, and from this the position of the particle at any time can be calculated.* Thus, in principle, the motion of the particle is completely predictable.

If the force acting on a particle is constant, then the acceleration is also constant. The motion is then given by the formulas we developed for motion with constant acceleration in Chapter 4 [Eqs. (4.20)–(4.24)]. If the acceleration is not constant, then formulas for the motion can be obtained by calculations using integration; we will learn to calculate integrals in Chapter 7, and we discuss their application to the equations of motion in Sections 2.7 and 12.4.

As a simple example of motion with a constant force, consider a rectangular box of mass m being pushed along a smooth, frictionless floor. Figure 5.36 shows the box and the man pushing it. The push can be represented by a vector **P** of magnitude P pointing at an angle θ with the horizontal direction. Besides this push, there are two other forces acting on the box: the weight **w** of the box pointing downward in the vertical direction, and the normal force of the floor **N** pointing upward in the vertical direction. Figure 5.37 is a "free-body" diagram showing these forces.

FIGURE 5.36 Man pushing a box.

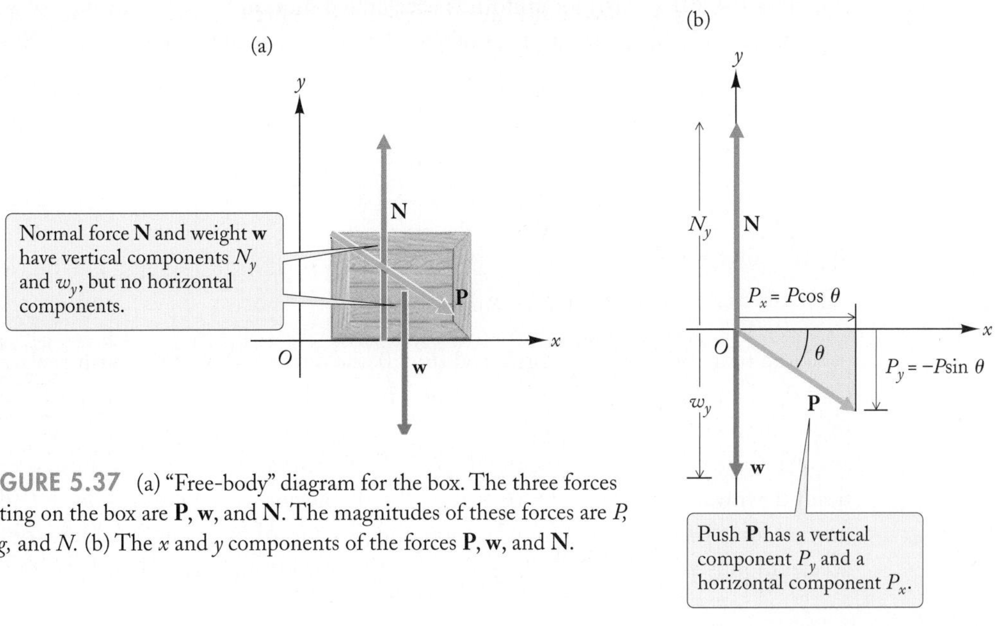

FIGURE 5.37 (a) "Free-body" diagram for the box. The three forces acting on the box are **P**, **w**, and **N**. The magnitudes of these forces are P, mg, and N. (b) The x and y components of the forces **P**, **w**, and **N**.

The net force on our box is the vector sum of **P**, **w**, and **N**; and the acceleration of the box is related to this net force by Newton's Second Law. Although the motion of the box is one-dimensional (horizontal, along the floor), the forces acting on it are not one-dimensional, and we must consider both the horizontal and the vertical components of these forces. For this purpose, it is convenient to take the x axis in the horizontal direction and the y axis in the vertical direction. Looking at Fig. 5.37b, we then find that the x and the y components of the individual forces are

$$P_x = P\cos\theta \qquad P_y = -P\sin\theta \tag{5.20}$$

$$w_x = 0 \qquad w_y = -mg \tag{5.21}$$

$$N_x = 0 \qquad N_y = N \tag{5.22}$$

Here N is the (unknown) magnitude of the normal force.
The components of the net force are

$$\begin{aligned} F_x &= P_x + w_x + N_x = P\cos\theta + 0 + 0 \\ &= P\cos\theta \end{aligned} \tag{5.23}$$

$$\begin{aligned} F_y &= P_y + w_y + N_y \\ &= -P\sin\theta - mg + N \end{aligned} \tag{5.24}$$

The x and y components of the equation of motion for the box are $ma_x = F_x$ and $ma_y = F_y$, from which we obtain the components of the acceleration:

$$a_x = \frac{F_x}{m} = \frac{P\cos\theta}{m} \tag{5.25}$$

$$a_y = \frac{F_y}{m} = \frac{-P\sin\theta - mg + N}{m} \tag{5.26}$$

Equation (5.25) says that the acceleration of the box along the floor is $(P\cos\theta)/m$. This determines the motion of the box along the floor, since the magnitude and direction (P and θ) of the push exerted by the man are assumed known. To find the velocity and the position at any time we need only substitute the acceleration a_x into our old equations [(4.20)–(4.24)] for uniformly accelerated motion.

Equation (5.26) can be used to evaluate the magnitude of the normal force. Since the motion is necessarily along the floor, the acceleration a_y in the direction perpendicular to the floor is zero; hence

$$0 = \frac{-P\sin\theta - mg + N}{m} \tag{5.27}$$

We can solve this for N, with the result

$$N = mg + P\sin\theta \tag{5.28}$$

Note that if the push is horizontal ($\theta = 0$ and $\sin\theta = 0$) or if the push is absent ($P = 0$), then the normal force is $N = mg$. Thus, under these conditions, the normal force simply balances the weight; this is, of course, exactly what we would expect for a box sitting on a floor when there is no push. Furthermore, note that if the push is vertically downward ($\theta = 90°$ and $\sin\theta = 1$, $\cos\theta = 0$), the normal force is $N = mg + P$ (the normal force balances the sum of weight and downward push), and the horizontal acceleration is zero. Finally, if the push has an upward component, then $P_y = P\sin\theta$

has become positive (the angle θ is above the x axis in Fig. 5.37). If this upward push is so large that it exceeds the weight (when $P\sin\theta > mg$), then the box no longer contacts the floor ($N = 0$) and it accelerates upward with $a_y = (P\sin\theta - mg)/m$.

EXAMPLE 9 Figure 5.38 shows a block of mass m sliding down a smooth, frictionless plane, or ramp, inclined at an angle θ with respect to the horizontal direction. Find the acceleration of the block along the inclined plane. Find the magnitude of the normal force that the plane exerts on the block.

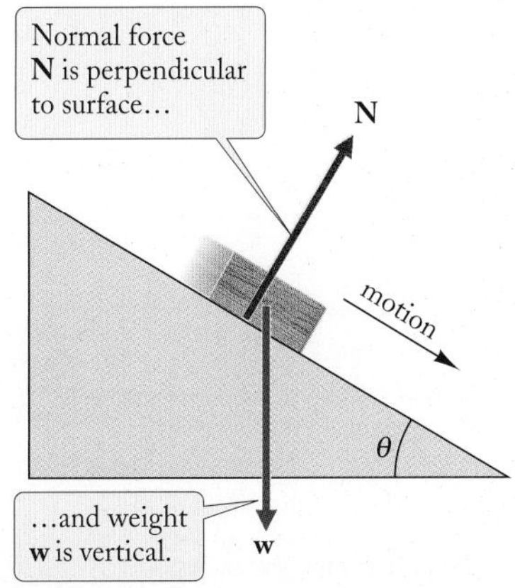

FIGURE 5.38 Block sliding down an inclined plane.

SOLUTION: There are two forces acting on the block: the weight **w** pointing vertically downward and the normal force **N** pointing in a direction perpendicular to the inclined plane. Figure 5.39a shows these two forces on a "free-body" diagram. The net force acting on the block is the vector sum of **w** and **N**. For the calculation of the components of these forces, it is convenient to take the x axis parallel to the inclined plane and the y axis perpendicular to it; this simplifies the calculation of the motion, since the velocity and the acceleration are then entirely along the chosen x axis. With this choice of axes, we find that the components of the two forces are (see Fig. 5.39b)

$$N_x = 0 \qquad N_y = N \tag{5.29}$$

$$w_x = mg\sin\theta \qquad w_y = -mg\cos\theta \tag{5.30}$$

and the components of the net force are

$$F_x = N_x + w_x = mg\sin\theta \tag{5.31}$$

$$F_y = N_y + w_y = N - mg\cos\theta \tag{5.32}$$

The equation of motion then gives us the corresponding components of the acceleration of the block:

$$a_x = \frac{F_x}{m} = \frac{mg\sin\theta}{m} = g\sin\theta \tag{5.33}$$

$$a_y = \frac{F_y}{m} = \frac{N - mg\cos\theta}{m} \tag{5.34}$$

Equation (5.33) tells us that the acceleration of the block along the inclined plane is $g\sin\theta$. In the case of a horizontal plane ($\theta = 0$ and $\sin\theta = 0$), there is no acceleration. In the case of a vertical plane ($\theta = 90°$ and $\sin\theta = 1$), the acceleration is $a_x = g$, which is the acceleration of free fall (with the x axis directed downward). Both of these extreme cases are as we would expect them to be.

Equation (5.34) can be used to evaluate the normal force. Since the motion is along the plane, the acceleration perpendicular to the plane must be identically zero, and thus

$$0 = \frac{N - mg\cos\theta}{m} \tag{5.35}$$

From this we find

$$N = mg\cos\theta \tag{5.36}$$

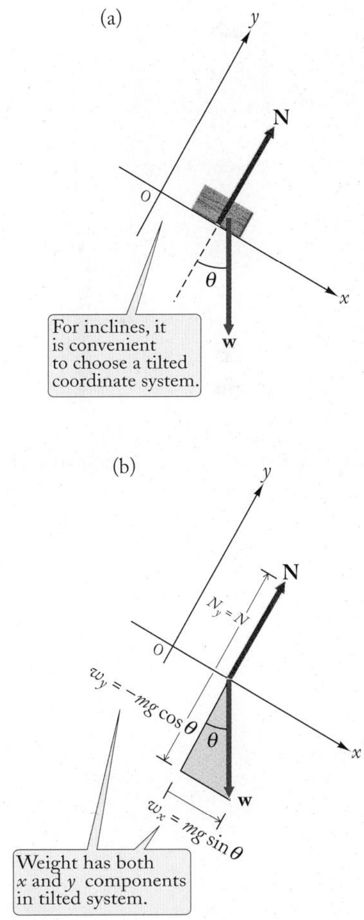

FIGURE 5.39 (a) "Free-body" diagram showing the forces **N** and **w.** (b) The x and the y components of the forces **N** and **w.**

COMMENT: Note that in the case of a horizontal plane ($\theta = 0$), the normal force has a magnitude mg, that is, the magnitude of the normal force matches the weight; in the case of a vertical plane ($\theta = 90°$), the normal force vanishes. This is reasonable, and we might have expected it.

Concepts in Context

EXAMPLE 10 A passenger elevator consists of an elevator cage of 1000 kg (empty) and a counterweight of 1100 kg connected by a cable running over a large pulley (see Fig. 5.40). Neglect the masses of the cable and of the pulley. (a) What is the upward acceleration of the elevator cage if the pulley is permitted to run freely, without friction? (b) What is the tension in the cable? (c) What are the tensions in the cable if the pulley is locked (by means of a brake) so that the elevator remains stationary?

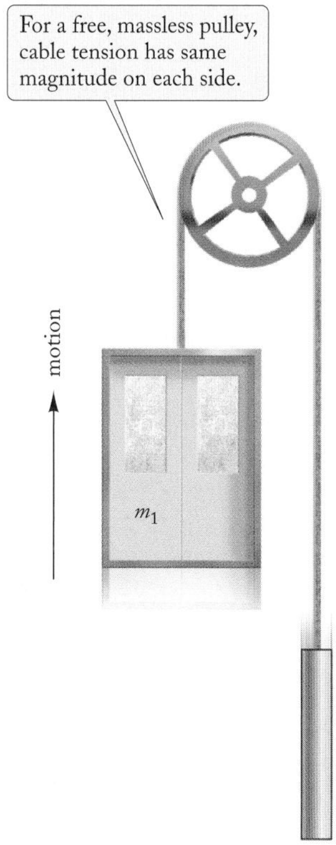

FIGURE 5.40 Elevator with counterweight.

SOLUTION: (a) Since the elevator cage and the counterweight are linked by the cable, it is necessary to solve the equations of motion for these two bodies simultaneously. For the free pulley, the cable merely transmits the tension from one body to the other, without change of its magnitude; consequently, the upward tension forces exerted by the ends of the cable on each body are exactly equal.

Figure 5.41 shows "free-body" diagrams for the elevator cage and the counterweight. The masses of these two bodies are designated by m_1 and m_2, respectively. For a system consisting of several bodies, such as the system of two bodies we are dealing with here, the "free-body" diagrams are especially helpful, since they permit us to view each body in isolation, and they give us a clear picture of what happens to each individual body. The vector $\mathbf{T}$ in Fig. 5.41 represents the tension force, and $\mathbf{w}_1$ and $\mathbf{w}_2$ represent the weights. Only vertical forces are present. With the y axis in the vertical direction, as indicated in Fig. 5.41, the y component of the force acting on m_1 is $F_1 = T - w_1$, and the y component of the force on m_2 is $F_2 = T - w_2$. Hence, the equation for the vertical motion of each mass is

$$m_1 a_1 = F_1 = T - w_1 \tag{5.37}$$

$$m_2 a_2 = F_2 = T - w_2 \tag{5.38}$$

where the forces and the accelerations are regarded as positive when directed upward. Since the two masses are tied together by a fixed length of cable, their accelerations a_1 and a_2 are always of the same magnitudes and in opposite directions; that is,

$$a_1 = -a_2 \tag{5.39}$$

With this equation and with $w_1 = m_1 g$ and $w_2 = m_2 g$, we obtain from Eqs. (5.37) and (5.38)

$$m_1 a_1 = T - m_1 g \tag{5.40}$$

$$-m_2 a_1 = T - m_2 g \tag{5.41}$$

These are two simultaneous equations for the two unknowns a_1 and T. To solve these equations first for a_1, we can eliminate T by subtracting each side of the second equation from each side of the first equation:

$$m_1 a_1 - (-m_2 a_1) = T - m_1 g - (T - m_2 g) \tag{5.42}$$

So the unknown T cancels out, leaving us with an equation for a_1:

$$m_1 a_1 + m_2 a_1 = -m_1 g + m_2 g \tag{5.43}$$

We can solve this equation for a_1 by rearranging:

$$a_1 = \frac{m_2 - m_1}{m_1 + m_2} g \tag{5.44}$$

This equation tells us the acceleration. With $m_1 = 1000$ kg and $m_2 = 1100$ kg, we find

$$a_1 = \frac{1100\ \text{kg} - 1000\ \text{kg}}{1100\ \text{kg} + 1000\ \text{kg}} g = \frac{100}{2100} g$$

$$= 0.0476g = 0.0476 \times 9.81\ \text{m/s}^2 = 0.467\ \text{m/s}^2 \tag{5.45}$$

The positive value of this acceleration indicates that the elevator cage accelerates upward, as we might have expected, since it has the smaller mass.

(b) Next, we must find the tension in the cable. Substituting the result for a_1 into Eq. (5.40), we obtain an equation for T:

$$m_1 \frac{m_2 - m_1}{m_1 + m_2} g = T - m_1 g \tag{5.46}$$

which leads to

$$T = m_1 \frac{m_2 - m_1}{m_1 + m_2} g + m_1 g = m_1 g \times \left(\frac{m_2 - m_1}{m_1 + m_2} + 1 \right)$$

$$= \frac{2m_1 m_2 g}{m_1 + m_2} \tag{5.47}$$

This tells us the tension in the cable; numerically,

$$T = \frac{2m_1 m_2 g}{m_1 + m_2} = \frac{2 \times 1000\ \text{kg} \times 1100\ \text{kg} \times 9.81\ \text{m/s}^2}{1000\ \text{kg} + 1100\ \text{kg}}$$

$$= 1.03 \times 10^4\ \text{N} \tag{5.48}$$

(c) If the pulley is locked and the elevator is stationary, the tension in the cable on either side of the pulley must match the weight hanging on that side. Thus

$$T_1 = w_1 = m_1 g = 1000\ \text{kg} \times 9.81\ \text{m/s}^2 = 9.81 \times 10^3\ \text{N} \tag{5.49}$$

and

$$T_2 = w_2 = m_2 g = 1100\ \text{kg} \times 9.81\ \text{m/s}^2 = 1.08 \times 10^4\ \text{N} \tag{5.50}$$

COMMENT: As we might have expected, Eq. (5.44) shows that if the two masses m_1 and m_2 are equal, the acceleration is zero—the two masses are then in equilibrium, and they either remain at rest or move with uniform velocity (until the cable runs out).

Note that when the two masses are unequal and the pulley is locked, the tensions in the two parts of the cable are *not* equal. This is because the locked pulley exerts extra friction forces on the cable, and the two portions of the cable now behave as though they were suspended independently.

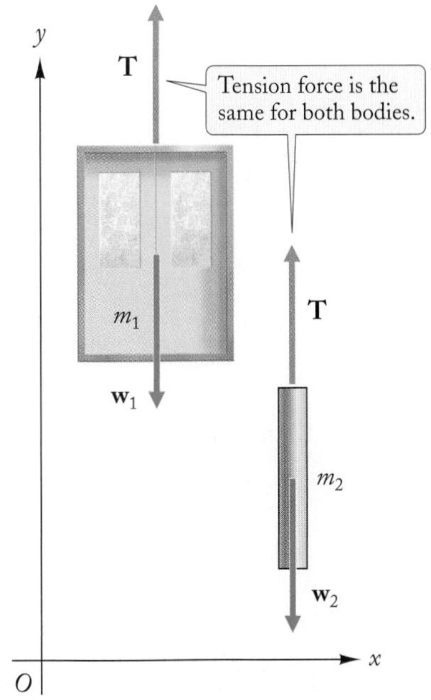

FIGURE 5.41 Separate "free-body" diagrams for the elevator and the counterweight.

EXAMPLE 11 The acceleration of a large block of mass m_1 down a frictionless ramp is to be kept small by using a hanging counterweight of mass m_2; the two are connected over a light, frictionless pulley by a light rope as shown in Fig. 5.42a. The angle of the incline is $\theta = 15°$. If the desired acceleration is to be one-hundredth of a standard g, what should the mass m_2 of the counterweight be?

SOLUTION: The "free-body" diagrams for the two masses are shown in Fig. 5.42b. It is convenient to use tilted coordinate axes for the mass m_1 on the incline, as in Example 9. The forces on the mass m_1 are the same as in Example 9, except that there is, additionally, the tension **T** from the rope. Accordingly, the x and the y components of the equation of motion of the mass m_1 are

$$m_1 a_{1,x} = m_1 g \sin\theta - T \tag{5.51}$$

$$m_1 a_{1,y} = N - m_1 g \cos\theta$$

The question does not ask about the normal force, so we will not need the second equation. For the hanging mass m_2, we use an upward $+y$ axis, and we have only vertical forces, so

$$m_2 a_{2,y} = T - m_2 g \tag{5.52}$$

As in Example 10, note that the linked accelerations must have the same magnitude, since they are connected by a taught rope. For our axis directions, this gives

$$a_{1,x} = a_{2,y} \tag{5.53}$$

Note that the pulley has linked motion in two different directions. The sign in Eq. (5.53) is chosen as positive, since motion of m_1 along the positive x axis results in motion of m_2 along the positive y axis. Using Eq. (5.53) in Eq. (5.52), and then adding Eq. (5.52) and Eq. (5.51), we can eliminate the tension T:

$$m_1 a_{1,x} + m_2 a_{1,x} = m_1 g \sin\theta - m_2 g$$

Collecting the m_2 terms on the left side, we obtain

$$m_2(a_{1,x} + g) = m_1 g \sin\theta - m_1 a_{1,x}$$

which we can solve for m_2:

$$m_2 = m_1 \frac{g \sin\theta - a_{1,x}}{a_{1,x} + g}$$

Finally, we substitute the desired value $a_{1,x} = 0.01g$, cancel the common factor of g, and evaluate the result for $\theta = 15°$:

$$\begin{aligned} m_2 &= m_1 \frac{g \sin\theta - 0.01g}{0.01g + g} \\ &= m_1 \frac{\sin\theta - 0.01}{1.01} \\ &= m_1 \frac{\sin 15° - 0.01}{1.01} = m_1 \frac{0.26 - 0.01}{1.01} \\ &= 0.25 m_1 \end{aligned}$$

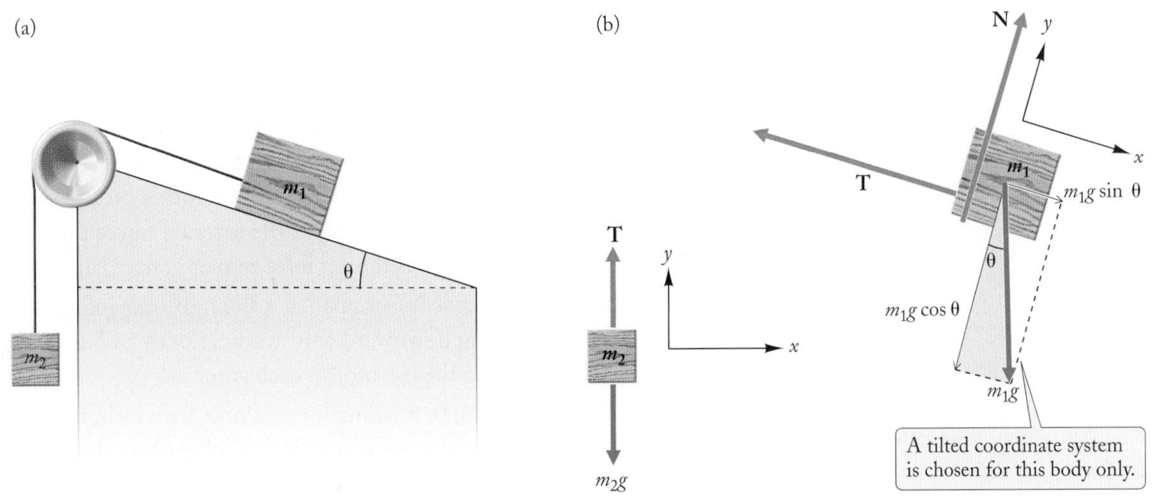

FIGURE 5.42 (a) Slowly accelerating mass m_1 on incline with counterweight m_2. (b) "Free-body" diagrams for the two masses.

PHYSICS IN PRACTICE ELEVATORS

Concepts — in — Context

The counterweight of an elevator usually consists of a concrete block, sliding within guide rails attached to the side of the elevator shaft. The mass of the counterweight is chosen to equal the mass of the elevator plus one-half of the mass of the expected average payload. With this choice of mass, the elevator–counterweight system will be nearly in equilibrium most of the time, and the extra force that the motor has to supply to move the elevator up (or down) will be minimized.

In Example 10, we neglected the mass of the elevator cable. For the sake of safety, modern elevators use several cables strung in parallel. In the elevators of skyscrapers, the total mass of the cables is considerable, and it often exceeds the mass of the elevator cage. If we take the mass of the cable into account, the acceleration of a freely running elevator depends on how much cable hangs on the elevator side and how much on the counterweight side. Since the lengths of these two segments of cable change as a function of the position of the elevator, the acceleration also changes with position.

Elevators have braking systems that would prevent their fall even if the cables were to break. The braking system consists of powerful jaws with braking pads that grip the guide rails (Figure 1). The jaws are triggered automatically if the speed of the elevator exceeds a critical value. Some elevators also have an air cushion at the bottom of the shaft. This is a downward extension of the shaft, which fits tightly around the sides of the elevator. If the elevator falls into this tightly fitting shaft, it will be cushioned by the air that it traps and compresses in the shaft. To prevent a bounce of the elevator, small vents allow the compressed air to leak out gradually.

FIGURE 1 Elevator brake assembly.

Checkup 5.6

QUESTION 1: A skydiver jumps out of a helicopter. At first, she falls freely, with the standard downward acceleration g. After a while, she reaches terminal speed and continues to fall with uniform velocity. Make a list of the external forces on the skydiver during the first part of the motion, and make a list of the external forces on the skydiver during the second part of the motion. Is the net force zero or nonzero in each case?

QUESTION 2: An elevator of 1000 kg is (a) stopped at a floor, (b) moving upward at a steady velocity of 5 m/s, (c) moving downward at a steady velocity of 5 m/s. What is the tension in the cable supporting the elevator in each case?

QUESTION 3: Figure 5.36 shows a man pushing a box along a smooth, frictionless floor. Suppose that instead of pushing downward on the rear of the box, the man pulls

PROBLEM-SOLVING TECHNIQUES "FREE-BODY" DIAGRAMS

From the examples analyzed in this section, we see that the solution of a problem of motion with several forces acting on a body often proceeds in a sequence of steps.

1 The first step is always a careful enumeration of all the forces. Make a complete list of these forces, and label each with a vector symbol.

2 Identify the body whose motion or whose equilibrium is to be investigated and draw a separate "free-body" diagram for this body in which each force is represented by an arrow labeled by a vector symbol. Remember that the only forces to be included in the "free-body" diagram are the forces that act *on* the body, not the forces exerted *by* the body. If there are several separate bodies in the problem (as in Examples 10 and 11), then you need to draw a separate "free-body" diagram for each. When drawing the arrows for the forces, try to draw the lengths of the arrows in proportion to the magnitudes of the forces; this will help you to see what the direction of the resultant is. (Do *not* include an "*ma*" force in the "free-body" diagram; the acceleration is caused by the resultant of several forces already included among the pushes and pulls displayed in the diagram.)

3 Then draw coordinate axes on each diagram, preferably placing one of the axes along the direction of motion. If the motion proceeds along an inclined plane, it is convenient to use tilted coordinate axes, with one axis along the plane, as in Fig. 5.39b. It is even sometimes convenient to use a different orientation of the axes for different bodies, as in Example 11, where one body was on an incline and attached by a rope to one that was hanging straight down.

4 Next, examine the components of the individual forces and the components of the net force. Remember that the signs of the components correspond to the directions of the force vectors. Remember that for freely moving massless pulleys and massless cables, the magnitude (but not necessarily the direction) of the tension in the cable is the same at every point.

5 Then apply Newton's Second Law for the components of the net force ($F_{\text{net},x} = ma_x$, $F_{\text{net},y} = ma_y$), and calculate the components of the acceleration (if the acceleration is unknown), or the components of some force (if some force is unknown). If there are several separate bodies in the problem, you need to apply Newton's Law separately for each. Another relation is available for bodies attached by a cable: the magnitude of the acceleration of each is the same (the direction may be opposite, as in Eq. (5.39) in Example 10, or even along different axes, as in Eq. (5.53) of Example 11).

6 As in the kinematics problems of Chapters 2 and 3, it is a good idea to solve for the unknowns algebraically, and substitute numbers only as a last step. This makes it easier to spot and correct mistakes, and it also makes it possible to check whether the final result behaves as you might have expected in various limiting cases (for instance, see the comments attached to Examples 9 and 10).

7 When substituting numbers, also substitute the units. Use the conversion $1\ \text{N} = 1\ \text{kg·m/s}^2$ where necessary. This should automatically result in the correct units for the final result.

upward at the front of the box, at the same angle θ and with the same magnitude of force. Does this change the acceleration? The normal force?

QUESTION 4: Consider a block placed on a smooth (frictionless) inclined plane. With your hands, you push against this block horizontally with a force of 100 N, and this barely holds the block stationary. (a) What must be the magnitude of your push if you want to keep the block moving at constant speed up the plane? Down the plane? (b) If instead of pushing horizontally, you push parallel to the surface of the plane, will your push have to be larger or smaller? To calculate how much larger or smaller, what do you need to know?

QUESTION 5: Which of the "free-body" diagrams in Fig. 5.43 respectively corresponds to each of the following bodies: (1) a book lying on a flat table, (2) a box on the floor with another box on top, and (3) a lamp hanging by a cord from a ceiling?

(A) a, b, c (B) a, c, b (C) b, a, c (D) b, c, a (E) c, a, b

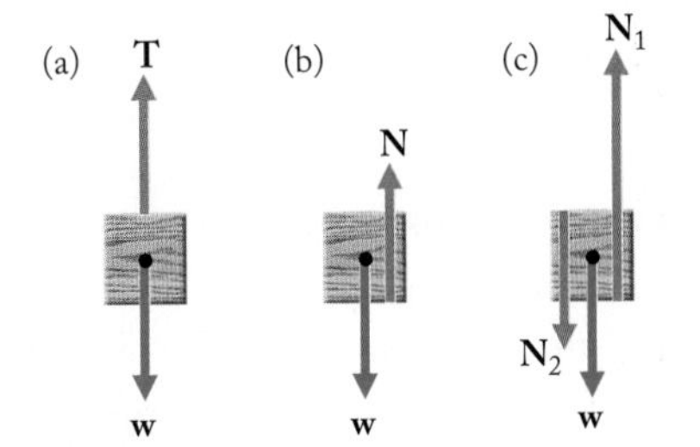

FIGURE 5.43 Three "free-body" diagrams.

SUMMARY

PHYSICS IN PRACTICE Elevators (page 157)

PROBLEM-SOLVING TECHNIQUES "Free-Body" Diagrams (page 158)

NEWTON'S FIRST LAW In an inertial reference frame, a body at rest remains at rest and a body in motion continues to move at constant velocity unless acted upon by a net external force.

NEWTON'S SECOND LAW $m\mathbf{a} = \mathbf{F}_{net}$ (5.2)

UNIT OF FORCE 1 newton = 1 N = 1 kg·m/s^2 (5.6)

DEFINITION OF MASS $\frac{m}{m_s} = \frac{a_s}{a}$ (5.3)

SUPERPOSITION OF FORCES $\mathbf{F}_{net} = \mathbf{F}_1 + \mathbf{F}_2 + \mathbf{F}_3 + \cdots$ (5.8)

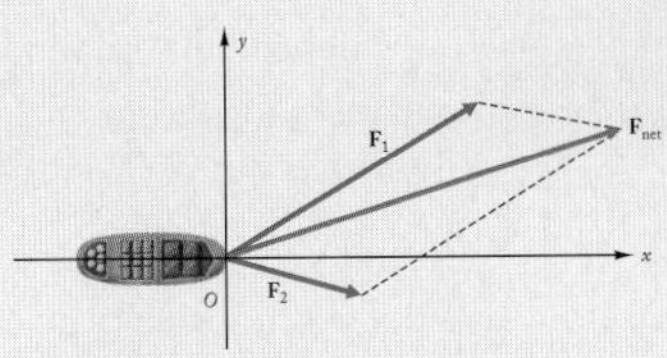

WEIGHT Magnitude: $w = mg$ (5.16)

Direction: downward

NORMAL FORCE The force provided by a surface that prevents interpenetration of solid bodies. The normal force has a direction perpendicularly outward from the surface, and, if the surface is not accelerating, a magnitude that balances the net force pushing perpendicularly toward the surface.

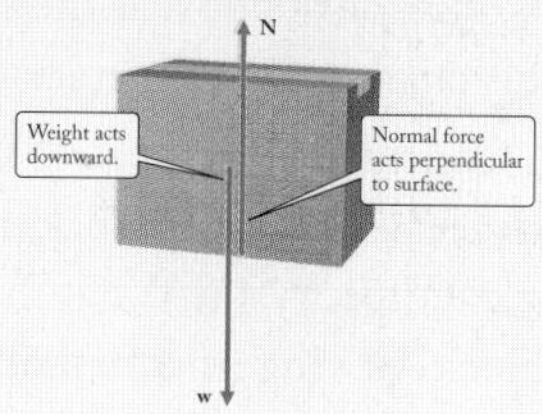

NEWTON'S THIRD LAW Whenever a body exerts a force on another body, the latter exerts a force of equal magnitude and opposite direction on the former. These *action–reaction pairs* of forces do not "cancel," since they *act on different bodies*.

$$\mathbf{F}_{1\text{ on }2} = -\mathbf{F}_{2\text{ on }1} \tag{5.17}$$

"FREE-BODY" DIAGRAM A diagram showing a body in isolation and each of the external force vectors acting on the body.

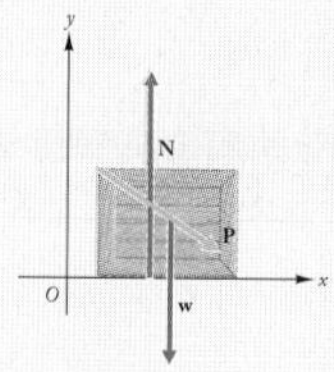

TENSION The force with which a string, rope, or cable pulls on what is attached to it. For a massless string, rope, or cable, the magnitude of the tension is the same at each point, even when passing around a massless, freely moving pulley.

QUESTIONS FOR DISCUSSION

1. If a glass stands on a table on top of a sheet of paper, you can remove the paper without touching the glass by jerking the paper away very sharply. Explain why the glass more or less stays put.
2. Make a critical assessment of the following statement: An automobile is a device for pushing the air out of the way of the passenger so that his body can continue to its destination in its natural state of motion at uniform velocity.
3. Does the mass of a body depend on the frame of reference from which we observe the body? Answer this by appealing to the definition of mass.
4. Suppose that a (strange) body has negative mass. Suppose that you tie this body to a body of positive mass of the same magnitude by means of a stretched rubber band. Describe the motion of the two bodies.
5. Does the magnitude or the direction of a force depend on the frame of reference?
6. A fisherman wants to reel in a large dead shark hooked on a thin fishing line. If he jerks the line, it will break; but if he reels it in very gradually and smoothly, it will hold. Explain.
7. If a body crashes into a water surface at high speed, the impact is almost as hard as on a solid surface. Explain.
8. Moving downwind, a sailboat can go no faster than the wind. Moving across the wind, a sailboat can go faster than the wind. How is this possible? (Hint: What are the horizontal forces on the sail and on the keel of a sailboat?)
9. A boy and girl are engaged in a tug-of-war. Draw a diagram showing the horizontal forces (a) on the boy, (b) on the girl, and (c) on the rope. Which of these forces are action–reaction pairs?

FIGURE 5.44 The Magdeburg hemispheres.

10. In an experiment performed in 1654, Otto von Guericke, mayor of Magdeburg and inventor of the air pump, gave a demonstration of air pressure before Emperor Ferdinand. He had two teams of 15 horses each pull in opposite directions on two evacuated hemispheres held together by nothing but air pressure. The horses failed to pull these hemispheres apart (Fig. 5.44). If each horse exerted a pull of 3000 N, what was the tension in the harness attached to each hemisphere? If the harness attached to one of the hemispheres had simply been tied to a stout tree, what would have been the tension exerted by a single team of horses hitched to the other harness? What would have been the tension exerted by the two teams of horses hitched in series to the other harness? Can you guess why von Guericke hitched up his horses in the way he did?
11. In a tug-of-war, two teams of children pull on a rope (Fig. 5.45). Is the tension constant along the entire length of the rope? If not, along what portion of the rope is it constant?

FIGURE 5.45 Two teams of children in a tug-of-war.

12. If you carry a spring balance from London to Hong Kong, do you have to recalibrate it? If you carry a beam balance?
13. When you stretch a rope horizontally between two fixed points, it always sags a little, no matter how great the tension. Why?
14. What are the forces on a soaring bird? How can the bird gain altitude without flapping its wings?
15. An automobile is parked on a street. (a) Draw a "free-body" diagram showing the forces acting on the automobile. What is the net force? (b) Draw a "free-body" diagram showing the forces that the automobile exerts on the Earth. Which of the forces in the diagrams (a) and (b) are action–reaction pairs?
16. A ship sits in calm water. What are the forces acting on the ship? Draw a "free-body" diagram for the ship.
17. Some old-time roofers claim that when walking on a rotten roof, it is important to "walk with a light step so that your full weight doesn't rest on the roof." Can you walk on a roof with less than your full weight? What is the advantage of a light step?
18. In a tug-of-war on sloping ground the party on the low side has the advantage. Why?
19. The label on a package of sugar claims that the weight of the contents is "1 lb or 454 g." What is wrong with this statement?
20. A physicist stands on a bathroom scale in an elevator. When the elevator is stationary, the scale reads 73 kg. Describe qualitatively how the reading of the scale will fluctuate while the elevator makes a trip to a higher floor.
21. How could you use a pendulum suspended from the roof of your automobile to measure its acceleration?
22. When you are standing on the Earth, your feet exert a force (push) against the surface. Why does the Earth not accelerate away from you?
23. When an automobile accelerates on a level road, the force that produces this acceleration is the push of the road on the wheels. If so, why does the automobile need an engine?
24. You are in a small boat in the middle of a calm lake. You have no oars, and you cannot put your hands in the water because the lake is full of piranhas. The boat carries a large load of coconuts. How can you get to the shore?
25. On a windless day, a sailor puts an electric fan, powered by a battery, on the stern of his boat and blows a stream of air into the sail. Will the boat move forward?
26. You are inside a ship that is trying to make headway against the strong current of a river. Without looking at the shore or other outside markers, is there any way you can tell whether the ship is making any progress?

PROBLEMS

5.2 Newton's Second Law†

1. According to the *Guinness Book of Records,* the heaviest man ever had a confirmed mass of 975 pounds. Express this in kilograms.
2. The hydrogen atom consists of one proton and one electron. What is the mass of one hydrogen atom? (See Table 5.2.) How many hydrogen atoms are there in 1.0 kg of hydrogen gas?
3. The oxygen atom consists of 8 protons, 8 neutrons, and 8 electrons. What is the mass of one oxygen atom? How many oxygen atoms are there in 1.0 kg of oxygen gas?
4. A boy and a girl are engaged in a tug-of-war while standing on the slippery (frictionless) surface of a sheet of ice (see Fig. 5.46). While they are pulling on the rope, the instantaneous acceleration of the boy is 7.0 m/s^2 toward the girl, and the instantaneous acceleration of the girl is 8.2 m/s^2 toward the boy. The mass of the boy is 50 kg. What is the mass of the girl?

FIGURE 5.46 A boy and a girl in a tug-of-war on ice.

5. On a flat road, a Maserati sports car can accelerate from 0 to 80 km/h (0 to 50 mi/h) in 5.8 s. The mass of the car is 1620 kg. What are the average acceleration and the average force on the car?
6. The Grumman F-14B fighter plane has a mass of 16 000 kg, and its engines develop a thrust of 2.7×10^5 N when at full power. What is the maximum horizontal acceleration that this plane can achieve? Ignore friction.
7. A woman of 57 kg is held firmly in the seat of her automobile by a lap-and-shoulder seat belt. During a collision, the automobile decelerates from 50 to 0 km/h in 0.12 s. What is the average horizontal force that the seat belt exerts on the woman? Compare the force with the weight of the woman.
8. A heavy freight train has a total mass of 16 000 metric tons. The locomotive exerts a pull of 670 000 N on this train. What is the acceleration? How long does it take to increase the speed from 0 to 50 km/h?
9. With brakes fully applied, a 1500-kg automobile decelerates at the rate of 8.0 m/s^2 on a flat road. What is the braking force acting on the automobile? Draw a "free-body" diagram showing the direction of motion of the automobile and the direction of the braking force.
10. Consider the impact of an automobile on a barrier. The initial speed is 50 km/h and the automobile comes to rest within a distance of 0.40 m, with constant deceleration. If the mass of the automobile is 1400 kg, what is the force acting on the automobile during the deceleration?
11. In a crash at the Silverstone circuit in England, a race-car driver suffered more than 30 fractures and dislocations and several heart stoppages after a deceleration from 174 km/h to 0 km/h within a distance of about 66 cm. If the deceleration was constant during the crash and the mass of the driver was 75 kg, what were the deceleration and force on the driver?
12. The projectile fired by the gun described in Problem 52 in Chapter 2 has a mass of 45 kg. What is the force on this projectile as it moves along the barrel?
13. A 70-g racquetball is accelerated from 0 to 30 m/s during an impact lasting 0.060 s. What is the average force experienced by the ball?
14. A Roman candle (a pyrotechnic device) of mass 35 g accelerates vertically from 0 to 22 m/s in 0.20 s. What is the average acceleration? What is the average force on the firework during this time? Neglect any loss in mass.
15. When the engine of a 240-kg motorboat is shut off, the boat slows from 15.0 m/s to 10.0 m/s in 1.2 s, and then from 10.0 m/s to 5.0 m/s in 2.1 s. What is the average acceleration during each of these intervals? What average frictional force does the water provide during each interval?
16. A tennis ball of mass 57 g is initially at rest. While being hit, it experiences an average force of 55 N during a 0.13-s interval. What is its final velocity?
17. An astronaut (with spacesuit) of mass 95 kg is tethered to a 750-kg satellite. By pulling on the tether, she accelerates toward the satellite at 0.50 m/s^2. What is the acceleration of the satellite toward the astronaut?

††*18. Figure 2.27 shows the plot of velocity vs. time for a Triumph sports car coasting along with its gears in neutral. The mass of the car is 1160 kg. From the values of the deceleration at the times $t = 0, 10, 20, 30,$ and 40 s [see Problem 39(b) in Chapter 2], calculate the friction force that the car experiences at these times. Make a plot of friction force vs. velocity.

*19. Table 2.3 gives the velocity of a projectile as a function of time. The projectile slows down because of the friction force exerted by the air. For the first 0.30-s time interval and for the last 0.20-s time interval, calculate the average friction force.

*20. A proton moving in an electric field has an equation of motion

$$\mathbf{r} = (5.0 \times 10^4 t)\mathbf{i} + (2.0 \times 10^4 t - 2.0 \times 10^5 t^2)j - (4.0 \times 10^5 t^2)\mathbf{k}$$

where distance is measured in meters and time in seconds. The proton has a mass of 1.7×10^{-27} kg. What are the components of the force acting on this proton? What is the magnitude of the force?

† For help, see Online Concept Tutorial 6 at www.wwnorton.com/physics

†† Due to the difficulty of accurately drawing tangents, answers for this problem that differ by up to 10% are acceptable.

*21. You launch a stone of mass 40 g horizontally with a slingshot. During the launch, the position of the stone is given by

$$x = x_0[1 - \cos(bt)]$$

where $x_0 = 30$ cm is the position of the stone at the end of the launch and $b = 4\pi\ \text{s}^{-1}$ is a constant. The launch begins at time $t = 0$ and ends at $t = 0.125$ s. Ignore any vertical motion. During the launch, what is the velocity of the stone as a function of time? What are the acceleration and the force as functions of time? Rewrite the force in terms of the position x.

5.3 The Combination of Forces†

22. While braking, an automobile of mass 1200 kg decelerates along a level road at 7.8 m/s^2. What is the horizontal force that the road exerts on each wheel of the automobile? Assume all wheels contribute equally to the braking. Ignore the friction of the air.

23. In 1978, in an accident at a school in Harrisburg, Pennsylvania, several children lost parts of their fingers when a nylon rope suddenly snapped during a giant tug-of-war among 2300 children. The rope was known to have a breaking tension of 58000 N. Each child can exert a pull of approximately 130 N. Was it safe to employ this rope in this tug-of-war?

24. An 1800-kg barge is pulled via cables by two donkeys on opposite riverbanks. The (horizontal) cables each make an angle of 30° with the direction of motion of the barge. At low speed, ignore friction and determine the acceleration of the barge when each donkey exerts a force of 460 N on a cable.

25. Three rescuers are pulling horizontally on a safety net to keep it taut. One pulls northward with a force of 270 N; the second pulls in a direction 30° south of west with a force of 240 N. In which direction and with what force must the third pull to keep the net stationary?

26. Two soccer players kick a ball at the same instant. During the kick, one applies a force of 25 N in a direction 30° east of north and the other a force of 35 N in a direction 30° east of south. What are the magnitude and direction of the net force?

27. An ocean current applies a force of 2500 N to a 1400-kg sailboat in a direction 15° east of north. The wind applies a force of 3200 N in a direction 30° east of north. What are the magnitude and direction of the resulting acceleration?

28. A 5.0-kg mass has the following forces on it:

$$\mathbf{F}_1 = (4.0\ \text{N})\mathbf{i} + (3.0\ \text{N})\mathbf{j}$$

$$\mathbf{F}_2 = (2.0\ \text{N})\mathbf{i} - (5.0\ \text{N})\mathbf{j}$$

Find the magnitude and direction of the acceleration of this mass.

*29. The Earth exerts a gravitational pull of 2.0×10^{20} N on the Moon; the Sun exerts a gravitational pull of 4.3×10^{20} N on the Moon. What is the net force on the Moon when the angular separation between the Earth and the Sun is 90° as seen from the Moon?

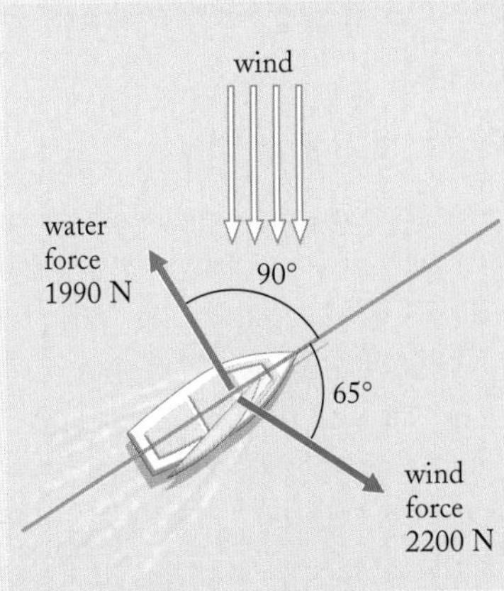

FIGURE 5.47 Forces on a sailboat. The angles are measured relative to the line of motion.

*30. A sailboat is propelled through the water by the combined action of two forces: the push ("lift") of the wind on the sail and the push of the water on the keel. Figure 5.47 shows the magnitudes and the directions of these forces acting on a medium-sized sailboat (this oversimplified diagram does not include the drag of wind and water). What is the resultant of the forces in Fig. 5.47?

*31. A boat is tied to a dock by four (horizontal) ropes. Two ropes, with a tension of 260 N each, are at right angles to the dock. Two other ropes, with a tension of 360 N each, are at an angle of 20° with the dock (Fig. 5.48). What is the resultant of these forces?

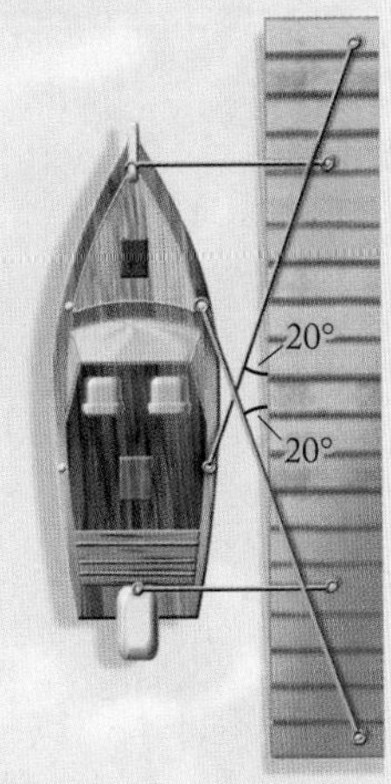

FIGURE 5.48 Ropes holding a boat at a dock.

**32. In a tug-of-war, a jeep of mass 1400 kg and a tractor of mass 2000 kg pull on a horizontal rope in opposite directions. At one instant, the tractor pulls on the rope with a force of 1.50×10^4 N while its wheels push horizontally against the ground with a force of 1.60×10^4 N. Calculate the instantaneous accelerations of the tractor and of the jeep; calculate the horizontal push of the wheels of the jeep. Assume the rope does not stretch or break.

† For help, see Online Concept Tutorial 7 at www.wwnorton.com/physics

5.4 Weight; Contact Force and Normal Force†
5.5 Newton's Third Law†

33. What is the mass of a laptop computer that weighs 25 N? What is the weight of a dictionary of mass 3.5 kg?

34. The surface gravity on Pluto is $0.045g$. What is the weight of a 60-kg woman on Pluto?

35. A man weighs 750 N on the Earth. The surface gravities of Mars and Jupiter are $0.38g$ and $2.53g$, respectively. What is the mass of the man on each of these planets? What is his weight on each planet?

36. What is the weight (in pounds-force) of a 1-lb bag of sugar in New York ($g = 9.803\ \text{m/s}^2$)? In Hong Kong ($g = 9.788\ \text{m/s}^2$)? In Quito ($g = 9.780\ \text{m/s}^2$)?

37. A bar of gold of mass 500.00 g is transported from Paris ($g = 9.8094\ \text{m/s}^2$) to San Francisco ($g = 9.7996\ \text{m/s}^2$).
    (a) What is the decrease of the weight of the gold? Express your answer as a fraction of the initial weight.
    (b) Does the decrease of weight mean that the bar of gold is worth less in San Francisco?

38. A woman stands on a chair. Her mass is 60 kg, and the mass of the chair is 20 kg. What is the force that the chair exerts on the woman? What is the force that the floor exerts on the chair?

39. A chandelier of 10 kg hangs from a cord attached to the ceiling, and a second chandelier of 3.0 kg hangs from a cord below the first (see Fig. 5.49). Draw the "free-body" diagram for the first chandelier and the "free-body" diagram for the second chandelier. Find the tension in each of the cords.

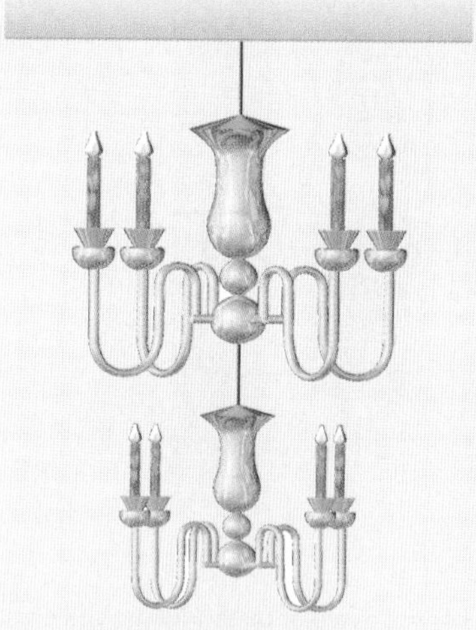

FIGURE 5.49 Two chandeliers.

40. A mass $m = 200$ kg hangs from a horizontal ceiling by two cables, one of length 3.0 m and the other 4.0 m; the two cables subtend an angle of 90° at the mass. What is the tension in each cable?

† For help, see Online Concept Tutorial 6 and 7 at www.wwnorton.com/physics

41. You hold a cable from which hangs a first mass m_1. A second cable below it connects to a second hanging mass m_2, and a third cable below the second mass connects to a third hanging mass m_3. If you hold the system stationary, what upward force F do you apply? What is the tension in the first cable? What is the tension in the second cable? In the third? Assume massless cables.

42. A horizontal force F holds a block on a frictionless inclined plane in equilibrium. If the block has a mass $m = 5.0$ kg and the incline makes an angle of 50° with the horizontal, what is the value of F? What is the normal force exerted by the incline on the block?

43. In a bosun's chair, a cable from a seat runs up over a pulley and back down. If a sailor of mass M sits on the seat of the otherwise massless system, with what force must a second sailor pull downward on the free end of the cable to get the first sailor moving? If, instead, the seated sailor pulls, what force must he apply to pull himself upward?

44. A locomotive pulls a train consisting of three equal boxcars with constant acceleration along a straight, frictionless track. Suppose that the tension in the coupling between the locomotive and the first boxcar is 12 000 N. What is the tension in the coupling between the first and the second boxcar? The second and the third? Does the answer depend on the absence of friction?

*45. A long, thick cable of diameter d and density ρ is hanging vertically down the side of a building. The length of the cable is l. What tension does the weight of the cable produce at its upper end? At its midpoint?

46. A small truck of 2800 kg collides with an initially stationary automobile of 1200 kg. The acceleration of the truck during the collision is $-500\ \text{m/s}^2$, and the collision lasts for 0.20 s. What is the acceleration of the automobile? What is the speed of the automobile after the collision? Assume that the frictional force due to the road can be neglected during the collision.

*47. Two heavy boxes of masses 20 kg and 30 kg sit on a smooth, frictionless surface. The boxes are in contact, and a horizontal force of 60 N pushes horizontally against the smaller box (Fig. 5.50). What is the acceleration of the two boxes? What is the force that the smaller box exerts on the larger box? What is the force that the larger box exerts on the smaller box?

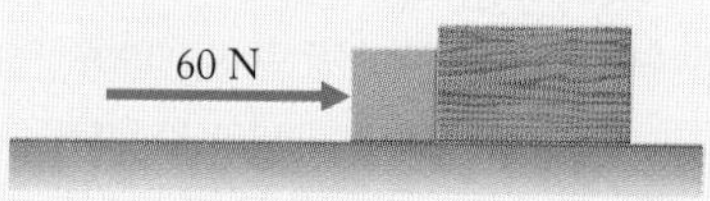

FIGURE 5.50 Two boxes in contact.

*48. An archer pulls the string of her bow back with her hand with a force of 180 N. If the two halves of the string above and below her hand make an angle of 120° with each other, what is the tension in each half of the string?

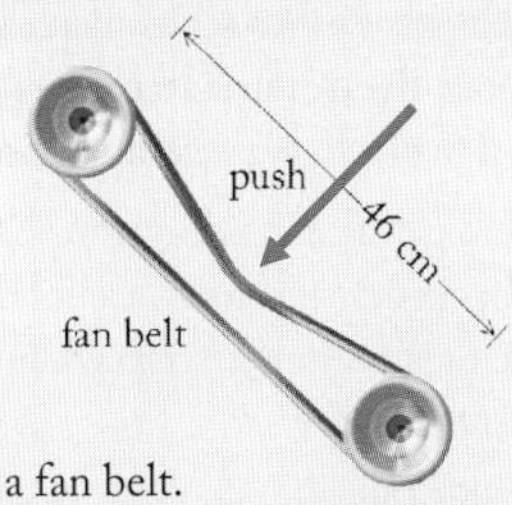

FIGURE 5.51 Push on a fan belt.

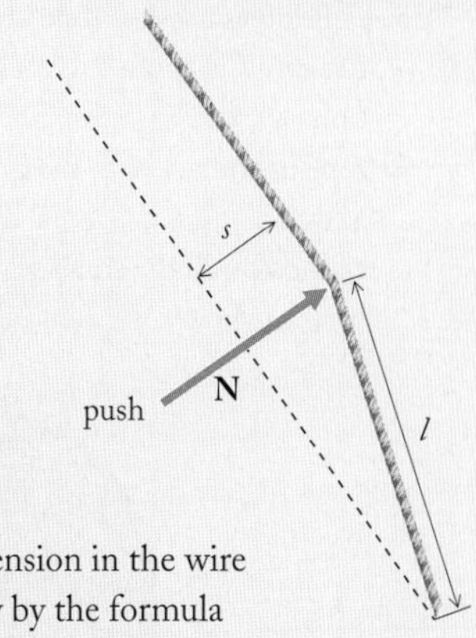

FIGURE 5.54 Push on a wire rope.

*49. A mechanic tests the tension in a fan belt by pushing against it with his thumb (Fig. 5.51). The force of the push is 130 N, and it is applied to the midpoint of a segment of belt 46 cm long. The lateral displacement of the belt is 2.5 cm. What is the tension in the belt (while the mechanic is pushing)?

*50. On a sailboat, a rope holding the foresail passes through a block (a pulley) and is made fast on the other side to a cleat (Fig. 5.52). The two parts of the rope make an angle of 140° with each other. The sail pulls on the rope with a force of 1.2×10^4 N. What is the force that the rope exerts on the block?

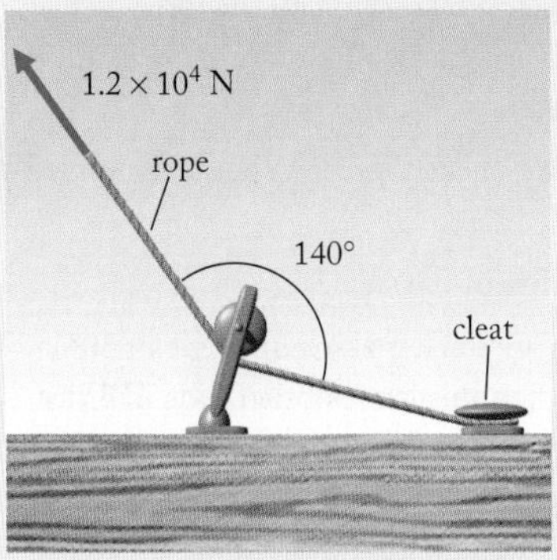

FIGURE 5.52 The left end of the rope is attached to the sail, and the right end is attached to a cleat.

*51. In order to pull an automobile out of the mud in which it is stuck, the driver stretches a rope tautly from the front end of the automobile to a stout tree. He then pushes sideways against the rope at the midpoint (see Fig. 5.53). When he pushes with a force of 900 N, the angle between the two halves of the rope on his right and his left is 170° . What is the tension in the rope under these conditions?

**52. A sailor tests the tension in a wire rope holding up a mast by pushing against the rope with his hand at a distance l from the lower end of the rope. When he exerts a transverse push N, the wire rope suffers a transverse displacement s (Fig. 5.54).

(a) Show that for $s << l$ the tension in the wire rope is given approximately by the formula

$$T = Nl/s$$

In your calculation, assume that the distance to the upper end of the rope is effectively infinite; i.e., the total length of the rope is much larger than l.

(b) What is the tension in the rope when it suffers a transverse displacement of 2.0 cm under a force of 150 N applied at a distance of 1.5 m from the lower end?

*53. On a windy day, a small tethered balloon is held by a long string making an angle of 70° with the ground. The vertical buoyant force on the balloon (exerted by the air) is 67 N. During a sudden gust of wind, the (horizontal) force of the wind is 200 N; the tension in the string is 130 N. What are the magnitude and direction of the force on the balloon?

*54. A horse, walking along the bank of a canal, pulls a barge. The horse exerts a pull of 300 N on the barge at an angle of 30° (Fig. 5.55). The bargeman relies on the rudder to steer the barge on a straight course parallel to the bank. What transverse force (perpendicular to the bank) must the rudder exert on the barge?

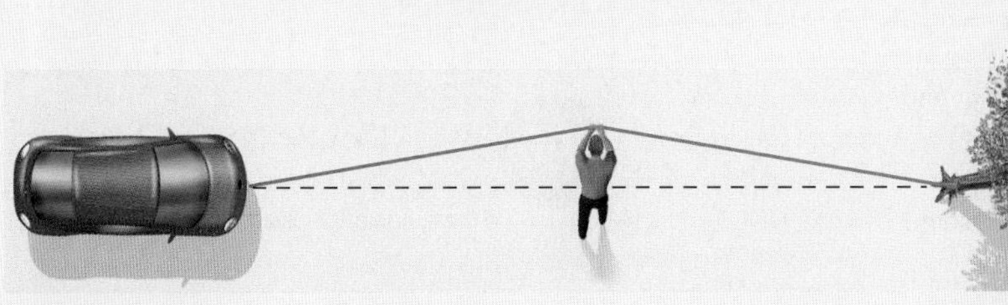

FIGURE 5.53 The rope is stretched between the automobile and a tree. The man pushes at the midpoint.

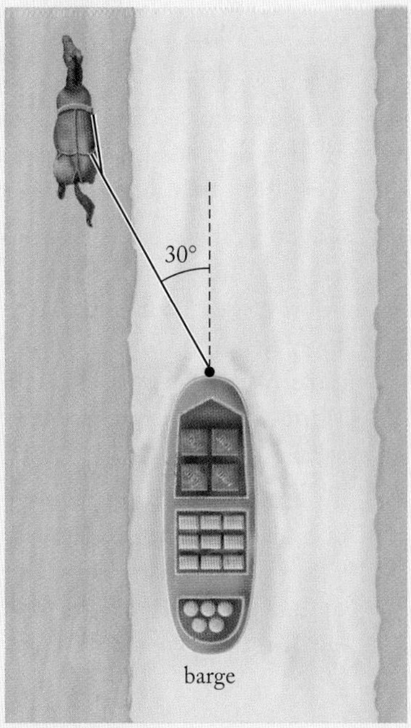

FIGURE 5.55 Horse pulling a barge.

**55. A flexible massless rope is placed over a cylinder of radius R. A tension T is applied to each end of the rope, which remains stationary (see Fig. 5.56). Show that each small segment $d\theta$ of the rope in contact with the cylinder pushes against the cylinder with a force $T\,d\theta$ in the radial direction. By integration of the forces exerted by all the small segments, show that the net vertical force on the cylinder is $2T$ and the net horizontal force is zero.

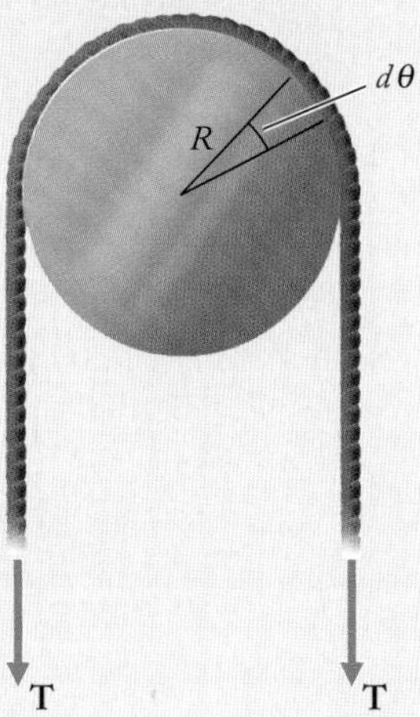

FIGURE 5.56 Rope and cylinder.

5.6 Motion with a Constant Force†

56. An elevator accelerates upward at 1.8 m/s^2. What is the normal force on the feet of an 80-kg passenger standing in the elevator? By how much does this force exceed his weight?

57. A parachutist of mass 80 kg approaches the ground at 5.0 m/s. Suppose that when he hits the ground, he decelerates at a constant rate (while his legs buckle under him) over a distance of 1.0 m. What is the force the ground exerts on his feet during the deceleration?

58. If the elevator described in Example 10 carries four passengers of 70 kg each, what speed will the elevator attain running down freely from a height of 10 m, starting from rest?

59. You lift a cable attached to a first mass m_1. A second cable below it connects to a second mass m_2, and a third cable below the second mass connects to a third mass m_3. If you apply an upward force F, what is the tension in the first cable? What is the tension in the second cable? In the third? What is the acceleration of the system? Assume massless cables.

60. Two adjacent blocks of mass $m_1 = 3.0$ kg and $m_2 = 4.0$ kg are on a frictionless surface. A force of 6.0 N is applied to m_1, and a force of 4.0 N is applied to m_2; these antiparallel forces squeeze the blocks together. What is the force due to m_1 on m_2? Due to m_2 on m_1? What is the acceleration of the system?

61. A 220-lb man stands on a scale in an elevator. What does the scale read when the elevator accelerates upward at 1.6 m/s^2? What does it read when accelerating downward at the same rate?

62. You are on an elevator holding your backpack by a single loose strap. The backpack and its contents have a mass of 9.5 kg. What is the tension in the strap when the elevator accelerates upward at 1.9 m/s^2?

63. Each of a pair of dice hangs from the rearview mirror of a car by a string. When the car accelerates forward at 2.5 m/s^2, the strings make an angle with the vertical. What is the angle? If each die has a mass of 25 g, what is the tension in the string?

64. A horizontal force $F = 25.0$ N attempts to push a block up a frictionless inclined plane. If the block has a mass $m = 3.50$ kg and the incline makes an angle of 50.0° with the horizontal, does the force succeed? What is the acceleration of the block along the incline?

65. A boy on a skateboard rolls down a hill of slope 1:5. What is his acceleration? What speed will he reach after rolling for 50 m? Ignore friction.

66. A skier of mass 75 kg is sliding down a frictionless hillside inclined at 35° to the horizontal.

(a) Draw a "free-body" diagram showing all the forces acting on the skier (regarded as a particle); draw a separate diagram showing the resultant of these forces.

(b) What is the magnitude of each force? What is the magnitude of the resultant?

(c) What is the acceleration of the skier?

67. A bobsled slides down an icy track making an angle of 30° with the horizontal. How far must the bobsled slide in order to attain a speed of 90 km/h if initially at rest? When will it attain this speed? Assume that the motion is frictionless.

68. A man carrying a 20-kg sack on his shoulder rides in an elevator. What is the force the sack exerts on his shoulder when the elevator is accelerating upward at 2.0 m/s^2?

*69. Figure 5.57 shows a spherical ball hanging on a string on a smooth, frictionless wall. The mass of the ball is m, its radius is R, and the length of the string is l. Draw a "free-body" diagram with all the forces acting on the ball. Find the normal force between the ball and the wall. Show that $N \to 0$ as $l \to \infty$.

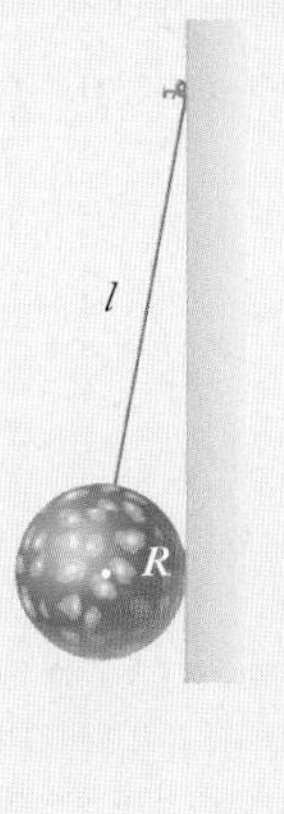

FIGURE 5.57 Ball on string against wall.

† For help, see Online Concept Tutorial 7 at www.wwnorton.com/physics

70. Figure 5.58 shows two masses hanging from a string running over a pulley. Such a device can be used to measure the acceleration of gravity; it is then called Atwood's machine. If the masses are nearly equal, then the acceleration a of the masses will be much smaller than g; that makes it convenient to measure a and then to calculate g by means of Eq. (5.44). Suppose that an experimenter using masses $m_1 = 400.0$ g and $m_2 = 402.0$ g finds that the masses move a distance of 0.50 m in 6.4 s starting from rest. What value of g does this imply? Assume the pulley is massless.

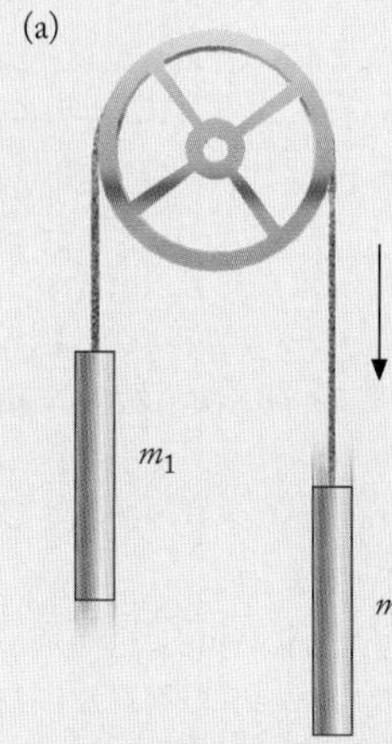

FIGURE 5.58 Two masses and a pulley.

*71. A woman pushes horizontally on a wooden box of mass 60 kg sitting on a frictionless ramp inclined at an angle of 30° (see Fig. 5.59).

(a) Draw the "free-body" diagram for the box.

(b) Calculate the magnitudes of all the forces acting on the box under the assumption that the box is at rest or in uniform motion along the ramp.

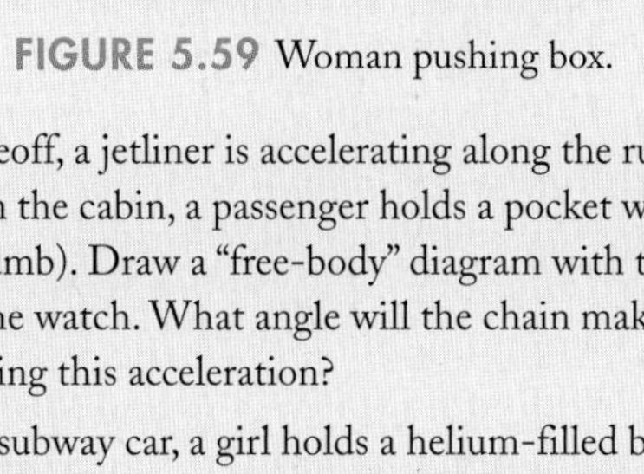

FIGURE 5.59 Woman pushing box.

*72. During takeoff, a jetliner is accelerating along the runway at 1.2 m/s^2. In the cabin, a passenger holds a pocket watch by a chain (a plumb). Draw a "free-body" diagram with the forces acting on the watch. What angle will the chain make with the vertical during this acceleration?

*73. In a closed subway car, a girl holds a helium-filled balloon by a string. While the car is traveling at constant velocity, the string of the balloon is exactly vertical. (Hint: For constant velocity, a buoyant force acts upward, but for constant deceleration, the buoyant force becomes tilted, but remains antiparallel to the tension.)

(a) While the subway car is braking, will the string be inclined forward or backward relative to the car?

(b) Suppose that the string is inclined at an angle of 20° with the vertical and remains there. What is the acceleration of the car?

**74. A string passes over a frictionless, massless pulley attached to the ceiling (see Fig. 5.60). A mass m_1 hangs from one end of this string, and a second massless, frictionless pulley hangs from the other end. A second string passes over the second pulley, and a mass m_2 hangs from one end of the string, whereas the other end is attached firmly to the ground. Draw separate "free-body" diagrams for the mass m_1, the second pulley, and the mass m_2. Find the accelerations of the mass m_1, the second pulley, and the mass m_2.

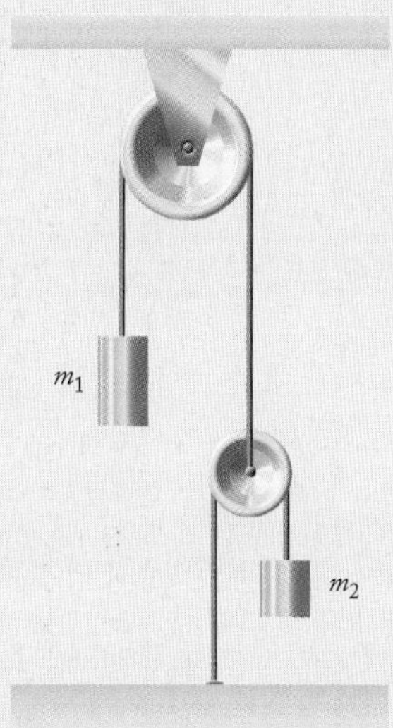

FIGURE 5.60 Two masses and two pulleys.

**75. A particle sliding down a frictionless ramp is to attain a given *horizontal* displacement Δx in a minimum amount of time. What is the best angle for the ramp? What is the minimum time?

**76. A mass m_1 hangs from one end of a string passing over a frictionless, massless pulley. A second frictionless, massless pulley hangs from the other end of the string (Fig. 5.61). Masses m_2 and m_3 hang from a second string passing over this second pulley. Find the acceleration of the three masses, and find the tensions in the two strings.

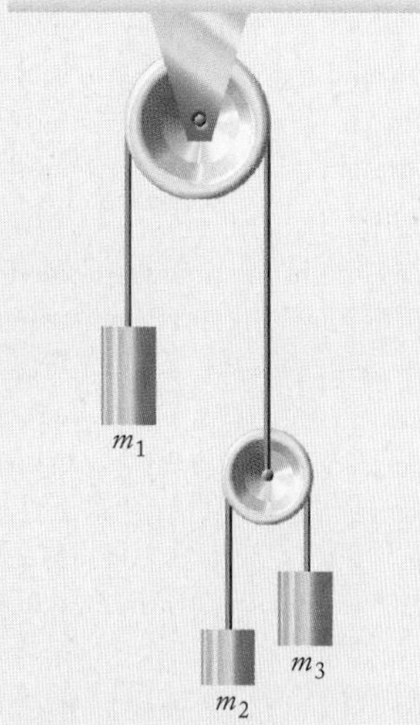

FIGURE 5.61 Three masses and two pulleys.

**77. A flexible cable of length l and mass m hangs over a small pulley. Initially, the cable is at rest, the length of the cable hanging on one side is more than $l/2$ by x_0, and the length of the cable hanging on the other side is less than $l/2$ by x_0 (see Fig. 5.62). What is the acceleration of the cable as a function of the distance x measured from the position of equal lengths? What is the position of the end of the cable as a function of time? (Hint: The differential equation $d^2x/dt^2 = kx$ has the solution $x = Ae^{\sqrt{k}t}$.)

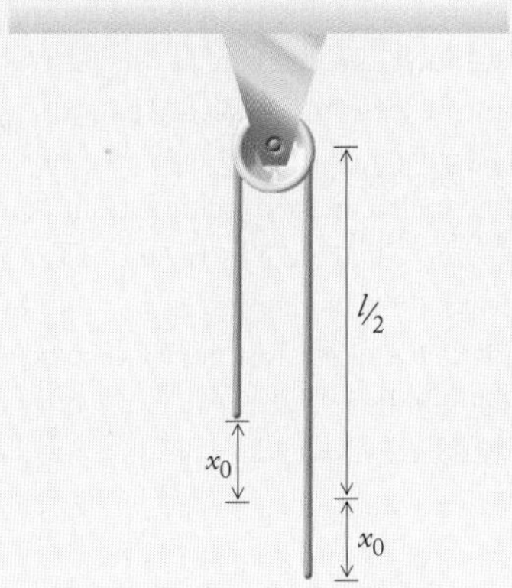

FIGURE 5.62 Massive cable and pulley.

REVIEW PROBLEMS

78. A 2000-kg truck, equipped with a winch, is trying to pull a car across a slippery (frictionless) sheet of ice by reeling in a cable attached to the car (see Fig. 5.63). Suppose that the truck is also on the sheet of ice, so the truck and the car both slip and accelerate toward each other with instantaneous accelerations of 1.2 m/s^2 and 2.5 m/s^2, respectively. What is the mass of the car?

FIGURE 5.63 Car and tow truck on frictionless surface.

79. Pushing with both hands, a sailor standing on a pier exerts a horizontal force of 270 N on a destroyer of 3400 metric tons. Assuming that the mooring ropes do not interfere and that the water offers no resistance, what is the acceleration of the ship? How far does the ship move in 60 s?

80. The speed of a projectile traveling horizontally and slowing down under the influence of air friction can be approximately represented by

$$v = 655.9 - 61.14t + 3.260t^2$$

where v is measured in meters per second and t in seconds; the mass of the projectile is 45.36 kg.

(a) What is the acceleration as a function of time?

(b) What is the force of air friction as a function of time? Can these formulas for the speed and the force remain valid when the speed becomes low ($v \approx 0$)?

81. Two forces $\mathbf{F}_1$ and $\mathbf{F}_2$ act on a particle of mass 6.0 kg. The forces are

$$\mathbf{F}_1 = 2\mathbf{i} - 5\mathbf{j} + 3\mathbf{k}$$

$$\mathbf{F}_2 = -4\mathbf{i} + 8\mathbf{j} + \mathbf{k}$$

where the force is measured in newtons.

(a) What is the net force vector?

(b) What is the acceleration vector of the particle, and what is the magnitude of the acceleration?

82. A diver of mass 75 kg is in free fall after jumping off a high platform.

(a) What is the force that the Earth exerts on the diver? What is the force that the diver exerts on the Earth?

(b) What is the acceleration of the diver? What is the acceleration of the Earth?

83. During a storm, a 2500-kg sailboat is anchored in a 10-m-deep harbor. The wind pushes against the boat with a horizontal force of 7000 N. The anchor rope that holds the boat in place is 50 m long and is stretched straight between the anchor and the boat (see Fig. 5.64).

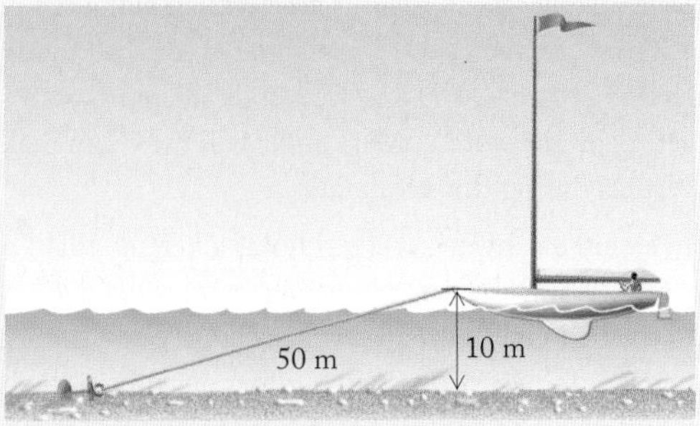

FIGURE 5.64 Anchored boat.

(a) Draw the "free-body" diagram for the boat. Be sure to include the upward force (buoyant force) that the water exerts on the boat, keeping it afloat.

(b) Calculate the tension in the anchor rope.

(c) Calculate the upward force (buoyant force) exerted by the water.

84. A box of mass 25 kg sits on a smooth, frictionless table. You push down on the box at an angle of 30° with a force of 80 N (see Fig. 5.65).

(a) Draw a "free-body" diagram for the box; include all the forces that act on the box.

(b) What is the acceleration of the box?

(c) What is the normal force that the table exerts on the box?

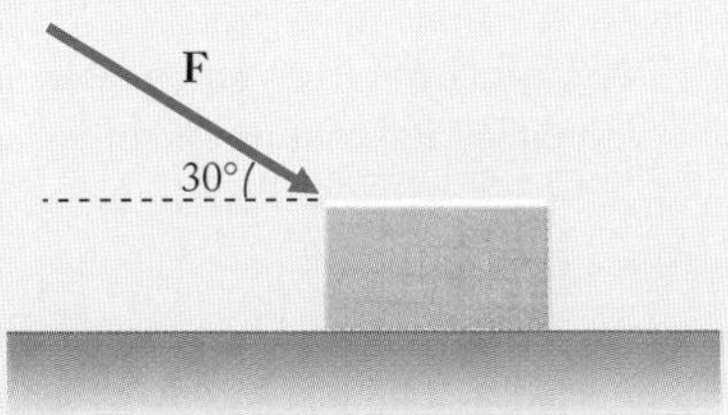

FIGURE 5.65 Pushing on a box.

85. A boy and a girl are engaged in a tug-of-war. Suppose that they are evenly matched, and neither moves. The boy pulls toward the left on the rope with a force of 250 N, and the girl pulls toward the right with a force of 250 N.

(a) Draw separate "free-body" diagrams for the boy, the girl, and the rope. In each of these diagrams include all the appropriate *horizontal* forces (ignore the vertical forces).

(b) What is the force that the ground exerts on the boy? On the girl?

(c) What is the tension in the rope?

(d) The girl ties her end of the rope to a stout tree and walks away, while the boy continues pulling as before. What is the tension in the rope now?

86. A long freight train consists of 250 cars each of mass 64 metric tons. The pull of the locomotive accelerates this train at the rate of 0.043 m/s^2 along a level track. What is the tension in the coupling that holds the first car to the locomotive? What is the tension in the coupling that holds the last car to the next-to-last car? Ignore friction.

87. The world's steepest railroad track, found in Guatemala, has a slope of 1:11 (see Fig. 5.66). A boxcar of 20 metric tons is being pulled up this track.

(a) Draw a "free-body" diagram for the boxcar.

(b) What force (along the track) is required to move the boxcar up the track at constant speed? Ignore friction and treat the motion as particle motion.

(c) What force is required to move the boxcar down the track at constant speed?

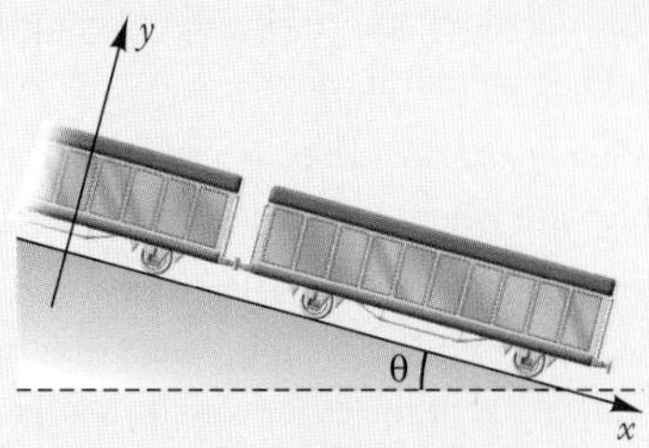

FIGURE 5.66 Boxcar on a steep railroad track.

88. To drag a heavy log of mass 500 kg up a ramp inclined at 30° to the horizontal, you attach the log to a cable that runs over a pulley at the top of the ramp, and you attach a 300-kg counterweight at the other end (see Fig. 5.67). Assume the log moves without friction.

(a) What is the acceleration of the log up the ramp?

(b) Could you use a smaller counterweight to move the log? How much smaller?

FIGURE 5.67 A log on a ramp with pulley and counterweight.

89. A 60-kg woman stands on a bathroom scale placed on the floor of an elevator.

(a) What does the scale read when the elevator is at rest?

(b) What does the scale read when the elevator is accelerating upward at 1.8 m/s^2?

(c) What does the scale read when the elevator is moving upward with constant velocity?

(d) What does the scale read if the cable of the elevator is cut (and the brakes have not yet engaged), so that the elevator is in free fall?

90. A crate of mass 2000 kg is hanging from a crane at the end of a cable 12 m long. If we attach a horizontal rope to this crate and gradually apply a pull of 1800 N, what angle will the cable finally make with the vertical?

*91. While a train is moving up a track of slope 1:10 at 50 km/h, the last car of the train suddenly becomes uncoupled. The car continues to roll up the slope for a while, then stops and rolls back.

(a) What is the deceleration of the car while it continues to roll up the slope? Assume there is no friction.

(b) How far along the track will the car roll before it stops and begins to roll back?

(c) What speed does the car attain when it has rolled back to the place where it first decoupled?

*92. An elevator consists of an elevator cage and a counterweight attached to the ends of a cable which runs over a pulley (see Fig. 5.68). The mass of the cage (with its load) is 1200 kg, and the mass of the counterweight is 1000 kg. Suppose that the elevator is moving upward at 1.5 m/s when its motor fails. The elevator then continues to coast upward until it stops and begins to fall down. Assume there is no friction, and assume the emergency brake of the elevator does not engage until the elevator begins to fall down. The brake locks the elevator cage to its guide rails.

(a) What is the deceleration of the elevator? How long does it take to stop?

(b) What is the tension in the cable when the elevator is coasting up?

(c) What is the tension in the cable when the emergency brake has engaged and the elevator is stopped?

FIGURE 5.68 Elevator with counterweight.

93. A mass m_1 sits on a horizontal, frictionless table, and is connected by a light string to a second, hanging mass m_2, as shown in Fig 5.69. (a) Find the acceleration of the system. (b) Find the tension in the string.

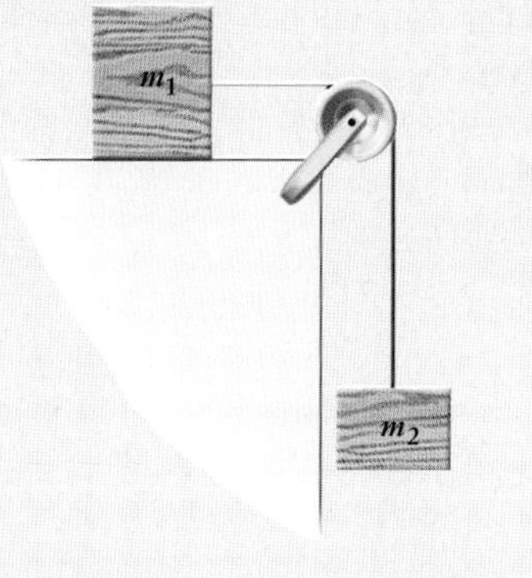

FIGURE 5.69 Mass m_1 on table with pulley and counterweight m_2.

94. Two masses m_1 and m_2 sit on a horizontal, frictionless table and are connected by a light string. Another light string connects the mass m_2 to a third, hanging mass m_3, as shown in Fig. 5.70.

(a) Find the acceleration of the system.

(b) Find the tension in each of the two strings.

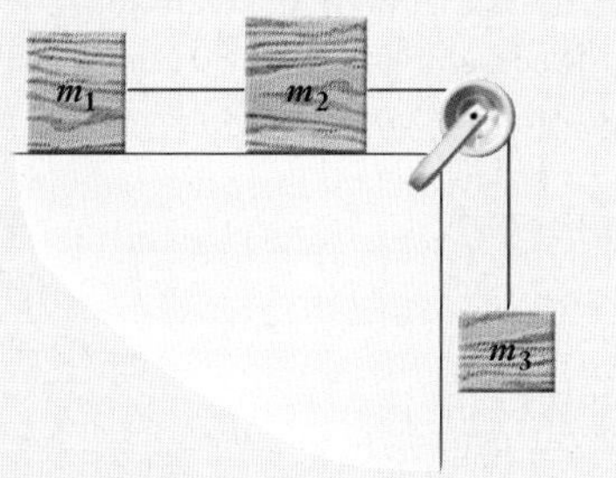

FIGURE 5.70 Two masses on table with pulley and counterweight.

Answers to Checkups

Checkup 5.1

1. No; the street exerts a friction force on the car's tires (the tires are both a source of friction and transfer friction forces from the axle to the road), and so an external pushing force is needed to counter it to maintain zero net external force.

2. No; the surface exerts friction forces on the ball that both roll and slow the ball. Thus, the ball does not have zero net external force on it, and so Newton's First Law does not apply.

3. Yes, the reference frame of a car traveling at constant speed along a straight, level road is an inertial reference frame (as

usual, we neglect the Earth's rotation). A car rounding a curve is accelerating centripetally, and so is not an inertial reference frame; similarly, a car braking on a level road is decelerating, and so is not an inertial reference frame.

4. Yes—while descending at constant speed, the elevator is inertial, since it moves at constant speed relative to an inertial reference frame, the Earth.
5. No—the diver is accelerating downward with $a_y = -g$, and so the reference frame of the diver is not an inertial reference frame.
6. (B) The reference frame of a bird descending at constant velocity. The other reference frames all involve accelerated motion; only the bird moves at constant velocity.

Checkup 5.2

1. You could push or pull each one horizontally with the same force; the one that accelerates more slowly has the greater mass.
2. In each case, the force is in the same direction, in the automobile's forward direction. The forces are merely applied to different points in each case.
3. No, the force is not an external force if the sailor is standing in the sailboat. The sailor's push on the mast is balanced by the push of his feet on the boat; wholly internal forces always cancel.
4. (C) 8, 2, 4. We assume that by "push on it equally" we mean each of the two men applies the same force. If two men push on the bobsled, the force is twice as large. By Newton's Second Law, $m\mathbf{a} = \mathbf{F}$, if the force is twice as large and the mass is the same, the acceleration will be twice as large, or 8 m/s^2. If the mass is twice as large, then the acceleration provided by a single man's force will be half as large, or 2 m/s^2; if both men push the doubled mass, then the doubling of both the force and the mass will leave the acceleration the same at 4 m/s^2.

Checkup 5.3

1. You can conclude that the net force is zero, since constant velocity means zero acceleration, and zero acceleration requires zero *net* force (Newton's Second Law). However, you cannot conclude that there are no forces: in this case of the parachutist, the downward force of gravity is balanced by the upward force of air friction.
2. When the elevator is at rest, there is zero net force on it. Only when the elevator is accelerating is there a net force; when accelerating upward, the net force is upward. When the elevator ascends at constant speed (zero acceleration), there is again zero net force on it.
3. When the subway car accelerates (and causes you to do so), the net external force on you is forward (even though you feel yourself being pushed backward in the reference frame of the car, this is merely your inertia trying to keep you at rest). When the car travels at constant velocity (zero acceleration), the net force on you is then zero ($ma = F$). During braking (deceleration), a negative acceleration means the net force on you is toward the rear of the car (even though you feel yourself lurch forward in the reference frame of the car, this is your inertia trying to maintain a constant velocity).
4. If both tugboats pull directly forward, then the magnitudes of their forces simply add (the two vectors are parallel), so that the net force is 2.5×10^5 N + 1.0×10^5 N = 3.5×10^5 N. If one pulls from the front and the other pushes from the rear, the two force vectors are still parallel, so the net force is again 3.5×10^5 N.
5. (B) 30° north of west. Since the two forces are of equal magnitude, the net force, and thus the acceleration, will be in a direction halfway between the two. Halfway between 30° west of north and west (which is 90° west of north) is 60° west of north, the same as 30° north of west.

Checkup 5.4

1. The force due to gravity is the stone's weight and, whether the stone is traveling up, or momentarily at rest, or traveling down, is always equal to $F = w = mg = 1.0 \text{ kg} \times 9.8 \text{ m/s}^2 = 9.8$ N in the downward direction.
2. No, the star has mass (inertia), and so will offer resistance to any attempt to accelerate it. That it is weightless reflects the fact that there are no external bodies exerting any measurable gravitational force on it.
3. Since weight is $w = mg$, your weight would be largest where the local value of the acceleration due to gravity, g, is largest, namely, Jupiter; your weight is smallest where g is smallest, on the Moon.
4. Apparent weightlessness will begin when the engines cut off (no external force from the engines' thrust); this is when the spacecraft begins coasting, that is, when free fall begins.
5. For the first book alone, the table exerts a normal force that balances the weight of the book, and so is 50 N upward. If a second, identical book is placed on top of the first, then the first book experiences two downward forces: its own weight (50 N) and a normal force on its top surface from the second book (50 N). Thus, the table exerts a normal force that balances both of these forces, and so is now 100 N upward.
6. (D) Smaller; larger. As the upward-traveling elevator decelerates, the passenger experiences a downward acceleration. Since her weight (a downward force) does not change, the normal force of the floor on her feet (an upward force) must be smaller than her weight to have a net downward force (and thus a net downward acceleration). For the downward-traveling elevator, the deceleration implies an upward acceleration. Thus the normal force (upward) is now larger than the passenger's weight.

Checkup 5.5

1. For an apple in free fall, your "free-body" diagram should show only the apple with its weight **w** as the only force acting on it (downward, acting at its center of mass). For an apple resting on a table, your diagram should show two forces: the apple's weight **w** acting downward at its center of mass, and the normal force of the table **N,** acting upward on the bottom of the apple; these two forces have the same magnitude (the two force vectors have the same length).
2. The reaction force to the apple's weight is the pull of gravity exerted by the apple on the Earth. The reaction force to the upward pull of the tree branch is the downward pull exerted by the apple on the tree branch.
3. During acceleration, the external force acts in the forward direction; while traveling at constant velocity, there is zero net external force on the car (no direction); while braking, the external force acts rearward on the car. The ground (the road) exerts these external forces on the tires of the car.
4. In both the cases of the supertanker and the rowboat, the vessels push backward against your feet with the same reaction force of 400 N. Because the supertanker has a much larger mass, your push of 400 N causes only a negligible acceleration, while for the much less massive rowboat, the same push of 400 N will cause a noticeable acceleration ($a = F/m$).
5. The top book has two forces acting upon it: its weight $\mathbf{w}_1$ downward and a normal force $\mathbf{N}_1$ of the lower book acting upward on the upper book; these two forces have equal magnitudes, since the book is not accelerating. The lower book has three forces acting on it: its downward weight $\mathbf{w}_2$, a downward normal force from the top book $\mathbf{N}_2$, and an upward normal force from the table $\mathbf{N}_3$; the two downward forces sum to the same magnitude as the upward force. Only the two normal forces of each book on the other are an action–reaction pair; that is, $\mathbf{N}_2 = -\mathbf{N}_1$.
6. If the man pulls with a force of 150 N on the rope, the rope pulls back on the man with the same force, and this is the tension in the rope, 150 N; the same tension acts at the woman's end. If the woman ties her end to a tree, the tension continues to be the same, since it still forms an action–reaction pair with the man's pull.
7. (D) 98 N; 98 N; 98 N. The tension T in the first, lower rope forms an action–reaction pair with the weight of the 10-kg mass, and so is $T = -w = -(-mg) = 10\ \text{kg} \times 9.8\ \text{m/s}^2 = 98\ \text{N}$. Since the spring scale and the ropes are assumed massless, this tension is transmited in full to the second (upper) rope, which then also has a tension of 98 N. The spring scale is set up to measure weight; since there is no net acceleration, it measures the actual weight of 98 N (we will examine the behavior of springs in detail in the next chapter).

Checkup 5.6

1. In the first part of the motion (low speed), friction is negligible ("she falls freely"), so the only force on the skydiver is the downward force due to her weight, $w = mg$. This is then the net force, and is nonzero. Later, reaching a uniform, terminal velocity implies zero acceleration; thus there is also zero net force. At that time, the force due to her weight (downward) is balanced by the force of air friction (upward).
2. In each case (zero or uniform velocity), there is zero acceleration, so there is zero net force. Thus, the upward tension exactly balances the weight, and the magnitude of the tension is $T = mg = 1000\ \text{kg} \times 9.81\ \text{m/s}^2 = 9810\ \text{N}$ in each case.
3. The acceleration may be changed, and the normal force is certainly changed. By pulling upward on the front, the vertical component of the force from the man is now opposite to its direction when pushing downward. If the vertical component of the pull is less than the weight, the normal force is reduced to a value equal to the difference between the weight and the vertical pull, and the vertical acceleration remains zero; if the vertical pull is greater than the weight, the normal force is zero, and the box accelerates vertically off the floor. In either case, the horizontal acceleration remains the same.
4. (a) Moving at constant speed up or down the plane implies no acceleration, so the force remains zero, and the horizontal push needed is the same as when the block was stationary, 100 N in each case. (b) If you push along the incline instead of horizontally, then more of your total push acts parallel to the motion (none merely opposes the normal force), so your push will have to be smaller to maintain zero net force. To calculate the value of push needed to balance the component of the weight along the incline, you would need to know the mass of the block and the angle θ of the incline [see Eq. (5.33)].
5. (D) b, c, a. For (1), the book diagram should have a weight vector downward, acting at the center of mass, and the normal force vector upward, on the surface in contact with the table; these two vectors are equal and opposite, as in (b). For (2), the box should also have a weight downward at the center of mass and an upward normal force on the bottom surface; in addition, there is another normal force from the second box downward on the top surface, as in (c). For (3), the lamp diagram should have a weight vector downward, acting at the center of mass, and a tension vector upward, acting at the point where the cord attaches to the lamp; the two vectors are equal and opposite, as in (a).

CHAPTER

6

Further Applications of Newton's Laws

Concepts — in — Context

CONCEPTS IN CONTEXT

Automobiles rely on the friction between the road and the tires to accelerate and to stop. We will see that one of two types of contact friction, kinetic or static, is involved. To see how these friction forces affect linear and circular motion, we ask:

? In an emergency, an automobile brakes with locked and skidding wheels. What deceleration can be achieved? (Example 1, page 176)

? What is the steepest slope of a street on which an automobile can rest without slipping? (Example 4, page 179)

? When braking without skidding, what maximum deceleration can be achieved? (Example 5, page 180)

? How quickly can a racing car round a curve without skidding sideways? (Example 10, page 186)

? How does a banked curve help to avoid skidding? (Example 11, page 186)

To find a solution of the equation of motion means to find a force **F** and a corresponding acceleration **a** such that Newton's equation $m\mathbf{a} = \mathbf{F}$ is satisfied. For a physicist, the typical problem involves a known force and an unknown motion; for example, the physicist knows the forces between the planets and the Sun, and she seeks to calculate the motion of these bodies. But for an engineer, the reverse problem with a known motion and an unknown force is often of practical importance; for example, the engineer knows that a train is to round a given curve at 60 km/h, and he seeks to calculate the forces that the track and the wheels must withstand. A special problem with a known motion is the problem of statics; here we know that the body is at rest (zero velocity and zero acceleration), and we wish to compute the forces that will maintain this condition of equilibrium. Thus, depending on the circumstances, we can regard either the right side or the left side of the equation $m\mathbf{a} = \mathbf{F}$ as an unknown that is to be calculated from what we know about the other side.

In the preceding chapter we found some solutions of the equation of motion with simple, constant forces, such as the weight and constant pushes or pulls. In this chapter we will examine further solutions of the equation of motion, and we will examine other, more complicated forces, such as friction and the forces exerted by springs.

Online Concept 8 Tutorial

6.1 FRICTION

Friction forces, which we have ignored up to now, play an important role in our environment and provide us with many interesting examples of motion with constant force. For instance, if the driver of a moving automobile suddenly slams on the brakes, the wheels will lock and begin to skid on the pavement. The skidding wheels experience an (approximately) constant friction force that opposes the motion and decelerates the automobile at an (approximately) constant rate of, say, 8 m/s^2. The magnitude of the friction force depends on the characteristics of the tires and the pavement; besides, the heavy friction of rubber wheels on a typical pavement is accompanied by abrasion of the wheels, which introduces additional complications.

For the sake of simplicity, let us focus on an idealized case of friction, involving a solid block of metal sliding on a flat surface of metal. Figure 6.1 shows a block of steel, in the shape of a brick, sliding on a tabletop of steel. If we give the block some initial velocity and then let it coast, friction will decelerate it. The forces acting on the block are the weight **w**, the normal force **N**, and the friction force **f**. The weight **w** acts downward with a magnitude mg. The normal force **N** exerted by the table on the block acts upward; the magnitude of this normal force must be mg, so that it balances the weight. The friction force **f** exerted by the table on the block acts horizontally, parallel to the tabletop, in a direction opposite to the motion. This force, like the normal force, is a contact force which acts over the entire bottom surface of the block; however, in Fig. 6.1 it is shown as though acting at the center of the surface.

LEONARDO da VINCI (1452–1519) *Italian artist, engineer, and scientist. Famous for his brilliant achievements in painting, sculpture, and architecture. Leonardo also made pioneering contributions to science. But Leonardo's investigations of friction were forgotten, and the laws of friction were rediscovered 200 years later by Guillaume Amontons, a French physicist.*

The friction force arises from adhesion between the two pieces of metal: the atoms in the block form bonds with the atoms in the tabletop, and when the block slides, these bonds are continually ruptured and formed again. The macroscopic friction force represents the effort required to rupture the microscopic bonds. Although at the microscopic level the phenomenon of friction is very complicated, at the macroscopic level the resulting friction force can often be described adequately by a simple empirical law, first enunciated by Leonardo da Vinci:

The magnitude of the friction force between unlubricated, dry surfaces sliding one over the other is proportional to the magnitude of the normal force acting on the surfaces and is independent of the area of contact and of the relative speed.

Friction involving surfaces in relative motion is called **sliding friction**, or **kinetic friction**. According to the above law, the magnitude of the force of kinetic friction can be written mathematically as

force of kinetic friction

$$f_k = \mu_k N \tag{6.1}$$

where μ_k is the **coefficient of kinetic friction**, a constant characteristic of the material involved. Table 6.1 lists typical friction coefficients for various materials.

Note that Eq. (6.1) states that the magnitudes of the friction force and the normal force are proportional. The *directions* of these forces are, however, quite different: the normal force $\mathbf{N}$ is perpendicular to the surface of contact, whereas the friction force $\mathbf{f}_k$ is parallel to this surface, in a direction opposite to that of the motion.

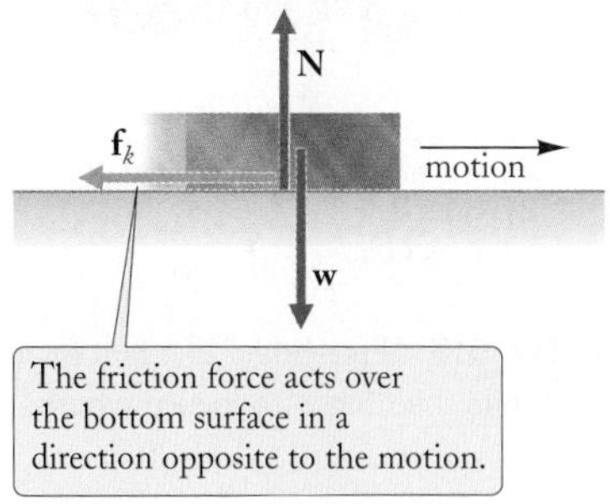

FIGURE 6.1 Forces on a block sliding on a plate.

The above simple "law" of friction lacks the general validity of, say, Newton's laws. It is only approximately valid, and it is phenomenological, which means that it is merely a descriptive summary of empirical observations which does not rest on any detailed theoretical understanding of the mechanism that causes friction. Deviations from this simple law occur at high speeds and at low speeds. However, we can ignore these deviations in many everyday engineering problems in which the speeds are not extreme. The simple friction law is then a reasonably good approximation for a wide range of materials, and it is at its best for metals sliding on metals.

The fact that the friction force is independent of the area of contact means that the friction force of the block sliding on the tabletop is the same whether the block slides on a large face or on one of the small faces (see Fig. 6.2). This may seem surprising at first—we might expect the friction force to be larger when the block slides on the larger face, with more area in contact with the tabletop. However, the normal force is then distributed over a larger area, and is therefore less effective in pressing the atoms together; and the net result is that the friction force is independent of the area of contact.

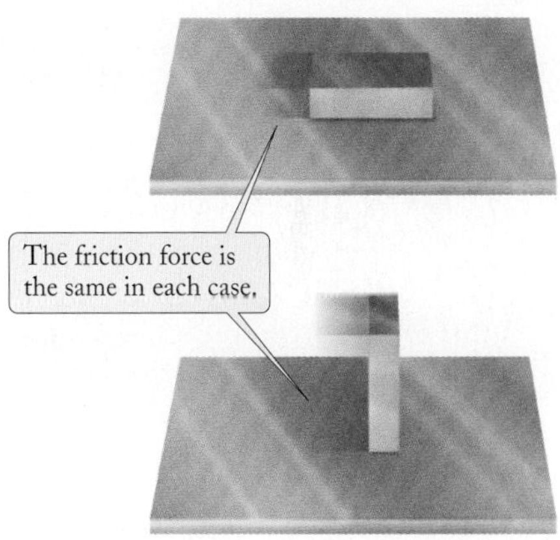

FIGURE 6.2 Steel block on a steel plate, sliding on a large face or on a small face.

TABLE 6.1 KINETIC AND STATIC FRICTION COEFFICIENTS[a]

MATERIALS	μ_k	μ_s
Steel on steel	0.6	0.7
Steel on lead	0.9	0.9
Steel on copper	0.4	0.5
Copper on cast iron	0.3	1.1
Copper on glass	0.5	0.7
Waxed ski on snow		
at −10°C	0.2	—
at 0°C	0.05	—
Rubber on concrete	≈1	≈1

[a] The friction coefficient depends on the condition of the surfaces. The values in this table are typical for dry surfaces but not entirely reliable.

Concepts in Context

EXAMPLE 1 Suppose that the coefficient of kinetic friction of the hard rubber of an automobile tire sliding on the pavement of a street is $\mu_k = 0.8$. What is the deceleration of an automobile on a flat street if the driver brakes sharply, so all the wheels are locked and skidding? (Assume the vehicle is an economy model without an antilock braking system.)

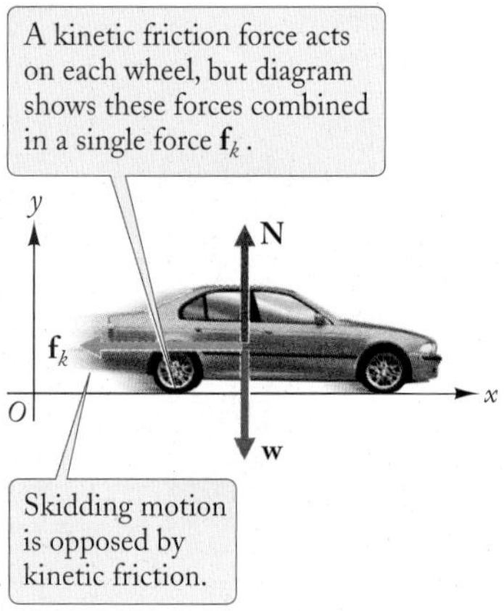

FIGURE 6.3 "Free-body" diagram for an automobile skidding with locked wheels.

SOLUTION: Figure 6.3 shows the "free-body" diagram with all the forces on the automobile. These forces are the weight **w**, the normal force **N** exerted by the street, and the friction force $\mathbf{f}_k$. The normal force must balance the weight; hence the magnitude of the normal force is the same as the magnitude of the weight, or $N = w = mg$. According to Eq. (6.1), the magnitude of the friction force is then

$$f_k = \mu_k N = 0.8 \times mg$$

Since this friction force is the only horizontal force on the automobile, the deceleration of the automobile along the street is

$$a_x = -\frac{f_k}{m} = -\frac{0.8 \times mg}{m} = -0.8 \times g = -0.8 \times 9.8\ \text{m/s}^2$$
$$= -8\ \text{m/s}^2$$

COMMENT: The normal forces and the friction forces act on all the four wheels of the automobile; but in Fig. 6.3 (and in other "free-body" diagrams in this chapter) these forces have been combined into a net force **N** and a net friction force $\mathrm{f_k}$, which, for convenience, are shown as though acting at the center of the automobile. To the extent that the motion is treated as purely translational motion (that is, particle motion), it makes no difference at what point of the automobile the forces act. Later, in Chapter 13, we will study how forces affect the rotational motion of bodies, and it will then become important to keep track of the exact point at which each force acts.

EXAMPLE 2 A ship is launched toward the water on a slipway making an angle of 5° with the horizontal direction (see Fig. 6.4). The coefficient of kinetic friction between the bottom of the ship and the slipway is $\mu_k = 0.08$. What is the acceleration of the ship along the slipway? What is the speed of the ship after accelerating from rest through a distance of 120 m down the slipway to the water?

SOLUTION: Figure 6.4b is the "free-body" diagram for the ship. The forces shown are the weight **w**, the normal force exerted by the slipway **N**, and the friction force $\mathbf{f}_k$. The magnitude of the weight is $w = mg$.

Since there is no motion in the direction perpendicular to the slipway, we find, as in Eq. (5.36), that the normal force is

$$N = mg\cos\theta$$

and the magnitude of the friction force is

$$f_k = \mu_k N = \mu_k mg\cos\theta \tag{6.2}$$

With the x axis parallel to the slipway, the x component of the weight is (see Fig. 6.4c)

$$w_x = mg\sin\theta$$

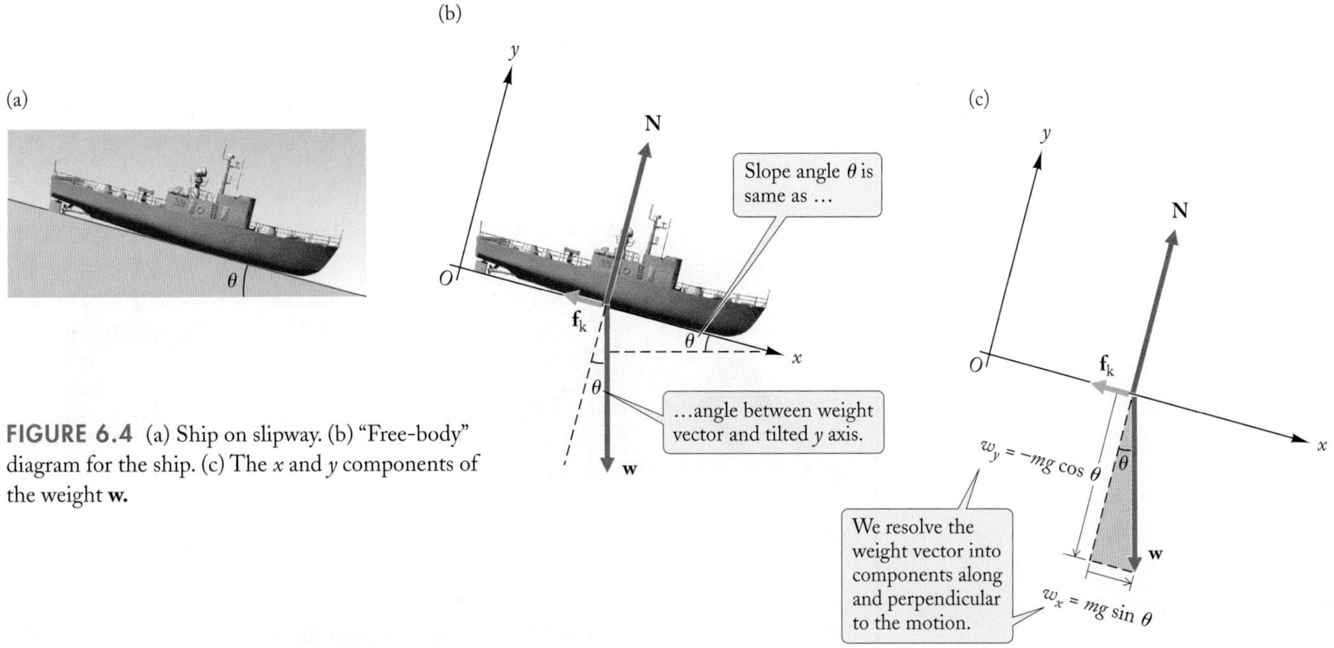

FIGURE 6.4 (a) Ship on slipway. (b) "Free-body" diagram for the ship. (c) The x and y components of the weight **w**.

The x component of the net force is then

$$F_x = w_x - f_k = mg\sin\theta - \mu_k mg\cos\theta$$

Hence the acceleration of the ship along the slipway is

$$a_x = \frac{F_x}{m} = \frac{mg\sin\theta - \mu_k mg\cos\theta}{m} = (\sin\theta - \mu_k\cos\theta)g \qquad (6.3)$$

Note that in this equation the mass has canceled—the acceleration is the same for a large ship and a small ship. With $\theta = 5°$ and $\mu_k = 0.08$, Eq. (6.3) gives

$$a_x = (\sin 5° - 0.08 \times \cos 5°) \times 9.81\ \text{m/s}^2 = 0.07\ \text{m/s}^2$$

From kinematics, Eqs. (4.20)–(4.24), we know that the velocity and displacement after constant acceleration from rest ($v_{0x} = 0$) will be

$$v_x = a_x t$$

$$x - x_0 = \tfrac{1}{2}a_x t^2$$

We can solve for the time t in the second equation:

$$t = \sqrt{\frac{2(x - x_0)}{a_x}}$$

and substitute it into the first:

$$v_x = a_x \times \sqrt{\frac{2(x - x_0)}{a_x}} = \sqrt{2(x - x_0)a_x}$$

$$= \sqrt{2 \times 120\ \text{m} \times 0.07\ \text{m/s}^2} = 4\ \text{m/s}$$

This is the speed of the ship as it enters the water.

EXAMPLE 3 A man pushes a heavy crate over a floor. The man pushes downward and forward, so his push makes an angle of 30° with the horizontal (Fig. 6.5a). The mass of the crate is 60 kg, and the coefficient of sliding friction is $\mu_k = 0.50$. What force must the man exert to keep the crate moving at uniform velocity?

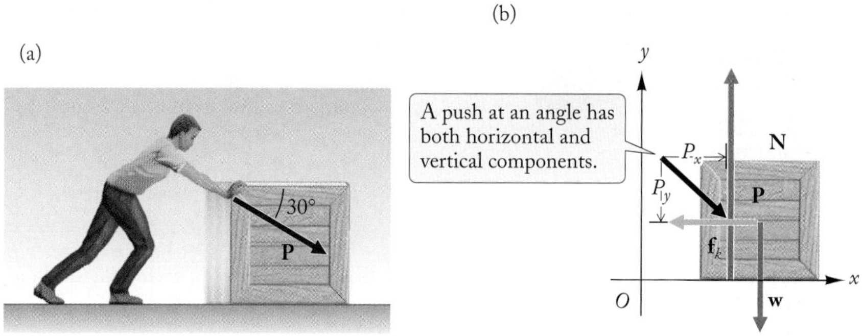

FIGURE 6.5 (a) Man pushing a crate. (b) "Free-body" diagram for the crate.

SOLUTION: Figure 6.5b is a "free-body" diagram for the crate. The forces on the crate are the push **P** of the man, the weight **w**, the normal force **N**, and the friction force $\mathbf{f}_k$. Note that because the man pushes the crate down against the floor, the magnitude of the normal force is not equal to *mg;* we will have to treat the magnitude of the normal force as unknown. Taking the *x* axis horizontal and the *y* axis vertical, we see from Fig. 6.5b that the *x* and *y* components of the forces are (see also Fig. 5.37)

$$P_x = P\cos 30^\circ \qquad P_y = -P\sin 30^\circ$$
$$w_x = 0 \qquad w_y = -mg$$
$$N_x = 0 \qquad N_y = N$$
$$f_{k,x} = -\mu_k N \qquad f_{k,y} = 0$$

Since the acceleration of the crate is zero in both the *x* and the *y* directions, the net force in each of these directions must be zero:

$$P\cos 30^\circ + 0 + 0 - \mu_k N = 0$$
$$-P\sin 30^\circ - mg + N + 0 = 0$$

These are two equations for the two unknowns *P* and *N*. By multiplying the second equation by μ_k and then adding the resulting equation to the first, we can eliminate *N*, and we find an equation for *P*:

$$P\cos 30^\circ - \mu_k P\sin 30^\circ - \mu_k mg = 0$$

Solving this for *P*, we find

$$P = \frac{\mu_k mg}{\cos 30^\circ - \mu_k \sin 30^\circ} = \frac{0.50 \times 60\text{ kg} \times 9.81\text{ m/s}^2}{\cos 30^\circ - 0.50 \times \sin 30^\circ} \qquad (6.4)$$
$$= 4.8 \times 10^2\text{ N}$$

Friction forces also act between two surfaces at rest. If we exert a force against the side of, say, a steel block initially at rest on a steel tabletop, the block will not move unless the force is sufficiently large to overcome the friction that holds it in place. Friction between surfaces at rest is called **static friction**. The maximum magnitude of the static friction force, that is, the magnitude that this force attains when the lateral push is just about to start the motion, can be described by an empirical law quite similar to that for the kinetic friction force:

The magnitude of the maximum static friction force between unlubricated, dry surfaces at rest with respect to each other is proportional to the magnitude of the normal force and independent of the area of contact.

Mathematically,

$$f_{s,\max} = \mu_s N \tag{6.5}$$

force of static friction

Here μ_s is a constant of proportionality, called the **coefficient of static friction**, which depends on the material. The direction of the static friction force is parallel to the surface, so as to oppose the total lateral push that tries to move the body (like the force **F** in Fig. 6.6).

The force in Eq. (6.5) is labeled with the subscript "max" because it represents the largest friction force that the surfaces can support without beginning to slide; in other words, $f_{s,\max}$ is the force at the "breakaway" point, when the lateral push is just about to start the motion. Of course, *if the lateral push is less than this critical value, then the static friction force f_s is less than $f_{s,\max}$ and its magnitude exactly matches the magnitude of the total lateral push.* This makes the net force on the block zero, as required if the block is to remain at rest.

Table 6.1 includes some typical values of the coefficient of static friction. *For most materials μ_s is larger than μ_k,* and therefore the maximum static friction force is larger than the kinetic friction force. This implies that if the lateral push applied to the block is large enough to overcome the static friction and start the block moving, it will more than compensate for the subsequent, smaller kinetic friction, and it will therefore accelerate the block continuously.

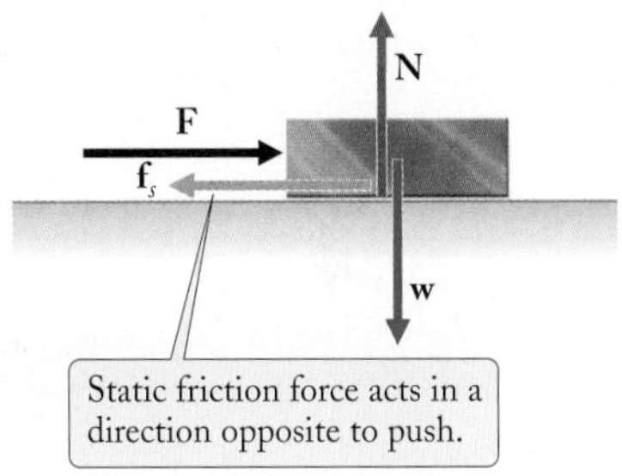

FIGURE 6.6 Forces on a steel block at rest on a steel plate. The friction force $\mathbf{f}_s$ has the same magnitude as the force **F**.

EXAMPLE 4 The coefficient of static friction of the rubber of an automobile tire on a street surface is $\mu_s = 0.90$. What is the steepest slope of a street on which an automobile with such tires (and locked wheels) can rest without slipping?

Concepts — in — Context

SOLUTION: The "free-body" diagram is shown in Fig. 6.7. The angle θ is assumed to be at its maximum value, so that the friction force has its maximum value $f_{s,\max} = \mu_s N$. As in Example 2, $N = mg\cos\theta$, and hence $f_{s,\max} = \mu_s mg\cos\theta$. With the x axis parallel to the street surface, we then find that the x component of the net force is

$$F_x = w_x - f_{s,\max} = mg\sin\theta - \mu_s\, mg\cos\theta \tag{6.6}$$

This component of the force determines the motion along the street. If the automobile is to remain stationary, F_x must be zero:

$$0 = mg\sin\theta - \mu_s mg\cos\theta$$

Hence, dividing by mg,

$$\sin\theta = \mu_s \cos\theta$$

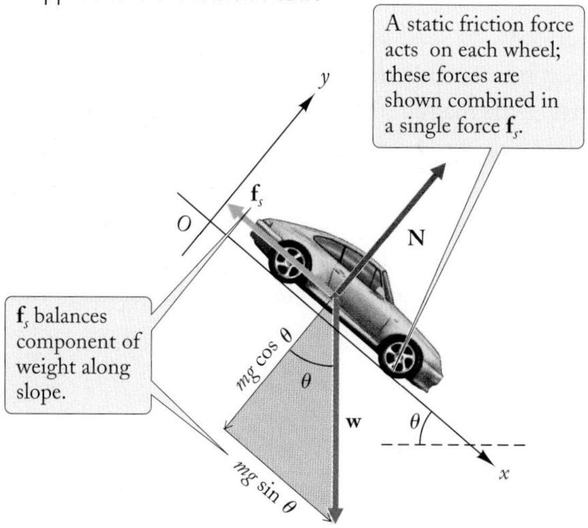

FIGURE 6.7 "Free-body" diagram for an automobile parked on a very steep street.

or, dividing this equation by cos θ,

$$\tan\theta = \mu_s \tag{6.7}$$

With $\mu_s = 0.90$, this gives $\tan\theta = 0.90$. Thus, the slope of the street is 0.90, or 9:10. (With a calculator, we find that the inverse tangent of 0.90 gives $\theta = 42°$ for the angle of the incline.)

Concepts — in — Context

EXAMPLE 5 An automobile is braking on a level road. What is the maximum deceleration that the automobile can achieve when it brakes without skidding? As in the preceding example, assume that the tires of the automobile have a coefficient of static friction $\mu_s = 0.9$.

SOLUTION: Figure 6.8 shows the "free-body" diagram. If wheels are rolling without skidding, their rubber surface does *not* slide on the street surface; that is, the point of contact between the rolling wheel and the street is instantaneously at rest on the street (you can easily convince yourself of this by rolling any round object on a tabletop). Since there is no sliding, the relevant friction force is the *static* friction force. The maximum value of this force is

$$f_{s,\text{max}} = \mu_s N \tag{6.8}$$

Here the magnitude of the normal force N is simply mg, since the normal force must balance the weight. The deceleration is then given by

$$a_x = -\frac{f_{s,\text{max}}}{m} = -\frac{\mu_s mg}{m} = -\mu_s g \tag{6.9}$$

which yields

$$a_x = -0.9 \times 9.8\ \text{m/s}^2 = -9\ \text{m/s}^2$$

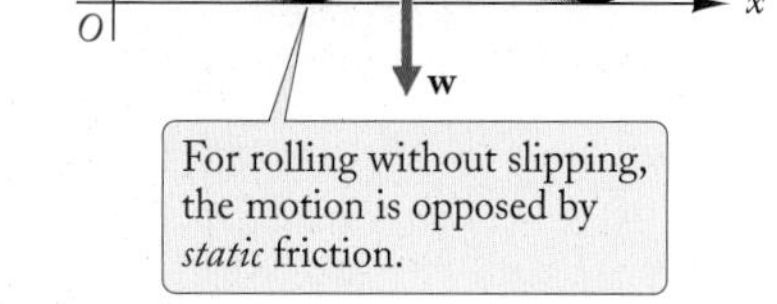

FIGURE 6.8 "Free-body" diagram for an automobile braking without skidding.

COMMENTS: Note that this is a larger deceleration than in Example 1, where the automobile skidded, because the coefficient of static friction is larger than that of kinetic friction. A car equipped with an antilock braking system (ABS) can exploit this difference and avoid skidding by rapid, repeated application of the brakes. The ABS also permits the driver to maintain directional control during rapid braking.

Note that throughout this section we have dealt only with the friction forces that act between solid surfaces in contact. There are also other friction forces that act when a solid body moves through a liquid or a gas, for instance, the friction experienced by an automobile moving through air. The magnitude of these **drag forces** can depend in a complicated way on the shape of the body, the speed of its motion, and the properties of the liquid or gas.

At low speeds in a liquid or gas, the friction force which opposes motion through the medium is very nearly proportional to the speed; this low-speed friction is due to the viscocity, or stickiness, of the medium; we explore such **viscous forces** in Problems 25 and 26. For automobiles or projectiles, the force due to **air resistance** at moderate to high speeds varies instead in direct proportion to the *square* of speed. The magnitude of the force can be written

$$f_{\text{air}} = \tfrac{1}{2}C\rho A v^2 \qquad (6.10)$$

force due to air resistance

Here, C is a dimensionless constant related to the shape of the body (its "aerodynamic design"), ρ is the density of air ($\rho = 1.3\ \text{kg/m}^3$ for dry air), and A is the cross-sectional area, perpendicular to the motion, of the automobile. For modern automobiles, values of C are in the range 0.3–0.5.

EXAMPLE 6 A manufacturer quotes an aerodynamic constant $C = 0.30$ for an automobile of mass 900 kg and cross-sectional area $A = 2.8\ \text{m}^2$. If the driver were to coast (in neutral) down a long hill with a slope of 8.0° (see Fig. 6.9a), what would be the terminal velocity? Recall from Section 2.6 that terminal velocity is the constant final velocity of a body moving under the combined influence of gravity and air resistance. Assume that air resistance is the only source of friction.

SOLUTION: We need only consider the components of the forces parallel to the motion. From Fig. 6.9b and Eq. (6.10), these are

$$w_x = mg\sin\theta \quad \text{and} \quad f_{\text{air},x} = -\tfrac{1}{2}C\rho A v^2$$

When terminal (constant) velocity is reached, the two forces balance and the acceleration is zero. Thus,

$$\tfrac{1}{2}C\rho A v^2 = mg\sin\theta$$

Solving for the velocity, we find

$$v = \sqrt{\frac{2\,mg\sin\theta}{C\rho A}}$$

Inserting the values given in the problem, plus the air density $\rho = 1.3\ \text{kg/m}^3$ from above, we have

$$v = \sqrt{\frac{2 \times 900\ \text{kg} \times 9.81\ \text{m/s}^2 \times \sin 8.0^\circ}{0.30 \times 1.3\ \text{kg/m}^3 \times 2.8\ \text{m}^2}}$$

$$= 47\ \text{m/s}$$

This is the same as

$$v = 47\ \text{m/s} \times \frac{1\ \text{km}}{1000\ \text{m}} \times \frac{3600\ \text{s}}{1\ \text{h}} = 170\ \text{km/h}$$

On such a slope, considerable speed can be attained before air resistance limits the motion.

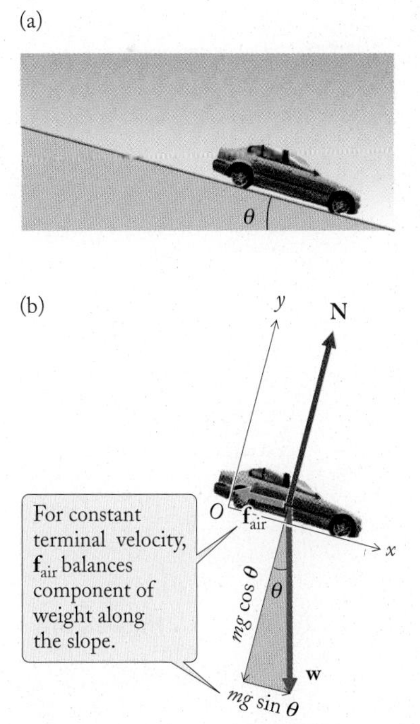

FIGURE 6.9 (a) An automobile coasting down a hill. (b) "Free-body" diagram for the automobile, including air resistance.

✓ Checkup 6.1

QUESTION 1: To overcome friction and to keep a book moving at constant speed on a flat table, you must push with a horizontal force of 10 N. If another, equal book is piled on the first, what horizontal force will be required to keep the books moving at constant speed? What if instead of piling the second book on top of the first you place it on the table in front of the first, so the first book pushes the second?

QUESTION 2: A block is sliding on an inclined plane, with friction. The block coasts up the plane, stops, and then slides down. What is the direction of the friction force while the block is coasting up the plane? While the block is sliding down the plane? Draw "free-body" diagrams for the block in these two cases.

QUESTION 3: You exert a lateral push on a brick resting on a surface with friction. If the brick remains at rest, you can conclude that the friction force is (a) equal to $\mu_s N$, (b) equal to $2\mu_s N$, or (c) smaller than or equal to $\mu_s N$.

QUESTION 4: A box sits on the floor of a delivery truck. If the truck brakes sharply, the box slides forward, but if the truck brakes gently, the box does not slide. Explain.

QUESTION 5: If the automobile of Example 6 had an extremely optimized aerodynamic constant of $C = 0.15$, by what factor would its terminal velocity increase?

QUESTION 6: A box is sliding down a ramp. Consider the mass m of the box, the friction coefficient μ_k, and the angle θ of the ramp. On which of these does the acceleration of the box depend?

(A) m and μ_k only
(B) m and θ only
(C) μ_k and θ only
(D) m, μ_k, and θ

6.2 RESTORING FORCE OF A SPRING; HOOKE'S LAW

A body is said to be **elastic** if it suffers a deformation when subjected to a stretching or compressing force, and if it returns to its original shape when the force is removed. For example, suitable forces can stretch a coil spring or a rubber band, and they can bend a flexible rod or a beam of metal or wood. Even bodies normally regarded as rigid, such as the balls of a ball bearing made of hardened steel, are somewhat elastic, and they experience slight deformations; but these deformations can be neglected unless the force is extremely large.

The force with which a body resists deformation is called its **restoring force**. If we stretch a spring by pulling with a hand at one end (see Fig. 6.10), we can feel the restoring force opposing our pull.

Under static conditions, the restoring force with which an elastic body opposes whatever pulls on it often obeys a simple empirical law known as **Hooke's Law**:

Hooke's Law

The magnitude of the restoring force is directly proportional to the deformation.

This law, like the law for friction, is not a general law of physics—the exact restoring force produced by the deformation of an elastic body depends in a complicated way on the shape of the body and on the detailed properties of the material of the body. Hooke's Law is only an approximate description of the restoring force. However, it is often a quite good approximation, provided the restoring force and the deformation are small.

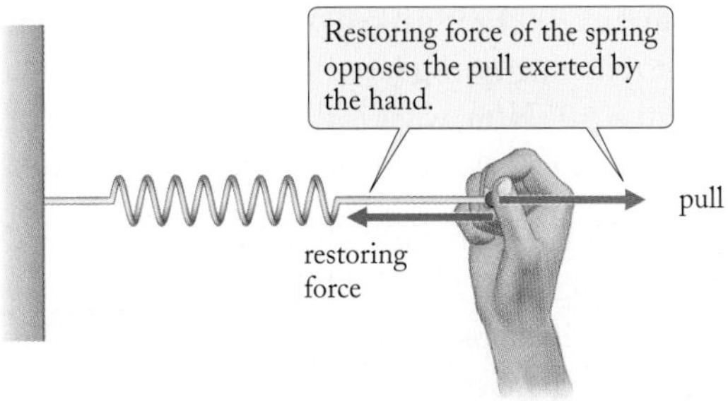

FIGURE 6.10 Restoring force of a stretched spring. The more the spring is stretched, the stronger the restoring force.

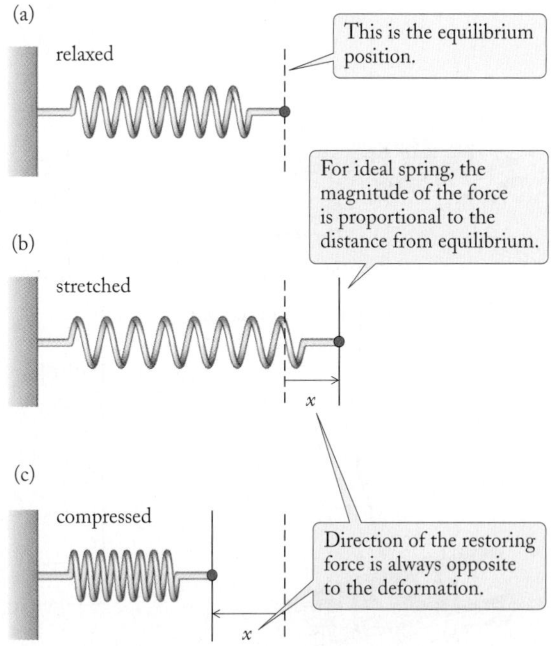

FIGURE 6.11 (a) Spring, relaxed. (b) Spring, stretched by a length x. (c) Spring, compressed by a length x.

As an example, consider a coil spring. Figure 6.11a shows such a spring in its relaxed state; it is loosely coiled and can therefore be compressed as well as stretched. Suppose we attach the left end of the spring to a wall and we apply a stretching or compressing force to the right end. Under the influence of this force, the spring will settle into a new equilibrium configuration such that the restoring force exactly balances the externally applied force. We can measure the deformation of the spring by the displacement that the right end undergoes relative to its initial position. In Fig. 6.11b, this displacement is denoted by x. Clearly, x is simply the change in the length of the spring. A positive value of x corresponds to an elongation, or stretching, of the spring, and a negative value of x corresponds to a compression.

Expressed mathematically, Hooke's Law then says that the restoring force opposes and is directly proportional to the displacement x:

$$F = -kx \tag{6.11}$$

restoring force of a spring

The constant of proportionality k is the **spring constant**; it is a positive number characteristic of the spring. The spring constant is a measure of the stiffness of the spring—a stiff spring has a high value of k, and a soft spring has a low value of k. The units for the spring constant are newtons per meter (N/m). The negative sign in Eq. (6.11) indicates that the restoring force opposes the deformation; if the spring in Fig. 6.11 is elongated (positive x), then the restoring force is negative and opposes the external stretching force; if the spring is compressed (negative x), then the restoring force is positive and opposes the external compressing force.

EXAMPLE 7 The manufacturer's specifications for the coil spring for the front suspension of a Triumph sports car call for a spring with a relaxed length of 0.316 m, and a length of 0.205 m when under a load of 399 kg. What is the spring constant?

SOLUTION: The weight of 399 kg is $w = mg = 399\ \text{kg} \times 9.81\ \text{m/s}^2 = 3.91 \times 10^3$ N. The magnitude of the restoring force that will balance this weight must then also be 3.91×10^3 N. For the given relaxed and compressed lengths, the corresponding change of length is $x = 0.205\ \text{m} - 0.316\ \text{m} = -0.111$ m. Hence, Eq. (6.11) gives us

$$k = -\frac{F}{x} = -\frac{3.91 \times 10^3\ \text{N}}{-0.111\ \text{m}} = 3.53 \times 10^4\ \text{N/m}$$

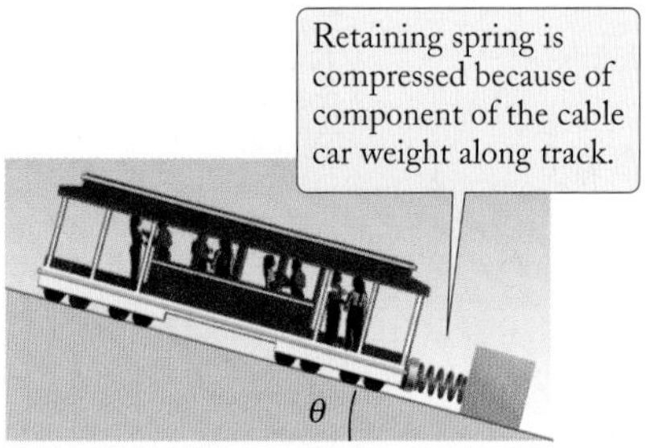

FIGURE 6.12 A cable car and retaining spring at a terminal station.

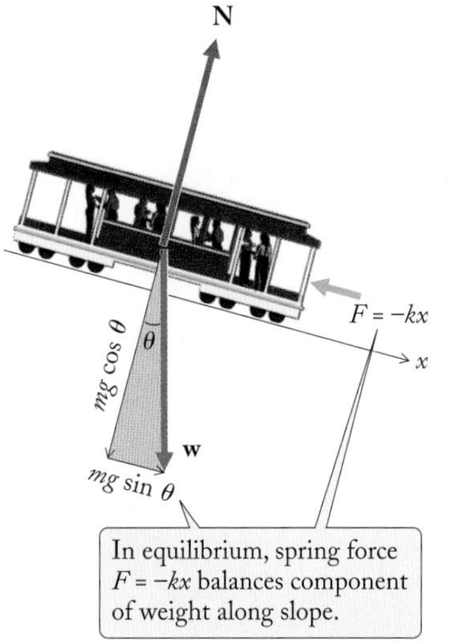

FIGURE 6.13 "Free-body" diagram for cable car parked against a retaining spring.

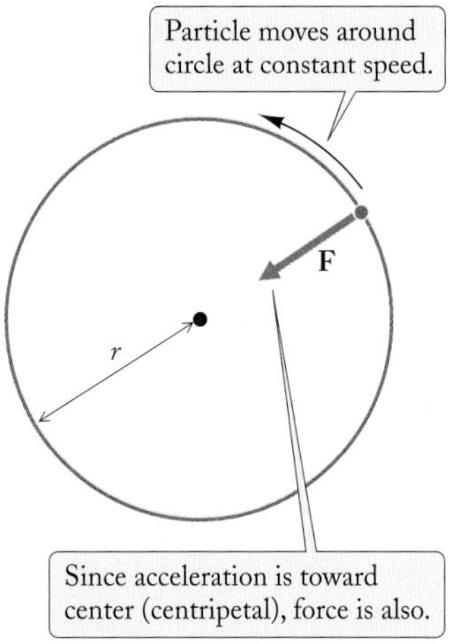

FIGURE 6.14 A particle in uniform circular motion. The force that acts on the particle is directed toward the center of the circle.

EXAMPLE 8 A cable car of mass 1200 kg is parked on a slope of 20°, compressing a giant retaining spring (see Fig. 6.12) to a length of 0.75 m. If the spring constant is 2.0×10^4 N/m, what is the length of the spring when relaxed? Neglect friction.

SOLUTION: A "free-body" diagram for the cable car is shown in Fig. 6.13, with the x axis parallel to the incline. The spring force $F = -kx$ balances the component of the cable car's weight that is parallel to the slope, $w_x = mg\sin\theta$. For the stationary cable car, the x component of the net force must be zero:

$$0 = mg\sin\theta - kx$$

This implies the length has been compressed by

$$x = \frac{mg\sin\theta}{k} = \frac{1200\text{ kg} \times 9.81\text{ m/s}^2 \times \sin 20°}{2.0 \times 10^4\text{ N/m}}$$
$$= 0.20\text{ m}$$

Thus the relaxed length of the spring is 0.75 m + 0.20 m = 0.95 m.

✔ Checkup 6.2

QUESTION 1: Nylon strings and ropes are elastic when stretched. Suppose that a mountain climber hanging from a long nylon rope stretches it by 20 cm. If two mountain climbers of equal mass hang from this rope, by how much will the rope stretch?

QUESTION 2: A spring is attached horizontally to a mass $m = 1.0$ kg, which sits on a table. The spring stretches an amount $x = 0.10$ m before the mass starts to move. If the spring constant is $k = 50$ N/m, what is the coefficient of static friction between the mass and the table?

QUESTION 3: A force F compresses a first spring by an amount x_1. If a second spring has twice the spring constant and a force $\frac{1}{2}F$ compresses it by an amount x_2, then the ratio x_2/x_1 is:

(A) 4 (B) 2 (C) 1 (D) $\frac{1}{2}$ (E) $\frac{1}{4}$

6.3 FORCE FOR UNIFORM CIRCULAR MOTION

All the examples of applications of Newton's laws we have examined so far involved particles moving along straight lines. But Newton's laws are also valid for motion along curved paths, for instance, motion with uniform speed along a circular path. As we saw in Section 4.5, such uniform circular motion is accelerated motion with a centripetal acceleration. If the motion proceeds with speed v around a circle of radius r, Eq. (4.49) tells us that the magnitude of the centripetal acceleration is

$$a = \frac{v^2}{r} \tag{6.12}$$

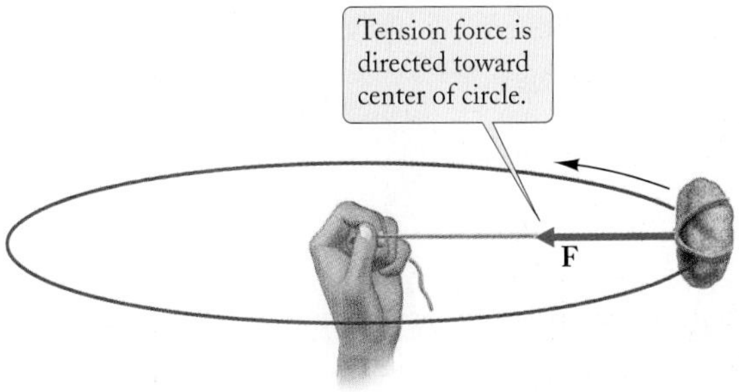

FIGURE 6.15 A stone being whirled around a circle. The string must exert a pull toward the center of the circle to produce a centripetal acceleration and to keep the stone in uniform circular motion.

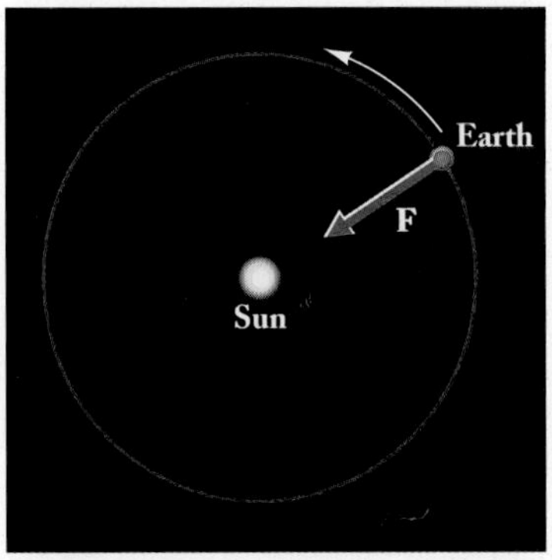

FIGURE 6.16 The Earth in its circular orbit around the Sun. The gravitational pull of the Sun produces the centripetal acceleration.

According to Newton's Second Law, this acceleration must be caused by a net force having the same direction as that of the acceleration; that is, *the direction of the net force must be centripetal, toward the center of the circle* (see Fig. 6.14). Such a force directed toward the center is called a **centripetal force**. For instance, the centripetal acceleration of a stone being whirled around a circle at the end of a string is caused by the pull of the string toward the center of the circle (see Fig. 6.15), and the centripetal acceleration of the Earth moving in its orbit around the Sun is caused by the gravitational pull toward the Sun (see Fig. 6.16).

The magnitude of the centripetal force required to maintain uniform circular motion is

$$F = ma = \frac{mv^2}{r} \tag{6.13}$$

centripetal force for circular motion

Note that this equation does not tell us how the force is produced. It is not a law of force, such as the law of friction or Hooke's Law, that tells us how to relate the force to the characteristics of the materials involved in producing the force. Instead, Eq. (6.13) merely tells us what magnitude of force we must produce, somehow, to keep the body in circular motion. For instance, to keep a stone whirling around a circle at the end of a string, we must exert this force with the string. If the string suddenly breaks, the stone will fly off in the direction of its instantaneous velocity, that is, in the direction of a tangent to the circle.

Equation (6.13) can be used to calculate the magnitude of the centripetal force required if the speed of the motion is known, or it can be used to calculate the speed if the force is known. The following examples illustrate such calculations with different kinds of forces.

EXAMPLE 9 In the hammer throw, an athlete launches a "hammer" consisting of a heavy metal ball attached to a handle by a steel cable (see Fig. 6.17). Just before launching the hammer, the athlete swings it around several times in a circle. The mass of the ball is 7.3 kg, and the distance from the hammer to the center of its circular motion is 1.9 m (including some length from the athlete's arms; see Fig. 6.18). The speed of the hammer is 27 m/s. What is the centripetal force that the athlete must exert with his arms to keep the hammer moving in its circle?

FIGURE 6.17 Hammer throw.

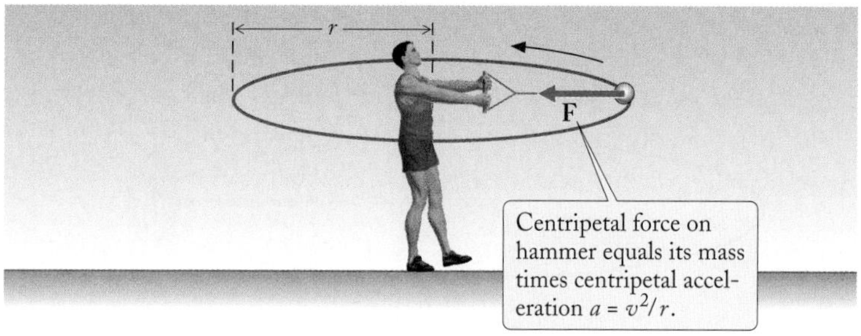

FIGURE 6.18 Circular motion of the hammer.

SOLUTION: According to Eq. (6.13), the magnitude of the force must be

$$F = ma = \frac{mv^2}{r} = \frac{7.3\ \text{kg} \times (27\ \text{m/s})^2}{1.9\ \text{m}} = 2.8 \times 10^3\ \text{N}$$

This is a rather large force! Hammer throwing requires great physical strength, and hammer throwers must be of hefty build.

Concepts in Context

EXAMPLE 10 What is the maximum speed with which an automobile can round a curve of radius 100 m without skidding sideways? Assume that the road is flat and that the coefficient of static friction between the tires and road surface is $\mu_s = 0.80$.

SOLUTION: The "free-body" diagram for the automobile is given in Fig. 6.19. The forces on the automobile are the weight **w**, the normal force **N**, and the friction force $\mathbf{f}_s$. The weight balances the normal force; that is, $N = mg$. The horizontal friction force must provide the centripetal force; hence the magnitude of the friction force must be

$$f_s = ma = \frac{mv^2}{r} \tag{6.14}$$

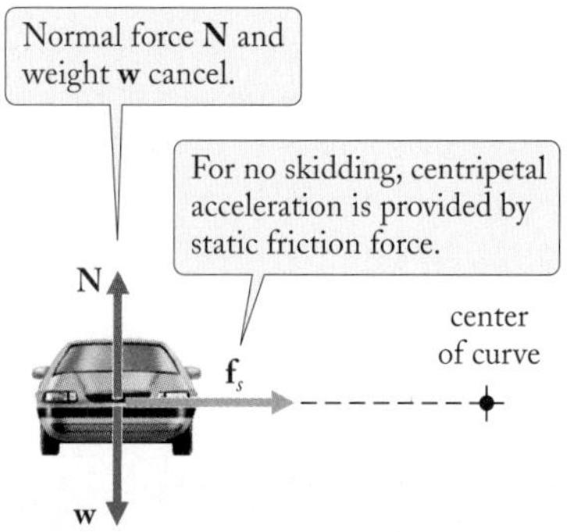

FIGURE 6.19 "Free-body" diagram for an automobile rounding a curve.

The friction is *static* because, by assumption, there is no lateral slippage. At the maximum possible speed, the friction force has its maximum value $f_s = f_{s,\text{max}} = \mu_s N = \mu_s mg$, and consequently

$$\mu_s mg = \frac{mv^2}{r} \tag{6.15}$$

We can cancel the masses on both sides of this equation and then multiply both sides by r. Taking the square root of both sides then yields

$$v = \sqrt{\mu_s gr}$$
$$= \sqrt{0.80 \times 9.81\ \text{m/s}^2 \times 100\ \text{m}} = 28\ \text{m/s} \tag{6.16}$$

This is about 100 km/h.

Concepts in Context

EXAMPLE 11 At a speedway in Texas, a curve of radius 500 m is banked at an angle of 22° (see Fig. 6.20). If the driver of a racing car does not wish to rely on lateral friction, at what speed should he take this curve?

SOLUTION: The "free-body" diagram for the car is shown in Fig. 6.21a. Lateral (sideways) friction is assumed absent, and hence the normal force **N** and the weight **w** are the only forces acting on the racing car perpendicular to its motion.[1] The resultant of these forces must play the role of centripetal force. Hence the resultant must be horizontal, as in Fig. 6.21b. From this figure we see that the magnitude of the resultant is $F = w \tan\theta$, which must coincide with the magnitude of the centripetal force:

$$w \tan\theta = ma = \frac{mv^2}{r} \tag{6.17}$$

or

$$mg \tan\theta = \frac{mv^2}{r}$$

which yields

$$\begin{aligned} v &= \sqrt{rg \tan\theta} \\ &= \sqrt{500\ \text{m} \times 9.81\ \text{m/s}^2 \times \tan 22°} = 45\ \text{m/s} \end{aligned}$$

This is 160 km/h. If the car goes faster than this, it will tend to skid up the embankment; if it goes slower than this, it will tend to skid down the embankment unless friction holds it there.

FIGURE 6.20 Racing car on a banked curve.

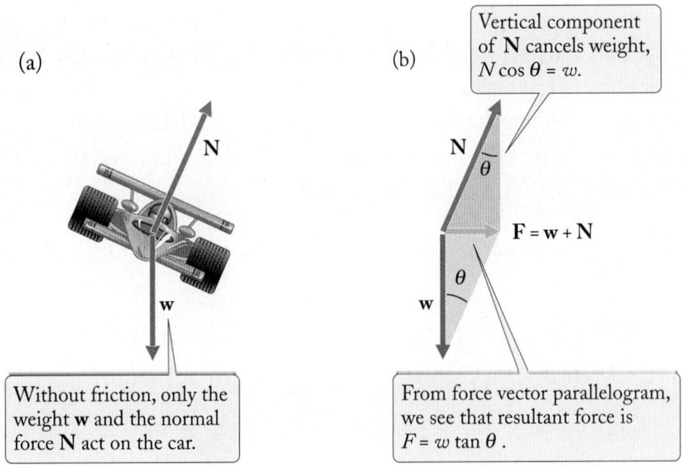

FIGURE 6.21 (a) "Free-body" diagram for the car rounding a banked curve. (b) The resultant **F** of the forces **N** and **w**.

EXAMPLE 12 A pilot in a fast jet aircraft loops the loop (see Fig. 6.22). The radius of the loop is 400 m, and the aircraft has a speed of 150 m/s when it passes through the bottom of the loop. What is the apparent weight that the pilot feels; in other words, what is the force with which she presses against her chair? Express the answer as a multiple of her normal weight.

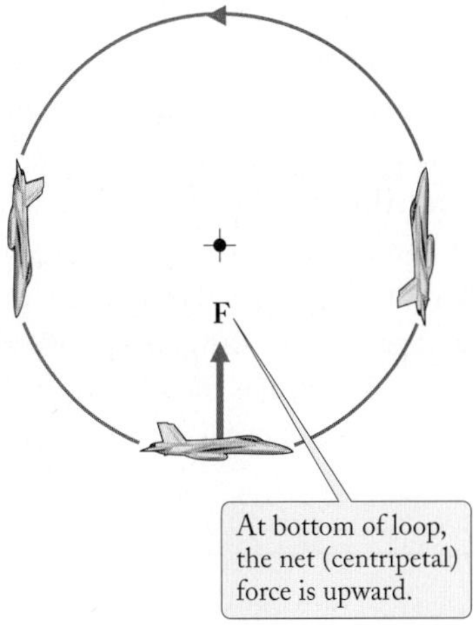

FIGURE 6.22 Jet aircraft looping the loop.

[1] Air resistance also acts on the car, but it is compensated by the propulsive force that the wheels produce, by reaction, with the *forward* friction force from the road.

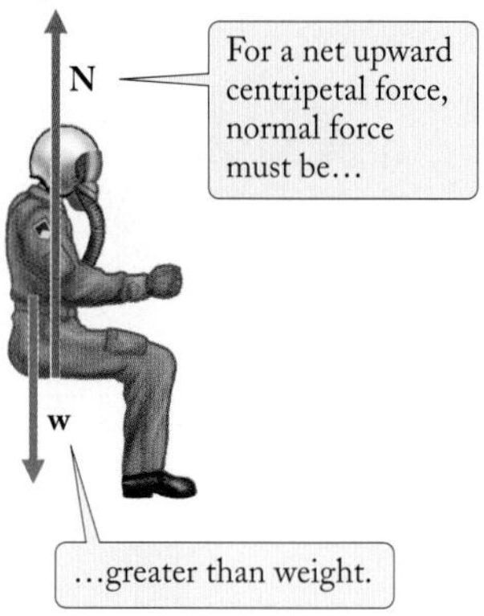

FIGURE 6.23 "Free-body" diagram for jet pilot at the bottom of the loop. The forces acting on the pilot are her weight **w** and the normal force **N** exerted by the chair.

SOLUTION: Figure 6.23 shows a "free-body" diagram for the pilot at the bottom of the loop. The forces acting on her are the (true) weight **w** and the normal force **N** exerted by the chair. The net vertical upward force is $N - mg$, and this must provide the centripetal acceleration:

$$N - mg = ma = \frac{mv^2}{r} \tag{6.18}$$

Note that here we used the the familiar formula v^2/r for the centripetal acceleration, even though the speed is not constant (the speed of the aircraft increases somewhat as it goes down the loop, and the speed decreases as it goes up the loop). Such a change of speed along the loop implies that there can be an extra acceleration *along* the loop. But this extra acceleration does not affect the centripetal acceleration—the two accelerations are at right angles, and they are independent. For the purposes of this problem, we do not need to pay any attention to the extra tangential acceleration along the loop at any point.

Solving Eq. (6.18) for N, we obtain

PHYSICS IN PRACTICE ULTRACENTRIFUGES

The operation of centrifuges and ultracentrifuges hinges on the effective increase of weight associated with circular motion. For instance, consider a test tube with some liquid that is being spun in a horizontal circle in a centrifuge (see Fig. 1). Suppose the liquid contains some particles in suspension; for instance, the liquid might be blood, consisting of a suspension of red blood cells (and other corpuscles) in liquid blood plasma. The radial force required to keep a corpuscle of mass m suspended in a fixed position relative to the test tube is mv^2/r in the centripetal direction. This centripetal force on the corpuscle has to be exerted by the liquid. By reaction, the corpuscle exerts a force of equal magnitude mv^2/r on the liquid, in the outward, or centrifugal, direction. These action and reaction forces are as though the particle were at rest but had an apparent weight mv^2/r in the outward, or centrifugal, direction. Note that the apparent weight is proportional to the mass, like a true weight. This apparent weight is called the **centrifugal force**. (Besides the apparent weight mv^2/r, there is also a true weight mg, in the vertical downward direction; but in a high-speed centrifuge, the true weight is negligible compared with the apparent weight.) Since the corpuscle is more dense than the plasma, its large apparent weight will enable it to shoulder the plasma aside, and to settle quickly against the outermost wall (the "bottom") of the test tube. Thus, the centrifuge accomplishes a quick segregation of the contents of the test tube into layers of different density (Fig. 2), just as though the test tube had been subjected to a manifold enhancement of gravity. Ultracentrifuges (see Fig. 3) spin-

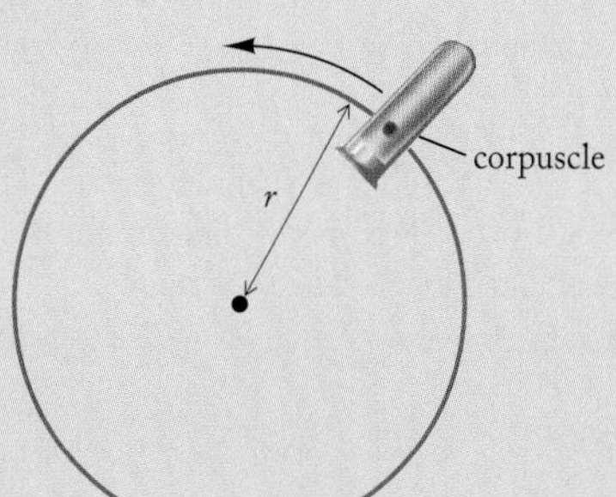

FIGURE 1 The test tube rotates around a circle of radius r. A corpuscle of mass m is suspended in the liquid in the test tube. A centripetal force mv^2/r is required to keep this particle moving on the circular path.

$$N = mg + \frac{mv^2}{r} = mg\left(1 + \frac{v^2}{gr}\right) \tag{6.19}$$

With $v = 150$ m/s and $r = 400$ m, we then find

$$N = mg\left(1 + \frac{(150\ \text{m/s})^2}{9.81\ \text{m/s}^2 \times 400\ \text{m}}\right) = mg \times 6.7$$

This is the force with which the chair presses against the pilot, and it is therefore the apparent weight that the pilot feels. The apparent weight equals the true weight multiplied by a factor of 6.7. In these circumstances, the pilot would say she is "pulling 6.7 g's," because she feels as though gravity had been magnified by a factor of 6.7.

COMMENT: This example shows that *the centripetal acceleration can generate an effective increase of weight*—the apparent weight in the reference frame of a body in circular motion can be much larger than its normal weight.

ning at up to 100 000 revolutions per minute generate apparent centrifugal weights of up to 500 000 times the true weight; they are used in chemical and biochemical research. Special ultracentrifuges can separate even different isotopes of chemical elements, such as the different isotopes of uranium.

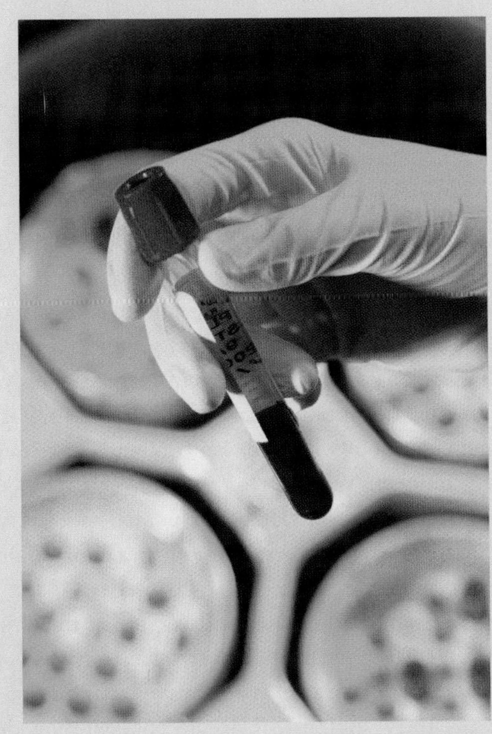

FIGURE 2 Test tube with a blood sample after centrifuging.

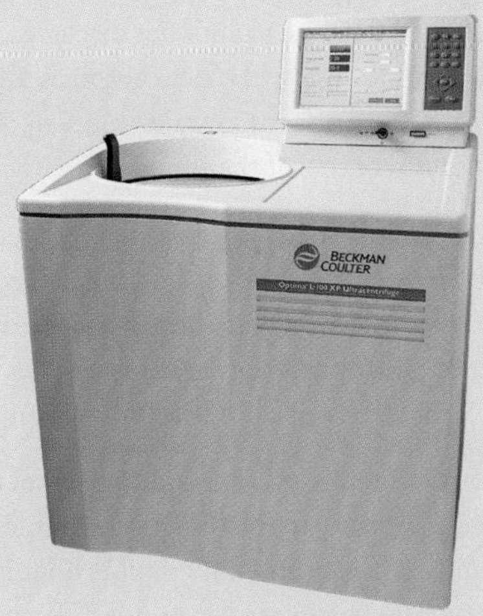

FIGURE 3 An ultracentrifuge.

Checkup 6.3

QUESTION 1: A stone is being whirled around a circle at the end of a string when the string suddenly breaks. Describe the motion of the stone after the string breaks; ignore gravity.

QUESTION 2: At an intersection, a motorcycle makes a right turn at constant speed. During this turn the motorcycle travels along a 90° arc of a circle. What is the direction of the acceleration of the motorcycle during this turn?

QUESTION 3: A car moves at constant speed along a road leading over a small hill with a spherical top. What is the direction of the acceleration of the car when at the top of this hill?

QUESTION 4: In Example 12, for the aircraft looping the loop, does the chair exert a centripetal or a centrifugal force on the pilot? Does the pilot exert a centripetal or a centrifugal force on the chair? What is the direction of the pilot's apparent, increased weight at the instant the aircraft passes through the bottom of the loop? Does the direction of the apparent weight change as the aircraft climbs up the loop?

QUESTION 5: Two cars travel around a traffic circle in adjacent (outer and inner) lanes. If the two cars travel at the same constant speed, which completes the circle first? Which has the larger acceleration?

(A) Outer; outer (B) Inner; outer
(C) Outer; inner (D) Inner; inner

PROBLEM-SOLVING TECHNIQUES: FRICTION FORCES AND CENTRIPETAL FORCES

The problems involving applications of Newton's laws in this chapter can be solved by the techniques discussed in the preceding chapter. In dealing with friction forces and with the centripetal force for uniform circular motion, pay special attention to the directions of the forces.

1 The magnitude of the sliding friction force is proportional to the magnitude of the normal force, but the direction is not the direction of the normal force. Instead, the sliding friction force is always parallel to the sliding surfaces, opposite to the direction of motion.

2 The static friction force is also always parallel to the sliding surfaces, opposite to the direction in which the body tends to move. If you have any doubts about the direction of the static friction force, pretend that the friction is absent, and ask yourself in what direction the body would then move; the static friction force is in the opposite direction.

3 Uniform circular motion requires a force toward the center of the circle, that is, a centripetal force. When preparing a "free-body" diagram for a body in uniform circular motion, include all the pushes and pulls acting on the moving body, but do *not* include a "centripetal mv^2/r force." This would be a mistake, like including an "ma force" in the "free-body" diagram for a body with some kind of translational motion. The quantity mv^2/r is not a force; it is merely the product of mass and centripetal acceleration. This acceleration is caused by one force or by the resultant of several forces already included among the pushes and pulls displayed in the "free-body" diagram. For instance, in Example 11 the resultant force is $w \tan\theta$, in Example 12 the resultant force is $N - mg$, and these resultants equal mv^2/r by Newton's Second Law [see Eqs. (6.17) and (6.18)]. To prevent confusion, do not include the resultant in the "free-body" diagram for a body in uniform circular motion. Instead, draw the resultant on a separate diagram (see Fig. 6.21b).

6.4 THE FOUR FUNDAMENTAL FORCES

In everyday experience we encounter an enormous variety of forces: the gravity of the Earth that pulls all bodies downward, contact forces between rigid bodies that resist their interpenetration, friction forces that resist the motion of a surface sliding over another surface, elastic forces that oppose the deformation of springs and beams, pressure forces exerted by air or water on bodies immersed in them, adhesive forces exerted by a layer of glue bonding two surfaces, electrostatic forces between two electrified bodies, magnetic forces between the poles of magnets, and so on.

Besides these forces that act in the macroscopic world of everyday experience, there are many others that act in the microscopic world of atomic and nuclear physics. There are intermolecular forces that attract or repel molecules to or from each other, interatomic forces that bind atoms into molecules or repel them if they come too close to each other, atomic forces within the atom that hold its parts together, nuclear forces that act on the parts of the nucleus, and even more esoteric forces which act during radioactive decay or act only for a brief instant when subnuclear particles are made to suffer violent collisions in high-energy experiments performed in accelerator laboratories.

Yet, at the fundamental level, this bewildering variety of forces involves only four different kinds of forces. The four fundamental forces are the gravitational force, the electromagnetic force, the "strong" force, and the "weak" force.

gravitational force

The **gravitational force** is a mutual attraction between all masses. Gravitation is the weakest of the four forces. The gravitational attraction between two masses of, say, 1 kg placed next to each other is so small that it is detectable only with extremely sensitive equipment. On the surface of the Earth, we feel the force of gravity only because the mass of the Earth is very large. We discuss gravity further in Chapter 9.

electromagnetic force

The **electromagnetic force** is an attraction or repulsion between electric charges. The electric and the magnetic forces, once considered to be separate, are now grouped together because they are closely related: the magnetic force is nothing but an extra electric force that acts whenever charges are in motion. Of all the forces, the electric force plays the largest role in our lives. With the exception of the Earth's gravity, every force in our immediate macroscopic environment is electric. Contact forces between rigid bodies, elastic forces, pressure forces, adhesive forces, friction forces, etc., are nothing but electric forces between charged particles in the atoms of one body and those in the atoms of another. Electricity and magnetism are the subject of Chapters 22–33.

"strong" force

The **"strong" force** acts mainly within the nuclei of atoms. It plays the role of a nuclear glue that prevents the protons and neutrons of the nucleus from flying apart. This nuclear force is called "strong" because it is the strongest of the four forces. It can be either attractive or repulsive: the strong force will push protons and neutrons apart if they come too near each other, and it will pull them together if they begin to drift too far apart. We will examine nuclei in Chapter 40.

"weak" force

Finally, the **"weak" force** manifests itself only in certain reactions among elementary particles. Most of the reactions caused by the weak force are radioactive-decay reactions; they involve the spontaneous breakup of a particle into several other particles (we will discuss this in Chapters 40 and 41). This force is called "weak" because it is weak compared with the "strong" force and the electromagnetic force.

SUMMARY

PHYSICS IN PRACTICE Ultracentrifuges (page 188)

PROBLEM-SOLVING TECHNIQUES Friction Forces and Centripetal Forces (page 190)

KINETIC FRICTION FORCE (Direction *opposes* motion.)

$$f_k = \mu_k N \quad (6.1)$$

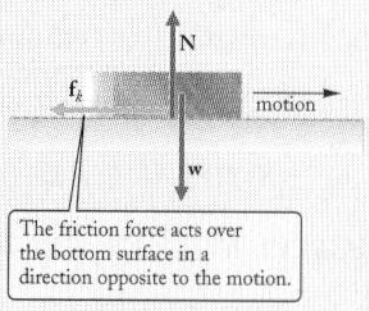

STATIC FRICTION FORCE (Direction opposes force which tries to move body; magnitude varies in response to applied force.)

$$f_{s,\max} = \mu_s N \quad (6.5)$$

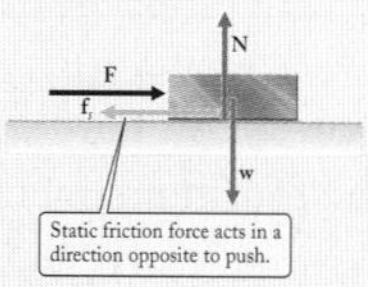

RESTORING FORCE OF A SPRING (HOOKE'S LAW) (Direction is toward relaxed position; x is measured from relaxed position.)

$$F = -kx \quad (6.11)$$

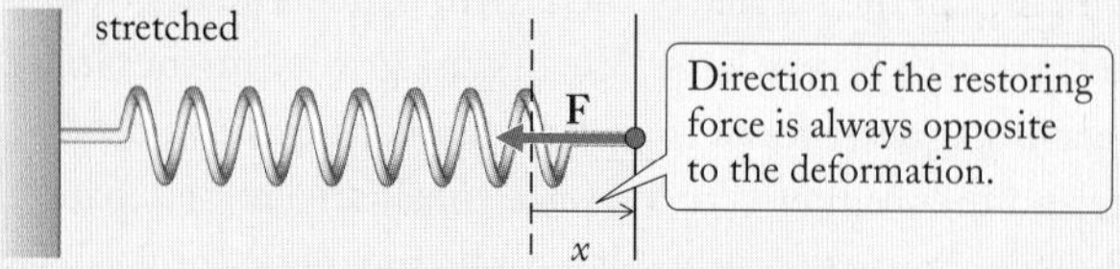

FORCE DUE TO AIR RESISTANCE At high speed v, where C is a dimensionless aerodynamic constant, ρ is the density of air, and A is the cross-sectional area.

$$f_{\text{air}} = \tfrac{1}{2} C \rho A v^2$$

FORCE REQUIRED FOR UNIFORM CIRCULAR MOTION (Direction is centripetal.)

$$F = \frac{mv^2}{r} \quad (6.13)$$

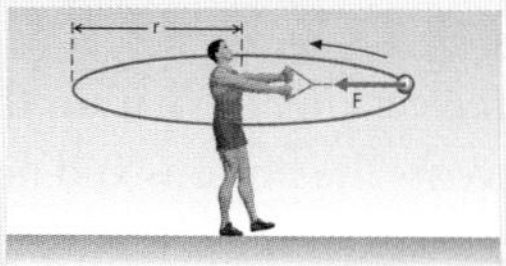

THE FOUR FUNDAMENTAL FORCES
Gravitational, "weak," electromagnetic, "strong"

QUESTIONS FOR DISCUSSION

1. According to the adherents of parapsychology, some people are endowed with the supernormal power of psychokinesis, e.g., spoon-bending-at-a-distance via mysterious psychic forces emanating from the brain. Physicists are confident that the only forces acting between pieces of matter are those listed in Section 6.4, none of which are implicated in psychokinesis. Given that the brain is nothing but a (very complicated) piece of matter, what conclusions can a physicist draw about psychokinesis?
2. If you carry a spring balance from London to Hong Kong, do you have to recalibrate it? If you carry a beam balance?
3. When you stretch a rope horizontally between two fixed points, it always sags a little, no matter how great the tension. Why?
4. What are the forces on a soaring bird? How can the bird gain altitude without flapping its wings?
5. How could you use a pendulum suspended from the roof of your automobile to measure its acceleration?
6. When an airplane flies along a parabolic path similar to that of a projectile, the passengers experience a sensation of weightlessness. How would the airplane have to fly to give the passengers a sensation of enhanced weight?
7. A frictionless chain hangs over two adjoining inclined planes (Fig. 6.24a). Prove the chain is in equilibrium, i.e., the chain will not slip to the left or to the right. [Hint: One method of proof, due to the seventeenth-century engineer and mathematician Simon Stevin, asks you to pretend that an extra piece of chain is hung from the ends of the original chain (Fig. 6.24b). This makes it possible to conclude that the original chain cannot slip.]

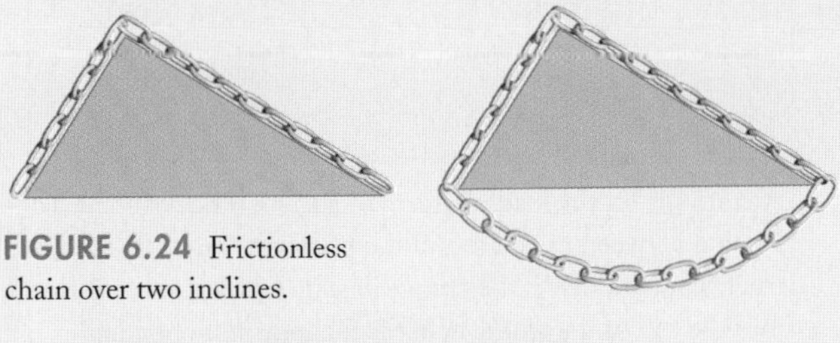

FIGURE 6.24 Frictionless chain over two inclines.

8. Seen from a reference frame moving with the wave, the motion of a surfer is analogous to the motion of a skier down a mountain.[2] If the wave were to last forever, could the surfer ride it forever? In order to stay on the wave as long as possible, in what direction should the surfer ski the wave?
9. Excessive polishing of the surfaces of a block of metal increases its friction. Explain.
10. Some drivers like to spin the wheels of their automobiles for a quick start. Does this give them greater acceleration? (Hint: $\mu_s > \mu_k$.)
11. Cross-country skiers like to use a ski wax that gives their skis a large coefficient of static friction, but a low coefficient of kinetic friction. Why is this useful? How do "waxless" skis achieve the same effect?
12. Designers of locomotives usually reckon that the maximum force available for moving the train ("tractive force") is one-fourth or one-fifth of the weight resting on the drive wheels of the locomotive. What value of the friction coefficient between the wheels and the track does this implicitly assume?
13. When an automobile with rear-wheel drive accelerates from rest, the maximum acceleration that it can attain is less than the maximum deceleration that it can attain while braking. Why? (Hint: Which wheels of the automobile are involved in acceleration? In braking?)
14. Can you think of some materials with $\mu_k > 1$?
15. For a given initial speed, the stopping distance of a train is much longer than that of a truck. Why?
16. Why does the traction on snow or ice of an automobile with rear-wheel drive improve when you place extra weight over the rear wheels?
17. Why are wet streets slippery?
18. In order to stop an automobile on a slippery street in the shortest distance, it is best to brake as hard as possible without initiating a skid. Why does skidding lengthen the stopping distance? (Hint: $\mu_s > \mu_k$.)
19. Suppose that in a panic stop, a driver locks the wheels of his automobile and leaves skid marks on the pavement. How can you deduce his initial speed from the length of the skid marks?
20. Hot-rod drivers in drag races find it advantageous to spin their wheels very fast at the start so as to burn and melt the rubber on their tires (Fig. 6.25). How does this help them to attain a larger acceleration than expected from the static coefficient of friction?

FIGURE 6.25 Drag racer at the start of the race.

[2] There is, however, one complication: surf waves grow higher as they approach the beach. Ignore this complication.

21. A curve on a highway consists of a quarter circle connecting two straight segments. If this curve is banked perfectly for motion at some given speed, can it be joined to the straight segments without a bump? How could you design a curve that is banked perfectly along its entire length and merges smoothly into straight segments without any bump?

22. Automobiles with rear engines (such as the old VW "Beetle") tend to oversteer; that is, in a curve the rear end tends to swing toward the outside of the curve, turning the car excessively into the curve. Explain.

23. When rounding a curve in your automobile, you get the impression that a force tries to pull you toward the outside of the curve. Is there such a force?

24. If the Earth were to stop spinning (other things remaining equal), the value of g at all points of the surface except the poles would become slightly larger. Why?

25. (a) If a pilot in a fast aircraft very suddenly pulls out of a dive (Fig. 6.26a), he will suffer blackout caused by loss of blood pressure in the brain. If he suddenly begins a dive while climbing (Fig. 6.26b), he will suffer *redout* caused by excessive blood pressure in the brain. Explain.

    (b) A pilot wearing a G suit—a tightly fitting garment that squeezes the tissues of the legs and abdomen—can tolerate $8g$ while pulling out of a dive (Fig. 6.26c). How does this G suit prevent blackout? A pilot can tolerate no more than $-2g$ while beginning a dive. Why does the G suit not help against redout?

26. While rounding a curve at high speed, a motorcycle rider leans the motorcycle toward the center of the curve. Why?

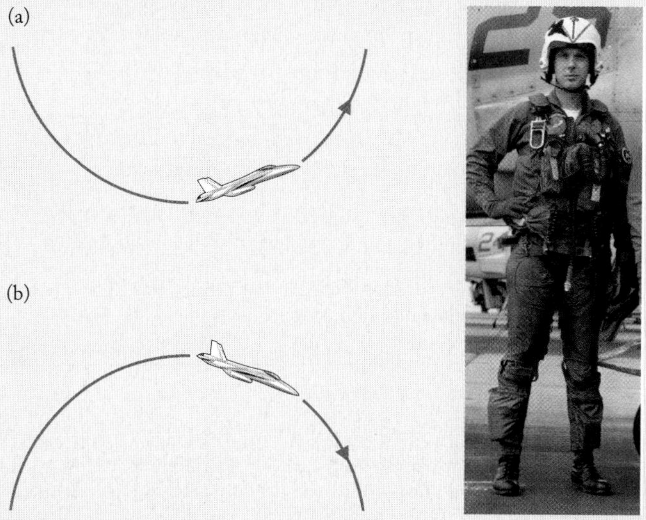

FIGURE 6.26 (a) Aircraft pulling out of a dive. (b) Aircraft beginning a dive. (c) Pilot wearing a G suit.

PROBLEMS

6.1 Friction†

1. The ancient Egyptians moved large stones by dragging them across the sand in sleds. How many Egyptians were needed to drag an obelisk of 700 metric tons? Assume that $\mu_k = 0.30$ for the sled on sand and that each Egyptian exerted a horizontal force of 360 N.

2. The base of a winch is bolted to a mounting plate with four bolts. The base and the mounting plate are flat surfaces made of steel; the friction coefficient of these surfaces in contact is $\mu_s =$ 0.40. The bolts provide a normal force of 2700 N each. What maximum static friction force will act between the steel surfaces and help oppose lateral slippage of the winch on its base?

3. According to tests performed by the manufacturer, an automobile with an initial speed of 65 km/h has a stopping distance of 20 m on a level road. Assuming that no skidding occurs during braking, what is the value of μ_s between the wheels and the road required to achieve this stopping distance?

4. A crate sits on the load platform of a truck. The coefficient of friction between the crate and the platform is $\mu_s = 0.40$. If the truck stops suddenly, the crate will slide forward and crash into the cab of the truck. What is the maximum braking deceleration that the truck may have if the crate is to stay put?

5. When braking (without skidding) on a dry road, the stopping distance of a sports car with a high initial speed is 38 m. What would have been the stopping distance of the same car with the same initial speed on an icy road? Assume that $\mu_s = 0.85$ for the dry road and $\mu_s = 0.20$ for the icy road.

6. In a remarkable accident on motorway M1 (in England), a Jaguar car initially speeding "in excess of 100 mph" skidded 290 m before coming to a rest. Assuming that the wheels were completely locked during the skid and that the coefficient of kinetic friction between the wheels and the road was 0.80, find the initial speed.

† For help, see Online Concept Tutorial 8 at www.wwnorton.com/physics

7. Because of a failure of its landing gear, an airplane has to make a belly landing on the runway of an airport. The landing speed of the airplane is 90 km/h, and the coefficient of kinetic friction between the belly of the airplane and the runway is $\mu_k = 0.60$. How far will the airplane slide along the runway?

8. A child slides down a playground slide; the coefficient of kinetic friction is $\mu_k = 0.15$, and the angle that the slide makes with the horizontal is 30°. She begins from rest and slides through a vertical height of 3.5 m. With what speed does she exit the slide?

9. A baseball player sprinting at 4.5 m/s begins to slide with his body flat on the ground when 2.8 m from home base. The coefficient of friction between the player's uniform and the ground is 0.30. Does he make it home? If so, what is his speed as he reaches home?

10. For microscopic objects, friction can be overwhelming. For example, tiny silicon microstructures (see Fig. 6.27) can become stuck when in contact. If a silicon cube 10 μm on each side with density 2.33 g/cm^3 requires a horizontal force of 0.50×10^{-9} N to begin sliding on a horizontal silicon surface, what is the effective coefficient of static friction?

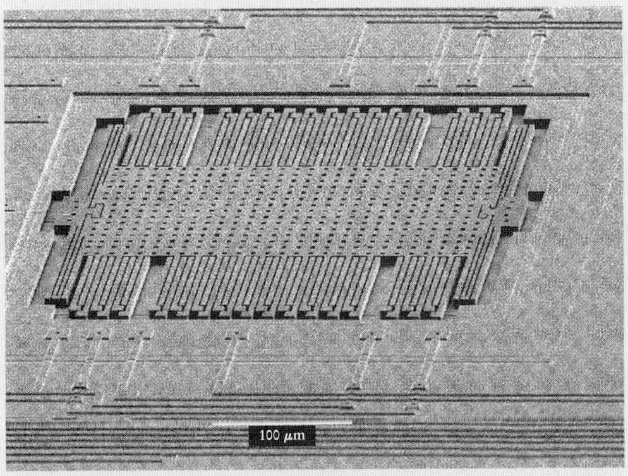

FIGURE 6.27 Micromachined silicon structures (an automobile accelerometer).

11. While braking in an emergency a train traveling at 35 m/s screeches to a halt with all wheels locked. The constant deceleration takes 7.5 s to complete. What is the coefficient of kinetic friction between the train and the tracks?

12. An automobile has an aerodynamic constant $C = 0.35$ and a cross-sectional area of 3.4 m^2. To balance air resistance, what force must be provided when traveling at 20 m/s? At 40 m/s?

13. The driver of the automobile in Example 6 is traveling on a flat road at 25 m/s. Considering only air resistance, what forward friction force must the road provide if the driver wants to begin accelerating at 2.0 m/s^2?

14. A Ping-Pong ball has an aerodynamic constant $C = 0.51$, a mass of 2.5 g, and a radius of 1.6 cm. What is its terminal velocity when dropped?

15. A falling golf ball (mass 45 g, radius 20 mm) reaches a high terminal speed of 45 m/s. What is the value of the aerodynamic constant C for this dimpled sphere?

16. A sky surfer (see Fig. 6.28) has a mass of 70 kg; in the position shown, the product of his aerodynamic constant and cross-sectional area is $CA = 0.42$ m^2. What is this surfer's terminal speed?

FIGURE 6.28 A sky surfer.

*17. A girl pulls a sled along a level dirt road by means of a rope attached to the front of the sled (Fig. 6.29). The mass of the sled is 40 kg, the coefficient of kinetic friction is $\mu_k = 0.60$, and the angle between the rope and the road is 30°. What pull must the girl exert to move the sled at constant velocity?

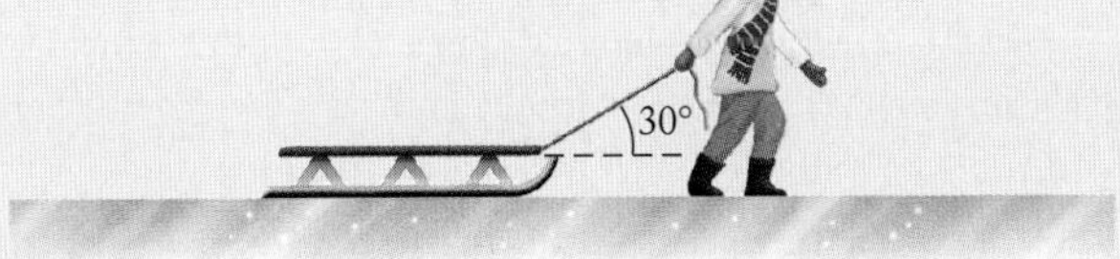

FIGURE 6.29 Pulling a sled.

*18. The "Texas" locomotives of the old T&P railways had a mass of 200 000 kg, of which 136 000 kg rested on the driving wheels. What maximum acceleration could such a locomotive attain (without slipping) when pulling a train of 100 boxcars of mass 18 000 kg each on a level track? Assume that the coefficient of static friction between the driving wheels and the track is 0.25.

*19. During braking, a truck has a steady deceleration of 7.0 m/s^2. A box sits on the platform of this truck. The box begins to slide when the braking begins and, after sliding a distance of 2.0 m (relative to the truck), it hits the cab of the truck. With what speed (relative to the truck) does the box hit? The coefficient of kinetic friction for the box is $\mu_k = 0.50$.

*20. The Schleicher ASW-22 is a high-performance sailplane of a wingspan of 24 m and a mass of 750 kg (including the pilot). At a forward speed (airspeed) of 35 knots, the sink rate, or the rate of descent, of this sailplane is 0.46 m/s. Draw a "free-body" diagram showing the forces on the plane. What is the friction force (antiparallel to the direction of motion) exerted on the plane by air resistance under these conditions? What is the lift force (perpendicular to the direction of motion) generated by air streaming past the wings?

*21. The friction force (including air friction and rolling friction) acting on an automobile traveling at 65 km/h amounts to 500 N. What slope must a road have if the automobile is to roll down this road at a constant speed of 65 km/h (with its gears in neutral)? The mass of the automobile is 1.5×10^3 kg.

*22. In a downhill race, a skier slides down a 40° slope. Starting from rest, how far must he slide down the slope in order to reach a speed of 130 km/h? How many seconds does it take him to reach this speed? The friction coefficient between his skis and the snow is $\mu_k = 0.10$. Ignore the resistance offered by the air.

*23. To measure the coefficient of static friction of a block of plastic on a plate of steel, an experimenter places the block on the plate and then gradually tilts the plate. The block suddenly begins to slide when the plate makes an angle of 38° with the horizontal. What is the value of μ_s?

*24. A solid steel ball bearing of radius 0.25 cm falling in air has a terminal speed of 88 m/s. What is the terminal speed of a solid steel ball of radius 5.0 cm?

*25. At very low speeds, the resistance to motion offered by a liquid or gas is nearly proportional to velocity (instead of the square of the velocity); such a *viscous drag force* opposes the motion and can be written $\mathbf{f}_{\text{viscous}} = -b\mathbf{v}$, where the constant of proportionality is known as the viscous drag coefficient. A tiny spherical metal particle of mass 3.9×10^{-6} g (neglect buoyancy effects) falling though oil has a drag coefficient $b = 2.8 \times 10^{-5}$ kg/s. What is its terminal speed?

**26. Show that the speed as a function of time of a particle falling from rest under the influence of gravity and a viscous force of the form $\mathbf{f}_{\text{viscous}} = -b\mathbf{v}$ (see Problem 25) is given by

$$v = \frac{mg}{b}(1 - e^{-bt/m})$$

[Hint: Integrate Newton's law in the form $m(dv/dt) = mg - bv$.] What is the value of the characteristic time $t = m/b$ for the particle in Problem 25?

**27. On a level road, the stopping distance for an automobile is 35 m for an initial speed of 90 km/h. What is the stopping distance of the same automobile on a similar road with a downhill slope of 1:10?

**28. Two masses, of 2.0 kg each, connected by a string slide down a ramp making an angle of 50° with the horizontal (Fig. 6.30). The mass m_1 has a coefficient of kinetic friction 0.60, and the mass m_2 has a coefficient of kinetic friction 0.40. Find the acceleration of the masses and the tension in the string.

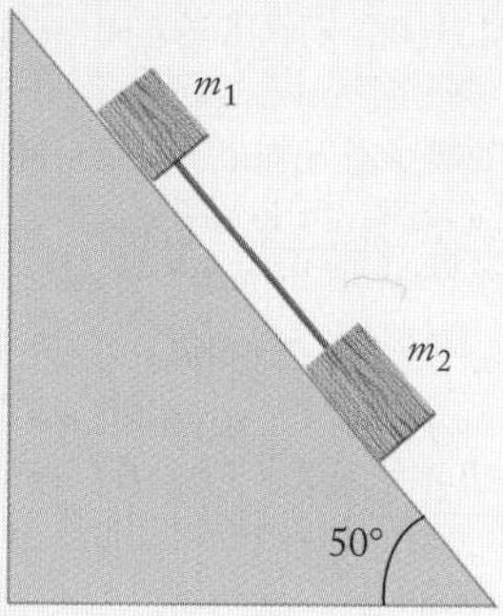

FIGURE 6.30 Two masses connected by a string sliding down a ramp.

**29. You are holding a book against a wall by pushing with your hand. Your push makes an angle of θ with the wall (see Fig. 6.31). The mass of the book is m, and the coefficient of static friction between the book and the wall is μ_s.

(a) Draw the "free-body" diagram for the book.

(b) Calculate the magnitude of the push you must exert to (barely) hold the book stationary.

(c) For what value of the angle θ is the magnitude of the required push as small as possible? What is the magnitude of the smallest possible push?

(d) If you push at an angle larger than 90°, you must push very hard to hold the book in place. For what value of the angle will it become impossible to hold the book in place?

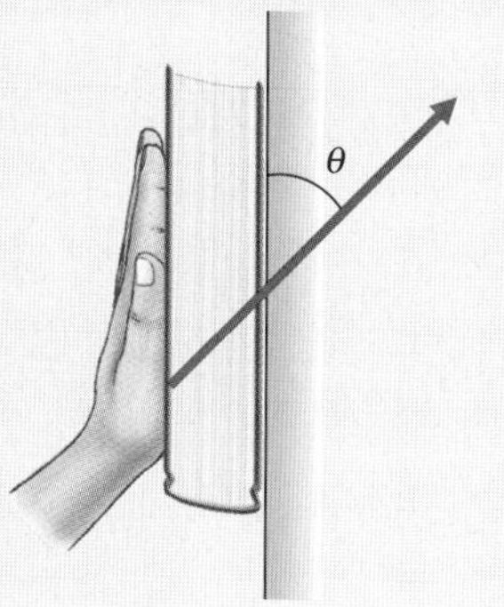

FIGURE 6.31 Pushing a book.

**30. A box is being pulled along a level floor at constant velocity by means of a rope attached to the front end of the box. The rope makes an angle θ with the horizontal. Show that for a given mass m of the box and a given coefficient of kinetic friction μ_k, the tension required in the rope is minimum if $\tan\theta = \mu_k$. What is the tension in the rope when at this optimum angle?

**31. Consider the man pushing the crate described in Example 3. Assume that instead of pushing down at an angle of 30°, he pushes down at an angle θ. Show that he will not be able to keep the crate moving if θ is larger than $\tan^{-1}(1/\mu_k)$.

32. A block of mass m_1 sits on top of a larger block of mass m_2 which sits on a flat surface (Fig. 6.32). The coefficient of kinetic friction between the upper and lower blocks is μ_1, and that between the lower block and the flat surface is μ_2. A horizontal force **F pushes against the upper block, causing it to slide; the friction force between the blocks then causes the lower block to slide also. Find the acceleration of the upper block and the acceleration of the lower block.

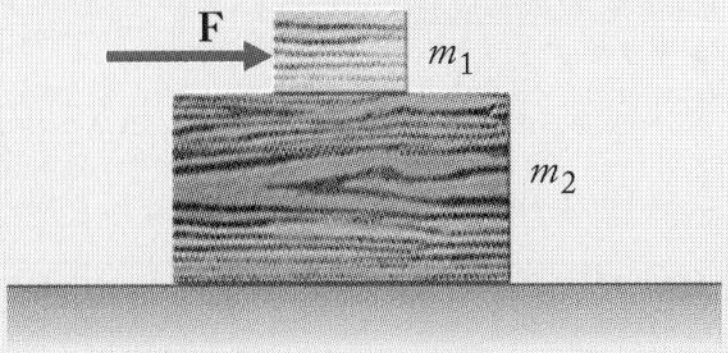

FIGURE 6.32 Block on block on surface.

**33. Two masses $m_1 = 1.5$ kg and $m_2 = 3.0$ kg are connected by a thin string running over a massless pulley. One of the masses hangs from the string; the other mass slides on a 35° ramp with a coefficient of kinetic friction $\mu_k = 0.40$ (Fig. 6.33). What is the acceleration of the masses?

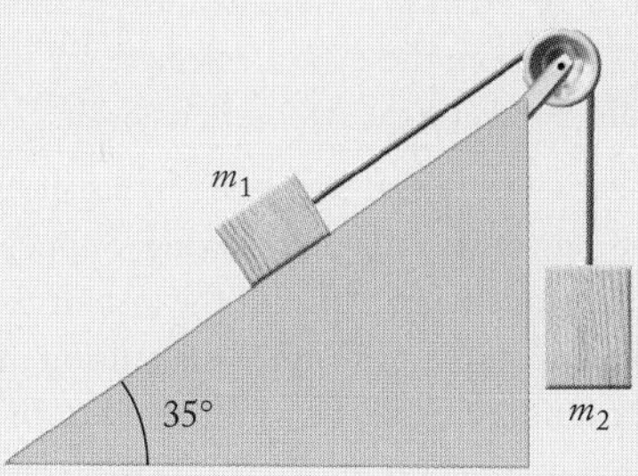

FIGURE 6.33 Two masses, an incline, and a pulley.

**34. A man pulls a sled up a ramp by means of a rope attached to the front of the sled (Fig. 6.34). The mass of the sled is 80 kg, the coefficient of kinetic friction between the sled and the ramp is $\mu_k = 0.70$, the angle between the ramp and the horizontal is 25°, and the angle between the rope and the ramp is 35°. What pull must the man exert to keep the sled moving at constant velocity?

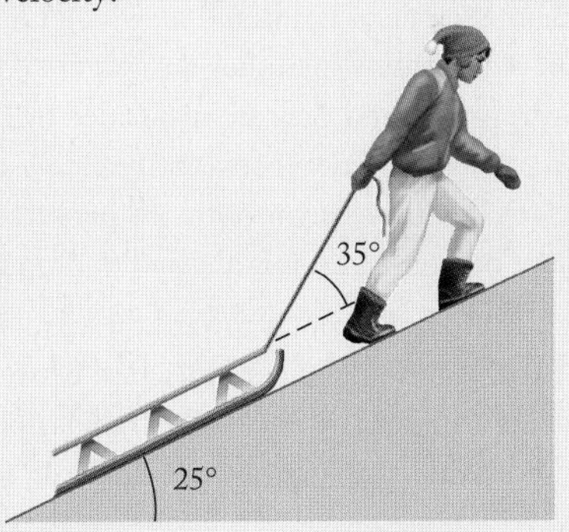

FIGURE 6.34 Pulling a sled up an incline.

**35. Two blocks of masses m_1 and m_2 are sliding down an inclined plane making an angle θ with the horizontal. The leading block has a coefficient of kinetic friction μ_k; the trailing block has a coefficient of kinetic friction $2\mu_k$. A string connects the two blocks; this string makes an angle ϕ with the ramp (Fig. 6.35). Find the tension in the string.

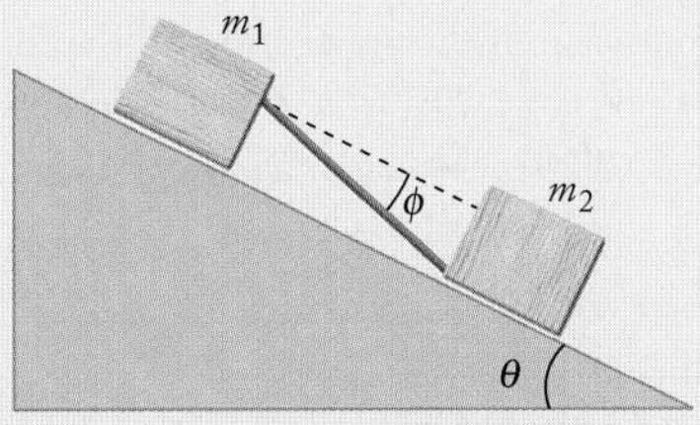

FIGURE 6.35 Two blocks connected by a slanted string sliding down an inclined plane.

**36. A man of 75 kg is pushing a heavy box along a flat floor. The coefficient of sliding friction between the floor and the box is 0.20, and the coefficient of static friction between the man's shoes and the floor is 0.80.

(a) If the man pushes downward on the box at an angle of 30° (see Fig. 6.36a), what is the maximum mass of the box he can move?

(b) If the man pushes upward on the box at an angle of 30° (see Fig. 6.36b), what is the maximum mass of the box he can move?

FIGURE 6.36 Pushing a box.

6.2 Restoring Force of a Spring; Hooke's Law

37. A spring with a force constant $k = 150$ N/m has a relaxed length of 0.15 m. What force must you exert to stretch this spring to twice its length? What force must you exert to compress this spring to one-half its length?

38. Attempting to measure the force constant of a spring, an experimenter clamps the upper end of the spring in a vise and suspends a mass of 1.5 kg from the lower end. This stretches the spring by 0.20 m. What is the force constant of the spring?

39. A rubber band of relaxed length 6.3 cm stretches to 10.2 cm under a force of 1.0 N, and to 15.5 cm under 2.0 N. Does this rubber band obey Hooke's Law?

40. A cantilever (such as a diving board) can be regarded as a spring. When a 70-kg diver stands on the edge of the board, it deflects downward by 16 cm. What is the effective spring constant of this diving board?

41. When pulled back to launch a ball in a pinball machine, a spring is compressed by 7.0 cm. This requires a pull of 3.5 N. What is the spring constant of the pinball spring?

42. Atomic-force microscopes (AFMs) use tiny cantilevers which act like springs; one manufacturer quotes a value of $k = 4.8 \times 10^{-2}$ N/m for the cantilever's spring constant. When the "spring" moves over an atom, it is compressed a distance of 2.0×10^{-11} m. What is the value of the atomic force in this case?

43. Retractable ballpoint pens contain a spring. When a mass of 250 g is placed on top of such a spring, the spring compresses by 2.8 mm. What is the spring constant of this spring?

44. When a 75-kg bungee jumper hangs from a bungee cord, it stretches by 2.9 m. What is the spring constant of the bungee cord?

45. Solid materials can act much like springs. Consider a steel cable with radius 2.0 cm and length 20 m, which has a spring constant of 1.4×10^7 N/m. If a 1500-kg elevator car is hung from this cable, how much does the cable stretch?

*46. Suppose that a uniform spring with a constant $k = 120$ N/m is cut into two pieces, one twice as long as the other. What are the spring constants of the two pieces?

*47. Show that if two springs, of constants k_1 and k_2, are connected in parallel (Fig. 6.37), the net spring constant k of the combination is given by

$$k = k_1 + k_2$$

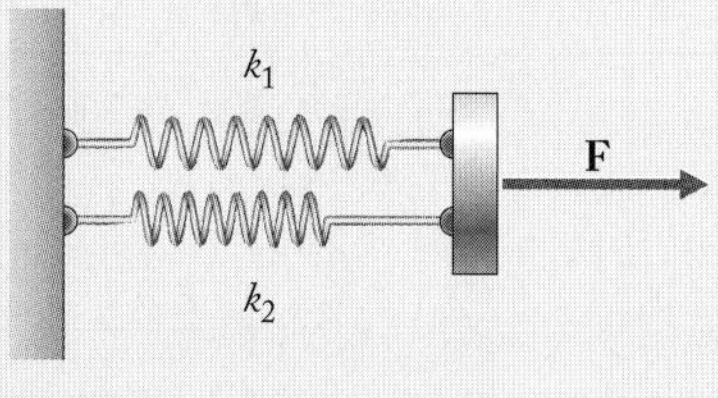

FIGURE 6.37 Springs acting in parallel.

*48. Show that if two springs, of constants k_1 and k_2, are connected in series (Fig. 6.38), the net spring constant k of the combination is given by

$$\frac{1}{k} = \frac{1}{k_1} + \frac{1}{k_2}$$

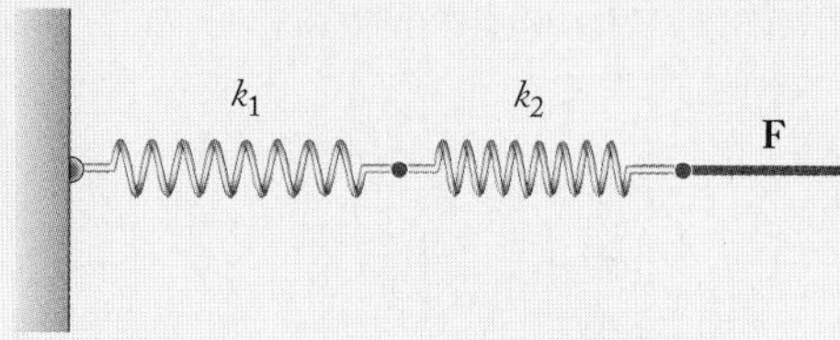

FIGURE 6.38 Springs acting in series.

6.3 Force for Uniform Circular Motion

49. A man of 80 kg is standing in the cabin of a Ferris wheel of radius 30 m rotating at 1.0 rev/min. What is the force that the feet of the man exert on the floor of the cabin when he reaches the highest point? The lowest point?

50. The Moon moves around the Earth in a circular orbit of radius 3.8×10^8 m in 27 days. The mass of the Moon is 7.3×10^{22} kg. From these data, calculate the magnitude of the force required to keep the Moon in its orbit.

51. A swing consists of a seat supported by a pair of ropes 5.0 m long. A 60-kg woman sits in the swing. Suppose that the speed of the woman is 5.0 m/s at the instant the swing goes through its lowest point. What is the tension in each of the two ropes? Ignore the masses of the seat and of the ropes.

52. A few copper coins are lying on the (flat) dashboard of an automobile. The coefficient of static friction between the copper and the dashboard is 0.50. Suppose the automobile rounds a curve of radius 90 m. At what speed of the automobile will the coins begin to slide? The curve is *not* banked.

53. A curve of radius 400 m has been designed with a banking angle such that an automobile moving at 75 km/h does not have to rely on friction to stay in the curve. What is the banking angle?

54. In an amusement park ride called "Drop Out," riders are spun in a horizontal circle of radius 6.0 m, which pins their backs against an outer wall. When they are spinning quickly enough, the floor drops out, and they are suspended by friction. If the coefficient of static friction between the riders and the wall is as small as 0.25, how many revolutions per second must the ride achieve before the floor is allowed to drop out?

55. An ant walks from the center toward the edge of a turntable of radius 15 cm. If the coefficient of friction between the ant's feet and the turntable is 0.30, at what radius does the ant begin to slide when the turntable rotates at 45 revolutions per minute?

56. Public skateboarding parks often include a well in the shape of a half cylinder (or "half-pipe"). The skateboarder's path traces out a semicircular arc, with the midpoint of the arc at the lowest point (Fig. 6.39). When starting from rest at one of the upper edges of the arc (at height h), the horizontal speed attained at the bottom can be shown to be the same as for vertical free fall, $v = \sqrt{2gh}$. At the bottom point, by what factor does the normal force from the skateboard on the skateboarder's feet exceed his weight?

FIGURE 6.39 Skateboard half-pipe.

57. A geosynchronous satellite orbits the Earth once per day; this requires an orbital radius of 4.23×10^4 km. From these data, deduce the weight of a (stationary) 1-kg mass at this distance.

58. A jet traveling at 140 m/s makes a turn of radius 6.0 km. What bank angle should the pilot use for the turn so that a passenger does not feel any lateral force?

*59. A rider of a swing carousel initially sits on a seat suspended vertically by a 7.0-m cable from a point 3.0 m from the center axis. When the carousel rotates, the seat swings outward to its equilibrium angle. If the seat speed during this rotation is 6.0 m/s, what is the angle that the cable makes with the vertical?

*60. Two identical automobiles enter a curve side by side, one traveling on the inside lane, the other on the outside. The curve is an arc of a circle, and it is unbanked. Each automobile travels through the curve at the maximum speed tolerated without skidding. Which automobile has a higher speed? Which automobile emerges from the curve first? Prove your answer.

*61. The highest part of a road over the top of a hill follows an arc of a vertical circle of radius 50 m. With what minimum speed must you drive an automobile along this road if its wheels are to lose contact with the road at the top of the hill?

*62. A woman holds a pail full of water by the handle and whirls it around a vertical circle at constant speed. The radius of this circle is 0.90 m. What is the minimum speed that the pail must have at the top of its circular motion if the water is not to spill out of the upside-down pail?

*63. In ice speedway races, motorcycles run at a high speed on an ice-covered track and are kept from skidding by long spikes on their wheels. Suppose that a motorcycle runs around a curve of radius 30 m at a speed of 96 km/h. What is the angle of inclination of the force exerted by the track on the wheels?

*64. An automobile traveling at speed v on a level surface approaches a brick wall (Fig. 6.40). When the automobile is at a distance d from the wall, the driver suddenly realizes that he must either brake or turn. If the coefficient of static friction between the tires and the surface is μ_s, what is the minimum distance that the driver needs to stop (without turning)? What is the minimum distance that the driver needs to complete a 90° turn (without braking)? What is the safest tactic for the driver?

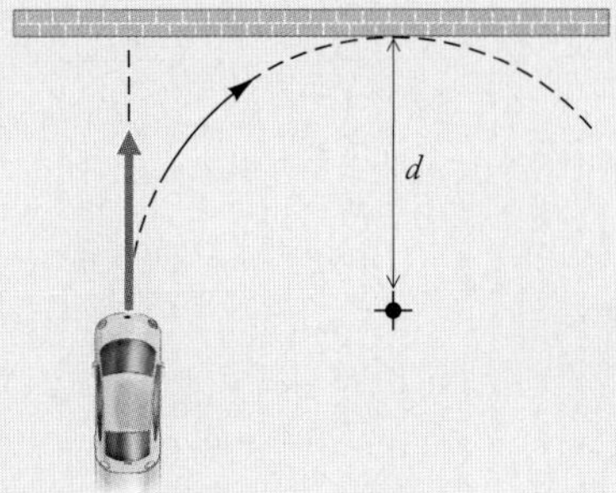

FIGURE 6.40 Automobile approaching a brick wall.

*65. A mass is attached to the lower end of a string of length l; the upper end of the string is held fixed. Suppose that the string initially makes an angle θ with the vertical. With what horizontal velocity must we launch the mass so that it continues to travel at constant speed along a horizontal circular path under the influence of the combined forces of the tension of the string and gravity? This device is called a **conical pendulum** (Fig. 6.41).

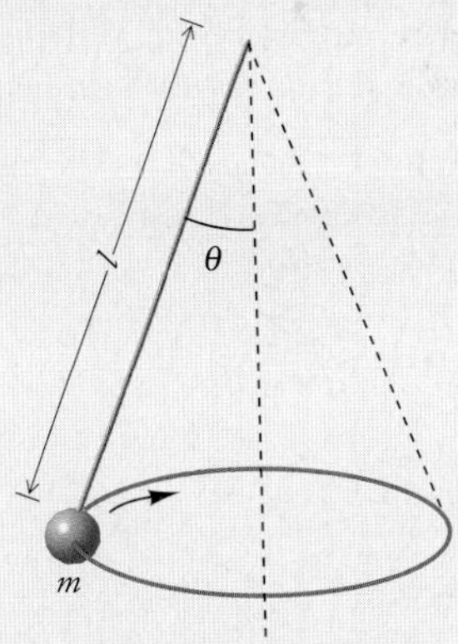

FIGURE 6.41 Mass suspended from a string swinging around a circle (conical pendulum).

*66. An automobile of mass 1200 kg rounds a curve at a speed of 25 m/s. The radius of the curve is 400 m, and its banking angle is 6.0°. What is the magnitude of the normal force on the automobile? The friction force?

*67. An airplane flies in a horizontal circular path at 320 km/h. Looking at the horizon, a passenger notices that the angle of bank of the airplane is 30°. What radius of the circular path can the passenger deduce from these data? [Hint: The force exerted by the air on the wings (lift) is perpendicular to the wings.]

*68. A mass m_1 slides on a smooth, frictionless table. The mass is constrained to move in a circle by a string that passes through a hole in the center of the table and is attached to a second mass m_2 hanging below the table. The second mass swings in a circle, so the string makes an angle θ with the vertical (see Fig. 6.42). The two masses move around their circles in unison, so they are always at diametrically opposite points from the hole. If the radius of the circular motion of the first mass is r_1, what must be the radius of the circular motion of the second mass?

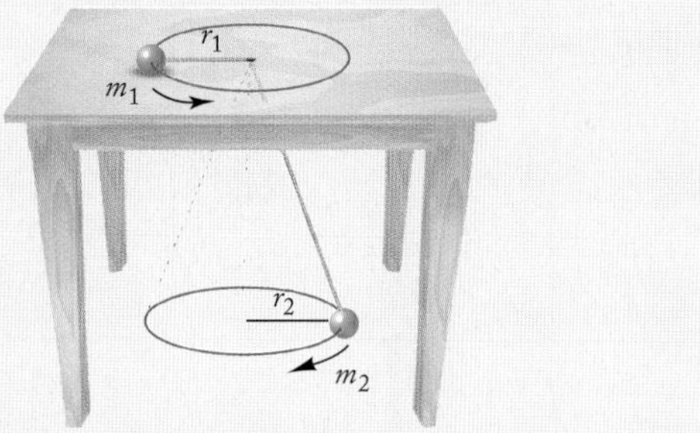

FIGURE 6.42 Mass on table and hanging mass are both in circular motion.

**69. Two masses m_1 and m_2 hang at the ends of a string that passes over a small pulley. The masses swing along circular arcs of equal radii and $m_1 > m_2$. Find a relation between m_1, m_2, and the initial angles (theta)$_1$ and (theta)$_2$ of swing, where the masses are at rest, if both ends of the string are to be under the same tension, so that the string is in equilibrium. Assume that the two masses later reach the bottom in equal times, so that the mass m_1 is moving faster there. Will the string be in equilibrium just before the masses reach the bottom?

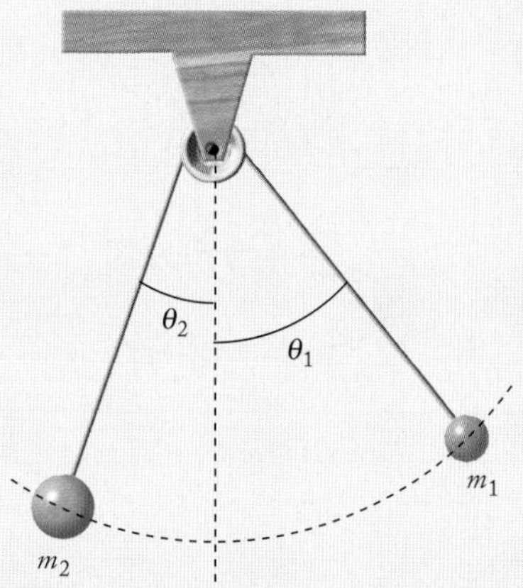

FIGURE 6.43 Masses hanging over pulley.

**70. A flexible drive belt runs over a flywheel turning freely on a frictionless axle (see Fig. 6.44). The mass per unit length of the drive belt is σ, and the tension in the drive belt is T. The speed of the drive belt is v. Show that each small segment $d\theta$

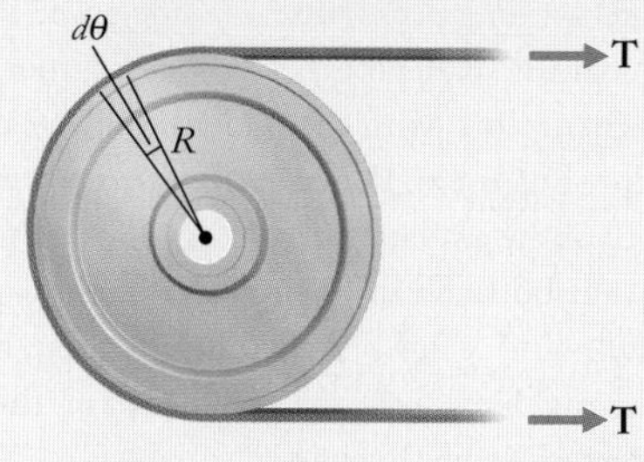

FIGURE 6.44 Belt over flywheel.

of the drive belt exerts a radial force $(T - \sigma v^2)\,d\theta$ on the flywheel. For what value of v is this force zero?

**71. A circle of rope of mass m and radius r is spinning about its center so each point of the rope has a speed v. Calculate the tension in the rope.

**72. The rotor of a helicopter consists of two blades 180° apart. Each blade has a mass of 140 kg and a length of 3.6 m. What is the tension in each blade at the hub when rotating at 320 rev/min? Pretend that each blade is a uniform thin rod.

**73. Assume that the Earth is a sphere and that the force of gravity (mg) points precisely toward the center of the Earth. Taking into account the rotation of the Earth about its axis, calculate the angle between the direction of a plumb line and the direction of the Earth's radius as a function of latitude. What is this deviation angle at a latitude of 45°?

**74. A curve of radius 120 m is banked at an angle of 10°. If an automobile with wheels with $\mu_s = 0.90$ is to round this curve without skidding, what is the maximum permissible speed?

**75. Figure 6.45 shows a pendulum hanging from the edge of a horizontal disk which rotates around its axis at a constant rate. The angle α that the rotating pendulum makes with the vertical increases with the speed of rotation, and can therefore be used as an indicator of this speed. Find a formula for the speed v_0 of the edge of the disk in terms of the angle α, the radius R of the disk, and the length l of the pendulum. If $R = 0.20$ m and $l = 0.30$ m, what is the speed when $\alpha = 45°$?

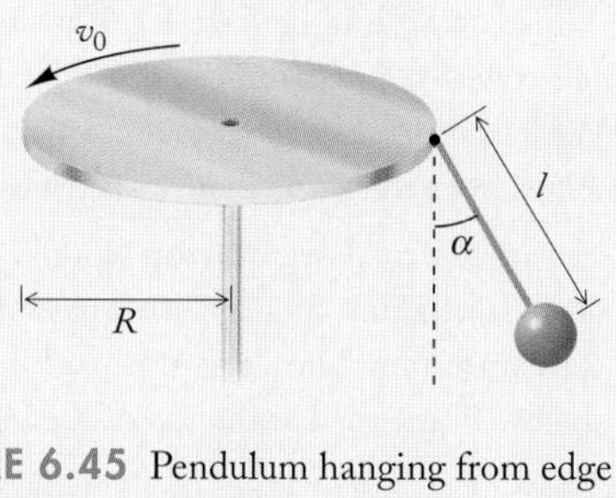

FIGURE 6.45 Pendulum hanging from edge of disk.

REVIEW PROBLEMS

76. At liftoff, the Saturn V rocket used for the Apollo missions has a mass of 2.45×10^6 kg.
    (a) What is the minimum thrust that the rocket engines must develop to achieve liftoff?
    (b) The actual thrust that the engines develop is 3.3×10^7 N. What is the vertical acceleration of the rocket at liftoff?
    (c) At burnout, the rocket has spent its fuel, and its remaining mass is 0.75×10^6 kg. What is the acceleration just before burnout? Assume that the motion is still vertical and that the strength of gravity is the same as when the rocket is on the ground.
77. If the coefficient of static friction between the tires of an automobile and the road is $\mu_s = 0.80$, what is the minimum distance the automobile needs in order to stop without skidding from an initial speed of 90 km/h? How long does it take to stop?
78. Suppose that the last car of a train becomes uncoupled while the train is moving upward on a slope of 1:6 at a speed of 48 km/h.
    (a) What is the deceleration of the car? Ignore friction.
    (b) How far does the car coast up the slope before it stops?
79. A 40-kg crate falls off a truck traveling at 80 km/h on a level road. The crate slides along the road and gradually comes to a halt. The coefficient of kinetic friction between the crate and the road is 0.80.
    (a) Draw a "free-body" diagram for the crate sliding on the road.
    (b) What is the normal force the road exerts on the crate?
    (c) What is the friction force the road exerts on the crate?
    (d) What is the weight force on the crate? What is the net force on the crate?
    (e) What is the deceleration of the crate? How far does the crate slide before coming to a halt?
80. A 2.0-kg box rests on an inclined plane which makes an angle of 30° with the horizontal. The coefficient of static friction between the box and the plane is 0.90.
    (a) Draw a "free-body" diagram for the box.
    (b) What is the normal force the inclined plane exerts on the box?
    (c) What is the friction force the inclined plane exerts on the box?
    (d) What is the net force the inclined plane exerts on the box? What is the direction of this force?
81. The body of an automobile is held above the axles of the wheels by means of four springs, one near each wheel. Assume that the springs are vertical and that the forces on all the springs are the same. The mass of the body of the automobile is 1200 kg, and the spring constant of each spring is 2.0×10^4 N/m. When the automobile is stationary on a level road, how far are the springs compressed from their relaxed length?

*82. A block of wood rests on a sheet of paper lying on a table. The coefficient of static friction between the block and the paper is $\mu_s = 0.70$, and that between the paper and the table is $\mu_s = 0.50$. If you tilt the table, at what angle will the block begin to move?

*83. Two blocks of masses m_1 and m_2 are connected by a string. One block slides on a table, and the other hangs from the string, which passes over a pulley (see Fig. 6.46). The coefficient of sliding friction between the first block and the table is $\mu_k = 0.20$. What is the acceleration of the blocks?

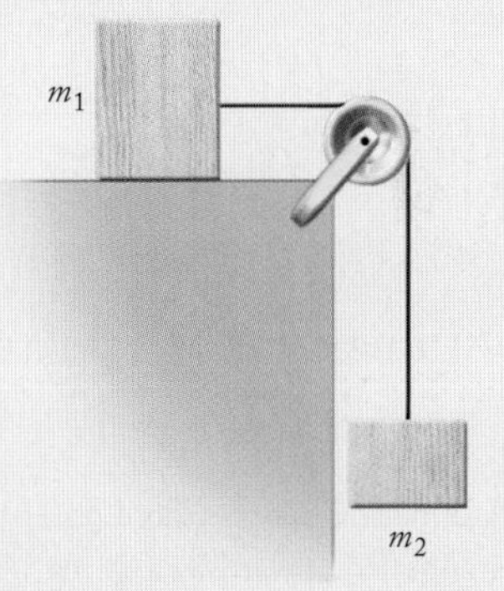

FIGURE 6.46 Mass on table, pulley, and hanging mass.

*84. A man of mass 75 kg is pushing a heavy box on a flat floor. The coefficient of sliding friction between the floor and the box is 0.20, and the coefficient of static friction between the man's shoes and the floor is 0.80. If the man pushes horizontally (see Fig. 6.47), what is the maximum mass of the box he can move?

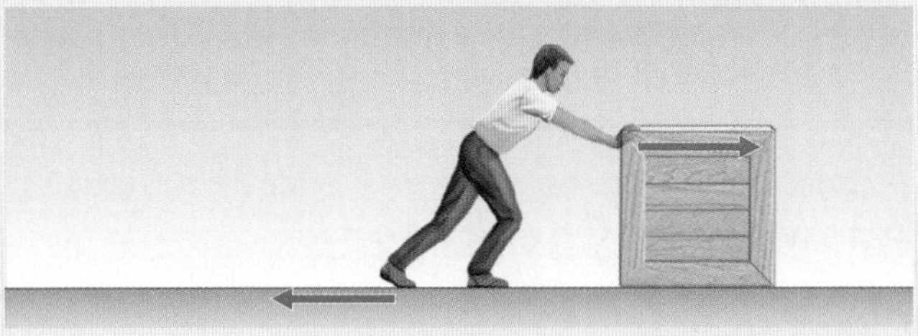

FIGURE 6.47 Pushing a box.

*85. Two springs of constants 2.0×10^3 N/m and 3.0×10^3 N/m are connected in tandem, and a mass of 5.0 kg hangs vertically from the bottom of the lower spring. By what amount does the mass stretch the combined spring? Each individual spring?

*86. A block of mass 1.5 kg is placed on a flat surface, and it is being pulled horizontally by a spring with a spring constant 1.2×10^3 N/m (see Fig. 6.48). The coefficient of static friction between the block and the table is $\mu_s = 0.60$, and the coefficient of sliding friction is $\mu_k = 0.40$.

(a) By what amount must the spring be stretched to start the block moving?

(b) What is the acceleration of the block if the stretch of the spring is maintained at a constant value equal to that required to start the motion?

(c) By what amount must the spring be stretched to keep the mass moving at constant speed?

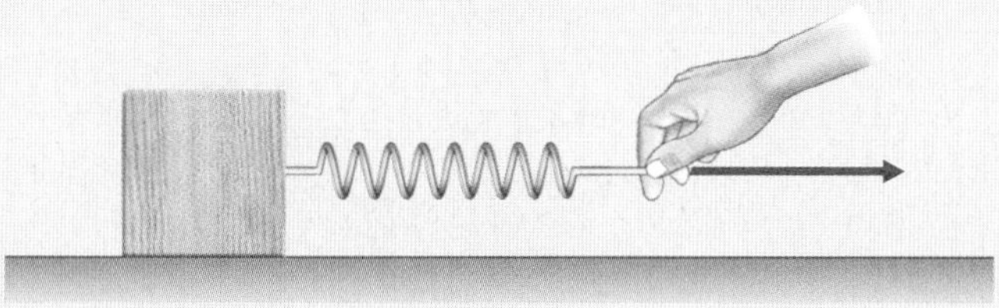

FIGURE 6.48 Mass pulled by spring.

*87. A block of mass 1.5 kg is placed on a plane inclined at 30°, and it is being pulled upward by a spring with a spring constant 1.2×10^3 N/m (see Fig. 6.49). The direction of pull of the spring is parallel to the inclined plane. The coefficient of static friction between the block and the inclined plane is $\mu_s = 0.60$, and the coefficient of sliding friction is $\mu_k = 0.40$.

(a) By what amount must the spring be stretched to start the block moving?

(b) What is the acceleration of the block if the stretch of the spring is maintained at a constant value equal to that required to start the motion?

(c) By what amount must the spring be stretched to keep the mass moving at constant speed?

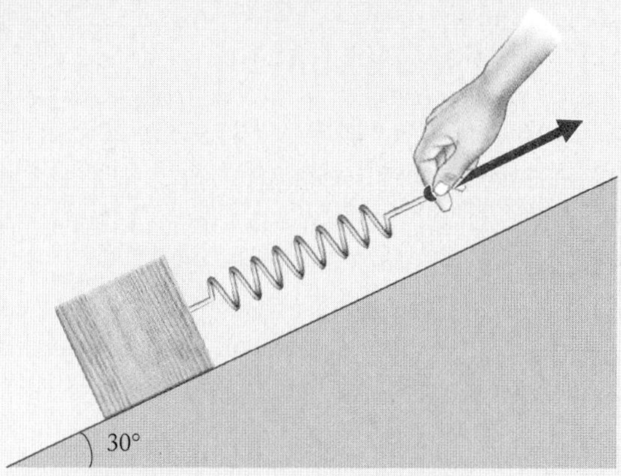

FIGURE 6.49 Block on incline pulled by spring.

*88. A mass m_1 slides on a smooth, frictionless table. The mass is constrained to move in a circle by a string that passes through a hole in the center of the table and is attached to a second mass m_2 hanging vertically below the table (Fig. 6.50). If the radius of the circular motion of the first mass is r, what must be its speed?

FIGURE 6.50 Mass in circular motion and hanging mass.

89. An automobile enters a curve of radius 45 m at 70 km/h. Will the automobile skid? The curve is not banked, and the coefficient of static friction between the wheels and the road is 0.80.

*90. A stone of 0.90 kg attached to a rod is being whirled around a vertical circle of radius 0.92 m. Assume that during this motion the speed of the stone is constant. If at the top of the circle the tension in the rod is (just about) zero, what is the tension in the rod at the bottom of the circle?

Answers to Checkups

Checkup 6.1

1. The weight of the second book results in a normal force between the first book and the table that is twice as large, so the friction force, and thus the horizontal push to overcome it, will be twice as large, or 20 N. If the first book pushes the second, then the friction force of the second book on the first adds to the friction force of the first to require a push also twice as large as the original, or 20 N.

2. While the block coasts up the incline, the friction, which always opposes the *motion,* is directed down the plane (the corresponding "free-body" diagram would have the weight component $mg \sin \theta$ and the friction $\mathbf{f}_k$ both pointing down the incline, the normal force **N** perpendicularly out from the incline, and the weight component $mg \cos \theta$ balancing the normal force). When the block slides down the incline, the friction force now points up the incline, again opposing the motion (except for this one reversed vector, the "free-body" diagram is the same).

3. For static friction, $\mu_s N$ is the *maximum* friction force, beyond which a body will begin to move. Since the brick remains at rest, the friction force can be any value less than or equal to $\mu_s N$, and so the correct answer is (c).

4. The static friction force can be no larger than $f_{s,\max} = \mu_s N$; thus the box will not slide if the deceleration is less than is $f_{s,\max}/m$, and will slide if the deceleration is larger than that value.

5. For automotive speeds, the friction force due to air resistance is proportional to the aerodynamic constant C and varies as the square of the speed [Eq. (6.10)], so for the same force (same road slope), the speed varies inversely with the square root of C. Thus, since C is reduced by a factor of 2 (from $C = 0.30$ to $C = 0.15$), the speed will increase by a factor of $\sqrt{2} \approx 1.4$.

6. (C) μ_k and θ only. The motion of the box down the ramp is the same as that of the ship down the slipway in Example 2; from Eq. (6.3), we see $a_x = (\sin\theta - \mu_k \cos\theta)g$. Thus, the acceleration does not depend on the mass, but depends on μ_k and θ. However, the friction force $f_k = \mu_k N = \mu_k mg \cos\theta$ does depend on all three quantities.

Checkup 6.2

1. The weight of the climbers has doubled, so the spring force will double to balance the weight. Since $F = -kx$, where k is constant, the amount of stretch x will double to 40 cm.

2. The mass starts to move when the magnitude of the spring force equals the maximum static friction force, or $kx = \mu_s N = \mu_s mg$. Thus, $\mu_s = kx/mg = (50\text{ N/m} \times 0.10\text{ m})/(1.0\text{ kg} \times 9.8\text{ m/s}^2) = 0.51$.

3. (E) $\frac{1}{4}$. The relationship between force, spring constant, and distance is $F = -kx$. So for the first spring, $x_1 = -F/k$. If k is twice as large and F half as large, then for the second spring $x_2 = -(F/2)/(2k) = -\frac{1}{4}F/k$. Comparing, we have $x_2/x_1 = \frac{1}{4}$.

Checkup 6.3

1. Once the string breaks, the stone is in free motion; by Newton's First Law, the stone will continue with uniform velocity in a straight line in the direction of its motion when the centripetal force from the string was removed. That direction is tangent to the circle.

2. The acceleration is centripetal, and so is always to the right of the motorcyclist, perpendicular to the instantaneous direction of motion.

3. At the top, the centripetal acceleration points straight downward, toward the center of the circle of motion.

4. The chair exerts a force upward on the pilot, toward the center of the circle, and so that force is centripetal. The pilot exerts a force downward on the chair, away from the center of the circle, and so that force is centrifugal. At the bottom of the loop, the apparent weight is downward (see Example 12). The apparent weight (if defined as the perpendicular force on the seat of the chair) changes direction as the chair swings around, varying in magnitude from $mv^2/r + mg$ at the bottom to $mv^2/r - mg$ at the top. There is an additional force parallel to the seat of the chair at other points (a tangential force), when the weight has a component parallel to the seat of the chair; there is also a tangential force on the chair if the speed v changes around the loop.

5. (D) Inner; inner. Since both cars have the same constant speed, the driver on the inner lane (the circle of smaller radius r) will have less distance ($2\pi r$) to travel, and so will finish first. That inner car also has the larger acceleration, since it has the smaller r, and $a = v^2/r$.

CHAPTER

7 Work and Energy

CONCEPTS IN CONTEXT

Concepts *in* Context

The high-speed and high-acceleration thrills of a roller coaster are made possible by the force of gravity. We will see that gravity does work on the roller-coaster car while it descends, increasing its kinetic energy.

To see how energy considerations provide powerful approaches for understanding and predicting motion, we will ask:

? What is the work done by gravity when the roller-coaster car descends along an incline? (Example 3, page 209)

? As a roller-coaster car travels up to a peak, over it, and then down again, does gravity do work? Does the normal force? (Checkup 7.1, question 1, page 210)

? For a complex, curving descent, how can the final speed be determined in a simple way? (Example 8, page 222; and Checkup 7.4, question 1, page 224)

Conservation laws play an important role in physics. Such laws assert that some quantity is conserved, which means that the quantity remains constant even when particles or bodies suffer drastic changes involving motions, collisions, and reactions. One familiar example of a conservation law is the conservation of mass. Expressed in its simplest form, this law asserts that the mass of a given particle remains constant, regardless of how the particle moves and interacts with other particles or other bodies. In the preceding two chapters we took this conservation law for granted, and we treated the particle mass appearing in Newton's Second Law ($m\mathbf{a} = \mathbf{F}$) as a constant, time-independent quantity. More generally, the sum of all the masses of the particles or bodies in a system remains constant, even when the bodies suffer transformations and reactions. In everyday life and in commercial and industrial operations, we always rely implicitly on the conservation of mass. For instance, in the chemical plants that reprocess the uranium fuel for nuclear reactors, the batches of uranium compounds are carefully weighed at several checkpoints during the reprocessing operation to ensure that none of the uranium is diverted for nefarious purposes. This procedure would make no sense if mass were not conserved, if the net mass of a batch could increase or decrease spontaneously.

This chapter and the next deal with the conservation of energy. This conservation law is one of the most fundamental laws of nature. Although we will derive this law from Newton's laws, it is actually much more general than Newton's laws, and it remains valid even when we step outside of the realm of Newtonian physics and enter the realm of relativistic physics or atomic physics, where Newton's laws fail. No violation of the law of conservation of energy has ever been discovered.

In mechanics, *we can use the conservation law for energy to deduce some features of the motion of a particle or of a system of particles* when it is undesirable or too difficult to calculate the full details of the motion from Newton's Second Law. This is especially helpful in those cases where the forces are not known exactly; we will see some examples of this kind in Chapter 11.

But before we can deal with energy and its conservation, we must introduce the concept of work. Energy and work are closely related. We will see that the work done by the net force on a body is equal to the change of the kinetic energy (the energy of motion) of the body.

7.1 WORK

Online Concept Tutorial 9

To introduce the definition of work done by a force, we begin with the simple case of motion along a straight line, with the force along the line of motion, and then we will generalize to the case of motion along some arbitrary curved path, with the force in some arbitrary direction at each point. Consider a particle moving along such a straight line, say, the x axis, and suppose that a constant force F_x, directed along the same straight line, acts on the particle. Then the **work done by the force** F_x *on the particle as it moves some given distance is defined as the product of the force and the displacement* Δx:

$$W = F_x \, \Delta x \tag{7.1}$$

work done by one constant force

This rigorous definition of work is consistent with our intuitive notion of what constitutes "work." For example, the particle might be a stalled automobile that you are pushing along a road (see Fig. 7.1). Then the work that you perform is proportional to the magnitude of the force you have to exert, and it is also proportional to the distance you move the automobile.

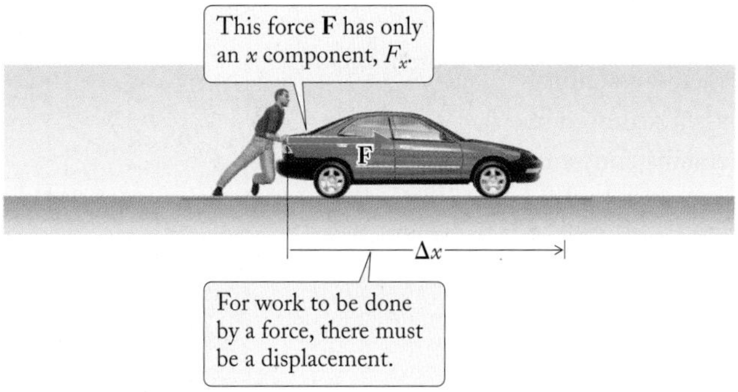

FIGURE 7.1 You do work while pushing an automobile along a road with a horizontal force **F**.

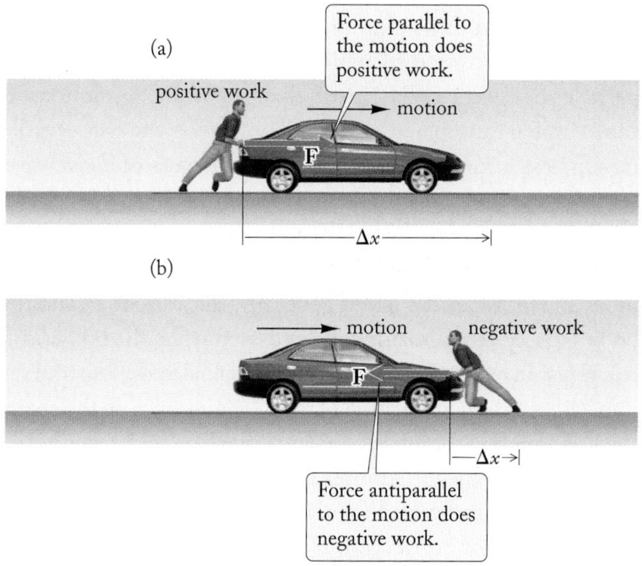

FIGURE 7.2 (a) The work you do on the automobile is positive if you push in the direction of motion. (b) The work you do on the automobile is negative if you push in the direction opposite to the motion.

Note that in Eq. (7.1), F_x is reckoned as positive if the force is in the positive x direction and negative if in the negative x direction. The subscript x on the force helps us to remember that F_x has a magnitude and a sign; in fact, F_x is the x component of the force, and this x component can be positive or negative. According to Eq. (7.1), *the work is positive if the force and the displacement are in the same direction* (both positive, or both negative), *and the work is negative if they are in opposite directions* (one positive, the other negative). When pushing the automobile, you do positive work on the automobile if you push in the direction of the motion, so your push tends to accelerate the automobile (Fig. 7.2a); but you do negative work on the automobile (it does work on you) if you push in the direction opposite to the motion, so your push tends to decelerate the automobile (Fig. 7.2b).

Equation (7.1) gives the work done by one of the forces acting on the particle. If several forces act, then Eq. (7.1) can be used to calculate the work done by each force. If we add the amounts of work done by all the forces acting on the particle, we obtain the net amount of work done by all these forces together. This net amount of work can be directly calculated from the net force:

$$W = F_{\text{net},x}\,\Delta x$$

In the SI system, *the unit of work is the* **joule** (J), which is the work done by a force of 1 N during a displacement of 1 m. Thus,

$$1\text{ joule} = 1\text{ J} = 1\text{ N}\cdot\text{m}$$

EXAMPLE 1 Suppose you push your stalled automobile along a straight road (see Fig. 7.1). If the force required to overcome friction and to keep the automobile moving at constant speed is 500 N, how much work must you do to push the automobile 30 m?

SOLUTION: With $F_x = 500$ N and $\Delta x = 30$ m, Eq. (7.1) gives

$$W = F_x\,\Delta x = 500\text{ N} \times 30\text{ m} = 15\,000\text{ J} \tag{7.2}$$

JAMES PRESCOTT JOULE (1818–1889) *English physicist. He established experimentally that heat is a form of mechanical energy, and he made the first direct measurement of the mechanical equivalent of heat. By a series of meticulous mechanical, thermal, and electrical experiments, Joule provided empirical proof of the general law of conservation of energy.*

EXAMPLE 2 A 1000-kg elevator cage descends 400 m within a skyscraper. (a) What is the work done by gravity on the elevator cage during this displacement? (b) Assuming that the elevator cage descends at constant velocity, what is the work done by the tension of the suspension cable?

SOLUTION: (a) With the x axis arranged vertically upward (see Fig. 7.3), the displacement is negative, $\Delta x = -400$ m; and the x component of the weight is also negative, $w_x = -mg = -1000\text{ kg} \times 9.81\text{ m/s}^2 = -9810$ N. Hence by the definition (7.1), the work done by the weight is

$$W = w_x\,\Delta x = (-9810\text{ N}) \times (-400\text{ m}) = 3.92 \times 10^6\text{ J} \tag{7.3}$$

(b) For motion at constant velocity, the tension force must exactly balance the weight, so the net force $F_{\text{net},x}$ is zero. Therefore, the tension force of the cable has the same magnitude as the weight, but the opposite direction:

$$T_x = +mg = 9810\text{ N}$$

The work done by this force is then

$$W = T_x\,\Delta x = 9810\text{ N} \times (-400\text{ m}) = -3.92 \times 10^6\text{ J} \tag{7.4}$$

This work is negative because the tension force and the displacement are in opposite directions. Gravity does work on the elevator cage, and the elevator cage does work on the cable.

COMMENTS: (a) Note that the work done by gravity is completely independent of the details of the motion; the work depends on the total vertical displacement and on the weight, but not on the velocity or the acceleration of the motion. (b) Note that the work done by the tension is exactly the negative of the work done by gravity, and thus the net work done by both forces together is zero (we can also see this by examining the work done by the net force; since the net force $F_{\text{net},x} = w_x + T_x$ is zero, the net work $W = F_{\text{net},x}\,\Delta x$ is zero). However, the result (7.4) for the work done by the tension depends implicitly on the assumptions made about the motion. Only for unaccelerated motion does the tension force remain constant at 9810 N. For instance, if the elevator cage were allowed to fall freely with the acceleration of gravity, then the tension would be zero; the work done by the tension would then also be zero, whereas the work done by gravity would still be 3.92×10^6 J.

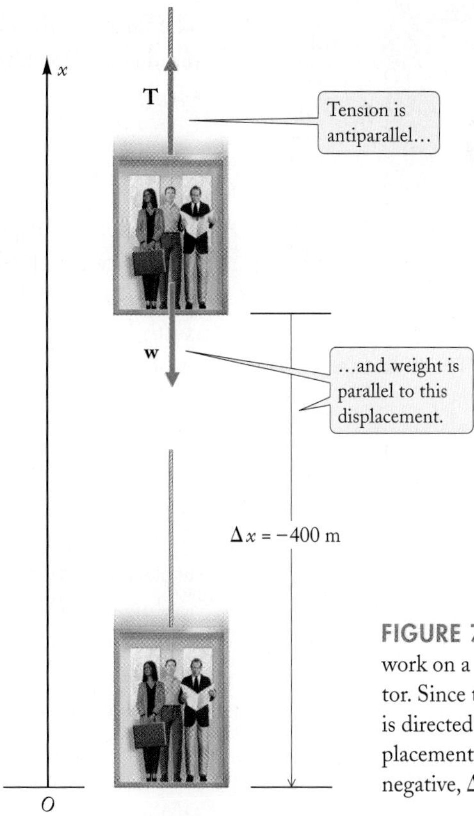

FIGURE 7.3 Gravity does work on a descending elevator. Since the positive x axis is directed upward, the displacement of the elevator is negative, $\Delta x = -400$ m.

Although the rigorous definition of work given in Eq. (7.1) agrees to some extent with our intuitive notion of what constitutes "work," the rigorous definition clashes with our

FIGURE 7.4 Man holding a ball. The displacement of the ball is zero; hence the work done on the ball is zero.

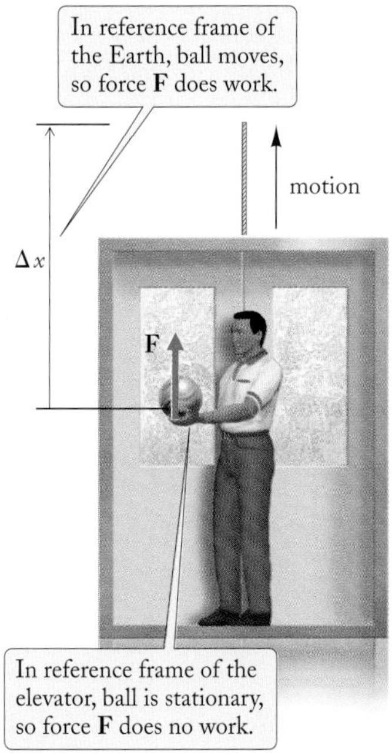

FIGURE 7.5 The man holding the ball rides in an elevator. The work done depends on the reference frame.

intuition in some instances. For example, consider a man holding a bowling ball in a fixed position in his outstretched hand (see Fig. 7.4). Our intuition suggests that the man does work—yet Eq. (7.1) indicates that no work is done on the ball, since the ball does not move and the displacement Δx is zero. The resolution of this conflict hinges on the observation that, although the man does no work *on the ball*, he does work *within his own muscles* and, consequently, grows tired of holding the ball. A contracted muscle is never in a state of complete rest; within it, atoms, cells, and muscle fibers engage in complicated chemical and mechanical processes that involve motion and work. This means that work is done, and wasted, internally within the muscle, while no work is done externally on the bone to which the muscle is attached or on the bowling ball supported by the bone.

Another conflict between our intuition and the rigorous definition of work arises when we consider a body in motion. Suppose that the man with the bowling ball in his hand rides in an elevator moving upward at constant velocity (Fig. 7.5). In this case, the displacement is not zero, and the force (push) exerted by the hand on the ball does work—the displacement and the force are in the same direction, and consequently the man continuously does positive work on the ball. Nevertheless, to the man the ball feels no different when riding in the elevator than when standing on the ground. This example illustrates that *the amount of work done on a body depends on the reference frame.* In the reference frame of the ground, the ball is moving upward and work is done on it; in the reference frame of the elevator, the ball is at rest, and no work is done on it. The lesson we learn from this is that before proceeding with a calculation of work, we must be careful to specify the reference frame.

If the motion of the particle and the force are not along the same line, then the simple definition of work given in Eq. (7.1) must be generalized. Consider a particle moving along some arbitrary curved path, and suppose that the force that acts on the particle is constant (we will consider forces that are not constant in the next section). The force can then be represented by a vector **F** (see Fig. 7.6a) that is constant in magnitude and direction. *The work done by this constant force during a (vector) displacement* **s** *is defined as*

$$W = Fs\cos\theta \tag{7.5}$$

where F is the magnitude of the force, s is the length of the displacement, and θ is the angle between the direction of the force and the direction of the displacement. Both F and s in Eq. (7.5) are positive; the correct sign for the work is provided by the factor $\cos\theta$. The work done by the force **F** is positive if the angle between the force and the displacement is less than 90°, and it is negative if this angle is more than 90°.

As shown in Fig. 7.6b, the expression (7.5) can be regarded as the magnitude of the displacement (s) multiplied by the component of the force along the direction of the displacement ($F\cos\theta$). If the force is parallel to the direction of the displacement ($\theta = 0$ and $\cos\theta = 1$), then the work is simply Fs; this coincides with the case of motion along a straight line [see Eq. (7.1)]. If the force is perpendicular to the direction of the displacement ($\theta = 90°$ and $\cos\theta = 0$), then the work vanishes. For instance, if a woman holding a bowling ball walks along a level road at constant speed, she does not do any work on the ball, since the force she exerts on the ball is perpendicular to the direction of motion (Fig. 7.7a). However, if the woman climbs up some stairs while holding the ball, then she does work on the ball, since now the force she exerts has a component along the direction of motion (Fig. 7.7b).

For two arbitrary vectors **A** and **B**, the product of their magnitudes and the cosine of the angle between them is called the **dot product** (or **scalar product**) of the vec-

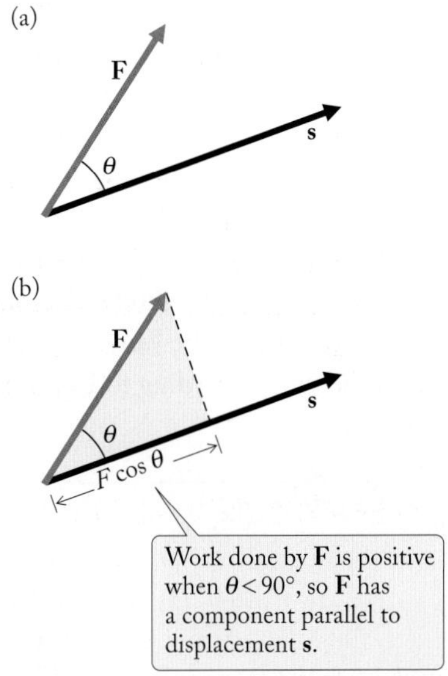

FIGURE 7.6 (a) A constant force **F** acts during a displacement **s**. The force makes an angle θ with the displacement. (b) The component of the force along the direction of the displacement is $F \cos \theta$.

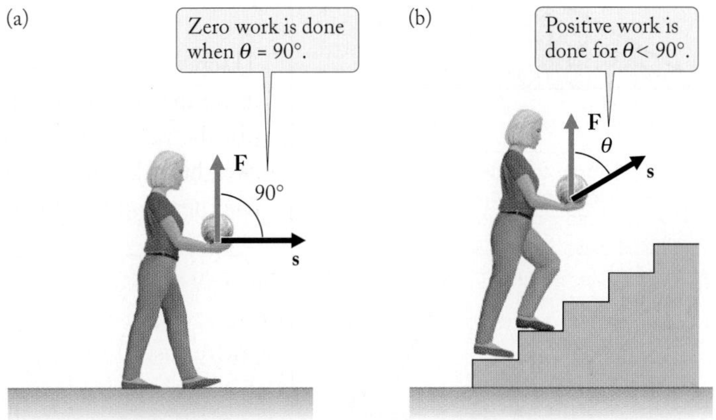

FIGURE 7.7 (a) The force exerted by the woman is perpendicular to the displacement. (b) The force exerted by the woman is now not perpendicular to the displacement.

tors (see Section 3.4). The standard notation for the dot product consists of the two vector symbols separated by a dot:

dot product (scalar product)

$$\mathbf{A} \cdot \mathbf{B} = AB \cos \theta \tag{7.6}$$

Accordingly, the expression (7.5) for the work can be written as the dot product of the force vector **F** and displacement vector **s**,

$$W = \mathbf{F} \cdot \mathbf{s} \tag{7.7}$$

In Section 3.4, we found that the dot product is also equal to the sum of the products of the corresponding components of the two vectors, or

$$\mathbf{A} \cdot \mathbf{B} = A_x B_x + A_y B_y + A_z B_z \tag{7.8}$$

If the components of **F** are F_x, F_y, and F_z and those of **s** are Δx, Δy, and Δz, then the second version of the dot product means that the work can be written

$$W = F_x \,\Delta x + F_y \,\Delta y + F_z \,\Delta z \tag{7.9}$$

Note that although this equation expresses the work as a sum of contributions from the x, y, and z components of the force and the displacement, the work does not have separate components. The three terms on the right are merely three terms in a sum. Work is a single-component, scalar quantity, not a vector quantity.

EXAMPLE 3 A roller-coaster car of mass m glides down to the bottom of a straight section of inclined track from a height h. (a) What is the work done by gravity on the car? (b) What is the work done by the normal force? Treat the motion as particle motion.

Concepts — in — Context

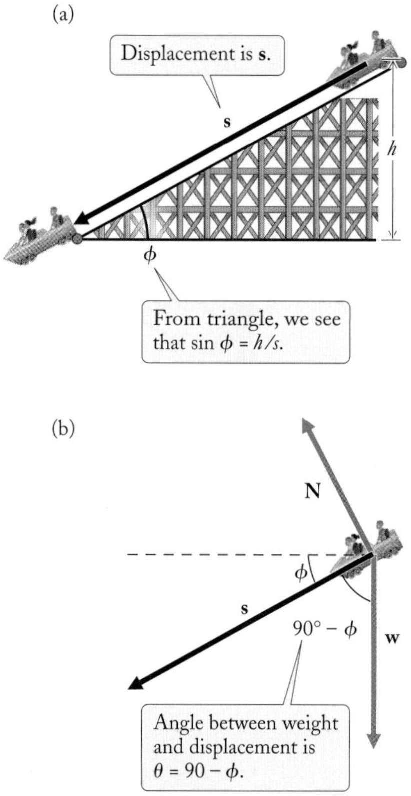

FIGURE 7.8 (a) A roller-coaster car undergoing a displacement along an inclined plane. (b) "Free-body" diagram showing the weight, the normal force, and the displacement of the car.

SOLUTION: (a) Figure 7.8a shows the inclined track. The roller-coaster car moves down the full length of this track. By inspection of the right triangle formed by the incline and the ground, we see that the displacement of the car has a magnitude

$$s = \frac{h}{\sin\phi} \tag{7.10}$$

[Here we use the label ϕ (Greek phi) for the angle of the incline to distinguish it from the angle θ appearing in Eq. (7.5).] Figure 7.8b shows a "free-body" diagram for the car; the forces acting on it are the normal force **N** and the weight **w**. The weight makes an angle $\theta = 90° - \phi$ with the displacement. According to Eq. (7.5), we then find that the work W done by the weight **w** is

$$W = ws\cos\theta = mg \times \frac{h}{\sin\phi} \times \cos(90° - \phi)$$

Since $\cos(90° - \phi) = \sin\phi$, the work is

$$W = mg \times \frac{h}{\sin\phi} \times \sin\phi = mgh \tag{7.11}$$

Alternatively, we can use components to calculate the work. For example, if we choose the x axis horizontal and the y axis vertical, the motion is two-dimensional, and we need to consider x and y components. The components of the weight are $w_x = 0$ and $w_y = -mg$. According to Eq. (7.9), the work done by the weight is then

$$W = w_x\,\Delta x + w_y\,\Delta y = 0 \times \Delta x + (-mg) \times \Delta y = 0 + (-mg) \times (-h) = mgh$$

Of course, this alternative calculation agrees with Eq. (7.11).

(b) The work done by the normal force is zero, since this force makes an angle of 90° with the displacement.

COMMENTS: (a) Note that the result (7.11) for the work done by the weight is independent of the angle of the incline—it depends only on the change of height, not on the angle or the length of the inclined plane. (b) Note that the result of zero work for the normal force is quite general. The normal force **N** acting on any body rolling or sliding on any kind of fixed surface never does work on the body, since this force is always perpendicular to the displacement.

✔ Checkup 7.1

Concepts — in — Context

QUESTION 1: Consider a frictionless roller-coaster car traveling up to, over, and down from a peak. The forces on the car are its weight and the normal force of the tracks. Does the normal force of the tracks perform work on the car? Does the weight?

QUESTION 2: While cutting a log with a saw, you push the saw forward, then pull backward, etc. Do you do positive or negative work on the saw while pushing it forward? While pulling it backward?

QUESTION 3: While walking her large dog on a leash, a woman holds the dog back to a steady pace. Does the dog's pull do positive or negative work on the woman? Does the woman's pull do positive or negative work on the dog?

QUESTION 4: You are trying to stop a moving cart by pushing against its front end. Do you do positive or negative work on the cart? What if you pull on the rear end?

QUESTION 5: You are whirling a stone tied to a string around a circle. Does the tension of the string do any work on the stone?

QUESTION 6: Figure 7.9 shows several equal-magnitude forces **F** and displacements **s**. For which of these is the work positive? Negative? Zero? For which of these is the work largest?

QUESTION 7: To calculate the work performed by a known constant force **F** acting on a particle, which two of the following do you need to know? (1) The mass of the particle; (2) the acceleration; (3) the speed; (4) the displacement; (5) the angle between the force and the displacement.

(A) 1 and 2 (B) 1 and 5 (C) 2 and 3
(D) 3 and 5 (E) 4 and 5

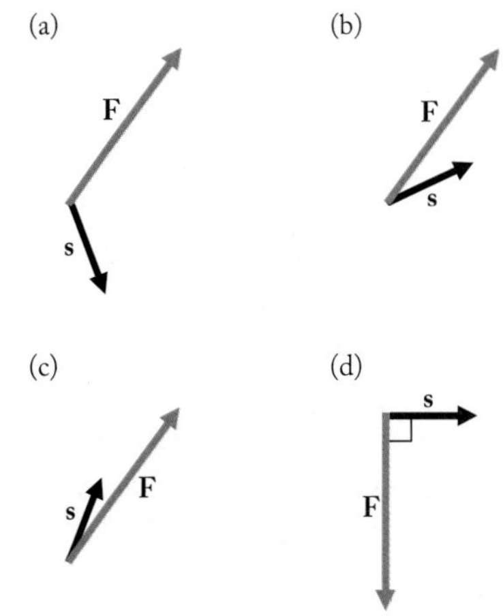

FIGURE 7.9 Several equal-magnitude forces and displacements.

7.2 WORK FOR A VARIABLE FORCE

Online Concept Tutorial 9

The definition of work in the preceding section assumed that the force was constant (in magnitude and in direction). But many forces are not constant, and we need to refine our definition of work so we can deal with such forces. For example, suppose that you push a stalled automobile along a straight road, and suppose that the force you exert is not constant—as you move along the road, you sometimes push harder and sometimes less hard. Figure 7.10 shows how the force might vary with position. (The reason why you sometimes push harder is irrelevant—maybe the automobile passes through a muddy portion of the road and requires more of a push, or maybe you get impatient and want to hurry the automobile along; all that is relevant for the calculation of the work is the value of the force at different positions, as shown in the plot.)

Such a variable force can be expressed as a function of position:

$$F_x = F_x(x)$$

(here the subscript indicates the x component of the force, and the x in parentheses indicates that this component is a function of x; that is, it varies with x, as shown in the diagram). To evaluate the work done by this variable force on the automobile, or on a particle, during a displacement from $x = a$ to $x = b$, we divide the total displacement into a large number of small intervals, each of length Δx (see Fig. 7.11). The beginnings and ends of these intervals are located at $x_0, x_1, x_2, \ldots, x_n$, where the first location x_0 coincides with a and the last location x_n coincides with b. Within each of the small intervals, the force can be regarded as approximately constant—within the interval x_{i-1} to x_i (where $i = 1$, or 2, or 3, . . . , or n), the force is approximately $F_x(x_i)$. This approximation is at its best if we select Δx to be very small. The work done by this force as the particle moves from x_{i-1} to x_i is then

$$W_i = F_x(x_i)\,\Delta x \tag{7.12}$$

and the total work done as the particle moves from a to b is simply the sum of all the small amounts of work associated with the small intervals:

$$W = \sum_{i=1}^{n} W_i = \sum_{i=1}^{n} F_x(x_i)\,\Delta x \tag{7.13}$$

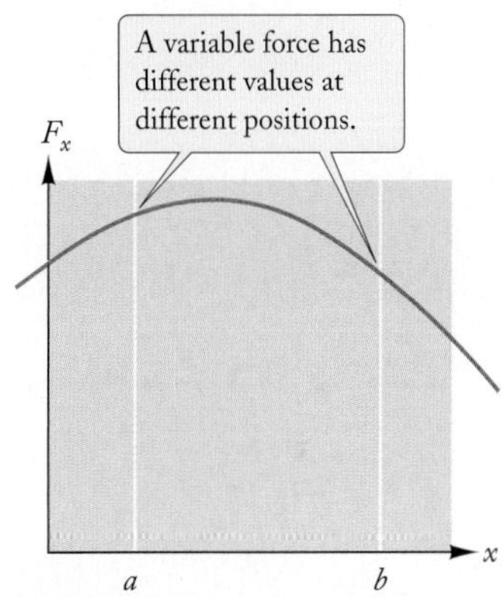

FIGURE 7.10 Plot of F_x vs. x for a force that varies with position.

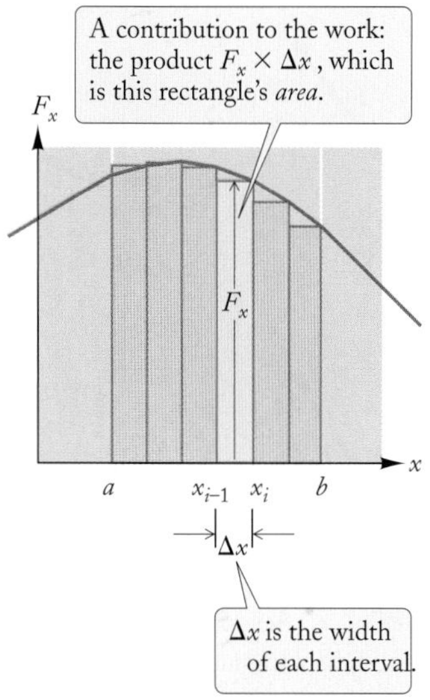

FIGURE 7.11 The curved plot of F_x vs. x has been approximated by a series of horizontal and vertical steps. This is a good approximation if Δx is very small.

Note that each of the terms $F_x(x_i)\,\Delta x$ in the sum is the area of a rectangle of height $F_x(x_i)$ and width Δx, highlighted in color in Fig. 7.11. Thus, Eq. (7.13) gives the sum of all the rectangular areas shown in Fig. 7.11.

Equation (7.13) is only an approximation for the work. In order to improve this approximation, we must use a smaller interval Δx. In the limiting case $\Delta x \to 0$ (and $n \to \infty$), the width of each rectangle approaches zero and the number of rectangles approaches infinity, so we obtain an exact expression for the work. Thus, the exact definition for the work done by a variable force is

$$W = \lim_{\Delta x \to 0} \sum_{i=1}^{n} F_x(x_i)\,\Delta x$$

This expression is called the **integral** of the function $F_x(x)$ between the limits a and b. The usual notation for this integral is

$$W = \int_a^b F_x(x)\,dx \tag{7.14}$$

where the symbol $\int$ is called the integral sign and the function $F_x(x)$ is called the integrand. The quantity (7.14) is equal to *the area bounded by the curve representing* $F_x(x)$, *the* x *axis, and the vertical lines* $x = a$ *and* $x = b$ in Fig. 7.12. More generally, for a curve that has some portions above the x axis and some portions below, the quantity (7.14) is the net area bounded by the curve above and below the x axis, with areas above the x axis being reckoned as positive and areas below the x axis as negative.

We will also need to consider arbitrarily small contributions to the work. From Eq. (7.12), the infinitesimal work dW done by the force $F_x(x)$ when acting over an infinitesimal displacement dx is

$$dW = F_x(x)\,dx \tag{7.15}$$

We will see later that the form (7.15) is useful for calculations of particular quantities, such as power or torque.

Finally, if the force is variable and the motion is in more than one dimension, the work can be obtained by generalizing Eq. (7.7):

work done by a variable force

$$W = \int \mathbf{F} \cdot d\mathbf{s} \tag{7.16}$$

To evaluate Eq. (7.16), it is often easiest to express the integral as the sum of three integrals, similar to the form of Eq. (7.9). For now, we consider the use of Eq. (7.14) to determine the total work done by a variable force as it acts over some distance in one dimension.

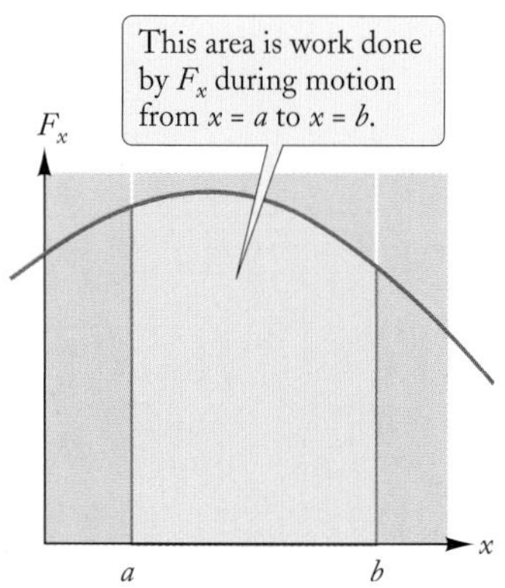

FIGURE 7.12 The integral $\int_a^b F_x(x)dx$ is the area (colored) under the curve representing $F_x(x)$ between $x = a$ and $x = b$.

EXAMPLE 4 A spring exerts a restoring force $F_x(x) = -kx$ on a particle attached to it (compare Section 6.2). What is the work done by the spring on the particle when it moves from $x = a$ to $x = b$?

SOLUTION: By Eq. (7.14), the work is the integral

$$W = \int_a^b F_x(x)\,dx = \int_a^b (-kx)\,dx$$

To evaluate this integral, we rely on a result from calculus (see the Math Help box on integrals) which states that the integral between a and b of the function x is the difference between the values of $\frac{1}{2}x^2$ at $x = b$ and $x = a$:

$$\int_a^b x\,dx = \tfrac{1}{2}x^2\Big|_a^b = \tfrac{1}{2}(b^2 - a^2)$$

where the vertical line | means that we evaluate the preceding function at the upper limit and then subtract its value at the lower limit. Since the constant $-k$ is just a multiplicative factor, we may pull it outside the integral and obtain for the work

$$W = \int_a^b (-kx)\,dx = -k\int_a^b x\,dx = -\tfrac{1}{2}k(b^2 - a^2) \qquad (7.17)$$

This result can also be obtained by calculating the area in a plot of force vs. position. Figure 7.13 shows the force $F(x) = -kx$ as a function of x. The area of the quadrilateral $aQPb$ that represents the work W is the difference between the areas of the two triangles OPb and OQa. The triangular area above the $F_x(x)$ curve between the origin and $x = b$ is $\frac{1}{2}$ [base] × [height] = $\frac{1}{2}b \times kb = \frac{1}{2}kb^2$. Likewise, the triangular area between the origin and $x = a$ is $\frac{1}{2}ka^2$. The difference between these areas is $\frac{1}{2}k(b^2 - a^2)$. Taking into account that areas below the x axis must be reckoned as negative, we see that the work W is $W = -\frac{1}{2}k(b^2 - a^2)$, in agreement with Eq. (7.17).

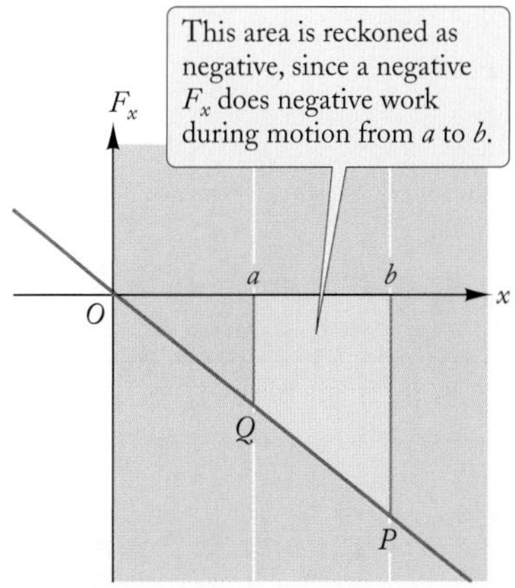

FIGURE 7.13 The plot of the force $F = -kx$ is a straight line. The work done by the force as the particle moves from a to b equals the (colored) quadrilateral area $aQPb$ under this plot.

MATH HELP INTEGRALS

The following are some theorems for integrals that we will frequently use.

The integral of a constant times a function is the constant times the integral of the function:

$$\int_a^b cf(x)\,dx = c\int_a^b f(x)\,dx$$

The integral of the sum of two functions is the sum of the integrals:

$$\int_a^b [f(x) + g(x)]\,dx = \int_a^b f(x)\,dx + \int_a^b g(x)\,dx$$

The integral of the function x^n (for $n \neq -1$) is

$$\int_a^b x^n\,dx = \frac{1}{n+1}x^{n+1}\Big|_a^b = \frac{1}{n+1}(b^{n+1} - a^{n+1})$$

In tables of integrals, this is usually written in the compact notation

$$\int x^n\,dx = \frac{x^{n+1}}{n+1} \qquad (\text{for } n \neq -1)$$

where it is understood that the right side is to be evaluated at the upper and at the lower limits of integration and then subtracted.

In a similar compact notation, here are a few more integrals of widely used functions (the quantity k is any constant):

$$\int \frac{1}{x}\,dx = \ln x$$

$$\int e^{kx}\,dx = \frac{1}{k}e^{kx}$$

$$\int \sin(kx)\,dx = -\frac{1}{k}\cos(kx)$$

$$\int \cos(kx)\,dx = \frac{1}{k}\sin(kx)$$

Appendix 4 gives more information on integrals.

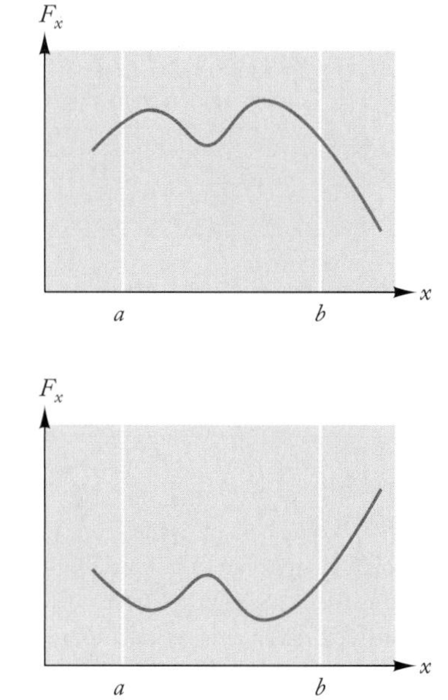

FIGURE 7.14 Two examples of plots of variable forces.

Checkup 7.2

QUESTION 1: Figure 7.14 shows two plots of variable forces acting on two particles. Which of these forces will perform more work during a displacement from a to b?

QUESTION 2: Suppose that a spring exerts a force $F_x(x) = -kx$ on a particle. What is the work done by the spring as the particle moves from $x = -b$ to $x = +b$?

QUESTION 3: What is the work that *you* must do to pull the end of the spring described in Example 4 from $x = a$ to $x = b$?

QUESTION 4: An amount of work W is performed to stretch a spring by a distance d from equilibrium. How much work is performed to further stretch the spring from d to $2d$?

(A) $\frac{1}{2}W$ (B) W (C) $2W$ (D) $3W$ (E) $4W$

7.3 KINETIC ENERGY

In everyday language, energy means a capacity for vigorous activities and hard work. Likewise, in the language of physics, energy is a capacity for performing work. Energy is "stored" work, or latent work, which can be converted into actual work under suitable conditions. *A body in motion has energy of motion, or kinetic energy.* For instance, a speeding arrow has kinetic energy that will be converted into work when the arrow strikes a target, such as a the trunk of a tree. The tip of the arrow then performs work on the wood, prying apart and cutting the wood fibers. The arrow continues to perform work and to penetrate the wood for a few centimeters, until all of its kinetic energy has been exhausted. A high-speed arrow has a deeper penetration and delivers a larger amount of work to the target than a low-speed arrow. Thus, we see that the kinetic energy of the arrow, or the kinetic energy of any kind of particle, must be larger if the speed is larger.

We now examine how work performed by or on a particle is related to changes of the speed of the particle. For clarity, we consider the work done *on* a particle by the net external force F_{net} acting on it (rather than the work done *by* the particle). When the force F_{net} acts on the particle, it accelerates the particle; if the acceleration has a component along the direction of motion of the particle, it will result in a change of the speed of the particle. The force does work on the particle and "stores" this work in the particle; or, if this force decelerates the particle, it does negative work on the particle and removes "stored" work.

We can establish an important identity between the work done by the net force and the change of speed it produces. Let us do this for the simple case of a particle moving along a straight line (see Fig. 7.15). If this straight line coincides with the x axis, then the work done by the net force $F_{\text{net},x}$ during a displacement from x_1 to x_2 is

$$W = \int_{x_1}^{x_2} F_{\text{net},x}\, dx \tag{7.18}$$

By Newton's Second Law, the net force equals the mass m times the acceleration $a = dv/dt$, and therefore the integral equals

$$\int_{x_1}^{x_2} F_{\text{net},x}\, dx = \int_{x_1}^{x_2} ma\, dx = m \int_{x_1}^{x_2} \frac{dv}{dt}\, dx \tag{7.19}$$

The velocity v is a function of time; but in the integral (7.19) it is better to regard the velocity as a function of x, and to rewrite the integrand as follows:

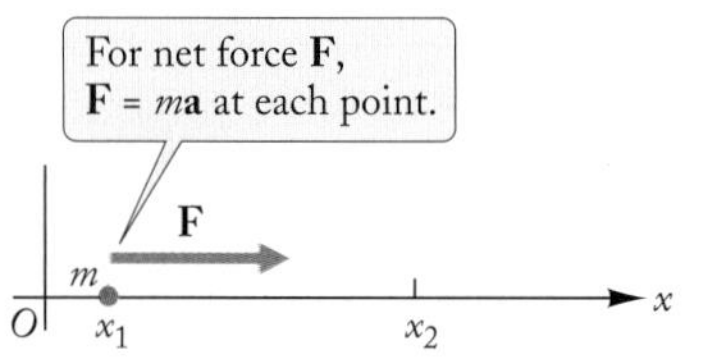

FIGURE 7.15 A particle moves on a straight line from x_1 to x_2 while a force **F** acts on it.

$$\frac{dv}{dt} = \frac{dv}{dx}\frac{dx}{dt} = \frac{dv}{dx}v = v\frac{dv}{dx} \tag{7.20}$$

Consequently, the work becomes

$$\begin{aligned} m\int_{x_1}^{x_2} \frac{dv}{dt}\,dx &= m\int_{x_1}^{x_2} v\frac{dv}{dx}\,dx = m\int_{v_1}^{v_2} v\,dv = m\tfrac{1}{2}v^2\Big|_{v_1}^{v_2} \\ &= \tfrac{1}{2}mv_2^2 - \tfrac{1}{2}mv_1^2 \end{aligned} \tag{7.21}$$

or

$$W = \tfrac{1}{2}mv_2^2 - \tfrac{1}{2}mv_1^2 \tag{7.22}$$

This shows that the change in the square of the speed is proportional to the work done by the force.

Although we have here obtained the result (7.22) for the simple case of motion along a straight line, it can be shown that the same result is valid for motion along a curve, in three dimensions.

According to Eq. (7.22), whenever we perform positive work on the particle, we increase the "amount of $\frac{1}{2}mv^2$" in the particle; and whenever we perform negative work on the particle (that is, when we let the particle perform work on us), we decrease the "amount of $\frac{1}{2}mv^2$" in the particle. Thus, *the quantity $\frac{1}{2}mv^2$ is the amount of work stored in the particle, or the* **kinetic energy** *of the particle.* We represent the kinetic energy by the symbol K:

$$K = \tfrac{1}{2}mv^2 \tag{7.23}$$

kinetic energy

With this notation, Eq. (7.22) states that *the change of kinetic energy equals the net work done on the particle;* that is,

$$K_2 - K_1 = W \tag{7.24}$$

or

$$\Delta K = W \tag{7.25}$$

work–energy theorem

This result is called the **work–energy theorem**. Keep in mind that the work in Eqs. (7.22), (7.24), and (7.25) must be evaluated with the *net* force; that is, all the forces that do work on the particle must be included in the calculation.

When a force does positive work on a particle initially at rest, the kinetic energy of the particle increases. The particle then has a capacity to do work: if the moving particle subsequently is allowed to push against some obstacle, then this obstacle does negative work on the particle and simultaneously the particle does positive work on the obstacle. When the particle does work, its kinetic energy decreases. The total amount of work the particle can deliver to the obstacle is equal to its kinetic energy. Thus, *the kinetic energy represents the capacity of a particle to do work by virtue of its speed.*

The acquisition of kinetic energy through work and the subsequent production of work by this kinetic energy are neatly illustrated in the operation of a waterwheel driven by falling water. In a flour mill of an old Spanish Colonial design, the water runs down from a reservoir in a steep, open channel (see Fig. 7.16). The motion of the water particles is essentially that of particles sliding down an inclined plane. If we

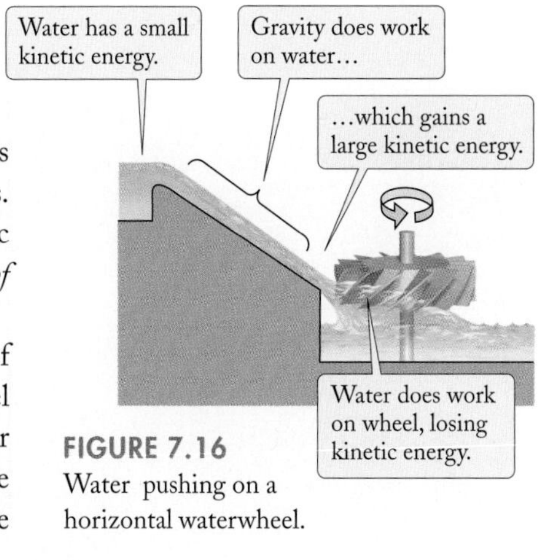

FIGURE 7.16 Water pushing on a horizontal waterwheel.

ignore friction, then the only force that does work on the water particles is gravity. This work is positive, so the kinetic energy of the water increases and it attains a maximum value at the lower end of the channel (where its speed is greatest). The stream of water emerges from this channel with high kinetic energy and hits the blades of the waterwheel. The water pushes on the wheel, turns it, and gives up its kinetic energy while doing work—and the wheel runs the millstones and does useful work on them. Thus, the work that gravity does on the descending water is ultimately converted into useful work, with the kinetic energy playing an intermediate role in this process.

The unit of kinetic energy is the joule, the same as the unit of work. Table 7.1 lists some typical kinetic energies.

FIGURE 7.17 Pitcher throwing a ball. The ball leaves his hand with a speed of 30 m/s.

EXAMPLE 5 During a baseball game, the pitcher throws the ball with a speed of 30 m/s (Fig. 7.17). The mass of the ball is 0.15 kg. What is the kinetic energy of the ball when it leaves his hand? How much work did his hand do on the ball during the throw?

SOLUTION: The final speed of the ball, when it leave the hand at the end of the throwing motion, is $v_2 = 30$ m/s. The final kinetic energy of the ball is

$$K_2 = \tfrac{1}{2}mv_2^2 = \tfrac{1}{2} \times 0.15 \text{ kg} \times (30 \text{ m/s})^2 = 68 \text{ J} \tag{7.26}$$

According to the work–energy theorem [Eq. (7.25)], the work done by the hand on the ball equals the change of kinetic energy. Since the initial kinetic energy at the beginning of the throwing motion is zero ($v_1 = 0$), the change of kinetic energy equals the final kinetic energy, and the work is

$$W = K_2 - K_1 = 68 \text{ J} - 0 = 68 \text{ J}$$

Note that for this calculation of the work we did not need to know the (complicated) details of how the force varies during the throwing motion. The work–energy theorem gives us the answer directly.

TABLE 7.1 SOME KINETIC ENERGIES

Orbital motion of Earth	2.6×10^{33} J
Ship *Queen Elizabeth* (at cruising speed)	9×10^{9} J
Jet airliner (Boeing 747 at maximum speed)	7×10^{9} J
Automobile (at 90 km/h)	5×10^{5} J
Rifle bullet	4×10^{3} J
Person walking	60 J
Falling raindrop	4×10^{-5} J
Proton from large accelerator (Fermilab)	1.6×10^{-7} J
Electron in atom (hydrogen)	2.2×10^{-18} J
Air molecule (at room temperature)	6.2×10^{-21} J

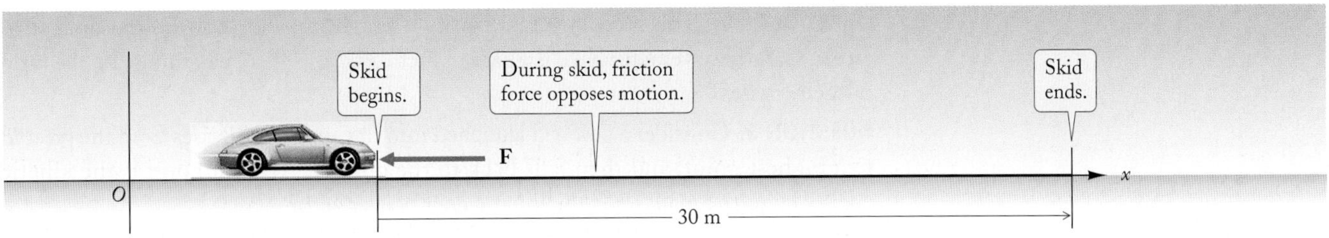

FIGURE 7.18 Automobile skidding on a street.

EXAMPLE 6 While trying to stop his automobile on a flat street, a drunk driver steps too hard on the brake pedal and begins to skid. He skids for 30 m with all wheels locked, leaving skid marks on the pavement, before he releases the brake pedal and permits the wheels to resume rolling (see Fig. 7.18). How much kinetic energy does the automobile lose to friction during this skid? If you find skid marks of 30 m on the pavement, what can you conclude about the initial speed of the automobile? The mass of the automobile is 1100 kg, and the coefficient of sliding friction between the wheels and the street is $\mu_k = 0.90$.

SOLUTION: The magnitude of the sliding friction force is $f_k = \mu_k N = \mu_k mg$. With the x axis along the direction of motion, the x component of this friction force is negative:

$$F_x = -\mu_k mg$$

Since the force is constant, the work done by this force is

$$W = F_x \, \Delta x = -\mu_k mg \times \Delta x$$

$$= -0.90 \times 1100 \text{ kg} \times 9.81 \text{ m/s}^2 \times 30 \text{ m} = -2.9 \times 10^5 \text{ J}$$

According to the work–energy theorem, this work equals the change of kinetic energy:

$$\Delta K = W = -2.9 \times 10^5 \text{ J}$$

Since the kinetic energy of the automobile decreases by 2.9×10^5 J, its initial kinetic energy must have been at least 2.9×10^5 J. Hence its initial speed must have been at least large enough to provide this kinetic energy; that is,

$$\tfrac{1}{2}mv_1^2 \geq 2.9 \times 10^5 \text{ J}$$

and so

$$v_1 \geq \sqrt{\frac{2 \times 2.9 \times 10^5 \text{ J}}{m}} = \sqrt{\frac{2 \times 2.9 \times 10^5 \text{ J}}{1100 \text{ kg}}} = 23 \text{ m/s} = 83 \text{ km/h}$$

✔ Checkup 7.3

QUESTION 1: Two automobiles of equal masses travel in opposite directions. Can they have equal kinetic energies?

QUESTION 2: A car is traveling at 80 km/h on a highway, and a truck is traveling at 40 km/h. Can these vehicles have the same kinetic energy? If so, what must be the ratio of their masses?

QUESTION 3: Consider a golf ball launched into the air. The ball rises from the ground to a highest point, and then falls back to the ground. At what point is the kinetic energy largest? Smallest? Is the kinetic energy ever zero?

QUESTION 4: A horse is dragging a sled at steady speed along a rough surface, with friction. The horse does work on the sled, but the kinetic energy of the sled does not increase. Does this contradict the work–energy theorem?

QUESTION 5: If you increase the speed of your car by a factor of 3, from 20 km/h to 60 km/h, by what factor do you change the kinetic energy?

(A) $\frac{1}{9}$ (B) $\frac{1}{3}$ (C) 1 (D) 3 (E) 9

PROBLEM-SOLVING TECHNIQUES — CALCULATION OF WORK

In calculations of the work done by a force acting on a body, keep in mind that

- A force that has a component in the direction of the displacement does positive work; a force that has a component in the direction opposite to the displacement does negative work.
- A force perpendicular to the displacement does no work [examples: the normal force acting on a body sliding on a surface, the centripetal force acting on a body in circular motion (uniform or not)].
- For a constant force, the work can be calculated either from Eq. (7.5) or from Eq. (7.9); use the former if you know the magnitude of the force and the angle, and use the latter if you know the components.
- For a variable force, the calculation of the work involves integration along the path [Eq. (7.14)]; also, Eq. (7.15) can be used for the work during an infinitesimal displacement.
- The work–energy theorem is valid only if the work is calculated with the *net* force. When two of the three quantities (work done, initial kinetic energy, and final kinetic energy) are known, the theorem can be applied to determine the third: $W = K_2 - K_1$.

Online Concept Tutorial 9

7.4 GRAVITATIONAL POTENTIAL ENERGY

As we saw in the preceding section, the kinetic energy represents the capacity of a particle to do work by virtue of its speed. We will now become acquainted with another form of energy that represents the capacity of the particle to do work by virtue of its position in space. This is the **potential energy**. In this section, we will examine the special case of gravitational potential energy for a particle moving under the influence of the constant gravitational force near the surface of the Earth, and we will formulate a law of conservation of energy for such a particle. In the next chapter we will examine other cases of potential energy and formulate the General Law of Conservation of Energy.

The gravitational potential energy represents the capacity of the particle to do work by virtue of its height above the surface of the Earth. When we lift a particle to some height above the surface, we have to do work against gravity, and we thereby store work in the particle. Thus, a particle high above the surface is endowed with a large amount of latent work, which can be exploited and converted into actual work by allowing the particle to push against some obstacle as it descends. A good example of such an exploitation of gravitational potential energy is found in a grandfather clock, where a weight hanging on a cord drives the wheel of the clock (Fig. 7.19). The weight does work on the wheel, and gradually converts all of its gravitational potential energy into work as it descends (in a typical grandfather clock, the weight takes about a week to sink down from the top to the bottom, and you must then rewind the clock, by lifting the weight).

FIGURE 7.19 The descending weights of the grandfather clock pull on the cords and do work on the wheel of the clock.

To obtain a general expression for the gravitational potential energy of a particle moving on a straight or a curving path, we first consider a particle moving on an inclined plane. According to Eq. (7.11), when a particle of mass m descends a distance h along an inclined plane, the work done by gravity is

$$W = mgh \tag{7.27}$$

As already remarked on in Example 3, this result is independent of the angle of inclination of the plane—it depends only on the change of height. More generally, for a curved path, the result is independent of the shape of the path that the particle follows from its starting point to its endpoint. For instance, the curved path and the straight sloping path in Fig. 7.20a lead to exactly the same result (7.27) for the work done by gravity. To recognize this, we simply approximate the curved path by small straight segments (see Fig. 7.20b). Each such small segment can be regarded as a small inclined plane, and therefore the work is mg times the small change of height. The net amount of work for all the small segments taken together is then mg times the net change of height, in agreement with Eq. (7.27).

If the vertical coordinate of the starting point is y_1 and the vertical coordinate of the endpoint is y_2 (see Fig. 7.20), then $h = y_1 - y_2$ and Eq. (7.27) becomes

$$W = mg(y_1 - y_2) \quad \text{or} \quad W = -(mgy_2 - mgy_1) \tag{7.28}$$

According to Eq. (7.28), whenever gravity performs positive work on the particle ($y_1 > y_2$, a descending particle), the "amount of mgy" of the particle decreases; and

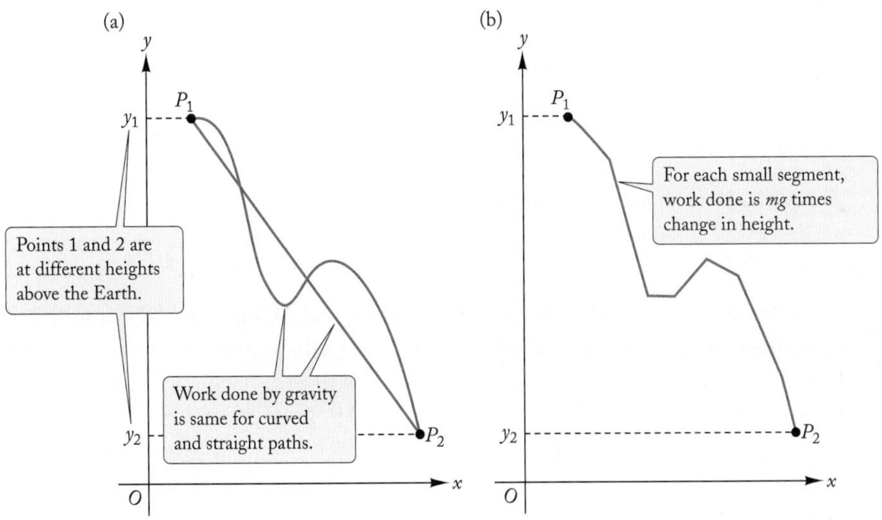

FIGURE 7.20 (a) A curved path (red) and a straight path (blue) from point P_1 to point P_2. (b) The curved path can be approximated by short straight segments.

whenever gravity performs negative work on the particle ($y_1 < y_2$, an ascending particle), the "amount of mgy" increases. Thus, *the quantity mgy represents the amount of stored, or latent, gravitational work; that is, it represents the gravitational potential energy.* We will adopt the notation U for the **gravitational potential energy**:

gravitational potential energy

$$U = mgy \tag{7.29}$$

This potential energy is directly proportional to the height y, and it has been chosen to be zero at $y = 0$ (see Fig. 7.21).

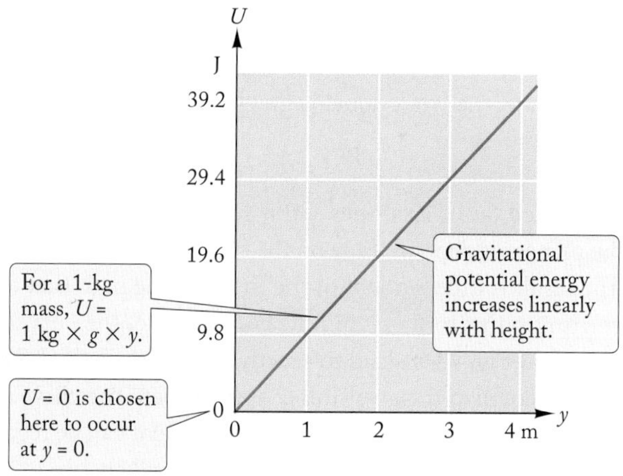

FIGURE 7.21 Plot of the gravitational potential energy of a mass of 1 kg as a function of height y.

In terms of the gravitational potential energy, Eq. (7.28) for the work done by gravity becomes

$$W = -U_2 + U_1 \tag{7.30}$$

Since $\Delta U = U_2 - U_1$ is the change in potential energy, Eq. (7.30) says that the work equals the negative of the change in potential energy,

$$W = -\Delta U \tag{7.31}$$

EXAMPLE 7 What is the kinetic energy and what is the gravitational potential energy (relative to the ground) of a jet airliner of mass 73 000 kg cruising at 240 m/s at an altitude of 9000 m?

SOLUTION: The kinetic energy is

$$K = \tfrac{1}{2}mv^2 = \tfrac{1}{2} \times 7.3 \times 10^4\ \text{kg} \times (240\ \text{m/s})^2 = 2.1 \times 10^9\ \text{J}$$

The gravitational potential energy is $U = mgy$. If we measure the y coordinate from the ground level, then $y = 9000$ m for our airliner, and

$$U = mgy = 7.3 \times 10^4\ \text{kg} \times 9.81\ \text{m/s}^2 \times 9.0 \times 10^3\ \text{m} = 6.4 \times 10^9\ \text{J}$$

We see that the airliner has about three times more potential energy than kinetic energy.

If we let the particle push or pull on some obstacle (such as the wheel of the grandfather clock) during its descent from y_1 to y_2, then the total amount of work that we can extract during this descent is equal to the work done by gravity; that is, it is equal to $-U_2 + U_1 = -(U_2 - U_1) = -\Delta U$, or the negative of the change of potential energy. Of course, the work extracted in this way really arises from the Earth's gravity—the particle can do work on the obstacle because gravity is doing work on the particle. Hence *the gravitational potential energy is really a joint property of the particle and the Earth; it is a property of the configuration of the particle–Earth system.*

If the only force acting on the particle is gravity, then by combining Eqs. (7.24) and (7.30) we can obtain a relation between potential energy and kinetic energy. According to Eq. (7.24), the change in kinetic energy equals the work, or $K_2 - K_1 = W$; and according to Eq. (7.30), the negative of the change in potential energy also equals the work: $W = -U_2 + U_1$. Hence the change in kinetic energy must equal the negative of the change in potential energy:

$$K_2 - K_1 = -U_2 + U_1$$

We can rewrite this as follows:

$$K_2 + U_2 = K_1 + U_1 \tag{7.32}$$

This equality indicates that the quantity $K + U$ is a constant of the motion; that is, it has the same value at the endpoint as it had at the starting point. We can express this as

$$K + U = [\text{constant}] \tag{7.33}$$

The sum of the kinetic and potential energies is called the **mechanical energy** *of the particle.* It is usually designated by the symbol E:

$$E = K + U \tag{7.34}$$

This energy represents the total capacity of the particle to do work by virtue of both its speed and its position.

Equation (7.33) shows that if the only force acting on the particle is gravity, then the mechanical energy remains constant:

$$E = K + U = [\text{constant}] \tag{7.35}$$

This is the **Law of Conservation of Mechanical Energy**.

Since the sum of the potential and kinetic energies must remain constant during the motion, an increase in one must be compensated by a decrease in the other; this means that *during the motion, kinetic energy is converted into potential energy and vice versa.* For instance, if we throw a baseball straight upward from ground level ($y = 0$), the initial kinetic energy is large and the initial potential energy is zero. As the baseball rises, its potential energy increases and, correspondingly, its kinetic energy decreases, so as to keep the sum of the kinetic and potential energies constant. When the baseball reaches its maximum height, its potential energy has the largest value, and the kinetic energy is (instantaneously) zero. As the baseball falls, its potential energy decreases, and its kinetic energy increases (see Fig. 7.22).

Apart from its practical significance in terms of work, the mechanical energy is very helpful in the study of the motion of a particle. If we make use of the formulas for K and U, Eq. (7.35) becomes

$$E = \tfrac{1}{2}mv^2 + mgy = [\text{constant}] \tag{7.36}$$

CHRISTIAAN HUYGENS (1629–1695) *Dutch mathematician and physicist. He invented the pendulum clock, made improvements in the manufacture of telescope lenses, and discovered the rings of Saturn. Huygens investigated the theory of collisions of elastic bodies and the theory of oscillations of the pendulum, and he stated the Law of Conservation of Mechanical Energy for motion under the influence of gravity.*

mechanical energy

Law of Conservation of Mechanical Energy

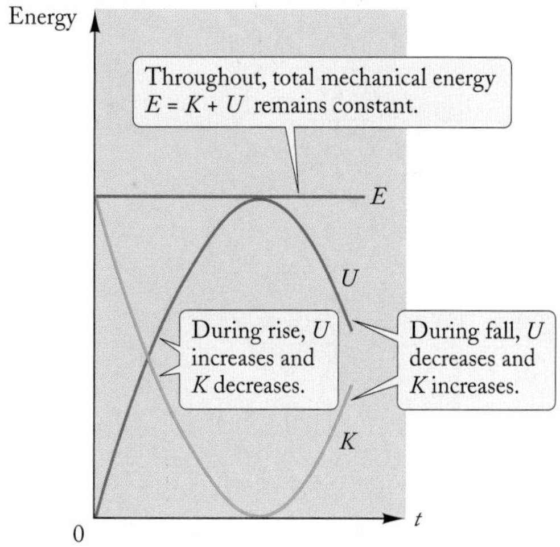

FIGURE 7.22 Kinetic energy K, potential energy U, and mechanical energy $E = K + U$ as functions of time during the upward and downward motions of a baseball.

This shows explicitly how the baseball, or any other particle moving under the influence of gravity, trades speed for height during the motion: whenever y increases, v must decrease (and conversely) so as to keep the sum of the two terms on the left side of Eq. (7.36) constant.

If we consider the vertical positions (y_1 and y_2) and speeds (v_1 and v_2) at two different times, we can equate the total mechanical energy at those two times:

$$\tfrac{1}{2}mv_1^2 + mgy_1 = \tfrac{1}{2}mv_2^2 + mgy_2$$

Rearranging, we immediately obtain

$$-g\,\Delta y = \tfrac{1}{2}(v_2^2 - v_1^2) \tag{7.37}$$

where $\Delta y = y_2 - y_1$. We recognize Eq. (7.37) as the same form that we obtained when studying the equations of motion [see Eq. (2.29)]. Here, however, the result follows directly from conservation of mechanical energy; we did not need to determine the detailed time dependence of the motion.

An important aspect of Eq. (7.36) is that it is valid not only for a particle in free fall (a projectile), but also for a particle sliding on a surface or a track of arbitrary shape, provided that there is no friction. Of course, under these conditions, besides the gravitational force there also acts the normal force; but this force does no work, and hence does not affect Eq. (7.28), or any of the equations following after it. The next example illustrates how these results can be applied to simplify the study of fairly complicated motions, which would be extremely difficult to investigate by direct calculation with Newton's Second Law. This example gives us a glimpse of the elegance and power of the Law of Conservation of Mechanical Energy.

Concepts — in — Context

EXAMPLE 8 A roller-coaster car descends 38 m from its highest point to its lowest. Suppose that the car, initially at rest at the highest point, rolls down this track without friction. What speed will the car attain at the lowest point? Treat the motion as particle motion.

SOLUTION: The coordinates of the highest and the lowest points are $y_1 = 38$ m and $y_2 = 0$, respectively (see Fig. 7.23). According to Eq. (7.36), the energy at the start of the motion for a car initially at rest is

$$E = \tfrac{1}{2}mv_1^2 + mgy_1 = 0 + mgy_1 \tag{7.38}$$

and the energy at the end of the motion is

$$E = \tfrac{1}{2}mv_2^2 + mgy_2 = \tfrac{1}{2}mv_2^2 + 0 \tag{7.39}$$

The conservation of energy implies that the right sides of Eqs. (7.38) and (7.39) are equal:

$$\tfrac{1}{2}mv_2^2 = mgy_1 \tag{7.40}$$

Solving this for v_2, we find

$$v_2 = \sqrt{2gy_1} \tag{7.41}$$

(a)

(b)

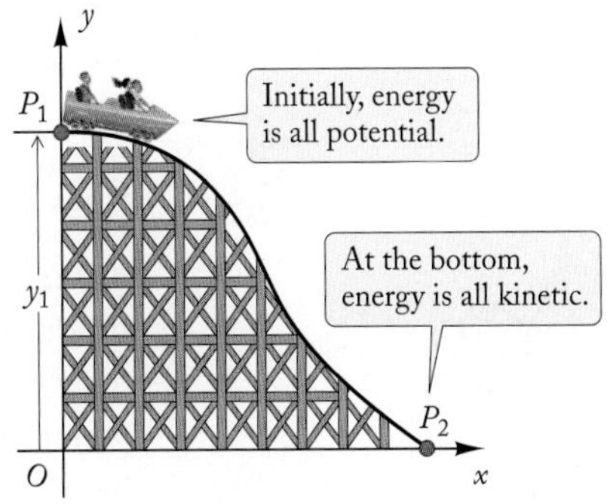

FIGURE 7.23 (a) A roller coaster. (b) Profile of a roller coaster. The roller-coaster car descends from P_1 to P_2.

which gives

$$v_2 = \sqrt{2 \times 9.81\ \text{m/s}^2 \times 38\ \text{m}} = 27\ \text{m/s}$$

Note that according to Eq. (7.41) the final velocity is independent of the mass of the car; since both the kinetic energy and the gravitational potential energy are proportional to mass, the mass cancels in this calculation.

COMMENT: This example illustrates how energy conservation can be exploited to answer a question about motion. To obtain the final speed by direct computation of forces and accelerations would have been extremely difficult—it would have required detailed knowledge of the shape of the path down the hill. With the Law of Conservation of Energy we can bypass these complications.

PROBLEM-SOLVING TECHNIQUES — ENERGY CONSERVATION IN ANALYSIS OF MOTION

As illustrated by the preceding example, the use of energy conservation in a problem of motion typically involves three steps:

1. First write an expression for the energy at one point of the motion [Eq. (7.38)].
2. Then write an expression for the energy at another point [Eq. (7.39)].
3. And then rely on energy conservation to equate the two expressions [Eq. (7.40)]. This yields one equation, which can be solved for the unknown final speed or the unknown final position (if the final speed is known).

Note that the value of the gravitational potential energy $U = mgy$ depends on the level from which you measure the y coordinate. However, the *change* in the potential energy does not depend on the choice of this level, and therefore any choice will lead to the same result for the change of kinetic energy. Thus, you can make any choice of zero level, but you must continue to use this choice throughout the entire calculation. You will usually find it convenient to place the zero level for the y coordinate either at the final position of the particle (as in the preceding example), or at the initial position, or at some other distinctive height, such as the bottom of a hill or the ground floor of a building. And always remember that the formula $U = mgy$ for the gravitational potential energy assumes that the y axis is directed vertically upward.

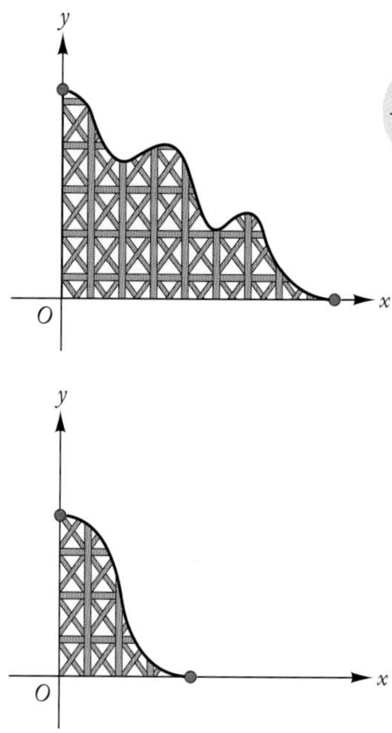

FIGURE 7.24 Two roller-coaster profiles. The two plots have the same vertical scale.

Checkup 7.4

Concepts — in — Context

QUESTION 1: Figure 7.24 shows two roller coasters in profile. Cars are released at the top of each, from rest. Which, if either, of these roller coasters gives the car a larger speed at the bottom? Neglect friction.

QUESTION 2: A piano is being moved from the second floor of one house to the second floor of another, nearby house. Describe the changes in the gravitational potential energy of the piano during this move.

QUESTION 3: A skidding truck slides down a mountain road, at constant speed. Is the mechanical energy $E = K + U$ conserved?

QUESTION 4: At an amusement park, a girl jumps off a high tower and lands in a pool. Meanwhile, a boy slides down a (frictionless) water slide that also takes him from the tower into the pool. Who reaches the pool with the higher speed? Who reaches the pool first?

QUESTION 5: A bicyclist rolls down a hill without braking, starting at the top, from rest. A second bicyclist rolls down the same hill, starting at one-half the height, from rest. By what factor will the speed of the first bicyclist be larger than that of the second, at the bottom? Ignore friction.

(A) $\sqrt{2}$ (B) 2 (C) $2\sqrt{2}$ (D) 4

SUMMARY

MATH HELP Integrals (page 213)

PROBLEM-SOLVING TECHNIQUES Calculation of Work (page 218)

PROBLEM-SOLVING TECHNIQUES Energy Conservation in Analysis of Motion (page 223)

SI UNIT OF WORK (Unit of energy) joule = J = N·m

WORK DONE BY A CONSTANT FORCE

Parallel to a displacement Δx $W = F_x \, \Delta x$ (7.1)

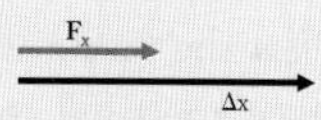

Not parallel to a displacement **s** $W = Fs\cos\theta = \mathbf{F} \cdot \mathbf{s}$ (7.6; 7.5)

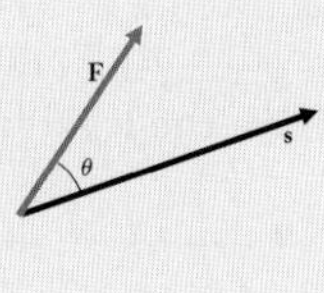

DOT PRODUCT (OR SCALAR PRODUCT) OF TWO VECTORS	$\mathbf{A} \cdot \mathbf{B} = AB\cos\theta = A_xB_x + A_yB_y + A_zB_z$	**(7.6; 7.8)**
WORK DONE BY CONSTANT GRAVITATIONAL FORCE (Descending from a height h)	$W = mgh$	**(7.11)**
WORK DONE BY A VARIABLE FORCE In one dimension	$W = \int_a^b F_x(x)\,dx$	**(7.15)**

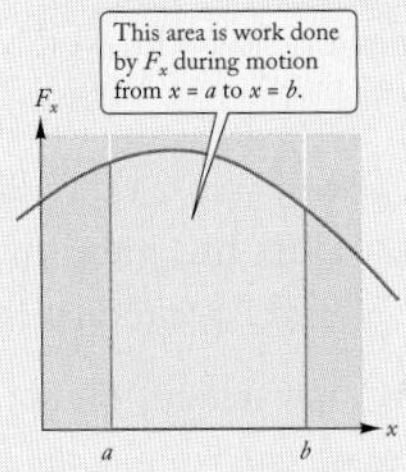

In two or three dimensions	$W = \int \mathbf{F} \cdot d\mathbf{s}$	**(7.16)**
WORK DONE BY A SPRING (Moving from $x = a$ to $x = b$)	$W = -\frac{1}{2}k(b^2 - a^2)$	**(7.17)**
KINETIC ENERGY	$K = \frac{1}{2}mv^2$	**(7.23)**
WORK–ENERGY THEOREM	$\Delta K = W$	**(7.25)**
GRAVITATIONAL POTENTIAL ENERGY	$U = mgy$	**(7.29)**
RELATION BETWEEN WORK AND CHANGE IN POTENTIAL ENERGY	$W = -\Delta U$	**(7.31)**
MECHANICAL ENERGY	$E = K + U$	**(7.34)**
CONSERVATION OF MECHANICAL ENERGY	$E = K + U = [\text{constant}]$	**(7.35)**
CONSERVATION OF MECHANICAL ENERGY AT TWO POINTS	$\frac{1}{2}mv_1^2 + mgy_1 = \frac{1}{2}mv_2^2 + mgy_2$	**(7.37)**

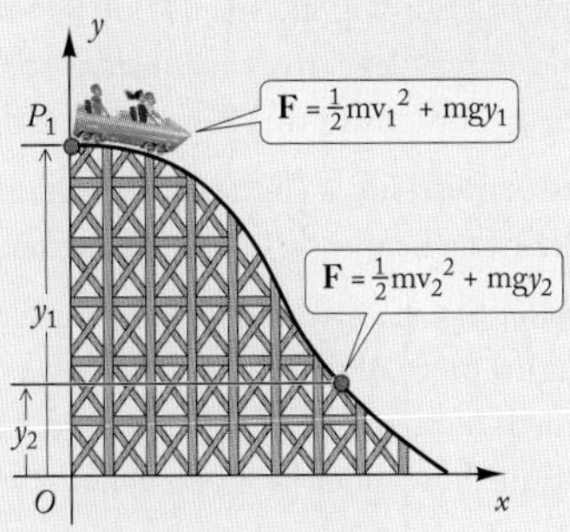

QUESTIONS FOR DISCUSSION

1. Does the work of a force on a body depend on the frame of reference in which it is calculated? Give some examples.
2. Does your body do work (external or internal) when standing at rest? When walking steadily along a level road?
3. Consider a pendulum swinging back and forth. During what part of the motion does the weight do positive work? Negative work?
4. Since $v^2 = v_x^2 + v_y^2 + v_z^2$, Eq.(7.23) implies $K = \frac{1}{2}mv_x^2 + \frac{1}{2}mv_y^2 + \frac{1}{2}mv_z^2$. Does this mean that the kinetic energy has x, y, and z components?
5. Consider a woman steadily climbing a flight of stairs. The external forces on the woman are her weight and the normal force of the stairs against her feet. During the climb, the weight does negative work, while the normal force does no work. Under these conditions how can the kinetic energy of the woman remain constant? (Hint: The entire woman cannot be regarded as a particle, since her legs are not rigid; but the upper part of her body can be regarded as a particle, since it is rigid. What is the force of her legs against the upper part of her body? Does this force do work?)
6. An automobile increases its speed from 80 to 88 km/h. What is the percentage of increase in kinetic energy? What is the percentage of reduction of travel time for a given distance?
7. Two blocks in contact slide past one another and exert friction forces on one another. Can the friction force *increase* the kinetic energy of one block? Of both? Does there exist a reference frame in which the friction force decreases the kinetic energy of both blocks?
8. When an automobile with rear-wheel drive is accelerating on, say, a level road, the horizontal force of the road on the rear wheels does not give the automobile any energy because the point of application of this force (point of contact of wheel on ground) is instantaneously at rest if the wheel is not slipping. What force gives the body of the automobile energy? Where does this energy come from? (Hint: Consider the force that the rear axle exerts against its bearings.)
9. Why do elevators have counterweights? (See Fig. 5.40.)
10. A parachutist jumps out of an airplane, opens a parachute, and lands safely on the ground. Is the mechanical energy for this motion conserved?
11. If you release a tennis ball at some height above a hard floor, it will bounce up and down several times, with a gradually decreasing amplitude. Where does the ball suffer a loss of mechanical energy?
12. Two ramps, one steeper than the other, lead from the floor to a loading platform (Fig. 7.25). It takes more force to push a (frictionless) box up the steeper ramp. Does this mean it takes more work to raise the box from the floor to the platform?

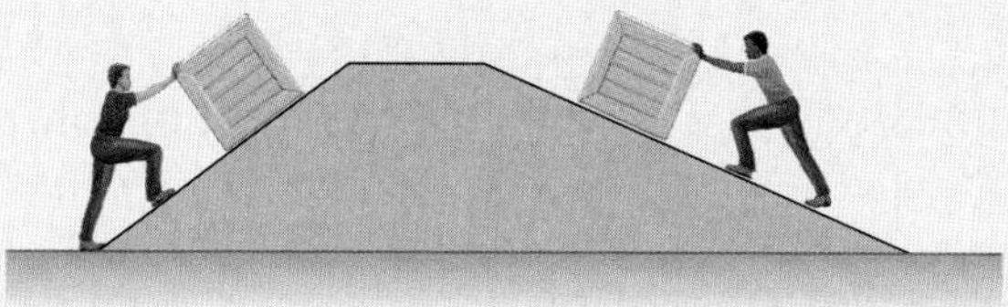

FIGURE 7.25 Two ramps of different steepness.

13. Consider the two ramps described in the preceding question. Taking friction into account, which ramp requires less work for raising a box from the floor to the platform?
14. A stone is tied to a string. Can you whirl this stone in a vertical circle with constant speed? Can you whirl this stone with constant energy? For each of these two cases, describe how you must move your hand.

PROBLEMS

7.1 Work†

1. If it takes a horizontal force of 300 N to push a stalled automobile along a level road at constant speed, how much work must you do to push this automobile a distance of 5.0 m?
2. In an overhead lift, a champion weight lifter raises 254 kg from the floor to a height of 1.98 m. How much work does he do?
3. Suppose that the force required to push a saw back and forth through a piece of wood is 35 N. If you push this saw back and forth 30 times, moving it forward 12 cm and back 12 cm each time, how much work do you do?
4. It requires 2200 J of work to lift a 15-kg bucket of water from the bottom of a well to the top. How deep is the well?
5. A child drags a 20-kg box across a lawn for 10 m and along a sidewalk for 30 m; the coefficient of friction is 0.25 for the first part of the trip and 0.55 for the second. If the child always pulls horizontally, how much work does the child do on the box?
6. A man moves a vacuum cleaner 1.0 m forward and 1.0 m back 300 times while cleaning a floor, applying a force of 40 N during each motion. The pushes and pulls make an angle of 60° with the horizontal. How much work does the man do on the vacuum cleaner?

† For help, see Online Concept Tutorial 9 at www.wwnorton.com/physics

7. A record for stair climbing was achieved by a man who raced up the 1600 steps of the Empire State Building to a height of 320 m in 10 min 59 s. If his average mass was 75 kg, how much work did he do against gravity? At what average rate (in J/s) did he do this work?

8. Suppose you push on a block sliding on a table. Your push has a magnitude of 50 N and makes a downward angle of 60° with the direction of motion. What is the work you do on the block while the block moves a distance of 1.6 m?

9. Consider the barge being pulled by two tugboats, as described in Example 4 of Chapter 5. The pull of the first tugboat is 2.5×10^5 N at 30° to the left, and the pull of the second tugboat is 1.0×10^5 N at 15° to the right (see Fig. 7.26). What is the work done by each tugboat on the barge while the barge moves 100 m forward (in the direction of the x axis in Fig. 7.26)? What is the total work done by both tugboats on the barge?

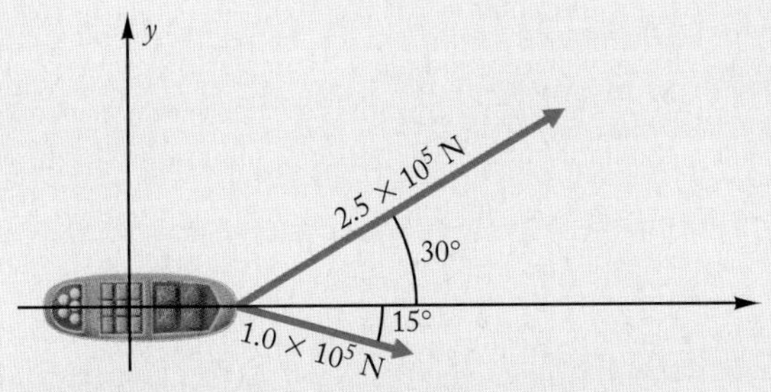

FIGURE 7.26 A barge pulled by two tugboats.

10. A 2.0-kg stone thrown upward reaches a height of 4.0 m at a horizontal distance of 6.0 m from the point of launch. What is the work done by gravity during this displacement?

*11. A man pushes a heavy box up an inclined ramp making an angle of 30° with the horizontal. The mass of the box is 60 kg, and the coefficient of kinetic friction between the box and the ramp is 0.45. How much work must the man do to push the box to a height of 2.5 m at constant speed? Assume that the man pushes on the box in a direction parallel to the surface of the ramp.

12. The driver of a 1200-kg automobile notices that, with its gears in neutral, it will roll downhill at a constant speed of 110 km/h on a road of slope 1:20. Draw a "free-body" diagram for the automobile, showing the force of gravity, the normal force (exerted by the road), and the friction force (exerted by the road and by air resistance). What is the magnitude of the friction force on the automobile under these conditions? What is the work done by the friction force while the automobile travels 1.0 km down the road?

13. Driving an automobile down a slippery, steep hill, a driver brakes and skids at constant speed for 10 m. If the automobile mass is 1700 kg and the angle of slope of the hill is 25°, how much work does gravity do on the car during the skid? How much work does friction do on the car?

14. The automobile in Example 6 of Chapter 6 is traveling on a flat road. For a trip of length 250 km, what is the total work done against air friction when traveling at 20 m/s? At 30 m/s?

15. A constant force of 25 N is applied to a body while it moves along a straight path for 12 m. The force does 175 J of work on the body. What is the angle between the force and the path of the body?

*16. A strong, steady wind provides a force of 150 N in a direction 30° east of north on a pedestrian. If the pedestrian walks first 100 m north and then 200 m east, what is the total work done by the wind?

*17. A man pulls a cart along a level road by means of a short rope stretched over his shoulder and attached to the front end of the cart. The friction force that opposes the motion of the cart is 250 N.

(a) If the rope is attached to the cart at shoulder height, how much work must the man do to pull the cart 50 m at constant speed?

(b) If the rope is attached to the cart below shoulder height so it makes an angle of 30° with the horizontal, what is the tension in the rope? How much work must the man now do to pull the cart 50 m? Assume that enough mass was added so the friction force is unchanged.

*18. A particle moves in the $x-y$ plane from the origin $x = 0$, $y = 0$ to the point $x = 2$, $y = -1$ while under the influence of a force $\mathbf{F} = 3\mathbf{i} + 2\mathbf{j}$. How much work does this force do on the particle during this motion? The distances are measured in meters and the force in newtons.

*19. An elevator consists of an elevator cage and a counterweight attached to the ends of a cable that runs over a pulley (Fig. 7.27). The mass of the cage (with its load) is 1200 kg, and the mass of the counterweight is 1000 kg. The elevator is driven by an electric motor attached to the pulley. Suppose that the elevator is initially at rest on the first floor of the building and the motor makes the elevator accelerate upward at the rate of 1.5 m/s^2.

(a) What is the tension in the part of the cable attached to the elevator cage? What is the tension in the part of the cable attached to the counterweight?

(b) The acceleration lasts exactly 1.0 s. How much work has the electric motor done in this interval? Ignore friction forces and ignore the mass of the pulley.

(c) After the acceleration interval of 1.0 s, the motor pulls the elevator upward at constant speed until it reaches the third floor, exactly 10.0 m above the first floor. What is the total amount of work that the motor has done up to this point?

FIGURE 7.27 Elevator cage and counterweight.

*20. By means of a towrope, a girl pulls a sled loaded with firewood along a level, icy road. The coefficient of friction between the sled and the road is $\mu_k = 0.10$, and the mass of the sled plus its load is 150 kg. The towrope is attached to the front end of the sled and makes an angle of 30° with the horizontal. How much work must the girl do on the sled to pull it 1.0 km at constant speed?

*21. During a storm, a sailboat is anchored in a 10-m-deep harbor. The wind pushes against the boat with a steady horizontal force of 7000 N.

(a) The anchor rope that holds the boat in place is 50 m long and is stretched straight between the anchor and the boat (Fig. 7.28a). What is the tension in the rope?

(b) How much work must the crew of the sailboat do to pull in 30 m of the anchor rope, bringing the boat nearer to the anchor (Fig. 7.28b)? What is the tension in the rope when the boat is in this new position?

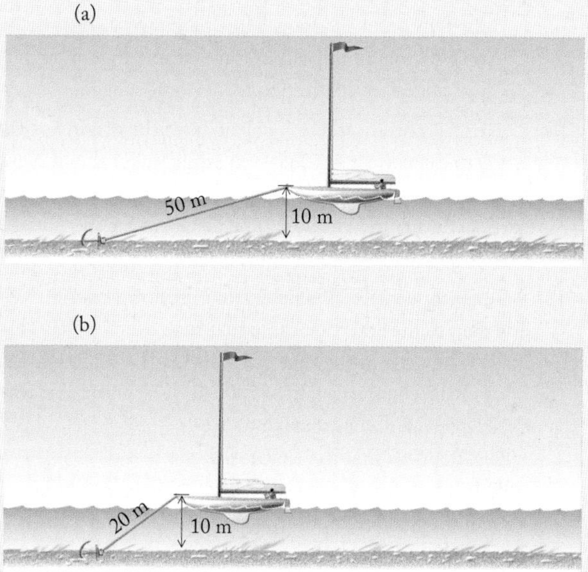

FIGURE 7.28 A sailboat at anchor.

7.2 Work for a Variable Force[†]

22. The spring used in the front suspension of a Triumph sports car has a spring constant $k = 3.5 \times 10^4$ N/m. How much work must you do to compress this spring by 0.10 m from its relaxed condition? How much more work must you do to compress the spring a further 0.10 m?

23. A particle moving along the x axis is subjected to a force F_x that depends on position as shown in the plot in Fig. 7.29. From this plot, find the work done by the force as the particle moves from $x = 0$ to $x = 8.0$ m.

24. A 250-g object is hung from a vertical spring, stretching it 18 cm below its original equilibrium position. How much work was done by gravity on the object? By the spring?

[†] For help, see Online Concept Tutorial 9 at www.wwnorton.com/physics

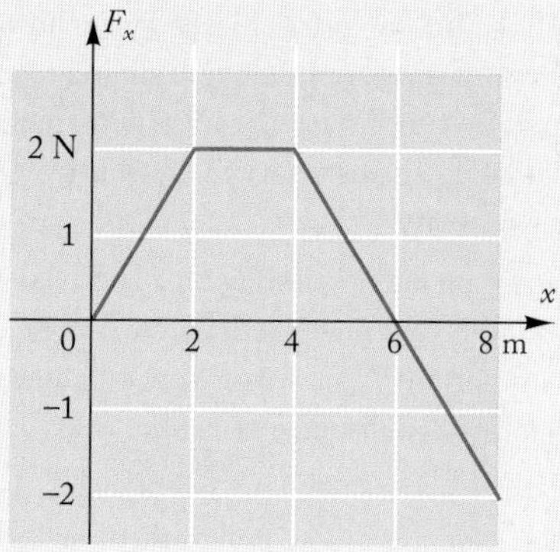

FIGURE 7.29 Position-dependent force.

25. When an ideal, horizontal spring is at equilibrium, a mass attached to its end is at $x = 0$. If the spring constant is 440 N/m, how much work does the spring do on the mass if the mass moves from $x = -0.20$ m to $x = +0.40$ m?

26. The spring on one kind of mousetrap has a spring constant of 4500 N/m. How much work is done to set the trap, by stretching the spring 2.7 cm from equilibrium?

*27. To stretch a spring a distance d from equilibrium takes an amount W_0 of work. How much work does it take to stretch the spring from d to $2d$ from equilibrium? From Nd to $(N + 1)d$ from equilibrium?

*28. A particular spring is not ideal; for a distance x from equilibrium, the spring exerts a force $F_x = -6x - 2x^3$, where x is in meters and F_x is in newtons. Compared with an ideal spring with a spring constant $k = 6.0$ N/m, by what factor does the work done by the nonideal spring exceed that done by the ideal spring when moving from $x = 0$ to $x = 0.50$ m? From $x = 1.0$ m to $x = 1.5$ m? From $x = 2.0$ m to $x = 2.5$ m?

*29. The ends of a relaxed spring of length l and force constant k are attached to two points on two walls separated by a distance l.

(a) How much work must you do to push the midpoint of the spring up or down a distance y (see Fig. 7.30)?

(b) How much force must you exert to hold the spring in this configuration?

*30. A particle moves along the x axis from $x = 0$ to $x = 2.0$ m. A force $F_x(x) = 2x^2 + 8x$ acts on the particle (the distance x is measured in meters, and the force in newtons). Calculate the work done by the force $F_x(x)$ during this motion.

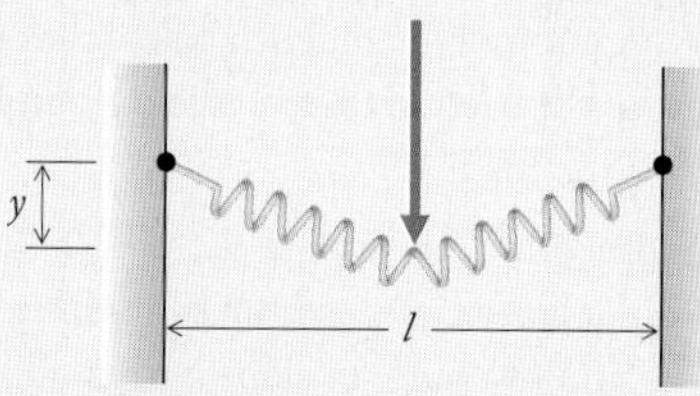

FIGURE 7.30 The midpoint of the spring has been pushed down a distance y. When the spring is relaxed, its length matches the distance l between the walls.

*31. Suppose that the force acting on a particle is a function of position; the force has components $F_x = 4x^2 + 1$, $F_y = 2x$, $F_z = 0$, where the force is measured in newtons and distance in meters. What is the work done by the force if the particle moves on a straight line from $x = 0$, $y = 0$, $z = 0$ to $x = 2.0$ m, $y = 2.0$ m, $z = 0$?

*32. A horse pulls a sled along a snow-covered curved ramp. Seen from the side, the surface of the ramp follows an arc of a circle of radius R (Fig. 7.31). The pull of the horse is always parallel to this surface. The mass of the sled is m, and the coefficient of sliding friction between the sled and the surface is μ_k. How much work must the horse do on the sled to pull it to a height $(1 - \sqrt{2}/2)R$, corresponding to an angle of 45° along the circle (Fig. 7.31)? How does this compare with the amount of work required to pull the sled from the same starting point to the same height along a straight ramp inclined at 22.5°?

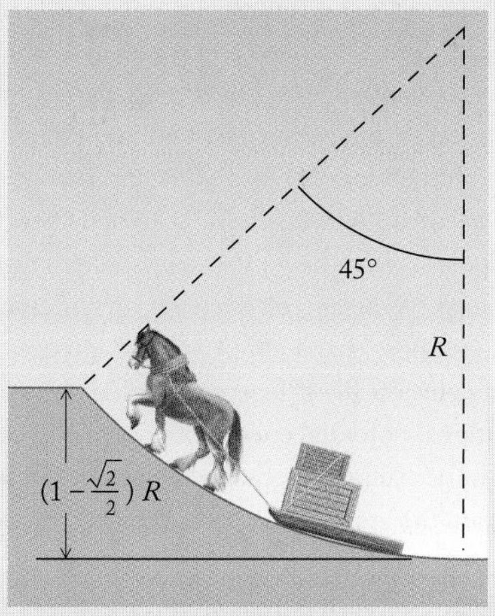

FIGURE 7.31 A horse pulling a sled along a curved ramp.

**33. The force between two inert gas atoms is often described by a function of the form

$$F_x = Ax^{-13} - Bx^{-7}$$

where A and B are positive constants and x is the distance between the atoms. Answer in terms of A and B.

(a) What is the equilibrium separation?

(b) What is the work done if the atoms are moved from their equilibrium separation to a very large distance apart?

7.3 Kinetic Energy

34. In a serve, a champion tennis player sends the ball flying at 160 km/h. The mass of the ball is 60 g. What is the kinetic energy of the ball?

35. Calculate the kinetic energy that the Earth has owing to its motion around the Sun.

36. The electron in a hydrogen atom has a speed of 2.2×10^6 m/s. What is the kinetic energy of this electron?

37. The fastest skier is Graham Wilkie, who attained 212.52 km/h on a steep slope at Les Arcs, France. The fastest runner is Robert Hayes, who briefly attained 44.88 km/h on a level track. Assume that the skier and the runner each have a mass of 75 kg. What is the kinetic energy of each? By what factor is the kinetic energy of the skier larger than that of the runner?

38. The Skylab satellite disintegrated when it reentered the atmosphere. Among the pieces that crashed down on the surface of the Earth, one of the heaviest was a lead-lined film vault of 1770 kg that had an estimated impact speed of 120 m/s on the surface. What was its kinetic energy? How many kilograms of TNT would we have to explode to release the same amount of energy? (One kilogram of TNT releases 4.6×10^6 J.)

39. An automobile of mass 1600 kg is traveling along a straight road at 80 km/h.

(a) What is the kinetic energy of this automobile in the reference frame of the ground?

(b) What is the kinetic energy in the reference frame of a motorcycle traveling in the same direction at 60 km/h?

(c) What is the kinetic energy in the reference frame of a truck traveling in the opposite direction at 60 km/h?

40. According to statistical data, the probability that an occupant of an automobile suffers lethal injury when involved in a crash is proportional to the square of the speed of the automobile. At a speed of 80 km/h, the probability is approximately 3%. What are the probabilities at 95 km/h, 110 km/h, and 125 km/h?

41. For the projectile described in Problem 47 of Chapter 2, calculate the initial kinetic energy ($t = 0$) and calculate the final kinetic energy ($t = 3.0$ s). How much energy does the projectile lose to friction in 3.0 s?

42. Compare the kinetic energy of a 15-g bullet fired at 630 m/s with that of a 15-kg bowling ball released at 6.3 m/s.

43. Compare the kinetic energy of a golf ball ($m = 45$ g) falling at a terminal velocity of 45 m/s with that of a person (75 kg) walking at 1.0 m/s.

44. A child's toy horizontally launches a 20-g ball using a spring that was originally compressed 8.0 cm. The spring constant is 30 N/m. What is the work done by the spring moving the ball from its compressed point to its relaxed position, where the ball is released? What is the kinetic energy of the ball at launch? What is the speed of the ball?

45. A mass of 150 g is held by a horizontal spring of spring constant 20 N/m. It is displaced from its equilibrium position and released from rest. As it passes through equilibrium, its speed is 5.0 m/s. For the motion from the release position to the equilibrium position, what is the work done by the spring? What was the initial displacement?

46. A 60-kg hockey player gets moving by pushing on the rink wall with a force of 500 N. The force is in effect while the skater extends his arms 0.50 m. What is the player's kinetic energy after the push? The player's speed?

47. A 1300-kg communication satellite has a speed of 3.1 km/s. What is its kinetic energy?

48. Suppose you throw a stone straight up so it reaches a maximum height h. At what height does the stone have one-half its initial kinetic energy?

49. The velocity of small bullets can be roughly measured with ballistic putty. When the bullet strikes a slab of putty, it penetrates a distance that is roughly proportional to the kinetic energy. Suppose that a bullet of velocity 160 m/s penetrates 0.80 cm into the putty and a second, identical bullet fired from a more powerful gun penetrates 1.2 cm. What is the velocity of the second bullet?

50. A particle moving along the x axis is subject to a force

$$F_x = -ax + bx^3$$

where a and b are constants.

(a) How much work does this force do as the particle moves from x_1 to x_2?

(b) If this is the only force acting on the particle, what is the change of kinetic energy during this motion?

*51. In the "tapping mode" used in atomic-force microscopes, a tip on a cantilever taps against the atoms of a surface to be studied. The cantilever acts as a spring of spring constant 2.5×10^{-2} N/m. The tip is initially displaced away from equilibrium by 3.0×10^{-8} m; it accelerates toward the surface, passes through the relaxed spring position, begins to slow down, and strikes the surface as the displacement approaches 2.5×10^{-8} m. What kinetic energy does the tip have just before striking the surface?

*52. With the brakes fully applied, a 1500-kg automobile decelerates at the rate of 8.0 m/s^2.

(a) What is the braking force acting on the automobile?

(b) If the initial speed is 90 km/h, what is the stopping distance?

(c) What is the work done by the braking force in bringing the automobile to a stop from 90 km/h?

(d) What is the change in the kinetic energy of the automobile?

*53. A box of mass 40 kg is initially at rest on a flat floor. The coefficient of kinetic friction between the box and the floor is $\mu_k = 0.60$. A woman pushes horizontally against the box with a force of 250 N until the box attains a speed of 2.0 m/s.

(a) What is the change of kinetic energy of the box?

(b) What is the work done by the friction force on the box?

(c) What is the work done by the woman on the box?

7.4 Gravitational Potential Energy†

54. It has been reported that at Cherbourg, France, waves smashing on the coast lifted a boulder of 3200 kg over a 6.0-m wall. What minimum energy must the waves have given to the boulder?

55. A 75-kg man walks up the stairs from the first to the third floor of a building, a height of 10 m. How much work does he do against gravity? Compare your answer with the food energy he acquires by eating an apple (see Table 8.1).

† For help, see Online Concept Tutorial 9 at www.wwnorton.com/physics

56. What is the kinetic energy and what is the gravitational potential energy (relative to the ground) of a goose of mass 6.0 kg soaring at 30 km/h at a height of 90 m?

57. Surplus energy from an electric power plant can be temporarily stored as gravitational energy by using this surplus energy to pump water from a river into a reservoir at some altitude above the level of the river. If the reservoir is 250 m above the level of the river, how much water (in cubic meters) must we pump in order to store 2.0×10^{13} J?

58. The track of a cable car on Telegraph Hill in San Francisco rises more than 60 m from its lowest point. Suppose that a car is ascending at 13 km/h along the track when it breaks away from its cable at a height of exactly 60 m. It will then coast up the hill some extra distance, stop, and begin to race down the hill. What speed does the car attain at the lowest point of the track? Ignore friction.

59. In pole vaulting, the jumper achieves great height by converting her kinetic energy of running into gravitational potential energy (Fig. 7.32). The pole plays an intermediate role in this process. When the jumper leaves the ground, part of her translational kinetic energy has been converted into kinetic energy of rotation (with the foot of the pole as the center of rotation) and part has been converted into elastic potential energy of deformation of the pole. When the jumper reaches her highest point, all of this energy has been converted into gravitational potential energy. Suppose that a jumper runs at a speed of 10 m/s. If the jumper converts all of the corresponding kinetic energy into gravitational potential energy, how high will her center of mass rise? The actual height reached by pole vaulters is 5.7 m (measured from the ground). Is this consistent with your calculation?

FIGURE 7.32 A pole vaulter.

60. Because of brake failure, a bicycle with its rider careens down a steep hill 45 m high. If the bicycle starts from rest and there is no friction, what is the final speed attained at the bottom of the hill?

61. Under suitable conditions, an avalanche can reach extremely great speeds because the snow rides down the mountain on a cushion of trapped air that makes the sliding motion nearly frictionless. Suppose that a mass of 2.0×10^7 kg of snow breaks loose from a mountain and slides down into a valley 500 m below the starting point. What is the speed of the snow when it hits the valley? What is its kinetic energy? The explo-

sion of 1 short ton (2000 lb) of TNT releases 4.2×10^9 J. How many tons of TNT release the same energy as the avalanche?

62. A parachutist of mass 60 kg jumps out of an airplane at an altitude of 800 m. Her parachute opens and she lands on the ground with a speed of 5.0 m/s. How much energy has been lost to air friction in this jump?

63. A block released from rest slides down to the bottom of a plane of incline 15° from a height of 1.5 m; the block attains a speed of 3.5 m/s at the bottom. By considering the work done by gravity and the frictional force, determine the coefficient of friction.

64. A bobsled run leading down a hill at Lake Placid, New York, descends 148 m from its highest to its lowest point. Suppose that a bobsled slides down this hill without friction. What speed will the bobsled attain at the lowest point?

65. A 2.5-g Ping-Pong ball is dropped from a window and strikes the ground 20 m below with a speed of 9.0 m/s. What fraction of its initial potential energy was lost to air friction?

66. A roller coaster begins at rest from a first peak, descends a vertical distance of 45 m, and then ascends a second peak, cresting the peak with a speed of 15 m/s. How high is the second peak? Ignore friction.

67. A skateboarder starts from rest and descends a ramp through a vertical distance of 5.5 m; he then ascends a hill through a vertical distance of 2.5 m and subsequently coasts on a level surface. What is his coasting speed? Ignore friction.

*68. In some barge canals built in the nineteenth century, barges were slowly lifted from a low level of the canal to a higher level by means of wheeled carriages. In a French canal, barges of 70 metric tons were placed on a carriage of 35 tons that was pulled, by a wire rope, to a height of 12 m along an inclined track 500 m long.

(a) What was the tension in the wire rope?

(b) How much work was done to lift the barge and carriage?

(c) If the cable had broken just as the carriage reached the top, what would have been the final speed of the carriage when it crashed at the bottom?

*69. A wrecking ball of mass 600 kg hangs from a crane by a cable of length 10 m. If this wrecking ball is released from an angle of 35°, what will be its kinetic energy when it swings through the lowest point of its arc?

*70. Consider a stone thrown vertically upward. If we take air friction into account, we see that $\frac{1}{2}mv^2 + mgy$ must decrease as a function of time. From this, prove that the stone will take longer for the downward motion than for the upward motion.

*71. A stone of mass 0.90 kg attached to a string swings around a vertical circle of radius 0.92 m. Assume that during this motion the energy (kinetic plus potential) of the stone is constant. If, at the top of the circle, the tension in the string is (just about) zero, what is the tension in the string at the bottom of the circle?

*72. A center fielder throws a baseball of mass 0.17 kg with an initial speed of 28 m/s and an elevation angle of 30°. What is the kinetic energy and what is the potential energy of the baseball when it reaches the highest point of its trajectory? Ignore friction.

*73. A jet aircraft looping the loop (see Problem 70 in Chapter 4) flies along a vertical circle of diameter 1000 m with a speed of 620 km/h at the bottom of the circle and a speed of 350 km/h at the top of the circle. The change of speed is due mainly to the downward pull of gravity. For the given speed at the bottom of the circle, what speed would you expect at the top of the circle if the thrust of the aircraft's engine exactly balances the friction force of air (as in the case for level flight)?

*74. A pendulum consists of a mass hanging from a string of length 1.0 m attached to the ceiling. Suppose that this pendulum is initially held at an angle of 30° with the vertical (see Fig. 7.33) and then released. What is the speed with which the mass swings through its lowest point? At what angle will the mass have one-half of this speed?

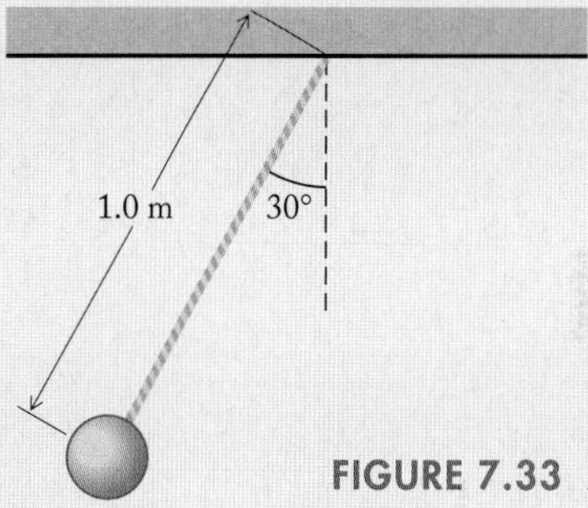

FIGURE 7.33 A pendulum.

**75. A stone is tied to a string of length R. A man whirls this stone in a vertical circle. Assume that the energy of the stone remains constant as it moves around the circle. Show that if the string is to remain taut at the top of the circle, the speed of the stone at the bottom of circle must be at least $\sqrt{5gR}$.

**76. In a loop coaster at an amusement park, cars roll along a track that is bent in a full vertical loop (Fig. 7.34). If the upper portion of the track is an arc of a circle of radius $R = 10$ m, what is the minimum speed that a car must have at the top of the loop if it is not to fall off? If the highest point of the loop has a height $h = 40$ m, what is the minimum speed with which the car must enter the loop at its bottom? Ignore friction.

FIGURE 7.34 A roller coaster with a full loop.

**77. You are to design a roller coaster in which cars start from rest at a height $h = 30$ m, roll down into a valley, and then up a mountain (Fig. 7.35).

(a) What is the speed of the cars at the bottom of the valley?

(b) If the passengers are to feel $8g$ at the bottom of the valley, what must be the radius R of the arc of the circle that fits the bottom of the valley?

(c) The top of the next mountain is an arc of a circle of the same radius R. If the passengers are to feel $0g$ at the top of this mountain, what must be its height h'?

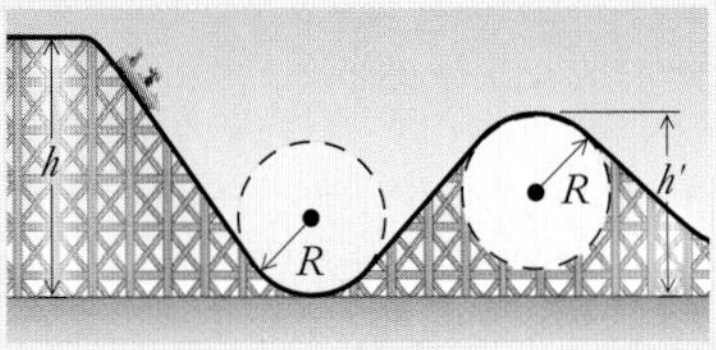

FIGURE 7.35 Profile of a roller coaster.

**78. One portion of the track of a toy roller coaster is bent into a full vertical circle of radius R. A small cart rolling on the track enters the bottom of the circle with a speed $2\sqrt{gR}$. Show that this cart will fall off the track before it reaches the top of the circle, and find the (angular) position at which the cart loses contact with the track.

**79. A particle initially sits on top of a large, smooth sphere of radius R (Fig. 7.36). The particle begins to slide down the sphere, without friction. At what angular position θ will the particle lose contact with the surface of the sphere? Where will the particle land on the ground?

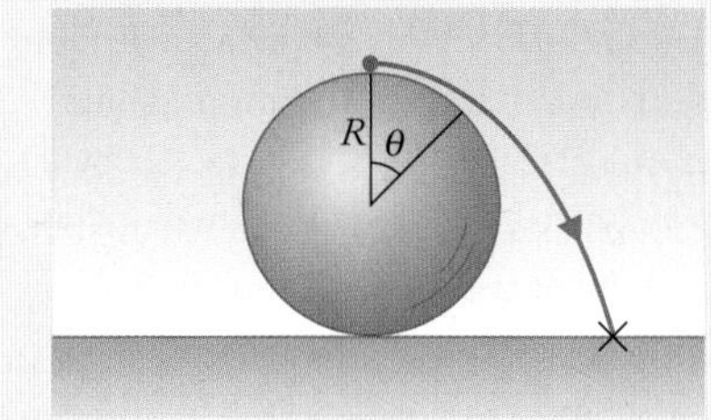

FIGURE 7.36 Particle sliding down a sphere.

REVIEW PROBLEMS

80. An apple falls down 35 m from the fifth floor of an apartment building to the street. The mass of the apple is 0.20 kg. How much work does gravity do on the apple during this fall?

81. A woman pulls a sled by a rope. The rope makes an upward angle of 45° with the ground, and the woman exerts a pull of 150 N on the rope. How much work does the woman do if she pulls this sled 20 m?

82. A man pushes a crate along a flat concrete floor. The mass of the crate is 120 kg, and the coefficient of friction between the crate and the floor is $\mu_k = 0.50$. How much work does the man do if, pushing horizontally, he moves the crate 15 m at constant speed?

83. A 1500-kg automobile is traveling at 20 m/s on a level road. How much work must be done on the automobile to accelerate it from 20 m/s to 25 m/s? From 25 m/s to 30 m/s?

84. A woman slowly lifts a 20-kg package of books from the floor to a height of 1.8 m, and then slowly returns it to the floor. How much work does she do on the package while lifting? How much work does she do on the package while lowering? What is the total work she does on the package? For the information given, can you tell how much work she expends internally in her muscles, that is, how many calories she expends?

85. An automobile of 1200 kg is traveling at 25 m/s when the driver suddenly applies the brakes so as to lock the wheels and cause the automobile to skid to a stop. The coefficient of sliding friction between the tires and the road is 0.90.

(a) What is the deceleration of the automobile, and what is the stopping distance?

(b) What is the friction force of the road on the wheels, and what is the amount of work that this friction force does during the stopping process?

*86. A golf ball of mass 50 g released from a a height of 1.5 m above a concrete floor bounces back to a height of 1.0 m.

(a) What is the kinetic energy of the ball just before contact with the floor begins? Ignore air friction.

(b) What is the kinetic energy of the ball just after contact with the floor ends?

(c) What is the loss of energy during contact with the floor?

87. A small aircraft of mass 1200 kg is cruising at 250 km/h at an altitude of 2000 m.

(a) What is the gravitational potential energy (relative to the ground), and what is the kinetic energy of the aircraft?

(b) If the pilot puts the aircraft into a dive, what will be the gravitational potential energy, what will be the kinetic energy, and what will be the speed when the aircraft reaches an altitude of 1500 m? Assume that the engine of the aircraft compensates the friction force of air, so the aircraft is effectively in free fall.

88. In a roller coaster, a car starts from rest on the top of a 30-m-high mountain. It rolls down into a valley, and then up a 20-m-high mountain. What is the speed of the car at the bottom of the valley, at ground level? What is the speed of the car at the top of the second mountain?

*89. In a compound bow (see Fig. 7.37), the pull of the limbs of the bow is communicated to the arrow by an arrangement of strings and pulleys that ensures that the force of the string against the arrow remains roughly constant while you pull the arrow back in preparation for letting it fly (in an ordinary bow, the force of the string increases as you pull back, which makes it difficult to continue pulling). A typical compound bow provides a steady force of 300 N. Suppose you pull an arrow of 0.020 kg back 0.50 m against this force.

(a) What is the work you do?

(b) When you release the arrow, what is the kinetic energy with which it leaves the bow?

(c) What is the speed of the arrow?

(d) How far will this arrow fly when launched with an elevation angle of 45°? Ignore friction and assume that the heights of the launch and impact points are the same.

(e) With what speed will it hit the target?

FIGURE 7.37 A compound bow.

*90. A large stone-throwing engine designed by Archimedes could throw a 77-kg stone over a range of 180 m. Assume that the stone is thrown at an initial angle of 45° with the horizontal.

(a) Calculate the initial kinetic energy of this stone.

(b) Calculate the kinetic energy of the stone at the highest point of its trajectory.

*91. The luge track at Lillehammer, the site of the 1994 Olympics, starts at a height of 350 m and finishes at 240 m. Suppose that a luger of 95 kg, including the sled starts from rest and reaches the finish at 130 km/h. How much energy has been lost to friction against the ice and the air?

*92. A pendulum consists of a mass m tied to one end of a string of length l. The other end of the string is attached to a fixed point on the ceiling. Suppose that the pendulum is initially held at an angle of 90° with the vertical. If the pendulum is released from this point, what will be the speed of the pendulum at the instant it passes through its lowest point? What will be the tension in the string at this instant?

*93. A roller coaster near St. Louis is 34 m high at its highest point.

(a) What is the maximum speed that the car can attain by rolling down from the highest point if initially at rest? Ignore friction.

(b) Some people claim that cars reach a maximum speed of 100 km/h. If this is true, what must be the initial speed of the car at the highest point?

*94. At a swimming pool, a water slide starts at a height of 6.0 m and ends at a height of 1.0 m above the water level with a short horizontal segment (see Fig. 7.38). A girl slides down the water slide.

(a) What is her speed at the bottom of the slide?

(b) How far from the slide does she land in the water?

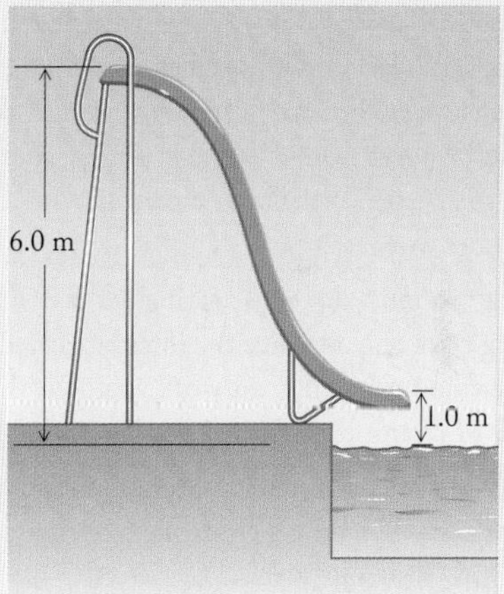

FIGURE 7.38 A water slide.

Answers to Checkups

Checkup 7.1

1. The normal force, which is perpendicular to the motion, does no work. The weight of the roller-coaster car does negative work as the car travels upward, and positive work as the car moves downward, since it has a component against or along the motion, respectively. At the peak, the work done by the weight is zero, since the weight is then perpendicular to the displacement.

2. For both the pushing and the pulling, the force is in the same direction as the displacement (you push when the saw moves forward and pull when the saw moves backward); thus, the work is positive in both cases.

3. The dog's pull is in the same direction as the displacement, and thus does positive work on the woman. The woman's pull is in the opposite direction to the displacement, and thus does negative work on the dog.

4. In each case, the force is opposite to the displacement (whether pushing against the front or pulling on the rear, the force is rearward), and so negative work is done on the cart in both cases.
5. No. The tension provides a centripetal acceleration, which is perpendicular to the (tangential) motion, and thus does no work.
6. The work is positive in (b) and (c), where the angle between the force and displacement is less than 90°; the work is negative in (a), where the angle is greater than 90°. The work is zero in (d), where the force is perpendicular to the displacement. The work is largest when the force is most nearly parallel to the displacement; for force vectors (and displacement vectors) of equal magnitude, this occurs in (c).
7. (E) 4 and 5. To calculate the work done by a constant force, $W = Fs\cos\theta$, you do not need to know the mass, acceleration, or speed. You do need to know the force, the displacement, and the angle between the two.

Checkup 7.2

1. The work done by a variable force is equal to the area under the $F(x)$ vs. x curve. Assuming the two plots are drawn to the same vertical scale, for a displacement from a to b, the upper plot clearly has a greater area between the $F(x)$ curve and the x axis.
2. If we consider a plot such as Fig. 7.13 and imagine extending the curve to the left to $x = -b$ [where $F(x) = +kb$], then we see that *positive* work is done on the particle as it moves from $x = -b$ to $x = 0$ [where the area between the $F(x)$ curve and the x axis is *above* the x axis]. *Negative* work is done on the particle as it moves from $x = 0$ to $x = +b$ [where the area between the $F(x)$ curve and the x axis is *below* the origin]. Thus the net work is zero.
3. The work you must do on the spring is the opposite of what the spring does on you, since the forces involved are an action–reaction pair. Thus the work you do is the negative of the result of Example 4, or $W = +\frac{1}{2}k(b^2 - a^2)$.
4. (D) $3W$. The work to stretch from equilibrium is $\frac{1}{2}kx^2$, so the first stretch requires $W = \frac{1}{2}kd^2$. The second stretch requires work $W' = \frac{1}{2}kx^2\big|_d^{2d} = \frac{1}{2}k(2d)^2 - \frac{1}{2}kd^2 = 4W - W = 3W$.

Checkup 7.3

1. Yes—the kinetic energy, $K = \frac{1}{2}mv^2$, depends only on the square of the speed, and not on the direction of the velocity. Thus if the two equal masses have the same speed, they have the same kinetic energy.
2. Yes, the kinetic energies can be equal. Since the kinetic energy is proportional to mass and proportional to the square of the speed ($K = \frac{1}{2}mv^2$), if the car has twice the speed of the truck (a factor of 4 contribution to the kinetic energy), then the kinetic energies can be equal if the truck has 4 times the mass of the car.
3. The kinetic energy of the golf ball is largest at the beginning (and end, if we neglect air resistance) of the trajectory; at higher points, the force of gravity has slowed the ball down. The kinetic energy is smallest at the top of the trajectory, where there is only a horizontal contribution to the speed ($v = \sqrt{v_x^2 + v_y^2}$). The kinetic energy is not zero while the ball is in the air (unless the ball was accidentally launched vertically; in that case, the kinetic energy would be zero at the top of the trajectory).
4. No. For the work–energy theorem to apply, one must consider the *net* external force on the sled. If traveling at constant velocity (zero acceleration), the total force must be zero (the horse's pull does positive work and is canceled by the friction force, which does negative work), and so the total work done on the sled is zero. Thus there is no change in kinetic energy.
5. (E) 9. The kinetic energy, $K = \frac{1}{2}mv^2$, is proportional to the square of the speed; thus increasing the speed by a factor of 3 increases the kinetic energy by a factor of 9.

Checkup 7.4

1. As in Example 8, the velocity at the bottom depends only on the height of release (the cars do not even have to have the same mass!); thus, the upper roller coaster will provide the larger speed at the bottom, since Δy is greater.
2. The gravitational potential energy U decreases as the piano is brought to street level from the first house; U remains constant during the trip to the nearby house (assuming travel over flat ground); then, the gravitational potential energy increases back to its original value as the piano is brought up to the second floor of the second house (assuming similar houses).
3. No. At constant speed, K is constant; since U decreases as the truck moves down, $E = K + U$ decreases also, and so is not conserved.
4. Since both the girl and the boy change height by the same amount, they both reach the pool with the same speed (at *any* vertical height, they have the same speed, but the boy's velocity has a horizontal component, so his vertical velocity is slower than that of the girl). Since the girl's velocity is all vertical, a larger vertical velocity implies that she reaches the pool first.
5. (A) $\sqrt{2}$. As in Example 8, the speed at the bottom (starting from rest) is proportional to the square root of the initial height. Thus, for twice the height, the speed of the first bicyclist will be $\sqrt{2}$ times as large at the bottom.

Conservation of Energy

CHAPTER

8

CONCEPTS IN CONTEXT

Concepts — *in* — Context

The two orange areas in the middle of this satellite image are the reservoirs of the hydroelectric pumped-storage plant on Brown Mountain in New York State. When full, the upper reservoir (at right) holds 19 million cubic meters of water. This reservoir is linked to the lower reservoir at the base, part of the Schoharie Creek, by a 320-m vertical shaft bored through the mountain. The water flowing out of this shaft drives four large turbines that generate electric power. During periods of low demand, the turbines are operated in reverse, so they pump water back into the upper reservoir.

With the concepts developed in this chapter we can address questions such as:

? How do pumped-storage power plants complement other power plants? (Physics in Practice: Hydroelectric Pumped Storage, page 242)

? What is the speed of water spurting out of the shaft at the bottom? (Example 3, page 242)

? How much gravitational potential energy is stored in the upper reservoir, and how much available electric energy does this represent? (Example 5, page 249)

? When generating power at its maximum capacity, at what rate does the power plant remove water from the upper reservoir? How many hours can it run? (Example 10, page 257)

In the preceding chapter we found how to formulate a law of conservation of mechanical energy for a particle moving under the influence of the Earth's gravity. Now we will seek to formulate the law of conservation of mechanical energy when other forces act on the particle—such as the force exerted by a spring—and we will state the general law of conservation of energy. As in the case of motion under the influence of gravity, the conservation law permits us to deduce some features of the motion without having to deal with Newton's Second Law.

Online Concept Tutorial 10

8.1 POTENTIAL ENERGY OF A CONSERVATIVE FORCE

JOSEPH LOUIS, COMTE LAGRANGE (1736–1813) *French mathematician and theoretical astronomer. In his elegant mathematical treatise* Analytical Mechanics, *Lagrange formulated Newtonian mechanics in the language of advanced mathematics and introduced the general definition of the potential-energy function. Lagrange is also known for his calculations of the motion of planets and for his influential role in securing the adoption of the metric system of units.*

To formulate the law of conservation of energy for a particle moving under the influence of gravity, we began with the work–energy theorem [see Eq. (7.24)],

$$K_2 - K_1 = W \tag{8.1}$$

We then expressed the work W as a difference of two potential energies [see Eq. (7.30)],

$$W = -U_2 + U_1 \tag{8.2}$$

This gave us

$$K_2 - K_1 = -U_2 + U_1$$

from which we immediately found the conservation law for the sum of the kinetic and potential energies, $K_2 + U_2 = K_1 + U_1$, or

$$E = K + U = [\text{constant}] \tag{8.3}$$

As an illustration of this general procedure for the construction of the conservation law for mechanical energy, let us deal with the case of a particle moving under the influence of the elastic force exerted by a spring attached to the particle. If the particle moves along the x axis and the spring lies along this axis, the force has only an x component F_x, which is a function of position:

$$F_x(x) = -kx \tag{8.4}$$

Here, as in Section 6.2, the displacement x is measured from the relaxed position of the spring. The crucial step in the construction of the conservation law is to express the work W as a difference of two potential energies. For this purpose, we take advantage of the result established in Section 7.2 [see Eq. (7.17)], according to which the work done by the spring force during a displacement from x_1 to x_2 is

$$W = \tfrac{1}{2}kx_1^2 - \tfrac{1}{2}kx_2^2 \tag{8.5}$$

This shows that if we identify the **elastic potential energy of the spring** as

potential energy of spring

$$U = \tfrac{1}{2}kx^2 \tag{8.6}$$

then the work is, indeed, the difference between two potential energies $U_1 = \frac{1}{2}kx_1^2$ and $U_2 = \frac{1}{2}kx_2^2$. According to Eq. (8.6), the potential energy of the spring is proportional to the square of the displacement. Figure 8.1 gives a plot of this elastic potential energy.

The potential energy of the spring represents the capacity of the spring to do work by virtue of its deformation. When we compress a spring, we store latent work in it, which we can recover at a later time by letting the spring push against something. An old-fashioned watch, operated by a wound spring, illustrates this storage of energy in a spring (however, the springs in watches are not coil springs, but spiral springs, which are compressed by turning the knob of the watch).

As in the case of the particle moving under the influence of gravity, we conclude that for the particle moving under the influence of the spring force, the sum of the kinetic and elastic potential energies is constant,

$$E = K + U = \frac{1}{2}mv^2 + \frac{1}{2}kx^2 = [\text{constant}] \tag{8.7}$$

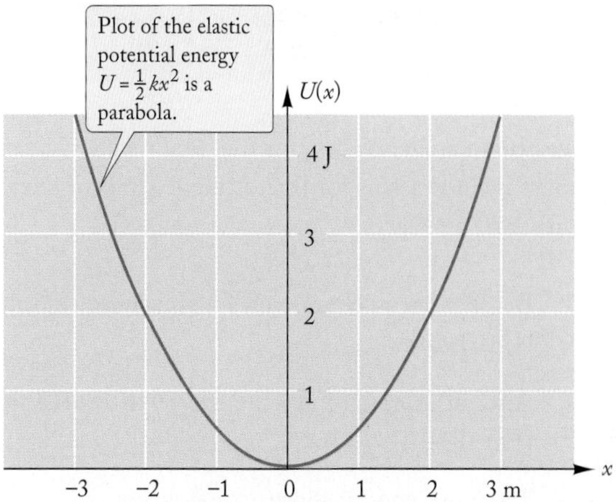

FIGURE 8.1 Plot of the potential energy of a spring as a function of the displacement x. In this plot, the spring constant is $k = 1$ N/m.

This equation gives us some information about the general features of the motion; it shows how the particle trades speed for an increase in the distance from the relaxed position of the spring. For instance, an increase of the magnitude of x requires a decrease of the speed v so as to keep the sum $\frac{1}{2}mv^2 + \frac{1}{2}kx^2$ constant.

EXAMPLE 1 A child's toy gun shoots a dart by means of a compressed spring. The constant of the spring is $k = 320$ N/m, and the mass of the dart is 8.0 g. Before shooting, the spring is compressed by 6.0 cm, and the dart is placed in contact with the spring (see Fig. 8.2). The spring is then released. What will be the speed of the dart when the spring reaches its relaxed position?

SOLUTION: The dart can be regarded as a particle moving under the influence of a force $F_x = -kx$, with a potential energy $U = \frac{1}{2}kx^2$. Taking the positive x axis along the direction of motion, the initial value of x is negative ($x_1 = -6.0$ cm); also, the initial speed is zero. According to Eq. (8.7), the initial energy is

$$E = \frac{1}{2}mv_1^2 + \frac{1}{2}kx_1^2 = 0 + \frac{1}{2}kx_1^2 \tag{8.8}$$

When the spring reaches its relaxed position ($x_2 = 0$), the energy will be

$$E = \frac{1}{2}mv_2^2 + \frac{1}{2}kx_2^2 = \frac{1}{2}mv_2^2 + 0 \tag{8.9}$$

Conservation of energy demands that the right sides of Eqs. (8.8) and (8.9) be equal:

$$\frac{1}{2}mv_2^2 = \frac{1}{2}kx_1^2 \tag{8.10}$$

If we cancel the factors of $\frac{1}{2}$ in this equation, divide both sides by m, and take the square root of both sides, we find that the speed of the dart as it leaves the spring at $x_2 = 0$ is

$$v_2 = \sqrt{\frac{k}{m}x_1^2} = \sqrt{\frac{320\ \text{N/m}}{0.0080\ \text{kg}} \times (-0.060\ \text{m})^2} = 12\ \text{m/s} \tag{8.11}$$

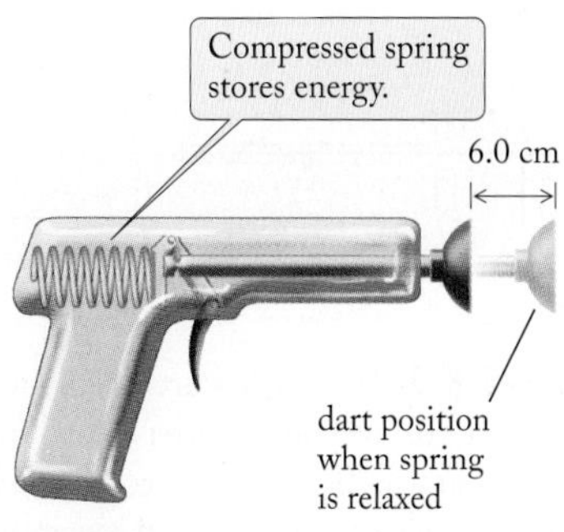

FIGURE 8.2 A toy gun. The spring is initially compressed 6.0 cm.

PROBLEM-SOLVING TECHNIQUES ENERGY CONSERVATION

To obtain an expression for the total mechanical energy, you must include terms for the different kinds of energy that are present:

1. Begin with an expression for the energy at one point [Eq. (8.8)].
2. And an expression for the energy at another point [Eq. (8.9)].
3. Then use energy conservation to equate these expressions [Eq. (8.10)].

With the appropriate expression for the mechanical energy, you can apply energy conservation to solve some problems of motion. As illustrated in the preceding example, this involves the three steps outlined in Section 7.4 and 8.1.

CONTRIBUTIONS TO THE MECHANICAL ENERGY

KIND OF ENERGY	APPLICABLE IF	CONTRIBUTION TO TOTAL MECHANICAL ENERGY
Kinetic energy	Particle is in motion	$K = \frac{1}{2}mv^2$
Gravitational potential energy	Particle is moving up or down near the Earth's surface	$U = mgy$
Elastic potential energy	Particle is subject to a spring force	$U = \frac{1}{2}kx^2$

To formulate the law of conservation of mechanical energy for a particle moving under the influence of some other force, we want to imitate the above construction. We will be able to do this if, and only if, the work performed by this force can be expressed as a difference between two potential energies, that is,

$$W = -U_2 + U_1 \tag{8.12}$$

If the force meets this requirement (and therefore permits the construction of a conservation law), the force is called **conservative.** Thus, the force of gravity and the force of a spring are conservative forces. Note that for any such force, the work done when the particle starts at the point x_1 and *returns* to the same point is necessarily zero, since, with $x_2 = x_1$, Eq. (8.12) implies

$$W = -U_1 + U_1 = 0 \tag{8.13}$$

This simply means that for a round trip that starts and ends at x_1, the work the force does during the outward portion of the trip is exactly the negative of the work the force does during the return portion of the trip, and therefore the net work for the round trip is zero (see Fig. 8.3). Thus, the energy supplied by the force is recoverable: the energy supplied by the force during motion in one direction is restored during the return motion in the opposite direction. For instance, when a particle moves downward from some starting point, gravity performs positive work; and when the particle moves upward, returning to its starting point, gravity performs negative work of a magnitude exactly equal to that of the positive work.

The requirement of zero work for a round trip can be used to discriminate between conservative and nonconservative forces. Friction is an example of a nonconservative force. If we slide a metal block through some distance along a table and then slide the block back to its starting point, the net work is not zero. The work performed by the friction force during the outward portion of the motion is negative, and the work performed by the friction force during the return portion of the trip is also negative—the friction

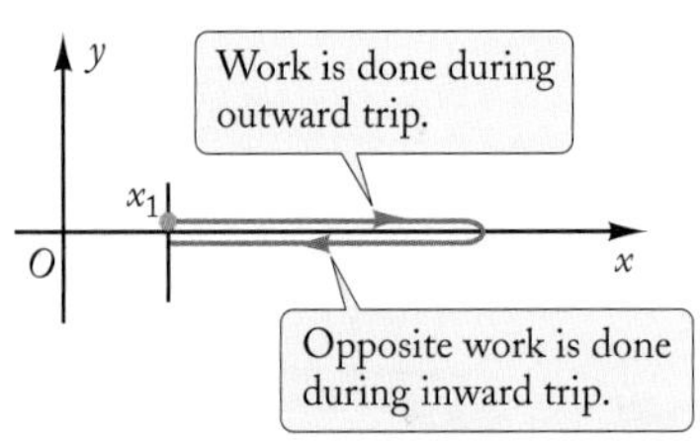

FIGURE 8.3 A particle starts at a point x_1 and returns to the point x_1 after completing some round trip. If the force is conservative, the work done is zero, because the work for the outward portion of the trip is opposite to that for the inward portion.

force always opposes the motion, and the work done by the friction force is always negative. Thus, the work done by the friction force cannot be expressed as a difference between two potential energies, and we cannot formulate a law of conservation of mechanical energy if friction forces are acting. However, as we will see in Section 8.3, we can formulate a more general law of conservation of energy, involving kinds of energy other than mechanical, which remains valid even when there is friction.

In the case of one-dimensional motion, a force is conservative whenever it can be expressed as an explicit function of position, $F_x = F_x(x)$. (Note that the friction force does *not* fit this criterion; the sign of the friction force depends on the direction of motion, and therefore the friction force is not uniquely determined by the position x.) For any such force $F_x(x)$, we can construct the potential energy function by integration. We take a point x_0 as reference point at which the potential energy is zero. The potential energy at any other point x is constructed by evaluating an integral (in the following equations, the integration variables are indicated by primes to distinguish them from the upper limits of integration):

$$U(x) = -\int_{x_0}^{x} F_x(x')\,dx' \tag{8.14}$$

potential energy as integral of force

To check that this construction agrees with Eq. (8.12), we examine $U_1 - U_2$:

$$U_1 - U_2 = U(x_1) - U(x_2) = -\int_{x_0}^{x_1} F_x(x')\,dx' + \int_{x_0}^{x_2} F_x(x')\,dx'$$

By one of the basic rules for integrals (see Appendix 4), the integral changes sign when we reverse the limits of integration. Hence

$$U_1 - U_2 = \int_{x_1}^{x_0} F_x(x')\,dx' + \int_{x_0}^{x_2} F_x(x')\,dx'$$

And by another basic rule, the sum of an integral from x_1 to x_0 and an integral from x_0 to x_2 is equal to a single integral from x_1 to x_2. Thus

$$U_1 - U_2 = \int_{x_1}^{x_2} F_x(x')\,dx' \tag{8.15}$$

Here the right side is exactly the work done by the force as the particle moves from x_1 to x_2, in agreement with Eq. (8.12). This confirms that our construction of the potential energy is correct.

In the special case of the spring force $F_x(x) = -kx$, our general construction (8.14) of the potential energy immediately yields the result (8.6), provided we take $x_0 = 0$.

For a particle moving under the influence of any conservative force, the total mechanical energy is the sum of the kinetic energy and the potential energy; as before, this total mechanical energy is conserved:

$$E = K + U = [\text{constant}] \tag{8.16}$$

or

$$E = \tfrac{1}{2}mv^2 + U = [\text{constant}] \tag{8.17}$$

conservation of mechanical energy

EXAMPLE 2 As we will see in later chapters, the **inverse-square force** plays a large role in physics—gravitational forces are inverse square, and electric forces are inverse square. If we consider a particle that can move in only one dimension along the positive x axis, this force has the form

$$F_x(x) = \frac{A}{x^2} \qquad (8.18)$$

where A is a constant. The point $x = 0$ is called the **center of force.** If A is positive, the force is repulsive (F_x is positive, and the force therefore pushes a particle on the positive x axis away from the center of force); if A is negative, the force is attractive (F_x is negative, and the force pulls the particle toward the center of force). The magnitude of the force is very large near $x = 0$, and it decreases as the distance from this point increases (Figs. 8.4a and b). What is the potential energy for this force?

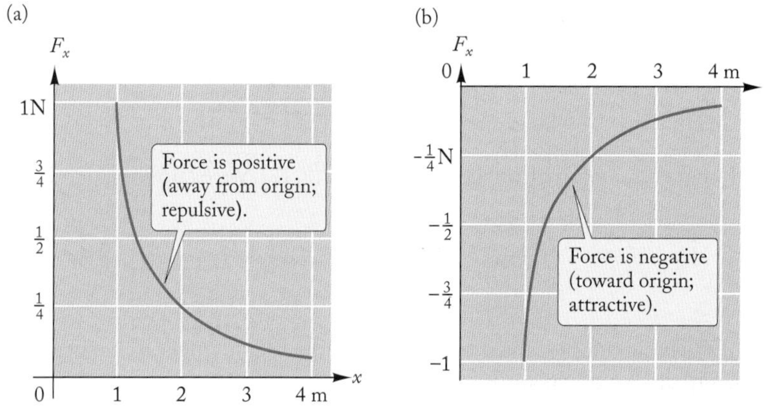

FIGURE 8.4 The inverse-square force A/x^2 as a function of x, (a) for a positive value of A (repulsive force; $A = 1\ \text{N·m}^2$) and (b) for a negative value of A (attractive force; $A = -1\ \text{N·m}^2$).

SOLUTION: According to Eq. (8.14),

$$U(x) = -\int_{x_0}^{x} \frac{A}{x'^2}\,dx'$$

In the compact notation of tables of integrals, $\int (1/x'^2)\,dx' = -1/x'$. Hence

$$U(x) = -\left[-\frac{A}{x'}\right]_{x_0}^{x} = -\left[-\frac{A}{x} - \left(-\frac{A}{x_0}\right)\right] = \frac{A}{x} - \frac{A}{x_0}$$

It is usually convenient to take $x_0 = \infty$ as the reference point, with $U_0 = 0$ at $x = \infty$. With this choice,

$$U(x) = \frac{A}{x} \qquad (8.19)$$

COMMENT: Note that for a repulsive force ($A > 0$), the potential energy decreases with x (see Fig. 8.5a), and for an attractive force ($A < 0$), the potential energy increases with x (the potential energy is large and negative near $x = 0$, and it increases toward zero as x increases; see Fig. 8.5b).

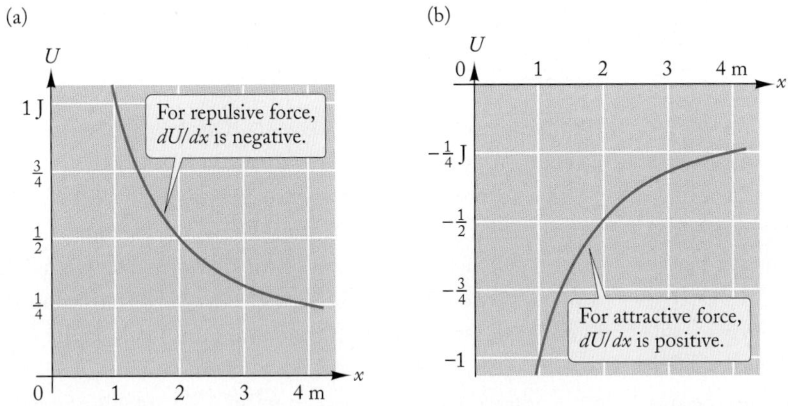

FIGURE 8.5 The potential A/x as a function of x, (a) for a positive value of A and (b) for a negative value of A.

For both the spring force and the inverse-square force, the force can be expressed in terms of the potential energy as $F_x(x) = -dU/dx$; that is, the force is the negative of the derivative of the potential energy. This relationship holds generally, for any kind of conservative force. We can see this by examining the change in potential energy produced by a small displacement dx. From Eq. (8.12) we see that if the points x_1 and x_2 are separated by a small distance $dx = x_2 - x_1$, then [see also Eq. (7.15)]

$$dU = U_2 - U_1 = -dW = -F_x\,dx \tag{8.20}$$

and if we divide this by dx, we obtain

$$F_x = -\frac{dU}{dx} \tag{8.21}$$

force as derivative of potential

This relation gives us a quick way to calculate the force if the potential energy is known.

From Eq. (8.21) we see that the force F_x is positive wherever the potential is a decreasing function of x, that is, wherever the derivative dU/dx is negative. Conversely, the force F_x is negative wherever the potential is an increasing function of x, that is, wherever the derivative dU/dx is positive. This is in agreement with the result we found for repulsive and attractive forces in Example 2.

Although in this section we have focused on one-dimensional motion, the criterion of zero work for a round trip is also valid for conservative forces in two or three dimensions. In one dimension, the path for a round trip is necessarily back and forth along a straight line; in two or three dimensions, the path can be of any shape, provided it forms a closed loop that starts and ends at the same point.

Furthermore, the law of conservation of mechanical energy is valid not only for the motion of a single particle, but also for the motion of more general systems, such as systems consisting of solids, liquids, or gases. When applying the conservation law to the kinetic and potential energies of such bodies, it may be necessary to take into account other forms of energy, such as the heat produced by friction and stored in the bodies (see Section 8.3). However, if such other forms of energy stored in the bodies are constant, then we can ignore them in our examination of the motion, as illustrated in the following example of the motion of water in a pipe.

PHYSICS IN PRACTICE HYDROELECTRIC PUMPED STORAGE

Concepts — in — Context

The demand for electric energy by industrial and commercial users is high during working hours, but low during nights and on weekends. For maximum efficiency, electric power companies prefer to run their large nuclear or coal-fired power plants at a steady, full output for 24 hours a day, 7 days a week. Thus electric power companies often have a surplus of electric energy available at night and on weekends, and they often have a deficit of energy during peak-demand times, which requires them to purchase energy from neighboring power companies. Hydroelectric pumped-storage plants help to deal with this mismatch between a fluctuating demand and a steady supply. A hydroelectric pumped-storage plant is similar to an ordinary hydroelectric power plant. It consists of an upper water reservoir and a lower water reservoir, typically separated by a few hundred meters in height. Large pipes (penstocks) connect the upper reservoir to turbines placed at the level of the lower reservoir. The water spurting out of the pipes drives the turbines, which drive electric generators. However, in contrast to an ordinary hydroelectric plant, the pumped-storage plant can be operated in reverse. The electric generators then act as electric motors which drive the turbines in reverse, and thereby pump water from the lower reservoir into the upper reservoir. At peak-demand times the hydroelectric storage plant is used for the generation of electric energy—it converts the gravitational potential energy of the water into electric energy. At low-demand times, the hydroelectric storage plant is used to absorb electric energy—it converts surplus electric energy into gravitational potential energy of the water. This gravitational potential energy can then be held in storage until needed.

The chapter photo shows the reservoirs of a large hydroelectric pumped-storage plant on Brown Mountain in New York State. The upper reservoir on top of the mountain is linked to the lower reservoir at the base by a vertical shaft of more than 320 m bored through the mountain. Each of the four reversible pump/turbines (see Fig. 8.17) and motor/generators in the powerhouse at the base (see Fig. 1) is capable of generating 260 MW of electric power. The upper reservoir holds $1.9 \times 10^7\ m^3$ of water, which is enough to run the generators at full power for about half a day.

FIGURE 1 Powerhouse at the lower reservoir of the Brown Mountain hydroelectric pumped-storage plant.

Concepts — in — Context

EXAMPLE 3 At the Brown Mountain hydroelectric storage plant, water from the upper reservoir flows down a pipe in a long vertical shaft (Fig. 8.6). The pipe ends 330 m below the water level of the (full) upper reservoir. Calculate the speed with which the water emerges from the bottom of the pipe. Consider two cases: (a) the bottom of the pipe is wide open, so the pipe does not impede the downward motion of the water; and (b) the bottom of the pipe is closed except for a small hole through which water spurts out. Ignore frictional losses in the motion of the water.

SOLUTION: (a) If the pipe is wide open at the bottom, any parcel of water simply falls freely along the full length of the pipe. Thus, the pipe plays no role at all in the motion of the water, and the speed attained by the water is the same as for a reservoir suspended in midair with water spilling out and falling freely through a height $h = 330$ m. For such free-fall motion, the final speed v can be obtained either from the equations for uniformly accelerated motion [from Eq. (2.29)] or from energy conservation [see Eq. (7.41)]. The result is

$$v = \sqrt{2gh} = \sqrt{2 \times 9.81\ \text{m/s}^2 \times 330\ \text{m}} = 80\ \text{m/s}$$

(b) For a closed pipe with a small hole, the motion of a parcel of water from the top of the upper reservoir to the hole at the bottom of the pipe is complicated and unknown. However, we can find the final speed of the water by relying on the law of energy conservation as applied to the system consisting of the entire volume of water in the reservoir and the pipe. For this purpose, we must examine the kinetic and the potential energy of the water. The water spurting out at the bottom has a large kinetic energy but a low potential energy. In contrast, the water at the top of the upper reservoir has a high potential energy, but next to no kinetic energy (while the water spurts out at the bottom, the water level in the reservoir gradually decreases; but the speed of this downward motion of the water level is very small if the reservoir is large, and this speed can be ignored compared with the large speed of the spurting water).

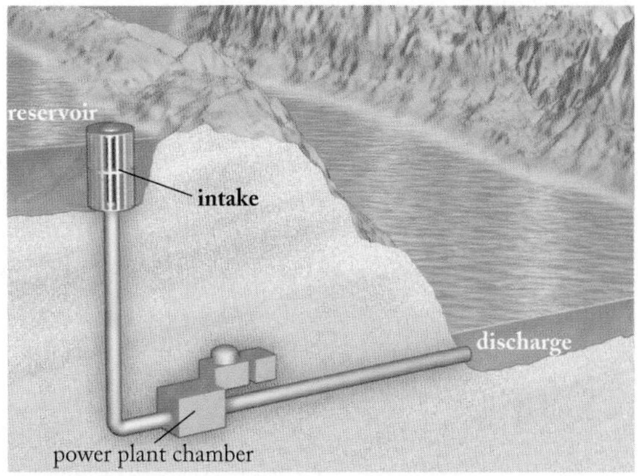

FIGURE 8.6 Cross-sectional view of hydroelectric pumped-storage power plant.

Consider, then, the energy changes that occur when a mass m of water, say, 1 kg of water, spurts out at the bottom of the pipe while, simultaneously, the water level of the upper reservoir decreases slightly. As concerns the energy balance, this effectively amounts to the removal of the potential energy of 1 kg from the top of the reservoir and the addition of the kinetic energy of 1 kg at the bottom of the pipe. All the water at intermediate locations, in the pipe and the reservoir, has the same energy it had before. Thus, energy conservation demands that the kinetic energy of the mass m of water emerging at the bottom be equal to the potential energy of a mass m at the top:

$$\tfrac{1}{2}mv^2 = mgh$$

This again gives

$$v = \sqrt{2gh} = 80\ \text{m/s}$$

that is, the same result as in part (a).

COMMENT: Note that the way the water acquires the final speed of 80 m/s in the cases (a) and (b) is quite different. In case (a), the water accelerates down the pipe with the uniform free-fall acceleration g. In case (b), the water flows down the pipe at a slow and nearly constant speed, and accelerates (strongly) only at the last moment, as it approaches the hole at the bottom. However, energy conservation demands that the result for the final speed of the emerging water be the same in both cases.

✓ Checkup 8.1

QUESTION 1: The potential energy corresponding to the spring force $F = -kx$ is $U = \frac{1}{2}kx^2$. Suppose that some new kind of force has a potential energy $U = -\frac{1}{2}kx^2$. How does this new kind of force differ from the spring force?

QUESTION 2: A particle moves along the positive x axis under the influence of a conservative force. Suppose that the potential energy of this force is as shown in Fig. 8.5a. Is the force directed along the positive x direction or the negative x direction?

QUESTION 3: Suppose that the force acting on a particle is given by the function $F_x = ax^3 + bx^2$, where a and b are constants. How do we know that the work done by this force during a round trip from, say, $x = 1$ back to $x = 1$ is zero?

QUESTION 4: Is the equation $W = U_1 - U_2$ valid for the work done by every kind of force? Is the equation $W = K_2 - K_1$ valid for the work done by each individual force acting on a particle?

(A) Yes; yes (B) Yes; no (C) No; yes (D) No; no

8.2 THE CURVE OF POTENTIAL ENERGY

If a particle of some given energy is moving in one dimension under the influence of a conservative force, then Eq. (8.17) permits us to calculate the speed of the particle as a function of position. Suppose that the potential energy is some known function $U = U(x)$; then Eq. (8.17) states

$$E = \tfrac{1}{2}mv^2 + U(x) \tag{8.22}$$

or, rearranging,

$$v^2 = \frac{2}{m}[E - U(x)] \tag{8.23}$$

Since the left side of this equation is never negative, we can immediately conclude that the particle must always remain within a range of values of x for which $U(x) \le E$. If $U(x)$ is increasing and the particle reaches a point at which $U(x) = E$, then $v = 0$; that is, the particle will stop at this point, and its motion will reverse. Such a point is called a **turning point** of the motion.

According to Eq. (8.23), v^2 is directly proportional to $E - U(x)$; thus, v^2 is large wherever the difference between E and $U(x)$ is large. We can therefore gain some insights into the qualitative features of the motion by drawing a graph of potential energy as a function of x on which it is possible to display the difference between E and $U(x)$. Such a graph of $U(x)$ vs. x is called the **curve of potential energy**. For example, Fig. 8.7 shows the curve of potential energy for an atom in a diatomic molecule. Treating the atom as a particle, we can indicate the value of the energy of the particle by a horizontal line in the graph (the red line in Fig. 8.7). We call this horizontal line the **energy level** of the particle. At any point x, we can then see the difference between E and $U(x)$ at a glance; according to Eq. (8.23), this tells us v^2. For instance, suppose that a particle has an energy $E = E_1$. Figure 8.7 shows this energy level. The particle has maximum speed at the point $x = x_0$, where the separation between the energy level and the potential-energy curve is maximum. The speed gradually decreases as the particle moves, say, toward the right. The potential-energy curve intersects the energy level at $x = a$; at this point the speed of the particle will reach zero, so this point is a turning point of the motion. The particle then moves toward the left, again attaining the same greatest speed at $x = x_0$. The speed gradually decreases as the particle continues to move toward the left, and the speed reaches zero at $x = a'$, the second turning point of the motion. Here the particle begins to move toward the right, and so on. Thus the particle continues to move back and forth between the two turning points—the particle is confined between the two turning points. The regions $x > a$ and $x < a'$ are forbidden regions;

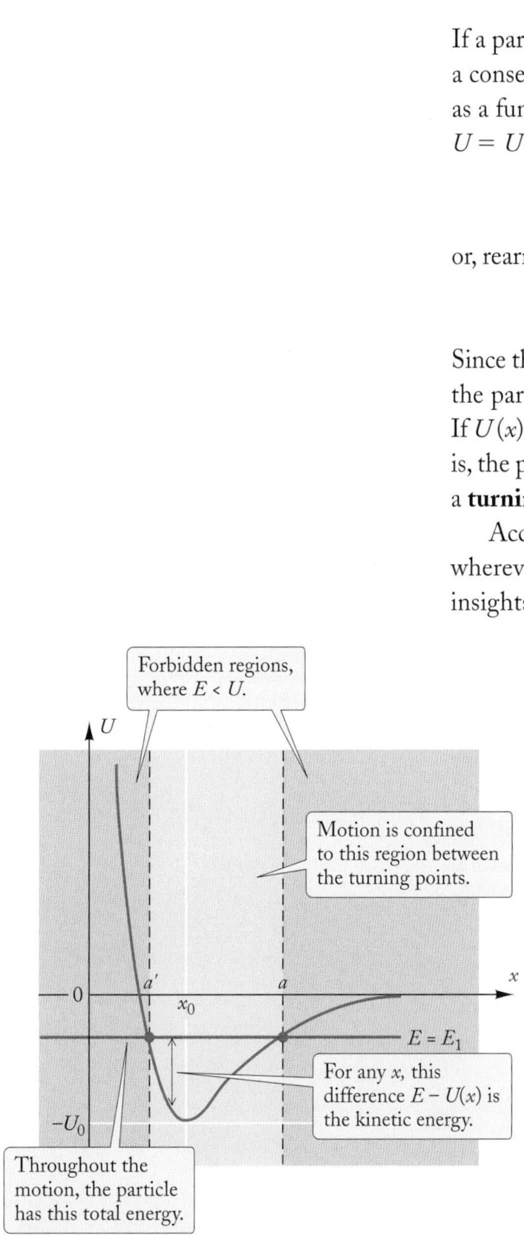

FIGURE 8.7 Potential-energy curve for an atom in a diatomic molecule. The horizontal line (red) is the energy level. The turning points are at $x = a$ and at $x = a'$.

only the region $a' \leq x \leq a$ is permitted. The particle is said to be in a **bound orbit.** The motion is periodic, that is, repeats again and again whenever the particle returns to its starting point.

The location of the turning points depends on the energy. For a particle with a lower energy level, the turning points are closer together. The lowest possible energy level intersects the potential-energy curve at its minimum (see $E = -U_0$ in Fig. 8.8); the two turning points then merge into the single point $x = x_0$. A particle with this lowest possible energy cannot move at all—it remains stationary at $x = x_0$. Note that the potential-energy curve has zero slope at $x = x_0$; this corresponds to zero force, $F_x = -dU/dx = 0$. A point such as $x = x_0$, where the force is zero, is called an **equilibrium point**. The point $x = x_0$ in Fig. 8.8 is a **stable** equilibrium point, since, after a small displacement, the force pushes the particle back toward that point. In contrast, at an **unstable** equilibrium point, after a small displacement, the force pushes the particle away from the point (see the point x_1 for the potential-energy curve shown in Fig. 8.9); and at a **neutral** equilibrium point no force acts nearby (see the point x_2 in Fig. 8.9). Equivalently, since the force is zero at an equilibrium point, the stable, unstable, and neutral equilibrium points correspond to negative, positive, or zero changes in the force with increasing x, that is, to negative, positive, or zero values of dF_x/dx. But $dF_x/dx = -d^2U/dx^2$, so the stable and unstable equilibrium points respectively correspond to positive and negative second derivatives of the function $U(x)$; in the former case the plot of $U(x)$ curves upward, and in the latter, downward (see Fig. 8.9).

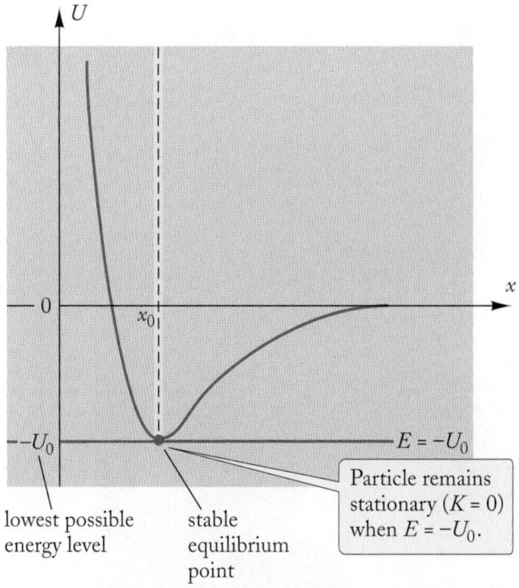

FIGURE 8.8 The energy level (red) coincides with the minimum of the potential-energy curve.

In Fig. 8.7, the right side of the potential-energy curve never rises above $U = 0$. Consequently, if the energy level is above this value (for instance, $E = E_2$; see Fig. 8.10), then there is only one single turning point on the left, and no turning point on the right. A particle with energy E_2 will continue to move toward the right forever; it is not confined. Such a particle is said to be in an **unbound orbit**.

The above qualitative analysis based on the curve of potential energy cannot tell us the details of the motion such as, say, the travel time from one point to another. But the qualitative analysis is useful because it gives us a quick survey of the types of motion that are possible for different values of the energy.

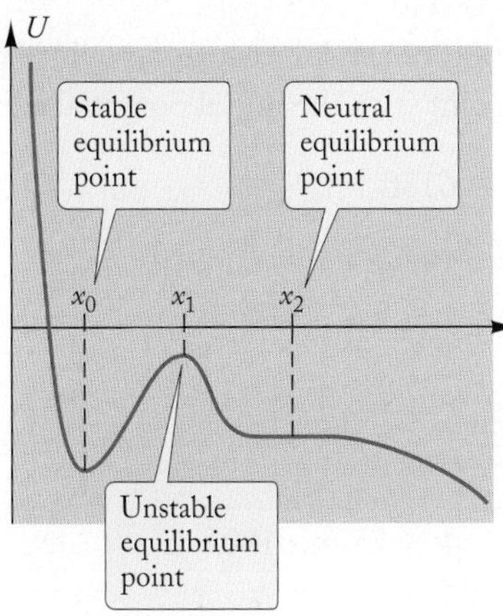

FIGURE 8.9 Types of equilibrium points. At the stable, unstable, and neutral equilibrium points, respectively, the potential-energy curve has a minimum, has a maximum, or is flat.

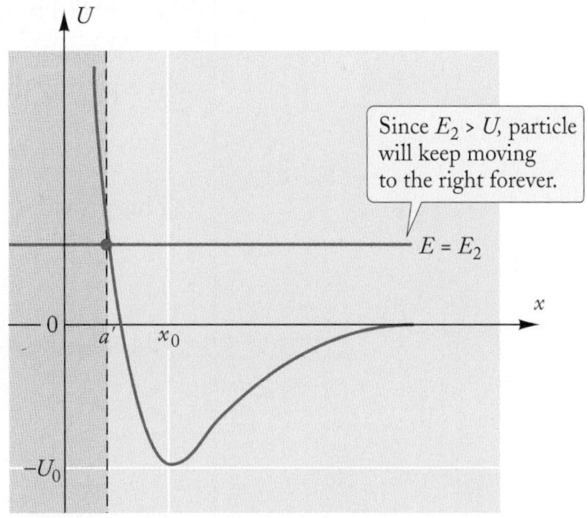

FIGURE 8.10 The energy level (red) is above the maximum height the potential-energy curve attains at its right. There is only one turning point, at $x = a'$.

FIGURE 8.11 Bungee jumping.

EXAMPLE 4 Some fanatics, in search of dangerous thrills, jump off high bridges or towers with bungee cords (long rubber cords) tied to their ankles (Fig. 8.11). Consider a jumper of mass 70 kg, with a 9.0-m cord tied to his ankles. When stretched, this cord may be treated as a spring, of spring constant 150 N/m. Plot the potential-energy curve for the jumper, and from this curve estimate the turning point of the motion, that is, the point at which the stretched cord stops the downward motion of the jumper.

SOLUTION: It is convenient to arrange the x axis vertically upward, with the origin at the point where the rubber cord becomes taut, that is, 9.0 m below the jump-off point (see Fig. 8.12a). The potential-energy function then consists of two pieces. For $x > 0$, the rubber cord is slack, and the potential energy is purely gravitational:

$$U = mgx \qquad \text{for } x > 0$$

For $x < 0$, the rubber cord is stretched, and the potential energy is a sum of gravitational and elastic potential energies:

$$U = mgx + \tfrac{1}{2}kx^2 \quad \text{for } x < 0$$

With the numbers specified for this problem,

$$\begin{aligned} U &= 70\text{ kg} \times 9.81\text{ m/s}^2 \times x \\ &= 687x \qquad \text{for } x > 0 \end{aligned} \tag{8.24}$$

and

$$\begin{aligned} U &= 70\text{ kg} \times 9.81\text{ m/s}^2 \times x + \tfrac{1}{2} \times 150\text{ N/m} \times x^2 \\ &= 687x + 75x^2 \qquad \text{for } x < 0 \end{aligned} \tag{8.25}$$

where x is in meters and U in joules. Figure 8.12b gives the plot of the curve of potential energy, according to Eqs. (8.24) and (8.25).

At the jump-off point $x = +9.0$ m, the potential energy is $U = 687x = 687 \times 9.0\text{ J} = 6180\text{ J}$. The red line in Fig. 8.12b indicates this energy level. The left intersection of the red line with the curve indicates the turning point at the lower end of the motion. By inspection of the plot, we see that this turning point is at $x \approx -15$ m. Thus, the jumper falls a total distance of 9.0 m + 15 m = 24 m before his downward motion is arrested.

We can accurately calculate the position of the lower turning point ($x < 0$) by equating the potential energy at that point with the initial potential energy:

$$687x + 75x^2 = 6180\text{ J}$$

This provides a quadratic equation of the form $ax^2 + bx + c = 0$:

$$75x^2 + 687x - 6180 = 0$$

This has the standard solution $x = (-b \pm \sqrt{b^2 - 4ac})/2a$, or

$$\begin{aligned} x &= \frac{-687 \pm \sqrt{(687)^2 + 4 \times 75 \times 6180}}{2 \times 75} \\ &= -14.7\text{ m} \approx -15\text{ m} \end{aligned}$$

in agreement with our graphical result. Here we have chosen the negative solution, since we are solving for x at the lower turning point using the form (8.25), which is valid only for $x < 0$.

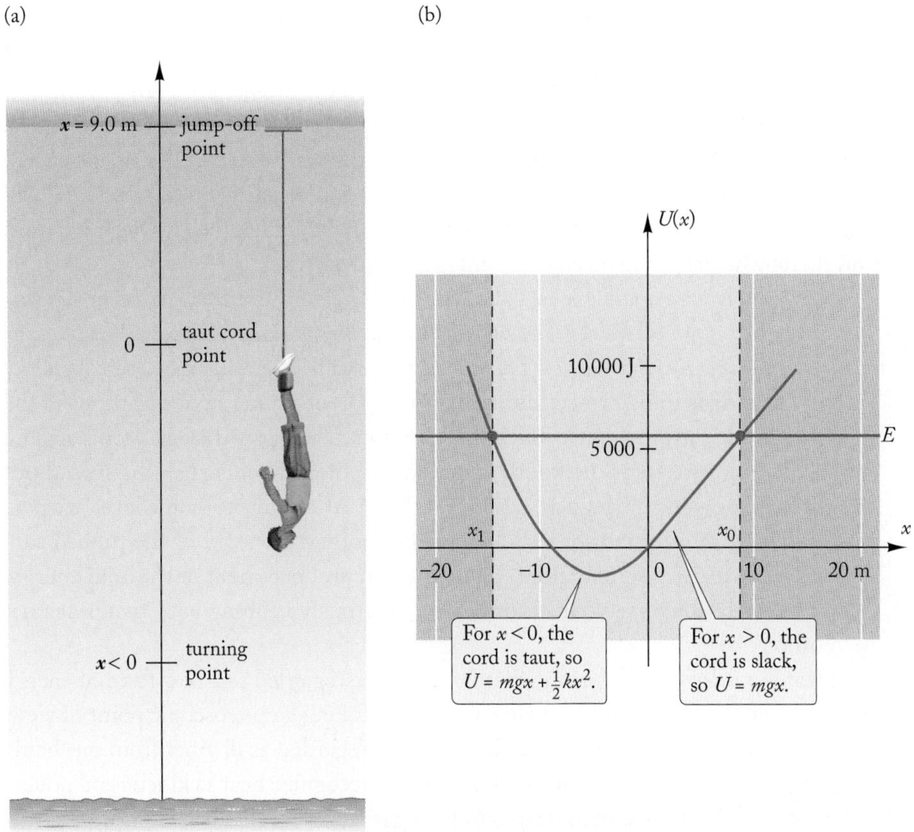

FIGURE 8.12 (a) The origin for the x coordinate is at the point where the rubber cord becomes taut. The jump-off point is at $x = 9.0$ m, and the turning point is at some negative value of x. (b) Curve of potential energy for the bungee jumper. The red line indicates the energy level. This line intersects the curve at approximately $x = -15$ m. This is the turning point for the jumper.

COMMENTS: If there were no friction, the motion would reverse, and the jumper would ascend to the bridge and bang against it. However, like a bouncing ball, the rubber cord has some energy loss due to friction within the material, and the jumper will not bounce back as high as the starting point.

Bungee jumping is a dangerous stunt. The human body has poor tolerance to deceleration in the head-down position. The pooling of blood in the head can lead to loss of consciousness ("redout"), rupture of blood vessels, eye damage, and temporary blindness. And in several instances, jumpers were killed by smashing their heads into the ground or by becoming entangled in their cords during the fall.

Checkup 8.2

QUESTION 1: A particle moving in one dimension under the influence of a given conservative force has either no turning point, one turning point, or two turning points, depending on the energy. Does the number of turning points increase or decrease with the energy? Is there any conceivable value of the energy that will result in three turning points?

QUESTION 2: By examining the curve of potential energy in Fig. 8.12, estimate at what points the bungee jumper attains his maximum downward speed and his maximum acceleration.

QUESTION 3: A particle moving under the influence of the spring force has a positive energy $E = 50$ J. How many turning points are there for this particle?

(A) 1 (B) 2 (C) 3 (D) 0

Online Concept 10 Tutorial

8.3 OTHER FORMS OF ENERGY

If the forces acting on a particle are conservative, then the mechanical energy of the particle is conserved. But if some of the forces acting on the particle are not conservative, then the mechanical energy of the particle—consisting of the sum of the kinetic energy and the net potential energy of all the conservative forces acting on the particle—will not remain constant. For instance, if friction forces are acting, they do negative work and thereby decrease the mechanical energy of the particle.

However, it is a remarkable fact about our physical universe that *whenever mechanical energy is lost by a particle or some other body, this energy never disappears—it is merely changed into other forms of energy.* Thus, in the case of friction, the mechanical energy lost by the body is transformed into kinetic and potential energy of the atoms in the body and in the surface against which it is rubbing. The energy that the atoms acquire in the rubbing process is disorderly kinetic and potential energy—it is spread out among the atoms in an irregular, random fashion. At the macroscopic level, we perceive the increase of the disorderly kinetic and potential energy of the rubbed surfaces as an increase of temperature. Thus, friction produces **heat** or **thermal energy.** (You can easily convince yourself of this by vigorously rubbing your hands against each other.)

Heat is a form of energy, but whether it is to be regarded as a new form of energy or not depends on what point of view we adopt. Taking a macroscopic point of view, we ignore the atomic motions; then heat is to be regarded as distinct from mechanical energy. Taking a microscopic point of view, we recognize heat as kinetic and potential energy of the atoms; then heat is to be regarded as mechanical energy. (We will further discuss heat in Chapter 20.)

Chemical energy and nuclear energy are two other forms of energy. The former is kinetic and potential energy of the electrons within the atoms; the latter is kinetic and potential energy of the protons and neutrons within the nuclei of atoms. As in the case of heat, whether these are to be regarded as new forms of energy depends on the point of view.

Electric and magnetic energy are forms of energy associated with electric charges and with light and radio waves. (We will examine these forms of energy in Chapters 25 and 31.)

Table 8.1 lists some examples of different forms of energy. All the energies in Table 8.1 are expressed in joules, the SI unit of energy. However, for reasons of tradition and convenience, some other energy units are often used in specialized areas of physics and engineering.

The energy of atomic and subatomic particles is usually measured in **electron-volts** (eV), where

$$1 \text{ electron-volt} = 1 \text{ eV} = 1.60 \times 10^{-19} \text{ J} \tag{8.26}$$

Electrons in atoms typically have kinetic and potential energies of a few eV.

The energy supplied by electric power plants is usually measured in **kilowatt-hours** (kW·h), where

$$1 \text{ kilowatt-hour} = 1 \text{ kW·h} = 3.60 \times 10^{6} \text{ J} \tag{8.27}$$

The electric energy used by appliances such as vacuum cleaners, hair dryers, or toasters during one hour of operation is typically 1 kilowatt-hour.

And the thermal energy supplied by the combustion of fuels is often expressed in **kilocalories** (kcal):

HERMANN VON HELMHOLTZ (1821–1894) *Prussian surgeon, biologist, mathematician, and physicist. His scientific contributions ranged from the invention of the ophthalmoscope and studies of the physiology and physics of vision and hearing to the measurement of the speed of light and studies in theoretical mechanics. Helmholtz formulated the general Law of Conservation of Energy, treating it as a consequence of the basic laws of mechanics and electricity.*

TABLE 8.1 SOME ENERGIES

Nuclear fuel in the Sun	1×10^{45} J
Explosion of a supernova	1×10^{44} J
Fossil fuel available on Earth	2.0×10^{23} J
Yearly energy expenditure of the United States (a)	8×10^{19} J
Volcanic explosion (Krakatoa)	6×10^{18} J
Annihilation of 1 kg of matter–antimatter	9.0×10^{16} J
Explosion of thermonuclear bomb (1 megaton)	4.2×10^{15} J
Gravitational potential energy of airliner (Boeing 747 at 10 000 m)	2×10^{10} J
Combustion of 1 gal of gasoline (b)	1.3×10^{8} J
Daily food intake of man (3000 kcal)	1.3×10^{7} J
Explosion of 1 kg of TNT	4.6×10^{6} J
Metabolization of one apple (110 kcal)	4.6×10^{5} J
One push-up (c)	3×10^{2} J
Fission of one uranium nucleus	3.2×10^{-11} J
Energy of ionization of hydrogen atom	2.2×10^{-18} J

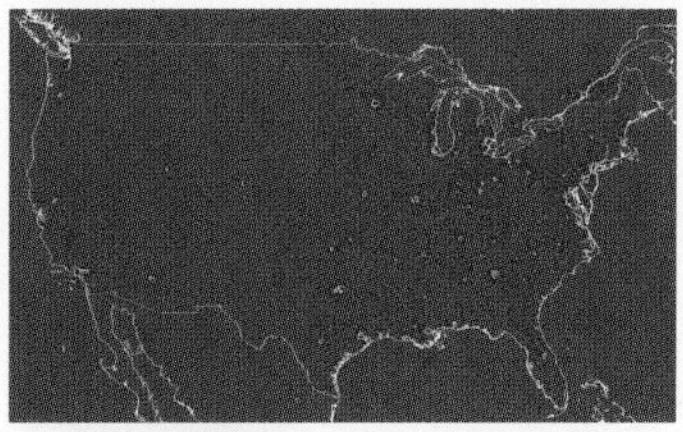
(a)

(b)

(c)

$$1 \text{ kilocalorie} = 1 \text{ kcal} = 4.187 \times 10^{3} \text{ J} \tag{8.28}$$

or in **British thermal units** (Btu):

$$1 \text{ Btu} = 1.055 \times 10^{3} \text{ J} \tag{8.29}$$

We will learn more about these units in later chapters.

All these forms of energy can be transformed into one another. For example, in an internal combustion engine, chemical energy of the fuel is transformed into heat and kinetic energy; in a hydroelectric power station, gravitational potential energy of the water is transformed into electric energy; in a nuclear reactor, nuclear energy is transformed into heat, light, kinetic energy, etc. However, in any such transformation process, the sum of all the energies of all the pieces of matter involved in the process remains constant: *the form of the energy changes, but the total amount of energy does not change.* This is the general **Law of Conservation of Energy**.

law of conservation of energy

EXAMPLE 5 At the Brown Mountain hydroelectric pumped-storage plant, the average height of the water in the upper reservoir is 320 m above the lower reservoir, and the upper reservoir holds 1.9×10^{7} m^3 of water. Expressed in kW·h, what is the gravitational potential energy available for conversion into electric energy?

Concepts in Context

SOLUTION: A cubic meter of water has a mass of 1000 kg. Hence the total mass of water is 1.9×10^{10} kg, and the gravitational potential energy is

$$U = mgh = 1.9 \times 10^{10} \text{ kg} \times 9.81 \text{ m/s}^2 \times 320 \text{ m} = 6.0 \times 10^{13} \text{ J}$$

Expressed in kW·h , this amounts to

$$6.0 \times 10^{13}\ \text{J} \times \frac{1\ \text{kW·h}}{3.6 \times 10^{6}\ \text{J}} = 1.7 \times 10^{7}\ \text{kW·h}$$

(The actual electric energy that can be generated is about 30% less than that, because of frictional losses during the conversion from one form of energy to the other. These frictional losses result in the generation of heat.)

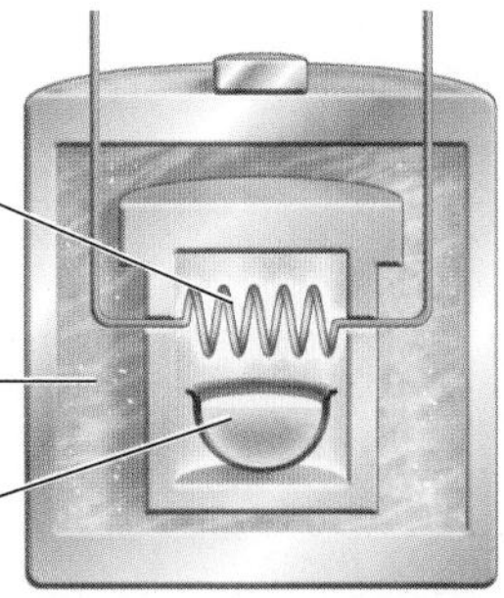

FIGURE 8.13 A bomb calorimeter. The sample is ignited electrically, by a glowing wire.

EXAMPLE 6 The "calorie" used by dietitians to express the energy equivalents of different foods is actually a kilocalorie, or a "large" calorie. To measure the energy equivalent of some kind of food—for instance, sugar—a sample is placed in a bomb calorimeter, a closed vessel filled with oxygen at high pressure (see Fig. 8.13). The sample is ignited and burned completely (complete oxidation). The number of calories released in this chemical reaction—for instance, 4.1 kcal for 1.0 g of sugar—tells us the maximum amount of energy that can be extracted from this food. The human body does not necessarily "burn" food quite as completely, and the muscles do not convert all of the available chemical energy into mechanical energy. However, energy conservation tells us that from one gram of sugar the body cannot produce more than 4.1 kcal of mechanical work.

If you eat one spoonful (4.0 g) of sugar, what is the maximum height to which this permits you to climb stairs? Assume your mass is 70 kg.

SOLUTION: Since 1.0 g of sugar releases 4.1 kcal of energy, the energy equivalent of 4.0 g of sugar is

$$4.0 \times 4.1\ \text{kcal} = 16.4\ \text{kcal} = 16.4\ \text{kcal} \times 4.18 \times 10^{3}\ \text{J/kcal} = 6.9 \times 10^{4}\ \text{J}$$

When you climb the stairs to a height y, this energy becomes gravitational potential energy:

$$mgy = 6.9 \times 10^{4}\ \text{J}$$

from which

$$y = \frac{6.9 \times 10^{4}\ \text{J}}{mg} = \frac{6.9 \times 10^{4}\ \text{J}}{70\ \text{kg} \times 9.81\ \text{m/s}^2} = 100\ \text{m}$$

In practice, because of the limited efficiency of your body, only about 20% of the chemical energy of food is converted into mechanical energy; thus, the actual height you can climb is only about 20 m. (Because of the strong musculature of the human leg, stair climbing is one of your most efficient activities; other physical activities are considerably less efficient in converting chemical energy into mechanical energy.)

✓ Checkup 8.3

QUESTION 1: A parachutist descends at uniform speed. Is the mechanical energy conserved? What happens to the lost mechanical energy?

QUESTION 2: You fire a bullet from a rifle. The increase of kinetic energy of the bullet upon firing must be accompanied by a decrease of some other kind of energy. What energy decreases?

QUESTION 3: A truck travels at constant speed down a road leading from a mountain peak to a valley. What happens to the gravitational potential energy of the truck? How is it dissipated?

QUESTION 4: When you apply the brakes and stop a moving automobile, what happens to the kinetic energy?

(A) Kinetic energy is converted to gravitational potential energy.
(B) Kinetic energy is converted to elastic potential energy.
(C) Kinetic energy is converted to heat due to frictional forces.
(D) Kinetic energy is converted to chemical energy.

8.4 MASS AND ENERGY

One of the great discoveries made by Albert Einstein early in the twentieth century is that energy can be transformed into mass, and mass can be transformed into energy. Thus, *mass is a form of energy*. The amount of energy contained in an amount m of mass is given by Einstein's famous formula

$$E = mc^2 \tag{8.30}$$

energy–mass relation

where c is the speed of light, $c = 3.00 \times 10^8$ m/s. This formula is a consequence of Einstein's relativistic physics. It cannot be obtained from Newton's physics, and its theoretical justification will have to wait until we study the theory of relativity in Chapter 36.

The most spectacular demonstration of Einstein's mass–energy formula is found in the annihilation of matter and antimatter (as we will see in Chapter 41, particles of antimatter are similar to the particles of ordinary matter, except that they have opposite electric charge). If a proton collides with an antiproton, or an electron with an antielectron, the two colliding particles react violently, and they annihilate each other in an explosion that generates an intense flash of very energetic light. According to Eq. (8.30), the annihilation of just 1000 kg of matter and antimatter (500 kg of each) would release an amount of energy

$$E = mc^2 = 1000 \text{ kg} \times (3.00 \times 10^8 \text{ m/s})^2 = 9.0 \times 10^{19} \text{ J} \tag{8.31}$$

This is enough energy to satisfy the requirements of the United States for a full year. Unfortunately, antimatter is not readily available in large amounts. On Earth, antiparticles can be obtained only from reactions induced by the impact of beams of high-energy particles on a target. These collisions occasionally result in the creation of a particle–antiparticle pair. Such pair creation is the reverse of pair annihilation. The creation process transforms some of the kinetic energy of the collision into mass, and a subsequent annihilation merely gives back the original energy.

But the relationship between energy and mass in Eq. (8.30) also has another aspect. *Energy has mass*. Whenever the energy of a body is changed, its mass (and weight) are changed. The change in mass that accompanies a given change of energy is

mass–energy relation

$$\Delta m = \frac{\Delta E}{c^2} \tag{8.32}$$

For instance, if the kinetic energy of a body increases, its mass (and weight) increase. At speeds small compared with the speed of light, the mass increment is not noticeable. But when a body approaches the speed of light, the mass increase becomes very large. The high-energy electrons produced at the Stanford Linear Accelerator provide an extreme example of this effect: these electrons have a speed of 99.999 999 97% of the speed of light, and their mass is 44 000 times the mass of electrons at rest!

The fact that energy has mass indicates that energy is a form of mass. Conversely, as we have seen above, mass is a form of energy. Hence mass and energy must be regarded as two aspects of the same thing. The laws of conservation of mass and conservation of energy are therefore not two independent laws—each implies the other. For example, consider the fission reaction of uranium inside the reactor vessel of a nuclear power plant. The complete fission of 1.0 kg of uranium yields an energy of 8.2×10^{13} J. The reaction conserves energy—it merely transforms nuclear energy into heat, light, and kinetic energy, but does not change the total amount of energy. The reaction also conserves mass—if the reactor vessel is hermetically sealed and thermally insulated from its environment, then the reaction does not change the mass of the contents of the vessel. However, if we open the vessel during or after the reaction and let some of the heat and light escape, then the mass of the residue will not match the mass of the original amount of uranium. The mass of the residues will be about 0.1% smaller than the original mass of the uranium. This mass defect represents the mass carried away by the energy that escapes. Thus, the nuclear fission reactions merely transform energy into new forms of energy and mass into new forms of mass. In this regard, a nuclear reaction is not fundamentally different from a chemical reaction. The mass of the residues in a chemical reaction that releases heat (exothermic reaction) is slightly less than the original mass. The heat released in such a chemical reaction carries away some mass, but, in contrast to a nuclear reaction, this amount of mass is so small as to be quite immeasurable.

EXAMPLE 7 As an example of the small mass loss in a chemical reaction, consider the binding energy of the electron in the hydrogen atom (one proton and one electron), which is 13.6 eV. What is the fractional mass loss when an electron is captured by a proton and the binding energy is allowed to escape?

SOLUTION: In joules, the binding energy is $13.6 \text{ eV} \times 1.60 \times 10^{-19}$ J/eV = 2.18×10^{-18} J. The mass loss corresponding to this binding energy is

$$\Delta m = \frac{\Delta E}{c^2} = \frac{2.18 \times 10^{-18} \text{ J}}{(3.00 \times 10^8 \text{ m/s})^2} = 2.42 \times 10^{-35} \text{ kg}$$

Since the mass of a proton and electron together is 1.67×10^{-27} kg (see Table 5.2), the fractional mass loss is

$$\frac{\Delta m}{m} = \frac{2.42 \times 10^{-35} \text{ kg}}{1.67 \times 10^{-27} \text{ kg}} = 1.45 \times 10^{-8}$$

This is about a millionth of one percent.

Checkup 8.4

QUESTION 1: The Sun radiates heat and light. Does the Sun consequently suffer a loss of mass?

QUESTION 2: In the annihilation of matter and antimatter, a particle and an antiparticle—such as a proton and an antiproton, or an electron and an antielectron—disappear explosively upon contact, giving rise to an intense flash of light. Is energy conserved in this reaction? Is mass conserved?

QUESTION 3: You heat a potful of water to the boiling point. If the pot is sealed so no water molecules can escape, then, compared with the cold water, the mass of the boiling water will:

(A) Increase (B) Decrease (C) Remain the same

8.5 POWER

Online Concept Tutorial 10

When we use an automobile engine to move a car up a hill or when we use an electric motor to lift an elevator cage, the important characteristic of the engine is not how much force it can exert, but rather how much work it can perform in a given amount of time. The force is only of secondary importance, because by shifting to a low gear we can make sure that even a "weak" engine exerts enough force on the wheels to propel the automobile uphill. But the work performed in a given amount of time, or the rate of work, is crucial, since it determines how fast the engine can propel the car up the hill. While the car moves uphill, the gravitational force takes energy from the car; that is, it performs negative work on the car. To keep the car moving, the engine must perform an equal amount of positive work. If the engine is able to perform this work at a fast rate, it can propel the car uphill at a fast speed.

The rate at which a force does work on a body is called the **power** *delivered by the force.* If the force does an amount of work W in an interval of time Δt, then the **average power** is the ratio of W and Δt:

$$\overline{P} = \frac{W}{\Delta t} \tag{8.33}$$

average power

The **instantaneous power** is defined by a procedure analogous to that involved in the definition of the instantaneous velocity. We consider the small amount of work dW done in the small interval of time dt and take the ratio of these small quantities:

$$P = \frac{dW}{dt} \tag{8.34}$$

instantaneous power

According to these definitions, the engine of your automobile delivers high power if it performs a large amount of work on the wheels (or, rather, the driveshaft) in a short time. The maximum power delivered by the engine determines the maximum speed of which this automobile is capable, since at high speed the automobile loses energy to air resistance at a prodigious rate, and this loss has to be made good by the engine. You might also expect that the power of the engine determines the maximum acceleration of which the automobile is capable. But the acceleration is determined

JAMES WATT (1736–1819) *Scottish inventor and engineer. He modified and improved an earlier steam engine and founded the first factory constructing steam engines. Watt introduced the* horsepower *as a unit of mechanical power.*

by the maximum force exerted by the engine on the wheels, and this is not directly related to the power as defined above.

The SI unit of power is the **watt** (W), which is the rate of work of one joule per second:

$$1 \text{ watt} = 1 \text{ W} = 1 \text{ J/s}$$

In engineering practice, power is often measured in **horsepower** (hp) units, where

$$1 \text{ horsepower} = 1 \text{ hp} = 746 \text{ W} \tag{8.35}$$

This is roughly the rate at which a (very strong) horse can do work.

Note that multiplication of a unit of power by a unit of time gives a unit of energy. An example of this is the kilowatt-hour (kW·h), already mentioned in Section 8.3:

$$\begin{aligned}1 \text{ kilowatt-hour} &= 1 \text{ kW·h} = 1 \text{ kW} \times 1 \text{ h} = 1000 \text{ W} \times 3600 \text{ s} \\ &= 3.6 \times 10^6 \text{ J}\end{aligned} \tag{8.36}$$

This unit is commonly used to measure the electric energy delivered to homes and factories.

For a constant (or average) power P delivered to a body during a time Δt, the work ΔW delivered is the rate times the time [see Eq. (8.33)]:

$$W = P\,\Delta t \tag{8.37}$$

If the rate of doing work P varies with time, then the total work W done between a time t_1 and another time t_2 is the sum of the infinitesimal $P\,\Delta t$ contributions; that is, the work done is the integral of the power over time:

$$W = \int dW = \int_{t_1}^{t_2} P\,dt \tag{8.38}$$

EXAMPLE 8 An elevator cage has a mass of 1000 kg. How many horsepower must the motor deliver to the elevator if it is to raise the elevator cage at the rate of 2.0 m/s? The elevator has no counterweight (see Fig. 8.14).

SOLUTION: The weight of the elevator is $w = mg = 1000 \text{ kg} \times 9.81 \text{ m/s}^2 \approx 9800$ N. By means of the elevator cable, the motor must exert an upward force equal to the weight to raise the elevator at a steady speed. If the elevator moves up a distance Δy, the work done by the force is

$$W = F\,\Delta y \tag{8.39}$$

To obtain the power, or the rate of work, we must divide this by the time interval Δt:

$$P = \frac{\Delta W}{\Delta t} = \frac{F\,\Delta y}{\Delta t} = F\frac{\Delta y}{\Delta t} = Fv \tag{8.40}$$

where $v = \Delta y/\Delta t$ is the speed of the elevator. With $F = 9800$ N and $v = 2.0$ m/s, we find

$$P = Fv = 9800 \text{ N} \times 2.0 \text{ m/s} = 2.0 \times 10^4 \text{ W}$$

Since 1 hp = 746 W [see Eq. (8.35)], this equals

$$P = 2.0 \times 10^4 \text{ W} \times \frac{1 \text{ hp}}{746 \text{ W}} = 27 \text{ hp}$$

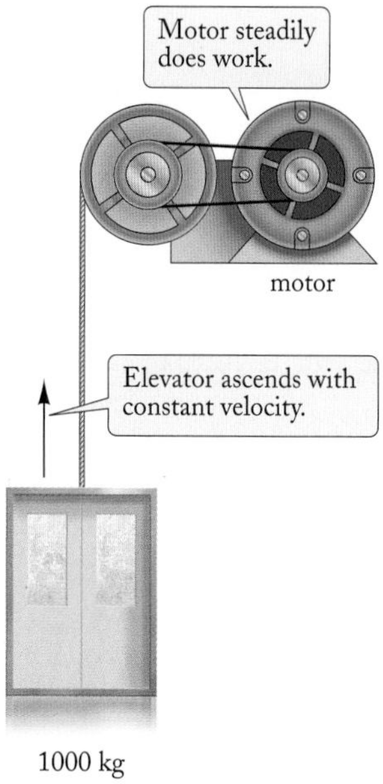

FIGURE 8.14 Elevator cage and motor.

Equation (8.40) is a special instance of a simple formula, which expresses the instantaneous power as the scalar product of force and velocity. To see this, consider that when a body suffers a small displacement $d\mathbf{s}$, the force $\mathbf{F}$ acting on the body will perform an amount of work

$$dW = \mathbf{F} \cdot d\mathbf{s} \tag{8.41}$$

or

$$dW = F\, ds \cos\theta$$

where θ is the angle between the direction of the force and the direction of the displacement (see Fig. 8.15). The instantaneous power delivered by this force is then

$$P = \frac{dW}{dt} = F\frac{ds}{dt}\cos\theta \tag{8.42}$$

Since ds/dt is the speed v, this expression for the power equals

$$P = Fv\cos\theta \tag{8.43}$$

or

$$P = \mathbf{F} \cdot \mathbf{v} \tag{8.44}$$

power delivered by a force

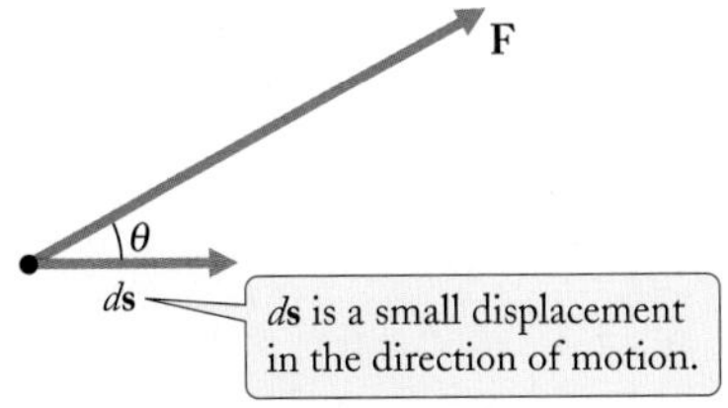

FIGURE 8.15 The force **F** makes an angle θ with the displacement $d\mathbf{s}$.

EXAMPLE 9 A horse pulls a sled up a steep snow-covered street of slope 1:7 (see Fig. 8.16a). The sled has a mass of 300 kg, and the coefficient of sliding friction between the sled and the snow is 0.12. If the horse pulls parallel to the surface of the street and delivers a power of 1.0 hp, what is the maximum (constant) speed with which the horse can pull the sled uphill? What fraction of the horse's power is expended against gravity? What fraction against friction?

SOLUTION: Figure 8.16b is a "free-body" diagram for the sled, showing the weight ($w = mg$), the normal force ($N = mg\cos\phi$), the friction force ($f_k = \mu_k N$), and the pull of the horse (T). With the x axis along the street and the y axis at right angles to the street, the components of these forces are

$$w_x = -mg\sin\phi \qquad w_y = -mg\cos\phi$$

$$N_x = 0 \qquad N_y = mg\cos\phi$$

$$f_{k,x} = -\mu_k mg\cos\phi \qquad f_{k,y} = 0$$

$$T_x = T \qquad T_y = 0$$

Since the acceleration along the street is zero (constant speed), the sum of the x components of these forces must be zero:

$$-mg\sin\phi + 0 - \mu_k mg\cos\phi + T = 0 \tag{8.45}$$

We can solve this equation for the pull of the horse:

$$T = mg\sin\phi + \mu_k mg\cos\phi \tag{8.46}$$

This simply says that the pull of the horse must balance the component of the weight along the street plus the friction force. The direction of this pull is parallel to the direction of motion of the sled. Hence, in Eq. (8.43), $\theta = 0$, and the power delivered by the horse is

$$P = Tv = (mg\sin\phi + \mu_k mg\cos\phi)v \tag{8.47}$$

Solving this equation for v, we find

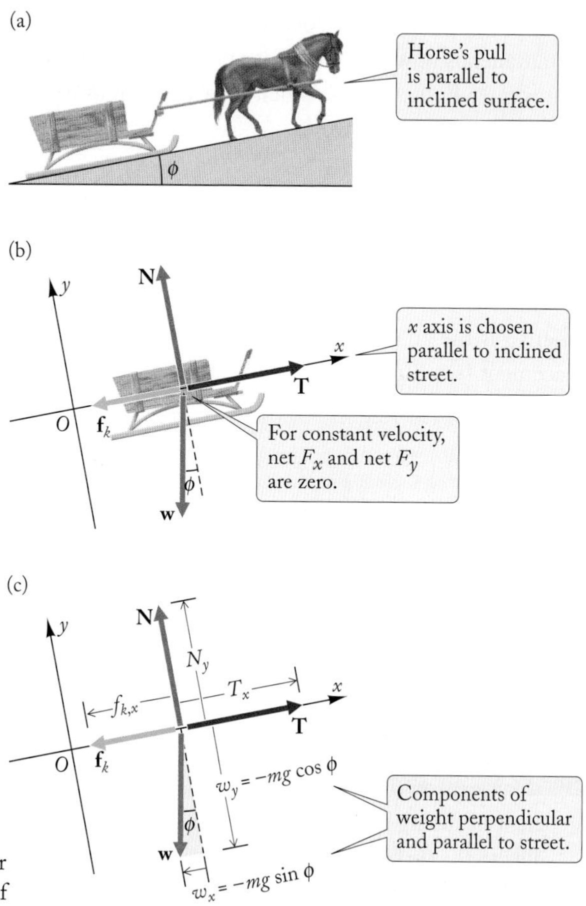

FIGURE 8.16 (a) Horse dragging a sled up a street. (b) "Free-body" diagram for the sled. (c) Components of the forces.

$$v = \frac{P}{mg\sin\phi + \mu_k mg\cos\phi} = \frac{P}{mg(\sin\phi + \mu_k\cos\phi)} \tag{8.48}$$

For a slope of 1:7, the tangent of the angle of inclination is $\tan\phi = 1/7$, and, using a calculator, the inverse tangent of 1/7 gives $\phi = 8.1°$. Hence

$$v = \frac{746\text{ W}}{300\text{ kg} \times 9.81\text{ m/s}^2 \times (\sin 8.1° + 0.12\cos 8.1°)}$$
$$= 0.98\text{ m/s}$$

The weight of the sled makes an angle of $90.0° + 8.1° = 98.1°$ with the direction of motion (see Fig. 8.16b). The power exerted by the weight of the sled is given by Eq. (8.43), with $F = mg$ and $\cos\theta = \cos 98.1°$:

$$P_{\text{weight}} = mgv\cos 98.1° = 300\text{ kg} \times 9.81\text{ m/s}^2 \times 0.98\text{ m/s} \times \cos 98.1°$$
$$= -406\text{ W} = -0.54\text{ hp}$$

Since the total power is 1.0 hp, this says that 54% of the horse's power is expended against gravity and, consequently, the remaining 46% against friction. The friction portion can also be calculated directly. The friction force acts opposite to the velocity ($\cos\theta = -1$), and so the power exerted is negative:

$$P_{\text{friction}} = -f_k v = -\mu_k mg\cos\phi\, v = -0.12 \times 300\text{ kg} \times 9.81\text{ m/s}^2 \times \cos 8.1° \times 0.98\text{ m/s}$$
$$= -343\text{ W} = -0.46\text{ hp}$$

The above equations all refer to *mechanical* power. In general, *power is the rate at which energy is transferred from one form of energy to another or the rate at which energy is transported from one place to another.* For instance, an automobile engine converts chemical energy of fuel into mechanical energy and thermal energy. A nuclear power plant converts nuclear energy into electric energy and thermal energy. And a high-voltage power line transports electric energy from one place to another. Table 8.2 gives some examples of different kinds of power.

TABLE 8.2 SOME POWERS

Light and heat emitted by the Sun	3.9×10^{26} W
Mechanical power generated by hurricane (a)	2×10^{13} W
Total power used in United States (average)	2×10^{12} W
Large electric power plant	$\approx 10^{9}$ W
Jet airliner engines (Boeing 747) (b)	2.1×10^{8} W
Automobile engine	1.5×10^{5} W
Solar light and heat per square meter at Earth	1.4×10^{3} W
Electricity used by toaster	1×10^{3} W
Work output of man (athlete at maximum)	2×10^{2} W
Electricity used by light bulb	1×10^{2} W
Basal metabolic rate for man (average)	88 W
Heat and work output of bumblebee (in flight) (c)	2×10^{-2} W
Atom radiating light	$\approx 10^{-10}$ W

(a)

(b)

(c)

In the previous example, part of the horse's work was converted into heat by the friction between the sled and the snow, and part was converted into gravitational potential energy. In the following example, gravitational potential energy is converted into electric energy.

EXAMPLE 10 Each of the four generators (Fig. 8.17) of the Brown Mountain hydroelectric plant generates 260 MW of electric power. When generating this power, at what rate does the power plant take water from the upper reservoir? How long does a full reservoir last? See the data in Example 5.

Concepts in Context

SOLUTION: We will assume that all of the potential energy of the water in the upper reservoir, at a height of 320 m, is converted into electric energy. The electric

FIGURE 8.17 Turbine generator at Brown Mountain hydroelectric plant, shown during installation.

power $P = 4 \times 260 \times 10^6$ W $= 1.0 \times 10^9$ W must then equal the negative of the rate of change of the potential energy (see Eq. 7.31):

$$P = -\frac{dU}{dt} = -\frac{dm}{dt}gh$$

from which we obtain the rate of change of mass,

$$\frac{dm}{dt} = -\frac{P}{gh} = -\frac{1.0 \times 10^9 \text{ W}}{9.81 \text{ m/s}^2 \times 320 \text{ m}} = -3.3 \times 10^5 \text{ kg/s}$$

Expressed as a volume of water, this amounts to an outflow of 330 m^3 per second. At this rate, the 1.9×10^7 m^3 of water in the reservoir will last for

$$\frac{1.9 \times 10^7 \text{m}^3}{330 \text{ m}^3\text{/s}} = 5.7 \times 10^4 \text{ s} = 16 \text{ h}$$

As mentioned in Example 5, there are also some frictional losses. As a result, the reservoir will actually be depleted about 30% faster than this, that is, in a bit less than half a day.

✔ Checkup 8.5

QUESTION 1: (a) You trot along a flat road carrying a backpack. Do you deliver power to the pack? (b) You trot uphill. Do you deliver power to the pack? (c) You trot downhill. Do you deliver power to the pack? Does the pack deliver power to you?

QUESTION 2: To reach a mountaintop, you have a choice between a short, steep road or a longer, less steep road. Apart from frictional losses, is the energy you have to expend in walking up these two roads the same? Why does the steeper road require more of an effort?

QUESTION 3: In order to keep a 26-m motor yacht moving at 88 km/h, its engines must supply about 5000 hp. What happens to this power?

QUESTION 4: Two cars are traveling up a sloping road, each at a constant speed. The second car has twice the mass and twice the speed of the first car. What is the ratio of the power delivered by the second car engine to that delivered by the first? Ignore friction and other losses.

(A) 1 (B) 2 (C) 4 (D) 8 (E) 16

SUMMARY

PROBLEM-SOLVING TECHNIQUES Energy Conservation (page 238)

PHYSICS IN PRACTICE Hydroelectric Pumped Storage (page 242)

CONSERVATIVE FORCE The work done by the force is zero for any round trip.

WORK DONE BY A CONSERVATIVE FORCE $W = -U_2 + U_1 = -\Delta U$ (8.2)

POTENTIAL ENERGY OF A SPRING $U = \frac{1}{2}kx^2$ (8.6)

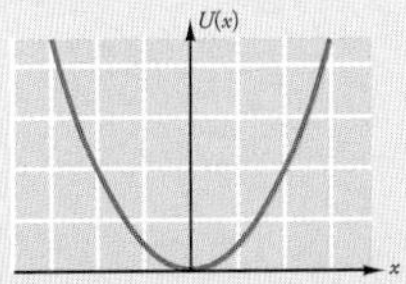

CONTRIBUTIONS TO THE MECHANICAL ENERGY

- Kinetic energy: $K = \frac{1}{2}mv^2$ (for motion)
- Gravitational potential energy: $U = mgy$ (near Earth's surface)
- Elastic potential energy: $U = \frac{1}{2}kx^2$ (for a spring)

POTENTIAL ENERGY AS INTEGRAL OF FORCE $U(x) = -\int_{x_0}^{x} F_x(x')\,dx'$ (8.14)

POTENTIAL OF INVERSE-SQUARE FORCE If $F(x) = \frac{A}{x^2}$ then $U(x) = \frac{A}{x}$ (8.19)

(for $x > 0$; attractive for $A < 0$, repulsive for $A > 0$.)

FORCE AS DERIVATIVE OF POTENTIAL ENERGY $F_x = -\frac{dU}{dx}$ (8.21)

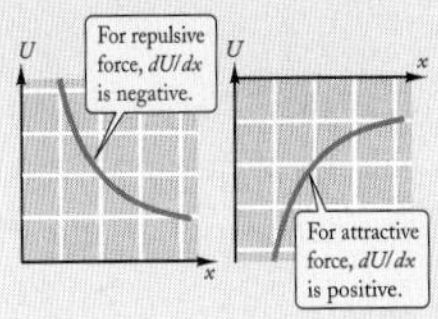

CONSERVATION OF MECHANICAL ENERGY $E = \frac{1}{2}mv^2 + U = [\text{constant}]$ (8.17)

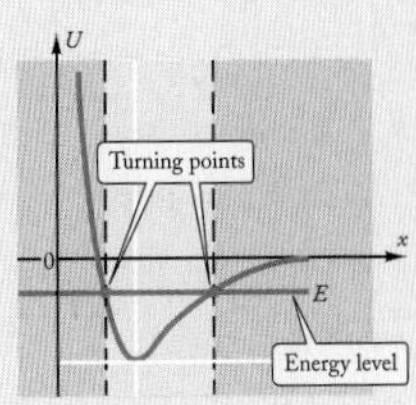

MASS IS A FORM OF ENERGY $E = mc^2$ (8.30)

ENERGY HAS MASS $\Delta m = \frac{\Delta E}{c^2}$ (8.32)

SI UNIT OF POWER	1 watt = 1 W = 1 J/s	
AVERAGE POWER	$\overline{P} = \dfrac{\Delta W}{\Delta t}$	(8.33)
INSTANTANEOUS POWER	$P = \dfrac{dW}{dt}$	(8.34)
MECHANICAL POWER DELIVERED BY A FORCE	$P = \mathbf{F}\cdot\mathbf{v} = Fv\cos\theta$	(8.43; 8.44)
WORK DONE AT CONSTANT POWER	$W = P\Delta t$	(8.37)
WORK DONE WITH TIME-DEPENDENT POWER	$W = \displaystyle\int_{t_1}^{t_2} P\,dt$	(8.38)

QUESTIONS FOR DISCUSSION

1. A body slides on a smooth horizontal plane. Is the normal force of the plane on the body a conservative force? Can we define a potential energy for this force according to the recipe in Section 8.1?
2. If you stretch a spring so far that it suffers a permanent deformation, is the force exerted by the spring during this operation conservative?
3. Is there any frictional dissipation of mechanical energy in the motion of the planets of the Solar System or in the motion of their satellites? (Hint: Consider the tides.)
4. What happens to the kinetic energy of an automobile during braking without skidding? With skidding?
5. An automobile travels down a road leading from a mountain peak to a valley. What happens to the gravitational potential energy of the automobile? How is it dissipated?
6. Suppose you wind up a watch and then place it into a beaker full of nitric acid and let it dissolve. What happens to the potential energy stored in the spring of the watch?
7. News reporters commonly speak of "energy consumption." Is it accurate to say that energy is *consumed*? Would it be more accurate to say that energy is *dissipated*?
8. The explosive yield of thermonuclear bombs (Fig. 8.18) is usually reported in kilotons or megatons of TNT. Would the explosion of a 1-megaton hydrogen bomb really produce the same effects as the explosion of 1 megaton of TNT (a mountain of TNT more than a hundred meters high)?
9. When you heat a potful of water, does its mass increase?
10. Since mass is a form of energy, why don't we measure mass in the same units as energy? How could we do this?
11. In order to travel at 130 km/h, an automobile of average size needs an engine delivering about 40 hp to overcome the effects of air friction, road friction, and internal friction (in the transmission and drive train). Why do most drivers think they need an engine of 150 or 200 hp?

FIGURE 8.18 A thermonuclear explosion.

PROBLEMS

8.1 Potential Energy of a Conservative Force†

1. The spring used in the front suspension of a Triumph sports car has a spring constant $k = 3.5 \times 10^4$ N/m. How far must we compress this spring to store a potential energy of 100 J?

2. A particle moves along the x axis under the influence of a variable force $F_x = 2x^3 + 1$ (where force is measured in newtons and distance in meters). Show that this force is conservative; that is, show that for any back-and-forth motion that starts and ends at the same place (round trip), the work done by the force is zero.

3. Consider a force that is a function of the velocity of the particle (and is not perpendicular to the velocity). Show that the work for a round trip along a closed path can then be different from zero.

4. The force acting on a particle moving along the x axis is given by the formula $F_x = K/x^4$, where K is a constant. Find the corresponding potential-energy function. Assume that $U(x) = 0$ for $x = \infty$.

5. A 50-g particle moving along the x axis experiences a force $F_x = -Ax^3$, where $A = 50$ N/m^3. Find the corresponding potential-energy function. If the particle is released from rest at $x = 0.50$ m, what is its speed as it passes the origin?

6. The force on a particle confined to move along the positive x axis is constant, $F_x = -F_0$, where $F_0 = 25$ N. Find the corresponding potential-energy function. Assume $U(x) = 0$ at $x = 0$.

7. A particular spring is not ideal; for a distance x from equilibrium, the spring exerts a force $F_x = -2x - x^3$, where x is in meters and F_x is in newtons. What is the potential-energy function for this spring? How much energy is stored in the spring when it is stretched 1.0 m? 2.0 m? 3.0 m?

8. The force on a particle moving along the x axis is given by

$$F_x = \begin{cases} F_0 & x \leq -a \\ 0 & -a < x < a \\ -F_0 & x \geq a \end{cases}$$

where F_0 is a constant. What is the potential-energy function for this force? Assume $U(x) = 0$ for $x = 0$.

9. Consider a particle moving in a region where the potential energy is given by $U = 2x^2 + x^4$, where U is in joules and x is in meters. What is the position-dependent force on this particle?

10. The force on an electron in a particular region of space is given by $\mathbf{F} = F_0 \sin(ax)\,\mathbf{i}$, where F_0 and a are constants (this force is achieved with two oppositely directed laser beams). What is the corresponding potential-energy function?

*11. A bow may be regarded mathematically as a spring. The archer stretches this "spring" and then suddenly releases it so that the bowstring pushes against the arrow. Suppose that when the archer stretches the "spring" 0.52 m, he must exert a force of 160 N to hold the arrow in this position. If he now releases the arrow, what will be the speed of the arrow when the "spring" reaches its equilibrium position? The mass of the arrow is 0.020 kg. Pretend that the "spring" is massless.

*12. A mass m hangs on a vertical spring of a spring constant k.

(a) How far will this hanging mass have stretched the spring from its relaxed length?

(b) If you now push up on the mass and lift it until the spring reaches its relaxed length, how much work will you have done against gravity? Against the spring?

*13. A particle moving in the x–y plane experiences a conservative force

$$\mathbf{F} = by\mathbf{i} + bx\mathbf{j}$$

where b is a constant.

(a) What is the work done by this force as the particle moves from $x_1 = 0, y_1 = 0$ to $x_2 = x, y_2 = y$? (Hint: Use a path from the origin to the point x_2, y_2 consisting of a segment parallel to the x axis and a segment parallel to the y axis.)

(b) What is the potential energy associated with this force? Assume that the potential energy is zero when the particle is at the origin.

*14. The four wheels of an automobile of mass 1200 kg are suspended below the body by vertical springs of constant $k = 7.0 \times 10^4$ N/m. If the forces on all wheels are the same, what will be the maximum instantaneous deformation of the springs if the automobile is lifted by a crane and dropped on the street from a height of 0.80 m?

*15. A rope can be regarded as a long spring; when under tension, it stretches and stores elastic potential energy. Consider a nylon rope similar to that which snapped during a giant tug-of-war at a school in Harrisburg, Pennsylvania (see Problem 23 of Chapter 5). Under a tension of 58000 N (applied at its ends), the rope of initial length 300 m stretches to 390 m. What is the elastic energy stored in the rope at this tension? What happens to this energy when the rope breaks?

*16. Among the safety features on elevator cages are spring-loaded brake pads which grip the guide rail if the elevator cable should break. Suppose that an elevator cage of 2000 kg has two such brake pads, arranged to press against opposite sides of the guide rail, each with a force of 1.0×10^5 N. The friction coefficient for the brake pads sliding on the guide rail is 0.15. Assume that the elevator cage is falling freely with an initial speed of 10 m/s when the brake pads come into action. How long will the elevator cage take to stop? How far will it travel? How much energy is dissipated by friction?

† For help, see Online Concept Tutorial 10 at www.ww norton.com/physics

*17. The force between two inert-gas atoms is often described by a function of the form

$$F_x = Ax^{-13} - Bx^{-7}$$

where A and B are positive constants and x is the distance between the atoms. What is the corresponding potential-energy function, called the **Lennard–Jones potential**?

*18. A particle moving in three dimensions is confined by a force $\mathbf{F} = -k(x\mathbf{i} + y\mathbf{j} + z\mathbf{k})$, where k is a constant. What is the work required to move the particle from the origin to a point $\mathbf{r} = x\mathbf{i} + y\mathbf{j} + z\mathbf{k}$? What is the potential-energy function?

**19. Mountain climbers use nylon safety rope whose elasticity plays an important role in cushioning the sharp jerk if a climber falls and is suddenly stopped by the rope.

(a) Suppose that a climber of 80 kg attached to a 10-m rope falls freely from a height of 10 m above to a height of 10 m below the point at which the rope is anchored to a vertical wall of rock. Treating the rope as a spring with $k = 4.9 \times 10^3$ N/m (which is the appropriate value for a braided nylon rope of 9.2 mm diameter), calculate the maximum force that the rope exerts on the climber during stopping.

(b) Repeat the calculations for a rope of 5.0 m and an initial height of 5.0 m. Assume that this second rope is made of the same material as the first, and remember to take into account the change in the spring constant due to the change in length. Compare your results for (a) and (b) and comment on the advantages and disadvantages of long ropes vs. short ropes.

**20. A package is dropped on a horizontal conveyor belt (Fig. 8.19). The mass of the package is m, the speed of the conveyor belt is v, and the coefficient of kinetic friction for the package on the belt is μ_k. For what length of time will the package slide on the belt? How far will it move in this time? How much energy is dissipated by friction? How much energy does the belt supply to the package (including the energy dissipated by friction)?

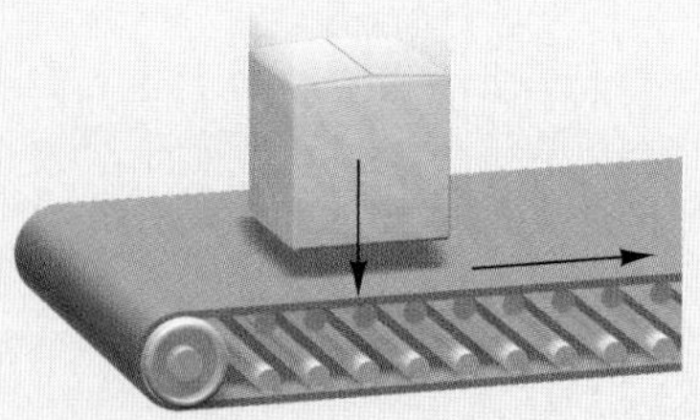

FIGURE 8.19 Package dropped on a conveyor belt

21. The potential energy of a particle moving in the x–y plane is $U = a/(x^2 + y^2)^{1/2}$, where a is a constant. What is the force on the particle? Draw a diagram showing the particle at the position x, y and the force vector.

22. The potential energy of a particle moving along the x axis is $U(x) = K/x^2$, where K is a constant. What is the corresponding force acting on the particle?

23. According to theoretical calculations, the potential energy of two quarks (see the Prelude) separated by a distance r is $U = \eta r$, where $\eta = 1.18 \times 10^{24}$ eV/m. What is the force between the two quarks? Express your answer in newtons.

8.2 The Curve of Potential Energy

24. The potential energy of a particle moving along the x axis is $U(x) = 2x^4 - x^2$, where x is measured in meters and the energy is measured in joules.

(a) Plot the potential energy as a function of x.

(b) Where are the possible equilibrium points?

(c) Suppose that $E = -0.050$ J. What are the turning points of the motion?

(d) Suppose that $E = 1.0$ J. What are the turning points of the motion?

25. In Example 4, we determined the turning point for a bungee jump graphically and numerically. Use the data given in this example for the following calculations.

(a) At what point does the jumper attain maximum speed? Calculate this maximum speed.

(b) At what point does the jumper attain maximum acceleration? Calculate this maximum acceleration.

26. The potential energy of one of the atoms in the hydrogen molecule is

$$U(x) = U_0\left[e^{-2(x-x_0)/b} - 2e^{-(x-x_0)/b}\right]$$

with $U_0 = 2.36$ eV, $x_0 = 0.037$ nm, and b = 0.034 nm.[2] Under the influence of the force corresponding to this potential, the atom moves back and forth along the x axis within certain limits. If the energy of the atom is $E = -1.15$ eV, what will be the turning points of the motion; i.e., at what positions x will the kinetic energy be zero? [Hint: Solve this problem graphically by making a careful plot of $U(x)$; from your plot find the values of x that yield $U(x) = -1.15$ eV.]

27. Suppose that the potential energy of a particle moving along the x axis is

$$U(x) = \frac{b}{x^2} - \frac{2c}{x}$$

where b and c are positive constants.

(a) Plot $U(x)$ as a function of x; assume $b = c = 1$ for this purpose. Where is the equilibrium point?

(b) Suppose the energy of the particle is $E = -\frac{1}{2}c^2/b$. Find the turning points of the motion.

(c) Suppose that the energy of the particle is $E = \frac{1}{2}c^2/b$. Find the turning points of the motion. How many turning points are there in this case?

[2] These values of U_0, x_0, and b are half as large as those usually quoted, because we are looking at the motion of *one* atom relative to the center of the molecule.

28. A particle moves along the x axis under the influence of a conservative force with a potential energy $U(x)$. Figure 8.20 shows the plot of $U(x)$ vs. x. Figure 8.20 shows several alternative energy levels for the particle: $E = E_1$, $E = E_2$, and $E = E_3$. Assume that the particle is initially at $x = 1$ m. For each of the three alternative energies, describe the motion qualitatively, answering the following questions:
    (a) Roughly, where are the turning points (right and left)?
    (b) Where is the speed of the particle maximum? Where is the speed minimum?
    (c) Is the orbit bound or unbound?

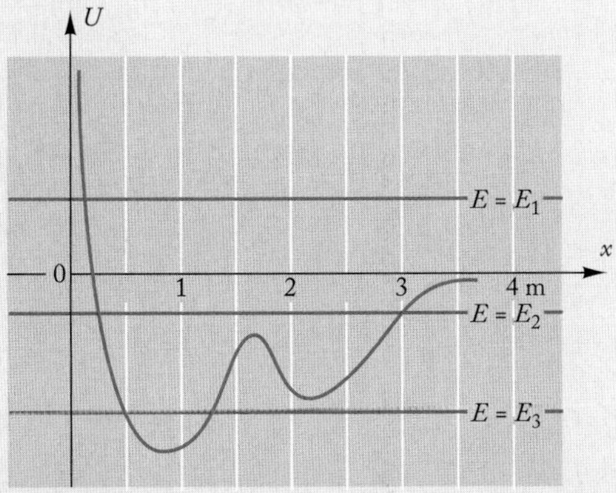

FIGURE 8.20 Plot of $U(x)$ vs. x.

29. A particle moving along the x axis experiences a potential of the form $U(x) = A|x|$, where A is a constant. A particle of mass m has speed v at the origin. Where are the turning points of its motion?

30. A particle initially at the origin moves in a potential of the form $U(x) = -U_0 \cos(ax)$, where U_0 and a are constants. What is the lowest energy the particle may have? If the energy of the particle is $E = 0$ and the particle is initially at $x = 0$, what are the turning points of the motion? For what energies is the particle motion unbound?

31. The potential energy of a particle moving along the x axis is $U(x) = -U_0/[1 + (x/a)^2]$, where $U_0 = 2.0$ J and $a = 1.0$ m. Sketch this function for $-3\text{ m} \leq x \leq 3\text{ m}$. What are the turning points for a particle with energy $E = -1.0$ J? For what energies is the particle unbound?

*32. Consider a particle moving in a region where the potential energy is given by $U = 2x^2 + x^4$, where U is in joules and x is in meters. Where are the turning points for a particle with total mechanical energy $E = 1.0$ J? with $E = 2.0$ J?

*33. The potential-energy function (Lennard–Jones potential) for two argon atoms as a function of their separation x is given by $U(x) = Cx^{-12} - Dx^{-6}$, where $C = 1.59 \times 10^{-24}\ \text{J}\cdot(\text{nm})^{12}$ and $D = 1.03 \times 10^{-21}\ \text{J}\cdot(\text{nm})^6$. (Recall that 1 nm = 10^{-9} m.)
    (a) What is their equilibrium separation in nanometers (nm)?
    (b) What is the lowest possible energy?
    (c) What are the turning points for a particle with energy $E = -2.0 \times 10^{-21}$ J?

8.3 Other Forms of Energy†

34. Express the last two entries in Table 8.1 in electron-volts.

35. The chemical formula for TNT is $CH_3C_5H_2(NO_2)_3$. The explosion of 1 kg of TNT releases 4.6×10^6 J. Calculate the energy released per molecule of TNT. Express your answer in electron-volts.

36. Using the data of Table 8.1, calculate the amount of gasoline that would be required if all the energy requirements of the United States were to be met by the consumption of gasoline. How many gallons per day would have to be consumed?

37. The following table lists the fuel consumption and the passenger capacity of several vehicles. Assume that the energy content of the fuel is that of gasoline (see Table 8.1). Calculate the amount of energy used by each vehicle per passenger per mile. Which is the most energy-efficient vehicle? The least energy-efficient?

VEHICLE	PASSENGER CAPACITY	FUEL CONSUMPTION
Motorcycle	1	60 mi/gal
Snowmobile	1	12
Automobile	4	12
Intercity bus	45	5
Concorde SST	110	0.12
Jetliner	360	0.1

38. The energy released by the metabolization of fat is about 9000 kcal per kg of fat. While jogging on a level road, you use 750 kcal/h. How long do you need to jog to eliminate 1.0 kg of fat?

39. A 12-ounce can of soda typically contains 150 kcal of food energy (150 food "calories"). If your body uses one-fifth of this to climb stairs, how high does one soda enable you to climb?

40. A large household may use as much as 3000 kilowatt-hours of energy during a hot summer month. Express this amount of energy in joules.

41. On food labels in Europe, energy content is typically listed in kilojoules (kJ) instead of kcal (food "calories"). Express a daily intake of 2500 kcal in kJ.

*42. When a humpback whale breaches, or jumps out of the water (see Fig. 8.21), it typically leaves the water at an angle of about 70° at high speed and sometimes attains a height of 3 m,

† For help, see Online Concept Tutorial 10 at www.ww norton.com/physics

FIGURE 8.21 A whale breaching.

measured from the water surface to the center of the whale. For a rough estimate of the energy requirements for such a breach, we can treat the translational motion of the whale as that of a particle moving from the surface of the water upward to a height of 3.0 m (for a more accurate calculation, we would have to take into account the buoyancy of the whale, which assists it in getting out of the water, but let us ignore this). What is the initial speed of the whale when it emerges from the water? What is the initial kinetic energy of a whale of 33 metric tons? Express the energy in kilocalories.

*43. The following table gives the rate of energy dissipation by a man engaged in diverse activities; the energies are given per kilogram of body mass:

RATE OF ENERGY DISSIPATION OF A MAN (PER kg OF BODY MASS)

Standing	1.3 kcal/(kg·h)
Walking (5 km/h)	3.3
Running (8 km/h)	8.2
Running (16 km/h)	15.2

Suppose the man wants to travel a distance of 2.5 km in one-half hour. He can walk this distance in exactly half an hour, or run slow and then stand still until the half hour is up, or run fast and then stand still until the half hour is up. What is the energy per kg of body mass dissipated in each case? Which program uses the most energy? Which the least?

8.4 Mass and Energy

44. The atomic bomb dropped on Hiroshima had an explosive energy equivalent to that of 20 000 tons of TNT, or 8.4×10^{13} J. How many kilograms of mass must have been converted into energy in this explosion?

45. How much energy is released by the annihilation of one proton and one antiproton (both initially at rest)? Express your answer in electron-volts.

46. How much energy is released by the annihilation of one electron and one antielectron (both initially at rest)? Express your answer in electron-volts.

47. The mass of the Sun is 2×10^{30} kg. The thermal energy in the Sun is about 2×10^{41} J. How much does the thermal energy contribute to the mass of the Sun?

48. The masses of the proton, electron, and neutron are 1.672623×10^{-27} kg, 9.11×10^{-31} kg, and 1.674929×10^{-27} kg, respectively. If a neutron decays into a proton and an electron, how much energy is released (other than the energy of the mass of the proton and electron)? Compare this extra energy with the energy of the mass of the electron.

49. Express the mass energy of the electron in keV. Express the mass energy of the proton in MeV.

50. A typical household may use approximately 1000 kilowatt-hours of energy per month. What is the equivalent amount of rest mass?

51. Combustion of one gallon of gasoline releases 1.3×10^{8} J of energy. How much mass is converted to energy? Compare this with 2.8 kg, the mass of one gallon of gasoline.

52. A small silicon particle of diameter 0.20 micrometers has a mass of 9.8×10^{-18} kg. What is the mass energy of such a "nanoparticle" (in J)?

*53. In a high-speed collision between an electron and an antielectron, the two particles can annihilate and create a proton and an antiproton. The reaction

$$e + \bar{e} \rightarrow p + \bar{p}$$

converts the mass energy and kinetic energy of the electron and antielectron into the mass energy of the proton and the antiproton. Assume that the electron and the antielectron collide head-on with opposite velocities of equal magnitudes and that the proton and the antiproton are at rest immediately after the reaction. Calculate the kinetic energy of the electron required for this reaction; express your answer in electron-volts.

8.5 Power†

54. For an automobile traveling at a steady speed of 65 km/h, the friction of the air and the rolling friction of the ground on the wheels provide a total external friction force of 500 N. What power must the engine supply to keep the automobile moving? At what rate does the friction force remove momentum from the automobile?

55. In 1979, B. Allen flew a very lightweight propeller airplane across the English Channel. His legs, pushing bicycle pedals, supplied the power to turn the propeller. To keep the airplane flying, he had to supply about 0.30 hp. How much energy did he supply for the full flight lasting 2 h 49 min? Express your answer in kilocalories.

† For help, see Online Concept Tutorial 10 at www.ww norton.com/physics

56. The ancient Egyptians and Romans relied on slaves as a source of mechanical power. One slave, working desperately by turning a crank, could deliver about 200 W of mechanical power (at this power the slave would not last long). How many slaves would be needed to match the output of a modern automobile engine (150 hp)? How many slaves would an ancient Egyptian have to own in order to command the same amount of power as the average per capita power used by residents of the United States (14 kW)?

57. An electric clock uses 2.0 W of electric power. How much electric energy (in kilowatt-hours) does this clock use in 1 year? What happens to this electric energy?

58. While an automobile is cruising at a steady speed of 65 km/h, its engine delivers a mechanical power of 20 hp. How much energy does the engine deliver per hour?

59. A large windmill delivers 10 kW of mechanical power. How much energy does the windmill deliver in a working day of 8 hours?

60. The heating unit of a medium-sized house produces 170 000 Btu/h. Is this larger or smaller than the power produced by a typical automobile engine of 150 hp?

61. The heart of a resting person delivers a mechanical power of about 1.1 W for pumping blood. Express this power in hp. How much work does the heart do on the blood per day? Express this work in kcal.

62. The lasers to be used for controlled fusion experiments at the National Ignition Facility at the Lawrence Livermore Laboratory will deliver a power of 2.0×10^{15} W, a thousand times the output of all the power stations in the United States, in a brief pulse lasting 1.0×10^{-9} s. What is the energy in this laser pulse? How does it compare with the energy output of all the power stations in the United States in one day?

63. During the seven months of the cold season in the Northeastern United States, a medium-sized house requires about 1.0×10^8 Btu of heat to keep warm. A typical furnace delivers 1.3×10^5 Btu of heat per gallon of fuel oil.

    (a) How many gallons of fuel oil does the house consume during the cold season?

    (b) What is the average power delivered by the furnace?

64. Experiments on animal muscle tissue indicate that it can produce up to 100 watts of power per kilogram. A 600-kg horse has about 180 kg of muscle tissue attached to the legs in such a way that it contributes to the external work the horse performs while pulling a load. Accordingly, what is the theoretical prediction for the maximum power delivered by a horse? In trials, the actual maximum power that a horse can deliver in a short spurt was found to be about 12 hp. How does this compare with the theoretical prediction?

65. If a 60-W light bulb is left on for 24 hours each day, how many kilowatt-hours of electric power does it use in one year? If the electric energy costs you 15 cents per kilowatt-hour, what is your cost for one year?

66. Nineteenth-century English engineers reckoned that a laborer turning a crank can do steady work at the rate of 5000 ft·lbf/min. Suppose that four laborers working a manual crane attempt to lift a load of 9.0 short tons (1 short ton = 2000 lb). If there is no friction, what is the rate at which they can lift this load? How long will it take them to lift the load 15 ft?

67. The driver of an automobile traveling on a straight road at 80 km/h pushes forward with his hands on the steering wheel with a force of 50 N. What is the rate at which his hands do work on the steering wheel in the reference frame of the ground? In the reference frame of the automobile?

68. An automobile with a 100-hp engine has a top speed of 160 km/h. When at this top speed, what is the friction force (from air and road) acting on the automobile?

69. A horse walks along the bank of a canal and pulls a barge by means of a long horizontal towrope making an angle of 35° with the bank. The horse walks at the rate of 5.0 km/h, and the tension in the rope is 400 N. What horsepower does the horse deliver?

70. A 900-kg automobile accelerates from 0 to 80 km/h in 7.6 s. What are the initial and the final translational kinetic energies of the automobile? What is the average power delivered by the engine in this time interval? Express your answer in horsepower.

71. A six-cylinder internal combustion engine, such as used in an automobile, delivers an average power of 150 hp while running at 3000 rev/min. Each of the cylinders fires once every two revolutions. How much energy does each cylinder deliver each time it fires?

72. In Chapter 6, we saw that an automobile must overcome the force of air resistance, $f_{air} = \frac{1}{2}\rho C A v^2$. For the automobile of Example 6 of Chapter 6 ($C = 0.30$, $A = 2.8\ m^2$, and $\rho = 1.3\ kg/m^3$), calculate the power dissipation due to air resistance when traveling at 30 km/h and when traveling at 90 km/h. What is the difference in the total energy supplied to overcome air friction for a 300-km trip at 30km/h? For a 300-km trip at 90 km/h?

73. A constant force of 40 N is applied to a body as the body moves uniformly at a speed of 3.5 m/s. The force does work on the body at a rate of 90 W. What is the angle between the force and the direction of motion of the body?

74. An electric motor takes 1.0 s to get up to speed; during this time, the power supplied by the motor varies with time according to $P = P_1 + (P_0 - P_1)(t - 1)^2$, where t is in seconds, $P_0 = 1.50$ kW, and $P_1 = 0.75$ kW. What is the total energy supplied for the time period $0 \leq t \leq 1$ s?

75. A constant force $\mathbf{F} = (6.0\ N)\mathbf{i} + (8.0\ N)\mathbf{j}$ acts on a particle. At what instantaneous rate is this force doing work on a particle with velocity $\mathbf{v} = (3.0\ m/s)\mathbf{i} - (2.5\ m/s)\mathbf{j}$?

76. An automobile engine typically has an efficiency of about 25%; i.e., it converts about 25% of the chemical energy available in gasoline into mechanical energy. Suppose that an automobile engine has a mechanical output of 110 hp. At what

rate (in gallons per hour) will this engine consume gasoline? See Table 8.1 for the energy content in gasoline.

77. The takeoff speed of a DC-3 airplane is 100 km/h. Starting from rest, the airplane takes 10 s to reach this speed. The mass of the (loaded) airplane is 11 000 kg. What is the average power delivered by the engines to the airplane during takeoff?

78. The Sun emits energy in the form of radiant heat and light at the rate of 3.9×10^{26} W. At what rate does this energy carry away mass from the Sun? How much mass does this amount to in 1 year?

79. The energy of sunlight arriving at the surface of the Earth amounts to about 1.0 kW per square meter of surface (facing the Sun). If all of the energy incident on a collector of sunlight could be converted into useful energy, how many square meters of collector area would we need to satisfy all of the energy demands in the United States? See Table 8.1 for the energy expenditure of the United States.

80. Equations (2.11) and (2.16) give the velocity and the acceleration of an accelerating Maserati sports car as a function of time. The mass of this automobile is 1770 kg. What is the instantaneous power delivered by the engine to the automobile? Plot the instantaneous power as a function of time in the time interval from 0 to 10 s. At what time is the power maximum?

81. The ship *Globtik Tokyo*, a supertanker, has a mass of 650 000 metric tons when fully loaded.
   (a) What is the kinetic energy of the ship when her speed is 26 km/h?
   (b) The engines of the ship deliver a power of 44 000 hp. According to the energy requirements, how long a time does it take the ship to reach a speed of 26 km/h, starting from rest? Make the assumption that 50% of the engine power goes into friction or into stirring up the water and 50% remains available for the translational motion of the ship.
   (c) How long a time does it take the ship to stop from an initial speed of 26 km/h if her engines are put in reverse? Estimate roughly how far the ship will travel during this time.

82. At Niagara Falls, 6200 m^3 per second of water falls down a height of 49 m.
   (a) What is the rate (in watts) at which gravitational potential energy is dissipated by the falling water?
   (b) What is the amount of energy (in kilowatt-hours) wasted in 1 year?
   (c) Power companies get paid about 5 cents per kilowatt-hour of electric energy. If all the gravitational potential energy wasted in Niagara Falls could be converted into electric energy, how much money would this be worth?

83. The movement of a grandfather clock is driven by a 5.0-kg weight which drops a distance of 1.5 m in the course of a week. What is the power delivered by the weight to the movement?

84. A 27 000-kg truck has a 550-hp engine. What is the maximum speed with which this truck can move up a 10° slope?

*85. Consider a "windmill ship," which extracts mechanical energy from the wind by means of a large windmill mounted on the deck (see Fig. 8.22). The windmill generates electric power, which is fed into a large electric motor, which propels the ship. The mechanical efficiency of the windmill is 70% (that is, it removes 70% of the kinetic energy of the wind and transforms it into rotational energy of its blades). The efficiency of the electric generator attached to the windmill is 90%, and the efficiency of the electric motor connected to the generator is also 90%. We want the electric motor to deliver 20 000 hp in a (relative) wind of 40 km/h. What size windmill do we need? The density of air is 1.29 kg/m^3.

FIGURE 8.22
A "windmill ship."

*86. An electric water pump is rated at 15 hp. If this water pump is to lift water to a height of 30 m, how many kilograms of water can it lift per second? How many liters? Neglect the kinetic energy of the water.

*87. The engines of the Sikorski Blackhawk helicopter generate 3080 hp of mechanical power, and the maximum takeoff mass of this helicopter is 7400 kg. Suppose that this helicopter is climbing vertically at a steady rate of 5.0 m/s.
   (a) What is the power that the engines deliver to the body of the helicopter?
   (b) What is the power that the engines deliver to the air (by friction and by the work that the rotors of the helicopter perform on the air)?

*88. In order to overcome air friction and other mechanical friction, an automobile of mass 1500 kg requires a power of 20 hp from its engine to travel at 64 km/h on a level road. Assuming the friction remains the same, what power does the same automobile require to travel uphill on an incline of slope 1:10 at the same speed? Downhill on the same incline at the same speed?

*89. With the gears in neutral, an automobile rolling down a long incline of slope 1:10 reaches a terminal speed of 95 km/h. At this speed the rate of decrease of the gravitational potential energy matches the power required to overcome air friction and other mechanical friction. What power (in horsepower) must the engine of this automobile deliver to drive it at 95 km/h on a level road? The mass of the automobile is 1500 kg.

*90. The power supplied to an electric circuit decreases exponentially with time according to $P = P_0 e^{-t/\tau}$, where $P_0 =$ 2.0 W and $\tau = 5.0$ s are constants. What is the total energy supplied to the circuit during the time interval $0 \leq t \leq 5.0$ s? During $0 \leq t \leq \infty$?

*91. Each of the two Wright Cyclone engines on a DC-3 airplane generates a power of 850 hp. The mass of the loaded plane is 10 900 kg. The plane can climb at the rate of 260 m/min. When the plane is climbing at this rate, what percentage of the engine power is used to do work against gravity?

*92. A fountain shoots a stream of water 10 m up in the air. The base of the stream is 10 cm across. What power is expended to send the water to this height?

*93. The record of 203.1 km/h for speed skiing set by Franz Weber at Velocity Peak in Colorado was achieved on a mountain slope inclined downward at 51°. At this speed, the force of friction (air and sliding friction) balances the pull of gravity along the slope, so the motion proceeds at constant velocity.

(a) What is the rate at which gravity does work on the skier? Assume that the mass of the skier is 75 kg.

(b) What is the rate at which sliding friction does work? Assume that the coefficient of friction is $\mu_k = 0.03$.

(c) What is the rate at which air friction does work?

*94. A windmill for the generation of electric power has a propeller of diameter 1.8 m. In a wind of 40 km/h, this windmill delivers 200 W of electric power.

(a) At this wind speed, what is the rate at which the air carries kinetic energy through the circular area swept out by the propeller? The density of air is 1.29 kg/m^3.

(b) What percentage of the kinetic energy of the air passing through this area is converted into electric energy?

*95. A small electric kitchen fan blows 8.5 m^3/min of air at a speed of 5.0 m/s out of the kitchen. The density of air is 1.3 kg/m^3. What electric power must the fan consume to give the ejected air the required kinetic energy?

*96. The final portion of the Tennessee River has a downward slope of 0.074 m per kilometer. The rate of flow of water in the river is 280 m^3/s. Assume that the speed of the water is constant along the river. How much power is dissipated by friction of the water against the riverbed per kilometer?

*97. Off the coast of Florida, the Gulf Stream has a speed of 4.6 km/h and a rate of flow of 2.2×10^3 km^3/day. At what rate is kinetic energy flowing past the coast? If all this kinetic energy could be converted into electric power, how many kilowatts would it amount to?

*98. Figure 8.23 shows an overshot waterwheel, in which water flowing onto the top of the wheel fills buckets whose weight causes the wheel to turn. The water descends in the buckets to the bottom, and there it is spilled out, so the ascending buckets are always empty. If in a waterwheel of diameter 10 m the amount of water carried down by the wheel is 20 liters per second (or 20 kg per second), what is the mechanical power that the descending water delivers to the wheel? Assume that the water flowing onto the top of the wheel has roughly the same speed as the wheel and exerts no horizontal push on the wheel. [Hint: The kinetic energy of the water is the same when the water enters the bucket and when it spills out (since the speed of the bucket is constant); hence the kinetic energy of the water does not affect the answer.]

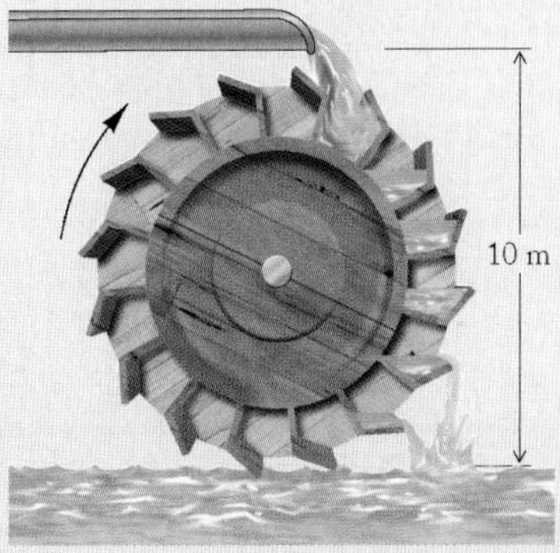

FIGURE 8.23 An overshot waterwheel.

*99. Suppose that in the undershot waterwheel shown in Fig. 8.24, the stream of water against the blades of the wheel has a speed of 15 m/s, and the amount of water is 30 liters per second (or 30 kg per second). If the water gives all of its kinetic energy to the blades (and then drips away with zero horizontal speed), how much mechanical power does the water deliver to the wheel?

FIGURE 8.24 An undershot waterwheel.

*100. (a) With its engines switched off, a small two-engine airplane of mass 1100 kg glides downward at an angle of 13° at a speed of 90 knots. Under these conditions, the weight of the plane, the lift force (perpendicular to the direction of motion) generated by air flowing over the wings, and the frictional force (opposite to the direction of motion) exerted by the air are in balance. Draw a "free-body" diagram for these forces, and calculate their magnitudes.

(b) Suppose that with its engine switched on, the plane climbs at an upward angle of 13° at a speed of 90 knots. Draw a "free-body" diagram for the forces acting on the airplane under these conditions; include the push that the air exerts on the propeller. Calculate the magnitudes of all the forces.

(c) Calculate the power that the engine must deliver to compensate for the rate of increase of the potential energy of the plane and the power lost to friction. For a typical small plane of 1100 kg, the actual engine power required for such a climb of 13° is about 400 hp. Explain the discrepancy between your result and the actual engine power. (Hint: What does the propeller do to the air?)

*101. The reaction that supplies the Sun with energy is

$$H + H + H + H \rightarrow He + [\text{energy}]$$

(The reaction involves several intermediate steps, but this need not concern us now.) The mass of the hydrogen (H) atom is 1.008 13 u, and that of the helium (He) atom is 4.003 88 u.

(a) How much energy is released in the reaction of four hydrogen atoms (by the conversion of mass into energy)?

(b) How much energy is released in the reaction of 1.0 kg of hydrogen atoms?

(c) The Sun releases energy at the rate of 3.9×10^{26} W. At what rate (in kg/s) does the Sun consume hydrogen?

(d) The Sun contains about 1.5×10^{30} kg of hydrogen. If it continues to consume hydrogen at the same rate, how long will the hydrogen last?

REVIEW PROBLEMS

102. A particle moves along the x axis under the influence of a variable force $F_x = 5x^2 + 3x$ (where force is measured in newtons and distance in meters).

(a) What is the potential energy associated with this force? Assume that $U(x) = 0$ at $x = 0$.

(b) How much work does the force do on a particle that moves from $x = 0$ to $x = 2.0$ m?

*103. A particle is subjected to a force that depends on position as follows:

$$\mathbf{F} = 4\mathbf{i} + 2x\mathbf{j}$$

where the force is measured in newtons and the distance in meters.

(a) Calculate the work done by this force as the particle moves from the origin to the point $x = 1.0$ m, $y = 1.0$ m along the straight path I shown in Fig. 8.25.

(b) Calculate the work done by this force if the particle returns from the point $x = 1.0$ m, $y = 1.0$ m to the origin along the path II consisting of a horizontal and a vertical segment (see Fig. 8.25). Is the force conservative?

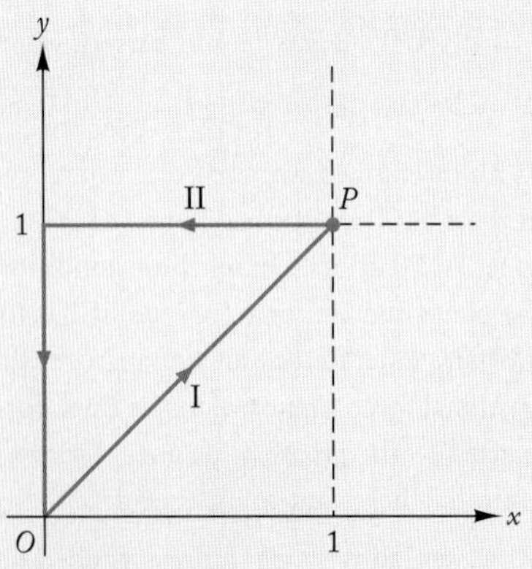

FIGURE 8.25 Outward and return paths of a particle.

*104. A 3.0-kg block sliding on a horizontal surface is accelerated by a compressed spring. At first, the block slides without friction. But after leaving the spring, the block travels over a new portion of the surface, with a coefficient of friction 0.20, for a distance of 8.0 m before coming to rest (see Fig. 8.26). The force constant of the spring is 120 N/m.

(a) What was the maximum kinetic energy of the block?

(b) How far was the spring compressed before being released?

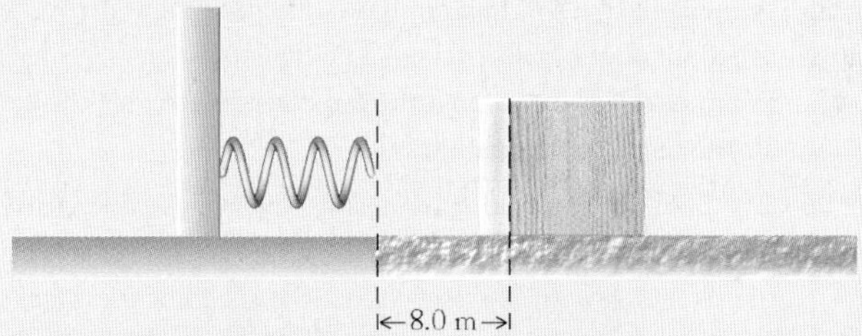

FIGURE 8.26 Block released from a spring.

105. The ancient Egyptians moved large stones by dragging them across the sand in sleds (Fig. 8.27). Suppose that 6000 Egyptians are dragging a sled with a coefficient of sliding friction $\mu_k = 0.30$ along a level surface of sand. Each Egyptian can exert a force of 360 N, and each can deliver a mechanical power of 0.20 hp.

(a) What is the maximum weight they can move at constant speed?

(b) What is the maximum speed with which they can move this weight?

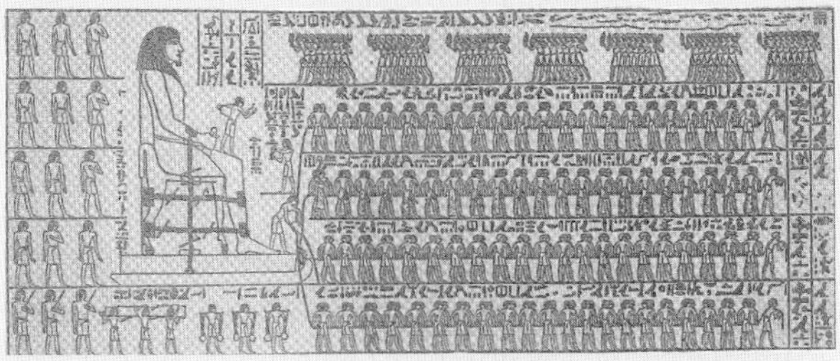

FIGURE 8.27 Ancient Egyptian wall mural from 1900 B.C.

106. In a braking test, a 990-kg automobile takes 2.1 s to come to a full stop from an initial speed of 60 km/h. What is the amount of energy dissipated in the brakes? What is the average power dissipated in the brakes? Ignore external friction in your calculation and express the power in horsepower.

107. In a waterfall on the Alto Paraná river (between Brazil and Paraguay), the height of fall is 33 m and the average rate of flow is 13 000 m^3 of water per second. What is the power dissipated by this waterfall?

108. When jogging at 12 km/h on a level road, a 70-kg man uses 750 kcal/h. How many kilocalories per hour does he require when jogging up a 1:10 incline at the same speed? Assume that the frictional losses are independent of the value of the slope.

109. Consider a projectile traveling horizontally and slowing down under the influence of air resistance, as described in Problems 47 and 48 of Chapter 2. The mass of this projectile is 45.36 kg, and the speed as a function of time is

$$v = 655.9 - 61.1t + 3.26t^2$$

where speed is in m/s and time in seconds.

(a) What is the instantaneous power removed from the projectile by air resistance?

(b) What is the kinetic energy at time $t = 0$? At time $t = 3.00$ s?

(c) What is the average power for the time interval from 0 to 3.00 s?

110. A woman exercising on a rowing machine pulls the oars back once per second. During such a pull, each hand moves 0.50 m while exerting an average force of 100 N.

(a) What is the work the woman does during each stroke (with both hands)?

(b) What is the average power the woman delivers to the oars?

111. The world's tallest staircase, of 2570 steps, is located in the CN tower in Toronto. It reaches a height of 457 m. Estimate how long it would take an athlete to climb this staircase. The athlete has a mass of 75 kg, and his leg muscle can deliver a power of 200 W.

112. A pump placed on the shore of a pond takes in 0.80 kg of water per second and squirts it out of a nozzle at 50 m/s. What mechanical power does this pump supply to the water?

113. The hydroelectric pumped-storage plant in Northfield, Massachusetts, has a reservoir holding 2.2×10^7 m^3 of water on top of a mountain. The water flows 270 m vertically down the mountain in pipes and drives turbines connected to electric generators.

(a) How much electric energy, in kW·h, can this storage plant generate with the water available in the reservoir?

(b) In order to generate 1000 MW of electric energy, at what rate, in m^3/s, must this storage plant withdraw water from the reservoir?

114. A 50-kg circus clown is launched vertically from a spring-loaded cannon using a spring with spring constant 3500 N/m. The clown attains a height of 4.0 m above the initial position (when the spring was compressed).

(a) How far was the spring compressed before launch?

(b) What was the maximum acceleration of the clown during launch?

(c) What was the maximum speed of the clown?

Answers to Checkups

Checkup 8.1

1. The force can be obtained from $F_x = -dU/dx = +kx$. Thus, the force is positive for positive x and negative for negative x; that is, the new force is repulsive (it pushes a particle away from $x = 0$), whereas the spring force is attractive.

2. The potential energy shown in Fig. 8.5a has a negative slope as a function of x. By Eq. (8.21), the force is the negative of the slope of the potential; the negative of a negative is positive, and so the force is directed along the positive x direction.

3. A force is always conservative if the force is an explicit function of position x. In that case, a potential-energy function can always be constructed by integration of the force according to Eq. (8.14).

4. (D) No; no. The work done is equal to the negative of the change in potential energy only for conservative forces. The work done is equal to the change in kinetic energy only for the *net* force acting on a particle.

Checkup 8.2

1. The number of turning points must decrease with increasing energy [we do not consider a stationary point of stable equilibrium (Fig. 8.8), since the particle is moving]. Consider the potential of Fig. 8.7: for small energies, the particle will move back and forth (two turning points); for somewhat higher energy, the particle will move back from the left end but escape from the right end (one turning point). Unless $U = \infty$, for sufficiently high energy the particle could escape from the left end also (no turning point). In one dimension, there cannot be more than two turning points, although the two turning points will of course be different for different energies.
2. The maximum speed corresponds to the deepest part of the curve (maximum kinetic energy, $K = E - U$); from the figure, this occurs at $x \approx -6$ m. The maximum acceleration and force ($F = -dU/dx$) occurs where the slope is largest; for the bungee jumper, this is at $x \approx -15$ m.
3. (B) 2. The potential-energy curve of the spring force is a simple parabola (Fig. 8.1), so there are two turning points for any positive energy.

Checkup 8.3

1. No. Gravitational potential energy is lost as the parachutist descends (at uniform speed, there is no change in kinetic energy). From a macroscopic viewpoint, the energy lost due to friction with the air is converted into heat.
2. The energy comes from a decrease in the chemical energy of the exploding gunpowder; microscopically, such chemical energy comes from changes in the kinetic and potential energy of electrons in the atoms and bonds of the elements involved.
3. The energy is converted to heat due to frictional forces; these may include friction in the engine, brakes, tires, and road, as well as air friction.
4. (C) The kinetic energy is converted into heat due to frictional forces, mostly in the brakes (brake pads rub against drums or disks), partly where the tires contact the road, and some from air friction. All the heat is eventually transferred to the air as the brakes, tires, and road cool.

Checkup 8.4

1. Yes; the Sun continually loses mass in the form of heat and light, as well as by emitting particles with mass.
2. Energy and mass are both conserved; the original rest mass is converted to the energy of the light (electromagnetic radiation), and this light carries away mass as well as energy.
3. (A) Increase. The mass of the water will increase by the usual $\Delta m = \Delta E/c^2$, where ΔE is the increase in thermal energy of the water.

Checkup 8.5

1. (a) No; there is no force parallel to the motion, so there is no work done and no power expended. (b) Trotting uphill, you deliver power at a rate $P = Fv = mg \sin \phi \, v$, where m is the mass of the pack, ϕ is the angle of the incline, and v is the speed along the incline. (c) Trotting downhill, the component of $\mathbf{F}$ along $\mathbf{v}$ is negative, so you do negative work on the backpack; that is, the backpack delivers power to you.
2. Yes, the energy you have to expend is mgh, whichever slope of road you take. The steeper road requires more of an effort, since, for example, for the same walking speed, the force is more nearly parallel to the velocity, and so the power expended, $P = \mathbf{F} \cdot \mathbf{v}$, is greater.
3. Some of the power is lost as heat, due to the friction force between the boat and the water; some of the energy is converted into a more macroscopic kinetic energy of the water, by the generation of water waves.
4. (C) 4. The power is equal to the force times the speed. At the same speed, a car with twice the mass will require twice the power to move against gravity; if that car is also traveling at twice the speed, it will then require four times as much power (ignoring other losses).

CHAPTER

9

Gravitation

CONCEPTS IN CONTEXT

Concepts in Context

Hundreds of artificial satellites have been placed in orbit around the Earth, such as this Syncom communications satellite shown just after launch from the Space Shuttle.

With the concepts we will develop in this chapter, we can answer various questions about artificial satellites:

? Communications satellites and weather satellites are placed in high-altitude "geosynchronous" orbits that permit them to keep in step with the rotation of the Earth, so the satellite always remains at a fixed point above the equator. What is the radius of such a geosynchronous orbit? (Example 6, page 279; and Physics in Practice: Communications Satellites and Weather Satellites, page 281)

? Surveillance satellites and spacecraft such as the Space Shuttle usually operate in low-altitude orbits, just above the Earth's atmosphere. How quickly does such a satellite circle the Earth? (Example 7, page 280)

? The Syncom satellite was carried by the Space Shuttle to a low-altitude orbit, and then it used its own booster rocket to lift itself to the high-altitude geosynchronous orbit. What is the increase of mechanical energy (kinetic and gravitational) of the satellite during this transfer from one orbit to another? (Example 9, page 290)

Within the Solar System, planets orbit around the Sun, and satellites orbit around the planets. These circular, or nearly circular, motions require a centripetal force pulling the planets toward the Sun and the satellites toward the planets. It was Newton's great discovery that this interplanetary force holding the celestial bodies in their orbits is of the same kind as the force of gravity that causes apples, and other things, to fall downward near the surface of the Earth. Newton found that a single formula, his Law of Universal Gravitation, encompasses both the gravitational forces acting between celestial bodies and the gravitational force acting on bodies near the surface of the Earth.

By the nineteenth century, Newton's theory of gravitation had proved itself so trustworthy that when astronomers noticed an irregularity in the motion of Uranus, they could not bring themselves to believe that the theory was at fault. Instead, they suspected that a new, unknown planet caused these irregularities by its gravitational pull on Uranus. The astronomers J. C. Adams and U. J. J. Leverrier proceeded to calculate the expected position of this hypothetical planet—and the new planet was immediately found at just about the expected position. This discovery of a new planet, later named Neptune, was a spectacular success of Newton's theory of gravitation. Newton's theory remains one of the most accurate and successful theories in all of physics, and in all of science.

In this chapter, we will examine Newton's Law of Universal Gravitation; we will see how it includes the familiar gravitational force near the Earth's surface. We will also examine circular and elliptical orbits of planets and satellites, and we will become familiar with Kepler's laws describing these orbits. Finally, we will discuss gravitational potential energy and apply energy conservation to orbital motion.

Online Concept Tutorial 11

9.1 NEWTON'S LAW OF UNIVERSAL GRAVITATION

Newton proposed that just as the Earth gravitationally attracts bodies placed near its surface and causes them to fall downward, the Earth also attracts more distant bodies, such as the Moon, or the Sun, or other planets. In turn, the Earth is gravitationally attracted by all these bodies. More generally, *every* body in the Universe attracts *every* other body with a gravitational force that depends on their masses and on their distances. The gravitational force that two bodies exert on each other is large if their masses are large, and small if their masses are small. The gravitational force decreases if we increase the distance between the bodies. The **Law of Universal Gravitation** formulated by Newton can be stated most easily for the case of particles:

> *Every particle attracts every other particle with a force directly proportional to the product of their masses and inversely proportional to the square of the distance between them.*

Expressed mathematically, the magnitude of the gravitational force that two particles of masses M and m separated by a distance r exert on each other is

Law of Universal Gravitation

$$F = \frac{GMm}{r^2} \tag{9.1}$$

where G is a universal constant of proportionality, the same for all pairs of particles.

The direction of the force on each particle is directly toward the other particle. Figure 9.1 shows the directions of the forces on each particle. Note that the two forces are of equal magnitudes and opposite directions; they form an action–reaction pair, as required by Newton's Third Law.

The constant G is known as the **gravitational constant**. In SI units its value is approximately given by

$$G = 6.67 \times 10^{-11}\,\text{N}\cdot\text{m}^2/\text{kg}^2 \qquad (9.2)$$

gravitational constant

The gravitational force of Eq. (9.1) is an inverse-square force: it decreases by a factor of 4 when the distance increases by a factor of 2, it decreases by a factor of 9 when the distance increases by a factor of 3, and so on. Figure 9.2 is a plot of the magnitude of the gravitational force as a function of the distance. Although the force decreases with distance, it never quite reaches zero. Thus, every particle in the universe continually attracts every other particle at least a little bit, even if the distance between the particles is very, very large.

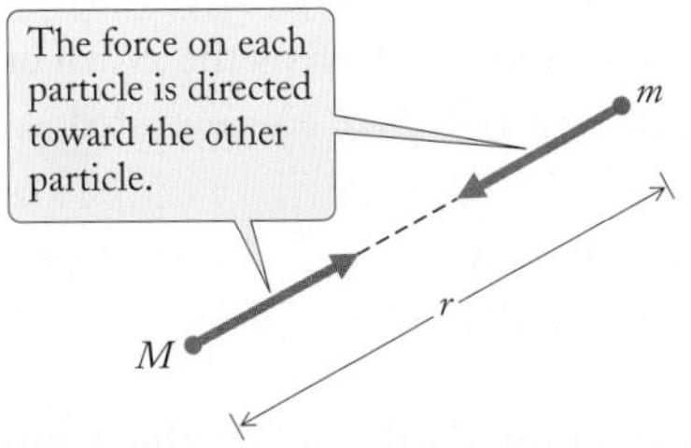

FIGURE 9.1 Two particles attract each other gravitationally. The forces are of equal magnitudes and of opposite directions.

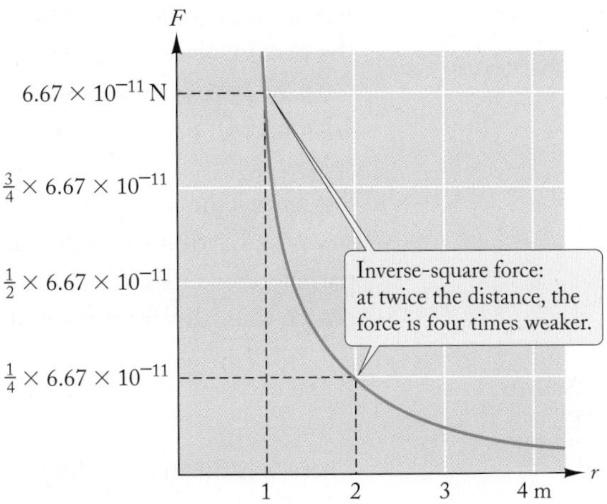

FIGURE 9.2 Magnitude of the gravitational force exerted by a particle of mass 1 kg on another particle of mass 1 kg.

EXAMPLE 1 What is the gravitational force between a 70-kg man and a 70-kg woman separated by a distance of 10 m? Treat both masses as particles.

SOLUTION: From Eq. (9.1),

$$F = \frac{GMm}{r^2}$$

$$= \frac{6.67 \times 10^{-11}\ \text{N}\cdot\text{m}^2/\text{kg}^2 \times 70\ \text{kg} \times 70\ \text{kg}}{(10\ \text{m})^2}$$

$$= 3.3 \times 10^{-9}\ \text{N}$$

This is a very small force, but as we will see in the next section, the measurement of such small forces is not beyond the reach of sensitive instruments.

The gravitational force does not require any contact between the interacting particles. In reaching from one remote particle to another, the gravitational force somehow bridges the empty space between the particles. This is called **action-at-a-distance**.

It is also quite remarkable that the gravitational force between two particles is unaffected by the presence of intervening masses. For example, a particle in Washington attracts a particle in Beijing with exactly the force given by Eq. (9.1), even though all of the bulk of the Earth lies between Washington and Beijing. This means that it is impossible to shield a particle from the gravitational attraction of another particle.

Since the gravitational attraction between two particles is completely independent of the presence of other particles, it follows that the net gravitational force between two bodies (e.g., the Earth and the Moon or the Earth and an apple) is merely the vector sum of the individual forces between all the particles making up the bodies—that is, the gravitational force obeys the principle of linear superposition of forces (see Section 5.3). As a consequence of this simple vector summation of the gravitational forces of the individual particles in a body, it can be shown that *the net gravitational force between two spherical bodies acts just as though each body were concentrated at the center of its respective sphere.* This result is known as **Newton's theorem**. The proof of Newton's theorem involves a somewhat tedious summation. Later, in the context of electrostatic force, we provide a much simpler derivation of Newton's theorem using Gauss' Law (see Chapter 24). Since the Sun, the planets, and most of their satellites are almost exactly spherical, this important theorem permits us to treat all these celestial bodies as pointlike particles in all calculations concerning their gravitational attractions. For instance, since the Earth is (nearly) spherical, the gravitational force exerted by the Earth on a particle above its surface is as though the mass of the Earth were concentrated at its center; thus, this force has a magnitude

Newton's theorem

$$F = \frac{GM_E m}{r^2} \tag{9.3}$$

where m is the mass of the particle, M_E is the mass of the Earth, and r is the distance from the center of the Earth (see Fig. 9.3).

If the particle is at the surface of the Earth, at a radius $r = R_E$, then Eq. (9.3) gives a force

$$F = \frac{GM_E m}{R_E^2} \tag{9.4}$$

The corresponding acceleration of the mass m is

$$a = \frac{F}{m} = \frac{GM_E}{R_E^2} \tag{9.5}$$

But this acceleration is what we usually call the acceleration of free fall; and usually designate by g. Thus, g is related to the mass and the radius of the Earth,

$$g = \frac{GM_E}{R_E^2} \tag{9.6}$$

This equation establishes the connection between the ordinary force of gravity we experience at the surface of the Earth and Newton's Law of Universal Gravitation. Notice that g is only approximately constant. Small changes in height near the Earth's surface have little effect on the value given by Eq. (9.6), since $R_E \approx 6.4 \times 10^6$ m is so large. But for a large altitude h above the Earth's surface, we must replace R_E with $R_E + h$ in Eq. (9.6), and appreciable changes in g can occur.

Note that an equation analogous to Eq. (9.6) relates the acceleration of free fall at the surface of any (spherical) celestial body to the mass and the radius of that body.

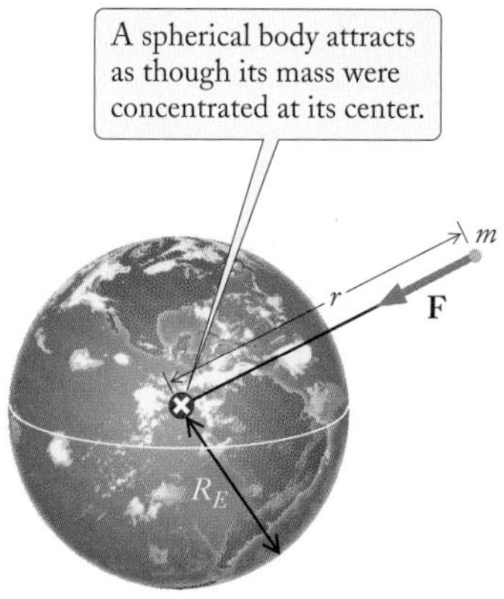

FIGURE 9.3 The gravitational force exerted by the Earth on a particle is directed toward the center of the Earth.

For example, we can calculate the acceleration of free fall on the surface of the Moon from its mass and radius.

EXAMPLE 2 The mass of the Moon is 7.35×10^{22} kg, and its radius is 1.74×10^{6} m. Calculate the acceleration of free fall on the Moon and compare with acceleration of free fall on the Earth.

SOLUTION: For the Moon, the formula analogous to Eq. (9.6) is

$$g_{\text{Moon}} = \frac{GM_{\text{Moon}}}{R_{\text{Moon}}^2} = \frac{6.67 \times 10^{-11}\,\text{N}\cdot\text{m}^2/\text{kg}^2 \times 7.35 \times 10^{22}\,\text{kg}}{(1.74 \times 10^{6}\,\text{m})^2}$$

$$= 1.62\ \text{m/s}^2$$

This is about 1/6 the acceleration of free fall on the surface of the Earth ($g = 9.81\ \text{m/s}^2$). If you can jump upward to a height of one-half meter on the Earth, then this same jump will take you to a height of 3 meters on the Moon!

EXAMPLE 3 The masses of the Sun, Earth, and Moon are 1.99×10^{30} kg, 5.98×10^{24} kg, and 7.35×10^{22} kg, respectively. Assume that the location of the Moon is such that the angle subtended by the lines from the Moon to the Sun and from the Moon to the Earth is 45.0°, as shown in Fig. 9.4a. What is the net force on the Moon due to the gravitational forces of the Sun and Earth? The Moon is 1.50×10^{11} m from the Sun and 3.84×10^{8} m from the Earth.

SOLUTION: Before finding the resultant force, we first find the magnitudes of the individual forces. The magnitude of the force due to the Sun on the Moon is

$$F_{SM} = \frac{GM_S M_M}{R_{SM}^2}$$

$$= \frac{6.67 \times 10^{-11}\,\text{N}\cdot\text{m}^2/\text{kg}^2 \times 1.99 \times 10^{30}\,\text{kg} \times 7.35 \times 10^{22}\,\text{kg}}{(1.50 \times 10^{11}\,\text{m})^2}$$

$$= 4.34 \times 10^{20}\,\text{N}$$

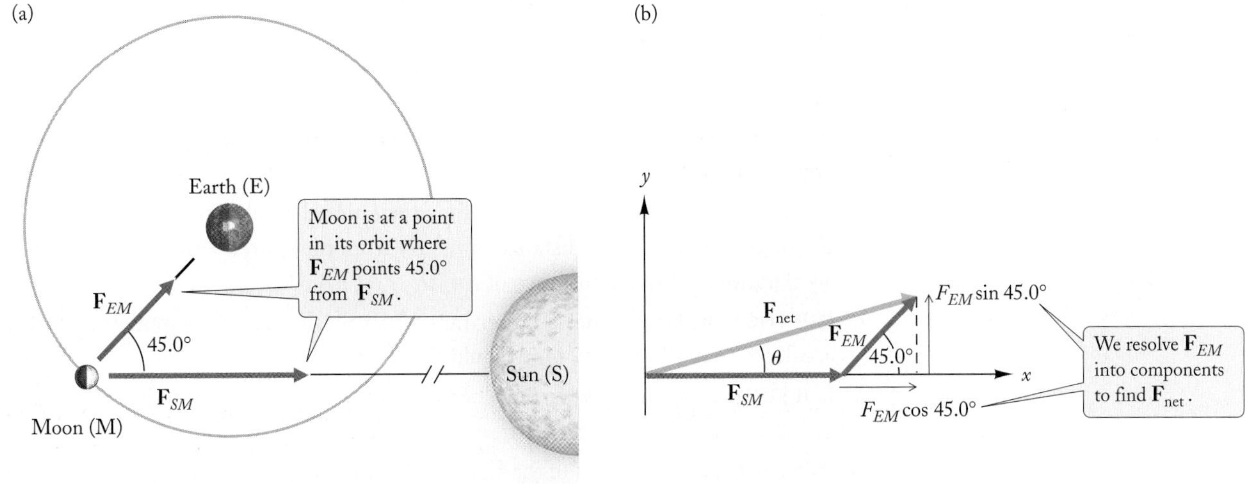

FIGURE 9.4 (a) Each of the gravitational forces on the Moon is directed toward the body producing the force. (b) Vector addition of the two forces.

The magnitude of the force due to the Earth on the Moon is

$$F_{EM} = \frac{GM_E M_M}{R_{EM}^2}$$

$$= \frac{6.67 \times 10^{-11}\,\text{N·m}^2/\text{kg}^2 \times 5.98 \times 10^{24}\,\text{kg} \times 7.35 \times 10^{22}\,\text{kg}}{(3.84 \times 10^8\,\text{m})^2}$$

$$= 1.99 \times 10^{20}\,\text{N}$$

The direction of each force on the Moon is toward the body producing that force, as indicated in Fig. 9.4a. We choose the x axis along the Moon–Sun direction and add the two forces vectorially as shown in Fig. 9.4b. By resolving $\mathbf{F}_{EM}$ into components, we see that the resultant force $\mathbf{F}_{\text{net}}$ has x component

$$F_{\text{net},x} = F_{SM} + F_{EM} \cos 45.0°$$

$$= 4.34 \times 10^{20}\,\text{N} + 1.99 \times 10^{20}\,\text{N} \times \cos 45.0° = 5.75 \times 10^{20}\,\text{N}$$

and y component

$$F_{\text{net},y} = F_{EM} \sin 45.0° = 1.99 \times 10^{20}\,\text{N} \times \sin 45.0° = 1.41 \times 10^{20}\,\text{N}$$

Thus the resultant force has magnitude

$$F_{\text{net}} = \sqrt{F_{\text{net},x}^2 + F_{\text{net},y}^2}$$

$$= \sqrt{(5.75 \times 10^{20}\,\text{N})^2 + (1.41 \times 10^{20}\,\text{N})^2} = 5.92 \times 10^{20}\,\text{N}$$

The direction of $\mathbf{F}_{\text{net}}$ is given by $\tan\theta = \dfrac{F_{\text{net},y}}{F_{\text{net},x}} = \dfrac{1.41 \times 10^{20}\,\text{N}}{5.75 \times 10^{20}\,\text{N}} = 0.245.$

With a calculator, we find that the inverse tangent of 0.238 is

$$\theta = 13.8°$$

✔ Checkup 9.1

QUESTION 1: Neptune is about 30 times as far away from the Sun as the Earth. Compare the gravitational force that the Sun exerts on a 1-kg piece of Neptune with the force it exerts on a 1-kg piece of the Earth. By what factor do these forces differ?

QUESTION 2: Saturn is about 10 times as far away from the Sun as the Earth, and its mass is about 100 times as large as the mass of the Earth. Is the force that the Sun exerts on Saturn larger, smaller, or about equal to the force it exerts on the Earth? Is the acceleration of Saturn toward the Sun larger, smaller, or about equal to the acceleration of the Earth?

QUESTION 3: Equation (9.6) gives the gravitational acceleration at the surface of the Earth, that is, at a radial distance of $r = R_E$ from the center. What is the gravitational acceleration at a radial distance of $r = 2R_E$? At $r = 3R_E$?

QUESTION 4: Uranus has a larger mass than the Earth, but a smaller gravitational acceleration at its surface. How could this be possible?

QUESTION 5: Consider a particle located at the exact center of the Earth. What is the gravitational force that the Earth exerts on this particle?

QUESTION 6: If the radius of the Earth were twice as large as it is but the mass remained unchanged, what would be the gravitational acceleration at its surface?

(A) $\frac{1}{8}g$ (B) $\frac{1}{4}g$ (C) g (D) $4g$ (E) $8g$

9.2 THE MEASUREMENT OF G

The gravitational constant G is rather difficult to measure with precision. The trouble is that gravitational forces between masses of laboratory size are extremely small, and thus a very delicate apparatus is needed to detect these forces. Measurements of G are usually done with a **Cavendish torsion balance** (see Fig. 9.5). Two equal, small spherical masses m and m' are attached to a lightweight horizontal beam which is suspended at its middle by a thin vertical fiber. When the beam is left undisturbed, it will settle into an equilibrium position such that the fiber is completely untwisted. If two equal, large masses M and M' are brought near the small masses, the gravitational attraction between each small mass and the neighboring large mass tends to rotate the beam clockwise (as seen from above). The twist of the fiber opposes this rotation, and the net result is that the beam settles into a new equilibrium position in which the force on the beam generated by the gravitational attraction between the masses is exactly balanced by the force exerted by the twisted fiber. The gravitational constant can then be calculated from the measured values of the angular displacement between the two equilibrium positions, the values of the masses, their distances, and the characteristics of the fiber.

HENRY CAVENDISH (1731–1810)
English experimental physicist and chemist. His torsion balance for the absolute measurement of the gravitational force was based on an earlier design used by Coulomb for the measurement of the electric force.

Note that the mass of the Earth can be calculated from Eq. (9.6) using the known values of G, R_E, and g:

$$M_E = \frac{R_E^2 g}{G} = \frac{(6.38 \times 10^6 \text{ m})^2 \times 9.81 \text{ m/s}^2}{6.67 \times 10^{-11} \text{ N}\cdot\text{m}^2/\text{kg}^2}$$

$$= 5.98 \times 10^{24} \text{ kg}$$

This calculation would seem to be a rather roundabout way to arrive at the mass of the Earth, but there is no direct route, since we cannot place the Earth on a balance. Because the calculation requires a prior measurement of the value of G, the Cavendish experiment has often been described figuratively as "weighing the Earth."

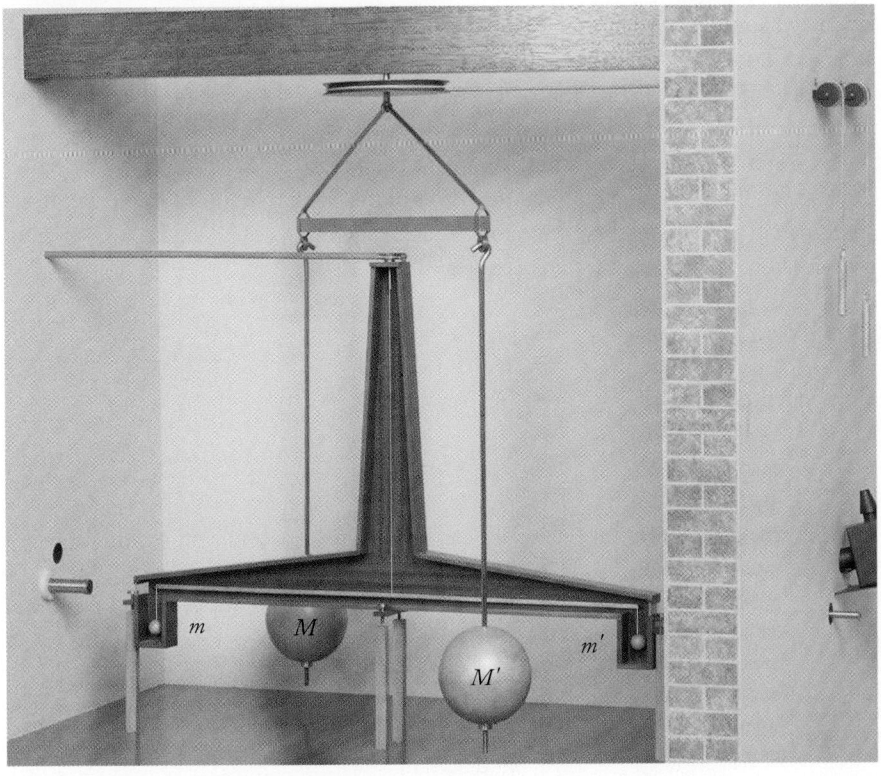

FIGURE 9.5 Model of large torsion balance used by Cavendish. The small masses m, m' hang from the ends of a horizontal beam which is suspended at its middle by a thin vertical fiber.

Checkup 9.2

QUESTION 1: Why don't we determine G by measuring the (fairly large) force between the Earth and a mass of, say, 1 kg?

QUESTION 2: Large mountains produce a (small) deflection of a plumb bob suspended nearby. Could we use this effect to determine G?

(A) Yes (B) No

9.3 CIRCULAR ORBITS

Online Concept Tutorial 11

The gravitational force is responsible for holding the Solar System together; it makes the planets orbit around the Sun, and it makes the satellites orbit around the planets. Although the mutual gravitational forces of the Sun on a planet and of the planet on the Sun are of equal magnitudes, the mass of the Sun is more than a thousand times as large as the mass of even the largest planet, and hence its acceleration is much smaller. It is therefore an excellent approximation to regard the Sun as fixed and immovable, and it then only remains to investigate the motion of the planet. If we designate the masses of the Sun and the planet by M_S and m, respectively, and their center-to-center separation by r, then the magnitude of the gravitational force on the planet is

$$F = \frac{GM_S m}{r^2} \tag{9.7}$$

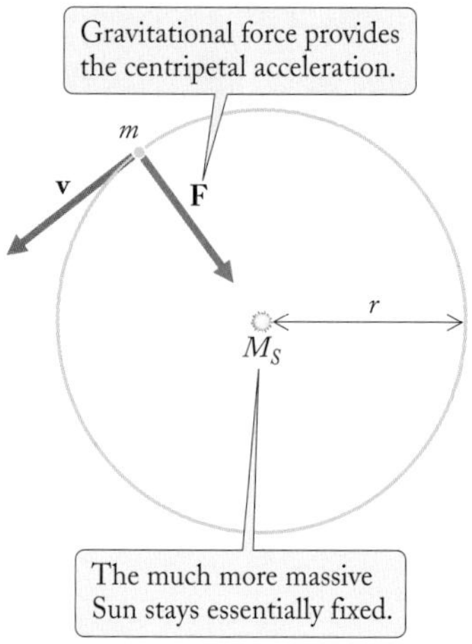

FIGURE 9.6 Circular orbit of a planet around the Sun.

This force points toward the center of the Sun; that is, the center of the Sun is the center of force (see Fig. 9.6). For a particle moving under the influence of such a central force, the simplest orbital motion is uniform circular motion, with the gravitational force acting as centripetal force. The motion of the planets in our Solar System is somewhat more complicated than that—as we will see in the next section, the planets move along ellipses, instead of circles. However, none of these planetary ellipses deviates very much from a circle, and as a first approximation we can pretend that the planetary orbits are circles.

By combining the expression (9.7) for the centripetal force with Newton's Second Law we can find a relation between the radius of the circular orbit and the speed. If the speed of the planet is v, then the centripetal acceleration is v^2/r [see Eq. (4.49)], and the equation of motion, $ma = F$, becomes

$$\frac{mv^2}{r} = F \tag{9.8}$$

Consequently,

$$\frac{mv^2}{r} = \frac{GM_S m}{r^2} \tag{9.9}$$

We can cancel a factor of m and a factor of $1/r$, in this equation, and we obtain

$$v^2 = \frac{GM_S}{r}$$

or

speed for circular orbit

$$v = \sqrt{\frac{GM_S}{r}} \tag{9.10}$$

EXAMPLE 4 The mass of the Sun is 1.99×10^{30} kg, and the radius of the Earth's orbit around the Sun is 1.5×10^{11} m. From this, calculate the orbital speed of the Earth.

SOLUTION: According to Eq. (9.10), the orbital speed is

$$v = \sqrt{\frac{GM_S}{r}} = \sqrt{\frac{6.67 \times 10^{-11}\,\text{N}\cdot\text{m}^2/\text{kg}^2 \times 1.99 \times 10^{30}\,\text{kg}}{1.5 \times 10^{11}\,\text{m}}}$$

$$= 3.0 \times 10^4\,\text{m/s} = 30\,\text{km/s}$$

NICHOLAS COPERNICUS (1473–1543)
Polish astronomer. In his book De Revolutionibus Orbium Coelestium *he formulated the heliocentric system for the description of the motion of the planets, according to which the Sun is immovable and the planets orbit around it.*

The time a planet takes to travel once around the Sun, or the time for one revolution, is called the **period** *of the planet.* We will designate the period by T. The speed of the planet is equal to the circumference $2\pi r$ of the orbit divided by the time T:

$$v = \frac{2\pi r}{T} \tag{9.11}$$

With this expression for the speed, the square of Eq. (9.10) becomes

$$\frac{4\pi^2 r^2}{T^2} = \frac{GM_S}{r} \tag{9.12}$$

which can be rearranged to read

period for circular orbit

$$T^2 = \frac{4\pi^2}{GM_S} r^3 \tag{9.13}$$

This says that *the square of the period is proportional to the cube of the radius of the orbit*, with a constant of proportionality depending on the mass of the central body.

EXAMPLE 5 Both Venus and the Earth have approximately circular orbits around the Sun. The period of the orbital motion of Venus is 0.615 year, and the period of the Earth is 1 year. According to Eq. (9.13), by what factor do the sizes of the two orbits differ?

SOLUTION: If we take the cube root of both sides of Eq. (9.13), we see that the orbital radius is proportional to the 2/3 power of the period. Hence we can set up the following proportion for the orbital radii of the Earth and Venus:

$$\frac{r_E}{r_V} = \frac{T_E^{2/3}}{T_V^{2/3}} = \frac{(1\ \text{year})^{2/3}}{(0.615\ \text{year})^{2/3}} = 1.38 \tag{9.14}$$

An equation analogous to Eq. (9.13) also applies to the circular motion of a moon or artificial satellite around a planet. In this case, the planet plays the role of the central body and, in Eq. (9.13), its mass replaces the mass of the Sun.

EXAMPLE 6 A communications satellite is in a circular orbit around the Earth, in the equatorial plane. The period of the orbit of such a satellite is exactly 1 day, so that the satellite always hovers in a fixed position relative to the rotating Earth. What must be the radius of such a "geosynchronous," or "geostationary," orbit?

Concepts in Context

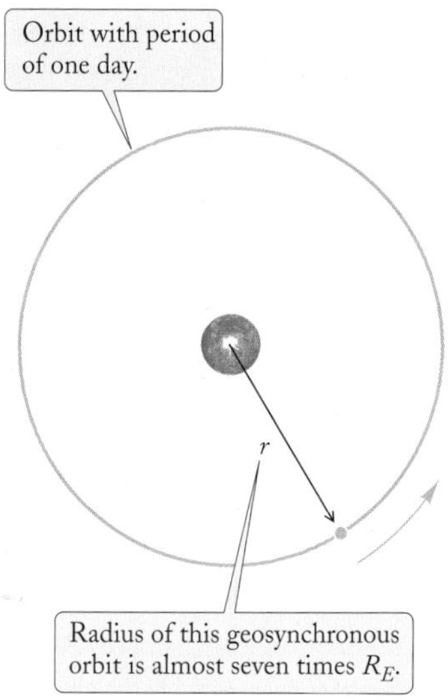

FIGURE 9.7 Orbit of a "geostationary" satellite around the Earth.

SOLUTION: Since the central body is the Earth, the equation analogous to Eq. (9.13) is

$$T^2 = \frac{4\pi^2}{GM_E} r^3 \tag{9.15}$$

or

$$r^3 = \frac{GM_E T^2}{4\pi^2} \tag{9.16}$$

Taking the cube root of both sides of this equation, we find

$$r = \left(\frac{GM_E T^2}{4\pi^2}\right)^{1/3}$$

$$= \left(\frac{6.67 \times 10^{-11}\,\mathrm{N{\cdot}m^2/kg^2} \times 5.98 \times 10^{24}\,\mathrm{kg} \times (24 \times 60 \times 60\,\mathrm{s})^2}{4\pi^2}\right)^{1/3}$$

$$= 4.23 \times 10^7\,\mathrm{m} \tag{9.17}$$

The orbit is shown in Fig. 9.7, which is drawn to scale. The radius of the orbit is about 6.6 times the radius of the Earth.

Concepts in Context

EXAMPLE 7 Surveillance satellites and spacecraft such as the Space Shuttle (Fig. 9.8) often operate in low-altitude orbits quite near the Earth, just above the atmosphere. Such orbits can have a radius as small as $r_{low} = 6.6 \times 10^6$ m; this is less than one-sixth of the geostationary orbit radius $r_{geo} = 4.23 \times 10^7$ m. Calculate how often the low-altitude satellites and spacecraft circle the Earth.

SOLUTION: Taking the square root of both sides of Eq. (9.13), we see that the period is proportional to the 3/2 power of the orbital radius. Hence we can set up the following proportion for the orbital periods:

$$\frac{T_{low}}{T_{geo}} = \left(\frac{r_{low}}{r_{geo}}\right)^{3/2} = \left(\frac{6.6 \times 10^6\,\mathrm{m}}{4.23 \times 10^7\,\mathrm{m}}\right)^{3/2}$$

$$= 0.062$$

or, since the geostationary period T_{geo} is one day, or 24 h,

$$T_{low} = 0.062 \times 24\,\mathrm{h} = 1.5\,\mathrm{h}$$

Thus such "fly-bys" occur quite frequently: 16 times per day.

FIGURE 9.8 The Space Shuttle in orbit with its cargo bay open.

Checkup 9.3

QUESTION 1: The orbit of the geostationary satellite illustrated in Fig. 9.7 is in the equatorial plane, and the satellite is stationary above a point on the Earth's equator. Why can't we keep a satellite stationary above a point that is not on the equator, say, above San Francisco?

PHYSICS IN PRACTICE — COMMUNICATIONS SATELLITES AND WEATHER SATELLITES

Concepts — in — Context

More than a hundred communications and weather satellites have been placed in geostationary orbits. The communications satellites use radio signals to relay telephone and TV signals from one point on the Earth to another. The weather satellites capture pictures of the cloud patterns and measure the heights of clouds, wind speeds, atmospheric and ground temperatures, and moisture in the atmosphere. These observations are especially useful for monitoring weather conditions over the oceans, where there are few observation stations at ground level. Data collected by weather satellites permit early detection of dangerous tropical storms (hurricanes, typhoons) and forecasting of the tracks and the strengths of these storms.

The launch vehicle for these satellites usually consists of a two-stage rocket, which carries the satellite to a low-altitude orbit. A small rocket motor attached to the satellite is then used to lift the satellite from the low-altitude orbit to the high-altitude geostationary orbit. Alternatively, the satellite can be ferried to the low-altitude orbit by the Space Shuttle.

At the high altitude of the geostationary orbit there is no atmospheric drag, and a satellite placed in such an orbit will continue to orbit the Earth indefinitely. However, the orbital motion of the satellite is disturbed by the Moon and the Sun, and it is also affected by the nonspherical shape of the Earth, which produces deviations from the ideal uniform centripetal force. These disturbances cause the satellite to drift from its geostationary position. This requires an adjustment of the orbit every few weeks, which is done with small control nozzles on the satellite. Typically, a satellite carries enough propellant to operate its control nozzles for 10 years, by which time other components in the satellite will also have worn out, or will have been superseded by new technology, so it becomes desirable to switch the satellite off, and replace it by a new model.

Communications satellites contain a radio receiver and a transmitter connected to dish antennas aimed at radio stations on the ground. The signal received from one station on the ground is amplified by the satellite, and then this amplified signal is retransmitted to the other station (the satellite acts as a transponder).

FIGURE 1 Astronauts handle an INTELSAT communications satellite.

Figure 1 shows a recent model of the INTELSAT series of communications satellites. This satellite has a length of 5.2 m, a diameter of 3.6 m, and a mass of 2240 kg. It is powered by solar panels that convert the energy of sunlight into electricity, delivering 2300 watts of power. It contains 50 transponders and is capable of handling 40 000 telephone circuits simultaneously.

For intercontinental communications, three groups of INTELSAT satellites are deployed at geostationary positions over the Atlantic, Pacific, and Indian Oceans. But communications satellites are also cost-effective for communications over shorter ranges, when there is a shortage of telephone cables. Many countries have launched communications satellites to handle telephone traffic within their borders. Communications satellites also relay TV transmissions. A small dish antenna connected to an amplifier permits home television sets to pick up a multitude of TV channels from these satellites.

QUESTION 2: The period of the orbital motion of the Moon around the Earth is 27 days. If the orbit of the Moon were twice as large as it is, what would be the period of its motion?

QUESTION 3: The mass of a planet can be determined by observing the period of a moon in a circular orbit around the planet. For such a mass determination, which of the following do we need: the period, the radius of the moon's orbit, the mass of the moon, the radius of the planet?

QUESTION 4: The radius of the orbit of Saturn around the Sun is about 10 times the radius of the orbit of the Earth. Accordingly, what must be the approximate period of its orbital motion?

(A) 1000 yr (B) 100 yr (C) 30 yr (D) 10 yr (E) 3 yr

Online Concept Tutorial 12

9.4 ELLIPTICAL ORBITS; KEPLER'S LAWS

Although the orbits of the planets around the Sun are approximately circular, none of these orbits are *exactly* circular. We will not attempt the general solution of the equation of motion for such noncircular orbits. A complete calculation shows that with the inverse-square force of Eq. (9.1), the planetary orbits are ellipses. This is **Kepler's First Law**:

Kepler's First Law

The orbits of the planets are ellipses with the Sun at one focus.

Figure 9.9 shows an elliptical planetary orbit (for the sake of clarity, the elongation of this ellipse has been exaggerated; actual planetary orbits have only very small elongations). The point closest to the Sun is called the **perihelion**; the point farthest from the Sun is called the **aphelion**. The sum of the perihelion and the aphelion distances is the **major axis** of the ellipse. The distance from the center of the ellipse to the perihelion (or aphelion) is the **semimajor axis**; this distance equals the average of the perihelion and aphelion distances.

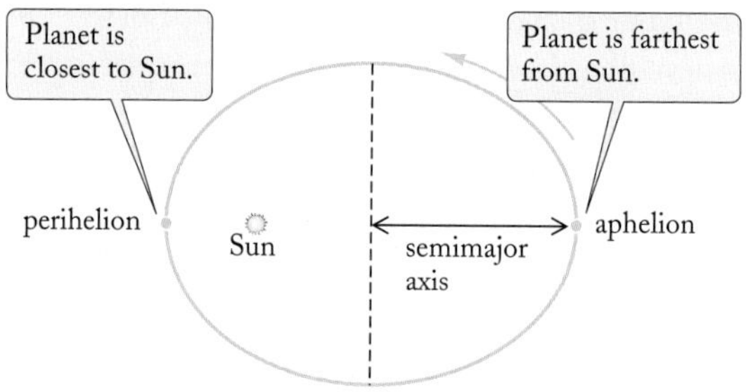

FIGURE 9.9 Orbit of a planet around the Sun. The orbit is an ellipse, with the Sun at one focus.

Kepler originally discovered his First Law and his other two laws (see below) early in the seventeenth century, by direct analysis of the available observational data on planetary motions. Hence, Kepler's laws were originally purely phenomenological statements; that is, they described the phenomenon of planetary motion but did not explain its causes. The explanation came only later, when Newton laid down his laws of motion and his Law of Universal Gravitation and deduced the features of planetary motion from these fundamental laws.

Kepler's Second Law describes the variation in the speed of the motion:

Kepler's Second Law

The radial line segment from the Sun to the planet sweeps out equal areas in equal times.

MATH HELP ELLIPSES

An ellipse is defined geometrically by the condition that the sum of the distance from one focus of the ellipse and the distance from the other focus is the same for all points on the ellipse. This geometrical condition leads to a simple method for the construction of an ellipse: Stick pins into the two foci and tie a length of string to these points. Stretch the string taut to the tip of a pencil, and move this pencil around the foci while keeping the string taut (see Fig. 1a).

An ellipse can also be constructed by slicing a cone obliquely (see Fig. 1b). Because of this, an ellipse is said to be a conic section.

The largest diameter of the ellipse is called the major axis, and the smallest diameter is called the minor axis. The semimajor axis and the semiminor axis are one-half of these diameters, respectively (see Fig. 1c).

If the semimajor axis of length a is along the x axis and the semiminor axis of length b is along the y axis, then the x and y coordinates of an ellipse centered on the origin satisfy

$$\frac{x^2}{a^2} + \frac{y^2}{b^2} = 1$$

The foci are on the major axis at a distance f from the origin given by

$$f = \sqrt{a^2 - b^2}$$

The separation between a planet and the Sun is $a - f$ at perihelion and is $a + f$ at aphelion.

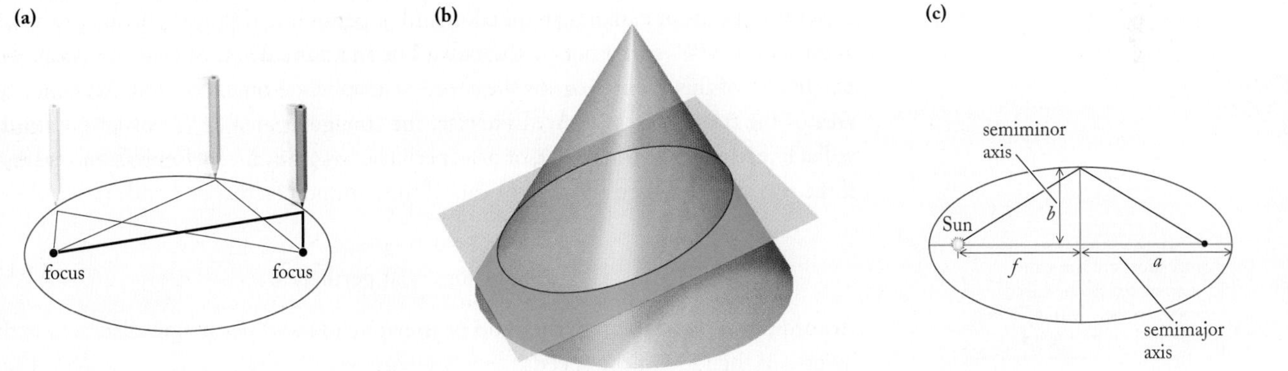

FIGURE 1 (a) Constructing an ellipse. (b) Ellipse as a conic section. (c) Focal distance f, semimajor axis a, and semiminor axis b of an ellipse.

Figure 9.10 illustrates this law. The two colored areas are equal, and the planet takes equal times to move from P to P' and from Q to Q'. According to Fig. 9.10, the speed of the planet is larger when it is near the Sun (at Q) than when it is far from the Sun (at P).

Kepler's Second Law, also called the law of areas, is a direct consequence of the central direction of the gravitational force. We can prove this law by a simple geometrical argument. Consider three successive positions P, P', P'' on the orbit, separated by a relatively small distance. Suppose that the time intervals between P, P' and between P', P'' are equal—say, each of the two intervals is one second. Figure 9.11 shows the positions P, P', P''. Between these positions the curved orbit can be approximated by straight line segments PP' and $P'P''$. Since the time intervals are one unit of time (1 second), the lengths of the segments PP' and $P'P''$ are in proportion to the

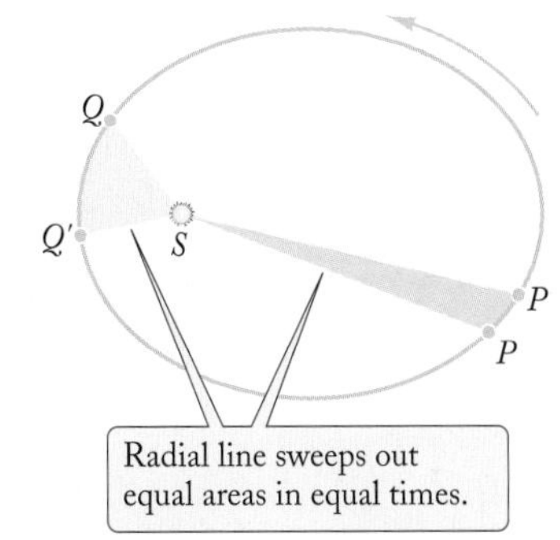

FIGURE 9.10 For equal time intervals, the areas SQQ' and SPP' are equal. The distance QQ' is larger than the distance PP'.

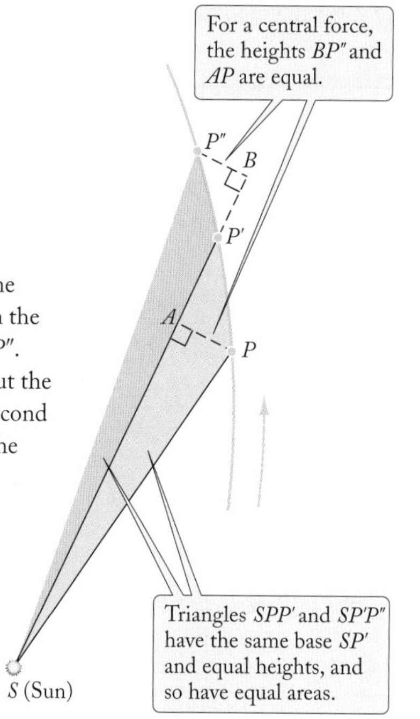

FIGURE 9.11 In one second the planet travels from P to P', and in the next second it travels from P' to P''. The radial line segment sweeps out the triangular area SPP' in the first second and the triangular area $SP'P''$ in the next second.

average velocities in the two time intervals. The velocities differ because the gravitational force causes an acceleration. However, since the direction of the force is toward the center, parallel to the radius, the component of the velocity perpendicular to the radius cannot change. The component of the velocity perpendicular to the radius is represented by the line segment PA for the first time interval, and it is represented by BP'' for the second time interval. These line segments perpendicular to the radius are, respectively, the heights of the triangles SPP' and $SP'P''$ (see Fig. 9.11). Since these heights are equal and since both triangles have the same base SP', their areas must be equal. Thus, the areas swept out by the radial line in the two time intervals must be equal, as asserted by Kepler's Second Law. Note that this geometrical argument depends only on the fact that the force is directed toward a center; it does not depend on the magnitude of the force. This means that Kepler's Second Law is valid not only for planetary motion, but also for motion with any kind of central force.

Let us explore what Kepler's Second Law has to say about the speeds of a planet at aphelion and at perihelion. Figure 9.12 shows the triangular area SPP' swept out by the radial line in a time Δt at, or near, aphelion. The height PP' of this triangle equals the speed v_1 at aphelion times the time Δt; hence the area of the triangle is $\frac{1}{2}r_1 v_1\,\Delta t$. Likewise, the triangular area SQQ' swept out by the radial line in an equal time Δt at, or near, perihelion is $\frac{1}{2}r_2 v_2\,\Delta t$. By Kepler's Second Law these two areas must be equal; if we cancel the common factors of $\frac{1}{2}$ and Δt, we obtain

$$\underset{\text{at aphelion}}{r_1 v_1} = \underset{\text{at perihelion}}{r_2 v_2} \tag{9.18}$$

According to this equation, the ratio of the aphelion and perihelion distances is the inverse of the ratio of the speeds.

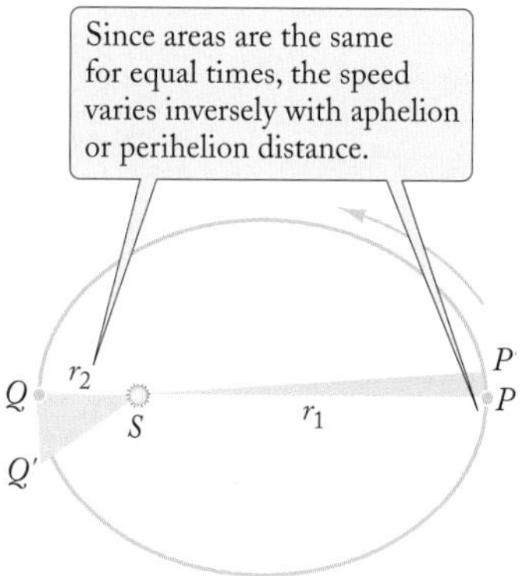

FIGURE 9.12 Triangular area SPP' swept out in one interval Δt after aphelion, and triangular area SQQ' swept out in an identical interval Δt after perihelion.

EXAMPLE 8 The perihelion and aphelion distances for Mercury are 45.9×10^9 m and 69.8×10^9 m, respectively. The speed of Mercury at aphelion is 3.88×10^4 m/s. What is the speed at perihelion?

SOLUTION: From Eq. (9.18),

$$v_2 = \frac{r_1}{r_2}v_1 = \frac{69.8 \times 10^9\ \text{m}}{45.9 \times 10^9\ \text{m}} \times 3.88 \times 10^4\ \text{m/s}$$
$$= 5.90 \times 10^4\ \text{m/s}$$

In Chapter 13 we will become acquainted with the **angular momentum** L, which, for a planet at aphelion or perihelion, is equal to the product rmv. By multiplying both sides of Eq. (9.18) by the mass of the planet m, we see that $r_1 m v_1 = r_2 m v_2$; that is, the angular momentum at aphelion equals the angular momentum at perihelion. Thus, Kepler's Second Law can be regarded as a consequence of a conservation law for angular momentum. We will see that angular momentum is conserved when a particle is under the influence of any central force.

Kepler's Third Law relates the period of the orbit to the size of the orbit:

The square of the period is proportional to the cube of the semimajor axis of the planetary orbit.

Kepler's Third Law

This Third Law, or law of periods, is nothing but a generalization of Eq. (9.13) to elliptical orbits.

Table 9.1 lists the orbital data for the planets of the Solar System. The mean distance listed in this table is defined as the average of the perihelion and aphelion distances; that is, it is the semimajor axis of the ellipse. The difference between the perihelion and aphelion distances gives an indication of the elongation of the ellipse.

TABLE 9.1 THE PLANETS

PLANET (a)	MASS	MEAN DISTANCE FROM SUN (SEMIMAJOR AXIS)	PERIHELION DISTANCE	APHELION DISTANCE	PERIOD
Mercury	3.30×10^{23} kg	57.9×10^6 km	45.9×10^6 km	69.8×10^6 km	0.241 yr
Venus	4.87×10^{24}	108	107	109	0.615
Earth	5.98×10^{24}	150	147	152	1.00
Mars	6.42×10^{23}	228	207	249	1.88
Jupiter	1.90×10^{27}	778	740	816	11.9
Saturn	5.67×10^{26}	1430	1350	1510	29.5
Uranus	8.70×10^{25}	2870	2730	3010	84.0
Neptune	1.03×10^{26}	4500	4460	4540	165
Pluto	1.50×10^{22}	5890	4410	7360	248

(a) A photomontage of the planets in sequence from Mercury (top left, partly hidden) to Pluto (bottom left).

JOHANNES KEPLER (1571–1630) *German astronomer and mathematician. Kepler relied on the theoretical framework of the Copernican system, and he extracted his three laws by a meticulous analysis of the observational data on planetary motions collected by the great Danish astronomer Tycho Brahe.*

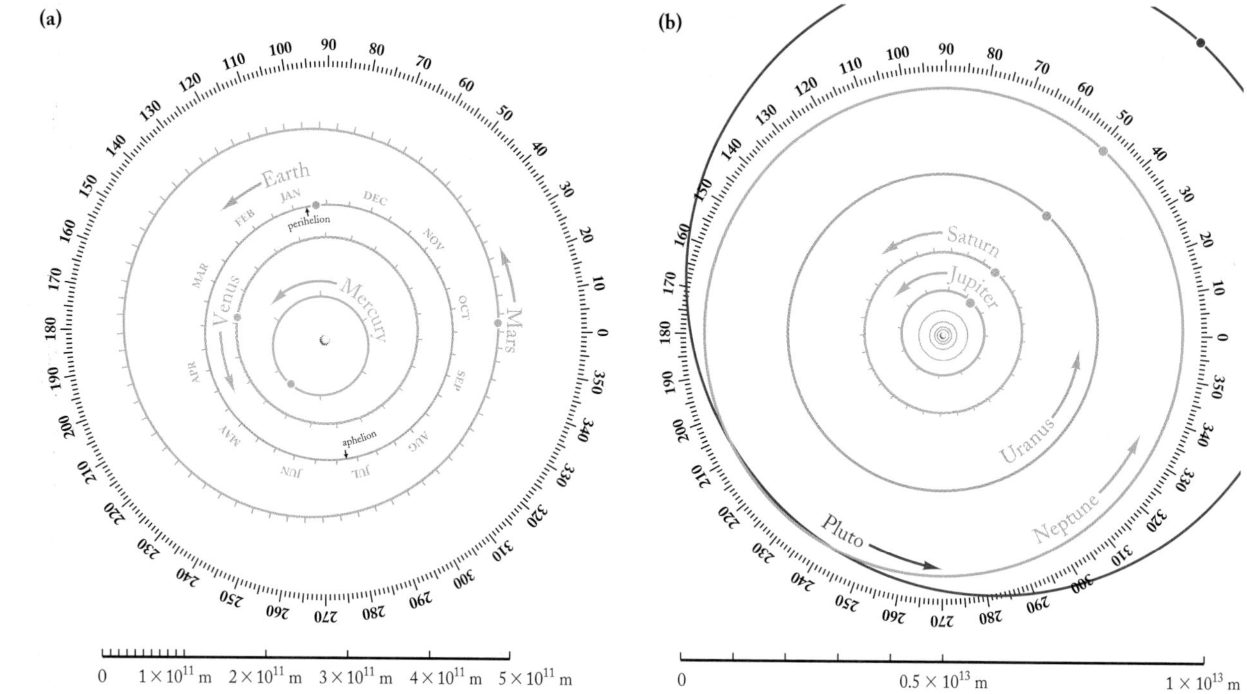

FIGURE 9.13 (a) Orbits of Mercury, Venus, Earth, and Mars. Elliptical orbits can appear quite circular, even when the focus is noticeably off-center, as with Mercury and Mars. The colored dots indicate the positions of these planets on January 1, 2000. The tick marks indicate the positions at intervals of 10 days. (b) Orbits of Jupiter, Saturn, Uranus, and Neptune, and a portion of the orbit of Pluto. The tick marks for Jupiter and Saturn indicate the positions at intervals of 1 year.

Figure 9.13a shows the orbits of the planets Mercury, Venus, Earth, and Mars on scale diagrams. The orbits of Saturn, Jupiter, Uranus, and Neptune and part of the orbit of Pluto are shown in Fig. 9.13b. Inspection of these diagrams reveals that the orbits of Mercury, Mars, and Pluto are noticeably different from circles.[1]

Kepler's three laws apply not only to planets, but also to satellites and to comets. For example, Fig. 9.14 shows the orbits of a few of the many artificial Earth satellites. All these orbits are ellipses. For Earth orbits, the point closest to the Earth is called **perigee**; the point farthest from Earth is called **apogee**. The early artificial satellites were quite small, with masses below 100 kg (see Fig. 9.15). Nowadays, satellites with masses

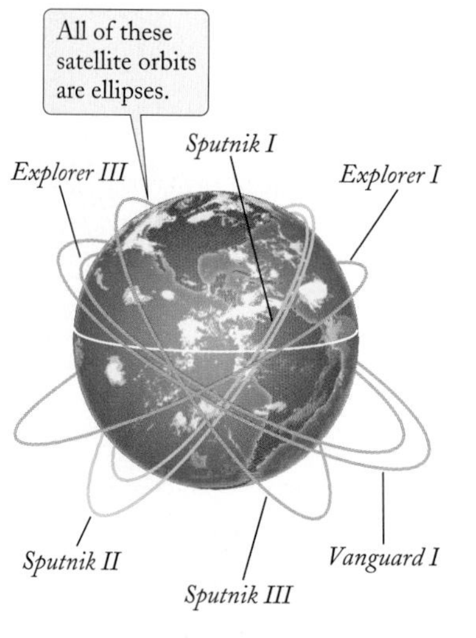

FIGURE 9.14 Orbits of the first artificial Earth satellites. See Table 9.3 for more data.

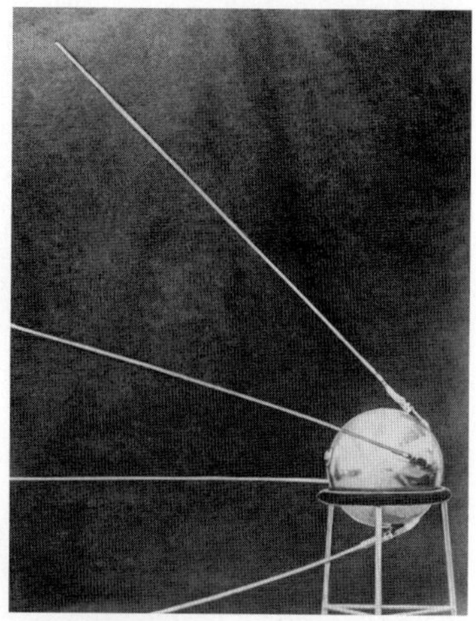

FIGURE 9.15 *Sputnik I*, the first artificial Earth satellite. This satellite had a mass of 83 kg.

[1] Pluto has recently been reclassified by the International Astronomical Union as a dwarf planet, in the same category as Ceres and 2003 UB_{313} (X ena).

of several tons are not unusual. All of the early artificial satellites burned up in the atmosphere after a few months or a few years because they were not sufficiently far from the Earth to avoid the effects of residual atmospheric friction.

Kepler's laws also apply to the motion of a projectile near the Earth. For instance, Fig. 9.16 shows the trajectory of an intercontinental ballistic missile (ICBM). During most of its trajectory, the only force acting on the missile is the gravity of the Earth; the thrust of the engines and the friction of the atmosphere act only during the relatively short initial and final segments of the trajectory (on the scale of Fig. 9.16, these initial and final segments of the trajectory are too small to be noticed). The trajectory is a portion of an elliptical orbit cut short by the impact on the Earth. Likewise, the motion of an ordinary low-altitude projectile, such as a cannonball, is also a portion of an elliptical orbit (if we ignore atmospheric friction). In Chapter 4 we made the near-Earth approximation that gravity was constant in magnitude and direction; with these approximations we found that the orbit of a projectile was a parabola. Although the exact orbit of a projectile is an ellipse, the parabola approximates this ellipse quite well over the relatively short distance involved in ordinary projectile motion; deviations do become noticeable for long-range trajectories (see Fig. 9.17).

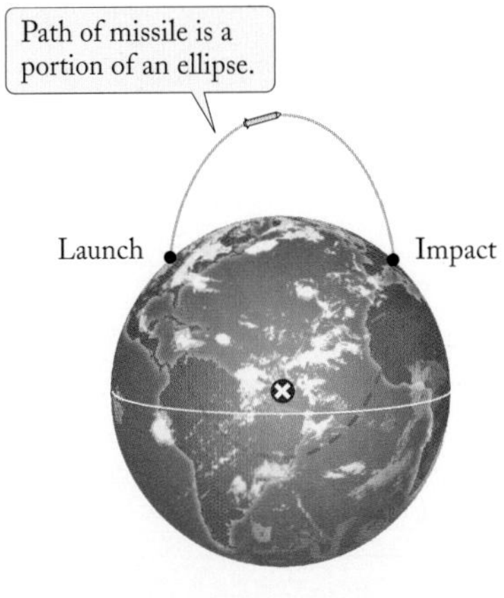

FIGURE 9.16 Orbit of an intercontinental ballistic missile (ICBM). The elongation of the ellipse and the height of the orbit are exaggerated.

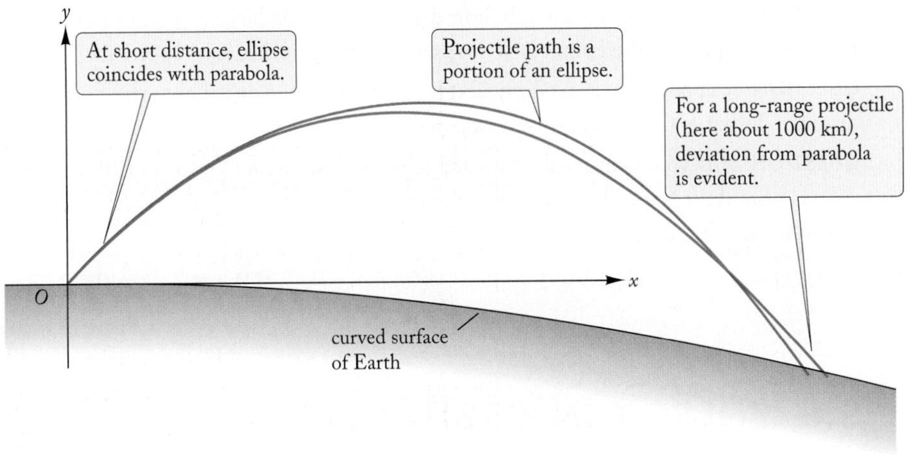

FIGURE 9.17 The parabola (blue curve) approximates the ellipse (red curve) for short distances.

The connection between projectile motion and orbital motion was neatly illustrated by Newton by means of an imaginary experiment, or what today we would call a *Gedankenexperiment.*[1] Newton proposed to fire a cannonball horizontally from a gun emplaced on a high mountain (see Fig. 9.18). If the muzzle velocity is fairly low, the cannonball will arc toward the Earth and strike near the base of the mountain. The trajectory is a segment of a parabola, or, more precisely, a segment of an ellipse. If we increase the muzzle velocity, the cannonball will describe larger and larger arcs. Finally, if the muzzle velocity is just large enough, the rate at which the trajectory curves downward is precisely matched by the curvature of the surface of the Earth—the cannonball never hits the Earth and keeps on falling forever while moving in a circular orbit. This example makes it very clear that orbital motion is free-fall motion.

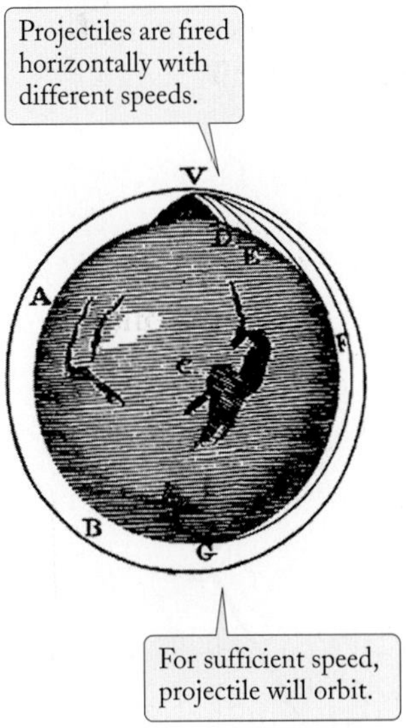

FIGURE 9.18 This drawing from Newton's *Principia* illustrates an imaginary experiment with a cannonball fired from a gun on a high mountain. For a sufficiently large muzzle velocity, the trajectory of the cannonball is a circular orbit.

[1] *Gedankenexperiment* is German for "thought experiment." This word is used by physicists for an imaginary experiment that can be done in principle, but that has never been done in practice, and whose outcome can be discovered by thought.

Finally, we note that in our mathematical description of planetary motion we have neglected the gravitational forces that the planets exert on one another. These forces are much smaller than the force exerted by the Sun, but in a precise calculation the vector sum of all the forces must be taken into account. The net force on any planet then depends on the positions of all the other planets. This means that the motions of all the planets are coupled together, and the calculation of the motion of one planet requires the simultaneous calculation of the motions of all the other planets. This makes the precise mathematical treatment of planetary motion extremely complicated. Kepler's simple laws neglect the complications introduced by the interplanetary forces; these laws therefore do not provide an exact description of planetary motions, but only a very good first approximation.

✓ Checkup 9.4

QUESTION 1: Suppose that the gravitational force were an inverse-cube force, instead of an inverse-square force. Would Kepler's Second Law remain valid? Would Kepler's Third Law remain valid?

QUESTION 2: A comet has an aphelion distance twice as large as its perihelion distance. If the speed of the comet is 40 km/s at perihelion, what is its speed at aphelion?

QUESTION 3: A comet has an elliptical orbit of semimajor axis equal to the Earth–Sun distance. What is the period of such a comet?

QUESTION 4: If you want to place an artificial satellite in an elliptical orbit of period 8 years around the Sun, what must be the semimajor axis of this ellipse? (Answer in units of the Earth–Sun distance.)

(A) 64 (B) $16\sqrt{2}$ (C) 8 (D) 4 (E) 2

9.5 ENERGY IN ORBITAL MOTION

The gravitational force is a conservative force; that is, the work done by this force on a particle moving from some point P_1 to some other point P_2 can be expressed as a difference between two potential energies, and the work done on any round trip starting and ending at some given point is zero. To construct the potential energy, we proceed as in Section 8.1: we calculate the work done by the gravitational force as the particle moves from point P_1 to point P_2, and we seek to express this work as a difference of two terms. In Fig. 9.19, the points P_1 and P_2 are at distances r_1 and r_2, respectively, from the central mass. To calculate the work, we must take into account that the force is a function of the distance; that is, the force is variable. From Section 7.2, we know that for such a variable force, the work is the integral of the force over the distance. If we place the x axis along the line connecting P_1 and P_2 (see Fig. 9.19), then the force can be expressed as

$$F_x = -\frac{GMm}{x^2}$$

and the work is

$$W = \int_{P_1}^{P_2} F_x(x)\,dx = \int_{r_1}^{r_2}\left(-\frac{GMm}{x^2}\right)dx$$

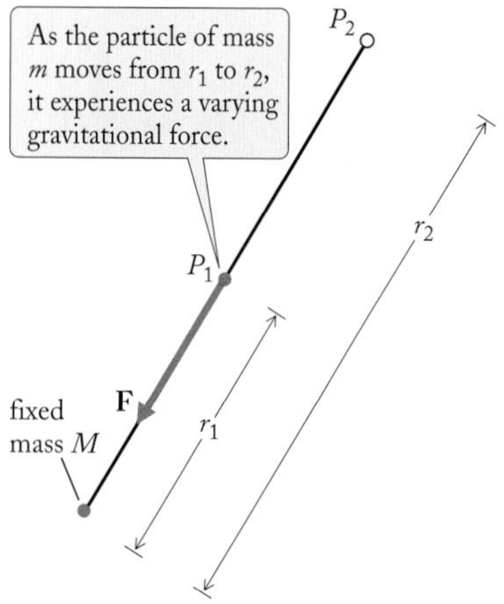

FIGURE 9.19 Two points P_1 and P_2 at distances r_1 and r_2 from the central mass.

We already have evaluated this kind of integral in Example 2 of Chapter 8 (in the case of the gravitational force, the constant A in that example is $A = -GMm$). The result of the integration is

$$W = \left.\frac{GMm}{x}\right|_{r_1}^{r_2} = \frac{GMm}{r_2} - \frac{GMm}{r_1} \tag{9.19}$$

As expected, this result shows that the work is the difference between two potential energies. Accordingly, we can identify the **gravitational potential energy** as

$$U = -\frac{GMm}{r} \tag{9.20}$$

gravitational potential energy

Note that in this calculation of the gravitational potential energy we assumed that the points P_1 and P_2 lie on the same radius (see Fig. 9.19). However, Eq. (9.19) is valid in general, even if P_1 and P_2 do not lie on the same radial line. We can see this by introducing an intermediate point Q, which is on the radial line of P_1 but at the radial distance of P_2 (see Fig. 9.20). To move the particle from P_1 to P_2, we first move it from P_1 to Q along the radial line; this takes the amount of work given by Eq. (9.19). We then move the particle from Q to P_2, along the circular arc of radius r_2; this costs no work, since such a displacement is perpendicular to the force. Any more general path can be constructed from small radial segments and small arcs of circles, and so Eq. (9.19) is true in general.

The potential energy (9.20) is always negative, and its magnitude is inversely proportional to r. Figure 9.21 gives a plot of this potential energy as a function of distance. If the distance r is small, the potential energy is low (the potential energy is much below zero); if the distance r is large, the potential energy is higher (the potential energy is still negative, but not so much below zero). Thus, the potential energy *increases* with distance; it increases from a large negative value to a smaller negative value or to zero. Such an increase of potential energy with distance is characteristic of an attractive force. For instance, if we want to lift a communications satellite from a low initial orbit (just above the Earth's atmosphere) into a high final orbit (such as the geostationary orbit described in Example 6), we must do work on this satellite (by means of a rocket engine). The work we do while lifting the satellite increases the gravitational potential energy from a large negative value (much below zero) to a smaller negative value (not so much below zero).

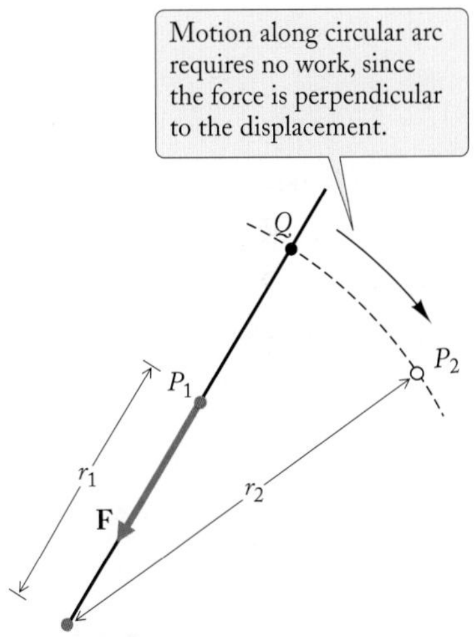

FIGURE 9.20 Two points P_1 and P_2 at distances r_1 and r_2 in different directions. The particle moves from P_1 to Q and then from Q to P_2.

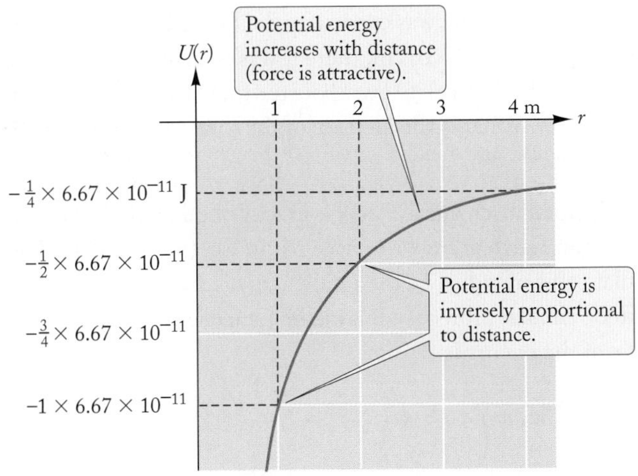

FIGURE 9.21 Gravitational potential energy for a particle of mass 1 kg gravitationally attracted by another particle of mass 1 kg.

The total mechanical energy is the sum of the potential energy and the kinetic energy. Since we are assuming that the mass M is stationary, the kinetic energy is entirely due to the motion of the mass m, and the Law of Conservation of Energy takes the form

Law of Conservation of Energy

$$E = K + U = \tfrac{1}{2}mv^2 - \frac{GMm}{r} = [\text{constant}] \tag{9.21}$$

If the only force acting on the body is the gravitational force (no rocket engine or other external force!), then this total energy remains constant during the motion. For instance, the energy (9.21) is constant for a planet orbiting the Sun, and for a satellite or a spacecraft (with rocket engines shut off) orbiting the Earth. As we saw in Chapter 8, examination of the energy reveals some general features of the motion. Equation (9.21) shows how the orbiting body trades distance ("height") for speed; it implies that if r decreases, v must increase, so that the sum of the two terms $\frac{1}{2}mv^2$ and $-GMm/r$ remains constant. Conversely, if r increases, v must decrease.

Let us now investigate the possible orbits around, say, the Sun from the point of view of their energy. For a circular orbit, we saw in Eq. (9.10) that the orbital speed is

$$v = \sqrt{\frac{GM_S}{r}} \tag{9.22}$$

and so the kinetic energy is

$$K = \tfrac{1}{2}mv^2 = \frac{GM_S m}{2r} \tag{9.23}$$

Hence the total energy is

$$E = K + U = \tfrac{1}{2}mv^2 - \frac{GM_S m}{r} = \frac{GM_S m}{2r} - \frac{GM_S m}{r}$$

or

$$E = -\frac{GM_S m}{2r} \tag{9.24}$$

Consequently, the total energy for a circular orbit is negative and is exactly one-half of the potential energy.

Concepts in Context

EXAMPLE 9 The 1300-kg Syncom communications satellite was placed in its high-altitude geosynchronous orbit of radius 4.23×10^7 m in two steps. First the satellite was carried by the Space Shuttle to a low-altitude circular orbit of radius 6.65×10^6 m; there it was released from the cargo bay of the Space Shuttle, and it used its own booster rocket to lift itself to the high-altitude circular orbit. What is the increase of the total mechanical energy during this change of orbit?

SOLUTION: The total mechanical energy is exactly one-half of the potential energy [Eq. (9.24)]. For an Earth orbit, we replace M_S in Eq. (9.24) by M_E. For the low-altitude circular orbit of radius r_1, the total energy is $E_1 = -GM_Em/2r_1$, and for the high-altitude circular orbit of radius r_2, the total energy is $E_2 = -GM_Em/2r_2$. So the change of the energy is

$$E_2 - E_1 = -\frac{GM_E m}{2}\left(\frac{1}{r_2} - \frac{1}{r_1}\right)$$

$$= -\frac{6.67 \times 10^{-11}\,\text{N}\cdot\text{m}^2/\text{kg}^2 \times 5.98 \times 10^{24}\,\text{kg} \times 1300\,\text{kg}}{2}$$

$$\times \left(\frac{1}{4.23 \times 10^7\,\text{m}} - \frac{1}{6.65 \times 10^6\,\text{m}}\right)$$

$$= 3.29 \times 10^{10}\,\text{J}$$

This energy was supplied by the booster rocket of the satellite.

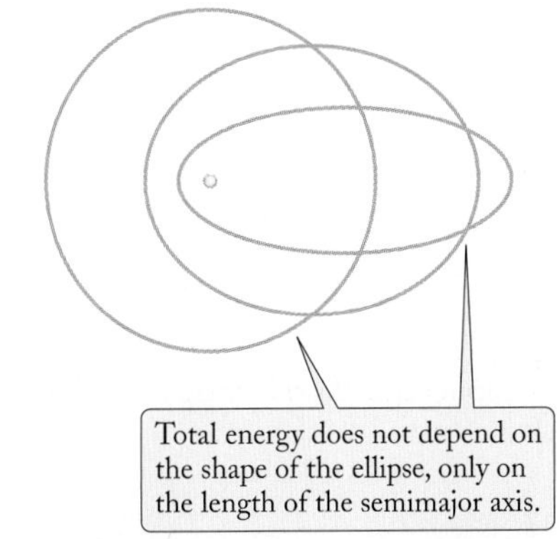

FIGURE 9.22 Orbits of the same total energy. All these orbits have the same semimajor axis.

For an elliptical orbit, the total energy is also negative. It can be demonstrated that the energy can still be written in the form of Eq. (9.24), but the quantity r must be taken equal to the semimajor axis of the ellipse. The total energy of the orbit does not depend on the shape of the ellipse, but only on its larger overall dimension. Figure 9.22 shows several orbits of different shapes but with exactly the same total energy.

From Eq. (9.24) we see that if the energy is nearly zero, then the size of the orbit is very large (note that $E \to 0$ as $r \to \infty$). Such orbits are characteristic of comets, many of which have elliptical orbits that extend far beyond the edge of the Solar System (see Fig. 9.23). If the energy is exactly zero, then the "ellipse" extends all the way to infinity and never closes; such an "open ellipse" is actually a parabola (see Fig. 9.24).

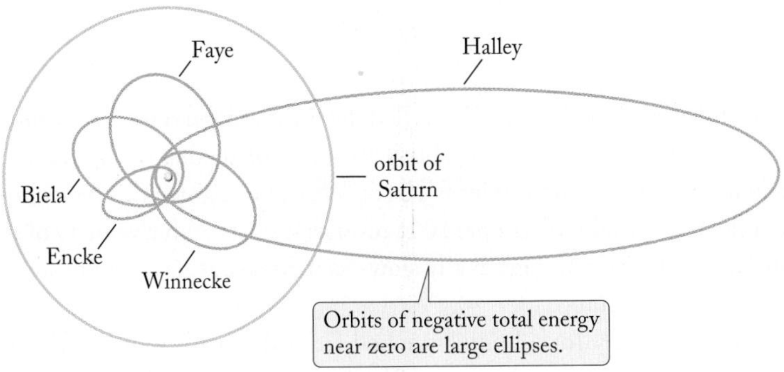

FIGURE 9.23 Orbits of some periodic comets.

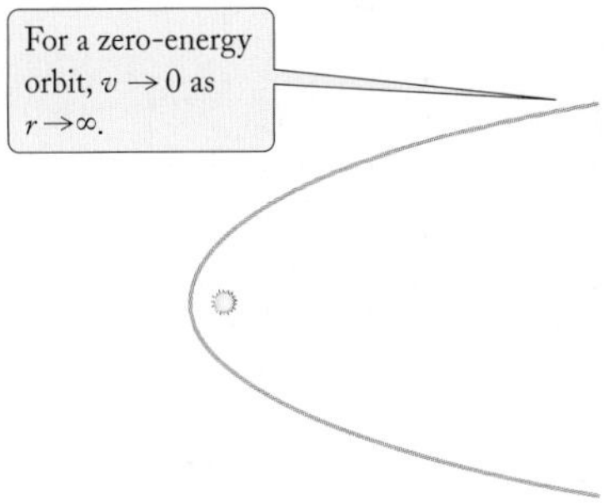

FIGURE 9.24 Orbit of zero energy—a parabola.

Equation (9.21) indicates that if the energy is zero, the comet will reach infinite distance with zero velocity (if $r = \infty$, then $v = 0$). By considering the reverse of this motion, we see that a comet of zero energy, initially at very large distance from the Sun, will fall along this type of parabolic orbit.

If the energy is positive, then the orbit again extends all the way to infinity and again fails to close; such an open orbit is a hyperbola. The comet will then reach infinite distance with some nonzero velocity and continue moving along a straight line (see Fig. 9.25).

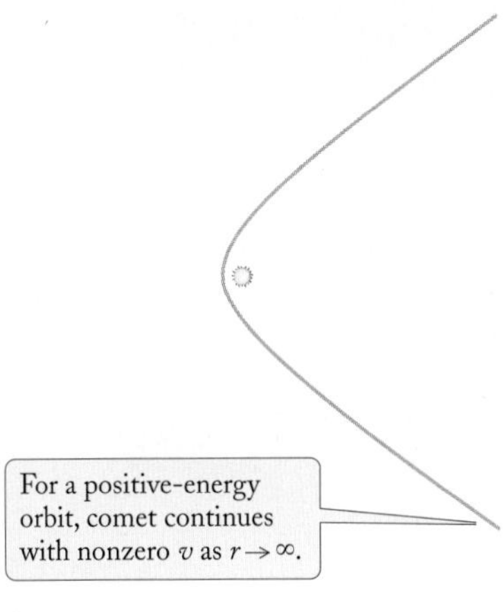

FIGURE 9.25 Orbit of positive energy—a hyperbola.

EXAMPLE 10 A meteoroid (a chunk of rock) is initially at rest in interplanetary space at a large distance from the Sun. Under the influence of gravity, the meteoroid begins to fall toward the Sun along a straight radial line. With what speed does it strike the Sun? The radius of the Sun is 6.96×10^8 m.

SOLUTION: The energy of the meteoroid is

$$E = K + U = \tfrac{1}{2}mv^2 - \frac{GM_S m}{r} = [\text{constant}] \tag{9.25}$$

Initially, both the kinetic and potential energies are zero ($v = 0$ and $r \approx \infty$). Hence at any later time

$$\tfrac{1}{2}mv^2 - \frac{GM_S m}{r} = 0$$

or

$$\tfrac{1}{2}mv^2 = \frac{GM_S m}{r} \qquad (9.26)$$

If we cancel a factor of m and multiply by 2 on both sides of this equation, take the square root of both sides, and substitute $r = R_S$ for the impact on the Sun's surface, we find the speed at the moment of impact:

impact speed and escape velocity

$$v = \sqrt{\frac{2GM_S}{R_S}} \qquad (9.27)$$

With $M_S = 1.99 \times 10^{30}$ kg (see Example 4) and $R_S = 6.96 \times 10^8$ m, we obtain

$$v = \sqrt{\frac{2 \times 6.67 \times 10^{-11}\,\text{N}\cdot\text{m}^2/\text{kg}^2 \times 1.99 \times 10^{30}\,\text{kg}}{6.96 \times 10^8\,\text{m}}}$$
$$= 6.18 \times 10^5\ \text{m/s} = 618\ \text{km/s}$$

The quantity given by Eq. (9.27) is called the Sun's **escape velocity** because it is the minimum initial velocity with which a body must be launched upward from the surface of the Sun if it is to escape and never fall back. We can recognize this by looking at the motion of the meteoroid in Example 10 in reverse: it starts with a velocity of 618 km/s at the surface of the Sun and gradually slows as it rises, but never quite stops until it reaches a very large distance ($r \approx \infty$).

The escape velocity for a body launched from the surface of the Earth can be calculated from a formula analogous to Eq. (9.27), provided that we ignore atmospheric friction and the pull of the Sun on the body. Atmospheric friction will be absent if we launch the body from just above the atmosphere. The pull of the Sun has only a small effect on the velocity of escape from the Earth if we contemplate a body that "escapes" to a distance of, say, $r = 100R_E$ or $200R_E$ rather than $r = \infty$, where we would also have to consider escape from the Sun. For such a motion, the displacement relative to the Sun can be neglected, and the escape velocity v is approximately $\sqrt{2GM_E/R_E} \approx 11.2$ km/s.

Note that the direction in which the escaping body is launched is immaterial—the body will succeed in its escape whenever the direction of launch is above the horizon. Of course, the path that the body takes will depend on the direction of launch (see Fig. 9.26).

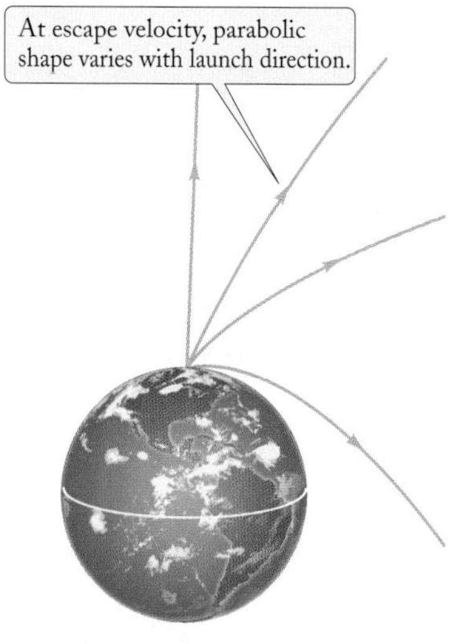

FIGURE 9.26 Different orbits with the same starting point and initial speed. All these orbits are segments of parabolas.

Checkup 9.5

QUESTION 1: An artificial satellite is initially in a circular orbit of fairly low altitude around the Earth. Because of friction with the residual atmosphere, the satellite loses some energy and enters a circular orbit of smaller radius. The speed of the satellite will then be *larger* in the new orbit. How can friction result in an increase of kinetic energy?

QUESTION 2: Does Kepler's Second Law apply to parabolic and hyperbolic orbits?

QUESTION 3: Suppose that we launch a body horizontally from the surface of the Earth, with a velocity exactly equal to the escape velocity of 11.2 km/s. What kind of orbit will this body have? Ignore atmospheric friction.

QUESTION 4: Uranus has a smaller gravitational acceleration at its surface than the Earth. Can you conclude that the escape velocity from its surface is smaller than from the Earth's surface?

QUESTION 5: Suppose that three comets, I, II, and III, approach the Sun. At the instant they cross the Earth's orbit, comet I has a speed of 42 km/s, comet II has a larger speed, and comet III a smaller speed. Given that the orbit of comet I is parabolic, what are the kinds of orbit for comets II and III, respectively?

(A) Elliptical; hyperbolic (B) Elliptical; parabolic (C) Hyperbolic; elliptical
(D) Hyperbolic; parabolic (E) Parabolic; elliptical

SUMMARY

PHYSICS IN PRACTICE Communication Satellites and Weather Satellites **(page 281)**

MATH HELP Ellipses **(page 283)**

LAW OF UNIVERSAL GRAVITATION **(9.1)**

Magnitude:

$$F = \frac{GMm}{r^2}$$

Direction: The force on each mass is directed toward the other mass.

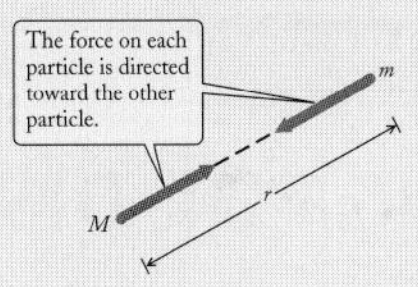

GRAVITATIONAL CONSTANT **(9.2)**

$$G = 6.67 \times 10^{-11}\,\text{N}\cdot\text{m}^2/\text{kg}^2$$

ACCELERATION OF FREE FALL ON EARTH **(9.6)**

$$g = \frac{GM_E}{R_E^2}$$

SPEED FOR CIRCULAR ORBIT AROUND SUN **(9.10)**

$$v = \sqrt{\frac{GM_S}{r}}$$

PERIOD OF ORBIT AROUND SUN

$$T^2 = \frac{4\pi^2}{GM_S} r^3$$

KEPLER'S FIRST LAW The orbits of the planets are ellipses with the Sun at one focus.

KEPLER'S SECOND LAW The radial line segment from the Sun to a planet sweeps out equal areas in equal times.

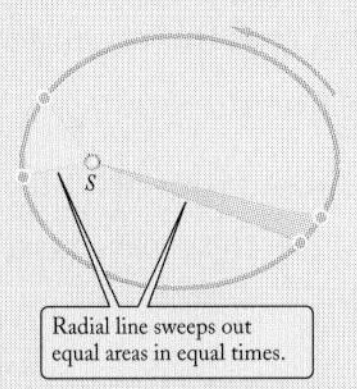

KEPLER'S THIRD LAW The square of the period is proportional to the cube of the semimajor axis of a planetary orbit.

GRAVITATIONAL POTENTIAL ENERGY

$$U = -\frac{GMm}{r} \tag{9.20}$$

CONSERVATION OF ENERGY

$$E = K + U = \tfrac{1}{2}mv^2 - \frac{GMm}{r} = [\text{constant}] \tag{9.21}$$

ENERGY FOR A CIRCULAR ORBIT AROUND THE SUN (Also the energy for an elliptical orbit with semimajor axis r.)

$$E = -\frac{GM_S m}{2r} \tag{9.24}$$

SHAPES OF ORBITS For total mechanical energy E,

$E < 0$ elliptical orbit
$E = 0$ parabolic trajectory
$E > 0$ hyperbolic trajectory

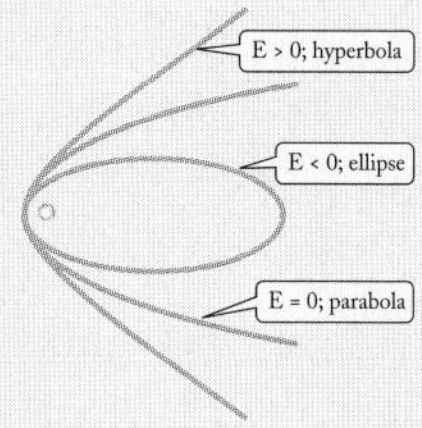

ESCAPE VELOCITY FROM EARTH

$$v = \sqrt{\frac{2GM_E}{R_E}} \tag{9.27}$$

QUESTIONS FOR DISCUSSION

1. Can you directly feel the gravitational pull of the Earth with your sense organs? (Hint: Would you feel anything if you were in free fall?)
2. According to a tale told by Professor R. Lichtenstein, some apple trees growing in the mountains of Tibet produce apples of negative mass. In what direction would such an apple fall if it fell off its tree? How would such an apple hang on the tree?
3. Eclipses of the Moon can occur only at full Moon. Eclipses of the Sun can occur only at new Moon. Why?
4. Explain why the sidereal day (the time of rotation of the Earth relative to the stars, or 23 h 56 min 4 s) is shorter than the mean solar day (the time between successive passages of the Sun over a given meridian, or 24 h). (Hint: The rotation of the Earth around its axis and the revolution of the Earth around the Sun are in the same direction.)
5. Suppose that an airplane flies around the Earth along the equator. If this airplane flew *very* fast, it would not need wings to support itself. Why not?
6. The mass of Pluto was not known until 1978 when a moon of Pluto was finally discovered. How did the discovery of this moon help?
7. It is easier to launch an Earth satellite into an eastward orbit than into a westward orbit. Why?
8. Would it be advantageous to launch rockets into space from a pad at very high altitude on a mountain? Why has this not been done?
9. Describe how you would play squash on a small, round asteroid (with no enclosing wall). What rules of the game would you want to lay down?
10. According to an NBC news report of April 5, 1983, a communications satellite launched from the Space Shuttle went into an orbit as shown in Fig. 9.27. Is this believable?

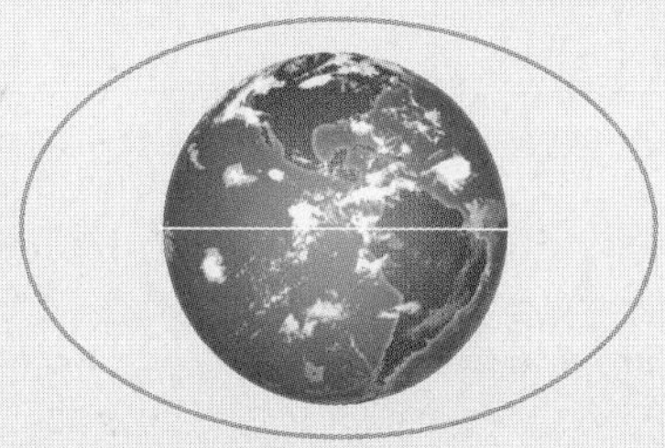

FIGURE 9.27 Proposed orbit for a communications satellite.

11. Does the radial line from the Sun to Mars sweep out area at the same rate as the radial line from the Sun to the Earth?

12. Why were the Apollo astronauts able to jump much higher on the Moon than on the Earth (Fig. 9.28)? If they had landed on a small asteroid, could they have launched themselves into a parabolic or hyperbolic orbit by a jump?

FIGURE 9.28 The jump of the astronaut.

13. The Earth reaches perihelion on January 3 and aphelion on July 6. Why is it not warmer in January than in July?

14. When the Apollo astronauts were orbiting around the Moon at low altitude, they detected several mass concentrations ("mascons") below the lunar surface. What is the effect of a mascon on the orbital motion?

15. An astronaut in a circular orbit above the Earth wants to take his spacecraft into a new circular orbit of larger radius. Give him instructions on how to do this.

16. A Russian and an American astronaut are in two separate spacecraft in the same circular orbit around the Earth. The Russian is slightly behind the American, and he wants to overtake him. The Russian fires his thrusters in the *forward* direction, braking for a brief instant. This changes his orbit into an ellipse. One orbital period later, the astronauts return to the vicinity of their initial positions, but the Russian is now ahead of the American. He then fires his thrusters in the *backward* direction. This restores his orbit to the original circle. Carefully explain the steps of this maneuver, drawing diagrams of the orbits.

17. The gravitational force that a hollow spherical shell of uniformly distributed mass exerts on a particle in its interior is zero. Does this mean that such a shell acts as a gravity shield?

18. Consider an astronaut launched in a rocket from the surface of the Earth and then placed in a circular orbit around the Earth. Describe the astronaut's weight (measured in an inertial reference frame) at different times during this trip. Describe the astronaut's *apparent* weight (measured in his own reference frame) at different times.

19. Several of our astronauts suffered severe motion sickness while under conditions of apparent weightlessness. Since the astronauts were not being tossed about (as in an airplane or a ship in a storm), what caused this motion sickness? What other difficulties does an astronaut face in daily life under conditions of weightlessness?

20. An astronaut on the International Space Station lights a candle. Will the candle burn like a candle on Earth?

21. Astrology is an ancient superstition according to which the planets influence phenomena on the Earth. The only force that can reach over the large distances between the planets and act on pieces of matter on the Earth is gravitation (planets do not have electric charge, and they therefore do not exert electric forces; some planets do have magnetism, but their magnetic forces are too weak to reach the Earth). Given that the Earth is in free fall under the action of the net gravitational force of the planets and the Sun, is there any way that the gravitational forces of the planets can affect what happens on the Earth?

PROBLEMS

9.1 Newton's Law of Universal Gravitation
9.2 The Measurement of G†

1. Two supertankers, each with a mass of 700 000 metric tons, are separated by a distance of 2.0 km. What is the gravitational force that each exerts on the other? Treat them as particles.

2. What is the gravitational force between two protons separated by a distance equal to their diameter, 2.0×10^{-15} m?

3. Somewhere between the Earth and the Moon there is a point where the gravitational pull of the Earth on a particle exactly balances that of the Moon. At what distance from the Earth is this point?

4. Calculate the value of the acceleration of gravity at the surface of Venus, Mercury, and Mars. Use the data on planetary masses and radii given in the table printed inside the book cover.

5. What is the magnitude of the gravitational force that the Sun exerts on you? What is the magnitude of the gravitational force that the Moon exerts on you? The masses of the Sun and the Moon and their distances are given inside the book cover; assume that your mass is 70 kg. Compare these forces with

† For help, see Online Concept Tutorial 11 at www.wwnorton.com/physics

your weight. Why don't you feel these forces? (Hint: You and the Earth are in free fall toward the Sun and the Moon.)

6. Calculate the gravitational force between our Galaxy and the Andromeda galaxy. Their masses are 2.0×10^{11} and 3.0×10^{11} times the mass of the Sun, respectively, and their separation is 2.2×10^6 light-years. Treat both galaxies as point masses.

7. The nearest star is Alpha Centauri, at a distance of 4.4 light-years from us. The mass of this star is 2.0×10^{30} kg. Compare the gravitational force exerted by Alpha Centauri on the Sun with the gravitational force that the Earth exerts on the Sun. Which force is stronger?

8. What is the magnitude of the gravitational attraction the Sun exerts on the Moon? What is the magnitude of the gravitational attraction the Earth exerts on the Moon? Suppose that the three bodies are aligned, with the Earth between the Sun and the Moon (at full moon). What is the direction of the net force acting on the Moon? Suppose that the three bodies are aligned with the Moon between the Earth and the Sun (at new moon). What is the direction of the net force acting on the Moon?

9. Calculate the value of the acceleration due to gravity at the surfaces of Jupiter, Saturn, and Uranus. Use the values of the planetary masses and radii given in the table printed inside the book cover.

10. Somewhere between the Earth and the Sun is a point where the gravitational attraction of the Earth exactly balances that of the Sun. At what fraction of the Earth–Sun distance does this occur?

11. Compare the weight of a 1-kg mass at the Earth's surface with the gravitational force between our Sun and another star of the same mass located at the far end of our galaxy, about 5×10^{20} m away.

12. Each of two adjacent 1.5-kg spheres hangs from a ceiling by a string. The center-to-center distance of the spheres is 8.0 cm. What (small) angle does each string make with the vertical?

13. A 7.0-kg mass is on the x axis at $x = 3.0$ m, and a 4.0-kg mass is on the y axis at $y = 2.0$ m. What is the resultant gravitational force (magnitude and direction) due to these two masses on a third mass of 3.0 kg located at the origin?

14. Three equal masses m are located at the vertices of an equilateral triangle of side a. What is the magnitude of the net gravitational force on each mass due to the other two?

15. Find the acceleration of the Moon due to the pull of the Earth. Express your result in units of the standard g.

16. If a "tower to the sky" of height 2000 km above the Earth's surface could be built, what would be your weight when standing at the top? Assume the tower is located at the South Pole. Express your answer in terms of your weight at the Earth's surface.

17. It has been suggested that strong tidal forces on Io, a moon of Jupiter, could be responsible for the dramatic volcanic activity observed there by Voyager spacecraft. Compare the difference in gravitational accelerations on the near and far surfaces of Io (due to Jupiter) with the difference in accelerations on the near and far side of the Earth (due to the Moon), both as absolute accelerations and as a fraction of the surface g. Io has a mass of 8.9×10^{22} kg and a radius of 1820 km, and is 422×10^3 km from the center of Jupiter.

*18. Suppose that the Earth, Sun, and Moon are located at the vertices of a right triangle, with the Moon located at the right angle (at first or last quarter moon; see Fig. 9.29). Find the magnitude and direction of the sum of the gravitational forces exerted by the Earth and the Sun on the Moon.

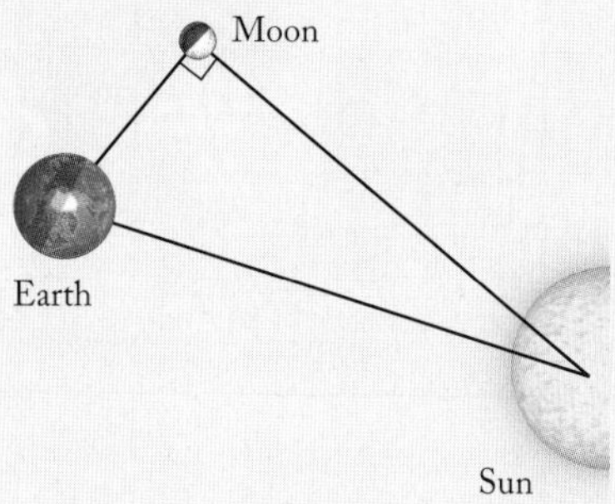

FIGURE 9.29 Earth, Moon, and Sun.

*19. Mimas, a small moon of Saturn, has a mass of 3.8×10^{19} kg and a diameter of 500 km. What is the maximum equatorial velocity with which we can make this moon rotate about its axis if pieces of loose rock sitting on its surface at its equator are not to fly off?

9.3 Circular Orbits†

20. The *Midas II* spy satellite was launched into a circular orbit at a height of 500 km above the surface of the Earth. Calculate the orbital period and the orbital speed of this satellite.

21. Consider the communications satellite described in Example 6. What is the speed of this satellite?

22. Calculate the orbital speed of Venus from the data given in Example 5.

23. The Sun is moving in a circular orbit around the center of our Galaxy. The radius of this orbit is 3×10^4 light-years. Calculate the period of the orbital motion and calculate the orbital speed of the Sun. The mass of our Galaxy is 4×10^{41} kg, and all of this mass can be regarded as concentrated at the center of the Galaxy.

24. Table 9.2 lists some of the moons of Saturn. Their orbits are circular.

    (a) From the information given, calculate the periods and orbital speeds of all these moons.

    (b) Calculate the mass of Saturn.

† For help, see Online Concept Tutorial 11 at www.wwnorton.com/physics

TABLE 9.2 SOME MOONS OF SATURN

MOON	DISTANCE FROM SATURN	PERIOD	ORBITAL SPEED
Tethys (Fig. 9.30)	2.95×10^5 km	1.89 days	—
Dione	3.77	—	—
Rhea	5.27	—	—
Titan	12.22	—	—
Iapetus	35.60	—	—

FIGURE 9.30 Tethys, one of the moons of Saturn.

25. Before clocks with long-term accuracy were constructed, it was proposed that navigators at sea should use the motion of the moons of Jupiter as a clock. The moons Io, Europa, and Ganymede have orbital radii of 422×10^3, 671×10^3, and 1070×10^3 km, respectively. What are the periods of the orbits of these moons? The mass of Jupiter is 1.90×10^{27} kg.

26. A satellite is to be put into an equatorial orbit with an orbital period of 12 hours. What is the radius of the orbit? What is the orbital speed? How many times a day will the satellite be over the same point on the equator if the satellite orbits in the same direction as the Earth's rotation? If it orbits in the opposite direction?

27. An asteroid is in a circular orbit at a distance of two solar diameters from the center of the Sun. What is its orbital period in days?

28. The Sun rotates approximately every 26 days. What is the radius of a "heliosynchronous" orbit, that is, an orbit that stays over the same spot of the Sun?

29. The Apollo command module orbited the Moon while the lunar excursion module visited the surface. If the orbit had a radius of 2.0×10^6 m, how many times per (Earth) day did the command module fly over the excursion module?

30. A Jupiter-sized planet orbits the star 55 Cancri with an orbital radius of 8.2×10^{11} m (see Fig. 9.31). The orbital period of this planet is 13 yr. What is the mass of the star 55 Cancri? How does this compare with the mass of the Sun?

*31. The *Discoverer II* satellite had an approximately circular orbit passing over both poles of the Earth. The radius of the orbit was about 6.67×10^3 km. Taking the rotation of the Earth into account, if the satellite passed over New York City at one instant, over what point of the United States would it pass after completing one more orbit?

*32. The binary star system PSR 1913+16 consists of two neutron stars orbiting with a period of 7.75 h about their center of mass, which is at the midpoint between the stars. Assume that the stars have equal masses and that their orbits are circular with a radius of 8.67×10^8 m.

(a) What are the masses of the stars?

(b) What are their speeds?

*33. Figure 9.32 shows two stars orbiting about their common center of mass in the binary system Krüger 60. The center of mass is at a point between the stars such that the distances of the stars from this point are in the inverse ratio of their masses. Measure the sizes of their orbits and determine the ratio of their masses.

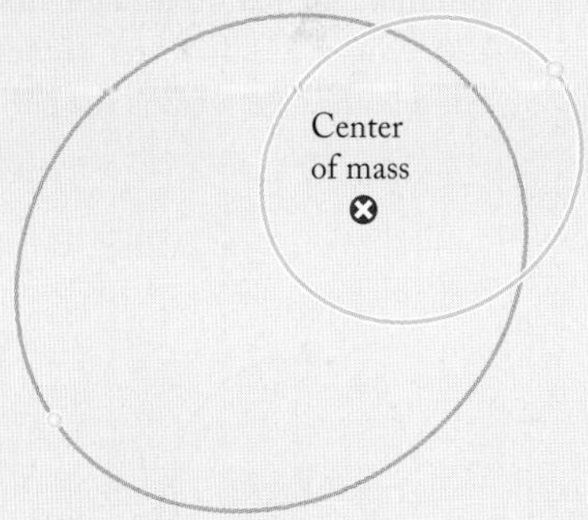

FIGURE 9.32 The orbits of the two stars in the binary system Krüger 60. Each ellipse has its focus at the center of mass.

FIGURE 9.31 (a) The Solar System and (b) the 55 Cancri system.

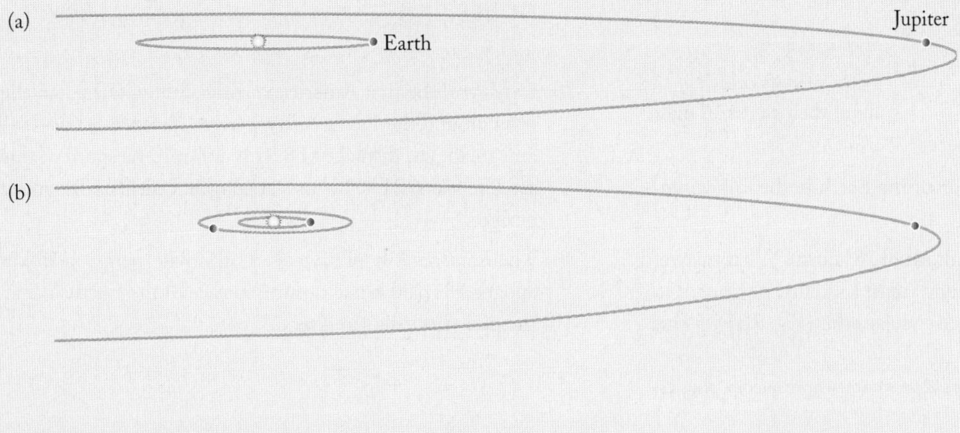

**34. A binary star system consists of two stars of masses m_1 and m_2 orbiting about each other. Suppose that the orbits of the stars are circles of radii r_1 and r_2 centered on the center of mass (Fig. 9.33). The center of mass is a point between the stars such that the radii r_1 and r_2 are in the ratio $r_1/r_2 = m_2/m_1$. Show that the period of the orbital motion is given by

$$T^2 = \frac{4\pi^2}{G(m_1 + m_2)}(r_1 + r_2)^3$$

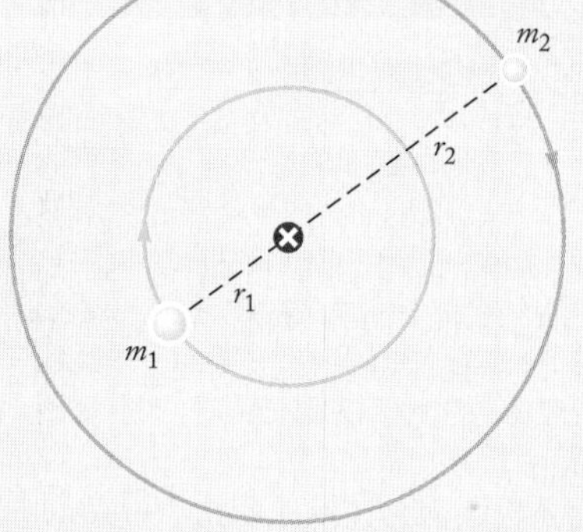

FIGURE 9.33 A binary star system. The orbits are circles about the center of mass.

*35. The binary system Cygnus X-1 consists of two stars orbiting about their center of mass under the influence of their mutual gravitational forces. The orbital period of the motion is 5.6 days. One of the stars is a supergiant with a mass 25 times the mass of the Sun. The other star is believed to be a black hole with a mass about 10 times the mass of the Sun. From the information given, determine the distance between the stars; assume that the orbits of both stars are circular. (Hint: See Problem 34.)

**36. A hypothetical triple star system consists of three stars orbiting about each other. For the sake of simplicity, assume that all three stars have equal masses and that they move along a common circular orbit maintaining an angular separation of 120° (Fig. 9.34). In terms of the mass M of each star and the orbital radius R, what is the period of the motion?

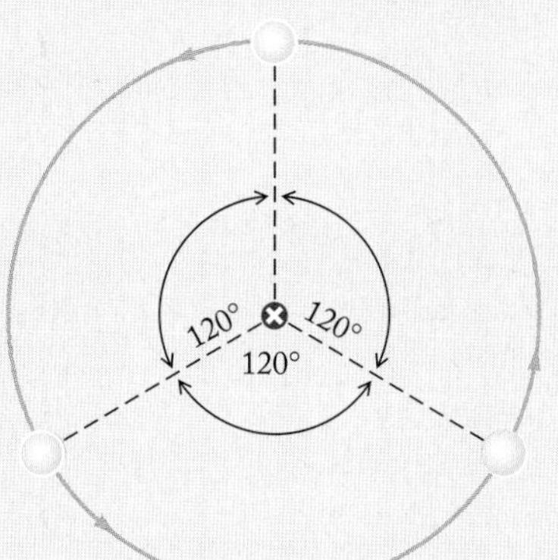

FIGURE 9.34 Three identical stars orbiting about their center of mass.

**37. Take into account the rotation of the Earth in the following problem:

(a) Cape Canaveral is at a latitude of 28° north. What eastward speed (relative to the ground) must a satellite be given if it is to achieve a low-altitude circular orbit (Fig. 9.35)? What westward speed must the satellite be given if it is to travel along the same orbit in the opposite direction? For the purpose of this problem, pretend that "low altitude" means essentially "zero altitude."

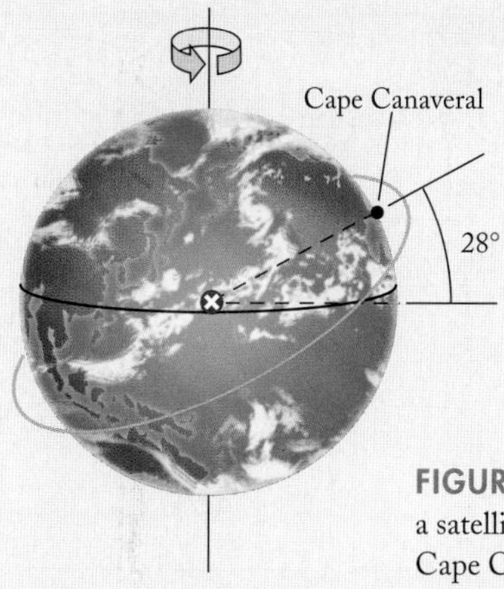

FIGURE 9.35 Orbit of a satellite launched from Cape Canaveral.

(b) Suppose that the satellite has a mass of 14.0 kg. What kinetic energy must the launch vehicle give to the satellite for an eastward orbit? For a westward orbit?

9.4 Elliptical Orbits; Kepler's Laws†

38. Halley's comet (Fig. 9.36) orbits the Sun in an elliptical orbit (the comet reached perihelion in 1986). When the comet is at perihelion, its distance from the Sun is 8.78×10^{10} m, and its speed is 5.45×10^4 m/s. When the comet is at aphelion, its distance is 5.28×10^{12} m. What is the speed at aphelion?

FIGURE 9.36 Halley's comet photographed in 1986.

39. *Explorer I*, the first American artificial satellite, had an elliptical orbit around the Earth with a perigee distance of 6.74×10^6 m and an apogee distance of 8.91×10^6 m. The speed of this satellite was 6.21×10^3 m/s at apogee. Calculate the speed at perigee.

40. The *Explorer X* satellite had an orbit with perigee 175 km and apogee 181,200 km above the surface of the Earth. What was the period of this satellite?

† For help, see Online Concept Tutorial 12 at www.wwnorton.com/physics

TABLE 9.3 THE FIRST ARTIFICIAL EARTH SATELLITES

SATELLITE	MASS	MEAN DISTANCE FROM CENTER OF EARTH (SEMIMAJOR AXIS)	PERIGEE DISTANCE	APOGEE DISTANCE	PERIOD
Sputnik I	83 kg	6.97×10^3 km	6.60×10^3 km	7.33×10^3 km	96.2 min
Sputnik II	3000	7.33	6.61	8.05	104
Explorer I	14	7.83	6.74	8.91	115
Vanguard I	1.5	8.68	7.02	10.3	134
Explorer III	14	7.91	6.65	9.17	116
Sputnik III	1320	7.42	6.59	8.25	106

41. Calculate the orbital periods of *Sputnik I* and *Explorer I* from their apogee and perigee distances in Table 9.3.

42. The aphelion distance for Saturn is 1510×10^6 km; its perihelion distance is 1350×10^6 km. By Kepler's First Law, the Sun is at one focus of this ellipse. How far from the Sun is the other focus? How does this compare with the orbital radius of Mercury?

43. The comet Hale–Bopp was spectacularly visible in the spring of 1997 (see Fig. 9.37) and may be the most viewed comet in history. Its perihelion distance was 137×10^6 km, and its orbital period is 2380 yr. What is its aphelion distance? How does this compare with the mean distance of Pluto from the Sun?

FIGURE 9.37 Comet Hale–Bopp photographed in 1997.

44. The orbit of the Earth deviates slightly from circular: at aphelion, the Earth–Sun distance is 1.52×10^8 km, and at perihelion it is 1.47×10^8 km. By what factor is the speed of the Earth at perihelion greater than the speed at aphelion?

9.5 Energy in Orbital Motion

45. The *Voskhod I* satellite, which carried Yuri Gagarin into space in 1961, had a mass of 4.7×10^3 kg. The radius of the orbit was (approximately) 6.6×10^3 km. What were the orbital speed and the orbital energy of this satellite?

46. What is the kinetic energy and what is the gravitational potential energy for the orbital motion of the Earth around the Sun? What is the total energy?

47. Compare the escape velocity given by Eq. (9.27) with the velocity required for a circular orbit of radius R_S, according to Eq. (9.10). By what factor is the escape velocity larger than the velocity for the circular orbit?

48. In July of 1994, fragments of the comet Shoemaker–Levy struck Jupiter.
    (a) What is the impact speed (equal to the escape speed) for a fragment falling on the surface of Jupiter?
    (b) What is the kinetic energy at impact for a fragment of 1.0×10^{10} kg? Express this energy as an equivalent number of short tons of TNT (the explosion of 1 short ton, or 2000 lb, of TNT releases 4.2×10^9 J).

49. A 1.0-kg mass is in the same orbit around the Earth as the Moon (but far from the Moon). What is the kinetic energy for this orbit? The gravitational potential energy? The total energy?

50. The boosters on a satellite in geosynchronous orbit accidentally fire for a prolonged period. At the instant this "burn" ends, the velocity is parallel to the original tangential direction, but the satellite has been slowed to one-half of its original speed. The satellite is thus at apogee of its new orbit. What is the perigee distance for such an orbit? What happens to the satellite?

51. A black hole is so dense that even light cannot escape its gravitational pull. Assume that all of the mass of the Earth is compressed in a sphere of radius R. How small must R be so the escape speed is the speed of light?

52. The spectacular comet Hale–Bopp (Fig. 9.37), most visible in 1997, entered the Solar System in an elliptical orbit with period 4206 yr. However, after a close encounter with Jupiter on its inbound path, it continues on a new elliptical orbit with a period of 2380 yr. By what fraction did the encounter with Jupiter change the energy of Hale–Bopp's orbit?

53. The typical speed of nitrogen molecules at a temperature of 117°C, the temperature of the Moon's surface at "noon," is 600 m/s; some molecules move slower, others faster. What fraction of the escape velocity from the Moon is this? Can you guess why the Moon has not retained an atmosphere?

*54. The Andromeda galaxy is at a distance of 2.1×10^{22} m from our Galaxy. The mass of the Andromeda is 6.0×10^{41} kg, and the mass of our Galaxy is 4.0×10^{41} kg.

(a) Gravity accelerates the galaxies toward each other. As reckoned in an inertial reference frame, what is the acceleration of the Andromeda galaxy? What is the acceleration of our Galaxy? Treat both galaxies as point particles.

(b) The speed of the Andromeda galaxy *relative to our Galaxy* is 266 km/s. What is the speed of the Andromeda and what is the speed of our Galaxy *relative to the center of mass* of the two galaxies? The center of mass is at a point between the galaxies such that the distances of the galaxies from this point are in the inverse ratios of their masses.

(c) What is the kinetic energy of each galaxy relative to the center of mass? What is the total energy (kinetic and potential) of the system of the two galaxies? Will the two galaxies eventually escape from each other?

*55. Neglect the gravity of the Moon, neglect atmospheric friction, and neglect the rotational velocity of the Earth in the following problem. A long time ago, Jules Verne, in his book *From Earth to the Moon* (1865), suggested sending an expedition to the Moon by means of a projectile fired from a gigantic gun.

(a) With what muzzle speed must a projectile be fired vertically from a gun on the surface of the Earth if it is to (barely) reach the distance of the Moon?

(b) Suppose that the projectile has a mass of 2000 kg. What energy must the gun deliver to the projectile? The explosion of 1 short ton (2000 lb) of TNT releases 4.2×10^{9} J. How many tons of TNT are required for firing this gun?

(c) If the gun barrel is 500 m long, what must be the average acceleration of the projectile during firing?

*56. An artificial satellite of 1300 kg made of aluminum is in a circular orbit at a height of 100 km above the surface of the Earth. Atmospheric friction removes energy from the satellite and causes it to spiral downward so that it ultimately crashes into the ground.

(a) What is the initial orbital energy (gravitational plus kinetic) of the satellite? What is the final energy when the satellite comes to rest on the ground? What is the energy change?

(b) Suppose that all of this energy is absorbed in the form of heat by the material of the satellite. Is this enough heat to melt the material of the satellite? To vaporize it? The heats of fusion and of vaporization of aluminum are given in Table 20.4.

*57. According to one theory, glassy meteorites (tektites) found on the surface of the Earth originate in volcanic eruptions on the Moon. With what minimum speed must a volcano on the Moon eject a stone if it is to reach the Earth? With what speed will this stone strike the surface of the Earth? In this problem ignore the orbital motion of the Moon around the Earth; use the data for the Earth–Moon system listed in the tables printed inside the book cover. (Hint: When the rock reaches the intermediate point where the gravitational pulls of the Moon and the Earth cancel out, it must have zero velocity.)

*58. A spacecraft is launched with some initial velocity toward the Moon from 300 km above the surface of the Earth.

(a) What is the minimum initial speed required if the spacecraft is to coast all the way to the Moon without using its rocket motors? For this problem pretend that the Moon does not move relative to the Earth. The masses and radii of the Earth and the Moon and their distance are listed in the tables printed inside the book cover. (Hint: When the spacecraft reaches the point in space where the gravitational pulls of the Earth and the Moon cancel, it must have zero velocity.)

(b) With what speed will the spacecraft strike the Moon?

59. The Pons–Brooks comet had a speed of 47.30 km/s when it reached its perihelion point, 1.160×10^{8} km from the Sun. Is the orbit of this comet elliptical, parabolic, or hyperbolic?

*60. At a radial distance of 2.00×10^{7} m from the center of the Earth, three artificial satellites (I, II, III) are ejected from a rocket. The three satellites I, II, III are given initial speeds of 5.47 km/s, 4.47 km/s, and 3.47 km/s, respectively; the initial velocities are all in the tangential direction.

(a) Which of the satellites I, II, III will have a circular orbit? Which will have elliptical orbits? Explain your answer.

(b) Draw the circular orbit. Also, superimposed on the same diagram, draw the elliptical orbits of the other satellites; label the orbits with the names of the satellites. (Note: You need not calculate the exact sizes of the ellipses, but your diagram should show where the ellipses are larger or smaller than the circle.)

*61. (a) Since the Moon (*our* Moon) has no atmosphere, it is possible to place an artificial satellite in a circular orbit that skims along the surface of the Moon (provided that the satellite does not hit any mountains!). Suppose that such a satellite is to be launched from the *surface* of the Moon by means of a gun that shoots the satellite in a horizontal direction. With what velocity must the satellite be shot out from the gun? How long does the satellite take to go once around the Moon?

(b) Suppose that a satellite is shot from the gun with a horizontal velocity of 2.00 km/s. Make a rough sketch showing the Moon and the shape of the satellite's orbit; indicate the position of the gun on your sketch.

(c) Suppose that a satellite is shot from the gun with a horizontal velocity of 3.00 km/s. Make a rough sketch showing the Moon and the shape of the satellite's orbit. Is this a closed orbit?

*62. According to an estimate, a large crater on Wilkes Land, Antarctica, was produced by the impact of a 1.2×10^{13}-kg meteoroid incident on the surface of the Earth at 70 000 km/h. What was the speed of this meteoroid relative to the Earth when it was at a "large" distance from the Earth?

*63. An experienced baseball player can throw a ball with a speed of 140 km/h. Suppose that an astronaut standing on Mimas, a small moon of Saturn of mass 3.76×10^{19} kg and radius 195 km, throws a ball with this speed.

(a) If the astronaut throws the ball horizontally, will it orbit around Mimas?

(b) If the astronaut throws the ball vertically, how high will it rise?

*64. An electromagnetic launcher, or rail gun, accelerates a projectile by means of magnetic fields. According to some calculations, it may be possible to attain muzzle speeds as large as 15 km/s with such a device. Suppose that a projectile is launched upward from the surface of the Earth with this speed; ignore air resistance.

(a) Will the projectile escape permanently from the Earth?

(b) Can the projectile escape permanently from the Solar System? (Hint: Take into account the speed of 30 km/s of the Earth around the Sun.)

*65. *Sputnik I*, the first Russian satellite (1957), had a mass of 83.5 kg; its orbit reached perigee at a height of 225 km and apogee at 959 km. *Explorer I*, the first American satellite (1958), had a mass of 14.1 kg; its orbit reached perigee at a height of 368 km and apogee at 2540 km. What was the orbital energy of these satellites?

*66. The orbits of most meteoroids around the Sun are nearly parabolic.

(a) With what speed will a meteoroid reach a distance from the Sun equal to the distance of the Earth from the Sun? (Hint: In a parabolic orbit the speed at any radius equals the escape velocity at the radius. Why?)

(b) Taking into account the Earth's orbital speed, what will be the speed of the meteoroid *relative to the Earth* in a head-on collision with the Earth? In an overtaking collision? Ignore the effect of the gravitational pull of the Earth on the meteoroid.

*67. Calculate the perihelion and the aphelion speeds of Encke's comet. The perihelion and aphelion distances of this comet are 5.06×10^7 km and 61.25×10^7 km. (Hint: Consider the total energy of the orbit.)

*68. The *Explorer XII* satellite was given a tangential velocity of 10.39 km/s when at perigee at a height of 457 km above the Earth. Calculate the height of the apogee. (Hint: Consider the total energy of the orbit.)

**69. Prove that the orbital energy of a planet or a comet in an elliptical orbit around the Sun can be expressed as

$$E = -\frac{GM_S m}{r_1 + r_2}$$

where r_1 and r_2 are, respectively, the perihelion and aphelion distances. [Hint: Use the conservation of energy and the conservation of angular momentum ($r_1 v_1 = r_2 v_2$) at perihelion and at aphelion to solve for v_1^2 and v_2^2 in terms of r_1 and r_2.]

*70. Suppose that a comet is originally at rest at a distance r_1 from the Sun. Under the influence of the gravitational pull, the comet falls radially toward the Sun. Show that the time it takes to reach a radius r_2 is

$$t = -\int_{r_1}^{r_2} \frac{dr}{\sqrt{2GM_S/r - 2GM_S/r_1}}$$

*71. Suppose that a projectile is fired horizontally from the surface of the Moon with an initial speed of 2.0 km/s. Roughly sketch the orbit of the projectile. What maximum height will this projectile reach? What will be its speed when it reaches maximum height?

**72. The Earth has an orbit of radius 1.50×10^8 km around the Sun; Mars has an orbit of radius 2.28×10^8 km. In order to send a spacecraft from the Earth to Mars, it is convenient to launch the spacecraft into an elliptical orbit whose perihelion coincides with the orbit of the Earth and whose aphelion coincides with the orbit of Mars (Fig. 9.38); this orbit requires the least amount of energy for a trip to Mars.

(a) To achieve such an orbit, with what speed (relative to the Earth) must the spacecraft be launched? Ignore the pull of the gravity of the Earth and Mars on the spacecraft.

(b) With what speed (relative to Mars) does the spacecraft approach Mars at the aphelion point? Assume that Mars actually is at the aphelion point when the spacecraft arrives.

(c) How long does the trip from Earth to Mars take?

(d) Where must Mars be (in relation to the Earth) at the instant the spacecraft is launched? Where will the Earth be when the spacecraft arrives at its destination? Draw a diagram showing the relative positions of Earth and Mars at these two times.

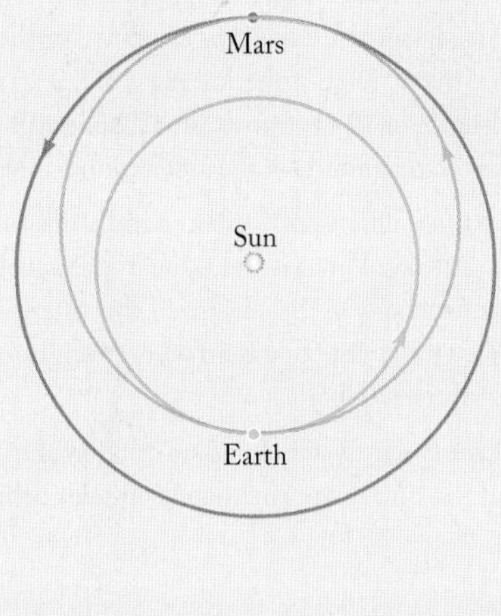

FIGURE 9.38 Orbit for a spacecraft on a trip to Mars.

**73. Repeat the calculations of Problem 72 for the case of a spacecraft launched on a trip to Venus. The orbit of Venus has a radius of 1.08×10^8 km.

**74. If a spacecraft, or some other body, approaches a moving planet on a hyperbolic orbit, it can gain some energy from the motion of the planet and emerge with a larger speed than it had initially. This slingshot effect has been used to boost the speeds of the two Voyager spacecraft as they passed near Jupiter. Suppose that the line of approach of the satellite makes an angle θ with the line of motion of the planet and the line of recession of the spacecraft is parallel to the line of motion of the planet (Fig. 9.39; the planet can be regarded as moving on a straight line during the time interval in question). The speed of the planet is u, and the initial speed of the spacecraft is v (in the reference frame of the Sun).

(a) Show that the final speed of the spacecraft is

$$v' = u + \sqrt{v^2 + u^2 - 2uv\cos\theta}$$

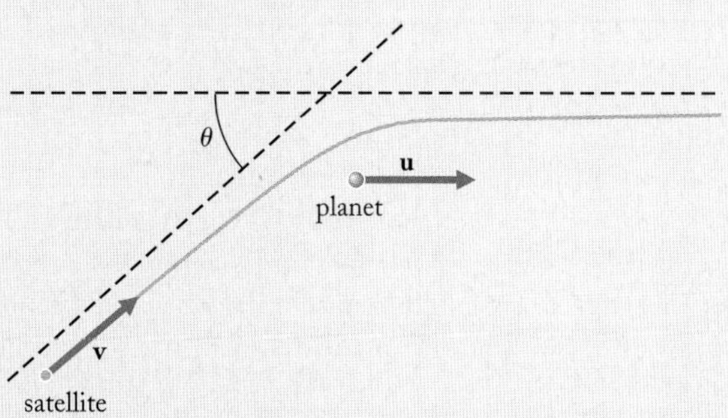

FIGURE 9.39 Trajectory of a spacecraft passing by a planet.

(b) Show that the spacecraft will not gain any speed in this encounter if $\theta = 0$, and show that the spacecraft will gain maximum speed if $\theta = 180°$.

(c) If a spacecraft with $v = 3.0$ km/s approaches Jupiter at an angle of $\theta = 20°$, what will be its final speed?

75. According to one design studied by NASA, a large space colony in orbit around the Earth would consist of a torus of diameter 1.8 km, looking somewhat like a gigantic wheel (see Fig. 9.40). In order to generate artificial gravity of 1g, how fast must this space colony rotate about its axis?

FIGURE 9.40 A rotating space station.

REVIEW PROBLEMS

76. Calculate the gravitational force that the Earth exerts on an astronaut of mass 75 kg in a space capsule at a height of 100 km above the surface of the Earth. Compare with the gravitational force that this astronaut would experience if on the surface of the Earth.

77. The masses used in the Cavendish experiment typically are a few kilograms for the large masses and a few tens of grams for the small masses. Suppose that a "large" spherical mass of 8.0 kg is at a center-to-center distance of 10 cm from a "small" spherical mass of 30 g. What is the magnitude of the gravitational force?

78. The asteroid Ceres has a diameter of 1100 km and a mass of (approximately) 7×10^{20} kg. What is the value of the acceleration of gravity at its surface? On the surface of this asteroid, what would be the weight (in lbf) of a man whose weight on the surface of the Earth is 170 lbf?

79. The asteroid belt of the Solar System consists of chunks of rock orbiting around the Sun in approximately circular orbits. The mean distance of the asteroid belt from the Sun is about 2.9 times the distance of the Earth. What is the mean period of the orbital motion of the asteroids?

80. Imagine that somewhere in interstellar space a small pebble is in a circular orbit around a spherical asteroid of mass 1000 kg. If the radius of the circular orbit is 1.0 km, what is the period of the motion?

81. Europa (Fig. 9.41) is a moon of Jupiter. Astronomical observations show that this moon is in a circular orbit of radius 6.71×10^8 m with a period of 3.55 days. From these data deduce the mass of Jupiter.

82. Observations with the Hubble Space Telescope have revealed that at the center of the galaxy M87, gas orbits around a very massive compact object, believed to be a black hole. The measurements show that gas clouds in a circular orbit of radius 250 light-years have an orbital speed of 530 km/s. From this information, deduce the mass of the black hole.

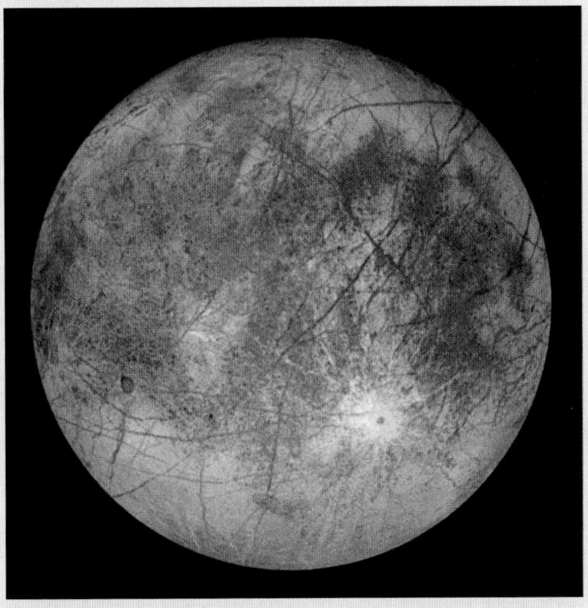

FIGURE 9.41 Europa, one of the moons of Jupiter.

83. Consider a space station in a circular orbit at an altitude of 400 km around the Earth and a piece of debris, left over from, say, the disintegration of a rocket, in an orbit of the same radius but of opposite direction.
    (a) What is the speed of the debris relative to the space station when they pass?
    (b) If the debris hit the spacecraft, it would penetrate the space station with catastrophic consequences for the crew. Penetration depends on the kinetic energy of the debris. What must be the mass of a piece of debris if it is to have an impact energy of 4.6×10^5 J, which corresponds to the explosion of 100 g of TNT?

84. *Vanguard I*, the second American artificial satellite (Fig. 9.42), moved in an elliptical orbit around the Earth with a perigee distance of 7.02×10^6 m and an apogee distance of 10.3×10^6 m. At perigee, the speed of this satellite was 8.22×10^3 m/s. What was the speed at apogee?

*85. The motor of a Scout rocket uses up all its fuel and stops when the rocket is at an altitude of 200 km above the surface of the Earth and is moving vertically at 8.50 km/s. How high will this rocket rise? Neglect any residual atmospheric friction.

*86. An astronaut in a spacecraft in a circular orbit around the Earth wants to get rid of a defective solar panel that he has detached from the spacecraft. He hits the panel with a blast from the steering rocket of the spacecraft, giving it an increment of velocity. This sends the solar panel into an elliptical orbit.
    (a) Sketch the circular orbit of the spacecraft and the elliptical orbit of the solar panel if the velocity increment is parallel to the velocity of the spacecraft and if it is antiparallel.
    (b) If the ratio of the semimajor axis of the ellipse to the radius of the circle has a special value, it is possible for the panel to meet with the spacecraft again after several orbits. What are these special values of the ratio?

*87. A communications satellite of mass 700 kg is placed in a circular orbit of radius 4.23×10^7 m around the Earth.
    (a) What is the total orbital energy of this satellite?
    (b) How much extra energy would we have to give this satellite to put it into a parabolic orbit that permits it to escape to infinite distance from the Earth?

88. What is the escape velocity for a projectile launched from the surface of our Moon?

FIGURE 9.42 The *Vanguard I* satellite.

Answers to Checkups

Checkup 9.1

1. The gravitational force varies inversely with the square of the distance, so the force will be $(30)^2 = 900$ times weaker for a 1-kg piece of Neptune than for a 1-kg piece of the Earth.
2. The gravitational force varies in proportion to the mass and in inverse proportion to the square of the distance, so the 100-times-larger mass for Saturn cancels the 10-times-larger distance; thus, the gravitational force that the Sun exerts on Saturn is about equal to that on the Earth. The acceleration is $a = F/m$, and so is about 100 times smaller for Saturn.
3. The acceleration varies inversely with the square of the distance, and so is $\frac{1}{4}g$ at $r = 2R_E$, and is $\frac{1}{9}g$ at $r = 3R_E$.

4. Since the acceleration at a planet's surface is $a = GM/R^2$, a larger mass M and a smaller gravitational acceleration a are possible only because the radius R of Uranus is sufficiently larger than that of the Earth.

5. At the exact center of the Earth, a particle would be equally attracted in all directions, and so would experience zero net force.

6. (B) $\frac{1}{4}g$. The acceleration at the surface is $a = GM_E/R_E^2$, so a doubled radius would result in an acceleration one-fourth as large, or $\frac{1}{4}g$.

Checkup 9.2

1. To determine G by measuring the force between the Earth and some known mass, we would also have to know the mass of the Earth; we have no independent way of determining the mass of the Earth.

2. (A) Yes. If we knew the mass of the mountain (and the spatial distribution of such mass), then we could determine the gravitational force from the plumb bob's deflection, and thus G.

Checkup 9.3

1. An orbit that is a circle at the latitude of San Francisco is impossible, since the center of every orbit must coincide with the center of the Earth.

2. The period is proportional to the 3/2 power of the radius of the orbit, so for a doubled radius, the period of the Moon would become $2^{3/2} \times 27$ days ≈ 76 days.

3. As in Eq. (9.13), we need only know the period and radius of the moon's orbit to determine the mass of the planet.

4. (C) 30 yr. The period is proportional to the 3/2 power of the radius of the orbit, so the period of Saturn's motion is $10^{3/2} \times 1$ yr ≈ 30 yr.

Checkup 9.4

1. Kepler's Second Law would remain valid, since it depends only on the central nature of the force, and otherwise not on any particular form of the force. Kepler's Third Law, however, like the law of periods, Eq. (9.13), depends on the inverse-square nature of the force. If we were to perform a similar derivation to that preceding Eq. (9.13) for an inverse-cube force, we would find that the period was proportional to the square of the radius.

2. As in Eq. (9.18), the speeds vary inversely with the distances, so for an aphelion distance twice as large as the perihelion distance, the speed at aphelion will be half as large as the speed at perihelion, or will be 20 km/s.

3. According to Kepler's Third Law, the period must be exactly one year. This is so because both the Earth's orbit (nearly circular; the semimajor axis of a circle is its radius) and the comet's orbit have the same semimajor axis, and both orbit the same central body, the Sun.

4. (D) 4. Kepler's Third Law states that the square of the period is proportional to the cube of the semimajor axis of the orbit, so to make the period 8 times as large as the Earth's period would make the cube of the semimajor axis 64 times as large; thus the semimajor axis would be $64^{1/3} = 4$ times as large as the Earth–Sun distance.

Checkup 9.5

1. For a circular orbit, we found that the magnitude of the (negative) potential energy is twice the size of the kinetic energy. Thus the potential energy decreases so much for the lower orbit (it becomes more negative) that the kinetic energy can increase and energy can be lost to friction.

2. Yes—our derivation of the law depended only on the central nature of the force, not on any particular type of orbit (or even any particular form of the central force).

3. If we ignore air friction (and the body does not encounter any obstacles), then the body will escape the Earth's influence in a parabolic "orbit," since the escape velocity provides for zero net energy. The orbit would be similarly parabolic if we launched the body at any angle (except straight up, although that resulting linear path can be considered a special case of the parabola). Ultimately, far from the Earth's influence, the path would be modified by the Sun.

4. No. The gravitational acceleration is $g = GM/R^2$, whereas the escape velocity depends on the gravitational potential energy, which is proportional to M/R. For example, a body with twice the mass and twice the radius of the Earth would have half the gravitational acceleration at the surface, but would have the same escape velocity.

5. (C) Hyperbolic; elliptical. Recall that a parabolic orbit is a zero-energy orbit, where the comet can just barely escape to infinity. The energy of comet II must be positive, since it has a larger speed (a greater kinetic energy, but the same potential energy as it crosses the Earth's orbit); we found that a positive-energy orbit is a hyperbola. Similarly, the energy of comet III must be negative, since it has a smaller speed; negative-energy orbits are ellipses, with a semimajor axis given by Eq. (9.24).

CHAPTER

10

Systems of Particles

Concepts — in — Context

CONCEPTS IN CONTEXT

While this high jumper is passing over the bar, he bends backward and keeps his extremities below the level of the bar. This means that the average height of his body parts is less than if he were to keep his body straight, and he requires less energy to pass over the bar.

The concepts introduced in this chapter permit us to examine in detail several aspects of the motion of the jumper:

? The body of the jumper is a system of particles. Where is the average position of the mass of this system of particles when the body is in a straight configuration? How does this change when the jumper reconfigures his extremities? (Example 8, part (a), page 322)

? What is the gravitational potential energy of a system of particles, and how much does the jumper reduce his potential energy by bending his body? (Page 321 in Section 10.2 and Example 8, part (b), page 322)

? What is equation of motion of a system of particles, and to what extent does the translational motion of a jumper resemble projectile motion? (Page 324 in Section 10.3)

So far we have dealt almost exclusively with the motion of a single particle. Now we will begin to study systems of particles interacting with each other via some forces. This means we must examine, and solve, the equations of motion of all these particles simultaneously.

Since chunks of ordinary matter are made of particles—electrons, protons, and neutrons—all the macroscopic bodies that we encounter in our everyday environment are in fact many-particle systems containing a very large number of particles. However, for most practical purposes, it is not desirable to adopt such an extreme microscopic point of view, and in the preceding chapters we treated the motion of a macroscopic body, such as an automobile, as motion of a particle. Likewise, in dealing with a system consisting of several macroscopic bodies, we will often find it convenient to treat each of these bodies as a particle and ignore the internal structure of the bodies. For example, when investigating a collision between two automobiles, we may find it convenient to pretend that each of the automobiles is a particle—we then regard the colliding automobiles as a system of two particles which exert forces on each other when in contact. And when investigating the Solar System, we may find it convenient to pretend that each planet and each satellite is a particle—we then regard the Solar System as a system of such planet and satellite particles loosely held together by gravitation and orbiting around the Sun and around each other.

The equations of motion of a system of several particles are often hard, and sometimes impossible, to solve. It is therefore necessary to make the most of any information that can be extracted from the general conservation laws. In the following sections we will become familiar with the *momentum* vector, and we will see how the laws of conservation of momentum and of energy apply to a system of particles.

10.1 MOMENTUM

Newton's laws can be expressed very neatly in terms of **momentum,** a vector quantity of great importance in physics. *The momentum of a single particle is defined as the product of the mass and the velocity of the particle:*[1]

momentum of a particle

$$\mathbf{p} = m\mathbf{v} \tag{10.1}$$

Thus, the momentum $\mathbf{p}$ is a vector that has the same direction as the velocity vector, but a magnitude that is m times the magnitude of the velocity. The SI unit of momentum is kg·m/s; this is the momentum of a mass of 1 kg when moving at 1 m/s.

The mathematical definition of momentum is consistent with our intuitive, everyday notion of "momentum." If two cars have equal masses but one has twice the velocity of the other, it has twice the momentum. And if a truck has three times the mass of a car and the same velocity, it has three times the momentum. During the nineteenth century physicists argued whether momentum or kinetic energy was the best measure of the "amount of motion" in a body. They finally decided that the answer

[1] The momentum $\mathbf{p} = m\mathbf{v}$ is sometimes referred to as *linear momentum* to distinguish it from *angular momentum,* discussed in Chapter 13.

depends on the context—as we will see in the examples in this chapter and the next, sometimes momentum is the most relevant quantity, sometimes energy is, and sometimes both are relevant.

Newton's First Law states that, in the absence of external forces, the velocity of a particle remains constant. Expressed in terms of momentum, *the First Law therefore states that the momentum remains constant:*

$$\mathbf{p} = [\text{constant}] \qquad \text{(no external forces)} \tag{10.2}$$

First Law in terms of momentum

Thus, we can say that the momentum of the particle is conserved. Of course, we could equally well say that the velocity of this particle is conserved; but the deeper significance of momentum will emerge when we study the motion of a system of several particles exerting forces on one another. We will find that the total momentum of such a system is conserved—any momentum lost by one particle is compensated by a momentum gain of some other particle or particles.

To express the Second Law in terms of momentum, we note that since the mass is constant, the time derivative of Eq. (10.1) is

$$\frac{d\mathbf{p}}{dt} = m\frac{d\mathbf{v}}{dt}$$

or

$$\frac{d\mathbf{p}}{dt} = m\mathbf{a}$$

But, according to Newton's Second Law, $m\mathbf{a}$ equals the force; hence, *the rate of change of the momentum with respect to time equals the force:*

$$\frac{d\mathbf{p}}{dt} = \mathbf{F} \tag{10.3}$$

Second Law in terms of momentum

This equation gives the Second Law a concise and elegant form.

EXAMPLE 1 A tennis player smashes a ball of mass 0.060 kg at a vertical wall. The ball hits the wall perpendicularly with a speed of 40 m/s and bounces straight back with the same speed. What is the change of momentum of the ball during the impact?

SOLUTION: Take the positive x axis along the direction of the initial motion of the ball (see Fig. 10.1a). The momentum of the ball before impact is then in the positive direction, and the x component of the momentum is

$$p_x = mv_x = 0.060\ \text{kg} \times 40\ \text{m/s} = 2.4\ \text{kg}\cdot\text{m/s}$$

The momentum of the ball after impact has the same magnitude but the opposite direction:

$$p'_x = -2.4\ \text{kg}\cdot\text{m/s}$$

(Throughout this chapter, the primes on mathematical quantities indicate that these quantities are evaluated *after* the collision.) The change of momentum is

$$\Delta p_x = p'_x - p_x = -2.4\ \text{kg}\cdot\text{m/s} - 2.4\ \text{kg}\cdot\text{m/s} = -4.8\ \text{kg}\cdot\text{m/s}$$

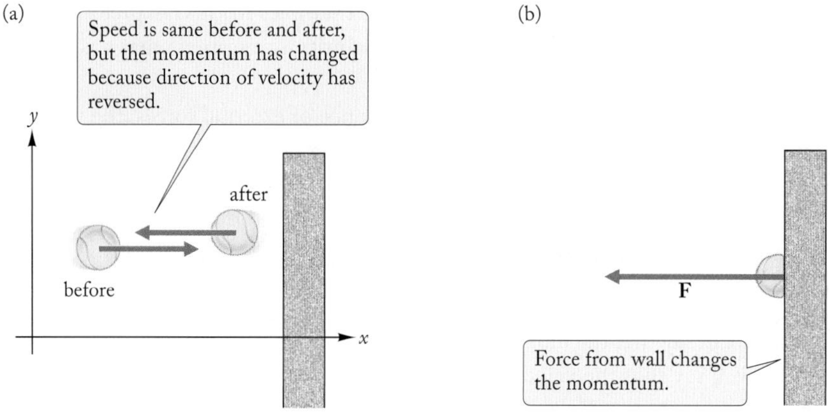

FIGURE 10.1 (a) A tennis ball bounces off a wall. (b) At the instant of impact, the wall exerts a large force on the ball.

This change of momentum is produced by the (large) force that acts on the ball during impact on the wall (see Fig. 10.1b). The change of momentum is negative because the force is negative (the force is in the negative x direction, opposite to the direction of the initial motion).

We can also express Newton's Third Law in terms of momentum. Since the action force is exactly opposite to the reaction force, the rate of change of momentum generated by the action force on one body is exactly opposite to the rate of change of momentum generated by the reaction force on the other body. Hence, we can state the Third Law as follows:

Third Law in terms of momentum

Whenever two bodies exert forces on each other, the resulting changes of momentum are of equal magnitudes and of opposite directions.

This balance in the changes of momentum leads us to a general law of conservation of the total momentum for a system of particles.

The total momentum of a system of n particles is simply the (vector) sum of all the individual momenta of all the particles. Thus, if $\mathbf{p}_1 = m_1\mathbf{v}_1$, $\mathbf{p}_2 = m_2\mathbf{v}_2$, . . . , and $\mathbf{p}_n = m_n\mathbf{v}_n$ are the individual momenta of the particles, then the total momentum is

momentum of a system of particles

$$\mathbf{P} = \mathbf{p}_1 + \mathbf{p}_2 + \cdots + \mathbf{p}_n \tag{10.4}$$

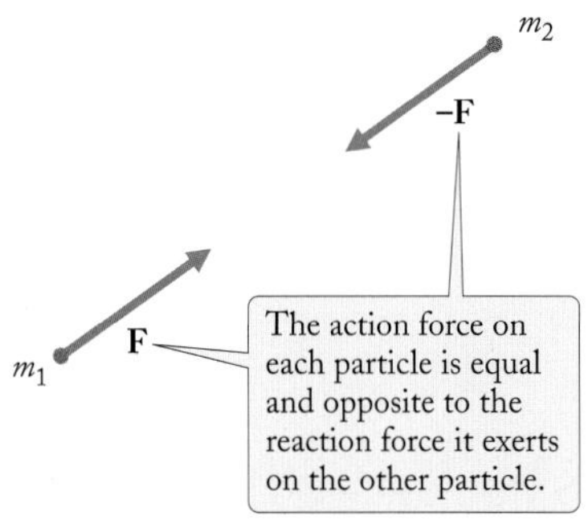

FIGURE 10.2 Two particles exerting mutual forces on each other. The net change of momentum of the isolated particle pair is zero.

The simplest of all many-particle systems consists of just two particles exerting some mutual forces on one another (see Fig. 10.2). Let us assume that the two particles are isolated from the rest of the Universe so that, apart from their mutual forces, they experience no extra forces of any kind. According to the above formulation of the Third Law, the rates of change of $\mathbf{p}_1$ and $\mathbf{p}_2$ are then exactly opposite:

$$\frac{d\mathbf{p}_1}{dt} = -\frac{d\mathbf{p}_2}{dt}$$

The rate of change of the sum $\mathbf{p}_1 + \mathbf{p}_2$ is therefore zero, since the rate of change of the first term in this sum is canceled by the rate of change of the second term:

$$\frac{d(\mathbf{p}_1 + \mathbf{p}_2)}{dt} = 0$$

This means that the sum $\mathbf{p}_1 + \mathbf{p}_2$ is a constant of the motion:

$$\mathbf{p}_1 + \mathbf{p}_2 = [\text{constant}] \tag{10.5}$$

momentum conservation for two particles

This is the **Law of Conservation of Momentum.** Note that Newton's Third Law is an essential ingredient for establishing the conservation of momentum: the total momentum is constant because the equality of action and reaction keeps the momentum changes of the two particles exactly equal in magnitude but opposite in direction—the particles merely exchange some momentum by means of their mutual forces. Thus, for our particles, the total momentum $\mathbf{P}$ at some instant equals the total momentum $\mathbf{P}'$ at some other instant, so

$$\mathbf{P} = \mathbf{P}'$$

Conservation of momentum is a powerful tool which permits us to calculate some general features of the motion even when we are ignorant of the detailed properties of the interparticle forces. The following examples illustrate how we can use conservation of momentum to solve some problems of motion.

EXAMPLE 2 A gun used onboard an eighteenth-century warship is mounted on a carriage which allows the gun to roll back each time it is fired (Fig. 10.3). The mass of the gun, including the carriage, is 2200 kg. The gun fires a 6.0-kg shot horizontally with a velocity of 500 m/s. What is the recoil velocity of the gun?

SOLUTION: The total momentum of the shot plus the gun must be the same before the firing and just after the firing. Before, the total momentum is zero (Fig. 10.3a):

$$\mathbf{P} = 0$$

After, the (horizontal) velocity of the shot is $\mathbf{v}_1'$, and the velocity of the gun is $\mathbf{v}_2'$ (as above, the primes on mathematical quantities indicate that these are evaluated *after* the firing); hence the total momentum is

$$\mathbf{P}' = m_1\mathbf{v}_1' + m_2\mathbf{v}_2'$$

where $m_1 = 6.0$ kg is the mass of the shot and $m_2 = 2200$ kg is the mass of the gun (including the carriage). Thus, momentum conservation tells us

$$0 = m_1\mathbf{v}_1' + m_2\mathbf{v}_2'$$

or

$$\mathbf{v}_2' = -\frac{m_1}{m_2}\mathbf{v}_1'$$

The negative sign indicates that $\mathbf{v}_2'$, the recoil velocity of the gun, is opposite to the velocity of the shot and has a magnitude

$$v_2' = \frac{m_1}{m_2}v_1'$$
$$= \frac{6.0\ \text{kg}}{2200\ \text{kg}} \times 500\ \text{m/s} = 1.4\ \text{m/s}$$

COMMENTS: Note that the final velocities are in the inverse ratio of the masses: the shot emerges with a large velocity, and the gun rolls back with a low velocity.

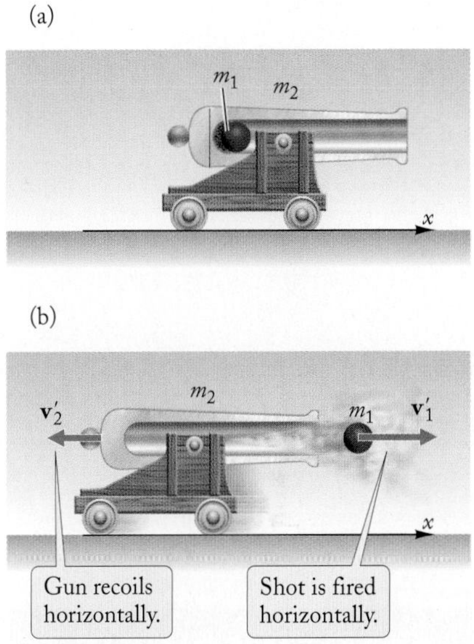

FIGURE 10.3 (a) Initially, the gun and the shot are at rest. (b) After the firing, the gun recoils toward the left (the velocity $\mathbf{v}_2'$ of the gun is negative).

This is a direct consequence of the equality of the magnitudes of the action and reaction forces that act on the shot and the gun during the firing. The force gives the shot (of small mass) a large acceleration, and the reaction force gives the gun (of large mass) a small acceleration.

In this calculation we neglected the mass and momentum of the gases released in the explosion of the gunpowder. This extra momentum increases the recoil velocity somewhat.

EXAMPLE 3 An automobile of mass 1500 kg traveling at 24 m/s crashes into a similar parked automobile. The two automobiles remain joined together after the collision. What is the velocity of the wreck immediately after the collision? Neglect friction against the road, since this force is insignificant compared with the large mutual forces that the automobiles exert on each other.

SOLUTION: Under the assumptions of the problem, the only horizontal forces are the mutual forces of one automobile on the other. Thus, momentum conservation applies to the horizontal component of the momentum: the value of this component must be the same before and after the collision. Before the collision, the (horizontal) velocity of the moving automobile is $v_1 = 24$ m/s and that of the other is $v_2 = 0$. With the x axis along the direction of motion (see Fig. 10.4), the total momentum is therefore

$$P_x = m_1 v_1 + m_2 v_2 = m_1 v_1$$

After the collision, both automobiles have the same velocity (see Fig. 10.4b). We will designate the velocities of the automobiles after the collision by v_1' and v_2', respectively. We can write $v_1' = v_2' = v'$ (the automobiles have a common v', since they remain joined), so the total momentum is

$$P_x' = m_1 v_1' + m_2 v_2' = (m_1 + m_2)v'$$

PROBLEM-SOLVING TECHNIQUES — CONSERVATION OF MOMENTUM

Note that the solution of these examples involves three steps similar to those we used in examples of energy conservation:

1. First write an expression for the total momentum **P** before the firing of the gun or the collision of the automobiles.
2. Then write an expression for the total momentum **P**′ after the firing or the collision.
3. And then use momentum conservation to equate these expressions.

However, in contrast to energy conservation, you must keep in mind that momentum conservation applies to the components of the momentum—*the x, y, and z components of the momentum are conserved separately.* Thus, before writing the expressions for the momentum, you need to select coordinate axes and decide which components of the momentum you want to examine. If the motion is one-dimensional, place one axis along the direction of motion, such as the x axis in the above examples. It then suffices to examine the x component of the momentum. However, sometimes it is necessary to examine two components of the momentum (or, rarely, three); then two (or three) equations result. When writing the components of the momentum, pay attention to the signs; the component is positive if the motion is along the direction of the axis, negative if the motion is opposite to the direction of the axis.

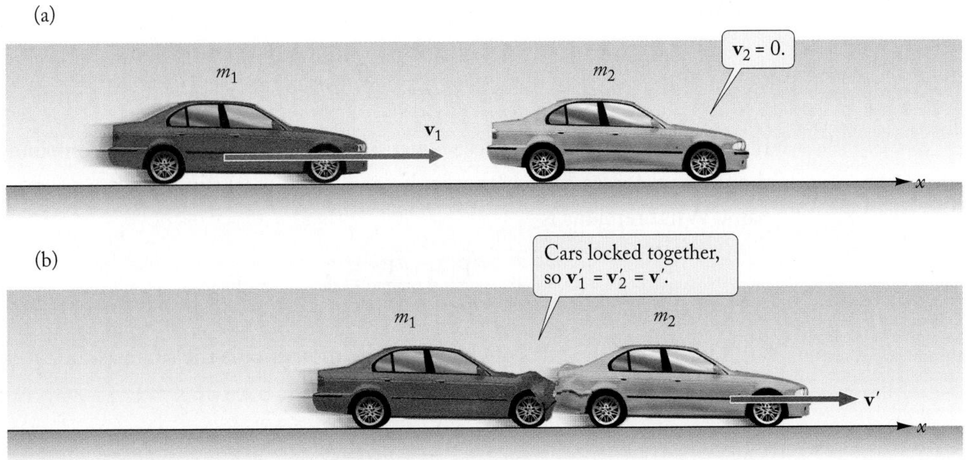

FIGURE 10.4 (a) Initially, the red automobile has a speed of 24 m/s, and the blue automobile is at rest. (b) After the collision, both automobiles are in motion with velocity $\mathbf{v}'$.

By momentum conservation, the momenta P_x and P'_x before and after the collision must be equal:

$$m_1 v_1 = (m_1 + m_2)v' \tag{10.6}$$

When we solve this for the velocity of the wreck v', we find

$$\begin{aligned} v' &= \frac{m_1 v_1}{m_1 + m_2} \\ &= \frac{1500\,\text{kg} \times 24\,\text{m/s}}{1500\,\text{kg} + 1500\,\text{kg}} = 12\,\text{m/s} \end{aligned} \tag{10.7}$$

The forces acting during the firing of the gun or the collision of the automobiles are quite complicated, but momentum conservation permits us to bypass these complications and directly obtain the answer for the final velocities. Incidentally: It is easy to check that kinetic energy is *not* conserved in these examples. During the firing of the gun, kinetic energy is supplied to the shot and the gun by the explosive combustion of the gunpowder, and during the collision of the automobiles, some kinetic energy is used up to produce changes in the shapes of the automobiles.

The conservation law for momentum depends on the absence of "extra" forces. If the particles are not isolated from the rest of the Universe, then besides the mutual forces exerted by one particle on the other, there are also forces exerted by other bodies not belonging to the particle system. The former forces are called **internal forces** of the system and the latter **external forces.** For instance, for the colliding automobiles of Example 3 the gravity of the Earth, the normal force of the road, and the friction of the road are external forces. In Example 3 we ignored these external forces, because gravity and the normal force cancel each other, and the friction force can be neglected in comparison with the much larger impact force that the automobiles exert on each other. But if the external forces are significant, we must take them into account, and we must modify Eq. (10.5). If the internal force on particle 1 is $\mathbf{F}_{1,\text{int}}$ and the external force is $\mathbf{F}_{1,\text{ext}}$, then the total force on particle 1 is $\mathbf{F}_{1,\text{int}} + \mathbf{F}_{1,\text{ext}}$ and its equation of motion will be

internal forces and external forces

$$\frac{d\mathbf{p}_1}{dt} = \mathbf{F}_{1,\text{int}} + \mathbf{F}_{1,\text{ext}} \tag{10.8}$$

Likewise

$$\frac{d\mathbf{p}_2}{dt} = \mathbf{F}_{2,\text{int}} + \mathbf{F}_{2,\text{ext}} \tag{10.9}$$

If we add the left sides of these equations and the right sides, the contributions from the internal forces cancel (that is, $\mathbf{F}_{1,\text{int}} + \mathbf{F}_{2,\text{int}} = 0$), since they are action–reaction pairs. What remains is

$$\frac{d\mathbf{p}_1}{dt} + \frac{d\mathbf{p}_2}{dt} = \mathbf{F}_{1,\text{ext}} + \mathbf{F}_{2,\text{ext}} \tag{10.10}$$

The sum of the rates of change of the momenta is the same as the rate of change of the sum of the momenta; hence,

$$\frac{d(\mathbf{p}_1 + \mathbf{p}_2)}{dt} = \mathbf{F}_{1,\text{ext}} + \mathbf{F}_{2,\text{ext}} \tag{10.11}$$

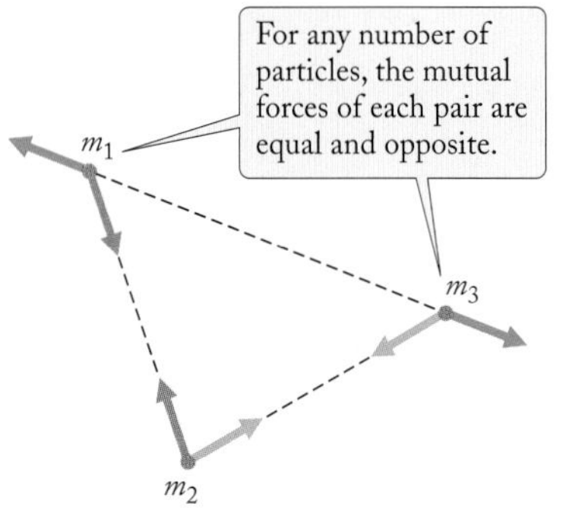

FIGURE 10.5 Three particles exerting forces on each other. As in the case of two particles, the mutual forces between pairs of particles merely exchange momentum between them.

The sum $\mathbf{P} = \mathbf{p}_1 + \mathbf{p}_2$ is the total momentum, and the sum $\mathbf{F}_{1,\text{ext}} + \mathbf{F}_{2,\text{ext}}$ is the total external force on the particle system. Thus, Eq. (10.11) states that the rate of change of the total momentum of the two-particle system equals the total *external* force.

For a system containing more than two particles, we can obtain similar results. If the system is isolated so that there are no external forces, then the mutual interparticle forces acting between pairs of particles merely transfer momentum from one particle of the pair to the other, just as in the case of two particles. Since all the internal forces necessarily arise from such forces between pairs of particles, these internal forces cannot change the total momentum. For example, Fig. 10.5 shows three isolated particles exerting forces on one another. Consider particle 1; the mutual forces between particles 1 and 2 exchange momentum between these two, while the mutual forces between particles 1 and 3 exhange momentum between those two. But none of these momentum transfers will change the total momentum. The same holds for particles 2 and 3. Consequently, the total momentum is constant. More generally, for an isolated system of n particles, the total momentum $\mathbf{P} = \mathbf{p}_1 + \mathbf{p}_2 + \cdots + \mathbf{p}_n$ obeys the conservation law

momentum conservation for a system of particles

$$\mathbf{P} = [\text{constant}] \qquad \text{(no external forces)} \tag{10.12}$$

If, besides the internal forces, there are external forces, then the latter will change the momentum. The rate of change can be calculated in essentially the same way as for the two-particle system, and again, the rate of change of the total momentum is equal to the total external force. We can write this as

Second Law for a system of particles

$$\frac{d\mathbf{P}}{dt} = \mathbf{F}_{\text{ext}} \tag{10.13}$$

where $\mathbf{F}_{\text{ext}} = \mathbf{F}_{1,\text{ext}} + \mathbf{F}_{2,\text{ext}} + \cdots + \mathbf{F}_{n,\text{ext}}$ is the total external force acting on the system.

Equations (10.12) and (10.13) have exactly the same mathematical form as Eqs. (10.2) and (10.3), and they may be regarded as the generalizations for a system of particles of Newton's First and Second Laws. As we will see in Section 10.3, Eq. (10.13) is an equation of motion for the system of particles—it determines the overall translational motion of the system.

Checkup 10.1

QUESTION 1: An automobile and a truck have equal momenta. Which has the larger speed? Which has the larger kinetic energy?

QUESTION 2: An automobile and a truck are traveling along a street in opposite directions. Can they have the same momentum? The same kinetic energy?

QUESTION 3: A rubber ball, dropped on a concrete floor, bounces up with reversed velocity. Is the momentum before the impact the same as after the impact?

QUESTION 4: Is the net momentum of the Sun and all the planets and moons of the Solar System constant? Is the net kinetic energy constant?

QUESTION 5: Consider two automobiles of equal masses m and equal speeds v. (a) If both automobiles are moving southward on a street, what are the total kinetic energy and the total momentum of this system of two automobiles? (b) If one automobile is moving southward and one northward? (c) If one automobile is moving southward and one eastward?

QUESTION 6: An automobile and a truck have equal kinetic energies. Which has the larger speed? Which has the larger momentum? Assume that the truck has the larger mass.

(A) Truck; truck
(B) Truck; automobile
(C) Automobile; truck
(D) Automobile; automobile

10.2 CENTER OF MASS

In our study of kinematics and dynamics in the preceding chapters we always ignored the size of the bodies; even when analyzing the motion of a large body—an automobile or a ship—we pretended that the motion could be treated as particle motion, position being described by means of some reference point marked on the body. In reality, large bodies are systems of particles, and their motion obeys Eq. (10.13) for a system of particles. This equation can be converted into an equation of motion containing just one acceleration rather than the rate of change of momentum of the entire system, by taking as reference point the **center of mass** of the body. The equation that describes the motion of this special point has the same mathematical form as the equation of motion of a particle; that is, the motion of the center of mass mimics particle motion (see, for example, Fig. 10.6).

center of mass

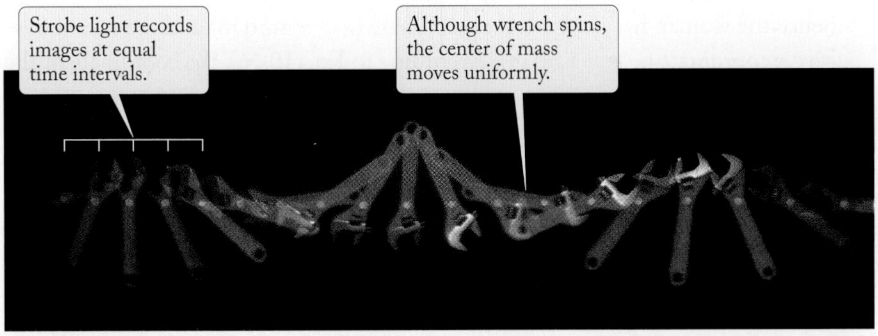

FIGURE 10.6 A wrench moving freely in the absence of external forces. The center of mass, marked with a dot, moves with uniform velocity, along a straight line (you can check this by laying a ruler along the dots).

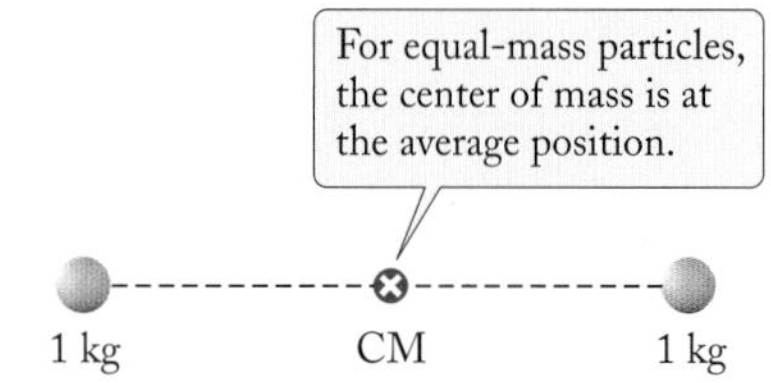

FIGURE 10.7 Two particles of equal masses, and their center of mass.

The position of the center of mass is merely the average position of the mass of the system. For instance, if the system consists of two particles, each of mass 1 kg, then the center of mass is halfway between them (see Fig. 10.7). In any system consisting of n particles of equal masses—such as a piece of pure metal with atoms of only one kind—the x coordinate of the center of mass is simply the sum of the x coordinates of all the particles divided by the number of particles,

$$x_{CM} = \frac{x_1 + x_2 + \cdots + x_n}{n} \quad \text{(for equal-mass particles)} \tag{10.14}$$

Similar equations apply to the y and the z coordinates, if the particles of the system are distributed over a three-dimensional region. The three coordinate equations can be expressed concisely in terms of position vectors:

$$\mathbf{r}_{CM} = \frac{\mathbf{r}_1 + \mathbf{r}_2 + \cdots + \mathbf{r}_n}{n} \quad \text{(for equal-mass particles)} \tag{10.15}$$

If the system consists of particles of unequal mass, then the position of the center of mass can be calculated by first subdividing the particles into fragments of equal mass. For instance, if the system consists of two particles, the first of mass 2 kg and the second of 1 kg, then we can pretend that we have *three* particles of equal masses 1 kg, two of which are located at the same position. The coordinate of the center of mass is then

$$x_{CM} = \frac{x_1 + x_1 + x_2}{3}$$

We can also write this in the equivalent form

$$x_{CM} = \frac{m_1 x_1 + m_2 x_2}{m_1 + m_2} \tag{10.16}$$

where $m_1 = 2$ kg and $m_2 = 1$ kg. The formula (10.16) is actually valid for any values of the masses m_1 and m_2. The formula simply asserts that in the average position, the position of particle 1 is included m_1 times and the position of particle 2 is included m_2 times—that is, the number of times each particle is included in the average is directly proportional to its mass.

EXAMPLE 4 A 50-kg woman and an 80-kg man sit on the two ends of a seesaw of length 3.00 m (see Fig. 10.8). Treating them as particles, and ignoring the mass of the seesaw, find the center of mass of this system.

SOLUTION: In Fig. 10.8, the origin of coordinates is at the center of the seesaw; hence the woman has a negative x coordinate ($x = -1.50$ m) and the man a positive x coordinate ($x = +1.50$ m). According to Eq. (10.16), the coordinate of the center of mass is

$$x_{CM} = \frac{m_1 x_1 + m_2 x_2}{m_1 + m_2} = \frac{50\,\text{kg} \times (-1.50\,\text{m}) + 80\,\text{kg} \times 1.50\,\text{m}}{50\,\text{kg} + 80\,\text{kg}}$$

$$= 0.35\,\text{m}$$

COMMENT: Note that the distance of the woman from the center of mass is 1.50 m + 0.35 m = 1.85 m, and the distance of the man from the center of mass is 1.50 m

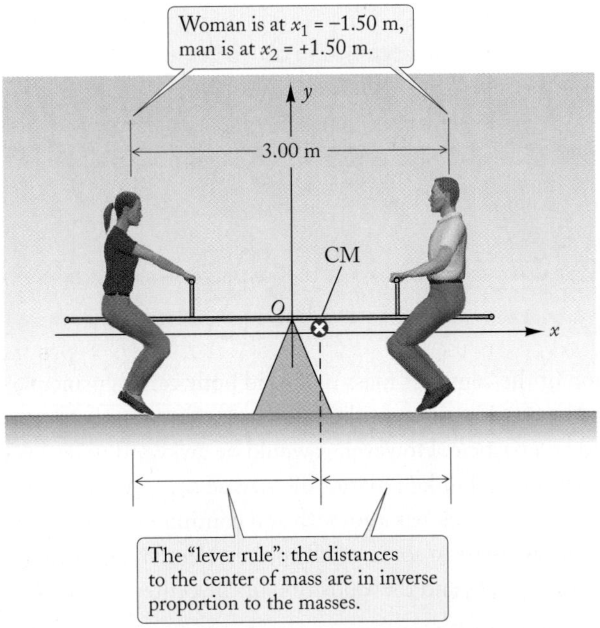

FIGURE 10.8 A woman and a man on a seesaw.

− 0.35 m = 1.15 m. The ratio of these distances is 1.6, which coincides with the inverse of the ratio of the masses, 50/80 = 1/1.6. This "lever rule" is quite general: the position of the center of mass of two particles divides the line segment connecting them in the ratio $m_1{:}m_2$, with the smaller length segment nearer to the larger mass.

If the system consists of n particles of different masses $m_1, m_2, \ldots, m_n$, then we apply the same prescription: the number of times each particle is included in the average is in direct proportion to its mass; the exact factor by which each particle's coordinate is multiplied is that particle's fraction of the total mass. This gives the following general expression for the coordinate of the center of mass:

$$x_{\text{CM}} = \frac{m_1x_1 + m_2x_2 + \cdots + m_nx_n}{m_1 + m_2 + \cdots + m_n} \tag{10.17}$$

or

$$x_{\text{CM}} = \frac{m_1x_1 + m_2x_2 + \cdots + m_nx_n}{M} \tag{10.18}$$

where M is the total mass of the system, $M = m_1 + m_2 + \cdots + m_n$. Similar formulas apply to the y and the z coordinates, if the particles of the system are distributed over a three-dimensional region:

$$y_{\text{CM}} = \frac{m_1y_1 + m_2y_2 + \cdots + m_ny_n}{M} \tag{10.19}$$

$$z_{\text{CM}} = \frac{m_1z_1 + m_2z_2 + \cdots + m_nz_n}{M} \tag{10.20}$$

By introducing the standard notation Σ for a summation of n terms, we can express these formulas more concisely as

coordinates of center of mass

$$x_{CM} = \frac{1}{M}\sum_{i=1}^{n} m_i x_i \tag{10.21}$$

$$y_{CM} = \frac{1}{M}\sum_{i=1}^{n} m_i y_i \tag{10.22}$$

$$z_{CM} = \frac{1}{M}\sum_{i=1}^{n} m_i z_i \tag{10.23}$$

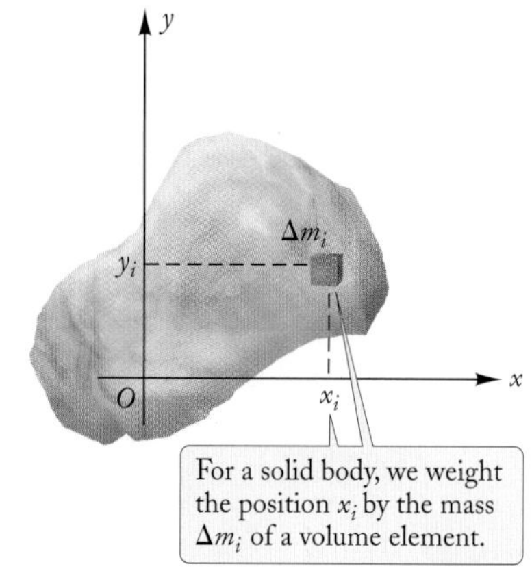

FIGURE 10.9 A small volume element of the body at position x_i has a mass Δm_i.

The position of the center of mass of a solid body can, in principle, be calculated from Eqs. (10.21)–(10.23), since a solid body is a collection of atoms, each of which can be regarded as a particle. However, it would be awkward to deal with the 10^{23} or so atoms that make up a chunk of matter the size of, say, a coin. It is more convenient to pretend that matter in bulk has a smooth and continuous distribution of mass over its entire volume. The mass in some small volume element at position x_i in the body is then Δm_i (see Fig. 10.9), and the x position of the center of mass is

$$x_{CM} = \frac{1}{M}\sum_{i=1}^{n} x_i \Delta m_i \tag{10.24}$$

In the limiting case of $\Delta m_i \to 0$ (and $n \to \infty$), this sum becomes an integral:

$$x_{CM} = \frac{1}{M}\int x\,dm \tag{10.25}$$

Similar expressions are valid for the y and z positions of the center of mass:

$$y_{CM} = \frac{1}{M}\int y\,dm \tag{10.26}$$

$$z_{CM} = \frac{1}{M}\int z\,dm \tag{10.27}$$

Thus, the position of the center of mass is the average position of all the mass elements making up the body.

For a body of uniform density, the amount of mass dm in any given volume element dV is directly proportional to the amount of volume. *For a uniform-density body, the position of the center of mass is simply the average position of all the volume elements of the body* (in mathematics, this is called the **centroid** of the volume). If the body has a symmetric shape, this average position will often be obvious by inspection. For instance, a sphere of uniform density, or a ring, or a circular plate, or a cylinder, or a parallelepiped will have its center of mass at the geometrical center (see Fig. 10.10). But for a less symmetric body, the center of mass must often be calculated, either by considering parts of the body (as in the next example) or by integrating over the entire body (as in the two subsequent examples).

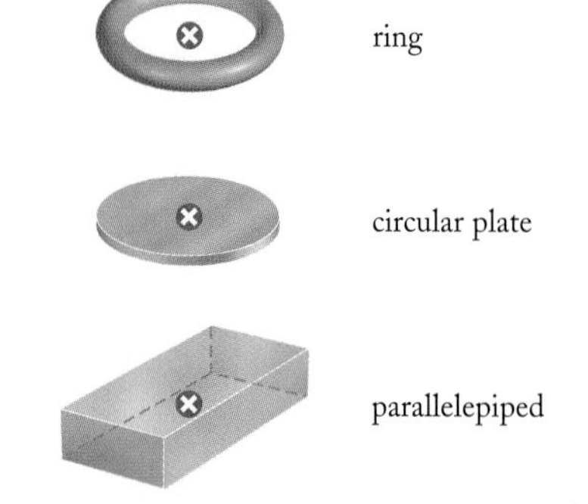

FIGURE 10.10 Several bodies for which the center of mass coincides with the geometrical center.

EXAMPLE 5 A meterstick of aluminum is bent at its midpoint so that the two halves are at right angles (see Fig. 10.11). Where is the center of mass of this bent stick?

SOLUTION: We can regard the bent stick as consisting of two straight pieces, each of 0.500 m. The centers of mass of these straight pieces are at their midpoints,

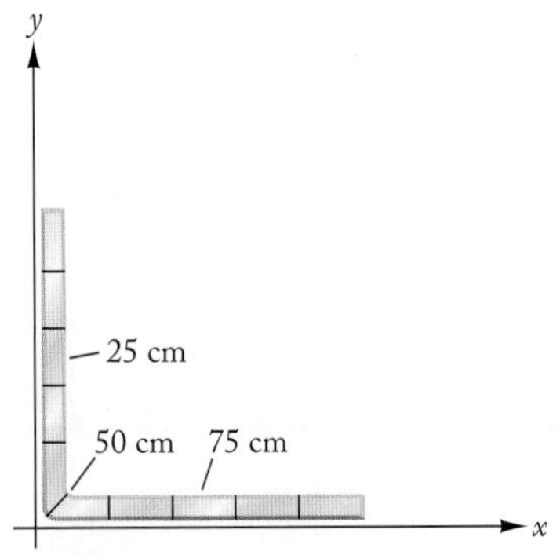

FIGURE 10.11 A meterstick, bent through 90° at its midpoint.

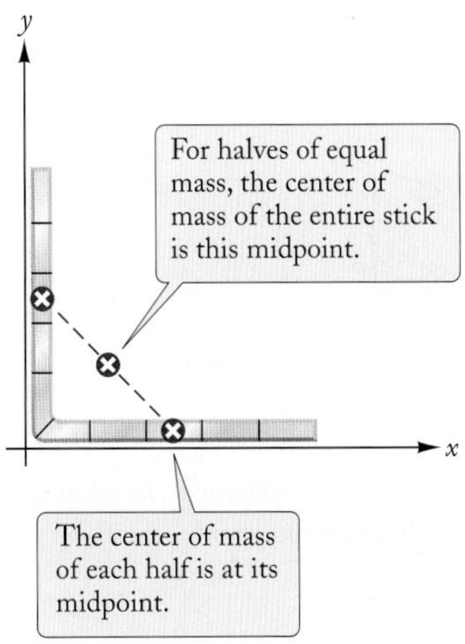

FIGURE 10.12 The center of mass of the bent meterstick is at the midpoint of the line connecting the centers of the halves. The coordinates x_{CM} and y_{CM} of this midpoint are one-half of the distances to the centers of mass of the horizontal and vertical sides—that is, 0.125 m each.

0.250 m from their ends (see Fig. 10.12). The center of mass of the entire stick is the average position of the centers of mass of the two halves. With the coordinate axes arranged as in Fig. 10.12, the x coordinate of the center of mass is, according to Eq. (10.14),

$$x_{CM} = \frac{0.250 \text{ m} + 0}{2} = 0.125 \text{ m} \tag{10.28}$$

Likewise, the y coordinate is

$$y_{CM} = \frac{0.250 \text{ m} + 0}{2} = 0.125 \text{ m}$$

Note that the center of mass of this bent stick is *outside* the stick; that is, it is not in the volume of the stick (see Fig. 10.12).

EXAMPLE 6 Figure 10.13 shows a mobile by Alexander Calder, which contains a uniform sheet of steel, in the shape of a triangle, suspended at its center of mass. Where is the center of mass of a right triangle of perpendicular sides a and b?

SOLUTION: Figure 10.14 shows the triangle positioned with a vertex at the origin and its right angle at a distance b along the x axis. To calculate the x coordinate of the center of mass, we need to sum mass contributions dm at each value of x; one such contribution is the vertical strip in Fig. 10.14, which has a height $y = (a/b)x$ and a width dx. Since the sheet is uniform, the strip has a fraction of the total mass M equal to the strip's area $y\,dx = (a/b)x\,dx$ divided by the total area $\frac{1}{2}ab$:

$$\frac{dm}{M} = \frac{(a/b)x\,dx}{\frac{1}{2}ab}$$

or

$$dm = M\frac{2x}{b^2}\,dx$$

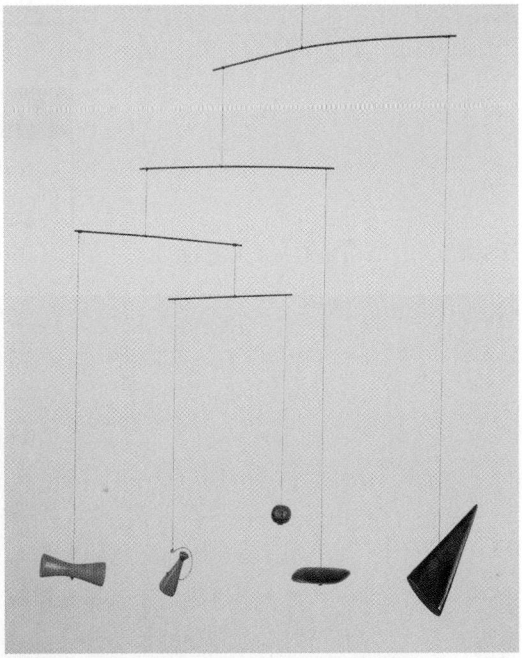

FIGURE 10.13 This mobile by Alexander Calder contains a triangle suspended above its center of mass.

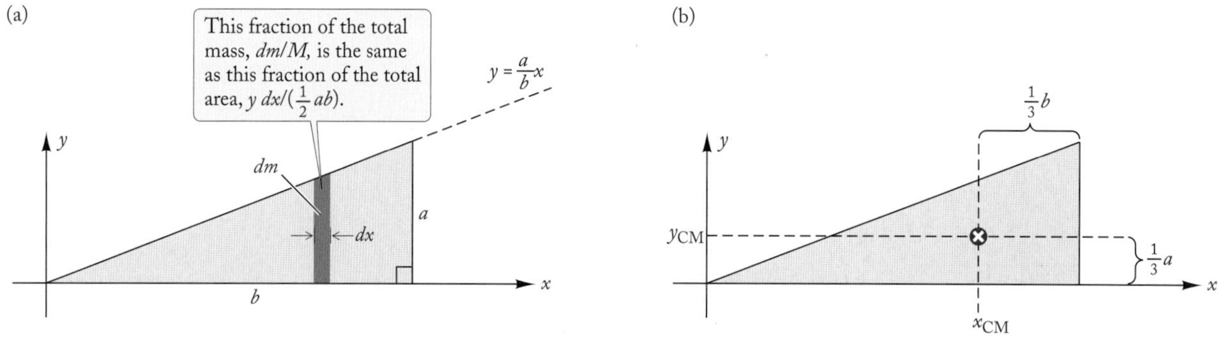

FIGURE 10.14 (a) A right triangle, with mass element dm of height y and width dx. (b) The center of mass is one-third of the distance from the right angle along sides a and b.

We integrate this in Eq. (10.25) for x_{CM} and sum the contributions from $x = 0$ to $x = b$:

$$x_{CM} = \frac{1}{M}\int x\,dm = \frac{1}{M}\int_0^b xM\frac{2x}{b^2}\,dx$$

$$= \frac{2}{b^2}\int_0^b x^2\,dx = \frac{2}{b^2}\frac{1}{3}x^3\Big|_0^b$$

$$= \frac{2}{b^2}\frac{1}{3}(b^3 - 0) = \frac{2}{3}b$$

So the center of mass is two-thirds of the distance toward the right angle. Performing a similar calculation for y_{CM} yields $y_{CM} = \frac{1}{3}a$. Thus each of x_{CM} and y_{CM} is a distance *away* from the right angle equal to one-third of the length of the corresponding side (see Fig. 10.14b).

EXAMPLE 7 The Great Pyramid at Giza (see Fig. 10.15) has a height of 147 m and a square base. Assuming that the entire volume is completely filled with stone of uniform density, find its center of mass.

SOLUTION: Because of symmetry, the center of mass must be on the vertical line through the apex. For convenience, we place the y axis along this line, and we arrange this axis downward, with origin at the apex. We must then find where the center of mass is on this y axis. Figure 10.16a shows a cross section through the pyramid, looking parallel to two sides. The half-angle at the apex is ϕ. By examination of the colored triangle, we see that at a height y (measured from the apex) the half-width is $x = y\tan\phi$ and the full width is $2x = 2y\tan\phi$. A horizontal slice through the pyramid at this height is a square measuring $2x \times 2x$ (see Fig. 10.16b). The volume of a horizontal slab of thickness dy at this height y is therefore $dV = (2x)^2\,dy = (2y\tan\phi)^2\,dy$. If we represent the uniform density of the stone by ρ (the Greek letter *rho*), the proportionality between mass and volume can be written

$$dm = \rho dV$$

FIGURE 10.15 The Great Pyramid.

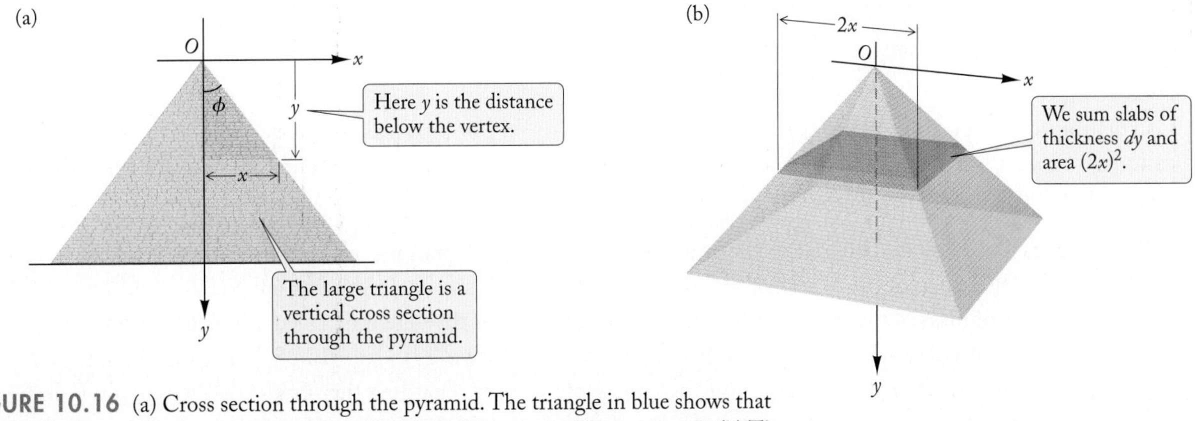

FIGURE 10.16 (a) Cross section through the pyramid. The triangle in blue shows that at a height y measured from the apex, the half-width of the pyramid is $x = y \tan\phi$. (b) The thin horizontal slab indicated in red is a square measuring $2x \times 2x$ with a thickness dy.

Thus the mass of the slab of thickness dy at this height y is

$$dm = \rho dV = \rho(2y\tan\phi)^2 dy = 4\rho(\tan^2\phi)y^2\,dy$$

Equation (10.26) then gives us the y coordinate of the center of mass:

$$y_{CM} = \frac{1}{M}\int y\,dm = \frac{1}{M}\int 4\rho(\tan^2\phi)y^3\,dy \tag{10.29}$$

The total mass is

$$M = \int dm = \int 4\rho(\tan^2\phi)y^2\,dy \tag{10.30}$$

When we substitute Eq. (10.30) into Eq. (10.29), the common factor $4\rho\tan^2\phi$ cancels, leaving

$$y_{CM} = \frac{\int y^3\,dy}{\int y^2\,dy} \tag{10.31}$$

As we sum the square slabs of thickness dy in both of these integrals, the integration runs from $y = 0$ at the top of the pyramid to $y = h$ at the bottom, where h is the height of the pyramid. Evaluation of these integrals yields

$$\int_0^h y^3\,dy = \left.\frac{y^4}{4}\right|_0^h = \frac{h^4}{4}$$

$$\int_0^h y^2\,dy = \left.\frac{y^3}{3}\right|_0^h = \frac{h^3}{3}$$

The y coordinate of the center of mass is therefore

$$y_{CM} = \frac{h^4/4}{h^3/3} = \frac{3}{4}h$$

This means that the center of mass is $3/4 \times 147$ m below the apex; that is, it is $1/4 \times 147$ m $= 37$ m above the ground.

PROBLEM-SOLVING TECHNIQUES — CENTER OF MASS

Calculations of the position of the center of mass of a body can often be simplified by exploiting the shape or the symmetry of the body.

- Sometimes it is profitable to treat the body as consisting of several parts and to begin by calculating the positions of the centers of mass of these parts (as in the example of the bent meterstick). Each part can then be treated as a particle located at its center of mass, and the center of mass of the entire body is then the center of mass of this system of particles, which can be calculated by the sums, Eqs. (10.18)–(10.20).
- If the body or some part of it has symmetry, the position of the center of mass will often be obvious by inspection. For instance, in the example of the bent meterstick, it is obvious that the center of mass of each half is at its center.
- Geometrical arguments can sometimes replace algebraic calculations of the coordinates of the center of mass. For instance, in the example of the bent meterstick, instead of the algebraic calculations of the coordinates [such as for x_{CM} in Eq. (10.28)], the coordinates can be obtained by regarding the stick as consisting of two straight pieces with known centers of mass; then the coordinates of the overall center of mass can be found from the geometry of a diagram, such as Fig. 10.12.

PHYSICS IN PRACTICE — CENTER OF MASS AND STABILITY

In the design of ships, engineers need to ensure that the position of the center of mass is low in the ship, to enhance the stability. If the center of mass is high, the ship is top-heavy and liable to tip over. Ships often carry ballast at the bottom of the hull to lower the center of mass. Many ships have been lost because of insufficient ballast or because of an unexpected shifting of the ballast. For instance, in 1628, the Swedish ship *Vasa* (see Fig. 1), the pride and joy of the Swedish navy and King Gustavus II Adolphus, capsized and sank on its maiden voyage when struck by a gust of wind, just barely out of harbor. It carried an excessive number of heavy guns on its upper decks, which made it top-heavy; and it should have carried more ballast to lower its center of mass.

The position of the center of mass is also crucial in the design of automobiles. A top-heavy automobile, such as an SUV, will tend to roll over when speeding around a sharp curve. High-performance automobiles, such as the Maserati shown in Fig. 2, have a very low profile, with the engine and transmission slung low in the body, so the center of mass is as low as possible and the automobile hugs the ground.

FIG. 1 The Swedish ship *Vasa*.

FIG. 2 A Maserati sports car.

The position of the center of mass enters into the calculation of the gravitational potential energy of an extended body located near the surface of the Earth. According to Eq. (7.29), the potential energy of a single particle of mass m at a height y above the ground is mgy. For a system of particles, the total gravitational potential energy is then

$$\begin{aligned} U &= m_1gy_1 + m_2gy_2 + \cdots + m_ngy_n \\ &= (m_1y_1 + m_2y_2 + \cdots + m_ny_n)g \end{aligned} \tag{10.32}$$

Comparison with Eq. (10.19) shows that the quantity in parentheses is My_{CM}. Hence, Eq. (10.32) becomes

$$U = Mgy_{CM} \tag{10.33}$$

potential energy in terms of height of center of mass

This expression for the gravitational potential energy of a system near the Earth's surface has the same mathematical form as for a single particle—it is as though the entire mass of the system were located at the center of mass.

Concepts — in — Context

For a human body standing upright, the position of the center of mass is in the middle of the trunk, at about the height of the navel. This is therefore the height to be used in the calculation of the gravitational potential energy of the body. However, if the body adopts any bent position, the center of mass shifts.

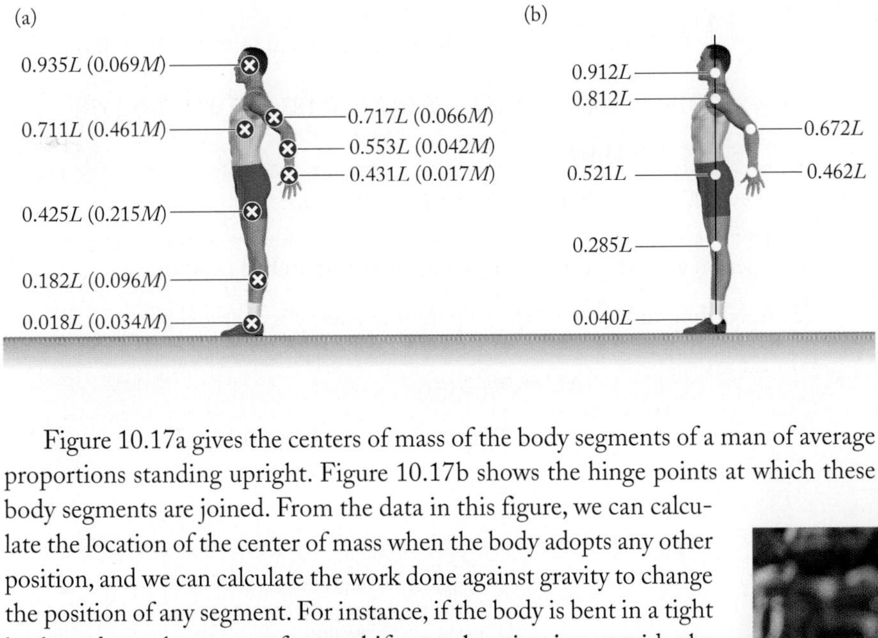

FIGURE 10.17 (a) Centers of mass of the body segments of an average male of mass M and height L standing upright. The numbers give the heights of the centers of mass of the body segments from the floor and (in parentheses) the masses of the body segments; right and left limbs are shown combined. (b) Hinge points of the body. The numbers give the heights of the joints from the floor.

Figure 10.17a gives the centers of mass of the body segments of a man of average proportions standing upright. Figure 10.17b shows the hinge points at which these body segments are joined. From the data in this figure, we can calculate the location of the center of mass when the body adopts any other position, and we can calculate the work done against gravity to change the position of any segment. For instance, if the body is bent in a tight backward arc, the center of mass shifts to a location just outside the body, about 10 cm below the middle of the trunk. Olympic jumpers (see Fig. 10.18) take advantage of this shift of the center of mass to make the most of the gravitational potential energy they can supply for a high jump. By adopting a bent position as they pass over the bar, they raise their trunk above the center of mass, so the trunk passes over the bar while the center of mass can pass *below* the bar. By this trick, the jumper raises the center of her trunk by about 10 cm relative to the center of mass, and she gains extra height without expending extra energy.

FIGURE 10.18 High jumper passing over the bar.

Concepts — in — Context

EXAMPLE 8 Suppose a man of average proportions performs a high jump, while arching his back (see the chapter opening photo). At the peak of his jump, his torso is approximately horizontal; his thighs, arms, and head make an angle of 45° with the horizontal; and his lower legs are vertical, as shown in Fig. 10.19b. (a) How much is his center of mass shifted downward compared with a man who goes over the pole horizontally (Fig. 10.19a)? (b) How much is his potential energy reduced? Assume the mass of the jumper is $M = 73$ kg and his height $L = 1.75$ m.

SOLUTION: (a) In Fig. 10.19a, the center of mass of the horizontal body is at $y = 0$, since each segment is essentially at $y = 0$. In Fig. 10.20, we have used the relative locations of the hinge points and centers of mass from Fig. 10.17 to determine the vertical position of each body segment in the arched-back position. For example, the center of mass of the thigh is at a distance $0.521L - 0.425L = 0.096L$ from the hip joint, and so is at a vertical distance $0.096L \times \sin 45° = 0.068L$ below $y = 0$. Similarly, we can determine that the centers of mass of the lower legs, the feet, the head, the upper arms, the forearms, and the hands are at $y = -0.270L$, $-0.434L$, $-0.016L$, $-0.067L$, $-0.183L$, and $-0.269L$, respectively. From Fig. 10.17, the masses of all seven segments are $0.215M$, $0.096M$, $0.034M$, $0.069M$, $0.066M$, $0.042M$, and $0.017M$, respectively. The torso, of mass $0.461M$, is again at $y = 0$. Thus, using Eq. (10.19) or (10.22), the arched-back center of mass is at

$$
\begin{aligned}
y_{CM} &= \frac{1}{M}\sum_{i=1}^{n} m_i y_i \\
&= -\frac{1}{M}(0.215 \times 0.068 + 0.096 \times 0.270 + 0.034 \times 0.434 + 0.069 \\
&\quad \times 0.016 + 0.066 \times 0.067 + 0.042 \times 0.183 + 0.017 \times 0.269 \\
&\quad + 0 \times 0.461)ML \\
&= -0.073L = -0.073 \times 1.75 \text{ m} = -0.13 \text{ m}
\end{aligned}
$$

Thus a height advantage of 13 cm is gained in this arched position.

(b) According to Eq. (10.33), the potential energy is changed by

$$
\begin{aligned}
\Delta U &= Mg\,\Delta y_{CM} \\
&= 73 \text{ kg} \times 9.81 \text{ m/s}^2 \times (-0.13 \text{ m}) \qquad (10.34) \\
&= -93 \text{ J}
\end{aligned}
$$

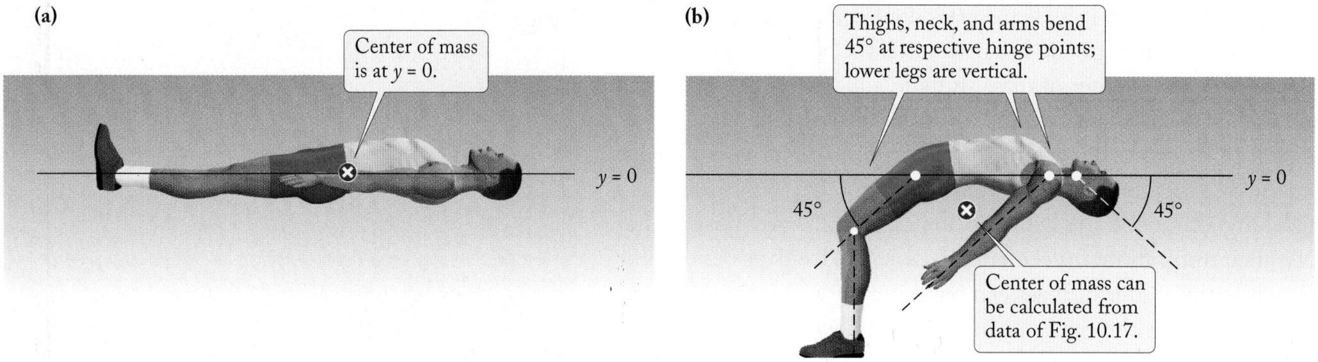

FIGURE 10.19 (a) Horizontal position. (b) High jumper in arched-back position.

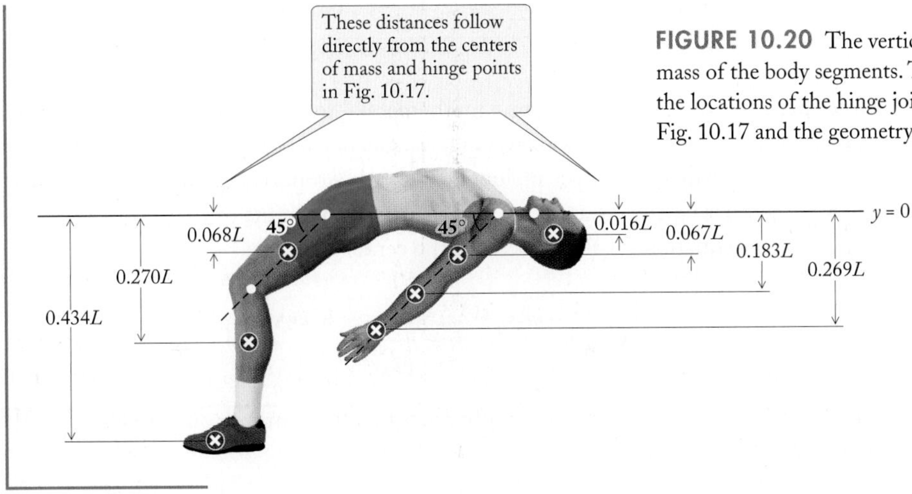

FIGURE 10.20 The vertical positions of the centers of mass of the body segments. These are determined from the locations of the hinge joints and centers of mass in Fig. 10.17 and the geometry of the arched-back position.

Checkup 10.2

QUESTION 1: Roughly where is the center of mass of the snake shown in Fig. 10.21a?

QUESTION 2: Roughly where is the center of mass of the horseshoe shown in Fig. 10.21b?

QUESTION 3: Is it possible for the center of mass of a body to be above the highest part of the body?

QUESTION 4: A sailboat has a keel with a heavy lead bulb at the bottom. If the bulb falls off, the center of mass of the sailboat:

(A) Remains at the same position (B) Shifts downward
(C) Shifts upward

(a)

(b)

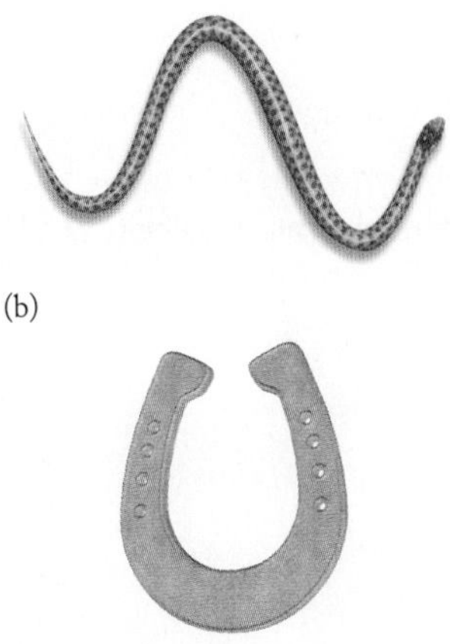

FIGURE 10.21 (a) A snake. (b) A horseshoe.

10.3 THE MOTION OF THE CENTER OF MASS

When the particles in a system move, often so does the center of mass. We will now obtain an equation for the motion of the center of mass, an equation which relates the acceleration of the center of mass to the external force. This equation will permit us to calculate the overall translational motion of a system of particles.

According to Eq. (10.18), if the x components of positions of the respective particles change by $dx_1, dx_2, \ldots, dx_n$, then the position of the center of mass changes by

$$dx_{CM} = \frac{1}{M}(m_1 dx_1 + m_2 dx_2 + \cdots + m_n dx_n) \tag{10.35}$$

Dividing this by the time dt taken for these changes of position, we obtain

$$\frac{dx_{CM}}{dt} = \frac{1}{M}\left(m_1 \frac{dx_1}{dt} + m_2 \frac{dx_2}{dt} + \cdots + m_n \frac{dx_n}{dt}\right) \tag{10.36}$$

The left side of this equation is the x component of the velocity of the center of mass, and the rates of change on the right side are the x components of the velocities of the individual particles; thus

$$v_{x,\text{CM}} = \frac{m_1 v_{x,1} + m_2 v_{x,2} + \cdots + m_n v_{x,n}}{M}$$

Note that this equation has the same mathematical form as Eq. (10.18); that is, the velocity of the center of mass is an average over the particle velocities, and the number of times each particle velocity is included is directly proportional to its mass.

Since similar equations apply to the y and z components of the velocity, we can write a vector equation for the velocity of the center of mass:

velocity of the center of mass

$$\mathbf{v}_{\text{CM}} = \frac{m_1 \mathbf{v}_1 + m_2 \mathbf{v}_2 + \cdots + m_n \mathbf{v}_n}{M} \tag{10.37}$$

The quantity in the numerator is simply the total momentum [compare Eq. (10.1)]; hence Eq. (10.37) says

$$\mathbf{v}_{\text{CM}} = \frac{\mathbf{P}}{M} \tag{10.38}$$

or

momentum in terms of velocity of CM

$$\mathbf{P} = M\mathbf{v}_{\text{CM}} \tag{10.39}$$

This equation expresses the total momentum of a system of particles as the product of the total mass and the velocity of the center of mass. Obviously, this equation is analogous to the familiar equation $\mathbf{p} = m\mathbf{v}$ for the momentum of a single particle.

We know, from Eq. (10.13), that the rate of change of the total momentum equals the net external force on the system,

$$\frac{d\mathbf{P}}{dt} = \mathbf{F}_{\text{ext}}$$

If we substitute $\mathbf{P} = M\mathbf{v}_{\text{CM}}$ and take into account that the mass is constant, we find

$$\frac{d\mathbf{P}}{dt} = \frac{d}{dt}(M\mathbf{v}_{\text{CM}}) = M\frac{d\mathbf{v}_{\text{CM}}}{dt} = M\mathbf{a}_{\text{CM}}$$

and consequently

motion of center of mass

$$M\mathbf{a}_{\text{CM}} = \mathbf{F}_{\text{ext}} \tag{10.40}$$

This equation for a system of particles is the analog of Newton's equation for motion for a single particle. The equation asserts that the center of mass moves as though it were a particle of mass M under the influence of a force $\mathbf{F}_{\text{ext}}$.

Concepts in Context

This result justifies some of the approximations we made in previous chapters. For instance, in Example 9 of Chapter 2 we treated a diver falling from a cliff as a particle. Equation (10.40) shows that this treatment is legitimate: the center of mass of the diver, under the influence of the external force (gravity), moves with a downward acceleration g, just as though it were a freely falling particle. Likewise, after a high jumper leaves the ground, his center of mass moves along a parabolic trajectory, as though it were a projectile, and the shape and height of this parabolic trajectory is unaffected by any contortions the high jumper might perform while in flight. From Chapter 4, we know that the initial vertical velocity v_y determines the maximum height h of the center of mass; that is, $v_y = \sqrt{2gh}$. The contortions of the jumper enable his body to pass over a bar roughly 10 cm above the maximum height of the center of mass.

If the net external force vanishes, then the acceleration of the center of mass vanishes; hence the center of mass remains at rest or it moves with uniform velocity.

FIGURE 10.22 Astronaut on a "space walk" during the *Gemini 4* mission.

EXAMPLE 9 During a "space walk," an astronaut floats in space 8.0 m from his spacecraft orbiting the Earth. He is tethered to the spacecraft by a long umbilical cord (see Fig. 10.22); to return, he pulls himself in by this cord. How far does the spacecraft move toward him? The mass of the spacecraft is 3500 kg, and the mass of the astronaut, including his space suit, is 110 kg.

SOLUTION: In the reference frame of the orbiting (freely falling) astronaut and spacecraft, each is effectively weightless; that is, the external force on the system is effectively zero. The only forces in the system are the forces exerted when the astronaut pulls on the cord; these forces are internal. The forces exerted by the cord on the spacecraft and on the astronaut during the pulling in are of equal magnitudes and opposite directions; the astronaut is pulled toward the spacecraft, and the spacecraft is pulled toward the astronaut. In the absence of external forces, the center of mass of the astronaut–spacecraft system remains at rest. Thus, the spacecraft and the astronaut both move toward the center of mass, and there they meet.

With the x axis as in Fig. 10.23, the x coordinate of the center of mass is

$$x_{CM} = \frac{m_1 x_1 + m_2 x_2}{m_1 + m_2} \tag{10.41}$$

where $m_1 = 3500$ kg is the mass of the spacecraft and $m_2 = 110$ kg is the mass of the astronaut. Strictly, the coordinates x_1 and x_2 of the spacecraft and of the astronaut should correspond to the centers of mass of these bodies, but, for the sake of simplicity, we neglect their size and treat both as particles. The initial values of the coordinates are $x_1 = 0$ and $x_2 = 8.0$ m; hence

$$x_{CM} = \frac{0 + 110 \text{ kg} \times 8.0 \text{ m}}{3500 \text{ kg} + 110 \text{ kg}} = 0.24 \text{ m}$$

During the pulling in, the spacecraft will move from $x_1 = 0$ to $x_1 = 0.24$ m; simultaneously, the astronaut will move from $x_2 = 8.0$ m to $x_2 = 0.24$ m.

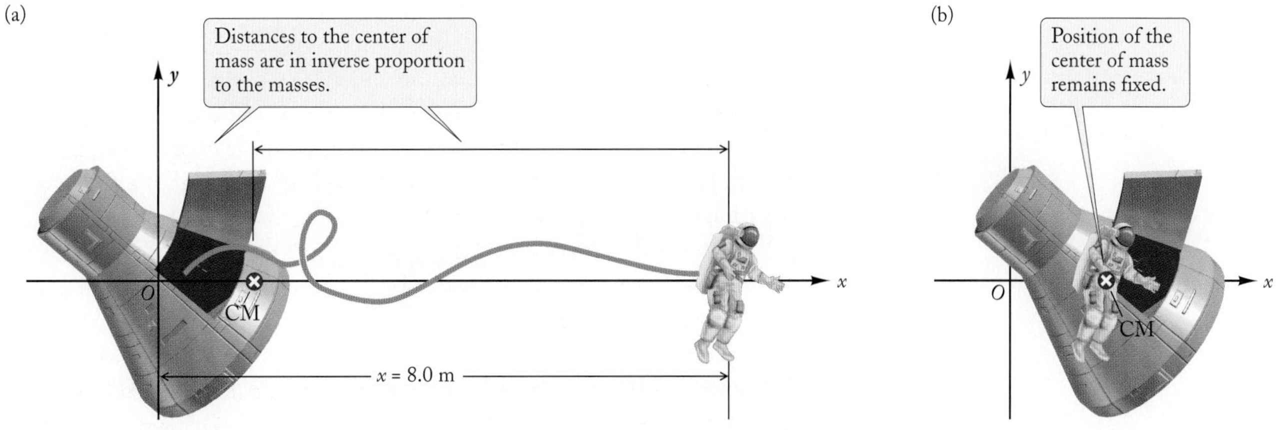

FIGURE 10.23 (a) Initial position of the astronaut and the spacecraft. The center of mass is between them. (b) Final position of the astronaut and the spacecraft. They are both at the center of mass.

COMMENT: The distances moved by the astronaut and by the spacecraft are in the inverse ratio of their masses. The astronaut (of small mass) moves a large distance, and the spacecraft (of large mass) moves a smaller distance. This is the result of the accelerations that the pull of the cord gives to these bodies: with forces of equal magnitudes, the accelerations of the astronaut and spacecraft are in the inverse ratio of their masses. However, our method of calculation based on the fixed position of the center of mass gives us the final positions directly, without any need to examine accelerations.

EXAMPLE 10 A projectile is launched at some angle θ with respect to the horizontal, $0° < \theta < 90°$. Just as it reaches its peak, it explodes into two pieces. The explosion causes a first, rear piece to come to a momentary stop, and it simply drops, striking the ground directly below the peak position. The explosion also causes the speed of the second piece to increase, and it hits the ground a distance five times further from the launch point than the first piece (see Fig. 10.24). If the original projectile had a mass of 12.0 kg, what are the masses of the pieces?

SOLUTION: Because the explosion does not produce external forces, the center of mass continues on its original path, a parabolic trajectory which strikes the ground at the range x_{max}, given by Eq. (4.43). The peak of the parabolic trajectory occurs at half this distance; thus the first piece, of some mass m_1, hits the ground a distance $\frac{1}{2}x_{max}$ from the launch point. We are also told that the second piece, of mass m_2, hits the ground a distance $5 \times \frac{1}{2}x_{max}$ from the launch point. The two pieces will reach the ground at the same instant, since this explosion affected only each piece's horizontal momentum. If we take our origin at the launch point, the x component of the center of mass is thus

$$x_{CM} = x_{max} = \frac{m_1x_1 + m_2x_2}{m_1 + m_2} = \frac{m_1x_{max}/2 + 5m_2x_{max}/2}{m_1 + m_2}$$

We can divide both sides of this equation by x_{max} and rearrange to obtain

$$m_1 = 3m_2$$

Since we know the total mass is $m_1 + m_2 = 12.0$ kg, or $4m_2 = 12.0$ kg, we obtain

$$m_1 = 9.0 \text{ kg} \qquad \text{and} \qquad m_2 = 3.0 \text{ kg}$$

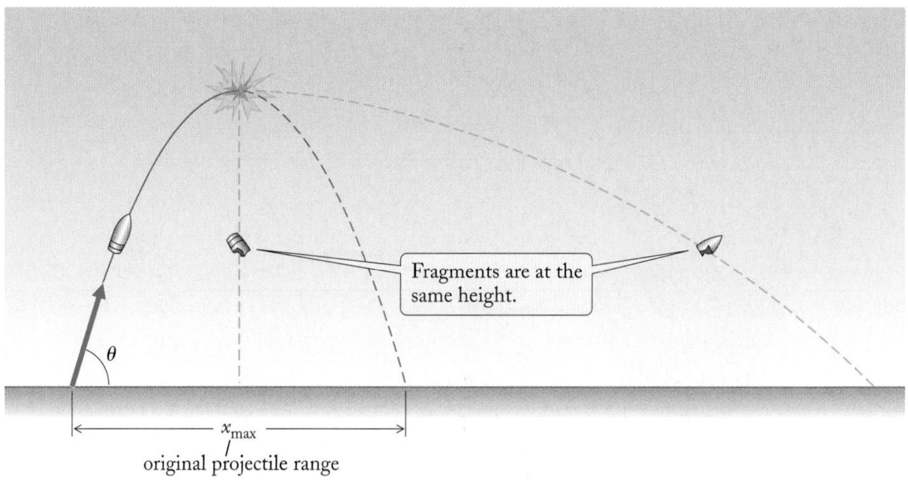

FIGURE 10.24 A projectile explodes at its apex. The rear fragment simply drops, and the forward piece lands five times further from the launch point.

COMMENT: Note that to relate both points of impact to the center of mass, we had to know that the impacts occurred at the *same instant*; we must always use the coordinates of a system of particles at a particular instant when calculating the center of mass.

✓ Checkup 10.3

QUESTION 1: When you crawl from the rear end of a canoe to the front end, the boat moves backward relative to the water. Explain.

QUESTION 2: You are locked inside a boxcar placed on frictionless wheels on railroad tracks. If you walk from the rear end of the boxcar to the front end, the boxcar rolls backward. Is it possible for you to make the boxcar roll a distance longer than its length?

QUESTION 3: You drop a handful of marbles on a smooth floor, and they bang into each other and roll away in all directions. What can you say about the motion of the center of mass of the marbles after the impact on the floor?

QUESTION 4: An automobile is traveling north at 25 m/s. A truck with twice the mass of the automobile is heading south at 20 m/s. What is the velocity of the center of mass of the two vehicles?

(A) 0 (B) 5 m/s south (C) 5 m/s north
(D) 10 m/s south (E) 10 m/s north

10.4 ENERGY OF A SYSTEM OF PARTICLES

The total kinetic energy of a system of particles is simply the sum of the individual kinetic energies of all the particles,

$$K = \tfrac{1}{2}m_1v_1^2 + \tfrac{1}{2}m_2v_2^2 + \cdots + \tfrac{1}{2}m_nv_n^2 \qquad (10.42)$$

kinetic energy of a system of particles

Since Eq. (10.39) for the momentum of a system of particles resembles the expression for the momentum of a single particle, we might be tempted to guess that the equation for the kinetic energy for a system of particles also can be expressed in the form of the translational kinetic energy of the center of mass $\frac{1}{2}Mv_{\text{CM}}^2$, resembling the kinetic energy of a single particle. But this is wrong! The total kinetic energy of a system of particles is usually larger than $\frac{1}{2}Mv_{\text{CM}}^2$. We can see this in the following simple example: Consider two automobiles of equal masses moving toward each other at equal speeds. The velocity of the center of mass is then zero, and consequently $\frac{1}{2}Mv_{\text{CM}}^2 = 0$. However, since each automobile has a positive kinetic energy, the total kinetic energy is *not* zero.

If the internal and external forces acting on a system of particles are conservative, then the system will have a potential energy. We saw above that for the specific example of the gravitational potential energy near the Earth's surface, the potential energy of the system took the same form as for a single particle, $U = Mgy_{\text{CM}}$ [see Eq. (10.33)]. But this form is a result of the particular force (uniform and proportional to mass); in general, the potential energy for a system does not have the same form as for a single particle. Unless we specify all of the forces, we cannot write down an explicit formula for the potential energy; but in any case, this potential energy will be some function of the positions of all the particles. The total mechanical energy is the sum of the total kinetic energy [Eq. (10.42)] and the total potential energy. This total energy will be conserved during the motion of the system of particles. Note that in reckoning the total potential energy of the system, we must include the potential energy of both the external forces and the internal forces. We know that the internal forces do not contribute to the changes of total momentum of the system, but these internal forces, and their potential energies, contribute to the total energy. For instance, if two particles are falling toward each other under the influence of their mutual gravitational attraction, the momentum gained by one particle is balanced by momentum lost by the other, but the kinetic energy gained by one particle is *not* balanced by kinetic energy lost by the other—both particles gain kinetic energy. In this example the gravitational attraction plays the role of an internal force in the system, and the gain of kinetic energy is due to a loss of mutual gravitational potential energy.

✔ Checkup 10.4

QUESTION 1: Consider a system consisting of two automobiles of equal mass. Initially, the automobiles have velocities of equal magnitudes in opposite directions. Suppose the automobiles collide head-on. Is the kinetic energy conserved?

QUESTION 2: The Solar System consists of the Sun, nine planets, and their moons. Is the total energy of this system conserved? Is the kinetic energy conserved? Is the potential energy conserved?

QUESTION 3: Two equal masses on a frictionless horizontal surface are connected by a spring. Each is given a brief push in a different direction. During the subsequent motion, which of the following remain(s) constant? ($\mathbf{P}$ = total momentum; K = total kinetic energy; U = total potential energy.)

(A) $\mathbf{P}$ only (B) $\mathbf{P}$ and K (C) $\mathbf{P}$ and U
(D) K and U (E) $\mathbf{P}$, K, and U

SUMMARY

PROBLEM-SOLVING TECHNIQUES Conservation of Momentum		(page 310)
PROBLEM-SOLVING TECHNIQUES Center of Mass		(page 320)
PHYSICS IN PRACTICE Center of Mass and Stability		(page 320)
MOMENTUM OF A PARTICLE	$\mathbf{p} = m\mathbf{v}$	(10.1)
MOMENTUM OF A SYSTEM OF PARTICLES	$\mathbf{P} = \mathbf{p}_1 + \mathbf{p}_2 + \cdots + \mathbf{p}_n$	(10.4)
RATE OF CHANGE OF MOMENTUM	$\frac{d\mathbf{P}}{dt} = \mathbf{F}_{\text{ext}}$	(10.13)
CONSERVATION OF MOMENTUM (in the absence of external forces)	$\mathbf{P} = [\text{constant}]$	(10.12)

CENTER OF MASS
(Using $M = m_1 + m_2 + \cdots + m_n$)

$$x_{\mathrm{CM}} = \frac{1}{M}(m_1x_1 + m_2x_2 + \cdots + m_nx_n) \quad (10.18)$$

$$y_{\mathrm{CM}} = \frac{1}{M}(m_1y_1 + m_2y_2 + \cdots + m_ny_n) \quad (10.19)$$

$$z_{\mathrm{CM}} = \frac{1}{M}(m_1z_1 + m_2z_2 + \cdots + m_nz_n) \quad (10.20)$$

CENTER OF MASS OF CONTINUOUS DISTRIBUTION OF MASS

where $dm = \rho dV$ (ρ is density and dV is a volume element).

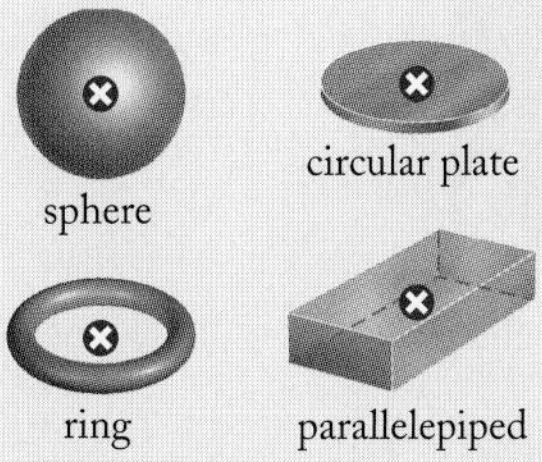

$$x_{\mathrm{CM}} = \frac{1}{M}\int x\,dm \quad (10.25)$$

$$y_{\mathrm{CM}} = \frac{1}{M}\int y\,dm \quad (10.26)$$

$$z_{\mathrm{CM}} = \frac{1}{M}\int z\,dm \quad (10.27)$$

VELOCITY OF THE CENTER OF MASS

$$\mathbf{v}_{\mathrm{CM}} = \frac{m_1\mathbf{v}_1 + m_2\mathbf{v}_2 + \cdots + m_n\mathbf{v}_n}{M} \quad (10.37)$$

MOMENTUM OF A SYSTEM OF PARTICLES

$$\mathbf{P} = M\mathbf{v}_{\mathrm{CM}} \quad (10.39)$$

MOTION OF THE CENTER OF MASS

$$M\mathbf{a}_{\mathrm{CM}} = \mathbf{F}_{\mathrm{ext}} \quad (10.40)$$

GRAVITATIONAL POTENTIAL ENERGY OF A SYSTEM OF PARTICLES (near the Earth's surface)

$$U = Mgy_{\mathrm{CM}} \quad (10.33)$$

KINETIC ENERGY OF A SYSTEM OF PARTICLES

$$K = \tfrac{1}{2}m_1v_1^2 + \tfrac{1}{2}m_2v_2^2 + \cdots + \tfrac{1}{2}m_nv_n^2 \quad (10.42)$$

QUESTIONS FOR DISCUSSION

1. When the nozzle of a fire hose discharges a large amount of water at high speed, several strong firefighters are needed to hold the nozzle steady. Explain.
2. When firing a shotgun, a hunter always presses it tightly against his shoulder. Why?
3. As described in Example 2, guns onboard eighteenth-century warships were often mounted on carriages (see Fig. 10.3). What was the advantage of this arrangement?
4. Hollywood movies often show a man being knocked over by the impact of a bullet while the man who shot the bullet remains standing, quite undisturbed. Is this reasonable?
5. Where is the center of mass of this book when it is closed? Mark the center of mass with a cross.
6. Roughly, where is the center of mass of this book when it is open, as it is at this moment?

7. A fountain shoots a stream of water up into the air (Fig. 10.25). Roughly, where is the center of mass of the water that is in the air at one instant? Is the center of mass higher or lower than the middle height?

FIGURE 10.25 Stream of water from a fountain.

8. Consider the moving wrench shown in Fig. 10.6. If the center of mass on this wrench had not been marked, how could you have found it by inspection of this photograph?
9. Is it possible to propel a sailboat by mounting a fan on the deck and blowing air on the sail? Is it better to mount the fan on the stern and blow air toward the rear?
10. Cyrano de Bergerac's sixth method for propelling himself to the Moon was as follows: "Seated on an iron plate, to hurl a magnet in the air—the iron follows—I catch the magnet—throw again—and so proceed indefinitely." What is wrong with this method (other than the magnet's insufficient pull)?
11. Within the Mexican jumping bean, a small insect larva jumps up and down. How does this lift the bean off the table?
12. Answer the following question, sent by a reader to the *New York Times:*

    A state trooper pulls a truck driver into the weigh station to see if he's overloaded. As the vehicle rolls onto the scales, the driver jumps out and starts beating on the truck box with a club. A bystander asks what he's doing. The trucker says: "I've got five tons of canaries in here. I know I'm overloaded. But if I can keep them flying I'll be OK." If the canaries are flying in that enclosed box, will the truck really weigh any less than if they're on the perch?

13. An elephant jumps off a cliff. Does the Earth move upward while the elephant falls?
14. A juggler stands on a balance, juggling five balls (Fig. 10.26). On the average, will the balance register the weight of the juggler plus the weight of the five balls? More than that? Less?

FIGURE 10.26 Juggler on a balance.

15. Suppose you fill a rubber balloon with air and then release it so that the air spurts out of the nozzle. The balloon will fly across the room. Explain.
16. The combustion chamber of a rocket engine is closed at the front and at the sides, but it is open at the rear (Fig. 10.27). Explain how the pressure of the gas on the walls of this combustion chamber gives a net forward force that propels the rocket.

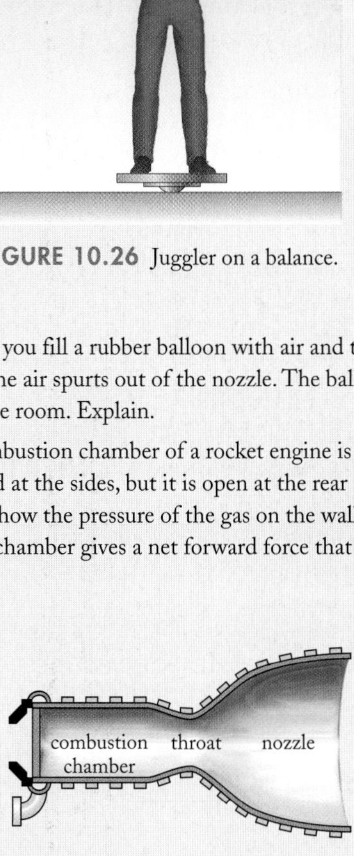

FIGURE 10.27 Combustion chamber of a rocket engine.

PROBLEMS

10.1 Momentum

1. What is the momentum of a rifle bullet of mass 15 g and speed 600 m/s? An arrow of mass 40 g and speed 80 m/s?
2. What is the momentum of an automobile of mass 900 kg moving at 65 km/h? If a truck of mass 7200 kg is to have the same momentum as the automobile, what must be its speed?
3. Using the entries listed in Tables 1.7 and 2.1, find the magnitude of the momentum for each of the following: Earth moving around the Sun, jet airliner at maximum airspeed, automobile at 55 mi/h, man walking, electron moving around a nucleus.
4. The push that a bullet exerts during impact on a target depends on the momentum of the bullet. A Remington .244

rifle, used for hunting deer, fires a bullet of 90 grains (1 grain is $\frac{1}{7000}$ lb) with a speed of 975 m/s. A Remington .35 rifle fires a bullet of 200 grains with a speed of 674 m/s. What is the momentum of each bullet?

5. An electron, of mass 9.1×10^{-31} kg, is moving in the x–y plane; its speed is 2.0×10^5 m/s, and its direction of motion makes an angle of 25° with the x axis. What are the components of the momentum of the electron?

6. A skydiver of mass 75 kg is in free fall. What is the rate of change of his momentum? Ignore friction.

7. A soccer player kicks a ball and sends it flying with an initial speed of 26 m/s at an upward angle of 30°. The mass of the ball is 0.43 kg. Ignore friction.
   (a) What is the initial momentum of the ball?
   (b) What is the momentum when the ball reaches maximum height on its trajectory?
   (c) What is the momentum when the ball returns to the ground? Is this final momentum the same as the initial momentum?

8. The Earth moves around the Sun in a circle of radius 1.5×10^{11} m at a speed of 3.0×10^4 m/s. The mass of the Earth is 6.0×10^{24} kg. Calculate the magnitude of the rate of change of the momentum of the Earth from these data. (Hint: The magnitude of the momentum does not change, but the direction does.)

9. A 1.0-kg mass is released from rest and falls freely. How much momentum does it acquire after one second? After ten seconds?

10. A 55-kg woman in a 20-kg rowboat throws a 3.0-kg life preserver with a horizontal velocity of 5.0 m/s. What is the recoil velocity of the woman and rowboat?

11. A 90-kg man dives from a 20-kg boat with an initial horizontal velocity of 2.0 m/s (relative to the water). What is the initial recoil velocity of the boat? (Neglect water friction.)

12. A hydrogen atom (mass 1.67×10^{-27} kg) at rest can emit a photon (a particle of light) with maximum momentum 7.25×10^{-27} kg·m/s. What is the maximum recoil velocity of the hydrogen atom?

13. Calculate the change of the kinetic energy in the collision between the two automobiles described in Example 3.

14. A rifle of 10 kg lying on a smooth table discharges accidentally and fires a bullet of mass 15 g with a muzzle speed of 650 m/s. What is the recoil velocity of the rifle? What is the kinetic energy of the bullet, and what is the recoil kinetic energy of the rifle?

15. A typical warship built around 1800 (such as the USS *Constitution*) carried 15 long guns on each side. The guns fired a shot of 11 kg with a muzzle speed of about 490 m/s. The mass of the ship was about 4000 metric tons. Suppose that all of the 15 guns on one side of the ship are fired (almost) simultaneously in a horizontal direction at right angle to the ship. What is the recoil velocity of the ship? Ignore the resistance offered by the water.

16. Two automobiles, moving at 65 km/h in opposite directions, collide head-on. One automobile has a mass of 700 kg; the other, a mass of 1500 kg. After the collision, both remain joined together. What is the velocity of the wreck? What is the change of the velocity of each automobile during the collision?

17. The nucleus of an atom of radium (mass 3.77×10^{-25} kg) suddenly ejects an alpha particle (mass 6.68×10^{-27} kg) of an energy of 7.26×10^{-16} J. What is the velocity of the recoil of the nucleus? What is the kinetic energy of the recoil?

18. A lion of mass 120 kg leaps at a hunter with a horizontal velocity of 12 m/s. The hunter has an automatic rifle firing bullets of mass 15 g with a muzzle speed of 630 m/s, and he attempts to stop the lion in midair. How many bullets would the hunter have to fire into the lion to stop its horizontal motion? Assume the bullets stick inside the lion.

*19. Find the recoil velocity for the gun described in Example 2 if the gun is fired with an elevation angle of 20°.

*20. Consider the collision between the moving and the initially stationary automobiles described in Example 3. In this example we neglected effects of the friction force exerted by the road during the collision. Suppose that the collision lasts for 0.020 s, and suppose that during this time interval the joined automobiles are sliding with locked wheels on the pavement with a coefficient of friction $\mu_k = 0.90$. What change of momentum and what change of speed does the friction force produce in the joined automobiles in the interval of 0.020 s? Is this change of speed significant?

*21. A Maxim machine gun fires 450 bullets per minute. Each bullet has a mass of 14 g and a velocity of 630 m/s.
   (a) What is the average force that the impact of these bullets exerts on a target? Assume that the bullets penetrate the target and remain embedded in it.
   (b) What is the average rate at which the bullets deliver their energy to the target?

*22. An owl flies parallel to the ground and grabs a stationary mouse with its talons. The mass of the owl is 250 g, and that of the mouse is 50 g. If the owl's speed was 4.0 m/s before grabbing the mouse, what is its speed just after the capture?

*23. A particle moves along the x axis under the influence of a time-dependent force of the form $F_x = 2.0t + 3.0t^2$, where F_x is in newtons and t is in seconds. What is the change in momentum of the particle between $t = 0$ and $t = 5.0$ s? [Hint: Rewrite Eq. (10.3) as $dp_x = F_x\,dt$ and integrate.]

*24. A vase falls off a table and hits a smooth floor, shattering into three fragments of equal mass which move away horizontally along the floor. Two of the fragments leave the point of impact with velocities of equal magnitudes v at right angles. What are the magnitude and direction of the horizontal velocity of the third fragment? (Hint: The x and y components of the momentum are conserved separately.)

*25. The nucleus of an atom of radioactive copper undergoing beta decay simultaneously emits an electron and a neutrino. The momentum of the electron is 2.64×10^{-22} kg·m/s, that of the neutrino is 1.97×10^{-22} kg·m/s, and the angle between their directions of motion is 30.0°. The mass of the residual nucleus is 63.9 u. What is the recoil velocity of the nucleus? (Hint: The x and y components of the momentum are conserved separately.)

*26. The solar wind sweeping past the Earth consists of a stream of particles, mainly hydrogen ions of mass 1.7×10^{-27} kg. There are about 1.0×10^{7} ions per cubic meter, and their speed is 4.0×10^{5} m/s. What force does the impact of the solar wind exert on an artificial Earth satellite that has an area of 1.0 m^2 facing the wind? Assume that upon impact the ions at first stick to the surface of the satellite.

*27. The record for the heaviest rainfall is held by Unionville, Maryland, where 3.12 cm of rain (1.23 in.) fell in an interval of 1.0 min. Assuming that the impact velocity of the raindrops on the ground was 10 m/s, what must have been the average impact force on each square meter of ground during this rainfall?

*28. An automobile is traveling at a speed of 80 km/h through heavy rain. The raindrops are falling vertically at 10 m/s, and there are 7.0×10^{-4} kg of raindrops in each cubic meter of air. For the following calculation assume that the automobile has the shape of a rectangular box 2.0 m wide, 1.5 m high, and 4.0 m long.

(a) At what rate (in kg/s) do the raindrops strike the front and top of the automobile?

(b) Assume that when a raindrop hits, it initially sticks to the automobile, although it falls off later. At what rate does the automobile give momentum to the raindrops? What is the horizontal drag force that the impact of the raindrops exerts on the automobile?

*29. A spaceship of frontal area 25 m^2 passes through a cloud of interstellar dust at a speed of 1.0×10^{6} m/s. The density of dust is 2.0×10^{-18} kg/m^3. If all the particles of dust that impact on the spaceship stick to it, find the average decelerating force that the impact of the dust exerts on the spaceship.

**30. A basketball player jumps straight up to launch a long jump shot at an angle of 45° with the horizontal and a speed of 15 m/s. The 75-kg player is momentarily at rest at the top of his jump just before the shot is released, with his feet 0.80 m above the floor. (a) What is the player's velocity immediately after the shot is released? (b) How far from his original position does he land? Treat the player as a point particle. The mass of a basketball is 0.62 kg.

**31. A gun mounted on a cart fires bullets of mass m in the backward direction with a horizontal muzzle velocity u. The initial mass of the cart, including the mass of the gun and the mass of the ammunition, is M, and the initial velocity of the cart is zero. What is the velocity of the cart after firing n bullets? Assume that the cart moves without friction, and ignore the mass of the gunpowder.

10.2 Center of Mass

32. A penny coin lies on a table at a distance of 20 cm from a stack of three penny coins. Where is the center of mass of the system of four coins?

33. A 59-kg woman and a 73-kg man sit on a seesaw, 3.5 m long. Where is their center of mass? Neglect the mass of the seesaw.

34. Consider the system Earth–Moon; use the data in the table printed inside the book cover. How far from the center of the Earth is the center of mass of this system?

35. Consider the Sun and the planet Jupiter as a two-particle system. How far from the center of the Sun is the center of mass of this system? Express your result as a multiple of the radius of the Sun. (Use the data inside the cover of this book.)

36. Two bricks are adjacent, and a third brick is positioned symmetrically above them, as shown in Fig. 10.28. Where is the center of mass of the three bricks?

FIGURE 10.28 Three bricks.

*37. Where is the center of mass of a uniform sheet in the shape of an isosceles triangle? Assume that the height of the triangle is h when the unequal side is the base.

*38. Consider a pyramid with height h and a triangular base. Where is its center of mass?

*39. In order to balance the wheel of an automobile, a mechanic attaches a piece of lead alloy to the rim of the wheel. The mechanic finds that if he attaches a piece of 40 g at a distance of 20 cm from the center of the wheel of 30 kg, the wheel is perfectly balanced; that is, the center of the wheel coincides with the center of mass. How far from the center of the wheel was the center of mass before the mechanic balanced the wheel?

*40. The distance between the oxygen and each of the hydrogen atoms in a water (H_2O) molecule is 0.0958 nm; the angle between the two oxygen–hydrogen bonds is 105° (Fig. 10.29). Treating the atoms as particles, find the center of mass.

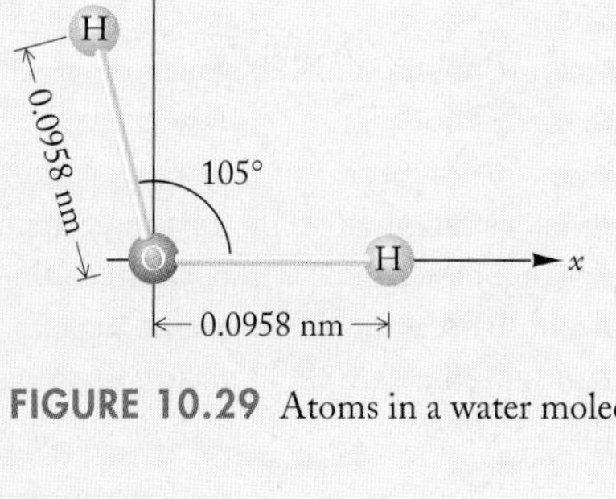

FIGURE 10.29 Atoms in a water molecule.

*41. Figure 10.30 shows the shape of a nitric acid (HNO_3) molecule and its dimensions. Treating the atoms as particles, find the center of mass of this molecule.

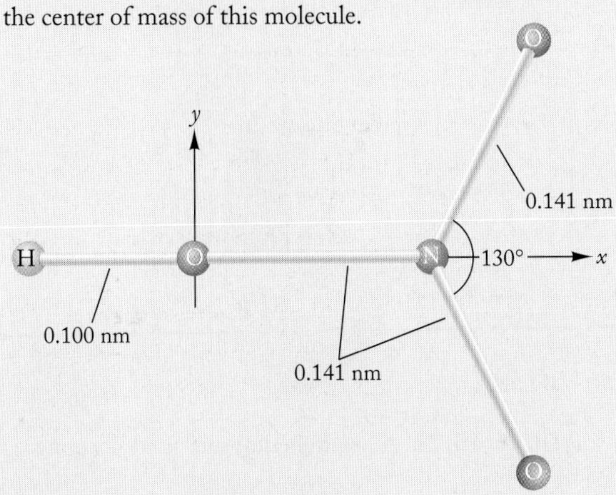

FIGURE 10.30 Atoms in a nitric acid molecule.

*42. Figure 9.13a shows the positions of the three inner planets (Mercury, Venus, and Earth) on January 1, 2000. Measure angles and distances off this figure and find the center of mass of the system of these planets (ignore the Sun). The masses of the planets are listed in Table 9.1.

*43. The Local Group of galaxies consists of our Galaxy and its nearest neighbors. The masses of the most important members of the Local Group are as follows (in multiples of the mass of the Sun): our Galaxy, 2×10^{11}; the Andromeda galaxy, 3×10^{11}; the Large Magellanic Cloud, 2.5×10^{10}; and NGC598, 8×10^{9}. The x, y, z coordinates of these galaxies are, respectively, as follows (in thousands of light-years): (0, 0, 0); (1640, 290, 1440), (8.5, 56.7, −149), and (1830, 766, 1170). Find the coordinates of the center of mass of the Local Group. Treat all the galaxies as point masses.

*44. A thin, uniform rod is bent in the shape of a semicircle of radius R (see Fig. 10.31). Where is the center of mass of this rod?

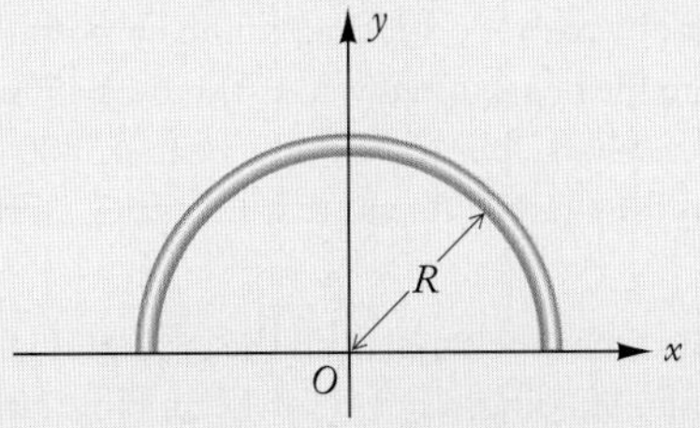

FIGURE 10.31 A rod bent in a semicircle.

*45. Three uniform square pieces of sheet metal are joined along their edges so as to form three of the sides of a cube (Fig. 10.32). The dimensions of the squares are $L \times L$. Where is the center of mass of the joined squares?

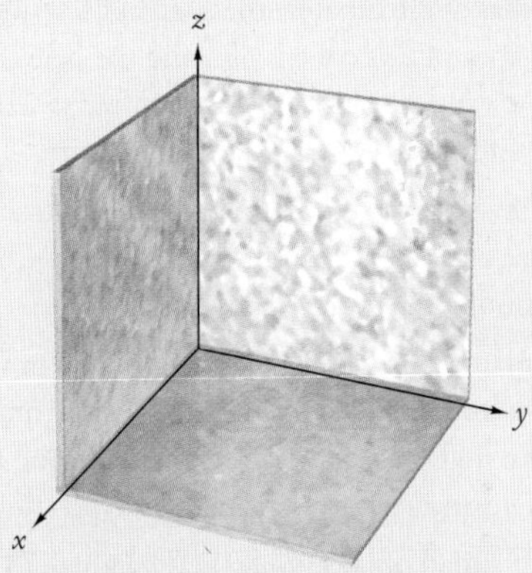

FIGURE 10.32 Three square pieces of sheet metal joined together at their edges.

*46. A box made of plywood has the shape of a cube measuring $L \times L \times L$. The top of the box is missing. Where is the center of mass of the open box?

*47. A cube of iron has dimensions $L \times L \times L$. A hole of radius $\frac{1}{4}L$ has been drilled all the way through the cube so that one side of the hole is tangent to the middle of one face along its entire length (Fig. 10.33). Where is the center of mass of the drilled cube?

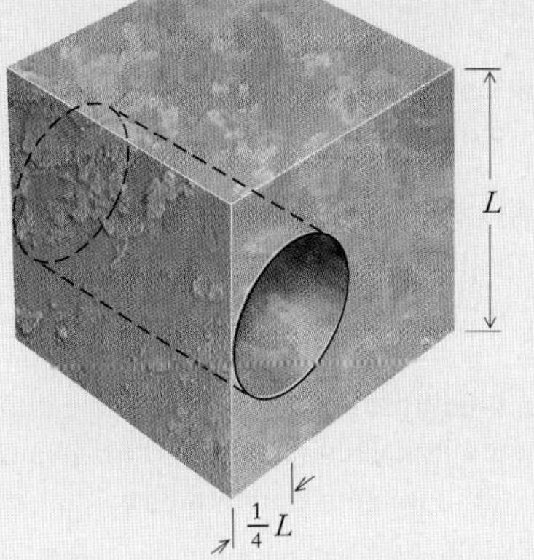

FIGURE 10.33 Iron cube with a hole.

*48. A semicircle of uniform sheet metal has radius R (Fig. 10.34). Find the center of mass.

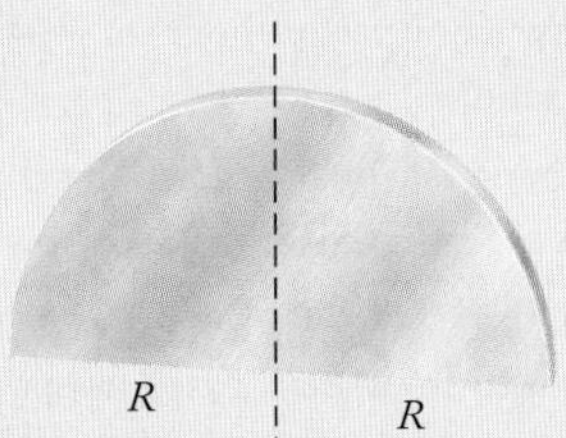

FIGURE 10.34 Semicircle of sheet metal.

*49. Mount Fuji has approximately the shape of a cone. The half-angle at the apex of this cone is 65°, and the height of the apex is 3800 m. At what height is the center of mass? Assume that the material in Mount Fuji has uniform density.

*50. Show that the center of mass of a uniform flat triangular plate is at the point of intersection of the lines drawn from the vertices to the midpoints of the opposite sides.

*51. Consider a man of mass 80 kg and height 1.70 m with the mass distribution described in Fig. 10.17. How much work does this man do to raise his arms from a hanging position to a horizontal position? To a vertically raised position?

*52. Suppose that a man of mass 75 kg and height 1.75 m runs in place, raising his legs high, as in Fig. 10.35. If he runs at the rate of 80 steps per minutes for each leg (160 total per minute), what power does he expend in raising his legs?

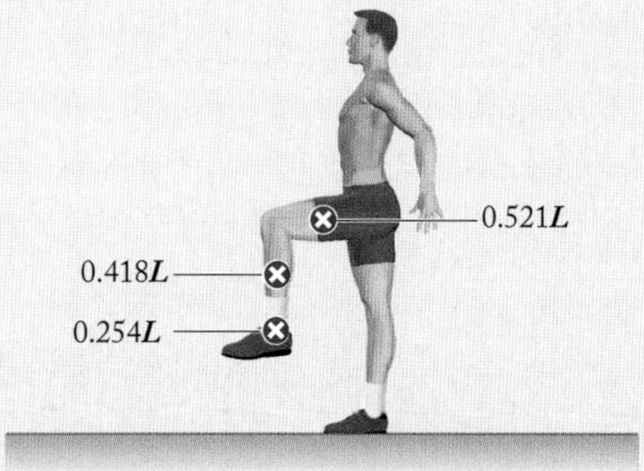

FIGURE 10.35 Man with raised leg.

*53. A lock on the Champlain Canal is 73 m long and 9.2 m wide; the lock has a lift of 3.7 m—that is, the difference between the water levels of the canal on one side of the lock and on the other side is 3.7 m. How much gravitational potential energy is wasted each time the lock goes through one cycle (involving the filling of the lock with water from the high level and then the spilling of this water to the low level)?

*54. The Great Pyramid at Giza has a mass of 6.6×10^6 metric tons and a height of 147 m (see Example 7). Assume that the mass is uniformly distributed over the volume of the pyramid.

(a) How much work must the ancient Egyptian laborers have done against gravity to pile up the stones in the pyramid?

(b) If each laborer delivered work at an average rate of 4.0×10^5 J/h, how many person-hours of work have been stored in this pyramid?

**55. A thin hemispherical shell of uniform thickness is suspended from a point above its center of mass as shown in Fig. 10.36. Where is that center of mass?

**56. Suppose that water drops are released from a point at the edge of a roof with a constant time interval Δt between one water drop and the next. The drops fall a distance l to the ground. If Δt is very short (so the number of drops falling though the air at any given instant is very large), show that the center of mass of the falling drops is at a height of $\frac{2}{3}l$ above the ground.

FIGURE 10.36 A hemispherical shell used as a gong.

From this, deduce that the time-average height of a projectile released from the ground and returning to the ground is $\frac{2}{3}$ of its maximum height. (This theorem is useful in the calculation of the average air pressure and air resistance encountered by a projectile.)

10.3 The Motion of the Center of Mass

57. A proton of kinetic energy 1.6×10^{-13} J is moving toward a proton at rest. What is the velocity of the center of mass of the system?

58. In a molecule, the atoms usually execute a rapid vibrational motion about their equilibrium position. Suppose that in an isolated potassium bromide (KBr) molecule the speed of the potassium atom is 5.0×10^3 m/s at one instant (relative to the center of mass). What is the speed of the bromine atom at the same instant?

59. A fisherman in a boat catches a great white shark with a harpoon. The shark struggles for a while and then becomes limp when at a distance of 300 m from the boat. The fisherman pulls the shark by the rope attached to the harpoon. During this operation, the boat (initially at rest) moves 45 m in the direction of the shark. The mass of the boat is 5400 kg. What is the mass of the shark? Pretend that the water exerts no friction.

60. A 75-kg man climbs the stairs from the ground floor to the fourth floor of a building, a height of 15 m. How far does the Earth recoil in the opposite direction as the man climbs?

61. A 6000-kg truck stands on the deck of an 80000-kg ferryboat. Initially the ferry is at rest and the truck is located at its front end. If the truck now drives 15 m along the deck toward the rear of the ferry, how far will the ferry move forward relative to the water? Pretend that the water has no effect on the motion.

62. While moving horizontally at 5.0×10^3 m/s at an altitude of 2.5×10^4 m, a ballistic missile explodes and breaks apart into

two fragments of equal mass which fall freely. One of the fragments has zero speed immediately after the explosion and lands on the ground directly below the point of the explosion. Where does the other fragment land? Ignore the friction of air.

63. A 15-g bullet moving at 260 m/s is fired at a 2.5-kg block of wood. What is the velocity of the center of mass of the bullet–block system?

64. A 60-kg woman and a 90-kg man walk toward each other, each moving with speed v relative to the ground. What is the velocity of their center of mass?

65. A projectile of mass M reaches the peak of its motion a horizontal distance D from the launch point. At its peak, it explodes into three equal fragments. One fragment returns directly to the launch point, and one lands a distance $2D$ from the launch point, at a point in the same plane as the initial motion. Where does the third fragment land?

*66. A projectile is launched with speed v_0 at an angle of θ with respect to the horizontal. At the peak of its motion, it explodes into two pieces of equal mass, which continue to move in the original plane of motion. One piece strikes the ground a horizontal distance D further from the launch point than the point directly below the explosion at a time $t < v_0 \sin\theta/g$ after the explosion. How high does the other piece go? Where does the other piece land? Answer in terms of v_0, θ, D, and t.

**67. Figure 9.13a shows the positions of the three inner planets (Mercury, Venus, Earth) on January 1, 2000. Measuring angles off this figure and using the data on masses, orbital radii, and periods given in Table 9.1, find the velocity of the center of mass of this system of three planets.

10.4 Energy of a System of Particles

68. Two automobiles, each of mass 1500 kg, travel in the same direction along a straight road. The speed of one automobile is 25 m/s, and the speed of the other automobile is 15 m/s. If we regard these automobiles as a system of two particles, what is the translational kinetic energy of the center of mass? What is the total kinetic energy?

69. Repeat the calculation of Problem 68 if the two automobiles travel in *opposite* directions.

70. A projectile of 45 kg fired from a gun has a speed of 640 m/s. The projectile explodes in flight, breaking apart into a fragment of 32 kg and a fragment of 13 kg (we assume that no mass is dispersed in the explosion). Both fragments move along the original direction of motion. The speed of the first fragment is 450 m/s and that of the second is 1050 m/s.

(a) Calculate the translational kinetic energy of the center of mass motion before the explosion.

(b) Calculate the translational kinetic energy of the center of mass motion after the explosion. Calculate the total kinetic energy. Where does the extra kinetic energy come from?

71. Consider the automobile collision described in Problem 16. What is the translational kinetic energy of the center of mass motion before the collision? What is the total kinetic energy before and after the collision?

72. Two isolated point masses m_1 and m_2 are connected by a spring. The masses attain their maximum speeds at the same instant. A short time later both masses are stationary. The maximum speed of the first mass is v_1. What is the maximum speed of the second mass? When the masses are stationary, what is the energy stored in the spring?

73. The typical speed of a helium atom in helium gas at room temperature is 1.4 km/s; that of an oxygen molecule (O_2) in oxygen gas is close to 500 m/s. Find the total kinetic energy of one mole of helium atoms and that of one mole of oxygen molecules.

*74. Two automobiles, each of mass $M/2$ and speed v, drive around a one-lane traffic circle. What is the total kinetic energy of the two-car system? What is the quantity $\frac{1}{2}Mv_{CM}^2$ if the automobiles are (a) on opposite sides of the traffic circle, (b) one-quarter of the circle apart, and (c) locked together?

*75. Consider the Sun and Jupiter to be a two-particle system, orbiting around the center of mass. Find the ratio of the kinetic energy of the Sun to that of Jupiter. (Use the data inside the book cover.)

*76. The typical speed of the vibrational motion of the iron atoms in a piece of iron at room temperature is 360 m/s. What is the total kinetic energy of a 1.0-kg chunk of iron?

REVIEW PROBLEMS

77. A hunter on skates on a smooth sheet of ice shoots 10 bullets at a target at the shore. Each bullet has a mass of 15 g and a speed of 600 m/s. The hunter has a mass of 80 kg. What recoil speed does he acquire?

78. Grain is being loaded into an almost full railroad car from an overhead chute (see Fig. 10.37). If 500 kg per second falls freely from a height of 4.0 m to the top of the car, what downward push does the impact of the grain exert on the car?

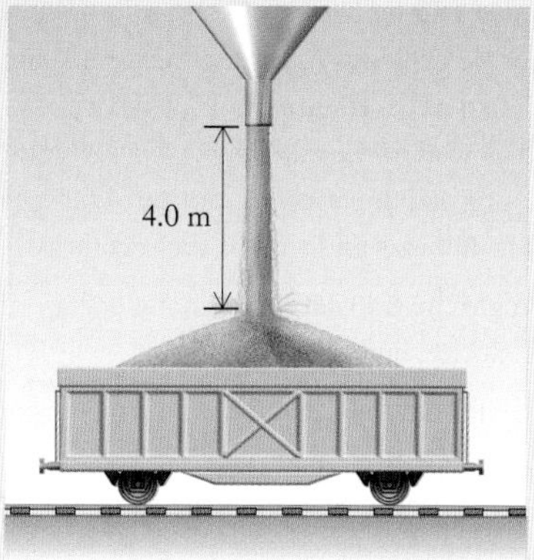

FIGURE 10.37 Grain from a chute falls into a railroad car.

79. A boy and a girl are engaged in a tug-of-war on smooth, frictionless ice. The mass of the boy is 40 kg, and that of the girl is 30 kg; their separation is initially 4.0 m. Each pulls with a force of 200 N on the rope. What is the acceleration of each? If they keep pulling, where will they meet?

80. An automobile of 1200 kg and an automobile of 1500 kg are traveling in the same direction on a straight road. The speeds of the two automobiles are 60 km/h and 80 km/h, respectively. What is the velocity of the center of mass of the two-automobile system?

81. An automobile traveling 40 km/h collides head-on with a truck which has 5 times the mass of the automobile. The wreck remains at rest after the collision. Deduce the speed of the truck.

82. The nozzle of a fire hose ejects 800 liters of water per minute at a speed of 26 m/s. Estimate the recoil force on the nozzle. By yourself, can you hold this nozzle steady in your hands?

83. The distance between the centers of the atoms of potassium and bromine in the potassium bromide (KBr) molecule is 0.282 nm (Fig. 10.38). Treating the atoms as particles, find the center of mass.

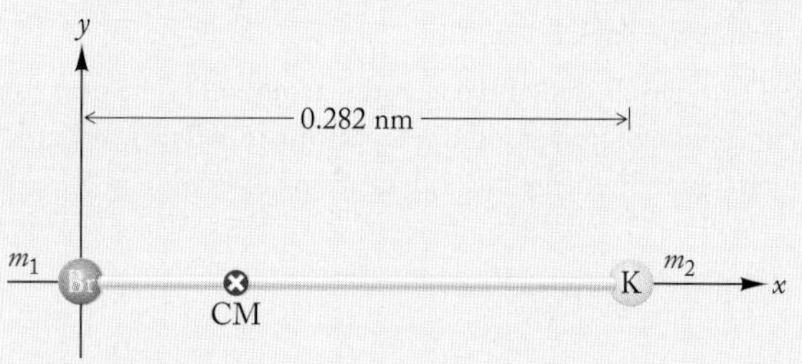

FIGURE 10.38 Atoms in a potassium bromide molecule.

84. A tugboat of mass 400 metric tons and a ship of 28 000 metric tons are joined by a long towrope of 400 m. Both vessels are initially at rest in the water. If the tugboat reels in 200 m of towrope, how far does the ship move relative to the water? The tugboat? Ignore the resistance that the water offers to the motion.

85. A cat stands on a plank of balsa wood floating in water. The mass of the cat is 3.5 kg, and the mass of the balsa is 5.0 kg. If the cat walks 1.0 m along the plank, how far does she move in relation to the water?

86. Three firefighters of equal masses are climbing a long ladder. When the first firefighter is 20 m up the ladder, the second is 15 m up, and the third is 5 m up. Where is the center of mass of the three firefighters?

87. Four identical books are arranged on the vertices of an equilateral triangle of side 1.0 m. Two of the books are together at one vertex of the triangle, and the other two are at the other two vertices. Where is the center of mass of this arrangement?

88. Three identical metersticks are arranged to form the letter U. Where is the center of mass of this system?

89. Two uniform squares of sheet metal of dimensions $L \times L$ are joined at right angles along one edge (see Fig. 10.39). One of the squares has twice the mass of the other. Find the center of mass of the combined squares.

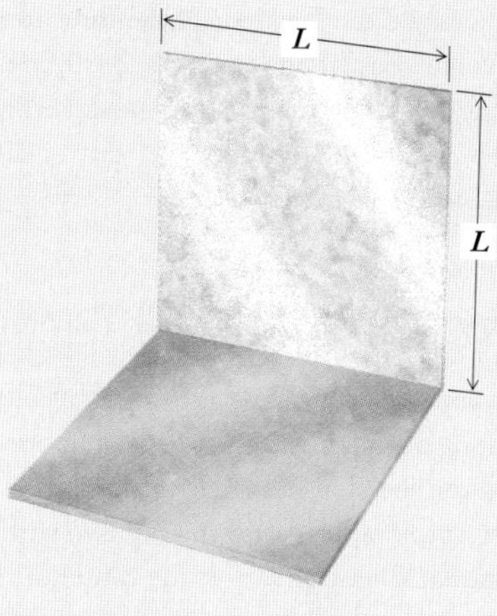

FIGURE 10.39 Two square pieces of sheet metal joined along one edge.

*90. Find the center of mass of a uniform solid hemisphere of radius R.

Answers to Checkups

Checkup 10.1

1. We assume the usual case that the truck has a larger mass than the automobile. Then the equality of their momenta (mv) implies that the automobile has a larger speed (the ratio of the velocities will be the inverse of the ratio of the masses). Since the kinetic energy is $\frac{1}{2}mv^2 = \frac{1}{2} \times mv \times v$ and the momenta (mv) are equal, then the vehicle with the larger speed, the automobile, will also have the larger kinetic energy.
2. They cannot have the same momentum, since the signs of their momenta will be opposite. They can have the same kinetic energy, since that depends only on the speed: $K = \frac{1}{2}mv^2$.
3. No—the momentum, like the velocity, is also reversed, and it has the opposite sign after the impact.
4. Yes—for practical purposes, the Solar System is essentially an isolated system, and so the net momentum is constant. The net kinetic energy is not constant, since during motion, kinetic energy is converted to potential energy and vice versa.
5. (a) For any directions, the total kinetic energy is $\frac{1}{2}mv^2 + \frac{1}{2}mv^2 = mv^2$. For parallel motion (both southward), the total momentum is $mv + mv = 2mv$ southward. (b) The total kinetic energy is again mv^2, while for antiparallel motion, the total momentum is $mv - mv = 0$ (no direction). (c) The total kinetic energy is again mv^2, while for perpendicular motion, the total momentum has magnitude $\sqrt{(mv)^2 + (mv)^2} = \sqrt{2}mv$ and is directed 45° south of east.
6. (C) Automobile; truck. The truck has a larger mass M than the automobile mass m. Let the truck speed be V and the automobile speed be v. The equal kinetic energies $(\frac{1}{2}MV^2 = \frac{1}{2}mv^2)$ then imply that the automobile will have the larger speed $v = (M/m)^{1/2}V$. If we substitute one power of this v into the kinetic energy equality and cancel a factor of $\frac{1}{2}V$, we find $MV = (M/m)^{1/2}mv$; thus, the truck momentum is larger.

Checkup 10.2

1. Consider the average position of the mass distribution. For the curved snake shown in the figure, the center of mass is at a point in the space below the top arc, perhaps slightly below center (because of the *two* bottom arcs) and slightly to the right of center (because of the head).
2. Consider the average position of the mass distribution. For the horseshoe, the center of mass is below center in the space in the middle of the arc, along the vertical line of symmetry, at a point well away from the open end.
3. No. The center of mass is a weighted average of position; such an average can never be greater than all of the positions averaged.
4. (C) Shifts upward. If the heavy mass at the bottom falls off, the center of mass is higher. When the center of mass of the sailboat is too high, it is top-heavy, and prone to tip over.

Checkup 10.3

1. For the (isolated) system of person plus canoe, with no initial motion, the center of mass stays at the same fixed position as you begin and continue your crawl. Thus as your mass moves from the rear to the front, the boat moves backward a sufficient distance to keep the center of mass of the combined system fixed.
2. No. In the extreme case where the boxcar has zero mass, you remain fixed relative to the ground, and the boxcar rolls a distance equal to its length as you walk from the rear to the front. If the boxcar has appreciable mass, you will move toward your common center of mass, which will be a distance less than the length of the boxcar.
3. If the marbles were dropped vertically, then the center of mass remains fixed at the point of impact with the floor, even though the marbles scatter in all directions.
4. (B) 5 m/s south. Using Eq. (10.37) with positive velocity northward, $v_{CM} = (M \times 25\text{ m/s} - 2M \times 20\text{ m/s})/(3M) = (-15\text{ m/s})/3 = -5\text{ m/s}$.

Checkup 10.4

1. No. The initial kinetic energy is large, and the final kinetic energy is small or zero; the energy is transformed into other forms: elastic energy (deformation of automobile parts), friction, sound, and heat.
2. The total energy is conserved, if we consider only gravitational potential energy and kinetic energy (in actuality, some other energy is lost, for example, as the Sun's light is radiated away into space). Neither kinetic nor potential energy is separately conserved; these two are traded back and forth, for example, as the planets move in their elliptical orbits.
3. (A) **P** only. Since there is no net external force, the total momentum of such an isolated system is always simply conserved. However, the spring will stretch and compress during the motion, trading kinetic for potential energy, so K and U will not remain constant.

CHAPTER

11 Collisions

CONCEPTS IN CONTEXT

Concepts in Context

In this crash test, the automobile was towed at a speed of 56 km/h (35 mi/h) and then crashed into a rigid concrete barrier. Anthropomorphic dummies that simulate human bodies are used for evaluation of injuries that would be sustained by driver and passengers. Accelerometers installed on the body of the automobile and the bodies of the dummies permit calculation of impact forces.

With the concepts of this chapter we can answer questions such as:

? What is the average force on the front of an automobile during impact? (Example 1, page 341)

? What is force on the head of a dummy during a collision with the windshield or the steering wheel? (Example 2, page 341)

? How do seat belts and air bags protect occupants of an automobile in a crash? (Physics in Practice: Automobile Collisions, page 343)

? How does the stiffness of the front end of an automobile affect the safety of its occupants in a collision? (Checkup 11.1, question 1, page 344)

? In a two-car collision, how are the initial velocities related to the final direction of motion? (Example 7, page 351)

The collision between two bodies—an automobile and a solid wall, a ship and an iceberg, a molecule of oxygen and a molecule of nitrogen—involves a violent change of the motion, a change brought about by very strong forces that begin to act suddenly when the bodies come into contact, last a short time, and then cease just as suddenly when the bodies separate. The forces that act during a collision are usually rather complicated, so their complete theoretical description is impossible (e.g., in an automobile collision) or at least very difficult (e.g., in a collision between subatomic particles). However, even without exact knowledge of the details of the forces, we can make some predictions about the collision by taking advantage of the general laws of conservation of momentum and energy we studied in the preceding chapters. In the following sections we will see what constraints these laws impose on the motion of the colliding bodies.

The study of collisions is an important tool in engineering and physics. In automobile collision and safety studies, engineers routinely subject vehicles to crash tests. Collisions are also essential for the experimental investigation of atoms, nuclei, and elementary particles. All subatomic bodies are too small to be made visible with any kind of microscope. Just as you might use a stick to feel your way around a dark cave, a physicist who cannot see the interior of an atom uses probes to "feel" for subatomic structures. The probe used by physicists in the exploration of subatomic structures is simply a stream of fast-moving particles—electrons, protons, alpha particles (helium nuclei), or others. These projectiles are aimed at a target containing a sample of the atoms, nuclei, or elementary particles under investigation. From the manner in which the projectiles collide and react with the target, physicists can deduce some of the properties of the subatomic structures in the target. Similarly, materials scientists, chemists, and engineers deduce the structure and composition of solids and liquids by bombarding such materials with particles and examining the results of such collisions.

11.1 IMPULSIVE FORCES

Online Concept Tutorial 13

The force that two colliding bodies exert on one another acts for only a short time, giving a brief but strong push. Such a force that acts for only a short time is called an **impulsive force.** During the collision, the impulsive force is much stronger than any other forces that may be present; consequently the impulsive force produces a large change in the motion while the other forces produce only small and insignificant changes. For instance, during the automobile collision shown in Fig. 11.1, the only important force on the automobile is the push of the wall on its front end; the effects produced by gravity and by the friction force of the road during the collision are insignificant.

Suppose the collision lasts some short time Δt, say, from $t = 0$ to $t = \Delta t$, and that during this time an impulsive force $\mathbf{F}$ acts on one of the colliding bodies. This force is zero before $t = 0$ and it is zero after $t = \Delta t$, but it is large between these times. For example, Fig. 11.2 shows a plot of the force experienced by an automobile in a collision with a solid wall lasting 0.120 s. The force is zero before $t = 0$ and after $t = 0.120$ s, and varies in a complicated way between these times.

*The **impulse** delivered by such a force to the body is defined as the integral of the force over time:*

$$\mathbf{I} = \int_0^{\Delta t} \mathbf{F}\,dt \qquad (11.1)$$

impulse

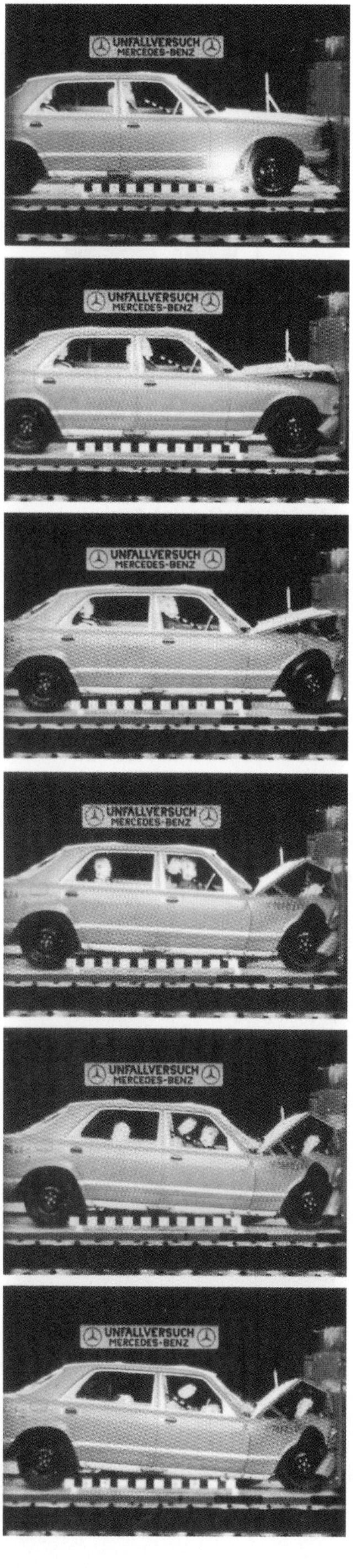

FIGURE 11.1 Crash test of a Mercedes–Benz automobile. The photographs show an impact at 49 km/h into a rigid barrier. The first photograph was taken 5×10^{-3} s after the initial contact; the others were taken at intervals of 20×10^{-3} s. The automobile remains in contact with the barrier for 0.120 s; it then recoils from the barrier with a speed of 4.7 km/h. The checkered bar on the ground has a length of 2 m.

According to this equation, the x component of the impulse for the force plotted in Fig. 11.2 is the area between the curve $F_x(t)$ and the t axis.

The SI units of impulse are N·s, or kg·m/s; these units are the same as those for momentum.

By means of the equation of motion, $\mathbf{F} = d\mathbf{p}/dt$, we can transform Eq. (11.1) into

$$\mathbf{I} = \int_0^{\Delta t} \mathbf{F}\,dt = \int_0^{\Delta t} \frac{d\mathbf{p}}{dt}\,dt = \int d\mathbf{p} = \mathbf{p}' - \mathbf{p} \qquad (11.2)$$

where $\mathbf{p}$ is the momentum of the body before the collision (at time 0) and $\mathbf{p}'$ is the momentum after the collision (at time $t = \Delta t$). Thus, the impulse of a force is simply equal to the momentum change produced by this force. This equality of impulse and momentum change is sometimes referred to as the *impulse–momentum relation*. However, since the force acting during a collision is usually not known in detail, Eq. (11.2) is not very helpful for calculating momentum changes. It is often best to apply Eq. (11.2) in reverse, for calculating the time-average force from the known momentum change. This time-average force is defined by

$$\overline{\mathbf{F}} = \frac{1}{\Delta t}\int_0^{\Delta t} \mathbf{F}\,dt \qquad (11.3)$$

In a plot of force vs. time, such as shown in Fig. 11.2, the time-average force simply represents the mean height of the function above the t axis; this mean height is shown

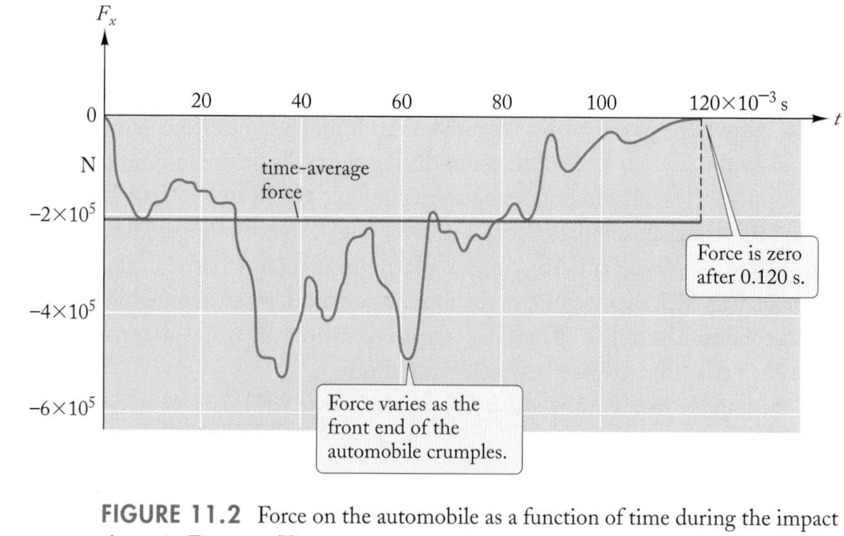

FIGURE 11.2 Force on the automobile as a function of time during the impact shown in Fig. 11.1. The colored horizontal line indicates the time-average force. (Calculated from data supplied by Mercedes–Benz of North America, Inc.)

by the red horizontal line in Fig. 11.2. According to Eq. (11.2), we can write the time-average force as

$$\overline{\mathbf{F}} = \frac{1}{\Delta t}\mathbf{I} = \frac{1}{\Delta t}(\mathbf{p}' - \mathbf{p}) \tag{11.4}$$

average force in collision

This relation gives a quick estimate of the average magnitude of the impulsive force acting on the body if the duration of the collision and the momentum change are known.

EXAMPLE 1 The collision between the automobile and the barrier shown in Fig. 11.1 lasts 0.120 s. The mass of the automobile is 1700 kg, and the initial and final velocities in the horizontal direction are $v_x = 13.6$ m/s and $v'_x = -1.3$ m/s, respectively (the final velocity is negative because the automobile recoils, or bounces back from the barrier). From these data, evaluate the average force that acts on the automobile during the collision. Evaluate the average force that acts on the barrier.

Concepts — in — Context

SOLUTION: With the x axis along the direction of the initial motion, the change of momentum is

$$\begin{aligned} p'_x - p_x &= mv'_x - mv_x \\ &= 1700 \text{ kg} \times (-1.3 \text{ m/s}) - 1700 \text{ kg} \times 13.6 \text{ m/s} \\ &= -2.53 \times 10^4 \text{ kg·m/s} \end{aligned}$$

According to Eq. (11.4), the average force is then

$$\begin{aligned} \overline{F}_x &= \frac{p'_x - p_x}{\Delta t} \\ &= \frac{-2.53 \times 10^4 \text{ kg·m/s}}{0.120 \text{ s}} = -2.11 \times 10^5 \text{ N} \end{aligned} \tag{11.5}$$

Since the mutual forces on two bodies engaged in a collision are an action–reaction pair, the forces on the automobile and on the barrier are of equal magnitudes and of opposite directions. Thus, the average force on the barrier is $\overline{F}_x = +2.11 \times 10^5$ N. This is quite a large force—it equals the weight of about 2×10^4 kg, or 20 tons.

EXAMPLE 2 When an automobile collides with an obstacle and suddenly stops, a passenger not restrained by a seat belt will not stop simultaneously with the automobile, but instead will continue traveling at nearly constant speed until he or she hits the dashboard and the windshield. The collision of the passenger's head with the windshield often results in severe or fatal injuries. In crash tests, dummies with masses, shapes, and joints simulating human bodies are used to determine likely injuries. Consider a dummy head striking a windshield at 15 m/s (54 km/h) and stopping in a time of 0.015 s (this time is considerably shorter than the time of about 0.12 s for stopping the automobile because the front end of the automobile crumples gradually and cushions the collision to some extent; there is no such cushioning for the head striking the windshield). What is the average force on the head during impact on the windshield? What is the average deceleration? Treat the head as a body of mass 5.0 kg, moving independently of the neck and trunk.

Concepts — in — Context

SOLUTION: The initial momentum of the head is

$$p_x = mv_x = 5.0 \text{ kg} \times 15 \text{ m/s} = 75 \text{ kg·m/s}$$

When the head stops, the final momentum is zero. Hence the average force is

$$\overline{F}_x = \frac{p'_x - p_x}{\Delta t} = -\frac{p_x}{\Delta t}$$

$$= -\frac{75 \text{ kg·m/s}}{0.015 \text{ s}} = -5.0 \times 10^3 \text{ N}$$

The average acceleration is

$$\overline{a}_x = \frac{\overline{F}_x}{m} = -\frac{5.0 \times 10^3 \text{ N}}{5.0 \text{ kg}} = -1.0 \times 10^3 \text{ m/s}^2$$

which is about 100 standard *g*'s!

Often it is not possible to calculate the motion of the colliding bodies by direct solution of Newton's equation of motion because the impulsive forces that act during the collision are not known in sufficient detail. We must then glean whatever information we can from the general laws of conservation of momentum and energy, which do not depend on the details of these forces. In some simple instances, these general laws permit the deduction of the motion after the collision from what is known about the motion before the collision.

In all collisions between two or more particles, the total momentum of the system is conserved. Whether or not the mechanical energy is conserved depends on the character of the forces that act between the particles. *A collision in which the total kinetic energy before and after the collision is the same is called* **elastic.** (This usage of the word *elastic* is consistent with the usage we encountered previously when discussing the restoring force of a deformable body in Section 6.2. For example, if the colliding bodies exert a force on each other by means of a massless elastic spring placed between them, then the kinetic energy before and after the collision will indeed be the same—that is, the collision will be elastic.) Collisions between macroscopic bodies are usually not elastic—during the collision some of the kinetic energy is transformed into heat by the internal friction forces and some is used up in doing work to change the internal configuration of the bodies. For example, the automobile collision shown in Fig. 11.1 is highly *inelastic;* almost the entire initial kinetic energy is used up in doing work on the automobile parts, changing their shape. On the other hand, the collision of a "Super Ball" and a hard wall or the collision of two billiard balls comes pretty close to being elastic—that is, the kinetic energies before and after the collision are almost the same.

elastic collision

Collisions between "elementary" particles—such as electrons, protons, and neutrons—are often elastic. These particles have no internal friction forces which could dissipate kinetic energy. A collision between such particles can be inelastic only if it involves the creation of new particles; such new particles may arise either by conversion of some of the available kinetic energy into mass or else by transmutation of the old particles by means of a change of their internal structure.

EXAMPLE 3 A Super Ball, made of a rubberlike plastic, is thrown against a hard, smooth wall. The ball strikes the wall from a perpendicular direction with speed v. Assuming that the collision is elastic, find the speed of the ball after the collision.

PHYSICS IN PRACTICE AUTOMOBILE COLLISIONS

Concepts — in — Context

We can fully appreciate the effects of the secondary impact on the human body if we compare the impact speeds of a human body on the dashboard or the windshield with the speed attained by a body in free fall from some height. The impact of the head on the windshield at 15 m/s is equivalent to falling four floors down from an apartment building and landing headfirst on a hard surface. Our intuition tells us that this is likely to be fatal. Since our intuition about the dangers of heights is much better than our intuition about the dangers of speeds, it is often instructive to compare impact speeds with equivalent heights of fall. The table lists impact speeds and equivalent heights, expressed as the number of floors the body has to fall down to acquire the same speed.

The number of fatalities in automobile collisions has been reduced by the use of air bags. The air bag helps by cushioning the impact over a longer time, reducing the time-average force. To be effective, the air bag must inflate quickly, before the passenger reaches it, typically in about 10 milliseconds. Because of this, a passenger, especially a child, too near an air bag prior to inflation can be injured or killed by the impulse from the inflation. But for a properly seated adult passenger, the inflated air bag cushions the passenger, reducing the severity of injuries.

COMPARISON OF IMPACT SPEEDS AND HEIGHTS OF FALL

SPEED	SPEED	EQUIVALENT HEIGHT (NUMBER OF FLOORS)[a]
15 km/h	9 mi/h	$\frac{1}{3}$
30	19	1
45	28	3
60	37	5
75	47	8
90	56	11
105	65	15

[a]Each floor is 2.9 m.

However, the impact can still be fatal—you wouldn't expect to survive a jump from an 11-floor building onto an air mattress.

For maximum protection, a seat belt should always be worn even in vehicles equipped with air bags. In lateral collisions, in repeated collisions (such as in car pileups), and in rollovers, an air bag is of little help, and a seat belt is essential. The effectiveness of seat belts is well demonstrated by the experiences of race car drivers. Race car drivers wear lap belts and crossed shoulder belts. Even in spectacular crashes at very high speeds (see the figure), the drivers rarely suffer severe injuries.

In a race at the California Speedway in October 2000, a car flips over and breaks in half after a crash, but the driver, Luis Diaz, walks away from the wreck.

SOLUTION: The only horizontal force on the ball is the normal force exerted by the wall; this force reverses the motion of the ball (see Fig. 11.3). Since the wall is very massive, the reaction force of the ball on the wall will not give the wall any appreciable velocity. Hence the kinetic energy of the system, both before and after the collision, is merely the kinetic energy of the ball. Conservation of this

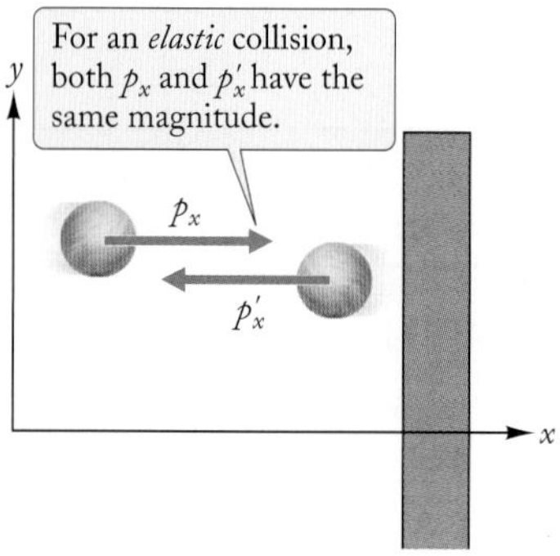

FIGURE 11.3 The initial momentum p_x of the ball is positive; the final momentum p'_x is negative.

kinetic energy then requires that the ball rebound with a speed v equal to the incident speed.

Note that although the kinetic energy of the ball is the same before and after the collision, the momentum is not the same before and after (see also Example 1 of Chapter 10). If the x axis is in the direction of the initial motion, then the momentum of the ball before the collision is $p_x = mv$, and after the collision it is $p'_x = -mv$. Hence the change of momentum is $p'_x - p_x = -2mv$. The wall suffers an equal and opposite momentum change of $+2mv$, so that the total momentum of the system is conserved. The wall can acquire the momentum $2mv$ without acquiring any appreciable velocity because its mass is large and it is attached to a building of even larger mass.

Concepts — in — Context

✓ Checkup 11.1

QUESTION 1: In order to protect the occupants of an automobile in a collision, is it better to make the front end of the automobile very hard (a solid block of steel) or fairly soft and crushable?

QUESTION 2: If a golf ball and a steel ball of the same mass strike a concrete floor with equal speeds, which will exert the larger average force on the floor?

QUESTION 3: You drop a Super Ball on a hard, smooth floor from a height of 1 m. If the collision is elastic, how high will the ball bounce up?

QUESTION 4: A child throws a wad of chewing gum against a wall, and it sticks. Is this an elastic collision?

QUESTION 5: A 3000-kg truck collides with a 1000-kg car. During this collision the average force exerted by the truck on the car is 3×10^6 N in an eastward direction. What is the magnitude of the average force exerted by the car on the truck?

(A) 0 (B) 1×10^6 N (C) 3×10^6 N (D) 9×10^6 N

Online Concept Tutorial 13 Online Concept Tutorial 14

11.2 ELASTIC COLLISIONS IN ONE DIMENSION

The collision of two boxcars on a railroad track is an example of a collision on a straight line. More generally, the collision of any two bodies that approach head-on and recoil along their original line of motion is a collision along a straight line. Such collisions will occur only under exceptional circumstances; nevertheless, we find it instructive to study such collisons because they display in a simple way some of the broad features of more complicated collisions.

In an elastic collision of two particles moving along a straight line, the laws of conservation of momentum and energy completely determine the final velocities in terms of the initial velocities. In the following calculations, we will assume that one particle (the "projectile") is initially in motion and the other (the "target") is initially at rest.

Figure 11.4a shows the particles before the collision, and Fig. 11.4b shows them after; the x axis is along the direction of motion. We will designate the x components of the velocity of particle 1 and particle 2 before the collision by v_1 and v_2, respectively. We will designate the x components of these velocities after the collision by v'_1 and v'_2.

Particle 2 is the target, initially at rest, so $v_2 = 0$. Particle 1 is the projectile. The initial momentum is therefore simply the momentum m_1v_1 of particle 1. The final momentum, after the collision, is $m_1v_1' + m_2v_2'$. Conservation of momentum then tells us that

$$m_1v_1 = m_1v_1' + m_2v_2' \tag{11.6}$$

The initial kinetic energy is $\frac{1}{2}m_1v_1^2$, and the final kinetic energy is $\frac{1}{2}m_1v_1'^2 + \frac{1}{2}m_2v_2'^2$. Since this collision is *elastic,* conservation of kinetic energy[1] tell us that

$$\tfrac{1}{2}m_1v_1^2 = \tfrac{1}{2}m_1v_1'^2 + \tfrac{1}{2}m_2v_2'^2 \tag{11.7}$$

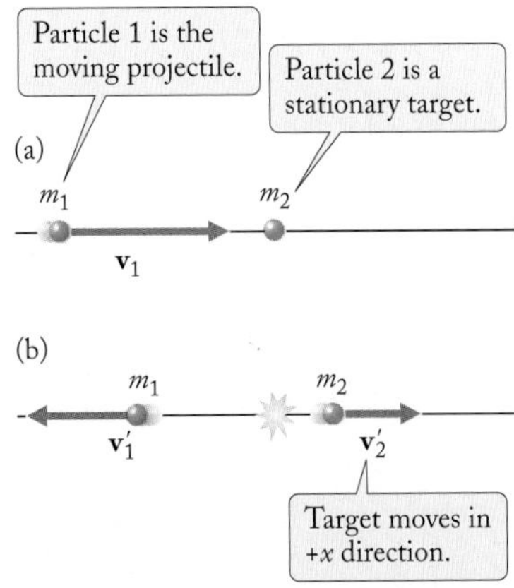

FIGURE 11.4 (a) Before the collision, particle 2 is at rest, and particle 1 has velocity $\mathbf{v}_1$. (b) After the collision, particle 1 has velocity $\mathbf{v}_1'$, and particle 2 has velocity $\mathbf{v}_2'$.

In these equations, we can regard the initial velocities v_1 and v_2 as known, and the final velocities v_1' and v_2' as unknown. We therefore want to solve these equations for the unknown quantities. For this purpose, it is convenient to rearrange the two equations somewhat. If we subtract m_1v_1' from both sides of Eq. (11.6), we obtain

$$m_1(v_1 - v_1') = m_2v_2' \tag{11.8}$$

If we multiply both sides of Eq. (11.7) by 2 and subtract $m_1v_1'^2$ from both sides, we obtain

$$m_1(v_1^2 - v_1'^2) = m_2v_2'^2 \tag{11.9}$$

With the identity $v_1^2 - v_1'^2 = (v_1 - v_1')(v_1 + v_1')$, this becomes

$$m_1(v_1 - v_1')(v_1 + v_1') = m_2v_2'^2 \tag{11.10}$$

Now divide Eq. (11.10) by Eq. (11.8)—that is, divide the left side of Eq. (11.10) by the left side of Eq. (11.8) and the right side of Eq. (11.10) by the right side of Eq. (11.8). The result is

$$v_1 + v_1' = v_2' \tag{11.11}$$

This trick gets rid of the bothersome squares in Eq. (11.7) and leaves us with two equations—Eqs. (11.8) and (11.11)—without squares. To complete the solution for our unknowns, we take the value $v_2' = v_1 + v_1'$ given by Eq. (11.11) and substitute it into the right side of Eq. (11.8):

$$m_1(v_1 - v_1') = m_2(v_1 + v_1') \tag{11.12}$$

We can solve this immediately for the unknown v_1', with the result

$$v_1' = \frac{m_1 - m_2}{m_1 + m_2}v_1 \tag{11.13}$$

final projectile velocity in elastic collision

Finally, we substitute this value of v_1' into the expression from Eq. (11.11), $v_2' = v_1 + v_1'$, and we find

$$v_2' = v_1 + \frac{m_1 - m_2}{m_1 + m_2}v_1 = \frac{(m_1 + m_2)v_1 + (m_1 - m_2)v_1}{m_1 + m_2}$$

[1] In the context of elastic collisions, "conservation of kinetic energy" is taken to mean that the kinetic energy is the same before and after the collision; duing the collision, when the particles are interacting, what is conserved is not the kinetic energy itself, but the sum of kinetic and potential energies.

or

final target velocity in elastic collision

$$v_2' = \frac{2m_1}{m_1 + m_2} v_1 \tag{11.14}$$

Equations (11.13) and (11.14) give us the final velocities v_1' and v_2' in terms of the initial velocity v_1.

EXAMPLE 4 An empty boxcar of mass $m_1 = 20$ metric tons rolling on a straight track at 5.0 m/s collides with a loaded stationary boxcar of mass $m_2 = 65$ metric tons (see Fig. 11.5). Assuming that the cars bounce off each other elastically, find the velocities after the collision.

SOLUTION: With $m_1 = 20$ tons and $m_2 = 65$ tons, Eqs. (11.13) and (11.14) yield

$$v_1' = \frac{20\text{ tons} - 65\text{ tons}}{20\text{ tons} + 65\text{ tons}} \times 5.0\text{ m/s} = -2.6\text{ m/s}$$

$$v_2' = \frac{2 \times 20\text{ tons}}{20\text{ tons} + 65\text{ tons}} \times 5.0\text{ m/s} = 2.4\text{ m/s}$$

Thus, boxcar 2 acquires a speed of 2.4 m/s, and boxcar 1 recoils with a speed of 2.6 m/s (note the negative sign of v_1').

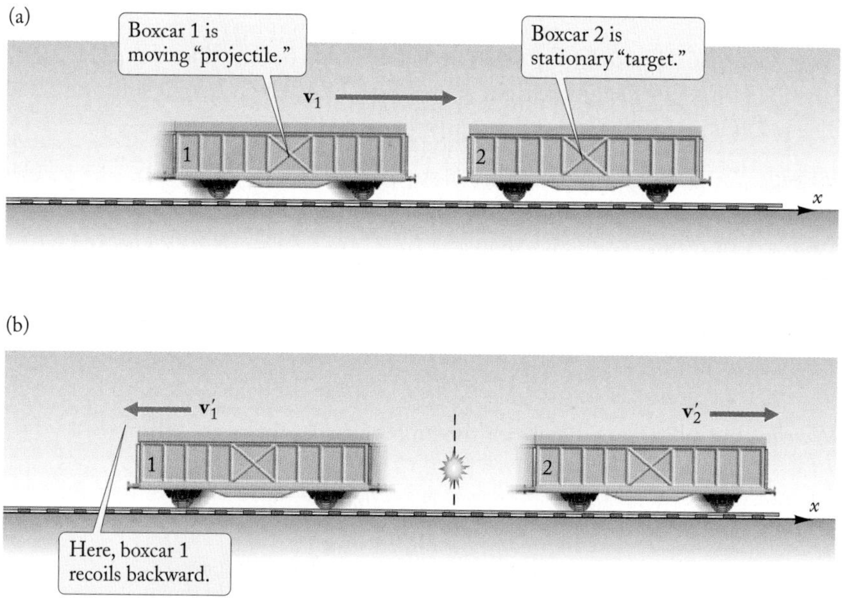

FIGURE 11.5 (a) Initially, boxcar 1 is moving toward the right, and boxcar 2 is stationary. (b) After the collision, boxcar 1 is moving toward the left, and boxcar 2 is moving toward the right.

Note that if the mass of the target is much larger than the mass of the projectile, then m_1 can be neglected compared with m_2. Equation (11.13) then becomes

$$v_1' \approx -\frac{m_2}{m_2} v_1 = -v_1 \tag{11.15}$$

and Eq. (11.14) becomes

$$v_2' \approx \frac{2m_1}{m_2} v_1 \approx 0 \tag{11.16}$$

This means the projectile bounces off the target with a reversed velocity and the target remains nearly stationary (as in the case of the Super Ball bouncing off the wall; see Example 3).

Conversely, if the mass of the projectile is much larger than the mass of the target, then m_2 can be neglected compared with m_1, and Eqs. (11.13) and (11.14) become

$$v_1' \approx \frac{m_1}{m_1} v_1 = v_1 \tag{11.17}$$

and

$$v_2' \approx \frac{2m_1}{m_1} v_1 = 2v_1 \tag{11.18}$$

This means that the projectile plows along with unchanged velocity and the target bounces off with *twice* the speed of the incident projectile. For example, when a (heavy) golf club strikes a golf ball, the ball bounces away at twice the speed of the club (see Fig. 11.6).

Also, if the two masses are equal, Eqs. (11.13) and (11.14) give

$$v_1' = 0 \qquad \text{and} \qquad v_2' = v_1$$

Thus, the projectile stops and the target moves off with the projectile's initial speed. This is common in a head-on collision in billiards, and is also realized in certain pendulum toys (see Discussion Question 9 at the end of the chapter).

Finally, if both particles involved in a one-dimensional elastic collision are initially moving ($v_1 \neq 0$ and $v_2 \neq 0$), conservation of the total momentum and the total kinetic energy can again be applied to uniquely determine the final velocities. The results are more complicated, but they are obtained in the same manner as in the stationary target case above.

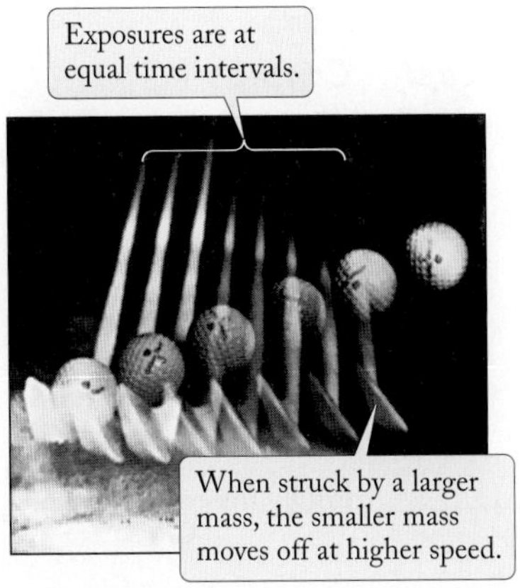

FIGURE 11.6 Impact of club on golf ball. By inspection of this multiple-exposure photograph, we see that the speed of the ball is larger than the initial speed of the club.

✔ Checkup 11.2

In the following questions assume that a projectile traveling in the direction of the positive x axis strikes a stationary target head-on and the collision is elastic.

QUESTION 1: Under what conditions will the velocity of the projectile be positive after the collision? Negative?

QUESTION 2: Can the speed of recoil of the target ever exceed twice the speed of the incident projectile?

QUESTION 3: For an elastic collision, the kinetic energies before and after the collision are the same. Is the kinetic energy *during* the collision also the same?

QUESTION 4: A marble with velocity v_1 strikes a stationary, identical marble elastically and head-on. The final velocities of the shot and struck marbles are, respectively:

(A) $\frac{1}{2}v_1$; $\frac{1}{2}v_1$ (B) v_1; $2v_1$ (C) $-v_1$; 0

(D) $-v_1$; $2v_1$ (E) 0; v_1

Online Concept Tutorial 13 Online Concept Tutorial 14

11.3 INELASTIC COLLISIONS IN ONE DIMENSION

If the collision is inelastic, kinetic energy is not conserved, and then the only conservation law that is applicable is the conservation of momentum. This, by itself, is insufficient to calculate the velocities of both particles after the collision. Thus, for most inelastic collisions, one of the final velocities must be measured in order for momentum conservation to provide the other. Alternatively, we must have some independent knowledge of the amount of kinetic energy lost. However, *if the collision is* **totally inelastic,** *so a maximum amount of kinetic energy is lost, then the common velocity of both particles after the collision can be calculated.*

totally inelastic collision

In a totally inelastic collision, the particles do not bounce off each other at all; instead, *the particles stick together,* like two automobiles that form a single mass of interlocking wreckage after a collision, or two railroad boxcars that couple together. Under these conditions, the velocities of both particles must coincide with the velocity of the center of mass. But the velocity of the center of mass *after* the collision is the same as the velocity of the center of mass *before* the collision, because there are no external forces and the acceleration of the center of mass is zero [see Eq. (10.40)]. We again consider a stationary target, so that before the collision the velocity of the target particle is zero ($v_2 = 0$) and the general equation [Eq. (10.37)] for the velocity of the center of mass yields

$$v_{CM} = \frac{m_1 v_1}{m_1 + m_2} \tag{11.19}$$

This must then be the final velocity of both particles after a totally inelastic collision:

final velocities in totally inelastic collision with stationary target

$$v'_1 = v'_2 = v_{CM} = \frac{m_1 v_1}{m_1 + m_2} \tag{11.20}$$

We have already come across an instance of this formula in Example 3 of Chapter 10.

EXAMPLE 5 Suppose that the two boxcars of Example 4 couple during the collision and remain locked together (see Fig. 11.7). What is the velocity of the combination after the collision? How much kinetic energy is dissipated during the collision?

SOLUTION: Since the boxcars remain locked together, this is a totally inelastic collision. With $m_1 = 20$ tons, $m_2 = 65$ tons, and $v_1 = 5.0$ m/s, Eq. (11.19) gives us the velocity of the center of mass:

$$v_{CM} = \frac{m_1 v_1}{m_1 + m_2} = \frac{20 \text{ tons} \times 5.0 \text{ m/s}}{20 \text{ tons} + 65 \text{ tons}} = 1.2 \text{ m/s}$$

and this must be the velocity of the coupled cars after the collision.

The kinetic energy before the collision is that of the moving boxcar,

$$\tfrac{1}{2}m_1 v_1^2 = \tfrac{1}{2} \times 20000 \text{ kg} \times (5.0 \text{ m/s})^2 = 2.5 \times 10^5 \text{ J}$$

and the kinetic energy after the collision is that of the two coupled boxcars,

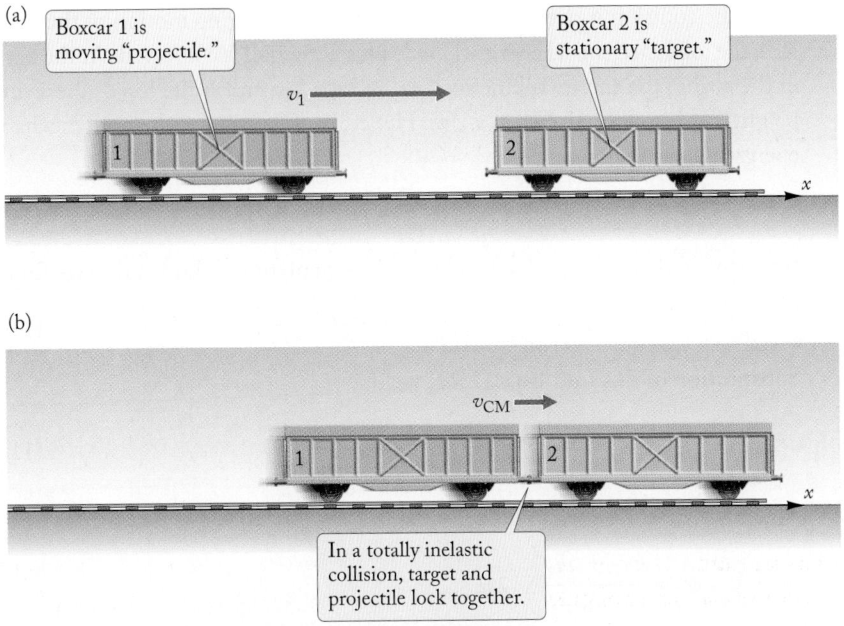

FIGURE 11.7 (a) Initially, boxcar 1 is moving toward the right, and boxcar 2 is stationary, as in Fig. 11.5. (b) After the collision, the boxcars remain locked together. Their common velocity must be the velocity of the center of mass.

$$\tfrac{1}{2}m_1v_{CM}^2 + \tfrac{1}{2}m_2v_{CM}^2 = \tfrac{1}{2}(m_1 + m_2)v_{CM}^2$$
$$= \tfrac{1}{2}(20000 \text{ kg} + 65000 \text{ kg}) \times (1.2 \text{ m/s})^2 = 0.61 \times 10^5 \text{ J}$$

Thus, the loss of kinetic energy is

$$2.5 \times 10^5 \text{ J} - 0.61 \times 10^5 \text{ J} = 1.9 \times 10^5 \text{ J} \tag{11.21}$$

This energy is absorbed by friction in the bumpers during the coupling of the boxcars.

EXAMPLE 6 Figure 11.8a shows a **ballistic pendulum,** a device once commonly used to measure the speeds of bullets. The pendulum consists of a large block of wood of mass m_2 suspended from thin wires. Initially, the pendulum is at rest. The bullet, of mass m_1, strikes the block horizontally and remains stuck in it. The impact of the bullet puts the block in motion, causing it to swing upward to a height h (see Fig. 11.8b), where it momentarily stops. In a test of a Springfield rifle firing a bullet of 9.7 g, a ballistic pendulum of 4.0 kg swings up to a height of 19 cm. What was the speed of the bullet before impact?

SOLUTION: The collision of the bullet with the wood is totally inelastic. Hence, immediately after the collision, bullet and block move horizontally with the velocity of the center of mass:

$$v_{CM} = \frac{m_1v_1}{m_1 + m_2} \tag{11.22}$$

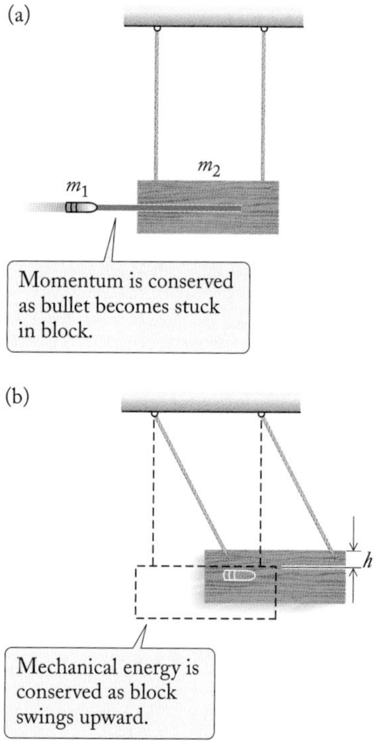

FIGURE 11.8 (a) Before the bullet strikes, the block of wood is at rest. (b) After the bullet strikes, the block, with the embedded bullet, moves toward the right and swings upward to a height h.

After the collision is over, during the subsequent swinging motion of the pendulum, the total mechanical energy (kinetic plus potential) is conserved. At the bottom of the swing, the energy is kinetic, $\frac{1}{2}(m_1 + m_2)v_{CM}^2$; and at the top of the swing at height h, it is potential, $(m_1 + m_2)gh$. Hence, conservation of the total mechanical energy tells us that

$$\tfrac{1}{2}(m_1 + m_2)v_{CM}^2 = (m_1 + m_2)gh \tag{11.23}$$

If we divide this by $(m_1 + m_2)$ and take the square root of both sides, we find

$$v_{CM} = \sqrt{2gh} \tag{11.24}$$

Substitution of this into Eq. (11.22) yields

$$\sqrt{2gh} = \frac{m_1 v_1}{m_1 + m_2} \tag{11.25}$$

which we can solve for v_1, with the result

$$\begin{aligned} v_1 &= \frac{m_1 + m_2}{m_1}\sqrt{2gh} \\ &= \frac{0.0097\text{ kg} + 4.0\text{ kg}}{0.0097\text{ kg}}\sqrt{2 \times 9.81\text{ m/s}^2 \times 0.19\text{ m}} \\ &= 800\text{ m/s} \end{aligned} \tag{11.26}$$

COMMENT: Note that during the collision, momentum is conserved but not kinetic energy (the collision is totally inelastic); and that during the swinging motion, the total mechanical energy is conserved, but not momentum (the swinging motion proceeds under the influence of the "external" forces of gravity and the tensions in the wires).

✔ Checkup 11.3

QUESTION 1: In a totally inelastic collision, do both particles lose kinetic energy?

QUESTION 2: Consider a collision between two particles of equal masses and of opposite velocities. What is the velocity after this collision if the collision is totally inelastic? If the collision is elastic?

QUESTION 3: Under what conditions is the velocity of the particles after a totally inelastic collision equal to one-half the velocity of the incident projectile? (Assume a stationary target.)

QUESTION 4: Does the length of the suspension wires affect the operation of the ballistic pendulum described in Example 6?

QUESTION 5: A particle is traveling in the positive x direction with speed v. A second particle with one-half the mass of the first is traveling in the opposite direction with the same speed. The two experience a totally inelastic collision. The final x component of the velocity is:

(A) 0 (B) $\frac{1}{3}v$ (C) $\frac{1}{2}v$ (D) $\frac{2}{3}v$ (F) v

11.4 COLLISIONS IN TWO AND THREE DIMENSIONS

In the previous sections, we have focused on collisions on a straight line, in one dimension. Collisions in two or three dimensions are more difficult to analyze, because the conservation laws for momentum and energy do not provide sufficient information to determine the final velocities completely in terms of the initial velocities. Momentum is always conserved during a collision, and this conservation provides one equation for each of the x, y, and z directions. If it is known that the collision is totally elastic, then conservation of the total kinetic energy provides another equation. However, these are not enough to determine the three final velocity components for each and every particle. Some information concerning the final velocities must also be known or measured.

The case of totally inelastic collisions is an exception: in this case, the conservation of momentum determines the outcome completely, even in two or three dimensions. The particles stick together, and their final velocities coincide with the velocity of the center of mass, as illustrated by the following example. The subsequent example explores a case where the solution exploits some knowledge of the final velocities.

EXAMPLE 7 A red automobile of mass 1100 kg and a green automobile of mass 1300 kg collide at an intersection. Just before this collision, the red automobile was traveling due east at 34 m/s, and the green automobile was traveling due north at 15 m/s (see Fig. 11.9). After the collision, the wrecked automobiles remain joined together, and they skid on the pavement with locked wheels. What is the direction of the skid?

Concepts in Context

SOLUTION: The final velocity of the wreck coincides with the final velocity of the center of mass, which is the same as the initial velocity of the center of mass. According to Eq. (10.37), this velocity is

$$\mathbf{v}_{CM} = \frac{m_1\mathbf{v}_1 + m_2\mathbf{v}_2}{m_1 + m_2} \tag{11.27}$$

final velocity in totally inelastic collision

With the x axis eastward and the y axis northward, the initial velocity $\mathbf{v}_1$ of the red automobile has an x component but no y component, and the initial velocity $\mathbf{v}_2$ of the green automobile has a y component but no x component. Hence the x component of $\mathbf{v}_{CM}$ is

$$v_{CM,x} = \frac{m_1 v_1}{m_1 + m_2} = \frac{1100\text{ kg} \times 34\text{ m/s}}{1100\text{ kg} + 1300\text{ kg}} = 16\text{ m/s}$$

and the y component of $\mathbf{v}_{CM}$ is

$$v_{CM,y} = \frac{m_2 v_2}{m_1 + m_2} = \frac{1300\text{ kg} \times 15\text{ m/s}}{1100\text{ kg} + 1300\text{ kg}} = 8.1\text{ m/s}$$

The angle between the direction of this velocity and the x axis is given by

$$\tan\theta = \frac{v_{CM,y}}{v_{CM,x}} = \frac{8.1\text{ m/s}}{16\text{ m/s}} = 0.51$$

from which

$$\theta = 27°$$

Since the x axis is eastward, this is 27° north of east.

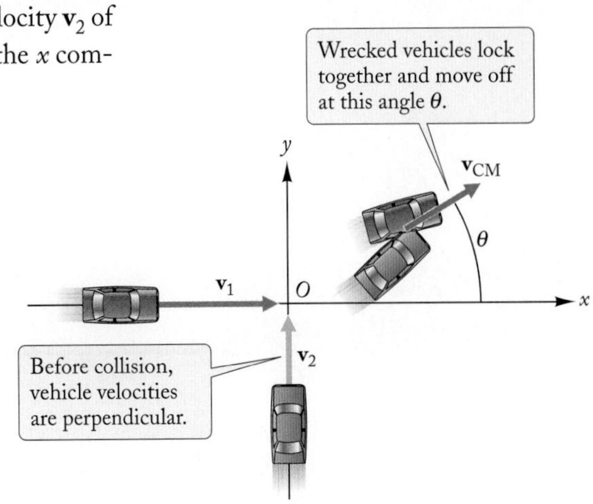

FIGURE 11.9 An automobile collision. Before the collision, the velocities of the automobiles were $\mathbf{v}_1$ and $\mathbf{v}_2$. After the collision, both velocities are $\mathbf{v}_{CM}$.

EXAMPLE 8 In an atomic collision experiment, or "scattering" experiment, a helium ion of mass $m_1 = 4.0$ u with speed $v_1 = 1200$ m/s strikes an oxygen (O_2) molecule of mass $m_2 = 32$ u which is initially at rest (see Fig. 11.10a). The helium ion exits the collision at 90° from its incident direction with one-fourth of its original kinetic energy. What is the recoil speed of the oxygen molecule? What fraction of the total kinetic energy is lost during the collision? [This energy is lost to the internal (vibrational and rotational) motions of the oxygen molecule.]

SOLUTION: In the absence of external forces, momentum is always conserved. If we choose the direction of incident motion along the x axis, then for 90° scattering, we can choose the direction in which the helium ion exits (the direction of $\mathbf{v}_1'$) to be along the y axis (see Fig. 11.10b). Conservation of momentum in the two directions then requires

$$\text{for } x \text{ direction: } m_1 v_1 = m_2 v_{2x}'$$

$$\text{for } y \text{ direction: } 0 = m_1 v_1' + m_2 v_{2y}'$$

Since the helium ion exits with one-fourth of its initial kinetic energy,

$$\tfrac{1}{2} m_1 v_1'^2 = \tfrac{1}{4} \times \tfrac{1}{2} m_1 v_1^2$$

or

$$v_1' = \tfrac{1}{2} v_1$$

Substituting this v_1' and the given $m_2 = 8m_1$ into the x and y components of the momentum gives for the velocity of the oxygen molecule:

$$v_{2x}' = \frac{m_1}{m_2} v_1 = \frac{1}{8} v_1 = \frac{1}{8} \times 1200 \text{ m/s} = 150 \text{ m/s}$$

$$v_{2y}' = -\frac{m_1}{m_2} v_1' = -\frac{1}{8} \times \frac{1}{2} v_1 = -\frac{1}{16} \times 1200 \text{ m/s} = -75 \text{ m/s}$$

The speed of recoil of the oxygen molecule is thus

$$v_2' = \sqrt{v_{2x}'^2 + v_{2y}'^2} = \sqrt{(150 \text{ m/s})^2 + (-75 \text{ m/s})^2} = 170 \text{ m/s}$$

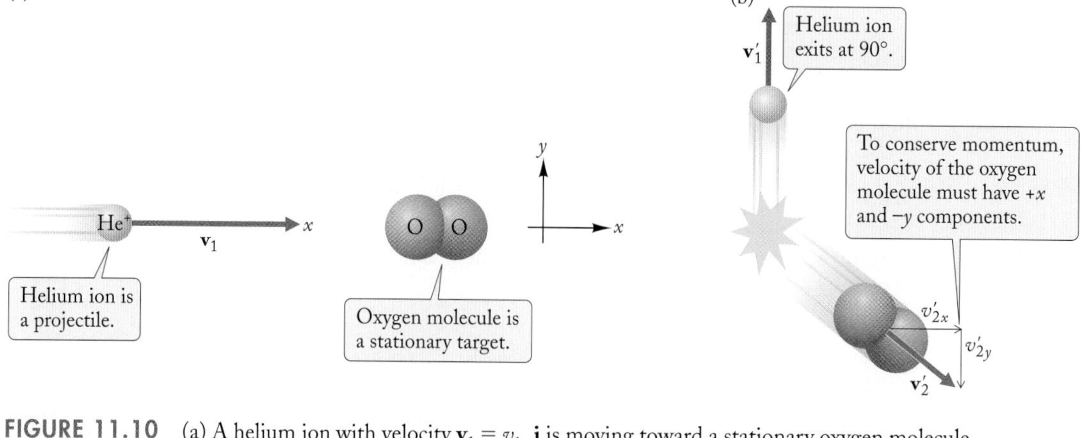

FIGURE 11.10 (a) A helium ion with velocity $\mathbf{v}_1 = v_{1x}\mathbf{i}$ is moving toward a stationary oxygen molecule. (b) After the collision, the helium ion exits perpendicular to its incident direction with velocity $\mathbf{v}_1'$, while the oxygen molecule acquires a velocity $\mathbf{v}_2' = v_{2x}'\mathbf{i} + v_{2y}'\mathbf{j}$.

The fraction of kinetic energy lost is the amount of kinetic energy lost divided by the original kinetic energy:

$$[\text{fraction lost}] = \frac{K - K'}{K} = \frac{\frac{1}{2}m_1v_1^2 - \left(\frac{1}{4} \times \frac{1}{2}m_1v_1^2 + \frac{1}{2}m_2v_2'^2\right)}{\frac{1}{2}m_1v_1^2}$$

$$= 1 - \frac{1}{4} - \frac{m_2}{m_1}\frac{v_2'^2}{v_1^2} = \frac{3}{4} - \frac{32}{4.0} \times \frac{(170\ \text{m/s})^2}{(1200\ \text{m/s})^2}$$

$$= 0.59$$

Thus, about 59% of the helium ion's initial kinetic energy is lost to internal motions of the oxygen molecule during the collision.

✓ Checkup 11.4

QUESTION 1: A car traveling south collides with and becomes entangled with a car of the same mass and speed heading west. In what direction does the wreckage emerge from the collision?

QUESTION 2: An object at rest explodes into three pieces; one travels due west and another due north. In which quadrant of directions does the third piece travel?

(A) Northeast (B) Southeast (C) Southwest (D) Northwest

PROBLEM-SOLVING TECHNIQUES: CONSERVATION OF ENERGY AND MOMENTUM IN COLLISIONS

For solving problems involving collisions, it is essential to know what conservation laws are applicable. The following table summarizes the conservation laws applicable for different collisions:

TYPE OF COLLISION	CONSERVATION OF KINETIC ENERGY	CONSERVATION OF MOMENTUM	COMMENTS
Elastic	Yes	Yes	For a one-dimensional collision, energy and momentum conservation determine the final velocities in terms of the initial velocities. For a 2- or 3-dimensional collision, there is not enough information in the initial velocities alone to determine the final velocities uniquely.
Totally inelastic	No	Yes	The two colliding bodies stick together, and momentum conservation determines the final velocities (in 1, 2, and 3 dimensions).
Inelastic	No	Yes	If the collision is not totally inelastic, there is not enough information in the initial velocities alone to determine the final velocities. Some information about the energy loss and final velocities must also be known.

SUMMARY

PHYSICS IN PRACTICE Automobile Collisions **(page 343)**

PROBLEM-SOLVING TECHNIQUES **(page 353)**
Conservation of Energy and Momentum in Collisions

IMPULSE

$$\mathbf{I} = \int_0^{\Delta t} \mathbf{F}\,dt = \mathbf{p}' - \mathbf{p} \qquad (11.1)$$

AVERAGE FORCE IN COLLISION

$$\overline{\mathbf{F}} = \frac{\mathbf{p}' - \mathbf{p}}{\Delta t} \qquad (11.4)$$

ALL COLLISIONS
The total momentum is conserved.

$$m_1\mathbf{v}_1 + m_2\mathbf{v}_2 + \cdots = m_1\mathbf{v}_1' + m_2\mathbf{v}_2' + \cdots \qquad (11.6)$$

ELASTIC COLLISION
The total kinetic energy is conserved.

$$\tfrac{1}{2}m_1v_1^2 + \tfrac{1}{2}m_2v_2^2 + \cdots = \tfrac{1}{2}m_1v_1'^2 + \tfrac{1}{2}m_2v_2'^2 + \cdots \qquad (11.7)$$

VELOCITIES IN ONE-DIMENSIONAL ELASTIC COLLISION WITH STATIONARY TARGET

Before: $v_1 \neq 0, \quad v_2 = 0$

After:

$$v_1' = \frac{m_1 - m_2}{m_1 + m_2}v_1 \qquad (11.13)$$

$$v_2' = \frac{2m_1}{m_1 + m_2}v_1 \qquad (11.14)$$

INELASTIC COLLISION
Kinetic energy is not conserved.

TOTALLY INELASTIC COLLISION
The colliding particles stick together.

VELOCITIES IN TOTALLY INELASTIC COLLISION
(1, 2, or 3 dimensions).

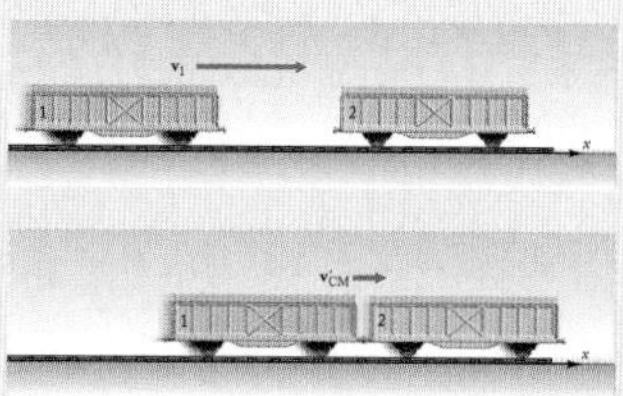

Before: $\mathbf{v}_1$ and $\mathbf{v}_2$

After:

$$\mathbf{v}_1' = \mathbf{v}_2' = \mathbf{v}_{CM} = \frac{m_1\mathbf{v}_1 + m_2\mathbf{v}_2}{m_1 + m_2} \qquad (11.27)$$

QUESTIONS FOR DISCUSSION

1. According to the data given in Example 1, what percentage of the initial kinetic energy does the automobile have after the collision?
2. A (foolish) stuntman wants to jump out of an airplane at high altitude without a parachute. He plans to jump while tightly encased in a strong safe which can withstand the impact on the ground. How would you convince the stuntman to abandon this project?
3. In the crash test shown in the photographs of Fig. 11.1, anthropomorphic dummies were riding in the automobile. These dummies were (partially) restrained by seat belts, which limited their motion relative to the automobile. How would the motion of the dummies have differed from that shown in these photographs if they had not been restrained by seat belts?
4. For the sake of safety, would it be desirable to design automobiles so that their collisions are elastic or inelastic?
5. Two automobiles have collided at a north–south east–west intersection. The skid marks their tires made after the collision point roughly northwest. One driver claims he was traveling west; the other driver claims he was traveling south. Who is lying?
6. Statistics show that, on the average, the occupants of a heavy ("full-size") automobile are more likely to survive a crash than those of a light ("compact") automobile. Why would you expect this to be true?
7. In Joseph Conrad's tale "Gaspar Ruiz", the hero ties a cannon to his back and, hugging the ground on all fours, fires several shots at the gate of a fort. How does the momentum absorbed by Ruiz compare with that absorbed by the gate? How does the energy absorbed by Ruiz compare with that absorbed by the gate?
8. Give an example of a collision between two bodies in which *all* of the kinetic energy is lost to inelastic processes.
9. Explain the operation of the five-pendulum toy, called Newton's cradle, shown in Fig. 11.11.
10. In order to split a log with a small ax, you need a greater impact speed than you would need with a large ax. Why? If the energy required to split the log is the same in both cases, why is it more tiring to use the small ax? (Hint: Think about the kinetic energy of your arms.)
11. If you throw an (elastic) baseball at an approaching train, the ball will bounce back at you with an increased speed. Explain.
12. You are investigating the collision of two automobiles at an intersection. The automobiles remained joined together after this collision, and their wheels made measurable skid marks on the pavement before they came to rest. Assume that during skidding all the wheels remained locked so that the deceleration was entirely due to sliding friction. You know the direction of motion of the automobiles before the collision (drivers are likely to be honest about this), but you do not know the speeds (drivers are likely to be dishonest about this). What do you have to measure at the scene of the accident to calculate the speeds of both the automobiles before the collision?
13. You are sitting in your car, stopped at an intersection. You notice another car approaching from behind, and you notice this car is not slowing down and is going to ram you. Because the time to impact is short, you have only two choices: push hard on your brake, or take your foot off the brake and give your car freedom to roll. Which of these tactics will minimize damage to yourself? Which will minimize damage to your car? Which will minimize damage to the other car?

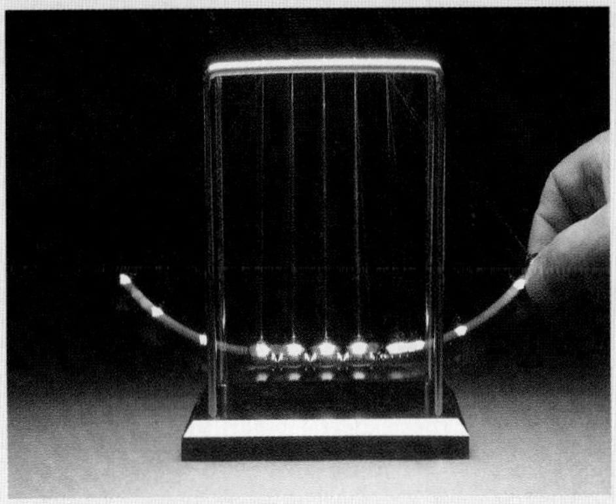

FIGURE 11.11 Newton's cradle.

PROBLEMS

11.1 Impulsive Forces†

1. A stuntman of mass 77 kg "belly-flops" on a shallow pool of water from a height of 11 m. When he hits the pool, he comes to rest in about 0.050 s. What is the impulse that the water and the bottom of the pool deliver to his body during this time interval? What is the time-average force?
2. A large ship of 7.0×10^5 metric tons steaming at 20 km/h runs aground on a reef, which brings it to a halt in 5.0 s. What is the impulse delivered to the ship? What is the average force on the ship? What is the average deceleration?

† For help, see Online Concept Tutorial 13 at www.wwnorton.com/physics

3. The photographs in Fig. 11.1 show the impact of an automobile on a rigid wall.

   (a) Measure the positions of the automobile on these photographs and calculate the average velocity for each of the 20×10^{-3}-s intervals between one photograph and the next; calculate the average acceleration for each time interval from the change between one average velocity and the next.

   (b) The mass of this automobile is 1700 kg. Calculate the average force for each time interval.

   (c) Make a plot of this force as a function of time and find the impulse by estimating the area under this curve.

4. The "land divers" of Pentecost Island (Vanuatu) jump from platforms 21 m high. Long liana vines tied to their ankles jerk them to a halt just short of the ground. If the pull of the liana takes 0.020 s to halt the diver, what is the average acceleration of the diver during this time interval? If the mass of the diver is 64 kg, what is the corresponding average force on his ankles?

5. A shotgun fires a slug of lead of mass 28 g with a muzzle velocity of 450 m/s. The slug acquires this velocity while it accelerates along the barrel of the shotgun, which is 70 cm long.

   (a) What is the impulse the shotgun gives the slug?

   (b) Estimate the average impulsive force; assume constant acceleration of the slug along the barrel.

6. A rule of thumb for automobile collisions against a rigid barrier is that the collision lasts about 0.11 s, for any initial speed and for any model of automobile (for instance, the collision illustrated in Fig. 11.1 lasted 0.120 s, in rough agreement with this rule of thumb). Accordingly, the deceleration experienced by an automobile during a collision is directly proportional to the change of velocity Δv (with a constant factor of proportionality), and therefore Δv can be regarded as a measure of the severity of the collision.

   (a) If the collision lasts 0.11 s, what is the average deceleration experienced by an automobile in an impact on a rigid barrier at 55 km/h? 65 km/h? 75 km/h?

   (b) For each of these speeds, what is the crush distance of the front end of the automobile? Assume constant deceleration for this calculation.

   (c) For each of these speeds, what is the average force the seat belt must exert to hold a driver of 75 kg in his seat during the impact?

7. Suppose that a seat-belted mother riding in an automobile holds a 10-kg baby in her arms. The automobile crashes and decelerates from 50 km/h to 0 in 0.10 s. What average force would the mother have to exert on the baby to hold it? Do you think she can do this?

8. In a test, an air force volunteer belted in a chair placed on a rocket sled was decelerated from 143 km/h to 0 in a distance of 5.5 m. Assume that the mass of the volunteer was 75 kg, and assume that the deceleration was uniform. What was the deceleration? What impulse did the seat belt deliver to the volunteer? What time-average force did the seat belt exert?

9. Assume that the Super Ball of Example 3 has a mass of 60 g and is initially traveling with speed 15 m/s. For simplicity, assume that the acceleration is constant while the ball is in contact with the wall. After touching the wall, the center of mass of the Super Ball moves 0.50 cm toward the wall, and then moves the same distance away to complete the bounce. What is the impulse delivered by the wall? What is the time-average force?

10. A 0.50-kg hammerhead moving at 2.0 m/s strikes a board and stops in 0.020 s. What is the impulse delivered to the board? What is the time-average force?

11. A soccer player applies an average force of 180 N during a kick. The kick accelerates a 0.45-kg soccer ball from rest to a speed of 18 m/s. What is the impulse imparted to the ball? What is the collision time?

12. When an egg ($m = 50$ g) strikes a hard surface, the collision lasts about 0.020 s. The egg will break when the average force during impact exceeds 3.0 N. From what minimum height will a dropped egg break?

*13. The net force on a body varies with time according to $F_x = 3.0t + 0.5t^2$, where F_x is in newtons and t is in seconds. What is the impulse imparted to the body during the time interval $0 \le t \le 3.0$ s?

*14. Suppose that in a baseball game, the batter succeeds in hitting the baseball thrown toward him by the pitcher. Suppose that just before the bat hits, the ball is moving toward the batter horizontally with a speed of 35 m/s; and that after the bat has hit, the ball is moving away from the batter and upward at an angle of 50° and finally lands on the ground 110 m away. The mass of the ball is 0.15 kg. From this information, calculate the magnitude and direction of the impulse the ball receives in the collision with the bat. Neglect air friction and neglect the initial height of the ball above the ground.

*15. Bobsleds racing down a bobsled run often suffer glancing collisions with the vertical walls enclosing the run. Suppose that a bobsled of 600 kg traveling at 120 km/h approaches a wall at an angle of 3.0° and bounces off at the same angle. Subsequent inspection of the wall shows that the side of the bobsled made a scratch mark of length 2.5 m along the wall. From these data, calculate the time interval the bobsled was in contact with the wall, and calculate the average magnitude of the force that acted on the side of the bobsled during the collision.

11.2 Elastic Collisions in One Dimension†

16. A particle moving at 10 m/s along the x axis collides elastically with another particle moving at 5.0 m/s in the *same* direction along the x axis. The particles have equal masses. What are their speeds after this collision?

17. In a lecture demonstration, two masses collide elastically on a frictionless air track. The moving mass (projectile) is 60 g,

† For help, see Online Concept Tutorial 13 and 14 at www.wwnorton.com/physics

and the initially stationary mass (target) is 120 g. The initial velocity of the projectile is 0.80 m/s.

(a) What is the velocity of each mass after the collision?

(b) What is the kinetic energy of each mass before the collision? After the collision?

18. A target sometimes used for target shooting with small bullets consists of a steel disk hanging on a rod which is free to swing on a pivot (in essence, a pendulum). The collision of the bullet with the steel disk is not elastic and not totally inelastic, but somewhere between these extremes. Suppose that a .22-caliber bullet of 15 g and initial speed 600 m/s strikes such a target of mass 40 g. With what velocity would this bullet bounce back (ricochet) if the collision were elastic? Assume that the disk acts like a free particle during the collision.

19. The impact of the head of a golf club on a golf ball can be approximately regarded as an elastic collision. The mass of the head of the golf club is 0.15 kg, and that of the ball is 0.045 kg. If the ball is to acquire a speed of 60 m/s in the collision, what must be the speed of the club before impact?

20. Suppose that a neutron in a nuclear reactor initially has an energy of 4.8×10^{-13} J. How many head-on collisions with carbon nuclei at rest must this neutron make before its energy is reduced to 1.6×10^{-19} J? The collisions are elastic.

21. The impact of a hammer on a nail can be regarded as an elastic collision between the head of the hammer and the nail. Suppose that the mass of the head of the hammer is 0.50 kg and it strikes a nail of mass 12 g with an impact speed of 5.0 m/s. How much energy does the nail acquire in this collision?

22. Consider two coins: a quarter of mass 5.6 g and a dime of mass 2.3 g. If one is sliding at 2.0 m/s on a frictionless surface and hits the other head-on, find the final velocities when either (a) the quarter or (b) the dime is the stationary target. Assume the collision is elastic.

23. Using a straw, a child shoots a series of small balls of mass 1.0 g with speed v at a block of mass 40 g on a frictionless surface. If the small balls elastically collide head-on with the block, how fast will the block be moving after five strikes?

24. A projectile of unknown mass and speed strikes a ball of mass $m =$ 0.15 kg initially at rest. The collision is head-on and elastic. The ball moves off at 1.50 m/s, and the projectile continues in its original direction at 0.50 m/s. What is the mass of the projectile? What was its original speed?

25. A marble of unknown mass m is shot at a larger marble of known mass M, initially at rest in the center of a circle. The collision is head-on and elastic. The smaller marble bounces backward and exits the circle in one-third of the time that it takes the larger marble to do so. What is the mass of the smaller marble? Neglect any rolling motion.

26. In materials science, **Rutherford backscattering** is used to determine the composition of materials. In such an experiment, alpha particles (helium nuclei, mass 4.0 u) of typical kinetic energy 1.6×10^{-13} J strike target nuclei at rest. The collisions are elastic and head-on. What is the recoil kinetic energy of the alpha particle when the target is (a) silicon ($m = 28$ u) and (b) copper ($m = 63$ u)?

*27. An automobile traveling at 60 km/h bumps into the rear of another automobile traveling at 55 km/h in the *same* direction. The mass of the first automobile is 1200 kg, and the mass of the second automobile is 1000 kg. If the collision is elastic, find the velocities of both automobiles immediately after this collision. (Hint: Solve this problem in a reference frame moving with a velocity equal to the initial velocity of one of the automobiles.)

*28. A projectile of 45 kg has a muzzle speed of 656.6 m/s when fired horizontally from a gun held in a rigid support (no recoil). What will be the muzzle speed (relative to the ground) of the same projectile when fired from a gun that is free to recoil? The mass of the gun is 6.6×10^3 kg. (Hint: The kinetic energy of the gun–projectile system is the same in both cases.)

*29. On a smooth, frictionless table, a billiard ball of velocity v is moving toward two other aligned billiard balls in contact (Fig. 11.12). What will be the velocity of each ball after impact? Assume that all balls have the same mass and that the collisions are elastic. Ignore any rotation of the balls. (Hint: Treat this as two successive collisions.)

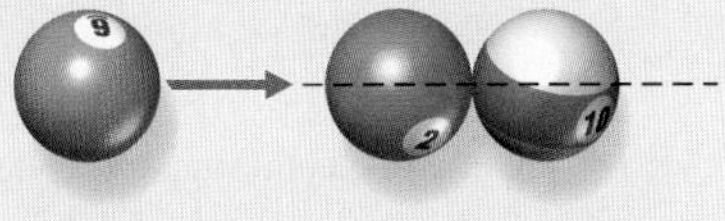

FIGURE 11.12 Three billiard balls along a line.

*30. Repeat Problem 29 but assume that the middle ball has twice the mass of each of the others.

*31. Two small balls are suspended side by side from two strings of length l so that they touch when in their equilibrium position (see Fig. 11.13). Their masses are m and $2m$, respectively. If the left ball (of mass m) is pulled aside and released from a height h, it will swing down and collide with the right ball (of mass $2m$) at the lowest point. Assume the collision is elastic.

(a) How high will each ball swing after the collision?

(b) Both balls again swing down, and they collide once more at the lowest point. How high will each swing after this second collision?

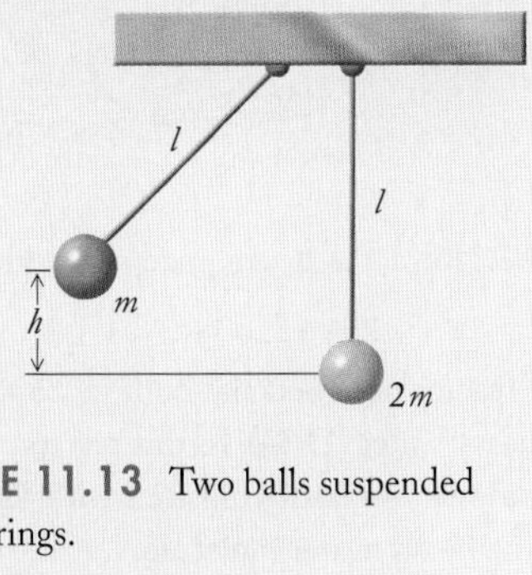

FIGURE 11.13 Two balls suspended from strings.

*32. If a spacecraft, or some other body, approaches a planet at fairly high speed at a suitable angle, it will whip around the planet and recede in a direction almost opposite to the initial direction of motion (Fig. 11.14). This can be regarded approximately as a one-dimensional "collision" between the satellite and the planet; the collision is elastic. In such a collision the satellite will gain kinetic energy from the planet, provided that it approaches the planet along a direction opposite to the direction of the planet's motion. This slingshot effect has been used to boost the speed of both Voyager spacecraft as they passed near Jupiter. Consider the head-on "collision" of a satellite of initial speed 10 km/s with the planet Jupiter, which has a speed of 13 km/s. (The speeds are measured in the reference frame of the Sun.) What is the maximum gain of speed that the satellite can achieve?

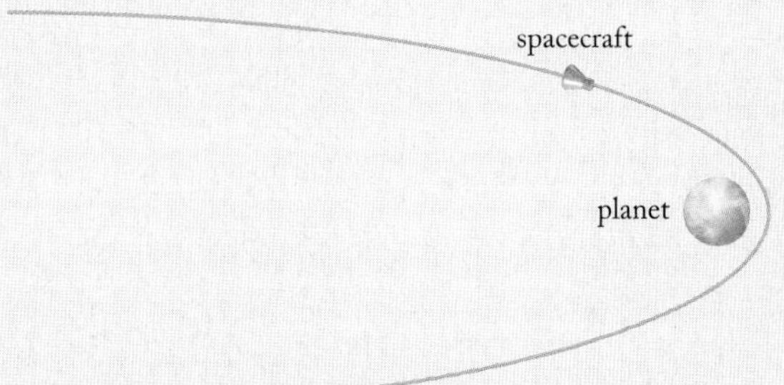

FIGURE 11.14 Spacecraft "colliding" with planet.

**33. A turbine wheel with curved blades is driven by a high-velocity stream of water that impinges on the blades and bounces off (Fig. 11.15). Under ideal conditions the velocity of the water particles after the collision with the blade is exactly zero, so that all of the kinetic energy of the water is transferred to the turbine wheel. If the speed of the water particles is 27 m/s, what is the ideal speed of the turbine blade? (Hint: Treat the collision of a water particle and the blade as a one-dimensional elastic collision.)

FIGURE 11.15 An undershot turbine wheel.

*34. A nuclear reactor designed and built in Canada (CANDU) contains heavy water (D_2O). In this reactor, the fast neutrons are slowed down by elastic collisions with the deuterium nuclei of the heavy-water molecule.

(a) By what factor will the speed of the neutron be reduced in a head-on collision with a deuterium nucleus? The mass of this nucleus is 2.01 u.

(b) After how many head-on collisions with deuterium nuclei will the speed be reduced by the same factor as in a single head-on collision with a proton?

**35. Because of brake failure, an automobile parked on a hill of slope 1:10 rolls 12 m downhill and strikes a parked automobile. The mass of the first automobile is 1400 kg, and the mass of the second automobile is 800 kg. Assume that the first automobile rolls without friction and that the collision is elastic.

(a) What are the velocities of both automobiles immediately after the collision?

(b) After the collision, the first automobile continues to roll downhill, with acceleration, and the second automobile skids downhill, with deceleration. Assume that the second automobile skids with all its wheels locked, with a coefficient of sliding friction 0.90. At what time after the first collision will the automobiles have another collision, and how far from the initial collision?

**36. (a) Show that for an elastic one-dimensional collision the relative velocity reverses during the collision; that is, show that $v_1' - v_2' = -v_1$ (for $v_2 = 0$).

(b) For a partially inelastic collision the relative velocity after the collision will have a smaller magnitude than the relative velocity before the collision. We can express this mathematically as $v_1' - v_2' = -ev_1$, where $e < 1$ is called the **coefficient of restitution.** For some kinds of bodies, the coefficient e is a constant, independent of v_1 and v_2. Show that in this case the final kinetic energy of the motion relative to the center of mass is less than the initial kinetic energy of this motion by a factor of e^2, that is, that $K' = e^2K$.

(c) Derive formulas analogous to Eqs. (11.13) and (11.14) for the velocities v_1' and v_2' in terms of v_1.

11.3 Inelastic Collisions in One Dimension†

37. In karate, the fighter makes the hand collide at high speed with the target; this collision is inelastic, and a large portion of the kinetic energy of the hand becomes available to do damage in the target. According to a crude estimate, the energy required to break a concrete block (28 cm × 15 cm × 1.9 cm supported only at its short edges) is of the order of 10 J. Suppose the fighter delivers a downward hammer-fist strike with a speed of 12 m/s to such a concrete block. In principle, is there enough energy to break the block? Assume that the fist has a mass of 0.4 kg.

38. According to a tall tale told by Baron Münchhausen, on one occasion, while cannon shots were being exchanged between a

† For help, see Online Concept Tutorial 13 and 14 at www.wwnorton.com/physics

besieged city and the enemy camp, he jumped on a cannonball as it was being fired from the city, rode the cannonball toward the enemy camp, and then, in midair, jumped onto an enemy cannonball and rode back to the city. The collision of Münchhausen and the enemy cannonball must have been inelastic, since he held on to it. Suppose that his speed just before hitting the enemy cannonball was 150 m/s southward and the speed of the enemy cannonball was 300 m/s northward. The mass of Münchhausen was 90 kg, and the mass of the enemy cannonball was 20 kg. What must have been the speed just after the collision? Do you think he made it back to the city?

39. As described in Problem 6, the change of velocity Δv of an automobile during a collision is a measure of the severity of the collision. Suppose that an automobile moving with an initial speed of 15 m/s collides with (a) an automobile of equal mass initially at rest, (b) an automobile of equal mass initially moving in the opposite direction at 15 m/s, or (c) a stationary rigid barrier. Assume that the collision is totally inelastic. What is Δv in each case?

40. A 25-kg boy on a 10-kg sled is coasting at 3.0 m/s on level ice toward his 30-kg sister. The girl jumps vertically and lands on her brother's back. What is the final speed of the siblings and sled? Neglect friction.

41. A 75-kg woman and a 65-kg man face each other on a frictionless ice pond. The woman holds a 5.0-kg "medicine ball." The woman throws the ball to the man with a horizontal velocity of 2.5 m/s relative to the ice. What is her recoil velocity? What is the man's velocity after catching the ball? The man then throws the ball horizontally to the woman at 3.0 m/s relative to himself at the instant before release. What is his final velocity? What is the woman's final velocity after catching it?

42. A 16-u oxygen atom traveling at 600 m/s collides head-on with another oxygen atom at rest. The two join and form an oxygen molecule. With what speed does the molecule move? What fraction of the original translational kinetic energy is transferred to internal energy of the molecule?

*43. A circus clown in a cannon is shot vertically upward with an initial speed of 12 m/s. After ascending 3.5 m, she collides with and grabs a performer sitting still on a trapeze. They ascend together and then fall. What is their speed when they reach the original launch height? The clown and trapeze artist have the same mass.

*44. As described in Problem 6, the change in velocity Δv of an automobile during a collision is a measure of the severity of the collision. For a collision between two automobiles of equal masses, Δv has the same magnitude for each automobile. But for a collision between automobiles of different masses, Δv is larger for the automobile of smaller mass. Suppose that an automobile of 800 kg moving with an initial speed of 15 m/s collides with (a) an automobile of 1400 kg initially at rest, (b) an automobile of 1400 kg initially moving in the opposite direction at 15 m/s, or (c) a stationary rigid barrier. Assume that the collision is totally inelastic. What is Δv in each case for each participating automobile?

*45. Two automobiles of 540 and 1400 kg collide head-on while moving at 80 km/h in opposite directions. After the collision the automobiles remain locked together.

(a) Find the velocity of the wreck immediately after the collision.

(b) Find the kinetic energy of the two-automobile system before and after the collision.

(c) The front end of each automobile crumples by 0.60 m during the collision. Find the acceleration (relative to the ground) of the passenger compartment of each automobile; make the assumption that these accelerations are constant during the collision.

*46. A speeding automobile strikes the rear of a parked automobile. After the impact the two automobiles remain locked together, and they skid along the pavement with all their wheels locked. An investigation of this accident establishes that the length of the skid marks made by the automobiles after the impact was 18 m; the mass of the moving automobile was 2200 kg and that of the parked automobile was 1400 kg, and the coefficient of sliding friction between the wheels and the pavement was 0.95.

(a) What was the speed of the two automobiles immediately after impact?

(b) What was the speed of the moving automobiles before impact?

*47. A proton of energy 8.0×10^{-13} J collides head-on with a proton at rest. How much energy is available for inelastic reactions between these protons?

*48. According to test procedures laid down by the National Highway Traffic Safety Administration, a stationary barrier (of very large mass) and a towed automobile are used for tests of front impacts (Fig. 11.16a), but a moving barrier of 1800 kg and a stationary, unbraked automobile are used for tests of rear impacts (Fig. 11.16b). Explain how this test with the moving barrier and the stationary automobile could be replaced by an equivalent test with a stationary barrier and an automobile towed *backward* at some appropriate speed. If the automobile has a mass of 1400 kg and the moving barrier has a speed of 8 km/h, what is the appropriate equivalent speed of the moving automobile towed backward to the stationary barrier? Assume the collision is inelastic.

FIGURE 11.16 (a) Test procedure for front impact. (b) Test procedure for rear impact.

*49. Regard the two automobiles described in Example 7 as a system of two particles.

(a) What is the translational kinetic energy of the center of mass before the collision? After the collision?

(b) What is the total kinetic energy before the collision? After the collision?

*50. A cat crouches on the floor, at a distance of 1.2 m from a desk chair of height 0.45 m. The cat jumps onto the chair, landing with zero vertical velocity (this is standard procedure for cat jumps). The desk chair has frictionless coasters and rolls away when the cat lands. The mass of the cat is 4.5 kg, and the mass of the chair is 12 kg. What is the speed of recoil of the chair and cat?

*51. A crude but simple method for measuring the speed of a bullet is to shoot the bullet horizontally into a block of wood resting on a table. The block of wood will then slide until its kinetic energy is expended against the friction of the surface of the table. Suppose that a 3.0-kg block of wood slides a distance of 6.0 cm after it is struck by a bullet of 12 g. If the coefficient of sliding friction for the wood on the table is 0.60, what impact speed can you deduce for the bullet?

*52. Another way (not recommended) to measure the speed of bullets with a ballistic pendulum is to shoot a steady stream of bullets into the pendulum, which will push it aside and hold it in a rough equilibrium position at some angle. The speed can be calculated from this equilibrium angle. Suppose that you shoot .22-caliber bullets of mass 15 g into a 4.0-kg ballistic pendulum at the rate of 2 per second. You find that the equilibrium angle is 24°. What is the speed of the bullets?

*53. You shoot a .22-caliber bullet through a piece of wood sitting on a table. The piece of wood acquires a speed of 8.0 m/s, and the bullet emerges with a reduced speed. The mass of the bullet is 15 g, and its initial speed is 600 m/s; the mass of the piece of wood is 300 g.

(a) What is the change of velocity of the bullet?

(b) What is the change of kinetic energy of the bullet?

(c) What is the change of kinetic energy of the wood?

(d) Account for the missing kinetic energy.

*54. An automobile traveling at 50 km/h strikes the rear of a parked automobile. After the collision, the two automobiles remain joined together. The parked automobile skids with all its wheels locked, but the other automobile rolls with negligible friction. The mass of each automobile is 1300 kg, and the coefficient of sliding friction between the locked wheels and the pavement is 0.90. How far do the joined automobiles move before they stop? How long do they take to stop?

*55. You can make a fairly accurate measurement of the speed of a bullet by shooting it horizontally into a block of wood sitting on a fence. The collision of the bullet and the block is inelastic, and the block will fall off the fence and land on the ground at some distance from the bottom of the fence. The speed of the bullet is proportional to this distance. Suppose that a bullet of mass 15 g fired into a block of 4.0 kg sitting on a 1.8-m fence causes the block to land 1.4 m from the bottom of the fence. Calculate the speed of the bullet.

11.4 Collisions in Two and Three Dimensions

56. A cheetah intercepts a gazelle on the run, and grabs it (a totally inelastic collision). Just before this collision, the gazelle was running due north at 20 m/s, and the cheetah was running on an intercepting course of 45° east of north at 22 m/s. The mass of the cheetah is 60 kg, and the mass of the gazelle is 50 kg. What are the magnitude and the direction of the velocity of the entangled animals at the instant after the collision?

57. Two hydrogen atoms ($m = 1.0$ u) with equal speeds, initially traveling in perpendicular directions, collide and join together to form a hydrogen molecule. If 6.1×10^{-22} J of the initial kinetic energy is transferred to internal energy in the collision, what was the initial speed of the atoms?

*58. Two automobiles of equal masses collide at an intersection. One was traveling eastward and the other northward. After the collision, they remain joined together and skid, with locked wheels, before coming to rest. The length of the skid marks is 18 m, and the coefficient of friction between the locked wheels and the pavement is 0.80. Each driver claims his speed was less than 14 m/s (50 km/h) before the collision. Prove that at least one driver is lying.

*59. Two hockey players (see Fig. 11.17) of mass 80 kg collide while skating at 7.0 m/s. The angle between their initial directions of motion is 130°.

(a) Suppose that the players remain entangled and that the collision is totally inelastic. What is their velocity immediately after collision?

(b) Suppose that the collision lasts 0.080 s. What is the magnitude of the average acceleration of each player during the collision?

FIGURE 11.17 Collision of two hockey players.

*60. On July 27, 1956, the ships *Andrea Doria* (40 000 metric tons) and *Stockholm* (20 000 metric tons) collided in the fog south of Nantucket and remained locked together (for a while). Immediately before the collision the velocity of the *Andrea Doria* was 22 knots at 15° east of south and that of the *Stockholm* was 19 knots at 48° east of south (1 knot = 1 nmi/h = 1.85 km/h).

(a) Calculate the velocity (magnitude and direction) of the combined wreck immediately after the collision.

(b) Find the amount of kinetic energy that was converted into other forms of energy by inelastic processes during the collision.

(c) The large amount of energy absorbed by inelastic processes accounts for the heavy damage to both ships. How many kilograms of TNT would have to be exploded to obtain the same amount of energy as was absorbed by inelastic processes in the collision? The explosion of 1 kg of TNT releases 4.6×10^6 J.

*61. Your automobile of mass $m_1 = 900$ kg collides at a traffic circle with another automobile of mass $m_2 = 1200$ kg. Just before the collision, your automobile was moving due east and the other automobile was moving 40° south of east. After the collision the two automobiles remain entangled while they skid, with locked wheels, until coming to rest. Your speed before the collision was 14 m/s. The length of the skid marks is 17.4 m, and the coefficient of kinetic friction between the tires and the pavement is 0.85. Calculate the speed of the other automobile before the collision.

*62. Two billiard balls are placed in contact on a smooth, frictionless table. A third ball moves toward this pair with velocity v in the direction shown in Fig. 11.18. What will be the velocity (magnitude and direction) of the three balls after the collision? The balls are identical and the collisions are elastic.

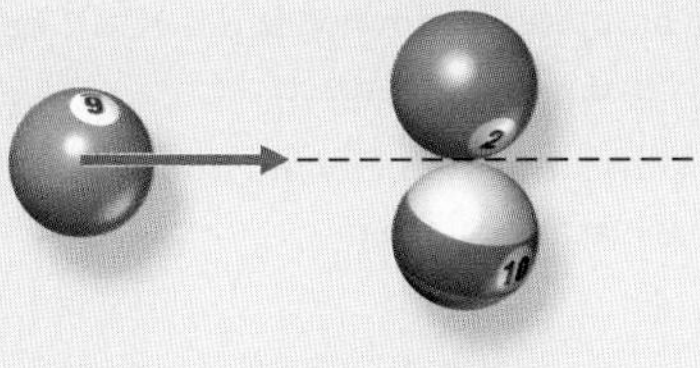

FIGURE 11.18 Three billiard balls.

*63. A billiard ball of mass m and radius R moving with speed v on a smooth, frictionless table collides elastically with an identical stationary billiard ball glued firmly to the surface of the table.

(a) Find a formula for the angular deflection suffered by the moving billiard ball as a function of the impact parameter b (defined in Fig. 11.19). Assume the billiard balls are very smooth so that the force during contact is entirely along the center-to-center line of the balls.

(b) Find a formula for the magnitude of the momentum change suffered by the billiard ball.

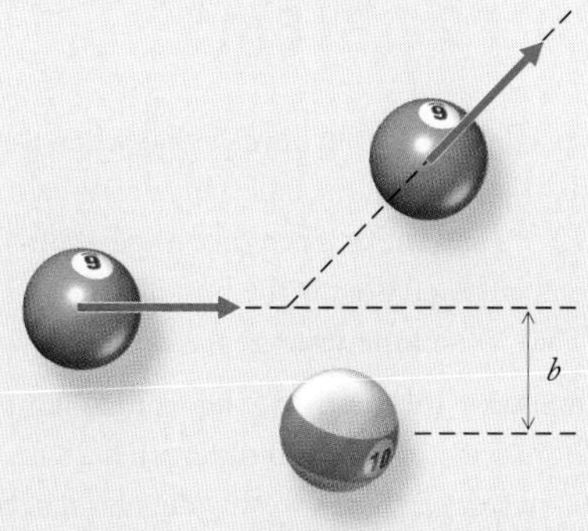

FIGURE 11.19 Two billiard balls.

*64. A coin of mass m slides along a table with speed v and elastically collides with a second, identical coin at rest. The first coin is deflected 60° from its original direction. What are the speeds of each of the two coins after the collision? At what angle does the second coin exit the collision?

*65. In a head-on elastic collision between a projectile and a stationary target of equal mass, we saw that the projectile stops. Show that if such a collision is not head-on, then the projectile and target final velocities are perpendicular (see Fig. 11.20). (Hint: Square the conservation of momentum equation, using $p^2 = \mathbf{p} \cdot \mathbf{p}$, and compare the resulting equation with the energy conservation equation.)

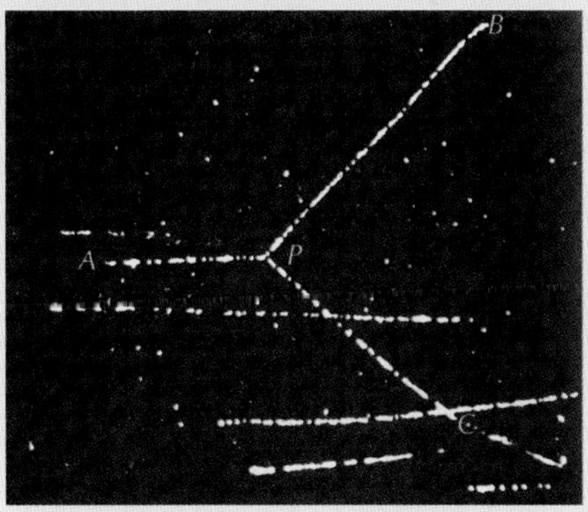

FIGURE 11.20 Elastic collision between two protons. The final velocities of the protons are perpendicular.

*66. In an elastic collision in two dimensions, the projectile has twice the mass of the stationary target. After the collision, the target moves off with three times the final speed of the projectile. Find the angle between the two final directions of motion.

REVIEW PROBLEMS

67. High-speed photography reveals that when a golf club hits a golf ball, the club and the ball typically remain in contact for 1.0×10^{-3} s and the ball acquires a speed of 70 m/s. The mass of the ball is 45 g. What is the impulse the club delivers to the ball? What is the time-average force?

68. In a remarkable incident, a 52-kg woman jumped from the 10th floor of an apartment building, fell 28 m, and landed on her side on soft earth in a freshly dug garden. She fractured her wrist and rib, but remained conscious and fully alert, and recovered completely after some time in a hospital. The earth was depressed 15 cm by her impact.
   (a) What was her impact speed?
   (b) Assuming constant deceleration upon contact with the ground, what was her deceleration?
   (c) What was the force of the ground on her body during deceleration?

69. An automobile approaching an intersection at 10 km/h bumps into the rear of another automobile standing at the intersection with its brakes off and its gears in neutral. The mass of the moving automobile is 1200 kg, and that of the stationary automobile is 700 kg. If the collision is elastic, find the velocities of both automobiles after the collision.

70. It has been reported (fallaciously) that the deer botfly can attain a maximum airspeed of 1318 km/h, that is, 366 m/s. Suppose that such a fly, buzzing along at this speed, strikes a stationary hummingbird and remains stuck in it. What will be the recoil velocity of the hummingbird? The mass of the fly is 2 g; the mass of the hummingbird is 50 g.

*71. A proton of energy 8.0×10^{-13} J collides head-on with a proton of energy 4.0×10^{-13} J moving in the opposite direction. How much energy is available for inelastic reactions between these protons?

*72. When a baseball bat strikes a ball, the impact can be approximately regarded as an elastic collision (the hands of the hitter have little effect on the short time the bat and the ball are in contact). Suppose that a bat of 0.85 kg moving horizontally at 30 m/s encounters a ball of 0.15 kg moving at 40 m/s in the opposite direction. We cannot directly apply the results of Section 11.2 to this collision, since *both* particles are in motion before collision ($\mathbf{v}_1 = 40$ m/s and $\mathbf{v}_2 = -30$ m/s). However, we can apply these results if we use a reference frame that moves at a velocity $\mathbf{V}_0 = -30$ m/s in the direction of the initial motion of the bat; in this reference frame, the initial velocity of the bat is zero ($\mathbf{v}_2 = 0$)
   (a) What is the initial velocity of the ball in this reference frame?
   (b) What are the final velocities of the ball and the bat, just after the collision?
   (c) What are these final velocities in the reference frame of the ground?

*73. A boy throws a baseball at another baseball sitting on a 1.5-m-high fence. The collision of the balls is elastic. The thrown ball moves horizontally at 20 m/s just before the head-on collision.
   (a) What are the velocities of the two balls just after the collision?
   (b) Where do the two balls land on the ground?

74. An automobile of 1200 kg traveling at 45 km/h strikes a moose of 400 kg standing on the road. Assume that the collision is totally inelastic (the moose remains draped over the front end of the automobile). What is the speed of the automobile immediately after this collision?

75. A ship of 3.0×10^4 metric tons steaming at 40 km/h strikes an iceberg of 8.0×10^5 metric tons. If the collision is totally inelastic, what fraction of the initial kinetic energy of the ship is converted into inelastic energy? What fraction remains as kinetic energy of the ship–iceberg system? Ignore the effect of the water on the motion of the ship and iceberg.

*76. When William Tell shot the apple off his son's head, the arrow remained stuck in the apple, which means the collision between the arrow and apple was totally inelastic. Suppose that the velocity of the arrow was horizontal at 80 m/s before it hit, the mass of the arrow was 40 g, and the mass of the apple was 200 g. Suppose Tell's son was 1.40 m high.
   (a) Calculate the velocity of the apple and arrow immediately after the collision.
   (b) Calculate how far behind the son the apple and arrow landed on the ground.

*77. Meteor Crater in Arizona (Fig. 11.21), a hole 180 m deep and 1300 m across, was gouged in the surface of the Earth by the impact of a large meteorite. The mass and speed of this meteorite have been estimated at 2.0×10^9 kg and 10 km/s, respectively, before impact.
   (a) What recoil velocity did the Earth acquire during this (inelastic) collision?

FIGURE 11.21 Meteor Crater in Arizona.

(b) How much kinetic energy was released for inelastic processes during the collision? Express this energy in the equivalent of tons of TNT; 1 ton of TNT releases 4.2×10^9 J upon explosion.

(c) Estimate the magnitude of the impulsive force.

*78. A black automobile smashes into the rear of a white automobile stopped at a stop sign. You investigate this collision and find that before the collision, the black automobile made skid marks 5.0 m long; after the collision the black automobile made skid marks 1.0 m long (in the same direction as the initial direction of motion), and the white automobile made skid marks 2.0 m long. Both automobiles made these skid marks with all their wheels. The mass of the black automobile is 1400 kg, and the mass of the white automobile is 800 kg. The coefficient of sliding friction between the wheels and the pavement is 0.90. From these data, deduce the speed of the black automobile just before the collision, and the speed before it started to brake.

*79. (a) Two identical small steel balls are suspended from strings of length l so they touch when hanging straight down, in their equilibrium position (Fig. 11.22). If we pull one of the balls back until its string makes an angle θ with the vertical and then let it go, it will collide elastically with the other ball. How high will the other ball rise?

(b) Suppose that instead of steel balls we use putty balls. They will then collide inelastically and remain stuck together. How high will the balls rise?

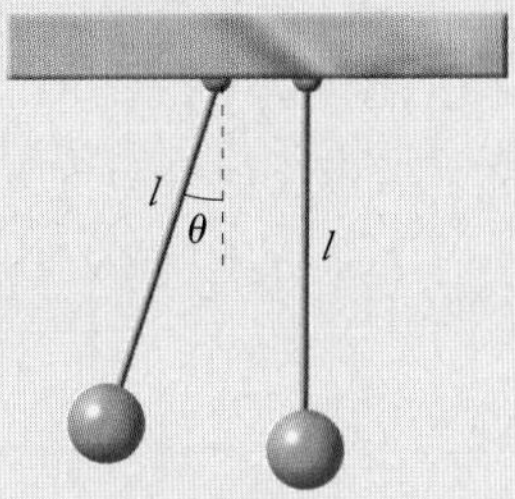

FIGURE 11.22 Two balls suspended from strings.

*80. While in flight, a peregrine falcon spots a pigeon flying 40 m below. The falcon closes its wings and, in free fall, dives on the pigeon and grabs it (a totally inelastic collision). The mass of the falcon is 1.5 kg, and the mass of the pigeon is 0.40 kg. Suppose that the velocity of the pigeon before this collision is horizontal, at 15 m/s, and the velocity of the falcon is vertical, equal to the free-fall velocity. What is the velocity (magnitude and direction) of both birds after the collision?

*81. On a freeway, a truck of 3500 kg collides with an automobile of 1500 kg that is trying to cut diagonally across the path of the truck. Just before the collision, the truck was traveling due north at 70 km/h, and the automobile was traveling at 30° west of north at 100 km/h. After the collision, the vehicles remain joined together.

(a) What is the velocity (magnitude and direction) of the joined vehicles immediately after the collision?

(b) How much kinetic energy is lost during the collision?

*82. Two asteroids of 1.0×10^7 kg and 8.0×10^7 kg, respectively, are initially at rest in interstellar space separated by a large distance. Their mutual gravitational attraction then causes them to fall toward each other on a straight line. Assume the asteroids are spheres of radius 100 m and 200 m, respectively.

(a) What is the velocity of each asteroid just before they hit? What is the kinetic energy of each? What is the total kinetic energy?

(b) The collision is totally inelastic. What is the velocity of the joined asteroids after they hit?

Answers to Checkups

Checkup 11.1

1. The front end should be soft and crushable to protect automobile occupants in a collision; this will spread the momentum change over a longer time, lowering the force experienced by the occupants.

2. The steel ball will exert a larger force, because it is less deformable than the golf ball. Thus, although the change in momentum (the impulse **I**) can be the same, the steel ball is in contact for a shorter time Δt, and so exerts a greater average force during that time ($\overline{\mathbf{F}} = \mathbf{I}/\Delta t$).

3. Because the collision is elastic, the ball will rebound with the same kinetic energy; as this energy gets converted to potential energy, the ball will rise up to the same height, 1 m, from which it was dropped before stopping.

4. No. Since the wad of gum stopped, kinetic energy was lost, and thus the collision was not elastic. We will later refer to such a collision (when the bodies stick together) as totally *in*elastic.

5. (C) 3×10^6 N. In a collision, each vehicle exerts an equal-magnitude, but opposite-direction, force on the other (an action–reaction pair), so the force exerted by the car on the truck is 3×10^6 N westward.

Checkup 11.2

1. As in the cases just discussed, and as in Eq. (11.13), where the projectile's final velocity is proportional to $m_1 - m_2$, the velocity of the projectile will be positive when it is more massive than the target ($m_1 > m_2$), and it will be negative when it is less massive than the target ($m_1 < m_2$).
2. No. As we saw in the cases just discussed, the speed of recoil of a massive target is very small; in the limit of a very light target, the speed approaches twice the speed of the projectile. For any values of m_1 and m_2, the final speed of the target (v_2'), given by Eq. (11.14), cannot exceed twice the projectile speed (v_1).
3. No; for instance, in the collision of the Super Ball and the wall, the ball is instantaneously at rest before it bounces back. The kinetic energy is transformed into elastic energy momentarily, and then converted back into kinetic energy.
4. (E) 0; v_1. As discussed above, when the masses of the target and projectile are identical, the speed of the projectile is zero after the collision [since we have $m_1 - m_2 = 0$ in Eq. (11.13)]. For identical masses, the target speed is equal to the initial speed of the projectile, v_1 [since we have $m_1 = m_2$ in Eq. (11.14)].

Checkup 11.3

1. No; for example, if the target is initially at rest, it gains kinetic energy.
2. Two particles of equal mass and opposite velocities have zero net momentum. Thus, in a totally inelastic collision, the composite particle has zero momentum, and thus zero velocity. In an elastic collision, the total kinetic energy is unchanged; since the net momentum is zero, the particles must again have opposite velocities. If we ignore the possibility that the particles might have passed through each other, then this means that their velocities were reversed by the collision.
3. The velocity of the joined particles after a totally inelastic collision is the velocity of the center of mass, $v_{CM} = m_1v_1/(m_1 + m_2)$; this is equal to one-half of the velocity of the incident projectile when the masses of the target and projectile are equal, or $m_1 = m_2$.
4. No, assuming the wires are long enough to permit the upward motion of the pendulum to the maximum height h.
5. (B) $\frac{1}{3}v$. Momentum is conserved, so equating the initial and final momenta, we have $mv - (m/2)v = (\frac{3}{2}m)v'$, which implies $v' = \frac{1}{3}v$.

Checkup 11.4

1. Because the cars have equal mass and speed, the total momentum before and after this totally inelastic collision is directed due southwest.
2. (B) Southeast. This explosion is like a three-particle totally inelastic collision in reverse. Since the total momentum before the "collision" (explosion) is zero, so must it be afterward: the third particle must have momentum components which cancel the northward and westward momentum contributions of the other two particles; thus, the third particle travels in the southeast quadrant of directions.

CHAPTER

12

Rotation of a Rigid Body

Concepts — in — Context

CONCEPTS IN CONTEXT

This large centrifuge at the Sandia National Laboratory is used for testing the behavior of components of rockets, satellites, and reentry vehicles when subjected to high accelerations. The components to be tested are placed in a compartment in one arm of this centrifuge; the opposite arm holds a counterweight. The arms rotate at up to 175 revolutions per minute, and they generate a centripetal acceleration of up to 300*g*.

The concepts of this chapter permit us to answer several questions about this centrifuge:

? How are the speed and the centripetal acceleration at the end of an arm related to the rate of rotation? (Example 5, page 373)

? How do we determine the resistance that the centrifuge offers to changes in its rotational motion? (Example 12, part (a), page 383)

? How is the kinetic energy of the centrifuge arms related to the rate of rotation? (Example 12, part (b), page 383)

rigid body

A *body is* **rigid** *if the particles in the body do not move relative to one another.* Thus, the body has a fixed shape, and all its parts have a fixed position relative to one another. A hammer is a rigid body, and so is a baseball bat. A baseball is not rigid—when struck a blow by the bat, the ball suffers a substantial deformation; that is, different parts of the ball move relative to one another. However, the baseball can be regarded as a rigid body while it flies through the air—the air resistance is not sufficiently large to produce an appreciable deformation of the ball. This example indicates that whether a body can be regarded as rigid depends on the circumstances. No body is absolutely rigid; when subjected to a sufficiently large force, any body will suffer some deformation or perhaps even break into several pieces. In this chapter, we will ignore such deformations produced by the forces acting on bodies. We will examine the motion of bodies under the assumption that rigidity is a good approximation.

12.1 MOTION OF A RIGID BODY

A rigid body can simultaneously have two kinds of motion: it can change its position in space, and it can change its orientation in space. Change of position is translational motion; as we saw in Chapter 10, this motion can be conveniently described as motion of the center of mass. Change in orientation is rotational motion; that is, it is rotation about some axis.

As an example, consider the motion of a hammer thrown upward (see Fig. 12.1). The orientation of the hammer changes relative to fixed coordinates attached to the ground. Instantaneously, the hammer rotates about a horizontal axis, say, a horizontal axis that passes through the center of mass. In Fig. 12.1, this horizontal axis sticks out of the plane of the page and moves upward with the center of mass. The complete motion can then be described as a rotation of the hammer about this axis and a simultaneous translation of the axis along a parabolic path.

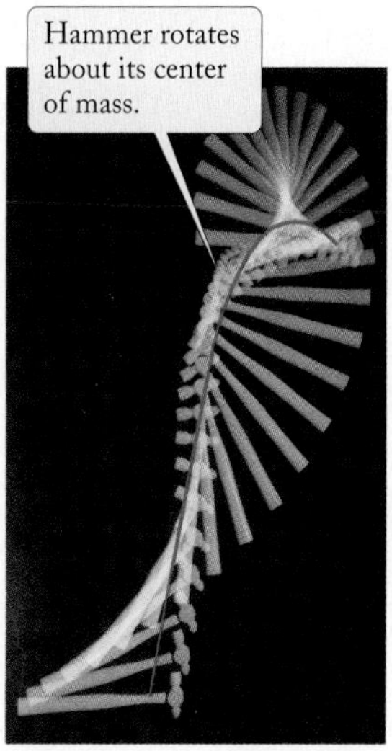

FIGURE 12.1 A hammer in free fall under the influence of gravity. The center of mass of the hammer moves with constant vertical acceleration g, just like a particle in free fall.

In this example of the thrown hammer, the axis of rotation always remains horizontal, out of the plane of the page. In the general case of motion of a rigid body, the axis of rotation can have any direction and can also change its direction. To describe such complicated motion, it is convenient to separate the rotation into three components along three perpendicular axes. The three components of rotation are illustrated by the motion of an aircraft (see Fig. 12.2): the aircraft can turn left or right (yaw), it can tilt to the left or the right (roll), and it can tilt its nose up or down (pitch). However, in the following sections we will usually not deal with this general case of rotation with three components; we will mostly deal only with the simple case of rotation about a fixed axis, such as the rotational motion of a fan, a roulette wheel, a compact disc, a swinging door, or a merry-go-round (see Fig. 12.3).

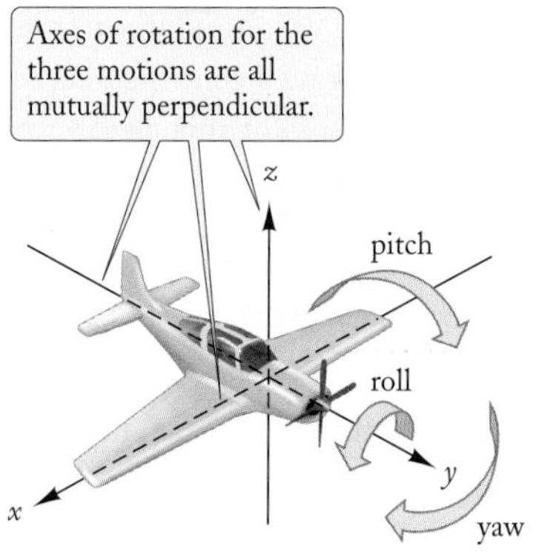

FIGURE 12.2 Pitch, roll, and yaw motions of an aircraft.

(a) (b) (c) (d)

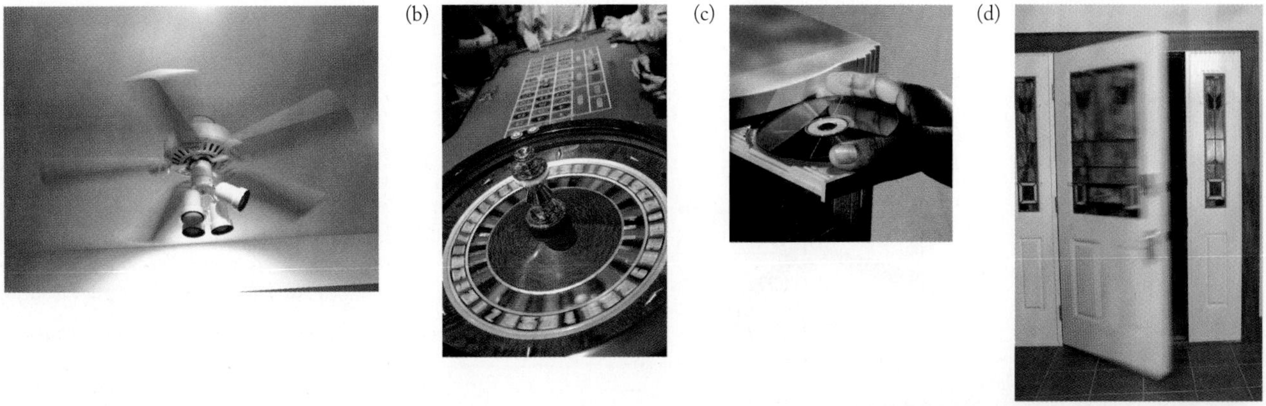

(e)

FIGURE 12.3 Some examples of rotational motion with a fixed axis (a) fan, (b) roulette wheel, (c) compact disc, (d) swinging door, (e) merry-go-round).

✔ Checkup 12.1

QUESTION 1: Characterize the following motions as translational, rotational, or both: swinging motion of door, motion of wheel of train, motion of propeller of airplane while in level flight.

QUESTION 2: Suppose that instead of selecting an axis through the center of mass of the hammer in Fig. 12.1, we select a parallel axis through the end of the handle. Can the motion still be described as rotation about this axis and a simultaneous translation of the axis along some path? Is this path parabolic?

QUESTION 3: Under what conditions will the passenger compartment of an automobile exhibit (limited) rolling, pitching, and turning motions?

QUESTION 4: Which of the rotating bodies in Fig. 12.3 does *not* rotate about an axis through its center of mass?

(A) Fan (B) Roulette wheel (C) Compact disc
(D) Swinging door (E) Merry-go-round

12.2 ROTATION ABOUT A FIXED AXIS

Online Concept Tutorial 15

Figure 12.4 shows a rigid body rotating about a fixed axis, which coincides with the z axis. During this rotational motion, each point of the body remains at a given distance from this axis and moves along a circle centered on the axis. To describe the orientation of the body at any instant, we select one particle in the body and use it as a reference point; any particle can serve as reference point, provided that it is not on the axis of rotation. The circular motion of this reference particle (labeled P in Fig. 12.4) is then representative of the rotational motion of the entire body, and the angular position of this particle is representative of the angular orientation of the entire body.

Figure 12.5 shows the rotating rigid body as seen from along the axis of rotation. The coordinates in Fig. 12.5 have been chosen so the z axis coincides with the axis of rotation, whereas the x and y axes are in the plane of the circle traced out by the motion of the reference particle. The angular position of the reference particle—and hence the angular orientation of the entire rigid body—can be described by the position angle ϕ (the Greek

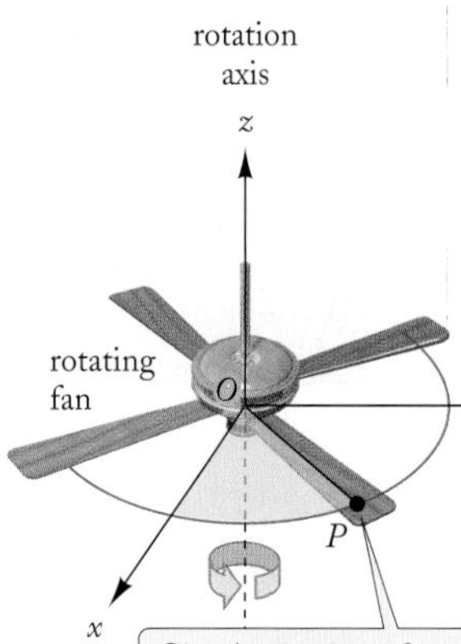

FIGURE 12.4 The four blades of this fan are a rigid body rotating about a fixed axis, which coincides with the z axis. The reference particle P in this rigid body moves along a circle around this axis.

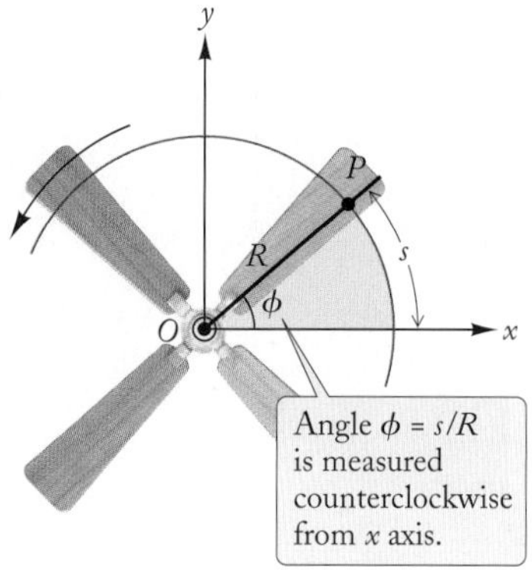

FIGURE 12.5 Motion of a reference particle P in the rigid body rotating about a fixed axis. The axis is indicated by the circled dot O. The radius of the circle traced out by the motion of the reference particle is R.

letter *phi*) between the radial line OP and the x axis. Conventionally, the angle ϕ is taken as positive when reckoned in a counterclockwise direction (as in Fig. 12.5). We will usually measure this position angle in radians, rather than degrees. By definition, *the* **angle ϕ in radians** *is the length s of the circular arc divided by the radius R,* or

angle in radians

$$\phi = \frac{s}{R} \tag{12.1}$$

In Fig. 12.5, the length s is the distance traveled by the reference particle from the x axis to the point P. Note that if the length s is the circumference of a full circle, then $s = 2\pi R$, and $\phi = s/R = 2\pi R/R = 2\pi$. Thus, there are 2π radians in a full circle; that is, there are 2π radians in 360°:

$$2\pi \text{ radians} = 360°$$

Accordingly, 1 radian equals $360°/2\pi$, or

$$1 \text{ radian} = 57.3°$$

EXAMPLE 1 The accuracy of the guidance system of the Hubble Space Telescope is such that if the telescope were sitting in New York, the guidance system could aim at a dime placed on top of the Washington Monument, at a distance of 320 km. The width of a dime is 1.8 cm. What angle does the dime subtend when seen from New York?

SOLUTION: Figure 12.6 shows the circular arc subtended by the dime. The radius of the circle is 320 km. For a small angle, such as in this figure, the length s of the arc from one side of the dime to the other is approximately the same as the length of the straight line from one side to the other, which is the width of the dime. Hence the angle in radians is

$$\phi = \frac{s}{R} = \frac{1.8 \times 10^{-2}\text{ m}}{3.2 \times 10^{5}\text{ m}} = 5.6 \times 10^{-8}\text{ radian}$$

Expressed in degrees, this becomes

$$\phi = 5.6 \times 10^{-8}\text{ radian} \times \frac{360°}{2\pi \text{ radians}} = 3.2 \times 10^{-6}\text{ degree}$$

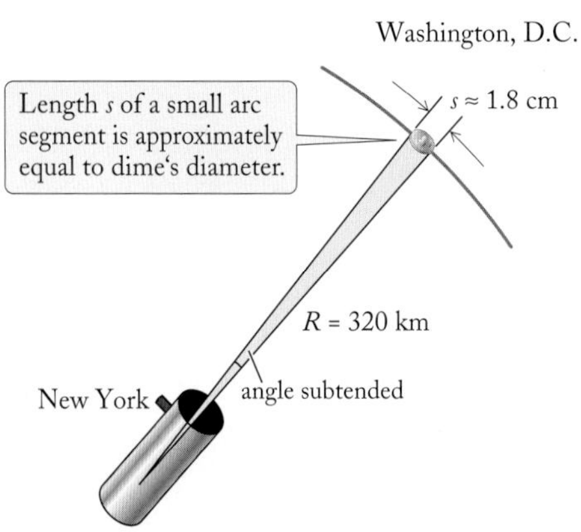

FIGURE 12.6 A dime placed at a distance of 320 km from the telescope. The length $s = 1.8$ cm is the diameter of the dime.

When a rigid body rotates, the position angle ϕ changes in time. The body then has an **angular velocity** ω (the Greek letter *omega*). The definition of the angular velocity for rotational motion is mathematically analogous to the definition of velocity for translational motion (see Sections 2.2 and 2.3). The **average angular velocity** $\overline{\omega}$ is defined as

$$\overline{\omega} = \frac{\Delta\phi}{\Delta t} \tag{12.2}$$

average angular velocity

where $\Delta\phi$ is the change in the angular position and Δt the corresponding change in time. The **instantaneous angular velocity** is defined as

$$\omega = \frac{d\phi}{dt} \tag{12.3}$$

instantaneous angular velocity

According to these definitions, the angular velocity is the rate of change of the angle with time. The unit of angular velocity is the radian per second (1 radian/s). The radian is the ratio of two lengths [compare Eq. (12.1)], and hence it is a pure number; thus, 1 radian/s is the same thing as 1/s. However, to prevent confusion, it is often useful to retain the vacuous label *radian* as a reminder that angular motion is involved. Table 12.1 gives some examples of angular velocities.

If the body rotates with constant angular velocity, then we can also measure the rate of rotation in terms of the ordinary **frequency** f, or the number of revolutions per second. Since each complete revolution involves a change of ϕ by 2π radians, the frequency of revolution is smaller than the angular velocity by a factor of 2π:

$$f = \frac{\omega}{2\pi} \tag{12.4}$$

frequency

This expresses the frequency in terms of the angular velocity. The unit of rotational frequency is the revolution per second (1 rev/s). Like the radian, the revolution is a pure number, and hence 1 rev/s is the same thing as 1/s. But we will keep the label *rev* to prevent confusion between rev/s and radian/s.

As in the case of planetary motion, the time per revolution is called the **period** of the motion. If the number of revolutions per second is f, then the time per revolution is $1/f$, that is,

$$T = \frac{1}{f} \tag{12.5}$$

period of motion

TABLE 12.1 SOME ANGULAR VELOCITIES

Computer hard disk	8×10^2 radians/s	Helicopter rotor	40 radians/s
Circular saw	7×10^2 radians/s	Compact disc (outer track)	22 radians/s
Electric blender blades	5×10^2 radians/s	Phonograph turntable	3.5 radians/s
Jet engine	4×10^2 radians/s	Neutron star (pulsar) rotation	0.1 radian/s
Airplane propeller	3×10^2 radians/s	Earth rotation	7.3×10^{-5} radian/s
Automobile engine	2×10^2 radians/s	Earth revolution about Sun	2.0×10^{-7} radian/s
Small fan	60 radians/s		

EXAMPLE 2 The rotational frequency of machinery is often expressed in revolutions per minute, or rpm. A typical ceiling fan on medium speed rotates at 150 rpm. What is the frequency of revolution? What is the angular velocity? What is the period of the motion?

SOLUTION: Each minute is 60.0 s; hence 150 revolutions per minute amounts to 150 revolutions in 60.0 s; so

$$f = \frac{150 \text{ rev}}{60.0 \text{ s}} = 2.50 \text{ rev/s}$$

Since each revolution comprises 2π radians, the angular velocity is

$$\omega = 2\pi f = 2\pi \times 2.50 \text{ rev/s} = 15.7 \text{ radians/s}$$

Note that here we have dropped a label *rev* in the third step and inserted a label *radians;* as remarked above, these labels merely serve to prevent confusion, and they can be inserted and dropped at will once they have served their purpose.

The period of the motion is

$$T = \frac{1}{f} = \frac{1}{2.50 \text{ rev/s}} = 0.400 \text{ s}$$

One complete revolution takes two-fifths of a second.

If the angular velocity of a rigid body is changing, the body has an **angular acceleration** α (the Greek letter *alpha*). The rotational motion of a ceiling fan that is gradually building up speed immediately after being turned on is an example of accelerated rotational motion. The mathematical definition of the **average angular acceleration** is, again, analogous to the definition of acceleration for translational motion. If the angular velocity changes by $\Delta\omega$ in a time Δt, then the average angular acceleration is

average angular acceleration

$$\overline{\alpha} = \frac{\Delta\omega}{\Delta t} \tag{12.6}$$

and the **instantaneous angular acceleration** is

instantaneous angular acceleration

$$\alpha = \frac{d\omega}{dt} \tag{12.7}$$

Thus, the angular acceleration is the rate of change of the angular velocity. The unit of angular acceleration is the radian per second per second, or radian per second squared (1 radian/s^2).

Since the angular velocity ω is the rate of change of the angular position ϕ [see Eq. (12.3)], the angular acceleration given by Eq. (12.7) can also be written

$$\alpha = \frac{d^2\phi}{dt^2} \tag{12.8}$$

Equations (12.3) and (12.7) give the angular velocity and acceleration of the rigid body; that is, they give the angular velocity and acceleration of every particle in the body. It is interesting to focus on one of the particles and evaluate its *translational* speed and acceleration as it moves along its circular path around the axis of rotation of the rigid body. If the particle is at a distance R from the axis of rotation (see Fig. 12.7), then the length along the circular path of the particle is, according to the definition of angle, Eq. (12.1),

$$s = \phi R \tag{12.9}$$

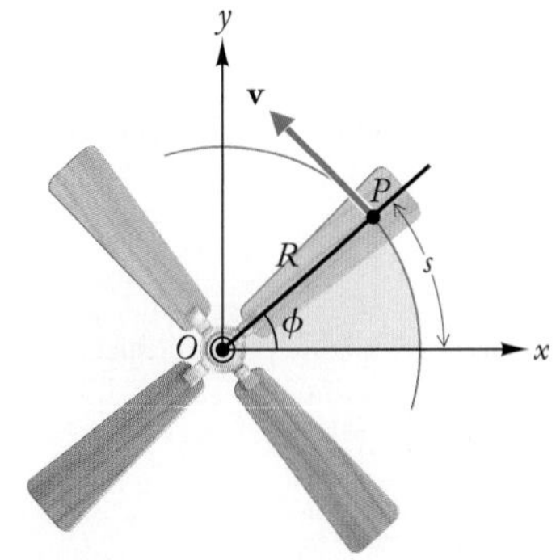

FIGURE 12.7 The instantaneous translational velocity of a particle in a rotating rigid body is tangent to the circular path.

Since R is a constant, the rate of change of s is entirely due to the rate of change of ϕ, so

$$\frac{ds}{dt} = \frac{d\phi}{dt}R \tag{12.10}$$

Here ds/dt is the translational speed v with which the particle moves along its circular path, and $d\phi/dt$ is the angular velocity ω; hence Eq. (12.10) is equivalent to

$$v = \omega R \tag{12.11}$$

translational speed in circular motion

This shows that the translational speed of the particle along its circular path around the axis is directly proportional to the radius: the farther a particle in the rigid body is from the axis, the faster it moves. We can understand this by comparing the motions of two particles, one on a circle of large radius R_1, and the other on a circle of smaller radius R_2 (see Fig. 12.8). For each revolution of the rigid body, both of these particles complete one trip around their circles. But the particle on the larger circle has to travel a larger distance, and hence must move with a larger speed.

For a particle at a given R, the translational speed is constant if the angular velocity is constant. This speed is the distance around the circular path (the circumference) divided by the time for one revolution (the period), or

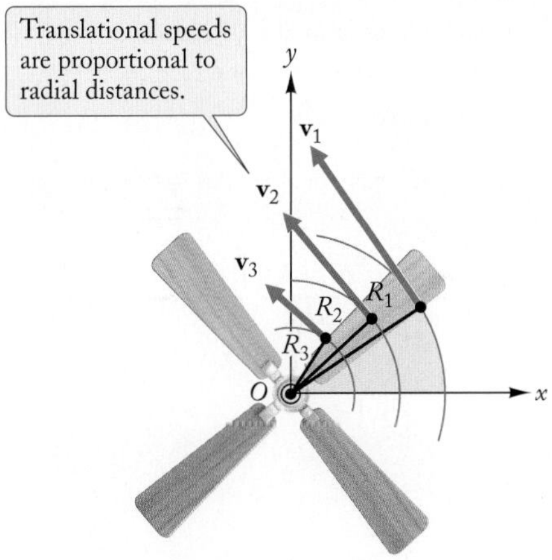

FIGURE 12.8 Several particles in a rigid body rotating about a fixed axis and their velocities.

$$v = \frac{2\pi R}{T} \quad \text{(constant speed)} \tag{12.12}$$

Since $2\pi/T = 2\pi f = \omega$, Eq. (12.12) can be obtained from Eq. (12.11).

If v is changing, it also follows from Eq. (12.11) that the rate of change of v is proportional to the rate of change of ω:

$$\frac{dv}{dt} = \frac{d\omega}{dt}R$$

A rate of change of the speed along the circle implies that the particle has an acceleration along the circle, called a **tangential acceleration.** According to the last equation, this tangential acceleration is

$$a_{\text{tangential}} = \alpha R \tag{12.13}$$

tangential acceleration

Note that, besides this tangential acceleration directed along the circle, the particle also has a **centripetal acceleration** directed toward the center of the circle. From Section 4.5, we know that the centripetal acceleration for uniform circular motion is

$$a_{\text{centripetal}} = \frac{v^2}{R} \tag{12.14}$$

With $v = \omega R$, this becomes

centripetal acceleration

$$a_{\text{centripetal}} = \omega^2 R \tag{12.15}$$

The net translational acceleration of the particle is the vector sum of the tangential and the centripetal accelerations, which are perpendicular (see Fig. 12.9); thus, the magnitude of the net acceleration is

$$a_{\text{net}} = \sqrt{a_{\text{tangential}}^2 + a_{\text{centripetal}}^2} \tag{12.16}$$

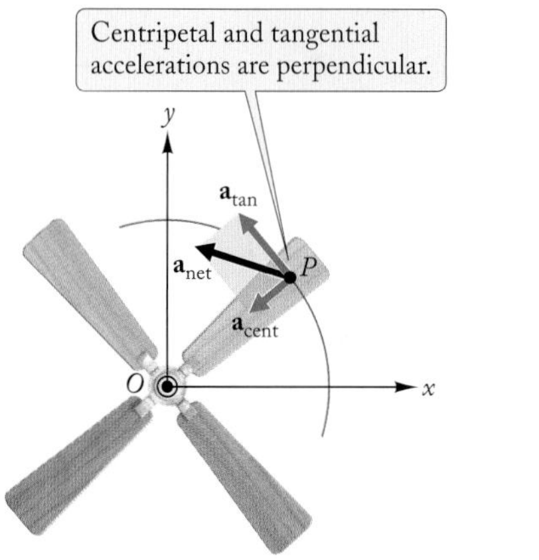

FIGURE 12.9 A particle in a rotating rigid body with an angular acceleration has both a centripetal acceleration $\mathbf{a}_{\text{centripetal}}$ and a tangential acceleration $\mathbf{a}_{\text{tangential}}$. The net instantaneous translational acceleration $\mathbf{a}_{\text{net}}$ is then the vector sum of $\mathbf{a}_{\text{centripetal}}$ and $\mathbf{a}_{\text{tangential}}$.

Although we have here introduced the concept of tangential acceleration in the context of the rotational motion of a rigid body, this concept is also applicable to the translational motion of a particle along a circular path or any curved path. For instance, consider an automobile (regarded as a particle) traveling around a curve. If the driver steps on the accelerator (or on the brake), the automobile will suffer a change of speed as it travels around the curve. It will then have both a tangential and a centripetal acceleration.

EXAMPLE 3 The blade of a circular saw is initially rotating at 7000 revolutions per minute. Then the motor is switched off, and the blade coasts to a stop in 8.0 s. What is the average angular acceleration?

SOLUTION: In radians per second, 7000 rev/min corresponds to an initial angular velocity $\omega_1 = 7000 \times 2\pi$ radians/min, or

$$\omega_1 = \frac{7000 \times 2\pi \text{ radians}}{60 \text{ s}} = 7.3 \times 10^2 \text{ radians/s}$$

The final angular velocity is $\omega_2 = 0$. Hence the average angular acceleration is

$$\bar{\alpha} = \frac{\Delta\omega}{\Delta t} = \frac{\omega_2 - \omega_1}{t_2 - t_1} = \frac{0 - 7.3 \times 10^2 \text{ radians/s}}{8.0 \text{ s} - 0}$$
$$= -91 \text{ radians/s}^2$$

FIGURE 12.10 Rotating wheel of the automobile as viewed in the reference frame of the automobile. The ground moves toward the left at speed v.

EXAMPLE 4 An automobile accelerates uniformly from 0 to 80 km/h in 6.0 s. The wheels of the automobile have a radius of 0.30 m. What is the angular acceleration of the wheels? Assume that the wheels roll without slipping.

SOLUTION: The translational acceleration of the automobile is

$$a = \frac{v - v_0}{t} = \frac{80 \text{ km/h}}{6.0 \text{ s}} = \frac{(80 \text{ km/h}) \times (1000 \text{ m/1 km}) \times (1 \text{ h/3600 s})}{6.0 \text{ s}}$$
$$= 3.7 \text{ m/s}^2$$

The angular acceleration of the wheel is related to this translational acceleration by $a = \alpha R$, the same relation as Eq. (12.13). We can establish this relationship most conveniently by viewing the motion of the wheel in the reference frame of the automobile (see Fig. 12.10). In this reference frame, the ground is moving backward at speed v, and the bottom point of the rotating wheel is moving backward at the tangential speed ωR. Since the wheel is supposed to move without slipping, the speed v of the ground must match the tangential speed of the bottom point of the wheel; that is, $v = \omega R$. This proportionality of v and ω implies the same proportionality of the accelerations a and α, and therefore establishes the relationship $a = \alpha R$.

The angular acceleration of the wheel is then

$$\alpha = \frac{a}{R} = \frac{3.7\ \text{m/s}^2}{0.30\ \text{m}} = 12\ \text{radians/s}^2$$

EXAMPLE 5 The large centrifuge shown in the chapter photo has an arm of length 8.8 m. When rotating at 175 revolutions per minute, what is the speed of the end of this arm, and what is the centripetal acceleration?

Concepts — in — Context

SOLUTION: 175 rpm amounts to 175/60 = 2.9 revolutions per second. The corresponding angular velocity is

$$\omega = 2\pi f = 2\pi \times 2.9\ \text{rev/s} = 18\ \text{radians/s}$$

According to Eq. (12.11), the speed at a radius $R = 8.8$ m is

$$v = \omega R = 18\ \text{radians/s} \times 8.8\ \text{m} = 1.6 \times 10^2\ \text{m/s}$$

and according to Eq. (12.15), the centripetal acceleration is

$$a_{\text{centripetal}} = \omega^2 R = (18\ \text{radians/s})^2 \times 8.8\ \text{m} = 2.9 \times 10^3\ \text{m/s}^2$$

This is almost 300 standard *g*'s!

Checkup 12.2

QUESTION 1: Consider a point P on the rim of a rotating, accelerating flywheel and a point Q near the center. Which point has the larger instantaneous speed? The larger instantaneous angular velocity? The larger angular acceleration? The larger tangential acceleration? The larger centripetal acceleration?

QUESTION 2: The Earth rotates steadily around its axis once per day. Do all points on the surface of the Earth have the same radius R for their circular motion? Do they all have the same angular velocity ω? The same speed v around the axis? The same centripetal acceleration $a_{\text{centripetal}}$? If not, which points have the largest R, ω, v, and $a_{\text{centripetal}}$?

QUESTION 3: A short segment of the track of a roller coaster can be approximated by a circle of suitable radius. If a (frictionless) roller-coaster car is passing through the highest point of the track, is there a centripetal acceleration? A tangential acceleration? What if the the roller coaster is some distance beyond the highest point?

QUESTION 4: Consider the motion of the hammer shown in Fig. 12.1. Taking into account only the rotational motion, which end of the hammer has the larger speed v around the axis? The larger centripetal acceleration $a_{\text{centripetal}}$?

(A) Head end; head end
(B) Head end; handle end
(C) Handle end; head end
(D) Handle end; handle end

12.3 MOTION WITH CONSTANT ANGULAR ACCELERATION

We will now examine the kinematic equations describing rotational motion for the special case of *constant* angular acceleration; these are mathematically analogous to the equations describing translational motion with constant acceleration (see Section 2.5), and they can be derived by the same methods. In the next section, we will develop an alternative method, based on integration, for obtaining the kinematic equations describing either angular or translational motion for the general case of accelerations with arbitrary time dependence.

If the rigid body rotates with a constant angular acceleration α, then the angular velocity increases at a constant rate, and after a time t has elapsed, the angular velocity will attain the value

constant angular acceleration: ω, α, and t

$$\omega = \omega_0 + \alpha t \tag{12.17}$$

where ω_0 is the initial value of the angular velocity at $t = 0$.

The angular position can be calculated from this angular velocity by the arguments used in Section 2.5 to calculate x from v [see Eqs. (2.17), (2.22), and (2.25)]. The result is

constant angular acceleration: ϕ, α, and t

$$\phi = \phi_0 + \omega_0 t + \tfrac{1}{2}\alpha t^2 \tag{12.18}$$

Furthermore, the arguments of Section 2.5 lead to an identity between acceleration, position, and velocity [see Eqs. (2.20)–(2.22)]:

constant angular acceleration: α, ϕ, and ω

$$\alpha(\phi - \phi_0) = \tfrac{1}{2}(\omega^2 - \omega_0^2) \tag{12.19}$$

Note that all these equations have exactly the same mathematical form as the equations of Section 2.5, with the angular position ϕ taking the place of the position x, the angular velocity ω taking the place of v, and the angular acceleration α taking the place of a. This analogy between rotational and translational quantities can serve as a useful mnemonic for remembering the equations for rotational motion. Table 12.2 displays analogous equations.

TABLE 12.2 ANALOGIES BETWEEN TRANSLATIONAL AND ROTATIONAL QUANTITIES

$v = \dfrac{dx}{dt}$	$\rightarrow$	$\omega = \dfrac{d\phi}{dt}$
$a = \dfrac{dv}{dt}$	$\rightarrow$	$\alpha = \dfrac{d\omega}{dt}$
$v = v_0 + at$	$\rightarrow$	$\omega = \omega_0 + \alpha t$
$x = x_0 + v_0 t + \frac{1}{2}at^2$	$\rightarrow$	$\phi = \phi_0 + \omega_0 t + \frac{1}{2}\alpha t^2$
$a(x - x_0) = \frac{1}{2}(v^2 - v_0^2)$	$\rightarrow$	$\alpha(\phi - \phi_0) = \frac{1}{2}(\omega^2 - \omega_0^2)$

EXAMPLE 6 The cable supporting an elevator runs over a wheel of radius 0.36 m (see Fig. 12.11). If the elevator begins from rest and ascends with an upward acceleration of 0.60 m/s^2, what is the angular acceleration of the wheel? How many turns does the wheel make if this accelerated motion lasts 5.0 s? Assume that the cable runs over the wheel without slipping.

SOLUTION: If there is no slipping, the speed of the cable must always coincide with the tangential speed of a point on the rim of the wheel. The acceleration $a = 0.60\ \text{m/s}^2$ of the cable must then coincide with the tangential acceleration of a point on the rim of the wheel:

$$a = a_{\text{tangential}} = \alpha R \tag{12.20}$$

where $R = 0.36$ m is the radius of the wheel. Hence

$$\alpha = \frac{a}{R} = \frac{0.60\ \text{m/s}^2}{0.36\ \text{m}} = 1.7\ \text{radians/s}^2$$

According to Eq. (12.18), the angular displacement in 5.0 s is

$$\begin{aligned}\phi - \phi_0 &= \omega_0 t + \tfrac{1}{2}\alpha t^2 \\ &= 0 + \tfrac{1}{2} \times 1.7\ \text{radians/s}^2 \times (5.0\ \text{s})^2 \\ &= 21\ \text{radians}\end{aligned}$$

Each revolution comprises 2π radians; thus, the number of turns the wheel makes is

$$[\text{number of turns}] = \frac{\phi - \phi_0}{2\pi} = \frac{21\ \text{radians}}{2\pi} = 3.3\ \text{revolutions}$$

FIGURE 12.11 Elevator supported by a cable that runs over a rotating wheel.

PROBLEM-SOLVING TECHNIQUES ANGULAR MOTION

The solution of kinematic problems about angular velocity and angular acceleration involves the same techniques as the problems about translational velocity and translational acceleration in Chapter 2. You might find it useful to review the procedures suggested on page 50.

Sometimes a problem contains a link between a rotational motion and a translational motion, such as the link between the rotational and translational motions of the wheels of an automobile (see Example 4) or the link between the translational motion of the elevator cable and the rotational motion of the wheel over which it runs (Example 6). If the body in contact with the rim of the wheel does not slip, the translational speed of this body equals the tangential speed of the contact point at the rim of the wheel; that is, $v = \omega R$ and $a = \alpha R$.

Keep in mind that although some of the equations in this chapter remain valid if the angular quantities are expressed in degrees, any equation that contains both angular quantities and distances (e.g., $v = \alpha R$) is valid only if the angular quantity is expressed in radians. To prevent mistakes, it is safest to express all angular quantities in radians; if degrees are required in the answer, convert from radians to degrees after completing your calculations.

✔ Checkup 12.3

QUESTION 1: Consider a point on the rim of the wheel shown in Fig. 12.11, (instantaneously) at the top of the wheel. What is the direction of the centripetal acceleration of this point? The tangential acceleration?

QUESTION 2: The wheel of a bicycle rolls on a flat road. Is the angular velocity constant if the translational velocity of the bicycle is constant? Is the angular acceleration constant if the translational acceleration of the bicycle is constant?

QUESTION 3: A grinding wheel accelerates uniformly for 3 seconds after being turned on. In the first second of motion, the wheel rotates 5 times. In the first two seconds of motion, the total number of revolutions is:

(A) 6 (B) 10 (C) 15 (D) 20 (E) 25

12.4 MOTION WITH TIME-DEPENDENT ANGULAR ACCELERATION

The equations of angular motion for the general case when the angular acceleration is a function of time are analogous to the corresponding equations of translational motion discussed in Section 2.7. Such equations are solved by integration. Integral calculus was discussed in detail in Chapter 7, and we now revisit the technique of integration of the equations of motion for the case of angular motion. To see how we can obtain kinematic solutions for nonconstant accelerations, consider the angular acceleration $\alpha = d\omega/dt$. We rearrange this relation and obtain

$$d\omega = \alpha\, dt$$

We can integrate this expression directly, for example, from the initial value of the angular velocity ω_0 at time $t = 0$, to some final value ω at time t (the integration variables are indicated by primes to distinguish them from the upper limits of integration):

$$\int_{\omega_0}^{\omega} d\omega' = \int_0^t \alpha\, dt'$$

$$\omega - \omega_0 = \int_0^t \alpha\, dt' \qquad (12.21)$$

This gives the angular velocity as a function of time:

angular velocity for time-dependent angular acceleration

$$\omega = \omega_0 + \int_0^t \alpha\, dt' \qquad (12.22)$$

Equation (12.22) enables us to calculate the angular velocity as a function of time for any angular acceleration that is a known function of time.

The angular position ϕ can be obtained in a similar manner:

$$d\phi = \omega\, dt$$

$$\int_{\phi_0}^{\phi} d\phi' = \int_0^t \omega\, dt'$$

angular position for time-dependent angular velocity

$$\phi - \phi_0 = \int_0^t \omega\, dt' \qquad (12.23)$$

In the special case of constant angular acceleration α, Eq. (12.22) gives us $\omega = \omega_0 + \alpha t$, which agrees with our previous result, Eq. (12.17). If we insert this into Eq. (12.23), we obtain $\phi - \phi_0 = \int_0^t (\omega_0 + \alpha t)dt = \omega_0 t + \frac{1}{2}\alpha t^2$, which agrees with our previous Eq. (12.18).

In the general case of a time-dependent angular acceleration α, we proceed in the same way: first, use Eq. (12.22) to find ω as a function of time, and then insert this function into Eq. (12.23) to find the angular position as a function of time, as in the following example.

EXAMPLE 7 When turned on, a motor rotates a circular saw wheel, beginning from rest, with an angular acceleration that has an initial value $\alpha_0 = 60$ radians/s^2 at $t = 0$ and decreases to zero acceleration during the interval $0 \leq t \leq 3.0$ s according to

$$\alpha = \alpha_0\left(1 - \frac{t}{3.0\,\text{s}}\right)$$

After $t = 3.0$ s, the motor maintains the wheel's angular velocity at a constant value. What is this final angular velocity? In the process of "getting up to speed," how many revolutions occur?

SOLUTION: The angular acceleration α is given as an explicit function of time. Since we are beginning from rest, the initial angular velocity is $\omega_0 = 0$, so Eq. (12.22) gives ω as a function of t:

$$\omega = \omega_0 + \int_0^t \alpha\,dt' = 0 + \int_0^t \alpha_0\left(1 - \frac{t'}{3.0\,\text{s}}\right)dt'$$

$$= \alpha_0\left(\int_0^t dt' - \frac{1}{3.0\,\text{s}}\int_0^t t'\,dt'\right) = \alpha_0\left(t'\Big|_0^t - \frac{1}{3.0\,\text{s}}\frac{t'^2}{2}\Big|_0^t\right)$$

$$= \alpha_0\left(t - \frac{t^2}{6.0\,\text{s}}\right) \tag{12.24}$$

where we have used the property that the integral of the sum is the sum of the integrals, and that $\int t^n\,dt = t^{n+1}/(n+1)$. At $t = 3.0$ s, this angular velocity reaches its final value of

$$\omega = 60\ \text{radians/s}^2 \times \left(3.0\,\text{s} - \frac{(3.0\,\text{s})^2}{6.0\,\text{s}}\right) = 90\ \text{radians/s}$$

To obtain the number of revolutions during the time of acceleration, we can calculate the change in angular position and divide by 2π. To do so, we must insert the time-dependent angular velocity obtained in Eq. (12.24) into Eq. (12.23):

$$\phi - \phi_0 = \int_0^t \omega\,dt' = \int_0^t \alpha_0\left(t' - \frac{t'^2}{6.0\,\text{s}}\right)dt'$$

$$= \alpha_0\left(\int_0^t t'\,dt' - \frac{1}{6.0\,\text{s}}\int_0^t t'^2\,dt'\right) = \alpha_0\left(\frac{t'^2}{2}\Big|_0^t - \frac{1}{6.0\,\text{s}}\frac{t'^3}{3}\Big|_0^t\right)$$

$$= \alpha_0\left(\frac{t^2}{2} - \frac{t^3}{18\,\text{s}}\right)$$

Evaluating this expression at $t = 3.0$ s, we find

$$\phi - \phi_0 = 60 \text{ radians/s}^2 \times \left(\frac{(3.0\,\text{s})^2}{2} - \frac{(3.0\text{ s})^3}{18\text{ s}} \right) = 180 \text{ radians}$$

Hence the number of revolutions during the acceleration is

$$[\text{number of revolutions}] = \frac{\phi - \phi_0}{2\pi} = \frac{180 \text{ radians}}{2\pi} = 29 \text{ revolutions} \quad (12.25)$$

As discussed in Section 2.7, similar integration techniques can be applied to determine any component of the translational velocity and the position when the time-dependent net force and, thus, the time-dependent translational acceleration are known. In Section 2.7 we also examined the case when the acceleration is a known function of the velocity; in that case, integration provides t as a function of v (and v_0), which can sometimes be inverted to find v as a function of t.

We saw in Chapters 7–9 that a conservation-of-energy approach is often the easiest way to determine the motion when the forces are known as a function of position. Now we have seen that direct integration of the equations of motion can be applied when the translational or angular acceleration is known as a function of time or of velocity.

✔ Checkup 12.4

QUESTION 1: Beginning from rest at $t = 0$, the angular velocity of a merry-go-round increases in proportion to the square root of the time t. By what factor is the angular position of the merry-go-round at $t = 4$ s greater than it was at $t = 1$ s?

QUESTION 2: A car on a circular roadway accelerates from rest beginning at $t = 0$, so that its angular acceleration increases in proportion to the time t. With what power of time does its centripetal acceleration increase?

(A) t (B) t^2 (C) t^3 (D) t^4 (E) t^5

12.5 KINETIC ENERGY OF ROTATION; MOMENT OF INERTIA

A rigid body is a system of particles, and as for any system of particles, the total kinetic energy of a rotating rigid body is simply the sum of the individual kinetic energies of all the particles (see Section 10.4). If the particles in the rigid body have masses m_1, m_2, m_3, . . . and speeds v_1, v_2, v_3, . . . , then the kinetic energy is

$$K = \tfrac{1}{2}m_1v_1^2 + \tfrac{1}{2}m_2v_2^2 + \tfrac{1}{2}m_3v_3^2 + \cdots \quad (12.26)$$

In a rigid body rotating about a given axis, all the particles move with the same angular velocity ω along circular paths. By Eq. (12.11), the speeds of the particles along their paths are proportional to their radial distances:

$$v_1 = R_1\omega, \quad v_2 = R_2\omega, \quad v_3 = R_3\omega, \quad \cdots \quad (12.27)$$

and hence the total kinetic energy is

$$K = \tfrac{1}{2}m_1R_1^2\omega^2 + \tfrac{1}{2}m_2R_2^2\omega^2 + \tfrac{1}{2}m_3R_3^2\omega^2 + \cdots$$

We can write this as

$$K = \tfrac{1}{2}I\omega^2 \qquad (12.28)$$

kinetic energy of rotation

where the quantity

$$I = m_1R_1^2 + m_2R_2^2 + m_3R_3^2 + \cdots \qquad (12.29)$$

moment of inertia

is called the **moment of inertia** of the rotating body about the given axis. The SI unit of moment of inertia is $\text{kg}\cdot\text{m}^2$.

Note that Eq. (12.28) has a mathematical form reminiscent of the familiar expression $\frac{1}{2}mv^2$ for the kinetic energy of a single particle—the moment of inertia replaces the mass, and the angular velocity replaces the translational velocity. As we will see in the next chapter, this analogy between moment of inertia and mass is of general validity. *The moment of inertia is a measure of the resistance that a body offers to changes in its rotational motion,* just as mass is a measure of the resistance that a body offers to changes in its translational motion.

Equation (12.29) shows that the moment of inertia—and consequently the kinetic energy for a given value of ω—is large if most of the mass of the body is at a large distance from the axis of rotation. This is very reasonable: for a given value of ω, particles at large distance from the axis move with high speeds, and therefore have large kinetic energies.

EXAMPLE 8 A 50-kg woman and an 80-kg man sit on a massless seesaw separated by 3.00 m (see Fig. 12.12). The seesaw rotates about a fulcrum (the point of support) placed at the center of mass of the system; the center of mass is 1.85 m from the woman and 1.15 m from the man, as obtained in Example 4 of Chapter 10. If the (instantaneous) angular velocity of the seesaw is 0.40 radian/s, calculate the kinetic energy. Treat both masses as particles.

SOLUTION: The moment of inertia for particles rotating about an axis depends only on the masses and their distances from the axis:

$$\begin{aligned} I &= m_1R_1^2 + m_2R_2^2 \\ &= 50\ \text{kg} \times (1.85\ \text{m})^2 + 80\ \text{kg} \times (1.15\ \text{m})^2 = 280\ \text{kg}\cdot\text{m}^2 \end{aligned} \qquad (12.30)$$

The kinetic energy for the rotational motion is

$$\begin{aligned} K &= \tfrac{1}{2}I\omega^2 \\ &= \tfrac{1}{2} \times 280\ \text{kg}\cdot\text{m}^2 \times (0.40\ \text{radian/s})^2 = 22\ \text{J} \end{aligned} \qquad (12.31)$$

This kinetic energy could equally well have been obtained by first calculating the individual speeds of the woman and the man ($v_1 = R_1\omega$, $v_2 = R_2\omega$) and then adding the corresponding individual kinetic energies.

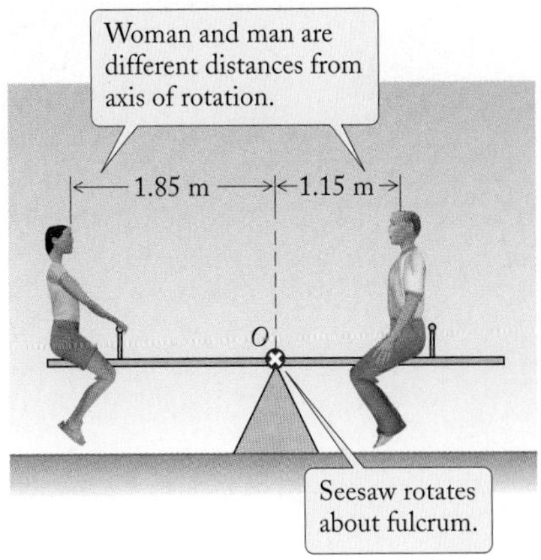

FIGURE 12.12 Woman and man on a seesaw.

If we regard the mass of a solid body as continuously distributed throughout its volume, then we can calculate the moment of inertia by the same method we used for the calculation of the center of mass: we subdivide the body into small mass elements and add the moments of inertia contributed by all these small amounts of mass. This leads to an approximation for the moment of inertia,

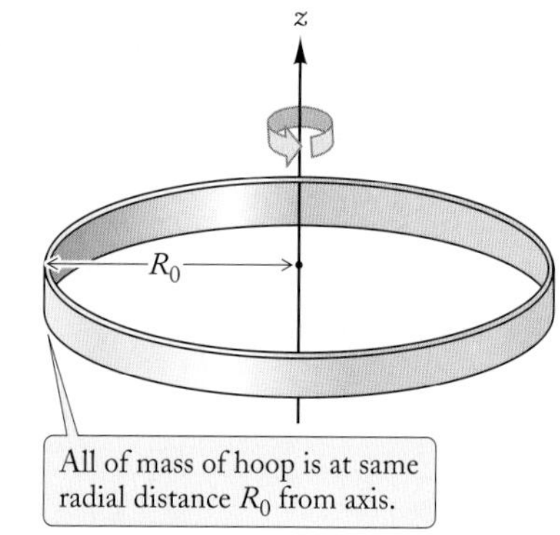

FIGURE 12.13 A thin hoop rotating about its axis of symmetry.

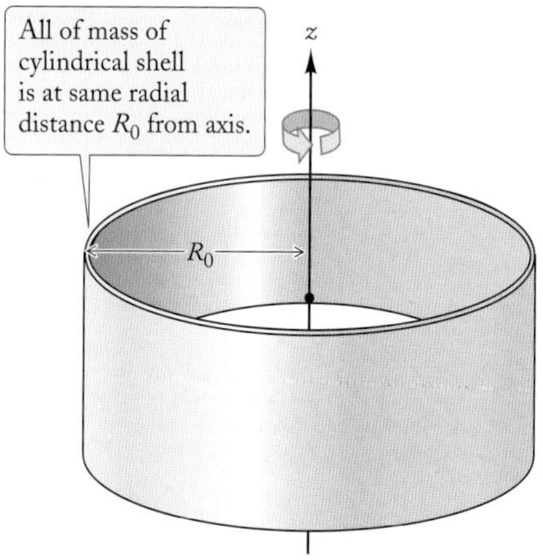

FIGURE 12.14 A thin cylindrical shell rotating about its axis of symmetry.

$$I \approx \sum_i R_i^2 \, \Delta m_i \tag{12.32}$$

where R_i is the radial distance of the mass element Δm_i from the axis of rotation. In the limit $\Delta m_i \to 0$, this approximation becomes exact, and the sum becomes an integral:

$$I = \int R^2 \, dm \tag{12.33}$$

In general, the calculation of the moment of inertia requires the evaluation of the integral (12.33). However, in a few exceptionally simple cases, it is possible to find the moment of inertia without performing this integration. For example, if the rigid body is a thin hoop (see Fig. 12.13) or a thin cylindrical shell (see Fig. 12.14) of radius R_0 rotating about its axis of symmetry, then *all* of the mass of the body is at the same distance from the axis of rotation—the moment of inertia is then simply the total mass M of the hoop or shell multiplied by its radius R_0 squared,

$$I = MR_0^2$$

If all of the mass is *not* at the same distance from the axis of rotation, then we must perform the integration (12.33); when summing the individual contributions, we usually write the small mass contribution as a mass per unit length times a small length, or as a mass per unit area times a small area, as in the following examples.

EXAMPLE 9 Find the moment of inertia of a uniform thin rod of length l and mass M rotating about an axis perpendicular to the rod and through its center.

SOLUTION: Figure 12.15 shows the rod lying along the x axis; the axis of rotation is the z axis. The rod extends from $x = -l/2$ to $x = +l/2$. Consider a small slice dx of the rod. The amount of mass within this slice is proportional to the length dx, and so is equal to the mass per unit length times this length:

$$dm = \frac{M}{l}\, dx$$

The square of the distance of the slice from the axis of rotation is $R^2 = x^2$, so Eq. (12.33) becomes

$$\begin{aligned} I &= \int R^2 \, dm = \int_{-l/2}^{+l/2} x^2 \frac{M}{l}\, dx = \frac{M}{l}\left(\frac{x^3}{3}\right)\Bigg|_{-l/2}^{+l/2} \\ &= \frac{M}{l} \times \frac{2(l/2)^3}{3} = \frac{1}{12} Ml^2 \end{aligned} \tag{12.34}$$

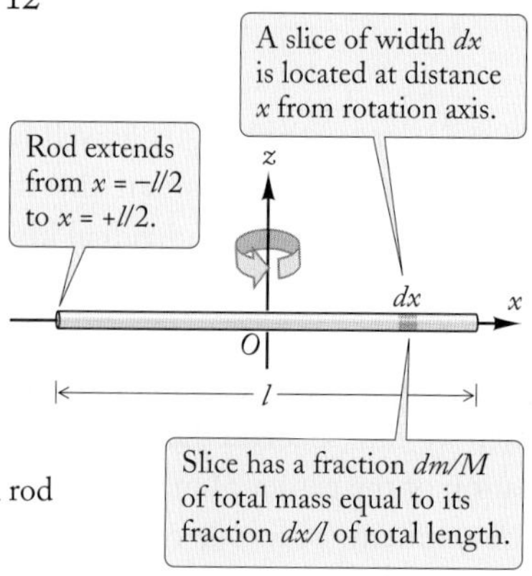

FIGURE 12.15 A thin rod rotating about its center.

EXAMPLE 10 Repeat the calculation of the preceding example for an axis of rotation through one end of the rod.

SOLUTION: Figure 12.16 shows the rod and the axis of rotation. The rod extends from $x = 0$ to $x = l$. Hence, instead of Eq. (12.34) we now obtain

$$I = \int_0^l x^2 \frac{M}{l}\, dx = \frac{M}{l}\left(\frac{x^3}{3}\right)\Big|_0^l = \frac{M}{l} \times \frac{l^3}{3} = \frac{1}{3}Ml^2 \qquad (12.35)$$

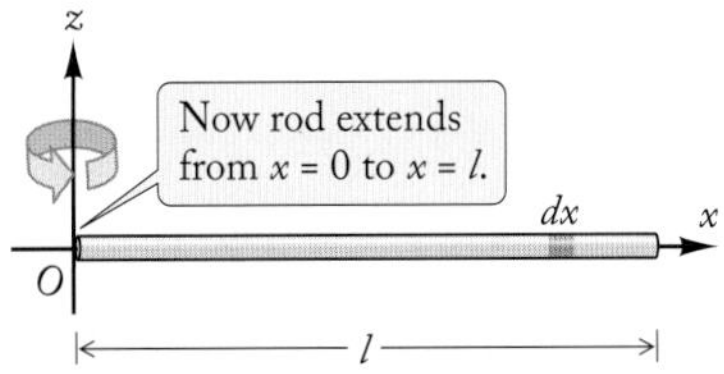

FIGURE 12.16 A thin rod rotating about its end.

EXAMPLE 11 Find the moment of inertia of a wide ring, or annulus, made of sheet metal of inner radius R_1, outer radius R_2, and mass M rotating about its axis of symmetry (see Fig. 12.17).

SOLUTION: The annulus can be regarded as made of a large number of thin concentric hoops fitting around one another. Figure 12.17 shows one such hoop, of radius R and width dR. All of the mass dm of this hoop is at the same radius R from the axis of rotation; hence the moment of inertia of the hoop is

$$dI = R^2\, dm$$

The area dA of the hoop is the product of its length (the perimeter $2\pi R$) and its width dR, so $dA = 2\pi R\, dR$. The mass dm of the hoop equals the product of this area and the mass per unit area of the sheet metal. Since the total area of the annulus is $\pi R_2^2 - \pi R_1^2$, the mass per unit area is $M/\pi(R_2^2 - R_1^2)$. The mass contributed by each hoop is the mass per unit area times its area:

$$dm = \frac{M}{\pi(R_2^2 - R_1^2)} \times 2\pi R\, dR = \frac{2M}{R_2^2 - R_1^2} R\, dR \qquad (12.36)$$

We sum the contributions dI from $R = R_1$ to $R = R_2$; hence

$$I = \int R^2\, dm = \frac{2M}{R_2^2 - R_1^2}\int_{R_1}^{R_2} R^3\, dR$$

$$= \frac{2M}{R_2^2 - R_1^2}\left(\frac{R^4}{4}\right)\Big|_{R_1}^{R_2} = \frac{M}{2(R_2^2 - R_1^2)} \times (R_2^4 - R_1^4)$$

$$= \frac{M}{2(R_2^2 - R_1^2)} \times (R_2^2 + R_1^2)(R_2^2 - R_1^2) = \frac{M}{2}(R_2^2 + R_1^2) \qquad (12.37)$$

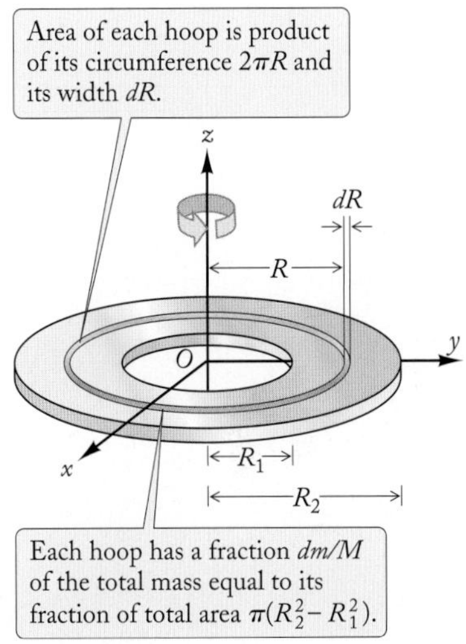

FIGURE 12.17 An annulus of sheet metal rotating about its axis of symmetry. The annulus can be regarded as made of a large number of concentric hoops. The hoop shown in the figure has radius R and width dR.

COMMENT: Note that for $R_1 = 0$, this becomes $I = MR_2^2/2$, which is the moment of inertia of a disk (see Table 12.3). And for $R_1 = R_2$, it becomes $I = MR_1^2$, which is the moment of inertia of a hoop. Note that the result (12.37) for a sheet also applies to a thick annulus or a thick cylindrical shell (rotating about the axis of symmetry).

Comparison of Eqs. (12.34) and (12.35) for the moment of inertia of a rod makes it clear that the value of the moment of inertia depends on the location of the axis of rotation. The moment of inertia is small if the axis passes through the center of mass, and large if it passes through the end of the rod. In the latter case, more of the mass of the rod is at a larger distance from the axis of rotation, which leads to a larger moment of inertia.

TABLE 12.3 SOME MOMENTS OF INERTIA

BODY		MOMENT OF INERTIA
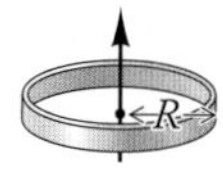	Thin hoop about symmetry axis	MR^2
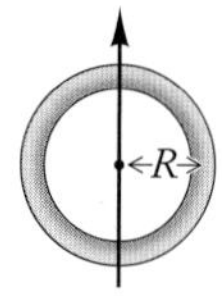	Thin hoop about diameter	$\frac{1}{2}MR^2$
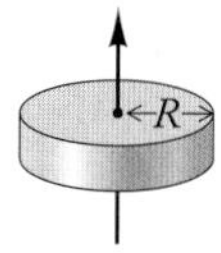	Disk or cylinder about symmetery axis	$\frac{1}{2}MR^2$
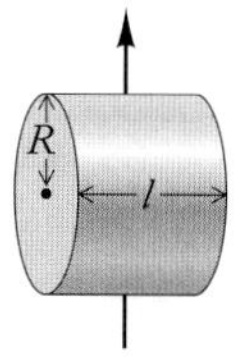	Cylinder about diameter through center	$\frac{1}{4}MR^2 + \frac{1}{12}Ml^2$
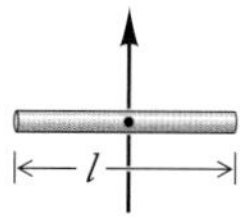	Thin rod about perpendicular axis through center	$\frac{1}{12}Ml^2$
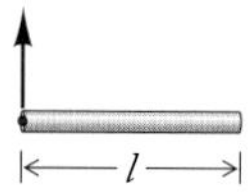	Thin rod about perpendicular axis through end	$\frac{1}{3}Ml^2$
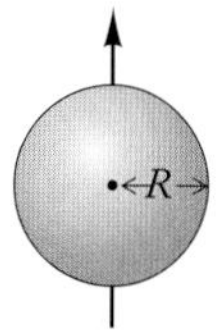	Sphere about diameter	$\frac{2}{5}MR^2$
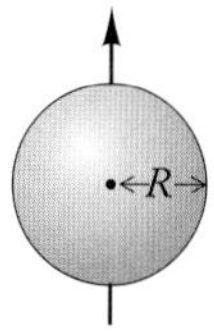	Thin spherical shell about diameter	$\frac{2}{3}MR^2$

It is possible to prove a theorem that relates the moment of inertia I_{CM} about an axis through the center of mass to the moment of inertia I about a parallel axis through some other point. This theorem, called the **parallel-axis theorem,** asserts that

parallel-axis theorem

$$I = I_{CM} + Md^2 \tag{12.38}$$

where M is the total mass of the body and d the distance between the two axes. We will not give the proof, but merely check that the theorem is consistent with our results for the moments of inertia of the rod rotating about an axis through the center $[I_{CM} = \frac{1}{12} Ml^2$; see Eq. (12.34)] and an axis through an end $[I = \frac{1}{3} Ml^2$; see Eq. (12.35)]. In this case, $d = l/2$, and the parallel-axis theorem asserts

$$\frac{1}{3}Ml^2 = \frac{1}{12}Ml^2 + M\left(\frac{l}{2}\right)^2 \tag{12.39}$$

which is identically true.

Note that it is a corollary of Eq. (12.38) that the moment of inertia about an axis passing through the center of mass is always smaller than that about any other parallel axis.

Table 12.3 lists the moments of inertia of a variety of rigid bodies about an axis through their center of mass; all the bodies are assumed to have uniform density.

EXAMPLE 12 The large centrifuge shown in the chapter photo carries the payload in a chamber in one arm and counterweights at the end of the opposite arm. The mass distribution depends on the choice of payload and the choice of counterweights. Figure 12.18 is a schematic diagram of the mass distribution attained with a particular choice of payload and counterweights. The payload arm (including the payload) has a mass of 1.8×10^3 kg uniformly distributed over a length of 8.8 m. The counterweight arm has a mass of 1.1×10^3 kg uniformly distributed over a length of 5.5 m, and it carries a counterweight of 8.6×10^3 kg at its end. (a) What is the moment of inertia of the centrifuge for this mass distribution? (b) What is the rotational kinetic energy when the centrifuge is rotating at 175 revolutions per minute?

Concepts in Context

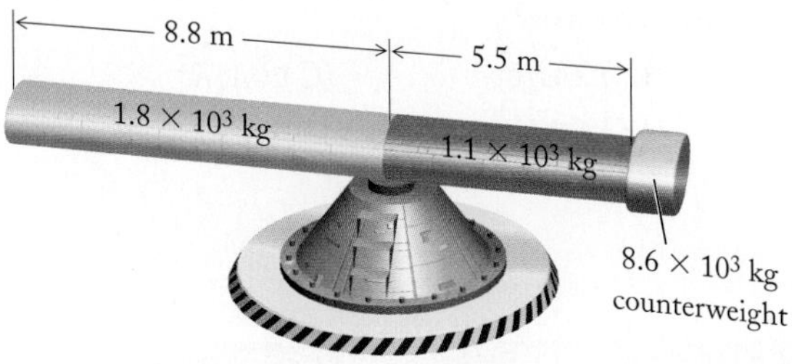

FIGURE 12.18 Centrifuge mass distribution.

SOLUTION: (a) The total moment of inertia is the sum of the moments of inertia of a rod of mass $m_1 = 1.8 \times 10^3$ kg, length $l_1 = 8.8$ m rotating about its end; a second rod of mass $m_2 = 1.1 \times 10^3$ kg, length $l_2 = 5.5$ m also rotating about its end; and a mass of $m = 8.6 \times 10^3$ kg at a radial distance of $R = 5.5$ m. The moments of inertia of the rods are given by Eq. (12.35), and the moment of inertia of the counterweight is mR^2. So the total moment of inertia is

$$\begin{aligned} I &= \tfrac{1}{3}m_1l_1^2 + \tfrac{1}{3}m_2l_2^2 + mR^2 \\ &= \tfrac{1}{3} \times 1.8 \times 10^3 \text{ kg} \times (8.8 \text{ m})^2 + \tfrac{1}{3} \times 1.1 \times 10^3 \text{ kg} \times (5.5 \text{ m})^2 \\ &\quad + 8.6 \times 10^3 \text{ kg} \times (5.5 \text{ m})^2 \\ &= 3.2 \times 10^5 \text{ kg}\cdot\text{m}^2 \end{aligned}$$

(b) At 175 revolutions per minute, the angular velocity is $\omega = 18$ radians/s (see Example 5), and the rotational kinetic energy is

$$\begin{aligned} K &= \tfrac{1}{2}I\omega^2 \\ &= \tfrac{1}{2} \times 3.2 \times 10^5 \text{ kg}\cdot\text{m}^2 \times (18 \text{ radians/s})^2 \\ &= 5.2 \times 10^7 \text{ J} \end{aligned}$$

(a)
(b)
(c)

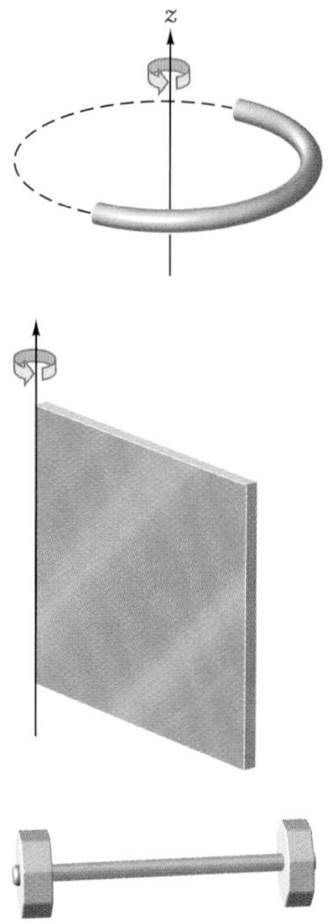

FIGURE 12.19 (a) A rod bent into an arc of a circle of radius R, rotating about its center of curvature. (b) A square plate rotating about an axis along one edge. (c) A dumbbell.

✔ Checkup 12.5

QUESTION 1: What is the moment of inertia of a rod of mass M bent into an arc of a circle of radius R when rotating about an axis through the center and perpendicular to the circle (see Fig. 12.19a)?

QUESTION 2: Consider a rod rotating about (a) an axis along the rod, (b) an axis perpendicular to the rod through its center, and (c) an axis perpendicular to the rod through its end. For which axis is the moment of inertia largest? Smallest?

QUESTION 3: What is the moment of inertia of a square plate of mass M and dimension $L \times L$ rotating about an axis along one of its edges (see Fig. 12.19b)? What is the moment of inertia if this square plate rotates about an axis through its center parallel to an edge?

QUESTION 4: A dumbbell consists of two particles of mass m each attached to the ends of a rigid, massless rod of length l (Fig. 12.19c). Assume the particles are point particles. What is the moment of inertia of this rigid body when rotating about an axis through the center and perpendicular to the rod? When rotating about a parallel axis through one end? Are these moments of inertia consistent with the parallel-axis theorem?

QUESTION 5: According to Table 12.3, the moment of inertia of a hoop about its symmetry axis is $I_{CM} = MR^2$. What is the moment of inertia if you twirl a large hoop around your finger, so that in essence it rotates about a point on the hoop, about an axis parallel to the symmetry axis?

(A) $5MR^2$ (B) $2MR^2$ (C) $\frac{3}{2}MR^2$.
(D) MR^2 (E) $\frac{1}{2}MR^2$.

SUMMARY

PROBLEM-SOLVING TECHNIQUES Angular Motion **(page 375)**

DEFINITION OF ANGLE (in radians)

$$\phi = \frac{[\text{arc length}]}{[\text{radius}]} = \frac{s}{R} \quad \textbf{(12.1)}$$

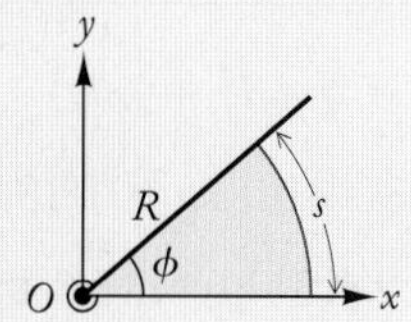

ANGLE CONVERSIONS

$$1 \text{ revolution} = 2\pi \text{ radians} = 360^\circ$$

AVERAGE ANGULAR VELOCITY

$$\bar{\omega} = \frac{\Delta\phi}{\Delta t} \quad \textbf{(12.2)}$$

INSTANTANEOUS ANGULAR VELOCITY

$$\omega = \frac{d\phi}{dt} \quad \textbf{(12.3)}$$

FREQUENCY

$$f = \frac{\omega}{2\pi} \tag{12.4}$$

PERIOD OF MOTION

$$T = \frac{1}{f} = \frac{2\pi}{\omega} \tag{12.5}$$

AVERAGE ANGULAR ACCELERATION

$$\overline{\alpha} = \frac{\Delta\omega}{\Delta t} \tag{12.6}$$

INSTANTANEOUS ANGULAR ACCELERATION

$$\alpha = \frac{d\omega}{dt} \tag{12.7}$$

SPEED OF PARTICLE ON ROTATING BODY

$$v = \omega R \tag{12.11}$$

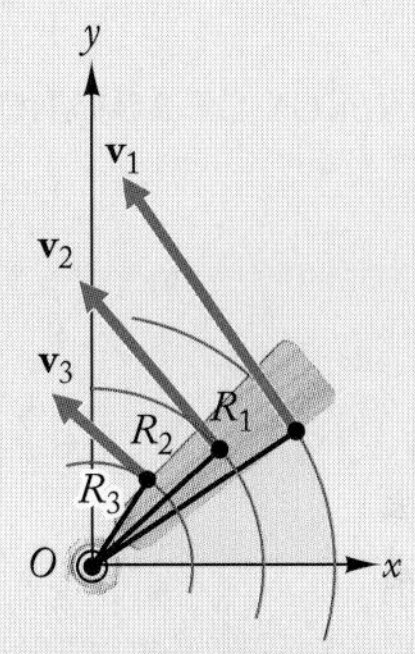

ACCELERATION OF PARTICLE ON ROTATING BODY

$$a_{\text{tangential}} = \alpha R, \tag{12.13}$$

$$a_{\text{centripetal}} = \omega^2 R \tag{12.15}$$

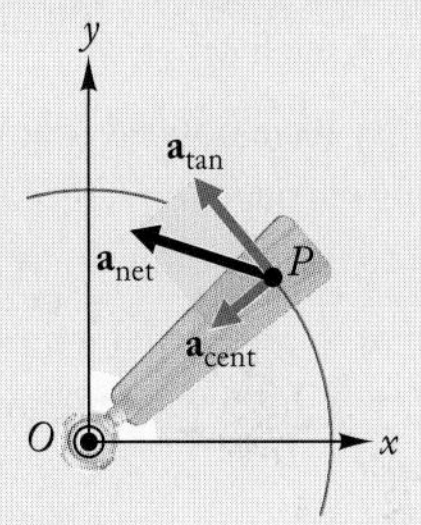

MOTION WITH CONSTANT ANGULAR ACCELERATION

$$\omega = \omega_0 + \alpha t \tag{12.17}$$

$$\phi = \phi_0 + \omega_0 t + \tfrac{1}{2}\alpha t^2 \tag{12.18}$$

$$\alpha(\phi - \phi_0) = \tfrac{1}{2}(\omega^2 - \omega_0^2) \tag{12.19}$$

MOTION WITH TIME-DEPENDENT ANGULAR ACCELERATION

$$\omega = \omega_0 + \int_0^t \alpha\, dt' \tag{12.22}$$

$$\phi = \phi_0 + \int_0^t \omega\, dt' \tag{12.23}$$

MOMENT OF INERTIA

where R_i is the radial distance of m_i from the axis of rotation.

$$I = m_1 R_1^2 + m_2 R_2^2 + m_3 R_3^3 + \cdots \tag{12.29}$$

MOMENT OF INERTIA OF RIGID BODY
(see also Table 12.3)
where R is the radial distance of the mass element dm from the axis of rotation; for uniformly distributed mass, dm is given by

$$I = \int R^2 dm$$

$$dm = \frac{M}{[\text{length}]} dx$$

$$dm = \frac{M}{[\text{area}]} dA, \qquad \textbf{(12.33)}$$

$$dm = \frac{M}{[\text{volume}]} dV$$

PARALLEL-AXIS THEOREM
where M is the total mass and d is the distance from CM axis.

$$I = I_{\text{CM}} + Md^2 \qquad \textbf{(12.38)}$$

KINETIC ENERGY OF ROTATION

$$K = \tfrac{1}{2} I\omega^2 \qquad \textbf{(12.28)}$$

QUESTIONS FOR DISCUSSION

1. A spinning flywheel in the shape of a disk suddenly shatters into many small fragments. Draw the trajectories of a few of these small fragments; assume that the fragments do not interfere with each other.
2. You may have noticed that in some old movies the wheels of moving carriages or stagecoaches seem to rotate backwards. How does this come about?
3. Relative to an inertial reference frame, what is your angular velocity right now about an axis passing through your center of mass?
4. Consider the wheel of an accelerating automobile. Draw the instantaneous acceleration vectors for a few points on the rim of the wheel.
5. The hands of a watch are small rectangles with a common axis passing through one end. The minute hand is long and thin; the hour hand is short and thicker. Assume both hands have the same mass. Which has the greater moment of inertia? Which has the greater kinetic energy and angular momentum?
6. What configuration and what axis would you choose to give your body the smallest possible moment of inertia? The greatest?
7. About what axis through the center of mass is the moment of inertia of this book largest? Smallest? (Assume the book is closed.)
8. A circular hoop made of thin wire has a radius R and mass M. About what axis perpendicular to the plane of the hoop must you rotate this hoop to obtain the minimum moment of inertia? What is the value of this minimum?
9. Automobile engines and other internal combustion engines have flywheels attached to their crankshafts. What is the purpose of these flywheels? (Hint: Each explosive combustion in one of the cylinders of such an engine gives a sudden push to the crankshaft. How would the crankshaft respond to this push if it had no flywheel?)
10. Suppose you pump a mass M of seawater into a pond on a hill at the equator. How does this change the moment of inertia of the Earth?

PROBLEMS

12.2 Rotation About A Fixed Axis†

1. The minute hand of a wall clock has a length of 20 cm. What is the angular velocity of this hand? What is the speed of the tip of this hand?

† For help, see Online Concept Tutorial 15 at www.wwnorton.com/physics

2. Quito is on the Earth's equator; New York is at latitude 41° north. What is the angular velocity of each city about the Earth's axis of rotation? What is the linear speed of each?
3. An automobile has wheels with a radius of 30 cm. What are the angular velocity (in radians per second) and the frequency (in revolutions per second) of the wheels when the automobile is traveling at 88 km/h?

4. In an experiment at the Oak Ridge Laboratory, a carbon fiber disk of 0.70 m in diameter was set spinning at 37 000 rev/min. What was the speed at the edge of this disk?

5. The rim of a phonograph record is at a distance of 15 cm from the center, and the rim of the paper label on the record is at a distance of 5 cm from the center.

   (a) When this record is rotating at $33\frac{1}{3}$ rev/min, what is the translational speed of a point on the rim of the record? The translational speed of a point on the rim of the paper label?

   (b) What are the centripetal accelerations of these points?

6. An electric drill rotates at 5000 rev/min. What is the frequency of rotation (in rev/s)? What is the time for one revolution? What is the angular velocity (in radians/s)?

7. An audio compact disk (CD) rotates at 210 rev/min when playing an outer track of radius 5.8 cm. What is the angular velocity in radians/s? What is the tangential speed of a point on the outer track? Because the CD has the same linear density of bits on each track, the drive maintains a constant tangential speed. What is the angular velocity (in radians/s) when playing an inner track of radius 2.3 cm? What is the corresponding rotational frequency (in rev/s)?

8. An automobile travels one-fourth of the way around a traffic circle in 4.5 s. The diameter of the traffic circle is 50 m. The automobile travels at constant speed. What is that speed? What is the angular velocity in radians/s?

9. When a pottery wheel motor is switched on, the wheel accelerates from rest to 90 rev/min in 5.0 s. What is its angular velocity at $t = 5.0$ s (in radians/s)? What is the linear speed of a piece of clay 10 cm from the center of the wheel at $t = 5.0$ s? What is its average angular acceleration during the acceleration?

10. A grinding wheel of radius 6.5 cm accelerates from rest to its operating speed of 3450 rev/min in 1.6 s. When up to speed, what is its angular velocity in radians/s? What is the linear speed at the edge of the wheel? What is its average angular acceleration during this 1.6 s? When turned off, it decelerates to a stop in 35 s. What is its average angular acceleration during this time?

11. When drilling holes, manufacturers stay close to a recommended linear cutting speed in order to maintain efficiency while avoiding overheating. The rotational speed of the drill thus depends on the diameter of the hole. For example, recommended linear cutting speeds are typically 20 m/min for steel and 100 m/min for aluminum. What is the corresponding rotational rate (in rev/s) when drilling a 3.0-mm-diameter hole in aluminum? When drilling a 2.5-cm-diameter hole in steel?

12. An electric blender accelerates from rest to 500 radians/s in 0.80 s. What is the average angular acceleration? What is the corresponding average tangential acceleration for a point on the tip of a blender blade a distance 3.0 cm from the axis? If this point has that tangential acceleration when the blender's angular velocity is 50 radians/s, what is the corresponding total acceleration of the point?

13. The angular position of a ceiling fan during the first two seconds after start-up is given by $\phi = C[t^2 - (t^3/4\text{ s})]$, where $C = 20/\text{s}^2$ and t is in seconds. What are the angular position, angular frequency, and angular acceleration at $t = 0$ s? At $t = 1.0$ s? At $t = 2.0$ s?

*14. An aircraft passes directly over you with a speed of 900 km/h at an altitude of 10 000 m. What is the angular velocity of the aircraft (relative to you) when directly overhead? Three minutes later?

*15. The outer edge of the grooved area of a long-playing record is at a radial distance of 14.6 cm from the center; the inner edge is at a radial distance of 6.35 cm. The record rotates at $33\frac{1}{3}$ rev/min. The needle of the pickup arm takes 25 min to play the record, and in that time interval it moves uniformly and radially from the outer edge to the inner edge. What is the radial speed of the needle? What is the speed of the outer edge relative to the needle? What is the speed of the inner edge relative to the needle?

*16. Consider the phonograph record described in Problem 15. What is the total length of the groove in which the needle travels?

12.3 Motion with Constant Angular Acceleration

17. The blade of a circular saw of diameter 20 cm accelerates uniformly from rest to 7000 rev/min in 1.2 s. What is the angular acceleration? How many revolutions will the blade have made by the time it reaches full speed?

18. A large ceiling fan has blades of radius 60 cm. When you switch this fan on, it takes 20 s to attain its final steady speed of 1.0 rev/s. Assume a constant angular acceleration.

    (a) What is the angular acceleration of the fan?

    (b) How many revolutions does it make in the first 20 s?

    (c) What is the distance covered by the tip of one blade in the first 20 s?

19. When you switch on a PC computer, the disk in the disk drive takes 5.0 s to reach its final steady speed of 7200 rev/min. What is the average angular acceleration?

20. When you turn off the motor, a phonograph turntable initially rotating at $33\frac{1}{3}$ rev/min makes 25 revolutions before it stops. Calculate the angular deceleration of this turntable; assume it is constant.

21. A large merry-go-round rotates at one revolution each 9.0 seconds. When shut off, it decelerates uniformly to a stop in 16 s. What is the angular acceleration? How many revolutions does the merry-go-round make during the deceleration?

22. A cat swipes at a spool of thread, which then rolls across the floor with an initial speed of 1.0 m/s. The spool decelerates uniformly to a stop 3.0 m from its initial position. The spool has a radius of 1.5 cm and rolls without slipping. What is the

initial angular velocity? Through what total angle does the spool rotate while slowing to a stop? What is the angular acceleration during this motion?

23. If you lift the lid of a washing machine during the rapid spin–dry cycle, the cycle stops (for safety), typically after 5.0 revolutions. If the clothes are spinning at 6.0 rev/s initially, what is their constant angular acceleration during the slowing motion? How long do they take to come to a stop?

24. A toy top initially spinning at 30 rev/s slows uniformly to a stop in 25 seconds. What is the angular acceleration during this motion? Through how many revolutions does the top turn while slowing to a stop?

*25. The rotation of the Earth is slowing down. In 1977, the Earth took 1.01 s longer to complete 365 rotations than in 1900. What was the average angular deceleration of the Earth in the time interval from 1900 to 1977?

*26. An automobile engine accelerates at a constant rate from 200 rev/min to 3000 rev/min in 7.0 s and then runs at constant speed.

(a) Find the angular velocity and the angular acceleration at $t = 0$ (just after acceleration begins) and at $t = 7.0$ s (just before acceleration ends).

(b) A flywheel with a radius of 18 cm is attached to the shaft of the engine. Calculate the tangential and the centripetal acceleration of a point on the rim of the flywheel at the times given above.

(c) What angle does the net acceleration vector make with the radius at $t = 0$ and at $t = 7.0$ s? Draw diagrams showing the wheel and the acceleration vector at these times.

12.4 Motion with Time-Dependent Angular Acceleration

27. A disk has an initial angular velocity of $\omega_0 = 8.0$ radians/s. At $t = 0$, it experiences a time-dependent angular acceleration given by $\alpha = Ct^2$, where $C = 0.25$ radian/s^4. What is the instantaneous angular velocity at $t = 3.0$ s? What is the change in angular position between $t = 0$ and $t = 1.0$ s?

28. A rigid body is initially at rest. Beginning at $t = 0$, it begins rotating, with an angular acceleration given by $\alpha = \alpha_0 \{1 - [t^2/(4\ \text{s}^2)]\}$ for $0 \leq t \leq 2.0$ s and $\alpha = 0$ thereafter. The initial value is $\alpha_0 = 20$ radians/s^2. What is the body's angular velocity after 1.0 s? After a long time? How many revolutions have occurred after 1.0 s?

*29. A sphere is initially rotating with angular velocity ω_0 in a viscous liquid. Friction causes an angular deceleration that is proportional to the instantaneous angular velocity, $\alpha = -A\omega$, where A is a constant. Show that the angular velocity as a function of time is given by

$$\omega = \omega_0 e^{-At}$$

12.5 Kinetic Energy of Rotation; Moment of Inertia

30. Find the moment of inertia of an orange of mass 300 g and diameter 9.0 cm. Treat the orange as a uniform sphere.

31. The original Ferris wheel built by George Ferris (see Fig. 12.20) had a radius of 38 m and a mass of 1.9×10^6 kg. Assume that all of the mass of the wheel was uniformly distributed along its rim. If the wheel was rotating at 0.050 rev/min, what was its kinetic energy?

FIGURE 12.20 The original Ferris wheel.

32. What is the moment of inertia of a broomstick of mass 0.50 kg, length 1.5 m, and diameter 2.5 cm about its longitudinal axis? About an axis at right angles to the broomstick, passing through its center?

33. According to spectroscopic measurements, the moment of inertia of an oxygen molecule about an axis through the center of mass and perpendicular to the line joining the atoms is 1.95×10^{-46} kg·m^2. The mass of an oxygen atom is 2.66×10^{-26} kg. What is the distance between the atoms? Treat the atoms as pointlike particles.

34. The moment of inertia of the Earth about its polar axis is $0.331 M_E R_E^2$, where M_E is the mass and R_E the equatorial radius. Why is the moment of inertia smaller than that of a sphere of uniform density? What would the radius of a sphere of uniform density have to be if its mass and moment of inertia are to coincide with those of the Earth?

35. Problem 41 in Chapter 10 gives the dimensions of a molecule of nitric acid (HNO_3). What is the moment of inertia of this molecule when rotating about the symmetry axis passing through the H, O, and N atoms? Treat the atoms as pointlike particles.

36. The water molecule has a shape shown in Fig. 12.21. The distance between the oxygen and the hydrogen atoms is d, and the angle between the hydrogen atoms is θ. From spectroscopic investigations it is known that the moment of inertia of the molecule is $1.93 \times 10^{-47}\ \text{kg}\cdot\text{m}^2$ for rotation about the axis AA' and $1.14 \times 10^{-47}\ \text{kg}\cdot\text{m}^2$ for rotation about the axis BB'. From this information and the known values of the masses of the atoms, determine the values of d and θ. Treat the atoms as pointlike.

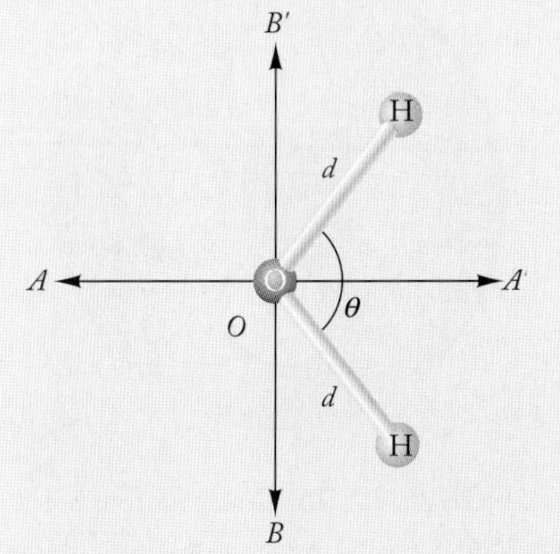

FIGURE 12.21 Atoms in a water molecule.

37. What is the moment of inertia (about the axis of symmetry) of a bicycle wheel of mass 4.0 kg, radius 0.33 m? Neglect the mass of the spokes.

38. An airplane propeller consists of three radial blades, each of length 1.2 m and mass 6.0 kg. What is the kinetic energy of this propeller when rotating at 2500 rev/min? Assume that each blade is (approximately) a uniform rod.

39. Estimate the moment of inertia of a human body spinning rigidly about its longitudinal axis. Treat the body as a uniform cylinder of mass 70 kg, length 1.7 m, and average diameter 23 cm.

40. Use the parallel-axis theorem to determine the moment of inertia of a solid disk or cylinder of mass M and radius R rotating about an axis parallel to its symmetry axis but tangent to its surface.

41. The moment of inertia of the Earth is approximately $0.331 M_E R_E^2$ (see also Problem 34). Calculate the rotational kinetic energy of the Earth.

42. Assume that a potter's kickwheel is a disk of radius 60 cm and mass 120 kg. What is its moment of inertia? What is its rotational kinetic energy when revolving at 2.0 rev/s?

43. A flywheel energy-storage system designed for the International Space Station has a maximum rotational rate of 53 000 rev/min. The cylindrical flywheel has a mass of 75 kg and a radius of 16 cm. For simplicity, assume the cylinder is solid and uniform. What is the moment of inertia of the flywheel? What is the maximum rotational kinetic energy stored in the flywheel?

*44. An empty beer can has a mass of 15 g, a length of 12 cm, and a radius of 3.3 cm. Find the moment of inertia of the can about its axis of symmetry. Assume that the can is a perfect cylinder of sheet metal with no ridges, indentations, or holes.

*45. Suppose that a supertanker transports 4.4×10^8 kg of oil from a storage tank in Venezuela (latitude 10° north) to a storage tank in Holland (latitude 53° north). What is the change of the moment of inertia of the Earth–oil system?

*46. A dumbbell consists of two uniform spheres of mass M and radius R joined by a thin rod of mass m. The distance between the centers of the sphers is l (Fig. 12.22). What is the moment of inertia of this device about an axis through the center of the rod perpendicular to the rod? About an axis along the rod?

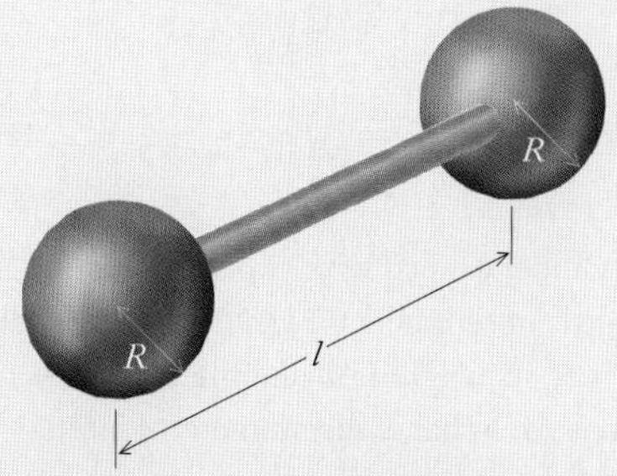

FIGURE 12.22 A dumbbell.

*47. Suppose that the Earth consists of a spherical core of mass $0.22M_E$ and radius $0.54R_E$ and a surrounding mantle (a spherical shell) of mass $0.78M_E$ and outer radius R_E. Suppose that the core is of uniform density and the mantle is also of uniform density. According to this simple model, what is the moment of inertia of the Earth? Express your answer as a multiple of $M_E R_E^2$.

*48. In order to increase her moment of inertia about a vertical axis, a spinning figure skater stretches her arms out horizontally; in order to reduce her moment of inertia, she brings her arms down vertically along her sides. Calculate the change of moment of inertia between these two configurations of the arms. Assume that each arm is a thin, uniform rod of length 0.60 m and mass 2.8 kg hinged at the shoulder a distance of 0.20 m from the axis of rotation.

*49. Find the moment of inertia of a thin rod of mass M and length L about an axis through the center inclined at an angle θ with respect to the rod.

*50. Given that the moment of inertia of a sphere about a diameter is $\frac{2}{5}MR^2$, show that the moment of inertia about an axis tangent to the surface is $\frac{7}{5}MR^2$.

*51. Find a formula for the moment of inertia of a uniform thin square plate (mass m, dimension $l \times l$) rotating about an axis that coincides with one of its edges.

*52. A conical shell has mass M, height h, and base radius R. Assume it is made from a thin sheet of uniform thickness. What is its moment of inertia about its symmetry axis?

*53. Suppose a peach of radius R and mass M consists of a spherical pit of radius $0.50R$ and mass $0.050M$ surrounded by a spherical shell of fruit of mass $0.95M$. What is the moment of inertia of the peach?

*54. Find the moment of inertia of the flywheel shown in Fig. 12.23 rotating about its axis. The flywheel is made of material of uniform thickness; its mass is M.

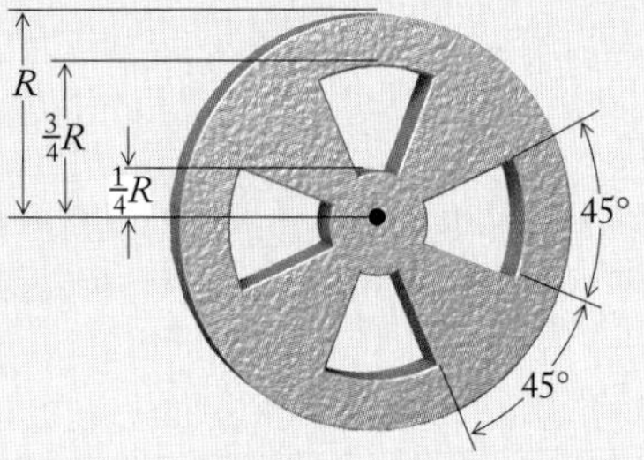

FIGURE 12.23 A flywheel.

*55. A solid cylinder capped with two solid hemispheres rotates about its axis of symmetry (Fig. 12.24). The radius of the cylinder is R, its height is h, and the total mass (hemispheres included) is M. What is the moment of inertia?

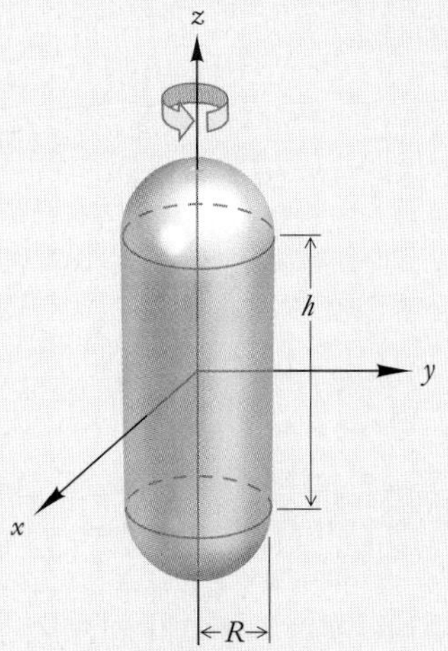

FIGURE 12.24 A solid cylinder capped with two solid hemispheres.

*56. A hole of radius r has been drilled in a circular, flat plate of radius R (Fig. 12.25). The center of the hole is at a distance d from the center of the circle. The mass of this body is M. Find the moment of inertia for rotation about an axis through the center of the circle, perpendicular to the plate.

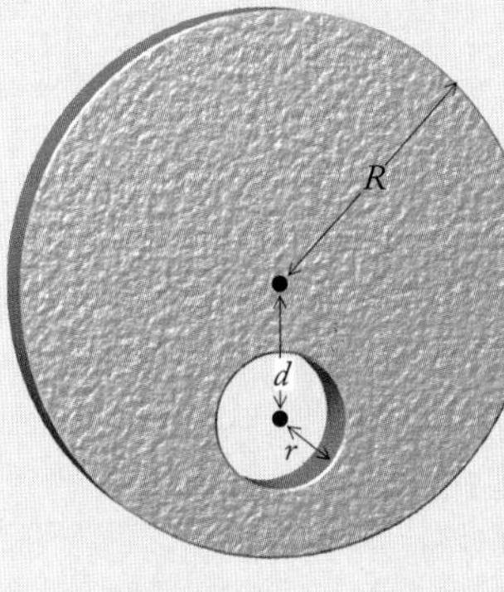

FIGURE 12.25 Circular plate with a hole.

*57. Derive a formula for the moment of inertia of a uniform spherical shell of mass M, inner radius R_1, outer radius R_2, rotating about a diameter.

*58. Find the moment of inertia of a flywheel of mass M made by cutting four large holes of radius r out of a uniform disk of radius R (Fig. 12.26). The holes are centered at a distance $R/2$ from the center of the flywheel.

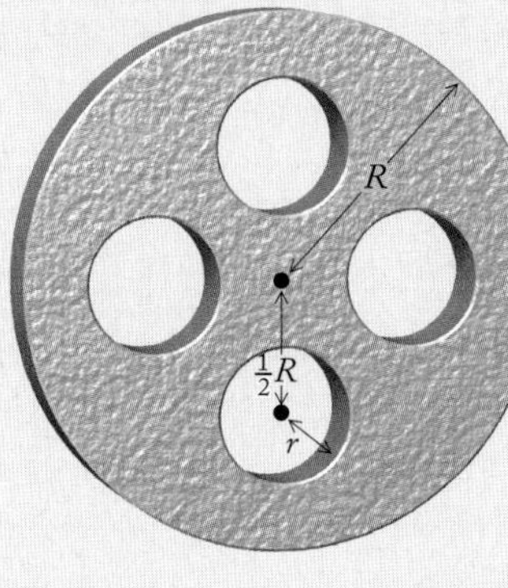

FIGURE 12.26 Disk with four holes.

*59. Show that the moment of inertia of a long, very thin cone (Fig. 12.27) about an axis through the apex and perpendicular to the centerline is $\frac{3}{5}Ml^2$, where M is the mass and l the height of the cone.

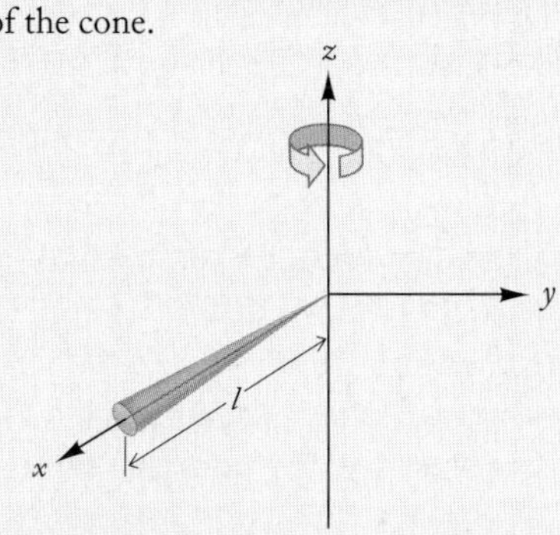

FIGURE 12.27 A long, thin cone rotating about its apex.

*60. The mass distribution within the Earth can be roughly approximated by several concentric spherical shells, each of constant density. The following table gives the outer and the inner radius of each shell and its mass (expressed as a fraction of the Earth's mass):

SHELL	OUTER RADIUS	INNER RADIUS	FRACTION OF MASS
1	6400 km	5400 km	0.28
2	5400	4400	0.25
3	4400	3400	0.16
4	3400	2400	0.20
5	2400	0	0.11

Use these data to calculate the moment of inertia of the Earth about its axis.

61. The drilling pipe of an oil rig is 2.0 km long and 15 cm in diameter, and it has a mass of 20 kg per meter of length. Assume that the wall of the pipe is very thin.

(a) What is the moment of inertia of this pipe rotating about its longitudinal axis?

(b) What is the kinetic energy when rotating at 1.0 rev/s?

62. Engineers have proposed that large flywheels be used for the temporary storage of surplus energy generated by electric power plants. A suitable flywheel would have a diameter of 3.6 m and a mass of 300 metric tons and would spin at 3000 rev/min. What is the kinetic energy of rotation of this flywheel? Give the answer in both joules and kilowatt-hours. Assume that the moment of inertia of the flywheel is that of a uniform disk.

63. An automobile of mass 1360 kg has wheels 76.2 cm in diameter of mass 27.2 kg each. Taking into account the rotational kinetic energy of the wheels about their axles, what is the total kinetic energy of the automobile when traveling at 80.0 km/h? What percentage of the kinetic energy belongs to the rotational motion of the wheels about their axles? Pretend that each wheel has a mass distribution equivalent to that of a uniform disk.

*64. The Oerlikon Electrogyro bus uses a flywheel to store energy for propelling the bus. At each bus stop, the bus is briefly connected to an electric power line, so that an electric motor on the bus can spin up the flywheel to 3000 rev/min. If the flywheel is a disk of radius 0.60 m and mass 1500 kg, and if the bus requires an average of 40 hp for propulsion at an average speed of 20 km/h, how far can it move with the energy stored in the rotating flywheel?

*65. Pulsars are rotating stars made almost entirely of neutrons closely packed together. The rate of rotation of most pulsars gradually decreases because rotational kinetic energy is gradually converted into other forms of energy by a variety of complicated "frictional" processes. Suppose that a pulsar of mass 1.5×10^{30} kg and radius 20 km is spinning at the rate of 2.1 rev/s and is slowing down at the rate of 1.0×10^{-15} rev/s^2. What is the rate (in joules per second, or watts) at which the rotational energy is decreasing? If this rate of decrease of the energy remains constant, how long will it take the pulsar to come to a stop? Treat the pulsar as a sphere of uniform density.

66. For the sake of directional stability, the bullet fired from a rifle is given a spin angular velocity about its axis by means of spiral grooves ("rifling") cut into the barrel. The bullet fired by a Lee–Enfield rifle is (approximately) a uniform cylinder of length 3.18 cm, diameter 0.790 cm, and mass 13.9 g. The bullet emerges form the muzzle with a translational velocity of 628 m/s and a spin angular velocity of 2.47×10^3 rev/s. What is the translational kinetic energy of the bullet? What is the rotational kinetic energy? What fraction of the total kinetic energy is rotational?

*67. Find a formula for the moment of inertia of a thin disk of mass M and radius R rotating about a diameter.

*68. Derive the formula for the moment of inertia of a thin hoop of mass M and radius R rotating about a diameter.

*69. Find a formula for the moment of inertia of a uniform thin square plate (mass M, dimension $l \times l$) rotating about an axis through the center and perpendicular to the plate.

*70. Find the moment of inertia of a uniform cube of mass M and edge l. Assume the axis of rotation passes through the center of the cube and is perpendicular to two of the faces.

*71. What is the moment of inertia of a thin, flat plate in the shape of a semicircle rotating about the straight side (Fig. 12.28)? The mass of the plate is M and the radius is R.

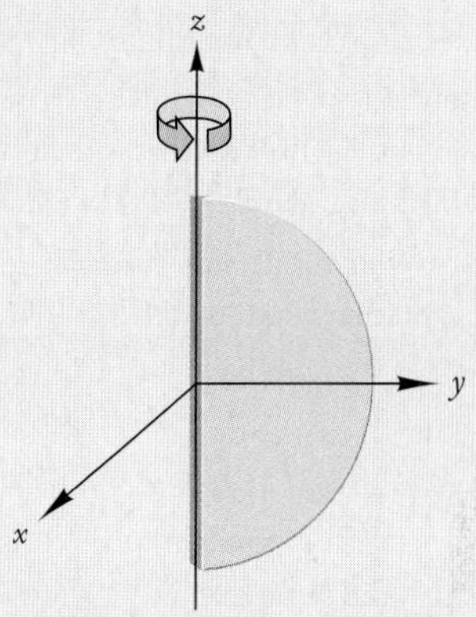

FIGURE 12.28 A semicircle rotating about its straight edge.

**72. Find the moment of inertia of the thin disk with two semicircular cutouts shown in Fig. 12.29 rotating about its axis. The disk is made of material of uniform thickness; its mass is M.

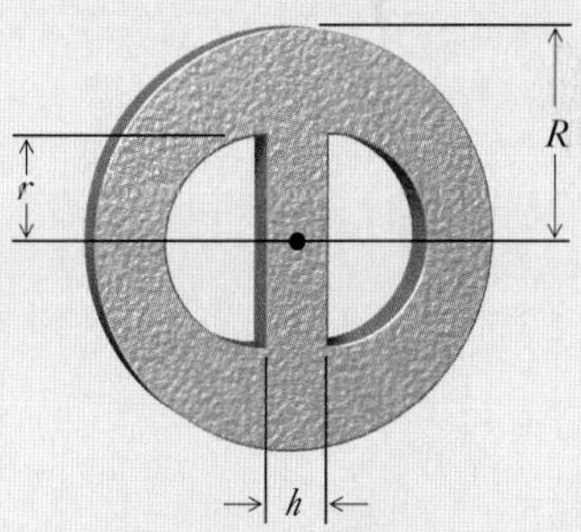

FIGURE 12.29 Disk with two semicircular cutouts.

**73. A cone of mass M has a height h and a base diameter R. Find its moment of inertia about its axis of symmetry.

**74. Derive the formula given in Table 12.3 for the moment of inertia of a sphere.

REVIEW PROBLEMS

75. An automobile has wheels of diameter 0.63 m. If the automobile is traveling at 80 km/h, what is the instantaneous velocity vector (relative to the ground) of a point at the top of the wheel? At the bottom? At the front?

76. The propeller of an airplane is turning at 2500 rev/min while the airplane is cruising at 200 km/h. The blades of the propeller are 1.5 m long. Taking into account both the rotational motion of the propeller and the translational motion of the aircraft, what is the velocity (magnitude and direction) of the tip of the propeller?

77. An automobile accelerates uniformly from 0 to 80 km/h in 6.0 s. The automobile has wheels of radius 30 cm. What is the angular acceleration of the wheels? What is their final angular velocity? How many turns do they make during the 6.0-s interval?

78. The minute hand of a wall clock is a rod of mass 5.0 g and length 15 cm rotating about one end. What is the rotational kinetic energy of the minute hand?

79. What is the kinetic energy of rotation of a phonograph record of mass 170 g and radius 15.2 cm rotating at $33\frac{1}{3}$ revolutions per minute? To give this phonograph record a translational kinetic energy of the same magnitude, how fast would you have to throw it?

80. The wheel of a wagon consists of a rim of mass 20 kg and eight spokes in the shape of rods of length 0.50 m and mass 0.80 kg each.

    (a) What is the moment of inertia of this wheel about its axle?

    (b) What is the kinetic energy of this wheel when rotating at 1.0 rev/s?

*81. A solid body consists of two uniform solid spheres of mass M and radius R welded together where they touch (see Fig. 12.30). What is the moment of inertia of this rigid body about the longitudinal axis through the center of the spheres? About the transverse axis through the point of contact?

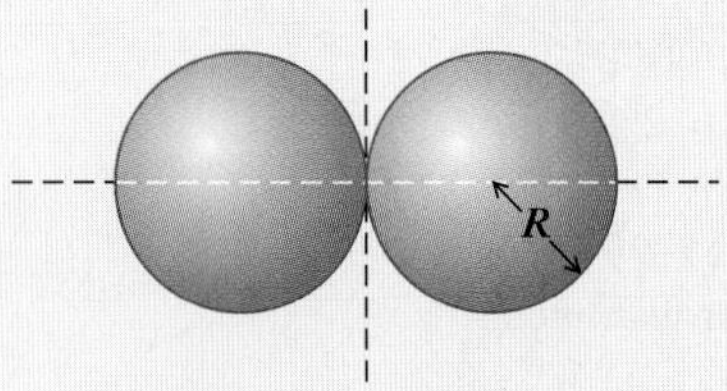

FIGURE 12.30 Two connected solid spheres.

*82. A .22-caliber bullet is a solid cylinder of length 7.0 mm and radius 2.7 mm capped at its front with a hemisphere of the same radius. The mass of the bullet is 15 g.

    (a) What is the moment of inertia of this bullet when rotating about its axis of symmetry?

    (b) What is the rotational kinetic energy of the bullet when rotating at 1.2×10^3 rev/s?

*83. Find the moment of inertia of the wheel shown in Fig. 12.31 rotating about its axis. The wheel is made of material of uniform thickness, its mass is M, and its radius is R. Treat the spokes as thin rods of length $R/2$ and width $R/12$.

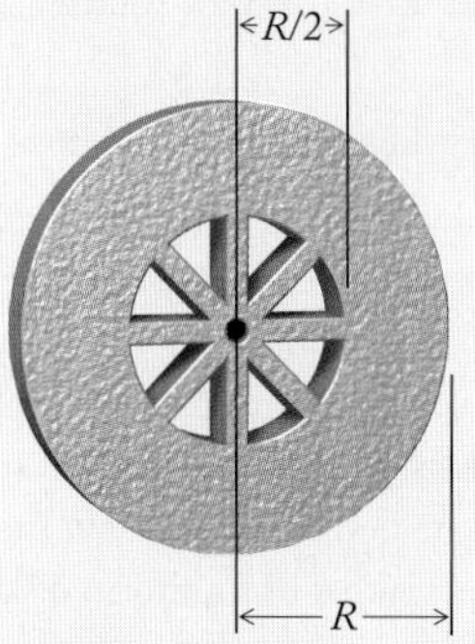

FIGURE 12.31 A wheel.

*84. The total kinetic energy of a rolling body is the sum of its translational kinetic energy $\frac{1}{2}Mv^2$ and its rotational kinetic energy $\frac{1}{2}I\omega^2$. Suppose that a cylinder, a sphere, and a pipe (a cylindrical shell) of equal masses 2.0 kg are rolling with equal speeds of 1.0 m/s. What is the total kinetic energy of each?

*85. A uniform solid cylinder is initially at rest at the top of a ramp of height 1.5 m. If the cylinder rolls down this ramp without slipping, what will be its speed at the bottom? (Hint: Use energy conservation. The kinetic energy of the cylinder at the bottom of the ramp is the sum of its translational kinetic energy $\frac{1}{2}Mv^2$ and its rotational kinetic energy $\frac{1}{2}I\omega^2$.)

**86. An airplane propeller (Fig. 12.32) is rotating at 3000 rev/min when one of the blades breaks off at the hub. Treat the blade as a rod, of length 1.2 m. The blade is horizontal and swinging upward at the instant it breaks.

    (a) What is the velocity (magnitude and direction) of the motion of the center of mass of the blade immediately after this instant?

    (b) What is the angular velocity of the rotational motion of the blade about its own center of mass?

    (c) Suppose that this happens while the aircraft is on the ground, with the hub of the propeller 2.4 m above the ground. How high above the ground does the center of mass of the broken propeller blade rise? Neglect air resistance.

FIGURE 12.32 An airplane propeller.

Answers to Checkups

Checkup 12.1

1. The swinging door executes only rotational motion about its (fixed) hinges. The motions of the wheel of a train and of the propeller of an airplane involve both rotational and translational motion; the wheel and propeller rotate as the vehicle moves through space.
2. Yes, the motion is describable as rotation about an axis and simultaneous translational motion. The rotational motion is rotation about an axis through the end of the hammer; the translational motion, however, is not along a parabolic path, but involves more complicated looping motion (see Fig. 12.1).
3. An automobile exhibits roll motion when driving on a banked surface; the auto is then tilted. Pitch motion can occur during sudden braking, when the front of the auto dives downward. Turning motion occurs whenever the auto is being driven around a curve (compare Fig. 12.2).
4. (D) Swinging door. The axis of rotation is through the hinges, along the edge of the door.

Checkup 12.2

1. The point P has the larger instantaneous speed (it travels through a greater distance per unit time). Both points have the same instantaneous angular velocity ω and the same angular acceleration α (as do all points on the same rigid body). Hence the point P has the larger tangential acceleration ($a_{\text{tangential}} = \alpha R$) and also the larger centripetal acceleration ($a_{\text{centripetal}} = \omega^2 R$).
2. The radius R for circular motion is the perpendicular distance from the axis of rotation, and so is equal to the Earth's radius only at the equator, and is increasingly smaller as one moves toward the poles; at a pole, R is zero. All points have the same angular velocity ω, as for any rigid body. The velocity is not the same for all points; since $v = \omega R$, v is largest at the equator. All points do not have the same centripetal acceleration; since $a_{\text{centripetal}} = \omega^2 R$, the centripetal acceleration is largest at the equator.
3. There is a centripetal acceleration; at the top of the arc, this is directed downward ($a_{\text{centripetal}} = v^2/R$). There is no tangential acceleration at the top (no forces act in this direction). Some distance beyond the highest point, there will be both a centripetal acceleration (since the car still moves along an arc) and a tangential acceleration (since now a component of the gravitational force is tangent to the path).
4. (D) Handle end; handle end. Since the rotation is about an axis through the center of mass (near the hammer head), the end of the handle is furthest from the axis. Thus both the speed $v = \omega R$ and the centripetal acceleration $a_{\text{centripetal}} = \omega^2 R$ are largest at the end of the handle, since R is largest there (and ω is a constant for all points on a rigid body).

Checkup 12.3

1. The centripetal acceleration always points toward the center of curvature of the circular arc of the problem; here, this is vertically down. The tangential acceleration points perpendicular to a radius at any point; since the elevator accelerates upward, the tangential acceleration at the top of the wheel points horizontally toward the left.
2. Yes to both. As long as there is no slipping, we have $\omega = v/R$ and $\alpha = a/R$, so the behavior of an angular quantity is the same as the corresponding translational quantity.
3. (D) 20. For constant acceleration and starting from rest, the angular position is $\phi = \frac{1}{2}\alpha t^2$. Since this is proportional to t^2, the angular position will be four times greater in twice the time. Thus the total number of revolutions in the first two seconds is $4 \times 5 = 20$.

Checkup 12.4

1. Since the angular velocity is proportional to $t^{1/2}$, the angular position, which is the integral of the angular velocity over time [Eq. (12.23)], will be proportional to $t^{1/2+1} = t^{3/2}$. Thus the angular position will be $4^{3/2} = 8$ times as large at $t = 4$ s as it was at $t = 1$ s.
2. (D) t^4. If the angular acceleration α increases in proportion to the time t, then the angular velocity $\omega = \int \alpha\, dt$ increases in proportion to t^2. The centripetal acceleration is given by $a_{\text{centripetal}} = v^2/R = \omega^2 R$, and so increases in proportion to the fourth power of the time.

Checkup 12.5

1. Since all of the mass M is at the same distance from the axis of rotation, the moment of inertia is simply $I = MR^2$.
2. Rotation about an axis perpendicular to the rod through its end gives the largest moment of inertia, since more mass is located at a greater distance from the axis of rotation. Rotation about an axis along the rod must give the smallest moment of inertia, since in this case all of the mass is very close to the axis.
3. About an axis along one edge or through its center parallel to one edge, the distribution of mass (relative to the axis of rotation) in each case is the same as for the corresponding rod (imagine viewing Fig. 12.19b from above, that is, along the axis of rotation). Thus the moment of inertia of the square about an axis along one edge is $I = \frac{1}{3}ML^2$; about an axis through its center parallel to one edge, it is $\frac{1}{12}ML^2$.
4. About an axis through the center, each particle is a distance $l/2$ from the axis, and so the moment of inertia is $I = m(l/2)^2 + m(l/2)^2 = \frac{1}{2}ml^2$. About an axis through one particle, one particle is a distance l from the axis and the other is at zero distance, so $I = ml^2 + 0 = ml^2$. Since we have shifted the axis by $d = l/2$ in the second case, we indeed have $I = I_{\text{CM}} + Md^2 = \frac{1}{2}ml^2 + (2m)(l/2)^2 = ml^2$, so the parallel-axis theorem is satisfied (notice we must use the total mass $M = 2m$).
5. (B) $2MR^2$. Since the axis is shifted by a distance $d = R$, the parallel-axis theorem gives $I = I_{\text{CM}} + Md^2 = MR^2 + MR^2 = 2MR^2$ for rotation about a point on the hoop.

CHAPTER

13 Dynamics of a Rigid Body

CONCEPTS IN CONTEXT

Concepts — in — Context

The *Gravity Probe B* satellite, containing four high-precision gyroscopes, was recently placed in orbit by a rocket. These gyroscopes are used for a delicate test of Einstein's theory of General Relativity. The rotor of one of these gyroscopes is shown here. It consists of a nearly perfect sphere of quartz, 3.8 cm in diameter, suspended electrically and spinning at 10 000 revolutions per minute.

Some of the questions we can address with the concepts developed in this chapter are:

? When initially placed in orbit, the rotor is at rest. What torque and what force are needed to spin up this gyroscope with a given angular acceleration? (Example 4, page 401)

? A rotating body, such as this rotor, has not only kinetic energy, but also an angular momentum, which is the rotational analog of the linear momentum introduced in Chapter 10. How is the angular momentum of the gyroscope expressed in terms of its angular velocity? (Example 8, page 406)

? The gyroscope is used like a compass, to establish a reference direction in space. How does a gyroscope maintain a fixed reference direction? (Physics in Practice: The Gyrocompass, page 414)

As we saw in Chapter 5, Newton's Second Law is the equation that determines the translational motion of a body. In this chapter, we will derive an equation that determines the rotational motion of a rigid body. Just as Newton's equation of motion gives us the translational acceleration and permits us to calculate the change in velocity and position, the analogous equation for rotational motion gives us the angular acceleration and permits us to calculate the change in angular velocity and angular position. The equation for rotational motion is not a new law of physics, distinct from Newton's three laws. Rather, it is a consequence of these laws.

13.1 WORK, ENERGY, AND POWER IN ROTATIONAL MOTION; TORQUE

We begin with a calculation of the work done by an external force on a rigid body constrained to rotate about a fixed axis. Figure 13.1 shows the body, with the axis of rotation perpendicular to the page. The force is applied at some point of the body at a distance R from the axis of rotation. For a start, we will assume that the force has no component parallel to the axis; any such component is of no interest in the present context since the body does not move in the direction parallel to the axis, and so a force parallel to the axis can do no work. In Fig. 13.1, the force is shown entirely in the plane of the page. The work done by this force during a small displacement of the point at which the force acts is the product of the force F, the displacement ds, and the cosine of the angle between the force and the displacement [see Eq. (7.5)]. The cosine of this angle is equal to the sine of the angle θ between the force and the radial line (see Fig. 13.1). Hence, we can write the work as

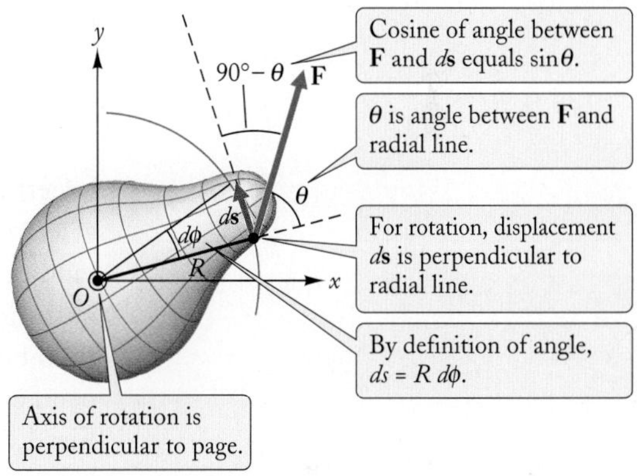

FIGURE 13.1 Force applied to a rigid body rotating about a fixed axis. As in Chapter 12, the axis of rotation is indicated by a circled dot. The force makes an angle θ with the radial line and an angle $90° - \theta$ with the instantaneous displacement $d\mathbf{s}$.

$$dW = F\,ds\,\sin\theta$$

If the body rotates through a small angle $d\phi$, the displacement is $ds = R\,d\phi$, and therefore

$$dW = FR\,d\phi\,\sin\theta \qquad (13.1)$$

The product $FR\sin\theta$ is called the **torque** of the force F, usually designated by the symbol τ (the Greek letter *tau*):

$$\tau = FR\sin\theta \qquad (13.2)$$

torque

With this notation, the work done by the force, or *the work done by the torque*, is simply

$$dW = \tau\,d\phi \qquad (13.3)$$

work done by torque

This is the rotational analog of the familiar equation $dW = F\,dx$ for work done in translational motion. The torque τ is analogous to the force F, and the angular displacement $d\phi$ is analogous to the translational displacement dx. The analogy between torque and force extends beyond the equation for the work. As we will see in the next section, a torque applied to a rigid body causes angular acceleration, just as a force applied to a particle causes translational acceleration.

According to Eq. (13.3), each contribution to the work is the product of the torque τ and the small angular displacement $d\phi$. Thus the total work done in rotating a body from an initial angle ϕ_1 to a final angle ϕ_2 is

$$W = \int dW = \int_{\phi_1}^{\phi_2} \tau \, d\phi \qquad (13.4)$$

In the special case of a constant torque, the torque may be brought outside the integral to obtain

$$W = \tau \int_{\phi_1}^{\phi_2} d\phi = \tau(\phi_2 - \phi_1)$$

or simply

work done by constant torque

$$W = \tau \, \Delta\phi \qquad (\text{for } \tau = \text{constant}) \qquad (13.5)$$

where $\Delta\phi = \phi_2 - \phi_1$ is the change in angular position during the time that the torque is applied. Equation (13.5) is analogous to the equation for the work done by a constant force on a body in one-dimensional translational motion, $W = F\,\Delta x$.

From Eq. (13.2), we see that the unit of torque is the unit of force multiplied by the unit of distance; *this SI unit of torque is the newton-meter* (N·m).

Note that according to Eq. (13.2), for a force of given magnitude, the torque is largest if the force acts at right angles to the radial line ($\theta = 90°$) and if the force acts at a large distance from the axis of rotation (large R). This dependence of the torque (and of the work) on the distance from the axis of rotation and on the angle of the push agrees with our everyday experience in pushing doors open or shut. A door is a rigid body, which rotates about a vertical axis through the hinges. If you push perpendicularly against the door, near the edge farthest from the hinge (largest R; see Fig. 13.2a), you produce a large torque, which does work on the door, increases its kinetic energy, and swings the door quickly on its hinges. If you push at a point near the hinge (small R; see Fig. 13.2b), the door responds more sluggishly. You produce a smaller torque, and you have to push harder to do the same amount of work and attain the same amount of kinetic energy and the same final angular velocity. Finally, if you push in a direction that is not perpendicular to the door (small θ; see Fig. 13.2c), the door again responds sluggishly, because the torque is small.

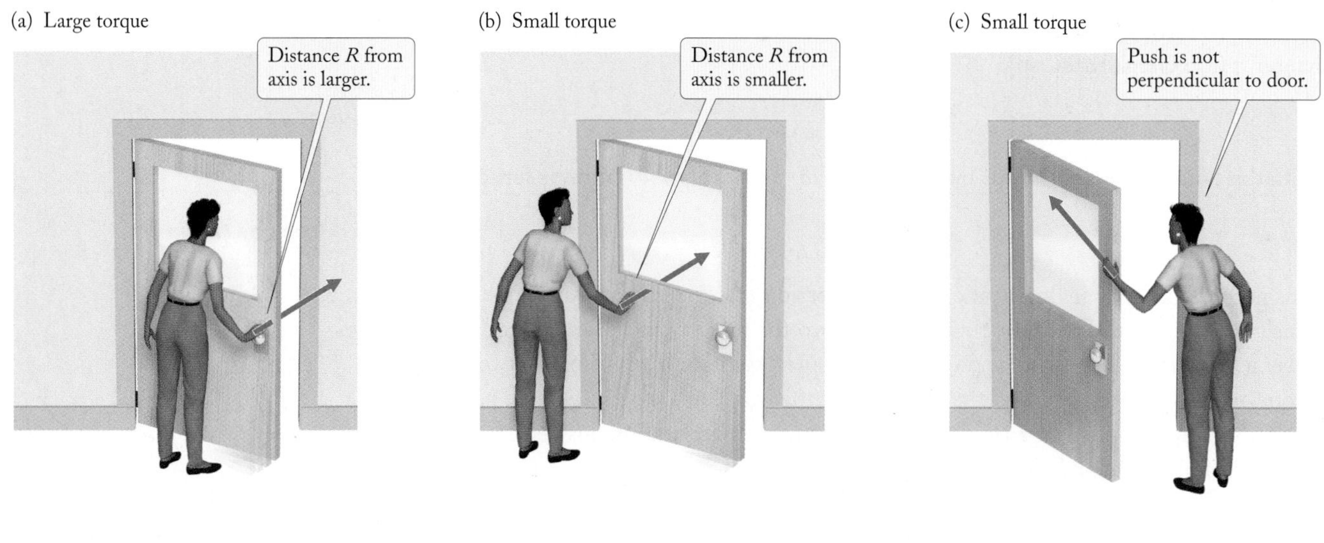

FIGURE 13.2 (a) A push against the door far from the hinge produces a large angular acceleration. (b) The same push near the hinge produces a small angular acceleration. (c) A push against the door at a small angle also produces a small angular acceleration.

EXAMPLE 1 Suppose that while opening a 1.0-m-wide door, you push against the edge farthest from the hinge, applying a force with a steady magnitude of 0.90 N at right angles to the surface of the door. How much work do you do on the door during an angular displacement of 30°?

SOLUTION: For a constant torque, the work is given by Eq. (13.5), $W = \tau\,\Delta\phi$. The definition of torque, Eq. (13.2), with $F = 0.90$ N, $R = 1.0$ m, and $\theta = 90°$, gives

$$\tau = FR \sin 90° = 0.90 \text{ N} \times 1.0 \text{ m} \times 1 = 0.90 \text{ N·m}$$

To evaluate the work, the angular displacement must be expressed in radians; $\Delta\phi = 30° \times (2\pi \text{ radians}/360°) = 0.52$ radian. Then

$$W = \tau\,\Delta\phi = 0.90 \text{ N·m} \times 0.52 \text{ radian}$$
$$= 0.47 \text{ J}$$

The equation for the power in rotational motion and the equations that express the work–energy theorem and the conservation law for energy in rotational motion are analogous to the equations we formulated for translational motion in Chapters 7 and 8. If we divide both sides of Eq. (13.3) by dt, we find *the instantaneous power delivered by the torque:*

$$P = \frac{dW}{dt} = \tau\frac{d\phi}{dt}$$

or

$$P = \tau\omega \qquad (13.6)$$

power delivered by torque

where $\omega = d\phi/dt$ is the angular velocity. Obviously, this equation is analogous to the equation $P = Fv$ obtained in Section 8.5 for the power in one-dimensional translational motion.

The work done by the torque changes the rotational kinetic energy of the body. Like the work–energy theorem for translational motion, the work–energy theorem for rotational motion says that the work done on the body by the external torque equals the change in rotational kinetic energy (the internal forces and torques in a rigid body do no net work):

$$W = K_2 - K_1 = \tfrac{1}{2}I\omega_2^2 - \tfrac{1}{2}I\omega_1^2 \qquad (13.7)$$

If the force acting on the body is conservative—such as the force of gravity or the force of a spring—then the work equals the negative of the change in potential energy, and Eq. (13.7) becomes

$$-U_2 + U_1 = \tfrac{1}{2}I\omega_2^2 - \tfrac{1}{2}I\omega_1^2 \qquad (13.8)$$

or

$$\tfrac{1}{2}I\omega_1^2 + U_1 = \tfrac{1}{2}I\omega_2^2 + U_2 \qquad (13.9)$$

This expresses the **conservation of energy in rotational motion**: the sum of the kinetic and potential energies is constant, that is,

$$E = \tfrac{1}{2}I\omega^2 + U = [\text{constant}] \qquad (13.10)$$

conservation of energy in rotational motion

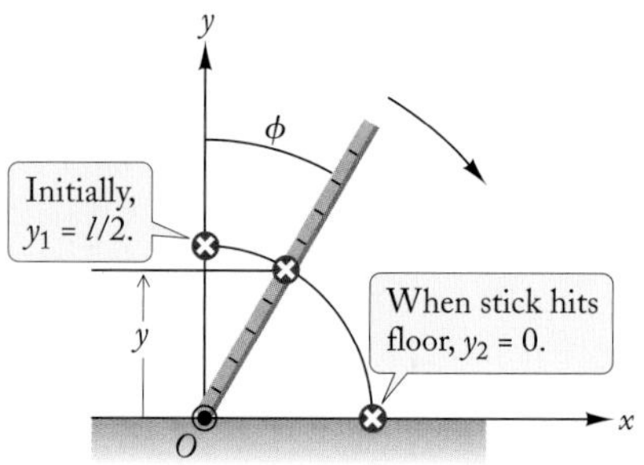

FIGURE 13.3 Meterstick rotating about its lower end.

EXAMPLE 2 A meterstick is initially standing vertically on the floor. If the meterstick falls over, with what angular velocity will it hit the floor? Assume that the end in contact with the floor does not slip.

SOLUTION: The motion of the meterstick is rotation about a fixed axis passing through the point of contact with the floor (see Fig. 13.3). The stick is a uniform rod of mass M and length $l = 1.0$ m. Its moment of inertia about the end is $Ml^2/3$ (see Table 12.3), and its rotational kinetic energy is therefore $\frac{1}{2}I\omega^2 = Ml^2\omega^2/6$. The gravitational potential energy is Mgy, where y is the height of the center of mass above the floor. When the meterstick is standing vertically, the initial angular velocity is $\omega_1 = 0$ and $y_1 = l/2$, so the total energy is

$$E = \tfrac{1}{6}Ml^2\omega_1^2 + Mgy_1 = 0 + Mgl/2 \tag{13.11}$$

Just before the meterstick hits the floor, the angular velocity is ω_2 and $y_2 = 0$. The energy is

$$E = \tfrac{1}{6}Ml^2\omega_2^2 + Mgy_2 = \tfrac{1}{6}Ml^2\omega_2^2 + 0 \tag{13.12}$$

Conservation of energy therefore implies

$$\tfrac{1}{6}Ml^2\omega_2^2 = Mgl/2$$

from which we obtain

$$\omega_2^2 = \frac{3g}{l} \tag{13.13}$$

Taking the square root of both sides, we find

$$\omega_2 = \sqrt{\frac{3g}{l}} = \sqrt{\frac{3 \times 9.81\,\text{m/s}^2}{1.0\text{ m}}} = 5.4\text{ radians/s}$$

EXAMPLE 3 At what instantaneous rate is gravity delivering energy to the meterstick of Example 2 just before it hits the floor? The mass of the meterstick is 0.15 kg.

SOLUTION: The rate of energy delivery is the power,

$$P = \tau\omega$$

From Example 2, we know $\omega = 5.4$ radians/s just before the stick hits the floor. At that instant, gravity acts perpendicular to the stick at the center of mass (in the next chapter we will see that the weight acts as if concentrated at the center of mass), a distance $R = l/2 = 0.50$ m from the end. So the torque exerted by gravity is

$$\tau = FR\sin\theta = mg\frac{l}{2}\sin 90^\circ = 0.15\text{ kg} \times 9.81\text{ m/s}^2 \times 0.50\text{ m} \times 1$$

$$= 0.74\text{ N}\cdot\text{m}$$

Thus the instantaneous power delivered by the torque due to gravity is

$$P = \tau\omega = 0.74\text{ N}\cdot\text{m} \times 5.4\text{ radians/s} = 4.0\text{ W}$$

Checkup 13.1

QUESTION 1: You are trying to tighten a bolt with a wrench. Where along the handle should you place your hand so you can exert maximum torque? In what direction should you push?

QUESTION 2: A force is being exerted against the rim of a freely rotating wheel, but the work done by this force is zero. What can you conclude about the direction of the force? What is the torque of the force?

QUESTION 3: Consider the meterstick falling over, as in Example 2. What is the torque that the weight exerts on the meterstick when it is in the upright, initial position? After the stick begins to fall over, the torque increases. When is the torque maximum?

QUESTION 4: Suppose you first push a door at its outer edge at right angles to the surface of the door with a force of magnitude F. Next you push the door at its center, again at right angles to the surface, with a force of magnitude $F/2$. In both cases you push the door as it moves through 30°. The ratio of the work done by the second push to the work done by the first push is:

(A) $\frac{1}{4}$ (B) $\frac{1}{2}$ (C) 1 (D) 2 (E) 4

13.2 THE EQUATION OF ROTATIONAL MOTION

Our intuition tells us that a torque acting on a wheel or some other body free to rotate about an axis will produce an angular acceleration. For instance, the push of your hand against a crank on a wheel (see Fig. 13.4) exerts a torque or "twist" that starts the wheel turning. The angular acceleration depends on the magnitude of your push on the crank and also on its direction (as well as on the inertia of the wheel). Your push will be most effective if exerted tangentially, at right angles to the radius (at $\theta = 90°$; see Fig. 13.4a). It will be less effective if exerted at a smaller or larger angle (see Fig. 13.4b). And it will be completely ineffective if exerted parallel to the radius (at $\theta = 0$ or 180°; see Fig. 13.4c)—such a push in the radial direction produces no rotation at all. These qualitative considerations are in agreement with the definition of torque,

$$\tau = FR\sin\theta \qquad (13.14)$$

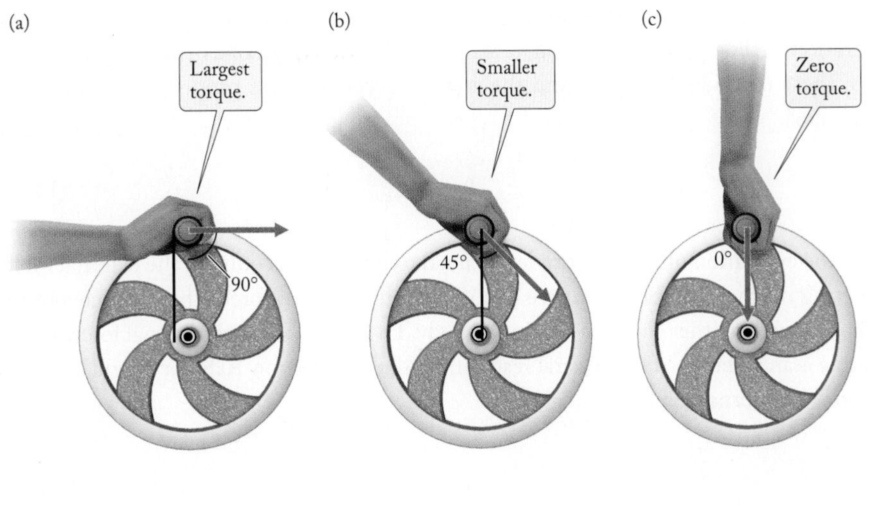

FIGURE 13.4 (a) A push at right angles to the radius is most effective in producing rotation. (b) A push at 45° is less effective. (c) A push parallel to the radius produces no rotation.

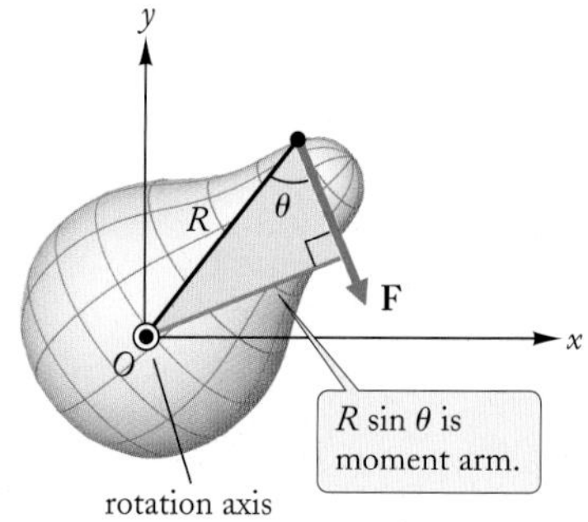

FIGURE 13.5 The distance between the center of rotation and the point of application of the force is R. The perpendicular distance between the center of rotation and the line of action of the force is $R \sin\theta$.

According to this equation, the torque provided by a force of a given magnitude F is maximum if the force is at right angles to the radius ($\theta = 90°$), and it is zero if the force is parallel to the radius ($\theta = 0°$ or $180°$).

The quantity $R \sin\theta$ appearing in Eq. (13.14) has a simple geometric interpretation: it is the perpendicular distance between the line of action of the force and the axis of rotation (see Fig. 13.5); this perpendicular distance is called the **moment arm of the force.** Hence, Eq. (13.14) states that the torque equals the magnitude of the force multiplied by the moment arm.

To find a quantitative relationship between torque and angular acceleration, we recall from Eq. (13.6) that the power delivered by a torque acting on a body is

$$\frac{dW}{dt} = \tau\omega \tag{13.15}$$

The work–energy theorem tells us that the work dW equals the change of kinetic energy in the small time interval dt. The small change in the kinetic energy is $dK = d(\frac{1}{2}I\omega^2) = \frac{1}{2}I \times 2\omega\, d\omega = I\omega\, d\omega$. Thus,

$$dW = I\omega\, d\omega \tag{13.16}$$

Inserting this into the left side of Eq. (13.15), we find

$$\frac{I\omega\, d\omega}{dt} = \tau\omega \tag{13.17}$$

Canceling the factor of ω on both sides of the equation, we obtain

$$I\frac{d\omega}{dt} = \tau \tag{13.18}$$

But $d\omega/dt$ is the angular acceleration α; hence

equation of rotational motion

$$I\alpha = \tau \tag{13.19}$$

This is the equation for rotational motion. As we might have expected, this equation says that *the angular acceleration is directly proportional to the torque.* Equation (13.19) is mathematically analogous to Newton's Second Law, $m\mathbf{a} = \mathbf{F}$, for the translational motion of a particle; the moment of inertia takes the place of the mass, the angular acceleration the place of the acceleration, and the torque the place of the force.

In our derivation of Eq. (13.19) we assumed that only one external force is acting on the rigid body. If several forces act, then each produces its own torque. If an individual torque would produce an angular acceleration in the rotational direction chosen as positive, it is reckoned as positive, and if a torque would produce an angular acceleration in the opposite direction, it is reckoned as negative. The net torque is the sum of these individual torques, and the angular acceleration is proportional to this net torque:

equation of rotational motion for net torque

$$I\alpha = \tau_{\text{net}} \tag{13.20}$$

In the evaluation of the net torque, we need to take into account all the external forces acting on the rigid body, but we can ignore the internal forces that particles in the body exert on other particles also in the body. The torques of such internal forces cancel (this is an instance of the general result mentioned in Section 10.4: for a rigid body, the work of internal forces cancels).

EXAMPLE 4 The rotor of the gyroscope of the *Gravity Probe B* experiment (see the chapter photo and Fig. 13.6) is a quartz sphere of diameter 3.8 cm and mass 7.61×10^{-2} kg. To start this sphere spinning, a stream of helium gas flowing in an equatorial channel in the surface of the housing is blown tangentially against the rotor. What torque must this stream of gas exert on the rotor to accelerate it uniformly from 0 to 10000 rpm (revolutions per minute) in 30 minutes? What force must it exert on the equator of the sphere?

Concepts — in — Context

FIGURE 13.6 A gyroscope sphere for *Gravity Probe B*.

SOLUTION: The final angular velocity is $2\pi \times 10000$ radians/60 s $= 1.05 \times 10^3$ radians/s, and therefore the angular acceleration is

$$\alpha = \frac{\omega_2 - \omega_1}{t_2 - t_1}$$

$$= \frac{1.05 \times 10^3 \text{ radians/s} - 0}{30 \times 60 \text{ s} - 0} = 0.582 \text{ radian/s}^2$$

The moment of inertia of the rotor is that of a sphere (see Table 12.3):

$$I = \tfrac{2}{5}MR^2$$

$$= \tfrac{2}{5} \times 7.61 \times 10^{-2} \text{ kg} \times (0.019 \text{ m})^2 = 1.1 \times 10^{-5} \text{ kg·m}^2$$

Hence the required torque is, according to Eq. (13.19),

$$\tau = I\alpha = 1.1 \times 10^{-5} \text{ kg·m}^2 \times 0.582 \text{ radian/s}^2$$

$$= 6.4 \times 10^{-6} \text{ N·m}$$

The driving force is along the equator of the rotor—that is, it is perpendicular to the radius—so $\sin\theta = 1$ and Eq. (13.2) reduces to $\tau = FR$, which yields

$$F = \frac{\tau}{R} = \frac{6.4 \times 10^{-6} \text{ N·m}}{0.019 \text{ m}} = 3.4 \times 10^{-4} \text{ N}$$

EXAMPLE 5 Two masses m_1 and m_2 are suspended from a string that runs, without slipping, over a pulley (see Fig. 13.7a). The pulley has a radius R and a moment of inertia I about its axle, and it rotates without friction. Find the accelerations of the masses.

SOLUTION: We have already found the motion of this system in Example 10 of Chapter 5, where the two masses were an elevator and its counterweight, and where we neglected the inertia of the pulley. Now we will take this inertia into account.

Figure 13.7c shows the "free-body" diagrams for the masses m_1 and m_2. In these diagrams, T_1 and T_2 are the tensions in the two parts of the string attached to the two masses. (Note that now T_1 and T_2 are not equal. For a pulley of zero moment of inertia, these tensions would be equal; but for a pulley of nonzero moment of inertia, a difference between T_1 and T_2 is required to produce the angular acceleration of the pulley.) If the acceleration of mass m_1 is a (reckoned as positive if upward), then the acceleration of mass m_2 is $-a$, and the equations of motion of the two masses are

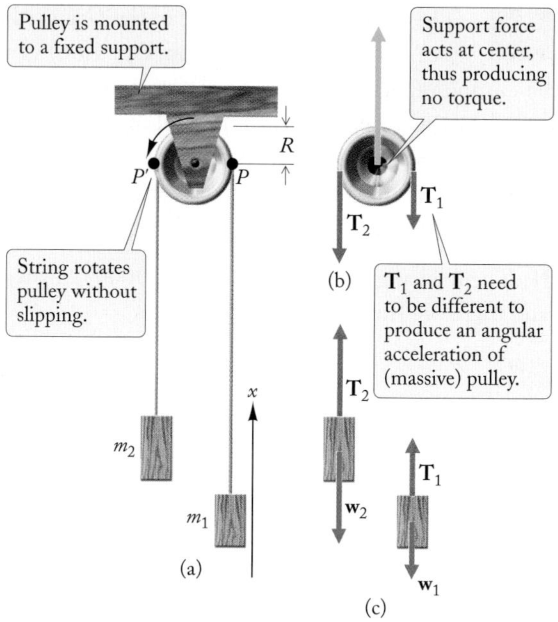

FIGURE 13.7 (a) Two masses m_1 and m_2 suspended from a string that runs over a pulley. (b) "Free-body" diagram for the pulley. (c) "Free-body" diagrams for the masses m_1 and m_2.

$$m_1 a = T_1 - m_1 g \tag{13.21}$$

$$-m_2 a = T_2 - m_2 g \tag{13.22}$$

Figure 13.7b shows the "free-body" diagram for the pulley. The tension forces act at the ends of the horizontal diameter (since the string does not slip, it behaves as though instantaneously attached to the pulley at the point of first contact; see points P and P' in Fig. 13.7a). The upward supporting force of the axle acts at the center of the pulley, and it generates no torque about the center of the pulley. The tensions act perpendicular to the radial direction, so $\sin\theta = 1$ in Eq. (13.2). Taking the positive direction of rotation as counterclockwise (to match the positive direction for the motion of mass m_1), we see that the tension forces T_1 and T_2 generate torques $-RT_1$ and RT_2 about the center. The equation of rotational motion of the pulley is

$$I\alpha = \tau_{\text{net}} = -RT_1 + RT_2 \tag{13.23}$$

The translational acceleration of each hanging portion of the string must match the instantaneous translational acceleration of the point of first contact (for the given condition of no slipping). Hence the translational acceleration a of the masses is related to the angular acceleration α by $a = \alpha R$, or $\alpha = a/R$ [see Eq. (12.13)]. Furthermore, according to Eqs. (13.21) and (13.22), $T_1 = m_1 g + m_1 a$ and $T_2 = m_2 g - m_2 a$. With these substitutions, Eq. (13.23) becomes

$$I(a/R) = -R(m_1 g + m_1 a) + R(m_2 g - m_2 a)$$

Solving this for a, we find

$$a = \frac{m_2 - m_1}{m_1 + m_2 + (I/R^2)}\, g \tag{13.24}$$

COMMENT: If the mass of the pulley is small, then I/R^2 can be neglected; with this approximation, Eq. (13.24) reduces to Eq. (5.44), which was obtained without taking into account the inertia of the pulley.

A device of this kind, called **Atwood's machine,** can be used to determine the value of g. For this purpose, it is best to use masses m_1 and m_2 that are nearly equal. Then a is much smaller than g and easier to measure; the value of g can be calculated from the measured value of a according to Eq. (13.24).

Atwood's machine

In some cases—for instance, the rolling motion of a wheel—the axis of rotation is in motion, perhaps accelerated motion, and is *not a fixed axis*. For such problems, some further arguments can be used to demonstrate that Eq. (13.19) remains valid for rotation about an axis in accelerated translational motion, *provided the axis passes through the center of mass of the rotating body*. When this condition is met, we can use the equation of rotational motion (13.19) as in the following examples.

EXAMPLE 6 An automobile with rear-wheel drive is accelerating at 4.0 m/s^2 along a straight road. Consider one of the front wheels of this automobile (see Fig. 13.8a). The axle pushes the wheel forward, providing an acceleration of 4.0 m/s^2. Simultaneously, the friction force of the road pushes the bottom of the wheel backward, providing a torque that gives the wheel an angular acceleration. The wheel has a radius of 0.38 m and a mass of 25 kg. Assume that the wheel is (approximately) a uniform disk, and assume it rolls without slipping. Find the backward force that the friction force exerts on the wheel, and find the forward force that the axle exerts on the wheel.

(a)

(b)

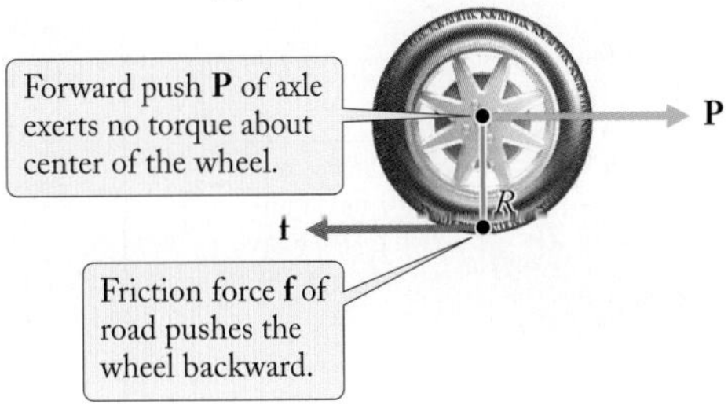

FIGURE 13.8 (a) Front wheel of an automobile. (b) "Free-body" diagram for the wheel. The friction force of the road pushes the wheel backward. The axle pushes the wheel forward.

SOLUTION: Figure 13.8b shows a "free-body" diagram of the wheel, with the horizontal forces acting on it (besides these horizontal forces, there are also a vertical downward push exerted by the axle and a vertical upward normal force exerted by the road; these forces exert no torque and cancel, so they need not concern us here). The forward push of the axle is P, and the rearward push of the ground is f. The force P, acting at the center of the wheel, exerts no torque; the force f, acting at the rim, exerts a torque Rf. Thus, the equation for the rotational motion of the wheel is

$$I\alpha = Rf$$

or, since $I = \frac{1}{2}MR^2$ for a uniform disk (see Table 12.3),

$$\tfrac{1}{2}MR\alpha = f$$

As we have seen in Example 4 of Chapter 12, the angular acceleration of a rolling wheel is related to the translational acceleration by $\alpha = a/R$. Hence

$$\tfrac{1}{2}Ma = f$$

from which

$$\begin{aligned} f &= \tfrac{1}{2}Ma = \tfrac{1}{2} \times 25 \text{ kg} \times 4.0 \text{ m/s}^2 \\ &= 50 \text{ N} \end{aligned}$$

To find the force P, we need to examine the equation for the translational motion. The net horizontal force is $F_{net} = P - f$. Hence the equation for the translational motion of the wheel is

$$Ma = P - f$$

from which

$$P = Ma + f = 25\ \text{kg} \times 4.0\ \text{m/s}^2 + 50\ \text{N}$$
$$= 150\ \text{N}$$

Thus, the force required to accelerate a rolling wheel is larger than the force required for a wheel that slips on a frictionless surface without rolling—for such a wheel the force would be only $Ma = 25\ \text{kg} \times 4.0\ \text{m/s}^2 = 100$ N. Here, the additional rotational inertia $\frac{1}{2}MR^2$ adds an additional amount $f = \frac{1}{2}Ma$ to the required force, so the total required force is $\frac{3}{2}$ that for sliding without rolling.

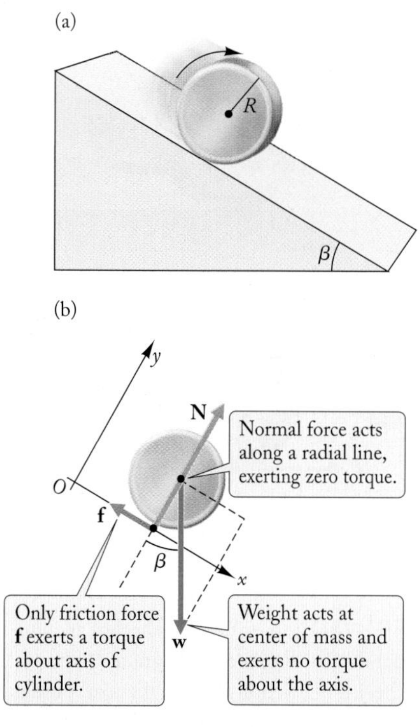

FIGURE 13.9 (a) A cylinder rolling down an inclined plane. (b) "Free-body" diagram for the cylinder.

EXAMPLE 7 A solid cylinder of mass M and radius R rolls down a sloping ramp that makes an angle β with the ground (see Fig. 13.9a). What is the acceleration of the cylinder? Assume that the cylinder is uniform and rolls without slipping.

SOLUTION: Figure 13.9b shows the "free-body" diagram for the cylinder. The forces on the cylinder are the normal force **N** exerted by the ramp, the friction force **f** exerted by the ramp, and the weight **w**. The friction force is exerted on the rim of the cylinder, and the weight is effectively exerted at the center of the cylinder (in the next chapter we will see that the weight can always be regarded as concentrated at the center of mass). As axis of rotation, we take the axis that passes through the center of the cylinder. The weight exerts no torque about this axis, and neither does the normal force (zero moment arm). Hence, the only force that exerts a torque is the friction force, and so

$$\tau = Rf$$

The equation of rotational motion is then

$$I\alpha = Rf$$

The moment of inertia of a uniform cylinder is the same as that of a disk, $I = \frac{1}{2}MR^2$. Furthermore, for rolling motion without slipping, $\alpha = a/R$. Hence

$$\frac{1}{2}MRa = Rf$$

or

$$a = \frac{2f}{M} \tag{13.25}$$

To evaluate the acceleration, we need to eliminate the friction force f from this equation. We can do this by appealing to the equation for the component of the translational motion along the ramp (the motion along the x direction in Fig. 13.9b).

The components of the forces along the ramp are $-f$ for the friction force and $Mg \sin \beta$ for the weight. Hence

$$Ma = Mg \sin \beta - f$$

or

$$f = Mg \sin \beta - Ma$$

Substituting this into Eq. (13.25), we find

$$a = 2g \sin \beta - 2a$$

which we can immediately solve for a:

$$a = \tfrac{2}{3} g \sin \beta$$

COMMENT: Note that the force $Mg \sin \beta$ along the ramp here produces an acceleration that is two-thirds of the acceleration that the cylinder would have if it were to slip down a frictionless ramp without rolling. This is consistent with the last example, where we saw that a force $\frac{3}{2}$ as large was required to produce a given acceleration. The same factor occurs in both cases, because both the disk and the cylinder have the same moment of inertia, $\frac{1}{2}MR^2$.

PROBLEM-SOLVING TECHNIQUES — TORQUES AND ROTATIONAL MOTION

The general techniques for the solution of problems of rotational motion are similar to the techniques we learned in Chapters 5 and 6 for translational motion.

1. The first step is always a careful enumeration of all the forces. Make a complete list of these forces, and label each with a vector symbol.
2. Identify the body whose motion or whose equilibrium is to be investigated and draw the "free-body" diagram showing the forces acting on this body. If there are several distinct bodies in the problem (as in Example 5), then you need to draw a separate "free-body" diagram for each. When drawing the arrows for the forces acting on a rotating body, be sure to draw the head or the tail of the arrow at the actual point of the body where the force acts, since this will be important for the calculation of the torque. Note that the weight acts at the center of mass (we will establish this in the next chapter).
3. Select which direction of rotation will be regarded as positive (for instance, in Example 5, we selected the counterclockwise direction of rotation as positive). If the problem involves joint rotational and translational motions, select coordinate axes for the translational motion, preferably placing one of the axes along the direction of motion.
4. Select an axis for the rotation of the rigid body, either an axis through the center of mass, or else a fixed axis (such as an axle or a pivot mounted on a support) about which the body is constrained to rotate. Calculate the torque of each force acting on the body about this center. Remember that the sign of the torque is positive or negative depending on whether it produces an angular acceleration in the positive or the negative direction of rotation.
5. Then apply the equation of rotational motion, $I\alpha = \tau$, to each rotating body, where τ is the net torque on a given body.
6. If the rigid body has a translational motion besides the rotational motion, apply Newton's Second Law, $\mathbf{F} = m\mathbf{a}$, for the translational motion (see Examples 5 and 6). For rolling without slipping, the translational and the rotational motions are related by $v = \omega R$ and $a = \alpha R$.
7. If there are several distinct bodies in the problem, you need to apply the equation of rotational motion or Newton's Second Law separately for each (see Example 5).

Checkup 13.2

QUESTION 1: Consider a meterstick falling over, as in Example 2. At what instant is the angular acceleration produced by the weight force maximum?

QUESTION 2: A rolling cylinder has both rotational kinetic energy (reckoned about its center of mass) and translational kinetic energy. Which is larger?

QUESTION 3: Consider the rolling cylinder of Example 7. When this cylinder reaches the bottom of the ramp, is its kinetic energy larger, smaller, or the same as that of a similar cylinder that slips down a frictionless ramp without rolling?

QUESTION 4: A sphere and a cylinder of equal masses roll down an inclined plane without slipping. Will they have equal kinetic energies when they reach the bottom? Which will get to the bottom first?

QUESTION 5: A thin hoop and a solid cylinder roll down an inclined plane without slipping. When they reach the bottom, the translational speed of the hoop is

(A) Less than that of the cylinder
(B) Greater than that of the cylinder
(C) Equal to that of the cylinder

13.3 ANGULAR MOMENTUM AND ITS CONSERVATION

In Chapter 10 we saw how to express the equation for the translational motion in terms of the momentum: the rate of change of the momentum equals the force ($dp_x/dt = F_x$). Likewise, we can express the equation for rotational motion in terms of **angular momentum.** *The angular momentum of a body rotating about a fixed axis is defined as the product of the moment of inertia and the angular velocity,*

angular momentum

$$L = I\omega \tag{13.26}$$

This equation for angular momentum is analogous to the equation $p = mv$ for translational momentum. *The SI unit of angular momentum is* kg·m²/s, which can also be written in the alternative form J·s. Table 13.1 gives some examples of typical values of angular momenta.

Concepts in Context

EXAMPLE 8 According to the data given in Example 4, what is the angular momentum of the rotor of the *Gravity Probe B* gyroscope when spinning at 10 000 revolutions per minute?

SOLUTION: From Example 4, the angular velocity is $\omega = 1.05 \times 10^3$ radians/s, and the moment of inertia is $I = 1.1 \times 10^{-5}$ kg·m². So

$$L = I\omega = 1.1 \times 10^{-5}\ \text{kg}\cdot\text{m}^2 \times 1.05 \times 10^3\ \text{radians/s}$$

$$= 1.2 \times 10^{-2}\ \text{kg}\cdot\text{m}^2/\text{s}$$

To express the equation for rotational motion in terms of angular momentum, we proceed as we did in the translational case. We note that if the change of angular velocity is $d\omega$, then $dL = I\,d\omega$. Dividing both sides of this relation by dt, we see

$$\frac{dL}{dt} = I\frac{d\omega}{dt}$$

If we compare this with Eq. (13.18), we see that the right side can be expressed as the torque, so

$$\frac{dL}{dt} = \tau \qquad (13.27)$$

equation of rotational motion in terms of angular momentum

This says that *the rate of change of angular momentum equals the torque.* Obviously, this equation is analogous to the equation $dp_x/dt = F_x$ for translational motion.

We now see that the analogy between rotational and translational quantities mentioned in Section 12.3 can be extended to angular momentum and momentum. Table 13.2 lists analogous quantities, including the quantities for work, power, and kinetic energy.

If there is no torque acting on the rotating body, $\tau = 0$ and therefore $dL/dt = 0$, which means that the angular momentum does not change:

$$L = [\text{constant}] \qquad (\text{when } \tau = 0) \qquad (13.28)$$

conservation of angular momentum

This is the **Law of Conservation of Angular Momentum.** Since $L = I\omega$, we can also write this law as

$$I\omega = [\text{constant}] \qquad (13.29)$$

TABLE 13.1 SOME ANGULAR MOMENTA

Orbital motion of Earth	2.7×10^{40} J·s
Rotation of Earth	5.8×10^{33} J·s
Helicopter rotor (320 rev/min)	5×10^{4} J·s
Automobile wheel (90 km/h)	1×10^{2} J·s
Electric fan	1 J·s
Frisbee	1×10^{-1} J·s
Toy gyroscope	1×10^{-2} J·s
Phonograph record (33.3 rev/min)	6×10^{-3} J·s
Compact disc (plating outer track)	2×10^{-3} J·s
Bullet fired from rifle	2×10^{-3} J·s
Orbital motion of electron in atom	1.05×10^{-34} J·s
Spin of electron	0.53×10^{-34} J·s

TABLE 13.2 FURTHER ANALOGIES BETWEEN 1D TRANSLATIONAL AND ROTATIONAL QUANTITIES

$dW = F\,dx$	$\rightarrow$	$dW = \tau\,d\phi$
$P = Fv$	$\rightarrow$	$P = \tau\omega$
$K = \frac{1}{2}mv^2$	$\rightarrow$	$K = \frac{1}{2}I\omega^2$
$ma = F$	$\rightarrow$	$I\alpha = \tau$
$p = mv$	$\rightarrow$	$L = I\omega$
$\frac{dp}{dt} = F$	$\rightarrow$	$\frac{dL}{dt} = \tau$

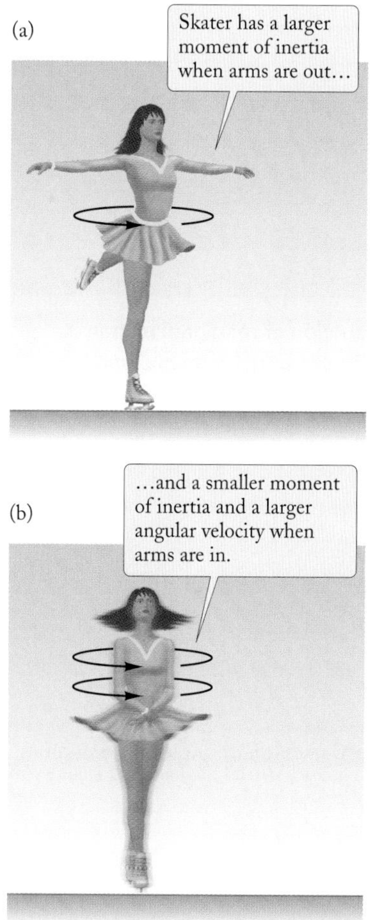

FIGURE 13.10 Figure skater performing a pirouette. (a) Arms extended. (b) Arms folded against body.

FIGURE 13.11 A figure skater whirling at high speed.

A pirouette performed by a figure skater on ice provides a nice illustration of the conservation of angular momentum. The skater begins the pirouette by spinning about her vertical axis with her arms extended horizontally (see Fig. 13.10a); in this configuration, the arms have a large moment of inertia. She then brings her arms close to her body (see Fig. 13.10b), suddenly decreasing her moment of inertia. Since the ice is nearly frictionless, the external torque on the skater is nearly zero, and therefore the angular momentum is conserved. According to Eq. (13.26), a decrease of I requires an increase of ω to keep the angular momentum constant. Thus, the change of configuration of her arms causes the skater to whirl around her vertical axis with a dramatic increase of angular velocity (see Fig. 13.11).

Like the law of conservation of translational momentum, the Law of Conservation of Angular Momentum is often useful in the solutions of problems in which the forces are not known in detail.

EXAMPLE 9 Suppose that a pottery wheel is spinning (with the motor disengaged) at 80 rev/min when a 6.0-kg ball of clay is suddenly dropped down on the center of the wheel (see Fig. 13.12). What is the angular velocity after the drop? Treat the ball of clay as a uniform sphere of radius 8.0 cm. The pottery wheel has a moment of inertia $I = 7.5 \times 10^{-2}$ kg·m^2. Ignore the (small) friction force in the axle of the turntable.

SOLUTION: Since there is no external torque on the system of pottery wheel and clay, the angular momentum of this system is conserved. The angular momentum before the drop is

$$L = I\omega \tag{13.30}$$

where ω is the initial angular velocity and I the moment of inertia of the pottery wheel. The angular momentum after the drop is

$$L' = I'\omega' \tag{13.31}$$

where ω' is the final angular velocity and I' the moment of inertia of pottery wheel and clay combined. Hence

$$I\omega = I'\omega' \tag{13.32}$$

from which we find

$$\omega' = \frac{I}{I'}\omega \tag{13.33}$$

The wheel is initially rotating with angular velocity

$$\omega = 2\pi \times f = 2\pi \times \frac{80 \text{ rev}}{60 \text{ s}} = 8.4 \text{ radians/s}$$

The moment of inertia of the pottery wheel is given,

$$I = 7.5 \times 10^{-2} \text{ kg·m}^2$$

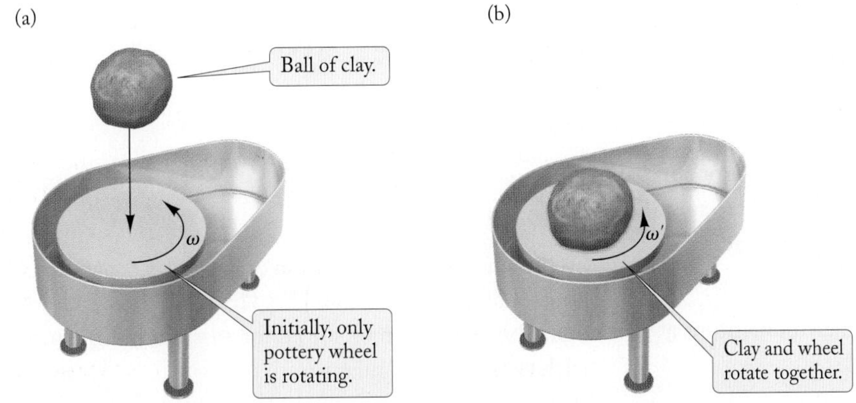

FIGURE 13.12 (a) A pottery wheel rotates with angular velocity ω; (b) when a ball of clay is dropped on the wheel, the angular velocity slows to ω'.

and the moment of inertia of the clay is that of a uniform sphere (see Table 12.3):

$$I_{\text{clay}} = \tfrac{2}{5}MR^2 = \tfrac{2}{5} \times 6.0 \text{ kg} \times (0.080 \text{ m})^2$$

$$= 1.5 \times 10^{-2} \text{ kg·m}^2$$

Accordingly,

$$\omega' = \frac{I}{I'}\omega = \frac{I}{I + I_{\text{clay}}}\omega$$

$$= \frac{7.5 \times 10^{-2} \text{ kg·m}^2}{7.5 \times 10^{-2} \text{ kg·m}^2 + 1.5 \times 10^{-2} \text{ kg·m}^2} \times 8.4 \text{ radians/s}$$

$$= 7.0 \text{ radians/s}$$

As already mentioned in Chapter 9, the Law of Conservation of Angular Momentum also applies to a single particle moving in an orbit under the influence of a central force. Such a force is always directed along the radial line, and it therefore exerts no torque. If the particle is moving along a circle of radius r with velocity v (see Fig. 13.13), its moment of inertia is mr^2 and its angular velocity is $\omega = v/r$. Hence $I\omega = mr^2 \times v/r = mvr$, and the angular momentum of the particle is

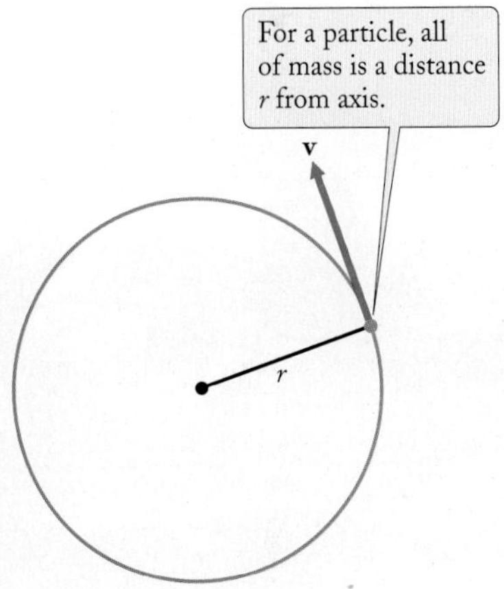

FIGURE 13.13 A particle moving with speed v along a circle of radius r. The moment of inertia of this particle with respect to the center of the circle is $I = mr^2$.

angular momentum for circular orbit

$$L = mvr \qquad \text{(circular orbit)} \qquad (13.34)$$

This formula is valid not only for a circular orbit, but also for the perihelion and aphelion points of an elliptical orbit, where the instantaneous velocity is perpendicular to the radius. In Chapter 9 we took advantage of the conservation of the angular momentum $L = mvr$ to compare the speeds of a planet at perihelion and at aphelion.

orbital angular momentum and spin angular momentum

The angular momentum defined by Eq. (13.34) is called the **orbital angular momentum** to distinguish it from **spin angular momentum** of a body rotating about its own axis. For instance, the Earth has both an orbital angular momentum (due to its motion around the Sun) and a spin angular momentum (due to its rotation about its own axis). Table 13.1 includes examples of both kinds of angular momentum.

PROBLEM-SOLVING TECHNIQUES — CONSERVATION OF ANGULAR MOMENTUM

The use of conservation of angular momentum in a problem involving rotational motion involves the familiar three steps we used with conservation of momentum or of energy in translational motion:

1. First write an expression for the angular momentum at one instant of the motion [Eq. (13.30)].
2. Then write an expression for the angular momentum at another instant [Eq. (13.31)].
3. And then rely on conservation of angular momentum to equate the two expressions [Eq. (13.32)]. This yields one equation, which can be solved for an unknown quantity, such as the final angular speed.

✔ Checkup 13.3

QUESTION 1: A hoop and a uniform disk have equal radii and equal masses. Both are spinning with equal angular speeds. Which has the larger angular momentum? By what factor?

QUESTION 2: Two automobiles of equal masses are traveling around a traffic circle side by side, with equal angular velocities. Which has the larger angular momentum?

QUESTION 3: You sit on a spinning stool with your legs tucked under the seat. You then stretch your legs outward. How does your angular velocity change?

QUESTION 4: Consider the spinning skater described in Fig. 13.10. While she brings her arms close to her body, does the rotational kinetic energy remain constant?

QUESTION 5: Three children sit on a tire swing (see Fig. 13.14), leaning backward as the wheel rotates about a vertical axis. What happens to the rotational frequency if the children sit up straight?

(A) Frequency increases (B) Frequency decreases
(C) Frequency remains constant

FIGURE 13.14 Children on a spinning tire swing.

13.4 TORQUE AND ANGULAR MOMENTUM AS VECTORS

The rotational motion of a rigid body about a fixed axis is analogous to one-dimensional translational motion. More generally, if the axis of rotation is not fixed but changes in direction, the motion becomes three-dimensional. A wobbling, spinning top provides an example of such a three-dimensional rotational motion. In this case, the torque and the angular momentum must be treated as vectors, analogous to the force vector and the momentum vector. The definitions of the torque vector and the angular-momentum vector involve the vector cross product that we introduced in Section 3.4. When a force **F** acts at some point with position vector **r**, *the resulting* **torque vector** *is the cross product of the position vector and the force vector:*

torque vector

$$\boldsymbol{\tau} = \mathbf{r} \times \mathbf{F} \tag{13.35}$$

According to the definition of the cross product, the magnitude of $\boldsymbol{\tau}$ is

$$\tau = rF \sin\theta \qquad (13.36)$$

and the direction of $\boldsymbol{\tau}$ is perpendicular to the force vector and the position vector, as specified by the right-hand rule (see Fig. 13.15). Note that since the position vector depends on the choice of origin, *the torque* also *depends on the choice of origin.* We will usually place the origin on some axis or some pivot, and the torque (13.35) is then reckoned in relation to this pivot. For instance, for rotation about a fixed axis, we place the origin on that axis, so **r** is in the plane of the circular motion of the point at which the force acts; then $r = R$, and Eq. (13.36) agrees with Eq. (13.2).

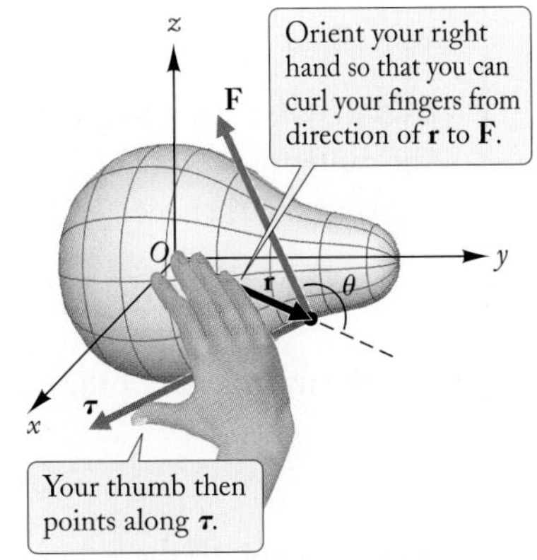

FIGURE 13.15 The torque vector $\boldsymbol{\tau}$ is perpendicular to the force **F** and the position vector **r**, in the direction specified by the right-hand rule: place the fingers of your right hand along the direction of **r** and curl toward **F** along the smaller angle between these vectors; your thumb will then point in the direction of $\mathbf{r} \times \mathbf{F}$.

The definition of the angular-momentum vector of a rigid body is based on the definition of the angular-momentum vector for a single particle. If a particle has translational momentum **p** at position **r**, then *its* **angular-momentum vector** *is defined as the cross product of the position vector and the momentum vector:*

$$\mathbf{L} = \mathbf{r} \times \mathbf{p} \qquad (13.37)$$

angular-momentum vector

As in the case of the torque, *the angular momentum vector depends on the choice of origin.* For instance, if the particle is moving along a circle, we place the origin at the center of the circle, so **r** and **p** are in the plane of the circular motion. Since the vectors **r** and **p** are perpendicular, the magnitude of their cross product is then $L = rp \sin 90° = rp = rmv$. By the right-hand rule, the direction of $\mathbf{r} \times \mathbf{p}$ is perpendicular to the plane of the circular motion, parallel to the axis of rotation. (see Fig. 13.16).

For a rigid body rotating about some (instantaneous) axis, the angular-momentum vector is defined as the sum of the angular-momentum vectors of all the particles in the body,

$$\mathbf{L} = \mathbf{r}_1 \times \mathbf{p}_1 + \mathbf{r}_2 \times \mathbf{p}_2 + \cdots \qquad (13.38)$$

As in the case of a single particle, the value of the angular momentum obtained from this formula depends on the choice of the origin of coordinates. For the calculation of the angular momentum of a rigid body rotating about a fixed axis, it is usually convenient to choose an origin on the axis of rotation.

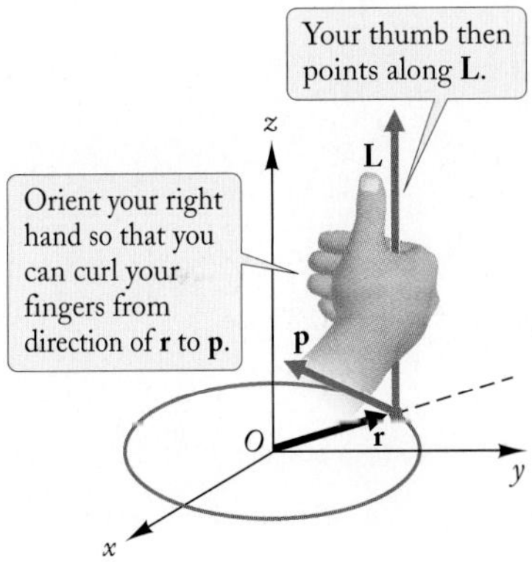

FIGURE 13.16 Angular-momentum vector for a particle.

EXAMPLE 10 Figure 13.17 shows a dumbbell, a rigid body consisting of two particles of mass m attached to the ends of a massless rigid rod of length $2r$. The body rotates with angular velocity ω about a perpendicular axis through the center of the rod. Find the angular momentum about this center.

SOLUTION: Each particle executes circular motion with speed $v = r\omega$. Hence the angular momentum of each has a magnitude $L = rmv = mr^2\omega$ (compare the case of a single particle, illustrated in Fig. 13.13). The direction of each angular-momentum vector is parallel to the axis of rotation (see Fig. 13.16). Thus the direction of the vector sum of the two angular-momentum vectors is also parallel to the axis of rotation, and its magnitude is

$$L = mr^2\omega + mr^2\omega = 2mr^2\omega$$

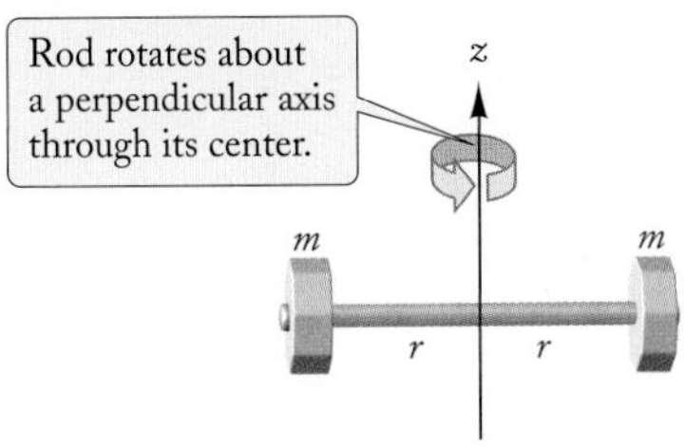

FIGURE 13.17 A rotating dumbbell.

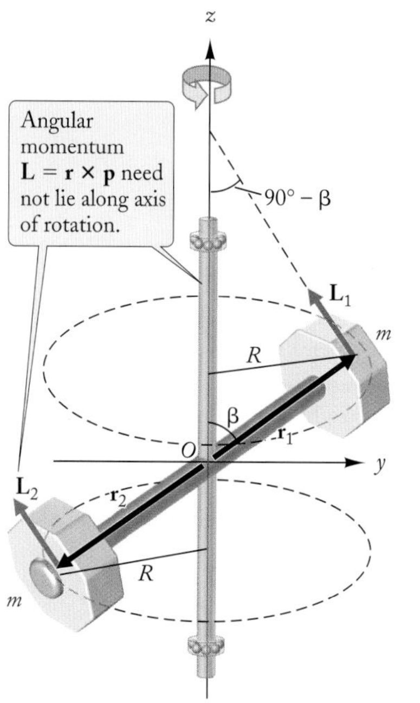

FIGURE 13.18 A rotating dumbbell oriented at an angle β with the axis of rotation.

EXAMPLE 11 Suppose that the rod of the dumbbell described in the preceding example is welded to an axle inclined at an angle β with respect to the rod. The dumbbell rotates with angular velocity ω about this axis, which is supported by fixed bearings (see Fig. 13.18). Find the angular momentum about an origin on the axis, at the center of mass.

SOLUTION: Each particle executes circular motion, but since the origin is not at the center of the circle, the angular momentum is not the same as in Example 10. The distance between each particle and the axis of rotation is

$$R = r \sin \beta$$

and the magnitude of the velocity of each particle is

$$v = \omega R = \omega r \sin \beta$$

The direction of the velocity is perpendicular to the position vector. Hence the angular-momentum vector of each mass has a magnitude

$$|\mathbf{L}_1| = |\mathbf{L}_2| = m|\mathbf{r} \times \mathbf{v}| = mrv = m\omega r^2 \sin \beta \qquad (13.39)$$

The direction of the angular-momentum vector of each mass is perpendicular to both the velocity and the position vectors, as specified by the right-hand rule. The angular-momentum vector of each mass is shown in Fig. 13.18; these vectors are parallel to each other, they are in the plane of the axis and the rod, and they make an angle of $90° - \beta$ with the axis. The total angular momentum is the vector sum of these individual angular momenta. This vector is in the same direction as the individual angular-momentum vectors, and it has a magnitude twice as large as either of those in Eq. (13.39):

$$L = 2m\omega r^2 \sin \beta \qquad (13.40)$$

As the body rotates, so does the angular-momentum vector, remaining in the plane of the axis and the rod. If at one instant the angular momentum lies in the z–y plane, a quarter of a cycle later it will lie in the z–x plane, etc.

COMMENT: Note that the z component of the angular momentum is

$$L_z = L \cos(90° - \beta) = 2m\omega r^2 \sin \beta \cos(90° - \beta) = 2m\omega r^2 \sin^2\beta$$

This can also be written as

$$L_z = 2m\omega R^2 \qquad (13.41)$$

where $R = r \sin \beta$ is the perpendicular distance between each mass and the axis of rotation. Since $2mR^2$ is simply the moment of inertia of the two particles about the z axis, Eq. (13.41) is the same as

$$L_z = I\omega \qquad (13.42)$$

As we will see below, this formula is of general validity for rotation around a fixed axis.

The preceding example shows that *the angular-momentum vector of a rotating body need not always lie along the axis of rotation.* However, if the body is symmetric about the axis of rotation, then the angular-momentum vector will lie along this axis. In such a symmetric body, each particle on one side of the axis has a counterpart on the other side of the axis, and when we add the angular-momentum vectors contributed by these two particles (or any other pair of particles), the resultant lies along the axis of rotation (see Fig. 13.19).

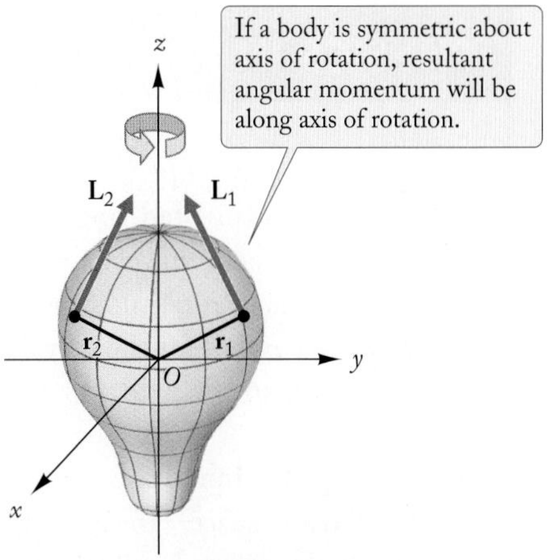

FIGURE 13.19 For a rotating symmetric body, the angular momentum is always along the axis of rotation.

Since Newton's Second Law for translational motion states that the rate of change of the momentum equals the force, the analogy between the equations for translational and rotational motion suggests that the rate of change of the angular momentum should equal the torque. It is easy to verify this for the case of a single particle. With the usual rule for the differentiation of a product,

$$\frac{d}{dt}\mathbf{L} = \frac{d}{dt}(\mathbf{r} \times \mathbf{p}) = \frac{d\mathbf{r}}{dt} \times \mathbf{p} + \mathbf{r} \times \frac{d\mathbf{p}}{dt} \qquad (13.43)$$

The first term on the right side is

$$\frac{d\mathbf{r}}{dt} \times \mathbf{p} = \mathbf{v} \times (m\mathbf{v}) = m(\mathbf{v} \times \mathbf{v}) = 0 \qquad (13.44)$$

This is zero because the cross product of a vector with itself is always zero. According to Newton's Second Law, the second term on the right side of Eq. (13.43) is

$$\mathbf{r} \times \frac{d\mathbf{p}}{dt} = \mathbf{r} \times \mathbf{F} \qquad (13.45)$$

where **F** is the force acting on the particle. Therefore, Eq. (13.43) becomes

$$\frac{d\mathbf{L}}{dt} = \mathbf{r} \times \mathbf{F} = \boldsymbol{\tau} \qquad (13.46)$$

In the case of a rigid body, the angular momentum is the sum of all the angular momenta of the particles in the body, and the rate of change of this total angular momentum can be shown to equal the net external torque:

equation of rotational motion for vector angular momentum

$$\frac{d\mathbf{L}}{dt} = \boldsymbol{\tau} \qquad (13.47)$$

This equation for the rate of change of the angular momentum of a rigid body is analogous to the equation $d\mathbf{p}/dt = \mathbf{F}$ for the rate of change of the translational momentum of a particle.

To compare the vector equation (13.47) with our earlier equation $I\alpha = \tau$, we must focus our attention on the component of the angular momentum along the axis of rotation, that is, the z axis. Figure 13.20 shows an arbitrary rigid body rotating about a fixed axis, which coincides with the z axis. As in Example 11, the angular-momentum vector of this body makes an angle with the axis of rotation. However, as we discussed in Example 11, the z component of the angular momentum of each particle in the rotating body is simply equal to its moment of inertia about the z axis multiplied by the angular velocity [see Eq. (13.42)]. Hence, when we sum the contributions of all the particles in the rotating body, we find that the z component of the net angular momentum of the entire rotating body equals the net moment of inertia of the entire body multiplied by the angular velocity. This establishes that the equation

$$L_z = I\omega \qquad (13.48)$$

is of general validity.

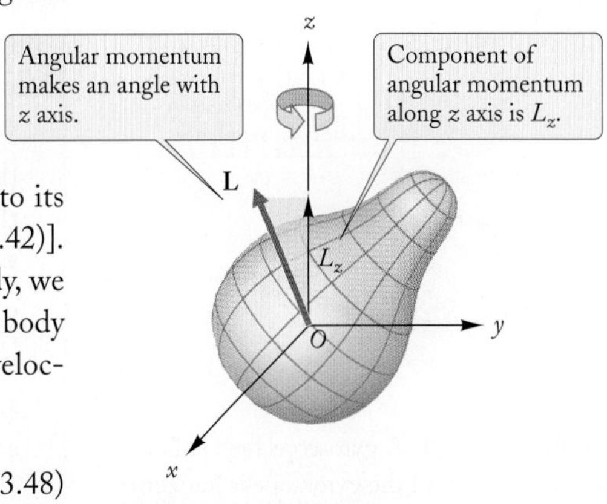

FIGURE 13.20 A body rotating about the z axis.

PHYSICS IN PRACTICE — THE GYROCOMPASS

Concepts — in — Context

A gyroscope is a flywheel suspended in gimbals (pivoted rings; see Fig. 1). The angular-momentum vector of the flywheel lies along its axis of rotation. Since there are no torques on this flywheel, except for the very small and negligible frictional torques in the pivots of the gimbals, the angular-momentum vector remains constant in both magnitude and direction. Hence the direction of the axis of spin remains fixed in space—the gyroscope can be carried about, its base can be twisted and turned in any way, and yet the axis always continues to point in its original direction. Thus, the gyroscope serves as a compass. High-precision gyroscopes are used in the inertial-guidance systems for ships, aircraft, rockets, and spacecraft (see Fig. 2). They provide an absolute reference direction relative to which the orientation of the vehicle can be established. In such applications, three gyroscopes aimed along mutually perpendicular axes define the absolute orientation of an x, y, z coordinate grid.

The best available high-precision gyroscopes, such as those used in the inertial-guidance system of the Hubble Space Telescope, are capable of maintaining a fixed reference direction with a deviation, or drift, of no more than 10 arc-seconds per hour. The special gyroscopes developed for the *Gravity Probe B* experiment are even better than that; their drift is less than 1 milliarcsecond per year!

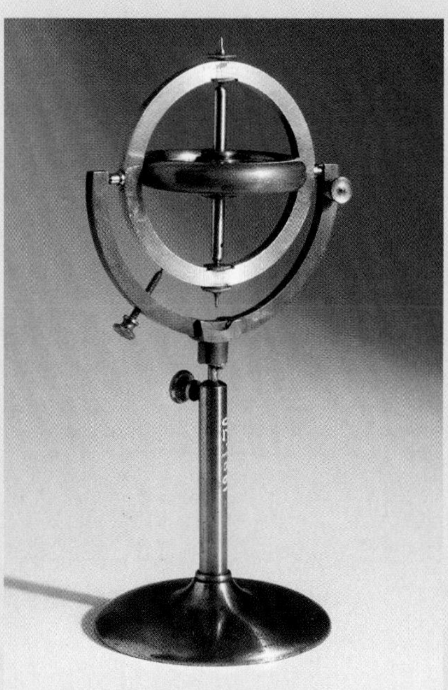

FIGURE 1 Gyroscope mounted in gimbals.

FIGURE 2 Internal-guidance system for an Atlas rocket. This system contains gyroscopes to sense the orientation of the rocket and accelerometers to measure the instantaneous acceleration. From these measurements, computers calculate the position of the rocket and guide it along the intended flight path.

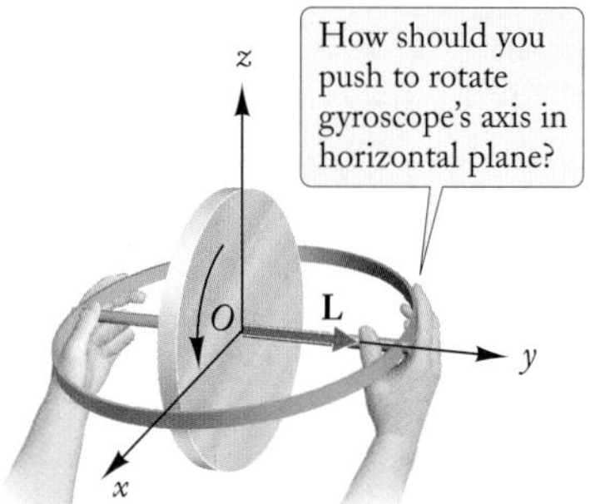

FIGURE 13.21 A gyroscope held in both hands. The axis of the gyroscope is horizontal, and the hands twist this axis sideways through an angle in the x–y plane.

EXAMPLE 12 You grasp the gimbals of a spinning gyroscope with both hands and you forcibly twist the axis of the gyroscope through an angle in the horizontal plane (see Fig. 13.21). If the angular momentum of the gyroscope spinning about its axis is 3.0×10^{-2} J·s, what are the magnitude and the direction of the torque you need to exert to twist the axis of the gyroscope at a constant rate through 90° in the horizontal plane in 1.0s?

SOLUTION: Figure 13.22a shows the angular-momentum vector **L** of the spinning gyroscope at an initial time and the new angular-momentum vector **L** + d**L** after you have turned the gyroscope through a small angle $d\beta$. From the figure, we see that d**L** is approximately perpendicular to **L**, and that the magnitude of d**L** is

$$dL = L\,d\beta$$

Hence

$$\frac{dL}{dt} = L\frac{d\beta}{dt} \tag{13.49}$$

According to Eq. (13.49), the magnitude of the torque is

$$\tau = \frac{dL}{dt} = L\frac{d\beta}{dt}$$

With $L = 3.0 \times 10^{-2}$ J·s and $d\beta/dt = (90°)/(1.0 \text{ s}) = \pi/2$ radians/s,

$$\tau = 3.0 \times 10^{-2} \text{ J·s} \times \frac{\pi}{2} \text{ radians/s} = 4.7 \times 10^{-2} \text{ N·m}$$

Since $\boldsymbol{\tau} = d\mathbf{L}/dt$, the direction of the torque vector $\boldsymbol{\tau}$ must be the direction of $d\mathbf{L}$; that is, the torque vector must be perpendicular to $\mathbf{L}$, or initially into the plane of the page (see Fig. 13.22b). To produce such a torque, your left hand must push up, and your right hand must pull down. This is contrary to intuition, which would suggest that to twist the axis in the horizontal plane, you should push forward with your right hand and pull back with your left! This surprising behavior also explains why a downward gravitational force causes the slow precession of a spinning top, as considered in the next example.

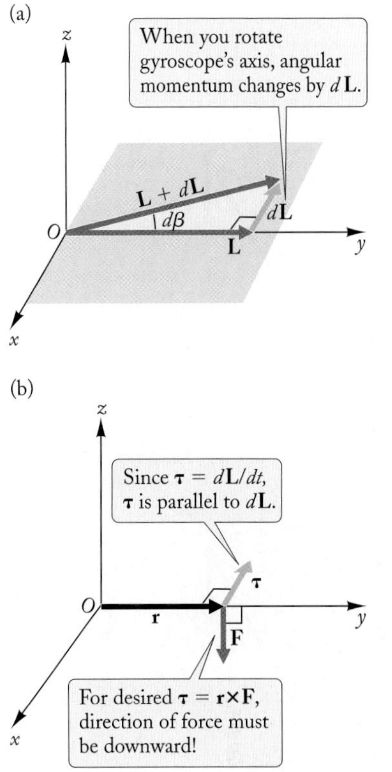

FIGURE 13.22 (a) $d\mathbf{L}$ is approximately perpendicular to $\mathbf{L}$, in the x–y plane. (b) The torque $\boldsymbol{\tau}$ is parallel to $d\mathbf{L}$, also in the x–y plane.

EXAMPLE 13 A toy top spins with angular momentum of magnitude L; the axis of rotation is inclined at an angle θ with respect to the vertical (see Fig. 13.23). The spinning top has mass M; its point of contact with the ground remains fixed, and its center of mass is a distance r from the point of contact. The top *precesses;* that is, its angular-momentum vector rotates about the vertical. Find the angular velocity Ω_p of this precessional motion. If a top has $r = 4.0$ cm and moment of inertia $I = MR^2/4$, where $R = 3.0$ cm, find the period of the precessional motion when the top is spinning at 250 radians/s.

SOLUTION: From Fig. 13.24a, we see that the weight, Mg, acting at the center of mass, produces a torque τ of magnitude

$$\tau = rMg\sin\theta \tag{13.50}$$

As in Example 12, the change in angular momentum $d\mathbf{L}$ will be parallel to the torque, since $\boldsymbol{\tau} = d\mathbf{L}/dt$. In a time dt, the top will precess though an angle $d\beta$ given by (see Fig. 13.24b)

$$d\beta = \frac{dL}{L\sin\theta}$$

Using $dL = \tau\, dt = rMg \sin\theta\, dt$, we thus have

$$d\beta = \frac{rMg\sin\theta\, dt}{L\sin\theta} = \frac{rMg}{L}dt$$

The precessional angular velocity is the rate of change of this angle:

$$\Omega_p = \frac{d\beta}{dt} = \frac{rMg}{L} \tag{13.51}$$

Thus the angular velocity of precession is independent of the tilt angle θ.

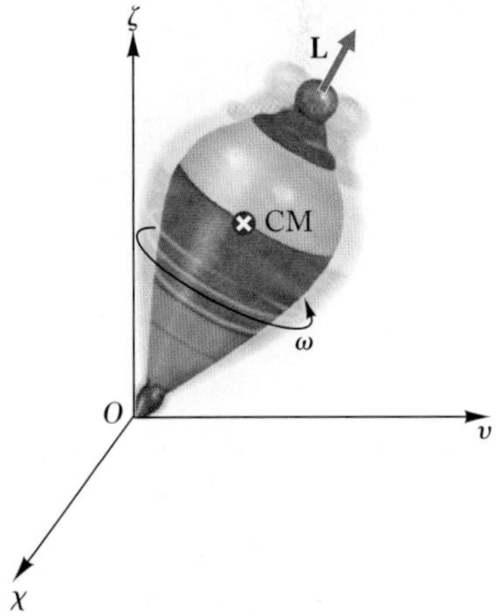

FIGURE 13.23 A tilted top spinning with angular velocity ω.

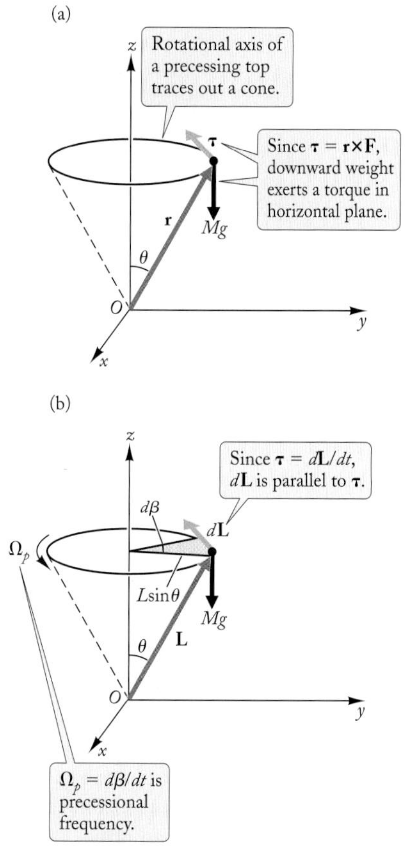

FIGURE 13.24 (a) The weight of the top, acting at the center of mass (a distance r from the point of contact), produces a torque perpendicular to $\mathbf{r}$ and to the weight. (b) The torque is parallel to $d\mathbf{L}$, which results in a slow precession around a vertical axis at an angular velocity Ω_p.

The period of the precession is related to the precessional angular velocity by

$$T = \frac{2\pi}{\Omega_p} = \frac{2\pi L}{rMg}$$

For the particular top described, we insert the angular momentum

$$L = I\omega = \frac{MR^2}{4}\omega$$

and obtain

$$T = \frac{2\pi MR^2\omega}{4rMg} = \frac{\pi R^2\omega}{2rg}$$

$$= \frac{\pi \times (0.030\,\text{m})^2 \times 250\,\text{radians/s}}{2 \times 0.040\,\text{m} \times 9.81\,\text{m/s}^2}$$

$$= 0.90\,\text{s}$$

Since this precessional period is proportional to ω, we see that as the spinning of the top slows down, the top will precess with a shorter period, that is, more quickly.

✔ Checkup 13.4

QUESTION 1: A particle has a nonzero position vector $\mathbf{r}$ and a nonzero momentum $\mathbf{p}$. Can the angular momentum of this particle be zero?

QUESTION 2: What is the angle between the momentum vector $\mathbf{p}$ and the angular-momentum vector $\mathbf{L}$ of a particle?

QUESTION 3: Suppose that instead of calculating the angular momentum of the dumbbell shown in Fig. 13.17 about the center, we calculate it about an origin on the z axis at some distance below the center. What are the directions of the individual angular-momentum vectors of the two masses m in this case? What is the direction of the total angular momentum?

QUESTION 4: Is a torque required to keep the dumbbell in Fig. 13.18 rotating around the z axis at constant angular velocity?

QUESTION 5: What is the direction of the angular-momentum vector of the rotating minute hand on your watch (calculated with respect to an origin at the center of the watch face)?

(A) In the direction that the minute hand points
(B) Antiparallel to the direction that the minute hand points
(C) In the plane of the watch face, but perpendicular to the minute hand
(D) Perpendicularly out of the face of the watch
(E) Perpendicularly into the face of the watch

SUMMARY

PROBLEM-SOLVING TECHNIQUES Torques and Rotational Motion **(page 405)**

PROBLEM-SOLVING TECHNIQUES Conservation of Angular Momentum **(page 410)**

PHYSICS IN PRACTICE The Gyrocompass **(page 414)**

TORQUE where θ is the angle between the force **F** and the radial line of length R.

$$\tau = FR \sin\theta \quad (13.3)$$

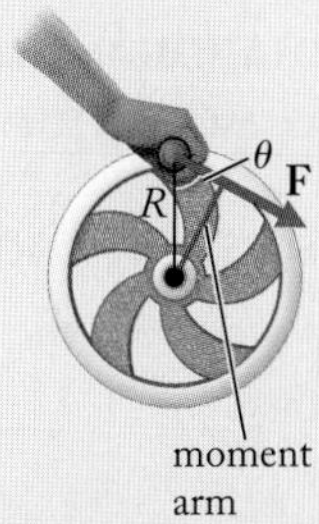

WORK DONE BY TORQUE

$$W = \int \tau \, d\phi \quad (13.4)$$

WORK DONE BY A CONSTANT TORQUE

$$W = \tau \Delta\phi \quad (13.5)$$

POWER DELIVERED BY TORQUE where ω is the angular velocity.

$$P = \tau\omega \quad (13.6)$$

CONSERVATION OF ENERGY IN ROTATIONAL MOTION

$$E = \tfrac{1}{2}I\omega^2 + U = [\text{constant}] \quad (13.10)$$

EQUATION OF ROTATIONAL MOTION (Fixed axis) where I is the moment of inertia and α is the angular acceleration.

$$I\alpha = \tau \quad (13.19)$$

ANGULAR MOMENTUM OF ROTATION

$$L = I\omega \quad (13.26)$$

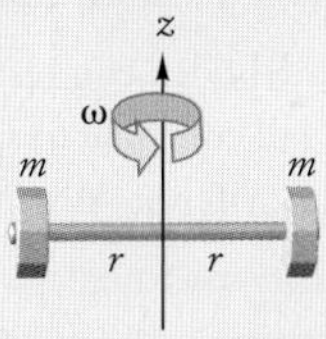

CONSERVATION OF ANGULAR MOMENTUM

$$I\omega = [\text{constant}] \quad (13.29)$$

ANGULAR MOMENTUM OF PARTICLE (In circular orbit)

$$L = mvr \quad (13.34)$$

ANGULAR MOMENTUM VECTOR

$$\mathbf{L} = \mathbf{r} \times \mathbf{p} \quad (13.37)$$

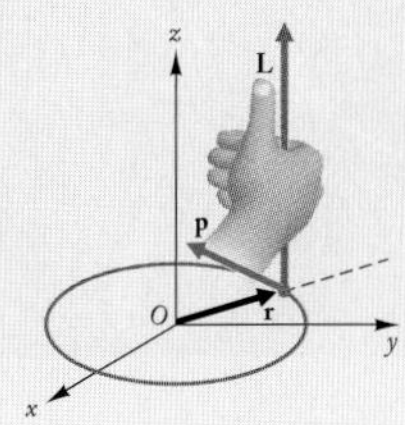

TORQUE VECTOR

$$\boldsymbol{\tau} = \mathbf{r} \times \mathbf{F} \tag{13.35}$$

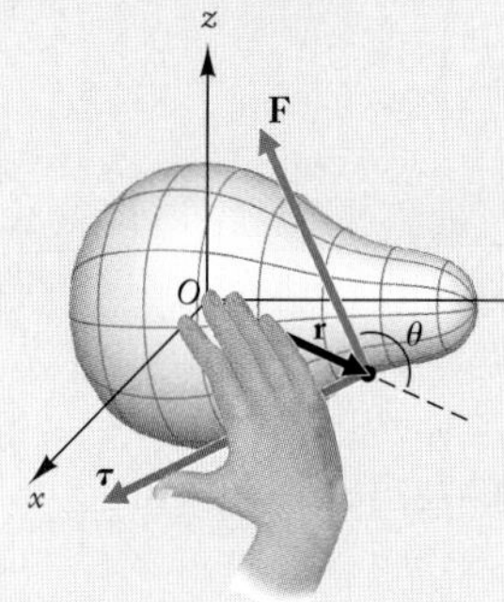

EQUATION OF ROTATIONAL MOTION FOR VECTOR ANGULAR MOMENTUM

$$\frac{d\mathbf{L}}{dt} = \boldsymbol{\tau} \tag{13.47}$$

GYROSCOPIC PRECESSION ANGULAR VELOCITY

where r is the distance from the point of contact to the center of mass.

$$\Omega_p = \frac{rMg}{L} \tag{13.51}$$

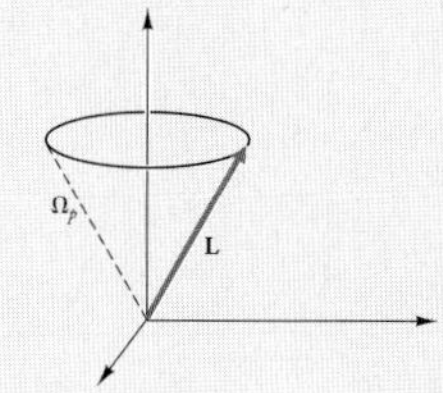

QUESTIONS FOR DISCUSSION

1. Suppose you push down on the rim of a stationary phonograph turntable. What is the direction of the torque you exert about the center of the turntable?
2. Many farmers have been injured when their tractors suddenly flipped over backward while pulling a heavy piece of farm equipment. Can you explain how this happens?
3. Rifle bullets are given a spin about their axis by spiral grooves ("rifling") in the barrel of the gun. What is the advantage of this?
4. You are standing on a frictionless turntable (like a phonograph turntable, but sturdier). How can you turn 180° without leaving the turntable or pushing against any exterior body?
5. If you give a hard-boiled egg resting on a table a twist with your fingers, it will continue to spin. If you try doing the same with a raw egg, it will not. Why?
6. A tightrope walker uses a balancing pole to keep steady (Fig. 13.25). How does this help?
7. Why do helicopters need a small vertical propeller on their tail?
8. The rate of rotation of the Earth is subject to small seasonal variations. Does this mean that angular momentum is not conserved?
9. Why does the front end of an automobile dip down when the automobile is braking sharply?
10. The friction of the tides against the ocean coasts and the ocean shallows is gradually slowing down the rotation of the Earth. What happens to the lost angular momentum?
11. An automobile is traveling on a straight road at 90 km/h. What is the speed, relative to the ground, of the lowermost point on one of its wheels? The topmost point? The midpoint?
12. A sphere and a hoop of equal masses roll down an inclined plane without slipping. Which will get to the bottom first? Will they have equal kinetic energy when they reach the bottom?
13. A yo-yo rests on a table (Fig. 13.26). If you pull the string horizontally, which way will it move? If you pull vertically?

FIGURE 13.25 A tightrope walker.

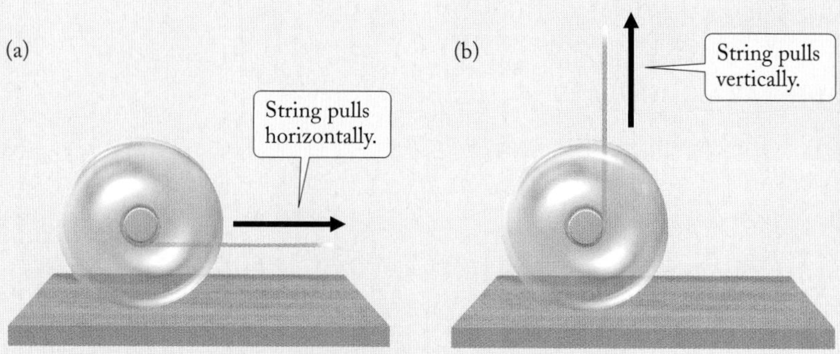

FIGURE 13.26 Yo-yo resting on a table. (a) String pulls horizontally. (b) String pulls vertically.

14. Stand a pencil vertically on its point on a table and let go. The pencil will topple over.
   (a) If the table is very smooth, the point of the pencil will slip in the direction opposite to that of the toppling. Why?
   (b) If the table is somewhat rough, or covered with a piece of paper, the point of the pencil will jump in the direction of the toppling. Why? (Hint: During the early stages of the toppling, friction holds the point of the pencil fixed; thus the pencil acquires horizontal momentum.)
15. An automobile travels at constant speed along a road consisting of two straight segments connected by a curve in the form of an arc of a circle. Taking the center of the circle as origin, what is the direction of the angular momentum of the automobile? Is the angular momentum constant as the automobile travels along this road?
16. Is the angular momentum of the orbital motion of a planet constant if we choose an origin of coordinates on the Sun?
17. A pendulum is swinging back and forth. Is the angular momentum of the pendulum bob constant?
18. What is the direction of the angular-momentum vector of the rotation of the Earth?
19. A bicycle is traveling east along a level road. What are the directions of the angular-momentum vectors of its wheels?

PROBLEMS

13.1 Work, Energy, and Power in Rotational Motion; Torque

1. The operating instructions for a small crane specify that when the boom is at an angle of 20° above the horizontal (Fig. 13.27), the maximum safe load for the crane is 500 kg. Assuming that this maximum load is determined by the maximum torque that the pivot can withstand, what is the maximum torque for 20° in terms of length R of the boom? What is the maximum safe load for 40°? For 60°?

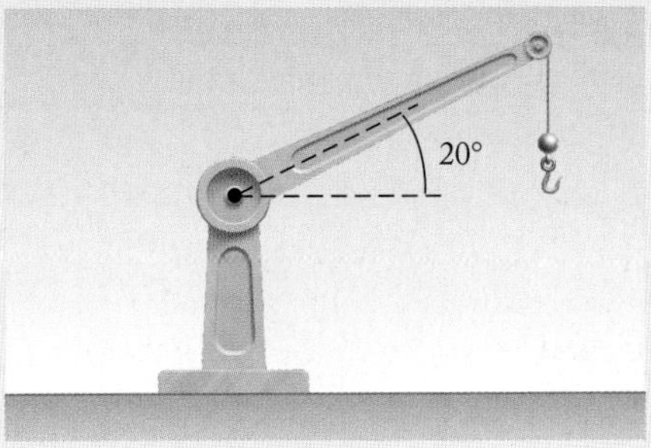

FIGURE 13.27 Small crane.

2. A simple manual winch consists of a drum of radius 4.0 cm to which is attached a handle of radius 25 cm (Fig. 13.28). When you turn the handle, the rope winds up on the drum and pulls the load. Suppose that the load carried by the rope is 2500 N. What force must you exert on the handle to hold this load?

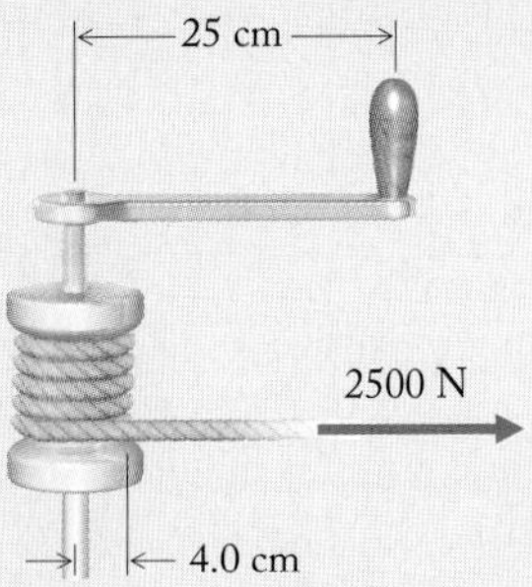

FIGURE 13.28 Manual winch.

3. The repair handbook for an automobile specifies that the cylinder-head bolts are to be tightened to a torque of 62 N·m. If a mechanic uses a wrench of length 20 cm on such a bolt, what perpendicular force must he exert on the end of this wrench to achieve the correct torque?
4. A 2.0-kg trout hangs from one end of a 2.0-m-long stiff fishing pole that the fisherman holds with one hand by the other end. If the pole is horizontal, what is the torque that the weight of the trout exerts about the end the fisherman holds? If the pole is tilted upward at an angle of 60°?
5. You hold a 10-kg book in your hand with your arm extended horizontally in front of you. What is the torque that the weight of this book exerts about your shoulder joint, at a distance of 0.60 m from the book?
6. If you bend over, so your trunk is horizontal, the weight of your trunk exerts a rather strong torque about the sacrum, where your backbone is pivoted on your pelvis. Assume that the mass of your trunk (including arms and head) is 48 kg, and that the weight effectively acts at a distance of 0.40 m from the sacrum. What is the torque that this weight exerts?
7. The engine of an automobile delivers a maximum torque of 203 N·m when running at 4600 rev/min, and it delivers a maximum power of 142 hp when running at 5750 rev/min. What power does the engine deliver when running at maximum torque? What torque does it deliver when running at maximum power?
8. The flywheel of a motor is connected to the flywheel of a pump by a drive belt (Fig. 13.29). The first flywheel has a radius R_1, and the second a radius R_2. While the motor wheel is rotating at a constant angular velocity ω_1, the tensions in the upper and the lower portions of the drive belt are T and T', respectively. Assume that the drive belt is massless.
   (a) What is the angular velocity of the pump wheel?
   (b) What is the torque of the drive belt on each wheel?

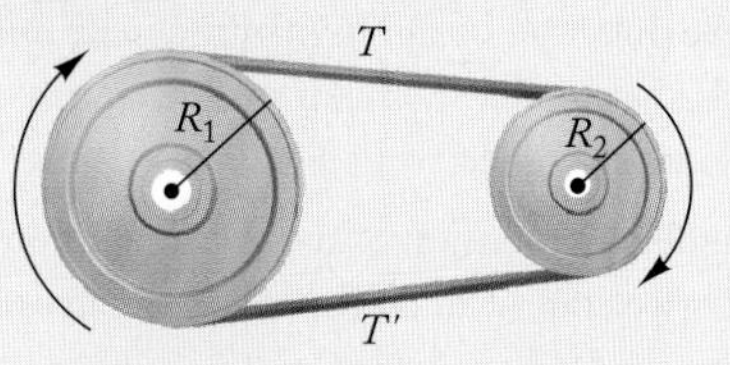

FIGURE 13.29 Motor and pump wheels connected by a drive belt.

(c) By taking the product of torque and angular velocity, calculate the power delivered by the motor to the drive belt, and the power removed by the pump from the drive belt. Are these powers equal?

9. The Wright Cyclone engine on a DC-3 airplane delivers a power of 850 hp with the propeller revolving steadily at 2100 rev/min. What is the torque exerted by air resistance on the propeller?

10. A woman on an exercise bicycle has to exert an (average) tangential push of 35 N on each pedal to keep the wheel turning at constant speed. Each pedal has a radial length of 0.18 m. If she pedals at the rate of 60 rev/min, what is the power she expends against the exercise bicycle? Express your answer in watts and in kilocalories per minute.

11. With what translational speed does the upper end of the meterstick in Example 2 hit the floor? If, instead of a 1.0-m stick, we use a 2.0-m stick, with what translational speed does it hit?

12. A ceiling fan uses 0.050 hp to maintain a rotational frequency of 150 rev/min. What torque does the motor exert?

13. The motor of a grinding wheel exerts a torque of 0.65 N·m to maintain an operating speed of 3450 rev/min. What power does the motor deliver?

14. From the human-body data of Fig. 10.17, calculate (a) the torque about the shoulder for an arm held horizontally and (b) the torque about the hip for a leg held horizontally.

15. A large grinding table is used to thin large batches of silicon wafers in the final stage of semiconductor manufacturing, a process called *backlap*. If the driving motor exerts a torque of 250 N·m while rotating the table 1200 times for one batch of wafers, how much work does the motor do?

16. Recently, a microfabricated torque sensor measured a torque as small as 7.5×10^{-24} N·m. If the torque is produced by a force applied perpendicular to the sensor at a distance of 25 μm from the axis of rotation, what is the smallest force that the sensor can detect?

*17. The angular position of a ceiling fan during the first two seconds after start-up is given by $\phi = Ct^2$, where $C =$ 7.5 radians/s^2 and t is in seconds. If the fan motor exerts a torque of 2.5 N·m, how much work has the motor done after $t = 1.0$ s? After $t = 2.0$ s?

*18. While braking, a 1500-kg automobile decelerates at the rate of 8.0 m/s^2. What is the magnitude of the braking force that the road exerts on the automobile? What torque does this force generate about the center of mass of the automobile? Will this torque tend to lift the front end of the automobile or tend to depress it? Assume that the center of mass of the automobile is 60 cm above the surface of the road.

*19. A tractor of mass 4500 kg has rear wheels of radius 0.80 m. What torque and what power must the engine supply to the rear axle to move the tractor up a road of slope 1:3 at a constant speed of 4.0 m/s?

*20. A bicycle and its rider have a mass of 90 kg. While accelerating from rest to 12 km/h, the rider turns the pedals through three full revolutions. What torque must the rider exert on the pedals? Assume that the torque is constant during the acceleration and ignore friction within the bicycle mechanism.

*21. A meterstick is held to a wall by a nail passing through the 60-cm mark (Fig. 13.30). The meterstick is free to swing about this nail, without friction. If the meterstick is released from an initial horizontal position, what angular velocity will it attain when it swings through the vertical position?

FIGURE 13.30 A meterstick.

*22. A uniform solid sphere of mass M and radius R hangs from a string of length $R/2$. Suppose the sphere is released from an initial position making an angle of 45° with the vertical (Fig. 13.31).

(a) Calculate the angular velocity of the sphere when it swings through the vertical position.

(b) Calculate the tension in the string at this instant.

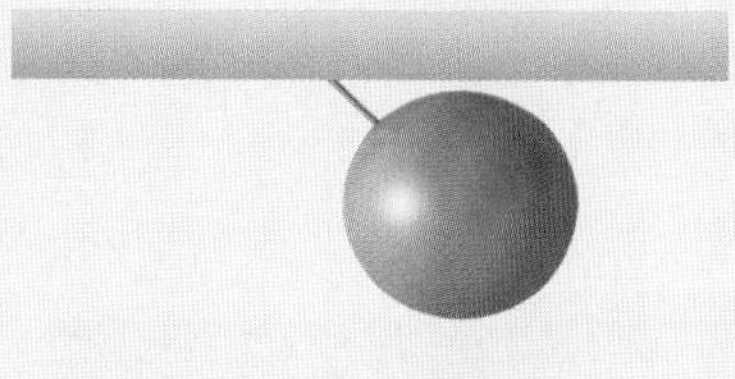

FIGURE 13.31 A hanging sphere.

*23. The maximum (positive) acceleration an automobile can achieve on a level road depends on the maximum torque the engine can deliver to the wheels.

(a) The engine of a Maserati sports car delivers a maximum torque of 441 N·m to the gearbox. The gearbox steps down the rate of revolution by a factor of 2.58; that is, whenever the engine makes 2.58 revolutions, the wheels make 1 revolution. What is the torque delivered to the wheels? Ignore frictional losses in the gearbox.

(b) The mass of the car (including fuel, driver, etc.) is 1770 kg, and the radius of its wheels is 0.30 m. What is the maximum acceleration? Ignore the moment of inertia of the wheels and frictional losses.

*24. An automobile of mass 1200 kg has four brake drums of diameter 25 cm. The brake drums are rigidly attached to the wheels of diameter 60 cm. The braking mechanism presses brake pads against the rim of each drum, and the friction between the pad and the rim generates a torque that slows the rotation of the wheel. Assume that all four wheels contribute equally to the braking. What torque must the brake pads exert on each drum in order to decelerate the automobile at 7.8 m/s^2? If the coefficient of friction between the pad and the drum is $\mu_k = 0.60$, what normal force must the brake pad exert on the rim of the drum? Ignore the masses of the wheels.

*25. In one of the cylinders of an automobile engine, the gas released by internal combustion pushes on the piston, which, in turn, pushes on the crankshaft by means of a piston rod (Fig. 13.32). If the crankshaft experiences a torque of 31 N·m and if the dimensions of the crankshaft and piston rod are as in Fig. 13.32, what must be the force of the gas on the piston when the crankshaft is in the horizontal position as in Fig. 13.32? Ignore friction, and ignore the masses of the piston and rod.

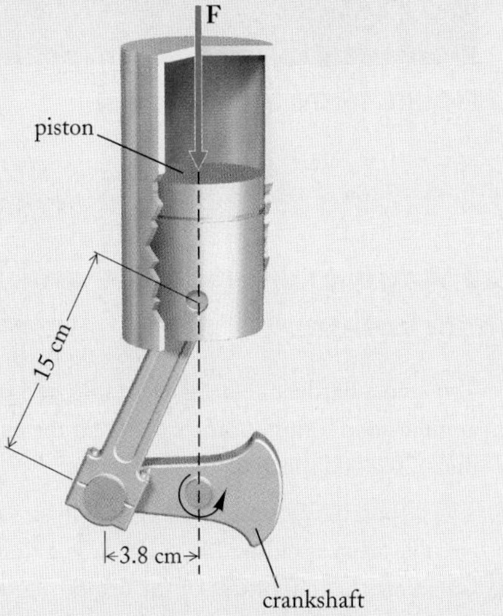

FIGURE 13.32 Automobile piston and crankshaft.

13.2 The Equation of Rotational Motion

26. While starting up a roulette wheel, the croupier exerts a torque of 100 N·m with his hand on the spokes of the wheel. What angular acceleration does this produce? Treat the wheel as a disk of mass 30 kg and radius 0.25 m.

27. The center span of a revolving drawbridge consists of a uniform steel girder of mass 300 metric tons and length 25 m. This girder can be regarded as a uniform thin rod. The bridge opens by rotating about a vertical axis through its center. What torque is required to open this bridge in 60 s? Assume that the bridge first accelerates uniformly through an angular interval of 45° and then the torque is reversed, so the bridge decelerates uniformly through an angular interval of 45° and comes to rest after rotating by 90°.

28. The original Ferris wheel, built by George Ferris, had a radius of 38 m and a mass of 1.9×10^6 kg. Assume that all of its mass was uniformly distributed along the rim of the wheel. If the wheel was initially rotating at 0.050 rev/min, what constant torque had to be applied to bring it to a full stop in 30 s? What force exerted on the rim of the wheel would have given such a torque?

29. The pulley of an Atwood machine for the measurement of g is a brass disk of mass 120 g. When using masses $m_1 = 0.4500$ kg and $m_2 = 0.4550$ kg, an experimenter finds that the larger mass descends 1.6 m in 8.0 s, starting from rest. What is the value of g?

30. A hula hoop rolls down a slope of 1:10 without slipping. What is the (linear) acceleration of the hoop?

31. A uniform cylinder rolls down a plane inclined at an angle θ with the horizontal. Show that if the cylinder rolls without slipping, the acceleration is $a = \frac{2}{3}g \sin\theta$.

32. The spare wheel of a truck, accidentally released on a straight road leading down a steep hill, rolls down the hill without slipping. The mass of the wheel is 60 kg, and its radius is 0.40 m; the mass distribution of the wheel is approximately that of a uniform disk. At the bottom of the hill, at a vertical distance of 120 m below the point of release, the wheel slams into a telephone booth. What is the total kinetic energy of the wheel just before impact? How much of this kinetic energy is translational energy of the center of mass of the wheel? How much is rotational kinetic energy about the center of mass? What is the speed of the wheel?

33. Galileo measured the acceleration of a sphere rolling down an inclined plane. Suppose that, starting from rest, the sphere takes 1.6 s to roll a distance of 3.00 m down a 20° inclined plane. What value of g can you deduce from this?

34. A yo-yo consists of a uniform disk with a string wound around the rim. The upper end of the string is held fixed. The yo-yo unwinds as it drops. What is its downward acceleration?

35. A man is trying to roll a barrel along a level street by pushing forward along its top rim. At the same time another man is pushing backward at the middle, with a force of equal magnitude F (see Fig. 13.33). The barrel rolls without slipping. Which way will the barrel roll? Find the magnitude and direction of the friction force at the point of contact with the street. The barrel is a uniform cylinder of mass M and radius R.

FIGURE 13.33 One man pushes horizontally at a cylinder's top; another pushes with equal force in the opposite direction at its middle. Which way does it roll?

36. An electric blender uniformly accelerates from rest beginning at $t = 0$; at $t = 0.50$ s, the blender has reached 250 radians/s and continues accelerating. If the rotating components have a moment of inertia of 2.0×10^{-4} kg·m^2, at what instantaneous rate is the motor delivering energy at $t = 0.50$ s?

37. A basketball is released from rest on a 15° incline. How many revolutions will the basketball undergo in 4.0 s? Assume the basketball is a thin spherical shell with a diameter of 23 cm, and that it rolls without slipping.

38. A 25-cm length of thin string is wound on the axle of a toy gyroscope that rotates in fixed bearings; the radius of the winding is 2.0 mm. If the string is pulled with a steady force of 5.0 N until completely unwound, how long does it take to complete the pull? What is the final angular velocity? The moment of inertia of the gyroscope (including axle) is 5.0×10^{-5} kg·m^2.

39. A phonograph turntable driven by an electric motor accelerates at a constant rate from 0 to 33.3 revolutions per minute in a time of 2.0 s. The turntable is a uniform disk of metal, of mass 1.2 kg and radius 15 cm. What torque is required to produce this acceleration? If the driving wheel makes contact with the turntable at its outer rim, what force must it exert?

*40. A bowling ball sits on the smooth floor of a subway car. If the car has a horizontal acceleration *a*, what is the acceleration of the ball? Assume that the ball rolls without slipping.

*41. A hoop rolls down an inclined ramp. The coefficient of static friction between the hoop and the ramp is μ_s. If the ramp is very steep, the hoop will slip while rolling. Show that the critical angle of inclination at which the hoop begins to slip is given by $\tan\theta = 2\mu_s$.

*42. A solid cylinder rolls down an inclined plane. The angle of inclination θ of the plane is large so that the cylinder slips while rolling. The coefficient of kinetic friction between the cylinder and the plane is μ_k. Find the rotational and translational accelerations of the cylinder. Show that the translational acceleration is the same as that of a block sliding down the plane.

**43. Suppose that a tow truck applies a horizontal force of 4000 N to the front end of an automobile similar to that described in Problem 63 of Chapter 12. Taking into account the rotational inertia of the wheels and ignoring frictional losses, what is the acceleration of the automobile? What is the percentage difference between this value of the acceleration and the value calculated by neglecting the rotational inertia of the wheels?

**44. A cart consists of a body and four wheels on frictionless axles. The body has a mass *m*. The wheels are uniform disks of mass *M* and radius *R*. Taking into account the moment of inertia of the wheels, find the acceleration of this cart if it rolls without slipping down an inclined plane making an angle θ with the horizontal.

**45. When the wheels of a landing airliner touch the runway, they are not rotating initially. The wheels first slide on the runway (and produce clouds of smoke and burn marks on the runway, which you may have noticed; see Fig. 13.34), until the sliding friction force has accelerated the wheels to the rotational speed required for rolling without slipping. From the following data, calculate how far the wheel of an airliner slips before it begins to roll without slipping: the wheel has a radius of 0.60 m and a mass of 160 kg, the normal force acting on the wheel is 2.0×10^5 N, the speed of the airliner is 200 km/h, and the coefficient of sliding friction for the wheel on the runway is 0.80. Treat the wheel as a uniform disk.

FIGURE 13.34 A landing airliner.

13.3 Angular Momentum and its Conservation

46. You spin a hard-boiled egg on a table, at 5.0 rev/s. What is the angular momentum of the egg? Treat the egg as a sphere of mass 70 g and mean diameter 5.0 cm.

47. The Moon moves around the Earth in an (approximately) circular orbit of radius 3.8×10^8 m in a time of 27.3 days. Calculate the magnitude of the orbital angular momentum of the Moon. Assume that the origin of coordinates is centered on the Earth.

48. At the Fermilab accelerator, protons of momentum 5.2×10^{-16} kg·m/s travel around a circular path of diameter 2.0 km. What is the orbital angular momentum of one of these protons? Assume that the origin is at the center of the circle.

49. Prior to launching a stone from a sling, a Bolivian native whirls the stone at 3.0 rev/s around a circle of radius 0.75 m. The mass of the stone is 0.15 kg. What is the angular momentum of the stone relative to the center of the circle?

50. A communications satellite of mass 100 kg is in a circular orbit of radius 4.22×10^7 m around the Earth. The orbit is in the equatorial plane of the Earth, and the satellite moves along it from west to east with a speed of 4.90×10^2 m/s. What is the magnitude of the angular momentum of this satellite?

51. According to Bohr's (oversimplified) theory, the electron in the hydrogen atom moves in one or another of several possible circular orbits around the nucleus. The radii and the orbital

velocities of the three smallest orbits are, respectively, 0.529×10^{-10} m, 2.18×10^6 m/s; 2.12×10^{-10} m, 1.09×10^6 m/s; and 4.76×10^{-10} m, 7.27×10^5 m/s. For each of these orbits calculate the orbital angular momentum of the electron, with the origin at the center. How do these angular momenta compare?

52. A high-speed meteoroid moves past the Earth along an (almost) straight line. The mass of the meteoroid is 150 kg, its speed relative to the Earth is 60 km/s, and its distance of closest approach to the center of the Earth is 1.2×10^4 km.
    (a) What is the angular momentum of the meteoroid in the reference frame of the Earth (origin at the center of the Earth)?
    (b) What is the angular momentum of the Earth in the reference frame of the meteoroid (origin at the center of the meteoroid)?

53. A train of mass 1500 metric tons runs along a straight track at 85 km/h. What is the angular momentum of the train about a point 50 m to the side of the track, left of the train? About a point on the track?

54. The electron in a hydrogen atom moves around the nucleus under the influence of the electric force of attraction, a central force pulling the electron toward the nucleus. According to the Bohr theory, one of the possible orbits of the electron is an ellipse of angular momentum $2\hbar$ with a distance of closest approach $(1 - 2\sqrt{2}/3)a_0$ and a distance of farthest recession $(1 + 2\sqrt{2}/3)a_0$, where $\hbar$ and a_0 are two atomic constants with the numerical values 1.05×10^{-34} kg·m²/s ("Planck's constant") and 5.3×10^{-11} m ("Bohr radius"), respectively. In terms of $\hbar$ and a_0, find the speed of the electron at the points of closest approach and farthest recession; then evaluate numerically.

55. According to a simple (but erroneous) model, the proton is a uniform rigid sphere of mass 1.67×10^{-27} kg and radius 1.0×10^{-15} m. The spin angular momentum of the proton is 5.3×10^{-35} J·s. According to this model, what is the angular velocity of rotation of the proton? What is the linear velocity of a point on its equator? What is the rotational kinetic energy? How does this rotational energy compare with the rest-mass energy mc^2?

56. What is the angular momentum of a Frisbee spinning at 20 rev/s about its axis of symmetry? Treat the Frisbee as a uniform disk of mass 200 g and radius 15 cm.

57. A phonograph turntable is a uniform disk of radius 15 cm and mass 1.4 kg. If this turntable accelerates from 0 rev/min to 78 rev/min in 2.5 s, what is the average rate of change of the angular momentum in this time interval?

58. The propeller shaft of a cargo ship has a diameter of 8.8 cm, a length of 27 m, and a mass of 1200 kg. What is the rotational kinetic energy of this propeller shaft when it is rotating at 200 rev/min? What is the angular momentum?

59. The Sun rotates about its axis with a period of about 25 days. Its moment of inertia is $0.20M_SR_S^2$, where M_S is its mass and R_S its radius. Calculate the angular momentum of rotation of the Sun. Calculate the total orbital angular momentum of all the planets; make the assumption that each planet moves in a circular orbit of radius equal to its mean distance from the Sun listed in Table 9.1. What percentage of the angular momentum of the Solar System is in the rotational motion of the Sun?

60. Suppose we measure the speed v_1 and the radial distance r_1 of a comet when it reaches perihelion. Use conservation of angular momentum and conservation of energy to determine the speed and the radius at aphelion.

61. A playground merry-go-round is rotating at 2.0 radians/s. Consider the merry-go-round to be a uniform disk of mass 20 kg and radius 1.5 m. A 25-kg child, moving along a radial line, jumps onto the edge of the merry-go-round. What is its new angular velocity? The child then kicks the ground until the merry-go-round (with the child) again rotates at 2.0 radians/s. If the child then walks radially inward, what will the angular velocity be when the child is 0.50 m from the center?

62. The moment of inertia of the Earth is approximately 0.331 $M_ER_E^2$. If an asteroid of mass 5.0×10^{18} kg moving at 150 km/s struck (and stuck in) the Earth's surface, by how long would the length of the day change? Assume the asteroid was traveling westward in the equatorial plane and struck the Earth's surface at 45°.

63. In a popular demonstration, a professor rotates on a stool at 0.50 rev/s, holding two 10-kg masses, each 1.0 m from the axis of rotation. If she pulls the weights inward until they are 10 cm from the axis, what is the new rotational frequency? Without the weights, the professor and stool have a moment of inertia of 6.0 kg·m² with arms extended and 4.0 kg·m² with arms pulled in.

64. In a demonstration, a bicycle wheel with moment of inertia 0.48 kg·m² is spun up to 18 radians/s, rotating about a vertical axis. A student holds the wheel while sitting on a rotatable stool. The student and stool are initially stationary and have a moment of inertia of 3.0 kg·m². If the student turns the bicycle wheel over so its axis points in the opposite direction, with what angular velocity will the student and stool rotate? For simplicity, assume the wheel is held overhead, so that the student, wheel, and stool all have the same axis of rotation.

65. A very heavy freight train made up of 250 cars has a total mass of 7700 metric tons. Suppose that such a train accelerates from 0 to 65 km/h on a track running exactly east from Quito, Ecuador (on the equator). The force that the engine exerts on the Earth will slow down the rotational motion of the Earth. By how much will the angular velocity of the Earth have decreased when the train reaches its final speed? Express your answer in revolutions per day. The moment of inertia of the Earth is $0.33M_ER_E^2$.

66. There are 1.1×10^8 automobiles in the United States, each of an average mass of 2000 kg. Suppose that one morning all these automobiles simultaneously start to move in an eastward direction and accelerate to a speed of 80 km/h.

(a) What total angular momentum about the axis of the Earth do all these automobiles contribute together? Assume that the automobiles travel at an average latitude of 40°.

(b) How much will the rate of rotation of the Earth change because of the action of these automobiles? Assume that the axis of the Earth remains fixed. The moment of inertia of the Earth is 8.1×10^{37} kg·m^2.

*67. Two artificial satellites of equal masses are in circular orbits of radii r_1 and r_2 around the Earth. The second has an orbit of larger radius than the first ($r_2 > r_1$). What is the speed of each? What is the angular momentum of each? Which has the larger speed? Which has the larger angular momentum?

*68. Consider the motion of the Earth around the Sun. Take as origin the point at which the Earth is today and treat the Earth as a particle.

(a) What is the angular momentum of the Earth about this origin today?

(b) What will be the angular momentum of the Earth about the same origin three months from now? Six months from now? Nine months from now? Is the angular momentum conserved?

*69. The friction of the tides on the coastal shallows and the ocean floors gradually slows down the rotation of the Earth. The period of rotation (length of a sidereal day) is gradually increasing by 0.0016 s per century. What is the angular deceleration (in radians/s^2) of the Earth? What is the rate of decrease of the rotational angular momentum? What is the rate of decrease of the rotational kinetic energy? The moment of inertia of the Earth about its axis is $0.331 M_E R_E^2$, where M_E is the mass of the Earth and R_E its equatorial radius.

*70. Phobos is a small moon of Mars. For the purposes of the following problem, assume that Phobos has a mass of 5.8×10^{15} kg and that it has a shape of a uniform sphere of radius 7.5×10^3 m. Suppose that a meteoroid strikes Phobos 5.0×10^3 m off center (Fig. 13.35) and remains stuck. If the momentum of the meteoroid was 3×10^{13} kg·m/s before impact and the mass of the meteoroid is negligible compared with the mass of Phobos, what is the change in the rotational angular velocity of Phobos?

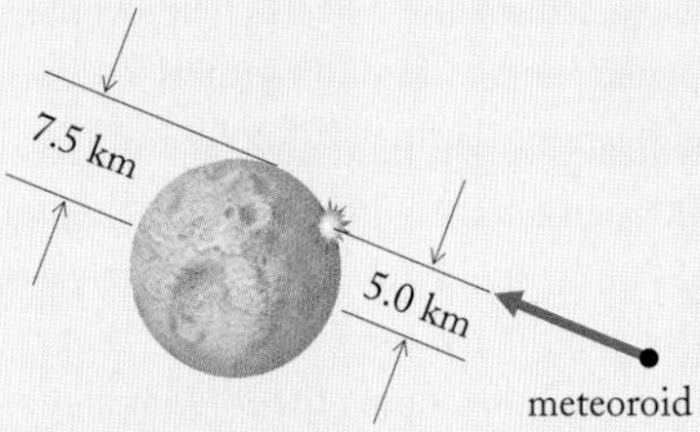

FIGURE 13.35 A meteoroid strikes Phobos.

*71. A woman stands in the middle of a small rowboat. The rowboat is floating freely and experiences no friction against the water. The woman is initially facing east. If she turns around 180° so that she faces west, through what angle will the rowboat turn? Assume that the woman performs her turning movement at constant angular velocity and that her moment of inertia remains constant during this movement. The moment of inertia of the rowboat about the vertical axis is 20 kg·m^2 and that of the woman is 0.80 kg·m^2.

*72. Two automobiles both of 1200 kg and both traveling at 30 km/h collide on a frictionless icy road. They were initially moving on parallel paths in opposite directions, with a center-to-center distance of 1.0 m (Fig. 13.36). In the collision, the automobiles lock together, forming a single body of wreckage; the moment of inertia of this body about its center of mass is 2.5×10^3 kg·m^2.

(a) Calculate the angular velocity of the wreck.

(b) Calculate the kinetic energy before the collision and after the collision. What is the change of kinetic energy?

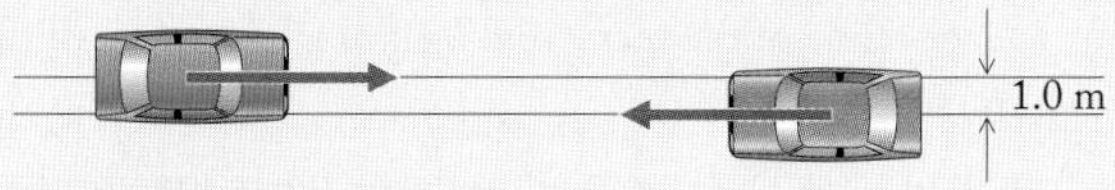

FIGURE 13.36 Two automobiles collide.

*73. In one experiment performed under weightless conditions in Skylab, the three astronauts ran around a path on the inside wall of the spacecraft so as to generate artificial gravity for their bodies (Fig. 13.37). Assume that the center of mass of each astronaut moves around a circle of radius 2.5 m; treat the astronauts as particles.

(a) With what speed must each astronaut run if the average normal force on his feet is to equal his normal weight (mg)?

(b) Suppose that before the astronauts begin to run, Skylab is floating in its orbit without rotating. When the astronauts begin to run clockwise, Skylab will begin to rotate counterclockwise. What will be the angular velocity of Skylab when the astronauts are running steadily with the speed calculated above? Assume that the mass of each astronaut is 70 kg and that the moment of inertia of Skylab about its longitudinal axis is 3×10^5 kg·m^2.

(c) How often must the astronauts run around the inside if they want Skylab to rotate through an angle of 30°?

FIGURE 13.37 Three astronauts about to start running around inside Skylab.

*74. A flywheel rotating freely on a shaft is suddenly coupled by means of a drive belt to a second flywheel sitting on a parallel shaft (Fig. 13.38). The initial angular velocity of the first flywheel is ω; that of the second is zero. The flywheels are uniform disks of masses M_1, M_2 and of radii R_1, R_2, respectively. The drive belt is massless and the shafts are frictionless.

(a) Calculate the final angular velocity of each flywheel.

(b) Calculate the kinetic energy lost during the coupling process. What happens to this energy?

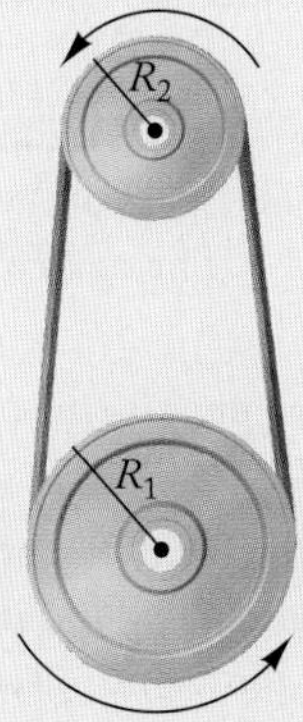

FIGURE 13.38 Two flywheels coupled by a drive belt.

*75. A thin rod of mass M and length l hangs from a pivot at its upper end. A ball of clay of mass m and of horizontal velocity v strikes the lower end at right angles and remains stuck (a totally inelastic collision). How high will the rod swing after this collision?

**76. If the melting of the polar ice caps were to raise the water level on the Earth by 10 m, by how much would the day be lengthened? Assume that the moment of inertia of the ice in the polar ice caps is negligible (they are very near the axis), and assume that the extra water spreads out uniformly over the entire surface of the Earth (that is, neglect the area of the continents compared with the area of the oceans). The moment of inertia of the Earth (now) is 8.1×10^{37} kg·m^2.

77. Consider a projectile of mass m launched with a speed v_0 at an elevation angle of 45°. If the launch point is the origin of coordinates, what is the angular momentum of the projectile at the instant of launch? At the instant it reaches maximum height? At the instant it strikes the ground? Is the angular momentum conserved in this motion with this choice of origin?

13.4 Torque and Angular Momentum as Vectors

78. Show that for a flat plate rotating about an axis perpendicular to the plate, the angular-momentum vector lies along the axis of rotation, even if the body is not symmetric.

79. A child's toy top consists of a uniform thin disk of radius 5.0 cm and mass 0.15 kg with a thin spike passing through its center. The lower part of the spike protrudes 6.0 cm from the disk. If you stand this top on its spike and start it spinning at 200 rev/s, what will be its precession frequency?

80. Suppose that the flywheel of a gyroscope is a uniform disk of mass 250 g and radius 3.0 cm. The distance of the center of this flywheel from the point of support is 4.0 cm. What is the precession frequency if the flywheel is spinning at 120 rev/s?

81. If a bicycle in forward motion begins to tilt to one side, the torque exerted by gravity will tend to turn the bicycle. Draw a diagram showing the angular momentum of a (slightly tilted) front wheel, the weight of the wheel, and the resulting torque. In which direction is the instantaneous change in angular momentum? Will this change make the tilt worse or better?

82. Slow precession can be used to determine a much more rapid rotational frequency. Consider a top made by inserting a small pin radially into a ball (a uniform sphere) of radius $R = 6.0$ cm. The pin extends 1.0 cm from the surface of the ball and supports the top. When set spinning, the top is observed to precess with a period of 0.75 s. What is the rotational frequency of the top?

*83. The wheel of an automobile has a mass of 25 kg and a diameter of 70 cm. Assume that the wheel can be regarded as a uniform disk.

(a) What is the angular momentum of the wheel when the automobile is traveling at 25 m/s (90 km/h) on a straight road?

(b) What is the rate of change of the angular momentum of the wheel when the automobile is traveling at the same speed along a curve of radius 80 m?

(c) For this rate of change of the angular momentum, what must be the torque on the wheel? Draw a diagram showing the path of the automobile, the angular-momentum vector of the wheel, and the torque vector.

*84. Consider the airplane propeller described in Problem 38 in Chapter 12. If the airplane is flying around a curve of radius 500 m at a speed of 360 km/h, what is the rate of change of the angular momentum of the propeller? What torque is required to change the angular momentum at this rate? Draw a diagram showing **L**, $d\mathbf{L}/dt$, and $\boldsymbol{\tau}$.

*85. A large flywheel designed for energy storage at a power plant has a moment of inertia of 5×10^5 kg·m^2 and spins at 3000 rev/min. Suppose that this flywheel is mounted on a horizontal axle oriented in the east–west direction. What are the magnitude and direction of its angular momentum? What is the rate of change of this angular momentum due to the rotational motion of the Earth and the consequent motion of the axle of the flywheel? What is the torque that the axle of the flywheel exerts against the bearings supporting it? If the bearings are at a distance of 0.60 m from the center of the flywheel on each side, what are the forces associated with this torque?

REVIEW PROBLEMS

86. A door is 0.80 m wide. What is the torque you exert about the axis passing through the hinges if you push against this door with a perpendicular force of 200 N at its middle? What is it if you push at the edge? A wind is blowing against the other side of the door and trying to push it open. Where should you push to keep the door closed?

87. An elevator of mass 900 kg is being lifted at constant speed by a cable wrapped around a wheel (see Fig. 13.39). The radius of the wheel is 0.35 m. What torque does the cable exert on the wheel?

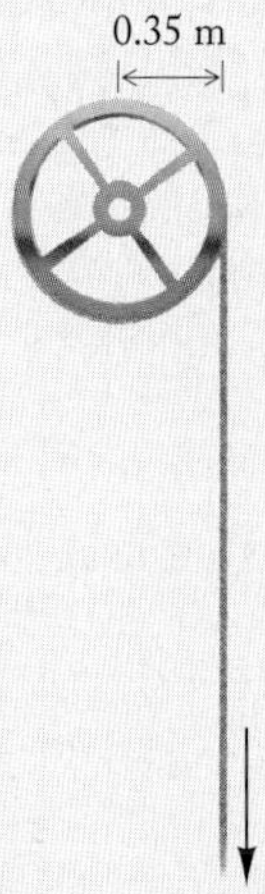

FIGURE 13.39 Elevator cable attached to a wheel.

88. Each of the two fuel turbopumps in the Space Shuttle delivers a power of 700 hp. The rotor of this pump rotates at 37 000 rev/min. What is the torque that the rotor exerts while pushing against the fuel?

89. A manual winch has a crank of length (radius) 0.25 m. If a laborer pushes against its handle tangentially with a force of 200 N, how much work does the laborer do while turning the crank through 10 revolutions?

90. A meterstick is initially standing vertically on the floor. If the meterstick falls over, with what angular velocity will it hit the floor? Assume that the end in contact with the floor experiences no friction and slips freely.

*91. A heavy hatch on a ship is made of a uniform plate of steel that measures 1.2 m × 1.2 m and has a mass of 400 kg. The hatch is hinged along one side; it is horizontal when closed, and it opens upward. A torsional spring assists in the opening of the hatch. The spring exerts a torque of 2.00×10^3 N·m when the hatch is horizontal and a torque of 0.30×10^3 N·m when the hatch is vertical; in the range of angles between horizontal and vertical, the torque decreases linearly (e.g., the torque is 1.15×10^3 N·m when the hatch is at 45°).

(a) At what angle will the hatch be in equilibrium so the spring exactly compensates the torque due to the weight?

(b) What minimum push must a sailor exert on the hatch to open it from the closed position? To close it from the open position? Assume that the sailor pushes perpendicularly on the hatch at the edge that is farthest from the hinge.

92. With your bicycle upside down on the ground, and the wheel free to rotate, you grasp the front wheel at the top and give it a horizontal push of 20 N. What is the instantaneous angular acceleration of the wheel? The wheel is a hoop of mass 4.0 kg and radius 0.33 m; ignore the mass of the spokes.

93. A toy top consists of a disk of radius 4.0 cm with a reinforced rim (a ring). The mass of the disk is 20 g, and the mass of the rim is 15 g. The mass of the pivot of this top is negligible.

(a) What is the moment of inertia of this top?

(b) When you give this top a twist and start it rotating at 100 rev/min on the floor, friction slows the top to a stop in 1.5 min. Assuming that the angular deceleration is uniform, what is the angular deceleration?

(c) What is the frictional torque on the top?

(d) What is the work done by the frictional torque?

94. The turntable of a record player is a uniform disk of radius 0.15 m and mass 1.2 kg. When in operation, it spins at $33\frac{1}{3}$ rev/min. If you switch the record player off, you find that the turntable coasts to a stop in 45 s.

(a) Calculate the frictional torque that acts on the turntable. Assume the torque is constant, that is, independent of the angular speed.

(b) Calculate the power that the motor of the record player must supply to keep the turntable in operation at $33\frac{1}{3}$ rev/min.

*95. A barrel of mass 200 kg and radius 0.50 m rolls down a 40° ramp without slipping. What is the value of the friction force acting at the point of contact between the barrel and the ramp? Treat the barrel as a cylinder of uniform density.

*96. A disk of mass M is free to rotate about a fixed horizontal axis. A string is wrapped around the rim of the disk, and a mass m is attached to this string (see Fig. 13.40). What is the downward acceleration of the mass?

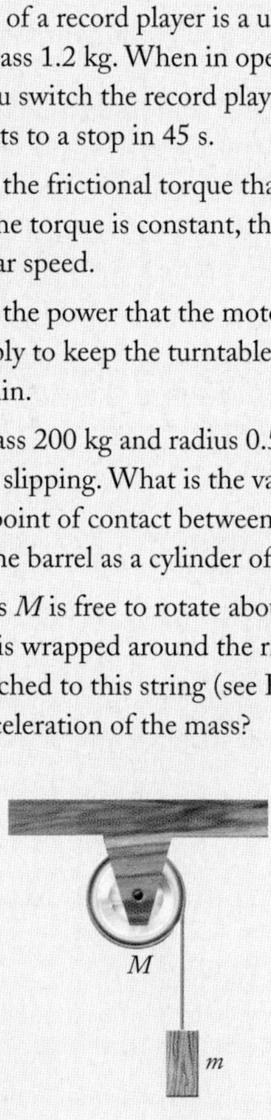

FIGURE 13.40 A mass m hanging from a disk.

*97. A hoop of mass M and radius R rolls down a sloping ramp that makes an angle of 30° with the ground. What is the acceleration of the hoop if it rolls without slipping?

*98. An automobile has the arrangement of the wheels shown in Fig. 13.41. The mass of this automobile is 1800 kg, the center of mass is at the midpoint of the rectangle formed by the wheels, and the moment of inertia about a vertical axis through the center of mass is 2200 kg·m^2. Suppose that during braking in an emergency, the left front and rear wheels lock and begin to skid while the right wheels continue to rotate just short of skidding. The coefficient of static friction between the wheels and the road is $\mu_s = 0.90$, and the coefficient of kinetic friction is $\mu_k = 0.50$. Calculate the instantaneous angular acceleration of the automobile about the vertical axis through the center of mass.

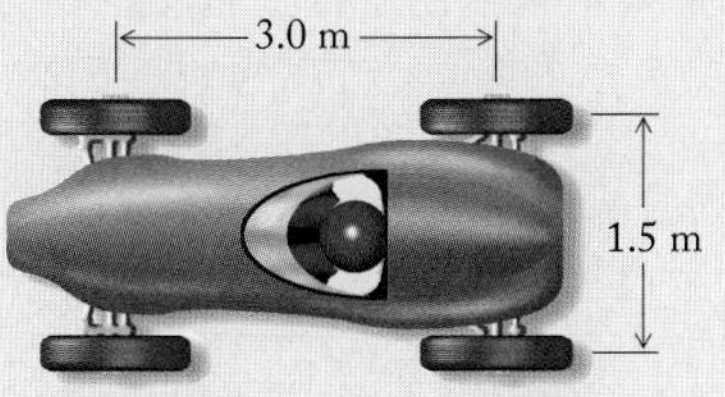

FIGURE 13.41 An automobile.

*99. Neutron stars, or pulsars, spin very quickly about their axes. Their high rate of spin is the result of the conservation of angular momentum during the formation of the neutron star by the gradual contraction (shrinking) of an initially normal star.

(a) Suppose that the initial star is similar to the Sun, with a radius of 7.0×10^8 m and a rate of rotation of 1.0 revolution per month. If this star contracts to a radius of 1.0×10^4 m, by what factor does the moment of inertia increase? Assume that the relative distribution of mass in the initial and the final stars is roughly the same.

(b) By what factor does the angular velocity increase? What is the final angular velocity?

*100. A rod of mass M and length l is lying on a flat, frictionless surface. A ball of putty of mass m and initial velocity v at right angles to the rod strikes the rod at a distance $l/4$ from the center (Fig. 13.42). The collision is inelastic, and the putty adheres to the rod.

(a) Where is the center of mass of the rod with adhering putty?

(b) What is the velocity of this center of mass after the collision?

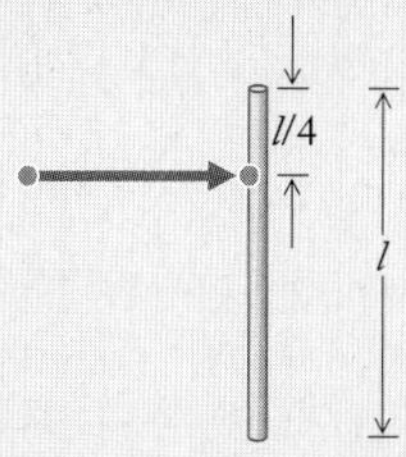

FIGURE 13.42 A ball of putty strikes a rod.

(c) What is the angular momentum about this center of mass? What is the moment of inertia, and what is the angular velocity?

*101. A communications satellite of mass 1000 kg is in a circular orbit of radius 4.22×10^7 m around the Earth. The orbit is in the equatorial plane of the Earth, and the satellite moves along it from west to east. What are the magnitude and the direction of the angular-momentum vector of this satellite?

*102. The spin angular momentum of the Earth has a magnitude of 5.9×10^{33} kg·m^2/s. Because of forces exerted by the Sun and the Moon, the spin angular momentum gradually changes direction, describing a cone of half-angle 23.5° (Fig. 13.43). The angular-momentum vector takes 26 000 years to swing once around this cone. What is the magnitude of the rate of change of the angular-momentum vector; that is, what is the value of $|d\mathbf{L}/dt|$?

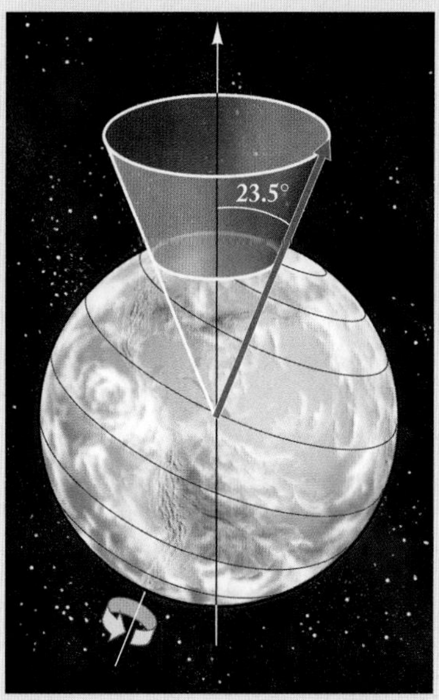

FIGURE 13.43 The precessing Earth.

Answers to Checkups

Checkup 13.1

1. You should place your hand at the end of the handle farthest from the bolt; this will provide the largest R in Eq. (13.2) and maximize the torque. Similarly, your push should be perpendicular to the wrench handle, in order to maximize $\sin\theta$ to the value $\sin 90° = 1$ in Eq. (13.2).
2. The direction must be toward the axis (along a radius), so that $\sin\theta = \sin 0° = 0$; thus both the torque and the work done will be zero.
3. Initially, when the stick is upright, the weight acts downward, along the radial direction, and so the torque is zero. As the stick falls, the weight (mg) and the point at which it acts ($R = y_{CM} = l/2$) remain constant. Only the angle between the force and the radial line changes; the sine of this angle is maximum just as the meterstick hits the floor (when $\sin\theta = \sin 90° = 1$), so the torque is maximum then.
4. (A) $\frac{1}{4}$. The work done is $W = \tau\,\Delta\phi = FR\sin\theta\,\Delta\phi$. In both cases, pushing at right angles implies $\sin\theta = 1$, and both angular displacements $\Delta\phi$ are the same. But with half the force applied at half the radius for the second push, the work will be one-fourth of that for the first push.

Checkup 13.2

1. The angular acceleration results from the torque exerted by gravity at the center of mass; this is maximum when the weight is perpendicular to the radial direction (when $\sin\theta = 1$). That occurs when the meterstick is horizontal, just before it hits the floor.
2. The translational kinetic energy is twice as large for a (uniform) rotating cylinder, because the rotational kinetic energy is $\frac{1}{2}I\omega^2 = \frac{1}{2} \times \frac{1}{2}MR^2 \times (v/R)^2 = \frac{1}{2} \times \frac{1}{2}Mv^2$.
3. The rolling cylinder's total kinetic energy is the same as for a slipping cylinder; in each case, it is equal to the change in potential energy Mgh. For the rolling cylinder, one-third of the total kinetic energy is rotational kinetic energy, and two-thirds is translational kinetic energy; thus, the rolling cylinder's translational speed is smaller when it reaches the bottom than that of a slipping cylinder (by a factor of $\sqrt{2/3}$).
4. The sphere and cylinder must have equal kinetic energies when they reach the bottom; each kinetic energy is equal to the change in potential energy Mgh. The sphere's moment of inertia is only $\frac{2}{5}MR^2$, compared with $\frac{1}{2}MR^2$ for the cylinder, so the sphere will achieve a higher speed and get to the bottom first.
5. (A) Less than that of the cylinder. For the thin hoop ($I = MR^2$), only one-half of its kinetic energy is translational; for the cylinder ($I = \frac{1}{2}Mr^2$), two-thirds of its kinetic energy will be translational. Since the total kinetic energy in each case will equal the change in potential energy (Mgh), the speed of the hoop will be smaller.

Checkup 13.3

1. Since the angular momentum is $L = I\omega$ and the angular speeds (ω) are equal, the hoop (with moment of inertia $I = MR^2$; see Table 12.3) has a larger angular momentum by a factor of 2 compared with the uniform disk (which has $I = \frac{1}{2}MR^2$).
2. Since the angular velocities are equal and the angular momentum is $L = I\omega$, the car with the larger moment of inertia $I = MR^2$ has the greater angular momentum. Since the masses are equal, this is the car on the outside, with the greater value of R.
3. Since there are no external torques on you, angular momentum $L = I\omega$ is conserved. Since you increase your moment of inertia I by stretching your legs outward (increasing R^2), your angular velocity ω must decrease.
4. No. Since angular momentum $L = I\omega$ is conserved and she decreases her moment of inertia I, her angular velocity ω increases. But her rotational kinetic energy is $K = \frac{1}{2}I\omega^2 = \frac{1}{2}I\omega \times \omega$. Since $I\omega$ is constant and ω increases, the kinetic energy increases. Thus the skater must do work to bring her arms close to her body.
5. (A) Frequency increases. The moment of inertia decreases when the children sit up, since more of their mass is closer to the axis. Since the angular momentum $L = I\omega$ is conserved, a smaller moment of inertia requires a larger angular frequency.

Checkup 13.4

1. Yes; since the angular-momentum vector is $\mathbf{L} = \mathbf{r} \times \mathbf{p}$, it will be zero when $\mathbf{r}$ and $\mathbf{p}$ are parallel (or antiparallel).
2. Since the angular-momentum vector is $\mathbf{L} = \mathbf{r} \times \mathbf{p}$, $\mathbf{L}$ is always perpendicular to $\mathbf{p}$; the angle between them is 90°.
3. The individual angular-momentum vectors will be inclined at an angle with respect to the z axis; each, however, will point toward the z axis, like the angular-momentum vectors $\mathbf{L}_1$ and $\mathbf{L}_2$ in Fig. 13.19. In this case, the horizontal components of the two angular-momentum vectors will cancel, and the total angular-momentum vector will point along the z axis.
4. Yes; the total angular momentum is changing as the dumbbell rotates about the z axis (because the direction of $\mathbf{L}$ is changing), so a torque is required to produce that change in angular momentum.
5. (E) Perpendicularly into the face of the watch. By the right-hand rule, with $\mathbf{r}$ pointing along the minute hand and $\mathbf{p}$ in the direction of motion, the clockwise rotation implies that the angular-momentum vector $\mathbf{L} = \mathbf{r} \times \mathbf{p}$ is perpendicularly *into* the face of the watch.

CHAPTER

14

Statics and Elasticity

CONCEPTS IN CONTEXT

Concepts — in — Context

Tower cranes are widely used at construction sites. The K-10000 tower crane shown here is the largest commercially available tower crane. Its central tower is 110 m high, and its long horizontal arm reaches out to 84 m. It can lift 120 tons at the end of the long arm, and more than twice as much at the middle of the long arm. The short arm holds a fixed counterweight of 100 tons (at the end, above the arm) and two additional mobile counterweights (below the arm). For the lift of a small load, the mobile counterweights are parked in the inboard position, near the central tower. For the lift of a large load, the mobile counterweights are moved outward to keep the crane in balance.

The concepts discussed in this chapter permit us to examine many aspects of the operation of such a crane:

? Where must the mobile counterweights be placed to keep the crane in balance for a given load? (Example 2, page 435)

? What is the tension in the tie-rod (stretched diagonally from the top of the tower to the end of the arm) that holds the short arm in place? (Example 3, page 435)

? What is the elongation of the lifting cable when subjected to a given load? (Example 8, page 448)

Engineers and architects concerned with the design of bridges, buildings, and other structures need to know under what conditions a body will remain at rest, even when forces act on it. For instance, the designer of a railroad bridge must make sure that the bridge will not tip over or break when a heavy train passes over it. *A body that remains at rest, even though several forces act on it, is said to be in equilibrium.* The branch of physics that studies the conditions for the equilibrium of a body is called **statics.** Statics is the oldest branch of physics. The ancient Egyptians, Greeks, and Romans had a good grasp of the basic principles of statics, as is evident from their construction of elegant arches for doorways and bridges. The oldest surviving physics textbook is a treatise on the statics of ships by Archimedes.

In the first three sections of this chapter, we will rely on the assumption that the "rigid" structural members—such as beams and columns—indeed remain rigid; that is, they do not deform. In essence, this means that we assume that the forces are not so large as to produce a significant bending or compression of the beams or columns. However, in the last section, we will take a brief look at the phenomenon of the elastic deformation of solid bodies when subjected to the action of large forces.

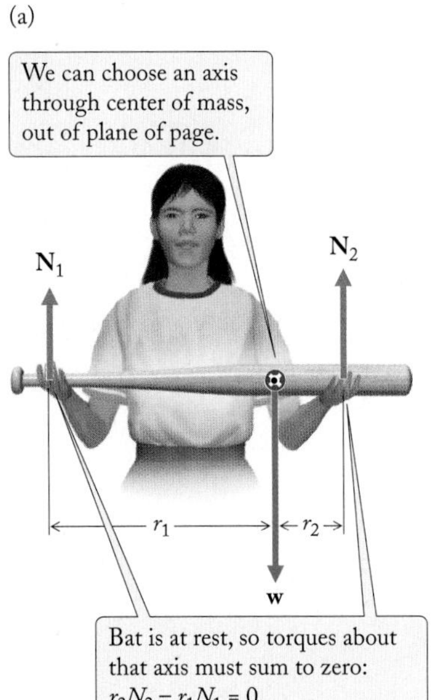

FIGURE 14.1 A baseball bat at rest in your hands. The external forces are the downward weight **w** and the upward pushes $\mathbf{N}_1$ and $\mathbf{N}_2$ of the right and left hands, respectively. These external forces add to zero. The external torques about any axis also add to zero. (a) Axis is through center of mass. (b) Axis is through left hand.

14.1 STATICS OF RIGID BODIES

If a rigid body is to remain at rest, its translational and rotational accelerations must be zero. Hence, the condition for the **static equilibrium** of a rigid body is that *the sum of external forces and the sum of external torques on the body must be zero.* This means that the forces and the torques are in balance; each force is compensated by some other force or forces, and each torque is compensated by some other torque or torques. For example, when a baseball bat rests in your hands (Fig. 14.1), the external forces on the bat are its (downward) weight **w** and the (upward) pushes $\mathbf{N}_1$ and $\mathbf{N}_2$ of your hands. If the bat is to remain at rest, the sum of these external forces must be zero—that is, $\mathbf{w} + \mathbf{N}_1 + \mathbf{N}_2 = 0$, or, in terms of magnitudes, $-w + N_1 + N_2 = 0$. Likewise, the sum of the torques of the external forces must be zero. Since the angular acceleration of the bat is zero about any axis of rotation whatsoever that we might choose in Fig. 14.1, the sum of torques must be zero about any such axis. For example, we might choose a horizontal axis of rotation through the center of mass of the bat, out of the plane of the page, as in Fig. 14.1a. With this choice of axis, the force $\mathbf{N}_2$ produces a counterclockwise torque r_2N_2 and the force $\mathbf{N}_1$ produces a clockwise torque r_1N_1, whereas the weight **w** (acting at the axis) produces no torque. The equilibrium condition for the torque is then $r_2N_2 - r_1N_1 = 0$. Alternatively, we might choose a horizontal axis of rotation through, say, the left hand, out of the plane of the page, as in Fig. 14.1b. With this choice, the force $\mathbf{N}_1$ produces a clockwise torque $(r_1 + r_2)N_1$, the weight produces a counterclockwise torque r_2w, and the force $\mathbf{N}_2$ produces no torque. The equilibrium condition for the torques is then $-(r_1 + r_2)N_1 + r_2w = 0$. With other choices of axis of rotation, we can generate many more equations than there are unknown forces or torques in a static equilibrium problem. However, the equations obtained with different choices of axis of rotation are related, and they can always be shown to be consistent.

From this discussion, we conclude that for the purposes of static equilibrium, *any line through the body or any line passing at some distance from the body can be thought of as a conceivable axis of rotation, and the torque about every such axis must be zero.* This means we have complete freedom in the choice of the axis of rotation, and we can make whatever choice seems convenient. With some practice, one learns to recognize which choice of axis will be most useful for the solution of a problem in statics.

The force of gravity plays an important role in many problems of statics. The force of gravity on a body is distributed over all parts of the body, each part being subjected to a force proportional to its mass. However, for the calculation of the torque exerted by gravity on a rigid body, *the entire gravitational force may be regarded as acting on the center of mass.* We relied on this rule in Fig. 14.1, where we assumed that the weight acts at the center of mass of the bat. The proof of this rule is easy: Suppose that we release some arbitrary rigid body and permit it to fall freely from an initial condition of rest. Since all the particles in the body fall at the same rate, the body will not change its orientation as it falls. If we consider an axis through the center of mass, the absence of angular acceleration implies that gravity does not generate any torque about the center of mass. Hence, if we want to simulate gravity by a single force acting at one point of the rigid body, that point will have to be the center of mass.

Given that in a rigid body the force of gravity effectively acts on the center of mass, we see that a rigid body supported by a single force acting at its center of mass or acting on the vertical line through its center of mass is in equilibrium, since the support force is then collinear with the effective force of gravity, and such collinear forces of equal magnitudes and opposite directions exert no net torque. This provides us with a simple method for the experimental determination of the center of mass of a body of complicated shape: Suspend the body from a string attached to a point on its surface (Fig. 14.2); the body will then settle into an equilibrium position such that the center of mass is on the vertical downward prolongation of the string (this vertical prolongation is marked dashed in Fig. 14.2). Next, suspend the body from a string attached at another point of its surface, and mark a new vertical downward prolongation of the string. The center of mass is then at the intersection of the new and the old prolongations of the string.

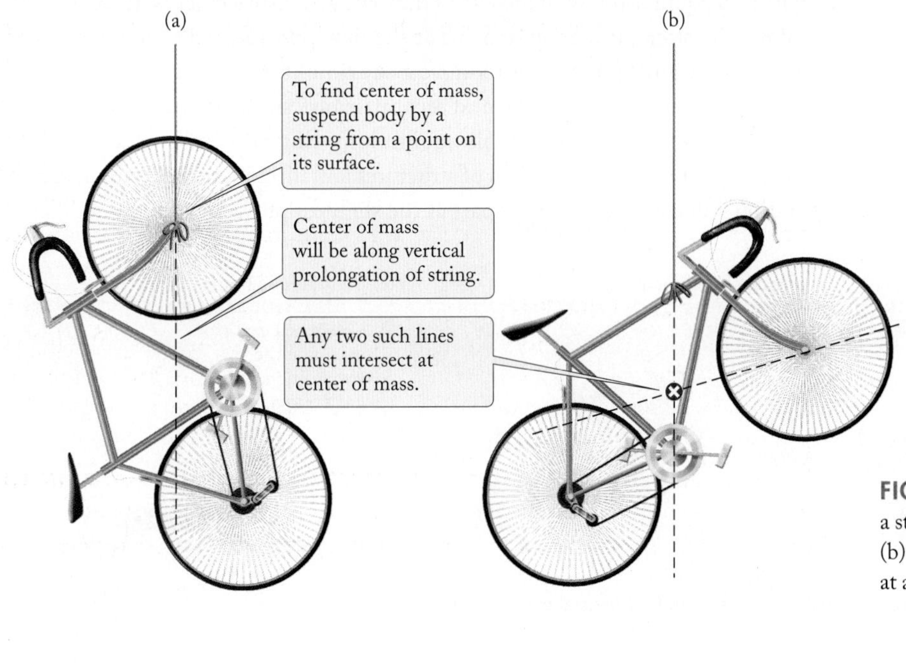

FIGURE 14.2 (a) Bicycle suspended by a string attached at a point on its "surface." (b) Bicycle suspended by a string attached at a different point.

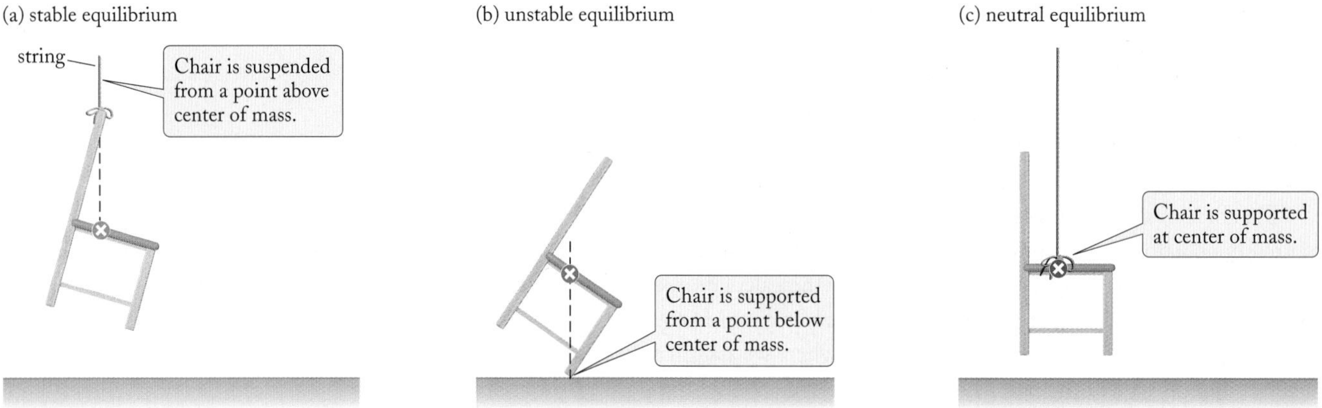

FIGURE 14.3 A body (a) in stable equilibrium; (b) in unstable equilibrium; (c) in neutral equilibrium.

A body suspended from a point above its center of mass, as in Fig. 14.3a, is in **stable equilibrium** (see also Section 8.2). If we turn this body through some angle, so the center of mass is no longer vertically below the point of support, the force of gravity and the supporting force will produce a torque that tends to return the body to the equilibrium position. In contrast, *if this body is supported by a single force applied at a point below the center of mass, as in Fig. 14.3b, the body is in* **unstable equilibrium.** If we turn the body ever so slightly, the force of gravity and the supporting force will produce a torque that tends to turn the body farther away from the equilibrium position—the body tends to topple over. Finally, *a body supported by a single force at its center of mass, as in Fig. 14.3c, is in* **neutral equilibrium.** If we turn such a body, it remains in equilibrium in its new position, and exhibits no tendency to return to its original position or to turn farther away.

Similar stability criteria apply to the translational motion of a body moving on a surface. A body is in stable equilibrium if it resists small disturbances and tends to return to its original position when the disturbance ceases. A car resting at the bottom of a dip in the road is an example of this kind of equilibrium; if we displace the car forward and then let go, the car rolls back to its original position. A body is in unstable equilibrium if it tends to move away from its original position when disturbed. A car resting on the top of a hill is an example of this second kind of equilibrium. If we displace the car forward, it continues to roll down the hill. A car resting on a flat street is in neutral equilibrium with respect to translational displacements. If we displace the car along the street, it merely remains at the new position, without any tendency to return to its original position or to move away from it (see Fig. 14.4).

The first four examples of the next section involve stable or neutral equilibrium; the next two examples involve unstable equilibrium. Engineers take great care to avoid unstable equilibrium in the design of structures and machinery, since an unstable configuration will collapse or come apart at the slightest provocation.

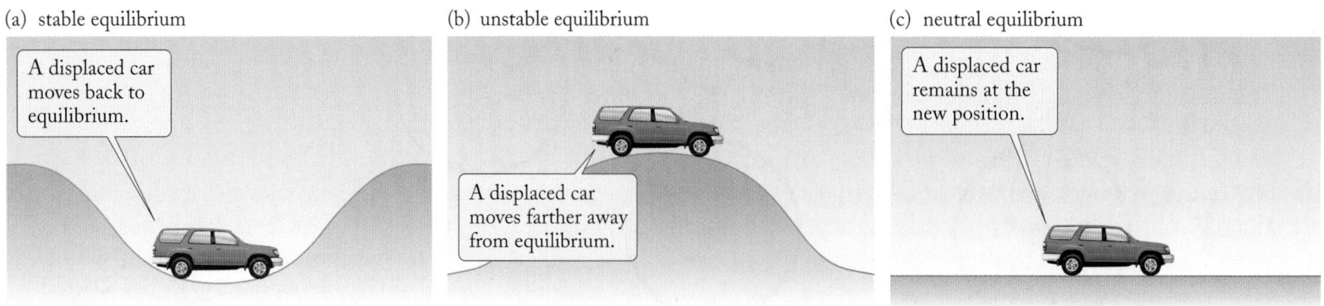

FIGURE 14.4 Stationary automobile in (a) stable, (b) unstable, and (c) neutral equilibrium.

Checkup 14.1

QUESTION 1: Is a cyclist balanced on an upright bicycle in stable or unstable equilibrium? Assume the cyclist sits rigidly, and makes no effort to avoid whatever might befall (see Fig. 14.5).

QUESTION 2: You sit in a swing, with your knees bent. If you now extend your legs fully, how will this change the equilibrium position of the swing and your body?

QUESTION 3: (a) You hold a fishing pole with both hands and point it straight up. Is the support force aligned with the weight? (b) You point the fishing pole horizontally. Is the support force aligned with the weight? Is there a single support force?

QUESTION 4: Consider a cone on a table (a) lying flat on its curved side, (b) standing on its base, (c) standing on its apex. Respectively, the equilibrium of each position is

(A) Stable, unstable, neutral
(B) Stable, neutral, unstable
(C) Unstable, stable, neutral
(D) Neutral, stable, unstable
(E) Neutral, unstable, stable

FIGURE 14.5 Is an upright bicycle in unstable equilibrium?

14.2 EXAMPLES OF STATIC EQUILIBRIUM

The following are some examples of solutions of problems in statics. In these examples, the conditions of a zero sum of external forces,

$$\mathbf{F}_1 + \mathbf{F}_2 + \mathbf{F}_3 + \cdots = 0 \tag{14.1}$$

and a zero sum of external torques,

$$\boldsymbol{\tau}_1 + \boldsymbol{\tau}_2 + \boldsymbol{\tau}_3 + \cdots = 0 \tag{14.2}$$

are used either to find the magnitudes of the forces that hold the body in equilibrium, or to find whether the body can achieve equilibrium at all.

EXAMPLE 1 A locomotive of mass 90 000 kg is one-third of the way across a bridge 90 m long. The bridge consists of a uniform iron girder of mass 900 000 kg, which is supported by two piers (see Fig. 14.6a). What is the load on each pier?

SOLUTION: The body whose equilibrium we want to investigate is the bridge. Figure 14.6b is a "free-body" diagram for the bridge, showing all the forces acting on it: the weight of the bridge, the downward push exerted by the locomotive, and the upward thrust exerted by each pier. The weight of the bridge can be regarded as acting at its center of mass. The bridge is static, and hence the net torque on the bridge reckoned about any point must be zero.

Let us first consider the torques about the point P_2, at the right pier. These torques are generated by the weight of the bridge acting at a distance of 45 m, the downward push of the locomotive acting at a distance of 30 m, and the upward thrust $\mathbf{F}_1$ of the pier at P_1 acting at a distance of 90 m (the upward thrust $\mathbf{F}_2$ has zero moment arm and generates no torque about P_2). The weight of the bridge is $m_{\text{bridge}}g = 9.0 \times 10^5$ kg $\times\ g$, and the downward push exerted by the locomotive equals its weight, $m_{\text{loc}}g = 9.0 \times 10^4$ kg $\times\ g$. Since each of the forces

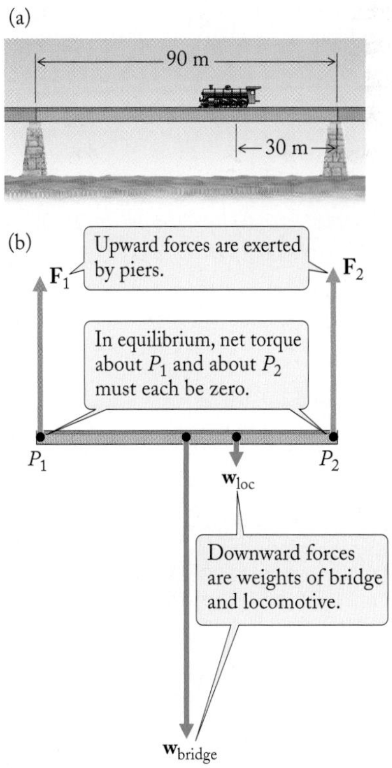

FIGURE 14.6 (a) Bridge with a locomotive on it. (b) "Free-body" diagram for the bridge.

acts at right angles to the (horizontal) line from P_2 to the point of application of the force, the magnitude of the torque $\tau = rF \sin 90°$ for each force is simply the product of the distance and the force, $\tau = rF$. According to the equilibrium condition, we must set the sum of the three torques equal to zero:

$$\tau_{\text{bridge}} + \tau_{\text{loc}} + \tau_{\text{pier}} = 0 \tag{14.3}$$

$$45 \text{ m} \times 9.0 \times 10^5 \text{ kg} \times g + 30 \text{ m} \times 9.0 \times 10^4 \text{ kg} \times g - 90 \text{ m} \times F_1 = 0 \tag{14.4}$$

Here, we have chosen to reckon the first two torques as positive, since they tend to produce counterclockwise rotation about P_2, and the last torque must then be reckoned as negative, since it tends to produce clockwise rotation. Equation (14.4) contains only the single unknown force F_1. Note that we were able to isolate this unknown force by evaluating the torques about P_2: the other unknown force F_2 is absent because it produces no torque about P_2. Solving this equation for the unknown F_1, we find

$$\begin{aligned} F_1 &= \frac{(45 \text{ m} \times 9.0 \times 10^5 \text{ kg} + 30 \text{ m} \times 9.0 \times 10^4 \text{ kg}) \times g}{90 \text{ m}} \\ &= 4.8 \times 10^5 \text{ kg} \times g \\ &= 4.8 \times 10^5 \text{ kg} \times 9.81 \text{ m/s}^2 = 4.7 \times 10^6 \text{ N} \end{aligned}$$

Next, consider the torques about the point P_1. These torques are generated by the weight of the bridge, the weight of the locomotive, and the upward thrust $\mathbf{F}_2$ at point P_2 (the upward thrust of $\mathbf{F}_1$ has zero moment arm and generates no torque about P_1). Setting the sum of these three torques about the point P_1 equal to zero, we obtain

$$-45 \text{ m} \times 9.0 \times 10^5 \text{ kg} \times g - 60 \text{ m} \times 9.0 \times 10^4 \text{ kg} \times g + 90 \text{ m} \times F_2 = 0$$

This equation contains only the single unknown force F_2 (the force F_1 is absent because it produces no torque about P_1). Solving for the unknown F_2, we find

$$F_2 = 5.0 \times 10^6 \text{ N}$$

The loads on the piers (the downward pushes of the bridge on the piers) are opposite to the forces $\mathbf{F}_1$ and $\mathbf{F}_2$ (these downward pushes of the bridge on the piers are the reaction forces corresponding to the upward thrusts of the piers on the bridge). Thus, the magnitudes of the loads are 4.7×10^6 N and 5.0×10^6 N, respectively.

COMMENT: Note that the net vertical upward force exerted by the piers is $F_1 + F_2 = 9.7 \times 10^6$ N. It is easy to check that this matches the sum of the weights of the bridge and the locomotive; thus, the condition for zero net vertical force, as required for translational static equilibrium, is automatically satisfied. This automatic result for the equilibrium of vertical forces came about because we used the condition for rotational equilibrium twice. Instead, we could have used the condition for rotational equilibrium once [Eq. (14.4)] and then evaluated F_2 by means of the condition for translational equilibrium [Eq. (14.1)]. The result for zero net torque about the point P_1 would then have emerged automatically.

Also note that instead of taking the bridge as the body whose equilibrium is to be investigated, we could have taken the bridge plus locomotive as a combined body. The downward push of the locomotive on the bridge would then not be an external force, and would not be included in the "free-body" diagram. Instead, the weight of the locomotive would be one of the external forces acting on the combined body and would have to be included in the "free-body" diagram. The vectors in Fig. 14.6b would therefore remain unchanged.

EXAMPLE 2 A large tower crane has a fixed counterweight of 100 tons at the end of its short arm, and it also has a mobile counterweight of 120 tons. The length of the short arm is 56 m, and the length of the long arm is 84 m; the total mass of both arms is 100 tons, and this mass is uniformly distributed along their combined length. The crane is lifting a load of 80 tons hanging at the end of the long arm. Where should the crane operator position the mobile counterweight to achieve a perfect balance of the crane, that is, a condition of zero (external) torque?

Concepts in Context

SOLUTION: To find the position of the counterweight, we consider the equilibrium condition for the entire crane (alternatively, we could consider the upper part of the crane, that is, the arms and the tie-rods that hold them rigid). Figure 14.7 is a "free-body" diagram for the crane. The external forces are the support force of the base and the weights of the load, the tower, the horizontal arms, the fixed counterweight, and the mobile counterweight. The weight of the arms acts at the center of mass of the combined arms. The total length of these arms is 84 m + 56 m = 140 m, and the center of mass is at the midpoint, 70 m from each end, that is, 14 m from the centerline of the tower.

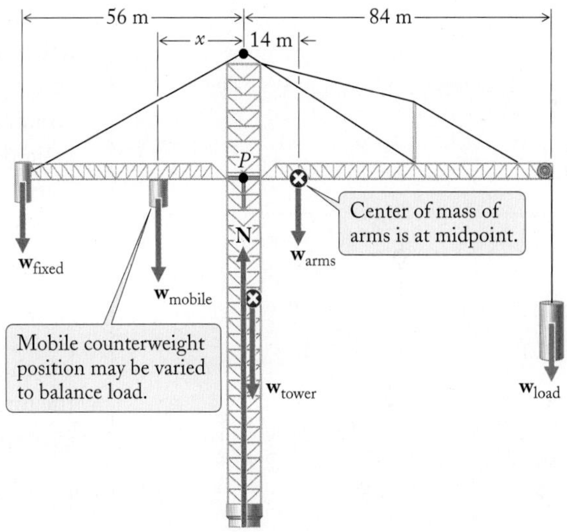

FIGURE 14.7 "Free-body" diagram of a tower crane. The crane is balanced, so that no torque is exerted by the base.

To examine the balance of torques, it is convenient to select the point P at the intersection of the arms and the midline of the tower. All the forces then act at right angles to the line from P to the point of application of the force, and the torque for each is simply the product of the distance and the force. The weight of the tower and the support force of the base do not generate any torques, since they act at zero distance. The equilibrium condition for the sum of the torques generated by the weights of the load, the arms, the fixed counterweight, and the mobile counterweight is

$$\tau_{load} + \tau_{arms} + \tau_{fixed} + \tau_{mobile} = 0 \tag{14.5}$$

Inserting the values of the weights and moment arms, we have

$$-84\text{ m} \times 80\text{ t} \times g - 14\text{ m} \times 100\text{ t} \times g + 56\text{ m} \times 100\text{ t} \times g$$
$$+ x \times 120\text{ t} \times g = 0$$

where we have again chosen to reckon counterclockwise torques as positive and clockwise torques as negative. When we solve this equation for x, we obtain

$$x = \frac{84\text{ m} \times 80\text{ t} + 14\text{ m} \times 100\text{ t} - 56\text{ m} \times 100\text{ t}}{120\text{ t}}$$
$$= 21\text{ m}$$

EXAMPLE 3 The short arm of the tower crane is held in place by a steel tie-rod stretched diagonally from the top of the tower to the end of the arm, as shown in Fig. 14.8a. The top part of the tower is 30 m high, and the short arm has a length of 56 m and a mass of 40 metric tons. The joint of the arm and the tower is somewhat flexible, so the joint acts as a pivot. Suppose

Concepts in Context

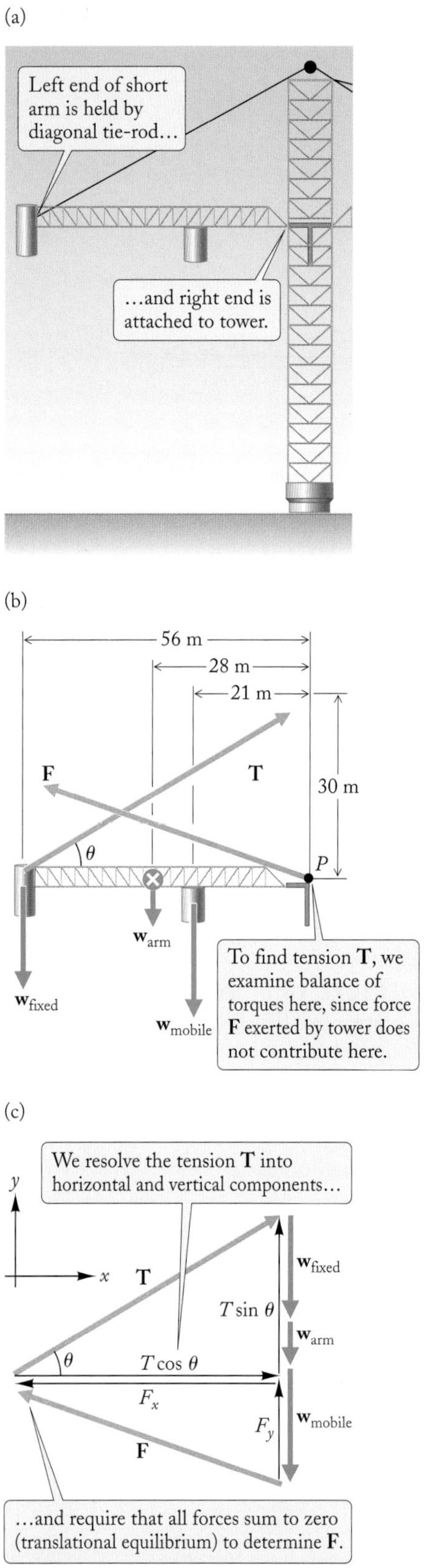

FIGURE 14.8 (a) Steel tie-rod supporting the short tower crane arm. (b) "Free-body" diagram for the short tower crane arm. (c) The x and y components of the forces.

that the counterweights are placed on the short arm as in the preceding example: the fixed counterweight of 100 metric tons is at the end of the arm, and the mobile counterweight of 120 metric tons is at a distance of 21 m from the centerline. (a) What is the tension in the tie-rod? (b) What is the force that the short arm exerts against the tower at the joint?

SOLUTION: Figure 14.8b is a "free-body" diagram of the short arm, displaying all the external forces acting on it. These forces are the weight $\mathbf{w}_{arm}$ of the arm, the weights of the counterweights $\mathbf{w}_{fixed}$ and $\mathbf{w}_{mobile}$, the tension **T** of the tie-rod, and the force **F** exerted by the tower at the joint. The force **F** is equal and opposite to the force that the short arm exerts against the tower. The weight of the arm acts at its center of mass, at a distance of 28 m from the centerline; the mobile counterweight acts at a distance of 21 m; and the fixed counterweight and the tension act at the end of the short arm, at a distance of 56 m.

(a) To find the tension **T**, it is convenient to examine the balance of torques about a point P that coincides with the joint. The force **F** does not generate any torque about this point, and hence the condition for the balance of the torques will contain **T** as the sole unknown. The weight of the short arm and the counterweights act at right angles to the line from P to the point of application of the force, so the torque for each is the product of the distance and the force. From Fig. 14.8b, we see that the tension acts at an angle θ, given by

$$\tan\theta = \frac{30\text{ m}}{56\text{ m}} = 0.54$$

which corresponds to $\theta = 28°$. With the same sign convention for the direction of the torques as in the preceding example, the equilibrium condition for the torques exerted by the weight of the arm, the counterweights, and the tension is then

$$28\text{ m} \times 40\text{ t} \times g + 21\text{ m} \times 120\text{ t} \times g + 56\text{ m} \times 100\text{ t} \times g - 56\text{ m} \times T \times \sin 28° = 0$$

We can solve this equation for T, with the result

$$T = \frac{28\text{ m} \times 40\text{ t} \times g + 21\text{ m} \times 120\text{ t} \times g + 56\text{ m} \times 100\text{ t} \times g}{56\text{ m} \times \sin 28°}$$

$$= 351\text{ t} \times g$$

$$= 351 \times 1000\text{ kg} \times 9.8\text{ m/s}^2 = 3.4 \times 10^6\text{ N}$$

(b) To find the components of the force **F** (Fig. 14.8c), we simply use the conditions for translational equilibrium: the sum of the horizontal components of all the forces and the sum of the vertical components of all the forces must each be zero. The weights of the short arm and the counterweights have vertical components, but no horizontal components. The tension force has a horizontal component $T\cos\theta$ and a vertical component $T\sin\theta$. Hence

$$3.4 \times 10^6\text{ N} \times \cos 28° + F_x = 0$$

and

$$3.4 \times 10^6\text{ N} \times \sin 28° - 40\text{ t} \times g - 120\text{ t} \times g - 100\text{ t} \times g + F_y = 0$$

When we solve these equations for F_x and F_y, we find

$$F_x = -3.0 \times 10^6\text{ N}$$

PROBLEM-SOLVING TECHNIQUES — STATIC EQUILIBRIUM

From the preceding examples we see that the steps in the solution of a problem of statics resemble the steps we employed in Chapter 5.

1 The first step is the selection of the body that is to obey the equilibrium conditions. The body may consist of a genuine rigid body (for instance, the bridge in Example 1), or it may consist of several pieces that act as a single rigid body for the purposes of the problem (for instance, the bridge plus the locomotive in Example 1). It is often helpful to mark the boundary of the selected rigid body with a distinctive color or with a heavy line; this makes it easier to recognize which forces are external and which internal.

2 Next, list all the external forces that act on this body, and display these forces on a "free-body" diagram.

3 If the forces have different directions, it is usually best to draw coordinate axes on the diagram and to resolve the forces into x and y components.

4 For each component, apply the static equilibrium condition for forces: the sum of forces is zero.

5 Make a choice of axis of rotation, calculate the torque of each force about this axis ($\tau = RF \sin \theta$), and apply the static equilibrium condition for torques: the sum of torques is zero. Establish and maintain a sign convention for torques; for example, for an axis pointing into the plane of the paper, counterclockwise torques to be positive and clockwise torques to be negative.

6 As mentioned in Section 14.1, any line can be thought of as an axis of rotation; and the torque about every such axis must be zero. You can make an unknown force disappear from the equation if you place the axis of rotation at the point of action or on the line of action of this force, so that this force has zero moment arm. Furthermore, as illustrated in Example 1, sometimes it is convenient to consider two different axes of rotation, and to examine the separate equilibrium conditions of the torques for each of these axes.

7 As recommended in Chapter 2, it is usually best to solve the equations algebraically for the unknown quantities, and to substitute numbers for the known quantities as a last step. But if the equations are messy, with a clutter of algebraic symbols, it may be convenient to substitute some of the numbers before proceeding with the solution of the equations.

and

$$F_y = 9.5 \times 10^5 \text{ N}$$

The x and y components of the force exerted by the short arm on the tower are therefore $+3.0 \times 10^6$ N and -9.5×10^5 N, respectively.

EXAMPLE 4 The bottom of a ladder rests on the floor, and the top rests against a wall (see Fig. 14.9a). If the coefficient of static friction between the ladder and the floor is $\mu_s = 0.40$ and the wall is frictionless, what is the maximum angle that the ladder can make with the wall without slipping?

SOLUTION: Figure 14.9b shows the "free-body" diagram for the ladder, with all the forces. The weight of the ladder acts downward at the center of mass. If the ladder is about to slip, the friction force at the floor has the maximum magnitude for a static friction force, that is,

$$f = \mu_s N_1 \tag{14.6}$$

If we reckon the torques about the point of contact with the floor, the normal force $\mathbf{N}_1$ and the friction force $\mathbf{f}$ exert no torques about this point, since their moment

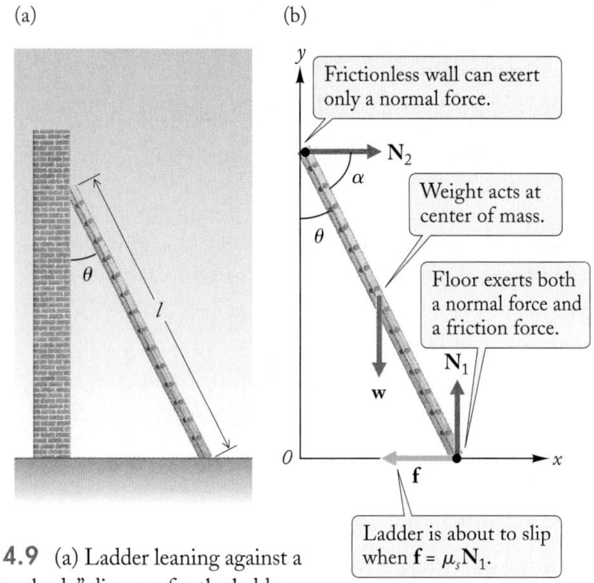

FIGURE 14.9 (a) Ladder leaning against a wall. (b) "Free-body" diagram for the ladder.

arms are zero. The weight $w = mg$ acting at the center of mass exerts a counterclockwise torque of magnitude $(l/2) \times mg \times \sin\theta$, and the normal force $\mathbf{N}_2$ of the wall exerts a clockwise torque of magnitude $l \times N_2 \times \sin\alpha$, where α is the angle between the ladder and the normal force (see Fig. 14.9b); since $\alpha = 90° - \theta$, the sine of α equals the cosine of θ, and the torque equals $l \times N_2 \times \cos\theta$. For equilibrium, the sum of these torques must be zero,

$$+\frac{l}{2}mg\sin\theta - lN_2\cos\theta = 0 \tag{14.7}$$

or, equivalently,

$$\frac{1}{2}mg\sin\theta = N_2\cos\theta \tag{14.8}$$

We collect the factors that depend on θ by dividing both sides of this equation by $\frac{1}{2}mg\cos\theta$, so

$$\frac{\sin\theta}{\cos\theta} = \frac{2N_2}{mg}$$

or, since $\sin\theta/\cos\theta = \tan\theta$,

$$\tan\theta = \frac{2N_2}{mg} \tag{14.9}$$

To evaluate the angle θ we still need to determine the unknown N_2. For this, we use the condition for translational equilibrium: the net vertical and the net horizontal forces must be zero, or

$$N_1 - mg = 0 \tag{14.10}$$

$$N_2 - \mu_s N_1 = 0 \tag{14.11}$$

From the first of these equations, $N_1 = mg$; therefore, from the second equation, $N_2 = \mu_s mg$. Inserting this into our expression (14.9) for the tangent of the angle θ, we obtain the final result

$$\tan\theta = \frac{2\,\mu_s mg}{mg} = 2\mu_s \qquad (14.12)$$

With $\mu_s = 0.40$, this yields $\tan\theta = 0.80$. With a calculator, we find that the angle with this tangent is

$$\theta = 39°$$

For any angle larger than this, equilibrium is impossible, because the maximum frictional force is not large enough to prevent slipping of the ladder.

EXAMPLE 5 A uniform rectangular box 2.0 m high, 1.0 m wide, and 1.0 m deep stands on a flat floor. You push the upper end of the box to one side and then release it (see Fig. 14.10a). At what angle of release will the box topple over on its side?

SOLUTION: The forces on the box when it has been released are as shown in the "free-body" diagram in Fig. 14.10b. Both the normal force **N** and the friction force **f** act at the bottom corner, which is the only point of contact of the box with the floor. The weight acts at the center of mass, which is at the center of the box.

Since the box rotates about the bottom corner, let us consider the torque about this point. The only force that produces a torque about the bottom corner is the weight. The weight acts at the center of mass; for a uniform box, this is at the center of the box. The torque exerted by the weight can be expressed as $d \times Mg$, where d is the perpendicular distance from the bottom corner to the vertical line through the center of mass (see Fig. 14.10b). This torque produces counterclockwise rotation if the center of mass is to the left of the bottom corner, and it produces clockwise rotation if the center of mass is to the right of the bottom corner. This means that in the former case, the box returns to its initial position, and in the latter case it topples over on its side. Thus, the critical angle beyond which the box will tip over corresponds to vertical alignment of the bottom corner and the center of the box (see Fig. 14.10c). This critical angle equals the angle between the side of the box and the diagonal. The tangent of this angle is the ratio of the width and the height of the box,

$$\tan\theta = \frac{0.50\text{ m}}{1.0\text{ m}} = 0.50$$

With our calculator we find that the critical angle is then

$$\theta = 27°$$

COMMENT: In this example we found that the box begins to topple over if its inclination is such that the center of mass is vertically aligned with the bottom corner. This is a special instance of the general rule that a rigid body resting on a surface (flat or otherwise) becomes unstable when its center of mass is vertically above the outermost point of support.

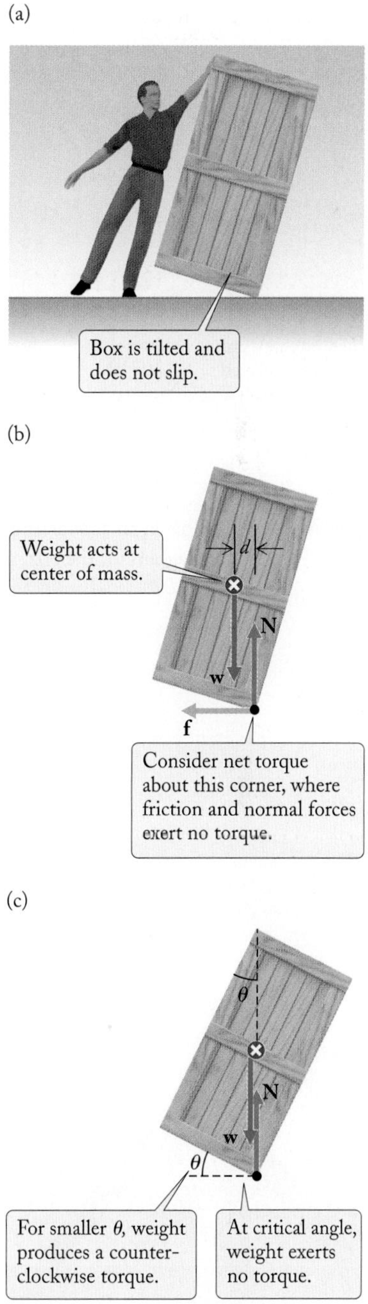

FIGURE 14.10 (a) Box standing on edge. (b) "Free-body" diagram for the box. (c) "Free-body" diagram if the box is tilted at the critical angle. The center of mass is directly above the edge.

EXAMPLE 6 A uniform rectangular box 2.0 m high, 1.0 m wide, and 1.0 m deep stands on the platform of a truck (Fig. 14.11a). What is the maximum forward acceleration of the truck that the box can withstand without toppling over? Assume that the coefficient of static friction is large enough that the box will topple over before it starts sliding.

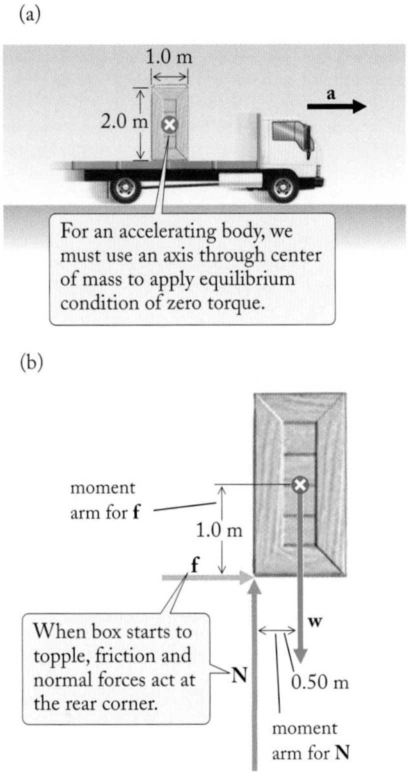

FIGURE 14.11 (a) Box on an accelerating truck. (b) "Free-body" diagram for the box.

SOLUTION: Strictly, this is not a problem of statics, since the translational motion is accelerated; however, the rotational motion involves a question of equilibrium and can be treated by the methods of this section. Under the conditions of the problem, the forces on the box are as shown in Fig. 14.11b. Both the normal force **N** and the friction force **f** act at the rear corner (when the box is about to topple, it makes contact with the platform only along the rear bottom edge). The weight acts at the center of mass; for a uniform box, this is at the center of the box, 1.0 m above and 0.50 m in front of the corner. Since the box is in accelerated motion, we have to be careful about the choice of axis for the calculation of the torque. As mentioned before Example 6 in Chapter 13, for an accelerated body, the equation of rotational motion (and the equilibrium condition of zero torque) is valid only for an axis through the center of mass. The forces that produce a torque about the center of mass are **N** and **f**, and each torque $\tau = RF \sin\theta$ may be expressed as the product of the force and the corresponding moment arm, $R \sin\theta$; the moment arms are the perpendicular distances shown in Fig. 14.11b. For an axis pointing into the page, the normal force tends to produce clockwise rotation and the frictional force counterclockwise; thus the condition of zero torque is

$$-0.50\text{ m} \times N + 1.0\text{ m} \times f = 0 \tag{14.13}$$

We can obtain expressions for f and N from the equations for the horizontal and vertical translational motions. The horizontal acceleration is a and the vertical acceleration is zero; accordingly, the horizontal and vertical components of Newton's Second Law are

$$f = ma$$
$$N - mg = 0$$

Inserting these expressions for f and for N into Eq. (14.13), we obtain

$$0.50\text{ m} \times mg - 1.0\text{ m} \times ma = 0$$

from which

$$a = 0.50g = 4.9\text{ m/s}^2$$

If the acceleration exceeds this value, rotational equilibrium fails, and the box topples.

Checkup 14.2

QUESTION 1: Why is it dangerous to climb a ladder that is leaning against a building at a large angle with the vertical? Why is it dangerous to climb a ladder that is leaning against a building at a small angle with the vertical?

QUESTION 2: Suppose that in Example 5 all the mass of the box is concentrated at the midpoint of the bottom surface, so the center of mass is at this midpoint. What is the critical angle at which such a box topples over on its side?

QUESTION 3: Two heavy pieces of lumber lean against each other, forming an A-frame (see Fig. 14.12). Qualitatively, how does the force that one piece of lumber exerts on the other at the tip of the A vary with the angle?

QUESTION 4: You hold a fishing pole steady, with one hand forward, pushing upward to support the pole, and the other hand further back, pushing downward to maintain

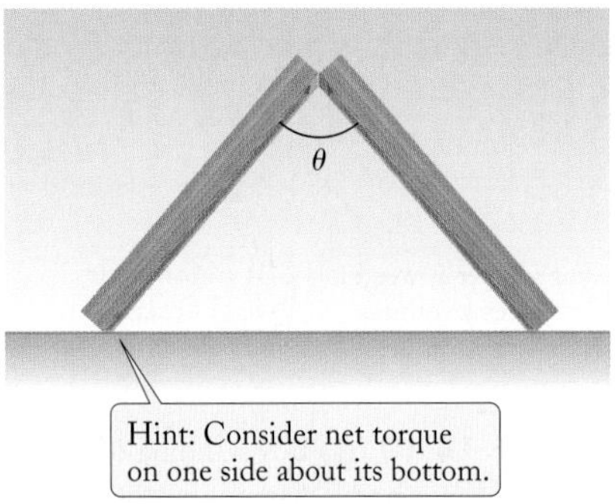

FIGURE 14.12 Two pieces of lumber forming an A-frame.

zero net torque. If a fish starts to pull downward on the far end of the pole, then to maintain equilibrium you must

(A) Increase the upward push and decrease the downward push
(B) Increase the upward push and increase the downward push
(C) Increase the upward push and keep the downward push the same

14.3 LEVERS AND PULLEYS

A lever consists of a rigid bar swinging on a pivot (see Fig. 14.13). If we apply a force at the long end, the short end of the bar pushes against a load with a larger force. Thus, the lever permits us to lift a larger load than we could with our bare hands. The relationship between the magnitudes of the forces at the ends follows from the condition for static equilibrium for the lever. Figure 14.13 shows the forces acting on the lever: the force **F** that we exert at one end, the force **F**′ exerted by the load at the other end, and the support force **S** exerted by the pivot point *P*. The net torque about the pivot point *P* must be zero. Since, for the arrangement shown in Fig. 14.13, the forces at the ends are at right angles to the distances l and l', the condition on the net torque is

$$Fl - F'l' = 0 \tag{14.14}$$

from which we find

$$\frac{F'}{F} = \frac{l}{l'} \tag{14.15}$$

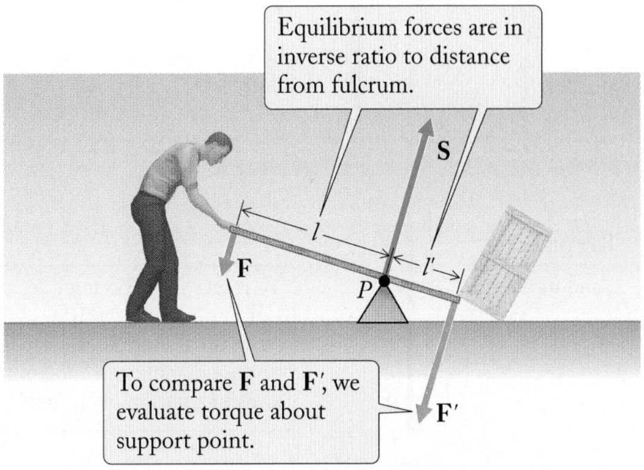

FIGURE 14.13 A lever. The vectors show the forces acting on the lever; **F** is our push, **F**′ is the push of the load, and **S** is the supporting force of the pivot. The force that the lever exerts on the load is of the same magnitude as **F**′, but of opposite direction.

mechanical advantage of lever

By Newton's Third Law, the force that the load exerts on the lever is equal in magnitude to the force that the lever exerts on the load (and of opposite direction). Hence Eq. (14.15) tells us the ratio of the magnitudes of the forces we exert and the lever exerts. *These forces are in the inverse ratio of the distances from the pivot point.* For a powerful lever, we must make the lever arm l as long as possible and the lever arm l' as short as possible. The ratio F'/F of the magnitudes of the force delivered by the lever and the force we must supply is called the **mechanical advantage.**

Apart from its application in the lifting of heavy loads, the principle of the lever finds application in many hand tools, such as pliers and bolt cutters. The handles of these tools are long, and the working ends are short, yielding an enhancement of the force exerted by the hand (see Fig. 14.14). A simple manual winch also relies on the principle of the lever. The handle of the winch is long, and the drum of the winch, which acts as the short lever arm, is small (see Fig. 14.15). The force the winch delivers to the rope attached to the drum is then larger than the force exerted by the hand pushing on the handle. Compound winches, used for trimming sails on sailboats, have internal sets of gears that provide a larger mechanical advantage; in essence, such compound winches stagger one winch within another, so the force ratio generated by one winch is further multiplied by the force ratio of the other.

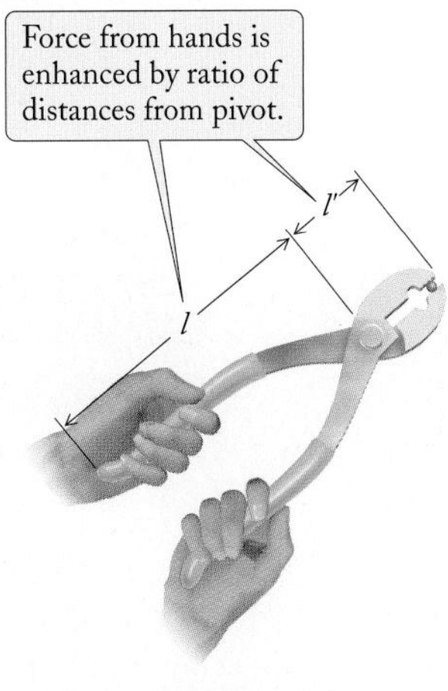

FIGURE 14.14 A pair of pliers serves as levers.

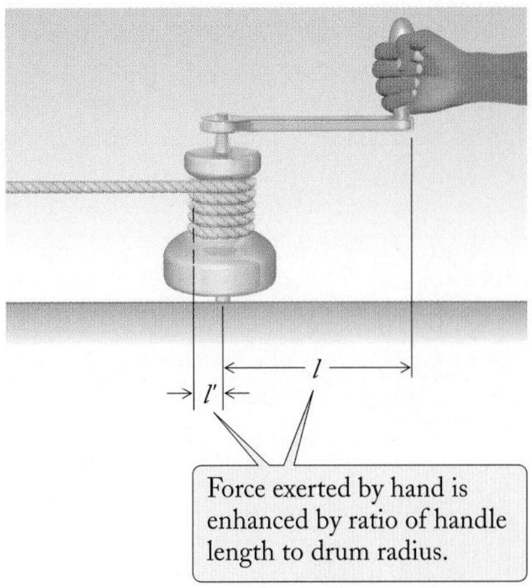

FIGURE 14.15 A manual winch.

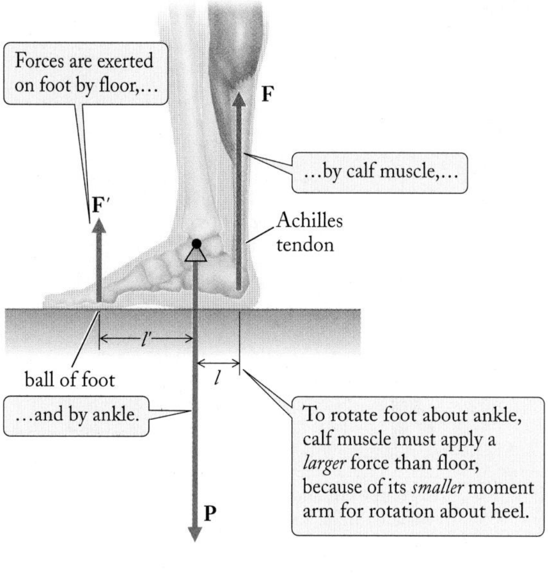

FIGURE 14.16 Bones of the foot acting as a lever.

In the human body, many bones play the role of levers that permit muscles or groups of muscles to support or to move the body. For example, Fig. 14.16 shows the bones of the foot; these act as a lever, hinged at the ankle. The rear end of this lever, at the heel, is tied to the muscles of the calf by the Achilles tendon, and the front end of the lever is in contact with the ground, at the ball of the foot. When the muscle contracts, it rotates the heel about the ankle and presses the ball of the foot against the ground, thereby lifting the entire body on tiptoe. Note that the muscle is attached to the short end of this lever—the muscle must provide a larger force than the force generated at the ball of the foot. At first sight, it would seem advantageous to install a longer projecting spur at the heel of the foot and attach the Achilles tendon to the end of this spur; but this would require that the contracting muscle move through a longer distance. Muscle is good at producing large forces, but not so good at contracting over long distances, and the attachment of the Achilles tendon represents the best compromise. In most of the levers found in the human skeleton, the muscle is attached to the short end of the lever.

Equation (14.15) is valid only if the forces are applied at right angles to the lever. A similar equation is valid if the forces are applied at some other angle, but instead of the lengths l and l' of the lever, we must substitute the lengths of the moment arms of the forces, that is, the perpendicular distances between the pivot point and the lines of action of the forces. These moment arms play the role of effective lengths of the lever.

EXAMPLE 7 When you bend over to pick up something from the floor, your backbone acts as a lever pivoted at the sacrum (see Fig. 14.17). The weight of the trunk pulls downward on this lever, and the muscles attached along the upper part of the backbone pull upward. The actual arrangement of the muscles is rather complicated, but for a simple mechanical model we can pretend that the muscles are equivalent to a string attached to the backbone at an angle of about 12° at a point beyond the center of mass (the other end of the "string" is attached to the pelvis). Assume that the mass of the trunk, including head and arms, is

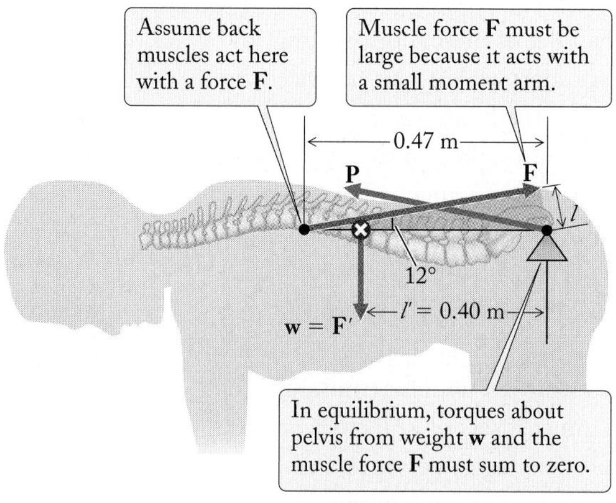

FIGURE 14.17 "Free-body" diagram for the backbone acting as lever. The forces on the backbone are the weight **w** of the trunk (including the weight of the backbone), the pull **F** of the muscles, and the thrust **P** of the pelvis acting as pivot.

48 kg, and that the dimensions are as shown in the diagram. What force must the muscles exert to balance the weight of the trunk when bent over horizontally?

SOLUTION: Figure 14.17 shows a "free-body" diagram for the backbone, with all the forces acting on it. Since the weight **w** of the trunk acts at right angles to the backbone, the lever arm for this weight is equal to the distance $l' = 0.40$ m between the pivot and the center of mass of the trunk. The lever arm for the muscle is the (small) distance l, which equals $l = 0.47 \text{ m} \times \sin 12° = 0.10$ m. According to Eq. (14.15), the force **F** exerted by the muscles then has magnitude

$$F = \frac{l'}{l}F' = \frac{l'}{l}w = \frac{l'}{l}Mg = \frac{0.40 \text{ m}}{0.10 \text{ m}} \times Mg = 4.0 \times Mg$$

$$= 4.0 \times 48 \text{ kg} \times 9.81 \text{ m/s}^2 = 1.9 \times 10^3 \text{ N}$$

This is a quite large force, 4.0 times larger than the weight of the trunk.

COMMENT: Bending over horizontally puts a severe stress on the muscles of the back. Furthermore, it puts an almost equally large compressional stress on the backbone, pulling it hard against the sacrum. The stresses are even larger if you try to lift a load from the floor while your body is bent over in this position. To avoid damage to the muscles and to the lumbosacral disk, it is best to lift by bending the knees, keeping the backbone vertical.

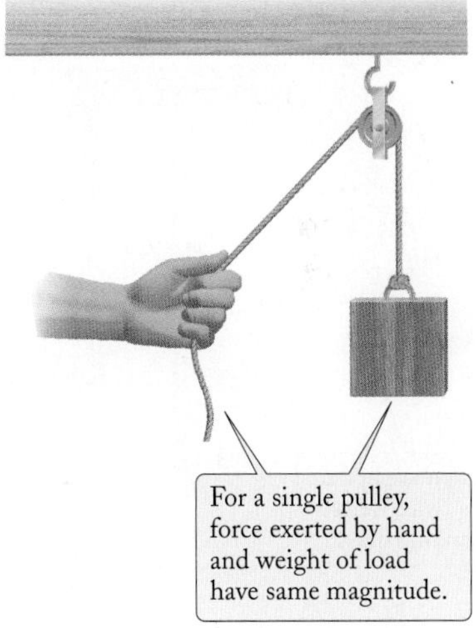

FIGURE 14.18 A single pulley.

Often, a force is applied to a load by means of a flexible rope, or a string. A pulley is then sometimes used to change the direction of the string or rope and the direction of the force exerted on the body. If the pulley is frictionless, the tension at each point of a flexible rope passing over the pulley is the same. For instance, if we want to lift a load with a rope passing over a single pulley attached to the ceiling (see Fig. 14.18), the force we must exert on the rope has the same magnitude as the weight of the load. Thus, there is no gain of mechanical advantage in such an arrangement of a single pulley; the only benefit is that it permits us to pull more comfortably than if we attempted to lift the load directly.

However, an arrangement of several pulleys linked together, called **block and tackle,** can provide a large gain of mechanical advantage. For example, consider the arrangement of three pulleys shown in Fig. 14.19a; the axles of the two upper pulleys are bolted together, and they are linked to each other and to the third pulley by a single rope. If the rope segments linking the pulleys are parallel and there is no friction, then the mechanical advantage of this arrangement is 3; that is, the magnitudes of the forces F and F' are in the ratio of 1 to 3. This can be most easily understood by drawing the "free-body" diagram for the lower portion of the pulley system, including the load (Fig. 14.19b). In this diagram, the three ropes leading upward have been cut off and replaced by the forces exerted on them by the external (upper) portions of the ropes. Since the tension is the same everywhere along the rope, the forces pulling upward on each of the three rope ends shown in the "free-body" diagram all have the same magnitude F, and thus the net upward force is $3F$.

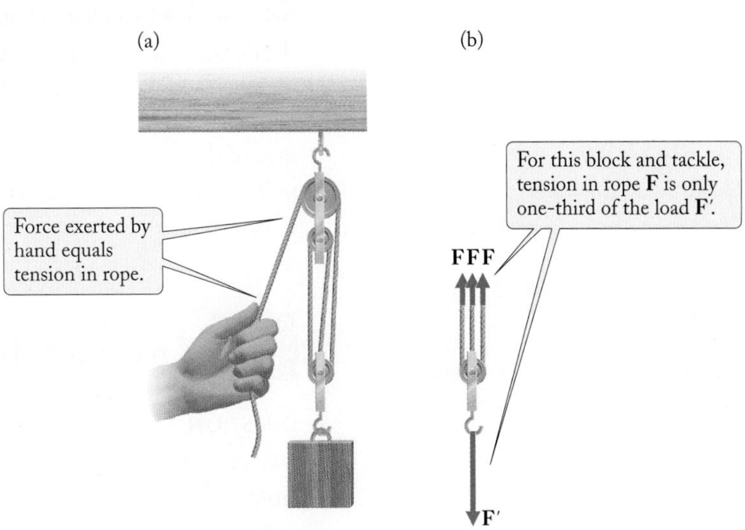

FIGURE 14.19 (a) Block and tackle. (b) "Free-body" diagram for the lower portion of the pulley system.

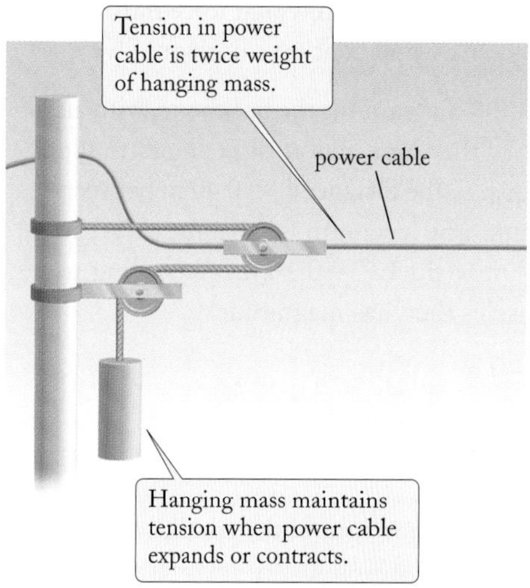

FIGURE 14.20 Block and tackle used for tensioning power line.

Block-and-tackle arrangements have many practical applications. For instance, they are used to provide the proper tension in overhead power cables for electric trains and trams (see Fig. 14.20); without such an arrangement, the cables would sag on warm days when thermal expansion increases their length, and they would stretch excessively tight and perhaps snap on cold days, when they contract. One common cause of power failures on cold winter nights is the snapping of power lines lacking such compensating pulleys.

Another practical application of block and tackle is found in the traction devices used in hospitals to immobilize and align fractured bones, especially leg bones. A typical arrangement is shown in Fig. 14.21; here the pull applied to the leg is twice as large as the magnitude of the weight attached on the lower end to the rope. Also, as in the case of the power line, the tension remains constant even if the leg moves.

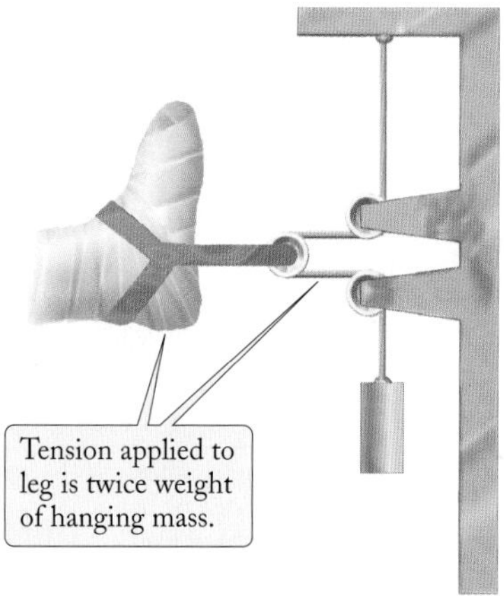

FIGURE 14.21 Block and tackle in traction apparatus for fractured leg.

The mechanical advantage provided by levers, arrangements of pulleys, or other devices can be calculated in a general and elegant way by appealing to the Law of Conservation of Energy. A lever merely transmits the work we supply at one end to the load at the other end. We can express this equality of work input and work output by

$$F'\Delta x' = F\Delta x \tag{14.16}$$

where Δx is the displacement of our hand and $\Delta x'$ the displacement of the load. According to this equation, the forces F' and F are in the inverse ratio of the displacements,

$$\frac{F'}{F} = \frac{\Delta x}{\Delta x'} \tag{14.17}$$

Consider, now, the rotation of the lever by a small angle (see Fig. 14.22). Since the two triangles included between the initial and final positions of the lever are similar, the distances Δx and $\Delta x'$ are in the same ratio as the lever arms l and l'; thus, we immediately recognize from Eq. (14.17) that the mechanical advantage of the lever is l/l'.

Likewise, we immediately recognize from Eq. (14.17) that the mechanical advantage of the arrangement of pulleys shown in Fig. 14.19 is 3, since whenever our hand pulls a length Δx of rope out of the upper pulley, the load moves upward by a distance of only $\Delta x/3$.

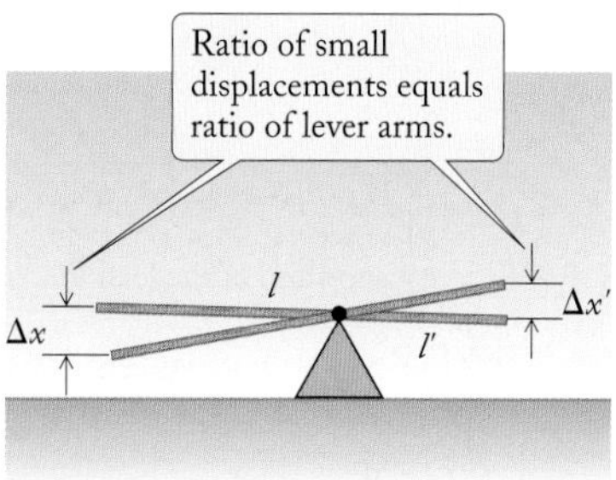

FIGURE 14.22 Rotation of lever by a small angle produces displacements Δx and $\Delta x'$ of the ends.

Checkup 14.3

QUESTION 1: Figure 14.23 shows two ways of using a lever. Which has the larger mechanical advantage?

QUESTION 2: Is Eq. (14.15) for the ratio of the forces F and F' on a lever valid if one or both of these forces are not perpendicular to the lever?

QUESTION 3: Suppose that the pulleys in a block and tackle are of different sizes. Does this affect the mechanical advantage?

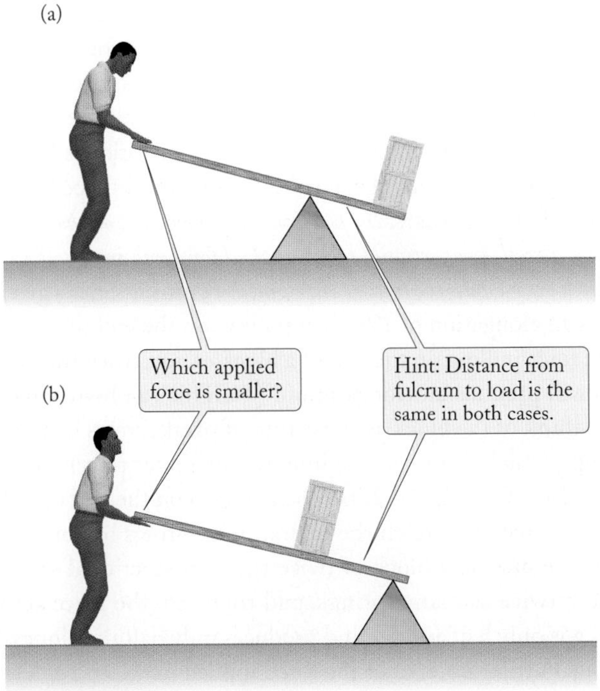

FIGURE 14.23
Two ways of using a lever.

QUESTION 4: A lever is used to lift a 100-kg rock. The distance from the rock to the fulcrum is roughly one-tenth of the distance from the fulcrum to the handle. If the rock has a mass of 100 kg, the downward force at the handle necessary to lift the rock is approximately:

(A) 1 N (B) 10 N (C) 100 N (D) 1000 N

14.4 ELASTICITY OF MATERIALS

In our examples of bridges, tower cranes, etc., we assumed that the bodies on which the forces act are rigid; that is, they do not deform. Although solid bodies, such as bars or blocks of steel, are nearly rigid, they are not exactly rigid, and they will deform by a noticeable amount if a large enough force is applied to them. A solid bar may be thought of as a very stiff spring. If the force is fairly small, this "spring" will suffer only an insignificant deformation, but if the force is large, it will suffer a noticeable deformation. Provided that the force and the deformation remain *within some limits, the deformation of a solid body is* **elastic,** *which means that the body returns to its original shape once the force ceases to act.* Such elastic deformations of a solid body usually obey Hooke's Law: the deformation is proportional to the force. But the constant of proportionality is small, giving a small deformation unless the force is large. The corresponding spring constant is thus very large, meaning that an appreciable deformation requires a large force.

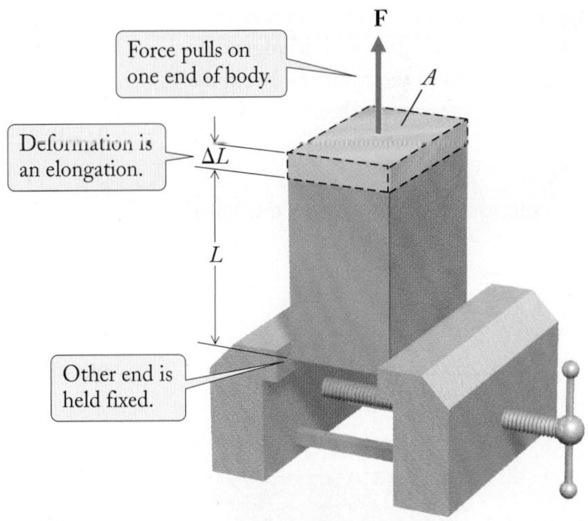

FIGURE 14.24 Tension applied to the end of a block of material causes elongation.

A solid block of material can suffer several kinds of deformation, depending on how the force is applied. If one end of the body is held fixed and the force pulls on the other end, the deformation is a simple **elongation** of the body (see Fig. 14.24). If one side of the body is held fixed and the force pushes tangentially along the other side, then the deformation is a **shear,** which changes the shape of the body from a rectangular parallelepiped to a rhomboidal parallelepiped (see Fig. 14.25a). During this deformation, the parallel layers of the body slide with respect to one another just as the pages of a book slide with respect to one another when we push along its cover (see Fig. 14.25b). If the force is applied from all sides simultaneously, by subjecting the body to the pressure of a fluid in which the body is immersed, then the deformation is a **compression** of the volume of the body, without any change of the geometrical shape (see Fig. 14.26).

In all of these cases, *the fractional deformation, or the percent deformation, is directly proportional to the applied force and inversely proportional to the area over which the force is distributed.* For instance, if a given force produces an elongation of 1% when pulling on the end of a block, then the same force pulling on the end of a block of, say, twice the cross-sectional area will produce an elongation of $\frac{1}{2}$%. This can be readily understood if we think of the block as consisting of parallel rows of atoms linked by springs, which represent the interatomic forces that hold the atoms in their places (see Fig. 14.27). When we pull on the end of the block with a given force, we stretch the interatomic springs by some amount; and when we pull on a block of twice the cross-sectional area, we have to stretch twice as many springs, and therefore the force acting on each spring is only half as large and produces only half the elongation in each spring. Furthermore, since the force applied to the end of a row of atoms is communicated to all the interatomic springs in that row, a given force produces a given elongation in each spring in a row. The net elongation of the block is the sum of the elongations of all the interatomic springs in the row, and hence the fractional elongation of the block is the same as the fractional elongation of each spring, regardless of the overall length of the block. For instance, if a block elongates by 0.1 mm when subjected to a given force, then a block of, say, twice the length will elongate by 0.2 mm when subjected to the same force.

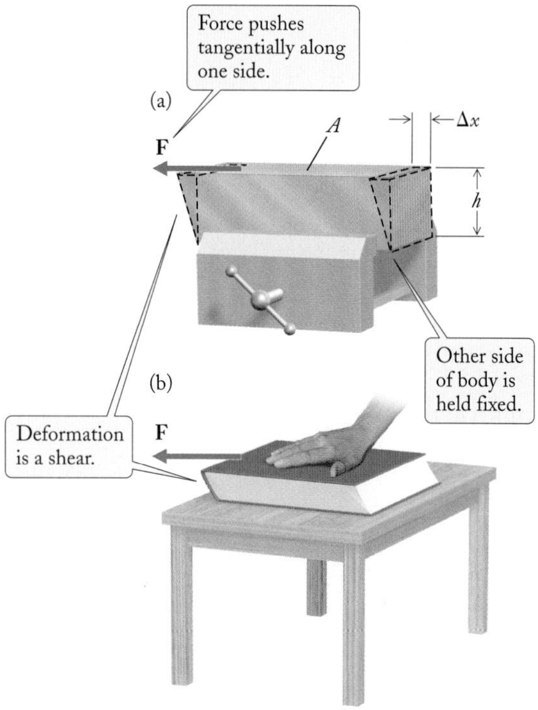

FIGURE 14.25 (a) Tangential force applied to the side of a block of material causes shear. (b) When such a tangential force is applied to the cover of a book, the pages slide past one another.

To express the relationships among elongation, force, and area mathematically, consider a block of initial length L and cross-sectional area A. If a force F pulls on the end of this block, the elongation is ΔL, and the fractional elongation is $\Delta L/L$. This fractional elongation is directly proportional to the force and inversely proportional to the area A:

elongation and Young's modulus

$$\frac{\Delta L}{L} = \frac{1}{Y}\frac{F}{A} \tag{14.18}$$

Here the quantity Y is the constant of proportionality. In Eq. (14.18) this constant written as $1/Y$, so it divides the right side, instead of multiplying it (this is analogous to writing Hooke's Law for a spring as $\Delta x = (1/k)\, F$, where Δx is the elongation

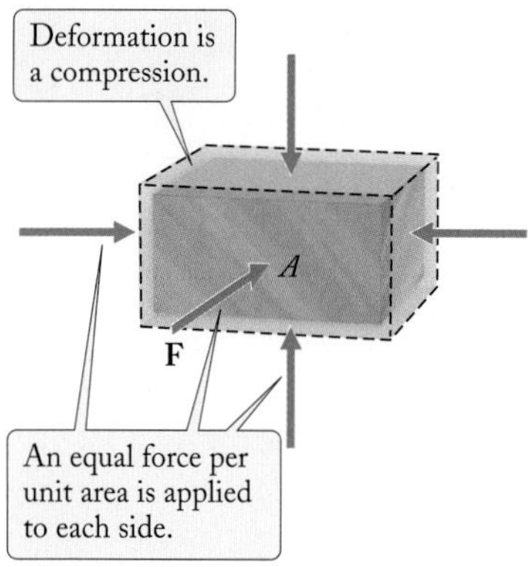

FIGURE 14.26 Pressure applied to all sides of a block of material causes compression.

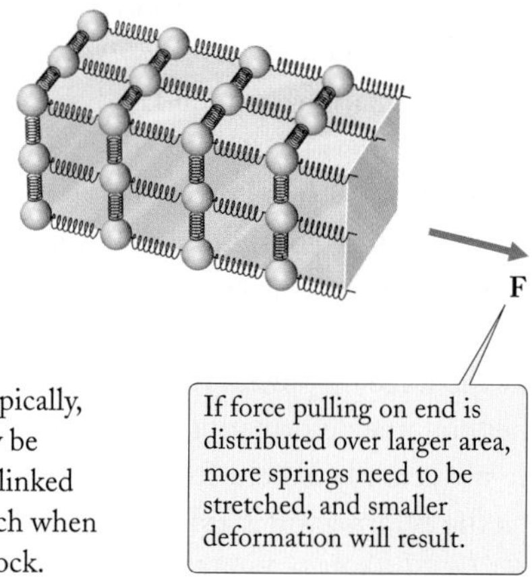

FIGURE 14.27 Microscopically, a block of solid material may be thought of as rows of atoms linked by springs. The springs stretch when a tension is applied to the block.

TABLE 14.1 ELASTIC MODULI OF SOME MATERIALS

MATERIAL	YOUNG'S MODULUS	SHEAR MODULUS	BULK MODULUS
Steel	22×10^{10} N/m^2	8.3×10^{10} N/m^2	16×10^{10} N/m^2
Cast iron	15	6.0	11
Brass	9.0	3.5	6.0
Aluminum	7.0	2.5	7.8
Bone (long)	3.2	1.2	3.1
Concrete	2	—	—
Lead	1.6	0.6	4.1
Nylon	0.36	0.12	0.59
Glycol	—	—	0.27
Water	—	—	0.22
Quartz	9.7(max)	3.1	3.6

produced by an applied force F). Thus, a stiff material, such as steel, that elongates by only a small amount has a large value of Y. The constant Y is called **Young's modulus.** Table 14.1 lists values of Young's moduli for a few solid materials. Note that if, instead of exerting a pull on the end of the block, we exert a push, then F in Eq. (14.18) must be reckoned as negative, and the change ΔL of length will then likewise be negative—the block becomes shorter.

In engineering language, *the fractional deformation is usually called the* **strain,** *and the force per unit area is called the* **stress.** In this terminology, Eq. (14.18) simply states that the strain is proportional to the stress.

This proportionality of strain and stress is also valid for shearing deformations and compressional deformations, provided we adopt a suitable definition of strain, or fractional deformation, for these cases. For shear, the fractional deformation is defined as the ratio of the sideways displacement Δx of the edge of the block to the height h of the block (see Fig. 14.25a). This fractional deformation is directly proportional to the force F and inversely proportional to the area A (note that the relevant area A is now the top area of the block, where the force is applied):

$$\frac{\Delta x}{h} = \frac{1}{S}\frac{F}{A} \tag{14.19}$$

shear and shear modulus

Here, the constant of proportionality S is called the **shear modulus.** Table 14.1 includes values of shear moduli of solids.

For compression, the fractional deformation is defined as the ratio of the change ΔV of the volume to the initial volume, and this fractional deformation is, again, proportional to the force F pressing on each face of the block and inversely proportional to the area A of that face:

$$\frac{\Delta V}{V} = -\frac{1}{B}\frac{F}{A} \tag{14.20}$$

compression and bulk modulus

In this equation, the minus sign indicates that ΔV is negative; that is, the volume decreases. The constant of proportionality B in the equation is called the **bulk modulus.**

Table 14.1 includes values of bulk moduli for solids. This table also includes values of bulk moduli for some liquids. The force per unit area, F/A, is also known as the **pressure:**

pressure

$$[\text{pressure}] = \frac{F}{A} \tag{14.21}$$

The formula (14.20) is equally valid for solids and for liquids—when we squeeze a liquid from all sides, it will suffer a compression. Note that Table 14.1 does not include values of Young's moduli and of shear moduli for liquids. *Elongation and shear stress are not supported by a liquid*—we can elongate or shear a "block" of liquid as much as we please without having to exert any significant force.

Concepts in Context

EXAMPLE 8 The lifting cable of a tower crane is made of steel, with a diameter of 5.0 cm. The length of this cable, from the ground to the horizontal arm, across the horizontal arm, and down to the load, is 160 m (Fig. 14.28). By how much does this cable stretch in excess of its initial length when carrying a load of 60 tons?

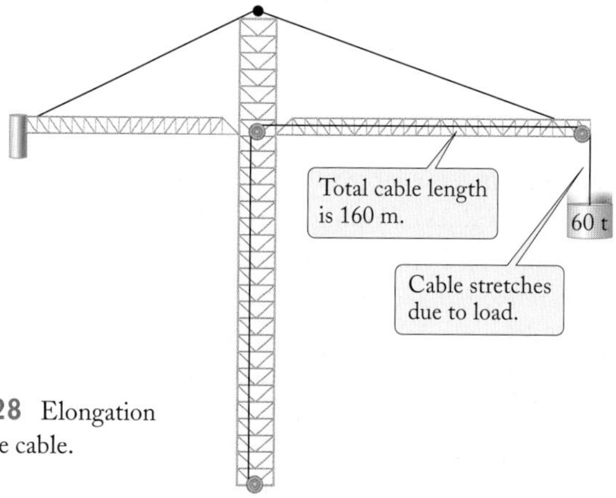

FIGURE 14.28 Elongation of a tower crane cable.

SOLUTION: The cross-sectional area of the cable is

$$A = \pi r^2 = \pi \times (0.025 \text{ m})^2 = 2.0 \times 10^{-3} \text{ m}^2$$

and the force per unit area is

$$\frac{F}{A} = \frac{(60\,000 \text{ kg} \times 9.81 \text{ m/s}^2)}{2.0 \times 10^{-3} \text{ m}^2} = 2.9 \times 10^8 \text{ N/m}^2$$

Since we are dealing with an elongation, the relevant elastic modulus is the Young's modulus. According to Table 14.1, the Young's modulus of steel is 22×10^{10} N/m^2. Hence Eq. (14.18) yields

$$\frac{\Delta L}{L} = \frac{1}{Y}\frac{F}{A} = \frac{1}{22 \times 10^{10} \text{ N/m}^2} \times 2.9 \times 10^8 \text{ N/m}^2$$
$$= 1.3 \times 10^{-3}$$

The change of length is therefore

$$\Delta L = 1.3 \times 10^{-3} \times L = 1.3 \times 10^{-3} \times 160 \text{ m}$$
$$= 0.21 \text{ m}$$

EXAMPLE 9 What pressure must you exert on a sample of water if you want to compress its volume by 0.10%?

SOLUTION: For volume compression, the relevant elastic modulus is the bulk modulus B. By Eq. (14.20), the pressure, or the force per unit area, is

$$\frac{F}{A} = -B\frac{\Delta V}{V}$$

For 0.10% compression, we want to achieve a fractional change of volume of $\Delta V/V = -0.0010$. Since the bulk modulus of water is 0.22×10^{10} N/m^2, the required pressure is

$$\frac{F}{A} = 0.22 \times 10^{10}\ \text{N/m}^2 \times 0.0010 = 2.2 \times 10^6\ \text{N/m}^2$$

The simple uniform deformations of elongation, shear, and compression described above require a rather special arrangement of forces. In general, the forces applied to a solid body will produce nonuniform elongation, shear, and compression. For instance, a beam supported at its ends and sagging in the middle because of its own weight or the weight of a load placed on it will elongate along its lower edge, and compress along its upper edge.

Finally, note that the formulas (14.18)–(14.20) are valid only as long as the deformation is reasonably small—a fraction of a percent or so. If the deformation is excessive, the material will be deformed beyond its elastic limit; that is, the material will suffer a permanent deformation and will *not* return to its original size and shape when the force ceases. If the deformation is even larger, the material will break apart or crumble. For instance, steel will break apart (see Fig. 14.29) if the tensile stress exceeds 5×10^8 N/m^2, or if the shearing stress exceeds 2.5×10^8 N/m^2, and it will crumble if the compressive stress exceeds 5×10^8 N/m^2.

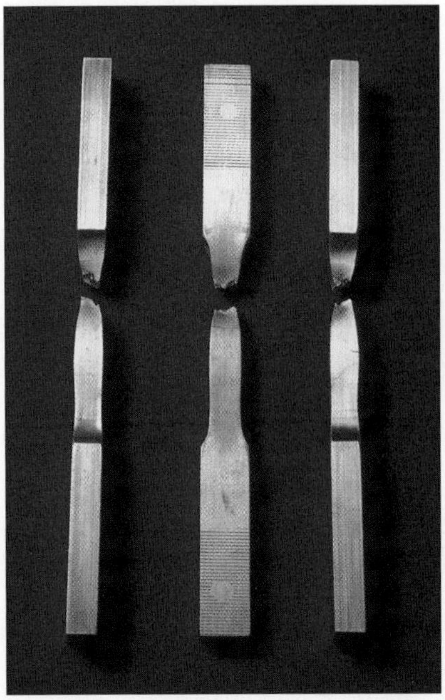

FIGURE 14.29 These rods of steel broke apart when a large tension was applied.

Checkup 14.4

QUESTION 1: When a tension of 70 N is applied to a piano wire of length 1.8 m, it stretches by 2.0 mm. If the same tension is applied to a similar piano wire of length 3.6 m, by how much will it stretch?

QUESTION 2: Is it conceivable that a long cable hanging vertically might snap under its own weight? If so, does the critical length of the cable depend on its diameter?

QUESTION 3: The bulk modulus of copper is about twice that of aluminum. Suppose that a copper and an aluminum sphere have exactly equal volumes at normal atmospheric pressure. Suppose that when subjected to a high pressure, the volume of the aluminum sphere shrinks by 0.01%. By what percentage will the copper sphere shrink at the same pressure?

QUESTION 4: While lifting a load, the steel cable of a crane stretches by 1 cm. If you want the cable to stretch by only 0.5 cm, by what factor must you increase its diameter?

(A) $\sqrt{2}$ (B) 2 (C) $2\sqrt{2}$ (D) 4

SUMMARY

PROBLEM-SOLVING TECHNIQUES Static Equilibrium (page 437)

STATIC EQUILIBRIUM The sums of the external forces and of the external torques on a rigid body are zero.

$$\mathbf{F}_1 + \mathbf{F}_2 + \mathbf{F}_3 + \cdots = 0 \quad (14.1)$$

$$\boldsymbol{\tau}_1 + \boldsymbol{\tau}_2 + \boldsymbol{\tau}_3 + \cdots = 0 \quad (14.2)$$

STATICS CALCULATION TECHNIQUE To eliminate an unknown force, evaluate torques about the point where that force acts (or about another point where the force has zero moment arm).

TORQUE DUE TO GRAVITY Gravity effectively acts at the center of mass.

MECHANICAL ADVANTAGE OF LEVER

$$\frac{F'}{F} = \frac{l}{l'} \quad (14.15)$$

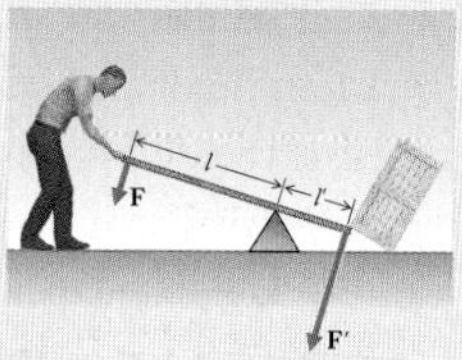

BLOCK AND TACKLE An arrangement of several pulleys that provides a mechanical advantage (equal to the ratio of the distance moved where the force is applied to the distance moved by the load).

PRESSURE

$$[\text{pressure}] = \frac{F}{A} \quad (14.21)$$

DEFORMATIONS OF ELASTIC MATERIAL

A is cross-sectional area.

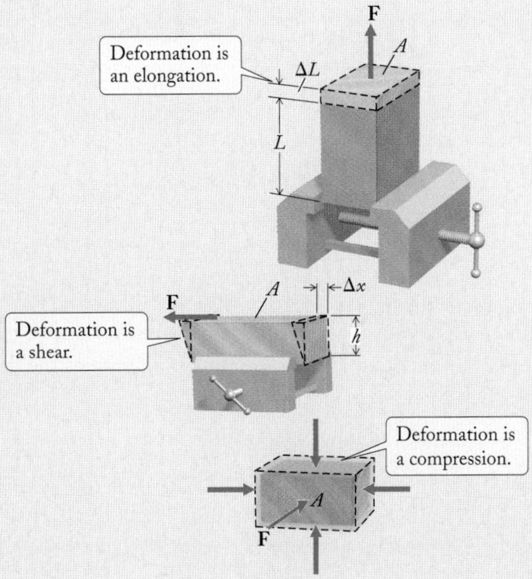

Elongation:
$$\frac{\Delta L}{L} = \frac{1}{Y}\frac{F}{A} \quad (Y = \text{Young's modulus}) \tag{14.18}$$

Shear:
$$\frac{\Delta x}{h} = \frac{1}{S}\frac{F}{A} \quad (S = \text{shear modulus}) \tag{14.19}$$

Compression:
$$\frac{\Delta V}{V} = -\frac{1}{B}\frac{F}{A} \quad (B = \text{bulk modulus}) \tag{14.20}$$

QUESTIONS FOR DISCUSSION

1. If the legs of a table are exactly the same length and if the floor is exactly flat, then the weight of the table will be equally distributed over all four legs. But if there are small deviations from exactness, then the weight will not be equally distributed. Is it possible for all of the weight to rest on three legs? On two?
2. List as many examples as you can of joints in the human skeleton that act as pivots for levers. Do any of these levers in the human skeleton have a mechanical advantage larger than 1?
3. Design a block and tackle with a mechanical advantage of 4, and another with a mechanical advantage of 5. If you connect these two arrangements in tandem, what mechanical advantage do you get?
4. Figure 14.30 shows a differential windlass consisting of two rigidly joined drums around which a rope is wound. A pulley holding a load hangs from this rope. Explain why this device gives a very large mechanical advantage if the radii of the two drums are nearly equal.
5. The collapse of several skywalks at the Hyatt Regency hotel in Kansas City on July 17, 1982, with the loss of 114 lives, was due to a defective design of the suspension system. Instead of suspending the beams of the skywalks directly from single, long steel rods anchored at the top of the building, some incompetent engineers decided to use several short steel rods joining the beams of each skywalk to those of the skywalk above (Fig. 14.31). Criticize this design, keeping in mind that the beams are made of a much weaker material than the rods.

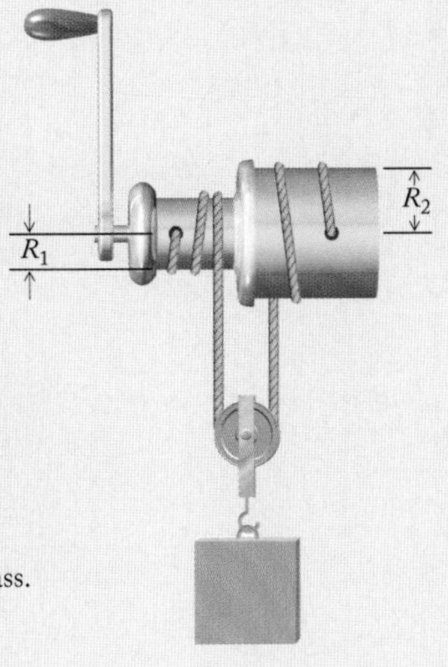

FIGURE 14.30 Differential windlass.

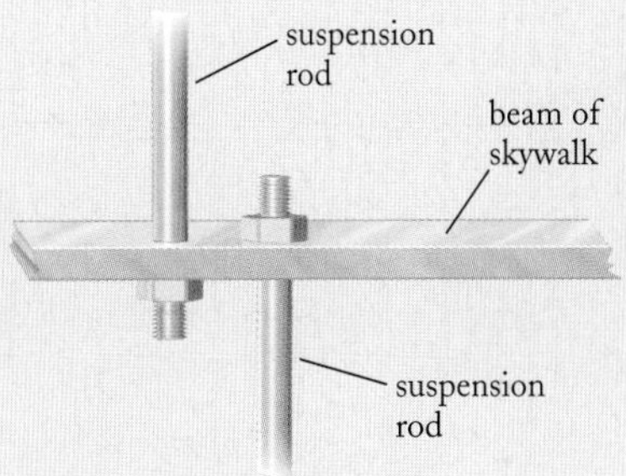

FIGURE 14.31 Beam of skywalk and suspension rods.

6. A steel rod is much less flexible than a woven steel rope of the same strength. Explain this.

7. A carpenter wants to support the (flat) roof of a building with horizontal beams of wood of rectangular cross section. To achieve maximum strength of the roof (least sag), should he install the beams with their narrow side up or with their wide side up?

8. The long bones in the limbs of vertebrates have the shape of hollow pipes. If the same amount of bone tissue had been assembled in a solid rod (of correspondingly smaller cross section), would the limb have been more rigid or less rigid?

PROBLEMS

4.2 Examples of Static Equilibrium

1. At a construction site, a laborer pushes horizontally against a large bucket full of concrete of total mass 600 kg suspended from a crane by a 20-m cable (see Fig. 14.32). What is the force the laborer has to exert to hold the bucket at a distance of 2.0 m from the vertical?

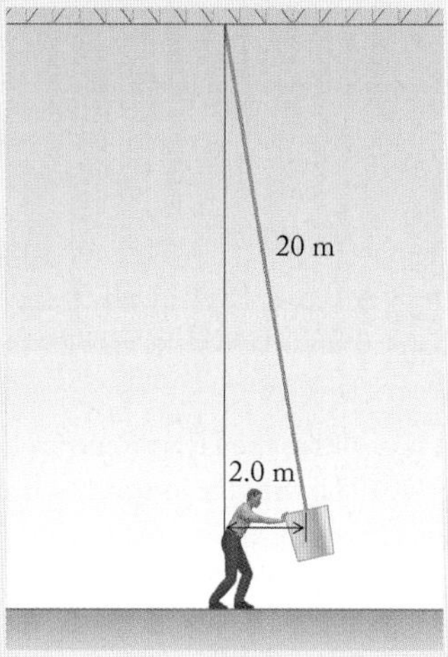

FIGURE 14.32 Bucket hanging from a cable.

2. You are holding a meterstick of 0.20 kg horizontally in one hand. Assume that your hand is wrapped around the last 10 cm of the stick (see Fig. 14.33), so the front edge of your hand exerts an upward force and the rear edge of your hand exerts a downward force. Calculate these forces.

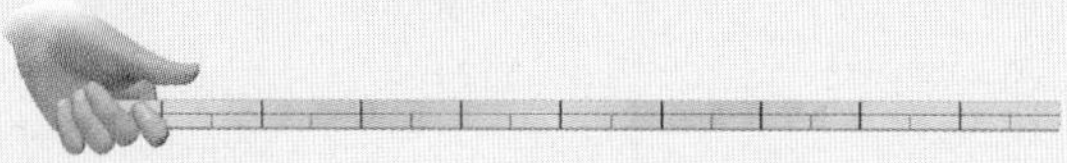

FIGURE 14.33 A meterstick held in a hand.

3. Consider the bridge with the locomotive described in Example 1 and suppose that, besides the first locomotive at 30 m from the right end, there is a second locomotive, also of 90 000 kg, at 80 m from the right end. What is the load on each pier in this case?

4. Repeat the calculations of Example 1 assuming that the bridge has a slope of 1:7, with the left end higher than the right.

5. In order to pull an automobile out of the mud in which it is stuck, the driver stretches a rope taut from the front end of the automobile to a stout tree. He then pushes sideways against the rope at the midpoint (see Fig. 14.34). When he pushes with a force of 900 N, the angle between the two halves of the rope on his right and left is 170°. What is the tension in the rope under these conditions?

6. A mountaineer is trying to cross a crevasse by means of a rope stretched from one side to the other (see Fig. 14.35). The mass of the mountaineer is 90 kg. If the two parts of the rope make angles of 40° and 20° with the horizontal, what are the tensions in the two parts?

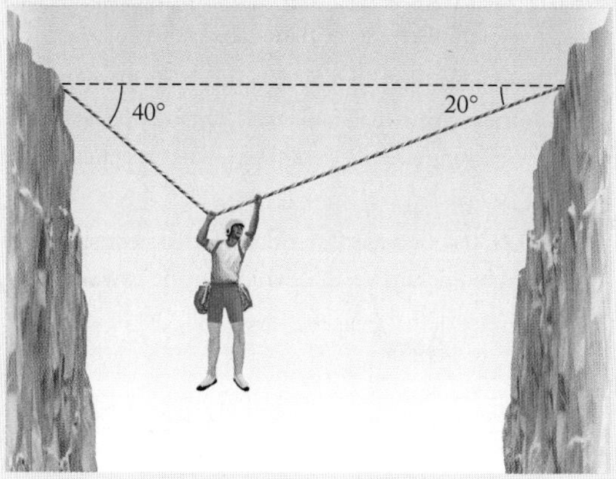

FIGURE 14.35 Mountaineer suspended from a rope.

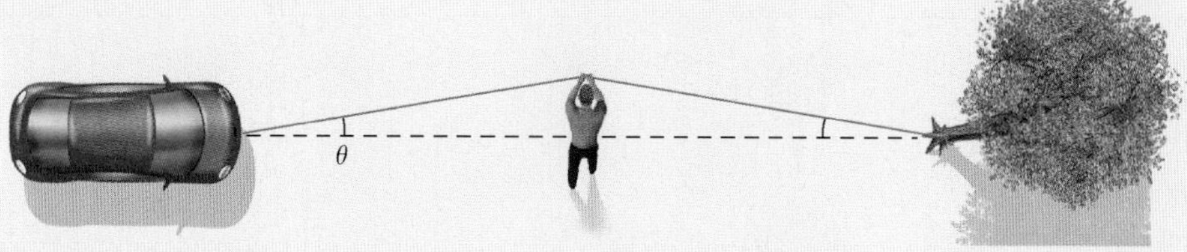

FIGURE 14.34 The rope is stretched between the automobile and a tree. The driver is pushing at the midpoint.

7. The plant of the foot of an average male is 26 cm, and the height of his center of mass above the floor is 1.03 m. When he is standing upright, the center of mass is vertically aligned with the ankle, 18 cm from the tip of the foot (see Fig. 14.36). Without losing his equilibrium, how far can the man lean forward or backward while keeping his body straight and his feet stiff and immobile?

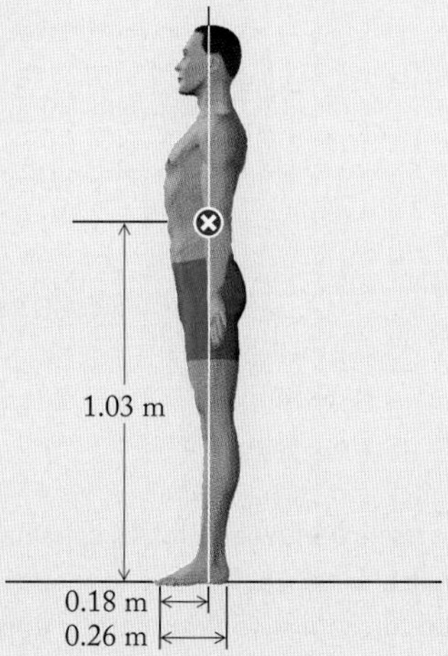

FIGURE 14.36 Man standing on stiff feet.

8. A 50-kg log of uniform thickness lies horizontally on the ground.
   (a) What vertical force must you exert on one end of the log to barely lift this end off the ground?
   (b) If you continue to exert a purely vertical force on the end of the log, what is the magnitude of the force required to hold the log at an angle of 30° to the ground? At an angle of 60°? At an angle of 85°?
   (c) If instead you exert a force at right angles to the length of the log, what is the magnitude of the force required to hold the log at an angle of 30° to the ground? At an angle of 60°? At an angle of 85°?

9. In an unequal-arm balance, the beam is pivoted at a point near one end. With such a balance, large loads can be balanced with small standard weights. Figure 14.37 shows such a balance with an arm of 50 cm swinging on a pivot 1.0 cm from one end. When a package of sugar is deposited in the balance pan, equilibrium is attained with a standard mass of 0.12 kg in the other pan. What is the mass of the sugar? Neglect the masses of the pans.

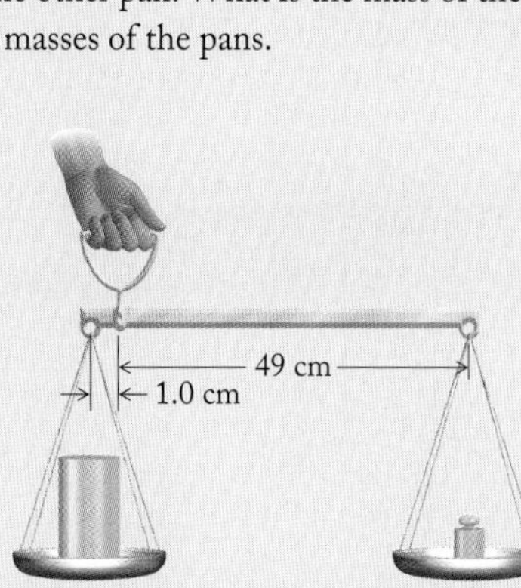

FIGURE 14.37 Unequal-arm balance.

10. One end of a uniform beam of mass 50 kg and length 3.0 m rests on the ground; the other end is held above the ground by a pivot placed 1.0 m from that end (see Fig. 14.38). An 80-kg man walks along the beam, from the low end toward the high end. How far beyond the pivot can the man walk before the high end of the beam swings down?

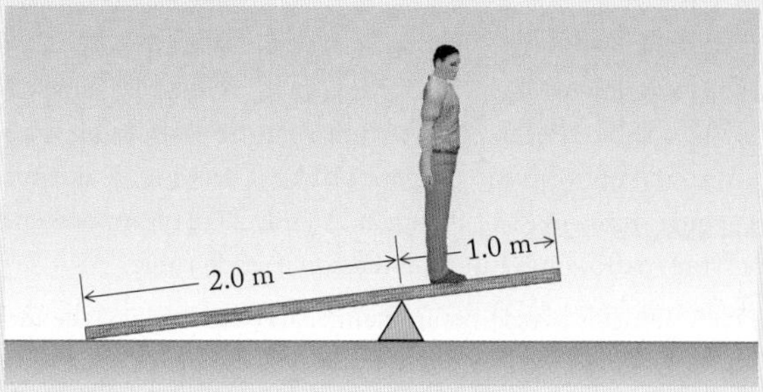

FIGURE 14.38 Man standing on a beam.

11. The mast of a sailboat is held by two steel cables attached as shown in Fig. 14.39. The front cable has a tension of 5.0×10^3 N. The mast is 10 m high. What is the tension in the rear cable? What force does the foot of the mast exert on the sailboat? Assume that the weight of the mast can be neglected and that the foot of the mast is hinged (and therefore exerts no torque).

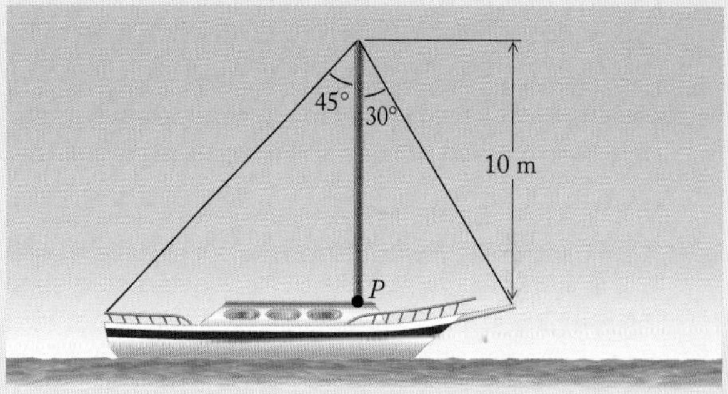

FIGURE 14.39 Steel cables staying a mast.

12. The center of mass of a 45-kg sofa is 0.30 m above its bottom, at its lateral midpoint. You lift one end of the 2.0-m-long sofa to a height of 1.0 m by applying a vertical force at the bottom of one end; the other end stays on the floor without slipping. What force do you apply? Compare this with the force you apply when a friend lifts the other end, also to 1.0 m, so that the weight is shared equally. Based on this, who has the easier task when a short and a tall person share a bulky load?

13. Suppose that you lift the lid of a chest. The lid is a uniform sheet of mass 12 kg, hinged at the rear. What is the smallest force you can apply at the front of the lid to hold it at an angle of 30° with the horizontal? At 60°?

14. A pole-vaulter holds a 4.5-m pole horizontally with her right hand at one end and her left hand 1.5 m from the same end.

The left hand applies an upward force and the right hand a downward force. If the mass of the pole is 3.0 kg, find those two forces.

15. A 50-kg diving board is 3.0 m long; it is a uniform beam, bolted down at one end and supported from below a distance 1.0 m from the same end. A 60-kg diver stands at the other end. Calculate the downward force at the bolted end and the upward support force.

16. A window washer's scaffolding is 12 m long; it is suspended by a cable at each end. Assume that the scaffolding is a horizontal uniform rod of mass 110 kg. The window washer (with gear) has a mass of 90 kg and stands 2.0 m from one end of the scaffolding. Find the tension in each cable.

17. A pencil is placed on an incline, and the angle of the incline is slowly increased. At what angle will the pencil start to roll? Assume the pencil has an exactly hexagonal cross section and does not slip.

18. A 10-kg ladder is 5.0 m long and rests against a frictionless wall, making an angle of 30° with the vertical. The coefficient of friction between the ladder and the ground is 0.35. A 60-kg painter begins to climb the ladder, standing vertically on each rung. How far up the ladder has the painter climbed when the ladder begins to slip?

19. Figure 14.40 shows the arrangement of wheels on a passenger engine of the Caledonian Railway. The numbers give the distances between the wheels in feet and the downward forces that each wheel exerts on the track in short tons (1 short ton = 2000 lbf; the numbers for the forces include both the right and left wheels). From the information given, find how far the center of mass of the engine is behind the front wheel.

20. A door made of a uniform piece of wood measures 1.0 m by 2.0 m and has a mass of 18 kg. The door is entirely supported by two hinges, one at the bottom corner and one at the top corner. Find the force (magnitude and direction) that the door exerts on each hinge. Assume that the *vertical* force on each hinge is the same.

21. You want to pick up a nearly massless rectangular cardboard box by grabbing its top and side between your forefinger and thumb (see Fig. 14.41). Show that this is impossible unless the coefficient of friction between your fingers and the box is at least 1.

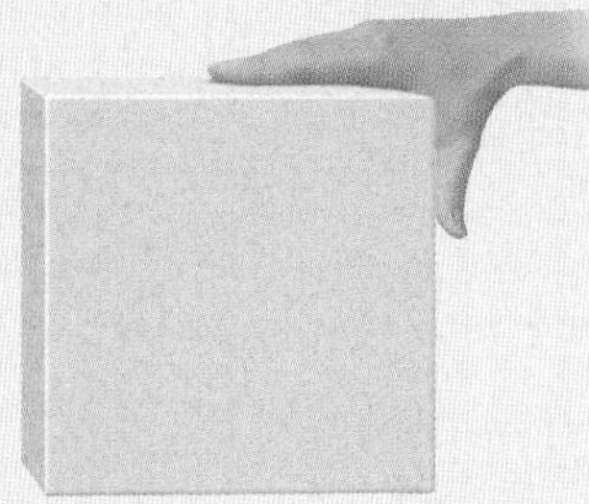

FIGURE 14.41 Box held in a hand.

22. A meterstick of wood of 0.40 kg is nailed to the wall at the 75-cm mark. If the stick is free to rotate about the nail, what horizontal force must you exert at the upper (short) end to deflect the stick 30° to one side?

*23. A wheel of mass M and radius R is to be pulled over a step of height h, where $R > h$. Assume that the pulling force is applied at the axis of the wheel. If the pull is horizontal, what force must be applied to barely begin moving? If the pull at the axis is instead in the direction that requires the least force to begin moving, what force must be applied? What is the new direction? (Hint: Consider the torques about the point of contact with the step.)

*24. Consider a heavy cable of diameter d and density ρ from which hangs a load of mass M. What is the tension in the cable as a function of the distance from the lower end?

*25. Figure 14.42 shows two methods for supporting the mast of a sailboat against the lateral force exerted by the pull of the sail. In Fig. 14.42a, the shrouds (wire ropes) are led directly to the top of the mast; in Fig. 14.42b, the shrouds are led around a rigid pair of spreaders. Suppose that the dimensions of the mast

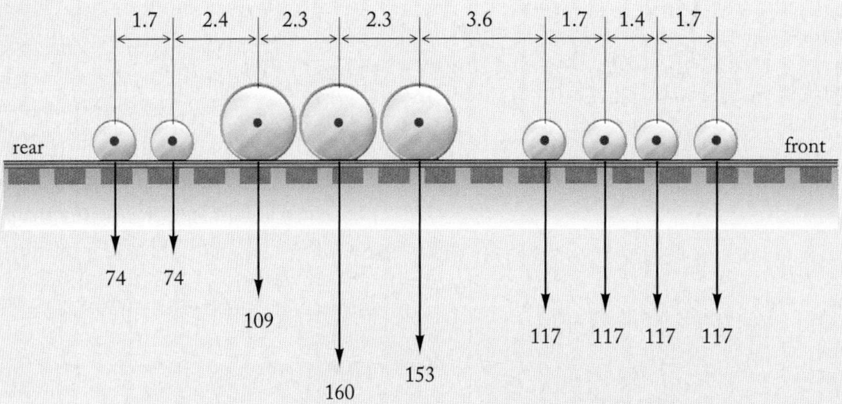

FIGURE 14.40 Wheels of a locomotive.

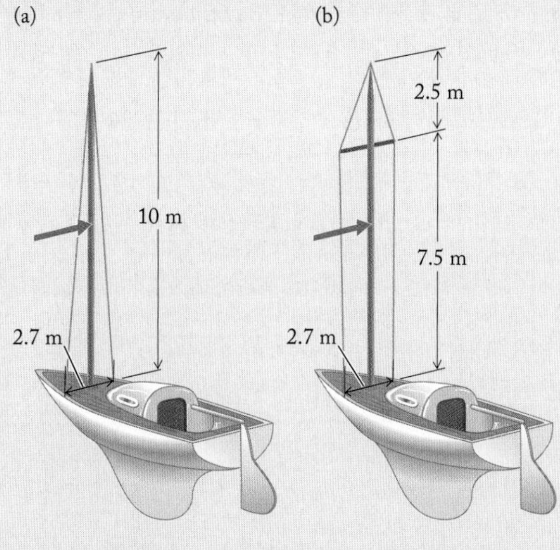

FIGURE 14.42 Two methods for supporting the mast of a boat.

and the boat are as indicated in this figure, and that the pull of the sail is equivalent to a horizontal force of 2400 N acting from the left at half the height of the mast. The foot of the mast permits the mast to tilt, so the only lateral support of the mast is that provided by the shrouds. What is the excess tension in the left shroud supporting the mast in case (a)? In case (b)? Which arrangement is preferable?

*26. A bowling ball of mass 10 kg rests in a groove with smooth, perpendicular walls, inclined at angles of 30° and 60° with the vertical, as shown in Fig. 14.43. Calculate the magnitudes of the normal forces at the points of contact.

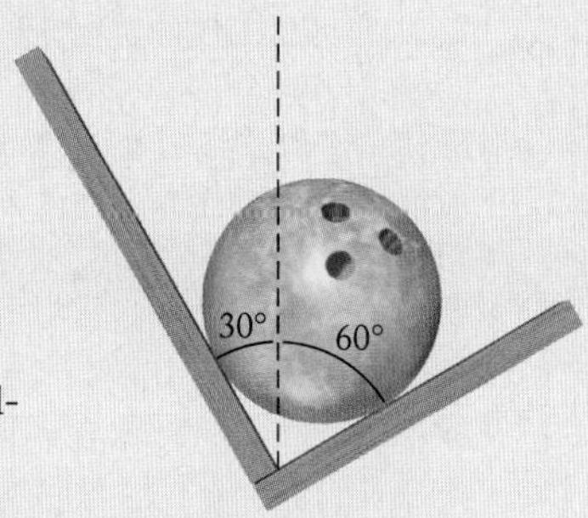

FIGURE 14.43 A bowling ball in a groove.

*27. A tetrahedral tripod consists of three massless legs (see Fig. 14.44). A mass M hangs from the apex of the (regular) tetrahedron. What are the compressional forces in the three legs?

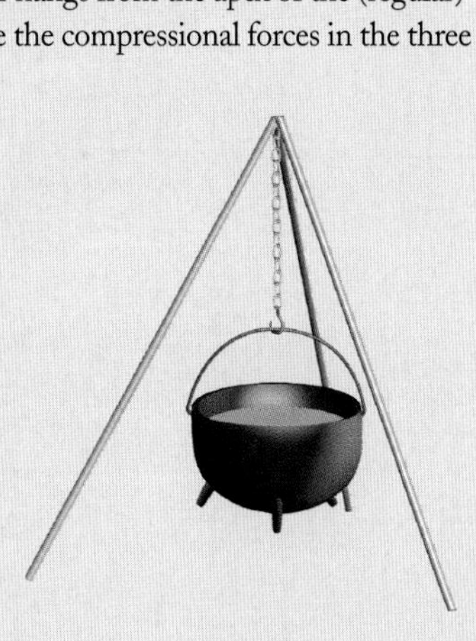

FIGURE 14.44 A tripod.

*28. A sailor is being transferred from one ship to another by means of a bosun's chair (see Fig. 14.45). The chair hangs from a roller riding on a rope strung between the two ships. The distance between the ships is d, and the rope has a length $1.2d$. The mass of the sailor plus the chair is m. If the sailor is at a (horizontal) distance of $0.25d$ from one ship, find the force that must be exerted on the pull rope to keep the sailor in equilibrium. Also find the tension in the long rope. Ignore the masses of the ropes.

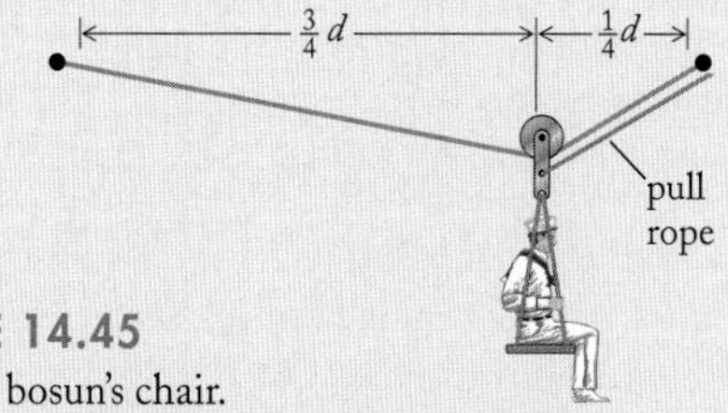

FIGURE 14.45 Sailor in bosun's chair.

*29. A uniform solid disk of mass M and radius R hangs from a string of length l attached to a smooth vertical wall (see Fig. 14.46). Calculate the tension in the string and the normal force acting at the point of contact of disk and wall.

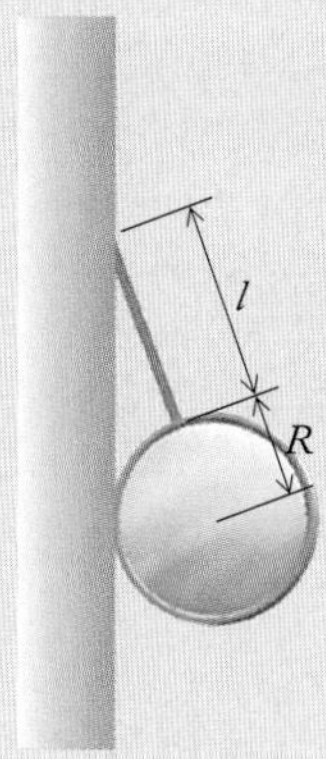

FIGURE 14.46 Disk hanging from string.

*30. Three traffic lamps of equal masses of 20 kg hang from a wire stretched between two telephone poles, 15 m apart (Fig. 14.47). The horizontal spacing of the traffic lamps is uniform. At each pole, the wire makes a downward angle of 10° with the horizontal line. Find the tensions in all the segments of wire, and find the distance of each lamp below the horizontal line.

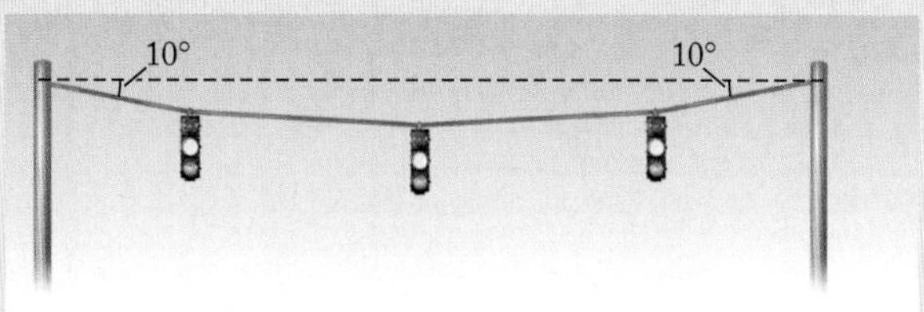

FIGURE 14.47 Three traffic lamps.

*31. Consider the ladder leaning against a wall described in Example 4. If the ladder makes an angle of 30° with the wall, how hard can you push down vertically on the top of the ladder with your hand before slipping begins?

*32. An automobile with a wheelbase (distance from the front wheels to the rear wheels) of 3.0 m has its center of mass at a point midway between the wheels at a height of 0.65 m above the road. When the automobile is on a level road, the force with which each wheel presses on the road is 3100 N. What is the normal force with which each wheel presses on the road when the automobile is standing on a steep road of slope 3:10 with all the wheels locked?

*33. A wooden box is filled with material of uniform density. The box (with its contents) has a mass of 80 kg; it is 0.60 m wide, 0.60 m deep, and 1.2 m high. The box stands on a level floor. By pushing against the box, you can tilt it over (Fig. 14.48). Assume that when you do this, one edge of the box remains in contact with the floor without sliding.

(a) Plot the gravitational potential energy of the box as a function of the angle θ between the bottom of the box and the floor.

(b) What is the critical angle beyond which the box will topple over when released?

(c) How much work must you do to push the box to this critical angle?

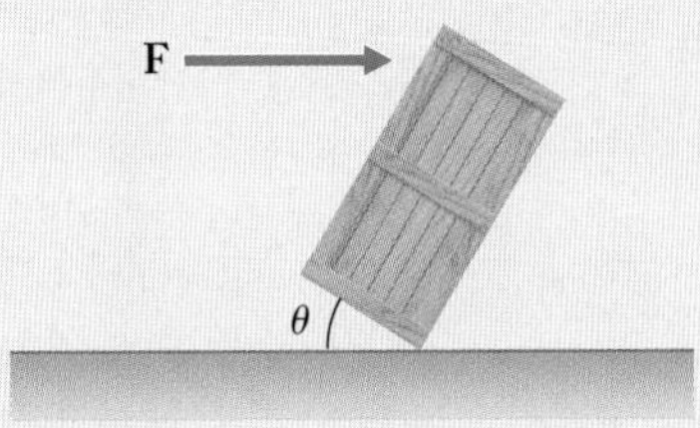

FIGURE 14.48 Tilted box.

*34. A meterstick of mass M hangs from a 1.5-m string tied to the meterstick at the 80-cm mark. If you push the bottom end of the meterstick to one side with a horizontal push of magnitude $Mg/2$, what will be the equilibrium angles of the meterstick and the string?

*35. Five identical books are to be stacked one on top of the other. Each book is to be shifted sideways by some variable amount, so as to form a curved leaning tower with maximum protrusion (see Fig. 14.49). How much must each book be shifted? What is the maximum protrusion? If you had an infinite number of books, what would be the limiting maximum protrusion? (Hint: Try this experimentally; start with the top book, and insert the others underneath, one by one.)

FIGURE 14.49 A stack of books.

**36. A wooden box, filled with a material of uniform density, stands on a concrete floor. The box has a mass of 75 kg and is 0.50 m wide, 0.50 m long, and 1.5 m high. The coefficient of friction between the box and the floor is $\mu_s = 0.80$. If you exert a (sufficiently strong) horizontal push against the side of the box, it will either topple over or start sliding without toppling over, depending on how high above the level of the floor you push. What is the maximum height at which you can push if you want the box to slide? What is the magnitude of the force you must exert to start the sliding?

*37. The left and right wheels of an automobile are separated by a transverse distance of $l = 1.5$ m. The center of mass of this automobile is $h = 0.60$ m above the ground. If the automobile is driven around a flat (no banking) curve of radius $R = 25$ m with an excessive speed, it will topple over sideways. What is the speed at which it will begin to topple? Express your answer in terms of l, h, and R; then evaluate numerically. Assume that the wheels do not skid.

*38. An automobile has a wheelbase (distance from front wheels to rear wheels) of 3.0 m. The center of mass of this automobile is at a height of 0.60 m above the ground. Suppose that this automobile has rear-wheel drive and that it is accelerating along a level road at 6.0 m/s^2. When the automobile is parked, 50% of its weight rests on the front wheels and 50% on the rear wheels. What is the weight distribution when it is accelerating? Pretend that the body of the automobile remains parallel to the road at all times.

*39. Consider a bicycle with only a front-wheel brake. During braking, what is the maximum deceleration that this bicycle can withstand without flipping over its front wheel? The center of mass of the bicycle with rider is 95 cm above the road and 70 cm behind the point of contact of the front wheel with the ground.

*40. A bicycle and its rider are traveling around a curve of radius 6.0 m at a constant speed of 20 km/h. What is the angle at which the rider must lean the bicycle toward the center of the curve (see Fig. 14.50)?

FIGURE 14.50 Bicycle traveling around curve.

**41. An automobile is braking on a flat, dry road with a coefficient of static friction of 0.90 between its wheels and the road. The wheelbase (the distance between the front and the rear wheels) is 3.0 m, and the center of mass is midway between the wheels, at a height of 0.60 m above the road.

(a) What is the deceleration if all four wheels are braked with the maximum force that avoids skidding?

(b) What is the deceleration if the rear-wheel brakes are disabled? Take into account that during braking, the normal force on the front wheels is larger than that on the rear wheels.

(c) What is the deceleration if the front-wheel brakes are disabled?

*42. A square framework of steel hangs from a crane by means of cables attached to the upper corners making an angle of 60° with each other (see Fig. 14.51). The framework is made of beams of uniform thickness joined (loosely) by pins at the corners, and its total mass is M. Find the tensions in the cables and the tensional and compressional forces in each beam at each of its two ends.

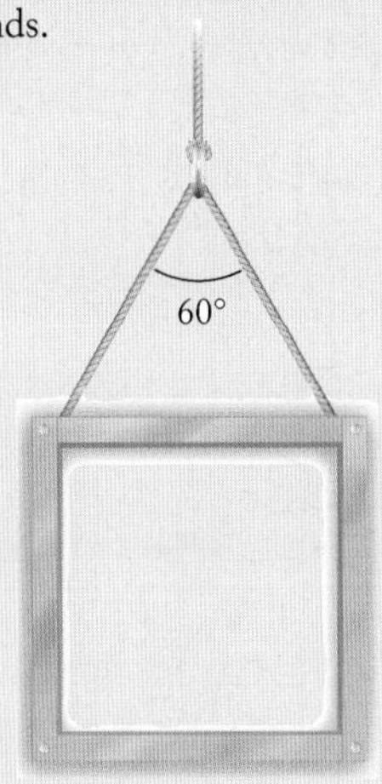

FIGURE 14.51 Hanging framework of beams.

**43. Two smooth balls of steel of mass m and radius R are sitting inside a tube of radius $1.5R$. The balls are in contact with the bottom of the tube and with the wall (at two points; see Fig. 14.52). Find the contact force at the bottom and at the two points on the wall.

FIGURE 14.52 Two balls in a tube.

**44. One end of a uniform beam of length L rests against a smooth, frictionless vertical wall, and the other end is held by a string of length $l = \frac{3}{2}L$ attached to the wall (see Fig. 14.53). What must be the angle of the beam with the wall if it is to remain at rest without slipping?

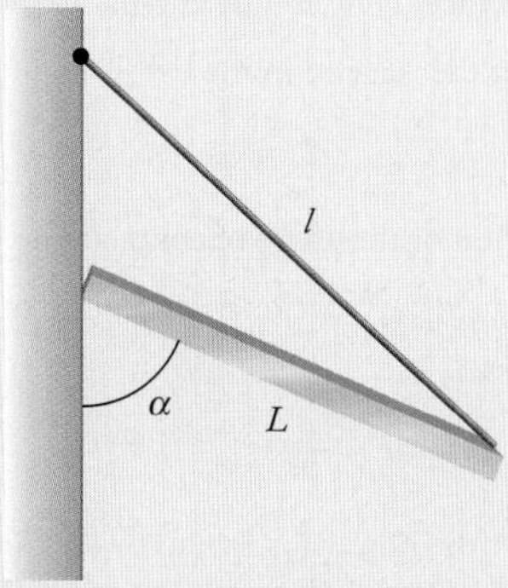

FIGURE 14.53 Beam, string, and wall.

**45. Two playing cards stand on a table leaning against each other so as to form an A-frame "roof." The frictional coefficient between the bottoms of the cards and the table is μ_s. What is the maximum angle that the cards can make with the vertical without slipping?

**46. A rope is draped over the round branch of a tree, and unequal masses m_1 and m_2 are attached to its ends. The coefficient of sliding friction for the rope on the branch is μ_k. What is the acceleration of the masses? Assume that the rope is massless. (Hint: For each small segment of the rope in contact with the branch, the small change in tension across the segment is equal to the friction force.)

**47. The flywheel of a motor is connected to the flywheel of an electric generator by a drive belt (Fig. 14.54). The flywheels are of equal size, each of radius R. While the flywheels are rotating, the tensions in the upper and the lower portions of the drive belt are T_1 and T_2, respectively, so the drive belt exerts a torque $\tau = (T_2 - T_1)R$ on the generator. The coefficient of static friction between each flywheel and the drive belt is μ_s. Assume that the tension in the drive belt is as low as possible with no slipping, and that the drive belt is massless. Show that under these conditions

$$T_1 = \frac{\tau}{R}\frac{1}{e^{\mu_s \pi} - 1}$$

$$T_2 = \frac{\tau}{R}\frac{1}{1 - e^{-\mu_s \pi}}$$

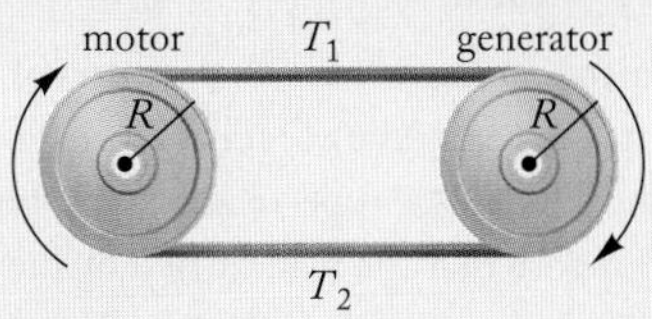

FIGURE 14.54 A drive belt connecting flywheels of a motor and a generator.

**48. A power brake invented by Lord Kelvin consists of a strong flexible belt wrapped once around a spinning flywheel (Fig. 14.55). One end of the belt is fixed to an overhead support; the other end carries a weight w. The coefficient of kinetic friction between the belt and the wheel is μ_k. The radius of the wheel is R, and its angular velocity is ω.

(a) Show that the tension in the belt is

$$T = we^{-\mu_k \theta}$$

as a function of the angle of contact (Fig. 14.55).

(b) Show that the net frictional torque the belt exerts on the flywheel is

$$\tau = wR\,(1 - e^{-2\pi\mu_k})$$

(c) Show that the power dissipated by friction is

$$P = wR\omega\,(1 - e^{-2\pi\mu_k})$$

FIGURE 14.55 Belt and flywheel.

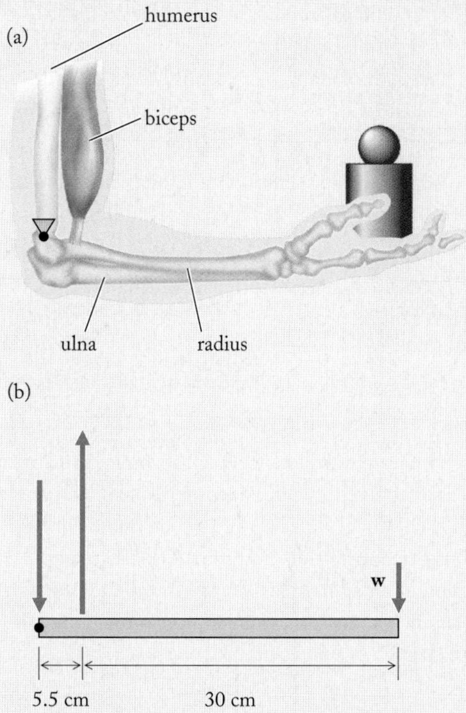

FIGURE 14.56 Forearm as lever.

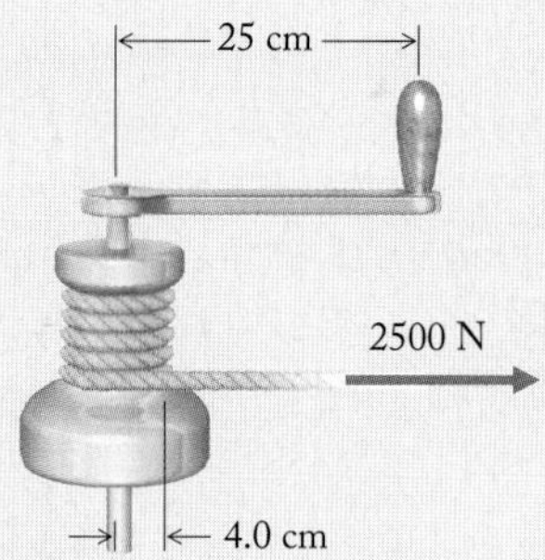

FIGURE 14.57 Manual winch.

14.3 Levers and Pulleys

49. The human forearm (including the hand) can be regarded as a lever pivoted at the joint of the elbow and pulled upward by the tendon of the biceps (Fig. 14.56a). The dimensions of this lever are given in Fig. 14.56b. Suppose that a load of 25 kg rests in the hand. What upward force must the biceps exert to keep the forearm horizontal? What is the downward force at the elbow joint? Neglect the weight of the forearm.

50. Repeat the preceding problem if, instead of being vertical, the upper arm is tilted, so as to make an angle of 135° with the (horizontal) forearm.

51. A simple manual winch consists of a drum of radius 4.0 cm to which is attached a handle of radius 25 cm (Fig. 14.57). When you turn the handle, the rope winds up on the drum and pulls the load. Suppose that the load carried by the rope is 2500 N. What force must you exert on the handle to hold this load?

52. The handle of a crowbar is 60 cm long; the short end is 4.0 cm from a bend, which acts as the fulcrum. If a 75-kg man leans on the handle with all his weight, how much mass can he lift at the short end?

53. A 60-kg woman sits 80 cm from the fulcrum of a 4.0-m-long seesaw. The woman's daughter pulls down on the other end of the seesaw. What minimum force must the child apply to hold her mother's end of the seesaw off the ground?

54. The fingers apply a force of 30 N at the handle of a pair of scissors, 4.0 cm from the hinge point. What force is available for cutting when the object to be cut is placed at the far end of the scissors, 12 cm from the hinge point? When the object is placed as close to the hinge point as possible, at a distance of 1.0 cm?

55. A laboratory microbalance has two weighing pans, one hanging 10 times farther away from the fulcrum than the other.

When an unknown mass is placed on the inner pan, the microbalance can measure changes in mass as small as 100 nanograms (1.0×10^{-7} g) and can measure masses up to 2.0 milligrams. What would you expect that the resolution and maximum load for the outer pan might be?

56. A man of 73 kg stands on one foot, resting all of his weight on the ball of the foot. As described in Section 14.3, the bones of the foot play the role of a lever. The short end of the lever (to the heel) measures 5.0 cm and the long end (to the ball of the foot) 14 cm. Calculate the force exerted by the Achilles tendon and the force at the ankle.

57. A rope hoist consists of four pulleys assembled in two pairs with rigid straps, with a rope wrapped around as shown in Fig. 14.58. A load of 300 kg hangs from the lower pair of pulleys. What tension must you apply to the rope to hold the load steady? Treat the pulleys and the rope as massless, and ignore any friction in the pulleys.

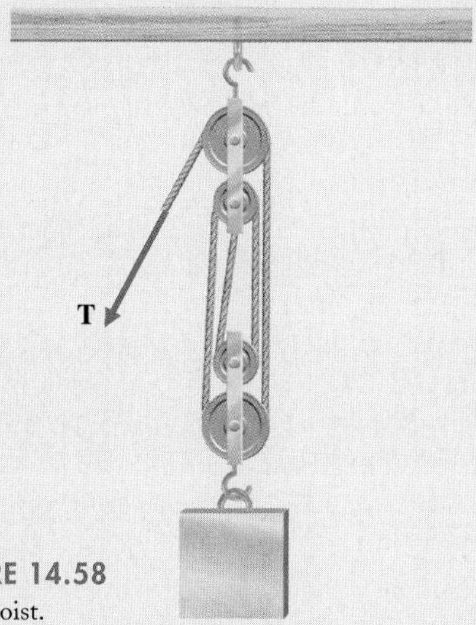

FIGURE 14.58 Rope hoist.

58. A parbuckle is a simple device used by laborers for raising or lowering a barrel or some other cylindrical object along a ramp. It consists of a loop of rope wrapped around the barrel (see Fig. 14.59). One end of the rope is tied to the top of the ramp, and the laborer pulls on the other end. Suppose that the laborer exerts a pull of 500 N on the rope, parallel to the ramp. What is the force that the rope exerts on the barrel? What is the mechanical advantage of the parbuckle?

FIGURE 14.59 Parbuckle used to move a barrel up a ramp.

59. Consider the differential windlass illustrated in Fig. 14.30. Calculate what clockwise torque must be applied to the handle to lift a load of mass m. What tangential force must be exerted on the handle? What is the mechanical advantage of this windlass?

60. Design a block and tackle with a mechanical advantage of 4, and another with a mechanical advantage of 5. If you connect these two arrangements in tandem, what mechanical advantage do you get?

*61. Figure 14.60 shows a compound bolt cutter. If the dimensions are as indicated in this figure, what is the mechanical advantage?

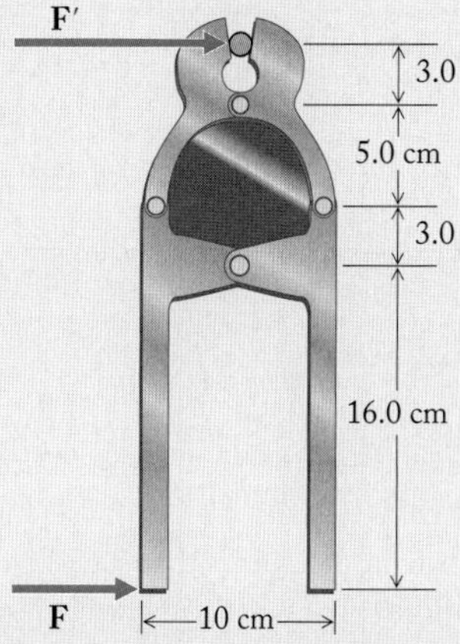

FIGURE 14.60 Compound bolt cutter.

*62. The drum of a winch is rigidly attached to a concentric large gear, which is driven by a small gear attached to a crank. The dimensions of the drum, the gears, and the crank are given in Fig. 14.61. What is the mechanical advantage of this geared winch?

FIGURE 14.61 Geared winch.

*63. The screw of a vise has a *pitch* of 4.0 mm; that is, it advances 4.0 mm when given one full turn. The handle of the vise is 25 cm long, measured from the screw to the end of the handle. What is the mechanical advantage when you push perpendicularly on the end of the handle?

*64. A scissors jack has the dimensions shown in Fig. 14.62. The screw of the jack has a pitch of 5.0 mm (as stated in the previous problem, this is the distance the screw advances when given one full turn). Suppose the scissors jack is partially extended, with an angle of 55° between its upper sides. What is the mechanical advantage provided by the jack?

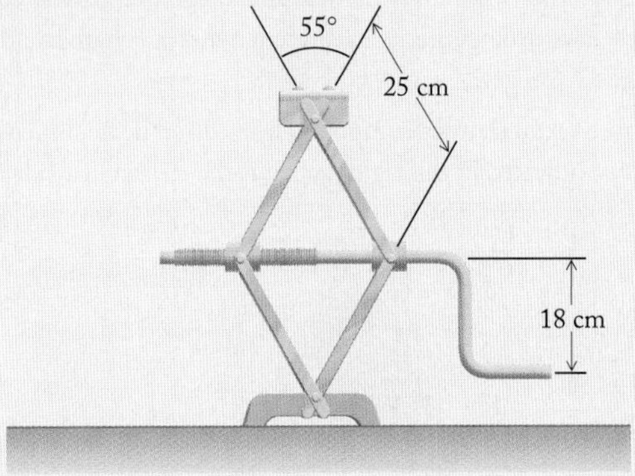

FIGURE 14.62 Scissors jack.

**65. Figure 14.63 shows a tensioning device used to tighten the rear stay of the mast of a sailboat. The block and tackle pulls down a rigid bar with two rollers that squeeze together the two branches of the split rear stay. If the angles are as given in the figure, what is the mechanical advantage?

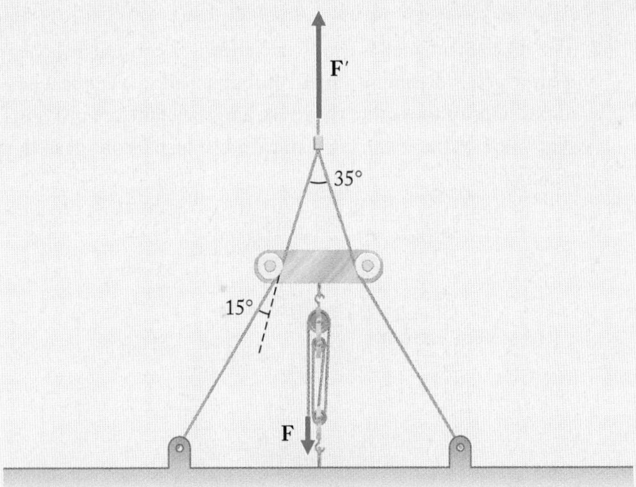

FIGURE 14.63 Tensioning device.

14.4 Elasticity of Materials

66. The anchor rope of a sailboat is a nylon rope of length 60 m and diameter 1.3 cm. While anchored during a storm, the sailboat momentarily pulls on this rope with a force of 1.8×10^4 N. How much does the rope stretch?

67. A piano wire of steel of length 1.8 m and radius 0.30 mm is subjected to a tension of 70 N by a weight attached to its lower end. By how much does this wire stretch in excess of its initial length?

68. An elastic cord is 5.0 m long and 1.0 cm in diameter and acts as a spring with spring constant 70 N/m. What is the Young's modulus for this material?

69. The piano wire described in Problem 67 can be regarded as a spring. What is the effective spring constant of this spring?

70. A simple hand-operated hydraulic press can generate a pressure of 6.0×10^9 N/m^2. If the system is used to compress a small volume of steel, what fraction of the original volume does the final volume of steel occupy?

71. A 10-m length of 1.0-mm-radius copper wire is stretched by holding one end fixed and pulling on the other end with a force of 150 N. What is the change of length? By briefly increasing the force to exceed the limit of elastic behavior (a fractional elongation of approximately 1.0%), the wire may be permanently deformed; this is often done in order to straighten out bends or kinks in a wire. Approximately what force is necessary?

72. A 0.50-mm-radius fishing line made of nylon is 100 m long when no forces are applied. A fish is hooked and pulls with a tension force of 250 N. What is the elongation?

73. In a skyscraper, an elevator is suspended from three equal, parallel 300-m-long steel cables, each of diameter 1.0 cm. How much do these cables stretch if the mass of the elevator is 1000 kg?

74. The length of the femur (thighbone) of a woman is 38 cm, and the average cross section is 10 cm^2. How much will the femur be compressed in length if the woman lifts another woman of 68 kg and carries her piggyback? Assume that, momentarily, all of the weight rests on one leg.

75. If the volume of a sphere subjected to an external pressure shrinks by 0.10%, what is the percent shrinkage of the radius? In general, show that the percent shrinkage of the volume equals three times the percent shrinkage of the radius, provided the shrinkage is small.

76. At the bottom of the Marianas Trench in the Pacific Ocean, at a depth of 10 900 m, the pressure is 1.24×10^8 N/m^2. What is the percent increase of the density of water at this depth as compared with the density at the surface?

77. A slab of stone of mass 1200 kg is attached to the wall of a building by two bolts of iron of diameter 1.5 cm (see Fig. 14.64). The distance between the wall and the slab of stone is 1.0 cm. Calculate by how much the bolts will sag downward because of the shear stress they are subjected to.

78. According to (somewhat oversimplified) theoretical considerations, the Young's modulus, the shear modulus, and the bulk modulus are related by

$$Y = \frac{9BS}{3B + S}$$

Check this for the first four materials listed in Table 14.1.

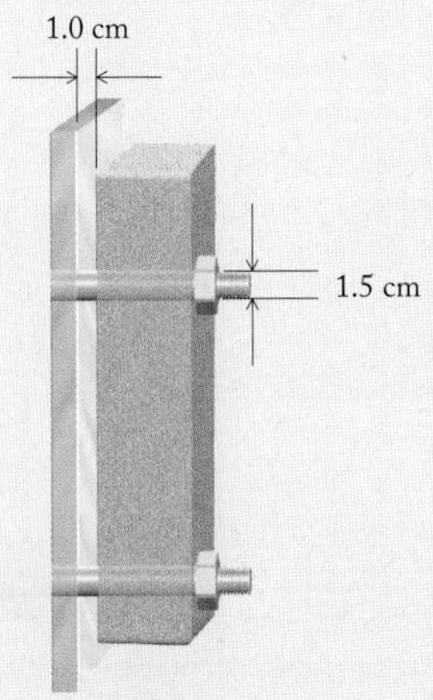

FIGURE 14.64 A slab of stone held by bolts.

79. A nylon rope of diameter 1.3 cm is to be spliced to a steel rope. If the steel rope is to have the same ultimate breaking strength as the nylon, what diameter should it have? The ultimate tensile strength is 2.0×10^9 N/m^2 for the steel and 3.2×10^8 N/m^2 for nylon.

*80. A rod of aluminum has a diameter of 1.000 002 cm. A ring of cast steel has an inner diameter of 1.000 000 cm. If the rod and the ring are placed in a liquid under high pressure, at what value of the pressure will the aluminum rod fit inside the steel ring?

*81. A heavy uniform beam of mass 8000 kg and length 2.0 m is suspended at one end by a nylon rope of diameter 2.5 cm and at the other end by a steel rope of diameter 0.64 cm. The ropes are tied together above the beam (see Fig. 14.65). The unstretched lengths of the ropes are 3.0 m each. What angle will the beam make with the horizontal?

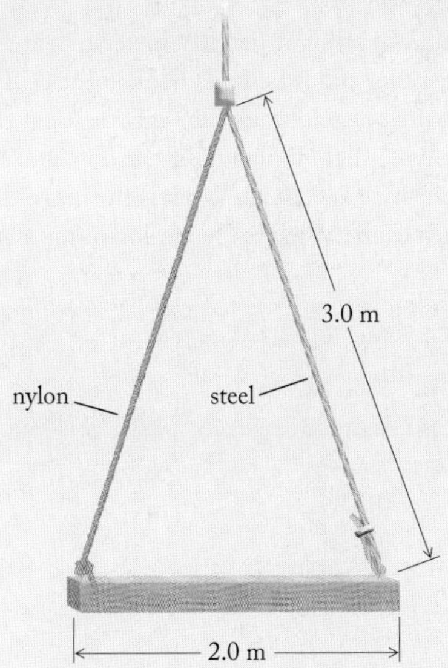

FIGURE 14.65 Beam hanging from two types of rope.

*82. A rod of cast iron is soldered to the upper edges of a plate of copper whose lower edge is held in a vise (see Fig. 14.66). The rod has a diameter of 4.0 cm and a length of 2.0 m. The copper plate measures 6.0 cm × 6.0 cm × 1.0 cm. If we pull the free end of the iron rod forward by 3.0 mm, what is the shear strain ($\Delta x/h$) of the copper plate?

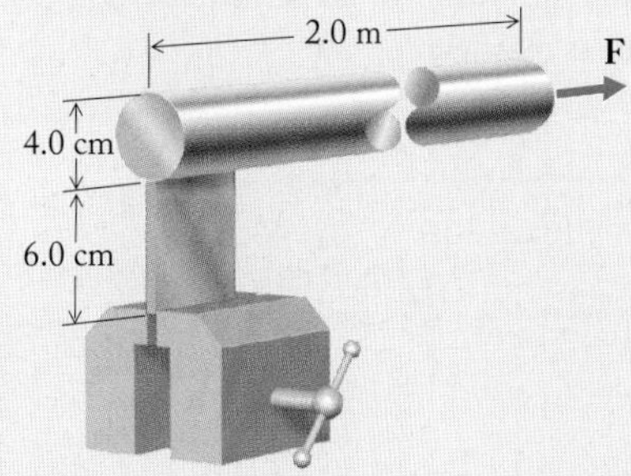

FIGURE 14.66 Iron rod and copper plate.

*83. When a bar of steel is heated, it expands in length by 0.0012% for each degree Celsius of temperature increase. If the length of the heated bar is to be reduced to its original value, a compressive stress must be applied to it. The compressive stress required to cancel the thermal expansion is called **thermal stress.** What is its value for a cylindrical bar of cast steel of cross section 4.0 cm^2 heated by 150°C?

*84. A power cable of copper is stretched straight between two fixed towers. If the temperature decreases, the cable tends to contract (compare Problem 83). The amount of contraction for a free copper cable or rod is 0.0017% per degree Celsius. Estimate what temperature decrease will cause the cable to snap. Pretend that the cable obeys Eq. (14.18) until it reaches its breaking point, which for copper occurs at a tensile stress of 2.4×10^8 N/m^2. Ignore the weight of the cable and the sag and stress produced by the weight.

**85. A meterstick of steel, of density 7.8×10^3 kg/m^3, is made to rotate about a perpendicular axis passing through its middle. What is the maximum angular velocity with which the stick can rotate if its center is to hold? Mild steel will break when the tensile stress exceeds 3.8×10^8 N/m^2.

**86. The wall of a pipe of diameter 60 cm is constructed of a sheet of steel of thickness 0.30 cm. The pipe is filled with water under high pressure. What is the maximum pressure, that is, force per unit area, that the pipe can withstand? See Problem 85 for data on mild steel.

**87. A hoop of aluminum of radius 40 cm is made to spin about its axis of symmetry at high speed. The density of aluminum is 2.7×10^3 kg/m^3, and the ultimate tensile breaking strength is 7.8×10^7 N/m^2. At what angular velocity will the hoop begin to break apart?

**88. A pipe of steel with a wall 0.40 cm thick and a diameter of 50 cm contains a liquid at a pressure of 2.0×10^4 N/m^2. How much will the diameter of the pipe expand due to this pressure?

REVIEW PROBLEMS

89. A traffic lamp of mass 25 kg hangs from a wire stretched between two posts. The traffic lamp hangs at the middle of the wire, and the two halves of the wire sag downward at an angle of 20° (see Fig. 14.67). What is the tension in the wire? Assume the wire is massless.

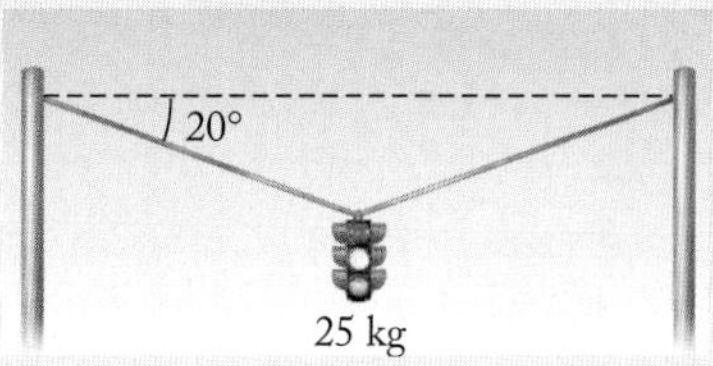

FIGURE 14.67 A traffic lamp.

90. A heavy shop sign hangs from a boom sticking out horizontally from a building (see Fig. 14.68). The boom is hinged at the building and is supported by a diagonal wire, making an angle of 45° with the boom. The mass of the sign is 50 kg, and the boom and the wire are massless. What is the tension in the wire? What is the force with which the end of the boom pushes against the building?

FIGURE 14.68 Sign hanging from a boom.

91. Figure 14.69 shows cargo hanging from the loading boom of a ship. If the boom is inclined at an angle of 30° and the cargo has a mass of 2500 kg, what is the tension in the upper cable? What is the compressional force in the boom? Neglect the mass of the boom.

92. Repeat the calculation of Problem 91, but assume that the mass of the boom is 800 kg, and that this mass is uniformly distributed along the length of the boom.

93. A tractor pulls a trailer along a street (see Fig. 14.70). The rear wheels, which are connected to the engine by means of the axle, have a radius of 0.60 m. Draw a "free-body" diagram for one of the rear wheels; be sure to include the forces and the torque exerted by the axle on the wheel, but neglect the weight of the wheel. If the tractor is to provide a pull of 8000 N (a pull of 4000 N from each rear wheel), what torque must the axle exert on each rear wheel?

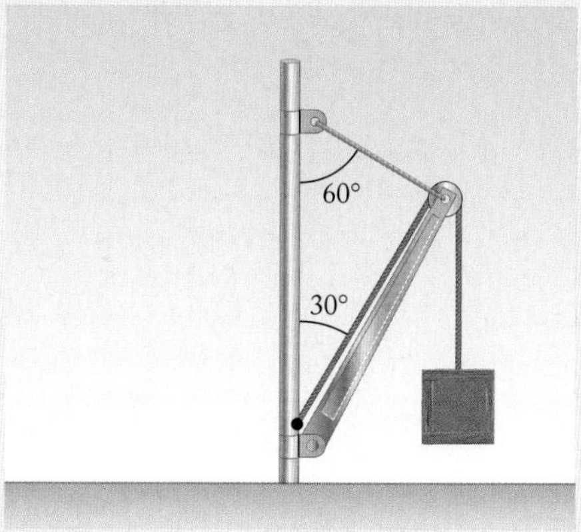

FIGURE 14.69 Cargo hanging from a boom.

FIGURE 14.70 Tractor pulling trailer.

94. One end of a string is tied to a meterstick at the 80-cm mark, and the other end is tied to a hook in the ceiling. You push against the bottom edge of the meterstick at the 30-cm mark, so the stick is held horizontally (see Fig. 14.71). The mass of the meterstick is 0.24 kg. What is the magnitude of the force you must exert? What is the tension in the string?

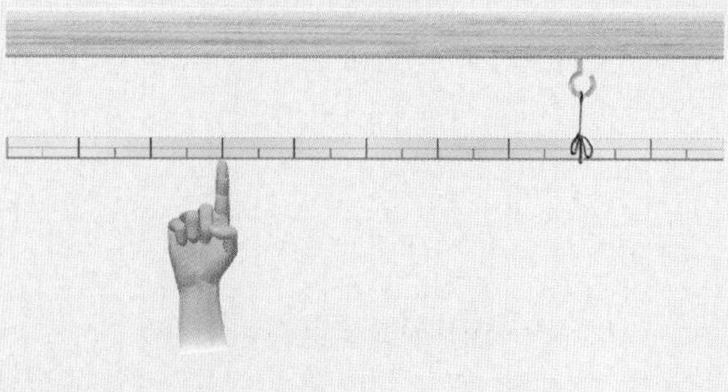

FIGURE 14.71 Meterstick tied to a hook.

95. A beam of steel hangs from a crane by means of cables attached to the upper corners of the beam making an angle of 60° with each other. The mass of the beam is M. Find the tensions in the cables and the compressional force in the beam.

96. Sheerlegs are sometimes used to suspend loads. They consist of two rigid beams leaning against each other, like the legs of the letter A (see Fig. 14.72). The load is suspended by a cable from the apex of the A. Suppose that a pair of sheerlegs, each at an angle of 30° with the vertical, are used to suspend an automobile engine of mass 400 kg. What is the compressional force in each leg? What are the horizontal and vertical forces that each leg exerts on the ground? Neglect the mass of the legs.

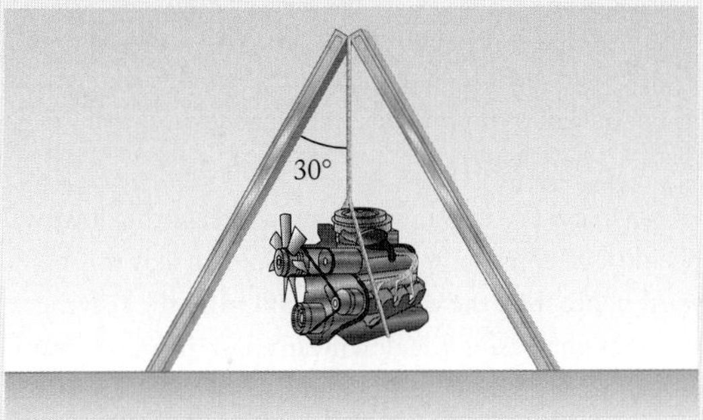

FIGURE 14.72 Sheerlegs supporting a load.

97. A 100-kg barrel is placed on a 30° ramp (see Fig. 14.73). What push, parallel to the ramp, must you exert against the middle of the barrel to keep it from rolling down? Assume that the friction between the barrel and the ramp prevents slipping of the barrel; that is, the barrel would roll without slipping if released.

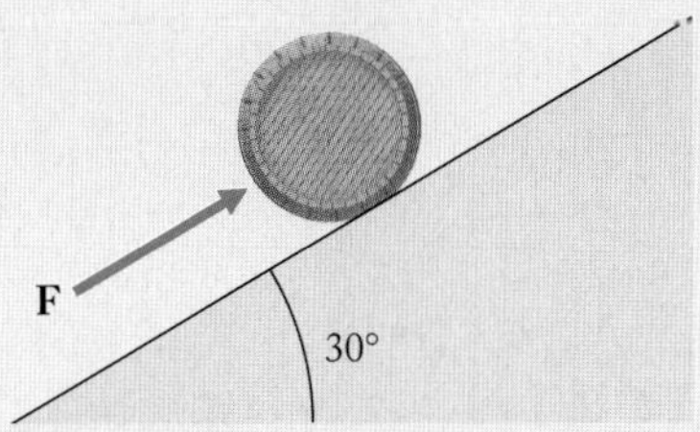

FIGURE 14.73 Barrel on a ramp.

98. Figure. 14.74 shows a pair of pliers and their dimensions. If you push against the handles of the pliers with a force of 200 N from each side, what is the force that the jaws of the pliers exert against each other?

99. To help his horses drag a heavy wagon up a hill, a teamster pushes forward at the top of one of the wheels (see Fig. 14.75). If he pushes with a force of 600 N, what forward force does he generate on the axle of the wagon? (Hint: The diameter of the wheel can be regarded as a lever pivoted at the ground.)

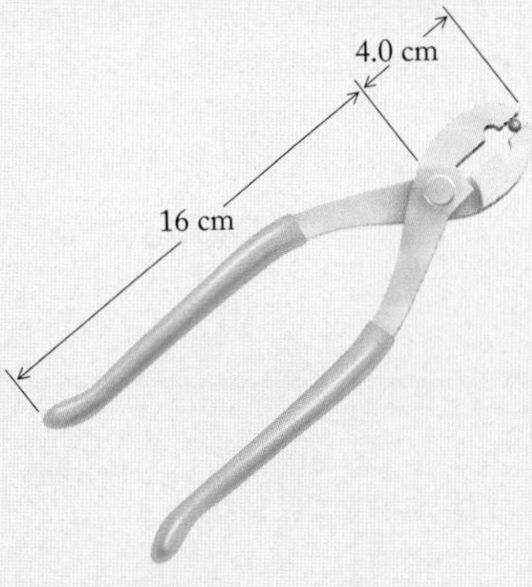

FIGURE 14.74 Pliers.

FIGURE 14.75 Teamster pushing on a wheel.

100. A flagpole points horizontally from a vertical wall. The pole is a uniform rod of mass M and length L. In addition to the pole mount at the wall end (which is hinged and exerts no torque), the pole is supported at its far end by a straight cable; the cable is attached to the wall a distance $L/2$ above the pole mount. What is the tension in the cable? What are the magnitude and direction of the force provided by the pole wall mount?

101. A wire stretches when subjected to a tension. This means that the wire can be regarded as a spring.

(a) Express the effective spring constant in terms of the length of the wire, its radius, and its Young's modulus.

(b) If a steel wire of length 2.0 m and radius 0.50 mm is to have the same spring constant as a steel wire of length 4.0 m, what must be the radius of the second wire?

102. If a steel rope and a nylon rope of equal lengths are to stretch by equal amounts when subjected to equal tensions, what must be the ratio of their diameters?

*103. A long rod of steel hangs straight down into a very deep mine shaft. For what length will the rod break off at the top because of its own weight? The density of mild steel is 7.8×10^3 kg/m^3, and its tensile stress for breaking is 3.8×10^8 N/m^2.

104. A rope of length 12 m consists of an upper half of nylon of diameter 1.9 cm spliced to a lower half of steel of diameter 0.95 cm. How much will this rope stretch if a mass of 4000 kg is suspended from it? The Young's modulus for steel rope is $19 \times 10^{10}\ N/m^2$.

105. Suppose you drop an aluminum sphere of radius 10 cm into the ocean and it sinks to a depth of 5000 m, where the pressure is $5.7 \times 10^7\ N/m^2$. Calculate by how much the diameter of this sphere will shrink.

Answers to Checkups

Checkup 14.1

1. If the bicyclist sits rigidly, the equilibrium is unstable: if tipped slightly, gravity will pull the bicycle and cyclist further over.

2. When you extend your legs while sitting on a swing, you are shifting your center of mass forward. To remain in equilibrium, the swing and your body will shift backward, and tilt, so as to keep your center of mass aligned below the point of support.

3. (a) Yes, the (vertical) support force is along the same line as the weight when holding the pole straight up (more precisely, it is slightly distributed around the edge of the pole). (b) No, the support force is provided by the more forward hand (which pushes up); an additional force from the rear hand pushes down, to balance the torques from the force of gravity and the support force.

4. (D) Neutral, stable, unstable. As our intuition might suggest, a cone on its side is in neutral equilibrium (after a small displacement, it remains on its side). A cone on its base is in stable equilibrium (after being tipped slightly, it will settle back on its base). Finally, a cone balanced on its apex is in unstable equilibrium (after being tipped slightly, the cone will fall over).

Checkup 14.2

1. When a ladder makes a large angle with the vertical, the weight of the ladder and the person climbing it exerts a large torque about the bottom, which can more easily overcome friction and make the ladder slip. When a ladder makes a small angle with the vertical, a person on the ladder can shift the center of mass to a point behind the bottom, causing the ladder to topple backward.

2. With the center of mass on the bottom, the box would have to be rotated 90° before toppling over. In that case, however, the box would then be on its side when it reaches the critical angle, where the center of mass is just above the support point.

3. For each side of the A, the force that one piece of lumber exerts on the other must exert a torque about the other's bottom that balances the torque due to the other's weight. Such a torque increases from zero when the pieces of lumber are vertical (when the tip of the A makes zero angle) to a maximum when the tip approaches 180°. Since the force exerted by one piece on the other acts with a smaller and smaller moment arm as the tip angle approaches 180°, the force must be very large as the tip angle approaches 180°.

4. (B) Increase the upward push and increase the downward push. If we consider the torques about an axis through the forward hand, then the downward pull from the fish must be balanced by increasing the downward push from the rear hand. The upward push of the forward hand must increase to balance those two increased downward forces.

Checkup 14.3

1. The arrangement shown in Fig. 14.23b has the larger ratio l/l', and thus has a greater mechanical advantage.

2. No. If, for example, the force F is not perpendicular to the lever, we must replace l by $l \sin\theta$, where θ is the angle between the force and the lever.

3. No. The pulley transmits tension to a different direction, independent of its size.

4. (C) 100 N. The weight of the rock is $w = mg = 100\ kg \times 9.8\ m/s^2 = 980\ N \approx 1000\ N$. The lever has a mechanical advantage of $l'/l \approx \frac{1}{10}$. So the force required to lift the rock is $F = (l'/l)F' = \frac{1}{10} \times 1000\ N = 100\ N$.

Checkup 14.4

1. The tension determines the *fractional* elongation [see Eq. (14.18)]; thus, for a piano wire of twice the length, the elongation will be twice as long, or 4.0 mm.

2. Yes, a cable can snap under its own weight (the downward weight below any point must be balanced by the upward tension at that point). Since the critical length for breaking is a condition of maximum tensile stress (a force per unit area), this depends on only the material and its mass density, not its area.

3. A material with a larger bulk modulus is stiffer, that is, its volume shrinks less in response to an applied pressure. We can rewrite Eq. (14.20) as $F/A = -B(\Delta V/V)$; thus, at constant pressure, a B that is larger by a given factor results in a fractional volume change that is smaller by the same factor. The volume of the copper sphere then shrinks by 0.005%.

4. (A) $\sqrt{2}$. Since the elongation is inversely proportional to the area of the elastic body [see Eq. (14.18)], if you want to decrease the elongation of a cable by a factor of 2, you must increase the cross-sectional area by a factor of 2; thus, you must increase the diameter by a factor of $\sqrt{2}$.

PART

2

Oscillations, Waves, and Fluids

CONTENTS

At 1.1 times the speed of sound, this T-38 training jet generates shock waves, or sonic booms, in the surrounding air. The shock waves are made visible by Schlieren photography, a special technique that detects changes in the density of air.

CHAPTER

15 Oscillations

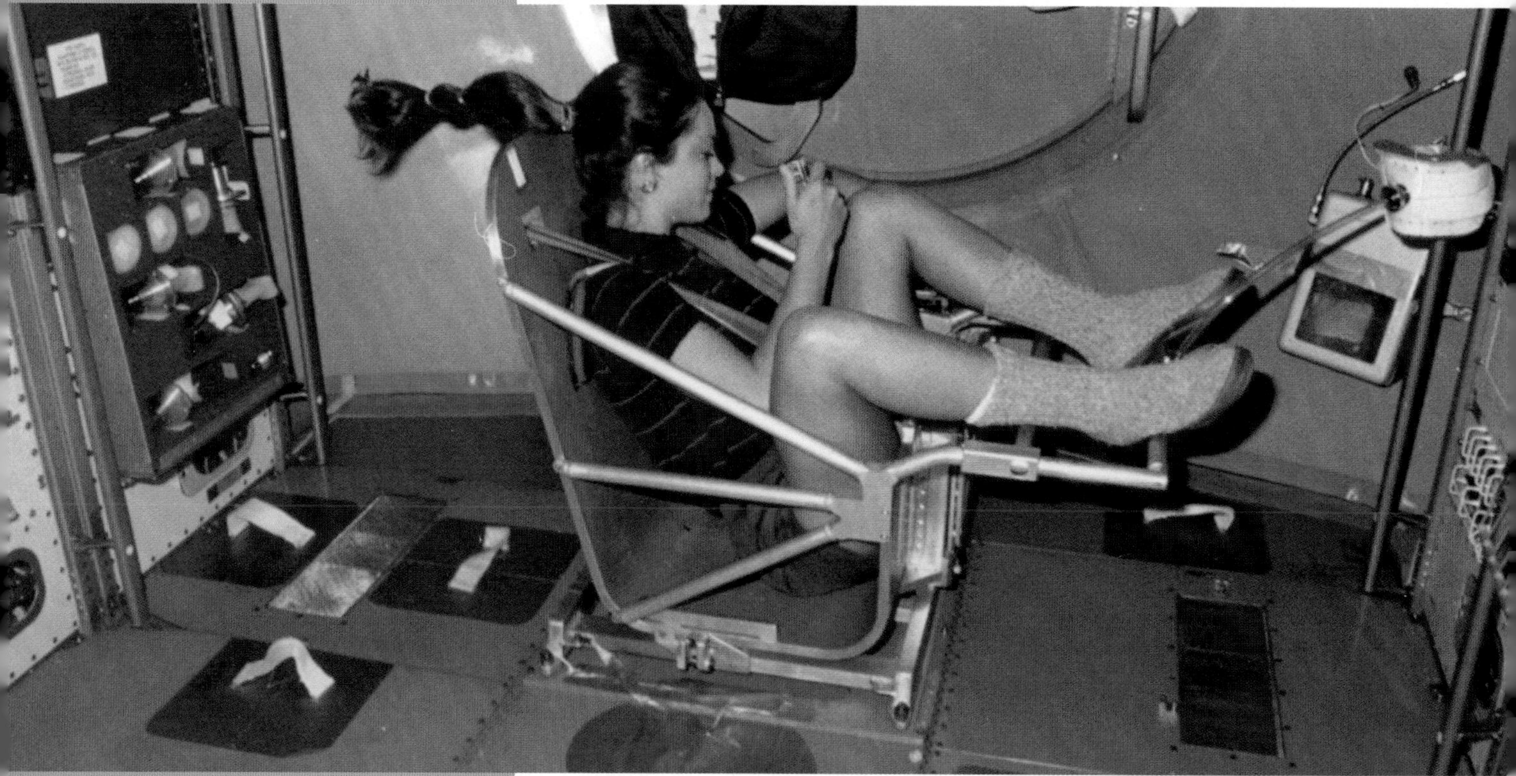

CONCEPTS IN CONTEXT

Concepts — *in* — Context

The body-mass measurement device shown is used aboard the International Space Station for the daily measurement of the masses of the astronauts. The device consists of a spring coupled to a chair into which the astronaut is strapped. Pushed by the spring, the chair with the astronaut oscillates back and forth. We will see in this chapter that the frequency of oscillation of the mass–spring system depends on the mass, and therefore the frequency can serve as an indicator of the mass of the astronaut.

While learning about oscillating systems, we will consider such questions as:

? When the spring pushes and pulls the astronaut, what is the position of the astronaut as a function of time? The velocity of the astronaut? (Example 4, page 478)

? What is the total mechanical energy of the astronaut–spring system? What are the kinetic and potential energies as the spring begins to push? At later times? (Example 5, page 482)

? Good oscillators have low friction. How do we measure the quality of an oscillator? (Example 10, page 490)

The motion of a particle or of a system of particles is **periodic,** *or* **cyclic,** *if it repeats again and again at regular intervals of time.* The orbital motion of a planet around the Sun, the uniform rotational motion of a carousel or of a circular saw blade, the back-and-forth motion of a piston in an automobile engine or in a water pump, the swinging motion of a pendulum bob in a grandfather clock, and the vibration of a guitar string are examples of periodic motions. *If the periodic motion is a back-and-forth motion along a straight or curved line, it is called an* **oscillation.** Thus, the motion of the piston is an oscillation, and so are the motion of the pendulum and the motion of the individual particles of the guitar string.

In this chapter we will examine in some detail the motion of a mass oscillating back and forth under the push and pull exerted by an ideal, massless spring. The equations that we will develop for the description of this mass–spring system are of great importance because analogous equations also occur in the description of all other oscillating systems. We will also examine some of these other oscillating systems, such as the pendulum.

15.1 SIMPLE HARMONIC MOTION

Online Concept Tutorial 16

Simple harmonic motion is a special kind of one-dimensional periodic motion. In any kind of one-dimensional periodic motion, the particle moves back and forth along a straight line, repeating the same motion again and again. In the special case of **simple harmonic motion,** *the particle's position can be expressed as a cosine or a sine function of time.* As we will see later, the motion of a mass oscillating back and forth under the push and pull of a spring is simple harmonic (Fig. 15.1a), and so is the motion of a pendulum bob swinging back and forth (provided the amplitude of swing is small; see Fig. 15.1b), and so is the up-and-down motion of the blade of a saber saw (Fig. 15.1c). However, in this first section we will merely deal with the mathematical description of simple harmonic motion, and we will postpone until the next section the question of what causes the motion.

As a numerical example of simple harmonic motion, suppose that the tip of the blade in Fig. 15.1c moves up and down between $x = -0.8$ cm and $x = +0.8$ cm (where

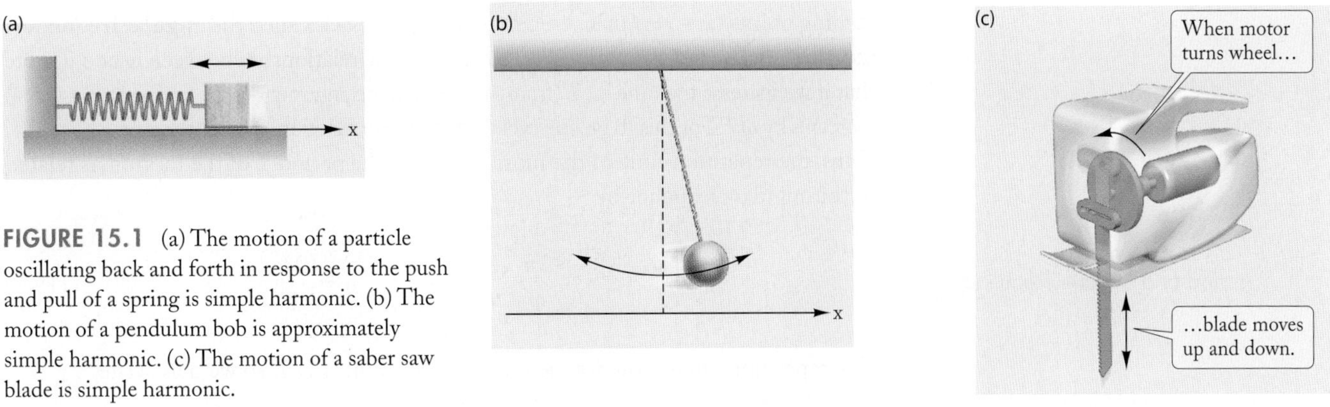

FIGURE 15.1 (a) The motion of a particle oscillating back and forth in response to the push and pull of a spring is simple harmonic. (b) The motion of a pendulum bob is approximately simple harmonic. (c) The motion of a saber saw blade is simple harmonic.

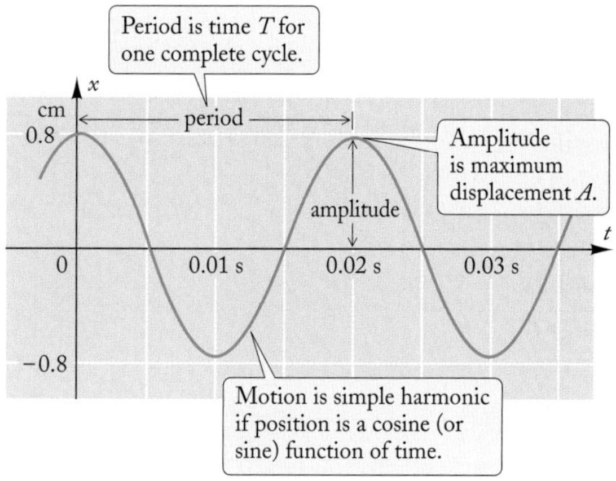

FIGURE 15.2 Plot of position vs. time for a case of simple harmonic motion up and down along the x axis.

the x axis is assumed to be vertical); further suppose that the blade completes 50 up-and-down cycles each second. Figure 15.2 gives a plot of the position of the tip of the blade as a function of time. The plot in Fig. 15.2 has the mathematical form of a cosine function of the time t,

$$x = 0.8\cos(100\pi t) \tag{15.1}$$

where it is assumed that distance is measured in centimeters and time in seconds, and it is assumed that the "angle" $100\pi t$ in the cosine function is reckoned in radians. [The factor 100π multiplying t in Eq. (15.1) has been selected so as to obtain exactly 50 complete cycles each second, which is typical for saber saws; we will see below in Eq. (15.5) how the factor multiplying t in Eq. (15.1) is related to the period of the motion.]

Cosines and sines are called **harmonic functions**, which is why we call the motion *harmonic*. For the harmonic motion plotted in Fig. 15.2, at $t = 0$, the blade tip is at its maximum upward displacement [evaluating Eq. (15.1) at $t = 0$, we have $\cos 0 = 1$, so $x = 0.8$ cm] and is just starting to move; at $t = 0.005$ s, it passes through the midpoint [since $\cos(100\pi \times 0.005) = \cos(\pi/2) = 0$, Eq. (15.1) gives $x = 0$]; at $t = 0.010$ s, it reaches maximum downward displacement [$\cos(\pi) = -1$, so $x = -0.8$ cm]; at $t = 0.015$ s, it again passes through midpoint. Finally, at $t = 0.020$ s, the tip returns to its maximum upward displacement, exactly as at $t = 0$—it has completed one cycle of the motion and is ready to begin the next cycle. Thus, the **period** T, or the repeat time of the motion (the number of seconds for one complete cycle of the motion), is

$$T = 0.020 \text{ s} \tag{15.2}$$

and the **frequency** f of the motion, or the rate of repetition of the motion (the number of cycles per second), is

$$f = \frac{1}{T} = \frac{1}{0.020 \text{ s}} = 50/\text{s} \tag{15.3}$$

The points $x = 0.8$ cm and $x = -0.8$ cm, at which the x coordinate attains its maximum and minimum values, are the **turning points** of the motion; and the point $x = 0$ is the midpoint.

Equation (15.1) is a special example of simple harmonic motion. More generally, the motion of a particle is simple harmonic if the dependence of position on time has the form of a cosine or a sine function, such as

simple harmonic motion

$$x = A\cos(\omega t + \delta) \tag{15.4}$$

The quantities A, ω, and δ are constants. The quantity A is called the **amplitude** of the motion; it is simply the distance between the midpoint ($x = 0$) and either of the turning points ($x = +A$ or $x = -A$). The quantity ω is called the **angular frequency**; its value is related to the period T. To establish the relationship between ω and T, note that if we increase the time by T (from t to $t + T$), the argument of the cosine in Eq. (15.4) increases by ωT. For this to be one cycle of the cosine function, we must require $\omega T = 2\pi$. Thus, the repetition time of the motion, that is, the period T of the motion, is related to the angular frequency by

period and angular frequency

$$T = \frac{2\pi}{\omega} \quad \text{or} \quad \omega = \frac{2\pi}{T} \tag{15.5}$$

The repetition rate, or the frequency of the motion, is $1/T$, so we may write

$$f = \frac{1}{T} = \frac{\omega}{2\pi} \qquad \text{or} \qquad \omega = 2\pi f \tag{15.6}$$

frequency and angular frequency

Note that the angular frequency ω and the frequency f differ by a factor of 2π, which corresponds to 2π radians = 1 cycle. The units of angular frequency are radians per second (radians/s). The units of frequency are cycles per second (cycles/s). Like the label *revolution* that we used in rev/s in rotational motion, the label *cycle* in cycle/s can be omitted in the course of a calculation, and so can the label *radian* in radian/s. But it is useful to retain these labels wherever there is a chance of confusion. The SI *unit of frequency is called the* **hertz** (Hz):

$$1 \text{ hertz} = 1 \text{ Hz} = 1 \text{ cycle/s} = 1/\text{s} \tag{15.7}$$

hertz (Hz)

For instance, in the example of the motion of the saber saw blade, the period of the motion is $T = 0.020$ s, the frequency is $f = 1/T = 1/(0.020 \text{ s}) = 50/\text{s} = 50$ Hz, and the angular frequency is

$$\omega = 2\pi f = 2\pi \times 50/\text{s} = 314 \text{ radians/s}$$

Here, in the last step of the calculation, the label *radians* has been inserted, so as to distinguish the angular frequency ω from the ordinary frequency f.

The argument $(\omega t + \delta)$ of the cosine function is called the **phase** of the oscillation, and the quantity δ is called the **phase constant.** This constant determines at what times the particle reaches the point of maximum displacement, when $\cos(\omega t + \delta) = 1$. One such instant is when

$$\omega t_{max} + \delta = 0$$

that is, when

$$t_{max} = -\delta/\omega \tag{15.8}$$

phase constant and time of maximum displacement

Hence the particle reaches the point of maximum displacement at a time δ/ω before $t = 0$ (see Fig. 15.3). Of course, the particle also passes through this point at periodic intervals before and after this time. If the phase constant is zero ($\delta = 0$), then the maximum displacement occurs at $t = 0$.

Note that the preceding equations connecting *angular frequency,* period, and frequency are formally the same as the equations connecting *angular velocity,* period, and frequency of uniform rotational motion [see Eqs. (12.4) and (12.5)]. This coincidence

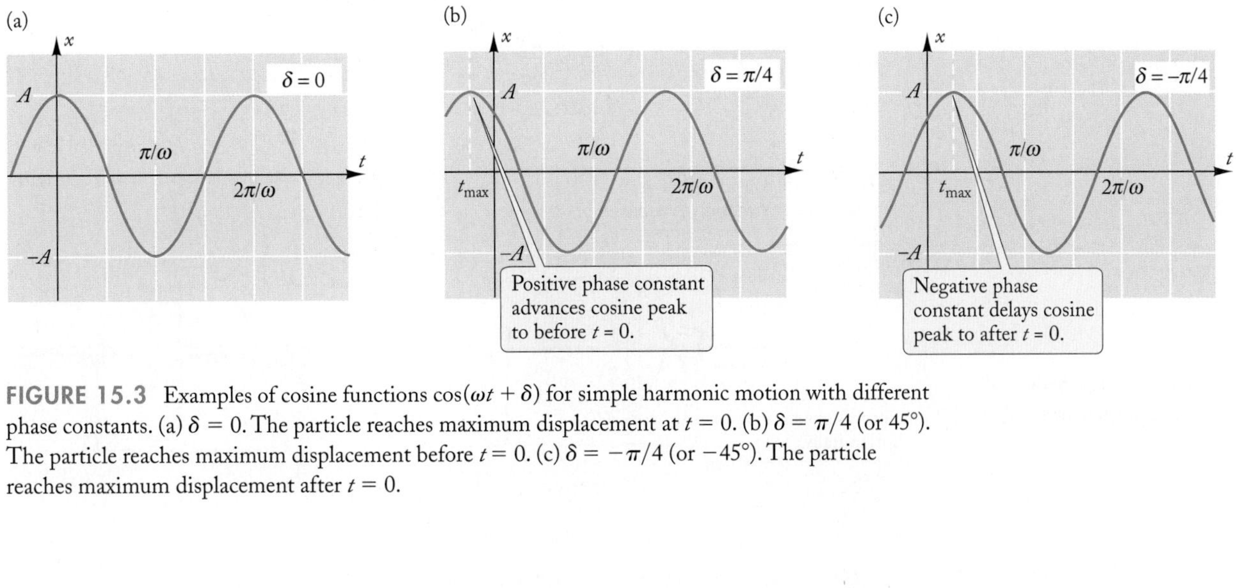

FIGURE 15.3 Examples of cosine functions $\cos(\omega t + \delta)$ for simple harmonic motion with different phase constants. (a) $\delta = 0$. The particle reaches maximum displacement at $t = 0$. (b) $\delta = \pi/4$ (or 45°). The particle reaches maximum displacement before $t = 0$. (c) $\delta = -\pi/4$ (or −45°). The particle reaches maximum displacement after $t = 0$.

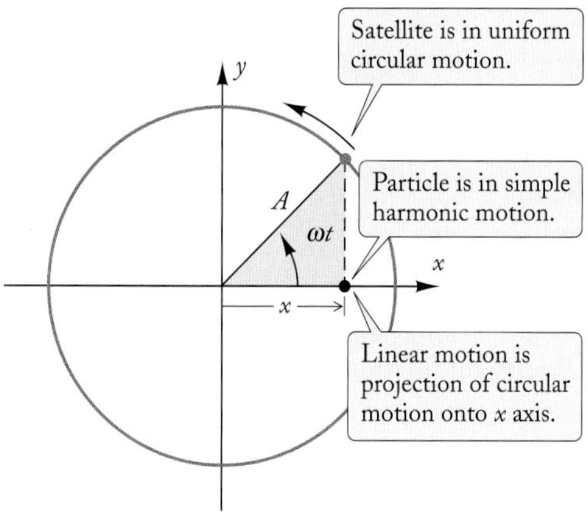

FIGURE 15.4 Particle oscillating along the x axis and satellite particle moving around reference circle. The particle and the satellite are always aligned vertically; that is, they have the same x coordinate.

arises from a special geometrical relationship between simple harmonic motion and uniform circular motion. Suppose that a particle moves with simple harmonic motion according to Eq. (15.4), with amplitude A and angular velocity ω; and consider a "satellite" particle that is constrained to move in uniform circular motion with angular velocity ω along a circle of radius A, centered on the midpoint of the harmonic motion, that is, centered on $x = 0$. Figure 15.4 shows this circle, called the **reference circle.** At time $t = 0$, both the particle and its satellite are on the x axis at $x = A$. After this time, the particle moves along the x axis, so its position is

$$x = A\cos(\omega t) \tag{15.9}$$

Meanwhile, the satellite moves around the circle, and its angular position is

$$\theta = \omega t$$

Now note that the x coordinate of the satellite is the adjacent side of the triangle shown in Fig. 15.4:

$$x_{\text{sat}} = A\cos\theta = A\cos(\omega t) \tag{15.10}$$

Comparing this with Eq. (15.9), we see that the x coordinate of the satellite always coincides with the x coordinate of the particle; that is, the particle and the satellite always have exactly the same x motion. This means that in Fig. 15.4 the satellite is always on that point of the reference circle directly above or directly below the particle.

This geometrical relationship between simple harmonic motion and uniform circular motion can be used to generate simple harmonic motion from uniform circular motion. Figure 15.5 shows a simple mechanism for accomplishing this by means of a slotted arm placed over a peg that is attached to a wheel in uniform circular motion. The slot is vertical, and the arm is constrained to move horizontally. The peg plays the role of "satellite," and the midpoint of the slot in the arm plays the role of "particle." The peg drags the arm left and right and makes it move with simple harmonic motion. A mechanism of this kind is used in electric saber saws and other devices to convert the rotational motion of an electric motor into the up-and-down motion of the saw blade or other moving component.

Finally, let us calculate the instantaneous velocity and instantaneous acceleration in simple harmonic motion. If the displacement is

$$x = A\cos(\omega t + \delta) \tag{15.11}$$

then differentiation of this displacement gives the velocity

$$v = \frac{dx}{dt} = -\omega A\sin(\omega t + \delta) \tag{15.12}$$

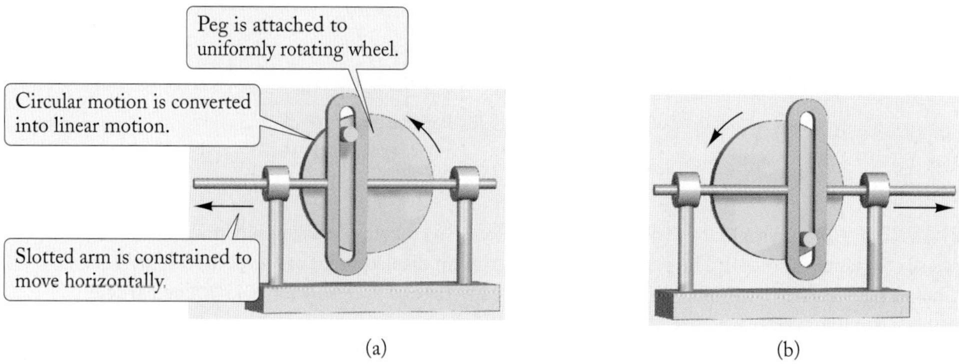

FIGURE 15.5 Rotating wheel with a peg driving a slotted arm back and forth.

MATH HELP DERIVATIVES OF TRIGONOMETRIC FUNCTIONS

Under the assumption that the argument of each trigonometric function is expressed in radians, the derivatives of the sine, cosine, and tangent are

$$\frac{d}{du}\sin bu = b\cos bu \qquad \frac{d}{du}\cos bu = -b\sin bu \qquad \frac{d}{du}\tan bu = b\sec^2 bu = \frac{b}{\cos^2 bu}$$

and differentiation of this velocity gives the acceleration

$$a = \frac{d^2x}{dt^2} = \frac{dv}{dt} = -\omega^2 A\cos(\omega t + \delta) \qquad (15.13)$$

Here we have used the standard formulas for the derivatives of the sine function and the cosine function (see Math Help: Derivatives of Trigonometric Functions). Bear in mind that the arguments of the sine and cosine functions in this chapter (and also the next) are always expressed in radians, as required for the validity of the standard formulas for derivatives.

As expected, the instantaneous velocity calculated from Eq. (15.12) is zero for $\omega t + \delta = 0$, when the particle is at the turning point. Furthermore, the instantaneous velocity attains a maximum magnitude of

maximum velocity

$$v_{\text{max}} = \omega A \qquad (15.14)$$

for $\omega t + \delta = \pi/2$, when the particle passes through the midpoint (note that the maximum magnitude of $\sin\omega t$ is 1).

Figure 15.6 shows a multiple-exposure photograph of the oscillations of a particle in simple harmonic motion. The picture illustrates the variations of speed in simple harmonic motion: the particle moves at low speed (smaller displacements between snapshots) near the turning points, and at high speed (larger displacements) near the midpoint.

The velocity (15.12) is a sine function, whereas the displacement (15.11) is a cosine function. When the cosine is at its maximum (say, $\cos 0 = 1$), the sine is small ($\sin 0 = 0$); when the cosine is small (say, $\cos\pi/2 = 0$), the sine is at its maximum

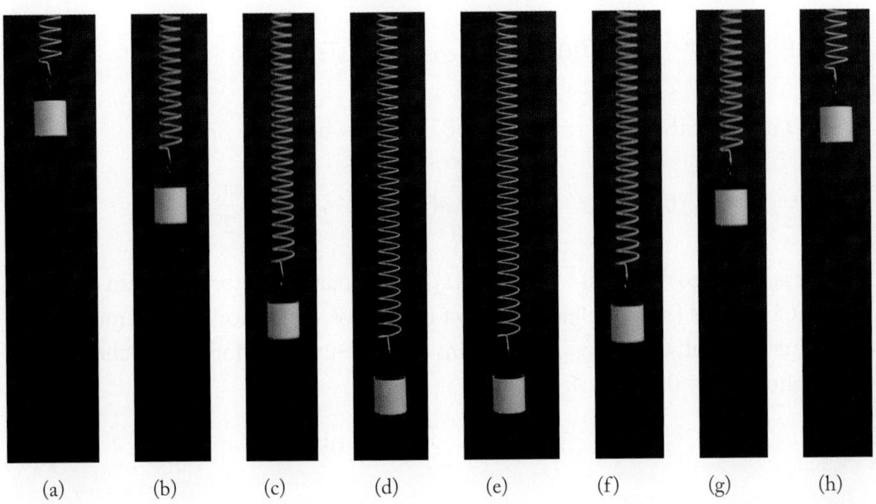

FIGURE 15.6 Sequence of snapshots at uniform time intervals of an oscillating mass on a spring (a-h). Note that the mass moves slowly at the extremes of its motion.

($\sin \pi/2 = 1$). Hence the displacement and the velocity are out of step—when one has a large magnitude, the other has a small magnitude, and vice versa. Figures 15.7a and b compare the velocity and the displacement for simple harmonic motion at different times. Graphically, the velocity is the slope of the position vs. time curve. When the position goes through a maximum or minimum, the slope is zero; when the position goes through zero, the magnitude of the slope is a maximum.

Comparison of Eqs. (15.11) and (15.13) shows that

acceleration in simple harmonic motion

$$\frac{d^2x}{dt^2} = -\omega^2 x \tag{15.15}$$

Thus, the acceleration is always proportional to the displacement x, but is in the opposite direction; see Fig. 15.7c. This proportionality is a characteristic feature of simple harmonic motion, a fact that will be useful in the next section. Even when a phenomenon does not involve motion along a line (for example, rotational motion or the behavior of electric circuits), harmonic behavior occurs whenever the second derivative of a quantity is proportional to the negative of that quantity, as in Eq. (15.15). The sine and cosine functions (or a combination of them) are the *only* functions that have this property.

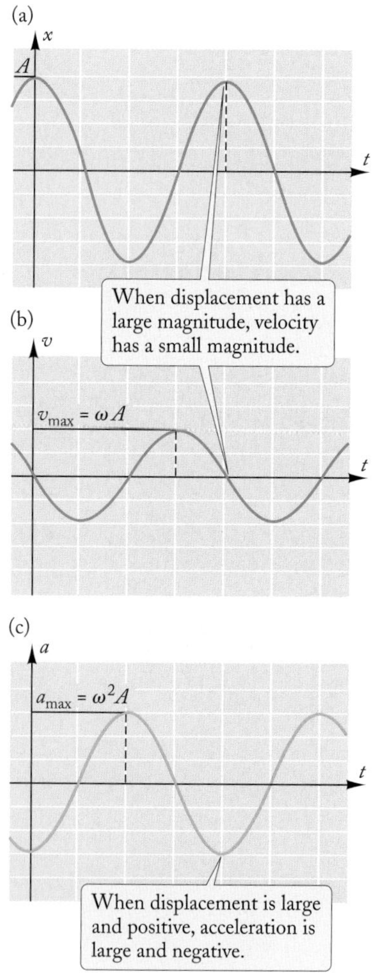

FIGURE 15.7 (a) Position, (b) velocity, and (c) acceleration of a particle in simple harmonic motion as functions of time.

EXAMPLE 1 Consider the blade of a saber saw moving up and down in simple harmonic motion with a frequency of 50.0 Hz, or an angular frequency of 314 radians/s. Suppose that the amplitude of the motion is 1.20 cm and that at time $t = 0$, the tip of the blade is at $x = 0$ and its velocity is positive. What is the equation describing the position of the tip of the blade as a function of time? How long does the blade take to travel from $x = 0$ to $x = 0.60$ cm? To 1.20 cm?

SOLUTION: The position as function of time is given by Eq. (15.4):

$$x = A\cos(\omega t + \delta)$$

with $\omega = 314$ radians/s and $A = 0.0120$ m. Since $x = 0$ at $t = 0$, we must adopt a value of δ such that $\cos\delta = 0$. The smallest values of δ that satisfy this condition are $\delta = \pi/2$ and $\delta = -\pi/2$ (other possible values of δ differ from these by $\pm 2\pi$, $\pm 4\pi$, etc.). From Eq. (15.12), we see that to obtain a positive value of v at $t = 0$, we need a *negative* value of δ; that is, $\delta = -\pi/2$. So the equation describing the motion is

$$x = (0.0120\text{ m})\cos\left[(314/\text{s})t - \frac{\pi}{2}\right]$$

The tip of the blade reaches $x = 0.0060$ m when

$$0.0060\text{ m} = (0.0120\text{ m})\cos\left[(314/\text{s})\,t - \frac{\pi}{2}\right]$$

that is, when $\cos[(314/\text{s})t - \pi/2] = (\frac{1}{2})$. With our calculator we obtain $\cos^{-1}\frac{1}{2} = -1.05$ radians (here, we have to select a negative sign, since the argument of the cosine is initially negative, and remains negative until the motion reaches the full amplitude, $x = 0.0120$ m). So

$$(314/\text{s})t - \frac{\pi}{2} = -1.05$$

from which

$$t = \frac{-1.05 + (\pi/2)}{314/\text{s}} = 0.0017 \text{ s}$$

To find when the tip of the blade reaches $x = 0.0120$ m, we can use Eq. (15.8), which gives

$$t = -\frac{\delta}{\omega} = -\frac{(-\pi/2)}{314/\text{s}} = 0.0050 \text{ s}$$

COMMENT: Note that the time taken to reach a distance of one-half of the amplitude is not one-half of the time taken to reach the full amplitude, because the motion does not proceed at constant speed.

EXAMPLE 2 In an atomic-force microscope (AFM), a cantilever beam with a sharp tip (Fig. 15.8a) oscillates near a surface. We can map the topography of a surface (see Fig. 15.8b) by slowly moving the tip laterally as it oscillates vertically, much like a blind person tapping a cane on the ground. The AFM tip shown in Fig. 15.8a oscillates with a period of 3.0×10^{-6} s. The tip moves up and down with amplitude 9.0×10^{-8} m. What is the maximum vertical acceleration of the tip? Its maximum vertical velocity?

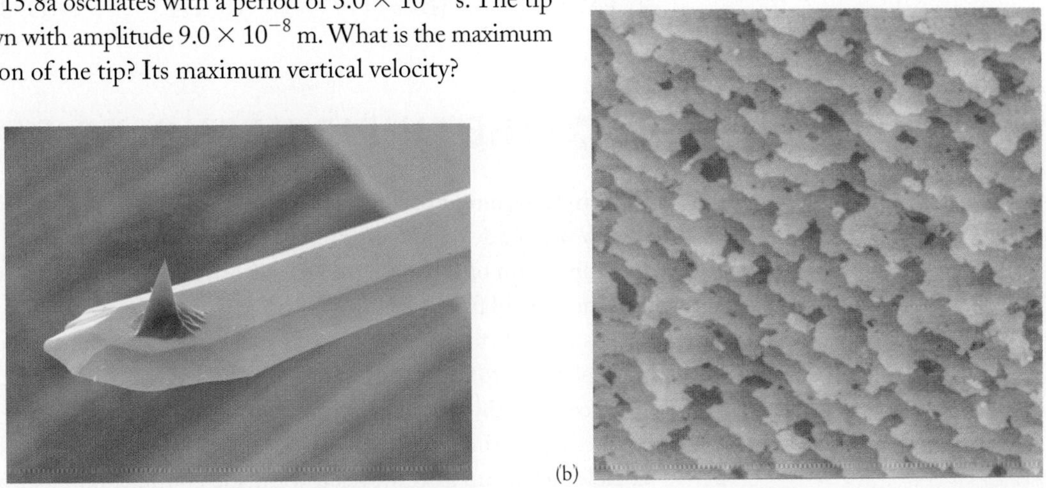

(a) (b)

FIGURE 15.8 (a) Atomic-force microscope (AFM) cantilever and tip. (b) AFM image of the surface of a crystal, obtained by scanning the vibrating tip across the surface. The area shown is 2 μm $\times$ 2 μm. The ragged terraces are single atomic "steps."

SOLUTION: As discussed above, the largest acceleration occurs at the point of maximum displacement. From Eq. (15.13) this maximum acceleration is [since the maximum value of $\cos(\omega t + \delta)$ is 1]

$$a_{\text{max}} = \omega^2 A \tag{15.16}$$

From Eq. (15.5) and the period $T = 3.0 \times 10^{-6}$ s, we obtain the angular frequency

$$\omega = \frac{2\pi}{T} = \frac{2\pi}{3.0 \times 10^{-6} \text{ s}} = 2.1 \times 10^6 \text{ radians/s}$$

Thus, with $A = 9.0 \times 10^{-8}$ m, the maximum acceleration is

$$a_{\text{max}} = \omega^2 A = (2.1 \times 10^6 \text{ radians/s})^2 \times 9.0 \times 10^{-8} \text{ m} = 4.0 \times 10^5 \text{ m/s}^2$$

This is more than 40000 standard g's, an enormous acceleration.

The maximum velocity is, from Eq. (15.12),

$$v_{\text{max}} = \omega A = 2.1 \times 10^6 \text{ radians/s} \times 9.0 \times 10^{-8} \text{ m} = 0.19 \text{ m/s}$$

Checkup 15.1

QUESTION 1: Is the rotational motion of the Earth about its axis periodic motion? Oscillatory motion?

QUESTION 2: For a particle with simple harmonic motion, at what point of the motion does the velocity attain maximum magnitude? Minimum magnitude?

QUESTION 3: For a particle with simple harmonic motion, at what point of the motion does the acceleration attain maximum magnitude? Minimum magnitude?

QUESTION 4: Two particles execute simple harmonic motion with the same amplitude. One particle has twice the frequency of the other. Compare their maximum velocities and accelerations.

QUESTION 5: Are the x coordinates of the particle and the satellite particle in Fig. 15.4 always the same? The y coordinates? The velocities? The x components of the velocities? The accelerations? The x components of the accelerations?

QUESTION 6: Suppose that a particle with simple harmonic motion passes through the equilibrium point ($x = 0$) at $t = 0$. In this case, which of the following is a possible value of the phase constant δ in $x = A\cos(\omega t + \delta)$?

(A) 0 (B) $\pi/4$ (C) $\pi/2$ (D) $3\pi/4$ (E) π

15.2 THE SIMPLE HARMONIC OSCILLATOR

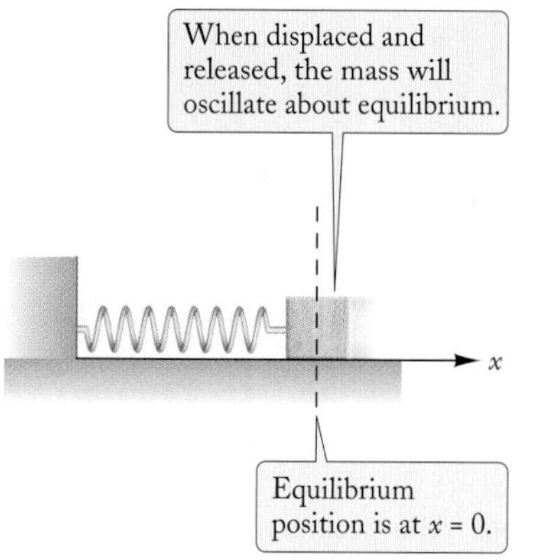

FIGURE 15.9 A mass attached to a spring slides back and forth on a frictionless surface. We regard the mass as a particle, whose position coincides with the center of the mass.

The **simple harmonic oscillator** *consists of a particle coupled to an ideal, massless spring that obeys Hooke's Law,* that is, a spring that provides a force proportional to the elongation or compression of the spring. One end of the spring is attached to the particle, and the other is held fixed (see Fig. 15.9). We will ignore gravity and friction, so the spring force is the only force acting on the particle. The system has an **equilibrium position** corresponding to the relaxed length of the spring. If the particle is initially at some distance from this equilibrium position (see Fig. 15.10), then the stretched spring supplies a restoring force that pulls the particle toward the equilibrium position. The particle speeds up as it moves toward the equilibrium position, and it overshoots the equilibrium position. Then, the particle begins to compress the spring and slows down, coming to rest at the other side of the equilibrium position, at a distance equal to its initial distance. The compressed spring then pushes the particle back toward the equilibrium position. The particle again speeds up, overshoots the equilibrium position, and so on. The result is that the particle oscillates back and forth about the equilibrium position—forever if there is no friction.

The great importance of the simple harmonic oscillator is that many physical systems are mathematically equivalent to simple harmonic oscillators; that is, these systems have an equation of motion of the same mathematical form as the simple harmonic oscillator. A pendulum, the balance wheel of a watch, a tuning fork, the air in an organ pipe, and the atoms in a diatomic molecule are systems of this kind; the restoring force and the inertia are of the same mathematical form in these systems as in the simple harmonic oscillator, and we can transcribe the general mathematical results directly from the latter to the former.

To obtain the equation of motion of the simple harmonic oscillator, we begin with Hooke's Law for the restoring force exerted by the spring on the particle [compare Eq. (6.11)]:

$$F = -kx \tag{15.17}$$

Here the displacement x is measured from the equilibrium position, which corresponds to $x = 0$. The constant k is the spring constant. Note that the force is negative if x is positive (stretched spring; see Fig. 15.10a); and the force is positive if the displacement is negative (compressed spring; see Fig. 15.10b).

With the force as given by Eq. (15.16), the equation of motion of the particle is

$$m\frac{d^2x}{dt^2} = -kx \tag{15.18}$$

equation of motion for simple harmonic oscillator

This equation says that the acceleration of the particle is always proportional to the distance x, but is in the opposite direction. We now recall, from Eq. (15.15), that such a proportionality of acceleration and distance is characteristic of simple harmonic motion, and we therefore can immediately conclude that the motion of a particle coupled to a spring must be simple harmonic motion. By comparing Eqs. (15.18) and (15.15), we see that these equations become identical if

$$\omega^2 = \frac{k}{m}$$

and we therefore see that the angular frequency ω of the oscillation of the particle on a spring is

$$\omega = \sqrt{\frac{k}{m}} \tag{15.19}$$

Consequently, the frequency and the period are

$$f = \frac{\omega}{2\pi} = \frac{1}{2\pi}\sqrt{\frac{k}{m}} \tag{15.20}$$

angular frequency, frequency, and period for simple harmonic oscillator

and

$$T = \frac{1}{f} = 2\pi\sqrt{\frac{m}{k}} \tag{15.21}$$

With the value (15.19) for the angular frequency, the expression (15.4) for the position as a function of time becomes

$$x = A\cos\left(\sqrt{\frac{k}{m}}\,t + \delta\right) \tag{15.22}$$

According to Eq. (15.20) the frequency of the motion of the simple harmonic oscillator depends *only* on the spring constant and on the mass. *The frequency of the oscillator is unaffected by the amplitude with which it has been set in motion*—if the oscillator has a frequency of, say, 2 Hz when oscillating with a small amplitude, then it also has a frequency of 2 Hz when oscillating with a large amplitude. This property of the oscillator is called **isochronism**.

Note that the period is long if the mass is large and the spring constant is small. This is as expected, since in each period the spring must accelerate and decelerate the mass, and a weak spring will give a large mass only little acceleration.

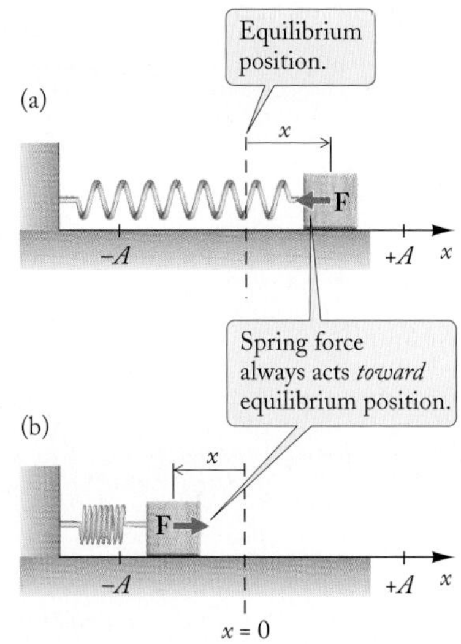

FIGURE 15.10 (a) Positive displacement of the particle; the force is negative. (b) Negative displacement of the particle; the force is positive.

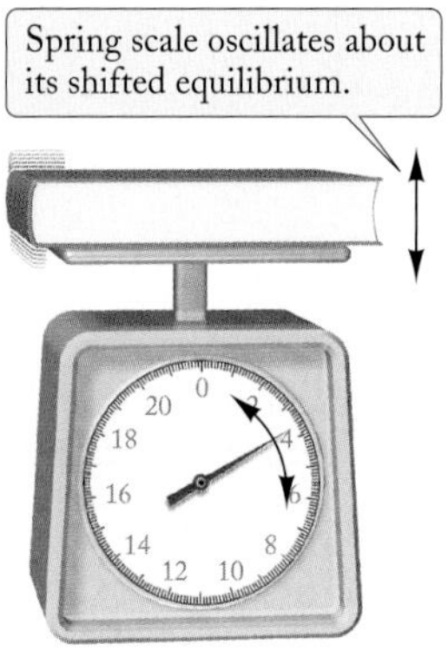

FIGURE 15.11 A heavy book on a spring scale oscillates up and down.

EXAMPLE 3 When you place a heavy encyclopedia, of mass 8 kg, on a kitchen scale (a spring scale; see Fig. 15.11), you notice that before coming to equilibrium, the pointer of the scale oscillates back and forth around the equilibrium position a few times with a period of 0.4 s. What is the effective spring constant of the internal spring of the kitchen scale? (Neglect other masses in the scale.)

SOLUTION: The mass of 8 kg in conjunction with the internal spring of the scale forms a mass-and-spring system, to which we can apply Eq. (15.21). If we square both sides of this equation, we obtain

$$T^2 = 4\pi^2 \frac{m}{k}$$

which gives us

$$k = 4\pi^2 \frac{m}{T^2} \qquad (15.23)$$

With $m = 8$ kg and $T = 0.4$ s, this becomes

$$k = 4\pi^2 \times \frac{8 \text{ kg}}{(0.4 \text{ s})^2} = 2 \times 10^3 \text{ N/m}$$

COMMENT: In this example, there is not only the force of the spring acting on the mass, but also the force of gravity on the mass (the weight) and friction forces. The force of gravity determines where the spring will reach equilibrium, but this force has no direct effect on the frequency of oscillation around equilibrium. The friction forces cause the oscillations to stop after a few cycles, but only slightly reduce the frequency (see Section 15.5). For negligible friction, the frequency depends exclusively on the mass and the spring constant.

Concepts — in — Context

EXAMPLE 4 Suppose that the astronaut in the chapter photo has a mass of 58 kg, including the chair device to which she is attached. She and the chair move under the influence of the force of a spring with $k = 2.1 \times 10^3$ N/m. There are no other forces acting. Consider the motion to be along the x axis, with the equilibrium point at $x = 0$. Suppose that at $t = 0$, she is (instantaneously) at rest at $x = 0.20$ m. Where will she be at $t = 0.10$ s? At $t = 0.20$ s? What will her velocity be when she passes through the equilibrium point?

SOLUTION: Since the astronaut is initially at rest at $x = 0.20$ m, this must be one of the turning points of the motion; thus, the amplitude of the motion must be $A = 0.20$ m. Furthermore, since at $t = 0$ the astronaut is at the turning point, the phase constant $\delta = 0$ [see Eq. (15.8)]. Consequently, at time $t = 0.10$ s, the position of the astronaut will be

$$x = A\cos\omega t = 0.20 \text{ m} \times \cos(\omega \times 0.10 \text{ s})$$

To evaluate this, we need the angular frequency of the oscillation. By Eq. (15.19) this is

$$\omega = \sqrt{\frac{k}{m}} = \sqrt{\frac{2.1 \times 10^3 \text{ N/m}}{58 \text{ kg}}} = 6.0 \text{ radians/s}$$

With this value of ω,

$$x = 0.20 \text{ m} \times \cos(6.0 \text{ radians/s} \times 0.10 \text{ s})$$
$$= 0.20 \text{ m} \times \cos(0.60 \text{ radian}) = 0.20 \text{ m} \times 0.83 = 0.17 \text{ m}$$

Likewise, at time $t = 0.20$ s, the position will be

$$x = A\cos(\omega t) = 0.20 \text{ m} \times \cos(6.0 \text{ radians/s} \times 0.20 \text{ s})$$
$$= 0.20 \text{ m} \times \cos(1.2 \text{ radian}) = 0.20 \text{ m} \times 0.36 = 0.072 \text{ m}$$

The astronaut passes through the equilibrium point when $\omega t = \pi/2$ (which makes $\cos \omega t = 0$). To find her velocity when she passes through the equilibrium point, we take the derivative of x with respect to t, and then evaluate the resulting expression at $\omega t = \pi/2$:

$$v = \frac{dx}{dt} = \frac{d}{dt}(A\cos\omega t) = -\omega A \sin\omega t \tag{15.24}$$
$$= -6.0 \text{ radians/s} \times 0.20 \text{ m} \times \sin(\pi/2) = -1.2 \text{ m/s}$$

Simple harmonic oscillators are used as the timekeeping element in modern watches. These watches use a quartz crystal as a spring-and-mass system. The crystal is elastic, with a high Young's modulus, and it therefore acts as a very stiff spring. The mass is not attached as a lump to the end of this spring, but it is uniformly distributed over the volume of the crystal (hence this spring–mass system is said to be "distributed," in contrast to a "lumped" system with separate springs and masses). The crystal is set into vibration by electric impulses, instead of mechanical pushes. The electric circuits attached to the crystal not only keep it vibrating, but also sense the frequency of vibration and control the display on the face of the clock.

The advantage of the quartz crystal as a timekeeping element is that the vibrations of the crystal are extremely stable, because any accelerations from bumping the watch are completely negligible compared with the immense accelerations of the oscillating masses in the crystal. Ordinary quartz clocks are accurate to within a few seconds per month; high-precision clocks are accurate to within 10^{-5} s per month.

✔ Checkup 15.2

QUESTION 1: For a particle with simple harmonic motion, at what point of the motion does the force on the particle attain maximum magnitude? Minimum magnitude?

QUESTION 2: Suppose we replace the particle in a simple harmonic oscillator by a particle of twice the mass. How does this alter the frequency of oscillation?

QUESTION 3: If we suddenly cut the spring of a simple harmonic oscillator when the particle is at the equilibrium point ($x = 0$), what is the subsequent motion of the particle? If we suddenly cut the spring when the particle is at maximum displacement ($x = A$)?

QUESTION 4: Suppose we replace the spring in a simple harmonic oscillator by a stronger spring, with twice the spring constant. What is the ratio of the new period of oscillation to the original period?

(A) 1/2 (B) $1/\sqrt{2}$. (C) 1 (D) $\sqrt{2}$. (E) 2

15.3 KINETIC ENERGY AND POTENTIAL ENERGY

We know from Section 8.1 that the force exerted by a spring is a conservative force, for which we can construct a potential energy. With this potential energy, we can formulate a law of conservation of the mechanical energy: the sum of the kinetic energy and the potential energy is a constant; that is,

$$E = K + U = [\text{constant}] \tag{15.25}$$

In this section we will see how to calculate the kinetic energy and the potential energy of the simple harmonic oscillator at each instant of time, and we will verify explicitly that the sum of these energies is constant.

The kinetic energy of a moving particle is

$$K = \tfrac{1}{2}mv^2 \tag{15.26}$$

For simple harmonic motion, the speed is given by Eq. (15.12), and the kinetic energy becomes

$$\begin{aligned} K &= \tfrac{1}{2}mv^2 = \tfrac{1}{2}m[-\omega A \sin(\omega t + \delta)]^2 \\ &= \tfrac{1}{2}m\omega^2 A^2 \sin^2(\omega t + \delta) \end{aligned} \tag{15.27}$$

Since $m\omega^2 = k$ [see Eq. (15.18)], we can also write this as

$$K = \tfrac{1}{2}kA^2\sin^2(\omega t + \delta) \tag{15.28}$$

The potential energy associated with the force $F = -kx$ is [see Eq. (8.6)]

$$U = \tfrac{1}{2}kx^2 \tag{15.29}$$

For simple harmonic motion, with $x = A\cos(\omega t + \delta)$, this becomes

$$U = \tfrac{1}{2}kA^2\cos^2(\omega t + \delta) \tag{15.30}$$

The kinetic energy and the potential energy both depend on time. According to Eqs. (15.28) and (15.30), each oscillates between a minimum value of zero and a maximum value of $\frac{1}{2}kA^2$. Figure 15.12 plots the oscillations of the kinetic energy and the potential energy as functions of time; for simplicity, we set the phase constant at $\delta = 0$. At the initial time $t = 0$, the particle is at maximum distance from the equilibrium point and its instantaneous speed is zero; thus, the potential energy is at its maximum value, and the kinetic energy is zero. A quarter of a cycle later, the particle passes through the equilibrium point and attains its maximum speed; thus, the kinetic energy is at its maximum value and the potential energy is zero. Thus energy is traded back and forth between potential energy and kinetic energy.

Since the force $F = -kx$ is conservative, the total mechanical energy $E = K + U$ is a constant of the motion. To verify this conservation law for the energy explicitly, we take the sum of Eqs. (15.28) and (15.30),

$$\begin{aligned} E &= K + U \\ &= \tfrac{1}{2}kA^2\sin^2(\omega t + \delta) + \tfrac{1}{2}kA^2\cos^2(\omega t + \delta) \\ &= \tfrac{1}{2}kA^2[\sin^2(\omega t + \delta) + \cos^2(\omega t + \delta)] \end{aligned} \tag{15.31}$$

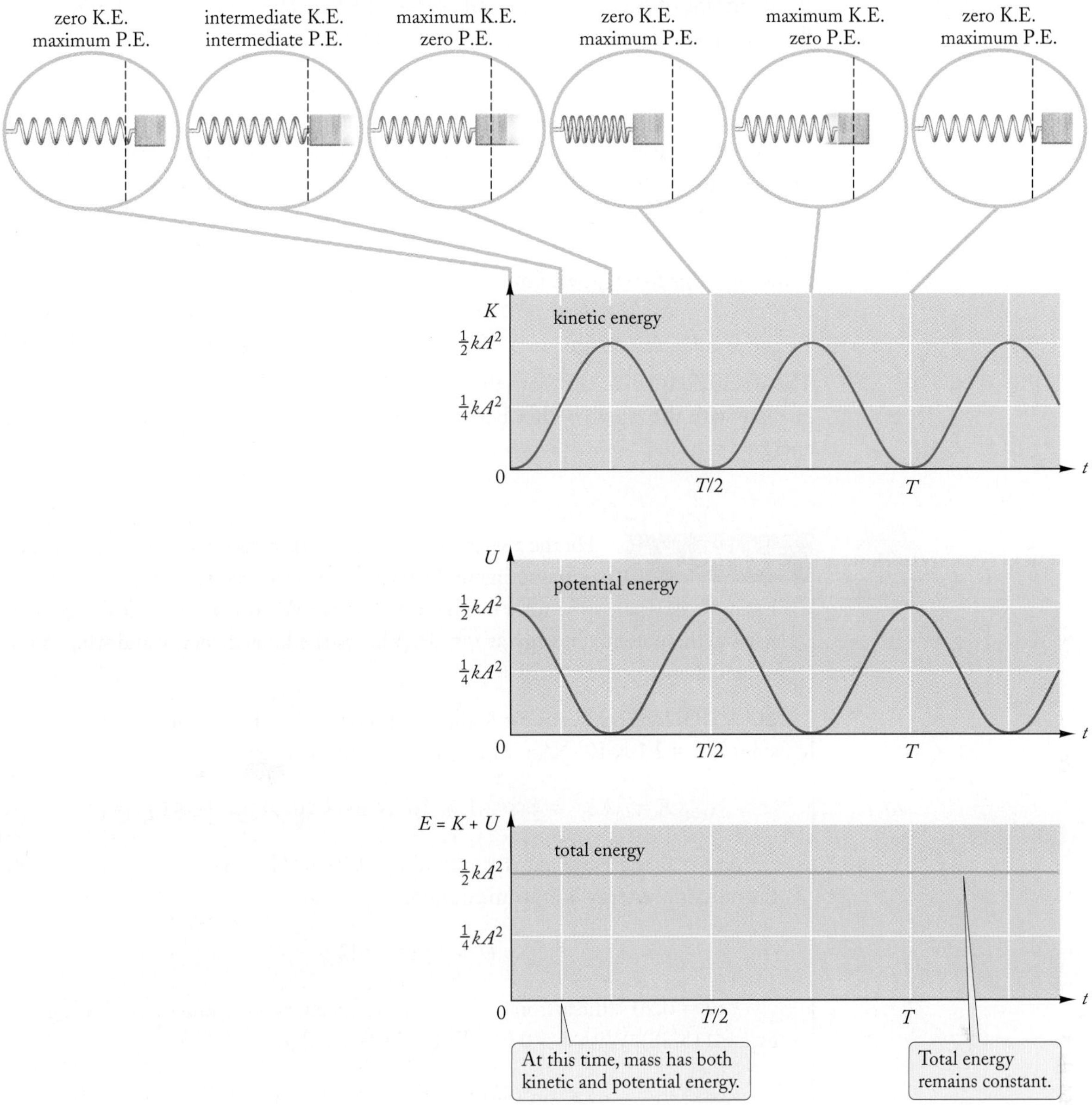

FIGURE 15.12 Kinetic energy and potential energy of a simple harmonic oscillator as a function of time.

We can simplify this expression if we use the trigonometric identity $\sin^2\theta + \cos^2\theta = 1$, which is valid for any angle θ. With this identity, we find that the right side of Eq. (15.31) is simply $\frac{1}{2}kA^2$:

$$E = \tfrac{1}{2}kA^2 \qquad (15.32)$$

energy of simple harmonic oscillator

This shows that *the energy of the motion is constant and is proportional to the square of the amplitude of oscillation.*

By means of Eq. (15.32), we can express the maximum displacement in terms of the energy. For this, we need only solve Eq. (15.32) for A:

$$x_{max} = A = \sqrt{2E/k} \tag{15.33}$$

Likewise, we can express the maximum speed in terms of the energy. For this, we note that when the particle passes through the equilibrium point, the energy is purely kinetic:

$$E = \tfrac{1}{2}mv_{max}^2 \tag{15.34}$$

If we solve this for v_{max}, we find

$$v_{max} = \sqrt{2E/m} \tag{15.35}$$

These equations tell us that both the maximum displacement and the maximum speed increase with the energy—they both increase in proportion to the square root of the energy.

Concepts — in — Context

EXAMPLE 5 For the 58-kg astronaut (with chair) moving under the influence of the spring in the body-mass measurement device described in Example 4, what is the total mechanical energy? What is the kinetic energy and what is the potential energy at $t = 0$? What is the kinetic energy and what is the potential energy at $t = 0.20$ s?

SOLUTION: From Example 4, the amplitude is $A = 0.20$ m and the spring constant is $k = 2.1 \times 10^3$ N/m. The total mechanical energy is

$$E = \tfrac{1}{2}kA^2 = \tfrac{1}{2} \times 2.1 \times 10^3\ \text{N/m} \times (0.20\ \text{m})^2 = 42\ \text{J}$$

At $t = 0$, the astronaut is at rest at $x = 0.20$ m. The kinetic energy is zero and the potential energy is at its maximum,

$$U = \tfrac{1}{2}kA^2 = 42\ \text{J}$$

At $t = 0.20$ s, the astronaut has nonzero speed, and the kinetic energy is given by Eq. (15.28). With $\delta = 0$ (see Example 4), we find

$$\begin{aligned} K &= \tfrac{1}{2}mv^2 = \tfrac{1}{2}kA^2\sin^2(\omega t) \\ &= \tfrac{1}{2} \times 2.1 \times 10^3\ \text{N/m} \times (0.20\ \text{m})^2 \\ &\qquad \times \sin^2(6.0\ \text{radians/s} \times 0.20\ \text{s}) \\ &= 36\ \text{J} \end{aligned} \tag{15.36}$$

The potential energy is given by Eq. (15.30), again with $\delta = 0$:

$$\begin{aligned} U &= \tfrac{1}{2}kA^2\cos^2(\omega t) \\ &= \tfrac{1}{2} \times 2.1 \times 10^3\ \text{N/m} \times (0.20\ \text{m})^2 \\ &\qquad \times \cos^2(6.0\ \text{radians/s} \times 0.20\ \text{s}) \\ &= 6\ \text{J} \end{aligned} \tag{15.37}$$

COMMENT: Note that the sum of the kinetic and potential energies is $K + U = 36\ \text{J} + 6\ \text{J} = 42\ \text{J}$, which agrees with our result for the total mechanical energy.

EXAMPLE 6 The hydrogen molecule (H_2) may be regarded as two particles joined by a spring (see Fig. 15.13). The center of the spring is the center of mass of the molecule. This point can be assumed to remain fixed, so this molecule consists of two identical simple harmonic oscillators vibrating in opposite directions. The spring constant for each of these oscillators is 1.13×10^3 N/m, and the mass of each hydrogen atom is 1.67×10^{-27} kg. Find the frequency of vibration in hertz. Suppose that the total vibrational energy of the molecule is 1.3×10^{-19} J. Find the corresponding amplitude of oscillation and the maximum speed.

SOLUTION: The frequency is given by Eq. (15.20):

$$f = \frac{1}{2p}\sqrt{\frac{k}{m}} = \frac{1}{2\pi}\left(\frac{1.13 \times 10^3 \text{ N/m}}{1.67 \times 10^{-27} \text{ kg}}\right)^{1/2} = 1.31 \times 10^{14} \text{ Hz}$$

Thus molecular vibrational frequencies can be quite high, about a hundred thousand billion cycles per second.

Each atom has half the total energy of the molecule; thus, the energy per atom is

$$E = \tfrac{1}{2} \times 1.3 \times 10^{-19} \text{ J} = 6.5 \times 10^{-20} \text{ J}$$

According to Eqs. (15.33) and (15.35), the amplitude of oscillation and the maximum speed of each atom are then

$$x_{\max} = \sqrt{\frac{2E}{k}} = \sqrt{\frac{2 \times 6.5 \times 10^{-20} \text{ J}}{1.13 \times 10^3 \text{ N/m}}} = 1.1 \times 10^{-11} \text{ m}$$

and

$$v_{\max} = \sqrt{\frac{2E}{m}} = \sqrt{\frac{2 \times 6.5 \times 10^{-20} \text{ J}}{1.67 \times 10^{-27} \text{ kg}}} = 8.8 \times 10^3 \text{ m/s}$$

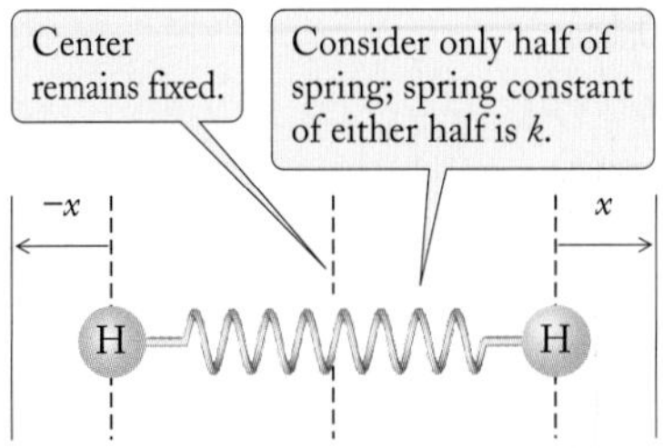

FIGURE 15.13 A hydrogen molecule, represented as two particles joined by a spring. The particles move symmetrically relative to the center of mass.

✔ Checkup 15.3

QUESTION 1: Two harmonic oscillators have equal masses and spring constants. One of them oscillates with twice the amplitude of the other. Compare the energies and compare the maximum speeds attained by the particles.

QUESTION 2: Two harmonic oscillators have equal spring constants and amplitudes of oscillation. One has twice the mass of the other. Compare the energies and the maximum speeds attained by the particles.

QUESTION 3: The period of a simple harmonic oscillator is 8.0 s. Suppose that at some time the energy is purely kinetic. At what later time will it be purely potential? At what later time again purely kinetic?

QUESTION 4: If the particle in a simple harmonic oscillator experiences a frictional force (say, air resistance), is the energy constant? Is the amplitude A constant?

QUESTION 5: The mass, frequency, and amplitude of one oscillator are each twice that of a second oscillator. What is the ratio of their stored energies, E_1/E_2?

(A) 2 (B) 4 (C) 8 (D) 16 (E) 32

Online Concept Tutorial 17

15.4 THE SIMPLE PENDULUM

A **simple pendulum** consists of a bob (a mass) suspended by a string or a rod from some fixed point (see Fig. 15.14). The bob is assumed to behave like a particle of mass m, and the string is assumed to be massless. Gravity acting on the bob provides a restoring force. When in equilibrium, the pendulum hangs vertically, just like a plumb line. When released at some angle with the vertical, the pendulum will swing back and forth along an arc of circle (see Fig. 15.15). The motion is two-dimensional; however, the position of the pendulum can be completely described by a single parameter: the angle θ between the string and the vertical (see Fig. 15.14). We will reckon this angle as positive on the right side of the vertical, and as negative on the left side.

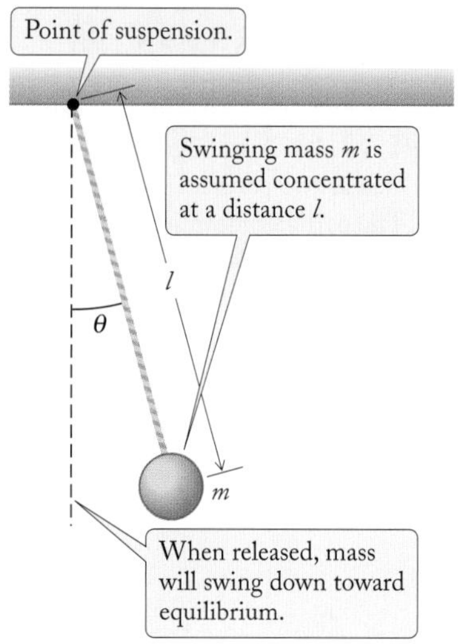

FIGURE 15.14 A pendulum swinging about a fixed suspension point. The angle θ is reckoned as positive if the deflection of the pendulum is toward the right, as in this figure.

Since the bob and the string swing as a rigid unit, the motion can be regarded as rotation about a horizontal axis through the point of suspension, and the equation of motion is that of a rigid body [see Eq. (13.19)]:

$$I\alpha = \tau \tag{15.38}$$

Here the moment of inertia I and the torque τ are reckoned about the horizontal axis through the point of suspension, and α is the angular acceleration.

Figure 15.16 shows the "free-body" diagram for the string–bob system with all the external forces. These external forces are the weight $\mathbf{w}$ of magnitude $w = mg$ acting on the mass m and the suspension force $\mathbf{S}$ acting on the string at the point of support. The suspension force exerts no torque, since its point of application is on the axis of rotation (its moment arm is zero). The weight exerts a torque [see Eq. (13.3)]

$$\tau = -mgl\sin\theta \tag{15.39}$$

where l is the length of the pendulum, measured from the point of suspension to the center of the bob. The minus sign in Eq. (15.39) indicates that this is a restoring torque, which tends to pull the pendulum toward its equilibrium position.

The moment of inertia I of the string–bob system is simply that of a particle of mass m at a distance l from the axis of rotation:

$$I = ml^2$$

FIGURE 15.15 Stroboscopic photograph of a swinging pendulum. The pendulum moves slowly at the extremes of its motion.

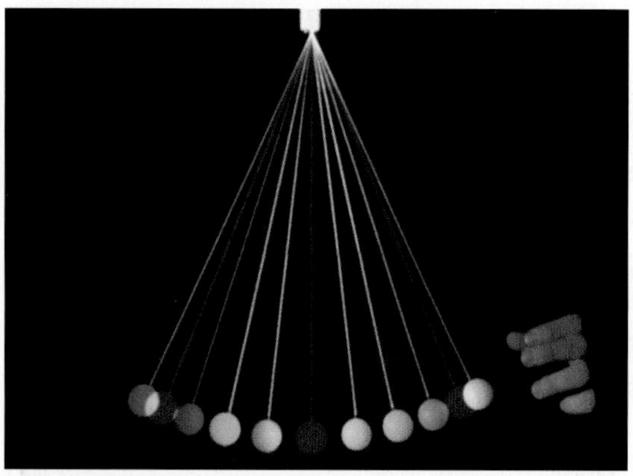

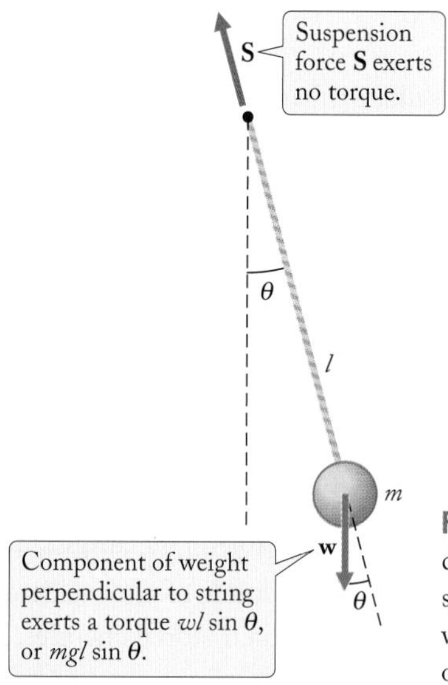

FIGURE 15.16 "Free-body" diagram for the string–bob system. The torque exerted by the weight $\mathbf{w}$ has magnitude $wl\sin\theta$, or $mgl\sin\theta$.

Hence the equation of rotational motion (15.38) becomes

$$ml^2\alpha = -mgl\sin\theta \tag{15.40}$$

or

$$\alpha = -\frac{g}{l}\sin\theta \tag{15.41}$$

We will solve this equation of motion only in the special case of small oscillations about the equilibrium position. If θ is small, we can make the approximation

$$\sin\theta \approx \theta \tag{15.42}$$

where the angle is measured in radians (see Math Help: Small-Angle Approximations for Sine, Cosine, and Tangent; and see Fig. 15.17).

With this approximation, the equation of motion becomes

$$\alpha = -\frac{g}{l}\theta \tag{15.43}$$

or, since the angular acceleration is $\alpha = d^2\theta/dt^2$,

$$\frac{d^2\theta}{dt^2} = -\frac{g}{l}\theta \tag{15.44}$$

This equation has the same mathematical form as Eq. (15.17). Comparing these two equations, we see that the angle θ replaces the distance x, the angular acceleration replaces the linear acceleration, l replaces m, and g replaces k. Hence the angular motion is simple harmonic. Making the appropriate replacements in Eq. (15.4), we find that the motion is described by the equation

$$\theta = A\cos(\omega t + \delta) \tag{15.45}$$

with an angular frequency [compare Eqs. (15.19) and (15.44)]

$$\omega = \sqrt{\frac{g}{l}} \tag{15.46}$$

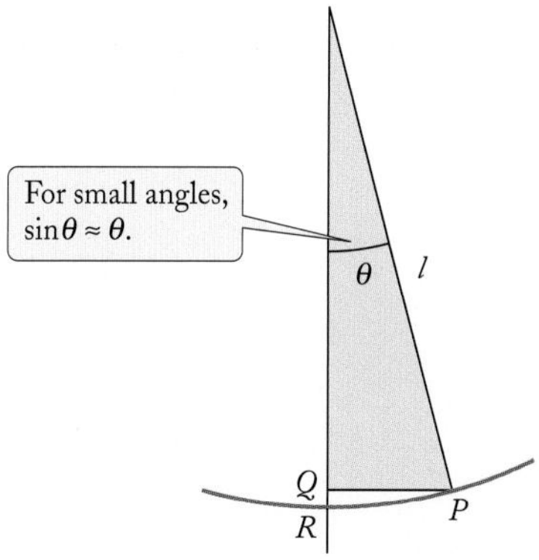

FIGURE 15.17 If the angle θ is small, the length of the straight line PQ is approximately the same as the length of the circular arc PR.

MATH HELP — SMALL-ANGLE APPROXIMATIONS FOR SINE, COSINE, AND TANGENT

With the assumption that an angle θ is expressed in radians and that this angle is small, the trigonometric functions have the simple approximations

$$\sin\theta \approx \theta$$

$$\cos\theta \approx 1 - \theta^2/2$$

$$\tan\theta \approx \theta$$

To understand how these approximations come about, consider the small angle θ shown in Fig. 15.17. The sine of this angle is $\sin\theta = PQ/l$. If θ is small, the length of the straight line PQ is approximately the same as the length of the curved circular arc PR (for small angles, the curved arc is almost a straight line). Thus, $\sin\theta \approx PR/l$. But the ratio PR/l is the definition of the angle θ expressed in radians, so $\sin\theta \approx \theta$. Similar arguments give the above approximations for the cosine and the tangent. These approximations are usually satisfactory if θ is less than about 0.2 radians, or about 10°.

The frequency and the period of the pendulum are then

angular frequency, frequency, and period for simple pendulum

$$f = \frac{\omega}{2\pi} = \frac{1}{2\pi}\sqrt{\frac{g}{l}}$$

and

$$T = \frac{1}{f} = \frac{2\pi}{\omega} = 2\pi\sqrt{\frac{l}{g}} \qquad (15.47)$$

Note that these expressions for the frequency and the period depend only on the length of the pendulum and on the acceleration of gravity; they do not depend on the mass of the pendulum bob or on the amplitude of oscillation (but, of course, our calculation depends on the assumption that the angle θ, and thus the amplitude of motion, is small).

Like the simple harmonic oscillator, the pendulum has the property of isochronism—*its frequency is (approximately) independent of the amplitude with which it is swinging.* This property can be easily verified by swinging two pendulums of equal lengths side by side, with different amplitudes. The pendulums will continue to swing in step for a long while.

FIGURE 15.18 Woman on a swing.

EXAMPLE 7 A woman sits in a swing of length 3.0 m (see Fig. 15.18). What is the period of oscillation of this swing?

SOLUTION: We can regard the swing as a pendulum of an approximate length 3.0 m. From Eq. (15.47) we then find

$$T = 2\pi\sqrt{\frac{l}{g}} = 2\pi\sqrt{\frac{3.0\text{ m}}{9.81\text{ m/s}^2}} = 3.5\text{ s}$$

EXAMPLE 8 The "seconds" pendulum in a pendulum clock built for an astronomical observatory has a period of exactly 2.0 s, so each one-way motion of the pendulum takes exactly 1.0 s. What is the length of such a "seconds" pendulum at a place where the acceleration of gravity is $g = 9.81\text{ m/s}^2$? At a place where the acceleration of gravity is 9.79 m/s^2?

SOLUTION: If we square both sides of Eq. (15.47) and then solve for the length l, we find

$$l = \left(\frac{T}{2\pi}\right)^2 g$$

With $g = 9.81\text{ m/s}^2$ and the known period $T = 2.0$ s, this gives

$$l = \left(\frac{2.0\text{ s}}{2\pi}\right)^2 \times 9.81\text{ m/s}^2 = 0.994\text{ m}$$

With $g = 9.79\text{ m/s}^2$, it gives

$$l = \left(\frac{2.0\text{ s}}{2\pi}\right)^2 \times 9.79\text{ m/s}^2 = 0.992\text{ m}$$

FIGURE 15.19 This electromechanical clock, regulated by a pendulum, served as the U.S. frequency standard in the 1920s. Its master pendulum is enclosed in the canister at right.

The most familiar application of pendulums is the construction of pendulum clocks. Up to about 1950, the most accurate clocks were pendulum clocks of a special design, which were kept inside airtight flasks placed in deep cellars to protect them from disturbances caused by variations of the atmospheric pressure and temperature (see Fig. 15.19). The best of these high-precision pendulum clocks were accurate to within a few thousandths of a second per day. Later, such pendulums were superseded by quartz clocks (see Section 15.2) and then by atomic clocks (see Section 1.3).

Another important application of pendulums is the measurement of the acceleration of gravity g. For this purpose it is necessary only to time the swings of a pendulum of known length; the value of g can then be calculated from Eq. (15.47). The pendulums used for precise determinations of g usually consist of a solid bar swinging about a knife edge at one end, instead of a bob on a string. Such a pendulum consisting of a swinging rigid body is called a **physical pendulum**; its period is related to its size and shape.

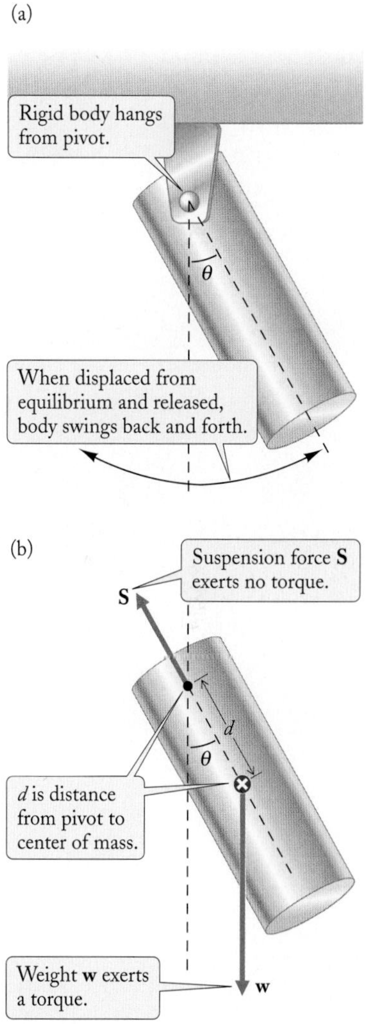

FIGURE 15.20 (a) A physical pendulum consisting of a rigid body swinging about a point of suspension. (b) "Free-body" diagram for the physical pendulum. The weight acts at the center of mass.

EXAMPLE 9 A physical pendulum has a moment of inertia I about its point of suspension, and its center of mass is at a distance d from this point (see Fig. 15.20a). Find the period of this pendulum.

SOLUTION: Figure 15.20b shows the "free-body" diagram for the pendulum. The suspension force **S** has zero moment arm about the pivot, and so exerts no torque. The weight acts at the center of mass, at a distance of d from the point of suspension, and it exerts a torque [see Eq. (13.3)]

$$\tau = -mgd\sin\theta$$

Hence the equation of rotational motion (15.38) is

$$I\alpha = -mgd\sin\theta$$

where $\alpha = d^2\theta/dt^2$ is the angular acceleration for the rotational motion. With the usual small-angle approximation $\sin\theta \approx \theta$, this becomes

$$\frac{d^2\theta}{dt^2} = -\frac{mgd}{I}\theta$$

As in the case of the simple pendulum, we compare this with Eq. (15.17). Since the second time derivative of θ is proportional to the negative of θ, the motion will again be simple harmonic. Hence the angular frequency of oscillation is

$$\omega = \sqrt{\frac{mgd}{I}} \tag{15.48}$$

and the period is

$$T = \frac{2\pi}{\omega} = 2\pi\sqrt{\frac{I}{mgd}} \tag{15.49}$$

COMMENT: Note that for a simple pendulum, the moment of inertia about the point of suspension is $I = ml^2$ and the distance of the center of mass from this point is $d = l$. Accordingly, Eq. (15.49) yields $T = 2\pi\sqrt{ml^2/mgl} = 2\pi\sqrt{l/g}$, which shows that the formula for the period of the simple pendulum is a special case of the general formula for the physical pendulum.

Finally, we must emphasize that the approximation contained in Eq. (15.43) is valid only for small angles. If the amplitude of oscillation of a pendulum is more than a few degrees—say, more than 10°—the approximation (15.43) begins to fail, and the motion of the pendulum begins to deviate from simple harmonic motion. At large amplitudes, the period of the pendulum depends on the amplitude—the larger the amplitude, the larger the period. For instance, a pendulum oscillating with an amplitude of 30° has a period 1.7% longer than the value given by Eq. (15.47).

✔ Checkup 15.4

QUESTION 1: If we shorten the string of a pendulum to half its original length, what is the alteration of the period? The frequency?

QUESTION 2: Two pendulums have equal lengths, but one has 3 times the mass of the other. If we want the energies of oscillation to be the same, how much larger must we make the amplitude of oscillation of the less massive pendulum?

QUESTION 3: A uniform metal rod of length l hangs from one end and oscillates with small amplitude. Such a rod, rotating about one end, has moment of inertia $I = \frac{1}{3}ml^2$ (Table 12.3). What is ω, the angular frequency of oscillation?

(A) $\sqrt{g/l}$. (B) $\sqrt{3g/2l}$. (C) $\sqrt{3g/l}$. (D) $\sqrt{6g/l}$.

15.5 DAMPED OSCILLATIONS AND FORCED OSCILLATIONS

So far we have proceeded on the assumption that the only force acting on a simple harmonic oscillator or a pendulum is the restoring force $F = -kx$ or the restoring torque $\tau = -mgl\sin\theta$. However, in a real oscillator or a real pendulum, there is always some extra force caused by friction. For instance, if the pendulum starts its swinging motion with some initial amplitude, the friction against the air and against the point of support will gradually brake the pendulum, reducing its amplitude of oscillation. Although good oscillators have low friction, sometimes more friction is desirable for damping out unwanted oscillations, as with the kitchen scale of Example 3, so that a steady, equilibrium position can be attained.

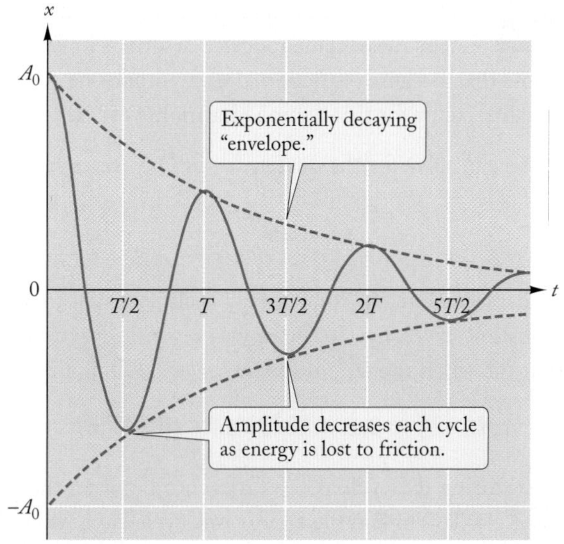

FIGURE 15.21 Plot of position vs. time for a particle with damped harmonic motion.

If the friction force is proportional to the velocity, the equation of motion becomes

$$m\frac{d^2x}{dt^2} = -kx - b\frac{dx}{dt} \tag{15.50}$$

where b is called the friction constant, or the damping constant. Figure 15.21 is a plot of the position as a function of time for a harmonic oscillator with fairly strong friction. The amplitude of oscillation suffers a noticeable decrease from one cycle to the next. Such a gradually decreasing oscillation is called **damped harmonic motion**. The oscillation amplitude decreases exponentially with time, as indicated by the dashed line in Fig. 15.21. Increasing the friction shortens the time it takes for the amplitude to decrease, and slows the frequency of oscillation somewhat. If the damping is very large, a displaced "oscillator" merely moves back to its equilibrium position, without oscillating. In Section 32.6, we will examine the damped harmonic oscillator in detail.

Since the oscillator must do work against the friction, the mechanical energy gradually decreases. The energy loss per cycle is a constant fraction of the energy E that the oscillator has at the beginning of the cycle. If we represent the energy loss per cycle by ΔE, then ΔE is proportional to E:

$$\Delta E = \left(\frac{2\pi}{Q}\right)E \tag{15.51}$$

Q **of oscillator**

Here, the constant of proportionality has been written in the somewhat complicated form $2\pi/Q$, which is the form usually adopted in engineering. The quantity Q is called the **quality factor** of the oscillator. In terms of the damping constant b,

$$Q = \frac{\sqrt{km}}{b} \tag{15.52}$$

An oscillator with low friction has a high value of Q, and a small energy loss per cycle; an oscillator with high friction has a low value of Q, and a large energy loss per cycle. The value of Q roughly coincides with the number of cycles the oscillator completes before the oscillations damp away significantly. Mechanical oscillators of low friction, such as tuning forks or piano strings, have Q values of a few thousand; that is, they "ring" for a few thousand cycles before their oscillations fade noticeably.

Concepts — in — Context

EXAMPLE 10 The maximum displacement from equilibrium of the body-mass measurement device described in Examples 4 and 5 was 0.200 m. Suppose that, because of friction, the amplitude one cycle later is 0.185 m. What is the quality factor for this damped harmonic oscillator?

SOLUTION: We can solve for the quality factor Q by rearranging Eq. (15.51):

$$Q = 2\pi \frac{E}{\Delta E}$$

At maximum displacement, the total energy is all potential energy, so $E = \frac{1}{2}kA^2$. The spring constant $k = 2.1 \times 10^3$ N/m was given in Example 4. We found in Example 5 that the when the amplitude was $A = 0.200$ m, the energy stored was

$$E = \tfrac{1}{2}kA^2 = \tfrac{1}{2} \times 2.1 \times 10^3 \text{ N/m} \times (0.200 \text{ m})^2 = 42 \text{ J}$$

The energy lost during the cycle is the difference between the energy when the amplitude was $A = 0.200$ m and the energy one cycle later, when the amplitude is $A' = 0.185$ m:

$$\Delta E = \tfrac{1}{2}kA^2 - \tfrac{1}{2}kA'^2 = \tfrac{1}{2}k(A^2 - A'^2)$$

$$= \tfrac{1}{2} \times 2.1 \times 10^3 \text{ N/m} \times [(0.200 \text{ m})^2 - (0.185 \text{ m})^2] = 6.1 \text{ J}$$

Hence the quality factor is

$$Q = 2\pi \frac{E}{\Delta E} = 2\pi \times \frac{42 \text{ J}}{6.1 \text{ J}} = 43$$

To maintain the oscillations of a damped harmonic oscillator at a constant level, it is necessary to exert a periodic force on the oscillator, so the energy fed into the oscillator by this extra force compensates for the energy lost to friction. An extra force is also needed to start the oscillations of any oscillator, damped or not, by supplying the initial energy for the motion. Any such extra force exerted on an oscillator is called a **driving force**. A familiar example of a driving force is the "pumping" force that you must exert on a playground swing (a pendulum) to start it moving and to keep it moving at a constant amplitude. This is an example of a *periodic* driving force.

With the addition of a harmonic driving force of amplitude F_0 and angular frequency ω, the equation of motion (15.50) becomes

$$m\frac{d^2x}{dt^2} = -kx - b\frac{dx}{dt} + F_0 \cos \omega t \qquad (15.53)$$

If the frequency ω of the driving force coincides with the frequency ω_0 of the natural oscillations of the oscillator, then even a quite small driving force can gradually build up large amplitudes. Under these conditions the driving force steadily feeds energy into the oscillations, and the amplitude of these grows until the friction becomes so large that it inhibits further growth. The ultimate amplitude reached depends on the amount of friction; in an oscillator of low friction, or high Q, this ultimate amplitude can be extremely large. The buildup of a large amplitude by the action of a driving force in tune with the natural frequency of an oscillator is called **resonance**. Figure 15.22 shows the value of the final amplitude of oscillation attained as a function of the frequency of

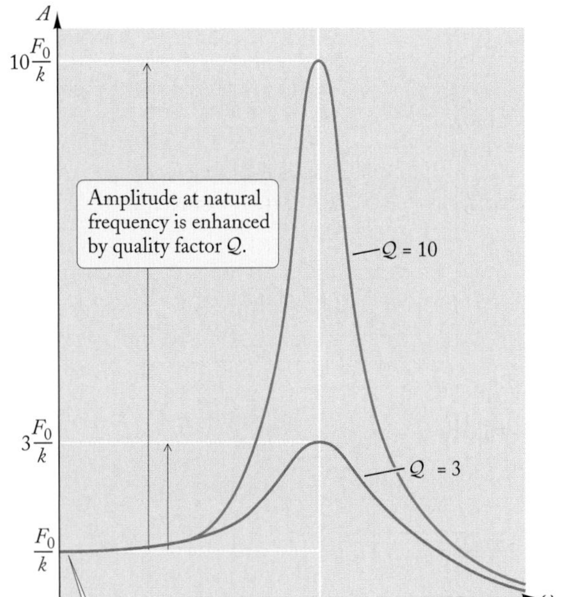

FIGURE 15.22 Amplitude of a forced damped harmonic oscillator as a function of the frequency of the oscillating force.

the harmonic driving force for two mass-and-spring systems with the same natural angular frequency $\omega_0 = \sqrt{k/m}$ but different values of Q, the quality factor. Notice that large amplitudes occur over a range of driving frequencies, and that some enhancement over the static Hooke's-Law displacement $x = -F_0/k$ occurs for any frequency of forced oscillation near or below the natural frequency ω_0. If the oscillator is forced precisely at resonance, the amplitude can be shown to take the value

$$A = \frac{F_0}{k} Q \tag{15.54}$$

amplitude at resonance of damped, driven harmonic oscillator

This is simply the magnitude of the static displacement $x = -F_0/k$ multiplied by Q; thus the quality factor is equivalent to an amplitude enhancement factor for a system at resonance.

The phenomenon of resonance plays a crucial role in many pieces of industrial machinery—if one vibrating part of a machine is driven at resonance by a perturbing force originating from some other part, then the amplitude of oscillation can build up to violent levels and shake the machine apart. Such dangerous resonance effects can occur not only in moving pieces of machinery, but also in structures that are normally regarded as static. In a famous accident that took place in 1850 in Angers, France, the stomping of 487 soldiers marching over a suspension bridge excited a resonant swinging motion of the bridge; the motion quickly rose to a disastrous level and broke the bridge apart, causing the death of 226 of the soldiers (Fig. 15.23).

FIGURE 15.23 Resonance disaster: the collapse of the bridge at Angers, as illustrated in a contemporary newspaper.

✔ Checkup 15.5

QUESTION 1: Suppose that the driving force has a frequency half as large as the frequency of the oscillator. Would you expect a buildup of oscillations by resonance?

QUESTION 2: Suppose that the driving force has a frequency twice as large as the frequency of the oscillator. Would you expect a buildup of oscillations by resonance?

QUESTION 3: Suppose that a bell has a high Q (it continues to ring for a long time after you strike it). If you rest your hand against the bell after striking it, how does this alter the Q?

QUESTION 4: An oscillator begins with 1.00 J of mechanical energy. After 10 oscillations, the energy stored has dropped to 0.90 J. What is the approximate Q of the system?

(A) 6.3 (B) 10 (C) 63 (D) 100 (E) 630

PHYSICS IN PRACTICE CHAOS

The motions we examined in this chapter and in the preceding chapters are either periodic or else regular in some other sense. For instance, the motion of a simple harmonic oscillator and the motion of a planet around the Sun are periodic; and the motion of a particle under the influence of a constant force is highly regular, proceeding with constant acceleration. But there also exist mechanical systems with highly irregular motions, without periodicity, and with a pathological sensitivity to small changes in initial conditions, so a small change of the initial velocity or position quickly leads to very large changes in the motion. Such motions are called **chaotic**.

An example of a system with chaotic motion is the double-well oscillator, in which the attractive force $-kx$ of the simple harmonic oscillator is replaced by a sum of a repulsive force $+kx$ and an attractive force $-ax^3$. Such an oscillator can be constructed by clamping a leaf spring in a vertical position and attaching a fairly large mass to its top (see Fig. 1). The central position is an equilibrium position, but it is unstable—the leaf spring will flop sideways either to the left or to the right, attaining a bent equilibrium position, which is stable. If disturbed, it can oscillate about this left or right equilibrium position. The potential energy for this system has a minimum at the left equilibrium position, a minimum at the right equilibrium position, and a maximum at the vertical position in between; that is, the curve of potential energy has two wells and a hump in between. The equation of motion for this system cannot be solved exactly, but it can be solved numerically by a computer program that calculates derivatives by evaluating changes in the position in small time increments.

FIGURE 1 (a) Leaf spring clamped at the lower end with a mass attached at the upper end. (b) The leaf spring flops to the left (blue) or to the right (red), and it can oscillate about these bent equilibrium positions.

Figure 2 shows two numerical solutions for a double-well oscillator that includes both a frictional damping force and a periodic driving force (such as discussed in Section 15.5). Note that at first the oscillator moves erratically—it sometimes oscillates in the left well, then in the right well, then back again, etc. Also note that although the initial conditions for the two solutions barely differ at all and the motions are initially almost indistinguishable, they soon begin to differ drastically. Finally, one of the solutions settles down in the left well (blue), and the other solution settles down in the right well (red), and they then continue to oscillate in a steady manner about the left or the right equilibrium position with a frequency equal to that of the driving force. The steady modes of oscillations in the left and the right wells are called **attractors**, because the motion tends to settle into these modes.

The erratic motion that precedes the steady oscillations is an instance of chaos. How long the chaos lasts depends on the initial conditions and on the strengths of the driving force and the damping. For some values of these parameters, the chaos lasts forever. We can prepare plots such as those in Fig. 2 for a wide variety of initial conditions, and in each case examine whether the oscillator settles into the left or the right well. Figure 3 is a color-coded diagram that summarizes the results of 900×900 such calculations, with different initial positions and velocities. The initial positions are plotted horizontally and the initial velocities vertically; the color indicates where the oscillator settles: blue for the left well, and red for the right well. The solid blue and red zones indicate that for all initial conditions in these zones, the oscillator settles in the same final steady state of oscillation. But the other regions of the diagram, with fine striations of intermingled red and blue points, are characteristic of chaos. A very minor change in initial conditions takes us from a blue point to a red point (or vice versa), which means the final motion depends sensitively on small changes. The striations in Fig. 3 have a **fractal** character—if we examine any small patch in the striated zone at higher magnification, we find striations within striations within striations.

The chaotic behavior implies that although in principle the motion can be calculated from the initial conditions, in practice the motion is not predictable, except for a short time. Any small uncertainty in the initial conditions or any small uncertainty introduced by round-off errors in the numerical calculation will make it impossible to decide whether the initial conditions fall on a blue or a red dot—which means we can't decide whether the oscillator will ultimately settle into steady oscillation on the left or on the right.

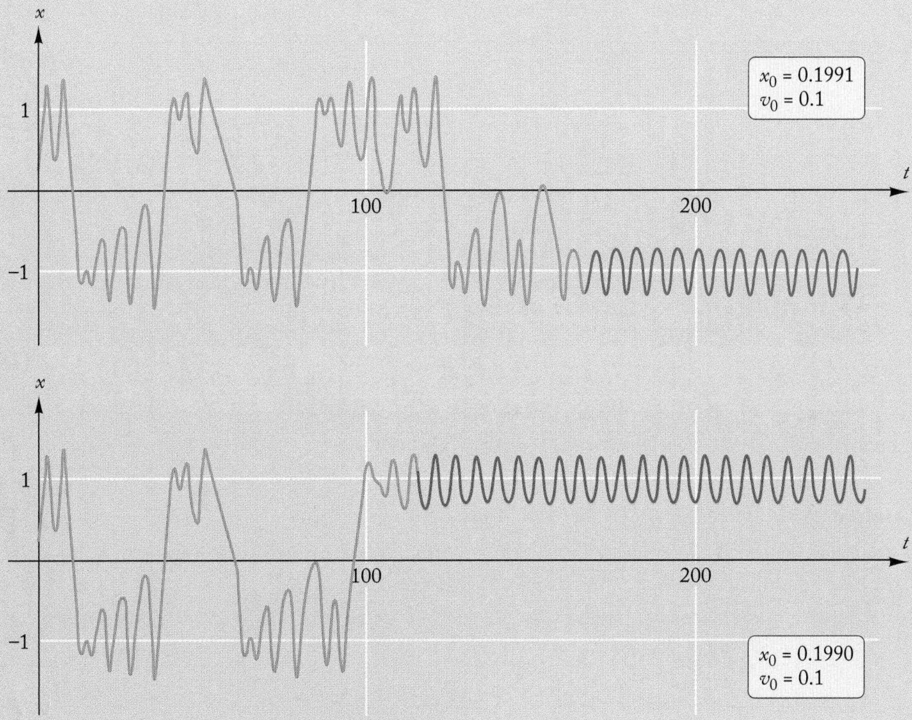

FIGURE 2 Motions for two slightly different initial conditions of the oscillator. At first the motion is chaotic (gray), but it ultimately settles into periodic oscillations about the left equilibrium position (blue) or the right equilibrium position (red). (For these plots, all the constants in the equation of motion were set equal to 1, except the damping constant and the strength of the driving force, which were set equal to 0.25.)

FIGURE 3 Plot of 900 × 900 initial positions (horizontal coordinate) and initial velocities (vertical coordinate). The color of each dot is blue if the oscillator ultimately settles into periodic oscillations about the left equilibrium position, red if about the right.

SUMMARY

MATH HELP Derivatives of trigonometric functions **(page 473)**

MATH HELP Small-angle approximations for sine, cosine, and tangent **(page 485)**

PHYSICS IN PRACTICE Chaos **(page 492)**

SIMPLE HARMONIC MOTION

where A is the amplitude $x = 0$; ω is the angular frequency, and δ is the phase constant.

$$x = A\cos(\omega t + \delta) \qquad (15.4)$$

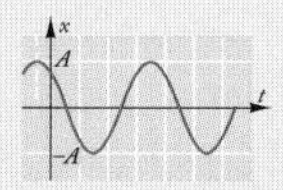

PERIOD (time for one cycle)

$$T = 2\pi/\omega \qquad (15.5)$$

FREQUENCY (number of cycles per second)

$$f = 1/T = \omega/2\pi \qquad (15.6)$$

PHASE CONSTANT AND TIME OF MAXIMUM DISPLACEMENT.

$$\delta = -\omega t_{max} \qquad (15.8)$$

MAXIMUM VELOCITY

$$v_{max} = \omega A \qquad (15.14)$$

MAXIMUM ACCELERATION

$$a_{max} = \omega^2 A \qquad (15.16)$$

EQUATION OF MOTION OF SIMPLE HARMONIC OSCILLATOR

where k is the spring constant.

$$m\frac{d^2x}{dt^2} = -kx \qquad (15.18)$$

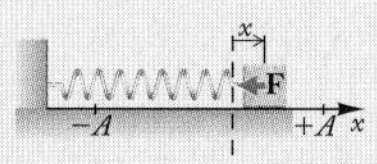

ANGULAR FREQUENCY AND PERIOD OF SIMPLE HARMONIC OSCILLATOR

$$\omega = \sqrt{\frac{k}{m}} \qquad T = 2\pi\sqrt{\frac{m}{k}} \qquad (15.19, 15.21)$$

ENERGY OF SIMPLE HARMONIC OSCILLATOR

$$E = \tfrac{1}{2}kA^2 = \tfrac{1}{2}m\omega^2 A^2 \qquad (15.32)$$

ANGULAR FREQUENCY AND PERIOD OF SIMPLE PENDULUM

$$\omega = \sqrt{\frac{g}{l}} \qquad T = 2\pi\sqrt{\frac{l}{g}} \qquad (15.46, 15.47)$$

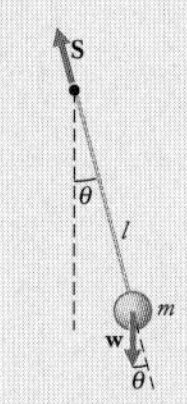

ANGULAR FREQUENCY AND PERIOD OF PHYSICAL PENDULUM

where I is the moment of inertia of the pendulum.

$$\omega = \sqrt{\frac{mgd}{I}} \qquad T = 2\pi\sqrt{\frac{I}{mgd}} \qquad (15.48, 15.49)$$

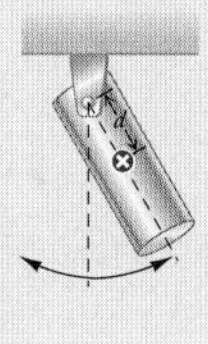

ENERGY LOSS PER CYCLE OF DAMPED OSCILLATOR where Q is the quality factor.

$$\Delta E = -\frac{2\pi}{Q} E \tag{15.51}$$

AMPLITUDE AT RESONANCE OF DAMPED HARMONIC OSCILLATOR where F_0 is the amplitude of a harmonic driving force.

$$A = \frac{F_0}{k} Q \tag{15.54}$$

QUESTIONS FOR DISCUSSION

1. Is the motion of the piston of an automobile engine simple harmonic motion? How does it differ from simple harmonic motion?
2. In our calculation of the frequency of the simple harmonic oscillator, we ignored the mass of the spring. Qualitatively, how does the mass of the spring affect the frequency?
3. A grandfather clock is regulated by a pendulum. If the clock is running late, how must we adjust the length of the pendulum?
4. Figure 15.24 shows the escapement of a pendulum clock, i.e., the linkage that permits the pendulum to control the rotation of the wheels of the clock. Explain how the wheel turns as the pendulum swings.

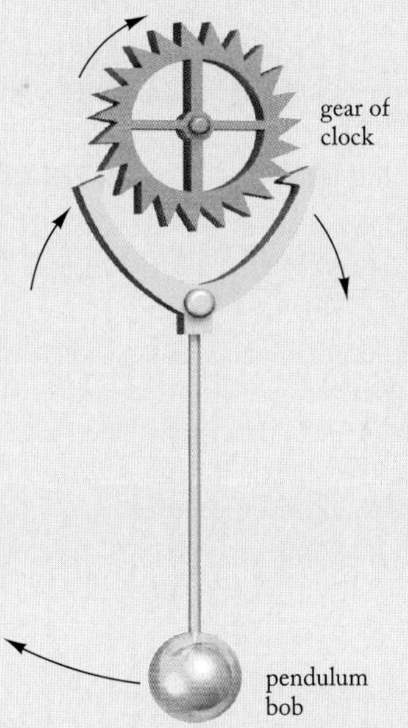

FIGURE 15.24 Escapement mechanism of a pendulum clock. At the instant shown, the tooth at the left has escaped from the left arm, and the tooth on the right is pushing against the right arm.

5. Would a pendulum clock keep good time on a ship?
6. Galileo claimed that the oscillators of a pendulum are isochronous, even for an amplitude of oscillation as large as 30°. What is your opinion of this claim?
7. Why would you expect a pendulum oscillating with an amplitude of nearly (but not quite) 180° to have a very long period?
8. Can a pendulum oscillate with an amplitude of more than 180°?
9. Figure 15.25 shows a "tilted pendulum" designed by Christiaan Huygens in the seventeenth century. When the pendulum is tilted, its period is longer than when the pendulum is vertical. Explain.

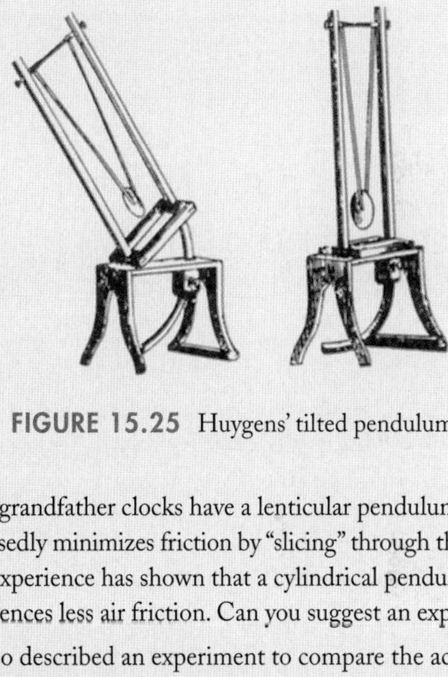

FIGURE 15.25 Huygens' tilted pendulum.

10. Most grandfather clocks have a lenticular pendulum bob which supposedly minimizes friction by "slicing" through the air. However, experience has shown that a cylindrical pendulum bob experiences less air friction. Can you suggest an explanation?
11. Galileo described an experiment to compare the acceleration of gravity of lead and of cork:

> I took two balls, one of lead and one of cork, the former being more than a hundred times as heavy as the latter, and suspended them from two equal thin strings, each four or five bracchia long. Pulling each ball aside from the vertical, I released them at the same instant, and they, falling along the circumferences of the circles having the strings as radii, passed through the vertical and returned along the same path. This free oscillation, repeated more than a hundred times, showed clearly that the heavy body kept time with the light body so well that neither in a hundred oscillations, nor in a thousand, will the former anticipate the latter by even an instant, so perfectly do they keep step.

Since air friction affects the cork ball much more than the lead ball, do you think Galileo's results are credible?

Newton reported a more careful experiment that avoided the inequality of friction:

> I tried the thing in gold, silver, lead, glass, sand, common salt, wood, water, and wheat. I provided two equal wooden boxes. I filled the one with wood, and suspended an equal weight of gold (as exactly as I could) in the centre of oscillation of the other. The boxes, hung by equal threads of 11 feet, made a couple of pendulums perfectly equal in weight and figure . . . and, placing the one by the other, I observed them to play together forwards and backwards for a long while, with equal vibrations. . . . And by these experiments, in bodies of the same weight, one could have discovered a difference of matter less than the thousandth part of the whole.

Explain how Newton's experiment was better than Galileo's.

12. A simple pendulum hangs below a table, with its string through a small hole in the tabletop. Suppose you gradually pull the string while the pendulum is swinging. What happens to the frequency of oscillation? To the (angular) amplitude?

13. Shorter people have a shorter length of stride, but a higher rate of step when walking "naturally." Explain.

14. A girl sits on a swing whose ropes are 1.5 m long. Is this a simple pendulum or a physical pendulum?

15. A simple pendulum consists of a particle of mass m attached to a string of length l. A physical pendulum consists of a body of mass m attached to a string in such a way that the center of mass is at a distance l from the point of support. Which pendulum has the shorter period?

16. Suppose that the spring in the front-wheel suspension of an automobile has a natural frequency of oscillation equal to the frequency of rotation of the wheel at, say, 80 km/h. Why is this bad?

17. When marching soldiers are about to cross a bridge, they break step. Why?

PROBLEMS

15.1 Simple Harmonic Motion[†]

1. A particle moves as follows as a function of time:

$$x = 3.0\cos(2.0t)$$

where distance is measured in meters and time in seconds.

(a) What is the amplitude of this simple harmonic motion? The frequency? The angular frequency? The period?

(b) At what time does the particle reach the midpoint, $x = 0$? The turning point?

2. A particle is performing simple harmonic motion along the x axis according to the equation

$$x = 0.6\cos\left(\frac{\pi t}{2}\right)$$

where the distance is measured in meters and the time in seconds.

(a) Calculate the position x of the particle at $t = 0$, $t = 0.50$ s, and $t = 1.00$ s.

(b) Calculate the instantaneous velocity of the particle at these times.

(c) Calculate the instantaneous acceleration of the particle at these times.

3. A particle moves back and forth along the x axis between the points $x = 0.20$ m and $x = -0.20$ m. The period of the motion is 1.2 s, and it is simple harmonic. At the time $t = 0$, the particle is at $x = 0.20$ m and its velocity is zero.

(a) What is the frequency of the motion? The angular frequency?

(b) What is the amplitude of the motion?

(c) At what time will the particle reach the point $x = 0$? At what time will it reach the point $x = -0.10$ m?

(d) What is the speed of the particle when it is at $x = 0$? What is the speed of the particle when it reaches the point $x = -0.10$ m?

4. Suppose that the peg on the rotating wheel illustrated in Fig. 15.5 is located at a radius of 4.0 cm. The wheel turns at a rate of 600 rev/min. What is the amplitude of the simple harmonic motion of the slotted arm? What are the period, the frequency, and the angular frequency?

5. Consider that the particle in Fig. 15.4 is executing simple harmonic motion according to Eq. (15.1).

(a) What is the speed of the satellite for this case?

(b) At $t = 0.050$ s, the particle is at the midpoint and its instantaneous velocity is parallel to that of the satellite. What is the speed of the particle? How does it compare with the speed of the satellite?

6. A given point on a guitar string executes simple harmonic motion with a frequency of 440 Hz and an amplitude of 1.2 mm. What is the maximum speed of this motion? The maximum acceleration?

7. A piston in a windmill-driven water pump is in simple harmonic motion. The motion has an amplitude of 50 cm and the mass of the piston is 6.0 kg. Find the maximum net force on

[†]For help, see Online Concept Tutorial 16 at www.wwnorton.com/physics

the piston when it oscillates 80 times per minute. Find the maximum velocity.

8. A particle moves in simple harmonic motion according to $x = A\cos(\omega t + \delta)$. At $t = 0$, the particle is at $x = 0$ with initial velocity $v_0 > 0$. What is the phase constant δ?

9. The position of a body can be described by $x = A\cos(\omega t + \delta)$. The angular frequency ω, the initial position x_0, and the initial velocity v_0 are known. Find the amplitude A and the phase constant δ in terms of ω, x_0, and v_0.

10. The central part of a piano string oscillates at 261.7 Hz with an amplitude of 3.0 mm. What is the angular frequency of the motion? The period? What is the maximum velocity? What is the maximum acceleration?

11. In a modern nonlinear dynamics experiment, small beads (spheres) are vibrated on a plate; when the beads start to move, interesting patterns form (see Fig. 15.26). If the plate vibrates at 250 Hz, for what amplitude of motion will the beads start to lift off? (Hint: This will occur when the maximum acceleration of the plate equals $g = 9.81\ \text{m/s}^2$.)

FIGURE 15.26 Oscillating beads.

12. A mass moves in a circle of radius 10 cm, centered on the origin in the x–y plane, with an angular velocity of $\pi/4$ radian/s. At $t = 0$, the mass is on the positive x axis. What are the x components of the position, velocity, and acceleration of the mass at $t = 1.0$ s? At $t = 2.0$ s?

13. A particle executes simple harmonic motion. Its displacement is given by $x = A\cos(\omega t + \delta)$, where as usual, the amplitude A is a positive constant. At $t = 0$, the particle is at the origin and moving in the positive x direction. What is the appropriate choice of the phase constant δ in this case?

*14. Experience shows that from one-third to one-half of the passengers in an airliner can be expected to suffer motion sickness if the airliner bounces up and down with a peak acceleration of 0.4 g and a frequency of about 0.3 Hz. Assume that this up-and-down motion is simple harmonic. What is the amplitude of the motion?

*15. The frequency of a mass attached to a spring is 3.0 Hz. At time $t = 0$, the mass has an initial displacement of 0.20 m and an initial velocity of 4.0 m/s.

(a) What is the position of the mass as a function of time?

(b) When will the mass first reach a turning point? What will be its acceleration at that time?

15.2 The Simple Harmonic Oscillator

16. A man of mass 70 kg is bouncing up and down on a pogo stick (see Fig. 15.27). He finds that if he holds himself rigid and lets the stick do the bouncing (after getting it started), the period of the up-and-down motion is 0.70 s. What is the spring constant of the spring in the pogo stick? Assume that the bottom of the stick remains in touch with the floor and ignore the mass of the stick.

FIGURE 15.27 Man on pogo stick.

17. The cable described in Example 8 in Chapter 14 can be regarded as a spring. What is the effective spring constant of this spring? What is the frequency of oscillation when a mass of 7.1×10^3 kg is attached to the lower end of the cable and allowed to oscillate up and down? Neglect the mass of the cable in your calculation.

18. A simple harmonic oscillator consists of a mass sliding on a frictionless surface under the influence of a force exerted by a spring connected to the mass. The frequency of this harmonic oscillator is 8.0 Hz. If we connect a second, identical spring to the mass, parallel to the first spring, what will be the new frequency of oscillation?

19. The body of an automobile of mass 1100 kg is supported by four vertical springs attached to the axles of the wheels. In order to test the suspension, a man pushes down on the body of the automobile and then suddenly releases it. The body rocks up and down with a period of 0.75 s. What is the spring constant of each of the springs? Assume that all the springs are identical and that the compressional force on each spring

is the same; also assume that the shock absorbers of the automobile are completely worn out so that they do not affect the oscillation frequency.

20. Deuterium (D) is an isotope of hydrogen. The mass of the deuterium atom is 1.998 times larger than the mass of the hydrogen atom. Given that the frequency of vibration of the H_2 molecule is 1.31×10^{14} Hz (see Example 6), calculate the frequency of vibration of the D_2 molecule. Assume the "spring" connecting the atoms is the same in H_2 and D_2.

*21. Calculate the frequency of vibration of the HD molecule consisting of one atom of hydrogen and one of deuterium. See Problem 20 for necessary data. (Hint: The center of mass is stationary.)

22. A mass attached to a spring oscillates with an amplitude of 15 cm; the spring constant is $k = 20$ N/m. When the position is half the maximum value, the mass moves with velocity $v = 25$ cm/s. Determine the period of the motion. Find the value of the mass.

23. A mass of 150 g is attached to a spring of constant $k = 8.0$ N/m and oscillates without friction. The mass is displaced 20 cm from equilibrium and, at $t = 0$, is released from rest. If the position as a function of time is written $x = A\cos(\omega t + \delta)$, determine the values of A, ω, and δ. What is the maximum velocity of the mass? Its maximum acceleration?

24. The equilibrium position of the bottom end of a light, hanging spring shifts downward by 15 cm when a 200-g mass is hung from it. The mass is then displaced an additional 5.0 cm and released. What is the period of motion?

25. A *thickness monitor* is a laboratory instrument used to determine the thickness of a thin film that is deposited on the surface of a quartz crystal. We may treat the crystal as a spring-and-mass system with $k = 6.0 \times 10^5$ N/m and $m = 0.50$ g. What is the frequency of oscillation of this system? This frequency changes slightly as mass is added to the crystal. If the frequency decreases 0.010%, how much mass was deposited? If the area of the crystal is 2.0 cm^2 and the mass density of the film material is 7.5 g/cm^2, how thick was the deposited film?

*26. A thin metal rod is attached to the ceiling and a mass $M = 15$ kg is attached to the bottom of the rod. The rod is 2.0 m long and has a 9.0-mm^2 cross-sectional area. Regard the rod as a (stiff) spring. If the Young's modulus of the rod material is 22×10^{10} N/m^2, what is its spring constant (for small elongations and compressions)? If the mass is displaced vertically, what is its frequency of oscillation (in Hz)? Neglect the mass of the rod.

*27. A mass $m = 2.5$ kg hangs from the ceiling by a spring with $k = 90$ N/m. Initially, the spring is in its unstretched configuration and the mass is held at rest by your hand. If, at time $t = 0$, you release the mass, what will be its position as a function of time?

*28. The wheel of a sports car is suspended below the body of the car by a vertical spring with a spring constant 1.1×10^4 N/m. The mass of the wheel is 14 kg, and the diameter of the wheel is 61 cm.

(a) What is the frequency of up-and-down oscillations of the wheel? Regard the wheel as a mass on one end of a spring, and regard the body of the car as a fixed support for the other end of the spring.

(b) Suppose that the wheel is slightly out of round, having a bump on one side. As the wheel rolls on the street, it receives a periodic push each time the bump comes in contact with the street. At what speed of the translational motion of the car will the frequency of this push coincide with the natural frequency of the up-and-down oscillations of the wheel? What will happen to the car at this speed? (Note: This problem is not quite realistic because the elasticity of the tire also contributes a restoring force to the up-and-down motion of the wheel.)

*29. A mass m slides on a frictionless plane inclined at an angle θ with the horizontal. The mass is attached to a spring, parallel to the plane (Fig. 15.28); the spring constant is k. How much is the spring stretched at equilibrium? What is the frequency of the oscillations of the mass up and down on the plane?

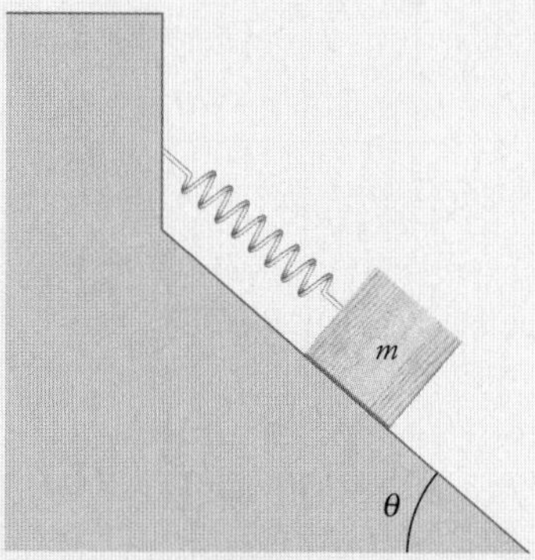

FIGURE 15.28 Mass sliding on a frictionless inclined plane.

**30. Two identical masses slide with one-dimensional motion on a frictionless plane under the influence of three identical springs attached as shown in Fig. 15.29. The magnitude of each mass is m, and the spring constant of each spring is k.

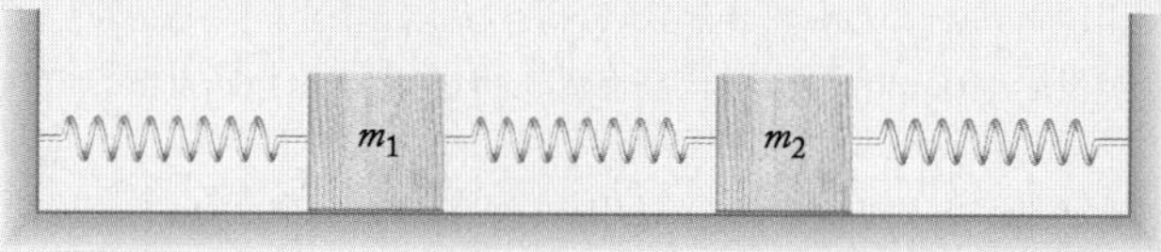

FIGURE 15.29 Two masses sliding on a frictionless plane.

(a) Suppose that at time $t = 0$, the masses are at their equilibrium positions and their instantaneous velocities are $v_1 = -v_2$. Find the position of each mass as a function of time. What is the frequency of the motion?

(b) Suppose that at time $t = 0$, the masses are at their equilibrium positions and their instantaneous velocities are $v_1 = v_2$. Find the position of each mass as a function of time. What is the frequency of the motion?

**31. A cart consists of a body and four wheels on frictionless axles. The body has a mass m. The wheels are uniform disks of mass M and radius R. The cart rolls, without slipping, back and forth on a horizontal plane under the influence of a spring attached to one end of the cart (Fig. 15.30). The spring constant is k. Taking into account the moment of inertia of the wheels, find a formula for the frequency of the back-and-forth motion of the cart.

FIGURE 15.30 A cart attached to a spring.

15.3 Kinetic Energy and Potential Energy

32. Suppose that a particle of mass 0.24 kg acted upon by a spring undergoes simple harmonic motion with the parameters given in Problem 3.
    (a) What is the total energy of this motion?
    (b) At what time is the kinetic energy zero? At what time is the potential energy zero?
    (c) At what time is the kinetic energy equal to the potential energy?
33. A mass of 8.0 kg is attached to a spring and oscillates with an amplitude of 0.25 m and a frequency of 0.60 Hz. What is the energy of the motion?
34. A simple harmonic oscillator consists of a mass of 2.0 kg sliding back and forth along a horizontal frictionless track while pushed and pulled by a spring with $k = 8.0 \times 10^2$ N/m. Suppose that when the mass is at the equilibrium point, it has an instantaneous speed of 3.0 m/s. What is the energy of this harmonic oscillator? What is the amplitude of oscillation?
35. A simple harmonic oscillator of mass 0.60 kg oscillates with a frequency of 3.0 Hz and an amplitude of 0.15 m. Suppose that, while the mass is instantaneously at rest at its turning point, we quickly attach another mass of 0.60 kg to it. How does this change the amplitude of the motion? The frequency? The energy? The maximum speed? The maximum acceleration?

*36. The separation between the equilibrium positions of the two atoms of a hydrogen molecule is 1.0×10^{-10} m. Using the data given in Example 6, calculate the value of the vibrational energy that corresponds to an amplitude of vibration of 0.5×10^{-10} m for each atom. Is it valid to treat the motion as small oscillation if the energy has this value?

37. A 500-g mass is connected to a spring and executes simple harmonic motion. The period of the motion is 1.5 s, and the total mechanical energy of the system is 0.50 J. Find the amplitude of motion.

38. A mass oscillates on a spring. At the points in a cycle when the kinetic energy is one-half of the potential energy, the displacement from equilibrium is 15 cm and the instantaneous velocity is ±25 cm/s. What is the period of the motion?

*39. One end of a horizontal spring of constant k is fixed and the other end is attached to a mass m on a frictionless surface. The spring is initially in its equilibrium position. At $t = 0$, a force F, constant thereafter, is applied in the direction of elongation of the spring. Sometime later, the mass has moved a distance d in the direction of the force. What is the kinetic energy at that time?

*40. A mass of 3.0 kg sliding along a frictionless floor at 2.0 m/s strikes and compresses a spring of constant $k = 300$ N/m. The spring stops the mass. How far does the mass travel while being slowed by the spring? How long does the mass take to stop?

*41. Two masses m_1 and m_2 are joined by a spring of spring constant k. Show that the frequency of vibration of these masses along the line connecting them is

$$\omega = \sqrt{\frac{k(m_1 + m_2)}{m_1 m_2}}$$

(Hint: The center of mass remains at rest.)

*42. Although it is usually a good approximation to neglect the mass of a spring, sometimes this mass must be taken into account. Suppose that a uniform spring has a relaxed length l and a mass m'; a mass m is attached to the end of the spring. The mass m' is uniformly distributed along the spring. Suppose that if the moving end of the spring has a speed v, all other points of the spring have speed directly proportional to their distance from the fixed end; for instance, a point midway between the moving and the fixed end has a speed $\frac{1}{2}v$.
    (a) Show that the kinetic energy in the spring is $\frac{1}{6}m'v^2$ and that the kinetic energy of the mass m and the spring is
    $$K = \tfrac{1}{2}mv^2 + \tfrac{1}{6}m'v^2 = \tfrac{1}{2}(m + \tfrac{1}{3}m')v^2$$
    Consequently, the effective mass of the combination is $m + \frac{1}{3}m'$.
    (b) Show that the frequency of oscillation is $\omega = \sqrt{k/(m + \frac{1}{3}m')}$.
    (c) Suppose that a spring has a mass of 0.05 kg. The frequency of oscillation of a 4.0-kg mass attached to this spring will then be somewhat smaller than calculated for a massless spring. How much smaller? Express your answer as a percentage of the value obtained for a massless spring.

15.4 The Simple Pendulum[†]

43. The longest pendulum in existence is a 27-m Foucault pendulum in Portland, Oregon. What is the period of this pendulum?

[†]For help, see Online Concept Tutorial 17 at www.wwnorton.com/physics

44. At a construction site, a bucket full of concrete hangs from a crane. You observe that the bucket slowly swings back and forth, 8.0 times per minute. What is the length of the cable from which the bucket hangs?

45. The elevator cage of a skyscraper hangs from a 300-m-long steel cable. The elevator cage is guided within the elevator shaft by railings. If we remove these railings and we let the elevator cage swing from side to side (with small amplitude), what is its period of oscillation?

46. On the Earth, a pendulum of length 0.994 m has a period of 2.00 s (compare Example 8). If we take this pendulum to the surface of Jupiter, where $g = 24.8\ \text{m/s}^2$, what will be its period?

47. A mass suspended from a parachute descending at constant velocity can be regarded as a pendulum. What is the frequency of the pendulum oscillations of a human body suspended 7.0 m below a parachute?

48. A "seconds" pendulum is a pendulum that has a period of exactly 2.0 s; each one-way swing of the pendulum therefore takes exactly 1.0 s. What is the length of the seconds pendulum in Paris ($g = 9.809\ \text{m/s}^2$), Buenos Aires ($g = 9.797\ \text{m/s}^2$), and Washington, D.C. ($g = 9.801\ \text{m/s}^2$)?

49. A grandfather clock controlled by a pendulum of length 0.9932 m keeps good time in New York ($g = 9.803\ \text{m/s}^2$).

(a) If we take this clock to Austin, Texas ($g = 9.793\ \text{m/s}^2$), how many minutes per day will it fall behind?

(b) In order to adjust the clock, by how many millimeters must we shorten the pendulum?

50. The pendulum of a grandfather clock has a length of 0.994 m. If the clock runs late by 1.0 minute per day, how much must you shorten the pendulum to make it run on time?

51. A small model of a 10-story construction crane used on a Hollywood movie set should appear realistic in motion. To make it look large, the mass hanging from the crane "cable" (actually, a rod) is constrained to oscillate with a period of 10 s. How long does this make the cable seem?

52. An astronaut lands on an asteroid and sets up a pendulum that has a period of 1.0 s on Earth. She finds that the pendulum has a period of 89 s on the asteroid. What is the local value of the acceleration due to gravity on the asteroid?

53. A circular painting is 2.00 m in diameter and has uniform thickness. It hangs on a wall, suspended by a nail 10 cm from the top edge. If it is pushed slightly, what is the period of small oscillations of the painting?

54. A hula hoop (a thin, uniform toy hoop) of radius 1.0 m hangs over a nail. If it is set to swinging with small amplitude, what is the period of motion?

*55. A **torsional oscillator** consists of a horizontal uniform disk of mass M and radius R attached at its center to the end of a massless vertical fiber. Some such oscillators can execute simple harmonic (twisting) motion with very large amplitudes (amplitudes greater than one rotation are possible). The restoring torque of the fiber is proportional to the angular rotation; that is, $\tau = -\kappa\theta$, where κ is called the torsional constant of the system.

(a) Find the angular frequency of oscillation in terms of M, R, and κ.

(b) If the disk is turned through an initial angle of θ_0 and released, what is the maximum rotational angular velocity of the subsequent motion?

(c) For what value of θ_0 do the answers to (a) and (b) have the same value?

*56. The balance wheel in a clock is a torsional oscillator with a period of 0.50 s. The restoring torque of the wheel spring is $\tau = -\kappa\theta$, where κ is the torsional constant. If the wheel is essentially a hoop, that is, all of its mass $m = 8.0$ g is concentrated at its radius $R = 1.0$ cm, what is the value of κ (in N·m/radian)?

*57. A pendulum hangs from an inclined wall (see Fig. 15.31). Suppose that this pendulum is released at an initial angle of 10° and it bounces off the wall elastically when it reaches an angle of −5°. What is the period of this pendulum?

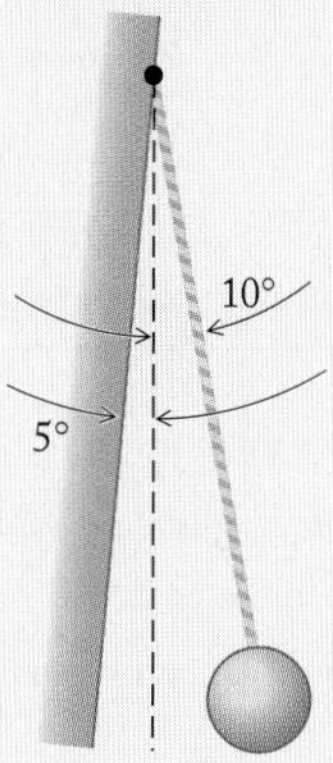

FIGURE 15.31 Pendulum hanging from an inclined wall.

*58. The pendulum of a pendulum clock consists of a rod of length 0.99 m with a bob of mass 0.40 kg. The pendulum bob swings back and forth along an arc of length 20 cm.

(a) What are the maximum velocity and the maximum acceleration of the pendulum bob along the arc?

(b) What is the force that the pendulum exerts on its support when it is at the midpoint of its swing? At the endpoint? Neglect the mass of the rod in your calculations.

*59. The pendulum of a regular clock consists of a mass of 120 g at the end of a (massless) wooden stick of length 44 cm.

(a) What is the total energy (kinetic plus potential) of this pendulum when oscillating with an amplitude of 4°?

(b) What is the speed of the mass when at its lowest point?

60. At the National Institute of Standards and Technology in Gaithersburg, Maryland, the value of the acceleration of gravity is $9.80095\ \text{m/s}^2$. Suppose that at this location a very precise physical pendulum, designed for measurements of the acceleration of gravity, has a period of 2.10356 s. If we take this pendulum to a new location at the U.S. Coast and Geodetic Survey, in nearby Washington, D.C., it has a period of 2.10354 s. What is the value of the acceleration of gravity at this new location? What is the percentage change of the acceleration between the two locations?

61. Consider a meterstick swinging about a pivot through its upper end. What is the period of oscillation of this physical pendulum?

*62. A pendulum consists of a brass rod with a brass cylinder attached to the end (Fig. 15.32). The diameter of the rod is 1.00 cm and its length is 90.00 cm; the diameter of the cylinder is 6.00 cm and its length is 20.00 cm. What is the period of this pendulum?

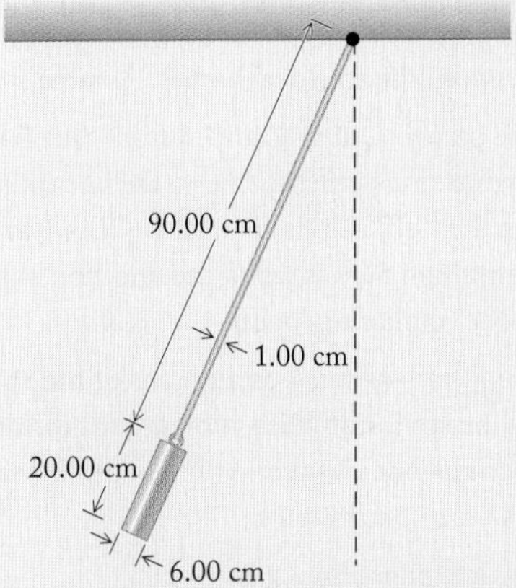

FIGURE 15.32 A physical pendulum.

*63. To test that the acceleration of gravity is the same for a piece of iron and a piece of brass, an experimenter takes a pendulum of length 1.800 m with an iron bob and another pendulum of the same length with a brass bob and starts them swinging in unison. After swinging for 12.00 min, the two pendulums are no more than one-quarter of a (one-way) swing out of step. What is the largest difference between the values of g for iron and brass consistent with these data? Express your answer as a fractional difference.

*64. Calculate the natural period of the swinging motion of a human leg. Treat the leg as a rigid physical pendulum with an axis at the hip joint. Pretend that the mass distribution of the leg can be approximated as two rods joined rigidly end to end. The upper rod (thigh) has a mass of 6.8 kg and a length of 43 cm; the lower rod (shin plus foot) has a mass of 4.1 kg and a length of 46 cm. Using a watch, measure the period of the natural swinging motion of *your* leg when you are standing on one leg and letting the other dangle freely. Alternatively, measure the period of the swinging motion of your leg when you walk at a normal rate (this approximates the natural swinging motion). Compare with the calculated number.

*65. A hole has been drilled through a meterstick at the 30-cm mark and the meterstick has been hung on a wall by a nail passing through this hole. If the meterstick is given a push so that it swings about the nail, what is the period of the motion?

*66. A physical pendulum has the shape of a disk of radius R. The pendulum swings about an axis perpendicular to the plane of the disk at a distance l from the center of the disk.

(a) Show that the frequency of the oscillations of this pendulum is

$$\omega = \sqrt{\frac{gl}{\frac{1}{2}R^2 + l^2}}$$

(b) For what value of l is the frequency a maximum?

*67. A physical pendulum consists of a massless rod of length $2l$ rotating about an axis through its center. A mass m_1 is attached at the lower end of the rod, and a smaller mass m_2 at the upper end (see Fig. 15.33). What is the period of this pendulum?

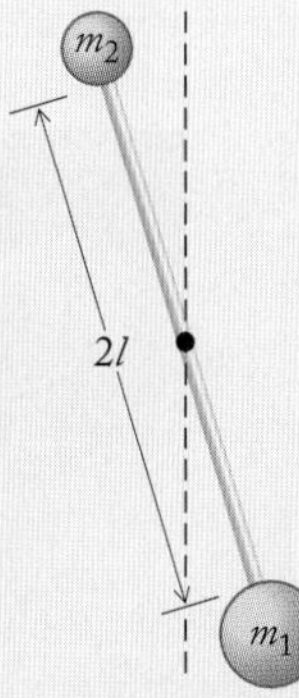

FIGURE 15.33 A physical pendulum with two bobs.

*68. Suppose that a physical pendulum consists of a thin rigid rod of mass m suspended at one end. Suppose that this rod has an initial position $\theta = 20°$ and an initial angular velocity $\omega = 0$. Calculate the force **F** that the support exerts on the pendulum at this initial instant (give horizontal and vertical components).

*69. The door of a house is made of wood of uniform thickness. The door has a mass of 27 kg and measures 1.90 m × 0.91 m. The door is held shut by a torsional spring with κ = 30 N·m/radian arranged so that it exerts a torque of 54 N·m when the door is fully open (at right angles to the wall of the house). What angular speed does the door attain if it slams shut from the fully open position? What linear speed does the edge of the door attain?

**70. Galileo claimed to have verified experimentally that a pendulum oscillating with an amplitude as large as 30° has the same period as a pendulum of identical length oscillating with a much smaller amplitude. Suppose that you let two pendulums of length 1.5 m oscillate for 10 min. Initially, the pendulums oscillate in step. If the amplitude of one of them is 30° and the amplitude of the other is 5°, by what fraction of a (one-way) swing will the pendulums be out of step at the end of the 10-min interval? What can you conclude about Galileo's claim?

**71. A thin vertical rod of steel is clamped at its lower end. When you push the upper end to one side, bending the rod, the upper end moves (approximately) along an arc of circle[2] of radius R and the rod opposes your push with a restoring force $F = -\kappa\theta$, where θ is the angular displacement and κ is a constant. If you attach a mass m to the upper end, what will be the frequency of small oscillations? For what value of m does the rod become unstable; that is, for what value of m is $\omega = 0$? Treat the rod as massless in your calculations. (Hint: Think of the rod as an inverted pendulum of length R, with an extra restoring force $-\kappa\theta$.)

[2] The radius R of the approximating (osculating) circle is somewhat shorter than the length of the rod.

**72. According to a proposal described in Problem 83 of Chapter 1, very fast trains could travel from one city to another in straight subterranean tunnels (see Fig. 15.34). For the following calculations, assume that the density of the Earth is constant, so the acceleration of gravity as a function of the radial distance r from the center of the Earth is $g = (GM/R^3)r$.

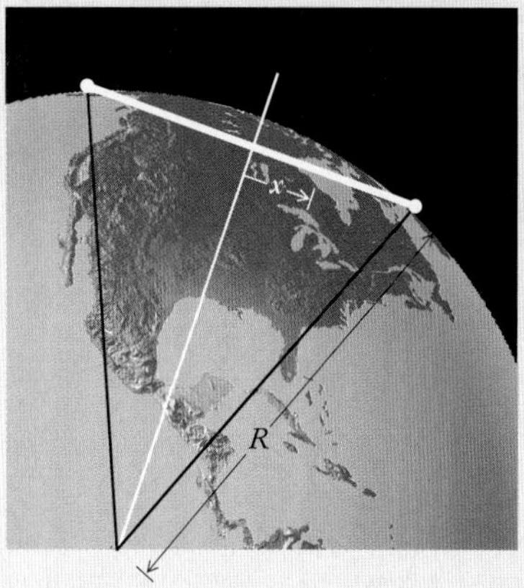

FIGURE 15.34 A straight tunnel connecting two points on the surface of the Earth.

(a) Show that the component of the acceleration of gravity along the track of the train is

$$g_x = -(GM/R^3)x$$

where x is measured from the midpoint of the track (see Fig. 15.34).

(b) Neglecting friction, show that the motion of the train along the track is simple harmonic motion with a period independent of the length of the track,

$$T = 2\pi\sqrt{\frac{R^3}{GM}}$$

(c) Starting from rest, how long would a train take to roll freely along its track from San Francisco to Washington, D.C.? What would be its maximum speed (at the midpoint)? Use the numbers you calculated in Problem 83 of Chapter 1 for the length and depth of the track.

**73. A physical pendulum consists of a long, thin cone suspended at its apex (Fig. 15.35). The height of the cone is l. What is the period of this pendulum?

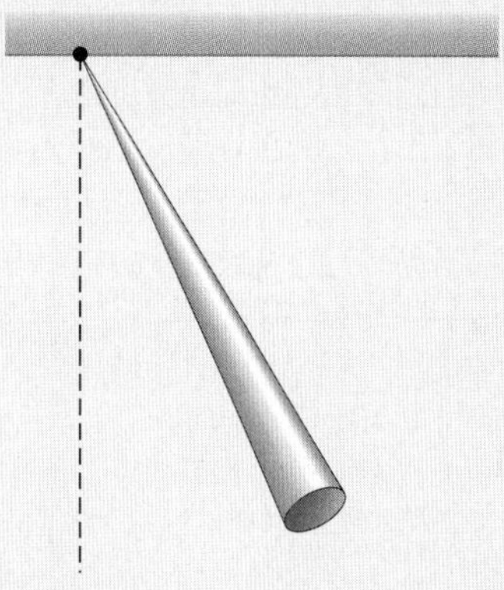

FIGURE 15.35
A long, thin cone.

**74. The net gravitational force on a particle placed midway between two equal spherical bodies is zero. However, if the particle is placed some distance away from this equilibrium point, then the gravitational force is not zero.

(a) Show that if the particle is at a distance x from the equilibrium point in a direction toward one of the bodies, then the force is approximately $4GMmx/r^3$, where M is the mass of each spherical body, m is the mass of the particle, and $2r$ is the distance between the spherical bodies. Assume $x \ll r$.

(b) Show that if the particle is at a distance x from the equilibrium point in a direction perpendicular to the line connecting the bodies, then the force is approximately $-2GMmx/r^3$, where the negative sign indicates that the direction of the force is toward the equilibrium point.

(c) What is the frequency of small oscillations of the mass m about the equilibrium point when moving in a direction perpendicular to the line connecting the bodies? Assume that the bodies remain stationary.

**75. The motion of a simple pendulum is given by

$$\theta = A\cos\left(\sqrt{\frac{g}{l}}\,t\right)$$

(a) Find the tension in the string of this pendulum; assume that $\theta \ll 1$. The mass of the suspended particle is m.

(b) The tension is a function of time. At what time is the tension maximum? What is the value of this maximum tension?

15.5 Damped Oscillations and Forced Oscillations

76. Roughly, what is the frequency of stomping of soldiers on the march? What must have been the resonant frequency of the bridge at Angers that broke when soldiers marched across it?

77. A pendulum of length 1.50 m is set swinging with an initial amplitude of 10°. After 12 min, friction has reduced the amplitude to 4°. What is the value of Q for this pendulum?

78. The pendulum of a grandfather clock has a length of 0.994 m and a mass of 1.2 kg.

(a) If the pendulum is set swinging, the friction of the air reduces its amplitude of oscillation by a factor of 2 in 13.0 min. What is the value of Q for this pendulum?

(b) If we want to keep this pendulum swinging at a constant amplitude of 8°, we must supply mechanical energy to it at a rate sufficient to make up for the frictional loss. What is the required mechanical power?

79. When a swing in motion is not being "pumped," the angular amplitude of oscillation decreases because of air and other friction. The motion of a 3.0-m-long swing decreases in amplitude from 12° to 10° after 5 complete cycles. What is the Q of the system? If the rider and seat are treated as a point mass with $m = 25$ kg, at what average rate is mechanical energy being dissipated?

80. A horizontal spring of constant k is attached to a mass m that slides on a slightly frictional floor. After the mass is displaced a distance A from equilibrium and released, the amplitude of oscillation decreases to $0.95A$ after 10 cycles. What is the Q of this system?

81. A harmonic force $F = F_0 \cos \omega t$, where $F_0 = 0.20$ N, is applied to a damped harmonic oscillator of spring constant $k = 15$ N/m and mass m, where $\omega = \sqrt{k/m}$. The amplitude of oscillation increases rapidly at first, and then settles to a constant value, $A = 40$ cm. What is the Q of the system? What would the amplitude be if the angular frequency of the force F had been much less than $\sqrt{k/m}$?

82. A microelectromechanical system (MEMS) consists of a microscopic silicon mechanical oscillator (see Fig. 15.36) with a spring constant $k = 5 \times 10^{-3}$ N/m. When it oscillates in a vacuum-sealed device (to remove air friction), the Q of such an oscillator is large: $Q = 5 \times 10^6$. What amplitude of motion will the oscillator attain if an oscillating force of amplitude 1×10^{-18} N (near the current limits of force detection) is applied?

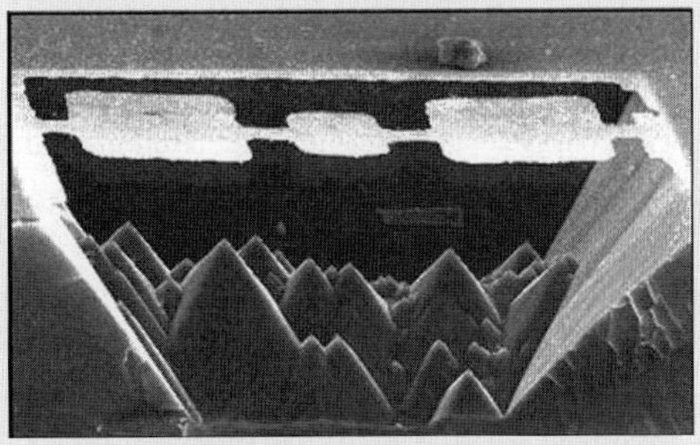

FIGURE 15.36 A microelectromechanical system (MEMS) oscillator, the silicon membrane structure suspended above the faceted silicon trench.

83. Using electron-beam lithography, engineers are attempting to fabricate nanoelectromechanical system (NEMS) oscillators with frequencies as high as 100 GHz (for communications and higher-speed computing). If the equivalent mass of such an oscillator is 1.0×10^{-18} g and a minimum amplitude of 0.10 nm is needed to detect an applied harmonic force of amplitude 1.0×10^{-10} N, what must the minimum Q of such an oscillator be?

*84. Consider the motion of the damped harmonic oscillator plotted in Fig. 15.21.

(a) According to this plot, what fraction of its amplitude does the oscillator lose in its first oscillation?

(b) What fraction of its energy does the oscillator lose in its first oscillation?

(c) According to Eq. (15.51), what is the value of Q for this oscillator?

*85. If you stand on one leg and let the other dangle freely back and forth starting at an initial amplitude of, say, 20° or 30°, the amplitude will decay to one-half of the initial amplitude after about four swings. Regarding the dangling leg as a damped oscillator, what value of Q can you deduce from this?

REVIEW PROBLEMS

86. A particle performs simple harmonic motion along the x axis with an amplitude of 0.20 m and a period of 0.80 s. At $t = 0$, the particle is at maximum distance from the origin; that is, $x = 0.20$ m.

(a) What is the equation that describes the position of the particle as a function of time?

(b) Calculate the position of the particle at $t = 0.10$ s, 0.20 s, 0.30 s, and 0.40 s.

87. In an electric saber saw, the rotational motion of the electric motor is converted into a back-and-forth motion of the saw blade by a mechanism similar to that shown in Fig. 15.5. Suppose the peg of the rotating wheel moves around a circle of diameter 3.0 cm at 4000 rev/min and thereby moves the slotted arm to which the saw blade is bolted. What are the amplitude and the frequency of the back-and-forth simple harmonic motion of the blade?

88. In response to a sound wave, the middle of your eardrum oscillates back and forth with a frequency of 4000 Hz and an amplitude of 1.0×10^{-5} m. What is the maximum speed of the eardrum?

89. Suppose that two particles are performing simple harmonic motion along the x axis with a period of 8.0 s. The first particle moves according to the equation

$$x = 0.30 \cos\left(\frac{\pi t}{4}\right)$$

and the second particle according to the equation

$$x' = 0.30 \sin\left(\frac{\pi t}{4}\right)$$

where the distance is measured in meters and the time in seconds.

(a) When does the first particle reach the midpoint? The turning point? Draw a diagram showing the particle and its satellite particle at these times.

(b) When does the second particle reach the midpoint? The turning point? Draw a diagram showing the particle and its satellite particle at these times.

(c) By some argument, establish that whenever the first particle passes through a point on the x axis, the second particle passes through this same point 2.0 s later.

90. A particle of 6.0 kg is executing simple harmonic motion along the x axis under the influence of a spring. The particle moves according to the equation

$$x = 0.20 \cos(3.0t)$$

where x is measured in meters and t in seconds.

(a) What is the frequency of the motion? What is the spring constant of the spring? What is the maximum speed of the motion?

(b) Suppose we replace the particle by a new particle of 2.0 kg (but we keep the same spring), and suppose we start the motion with the same amplitude of 0.20 m. What will be the new frequency of the motion? What will be the new maximum speed?

91. The motion of the piston in an automobile engine is approximately simple harmonic. Suppose that the piston travels back and forth over a distance of 8.50 cm and has a mass of 1.2 kg. What are its maximum acceleration and maximum speed if the engine is turning at its highest safe rate of 6000 rev/min? What is the maximum force on the piston?

92. A Small Mass Measurement Instrument (SMMI) was used in Skylab to measure the masses of biological samples, small animals, chemicals, and other such items used in life-sciences experiments while in orbit (see Fig. 15.37). The sample to be measured is strapped to a tray supported by leaf springs, and the mass is determined from the observed period of oscillation of the tray-and-mass. To calibrate this instrument, a test mass of 1.00 kg is first placed on the tray; the period of oscillation is then 1.08 s. Suppose that when the test mass is removed and an unknown sample is placed on the tray, the period becomes 1.78 s. What is the mass of the sample? Assume that the mass of the tray (and the straps) is 0.400 kg.

FIGURE 15.37 Small Mass Measurement Instrument.

93. A simple harmonic oscillator has a frequency of 1.5 Hz. What will happen to the frequency if we cut the spring in half and attach both halves to the mass so that both springs push jointly?

94. A physicist of 55 kg stands on a bathroom scale (a spring scale, with an internal spring). She observes that when she mounts the scale suddenly, the pointer of the scale first oscillates back and forth a few times with a frequency of 2.4 Hz.

(a) What value of the spring constant can she deduce from these data?

(b) If she then takes a child of 20 kg in her arms and again stands on the scale, what will be the new frequency of oscillation of the pointer?

95. Ropes used by mountain climbers are quite elastic, and they behave like springs. A rope of 10 m has a spring constant $k = 4.9 \times 10^3$ N/m. Suppose that a mountain climber of 80 kg hangs on this rope, which is stretched vertically down. What is the frequency of up-and-down oscillations of the mountain climber?

96. Consider a particle of mass m moving along the x axis under the influence of a spring of spring constant k. The equilibrium point is at $x = 0$, and the amplitude of the motion is A.

(a) At what point x is the kinetic energy of the particle equal to its potential energy?

(b) When the particle reaches the point $x = \frac{1}{2}A$, what fraction of its energy is potential, and what fraction is kinetic?

97. A simple harmonic oscillator consists of a mass of 3.0 kg sliding back and forth along a horizontal frictionless track while pushed and pulled by a spring with $k = 6.0 \times 10^2$ N/m. Suppose that initially the mass is released from rest at a distance of 0.25 m from the equilibrium point. What is the energy of this harmonic oscillator? What is the maximum speed it attains when passing through the equilibrium point?

98. A simple harmonic oscillator of mass 0.80 kg oscillates with a frequency of 2.0 Hz and an amplitude of 0.12 m. Suppose that, while the mass is instantaneously at rest at its turning point, we quickly shift the fixed end of the spring to a new fixed position, 0.12 m farther away from the mass. How does this change the amplitude of the motion? The frequency? The energy? The maximum speed? The maximum acceleration?

99. A pendulum has a length of 1.5 m. What is the period of this pendulum? If you wanted to construct a pendulum with exactly half this period, how long would it have to be?

100. An "interrupted" pendulum consists of a simple pendulum of length l that encounters a nail placed at a distance $\frac{3}{4}l$ below the point of support. If this pendulum is released from one side, it will begin to wrap around the nail as soon as it passes through the vertical position (Fig. 15.38). What is the period of this pendulum?

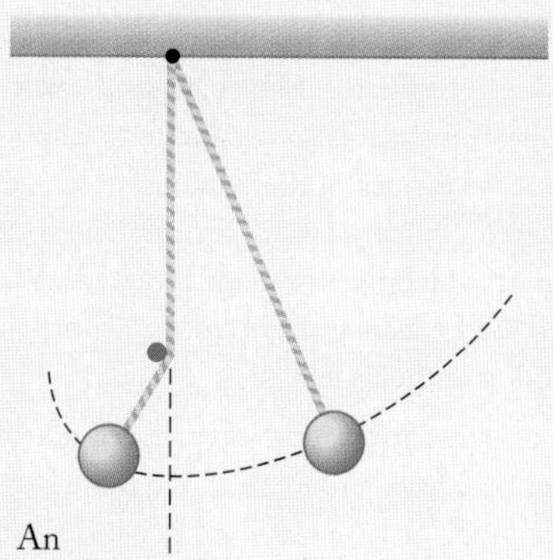

FIGURE 15.38 An "interrupted" pendulum.

101. A physical pendulum consists of a uniform spherical bob of mass M and radius R suspended from a massless string of length L (see Fig. 15.39). Taking into account the size of the bob, show that the period of small oscillations of this pendulum is

$$T = 2\pi\sqrt{\frac{\frac{2}{5}R^2 + (R + L)^2}{g(R + L)}}$$

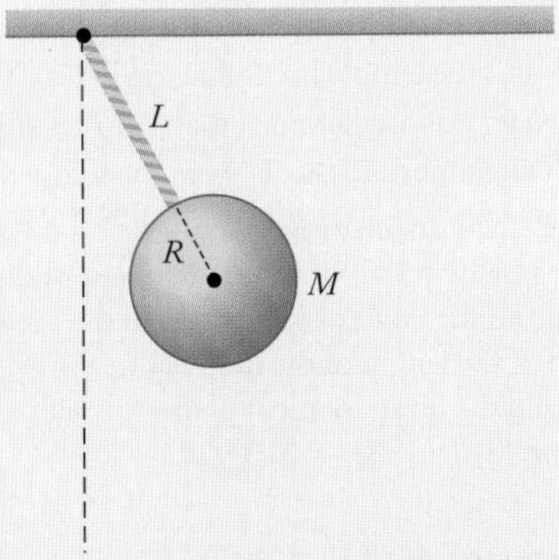

FIGURE 15.39 A physical pendulum with a large bob.

102. A uniform rod of length L is swinging about a pivot at a distance x from its center (see Fig. 15.40). Find the period of oscillation of this physical pendulum as a function of x. For what choice of x is the period shortest?

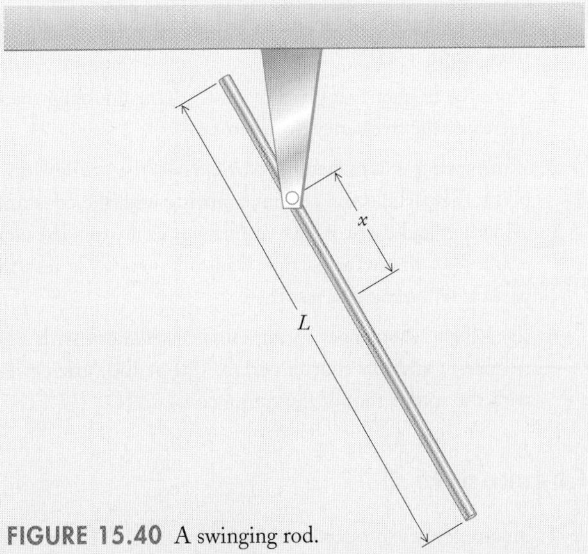

FIGURE 15.40 A swinging rod.

103. A swing of length 2.0 m hangs from a horizontal branch of a tree. With what frequency should you rock the branch to build up oscillations of the pendulum by resonance?

Answers to Checkups

Checkup 15.1

1. The Earth's rotational motion is periodic; it repeats with each daily cycle. It is not a back-and-forth motion along a line or arc, so it is not an oscillation.

2. The velocity attains its maximum magnitude at $x = 0$, that is, where the displacement is zero; the velocity attains its minimum magnitude at $x = \pm A$, that is, at the points of maximum displacement. This is because the displacement and velocity are 90° out of phase; if one is a cosine function, the other is a sine function [see Eqs. (15.11) and (15.12)].

3. The acceleration attains its maximum magnitude at $x = \pm A$, that is, at the point of maximum displacement from the origin; the acceleration attains its minimum magnitude at $x = 0$, that is, where the displacement is zero. This is because the displacement and acceleration are 180° out of phase; if one is a cosine function, the other is a negative cosine function [see Eqs. (15.11) and (15.13)].

4. If the maximum displacement is A, the maximum velocity is ωA [compare Eqs. (15.11) and (15.12)]. Thus, for the same amplitude, the particle with twice the frequency has twice the maximum velocity. Similarly, the maximum acceleration is $\omega^2 A$, so the particle with twice the frequency has 4 times the maximum acceleration.

5. As described in Section 15.1, the x coordinates of the particle and satellite are identical. Obviously, the y coordinates are not, since the particle is always at $y = 0$, while the satellite executes circular motion. The velocities are not the same, since the particle has zero y velocity, unlike the satellite. The x components of the velocities and accelerations are the same, since they are derived from the identical time dependence of the x coordinate. Since the particle is always at $y = 0$, the y components of the velocity and acceleration are not the same.

6. (C) $\pi/2$. If we insert $x = 0$ and $t = 0$ in $x = A\cos(\omega t + \delta)$ [Eq. (15.4)], then we see that $0 = \cos\delta$, which is true if $\delta = \pi/2$ or if $\delta = -\pi/2$. Of these two, only $\delta = \pi/2$ is listed.

Checkup 15.2

1. The force on the particle attains maximum magnitude at the extreme displacements (the turning points) of the motion, $x = \pm A$. The force on the particle attains the minimum magnitude of zero when the particle passes through the equilibrium point, $x = 0$.
2. Since the frequency is given by $\omega = \sqrt{k/m}$, doubling the mass decreases the frequency by a factor of $1/\sqrt{2}$.
3. If the spring is cut when the particle is at the equilibrium point, the particle will continue moving with the constant velocity it had there, $v_{max} = \omega A$. If we cut it when the particle is at $x = A$, where the particle is instantaneously at rest, the particle will remain at rest there.
4. (B) $1/\sqrt{2}$. A stronger spring causes oscillations with a higher frequency, and so a shorter period. The period varies inversely with the square root of the spring constant [Eq. (15.21)].

Checkup 15.3

1. Since energy is proportional to the square of the amplitude [Eq. (15.32)], the oscillator with twice the amplitude has 4 times the energy. Since the maximum speed is proportional to the amplitude, $v_{max} = \omega A$ [Eq. (15.34)], the oscillator with twice the amplitude also has twice the maximum speed.
2. Both oscillators have the same energy, since $E = \frac{1}{2}kA^2$. But the maximum speed is inversely proportional to the square root of the mass [Eq. (15.35)], so the particle with twice the mass has a smaller maximum speed by a factor of $1/\sqrt{2}$.
3. The energy is purely kinetic when the oscillator passes through equilibrium. The energy will be purely potential at maximum amplitude, which is one-quarter of a cycle later, or 2.0 s later. An oscillator passes through equilibrium twice each cycle (once in each direction), so the energy will be purely kinetic 4.0 s after the initial time, or another 2.0 s after the energy is purely potential.
4. Friction removes energy from the system, so the energy will decrease whenever the particle is moving, and will not remain constant. Since $E = \frac{1}{2}kA^2$, the amplitude A will also decrease each cycle due to friction.
5. (E) 32. The stored energy is $E = \frac{1}{2}kA^2$. But from Eq. (15.18), $k = m\omega^2$, so $E = \frac{1}{2}m\omega^2A^2$; if each of m, ω, and A increases by a factor of 2, then the energy increases by a factor of $2^5 = 32$.

Checkup 15.4

1. The period of the simple pendulum is proportional to the square root of the length ($T = 2\pi\sqrt{l/g}$), so the period of the shorter pendulum will be decreased by a factor of $1/\sqrt{2}$. The frequency is the inverse of the period ($f = 1/T$), and so will increase by $\sqrt{2}$.
2. Two pendulums of the same length have the same angular frequency of oscillation, since $\omega = \sqrt{g/l}$. But the energy of a pendulum is $\frac{1}{2}m\omega^2A^2$ (this formula is equally valid for the pendulum and the simple harmonic oscillater). Thus to have the same energy of oscillation, a mass 3 times smaller must move with an amplitude that is $\sqrt{3}$ times larger.
3. (B) $\sqrt{3g/2l}$. The angular frequency of such a physical pendulum is given by Eq. (15.48), $\omega = \sqrt{mgd/I}$. The distance d is measured from the point of suspension to the center of mass and thus is half of the length of the rod; that is, $d = l/2$. Inserting this value and the given moment of inertia yields $v = \sqrt{(mgl/2)/(ml^2/3)} = \sqrt{3g/2l}$.

Checkup 15.5

1. Yes, at least some slight buildup always occurs at frequencies below resonance. Figure 15.22 indicates that forced oscillations far below the resonant frequency approach an amplitude $A \approx F_0/k$, the magnitude of the static-force spring displacement. This occurs because a slowly varying force allows the (faster) mass–spring system to follow the force over time.
2. No. In this case the slowly responding oscillator cannot follow the oscillating force; it is as if the response is averaged nearly equally over the positive and negative force contributions.
3. Your hand provides friction; you remove energy from the bell. When ΔE, the energy lost per cycle, increases, the Q must decrease.
4. (E) 630. We can solve Eq. (15.51) for Q and obtain $Q = 2\pi E/\Delta E$. Since 0.10 J is lost in 10 cycles, about 0.010 J is lost each cycle. Thus $Q = 2\pi(1.0\text{ J})/(0.010\text{ J}) = 200\pi \approx 630$.

Waves

CHAPTER

16

CONCEPTS IN CONTEXT

At water parks, engineers have developed wave pools like this one, where water waves are generated by large alternating pumps at the deep end and propagate toward the shallow end. Design considerations include the following questions:

? How are the frequency and spacing of the waves related? (Example 2, page 511)

? What vertical acceleration does a swimmer feel as a wave passes by? (Example 3, page 513)

? The speed of a wave changes as it approaches the shallow end of the pool. Does its frequency change? Does its wavelength change? (Example 5, page 515)

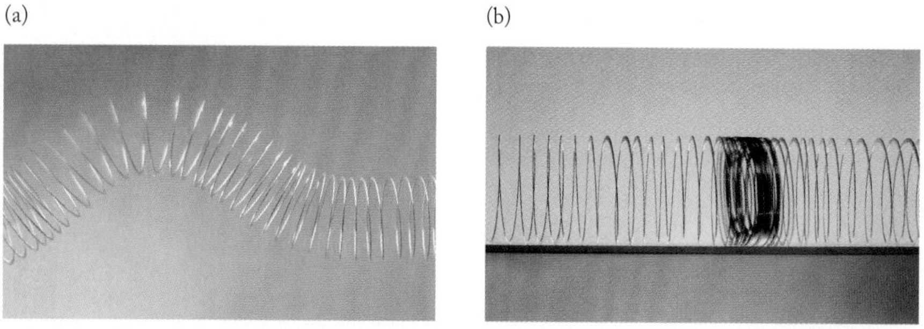

FIGURE 16.1 (a) A kink (a transverse deformation) traveling along a spring. (b) A compression (a longitudinal deformation) traveling along a spring.

A **wave** *is a vibrational, shaking motion in an elastic, deformable body.* The wave is initiated by some external force that acts on some part of the body and deforms it. The elastic restoring forces within the body communicate this initial disturbance from one part of the body to the next, adjacent part. The disturbance therefore gradually propagates along the elastic body. For instance, Fig. 16.1 shows a simple example of wave motion: a long spring, such as a Slinky, has been disturbed by a sudden up-and-down motion or a back-and-forth motion, which produced a kink (Fig. 16.1a) or a compressional deformation (Fig. 16.1b). The disturbance propagates along the spring as a wave pulse.

The elastic body in which the wave propagates is called the **medium.** Thus, a spring is the medium for the deformational waves illustrated in Figs. 16.1a and b, a stretched string is the medium for similar deformational "string" waves, water is the medium for water waves, air is the medium for sound waves, the crust of the Earth is the medium for seismic waves, and so on. When a wave propagates through a medium, *the particles in the medium vibrate back and forth, but the medium as a whole does not perform translational motion.* This is obvious in the case of a wave propagating on the spring, where we can see the spring vibrate and we know that the spring cannot travel anywhere, since it is held fixed at its ends. It is also obvious for a wave propagating on a stretched string, again held fixed at its ends. But the lack of motion of the medium as a whole is not so obvious for water waves—when watching ocean waves, we often gain the impression that the water travels with the wave, especially when the waves are large and when we see them crashing against a seawall or some other obstruction. But we can check that the water does not flow with the wave if we watch a chip of wood or some other flotsam on the water. Such a chip of wood only bobs up and down, and it rocks back and forth; it does not travel forward with the wave.

For the sake of simplicity, in this chapter we will concentrate on the motion of transverse waves on a stretched string. However, most of our mathematical results also apply to wave motion in other elastic bodies. In the next chapter we will examine some features of wave motion in air, that is, sound waves.

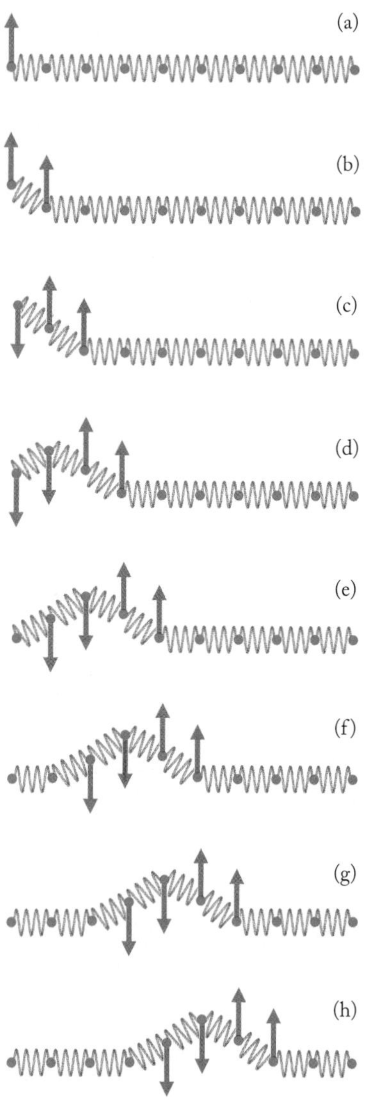

FIGURE 16.2 Particles joined by springs. A transverse disturbance propagates from left to right. The diagrams show snapshots at successive times. The particles move up and down.

16.1 TRANSVERSE AND LONGITUDINAL WAVE MOTION

To gain some qualitative understanding of the mechanism of wave motion, consider a tightly stretched elastic string, such as a long rubber cord. The elastic string may be regarded as a row of particles connected by small, massless springs. If we shake one end of the string up and down with a flick of the wrist, a disturbance travels along the row of particles. Figure 16.2 shows in detail how such a traveling disturbance comes about. Initially, the particles are at their equilibrium positions, evenly spaced along the string. When we jerk the first particle upward, it will pull the second particle upward, and this will pull the third, and so on. If we then jerk the first particle back to its original position, it will pull the second particle back, and this will likewise pull the third, and so on. As the motion is transmitted from one particle to the next particle, the disturbance propagates along the row of particles. Such a disturbance, in which the particles move at right angles to the direction of propagation of the disturbance, is called a **transverse** wave pulse.

Alternatively, we can generate a disturbance by suddenly pushing the first particle toward the second and, soon after, pulling it back. Figure 16.3 shows how such a compressional disturbance propagates along the row of particles. This kind of disturbance, in which the particles move back and forth along the direction of propagation of the disturbance, is called a **longitudinal** wave pulse.

Note that although the wave pulse travels along the full length of the string, the particles do not—they merely move back and forth around their equilibrium positions. Also note that in the region of the wave pulse, the string has kinetic energy (due to the back-and-forth motion of the particles) and potential energy (due to the deformation of the springs between the particles). Hence, a wave pulse traveling along the string carries energy with it—the wave transports energy from one end of the string to the other.

Wave motion in air, water, or any other medium displays the same general features. The waves are propagating disturbances in the medium communicated by pushes and pulls from one particle to the next. The waves transport energy without transporting particles. The waves are longitudinal, transverse, or both. A sound wave in air is longitudinal; the air molecules move forward and backward, parallel to the direction of propagation of the wave. But a wave on the surface of the ocean is both longitudinal and transverse; the water molecules move up and down and, simultaneously, forward and backward—the net result of these simultaneous motions is that each water molecule traces out an elliptical path. Seismic waves in the body of the Earth can be either longitudinal (P waves) or transverse (S waves). These two kinds of seismic waves have different speeds, and their relative intensities depend on the characteristics of the earthquake that generated them.

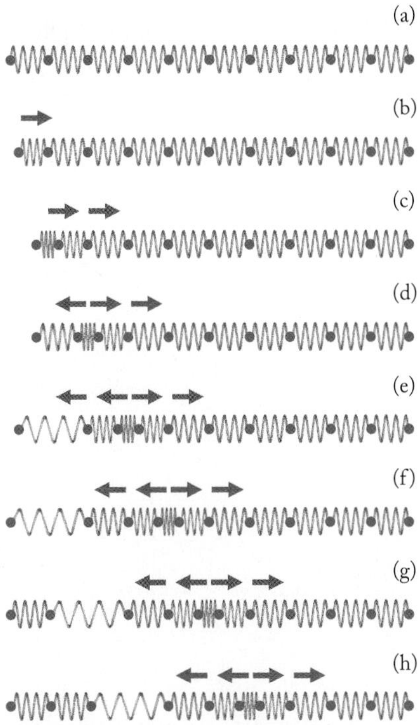

FIGURE 16.3 A longitudinal disturbance propagates from left to right. The particles move back and forth.

Checkup 16.1

QUESTION 1: You shake a baseball bat back and forth. Does this produce a wave motion in the bat? You shake a bowl of Jell-O. Does this produce a wave motion in the Jell-O?

QUESTION 2: You stretch a rubber cord from the porch of a house to a tree in the garden. How can you initiate a transverse wave in this cord? A longitudinal wave?

QUESTION 3: When a guitar player plucks the string of her guitar, does she produce a transverse wave or a longitudinal wave?

QUESTION 4: You have a long rod of steel and a hammer. How must you hit the end of the rod to generate a longitudinal wave along the rod? A transverse wave?

QUESTION 5: An ocean wave travels from the coast of Africa to Florida. Does this wave carry water from Africa to Florida? Does it carry energy?

(A) Yes; yes (B) Yes; no (C) No; no (D) No; yes

16.2 PERIODIC WAVES

periodic wave

wave crest and wave trough

If we shake the end of a long string up and down and we continue shaking it steadily, we will generate a **periodic wave** on the string. Such a wave can be regarded as consisting of a steady succession of positive (upward) and negative (downward) wave pulses, which repeat at regular intervals. Figure 16.4a shows a periodic wave at one instant of time. The high points of the wave are called the **wave crests**, and the low points are called the **wave troughs**. *The distance from one crest to the next or from one*

wavelength

trough to the next is called the **wavelength**, designated by the symbol λ (the Greek letter *lambda*). The wavelength is the repeat distance of the wave pattern—a shift of the wave pattern by one wavelength to the right (or the left) reproduces the original wave pattern.

period

With the passing of time, the wave crests and wave troughs travel toward the right at a speed v. As the wave travels, the entire wave pattern shifts toward the right; that is, the wave pattern (but not the string) performs a rigid translational motion. Figures 16.4b–h show the wave at successive instants of time. These pictures span one **period** of the wave; that is, they span the interval T of time required for the wave pattern to travel exactly one wavelength to the right. The period is the repeat time of the wave pattern—after one period, each wave crest or wave trough will have traveled to the position previously occupied by the adjacent wave crest or wave trough, and the wave will have attained exactly the same configuration as it had at the initial time.

Since in one period, the wave travels a distance equal to one wavelength, the ratio of wavelength to period must equal the **wave speed**,

wave speed

$$\frac{\lambda}{T} = v \tag{16.1}$$

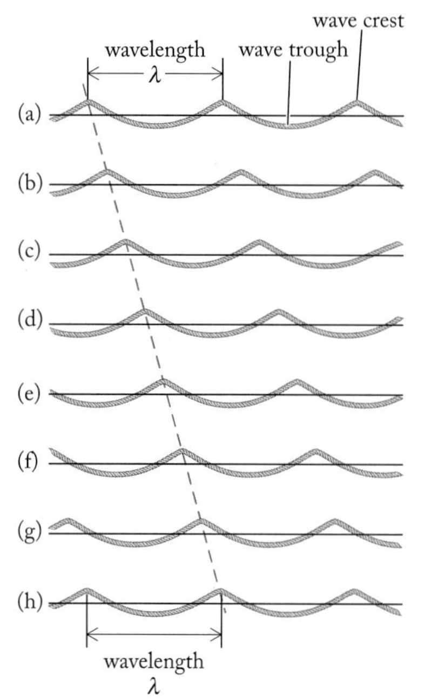

FIGURE 16.4 A periodic wave traveling to the right. The diagrams show snapshots of the wave at successive instants of time. The wave pattern (h) coincides with the wave pattern (a) because the wave has moved exactly one wavelength to the right.

As in the case of simple harmonic motion, we define the frequency of the wave as the inverse of the period:

$$f = \frac{1}{T} \tag{16.2}$$

The frequency of the wave is simply the number of wave crests arriving at some point on the string per second. The unit for the frequency is cycles per second, or Hz (hertz). For example, if the period of the wave is 0.1 s, then in one second there will be 10 wave crests arriving at some point on the string, and consequently the frequency of the wave is 10 cycles per second, or 10 Hz.

In terms of the frequency, Eq. (16.1) becomes

$$\lambda f = v \tag{16.3}$$

This equation permits us to calculate the frequency from the wavelength, or the wavelength from the frequency, provided we know the speed of the wave.

EXAMPLE 1 A long clothesline is stretched horizontally between two trees. While shaking this clothesline up and down near one end at the rate of 4.0 cycles per second, you observe that the wavelength of the waves you generate is 1.0 m. What is the speed of these waves? If the distant end of the clothesline is 10 m away, how long does a wave pulse take to return to you?

SOLUTION: From Eq. (16.3),

$$v = \lambda f = 1.0 \text{ m} \times 4.0 \text{ Hz} = 4.0 \text{ m/s}$$

To return to you, the wave pulse has to complete a round-trip distance of $d = 20$ m. Hence, the time required is

$$t = \frac{d}{v} = \frac{20 \text{ m}}{4.0 \text{ m/s}} = 5.0 \text{ s}$$

Concepts in Context

EXAMPLE 2 At the deep end of a wave pool (see the chapter photo), the speed of water waves is 5.2 m/s. In order to avoid an excessively rough ride, the frequency of the waves is kept low, at 0.40 Hz. What is the period of such waves? What is the distance beween wave crests?

SOLUTION: The period is the inverse of the frequency [Eq. (16.2)]:

$$T = \frac{1}{f} = \frac{1}{0.40 \text{ Hz}} = 2.5 \text{ s}$$

The distance between crests is the wavelength; from Eq. (16.1), we know the wavelength is the product of the wave speed and the period:

$$\lambda = vT = 5.2 \text{ m/s} \times 2.5 \text{ s} = 13 \text{ m}$$

harmonic wave

An important special case of a periodic wave is a **harmonic wave**. *This kind of wave has the shape of a harmonic function, that is, a sine curve or a cosine curve.* If we assume that a wave crest is at the origin at the initial time $t = 0$, the wavefunction is

$$y = A \cos kx \qquad \text{for the initial time } t = 0 \tag{16.4}$$

wave amplitude

The constant A, which represents the height of the wave crests (and the depth of the wave troughs), is called the **wave amplitude**, and the constant k is called the **wave number**. Note that here, as in the preceding chapter, the argument of the cosine function is supposed to be expressed in radians.

Figure 16.5 is a plot of the wavefunction (16.4). The wave crests (maxima) occur where $\cos kx = 1$, that is, at

$$kx = 0,\ 2\pi,\ 4\pi,\ 6\pi, \text{ etc.} \qquad \text{(maxima)} \tag{16.5}$$

and the wave troughs (minima) occur where $\cos kx = -1$, or at

$$kx = \pi,\ 3\pi,\ 5\pi, \text{ etc.} \qquad \text{(minima)} \tag{16.6}$$

From these equations we see that the distance from one crest to the next, or from one trough to the next, is $2\pi/k$. Thus the wavelength λ and the wave number k of a harmonic wave are related by

wavelength and wave number

$$\lambda = \frac{2\pi}{k} \qquad \text{or} \qquad k = \frac{2\pi}{\lambda} \tag{16.7}$$

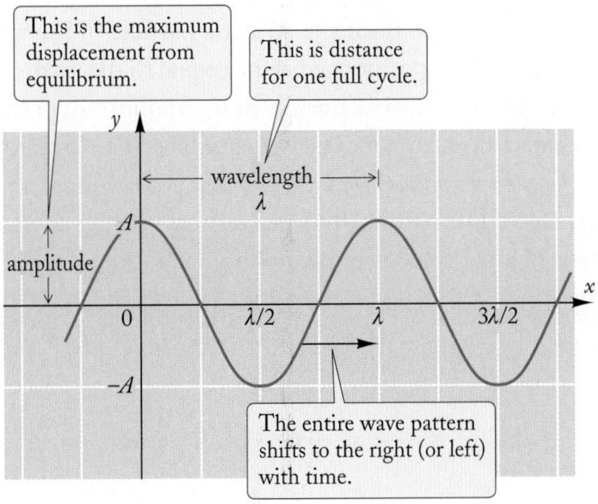

FIGURE 16.5 A harmonic wave.

Since the cosine function is periodic in intervals of 2π, we see that the relation $k = 2\pi/\lambda$ ensures that the value of the cosine function repeats whenever the position x in Eq. (16.4) changes by a distance equal to the wavelength λ.

At any later time, the harmonic wave will have traveled some distance to the right or to the left. This means that the wave pattern plotted in Fig. 16.5 shifts some distance to the right or to the left. If the speed of the wave is v, the wave pattern shifts a distance vt in a time t. The initial wavefunction (16.4) must then be replaced by a new wavefunction in which the value of the argument is shifted by a distance vt. Thus, we must replace x in Eq. (16.4) by $x - vt$ or by $x + vt$, for a wave that travels to the right or to the left, respectively. The wavefunction at time t is then

harmonic traveling wave

$$y = A\cos[k(x - vt)] \quad \text{for a wave traveling in the positive } x \text{ direction} \tag{16.8}$$

or

$$y = A\cos[k(x + vt)] \quad \text{for a wave traveling in the negative } x \text{ direction} \tag{16.9}$$

Note that according to Eq. (16.4) there is a wave crest at $kx = 0$, and that according to Eq. (16.8) the corresponding wave crest is at $k(x - vt) = 0$. Hence, at time t, this wave crest is at $x - vt = 0$, or at $x = vt$, as expected for a wave traveling in the positive x direction. In consequence of the *negative sign* in Eq. (16.8), as the time t increases, x must also increase to stay on the crest of the wave; the wave thus travels in the *positive* x direction.

For a harmonic wave it is customary to introduce the **angular frequency**

angular frequency

$$\omega = 2\pi f = \frac{2\pi}{T} = \frac{2\pi v}{\lambda} = kv \tag{16.10}$$

In terms of the wavelength, period, wave number, and angular frequency, we can express the wavefunction (16.8) in the alternative forms

$$y = A\cos\left(2\pi\frac{x}{\lambda} - 2\pi\frac{t}{T}\right)$$

and

harmonic wavefunction

$$y = A\cos(kx - \omega t) \tag{16.11}$$

When the wave passes a point of the string, a particle in the string at this point moves up from its equilibrium position a distance equal to the wave amplitude A; then, half a cycle later, the particle moves down from its equilibrium position a distance A; and then, another half cycle later, it moves up again. Thus, *the particle executes simple harmonic motion of a frequency and an amplitude equal to the frequency and the amplitude of the wave* (see Fig. 16.6). The particle thus has vertical velocity and acceleration of the same form as those of a particle in simple harmonic motion:[1]

$$v_y = \frac{dy}{dt} = A\omega\sin(kx - \omega t) \tag{16.12}$$

$$a_y = \frac{dv_y}{dt} = -A\omega^2\cos(kx - \omega t) \tag{16.13}$$

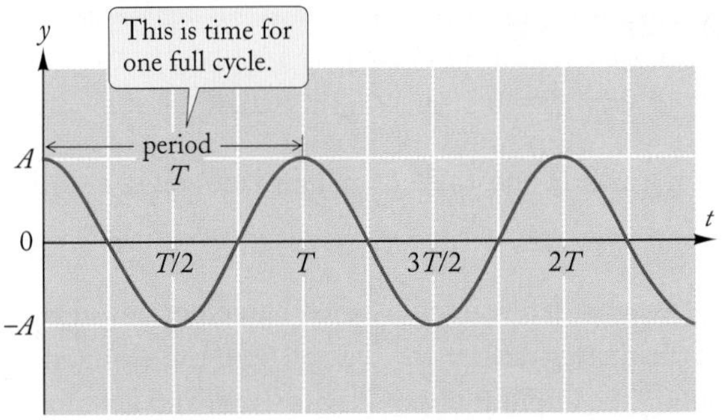

FIGURE 16.6 Each particle on the string executes simple harmonic motion with a period equal to the period of the wave, and an amplitude equal to the amplitude of the wave.

As mentioned previously, waves carry energy from one point to another. The kinetic energy of the particles involved in wave motion is proportional to the square of their velocity ($K = \frac{1}{2}mv^2$); combined with Eq. (16.12), this reveals a general feature of wave motion: *the energy stored in a wave is proportional to the square of the amplitude of the wave.*

Harmonic waves play a central role in the study of wave motion because, as we will see in the next section, any periodic wave of arbitrary shape can be regarded as a superposition, or sum, of several harmonic waves of suitably chosen amplitudes and wavelengths. Thus, if we understand the motion of harmonic waves, we understand the motion of any kind of periodic wave. Hereafter, we will concentrate on harmonic waves.

EXAMPLE 3 The waves near the deep end of the wave pool in the chapter photo have an amplitude of 0.50 m. For the 0.40-Hz waves described in Example 2, what is the maximum vertical speed of a swimmer floating on the surface? What is the maximum vertical acceleration experienced by that swimmer?

Concepts in Context

SOLUTION: According to Eq. (16.11), for a swimmer at a given horizontal position x_0, the vertical displacement is

$$y = A\cos(kx_0 - \omega t)$$

and therefore the vertical speed is, as in Eq. (16.12),

$$v_y = \frac{dy}{dt} = \frac{d}{dt}[A\cos(kx_0 - \omega t)] = A\omega\sin(kx_0 - \omega t)$$

This has a maximum magnitude $A\omega$. Since $\omega = 2\pi f$, the maximum vertical speed is

$$v_{y,\text{max}} = A \times 2\pi f = 0.50 \text{ m} \times 2\pi \times 0.40 \text{ Hz} = 1.3 \text{ m/s}$$

Similarly, according to Eq. (16.13), the vertical acceleration at x_0 is

$$a_y = \frac{dv_y}{dt} = \frac{d}{dt}[A\omega\sin(kx_0 - \omega t)] = -A\omega^2\cos(kx_0 - \omega t)$$

The maximum vertical acceleration is:

$$a_{y,\text{max}} = A\omega^2$$

[1] In this differentiation, x is held constant, since the particle remains at a fixed x position and only its y position changes with t. Mathematicians would use the *partial derivative* symbol $\partial y/\partial t$ to indicate that in the differentiation of the function $y = y(x, t)$ only t is differentiated, while x is held fixed.

Again using $\omega = 2\pi f$, this gives

$$a_{y,\max} = A(2\pi f)^2 = 0.50 \text{ m} \times (2\pi \times 0.40 \text{ Hz})^2 = 3.2 \text{ m/s}^2$$

This acceleration is nearly one-third of g; it provides quite a thrilling experience.

The speed of a wave is determined by the characteristics of the medium. For a string, the relevant characteristics are the tension F in the string, its mass M, and its length L. By examining the implications of Newton's Second Law for a short deformed segment of string being accelerated by the tension forces acting on the ends of the segment, we can demonstrate that the speed of the wave on a string is

speed of wave on a string

$$v = \sqrt{\frac{F}{M/L}} \tag{16.14}$$

To obtain this result, consider Fig. 16.7a, which shows a wave pulse on a string. We analyze the motion in a frame of reference moving to the right with the pulse; in that frame, the pulse is at rest, and the entire string moves to the left with the pulse speed v. Figure 16.7a shows one short segment ΔL in the pulse, which can be approximated as an arc of a circle with radius R, subtending a small angle $\Delta\theta = \Delta L/R$. The mass m of this segment is a fraction $\Delta L/L$ of the total mass M, $m = (\Delta L/L)M$; the centripetal acceleration is v^2/R. The radial components of the tension on either side of the segment provide the centripetal force; these sum to $F_{\text{net}} = 2F\sin(\Delta\theta/2) \approx F\Delta\theta = F\Delta L/R$ (see Fig. 16.7b). Newton's Second Law, $ma = F_{\text{net}}$, then implies

$$\frac{\Delta L}{L}M\frac{v^2}{R} = F\frac{\Delta L}{R} \tag{16.15}$$

Solving Eq. (16.15) for v yields Eq. (16.14).

The characteristic of the string that enters into Eq. (16.14) is the ratio of mass to length (M/L), or the mass per unit length. If we use a given kind of string, of some given thickness, the mass per unit length will be the same, regardless of whether we use a short length of string or a long length in our experiments. Note that the speed of the wave is large if the tension is large and the mass per unit length is small (a thin string).

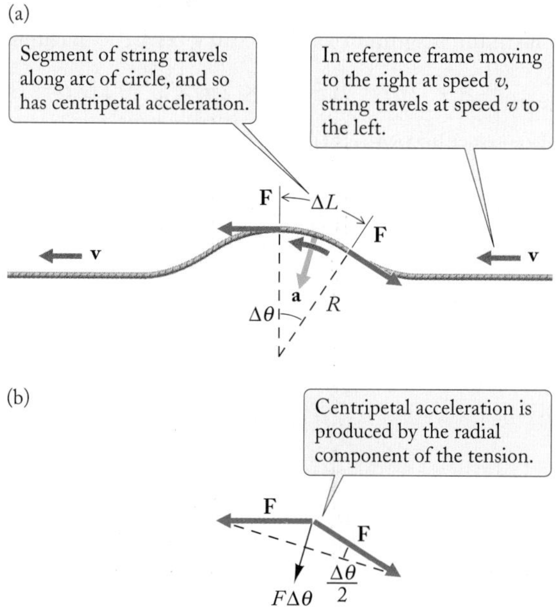

FIGURE 16.7 (a) A wave pulse moving to the right, viewed in a frame of reference moving with the pulse, so the string moves to the left at a speed v. For a small segment ΔL, the centripetal acceleration is related to the radial component of the tension forces **F** by the geometry shown. (b) Vector sum of the tension forces **F**.

This is intuitively reasonable, since a large tension can accelerate a small mass very quickly, and therefore produces a quick back-and-forth motion of the mass elements in the string; that is, it produces a high frequency of the motion. According to Eq. (16.3), for a given wavelength, a high frequency implies a large speed of the wave.

Of course, Eq. (16.14) is valid not only for strings, but also for tightly stretched wires and ropes.

EXAMPLE 4 A long piece of piano wire of a mass of 3.9×10^{-3} kg per meter is under a tension of 1.0×10^3 N. What is the speed of transverse waves on this wire? What is the wavelength of a harmonic wave on this wire if its frequency is 262 Hz?

SOLUTION: If the mass per unit length is 3.9×10^{-3} kg per meter, the ratio M/L in Eq. (16.14) is 3.9×10^{-3} kg/m, and hence the speed of the wave is

$$v = \sqrt{\frac{F}{M/L}} = \sqrt{\frac{1.0 \times 10^3 \text{ N}}{3.9 \times 10^{-3} \text{ kg/m}}} = 5.1 \times 10^2 \text{ m/s}$$

Consequently, with $f = 262 \text{ Hz} = 262 \text{ s}^{-1}$, the wavelength is

$$\lambda = \frac{v}{f} = \frac{510 \text{ m/s}}{262 \text{ s}^{-1}} = 1.9 \text{ m}$$

Although a wave on a string is a rather special case of wave motion, the mathematical description of other kinds of waves is similar to that of waves on a string. The instantaneous configuration of the wave can always be described by a plot of the wave disturbance vs. position, such as the plot in Fig. 16.6, but the vertical axis of the plot must be adapted to the physical properties of the wave. For instance, to describe a sound wave in air, we can plot the pressure disturbance produced by the wave vs. the position.

Our Eq. (16.14) for the wave speed applies only to waves on a string (or a wire, or a rope). But this equation exhibits a general feature of wave propagation: in broad terms, this equation states that the speed of the wave depends on the restoring force and on the inertia of the elastic medium in which the wave is propagating. This is true for all kinds of waves. In all cases, some force within the medium opposes its deformation—tension tends to keep the string straight, the pressure within a gas tends to keep the density of the gas uniform, gravity tends to keep the surface of the sea smooth, and so on. But if something provides an initial disturbance, then the restoring force will cause it to propagate, as in Fig. 16.2, with a speed depending on the magnitude of the restoring force and on the amount of inertia or, equivalently, the amount of mass in the medium. In general, the speed will be large if the restoring force is large and the amount of mass in the medium is small.

EXAMPLE 5 Gravity provides the restoring force for the water waves in a wave pool. A somewhat complicated analysis of the motion of water, including the requirement that the vertical velocity is zero at the bottom of the pool, yields for the wave speed in shallow water the approximate formula

Concepts in Context

$$v = \sqrt{gD}$$

where g is the acceleration of gravity and D is the depth of the water. This formula is valid when the depth is much less than the wavelength. For the shallow wave pool described in Examples 2 and 3, the speed at the deep end was 5.2 m/s. Recall that pumps produced waves at a rate of 0.40 Hz, and the resulting wavelength at the deep end was 13 m. What is the speed at the shallow end, where $D = 0.50$ m? What are the frequency and wavelength of the wave there?

SOLUTION: Using the given expression, the speed at the shallow end is

$$v = \sqrt{gD} = \sqrt{9.81\ \text{m/s}^2 \times 0.50\ \text{m}} = 2.2\ \text{m/s}$$

The frequency of the wave is unchanged. This is so because at any fixed location, water elements are pushed by adjacent elements; ultimately, this process begins at the water-pump source, which sets the overall frequency. To confirm this, we can consider a counterexample: if the frequency were to decrease away from the source, oscillations would continually "pile up" somewhere, since more would be steadily produced at the source than would pass another location. This is an impossible consequence; thus, the frequency must be the same at different locations.

The wavelength is, by Eq. (16.3),

$$\lambda = \frac{v}{f} = \frac{2.2\ \text{m/s}}{0.40\ \text{s}^{-1}} = 5.5\ \text{m}$$

Thus the wavelength is shorter in a region where the speed is slower.

Checkup 16.2

QUESTION 1: Are all periodic waves harmonic? Are all harmonic waves periodic?

QUESTION 2: Are the waves sketched in Fig. 16.4 harmonic?

QUESTION 3: Consider the piano wire described in Example 4. If we want to increase the wave speed by a factor of 2, by what factor must we increase the tension?

QUESTION 4: A steel wire stretched tightly across a room consists of two segments of the same material but of different diameters. If the first segment has a diameter half as large as the second segment, by what factor do the speeds of waves in the two segments differ?

QUESTION 5: A wave on a string has a wavelength of 30 cm and a frequency of 40 Hz. What is the frequency of a wave of wavelength 60 cm on this same string?

(A) 20 Hz (B) 40 Hz (C) 60 Hz (D) 80 Hz (E) 120 Hz

Online Concept Tutorial 18

16.3 THE SUPERPOSITION OF WAVES

Waves on a string and waves in other elastic bodies usually obey a **Superposition Principle:** *when two or more waves are present simultaneously in an elastic body, the resultant instantaneous displacement of a particle is the sum of the individual instantaneous displacements.* Such a superposition means that the waves do not interact; they have no effect on one another. Each wave propagates as though the other were not present, and the contribution that each makes to the displacement of a particle in the elastic body is as though the other were not present. For instance, if the sound waves from a violin and a flute reach us simultaneously, then each of these waves produces a displacement of the air molecules just as though it were acting alone, and the net displacement of the air molecules is simply the (vector) sum of these individual displacements.

For waves of low amplitude on a string and for sound waves of ordinary intensity in air, the Superposition Principle is very well satisfied. However, for waves of very large amplitude or intensity, the Superposition Principle fails. When a wave of very large amplitude is propagating on a string, it alters the tension of the string, and therefore affects the behavior of a second wave propagating on the same string. Likewise, a very intense sound wave (a shock wave, such as the loud bang from an explosion) produces significant alterations of the temperature and the pressure of the air, and therefore affects the behavior of a second wave propagating through this same region. However, we will not worry about such extreme conditions, and we will assume that the Superposition Principle is applicable.

As a first example of superposition, let us consider two waves propagating in the same direction with the same frequency and the same amplitude. If the wave crests and the wave troughs of the two waves coincide, the waves are said to be **in phase.** The superposition of these two waves yields a wave of twice the amplitude of the individual waves (see Fig. 16.8). *Such a reinforcement of one wave by another is called* **constructive interference.** If the wave crests of one wave match the wave troughs of the other, the waves are said to be **out of phase,** or to differ in phase by half a cycle. The superposition of these two waves yields a wave of zero amplitude (see Fig. 16.9). *Such a cancellation of one wave by another is called* **destructive interference.**[2]

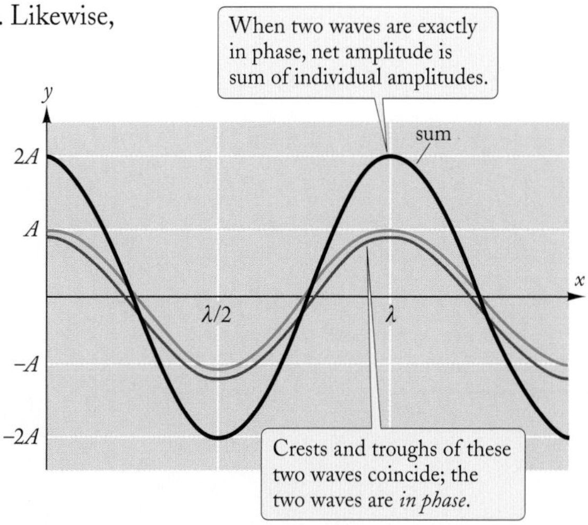

FIGURE 16.8 Constructive interference of two waves.

If the two waves are out of phase but their amplitudes are not equal, then their cancellation will not be total; some portion of the wave that has the larger amplitude will be left over (see Fig. 16.10). Similarly, if two equal-amplitude waves are not exactly in phase or exactly out of phase, then their sum will have an amplitude somewhere between zero and twice the amplitude of either wave. In Chapter 35, we will examine interference of waves with the same frequency in more detail.

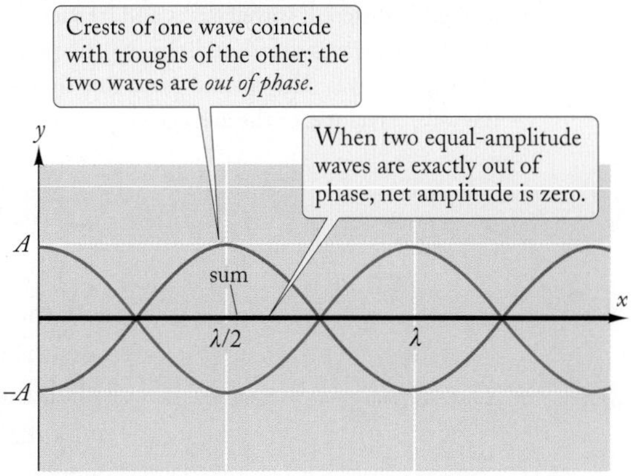

FIGURE 16.9 Destructive interference of two waves; the waves cancel everywhere.

For a much different example of superposition, let us consider two waves of the same amplitude, but slightly different frequencies and, therefore, slightly different wavelengths. Figure 16.11 shows the two waves at one instant of time and their superposition. At $x = 0$, the waves are in phase, and they interfere constructively, giving a large net amplitude. But farther along the x axis, the difference in wavelengths gradually causes the waves to acquire a phase difference. At the point P, the waves are out of phase by half a cycle, and they interfere destructively, giving a net amplitude of zero. Beyond this point, the phase difference exceeds one half cycle. At the point R the phase difference has grown to one cycle; but since a phase difference of one cycle means that the crests of the two waves coincide, they interfere constructively, again giving a large amplitude, and so on. Thus, the superposition of the two waves displays regularly alternating regions of constructive and destructive interference, that is, alternating regions of large amplitude and small amplitude (see Fig. 16.11b).

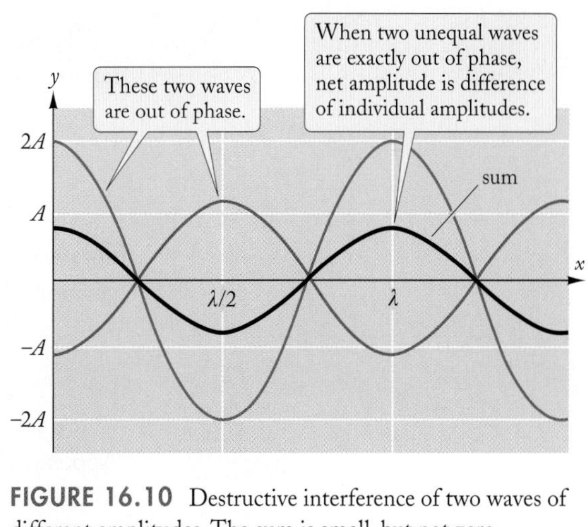

FIGURE 16.10 Destructive interference of two waves of different amplitudes. The sum is small, but not zero.

[2] This cancellation raises a question: If the waves cancel, what happens to the energy they carry? To answer this question we must examine in detail how the two waves were brought together. For instance, if the two waves were initially propagating on two separate strings that merge into a single string at a junction or knot, then the cancellation of the waves beyond the junction is necessarily associated with a strong backward reflection of the two incident waves at the junction, and the waves reflected backward from the junction carry away the missing energy.

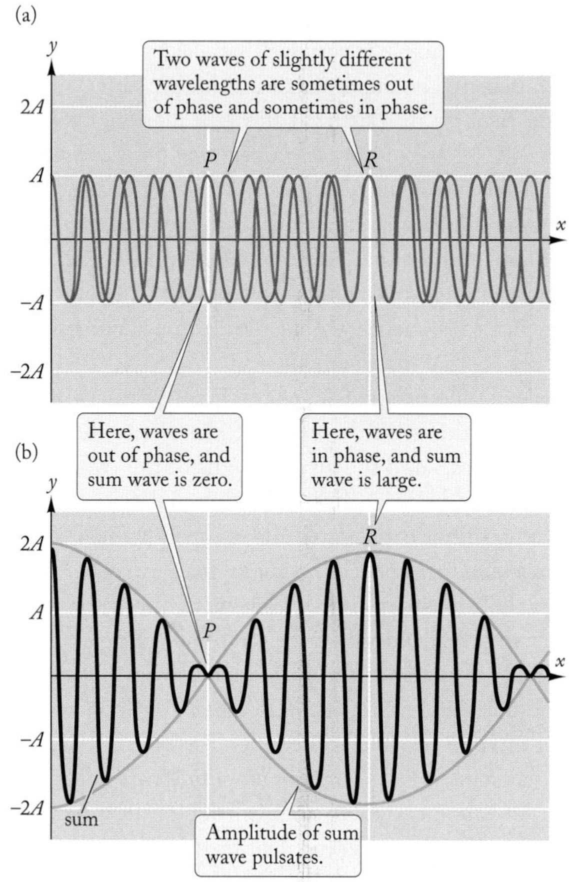

FIGURE 16.11 Superposition of two waves of slightly different wavelengths and frequencies. (a) The two waves before addition. (b) The sum of the two waves. The colored green line shows the wave envelope, or the average amplitude.

With the passing of time, the entire pattern in Fig. 16.11b moves toward the right with the wave velocity. This gives rise to the phenomenon of **beats.** At any given position, the amplitude of the wave pulsates—first the amplitude is large, then it becomes small, then again large, and so on. *The frequency with which the amplitude pulsates is called the* **beat frequency.** The beat frequency is simply the difference between the frequencies of the two waves:

beat frequency

$$f_{\text{beat}} = f_1 - f_2 \tag{16.16}$$

To establish this result, consider two traveling waves of the form (16.11), with different frequencies:

$$\begin{aligned} y_1 &= A\cos(k_1 x - \omega_1 t) \\ y_2 &= A\cos(k_2 x - \omega_2 t) \end{aligned} \tag{16.17}$$

For simplicity, let us consider these wavefunctions at a single point, $x = 0$. The superposition of these waves is the sum

$$y = y_1 + y_2 = A[\cos(\omega_1 t) + \cos(\omega_2 t)] \tag{16.18}$$

where we have used $\cos(-\theta) = \cos\theta$. If we apply the trigonometric identity (see Appendix 3)

$$\cos\theta_1 + \cos\theta_2 = 2\cos\left(\frac{\theta_1 - \theta_2}{2}\right)\cos\left(\frac{\theta_1 + \theta_2}{2}\right) \tag{16.19}$$

we obtain

$$y = 2A\cos\left[\frac{(\omega_1 - \omega_2)}{2}t\right]\cos\left[\frac{(\omega_1 + \omega_2)}{2}t\right] \tag{16.20}$$

The second cosine in Eq. (16.20) represents the rapid oscillations of the wave at the average angular frequency $(\omega_1 + \omega_2)/2$. The first cosine function in Eq. (16.20) represents the slow variation in the amplitude of that wave, producing the beats. Both the positive and negative parts of the slowly varying "envelope" in Fig. 16.11b produce large amplitudes; hence there are two beats per cycle of the function $\cos[(\omega_1 - \omega_2)t/2]$, and the angular beat frequency is $2 \times (\omega_1 - \omega_2)/2 = \omega_1 - \omega_2$. Thus the beat frequency is given by Eq. (16.16).

EXAMPLE 6 Suppose that two flutes generate sound waves of frequency 264 Hz and 262 Hz, respectively. What is the beat frequency?

SOLUTION: According to Eq. (16.16),

$$f_{\text{beat}} = f_1 - f_2 = 264\ \text{Hz} - 262\ \text{Hz} = 2\ \text{Hz}$$

Hence, a listener will hear a tone of average frequency 263 Hz, but with an amplitude pulsating 2 times per second.

Beats are a sensitive indication of small frequency differences, and they are very useful in the tuning of musical instruments. For example, when musicians want to bring two flutes in tune, they listen to the beats and, by trial and error, adjust one of the flutes so as to reduce the beat frequency; when the beats disappear entirely (zero beat frequency), the two flutes will be generating waves of exactly the same frequencies.

By the superposition of harmonic waves of different amplitudes and wavelengths, we can construct some rather complicated waveshapes. For example, Fig. 16.12 shows a periodic wave constructed by the superposition of three harmonic waves of wavelengths L, $L/3$, and $L/5$ whose amplitudes are in the ratio $1 : \frac{1}{3} : \frac{1}{5}$. It can be shown that any arbitrary periodic wave can be constructed by the superposition of a sufficiently large number of harmonic waves. If the desired periodic wave repeats each distance L, then the only harmonic waves needed in the sum are those with wavelengths equal to L divided by an integer. Since $f = v/\lambda$, the frequencies of the only harmonic waves needed are integer multiples of the frequency of the arbitrary periodic wave. This is called **Fourier's theorem.** As already mentioned in the preceding section, this theorem means that we can regard any arbitrary periodic wave as a sum of harmonic waves; and it means that the study of periodic waves is, in essence, the study of harmonic waves.

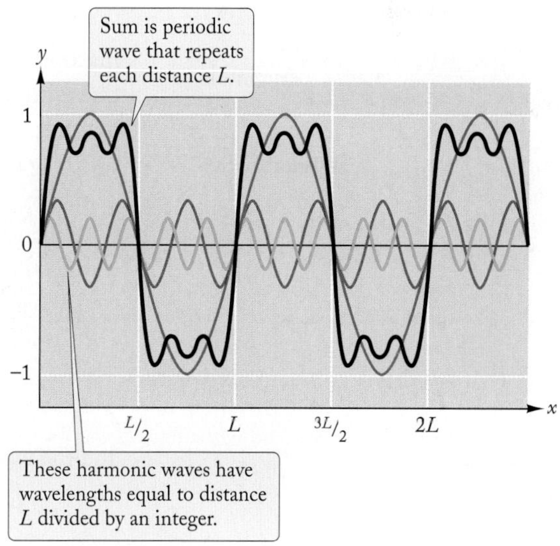

FIGURE 16.12 A wave constructed by superposition of three harmonic waves of wavelengths L (blue), $L/3$ (red), and $L/5$ (green). The harmonic waves are $y_{\text{blue}} = \sin(2\pi x/L)$, $y_{\text{red}} = \frac{1}{3}\sin[2\pi x/(L/3)]$, and $y_{\text{green}} = \frac{1}{5}\sin[2\pi x/(L/5)]$.

✔ Checkup 16.3

QUESTION 1: Consider a harmonic wave of small amplitude and short wavelength and a wave of much larger amplitude and much longer wavelength. What does the superposition of these two waves look like?

QUESTION 2: To check the tuning of a guitar, the lowest-pitch string with a finger on the fifth fret and the next string (open) are played simultaneously. If, for an out-of-tune guitar, the resulting tone pulsates 4 times per second and has a frequency of 107 Hz, at what frequencies are the two strings vibrating?

QUESTION 3: Two water waves of amplitudes 0.6 m and 0.8 m, respectively, arrive simultaneously at a buoy. What is the amplitude of the net wave if these two waves are in phase? If these two waves are out of phase?

(A) 0.2 m; 1.4 m (B) 0.2 m; 0.2 m (C) 1.0 m; 1.4 m
(D) 1.4 m; 0.2 m (E) 1.4 m; 1.0 m

Online Concept Tutorial 18

16.4 STANDING WAVES

Next, we want to consider the superposition of two waves of the same amplitude and the same frequency, but of opposite directions of propagation. Figure 16.13a shows the two traveling waves and their sum at an initial instant of time. At this instant, the waves are in phase, and their sum is a wave of twice the amplitude of each. At a slightly later time, one wave has moved to the right and the other to the left (see Fig. 16.13b). At the points *P, Q, R*, . . . the waves were initially zero; now, one wave has a positive value at these points, and the other wave has an equally large negative value. Thus, the sum of the two waves still yields a result of zero at these points. The two waves will continue to cancel at these points at all times (see Figs. 16.13c, d, and e). These points at which the sum of the waves is zero are called **nodes.** They are one-half wavelength apart.

node

Midway between the nodes we find points at which the sum of the two waves is maximum (positive or negative). These points are called **antinodes.** Figure 16.13 shows that with the passing of time, the height of the wave crests at the antinodes oscillates, but the positions of these wave crests remain fixed. Thus, *the superposition of two waves traveling in opposite directions is a* **standing wave.** This means that the net wave travels neither to the right nor to the left; its wave crests remain at fixed positions while the entire wave increases and decreases in unison. The frequency of this pulsation of the standing wave is the same as the frequency of the two underlying traveling waves.

antinode

standing wave

We can find the wavefunction for the standing wave by adding the wavefunctions of the individual traveling waves. The wave traveling to the right is

$$y_1 = A\cos(kx - \omega t) \tag{16.21}$$

and the wave traveling toward the left is

$$y_2 = A\cos(kx + \omega t) \tag{16.22}$$

The sum of these two waves is

$$y = y_1 + y_2 = A\cos(kx - \omega t) + A\cos(kx + \omega t) \tag{16.23}$$

To evaluate this sum, we again use the trigonometric identity (16.19)

$$\cos\theta_1 + \cos\theta_2 = 2\cos[\tfrac{1}{2}(\theta_1 - \theta_2)]\cos[\tfrac{1}{2}(\theta_1 + \theta_2)] \tag{16.24}$$

which, with $\theta_1 = kx - \omega t$ and $\theta_2 = kx + \omega t$, gives us

$$\begin{aligned} y &= 2\cos[\tfrac{1}{2}(kx - \omega t - kx - \omega t)]\cos[\tfrac{1}{2}(kx - \omega t + kx + \omega t)] \\ &= 2A\cos(-\omega t)\cos(kx) = 2A\cos(\omega t)\cos(kx) \end{aligned} \tag{16.25}$$

The time dependence in this wavefunction appears as an *overall factor* $\cos(\omega t)$, which shows that the entire wave indeed increases and decreases in unison. The dependence on x appears as an overall factor of $\cos(kx)$. This factor gives us the positions of the antinodes and nodes. The antinodes correspond to a maximum (positive or negative) value of $\cos(kx)$, which occurs at

$$kx = 0,\ \pi,\ 2\pi,\ 3\pi,\ldots \quad \text{for antinodes (maxima)} \tag{16.26}$$

The nodes correspond to a zero value of $\cos(kx)$, which occurs at

$$kx = \pi/2,\ 3\pi/2,\ 5\pi/2,\ldots \quad \text{for nodes (zeros)} \tag{16.27}$$

With $k = 2\pi/\lambda$, the condition for the antinodes becomes

$$x = 0,\ \lambda/2,\ \lambda,\ 3\lambda/2,\ldots \quad \text{for antinodes (maxima)} \tag{16.28}$$

and the condition for the nodes becomes

$$x = \lambda/4,\ 3\lambda/4,\ 5\lambda/4,\ldots \quad \text{for nodes (zeros)} \tag{16.29}$$

As expected from the discussion of Fig. 16.13, the nodes are one-half wavelength apart, and the antinodes are midway between the nodes.

If this standing wave is a wave on a string, then each particle of this string executes simple harmonic motion. However, in contrast to the case of a traveling wave, where the amplitudes of the harmonic oscillations of all the particles are the same, the amplitudes of oscillation now depend on position: the amplitude is maximum at the antinodes, and it is minimum (zero) at the nodes.

So far, in our discussion of the waves on a string, we have assumed that the string is very long, and we have ignored the endpoints of the string. When a traveling wave arrives at an endpoint, something drastic will have to happen to it: the wave will either have to be absorbed at the endpoint or it will have to be reflected, with a reversal of its direction of propagation. If the endpoint is a fixed point (the string is attached to a rigid support), then the endpoint cannot absorb the energy of the wave, and the wave will be completely reflected. This results in the simultaneous presence of two waves of equal amplitudes and opposite directions of travel; that is, it results in a standing wave. Reflection will also occur if the endpoint is free to move but cannot absorb energy; in this case, the endpoint is an antinode. For the case of a fixed endpoint, the reflected wave is inverted, as illustrated by the behavior of a reflected pulse in Fig. 16.14a; this inversion ensures that the endpoint is a node. For the case of a free endpoint, the reflected wave is not inverted, as illustrated in Fig. 16.14b.

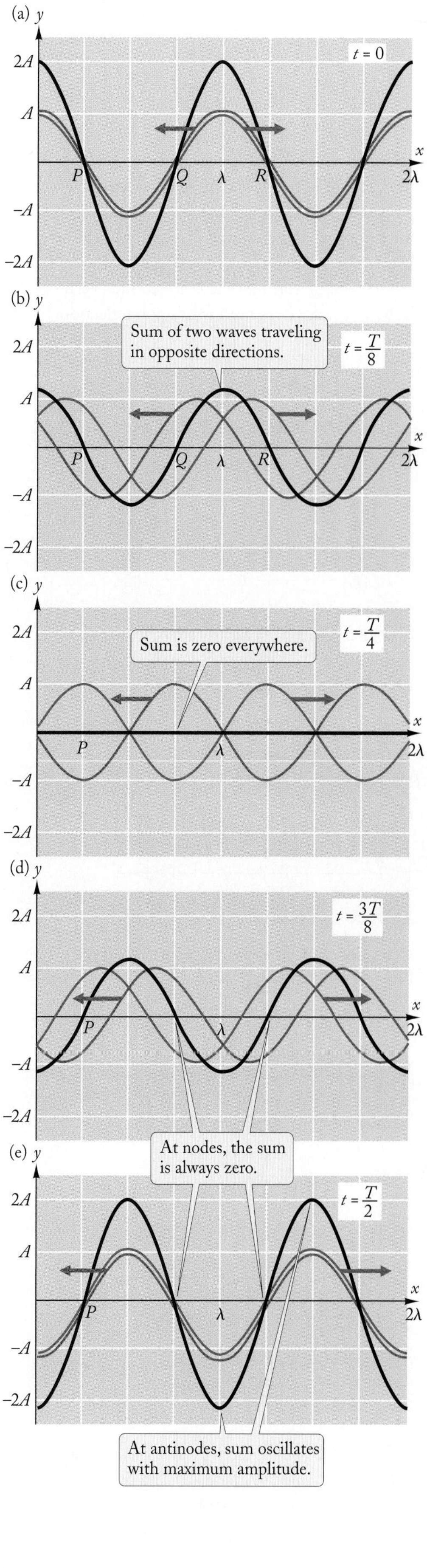

FIGURE 16.13 Superposition of two waves of the same amplitude but of opposite directions of propagation. (a) At $t = 0$ the waves are in phase. (b) At $t = \frac{1}{8}$ of a period $= \frac{1}{8}\,T$, one wave has moved $\frac{1}{8}$ of a wavelength to the right, and the other the same distance to the left. (c) At $t = \frac{1}{4}\,T$, one wave has moved $\frac{1}{4}$ of a wavelength to the right and the other the same distance to the left; the waves have therefore moved apart $\frac{1}{2}$ wavelength, and they are out of phase and they cancel everywhere. (d) At $t = \frac{3}{8}\,T$, the waves have moved $\frac{3}{8}$ of a wavelength to the right and the left, respectively. At $t = \frac{1}{2}\,T$, the waves are again in phase, and so on.

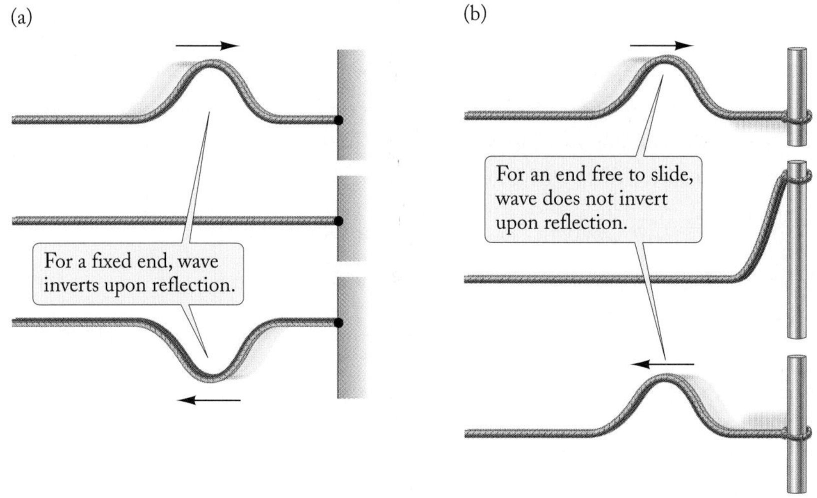

FIGURE 16.14 (a) A wave pulse approaches a fixed end of a string. The pulse is inverted upon reflection, as it must be for the superposition of the incoming and outgoing displacements to be zero at the endpoint. (b) A wave pulse approaching the end of a string that is free to slide vertically. The pulse does not invert upon reflection.

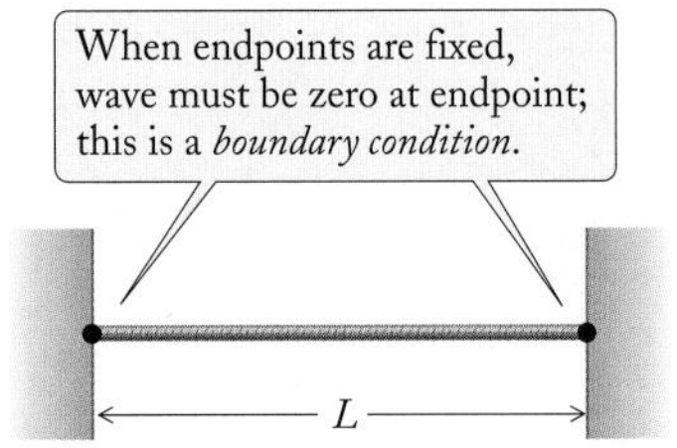

FIGURE 16.15 A tightly stretched string with fixed ends. The next figure shows possible standing waves on this string.

For a string with two fixed endpoints (see Fig. 16.15), the possible standing waves are subject to the restriction that the wave must be zero at each endpoint at all times. Such a restriction on what happens at the endpoints of a wave is called a **boundary condition.** Obviously, the boundary condition for our standing wave will be satisfied if the endpoints are nodes. Figure 16.16 shows possible standing waves on a string of some given length L; all these standing waves have nodes at the endpoints. *The possible standing-wave motions of the string shown in Fig. 16.16 are called the* **normal modes**. Figure 16.16a shows the **fundamental** mode; Fig. 16.16b, the **first overtone**; Fig. 16.16c, the **second overtone**. Figure 16.17 shows time-exposure photographs of a string with these modes.

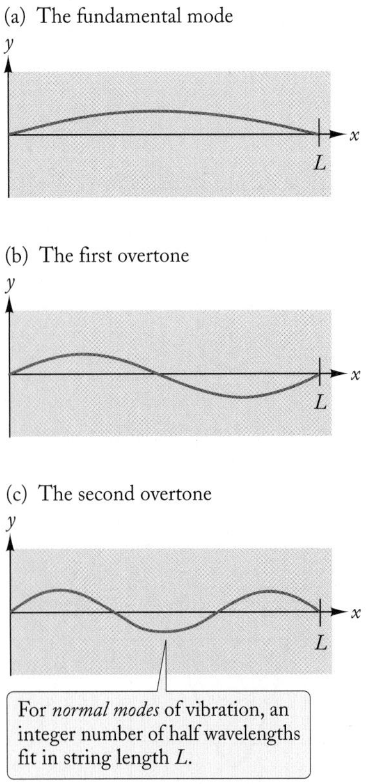

FIGURE 16.16 Standing waves on a string. (a) The fundamental mode. (b) The first overtone. (c) The second overtone.

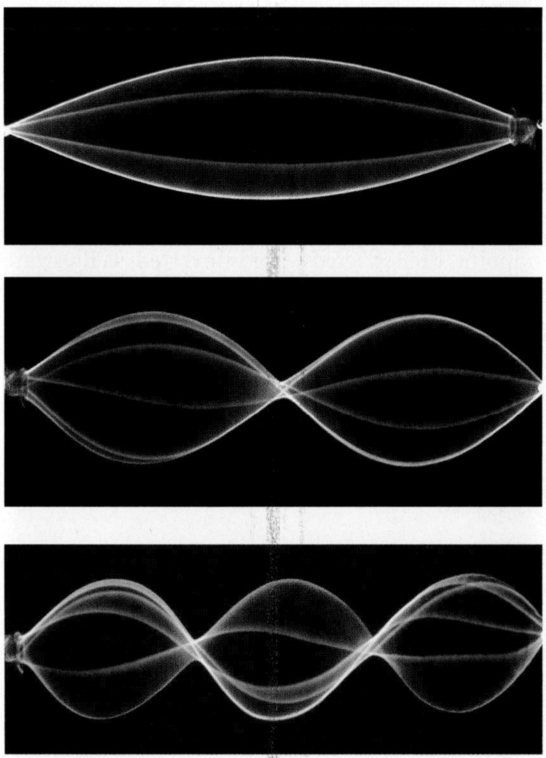

FIGURE 16.17 Stroboscopic photographs of standing waves on a string.

In all these modes, some integer number of half wavelengths exactly fits the length of the string. In the fundamental mode, one half wavelength fits the string; in the first overtone, two half wavelengths fit the string; in the second overtone, three half wavelengths fit the string; and so on. Thus, the normal modes occur when

$$L = \frac{\lambda}{2} \qquad L = 2\frac{\lambda}{2} \qquad L = 3\frac{\lambda}{2} \quad \cdots$$

and the wavelengths for the different modes are thus

$$\lambda_1 = 2L \qquad \lambda_2 = \frac{2L}{2} = L \qquad \lambda_3 = \frac{2L}{3} \quad \cdots \tag{16.30}$$

wavelengths of normal modes of string

The frequencies of oscillation of the modes are related to the wavelengths in the usual way ($f = v/\lambda$, where v is the velocity):

$$f_1 = \frac{v}{2L} \qquad f_2 = 2\frac{v}{2L} \qquad f_3 = 3\frac{v}{2L} \quad \cdots \tag{16.31}$$

The frequencies of these modes are called the normal frequencies, proper frequencies, or **eigenfrequencies** of the string. Note that all these frequencies are integer multiples of the **fundamental frequency**: the frequency of the first overtone is twice the frequency of the fundamental, the frequency of the second overtone is three times the frequency of the fundamental, and so on. The first overtone is also called the **second harmonic**; the second overtone is called the **third harmonic**; and so on.

eigenfrequencies

In general, any arbitrary motion of a freely vibrating string (with fixed endpoints) will be some superposition of several of the above normal modes. Which modes will be present in the superposition depends on how the motion is started. For instance, when a guitar player plucks a string on his guitar near the middle, he will excite the fundamental mode and also the second overtone and, to a lesser extent, some of the higher even-numbered overtones.

EXAMPLE 7 The low E string on the guitar vibrates with a frequency of 82.4 Hz when excited in its fundamental mode. What are the frequencies of the first, second, and third overtones of this string?

SOLUTION: According to Eq. (16.31), $f_2 = 2f_1$, $f_3 = 3f_1$, and $f_4 = 4f_1$. Hence the frequencies of the first, second, and third overtones are, respectively, 2×82.4 Hz = 165 Hz, 3×82.4 Hz = 247 Hz, and 4×82.4 Hz = 330 Hz.

The normal modes of vibration of a long, thin elastic rod or a beam fixed at both ends are mathematically similar to the normal modes of a string. However, such an elastic body can experience transverse deformations (like those of a string), compressional deformations, and rotational, or "torsional," deformations.

Recall from Chapter 15 that if an external driving force oscillates at the natural frequency of oscillation of a system, a large-amplitude oscillation can be built up, a phenomenon called **resonance.** Figure 16.18 shows a spectacular example of a torsional standing wave in the span of a bridge at Tacoma, Washington. This standing wave was excited by a wind blowing across the bridge, which generated a periodic succession of vortices, or regions of swirling motion in the air, with a frequency equal to

FIGURE 16.18 Standing wave on the deck of the Tacoma Narrows bridge, July 1, 1940. The bridge broke apart a short time after this picture was taken.

that of one of the normal modes of vibration of the span. Thus, the periodic generation of vortices was in resonance with the natural vibration of the bridge, and this gradually built up a large amplitude of vibration. The bridge vibrated for several hours, with increasing amplitude, and then broke apart.

✔ Checkup 16.4

QUESTION 1: Describe the third overtone and the fourth overtone for the wave on the string illustrated in Fig. 16.15.

QUESTION 2: Can the midpoint of a guitar string be an antinode? Can it be a node?

QUESTION 3: Is the number of nodes for a normal mode of a string fixed at both ends always larger than the number of antinodes? How much larger?

QUESTION 4: The lowest frequency on a guitar is obtained when the top string (the low E string) is played open, in its fundamental mode; the eighth harmonic of that frequency can be obtained on the bottom guitar string (the high E string) when it is constrained at its midpoint. Both strings have the same length. What is the ratio of the wave speed on the high E string to the wave speed on the low E string?

(A) 1 (B) 2 (C) 4 (D) 8 (E) 16

SUMMARY

WAVELENGTH, PERIOD, FREQUENCY, AND WAVE SPEED

$$\frac{\lambda}{T} = \lambda f = v \qquad (16.1, 16.3)$$

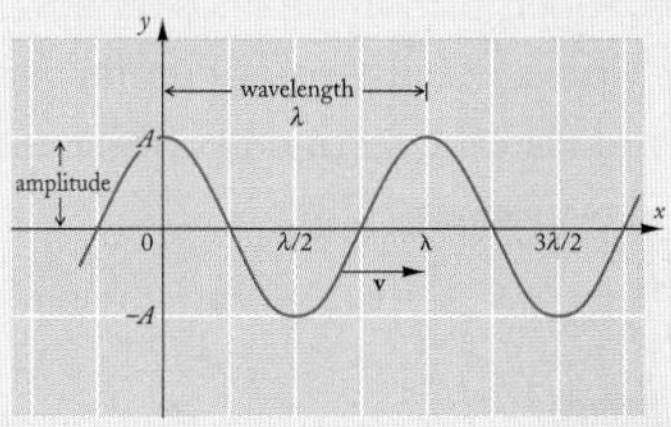

WAVE NUMBER

$$k = \frac{2\pi}{\lambda} \qquad (16.7)$$

ANGULAR FREQUENCY

$$\omega = 2\pi f \qquad (16.10)$$

HARMONIC WAVE

Wave traveling in the positive x direction:

$$y = A\cos(kx - \omega t) \qquad (16.8)$$

Wave traveling in the negative x direction:

$$y = A\cos(kx + \omega t) \qquad (16.9)$$

SPEED OF WAVE ON A STRING

$$v = \sqrt{\frac{F}{M/L}} \qquad (16.14)$$

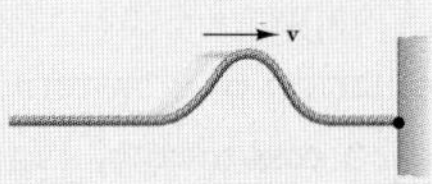

SUPERPOSITION PRINCIPLE FOR TWO OR MORE WAVES
The net instantaneous displacement is the sum of the individual instantaneous displacements.

CONSTRUCTIVE INTERFERENCE Waves meet crest to crest.

DESTRUCTIVE INTERFERENCE Waves meet crest to trough.

BEAT FREQUENCY $$f_{\text{beat}} = f_1 - f_2 \qquad (16.16)$$

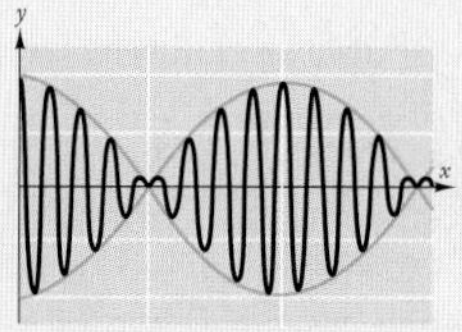

STANDING HARMONIC WAVE $$y = A\cos(kx)\cos(\omega t) \qquad (16.25)$$

NODE Point of zero oscillation.

ANTINODE Point of maximum oscillation.

WAVELENGTHS OF NORMAL MODES OF STRING
(of length L, fixed at both ends) $$\lambda_1 = 2L,\ \lambda_2 = L,\ \lambda_3 = 2L/3, \text{ etc.} \qquad (16.30)$$

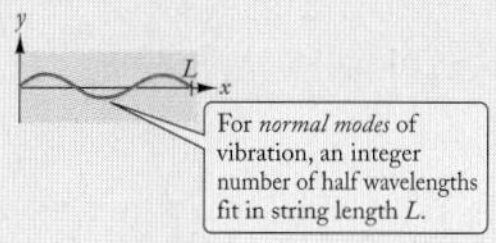

EIGENFREQUENCIES

$f_1 = v/2L$ fundamental mode (or first harmonic),
$f_2 = 2(v/2L)$ first overtone (or second harmonic),
$f_3 = 3(v/2L)$ second overtone (or third harmonic), etc.

QUESTIONS FOR DISCUSSION

1. You have a long, thin steel rod and a hammer. How must you hit the end of the rod to generate a longitudinal wave? A transverse wave?
2. Some people enjoy arranging long rows of dominoes on the floor, so the toppling of one domino triggers the toppling of all the others, by a chain reaction (see Fig. 16.19). The propagation of the disturbance along such a chain of dominoes has some of the properties of a wave pulse. In what way is it similar to a wave pulse? In what way is it different?

FIGURE 16.19 Toppling dominoes.

3. A wave pulse on a string transports energy. Does it also transport momentum? To answer this question, imagine a washer loosely encircling the string at some place; what happens to the washer when the wave pulse strikes it?
4. According to Eq. (16.14), the speed of a wave on a string increases by a factor of 2 if we increase the tension by a factor of 4. However, in the case of a rubber string, the speed increases by more than a factor of 2 if we increase the tension by a factor of 4. Why are rubber strings different?
5. A harmonic wave is traveling along a string. Where in this wave is the kinetic energy at maximum? The potential energy? The total energy?
6. Suppose that two strings of different densities are knotted together to make a single long string. If a wave pulse travels along the first string, what will happen to the wave pulse when it reaches the junction? (Hint: If the second string had the same density as the first string, the wave pulse would proceed

without interruption; if the second string were much denser than the first, the wave pulse would be totally reflected.)

7. Figure 16.13 shows a standing wave on a string. At time $t = T/4$, the amplitude of the wave is everywhere zero. Does this mean the wave has zero energy at this instant?
8. After an arrow has been shot from a bow, the bowstring will oscillate back and forth, forming a standing wave. Which of the overtones shown in Fig. 16.16 do you expect to be present?
9. In tuning a guitar or violin, by what means do you change the frequency of a string?
10. A mechanic can make a rough test of the tension on the spokes of a wire wheel (see Fig. 16.20) by striking the spokes with a wrench or a small hammer. A spoke under tension will ring, but a loose spoke will not. Explain.

FIGURE 16.20 Wire wheels on an automobile.

FIGURE 16.21 Frets on the neck of a guitar.

11. What is the purpose of the frets (see Fig. 16.21) on the neck of a guitar or a mandolin?
12. The strings of a guitar are made of wires of different thicknesses (the thickest strings are manufactured by wrapping copper or brass wire around a strand of steel). Why is it impractical to use wire of the same thickness for all the strings?

PROBLEMS

16.2 Periodic Waves

1. The speed of light waves is 3.0×10^8 m/s. The wavelengths of light waves range from 4.0×10^{-7} m (violet) to 7.0×10^{-7} m (red). What is the range of frequencies of these waves?
2. An ocean wave has a wavelength of 120 m and a period of 8.77 s. Calculate the frequency, angular frequency, wave number, and speed of this wave.
3. Figure 16.22 is a record of a tsunami that struck the coast of Mexico. Approximately, what was the frequency of this wave? What was its wavelength in the open sea? Assume that the speed of the wave in the open sea was 740 km/h.
4. In deep water (where the depth is much larger than the wavelength), the speed of waves is given by the formula $v = \sqrt{g\lambda/2\pi}$. Calculate the speed of short water waves, with $\lambda = 1.0$ m. Calculate the speed of long waves with $\lambda = 300$ m.

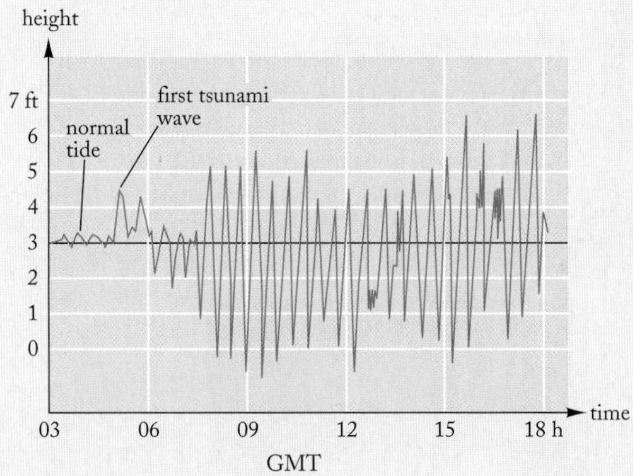

FIGURE 16.22 Height of a tsunami that struck the coast of Mexico.

5. In shallow water (where the depth is shorter than a wavelength), the speed of waves is given by the formula $v = \sqrt{gD}$, where D is the depth of the water.
   (a) Calculate the speed of water waves in a shallow pond with a depth of 2.0 m.
   (b) For ocean waves of extremely long wavelength, such as tidal waves, the oceanic basins can be treated as shallow ponds, since their depth is small compared with the wavelength. Calculate the speed of a tidal wave in the Pacific Ocean, where the mean depth is 4.3 km.
6. To determine the speed and the frequency of periodic waves on a lake, the owner of a motorboat first runs the boat in the direction of the waves, and he finds that when his boat keeps up with a wave crest the speed indicator shows 16 m/s. He then anchors the boat and finds that the waves make it bounce up and down 6.0 times per minute. What are the speed, frequency, and wavelength of the waves?
7. The speed of tidal waves in the Pacific is about 740 km/h.
   (a) How long does a tidal wave take to travel from Japan to California, a distance of 8000 km?
   (b) If the wavelength of the wave is 300 km, what is its frequency?

*8. In the open sea, a tsunami usually has an amplitude less than 30 cm and a wavelength longer than 80 km. Assume that the speed of the tsunami is 740 km/h. What are the maximum vertical velocity and acceleration that such a tsunami will give to a ship floating on the water? Will the crew of the ship notice the passing of the tsunami?

9. A transverse harmonic wave on a stretched string has an amplitude of 1.2 cm, a speed of 8.0 m/s, and a wavelength of 2.2 m.
   (a) What is the maximum transverse speed attained by a particle on the string? Does the particle attain this maximum speed when a wave crest passes the particle or at some other time?
   (b) What is the maximum transverse acceleration attained by a particle on the string? Does the particle attain this maximum acceleration when a wave crest passes the particle or at some other time?
10. When a periodic transverse wave travels along a clothesline, a ladybug sitting on the line experiences a maximum transverse velocity of 0.20 m/s and a maximum transverse acceleration of 4.0 m/s^2. Deduce the amplitude and the frequency of the wave.
11. A transverse wave travels along a stretched string with a wave speed of 14 m/s. A particle at a fixed location on this string oscillates up and down as follows as a function of time:

$$y = 0.020 \cos(9.0t)$$

    where the displacement y is measured in meters and the time t is measured in seconds.
   (a) What is the amplitude of the wave?
   (b) What is the frequency of the wave?
   (c) What is the wavelength of the wave?
12. Suppose that the function $y = 6.0 \times 10^{-3} \cos(20x + 4.0t + \pi/3)$ describes a wave on a long string (distance is measured in meters and time in seconds).
   (a) What are the amplitude, wavelength, wave number, frequency, angular frequency, direction of propagation, and speed of this wave?
   (b) At what time does this wave have a maximum at $x = 0$?
13. A harmonic wave on a string has an amplitude of 2.0 cm, a wavelength of 1.2 m, and a velocity of 6.0 m/s in the positive x direction. At time $t = 0$, this wave has a crest at $x = 0$.
   (a) What are the period, frequency, angular frequency, and wave number of this wave?
   (b) What is the mathematical equation describing this wave as a function of x and t?
14. Ocean waves smash into a breakwater at the rate of 12 per minute. The wavelength of these waves is 39 m. What is their speed?
15. The velocity of sound in freshwater at 15°C is 1440 m/s, and at 30°C it is 1530 m/s. Suppose that a sound wave of frequency 440 Hz penetrates from a layer of water at 30°C into a layer of water at 15°C. What will be the change in the wavelength? Assume that the frequency remains unchanged.
16. A light wave of frequency 5.5×10^{14} Hz penetrates from air into water. What is its wavelength in air? In water? The speed of light is 3.0×10^8 m/s in air and 2.3×10^8 m/s in water; assume that the frequency remains the same.
17. The National Ocean Survey has deployed buoys off the Atlantic coast to measure ocean waves. Such a buoy detects waves by the vertical acceleration that it experiences as it is lifted and lowered by the waves. In order to calibrate the device that measures the acceleration, scientists placed the buoy on a Ferris wheel at an amusement park. The vertical acceleration (as a function of time) of a buoy riding on a Ferris wheel of radius 6.1 m rotating at 6.0 rev/min is used to simulate the vertical acceleration of a buoy riding on a wave. What is the maximum vertical acceleration? What is the wavelength of the corresponding wave? Assume that the waves are in deep water and that the buoy always rides on the surface of the wave. In deep water, the wave speed is given by $v = \sqrt{g\lambda/2\pi}$, where g is the acceleration of free fall and λ is the wavelength."
18. The wavefunction for a wave on a string is

$$y = A\cos(kx - \omega t + \delta)$$

    where, as in Chapter 15, δ is a phase constant. If $A = 0.13$ m, $\omega = 20\pi$ radians/s, $k = 15\pi$ m^{-1}, and $\delta = \pi/4$, what is the speed of the wave? What is its wavelength? What is its frequency? What is the maximum transverse speed of a particle in the string? What is the vertical displacement of the string at $x = 0$ and $t = 0$?
19. Nine water wave crests and troughs pass a point in 15 s. If the horizontal distance between a crest and the nearest trough is 0.75 m, find the speed of the wave.
20. An astronaut wishes to measure gravitational acceleration. A 7.0-kg mass is suspended from a thin wire of length 2.0 m and

mass 6.0 g. A pulse travels the length of the wire in 33 ms. What is the local value of the acceleration due to gravity? (Neglect the mass of the wire when determining the tension.)

21. A copper wire is 100 m long and 0.50 mm in diameter. The wire is stretched to a tension of 75 N. How long does a wave pulse take to travel to the end of the wire?

22. A pendulum is made of a 2.0-kg mass hanging from a string of mass 5.0 g. For small amplitude, a complete pendulum oscillation takes 1.0 s. Find the speed of a transverse wave on the pendulum string.

23. After adjusting its tension to 150 N, an electrician taps a cable that is hanging between two utility poles. She notes that it takes 3.5 s for a pulse to travel to the next pole, 30 m away. What is the mass per unit length of the wire?

24. A piano wire of mass 35 g is 1.50 m long when unstretched, and has a cross-sectional area of 3.0 mm^2. The Young's modulus of the wire is 1.5×10^{11} N/m^2. If the wire is stretched 2.0 cm, find the speed of transverse waves on the wire.

25. A string has a length of 3.0 m and a mass of 12 g. If this string is subjected to a tension of 250 N, what is the speed of transverse waves?

26. A clothesline of length 10 m is stretched between a house and a tree. The clothesline is under a tension of 50 N, and it has a mass per unit length of 6.0×10^{-2} kg/m. How long does a wave pulse take to travel from the house to the tree and back?

27. A wire rope used to support a radio mast has a length of 20 m and a mass per unit length of 0.80 kg/m. When you give the wire rope a sharp blow at the lower end and generate a wave pulse, it takes 1.0 s for this wave pulse to travel to the upper end and to return. What is the tension in the wire rope?

28. A nylon rope of length 24 m is under a tension of 1.3×10^4 N. The total mass of this rope is 2.7 kg. If a wave pulse starts at one end of this rope, how long does it take to reach the other end?

*29. Ocean waves of wavelength 100 m have a speed of 6.2 m/s; ocean waves of wavelength 20 m have a speed of 2.8 m/s.[3] Suppose that a sudden storm at sea generates waves of all wavelengths. The long-wavelength waves travel fastest and reach the coast first. A fisherman standing on the coast first notices the arrival of 100-m waves; 10 hours later he notices the arrival of 20-m waves. How far is the storm from the coast?

*30. A motorboat is speeding at 12 m/s through a group of periodic ocean waves. When the motorboat travels in the same direction as the waves, it smashes into 6.5 waves per minute. When the motorboat is traveling in the direction opposite to the waves, it smashes into 30 waves per minute. Calculate the speed, frequency, and wavelength of the ocean waves.

*31. A string of mass per unit length μ is tied to a second string of mass per unit length μ'. A harmonic wave of speed v traveling along the first string reaches the junction and enters the second string. What will be the speed v' of this wave in the second string? Your answer should be a formula involving μ, μ', and v.

[3]These values are group velocities, or signal velocities.

*32. The maximum tensile stress that can be tolerated without breakage by steel is 5.0×10^8 N/m^2, and the density of steel is 7800 kg/m^3. If you apply a tension to a thin steel rod just barely less than the breaking tension, what is the speed of transverse waves on the rod? Does the answer depend on the diameter of the rod?

*33. A mass of 30 kg hangs from a string which, in turn, hangs from two other strings making an angle of 45° with each other (see Fig. 16.23). Each string has a length of 2.0 m, and the mass per unit length of each is 4.0×10^{-3} kg/m. How long does a wave pulse take to travel down from the upper end of this arrangement of strings to the mass?

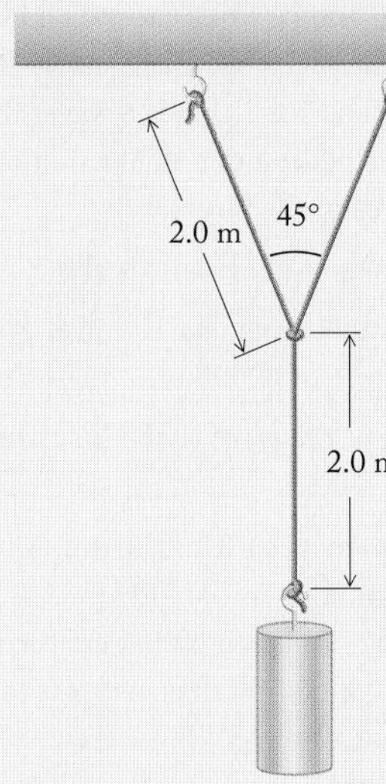

FIGURE 16.23 A mass hanging from an arrangement of strings.

*34. A steel wire of length 5.0 m and radius 0.30 mm is knotted to another steel wire of length 5.0 m and radius 0.10 mm. The wires are stretched with a tension of 150 N. How long does a transverse wave pulse take to travel the distance of 10 m from the beginning of the first wire to the end of the other? The density of steel is 7.8×10^3 kg/m^3.

**35. A long, uniform rope of length l hangs vertically. The only tension in the rope is that produced by its own weight. Show that, as a function of the distance z from the lower end of the rope, the speed of a transverse wave pulse of the rope is $\sqrt{gz}$. What is the time the wave pulse takes to travel from one end of the rope to the other?

**36. The speed of an ocean wave in shallow water is given by $v = \sqrt{gD}$, where D is the depth. A wave starts 50 m from shore, where the depth is 4.0 m. If the depth decreases linearly with distance as the wave approaches the shore, how long does it take for the wave to reach the shore?

**37. Suppose you take a loop of rope and make it rotate about its center at speed V. The centrifugal tendency of the segments of rope will then stretch it out along a circle of some radius R (see Fig. 16.24). What is the tension in the rope under these

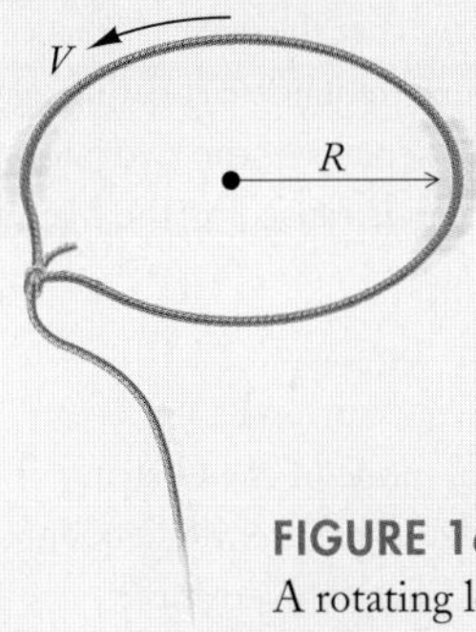

FIGURE 16.24
A rotating loop of rope.

conditions? Show that the speed of transverse waves on the rope (relative to the rope) coincides with the speed of rotation V.

**38. A flexible rope of length l and mass m hangs between two walls. The length of the rope is more than the distance between the walls (see Fig. 16.25), and the rope sags downward. At the ends, the rope makes an angle of α with the walls. At the middle, the rope approximately has the shape of an arc of a circle; the radius of the approximating (osculating) circle is R. What is the tension in the rope at its ends? What is the tension in the rope at its middle? What is the speed of transverse waves at the ends? At the middle?

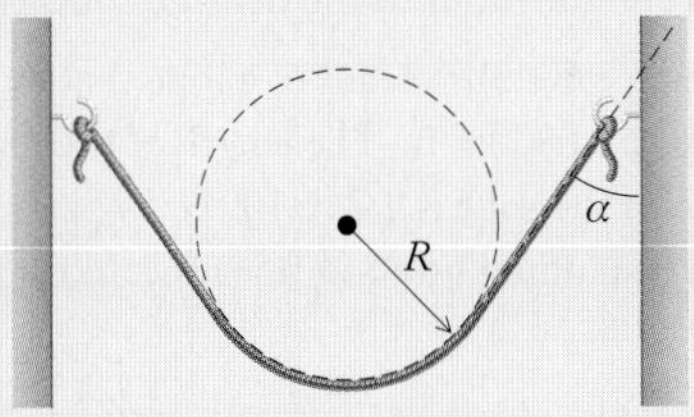

FIGURE 16.25 A rope hanging between two walls.

**39. The end of a long string of mass per unit length μ is knotted to the beginning of another long string of mass per unit length μ' (the tensions in these strings are equal). A harmonic wave travels along the first string toward the knot. The incident wave will be partially transmitted into the second string, and partially reflected. The frequencies of all these waves are the same. With the knot at $x = 0$, we can write the following expressions for the incident, reflected, and transmitted waves:

$$y_1 = A_{\text{in}} \cos(kx - \omega t)$$

$$y_2 = A_{\text{refl}} \cos(kx + \omega t)$$

$$y_3 = A_{\text{trans}} \cos(k'x - \omega t)$$

Show that

$$A_{\text{refl}} = \frac{k - k'}{k + k'} A_{\text{in}} = \frac{\sqrt{\mu} - \sqrt{\mu'}}{\sqrt{\mu} + \sqrt{\mu'}} A_{\text{in}}$$

$$A_{\text{trans}} = \frac{2k}{k + k'} A_{\text{in}} = \frac{2\sqrt{\mu}}{\sqrt{\mu} + \sqrt{\mu'}} A_{\text{in}}$$

[Hint: At $x = 0$, the displacement of the string must be continuous ($y_1 + y_2 = y_3$); if not, the string would break at the knot. Furthermore, the slope of the string must be continuous ($dy_1/dx + dy_2/dx = dy_3/dx$); if not, the string would have a kink and the (massless) knot would receive an infinite acceleration.]

16.3 The Superposition of Waves†

40. At one instant of time two transverse waves are traveling in the same direction along a stretched string. The instantaneous shapes of the wave are represented by

$$y = 0.020 \cos(4.0x) \quad \text{and} \quad y = 0.030 \cos(4.0x)$$

where the transverse displacements y and the position x are measured in meters.

(a) Are these waves in phase or out of phase?

(b) What are the amplitude and the wavelength of the net wave?

*41. At one instant of time two transverse waves are traveling in the same direction along a stretched string. The instantaneous shapes of the wave are represented by

$$y = 0.030 \cos(4.0x) \quad \text{and} \quad y = 0.030 \sin(4.0x)$$

where the transverse displacements y and the position x are measured in meters.

(a) What is the phase difference between these waves?

(b) Find the position x nearest the origin where the net wave has a wave crest. What is the amplitude of the net wave?

42. Three waves are traveling in the same direction; their individual amplitudes are 0.30 m, 0.50 m, and 0.80 m. What is the largest amplitude of the net wave that could occur? What is the smallest amplitude of the net wave that could occur, and how could that come about?

43. Two harmonic waves are described by

$$y_1 = A\cos(4x - 5t) \quad \text{and} \quad y_2 = 2A\cos(4x - 5t - \pi)$$

where $A = 6.0$ m, x is in meters, and t is in seconds. What are the amplitude, wavelength, and frequency of the superposition of these waves? At $x = 1.0$ m and $t = 1.0$ s, what is the net displacement?

44. At one point in space, two waves are described by

$$y_1 = A\cos(\omega_1 t) \quad \text{and} \quad y_2 = A\cos(\omega_2 t)$$

where $\omega_1 = 145$ radians/s and $\omega_2 = 152$ radians/s. When the two waves are superposed, how many beats are heard per second?

45. Two waves are described by

$$y_1 = A\cos(5x - 6t) \quad \text{and} \quad y_2 = A\cos(6x - 7t)$$

The two waves are superposed, and a snapshot of the resulting disturbance reveals a short-wavelength oscillation that gradually varies in amplitude over a longer length. What is the short

†For help, see Online Concept Tutorial 18 at www.wwnorton.com/physics

wavelength? What is the distance between the points where the amplitude of the short-wavelength oscillation goes to zero?

46. A periodic disturbance repeats every 2.0 m along a string (for example, a series of pulses). If we wish to describe this disturbance by a sum of harmonic waves, what are the only wavelengths that might be needed?

47. The wavefunctions for two transverse waves on a string are

$$y_1 = 0.030\cos(6.0x - 18t + 1.5)$$

$$y_2 = 0.030\cos(6.0x - 18t - 2.3)$$

where y and x are measured in meters and t in seconds.

(a) What is the amplitude of the sum of these waves?

(b) What is the transverse displacement of the string at $x = 0$ and $t = 0$?

*48. Consider the wavefunction $y = 3.0\cos(5.0x - 8.0t) + 4.0\sin(5.0x - 8.0t)$, which is a superposition of two wavefunctions expressed in some suitable units. Show that this wavefunction can be written in the form $y = A\cos(5.0x - 8.0t + \delta)$. What are the values of A and δ?

*49. A thin wire of length 1.0 m vibrates in a superposition of the fundamental mode and the second harmonic. The wavefunction is

$$y = 0.0060\sin\pi x\cos 400\pi t - 0.0040\sin 3\pi x\cos 1200\pi t$$

where y and x are measured in meters and t in seconds.

(a) What is the displacement at $x = 0.50$ m as a function of time?

(b) Plot this displacement as a function of time in the interval $0\text{ s} \le t \le 0.0050$ s.

50. Two ocean waves with $\lambda = 100$ m, $f = 0.125$ Hz and $\lambda = 90$ m, $f = 0.132$ Hz arrive at a seawall simultaneously. What is the beat frequency of these waves?

51. A guitar player attempts to tune her instrument perfectly with the help of a tuning fork. If the guitar player sounds the tuning fork and a string on her guitar simultaneously, she perceives beats at a frequency of 4.0 per second. The tuning fork is known to have a frequency of 294.0 Hz. What fractional increase (or decrease) of the tension of the guitar string is required to bring the guitar in tune with the tuning fork? From the available information, can you tell whether an increase or decrease of tension is required?

*52. Figure 16.26 shows the height of the tide at Pakhoi. These tides can be regarded as a wave. The shape of the curve in Fig. 16.26 indicates that the wave consists of two periodic waves of slightly different frequencies beating against each other. What are the frequencies of the two periodic waves? What are the periods? Which is caused by the Moon, which by the Sun?

16.4 Standing Waves†

53. The fundamental mode of the G string of a violin has a frequency of 196 Hz. What are the frequencies of the first, second, third, and fourth overtones?

54. Suppose that a vibrating mandolin string of length 0.34 m vibrates in a mode with five nodes (including the nodes at the ends) and four antinodes. What overtone is this? What is its wavelength?

55. A telegraph wire made of copper is stretched tightly between two telephone poles 50 m apart. The tension in the wire is 500 N, and the mass per unit length is 2.0×10^{-2} kg/m. What is the frequency of the fundamental mode? The first overtone?

56. A violin has four strings; all the strings have (approximately) equal tensions and lengths but they have different masses per unit length (kg/m), so that when excited in their fundamental modes they vibrate at different frequencies. The fundamental frequencies of the four strings are 196, 294, 440, and 659 Hz. What must be the ratios of the densities of the strings?

57. A car is being towed by means of a rope that has a mass per unit length of 0.080 kg/m. The length of the rope is 3.0 m, and the tension in the rope is 2.2×10^3 N. What is the eigenfrequency for a standing wave on this rope, in the fundamental mode?

58. You notice that a string fixed at both ends has a resonant frequency of 660 Hz, and also a resonant frequency of 440 Hz, but no resonant frequencies at any intermediate value. Identify what overtones these frequencies correspond to. Deduce the fundamental frequency of this string.

59. Two transverse harmonic waves are described by

$$y_1 = A\cos(\pi x - 3\pi t) \qquad \text{and} \qquad y_2 = A\cos(\pi x + 3\pi t)$$

where $A = 5.0$ m, x is in meters, and t is in seconds. What is the maximum amplitude of the superposition of these two waves at $x = 0.25$ m? What are the maximum transverse speed and acceleration at that point?

†For help, see Online Concept Tutorial 18 at www.wwnorton.com/physics

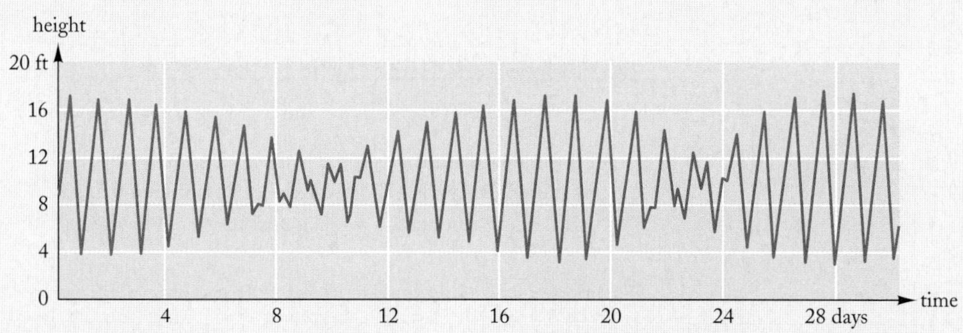

FIGURE 16.26 Height of the tide at Pakhoi as a function of time (because of exceptional local conditions, there is only one high and one low tide per day).

60. Several pulses of amplitude 3.0 cm are sent down a string that is attached to a pole at the far end. The pulses reflect there, and maintain their original amplitude. At the places where the forward and reflected pulses cross, what is the net amplitude if the string is fixed at the pole? If the string is free to slide up and down at the pole?

61. You pluck a guitar string upward near the center. The length of the string (fixed at both ends) is 65 cm. The wave speed is 70 m/s. How long does it take for the pulse you create to travel to the ends of the string and return to the center? When it returns, is the displacement now upward or downward? How many times per second does the string vibrate up and down?

62. If a human can hear up to 20000 Hz, how many overtones of a low A ($f = 27.5$ Hz) can be heard by the human ear?

63. A 15-g string is 10 m long and is fixed at both ends. If the tension in the string is 40 N, what are the first five eigenfrequencies of the string?

64. The Bay of Fundy in Nova Scotia, Canada, is known for its extreme tides (Fig. 16.27); these are "resonance" tides, because the natural period of oscillation in the bay is about 12 hours, nearly matching the tidal period. Assume that a one-quarter-wavelength standing wave just fits the 250-km length of the bay, and that the wave speed is $v = \sqrt{gD}$, where D is the average depth of the bay. From this information, find the average depth of the Bay of Fundy.

65. The fundamental mode of the G string on a mandolin has a frequency of 196 Hz. The length of this string is 0.34 m, and its mass per unit length is 4.0×10^{-3} kg/m. What is the tension of this string?

66. Some automobiles are equipped with wire wheels (see Fig. 16.20). The spokes of these wheels are made of short segments of thick wire installed under large tension. Suppose that one of these wires is 9.0 cm long, 0.40 cm in diameter, and under a tension of 2200 N. The wire is made of steel; the density of steel is 7.8 g/cm^3. To check the tension, a mechanic gives the spoke a light blow with a wrench near its middle. With what frequency will the spoke ring? Assume that the frequency is that of the fundamental mode.

67. A light wave of wavelength 5.0×10^{-7} m strikes a mirror perpendicularly. The reflection of the wave by the mirror makes a standing wave with a node at the mirror. At what distance from the mirror is the nearest antinode? The nearest node?

68. A wave on the surface of the sea with a wavelength of 3.0 m and a period of 4.4 s strikes a seawall oriented perpendicularly to its path. The reflection of the wave by the seawall sets up a standing wave. For such a wave, there is an antinode at the seawall. How far from the seawall will there be nodes?

*69. The D string of a violin vibrates in its fundamental mode with a frequency of 294 Hz and an amplitude of 2.0 mm. What are the maximum velocity and the maximum acceleration of the midpoint of the string?

*70. The middle C string of a piano is supposed to vibrate at 261.6 Hz when excited in its fundamental mode. A piano tuner finds that in a piano that has a tension of 900 N on this string, the frequency of vibration is too low (flat) by 15.0 Hz. How much must he increase the tension of the string to achieve the correct frequency?

*71. The wire rope supporting the mast of a sailboat from the rear is under a large tension. The rope has a length of 9.0 m and a mass per unit length of 0.22 kg/m.

(a) If a sailor pushes on the rope sideways at its midpoint with a force of 150 N, he can deflect it by 7.0 cm. What is the tension in the rope?

(b) If the sailor now plucks the rope near its midpoint, the rope will vibrate back and forth like a guitar string. What is the frequency of the fundamental mode?

*72. Many men enjoy singing in shower stalls because their voice resonates in the cavity of the shower stall. Consider a shower stall measuring 1.0 m × 1.0 m × 2.5 m. What are the four lowest resonant frequencies of standing sound waves in such a shower stall? The speed of sound is 331 m/s.

*73. A piano wire of length 1.5 m fixed at its end vibrates in its second overtone. The frequency of vibration is 440 Hz, and the amplitude at the midpoint of the wire is 0.40 mm. Express this standing wave as a superposition of traveling waves. What are the amplitudes and speeds of the traveling waves?

FIGURE 16.27 Resonance tides in the Bay of Fundy.

*74. Two strings are tied together and stretched across a room. The strings have equal lengths of 2.0 m, and their masses are 1.0 g and 4.0 g, respectively. Find the frequencies of the standing-wave modes on these tied strings if the knot between the strings is a node.

*75. Consider a string with a mass density that depends on position. One end of the string is at $x = 0$, the other at $x = L$; and the mass density increases linearly from one end to the other, that is, [mass density] $= A + Bx$, where A and B are constants.

(a) If the tension in the string is F, what is the speed of transverse waves as a function of x?

(b) A wave on such a string does not have a well-defined wavelength. However, it can be described approximately as having a wavelength that depends on position. If the frequency of the wave is f, what is the wavelength $\lambda(x)$?

(c) The condition for standing waves is then the usual condition: the number of wavelengths that fit within the length of the string must be 1/2, or 1, or 3/2, etc. Calculate the corresponding eigenfrequencies.

*76. (a) A long string is stretched along the x direction. Two transverse waves of equal wavelengths and equal frequencies travel simultaneously along this string. Suppose that one wave produces a displacement of the string in the y direction (see Fig. 16.28)

$$y = A\cos(kx - \omega t)$$

and the other wave produces a displacement of the string in the z direction (see Fig. 16.28)

$$z = A'\cos(kx - \omega t)$$

Show that the resulting motion of a particle on this string is back and forth along a line in the y–z plane. What is the angle of this line with respect to the y axis? The wave formed by a superposition of these y and z waves is called a wave of **linear polarization**, and the direction of the line of motion is called the direction of polarization.

(b) Now suppose that the two waves have the same amplitude but the wave in the z direction is a quarter of a cycle out of phase with the wave in the y direction, so

$$y = A\cos(kx - \omega t)$$

and

$$z = A\sin(kx - \omega t)$$

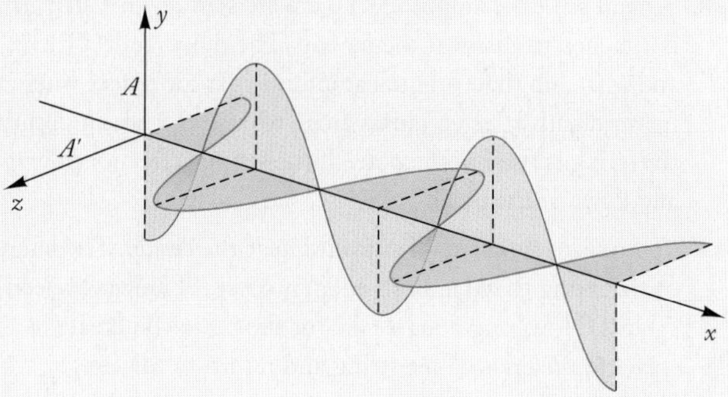

FIGURE 16.28 Displacements of a string in the y direction and the z direction.

Show that the resulting motion of a particle on the string is uniform circular motion. What are the radius, the frequency, the speed, and the centripetal acceleration of this circular motion? The wave formed by the superposition of these y and z waves is called a wave of **circular polarization.**

*77. Consider the superposition of two waves of the same frequency, opposite direction of propagation, and unequal amplitudes A_1 and A_2,

$$y_1 = A_1\cos(kx - \omega t)$$

$$y_2 = A_2\cos(kx + \omega t)$$

This superposition does not form a standing wave, but a wave with a modulated amplitude: the wave amplitude is large at some positions, and smaller at other positions. Show that the largest wave amplitude attained by the wave is $A_1 + A_2$. At what positions x is this large amplitude found? Show that the smallest wave amplitude attained by the wave is $|A_1 - A_2|$. At what positions x is this smallest amplitude found?

**78. A piano wire of length 0.18 m vibrates in its fundamental mode. The frequency of vibration is 494 Hz; the amplitude is 3.0×10^{-3} m. The mass per unit length of the wire is 2.2×10^{-3} kg/m. What is the energy of vibration of the entire wire? (Hint: Treat each small segment of the wire as a particle of mass dm with simple harmonic motion, and sum the energies of these simple harmonic motions.)

REVIEW PROBLEMS

79. A string is stretched along the x axis (horizontally), and it oscillates in the y direction (vertically). At one instant of time, a transverse traveling wave on this string is described by the mathematical formula

$$y = 0.030\cos(1.2x)$$

where y and x are measured in meters.

(a) What is the amplitude of the wave?

(b) What is the wavelength of this wave?

(c) Where are the first three wave crests and the first three wave troughs on the positive x axis nearest the origin?

*80. In the crust of the Earth, seismic waves of the P type have a speed of almost 5.0 km/s; waves of the S type have a speed of about 3.0 km/s. Suppose that after an earthquake, a seismometer placed at some distance first registers the arrival of P waves and 9.0 min later the arrival of S waves. What is the distance between the seismometer and the source of the waves?

81. A giant, freak wave encountered by a weather ship in the North Atlantic was 23.5 m high from trough to crest; its wavelength was 350 m and its period 15.0 s. Calculate the maximum vertical acceleration of the ship as the wave passed underneath; calculate the maximum vertical velocity. Assume that the motion of the ship was purely vertical.

*82. Many inhabitants of Tangshan, China, reported that during the catastrophic earthquake of July 28, 1976, they were thrown 2.0 m into the air as if hit by a "huge jolt from below."

(a) With what speed must a body be thrown upward to reach a height of 2.0 m?

(b) Assume that the vertical wave motion of the ground was simple harmonic with a frequency of 1.0 Hz. What amplitude of the vertical motion is required to generate a speed equal to that calculated in part (a)?

83. A harmonic transverse wave traveling on a tightly stretched wire has an amplitude of 0.020 m and a frequency of 100 Hz. What is the maximum speed attained by a particle on the string as this wave passes? What is the maximum acceleration? What is the time difference between the instant of maximum speed and the instant of maximum acceleration?

84. A passenger in an airplane flying over an anchored ship notices that ocean waves are smashing into the ship regularly at the rate of 10 per minute. He knows that ocean waves of this frequency have a speed of 9.4 m/s. He also notices that the length of the ship is the same as about three wavelengths. Deduce the length of the ship from this information.

85. While an anchored sailboat pulls on its anchor rope, the tension in the rope is 5.0×10^3 N. The anchor rope is nylon, of diameter 0.92 cm. The density of nylon is 1.1×10^3 kg/m^3. What is the mass per unit length for this rope? What is the speed of transverse waves on this rope?

86. Two strings are tied together and stretched across a room. The strings have equal lengths of 3.0 m, and their masses are 6.0 g and 9.0 g, respectively. If the tension in the strings is 200 N, what is the time a wave pulse takes to travel from the farthest end of one string to the farthest end of the other?

87. Three strings of identical material are tied together. They are under (different) tensions, and they make the angles indicated in Fig. 16.29.

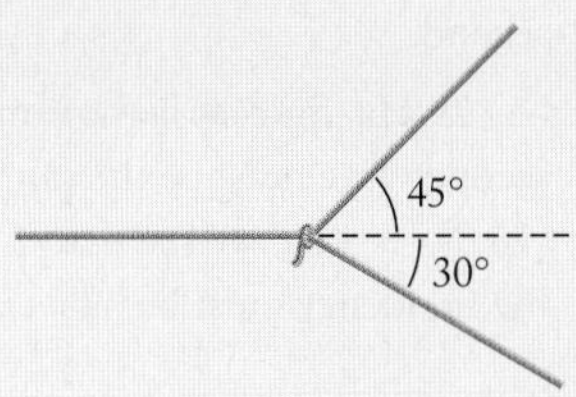

FIGURE 16.29 Three strings tied together.

(a) If the tension in the left string is T, what are the tensions in the other two?

(b) If the wave speed in the left string is 10 m/s, what are the speeds in the other two?

88. Consider two transverse harmonic waves of different wavelengths traveling in the same direction along a stretched string. At one instant of time, the shapes of the waves are given by

$$y = 0.012 \cos(3.0x) \qquad \text{and} \qquad y = 0.030 \cos(5.0x)$$

where the transverse displacements y and the position x are measured in meters.

(a) Is the superposition of these waves a harmonic wave? Is it a periodic wave?

(b) What is the wavelength of the net wave formed by the superposition of the two waves? What is the amplitude (the maximum transverse displacement) of the net wave?

89. Two cars, of identical make, have horns that generate sound waves of slightly different frequencies, 600 Hz and 612 Hz. What beat frequency do you hear if both of these cars are blowing their horns?

90. A standing wave on a string has the form

$$y = 0.020 \cos(15x) \cos(3.0t)$$

where distances are measured in meters and time in seconds.

(a) What are the amplitude and the frequency of this standing wave?

(b) What are the amplitudes, frequencies, and wavelengths of the two traveling waves whose superposition forms the standing wave?

(c) Where are the nodes and the antinodes of the standing wave?

(d) What are the maximum speed and the maximum acceleration of a particle on the string at one of the antinodes?

91. A uniform 20-m rope has a mass of 0.90 kg. The rope is hanging vertically from a support, so the only tension in the rope is that provided by its own weight.

(a) Find the speed of transverse waves in this rope, as a function of position along the rope. What is the speed at the top of the rope? At the midpoint? At the bottom?

(b) Find the time required for a wave pulse to travel from the top of the rope to the bottom.

92. An elevator of mass 2000 kg is hanging from a steel cable of length 60 m. The mass per unit length of this cable is 0.60 kg/m.

(a) What is the speed of transverse waves on this cable? (Neglect the mass of the cable when determining the tension.)

(b) What is the eigenfrequency of the fundamental mode on this cable? The first overtone? The second overtone?

93. A string of length L is fixed at one end and looped over a vertical frictionless rod at the other end (see Fig. 16.30). With

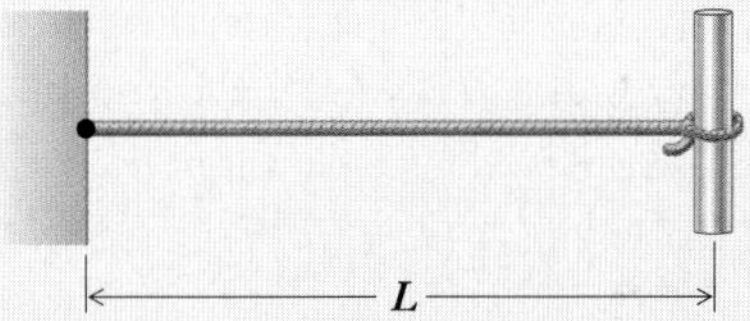

FIGURE 16.30 The left end of this string is fixed; the right end can slide up and down on the rod.

this arrangement the rod can maintain a tension in the string, but does not inhibit the up-and-down motion of the end of the string. Assume that the mass of the loop is negligible. What are the wavelengths of the standing waves of this string? What is the longest possible wavelength? (Hint: Since the loop is massless, the vertical forces that the tension in the string exerts on the loop must be zero at all times. This requires that the end of the string must be horizontal at all times.)

94. A mass of 15 kg hangs from two 6.0-m ropes, each of which leads downward at an angle of 30° with the vertical (Fig. 16.31). Each rope has a mass density $M/L = 0.012$ kg/m. Find the frequencies of those standing-wave modes for which the mass of 15 kg is at a node.

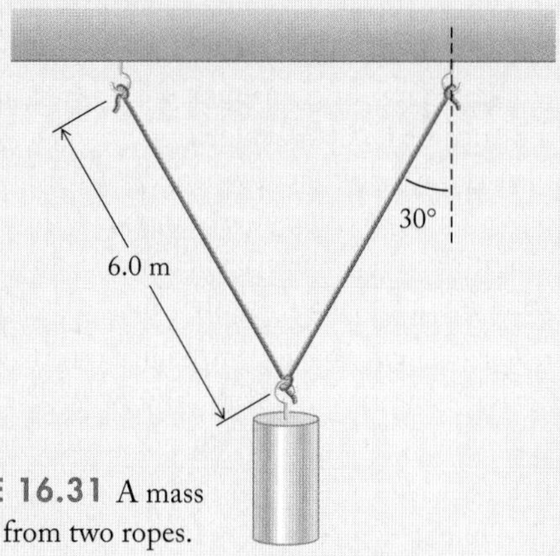

FIGURE 16.31 A mass hanging from two ropes.

Answers to Checkups

Checkup 16.1

1. The bat will remain essentially rigid under a small shaking motion, and so no appreciable wave motion is produced. The Jell-O (gelatin) is easily deformed, and so the shaking motion does produce a wave; the wiggling motion of the medium can be directly viewed.
2. A transverse wave can be generated by grabbing the cord and shaking it to the side and back (or up and down). A longitudinal wave can be generated by grabbing the cord and yanking it back and forth parallel to its length.
3. Such a wave on a string is transverse; except at the fixed endpoints, each point on the string moves up and down as it vibrates.
4. To generate a longitudinal wave, you must hit the end of the rod (causing a compression); to generate a transverse wave, you must hit the rod on its side (causing lateral motion).
5. (D) No; yes. The water wave does not carry the water from Africa to Florida; the particles merely oscillate about their equilibrium positions. The wave does carry energy from Africa to Florida; the motion of the arriving wave carries kinetic energy, and there is gravitational potential energy associated with the vertical displacements in water waves.

Checkup 16.2

1. All periodic waves are not harmonic; any shape of disturbance that repeats at regular intervals is periodic, but only a disturbance that takes the particular shape of a sine or cosine function is harmonic. However, all harmonic waves are periodic, since the sine and cosine functions repeat at regular intervals.
2. No. They are too pointy at the crests and flat in the troughs to be accurately described by a sine or a cosine function.
3. By Eq. (16.14), the speed is proportional to the square root of the tension; thus, if we want to increase the speed a factor of 2, we must increase the tension a factor of 4.
4. Since they are connected and stretched, they share the same tension. However, the first segment has a diameter half as large as the second; made of the same material, it must then have one-fourth of the cross-sectional area and thus one-fourth of the mass per unit length of the second. Since the speed is inversely proportional to the mass per unit length, the speed of waves in the first segment is twice that of the second.
5. (A) 20 Hz. Since it is the same string, the wave speed of the second wave must be the same as the first wave. Using Eq. (16.3), $v = \lambda f$, we see that if the wavelength is twice as long, the frequency must be half as large, or equal to 20 Hz.

Checkup 16.3

1. The superposition would look like the large-amplitude, long-wavelength wave, but with a quickly varying wiggle all along it due to the added small-amplitude, short-wavelength wave.

2. The superposition of the two waves vibrates at the average frequency, 107 Hz. The pulsating frequency of 4 Hz is the difference frequency (beats). So the two strings have frequencies of 105 Hz and 109 Hz.

3. (D) 1.4 m; 0.2 m. If the two waves are in phase, their amplitudes add, so that the net wave has amplitude 1.4 m. If the two waves are out of phase, their amplitudes subtract, giving a net wave of amplitude 0.2 m.

Checkup 16.4

1. For the third overtone, four half wavelengths fit in the string length L (two complete "cycles"). The fourth overtone has five half wavelengths in the string length L.

2. Yes; ordinarily, the midpoint of a guitar string is an antinode; the string primarily vibrates in its fundamental mode, similar to Fig. 16.16a. The midpoint (at the twelfth "fret") of the guitar string can also be a node; if constrained there (e.g., by holding a finger on it) while plucking the guitar string elsewhere, the string will vibrate primarily in its first overtone, similar to Fig. 16.16b.

3. Yes; if fixed at both ends, the number of nodes will always be one more than the number of antinodes. For example, the fundamental mode has two nodes (the ends) and one antinode (the center); each overtone adds one node and one antinode.

4. (C) 4. When the high E string is constrained, it has a node at its midpoint, and thus a half wavelength fits into half the length. The low E string vibrates in its fundamental mode, where one half wavelength fits into the full length. Since $v = \lambda f$, the high E string, with eight times the frequency and half the wavelength, has $8 \times \frac{1}{2} = 4$ times the speed.

CHAPTER

17 Sound

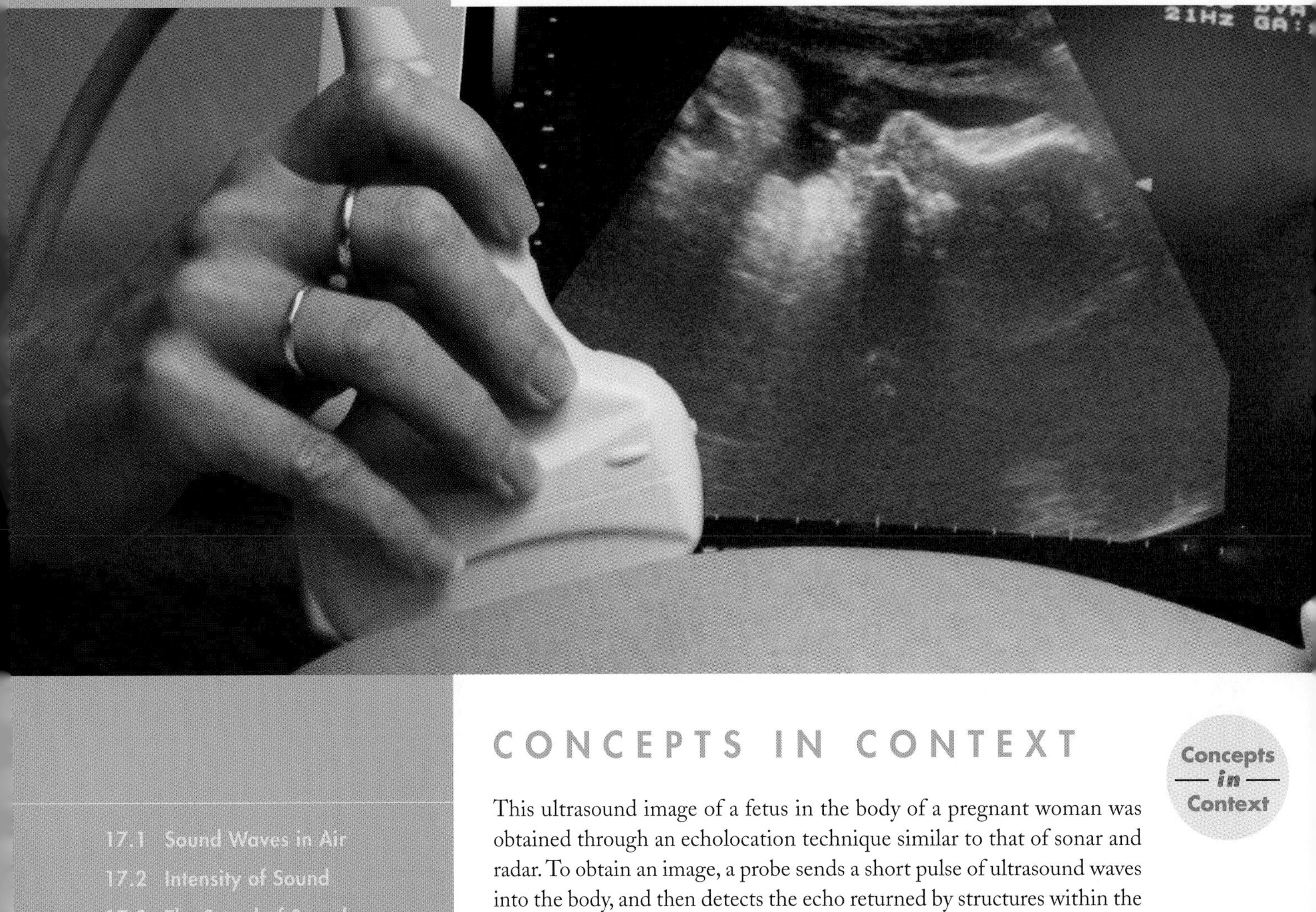

CONCEPTS IN CONTEXT

Concepts in Context

This ultrasound image of a fetus in the body of a pregnant woman was obtained through an echolocation technique similar to that of sonar and radar. To obtain an image, a probe sends a short pulse of ultrasound waves into the body, and then detects the echo returned by structures within the body; a computer analyzes the echoes and constructs the image. Such *sonography* is also widely used for examination of the heart.

As we learn about sound waves, we will consider such questions as:

? What are the frequencies of audible sound waves? Of ultrasound waves? (Section 17.1, page 539)

? How does ultrasound quantitatively determine the location of structures in the body? (Example 3, page 544)

? What determines the smallest size of a structure within the body that can be detected with ultrasound? (Example 8, page 555)

In the preceding chapter we dealt with waves propagating in one dimension, such as waves propagating on a string. These waves are confined; they can travel only in one direction (or the opposite direction) along the string. In this chapter we will deal with waves propagating in two or three dimensions. These waves spread out in all available directions, expanding as they move away from the source. For instance, Fig. 17.1a shows water waves spreading out in two dimensions on the surface of a pond from a pointlike disturbance caused by the impact of a pebble on the surface. And Fig. 17.2a shows sound waves spreading out in three dimensions from the earpiece of a telephone. For a simple diagrammatic representation of such two- or three-dimensional waves, we can use their **wave fronts**, that is, the locations of the wave crests at one instant of time. The wave fronts of the water waves are concentric circles (see Fig. 17.1b), and the wave fronts of the sound waves are concentric spherical shells (see Fig. 17.2b).

As time passes, the wave fronts spread outward, expanding as they move away from the source. This spreading of the waves is a characteristic feature of wave propagation in two or three dimensions. It implies that *the amplitude at a given wave front decreases as the wave front increases in size.* To understand this, consider the water waves in Fig. 17.1a. As a given circular wave front spreads outward, its circumference increases, and its energy is distributed along this larger circumference; thus, the amplitude at the wave front must decrease in accord with the decreased concentration of its energy. This decrease of the wave amplitude with distance is clearly visible in the water waves in Fig. 17.1a. Likewise, for the spherical sound wave spreading out from the telephone earpiece, the wave amplitude decreases with distance (but Fig. 17.2a fails to reveal this decrease, because the method used to make the sound waves visible is not sufficiently sensitive to the amplitude).

At a large distance from the source, the spherical wave fronts of a sound wave can be regarded as nearly flat, provided we concentrate our attention on a small region (see Fig. 17.3). Such *waves with flat, parallel wave fronts are called* **plane waves**.

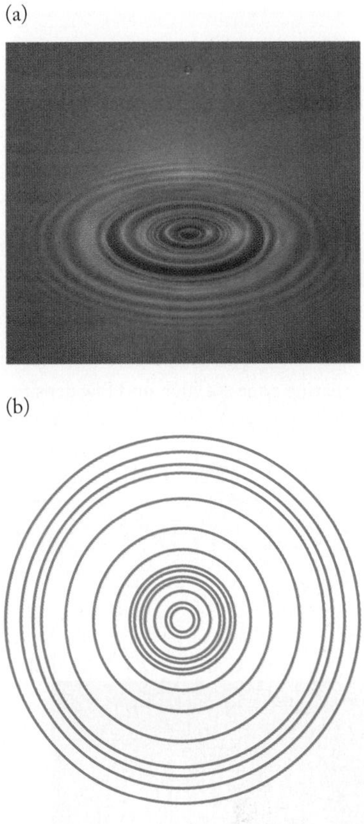

FIGURE 17.1 (a) Circular wave crests and wave troughs spreading on a water surface. (b) The wave fronts are concentric circles.

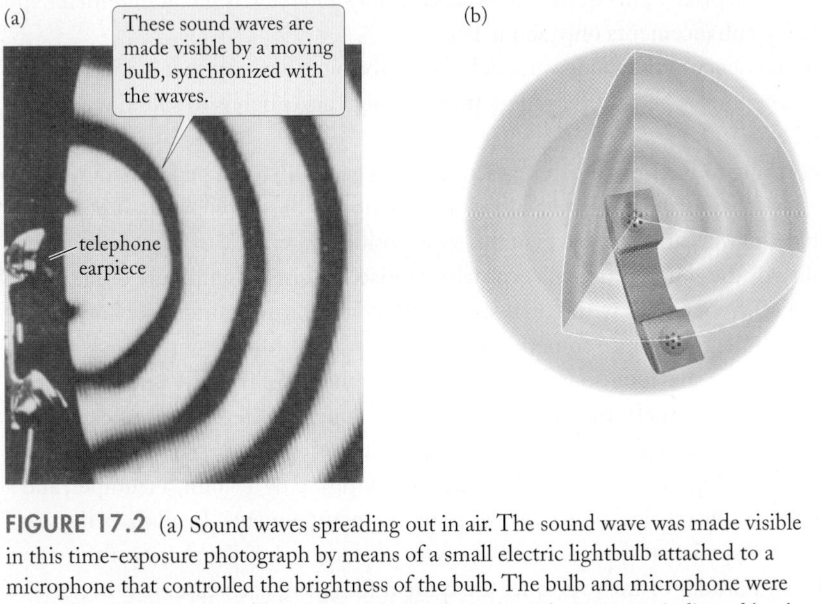

FIGURE 17.2 (a) Sound waves spreading out in air. The sound wave was made visible in this time-exposure photograph by means of a small electric lightbulb attached to a microphone that controlled the brightness of the bulb. The bulb and microphone were swept through the space in front of the telephone earpiece along arcs, as indicated by the fine pattern of ridges. The bulb and microphone were activated (gated) by an electronic circuit synchronized with the oscillations of the sound wave; this effectively produced a snapshot of the sound wave. (b) The wave fronts are concentric spherical shells.

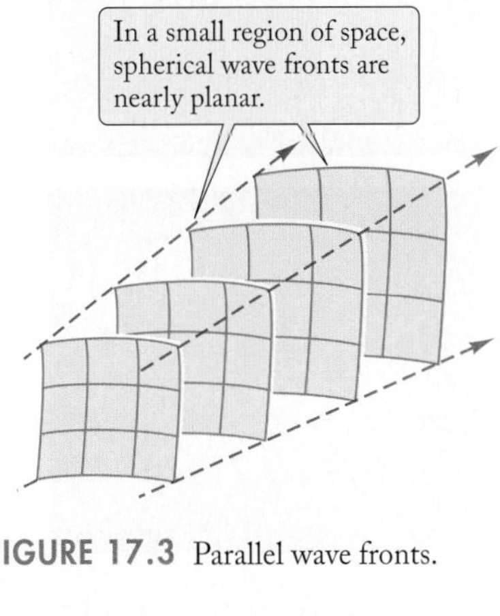

FIGURE 17.3 Parallel wave fronts.

17.1 SOUND WAVES IN AIR

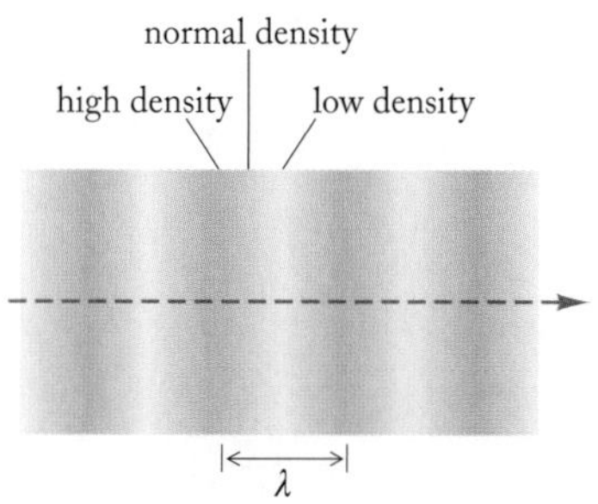

FIGURE 17.4 The sound wave consists of alternating zones of high and low density of air.

A sound wave in air consists of alternating zones of low density and high density (or, equivalently, zones of low pressure and of high pressure). Such zones of alternating density are generated by the vibrating diaphragm of a loudspeaker or the vibrating prong of a tuning fork, which exerts successive pushes on the air that is in contact with it. Figure 17.4 shows a sound wave in air at one instant of time. The alternating zones of low density and high density travel to the right, away from the source. However, although these density disturbances travel, the air as a whole does not travel—the air molecules merely oscillate back and forth.

The pushes of the loudspeaker or of the tuning fork on the air are longitudinal, and the sound wave itself is also **longitudinal**. The air molecules oscillate back and forth *along* the direction of propagation of the sound wave. The restoring force that drives these oscillations comes from the pressure of air. Wherever the density of molecules is higher than normal, the pressure also is higher than normal and pushes the molecules apart; wherever the density of molecules is lower than normal, the pressure also is lower than normal, and therefore the higher pressure of the adjacent regions pushes these molecules together. Thus, the pressure tends to keep the density uniform—it opposes the "deformation" of the air, and it gives air the elasticity required to permit the propagation of a wave.

Note that at the centers of the zones of high density in the wave, the molecules are instantaneously at rest while molecules from the right and the left have converged on them. At the centers of the zones of low density, the molecules also are instantaneously at rest, but molecules on the right and the left have moved away. Thus, *the zones of maximum and minimum density in the wave coincide with zones of zero displacement of the molecules. Conversely, the zones of zero density disturbance in the wave coincide with zones of maximum (positive or negative) displacement of the molecules.*

In Fig. 17.4, the density disturbances are shown much exaggerated. Even in an extremely intense sound wave, such as that produced by the engines of a jet airliner at takeoff, the displacements of the molecules are only about a tenth of a millimeter and the density enhancements only about 1%.

The frequency of the sound wave determines the pitch we hear; that is, it determines whether the tone is perceived as high or low by our ears (pitch is to sound what color is to light).

According to Fourier's theorem, mentioned in Section 16.3, a periodic sound wave of arbitrary shape can be regarded as a superposition of harmonic waves. The relative amplitudes of the harmonic waves in this superposition determine the perceived timbre, or quality, of the sound. Pure noise, or **white noise**, consists of a mixture of harmonic waves of all frequencies with equal strengths. White noise sounds like air rushing through a hole; to produce something like white noise, blow air out of your mouth, making a strong shushing sound (or turn up the volume on your TV set after selecting an inactive channel). In contrast, the musical tones emitted by a musical instrument consist of a mixture of just a few harmonic waves: the fundamental and its first few overtones. Figure 17.5 shows the waveforms emitted by a violin, a trumpet, and a clarinet when the musical note C is played on these instruments. In all cases the wave is periodic, repeating at the rate of 261.7 cycles per second; but the shapes of the waves and the amplitudes of the overtones are quite different in each case. It is because of this difference in the shapes of the waves that the ear can distinguish between diverse musical instruments.

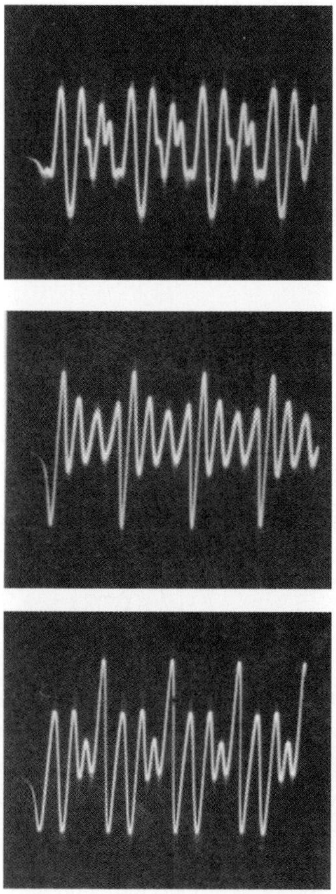

FIGURE 17.5 Waveforms emitted by (a) a violin, (b) a trumpet, and (c) a clarinet playing the note C. These pictures show the instantaneous wave amplitude as a function of time, displayed on an oscilloscope. The wave pattern repeats 261.7 times per second.

Table 17.1 lists the frequencies of the 12 notes of the chromatic musical scale. Other, nonchromatic scales are sometimes used, but we will not consider them here. The chromatic-scale frequencies are based on the system of equal temperament: successive frequencies in the scale differ by a factor of $2^{1/12} = 1.059$. The first entry in the table is middle C, with a frequency of 261.7 Hz (see Fig. 17.6). Any musical note not listed in the table can be obtained by multiplying or dividing the listed frequencies by a factor of 2, or 4, or 8, etc. Musical notes that differ by a factor of 2 in frequency are said to be separated by an **octave.** For example, C one octave above middle C has a frequency of 2×261.7 Hz $= 523.4$ Hz; C two octaves above middle C has a frequency of $2 \times 2 \times 261.7$ Hz $= 1046.8$ Hz, and so on. Incidentally: For a musician, the absolute values of these frequencies are not as important as the ratios of the frequencies. If an orchestra tunes its instruments so their middle C has a frequency of, say, 257 Hz, this will not do any noticeable harm to the music, provided that the frequencies of all the other notes are also decreased in proportion.

TABLE 17.1

THE CHROMATIC MUSICAL SCALE

NOTE	FREQUENCY[a]
C	261.7 Hz
C#	277.2
D	293.7
D#	311.2
E	329.7
F	349.2
F#	370.0
G	392.0
G#	415.3
A	440.0
A#	466.2
B	493.9

[a] Based on a frequency of 440 Hz for A.

The ear performs the task of converting the mechanical oscillations of a sound wave into electric nerve impulses. Thus, it is similar to a microphone, which also converts the mechanical oscillations of sound into electric signals. However, the ear is unmatched in its ability to accommodate a wide range of intensities of sound—the intensities of the faintest and the loudest sounds acceptable to the ear differ by a factor of 10^{12}!

The range of frequencies audible to the human ear extends from 20 Hz to 20000 Hz. These limits are somewhat variable; for instance, the ears of older people are less sensitive to high frequencies. Sound waves above 20000 Hz are called **ultrasound**; some animals—dogs, cats, bats, and dolphins—can hear these frequencies.

Concepts in Context

Ultrasonic waves of very high frequency do not propagate well in air—they are rapidly dissipated and absorbed by air molecules. However, these waves propagate readily through liquids and solids, and this property has been exploited in applications of ultrasound. The use of ultrasound for images such as the chapter photo avoids the damage that X rays might do to the very sensitive tissues of the fetus. The sonography "cameras" that produce such images employ ultrasound waves of a frequency of about 10^6 Hz. Further development of this technique has led to the construction of acoustic microscopes. Most of these devices employ ultrasound waves of a frequency in excess of 10^9 Hz to produce highly magnified images of small samples of materials. The wavelength of sound waves of such extremely high frequency is about 10^{-6} m, roughly the same as the wavelength of ordinary light waves. The micrographs made by experimental acoustic microscopes compare favorably with micrographs made by ordinary optical microscopes (see Fig. 17.7).

FIGURE 17.6 Middle C.

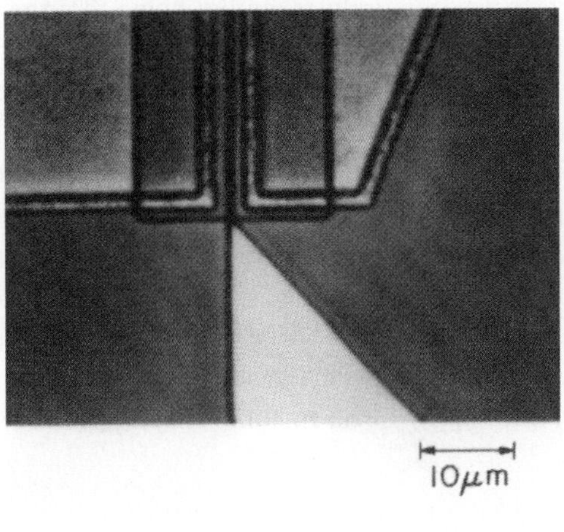

FIGURE 17.7 Acoustic micrograph of a portion of a transistor, at a magnification of 1000×. This picture was made with sound waves of 2.7×10^9 Hz.

Checkup 17.1

QUESTION 1: In the picture of the sound wave shown in Fig. 17.2, where is the amplitude of the wave largest? Where smallest?

QUESTION 2: An old-fashioned speaking tube, for communication between different floors of a house, consists of a pipe that serves as a conduit of sound waves. For a sound wave traveling in such a tube, do the wave fronts expand in size? Does the amplitude of the wave decrease (apart from frictional losses)?

QUESTION 3: Consider a standing sound wave in air. At a density node, is the amplitude of the pressure oscillation maximum or minimum? Is the displacement of the air molecules maximum or minimum?

QUESTION 4: A pianist simultaneously strikes middle C on the piano and C one octave above that. Which note has the shorter wavelength? By what factor?

QUESTION 5: Apart from frictional losses, the amplitude of a wave pulse traveling on a string remains constant, but a wave pulse traveling on the surface of a pond or in air decreases with distance. What causes this difference?

QUESTION 6: At a density antinode, is the amplitude of the pressure oscillation maximum or minimum? Is the displacement of the air molecules maximum or minimum?

(A) Maximum; maximum (B) Maximum; minimum
(C) Minimum; maximum (D) Minimum; minimum

17.2 INTENSITY OF SOUND

intensity

A sound wave is intense and loud if it has a large amplitude. However, the amplitude of a sound wave is hard to measure directly, and it is more convenient to reckon the intensity of a sound wave by the energy it carries. The **intensity** I of a sound wave is defined as the energy per unit time transported by this wave per square meter of wave front, that is, the power transported by this wave per unit area:

$$I = \frac{[\text{energy/time}]}{[\text{area}]} = \frac{[\text{power}]}{[\text{area}]} = \frac{P}{A} \tag{17.1}$$

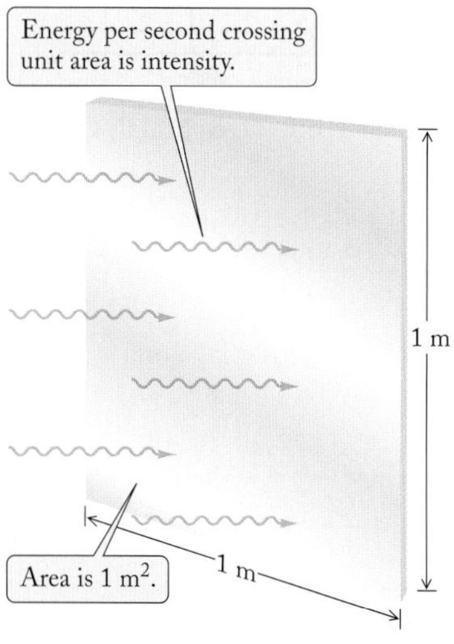

FIGURE 17.8 Area of 1 m^2 facing a wave. The intensity of the wave equals the energy incident on this area in 1 s.

Thus, to measure the intensity, we have to erect an area facing the wave, and we have to check how much energy the wave carries through this area per second (Fig. 17.8). Recall that the energy carried by a wave on a string was proportional to the square of the amplitude of the wave. Similarly, it can be shown that the intensity of a sound wave is proportional to the square of the pressure disturbance it produces in the air; equivalently, the intensity is proportional to the square of the density disturbance.

The unit of intensity is the watt per square meter (W/m^2). At a frequency of 1000 Hz, the minimum intensity audible to the human ear is about 2.5×10^{-12} W/m^2. This intensity is called the threshold of hearing. There is no upper limit for the audible intensity of sound; however, an intensity above 1 W/m^2 produces a painful sensation in the ear.

Note that since the eardrum has an area of about 4×10^{-5} m^2, the energy delivered per second by a sound wave of minimum intensity is only about 2.5×10^{-12} $J/m^2 \times 4 \times 10^{-5}$ $m^2 \approx 10^{-16}$ J; this is a very small amount of energy, and it testifies to the extreme sensitivity of the ear.

The intensity of sound is often expressed on a logarithmic scale called the intensity level. The unit of intensity level is the **decibel** (dB); like the radian, this unit is a pure number, without any dimensions of m, s, or kg. The definition of intensity level is as follows: We take an intensity of 1.0×10^{-12} W/m^2 as our standard of intensity, which corresponds to 0 dB.[1] An intensity 10 times as large corresponds to 10 dB; an intensity 100 times as large corresponds to 20 dB; an intensity 1000 times as large corresponds to 30 dB; and so on. This scale of intensity level is intended to agree with our subjective perception of the loudness of sounds. We tend to underestimate increments in the intensity of sound—our ears perceive a sound of 100×10^{-12} W/m^2 as only twice as loud as a sound of 10×10^{-12} W/m^2.

Mathematically, the relationship between the intensity in W/m^2 and the intensity level in dB is given by a formula involving a logarithm:

$$[\text{intensity level in dB}] = (10\ \text{dB}) \times \log_{10}\left(\frac{[\text{intensity in W/m}^2]}{1.0 \times 10^{-12}\ \text{W/m}^2}\right) \qquad (17.2)$$

intensity level in decibels

EXAMPLE 1 Express the threshold of hearing (2.5×10^{-12} W/m^2) and the threshold of pain (1.0 W/m^2) in decibels.

SOLUTION: According to Eq. (17.2), we find that the intensity level corresponding to 2.5×10^{-12} W/m^2 is

$$(10\ \text{dB}) \times \log_{10}\left(\frac{2.5 \times 10^{-12}\ \text{W/m}^2}{1.0 \times 10^{-12}\ \text{W/m}^2}\right) = (10\ \text{dB}) \times \log_{10}(2.5)$$

With our calculator, we obtain log(2.5) = 0.40 (on calculators, the base-10 logarithm "$\log_{10}$" usually appears simply as "log"), and hence the intensity level is (10 dB) × 0.40 = 4.0 dB.

Likewise, we find that the intensity level corresponding to 1.0 W/m^2 is

$$(10\ \text{dB}) \times \log_{10}\left(\frac{1.0\ \text{W/m}^2}{1.0 \times 10^{-12}\ \text{W/m}^2}\right) = (10\ \text{dB}) \times \log_{10}(10^{12})$$

Since $\log_{10}(10^{12}) = 12$, the intensity level is 120 dB.

Table 17.2 gives some examples of sounds of different intensities.

As a sound wave spreads out from its source, its intensity falls off because the area of the wave front grows larger, and therefore the wave energy per unit area grows smaller. Figure 17.9 helps to make this clear; it shows a spherical wave front at successive instants of time (such a spherical wave front is produced when a source emits waves uniformly in all directions). The wave front grows from an old radius r_1 to a new radius r_2; correspondingly, the area of the spherical wave front grows from $4\pi r_1^2$ to $4\pi r_2^2$. The total power carried by the wave front remains the same; hence, by the definition Eq. (17.1), the power per unit area, or the intensity, must be in inverse proportion to the area. If the intensity at r_1 is I_1 and the intensity at r_2 is I_2, then

$$I_1 \propto \frac{1}{4\pi r_1^2} \quad \text{and} \quad I_2 \propto \frac{1}{4\pi r_2^2} \qquad (17.3)$$

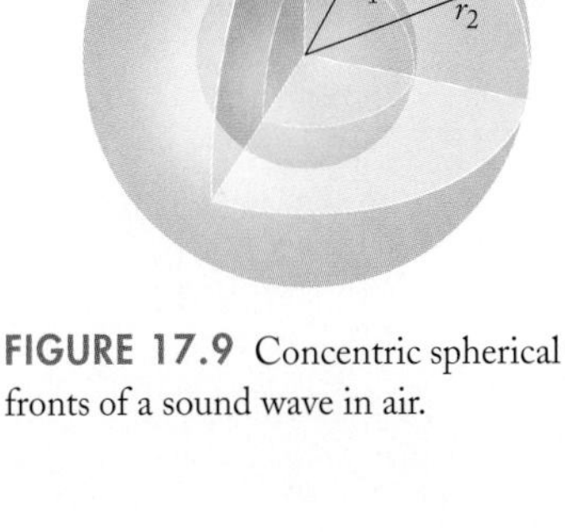

FIGURE 17.9 Concentric spherical wave fronts of a sound wave in air.

[1] More precisely, the standard of intensity, which corresponds to 0 dB, is 0.937×10^{-12} W/m^2. But it is conventional, and adequate for our purposes, to round this up to 1.0×10^{-12} W/m^2.

Taking the ratio of these two proportions, we obtain

intensity decrease with distance

$$\frac{I_2}{I_1} = \frac{r_1^2}{r_2^2} \tag{17.4}$$

This says that *the intensity of a spherical wave is inversely proportional to the square of the distance from the source.*

Although Fig. 17.9 shows the wave front spreading out uniformly in all directions, Eq. (17.4) remains valid if the wave is a beam, such as the beam emitted by a loudspeaker with a horn. Although the beam is aimed in some preferential direction, it spreads as it propagates, and the intensity falls off as the inverse square of the distance along any radial line within the beam.

TABLE 17.2 SOME SOUND INTENSITIES

SOUND	INTENSITY LEVEL	INTENSITY
Rupture of eardrum	160 dB	1.0×10^4 W/m^2
Jet engine (at 30 m)	130	10
Threshold of pain	120	1.0
Rock music	115	0.30
Thunder (loud)	110	0.10
Subway train (New York City)	100	1.0×10^{-2}
Heavy street traffic	70	1.0×10^{-5}
Normal conversation	60	1.0×10^{-6}
Whisper	20	1.0×10^{-10}
Normal breathing	10	1.0×10^{-11}
Threshold of hearing	4	2.5×10^{-12}

EXAMPLE 2 At a distance of 30 m from a jet engine, the intensity of sound is 10 W/m^2, and the intensity level is 130 dB. What are the intensity and the intensity level at a distance of 180 m?

SOLUTION: According to Eq. (17.4), when we increase the distance from r_1 = 30 m to r_2 = 180 m (see Fig. 17.10), we decrease the intensity by a factor r_1^2/r_2^2. Hence the intensity will be

$$I_2 = \frac{r_1^2}{r_2^2} I_1 = \frac{(30\text{ m})^2}{(180\text{ m})^2} I_1 = 2.8 \times 10^{-2} \times I_1 = 2.8 \times 10^{-2} \times 10\text{ W/m}^2$$
$$= 0.28\text{ W/m}^2$$

By definition (17.2), the corresponding intensity level is

$$(10\text{ dB}) \times \log_{10}\left(\frac{0.28\text{ W/m}^2}{1.0 \times 10^{-12}\text{ W/m}^2}\right) = (10\text{ dB}) \times \log_{10}(0.28 \times 10^{12})$$
$$= 114\text{ dB}$$

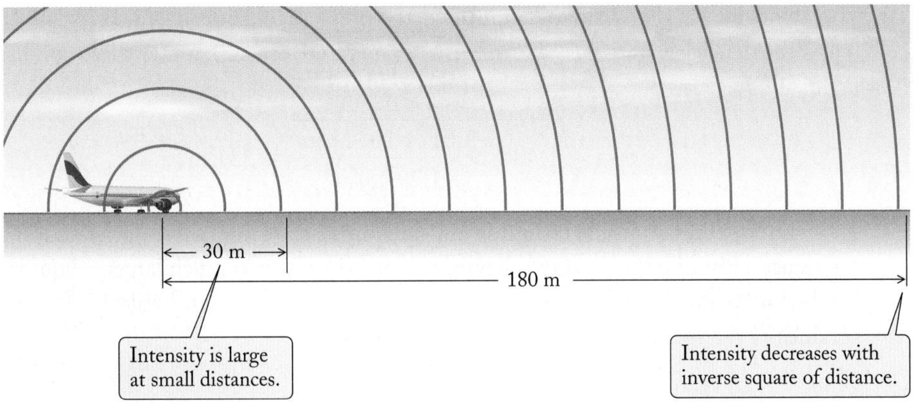

FIGURE 17.10 Sound wave spreading outward from a jet engine.

Checkup 17.2

QUESTION 1: According to Table 17.2, by how many dB do the intensity levels of the noise of thunder and of a subway train differ? By what factor do the intensities differ?

QUESTION 2: If the intensity of sound at a distance of 1 m in front of a loudspeaker is $10^{-5}\ \mathrm{W/m^2}$, what is the intensity at a distance of 10 m? 100 m?

QUESTION 3: One source emits a spherical sound wave, uniformly in all directions. A second source emits a hemispherical sound wave, uniformly only over all the directions *in front of* the source. Both sources produce the same total power. What is the ratio of the intensity at some distance from the first source to the intensity at twice that distance in front of the second source?

(A) $\frac{1}{4}$ (B) $\frac{1}{2}$ (C) 1 (D) 2 (E) 4

17.3 THE SPEED OF SOUND; STANDING WAVES

As in the case of a wave on a string, the speed of sound in air depends on the restoring force and on the amount of mass. Since the restoring force is proportional to the pressure, and since the amount of mass is proportional to the density of the air, we expect the speed of sound to depend on the pressure and on the density. A somewhat involved calculation shows that the theoretical formula for the speed of sound in air is

$$v = \sqrt{1.40\frac{p_0}{\rho_0}} \qquad (17.5)$$

where p_0 designates the air pressure and ρ_0 the mass density of the air. Note that this formula does display the expected dependence discussed in Section 16.2: the speed is the square root of an elastic property (the pressure, proportional to the force) divided by an inertial property (the density, proportional to the mass)—the speed is high if the pressure is large, and the speed is low if the density is large.

Under so-called standard conditions, the pressure of air is $p_0 = 1.01 \times 10^5\ \text{N/m}^2$ and the density is $\rho_0 = 1.29\ \text{kg/m}^3$ (at a temperature of 0°C); under these conditions Eq. (17.5) yields

$$v = \sqrt{1.40 \times \frac{1.01 \times 10^5\ \text{N/m}^2}{1.29\ \text{kg/m}^3}} = 331\ \text{m/s} \tag{17.6}$$

The speed of sound in liquids and in solids is considerably higher than in air because, although the density is larger, the restoring force is much larger—liquids and solids offer much more opposition to compression than gases. Table 17.3 gives the values of the speed of sound in some materials.

Concepts — in — Context

EXAMPLE 3 In sonography (see the chapter photo), a probe pressed against the skin is used to send a sound pulse into the human body, and the same probe then detects the echo reflected by structures in the body. (a) Suppose that a pulse from the probe takes 1.6×10^{-4} s to travel to a bone and return to the probe. How far is the bone from the probe? (b) The probe must emit pulses that are short enough that the emitted pulse is complete before the reflected pulse returns. If we wish to detect a structure as close as 1.0 cm, what is the maximum pulse duration? The speed of sound in body tissues is about the same as that in water, 1500 m/s.

SOLUTION: (a) The round-trip distance traveled by the sound pulse is

$$x = v\Delta t = 1500\ \text{m/s} \times 1.6 \times 10^{-4}\ \text{s} = 0.24\ \text{m} \tag{17.7}$$

The distance to the bone is one-half as large as the round-trip distance, 0.12 m.

(b) For a structure at 1.0 cm, the round-trip distance is 0.020 m. Thus the pulse must be complete before a time

$$t = \frac{\Delta x}{v} = \frac{2.0 \times 10^{-2}\ \text{m}}{1500\ \text{m/s}} = 13 \times 10^{-6}\ \text{s}$$

The maximum pulse duration is 13 microseconds.

TABLE 17.3

THE SPEED OF SOUND IN SOME MATERIALS

MATERIAL	v
Air	
0°C, 1 atm	331 m/s
20°C, 1 atm	344
100°C, 1 atm	386
Helium gas, 0°C, 1 atm	965
Water (distilled)	1497
Water (sea)	1531
Aluminum	5104
Iron	5130
Glass	5000–6000
Granite	6000

A simple method for the measurement of the speed of sound in air takes advantage of standing waves in a tube open at one end and closed at the other (see Fig. 17.11). The *standing sound wave in the column of air in this tube must then have a displacement node at the closed end, since the motion of the air is restricted by the wall at this end. The wave must have a displacement antinode at the open end.* This second condition is not so obvious, but can be understood by first considering the pressure variations in the wave. The pressure excess, relative to normal atmospheric pressure, is large where the displacement of the molecules is small, and small where the displacement of the molecules is large (see the discussion of Fig. 17.4). Thus, the pressure antinodes are displacement nodes, and the pressure nodes are displacement antinodes. At the open end of our tube, the pressure must remain approximately constant, because the open end is accessible to the atmosphere, and hence any incipient decrease or increase of the pressure would immediately lead to an inflow or outflow of air from the surrounding atmosphere, canceling the pressure change. Thus, the atmosphere behaves as a "reservoir" of approximately constant pressure, and the pressure excess of the standing wave must have a node at the open end of the tube. This implies that the open end is a displacement antinode.

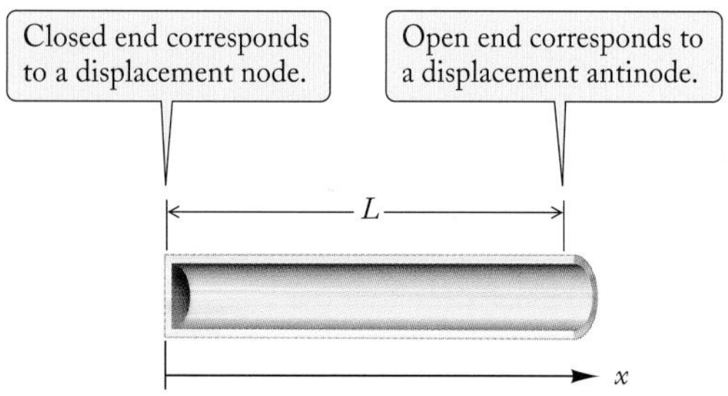

FIGURE 17.11 A tube open at one end and closed at the other.

With these boundary conditions for the closed end and the open end of the tube, the possible standing waves, or normal modes, are as shown in Fig. 17.12. If the length of the tube is L, then an odd number of quarter wavelengths must fit in the length L:

$$L = \frac{\lambda}{4}, \quad 3\frac{\lambda}{4}, \quad 5\frac{\lambda}{4}, \quad \ldots \tag{17.8}$$

so the wavelengths for these normal modes are

$$\lambda_1 = 4L, \quad \lambda_2 = \tfrac{4}{3}L, \quad \lambda_3 = \tfrac{4}{5}L, \quad \ldots \tag{17.9}$$

The eigenfrequencies of these normal modes are given by $f = v/\lambda$:

tube open at one end:

$$f_1 = \frac{v}{4L}, \quad f_2 = \frac{3v}{4L}, \quad f_3 = \frac{5v}{4L}, \quad \ldots \tag{17.10}$$

Note that the expressions (17.9) for the wavelengths of the normal modes of sound in a tube differ from the expressions (16.30) for the wavelengths of the normal modes of a string. This is due to the difference in boundary conditions: the tube has a node at one end and an antinode at the other end, whereas the string has nodes at both ends. For the tube closed at one end, we see from Eq. (17.10) that only odd-integer multiples (odd harmonics) of the fundamental frequency occur. In contrast, for a tube *open at both ends,* all integer harmonics occur. This is due to the boundary condition that the wave now has displacement antinodes (and pressure nodes) at *both* ends, and, like the string, requires that an integer number of half wavelengths must fit into the length L. The normal mode wavelengths are given by Eq. (16.30), and the eigenfrequencies are [see Eq. (16.31)]

tube open at both ends:

$$f_1 = \frac{v}{2L}, \quad f_2 = \frac{v}{L}, \quad f_3 = \frac{3v}{2L}, \quad \ldots \tag{17.11}$$

The speed of sound can be determined by means of Eq. (17.10) or Eq. (17.11) by measuring the resonant frequency of a tube of known length.

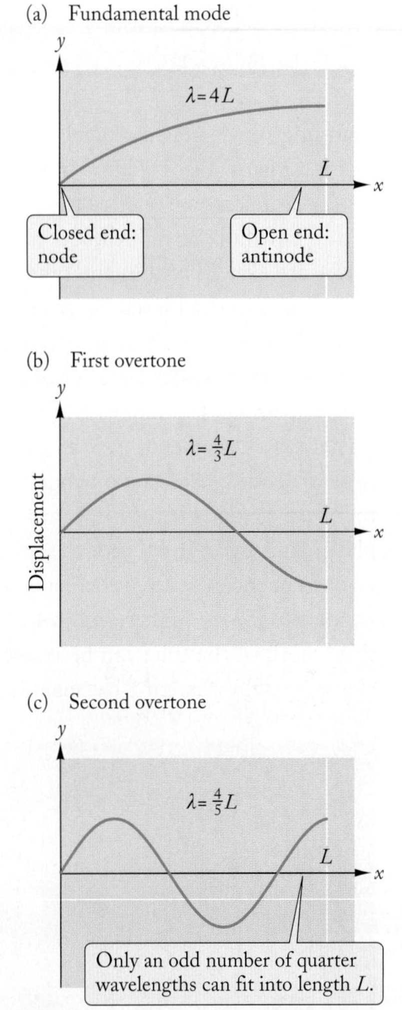

FIGURE 17.12 Possible standing waves in tube open at one end. (a) The fundamental mode. (b) The first overtone. (c) The second overtone.

✔ Checkup 17.3

QUESTION 1: When you watch a thunderstorm, you see the lightning first, and you hear the thunder afterward. Why is the thunder delayed?

QUESTION 2: Consider a tube of length L *closed* at both ends. What is the frequency of the fundamental mode of a sound wave in this tube?

QUESTION 3: For a violin, does the wavelength of the emerging sound match the wavelength of the standing wave on its string? Does the frequency of the emerging sound match the frequency of the standing wave on the string?

QUESTION 4: You have two tubes of equal lengths. The first is open at one end and closed at the other; the second is open at both ends. What is the ratio of the fundamental frequency of sound waves in the first tube to that in the second tube?

(A) $\frac{1}{4}$ (B) $\frac{1}{2}$ (C) 1 (D) 2 (E) 4

PHYSICS IN PRACTICE MUSICAL INSTRUMENTS

Standing waves play a crucial role in most musical instruments. Organs (Fig. 1), flutes, trumpets, trombones, and other wind instruments are essentially tubes open at their distant end with a blowhole or mouthpiece at the other end. Standing waves are excited within the tube by a stream of air blown across or into the blowhole or mouthpiece. In organs and flutes, the blowhole acts as an open end of the tube, and therefore the normal modes are those of a tube with *two* open ends. In trumpets and trombones, the lips of the player act approximately as a closed end, and the normal modes are those of a tube with one closed and one open end (however, the standing wave extends somewhat into the mouth cavity of the player, and the normal modes are quite complicated). The eigenfrequencies of the tube depend on its length. In many wind instruments—flutes, trumpets, French horns—the effective length of the tube can be varied by opening or closing valves, thereby changing the eigenfrequencies.

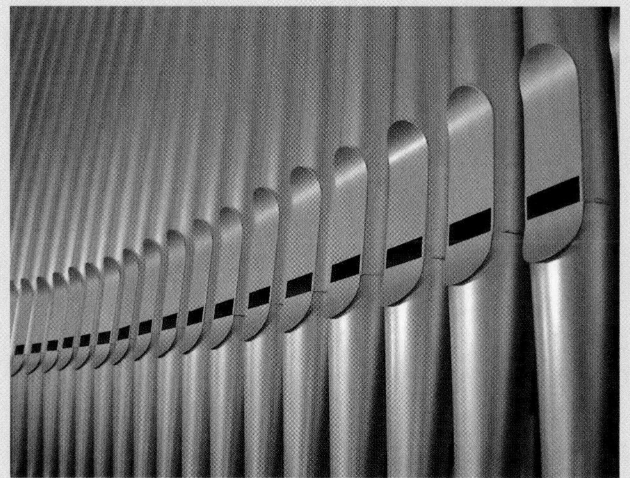

FIGURE 1 Organ pipes.

The excitation of vibrations in the tube of an organ or a flute by a steady stream of air blown across the blowhole arises from a rotational motion that develops behind the edges of the blowhole when the velocity of flow is high. As the air streams past the edge, it forms a vortex (see Fig. 2). This vortex soon breaks away from the edge and is replaced by another vortex, and another, and so on. The regular succession of vortices constitutes a vibration of the stream of air, and this excites standing waves in the tube of the organ by resonance.

The excitation of vibrations in the tube of a trumpet or trombone involves a different mechanism. These instruments have a cup-shaped mouthpiece, across which the player stretches his lips, which then behave somewhat like a pair of strings under tension, with a natural period of vibration. The vibration of the lips is triggered by the stream of air that the player blows out of his mouth. If the lips are initially close together and the gap between them is small, the pressure in the mouth builds up. This high pressure pushes the lips apart. But when the gap between them becomes wide, air rushes out and the pressure decreases. This permits the lips to snap back to their initial configuration. The gap between the lips therefore periodically widens and narrows, with a natural frequency that is determined by their tension. The periodic puffs of air produced by this vibration excite standing waves in the tube of the trumpet.

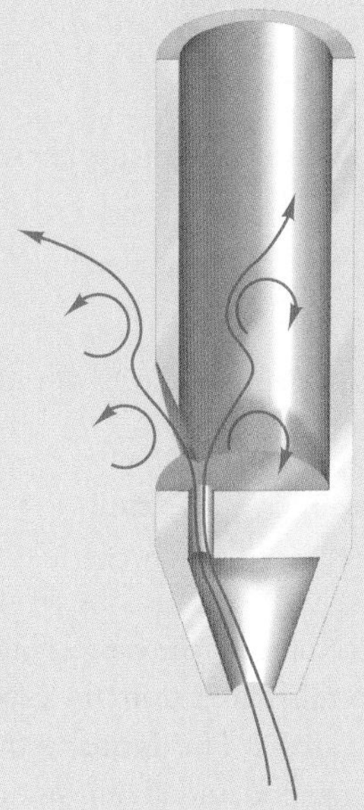

FIGURE 2 Vortices at the blowhole of an organ pipe.

Stringed instruments—violins, guitars, mandolins—use a resonant cavity to amplify and modify the sound produced by the string. The cavity is mechanically coupled to the string, and the vibrations of the latter excite resonant vibrations in the former. The resonant vibrations involve not only standing waves in the air in the cavity, but also standing waves in the solid material (wood) of the walls. Because the area of the body of, say, a violin is much larger than the area of its strings, the body pushes against much more air and radiates sound more efficiently than the strings (Fig. 3). Hence, most of the sound from a violin emerges from its body.

FIGURE 3 Vibrating violin.

17.4 THE DOPPLER EFFECT

Online Concept Tutorial 19

The speed of a sound wave in air is 331 m/s when measured in a reference frame at rest in the air. But when measured in a reference frame moving through the air, the speed of the sound wave will be larger or smaller, depending on the direction of motion of the reference frame. For example, if a train moving at 30 m/s approaches a stationary siren emitting sound waves (see Fig. 17.13), the speed of the sound waves relative to the train will be (331 + 30) m/s = 361 m/s. And if the train moves away from the siren, the speed of the sound waves relative to the train will be (331 − 30) m/s = 301 m/s.

The motion of the train affects not only the speed of the sound waves, but also their frequency. For instance, if the train approaches the siren, it runs head-on into the sound waves (see Fig. 17.13a) and hence encounters more wave fronts per second than if it were stationary; and if the train recedes from the siren, it runs with the sound waves (see Fig. 17.13b) and hence encounters fewer wave fronts per second. Consequently, *a receiver on the train will detect a higher frequency when approaching the siren, and a lower frequency when receding.* This frequency change caused by the motion of the receiver (or by motion of the source; see below) is called the **Doppler shift**.

To calculate the frequency shift, we note that in the reference frame of the air, we have the usual relation between the frequency, speed, and wavelength of the sound wave,

$$f = v/\lambda \tag{17.12}$$

and in the reference frame of the train, we have a corresponding relation

$$f' = v'/\lambda \tag{17.13}$$

The wavelengths in Eqs. (17.12) and (17.13) are exactly the same because the distance between the wave crests does not depend on the reference frame. Dividing Eq. (17.13) by Eq. (17.12), we obtain

$$\frac{f'}{f} = \frac{v'}{v} \tag{17.14}$$

We will designate by V_R the speed of the train acting as receiver of sound waves (R for receiver). In the reference frame of the train, the speed of sound is then $v' = v \pm V_R$, where the positive sign corresponds to motion of the train toward the source of sound

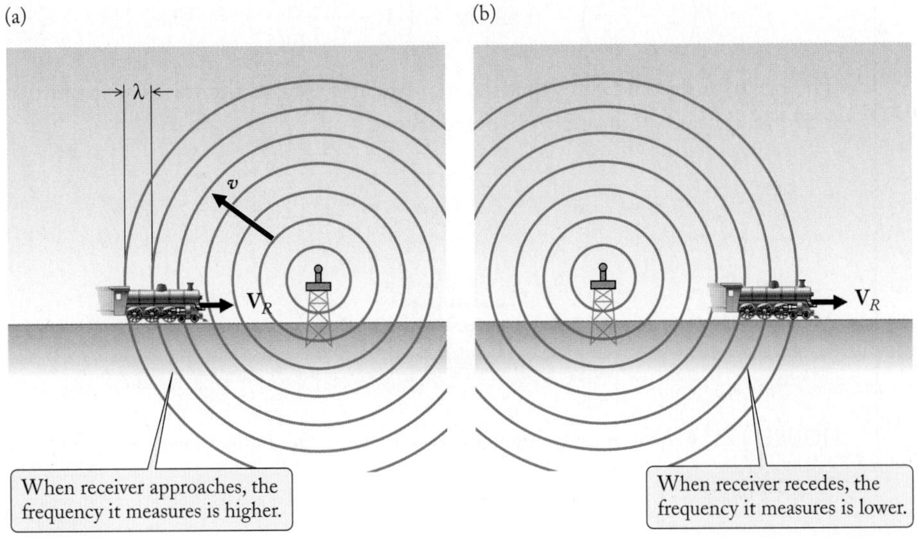

FIGURE 17.13 (a) Train approaching a siren. The train encounters *more* wave fronts per unit time than when stationary. (b) Train receding from a siren. The train encounters *fewer* wave fronts per unit time than when stationary.

and the negative sign to motion of the train away from the source of sound. With this expression for v', Eq. (17.14) yields

Doppler shift, moving receiver

$$f' = f\left(1 \pm \frac{V_R}{v}\right) \quad \begin{array}{l} + \text{ for approaching receiver} \\ - \text{ for receding receiver} \end{array} \tag{17.15}$$

EXAMPLE 4 Suppose that a stationary siren emits a tone of frequency 440 Hz as the train moves away from it at 30.0 m/s. What is the frequency received on the train?

SOLUTION: From Eq. (17.14),

$$f' = f\left(1 - \frac{V_R}{v}\right) = 440\ \text{Hz} \times \left(1 - \frac{30.0\ \text{m/s}}{331\ \text{m/s}}\right) = 400\ \text{Hz}$$

Note that if the receiver is moving away from the source at a speed equal to the speed of sound ($V_R = v$), then the frequency f' is zero; this simply means that the receiver is moving exactly with the waves, and therefore no wave fronts catch up with it. If the receiver is moving away at a speed greater than the speed of sound ($V_R > v$), then Eq. (17.15) gives a negative frequency; this means that the receiver overruns the wave fronts from behind. A receiver speed equal to or larger than the speed of sound can be achieved only by mounting the receiver on a supersonic aircraft, an arrangement of no practical interest. However, Eq. (17.15) applies not only to sound waves, but also to water waves and other kinds of waves. For water waves it is not at all hard to arrange for a "receiver" with a speed V_R in excess of the speed v of the waves.

EXAMPLE 5 A motorboat (see Fig. 17.14) speeding at 6.0 m/s is moving in the same direction as a group of water waves of frequency 0.62 Hz and speed 2.5 m/s (relative to the water). What is the frequency with which the wave crests pound on the motorboat?

SOLUTION: We again use Eq. (17.14):

$$f' = f\left(1 - \frac{V_R}{v}\right) = 0.62\ \text{Hz} \times \left(1 - \frac{6.0\ \text{m/s}}{2.5\ \text{m/s}}\right) = -0.87\ \text{Hz}$$

The negative sign indicates that the motorboat overtakes the waves at the rate of 0.87 Hz, that is, about one wave every second.

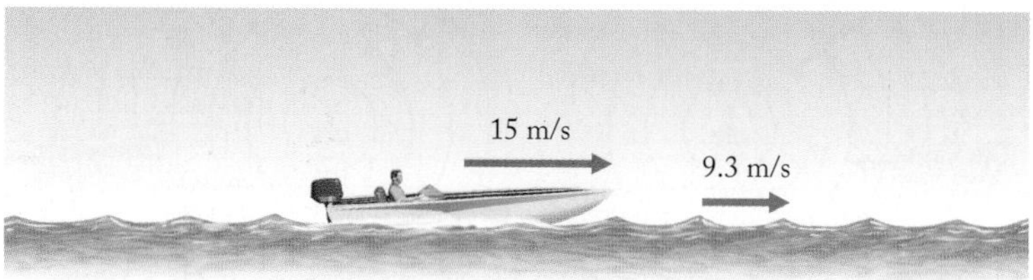

FIGURE 17.14 Motorboat overtaking waves on the surface of the water.

A shift between the frequency emitted by the source of sound waves (or other waves) and the frequency detected by the receiver will also occur if the emitter is in motion relative to the air and the receiver is at rest. For example, if a train approaching a railroad crossing blows a whistle, the successive wave fronts emitted by the whistle are centered at intervals along the path of the whistle, and they will be crowded together in the forward direction and spread apart in the rearward direction (see Fig. 17.15). Consequently, *a stationary receiver will detect a higher frequency when the emitter of sound approaches, and a lower frequency when the emitter of sound recedes.* As the emitter passes by the stationary receiver, the detected frequency suddenly changes from high to low; that is, the pitch of the train's whistle suddenly drops. This also explains the sudden drop in pitch of the hum of a car engine that you hear when standing next to a road as the car passes by you.

We will designate by V_E the speed of the emitter of sound (E for emitter). To calculate the frequency change produced by the motion of the emitter, we begin by noting that in the time $1/f$ corresponding to one period, the train travels a distance $(1/f)V_E$ and hence the wavelength is shortened or lengthened from its normal value λ to a new value $\lambda' = \lambda \mp V_E/f$, where the negative sign corresponds to motion of the emitter toward the receiver and the positive sign to motion away from the receiver. The new frequency is therefore

$$f' = \frac{v}{\lambda'} = \frac{v}{\lambda \mp (V_E/f)} = \frac{v}{(v/f) \mp (V_E/f)} \tag{17.16}$$

If we multiply both the numerator and the denominator of the right side of this equation by f/v, we obtain

$$f' = f\left[\frac{1}{1 \mp (V_E/v)}\right] \qquad \begin{matrix} - \text{ for approaching emitter} \\ + \text{ for receding emitter} \end{matrix} \tag{17.17}$$

Doppler shift, moving emitter

By combining the formulas (17.15) and (17.17) we can obtain a general formula for the Doppler shift when both emitter and receiver are in motion. In this case, the

CHRISTIAN DOPPLER (1803–1853)
Austrian physicist. After Doppler discovered his formula for the frequency shift of sound, he recognized that light from a moving source should also be subject to a frequency shift, resulting in a change of color of the received light. However, unless the speed of the source is extremely large, such changes of the color of light are too small to be perceived by the eye.

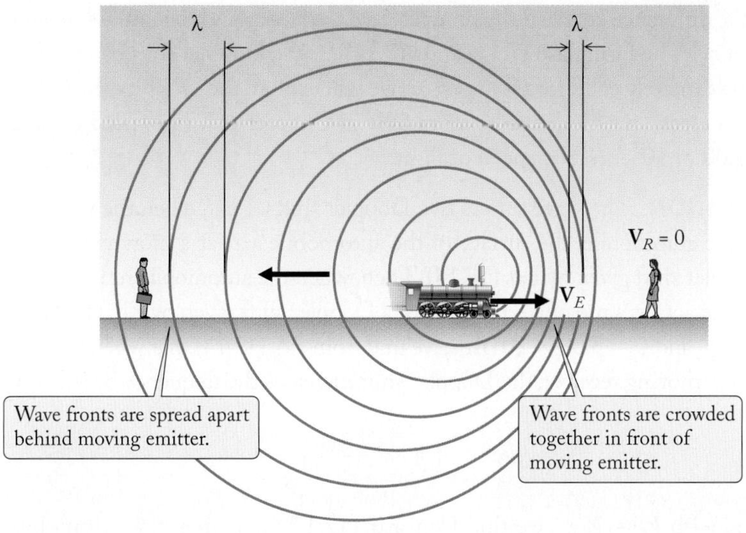

FIGURE 17.15 Train emitting sound waves while in motion. The wavelength ahead of the train is shorter and that behind the train is longer than when the train is stationary.

motion of the emitter causes a change of frequency by a factor of $1/[1 \mp (V_E/v)]$ and the motion of the receiver causes a further change of frequency by a factor of $[1 \pm (V_R/v)]$, leading to a final frequency

$$f' = f\left[\frac{1 \pm (V_R/v)}{1 \mp (V_E/v)}\right] \tag{17.18}$$

EXAMPLE 6 Suppose that the whistle of a train emits a tone of frequency 440 Hz as the train recedes from a stationary observer at 30.0 m/s. What frequency does the observer hear?

SOLUTION: According to Eq. (17.17), for a receding emitter,

$$f' = f\left[\frac{1}{1 + (V_E/v)}\right]$$

$$= 440 \text{ Hz} \times \frac{1}{1 + [(30.0 \text{ m/s})/(331 \text{ m/s})]} = 403 \text{ Hz}$$

COMMENT: If we compare the results of Examples 4 and 6, we see that motion of the emitter and motion of the receiver have nearly the same effect on the frequency—in both examples the frequency is decreased by about 10%. This symmetry of the Doppler shift has to do with the low speed of the motion. When the speed is low compared with the speed of sound, the effects of motion of the emitter and motion of the receiver are approximately the same; but when the speed is high, the effects of motion of the emitter and motion of the receiver are quite different.

EXAMPLE 7 Doppler radar units ("radar guns"), employed by police to measure the speeds of automobiles, consist of a transmitter of radar waves and a receiver. The transmitter sends a wave of frequency 8.00×10^9 Hz toward the target, and the receiver detects the reflected wave sent back by the target. Suppose that an automobile is approaching a radar unit at 100 km/h. What is the difference between the final received frequency and the initial transmitted frequency? The formulas (17.15) and (17.17) for the Doppler shift of sound waves are approximately valid for radar waves (we will explore the Doppler shift of radar and other electromagnetic waves more fully in Chapter 36). The speed of radar waves is 3.00×10^8 m/s, the speed of light.

SOLUTION: The wave suffers two Doppler shifts. First, when the wave is incident on the automobile, the surface of the automobile acts as a moving receiver, with a Doppler shift given by Eq. (17.15). Then, when the automobile reflects the wave, its surface acts as a moving emitter, with a Doppler shift given by Eq. (17.17). With $V_R = V_{\text{auto}}$ and $v = 3.00 \times 10^8$ m/s, we find from Eq. (17.14) that when the automobile acts as moving receiver, the Doppler shift increases the frequency by a factor of

$$\left(1 + \frac{V_{\text{auto}}}{v}\right)$$

And with $V_E = V_{\text{auto}}$, we find from Eq. (17.17) that when the automobile acts as moving emitter, the Doppler shift increases the frequency by an extra factor of

$$\frac{1}{1 - (V_{\text{auto}}/v)}$$

Multiplying the initial frequency by the product of these factors gives us the final frequency:

$$f' = f \times \left(1 + \frac{V_{\text{auto}}}{v}\right) \times \frac{1}{1 - (V_{\text{auto}}/v)} = f \times \frac{1 + (V_{\text{auto}}/v)}{1 - (V_{\text{auto}}/v)}$$

The difference between the final and the initial frequencies is

$$f' - f = f \times \left[\frac{1 + (V_{\text{auto}}/v)}{1 - (V_{\text{auto}}/v)} - 1\right] = f \times \frac{[1 + (V_{\text{auto}}/v)] - [1 - (V_{\text{auto}}/v)]}{1 - (V_{\text{auto}}/v)}$$

$$= f \times \frac{(2V_{\text{auto}}/v)}{1 - (V_{\text{auto}}/v)} \qquad (17.19)$$

The speed of the automobile is

$$V_{\text{auto}} = 100\text{ km/h} = 100\text{ km/h} \times \frac{1000\text{ m}}{1\text{ km}} \times \frac{1\text{ h}}{3600\text{ s}} = 27.8\text{ m/s}$$

Since V_{auto} is much, much smaller than v, we can here neglect the term V_{auto}/v in the denominator of Eq. (17.19), and we obtain the approximate result for the frequency shift

$$f' - f = f \times \frac{2V_{\text{auto}}}{v} = 8.00 \times 10^9\text{ Hz} \times \frac{2 \times 27.8\text{ m/s}}{3.00 \times 10^8\text{ m/s}}$$

$$= 1.48 \times 10^3\text{ Hz}$$

COMMENT: Although this is a very small shift (less than one part in a million), the electronics in a radar gun can accurately measure it by mixing the shifted frequency and the original frequency; this results in an easily detected difference frequency, as in the phenomenon of "beats" discussed in Chapter 16.

Finally, let us consider the case of an emitter, such as a fast aircraft, moving at a speed nearly equal to the speed of sound. If the aircraft emits sound of some frequency f, then Eq. (17.17) indicates that the frequency received at points just ahead of the aircraft is very large—in the limiting case of a speed V_E equal to the speed of sound, the frequency becomes infinite. This is because all the wave fronts are infinitely bunched together, and they all arrive at almost the same instant as the aircraft (see Fig. 17.16).

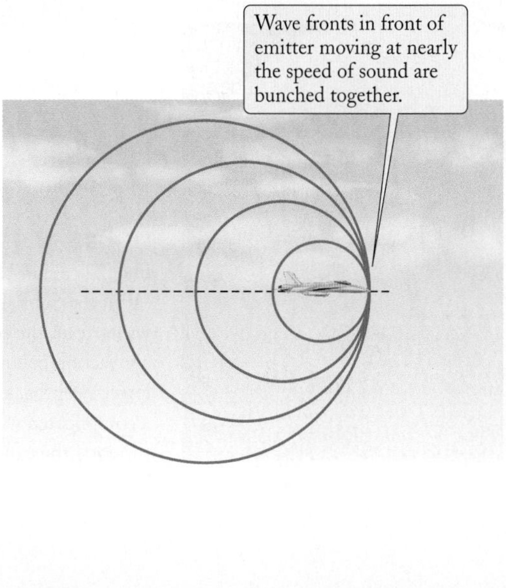

FIGURE 17.16 A subsonic aircraft at a speed very close to the speed of sound emitting sound waves.

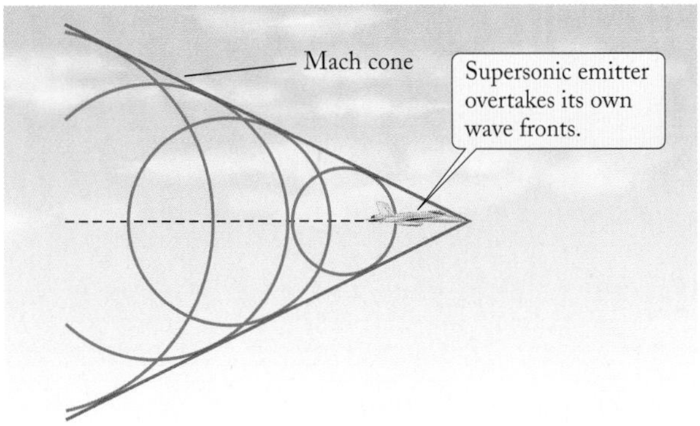

FIGURE 17.17 A supersonic aircraft emitting sound waves.

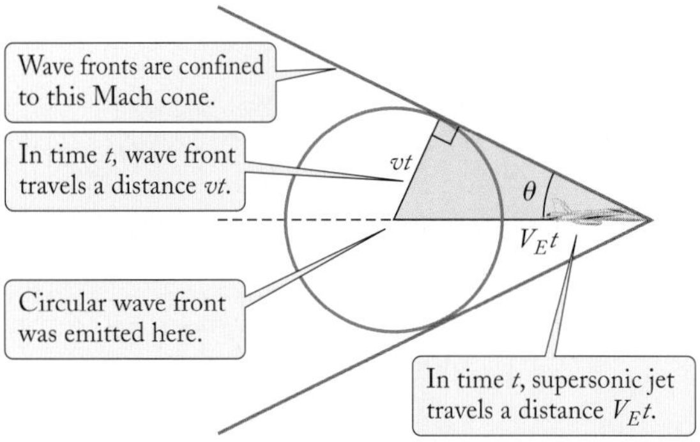

FIGURE 17.18 Mach cone. In a time t, the aircraft moves a distance $V_E t$ and the initial sound wave moves a distance vt.

If the speed of the aircraft exceeds the speed of sound, then the aircraft will overtake the wave fronts (see Fig. 17.17). In this case the sound is always confined to a conical region that has the aircraft at its apex and moves with the aircraft at the speed V_E; ahead of this region, the air has not yet been disturbed, although it will be disturbed when the aircraft and the cone move sufficiently far to the right. The cone is called the **Mach cone**.

The half-angle of the apex of the Mach cone is given by the formula

Mach cone

$$\sin\theta = v/V_E \tag{17.20}$$

This can readily be seen from Fig. 17.18, which shows the aircraft at a time t and the wave front that was emitted by the aircraft at time zero. In the time t the sound travels a distance vt, and, simultaneously, the aircraft travels a distance $V_E t$. Thus, the radius of the wave front is vt. This radius is the opposite side of a right triangle of hypotenuse $V_E t$. Consequently,

$$\sin\theta = \frac{vt}{V_E t} \tag{17.21}$$

which gives Eq. (17.20) if we cancel the factor t.

Any supersonic aircraft (or other body) will generate a Mach cone, regardless of whether or not it carries an artificial source of sound aboard (see Fig. 17.19). The motion of the body through the air creates a pressure disturbance that spreads out-

ERNST MACH (1838–1916)
Austrian philosopher and physicist. Mach obtained visual evidence for the Mach cone by photographing projectiles in flight. His book The Science of Mechanics *is a profound critical examination of the historical and logical foundations of mechanics.*

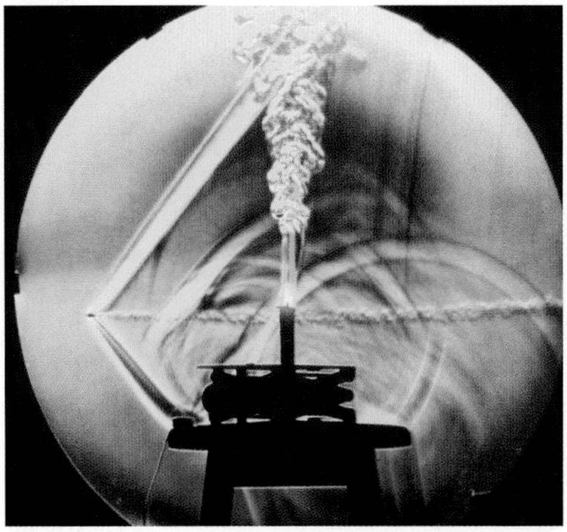

FIGURE 17.19 Schlieren photograph of a .22 caliber bullet passing a candle. With the Schlieren technique the shock waves from the bullet and turbulent convection column become visible. The lower sweptback shock wave reflects in a complicated way from the stool that supports the candle.

ward with the speed of sound and forms the cone. The cone trails behind the body much as a wake trails behind a ship. The sharp pressure disturbance at the surface of the cone is heard as a loud bang whenever the cone sweeps over the ear. This bang is called a **sonic boom**. For a large aircraft, such as the Concorde SST (Fig. 17.20), the noise level of the sonic boom reaches the pain threshold even if the aircraft is 20 km away. On a smaller scale, you can generate a sonic boom by cracking a whip. The crack of the whip occurs when the speed of the end of the whip through the air exceeds the speed of sound.

sonic boom

FIGURE 17.20 Concorde SST.

Checkup 17.4

QUESTION 1: With what speed must a train approach you if you are to hear its whistle at twice the frequency it has when stationary?

QUESTION 2: With what speed must you approach a stationary whistle if you are to hear it at twice the frequency it has when you are stationary?

QUESTION 3: Can the Doppler-shifted frequency of a moving source of sound ever be zero?

QUESTION 4: A car races around a traffic circle while steadily blowing its horn. Describe the changes of pitch you hear if (a) you stand in the center of the circle or (b) you stand at the rim of the circle.

QUESTION 5: The frequency of the sound from the turbines of an approaching jet plane is twice the value of the frequency as the plane recedes. At what fraction of the speed of sound is the plane moving?

(A) $\frac{1}{4}$ (B) $\frac{1}{3}$ (C) $1 - \sqrt{2}/2$ (D) $\frac{1}{2}$

17.5 DIFFRACTION

It is a characteristic feature of waves that they will deflect around the edges an obstacle placed in their path and penetrate into the "shadow" zone behind the obstacle. For example, Fig. 17.21 shows water waves striking a breakwater at the entrance of a harbor. The region directly behind the breakwater is out of the direct path of the waves, but nevertheless waves reach this region because each wave front spreads sideways once it has passed the entrance to the harbor. This lateral spreading of the wave fronts can be easily understood: The breakwater cuts a segment out of each wave front, and such a segment of wave front cannot just keep moving straight on as though nothing had happened—the end of the segment is a vertical wall of water where the breakwater has chopped off the wave front. The water at the ends will immediately begin to spill out sideways, producing a disturbance at the edge of the segment. This disturbance continues to spread out and gradually forms the curved wave fronts to the left and the right of the main beam of the wave.

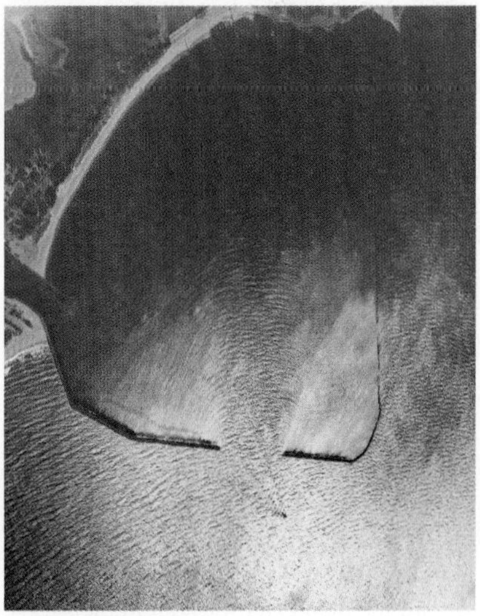

FIGURE 17.21 Ocean waves incident on a breakwater with an aperture.

Such a *deflection of waves at the edge of an obstacle is called* **diffraction**. We will perform a detailed analysis of diffraction phenomena in Chapter 35, when we examine the properties of light and other electromagnetic waves; here, we introduce a few of the main features of diffraction. It is a general rule for a wave passing through a gap that *the amount of diffraction increases with the ratio of wavelength to the size of the gap*. An increase of the wavelength (or a decrease of the size of the gap) makes the diffraction effects more pronounced; a decrease of the wavelength (or an increase of the size of the gap) makes

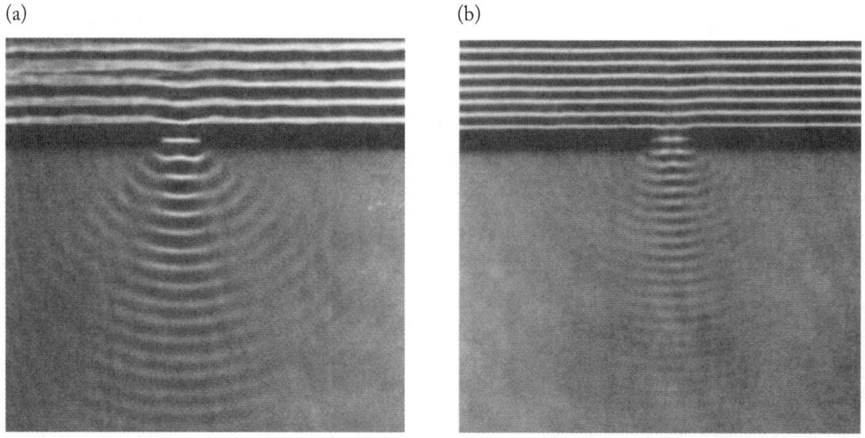

FIGURE 17.22 (a) Water waves of fairly long wavelength in a ripple tank exhibit strong diffraction when passing through an aperture. These waves spread out beyond the aperture in a fanlike pattern. (b) Water waves of shorter wavelength exhibit less diffraction.

the diffraction effects less pronounced. For instance, Fig. 17.22a shows diffraction of waves of relatively long wavelength by a small gap; the waves spread out very strongly, forming divergent, fanlike beams of concentric wave fronts. Figure 17.22b shows diffraction of waves of shorter wavelength; here the wave spreads out only slightly; most of the wave remains within a straight beam of nearly parallel wave fronts. Note that in Fig. 17.22b, the beam has a fairly well-defined edge—the region in the "shadow" of the barrier remains nearly undisturbed while the region facing the gap receives the full impact of the waves.

The fanlike beams of waves spreading out from the gap constitute a **diffraction pattern.** In Fig. 17.22a, the diffraction pattern consists of a central beam and two clearly recognizable secondary beams on each side. The beams are separated by nodal lines along which the wave amplitude is zero. Figures 17.23a and b show the diffraction patterns generated by a small island. Note that if the wavelength is large compared with the size of the island, then there exists no shadow zone; instead, the island merely produces some distortion of the waves.

Diffraction plays an important role in the propagation of sound. You can hear a person whose mouth is out of your direct line of sight—say, a person facing away from you or a person talking in an adjacent room—because the sound waves diffract through the open mouth or the open door and spread to fill the entire vicinity.

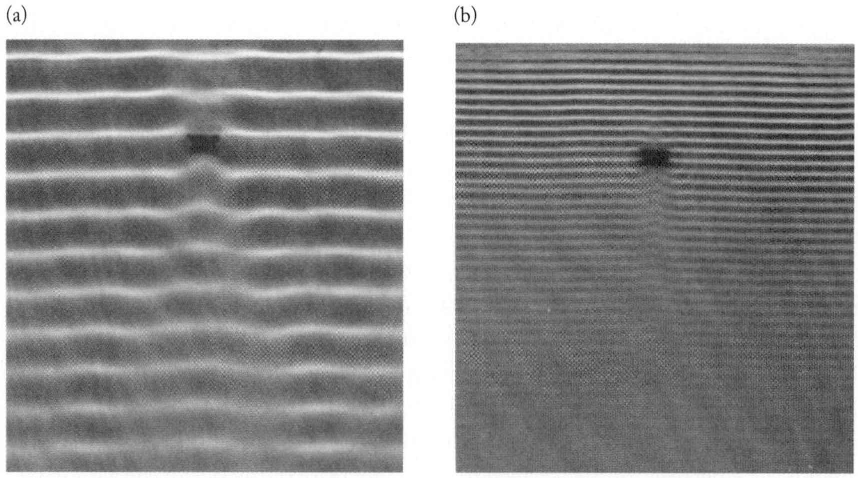

FIGURE 17.23 Diffraction by a small island. (a) Waves of long wavelength diffract around the island and spread into the shadow zone. (b) Waves of short wavelength do not spread into the shadow zone.

EXAMPLE 8 The minimum size of a structure that can be detected in an ultrasound image is limited by diffraction. With careful analysis of the echoes reflected from structures within the body, features as small as one-quarter of a wavelength can be imaged. Suppose that the probe uses ultrasound with a frequency of 2.0×10^6 Hz. What is the smallest feature that can be detected?

Concepts in Context

SOLUTION: Diffraction effects are determined by the size of the features of interest relative to the wavelength. From Example 3, the speed of sound in body tissues is about 1500 m/s, so the wavelength of ultrasound waves of frequency 2.0×10^6 Hz is

$$\lambda = \frac{v}{f} = \frac{1500 \text{ m/s}}{2.0 \times 10^6 \text{ Hz}} = 7.5 \times 10^{-4} \text{ m}$$

If a feature as small as one-quarter of a wavelength can be detected, the size of such a structure is

$$\frac{\lambda}{4} = \frac{7.5 \times 10^{-4} \text{ m}}{4} = 1.9 \times 10^{-4} \text{ m}$$

Thus features as small as 0.2 mm can be detected.

Checkup 17.5

QUESTION 1: If the water waves illustrated in Fig. 17.22 had a much longer wavelength, how would the distribution of waves beyond the barrier be different? If the waves had a much shorter wavelength?

QUESTION 2: Sound waves of frequency 1 kHz are traveling through a medium and strike an opening 1 m in width. The pattern of waves beyond the opening will be most spread out if the medium in which the sound waves are traveling is (hint: see Table 17.3.):

(A) Air (B) Helium gas (C) Water

SUMMARY

PHYSICS IN PRACTICE Musical instruments (page 546)

SOUND WAVE PROPERTIES Pressure (and density) antinodes correspond to displacement nodes (and vice versa).

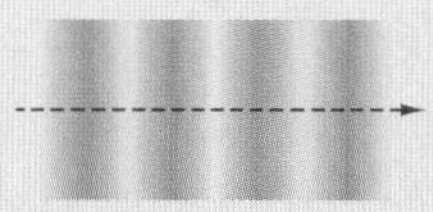

CHROMATIC MUSIC SCALE Successive notes are separated in frequency by a factor of $2^{1/12}$.

OCTAVE Musical notes differing by a factor of 2 in frequency are separated by an octave.

INTENSITY LEVEL (in dB)

$$(10\ \text{dB}) \times \log_{10}\left(\frac{[\text{intensity in W/m}^2]}{1.0 \times 10^{-12}\ \text{W/m}^2}\right) \qquad (17.2)$$

INTENSITY Transported energy per second (power) per unit area

$$I = \frac{P}{A} \qquad (17.1)$$

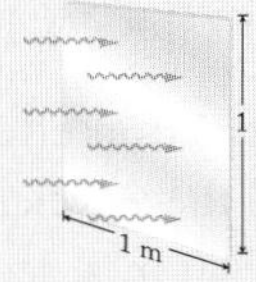

DECREASE OF INTENSITY OF SOUND WAVE WITH DISTANCE

$$I_2 = I_1\frac{r_1^2}{r_2^2} \qquad (17.4)$$

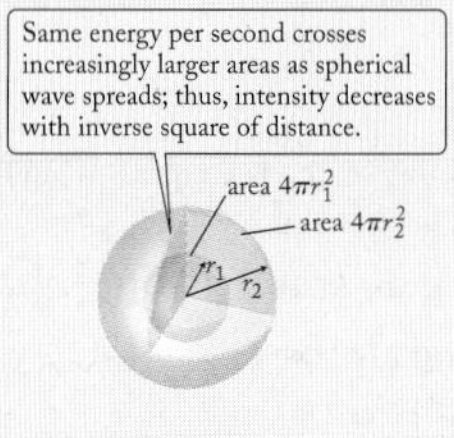

SPEED OF SOUND IN AIR
(With pressure p_0 and density ρ_0)

$$v = \sqrt{1.40\frac{p_0}{\rho_0}} \qquad (17.5)$$

SPEED OF SOUND IN AIR (At 0°C and 1 atm) 331 m/s

STANDING WAVES IN TUBE OPEN AT ONE END
An odd number of quarter wavelengths fits in length L.

$$f_1 = \frac{v}{4L},\ f_2 = \frac{3v}{4L},\ f_3 = \frac{5v}{4L},\ \ldots \qquad (17.10)$$

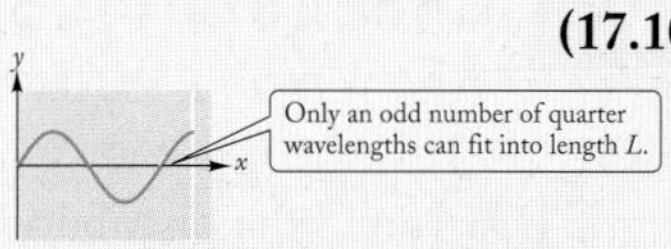

STANDING WAVES IN TUBE OPEN AT BOTH ENDS
An integer number of half wavelengths fits in length L.

$$f_1 = \frac{v}{2L},\ f_2 = \frac{v}{L},\ f_3 = \frac{3v}{2L},\ \ldots \qquad (17.11)$$

DOPPLER SHIFT (v is sound speed; V_R is receiver speed; V_E is emitter speed)
Use upper sign when approaching ($f' > f$); lower sign when receding ($f' < f$).

Moving receiver, stationary emitter

$$f' = f\left(1 \pm \frac{V_R}{v}\right) \qquad (17.15)$$

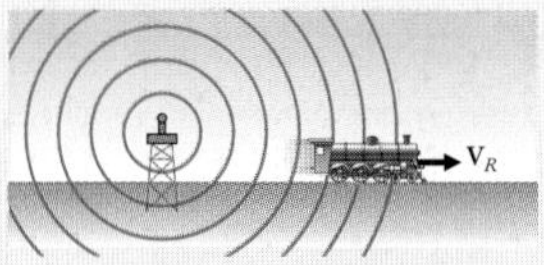

Moving emitter, stationary receiver

$$f' = f\left[\frac{1}{1 \mp (V_E/v)}\right] \qquad (17.17)$$

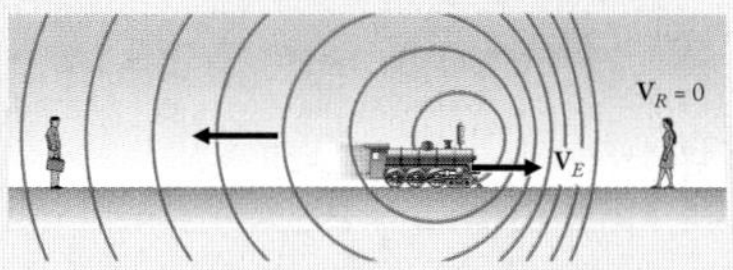

MACH CONE (For a supersonic emitter)

$$\sin\theta = \frac{v}{V_E} \qquad (17.20)$$

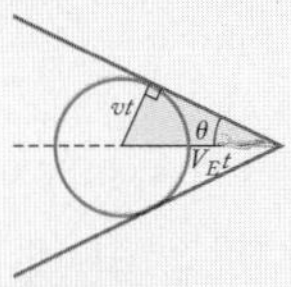

DIFFRACTION The spreading of waves at an obstacle or gap. Diffraction increases with the ratio of the wavelength to the size of the obstacle or gap.

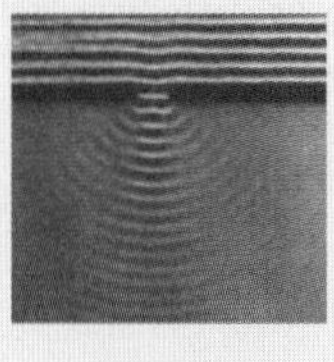

QUESTIONS FOR DISCUSSION

1. Could an astronaut be heard playing the violin while standing on the surface of the Moon?
2. A hobo can hear a very distant train by placing an ear against the rail. How does this help? (Hint: Ignoring frictional losses, how does the intensity of sound decrease with distance in air? In the rail?)
3. If you speak while standing in a corner with your face toward the wall, you will sometimes notice that your voice sounds unusually loud. Explain.
4. What happens to the frequency of musical notes if you play a $33\frac{1}{3}$-rpm record at 45 rpm?
5. Why does the wind whistle in the rigging of a ship or in the branches of a tree?
6. How does a flutist play different musical notes?
7. When inside a boat, you can often hear the engine noises of another boat much more loudly than when on deck. Can you guess why?
8. Many men like singing in the shower stall because the stall somehow enhances their voice. How does this happen? Would the effect be different for men and women?
9. According to a novel proposal for the reduction of engine noise inside aircraft cabins, loudspeakers installed along each side of the cabin are to cancel the noise by "antinoise," that is, sound waves of equal amplitude out of phase to the noise (see Fig. 17.24). The loudspeakers would be controlled by sensors and electronic circuits that detect the arriving engine noise and continuously adjust the amplitude and the phase of the required antinoise. Can such a noise cancellation system eliminate the noise throughout the cabin? In what part of the cabin would it be most effective? What happens to the energy in the arriving sound waves?

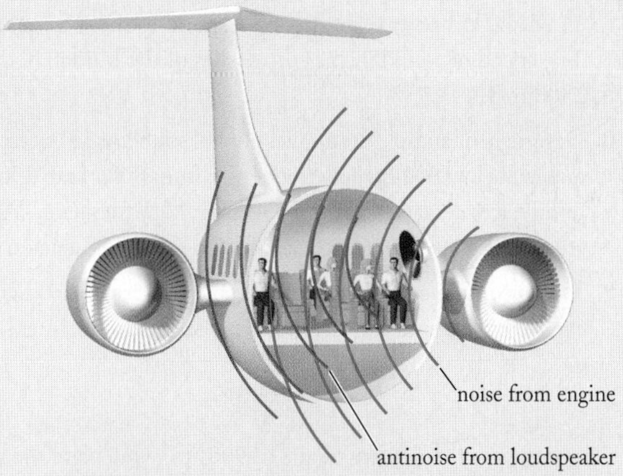

FIGURE 17.24 A system for the reduction of engine noise in an aircraft cabin.

10. The pipes that produce the lowest frequencies in a great organ are very long, usually 5.0 m. Why must they be so long? These pipes are also very thick. Why would a thin pipe give poor performance?
11. Some of the old European opera houses and concert halls renowned for their acoustic excellence have very irregular walls, heavily encrusted with an abundance of stucco ornamentation that reflects sound waves in almost all directions. How does the sound reaching a listener in such a hall differ from the sound reaching a listener in a modern concert hall with four flat, plain walls?
12. Electric guitars amplify the sound of the strings electronically. Do such guitars need a body?
13. Does the temperature of the air affect the pitch of a flute? A guitar?
14. The human auditory system is very sensitive to small differences between the arrival times of a sound signal at the right and left ears. Explain how this permits us to perceive the direction from which a sound signal arrives.
15. The depth finder (or "fish finder") on a boat sends a pulse of sound toward the bottom and measures the time an echo takes to return. The screen of the depth finder displays this echo time on a graph directly calibrated in distance units. Experienced operators can tell whether the bottom is clean rock or rock covered by a layer of mud, or whether a school of fish is swimming somewhere above the bottom. What echo times would you expect to see displayed on the screen of the depth finder in each of these instances?
16. The helmsman of a fast motorboat heading toward a cliff sounds his horn. A woman stands on the top of the cliff and listens. Compare the frequency of the horn, the frequency heard by the woman, and the frequency heard by the helmsman in the echo from the cliff. Which of these three is the highest frequency? Which is the lowest?
17. Two automobiles are speeding in opposite directions while sounding their horns. Describe the changes of pitch that each driver hears as they pass by one another.
18. A man is standing north of a woman while a strong wind is blowing from the south. If the man and the woman yell at each other, how does the wind affect the pitch of the voice of each as heard by the other?
19. A Concorde SST passing overhead at an altitude of 20 km produces a sonic boom with an intensity level of 120 dB lasting about half a second. How does this compare with some other loud noises? Would it be acceptable to let this aircraft make regular flights over populated areas?
20. Many people have reported seeing UFOs traveling through air noiselessly at speeds much greater than the speed of sound. If the UFO consisted of a solid impenetrable body, would you expect its motion to produce a sonic boom? What can you conclude from the absence of sonic booms?
21. When an ocean wave approaches a beach, its height increases. Why?

22. Occasionally ocean waves passing by a harbor entrance will excite very high standing waves ("seiches") within the harbor. Under what conditions will this happen?

23. Seismic waves of the S and P types have different speeds. Explain how a scientist at a seismometer station can take advantage of this difference in speed to determine the distance between his station and the point of origin of the waves.

24. The amplitude of an ocean wave initially decreases as the wave travels outward from its point of origin; but when the wave has traveled a quarter of the distance around the Earth, its amplitude *increases*. Explain how this comes about. (Hint: If the wave were to travel half the distance around the Earth, it would converge on a point, if no continents block its progress.)

25. Underground nuclear explosions generate seismic waves. How could you discriminate between the seismic waves received from such an explosion and the seismic waves from an earthquake? (Hint: Would you expect an explosion to produce mainly S waves or mainly P waves?)

26. Figure 17.25 shows a seismometer, an instrument used to detect and measure seismic waves. A vertical post is firmly set in the ground, and a large mass is suspended from it by a rigid horizontal beam and a diagonal wire. The beam ends in a sharp point that rests against the post; the beam is therefore free to swing in the horizontal plane. Describe how the beam will swing if the ground moves and tilts the post. For what direction of motion of the post is this seismometer most sensitive?

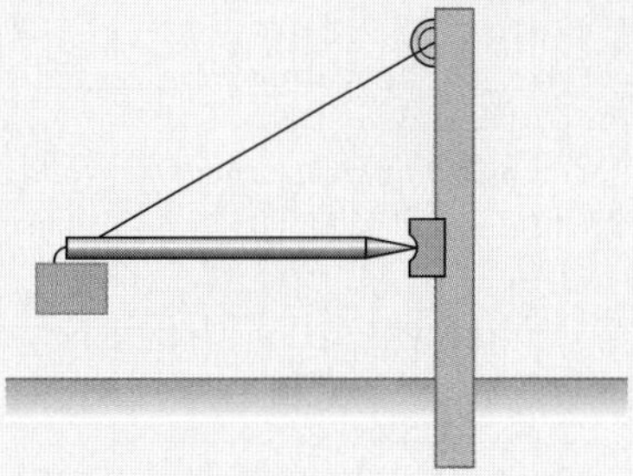

FIGURE 17.25 Seismometer.

27. If you are standing on the south side of a house, you can speak to a friend standing on the east side, out of sight around the corner. How do your sound waves reach into the shadow zone?

PROBLEMS[2]

17.1 Sound Waves in Air
17.2 Intensity of Sound

1. The range of frequencies audible to the human ear extends from 20 to 20000 Hz. What is the corresponding range of wavelengths?

2. The lowest musical note available on a piano is A, four octaves below that listed in Table 17.1; and the highest note available is C, four octaves above that listed in Table 17.1. What are the frequencies of these notes?

3. Both whales and elephants use low-frequency sound waves of a few hertz for communication. What is the wavelength of a whale song at 2.0 Hz (in water)? What is the wavelength of an elephant's rumble at 2.0 Hz (in air)?

4. Dogs can hear ultrasound of 40000 Hz; bats can hear ultrasound of 75000 Hz. What are the corresponding wavelengths?

5. Estimate the wavelength of the sound waves made visible in Fig. 17.2. Assume that the telephone handset is of standard size, like those found on pay telephones.

6. The lowest note that can be played on a guitar is E two octaves below the middle E listed in Table 17.1. This note is produced by the thickest string when vibrating in its fundamental mode. The length of the string is 0.62 m, and its mass per unit length is 5.4×10^{-3} kg/m. What must be the tension in this string?

7. Sound waves used for medical ultrasound scans of the soft tissues in the human body have frequencies in the range 0.80 to 15 MHz. What wavelengths correspond to these frequencies? The speed of sound in the soft tissues of the human body is 1500 m/s.

8. Designers of audio systems usually call sound waves of frequency below 800 Hz low frequency; from 800 Hz to 3500 Hz, middle frequency; and above 3500 Hz, high frequency. What are the wavelengths that correspond to these frequencies?

9. The maximum speed $v_{\max}$ acquired by particles of air when exposed to a sound wave of speed v is related to the intensity I by

$$I = \frac{v}{2}\rho_0 v_{\max}^2$$

where ρ_0 is the density of air, 1.29 kg/m^3. Calculate the maximum speed acquired by the particles in a sound wave of intensity 1.0×10^4 W/m^2.

*10. A violin has four strings, each of them 0.326 m long. When vibrating in their fundamental modes, the four strings have frequencies of 196, 294, 440, and 659 Hz, respectively.

[2] In all the problems assume that the speed of sound in air is 331 m/s, unless otherwise stated.

(a) What is the wavelength of the standing wave on each string? What is the wavelength of the sound wave generated by the string?

(b) What are the frequency and the wavelength of the first overtone on each string? What is the corresponding wavelength of the sound wave generated by each string?

(c) According to Table 17.1, to what musical tones do the frequencies calculated above correspond?

*11. A mandolin has strings 34.0 cm long fixed at their ends. When the mandolin player plucks one of these strings, exciting its fundamental mode, this string produces the musical note D (293.7 Hz; see Table 17.1). In order to produce other notes of the musical scale, the player shortens the string by holding a portion of the string against one or another of several frets (small transverse metal bars) placed underneath the string. The player shortens the string by one fret to produce the note D#, by two frets to produce the note E, by three frets to produce the note F, etc. Calculate the correct spacing between the successive frets of the mandolin for one complete octave. Assume that the string always vibrates in its fundamental mode, and assume that the tension in the string is always the same.

12. In general, the numerical values of the intensity of sound in W/m^2 and of the intensity level in dB are different, but at one value of the intensity they are equal. What is this value?

13. The noise level in a quiet automobile is 50 dB. Find the sound intensity in W/m^2.

14. The intensity level of sound near a loud rock band is 120 dB. What is the intensity level of sound near two such rock bands playing together?

15. The highest frequency typically detectable by humans is 20000 Hz. What note on the chromatic scale is closest to this frequency? How many octaves above the one listed in Table 17.1 is this frequency?

16. Assume that a particular loudspeaker emits sound waves equally in all directions; a total of 1.0 watt of power is in the sound waves. What is the intensity at a point 10 m from this source (in W/m^2)? What is the intensity level 20 m from this source (in dB)?

17. When 50 people are talking at once at a party, the intensity level is 70 dB. How much does the intensity level change when 25 people are talking?

18. A solo violinist generates an intensity level at the location of a listener of 60 dB. What is the intensity level there when 12 violinists play together?

19. A noisy machine produces an intensity level of 80 dB. What is the intensity level when two such machines operate at the same time?

20. The intensity level 50 m from an ambulance siren is 80 dB. What is the intensity level 1.0 m from the siren?

21. A sound source emits power equally in all directions. The intensity level 30 m from the source is 70 dB. What is the total sound power emitted by the source (in watts)?

*22. A loudspeaker receives 8.0 W of electric power from an audio amplifier and converts 3.0% of this power into sound waves. Assuming that the loudspeaker radiates the sound uniformly over a hemisphere (a vertical and horizontal angular spread of 180°), what will be the intensity and the intensity level at a distance of 10 m in front of the loudspeaker?

*23. An old-fashioned hearing trumpet has the shape of a flared funnel, with a diameter of 8.0 cm at its wide end and a diameter of 0.70 cm at its narrow end. Suppose that all of the sound energy that reaches the wide end is funneled into the narrow end. By what factor does this hearing trumpet increase the intensity of sound (measured in W/m^2)? By how many decibels does it increase the intensity level of sound?

17.3 The Speed of Sound; Standing Waves

24. Spectators at soccer matches often notice that they hear the sound of the impact of the ball on the player's foot (or head) sometime after seeing this impact. If a spectator notices that the delay time is about 0.50 s, how far is he from the player?

25. In the nineteenth century a signal gun was fired at noon at most harbors so that the navigators of the ships at anchor could set their chronometers. This method is somewhat inaccurate, because the sound signal takes some time to travel the distance from gun to ship. If this distance is 3.0 km, how long does the signal take to reach the ship? Can you suggest a better method for signaling noon?

26. In freshwater, sound travels at a speed of 1460 m/s. In air, sound travels at a speed of 331 m/s. Suppose that an explosive charge explodes on the surface of a lake. A woman with her head in the water hears the bang of the explosion and, lifting her head out of the water, hears the bang again 5.0 s later. How far is she from the site of the explosion?

27. While standing at some distance from a large stone cliff, you notice that if you clap your hands, an echo of the clap returns to you about 1.5 s later. What is the distance to the cliff?

28. The pitch of the vowels produced by the human voice is determined by the frequency of standing waves in several resonant cavities (larynx, pharynx, mouth, and nose). In an amusing demonstration experiment, a volunteer inhales helium gas and then speaks a few words. As long as his resonant cavities are filled with helium gas, the pitch of his voice will be much higher than normal. Given that the speed of sound in helium is about three times as large as in air, calculate the factor by which the eigenfrequencies of his resonant cavities will be higher than normal.

29. The Bay of Fundy (Nova Scotia) is about 250 km long. The speed of water waves of long wavelength in the bay is about 30 m/s.

(a) What are the frequency and the period of the fundamental mode of oscillation of the bay? Treat the bay as a long, narrow tube open at one end and closed at the other.

(b) The period of the tidal pull exerted by the Moon is about 12 h. Would you expect that the very large tidal oscillations (with heights of up to 15 m) observed in the Bay of Fundy are due to resonance?

30. You can estimate your distance from a bolt of lightning by counting the seconds between seeing the flash and hearing the thunder, and then dividing by 3 to obtain the distance in kilometers (or by 5 to obtain the distance in miles). Verify this rule.

31. The ultrasonic range finder on an automatic camera sends a pulse of sound to the target and determines the distance by the time an echo takes to return.

(a) If the range finder is to determine a distance of 50 cm with an error no larger than ± 2 cm, how accurately (in seconds) must it measure the travel time?

(b) If you aim this camera at an object placed beyond a sheet of glass (a window of a glass door), on what will the camera focus?

32. As described in Problem 31, the ultrasonic range finder on a camera sends a pulse of sound to the target and determines the distance by the time an echo takes to return. Suppose that, after waterproofing this camera somehow, you try to use it under water, in a swimming pool. If you aim the camera at a target 5.0 m away, what distance will the range finder indicate? The speed of sound in water is 1500 m/s.

33. The tube of a flute has a sliding joint that can be used to change the length, to tune the flute. Suppose that a flute has been tuned to perfect pitch while outdoors, where the temperature is 0°C and the speed of sound is 331 m/s. For this perfectly tuned flute, the frequency of middle C, which corresponds to the fundamental mode of the tube of the flute, is 261.7 Hz. Suppose that this flute is then taken indoors, where the temperature is 20°C and the speed of sound is 344 m/s. What will the frequency of the flute's middle C be now? To restore the flute to perfect pitch, how much must we increase the length of the tube of the flute? Express your answer as a percentage of the length.

34. The commonly accepted value for the speed of sound in dry air under standard conditions is 331.45 m/s. However, a scientist at the National Research Council of Canada recently discovered an error in the earliest determinations of the speed of sound, and he concluded that the correct value for the speed of sound is 331.29 m/s. What is the percent difference between the old and the new values? According to Eq. (17.5), what percent change of the pressure or of the density of the air will produce an equal change in the speed of sound?

35. Estimate the frequency of the sound waves made visible in Fig. 17.2.

36. A bat emits a pulse of ultrasound and detects an echo from a tree 0.20 s later. How far away is the tree?

37. A lightning flash is seen. Six seconds later, thunder is heard. How far away did the lightning occur?

38. Sound waves of wavelength 25.5 m travel in a piece of iron. The vibrations of the iron generate sound waves in the air nearby. What is the wavelength of the sound in air?

39. A scuba diver clangs her hammer on an underwater pipeline. Another diver, with his hand on the pipeline, feels a vibration from the clang and, 1.5 seconds later, hears the clang through the water. How far along the pipeline are the two divers separated? The speed of sound in the metal pipe is 5100 m/s.

40. To measure the level of liquid helium, scientists often use a thin tube that permits oscillations of the gas in the tube above the liquid. When the bottom end of the tube is barely immersed in liquid (a closed end) and the top end is essentially open, a frequency f of oscillation is detected. When the bottom end of the tube is lifted out of the liquid, so that it is also open, what frequency of oscillations do you expect to occur?

41. With your ear on an iron railroad rail, you hear the sound of a distant train whistle through the iron. Eight seconds later, you hear the same whistle through the air. How far away is the train?

42. The musical note A (440 Hz) is played on a flute in air. What frequency would be heard if the air were replaced with helium gas?

*43. A rock is dropped into a deep well, and the sound of it hitting water is heard 4.62 s after the drop. Take the speed of sound to be 331 m/s. How far down the well is the water?

*44. The mass per unit length of a steel wire of diameter 1.3 mm is 0.010 kg/m, and the yield strength, or maximum tension that the wire can withstand, is 3.6×10^3 N. Is it possible to apply enough tension to the wire so that the speed of a transverse wave on this wire exceeds the speed of sound in the steel of the wire, 5000 m/s?

*45. In the eighteenth century, members of the French Academy organized the first careful measurement of the speed of sound in air. To compensate for wind speed, they adopted a reciprocal method. Cannons were fired alternately at Montmartre (in Paris) and Montlhéry, 29.0 km apart. Observers at each station measured the time delay between the muzzle flash seen at the *other* station and the arrival of the sound. Show that from the measurements of these travel times t_1 and t_2 of sound in both directions, the speed of sound in still air can be calculated as follows, independently of the speed of the wind blowing from one station to the other:

$$v = \frac{d}{2}\left(\frac{1}{t_1} + \frac{1}{t_2}\right)$$

where d is the distance between the stations. Evaluate numerically for the measured travel times of 87.4 s and 84.8 s.

*46. Because the human auditory system is very sensitive to small differences between the arrival times of a sound signal at the right and the left ear, we can perceive the direction from which a sound signal arrives to within about 5°. Suppose that a source of sound (a ringing bell) is 10 m in front of and 5° to the left of a listener. What is the difference in the arrival times of sound signals at the left and right ears? The separation between the ears is about 15 cm.

*47. Consider a tube of length L open at both ends. Show that the eigenfrequencies of standing sound waves in this tube are

$$f_n = n\frac{v}{2L} \qquad n = 1, 2, 3, \ldots$$

Draw diagrams similar to those in Fig. 17.12 showing the displacement amplitude for each of the first four standing waves.

*48. The largest pipes in a great organ usually have a length of about 16 ft (4.8 m). These pipes are open at both ends so that a standing sound wave will have a displacement antinode at each end. What is the frequency of the fundamental mode of such a pipe?

*49. A flute can be regarded as a tube open at both ends. It will emit a musical note if the flutist excites a standing wave in the air column in the tube.

(a) The lowest musical note that can be played on a flute is C (261.7 Hz; see Table 17.1). What must be the length of the tube? Assume that the air column is vibrating in its fundamental mode (see Problem 47).

(b) In order to produce higher musical notes, the flutist opens valves arranged along the side of the tube. Since the holes in these valves are large, an open valve has the same effect as shortening the tube. The flutist opens one valve to play C#, two valves to play D, etc. Calculate the successive spacings between the valves of a flute for one complete octave. (The actual spacings used on flutes differ slightly from the results of this simple theoretical evaluation because the mouth cavity of the flutist also resonates and affects the frequency.)

50. The human ear canal is approximately 2.7 cm long. The canal can be regarded as a tube open at one end and closed at the other. What are the eigenfrequencies of standing waves in this tube? The ear is most sensitive at a frequency of about 3000 Hz. Would you expect that resonance plays a role in this?

*51. Consider a tube of length L closed at both ends. Show that the eigenfrequencies of standing sound waves in this tube are

$$f_n = n\frac{v}{2L} \qquad n = 1, 2, 3, \ldots$$

Draw diagrams similar to those in Fig. 17.12 showing the displacement amplitude for each of the first four standing waves.

*52. If you stand in the vicinity of a picket fence and clap your hands, you will notice that the sound waves reflected by the fence and reaching your ear are strongly reinforced at a selected wavelength; that is, the picket fence seems to ring with a musical tone. This selective reinforcement of sound waves occurs whenever the wavelength is such that waves reflected by different boards in the fence arrive at your ear in phase, giving constructive interference. Suppose that the picket fence consists of boards separated by a distance d and you stand in line with this fence (see Fig. 17.26). Show that in this case the condition for constructive interference for waves traveling from you to the boards and back to you is that the wavelength be equal to $2d$.

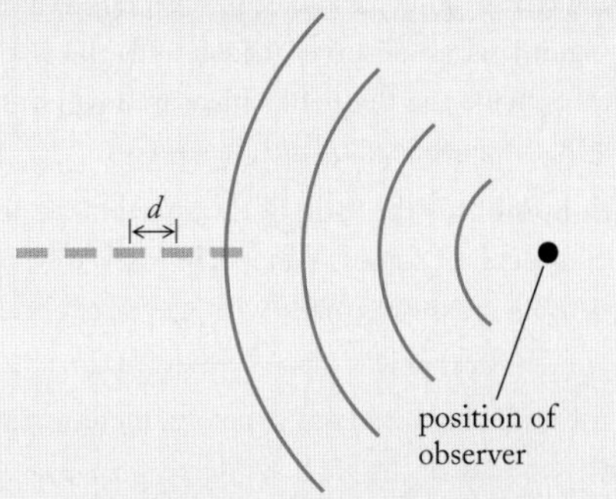

FIGURE 17.26 Picket-fence interference.

17.4 The Doppler Effect†

53. The horn of a stationary automobile emits a sound wave of 580 Hz. What frequency will you hear if you are driving toward this automobile at 80 km/h?

54. In an experiment performed shortly after Doppler proposed his theoretical formula for the frequency shift, several trumpeters were placed on a train and told to play a steady musical tone. As the train sped by, listeners standing on the side of the track judged the pitch of the tone received from the trumpeters. Suppose that the train had a speed of 60 km/h and that the trumpets on the train sounded the note of E (329.7 Hz; see Table 17.1). What was the frequency of the note perceived by a listener on the ground when the train was approaching? When the train was receding? Approximately to what musical notes do these Doppler-shifted frequencies correspond?

55. Ocean waves with a wavelength of 100 m have a period of 8.0 s. A motorboat, with a speed of 9.0 m/s, heads directly into the oncoming waves. What is the speed of the waves relative to the motorboat? With what frequency do wave crests hit the front of the motorboat?

56. The horn of an automobile emits a tone of frequency 520 Hz. What frequency will a pedestrian hear when the automobile is approaching at a speed of 85 km/h? Receding at the same speed?

57. An aircraft is flying at an altitude of 12 000 m. On the ground, a sonic boom is heard 18 s after the jet passes directly overhead. What is the speed of the aircraft?

58. While watching a high-speed chase, a stationary observer measures the frequency of the siren of an approaching police car to be 497 Hz; after the car passes, the same observer measures 395 Hz. What is the speed of the police car? What is the frequency it emits?

59. A bat flies toward a wall and emits a pulse of ultrasound of frequency 50 kHz. The echo received by the bat is Doppler-shifted 800 Hz toward higher frequency. How fast is the bat flying?

†For help, see Online Concept Tutorial 19 at www.wwnorton.com/physics

60. An airplane emits a tone as it flies past an observation tower. The frequency that an observer measures as the airplane approaches is twice the frequency measured when the airplane recedes. What is the speed of the airplane?

61. The **Mach number** is the ratio of the speed of an aircraft (or body) to the speed of sound. When a jet flies at Mach 2, what angle does the shock wave make with the direction of travel of the jet?

62. What is the angle of the Mach cone of a meteor traveling at 15 km/s?

*63. In movies, a person who dives from a high cliff often screams all the way down to the water. Assume that such a person emits a constant frequency of 440 Hz. What frequency will an observer at the top of the cliff hear after 3.0 seconds? Take into account the travel time of the sound.

*64. A car horn sounds a pure tone as the car approaches a wall. A stationary listener behind the car hears an average frequency of 250 Hz, which pulsates (beats) at 12 Hz. What is the speed of the car?

*65. Two automobiles are driving on the same road in opposite directions. The speed of the first automobile is 90.0 km/h, and that of the second is 60.0 km/h. The horns of both automobiles emit tones of frequency 524 Hz. Calculate the frequency that the driver of each automobile hears coming from the other automobile. Assume that there is no wind blowing along the road.

*66. Repeat Problem 65 under the assumption that a wind of 40.0 km/h blows along the road in the same direction as that of the faster automobile.

*67. A train approaches a mountain at a speed of 75 km/h. The train's engine sounds a whistle that emits a frequency of 420 Hz. What will be the frequency of the echo that the engineer hears reflected off the mountain?

*68. Suppose that a moving train carries a source of sound and also a receiver of sound so that both have the same velocity relative to the air. Show that in this case the Doppler shift due to motion of the source cancels the Doppler shift due to motion of the receiver—the frequency detected by the receiver is the same as the frequency generated by the source.

*69. The whistle on a train generates a tone of 440 Hz as the train approaches a station at 30 m/s. A wind blows at 20 m/s in the same direction as the motion of the train. What is the frequency that an observer standing at the station will hear?

70. Figure 17.19 shows the shock wave of a bullet speeding through air. Measure the angle of the Mach cone and calculate the speed of the bullet.

71. A rifle bullet has a speed of 674 m/s. What is the half-angle of the Mach cone generated by this bullet?

72. According to tradition, Superman flies faster than a speeding bullet. The speed of a typical bullet is 700 m/s, so let us guess that the speed of Superman is 800 m/s. At this speed, what is the half-angle of the Mach cone that Superman produces?

*73. You may have noticed that at the instant a fast (but subsonic) jet aircraft passes directly over your head, the sound it makes seems to come from a point behind the aircraft.

(a) Show that the direction from which the sound seems to come makes an angle θ with the vertical such that $\sin\theta = V_E/v$, where V_E is the speed of the aircraft and v the speed of sound.

(b) If you hear the sound from an angle of 30° behind the aircraft, what is the speed of the aircraft?

17.5 Diffraction

74. Because of diffraction, it can be difficult to detect reflected sound from objects smaller even than one full wavelength of the sound used. Using this one-wavelength criterion, what is the size of the smallest insect a bat can echolocate when it emits ultrasound of frequency 40 kHz?

75. To circumvent diffraction effects, an ultrasonic imaging apparatus uses sound of a wavelength smaller than the size of most objects to be imaged. A particular medical apparatus operates at 5.0 MHz and can detect structures as small as one-half of a wavelength. What is the size of the smallest object that can be detected?

76. Ultrasonic microscopes are used to study features in materials as small as one-quarter of the wavelength of sound in the material; at smaller length scales, diffraction obscures such details. If ultrasound of frequency 2.0×10^9 Hz is used to examine aluminum, what size features can be detected?

REVIEW PROBLEMS

77. The speed of sound waves in water is 1500 m/s. Dolphins emit ultrasound waves at 1.0×10^6 Hz. What is the wavelength of these waves? What is the wavelength if these waves penetrate from water into air?

78. When vibrating in its lowest mode, the C string of the cello produces the musical note C that is two octaves below middle C. The length of this string is 0.68 m.

(a) What is the frequency of this note? What is the wavelength of the sound wave?

(b) What is the frequency of the standing wave on the string, and what is its wavelength?

79. The sound waves used for medical ultrasound scans of the human body typically have a frequency of 10 MHz and an amplitude of 8.0×10^{-8} m (this is the amplitude of the

simple harmonic motion of the particles in the tissues through which the wave passes). When subjected to such a sound wave, what is the maximum speed of the motion of a particle? What is the maximum acceleration of the particle? Express the acceleration in standard g's ($1g = 9.81\ \text{m/s}^2$).

80. At a distance of 15 m from a pneumatic drill, the sound intensity is about $1.0 \times 10^{-3}\ \text{W/m}^2$. What is the intensity level in decibels?

81. In a screaming contest, a Japanese woman achieved 115 dB. How many such women would have to scream at you to bring you to the threshold of pain, 120 dB?

82. What is the energy incident per second on your eardrum (of diameter 7.0 mm) if exposed to a sound wave of 160 dB?

83. Suppose we turn the volume control of a loudspeaker up, and we increase the intensity of the sound reaching our ears by a factor of 2. What is the corresponding increase in intensity level, in dB?

84. Suppose that a whisper has an intensity level of 20 dB at a distance of 0.50 m from the speaker's mouth. At what distance will this whisper be below your threshold of hearing?

85. A bat can sense its distance from the wall of a cave (or whatever) by emitting a sharp ultrasonic pulse that is reflected by the wall. The bat can tell the distance from the time the echo takes to return.
   (a) If the bat is to determine the distance of a wall 10.0 m away with an error of less than ± 0.5 m, how accurately must it sense the time interval between emission and return of the pulse?
   (b) Suppose that a bat flies into a cave filled with methane (swamp gas). By what factor will this gas distort the bat's perception of distances? The speed of sound in methane is 432 m/s.

86. In order to measure the depth of a ravine, a physicist standing on a bridge drops a stone and counts the seconds between the instant he releases the stone and the instant he hears it strike some rocks at the bottom. If this time interval is 6.0 s, how deep is the ravine? Take into account the travel time of the sound signal, but ignore air friction.

87. The Concorde SST had a cruising speed of 2160 km/h.
   (a) What was the half-angle of the Mach cone generated by this aircraft?
   (b) If the aircraft passed directly over your head at an altitude of 12000 m, how long after this instant would the shock wave strike you?

88. The French TGV (*train à grande vitesse*; Fig. 17.27) attains a speed of 510 km/h. Suppose the train carries a whistle tuned to a frequency of 600 Hz. If it sounds this whistle, what is the frequency you hear standing alongside its track while the train is approaching? Receding?

FIGURE 17.27 The French TGV.

89. A man at the helm of a fast motorboat heading toward a cliff sounds his horn. The speed of the motorboat is 15 m/s, and the frequency of the sound emitted by the horn, when at rest, is 660 Hz. A woman stands on the top of the cliff and listens.
   (a) What is the frequency heard by the man?
   (b) What is the frequency heard by the woman?
   (c) What is the frequency heard by the man in the echo from the cliff?

Answers to Checkups

Checkup 17.1

1. The amplitude is largest near the earpiece of the telephone, where the sound waves have not spread appreciably; the amplitude is smallest far from the earpiece, where the energy in the wave has become distributed over a much larger area than near the earpiece.

2. No to both questions: because the sound wave is reflected from the inner wall of the tube, the wave fronts stay much the same size and maintain the same amplitude as they propagate along the tube.

3. At a density node, the pressure pattern also has a node; the density and pressure vary together. The displacement of the air molecules is maximum there.

4. The C that is an octave higher than the other has twice the frequency, and so a constant sound speed, $v = \lambda f$, requires that it has the shorter wavelength (by a factor of 2).

5. A wave pulse on the surface of a pond (or in air) is traveling in two (or three) dimensions, and so spreads out over a larger circumference (or a larger area), with a decreasing concentration of its energy. A wave pulse on a string travels in one dimension, and so does not spread.

6. (B) Maximum; minimum. At a density antinode, the pressure also experiences maximum amplitude oscillations. The air molecules experience minimum displacement there; thus a density and pressure antinode is a displacement node.

Checkup 17.2

1. From Table 17.2, the noise of thunder and of a subway train differ by 110 dB − 100 dB = 10 dB. The intensities differ by a factor of $0.1/(1.0 \times 10^{-2}) = 10$. Thus a change in intensity by a factor of 10 is equivalent to a change in intensity level by an increment of 10 dB.

2. The intensity decreases in proportion to the inverse of the square of the distance, so at 10 m it is 100 times smaller than at 1 m; thus it is 10^{-7} W/m^2 there. At 100 m, it is $(100)^2 = 10^4$ times smaller than at 1 m; thus it is 10^{-9} W/m^2 there.

3. (D) 2. Both sources emit the same power, but at any particular distance, that power is distributed over twice as much area for the first source, resulting in half the intensity. In both cases, the intensity will be inversely proportional to the square of the distance, and so will be another factor of 4 larger for the first source, resulting in an overall intensity ratio of $\frac{1}{2} \times 4 = 2$.

Checkup 17.3

1. The thunder is delayed compared with the lightning because the speed of sound is much less than the speed of light. The speed of sound is around 331 m/s, and the speed of light is roughly a million times faster; thus, the lightning flash reaches us almost instantly, and the delay of the thunderclap can be used to judge the listener's distance to the active cloud (about 3 seconds per kilometer, or 5 seconds per mile).

2. For a tube closed at both ends, each end must be a displacement node (a pressure antinode); two nodes (or antinodes) occur one-half wavelength apart. The fundamental mode thus contains a half wavelength in the length L, or $L = \lambda/2$. The fundamental frequency is then $f = v/\lambda = v/2L$.

3. The wavelengths do not match; the wavelength on the string is determined by the string length, while that in the air is determined by how far a disturbance propagates in one cycle (that is, it depends on the velocity of sound in air). The frequencies on the string and in the air do match; the oscillations of the string are mechanically transferred to the air, so the number of string oscillations per second matches the number of oscillations per second of the air.

4. (B) $\frac{1}{2}$. For the tube open at one end, a quarter wavelength must fit in the tube length; for the tube open at both ends, a half wavelength must fit. Thus the tube open at one end has twice the wavelength and, since $f = v/\lambda$, half the fundamental frequency of the tube open at both ends.

Checkup 17.4

1. For an emitter in motion, the increase in frequency is given by Eq. (17.17), where $f' = f/[1 - (V_E/v)]$ for an approaching emitter. Thus to hear twice the frequency, we require $f' = 2f$, which gives $\frac{1}{2} = 1 - (V_E/v)$, or $V_E = v/2$. Thus the train would have to approach at half the speed of sound!

2. For a receiver in motion, the increase in frequency is given by Eq. (17.15), where $f' = f[1 + (V_R/v)]$ for a receiver approaching the source of sound. Thus to hear twice the frequency, you must approach the whistle at the speed of sound, at $V_R = v$; this gives $f' = 2f$.

3. Not if the receiver remains stationary. This would require $V_E \to \infty$ for a receding emitter in Eq. (17.17).

4. (a) If you stand in the center of the circle, then there is no motion toward or away from you, and the pitch does not change. (b) If you stand at the rim of the circle, the pitch begins to increase when the car is diametrically opposite you (as the component of its velocity toward you increases from zero); the pitch continues to increase until the car is close to you (all of its velocity is toward you). The pitch then suddenly changes to its lowest value just after the car passes by. The pitch becomes less low until the car is again opposite you, when the pitch momentarily assumes its unshifted value.

5. (B) $\frac{1}{3}$. In both cases the emitter is moving, so the frequency change is given by Eq. (17.17). Since the frequency is twice as high when approaching, we have $1 + V_E/v = 2(1 - V_E/v)$, which implies $3V_E/v = 1$, or $V_E = \frac{1}{3}v$.

Checkup 17.5

1. If the waves had a much longer wavelength, diffraction effects would be greater, and the waves beyond the barrier would be even more spread out. If the waves had a much shorter wavelength, diffraction effects would be lessened, and the waves beyond the barrier would be less spread out, forming a more narrow "beam" of waves, similar to a shadow.

2. (C) Water. The waves are most spread out (diffraction effects are greatest) when the wavelength is longest. For waves of a given frequency, the wavelength is longest when the speed is largest ($v = \lambda f$). From Table 17.3, sound waves in water have a larger speed than in air or helium gas.

CHAPTER

18

Fluid Mechanics

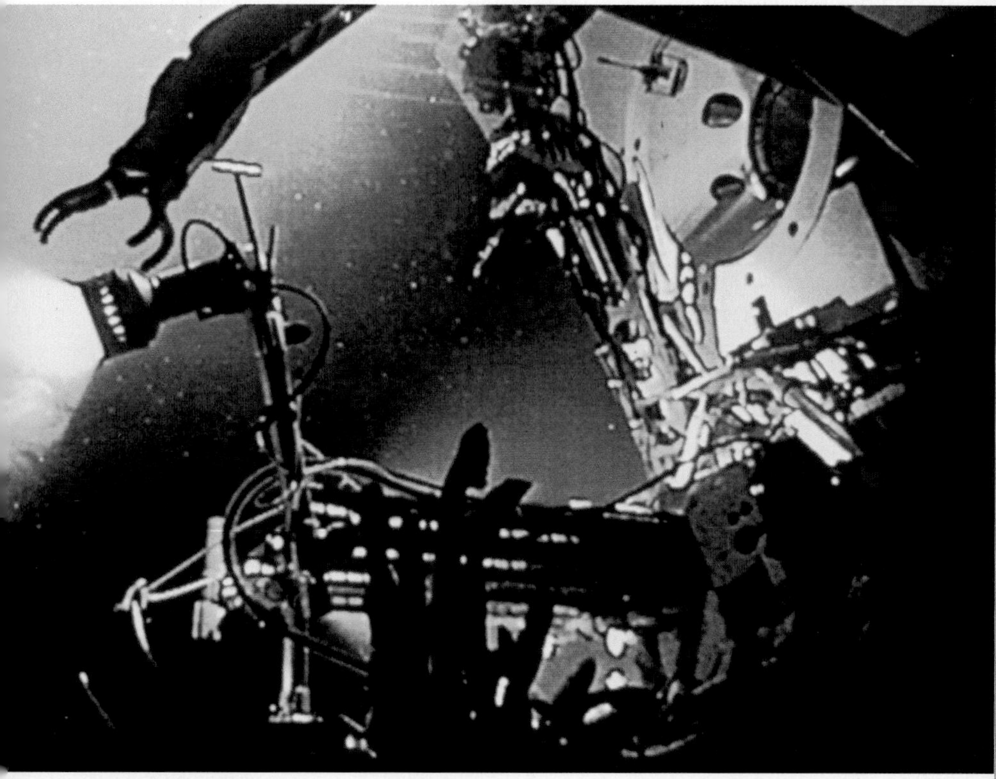

Concepts — *in* — Context

CONCEPTS IN CONTEXT

Deep-sea exploration demands submersibles of special design to withstand the immense underwater pressures and to provide maneuverability. The DSV (deep-sea vessel) *Alvin* was built in 1964. It carries a pilot and two passengers in a spherical cabin of thick titanium with several viewing ports. It was initially intended for dives of up to 1800 m, but it proved extremely successful in exploration and in salvage operations, and it is now certified for dives of up to 4500 m.

To appreciate the demands placed on such a submersible, we will consider these questions:

? What is the force that water pressure exerts on the viewing port of the vessel? (Example 4, page 574)

? What is the pressure at a depth of 4500 m? (Example 7, page 577)

? How does a submersible vessel change depth? (Section 18.5, page 580–581, and Example 11, page 582)

A **fluid** *is a system of particles loosely held together by their own cohesive forces or by the restraining forces exerted by the walls of a container.* Both liquids and gases are fluids—liquids are held together by their cohesive forces, and gases are held together by the restraining forces of a container (or, in the case of atmospheric air, by the weight of the atmosphere). In contrast to the particles in a rigid body, which are permanently locked into fixed positions, the particles in a fluid body are more or less free to wander about within the volume of the fluid body (see Fig. 18.1). A fluid will change its shape in response to external forces; for instance, a body of water or a body of air will change its shape in response to the forces exerted by gravity and by the container. The difference between these two kinds of fluids is that liquids are nearly incompressible, whereas gases are compressible. This means that a body of water has a constant volume independent of the container, whereas a body of air has a variable volume—the air always spreads out so as to entirely fill the container, and it can be made to expand or contract by increasing or decreasing the size of the container.

Although a fluid is a system of particles, the number of particles in, say, a cubic centimeter of water is so large that it is not feasible to describe the state of the fluid microscopically, in terms of the masses, positions, and velocities of all the individual particles in the system. Instead, we will describe the fluid in terms of its *density, velocity of flow,* and *pressure,* and we will see how the equations governing the statics and the dynamics of a fluid can be expressed in terms of these quantities.

In our discussion, we will neglect some other properties of fluids, such as the *viscosity* and the *surface tension*. Viscosity is an internal friction or stickiness within the fluid that offers resistance to its flow. For instance, honey is a fluid of high viscosity, whereas water is a fluid of fairly low viscosity. Surface tension is an elasticity of the exposed surface of fluid that tends to shrink this surface to a minimum area. Surface tension is responsible for the formation of drops of water in rain and for the beading of water splashed on a floor. Viscosity can be neglected when a fluid flows slowly, and surface tension can be ignored when the fluid has no exposed surface, as in the case of flow in completely filled pipes.

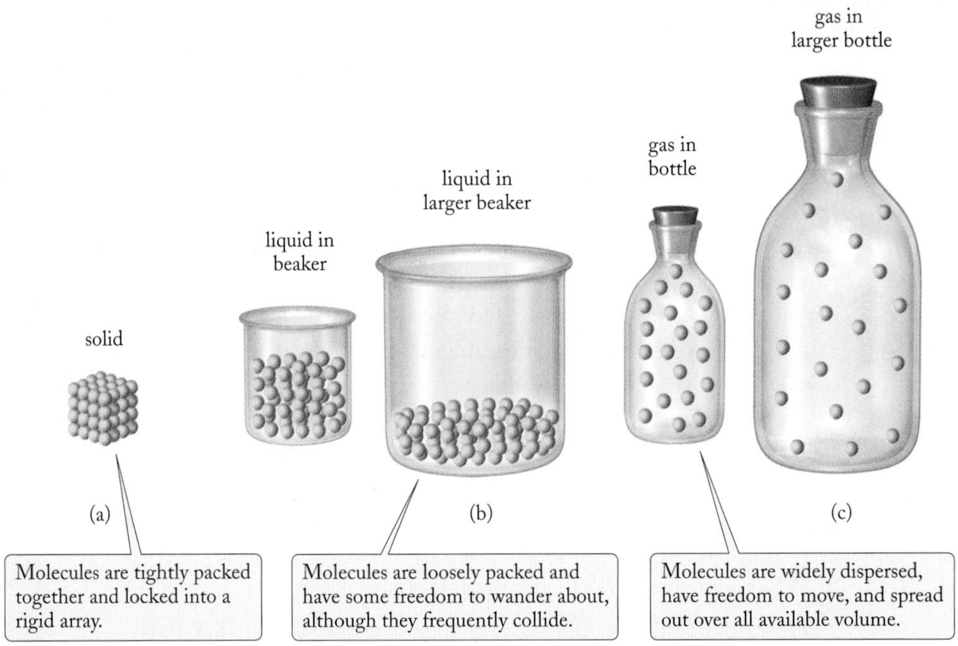

FIGURE 18.1 Molecules in (a) a solid, (b) a liquid, and (c) a gas.

18.1 DENSITY AND FLOW VELOCITY

Online Concept Tutorial 20

Density, velocity of flow, and pressure give us a macroscopic description of the fluid—they tell us the average behavior of the particles in regions within the fluid, and they can be measured with large-scale instruments. For example, if our instruments indicate that the flow velocity of water in a fire hose is 4 m/s, this does not mean that all the individual water molecules have this velocity. The water molecules have a high-speed thermal motion of about 900 m/s; they move in short zigzags because they frequently collide with one another. This thermal motion of a water molecule is random; the motion is as likely to be in a direction opposite to the flow as along the flow (see Fig. 18.2). The flow of water molecules along the fire hose at 4 m/s represents a slow drift superimposed on the much faster random zigzag motion. However, on a macroscopic scale, we notice only the drift and not the random small-scale motion—we notice only the *average* motion of the water molecules.

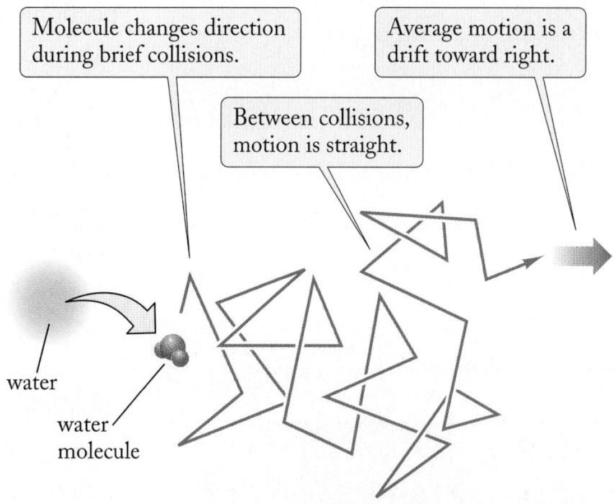

FIGURE 18.2 Motion of a molecule in water. The straight segments of the motion are typically 10^{-10} m long.

The **density** *is the amount of mass per unit volume.* Table 18.1 lists the densities of a few liquids and gases. The SI unit of density is the kilogram per cubic meter (kg/m^3). In the table, the density is designated by the customary symbol ρ, the Greek letter *rho*. Since density is the mass per unit volume, the total mass m in a volume V is the density times the volume:

$$m = \rho V \tag{18.1}$$

The densities of gases depend on the temperature and the pressure (this dependence will be discussed in Chapter 19); unless otherwise noted, the values of the densities listed in Table 18.1 are for standard temperature and standard pressure (0°C and 1 atm). The densities of liquids depend only slightly on pressure, but they do depend appreciably on temperature. For instance, water has a maximum density at about 4°C (a few degrees above its freezing temperature), and lowest density at 100°C (when it begins to boil).

TABLE 18.1 DENSITIES OF SOME FLUIDS[a]

LIQUID	ρ	GAS	ρ
Water		Air	
0°C	999.8 kg/m^3	0°C	1.29 kg/m^3
4°C	1000.0	20°C	1.20
20°C	998.2	Water vapor, 100°C	0.598
100°C	958.4	Hydrogen	0.0899
Seawater, 15°C	1025	Helium	0.178
Mercury	13600	Nitrogen	1.25
Sodium, liquid at 98°C	929	Oxygen	1.43
Texas crude oil, 15°C	875	Carbon dioxide	1.98
Gasoline, 15°C	739	Propane	2.01
Olive oil, 15°C	920		
Human blood, 37°C	1060		

[a] At 0°C and 1 atm, unless otherwise noted.

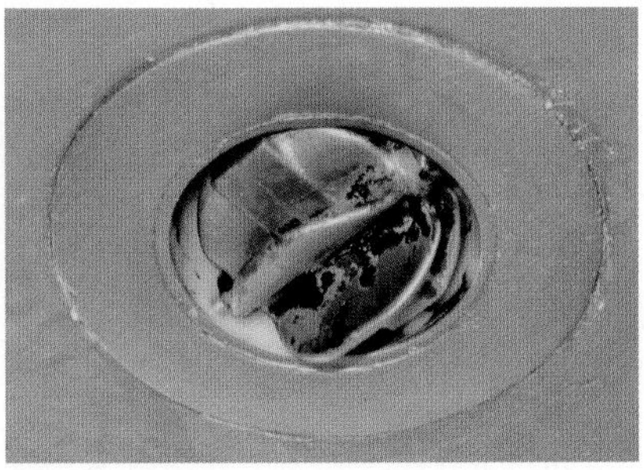

FIGURE 18.3 Paddle wheel of knot meter from a sailboat.

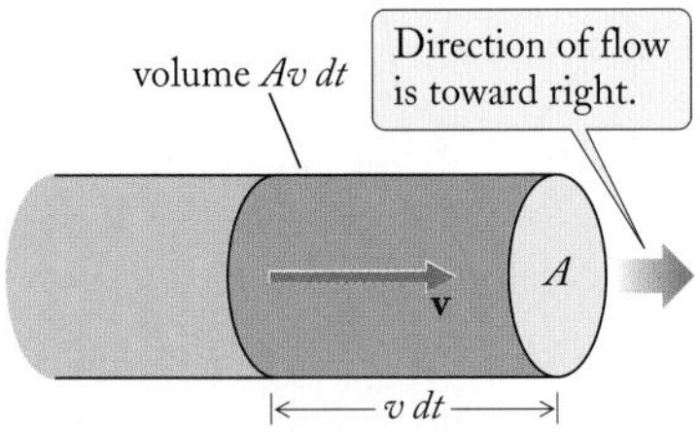

FIGURE 18.4 Flow of a fluid across an area A. In a time dt, the fluid within a distance $v\,dt$ reaches the area A, but fluid farther to the left does not.

The velocity of flow at some given point is the velocity of a small parcel of fluid passing this point. You can detect the flow velocity of the water at some point in a river by sprinkling bits of paper on the water. Quantitatively, the flow velocity can be measured by a small freely turning propeller or paddle wheel immersed in the fluid (Fig. 18.3). The motion of the fluid gives the paddle wheel a rotational speed directly proportional to the flow velocity, and the rotational speed indicator of the paddle wheel can be calibrated in terms of flow velocity.

From the magnitude v of the flow velocity of the fluid we can calculate the volume of fluid that flows across a unit area perpendicular to the direction of the velocity per unit time. Figure 18.4 shows a (stationary) area A perpendicular to the direction of flow and a volume of fluid about to cross this area. The fluid that crosses the area in a time dt is initially in a cylinder of base A and length $v\,dt$. The volume of fluid that crosses the area is therefore

$$dV = Av\,dt$$

and the volume that crosses per unit time is

$$\frac{dV}{dt} = Av \tag{18.2}$$

Here we assumed that the flow velocity v is constant over the area A. If the flow velocity in the fluid varies with position, then Eq. (18.2) is not valid for every area A, but only for a sufficiently small (infinitesimal) area A within which the flow velocity can be treated as constant.

EXAMPLE 1 The water in a fire hose of diameter 6.4 cm has a flow velocity of 4.0 m/s. At what rate does this hose deliver water? Give the answer in both cubic meters per second and kilograms per second.

SOLUTION: The radius of the hose is $r = 3.2\text{ cm} = 0.032\text{ m}$. The cross-sectional area of the hose is thus $A = \pi r^2 = \pi \times (0.032\text{ m})^2 = 0.0032\text{ m}^2$. Hence the rate of delivery is

$$\frac{dV}{dt} = Av = 0.0032\text{ m}^2 \times 4.0\text{ m/s} = 0.013\text{ m}^3\text{/s} \tag{18.3}$$

This gives the answer in terms of cubic meters per second. To find the rate of delivery in terms of kilograms per second, we must multiply dV/dt by the (constant) density ρ [see Eq. (18.1)]:

$$\frac{dm}{dt} = \rho\frac{dV}{dt} = 1000\text{ kg/m}^3 \times 0.013\text{ m}^3\text{/s} = 13\text{ kg/s} \tag{18.4}$$

✔ Checkup 18.1

QUESTION 1: According to Table 18.1, the mass of one cupful of mercury equals the mass of how many cupfuls of water?

QUESTION 2: What would be the rate at which the fire hose in Example 1 would deliver water if its diameter were half as large, that is, 3.2 cm instead of 6.4 cm?

QUESTION 3: The volume per second of water that flows through each of two pipes is the same. The flow velocity in the first pipe is one-quarter of that in the second pipe. What is the ratio of the radius of the first pipe to the radius of the second pipe?

(A) $\frac{1}{4}$ (B) $\frac{1}{2}$ (C) 1 (D) 2 (E) 4

18.2 INCOMPRESSIBLE STEADY FLOW; STREAMLINES

Online Concept 20 Tutorial

In most of the examples in this chapter we will deal with **steady flow**, *for which the velocity at any given point of space remains constant in time.* Thus, in steady flow, each small parcel of fluid that starts at any given point follows exactly the same path as a small parcel that passes through the same point at an earlier (or later) time. For example, Fig. 18.5 shows velocity vectors for the steady flow of water around a cylindrical obstacle, say, the flow of the water of a broad river around a cylindrical piling placed in the middle. The water enters the picture in a broad stream from the left, and disappears in a similar broad stream toward the right.

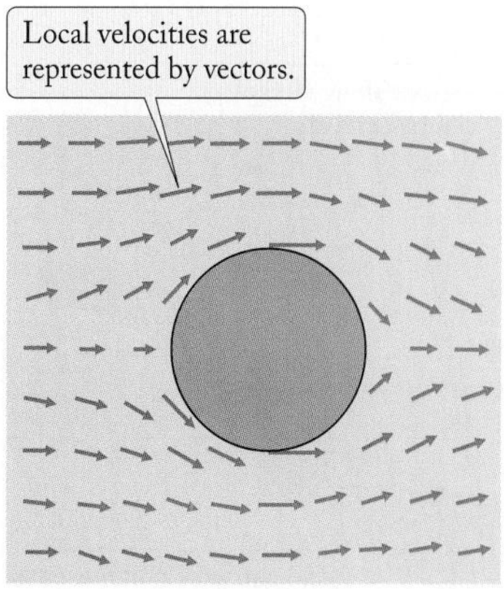

FIGURE 18.5 Velocity vectors for water flowing around a cylinder. The longest velocity vectors are found just above and just below the cylinder.

For the steady flow of an incompressible fluid, such as water, the picture of velocity vectors can be replaced by an alternative graphical representation. Suppose we focus our attention on a small volume of water, say, 1 mm^3 of water, and we observe the path of this 1 mm^3 from the source to the sink. *The path traced out by the small volume of fluid is called a* **streamline**. Neighboring small volumes will trace out neighboring streamlines. In Fig. 18.6 we show the pattern of streamlines for the same steady flow of water that we already represented in Fig. 18.5 by means of velocity vectors. The streamlines on the far left (and far right) of Fig. 18.6 are evenly spaced to indicate the uniform and parallel flow in this region.

The steady flow of an incompressible fluid is often called **streamline flow**. Note that streamlines never cross. A crossing of two streamlines would imply that a small parcel of water moving along one of these streamlines has to penetrate through a small parcel of water moving along the other streamline. This is impossible—it would lead to disruption of both the small parcels and to destruction of the steadiness of flow. Because the streamlines for steady incompressible flow never cross, such flow is also called **laminar flow**, which refers to the layered arrangement of the streamlines.

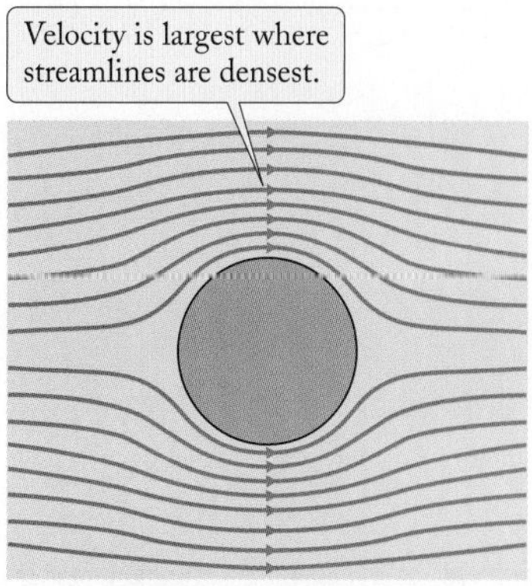

FIGURE 18.6 Streamlines for water flowing around a cylinder. The densest streamlines are found just above and just below the cylinder.

If we know the velocity of flow throughout the fluid, we can trace out the motion of small parcels of fluid and therefore construct the streamlines. But the converse is also true—if we know the streamlines, we can reconstruct the velocity of flow. We can do this by means of the following rule:

The direction of the velocity at any one point is tangent to the streamline, and the magnitude of the velocity is proportional to the density of streamlines.

The first part of this rule is self-evident, since the direction of motion of a small parcel of fluid is tangent to the streamline. To establish the second part, consider a bundle of streamlines forming a pipelike region, called a **stream tube**. Any fluid inside the stream tube will have to move along the tube; it cannot cross the surface of the tube because streamlines never cross. The tube therefore plays the same role as a pipe made of some impermeable material—it serves as a conduit for the fluid. If we consider a tube that is very narrow, so its cross-sectional area is very small, the velocity of flow will vary only along the length of the tube, and we can assume it will be the same at all points on a given cross-sectional area. For instance, on the area A_1 (see Fig. 18.7) the velocity is v_1, and on the area A_2 the velocity is v_2. In a time dt, Eq. (18.2) implies that the fluid volume that enters across the area A_1 is $dV_1 = v_1A_1\,dt$ and the fluid volume that leaves across the area A_2 is $dV_2 = v_2A_2\,dt$. The amount of fluid that enters must match the amount that leaves, since, under steady conditions, fluid cannot accumulate in the segment of tube between A_1 and A_2. Hence $dV_1 = dV_2$, and

$$v_1A_1\,dt = v_2A_2\,dt$$

or, canceling the factor dt on both sides of the equation,

continuity equation

$$v_1A_1 = v_2A_2 \tag{18.5}$$

This relation is called the **continuity equation**. It shows that along any stream tube, the speed of flow is inversely proportional to the cross-sectional area of the stream tube.

The density of streamlines inside the stream tube is the number of such lines divided by the cross-sectional area; since the number of streamlines entering A_1 is necessarily the same as that leaving A_2, the density of streamlines is inversely proportional to the cross-sectional area. This implies that the speed at any point in the fluid is directly proportional to the density of streamlines at that point. For example, in Fig. 18.6, the speed of the water is large at the top and bottom of the obstacle (large density of streamlines) and smaller to the left and right (smaller density of streamlines).

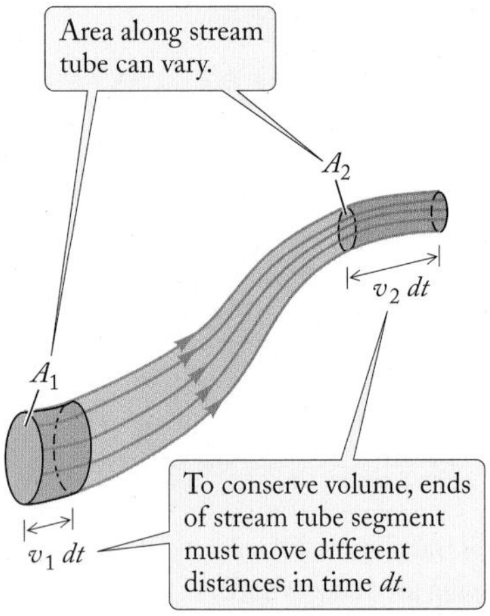

FIGURE 18.7 A stream tube.

In experiments on fluid flow, the streamlines of a fluid can be made directly visible by several clever techniques. If the fluid is water, we can place grains of dye at diverse points within the volume of water; the dye will then be carried along by the flow, and it will mark the streamlines. The photograph in Fig. 18.8 shows a pattern of streamlines made visible by this technique. The water emerges from a pointlike source on the left and disappears into a pointlike sink on the right. The colored streamers were created by small grains of potassium permanganate dissolving in the water.

If the fluid is air, we can make the streamlines visible by releasing smoke from small jets at diverse points within the flow of air. The photograph in Fig. 18.9 shows fine trails of smoke marking the streamlines in air flowing past a scale model of the wing of an airplane in a wind tunnel. The experimental investigation of such streamline patterns plays an important role in airplane design. Incidentally: Under some conditions, the flow of air can be regarded as nearly incompressible, provided that the speed of flow is well below the speed of sound (331 m/s). Although the air will suffer some changes of density in its flow around obstacles, the changes are usually small enough to be neglected.

Finally, Fig. 18.10 shows an example of **turbulent flow**. In the region behind the wing, the streamers of smoke become twisted and chaotic. This is due to the generation of vortices, or swirls of air, in this region. As the vortices form, grow, break away, and disappear in quick succession, the velocity of flow fluctuates violently. The flow of the fluid becomes unsteady and irregular. The formation of vortices and the onset of turbulence have to do with viscosity in the fluid (see Problem 73). It is a general rule that vortices and turbulence will develop in a fluid of given viscosity whenever the velocity of flow, the length of the flow, or both exceed a certain limit. We can see the transition from steady flow to turbulent flow in the ascending smoke trail from a cigarette (see Fig. 18.11). The flow starts out steady, with smoke particles moving along well-defined streamlines; but at some height above the cigarette, where the length of the flow exceeds the critical limit, the flow becomes turbulent.

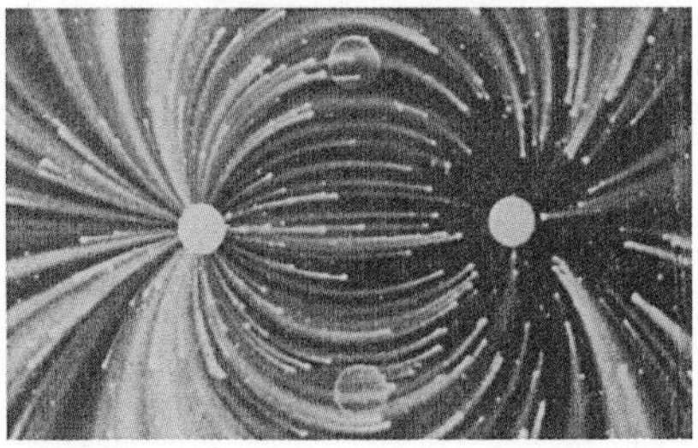

FIGURE 18.8 Streamlines in water flowing from a source (left) to a sink (right).

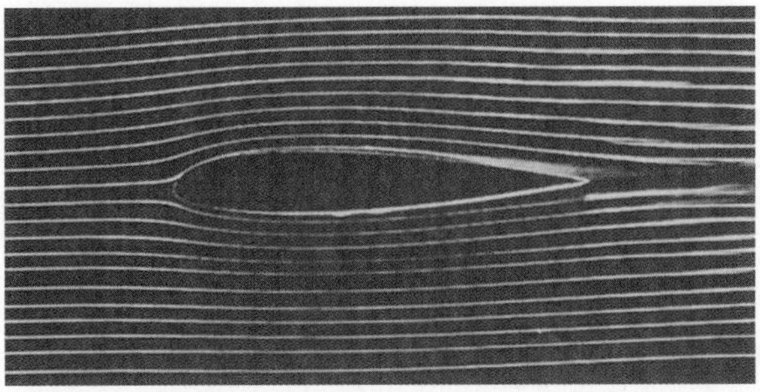

FIGURE 18.9 Fine trails of smoke indicate the streamlines in air flowing around the wing of an aircraft.

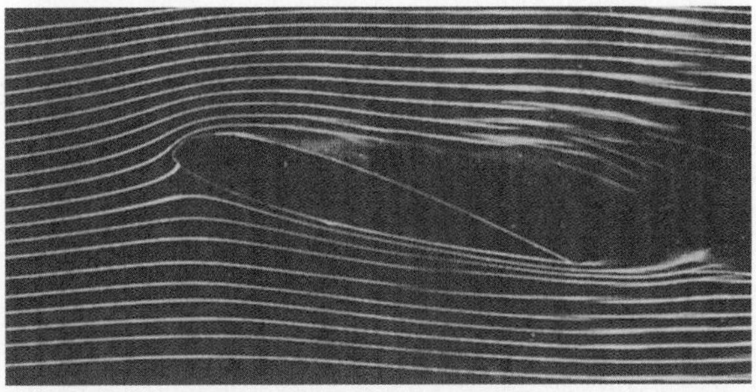

FIGURE 18.10 Here, the wing is in a partial stall, and the flow behind the wing has become turbulent.

FIGURE 18.11 Ascending smoke from a cigarette.

EXAMPLE 2 In the human circulatory system, the blood flows out of the heart via the aorta, which is connected to other arteries that branch out into a multitude of small capillaries (see Fig. 18.12). In the average adult, the aorta has a radius of 1.2 cm, and the speed of flow of the blood is 0.20 m/s. The radius of each capillary is about 3×10^{-6} m, and the number of open capillaries, under conditions of rest, is about 1×10^{10}. Calculate the speed of flow of the blood in the capillaries.

SOLUTION: The cross-sectional area of the aorta is

$$A_1 = \pi r_1^2 = \pi \times (0.012 \text{ m})^2 = 4.5 \times 10^{-4} \text{ m}^2$$

and the net cross-sectional area of all the capillaries is

$$A_2 = [\text{number of capillaries}] \times [\text{area of each}]$$

$$= 1 \times 10^{10} \times \pi r_2^2 = 1 \times 10^{10} \times \pi \times (3 \times 10^{-6} \text{ m})^2 = 3 \times 10^{-1} \text{ m}^2$$

From the continuity equation (18.5), with $v_1 = 0.20$ m/s, we then find that the speed of flow in the capillaries is

$$v_2 = \frac{A_1}{A_2} v_1 = \frac{4.5 \times 10^{-4} \text{ m}^2}{3 \times 10^{-1} \text{ m}^2} \times 0.20 \text{ m/s}$$

$$= 3 \times 10^{-4} \text{ m/s} = 0.3 \text{ mm/s}$$

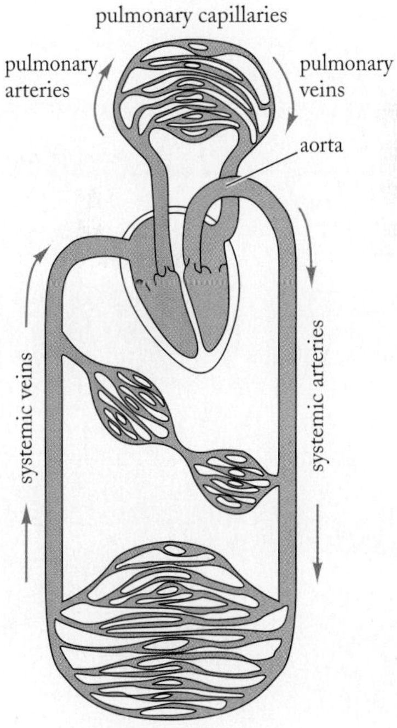

FIGURE 18.12 The human circulatory system.

FIGURE 18.13 The tall fountain at Fountain Hills, Arizona.

EXAMPLE 3 One of the world's tallest fountains (at Fountain Hills, Arizona; see Fig. 18.13) shoots water to a height of 170 m at the rate of 26 000 liters/min. If we ignore friction, the motion of small volumes of water is projectile motion. From this, it is easy to determine that the speed of flow must be 58 m/s at the base, and 37 m/s at a height of 100 m. Given these speeds, calculate the cross-sectional area of the water column at the base and at a height of 100 m.

SOLUTION: To find the cross-sectional area, we use Eq. (18.2),

$$A = \frac{1}{v}\frac{dV}{dt}$$

The rate of delivery of the fountain is (using 1 liter = 1000 cm^3 = 10^{-3} m^3)

$$\frac{dV}{dt} = \frac{2.6 \times 10^4 \text{ liters}}{1 \text{ min}} = \frac{26 \text{ m}^3}{60 \text{ s}} = 0.43 \text{ m}^3\text{/s}$$

Therefore, at the base,

$$A = \frac{1}{58 \text{ m/s}} \times 0.43 \text{ m}^3\text{/s} = 0.0075 \text{ m}^2 = 75 \text{ cm}^2$$

and at a height of 100 m

$$A = \frac{1}{37 \text{ m/s}} \times 0.43 \text{ m}^3\text{/s} = 0.012 \text{ m}^2 = 120 \text{ cm}^2$$

Note that the water column is narrow at the base and increasingly widens toward the top.

FIGURE 18.14 Water flowing down from a faucet.

✔ Checkup 18.2

QUESTION 1: Is the continuity equation (18.5) valid for a compressible fluid, that is, for a fluid whose density changes as it flows?

QUESTION 2: Does steady flow mean that the velocity of a parcel of fluid remains constant?

QUESTION 3: Consider a gradually broadening river flowing out into the sea. Compare the densities of the streamlines before the mouth, at the mouth, and beyond the mouth. Where is the density of streamlines largest and where smallest?

QUESTION 4: According to the calculations in Example 2, is the density of streamlines larger in the aorta or in the capillaries?

QUESTION 5: When water flows vertically downward out of a faucet, the cross-sectional area of the stream of water gradually narrows (see Fig. 18.14). Explain.

QUESTION 6: Figure 18.6 shows the streamlines of water flowing around a cylinder. Where is the speed of flow largest?

(A) At top and bottom of cylinder (B) At center of cylinder
(C) At left and right of cylinder

18.3 PRESSURE

The **pressure** within a fluid is defined as the force per unit area that a small volume of fluid exerts on an adjacent volume or on the adjacent wall of a container. Figure 18.15a shows two small adjacent cubical volumes of fluid that are within a larger volume of fluid surrounding them. The cube of fluid on the left presses against the cube on the right, and vice versa. Suppose that the magnitude of the perpendicular force between the two cubes is F and that the area of one face of one of the cubes is A; then the pressure p is defined as the magnitude F of the force divided by the area A:

$$p = \frac{F}{A} \tag{18.6}$$

According to this definition, pressure is simply the force per unit area. Note that, in contrast to the force, the pressure is a quantity *without direction,* that is, a scalar. We cannot associate a direction with the pressure, because each small volume exerts pressure forces in all directions perpendicular to its surface. For instance, each of the small cubes in Fig. 18.15a exerts pressure forces in all directions on the surrounding fluid (see Fig. 18.15b). We will see later that pressure can vary with position (for example, because of gravity); however, at any given location, the pressure is the same in all directions.

A simple mechanical device for the measurement of pressure consists of a hermetically sealed, evacuated cylindrical can with corrugated flexible bases (see Fig. 18.16). If this capsule is immersed in a fluid at high pressure, the bases will be compressed inward; if it is immersed in a fluid at low pressure, the bases will bulge outward. Thus, the deformation of the bases of the capsule serves as an indicator of the pressure, and a pointer linked to one base can be calibrated to read the pressure. This device is widely used in aneroid barometers for the measurement of atmospheric pressure.

In the SI system, the unit of pressure is the N/m^2, which has been given the name **pascal** (Pa),

$$1 \text{ Pa} = 1 \text{ N/m}^2 \tag{18.7}$$

Another unit in common use is the **atmosphere** (atm):

$$1 \text{ atm} = 1.01 \times 10^5 \text{ Pa} \tag{18.8}$$

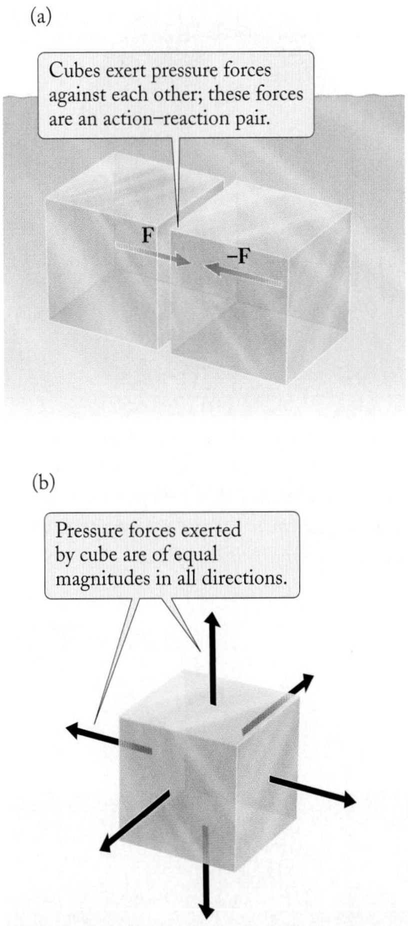

FIGURE 18.15 (a) Adjacent small cubes of fluid exerting pressure forces on each other. (b) Each small cube exerts pressure forces in all directions on the surrounding fluid.

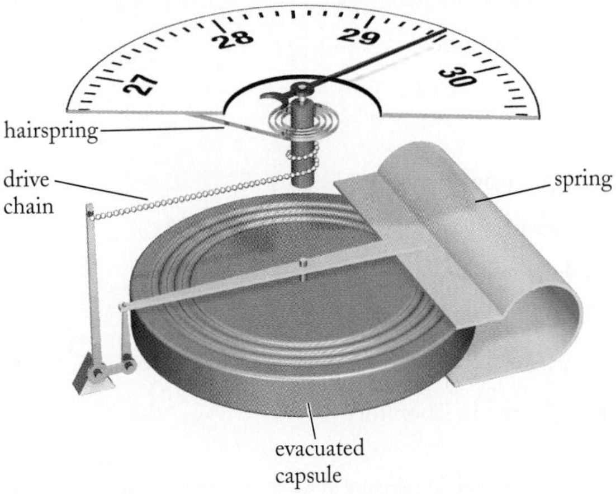

FIGURE 18.16 Device for the measurement of atmospheric pressure consisting of a hermetically sealed evacuated cylindrical capsule made of thin sheet metal. A spring holds the flexible bases apart, and prevents the atmospheric pressure from collapsing the capsule. The upper flexible base of the capsule is linked to a pointer that indicates the atmospheric pressure.

BLAISE PASCAL (1623–1662) *French scientist. He made important contributions to mathematics and is regarded as the founder of modern probability theory. In physics, he performed experiments on atmospheric pressure and on the equilibrium of fluids.*

The common unit of pressure in the British system is the **pound-force per square inch** (lbf/in.^{2}, abbreviated psi); in psi, one atmosphere is

$$1 \text{ atm} = 14.7 \text{ psi} \tag{18.9}$$

This is the average value of the pressure of air at sea level. Note that this is quite a large pressure. For instance, the force that the atmospheric pressure of 1 atm exerts on the palm of your hand, of approximate area 0.006 m^2, is

$$F = Ap = 0.006 \text{ m}^2 \times 1.01 \times 10^5 \text{ Pa} = 600 \text{ N} \tag{18.10}$$

This is roughly the weight of 60 kg, but you do not notice this pressure force because an equal pressure force of opposite direction acts on the back of your hand, leaving the hand in equilibrium (you don't even notice that these opposed forces squeeze your hand, because the external pressure forces directed against your skin are compensated by the internal pressure forces exerted by your body fluids).

Table 18.2 gives diverse examples of values of pressure.

Concepts in Context

EXAMPLE 4 The viewing ports of the DSV *Alvin* have a diameter of 30 cm (see Fig. 18.17). What is the force that water pressure exerts on the outside of such a viewing port when the *Alvin* is at a depth of 4500 m below the surface of the ocean, where the water pressure is 4.5×10^7 Pa?

SOLUTION: The area of the porthole is $A = \pi r^2 = \pi \times (0.15 \text{ m})^2 = 0.071 \text{ m}^2$. According to Eq. (18.6), the force is then

$$F = A \times p = 0.071 \text{ m}^2 \times 4.5 \times 10^7 \text{ Pa} = 3.2 \times 10^6 \text{ N}$$

This is a weight of about 320 tons!

FIGURE 18.17 Viewing ports of the DSV *Alvin*.

TABLE 18.2 SOME PRESSURES

Core of neutron star	1×10^{38} Pa
Center of Sun	2×10^{16}
Highest sustained pressure achieved in laboratory	5×10^{11}
Center of Earth	4×10^{11}
Bottom of Pacific Ocean (5.5-km depth)	6×10^7
Water in core of nuclear reactor	1.6×10^7
Overpressure[a] in automobile tire	2×10^5
Air at sea level	1.0×10^5
Overpressure at 7 km from 1-megaton explosion	3×10^4
Air in funnel of tornado	2×10^4
Overpressure in human heart	
Systolic	1.6×10^4
Diastolic	1.1×10^4
Lowest vacuum achieved in laboratory	10^{-14}

[a] The *overpressure* is the amount of pressure in excess of normal atmospheric pressure.

✓ Checkup 18.3

QUESTION 1: Consider a cube of solid concrete, 1 m × 1 m × 1 m, surrounded by air at 1 atm of pressure. What is the pressure force of air on its lower face? On its upper face? On its left face? On its right face?

QUESTION 2: A brick rests on a table. Does the atmospheric pressure force push this brick downward on the table?

QUESTION 3: A rubber balloon is filled with air and sealed tightly. Would you expect this balloon to change in size as the atmospheric pressure of air changes from one day to the next?

QUESTION 4: A closed, empty soda bottle, with a diameter of 10 cm at its base and 2 cm at its cap, is lying sideways on a desk in the air. What is the ratio of the value of the external pressure at its base to the value of the pressure at its cap?

(A) 25 (B) 5 (C) 1 (D) $\frac{1}{5}$ (E) $\frac{1}{25}$

18.4 PRESSURE IN A STATIC FLUID

A fluid is said to be in static equilibrium when the flow velocity is everywhere zero, that is, the fluid is at rest. An example of such a static fluid is the air in a closed room with no air currents. At first, we will neglect gravity and pretend that the only forces acting on the fluid are those exerted by the walls of the container. Under these conditions, the pressure at all points within the fluid must be the same. To see that this is so, consider two points 1 and 2, and imagine a long, thin parallelepiped of fluid with bases at these two points (see Fig. 18.18). The fluid outside the parallelepiped exerts pressure forces on the fluid inside the parallelepiped. The components of these forces along the long direction of the parallelepiped are entirely due to the forces on the bases at 1 and 2. If the parallelepiped of fluid is to remain static, these forces on the opposite bases must be equal in magnitude. Hence, the pressures at 1 and 2 must be equal. For example, the pressure of the air is the same at all points of a room—if the pressure is 1 atm in one corner of the room, it will be the same at any other point of the room.

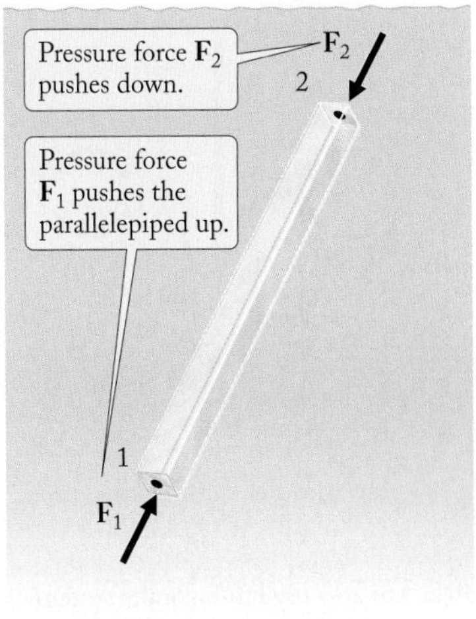

FIGURE 18.18 A long, thin parallelepiped of fluid within a static fluid.

The uniformity of pressure throughout a static fluid implies that if we apply a pressure to some part of the surface of a confined fluid by means of a piston or a weight pushing against the surface, then this pressure will be transmitted without change to all parts of the fluid. This rule for the transmission of pressure in a static fluid is called **Pascal's Principle**, and it finds widespread application in the design of hydraulic presses, jacks, and remote controls. Figure 18.19 is a schematic diagram illustrating the principle of a hydraulic press. The mechanism that generates a large force in such a press consists of two cylinders with pistons, one small and one large. The cylinders are filled with an incompressible fluid, and they are connected by a pipe. By pushing down on the small piston, we increase the pressure in the fluid; this increases the force on the large piston. Since the pressures on both pistons are the same, the forces on the pistons are in the ratio of the areas of their faces; thus, a small force on the piston in the small cylinder will generate a large force on the piston in the large cylinder. Figure 18.20 shows a hydraulic jack for an automobile. The pumping lever pushes on the small piston, and the hydraulic fluid communicates the

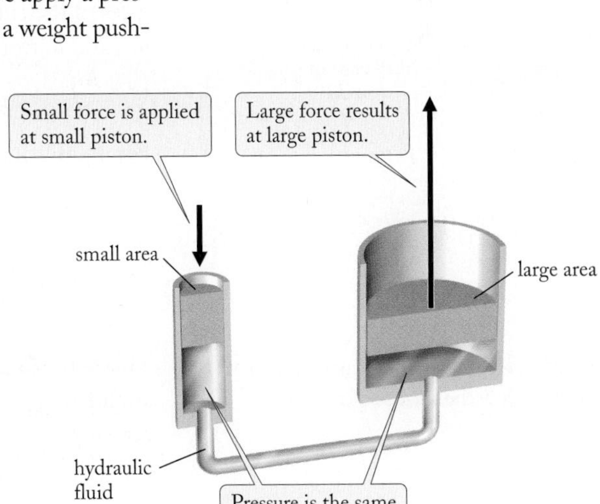

FIGURE 18.19 Mechanism of a hydraulic press (schematic).

FIGURE 18.20 Hydraulic car jack.

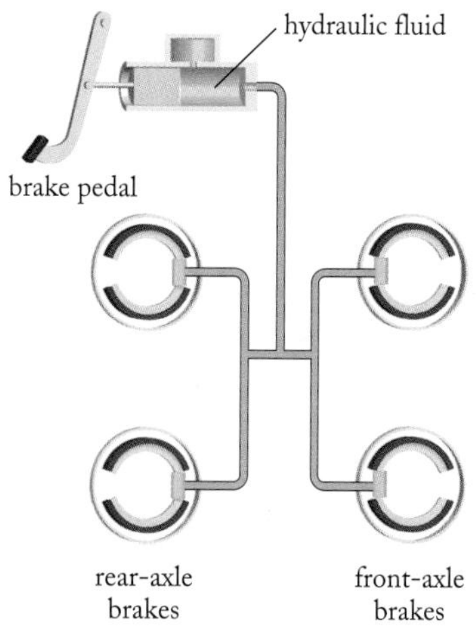

FIGURE 18.21 Hydraulic brake system.

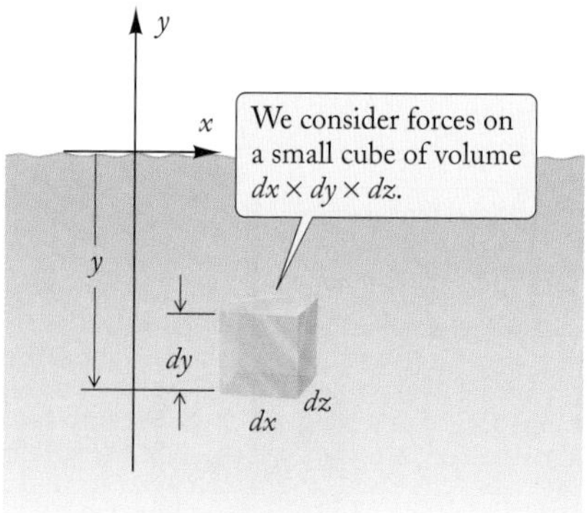

FIGURE 18.22 A small cube of fluid at depth y below the surface of the fluid.

resultant pressure to the large piston, which lifts the automobile. The brake systems and other control systems on automobiles, trucks, and aircraft also employ such arrangements of cylinders connected by pipes filled with hydraulic fluid (see Fig. 18.21).

EXAMPLE 5 The diameter of the small piston in Fig. 18.19 is 1.5 cm and that of the large piston is 7.0 cm. If you exert a force of 300 N on the small piston, what force will this generate on the large piston?

SOLUTION: The pressure is the same at all points in a static fluid (neglecting gravity), so $p_1 = p_2$ requires $F_1/A_1 = F_2/A_2$. Thus the forces are in the ratio of the areas of the pistons, and these areas are in the ratio of the squares of the diameters:

$$F_2 = F_1 \frac{A_2}{A_1}$$

$$= 300\ \text{N} \times \frac{(7.0\ \text{cm})^2}{(1.5\ \text{cm})^2} = 6500\ \text{N}$$

Next, we want to take into account the effect of gravity on the pressure in a fluid. For a static fluid subjected to gravity, such as the water of a calm lake, the pressure force at any given depth must support the weight of the overlying mass of fluid; consequently, the pressure must increase with depth. To derive a formula for the dependence of pressure on depth, consider the condition for the equilibrium of a small cubical volume of fluid. Figure 18.22 shows a small cube at some depth y below the surface of the fluid. As always, the y coordinate is reckoned as positive in the upward direction; hence the depth of the cube corresponds to some negative value of y. The dimensions of the cube are $dx \times dy \times dz$, and hence its weight is

$$g\,dm = g\rho\, dx \times dy \times dz$$

where ρ is the mass density, or mass per unit volume, of the fluid. The weight of the cube must be balanced by the vertical forces contributed by the pressure. Suppose that the pressure at the top of the cube is p; the pressure at the bottom of the cube is then some larger value $p + dp$, where dp represents the change of pressure in the interval dy (note that if dy is negative, dp will be positive). Since the area of the top and bottom faces of the cube is $dx \times dz$, the difference between the vertical pressure forces on the top and bottom of the cube is

$$p \times dx \times dz - (p + dp) \times dx \times dz = -dp \times dx \times dz$$

For equilibrium, this pressure force must balance the weight:

$$-dp \times dx \times dz = g\rho\, dx \times dy \times dz$$

or

$$dp = -g\rho\, dy \qquad (18.11)$$

This formula gives us the small change in pressure for a small increase in depth. For an incompressible fluid, ρ is constant, and Eq. (18.11) shows that the small change in pressure is proportional to the small change in y. Hence the total change in pressure must be proportional to the total change in y; that is,

pressure in incompressible fluid

$$p - p_0 = -\rho g y \qquad \text{(incompressible fluid)} \qquad (18.12)$$

Here p_0 is the pressure at the surface of the liquid, where $y = 0$. Note that depths below the surface are reckoned as negative in Eq. (18.12), and therefore the right side of the equation is positive, which leads to a positive value of $p - p_0$, that is, an increase of pressure with depth.

EXAMPLE 6 What is the pressure at a depth of 10 m below the surface of a lake? Assume that the pressure of air at the surface of the lake is 1.0 atm.

SOLUTION: With $p_0 = 1\text{ atm} = 1.01 \times 10^5\text{ Pa}$, $\rho = 1000\text{ kg/m}^3$, and $y = -10\text{ m}$, Eq. (18.12) gives

$$\begin{aligned} p &= p_0 - \rho g y \\ &= 1.01 \times 10^5\text{ Pa} + 1000\text{ kg/m}^3 \times 9.81\text{ m/s}^2 \times 10\text{ m} \qquad (18.13) \\ &= 1.99 \times 10^5\text{ Pa} = 2.0\text{ atm} \end{aligned}$$

Thus, the pressure is 1 atm at the surface of the lake and about 2 atm at a depth of 10 m; that is, the pressure increases by about 1 atm per 10 m of water, an easily remembered change with depth.

EXAMPLE 7 What is the pressure at a depth of 4500 m below the surface of the ocean, the maximum depth for which the DSV *Alvin* (see chapter photo) is certified? The density of seawater is 1025 kg/m^3.

Concepts in Context

SOLUTION: The calculation proceeds as in the preceding example:

$$\begin{aligned} p &= p_0 - \rho g y \\ &= 1.01 \times 10^5\text{ Pa} + 1025\text{ kg/m}^3 \times 9.81\text{ m/s}^2 \times 4500\text{ m} \\ &= 4.54 \times 10^7\text{ Pa} \end{aligned}$$

This is the value of the pressure used in Example 4. This is approximately 450 atm, in agreement with the expected pressure increase of about 1 atm for every 10 m.

Several simple instruments for the measurement of pressure make use of a column of liquid. Figure 18.23 shows a mercury **barometer** consisting of a tube of glass, about 1 m long, closed at the upper end and open at the lower end. The tube is filled with mercury, except for a small evacuated space at the top. The bottom of the tube is immersed in an open bowl filled with mercury. The atmospheric pressure acting on the exposed surface of mercury in the bowl prevents the mercury from flowing out of the tube. At the level of the exposed surface, the pressure exerted by the column of mercury is $\rho g h$ [see Eq. (18.12)], where $\rho = 1.36 \times 10^4\text{ kg/m}^3$ is the density of mercury and h the height of the mercury column. For equilibrium, this pressure must match the atmospheric pressure:

$$p_0 = \rho g h \qquad (18.14)$$

This equation permits a simple determination of the atmospheric pressure from a measurement of the height of the mercury column.

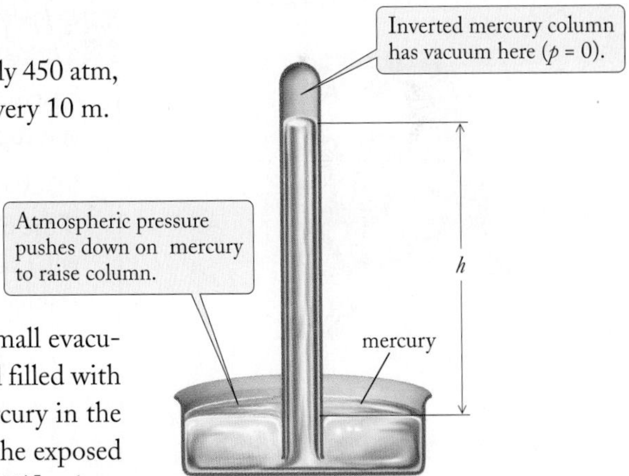

FIGURE 18.23 A mercury barometer.

In view of the direct correspondence of the atmospheric pressure and the height of the mercury column, the pressure is often quoted in terms of this height, usually expressed in millimeters of mercury (abbreviated mm-Hg, where Hg is the symbol for the element mercury). The average value of the **atmospheric pressure** at sea level is 760 mm-Hg, which by definition is one atmosphere (atm). Hence,

$$\begin{aligned} 1 \text{ atm} &= 760 \text{ mm-Hg} = \rho \times g \times 0.760 \text{ m} \\ &= 1.36 \times 10^4 \text{ kg/m}^3 \times 9.81 \text{ m/s}^2 \times 0.760 \text{ m} \\ &= 1.01 \times 10^5 \text{ N/m}^2 = 1.01 \times 10^5 \text{ Pa} \end{aligned} \tag{18.15}$$

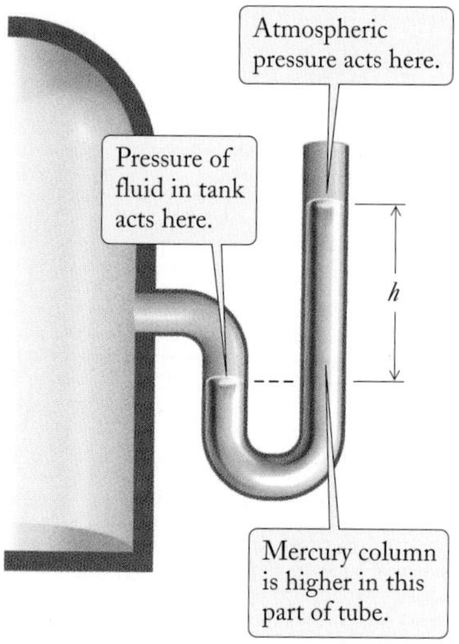

FIGURE 18.24 Open-tube manometer.

This value of the atmosphere has already been mentioned, in Eq. (18.8). The unit mm-Hg is also referred to as the **torr** (from Torricelli); that is, 1 torr = 1 mm-Hg. In British units, the atmospheric pressure is often quoted in inches of mercury; typical values are around 30 inches, since 760 mm × (1 inch)/(25.4 mm) = 29.9 inches.

Figure 18.24 shows an open-tube **manometer**, a device for the measurement of the pressure of a fluid, such as that contained in the tank shown on the left. The U-shaped tube contains mercury, or water, or oil. One side of the tube is in contact with the fluid in the tank; the other is in contact with the air. The fluid in the tank therefore presses down on one end of the mercury column and the air presses down on the other end. The difference h in the heights of the levels of mercury at the two ends gives the difference in the pressure at the two ends:

$$p - p_0 = \rho g h \tag{18.16}$$

Hence, this kind of manometer indicates the amount of pressure in the tank in excess of the atmospheric pressure. This excess is called the **overpressure**, or **gauge pressure**. It is well to keep in mind that many pressure gauges used in engineering practice are calibrated in terms of the overpressure rather than absolute pressure. For instance, the pressure gauges used for automobile tires read overpressure.

EXAMPLE 8 What is the change in atmospheric pressure between the basement of a house and the attic, at a height of 10 m above the basement? Express the result in mm-Hg. Assume that the density of air has its standard value 1.29 kg/m³.

SOLUTION: Although air is a compressible fluid, the change in its density is small if the change of altitude (and pressure) is small, as it is in the present example; we will examine the behavior for larger changes in the next chapter. Therefore, we can assume Eq. (18.12) is a good approximation:

$$\begin{aligned} p - p_0 &= -\rho g y \\ &= -1.29 \text{ kg/m}^3 \times 9.81 \text{ m/s}^2 \times 10 \text{ m} = -1.3 \times 10^2 \text{ Pa} \end{aligned} \tag{18.17}$$

Since 1.01×10^5 Pa equals 760 mm-Hg, -1.3×10^2 Pa equals

$$-1.3 \times 10^2 \text{ Pa} \times \frac{760 \text{ mm-Hg}}{1.01 \times 10^5 \text{ Pa}} = -0.95 \text{ mm-Hg}$$

Hence the pressure change is indeed small; the pressure decreases by about 1 mm-Hg for a 10-m increase in height in air. This decrease of pressure can be detected by carrying an ordinary barometer from the basement to the attic of the house.

EVANGELISTA TORRICELLI (1608–1647) *Italian physicist, mathematician, and inventor. At age 33 he assisted Galileo, serving as both researcher and secretary. He was the first to understand air pressure in terms of the weight of the atmosphere and to explain the limitations of suction pumps. His accomplishments included contributions to the development of integral calculus and the invention of the barometer.*

PHYSICS IN PRACTICE THE SPHYGMOMANOMETER

Among the many practical applications of manometers is the **sphygmomanometer** used to measure cardiac blood pressure. This consists of a manometer connected to an air sac in the form of a cuff (see the figures). The air sac is wrapped around the upper arm of the patient, and is then inflated by means of a hand pump until the pressure of the sac against the arm collapses the brachial artery and cuts off the blood flow. The air is then slowly allowed to leak out of the sac. When the pressure drops to a value equal to the systolic (maximum) cardiac pressure, blood will intermittently squirt through the artery with each heartbeat. This initiation of intermittent blood flow can be readily detected by listening to the noise of the rushing blood with a stethoscope placed just below the cuff. As the pressure drops further, the intervals of intermittent blood flow become longer; and when the pressure has dropped to a value equal to the minimum (diastolic) cardiac pressure, the blood flow becomes continuous. Thus, the onset of intermittent noises and the cessation of intermittent noises signal, respectively, the systolic and the diastolic cardiac pressures. Typical values of these pressures in healthy adults are 120 mm-Hg and 80 mm-Hg, usually reported in the abbreviated notation 120/80. Note that in order to obtain an accurate reading of the cardiac pressure, the cuff must be placed at the level of the heart; if it were placed lower (say, on a leg) or higher (say, on a raised arm), then there would be a pressure difference ρgh between the height of the heart and the height of the cuff. This pressure difference would amount to about 10 mm-Hg for every 10 cm of height difference—a significant discrepancy.

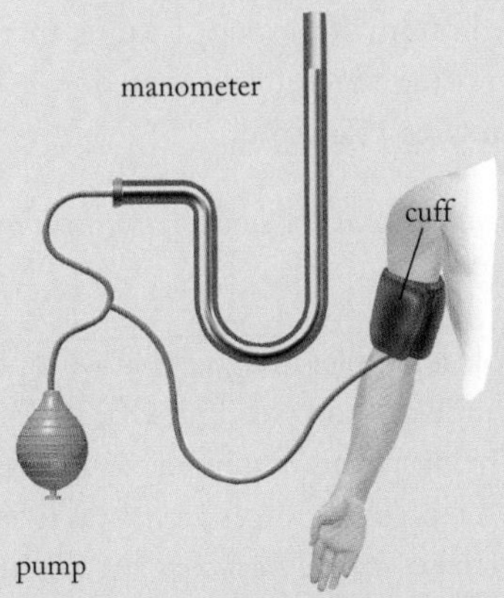

FIGURE 1 A sphygmomanometer.

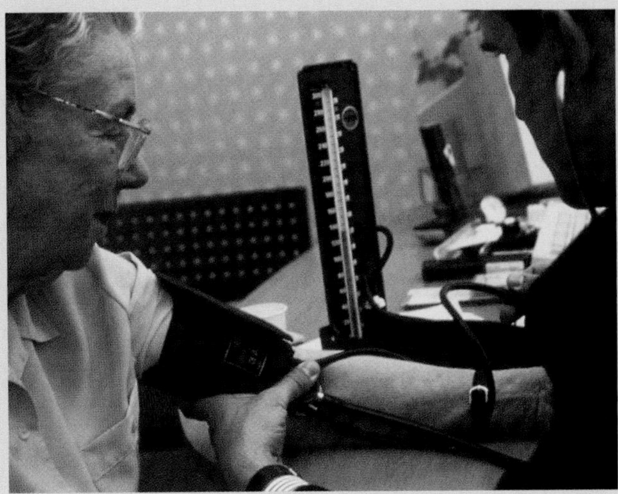

FIGURE 2

Checkup 18.4

QUESTION 1: Pipes connect the faucets in a house to the water main that supplies the house. Is the pressure in the faucets on the first floor the same as the pressure on the second floor?

QUESTION 2: You carry an inflated rubber balloon up a mountain. Would you expect this balloon to change in size?

QUESTION 3: Does the increase of pressure with depth [Eq. (18.12)] contradict Pascal's Principle?

QUESTION 4: A scuba diver descends to 10 m below the surface of the sea. What is the pressure inside her muscle tissues when she is at the surface? When she reaches 10 m?

(A) 0 atm, 0 atm (B) 0 atm, 1 atm
(C) 1 atm, 1 atm (D) 1 atm, 2 atm

18.5 ARCHIMEDES' PRINCIPLE

If you try to push a beach ball below the surface of the water, you notice that the water exerts a strong upward push on the ball. You can barely force the ball under, and if you release it, the ball pops out of the water with violence. The upward force that water or some other fluid exerts on a partially or totally immersed body is called the **buoyant force**. This force results from the pressure difference between the bottom and the top of the body. In a fluid in equilibrium under the influence of gravity, the pressure increases with depth; hence the pressure of the fluid at the bottom of the body is larger than that at the top of the body, and there is more force pressing the body up than down. The magnitude of the buoyant force is given by **Archimedes' Principle:**

Archimedes' Principle

The buoyant force on an immersed body has the same magnitude as the weight of the fluid displaced by the body.

The proof of this famous principle is simple. Imagine that we replace the immersed volume of the body by an equal volume of fluid (see Fig. 18.25). The volume of fluid will then be in static equilibrium. Obviously, this requires a balance between the weight of the fluid and the resultant of all the pressure forces acting on the surface enclosing this volume of fluid. But the pressure forces on the surface of the original immersed body are exactly the same as the pressure forces on the surface of the volume of fluid by which we have replaced it. Hence, the magnitude of the resultant of the pressure forces acting on the original body must equal the weight of the displaced fluid.

If a body of density less than that of water floats in water, equilibrium of the weight and of the buoyant force is achieved when the body is partially submerged—the weight of the body must match the weight of the water displaced by the submerged part of the body. Thus, the weight of a ship must match the weight of the water displaced by the submerged part of its hull. If we increase the weight of the ship, by loading more cargo, the hull will submerge more deeply until equilibrium is established (Fig. 18.26). If we increase the weight of the ship to such an extent that its average density exceeds that of water, equilibrium becomes impossible, and the ship sinks.

Concepts in Context

The human body has an average density slightly less than that of water; if inert, it floats with only the top of the head sticking out. But the equilibrium is quite delicate, and some people can make themselves sink by merely exhaling air, thereby reducing their chest volume and increasing their average density. Most species of bony fishes use a similar method to adjust their buoyancy; they have an internal swim bladder filled with gas,

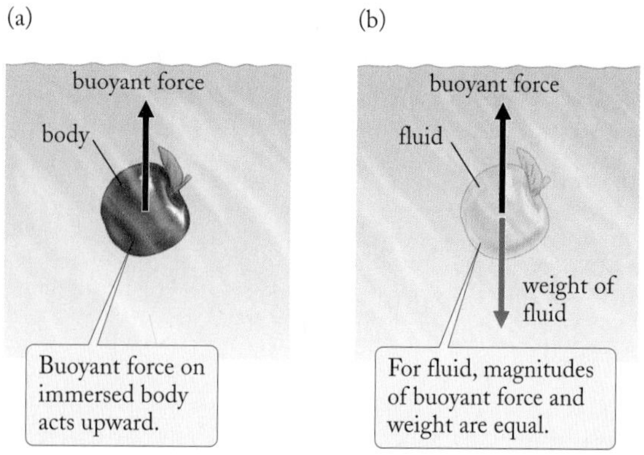

FIGURE 18.25 (a) A submerged body and the buoyant force that acts on this body. (b) "Free-body" diagram for a volume of fluid of the same shape as the body. For the volume of fluid, the buoyant force is balanced by the weight.

FIGURE 18.26 The markings along the hull of the ship indicate the immersion level.

and they preserve neutral buoyancy by making adjustments to the volume of this swim bladder. Submarines use much the same method to move down or up in the water. They have diving tanks ("ballast tanks") filled partially with water and partially with air. To dive, they pump air out of the tanks, into pressurized storage cylinders, and allow water to flood the tanks, thereby effectively increasing the mass and the density of the submarine. To surface, they blow high-pressure air into the tanks, and drive out the water.

ARCHIMEDES (287–212 B.C.) *Greek philosopher, mathematician, and physicist. He calculated an accurate value for π and obtained many other geometrical results, especially regarding the surface areas and volumes of curved bodies. He investigated the laws of the lever, the statics of fluids, and the static equilibrium of floating bodies. For these investigations, he relied on mathematical demonstrations, and he thereby initiated the mathematical analysis of physical phenomena, which is a cornerstone of theoretical physics.*

EXAMPLE 9 A chunk of ice floats in water (see Fig. 18.27). What percentage of the volume of ice will be above the level of the water? The density of ice is 917 kg/m^3.

FIGURE 18.27 Ice floating in water.

SOLUTION: If the mass of the chunk of ice is, say, $M = 1000$ kg, it must displace an amount of water of the same weight, that is, 1.000 m^3 of water. The volume of ice below the water level must then be 1.000 m^3. For this mass of ice, the total volume is

$$V = \frac{M}{\rho} = \frac{1000\text{ kg}}{917\text{ kg/m}^3} = 1.091\text{ m}^3$$

If the volume below water is 1.000 m^3, the volume above water is 1.091 m^3 − 1.000 m^3 = 0.091 m^3. The fraction of ice above the water level is therefore

$$\frac{\Delta V}{V} = \frac{0.091\text{ m}^3}{1.091\text{ m}^3} = 0.083$$

or 8.3%. Thus the expression "the tip of the iceberg" has come to mean a small part of the whole.

EXAMPLE 10 A hot-air balloon (Fig. 18.28) has a volume of 2.20×10^3 m^3. What is the buoyant force that the surrounding cold air exerts on the balloon? Assume that the density of the surrounding air is 1.29 kg/m^3.

SOLUTION: The mass M of the cold air displaced by the balloon volume V is

$$M = \rho V = 1.29\text{ kg/m}^3 \times 2.20 \times 10^3\text{ m}^3 = 2.84 \times 10^3\text{ kg}$$

The weight of this air is $W = Mg$. By Archimedes' Principle, this weight gives us the magnitude of the buoyant force,

$$F = Mg = 2.84 \times 10^3\text{ kg} \times 9.81\text{ m/s}^2 = 2.79 \times 10^4\text{ N}$$

If the balloon is to stay aloft, its weight (including the weight of the hot air inside it) must be less than or equal to the buoyant force of 2.79×10^4 N.

FIGURE 18.28 A hot-air balloon.

Concepts in Context

EXAMPLE 11 The deep-sea research diving vessel *Alvin* can operate at depths of 4500 m. With its ballast tanks empty, the vessel displaces a total volume of 16.8 m^3 and has a total mass of 17 000 kg. What minimum volume must the ballast tanks have to permit the vessel to dive?

SOLUTION: To dive, the vessel must be slightly heavier than the weight that provides neutral buoyancy when fully submerged. The vessel displaces a mass of seawater equal to

$$m = \rho V = 1025 \text{ kg/m}^3 \times 16.8 \text{ m}^3 = 17\,200 \text{ kg}$$

The additional mass required to achieve neutral buoyancy is thus 17 200 kg − 17 000 kg = 200 kg. Accordingly, the ballast tanks must permit the entry of at least 200 kg of water into the vessel. They must therefore have a volume

$$V = \frac{m}{\rho} = \frac{200 \text{ kg}}{1025 \text{ kg/m}^3} = 0.20 \text{ m}^3$$

Checkup 18.5

QUESTION 1: You completely immerse a sealed can of volume 1.0×10^{-3} m^3 in water. What is the buoyant force on the can?

QUESTION 2: If a ship is holed below the water line and it fills with water, it will sink. Explain in terms of Archimedes' Principle.

QUESTION 3: Some people claim that it is easier to stay afloat in seawater than in freshwater. Is this claim justified?

QUESTION 4: A rock has a density of 5000 kg/m^3. Will this rock float in water? Will it float in liquid mercury?

QUESTION 5: Suppose that in Example 9, the ice is floating in a glass of water that is filled to the brim, so that the ice above the water level sticks up above the rim of the glass. What happens when the ice melts?

(A) Some water overflows. (B) The water level drops below the rim.
(C) The water level remains at the rim.

Online Concept Tutorial 20

18.6 FLUID DYNAMICS; BERNOULLI'S EQUATION

When moving air encounters an obstacle that slows down its motion, the air exerts an extra pressure on the obstacle. You can feel the push of this extra pressure if you stand in a strong wind, or if you put a hand out of the window of a speeding car. The pressure changes that occur when air flows around obstacles or when water or some other fluid flows through pipes of varying cross sections can be calculated in a simple way by exploiting the conservation theorem for the mechanical energy. In this section, we will formulate the conservation theorem for energy for the special case of steady flow of an incompressible fluid without viscosity. This conservation theorem is called *Bernoulli's equation*.

We know from Section 18.2 that steady incompressible flow can be described by streamlines. As in the derivation of the equation of continuity, we consider a bundle of streamlines forming a thin stream tube. The fluid flows inside this tube as though the surface of the tube were an impermeable pipe. Figure 18.29a shows a segment of this "pipe"; this segment contains some mass of fluid. The left end of the mass of fluid is at height y_1 and has area A_1; the right end is at height y_2 and has area A_2. Figure 18.29b shows the same mass of fluid at a slightly later time—the fluid has moved toward the right. During this movement, the external pressures at the left and at the right ends of the segment exert forces and do some work on the mass of fluid. By energy conservation, this work must equal the change of kinetic and potential energy. To express this mathematically, we begin by calculating the work done by pressure. As the left end of the mass of fluid moves through a distance Δl_1, the work done by the pressure is the force $A_1 p_1$ multiplied by the distance Δl_1:

$$W_1 = A_1 p_1 \Delta l_1 \qquad (18.18)$$

Since the product $A_1 \Delta l_1$ is the volume ΔV vacated by the movement of the left end of the mass of fluid, we can also write this as

$$W_1 = p_1 \Delta V \qquad (18.19)$$

Likewise, the work done by the pressure at the right end is

$$W_2 = -p_2 \Delta V \qquad (18.20)$$

This is negative because the external force at the right end is opposite to the displacement. Note that the *same* volume ΔV appears in Eqs. (18.19) and (18.20)—the fluid is incompressible, and hence the volume vacated by the movement of the fluid at the left end must equal the volume newly occupied at the right end. The net work done by the pressure is then

$$\Delta W = W_1 + W_2 = p_1 \Delta V - p_2 \Delta V \qquad (18.21)$$

The change in kinetic and potential energy is entirely due to the changes at the ends of the mass of fluid; everywhere else, the shift of the fluid merely replaces fluid of some kinetic energy and potential energy with fluid of exactly the same kinetic and potential energy. The change at the ends involves replacing a mass Δm of fluid, of speed v_1 at height y_1, by an equal mass Δm, of speed v_2 at height y_2. The corresponding change of kinetic and potential energy is

$$\Delta K + \Delta U = \tfrac{1}{2}\Delta m\, v_2^2 - \tfrac{1}{2}\Delta m\, v_1^2 + \Delta m\, g y_2 - \Delta m\, g y_1 \qquad (18.22)$$

This change of mechanical energy must match the work done by the pressure:

$$\tfrac{1}{2}\Delta m\, v_2^2 - \tfrac{1}{2}\Delta m\, v_1^2 + \Delta m\, g y_2 - \Delta m\, g y_1 = p_1 \Delta V - p_2 \Delta V \qquad (18.23)$$

If we divide both sides of this equation by ΔV and we move all terms with subscript 2 to the left side and all terms with subscript 1 to the right side, we obtain

$$\frac{1}{2}\frac{\Delta m}{\Delta V} v_2^2 + \frac{\Delta m}{\Delta V} g y_2 + p_2 = \frac{1}{2}\frac{\Delta m}{\Delta V} v_1^2 + \frac{\Delta m}{\Delta V} g y_1 + p_1 \qquad (18.24)$$

or, since $\Delta m/\Delta V$ is the density ρ of the fluid,

$$\tfrac{1}{2}\rho v_2^2 + \rho g y_2 + p_2 = \tfrac{1}{2}\rho v_1^2 + \rho g y_1 + p_1 \qquad (18.25)$$

From this we see that the quantity $\frac{1}{2}\rho v^2 + \rho g y + p$ has the same value at different locations along the streamline, which means it is a constant of the motion:

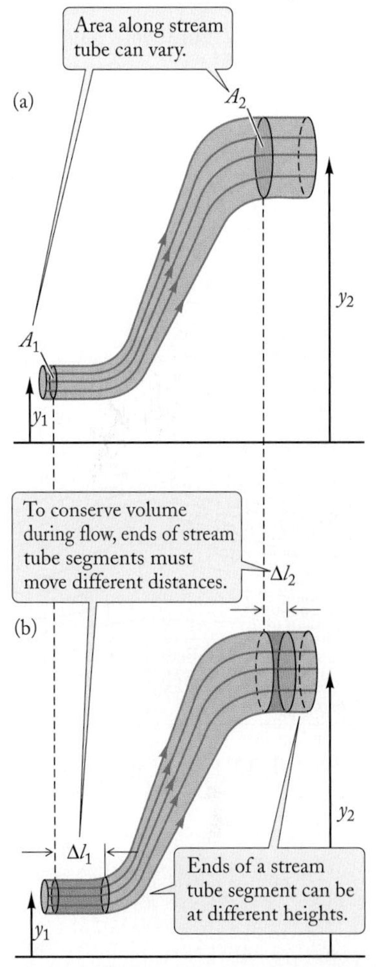

FIGURE 18.29 (a) A segment of a thin stream tube. The beginning and the end of the segment are marked by the dashed lines. (b) Motion of the fluid along the stream tube. The fluid enters the segment from the left and emerges on the right.

Bernoulli's equation

$$\tfrac{1}{2}\rho v^2 + \rho g y + p = [\text{constant}] \tag{18.26}$$

This is **Bernoulli's equation**. Note that $\frac{1}{2}\rho v^2$ is the density of kinetic energy (kinetic energy per unit volume) and $\rho g y$ is the density of potential energy (potential energy per unit volume). Hence Bernoulli's equation states that along any streamline, the sum of the density of kinetic energy, density of potential energy, and pressure is a constant. The kinetic and potential energy density terms correspond to the kinetic and potential energies in the Law of Conservation of Mechanical Energy for a particle [see Eq. (7.36)].

In the special case of a static fluid, with $v = 0$, Bernoulli's equation reduces to

$$p + \rho g y = [\text{constant}] \tag{18.27}$$

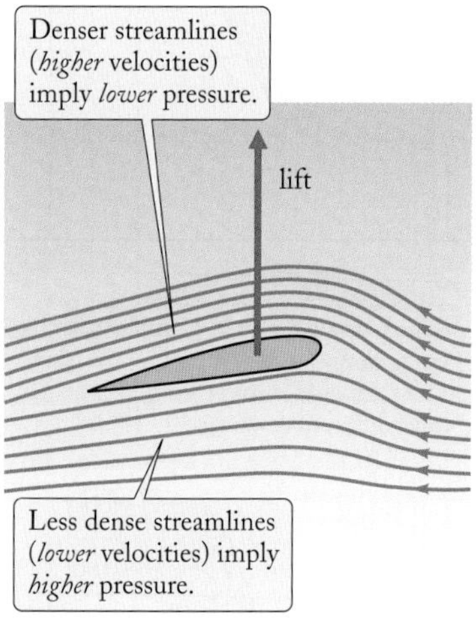

FIGURE 18.30 Flow of air around an airfoil.

This is equivalent to Eq. (18.12) for the pressure in a static fluid.

According to Bernoulli's equation (18.26), in any region in which the term $\rho g y$ is constant or approximately constant, *the pressure along any given streamline must decrease wherever the speed increases.* Intuitively, we might expect that where the speed is large, the pressure is large; but energy conservation demands exactly the opposite! The relation between pressure and speed plays an important role in the design of wings for airplanes. Figure 18.30 shows an airfoil and the streamlines of air flowing around it, as seen in the reference frame of the airplane. The shape of the airfoil has been designed so that along its upper part the speed is large (high density of streamlines) and along its lower part the speed is small (low density of streamlines). Consider now one streamline passing just over the airfoil and one just under it. At a large distance to the right, the fluid in all streamlines has the same pressure and the same speed. Bernoulli's equation applied to each of the two streamlines therefore tells us that in the region just above the airfoil, the pressure is low, and in the region just below the airfoil, the pressure is high. This leads to a net upward force, or **lift**, on the airfoil—this lift force supports the airplane in flight. As Fig. 18.30 shows, at a large distance to the left, the streamlines have a slight downward trend; what has happened is that the air has given some upward momentum to the airfoil and, in return, acquired an equal amount of downward momentum. Ultimately, the downward flow of air presses against the ground and transmits the weight of the airplane to the ground.

lift

To conclude this section, we will work out some examples in which Bernoulli's equation is applicable.

EXAMPLE 12 A water tank has a (small) hole near its bottom at a depth of 2.0 m from the top surface (see Fig. 18.31). What is the speed of the stream of water emerging from the hole?

SOLUTION: Qualitatively, one of the streamlines for the water flowing out of the tank will look as shown in Fig. 18.31. Since the hole is small, the water level at the top of the tank drops only very slowly; we can therefore take $v_1 = 0$ in Eq. (18.25). Furthermore, the pressures at the top and in the emerging stream of water are the same; both are equal to the atmospheric pressure p_0. Thus, $p_1 = p_2 = p_0$. With this, Eq. (18.25) becomes

$$\tfrac{1}{2}\rho v_2^2 + \rho g y_2 + p_0 = \rho g y_1 + p_0 \tag{18.28}$$

We can cancel the terms p_0, and we can move the term $\rho g y_2$ to the right side of the equation:

$$\tfrac{1}{2}\rho v_2^2 = \rho g y_1 - \rho g y_2 \tag{18.29}$$

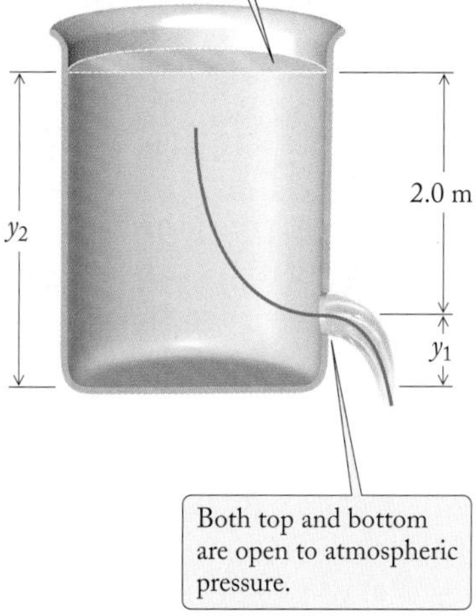

FIGURE 18.31 Streamline for a parcel of water flowing out of a tank.

If we divide both sides by $\frac{1}{2}\rho$ and extract the square root of both sides, we find

$$v_2 = \sqrt{2g(y_1 - y_2)} \tag{18.30}$$

With $y_1 - y_2 = 2.0$ m, the speed is

$$v_2 = \sqrt{2 \times 9.81 \text{ m/s}^2 \times 2.0 \text{ m}} = 6.3 \text{ m/s}$$

COMMENT: According to Eq. (18.30), the speed of the emerging water is exactly what it would be if the water were to fall freely through a height $y_1 - y_2$ [compare Eq. (2.29)]. This result is called **Torricelli's theorem**. It merely expresses conservation of energy: when a drop of water flows out at the bottom, the loss of potential energy of the water in the tank is equivalent to the removal of a drop of water from the top; the conversion of this potential energy into kinetic energy will give the drop the speed of free fall. We have already relied on this energy argument in Chapter 8, where we calculated the speed of water flowing out of a pipe in a hydroelectric storage plant.

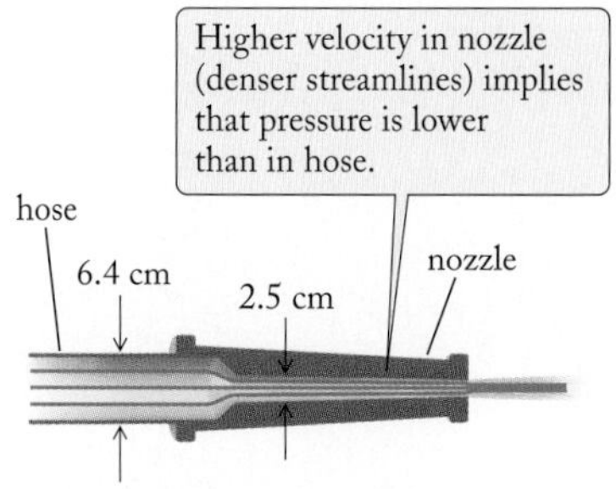

FIGURE 18.32 A fire-hose tip.

EXAMPLE 13 The overpressure in a fire hose of diameter 6.4 cm is 3.5×10^5 Pa, and the speed of flow is 4.0 m/s. The (horizontal) fire hose ends in a metal tip of diameter 2.5 cm (see Fig. 18.32). What are the overpressure and the velocity of water in the tip?

SOLUTION: The magnitudes of the velocities in the hose and in the tip are related to the cross-sectional areas by the continuity equation [Eq. (18.5)],

$$v_2 = v_1 \frac{A_1}{A_2} \tag{18.31}$$

The ratio of the cross-sectional areas is equal to the ratio of the squares of the diameters; hence the velocity in the tip is

$$v_2 = 4.0 \text{ m/s} \times \frac{(6.4 \text{ cm})^2}{(2.5 \text{ cm})^2} = 26 \text{ m/s}$$

The overpressure of the water in the tip can then be calculated from Eq. (18.25) with $y_1 = y_2 = 0$ for a horizontal hose:

$$\tfrac{1}{2}\rho v_2^2 + p_2 = \tfrac{1}{2}\rho v_1^2 + p_1$$

which yields

$$\begin{aligned} p_2 &= p_1 + \tfrac{1}{2}\rho v_1^2 - \tfrac{1}{2}\rho v_2^2 \\ &= 3.5 \times 10^5 \text{ Pa} + \tfrac{1}{2} \times 1000 \text{ kg/m}^3 \times (4.0 \text{ m/s})^2 \\ &\quad - \tfrac{1}{2} \times 1000 \text{ kg/m}^3 \times (26 \text{ m/s})^2 \\ &= 2 \times 10^4 \text{ Pa} \end{aligned} \tag{18.32}$$

DANIEL BERNOULLI (1700–1782) *Swiss physician, physicist, and mathematician. His great treatise* Hydrodynamica *included the equation named after him. Several other members of the Bernoulli family made memorable contributions to mathematics and physics.*

EXAMPLE 14 The **Venturi flowmeter** is a simple device that measures the velocity of a fluid flowing in a pipe. It consists of a constriction in the pipe with a cross-sectional area A_2 that is smaller than the cross-sectional area A_1 of the pipe itself (see Fig. 18.33). Small holes in the constriction and in

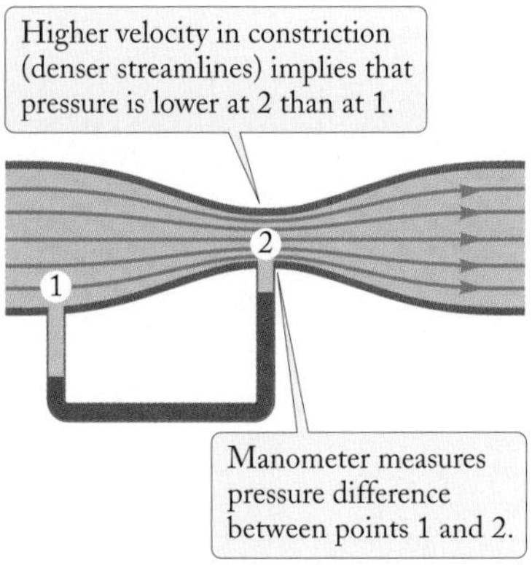

FIGURE 18.33 Venturi flowmeter.

the pipe permit the measurement of the pressures at these points by means of a manometer. Express the velocity of flow in terms of the pressure difference registered by the manometer.

SOLUTION: As in the preceding example, the magnitudes of the velocities at points 1 and 2 are related to the cross-sectional areas by the continuity equation,

$$v_2 = v_1\frac{A_1}{A_2} \tag{18.33}$$

With $y_1 = y_2$, Eq. (18.25) then again gives us

$$\tfrac{1}{2}\rho v_2^2 + p_2 = \tfrac{1}{2}\rho v_1^2 + p_1$$

From this we obtain a pressure difference, in which we then substitute for v_2 from Eq. (18.33):

$$\begin{aligned} p_1 - p_2 &= \tfrac{1}{2}\rho v_2^2 - \tfrac{1}{2}\rho v_1^2 \\ &= \frac{1}{2}\rho\left(v_1\frac{A_1}{A_2}\right)^2 - \frac{1}{2}\rho v_1^2 = \frac{1}{2}\rho v_1^2\left[\left(\frac{A_1}{A_2}\right)^2 - 1\right] \end{aligned} \tag{18.34}$$

Taking the square root of both sides of this equation and solving for v_1, we find

$$v_1 = \sqrt{\frac{2(p_1 - p_2)}{\rho[(A_1/A_2)^2 - 1]}} \tag{18.35}$$

This says that the flow velocity is proportional to the square root of the pressure difference.

PROBLEM-SOLVING TECHNIQUES BERNOULLI'S EQUATION

The application of Bernoulli's equation (18.26) to problems of fluid motion involves the same three steps we used in the application of energy conservation to problems of particle motion:

1 First evaluate the sum of the terms in Bernoulli's equation at one point of a streamline.

2 Then evaluate the sum of the terms in Bernoulli's equation at another point of the same streamline.

3 And then equate these two quantities.

When substituting the pressure into Bernoulli's equation, you can use either the absolute pressure (including atmospheric pressure) or the overpressure (excluding atmospheric pressure), but you must use the same kind of pressure on both sides of the equation. These two kinds of pressure differ by only a constant, and adding or subtracting the same constant on both sides of the equation does not affect its validity. Also remember that the pressure in Bernoulli's equation must be expressed in SI units (N/m^2, or Pa), not atmospheres or mm-Hg.

If two points in a fluid system are in contact with the atmosphere, then they are at essentially equal pressure (as in Example 12. However, if both points are not in contact with the atmosphere, then the pressure difference depends on the flow velocity (as in Examples 13 and 14) and the depth (as in Examples 6 and 7).

If the fluid moves through a pipe or channel of varying cross section, then the other general equation governing the flow is the continuity equation $v_1A_1 = v_2A_2$, which relates the cross-sectional area to the flow velocity. Most problems involving steady incompressible flow can be solved by combining Bernoulli's equation and the continuity equation.

Checkup 18.6

QUESTION 1: A fire hose of constant diameter carries water from a fire truck to the top of a burning house. According to Bernoulli's equation, does the pressure difference between the ends of this hose depend on the speed of flow?

QUESTION 2: Consider water flowing downhill in a shallow, open channel. Which of the three terms in Bernoulli's equation are constant?

QUESTION 3: A lawn sprinkler is connected to a faucet by a horizontal hose of constant diameter. The water emerges from the nozzle of the sprinkler in an arc. Which of the terms in Bernoulli's equation are constant during the motion of the water in the hose? During the motion after the water emerges from the nozzle?

QUESTION 4: An air hose is connected to the narrow end of a funnel. Air from the hose flows into the narrow end and out of the wide end of the funnel (Fig. 18.34a), and this can levitate a body (Fig. 18.34b). Where is the air pressure the lowest?

(A) At the narrow end of the funnel (B) At the wide end of the funnel
(C) Far from the funnel

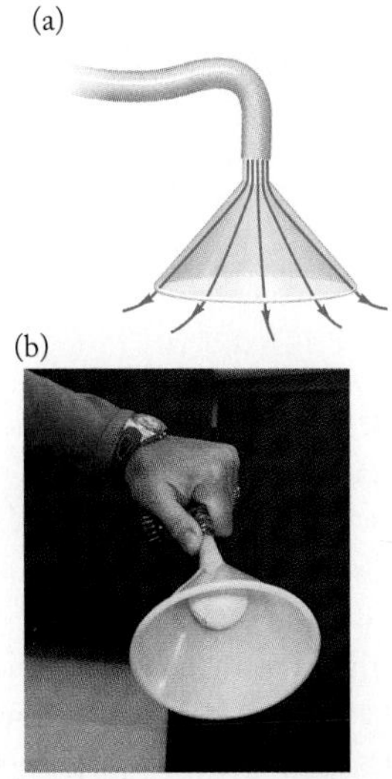

FIGURE 18.34 (a) Airflow through a funnel. (b) Levitation of a plastic ball.

SUMMARY

PHYSICS IN PRACTICE The sphygmomanometer **(page 579)**

PROBLEM-SOLVING TECHNIQUES Bernoulli's equation **(page 586)**

DENSITY, OR MASS PER UNIT VOLUME

$$\rho = \frac{m}{V} \qquad (18.1)$$

CONTINUITY EQUATION

$$v_1 A_1 = v_2 A_2 \qquad (18.5)$$

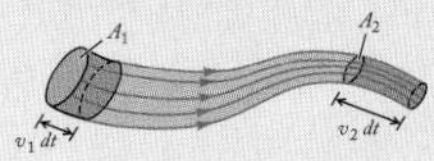

PRESSURE IS FORCE PER UNIT AREA

$$p = \frac{F}{A} \qquad (18.6)$$

UNIT OF PRESSURE

$$1 \text{ pascal} = 1 \text{ Pa} = 1 \text{ N/m}^2 \qquad (18.7)$$

ATMOSPHERIC PRESSURE

$$1 \text{ atm} = 760 \text{ mm-Hg} = 1.01 \times 10^5 \text{ Pa} \qquad (18.8)$$

HYDROSTATIC PRESSURE (FOR INCOMPRESSIBLE FLUID)
For water, this is very close to 1 atm for every 10 m of depth.

$$p - p_0 = -\rho g y \qquad (18.12)$$

ARCHIMEDES' PRINCIPLE The upward buoyant force has the same magnitude as the weight of the displaced fluid.

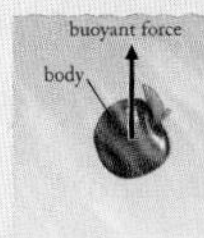

BERNOULLI'S EQUATION

$$\tfrac{1}{2}\rho v^2 + \rho g y + p = [\text{constant}] \qquad (18.26)$$

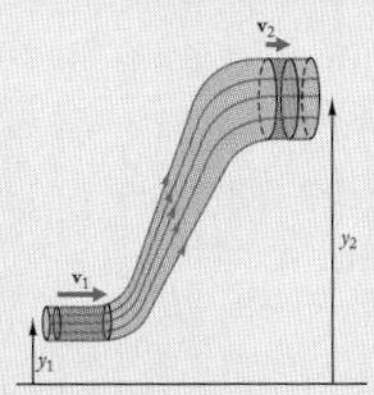

QUESTIONS FOR DISCUSSION

1. The sheet of water of a waterfall is thick at the top and thin at the bottom (see Fig. 18.35). Explain.

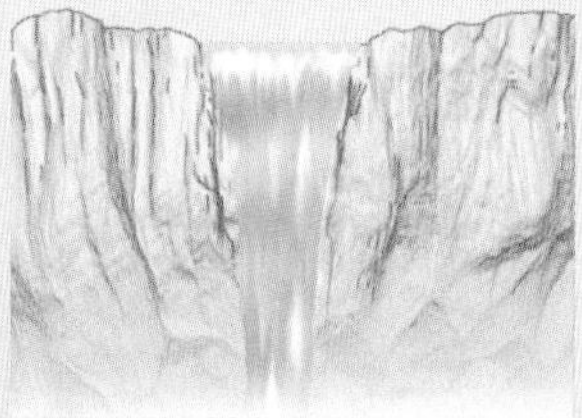

FIGURE 18.35 Sheet of water on a smooth waterfall, narrowing at the bottom.

2. During construction, three lanes of a four-lane highway are closed. Suppose that bumper-to-bumper traffic on the four lanes funnels into the single open lane. If the cars on the four approaching lanes proceed at a crawl, say, 15 km/h, what will be their speed when they proceed along the single lane? (Hint: Think of the continuity equation.)
3. If you place a block of wood on the bottom of a swimming pool, why does the pressure of the water not keep it there?
4. Explain what holds a suction cup on a smooth surface.
5. Newton gave the following description of his classic experiment with a rotating bucket:

   If a vessel, hung by a long cord, is so often turned about that the cord is strongly twisted, then filled with water, and held at rest together with the water; thereupon, by the sudden action of another force, it is whirled about the contrary way, and while the cord is untwisting itself, the vessel continues for some time in this motion; the surface of the water will at first be plain, as before the vessel began to move; but after that, the vessel, by gradually communicating its motion to the water, will make it begin sensibly to revolve, and recede by little and little from the middle, and ascend to the sides of the vessel, forming itself into a concave figure (as I have experienced), and the swifter the motion becomes, the higher will the water rise, till at last, performing its revolutions in the same times with the vessel, it becomes relatively at rest in it.

   Explain why hydrostatic equilibrium requires that the rotating water be higher at the rim of the bucket than at the center.
6. When a physician measures the blood pressure of a patient, he places the cuff on the arm, at the same vertical level as the heart. What would happen if he were to place the cuff around the leg?
7. Scuba divers have survived short intervals of free swimming at depths of 430 m. Why does the pressure of the water at this depth not crush them?
8. Figure 18.36 shows a glass vessel with vertical tubes of different shapes. Explain why the water level in all the tubes is the same.

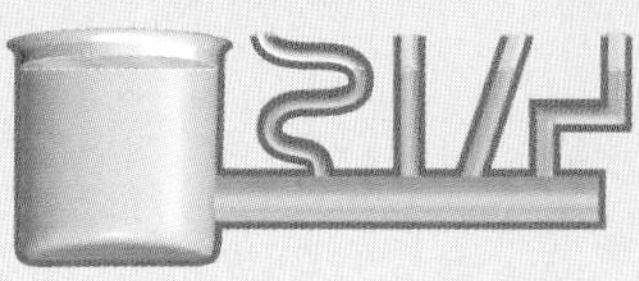

FIGURE 18.36 Glass vessel with vertical tubes.

9. In a celebrated experiment, Blaise Pascal attached a long metal funnel to a tight cask (Fig. 18.37). When he filled the cask by pouring water into this funnel, the cask burst. Explain.

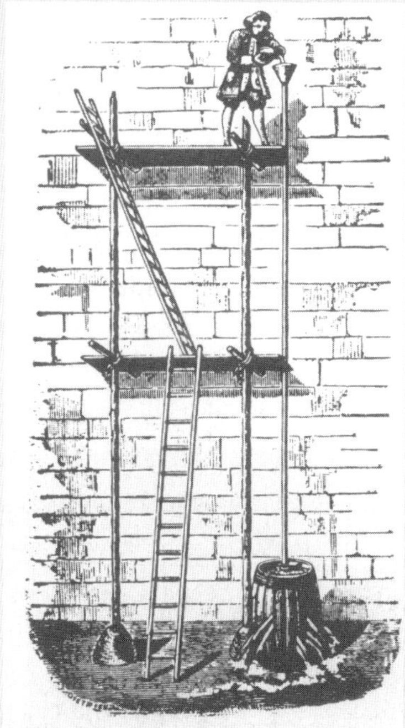

FIGURE 18.37 Blaise Pascal's experiment.

10. Face masks for scuba divers have two indentations at the bottom into which the diver can stick thumb and forefinger to pinch his nose shut. What is the purpose of this arrangement?
11. Figure 18.38 shows a grain silo held together by circumferential steel bands. Why has the farmer placed more bands near the bottom than near the top?

FIGURE 18.38 Grain silo.

12. Flooding on the low coasts of England and the Netherlands is most severe when the following three conditions are in coincidence: onshore wind, full or new moon, and low barometric pressure. Can you explain this?
13. Why do some men or women float better than others?

14. Will an ice cube float in a tub full of gasoline?

15. An ice cube floats in a glass full of water. Will the water level rise or fall when the ice cube melts?

16. The density of a solid body can be determined by first weighing the body in air and then weighing it again when it is immersed in water (Fig. 18.39). How can you deduce the density from these two measurements?

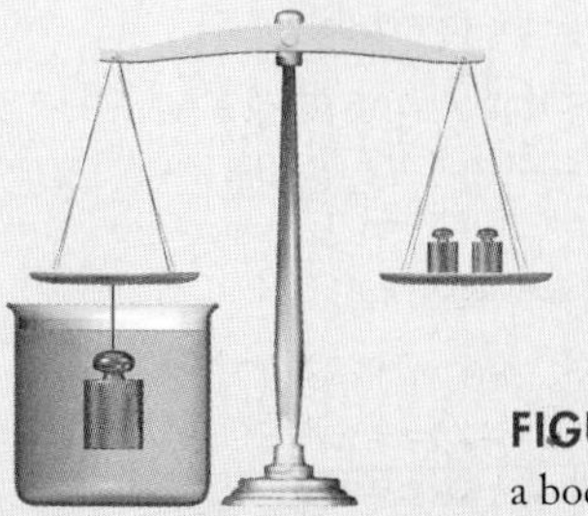

FIGURE 18.39 Weighing a body underwater.

17. A buoy floats on the water. Will the flotation level of the buoy change when the atmospheric pressure changes?

18. Will the water level in a canal lock rise or fall if a ship made of steel sinks in the lock? What if the ship is made of wood?

19. While training for the conditions of weightlessness they would encounter in an orbiting spacecraft, NASA astronauts were made to float submerged in a large water tank (Fig. 18.40). Small weights attached to their space suits gave the astronauts neutral buoyancy. To what extent does such simulated weightlessness imitate true weightlessness?

FIGURE 18.40 Astronauts floating underwater.

20. A girl standing in a subway car holds a helium balloon on a string. Which way will the balloon move when the car accelerates?

21. A slurry of wood chips is used in some tanning operations. What would happen to a man were he to fall into a vat filled with such a slurry?

22. The bathyscaphe *Trieste,* which set a record of 10917 m in a deep dive in the Marianas Trench in 1960, consists of a large tank filled with gasoline below which hangs a steel sphere for carrying the crew (Fig. 18.41). Can you guess the purpose of the tank of gasoline?

FIGURE 18.41 The bathyscaphe *Trieste*.

23. Gasoline vapor from, say, a small leak in the fuel system poses a very serious hazard in a motorboat, but only a minor hazard in an automobile. Explain. (Hint: Gasoline vapor is denser than air.)

24. Accidental release of carbon dioxide or of argon creates a much greater danger of suffocation than release of helium gas. Why?

25. How does a balloonist control the ascent and descent of a hot-air balloon?

26. When you release a bubble of air while underwater, the bubble grows in size as it ascends. Explain.

27. A "Cartesian diver" consists of a small inverted bottle floating inside a larger bottle whose mouth is covered by a rubber membrane (Fig. 18.42). By depressing the membrane, you can increase the water pressure in the large bottle. How does this affect the buoyancy of the small bottle?

FIGURE 18.42 A Cartesian diver.

28. To throw a curveball, a baseball pitcher gives the ball a spinning motion about a vertical axis. The air on the left and right sides of the ball will then be dragged along by the rotation and acquire slightly different speeds. Using Bernoulli's equation, explain how this creates a lateral deflecting force on the ball.

29. If you place a Ping-Pong ball in the jet of air from a hose aimed vertically upward, the Ping-Pong ball will be held in stable equilibrium within this jet. Explain this by means of Bernoulli's equation. (Hint: The speed of air is maximum at the center of the jet.)

PROBLEMS

18.1 Density and Flow Velocity
18.2 Incompressible Steady Flow; Streamlines†

1. The following table, taken from a firefighter's manual, lists the rate of flow of water (in liters per minute) through a fire hose of diameter 3.81 cm (1.5 in.) connected to a nozzle of given diameter; the listed rate of flow will maintain a pressure of 3.4 atm (50 lbf/in.^{2}) in the nozzle:

RATE OF FLOW FOR A 3.81-cm HOSE

RATE OF FLOW	NOZZLE DIAMETER	NOZZLE PRESSURE
95 liters/min	0.95 cm	3.4 atm
190	1.27	3.4
284	1.59	3.4

For each case calculate the speed of flow (in meters per second) in the hose and in the nozzle.

2. The average rate of flow of blood in the human aorta is 92 cm^3/s. The radius of the aorta is 1.2 cm. From this calculate the average speed of the blood in the aorta.
3. What pump power is required to shoot 26 000 liters/min of water to a height of 170 m, as in the tallest fountain?
4. Using a garden hose, you take 4.0 minutes to fill a water barrel. The inner diameter of the garden hose is 2.5 cm, and the capacity of the barrel is 250 liters. What is the flow velocity of the water in the hose?
5. Water flows in a 10-cm-diameter pipe with flow velocity 6.0 m/s. If the water encounters a constricted region of the pipe with diameter 7.0 cm, what is the flow velocity in that region?
6. During a heavy storm, a steady stream of water flows down from a 5.0-cm-diameter hole in a gutter. The downward speed of the water exiting the hole is 0.50 m/s. What is the speed of the flow just before it hits the ground, 4.0 m below the hole? What is the diameter of the flow there?
7. A patient receives 0.50 liter of intravenous fluid in 30 minutes. The tubing carrying the fluid has an inner diameter of 2.0 mm; the needle at the end of the tube has an inner diameter of 0.20 mm. What is the speed of flow of the fluid in the tubing? In the needle?
8. Water flows though a pipe of a certain radius at a rate of 10 kg/s; propane flows through a pipe with a radius 10 times larger than that of the water pipe, also at a rate of 10 kg/s. What is the ratio of the flow velocities of these fluids, $v_{\text{water}}/v_{\text{propane}}$?

†For help, see Online Concept Tutorial 20 at www.wwnorton.com/physics

*9. A fountain shoots a stream of water vertically upward. Assume that the stream is inclined very slightly to one side so that the descending water does not interfere with the ascending water. The upward velocity at the base of the column of water is 15 m/s.

  (a) How high will the water rise?

  (b) The diameter of the column of water is 7.0 cm at the base. What is the diameter at the height of 5.0 m? At the height of 10 m?

18.3 Pressure

10. A metallic can is filled with a carbonated soft drink. The can is cylindrical, of radius 3.2 cm. The (over)pressure of the carbon dioxide in the liquid is 0.8 atm. What is the force acting on one of the circular bases of the can?
11. You can use a barometer as an altimeter. Suppose that when you carry the barometer up a hill, its reading decreases by 8.0 mm-Hg. What is the height of the hill? Assume the density of air is 1.29 kg/m^3.
12. Within the funnel of a tornado, the air pressure is much lower than normal—about 0.20 atm as compared with the normal value of 1.00 atm. Suppose that such a tornado suddenly envelops a house; the air pressure inside the house is 1.00 atm and the pressure outside suddenly drops to 0.20 atm. This will cause the house to burst explosively. What is the net outward pressure force on a 12 m × 3.0 m wall of this house? Is the house likely to suffer less damage if all the windows and doors are open?
13. What is the downward force that air pressure (1 atm) exerts on the upper surface of a sheet of paper ($8\frac{1}{2}$ in. × 11 in.) lying on a table? Why does this force not squash the paper against the table?
14. At a distance of 7.0 km from a 1-megaton nuclear explosion, the blast wave has an overpressure of 3.0×10^4 N/m^2. Calculate the force that this blast wave exerts on the front of a standing man; the frontal area of the man is 0.70 m^2. (The actual force on a man exposed to the blast wave is larger than the result of this simple calculation because the blast wave will be reflected by the man, and this leads to a substantial increase in pressure.)
15. The overpressure in the tires of a 1300-kg automobile is 2.4 atm. If each tire supports one-fourth the weight of the automobile, what must be the area of each tire in contact with the ground? Pretend the tires are completely flexible.
16. The shape of the wing of an airplane is carefully designed so that, when the wing moves through air, a pressure difference develops between the bottom surface of the wing and the top surface; this supports the weight of the airplane. A fully loaded DC-3 airplane has a mass of 10 900 kg. The (bottom) surface area of its wings is 92 m^2. What is the average pressure

difference between the top and the bottom surfaces when the airplane is in flight?

17. Pressure gauges used on automobile tires read the overpressure, that is, the amount of pressure in excess of atmospheric pressure. If a tire has an overpressure of 2.4×10^5 N/m^2 on a day when the barometric pressure is 0.95 atm, what will be the overpressure when the barometric pressure increases to 1.01 atm? Assume that the volume and the temperature of the tire remain constant.

18. Commercial jetliners have pressurized cabins enabling them to carry passengers at a cruising altitude of 10000 m. The air pressure at this altitude is 0.28 atm. If the air pressure inside the jetliner is 1.00 atm, what is the net outward force on a 1.0 m $\times$ 2.0 m door in the wall of the cabin?

19. In 1654 Otto von Guericke, the inventor of the air pump, gave a public demonstration of air pressure (see Fig. 5.44). He took two hollow hemispheres of copper whose rims fitted tightly together and evacuated them with his air pump. Two teams of 15 horses each, pulling in opposite directions, were unable to separate these hemispheres. If the evacuated sphere had a radius of 40 cm and the pressure inside was nearly zero, what force would each team of horses have had to exert to pull the hemispheres apart?

20. A long bar of iron has a diameter of 10 cm and a length of 2.0 m. If the mass of the bar is 123 kg and the bar stands on one end, what pressure does it exert on the surface supporting it?

21. A 50-kg boy hangs from the ceiling by a (completely evacuated) suction cup. What is the minimum area the suction cup must have?

22. A large, cylindrical, aboveground cistern of negligible mass holds 50000 liters of water and has a radius of 2.0 m. When full, what overpressure does it exert on the ground?

23. A 60-kg woman stands with her weight evenly distributed over the total area of her shoe bottoms, 300 cm^2. What pressure do the shoes exert on the floor? Then she balances on the bottom of one circular heel of diameter 1.0 cm. What pressure does the heel exert on the floor?

24. The baggage compartment of the DC-10 airliner is under the floor of the passenger compartment. Both compartments are pressurized at a normal pressure of 1.0 atm. In a disastrous accident near Orly, France, in 1974, a faulty lock permitted the baggage compartment door to pop open in flight, depressurizing this compartment. The normal pressure in the passenger compartment then caused the floor to collapse, jamming the control cables. At the time the airliner was flying at an altitude of 3800 m, where the air pressure is 0.64 atm. What is the net pressure force on a 1.0 m $\times$ 1.0 m square of the floor?

25. A pencil sharpener is held to the surface of a desk by means of a rubber suction cup measuring 6.0 cm $\times$ 6.0 cm. The air pressure under the suction cup is zero and the air pressure above the suction cup is 1.0 atm.

    (a) What is the magnitude of the pressure force pushing the cup against the table?

    (b) If the coefficient of static friction between the rubber and the table is 0.90, what is the maximum transverse force the suction cup can withstand?

18.4 Pressure in a Static Fluid

26. Porpoises dive to a depth of 520 m. What is the water pressure at this depth?

27. A tanker is full of oil of density 880 kg/m^3. The flat bottom of the hull is at a depth of 26 m below the surface of the surrounding water. Inside the hull, oil is stored with a depth of 30 m (Fig. 18.43). What is the pressure of the water on the bottom of the hull? The pressure of the oil? What is the vertical pressure force on 1.0 m^2 of bottom?

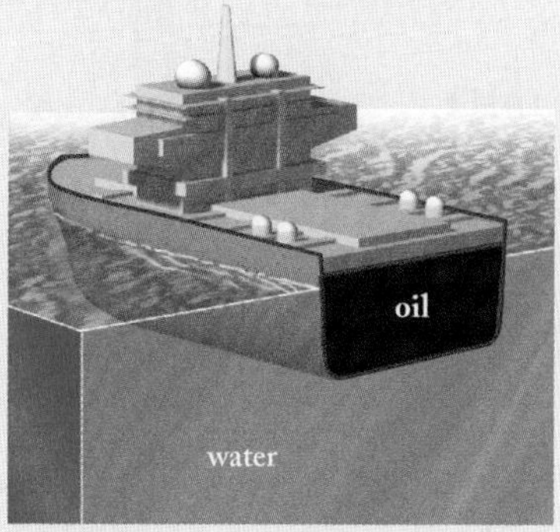

FIGURE 18.43 Cross section of an oil tanker.

28. (a) Calculate the mass of air in a column of base 1.00 m^2 extending from sea level to the top of the atmosphere. Assume that the pressure at sea level is 760 mm-Hg and that the value of the acceleration of gravity is 9.81 m/s^2, independent of height.

    (b) Multiply your result by the surface area of the Earth to find the total mass of the entire atmosphere.

29. Suppose that a zone of low atmospheric pressure (a "low") is at some place on the surface of the sea. The pressure at the center of the "low" is 64 mm-Hg less than the pressure at a large distance from the center. By how much will this cause the water level to rise at the center?

30. (a) Under normal conditions the human heart exerts a pressure of 120 mm-Hg on the arterial blood. What is the arterial blood pressure in the feet of a man standing upright? What is the blood pressure in the brain? The feet are 140 cm below the level of the heart; the brain is 40 cm above the level of the heart; the density of human blood is 1055 kg/m^3. Neglect the speed of flow of the blood; i.e., pretend it is at rest.

    (b) Under conditions of stress the human heart can exert a pressure of up to 190 mm-Hg. Suppose that an astronaut lands on the surface of a large planet where the acceleration of gravity is 61 m/s^2. Could the astronaut's heart maintain a positive blood pressure in his brain while he is standing upright? Could the astronaut survive?

31. A "suction" pump consists of a piston in a cylinder with a long pipe leading down into a well (Fig. 18.44). What is the maximum height to which such a pump can "suck" water?

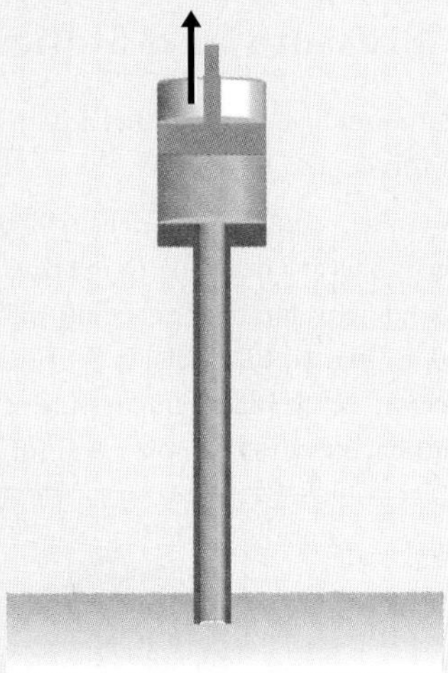

FIGURE 18.44 Suction pump.

32. On a day of high pressure, the barometer may read 790 mm-Hg; on a low-pressure day, it may drop to 730 mm-Hg. What are these pressures in units of atmospheres?

33. If a 2.0-m-high tank is filled with sand (with average density $\rho = 1.7\ \text{g/cm}^3$), what is the overpressure at the bottom of the tank?

34. A hydraulic lift for cars has a small piston with radius 1.0 cm and a large piston with radius 12 cm. What pressure (in atm) must be exerted on the small piston for the large piston to lift a car of mass 1500 kg?

35. A 30-m-high dam holds back a full reservoir of water. What is the overpressure at the base of the dam?

36. A tube is bent so that both ends are upward; the tube contains some water. Gasoline is poured into one end of the tube as shown in Fig. 18.45; the region with gasoline is 20 cm high. How much higher is the gasoline–air surface compared with the water–air surface?

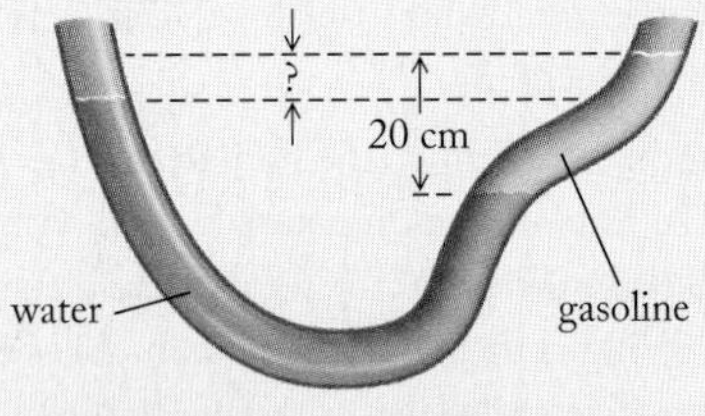

FIGURE 18.45 A tube with two liquids.

37. As any diver who has ever "belly-flopped" knows, water is a very incompressible fluid. From Example 7, the pressure at 4500 m below the surface of the sea has the value 4.54×10^7 Pa. What is the percent volume change of water at this immense pressure? [Hint: The fractional volume change was given in Eq. (14.20); the value of the bulk modulus of water appears in Table 14.1.]

38. A swimming pool is 4.0 m in depth; a swimmer at this depth feels discomfort in the ear. Calculate the net force on a 0.50-cm-diameter eardrum, assuming the inner surface of the eardrum remains at 1 atm pressure.

*39. A 30-m-high dam holds back a full reservoir of water. The width of the dam is 70 m. What is the total force that the water exerts on the dam? (Hint: Sum the forces $dF = p\,dA$ on thin horizontal strips of area dA of the dam.)

*40. A man whirls a bucket full of water around a vertical circle at the rate of 0.70 rev/s. The surface of the water is at a radial distance of 1.0 m from the center.

(a) What is the pressure difference between the surface of the water and a point 1.0 cm below the surface when the bucket is at the lowest point of the circle? You may assume that a point on the surface and a point 1.0 cm below have practically the same centripetal acceleration.

(b) What if the bucket is at the highest point of the circle?

*41. A test tube filled with water is being spun around in an ultracentrifuge with angular velocity ω. The test tube is lying along a radius, and the free surface of the water is at a radius r_0 (Fig. 18.46).

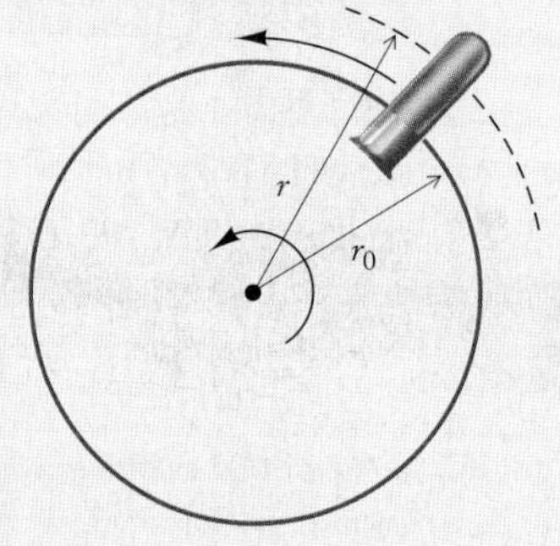

FIGURE 18.46 Test tube in an ultracentrifuge.

(a) Show that the pressure at radius r within the test tube is

$$p = \tfrac{1}{2}\rho\omega^2(r^2 - r_0^2)$$

where ρ is the density of the water. Ignore gravity and ignore atmospheric pressure.

(b) Suppose that $\omega = 3.8 \times 10^4$ radians/s and $r_0 = 10$ cm. What is the pressure at $r = 13$ cm?

*42. In a test centrifuge, a NASA scientist tolerated (suffered?) a sustained centripetal acceleration of 25 standard g's. Estimate the pressure difference between the front and back of his brain during the ordeal. Measure the relevant distance on your own head, and assume this distance was radial during the test.

*43. (a) Figure 18.47a shows a round conical flask filled with water of a depth h. The radius of the upper water surface is R_1, and that of the lower surface is R_2. What is the net force that the water exerts on the sides of the flask? On the bottom of the flask? What is the sum of these forces? Ignore atmospheric pressure.

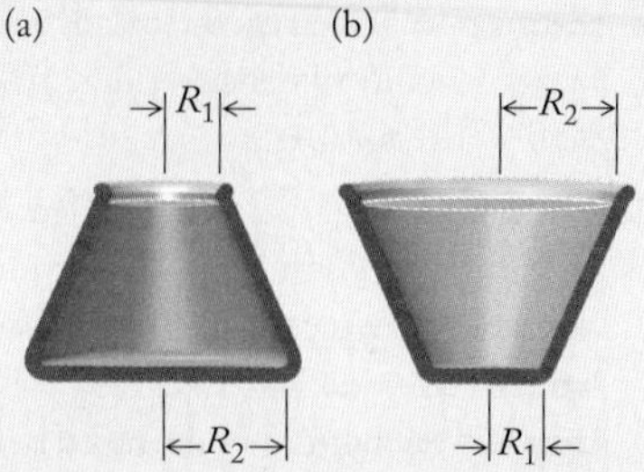

FIGURE 18.47 Conical flasks.

(b) Figure 18.47b shows another round conical flask, with the same radii. Answer the same questions for this flask.

18.5 Archimedes' Principle

44. What is the buoyant force on a human body of volume 7.4×10^{-2} m^3 when totally immersed in air? In water?

45. Icebergs commonly found floating in the North Atlantic (Fig. 18.48) are 30 m high (above the water) and 400 m $\times$ 400 m across. The density of ice is 920 kg/m^3, and the density of seawater is 1025 kg/m^3.
   (a) What is the total volume of such an iceberg (including the volume below the water)?
   (b) What is the total mass?

FIGURE 18.48 A large iceberg.

46. You can walk on water if you wear very large shoes shaped like boats. Calculate the length of the shoes that will support you; assume that each shoe is 30 cm $\times$ 30 cm in cross section.

47. A gasoline barrel, made of steel, has a mass of 20 kg when empty. The barrel is filled with 0.12 m^3 of gasoline with a density of 739 kg/m^3. Will the full barrel float in water? Neglect the volume of the steel.

48. A typical medium-size balloon used for scientific research (Fig. 18.49) is designed to attain an altitude of 40 km, at which altitude the helium in the balloon will have expanded to 570 000 m^3. What is the buoyant force on the balloon under these conditions? The density of air at this altitude is 4.3×10^{-3} kg/m^3.

49. A spherical balloon of mass 4.0 g is filled with helium gas of density 0.18 kg/m^3. If the balloon can barely lift a mass of 150 g off the ground, what is its radius?

50. When submerged in water, what is the buoyant force on a Ping-Pong ball (radius 2.0 cm)? A basketball (radius 11.4 cm)? A beach ball (radius 50 cm)?

FIGURE 18.49 A research balloon being inflated with helium.

51. A balloon is filled with helium gas of density 0.18 kg/m^3; the air around it is at standard density. If released, what is the initial acceleration of the balloon? (For this simple estimate, neglect the mass of the balloon and the inertia of the surrounding air.)

52. An egg sinks to the bottom of 1.000 liter of freshwater. Table salt is slowly dissolved in the water until 83 g has been added; the egg then becomes neutrally buoyant. If the salt has negligible effect on the volume of the water, what is the average density of the egg?

53. You bet your friend that he cannot completely submerge your beach ball, which has a diameter of 60 cm and negligible mass. He must balance on the beach ball, not jump on it. What is the smallest mass that your friend can have and be successful?

*54. A supertanker has a mass of 220 000 metric tons when empty and can carry up to 440 000 metric tons of oil when fully loaded. Assume that the shape of its hull is approximately that of a rectangular parallelepiped 380 m long, 60 m wide, and 40 m high (Fig. 18.50).

FIGURE 18.50 Empty and fully loaded supertanker.

(a) What is the draft of the empty tanker, that is, how deep is the hull submerged in the water? Assume the density of seawater is 1025 kg/m^3.

(b) What is the draft of the fully loaded tanker?

*55. A supertanker has a draft (submerged depth) of 30 m when in seawater (density ρ = 1025 kg/m^3). What will be the draft of this tanker when it enters a river estuary with freshwater (density ρ = 1000 kg/m^3)? Assume that the sides of the ship are vertical.

*56. The Raven S-66A hot-air balloon has a volume of 4000 m^3 and a height of 27 m. Fully loaded, its mass is 1400 kg. If the density of the air outside the balloon is 1.29 kg/m^3, what must be the density of the air inside the balloon to achieve liftoff? What is the inside overpressure at the top of the balloon? (Hint: The bottom of the balloon is open, and therefore at the same pressure as the exterior air. Treat the air outside and the air inside the balloon as fluids of uniform densities.)

*57. A round log of wood of density 600 kg/m^3 floats in water. The diameter of the log is 30 cm. How high does the upper surface of the log protrude from the water?

*58. A child's rubber balloon of mass 2.5 g is filled with helium gas of density 0.33 kg/m^3. The balloon is spherical, with a radius of 12 cm. A long cotton string with a mass of 2.0 g per meter hangs from the bottom of the balloon. Initially, the string lies loosely on the floor, but when the balloon ascends, it pulls the string upward and straightens it out. At what height will the balloon stop ascending, having reached equilibrium with the hanging portion of the string? Assume that the surrounding air has the standard density 1.29 kg/m^3.

*59. When a force accelerates a body immersed in a fluid, some of the fluid must also be accelerated, since it must be pushed out of the way of the body and flow around it. Thus, the force must overcome not only the inertia of the body, but also the inertia of the fluid pushed out of the way. It can be shown that for a spherical body completely immersed in a nonviscous fluid, the extra inertia is that of a mass of fluid half as large as the fluid displaced by the body.

(a) From this, deduce that the downward acceleration of a spherical body of density ρ falling through a fluid of density ρ' is

$$a = \frac{\rho - \rho'}{\rho + \frac{1}{2}\rho'}\,g$$

(b) Find the upward acceleration of an empty bubble in the fluid.

(c) What value would you have found for the acceleration of an empty bubble if you had not taken into account the extra inertia of the displaced fluid?

*60. When a body is weighed in air on an analytical balance, a correction must be made for the buoyancy contributed by the air. Suppose that the weights used in the balance are made of brass, of density 8.7 × 10^3 kg/m^3, and that the density of air is 1.3 kg/m^3. By what percentage must you increase (or decrease) the mass indicated by the balance in order to obtain the true mass of a body of density 1.0 × 10^3 kg/m^3? 5.0 × 10^3 kg/m^3? 10.0 × 10^3 kg/m^3?

*61. The bottom half of a tank is filled with water (ρ = 1.0 × 10^3 kg/m^3), and the top half is filled with oil (ρ = 8.5 × 10^2 kg/m^3). Suppose that a rectangular block of wood of mass 5.5 kg, 30 cm long, 20 cm wide, and 10 cm high is placed in this tank. How deep will the bottom of the block be submerged in the water?

**62. A hydrometer, the device used to determine the density of battery acid, wine, or other fluids, consists of a bulb with a long vertical stem (Fig. 18.51). The device floats in the liquid with the bulb submerged and the stem protruding above the surface. The density of the liquid is directly related to the length h of the protruding portion of the stem. Suppose that the bulb is a sphere of radius R and that the stem is a cylinder of radius R' and length l; the mass of both together is M. Derive a formula for the density of the liquid in terms of h, R, R', l, and M.

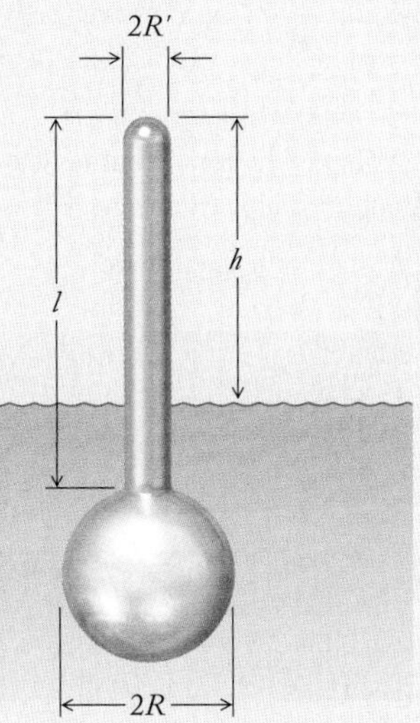

FIGURE 18.51 Hydrometer.

**63. A ship that displaces a weight of water equal to its own weight is in equilibrium with regard to vertical motion. If the ship were placed lower in the water, the buoyant force would exceed gravity and the ship would surge upward. If the ship were placed higher in the water, the buoyant force would be less than gravity and the ship would sink downward. Suppose that the sides of a ship are vertical above and below its normal waterline.

(a) Show that the frequency of small up-and-down oscillations of the ship is roughly $\omega = \sqrt{A\rho g/M}$, where A is the horizontal area bounded by the waterline, M is the mass of the ship, and ρ is the density of water. (This formula is only a rough approximation because, as the ship moves up and down, the water also has to move; hence the effective inertia is greater than M.)

(b) What is the frequency of such up-and-down oscillations for the fully loaded supertanker described in Problem 54?

**64. You drop a pencil, point down, into the water. If you release the pencil from a height of 4.0 cm (measured from the point to the water level), how far will it dive? Treat the pencil as a cylinder of radius 0.40 cm, length 19 cm, and mass 4.0 g. Ignore the frictional and inertial resistance of the water, but take into account the buoyant force.

18.6 Fluid Dynamics; Bernoulli's Equation[†]

65. A thin stream of water emerges vertically from a small hole on the side of a water pipe and ascends to a height of 1.2 m. What is the pressure inside the pipe? Assume the water inside the pipe is nearly static.

66. If you blow a thin stream of air with a speed of 7.0 m/s out of your mouth, what must be the overpressure in your mouth? Assume that the speed of the air in your mouth is (nearly) zero.

67. A pump has a horizontal intake pipe at a depth h below the surface of a lake. What is the maximum speed for steady flow of water into this pipe? (Hint: Inside the pipe the maximum speed of flow corresponds to zero pressure.)

68. The wind blows over the horizontal roof of a closed house at speed v. Find an expression for the difference in pressure inside and outside the house. For wind with speed $v = 20$ m/s, find the force on a square meter of the roof.

69. A small airplane has a total wing area of 8.0 m^2. If the air flowing above the wing has a speed of 130 m/s and the air below the wing a speed of 115 m/s, what is the net force on the wing?

70. A windmill has blades of length 20 m and extracts 5.0% of the kinetic energy of the wind that passes through the circle defined by the blade motion. If the wind blows at 5.0 m/s, how much power is extracted?

71. A certain fire extinguisher consists of a large sealed tank with a thin tube leading from the bottom of the tank to an exit nozzle; the air above the liquid in the tank is at an overpressure p_{tank} with respect to atmospheric pressure p_0. The exit nozzle is at the same level as the top of the liquid; both are a distance y above the bottom of the tank (see Fig. 18.52). The liquid has density ρ. Find an expression for the speed of the stream of liquid emerging from the nozzle.

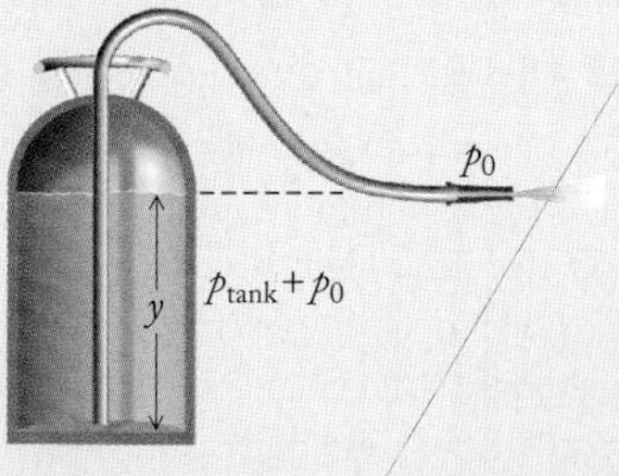

FIGURE 18.52 A fire extinguisher.

[†]For help, see Online Concept Tutorial 20 at www.wwnorton.com/physics

72. The friction experienced by real fluids is measured in terms of the **viscosity** η (the Greek letter *eta*), which for water at 20°C has the value $\eta = 1.0 \times 10^{-3}$ kg/m·s. Because of friction, the volume flow of fluid through a pipe requires a pressure difference Δp between the ends of the pipe. For a pipe of radius R and length L, the resulting volume flow is

$$\frac{dV}{dt} = \frac{\pi R^4}{8\eta L}\Delta p$$

Suppose that a 3.0-atm pressure difference is available. Calculate the volume flow rate for a 100-m length of pipe when the radius of the pipe is 1.0 cm (a garden hose) and when the radius is 20 cm (a water main).

73. As discussed in Section 18.2, laminar flow breaks down and the flow becomes turbulent when the velocity v of flow or the relevant dimension R of the flow becomes large. The transition to turbulence also depends on the viscosity η of the fluid (see Problem 72) and is related to a dimensionless quantity called the **Reynolds number** N. For flow through a tube of radius R, the Reynolds number is

$$N = \frac{2R\rho v}{\eta}$$

where ρ is the mass density of the fluid. Turbulence occurs when the Reynolds number gets large. Using $N = 2500$ as a threshold for turbulence, calculate the flow velocity at which turbulence will set in for water at 20°C when the tube radius is $R = 10$ cm.

*74. To fight a fire on the fourth floor of a building, firefighters want to use a hose of diameter 6.35 cm ($2\frac{1}{2}$ in.) to shoot 950 liters/min of water to a height of 12 m.

(a) With what minimum speed must the water leave the nozzle of the fire hose if it is to ascend 12 m?

(b) What pressure must the water have inside the fire hose? Ignore friction.

*75. Streams of water from fire hoses are sometimes used to disperse crowds. Suppose that a stream of water of diameter 2.5 cm emerging from a fire hose at a speed of 26 m/s impinges horizontally on a man. The collision of the water with the man is totally inelastic.

(a) What is the force that the stream of water exerts on the man; that is what is the rate at which the water delivers momentum to the man?

(b) What is the rate at which the water delivers energy?

*76. A **siphon** is an inverted U-shaped tube that is used to transfer liquid from a container at a high level to a container at a low level (Fig. 18.53).

(a) Using the lengths shown in Fig. 18.53 and the density ρ of the liquid, find a formula for the speed with which the liquid emerges from the lower end of the siphon. Assume that the containers are large, and that the lower end of the tube is *not* immersed in the container.

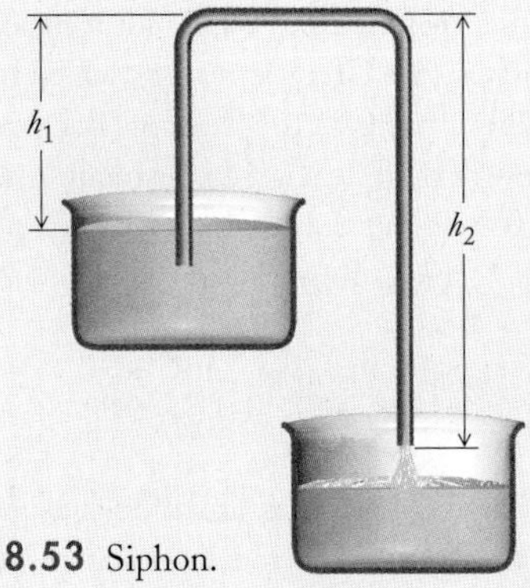

FIGURE 18.53 Siphon.

(b) Find the pressure at the highest point of the siphon.

(c) By setting this pressure equal to zero, find the maximum height h_1 with which the siphon can operate.

*77. The pump of a fire engine draws 1100 liters of water per minute from a pond with a water level 4.5 m below the pump and discharges this water into a fire hose of diameter 6.35 cm ($2\frac{1}{2}$ in.) at a pressure of 5.4 atm. In the absence of friction, what power (in hp) does this pump require?

*78. In the calculation of pressures in fire hoses, it is often necessary to take into account the frictional losses suffered by the water as it flows along the hose. Consider a horizontal hose of diameter 3.81 cm ($1\frac{1}{2}$ in.) and length 30 m carrying 380 liters/min of water. According to tables used by firefighters, a pressure of 6.1 atm at the upstream end of this hose will result in a pressure of only 3.4 atm at the downstream end; the loss of 2.7 atm is attributed to friction. At what rate (in hp) does friction remove energy from the water?

*79. A Venturi flowmeter in a water main of diameter 30 cm has a constriction of diameter 10 cm. Vertical pipes are connected to the water main and to the constriction (Fig. 18.54); these pipes are open at their upper ends, and the water level within them indicates the pressure at their lower ends. Suppose that the difference in the water levels in these two pipes is 3.0 m. What is the velocity of flow in the water main? What is the rate (in liters per second) at which water is delivered?

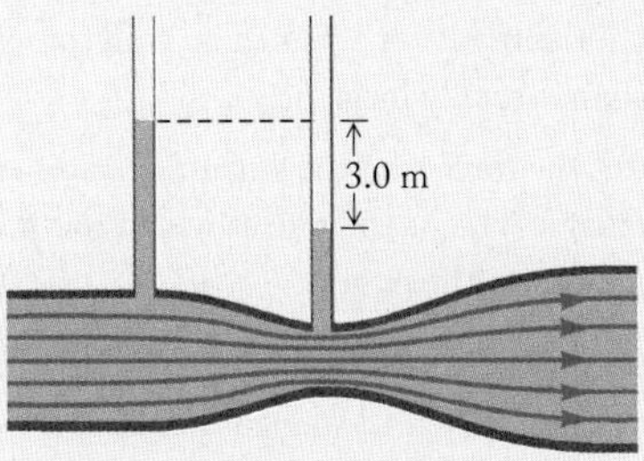

FIGURE 18.54 Venturi flowmeter.

*80. The **Pitot tube** is used for the measurement of the flow speeds of fluids, such as the flow speed of air past the fuselage of an airplane. It consists of a bent tube protruding into the airstream (Fig. 18.55) and another tube opening flush with the fuselage. The pressure difference between the air in the two tubes can be measured with a manometer. Show that in terms of the pressures p_1 and p_2 in the two tubes, the speed of airflow is

$$v = \sqrt{\frac{2(p_2 - p_1)}{\rho}}$$

where ρ is the density of air. (Hint: p_1 is simply the static air pressure. To find p_2, consider a streamline reaching the opening of the bent tube; at this point, the velocity of the air is zero.)

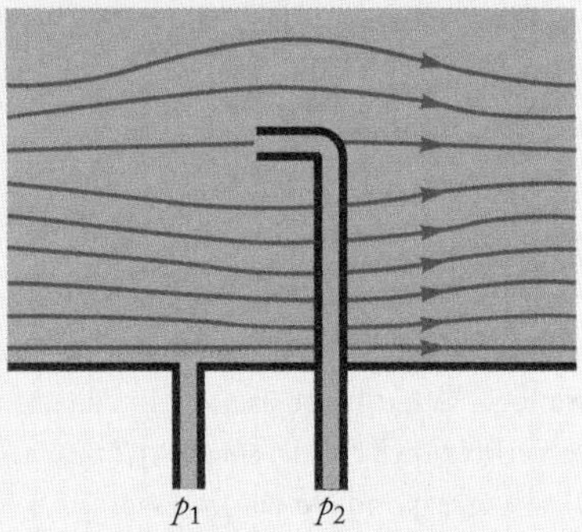

FIGURE 18.55 Pitot tube.

*81. The central front window of the cockpit of an airliner measures 30 cm × 30 cm. Estimate the force with which the air presses against this window when the airliner is flying at 900 km/h. [Hint: At the window, the air (almost) stops relative to the airliner.]

*82. A water tank filled to a height h has a small hole at the height z (see Fig. 18.56). Show that the stream of water emerging from this hole strikes the ground at a horizontal distance $2\sqrt{(h - z)z}$ from the base of the tank. What choice of z gives the largest horizontal distance? What is the largest horizontal distance?

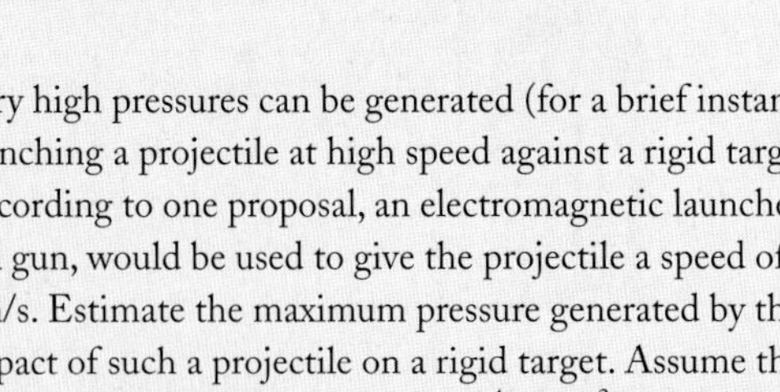

FIGURE 18.56 Tank with small hole.

*83. Very high pressures can be generated (for a brief instant) by launching a projectile at high speed against a rigid target. According to one proposal, an electromagnetic launcher, or rail gun, would be used to give the projectile a speed of 15 km/s. Estimate the maximum pressure generated by the impact of such a projectile on a rigid target. Assume that the density of the projectile is 1.0×10^4 kg/m^3. (Hint: At extreme pressures, the solid projectile will flow like a liquid of approximately constant density; hence Bernoulli's equation is approximately valid.)

**84. A cylindrical tank has a base area A and a height l; it is initially full of water. The tank has a small hole of area A' at its bottom. Calculate how long it will take all the water to flow out of this hole.

**85. A rectangular opening on the side of a tank has a width l. The top of the opening is at a depth h_1 below the surface of the water, and the bottom is at a depth h_2 (Fig. 18.57). Show that the volume of water that emerges from the opening in a unit time is

$$\tfrac{2}{3} l\sqrt{2g}\,(h_2^{3/2} - h_1^{3/2})$$

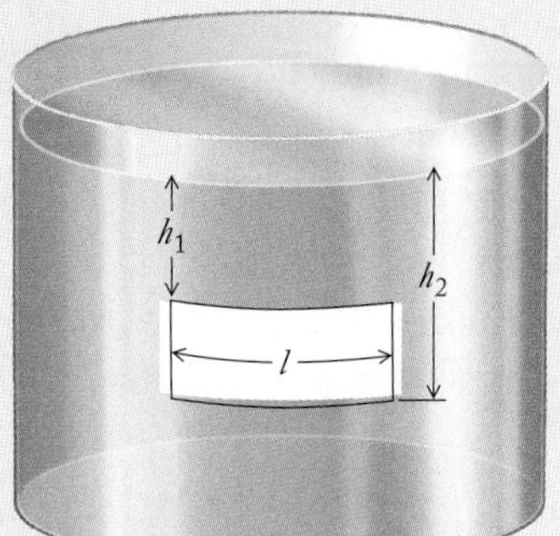

FIGURE 18.57 Tank with rectangular opening.

REVIEW PROBLEMS

86. (a) What is the arterial systolic pressure at your wrist if your arm is hanging straight down? Assume that the pressure is 120 mm-Hg at heart level, and that the wrist is 23 cm below this level.

    (b) What is the pressure if you raise your arm straight up? Assume that the wrist is then 78 cm above heart level.

87. Figure 18.58 shows a device for the measurement of venous pressure. A hypodermic needle is inserted into the vein, and saline solution from the syringe is pushed into the vertical manometer, until the hydrostatic pressure of the column of saline solution matches the pressure of the blood (there is then no blood flow or saline flow through the needle). If the height of the equilibrium column of saline is 103 mm, what is the blood pressure, in mm-Hg? The density of saline is the same as that of blood.

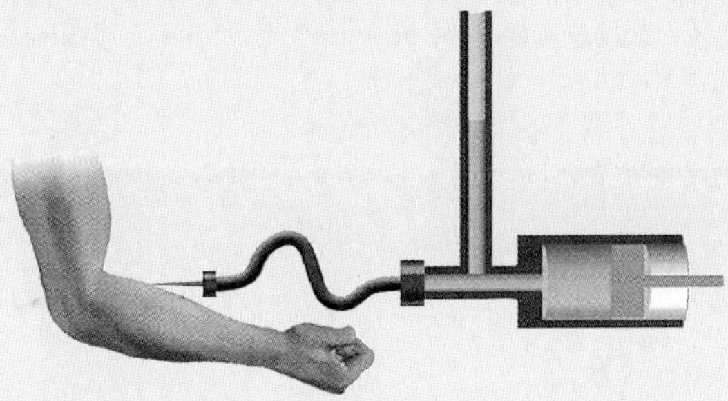

FIGURE 18.58 Device for the measurement of venous pressure.

88. A deep bowl has a layer of olive oil floating on top of a layer of water. The thickness of the layer of olive oil is 6.0 cm, and the thickness of the layer of water is 5.0 cm. The density of the olive oil is 918 kg/m^3. What is the pressure at the bottom of the bowl?

89. A diver attempts to breathe through a long snorkel, that is, a tube connecting his mouth to a float on the surface of the water (see Fig. 18.59). If the diver stays near the surface, he can breathe through this snorkel. But suppose the diver descends to a depth of 2.0 m. What is the water pressure on the outside of his chest? What is the air pressure on the inside of his chest? What is the net force on the front of his chest, of area 0.10 m^2? Can he breathe? (Hint: Could you breathe if two hefty football players were sitting on your chest?)

FIGURE 18.59 Diver with snorkel.

90. A submersible pump installed at the bottom of a well supplies water for a house. The pump is 30 m below the water level, and the house is 5 m above the water level. The house requires 20 liters of water per minute, at an overpressure of 3.0×10^5 N/m^2.

    (a) To supply this water, what power must the pump deliver in doing work against gravity?

    (b) What power must the pump deliver in doing work against the overpressure?

    (c) What is the total power required? Is a $\frac{1}{2}$-hp pump adequate for this purpose?

91. A large slick of very viscous oil dumped by a tanker floats on the surface of the sea. The density of the oil is 950 kg/m^3. If the slick is 10 cm thick at its center, how high is the center above the normal level of the sea?

92. A rectangular block of wood has a density of 600 kg/m^3. What fraction of this block will be submerged if the block floats in water?

93. A rectangular barge measures 5.0 m wide, 20 m long, and 3.0 m deep. Its mass is 5.0×10^4 kg when empty. What is the maximum load that this barge can carry, before the water swamps its deck?

94. A water trough is 1.0 m deep, 1.0 m wide, and 2.0 m long.
   (a) You slip a log of mass 200 kg and density 500 kg/m^3 into this trough. How much will the water level in the trough rise?
   (b) You slip a rock of mass 200 kg and density 2000 kg/m^3 into the trough. How much will the water level rise?

95. The mean density of the tissues in an average human body is 1.071×10^3 kg/m^3 (this value of the density excludes the volume of the lungs; that is, the air in the lungs is not counted as part of the body). What volume of air must a man of mass 80 kg take into his lungs if he wants to remain (barely) afloat in water?

96. A "gold" bracelet is made of a gold and copper alloy. The bracelet has a mass of 0.900 kg. To determine the amount of gold in the alloy, you weigh the bracelet when immersed in water. Its weight is then 0.820 N. What is the percentage of gold in the alloy? The density of gold is 19 300 kg/m^3, and the density of copper is 8960 kg/m^3.

97. In an experiment to measure the average density of the tissues of the human body, a man was first weighed in air, and then he was weighed while immersed in water. In air, his weight was 83.20×9.81 N; in water (after he expelled as much air from his lungs as he could, which caused him to sink), his weight was 4.30×9.81 N. The estimated residual volume of air in his lungs was 1.20 liters. What density do you deduce from these data?

98. An airplane has a wing of area (bottom area) 92 m^2. When in flight, the speed of the air along the bottom of the wing is 85 m/s, and along the top of the wing it is 95 m/s. Approximately what lift force does this wing provide? The density of air is 1.29 kg/m^3.

99. A tank full of water, 2.0 m deep, has a circular opening of radius 1.0 cm at its bottom (see Fig. 18.60). What is the rate (in m^3/s) at which water flows out of this opening?

FIGURE 18.60 Tank with circular opening.

100. Hurricanes with the lowest central barometric pressures have the highest wind speeds. Observations show that the square of the wind speed in a hurricane is roughly proportional to the difference between the barometric pressure outside the hurricane and the barometric pressure in the hurricane. Show that this proportionality is expected from Bernoulli's equation. (Hint: Consider a streamline of air that starts outside the hurricane and gradually spirals in toward the "eye.")

101. Suppose that a brisk wind of 18 m/s is blowing at your house from the north. The outside pressure is 1.0 atm.
   (a) If there were an open window on the north side, what would be the pressure inside your house? (Hint: Consider a streamline that enters the room and reaches a wall, where the speed of flow is zero.)
   (b) If, instead, there were an open window on the east side, would you expect the pressure to be larger or smaller than the atmospheric pressure?

Answers to Checkups

Checkup 18.1

1. From Table 18.1, we see that mercury has a density of 13 600 kg/m^3, that is, 13.6 times the density of water. Hence you need 13.6 cups of water to equal the mass of one cup of mercury.

2. For half the diameter, the cross-sectional area would be one-quarter as large ($A = \pi r^2$), so the rate would be $(1/4) \times 0.012\ \text{m}^2 \times 4.0\ \text{m/s} = 0.0032\ \text{m}^3\text{/s}$, or 3.2 kg/s.

3. (D) 2. The volume flow is given by Eq. (18.2), $dV/dt = Av$, so the product Av will be the same for each fluid. The pipe with one-quarter the flow velocity ratio thus has 4 times the area. Area is proportional to the square of the radius, so the first pipe has 2 times the radius of the second.

Checkup 18.2

1. No. If the fluid density can change, then matching the amounts of fluid that enter and leave a stream tube must include a density factor; this would give a more general continuity equation for compressible flow, $\rho_1 A_1 v_1 = \rho_2 A_2 v_2$.

2. No, a parcel of fluid may change velocity as it moves along to different locations. Steady flow means that as subsequent parcels of fluid pass a given point in space, they will each have the same velocity at that point.

3. The density of streamlines will be largest before the mouth (where the river is narrower), as required by the continuity equation (18.5); similarly, the density of streamlines will be

smallest out in the sea, beyond the mouth, where the flow has spread over a larger area.

4. From Example 2, the flow velocity is much larger in the aorta than in the capillaries; thus, the density of streamlines is larger in the aorta.
5. As the water falls, its velocity increases because of the acceleration of gravity. By the continuity equation (18.5), if the velocity increases, the cross-sectional area of the flow must decrease.
6. (A) At top and bottom of cylinder. The speed is the largest where the density of streamlines is greatest; from Fig. 18.6, thus is at the top and bottom of the cylinder.

Checkup 18.3

1. The pressure on each face from the air is one atmosphere; the value of the pressure force is this 1.01×10^5 N.
2. No. There is a large downward atmospheric pressure force on the upper face of the brick, but it is balanced by the upward atmospheric force on the lower surface of the brick, since there are many small pockets of air between the brick and the table.
3. Yes. If the balloon does not leak, it will expand on days of low atmospheric pressure and be compressed on high-pressure days.
4. (C) 1. The *pressure* is the same everywhere in any small region of the air, and so is the same at the base and at the cap of the bottle. (However, the *force* due to this pressure varies in proportion to the area, and so the force from the atmosphere on the base surface is 25 times the force on the bottle's cap.)

Checkup 18.4

1. No, the pressure is higher on the first floor, because of the weight of the water in the pipes going to the second floor. As we saw in Example 6, water pressure increases 1 atm for every 10 m of depth; for a typical distance between floors of about 3 m, this means the water pressure is about $\frac{1}{3}$ atm lower for each successive floor upward. In the upper floors of tall buildings, pumps are used to restore adequate water pressure.
2. Yes. As with any fluid, atmospheric air pressure decreases with height. As you climb the mountain, the decreased external pressure will allow the balloon to expand until the balloon's combined internal pressure and elastic forces reach equilibrium with the external pressure force.
3. No, Pascal's Principle deals with that part of the pressure not resulting from gravity. Gravity determines the pressure *difference* between different parts of a fluid that are at different heights [Eq. (18.12)]; Pascal's Principle states that a pressure applied at one point will change the pressure by the same amount throughout the fluid.
4. (D) 1 atm; 2 atm. The pressure in the muscle tissues balances the external pressure to maintain equilibrium. Thus, at the surface, the pressure in her muscle tissue is 1 atm; at a depth of 10 m, the pressure has increased to 2 atm, as in Example 6.

Checkup 18.5

1. By Archimedes' Principle, the buoyant force is equal to the weight of the fluid displaced. This is $W = Mg = \rho Vg$ $= 1000\ \text{kg/m}^3 \times 1.0 \times 10^{-3}\ \text{m}^3 \times 9.81\ \text{m/s}^2 = 9.8$ N.
2. The ship still experiences the same buoyant force, but the water that gets in through the hole acts like extra cargo; if not limited, it increases the weight of the ship until the buoyant force is no longer sufficient to balance the weight.
3. Yes. Seawater is more dense than freshwater (by almost 3%; see Table 18.1), so for a given body weight in equilibrium, a smaller volume of fluid is displaced, and more of the body floats above the water level.
4. The rock will not float in water; since water has a density of $1000\ \text{kg/m}^3$, the buoyant force will be smaller than the weight of the rock even for total submersion. The rock will float in liquid mercury; because mercury has a much higher density of $13\,600\ \text{kg/m}^3$, most of the rock will be above the mercury surface level.
5. (C) The water level remains at the rim. The weight of the ice cube equals the weight of the displaced water. When the ice melts to become water, this weight will thus precisely fill the volume previously displaced.

Checkup 18.6

1. No. Since the diameter is constant, the continuity equation (18.5) implies that the speed of flow does not change; according to Bernoulli's equation, the pressure difference between the ends depends on the difference in the square of the speeds, which is zero and thus independent of the speed of flow.
2. The channel is open to the atmosphere, so the pressure is the same at both ends; it is the only constant term. The height y will be different at the two ends (different potential-energy densities), and the speed will correspondingly change because of acceleration down the channel (different kinetic-energy densities).
3. Since the hose is horizontal, the height does not change, and the potential-energy term is constant. Since the diameter is constant, the speed does not change [continuity equation (18.5)], and so the kinetic-energy term is constant. Bernoulli's equation thus implies that the pressure is also constant: all three terms are constant in the hose. Outside the hose, the water is in projectile motion, but the pressure at every point is atmospheric pressure; thus $p = p_0$ is constant, but the kinetic- and potential-energy terms vary during the motion.
4. (A) At the narrow end of the funnel. Since the narrow end has the smaller area, the flow velocity is largest there [continuity equation (18.5)]. Bernoulli's equation tells us that where the velocity is largest, the pressure is lowest. This low pressure in an inverted funnel can be used to levitate an object, as in Fig. 18.34b.

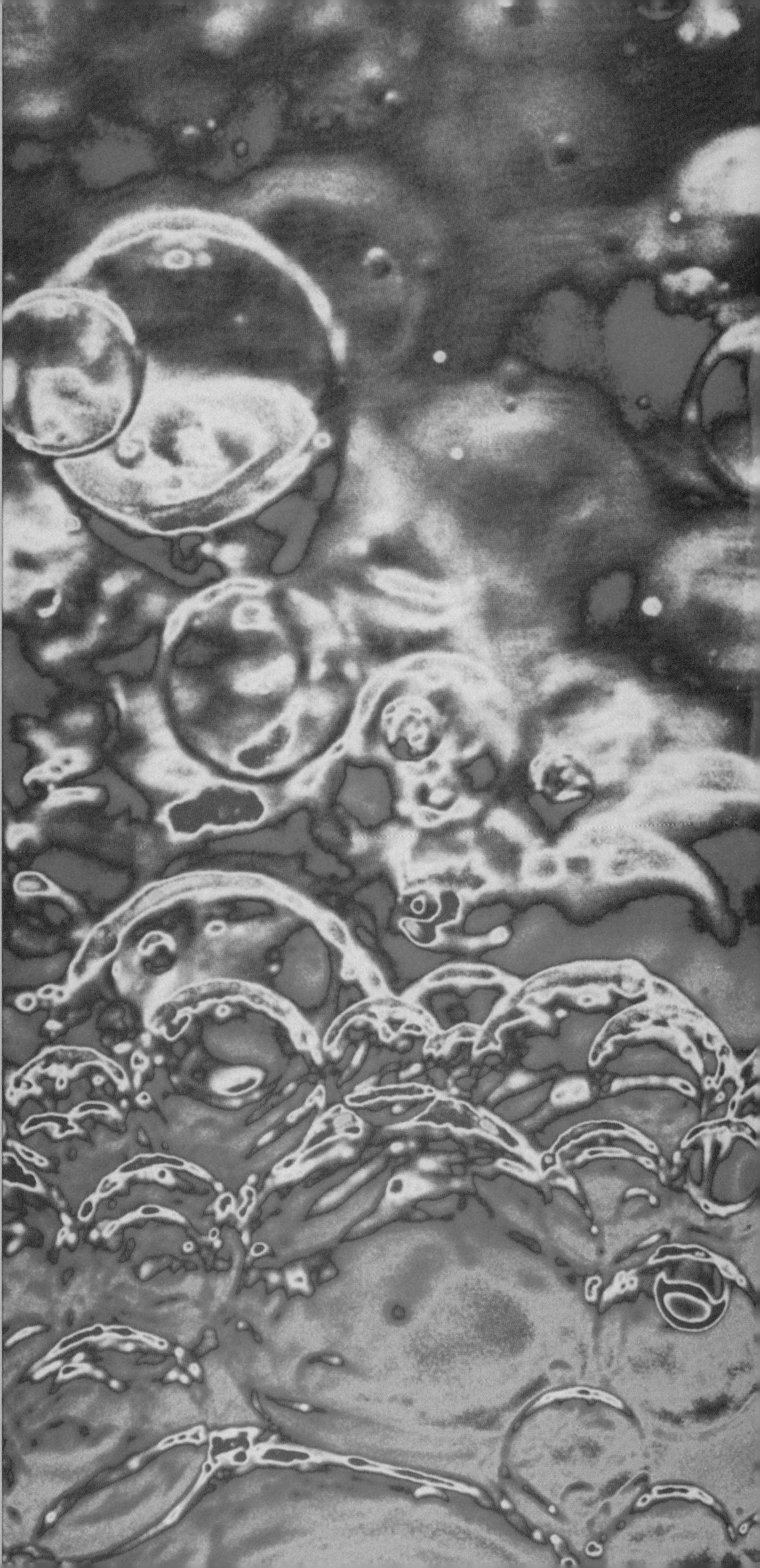

PART

3

Temperature, Heat, and Thermodynamics

CONTENTS

A liquid boils when bubbles of vapor form throughout its volume. The nature of the liquid and the atmospheric pressure determine the temperature required to form vapor. Pure water boils at 100° C at standard atmospheric pressure.

CHAPTER

19 The Ideal Gas

CONCEPTS IN CONTEXT

Concepts — in — Context

These spectacular hot-air balloons rely on buoyancy to fly: the warmed air in the balloon is less dense than the surrounding cold air.

To appreciate the properties of a gas that permit buoyancy, we will consider questions such as:

? How much air is in the balloon when cold? (Example 1, page 605)

? How much mass can the balloon lift when its air is heated to 50°C? (Example 6, page 612)

? How much energy (and fuel) does it take to heat the gas? (Example 9, page 618)

The particles of a gas—that is, the individual atoms or molecules of the gas—are well separated, and they fly about quite independently. The gas would disperse if it were not restrained by the forces exerted by the wall of the container. The gas molecules collide with each other, but these collisions do not alter the average speeds and average distribution of the gas molecules in the container. For most purposes, we can ignore these intermolecular collisions and regard the gas as a system of free particles. The motion of free gas particles is then very simple: they move with uniform velocity on straight lines, except when they collide with the walls of the container (Fig. 19.1). However, in spite of this simplicity of the motion of the gas particles, we cannot keep track of their motion in detail, because there are too many of them. For instance, one cubic centimeter of air contains about 2.7×10^{19} molecules. We have no way of ascertaining the initial position and velocity of each individual molecule; and even if we had, the calculation of the simultaneous motions of such an enormous number of molecules is far beyond the capabilities of even the fastest conceivable computer.

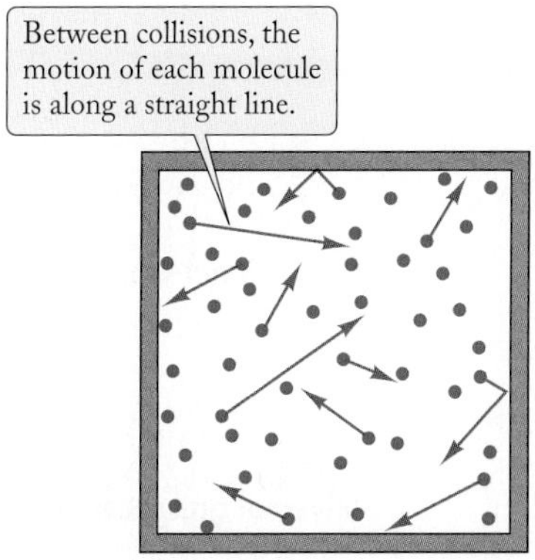

FIGURE 19.1 Random motion of the molecules of gas in a container.

In the absence of a microscopic description involving the individual positions and velocities of the molecules of the gas, we must be satisfied with a macroscopic description involving just a few parameters that characterize the average conditions in the volume of gas and that can be measured with large-scale laboratory instruments. We have already adopted such a macroscopic description for fluids in the preceding chapter. For a gas, the relevant macroscopic parameters characterizing the average conditions are *the mass, the number of moles, the volume, the density, the pressure, and the temperature.*

In this chapter we will study the macroscopic properties of gases, and we will see how these macroscopic properties are related to the average microscopic properties of the molecules of the gas.

19.1 THE IDEAL-GAS LAW

Online Concept Tutorial 21

The pressure, volume, and temperature of a gas obey some simple laws. Before we state these laws, let us recall the definition of pressure from the preceding chapter. Imagine that the gas is divided into small adjacent cubical volumes. The pressure is the force that one of these cubes exerts on an adjacent cube, or on an adjacent wall, divided by the area of one face of the cube (see Fig. 18.15); that is, the pressure is the force per unit area:

$$p = \frac{F}{A} \tag{19.1}$$

pressure

As we know from Section 18.4, the pressure is the same throughout the entire volume of a container of gas. (Within a container of gas, gravity causes a small decrease of pressure from the bottom to the top, but this decrease of pressure can usually be neglected.)

Consider now a given amount of gas, say, n moles of gas. We saw in Chapter 1 that a mole of any chemical element (or chemical compound) is the amount of matter that contains exactly as many atoms (or molecules) as there are atoms in 12 g of carbon. The "atomic mass" of a chemical element (or the "molecular mass" of a compound) is the mass of 1 mole. Thus, according to the table of "atomic masses" (see Appendix 8), 1 mole of carbon has a mass of 12.0 g, 1 mole of oxygen molecules (O_2) has a mass of 32.0 g, 1 mole of nitrogen molecules (N_2) has a mass of 28.0 g, and so on. For a mixture, such as air (consisting of 76% nitrogen, 23% oxygen, and 1% argon by mass), the mass of 1 mole can be obtained by adding the masses of suitable fractions of moles of the constituents; 1 mole of air has a mass of 29.0 g.

Suppose we place these n moles of gas in a container of volume V at a temperature T. The gas will then exert a pressure p. Experiments show that—to a good approximation—*the pressure p, the volume V, and the temperature T of the n moles of gas are related by the* **Ideal-Gas Law:**

Ideal-Gas Law

$$pV = nRT \tag{19.2}$$

Here R is the **universal gas constant**, with the value

universal gas constant

$$R = 8.31 \text{ J/mole}\cdot\text{K} \tag{19.3}$$

From the Ideal-Gas Law, we can calculate one of the three quantities that characterize the state of the gas (pressure, temperature, volume) if the other two are known.

The temperature in Eq. (19.2) is measured on the **absolute temperature scale,** and the SI unit of temperature is the **kelvin,** abbreviated K. We have not previously given the definition of temperature because Eq. (19.2) plays a dual role: it is a law of physics and also serves for the definition of temperature. This is by now a familiar story—in Chapter 5 we already saw that Newton's Second Law is a law of physics and also serves as a definition of mass.

We will give the details concerning the definition of temperature in the next section. For now it will suffice to note that the freezing point of water corresponds to a temperature of of 273.15 K, and the boiling point of water corresponds to 373.15 K; hence, there is an interval of exactly 100 K between the freezing and the boiling points. The zero of temperature on the absolute scale is the absolute zero, $T = 0$ K. According to Eq. (19.2), the pressure of the gas vanishes at this point. Actually, the gas will liquefy or even solidify before the absolute zero of temperature can be reached; when this happens, Eq. (19.2) becomes inapplicable.

The Ideal-Gas Law is a simple relation between the macroscopic parameters that characterize a gas. At normal densities and pressures, real gases obey this law quite well; but if a real gas is compressed to an excessively high density, then its behavior will deviate from this law. We briefly examine the effects of such extreme conditions in Problem 61, but elsewhere in this chapter we will neglect any deviations from Eq. (19.2). An **ideal gas** is a gas that obeys Eq. (19.2) exactly. The ideal gas is a limiting case of a real gas when the density and the pressure of the latter tend to zero. The ideal gas may be thought of as consisting of atoms of infinitesimal size, exerting no forces on each other or on the walls of the container, except for instantaneous impact forces exerted during collisions.

WILLIAM THOMSON, LORD KELVIN (1824–1907) *British physicist and engineer. Besides inventing the absolute temperature scale, he made many other contributions to the theory of heat. He was first to state the principle of dissipation of energy incorporated in the Second Law of Thermodynamics.*

The conditions of a temperature of 273 K and a pressure of 1 atm are called **standard temperature and pressure**, abbreviated **STP**. We can apply the Ideal-Gas Law to determine the volume of one mole of gas at STP. In SI units, a pressure of 1 atm is 1.01×10^5 N/m^2 [see Eq. (18.8)]. For one mole ($n = 1$ mole), the Ideal-Gas Law then gives us the volume

$$V = \frac{nRT}{p} = \frac{1 \text{ mole} \times 8.31 \text{ J/mole}\cdot\text{K} \times 273 \text{ K}}{1.01 \times 10^5 \text{ N/m}^2}$$

$$= 2.24 \times 10^{-2} \text{ m}^3 = 22.4 \text{ liters}$$

In the last equality, we have used 1 liter = 1000 cm^3 = 10^{-3} m^3. Note that it makes no difference whether the gas in this calculation is air or something else—one mole of *any* gas at STP has a volume of 22.4 liters (see Fig. 19.2).

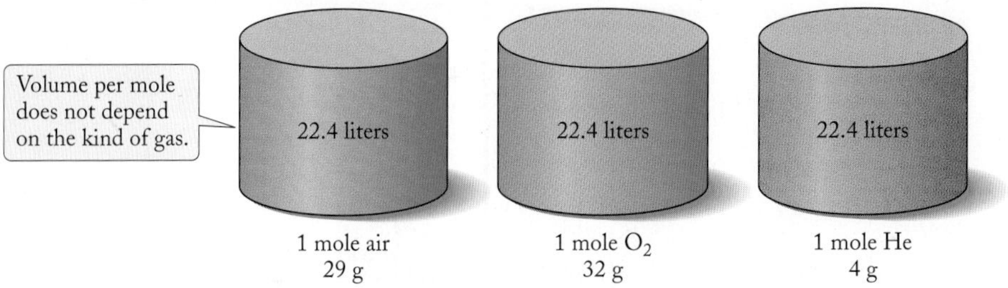

FIGURE 19.2 At a standard temperature of 273 K and pressure of 1 atm (STP conditions) a mole of any kind of gas occupies a volume of 22.4 liters.

EXAMPLE 1 The hot-air balloons pictured at the beginning of the chapter are large, with a typical volume of 2200 m^3. How many moles of gas does such a balloon contain when cold (at STP)? What mass of air does the balloon contain? The mass of 1 mole of air is 29.0 g.

Concepts in Context

SOLUTION: We found just above that one mole of any gas occupies 22.4 liters at STP. The volume of the balloon is 2200 $m^3 = 2.20 \times 10^6$ liters. Thus the number of moles n in the balloon is the volume in liters times the molar density in moles per liter:

$$n = 2.20 \times 10^6 \text{ liters} \times \frac{1 \text{ mole}}{22.4 \text{ liters}} = 9.82 \times 10^4 \text{ moles}$$

The mass of this number of moles is

$$m = 29.0 \text{ g/mole} \times 9.82 \times 10^4 \text{ moles} = 2.85 \times 10^6 \text{ g} = 2.85 \times 10^3 \text{ kg}$$

The balloon contains almost three metric tons of air!

EXAMPLE 2 Suppose you heat 1.00 kg of water and convert it into steam at the boiling temperature of water, 373 K, and at normal atmospheric pressure, 1.00 atm. What is the volume of the steam?

SOLUTION: The molecular mass of water (H_2O) is the sum of the "atomic masses" of two hydrogen atoms and one oxygen atom; that is, 1.0 g + 1.0 g + 16.0 g = 18.0 g. Thus, the number of moles of steam in 1.00 kg is

$$n = \frac{1.00 \text{ kg}}{18.0 \text{ g/mole}} = \frac{1000 \text{ g}}{18.0 \text{ g/mole}} = 55.6 \text{ moles}$$

With this, and with $p = 1.00$ atm $= 1.01 \times 10^5$ N/m^2, the Ideal-Gas Law tells us that the volume of the steam at 373 K is

$$V = \frac{nRT}{p} = \frac{55.6 \text{ moles} \times 8.31 \text{ J/mole}\cdot\text{K} \times 373 \text{ K}}{1.01 \times 10^5 \text{ N/m}^2}$$

$$= 1.71 \text{ m}^3$$

Thus when we boil 1 liter of cold water, it becomes over 1700 liters of steam.

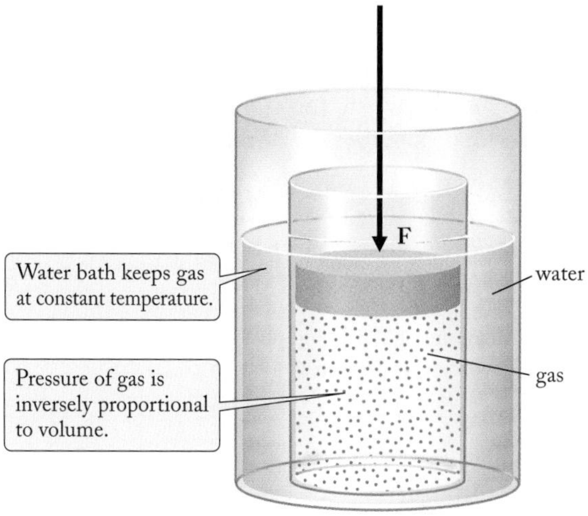

FIGURE 19.3 Compression of a gas at constant temperature by a force applied to the piston.

The Ideal-Gas Law incorporates two other laws: the **Law of Boyle** and the **Law of Charles and Gay-Lussac**. The Law of Boyle asserts that if the temperature is held constant, then the product of pressure and volume must remain constant as a given amount of gas is compressed or expanded:

$$pV = [\text{constant}] \qquad \text{for } T = [\text{constant}] \tag{19.4}$$

This law can be tested experimentally by placing a sample of gas in a cylinder with a movable piston, surrounded by some substance at a fixed temperature, say, a water bath (see Fig. 19.3). By moving the piston in or out, we can vary the volume V and we can check that the pressure then varies in inverse proportion to the volume, as demanded by Eq. (19.4).

The Law of Charles and Gay-Lussac asserts that if the pressure is held constant, the ratio of volume to temperature remains constant as a given amount of gas is heated or cooled:

$$\frac{V}{T} = [\text{constant}] \qquad \text{for } p = [\text{constant}] \tag{19.5}$$

ROBERT BOYLE (1627–1691) *English experimental physicist. He invented a new air pump, with which he performed the experiments on gases that led to discovery of the law named after him.*

This law can be tested with a similar cylinder–piston arrangement with a fixed weight mounted on the piston and a heat source placed below the cylinder (Fig. 19.4). The weight mounted on the piston, in conjunction with the weight of the piston, subjects the gas to a fixed pressure. By increasing or decreasing the temperature of the gas we then cause the gas to expand or contract, and we can check that the volume is proportional to the temperature, as demanded by Eq. (19.5). Together, the experimental tests of Eqs. (19.4) and (19.5) amount to a test of the Ideal-Gas Law.

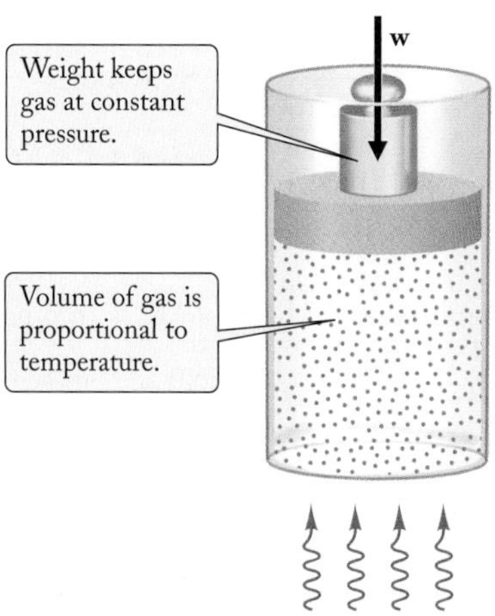

FIGURE 19.4 Expansion of a gas kept at constant pressure by a weight on the piston. The gas is being heated.

EXAMPLE 3 Early in the morning, at the beginning of a trip, the tires of an automobile are cold (280 K), and their air is at a pressure of 3.0 atm. Later in the day, after a long trip on hot pavements, the tires are hot (330 K). What is the pressure? Assume that the volume of the tires remains constant.

SOLUTION: At constant volume, the pressure is proportional to the temperature [see Eq. (19.2)]. Hence

$$p_2 = p_1 \times \frac{T_2}{T_1}$$

$$= 3.0 \text{ atm} \times \frac{330 \text{ K}}{280 \text{ K}} = 3.5 \text{ atm}$$

COMMENT: The pressure gauges for automobile tires are commonly calibrated to read **overpressure**, that is, the excess above atmospheric pressure (see also Section 18.4). Thus, a pressure gauge would read 2.0 atm in the morning, and 2.5 atm later in the day, if we assume that the atmospheric pressure remains constant at 1.0 atm (it rarely varies more than a few percent).

overpressure

The Ideal-Gas Law can also be written in terms of the number of molecules, instead of the number of moles. The number of molecules per mole is **Avogadro's number** N_A. As already mentioned in Chapter 1 [see Eq. (1.2)], the value of Avogadro's number is approximately

$$N_A = 6.02 \times 10^{23} \text{ molecules per mole} \tag{19.6}$$

Avogadro's number

Thus, if the number of moles is n, the number of molecules N is

$$N = N_A n \tag{19.7}$$

With this, the Ideal-Gas Law (19.2) becomes

$$pV = \frac{N}{N_A} RT \tag{19.8}$$

or

$$pV = NkT \tag{19.9}$$

Ideal-Gas Law in terms of number of molecules

where

$$k = \frac{R}{N_A} = \frac{8.31 \text{ J/mole}\cdot\text{K}}{6.02 \times 10^{23}\text{/mole}} = 1.38 \times 10^{-23} \text{ J/K} \tag{19.10}$$

Boltzmann's constant

The constant k is called **Boltzmann's constant**. As we will see, this constant tends to make an appearance in equations relating macroscopic quantities (such as p or V) to microscopic quantities (such as the number N of molecules).

EXAMPLE 4 (a) What is the number of molecules in 1.00 cm^3 of air at a temperature of 273 K and a pressure of 1.00 atm? (As in Example 1, these conditions are *standard temperature and pressure,* or STP.) (b) What is the mass density of air at STP? Recall from above that the mass of 1 mole of air is 29.0 g.

LUDWIG BOLTZMANN (1844–1906)
Austrian theoretical physicist. He made crucial contributions in the kinetic theory of gases and in statistical mechanics.

SOLUTION: (a) With $p = 1.00$ atm $= 1.01 \times 10^5$ N/m^2 and $V = 1.00$ cm^3 $= 1.00 \times 10^{-6}$ m^3, we can obtain the number of molecules N directly from the Ideal-Gas Law in the form (19.9):

$$N = \frac{pV}{kT} = \frac{1.01 \times 10^5\ \text{N/m}^2 \times 10^{-6}\ \text{m}^3}{1.38 \times 10^{-23}\ \text{J/K} \times 273\ \text{K}}$$

$$= 2.68 \times 10^{19}\ \text{molecules}$$

Note that this result is valid for any kind of gas—the number of molecules in 1 cm^3 of any kind of gas under STP conditions is 2.68×10^{19}, or about 27 billion billion!

(b) The mass density is the mass per unit volume. We can find the mass of our 1.00-cm^3 volume by first obtaining the number of moles n from the number of molecules:

$$n = \frac{N}{N_A} = \frac{2.68 \times 10^{19}\ \text{molecules}}{6.02 \times 10^{23}\ \text{molecules/mole}} = 4.45 \times 10^{-5}\ \text{moles}$$

Since the "molecular mass" is 29.0 grams per mole, the amount of mass in 1 cm^3 of air at STP is

$$4.45 \times 10^{-5}\ \text{moles} \times 29.0\ \text{g/mole} = 1.29 \times 10^{-3}\ \text{g} = 1.29 \times 10^{-6}\ \text{kg}$$

Thus the density of air is 1.29×10^{-3} g/cm^3, or 1.29 kg/m^3, in agreement with the value listed in Table 18.1.

PROBLEM-SOLVING TECHNIQUES — IDEAL-GAS LAW

When applying the Ideal-Gas Law to problems, care must be taken to use the correct units.

- The temperature T in the Ideal-Gas Law must be expressed in kelvins (K); the Ideal-Gas Law cannot be used in the forms (19.2) or (19.9) if the temperature is expressed in a different temperature scale, such as degrees celsius (°C) or degrees Fahrenheit (°F) (see Section 19.2).
- The pressure p in the Ideal-Gas Law must be the absolute pressure (not the overpressure), and it must be expressed in SI units (N/m^2, or pascals), not in atmospheres or mm-Hg.

✔ Checkup 19.1

QUESTION 1: If you heat the gas in a sealed jar from a temperature of 300 K to 600 K, by what factor does the pressure increase?

QUESTION 2: Suppose that the pressure of atmospheric air increases from 1.00 atm to 1.05 atm while the temperature remains constant. What happens to the density of air?

QUESTION 3: Suppose that the temperature of atmospheric air increases from 270 K to 290 K while the pressure remains constant. What happens to the density of air?

QUESTION 4: The air in the tires of an automobile is at an overpressure of 2 atm, but at the same temperature as the surrounding air. By what factor is the density of the air larger than the surrounding air?

QUESTION 5: A fixed amount of an ideal gas is in a variable-volume cylinder. The pressure is increased a factor of $\frac{3}{2}$, and the temperature is decreased by a factor of 2. The final volume occupied by the gas is what factor times the original volume?

(A) 3 (B) $\frac{4}{3}$ (C) 1 (D) $\frac{3}{4}$ (E) $\frac{1}{3}$

19.2 THE TEMPERATURE SCALE

As mentioned in the preceding section, the ideal gas can be used for the definition of temperature. For this purpose, we take a fixed amount of some gas, such as helium, and place it an airtight, nonexpanding container, such as a Pyrex glass bulb. According to Eq. (19.2), for a gas kept in such a constant volume, the pressure is directly proportional to the temperature. Thus, a simple measurement of pressure gives us the temperature.

To calibrate the scale of this thermometer, we must choose a standard reference temperature. The standard adopted in the SI system of units is the temperature of the **triple point of water**, that is, *the temperature at which water, ice, and water vapor coexist when placed in a closed vessel.* Figure 19.5 shows a triple-point cell used to achieve the standard temperature. This standard temperature has been assigned the value of 273.16 kelvins, or 273.16 K, which is ever so slightly above the freezing point of water. If the bulb of the gas thermometer is placed in thermal contact with this cell so that it attains a temperature of 273.16 K, it will read some pressure p_{triple}. If the bulb is then placed in thermal contact with some body at an unknown temperature T, it will read a pressure p that is greater or smaller than p_{triple} by some factor. The unknown temperature T is then greater or smaller than 273.16 K by this same factor; for instance, if the pressure p is half as large as p_{triple}, then $T = \frac{1}{2} \times 273.16$ K. *The temperature scale defined by this procedure is called the* **ideal-gas temperature scale**, or the **absolute temperature scale**.

When connecting a pressure gauge to the bulb of gas, we must take special precautions to ensure that the operation of the pressure gauge does not alter the volume available to the gas. Figure 19.6 shows a device designed for this purpose; this device is called a **constant-volume gas thermometer**. The pressure gauge used in this thermometer consists of a closed-tube manometer; one branch of the manometer is connected to the bulb of gas, and the other branch consists of a closed, evacuated tube. The difference h in the heights of the levels of mercury in these two branches is proportional to the pressure of the gas. The manometer is also connected to a mercury reservoir. During the operation of the thermometer, this reservoir must be raised or lowered so that the level of mercury in the left branch of the manometer tube always

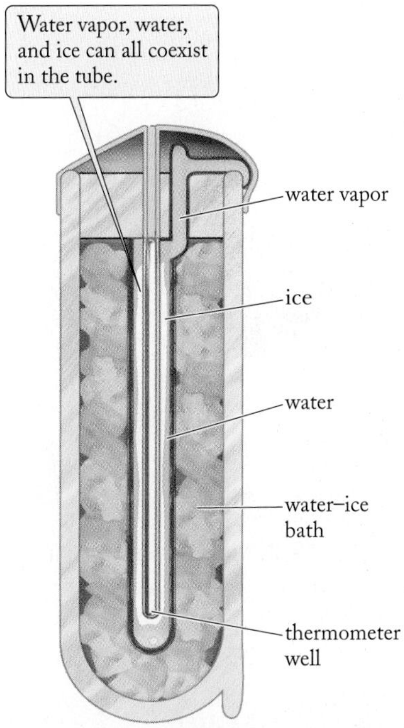

FIGURE 19.5 Triple-point cell of the National Institute of Standards and Technology. The inner tube (red) contains water, water vapor, and ice in equilibrium.

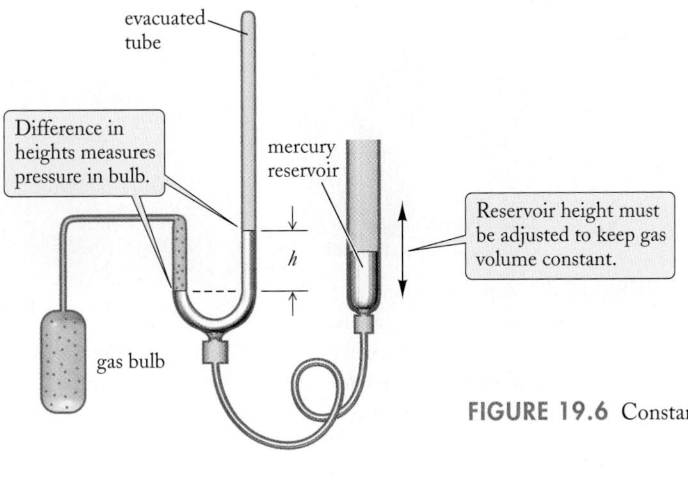

FIGURE 19.6 Constant-volume gas thermometer.

remains at a constant height; this keeps the gas in the bulb at a constant volume. The bulb of this thermometer may be put in contact with any body whose temperature we wish to measure, and the pressure registered by the manometer then gives us the absolute temperature.

Table 19.1 lists some examples of temperatures of diverse bodies.

For everyday and industrial use, the ideal-gas thermometer is somewhat inconvenient, and is often replaced by mercury-bulb thermometers, bimetallic strips, electrical-resistance thermometers, thermocouples, optical pyrometers, or color-strip thermometers (see Figs. 19.7–19.12). These must be calibrated in terms of the ideal-gas thermometer if they are to read absolute temperature.

FIGURE 19.7 Mercury-bulb thermometers. The thermal expansion of the mercury in the bulb causes it to rise in the thin capillary tube, indicating the temperature.

FIGURE 19.8 Bimetallic-strip thermometer. The thermometer contains a helix consisting of joined bands of different metals. With an increase of temperature, the bands expand by different amounts, which coils the helix and turns the pointer.

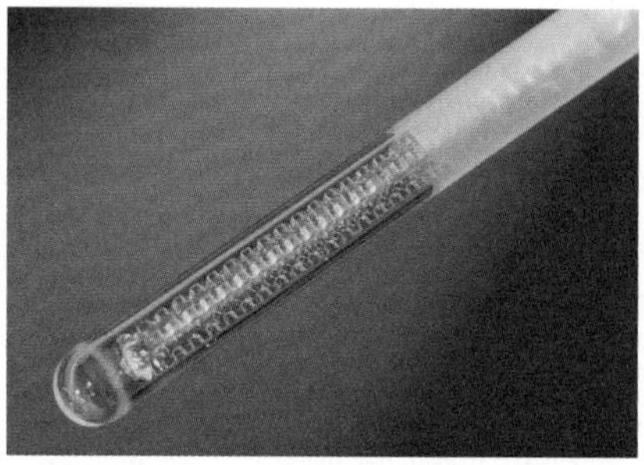

FIGURE 19.9 Platinum-resistance thermometer. The resistance that the fine coiled wire of platinum offers to an electric current serves as an indicator of the temperature (electrical resistance is discussed in Chapter 27).

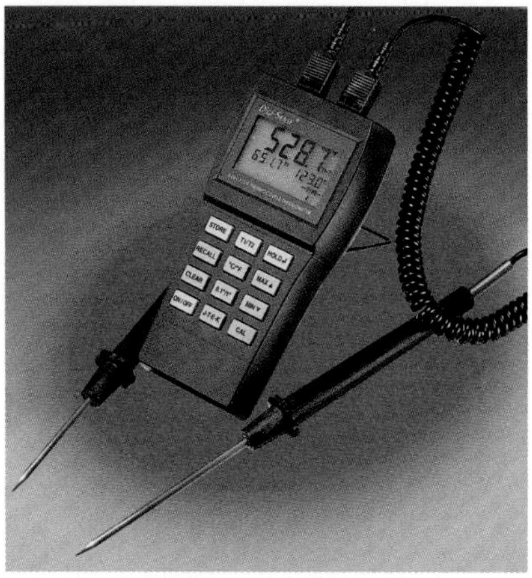

FIGURE 19.10 Thermocouple thermometer. A thermocouple consists of two wires of different metals, for example, one wire of platinum and one wire of platinum–rhodium alloy, joined at their ends. The other ends of the wires are at a reference temperature, usually 0°C; because of the temperature difference and the dissimilar materials, an electrical voltage develops between the other ends. Most digital thermometers use thermocouples.

FIGURE 19.11 Optical pyrometer. From the color distribution emitted by an incandescent material, the temperature can be determined (this effect is discussed in Chapter 37).

FIGURE 19.12 Color-strip thermometer. Material with different temperature-induced color transitions provides an indication of the temperature.

TABLE 19.1 SOME TEMPERATURES

	KELVIN TEMPERATURE	CELSIUS TEMPERATURE
Interior of hottest stars	10^9 K	10^9°C
Center of H-bomb explosion	10^8	10^8
Highest temperature attained in plasma in laboratory	6×10^7	6×10^7
Center of Sun	1.5×10^7	1.5×10^7
Surface of Sun (a)	4.5×10^3	4.2×10^3
Center of Earth	4×10^3	3.7×10^3
Acetylene flame	2.9×10^3	2.6×10^3
Melting of iron (b)	1.8×10^3	1.5×10^3
Melting of lead	6.0×10^2	3.3×10^2
Boiling of water	373	100
Human body (c)	310	37
Surface of Earth (average)	287	14
Freezing of water	273	0
Liquefaction of nitrogen	77	−196
Liquefaction of hydrogen	20	−253
Liquefaction of helium	4.2	−269
Interstellar space	3	−270
Lowest temperature attained in laboratory	3×10^{-8}	≈ -273.15

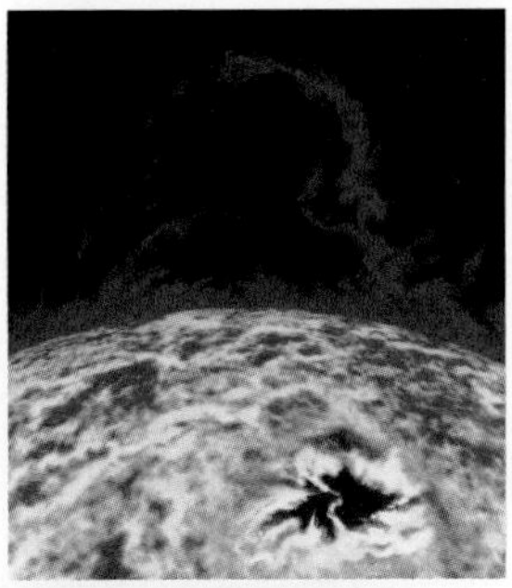
(a)

(b)

(c)

Although the absolute temperature scale is the only scale of fundamental significance, several other temperature scales are in practical use. The **Celsius scale** (formerly known as the **centigrade** scale) is shifted 273.15 K relative to the absolute scale:

$$T_C = T - 273.15°\text{C} \qquad (19.11)$$

Celsius scale

where the notation "°C" means "degrees Celsius." Note that on the Celsius scale, absolute zero is at −273.15°C. The triple point of water is then at 0.01°C, the freezing point at 0°C, and the boiling point at 100°C.

The **Fahrenheit scale** is shifted relative to the Celsius scale and, furthermore, uses degrees of smaller size, each degree Fahrenheit corresponding to $\frac{5}{9}$ degree Celsius:

$$T_F = \tfrac{9}{5}T_C + 32°\text{F} \quad \text{or} \quad T_C = \tfrac{5}{9}(T_F - 32°\text{F}) \qquad (19.12)$$

Fahrenheit scale

On this scale, the freezing point of water is at 32°F, the boiling point of water is at 212°F, and absolute zero is at −459.67°F. Figure 19.13 can be used for a rough conversion between the Fahrenheit and Celsius scales.

EXAMPLE 5 The normal temperature of the human body is 98.6°F. Express this in °C.

SOLUTION: According to Eq. (19.12)

$$T_C = \tfrac{5}{9}(T_F - 32°F) = \tfrac{5}{9}(98.6°F - 32°F) = 37°C$$

Concepts in Context

EXAMPLE 6 Hot-air balloons become buoyant when the gas inside is heated with a burner below the balloon. Recall from Example 1 that at STP, a typical balloon contained 2.85×10^3 kg of air. If the temperature of the gas is increased to 50°C (at constant pressure), what is the mass of the gas that remains in the balloon? What mass can be lifted by the balloon?

SOLUTION: We must convert to absolute temperature to use the Ideal-Gas Law, so a temperature of 50°C is equivalent to an absolute temperature of

$$T = T_C + 273.15 \text{ K} = 50 + 273.15 \text{ K}$$
$$= 323 \text{ K}$$

At constant pressure (and volume), the Ideal-Gas Law, $pV = nRT$, shows that the number of moles (and thus the mass) will be inversely proportional to temperature; hence the mass of air in the hot-air balloon is

$$m_{323\,\text{K}} = m_{\text{STP}} \times \frac{273 \text{ K}}{323 \text{ K}} = 2.85 \times 10^3 \text{ kg} \times \frac{273 \text{ K}}{323 \text{ K}} = 2.41 \times 10^3 \text{ kg}$$

The balloon will rise if its total mass (balloon, basket, equipment, and riders) is less than the difference between the displaced-air mass and the hot-air mass:

$$2.85 \times 10^3 \text{ kg} - 2.41 \times 10^3 \text{ kg} = 4.4 \times 10^2 \text{ kg} = 440 \text{ kg}$$

Thus the balloon can lift nearly one-half a metric ton.

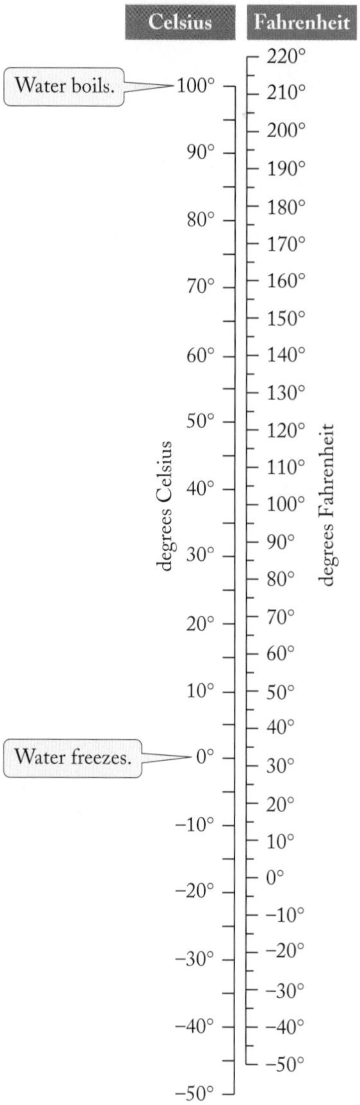

FIGURE 19.13 The correspondence between the Fahrenheit scale and the Celsius scale.

Checkup 19.2

QUESTION 1: Is a negative Fahrenheit temperature always a negative Celsius temperature? Is a negative Celsius temperature always a negative Fahrenheit temperature?

QUESTION 2: Three cups of water are at temperatures of 320 K, 20°C, and 90°F. Which is the hottest? Which the coldest?

QUESTION 3: Consider a cylindrical can filled with gas, closed off at the top by a piston. The weight of the piston then provides a constant force, which keeps the gas under a constant pressure. Explain how this device could be used as a "constant-pressure" gas thermometer.

QUESTION 4: To the nearest degree, at what temperature does the Celsius scale indicate the same numerical value of temperature as the Fahrenheit scale?

(A) −72°C (B) −50°C (C) −40°C (D) −26°C

19.3 KINETIC PRESSURE

The pressure of a gas against the walls of its container is due to the impacts of the molecules on the walls. We will now calculate this pressure by considering the average motion of the molecules of gas. This permits us to understand how a macroscopic property, such as the pressure, emerges from the microscopic behavior of individual molecules of the gas. In our calculation, we will assume that the container is a cube of side L, that the gas molecules collide only with the walls but not with each other, and that the collisions are elastic. These assumptions are not required, but they simplify the calculations.

Figure 19.14 shows the container filled with gas molecules. The motion of each molecule can be resolved into x, y, and z components. Consider one molecule, and consider the component of its motion in the x direction. The component of the velocity in this direction is $\pm v_x$, and the magnitude of this velocity remains constant, since the collisions with the wall are elastic. The time that the molecule takes to move from the face of the cube at $x = L$ to $x = 0$ and back to $x = L$ is the distance divided by the speed:

$$t = \frac{2L}{v_x} \tag{19.13}$$

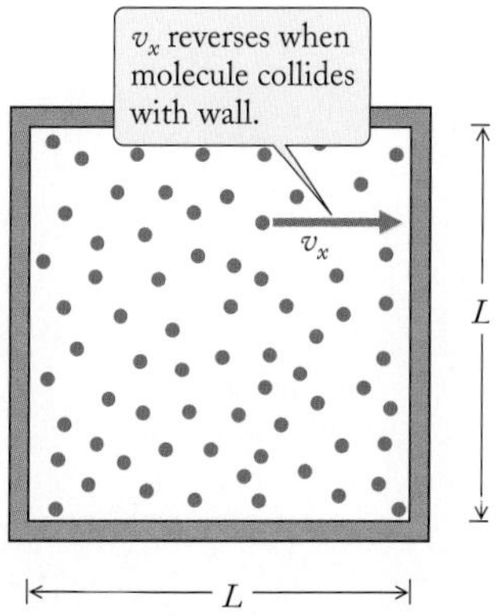

FIGURE 19.14 Gas molecules in a container.

This is therefore the time between one collision with the face at $x = L$ and the next collision with the same face. When the molecule strikes the face, its x velocity is reversed from $+v_x$ to $-v_x$. Hence, during each collision at $x = L$, the x momentum of the molecule changes from $+mv_x$ to $-mv_x$, a net change of $-2mv_x$, where m is the mass of the molecule. Thus each collision transfers a momentum $+2mv_x$ to the face at $x = L$. The average rate at which the molecule transfers momentum to the face at $x = L$ is then the momentum transfer per collision divided by the time between collisions:

$$\frac{2mv_x}{t} = \frac{2mv_x}{2L/v_x} = \frac{mv_x^2}{L} \tag{19.14}$$

This average rate of momentum transfer equals the average force that the impacts of this one molecule exert on the wall. To find the total force exerted by the impacts of all the N molecules, we must multiply the force given in Eq. (19.14) by N; and to find the pressure, we must divide the total force by the area L^2 of the face of the cube. This leads to a pressure

$$p = \frac{N}{L^2}\frac{mv_x^2}{L} \tag{19.15}$$

or, in terms of the volume $V = L^3$ of the gas,

$$p = \frac{Nmv_x^2}{V} \tag{19.16}$$

In this calculation we made the implicit assumption that all the molecules have the same velocity. This is, of course, not true; the molecules of the gas have a distribution of velocities—some have high velocities, and some have low velocities. To account for this spread of velocities, we must replace the force in Eq. (19.14) due to one given molecule by the average over all the molecules. Consequently, we must replace v_x^2 by an average over all the molecules in the container. We will designate the average by an overbar, $\overline{v_x^2}$. Equation (19.16) then becomes

$$p = \frac{Nm\overline{v_x^2}}{V} \tag{19.17}$$

To proceed further, we note that on the average, molecules are just as likely to move in the x, y, or z direction. Hence, the average values of v_x^2, v_y^2, and v_z^2 are equal,

$$\overline{v_x^2} = \overline{v_y^2} = \overline{v_z^2} \tag{19.18}$$

The sum of the squares of the components of the velocity is the square of the magnitude of the velocity,

$$\overline{v_x^2} + \overline{v_y^2} + \overline{v_z^2} = \overline{v^2} \tag{19.19}$$

Since all three terms on the left side of this equation are equal, each of them must equal $\frac{1}{3}\overline{v^2}$. We can then write Eq. (19.17) as

$$p = \frac{Nm\overline{v^2}}{3V} \tag{19.20}$$

Let us now compare this result for the pressure with the Ideal-Gas Law [Eq. (19.9)], according to which the pressure is

$$p = \frac{NkT}{V} \tag{19.21}$$

The agreement between these two expression (19.20) and (19.21) for the pressure demands

$$\frac{m\overline{v^2}}{3} = kT \tag{19.22}$$

This shows that the average square of the molecular speed is proportional to the temperature. *The square root of* $\overline{v^2}$ *is called the* **root-mean-square speed**, or the **rms speed**, and it is usually designated by v_{rms}. If we divide both sides of Eq. (19.22) by $m/3$ and extract the square root of both sides, we find

root-mean-square (rms) speed

$$v_{\text{rms}} = \sqrt{\overline{v^2}} = \sqrt{\frac{3kT}{m}} \tag{19.23}$$

This root-mean-square speed may be regarded as the typical speed of the molecules of the gas. Incidentally: There are other ways of calculating a typical speed; for example, we may want to calculate the average of all the molecular speeds, or the most probable of all the molecular speeds. The average and most probable speeds turn out to be somewhat less than the rms speed, but their calculation requires some further knowledge of the distribution of molecular speeds.

EXAMPLE 7 What is the root-mean-square speed of nitrogen molecules in air at 300 K? Of oxygen molecules?

SOLUTION: The "molecular mass" of N_2 molecules is 28.0 g, twice the "atomic mass" of nitrogen. Hence, the mass of one molecule is

$$m = \frac{28.0\text{ g}}{N_A} = \frac{28.0 \times 10^{-3}\text{ kg}}{6.02 \times 10^{23}} = 4.65 \times 10^{-26}\text{ kg}$$

and Eq. (19.23) yields

$$v_{\text{rms}} = \sqrt{\frac{3kT}{m}}$$

$$= \sqrt{\frac{3 \times 1.38 \times 10^{-23}\,\text{J/K} \times 300\,\text{K}}{4.65 \times 10^{-26}\,\text{kg}}} = 517\ \text{m/s} \tag{19.24}$$

For O_2 molecules the "molecular mass" is 32.0 g, twice the "atomic mass" of oxygen. By a similar calculation we find that the mass of one oxygen molecule is 5.32×10^{-26} kg, and that v_{rms} is 483 m/s.

COMMENT: Note that the rms speed of nitrogen molecules is slightly larger than that of oxygen molecules. In general, Eq. (19.23) shows that the rms speed is inversely proportional to the square root of the mass of the molecule—at a given temperature, the molecules of lowest mass have the highest speeds.

The distribution of molecular speeds in a sample of gas at some given temperature can be deduced by means of kinetic theory; it is called the **Maxwell distribution** of molecular speeds. We will not attempt to deduce this distribution here, but only examine some of its qualitative features. The distribution of speeds in a sample of gas is described mathematically by a distribution function N_v, which specifies the number of molecules per unit speed interval. Thus, if dv is a small interval of speeds centered on a given speed v, then the number dN of molecules that have speeds in the interval dv is

$$dN = N_v\, dv$$

Keep in mind that when dealing with the distribution of speeds, or the distribution of any other kind of physical quantity that has a continuous range of variation, it is not reasonable to ask how many molecules have a speed v, since it is unlikely that any molecule has a speed *exactly* equal to v. The only reasonable question is how many molecules have speeds in some specified interval of speeds.

Figure 19.15 shows plots of the function N_v for the Maxwell distribution of speeds at two different temperatures. As we can see from these plots, the molecular speeds are spread over broad ranges. The distribution functions fade away as $v \to 0$ and as $v \to \infty$; thus, there are few molecules near zero speed, and few molecules of very large speed. The peaks of the distribution functions indicate the **most probable speed**. Comparing the peaks at the two different temperatures, we see that the most probable speed increases with temperature. For the Maxwell distribution, the most probable speed and the average speed are somewhat lower than the rms speed; the most probable speed and the average speed turn out to be approximately $0.82v_{rms}$ and $0.92v_{rms}$ respectively.

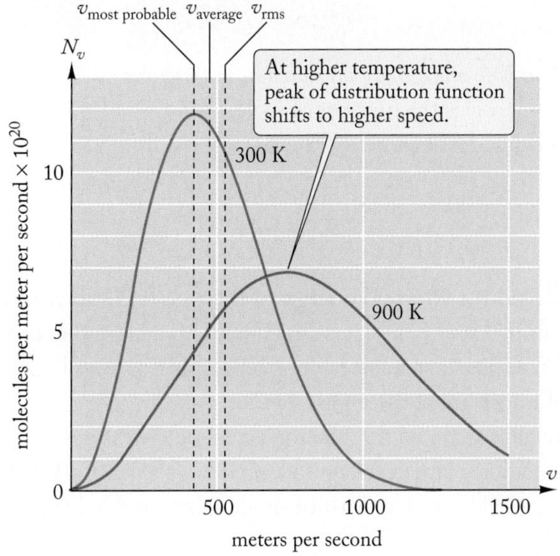

FIGURE 19.15 Maxwell distribution of speeds for one mole of N_2 gas molecules. The higher curve corresponds to a temperature of 300 K, the lower curve to 900 K. The most probable speed is indicated by the peak on the curve.

Checkup 19.3

QUESTION 1: The pressure of a gas on the walls of its container increases with temperature. Is this increase due to a higher rate of impacts on the walls, or a higher amount of momentum transferred to the wall per impact, or both?

QUESTION 2: If we increase the temperature of a gas from 300 K to 600 K, by what factor do we increase the rms speed of its molecules?

QUESTION 3: Is the rms speed of water molecules in air higher or lower than that of oxygen molecules?

QUESTION 4: Consider hydrogen (H_2) gas at a temperature of 50 K and oxygen gas (O_2) at 200 K, both at the same pressure $p = 1$ atm. What is the ratio of the mass per unit volume of the hydrogen gas to that of the oxygen gas?

(A) 64 (B) 4 (C) 1 (D) $\frac{1}{4}$ (E) $\frac{1}{64}$

19.4 THE ENERGY OF AN IDEAL GAS

Since a gas is a system of particles, its energy is the sum of the energies of all these particles. To calculate the energy of the gas, we begin with the kinetic energy of one molecule. According to Eq. (19.23), the average value of the square of the speed of a molecule in an ideal gas is

$$\overline{v^2} = \frac{3kT}{m}$$

Hence the average kinetic energy of a molecule of ideal gas is

$$\tfrac{1}{2}m\overline{v^2} = \tfrac{1}{2} \times 3kT = \tfrac{3}{2}kT \tag{19.25}$$

If we multiply this by the total number N of molecules, we obtain the total translational kinetic energy of all the molecules jointly. In an ideal gas, the molecules exert no forces on one another (they do not collide) and hence there is no intermolecular potential energy. The kinetic energy is then the total energy,

internal kinetic energy of monatomic gas

$$E = K = \tfrac{3}{2}NkT \tag{19.26}$$

This formula tells us how much energy is stored in the microscopic thermal motion of the gas. Since $Nk = nR$, where n is the number of moles [see Eqs. (19.7) and (19.10)], we can also write this formula for the energy as

$$E = \tfrac{3}{2}nRT \tag{19.27}$$

This energy is called the **thermal energy** of the gas. It is sometimes also called the **internal energy** of the gas, because it is stored (and hidden) in the microscopic motions in the interior of the gas, instead of being manifest in an overall macroscopic translational motion of the entire body of gas.

Note that we have assumed that the molecules behave as pointlike particles—each molecule has translational kinetic energy, but no internal energy. **Monatomic** gases, such as helium, argon, and krypton, consist of single atoms and behave in this way.

EXAMPLE 8 What is the thermal kinetic energy in 1.0 kg of helium gas at 0°C? How much extra energy must be supplied to this gas to increase its temperature to 60°C (at constant volume)?

SOLUTION: The "atomic mass" of helium is 4.0 g/mole; hence the number of moles in 1.0 kg of helium is

$$n = \frac{1.0 \text{ kg}}{4.0 \text{ g/mole}} = 250 \text{ moles}$$

According to Eq. (19.26), the energy at a temperature of 273 K is

$$\begin{aligned} E &= \tfrac{3}{2}nRT = \tfrac{3}{2} \times 250 \text{ moles} \times 8.31 \text{ J/mole·K} \times 273 \text{ K} \\ &= 8.5 \times 10^5 \text{ J} \end{aligned}$$

Since thermal energy changes in proportion to absolute temperature, the extra energy needed to increase the temperature by 60°C, or by 60 K, is

$$\begin{aligned} \Delta E &= \tfrac{3}{2}nR\Delta T = \tfrac{3}{2} \times 250 \text{ moles} \times 8.31 \text{ J/mole·K} \times 60 \text{ K} \\ &= 1.9 \times 10^5 \text{ J} \end{aligned}$$

Diatomic gases, which consist of two-atom molecules, such as N_2 and O_2, store an additional amount of energy in the internal motions of the atoms within each molecule. The molecules of these gases may be regarded as two pointlike particles rigidly connected together (a dumbbell; see Fig. 19.16). If such a molecule collides with another molecule or with the wall of the container, it will usually start rotating about its center of mass. We therefore expect that, on the average, an appreciable fraction of the energy of the gas will be in the form of this kind of rotational kinetic energy.

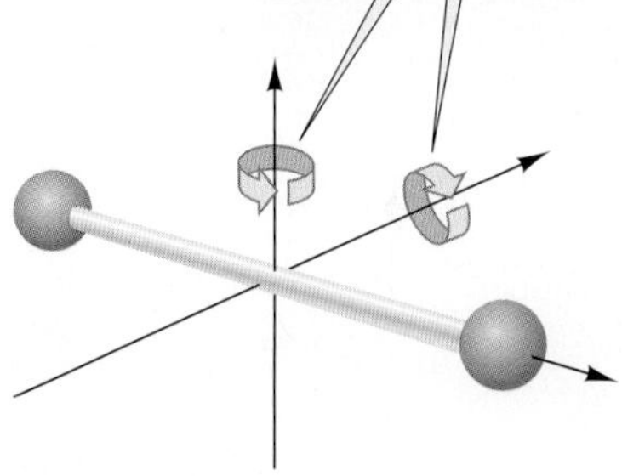

FIGURE 19.16 A diatomic molecule represented as two pointlike particles joined by a rod.

The molecule may rotate about either of the two axes through the center of mass perpendicular to the line joining the atoms (see Fig. 19.16). If the moments of inertia about these axes are I_1 and I_2, and if the corresponding angular velocities are ω_1 and ω_2, then the kinetic energy for these two rotations is [see Eq. (12.28)]

$$\tfrac{1}{2}I_1\omega_1^2 + \tfrac{1}{2}I_2\omega_2^2 \qquad (19.28)$$

The average of this kinetic energy is

$$\tfrac{1}{2}I_1\overline{\omega_1^2} + \tfrac{1}{2}I_2\overline{\omega_2^2} \qquad (19.29)$$

where, as in the preceding section, the overbars denote the average over all the molecules of the gas.

To discover the value of these average rotational energies, let us return to Eq. (19.22) and write it in terms of the x, y, and z components of velocity:

$$\tfrac{1}{2}m\overline{v_x^2} + \tfrac{1}{2}m\overline{v_y^2} + \tfrac{1}{2}m\overline{v_z^2} = \tfrac{3}{2}kT \qquad (19.30)$$

We know that the average x, y, and z speeds are equal. Hence, Eq. (19.30) asserts that the kinetic energy for each component of the motion has the value $\tfrac{1}{2}kT$:

$$\tfrac{1}{2}m\overline{v_x^2} = \tfrac{1}{2}kT, \quad \tfrac{1}{2}m\overline{v_y^2} = \tfrac{1}{2}kT, \quad \text{and} \quad \tfrac{1}{2}m\overline{v_z^2} = \tfrac{1}{2}kT \qquad (19.31)$$

It turns out that this is true not only for the components of the translational motion, but also for rotational and other motion. The general result is known as the **equipartition**

theorem. Applied to translational and rotational motion, the equipartition theorem states that

equipartition theorem

> ***Each translational or rotational component of the random thermal motion of a molecule has an average kinetic energy of $\frac{1}{2}kT$.***

We will not prove this theorem, but we will make use of it.

According to this theorem, each of the terms in Eq. (19.29) has a value $\frac{1}{2}kT$. The total average rotational energy is then kT, and when we add this to the translational kinetic energy $\frac{3}{2}kT$, we obtain the total kinetic energy for one molecule:

$$kT + \tfrac{3}{2}kT = \tfrac{5}{2}kT \tag{19.32}$$

The energy of all the molecules of the diatomic gas taken together is then

$$E = \tfrac{5}{2}NkT \tag{19.33}$$

Note that in this calculation we have ignored the possibility of rotation about the longitudinal axis of the molecule. This means we have ignored the rotation of the atoms about an axis through them, just as we have ignored this kind of rotation of the atoms in a monatomic gas. The reason why we have ignored this rotational degree of freedom for monatomic and diatomic gases has to do with their very small moment of inertia for these rotations. The full analysis lies beyond the realm of classical physics; it lies in the realm of quantum physics. There it is established that only certain energies are allowed, and that for much smaller moments of inertia, the allowed energies are much higher. It turns out that the typical available thermal energy $\frac{1}{2}kT$ is not nearly enough to attain these higher energies, and thus their contribution is "frozen out" at normally attainable temperatures. If, however, the molecule in question is **polyatomic and nonlinear** (like H_2O), then the third rotational degree of freedom would have the usual energy $\frac{1}{2}kT$ of random thermal motion, and the total rotational energy would be $\frac{3}{2}kT$, and the total energy would be $3kT$.

Similarly, we have ignored the vibrational motion of the atoms of the diatomic molecule. The interatomic forces do not really hold these atoms in a rigid embrace; rather, the forces act somewhat like springs (see Example 6 in Chapter 15), and they permit a restricted back-and-forth vibration of the atoms about their equilibrium positions. We have ignored the kinetic and potential energies associated with these vibrations because again the quantum-mechanically allowed energies are high. The vibration of atoms in a molecule can occur, however, if the temperature is rather high, 400°C or more. As we will see in Section 20.5, the energies calculated from Eqs. (19.27) and (19.33) actually agree quite well with experiments, provided we do not exceed this temperature limit.

Concepts — in — Context

EXAMPLE 9 We saw in Example 6 that a hot-air balloon achieves appreciable buoyancy when its air is heated from 273 K to 323 K. How much larger is the internal energy of the initial 9.82×10^4 moles of air when heated?

SOLUTION: Since air is 99% diatomic, it is a good approximation to treat it as a diatomic gas. Thus the energy change with temperature is given by Eq. (19.33):

$$\Delta E = \tfrac{5}{2}Nk\,\Delta T \tag{19.34}$$

We convert in the usual way from the number of molecules N to the number of moles n using [see Eqs. (19.7) and (19.10)]

$$Nk = nR \tag{19.35}$$

Substituting this and the given values into Eq. (19.34) gives

$$\Delta E = \tfrac{5}{2}nR\,\Delta T$$
$$= \tfrac{5}{2} \times 9.82 \times 10^4 \text{ moles} \times 8.31\ \text{J/mole}\cdot\text{K} \times (323\,\text{K} - 273\,\text{K})$$
$$= 1.0 \times 10^8\ \text{J}$$

This is comparable to the energy released by the combustion of 1 gallon of gasoline, 1.3×10^8 J (see Table 8.1).

COMMENT: The balloonist must supply not only this internal energy, but also the work required to push some of the expanding gas out of the balloon against the pressure of the atmosphere. We will consider some aspects of gas expansion in the next chapter. Also, after initially heating the air, the balloonist must supply some energy thereafter to maintain the temperature, since heat is conducted away through the surface of the balloon.

✔ Checkup 19.4

QUESTION 1: If we increase the temperature of a gas from 300 K to 400 K (at constant volume), by what factor do we increase the thermal energy?

QUESTION 2: Why does a diatomic gas have a higher thermal energy per mole than a monatomic gas?

QUESTION 3: In Example 8 we assumed that the gas was heated at constant volume. If, instead, we were to heat the gas at constant pressure (in a container equipped with a piston), we would have to supply *more* energy to achieve the same increase of temperature. Why?

QUESTION 4: You have one mole of each of three gases at $T = 400$ K. Place them in order of increasing total thermal energy: (1) diatomic oxygen (O_2); (2) monatomic neon (Ne); (3) polyatomic (and nonlinear) water vapor (H_2O).

(A) 1, 2, 3 (B) 1, 3, 2 (C) 2, 1, 3 (D) 2, 3, 1 (E) 3, 1, 2

SUMMARY

PROBLEM-SOLVING TECHNIQUES Ideal-Gas Law **(page 608)**

IDEAL-GAS LAW	$pV = nRT$ (n = number of moles)	**(19.2)**
	$pV = NkT$ (N = number of molecules)	**(19.9)**
UNIVERSAL GAS CONSTANT	$R = 8.31$ J/mole·K	**(19.3)**
BOLTZMANN'S CONSTANT	$k = 1.38 \times 10^{-23}$ J/K	**(19.10)**
k and R are related by Avogadro's number N_A.	$N_A k = R$	
STANDARD TEMPERATURE AND PRESSURE (STP)	$T = 273$ K	
	$p = 1\ \text{atm} = 1.01 \times 10^5$ Pa	

TEMPERATURE SCALES

Absolute (kelvin or K)	T	
Celsius (°C)	$T_C = T - 273.15$	**(19.12)**
Fahrenheit (°F)	$T_F = \frac{9}{5}T_C + 32$	

ROOT-MEAN-SQUARE SPEED

$$v_{rms} = \sqrt{\overline{v^2}} = \sqrt{\frac{3kT}{m}} \qquad \textbf{(19.23)}$$

INTERNAL KINETIC ENERGY OF IDEAL GAS

$$E = \tfrac{3}{2}NkT \qquad \text{(monatomic)} \qquad \textbf{(19.26)}$$

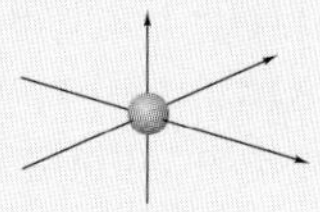

$$E = \tfrac{5}{2}NkT \qquad \text{(diatomic)} \qquad \textbf{(19.33)}$$

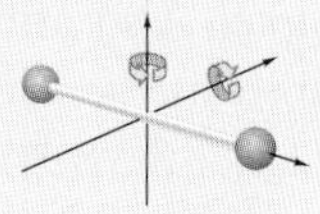

$$E = 3NkT \qquad \text{(nonlinear polyatomic)}$$

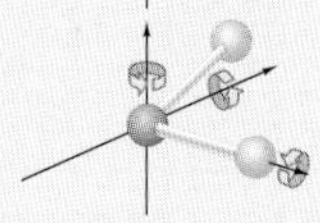

EQUIPARTITION THEOREM Each translational or rotational component of the random thermal motion of a molecule has an average kinetic energy of $\frac{1}{2}kT$.

QUESTIONS FOR DISCUSSION

1. Why do meteorologists usually measure the temperature in the shade rather than in the sun?
2. Why are there no negative temperatures on the absolute temperature scale?
3. The temperature of the ionized gas in the ionosphere of the Earth is about 2000 K, but the density of this gas is extremely low, only about 10^5 gas particles per cubic centimeter. If you were to place an ordinary mercury thermometer in the ionosphere, would it register 2000 K? Would it melt?
4. The temperature of intergalactic space is 3 K. How can empty space have temperature?
5. At the airport of La Paz, Bolivia, one of the highest in the world, pilots of aircraft find it preferable to take off early in the morning or late at night, when the air is very cold. Why?
6. If you release a rubber balloon filled with helium, it will rise to a height of a few thousand meters and then remain stationary. What determines the height reached? Is there an optimum pressure to which you should inflate the balloon to reach greatest height?
7. How can you use a barometer as an altimeter?
8. Explain why a real gas behaves like an ideal gas at low densities but not at high densities.
9. Helium and neon approach the behavior of an ideal gas more closely than do any other gases. Why would you expect this?
10. Ultrasound waves of extremely short wavelength cannot propagate in air. Why not?
11. Prove that it is impossible for all of the molecules in a gas to have the same speeds and to keep these speeds forever. (Hint: Consider an elastic collision between two molecules with the same speed. Will the speeds remain unchanged if the initial lines of motion are not parallel?)
12. If you increase the absolute temperature by a factor of 2, by what factor will you increase the average speed of the molecules of gas?
13. Air consists of a mixture of nitrogen (N_2), oxygen (O_2), and argon (Ar). Which of these molecules has the highest rms speed? The lowest?
14. Equipartition of energy applies not only to atoms and molecules, but also to macroscopic "particles" such as golf balls. If so, why do golf balls remain at rest on the ground instead of flying through the air like molecules?
15. If you open a bottle of perfume in one corner of a room, it takes a rather long time for the smell to reach the opposite corner (assuming that there are no air currents in the room). Explain why the smell spreads slowly, even though the typical speeds of perfume molecules are 300–400 m/s.

PROBLEMS

19.1 The Ideal-Gas Law†
19.2 The Temperature Scale†

1. Express the last six temperatures listed in Table 19.1 in terms of degrees Fahrenheit.
2. The hottest place on the Earth is Al-'Aziziyah, Libya, where the temperature has soared to 136.4°F. The coldest place is Vostok, Antarctica, where the temperature has plunged to −126.9°F. Express these temperatures in degrees Celsius and in kelvins.
3. A paper clip has a mass of 0.50 g. The paper clip is made of iron. How many atoms are in this paper clip?
4. What is the number of sodium and of chlorine atoms in one spoonful (10 g) of salt, NaCl?
5. Assume that air is 76% nitrogen and 24% oxygen by mass. What is the percent composition of air by number of molecules?
6. At an altitude of 160 km, the density of air is 1.5×10^{-9} kg/m^3 and the temperature is approximately 500 K. What is the pressure?
7. How much does the frequency of middle C (see Table 17.1) played on the flute change when the air temperature drops from 20°C to −10°C? [Hint: The speed of sound in air is given by Eq. (17.5).]
8. In the Middle Ages, physicians applied suction cups to the skin, to draw out "bad humors." The cups produced "suction" by means of hot air. Suppose that a hot suction cup, at a temperature of 85°C, is applied to the skin and its rim makes an airtight seal against the skin. The cup initially contains air at 80°C and at atmospheric pressure, 1.0 atm. What will be the underpressure generated in the cup (that is, the difference between the pressure in the cup and atmospheric pressure) when the cup and the air trapped inside it cool from 85°C to 30°C?
9. In summer when the temperature is 30°C, the overpressure within an automobile tire is 2.2 atm. What will be the overpressure within this tire in winter when the temperature is 0°C? Assume that no air is added to the tire and that no air leaks from the tire; assume that the volume of the tire remains constant and that the atmospheric pressure remains at 1.0 atm.
10. What is the number of oxygen molecules in 1.00 cm^3 of air at 273 K and 1.00 atm? Nitrogen molecules? What is the number of atoms?
11. A tank of a volume of 1.0 liter contains 1.0 g of nitrogen gas at 290 K. Another tank of equal volume at equal temperature contains 1.0 g of oxygen gas.
    (a) What is the pressure in each tank?
    (b) If we pump the oxygen gas into the nitrogen tank, what is the pressure produced by the mixture of the two gases? Assume that the temperature remains constant at 290 K.
12. Estimate the average distance between molecules of air at STP.
13. Repeat the calculation of Example 3 assuming that, because of the increase of pressure, the volume of the tire increases by 5%.
14. The storage tank of a small air compressor holds 0.30 m^3 of air at a pressure of 5.0 atm and a temperature of 20°C. How many moles of air is this?
15. On a warm day, the outdoor temperature is 35°C and the indoor temperature in an air-conditioned house is 21°C. What is the difference between the densities of the air outdoors and indoors? Assume the pressure is 1.00 atm.
16. The lowest pressure attained in a "vacuum" in a laboratory on the Earth is 1.0×10^{-16} atm. Assuming a temperature of 20°C, what is the number of molecules per cubic centimeter in this vacuum?
17. Clouds of interstellar hydrogen gas have densities of up to 1.0×10^{10} atoms/m^3 and temperatures of up to 1.0×10^4 K. What is the pressure in such a cloud?
18. The following table gives the pressure and density of the Earth's upper atmosphere as a function of altitude:

ALTITUDE	PRESSURE	DENSITY
20000 m	5600 Pa	9.2×10^{-2} kg/m^3
40000	320	4.3×10^{-3}
60000	28	3.8×10^{-4}
80000	1.3	2.5×10^{-5}

Calculate the temperature at each altitude. The mean molecular mass for air is 29.0 g.

19. What is the density (in kilograms per cubic meter) of helium gas at 1.0 atm at the temperature of boiling helium liquid (see Table 19.1)?
20. The volume of an automobile tire is 2.5×10^{-2} m^3. The pressure of the air in this tire is 3.0 atm and the temperature is 17°C. What is the mass of air? The mean molecular mass of air is 29.0 g.
21. When a bicycle tire is filled early in the morning at 22°C, the overpressure is 4.0 atm. What is the overpressure later in the day, when the temperature is 38°C? Assume that the atmospheric pressure of 1.0 atm is constant.
22. For gas storage, thick metal cylinders (Fig. 19.17) with an internal volume of 35 liters are used; typical cylinders can safely maintain a pressure of 180 atm. At 25°C, how many moles of gas can such a cylinder hold?

†For help, see Online Concept Tutorial 21 at www.wwnorton.com/physics

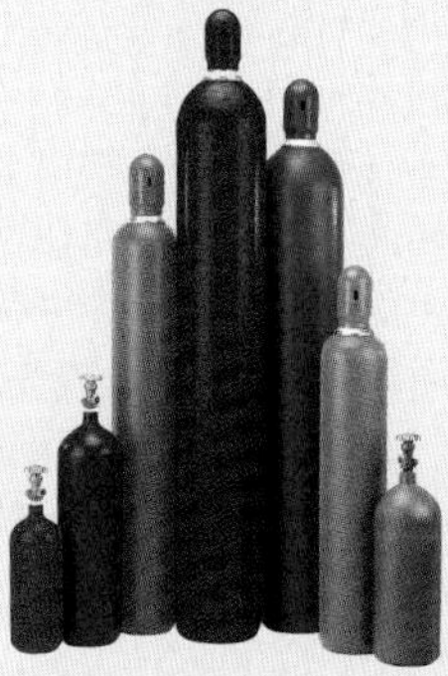

FIGURE 19.17 Gas cylinders.

23. Gas-storage cylinders that can maintain 400 atm pressure are commercially available. Treat nitrogen gas at such a high pressure approximately as an ideal gas. How does its mass density compare with that of liquid nitrogen? (Liquid nitrogen has a density of 800 kg/m^3.)
24. A tube of argon gas at STP is sealed and placed in an oven at 850°C. What is the pressure in the gas at that temperature?
25. Carbon dioxide can be pressurized to 56 atm in the gaseous state at 25°C. What is the mass density of such high-pressure carbon dioxide vapor?

*26. Suppose you pour 10 g of water into a 1.0-liter jar and seal it tightly. You then place the jar into an oven and heat it to 500°C (a dangerous thing to do!). What will be the pressure of the vaporized water?

*27. A scuba diver releases an air bubble of diameter 1.0 cm at a depth of 15 m below the surface of a lake. What will be the diameter of this bubble when it reaches the surface? Assume that the temperature of the bubble remains constant.

*28. The helium atom has a volume of about 3.0×10^{-30} m^3. What fraction of a volume of helium gas at STP is actually occupied by atoms?

29. A carbon dioxide (CO_2) fire extinguisher has an interior volume of 2.8×10^{-3} m^3. The extinguisher has a mass of 5.9 kg when empty and a mass of 8.2 kg when fully loaded with CO_2. At a temperature of 20°C, what is the pressure of CO_2 in the extinguisher?

*30. (a) When you heat the air in a house, some air escapes because the pressure inside the house must remain the same as the pressure outside. Suppose you heat the air from 10°C to 30°C. What fraction of the mass of air originally inside will escape?
(b) If the house were completely airtight, the pressure would have to increase as you heated the house. Suppose that the initial pressure inside the house is 1.00 atm. What is the final pressure? What force does this excess inside pressure exert on a window 1.0 m high and 1.0 m wide? Do you think the window can withstand this force?

*31. During the volcanic eruption of Mt. Pelée on the island of Martinique in 1902, a *nuée ardente* (burning cloud) of very hot gas and fine suspended ash rolled down the side of the volcano and killed the 30000 inhabitants of Saint-Pierre. The temperature in the cloud has been estimated at 700°C. Treat this cloud as a gas of high molecular mass. What must have been this molecular mass to make this cloud as dense as, or denser than, the surrounding air (at 20°C)?

*32. A typical hot-air balloon has a volume of 2200 m^3 and a mass of 730 kg (including balloon, gondola, four passengers, and a propane tank). Since the balloon is open at the bottom, the pressures of the internal and the external air are (approximately) equal. If the temperature of the external air is 20°C, what must be the minimum temperature of the internal air in the balloon to achieve liftoff? The density of the external air is 1.20 kg/m^3.

*33. A research balloon (Fig. 19.18) ascends to an altitude of 40 km and floats in equilibrium. The pressure (outside and also inside the balloon) is 3.2×10^2 N/m^2, and the temperature is -13°C. The volume of the balloon is 8.5×10^5 m^3, and it is filled with helium. What payload (including the mass of the fabric but excluding the helium) can this balloon carry? What was the volume of the balloon on the ground (at STP), before it was released?

FIGURE 19.18 A research balloon just after launch.

*34. A sunken ship of steel is to be raised by making the upper part of the hull airtight and then pumping compressed air into it while letting the water escape through holes in the bottom. The mass of the ship is 50000 metric tons, and it is at a depth of 60 m. How much compressed air (in kilograms) must be pumped into the ship? The temperature of the air and the water is 15°C.

*35. A diving bell is a cylinder closed at the top and open at the bottom; when it is immersed in the water, any air initially in the cylinder remains trapped in the cylinder. Suppose that such a diving bell, 2.0 m high and 1.5 m across, is initially full of air and is immersed to a depth of 15 m measured from water level to water level (see Fig. 19.19).

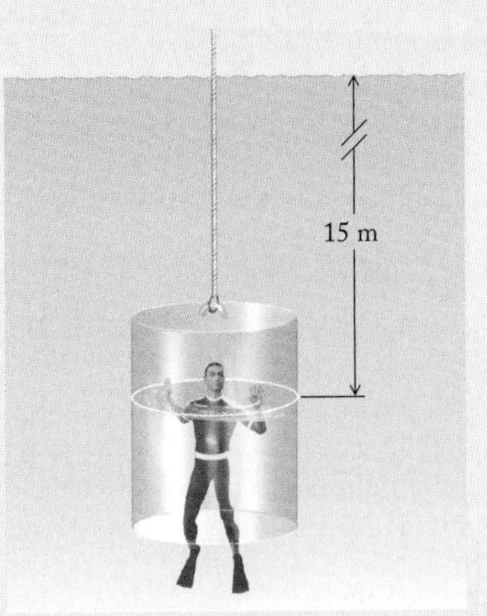

FIGURE 19.19 Submerged diving bell.

(a) How high will the water have risen within the diving bell?

(b) If compressed air is pumped into the bell, water will be expelled from the bell. How much air (in kilograms) must be pumped into the bell, and at what pressure, to get rid of all of the water? Assume that the temperature of the air is 15°C.

*36. At high altitudes, pilots and mountain climbers must breathe an enriched mixture containing more oxygen than the standard concentration of 23% found in ordinary air at sea level. At an altitude of 11000 m, the atmospheric pressure is 0.24 atm. What oxygen concentration is required at this altitude if with each breath the same number of oxygen molecules is to enter the lungs as for ordinary air at sea level?

*37. Air is 75.54% nitrogen (N_2), 23.1% oxygen (O_2), and 1.3% argon (Ar) by mass. From this information and from the molecular masses of N_2, O_2, and Ar, deduce the mean molecular mass of air.

*38. (a) The gas at the center of the Sun is 38% hydrogen and 62% helium at a temperature of 1.50×10^7 K and a density of 1.48×10^5 kg/m^3. What is the pressure?

(b) The gas at a distance of 20% of the solar radius from the center of the Sun is 71% hydrogen and 29% helium at a temperature of 9.0×10^6 K and a density of 3.6×10^4 kg/m^3. What is the pressure?

*39. The pressure change with height given in Eq. (18.12), $p - p_0 = -\rho g y$, is not valid for a compressible fluid, like an ideal gas, except for small changes dp and dy in p and y. Show that at constant temperature, this and the Ideal-Gas Law lead to the relation

$$\frac{dp}{p} = -\frac{Mg}{RT}dy$$

where M is the mass of one mole of the gas.

**40. Show that if the temperature in the atmosphere is independent of altitude, then the pressure as a function of altitude y is

$$p = p_0 e^{-mgy/kT}$$

where m is the average mass per molecule of air. See Problem 39. (This formula is applicable only for altitudes less than about 2 km; higher up, the temperature depends on the altitude.)

19.3 Kinetic Pressure

41. What is the rms speed of molecules of water vapor in air at 0°C?

42. What is the rms speed of hydrogen ions on the surface of the Sun, where the temperature is 4.5×10^3 K? At the center of the Sun, where the temperature is 1.5×10^7 K?

43. In Example 7 we calculated the rms speed of nitrogen and of oxygen molecules in air at 0°C. If we want to increase these rms speeds by a factor of 2, what temperature do we need?

44. Consider separate samples of nitrogen gas and oxygen gas. The temperature of the nitrogen gas is 20°C. What must be the temperature of the oxygen gas if the rms speed of the oxygen molecules is to equal the rms speed of the nitrogen molecules?

45. Free neutrons in the core of a nuclear reactor have a temperature of 400 K. What are the rms speed and the average kinetic energy of such neutrons?

46. The fireball of a 1.0-megaton nuclear explosion attains a temperature of 7000 K. It contains ionized gas and free electrons. What are the rms speed and the average kinetic energy of free electrons in the fireball? What are the rms speed and the average kinetic energy of nitrogen ions?

47. In order to achieve a novel state of matter known as a **Bose–Einstein condensate**, a gas of rubidium atoms is cooled to $T = 5.0 \times 10^{-5}$ K. What is the rms speed of a rubidium atom at this temperature?

48. According to Eq. (17.5) the speed of sound in air is $\sqrt{1.4p/\rho}$.

(a) Show by means of the Ideal-Gas Law that this expression equals $\sqrt{1.4kT/m}$, where m is the average mass per molecule of air.

(b) Show that, in terms of the rms speed, the latter expression equals $\sqrt{1.4/3}v_{rms}$, or $0.68v_{rms}$.

(c) Calculate the speed of sound in air at 0°C, 10°C, 20°C, and 30°C.

49. What is the average kinetic energy of an oxygen molecule in air at STP? A nitrogen molecule?

50. What is the rms speed of a helium atom at 0°C? At −269°C?

51. At the top of the stratosphere, at an altitude of 30 km, the temperature is −38°C. What is the rms speed of an oxygen molecule at −38°C? Of an ozone (O_3) molecule? What are the average kinetic energies for these molecules?

52. The rms speed of nitrogen molecules in air at some temperature is 493 m/s. What is the rms speed of hydrogen molecules in air at the same temperature?

53. One method for the separation of the rare isotope ^{235}U (used in nuclear bombs and reactors) from the abundant isotope ^{238}U relies on diffusion through porous membranes. Both isotopes are first made into a gas of uranium hexafluoride (UF_6). The molecules of $^{235}UF_6$ have a higher rms speed and they

will diffuse faster through a porous membrane than the molecules of $^{238}UF_6$. The "molecular masses" of $^{235}UF_6$ and $^{238}UF_6$ are 349 g and 352 g, respectively. What is the percent difference between their rms speeds at a given temperature?

54. Calculate the rms velocity of molecules of oxygen gas (O_2) and hydrogen gas (H_2) at room temperature; also, find the ratio of these two velocities.

55. By shaking containers of small grains or beads, some fluidlike behavior can be observed. If an evacuated container with a low density of small spheres of mass 1.0×10^{-9} g is shaken so the beads have an rms velocity of 2.0×10^{-2} m/s, what is the effective ideal-gas temperature of such a system?

56. Helium liquid at temperature 0.90 K (below its standard boiling point, achieved by **evaporative cooling**) is in equilibrium with 0.042 torr of helium gas pressure. What is the rms velocity of helium gas molecules at this temperature?

57. Using laser beams, physicists can cool a small amount of a gas of sodium atoms to extremely low temperature. Determine the rms speed of a sodium atom when such **laser cooling** results in a temperature of $T = 2.0 \times 10^{-4}$ K.

*58. For semiconductor fabrication, **ultrahigh vacuum** (UHV) is often needed to prevent surface contamination (Fig. 19.20). The time t for a surface to become appreciably contaminated by gas particles is inversely proportional to the particle flux $(N/V)v_{\mathrm{rms}}/(2\sqrt{3})$ and to the surface area per particle a^2:

$$t = \frac{2\sqrt{3}}{(N/V)v_{\mathrm{rms}}a^2}$$

where typically $a^2 = (3.0 \times 10^{-10}\ \mathrm{m})^2 = 9.0 \times 10^{-20}\ \mathrm{m}^2$. For room temperature, calculate this contamination time for hydrogen gas (a) at 1.0 atm, (b) at an "ordinary" vacuum pressure of 1.0×10^{-6} torr, and (c) at a UHV pressure of 1.0×10^{-11} torr.

FIGURE 19.20 An ultrahigh-vacuum (UHV) chamber.

59. Real gas molecules experience collisions with each other; the number of such collisions depends on the size of the gas molecule and the density N/V of the gas. The average distance the molecule travels before suffering one collision is called the **mean free path, denoted l. If we assume the molecules are spheres of radius R_0, a collision occurs when two molecules come within a distance $2R_0$ of each other. By considering the volume swept out by a moving molecule, show that the mean free path is given by

$$l = \frac{1}{4\pi R_0^2(N/V)}$$

(Hint: For one collision, equate the effective cylindrical volume of radius $2R_0$ swept out by one molecule with the average volume per molecule.)

*60. The average distance an atom in a fluid travels before suffering a collision with another atom is called the **mean free path** l, given by (see Problem 59)

$$l = \frac{1}{4\pi R_0^2(N/V)}$$

where R_0 is the radius of the (spherical) atom.
(a) Calculate the mean free path of a helium atom in helium gas with $R_0 = 1.3 \times 10^{-10}$ m under STP conditions.
(b) On the average, how many collisions does the atom make per second?
(c) How many collisions do all the atoms in 1.0 cm^3 of helium gas make per second?

*61. The behavior of a real gas deviates from the Ideal-Gas Law, particularly at high density. There are two primary effects: the nonzero size of molecules decreases the available volume, and the long-range attractive forces of molecules decrease the momentum transfer to container walls. These effects are included in the **van der Waals equation of state,**

$$\left[p + A\left(\frac{N}{V}\right)^2\right](V - NV_m) = nRT$$

where A is a constant that depends on the attractive interaction, N and V are the total number of molecules and the total volume, and V_m is an effective volume of a molecule. Neglecting the pressure effect (set $A = 0$), calculate the percentage volume correction for helium atoms with $V_m = 3.7 \times 10^{-29}$ m^3 at room temperature for (a) $p = 1.0$ atm and (b) $p = 1000$ atm.

*62. In our calculation of the pressure on the walls of a box (see Section 19.3) we have ignored gravity. If we take gravity into account, the pressure on the bottom of the box will be greater than that at the top. Show that the pressure difference is $p - p_0 = (N/V)mgL$. (Hint: When a molecule falls from the top to the bottom, its speed increases according to $v^2 - v_0^2 = 2gL$.)

19.4 The Energy of an Ideal Gas

63. What is the thermal kinetic energy in 1.0 kg of oxygen gas at 20°C? What fraction of this energy is translational?

64. Assume that air consists of the diatomic gases O_2 and N_2. How much must we increase the thermal energy of 1.0 kg of air in order to increase its temperature by 1.0°C?

65. What is the thermal kinetic energy of 1.0 mole of helium gas at 300 K? How much does this kinetic energy increase if we increase the temperature by 20 K? If we increase the pressure by 3.0 atm (at fixed temperature)?

66. A 1.0-liter vessel contains a monatomic gas under STP conditions. If 50 J of energy is added to the gas (at constant volume), what is the new temperature of the gas? The new pressure?

67. A gas cylinder contains 30 liters of diatomic nitrogen gas at 273 K and a pressure of 140 atm. If the temperature is increased to 300 K, how much will the internal energy of the gas increase?

*68. Water vapor consists of nonlinear polyatomic molecules, so all three rotational modes are excited at ordinary temperatures. What is the thermal kinetic energy of 1.0 mole of water vapor at 100°C?

*69. The two vibrational degrees of freedom (one kinetic, one potential) of diatomic hydrogen gas can be excited at very high temperatures (thousands of kelvins), resulting in seven total degrees of freedom. For one mole of hydrogen gas at such high temperatures, how much energy must be added to increase the temperature by 10 K?

*70. A container is divided into two equal compartments by a partition. One compartment is initially filled with helium at a temperature of 250 K; the other is filled with oxygen at a temperature of 310 K. Both gases are at the same pressure. If we remove the partition and allow the gases to mix, what will be their final temperature?

REVIEW PROBLEMS

71 What is the molecular mass of methanol, CH_3OH? What is the number of molecules in 1.0 kg of methanol?

72. A bicycle pump is a cylinder of diameter 2.5 cm and length 30 cm. Initially, when the piston is fully pulled out, the cylinder is filled with air at 25°C and 1.0 atm. If you slowly compress the air to half its initial volume, what force must you exert on the piston to hold it in the compressed position? Assume that the temperature of the air remains constant and that the valve is blocked, so no air escapes.

73. An airplane flies through air at a temperature of 5°C. The lift force generated by the flow of air over the wings is 1.2×10^3 N. What would be the lift force if the airplane were flying through air at a temperature of 35°C, other conditions remaining equal?

74. An oxygen cylinder for medical use contains oxygen at a pressure of 140 atm, at room temperature (20°C). The cylinder measures 20 cm in diameter and 110 cm in length. How many kilograms of oxygen does this cylinder contain? What will be the volume of the oxygen if it is allowed to expand slowly to normal atmospheric pressure, at room temperature?

75. The volume of air in the fully expanded human lungs is 5.0 liters. How many molecules are in the lungs? How many molecules of oxygen and how many of nitrogen? Assume that the air is 76% nitrogen and 24% oxygen by mass, at a temperature of 37°C and a pressure of 1.0 atm.

76. An inflatable life jacket, loosely inflated, provide a buoyant force of 50 N when completely immersed just below the surface of a lake. What buoyant force does this life jacket provide if you push it to a depth of 2.0 m below the surface? Assume that the initial pressure in the life jacket is 1.0 atm, and that the temperature of the air does not change when you push the life jacket down.

77. At the center of the Sun, the temperature is 1.5×10^7 K and the density is 1.5×10^5 kg/m^3.

(a) Assume that the material in the Sun is a mixture of equal numbers of hydrogen ions and free electrons. Find the number of particles per unit volume, and find the pressure.

(b) What would you have found for the pressure if you had assumed that the material of the Sun consists of hydrogen atoms (without free electrons)?

(c) What if you had assumed that the material in the Sun consists of H_2 molecules?

78. In a mercury barometer (see Fig. 19.21), the space at the upper end of the tube is supposed to be evacuated. Suppose that by some mistake, a barometer has some small amount of air in this space. A barometer with this defect always underreads the atmospheric pressure.

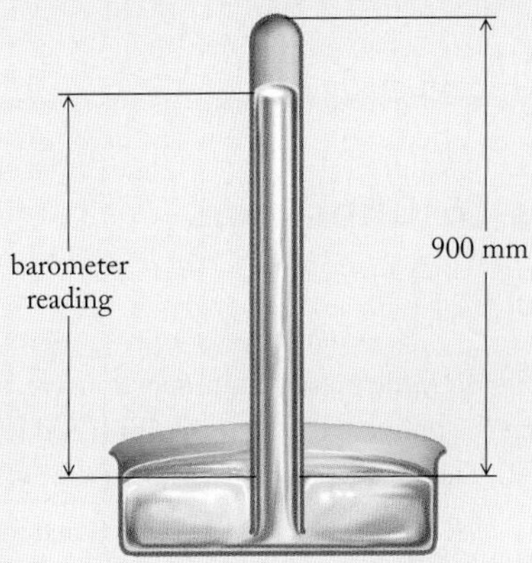

FIGURE 19.21 A mercury barometer.

(a) Suppose that when the actual atmospheric pressure is 760 mm-Hg, the barometer reads 750 mm-Hg. What will this barometer read when the actual pressure increases to 780 mm-Hg? The length of the barometer tube is 900 mm (see Fig. 19.21), and the temperature of the air remains constant.

(b) Suppose that while the pressure increases as in part (a), the temperature decreases from 300 K to 270 K. What will the barometer read in this case?

79. In one method for the determination of the average density of the tissues of the human body, the subject is locked in a hermetic chamber of known volume V_c (see Fig. 19.22) containing an unknown volume V of air at an initial pressure p. Then, by means of a small piston, the volume of this chamber is reduced by an amount ΔV. This causes an increase of pressure Δp (at constant temperature). Show that the volume of *air* in the chamber is approximately given by

$$V = \frac{p}{\Delta p}\Delta V$$

and that therefore the volume of the body of the subject is given by

$$V_s = V_c - \frac{p}{\Delta p}\Delta V$$

[Hint: $pV =$ [constant], and hence $pV = (p + \Delta p)(V - \Delta V)$.]

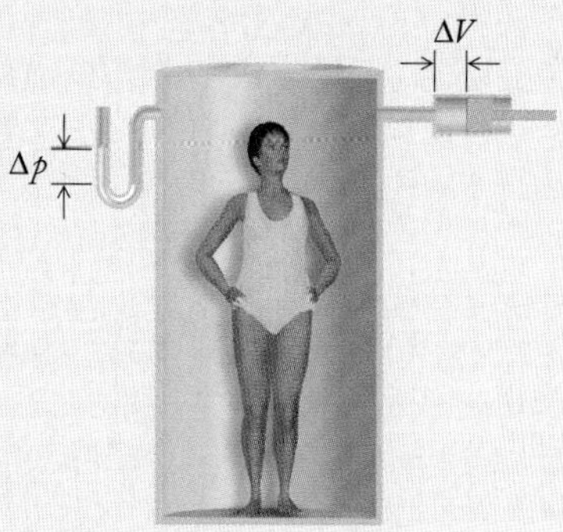

FIGURE 19.22 A volunteer locked in a hermetic chamber.

80. Estimate the number of impacts of air molecules per second on the palm of your hand (area 80 cm^2). Assume that the air is at 20°C and 1.0 atm, and assume that it consists entirely of nitrogen molecules.

81. What is the rms speed of molecules of water vapor in air at 0°C?

82. At a time of 100000 years after the first instant of the Big Bang at the beginning of the Universe, the temperature was about 1.0×10^4 K, and the density of the hot gas filling the Universe was about 1.0×10^{-16} kg/m^3. By mass, this gas consisted of 75% H atoms (not molecules) and 25% He atoms.

(a) Calculate the number of atoms per unit volume.

(b) Calculate the pressure contributed by these atoms.

(c) Calculate the rms speed of the H atoms and of the He atoms.

83. A sample of gas has some initial pressure p, volume V, and temperature T. By what factor does the rms speed of its molecules increase or decrease if we do one of the following:

(a) Increase the temperature to $2T$?

(b) Decrease the volume to $V/3$?

(c) Increase the volume to $4V$ and simultaneously decrease the pressure to $p/2$?

(d) Decrease the temperature to $T/4$ and simultaneously decrease the pressure to $p/3$?

84. At the center of the Sun, the temperature is 1.5×10^7 K. What are the rms speed and the average kinetic energy of hydrogen ions? What are the rms speed and the average kinetic energy of helium ions?

85. Two moles of H_2 gas react with 1.0 mole of O_2 gas to form 2.0 moles of water vapor. If the initial temperature of the H_2 and O_2 gases is 300 K, what final temperature must we give to the water vapor, if its thermal kinetic energy is to equal the initial kinetic energy?

Answers to Checkups

Checkup 19.1

1. Since $pV = nRT$, the pressure is proportional to the temperature; when the temperature increases from 300 K to 600 K (a factor of 2), the pressure also increases a factor of 2.

2. Since $pV = nRT$, the pressure p is proportional to the density n/V. So when the pressure increases a factor of 1.05, the density also increases a factor of 1.05.

3. Since $pV = nRT$, or $p = (n/V)RT$, when the temperature increases at constant pressure, the density must decrease by the same factor, here by a factor of 270/290.

4. Since $pV = nRT$, the density (n/V) is proportional to the absolute pressure. For an *overpressure* of 2 atm, the absolute pressure is 3 atm, so the density is a factor of 3 larger inside the tire than in the surrounding air.

5. (E) $\frac{1}{3}$. Since the Ideal-Gas Law, $pV = nRT$, says that volume is directly proportional to temperature and inversely proportional to pressure, an increase in pressure and a decrease in temperature will each cause the volume to decrease by the corresponding factor, so overall it will decrease by $\frac{2}{3} \times \frac{1}{2} = \frac{1}{3}$.

Checkup 19.2

1. Since 0°C corresponds to 32°F, any negative Fahrenheit temperature will be far below this and will always correspond to a negative Celsius temperature. But temperatures slightly below 0°C will be negative on the Celsius scale and positive on the Fahrenheit scale, so the answer to the second question is no.
2. To determine which are the extreme temperatures, convert each to a common unit. In degrees Celsius, the first is $T_C = 320\text{ K} - 273.15°\text{C} \approx 47°\text{C}$. From Fig. 19.13, we see that the last temperature is 90°F ≈ 32°C. Thus the first temperature (320 K) is the hottest, and the second temperature (20°C) is the coldest.
3. Since the thermometer is designed to maintain constant pressure, the volume of the gas would be proportional to temperature ($pV = nRT$). Thus the temperature could be determined by measuring the height of the piston, which determines the volume of the gas.
4. (C) −40°C. From Fig. 19.13, we see that the numerical values on the two scales coincide only at −40°C, a result that may also be obtained by equating the numerical values of T_C and T_F in Eq. (19.12).

Checkup 19.3

1. Both. For any given molecule, the root-mean-square speed will increase with temperature [Eq. (19.23)]. Thus it will make the round trip between impacts more quickly, increasing the rate of impacts in proportion to the speed. The transferred momentum per impact, $2mv$, will also increase in proportion to the speed. Thus the two effects contribute equally.
2. The rms speed is proportional to the square root of the temperature [Eq. (19.23)], so an increase in T from 300 K to 600 K (a factor of 2) increases v_{rms} a factor of $\sqrt{2}$.
3. The "molecular mass" of oxygen molecules, O_2, is $2 \times 16 = 32$, which is larger than that of water, H_2O, which is $(2 \times 1) + 16 = 18$. From Eq. (19.23), $v_{rms} = \sqrt{3\,kT/m}$, we see that the lighter molecule, water, will have the higher rms speed.
4. (D) $\frac{1}{4}$. The Ideal-Gas Law, $p = (N/V)kT$, indicates that for the same pressure, the number density (N/V) will change inversely with temperature; thus N/V will be $200/50 = 4$ times larger for hydrogen gas. But the mass per unit volume is the number density times the mass per molecule; the latter is $m_{H_2}/m_{O_2} = 2/32 = 1/16$ times as large for hydrogen. Thus the mass per unit volume ratio will be $4 \times (1/16) = 1/4$.

Checkup 19.4

1. The thermal energy [see Eq. (19.27) or (19.33)] is proportional to temperature, so an increase to 400 K from 300 K increases the thermal energy a factor of 4/3.
2. The diatomic molecule has two rotational components of motion that can store energy at ordinary temperatures, in addition to the usual three translational components of motion.
3. As the gas expanded at constant pressure, it would apply a force to the moving piston, thus doing work.
4. (C) 2, 1, 3. The total energy is lowest for the monatomic gas (2), which has only translational kinetic energy, $\frac{3}{2}nRT$. With two rotational components of motion, the diatomic gas (1) has the next higher energy, $\frac{5}{2}nRT$. Finally, a nonlinear, polyatomic gas (3) has three rotational components of motion and thus the highest total energy, $3nRT$.

CHAPTER

20 Heat

CONCEPTS IN CONTEXT

Concepts *in* Context

This image of the intensity of infrared energy emitted by a house is known as a *thermograph*; the colors indicate the different levels of heat loss through various parts of the building.

In this chapter, we will consider questions such as

? At what rate does thermal energy flow through a brick wall? (Example 6, page 639)

? How well does additional insulation reduce the flow of thermal energy through the wall? (Example 7, page 640)

? The flow of thermal energy through a window can be reduced by using two layers of glass with a layer of gas between them. What gas is best? (Checkup 20.5, Question 2, page 647)

In everyday language, heat is what makes things hot. When we place a kettle full of water on a stove, the water absorbs heat from the stove and becomes hot. But in the precise language of physics, what makes the water hot is **thermal energy**, *that is, the kinetic and potential energy of the random microscopic motions of molecules, atoms, ions, electrons, and other particles.* When the water is in contact with the hot stove, the atoms of the stove communicate some of their violent random microscopic motions to the water molecules. Thus, the thermal energy of the water molecules increases—they bounce around more violently than before. At the macroscopic level, such an increase of the energy or the random microscopic motions manifests itself as an increase of the temperature of the water.

In the language of physics, **heat** *is thermal energy transferred from a hotter body to a colder body.* The relationship of heat to thermal energy is analogous to the relationship of work to mechanical energy we studied in Chapter 7. Work done on a particle increases the mechanical energy of the particle. Thus, work is mechanical energy transferred by a force. Likewise, heat is thermal energy transferred by a temperature difference. This analogy between heat and work is not merely formal. In fact, heat can be regarded as microscopic work done by the particles in the hotter body on the particles in the colder body, and this microscopic work accomplishes the transfer of thermal energy. Although in a strict sense heat is a transfer of thermal energy, physicists sometimes use the word *heat* in a loose sense as a synonym for *thermal energy*. Thus, we speak of heat flow, heat storage, heat loss, etc., when there is a flow of thermal energy, storage of thermal energy, loss of thermal energy, etc.

The ambiguity in the usage of the word *heat* arises from historical roots. Until well into the nineteenth century, scientists did not have a clear understanding of the concept of energy, and they thought that heat was an invisible, weightless fluid, which they called "caloric." The first experiments to give conclusive evidence of the nature of heat were performed by Benjamin Thompson, Count Rumford, who showed that the mechanical energy lost in friction is converted into heat. You can verify such a frictional conversion of mechanical energy into heat by rubbing your hands against each other—a few seconds of rubbing produces a noticeable warming.

In this chapter, we will examine various effects caused by the application of heat in materials: increase of temperature, expansion of length or volume, thermal conduction, melting, and vaporization.

BENJAMIN THOMPSON, COUNT RUMFORD (1753–1814)
American–British scientist. On the basis of experimental observations that he collected while supervising the boring of cannon, Rumford argued against the prevailing view that heat is a substance, and he proposed that heat is nothing but the random microscopic motion of the particles within a body. **Robert von Mayer** *(1814–1878), German physician and physicist, calculated the mechanical equivalent of heat by comparing the work done on a gas during compression with the consequent increase of temperature. Finally,* **J. P. Joule** *(1818–1889) measured this quantity directly by means of his famous experiment.*

20.1 HEAT AS A FORM OF ENERGY TRANSFER

Online Concept 22 Tutorial

We examined the connection between random microscopic motion and temperature in the preceding chapter, where we saw that the increase of the kinetic energy of the random microscopic motions of the molecules in a gas is directly proportional to the increase of temperature [see Eq. (19.26)]. In a liquid or a solid, the kinetic energy of the random microscopic motions also increases with the temperature. Furthermore, the atoms and molecules in a liquid or a solid have potential energies associated with the forces they exert on one another; these potential energies also increase with temperature. Thus, the microscopic view of thermal energy as kinetic and potential energy of the random motions of atoms and molecules agrees with the intuitive notion that absorption of thermal energy should lead to an increase of temperature.

Long before physicists recognized that heat is the transfer of kinetic and potential energy of the random microscopic motion of atoms, they had defined heat in terms of

calorie (cal)

the temperature changes it produces in a body. A traditional, but non-SI, unit of heat is the **calorie** (cal), which was originally defined as the amount of heat needed to raise the temperature of 1 g of water by 1°C The kilocalorie is 1000 cal,

$$1 \text{ kcal} = 1000 \text{ cal} \tag{20.1}$$

Incidentally: The "calories" marked on some packages of food in grocery stores are actually kilocalories, also called large calories. Sometimes this is made more explicit by use of a capital letter: 1 Cal = 1 kcal.

In the British system of units, the unit of heat is the **British thermal unit** (Btu), which is the heat needed to raise the temperature of 1 lb of water by 1°F The relationship between this unit and the kcal is

British thermal unit (Btu)

$$1 \text{ Btu} = 0.252 \text{ kcal}$$

specific heat

The heat necessary to raise the temperature of 1 kg *of a material by* 1°C *is called the* **specific heat capacity**, *or the* **specific heat**, usually designated by the symbol c. Thus, water has a specific heat of

specific heat of water

$$c = 1.00 \frac{\text{kcal}}{\text{kg}\cdot{}^\circ\text{C}} \tag{20.2}$$

Table 20.1 lists the specific heats of some common substances. Note that water has a larger specific heat than all the other substances listed in Table 20.1. This means that, per kilogram, a temperature change in water requires more heat than an equal temperature change in these other substances. We might say that water has a large "thermal inertia"—it is capable of storing a large thermal energy with a small change of temperature. This makes water very useful for the storage and transport of thermal energy, for instance, in the heating system of a house (where water carries thermal energy from the boiler to the radiators) and in the cooling system of an automobile engine (where water carries thermal energy from the engine block to the radiator).

The specific heat of most substances varies slightly with temperature. For example, the specific heat of water varies by about 1% between 0°C and 100°C, reaching a minimum at 35°C.[1] Finally, the specific heat depends somewhat on the pressure to which the material is subjected during the heating. All the values listed in Table 20.1 were obtained at room temperature (20°C) and at a constant pressure of 1.0 atm.

The values in Table 20.1 give the amount of heat required to increase the temperature of 1 kg of a given substance by 1°C. For a mass m of this substance, the

TABLE 20.1 SOME SPECIFIC HEATS[a]

SUBSTANCE	c	
Aluminum	0.215 kcal/kg·°C	902 J/kg·°C
Brass	0.092	390
Copper	0.092	390
Iron, steel	0.106	445
Lead	0.031	130
Tin	0.054	230
Silver	0.056	240
Mercury	0.033	140
Water	1.000	4187
Seawater	0.93	3900
Ice, −10°C	0.530	2230
Ethyl alcohol	0.581	2430
Glycol	0.571	2390
Mineral oil	0.5	2000
Glass, thermometer	0.20	840
Marble	0.21	880
Granite	0.19	800

[a] At room temperature (20°C) and 1 atm, unless otherwise noted.

[1] This variation must be taken into acount for precise definitions: a calorie is the heat needed to raise the temperature of 1 g of water from 14.5°C to 15.5°C; a British thermal unit is the heat necessary to raise the temperature of 1 lb of water from 63°F to 64°F.

amount of heat Q and the increase of temperature ΔT are related by

$$Q = mc\Delta T \qquad (20.3)$$

heat and temperature change

This merely says that a large mass or a large temperature change requires more heat, in proportion to the mass or the temperature change.

EXAMPLE 1 You pour 0.10 kg of water at 20°C into an aluminum pot of 0.20 kg at the same temperature. How much heat must you supply to bring the water and the pot to a temperature of 100°C? (Neglect any heating of the environment.)

SOLUTION: The temperature change is $\Delta T = 80°\text{C}$. Hence the heat absorbed by the water is

$$Q_{\text{water}} = m_{\text{water}}\, c_{\text{water}}\, \Delta T = 0.10 \text{ kg} \times 1.00 \text{ kcal/(kg·°C)} \times 80°\text{C} = 8.0 \text{ kcal}$$

The specific heat of aluminum is $c_{\text{Al}} = 0.215$ kcal/(kg·°C) (see Table 20.1). Hence the heat absorbed by the aluminum is

$$Q_{\text{Al}} = m_{\text{Al}}\, c_{\text{Al}}\, \Delta T = 0.20 \text{ kg} \times 0.215 \text{ kcal/(kg·°C)} \times 80°\text{C} = 3.4 \text{ kcal}$$

The net heat absorbed by the water and the pot is then

$$Q_{\text{total}} = Q_{\text{water}} + Q_{\text{Al}} = 8.0 \text{ kcal} + 3.4 \text{ kcal} = 11.4 \text{ kcal}$$

Since heat is a form of work, it can be transformed into macroscopic mechanical work and vice versa. The transformation of heat into work is accomplished by a steam engine, a steam turbine, or a similar machine; we will examine the theory of such heat engines in the next chapter. The transformation of work into heat requires no special machinery—any kind of friction will convert work into heat. Since heat is a form of energy transfer, the calorie is a unit of energy, and it must be possible to express it in joules. The conversion factor between these units is called the **mechanical equivalent of heat**.

The traditional method for the measurement of the mechanical equivalent of heat is Joule's experiment. A set of falling weights drives a paddle wheel that churns the water in a bucket (see Fig. 20.1). The bucket is surrounded by insulation, so no heat can escape from it. The friction inherent in the churning raises the temperature of the water in the bucket by a measurable amount, converting the initial gravitational potential energy of the falling weights into a measurable amount of heat. The best available experimental results for this conversion of mechanical energy into heat give

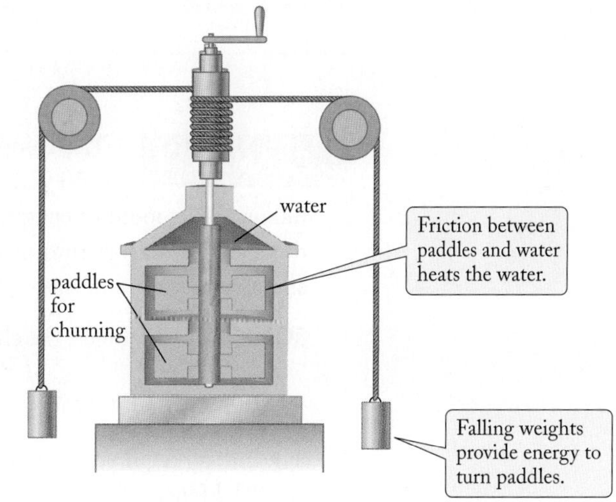

FIGURE 20.1 Joule's apparatus.

$$1 \text{ cal} = 4.187 \text{ J} \qquad (20.4)$$

mechanical equivalent of heat

for the mechanical equivalent of heat.

In the modern SI system of units, the calorie is taken as equal to 4.187 J by definition. This means that Joule's experiment is not needed anymore to find the mechanical equivalent of heat; instead, it is needed to determine the specific heat of water, which now must be regarded as a quantity to be measured experimentally. Hereafter we will mostly use joules to measure thermal energies and only occasionally revert to calories.

EXAMPLE 2 When an automobile is braking, the friction between the brake drums and the brake shoes converts translational kinetic energy into heat. If a 2000-kg automobile brakes from 25 m/s to 0 m/s, how much heat is generated in the brakes? If each of the four brake drums has a mass of 9.0 kg of iron of specific heat 450 J/(kg·°C), how much does the temperature of the brake drums rise? Assume that all the heat accumulates in the brake drums (there is not enough time for the heat to leak away into the air, and not much heat goes into the brake shoes).

SOLUTION: The initial kinetic energy of the automobile is

$$K = \tfrac{1}{2}mv^2 = \tfrac{1}{2} \times 2000\ \text{kg} \times (25\text{m/s})^2 = 6.3 \times 10^5\ \text{J}$$

The brakes convert this kinetic energy into heat. Each of the four brake drums absorbs one-fourth of the total, or $Q = \frac{1}{4} \times 6.3 \times 10^5$ J. Since the mass of each brake drum is 9.0 kg, the temperature increase of each brake drum is, according to Eq. (20.3),

$$\begin{aligned}\Delta T &= \frac{Q}{mc} \\ &= \frac{\frac{1}{4} \times 6.3 \times 10^5\ \text{J}}{9.0\ \text{kg} \times 450\ \text{J/(kg·°C)}} = 39\text{°C}\end{aligned} \tag{20.5}$$

EXAMPLE 3 The complete metabolization of one apple supplies 110 kcal (110 Cal) of chemical energy. How high can you climb up a hill on this amount of energy? Assume that your muscles can completely convert the chemical energy into mechanical energy and that there is no frictional loss, and assume that your mass is 75 kg.

SOLUTION: In joules, the chemical energy is

$$110\ \text{kcal} \times 4.187 \times 10^3\ \text{J/kcal} = 4.6 \times 10^5\ \text{J}$$

The energy required to lift a body of mass m to a height y is $\Delta U = mgy$ [see Eq. (7.29)]. Hence

$$mgy = 4.6 \times 10^5\ \text{J}$$

and

$$y = \frac{4.6 \times 10^5\ \text{J}}{75\ \text{kg} \times 9.81\ \text{m/s}^2} = 6.3 \times 10^2\ \text{m} = 630\ \text{m}$$

COMMENT: In practice, the height you can climb with the energy supplied by one apple is much less, maybe 100 m. Your metabolic system fails to extract all of the available chemical energy, your muscles fail to convert all of the extracted chemical energy into mechanical energy, and finally there are frictional losses.

Checkup 20.1

QUESTION 1: You mix 1.0 kg of water at 80°C with 1.0 kg of water at 20°C. What is the final temperature?

QUESTION 2: The specific heat of iron is about one-ninth that of water. By how many degrees will the temperature of 1 kg of iron increase if you supply 1 kcal of heat to it?

QUESTION 3: Consider the following process: A laborer pushes a heavy crate over a rough but level floor at constant speed. Friction heats the bottom of the crate. Does the laborer do work on the crate? Does he transfer (or remove) mechanical energy to (or from) the crate? Does he transfer thermal energy to the crate? Does the floor do work on the crate? Does the crate do work on the floor?

QUESTION 4: As in Joule's experiment, suppose a weight of 1000 N slowly falls a distance of 4.187 m. The falling weight turns a paddle that churns 10 kg of water. Assuming ideal conditions, what is the change in water temperature?

(A) 0.010°C (B) 0.10°C (C) 1.0°C (D) 10°C (E) 100°C

20.2 THERMAL EXPANSION OF SOLIDS AND LIQUIDS

As we saw in the preceding chapter, if the pressure is held constant, the volume of a given amount of gas will increase with the temperature [see Eq. (19.4)]. Such an increase of volume with temperature also occurs for solids and liquids; this phenomenon is called **thermal expansion**. However, the thermal expansion of solids and of liquids is much less than that of gases. For example, if we raise the temperature of a piece of iron by 100°C, we will increase its volume by only 0.36%. During the expansion, the solid retains its shape, but all its dimensions increase in proportion. Figure 20.2a illustrates the expansion of a piece of metal; for the sake of clarity, the expansion has been exaggerated. An expanding liquid does not retain its shape; the liquid will merely fill more of the container that holds it. Figure 20.2b illustrates the thermal expansion of a liquid.

thermal expansion

From a microscopic point of view, the thermal expansion of solids and liquids is due to the increase of thermal motion caused by the increase of temperature—in a solid the speed of the back-and-forth motions of the atoms about their equilibrium positions increases with temperature, and in a liquid the bouncing zigzag motions of the molecules increase with temperature. This increase of the random motions tends to push the atoms or molecules apart, and therefore leads to an increase of the volume of the solid or liquid.

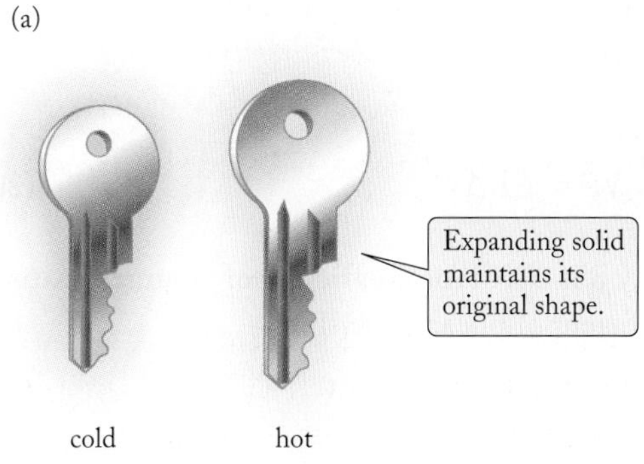

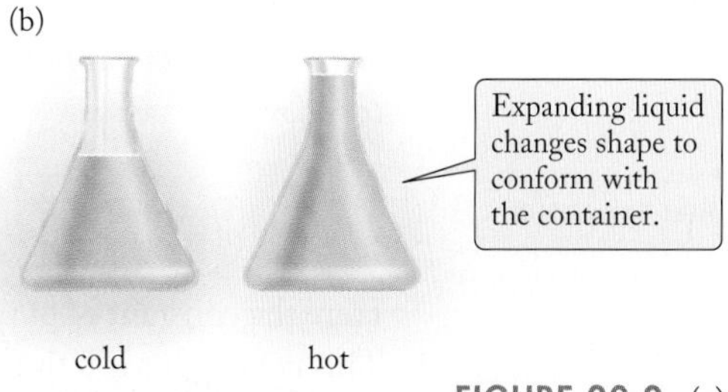

FIGURE 20.2 (a) Thermal expansion of a solid. (b) Thermal expansion of a liquid. The expansion of the flask has been neglected.

The thermal expansion of a solid can be best described mathematically by the increase in the linear dimensions of the solid (see Fig. 20.3). For most solids and for a broad range of temperatures near room temperature, *the increment ΔL in the length L is directly proportional to the increment of temperature and to the original length*:

linear expansion

$$\Delta L = \alpha L \Delta T \qquad (20.6)$$

The constant of proportionality α in this equation is called the **coefficient of linear expansion.** Table 20.2 lists the value of this coefficient for a few materials.

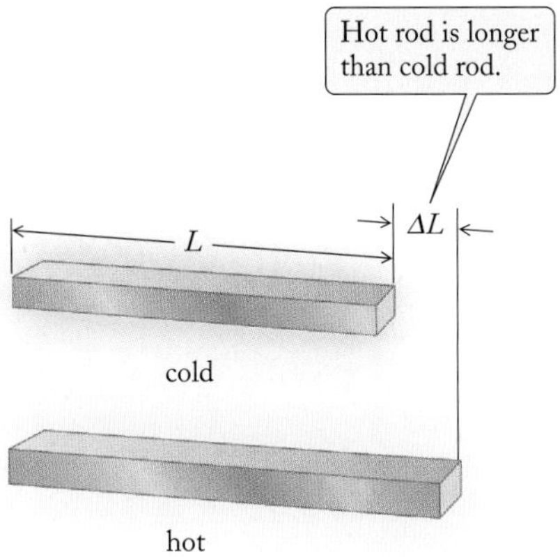

FIGURE 20.3 Increase of the length of a solid rod by thermal expansion. The initial length is L; the final length is $L + \Delta L$.

TABLE 20.2 COEFFICIENTS OF EXPANSION[a]

SOLIDS	α (linear)	LIQUIDS	β (volume)
Lead	29×10^{-6}/°C	Alcohol, ethyl (99%)	1.01×10^{-3}/°C
Aluminum	24	Carbon tetrachloride	1.18
Brass	19	Ether	1.51
Copper	17	Gasoline	0.95
Iron, steel	≈12	Glycerine	0.49
Concrete	≈12	Olive oil	0.68
Glass	9.0	Mercury	0.182
Pyrex	3.6		
Quartz, fused	0.50		

[a] At room temperature (20°C).

EXAMPLE 4 The highest tower in the world is the steel radio mast of Warsaw Radio in Poland, which has a height of 646 m. How much does its height increase between a cold winter day when the temperature is −35°C and a hot summer day when the temperature is +35°C?

SOLUTION: The increment of temperature is $\Delta T = 70$°C. With a value of $\alpha = 12 \times 10^{-6}$/°C for steel (see Table 20.2), we then find

$$\Delta L = \alpha L \Delta T = \frac{12 \times 10^{-6}}{°\text{C}} \times 646 \text{ m} \times 70°\text{C} = 0.54 \text{ m} \qquad (20.7)$$

Similarly, for temperatures near room temperature, for most materials *the increment in the volume of the solid is directly proportional to the increment of temperature and to the original volume*:

volume expansion

$$\Delta V = \beta V \Delta T \qquad (20.8)$$

Here the constant of proportionality β is called the **coefficient of volume expansion**. This coefficient is 3 times the coefficient of linear expansion:

$$\beta = 3\alpha \qquad (20.9)$$

To see how this relationship comes about, suppose that the solid has the shape of a cube of side L (see Fig. 20.4). The increment in the length of each side is ΔL [Eq. (20.6)], and treating this as a small (infinitesimal) quantity, the increment in the volume L^3 is

$$\begin{aligned} \Delta V &= \Delta(L^3) = 3L^2 \Delta L \\ &= 3L^2 \times \alpha L \Delta T \qquad (20.10) \\ &= 3\alpha L^3 \Delta T = 3\alpha V \Delta T \end{aligned}$$

Comparing this result for ΔV with Eq. (20.8), we see that, indeed, $\beta = 3\alpha$.

The increment in the volume of a liquid can be described by the same equation [Eq. (20.8)] as the increment in the volume of a solid. Table 20.2 gives values of coefficients of volume expansion for some liquids.

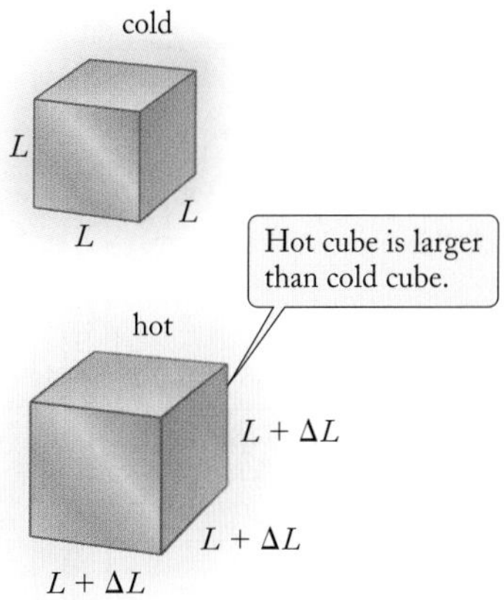

FIGURE 20.4 Thermal expansion of a solid cube. The initial volume is $L \times L \times L$; the final volume is $(L + \Delta L) \times (L + \Delta L) \times (L + \Delta L) = L^3 + 3L^2\Delta L + 3L(\Delta L)^2 + (\Delta L)^3 \approx L^3 + 3L^2\Delta L$.

Water has not been included in this table because its behavior is quite peculiar: from 0°C to 4°C, the volume *decreases* with increasing temperature, but not uniformly; above 4°C the volume increases with increasing temperature. Figure 20.5 gives the volume of 1 kg of water for temperatures ranging from 0°C to 10°C The strange behavior of the density of water at low temperatures can be traced to the crystal structure of ice. Water molecules have a rather angular shape that prevents a tight fit of these molecules; when they assemble in a solid, they adopt a very complicated crystal structure with large gaps. As a result, ice has a lower density than water—the density of ice is 917 kg/m^3, and the volume of 1 kg of ice is 1091 cm^3. At a temperature slightly above the freezing point, water is liquid, but some of the water molecules already have assembled themselves into microscopic (and short-lived) ice crystals; these microscopic ice crystals give the cold water an excess volume.

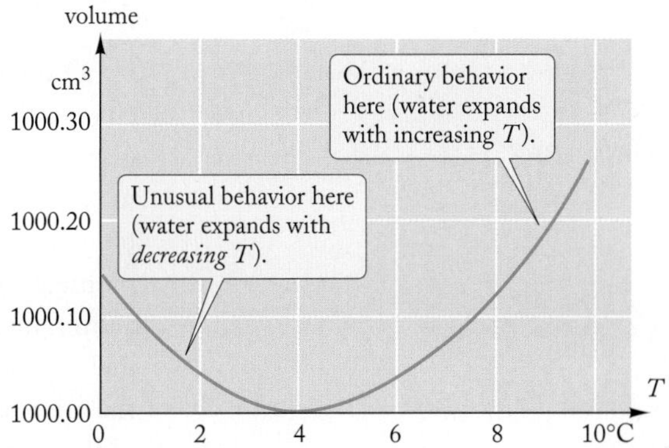

FIGURE 20.5 Volume of 1 kg of water as a function of temperature.

The maximum in the density of water at about 4°C has an important consequence for the ecology of lakes. In winter, the layer of water on the surface of the lake cools, becomes denser than the lower layer, and sinks to the bottom. This process continues until the temperature of the entire body of the lake reaches 4°C. Beyond this point, the cooling of the surface layer will make it *less dense* than the lower layers; thus, the surface layer stays in place, floating on the top of the lake. Ultimately, this surface layer freezes, becoming a solid sheet of ice while the body of the lake remains at 4°C. The sheet of ice inhibits the heat loss from the lake, especially if covered with an insulating blanket of snow. Besides, any further heat loss merely causes some thickening of the sheet of ice, without disturbing the deeper layers of water, which remain at a stable temperature of 4°C—fish and other aquatic life can survive the winter in this stable environment (Fig. 20.6).

FIGURE 20.6 Fish under ice. Water is unusual, solid ice is *less* dense than liquid water and floats.

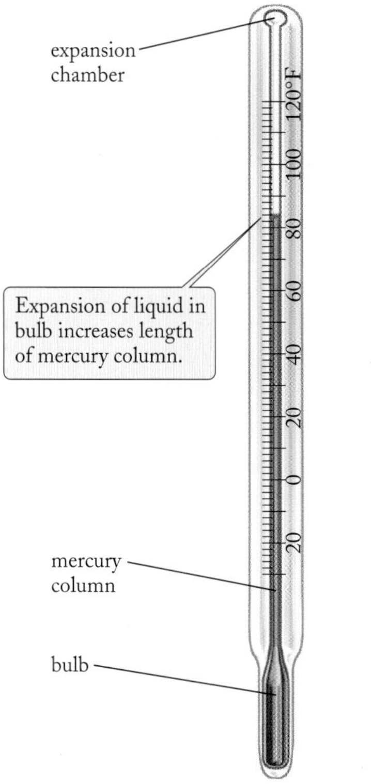

FIGURE 20.7 Mercury-bulb thermometer.

EXAMPLE 5 A glass vessel of volume 200 cm^3 is filled to the rim with mercury (atomic symbol Hg). How much of the mercury will overflow the vessel if we raise the temperature by 30.0°C?

SOLUTION: The volume of mercury will increase by

$$\Delta V_{Hg} = \beta_{Hg} V \Delta T$$
$$= 0.182 \times 10^{-3}/°C \times 200 \text{ cm}^3 \times 30.0°C = 1.09 \text{ cm}^3 \tag{20.11}$$

The volume of the glass vessel will also increase, just as though all of the vessel were filled with glass (as in Fig. 20.2, the hole, or cavity, in the vessel expands as though it were completely filled with glass); hence

$$\Delta V_{glass} = \beta_{glass} V \Delta T = 3\alpha_{glass} V \Delta T \tag{20.12}$$
$$= 3 \times 9.0 \times 10^{-6}/°C \times 200 \text{ cm}^3 \times 30.0°C = 0.16 \text{ cm}^3$$

The difference

$$1.09 \text{ cm}^3 - 0.16 \text{ cm}^3 = 0.93 \text{ cm}^3$$

is the volume of mercury that will overflow.

Ordinary thermometers and thermostats make use of thermal expansion to sense changes in temperature. The mercury-bulb thermometer (see Fig. 20.7) consists of a glass bulb filled with mercury connected to a thin capillary tube. Thermal expansion makes the mercury overflow into the capillary tube and increases the length of the mercury column; this length indicates the temperature. The bimetallic-strip thermometer (see Fig. 20.8) consists of two parallel strips of different metals—such as aluminum and iron—welded together and curled into a spiral. The differential thermal expansion increases the length of one side of the welded strip more than that of the other side. This causes the strip to curl up more tightly and rotates the upper end of the spiral relative to the lower end; a pointer attached to the upper end indicates the temperature.

PROBLEM-SOLVING TECHNIQUES — TEMPERATURE UNITS; THERMAL EXPANSION

- The temperature units in the tables of specific heats and coefficients of expansion in this chapter are degrees Celsius (°C). The temperature difference ΔT used in the calculations must therefore also be expressed in °C, not in °F. However, since the Celsius and the absolute temperature scales differ only by an additive constant (273.15°C), any temperature difference ΔT is the same in °C and in K, and these two units can be used interchangeably in any such temperature difference.
- If the linear dimensions of a solid body expand by, say, 0.001%, then the area of the body (or any part of the body) expands by about 0.002%, and the volume of the body expands by about 0.003%; that is, the linear, area, and volume fractional expansions are in a ratio of 1:2:3.
- During the thermal expansion of a solid body with holes or cavities, such as the glass vessel in Example 5, the hole or cavity expands just as if it were filled with the same solid material—if the solid body expands by, say, 0.003% in volume, the hole or cavity likewise expands by 0.003% in volume.

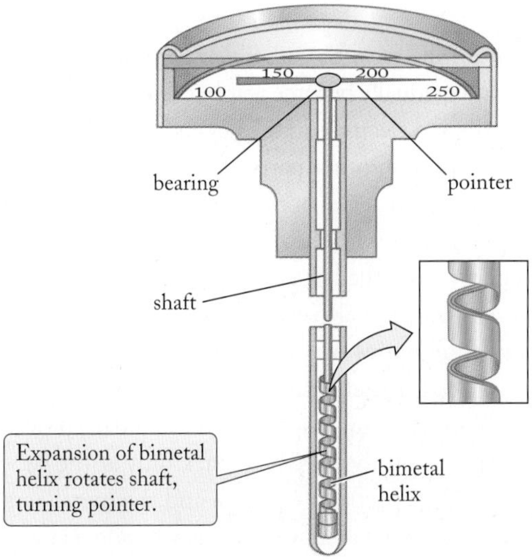

FIGURE 20.8 Bimetallic-strip thermometer.

FIGURE 20.9 Expansion joints in deck of a bridge.

Thermal expansion must be taken into account in the design of long structures, such as bridges or railroad tracks. The decks of bridges usually have several expansion joints with gaps (see Fig. 20.9) that permit changes of length and prevent the bridge from buckling. Likewise, gaps are left between the segments of rail in a railroad track; but if the temperature changes exceed the expectations of the designers, the results can be disastrous (see Fig. 20.10).

Incidentally: Our ability to erect large buildings and other structures out of reinforced concrete hinges on the fortuitous coincidence of the coefficients of expansion of iron and concrete (see Table 20.2). Reinforced concrete consists of iron rods in a concrete matrix. If the coefficients of expansion for these two materials were appreciably different, the daily and seasonal temperature changes would cause the iron rods to move relative to the concrete—ultimately, the iron rods would work loose, and the reinforcement would come to an end.

FIGURE 20.10 Buckling of railroad rails due to heat.

✓ Checkup 20.2

QUESTION 1: Suppose that when we heat an aluminum rod, its length increases by 0.02%. What is the corresponding percent increase of the volume of the rod?

QUESTION 2: According to Fig. 20.5, what is the percent decrease of the volume of 1 kg of water between 0°C and 4°C?

QUESTION 3: An engineer proposes to build a bridge out of concrete reinforced with aluminum rods. What is wrong with this proposal?

QUESTION 4: Consider the thermal expansion of the key illustrated in Fig. 20.2a. During the thermal expansion, does the size of the hole in this key increase, decrease, or stay the same?

(A) Increases (B) Decreases (C) Stays the same

20.3 THERMAL CONDUCTION

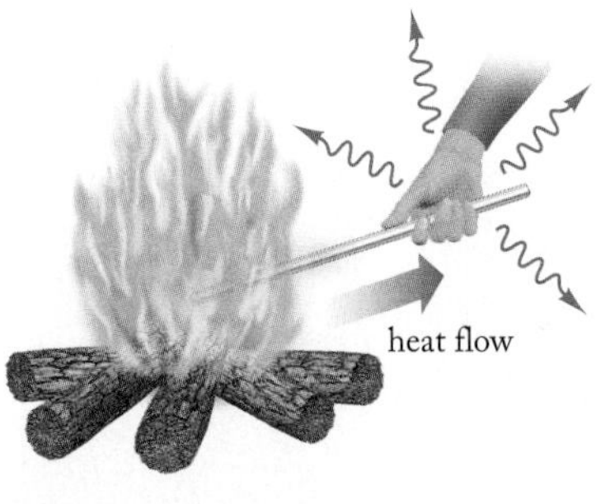

FIGURE 20.11 Heat flows from the hot end of the rod to the cold end.

If you stick one end of an iron rod into a fire and hold the other end in your hand (see Fig. 20.11), you will feel the end in your hand gradually become warmer. This is an example of heat transfer by **conduction**. The atoms and electrons in the hot end of the rod have greater kinetic and potential energies than those in other parts of the rod. In random collisions, these energetic atoms and electrons share some of their energy with their less energetic neighbors; these, in turn, share their energy with their neighbors, and so on. The result is a gradual diffusion of thermal energy from the hot end to the cold end.

Most metals are excellent conductors of heat, and also excellent conductors of electricity. The high thermal and electric conductivities of a metal are due to an abundance of "free" electrons within the volume of the metal; these are electrons that have become detached from their atoms—they move at high speeds, they wander all over the volume of the metal with little hindrance, and they are held back only at the surface of the metal. The free electrons behave like particles of a gas, and the metal acts like a bottle holding this gas. Typically, a free electron will move past a few hundred atoms before it suffers a collision. Because the electrons move such fairly large distances between collisions, they quickly transport energy from one end of a metallic rod to the other. The motion of the free electrons transports the thermal energy much more efficiently than does the back-and-forth vibrational motion of the atoms.

heat flow

Quantitatively, we can describe the transport of heat by the **heat flow**, *or the heat current; this is the amount of heat that passes by some given place on the rod per unit time.* We will use the symbol $\Delta Q/\Delta t$ for heat flow. In the SI system, the unit of heat flow is the joule per second (J/s); however, in practice, the calorie per second (cal/s) and the British thermal unit per hour (Btu/h) are also used.

Consider a rod of cross-sectional area A and length Δx (see Fig. 20.12). Assume that the cold end of the rod is kept at a constant temperature T_1 and the hot end at a constant temperature T_2, so the difference of temperature between the ends is $\Delta T = T_2 - T_1$. If the ends are kept at these constant temperatures for a while, then the temperatures at all other points of the rod will settle to final steady values. Under such steady-state conditions, the heat flow of the rod is found to be directly proportional to the temperature difference ΔT and to the cross-sectional area A, and inversely proportional to the length Δx:

heat conduction

$$\frac{\Delta Q}{\Delta t} = kA\frac{\Delta T}{\Delta x} \tag{20.13}$$

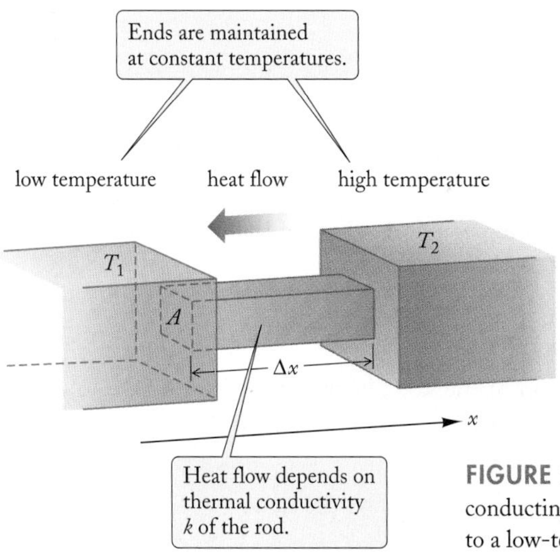

FIGURE 20.12 A rod of cross-sectional area A conducting heat from a high-temperature reservoir (T_2) to a low-temperature reservoir (T_1).

TABLE 20.3 SOME THERMAL CONDUCTIVITIES[a]

SUBSTANCE	k	
Silver	102 cal/(s·m·°C)	427 J/(s·m·°C)
Copper	95	398
Aluminum	57	237
Iron, cast	11	46
Steel	11	46
Lead	8.3	35
Ice, 0°C	0.3	1.3
Snow, 0°C, compact	0.05	0.2
Glass, crown	0.25	1.0
Porcelain	0.25	1.0
Concrete	0.2	0.8
Brick	0.15	0.63
Wood (pine, across grain)	0.03	0.13
Fiberglass (batten)	0.010	0.042
Down	0.005	0.02
Styrofoam	0.002	0.008

[a] At room temperature (20°C), unless otherwise noted.

The direction of the heat flow is, of course, from the hot end of the rod toward the cold end. The constant of proportionality k in our equation is called the **thermal conductivity**. Table 20.3 lists values of k for some materials.[2]

heat conduction

EXAMPLE 6 A house is built of bricks, with walls 20 cm thick. The wall in one of the rooms of this house measures 5.0 m × 3.0 m (see Fig. 20.13). What is the heat flow through this wall if the inside temperature is 21°C and the outside temperature −18°C?

Concepts in Context

SOLUTION: The temperature change across the wall is $\Delta T = 39°\text{C}$, the thickness of the wall is Δx = 20 cm = 0.20 m, and the area of the wall is 5.0 m × 3.0 m = 15 m². Hence, with $k = 0.63$ J/(s·m·°C) from Table 20.3, Eq. (20.13) gives

$$\frac{\Delta Q}{\Delta t} = kA\frac{\Delta T}{\Delta x}$$

$$= 0.63\ \text{J/(s·m·°C)} \times 15\ \text{m}^2 \times \frac{39°\text{C}}{0.20\ \text{m}} \qquad (20.14)$$

$$= 1.8 \times 10^3\ \text{J/s}$$

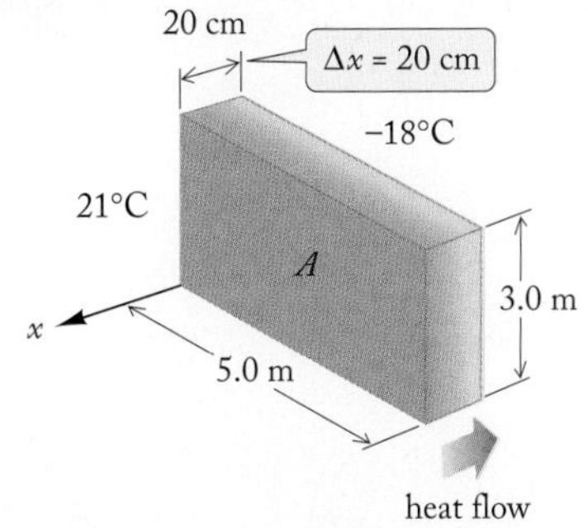

FIGURE 20.13 Heat flow through the brick wall of a house.

[2] The thermal conductivity must not be confused with the Boltzmann constant; both these quantities are designated with the same letter k, but they are not related.

Concepts — in — Context

EXAMPLE 7 To reduce the heat loss, the owner of the brick house described in Example 6 covers the brick wall with a 12-cm layer of fiberglass insulation (see Fig. 20.14). What is the heat loss now?

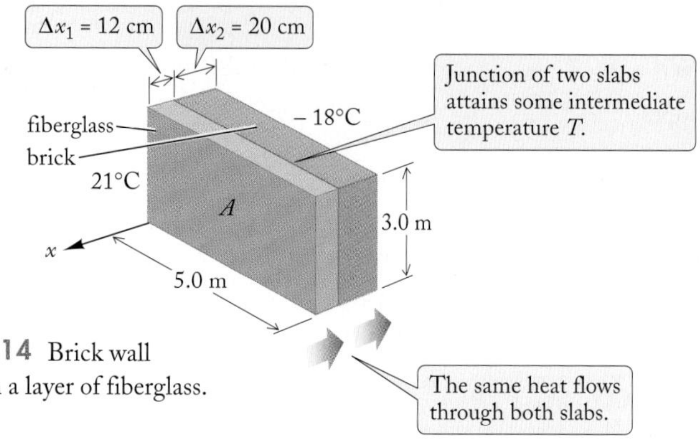

FIGURE 20.14 Brick wall insulated with a layer of fiberglass.

SOLUTION: The wall now consists of two layers. For steady-state conditions, heat does not accumulate anywhere; any heat that enters the first layer reaches the second, and hence the heat flow through each layer is the same. However, the temperature differences across the two layers are different, ΔT_1 for the brick layer and ΔT_2 for the fiberglass. The sum of these temperature differences must equal the net temperature change $\Delta T = 39°C$,

$$\Delta T_1 + \Delta T_2 = \Delta T \tag{20.15}$$

The heat flows through the brick and the fiberglass are, respectively,

$$\frac{\Delta Q_1}{\Delta t} = k_1 A \frac{\Delta T_1}{\Delta x_1} \quad \text{and} \quad \frac{\Delta Q_2}{\Delta t} = k_2 A \frac{\Delta T_2}{\Delta x_2}$$

Since, as discussed above, these heat flows must be equal,

$$k_1 A \frac{\Delta T_1}{\Delta x_1} = k_2 A \frac{\Delta T_2}{\Delta x_2} \tag{20.16}$$

Solving Eq. (20.16) for ΔT_2 and substituting into Eq. (20.15), we obtain

$$\Delta T_1 + \frac{k_1}{k_2}\frac{\Delta x_2}{\Delta x_1}\Delta T_1 = \Delta T$$

From this we find ΔT_1:

$$\Delta T_1 = \frac{\Delta T}{1 + (k_1 \Delta x_2)/(k_2 \Delta x_1)}$$

The heat flow is then

$$\frac{\Delta Q}{\Delta t} = k_1 A \frac{\Delta T_1}{\Delta x_1} = k_1 A \frac{\Delta T}{\Delta x_1[1 + (k_1 \Delta x_2)/(k_2 \Delta x_1)]}$$

or more simply

two-layer heat conduction

$$\frac{\Delta Q}{\Delta t} = A \frac{\Delta T}{\Delta x_1/k_1 + \Delta x_2/k_2} \tag{20.17}$$

With $k_1 = 0.63\ \text{J/(s·m·°C)}$ and $k_2 = 0.042\ \text{J/(s·m·°C)}$, the heat flow is

$$\frac{\Delta Q}{\Delta t} = A\frac{\Delta T}{\Delta x_1/k_1 + \Delta x_2/k_2}$$

$$= 15\text{ m}^2 \times \frac{39°\text{C}}{[0.20\text{ m}/(0.63\text{ J/s·m·°C})] + [0.12\text{ m}/(0.042\text{ J/s·m·°C})]}$$

$$= 1.8 \times 10^2\text{ J/s}$$

Thus, the extra insulation reduces the heat loss of this wall by a factor of 10.

COMMENT: From the denominator of Eq. (20.17), we see that the heat flow is inversely proportional to the sum $\Delta x_1/k_1+\Delta x_2/k_2$. In the description of home insulation, the quantity $\Delta x/k$ is commonly referred to as the **R value**, and it indicates the resistance that a layer offers to heat flow. Thus, for example, our 20-cm-thick brick wall has an R value of (0.20 m)/[0.63 J/(s·m·°C)] = 0.32 s·m^2·°C/J. In common engineering practice, a mixed system of units is often used, where R values are expressed in ft^2·°F·h/Btu; the conversion factor between these units is 1 s·m^2·°C/J = 5.54 ft^2·°F·h/Btu.

Besides conduction, there are two other mechanisms of heat transfer: convection and radiation. *In* **convection**, *the heat is stored in a moving fluid and is carried from one place to another by the motion of this fluid. In* **radiation**, *the heat is carried from one place to another by electromagnetic waves*—for example, light waves, infrared waves, or radio waves. All three mechanisms of heat transfer are neatly illustrated by the operation of a hot-water heating system in a house. In this system, the heat is carried from the boiler to the radiators in the rooms by means of water flowing in pipes (convection); the heat then diffuses through the metallic walls of the radiators (conduction) and finally spreads from the surface of the radiators into the volume of the room (radiation and also convection of the air heated by direct contact with the radiators).

convection and radiation

Radiation is the only mechanism of heat transfer that can carry heat through a vacuum; for instance, the heat of the Sun reaches the Earth by radiation. We will study thermal radiation in Chapter 37.

✔ Checkup 20.3

QUESTION 1: The walls of houses built in the Northeastern United States are commonly insulated with 6 in. (15 cm) of fiberglass insulation. If you wanted to achieve the same insulation with solid wood walls, what thickness of wood would you need?

QUESTION 2: A piece of ice at 0°C is in contact with a piece of steel. If heat is flowing from the ice into the steel, what can you say about the temperature of the steel?

QUESTION 3: Does each of the following changes increase or decrease the heat loss through the wall of the house calculated in Example 6: reduce wall thickness to 10 cm; increase wall area to 20 m^2; reduce external temperature to −20°C; reduce internal temperature to 19°C?

QUESTION 4: The following devices are used to deliver heat to the human body: hair dryer, heat lamp, hot-water bottle. By which respective mechanism does each transfer heat?

(A) Conduction, convection, radiation (B) Conduction, radiation, convection
(C) Convection, conduction, radiation (D) Convection, radiation, conduction
(E) Radiation, convection, conduction

Online Concept 22 Tutorial

20.4 CHANGES OF STATE

Heat absorbed by a body will not only increase the temperature, but it will also bring about a change of state from solid to liquid or from liquid to gas when the body reaches its melting point or its boiling point. At the melting temperature or the boiling temperature, the thermal motion of the atoms and molecules becomes so violent that the bonds holding them in the solid or liquid loosen or break. The loosening of the strong bonds in a solid transforms it into a liquid, and the breaking of the remaining weak bonds in a liquid transforms it into a gas.

While the body is melting or boiling, it absorbs some amount of heat without any increase of temperature. This heat represents the energy required to loosen and break the bonds that hold the atoms inside the solid or liquid. *The heat absorbed during the change of state is called the* **latent heat** *or the* **heat of transformation**, *and more specifically, the* **heat of fusion** *or the* **heat of vaporization**, for the change of state from solid to liquid or from liquid to gas, respectively. Table 20.4 lists the heats of fusion and vaporization for a few substances (at a pressure of 1 atm).

heat of transformation

The quantities listed in Table 20.4 depend on the pressure. The decrease of the boiling point of water with a decrease of atmospheric pressure is a phenomenon familiar to people living at high altitude; for instance, in Denver, Colorado, at an altitude of 1600 m, the mean pressure is 0.83 atm, and the boiling point of water is 95°C.

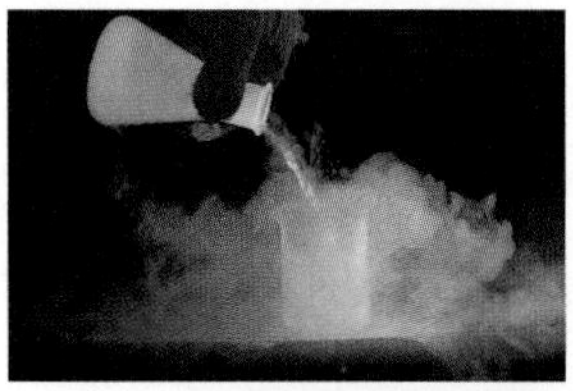
(a)

(b)

(c)

TABLE 20.4 HEATS OF FUSION AND VAPORIZATION[a]

SUBSTANCE	MELTING POINT	HEAT OF FUSION	BOILING POINT	HEAT OF VAPORIZATION
Water	0°C	3.34×10^5 J/kg	100°C	2.26×10^6 J/kg
Nitrogen (a)	−210	2.6×10^4	−196	2.00×10^5
Oxygen	−218	1.4×10^4	−183	2.1×10^5
Helium	—	—	−269	2.06×10^4
Hydrogen	−259	6.3×10^4	−253	4.5×10^5
Aluminum	660	3.99×10^5	2467	1.1×10^7
Copper	1083	2.05×10^5	2567	5.2×10^6
Iron (b)	1535	2.7×10^5	2750	6.8×10^6
Lead	328	2.9×10^4	1740	8.5×10^5
Tin	232	5.9×10^4	2270	1.9×10^6
Silver	962	9.9×10^4	2212	2.4×10^6
Tungsten	3410	1.8×10^5	5660	4.9×10^6
Mercury	−39	1.1×10^4	357	2.92×10^5
Carbon dioxide[b] (c)	−79	—	—	5.8×10^5

[a] At a pressure of 1 atm.
[b] Undergoes direct vaporization (sublimation) from solid to gas.

EXAMPLE 8 How many ice cubes (at 0°C) must be added to a bowl containing 1.00 liter of boiling water at 100°C so that the resulting mixture reaches a temperature of 40°C? Assume that each ice cube has a mass of 20 g and that the bowl and the environment do not exchange heat with the water, and assume that the average specific heat of water is 4.2×10^3 J/kg.

SOLUTION: Since all of the heat released by the water is absorbed by the ice, the amount of heat released by the hot water during cooling must equal the amount of heat absorbed by the ice during melting and during the subsequent heating of the molten ice from 0°C to 40°C. Thus we can write an expression for each of these amounts of heat and equate them.

The heat released by the hot water during cooling from 100°C to 40°C is, according to Eq. (20.3),

$$Q = mc\Delta T = 1.00 \text{ kg} \times 4.2 \times 10^3 \text{ J/(kg·°C)} \times 60\text{°C} = 2.5 \times 10^5 \text{ J}$$

If the total mass of ice is m, then, from Table 20.4, the heat absorbed by this mass during melting is $m \times 3.34 \times 10^5$ J/kg, and the heat absorbed during the subsequent heating from 0°C to 40°C is $m \times 4.2 \times 10^3 \text{ J/(kg·°C)} \times \Delta T = m \times 4.2 \times 10^3 \text{ J/(kg·°C)} \times 40\text{°C}$. Hence the total heat absorbed is

$$Q = m \times 3.34 \times 10^5 \text{ J/kg} + m \times 4.2 \times 10^3 \text{ J/(kg·°C)} \times 40\text{°C}$$

$$= m \times (3.34 \times 10^5 \text{ J/kg} + 1.7 \times 10^5 \text{ J/kg}) = m \times (5.0 \times 10^5 \text{ J/kg})$$

These amounts of heat must be equal:

$$m \times (5.0 \times 10^5 \text{ J/kg}) = 2.5 \times 10^5 \text{ J}$$

Solving this for m, we find that the mass of ice required is

$$m = 0.50 \text{ kg}$$

Since each ice cube has a mass of 0.020 kg, this is

$$\frac{0.50 \text{ kg}}{0.020 \text{ kg/ice cube}} = 25 \text{ ice cubes}$$

✓ Checkup 20.4

QUESTION 1: Which of the materials listed in Table 20.4 are gases at a temperature of −200°C? Which are liquids? Which are solids?

QUESTION 2: Which of the materials listed in Table 20.4 releases the largest amount of heat (per kg) when it freezes?

QUESTION 3: For the substances listed in Table 20.4, is more heat required for the melting or for the vaporization of a given mass of substance?

QUESTION 4: Place the following in increasing order of the amount of heat required: (a) vaporizing 1.0 kg of water; (b) melting 1.0 kg of ice; and (c) heating 1.0 kg of water from 0°C to 100°C.

(A) a, b, c (B) a, c, b (C) b, a, c (D) b, c, a (E) c, b, a

20.5 THE SPECIFIC HEAT OF A GAS

molar specific heat at constant volume and at constant pressure

If you heat a gas, the increase of temperature causes an increase of the pressure, and this tends to bring about an expansion of the gas. You can observe this expansion if you leave a loosely inflated beach ball or plastic bag in the sun; the beach ball soon becomes taut because the air warms and expands. The value of the specific heat of a gas depends on whether the container permits expansion during heating. If the container is perfectly rigid, the heating proceeds at constant volume (see Fig. 20.15). For gases it is customary to reckon the specific heat *per mole,* rather than per kilogram. The **molar specific heat at constant volume** is designated by C_V; it is the amount of heat needed to raise the temperature of 1 mole of gas by 1°C. The amount of heat Q required to increase the temperature of n moles by ΔT is proportional to n and to ΔT:

$$Q = nC_V \Delta T \tag{20.18}$$

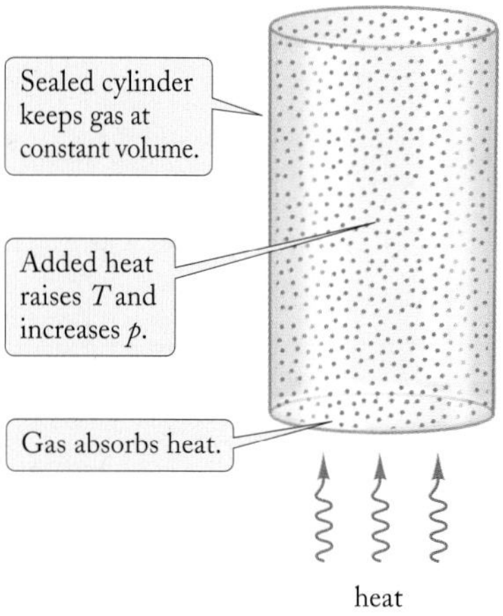

FIGURE 20.15 A gas kept at constant volume while being heated.

This equation resembles Eq. (20.3), but the number of moles n appears instead of the mass m, because we now are reckoning the specific heat per mole.

If the container is fitted with a vertical piston whose weight presses down on the gas, the heating proceeds at a constant pressure determined by the weight of the piston and its area A (see Fig. 20.16). The **molar specific heat at constant pressure** is designated by C_p. For n moles of gas, the heat absorbed and the temperature increase are related by

$$Q = nC_p \Delta T \tag{20.19}$$

We expect C_p to be larger than C_V because, if we supply some amount of heat to the gas in Fig. 20.16, only part of this heat will go into a temperature increase of the gas; the rest will be converted into work as the expanding gas lifts the piston. Let us calculate the difference between the two heat capacities.

At constant volume, the gas does no work. Hence all the heat absorbed will go into the energy of the gas; if the gas absorbs a small amount of heat dQ, the energy increment dE must match dQ:

$$dQ = dE \tag{20.20}$$

or, according to Eq. (20.18),

$$nC_V\, dT = dE \tag{20.21}$$

At constant pressure, the gas does work against the moving piston. Suppose that the piston is displaced a small (infinitesimal) distance dx (see Fig. 20.17). The force of the gas on the piston is pA and the work done by the gas, $dW = F\,dx$, is

$$dW = pA\,dx \tag{20.22}$$

The product $A\,dx$ is simply the small change dV of the volume of the gas. Hence

$$dW = p\,dV \tag{20.23}$$

The heat absorbed by the gas must provide both the energy increase of the gas and the work done by the gas:

$$dQ = dE + dW = dE + p\,dV \tag{20.24}$$

or, according to Eq. (20.19),

$$nC_p\, dT = dE + p\,dV \tag{20.25}$$

In an ideal gas, the energy E depends on the temperature only [see Eqs. (19.26) and (19.33)]; consequently, if the temperature increment at constant pressure has the same

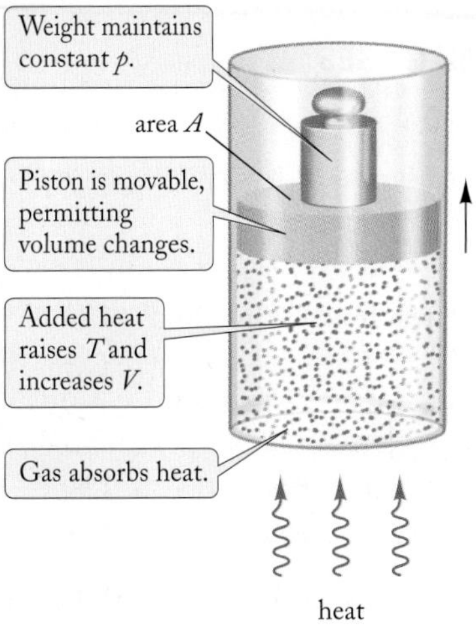

FIGURE 20.16 A gas kept at constant pressure while being heated. The pressure equals the atmospheric pressure plus the combined weight of the piston and the load divided by the area A.

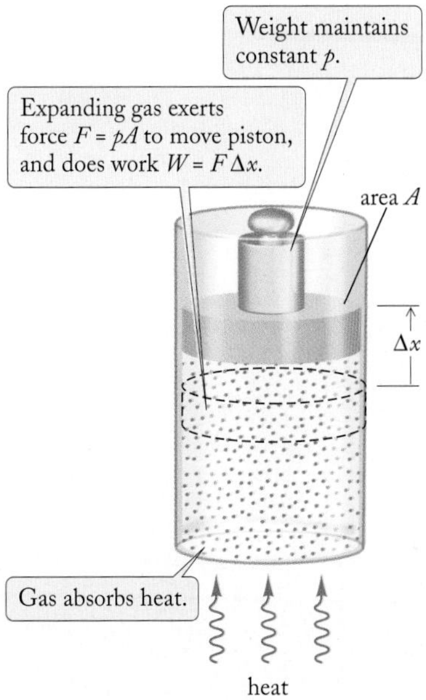

FIGURE 20.17 Expansion of a gas. During its expansion, the gas displaces the piston and performs work.

value as the temperature increment at constant volume, the increase dE of the energy must be the same. We can therefore insert the expression for dE given by Eq. (20.21) into Eq. (20.25) and obtain

$$nC_p\,dT = nC_V\,dT + p\,dV \tag{20.26}$$

From the Ideal-Gas Law [Eq. (19.2)], we find that, at constant pressure, the changes of volume and of temperature are related by

$$p\,dV = nR\,dT \tag{20.27}$$

With this, Eq. (20.26) becomes

$$nC_p\,dT = nC_V\,dT + nR\,dT \tag{20.28}$$

and, canceling the factors $n\,dT$ in this equation, we obtain the final result

$$C_p = C_V + R \tag{20.29}$$

molar specific heat relation

The numerical value of R is 8.31 J/K·mole, or 1.99 cal/K·mole. Hence, Eq. (20.29) shows that C_p is larger than C_V by about 8.3 J/K·mole—*it takes about* 8.3 *more joules to heat a mole of gas by* 1°C *at constant pressure than at constant volume.*

Note that although the above general argument did permit us to evaluate the difference between C_p and C_V, it does not permit us to find the individual values of C_p and C_V. For this, we must know something about the energy stored in the internal rotational or vibrational motions of the molecules. For a monatomic gas there is no such extra energy, and according to Eq. (19.27),

$$dE = \tfrac{3}{2}nR\,dT \tag{20.30}$$

By the definition of C_V [Eq. (20.21)], this leads to

$$C_V = \tfrac{3}{2}R \quad \text{(monatomic gas)} \tag{20.31}$$

and by Eq. (20.29)

$$C_p = C_V + R = \tfrac{3}{2}R + R = \tfrac{5}{2}R \quad \text{(monatomic gas)} \tag{20.32}$$

For a diatomic gas, the rotational energy of the molecules results in a larger value of E [see Eq. (19.33)], and consequently a larger value of C_V and C_p:

$$C_V = \tfrac{5}{2}R \quad \text{and} \quad C_p = \tfrac{7}{2}R \quad \text{(diatomic gas)} \tag{20.33}$$

TABLE 20.5 SPECIFIC HEATS OF SOME GASES[a]

GAS	C_V	C_p	$C_p - C_V$	$\gamma = C_p/C_V$
Monatomic				
Helium (He)	12.5 J/mole·K	20.8 J/mole·K	8.3 J/mol·K	1.66
Argon (Ar)	12.5	20.9	8.4	1.67
Diatomic				
Nitrogen (N_2)	20.8	29.1	8.3	1.40
Oxygen (O_2)	20.8	29.1	8.3	1.40
Carbon monoxide (CO)	20.7	29.1	8.4	1.41
Polyatomic				
Ammonia (NH_3)	27.3	35.8	8.5	1.31
Methane (CH_4)	27.1	35.5	8.4	1.31

[a] At STP.

Table 20.5 lists the values of the specific heats of some gases. In the cases of monatomic and diatomic gases, these values are in reasonable agreement with Eqs. (20.31)–(20.33), although there are some minor deviations because the gases are not quite ideal gases. Note that in all cases the difference $C_p - C_V$ agrees quite well with Eq. (20.29).

EXAMPLE 9 During a sunny day, the sunlight warms the ground, which in turn warms the air in contact with it. How many joules must the ground supply to heat an initial volume of 1.00 m^{3}of air from 0.0°C to 10.0°C? The atmospheric pressure is steady at 1.00 atm.

SOLUTION: Since the atmosphere surrounding the given amount of air provides a constant pressure, the relevant specific heat is C_p, the specific heat at constant pressure. Air is mostly N_2 and O_2, and Table 20.5 tells us that for N_2 and for O_2, $C_p = 29.1$ J/°C·mole. This must then also be the right value of C_p for any mixture of these two gases.

During the heating, the volume of air expands, but we can calculate the number of moles from the initial volume and temperature. For this calculation, we can either use the Ideal-Gas Law or, more simply, the known volume of 22.4 liters for one mole at STP (Example 1 in Chapter 19). By proportions, a volume of 1.00 m^3, or 1000 liters, contains n = (1000 liters)/(22.4 liters/mole) = 44.6 moles. Hence the amount of heat absorbed by the air is

$$Q = nC_p\,\Delta T = 44.6 \text{ moles} \times 29.1 \text{ J/°C·mole} \times 10.0\text{°C} = 1.30 \times 10^4 \text{ J} \tag{20.34}$$

This is a fairly small amount of heat (for comparison, if we wanted to heat a volume of 1 m^3 of water by 10°C, we would need to supply about 4×10^7 J).

Checkup 20.5

QUESTION 1: Why is the specific heat of a gas at constant pressure larger than the specific heat at constant volume?

QUESTION 2: The flow of thermal energy through a window can be reduced by using two layers of glass with a layer of gas between them. Under appropriate conditions, the thermal conductivity of a gas is proportional to the specific heat of the gas and to the rms speed of the gas moleules. Under such conditions, which of the gases in Table 20.5 would best reduce the flow of thermal energy?

Concepts in Context

QUESTION 3: A rubber balloon and a sealed glass jar contain equal amounts of air. For equal increases of temperature, which requires more heat? Is the difference in the specific heats 8.31 J/°C·mole, or is it smaller?

QUESTION 4: Under what conditions might the specific heat of a gas exceed C_V by more than 8.31 J/°C·mole?

QUESTION 5: Heat is added to one mole of air at constant pressure, resulting in a temperature increase of 100°C. If the same amount of heat is instead added at constant volume, what is the temperature increase?

(A) 50°C (B) 60°C (C) 71°C (D) 140°C (E) 167°C

20.6 ADIABATIC EXPANSION OF A GAS

If an amount of gas at high pressure and temperature is placed in a container fitted with a piston, the gas will push the piston outward and do work on it. Such a process of expansion against a piston converts thermal energy into useful mechanical energy—the temperature of the gas decreases as it delivers work to the piston. This process is at the core of the operation of steam engines, automobile engines, and other heat engines.

In this section we will investigate the equation for the expansion of a gas. We will assume that the gas is thermally insulated (see Fig. 20.18), so it neither receives heat from its environment nor loses any. The temperature change of the gas is then entirely due to the work that the gas does on its environment. Such *a process occurring without the exchange of heat with the environment is called* **adiabatic**.

If the volume of gas increases by a small amount dV, the work done by the gas on the piston is [see Eq. (20.23)]

$$dW = p\,dV \tag{20.35}$$

The heat absorbed is zero; hence the change of energy of the gas is entirely due to the work done by the gas [see also Eq. (20.24)]:

$$dE = -dW = -p\,dV \quad \text{(adiabatic process)} \tag{20.36}$$

The change of energy can also be expressed in terms of the temperature change [see Eq. (20.21) and the discussion after Eq. (20.25)]:

$$dE = nC_V\,dT \tag{20.37}$$

Combining Eqs. (20.36) and (20.37), we find

$$nC_V\,dT = -p\,dV \tag{20.38}$$

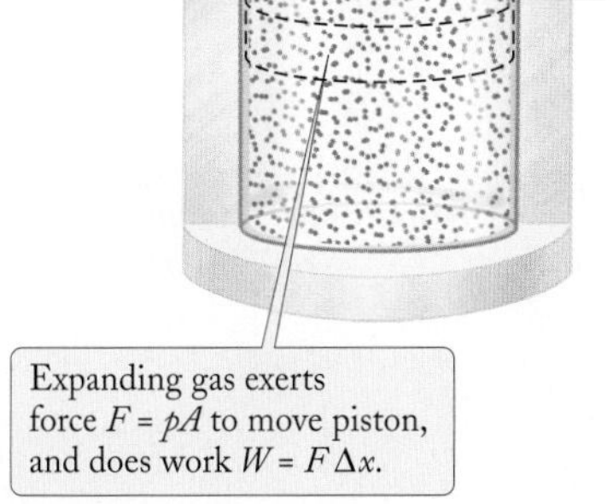

FIGURE 20.18 Expansion of a thermally insulated gas. The insulation prevents the escape of heat from the gas.

According to the Ideal-Gas Law, the pressure is $p = nRT/V$. Thus,

$$nC_V\,dT = -nRT\,\frac{dV}{V} \tag{20.39}$$

If we divide both sides of this equation by nTC_V, we obtain

temperature change in adiabatic expansion

$$\frac{dT}{T} = -\frac{R}{C_V}\frac{dV}{V} \tag{20.40}$$

This shows that the fractional decrease of temperature (dT/T) is directly proportional to the fractional increase of volume (dV/V). For monatomic gases, $C_V = \frac{3}{2}R$ [see Eq. (20.31)]; hence $R/C_V = \frac{2}{3}$ and, consequently, $dT/T = -\frac{2}{3}\,dV/V$. This means that a 3% increase of volume of the gas will produce a 2% decrease of the absolute temperature.

The simple proportionality implied by Eq. (20.40) is valid only for *small* changes of volume and temperature; for large changes, we must integrate Eq. (20.40). We first define the quantity γ (the Greek letter *gamma*) to be the ratio of specific heats at constant pressure and constant volume:

$$\gamma = \frac{C_p}{C_V} \tag{20.41}$$

From Eqs. (20.31)–(20.33), we see that for a monatomic ideal gas, $\gamma = \frac{5}{3} \approx 1.67$ and for a diatomic ideal gas, $\gamma = \frac{7}{5} = 1.40$. Measured values of γ are given in Table 20.5; these are quite close to the ideal values. Since $R = C_p - C_V$ [Eq. (20.29)], we see that

$$\frac{R}{C_V} = \frac{C_p}{C_V} - 1 = \gamma - 1 \tag{20.42}$$

We substitute this form into Eq. (20.40) and integrate from an initial temperature T_1 and volume V_1 to a final temperature T_2 and volume V_2:

$$\int_{T_1}^{T_2}\frac{dT}{T} = -(\gamma - 1)\int_{V_1}^{V_2}\frac{dV}{V} \tag{20.43}$$

The integrals of $1/T$ and $1/V$ are the natural logarithms $\ln T$ and $\ln V$, respectively:

$$\ln\left(\frac{T_2}{T_1}\right) = -(\gamma - 1)\ln\left(\frac{V_2}{V_1}\right) \tag{20.44}$$

where we have used the property of logarithms $\ln A - \ln B = \ln(A/B)$. Taking the exponential of both sides, using $\exp(\ln x) = x$, and rearranging, we obtain

temperature and volume in adiabatic process

$$T_1V_1^{\gamma-1} = T_2V_2^{\gamma-1} \tag{20.45}$$

Equation (20.45) can be used to relate any changes in temperature and volume during an adiabatic process. It is easy to see that if the volume increases, the temperature must drop.

The decrease of temperature of an expanding gas can be perceived quite readily when air is allowed to rush out of the valve of an automobile tire; this expanding air feels quite cool. This process is approximately adiabatic because the expanding air, although not insulated from its surroundings, expands so quickly that it does not have time to exchange heat with the surrounding atmospheric air. Conversely, the increase of temperature during the adiabatic compression of air can be perceived when operating a manual air pump. The compression of air in the barrel of the pump produces a noticeable warming of the pump.

Finally, we use the Ideal-Gas Law to eliminate T from Eq. (20.45); substituting $T = pV/nR$, we have

$$\frac{p_1 V_1}{nR} V_1^{\gamma-1} = \frac{p_2 V_2}{nR} V_2^{\gamma-1} \tag{20.46}$$

or simply

$$p_1 V_1^{\gamma} = p_2 V_2^{\gamma} \tag{20.47}$$

pressure and volume in adiabatic process

Intuitively, we expect the volume to decrease when the pressure increases. Equation (20.47) tells us quantitatively how pressure and volume vary in an adiabatic process.

EXAMPLE 10 In a diesel engine, the piston compresses the air–fuel mixture from an initial volume of 630 cm^3 to a final volume of 30.0 cm^3. The initial temperature of the mixture is 45°C. Assume that the compression occurs adiabatically. What is the final temperature? The value of γ for the air–fuel mixture is 1.37.

SOLUTION: For an adiabatic process, we use Eq. (20.45) to relate temperature and volume:

$$T_1 V_1^{\gamma-1} = T_2 V_2^{\gamma-1}$$

Solving for the final temperature T_2, we obtain

$$T_2 = T_1 \left(\frac{V_1}{V_2} \right)^{\gamma-1}$$

Inserting the initial temperature $T_1 = 45°\text{C} = 318\text{ K}$, the value $\gamma - 1 = 0.37$, and the given volumes, we find

$$T_2 = 318\text{ K} \times \left(\frac{630\text{ cm}^3}{30.0\text{ cm}^3} \right)^{0.37} = 981\text{ K} = 708°\text{C}$$

In a diesel engine, the high temperature resulting from nearly adiabatic compression triggers combustion without the need for the spark used for ignition in ordinary internal combustion engines.

✔ Checkup 20.6

QUESTION 1: When you switch on the heater in your room, the air warms and expands (and some air escapes from the room). Is this an adiabatic expansion?

QUESTION 2: The temperature on the tops of hills is usually lower than at the bottoms of valleys. In dry air, the temperature decrease with height is about 1°C per 100 m. This can be explained by examining the adiabatic behavior of parcels of air that drift upward (or downward). As a parcel drifts upward from the bottom of a valley to the top of a hill, what happens to its pressure, its volume, its temperature?

QUESTION 3: Three types of gases are initially at the same temperature. Each is adiabatically compressed to one-half of its original volume. Which type of gas attains the highest temperature?

(A) Monatomic (B) Diatomic (C) Polyatomic

SUMMARY

PROBLEM-SOLVING TECHNIQUES Temperature units; thermal expansion (page 636)

SPECIFIC HEAT OF WATER

$$c = 1.00 \text{ kcal/kg}\cdot{}^\circ\text{C} \quad (20.2)$$

HEAT Q FOR TEMPERATURE CHANGE ΔT OF A MASS m where c is the specific heat.

$$Q = mc\Delta T \quad (20.3)$$

MECHANICAL EQUIVALENT OF HEAT

$$1.00 \text{ cal} = 4.187 \text{ J} \quad (20.4)$$

THERMAL EXPANSION

where α and β are the linear and volume coefficients of thermal expansion, respectively.

$$\Delta L = \alpha L \,\Delta T \quad (20.6)$$

$$\Delta V = \beta V \,\Delta T \quad (20.8)$$

$$\beta = 3\alpha \quad (20.9)$$

HEAT CONDUCTION (energy per second)

where k is the thermal conductivity, A is the area that the heat flows across, ΔT is the temperature difference, and Δx is the length.

$$\frac{\Delta Q}{\Delta t} = kA\frac{\Delta T}{\Delta x} \quad (20.13)$$

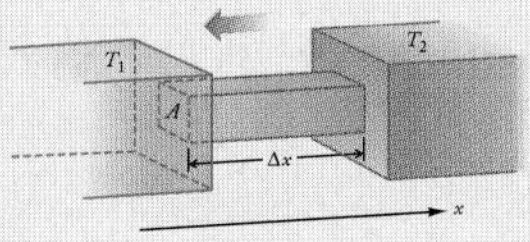

HEAT CONDUCTION THROUGH MULTIPLE SLABS

$$\frac{\Delta Q}{\Delta t} = A\frac{\Delta T}{(\Delta x_1/k_1) + (\Delta x_2/k_2) + \cdots} \quad (20.17)$$

CONVECTION Transport of heat by a moving fluid.

RADIATION Transport of heat by electromagnetic waves.

HEAT OF FUSION; HEAT OF VAPORIZATION The heat required to change the state of a material from solid to liquid or from liquid to gas, respectively.

MOLAR SPECIFIC HEATS AT CONSTANT VOLUME (per mole)

$$C_V = \tfrac{3}{2}R \quad \text{(monatomic ideal gas)} \tag{20.31}$$
$$C_V = \tfrac{5}{2}R \quad \text{(diatomic ideal gas)} \tag{20.33}$$

RELATION BETWEEN SPECIFIC HEATS OF A MOLE OF IDEAL GAS where C_p is the molar specific heat at constant pressure.

$$C_p = C_V + R \tag{20.29}$$

ADIABATIC EXPANSION OF GAS Using $\gamma = C_p/C_V$,

$$\frac{dT}{T} = -\frac{R}{C_V}\frac{dV}{V} \tag{20.40}$$

$$TV^{\gamma-1} = [\text{constant}] \tag{20.45}$$

$$pV^{\gamma} = [\text{constant}] \tag{20.47}$$

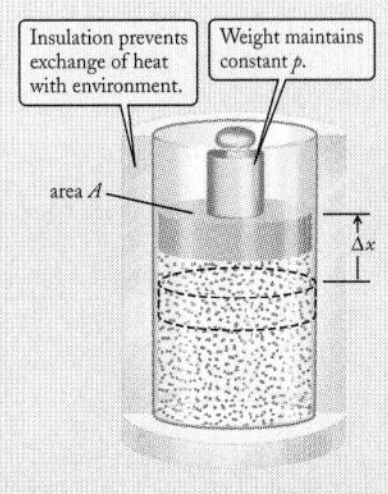

QUESTIONS FOR DISCUSSION

1. Can the body heat from a crowd of people produce a significant temperature increase in a room?
2. The expression "cold enough to freeze the balls off a brass monkey"originated aboard ships of the British Navy where cannonballs of lead were kept in brass racks ("monkeys"). Can you guess how the balls might fall off a "monkey"on a very cold day?
3. If the metal lid of a glass jar is stuck, it can usually be loosened by running hot water over the lid. Explain.
4. On hot days, bridges expand. How do bridge designers prevent this expansion from buckling the road?
5. At regular intervals, oil pipelines have lateral loops (shaped like a U; see Fig. 20.19). What is the purpose of these loops?

FIGURE 20.19 Loop in an oil pipeline.

6. When you heat soup in a metal pot, sometimes the soup rises at the rim of the pot and falls at the center. Explain.
7. A sheet of glass will crack if heated in one spot. Why?
8. When aluminum wiring is used in electrical circuits, special terminal connectors are required to hold the ends of the wires securely. If an ordinary brass screw were used to hold the end of an aluminum wire against a brass plate, what would be likely to happen during repeated heating and cooling of the circuit?
9. Suppose that a piece of metal and a piece of wood are at the same temperature. Why does the metal feel colder to the touch than the wood?
10. For lack of a better method, some nineteenth-century explorers in Africa and Asia measured altitude by sticking a thermometer into a pot of boiling water. Explain.
11. Can you guess why an alloy of two metals usually has a lower melting point than either pure metal?
12. In the cooling system of an automobile, how is the heat transferred from the combustion cylinder to the cooling water—by conduction, convection, or radiation? How is the heat transferred from the water in the engine to the water in the radiator? How is the heat transferred from the radiator to the air?
13. A fan installed near the ceiling of a room blows air down toward the floor. How does such a fan help to keep you cool in the summer and warm in the winter?
14. It is often said that an open fireplace sends more heat up the chimney than it delivers to the room. What is the mechanism

for the heat transport to the room? For the heat transport up the chimney?

15. A large fraction of the heat lost from a house escapes through the windows (Fig. 20.20). This heat is carried to the windowpane by convection—hot air at the top of the room descends along the windowpane, giving up its heat. Suppose that the windows are equipped with venetian blinds. In order to minimize the heat loss, should you close the blinds so that the slats are oriented down and away from the window or up and away?

FIGURE 20.20 Thermal photograph of a house. Bright regions indicate high heat loss.

16. Fiberglass insulation used in the walls of houses has a shiny layer of aluminum foil on one side. What is the purpose of this layer?
17. Is it possible to add heat to a system without changing its temperature? Give an example.
18. Why is boiling oil much more likely to cause severe burns on the skin than boiling water?
19. Would you expect the melting point of ice to increase or decrease with an increase of pressure?
20. A very cold ice cube, fresh out of the freezer, tends to stick to the skin of your fingers. Why?
21. What is likely to happen to the engine of an automobile if there is no antifreeze in the cooling system and the water freezes?
22. If an evacuated glass vessel, such as a TV tube, fractures and implodes, the fragments fly about with great violence. Where does the kinetic energy of these fragments come from?
23. When you boil water and convert it into water vapor, is the heat you supply equal to the change of the internal energy of the water?
24. A gas is in a cylinder fitted with a piston. Does it take more work to compress the gas at constant temperature or adiabatically?
25. According to the result of Section 11.2 [see Eqs. (11.13) and (11.14)], when a particle of small mass collides elastically with a body of very large mass, the particle gains kinetic energy if the body of large mass was approaching the particle before the collision. Using this result, explain how the collisions between the particles of gas and the moving piston lead to an increase of temperature during an adiabatic compression.
26. When a gas expands adiabatically, its temperature decreases. How could you take advantage of this effect to design a refrigerator?
27. A sample of ideal gas is initially confined in a bottle at some given temperature. If we break the bottle and let the gas expand freely into an evacuated chamber of larger volume, will the temperature of the gas change?

PROBLEMS

20.1 Heat as a Form of Energy Transfer†

1. The immersible electric heating element in a coffeemaker converts 620 W of electric power into heat. How long does this coffeemaker take to heat 1.0 liter of water from 20°C to 100°C? Assume that no heat is lost to the environment.
2. The body heat released by children in a school makes a contribution toward heating the building. How many kilowatts of heat do 1000 children release? Assume that the daily food intake of each child has a chemical energy of 2000 kcal and that this food is burned at a steady rate throughout the day.
3. In 1847 Joule attempted to measure the frictional heating of water in a waterfall near Chamonix in the French Alps. If the water falls 120 m and all of its gravitational energy is converted into thermal energy, how much does the temperature of the water increase? Actually, Joule found no increase of temperature because falling water cools by evaporation.
4. A nuclear power plant takes in 5.0×10^6 m^3 of cooling water per day from a river and exhausts 1200 megawatts of waste heat into this water. If the temperature of the inflowing water is 20°C, what is the temperature of the outflowing water?
5. Your metabolism extracts about 100 kcal of chemical energy from one apple. If you want to get rid of all this energy by jogging, how far must you jog? At a speed of 12 km/h, jogging requires about 750 kcal/h.
6. For basic subsistence a human body requires a diet with about 2000 kcal/day. Express this power in watts.
7. By turning a crank, you can do mechanical work at the steady rate of 0.15 hp. If the crank is connected to paddles churning 4.0 liters of water, how long must you churn the water to raise its temperature by 5.0°C?

† For help, see Online Concept Tutorial 22 at www.wwnorton.com/physics

8. A glass mug has a mass of 125 g when empty. It contains 180 g of coffee, and both the mug and the coffee are at 70°C. You add 15 g of cream at 5°C to the coffee. Assuming the mug–coffee system is isolated, find the common final temperature. Assume that coffee has the same specific heat as water and that the cream has a specific heat of 2900 J/kg·°C.

9. A cafeteria snack food label indicates that a "creme"-filled dessert contains 350 Calories (recall that 1 Calorie = 1000 cal = 1 kcal). How many steps would a person of mass 70 kg have to climb to do an amount of work equal to the energy contained in the snack food? Each step is 25 cm high.

10. A printing plate of mass 20 kg is made of lead. After casting, the solid lead cools to 90°C. It is then dropped into a 200-liter tank of water, initially at 25°C. What is the final equilibrium temperature of the lead and water?

11. Ten people can swim in a lap pool at the same time; each releases about 3.6×10^6 joules per hour. If the total volume of water in the pool is 600 m^3, what temperature change occurs over a 12-h period, assuming that the pool is fully utilized?

12. An iron frying pan becomes dangerously hot above 400°C. After a 600-W stove burner is turned on full, how long will it take for a 5.0-kg frying pan (initially at 25°C) to become this hot? Assume for simplicity that all the heat goes to the frying pan.

13. An industrial polishing apparatus generates 300 W of heat due to friction. The heat is carried away by a water flow of 2.5 liters per minute. How much warmer is the water leaving the polishing station than the water entering it?

*14. You can warm the surfaces of your hands by rubbing one against the other. If the coefficient of friction between your hands is 0.60 and if you press your hands together with a force of 60 N while rubbing them back and forth at an average speed of 0.50 m/s, at what rate (in joules per second) do you generate heat on the surfaces of your hands?

*15. Problem 96 of Chapter 8 gives the relevant numbers for frictional losses in the Tennessee River. If all the frictional heat were absorbed by the water and if there were no heat loss by evaporation, how much would the water temperature rise per kilometer?

*16. The first quantitative determination of the mechanical equivalent of heat was made by Robert von Mayer, who compared available data on the amount of mechanical work needed to compress a gas and the amount of heat generated during the compression. From this comparison, Mayer deduced that the energy required for warming 1.00 kg of water by 1.00°C is equivalent to the potential energy released when a mass of 1.00 kg falls from a height of 365 m. By what percent does Mayer's result differ from the modern result given by Eq. (20.4)?

*17. On a hot summer day, the use of air conditioners raises the consumption of electric power in New York City to 22 400 megawatts. All of this electric power ultimately produces heat. Compare the heat produced in this way with the solar heat incident on the city. Assume that the incident flux of solar energy is 1.0 kW/m^2 and the area of the city is 850 km^2. Would you expect that the consumption of electric power significantly increases the ambient temperature?

*18. A simple gadget for heating water for showers consists of a black plastic bag holding 10 liters of water. When hung in the sun, the bag absorbs heat. On a clear, sunny day, the power delivered by sunlight per unit area facing the Sun is 1.0×10^3 W/m^2. The bag has an area of 0.10 m^2 facing the Sun. How long does it take for the water to warm from 20°C to 50°C? Assume that the bag loses no heat.

*19. In the heating system of a house, an oil furnace heats the water in a boiler, and the water is pumped into pipes connected to radiators. The hot water releases some of its heat in the radiators, and it then returns to the boiler, to be reheated. The furnace delivers 1.8×10^8 J of thermal energy per hour to the water, and the water leaves the boiler at 88°C and returns at 77°C. What rate of flow of the water (in m^3/s) is required to achieve this?

*20. A solar collector consists of a flat plate that absorbs the heat of sunlight. A water pipe attached to the back of the plate carries away the absorbed heat (Fig. 20.21). Assume that the solar collector has an area of 4.0 m^2 facing the Sun and that the power per unit area delivered by sunlight is 1.0×10^3 W/m^2. What is the rate at which water must circulate through the pipe if the temperature of the water is to increase by 40°C as it passes through the collector?

*21. A fast-flowing stream of water in a horizontal channel strikes the bottom rim of an undershot waterwheel of radius 2.2 m (see Fig. 8.24). The water approaches the wheel with a speed of 5.0 m/s and leaves with a speed of 2.5 m/s; the amount of water passing through is 300 kg/s.

(a) At what rate does the water deliver angular momentum to the wheel? What torque does the water exert on the wheel?

(b) If the angular velocity of the wheel is 1.4 radians/s, what is the power delivered to the wheel?

(c) How much does the temperature of the water increase as it passes through the rim of the wheel?

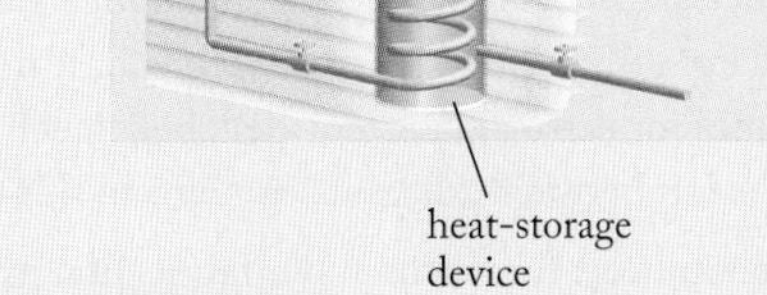

FIGURE 20.21 Collector of solar heat.

20.2 Thermal Expansion of Solids and Liquids

22. The tallest building in the United States is the Sears Tower in Chicago (Fig. 20.22), which is 443 m high. It is made of

concrete and steel. How much does its height change between a day when the temperature is 35°C and a day when the temperature is −29°C?

FIGURE 20.22 Sears Tower, Chicago.

23. The height of the Eiffel Tower is 321 m. What increase of temperature will lead to an increase of height by 10 cm?

24. Machinists use gauge blocks of steel as standards of length. A one-inch gauge block is supposed to have a length of 1 in., to within $\pm 10^{-6}$ in. In order to keep the length of the block within this tolerance, how precisely must the machinist control the temperature of the block?

25. A mechanic wants to place a sleeve (pipe) of copper around a rod of steel. At a temperature of 18°C the sleeve of copper has an inner diameter of 0.998 cm and the rod of steel has a diameter of 1.000 cm. To what temperature must the mechanic heat the copper to make it fit around the steel?

26. (a) Segments of steel railroad rails are laid end to end. In an old railroad, each segment is 18 m long. If they are originally laid at a temperature of −7°C, how much of a gap must be left between adjacent segments if they are to just barely touch at a temperature of 43°C?

    (b) In a modern railroad, each segment is continuously welded, typically 790 m long, for a smoother ride. A special expansion joint is used at each end. If no other allowance for expansion is provided, how much of a gap must be left between adjacent segments in this case?

27. A quartz photomask for silicon wafer fabrication must be positioned to within 1.0×10^{-7} m to match up with features from a previous quartz mask. If a wafer is 300 mm wide, what temperature change can be tolerated for accurate positioning across the entire wafer?

28. The outer walls of buildings include expansion joints (often filled with a soft caulking material). If the walls are made of concrete with expansion joints spaced every 10 m, how wide should the gaps be at low temperature to allow adequate expansion when the temperature rises 120°F?

29. Opticians immerse plastic eyeglass frames in a container of warm beads in order to expand the frames, permitting facile insertion of lenses. If the thermal expansion coefficient of such a plastic is 2.0×10^{-4}/°C and a lens is 0.75% larger than the opening into which it must be inserted, what minimum temperature increase of the plastic is required for easy insertion of the lens?

30. A particular experiment requires accurate positioning of a specimen at the end of a 1.5-m steel rod. In the experiment, the rod and sample are heated 200°C, while the surroundings stay at constant temperature. How far does the specimen move?

31. A 1.00-liter container made of glass is full to the brim with ethyl alcohol at −110°C (near its freezing point). How many cubic centimeters of alcohol overflow the container if the system is heated to +75°C (near the boiling point)?

32. An ordinary mercury thermometer consists of a glass bulb to which is attached a fine capillary tube. Given that the bulb has a volume of 0.20 cm^3 and that the capillary tube has a diameter of 7.0×10^{-3} cm, how far will the mercury column rise up the capillary tube for a temperature increase of 10°C? Ignore the expansion of the glass and ignore the expansion of the mercury in the capillary tube.

*33. Suppose you heat a 1.0-kg cube of iron from 20°C to 80°C while it is surrounded by air at a pressure of 1.0 atm. How much work does the iron do against the atmospheric pressure while expanding? Compare this work with the heat absorbed by the iron. (The density of iron is 7.9×10^3 kg/m^3.)

*34. When a solid expands, the increment of the area of one of its faces is directly proportional to the increment of temperature and to the original area. Show that the coefficient of proportionality for this expansion of area is 2 times the coefficient of linear expansion.

*35. A spring made of steel has a relaxed length of 0.316 m at a temperature of 20°C. By how much will the length of this spring increase if we heat it to 150°C? What compressional force must we apply to the hot spring to bring it back to its original length? The spring constant is 3.5×10^4 N/m.

*36. A wheel of metal has a moment of inertia I at some given temperature. Show that if the temperature increases by ΔT, the moment of inertia will increase by approximately $\Delta I = 2\alpha I \Delta T$.

*37. The pendulum (rod and bob) of a pendulum clock is made of brass.

    (a) What will be the fractional increment in the length of this pendulum if the temperature increases by 20°C? What will be the fractional increase in the period of the pendulum?

    (b) The pendulum clock keeps good time when its temperature is 15°C. How much time (in seconds per day) will the clock lose when its temperature is 35°C?

*38. (a) The density of gasoline is 730 kg/m^3 when the temperature is 0°C. What will be the density of gasoline when the temperature is 30°C?

(b) The price of gasoline is 60 cents per liter. What is the price per kilogram at 0°C? What is the price per kilogram at 30°C? Is it better to buy cold gasoline or warm gasoline?

**39. In order to compensate for deviations caused by temperature changes, a pendulum clock built during the nineteenth century for an astronomical observatory uses a large cylindrical glass tube filled with mercury as a pendulum bob. This tube is held by a brass rod and bracket (Fig. 20.23); the combined length of the rod and bracket is l (measured from the point of suspension of the pendulum). Neglecting the mass of the brass and the glass and neglecting the expansion of the glass, show that the height of the mercury in the glass tube must be

$$h = \left(\frac{2\alpha_{\text{brass}}}{\beta_{\text{mercury}}}\right) l$$

if the center of mass of the mercury is to remain at a fixed distance from the point of suspension, regardless of temperature.

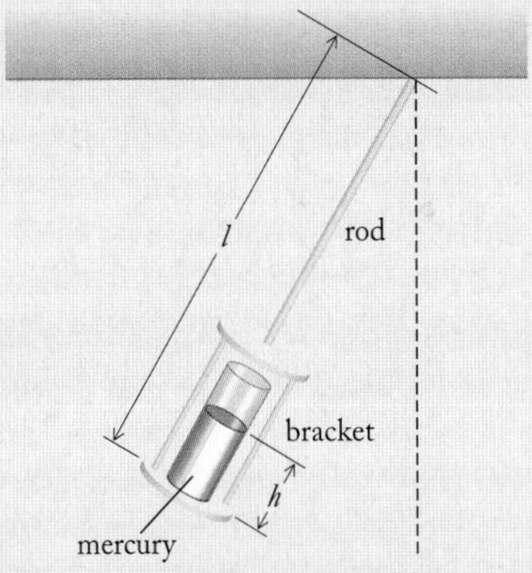

FIGURE 20.23 Pendulum with a temperature compensator.

20.3 Thermal Conduction

40. The walls of an igloo are made of compacted snow, 30 cm thick. What thickness of Styrofoam would provide the same insulation as the snow?

41. A pan of aluminum, filled with boiling water, sits on a hot plate. The bottom area of the pan is 300 cm^2, and the thickness of the aluminum is 0.10 cm. If the hot plate supplies 2000 W of heat to the bottom of the pan, what must be the temperature of the upper surface of the hot plate?

42. A rod of steel 0.70 cm in diameter is surrounded by a tight copper sleeve of inner diameter 0.70 cm and outer diameter 1.00 cm. What will be the heat flow along this compound rod if the temperature gradient along the rod is 50°C/cm? What fraction of the heat flows in the copper? What fraction in the steel?

43. A window in a room measures 1.0 m × 1.5 m. It consists of a single sheet of glass of thickness 2.5 mm. What is the heat flow through this window if the temperature difference between the inside surface of the glass and the outside is 39°C? Compare the heat loss through the window with the heat loss through the wall calculated in Example 6.

44. The bottom of a teakettle consists of a layer of stainless steel 0.050 cm thick welded to a layer of copper 0.030 cm thick. The area of the bottom of the kettle is 300 cm^2. The copper sits in contact with a hot plate at a temperature of 101.2°C, and the steel is covered with boiling water at 100.0°C. What is the rate of heat transfer through the bottom of the kettle from the hot plate to the water?

45. According to Eq. (20.13), the heat flow through a rod or slab of cross-sectional area A can be expressed as

$$\frac{\Delta Q}{\Delta t} = \frac{A\,\Delta T}{R}$$

where $R = \Delta x/k$ is called the thermal resistance, or the **R value** (see also the comments in Example 7). Since the heat flow is inversely proportional to the R value, a good insulator has a high R value.

(a) What is the R value of a slab of fiberglass insulation, 10 cm thick?

(b) In the United States, R values of commercially available insulation are commonly expressed in units of ft^2·°F·h/Btu. What is the R value of the 10-cm slab of fiberglass in these units?

46. An insulated coffee cup and lid contain coffee at 80°C. What is the initial heat flow per cm^2 of wall if the wall insulation is 1.0 cm thick and has a thermal conductivity of 0.080 J/s·m·°C? The ambient temperature is 22°C.

47. The water and the airspace in a tropical fish tank are maintained at 26°C by a heater when the temperature in the room is 18°C. If the walls, base, and lid of the fish tank are made of 3.0-mm-thick glass and the tank measures 80 cm × 50 cm × 30 cm, what average power (in watts) must be supplied by the heater?

48. What is the ratio of the heat flow though a glass door of thickness 3.0 mm to the heat flow through a wood door of thickness 25 mm? Each door has the same area and same temperature difference.

*49. Several slabs of different materials are piled one on top of another. All slabs have the same face area A, but their thicknesses and conductivities are Δx_i and k_i, respectively. Show that the heat flow through the pile of slabs is

$$\frac{\Delta Q}{\Delta t} = \frac{A\,\Delta T}{\sum_{i=1}^{n} \Delta x_i/k_i}$$

where ΔT is the temperature difference between the bottom of the first slab and the top of the last slab.

*50. A man has a skin area of 1.8 m^2; his skin temperature is 34°C. On a cold winter day, the man wears a whole-body suit insulated with down. The temperature of the outside surface of his suit is $-25°C$. If the man can stand a heat loss of no more than 4.0×10^5 J/h, what is the minimum thickness of down required for his suit?

*51. The end of a rod of copper 0.50 cm in diameter is welded to a rod of silver of half the diameter. Each rod is 6.0 cm long. What is the heat flow along these rods if the free end of the copper rod is in contact with boiling water and the free end of the silver rod is in contact with ice? What is the temperature of the junction? Assume that there is no heat loss through the lateral surfaces of the rods.

*52. A ceiling skylight uses a "thermal window" of area 2.0 m^2. The window consists of two sheets of 3.2-mm-thick glass, separated by a 2.0-mm-thick layer of argon gas. Take the thermal conductivity of argon to be 0.017 J/s·m·°C. The outside temperature is 38°C, and the inside temperature is 25°C. For such a horizontal gas layer, heated from above, convection can be neglected. Assume an overcast day so that radiation can also be neglected. Find the thermal energy entering the skylight per day. (Hint: See the discussion in Example 7 or use the result of Problem 49.)

**53. A cable used to carry electric power consists of a metallic conductor of radius r_1 encased in an insulator of inner radius r_1 and outer radius r_2. The metallic conductor is at a temperature T_1, and the outer surface of the insulator is at a temperature T_2. Show that the radial heat flow across the insulator is given by

$$\frac{\Delta Q}{\Delta t} = \frac{2\pi k(T_2 - T_1)l}{\ln(r_2/r_1)}$$

where k is the thermal conductivity of the insulator and l is the length of the cable.

20.4 Changes of State†

54. The icebox on a sailboat measures 60 cm × 60 cm × 60 cm. The contents of this icebox are to be kept at a temperature of 0°C for 4 days by the gradual melting of a block of ice of 20 kg, while the temperature of the outside of the box is 30°C. What minimum thickness of the Styrofoam insulation is required for the walls of the icebox?

*55. On a cold winter day, the water of a shallow pond is covered with a layer of ice 6.0 cm thick. The temperature of the air is $-20°C$, and the temperature of the water is 0°C. What is the (instantaneous) rate of growth of the thickness of the ice (in centimeters per hour)? Assume that the windchill keeps the top surface of the ice at exactly the temperature of the air, and assume that there is no heat transfer through the bottom of the pond.

*56. Suppose that the pond described in the preceding problem has a layer of compacted snow 3.0 cm thick on top of the ice. What is the rate of growth of the thickness of the ice?

†For help, see Online Concept Tutorial 22 at www.wwnorton.com/physics

57. Thunderstorms obtain their energy by condensing the water vapor contained in humid air. Suppose that a thunderstorm succeeds in condensing *all* the water vapor in 10.0 km^3 of air.

(a) How much heat does this release? Assume the air is initially at 100% humidity and that each cubic meter of air at 100% humidity (at 20°C and 1.0 atm) contains 1.74×10^{-2} kg of water vapor. The heat of vaporization of water is 2.45×10^6 J/kg at 20°C.

(b) The explosion of a nuclear bomb releases an energy of 8.0×10^{13} J. How many nuclear bombs does it take to make up the energy of one thunderstorm?

58. You place 1.0 kg of ice (at 0°C) in a pot and heat it until the ice melts and the water boils off, making steam. How much heat must you supply to achieve this?

59. Warm tea at 50°C is poured over ice to make iced tea. The specific heat of tea is the same as that of water. If 500 milliliters of tea are to be cooled to 5.0°C, how many grams of ice at 0.0°C are needed?

60. In magnets used for magnetic resonance imaging (MRI), both liquid nitrogen and liquid helium (which costs about 10 times more) are used; the magnet is precooled with liquid nitrogen to $-196°C$, and then cooled to $-269°C$ with liquid helium. Considering only the latent heat of vaporization, how many liters of liquid nitrogen are needed to cool 20 kg of copper from 20°C to $-196°C$? How many liters of liquid helium would be needed to do the same job? (In reality, some cooling is also provided by the warming of the vaporized gas.)

61. Large central air conditioners for entire homes are typically rated in "tons"; this is the mass of ice that would have to melt in one day to remove the same amount of heat. If such a "ton" corresponds to 907 kg, how much heat in joules is removed per day by a 3.5-ton air-conditioning system? What is this rate of energy removal in watts?

*62. A liquid can be cooled and even frozen by pumping with a vacuum pump above the liquid; some of the liquid is vaporized, leaving behind a colder liquid, and, eventually, a solid. If 1.0 kg of liquid nitrogen at $-196°C$ is pumped on, how much frozen nitrogen at $-210°C$ is obtained? The specific heat of liquid nitrogen is 2.0×10^3 J/kg·°C. Assume for simplicity that the heats of fusion and vaporization do not vary significantly with temperature or pressure.

*63. The vaporization of liquid nitrogen is used in a popular demonstration to make ice cream. Assume that cream has a specific heat of 2900 J/kg·°C and has the same heat of fusion and melting point as water, and that ice cream has a specific heat of 2200 J/kg·°C. If two kilograms of cream are initially at 10°C, how much liquid nitrogen must be vaporized to cool, freeze, and further cool the cream to $-10°C$?

*64. The heat of vaporization of water at 100°C and 1.0 atm is 2.26×10^6 J/kg. How much of this energy is due to the work the water vapor does against atmospheric pressure? What would this work be at a pressure of 0.10 atm? At (nearly) zero pressure?

*65. During a rainstorm lasting 2 days, 7.6 cm of rain fell over an area of 2.6×10^3 km^2.

(a) What is the total mass of the rain (in kilograms)?

(b) Suppose that the heat of vaporization of water in the rain clouds is 2.43×10^6 J/kg. How many calories of heat are released during formation of the total mass of rain by condensation of the water vapor in these clouds?

(c) Suppose that the rain clouds are at a height of 1500 m above the ground. What is the gravitational potential energy of the total mass of rain before it falls? Express your answer in joules and in calories.

(d) Suppose that the raindrops hit the ground with a speed of 10 m/s. What is the total kinetic energy of all the raindrops taken together? Express your answer in joules and in calories. Why does your answer to part (c) not agree with this?

*66. If you pour 0.50 kg of molten lead at 328°C into 2.5 liters of water at 20°C, what will be the final temperature of the water and the lead? The specific heat of (solid) lead has an average value of 140 J/kg·°C over the relevant temperature range.

*67. Suppose you drop a cube of titanium of mass 0.25 kg into a Dewar flask (a thermos bottle) full of liquid nitrogen at −196°C. The initial temperature of the titanium is 20°C. How many kilograms of nitrogen will boil off as the titanium cools from 20°C to −196°C? The specific heat of titanium is 340 J/kg·°C.

*68. While jogging on a level road, your body generates heat at the rate of 750 kcal/h. Assume that evaporation of sweat removes 50% of this heat, and convection and radiation the remainder. The evaporation of 1.0 kg (or 1.0 liter) of sweat requires 580 kcal. How many kilograms of sweat do you evaporate per hour?

*69. The Mediterranean loses a large volume of water by evaporation. The loss is made good, in part, by currents flowing into the Mediterranean through the straits joining it to the Atlantic Ocean and the Black Sea. Calculate the rate of evaporation (in km^3/h) of the Mediterranean on a clear summer day from the following data: the area of the Mediterranean is 2.9×10^6 km^2, the power per unit area supplied by sunlight is 1.0×10^3 W/m^2, and the heat of vaporization of water is 2.43×10^6 J/kg (at a temperature of 21°C). Assume that all the heat of sunlight is used for evaporation.

20.5 The Specific Heat of a Gas

70. A TV tube of glass with zero pressure inside and atmospheric pressure outside suddenly cracks and implodes. The volume of the tube is 2.5×10^{-2} m^3. During the implosion, the atmosphere does work on the fragments of the tube and on the layer of air immediately adjacent to the tube. This amount of work represents the energy released in the implosion. Calculate this energy. If all of this energy is acquired by the fragments of the glass, what will be the mean speed of the fragments? The total mass of the glass is 2.0 kg.

71. The rear end of an air conditioner dumps 1.2×10^7 J/h of waste heat into the air outside a building. A fan assists in the removal of this heat. The fan draws in 15 m^3/min of air at a temperature of 30°C and ejects this air after it has absorbed the waste heat. With what temperature does the air emerge?

72. What are the specific heats C_V and C_p for air consisting of 75% nitrogen, 24% oxygen, and 1% argon (by mass) at STP? Use the values for C_V and C_p of nitrogen, oxygen, and argon listed in Table 20.5.

73. Table 20.5 gives the specific heat C_V per mole for several gases. Calculate the specific heat per kilogram for each gas. Which gas has the highest value of the specific heat per kilogram? The lowest?

74. If we heat 1.00 kg of hydrogen gas from 0.0°C to 50.0°C in a cylinder with a piston keeping the gas at a constant pressure of 1.00 atm, we must supply 7.08×10^5 J of thermal energy. How much work does the gas deliver to the cylinder during this process? How many joules of thermal energy must we supply to heat the same amount of gas from 0.0°C to 50.0°C in a container of constant volume?

75. The theoretical expression for the speed of sound in a gas is $\sqrt{\gamma p/\rho}$ [compare Eq. (17.5)]. Calculate the speed of sound in helium gas at STP. See Table 20.5 for $\gamma = C_p/C_V$.

76. A helium balloon consists of a large bag loosely filled with 600 kg of helium at an initial temperature of 10°C. While exposed to the heat of the Sun, the helium gradually warms to a temperature of 30°C. The heating proceeds at a constant pressure of 1.0 atm. How much heat does the helium absorb during this temperature change?

77. As in Example 9, consider an initial volume of 1.0 m^3 of air that is warmed by sunlight from 0°C to 10°C at a constant pressure of 1.0 atm. What is the change in volume of this air? During its expansion, how much work does the air do against the pressure of the surrounding atmosphere?

78. A quartz tube contains one mole of helium gas at 20°C. The gas is heated at constant volume to 300°C. How much thermal energy is transferred to the gas? If the same amount of gas were heated at constant pressure, how much energy would be required?

79. A gas mixture is made consisting of 1.00 mole of helium gas and 2.00 moles of oxygen gas. What is the molar specific heat of this mixture at constant volume? At constant pressure?

*80. An air conditioner removes heat from the air of a room at the rate of 8.0×10^6 J/h. The room measures 5.0 m × 5.0 m × 2.5 m, and the pressure is constant at 1.0 atm.

(a) If the initial temperature of the air in the room is 30.0°C, how long does it take the air conditioner to reduce the temperature of the air by 5.0°C? Pretend that the mass of the air in the room is constant.

(b) As the air in the room cools, it contracts slightly and draws in some extra air from the outside; hence the mass of air is not exactly constant. Repeat your calculation taking into account this increase of the mass of air. Assume that the extra air enters with an initial temperature of 30°C. Does the result of your second calculation differ appreciably from that of your first calculation?

*81. On a winter day you inhale cold air at a temperature of −30°C and at 0% humidity. The amount of air you inhale is 0.45 kg per hour. Inside your body you warm and humidify the air; you then exhale the air at a temperature of 37°C and 100% relative humidity. At a temperature of 37°C, each kilogram of air at 100% relative humidity contains 0.041 kg of water vapor. How many calories are carried out of your body by the air that passes through your lungs in one hour? Take into account both the heat needed to warm the air at constant pressure and the heat needed to vaporize the moisture that the exhaled air carries out of your body. The specific heat of air at constant pressure is 1.0×10^3 J/kg·°C; the heat of vaporization of water at 37°C is 2.42×10^6 J/kg.

20.6 Adiabatic Expansion of a Gas

82. (a) Show that for a small adiabatic expansion,

$$\frac{dp}{p} = -\left(1 + \frac{R}{C_V}\right)\frac{dV}{V}$$

Since $R = C_p - C_V$, this can also be written in the form

$$\frac{dp}{p} = -\frac{C_p}{C_V}\frac{dV}{V}$$

(b) If an ideal monatomic gas adiabatically expands by 3%, what is the percent decrease of pressure?

*83. Under normal conditions, the temperature of air in the atmosphere decreases with altitude; that is, the air temperature at the top of a mountain is lower than that at the foot of a mountain. This temperature difference is maintained by winds that move the air from one place to another. When a wind carries a parcel of air up the side of a mountain, the parcel expands adiabatically and cools; when the wind carries the parcel of air down, it is compressed adiabatically and warms. Calculate the temperature difference between the bottom and the top of a mountain 100 m high. The temperature at the bottom is 20°C. The pressure at the bottom is 1.00 atm, and the pressure at the top is 0.988 atm. For air, the specific heats are those of nitrogen and oxygen (see Table 20.5).

84. Argon gas is compressed without heat loss to one-tenth of its initial volume. If the initial temperature is 25°C, what is the final temperature?

85. A fire extinguisher is filled with 1.0 kg of nitrogen gas at a pressure of 1.2×10^6 N/m^2 and a temperature of 20°C.

(a) What is the volume of this gas?

(b) If the gas is allowed to escape adiabatically against atmospheric pressure, what will be the volume and temperature of the expanded gas?

86. Suppose that you have a sample of oxygen gas at a pressure of 300 atm and a temperature of −29°C. If you suddenly (adiabatically) let this gas expand to a final pressure of 1.00 atm, what will be the final temperature? (This method was used by Cailletet in 1877 to liquefy oxygen; oxygen liquefies at −183°C.)

87. Air in an automobile tire is at an overpressure of 2.0 atm and a temperature of 20°C. The pressure outside the tire is at 1.0 atm. If you let some air escape through the valve, what will be the final temperature of the escaping air? Assume that the air expands adiabatically.

*88. By means of a hand pump, you inflate an automobile tire from 0.0 atm to 2.4 atm overpressure. The volume of the tire remains constant at 0.10 m^3. How much work must you do on the air with the pump? Assume that each stroke of the pump is an adiabatic process and that the air is initially at STP.

REVIEW PROBLEMS

89. In the cooling system of the engine of a boat, water is pumped from the outside through the engine and then returned to the outside. The engine produces 3.0 kW of waste heat, and the pump circulates 12 liters of water per minute through the engine. If all of the waste heat is carried away by the water, what is the increase in the temperature of the water?

90. A shallow pond has a depth of 50 cm. On a winter day, the initial temperature of the water is 6.0°C. If a snowstorm deposits a 50-cm layer of snow (which melts to form a 5.0-cm layer of water) on this pond, what will be the final temperature of the water? Assume that the initial temperature of the snow is 0.0°C, and that there is no heat exchange between the pond and the air or the ground.

91. A radiator, fed by hot water, is used to heat a room in a house. According to the manufacturer's specifications, the output of the radiator is 7.5×10^6 J/h when the inflowing water is at 88°C and the outflowing water at 77°C. What must be the rate of flow of the water through the radiator to achieve this output?

*92. The beam dump at the Stanford Linear Accelerator Center (SLAC) consists of a large tank with 12 m^3 of water into which the accelerated electrons can be aimed when they are not wanted elsewhere (Fig. 20.24). The beam carries 3.0×10^{14} electrons/s; the kinetic energy per electron is 3.2×10^{-9} J. In the beam dump this energy is converted into heat.

(a) What is the rate of production of heat?

(b) If the water in the tank is stagnant and does not lose any heat to the environment, what is the rate of increase of temperature of the water?

(c) To prevent overheating, cooling water is pumped through the tank at the rate of 2.0 m^3/min; this carries away the heat. If the temperature of the inflowing water is 20°C, what is the temperature of the outflowing water?

FIGURE 20.24 Beam dump at SLAC.

93. The largest ship is the supertanker *Seawise Giant,* with a length of 458 m and a beam of 69 m. By how many meters does this tanker expand in length and in width when it travels from the wintry North Atlantic (−20°C) to the hot Persian Gulf (+40°C)? By how many square meters does its deck area increase? Assume that the deck is approximately a rectangle, 458 m × 69 m.

94. The supertanker *Seawise Giant* (see also Problem 93) has an enclosed volume of 1.8×10^6 m^3. By how many cubic meters does its volume expand when it travels from the wintry North Atlantic (−20°C) to the hot Persian Gulf (+40°C)?

*95. A Styrofoam box, used for the transportation of medical supplies, is filled with dry ice (carbon dioxide) at a temperature of −79°C. The box measures 30 cm × 30 cm × 40 cm, and its walls are 4.0 cm thick. If the outside surface of the box is at a temperature of 20°C, what is the rate of loss of dry ice by vaporization?

*96. The wall of a house is to be built of a layer of wood, an adjacent layer of fiberglass insulation, and an adjacent layer of brick. The thickness of the wood is 1.2 cm, and that of the brick is 10 cm. The heat loss per square meter of wall is to be no more than 4.0×10^4 J/h when the temperatures inside and outside the house differ by 30°C. How thick must the fiberglass insulation be?

97. You dump a 0.30-kg chunk of dry ice (solid carbon dioxide) into a beaker containing a water–ice mixture. How much more ice will form in the beaker as the dry ice bubbles away, becoming carbon dioxide gas?

98. A blacksmith drops a 0.70-kg horseshoe of iron at a temperature of 1200°C into a bucket containing half a liter of water at an initial temperature of 30°C. How much of the water boils off? Assume that the bucket absorbs none of the heat.

99. How much heat must we supply to heat 1.0 mole of argon gas from 30°C to 100°C at constant volume? At constant pressure?

100. To discover whether an unknown gas is monatomic or diatomic, an experimenter takes a 2.0-liter sample of the gas at STP and heats this sample to 100°C at constant volume.

(a) The heat that the gas absorbs during this process is 180 J. Is the gas monatomic or diatomic? Assume it is an ideal gas.

(b) The experimenter weighs the sample and finds that its mass is 2.5 g. Can you tell what gas it is?

101. Suppose we heat 1.0 mole of oxygen gas at a constant pressure of 1.0 atm from 20°C to 80°C, and then cool it at a constant volume from 80°C back to 20°C.

(a) How much heat is absorbed by the gas during the first step?

(b) How much heat is released by the gas during the second step?

(c) What is the volume of the gas at the end of the first step? What is the pressure at the end of the second step?

(d) How much work does gas perform during the first step? During the second step?

Answers to Checkups

Checkup 20.1

1. Since the masses and heat capacities of the two entities being mixed are the same (and using the fact that the heat capacity of water is approximately independent of temperature), Eq. (20.3) and a common final temperature imply that the mixture will attain the average temperature 50°C.

2. Since 1 kcal would change the temperature of 1 kg of water by 1°C, Eq. (20.3) implies that for the same amount of heat and mass, but one-ninth the specific heat, the temperature change would be 9 times as large, or 9°C.

3. The laborer must apply a force during motion and so does work on the crate. This work transfers mechanical energy to the crate. The laborer does not transfer thermal energy to the crate. The floor does negative work on the crate because the floor exerts a frictional force in a direction opposite to the motion.

4. (B) 0.10°C. We equate the mechanical work done, $W = F \times \Delta y = 1000\text{ N} \times 4.187\text{ m} = 4187\text{ J}$, with the heat of Eq. (20.3), $Q = mc\Delta T$: $4187\text{ J} = 10\text{ kg} \times 4187\text{ J/(kg·°C)} \times \Delta T$; and we obtain $\Delta T = 0.10$°C.

Checkup 20.2

1. The fractional volume expansion is 3 times the fractional linear expansion; thus, for a 0.02% linear expansion, the corresponding volume expansion is 0.06%.
2. From Fig. 20.5, between 0°C and 4°C the volume decreases from 1000.14 cm^3 to 1000.00 cm^3, a change of 0.14/1000, or 0.014%.
3. The coefficients of expansion of aluminum and concrete do not match (from Table 20.2, aluminum expands twice as much as concrete), so the rods would loosen from the concrete, or cause the concrete to crack.
4. (A) Increases. A hole or cavity in a material expands just as though it were filled with the same material as that which surrounds the hole or cavity.

Checkup 20.3

1. From Table 20.3, the thermal conductivity of wood is 3 times that of fiberglass. Thus, to provide the same insulation, the solid wood walls would have to be 3 times as thick, or 18 in. (45 cm).
2. Since heat always flows from the hotter body to the colder body, you can assert that the temperature of the steel is below 0°C.
3. Reducing the wall thickness will increase the heat loss [smaller Δx in Eq. (20.13)], increasing the wall area will also increase the heat loss [increasing A in Eq. (20.13)], reducing the external temperature to −20°C will increase the heat loss (larger ΔT), and reducing the internal temperature to 19°C will decrease the heat loss (smaller ΔT).
4. (D) Convection, radiation, conduction. The hair dryer blows warm air; heat transfer by fluid motion is convection. The heat lamp radiates light and infrared radiation. The hot-water bottle is used in contact with the body, transferring heat by conduction.

Checkup 20.4

1. The gases at −200°C are those materials with boiling points *below* −200°C; in Table 20.4, these are helium and hydrogen. The liquids are those materials with boiling points above −200°C and melting points below −200°C; nitrogen and oxygen satisfy these conditions. The solids are all the other materials; these all have melting points above −200°C.
2. The heat released when freezing refers to the heat of fusion; for the materials in Table 20.4, this is largest for aluminum.
3. For all of the materials in Table 20.4, the heat of vaporization is greater than the heat of fusion; thus, more heat is required for vaporization than for melting.
4. (D) b, c, a. From Table 20.4, vaporization of 1.0 kg requires 2.3×10^6 J and melting requires 3.3×10^5 J. From Eq. (20.4), heating from 0°C to 100°C will require approximately 4.2×10^5 J. Thus, in increasing order, they are melting, heating, and vaporization.

Checkup 20.5

1. C_p is larger than C_V because work is done when heating at constant pressure: to maintain constant pressure, the volume must increase when heating ($pV = nRT$). A force (pressure times area) applied during the displacement (expanding walls or moving piston) represents work done by the gas.
2. From Table 20.5, both helium and argon have the lowest value of the specific heat. Of these, argon atoms have the greater mass and thus the lower rms speed [see Eq. (19.23)], and so argon has the lower thermal conductivity (and is used in practice).
3. The gas in the balloon will require more heat, since it is both changing temperature (internal energy) and expanding against atmospheric pressure (work done). For an expanding balloon, the difference in specific heats will be smaller than 8.31 J/C·mole, since the elastic properties of the balloon imply that the gas is actually expanding against *increasing* pressure (so the volume change, and the work done, will be smaller than at constant pressure).
4. If the gas expands against a *decreasing* pressure, then the volume change (and thus the work done) will be greater than at constant pressure. This occurs, for example, with an air bubble in water, expanding while it ascends.
5. (D) 140°C. Air is essentially diatomic, so $C_p = \frac{7}{2}R$ and $C_V = \frac{5}{2}R$ [Eq. (20.33)]. Because C_V is smaller than C_p, the same added heat results in a temperature increase larger by a factor $C_p/C_V = \frac{7}{2}/\frac{5}{2} = \frac{7}{5} = 1.4$. Thus the temperature increase at constant volume is $1.4 \times 100°C = 140°C$.

Checkup 20.6

1. No. For an adiabatic process, no heat is transferred to or from the environment. For the heated room, both the heat supplied by the heater and the escaping air carrying heat to the environment violate adiabatic conditions.
2. In a fluid, pressure decreases with height. So as a parcel of air drifts upward, its pressure decreases. Under adiabatic conditions, the volume of the parcel must therefore increase [Eq. (20.47)], and the temperature must decrease [Eq. (20.45)].
3. (A) Monatomic. From Eq. (20.45), $T_2/T_1 = (V_1/V_2)^{\gamma-1}$. From Table 20.5, the monatomic gas has the largest value of γ, and so for any $V_1 > V_2$ (compression) the monatomic gas attains the highest temperature.

CHAPTER

21

Thermodynamics

CONCEPTS IN CONTEXT

The operation of this steam-engine locomotive requires the transfer of heat in order to do work. As we learn about machines that generate mechanical energy, we can ask:

? How do we measure the efficiency of an engine, and how efficient is a typical steam engine? (Example 2, page 666)

? How efficient could an ideal steam engine be? (Example 4, page 671)

? How does a practical steam engine differ from an ideal engine in operation? (Section 21.2, page 671)

Thermodynamics *is the branch of physics that deals with the conversion of one form of energy into another, especially the conversion of heat into other forms of energy.* These conversions are governed by the two fundamental laws of thermodynamics. As we will see in this chapter, the first of these is essentially a general statement of the Law of Conservation of Energy, and the second is a statement about the maximum efficiency attainable in the conversion of heat into work.

Thermodynamics describes physical processes in terms of purely macroscopic parameters. Such a macroscopic, large-scale description is necessarily somewhat crude, since it overlooks the small-scale, microscopic phenomena, such as the collisions of molecules with the walls of the container we investigated in Section 19.3. However, in practical applications, a knowledge of the microscopic phenomena is often unnecessary. For instance, an engineer investigating the combustion of fuel in a rocket engine will find it satisfactory to deal with only such macroscopic quantities as temperature, pressure, density, and heat capacity, and ignore the microscopic behavior of the gases.

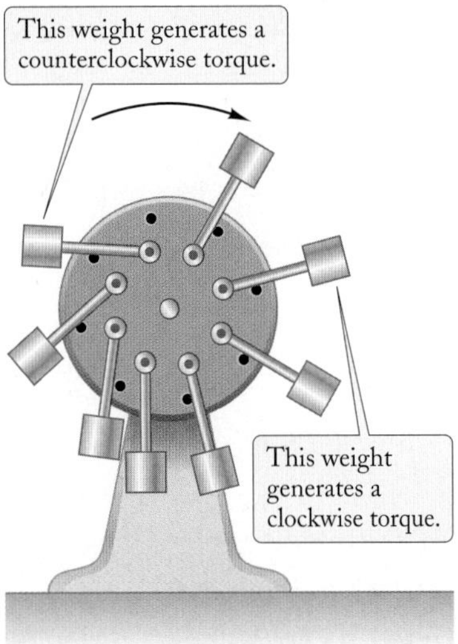

FIGURE 21.1 A hypothetical perpetual motion machine.

The development of steam engines for the industrial generation of (macroscopic) mechanical energy from heat motivated a careful examination of the theoretical principles underlying the operation of such engines, which led to the discovery of the Law of Conservation of Energy and to the recognition that heat is a form of energy transfer. Steam engines and other heat engines do not create energy; they merely convert thermal energy into mechanical energy, which can be used to perform useful work. For example, the steam engine of an old-fashioned locomotive converts thermal energy from the combustion of coal into mechanical energy, and the engine of an automobile converts thermal energy from the combustion of gasoline into mechanical energy.

Nineteenth-century engineers inaugurated the study of thermodynamics to discover what ultimate limitations the laws of physics impose on the operation of steam engines and other machines that generate mechanical energy. They soon established that perpetual motion machines, which earlier inventors had sought to build, are impossible. A **perpetual motion machine of the first kind** is a (hypothetical) device that supplies an endless output of work without any input of fuel or any other input of energy. Figure 21.1 shows a proposed design for such a machine. Weights are attached to the rim of a wheel by short, pivoted rods resting against pegs. With the rods in the position shown, there is an imbalance in the weight distribution causing a clockwise torque on the wheel; as the wheel turns, the rod coming to the top presumably flips over, maintaining the imbalance. This perpetual torque would not only keep the wheel turning, but would also continually deliver energy to the axle of the wheel. However, a detailed analysis demonstrates that the machine will not perform as intended—the wheel actually settles in a static equilibrium configuration such that the top rod just barely fails to flip over. The First Law of Thermodynamics, or the Law of Conservation of Energy, directly tells us of the failure of this machine: after one revolution of the wheel, the masses all return to their initial positions, their potential energy returns to its initial value, and they will not have delivered net energy to the motion of the wheel.

perpetual motion machine of the first kind

A **perpetual motion machine of the second kind** is a device that extracts thermal energy from some heat source, such as air or the water of the ocean, and converts it into mechanical energy. Such a device is not forbidden by conservation laws. The oceans are enormous reservoirs of thermal energy; if we could extract this thermal energy, a temperature drop of just 1°C of the oceans would supply the energy needs of the United States for the next 50 years. But, as we will see, the Second Law of Thermodynamics tells us that conversion of heat into work requires not only a heat

perpetual motion machine of the second kind

source, but also a heat sink. Heat flows out of a warm body only if there is a cooler body that can absorb heat. If we want heat to flow from the ocean into our machine, we must provide a low-temperature heat sink toward which the heat will tend to flow spontaneously. If no low-temperature sink is available, the extraction of heat from the ocean is impossible, and we cannot build a perpetual motion machine of the second kind.

21.1 THE FIRST LAW OF THERMODYNAMICS

Consider some amount of gas with a given initial volume V_1, pressure p_1, and temperature T_1. The gas is in a container fitted with a piston (see Fig. 21.2). Suppose we compress the gas to some smaller volume V_2 and also lower its temperature to some smaller value T_2; the pressure will then reach some new value p_2. (Such a compression and cooling process is of practical importance in the liquefaction of gases, say, oxygen or nitrogen—before the gas can be liquefied, it must be compressed and cooled.) Obviously, we can reach the new state V_2, p_2, and T_2 from the old state V_1, p_1, and T_1 in a variety of ways. For instance, we may first compress the gas and then cool it. Or else, we may first cool it and then compress it. Or we may go through small alternating steps of compressing and cooling. In order to compress the gas, we must do work on it [see Eq. (20.23)]; and in order to cool the gas, we must remove heat from it. The work done on or by the gas and the heat transferred from or to the gas result in a change of the internal energy of the gas. We can express this change of internal energy as

$$\Delta E = Q - W \tag{21.1}$$

where Q is the amount of heat transferred to the gas and W is the amount of work performed by the gas. Note the sign conventions in this equation: Q is positive if we add heat to the gas and negative if we remove heat; W is positive if the gas does work on us and negative if we do work on the gas (see Fig. 21.3).

The values of Q and W depend on the process. If we first compress the gas adiabatically and subsequently cool it at constant volume, then during the first step W is negative and Q zero; and during the second step W is zero and Q is negative. If we first cool the gas and then compress it, the values of W and Q will be quite different. Yet, it turns out that regardless of what sequence of operations we use to transform the gas from its initial state V_1, p_1, T_1 to its final state V_2, p_2, T_2, the net change ΔE in the internal energy is always the same: Q and W vary, but the sum of Q and $-W$ remains fixed. This is the **First Law of Thermodynamics**:

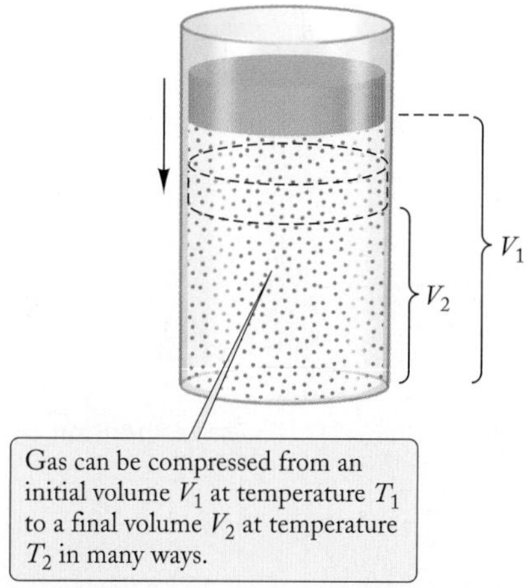

FIGURE 21.2 Compression of a gas by a piston.

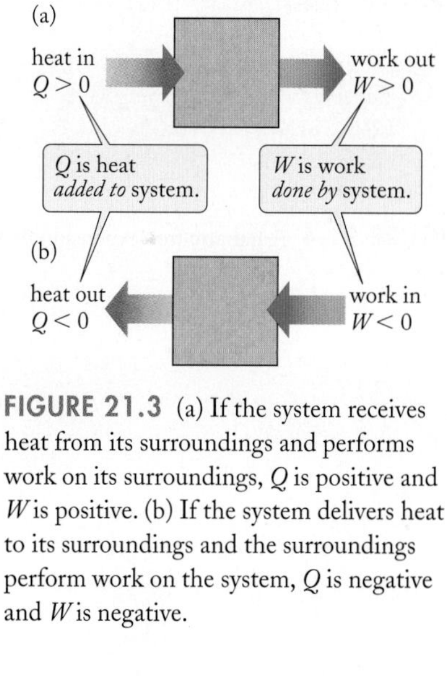

FIGURE 21.3 (a) If the system receives heat from its surroundings and performs work on its surroundings, Q is positive and W is positive. (b) If the system delivers heat to its surroundings and the surroundings perform work on the system, Q is negative and W is negative.

> ***When several alternative processes involving heat and work are available to change a system from an initial state characterized by given values of the macroscopic parameters to a final state characterized by new values of the macroscopic parameters, the amounts Q of heat and W of work depend on the process. But the change in the internal energy of the system***

$$\Delta E = Q - W \tag{21.2}$$

First Law of Thermodynamics

> ***has a fixed value which does not depend on the process.***

Note that the First Law tells us that the energy is conserved—the change of internal energy of the system equals the input of heat minus the output of work, that is, it equals the input of microscopic work minus the output of macroscopic work. But the First Law tells us more than that. If we describe a system in terms of the detailed microscopic positions and velocities of all its constituent particles, then energy conservation is uncontestable—it is a theorem of mechanics. But if we describe a system in terms of nothing but macroscopic parameters, then it is not at all obvious that we have available enough information to determine the energy and to formulate a conservation law. The First Law of Thermodynamics tells us that a knowledge of the macroscopic parameters is indeed sufficient to determine the energy of the system.

EXAMPLE 1 Some amount of gas at temperature T_1 is stored in a thermally insulated bottle. By means of a pipe with a valve, we connect this bottle to another insulated bottle which is evacuated (see Fig. 21.4). If we suddenly open the valve, the gas will rush from the first bottle into the second until the pressures are equalized. This is called a **free expansion** of the gas, because the gas expands without pushing against anything. What does the First Law say about the change of the internal energy of the gas in this process?

free expansion

SOLUTION: The use of thermally insulated bottles makes the expansion adiabatic, so the expansion process neither adds nor removes heat from the gas, that is, $Q = 0$. Furthermore, the expansion process involves no work (the gas does not push against any moving piston), that is, $W = 0$. Consequently, Eq. (21.2) tells us that the internal energy of the gas does not change:

$$\Delta E = Q - W = 0 \tag{21.3}$$

We can take this conclusion a step further if we assume that the gas behaves like an ideal gas. If so, the internal energy E depends on the temperature only [E is directly proportional to T; see Eqs. (19.26) and (19.33)]. Since the energy does not change, we can then conclude that the temperature does not change. Thus, in the free expansion of the gas, the temperature remains constant.

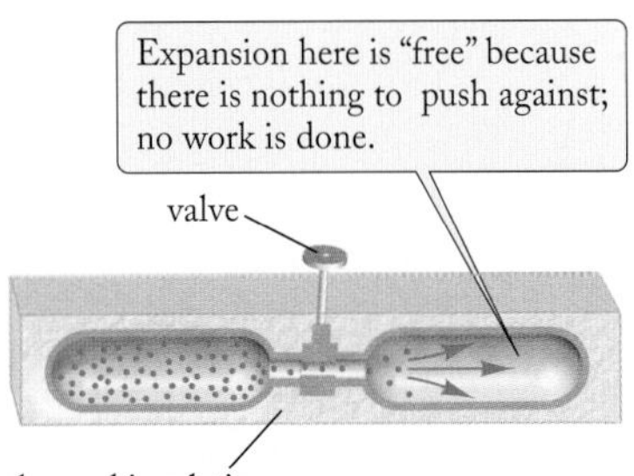

FIGURE 21.4 Adiabatic free expansion of a gas.

✔ Checkup 21.1

QUESTION 1: Consider some fixed amount of gas sealed in a rubber balloon. For this system, what are the signs of Q and W if the balloon is heated by sunlight and expands?

QUESTION 2: Consider some fixed amount of gas sealed in a rigid glass jar. For this system, what are the signs of Q and W if the jar is heated?

QUESTION 3. An athlete pedals an exercise bicycle until he gets hot and begins to sweat. If we regard the athlete as the system, what are the respective signs of Q and W for this exercise process?

(A) Positive, positive
(B) Positive, negative
(C) Negative, positive
(D) Negative, negative

21.2 HEAT ENGINES; THE CARNOT ENGINE

Online Concept Tutorial 23

Steam engines and automobile engines convert heat into mechanical energy. The steam engine in Fig. 21.5 obtains heat from the combustion of coal or oil in a boiler; the automobile engine (Fig. 21.6) obtains heat from the (explosive) combustion of gasoline in its cylinders. If an engine is to convert heat continually into mechanical energy, it must operate cyclically. At the end of each cycle, it must return to its initial configuration, so it can repeat the process of conversion of heat into work over and over again. Steam engines and automobile engines are obviously cyclic—after one (or sometimes two) revolutions of their crankshaft or flywheel, they return to their initial configuration. These engines are not 100% efficient. The condenser of the steam engine and the radiator and exhaust of the automobile engine eject a substantial amount of heat into the environment; this waste heat represents lost energy.

Any device that converts heat into work by means of a cyclic process is called a **heat engine**. The engine absorbs heat from a heat reservoir at high temperature, converts this heat partially into work, and ejects the remainder as waste heat into a reservoir at low temperature. In this context, a **heat reservoir** is simply a body that remains at constant temperature, even when heat is removed from or added to it. In practice, the high-temperature heat reservoir is often a boiler whose temperature is kept constant by the controlled combustion of some fuel, and the low-temperature reservoir is usually a condenser in contact with a body of water or in contact with the atmosphere of the Earth, whose large volume permits it to absorb the waste heat without appreciable change of temperature.

heat engine

heat reservoir

Figure 21.7 is a flowchart for the energy, showing the heat Q_1 flowing into the engine from the high-temperature reservoir, the heat Q_2 (waste heat) flowing out of the engine into the low-temperature reservoir, and the work generated. The work generated by the engine is the difference between Q_1 and Q_2,

$$W = Q_1 - Q_2 \tag{21.4}$$

FIGURE 21.5 A steam engine.

FIGURE 21.6 An automobile engine.

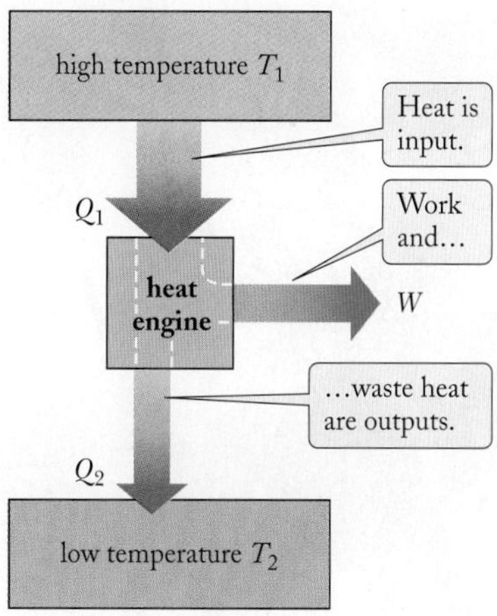

FIGURE 21.7 Flowchart for a heat engine. The center cube represents the heat engine.

The **efficiency** *of the engine is defined as the ratio of this work to the heat absorbed from the high-temperature reservoir*:

efficiency of heat engine

$$e = \frac{W}{Q_1} = \frac{Q_1 - Q_2}{Q_1} = 1 - \frac{Q_2}{Q_1} \tag{21.5}$$

This says that if $Q_2 = 0$ (no waste heat), then the efficiency would be $e = 1$, or 100%. If so, the engine would convert the high-temperature heat *totally* into work. As we will see later, this extreme efficiency is unattainable. Even under ideal conditions, the engine will produce some waste heat. It turns out that the efficiency of an ideal engine depends only on the temperatures of the heat reservoirs.[1]

Concepts — in — Context

EXAMPLE 2 The steam engine of a locomotive delivers 5.4×10^8 J of work per minute and receives 3.6×10^9 J of heat per minute from its boiler. What is the efficiency of this engine? How much heat is wasted per minute?

SOLUTION: From Eq. (21.5), the efficiency is the ratio of the work generated to the heat absorbed:

$$e = \frac{W}{Q_1} = \frac{5.4 \times 10^8 \text{ J}}{3.6 \times 10^9 \text{ J}} = 0.15 \tag{21.6}$$

Expressed in percent, this is 15%.

The wasted heat is the difference between the heat received and the work:

$$Q_2 = Q_1 - W = 3.6 \times 10^9 \text{ J} - 5.4 \times 10^8 \text{ J} = 3.1 \times 10^9 \text{ J} \tag{21.7}$$

FIGURE 21.8 Athlete on a stationary bicycle.

EXAMPLE 3 During strenuous bicycling on a stationary bicycle (see Fig. 21.8), an athlete delivers 220 W of mechanical power to the pedals of the bicycle and, simultaneously, generates 760 W of waste heat. What overall efficiency is implied by these data?

SOLUTION: A power of 1 W is equal to 1 J per second. Thus, in one second, the mechanical work delivered by the athlete is 220 J and the waste heat generated is 760 J. Accordingly, the chemical energy flowing into the athlete's muscles must be $Q_1 = Q_2 + W = 760 \text{ J} + 220 \text{ J} = 980 \text{ J}$, which implies an efficiency of

$$e = \frac{W}{Q_1} = \frac{220 \text{ J}}{980 \text{ J}} = 0.22$$

Thus the efficiency of this process is 22%.

[1] Note that according to the sign conventions summarized in Fig. 21.3, the heat flowing into the engine is supposed to be a positive quantity and the heat flowing out of the engine a negative quantity. But in discussions of the efficiency of engines, it is customary to treat both Q_1 and Q_2 as positive quantities, and an explicit minus sign therefore appears with Q_2 in Eq. (21.4); that is, $Q = Q_1$ for the absorbed heat, but $Q = -Q_2$ for the rejected heat.

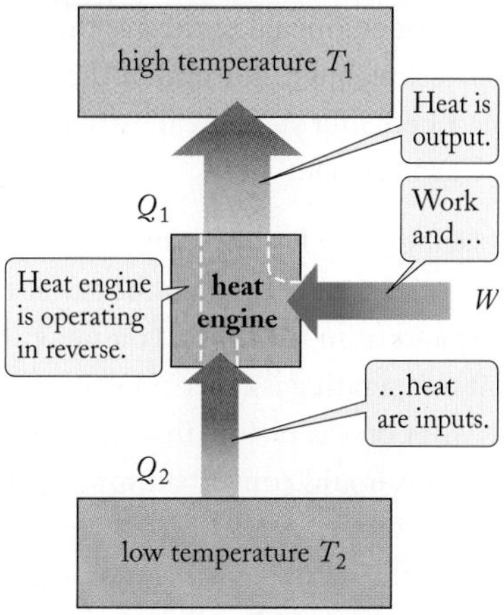

FIGURE 21.9 Flowchart for an engine operating in reverse. When the engine operates in the forward direction, it converts some amount of heat into work. When it operates in the reverse direction, it reconverts this amount of work into the original amount of heat.

We will now calculate the efficiency of an ideal heat engine that converts heat into work with maximum efficiency. As we will see in the next section, *for maximum efficiency*, the thermodynamic process within the engine should be **reversible**, *which means that the engine can, in principle, be operated in reverse and it then converts work into heat at the same rate as it converts heat into work when operating in the forward direction* (see Fig. 21.9).

reversible process

The simplest kind of reversible engine is the **Carnot engine**, consisting of some amount of ideal gas enclosed in a cylinder with a piston (see Fig. 21.10). We can alternately place the cylinder in thermal contact with a high-temperature reservoir (where it absorbs heat) or a low-temperature reservoir (where it dumps waste heat). The gas delivers work when it pushes the piston outward, and the gas absorbs work when we push the piston inward. To achieve reversibility with this engine, the motion of the piston must be sufficiently slow, so that the gas is always in an equilibrium configuration. If we were to give the piston a sudden motion, a pressure disturbance would travel through the gas, and the motion of this pressure disturbance could not be reversed by giving the piston a sudden motion in the opposite direction—this would merely create a second pressure disturbance. Furthermore, the temperature of the gas must coincide with the temperature of the heat reservoir during contact. If the gas were at, say, lower temperature than that of the heat reservoir with which it was in contact, heat would rush from the reservoir into the gas, and this flow of heat could not be reversed by any manipulation of the piston. In practice, we cannot attain exact reversibility; but it is nevertheless worthwhile to consider the ideal Carnot engine with exact reversibility, because this tells us what is the best we can hope for when attempting to convert heat into work.

Carnot engine

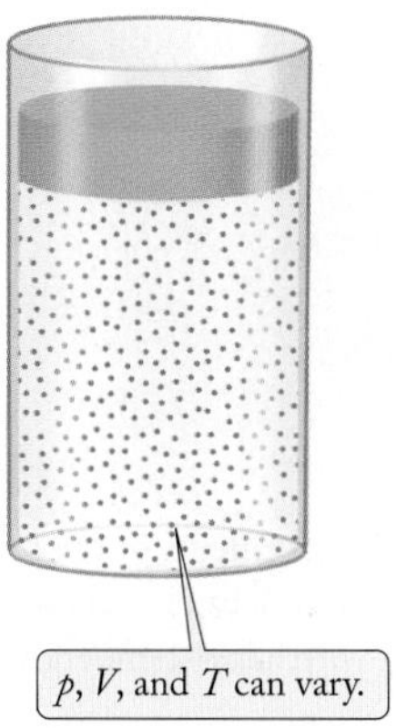

FIGURE 21.10 Carnot engine: a gas-filled cylinder with a piston.

SADI CARNOT (1796–1832) *French engineer and physicist. In his book* On the Motive Power of Heat *he formulated the theory of the conversion of heat into work.*

The operation of the Carnot engine takes the gas through a sequence of four steps with varying volume and pressure, but at the end of the last step the gas returns to its initial volume and pressure. These four steps are illustrated in Fig. 21.11. The sequence of four steps is called the **Carnot cycle**:

Carnot cycle

1. We begin the cycle by placing the cylinder in contact with the high-temperature heat reservoir, which maintains the temperature of the gas at the constant value T_1. The gas is now allowed to expand from the initial volume V_1 to a new volume V_2. Such an expansion at constant temperature is called an **isothermal expansion**. During the expansion, the gas does work on the piston, that is, the engine absorbs heat Q_1 from the high-temperature reservoir and converts it into work.
2. When the gas has reached volume V_2 and pressure p_2, we remove it from the heat reservoir and allow it to continue the expansion on its own, in thermal isolation. We know from Section 20.6 that such an expansion of a thermally isolated gas, which neither receives heat from its surroundings nor loses any, is called an **adiabatic expansion**. During this expansion, the temperature of the gas decreases, as we saw in Section 20.6.
3. When the temperature of the gas has decreased to the temperature T_2 of the low-temperature reservoir, we stop the piston and place the gas in contact with this low-temperature reservoir. The volume at this instant is V_3 and the pressure is p_3. We now begin to push the piston back toward its starting position, that is, we compress the gas isothermally. This means that the engine converts work to heat and ejects this heat Q_2 into the low-temperature reservoir.
4. When the gas has reached volume V_4 and pressure p_4, we remove it from the reservoir and continue to compress it adiabatically until the volume and the pressure return to their initial values. During this adiabatic compression, the temperature increases from T_2 to its initial value T_1.

isothermal expansion

adiabatic expansion

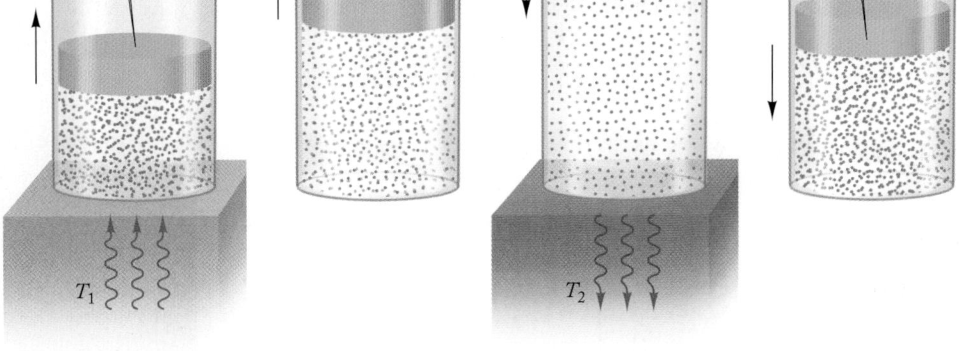

FIGURE 21.11 The Carnot cycle. The arrows indicate the displacements of the piston. (1) Isothermal expansion at temperature T_1 while in contact with a high-temperature heat reservoir. (2) Adiabatic expansion. (3) Isothermal compression at temperature T_2 while in contact with a low-temperature heat reservoir. (4) Adiabatic compression to the initial volume and pressure.

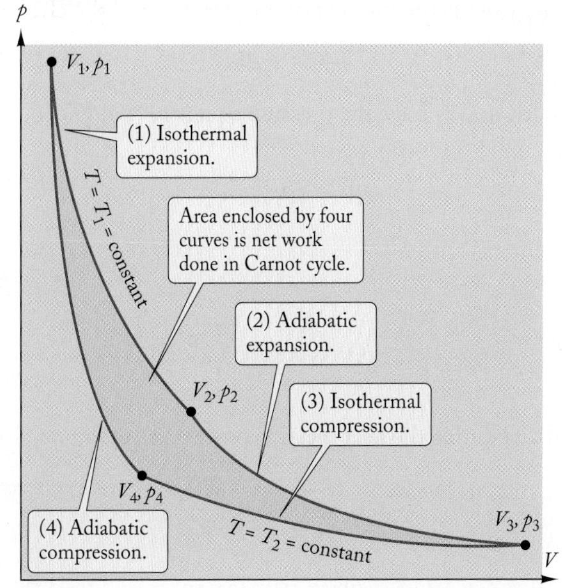

FIGURE 21.12 The Carnot cycle displayed on a p–V diagram.

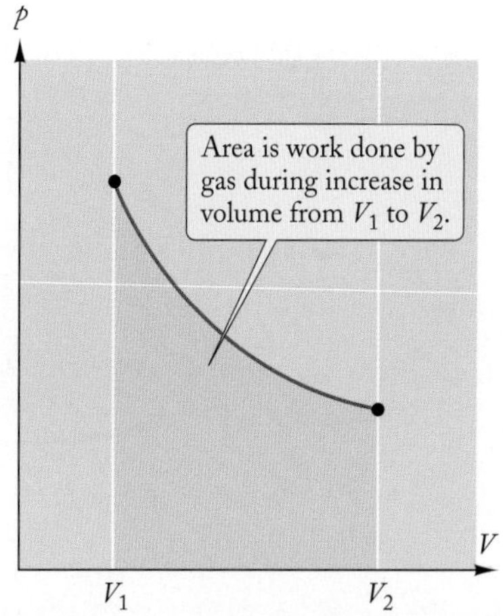

FIGURE 21.13 Expansion of a gas displayed on a p–V diagram. The work done by the gas is $W = \int dW = \int_{V_1}^{V_2} p\,dV$.

To describe the operation of this engine mathematically, it is convenient to use a p–V diagram (see Fig. 21.12). Each point in this diagram represents a possible state of the gas; p and V can be read directly from the diagram, and T can then be calculated from the ideal-gas equation. Since $dW = p\,dV$ is the work done by the gas during a small increase of volume, the work done by the gas during a finite increase from volume V_1 to volume V_2 is $W = \int dW = \int_{V_1}^{V_2} p\,dV$; this integral is the area under the curve in the p–V diagram (see Fig. 21.13). When the volume decreases, work is instead done on the gas, that is, the work done by the gas is negative. Hence the net work done in the Carnot cycle is the area enclosed between the four curves representing the four steps of the cycle (see the green area in Fig. 21.12). However, in the following calculation of the efficiency we will not have occasion to use this interpretation of the area in the p–V diagram. Instead, we will directly evaluate the heat Q_1 that the engine absorbs from the high-temperature reservoir in step (1) and the heat Q_2 that it ejects into the low-temperature reservoir in step (3), and then calculate the efficiency from Eq. (21.5).

The calculations of Q_1 and Q_2 are somewhat difficult because the volume and the pressure vary in a somewhat complicated way along the curved lines in Fig. 21.12. To simplify the calculations, let us consider the small (infinitesimal) Carnot cycle shown in Fig. 21.14. In this cycle, the pressure and the volume vary by only small amounts during each of the four steps (1)–(4); these small changes of pressure and of volume are represented by the straight lines drawn in Fig. 21.14. We will designate the changes of volume for the four steps of the cycle by dV_{12}, dV_{23}, dV_{34}, and dV_{41}; the first two of these changes of volume are positive (expansion), and the last two are negative (compression). Since the net change of volume for the entire cycle is zero, the changes of volume satisfy the identity

$$dV_{12} + dV_{23} + dV_{34} + dV_{41} = 0 \tag{21.8}$$

During the heat-absorption process (1), the temperature is constant; since the energy E of an ideal gas depends on the temperature only, we see that $dE = 0$ and, by Eq. (21.2), the amount of absorbed heat dQ_1 then equals the work dW done by the gas:

$$dQ_1 = dE + dW = dW \tag{21.9}$$

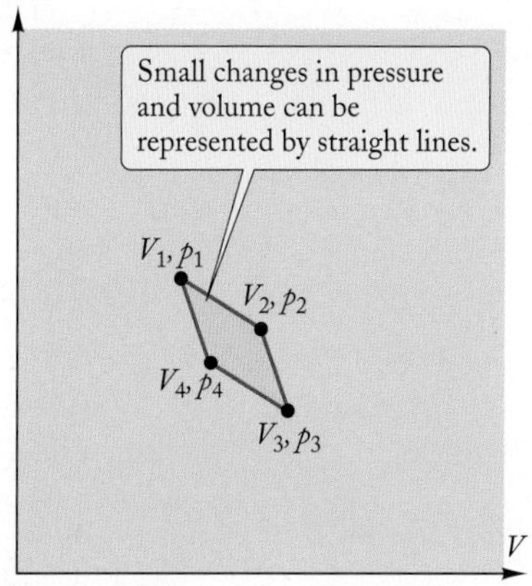

FIGURE 21.14 A small (infinitesimal) Carnot cycle, with small changes in temperature, volume, and pressure.

By Eq. (20.23), the work is the product of the pressure and the change of volume:

$$dW = p_1 dV_{12} \tag{21.10}$$

According to the Ideal-Gas Law, the pressure equals $p_1 = nRT_1/V_1$, and therefore

$$dW = nRT_1 \frac{dV_{12}}{V_1} \tag{21.11}$$

so

$$dQ_1 = nRT_1 \frac{dV_{12}}{V_1} \tag{21.12}$$

Likewise, we find that during the heat-ejection process (3), the amount of ejected heat is

$$dQ_2 = -nRT_2 \frac{dV_{34}}{V_1} \tag{21.13}$$

[Note that dV_{34} is negative; hence dQ_2 in Eq. (21.13) is positive.] For the calculation of the efficiency (21.5), we must evaluate the ratio dQ_2/dQ_1:

$$\frac{dQ_2}{dQ_1} = -\frac{nRT_2 dV_{34}}{nRT_1 dV_{12}} = -\frac{T_2}{T_1}\frac{dV_{34}}{dV_{12}} \tag{21.14}$$

To complete the evaluation of the right side of this equation, we need to know how the volume changes dV_{12} and dV_{34} are related. A relationship between these volume changes is given by Eq. (21.8), which contains not only the volume changes dV_{12} and dV_{34}, but also the volume changes dV_{23} and dV_{41}. The volume changes dV_{23} and dV_{41} correspond to the adiabatic processes (2) and (4), and we can exploit the results we obtained in Section 20.6 for the volume change in an adiabatic process. According to Eq. (20.40), the (small) volume change in an adiabatic process is directly proportional to the (small) temperature change. Since the temperature change in process (3) is simply the opposite of the temperature change in process (4), it follows that the volume changes dV_{23} and dV_{41} must also be opposites, that is, $dV_{23} = -dV_{41}$. We therefore see that the terms dV_{23} and dV_{41} in Eq. (21.8) cancel, leaving only

$$dV_{12} + dV_{34} = 0 \tag{21.15}$$

which implies that dV_{12} and dV_{34} are also opposites:

$$dV_{12} = -dV_{34} \tag{21.16}$$

With this result, the volume ratio on the right side of Eq. (21.14) is equal to -1, and the ratio of the amounts of heat is

$$\frac{dQ_2}{dQ_1} = \frac{T_2}{T_1} \tag{21.17}$$

This simply says that the amounts of heat are in direct proportion to the temperatures of the reservoirs.

With this simple expression for dQ_2/dQ_1, the formula (21.5) for the efficiency tells us that

efficiency of heat engine in terms of absolute temperatures

$$e = 1 - \frac{dQ_2}{dQ_1} = 1 - \frac{T_2}{T_1} \tag{21.18}$$

This shows that *the efficiency depends only on the temperatures of the heat reservoirs.* Although our derivation of this formula for the efficiency was based on the small Carnot cycle shown in Fig. 21.14, it can be demonstrated that this formula is also valid for the large Carnot cycle shown in Fig. 21.12, because such a large cycle can be regarded as a combination of many small cycles (in the same way that any given area in a plane can be approximated by a large number of small rectangles).

Note that an efficiency of $e = 1$ (or 100%) can be achieved only if $T_2 = 0$, that is, if the low-temperature reservoir is at the absolute zero of temperature. Unfortunately, we have no such absolutely cold reservoir available anywhere.

EXAMPLE 4 The boiler of a locomotive steam engine produces steam at a temperature of 500°C. The engine exhausts its waste heat into the atmosphere, where the temperature is 20°C. As in Example 2, the actual efficiency of this steam engine is 0.15. In order to see how much improvement on this efficiency is possible, compare this with the efficiency of a Carnot engine operating between the same temperatures.

Concepts — in — Context

SOLUTION: Our analysis of efficiency is described in terms of absolute temperature, so the releveant temperatures of operation are

$$T_1 = (500 + 273)\text{K} = 773 \text{ K} \quad \text{and} \quad T_2 = (20 + 273)\text{K} = 293 \text{ K}$$

According to Eq. (21.18), the efficiency of a Carnot engine operating between these temperatures is

$$e = 1 - \frac{T_2}{T_1} = 1 - \frac{293 \text{ K}}{773 \text{ K}} = 0.62$$

or 62%; thus the maximum theoretical efficiency is here more than 4 times the typical efficiency.

The basic steps in the operation of a practical steam engine resemble the steps of the Carnot cycle. The **steam engine** has a piston that performs a cyclic motion of expansion and contraction. However, in contrast to the Carnot engine, which uses gas as working fluid, practical steam engines use gas and liquid (steam and water) as working fluid, and the cycle of a steam engine differs from the Carnot cycle. Figure 21.15 is a schematic diagram of the main parts of a simple steam engine with boiler, cylinder, and condenser. The boiler produces hot, high-pressure steam, which enters the cylinder, pushes against the piston, expands, and does work. The piston is linked to a crankshaft, by means of which the steam engine drives some external machinery, and does useful external work. Upon completion of the expansion, the low-pressure, spent steam is exhausted from the cylinder, and the piston returns to its initial position; these operations are controlled by the opening and closing of valves. Meanwhile, the spent steam is sent to a condenser, where an external coolant (air or flowing water) condenses the steam into liquid water. This liquid water is pumped back to the boiler. Each completed circulation of the fluid through the circuit can be regarded as one complete cycle of operation of the steam engine. Such simple steam engines have efficiencies of only 5–18%. Most modern steam engines employ a turbine wheel instead of the cylinder and piston; large engines of this kind achieve efficiencies of up to 40%.

steam engine

Concepts — in — Context

For a description of the operation of an automobile engine see Physics in Practice: Efficiency of Automobiles on page 674.

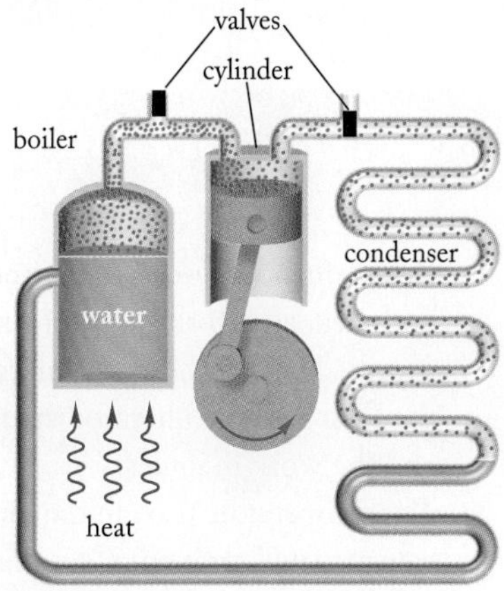

FIGURE 21.15 A steam engine (schematic diagram).

A Carnot engine can be operated in reverse, so the Carnot cycle begins with step (4) and ends with step (1). The Carnot engine then uses up work to transfer heat from the low-temperature reservoir to the high-temperature reservoir. This is the principle involved in the operation of **refrigerators**, air conditioners, and "heat pumps." The amount of work required to operate a Carnot engine in reverse can be calculated from Eq. (21.17); as stated above, this equation is valid not only for an infinitesimal Carnot cycle, but also for a finite Carnot cycle, for which the equation takes the form

refrigerator

$$\frac{Q_2}{Q_1} = \frac{T_2}{T_1} \tag{21.19}$$

EXAMPLE 5 Suppose a homeowner uses a Carnot engine operating in reverse as a heat pump to extract heat from the outside air and inject it into his home. If the outside temperature is $-10°C$ and the inside temperature is $20°C$, what is the amount of work that must be supplied to pump 1.00 J of heat from the outside to the inside? How does this compare with the direct conversion of energy to heat, as in an electric heater?

SOLUTION: For a Carnot engine, the ratio of the heats exchanged at the low-temperature reservoir and at the high-temperature reservoir is given by Eq. (21.19):

$$\frac{Q_2}{Q_1} = \frac{T_2}{T_1} = \frac{-10 + 273\ \text{K}}{20 + 273\ \text{K}} = \frac{263\ \text{K}}{293\ \text{K}} = 0.898$$

If $Q_2 = 1.00$ J, then $Q_1 = Q_2/0.898 = 1.11$ J. The difference $Q_1 - Q_2$ represents the work that must be supplied (see the flowchart in Fig. 21.16); hence the work is 0.11 J.

COMMENTS: By the expenditure of 0.11 J of work, the heat pump delivers a total of 1.11 J of heat into the house. Thus a small amount of work can bring a large amount of heat into the home. This is obviously a much more economical heating method than the direct expenditure of 1.11 J of fuel or electric energy in a conventional furnace or electric heater. Heat pumps achieve their highest efficiencies when the temperatures T_1 and T_2 are nearly equal, and they are widely used in mild climates (see Fig. 21.17).

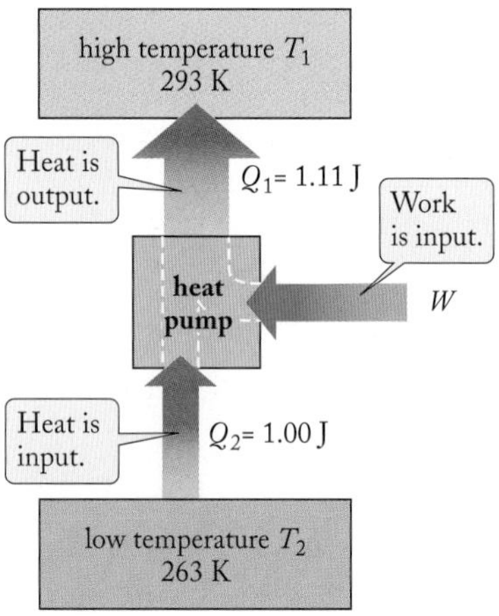

FIGURE 21.16 Flowchart for the operation of a heat pump.

Practical refrigerators use a gas and liquid refrigerant (such as R15, an environmentally safe refrigerant that has replaced Freon) as working fluid, and their cycle differs from the Carnot cycle. The fluid refrigerants have a boiling point near room

PROBLEM-SOLVING TECHNIQUES — THERMODYNAMIC CALCULATIONS

- Be careful about units in thermodynamic calculations. If both heat and work appear in an equation, then you must express both in the same energy units. For instance, in the First Law of Thermodynamics, express both the heat and the work in joules.
- The temperature used in the formula (21.18) for the efficiency of a Carnot engine is the absolute temperature, in kelvins. The formula is not valid if you express the temperature in °C or °F.
- To keep track of the amounts of heat and work, it is helpful to draw a flowchart, such as the flowchart for the heat engine in Fig. 21.7 or the flowchart for the heat pump in Fig. 21.16.

FIGURE 21.17 A heat pump installed outside a home.

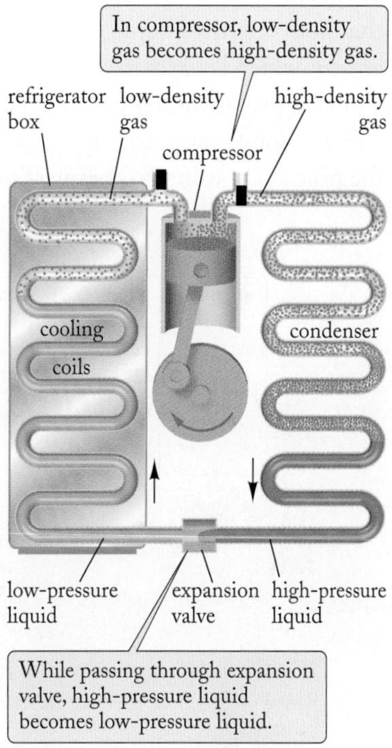

FIGURE 21.18 A refrigerator (schematic diagram).

temperature when at high pressure, but a boiling point below 0°C when at low pressure. Figure 21.18 is a schematic diagram of the parts of a practical refrigerator. Liquid refrigerant at low pressure enters the cooling coils in the refrigerator box and absorbs heat while evaporating into gas. This gas flows to the compressor, where its pressure and density are increased by the push of a piston. The high-pressure gas then circulates through the condenser coils, which are exposed to the atmospheric air (see Fig. 21.19). The gas loses its heat and condenses into liquid. This high-pressure liquid then passes through an expansion valve (a small orifice) where its pressure is reduced to match the low pressure in the cooling coils. This return of the fluid to the cooling coils completes the cycle.

An air conditioner employs a similar refrigeration cycle; and a "heat pump" is, in essence, an air conditioner turned around, so its cold end is outdoors and its warm end indoors.

coefficient of performance *CP*

The performance of a heat pump, an air conditioner, or a refrigerator is often rated using a **coefficient of performance** *CP*. A heat pump delivers heat to the high-temperature reservoir (the inside of a house), and so its coefficient of performance is defined as the ratio of the heat delivered to the work done: $CP = Q_1/W$. An air conditioner or refrigerator removes heat from a low-temperature reservoir; the coefficient of performance is then defined as the ratio of the heat removed to the work done: $CP = Q_2/W$. For practical devices, values of the coefficient of performance are usually greater than 1, often around 5.

FIGURE 21.19 Condenser coils of a refrigerator.

PHYSICS IN PRACTICE EFFICIENCY OF AUTOMOBILES

An automobile can be regarded as a thermodynamic engine that converts the chemical energy of fuel into mechanical energy, which is ultimately dissipated by friction. Figure 1 is a flowchart illustrating the successive energy conversions in an automobile propelled by a gasoline engine; Fig. 2 shows the piston's cycle: intake, compression, expansion, and exhaust.

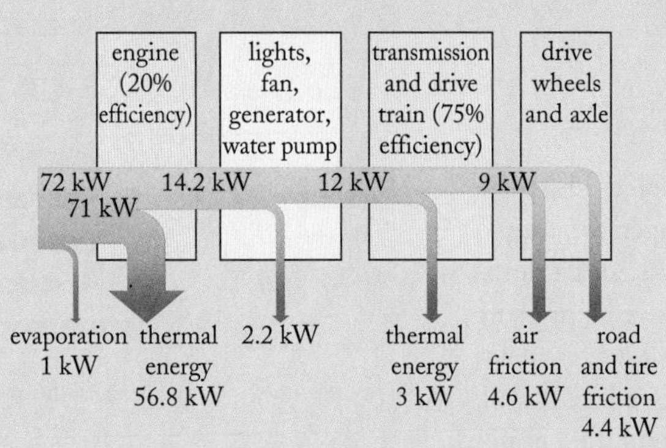

FIGURE 1 Energy conversions in an automobile. (Based on R.H. Romer, *An Introduction to Physics*.)

One liter of gasoline releases 3.4×10^7 J of chemical energy when burned with oxygen. While traveling at 65 km/h on a level road, the automobile described in this flowchart consumes 7.6 liters per hour, equivalent to a chemical energy of 2.6×10^8 J per hour, or 72 kW. The engine converts about 20% of this chemical energy into mechanical energy. It delivers most of this mechanical energy to the transmission and the axles of the automobile. The energy reaching the axles is in part fed into the body of the automobile (the axle pushes forward against its bearings, doing work on the body of the automobile), and in part it is fed into the wheels. The energy delivered to the body is dissipated into heat by air friction, and the energy delivered to the wheels is dissipated into heat by rolling friction against the road. Of the 72 kW of chemical energy entering the engine, only 14 kW emerges as propulsive power, much of which does useful work against the external friction forces of air and road. Most of the energy is dissipated in the conversion process. Note that the largest waste occurs in the engine itself—in gasoline engines most of the chemical energy is ejected into the exhaust gases. Diesel engines achieve a somewhat higher efficiency, because they burn their fuel more completely.

However, in part, the inefficiency of these engines is a consequence of the basic laws of thermodynamics. The gasoline engine is a heat engine—it converts the heat released by the burning fuel into mechanical energy. The efficiency of an ideal heat engine is given by Eq. (21.18). In the gasoline engine, the heat source is the hot gas produced by the combustion, and the heat sink is the air surrounding the engine. The complete combustion of hydrocarbon fuel gives the residual gas a temperature of about 2400 K. The air surrounding the engine typically has a temperature of 300 K. Hence the maximum attainable efficiency is

$$e = 1 - \frac{T_2}{T_1} = 1 - \frac{300\ \text{K}}{2400\ \text{K}} = 0.88$$

Gasoline engines and other engines that burn hydrocarbon fuels do not come anywhere near this maximum efficiency of an ideal engine, because they do not burn the fuel completely, they eject exhaust gases at a temperature much higher than 300 K, and they suffer from internal friction.

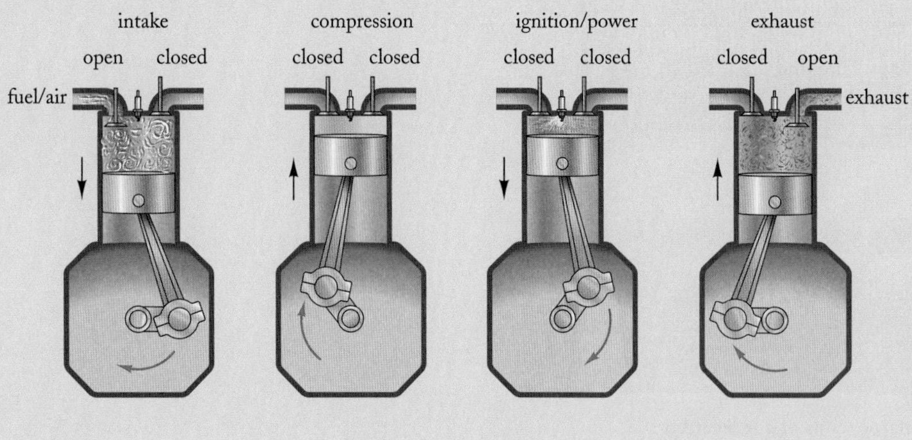

FIGURE 2 Intake, compression, expansion, and exhaust stages of an automobile engine cycle.

Checkup 21.2

QUESTION 1: What is the difference between an isothermal expansion and an adiabatic expansion?

QUESTION 2: Can the efficiency of an engine ever be larger than 1? Why or why not?

QUESTION 3: Is the operation of an automobile engine reversible?

QUESTION 4: Does the efficiency of a Carnot engine depend on the size of the cycle, that is, how far we let the piston in Fig. 21.11 travel within its cylinder?

QUESTION 5: Suppose we connect two Carnot engines together, so the work output of the first drives the second in reverse (see Figs. 21.7 and 21.9). What is the net effect?

(A) Net heat is added. (B) Net work is done. (C) There is no net effect.

21.3 THE SECOND LAW OF THERMODYNAMICS

As we saw in the preceding section, a Carnot engine operating with an ideal gas as working fluid has a limited efficiency—it fails to convert all of the heat into work and instead produces some waste heat. The **Second Law of Thermodynamics** asserts that this is a limitation from which all heat engines suffer. This law may be stated in several ways. As formulated by Lord Kelvin and Max Planck, this law simply states,

Second Law of Thermodynamics

An engine operating in a cycle cannot transform heat into work without some other effect on its surroundings.

It is an immediate corollary of this law that *the efficiency of any heat engine operating between two heat reservoirs of high and low temperature is never greater than the efficiency of a Carnot engine; furthermore, the efficiency of any reversible engine equals the efficiency of a Carnot engine.* The proof of these statements, known as **Carnot's theorem**, is by contradiction. Imagine that some heat engine is more efficient than a Carnot engine, so this engine converts heat from a reservoir at a high temperature into work and ejects only a small amount of waste heat into a reservoir at low temperature. We can then use the work output of this engine to drive a Carnot engine in reverse, and thereby pump the waste heat from the low-temperature reservoir to the high-temperature reservoir. By hypothesis, the given engine is more efficient than the Carnot engine; hence only part of its work output will be needed to drive the reversed Carnot engine and return all of the waste heat to the high-temperature reservoir. The remainder of the output constitutes available work (see Fig. 21.20 for a flowchart). The net effect of the joint operation of both engines is the complete conversion of heat into work, without waste heat, in contradiction to the Second Law. We can avoid this contradiction only if the efficiency of any engine is never greater than that of a Carnot engine.

Carnot's theorem

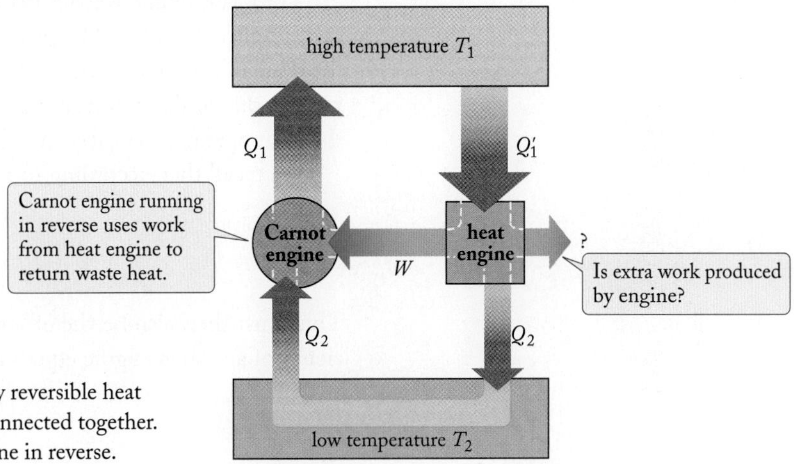

FIGURE 21.20 Flowchart for an arbitrary reversible heat engine (box) and a Carnot engine (circle) connected together. The arbitrary engine drives the Carnot engine in reverse.

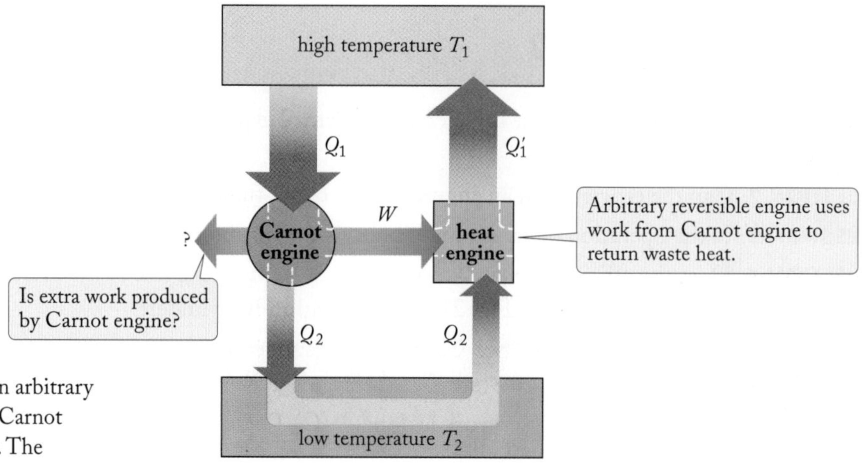

FIGURE 21.21 Flowchart for an arbitrary reversible heat engine (box) and a Carnot engine (circle) connected together. The Carnot engine drives the arbitrary engine in reverse.

To prove that the efficiency of any reversible engine equals that of a Carnot engine, we again consider the net effect of the joint operation of the two engines with the Carnot engine running in the forward direction and its work output driving the other reversible engine in the backward direction (see Fig. 21.21). By an argument similar to that given above, it now follows that the efficiency of the Carnot engine cannot be greater than that of the other engine. Thus, the efficiency of each engine can be no greater than that of the other; that is, they both must have exactly the same efficiency.

We recall that a perpetual motion machine of the second kind is an engine that takes heat energy from a reservoir and completely converts it into work. Thus, the Second Law asserts that no perpetual motion machine of the second kind can exist. Essentially, the operation of any heat engine hinges on the temperature difference between two heat reservoirs; heat tends to rush from the high-temperature reservoir to the low-temperature reservoir. By interposing a heat engine in the path of this rush of heat, we can force the heat to do useful work. The Second Law can be formulated in an alternative form which is based on this characteristic of the flow of heat. This formulation, due to Rudolf Clausius, states,

An engine operating in a cycle cannot transfer heat from a cold reservoir to a hot reservoir without some other effect on its surroundings.

The Clausius and Kelvin–Planck formulations of the Second Law are equivalent—each implies the other. The proof is, again, by contradiction, and relies on examining the result of a joint operation of two engines: a Carnot engine, and a second engine that violates one of the two formulations of the Second Law. It is then easy to show that the joint operation violates the other formulation of the Second Law.

We recall that, according to Eq. (21.18), the efficiency of a Carnot engine is

$$e = 1 - \frac{T_2}{T_1} \tag{21.20}$$

This must then also be the efficiency of any other reversible engine, since the efficiency of any such engine equals that of a Carnot engine.

Checkup 21.3

QUESTION 1: What happens to the efficiency of a Carnot engine if $T_2 = T_1$? What if $T_2 > T_1$?

QUESTION 2: A refrigerator transfers heat from a cold reservoir (the interior of the refrigerator) to a hot reservoir (the kitchen). Does this violate the Second Law of Thermodynamics?

QUESTION 3: What would happen if you were to try to operate a steam engine in a very hot environment, where the condenser is surrounded by air at about the same temperature as the boiler?

(A) With a warm environment, heating is easier, so the efficiency increases.
(B) Heating is easier, but heat flow decreases, so the efficiency is the same.
(C) The cycle could not be completed, since the steam could not condense.

21.4 ENTROPY

In the operation of a Carnot engine, or any other reversible engine, the ratio of the heats absorbed and ejected at the high- and low-temperature reservoirs is, according to Eq. (21.19).

$$\frac{Q_2}{Q_1} = \frac{T_2}{T_1} \tag{21.21}$$

We can rewrite this equation in a suggestive way by multiplying both sides by $-Q_1/T_2$ and then moving all terms to the left side:

$$\frac{Q_1}{T_1} - \frac{Q_2}{T_2} = 0 \tag{21.22}$$

If, as in Section 20.1, we regard the heat change as positive when heat enters the system and negative when it leaves the system, then $\Delta Q_1 = Q_1$ for the heat exchange with the high-temperature reservoir and $\Delta Q_2 = -Q_2$ for the heat exchange with the low-temperature reservoir. Thus, Eq. (21.22) says that, for a Carnot cycle, the sum of all the heat exchanges ΔQ divided by their respective temperatures is zero:

$$\frac{\Delta Q_1}{T_1} + \frac{\Delta Q_2}{T_2} = 0 \tag{21.23}$$

This result can also be shown to be valid for any other reversible cycle, because any other reversible cycle can be approximated by a large number of small Carnot cycles. A general reversible cycle may have a large number of successive heat exchanges, each at some different temperature. For such a cycle, the generalized version of Eq. (21.23) is

$$\frac{dQ_1}{T_1} + \frac{dQ_2}{T_2} + \frac{dQ_3}{T_3} + \frac{dQ_4}{T_4} + \cdots = 0$$

If we proceed to the limit of a large number of small (infinitesimal) reversible heat exchanges, this sum becomes an integral,

Clausius' theorem

$$\oint \frac{dQ}{T} = 0 \quad \text{(reversible cycle)} \tag{21.24}$$

This result is called **Clausius' theorem.**

entropy

The importance of Clausius' theorem is that it permits us to define a new physical quantity called the **entropy**, or more precisely, the entropy difference. Given two states A and B of a system, the entropy difference ΔS between them is defined as the sum of the heat exchanges divided by the temperature for some reversible process that takes the system from the initial state A to the final state B:

$$\Delta S = \int_A^B \frac{dQ}{T} \quad \text{(reversible process)} \tag{21.25}$$

Since there is usually a wide choice of different reversible processes that take the system from the given initial state to the given final state, the definition (21.25) would make no sense if these different processes gave different values to the sum (21.25). Clausius' theorem ensures that all these reversible processes lead to the same value for (21.25). To understand the connection between the definition of entropy and Clausius' theorem, consider the example of the ideal gas in the Carnot engine, for which different states are plotted in the p–V diagram in Fig. 21.12. If the initial state A corresponds to the point 1 in Fig. 21.12, and the final state B to the point 3, then one possible reversible path from 1 to 3 consists of the steps (1) and (2) of the Carnot cycle. The heat exchanged in step (1) is Q_1, and the heat exchanged in step (2) is zero; thus Eq. (21.25) yields

$$\Delta S = \frac{Q_1}{T_1} \tag{21.26}$$

for the entropy change between points 1 and 3. But another reversible path from 1 to 3 consists of step (4) *in reverse* followed by step (3) *in reverse*—this also takes the system from point 1 to point 3. The heat exchanged in step (4) is zero, and the heat exchanged in step (3) is Q_2. Hence Eq. (21.25) now yields

$$\Delta S = \frac{Q_2}{T_2} \tag{21.27}$$

But Eq. (21.22)—which is Clausius' theorem for the Carnot cycle—immediately tells us that the two expressions (21.26) and (21.27) for the entropy change are equal.

The SI unit of entropy is the unit of energy divided by the unit of temperature: joule per kelvin (J/K). An alternative unit is calories per kelvin (cal/K).

The entropy in thermodynamics plays a role somewhat analogous to that of the potential energy in mechanics. Just as the potential energy allows us to make some predictions about possible motions of a mechanical system, the entropy allows us to make some predictions about the possible behavior of a thermodynamic system. For instance, it can be shown from the Second Law of Thermodynamics that the entropy of a closed system—which is thermally and mechanically isolated from its surroundings—can never decrease. The entropy of such a system either remains constant (if only reversible processes are occurring in the system) or increases (if irreversible processes

RUDOLF CLAUSIUS (1822–1888)
German mathematical physicist. He was one of the creators of the science of thermodynamics. He contributed the concept of entropy, as well as the restatement of the Second Law of Thermodynamics.

are occurring). Thus an alternative formulation of the Second Law of Thermodynamics asserts that

The entropy of a closed system can never decrease.

The increase of entropy in closed systems is illustrated in the following examples.

EXAMPLE 6 A heat reservoir at a temperature of $T_1 = 400$ K is briefly put in thermal contact with a reservoir at $T_2 = 300$ K (see Fig. 21.22). If 1.00 J of heat flows from the hot reservoir to the cold reservoir, what is the change of the entropy of the system consisting of both reservoirs?

SOLUTION: Obviously, the heat flow from the hot to the cold reservoir is irreversible, and hence Eq. (21.25), which is restricted to reversible processes, is not directly applicable. To evaluate the entropy change by means of Eq. (21.25), we must first invent some process that reversibly takes the system from the initial state to the final state. We can imagine that 1.00 J flows from the hot reservoir into an auxiliary reservoir of a temperature just barely below 400 K, and that simultaneously 1.00 J flows from another auxiliary reservoir of a temperature just barely above 300 K into the cold reservoir. Then all processes are reversible, and Eq. (21.25) gives us the change of entropy:

$$\begin{aligned}\Delta S &= \frac{\Delta Q_1}{T_1} + \frac{\Delta Q_2}{T_2} = \frac{-1.00\text{ J}}{400\text{ K}} + \frac{1.00\text{ J}}{300\text{ K}} \\ &= 8.3 \times 10^{-4}\text{ J/K}\end{aligned} \tag{21.28}$$

COMMENT: Note that when the magnitudes of the heat transfers are equal, as they are in this example, the contributions to the entropy are inversely proportional to the temperature; that is, "hot heat" contributes a smaller entropy change than "cold heat."

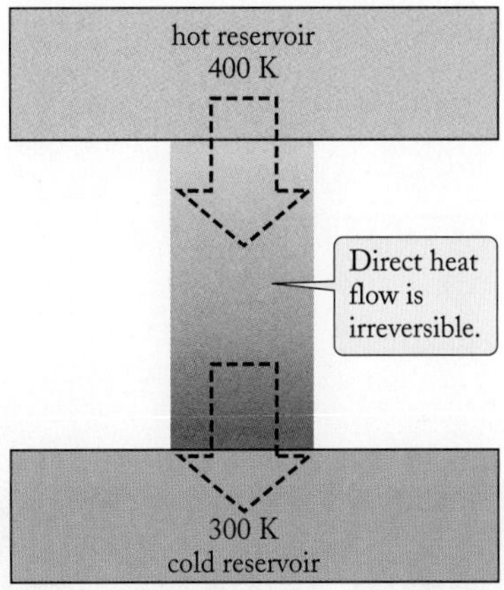

FIGURE 21.22 A hot reservoir (400 K) is in thermal contact with a colder reservoir (300 K). Heat flows through a conducting rod connecting the two reservoirs, from the hot reservoir into the colder reservoir.

EXAMPLE 7 A large stone, of mass 80 kg, slides down a hill of a vertical height of 100 m and is stopped by friction at the bottom. What is the increase of the entropy of the stone plus the surroundings? Assume that the temperature of the surroundings (hill and air) is 270 K.

SOLUTION: All of the initial mechanical energy of the stone is converted into heat:

$$\begin{aligned}Q = mgh &= 80\text{ kg} \times 9.81\text{ m/s}^2 \times 100\text{ m} \\ &= 7.8 \times 10^4\text{ J}\end{aligned}$$

This heat is delivered to the surroundings at a temperature of 270 K.

Clearly, the process described is irreversible (the heat generated could not be completely reconverted into mechanical work). To calculate the entropy with Eq. (21.25), we must imagine a reversible process that brings the stone down the hill and delivers the heat into the surroundings. In principle, we can use an elevator to let the stone down slowly without friction and extract work while removing the potential energy; and afterward we can use a heat reservoir of a temperature barely above 270 K to supply reversibly the correct amount of heat (7.8×10^4 J) to the stone's surroundings. The first of these steps makes no contribution to the entropy, and the second makes a contribution

$$\Delta S = \frac{\Delta Q}{T} = \frac{7.8 \times 10^4\ \text{J}}{270\ \text{K}} = 290\ \text{J/K} \tag{21.29}$$

From a microscopic point of view, the increase of entropy of a system is an increase of disorder. This is obvious in the preceding example of the conversion of mechanical energy into heat by friction. The translational motion of a macroscopic body is ordered energy—all the particles in the body move in the same direction with the same speed. Heat is disordered energy—the particles move in random directions with a mixture of speeds.

The increase of disorder also holds true in our other example of spontaneous flow of heat from a hot to a cold reservoir. When a given amount of heat is added to a reservoir of low temperature, it causes more disorder than when the same amount of heat is added to a reservoir of high temperature; in both reservoirs, the added heat generates extra random motion and extra disorder, but in the cold reservoir the percent increment of the random motions is larger than in the hot reservoir, and consequently the extra disorder is larger.

WALTHER HERMANN NERNST (1864–1941) *German physicist and chemist. He was a pioneer in physicial chemistry and received in Nobel Prize in Chemistry in 1920 for his discovery of the Third Law of Thermodynamics.*

The connection between entropy and disorder can be given a precise mathematical meaning, but the details would require a discussion of statistical mechanics and information theory—and here we will not go into this. We merely assert that the Second Law can be reformulated to say,

Processes in a closed system always tend to increase the amount of disorder.

Human activities on the Earth, like any other processes in nature, cause an increase of disorder. Of course, *some* of our activities result in an increase of *order* in some portion of the system. For instance, when we extract dispersed bits of metal from ores and assemble them into a watch, we are obviously increasing the order of the bits of metal. However, we can do this only by simultaneously generating disorder somewhere else—the smelting of ores and the machining of metals demand an input of energy that is converted into waste heat, and increases the disorder of the environment. The net result is always an increase of disorder. All of our activities depend on a supply of highly ordered energy in the form of chemical or nuclear fuels or light from the Sun that can "soak up" disorder while becoming degraded into heat. We continually convert useful, ordered energy into useless, disordered energy.

Third Law of Thermodynamics

We end this chapter with a brief statement of the **Third Law of Thermodynamics**. This law, as formulated by Walther Nernst, asserts that

The entropy of a system at absolute zero is zero.

We can understand this law in terms of the connection between entropy and disorder, mentioned above. As we lower the temperature of a system, we decrease the random thermal motions, and we decrease the disorder. According to classical theory, the random thermal motions cease completely as the temperature of the system approaches absolute zero. The system then tends to settle into a state of minimum disorder, that is, a state of minimum entropy.

Checkup 21.4

QUESTION 1: Two reservoirs are at temperatures of 300 K and 400 K, respectively. Each gains 100 J of heat from another reservoir. What is the entropy change in each?

QUESTION 2: Is the flowchart in Fig. 1 of Physics in Practice: Efficiency of Automobiles on page 674 consistent with the Second Law of Thermodynamics?

QUESTION 3: A pot full of hot water is placed in a cold room, and the pot gradually cools. Does the entropy of the water decrease? Does this violate the Second Law of Thermodynamics?

QUESTION 4: Which of the following processes results in an overall increase of entropy: (1) braking of an automobile, (2) gradual slowing of an automobile as it rolls uphill (without friction), (3) electric pump lifting water (without friction) from a well to a reservoir above, (4) firewood burning in a fireplace?

(A) 1 and 2 (B) 1 and 4 (C) 2 and 3 (D) 2 and 4 (E) 3 and 4

SUMMARY

PROBLEM-SOLVING TECHNIQUES Thermodynamic Calculations **(page 672)**

PHYSICS IN PRACTICE Efficiency of Automobiles **(page 674)**

FIRST LAW OF THERMODYNAMICS
where ΔE is the change in internal energy of the gas, Q is the heat added to the gas, and W is the work done by the gas.

$$\Delta E = Q - W \qquad (21.2)$$

EFFICIENCY OF HEAT ENGINE
where W is the work done by the engine, Q_1 is the heat absorbed from the high-temperature reservoir, and Q_2 is the waste heat.

$$e = \frac{W}{Q_1} = 1 - \frac{Q_2}{Q_1} \qquad (21.5)$$

CARNOT CYCLE (1) Isothermal expansion at T_1; (2) adiabatic expansion to T_2; (3) isothermal compression at T_2; (4) adiabatic compression to T_1.

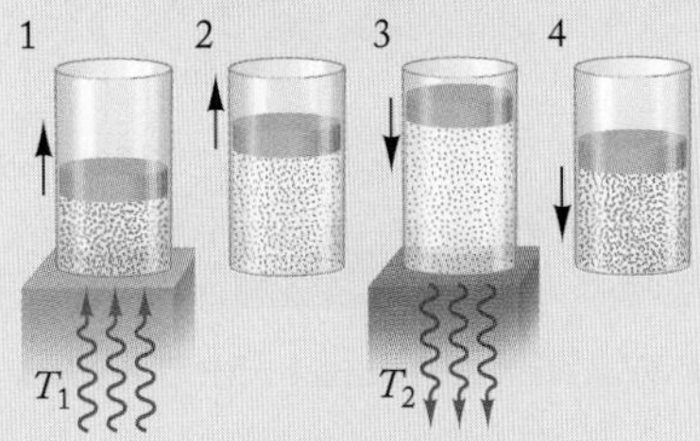

EFFICIENCY OF CARNOT ENGINE (OR ANY OTHER REVERSIBLE HEAT ENGINE) Where T_1 and T_2 are absolute temperatures.

$$e = 1 - \frac{T_2}{T_1} \qquad (21.18)$$

CARNOT'S THEOREM The efficiency of any engine cannot exceed that of a Carnot engine; the efficiency of any reversible engine equals that of a Carnot engine.

ENTROPY CHANGE (REVERSIBLE PROCESS)

$$\Delta S = \int_A^B \frac{dQ}{T} \tag{21.25}$$

SECOND LAW OF THERMODYNAMICS

- An engine operating in a cycle cannot transform heat into work without some other effect on its surroundings; or
- The entropy of a closed system can never decrease; or
- Processes in a closed system always tend to increase the amount of disorder.

THIRD LAW OF THERMODYNAMICS

The entropy is zero at absolute zero.

QUESTIONS FOR DISCUSSION

1. Figure 21.23 shows a perpetual motion machine designed by M. C. Escher. Exactly at what point is there a defect in this design?

FIGURE 21.23 Waterfall, lithograph by M. C. Escher.

2. An inventor proposes the following scheme for the propulsion of ships on the ocean without the input of energy: cover the hull of the ship with copper sheets and suspend an electrode of zinc in the water at some distance from the hull; since seawater is an electrolyte, the hull of the ship and the electrode will then act as the terminals of a battery, which can deliver energy to an electric motor propelling the ship. Will this scheme work? Does it violate the First Law of Thermodynamics?
3. The lowest-temperature heat sink available in nature is interstellar and intergalactic space, with a temperature of 3 K. Why don't we use this heat sink in the operation of a Carnot engine?
4. In some showrooms, salespeople demonstrate air conditioners by simply plugging them into an outlet without bothering to install them in a window or a wall. Does such an air conditioner cool the showroom or heat it?
5. Can mechanical energy be converted completely into heat? Give some examples.
6. If you leave the door of a refrigerator open, will it cool the kitchen?
7. Which of the following processes are irreversible, and which are reversible?
   (a) Burning a piece of paper.
   (b) Slow descent of an elevator attached to a perfectly balanced counterweight.
   (c) Breaking of a windowpane.
   (d) Exploding of a stick of dynamite.
   (e) Descent of a roller-coaster car from a hill, without friction.
8. If we measure the efficiency of a Carnot engine directly, we can use Eq. (21.20) to calculate the temperature of one of the heat reservoirs. What are the advantages and the disadvantages of such a thermodynamic determination of temperature?
9. In the absence of a heat sink, energy cannot be extracted from an ocean at uniform temperature. However, within the oceans there are small temperature differences between the warm water on the surface and the cooler water in the depths.

Design a Carnot engine that exploits this temperature difference.

10. Why are heat pumps used for heating houses in mild climates but not in very cold climates?
11. Does the Second Law of Thermodynamics forbid the spontaneous flow of heat between two bodies of equal temperature?
12. The inside of an automobile parked in the sun becomes much hotter than the surrounding air. Does this contradict the Second Law of Thermodynamics?
13. According to a story by George Gamow,[2] on one occasion Mr. Tompkins was drinking a cocktail when all of a sudden one small part at the surface of the liquid became hot and boiled with violence, releasing a cloud of steam, while the remainder of the liquid became cooler. Is this consistent with the First Law of Thermodynamics? With the Second Law?
14. A vessel is divided into two equal volumes by a partition. One of these volumes contains helium gas, and the other contains argon at the same temperature and pressure. If we remove the partition and allow the gases to mix, does the entropy of the system increase?
15. Does the motion of the planets around the Sun generate entropy?
16. Does static friction generate entropy?
17. Consider the process of emission of light by the surface of the Sun followed by absorption of this light by the surface of the Earth. Does this entail an increase in entropy?
18. Suppose that a box contains gas of extremely low density, say, only 50 molecules of gas altogether. The molecules move at random, and it is possible that once in a while all of the 50 molecules are simultaneously in the left half of the box, leaving the right half empty. At this instant the entropy of the gas is less than when the gas is more or less uniformly distributed throughout the box. Is this a violation of the Second Law of Thermodynamics? What are the implications for the range of validity of this law?

[2] G. Gamow, *Mr. Tompkins in Paperback.*

19. **Maxwell's demon** is a tiny hypothetical creature that can see individual molecules. The demon can make the heat flow from a cold body to a hot body as follows: Suppose that a box initially filled with gas at uniform temperature and pressure is divided into two equal volumes by a partition, equipped with a small door that is closed but can be opened by the demon (Fig. 21.24). Whenever a molecule of above-average speed approaches the door from the left, the demon quickly opens the door and lets it through. Whenever a molecule of below-average speed approaches from the right, the demon also lets it through. This selective action of the demon accumulates hot gas in the right volume, and cool gas in the left volume. Does this violate the Second Law of Thermodynamics? Do any of the activities of the demon involve an *increase* of entropy?

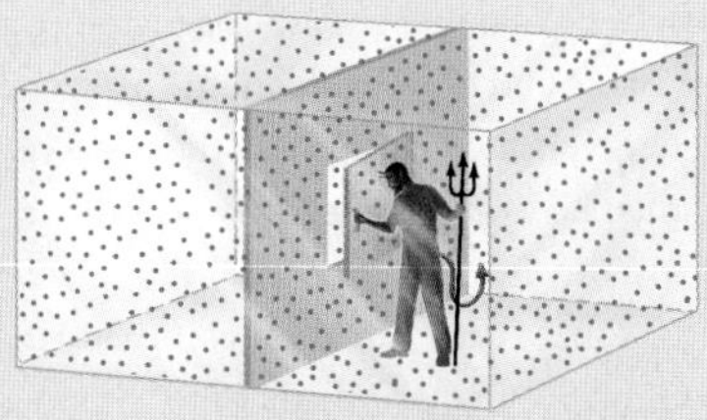

FIGURE 21.24 Maxwell's demon.

20. According to one cosmological model based on Einstein's theory of General Relativity, the Universe oscillates—it expands, then contracts, then expands, and then contracts, etc. If the Second Law of Thermodynamics is valid, can each cycle of oscillation be the same as the preceding cycle?
21. **Negentropy** is defined as the negative of the entropy ([negentropy] $= -S$). Explain the following statement: "In our everyday activities on the Earth, we do not consume energy, but we consume negentropy."
22. The amount of energy dissipated in the United States per year is 8×10^{19} J. Roughly, what is the entropy increase that results from this dissipation?
23. If you tidy up a messy room, you are producing a decrease of disorder. Does this violate the Second Law?

PROBLEMS

21.1 The First Law of Thermodynamics

1. We place a 1.00-mole sample of an ideal monatomic gas in a cylinder with a piston and we heat the gas so it expands and performs work against the piston. Suppose the temperature of the gas increases by 90°C while at the same time it performs 800 J of work. What is the change of the internal energy of the gas in this process? How much heat does the gas absorb during this process?
2. A sample of gas in a cylinder with a piston is in thermal contact with a heat reservoir at a temperature of 353 K. While keeping this gas at this constant temperature and at a constant pressure of 1.01×10^5 N/m^2, we permit the gas to expand by 1.50×10^{-5} m^3. How much heat does the gas absorb during this process?
3. A 4.0-liter sample of a monatomic ideal gas is initially at STP. We first heat this gas at constant volume, until its pressure is doubled. We then continue heating at constant pressure, and

allow the gas to expand until its volume is also doubled.

(a) What are the values of ΔE, Q, and W for the first step of this process?

(b) What are the values of ΔE, Q, and W for the second step of this process?

4. When pressurized air rushes out of the nozzle of a tire, the air must do work against the surrounding atmosphere while it expands. This means that the air suffers a loss of internal energy, and its temperature therefore drops (you can easily feel that the emerging air is quite cold). Suppose that each mole of emerging air does 1800 J of work, and suppose that the air rushes out so fast that it has no time to absorb heat from the surrounding atmosphere (adiabatic expansion). If the initial temperature of the air in the tire is 20°C, what is the temperature of the emerging air? Treat the air as an ideal diatomic gas.

5. In the cylinder of a diesel engine, the piston compresses the air–fuel mixture and does work on it. This work increases the internal energy of the mixture, and therefore heats it. The temperature attained by the compression is sufficient to ignite the mixture, without any need of a spark plug. How much work must the piston do on a parcel of gas, 0.0300 mole, to heat it from 40°C to 790°C? Assume that the compression is so fast that the gas loses no heat to the surroundings (adiabatic compression), and assume the gas behaves like an ideal diatomic gas.

6. A ball of lead of mass 0.25 kg drops from a height of 0.80 m, hits the floor, and remains there at rest. Assume that all the heat generated during the impact remains within the lead. What are the values of Q, W, and ΔE for the lead during this process? What is the increase of temperature of the lead?

7. A large closed bag of plastic contains 0.100 m^3 of an unknown ideal gas at an initial temperature of 10°C and at the same pressure as the surrounding atmosphere, 1.0 atm. You place this bag in the sun and let the gas warm up to 38°C and expand to 0.110 m^3. During this process, the gas absorbs 3500 J of heat. Assume the bag is large enough that the gas never strains against it, and therefore remains at a constant pressure of 1.0 atm.

(a) How many moles of gas are in the bag?

(b) What is the work done by the gas in the bag against the atmosphere during the expansion?

(c) What is the change in the internal energy of the gas in the bag?

(d) Is the gas a monatomic gas? A diatomic gas?

8. The vaporization of 1.00 kg of water converts 1.00 liter of liquid water at 100°C into a volume of water vapor at the same temperature. The water vapor occupies a larger volume than the water; hence, during vaporization, the water must do work on the surrounding atmosphere while it expands.

(a) How much work does 1.00 kg of water do while it expands into vapor, at a pressure of 1.00 atm? Given that the heat of vaporization of water is 2.26×10^6 J/kg, what is the change of internal energy of the water during vaporization?

(b) How much work does 1.00 kg of water do while it expands into vapor, at a pressure of 2.00 atm? Deduce the heat of vaporization of water at 2.00 atm (and 100°C). Assume that the change of the density of liquid water between 1.00 atm and 2.00 atm can be neglected in this calculation.

9. The melting of 1.00 kg of ice converts a volume of 1.091 liters of ice into 1.000 liter of liquid water at 0°C. The water occupies a smaller volume; hence, during melting, the surrounding atmosphere does work on the ice.

(a) How much work does the atmosphere do on 1.00 kg of ice while it melts at a pressure of 1.00 atm? Given that the heat of fusion of ice is 3.34×10^5 J/kg, what is the change of internal energy of the water during melting?

(b) How much work does the atmosphere do on 1.00 kg of ice while it melts at a pressure of 2.00 atm? Deduce the heat of fusion of ice at 2.00 atm (and 0°C). Assume that the change of the density of the ice and the water between 1.00 atm and 2.00 atm can be neglected in this calculation.

10. (a) For a small isothermal expansion, the work done by a gas is $dW = nRT\,dV/V$ [see Eq. (21.11)]. Integrate this to obtain the work done in a finite expansion from an initial volume V to a final volume V':

$$W = nRT \ln \frac{V'}{V}$$

(b) Suppose that 0.10 mole of helium gas absorbs 500 J of heat from a heat reservoir at 700 K. If the initial volume of the gas is 0.0020 m^3, what is the final volume of the gas?

11. The volume of a gas is decreased by a constant external pressure of 0.25 atm, changing from 2.0 liters to 0.50 liter. During this compression, 75 J of energy flows out of the gas. Find the work done by the gas and its change in internal energy.

12. A piston of area 90 cm^2 slowly compresses a gas, which remains at a constant pressure of 15 atm. The final position of the piston is 12 cm from its initial position, and the internal energy of the gas drops by 15 J. Is heat removed from or added to the gas in this process? Find this amount of heat and the work done by the gas.

13. An initial volume of 1.00 liter of ethyl alcohol (ethanol) is heated from 5°C to 30°C at a pressure of 1.00 atm. Using the data in Tables 20.1 and 20.2, find the heat added to the ethanol and the work done by the ethanol in this process. The density of ethanol at 20° is 789 kg/m^3.

14. In a hand pump, an initial volume of 80 cm^3 of air at 1.0 atm and 300 K is quickly (adiabatically) compressed to a volume of 20 cm^3. What is the change in internal energy just after compression? After a long time, the gas exchanges heat with its container and surroundings, and returns to 300 K, while remaining compressed to 20 cm^3. What is the final internal energy of the gas? What is the heat transferred?

*15. An ideal gas is kept at 50°C throughout its expansion from 2.0 liters to 10 liters. If the quantity of gas is 10 moles, how much work is done by the gas? (Hint: Recall that when the pressure changes with volume, the work done is $W = \int p\,dV$.)

*16. 2.0 moles of an ideal gas at 100°C expands at constant temperature to a volume that is 10 times the initial volume. How much heat was added to the gas?

*17. Five moles of an ideal gas does 2.0×10^4 J of work as the gas expands at constant temperature to a final volume of 0.10 m^3 and a final pressure of 1.0 atm. Determine the temperature of the gas, the initial volume, and the initial pressure.

*18. A submerged scuba diver, 10 m below the surface of the water, blows a bubble of air from the valve of her tank. The final volume of the bubble is 4.0 cm^3, and the temperature of the air is that of the water, 15°C. Calculate how much heat the bubble absorbs as it forms, and how much work it does on the surrounding water. Assume that the temperature of the air in the bubble remains equal to the water temperature. The pressure in the tank is very high, so the initial volume of the air is negligible compared with the final volume at the bubble.

**19. At a pressure of 1.0 atm, the heat of vaporization of water is 2.26×10^6 J/kg; this is the heat required to convert 1.0 kg of water at 100°C into water vapor at the same temperature. Given that the specific heat of water is c = 4187 J/kg·°C and that of water vapor is (approximately) c_p = 2010 J/kg·°C, use the First Law of Thermodynamics to calculate the heat of vaporization of water at 20°C. [Hint: Instead of directly converting water at 20°C into water vapor, we can first heat the water to 100°C, then vaporize it, and then cool the vapor to 20°C (without condensation). The net work done against atmospheric pressure is the same during this indirect process as during direct vaporization (why?); hence, according to the First Law, the heat absorbed during each process is also the same.]

**20. Calculate the heat of vaporization of water at a temperature of 140°C, at a pressure of 1.0 atm. Use the data and the method of calculation described in Problem 19.

21.2 Heat Engines; the Carnot Engine†

21. A coal burning power plant uses thermal energy at a rate of 850 megawatts and produces 300 megawatts of mechanical power for the generation of electricity. What is the efficiency of this power plant?

22. To produce 120 hp of mechanical power, an automobile engine requires a supply of heat of 4.40×10^5 J per second from combustion of the fuel. What is the efficiency of this engine?

23. While running up stairs at a (vertical) rate of 0.30 m/s, a man of 70 kg generates waste heat at a rate of 1300 J/s. What efficiency for the human body can you deduce from this?

24. Suppose that a heat engine takes 3.0×10^4 J of heat from the high-temperature reservoir to produce 2.0×10^4 J of mechanical work. What is the efficiency of this engine? How much waste heat does it produce?

†For help, see Online Concept Tutorial 23 at www.wwnorton.com/physics

25. A heat engine takes 2.0×10^7 J of heat from the high-temperature reservoir and ejects 8.0×10^6 J of heat into the low-temperature reservoir. How much work does it produce? What is its efficiency?

26. Electric motors convert electric energy into mechanical energy with an efficiency of 95%. If the electric current supplies a power of 3.0 kW to such an electric motor, what mechanical power will the motor produce? How much heat (in kW) will the motor produce?

27. An electric power plant consists of a coal-fired boiler that makes steam, a turbine, and an electric generator. The boiler delivers 90% of the heat of combustion of the coal to the steam; the turbine converts 50% of the heat of the steam into mechanical energy; and the electric generator converts 99% of this mechanical energy into electric energy. What is the overall efficiency of generation of electric power?

28. Although a sprinter running at a steady speed on level ground performs no external work (except for a small amount of work against air resistance), the sprinter performs a considerable amount of work to accelerate and decelerate his own limbs during each stride and to lift his limbs against gravity. The following table lists the mechanical power used by a sprinter for motion of limbs and body during a sprint (the power was calculated from data obtained from photographic analysis of the motion):

Acceleration of limbs	1.5 hp
Deceleration of limbs	0.67 hp
Work against gravity	0.1 hp
Speed changes of body	0.5 hp

What is the total mechanical power expended in these motions of the sprinter? According to a measurement of the oxygen consumption of the sprinter, his expenditure of chemical energy was 13 hp during the sprint. What is his efficiency of conversion of chemical energy into mechanical energy?

29. According to some naive speculations of the nineteenth century, the human body was supposed to be a heat engine in which the combustion of food produces body heat that is then somehow converted into mechanical work by the muscles. If this were true, what would be the maximum (ideal) efficiency of the human body? The temperature of the "hot" human body (the heat source) is 37°C, and the temperature of the heat sink is that of the surrounding air, about 20°C. How many kcal of food energy would you need to consume to climb up one flight of stairs, to a height of 3.0 m? (Actually, muscles convert chemical energy directly into work; they are not heat engines.)

30. The efficiency of striated muscle is typically 37%, that is, the muscle converts 37% of the chemical energy reaching it (in the form of glucose) into mechanical work, and converts the remainder into waste heat. If your biceps muscle performs mechanical work at the rate of 50 W, what is the rate at which it consumes chemical energy, and what is the rate at which it produces waste heat? Given that the oxidation of glucose

yields 3.7×10^3 kcal/kg, what is the rate at which this muscle consumes glucose?

31. Each of the two engines of a DC-3 airplane produces 1100 hp. The engines consume gasoline; the combustion of 1.0 kg of gasoline yields 4.4×10^7 J. If the efficiency of the engines is 20%, at what rate do the two engines consume gasoline?

32. A steam turbine in a power plant converts the thermal energy of hot steam into mechanical energy. The turbine consists of a high-pressure stage and a low-pressure stage operating in tandem. The steam first passes through the high-pressure stage and gives up some of its thermal energy. The waste heat from this stage is used to reheat the steam, which then enters the low-pressure stage, where it gives up some of its remaining thermal energy.
   (a) Draw a flowchart showing the two stages of the turbine and the heat and work inputs and outputs.
   (b) If the efficiency of the high-pressure stage is 40% and the efficiency of the low-pressure stage is 20%, what is the overall efficiency of the turbine?

33. In principle, nuclear reactions can achieve temperatures of the order of 10^{11} K. What is the efficiency of a Carnot engine taking in heat from such a nuclear reaction and exhausting waste heat at 300 K?

*34. A nuclear power plant generates 1000 megawatts of electric (or mechanical) power. If the efficiency of this plant is 33%, at what rate does the plant generate waste heat? If this waste heat is to be removed by passing water from a river through the plant, and if the water is to suffer a temperature increase of at most 8°C, how many cubic meters of water per second is required?

35. A Carnot engine operates between a high-temperature reservoir at 100°C and a low-temperature reservoir at 0°C. How much energy must the engine take from the high-temperature reservoir to produce 5.0×10^4 J of work? How much waste heat does it produce?

36. Consider a Carnot engine operating between heat reservoirs at 400 K and at 300 K. What is the efficiency of this Carnot engine? If we want to increase the efficiency by 10%, by how much must we increase the temperature of the high-temperature reservoir? Alternatively, by how much must we decrease the temperature of the low-temperature reservoir?

37. A heat engine absorbs 2500 J of heat from a hot reservoir and releases 1400 J of heat to a cold reservoir each second. Find the efficiency of the engine and its mechanical power output.

38. If a heat engine has an efficiency of 15%, does mechanical work at a rate of 12 kW, and releases 5.0×10^3 J of heat to a cold reservoir on each cycle, find the heat absorbed in each cycle and the time for one cycle.

39. A Texas home is cooled by a Carnot engine in reverse. The outside temperature is 104°F and the inside temperature is 75°F, and 5.0×10^6 J of heat must be removed each hour. What rate of work input does this Carnot air conditioner require?

40. When the hot reservoir of a Carnot engine is maintained at 350°C, the efficiency is 15%. If we keep the cold reservoir at constant temperature, to what temperature should the hot reservoir be increased to obtain an efficiency of 25%?

41. A Carnot heat pump brings heat from outside at 5°C into a house at 20°C. What is the ratio of the heat input to the house to the work input to the engine (the coefficient of performance) for this Carnot engine?

42. The Carnot engine can be used to determine an unknown temperature. Suppose that a Carnot engine operates between a heat reservoir at an unknown temperature and a heat sink consisting of a water and ice mixture at 0°C. If this Carnot engine is found to have an efficiency of $e = 0.322$, what temperature can you deduce for the heat reservoir? The temperature scale defined by means of a Carnot engine is called the **absolute thermodynamic temperature scale**.

43. The electric motor of your refrigerator uses 2.5×10^3 J of electric energy to remove 6.0×10^3 J of heat from the refrigerator compartment. How much heat does this refrigerator dump into your kitchen? What is its coefficient of performance?

44. Suppose we operate a Carnot engine in reverse to "pump" heat from a heat reservoir at 0°C into a heat reservoir at 100°C. How much mechanical work must we supply per joule of heat removed from the 0°C reservoir?

45. One mole of an ideal monatomic gas is enclosed in a cylinder with a piston and placed in contact with a heat reservoir of variable temperature. By suitable changes of volume and temperature, the gas is taken through the three-step cycle described by the p–V diagram illustrated in Fig. 21.25.

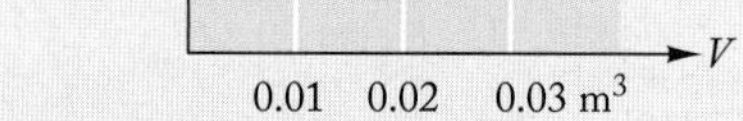

FIGURE 21.25 Three-step cycle.

   (a) In terms of the volumes and pressures given in Fig. 21.25, what is the net work done by the gas during this cycle?
   (b) What is the heat absorbed or ejected by the gas during steps 2 and 3?
   (c) What is the heat absorbed or ejected during step 1? [Hint: Use the results obtained in (a) and (b).]
   (d) What is the efficiency of this cycle?

46. One mole of an ideal monatomic gas is enclosed in a cylinder with a piston and placed in contact with a heat reservoir of

variable temperature. By suitable changes of volume and temperature, the gas is taken through the four-step cycle described by the p–V diagram illustrated in Fig. 21.26.

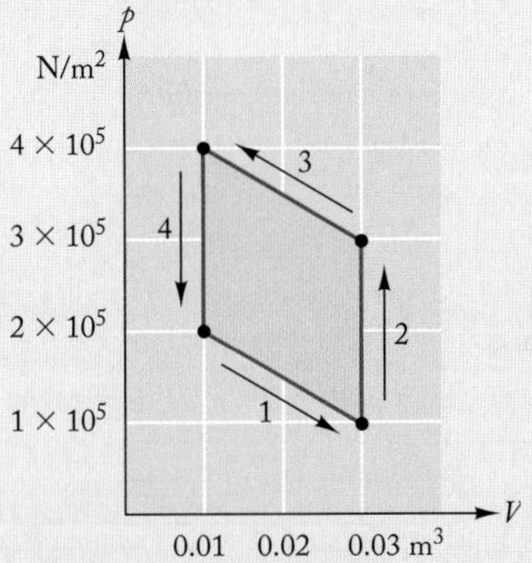

FIGURE 21.26 Four-step cycle.

(a) In terms of the volumes and pressures given in Fig. 21.26, what is the net work done by the gas during this cycle?

(b) What is the heat absorbed or ejected by the gas during steps 2 and 4?

(c) What is the heat absorbed or ejected during steps 1 and 3? (Hint: To find these amounts of heat, calculate W and ΔE for these steps.)

(d) What is the efficiency of this cycle?

47. A scheme for the extraction of energy from the oceans attempts to take advantage of the temperature difference between the upper and lower layers of ocean water. The temperature at the surface in tropical regions is about 25°C; the temperature at a depth of 300 m is about 5°C.

(a) What is the efficiency of a Carnot engine operating between these temperatures?

(b) If a power plant operating at the maximum theoretical efficiency generates 1.00 megawatt of mechanical power, at what rate does this power plant release waste heat?

(c) The power plant obtains the mechanical power and the waste heat from the surface water by cooling this water from 25°C to 5°C. At what rate must the power plant take in surface water?

48. In a nuclear power plant, the reactor produces steam at 520°C and the cooling tower eliminates waste heat into the atmosphere at 30°C. The power plant generates 500 megawatts of electric (or mechanical) power.

(a) If the efficiency is that of a Carnot engine, what is the rate of release of waste heat (in megawatts)?

(b) Actual efficiencies of nuclear power are about 33%. For this efficiency, what is the rate of release of heat?

49. An air conditioner removes 8.4×10^6 J/h of heat from a room at a temperature of 21°C and ejects this heat into the ambient air at a temperature of 27°C. This air conditioner requires 950 W of electric power.

(a) How much mechanical power would a Carnot engine, operating in reverse, require to remove this heat at the same rate?

(b) By what factor is the power required by the air conditioner larger than that required by the Carnot engine?

*50. An ice-making plant consists of a reversed Carnot engine extracting heat from a well-insulated box. The temperature in the icebox is −5°C and the temperature of the ambient air is 30°C. Water, of an initial temperature of 30°C, is placed in the icebox and allowed to freeze and to cool to −5°C. If the ice-making plant is to produce 10 000 kg of ice per day, what mechanical power is required by the Carnot engine?

*51. The boiler of a power plant supplies steam at 540°C to a turbine that generates mechanical power. The steam emerges from the turbine at 260°C and enters a steam engine that generates extra mechanical power. The steam is finally released into the atmosphere at a temperature of 38°C. Assume that the conversion of heat into work proceeds with the efficiency of a Carnot engine.

(a) What is the efficiency of the turbine? Of the steam engine?

(b) What is the net efficiency of both engines acting together? How does it compare with the efficiency of a single engine operating between 540°C and 38°C?

**52. One liter of water is initially at the same temperature as the surrounding air, 30°C. You wish to cool this water to 5°C by transferring heat from the water into the air. What is the minimum amount of work that you must supply in order to accomplish this? (Hint: Take a sequence of Carnot engines operating in reverse; use each engine to reduce the temperature by an infinitesimal amount.)

21.3 The Second Law of Thermodynamics
21.4 Entropy

53. What is the increase of entropy of 1.0 kg of water when it vaporizes at 100°C and 1.0 atm?

54. On a winter day heat leaks out of a house at the rate of 2.5×10^4 kcal/h. The temperature inside the house is 21°C and the temperature outside is −5°C. At what rate does this process produce entropy?

55. Consider the air conditioner described in Problem 49. Calculate the rate of increase of entropy contributed by the operation of this air conditioner.

56. Your body generates about 2000 kcal of heat per day. Estimate how much entropy you generate per day. Neglect the (small) amount of entropy that enters your body in the food you consume.

57. A steam engine operating between reservoirs at temperatures of 480°C and 27°C has an efficiency of 40%. The engine delivers 2000 hp of mechanical power. At what rate does this engine generate entropy?

58. Suppose that 1.0 kg of water freezes while at 0°C. What is the change of entropy of the water during this freezing process?

59. At a temperature of $-79°C$, solid carbon dioxide ("dry ice") transforms into a gas by sublimation (that is, direct vaporization from solid to gas). From the heat of transformation given in Table 20.4, calculate the increase of entropy per kilogram of carbon dioxide during sublimation.

60. Table 20.4 gives data for the melting and the vaporization of copper. What is the increase of entropy per kilogram of copper during melting? During vaporization?

61. Compare the increases of entropy during the melting of 1.0 kg of aluminum, iron, silver, and mercury (see Table 20.4 for data). Which of these has the largest change of entropy? Which has the smallest?

62. A parachutist of 80 kg descends at a constant speed of 5.0 m/s. What is the rate of increase of entropy of the parachute and the environment? The air temperature is 20°C.

63. For an automobile moving at a constant speed of 65 km/h on a level road, rolling friction, air friction, and friction in the drive train absorb a mechanical power of 12 kW. At what rate do these processes generate entropy? The temperature of the environment is 20°C.

64. An automobile of 2100 kg moving at 80 km/h brakes to a stop. In this process the kinetic energy of the automobile is first converted into thermal energy of the brake drums; this thermal energy later leaks away into the ambient air. Suppose that the temperature of the brake drums is 60°C when the automobile stops, and that the temperature of the air, and the final temperature of the brake drums, is 20°C.

(a) How much entropy is generated by the conversion of mechanical energy into thermal energy of the brake drums?

(b) How much extra entropy is generated as the heat leaks away into the air?

65. At Niagara Falls (Fig. 21.27), 5700 m^3/s of water falls through a vertical distance of 50 m, dissipating all of its gravitational energy. Calculate the rate of increase of entropy contributed by this falling water. The temperature of the environment is 20°C.

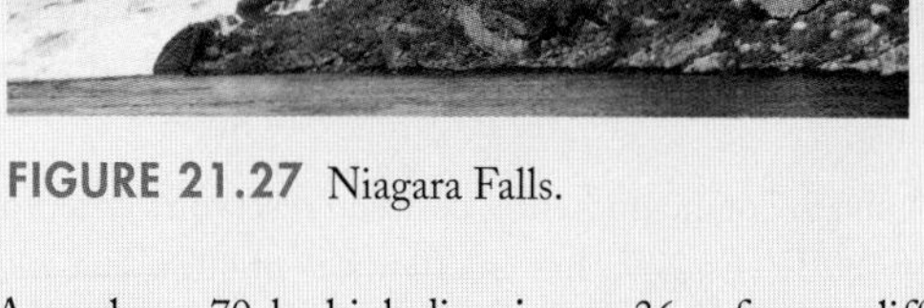

FIGURE 21.27 Niagara Falls.

66. At Acapulco, a 70-kg high diver jumps 36 m from a cliff into the sea. If the ambient temperature is 295 K, what is the net change in entropy for this process?

67. In Example 6 of Chapter 20, 1.8×10^3 J/s of heat is transferred through the wall of a house when the inside temperature is 21°C and the outside temperature is $-18°C$. At what rate is entropy generated?

68. The melting of lead at 328°C requires 2.9×10^4 J/kg. Determine the entropy change when 20 kg of solid lead at 328°C is melted. Is this change an increase or a decrease?

*69. What is the entropy change during the free expansion of 1.00 mole of an ideal gas from an initial volume of 1.00 liter into an evacuated volume of 1.00 additional liter? (Hint: See Example 1.)

*70. A 0.500-kg piece of silver is removed from an annealing furnace at 950°C and dropped into a bucket containing 5.00 kg of water initially at 20°C. If the heat stays within the silver–water system, calculate the entropy change when a common final temperature has been reached.

*71. (a) Consider a material with a constant specific heat capacity c. Show that if we gradually supply heat to a mass m of this material, increasing its temperature from T_1 to T_2, the increase of entropy of the mass m is

$$S_2 - S_1 = mc \ln\left(\frac{T_2}{T_1}\right)$$

(b) What is the increase of entropy in 1.0 kg of water that is heated from 20°C to 80°C?

*72. Consider the process of dissolving ice cubes in water as described in Example 8 of Chapter 20. What is the change of entropy of the system? (Hint: Use the formula derived in Problem 71.)

*73. You mix 1.0 liter of water at 20°C with 1.0 liter of water at 80°C. What is the increase of entropy?

*74. (a) Show that the increase of entropy of n moles of a gas heated at constant pressure from a temperature T_1 to a temperature T_2 is

$$\Delta S = nC_p \ln\frac{T_2}{T_1}$$

(b) Show that the increase of entropy of a gas heated at constant volume from a temperature T_1 to a temperature T_2 is

$$\Delta S = nC_V \ln\frac{T_2}{T_1}$$

*75. Consider the free expansion (see Example 1) of a sample of n moles of an ideal gas from an initial volume V to a final volume V'. Show that the increase of entropy in this process is

$$\Delta S = nR \ln\frac{V'}{V}$$

[Hint: The initial pressure is p and the final pressure is $p \times V/V'$. Replace the (irreversible) process of free expansion by an expansion at constant pressure p from the volume V to the volume V', followed by a reduction of pressure at constant volume V' from the initial pressure p to the final pressure $p \times V/V'$. The entropy changes for each of these two processes are given by the formulas in Problem 74.]

REVIEW PROBLEMS

76. A cylinder with a piston contains one mole of helium gas. Suppose we heat the cylinder and allow the gas to expand, so it performs work on the piston, at constant pressure. If during this expansion, the gas absorbs 3000 J of heat and performs 2000 J of work on the piston, what is the change ΔE in the internal energy of the gas? What is the change of its temperature?

77. A cylinder with a piston contains one mole of air (regarded as an ideal diatomic gas). The cylinder is insulated, so no heat can escape from it. If we push the piston inward and compress the gas, the temperature of the gas will increase. How much does the temperature increase if we do 2500 J of work on the gas during the compression?

78. One mole of an ideal monatomic gas in a cylinder with a piston has an initial volume V_1 and pressure p_1. We allow this gas to expand to a final (larger) volume V_2 and a final (lower) pressure p_2. Assume that on a p–V diagram, the expansion process is represented by a straight line (see Fig. 21.28).
(a) What is the work done by the gas?
(b) What is the change of energy of the gas?
(c) What is the heat absorbed by the gas?

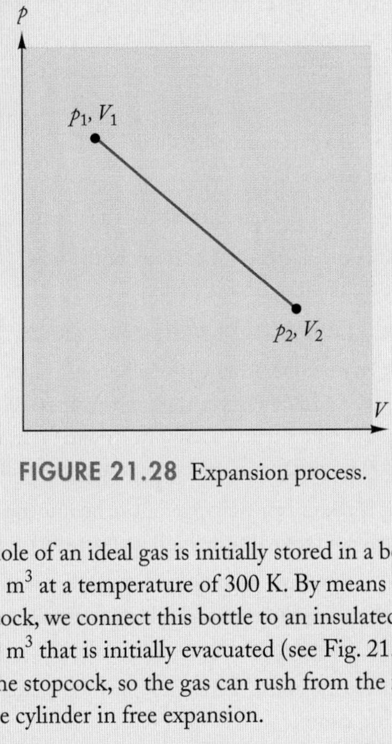

FIGURE 21.28 Expansion process.

79. 0.20 mole of an ideal gas is initially stored in a bottle of 0.0012 m^3 at a temperature of 300 K. By means of a pipe with a stopcock, we connect this bottle to an insulated cylinder of 0.0010 m^3 that is initially evacuated (see Fig. 21.29). We then open the stopcock, so the gas can rush from the first bottle into the cylinder in free expansion.
(a) What are the initial and final pressures of the gas?
(b) By pushing the piston of the cylinder inward, we now slowly compress the gas until all is expelled from the cylinder and its volume returns to the initial volume. During this compression, we keep the gas in contact with a heat reservoir at a temperature of 300 K. How much work must we do during this compression? How much heat is ejected into the heat reservoir?

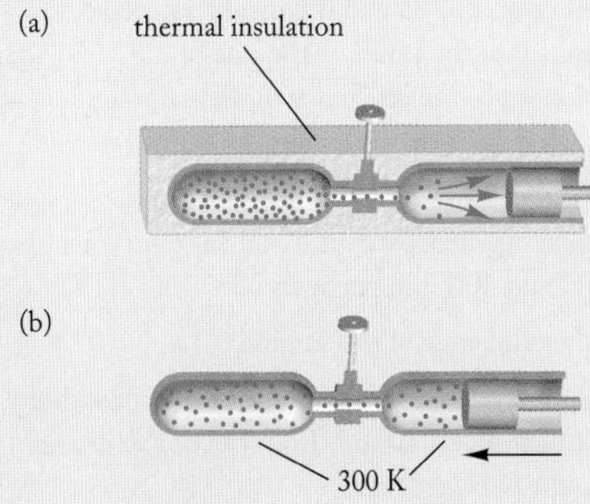

FIGURE 21.29 (a) Adiabatic free expansion. (b) Isothermal compression.

80. A tightly sealed plastic bag contains 4.0 liters of air at 0°C. The plastic bag is only loosely filled, and it exerts no compression on the air; the pressure of the air is therefore the pressure of the surrounding atmosphere, 1.0 atm. We place this plastic bag in the sun, and let it warm up to 60°C.
(a) What is the new volume of the air? Assume that the plastic bag remains loosely filled with air, so the pressure remains 1.0 atm.
(b) How much work did the air in the bag do on the surrounding atmosphere while it expanded? What is the change in the internal energy of the air? What is the amount of heat the air has absorbed? Assume that the air behaves like an ideal diatomic gas.

81. In an automobile proceeding at medium speed, the engine delivers 20 hp of mechanical power. The engine burns gasoline, which provides thermal energy at the rate of 6.3×10^4 J per second. What is the efficiency of the engine under these conditions? What is the rate at which the engine ejects waste heat?

82. In an experiment on the work efficiency of horses, a horse connected to an oxygen supply was made to do work on a treadmill (see Fig. 21.30). When the horse was delivering 869 watts of work to the treadmill, its rate of oxygen consumption was 10.8 liters per minute. The horse's metabolism yields 2.1×10^4 J of chemical energy per liter of oxygen consumed. Calculate the efficiency of conversion of chemical energy into external mechanical work.

FIGURE 21.30 Horse doing work on a treadmill.

83. On a hot day a house is kept cool by an air conditioner. The outside temperature is 32°C and the inside temperature is 21°C. Heat leaks into the house at the rate of 3.8×10^7 J/h. If the air conditioner has the efficiency of a Carnot engine, what is the mechanical power that it requires to hold the inside temperature constant?

84. A Carnot engine operates between two heat reservoirs of temperature 500°C and 30°C, respectively.
   (a) What is the efficiency of this engine?
   (b) If the engine generates 1.5×10^3 J of work, how many calories of heat does it absorb from the hot reservoir? Eject into the cold reservoir?

85. One mole of an ideal monatomic gas is enclosed in a cylinder with a piston and placed in contact with a heat reservoir of variable temperature. By suitable changes of volume and temperature, the gas is taken through the four-step cycle described by the p–V diagram illustrated in Fig. 21.31.
   (a) Describe how you must vary the temperature of the heat reservoir and the volume of the gas to accomplish each of the four steps. Begin at top left corner of the cycle and proceed clockwise.
   (b) From the volumes and pressures given in Fig. 21.31, calculate the work done by the gas during each step.
   (c) Calculate the heat absorbed or ejected by the gas during each step and verify that the net work done equals the net heat absorbed.
   (d) Calculate the efficiency of this cycle.

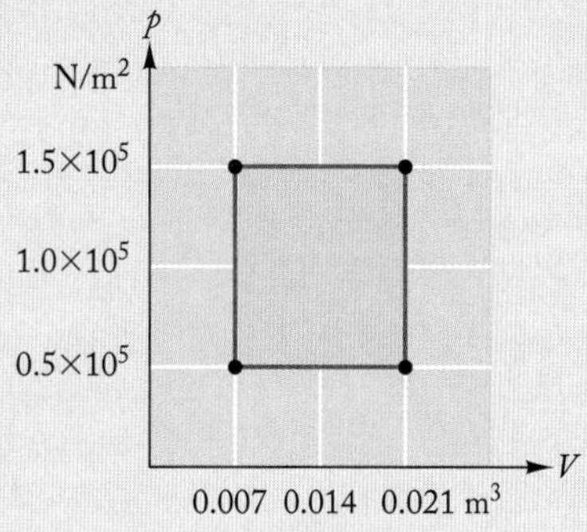

FIGURE 21.31 Four-step cycle.

86. A Carnot engine operating with an ideal gas uses a high-temperature reservoir at 600 K and a low-temperature reservoir at 200 K. In one cycle, the engine absorbs 1200 J from the high-temperature reservoir. What is the waste heat ejected into the low-temperature reservoir? What is the work produced by the engine during the cycle?

87. A geothermal power plant at Wairakei, New Zealand (Fig. 21.32), uses hot underground water at a temperature of 300°C as heat source and uses the atmosphere at a temperature of, say, 25°C as heat sink. What would be the efficiency of a Carnot engine operating between these temperatures? If this Carnot engine produces 10 000 kW of mechanical power, what power does it eject as waste heat?

FIGURE 21.32 Geothermal power plant at Wairakei, New Zealand.

88. A refrigerator is operated by a Carnot engine. The temperature inside the refrigerator is 4°C and the temperature of the surrounding room is 20°C. How much mechanical work must the refrigerator's electric motor perform to remove 5.0×10^3 J of heat from the inside of the refrigerator and dump it into the room?

89. Compare the increases of entropy during the vaporization of 1.0 kg of nitrogen, oxygen, and hydrogen (see Table 20.4 for data). Which of these has the largest change of entropy? Which the smallest?

90. You dump 1.0 kg of molten lead (at 328°C) into 1.0 kg of crushed ice (at 0°C).
   (a) What is the final temperature of the mixture?
   (b) What is the decrease of entropy of the lead? (Hint: Use the formula derived in Problem 71.)
   (c) What is the increase of entropy of the water?

91. A Carnot engine operating with helium gas (approximately an ideal gas) uses a high-temperature reservoir at 700 K and a low-temperature reservoir at 300 K. In one cycle, the engine absorbs 800 J from the high-temperature reservoir.
   (a) What is the entropy change of the helium gas while in thermal contact with the high-temperature reservoir?
   (b) What is the waste heat ejected, and what is the entropy change of the helium gas while in thermal contact with the low-temperature reservoir?
   (c) What are the entropy changes during the adiabatic expansion and contraction?

Answers to Checkups

Checkup 21.1

1. The sunlight heats the balloon of gas, so Q, the amount of heat added, is positive. The balloon does work as it pushes against atmospheric pressure, so W, the work done by the gas, is positive.
2. If the jar is heated, Q, the amount of heat added to the system, is positive. The work done by the gas, W, is zero, since the volume of the gas does not change appreciably. (However, because the jar will experience a slight thermal expansion, W will actually have a very small positive value.)
3. (C) Negative, positive. The athlete loses heat to the environment, so Q, the heat added to the athlete, is negative. The athlete does work to move the bicycle (against friction), so W, the work done by the system, is positive.

Checkup 21.2

1. An isothermal expansion proceeds at constant temperature; an adiabatic expansion proceeds without loss or gain of heat.
2. No. An efficiency greater than 1 would mean more energy is coming out than going in, and this is not possible.
3. No. The burning of fuel cannot be reversed by performing mechanical work on the crankshaft.
4. No. The efficiency of a Carnot engine depends only on the two temperatures of operation.
5. (C) There is no net effect. Since the Carnot cycle is reversible, using the work output of the first to drive the second in reverse results in no net effect.

Checkup 21.3

1. At $T_2 = T_1$, the efficiency is zero; no useful work can be extracted. For $T_2 > T_1$, the heat flow will be opposite the usual direction; that is, the engine will absorb heat from the reservoir at T_2, so the efficiency is $e = 1 - T_1/T_2$.
2. No. The refrigerator has the effect of removing mechanical or electrical energy from the environment (work input to the refrigerator is required for operation).
3. (C) The cycle could not be completed, since the steam could not condense. The steam could not condense if the "condenser" was at very high temperature; also, with the two temperatures nearly equal, the efficiency would approach zero.

Checkup 21.4

1. The entropy changes are given by $\Delta Q/T$ and so are 100 J/300 K $= \frac{1}{3}$ J/K for the first reservoir and 100 J/400 K $= \frac{1}{4}$ J/K for the second.
2. Yes. The flowchart shows the release of waste heat, that is, only a partial conversion of heat into work, and so is consistent with the Second Law.
3. Yes, the entropy of the water decreases as its temperature is lowered (heat is removed, order increases). This does not violate the Second Law, since the entropy of the surroundings increases (nearby air absorbs heat).
4. (B) 1 and 4. The braking of an automobile converts mechanical work into heat; this process is not reversible, and so (1) creates an increase in entropy. Both the slowing of an automobile uphill and the pumping of water upward (without friction) are reversible, since the mechanical energy can be completely recovered by reversing the process; thus, (2) and (3) do not result in an increase in entropy. The burning of firewood is a conversion of chemical energy into heat, which is irreversible, and so (4) increases the entropy.

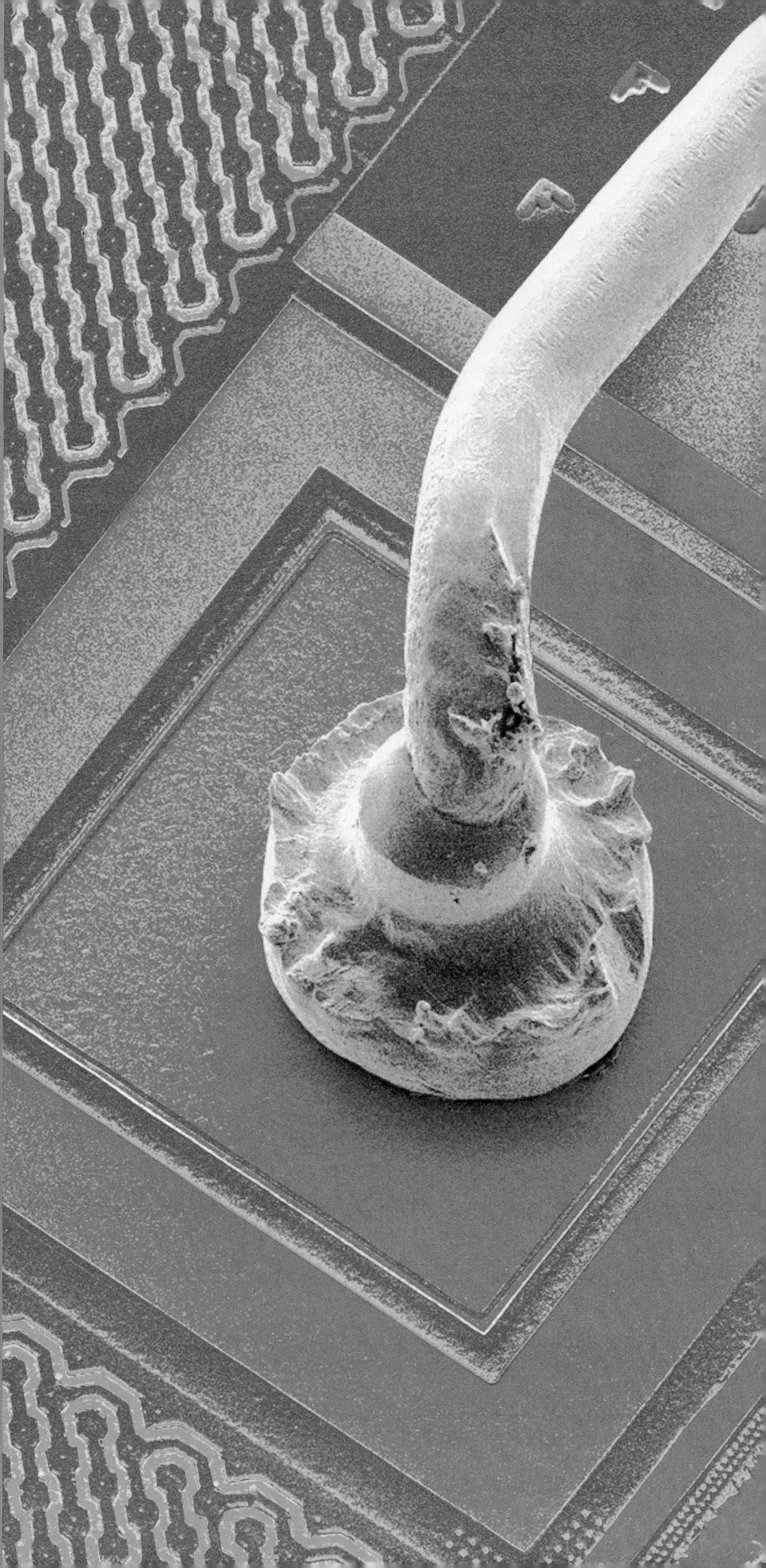

PART

4

Electricity and Magnetism

CONTENTS

A microchip contains precision-engineered circuits packed into a very small space. This gold micro-wire is one of several connections to the microchip, and it leads to larger pins that are used to plug the microchip into a circuit board, such as found inside a computer.

CHAPTER

22 Electric Force and Electric Charge

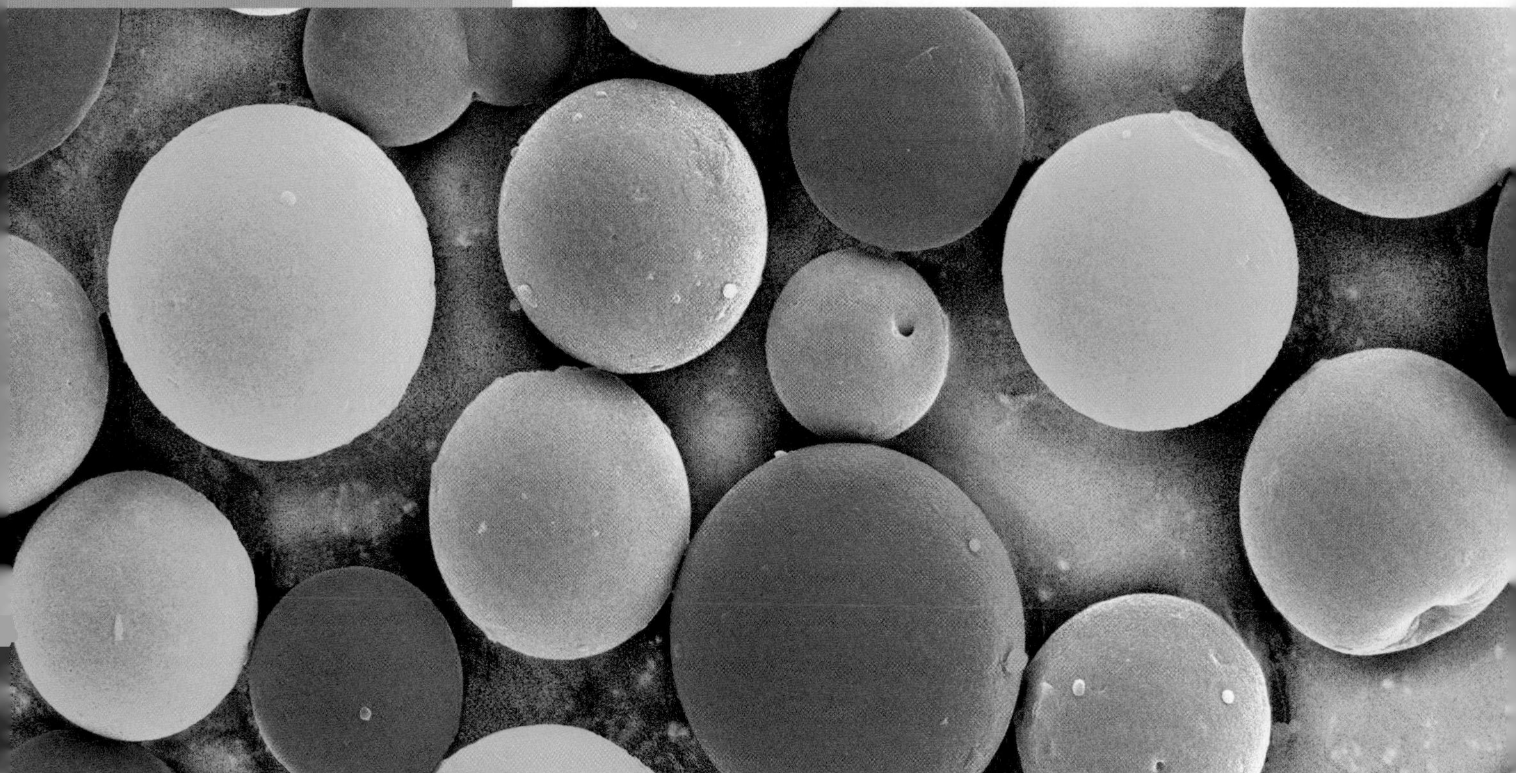

CONCEPTS IN CONTEXT

Concepts in Context

Photocopiers and laser printers use toner particles, such as these magenta, cyan, yellow, and black modern toner particles made spherical with a polymer coating. An attractive electric force holds the particles to the plate beneath them; the particles also exert repulsive electric forces on one another.

With the properties of electric force and electric charge introduced in this chapter, we can answer such questions as:

? What is the force that one toner particle exerts on another? (Example 4, page 701)

? What is the total force on a particle surrounded by several other particles? (Example 7, page 704)

? How are the toner particles transferred to form an image? (Physics in Practice: Xerography, page 709)

Our society is dependent on electricity. An electric power failure demonstrates our dependence—subways and elevators stop; traffic lights, streetlights, and the lights in our homes go out; refrigerators fail; food can't be cooked; homes can't be heated; radios, TVs, and computers can't be operated. But our dependence on electricity runs even deeper than our reliance on electrical machinery and gadgetry would suggest. Electricity is an essential ingredient in all the atoms in our bodies and in our environment. The forces that hold the parts of an atom together are electric forces. Furthermore, so are the forces that bind atoms in a molecule and hold these building blocks together in large-scale macroscopic structures, such as a rock, a tree, a human body, a skyscraper, or a supertanker. All the mechanical "contact" forces of everyday experience—the push of a hand against a door, the pull of an elevator cable, the pressure of water against the hull of a ship—are nothing but the combined electric forces of many atoms. Thus, our immediate environment is dominated by **electric forces**.

electric force

In the following chapters, we will study electric forces and their effects. For a start (Chapters 22–28), we will assume that the particles exerting these forces are at rest or moving only very slowly. The electric forces exerted under these conditions are called **electrostatic forces**. Later on (Chapters 29–31), we will consider the electric forces when the particles are moving with uniform velocity or nearly uniform velocity. Under these conditions the electric forces are modified—besides the electrostatic force there arises a **magnetic force**, which depends on the velocities of the particles. The combined electrostatic and magnetic forces are called **electromagnetic forces**. Finally, we will consider the forces exerted when the particles are moving with accelerated motion (Chapter 33). The electromagnetic forces are then further modified with a drastic consequence, that is, the emission of electromagnetic waves, such as light and radio waves.

magnetic force

Electricity was first discovered through friction. The ancient Greeks noticed that rods of amber (*elektron,* in Greek), when rubbed with a cloth or with fur, gave off sparks and attracted small bits of straw or feathers. You can easily duplicate this ancient discovery by rubbing a plastic comb on a shirt or a sweater; in the dark, you can then see a multitude of small sparks produced by this rubbing process, and the electrified comb will attract small bits of paper or lint. In the nineteenth century, practical applications of electricity were gradually developed, but it was only in the twentieth century that the pervasive presence of electric forces holding together all the matter of our environment was recognized.

22.1 THE ELECTROSTATIC FORCE

Online Concept Tutorial 25

Ordinary matter—solids, liquids, and gases—consists of atoms, each with a nucleus surrounded by a swarm of electrons. For example, Fig. 22.1 shows the structure of an atom of neon. At the center of this atom there is a nucleus made of ten protons and ten neutrons packed very tightly together—the diameter of the nucleus is only about 6×10^{-15} m. Moving around this nucleus are ten electrons; these electrons are confined to a roughly spherical region about 3×10^{-10} m across.

The atom somewhat resembles the Solar System, with the nucleus as Sun and the electrons as planets. In the Solar System, the force that holds a planet near the Sun is the gravitational force. In the atom the force that holds an electron near the nucleus is the electric force of attraction between the electron and the protons in the nucleus. This electric force resembles gravitation in that it decreases in proportion to the inverse square of the distance. But the electric force is much stronger than the gravitational force.

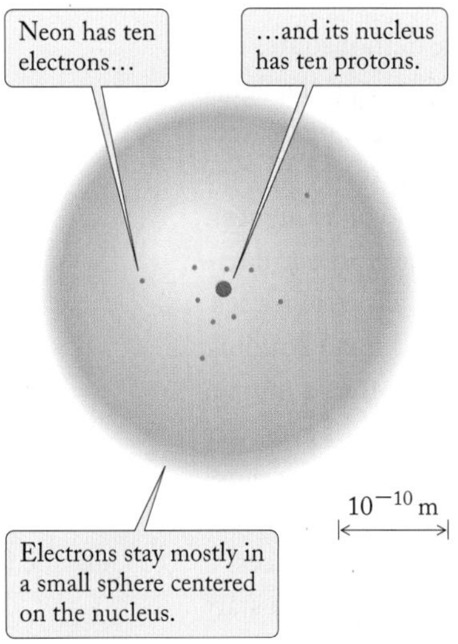

FIGURE 22.1 Neon atom. This drawing shows the electrons at one instant of time, as they would appear in a (hypothetical) extremely powerful microscope.

The electric attraction between an electron and a proton (at any given distance) is about 2×10^{39} times as strong as the gravitational attraction. Thus, the electric force is by far the strongest force felt by an electron in an atom.

The other great difference between the gravitational force and the electric force is that the gravitational force between two particles is always attractive, whereas electric forces can be attractive, repulsive, or zero, depending on what two particles we consider. Table 22.1 gives a qualitative summary of the electric forces between fundamental particles.

electric charge

Particles that exert electric forces are said to have an **electric charge**; particles that do not exert electric forces are said to have no electric charge. Thus, *electric charge is thought of as the source of electric force,* just as mass is the source of gravitational force. Electrons and protons have electric charge, but neutrons have no electric charge. The electron–proton force, the electron–electron force, and the proton–proton force all have the same magnitudes (for a given distance). Thus, the strengths of the electric forces associated with electrons and with protons are of equal magnitudes; that is, their electric charges are of equal magnitudes. For the mathematical formulation of the law of electric force, we assign a positive charge to the proton and a negative charge (of equal magnitude) to the electron. We designate these charges of the proton and the electron by $+e$ and $-e$, respectively. Table 22.2 summarizes these values of the charges.

In terms of these electric charges, we can then state that *the electric force between charges of like sign is repulsive and the electric force between charges of unlike sign is attractive.*

coulomb (C)

The numerical value of the charge e of the proton depends on the system of units. In the SI system of units, the electric charge is measured in **coulombs** (C), and the corresponding numerical values of the fundamental charges of the proton and of the electron are[1]

charge of proton and electron

$$e = 1.60 \times 10^{-19} \text{ C for proton}$$

$$-e = -1.60 \times 10^{-19} \text{ C for electron} \qquad (22.1)$$

TABLE 22.1

ELECTRIC FORCES (QUALITATIVE)

PARTICLES	FORCE
Electron and proton	Attractive
Electron and electron	Repulsive
Proton and proton	Repulsive
Neutron and anything	Zero

TABLE 22.2

ELECTRIC CHARGES OF PROTONS, ELECTRONS, AND NEUTRONS

PARTICLE	CHARGE
Proton, p	$+e$
Electron, e	$-e$
Neutron, n	0

[1]As with all physical constants, we have rounded these values to three significant figures. A more exact value can be found in Appendix 6 or inside the book covers.

In the SI system of units, the coulomb is defined in terms of a standard electric current, that is, a standard rate of flow of charge: one coulomb is the amount of electric charge that a current of one ampere delivers in one second. Unfortunately, the definition of the standard current involves the use of magnetic fields, and we will therefore have to postpone the question of the precise definition of ampere and coulomb to a later chapter.

One coulomb of charge represents a large number of fundamental charges; this is easily seen by taking the inverse of Eq. (22.1),

$$1\ \text{C} = \frac{1}{1.60 \times 10^{-19}}\, e = 6.25 \times 10^{18} \times e$$

Thus, a coulomb is more than 6 billion billion fundamental charges. Since the fundamental charge is so small, we can often ignore the discrete character of charge in practical and engineering applications of electricity, and we can treat macroscopic charge distributions as continuous. This is analogous to treating macroscopic mass distributions as continuous, even though, on a microscopic scale, the mass consists of discrete atoms.

The net electric charge of a body containing some number of electrons and protons is the (algebraic) sum of the electron and proton charges. For instance, the net electric charge of an atom containing equal numbers of electrons and protons is zero; that is, the atom is electrically neutral. Sometimes atoms lose an electron, and sometimes they gain an extra electron. *Such atoms with missing electrons or with extra electrons are called* **ions**. They have a net positive charge if they have lost electrons, and a net negative charge if they have gained electrons. The positive or negative charge on a macroscopic body—such as on a plastic comb electrified by rubbing—arises in the same way, from a deficiency or an excess of electrons.

ion

The electric forces between two neutral atoms tend to cancel; each electron in one atom is attracted by the protons in the nucleus of the other atom, and simultaneously it is repelled by the equal number of electrons of that atom. However, the cancellation of these electric attractive and repulsive forces among the electrons and the protons in the two atoms is sometimes not complete. For instance, the "contact" force between two atoms close together arises from an incomplete cancellation of the attractive and repulsive forces. The force between the atoms depends on the relative locations of the electrons and the nuclei. If the distributions of the electrons are somewhat distorted so, on the average, the electrons in one atom are closer to the electrons of the neighboring atom than to its nucleus, then the net force between these atoms will be repulsive. Figure 22.2a shows such a distortion that leads to a repulsive net force; the distortion may either be intrinsic to the structure of the atom or induced by the presence of the neighboring atom. Figure 22.2b shows a distortion that leads to an attractive force.

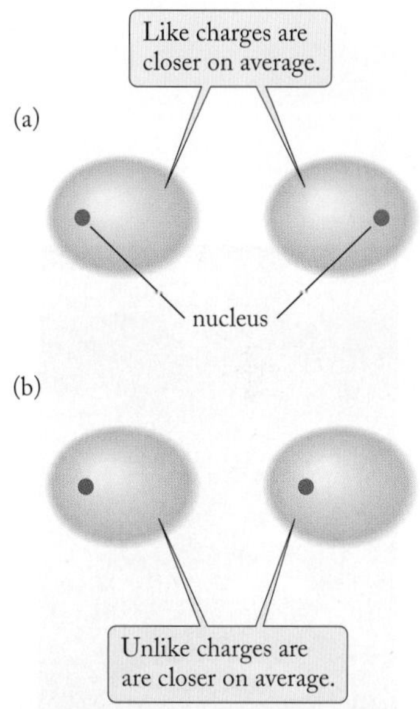

FIGURE 22.2 (a) Two neighboring distorted atoms. The colored regions represent the average distribution of the electrons. The electrons of the left atom are closer to most of the electrons of the right atom than to its nucleus. This results in a net repulsive force between the atoms. (b) The electrons of the left atom are closer to the nucleus of the right atom than to most of its electrons. This results in a net attractive force between the atoms.

Likewise, the electric forces between two neutral macroscopic bodies separated by some appreciable distance tend to cancel. For example, if the macroscopic bodies are a baseball and a tennis ball separated by a distance of 2 m, then each electron of the baseball is attracted by the protons of the tennis ball, but simultaneously it is repelled by the electrons of the tennis ball; and these forces cancel each other. Only when the surfaces of the two balls are very near one another ("touching") will the atoms in one surface exert a net electric force on those in the other surface.

This cancellation of the electric forces between neutral macroscopic bodies explains why we do not see large electric attractions or repulsions between the macroscopic bodies in our environment, even though the electric forces between individual electrons and protons are much stronger than the gravitational forces. Most such macroscopic bodies are electrically neutral, and they therefore will exert no net electric forces on each other, except for contact forces.

CHARLES AUGUSTIN de COULOMB (1736–1806) *French physicist, shown here with the torsion balance, which he invented. He established that the electric force between small charged balls obeys an inverse-square law.*

Checkup 22.1

QUESTION 1: The planets in the Solar System exert large gravitational forces on each other, but only insignificant electric forces. The electrons in an atom exert large electric forces on each other, but only insignificant gravitational forces. Explain this difference.

QUESTION 2: A stone of mass 1.0 kg rests on the ground. What is the net electric force that the ground exerts on the stone?

QUESTION 3: Suppose that in Fig. 22.2a, the atomic nuclei were displaced toward each other, instead of away from each other. Would the net electric force between the atoms be attractive or repulsive?

QUESTION 4: Six electrons are added to 1.0 coulomb of positive charge. The net charge is approximately

(A) 7.0 C (B) −5.0 C (C) 1.0 C (D) $-6e$ (E) $-5e$

22.2 COULOMB'S LAW

Online Concept 24 Tutorial

As already mentioned above, the electric force between two particles decreases with the inverse square of the distance, just as does the gravitational force. The dependence of the electric force on distance was discovered through experiments by Charles Augustin de Coulomb, who investigated the repulsion between small balls that he charged by a rubbing process. To measure the force between the balls, he used a delicate torsion balance (see Fig. 22.3) similar to the torsion balance later used by Henry Cavendish to measure gravitational forces. His experimental results are summarized in **Coulomb's Law**:

> *The magnitude of the electric force that a particle exerts on another particle is directly proportional to the product of their charges and inversely proportional to the square of the distance between them. The direction of the force is along the line joining the particles.*

Mathematically, the electric force F that a particle of charge q' exerts on a particle of charge q at a distance r is given by the formula

$$F = k \times \frac{q'q}{r^2} \tag{22.2}$$

where k is a constant of proportionality. This formula not only gives the magnitude of the force, but also the direction, if we *interpret a positive value of the force F as repulsive and a negative value as attractive.* For instance, in the case of the force exerted by a proton on an electron, the charges are $q' = e$ and $q = -e$, and the formula (22.2) yields

$$F = k \times \frac{e \times (-e)}{r^2} = -k\frac{e^2}{r^2} \tag{22.3}$$

which is negative, indicating attraction.

The electric force that the particle of charge q exerts on the particle of charge q' has the same magnitude as the force exerted by q' on q, but the opposite direction. These mutual forces are an action–reaction pair (see Fig. 22.4 for two examples).

FIGURE 22.3 Coulomb's torsion balance. The beam carries a small charged ball (a) at one end, and a counterweight on the other. A second small charged ball (b) is brought near the first ball. If the balls carry charges of equal signs, they repel each other, and the beam of the balance rotates.

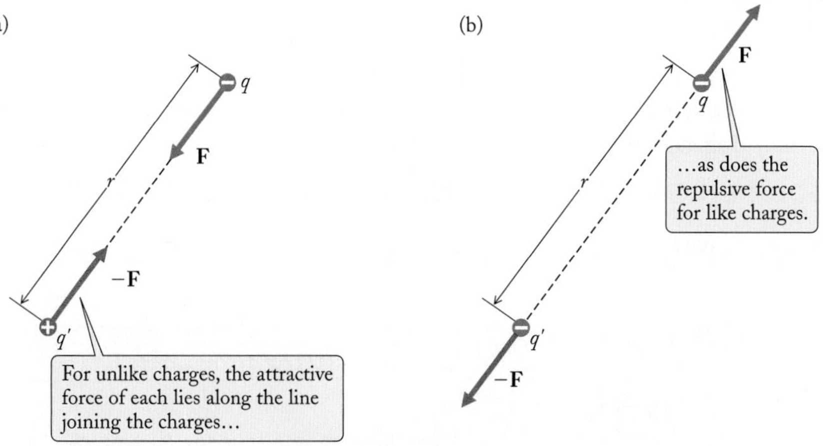

FIGURE 22.4 (a) Two charged particles q and q', one of which is positive and one of which is negative. (b) Two other particles q and q'; both are negative.

In the SI system of units, the constant of proportionality k in Coulomb's Law, called the **Coulomb constant**, or the electric force constant, has the value

$$k = 8.99 \times 10^9\ \text{N}\cdot\text{m}^2/\text{C}^2 \tag{22.4}$$

Coulomb constant

This constant is traditionally written in the more complicated but equivalent form

$$k = \frac{1}{4\pi\epsilon_0} \tag{22.5}$$

with

$$\epsilon_0 = 8.85 \times 10^{-12}\ \text{C}^2/(\text{N}\cdot\text{m}^2) \tag{22.6}$$

permittivity constant

The quantity ϵ_0 ("epsilon nought") is called the **electric constant** or the **permittivity constant.** In terms of the permittivity constant, Coulomb's Law for the force that a particle of charge q' exerts on a particle of charge q becomes

$$F = \frac{1}{4\pi\epsilon_0}\frac{q'q}{r^2} \tag{22.7}$$

Coulomb's Law

Remember that a positive value of F indicates a repulsive force directed along the line joining the charges, and a negative value indicates an attractive force. Although the expression (22.2), with the value of the constant given in Eq. (22.4), is most convenient for the numerical calculation of the Coulomb force, the somewhat more complicated expression (22.7) is widely used in physics and engineering. Of course, the two expressions are mathematically equivalent, and they give the same results.

Coulomb's Law applies to particles—electrons and protons—and also to any small charged bodies, provided that the sizes of these bodies are much smaller than the distances between them; such bodies are called **point charges**. Equation (22.2) obviously resembles Newton's Law for the gravitational force (see Section 9.1); the constant k is analogous to the gravitational constant G, and the electric charges are analogous to the gravitating masses.

point charge

EXAMPLE 1 Compare the magnitudes of the gravitational force of attraction and of the electric force of attraction between the electron and the proton in a hydrogen atom. According to Newtonian mechanics, what is the acceleration of the electron? Assume that the distance between these particles in a hydrogen atom is 5.3×10^{-11} m.

SOLUTION: From Chapter 9, the magnitude of the gravitational force is

$$F_g = G\frac{mM}{r^2}$$

$$= (6.67 \times 10^{-11}\,\text{N}\cdot\text{m}^2/\text{kg}^2) \times \frac{(9.11 \times 10^{-31}\,\text{kg})(1.67 \times 10^{-27}\,\text{kg})}{(5.3 \times 10^{-11}\,\text{m})^2}$$

$$= 3.6 \times 10^{-47}\,\text{N}$$

The magnitude of the electric force is

$$F_e = \frac{1}{4\pi\epsilon_0}\frac{e \times e}{r^2} = (8.99 \times 10^9\,\text{N}\cdot\text{m}^2/\text{C}^2) \times \frac{(1.60 \times 10^{-19}\,\text{C})^2}{(5.3 \times 10^{-11}\,\text{m})^2}$$

$$= 8.2 \times 10^{-8}\,\text{N}$$

The ratio of these forces is $(8.2 \times 10^{-8}\,\text{N})/(3.6 \times 10^{-47}\,\text{N}) = 2.3 \times 10^{39}$. Thus the electric force overwhelms the gravitational force.

Since the gravitational force is insignificant compared with the electric force, it can be neglected. The acceleration of the electron is then

$$a = \frac{F}{m} = \frac{8.2 \times 10^{-8}\,\text{N}}{9.11 \times 10^{-31}\,\text{kg}} = 9.0 \times 10^{22}\,\text{m/s}^2$$

This is a gigantic acceleration. If it occurred along the electron's motion instead of centripetally, such an acceleration could boost the electron's velocity close to one-third of the speed of light in only a femtosecond (10^{-15} s)!

COMMENTS: Note that for the ratio of the electric force and the gravitational force between the proton and electron, we would obtain the same immense value 2.3×10^{39} whatever the separation between the two particles, since both are inverse-square forces. Also notice that for the given atomic-scale distance, the electric force has a measurable value, the same as weighing an 8-microgram mass, whereas the gravitational force is far below the current limits of detection (the highest sensitivity attained by a measurement of force is near 10^{-20} N).

EXAMPLE 2 How much negative charge and how much positive charge are there on the electrons and the protons in a cup of water (0.25 kg)?

SOLUTION: The "molecular mass" of water is 18 g; hence, 250 g of water amounts to 250/18 moles. Each mole has 6.02×10^{23} molecules, giving $(250/18) \times 6.0 \times 10^{23}$ molecules in the cup. Each molecule consists of two hydrogen atoms (one electron apiece) and one oxygen atom (eight electrons). Thus, there are 10 electrons in each molecule, and the total negative charge on all the electrons together is

$$(250\,\text{g})(1\,\text{mole}/18\,\text{g})(6.02 \times 10^{23}\,\text{molecules/mole})$$
$$\times\,(10\,\text{electrons/molecule})(-1.60 \times 10^{-19}\,\text{C/electron})$$

$$= -1.3 \times 10^7 \text{ C}$$

The positive charge on the protons is the opposite of this.

EXAMPLE 3 What is the magnitude of the attractive force exerted by the electrons in a cup of water on the protons in a second cup of water at a distance of 10 m?

SOLUTION: According to the preceding example, the charge on the electrons in the cup is -1.3×10^7 C and the charge on the protons is $+1.3 \times 10^7$ C. If we treat both of these charges as point charges, the force on the protons is

$$F = \frac{1}{4\pi\epsilon_0}\frac{qq'}{r^2}$$

$$= (8.99 \times 10^9 \text{ N}\cdot\text{m}^2/\text{C}^2)\frac{(-1.3 \times 10^7 \text{ C})(1.3 \times 10^7 \text{ C})}{(10 \text{ m})^2}$$

$$= -1.5 \times 10^{22} \text{ N}$$

This is approximately the weight of a billion billion tons! This enormous attractive force on the protons is precisely canceled by an equally large repulsive force exerted by the protons in one cup on the protons in the other cup. Thus, the cups exert no net forces on each other.

EXAMPLE 4 Consider two toner particles separated by 1.2×10^{-5} m; each of the two particles has a negative charge of -3.0×10^{-14} C. What is the electric force that one particle exerts on the other? Treat the toner particles approximately as point particles.

Concepts in Context

SOLUTION: Since we may treat the particles approximately as point charges, the force that one particle exerts on the other is

$$F = \frac{1}{4\pi\epsilon_0}\frac{qq'}{r^2}$$

$$= (8.99 \times 10^9 \text{ N}\cdot\text{m}^2/\text{C}^2)\frac{(-3.0 \times 10^{-14} \text{ C})^2}{(1.2 \times 10^{-5} \text{ m})^2}$$

$$= +5.6 \times 10^{-8} \text{ N}$$

The positive sign reminds us that the force is repulsive, tending to push each particle directly away from the other, along the line joining them (see Fig. 22.4b). This mutual repulsion helps to keep the toner particles dispersed, so they do not clump up in one region.

EXAMPLE 5 A simple electroscope for the detection and measurement of electric charge consists of two small foil-covered cork balls of 1.5×10^{-4} kg each suspended by threads 10 cm long (see Fig. 22.5). When equal electric charges are placed on the balls, the electric repulsive force pushes them apart, and the angle between the threads indicates the magnitude of the electric

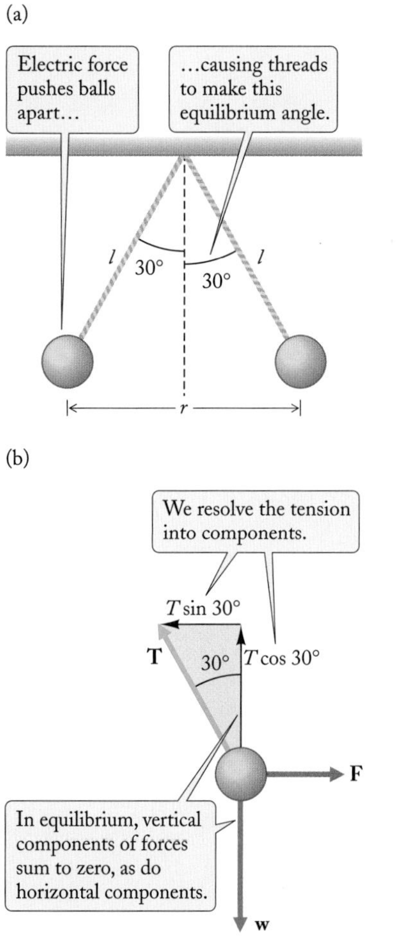

FIGURE 22.5 (a) Two equal charged balls suspended by threads. (b) "Free-body" diagram for the right ball.

charge. If the equilibrium angle between the threads is 60°, what is the magnitude of the charge?

SOLUTION: Figure 22.5b shows a "free-body" diagram for one of the balls. The electric force **F** acts along the line joining the two charges and is thus horizontal. In equilibrium, the vector sum of the electric repulsion **F**, the weight **w**, and the tension **T** of the thread must be zero. Accordingly, the horizontal component of the tension must balance the electric repulsion, and the vertical component of the tension must balance the weight:

$$F = T \sin 30°$$

$$mg = T \cos 30°$$

We can eliminate the tension from the problem by taking the ratio of these equations, yielding

$$F = mg \tan 30°$$

From Fig. 22.5a we see that the distance between the balls is $r = 2l \sin 30°$, so Coulomb's Law tells us

$$F = \frac{1}{4\pi\epsilon_0} \frac{q^2}{(2l \sin 30°)^2}$$

Equating these two expressions for F, we find

$$mg \tan 30° = \frac{1}{4\pi\epsilon_0} \frac{q^2}{(2l \sin 30°)^2}$$

and

$$q = \sqrt{4\pi\epsilon_0 mg \tan 30°} \times 2l \sin 30°$$

$$= \sqrt{(4\pi)(8.85 \times 10^{-12}\ \text{C}^2/\text{N}\cdot\text{m}^2)(1.5 \times 10^{-4}\ \text{kg})(9.81\ \text{m/s}^2)(\tan 30°)}$$

$$\times (2)(0.10\ \text{m})(\sin 30°)$$

$$= 3.1 \times 10^{-8}\ \text{C}$$

✓ Checkup 22.2

QUESTION 1: Suppose that the electric force between two charges is attractive. What can you conclude about the signs of these charges?

QUESTION 2: Suppose that the electric force between two charges separated by a distance of 1 m is 1×10^{-4} N. What will be the electric force if we increase the distance to 10 m? To 100 m?

QUESTION 3: Two balls, separated by some distance, carry equal electric charges and exert a repulsive electric force on each other. If we transfer a fraction of the electric charge of one ball to the other, will the electric force increase or decrease?

QUESTION 4: Two particles are separated by a distance of 3.0 m; each exerts an electric force of 1.0 N on the other. If one particle carries 10 times as much electric charge as the other, what is the magnitude of the smaller charge?

(A) 10 pC (B) 10 μC (C) 10 C (D) 10 kC

22.3 THE SUPERPOSITION OF ELECTRIC FORCES

Online Concept Tutorial 25

The electric force, like any other force, has a magnitude and a direction; that is, the electric force is a vector. According to Coulomb's Law, the *magnitude* of the electric force exerted by a point charge q' on a point charge q is

$$F = \frac{1}{4\pi\epsilon_0}\frac{qq'}{r^2} \tag{22.8}$$

The *direction* of this force is along the line from one charge to the other. As illustrated in Fig. 22.6, this force can be represented by a vector **F** pointing along the line from one charge to the other.

If several point charges $q_1, q_2, q_3, \ldots$ simultaneously exert electric forces on the charge q, then *the net force on q is obtained by taking the vector sum of the individual forces* (see Fig. 22.7). Thus, if the vectors representing the individual electric forces produced by $q_1, q_2, q_3, \ldots$ are $\mathbf{F}_1, \mathbf{F}_2, \mathbf{F}_3, \ldots$, respectively, then the net force is

Superposition Principle of electric forces

$$\mathbf{F} = \mathbf{F}_1 + \mathbf{F}_2 + \mathbf{F}_3 + \cdots \tag{22.9}$$

Equation (22.9) expresses the **Superposition Principle of electric forces.** According to Eq. (22.9), the force contributed by each charge is independent of the presence of the other charges. For instance, the charge q_2 does not affect the interaction of charge q_1 with q; it merely adds its own interaction with q. This simple combination law is an important empirical fact about electric forces. Since the contact forces of everyday experience, such as the normal force and the friction force, arise from electric forces between the atoms, they will likewise obey the Superposition Principle, and they can be combined with simple vector addition. Incidentally: The gravitational forces on the Earth and within the Solar System also obey the Superposition Principle. Thus, all the forces in our immediate environment obey this principle. We have already made much use of the Superposition Principle in our study of mechanics; now we recognize that *the superposition of mechanical forces, such as "contact" forces, hinges on the superposition of electric forces.*

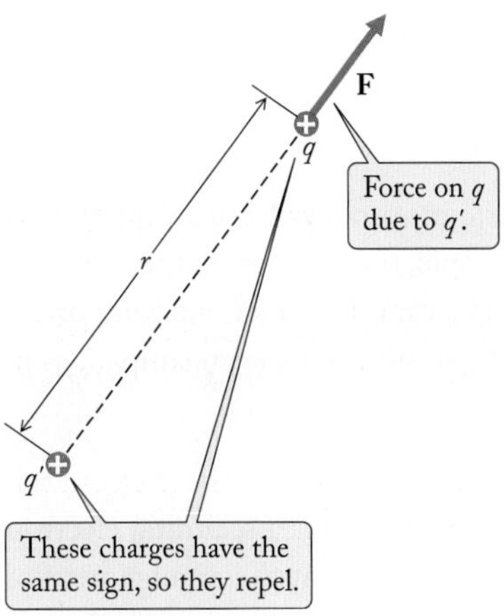

FIGURE 22.6 A charge q' exerts an electric force **F** on the charge q.

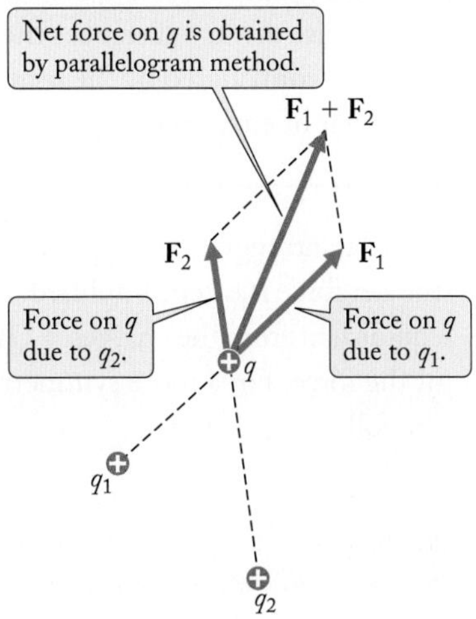

FIGURE 22.7 Two point charges q_1 and q_2 exert electric forces $\mathbf{F}_1$ and $\mathbf{F}_2$ on the point charge q. The net force on q is the vector sum of these forces.

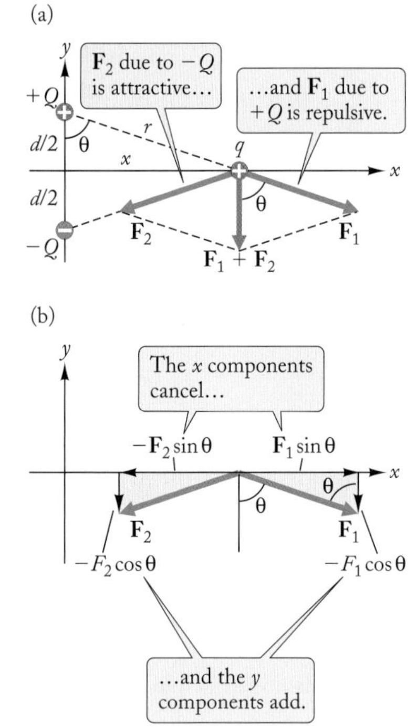

FIGURE 22.8 (a) The charges $+Q$ and $-Q$ exert forces $\mathbf{F}_1$ and $\mathbf{F}_2$ on the charge q. The net force is the vector sum $\mathbf{F}_1 + \mathbf{F}_2$. (b) The x and y components of $\mathbf{F}_1$ and $\mathbf{F}_2$.

EXAMPLE 6 Two point charges $+Q$ and $-Q$ are separated by a distance d, as shown in Fig. 22.8. A positive point charge q is equidistant from these charges, at a distance x from their midpoint. What is the electric force $\mathbf{F}$ on q?

SOLUTION: As illustrated in Fig. 22.8a, the charge $+Q$ produces a repulsive force on the charge q, and the charge $-Q$ produces an attractive force. Thus, the vector $\mathbf{F}_1$ points away from $+Q$, and the vector $\mathbf{F}_2$ points toward $-Q$. From the geometry of Fig. 22.8a, the distance from each of the charges $+Q$ and $-Q$ to the charge q is $r = \sqrt{x^2 + d^2/4}$. Hence, the *magnitudes* of the individual Coulomb forces exerted by $+Q$ and $-Q$ are

$$F_1 = F_2 = \frac{1}{4\pi\epsilon_0}\frac{qQ}{r^2} = \frac{1}{4\pi\epsilon_0}\frac{qQ}{x^2 + d^2/4} \quad \text{(magnitudes)} \tag{22.10}$$

From Fig. 22.8b we see that in the vector sum $\mathbf{F} = \mathbf{F}_1 + \mathbf{F}_2$, the horizontal components (x components) of $\mathbf{F}_1$ and $\mathbf{F}_2$ cancel, and the vertical components (y components) add, giving a net vertical component twice as large as each individual vertical component. Thus in this case the net force $\mathbf{F}$ has a y component but no x component. In terms of the angle θ shown in Fig. 22.8b, the y component of $\mathbf{F}_1$ is $-F_1 \cos\theta$, and the y component of $\mathbf{F}_2$ is $-F_2 \cos\theta$. Since these are equal, the net force is then

$$\begin{aligned} F = F_y &= -F_1\cos\theta - F_2\cos\theta = -2F_1\cos\theta \\ &= -2\frac{1}{4\pi\epsilon_0}\frac{qQ}{x^2 + d^2/4}\cos\theta \end{aligned} \tag{22.11}$$

From Fig. 22.8a, we see that $\cos\theta = \frac{1}{2}\, d/(x^2 + d^2/4)^{1/2}$, and therefore

$$F = F_y = -\frac{1}{4\pi\epsilon_0}\frac{qQd}{(x^2 + d^2/4)^{3/2}} \tag{22.12}$$

COMMENTS: Note that if the charge q is at a large distance from the two charges $\pm Q$, then d^2 can be neglected compared with x^2, so $(x^2 + d^2/4)^{3/2} \approx (x^2)^{3/2} = x^3$. The force F is then proportional to $1/x^3$; that is, the force decreases in proportion to the inverse cube of the distance. Thus, although the force contributed by each point charge $\pm Q$ is an inverse-square force, the net force has a different behavior, because at large distances the force contributed by one charge tends to cancel the force contributed by the other. We will further discuss this $1/x^3$ behavior far from a pair of equal and opposite charges in Chapter 23.

If an arrangement of point charges is symmetric in some way, the calculation of the net force is often simplified. In the previous example, the position of the charge q, equidistant from the charges $+Q$ and $-Q$, resulted in a cancellation of one component of the force. For a more symmetric arrangement, the result can be even simpler, as in the following example.

Concepts in Context

EXAMPLE 7 Similar to the chapter photo, Fig. 22.9a is a scanning electron micrograph of toner particles. We see that the arrangement of toner particles is nearly such that one is at the center of a hexagon and the other six are at the vertices of the hexagon. If the toner particles carry equal charges and are arranged precisely at the center and vertices of a hexagon, what is the net force on the central charge? Assume for simplicity that each particle acts as a point charge located at the center of the particle.

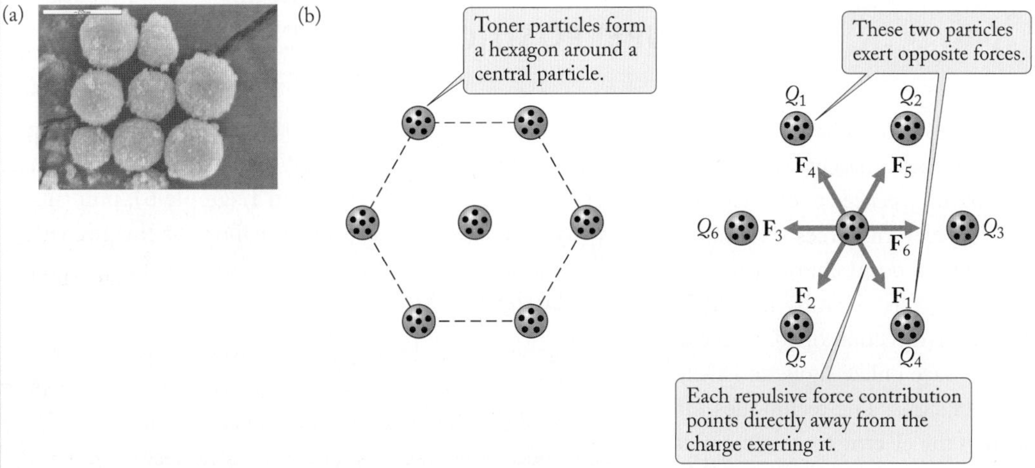

FIGURE 22.9 (a) Scanning electron micrograph of laser printer toner particles. (b) Electric forces on the central particle due to each of the other six.

SOLUTION: Like charges repel, so the force on the central particle due to any one of the other charges will point directly away from that charge. Since all charges are equal and the distance from the center to each vertex is the same, each of the force contributions from the outer six charges will have the same magnitude. Figure 22.9b shows the force vectors for the electric forces on the central particle due to each of the six surrounding toner particles. Inspecting this diagram, we see that the force vectors cancel pairwise; for example, $\mathbf{F}_1$ exactly balances $\mathbf{F}_4$. The net force on the central particle is zero.

Thus the mutual repulsion of the particles not only keeps them dispersed, as mentioned in Example 4, but also tends to keep them in equilibrium. Repulsive forces between particles, atoms, or other entities often result in stable, hexagonal structures, such as the examples shown in Fig. 22.10.

(a) (b) (c) (d)

FIGURE 22.10 Because of mutual repulsion, many types of objects tend to form hexagonal arrangements: (a) billiard balls, (b) microscopic polystyrene beads, (c) superconducting vortices, and (d) atoms in an oxide crystal.

Checkup 22.3

QUESTION 1: Suppose in Example 6 that instead of charges $\pm Q$, both charges Q were positive. What is the direction of the electric force $\mathbf{F}$ on the positive charge q in this case?

QUESTION 2: Three identical point charges are at the vertices of an equilateral triangle. A fourth, identical point charge is placed at the midpoint of one side of the triangle. As a result of the three electric force contributions from the vertex charges, the fourth charge

(A) Is in equilibrium and remains at rest
(B) Is pushed toward the center of the triangle
(C) Is pushed outside the triangle

PROBLEM-SOLVING TECHNIQUES — SUPERPOSITION OF ELECTRIC FORCES

- To find the total electric force on one charge due to an arrangement of several other point charges (such as the point charges $\pm Q$ of Example 6), you need to calculate the vector sum of the individual electric forces. The magnitude of the electric force on a point charge q due to each individual point charge q' is given by Eq. 22.7, $F = (1/4\pi\epsilon_0)\,qq'/r^2$; and the direction of this electric force is toward q' if q and q' have unlike signs, and away from q' if q and q' have like signs. The techniques for evaluating the vector sum of the electric force vectors of several charges are the same as for the sum of any other kind of vector.
- Recall the basic techniques for calculating vector sums: First, draw a careful diagram for the forces *on* a given charge q_1 *due to several other* charges $q_2, q_3, \ldots$, with each electric force vector along the line through the charge exerting the force and the charge acted upon. It may be easiest to put the tail of each contributing vector at the charge acted upon (as in Example 6), pointing away from the charge exerting the force for charges with like signs, and toward the charge exerting the force for unlike signs.
- Use geometry to decompose each vector into its x and y (and, if necessary, z) components. You can then obtain the total electric force vector by separately summing the x, y, and z components of the contributing vectors. You can obtain the magnitude of the total electric force in the usual way, from $F = \sqrt{F_x^2 + F_y^2 + F_z^2}$. In many examples of calculations of electric forces, some components of the force will cancel. If you select the x, y, and z axes judiciously, you can often achieve cancellation of all components except one (see Example 6).

22.4 CHARGE QUANTIZATION AND CONSERVATION

Not only electrons and protons exert electric forces on each other, but so do many other particles. The magnitudes of these electric forces are given by Eq. (22.7) with the appropriate values of the electric charges. Table 22.3 lists the electric charges of some particles; a more complete list will be found in Chapter 41. Antiparticles have electric charges that are opposite to those of the corresponding particles; for example, the antielectron (or positron) has charge $+e$, the antiproton has charge $-e$, the antineutron has charge 0, and so on.

All the known particles have charges that are some integer multiple of the fundamental charge; that is, the charges are always $0, \pm e, \pm 2e, \pm 3e$, etc. Why no other charges exist is a mystery for which classical physics offers no explanation. Since charges exist in discrete packets, we say that charge is **quantized**—the fundamental charge e is called the quantum of charge. Thus any amount of charge that is ever encountered is an integer multiple of the fundamental charge e. Of course, as discussed in Section 22.1, the extreme smallness of the fundamental charge permits us to treat macroscopic charge distributions as continuous.

charge quantization

charge conservation

The electric charge is a **conserved quantity**: *in any reaction involving charged particles, the total charges before and after the reaction are always the same.* For instance, here is an example of a reaction in which particles are destroyed, yet the net electric charge remains constant:

matter–antimatter annihilation:

$$\begin{array}{lccccc} & \text{[electron]} & + & \text{[antielectron]} & \rightarrow & 2\text{[photons]} \\ \textit{charges:} & -e & + & e & \rightarrow & 0 \end{array} \qquad (22.13)$$

The same charge conservation holds for any reaction. No reaction that creates or destroys any net electric charge has ever been discovered.

Charge is of course also conserved in chemical reactions. For instance, in a lead–acid battery (automobile battery), plates of lead and of lead dioxide are immersed in an electrolytic solution of sulfuric acid (Fig. 22.11). The reactions that take place on these plates involve sulfate ions (SO_4^{2-}, where the superscript 2− indicates an ion with two extra electrons) and hydrogen ions (H^+); the reactions release electrons at the lead plate, and they absorb electrons at the lead dioxide plate:

at lead plate:

$$\begin{array}{lll} & Pb + SO_4^{2-} \rightarrow PbSO_4 + 2[\text{electrons}] \\ \textit{charges}: & 0 + (-2e) \rightarrow 0 + (-2e) \end{array} \tag{22.14}$$

at lead dioxide plate:

$$\begin{array}{ll} & PbO_2 + 4H^+ + SO_4^{2-} + 2[\text{electrons}] \rightarrow PbSO_4 + 2H_2O \\ \textit{charges}: & 0 + 4e + (-2e) + (-2e) \rightarrow 0 + 0 \end{array} \tag{22.15}$$

The plates of such a battery are connected by an external circuit (a wire), and the electrons released by the reaction (22.14) travel from one plate to the other, forming an electric current (see Fig. 22.11).

TABLE 22.3

ELECTRIC CHARGES OF SOME PARTICLES

PARTICLE	CHARGE
Electron, e	$-e$
Muon, μ	$-e$
Pion, π^0	0
Pion, π^+	$+e$
Pion, π^-	$-e$
Proton, p	$+e$
Neutron, n	0
Neutrino, ν	0
Photon, γ	0
Delta, Δ^+	$+e$
Delta, Δ^{++}	$+2e$

EXAMPLE 8 A fully "charged" battery contains a large amount of sulfuric acid in the electrolytic solution (H_2SO_4 in the form of SO_4^{2-} ions and H^+ ions). As the battery delivers electric charge to the external circuit connecting its terminals, the amount of sulfuric acid in solution gradually decreases. Suppose that while discharging completely, the positive terminal of an automobile battery delivers an electric charge of 1.8×10^5 C through the external circuit. How many grams of sulfuric acid will be used up in this process?

SOLUTION: Let us first consider the number of electrons transferred, then the number of sulfate ions used up, and finally obtain the amount of sulfuric acid needed. Since the charge per electron is -1.6×10^{-19} C/electron, the number of electrons in -1.8×10^5 C is $(-1.8 \times 10^5\ \text{C})/(-1.6 \times 10^{-19}\ \text{C/electron}) = 1.1 \times 10^{24}$ electrons. According to the reactions (22.14) and (22.15), whenever two electrons are transferred from the lead to the lead dioxide plate, two sulfate ions are absorbed (one at each plate). Thus, 1.1×10^{24} sulfate ions will be used up, which means 1.1×10^{24} molecules of sulfuric acid will be used up, since sulfuric acid has one sulfate ion per molecule (H_2SO_4). The required number of moles of sulfuric acid is therefore $(1.1 \times 10^{24}\ \text{molecules})/(6.02 \times 10^{23}\ \text{molecules/mole}) = 1.9$ moles, and, since the molecular mass of sulfuric acid is 98 grams per mole, the required mass of sulfuric acid is $(1.9\ \text{mole}) \times (98\ \text{g/mole}) = 183$ g.

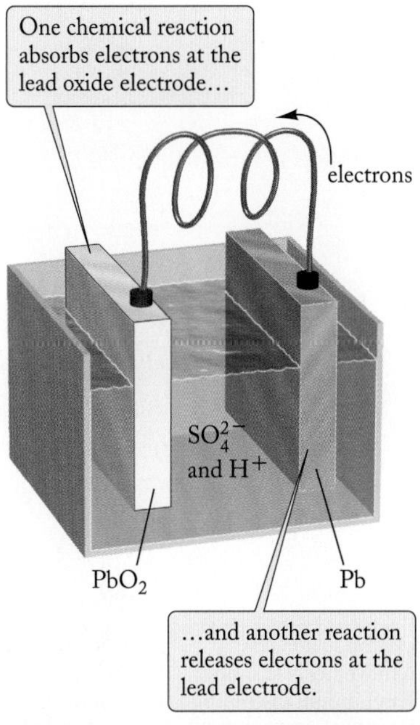

FIGURE 22.11 A lead–acid battery.

The conservation of electric charge in chemical reactions, such as the reactions (22.14) and (22.15), is a trivial consequence of the conservation of electrons and protons. All such reactions involve nothing but a rearrangement of the electrons and protons in the molecules; during this rearrangement, the numbers of electrons and of protons remain constant. Obviously, the net electric charge must then also remain constant.

The same argument applies to all macroscopic electric processes, such as the operation of electrostatic machines and generators, the flow of currents on wires, the storage of charge in capacitors, the electric discharge of thunderclouds, etc. All such processes involve nothing but a rearrangement of electrons and protons. Consequently, the net electric charge must remain constant.

✔ Checkup 22.4

QUESTION 1: If an atom loses two electrons, what is the electric charge of the resulting ion? If an atom loses three electrons?

QUESTION 2: Is it possible for a body to have an electric charge of 2.0×10^{-19} C? 3.2×10^{-19} C?

(A) Yes; yes (B) Yes; no (C) No; yes (D) No; no

Online Concept Tutorial 25

22.5 CONDUCTORS AND INSULATORS; CHARGING BY FRICTION OR BY INDUCTION

conductor

insulator

A **conductor**—such as copper, aluminum, or iron—*is a material that permits the motion of electric charges through its volume.* An **insulator**—such as glass, porcelain, rubber, or nylon—*is a material that does not readily permit the motion of electric charges.* Thus, when we place some electric charge on one end of a conductor, it immediately spreads out over the entire conductor until it finds an equilibrium distribution. (We will study the conditions for the equilibrium of electric charge on a conductor in Chapter 24. It turns out that when the charges finally reach equilibrium, they will all be located on the surface of the conductor.) In contrast, when we place some charge on one end of an insulator, it stays in place (see Fig. 22.12).

All metals are good conductors. They readily permit the motion of electric charges, and therefore metallic wires are widely used in electric circuits, such as the circuits in your home, where copper wires serve as conduits for the flow of electric charge. The motion of charge in a metal is due to the motion of electrons. *In a metal, some of the electrons of each atom are free,* that is, they are not bound to any particular atom although they are bound to the metal as a whole. The free electrons come from the outer parts of the atoms. The outer electrons of the atom are not very strongly attached and readily come loose; the inner electrons are firmly bound to the nucleus of the atom and are likely to stay put. The free electrons wander through the entire volume of the metal, suffering occasional collisions, but they experience a restraining force only when they encounter the surface of the metal. The electrons are held inside the metal in much the same way as particles of a gas are held inside a container—the particles of gas can wander through the volume of the container, but they are restrained by the walls. In view of this analogy, electrons in a metallic conductor are often said to form a free-electron gas. If one end of a metallic conductor has an excess or deficit of electrons, the motion of the free-electron gas will quickly distribute this excess or deficit to other parts of the metallic conductor.

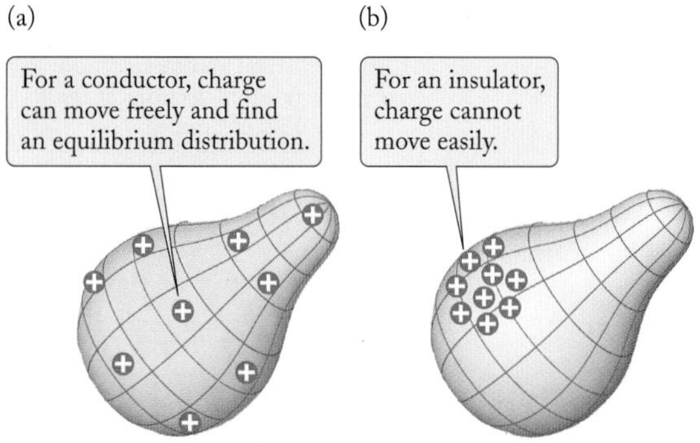

FIGURE 22.12 (a) Charge placed on a conductor spreads out over the entire conductor. (b) Charge placed on an insulator stays at its original place.

The charging of a body of metal is usually accomplished by the removal or the addition of electrons. A body will acquire a net positive charge if electrons are removed, and a net negative charge if electrons are added. Thus, positive charge on a body of metal is simply a deficit of electrons, and negative charge an excess of electrons.

PHYSICS IN PRACTICE XEROGRAPHY

Concepts in Context

The process of xerography (from Greek words meaning "dry writing"), commonly used in photocopiers and laser printers, exploits electric forces to put marks on paper. The figures illustrate the process. First, in Fig. 1a, the surface of an insulating cylindrical drum is positively charged, either by contact with a positively charged rod or by proximity to an electron-attracting wire ("corona wire"). The drum has a special surface containing selenium or some other photoconductive material having the unusual property that it becomes a conductor when exposed to light. This property permits the creation of a temporary "charge image" on the drum by exposure to light from an original document in a photocopier, or from microprocessor-controlled laser light in a laser printer. The charge from the exposed, conductive regions is neutralized by electrons from the metal drum below the photoconductive surface layer, and only the dark, insulating areas maintain their layer of positive charge (Fig. 1b). Next, small glass or plastic carrier particles, each with many negatively charged toner particles attached, are sprinkled onto the drum; the toner particles are attracted and stick to the regions of the drum where the positively charged image resides (Fig. 1c). We have seen in Examples 4 and 7 that the mutual repulsion of the toner particles maintains a more or less uniform dispersion throughout the dark image areas. The paper receiving the final image is positively charged and attracts the toner particles upon contact, creating a delicate powder-on-paper dry image (Fig. 1d). Finally, the toner particles are "fixed" (Fig. 1e) by means of a hot roller that melts the black plastic toner particles so they bond with the paper.

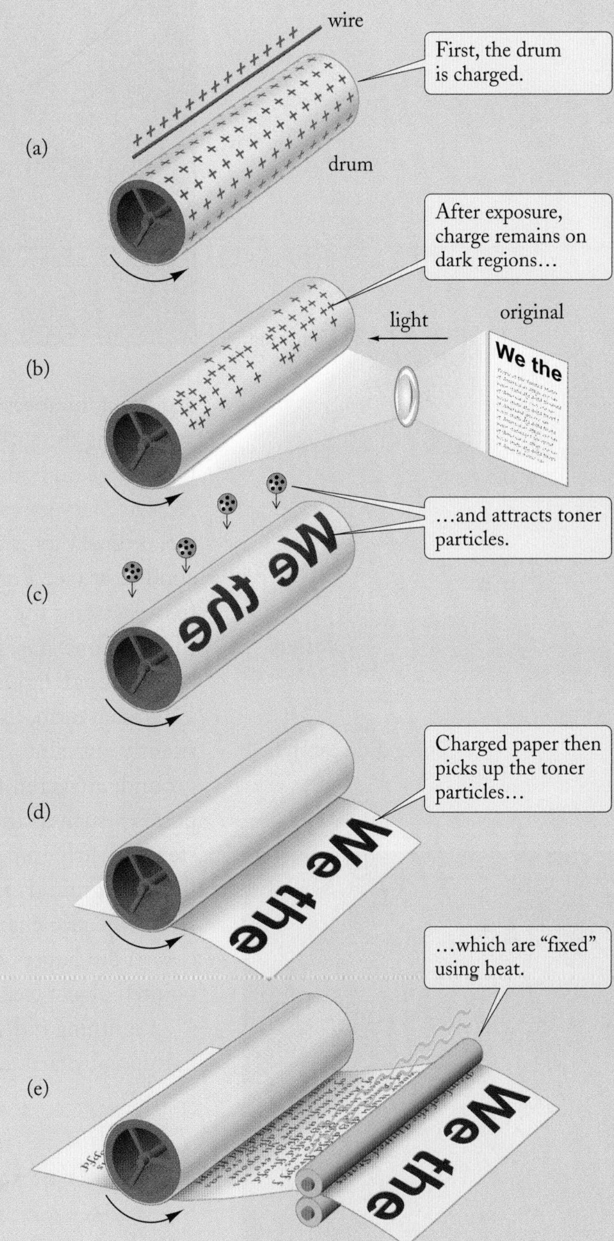

FIGURE 1 The xerographic process.

Liquids containing ions (atoms or molecules with missing electrons or with excess electrons) also are good conductors. For instance, a solution of common salt in water contains ions of Na^+ and Cl^-. The motion of charges through the liquid is due to the motion of these ions. *Liquid conductors with an abundance of ions are called* **electrolytes**.

electrolyte

Incidentally: Very pure distilled water is a poor conductor because it lacks ions. But ordinary water is a good conductor because it contains some ions contributed by dissolved impurities. The ubiquitous water in our environment makes many substances into conductors. For example, earth (ground) is a reasonably good conductor, mainly because of the presence of the water. Furthermore, on a humid day many insulators

FIGURE 22.13 Lightning.

acquire a microscopic surface film of water, and this permits electric charge to leak away along the surface of the insulator; thus, on humid days, it is difficult to store electric charge on bodies supported by insulators.

plasma

Ordinary gases are insulators, but ionized gases are good conductors. For example, ordinary air is an insulator, but the ionized air found in a lightning bolt is a good conductor (see Fig. 22.13). This ionized air contains a mixture of positive ions and free electrons; the motion of charge in such a mixture is due mainly to the motion of the electrons. *Such an ionized gas is called a* **plasma**.

corona discharge

Although lightning gives us the most spectacular evidence of electrical activity in a thunderstorm, the largest part of the electrical activity in a thunderstorm proceeds silently and almost imperceptibly in the form of **corona discharge**. Charge on the ground, attracted to opposite charge in a thundercloud, concentrates at any sharp points—such as the tips of the leaves of trees. The concentration of charge causes a steady, nearly imperceptible ionization of the air near the sharp points. The charge from the ground can then leak into this ionized air and flow toward the thundercloud, whose electric charge it gradually neutralizes. This kind of ionization of air is called corona discharge because it can sometimes be seen as a glowing halo surrounding the pointed object (see Fig. 22.14).

Lightning rods, invented by Benjamin Franklin, were originally intended to inhibit lightning by facilitating corona discharge. However, the traditional lightning rod, with

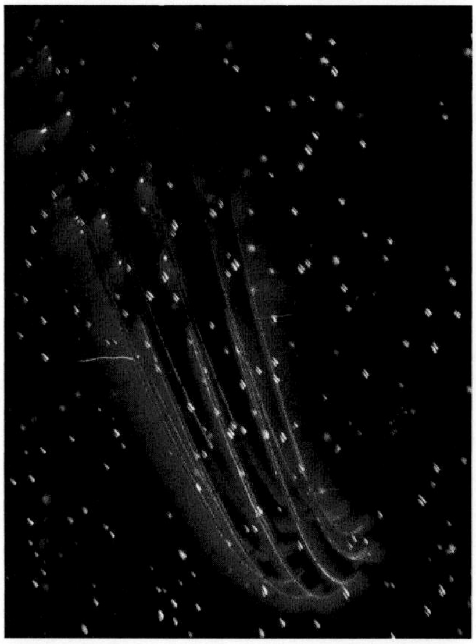

FIGURE 22.14 Corona discharge from power lines.

BENJAMIN FRANKLIN (1706–1790)
American scientist, statesman, and inventor. He is most often remembered for his hazardous experiments with a kite in a thunderstorm, which demonstrated that lightning is an electric phenomenon, and for his invention of lightning rods. Franklin also made other significant contributions to the experimental and theoretical studies of electricity, and he was admired and honored by the leading scientific associations in Europe. Among these contributions were his formulation of the Law of Conservation of Electric Charge and his introduction of the modern notation for plus and minus charges, which he regarded as an excess or deficiency of "electric fluid."

a single sharp point, cannot produce a sufficient amount of corona discharge; its main benefit is to conduct the electric charge of a lightning stroke harmlessly to the ground. Some lightning rods of modern design are much better at producing corona discharge. These dissipative lightning rods end in a multitude of sharp points (see Fig. 22.15), which can dissipate a substantial fraction of the total charge of a thundercloud passing overhead.

We will end this chapter with a brief discussion of two different ways in which a body may be charged. The first, frictional electricity, has already been mentioned in the introduction. The second, charging by induction, is a simple application of the Coulomb force that utilizes the properties of a conductor. We discuss each in turn.

Frictional electricity is quite common. It is easy to accumulate electric charge on a glass rod merely by rubbing it with a piece of silk. The silk becomes negatively charged, and the glass positively charged—the rubbing motion between the surfaces of the silk and the glass rips charges off one of these surfaces and makes them stick to the other, but the detailed mechanism is not well understood. It is believed that what is usually involved is a transfer of ions from one surface to the other. Contaminants residing on the rubbed surfaces play a crucial role in frictional electricity. If glass is rubbed with an absolutely clean piece of silk or other textile material, the glass becomes negatively charged rather than positively charged. Ordinary pieces of silk apparently have such a large amount of dirt on their surfaces that the charging process is dominated by the dirt rather than by the silk. Even air can act as contaminant for some surfaces; for instance, careful experiments on the rubbing of platinum with silk show that in vacuum the platinum becomes negatively charged, but in air it becomes positively charged.

The electric charge that can be accumulated on the surface of a body of ordinary size (a centimeter or more) by rubbing may be as much as 10^{-9} or 10^{-8} coulomb per square centimeter. If the charge concentration on a body is higher than that, it will cause an electric discharge into the surrounding air (a corona discharge).

FIGURE 22.15 Modern dissipative lightning rod. The rod ends in a tassel of fine wires, which provide many sharp points for corona discharge.

charging by induction

A second, quite different way to charge an object is called **charging by induction**. Once we have accumulated some charge, say, positive charge on a rod of glass, we can produce charges on other bodies by the induction process, as follows: First we bring the glass rod near a metallic body supported on an insulating stand (see Fig. 22.16a). The positive charge on the rod will then attract free electrons to the near side of the body and leave a deficit of free electrons on the far side; thus, the near side will acquire negative charge and the far side positive charge. If we next connect the far side to the ground with a wire, electrons will flow from the ground, attracted by the positive charge, which they neutralize (Fig. 22.16b). We then disconnect the wire; this leaves the metallic body with a net negative charge (Fig. 22.16c). When we finally withdraw the glass rod, the induced charge will remain on the metallic body, distributing itself over the entire surface, as in Fig. 22.16d. Thus the positive glass rod "induces" a charge distribution (and, ultimately, a net negative charge) on the metal sphere without actually making contact with the sphere.

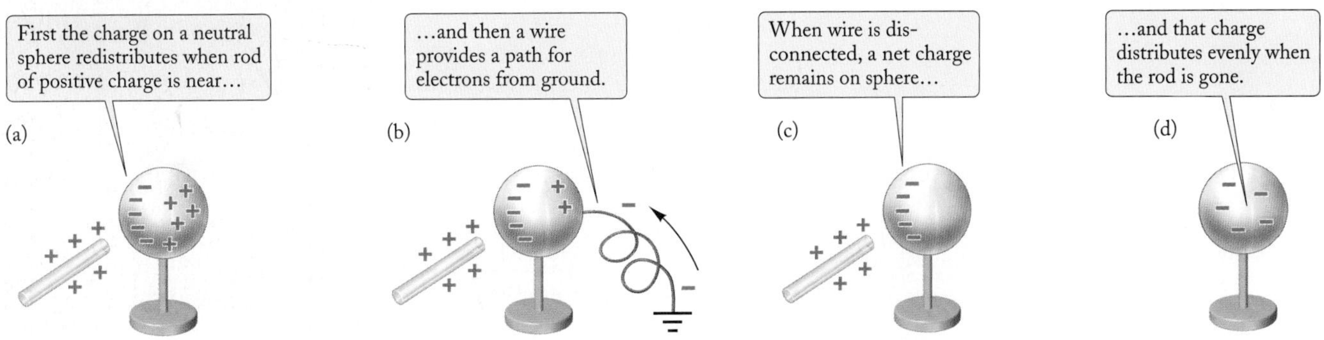

FIGURE 22.16 Charging by induction. (a) The positively charged glass rod induces a charge distribution on a neutral metallic sphere. (b) When the far side of the sphere is connected to the ground by means of a wire, electrons move from ground to the sphere and neutralize the positive charge. (c) When the grounding wire is removed, a net negative charge remains. (d) When the glass rod is moved away, the net negative charge spreads over the sphere.

Checkup 22.5

QUESTION 1: What is the difference between an ordinary liquid and an electrolyte? What is the difference between an ordinary gas and a plasma?

QUESTION 2: Figure 22.16 shows a conducting body mounted on an insulating stand. Suggest some material for making the conducting body and some material for the insulating stand.

QUESTION 3: Suppose you have two metallic balls of equal size, both like the one illustrated in Fig. 22.16d, one with a charge of $+1 \times 10^{-7}$ C, and the other with a charge of -3×10^{-7} C. If you touch them together, what will be the resulting charge on each ball?

(A) There will remain $+1 \times 10^{-7}$ C on one and -3×10^{-7} C on the other
(B) There will be 0 C on one and -2×10^{-7} C on the other
(C) There will be -1×10^{-7} C on each

SUMMARY

PROBLEM-SOLVING TECHNIQUES Superposition of Electric Forces **(page 706)**

PHYSICS IN PRACTICE Xerography **(page 709)**

ELECTRIC CHARGES May be positive, negative, or zero; like charges repel, unlike charges attract.

SI UNIT OF CHARGE 1 coulomb = 1 C

FUNDAMENTAL CHARGE, OR CHARGE OF PROTON $e = 1.60 \times 10^{-19}$ C **(22.1)**

CHARGE OF ELECTRON $-e = -1.60 \times 10^{-19}$ C **(22.1)**

COULOMB'S LAW Direction of Coulomb force is along the line joining the particles.

$$F = \frac{1}{4\pi\epsilon_0}\frac{qq'}{r^2} \quad (22.7)$$

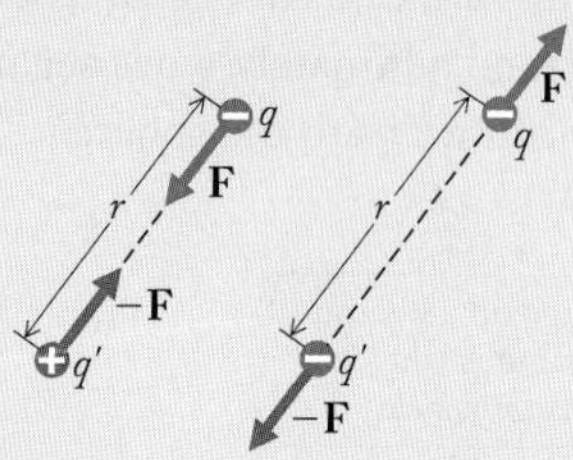

PERMITTIVITY CONSTANT (Electric constant) $\epsilon_0 = 8.85 \times 10^{-12}\ \text{C}^2/\text{N·m}^2$ **(22.6)**

COULOMB CONSTANT

$$k = \frac{1}{4\pi\epsilon_0} = 8.99 \times 10^9\ \text{N·m}^2/\text{C}^2 \quad (22.4, 22.5)$$

SUPERPOSITION PRINCIPLE Net force is vector sum of individual forces.

$$\mathbf{F} = \mathbf{F}_1 + \mathbf{F}_2 + \mathbf{F}_3 + \cdots \qquad (22.9)$$

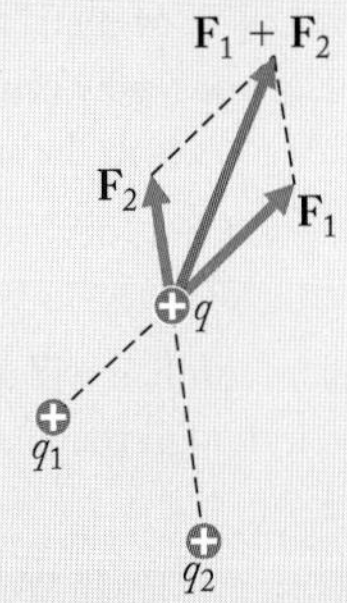

CHARGE CONSERVATION In any reaction or process, the net electric charge remains constant.

CHARGE QUANTIZATION Any charge is an integer multiple of the fundamental charge.

$$q = 0, \pm e, \pm 2e, \pm 3e, \cdots$$

ION An atom with net charge (missing or extra electrons).

ELECTROLYTE A liquid with many dissolved ions.

PLASMA A gas with many ionized atoms and free electrons.

CONDUCTOR Permits the motion of charge.

INSULATOR Does not permit the motion of charge.

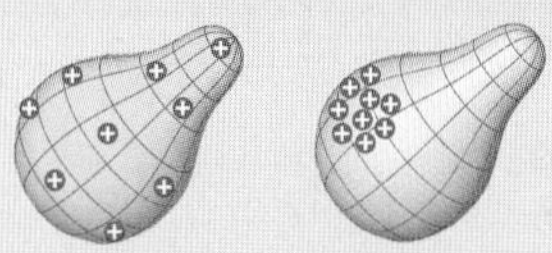

QUESTIONS FOR DISCUSSION

1. Suppose that the Sun has a positive electric charge and each of the planets a negative electric charge, and suppose that there is no gravitational force. In what way will the motions of the planets predicted by this "electric" model of the Solar System differ from the observed motions?
2. Describe how you would set up an experiment to determine whether the electric charges on an electron and a proton are exactly the same.
3. Assume that neutrons have a small amount of electric charge, say, positive charge. Discuss some of the consequences of this assumption for the behavior of matter.
4. If we were to assign a positive charge to the electron and a negative charge to the proton, would this affect the mathematical statement [Eq. (22.7)] of Coulomb's Law?
5. In the cgs, or Gaussian, system of units, Coulomb's Law is written as $F = qq'/r^2$. In terms of grams, centimeters, and seconds, what are the units of electric charge in this system?
6. Could we use the electric charge of an electron as an atomic standard of electric charge to define the coulomb? What would be the advantages and disadvantages of such a standard?
7. Besides electric charge, what other physical quantities are conserved in reactions among particles? Which of these quantities are quantized?
8. Since the free electrons in a piece of metal are free to move any which way, why don't they all fall to the bottom of the piece of metal under the influence of the pull of gravity?

9. If the surface of a piece of metal acts like a container in confining the free electrons, why can't we cause these electrons to spill out by drilling holes in the surface?
10. Water is a conductor, but (dry) snow is an insulator. How can this be?
11. If you rub a plastic comb, it will attract hairs or bits of paper (Fig. 22.17), even though they have no net electric charge. Explain.

FIGURE 22.17 The electric charge on this plastic comb attracts small bits of paper.

12. Some old-fashioned physics textbooks define positive electric charge as the kind of charge that accumulates on a glass rod when rubbed with silk. What is wrong with this definition?
13. When you rub your shoes on a carpet, you sometimes pick up enough electric charge to feel an electric shock if you subsequently touch a radiator or some other metallic body connected to the ground. Why is this more likely to happen in winter than in summer?
14. Some automobile operators hang a conducting strap on the underside of their automobiles, so that this strap drags on the street. What is the purpose of this arrangement?
15. An amount of electric charge has been deposited on a Ping-Pong ball. How can you find out whether the charge is positive or negative?
16. Two aluminum spheres of equal radii hang from the ceiling on insulating threads. You have a glass rod and a piece of silk. How can you give these two spheres exactly equal amounts of electric charge?

PROBLEMS

22.1 The Electrostatic Force†
22.2 Coulomb's Law‡

1. Within a typical thundercloud there are electric charges of −40 C and +40 C separated by a vertical distance of 5.0 km (Fig. 22.18). Treating these charges as pointlike, find the magnitude of the electric force of attraction between them.

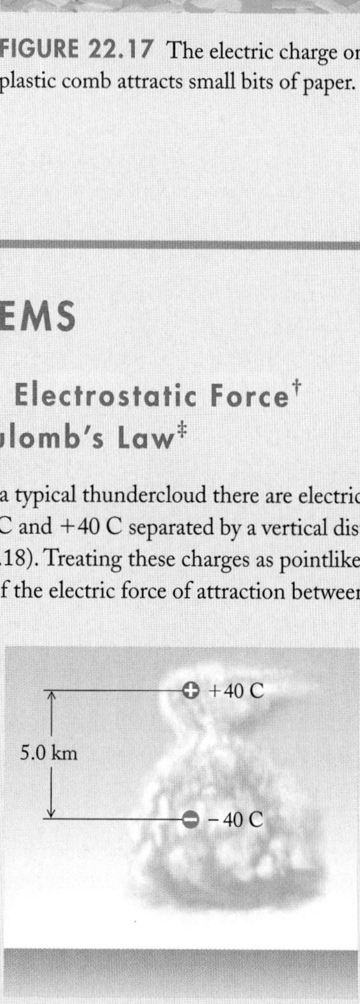

FIGURE 22.18 Charges in a thundercloud.

†For help, see Online Concept Tutorial 25 at www.wwnorton.com/physics
‡For help, see Online Concept Tutorial 24 at www.wwnorton.com/physics

2. A crystal of NaCl (common salt) consists of a regular arrangement of ions of Na^+ and Cl^-. The distance from one ion to its neighbor is 2.82×10^{-10} m. What is the magnitude of the electric force of attraction between the two ions? Treat the ions as point charges.
3. Suppose that the two protons in the nucleus of a helium atom are at a distance of 2.0×10^{-15} m from each other. What is the magnitude of the electric force of repulsion that they exert on each other? What would be the acceleration of each if this were the only force acting on them? Treat the protons as point charges.
4. An alpha particle (a helium nucleus with charge $+2e$) is launched at high speed toward a nucleus of uranium (charge $+92e$). What is the magnitude of the electric force on the alpha particle when it is at a distance of 5.0×10^{-14} m from the nucleus? What is the corresponding instantaneous acceleration of the alpha particle? Treat the alpha particle and the nucleus as point charges.
5. According to recent theoretical and experimental investigations, the subnuclear particles are made of quarks and of antiquarks (see Chapter 41). For example, a positive pion is made of a u quark and a d antiquark. The electric charge on the u quark is $\frac{2}{3}e$ and that on the d antiquark is $\frac{1}{3}e$. Treating the quarks as classical particles, calculate the electric force of repulsion between the quarks in the pion if the distance between them is 1.0×10^{-15} m.

6. How many electrons do you need to remove from an initially neutral bowling ball to give it a positive electric charge of 1.0×10^{-6} C?

7. A lightning stroke typically deposits -25 C on the ground. How many electrons is this?

8. The mass of the electron cannot be measured directly, since macroscopic amounts of mass always contain a combination of electrons, protons, and neutrons, never pure electrons. Instead, the mass is calculated from a measurement of the electric charge $-e$ of the electron and a measurement of the charge-to-mass ratio $-e/m_e$. The best values for these quantities are $-e = -1.602\,177 \times 10^{-19}$ C and $-e/m_e = -1.758\,820 \times 10^{11}$ C/kg. What best value of the mass of the electron can you deduce from this?

9. In an HCl molecule, the nuclei of the H and the Cl atoms, with charges $+e$ and $+17e$, respectively, are separated by a distance of 1.28×10^{-10} m. What is the electric force of repulsion between these nuclei?

10. Consider two protons separated by a distance of 1.0×10^{-12} m.
    (a) What is the gravitational force of attraction between the protons?
    (b) What is the electric force of repulsion between these protons? What is the ratio of the electric force and the gravitational force?
    (c) Consider a second pair of protons separated by a larger distance, so their electric repulsion matches the gravitational attraction of the other, closer pair calculated in part (a). How far apart would these protons have to be?

11. According to recent theoretical speculations, there might exist an elementary particle of mass 1.0×10^{-11} kg. If such a particle carries a charge e, what is the gravitational force and what is the electric force on an identical particle placed at a distance of 1.0×10^{-10} m?

12. In the lead atom, the nucleus has an electric charge $82e$. The innermost electron in this atom is typically at a distance of 6.5×10^{-13} m from the nucleus. What is the electric force that the nucleus exerts on such an electron? What is the acceleration that this force produces on the electron? Treat the electron as a classical particle.

13. The electric charge in one mole of protons is called **Faraday's constant.** What is its numerical value?

14. The electric charge flowing through an ordinary 115-volt, 150-watt lightbulb is 1.3 C/s. How many electrons per second does this amount to?

15. A maximum electric charge of 7.5×10^{-6} C can be placed on a metallic sphere of radius 15 cm before the surrounding air suffers electric breakdown (sparks). How many excess electrons (or missing electrons) does the sphere have when breakdown is about to occur?

16. How many electrons are in a paper clip of iron of mass 0.30 g?

17. Suppose that you remove all the electrons in a copper penny of mass 2.7 g and place them at a distance of 2.0 m from the remaining copper nuclei. What is the electric force of attraction on the electrons?

18. What is the number of electrons and of protons in a human body of mass 73 kg? The chemical composition of the body is roughly 70% oxygen, 20% carbon, and 10% hydrogen (by mass).

19. It is possible to dissolve 36 g of sodium chloride (table salt) in 100 g of water. By what factor does the number of electrons (or protons) in the solution exceed that of the water alone?

20. An introductory physics laboratory experiment uses two small spheres, each charged with -2.0×10^{-6} C of charge. What is the electric force between the spheres when they are 1.0 m apart?

21. The value of the Coulomb constant is defined to be exactly $k = 1/4\pi\epsilon_0 = 8.987\,551\,787 \times 10^9$ N·m²/C². What is the approximate difference (in percent) between the simple number 9.0×10^9 N·m²/C² and the exact value of the Coulomb constant? (For most calculations, the simple value suffices.)

22. From far away, any charge distribution with a net charge behaves more or less like a point charge. Consider two thin disks, each of radius 1.0 cm. Each disk has a charge per unit area of 2.5×10^{-8} C/m². What is the electric force between the disks when they are separated by 2.0 m?

*23. A long, linear organic molecule is initially 1.9 μm in length. An atom at each end is then singly ionized; overall, the molecule remains neutral. The two ionizations result in a length change of -1.2%. What is the effective spring constant for this molecule?

*24. Deimos is a small moon of Mars, with a mass of 2.0×10^{15} kg. Suppose that an electron is at a distance of 100 km from Deimos. What is the gravitational attraction acting on the electron? What negative electric charge would have to be placed on Deimos to balance this gravitational attraction? How many electron charges does this amount to? Treat the mass and the charge distribution as pointlike in your calculations.

*25. A small charge of -2.0×10^{-8} C is at the point $x = 2.0$ m, $y = 0$ on the x axis. A second small charge of -3.0×10^{-6} C is at the point $x = 0$, $y = -3.0$ m on the y axis (see Fig. 22.19). What is the electric force that the first charge exerts on the second? What is the force that the second charge exerts on the first? Express your answers as vectors, with x and y components.

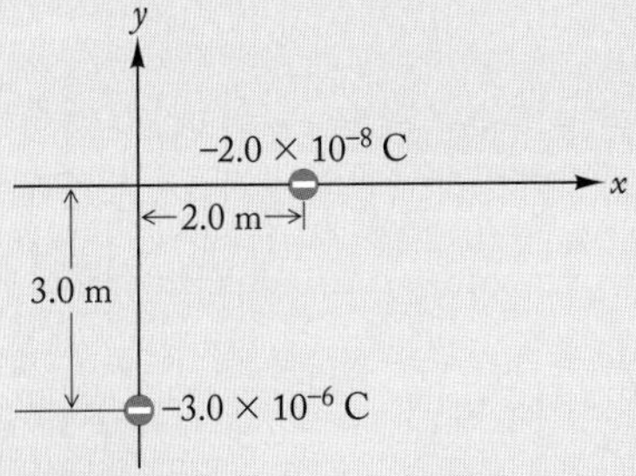

FIGURE 22.19 Two pointlike charges.

*26. Two tiny chips of plastic of masses 5.0×10^{-5} g are separated by a distance of 1.0 mm. Suppose that they carry equal and opposite electrostatic charges. What must the magnitude of the charge be if the electric attraction between them is to equal their weight?

*27. How many extra electrons would we have to place on the Earth and on the Moon so that the electric repulsion between these bodies cancels their gravitational attraction? Assume that the numbers of extra electrons on the Earth and on the Moon are in the same proportion as the radial dimensions of these bodies (6.38:1.74).

*28. At a place directly below a thundercloud, the induced electric charge on the surface of the Earth is $+1.0 \times 10^{-7}$ coulomb per square meter of surface. How many singly charged positive ions per square meter does this represent? The number of atoms on the surface of a solid is typically 2.0×10^{19} per square meter. What fraction of these atoms must be ions to account for the above electric charge?

*29. Although the best available experimental data are consistent with Coulomb's Law, they are also consistent with a modified Coulomb's Law of the form

$$F = \frac{1}{4\pi\epsilon_0} \frac{q_1 q_2}{r^2} e^{-r/r_0}$$

where r_0 is a constant with the dimensions of length and a numerical value that is known to be no less than 10^9 m and is probably much larger. Here, e is the base of the natural logarithms. Assuming that $r_0 = 1.0 \times 10^9$ m, what is the fractional deviation between Coulomb's Law and the modified Coulomb's Law for $r = 1.0$ m? For $r = 1.0 \times 10^4$ m? (Hint: Use the approximation $e^x \approx 1 + x$ for small x.)

*30. A proton is at the origin of coordinates. An electron is at the point $x = 4.0 \times 10^{-11}$ m, $y = 2.0 \times 10^{-11}$ m in the x–y plane (see Fig. 22.20). What are the x and y components of the electric force that the proton exerts on the electron? That the electron exerts on the proton?

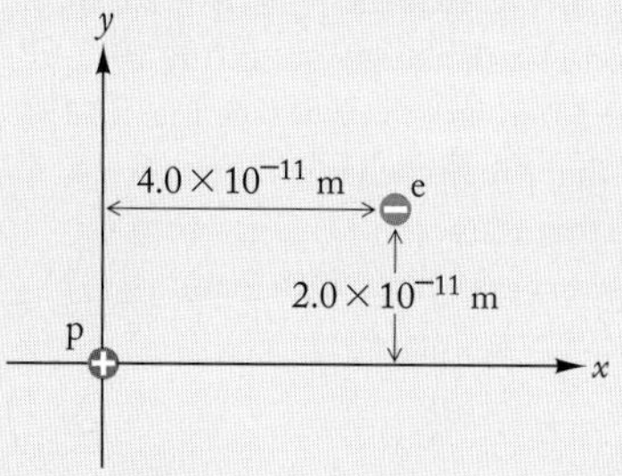

FIGURE 22.20 A proton and an electron.

*31. Precise experiments have established that the magnitudes of the electric charges of an electron and a proton are equal to within an experimental error of $\pm 10^{-21}e$ and that the electric charge of a neutron is zero to within $\pm 10^{-21}e$. Making the worst possible assumption about the combination of errors, what is the largest conceivable electric charge of an oxygen atom consisting of 8 electrons, 8 protons, and 8 neutrons? Treating the atoms as point particles, compare the electric force between two such oxygen atoms with the gravitational force between these atoms. Is the net force attractive or repulsive?

*32. Under the influence of the electric force of attraction, the electron in a hydrogen atom orbits around the proton on a circle of radius 5.3×10^{-11} m. What is the orbital speed? What is the orbital period?

22.3 The Superposition of Electric Forces†

33. Suppose that in Example 6 both charges Q are positive. What are the magnitude and direction of the electric force on the electric charge q in this case?

34. The distribution of electric charges in a thundercloud can be approximated by several pointlike charges placed at different heights. Suppose that a thundercloud has electric charges of +10 C, −40 C, and +40 C at altitudes of 2.0 km, 5.0 km, and 10 km, respectively (see Fig. 22.21). Treating these charges as pointlike, find the net electric force that the two charges of ±40 C exert on the charge of +10 C.

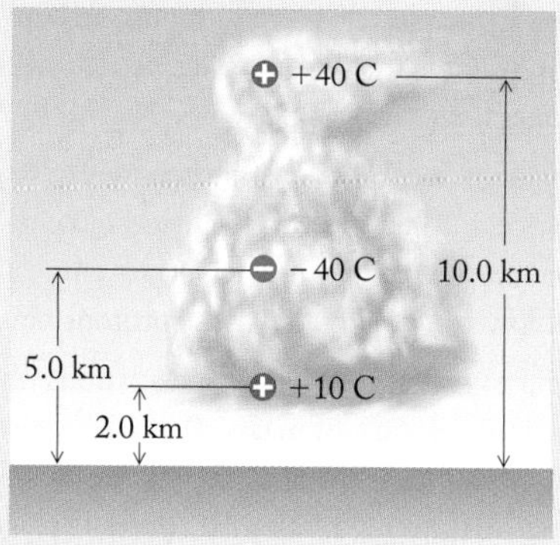

FIGURE 22.21 Charges in a thundercloud.

35. Figure 22.22 shows the arrangement of nuclear charges (positive charges) in an HCl molecule. The magnitudes of the H and Cl nuclear charges are e and $17e$, respectively, and the distance between them is 1.28×10^{-10} m. What is the net electric force that these charges exert on an electron placed 5.0×10^{-11} m above the H nucleus?

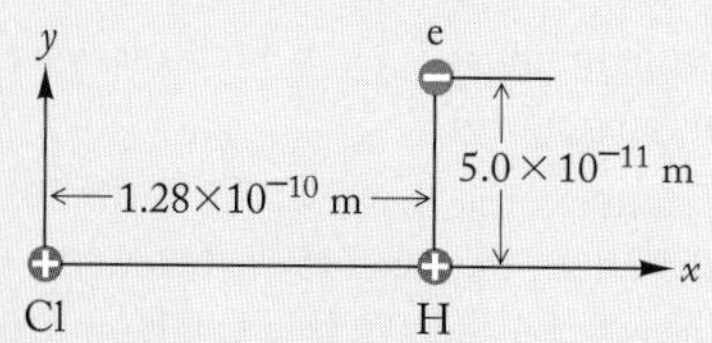

FIGURE 22.22 The positive nuclear charges in the chlorine and hydrogen atoms exert electric forces on an electron.

†For help, see Online Concept Tutorial 26 at www.wwnorton.com/physics

36. Five identical charges $+Q$ are at the vertices of a pentagon. What is the net electric force due to those five charges on a sixth charge $+q$ at the center of the pentagon?

37. Suppose that two balls of mass 2.5×10^{-4} kg each carry equal charges and are suspended by identical threads of length 10 cm, similar to Example 5; however, these threads are anchored at points 25 cm apart. If each thread makes an angle of 20° with the vertical, what is the charge on each ball?

*38. Point charges $+Q$ and $-2Q$ are separated by a distance d. A point charge q is equidistant from these charges, at a distance x from their midpoint (see Fig. 22.23). What is the electric force on q?

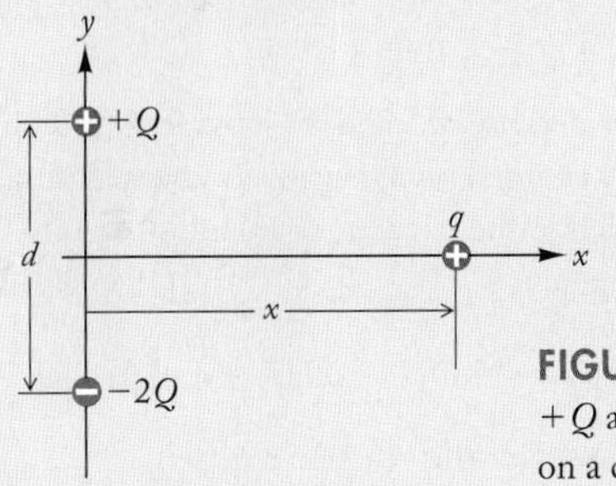

FIGURE 22.23 Charges $+Q$ and $-2Q$ exert forces on a charge q.

*39. Three positive point charges $+Q$ are placed at three corners of a square, and a negative point charge $-Q$ is placed at the fourth corner (see Fig. 22.24). The side of the square is L. Calculate the net electric force that the positive charges exert on the negative charge.

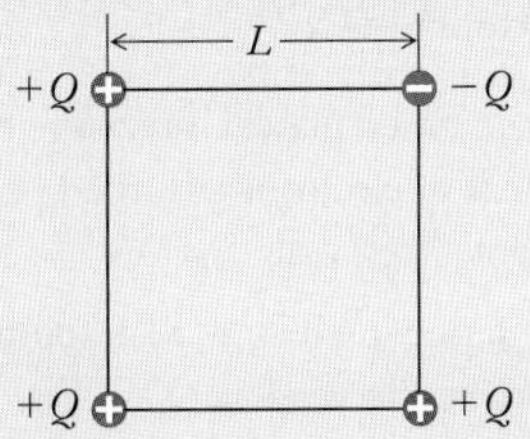

FIGURE 22.24 Three positive point charges and one negative point charge.

*40. Four point charges of $\pm Q$ are arranged on the corners of a square of side L as illustrated in Fig. 22.25. What is the net electric force that these charges exert on a point charge q placed at the center of the square?

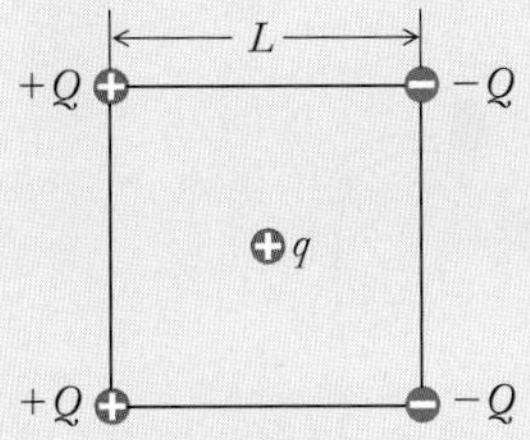

FIGURE 22.25 Five point charges.

*41. Figure 22.26 shows the approximate charge distribution in a thundercloud consisting of a pointlike charge of $+40$ C at a height of 10.0 km and a pointlike charge of -30 C at a height of 4.0 km. What is the force that these two charges exert on an electron located at a height of 10.0 km and at a horizontal distance of 4.0 km to the right of the charges?

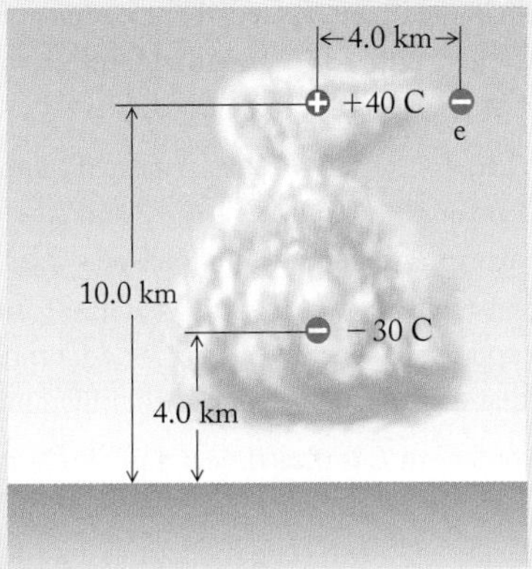

FIGURE 22.26 Charges in a thundercloud.

*42. Repeat the preceding problem but with the electron at a height of 7.0 km and at a horizontal distance of 4.0 km to the right.

*43. Suppose that two balls are suspended by identical threads of length 10 cm anchored at the same point, similar to Example 5, except that these balls have different masses and charges. When one ball carries a charge of $+2.0 \times 10^{-7}$ C and the other a charge of $+6.0 \times 10^{-8}$ C, the threads each make the same equilibrium angle of 25° with the vertical. What is the mass of each ball?

*44. Three charges ($+q$, $+q$, and $-q$) have equal magnitudes and are located at the vertices of an equilateral triangle. Find the magnitude of the total force on one of the positive charges due to the other two charges.

*45. Two equal charges $+Q$ are at two vertices of an equilateral triangle of side a; a third charge $-q$ is at the other vertex. A fourth charge q_0, located a distance $a/2$ outside the triangle along the perpendicular bisector of the $+Q$ charges (see Fig. 22.27), experiences zero net force. Find the value of the ratio q/Q.

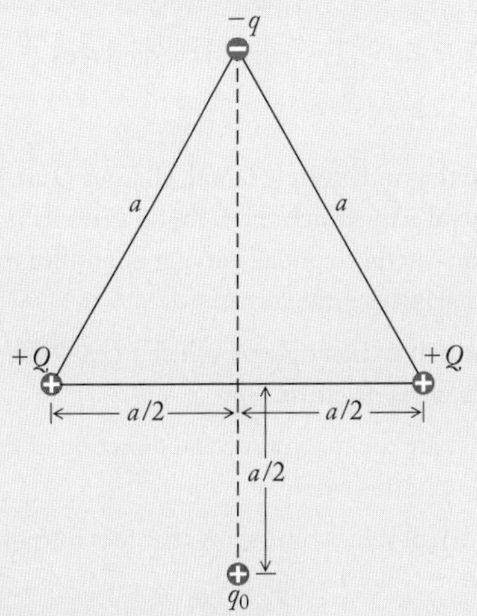

FIGURE 22.27 Three charges $+Q$, $+Q$, and $-q$ exert forces on a fourth charge q_0.

*46. Eight equal charges q are located at the corners of a cube of side a. Find the magnitude of the total force on one of the charges due to the other seven charges.

*47. Two point charges $+Q$ and $-Q$, separated a distance d (an **electric dipole**), are on the x axis at $x = +d/2$ and $x = -d/2$, respectively. Find an expression for the net force on a third charge $+q$ also on the x axis at $x > d/2$. Simplify your result and obtain the form of the approximate net force for $x \gg d$. Compare your result with that discussed in the Comment of Example 6.

*48. A thin rod of length L is placed near a point charge q (see Fig. 22.28), with the nearest end a distance d from the charge and oriented radially as shown. The rod carries a uniform distribution of charge λ coulombs per meter. Find the electric force that acts on the rod. (Hint: Sum the force contributions $dF = kq\,dq'/r^2$ due to each small charge $dq' = \lambda dx$ on the rod to obtain the total force $F = \int dF$.)

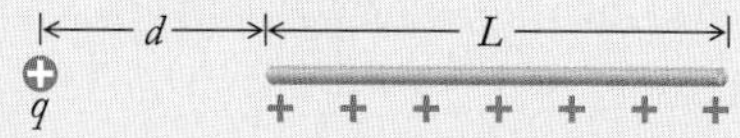

FIGURE 22.28 A charged rod and a point charge q.

**49. Four equal point charges $+Q$ are located at the vertices of a regular tetrahedron of side a. What is the force on one of the charges due to the other three?

**50. Two thin rods of length L carry equal charges Q uniformly distributed over their lengths. The rods are aligned, and their nearest ends are separated by a distance x (see Fig. 22.29). What is the electric force of repulsion between these two rods? (Hint: Sum the force contributions $dF = kdq_1\,dq_2/r^2$ due to each pair of small charges $dq_1 = (Q/L)\,dx_1$ and $dq_2 = (Q/L)\,dx_2$ to obtain the total force $F = \int\int dF$.)

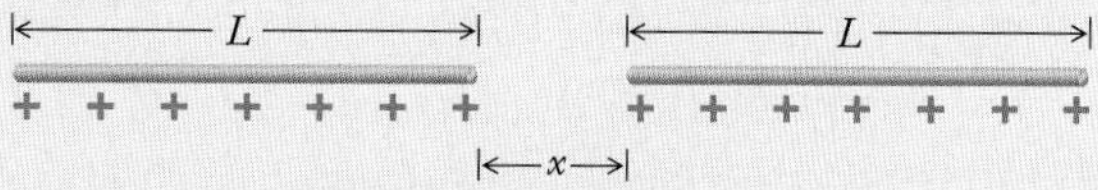

FIGURE 22.29 Two aligned charged rods.

**51. Suppose that in Example 6 both charges Q are positive. Find the value of x for which the force on q has its maximum value. (Hint: Be careful to put all varying quantities in terms of x before maximizing the force.)

**52. Suppose that in Example 6 both charges Q are positive but the other charge is negative, $-q$.

(a) What are the magnitude and direction of the electric force on $-q$ in this case?

(b) For small values of x, show that this force is proportional to x.

(c) For such a force, one expects simple harmonic motion. What is the period of such motion? (Assume that the charge $-q$ has a mass m.)

22.4 Charge Quantization and Conservation

53. Consider the following hypothetical reactions involving the collision between a high-energy proton (from an accelerator) and a stationary proton (in the nucleus of a hydrogen atom serving as target):

$$p + p \rightarrow n + n + \pi^+$$

$$p + p \rightarrow n + p + \pi^0$$

$$p + p \rightarrow n + p + \pi^+$$

$$p + p \rightarrow p + p + \pi^0 + \pi^0$$

$$p + p \rightarrow n + p + \pi^0 + \pi^-$$

where the symbols p, n, π^+, π^-, and π^0 stand for a proton, neutron, positively charged pion, negatively charged pion, and neutral pion. Which of these reactions are impossible?

54. Consider the reaction

$$Ni^{2+} + 4H_2O \rightarrow NiO_4^{2-} + 8H^+ + [\text{electrons}]$$

How many electrons does this reaction release?

55. Lithium ions are often dissolved for use in electrolytes. The reactions in a rechargeable lithium–cobalt (Li–Co) battery can be represented as

$$Li \rightarrow Li^+ + 1[\text{electron}]$$

at the electron-releasing lithium plate and

$$Co^{4+} + N[\text{electrons}] \rightarrow Co^{3+}$$

at the electron-absorbing cobalt-based plate. Use charge balance to determine the number N of electrons absorbed per cobalt atom during the reaction.

56. A spherical shell has net charge only on its inner and outer surfaces. The total charge on the entire shell is $Q_{total} = -1.0 \times 10^{-8}$ C. The charge on the inner surface of the shell is $Q_{inner} = +2.0 \times 10^{-8}$ C. What charge is on the outer surface of the shell?

*57. We can silver-plate a metallic object, such as a spoon, by immersing the spoon and a bar of silver (Ag) in a solution of silver nitrate ($AgNO_3$). If we then connect the spoon and the silver

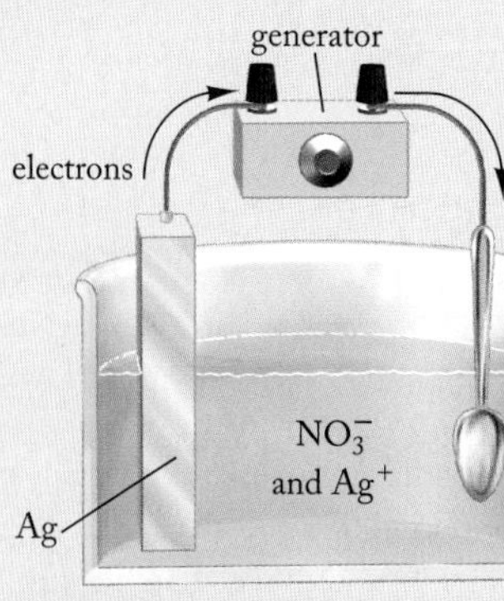

FIGURE 22.30 Silver-plating a spoon.

bar to an electric generator and make a current flow from one to the other, the following reactions will occur at the immersed surfaces (Fig. 22.30):

$$Ag^+ + [\text{electron}] \rightarrow Ag_{(\text{metal})}$$

$$Ag_{(\text{metal})} \rightarrow Ag^+ + [\text{electron}]$$

The first reaction deposits silver on the spoon, and the second removes silver from the silver bar. How many electrons must we make flow from the silver bar to the spoon in order to deposit 1.0 g of silver on the spoon?

REVIEW PROBLEMS

58. By rubbing a small glass ball against a small nylon ball, you deposit a charge of 6.0×10^{-11} C on the glass ball and a charge of -6.0×10^{-11} C on the nylon ball. If you then separate the balls to a distance of 20 cm, what is the attractive electric force between them?

59. Suppose that two grains of dust of equal masses each have a single electron charge. What must be the masses of the grains if their gravitational attraction is to balance their electric repulsion?

60. The arm in Coulomb's torsion balance was a rod with a charged ball at one end and a counterweight on the other (see Fig. 22.31). Suppose that the length of the arm in such a balance is 15 cm. Suppose that the probe is at a distance of 3.0 cm in a direction perpendicular to the arm. If the charged ball and the probe both carry $+2.0 \times 10^{-9}$ C, what is the torque exerted on the balance arm?

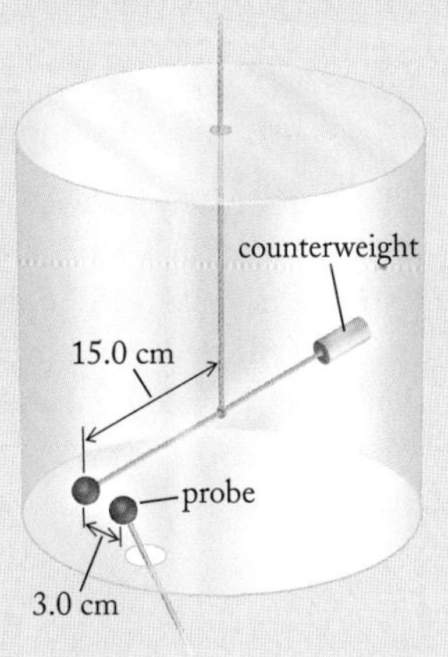

FIGURE 22.31 The arm in Coulomb's balance.

61. Three identical positive point charges $+Q$ are at the vertices of an equilateral triangle. A negative point charge is at the center of the triangle. The four charges are in equilibrium. What is the value of the negative charge?

62. Two small balls, each of mass 2.0×10^{-4} kg, carry opposite charges of equal magnitude. The balls are suspended by identical threads of length 10 cm, similar to Example 5; however, these threads are anchored at points 20 cm apart. If the equilibrium separation of the balls is 10 cm, what is the magnitude of the charge on each ball?

63. Water drops in thunderclouds carry electric charges. Suppose that two such drops are falling side by side separated by a horizontal distance of 1.0 cm. Each drop has a radius of 0.5 mm and carries a charge of 2.0×10^{-11} C. What is the electric repulsive force on each drop? What is the instantaneous horizontal acceleration of each?

64. Suppose that during a thunderstorm, the corona discharge from a dissipative lightning rod into the surrounding air amounts to 1.0×10^{-4} C of positive charge per second. If this discharge goes on more or less steadily for an hour, how much electric charge flows out of the lightning rod? How many electrons flow into the lightning rod from the surrounding air?

65. Two small balls of plastic carry equal charges of opposite signs and of unknown magnitudes. When the balls are separated by a distance of 18 cm, the attractive force between them is 0.30 N. What is the excess of electrons on one ball and the deficit of electrons on the other?

66. A different version of the electroscope discussed in Example 5 uses a fixed cork ball and a suspended cork ball (see Fig. 22.32). The mass of the suspended ball is 1.5×10^{-4} kg, and the length of the suspension thread is 10 cm. The fixed ball is

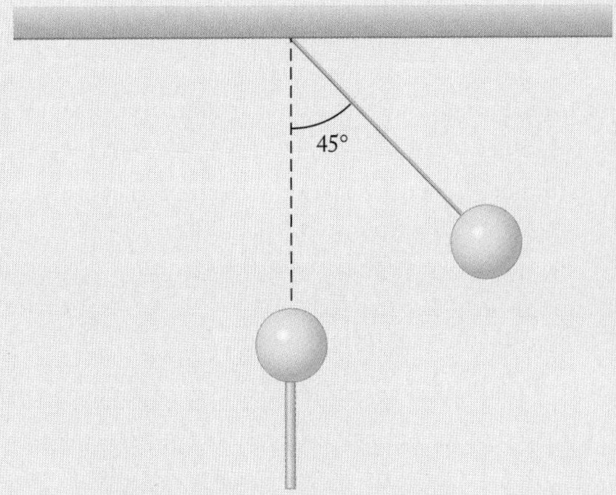

FIGURE 22.32 A charge suspended by a thread and a fixed charge.

located 10 cm directly below the point of suspension if the suspended ball. Assume that when equal electric charges are placed on the two balls, the electric repulsive force pushes the suspended ball up so its thread makes an angle of 45° with the vertical. What is the magnitude of the electric charge?

67. In each of the following decay reactions of elementary particles there is a missing particle. What is its electric charge?

$$n \rightarrow p + e + ?$$

$$\Delta^{++} \rightarrow p + \pi^{0} + ?$$

$$\Delta^{+} \rightarrow n + ?$$

$$\pi^{-} \rightarrow \mu^{-} + ?$$

Answers to Checkups

Checkup 22.1

1. Planets have large masses, and so exert large gravitational forces. However, planets have negligible net charge (if any); overall, they are electrically neutral. In an atom, on the other hand, the electrons have (relatively) large charges and (relatively) small masses.
2. The ground exerts the normal "contact" force on the stone, which is electric in origin; for a stone at rest, the magnitude of the contact force is the same as the gravitational force it cancels: $F = mg = (1.0\text{ kg})(9.8\text{ m/s}^2) = 9.8\text{ N}$.
3. The signs of the charges that are, on average, closest to each other determine the direction of the net force. Thus with two nuclei (like charges) closest to each other, the net electric force is repulsive. The overall effect is somewhat subtle, as it depends on summing four forces.
4. (C) 1.0 C. The addition of a few elementary charges will cause only negligible change (here, in the eighteenth decimal place) in the net charge of any macroscopic quantity of charge, such as one coulomb.

Checkup 22.2

1. For point charges, the two charges must have opposite signs.
2. Since the electric force varies in proportion to the inverse square of distance, $1/r^2$, the increase from 1 m to 10 m will decrease the force by a factor of 100 to 1×10^{-6} N, and the increase from 1 m to 100 m will result in a force of 1×10^{-8} N.
3. Consider any two equal charges q, with an electric force $F = kq^2/r^2$. If we transfer a charge δ from one to the other, the charges become $(q + \delta)$ and $(q - \delta)$, so the force becomes $F' = k(q + \delta)(q - \delta)/r^2$. But $(q + \delta)(q - \delta) = q^2 - \delta^2$, which is less than q^2, so the force decreases.
4. (B) 10 μC. For the given force and distance, and with $q' = 10q$, Coulomb's Law gives $1.0\text{ N} = F = kqq'/r^2 = (9.0 \times 10^9\text{ N}\cdot\text{m}^2/\text{C}^2) \times 10q^2/(3.0\text{ m})^2 = 1.0 \times 10^{10}\text{ N/C}^2 \times q^2$. Thus $q = \sqrt{1.0 \times 10^{-10}\text{ C}^2} = 1.0 \times 10^{-5}\text{ C} = 10\ \mu\text{C}$.

Checkup 22.3

1. In this case, the two forces on q are both away from the Q charges, so now the $+y$, $-y$ contributions cancel and the two $+x$ contributions add. Thus the total force is to the right, in the $+x$ direction (sketch a quick vector diagram to convince yourself of this).
2. (C) Is pushed outside the triangle. Like charges repel. The forces from the two charges at vertices adjacent to the fourth charge are equal but opposite and cancel. The net force is thus the same as the contribution from the charge at the opposite vertex; that repulsion will push the fourth charge away, out of the triangle.

Checkup 22.4

1. If an atom (which is neutral, in contrast to an ion) loses two electrons, it leaves behind an ion with a net positive charge of $+2e = +3.2 \times 10^{-19}$ C. If three electrons are lost, the remaining ion has a charge $+3e = +4.8 \times 10^{-19}$ C.
2. (C) No; yes. The charge of any body must be an integral multiple of the fundamental charge, such as e, $2e$, $3e = 1.6 \times 10^{-19}$ C, 3.2×10^{-19} C, 4.8×10^{-19} C. Thus the first charge given is impossible, while the second charge given, twice the fundamental charge, is possible—indeed, it is common.

Checkup 22.5

1. An electrolyte has an abundance of ions and is a good conductor; an ordinary liquid has few or no ions and is a poor conductor or an insulator. Similarly, a plasma has an abundance of ions and is a conductor, while an ordinary gas does not.
2. A good material for the conductor would be a metal such as aluminum or copper; for the stand, glass, as well as most rubber or plastic, would be a good insulator.
3. (C) There will be -1×10^{-7} C on each. Since the balls are metal, charge is free to move. When touching, the arrangement is symmetric (balls of equal size), and so the net charge will distribute evenly between the two balls. The net charge is $(+1 - 3) \times 10^{-7}\text{ C} = -2 \times 10^{-7}$ C, so the charge on each will be -1×10^{-7} C.

CHAPTER

23

The Electric Field

Concepts
— in —
Context

CONCEPTS IN CONTEXT

Lightning is a spectacular visual result of the electric forces generated by the charge distribution in a thundercloud. Electric forces accelerate electrons to high speeds; the electrons ionize atoms, resulting in an avalanche of moving charge, a lightning bolt. In this chapter, we introduce the *electric field,* which is the electric force per unit charge.

With this and other concepts developed in this chapter, we can investigate questions such as

? What is the approximate distribution of charge in a thundercloud, and what electric field does it produce some distance from the cloud? (Example 2, page 725)

? How does the charge distribution in a thundercloud affect the arrangement of charges on the (conducting) ground below the cloud? What is the resulting electric field at ground level? (Example 3, page 727)

? What is the electric force on an electron at ground level? What is the resulting acceleration of the electron? (Example 4, page 728)

action-at-a-distance

field

action-by-contact

Up to here we have taken the view that the gravitational forces and the electric forces between particles are **action-at-a-distance**, that is, a particle exerts a direct gravitational or electric force on another particle even when the particles are widely separated. Such an interpretation of gravitational and electric forces as a ghostly tug-of-war between distant bodies is suggested by Newton's Law of Gravitation and Coulomb's Law. However, according to the modern view, there is a physical entity that acts as mediator of force, conveying the force over the distance from one body to another. This entity is the **field**. A gravitating or electrically charged body generates a gravitational or electric field which permeates the (apparently) empty space around the body, and this field exerts pushes or pulls whenever it comes in contact with another body. Thus, fields convey forces from one body to another through **local action**, or **action-by-contact**.

In the present chapter, we will become acquainted with the electric field that conveys the electric force from one body to another body. We will first examine the electric field due to a point charge. Then, using a superposition of electric fields similar to the superposition of electric forces, we will consider the electric field generated by several point charges, and by continuous distributions of charges. We will also introduce the useful concept of lines of electric field. Finally, we will consider what motion occurs when the electric field acts on a charge.

23.1 THE ELECTRIC FIELD OF POINT CHARGES

Online Concept Tutorial 27

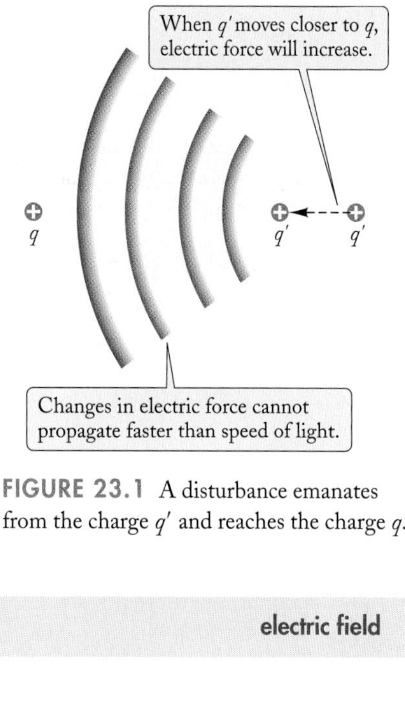

FIGURE 23.1 A disturbance emanates from the charge q' and reaches the charge q.

electric field

According to the naive interpretation of Coulomb's Law, the electric forces between charges are action-at-a-distance, that is, a charge q' exerts a direct force on a charge q even though these charges are separated by a large distance and are not touching. However, such an action-at-a-distance interpretation of electric forces leads to serious difficulties in the case of moving charges. Suppose that we suddenly move the charge q' somewhat nearer to the charge q; then the electric force on q has to increase, according to the inverse-square law. But the required increase cannot occur instantaneously—the increase can be regarded as a signal from q' to q, and it is a fundamental principle of physics, based on the theory of relativity, that no signal can propagate faster than the speed of light. This suggests that, when we suddenly move the charge q', some kind of physical disturbance propagates through space from q' to q and adjusts the electric force to the new increased value (see Fig. 23.1). Thus, *charges exert forces on one another by means of disturbances that they generate in the space surrounding them. These disturbances are called* **electric fields**.

Fields are a form of matter—they are endowed with energy and momentum (as we shall find in Chapter 33), and they therefore exist in a material sense. In the context of the above example, it is easy to see why the disturbance, or field, generated by the sudden displacement of q' must carry momentum: when we suddenly move q' toward q, the increase in the force that q' exerts on q will be delayed until a signal has had time to propagate from q' carrying the information regarding the changed position of q'. But the increase in the force that q exerts on q' does not suffer a similar delay. Since we have not moved q, we have not disturbed the way in which q causes electric forces on other charges, according to Coulomb's Law. This temporary deviation between the forces exerted by q' on q and by q on q' means that, when only the charges are considered, Newton's Third Law on the balance of action and reaction fails. Thus an extra entity, such as the field, is needed to take up the momentum and energy missing from the particles.

Although the above argument for the existence of the electric field arose by considering charges in motion, we will now adopt the very natural view that the forces on charges at rest involve the same mechanism. We suppose that each charge at rest generates a permanent, static disturbance in the space surrounding it, and that this disturbance exerts forces on other charges when they come in contact with the disturbance. Thus, we take the view that *the electric interaction between charges is action-by-contact: a charge q' generates an electric field which permeates the surrounding space and exerts forces on any other charges that it touches.* The electric field serves as the mediator of forces according to the scheme

$$\text{charge } q' \Rightarrow \text{electric field of charge } q' \Rightarrow \text{force on charge } q$$

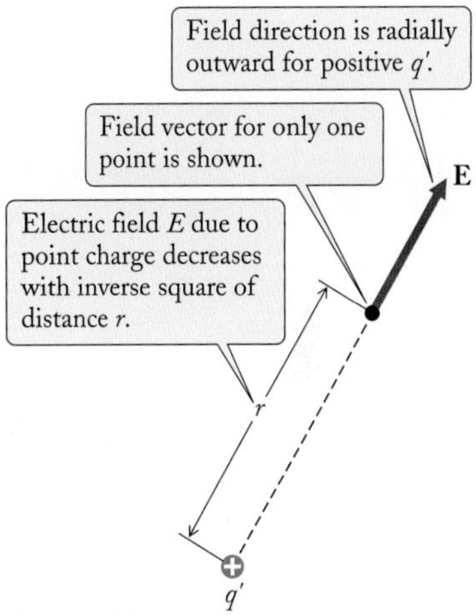

FIGURE 23.2 A charge q' generates an electric field **E** at some distance r. The direction of this electric field is along the radial line.

To translate this conceptual scheme into mathematical language, we start with Coulomb's Law for the force exerted by the charge q' on the charge q,

$$F = \frac{1}{4\pi\epsilon_0}\frac{qq'}{r^2}$$

and we separate this expression into a product of two factors: a factor q characteristic of the point charge on which the force is being exerted, and a factor q'/r^2 characteristic of the point charge that exerts the force at the distance r. We also include the constant of proportionality $1/4\pi\epsilon_0$ in the second factor, so the expression for the force becomes

$$F = q \times \left(\frac{1}{4\pi\epsilon_0}\frac{q'}{r^2}\right) \tag{23.1}$$

The second factor is defined to be the electric field generated by the point charge q'; we designate this electric field by E:

$$E = \frac{1}{4\pi\epsilon_0}\frac{q'}{r^2} \tag{23.2}$$

electric field of point charge

This says that the magnitude of the electric field of the point charge q' is directly proportional to the magnitude of this charge and inversely proportional to the square of the distance. The force that this electric field exerts on the charge q is then simply

$$F = qE \tag{23.3}$$

Like the force, the electric field is a vector, and it can be represented by an arrow (see Fig. 23.2). The direction of the electric field depends on the sign of the charge q'. *The electric field is directed radially outward if q' is positive, and radially inward if q' is negative.* In vector language, the force on the charge q becomes

$$\mathbf{F} = q\mathbf{E} \tag{23.4}$$

With this special case of the electric field of a point charge as guidance, we can proceed to the definition of the electric field for the general case of an arbitrary charge distribution, such as the charge distribution in a thundercloud, or the charge distribution on an electric power line. To find the electric field that the charge distribution generates

TABLE 23.1 SOME ELECTRIC FIELDS

At surface of pulsar	$\approx 10^{14}$ N/C
At orbit of electron in hydrogen atom	6×10^{11}
In X-ray tube	5×10^{6}
Electrical breakdown of air	3×10^{6}
In Van de Graaff accelerator	2×10^{6}
Within lightning bolt	$\approx 10^{4}$
Under thundercloud	1×10^{4}
Near radar transmitter (FPS-6)	7×10^{3}
In sunlight (rms)	1×10^{3}
In atmosphere (fair weather)	1×10^{2}
In beam of small laser (rms)	1×10^{2}
In fluorescent lighting tube	10
In radio wave	$\approx 10^{-1}$
Within household wiring	$\approx 3 \times 10^{-2}$

at a given position, we take a point charge q (a "test charge") and place it at that position. The charge q will then experience an electric force F. By dividing out the charge q, we isolate the factor characteristic of the charge distribution and we eliminate the factor characteristic of the test charge. We define the electric field **E** as the force **F** divided by the magnitude q of the charge:

electric field and electric force

$$\mathbf{E} = \frac{\mathbf{F}}{q} \qquad (23.5)$$

This means that the electric field is the force per unit charge. Note that with this definition, the electric field is independent of the magnitude of the test charge. The electric field is a very useful concept: for a given charge distribution, instead of calculating the force that the charges exert on another charge q at some point (which would be different for different values of q), the electric field tells us the force per unit charge on *any* charge placed at that point. Thus, the electric field **E** depends on the magnitudes and positions of the charges that produce the electric field, but it does not depend on the magnitude of the test charge q used to detect it.

newton per coulomb (N/C)

The SI unit of electric field is the **newton per coulomb** (N/C).[1] Table 23.1 gives the magnitudes of some typical electric fields.

The net electric field generated by any distribution of point charges can be calculated by forming the vector sum of the individual electric fields due to the point charges, where each individual electric field is given by Eq. (23.2). This procedure is justified by the Principle of Superposition for electric forces stated in Section 22.3. Thus, to find the total electric field at some point in space, you simply sum the individual vectors [with magnitude given by Eq. (23.2), and direction away from positive charges or toward negative charges].

[1] As we shall see in Chapter 25, newton/coulomb is the same thing as volt/meter.

EXAMPLE 1 Consider two charges $\pm Q$ of equal magnitudes and opposite signs, separated by a distance d. Such an arrangement of charges, similar to the ones exerting the force in Example 6 of Chapter 22, is called an **electric dipole**. Find the electric field at a point equidistant from the two charges, a distance x from their midpoint (see Fig. 23.3). What is the dependence on distance of this "dipole field" for $x \gg d$?

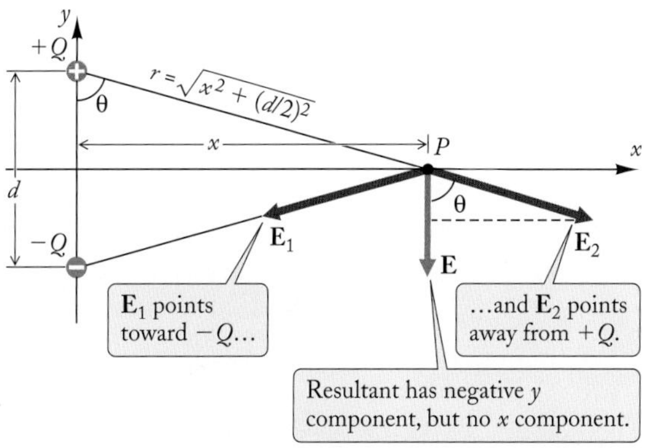

FIGURE 23.3 Two point charges $-Q$ and $+Q$ and a point P on the x axis. The net electric field $\mathbf{E}$ at the point P is the vector sum of the contributions $\mathbf{E}_1$ due to $-Q$ and $\mathbf{E}_2$ due to $+Q$.

SOLUTION: The *magnitude* of each of the two individual electric fields is $E_1 = E_2 = (1/4\pi\epsilon_0)Q/r^2$, where $r = \sqrt{x^2 + (d/2)^2}$. At the point P, the electric field due to the $+Q$ charge points away from this charge, and the electric field due to the $-Q$ charge points toward that charge, as shown in Fig. 23.3. As in the solution of Example 6 of Chapter 22, the x components of the electric field at the point P in Fig. 23.3 cancel, and the y components are equal, giving

$$E = E_y = -2E_1 \cos\theta = -2\frac{1}{4\pi\epsilon_0}\frac{Q}{x^2 + d^2/4}\frac{d/2}{\sqrt{x^2 + d^2/4}} \tag{23.6}$$

$$= -\frac{1}{4\pi\epsilon_0}\frac{Qd}{(x^2 + d^2/4)^{3/2}}$$

For $x \gg d$ we can neglect d^2 compared with x^2 and obtain

$$E = -\frac{1}{4\pi\epsilon_0}\frac{Qd}{x^3}$$

COMMENTS: Note that since we already calculated the force due to the charges $\pm Q$ on a charge q in Example 6 of Chapter 22, we could equally well solve this problem using the definition $\mathbf{E} = \mathbf{F}/q$ and obtain the same result. Also note that for large distances, the dependence of the dipole field on distance is $E \propto 1/x^3$. It turns out that this $1/x^3$ behavior is characteristic of the electric dipole field at large distances *in any direction*, not just along the perpendicular bisector. Finally, note that here we have calculated the field *generated* by an electric dipole, that is, the field that would act on a third charge when placed near the dipole. Later, in Section 23.5, we will examine the *response* of an electric dipole to an *external* field. Be careful not to confuse the two situations.

EXAMPLE 2 The distribution of electric charge in a thundercloud can be roughly approximated by several pointlike charges placed at different heights. Figure 23.4a shows such an approximate charge distribution consisting of a charge of +40 C at a height of 10.0 km and a charge of −30 C at a height of 4.0 km in the thundercloud. What are the horizontal and the vertical components of the electric field that these two charges produce at a point P at a height of 10.0 km and at a horizontal distance of 6.0 km to the right? What is the magnitude of this field at that point?

Concepts in Context

SOLUTION: At the point P, the charge $Q_1 = +40$ C produces an electric field that is directed away from this charge, and the charge $Q_2 = -30$ C produces an electric field that is directed toward that charge. The net electric field is the vector sum of these two individual electric fields.

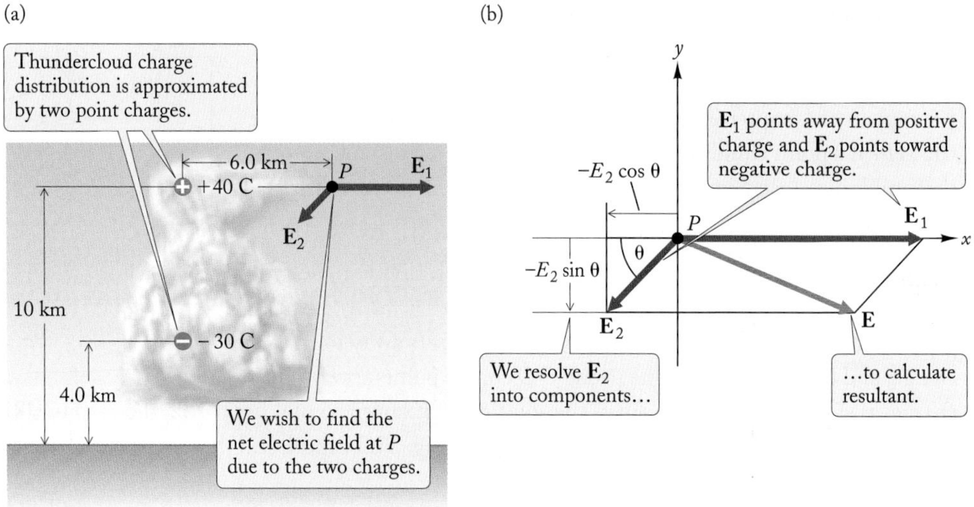

FIGURE 23.4 (a) Distribution of positive and negative electric charges in a thundercloud. (b) The net electric field is the vector sum of the individual electric fields $\mathbf{E}_1$ and $\mathbf{E}_2$.

If h represents the height difference between the charges and d the horizontal distance of the point P, then the radial distances from the charges Q_1 and Q_2 to P are, respectively, d and $\sqrt{d^2 + h^2}$. According to Eq. (23.2), the *magnitudes* of the two individual electric fields are

$$E_1 = \frac{1}{4\pi\epsilon_0}\frac{Q_1}{d^2} \quad \text{and} \quad E_2 = \frac{1}{4\pi\epsilon_0}\frac{|Q_2|}{d^2 + h^2}$$

From Fig. 23.4b we see that the horizontal components (x components) of the two individual electric fields are E_1 and $-E_2 \cos\theta$, and the vertical components (y components) are 0 and $-E_2 \sin\theta$, respectively. We therefore obtain the following components of the net electric field:

$$E_x = \frac{1}{4\pi\epsilon_0}\left(\frac{Q_1}{d^2} - \frac{|Q_2|}{d^2 + h^2}\cos\theta\right)$$

and

$$E_y = -\frac{1}{4\pi\epsilon_0}\frac{|Q_2|}{d^2 + h^2}\sin\theta$$

For this example, we have $d = h = 6.0$ km, and thus $\theta = 45°$, so we find

$$E_x = (9.0 \times 10^9\ \text{N}\cdot\text{m}^2/\text{C}^2)\left[\frac{40\ \text{C}}{(6.0 \times 10^3\ \text{m})^2} - \frac{30\ \text{C}}{2 \times (6.0 \times 10^3\ \text{m})^2}\cos 45°\right]$$

$$= 7.3 \times 10^3\ \text{N/C}$$

and

$$E_y = -(9.0 \times 10^9\ \text{N}\cdot\text{m}^2/\text{C}^2)\frac{30\ \text{C}}{2 \times (6.0 \times 10^3\ \text{m})^2}\sin 45°$$

$$= -2.7 \times 10^3\ \text{N/C}$$

The magnitude of the electric field is then

$$E = \sqrt{E_x^2 + E_y^2} = \sqrt{(7.3 \times 10^3 \text{ N/C})^2 + (-2.7 \times 10^3 \text{ N/C})^2}$$
$$= 7.8 \times 10^3 \text{ N/C}$$

EXAMPLE 3 Consider the thundercloud charge distribution of Example 2. To find the total electric field, we must take into account that the ground is a conductor and that the charges in the thundercloud induce charges in the ground. It can be shown that the effect of the charge induced by an above-ground point charge can be simulated by an opposite point charge the same distance *below ground*; these fictitious charges are called **image charges**. The resulting charge arrangement is shown in Fig. 23.5. By summing the electric fields due to the real charges in the thundercloud and due to the image charges, calculate the electric field at a point on the ground directly below the thundercloud charges.

Concepts in Context

image charge

SOLUTION: At the ground, the electric field due to the -30 C thundercloud charge is directed upward, toward this negative charge, and the electric field due to the $+30$ C below ground image charge is also upward, directed away from that positive charge. The charges are each 4.0 km from the ground, so these two electric field contributions have equal magnitudes and are both directed upward, which we choose as the positive y direction. Hence the sum is twice an individual contribution:

$$E_{y,\pm 30\text{ C}} = 2 \times \frac{1}{4\pi\epsilon_0}\frac{|Q|}{r^2} = 2 \times (9.0 \times 10^9 \text{ N}\cdot\text{m}^2/\text{C}^2)\frac{30 \text{ C}}{(4.0 \times 10^3 \text{ m})^2}$$
$$= +3.4 \times 10^4 \text{ N/C}$$

The electric field at the ground due to the $+40$ C thundercloud charge is directed downward, away from this positive charge, and the electric field due to the -40 C below ground image charge is also downward, directed toward that negative charge. These charges are 10 km from the ground, so the two electric field contributions again have equal magnitudes, but now are directed in the negative y direction. Hence

$$E_{y,\pm 40\text{ C}} = -2 \times \frac{1}{4\pi\epsilon_0}\frac{|Q|}{r^2} = -2 \times (9.0 \times 10^9 \text{ N}\cdot\text{m}^2/\text{C}^2)\frac{40 \text{ C}}{(1.0 \times 10^4 \text{ m})^2}$$
$$= -7.2 \times 10^3 \text{ N/C}$$

Since each of the ± 30 C and ± 40 C pairs of charges is an electric dipole, we could alternatively obtain these results by evaluating Eq. (23.6) in Example 1 directly between each pair of charges, at $x = 0$. In any case, the net electric field on the ground directly below the thundercloud charges is

$$E_y = E_{y,\pm 30\text{ C}} + E_{y,\pm 40\text{ C}} = +3.4 \times 10^4 \text{ N/C} - 7.2 \times 10^3 \text{ N/C}$$
$$= +2.7 \times 10^4 \text{ N/C}$$

The positive result indicates that the net electric field is directed vertically upward.

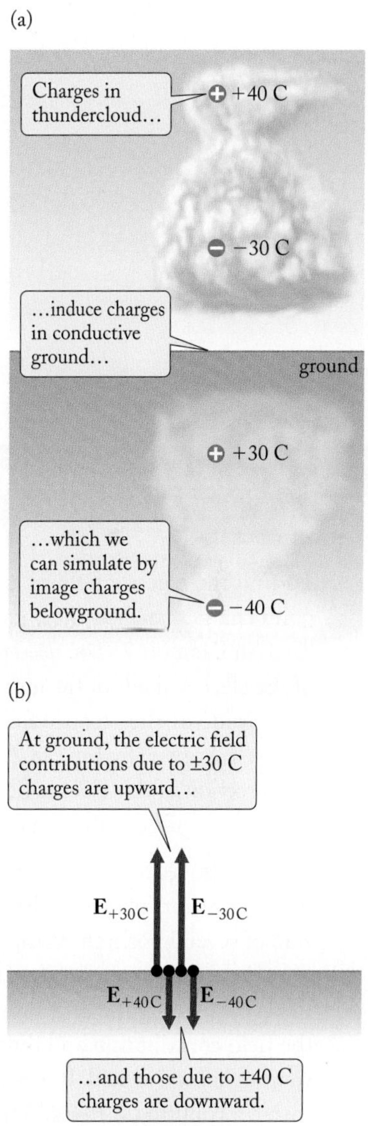

FIGURE 23.5 (a) Charges in a thundercloud and image charges below the ground. (b) Electric field contributions at ground level.

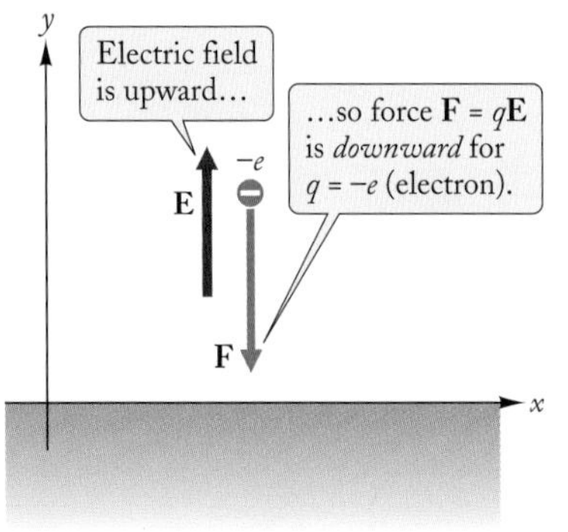

FIGURE 23.6 The upward electric field **E** exerts a downward force **F** on the electron.

EXAMPLE 4 An electron is near the ground, directly below the thundercloud charges of Example 3, in the net electric field of 2.7×10^4 N/C pointing vertically upward (see Fig. 23.6). What is the force on the electron, and what is its acceleration?

SOLUTION: By Eq. (23.4), the force on the electron equals the charge multiplied by the electric field. If the y axis is vertically upward, $E_y = 2.7 \times 10^4$ N/C, and

$$F_y = qE_y = -eE_y = (-1.60 \times 10^{-19}\text{ C})(2.7 \times 10^4\text{ N/C}) = -4.3 \times 10^{-15}\text{ N}$$

This force gives the electron an acceleration

$$a_y = \frac{F_y}{m_e} = \frac{-4.3 \times 10^{-15}\text{ N}}{9.11 \times 10^{-31}\text{ kg}} = -4.7 \times 10^{15}\text{ m/s}^2$$

This is an enormous acceleration.

COMMENTS: Note that, since the charge of the electron is negative, the direction of the force and of the acceleration is *opposite* to the direction of the electric field. Note also that a charge in the thundercloud is attracted to its own (opposite) image charge (the same image-charge attraction is responsible for the binding of free electrons within the volume of a metal). Also, note that in this calculation we have neglected the acceleration due to gravity; since this acceleration is much smaller than that due to the electric field, the neglect is justified.

PROBLEM-SOLVING TECHNIQUES ELECTRIC FIELDS

- To find the electric field of an arrangement of several point charges (such as the point charges of the thundercloud in Example 2), you need to calculate the vector sum of the electric fields of the individual point charges. The magnitude of the electric field of an individual point charge q' is given by Eq. (23.2), $E = (1/4\pi\epsilon_0)q'/r^2$; and the direction of this electric field is radially outward if q' is positive, and radially inward if q' is negative.
- The techniques for evaluating the vector sum of the electric field vectors of several charges are the same as for the sum of several position vectors or any other kind of vectors. First, draw a careful diagram, showing each electric field vector along the line through the charge producing the field contribution and through a point P where you want to know the field. It may be easiest to put the tail of each contributing vector at P (as in Example 1) and then use geometry to break each vector into its x and y (and, if necessary, z) components. You obtain the total electric field vector by separately summing the x, y, and z components of the contributing vectors. In many examples of calculations of electric fields, some components of the field will cancel. If you select the x, y, and z axes judiciously, you can often achieve cancellation of all components except one (see Example 1).
- When calculating the electric force $\mathbf{F} = q\mathbf{E}$ on charge q from the electric field **E** generated by a charge distribution, be sure to keep in mind the distinction between the charges that generate the electric field and the charge q that is placed in the resultant electric field (and experiences the force). The electric field of the latter charge is not to be included in the calculation of the electric field **E**; only the *other* charges contribute to the electric field that acts on the charge q.

Checkup 23.1

QUESTION 1: In the discussion leading up to Eq. (23.3), we began with the Coulomb force between two point charges q and q', and we found the electric field generated by the charge q'. If, instead, we had wanted to find the electric field generated by the charge q, what would have been the result?

QUESTION 2: The magnitude of the electric field at a distance of 1 m from a point charge is 1×10^5 N/C. What is the magnitude of the electric field at a distance of 2 m? 3 m?

QUESTION 3: Given that an electric field of 10 N/C has a northward direction and another electric field of 10 N/C has an eastward direction, what are the magnitude and direction of the superposition of these two electric fields?

QUESTION 4: Four equal positive charges are located at the corners of a square. What is the magnitude of the electric field at the center of the square?

QUESTION 5: Three equal positive charges are located at three corners of a square, and a negative charge is located at the fourth corner. What is the direction (if any) of the electric field at the center of the square?

(A) Toward the negative charge (B) No direction, since field is zero
(C) Away from the negative charge
(D) Perpendicular to the diagonal through the negative charge

23.2 THE ELECTRIC FIELD OF CONTINUOUS CHARGE DISTRIBUTIONS

We learned in Chapter 22 that charge is discrete, or quantized; that is, it always occurs in integer multiples of the fundamental charge e. However, in a description of the charge distribution on macroscopic bodies, the discrete nature of charge can often be ignored, and it is usually sufficient to treat the charge as a continuous "fluid" with a charge density (C/m^3) that varies more or less smoothly over the volume of the charged body. This is analogous to describing the mass distribution of a solid, a liquid, or a gas by a smooth mass density (kg/m^3), which ignores the fact that, on a microscopic scale, the mass is concentrated in atoms. A solid, a liquid, or a gas seems smooth because there are very many atoms in each cubic millimeter, and the distances between atoms are very small. Likewise, charge distributions placed on wires or other conductors seem smooth because they consist of very many electrons (or protons) in each cubic millimeter.

The Superposition Principle still holds for continuous charge distributions; we need merely sum the individual small contributions to the electric field from each small piece of charge. Suppose that a small region contains a charge dq and contributes a field of magnitude dE at some point where we want to know the total field. If this region is sufficiently small, we can treat the contribution as that of a point charge, with the usual Coulomb field:

$$dE = \frac{1}{4\pi\epsilon_0}\frac{dq}{r^2} \tag{23.7}$$

magnitude of electric field contribution

Next, we must sum these contributions to the electric field from the various small regions of charge. Mathematically, a sum over a continuous region means we integrate the contributions to get the total field. Since the electric field is a vector, we have to perform the integration for each component of **E**. For example, if θ is the angle that an electric field contribution makes with the x axis, then the x component of the contribution is $dE_x = dE \cos\theta$, that is,

x component of electric field contribution

$$dE_x = \frac{1}{4\pi\epsilon_0}\frac{\cos\theta}{r^2}dq \tag{23.8}$$

When integrating this over the charge distribution to obtain the total E_x, we must take into account that $\cos\theta$ and r^2 vary over the distribution. We can perform similar calculations for E_y and E_z and obtain the total field $\mathbf{E} = E_x\mathbf{i} + E_y\mathbf{j} + E_z\mathbf{k}$. In the general case, such a calculation can be very difficult, and numerical methods (and computers) often must be used to obtain accurate results. However, in some important examples, the integration is simple enough to be performed directly. In the examples that follow, such exact solutions provide significant insight into the behavior of the electric field due to continuous charge distributions.

In the examples below, we will encounter two of the three types of continuous charge distributions: a linear charge distribution (charge distributed on a string, a wire, or a thin rod) and a surface charge distribution (charge distributed over a flat or curved surface). In Chapter 24, we will also encounter the third type, a volume charge distribution (charge distributed throughout a three-dimensional volume). In all the cases of immediate interest, we will assume the charge distribution is *uniform*, that is, constant, over a specified region. For a uniform distribution, the small charge dq in a small linear region of length dL, or on a small piece of surface area dA, is equal to the length dL or the area dA multiplied by a constant:

$$\text{Linear:}\quad dq = \lambda\, dL$$

$$\text{Surface:}\quad dq = \sigma dA. \tag{23.9}$$

where λ (lambda) is the charge per unit length in coulombs per meter (C/m) and σ (sigma) is the charge per unit area in coulombs per square meter (C/m^2). For example, a rod of length L with a total charge Q uniformly distributed along its length has $\lambda = Q/L$, and a sheet of area A with a total charge Q uniformly distributed over its area has $\sigma = Q/A$. The more general possibility of a *nonuniform* distribution, for example, where the value of λ may vary along a line of charge, is considered in Problem 43. In the examples, we restrict ourselves to uniform distributions.

EXAMPLE 5 Charge is distributed uniformly along an infinitely long, straight thin rod. If the charge per unit length of the rod is λ, what is the electric field at some distance from the rod?

SOLUTION: Figure 23.7 shows the charged rod lying along the y axis. The point P at which the electric field is to be evaluated is on the x axis, at a perpendicular distance x from the rod. To find the electric field, we must perform an integration, regarding the rod as made up of many infinitesimal line elements, each of which can be treated as a point charge. Before proceeding with such an integration, it is helpful to determine the direction of the field by a qualitative symmetry argument. The electric field generated by the line element dy shown below the origin in Fig. 23.7 has both x and y components. Upon integration, the y component will cancel against the y component contributed by a line element lying at the same distance above the origin, and the total y component of the electric field will be zero. Therefore the net electric field will have only an x component.

The line element dy carries a charge $dQ = \lambda\, dy$. Since this line element can be treated as a point charge, it generates an electric field of magnitude

$$dE = \frac{1}{4\pi\epsilon_0}\frac{dQ}{r^2} = \frac{1}{4\pi\epsilon_0}\frac{\lambda\, dy}{r^2}$$

This electric field has an x component

$$dE_x = dE\cos\theta = \frac{1}{4\pi\epsilon_0}\frac{\lambda\, dy\cos\theta}{r^2}$$

The net electric field is then

$$E_x = \int_{-\infty}^{+\infty}\frac{1}{4\pi\epsilon_0}\frac{\lambda\, dy\cos\theta}{r^2}$$

As we vary the position of the contribution along y in this integral, the quantities r and θ also change. To evaluate this integral, we must express all variables in terms of a single variable. It turns out that the most convenient choice in this particular case is the angle θ. From Fig. 23.7,

$$y = x\tan\theta \quad\text{and}\quad dy = x\, d(\tan\theta) = x\sec^2\theta\, d\theta$$

Furthermore,

$$r = \frac{x}{\cos\theta} = x\sec\theta$$

With these substitutions, we obtain an integral over the angle θ. In terms of this angle, the limits of integration are $\theta = -90°$ and $\theta = 90°$, or, in radians, $\theta = -\pi/2$ and $\theta = \pi/2$, and the integral becomes

$$E_x = \frac{1}{4\pi\epsilon_0}\int_{-\pi/2}^{\pi/2}\frac{\lambda\cos\theta}{(x\sec\theta)^2}x\sec^2\theta\, d\theta = \frac{1}{4\pi\epsilon_0}\frac{\lambda}{x}\int_{-\pi/2}^{\pi/2}\cos\theta\, d\theta$$

Since

$$\int_{-\pi/2}^{\pi/2}\cos\theta\, d\theta = \sin\theta\Big|_{-\pi/2}^{\pi/2} = 2$$

we find

$$E_x = \frac{1}{2\pi\epsilon_0}\frac{\lambda}{x} \qquad (23.10)$$

The direction of the electric field is radially away from the rod (if λ is positive).

COMMENTS: The magnitude of this electric field decreases in inverse proportion to the distance from the rod (not the square of the distance). The result (23.10) is exact for an infinitely long rod, but it is also a reasonable approximation for a rod of finite length, provided the point P is near the rod and away from its ends (seen from such a point, the finite rod looks almost like an infinite rod).

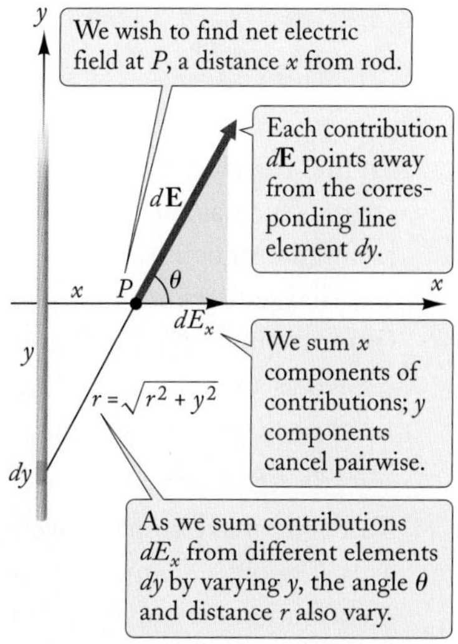

FIGURE 23.7 A long, thin, charged rod lies along the y axis. A small line element dy of this rod contributes an electric field $d\mathbf{E}$ at the point P.

PROBLEM-SOLVING TECHNIQUES ELECTRIC FIELD OF CHARGE DISTRIBUTION

- To find the electric field of a continuous distribution of charge (such as the charge distributed uniformly along an infinite straight rod in Example 5), you need to regard the charge distribution as a collection of many small (infinitesimal) charge elements, each of which can be treated as a point charge. The vector sum of the electric fields of all the small charge elements then involves integration over all these charge elements.
- To begin, make some convenient choice of coordinate axes. Choose a charge element dQ and look at a typical contribution $dE = k\,dQ/r^2$; then consider the x, y, and z components of the electric field contribution one by one, and try to decide if any component is zero because of cancellations brought about by the symmetry of the charge distribution. Any component that is not zero must be evaluated by integration over the small charge elements.
- The choice of variable of integration is often a delicate matter, since the integrand might become very simple for some choices of variable of integration, but very complicated and intractable for other choices. Thus, in Example 5, the integral is elementary if θ is used as variable, but it is much more complicated if y is used as variable. When faced with an apparently intractable integral, try substitutions of variables (or try finding the integral in a table, such as the one in Appendix 4).
- For some problems, the simplest approach is best; the variable of integration can remain the length element obtained when substituting $dQ = \lambda\,dx$. Be sure to specify the limits of integration and the quantity r in a manner consistent with the chosen variable of integration.
- Sometimes the electric field of a complicated charge distribution can be found by superposition of the individual electric fields of the separate pieces of the charge distribution. For example, the electric field of the two large charged sheets in Fig. 23.10 is the superposition of their individual electric fields, and the net electric field of two long lines of charge arranged parallel or arranged at some angle is the superposition of their individual electric fields.

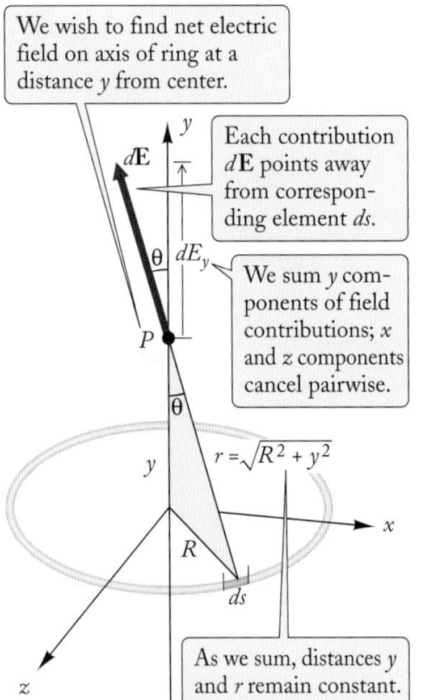

FIGURE 23.8 A small segment ds of a ring of charge contributes an electric field $d\mathbf{E}$.

EXAMPLE 6 An amount Q of positive charge is distributed uniformly along the circumference of a thin ring of radius R. What is the electric field on the axis of the ring at some distance from the center?

SOLUTION: Figure 23.8 shows the ring of charge and the point P at which the electric field is to be evaluated. This point P is on the y axis, at a distance y above the ring. We can regard the ring as made up of many small elements, each of which can be treated as a point charge. Figure 23.8 shows one of these elements, of length ds. The electric field dE contributed by this element points away from the element. To find the net electric field, we must evaluate the vector sum of all the small contributions to the electric field arising from all the small elements of the ring. Before evaluating this vector sum, let us determine the direction of the net electric field. The field contributed by the small element ds shown in Fig. 23.8 has both a horizontal and a vertical component. For any given small element ds of the ring, there is an equal small element on the opposite side of the ring's center that contributes an electric field of opposite horizontal component; consequently, the horizontal components cancel pairwise. The net electric field is therefore vertical.

The small element ds has a small charge dQ. The distance from this charge dQ to the point P is $r = \sqrt{R^2 + y^2}$. By Coulomb's Law, the magnitude of the electric field contributed by dQ is

$$dE = \frac{1}{4\pi\epsilon_0}\frac{dQ}{r^2} = \frac{1}{4\pi\epsilon_0}\frac{dQ}{R^2 + y^2}$$

We are interested only in the vertical component of this electric field, since we already know that the horizontal component makes no net contribution to the net electric field. The vertical component of $d\mathbf{E}$ is

$$dE_y = dE \cos\theta$$

where $\cos\theta = y/\sqrt{R^2 + y^2}$ (see Fig. 23.8). Consequently,

$$dE_y = \frac{1}{4\pi\epsilon_0}\frac{dQ}{R^2 + y^2}\frac{y}{\sqrt{R^2 + y^2}}$$

$$= \frac{1}{4\pi\epsilon_0}\frac{dQ\,y}{(R^2 + y^2)^{3/2}}$$

To find the total electric field, we must integrate around the ring:

$$E_y = \int dE_y = \frac{1}{4\pi\epsilon_0}\int\frac{dQ\,y}{(R^2 + y^2)^{3/2}}$$

Notice that the quantities y and R do not vary around the ring, so

$$E_y = \frac{1}{4\pi\epsilon_0}\frac{y}{(R^2 + y^2)^{3/2}}\int dQ$$

The integral of dQ around the ring is simply the net electric charge Q of the ring, so we obtain the result

$$E_y = \frac{1}{4\pi\epsilon_0}\frac{Qy}{(R^2 + y^2)^{3/2}} \tag{23.11}$$

COMMENTS: Note that if $y = 0$, then $E_y = 0$; that is, at the center of the ring of charge, the electric field is zero. That we could have deduced from symmetry, since contributions from opposite sides of the ring exactly cancel at the center of the ring. Note also the behavior far from the ring: if y is very much larger than R, then Eq. (23.11) becomes approximately

$$E_y \approx \frac{1}{4\pi\epsilon_0}\frac{Qy}{(y^2)^{3/2}} = \frac{1}{4\pi\epsilon_0}\frac{Q}{y^2}$$

which is the usual inverse-square electric field of a point charge. Thus, the ring behaves like a point charge whenever the distance y is very large.

EXAMPLE 7 Positive charge is uniformly distributed over an infinite flat horizontal sheet, such as a very large sheet of paper. Suppose that the amount of charge per unit area on this sheet is σ. Find the electric field in the space above and below the sheet.

SOLUTION: The sheet can be regarded as made up of many concentric rings. Figure 23.9 shows one of these rings, with radius R and width dR; this ring has

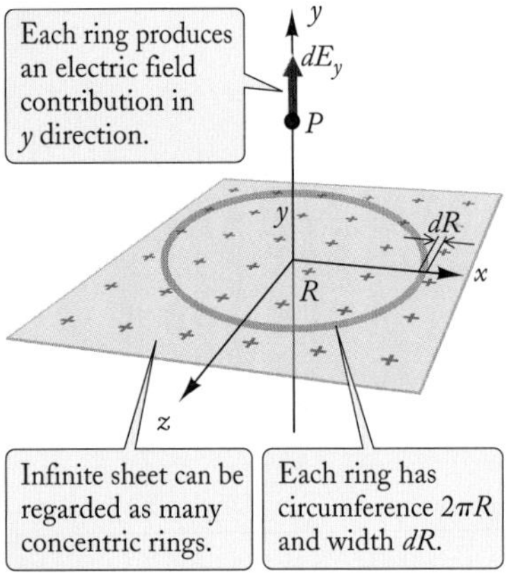

FIGURE 23.9 A very large sheet of charge lies in the x–z plane. A thin ring of charge within this sheet produces an electric field $d\mathbf{E}$.

an area dA = (length × width) = $(2\pi R)\,dR$ and, from Eq. (23.9), a charge $dQ = \sigma\,dA = \sigma \times 2\pi R\,dR$. According to Eq. (23.11), the ring produces an electric field

$$dE_y = \frac{1}{4\pi\epsilon_0}\frac{(2\pi\sigma R\,dR)y}{(R^2+y^2)^{3/2}}$$

The net electric field of the entire infinite sheet is then obtained by integrating over R, from $R = 0$ (the radius of the smallest ring) to $R = \infty$ (the radius of the largest ring):

$$E_y = \frac{\sigma y}{2\epsilon_0}\int_0^\infty \frac{R\,dR}{(R^2+y^2)^{3/2}}$$

With the substitutions $u = R^2$ and $du = 2R\,dR$, this becomes

$$E_y = \frac{\sigma y}{2\epsilon_0}\int_0^\infty \frac{\frac{1}{2}du}{(u+y^2)^{3/2}} = \frac{\sigma y}{2\epsilon_0}\left[-\frac{1}{(u+y^2)^{1/2}}\right]\Bigg|_0^\infty = \frac{\sigma y}{2\epsilon_0}\frac{1}{y}$$

or simply

electric field of flat sheet

$$E_y = \frac{\sigma}{2\epsilon_0} \tag{23.12}$$

COMMENTS: This electric field is proportional to the charge density, and it is constant, that is, it is independent of the distance from the sheet. Although the result (23.12) is strictly valid only for the case of an infinitely large sheet, it is also a good approximation for a sheet of finite size, provided we stay near the sheet and we stay away from the vicinity of the edges.

Large charged flat sheets or plates are often used in laboratories to generate uniform electric fields. In practice, two parallel charged sheets with charges of opposite signs are preferred over a single sheet. To find the field due to two sheets, consider the electric fields due to each sheet separately, and then apply the Superposition Principle to sum them. In the space between the sheets, the individual electric fields of the two

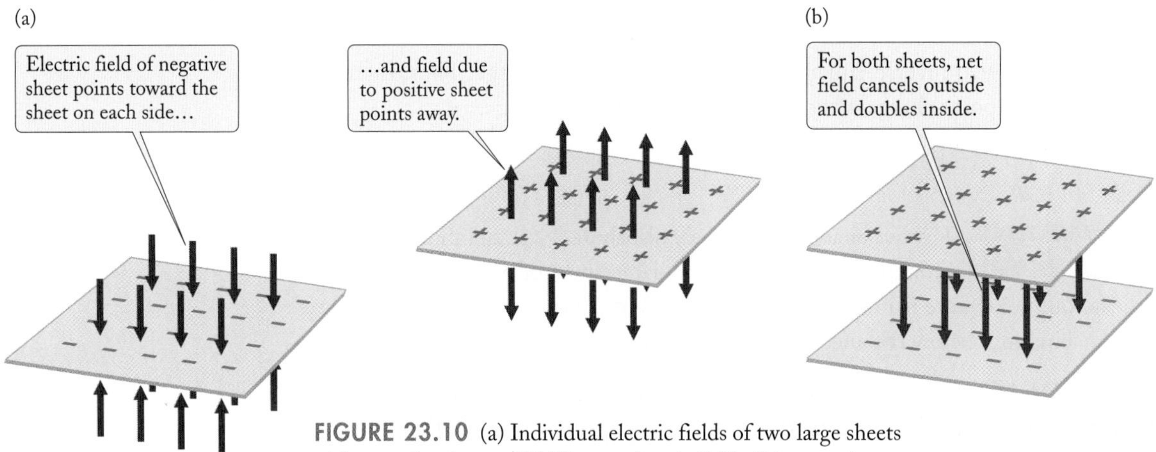

FIGURE 23.10 (a) Individual electric fields of two large sheets with opposite charges. (b) The net electric field of the two sheets.

PHYSICS IN PRACTICE ELECTROSTATIC PRECIPITATORS

Electrostatic precipitators are widely used to remove pollutants from the smoke produced by power plants, smelters, and other industrial installations. Before the installation of such cleaning devices, industrial smoke ejected huge amounts of pollutants into the atmosphere. For instance, in just one day, a large copper smelter would eject more than 3000 tons of sulfur compounds, 30 tons of arsenic compounds, and several tons of zinc, copper, lead, and antimony. The electrostatic precipitator consists of a metallic cylindrical chamber in which hangs a wire connected to a high-voltage generator (see Fig. 1). The high concentration of electric charge on the wire produces a strong electric field—as much as 10^6 N/C—in the space between the wire and the wall of the chamber. When dirty smoke, loaded with particles of pollutants, enters this strong electric field, it triggers a discharge, called a corona discharge, from the wire into the smoke. This discharge deposits negative electric charges on the particles of pollutants. The electric field then pushes the charged particles sideways and drives them against the wall, where they adhere. Every so often, a motorized mechanical hammer strikes and shakes the chamber, dislodging the accumulated layers of pollutants, which fall down into hoppers for collection and disposal.

Electrostatic precipitators used in industrial installations contain large arrays of many parallel plates and charged wires (see Fig. 2). Automatic control systems connected to the wires and the generators continually adjust the charges on the wires and the resulting electric fields to optimize the collection of pollutants. Precipitators can collect solid particles, such as fly ash, soot, or dust; they can also collect liquid droplets, such as sulfuric acid mist or tar fumes.

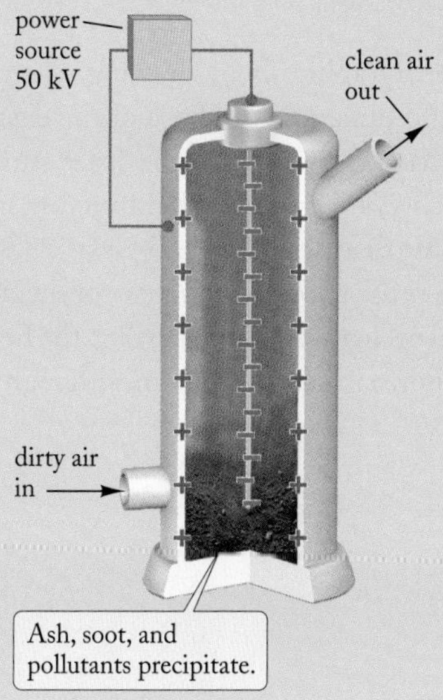

FIGURE 1 A schematic diagram of an electrostatic precipitator.

FIGURE 2 An electrostatic precipitator array.

sheets are in the same direction (see Fig. 23.10a); they therefore reinforce each other, giving a net field twice as large as the field of one sheet, that is,

$$E_y = \frac{\sigma}{\epsilon_0} \tag{23.13}$$

electric field between two oppositely charged flat sheets

In the space above or below the two sheets, the individual electric fields of the two sheets are in opposite directions; thus, these electric fields cancel, and the space above and below has zero electric field. Figure 23.10b shows the net electric field of the two sheets.

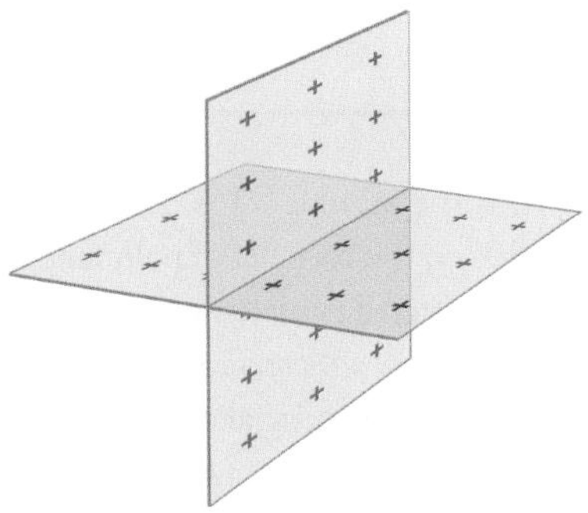

FIGURE 23.11 Two intersecting sheets of charge.

Checkup 23.2

QUESTION 1: Two infinitely large flat sheets with equal uniform positive charge distributions intersect at right angles (see Fig. 23.11). What is the direction of the electric field in each quadrant?

QUESTION 2: A small disk (approximate diameter 1 cm) has a charge Q uniformly distributed over its area A. What is the electric field far from this disk (e.g., meters away)? What is the field very close to its surface (e.g., 10^{-3} cm) and far from the edges?

QUESTION 3: The electric field at a distance of 1 m from a uniformly charged infinite sheet has the value $E = E_0$. What is the value of E at a distance of 2 m? At 4 m?

QUESTION 4: Suppose that two large parallel sheets have equal surface charge densities σ with the same sign, say, positive. What is the magnitude of the electric field in the space between the sheets? Outside the sheets?

(A) $\sigma/\epsilon_0, 0$ (B) $0, \sigma/\epsilon_0$ (C) $\sigma/\epsilon_0, \sigma/\epsilon_0$ (D) $\sigma/2\epsilon_0, \sigma/2\epsilon_0$ (E) $0, 0$

23.3 LINES OF ELECTRIC FIELD

The electric field can be represented graphically by drawing, at any given point of space, a vector whose magnitude and direction are those of the electric field at that point. Figure 23.12 shows the electric field vectors in the space surrounding a positive point charge, and Fig. 23.13 shows the electric field vectors in the space surrounding a negative point charge. Note that, as demanded by Eq. (23.8), the magnitude of these vectors decreases in inverse proportion to the square of the distance from the point charge. Also note that, to avoid overcrowding the drawing, the field vectors at only a few representative points can be shown in such a diagram, although the electric field has a value at every point in space.

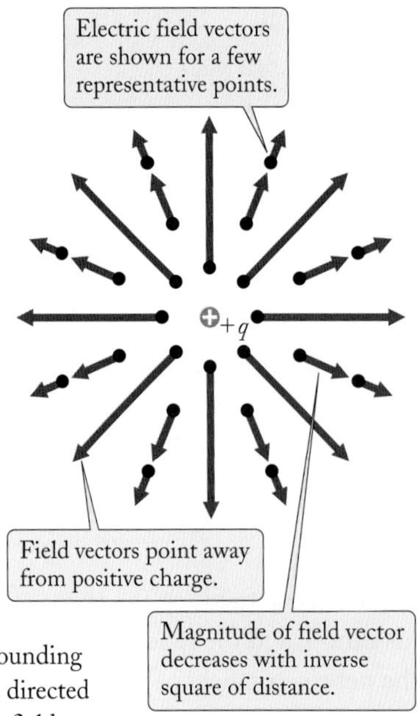

FIGURE 23.12 Electric field vectors surrounding a positive point charge. The field vectors are directed radially outward. The positions to which the field vectors belong are marked in black.

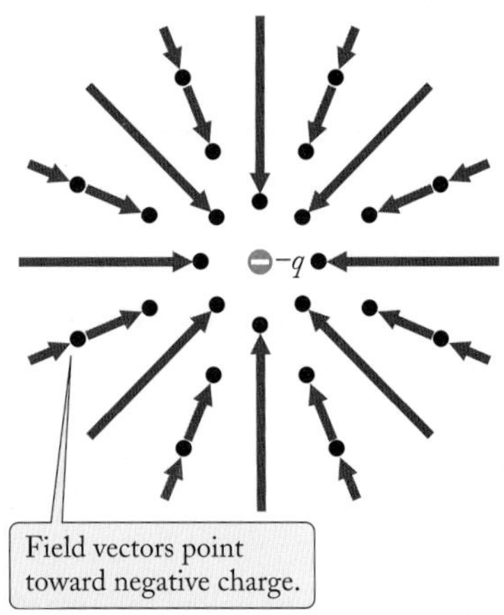

FIGURE 23.13 Electric field vectors surrounding a negative point charge. The field vectors are directed radially inward.

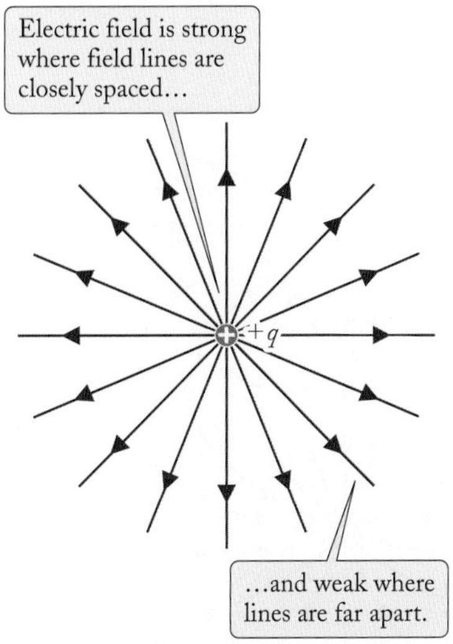

FIGURE 23.14 Electric field lines of a positive point charge. Note that in three dimensions the lines spread out in all three directions of space, whereas the diagram shows the lines spreading out in only the two directions within the page. This gives a misleading impression of the density of field lines as a function of distance. This limitation of two-dimensional representations should be kept in mind when looking at diagrams of field lines.

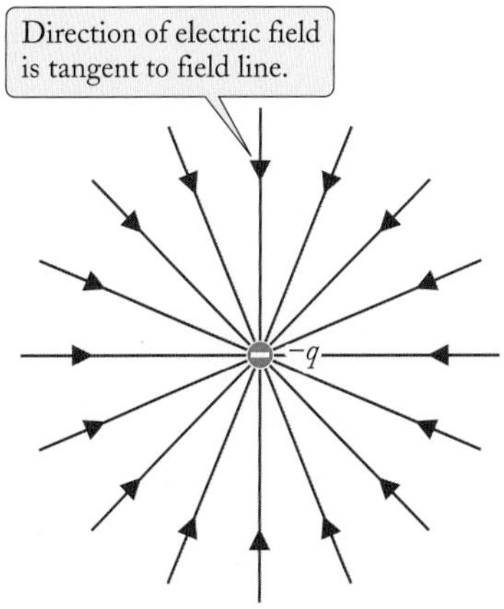

FIGURE 23.15 Electric field lines of a negative point charge.

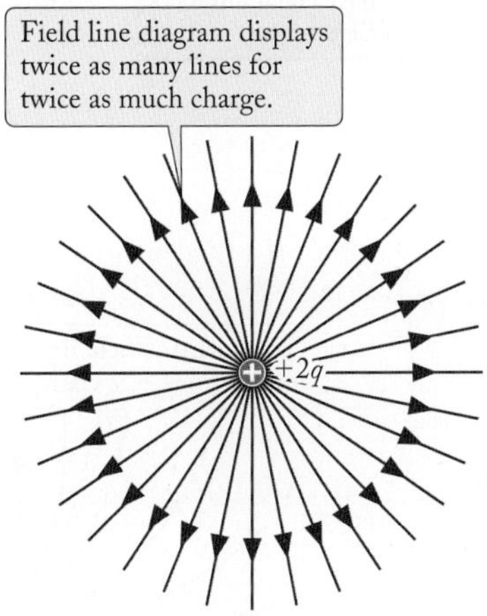

FIGURE 23.16 Electric field lines of a positive point charge twice as large as in Fig. 23.14.

field lines

Alternatively, the electric field can be represented graphically by **field lines**. These lines are drawn in such a way that, at any given point, the tangent to the line has the direction of the electric field, that is, the direction of the field vector. Furthermore, the density of lines is directly proportional to the magnitude of the electric field; that is, where the lines are closely spaced the electric field is strong, and where the lines are far apart the electric field is weak. Figure 23.14 shows the electric field lines of a positive point charge and Fig. 23.15 shows those of a negative point charge. The arrows on these lines indicate the direction of the electric field along each line.

When we draw a pattern of field lines, we must begin each line on a positive point charge and end on a negative point charge (or the line may extend to infinity). Since the magnitude of the electric field is directly proportional to the amount of electric charge, the number of field lines that we draw emerging from a positive point charge must be proportional to the amount of charge. Figure 23.16 shows the electric field lines of a positive point charge twice as large as that of Fig. 23.14. Figure 23.17 shows the field lines generated jointly by a positive and a negative charge of equal magnitudes (an electric dipole), and Fig 23.18 shows those of a pair of equal positive charges (note the equal number of field lines in each case). Figure 23.19 shows the field lines of a pair of unequal positive and negative charges. Note that there are 2 times as many field lines originating on the $+2q$ charge (32 lines) as there are ending on the $-q$ charge (16 lines). If desired, we can vary the number of field lines actually drawn, but we must

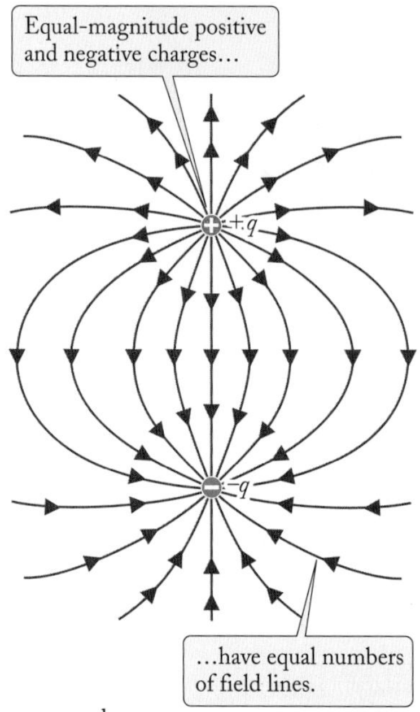

FIGURE 23.17 Field lines generated by positive and negative charges of equal magnitudes.

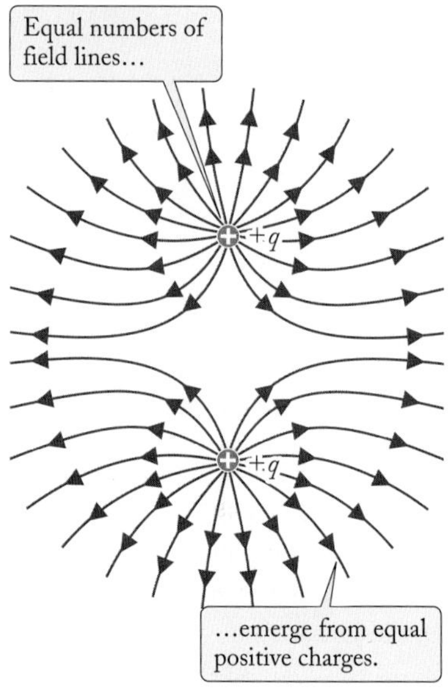

FIGURE 23.18 Field lines generated by two positive charges of equal magnitudes.

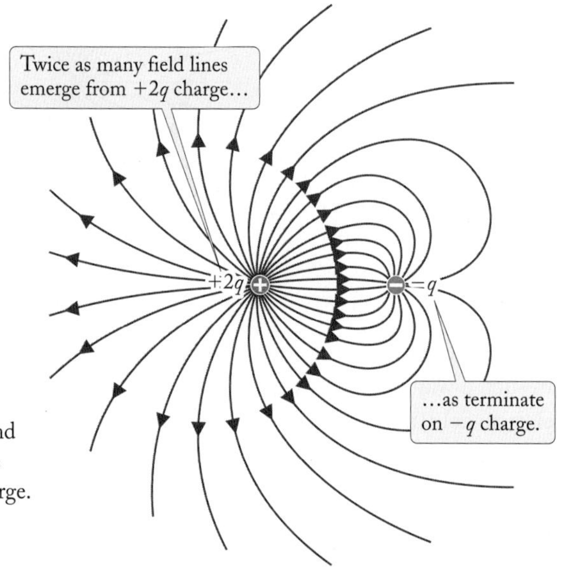

FIGURE 23.19 Field lines generated by positive and negative charges of unequal magnitudes. The positive charge has 2 times the magnitude of the negative charge.

be careful to maintain a fixed ratio of field lines, proportional to charge. It is necessary to keep such a normalization fixed through any given computation or series of drawings.

Note that the field lines start on positive charges and end on negative charges—the positive charges are sources of field lines and the negative charges are sinks. Also, note that the field lines never intersect (except where they start or end on point charges). If the lines ever were to intersect, the electric field would have *two* directions at the point of intersection; this is impossible.

The above pictures of field lines help us to develop some intuitive feeling for the spatial dependence of the electric fields surrounding diverse arrangements of electric charges. But we must refrain from thinking of the field lines as physical objects. The

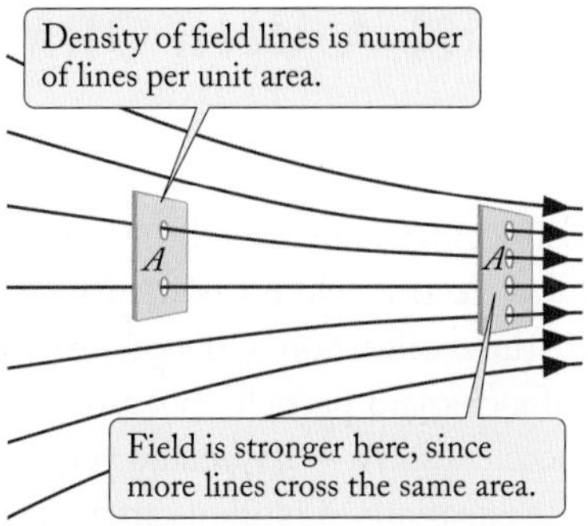

FIGURE 23.20 Two identical areas A intercept different numbers of field lines.

electric field is a physical entity, it is a form of matter; but the field lines are merely mathematical crutches to aid our imagination.

We said the electric field is proportional to the density of field lines. The precise definition is that the density of lines is the number of lines per unit area, that is, the number of lines intercepted by a small area A erected perpendicularly to the lines (see Fig. 23.20) divided by the magnitude of this area. So if a greater number of field lines cross a given amount of area A at one location than at a second location, the electric field is proportionally stronger at the first location.

The inverse-square law for the electric field of a point charge can be "derived" from the picture of field lines. By a simple argument, we can show that the density of lines necessarily obeys an inverse-square law. Since a point charge q is spherically symmetric and makes no distinction between one radial direction and another, symmetry tells us that the arrangement of field lines must also be spherically symmetric, and they must be uniformly distributed over all radial directions. At a distance r from the point charge, these lines are uniformly distributed over the area $A = 4\pi r^2$. For larger concentric spheres, the same number of field lines is distributed over increasingly larger areas; hence the density of lines decreases in proportion to the inverse square of the distance. This "derivation" of the inverse-square law from the picture of field lines is really no more than a consistency check—the electric field can be represented by field lines only because it is an inverse-square field; any other dependence on distance would make it impossible to draw continuous field lines that start and end only on charges.

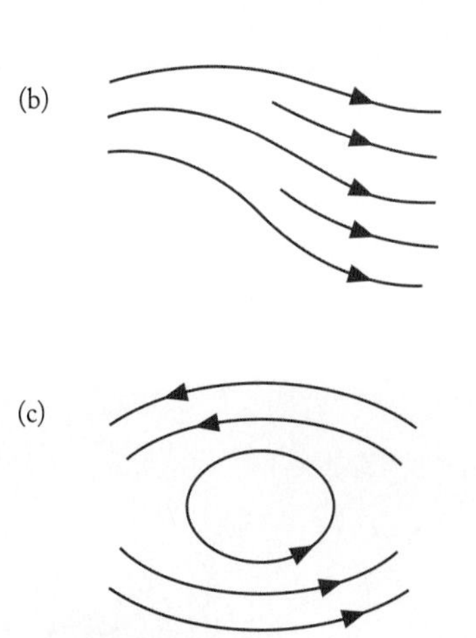

FIGURE 23.21 Hypothetical electric field lines; the patterns are incorrect if due to static charges beyond the edges of the diagram.

Checkup 23.3

QUESTION 1: Figure 23.21 shows diagrams of hypothetical (but incorrect) field lines corresponding to some static charge distributions, which are beyond the edge of the diagram. What is wrong with each of these diagrams of field lines?

QUESTION 2: A field line diagram shows a point charge with eight field lines emerging from it. Nearby in the same diagram is a metal sphere with four lines emerging and two lines entering the sphere. What can we say about the arrangement of charge on the sphere? The net charge on the sphere?

QUESTION 3: During a day of fair weather, the Earth has an atmospheric electric field that points vertically down. This electric field is due to charges distributed over the surface of the Earth. What must be the sign of these charges?

QUESTION 4: A field line diagram shows a +6.0 C point charge with eight field lines emerging from it. Nearby in the diagram is a short, charged rod with 12 field lines entering it. The charge on the rod is

(A) +4.0 C (B) −4.0 C (C) +9.0 C (D) −9.0 C (E) −12.0 C

23.4 MOTION IN A UNIFORM ELECTRIC FIELD

We have now discussed the electric field and the electric force at some length. A charged particle in a known electric field **E** experiences an electric force $\mathbf{F} = q\mathbf{E}$, and from this we can calculate the acceleration of the charged particle, as in Example 4. To find the actual motion of a charged particle (position **r** as a function of time t), we apply the definitions of acceleration $\mathbf{a} = d\mathbf{v}/dt$ and velocity $\mathbf{v} = d\mathbf{r}/dt$. As we learned in mechanics, solving these equations of motion can in general be quite difficult; however, the simplest case of a constant acceleration is easy. Recall that for a constant acceleration in two dimensions, the solution for the individual components of the velocity, Eqs. (4.20)–(4.21), gives

$$v_x = v_{0x} + a_x t \quad \text{and} \quad v_y = v_{0y} + a_y t \tag{23.14}$$

where v_{0x} and v_{0y} are the initial velocity components, $a_x = F_x/m = qE_x/m$, and $a_y = F_y/m = qE_y/m$. The solution for the components of the position, Eqs. (4.23)–(4.24), yields

$$x = x_0 + v_{0x}t + \tfrac{1}{2}a_x t^2 \quad \text{and} \quad y = y_0 + v_{0y}t + \tfrac{1}{2}a_y t^2 \tag{23.15}$$

where x_0 and y_0 specify the initial position. You might remember from Chapter 4 that in the case of a uniform gravitational field, the result was projectile motion, where the position of the particle traces out a parabola in two dimensions.

The same is true for uniform electric fields. As we saw above, two parallel sheets of charge with equal and opposite surface charge densities produce a region of constant electric field. This arrangement, typically attained with metal plates, is exploited in many electric devices. One difference with projectile motion is that electric charges can be positive or negative (and electric fields can point in any direction), so the parabolic path is not always concave down, as it is with gravity. Another feature is that, in practice, the charged particle's motion can be manipulated only in the limited region of space between the parallel sheets of charge.

To calculate the motion of a charged particle in a uniform **E** field, we use Eqs. (23.14) and (23.15). If the only motion is along the applied field, the problem is one-dimensional and simple. But if there is motion both along the field and perpendicular to it, two equations will probably need to be solved simultaneously. We can best simplify such problems if we choose our coordinate system with one axis parallel to the field; then, the motion along the other direction (perpendicular to the field) has zero acceleration.

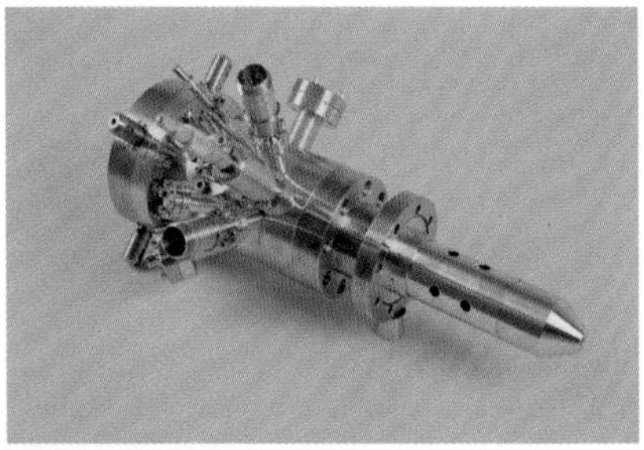

FIGURE 23.22 Ion gun from an ion-milling machine.

EXAMPLE 8 An ion milling machine (see Fig. 23.22) uses a beam of gallium ions ($m = 70$ u) to carve microstructures from a target. A region of uniform electric field between parallel sheets of charge is used for precise control of the beam direction. Singly ionized gallium atoms with an initially horizontal velocity of 1.8×10^4 m/s enter a 2.0-cm-long region of uniform electric field which points vertically upward as shown in Fig. 23.23. The ions are redirected by the field and exit the field region at the angle θ shown. If the field is set to a value of $E = 90$ N/C, what is the exit angle θ?

SOLUTION: A singly ionized atom has charge $q = +e$. It is convenient to choose the origin of our coordinate system at the initial ion position at the left end of the field region ($x_0 = 0$, $y_0 = 0$) and the x axis horizontal. The desired exit angle θ is

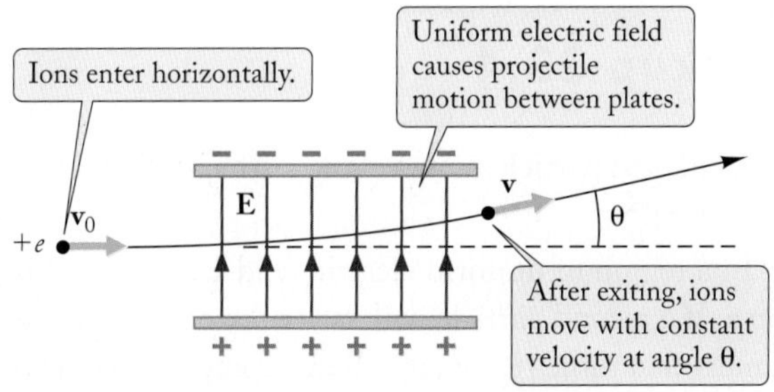

FIGURE 23.23 Motion of a charged particle; between the parallel plates is a region of uniform electric field.

given by the final direction of motion of the beam as it exits the electric field region:

$$\tan\theta = \frac{v_y}{v_x}$$

There is no horizontal electric field, so the x component of the acceleration is zero; Eq. (23.14) confirms that the x component of the velocity does not change, so

$$v_x = v_{0x}$$

The initial y component of the velocity is zero, so Eq. (23.14) for the y component of the velocity yields

$$v_y = a_y t$$

To obtain the time t that the ions spend in the region of uniform electric field, we rearrange Eq. (23.15) for the x component of the position, with $a_x = 0$:

$$t = \frac{x}{v_{0x}}$$

where x is the width of the field region. Inserting this expression for t in the previous equation, we have

$$v_y = a_y \frac{x}{v_{0x}}$$

Using $a_y = F_y/m = eE/m$, we find

$$v_y = \frac{eEx}{mv_{0x}}$$

We can now determine the exit angle:

$$\tan\theta = \frac{v_y}{v_x} = \frac{1}{v_{0x}}\frac{eEx}{mv_{0x}}$$
$$= \frac{eEx}{mv_{0x}^2}$$

Inserting the given numerical values and the value of the atomic mass unit, 1 u = 1.66×10^{-27} kg, we obtain

$$\tan\theta = \frac{eEx}{mv_{0x}^2} = \frac{1.60 \times 10^{-19}\ \text{C} \times 90\ \text{N/C} \times 2.0 \times 10^{-2}\ \text{m}}{70 \times 1.66 \times 10^{-27}\ \text{kg} \times (1.8 \times 10^4\ \text{m/s})^2}$$
$$= 7.6 \times 10^{-3}$$

Using a calculator, we take the inverse tangent of 7.6×10^{-3} and obtain

$$\theta = 0.44°$$

Checkup 23.4

QUESTION 1: A charged particle moves along a straight line. Does this mean that no electric field is present?

QUESTION 2: In a region of uniform electric field **E**, a charged particle experiences an acceleration **a**. If a second particle with twice the charge and twice the mass of the first particle enters that same region, it will experience an acceleration

(A) $\frac{1}{4}\mathbf{a}$ (B) $\frac{1}{2}\mathbf{a}$ (C) $\mathbf{a}$ (D) $2\mathbf{a}$ (E) $4\mathbf{a}$

23.5 ELECTRIC DIPOLE IN AN ELECTRIC FIELD

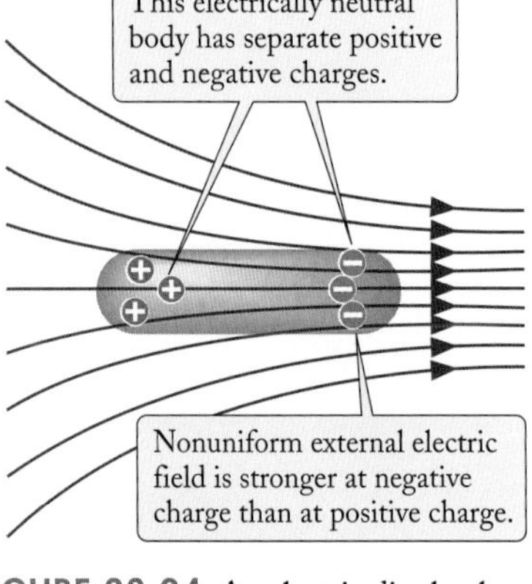

FIGURE 23.24 An electric dipole placed in a nonuniform electric field.

If an electrically neutral body is placed in a given electric field, we might expect that the body experiences no force. However, this expectation is not always realized. A neutral body may contain within it separate positive and negative charges (of equal magnitudes), and it is possible that the electric force on one of these charges is larger than that on the other; the body then experiences a net force. Such an imbalance of the forces on the positive and negative charges will happen if the electric field is stronger at the location of one kind of charge than at the location of the other. For example, in a nonuniform electric field, the neutral body shown in Fig. 23.24 with positive charges on one end and negative charges on the other end will be pushed to the left because the electric field that acts on the body is stronger at the location of the negative charges. Note that this electric field—indicated by the field lines in Fig. 23.24—is not the field generated by the body; rather it is an electric field generated by some other charges (not shown in Fig. 23.24). The electric field that acts on a body is usually called the **external field**; in contrast, the field generated by the body itself is called the **self-field**. The latter field exerts only internal forces within the body and does not contribute to the net force acting on the body from the outside.

If the external electric field is uniform, such as the electric field generated by the large flat charged sheets discussed in the preceding sections, then the forces on the positive and negative charges in a neutral body cancel, and there is no net force. However, there may still be a torque. Figure 23.25 shows a neutral body, with positive charge on one end and negative charge on the other end, in a uniform electric field. The body carries equal positive and negative charges $\pm Q$, with the average positions of these charges separated by a distance l. As we already mentioned in Example 1, such a body is called an **electric dipole**. Since the force on the positive charge in Fig. 23.25 is toward the right and the force on the negative charge is toward the left, there is a torque on the body. The torque τ of each force about the center of the body is $\tau = rF\sin\theta$, that is, the product of the distance from the center ($r = \frac{1}{2}l$), the magnitude of the force ($F = QE$), and the sine of the angle between the line from the center and the force ($\sin\theta$). Thus, the magnitude of the torque due to each force is $\tau = \frac{1}{2}lQE\sin\theta$, and the net torque of both forces together is

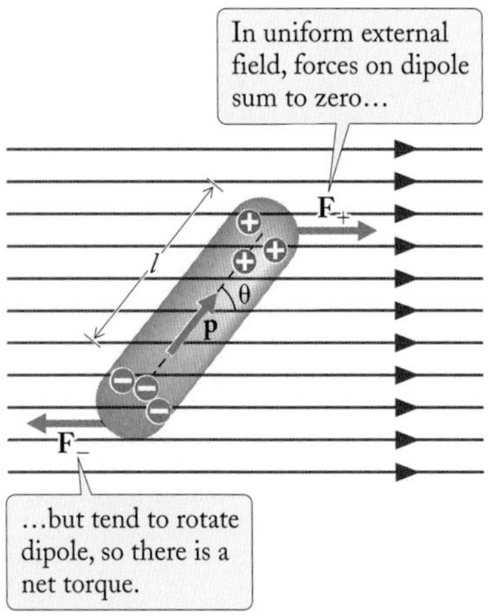

FIGURE 23.25 Electric dipole placed in a uniform electric field. The dipole moment vector **p** is directed from the negative charge to the positive charge.

$$\tau = -lQE\sin\theta \qquad (23.16)$$

The minus sign in this equation is a reminder that here the torque is clockwise, in the sense of negative angles. The torque tends to align the body with the electric field.

We can write Eq. (23.16) as

$$\tau = -pE \sin\theta \tag{23.17}$$

torque on electric dipole

where

$$p = lQ \tag{23.18}$$

dipole moment

The quantity p is called the **dipole moment** of the body; it is simply the magnitude of the charge at either end multiplied by the separation between the charges. The units of dipole moment are coulomb-meters (C·m). We can also introduce a **dipole moment vector p** whose magnitude is equal to the dipole moment p and whose direction is from the negative charge toward the positive charge. Then the torque vector can be simply expressed in terms of the vector **p** as

dipole moment vector p

$$\boldsymbol{\tau} = \mathbf{p} \times \mathbf{E} \tag{23.19}$$

where the direction of $\boldsymbol{\tau}$ is, as usual, the axis of rotation.

Corresponding to the torque (23.17), there exists a potential energy that equals the amount of work that *you* must do against the electric forces to twist the dipole through some angle. If we choose $\theta = 90°$ as a reference position (when the dipole is perpendicular to the field), then, from Eq. (13.4), the work done to rotate the dipole to some other angle θ is

$$W = \int_{90°}^{\theta} \tau\, d\theta = -pE \int_{90°}^{\theta} \sin\theta\, d\theta = pE\cos\theta \Big|_{90°}^{\theta} = pE(\cos\theta - 0)$$

$$= pE\cos\theta$$

The potential energy is the negative of the work done, $U = -W$, so

$$U = -pE\cos\theta \tag{23.20}$$

potential energy of dipole

In terms of the dipole moment vector **p**, the quantity $pE\cos\theta$ is $\mathbf{p}\cdot\mathbf{E}$, and the potential energy U becomes

$$U = -\mathbf{p}\cdot\mathbf{E} \tag{23.21}$$

The potential energy (23.20) or (23.21) has a minimum ($-pE$) when the dipole is oriented parallel to the electric field ($\theta = 0$), and it has a maximum ($+pE$) when the dipole is oriented antiparallel to the electric field ($\theta = 180°$). Figure 23.26 is a plot of the potential energy vs. the angle θ. The equilibrium orientation of the dipole is the parallel orientation, corresponding to the minimum of the potential energy.

Many asymmetric molecules have **permanent dipole moments** due to an excess of electrons on one end of the molecule and a corresponding deficit on the other. This means that the molecule has an accumulation of negative charge on one end and of positive charge on the other. For example, Fig. 23.27 shows the structure of a water molecule. In this molecule, the electrons tend to concentrate on the oxygen atom; in Fig. 23.27, the left side of the molecule is negatively charged, and the right side positively charged. Because of this separation of the average positions of the negative and positive charges, the molecule has a dipole moment. The direction of the dipole vector is from the negative side of the molecule toward the positive side. For a water molecule in water vapor, the dipole moment is $p = 6.1 \times 10^{-30}$ C·m.

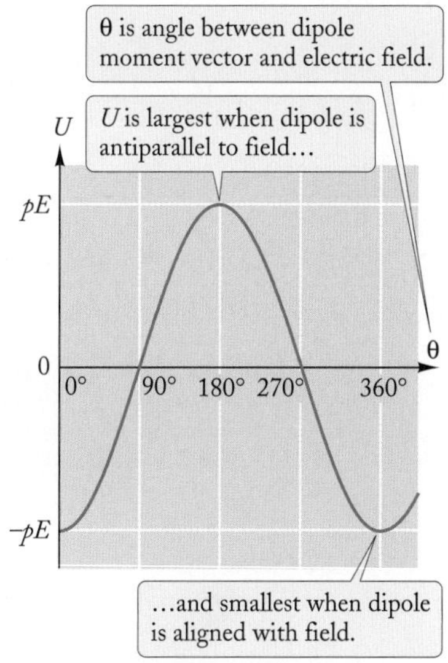

FIGURE 23.26 Potential energy of an electric dipole as a function of angle.

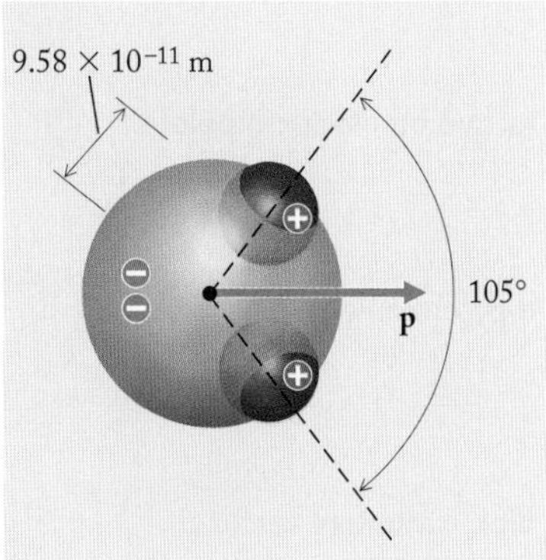

FIGURE 23.27 Arrangement of the atoms of hydrogen (red) and oxygen (blue) in a water molecule. The direction of the dipole moment vector **p** is from the negative side of the molecule toward the positive side.

EXAMPLE 9 A molecule of water is placed in an electric field of 2.0×10^5 N/C. What is the difference between the potential energies for the parallel and the antiparallel orientations of the molecule in this electric field?

SOLUTION: As given just above, the dipole moment of the water molecule is $p = 6.1 \times 10^{-30}$ C·m. When the dipole is parallel to the electric field, the angle θ in Eq. (23.20) is zero, and

$$U = -pE\cos 0° = (-6.1 \times 10^{-30}\ \text{C·m})(2.0 \times 10^5\ \text{N/C})(1)$$
$$= -1.2 \times 10^{-24}\ \text{J}$$

When the dipole is antiparallel to the electric field, the angle is 180°, and

$$U = -pE\cos 180° = (-6.1 \times 10^{-30}\ \text{C·m})(2.0 \times 10^5\ \text{N/C})(-1)$$
$$= 1.2 \times 10^{-24}\ \text{J}$$

Hence the energy difference between the parallel and antiparallel orientations is 2.4×10^{-24} J.

Molecules of atoms that do not have a permanent dipole moment may acquire a temporary dipole moment when placed in an electric field. The opposite electric forces on the positive and negative charges can distort the molecule and produce a charge separation. Such a dipole moment, which lasts only as long as the molecule is immersed in the electric field, is called an **induced dipole moment**. The magnitude of the induced dipole moment is usually directly proportional to the magnitude of the electric field causing it. This *linear* behavior of the induced dipole moment is useful in devices that store electric charge, discussed in Chapter 26.

induced dipole moment

✔ Checkup 23.5

QUESTION 1: The torque on a dipole in an electric field depends on the angle between the dipole and the electric field. For what angle is the torque maximum?

QUESTION 2: For what angle between the dipole and the electric field is the potential energy zero?

QUESTION 3: Consider a body with an induced dipole moment. If the body is in a nonuniform electric field (stronger in one region than in another), would you expect the body to be pulled into the region of stronger electric field or pushed into the region of weaker electric field?

QUESTION 4: The permanent electric dipole moment vector of a body is directed along the $+x$ axis. An electric field is applied in the $+y$ direction. The torque on the dipole will cause the dipole moment vector to rotate toward which direction?

(A) $-x$ (B) $+y$ (C) $-y$ (D) $+z$ (E) $-z$

SUMMARY

PROBLEM-SOLVING TECHNIQUES Electric Fields **(page 728)**

PROBLEM-SOLVING TECHNIQUES Electric Field of Charge Distribution **(page 732)**

PHYSICS IN PRACTICE Electrostatic Precipitators **(page 735)**

SUPERPOSITION PRINCIPLE Electric forces and electric fields produced by different charges or by different charge distributions combine by vector addition.

DEFINITION OF ELECTRIC FIELD

$$\mathbf{E} = \frac{\mathbf{F}}{q} \qquad (23.5)$$

SI UNIT OF ELECTRIC FIELD 1 newton/colomb = 1 N/C

ELECTRIC FIELD OF POINT CHARGE q'
Direction is radially outward for positive charge, radially inward for negative charge.

$$E = \frac{1}{4\pi\epsilon_0}\frac{q}{r^2} \qquad (23.2)$$

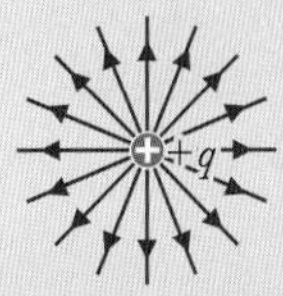

ELECTRIC FIELD OF A CONTINUOUS CHARGE DISTRIBUTION Magnitude of contribution:

$$dE = \frac{1}{4\pi\epsilon_0}\frac{dq}{r^2} \qquad (23.7)$$

The total field $\mathbf{E} = E_x\mathbf{i} + E_y\mathbf{j} + E_z\mathbf{k}$ is obtained by summing contributions to each component; for example $E_x = \int dE_x$, with:

$$dE_x = \cos\theta\, dE = \frac{1}{4\pi\epsilon_0}\cos\theta\frac{dq}{r^2} \qquad (23.8)$$

where θ is the angle between the electric field contribution and the x axis.

LINEAR AND SURFACE CHARGE DISTRIBUTIONS (23.9)

Linear: $dq = \lambda\, dL$ where λ is in coulombs per meter.

Surface: $dq = \sigma dA$ where σ is in coulombs per square meter.

ELECTRIC FIELD OF INFINITE, UNIFORMLY CHARGED THIN ROD
Direction is perpendicular to rod (outward for $+\lambda$, inward for $-\lambda$).

$$E = \frac{1}{2\pi\epsilon_0}\frac{\lambda}{r} \qquad (23.10)$$

ELECTRIC FIELD OF INFINITE, UNIFORMLY CHARGED FLAT SHEET
Direction is perpendicular to sheet (outward for $+\sigma$, inward for $-\sigma$).

$$E = \frac{\sigma}{2\epsilon_0} \qquad (23.12)$$

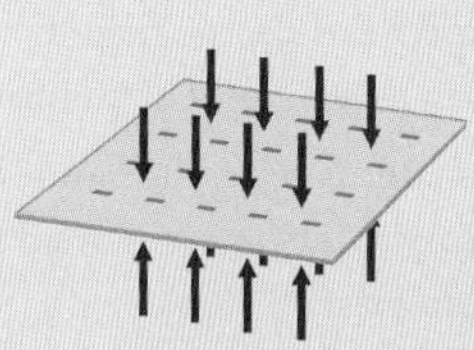

ELECTRIC FIELD OF A PAIR OF OPPOSITELY CHARGED, PARALLEL FLAT SHEETS

$$E = \frac{\sigma}{\epsilon_0} \quad \text{(between sheets)}$$

$$E = 0 \quad \text{(outside sheets)}$$

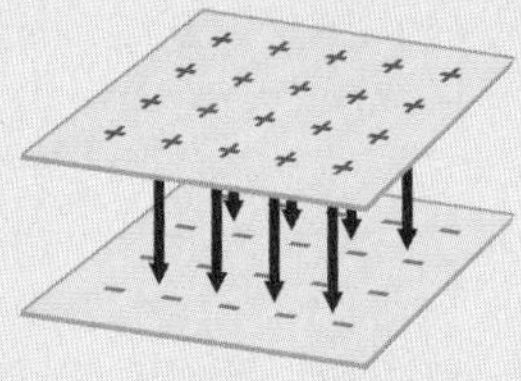

PROPERTIES OF ELECTRIC FIELD LINES

Lines are tangent to the electric field vector at any point.

Density of lines is proportional to the magnitude of the field.

Field lines do not cross.

Field lines start on positive charges and end on negative charges.

The number of field lines emerging from (terminating on) a positive (negative) charge is proportional to the charge.

MOTION IN UNIFORM E

$$\mathbf{a} = \frac{\mathbf{F}}{m} = \frac{q\mathbf{E}}{m} \tag{23.14}$$

$$\mathbf{v} = \mathbf{v}_0 + \mathbf{a}t$$

$$\mathbf{r} = \mathbf{r}_0 + \mathbf{v}_0 t + \tfrac{1}{2}\mathbf{a}t^2$$

ELECTRIC DIPOLE MOMENT

Direction of dipole moment vector **p** is from − to +.

$$p = lQ \tag{23.18}$$

TORQUE ON DIPOLE

$$\tau = -pE\sin\theta \quad \text{or} \quad \boldsymbol{\tau} = \mathbf{p} \times \mathbf{E} \tag{23.17, 23.19}$$

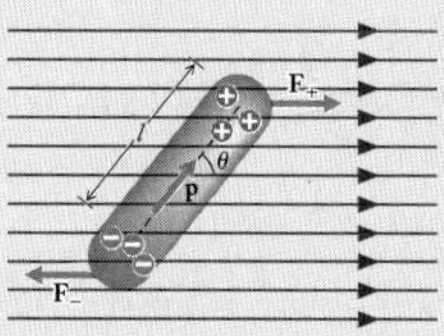

POTENTIAL ENERGY OF DIPOLE

$$U = -pE\cos\theta \quad \text{or} \quad U = -\mathbf{p}\cdot\mathbf{E} \tag{23.20, 23.21}$$

QUESTIONS FOR DISCUSSION

1. Does it make any difference whether the value of the charge q in the equation defining the electric field [Eq. (23.5)] is positive or negative?
2. How would you formally define a gravitational field vector? Is the unit of gravitational field the same as the unit of electric field? According to your definition, what are the magnitude and the direction of the gravitational field at the surface of the Earth?
3. A large, flat sheet measures $L \times L$; the sheet carries a uniform distribution of charge. Roughly how far from the center of the sheet would you expect the electric field to be markedly different from the uniform electric field of an infinitely large sheet?
4. If a positive point charge is released from rest in an electric field, will its trajectory coincide with a field line? What if the point charge has zero mass?
5. A **tube of force** is the volume enclosed by a bundle of adjacent field lines (Fig. 23.28). Such a tube of force is analogous to a stream tube in hydrodynamics, and the field lines are analogous to streamlines. Along such a tube of force, the magnitude of the electric field varies in inverse proportion to the cross-sectional area of the tube. Explain.

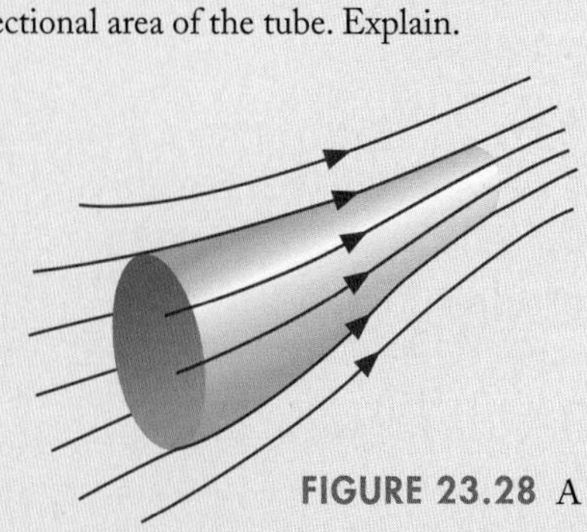

FIGURE 23.28 A tube of force.

6. A negative point charge $-q$ sits in front of a very large, flat sheet with a uniform distribution of positive charge. Make a rough sketch of the pattern of field lines. Is there any point where the electric field is zero?

7. A very long straight line of positive charge lies along the z axis. A very large flat sheet of positive charge lies in the x–y plane. Sketch a few of the field lines of the net electric field produced by both of these charge distributions acting together.

8. A large, flat, thick slab of insulator has positive charge uniformly distributed over its volume. Sketch the field lines on both sides and inside the slab; pay careful attention to the starting points of the field lines.

9. If our Universe is topologically closed, so that a straight line drawn in any direction returns on itself from the opposite direction, can the net charge in the Universe be different from zero?

10. How could you build a "compass" that indicates the direction of the electric field?

11. When a neutral metallic body, insulated from the ground, is placed in an electric field, it develops a charge separation, acquiring positive charge on one end and negative charge on the other. This means the body acquires an induced dipole moment. How is the direction of this dipole moment related to the direction of the electric field?

12. One electric dipole is at the origin, oriented parallel to the z axis. Another electric dipole is at some distance on the x axis. The electric field of the first dipole then exerts a torque on the second dipole. For what orientation of the second dipole is the potential energy minimum?

PROBLEMS

†23.1 The Electric Field of Point Charges

1. Electric fields as large as 3.4×10^5 N/C have been measured by airplanes flying through thunderclouds. What is the force on an electron exposed to such a field? What is the acceleration?

2. An electron moving through an electric field is observed to have an acceleration of 1.0×10^{16} m/s^2 in the x direction. What must be the magnitude and the direction of the electric field that produces this acceleration?

3. Example 2 gives the x and y components and the magnitude of the electric field near a thundercloud. What is the direction of this electric field?

4. The Earth has an atmospheric electric field. During days of fair weather (no thunderclouds), this atmospheric electric field has a strength of about 100 N/C and points down. Taking into account this electric field and also gravity, what will be the acceleration (magnitude and direction) of a grain of dust of mass 1.0×10^{-18} kg carrying a single *electron* charge?

5. **Millikan's experiment** measures the elementary charge e by the observation of the motion of small oil droplets in an electric field. The oil droplets are charged with one or several elementary charges, and if the (vertical) electric field has the right magnitude, the electric force on the droplet will balance its weight, holding the drop suspended in midair. Suppose that an oil droplet of radius 1.0×10^{-4} cm carries a single elementary charge. What electric field is required to balance the weight? The density of oil is 0.80 g/cm^3.

6. In an X-ray tube, electrons are exposed to an electric field of 8.0×10^5 N/C. What is the force on an electron? What is its acceleration?

7. According to a theoretical estimate, at the surface of a neutron star of mass 1.4×10^{30} kg and radius 1.0×10^4 m there is an electric field of magnitude 6.0×10^3 N/C pointing vertically up. Compare the electric force on a proton with the gravitational force on the proton.

8. A long hair, taken from a girl's braid, has a mass of 1.2×10^{-3} g. The hair carries a charge of 1.3×10^{-9} C distributed along its length. If we want to suspend this hair in midair, what (uniform) electric field do we need?

9. Electric breakdown (sparks) occurs in air if the electric field reaches 3.0×10^6 N/C. At this field strength, free electrons present in the atmosphere are quickly accelerated to such large speeds that upon impact on atoms they knock electrons off the atom and thereby generate an avalanche of electrons. How far must a free electron move under the influence of the above electric field if it is to attain a kinetic energy of 3.0×10^{-19} J (which is sufficient to produce ionization)?

10. The nuclei of the atoms in a chunk of metal lying on the surface of the Earth would fall to the bottom of the metal if their weight were the only force acting on them. Actually, within the interior of any metal exposed to gravity there exists a very small electric field that points vertically up. The corresponding electric force on a nucleus just balances the weight of the nucleus. Show that for a nucleus of atomic number Z and mass m, the required field has a magnitude $mg/(Ze)$. What is the numerical value of this electric field in a chunk of iron?

11. The hydrogen atom has a radius of 5.3×10^{-11} m. What is the magnitude of the electric field that the nucleus of the atom (a proton) produces at this radius?

†For help, see Online Concept Tutorial 27 at www.wwnorton.com/physics

12. What is the strength of the electric field at the surface of a uranium nucleus? The radius of the nucleus is 7.4×10^{-15} m and the electric charge is $92e$. For the purposes of this problem the electric charge may be regarded as concentrated at the center.

13. Figure 23.29 shows the arrangement of nuclear charges (positive charges) of a KBr molecule. Find the electric field that these charges produce at the center of mass at a distance of 9.3×10^{-11} m from the Br atom.

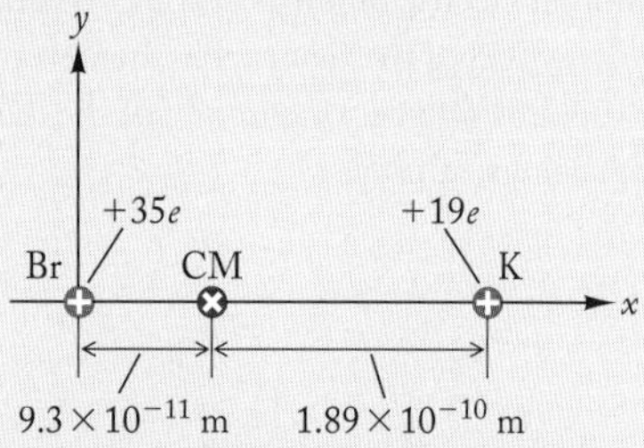

FIGURE 23.29 The positive (nuclear) charges in a KBr molecule.

14. Suppose that in a hydrogen atom the electron is (instantaneously) at a distance of 2.1×10^{-10} m from the proton. What is the net electric field that the electron and the proton produce jointly at a point midway between them?

*15. If a 1.0×10^{-10} C charge is placed on the x axis 0.15 m from the origin of a coordinate system (see Fig. 23.30), what is the magnitude of the electric field at a point 0.10 m up the y axis?

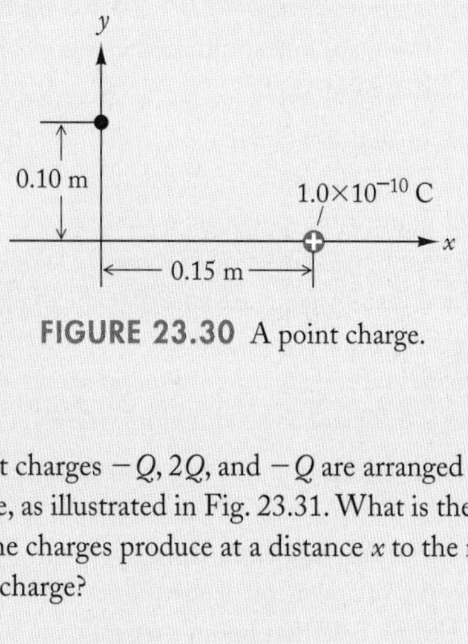

FIGURE 23.30 A point charge.

*16. Three point charges $-Q$, $2Q$, and $-Q$ are arranged on a straight line, as illustrated in Fig. 23.31. What is the electric field that the charges produce at a distance x to the right of the central charge?

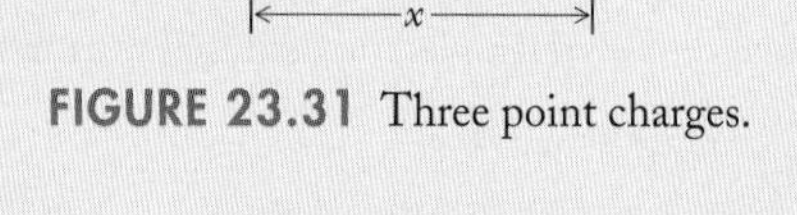

FIGURE 23.31 Three point charges.

*17. Four point charges $\pm Q$ are arranged at the corners of a square of side L as shown in Fig. 23.32. Consider the midpoint of each of the four sides of the square. At each midpoint, find the magnitude and the direction of the electric field.

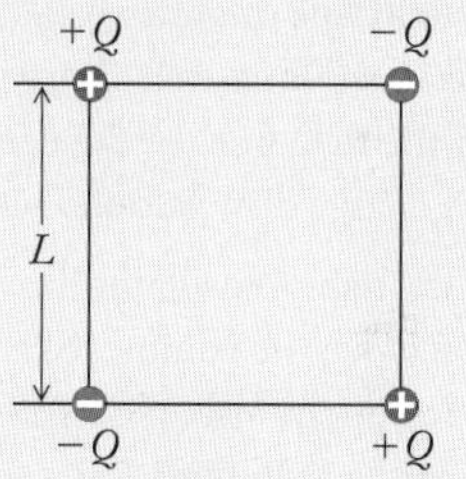

FIGURE 23.32 Four point charges.

*18. The distance between the oxygen nucleus and each of the hydrogen nuclei in an H_2O molecule is 9.58×10^{-11} m; the angle between the hydrogen atoms is 105° (Fig. 23.33). Find the electric field produced by the nuclear charges (positive charges) at the point P at a distance of 1.20×10^{-10} m to the right of the oxygen nucleus.

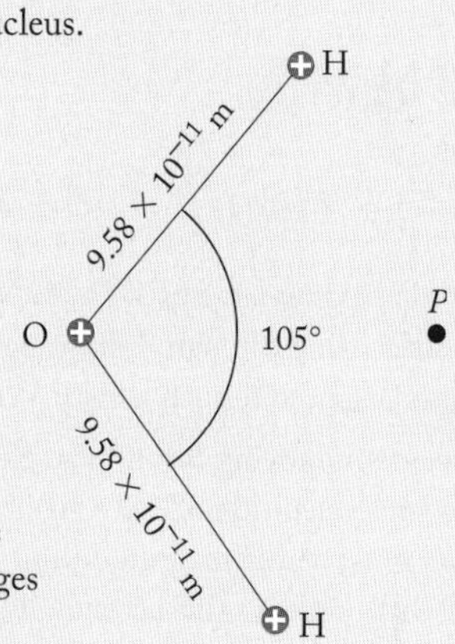

FIGURE 23.33 The positive (nuclear) charges in a water molecule.

*19. Figure 23.34 shows the charge distribution within a thundercloud. There is a charge of 40 C at a height of 10 km, -40 C at 5.0 km, and 10 C at 2.0 km. Treating these charges as pointlike, find the electric field (magnitude and direction) that they produce at a height of 8.0 km and a horizontal distance of 3.0 km.

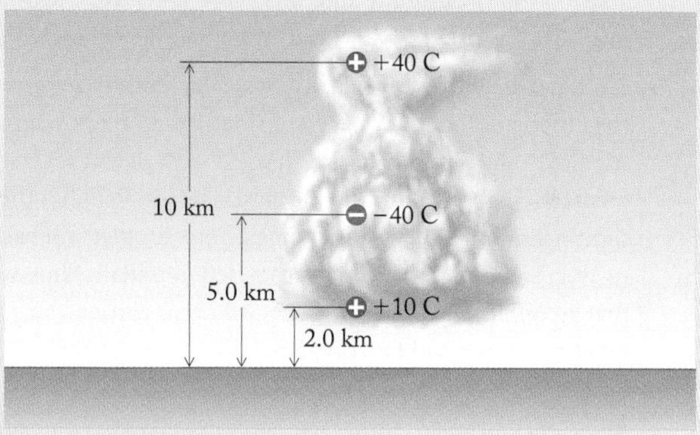

FIGURE 23.34 Charges in a thundercloud.

*20. Suppose that an airplane flies through the thundercloud described in Problem 19 at the 8.0-km level. Plot the magnitude of the electric field as a function of position along the path of the airplane; start with the airplane 10 km away from the thundercloud.

*21. For the four thundercloud and image charges of Fig. 23.5, calculate the magnitude of the electric field at horizontal distances of 2.0, 4.0, 6.0, 8.0, and 10 km from the point on the ground directly below the thundercloud charges. Plot the magnitude of the field vs. the horizontal distance.

*22. Consider eight of the ions of Cl^- and Na^+ on a crystal lattice of common salt. The ions are located at the vertices of a cube measuring 2.82×10^{-10} m on an edge (Fig. 23.35). Calculate the magnitude of the electric force that seven of these ions exert on the eighth.

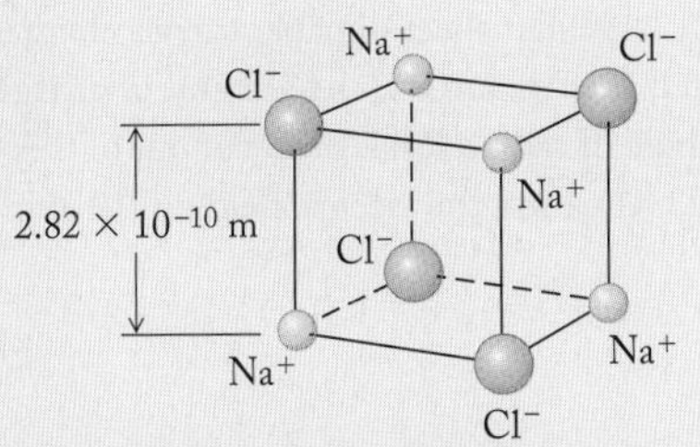

FIGURE 23.35 Ions of Cl^- and Na^+ in a salt crystal.

*23. For the three charges described in Problem 16, again consider the electric field at a point on the positive x axis. What is the approximate dependence of the electric field on the distance x for $x \gg d$? This is the **electric quadrupole** field. [Hint: For small $\delta = (d/x)$, use the approximation $(1 + \delta)^n \approx 1 + n\delta$.]

*24. Identical charges $+Q$ are at seven of the corners of a cube of side a. The eighth corner is empty. From symmetry considerations, what is the direction of the net electric field at the empty corner? Calculate the magnitude of the net electric field at the empty corner.

23.2 The Electric Field of Continuous Charge Distributions

25. (a) Equation (23.11) gives the electric field on the axis of a charged ring. Where is the strength of this electric field maximum?

(b) Roughly sketch the electric field in the space surrounding the ring.

26. You wish to generate a uniform electric field of 2.0×10^5 N/C in the space between two flat, parallel plates of metal placed face to face. The plates measure 0.30 cm × 0.30 cm. How much electric charge must you put on each plate? Assume that the gap between the plates is small so that the charge distribution and the electric field are approximately uniform, as for infinite plates.

27. A long, straight rod carries a uniform distribution of electric charge of 2.0×10^{-14} C per meter. What is the electric field at a perpendicular distance of 0.50 m from this rod? At 1.0 m? At 1.5 m?

*28. Each of two very long, straight rods carries a positive charge of 1.0×10^{-12} C per meter. One rod lies along the x axis, the other along the y axis. Find the electric field (magnitude and direction) at the point $x = 0.50$ m, $y = 0.20$ m.

*29. Two thin rods of length L, each with uniform linear charge distribution λ, form a cross. Find the electric field at a point a distance $L/2$ from each rod, in the plane of the cross.

*30. Consider an infinitely long line of charge of zero thickness, with a uniform linear charge distribution λ. Show that the tension in such a line is infinite (hence such a line is a physical impossibility). Show that the tension is also infinite in a line of charge of finite length.

*31. Each of two very long, straight, parallel rods carries a positive charge of λ coulombs per meter. The distance between the rods is d (Fig. 23.36). Find the electric field at a point equidistant from the rods, with a distance $2d$ from each rod. Draw a diagram showing the direction of the electric field.

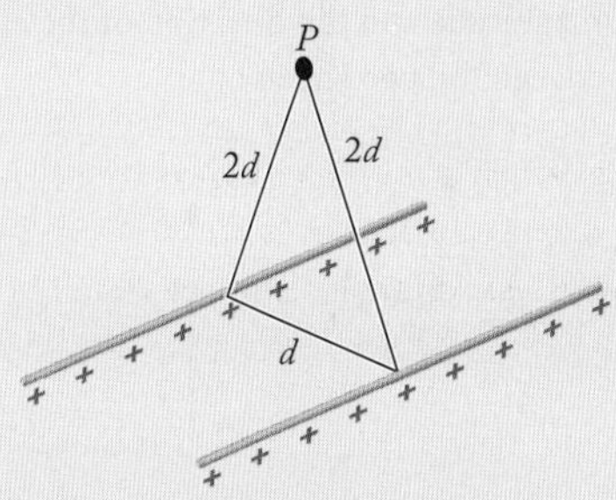

FIGURE 23.36 Two parallel charged rods.

*32. Two infinite lines of silk with uniform linear charge distribution λ lie along the x and the y axes, respectively. Find the electric field at a point with coordinates x, y, z; assume that $x > 0, y > 0, z > 0$.

*33. A semi-infinite line carrying a uniform charge distribution of λ coulombs per meter lies along the positive x axis from $x = 0$ to $x = \infty$. Find the components of the electric field at the point with coordinates x, y, with $z = 0$; assume $x > 0, y > 0$.

*34. A semi-infinite line with a uniform charge distribution of $+\lambda$ coulombs per meter lies along the positive x axis from $x = 0$ to $x = \infty$. Another semi-infinite line with a charge distribution of $-\lambda$ coulombs per meter lies along the negative x axis from $x = 0$ to $x = \infty$. Find the electric field at a point on the y axis.

*35. Electric charge is uniformly distributed over each of three large parallel sheets of paper (Fig. 23.37). The charges per unit area on the sheets are 2.0×10^{-6} C/m^2, 2.0×10^{-6} C/m^2, and -2.0×10^{-6} C/m^2, respectively. The distance between one sheet and the next is 1.0 cm. Find the strength of the electric field **E** above the sheets, below the sheets, and in the spaces between the sheets. Find the direction of **E** at each place.

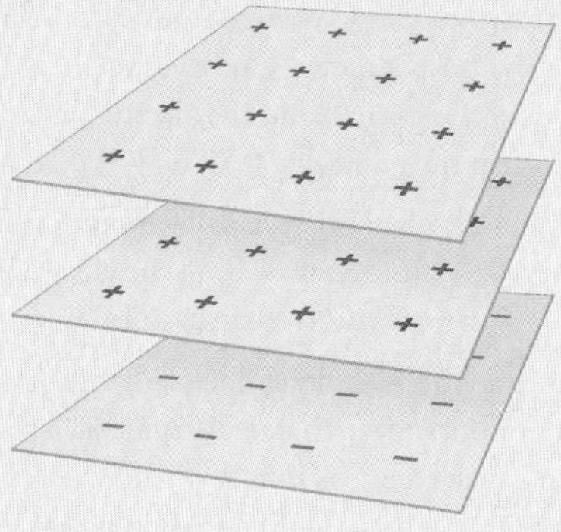

FIGURE 23.37 Three parallel charged sheets of paper.

*36. Each of two very large flat sheets of paper carries a uniform positive charge distribution of 3.0×10^{-4} C/m^2. The two sheets of paper intersect at an angle of 45° (Fig. 23.38). What are the magnitude and the direction of the electric field at a point between the two sheets?

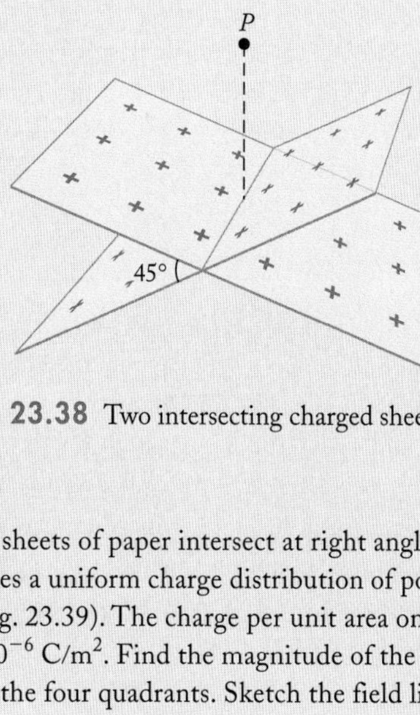

FIGURE 23.38 Two intersecting charged sheets of paper.

*37. Two large sheets of paper intersect at right angles. Each sheet carries a uniform charge distribution of positive charge (Fig. 23.39). The charge per unit area on the sheets is 3.0×10^{-6} C/m^2. Find the magnitude of the electric field in each of the four quadrants. Sketch the field lines in each quadrant.

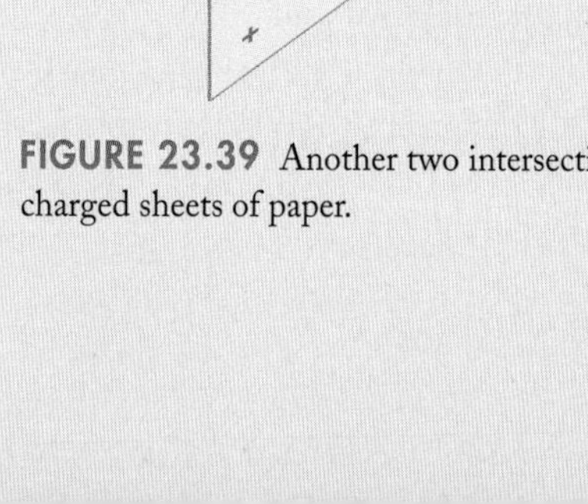

FIGURE 23.39 Another two intersecting charged sheets of paper.

*38. The electric field within a chunk of metal exposed to the Earth's gravity (see Problem 10) is due to a distribution of surface charge. Suppose that we have a slab of iron oriented horizontally (Fig. 23.40). What must be the surface charge densities of the upper and lower surfaces?

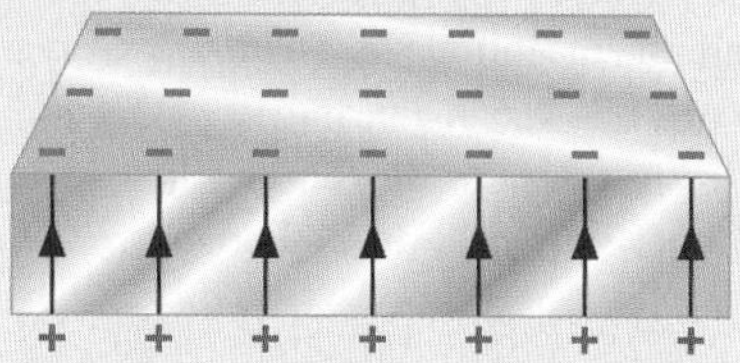

FIGURE 23.40 A horizontal slab of iron.

*39. A very large, flat sheet of paper carries charge uniformly distributed over its surface; the amount of charge per unit area is σ. A hole of radius R has been cut out of this paper (see Fig. 23.41). Find the electric field on the axis of the hole.

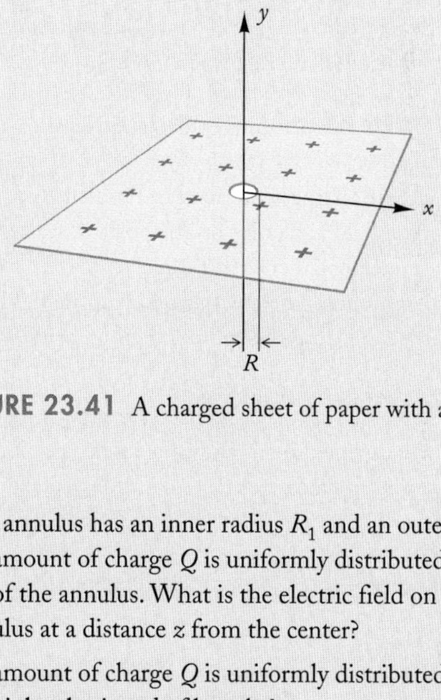

FIGURE 23.41 A charged sheet of paper with a hole.

*40. A paper annulus has an inner radius R_1 and an outer radius R_2. An amount of charge Q is uniformly distributed over the surface of the annulus. What is the electric field on the axis of the annulus at a distance z from the center?

*41. A total amount of charge Q is uniformly distributed along a thin, straight plastic rod of length l.

(a) Find the electric field at the point P, at a distance x from one end of the rod (Fig. 23.42).

(b) Find the electric field at point P', at a distance y from the midpoint of the rod (see Fig. 23.42).

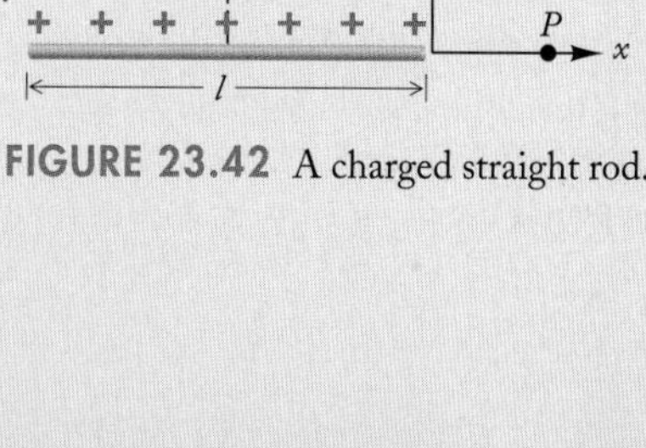

FIGURE 23.42 A charged straight rod.

*42. Two infinite parallel rods are separated a distance $2d$. One carries a uniformly distributed positive charge λ, the other an opposite charge $-\lambda$. The rods are parallel to the z axis and intersect the x–y plane at $x = 0, y = \pm d$. Find the electric field for a point on the positive x axis. How does E behave for $x \gg d$?

*43. Consider the possibility of a **nonuniform charge distribution**. Specifically, consider the line segment of charge of Fig. 23.42, but now assume the total charge Q is distributed unevenly along the length l, with charge density λ that decreases linearly from left to right according to $\lambda = -\lambda_0 x/l$, where λ_0 is a constant, $x = -l$ is at the left end of the rod, and $x = 0$ is at the right end of the rod, as in Fig. 23.42. Note that a small charge dQ in a length element dx is still given by $dQ = \lambda\, dx$.

(a) By summing the charge along the rod, $Q = \int dQ$, determine the value of λ_0 (in terms of Q and l).

(b) Calculate the electric field E at the point P a distance x from the end of the rod (on the positive x axis).

(c) Check the behavior of your result for $x \gg l$. [Hint: You will need to expand the expressions from (b)].

*44. A thin plastic rod is bent so that it has the shape of a semicircle of radius R (Fig. 23.43). An amount of charge Q is uniformly distributed along the rod. What is the electric field at the center of the circle?

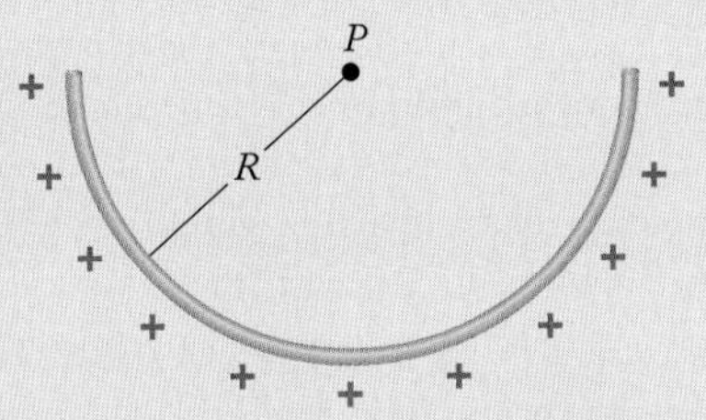

FIGURE 23.43 A charged semicircular rod.

*45. A Plexiglas square of dimension $l \times l$ has a uniform charge density of magnitude λ coulombs per meter along its edges. Two of the edges are positive and two are negative (Fig. 23.44). Find the electric field at the center of the square.

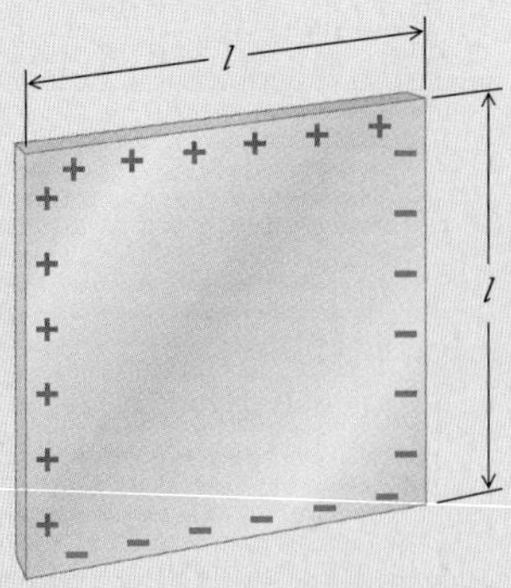

FIGURE 23.44 A square, with charge along its edges.

*46. A cylindrical Plexiglas tube of length l, radius R carries a charge Q uniformly distributed over its surface. Find the electric field on the axis of the tube at one of its ends. [Hint: Sum field contributions from ringlike elements of the tube; see Eq. (23.11) from Example 6.]

*47. A thin, semi-infinite rod with a uniform charge distribution of λ coulombs per meter lies along the positive x axis from $x = 0$ to $x = \infty$; a similar rod lies along the positive y axis from $y = 0$ to $y = \infty$ (see Fig. 23.45). Calculate the electric field at a point in the x–y plane in the first quadrant.

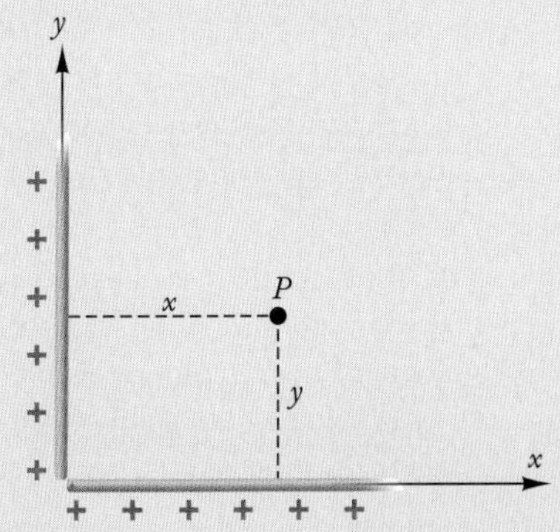

FIGURE 23.45 Two charged straight rods.

*48. (a) Consider an infinitely long rod of radius R with charge uniformly distributed over its volume at the rate of λ coulombs per meter of length. Show that the tension in such a rod is infinite.

(b) Consider a rod of finite length L; assume $R \ll L$. Estimate the tension in such a rod, at some point near its middle. Do not bother with an exact calculation, but try to discover the correct dependence on λ and R. What is the estimated tension for a long rod of radius 1.0 mm with a charge of 1.0×10^{-7} C/m?

**49. Two thin, semi-infinite rods lie in the same plane. They make an angle of 45° with each other, and they are joined by another thin rod bent along an arc of circle of radius R, with center P (see Fig. 23.46). All the rods carry a uniform charge distribution of λ coulombs per meter. Find the electric field at the point P.

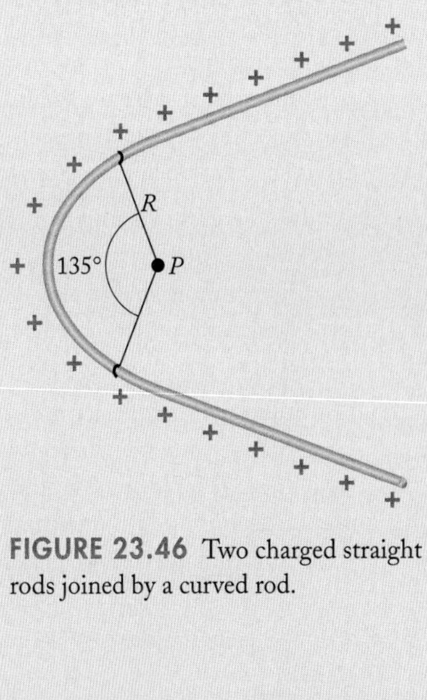

FIGURE 23.46 Two charged straight rods joined by a curved rod.

**50. A square of paper measuring $l \times l$ carries an amount of charge Q uniformly distributed over its surface. The square lies in the x–y plane with its center at the origin and its sides parallel to the x and y axes. Find the electric field at a point on the y axis; assume that the point is outside the square. (Hint: Sum the contributions from each line segment of charge along the square.)

†23.3 Lines of Electric Field

51. A positive charge $+2q$ and a negative charge $-3q$ are separated by a distance d. Sketch the field lines of the electric field produced jointly by these charges in a plane containing the two charges.

52. A very long, thin rod carries $+Q$ coulombs of charge uniformly distributed in each meter of its length, and a point charge $-Q$ is located one meter from the rod. Sketch the field lines of the electric field produced jointly by these charges in the plane containing the rod and the point charge.

53. Sketch the electric field lines due to a finite rod of length L which has a charge Q uniformly distributed over its length. (Hint: Consider the field both very close to the rod, where it looks like an infinite rod, and very far from the rod, where it looks like a point charge.)

54. Three positive charges are at the vertices of an equilateral triangle. Sketch the field line pattern in the plane of the triangle.

55. Two positive charges $+Q$ are at two of the vertices of an equilateral triangle; at the third vertex is a negative charge $-Q$. Sketch the field line pattern in the plane of the triangle.

56. Sketch the electric field line pattern for an infinite row of point charges $+q$, each separated by a distance d from its two nearest neighbors.

23.4 Motion in a Uniform Electric Field

57. The electric field in the electron gun of a TV tube is supposed to accelerate electrons uniformly from 0 to 3.3×10^7 m/s within a distance of 1.0 cm. What electric field is required?

58. An electron is initially at rest, 0.10 cm from a very large disk which carries a uniform surface charge density $\sigma = +3.0 \times 10^{-8}$ C/m^2. How long does it take the electron to strike the plate? What is its speed just before striking the plate?

59. A proton, initially 4.0 cm from a uniformly charged sheet with $\sigma = 5.0 \times 10^{-8}$ C/m^2, is fired with initial speed v_0 directly toward the sheet. The proton slows down and momentarily stops a distance 0.15 cm from the sheet before recoiling away from the sheet. What was the value of v_0?

60. In a **cathode-ray tube**, a beam of electrons (the cathode ray) is deflected in a region of electric field on its way to a fluorescent screen, as shown in Fig. 23.47. Consider the parallel-plate arrangement in the figure, and assume that the electric field $E = 400$ N/C is uniform between the plates and that $E = 0$ outside the plates. The beam of electrons is injected horizontally with velocity $v_0 = 5.0 \times 10^6$ m/s. If the width of the plates is $L = 4.0$ cm, by what vertical distance y_1 is the beam deflected upon just exiting the plates? If the distance from the end of the plates to the screen is $D = 12.0$ cm, what is the total vertical deflection $y = y_1 + y_2$ upon reaching the screen?

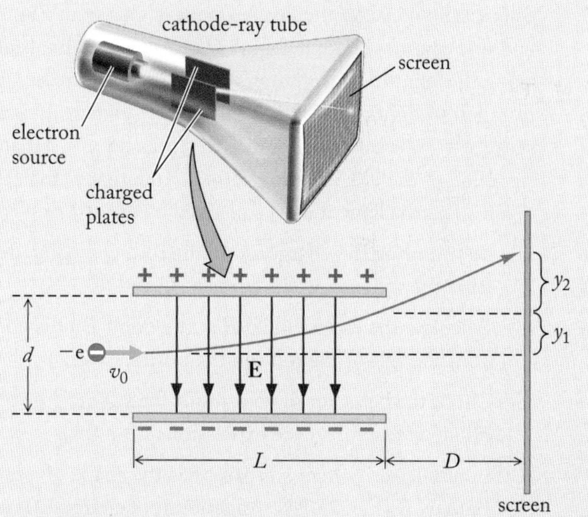

FIGURE 23.47 Schematic diagram of a cathode-ray tube.

61. In the cathode-ray tube of Fig. 23.47, a beam of electrons is injected horizontally into the exact center of the parallel-plate region. If $L = 3.0$ cm, $d = 0.20$ cm, and each plate carries a uniform charge density of magnitude $\sigma = 1.5 \times 10^{-7}$ C/m^2, what minimum initial speed v_0 must the electrons have to ensure that they do not strike the upper plate?

62. Consider the parallel plates of Fig. 23.48. Assume the injected particles are electrons with initial speed $v_0 = 4.0 \times 10^6$ m/s and the angle $\theta = 35°$. If the uniform vertical electric field has the value $E = 3000$ N/C, at what horizontal distance will the electrons strike the lower plate?

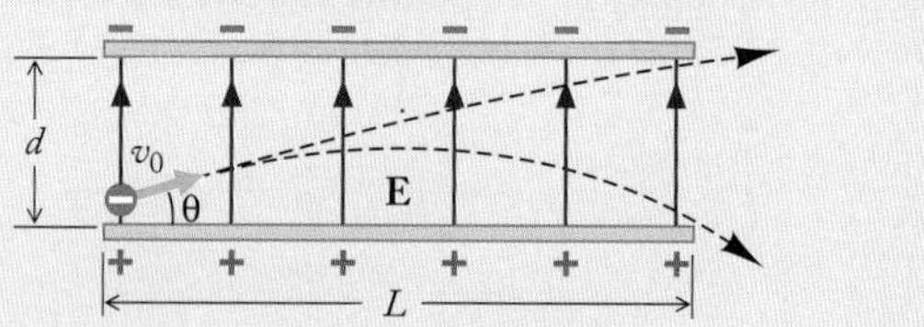

FIGURE 23.48 Charge injected at an angle between plates.

*63. Consider the parallel plates of Fig. 23.48. Assume that the injected particles are polystyrene beads, each of mass 2.0×10^{-11} g; the beads carry negative charges of hundreds to thousands of elementary charges. The uniform electric field between the plates is $E = 900$ N/C, directed vertically upward. The uniform field region is $L = 20$ cm long and $d = 1.0$ cm high. The balls have initial velocity $v_0 = 3.0$ m/s and enter the corner of the field region at $\theta = 5.0°$. For what numbers of elementary charges will the balls exit the field region without striking one of the plates?

†For help, see Online Concept Tutorial 27 at www.wwnorton.com/physics

23.5 Electric Dipole in an Electric Field

64. A small, straight bit of thread has a charge of $+1.0 \times 10^{-14}$ C at one end and a charge of -1.0×10^{-14} C at the other. The length of the thread is 2.0 mm.

    (a) What is the dipole moment?

    (b) What is the torque on this thread if it is placed in an electric field of 6.0×10^5 N/C at right angles to the field?

65. The two charges of ± 40 C in the thundercloud of Fig. 23.34 form a dipole. What is the dipole moment?

66. In a hydrogen atom, the electron is at a distance of 5.3×10^{-11} m from the proton.

    (a) What is the instantaneous dipole moment of this system?

    (b) Taking into account that the electron moves around the proton on a circular orbit, what is the time-average dipole moment of this system?

67. (a) Pretend that the HCl molecule consists of (pointlike) ions of H^+ and Cl^- separated by a distance of 1.0×10^{-10} m. If so, what would be the dipole moment of this system?

    (b) The observed dipole moment is 3.4×10^{-30} C·m. Can you suggest a reason for this discrepancy?

68. The dipole moment of an HCl molecule is 3.4×10^{-30} C·m. Calculate the magnitude of the torque that an electric field of 2.0×10^6 N/C exerts on this molecule when the angle between the electric field and the longitudinal axis of the molecule is 45°.

REVIEW PROBLEMS

69. To measure the magnitude of a horizontal electric field, an experimenter attaches a small charged cork ball to a string and suspends this device in the electric field. The electric force pushes the ball to one side, and the ball attains equilibrium when the string makes an angle of 35° with the vertical (see Fig. 23.49). The mass of the ball is 3.0×10^{-5} kg, and the charge on the ball is 4.0×10^{-7} C. What is the magnitude of the electric field?

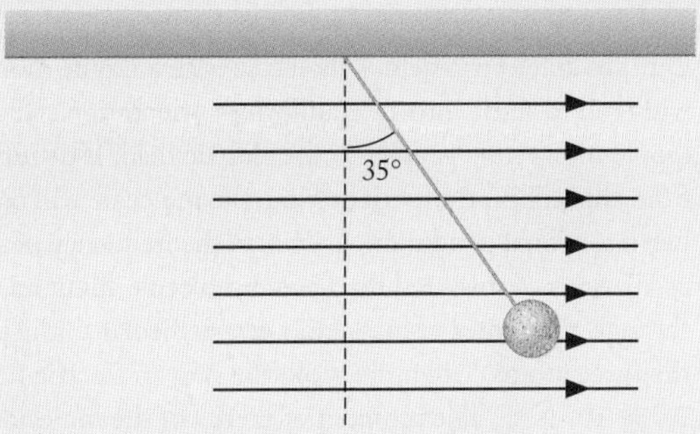

FIGURE 23.49 Cork ball suspended in an electric field.

70. When placed in a given uniform electric field, a proton develops an acceleration of 2.0×10^{10} m/s^2 in the eastward direction. What acceleration will be developed by an electron when placed in the same electric field?

71. A large sheet of paper with a uniform charge distribution of 2.0×10^{-4} C/m^2 is pierced perpendicularly by a very long straight rod with a uniform linear charge distribution of 3.0×10^{-7} C/m (see Fig. 23.50). Find the magnitude and direction of the electric field at a point above the sheet distance of 1.0 cm from the rod.

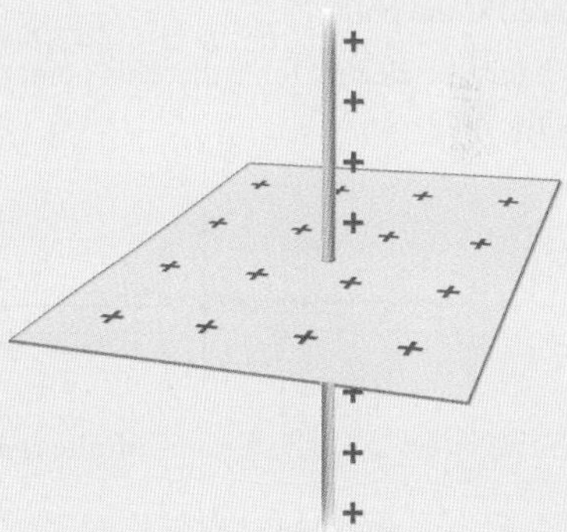

FIGURE 23.50 A charged sheet of paper and a charged rod.

72. Consider two parallel sheets with uniform, opposite charge distributions, as in Fig. 23.10. Suppose that the magnitude of the charge density on each sheet is 2.0×10^{-5} C/m^2. The upper sheet is positive and the lower is negative.

    (a) What is the magnitude of the electric field in the region between the sheets?

    (b) You now insert another large parallel sheet with positive charge density 1.0×10^{-5} C/m^2 in the space between the two charged sheets. What is the magnitude of the electric field in the space immediately above the middle sheet? Below the middle sheet?

73. Consider the charge distribution within the thundercloud illustrated in Fig. 23.4a. Find the electric field (magnitude and direction) that this charge distribution produces at a height of 10 km and a horizontal distance of 5.0 km to the right of the charges.

*74. Three charges $\pm Q$ are arranged at the vertices of an equilateral triangle of side L as shown in Fig. 23.51. Calculate the electric field that these charges produce at the center of the triangle.

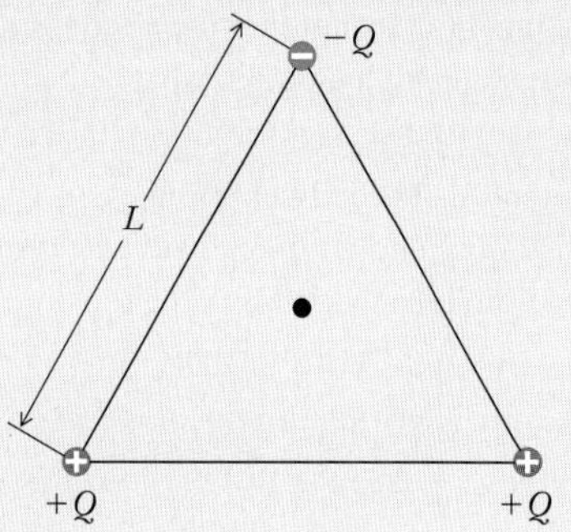

FIGURE 23.51 Three point charges.

*75. A semi-infinite thin rod with uniform linear charge density λ lies along the negative x axis, from $x = 0$ to $x = -\infty$ (see Fig. 23.52).

(a) Find the electric field at a point P a distance d on the positive x axis.

(b) What is the force that this field exerts on a charge q placed at the point P?

(c) What is the electric force that this point charge q exerts on the rod?

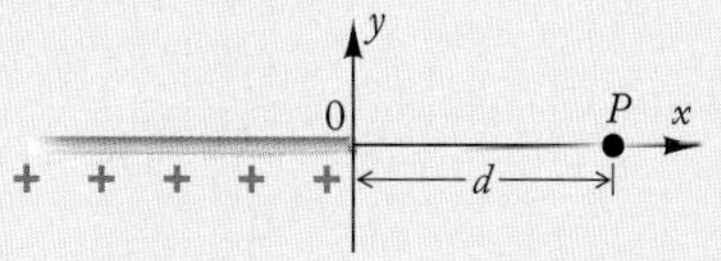

FIGURE 23.52 A semi-infinite charged rod.

*76. Two infinite thin rods intersect at an angle α (see Fig. 23.53). Each charge carries a uniform linear charge distribution λ. Find the electric field on the bisector of the angle α.

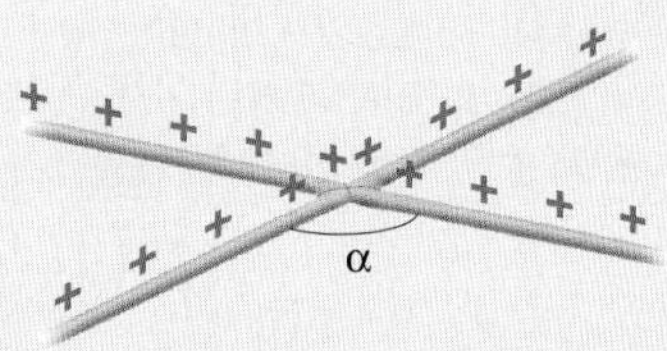

FIGURE 23.53 Two intersecting charged rods.

*77. Three point charges $-Q$, $2Q$, and $-Q$ are arranged on a straight line, as illustrated in Fig. 23.54. What is the electric field that these charges produce at a distance y above the central charge?

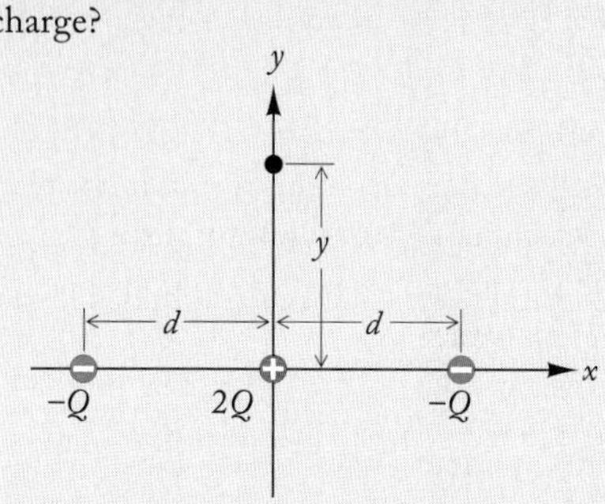

FIGURE 23.54 Three point charges.

*78. Three thin glass rods carry charges Q, Q, and $-Q$, respectively. The length of each rod is l, and the charge is uniformly distributed along each rod. The rods form an equilateral triangle. Calculate the electric field at the center of the triangle.

*79. A paper disk of radius R carries an amount of charge Q uniformly distributed over its surface. Find the electric field at a point on the axis of the disk at a distance z from the center. Show that in the limiting case $z \gg R$, the result reduces to that for a point charge. Hint: To estimate the repulsive force between the right and the left halves of the rod, divide each half into a short segment of whose length is a few times R (say nR) and a long segment of length L/2 − nR. The repulsive force between the long segments can be calculated as in Problem 22.50. Estimate the repulsive force contributed by the short segments by treating them as lumps of charge located at the centers of the segments, and show that this contribution is small if $L \gg R$.)

*80. The dipole moment of the water molecule is 6.1×10^{-30} C·m. In an electric field, a molecule with a dipole moment will tend to settle into an equilibrium orientation such that the dipole moment is parallel to the electric field. If disturbed from this equilibrium orientation, the molecule will oscillate like a torsional pendulum. Calculate the frequency of small oscillations of this kind for a water molecule about an axis through the center of mass (and perpendicular to the plane of the three atoms) when the molecule is in an electric field of 2.0×10^6 N/C. The moment of inertia of the molecule about this axis is 1.93×10^{-47} kg·m^2.

Answers to Checkups

Checkup 23.1

1. In this case, the test charge is q', and the charge producing the field is q. So the magnitude of the field due to q is $E = F/q'$ (the force divided by the test charge), or $E = (1/4\pi\epsilon_0)q/r^2$. The direction is radially away from q for positive q, or radially toward q for negative q.

2. Since the field due to a point charge decreases with the inverse square of the distance, the field at 2 m is one-fourth as large, or $(1/4) \times 10^5$ N/C. Similarly, at 3 m, the field is $(1/9) \times 10^5$ N/C.

3. Since these field contributions are in perpendicular directions, the resultant magnitude is $E = \sqrt{10^2 + 10^2}$ N/C $= \sqrt{2} \times 10$ N/C $= 14$ N/C; since the two perpendicular components are equal, the direction of the resultant is at 45°, i.e., precisely in the northeast direction.

4. At the center of the square, the contribution to the electric field from each point charge points away from that point charge. Since the charges are equal and are equidistant from the center, the electric fields due to charges diagonally across from each other exactly cancel, and the net field is zero.

5. (A) Toward the negative charge. The fields due to the two equal positive charges that are diagonally across from one another cancel at the center of the square. However, the electric field at the center of the square due to the remaining positive charge points away from it, and the electric field due to the negative charge points toward it; these two contributions are in the same direction, directly toward the negative charge.

Checkup 23.2

1. The contribution to the electric field from each positive sheet of charge points perpendicularly out from the corresponding sheet and is independent of distance from the sheet. For two perpendicular equivalent sheets, these two vectors are equal in magnitude and at right angles; they add to produce a field at 45° to the sheets (with magnitude $\sqrt{2}$ times as large as each individual sheet field).

2. As we saw in the examples, far away from a charge distribution, the field is essentially that of a point charge, in this case $E = (1/4\pi\epsilon_0)Q/r^2$. Very close to the surface, the disk looks like an infinite plane of charge, with field $E = \sigma/2\epsilon_0 = (Q/A)/2\epsilon_0$.

3. For an infinite sheet (Example 7), the electric field is independent of distance. So $E = E_0$ at both 2 m and 4 m from the sheet.

4. (B) 0; σ/ϵ_0. Since each sheet by itself produces a field that points perpendicularly away from the sheet on each side, the total field is zero in between the two sheets, where the two equal-magnitude contributions oppose. Outside the sheets, the two contributions are parallel, so the field is twice as strong as that due to either sheet alone. Outside, the sheets behave like a single sheet with twice the charge density.

Checkup 23.3

1. In Fig. 23.21a, there are electric field lines that cross; since the direction of the electric field is tangent to the field lines, crossing field lines (two directions at one point) are not allowed in field line diagrams. Figure 23.21b shows field lines beginning in empty space; a field line must begin and end on charges (or at infinity). Last, Fig. 23.21c shows a field line looping around and closing on itself; in electrostatics, field lines must have a beginning and an end.

2. The sphere has some positive charge (in the region where field lines are emerging) and some negative charge (where field lines are entering). Since the net number of field lines is two lines emerging (four emerging minus two entering), the net charge on the sphere must be one-quarter the value of the point charge (which had eight lines emerging).

3. Since the field is pointing downward, the field lines terminate at the Earth's surface, and the Earth's surface charge must be negative.

4. (D) -9.0 C. Since field lines only enter the rod, its charge is negative. Since there are 12 lines entering the rod, compared with 8 lines emerging from the point charge, the magnitude of the charge on the rod is $12/8 = 3/2$ times that of the point charge. Thus the charge on the rod is $-(3/2) \times 6.0$ C $= -9.0$ C.

Checkup 23.4

1. No. If the motion is at constant velocity, then there is no net force and thus no electric field; but if the straight-line motion is accelerated, then there is an electric field directed along the line of motion.

2. (C) **a**. The acceleration of a charged particle in an electric field **E** is $\mathbf{a} = \mathbf{F}/m = q\mathbf{E}/m$. So if **E** is the same and the charge q and mass m of the particle both double, then the charge-to-mass ratio q/m remains the same, and the acceleration is unchanged.

Checkup 23.5

1. Since the torque is proportional to the sine of the angle between the field and the dipole moment ($\sin\theta$), it is maximum for $\theta = 90°$. That is, there is a maximum torque when the dipole moment is perpendicular to the field.

2. Since the potential energy of a dipole is defined by $U = -pE\cos\theta$, it is zero for $\theta = 90°$ (or $\theta = 270°$), when the dipole is perpendicular to the field.

3. For an induced dipole moment, the dipole moment is parallel to the applied field. For the equal-magnitude charges of a dipole, the greater force occurs where the field is stronger: if this is at the positive charge, the net force will be parallel to the field; if this is at the negative charge, the net force will be antiparallel to the field. In either case, the net force is toward the region of stronger field.

4. (B) $+y$. The torque tends to align the dipole moment of a body with the electric field, here the $+y$ direction.

CHAPTER

24

Gauss' Law

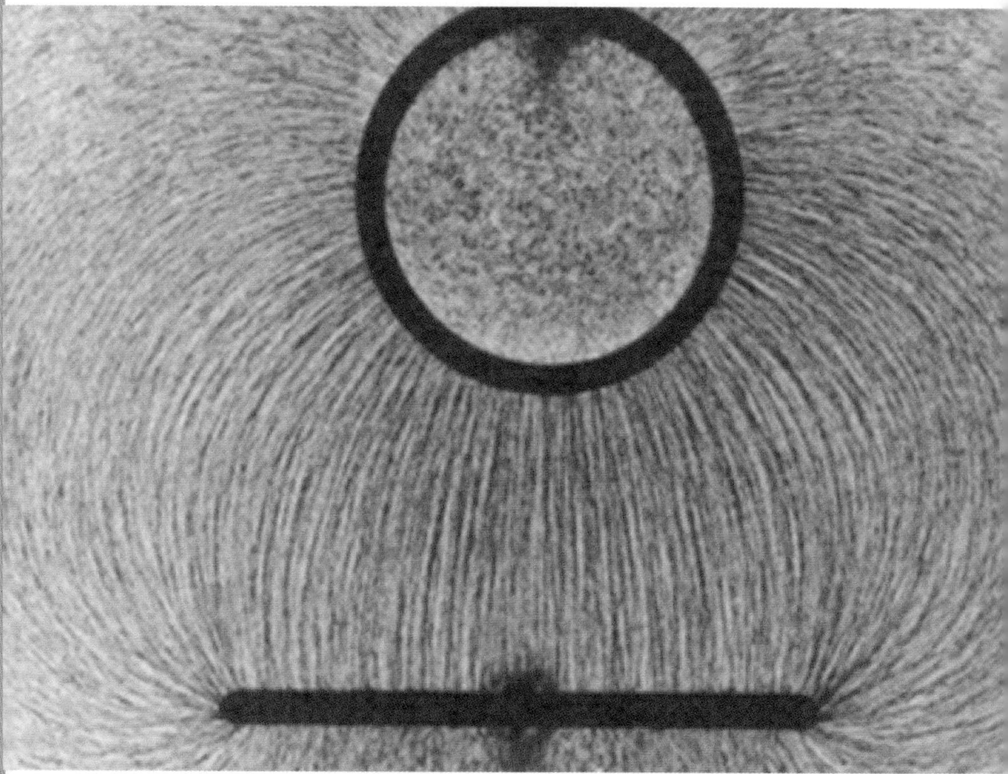

CONCEPTS IN CONTEXT

Concepts in Context

This photograph illustrates bits of thread dispersed in oil to reveal the pattern of the electric field generated by the electric charges residing on two parallel conductors. The bits of thread act as electric dipoles, which tend to align with the local electric field. The charge distribution in this photograph is highly symmetric, and correspondingly, the electric field pattern is symmetric.

We can exploit symmetry to facilitate the calculation of electric fields, which will permit us to answer the following questions:

? What is the electric field due to a spherically symmetric distribution of charge? (Examples 4 and 5, pages 764–766)

? What is the electric field near a flat sheet of charge, or near a flat conducting slab? (Examples 9 and 11, pages 770 and 775)

? What is the electric field inside a conductor in electrostatic equilibrium, and what does this tell us about the distribution of charge on the conductor? (Section 24.5, page 774, and Example 12, page 776)

The electric field of any charge distribution can be calculated by means of Coulomb's Law, as in the examples of the preceding chapter. The calculation is fairly simple if the charge distribution consists of an arrangement of just a few point charges. But as we saw in Section 23.2, this method often becomes complicated and tedious if the charge is distributed over a surface or volume, when we have to perform a difficult summation or integration of all the individual electric field vectors of the small pieces of the charge distribution. Fortunately, there is another method for calculating the electric field; this method relies on a theorem called Gauss' Law. This law is a consequence of Coulomb's Law, and therefore it contains no new physics. It does, however, contain some new mathematics, which supplies an elegant shortcut for calculating the electric field, provided that the charge distribution has a high degree of symmetry. This means that whereas Gauss' Law is always valid, it does not ease every calculation, but when it does, it works wonders.

In this chapter we will first introduce the concept of electric flux; we will then present Gauss' Law and show how it makes finding the electric field easy in many situations. We will also discuss how charge can be distributed on metals and insulators, and how such knowledge of allowed charge distributions helps in applying Gauss' Law.

24.1 ELECTRIC FLUX

Online Concept Tutorial 28

Consider a mathematical (that is, imagined) surface in the shape of a rectangle of area A. Suppose that this surface is immersed in a constant electric field **E** (see Fig. 24.1a). This electric field makes some angle with the surface; the electric field vector has a component tangential, or parallel, to the surface and a component normal, or perpendicular, to the surface. The **electric flux** Φ_E *through the surface is defined as the product of the area A and the normal component of the electric field.* If we designate this perpendicular component of the electric field by $E_\perp$, the electric flux is

$$\Phi_E = E_\perp A \tag{24.1}$$

The SI unit for electric flux is the unit of electric field times the unit of area, that is, $(\text{N/C})\cdot\text{m}^2$, or $\text{N}\cdot\text{m}^2/\text{C}$.

The normal component $E_\perp$ can also be expressed as $E \cos\theta$, where θ is the angle between **E** and a perpendicular to the surface (see Fig. 24.1a). With this expression for $E_\perp$, Eq. (24.1) becomes

$$\Phi_E = EA\cos\theta \tag{24.2}$$

electric flux Φ_E

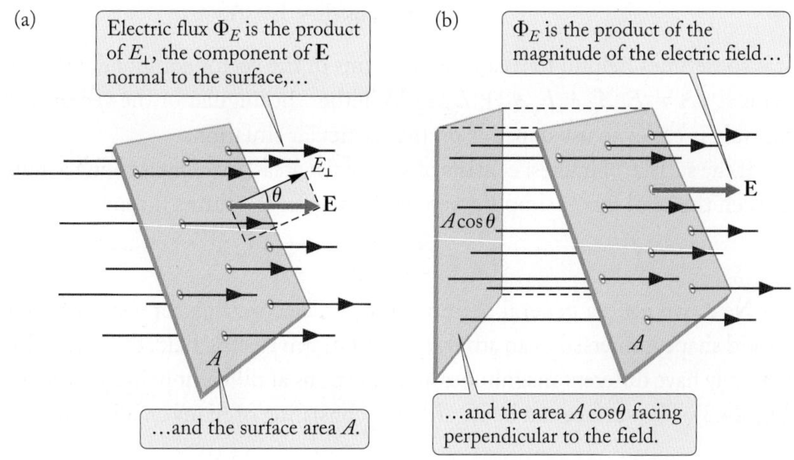

FIGURE 24.1 (a) Flat rectangular surface immersed in a uniform electric field. The perpendicular to the surface makes an angle θ with the field lines. The normal component of the electric field is $E_\perp = E\cos\theta$. The electric flux is $\Phi = E_\perp A$. In terms of field lines, the electric flux is proportional to the number of lines intercepted by the area. (b) The area $A\cos\theta$ is the projection of the area A onto a plane perpendicular to the electric field.

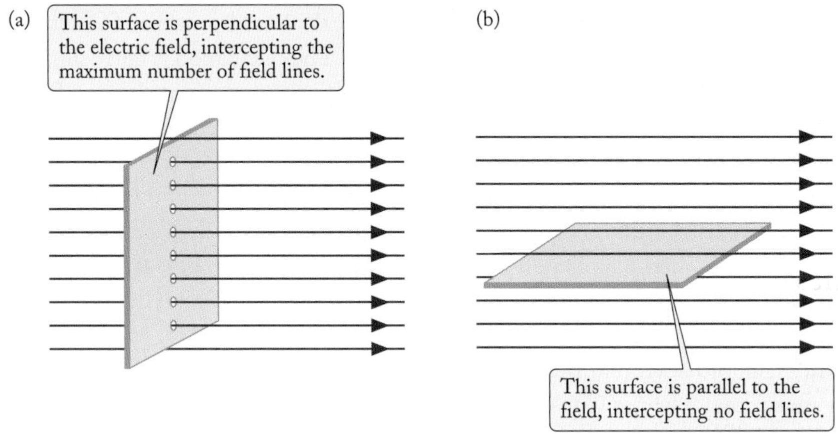

FIGURE 24.2 (a) A surface of area A is perpendicular to the electric field lines. It intercepts an electric flux $\Phi = EA$. (b) The surface is parallel to the field lines. It intercepts no electric flux.

The product $A \cos \theta$ can be interpreted as the projection of the area A onto a plane perpendicular to the electric field; that is, $A \cos \theta$ can be regarded as that portion of the area that effectively faces the electric field, as shown in Fig. 24.1b. Thus, Eq. (24.2) also states that the flux is the magnitude E of the electric field multiplied by the area facing perpendicular to the electric field. If we think in terms of the field lines introduced in Section 23.3, then the electric flux through an area A is proportional to the number of field lines intercepted by that area.

Note that if $\theta = 0$, the flux is simply $\Phi_E = EA$; in this case, the area A is exactly face on to the electric field, and it intercepts the maximum number of field lines (see Fig. 24.2a). On the other hand, if $\theta = 90°$, then the flux given by Eq. (24.2) is zero; in that case, the area A is parallel to the electric field, and all the field lines skim by the area without crossing it (see Fig. 24.2b). This dependence of flux, or the intercepted number of field lines, on the angle of orientation of the area can be understood by a simple analogy: Think of the electric field lines as analogous to the trajectories of falling raindrops, and think of the area as a sheet of paper. The flux is the flow of raindrops intercepted by the paper (*flux* comes from the Latin word for a flow). If you hold the paper face on to the rain, it will intercept a maximum number of drops; if you hold it parallel to the rain, it will intercept no drops. At intermediate angles, the number of drops will be proportional to $\cos \theta$, as in Eq. (24.2).

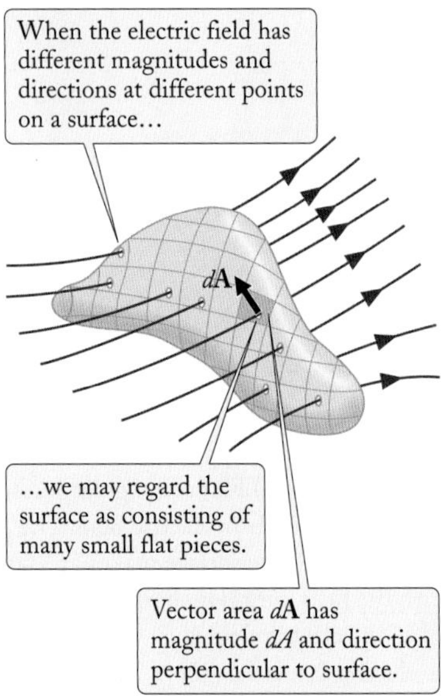

FIGURE 24.3 An arbitrary surface immersed in an arbitrary, nonuniform electric field.

Equation (24.1) is mathematically equivalent to the dot product, or scalar product, of two vectors: the electric field vector $\mathbf{E}$ and the "vector area" $\mathbf{A}$. The vector $\mathbf{A}$ is defined with a magnitude equal to the ordinary surface area A and a direction perpendicular to the surface; that is, $\mathbf{A} = A\,\hat{\mathbf{n}}$, where $\hat{\mathbf{n}}$ is a normal (perpendicular) unit vector. With this definition, the electric flux is

$$\Phi_E = EA \cos \theta = \mathbf{E} \cdot \mathbf{A} \tag{24.3}$$

It is sometimes helpful to recall that in terms of the vector components the dot product is $\mathbf{E} \cdot \mathbf{A} = E_x A_x + E_y A_y + E_z A_z$. Whether the angular or the vector-component formula is easier to use depends on the particular situation.

If the surface of interest consists of several flat areas, each intercepting a uniform field $\mathbf{E}$, then the total flux is simply given by the sum of the fluxes through each flat area:

$$\Phi_E = \sum E_{\perp} A = \sum \mathbf{E} \cdot \mathbf{A} \tag{24.4}$$

Next, we need a general definition of the electric flux for a surface of arbitrary, curved shape immersed in an arbitrary, nonuniform electric field, that is, an electric field that may have different magnitudes and directions at different points on the surface (see Fig. 24.3). The surface can be regarded as consisting of many small (infinitesimal) flat pieces. For a small piece with a vector area $d\mathbf{A}$ in an electric field $\mathbf{E}$, the contribution to the electric flux is $\mathbf{E} \cdot d\mathbf{A}$. If we take the sum, or integral, of all these small amounts of electric flux for all the small pieces of the surface, we obtain the electric flux through the entire surface,

$$\Phi_E = \int \mathbf{E} \cdot d\mathbf{A} = \int E_\perp \, dA \tag{24.5}$$

where again $E_\perp$ is the component of $\mathbf{E}$ perpendicular to the surface.

Keep in mind that the electric flux is proportional to the number of field lines intercepted by the surface. Note that whereas lines going through the surface from one side make a positive contribution to the flux, lines going through from the other side make a negative contribution (see Fig. 24.4 for an example). However, for the *open* surfaces thus far considered, the reference direction (the side for which outward flux is positive) can be chosen either one way or the other.

The definition of flux is also valid for a *closed* surface, such as the surface of a sphere or the surface of a cube. The electric flux through such a surface is

$$\Phi_E = \oint \mathbf{E} \cdot d\mathbf{A} = \oint E_\perp \, dA \tag{24.6}$$

where the circle on the integral sign merely indicates a closed surface. For the case of closed surfaces, there is a sign convention: *the normal component $E_\perp$ is reckoned as positive if the direction of the electric field* **E** *is outward from the surface, and negative if* **E** *is inward, into the surface.* Thus, the electric flux through a closed surface is proportional to the *net* number of field lines emerging from the surface. The electric flux will be positive or negative depending on whether more field lines emerge from the surface or enter the surface.

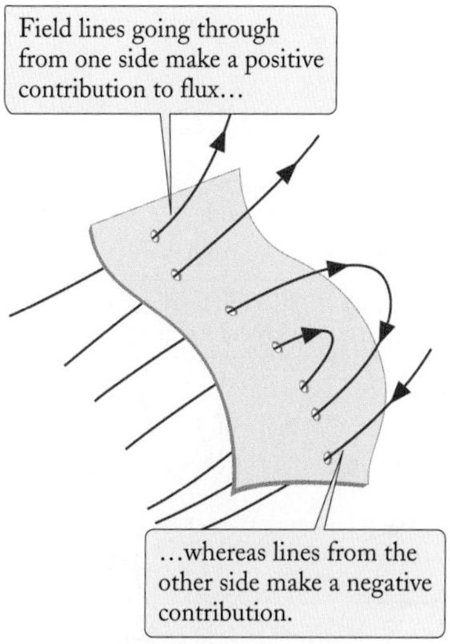

FIGURE 24.4 Arbitrary surface immersed in an arbitrary, nonuniform electric field. Although seven field lines cross the surface, four are in one direction and three are in the other; thus the net flux through the surface is one-seventh the flux one would have if all the field lines crossed in the same direction.

EXAMPLE 1 A flat sheet of paper measuring 22 cm × 28 cm is placed in a uniform electric field of 100 N/C. What is the flux through the paper if the paper makes an angle of 90° with the electric field? If the paper makes an angle of 30°? (See Fig. 24.5.)

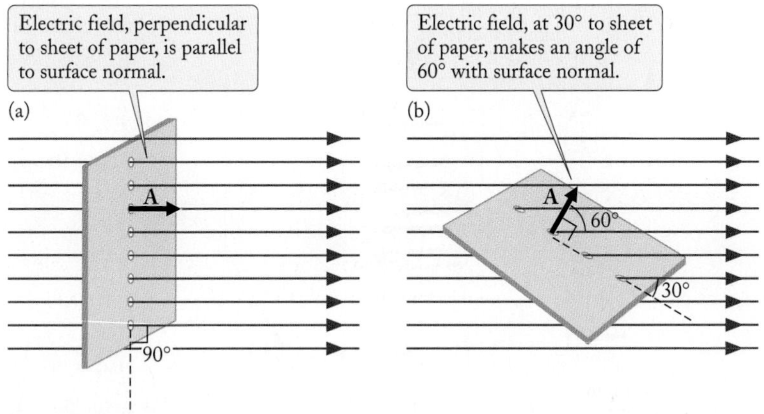

FIGURE 24.5 A sheet of paper placed in a uniform electric field (a) at an angle of 90° with the electric field and (b) at an angle of 30°. The black arrow **A** indicates the perpendicular to the surface.

SOLUTION: For a uniform field and a flat area, the flux is merely the product $\Phi_E = EA\cos\theta$, and since an angle of 90° between the sheet of paper and the electric field means $\theta = 0$ in Eq. (24.2) See Fig. 24.5a,

$$\Phi_E = EA\cos\theta = (100\ \text{N/C}) \times (0.22\ \text{m} \times 0.28\ \text{m}) \times 1 = 6.2\ \text{N}\cdot\text{m}^2/\text{C}$$

An angle of 30° between the paper and the electric field means $\theta = 60°$ (see Fig. 24.5b), so

$$\Phi_E = EA\cos\theta = (100\ \text{N/C}) \times (0.22\ \text{m} \times 0.28\ \text{m}) \times \cos 60° = 3.1\ \text{N}\cdot\text{m}^2/\text{C}$$

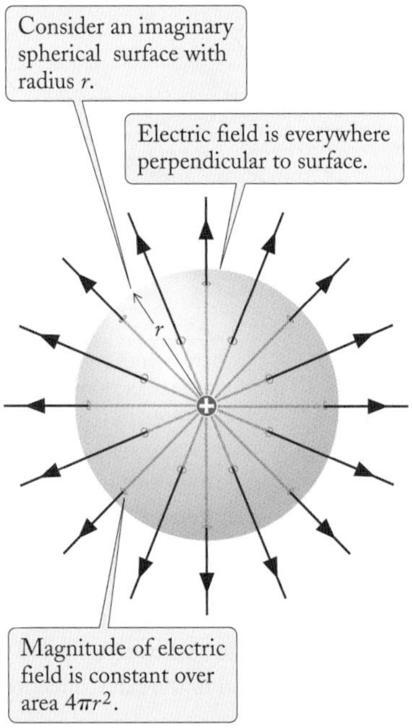

FIGURE 24.6 A spherical surface with a point charge q at its center.

EXAMPLE 2 Consider an imaginary sphere of radius r which has a positive point charge q at its center (see Fig. 24.6). What electric flux does the electric field of this point charge produce through the surface of the sphere?

SOLUTION: The positive point charge q produces a radial, outward electric field of magnitude $E = q/(4\pi\epsilon_0 r^2)$ that crosses the spherical surface everywhere perpendicularly. The area of such a sphere is $4\pi r^2$, so

$$\Phi_E = EA\cos\theta = EA = \frac{q}{4\pi\epsilon_0 r^2} \times 4\pi r^2 = \frac{q}{\epsilon_0} \tag{24.7}$$

COMMENT: We will see in the next section that this simple result is generally true; indeed, the same result, $\Phi_E = q/\epsilon_0$, would have been obtained for *any* surface enclosing the charge q, and for *any* way that the enclosed charge q was distributed. But in this example we see explicitly and simply how the result comes from the inverse-square nature of the electric field, which exactly cancels the surface area $4\pi r^2$.

EXAMPLE 3 Consider a cube of side $a = 2.0$ m with one corner at $(x, y, z) = (1.0\ \text{m}, 0, 0)$ and its sides parallel to the axes as shown in Fig. 24.7a. The cube is in a region where the electric field points everywhere in

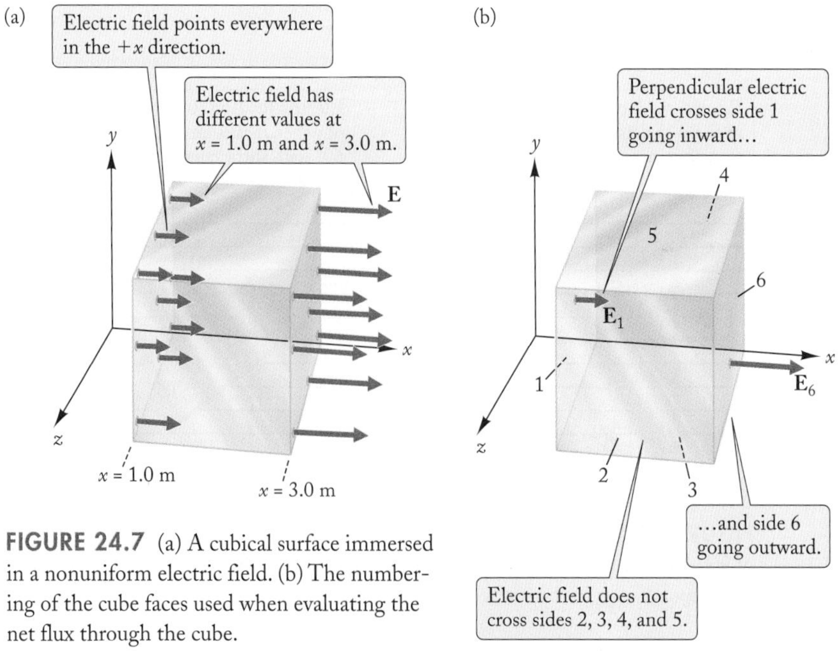

FIGURE 24.7 (a) A cubical surface immersed in a nonuniform electric field. (b) The numbering of the cube faces used when evaluating the net flux through the cube.

the $+x$ direction. The magnitude of this electric field varies as a function of x only, and has the values $E_x = 5.0$ N/C at $x = 1.0$ m and $E_x = 15$ N/C at $x = 3.0$ m. What is the electric flux through the cube?

SOLUTION: The total flux is the sum of the fluxes through the individual faces of the cube, labeled 1 through 6 in Fig. 24.7b. The area of each face is $A = a^2 = (2.0\text{ m})^2 = 4.0\text{ m}^2$. Note that for sides 2, 3, 4, and 5, the field skims along the surfaces without crossing them. Thus

$$\begin{aligned}\Phi_E = \oint E_\perp\, dA &= \Phi_{E1} + \Phi_{E2} + \Phi_{E3} + \Phi_{E4} + \Phi_{E5} + \Phi_{E6} \\ &= \Phi_{E1} + 0 + 0 + 0 + 0 + \Phi_{E6} \qquad (24.8) \\ &= (E_\perp A)_1 + (E_\perp A)_6\end{aligned}$$

Since the direction of **E** at side 1 of the cube is inward, the flux for side 1 is negative, whereas the flux for side 6 is positive. Also, the electric field is perpendicular to both areas, so the net flux through the cube is

$$\Phi_E = -(5.0\text{ N/C}) \times (4.0\text{ m}^2) + (15\text{ N/C}) \times (4.0\text{ m}^2) = 40\text{ N}\cdot\text{m}^2/\text{C} \qquad (24.9)$$

Checkup 24.1

QUESTION 1: Consider Fig. 24.1a. For the surface with the orientation shown in that figure, the flux has some particular value. Would the flux be larger or smaller if the surface were horizontal? Vertical?

QUESTION 2: Would the answer to Example 2 change if instead of the spherical surface we had the same charge inside a cube, as in Fig. 24.8? Inside a cylinder?

QUESTION 3: Consider the point charge and surface of Fig. 24.9. The net flux through this surface is

(A) Positive (B) Negative (C) Zero

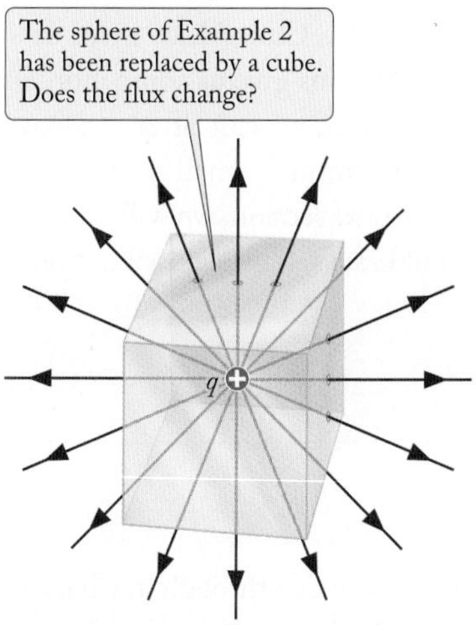

FIGURE 24.8 A cubical surface encloses a point charge q.

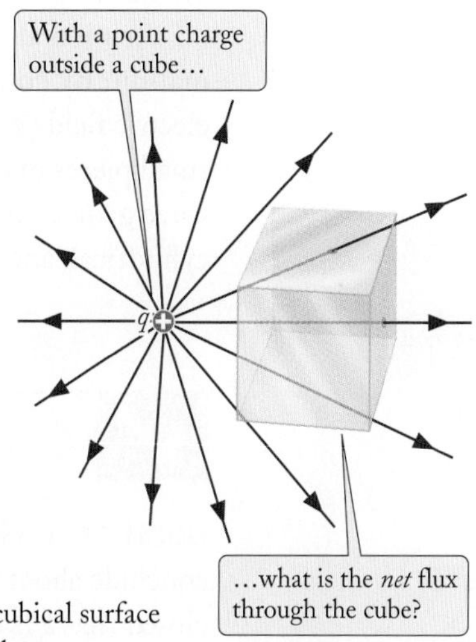

FIGURE 24.9 A cubical surface adjacent to a point charge q.

Online Concept 29 Tutorial

24.2 GAUSS' LAW

We saw in Example 2 that a point charge q enclosed within a surface produces an electric flux $\Phi_E = q/\epsilon_0$ through the surface. This result can be generalized for an arbitrary charge distribution enclosed within an arbitrary surface. The general result is known as **Gauss' Law**:

> *If an arbitrary closed surface has a net electric charge Q_{inside} within it, then the electric flux through the surface is $Q_{\text{inside}}/\epsilon_0$, that is,*

Gauss' Law

$$\Phi_E = \oint \mathbf{E} \cdot d\mathbf{A} = \frac{Q_{\text{inside}}}{\epsilon_0} \quad \text{or} \quad \Phi_E = \oint E_\perp \, dA = \frac{Q_{\text{inside}}}{\epsilon_0} \tag{24.10}$$

Gaussian surface

Gauss' Law says that the *total* electric flux through *any* closed surface is determined solely by the amount of charge *inside* the surface. The surface, called a **Gaussian surface**, can be a real one, but it is often an imaginary surface in space, chosen to pass through some point of interest. In general, as above in Figs. 24.8 and 24.9, a charge inside the surface generates a net flux through the surface, whereas any charge outside the surface makes zero contribution to the net flux through the surface.

The proof of Gauss' Law is easy, if we exploit the concept of field lines. The electric field appearing in Eq. (24.10) is the sum of individual electric fields of point charges, some *outside* the closed surface and some inside it. The individual electric field of a charge outside the closed surface generates no net flux through the closed surface—any such field line either does not touch the surface or else enters it at one point and leaves at another; neither case makes any net contribution to the flux. However, the individual electric field of a point charge within the surface does contribute to the flux. For example, a positive point charge q has a number of outward field lines proportional to its charge q, all of which will pierce the closed surface—such a charge thus contributes a flux proportional to q. We saw in Example 2 that the constant of proportionality between a flux and the charge producing it is $1/\epsilon_0$. Taking into account that positive charges generate positive flux and negative charges negative flux, we see that the net flux, proportional to the net number of field lines piercing the surface, is equal to the net charge divided by ϵ_0. This completes the proof of Gauss' Law.

Gauss' Law can be used to calculate the electric field, provided that the charge distribution has a high degree of symmetry. Essentially, Gauss' Law gives us some information about the electric field. Symmetry conditions give us some more information. By combining these two kinds of information, we can often evaluate the electric field easily, without the laborious process of summing small contributions of small pieces of charge using Coulomb's Law. In the next section, we will examine the three geometries where such an easy application of Gauss' Law is possible: spherical, cylindrical, and planar.

✓ Checkup 24.2

QUESTION 1: Given that a closed surface has zero electric flux through it, what can you conclude about the electric charge inside the surface? What if the surface is not quite closed (has a hole)?

QUESTION 2: Consider a closed surface that surrounds both of the charges in Fig. 24.10. What is the flux through this surface?

QUESTION 3: What is the charge enclosed by the cube of Example 3?

QUESTION 4: A sphere of radius R is in a region of uniform electric field E. The net flux through the surface of the sphere is

(A) $E \times \pi R^2$ (B) $E \times 2\pi R^2$ (C) $E \times 4\pi R^2$ (D) $E \times \frac{4}{3}\pi R^3$ (E) Zero

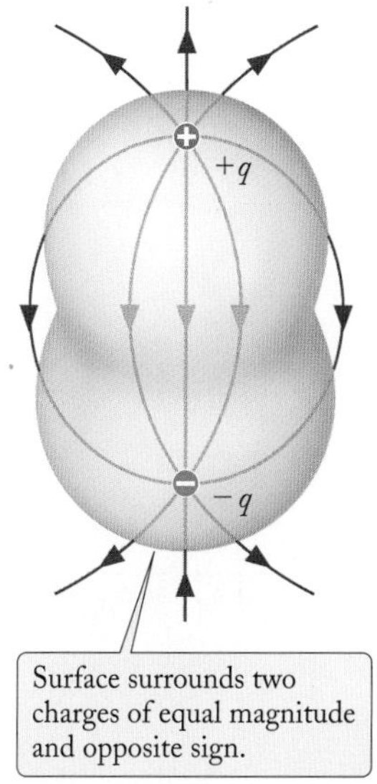

FIGURE 24.10 Field lines generated by positive and negative charges of equal magnitudes.

24.3 APPLICATIONS OF GAUSS' LAW

Gauss' Law relates the normal component $E_\perp$ of the electric field on a closed surface to the total charge inside the surface. To apply Gauss' Law in each of the three geometries with high symmetry (spherical, cylindrical, and planar), we will first choose a spherical or cylindrical surface for which the flux on the left side of Gauss' Law can be rewritten as the simple product EA, where E is the electric field at a desired point and A is a known area containing that point. We will then need to calculate Q_{inside}, the charge inside our surface. Finally, we can equate the flux EA with $Q_{\text{inside}}/\epsilon_0$ and solve for the desired quantity (usually E). To see how to calculate Q_{inside} for our surface, let us consider useful ways to quantify how charge may be distributed.

Macroscopically, the smallness of each elementary charge means that charge distributions can be viewed as *continuous* (even 1 μC contains several thousand billion elementary charges). We already saw two kinds of charge distributions in the previous chapter: a linear charge distribution, and a surface charge distribution. In the examples below, we will also encounter the third kind, a volume charge distribution. In the cases of immediate interest, we will again assume the charge distribution is *uniform*, that is, constant, over a specified region. For a uniform distribution, the charge q on a line of length L, or on a surface area A, or in a volume V, is proportional to the length, or the area, or the volume:

$$\begin{aligned} &\text{Line:} \quad && q = \lambda L \\ &\text{Surface:} \quad && q = \sigma A \\ &\text{Volume:} \quad && q = \rho V \end{aligned} \qquad (24.11)$$

where λ (lambda) is the charge per unit length in coulombs per meter, or C/m; σ (sigma) is the charge per unit area in coulombs per square meter, or C/m^2; and ρ (rho) is the charge per unit volume in coulombs per cubic meter, or C/m^3. For instance, a volume of 1 cm^3 with a charge of 1 μC uniformly distributed throughout it has $\rho = 1\ \mu\text{C/cm}^3 = 1\ \text{C/m}^3$. The more general possibility of a *nonuniform* distribution, where the value of ρ may vary as a function of position, is discussed in Problems 39–44. In the examples, we restrict ourselves to uniform distributions.

We will now examine some calculations of the electric field **E** in each of the three high-symmetry geometries: spherical, cylindrical, and planar; with these as background, Gauss' Law can then be applied to an enormous number of problems. We begin with two examples in spherical symmetry.

KARL FRIEDRICH GAUSS (1777–1855) *German mathematician, physicist, and astronomer. Gauss was an indefatigable calculator, and he loved to perform enormously complicated computations, which today would be regarded as impossible without an electronic computer. He developed many new methods for calculations in celestial mechanics; later he became interested in electric and magnetic phenomena, which he researched in collaboration with Wilhelm Eduard Weber (1804–1891).*

Spherical Symmetry

Charge distributions with spherical symmetry include points, spheres, thin or thick spherical shells, etc., and concentric layers of such objects.

Concepts — in — Context

EXAMPLE 4 Use Gauss' Law to deduce the electric field of a point charge.

SOLUTION: Although we already know the solution (Coulomb's Law, $E = kq/r^2$), it is instructive to obtain it from Gauss' Law. We begin with a symmetry argument: Since the point charge is spherically symmetric, so must be the electric field. This spherical symmetry requires that the magnitude of the electric field be the same at all points at equal distances from the center, that is, at all points on a spherical surface centered on the charge, since otherwise the electric field would make an unacceptable distinction between such points. Furthermore, the spherical symmetry requires that the direction of the electric field be rotationally symmetric around any radial line extending outward from the charge. The only direction of the electric field consistent with this requirement is *radial*, since any nonradial component would make an unacceptable distinction between different directions around this line.

Now consider a spherical Gaussian surface of radius r centered on the charge q, as shown in Fig. 24.6. According to the symmetry arguments, at all points on this surface, the electric field must be radial (radially outward for a positive charge and radially inward for a negative charge) and constant (for a given value of r). On this Gaussian surface, the normal component of the electric field is the same as the total field, $E_\perp = E$; and the flux is therefore

$$\oint E_\perp \, dA = EA = E \times 4\pi r^2 \tag{24.12}$$

The charge inside is merely the point charge q, so Gauss' Law tells us

$$E \times 4\pi r^2 = \frac{Q_{\text{inside}}}{\epsilon_0} = \frac{q}{\epsilon_0} \tag{24.13}$$

from which we immediately obtain Coulomb's Law:

$$E = \frac{1}{4\pi\epsilon_0}\frac{q}{r^2} \tag{24.14}$$

COMMENT: Notice that the same argument holds for *any* spherically symmetric charge distribution, provided $q = Q_{\text{inside}}$. Hence the general result: the electric field E at a radius r due to *any* spherically symmetric charge distribution is the same as if all the charge *inside the shell of radius r* were concentrated at a point charge in the center.

Concepts — in — Context

EXAMPLE 5 A spherical region of radius R has a total charge Q distributed uniformly thougout the volume of this region. (a) What is the electric field at points inside the sphere? (b) What is the electric field at points outside the sphere?

SOLUTION: Since there is charge throughout the volume of the sphere, the electric field lines of this charge distribution start within the volume of the spherical region, at varying distances from the center (see Fig. 24.11). Outside the sphere of charge, the number of electric field lines remains constant.

(a) Once again, since the charge distribution is spherically symmetric, the electric field must be radial and of constant magnitude over any concentric spherical surface of radius r. To find the magnitude of the electric field E at a radius

$r < R$ inside the charge distribution, consider the spherical Gaussian surface of Fig. 24.12a. As always in spherical symmetry, $E_\perp = E$ over the surface, so the flux through such a sphere is again the product $E \times 4\pi r^2$, and Gauss' Law again becomes

$$E \times 4\pi r^2 = \frac{Q_{\text{inside}}}{\epsilon_0} \tag{24.15}$$

To express Q_{inside} in terms of the given quantities, we use the fact that the charge is uniformly distributed throughout the specified volume. Thus for $r < R$, the charge inside is

$$Q_{\text{inside}} = \rho V_{\text{inside}} \tag{24.16}$$

where the volume charge density ρ (charge per unit volume) is determined by the total charge Q and total volume V of the *entire* charged sphere:

$$\rho = \frac{Q}{V} = \frac{Q}{\frac{4}{3}\pi R^3} \tag{24.17}$$

In Eq. (24.16), the quantity V_{inside} is the volume inside our Gaussian surface of radius r:

$$V_{\text{inside}} = \tfrac{4}{3}\pi r^3$$

Substituting this in Eq. (24.16), we find

$$Q_{\text{inside}} = \rho V_{\text{inside}} = \frac{Q}{\frac{4}{3}\pi R^3} \cdot \tfrac{4}{3}\pi r^3 = Q\frac{r^3}{R^3}$$

and Gauss' Law then allows us to solve for E:

$$E \times 4\pi r^2 = \frac{Q_{\text{inside}}}{\epsilon_0} = \frac{Q}{\epsilon_0}\frac{r^3}{R^3}$$

$$E = \frac{Q}{4\pi\epsilon_0}\frac{r}{R^3} \quad (r \le R) \tag{24.18}$$

Note that this electric field inside a uniformly charged sphere increases in direct proportion to the radius r. It is zero at the center, and reaches its maximum value at the edge of the charge distribution ($r = R$), where $E = (1/4\pi\epsilon_0)Q/R^2$.

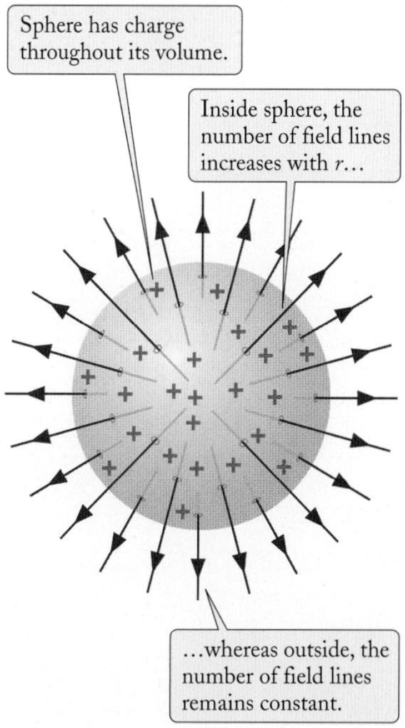

FIGURE 24.11 A sphere of radius R with a uniform volume distribution of positive charge. The electric field lines start on positive charges, at different radii. The electric field is everywhere radially outward.

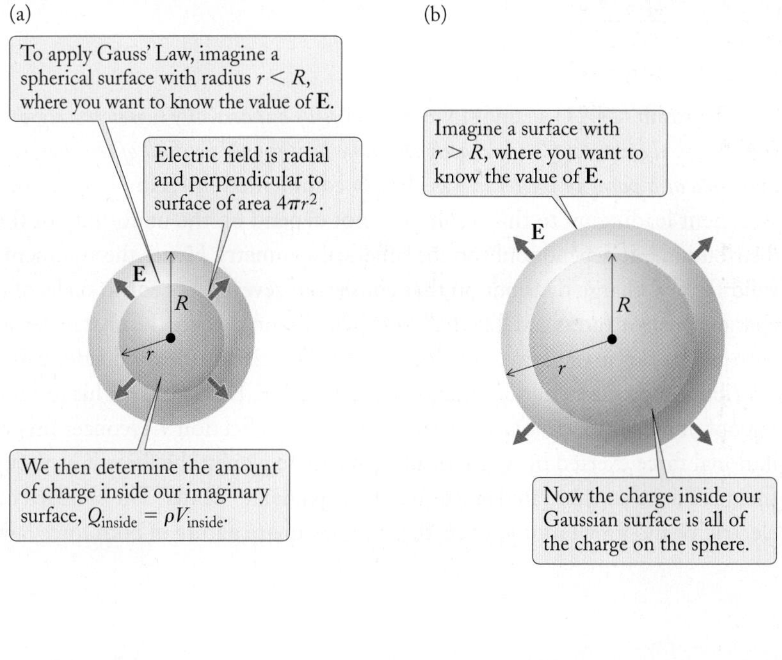

FIGURE 24.12 (a) An imaginary Gaussian spherical surface of radius $r < R$ (green) inside a uniformly charged sphere of radius R (tan). (b) Same charged sphere as (a), but now the imaginary spherical surface has $r > R$.

(b) To find the electric field outside the sphere, we again take a spherical Gaussian surface of radius r, but now r is larger than R (see Fig. 24.12b). With the usual spherical symmetry argument, Gauss' Law again has the form

$$E \times 4\pi r^2 = \frac{Q_{\text{inside}}}{\epsilon_0} \tag{24.19}$$

For any $r \geq R$, the charge inside the Gaussian surface is now equal to Q, the total charge. Thus,

$$E \times 4\pi r^2 = \frac{Q}{\epsilon_0} \tag{24.20}$$

and

$$E = \frac{1}{4\pi\epsilon_0}\frac{Q}{r^2} \quad (r \geq R) \tag{24.21}$$

Figure 24.13 is a plot of the electric field as a function of distance r from the center. Notice that it increases linearly inside the sphere, but decreases in proportion to $1/r^2$ outside the distribution.

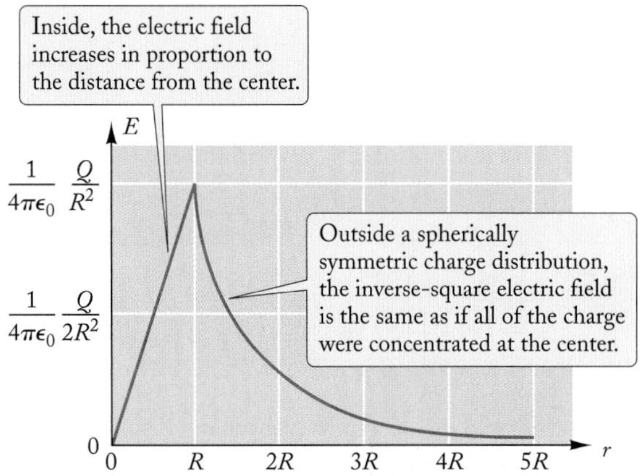

FIGURE 24.13 Magnitude of the electric field as a function of radial distance r for a uniformly charged sphere.

The result (24.21) again implies that *outside a spherically symmetric region containing charge, the electric field is exactly the same as it would be if all of the charge were concentrated at a point in the center* (as already commented in Example 4). Note that the argument leading up to this result does not depend on the uniformity of the charge distribution—it depends only on the spherical symmetry. Hence the argument remains valid for any charge distribution that consists of several concentric shells of different charge densities (arranged like the layers of an onion), where the charge density varies with radius but is the same in all directions. The electric field outside such a charge distribution precisely mimics that of a point charge, Eq. (24.21). This result provides a proof of Newton's famous theorem, mentioned in Section 9.1, concerning the gravitational force exerted by a spherically symmetric planet: the force is as though all of the mass were concentrated at a point at the planetary center. The similarity between electricity and gravitation reflects the inverse-square nature of both force laws.

EXAMPLE 6 A uranium nucleus (charge $+92e$) is (approximately) a uniformly charged sphere of radius 7.4×10^{-15} m. What is the electric field inside a uranium nucleus at a point halfway from the center to the surface? What is the radial electric force if such a field acts on a proton at the same point?

SOLUTION: For the field halfway inside a uniformly charged sphere, we can use Eq. (24.18) with $r = R/2$:

$$E = \frac{Q}{4\pi\epsilon_0}\frac{r}{R^3} = \frac{92e}{4\pi\epsilon_0}\frac{1}{2R^2}$$

$$= \frac{(92)(1.60 \times 10^{-19}\ \text{C})(9.0 \times 10^9\ \text{N}\cdot\text{m}^2/\text{C}^2)}{(2)(7.4 \times 10^{-15}\ \text{m})^2}$$

$$= 1.2 \times 10^{21}\ \text{N/C}$$

The force exerted on a proton by such a field is

$$F = eE = (1.6 \times 10^{-19}\ \text{C})\,(1.2 \times 10^{21}\ \text{N/C}) = 192\ \text{N}$$

In the nucleus of an atom, this large repulsive force is more than compensated by an even larger binding force—the "strong" force—that holds the nucleus together.

Cylindrical Symmetry

Charge distributions with cylindrical symmetry include infinitely long lines, rods (solid cylinders), tubes (cylindrical shells), etc., and concentric layers of such objects.

EXAMPLE 7 Using Gauss' Law, find the electric field of an infinitely long, thin straight rod of charge, with a uniform linear charge density λ.

SOLUTION: We calculated this electric field in Chapter 23 by direct integration of the electric fields of all the infinitesimal charge contributions, a tedious process. Now we will see that the calculation is much simpler by Gauss' Law. First we need to determine the direction of the electric field. The only direction consistent with a cylindrically symmetric charge distribution is *radial* (see Fig. 24.14); electric field lines must originate on charges in the rod, and if the electric field had any nonradial direction, it would make an unacceptable distinction between the upper and lower portions of the rod. Furthermore, the rotational symmetry around the rod dictates that the electric field has the same magnitude at any point on the surface of a cylinder concentric with the rod: otherwise, it would make an unacceptable distinction between different positions along or around the rod.

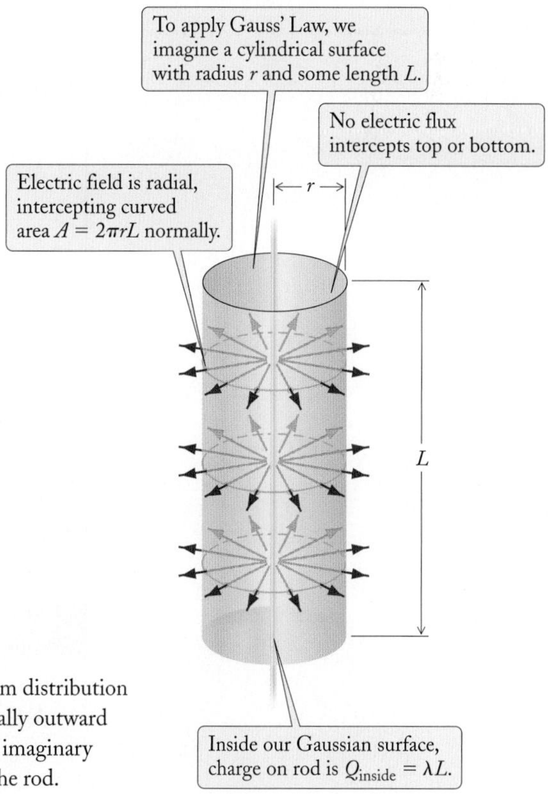

FIGURE 24.14 A very long rod with a uniform distribution of positive charge (blue). The field lines are radially outward and are uniformly distributed along the rod. An imaginary cylindrical Gaussian surface (green) surrounds the rod.

For cylindrical problems, we take for our Gaussian surface an imaginary cylindrical surface of radius r (where we want to know E) and of some arbitrary length L. The closed cylindrical surface consists of two circles (top and bottom) and a curved side. Since the electric field is radial, it is parallel to the top and bottom circular surfaces, so no electric flux is intercepted by these surfaces. Thus the top and bottom ends of the cylinder make zero contribution to the flux in cylindrical symmetry. Also, the radial electric field is constant and normal over the curved side of the cylindrical surface, which has area $A = 2\pi rL$ (the circumference multiplied by the height). The total flux is

$$\begin{aligned}\oint E_\perp\, dA &= \int_{\text{top}} E_\perp\, dA + \int_{\text{bottom}} E_\perp\, dA + \int_{\text{side}} E_\perp\, dA \\ &= 0 + 0 + \int_{\text{side}} E_\perp\, dA \\ &= E_\perp \int_{\text{side}} dA = E_\perp A_{\text{side}} = E_\perp \times 2\pi rL\end{aligned}$$

So for all problems with cylindrical symmetry, Gauss' Law may always be written

$$E \times 2\pi rL = \frac{Q_{\text{inside}}}{\epsilon_0} \tag{24.22}$$

For the simple uniform linear distribution, the charge enclosed by our surface is the charge per unit length multiplied by the length:

$$Q_{\text{inside}} = \lambda L$$

Hence Gauss' Law implies

$$E \times 2\pi rL = \frac{\lambda L}{\epsilon_0}$$

and, solving for E, we find

$$E = \frac{\lambda}{2\pi\epsilon_0 r} \tag{24.23}$$

which agrees with the result from Chapter 23, Eq. (23.10).

COMMENT: Notice that the same relation (24.23) holds for *any* cylindrically symmetric charge distribution, provided λ represents the total charge per unit length inside a cylinder of radius r where we want to know the value of E. Thus we see another general result: the E field at a radius r due to *any* cylindrically symmetric charge distribution is the same as if all the charge *inside the cylinder of radius r* were concentrated on a line of charge along the axis.

EXAMPLE 8 A thick, insulating cylindrical shell with inner radius a and outer radius b has charge distributed thoughout its volume, with a uniform volume charge density ρ, as shown in Fig. 24.15. Find the electric field in the three regions (a) $r \leq a$, (b) $a \leq r \leq b$, and (c) $r \geq b$.

SOLUTION: As just discussed, in cylindrical symmetry the electric field must be radial and of constant magnitude for any particular radius r. Thus Gauss' Law in cylindrical symmetry always becomes

$$E \times 2\pi r L = \frac{Q_{\text{inside}}}{\epsilon_0} \qquad (24.24)$$

(a) For $r \leq a$, consider the cylindrical Gaussian surface shown in green in Fig. 24.15a. Since it is contained completely inside the cylindrical "hole" within the shell, it encloses no charge. Thus $Q_{\text{inside}} = 0$ and $E \times 2\pi r L = 0$, or

$$E = 0 \qquad (r \leq a) \qquad (24.25)$$

(b) For $a \leq r \leq b$, consider a cylindrical Gaussian surface with its curved part inside the shell, as in Fig. 24.15b. In terms of the given volume charge density ρ,

$$Q_{\text{inside}} = \rho V_{\text{inside}}$$

but in calculating V_{inside}, we must include only that volume that *actually contains charge* (and is inside the radius r). The volume of our Gaussian cylinder is the area πr^2 of its end times its length L; if we subtract the volume that is empty, we have

$$V_{\text{inside}} = \pi r^2 L - \pi a^2 L = \pi L(r^2 - a^2)$$

so Gauss' Law becomes

$$E \times 2\pi r L = \frac{\rho V_{\text{inside}}}{\epsilon_0} = \frac{\rho \pi L(r^2 - a^2)}{\epsilon_0}$$

and from this we find

$$E = \frac{\rho}{2\epsilon_0}\left(r - \frac{a^2}{r}\right) \qquad (a \leq r \leq b) \qquad (24.26)$$

(c) For $r \geq b$, the cylindrical Gaussian surface is outside the shell (see Fig. 24.15c). The charge inside is obtained using *only the volume that contains charge* (this is the volume of the charged shell, since we exclude the empty volume inside and outside the shell):

$$Q_{\text{inside}} = \rho V_{\text{inside}} = \rho(\pi b^2 L - \pi a^2 L)$$

so Gauss' Law gives us

$$E \times 2\pi r L = \frac{\rho \pi L(b^2 - a^2)}{\epsilon_0}$$

and

$$E = \frac{\rho(b^2 - a^2)}{2\epsilon_0 r} \qquad (r \geq b) \qquad (24.27)$$

Note the $1/r$ dependence for this electric field outside the cylindrically symmetric charge distribution. Indeed, Eq. (24.27) is identical to the result (24.23) for a line of charge, $E = \lambda/2\pi\epsilon_0 r$, where now the charge per unit length is the charge per unit volume times the cross-sectional area of the shell, that is, $\lambda = \rho(\pi b^2 - \pi a^2)$.

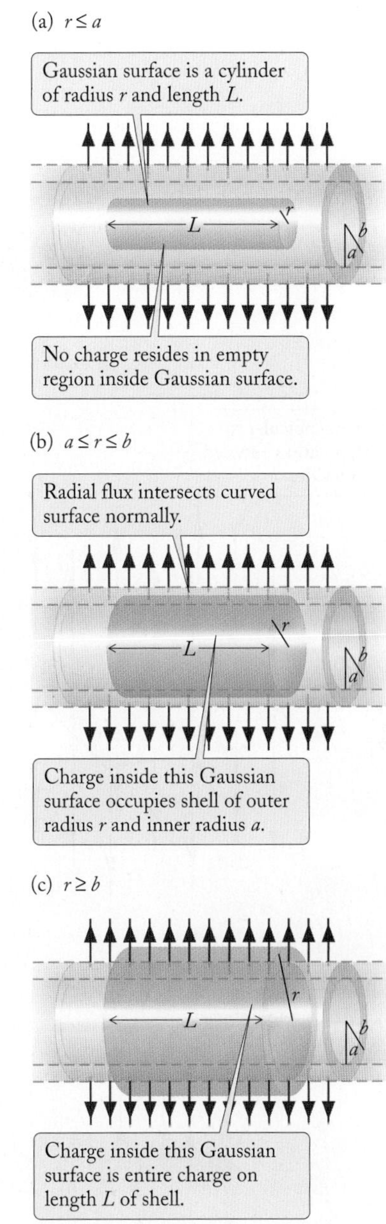

FIGURE 24.15 (a) An infinite cylindrical shell (tan), with inner radius a and outer radius b, with a uniform volume distribution of positive charge. The field lines are radially outward. To find the electric field at different radii, we consider imaginary Gaussian cylindrical surfaces (green) of different radii, (a) Gaussian surface of radius $r \leq a$. (b) Gaussian surface of radius $a \leq r \leq b$. (c) Gaussian surface of radius $r \geq b$.

Planar Symmetry

Charge distributions with planar symmetry include infinite sheets and slabs of charge and symmetric layers of such sheets or slabs.

Concepts in Context

EXAMPLE 9 Using Gauss' Law, find the electric field of a very large uniform sheet of charge with σ coulombs per square meter.

SOLUTION: This is another instance of an electric field we have already calculated by direct integration in Chapter 23. Again, the calculation is much simpler by Gauss' Law. Symmetry tells us that the electric field is everywhere perpendicular to the sheet of charge and has a constant magnitude over any particular surface parallel to the sheet (see Fig. 24.16a). We take a Gaussian surface in the shape of a cylinder of base area A that extends symmetrically on either side of the sheet of charge (see Fig. 24.16b). In this geometry, the electric field is parallel to the curved surface of the cylinder, which therefore intercepts no flux. On each of the two circular ends, the electric field is perpendicular to the surface. Thus the flux through each end is EA. The total flux is

$$\oint E_\perp\, dA = \int_{\text{bottom}} E_\perp\, dA + \int_{\text{curved part}} E_\perp\, dA + \int_{\text{top}} E_\perp\, dA$$

$$= EA + 0 + EA = 2EA$$

so Gauss' Law for this planar geometry becomes

$$2EA = \frac{Q_{\text{inside}}}{\epsilon_0} \tag{24.28}$$

The amount of charge inside the Gaussian surface is the charge per unit area multiplied by the area of the sheet enclosed by the surface, or $Q_{\text{inside}} = \sigma A$. So Gauss' Law gives

$$2EA = \frac{\sigma A}{\epsilon_0}$$

Canceling the common factor of area A, we find

$$E = \frac{\sigma}{2\epsilon_0} \tag{24.29}$$

This result, of course, agrees with the result obtained in Section 23.2.

COMMENTS: In this geometry, we could have used a cubical or rectangular box as our Gaussian surface, instead of a cylinder. The shape of the ends is irrelevant; the crucial features of the Gaussian surface are that its ends must be parallel to the sheet, the surface around it must be perpendicular to the sheet, and it should be symmetric on either side of the sheet. Note that outside the planar charge distribution, the magnitude of the electric field is independent of the distance from the sheet. For each type of symmetry, it is useful to remember the behavior of the electric field outside the charge distribution: spherical, $E \propto 1/r^2$; cylindrical, $E \propto 1/r$; and planar, $E \propto$ constant.

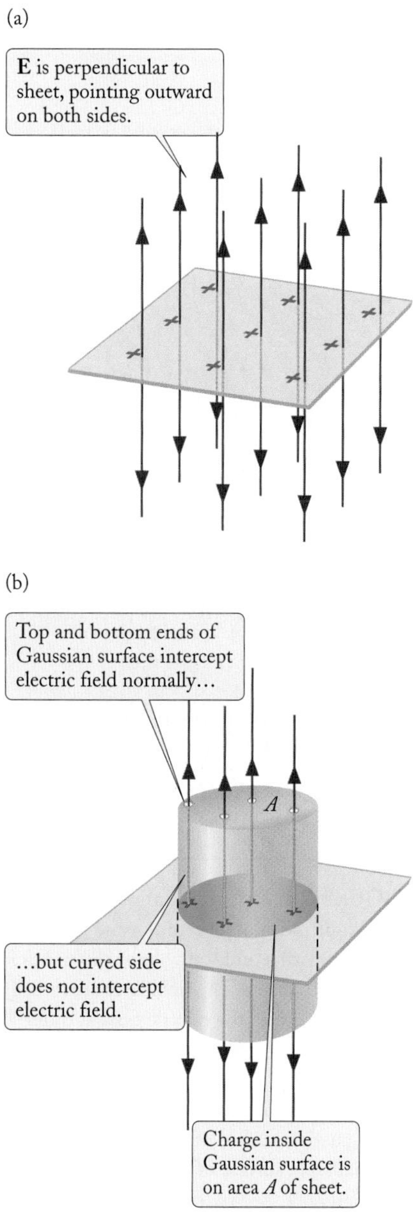

FIGURE 24.16 (a) A very large sheet of uniformly distributed positive charge. The field lines are uniformly distributed. (b) A cylindrical Gaussian surface; the area of each end is A.

EXAMPLE 10 An infinite, insulating slab of thickness d has a uniform volume charge density ρ, as shown in Fig. 24.17. Find the electric field inside and outside the slab.

SOLUTION: With $x = 0$ at the center of the slab, "inside" the slab means $|x| < d/2$ and "outside" means $|x| > d/2$. Consider the Gaussian surfaces shown in Fig. 24.17. To obtain E inside the slab, we use the smaller cylinder to evaluate $Q_{\text{inside}} = \rho V_{\text{inside}}$. Since its volume is V_{inside} = [length] × [base area] = $2x\,A$, Gauss' Law becomes

$$2EA = \frac{Q_{\text{inside}}}{\epsilon_0} = \frac{\rho \times 2x\,A}{\epsilon_0}$$

and solving for E we obtain

$$E = \frac{\rho x}{\epsilon_0} \quad (|x| \le d/2) \tag{24.30}$$

Thus the E field is zero at the center of the slab and increases linearly with distance inside the slab.

Outside the slab, we use the larger Gaussian cylinder. The amount of volume inside it *which contains charge* is $V_{\text{inside}} = d \times A$, so Gauss' Law implies

$$2EA = \frac{Q_{\text{inside}}}{\epsilon_0} = \frac{\rho \times d \times A}{\epsilon_0}$$

and

$$E = \frac{\rho d}{2\epsilon_0} \quad (|x| \ge d/2) \tag{24.31}$$

Note that this constant electric field is the same as for a sheet of charge ($E = \sigma/2\epsilon_0$) with surface charge density $\sigma = \rho d$, as it must be.

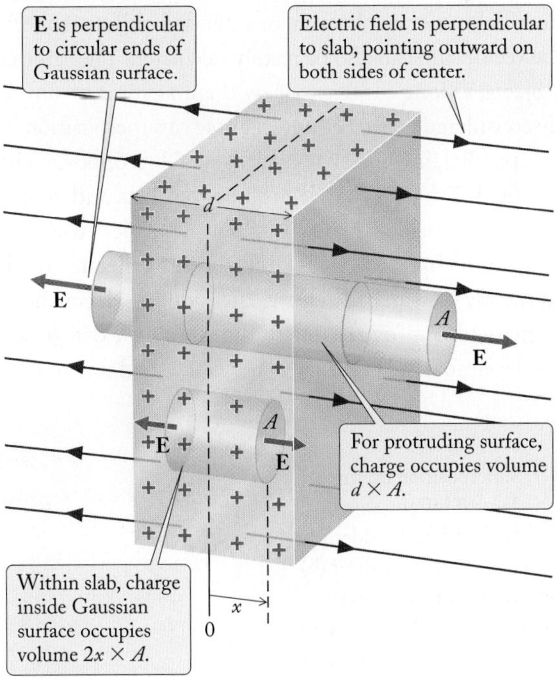

FIGURE 24.17 An infinite slab of thickness d with a uniform volume distribution of positive charge. A small, imaginary Gaussian cylinder with end area A is used to find the field inside the slab, a distance x from the center; a longer Gaussian cylinder with end area A is used to find the field outside the slab.

Checkup 24.3

QUESTION 1: A charge distribution consists of four equal charges at the corners of a square. Does this arrangement have enough symmetry so you can use Gauss' Law to calculate the electric field at some distance from the square?

QUESTION 2: A long wire of radius 10^{-5} m carries a uniform linear charge distribution. Another long wire of radius 10^{-6} m carries a linear charge distribution of the same magnitude. Which wire produces a larger electric field at its surface?

QUESTION 3: A charge Q is distributed uniformly over the volume of a sphere of radius R. What is the electric field at the center of the sphere? Where is the electric field strongest? What is the magnitude of this strongest electric field?

QUESTION 4: A charge Q is distributed over the the volume of a sphere of radius R. The charge distribution consists of several concentric shells of different charge densities; although the charge density varies with radius, it is the same in all directions. What is the electric field at the center of this sphere? At the surface?

QUESTION 5: The electric field at some distance from a long wire carrying a uniform charge distribution is 1.0×10^5 N/C. What is the electric field at twice the distance?

(A) 2.5×10^4 N/C (B) 5.0×10^4 N/C (C) 1.0×10^5 N/C
(D) 2.0×10^5 N/C (E) 4.0×10^5 N/C

24.4 SUPERPOSITION OF ELECTRIC FIELDS

It is worthwhile to emphasize that the Superposition Principle holds also for extended charge distributions, the same way it did for point charges. That is, the electric field at any point due to several charge distributions is simply the vector sum of the electric fields of the individual charge distributions. With point charges, this is a straightforward sum of Coulomb fields. For a collection of extended objects (spheres, cylinders, shells, planes, etc.), the vector sum can also be readily calculated. The same can be true for more complicated objects, as long as we can view them as sums of simpler objects.

We already considered an important example of superposition in Fig. 23.10, where we saw that two parallel, infinite sheets with equal but opposite charge densities produce an electric field $E = \sigma/\epsilon_0$ between the two sheets and zero field outside them. For another example, consider two parallel, infinite sheets with equal positive charge densities, as shown in Fig. 24.18a. To find the total electric field in each region of space, we need merely consider the electric field due to either sheet separately. Figure 24.18b shows the field $E = \sigma/2\epsilon_0$ due to the left sheet, which points away from the sheet on each side; similarly, Fig. 24.18c shows the field due to the right sheet, which

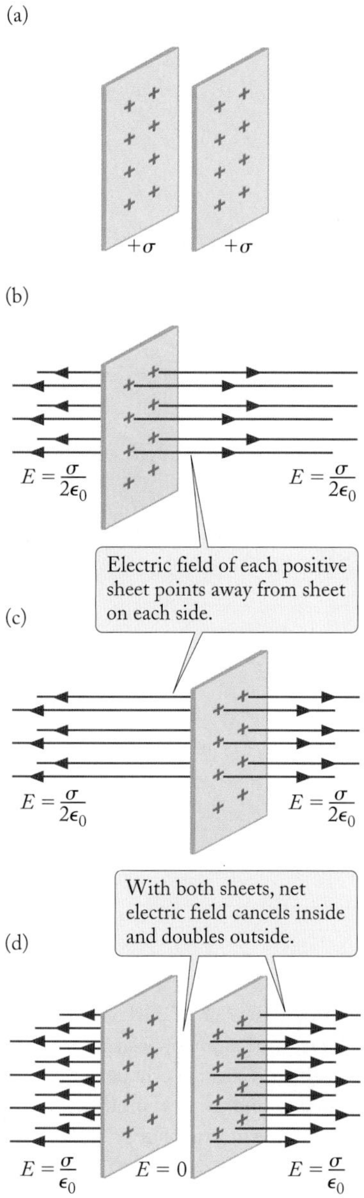

FIGURE 24.18 Superposition example. (a) Two positively charged infinite sheets. The total field can be found by considering the individual fields due to the left sheet (b) and the right sheet (c) separately. In each region of space, one can sum the fields due to the individual sheets to get the total field (d).

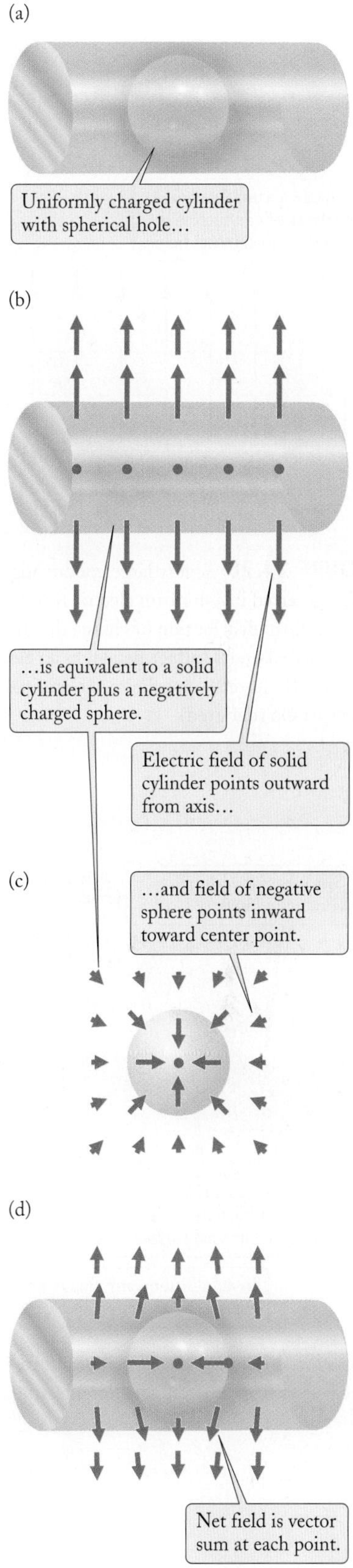

FIGURE 24.19 Superposition example. (a) A solid, infinite cylinder with a uniform volume distribution of positive charge, with a spherical hole. The field due to such an object with missing volume can be found by summing the field due to a solid cylinder (b) plus a field due to an oppositely charged sphere (c). At each point, the net field (d) is the vector sum of these two fields.

points away from it on each side. When we sum the fields due to the two sheets, as in Fig. 24.18d, we see immediately that $E = 0$ between the two sheets, whereas $E = \sigma/2\epsilon_0 + \sigma/2\epsilon_0 = \sigma/\epsilon_0$ to the left and right of both sheets. These examples of sheets with charge densities of equal magnitudes lead to simple doublings or cancellations of the field. However, the Superposition Principle can just as easily be applied to any number of sheets with any values of uniform surface charge densities.

Another example is shown in Fig. 24.19a. Here we have a solid cylinder with uniform volume charge distribution ρ that has a spherical cavity inside of it. Such a hole may be viewed as consisting of equal amounts of positive and negative charge (zero net charge in the hole), so that the electric field at any point is the sum of the electric field due to a solid cylinder (no hole) with charge density ρ and that due to a solid sphere of charge density $-\rho$. Of course, at any position in space, we must take the vector sum of the cylindrical (radially outward from the axis, Fig. 24.19b) and spherical (radially toward the center point, Fig. 24.19c) fields. Such a vector sum is indicated schematically at various points in Fig. 24.19d. Note that for such an object, the highly symmetric field due to each "piece" is a simple application of Gauss' Law, whereas the vector sum of the two contributions is less symmetric and must be calculated for each point.

✓ Checkup 24.4

QUESTION 1: Suppose that the surface of the spherical cavity in Fig. 24.19 has a uniform surface charge density σ. Can the electric field still be calculated from a superposition of simple high-symmetry applications of Gauss' Law?

QUESTION 2: Assume that the cavity in Fig. 24.19 is cubic instead of spherical. Can the electric field still be calculated from a superposition of simple high-symmetry applications of Gauss' Law?

QUESTION 3: Two parallel infinite sheets carry positive uniform surface charge densities σ and 2σ. What is the magnitude of the electric field between the sheets? Outside both sheets?

(A) σ/ϵ_0; $3\sigma/\epsilon_0$ (B) $\sigma/2\epsilon_0$; $3\sigma/2\epsilon_0$ (C) $3\sigma/2\epsilon_0$; $\sigma/2\epsilon_0$
(D) σ/ϵ_0; 0 (E) 0; σ/ϵ_0

24.5 CONDUCTORS AND ELECTRIC FIELDS

Concepts in Context

In insulators (or *dielectrics*), electrons and ions in atoms are firmly bound to each other, and there are no free, mobile charges. Extra charges can be placed on an insulator in an arbitrary arrangement; these charges also stay put and do not move. However, as we pointed out in Section 22.5, in conductors the charges can move. For instance, in metals—such as copper, silver, or aluminum—some of the electrons are free, that is, they can move about without restraint within the volume of the metal. If such a conductor is immersed in an electric field, the free electrons move in response to the electric force. As an excess of electrons accumulates on one part of the conductor, a deficit of electrons will appear on another part of the conductor; thus, negative and positive charges are induced on the conductor. Within the volume of the conductor, the electric field of the induced charges tends to cancel the external electric field in which the conductor was originally immersed (see Fig. 24.20). The accumulation of negative and positive charges on the surfaces of the conductor continues until the electric field generated by these charges exactly cancels the original electric field that initiated the motion of the electrons. Consequently, *when the charge distribution on a conductor reaches electrostatic equilibrium, the net electric field within the material of the conductor is exactly zero.* The proof of this statement is by contradiction: if the electric field were different from zero, the free electrons would continue to move, and the charge distribution would not (yet) be in equilibrium. For a good conductor (copper, aluminum, etc.), the equilibrium is reached quickly, in a very small fraction of a second.

Furthermore, *for a conductor in electrostatic equilibrium, any (extra) electric charge deposited on the conductor resides on the surfaces of the conductor.* We can prove this using Gauss' Law: Consider a small Gaussian surface *inside* the conducting material (see Fig. 24.21). Since at equilibrium $\mathbf{E} = 0$ everywhere in this material, the left side of Gauss' Law vanishes and therefore the right side must also vanish—which means the charge enclosed by *any* arbitrary surface inside the conducting material is zero. Obviously, if the charges are not inside the conductor, they must be on the surfaces.

Finally, we can say something about the electric field just outside a conductor: *the electric field at the surface of a conductor in electrostatic equilibrium is perpendicular to the surface.* The proof is again by contradiction: if the electric field had a component tangential to the surface, the free electrons would move along the surface, and the charge distribution would not be in equilibrium.

Note that this argument does not exclude an electric field perpendicular to the surface of the conductor; such an electric field merely pushes the free electrons against the surface, where they are held in equilibrium by the combination of the force exerted by the electric field and the restraining force of the surface of the conductor. Figure 24.22

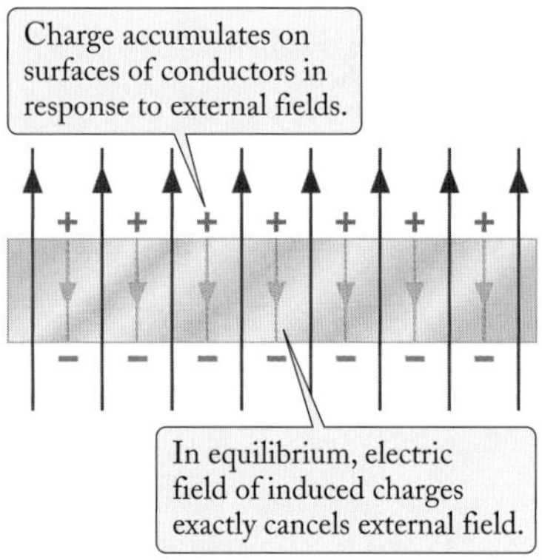

FIGURE 24.20 A very large conducting slab immersed in a uniform electric field. The uniform distribution of charge that has accumulated on its surface generates an electric field (light red) opposite to the original electric field (dark red).

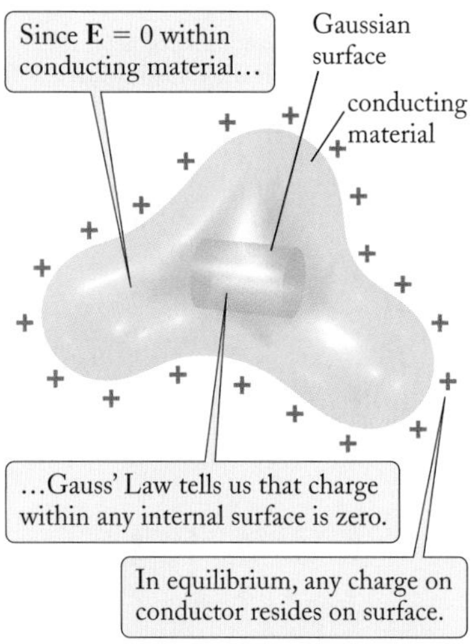

FIGURE 24.21 A closed Gaussian surface (green) inside a volume of conducting material.

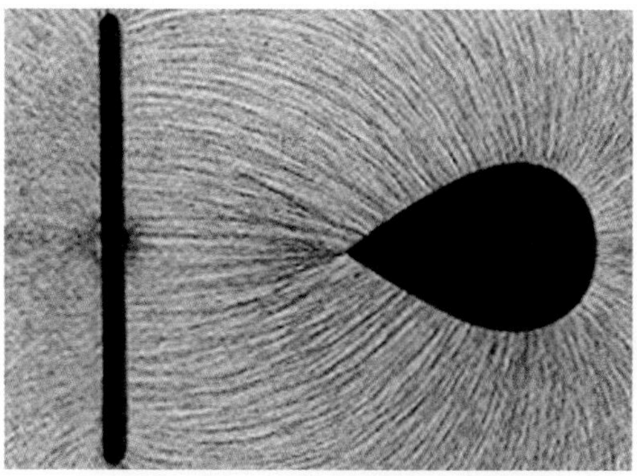

FIGURE 24.22 Field lines in the space surrounding charged conductors made visible by small bits of thread suspended in oil.

displays an experimental demonstration of electric fields perpendicular to the surfaces of conductors. The flat plate and the cylinder in this figure are conductors, and we see that the the field lines meet the surfaces at right angles.

EXAMPLE 11 Find the electric field above a very large, flat conducting slab (see Fig. 24.23) on which there is a surface charge density of σ coulombs per square meter.

Concepts in Context

SOLUTION: In view of the symmetry of the charge configuration, the electric field will be perpendicular to the conducting slab and will have a constant magnitude over any plane parallel to the conducting slab. As a Gaussian surface, take the cylinder shown in Fig. 24.23, with a base area A. The upper base contributes an amount EA to the flux. The lower base, inside the conductor, does not contribute to the flux, since the electric field is zero in this region. Finally, the curved surface does not contribute to the flux, since any electric field is parallel to this surface. The charge within the Gaussian surface is $Q_{\text{inside}} = \sigma A$. Hence, from Gauss' Law,

$$\oint E_{\perp}\, dA = EA = \frac{Q_{\text{inside}}}{\epsilon_0} = \frac{\sigma A}{\epsilon_0} \tag{24.32}$$

we obtain

$$E = \frac{\sigma}{\epsilon_0} \tag{24.33}$$

This is a constant electric field, independent of the distance from the conducting slab. This result is strictly valid for a slab of infinite extent, but it is approximately valid for the electric field near a slab of finite extent.

COMMENTS: Note that, for a given value of the surface charge density σ, the field generated by a conducting slab is twice as strong as the field generated by a sheet of charge [see Eq. (24.29)]. The reason is clear from a comparison of Figs. 24.16 and 24.23. In the latter case, the charges produce flux only on one side of the surface; in the former case, the flux is distributed over both sides. How is this possible? In the conducting case, there must always be another surface charge distribution that produces a field which cancels that due to the surface shown in Fig. 24.23. For example, a charged, conducting infinite slab can be considered as two charged sheets, such as the two we considered in Fig. 24.18.

Over a small region *near the surface,* any smooth, curved conducting surface can be approximated by a flat surface. Hence we can use the formula (24.33) to find the electric field in a region very near *any* smooth curved conducting surface (but $E = \sigma/\epsilon_0$ is *not* a good approximation near sharp edges). The formula is valid even if σ is not constant—if σ depends on position along the surface, then the electric field also depends on position.

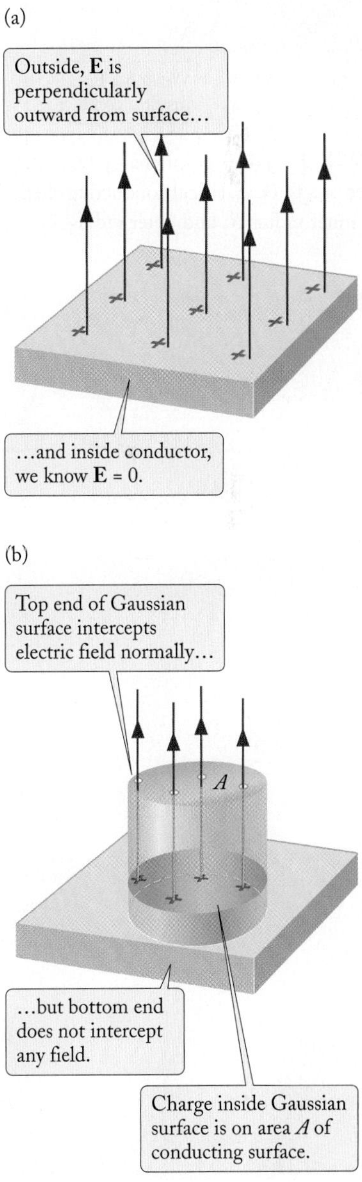

FIGURE 24.23 (a) A large, flat conducting slab with a uniform distribution of charge on its surface. (b) The cylindrical Gaussian surface has one base outside the conductor and the other base in the conductor.

Concepts — in — Context

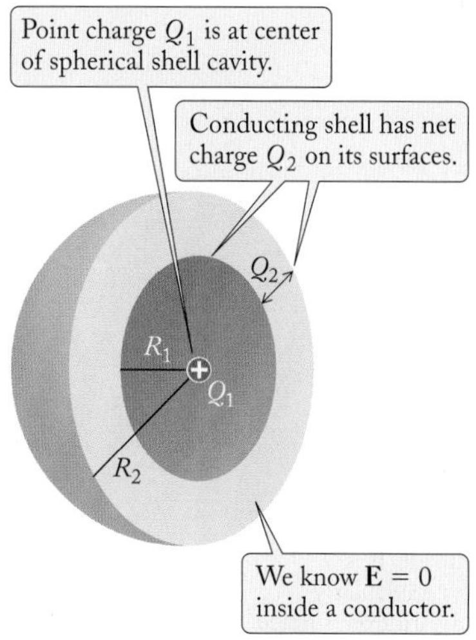

FIGURE 24.24 A point charge Q_1 at the center of a thick spherical conducting shell with inner radius R_1 and outer radius R_2.

EXAMPLE 12 A point charge Q_1 is at the center of a spherical conducting shell of inner radius R_1 and outer radius R_2, as shown in Fig. 24.24. The shell has a net charge Q_2 on its surfaces. Find the electric field in the three regions $r < R_1$, $R_1 < r < R_2$, and $r > R_2$. How much charge is on the inner surface of the shell? The outer surface?

SOLUTION: For any spherical Gaussian surface of radius $r < R_1$, only the point charge is inside the surface, so $Q_{\text{inside}} = Q_1$. Gauss' Law in spherical symmetry gives us $E \times 4\pi r^2 = Q_1/\epsilon_0$ and

$$E = \frac{Q_1}{4\pi\epsilon_0 r^2} \quad (r < R_1) \tag{24.34}$$

The region $R_1 < r < R_2$ is inside the conducting material, so we know immediately

$$E = 0 \quad (R_1 < r < R_2) \tag{24.35}$$

For a spherical Gaussian surface of radius $r > R_2$, the enclosed charge is $Q_1 + Q_2$, so Gauss' Law gives us

$$E = \frac{Q_1 + Q_2}{4\pi\epsilon_0 r^2} \quad (r > R_2) \tag{24.36}$$

To determine the charge on the inner surface, we use Gauss' Law in reverse: since $E = 0$ in the conducting material, there must be zero net charge inside any Gaussian sphere in the conducting material ($R_1 < r < R_2$). Since a point charge Q_1 is at the center of the spherical shell, the charge on the inner surface of the shell must be the opposite of Q_1:

$$Q_{\text{inner surface}} = -Q_1 \tag{24.37}$$

We say the point charge *induces* an equal but opposite charge on the inner surface of the conductor. To find the charge on the outer surface, we merely use conservation of charge:

$$Q_{\text{outer surface}} = Q_2 - Q_{\text{inner surface}} = Q_2 - (-Q_1) = Q_2 + Q_1 \tag{24.38}$$

The electrostatic requirement that $E = 0$ in a conducting material can lead to induced charges on any number of concentric conducting shells in either spherical or cylindrical geometry. We can always consider a Gaussian surface within the conducting material of a particular shell, and the requirement of zero net charge inside the Gaussian surface implies that the charge on the inner surface of that shell is exactly opposite to any net charge interior to that shell. Such interior charges can reside on any number of other conducting or insulating shells, or on a central sphere or point charge (spherical geometry) or a central rod or line of charge (cylindrical geometry).

For a planar geometry, there is a difference: an infinite sheet with surface charge density σ will induce a charge density of only $-\sigma/2$ on the near face of a parallel, conducting slab; an opposite charge density $+\sigma/2$ must also be induced on the far face of the slab. In the spherical and cylindrical shells, this "outer" charge does not contribute to the field inside the conductor, but in the planar slab the contribution is equal to that of the other face. Finally, we saw in Example 12 that the value of the charge on the outer surface of any conducting spherical or cylindrical shell is equal to the net charge on both surfaces of the shell minus the value of any induced charge on its inner surface. For a planar conducting slab, however, any net charge on the surfaces divides equally between the two faces of the slab.

PROBLEM-SOLVING TECHNIQUES — GAUSS' LAW FOR CHARGE DISTRIBUTIONS WITH SYMMETRY

As in Examples 4 through 12, the calculation of the electric field is readily possible by means of Gauss' Law only if the charge distribution has high symmetry (spherical, cylindrical, or planar). If so, you use Gauss' Law,

$$\oint E_\perp \, dA = \frac{Q_{\text{inside}}}{\epsilon_0}$$

in the following way:

- Determine the *direction* of the electric field from symmetry: *radial* for spherical and cylindrical geometries; *normal* for planar geometries.
- Choose an (imaginary) Gaussian surface in the region where you want to know the electric field: a sphere of radius r, a cylinder of radius r and length L, or a cylinder of base area A, as in the examples.
- Rewrite the left side of Gauss' Law (the electric flux) as the product [electric field] × [area] in one of the following ways:

$$\oint E_\perp \, dA = \begin{cases} E \times 4\pi r^2 & \text{(spherical)} \\ E \times 2\pi r L & \text{(cylindrical)} \\ 2EA & \text{(planar, both ends in field)} \\ EA & \text{(planar, one end in field)} \end{cases}$$

- For the right side of Gauss' Law, find Q_{inside} for your surface. Be sure to include only the charge *inside* the surface. You may need to express the enclosed charge in terms of the appropriate charge density (see Examples 4 through 12). For uniform densities, use

Line: $q = \lambda L$

Surface: $q = \sigma A$

Volume: $q = \rho V$

- Apply Gauss' Law: equate the appropriate expression (for $\oint E_\perp \, dA$) with $Q_{\text{inside}}/\epsilon_0$.
- Solve the resulting equation for the electric field E.
- If the problem involves conductors (metal) in electrostatic equilibrium, you have additional information (Section 24.5): $E = 0$ inside the conducting material, and any net charge resides on the surfaces of the conductors.

✓ Checkup 24.5

QUESTION 1: Can there be an electric field within the material of a conductor in electrostatic equilibrium?

QUESTION 2: A uniform horizontal electric field has horizontal field lines. Suppose we shove an uncharged copper ball into this electric field. Will this alter the direction of the field lines?

QUESTION 3: A large, horizontal sheet of paper carries a uniform charge distribution of 10^{-6} C/m^2. A large, horizontal aluminum plate carries a uniform charge distribution of 10^{-6} C/m^2 equally divided between its upper side and its lower side (0.5×10^{-6} C/m^2 on each side). Compare the electric fields produced above and below the paper sheet and the aluminum plate.

QUESTION 4: A point charge Q is at the center of an uncharged spherical conducting shell. How much charge is on the inner surface of the shell? The outer surface?

(A) 0; 0 (B) Q; 0 (C) $-Q$; 0 (D) $-Q$; Q (E) Q; $-Q$

SUMMARY

PROBLEM-SOLVING TECHNIQUES Gauss' Law for Charge Distributions with Symmetry (page 777)

ELECTRIC FLUX THROUGH AN OPEN SURFACE

$$\Phi_E = \mathbf{E} \cdot \mathbf{A} = E_\perp A = EA\cos\theta \quad \text{(flat surface, uniform } \mathbf{E}\text{)} \qquad (24.1)$$

$$\Phi_E = \int \mathbf{E} \cdot d\mathbf{A} = \int E_\perp \, dA \quad \text{(arbitrary surface, varying } \mathbf{E}\text{)} \qquad (24.5)$$

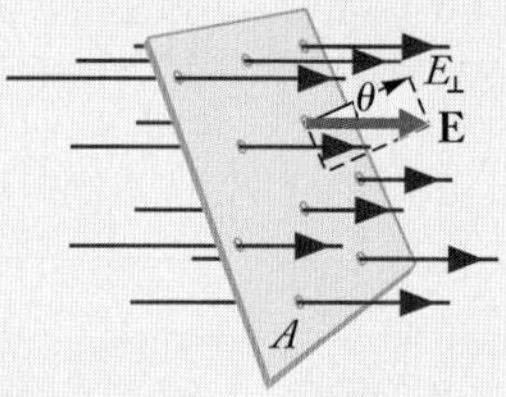

ELECTRIC FLUX THROUGH A CLOSED SURFACE
(positive for **E** outward)

$$\oint \mathbf{E} \cdot d\mathbf{A} = \oint E_\perp \, dA \qquad (24.6)$$

GAUSS' LAW

$$\oint E_\perp \, dA = \frac{Q_{\text{inside}}}{\epsilon_0} \qquad (24.10)$$

where

$$\oint E_\perp \, dA = \begin{cases} E \times 4\pi r^2 & \text{(spherical)} \\ E \times 2\pi r L & \text{(cylindrical)} \\ 2EA & \text{(planar, both ends in field)} \\ EA & \text{(planar, one end in field)} \end{cases}$$

UNIFORM CHARGE DISTRIBUTIONS

Line: $q = \lambda L$

Surface: $q = \sigma A$ (24.11)

Volume: $q = \rho V$

CONDUCTORS IN ELECTROSTATIC EQUILIBRIUM

The electric field within the conducting material is zero. Any charge resides on the surface(s). The electric field at the surface is perpendicular and of magnitude $E = \sigma/\epsilon_0$

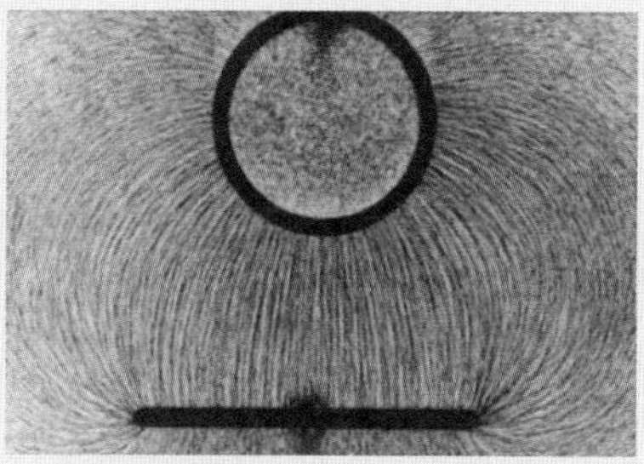

QUESTIONS FOR DISCUSSION

1. A point charge Q is inside a spherical Gaussian surface, which is enclosed in a larger, cubical Gaussian surface. Compare the fluxes through these two surfaces.

2. Figure 24.25 shows a Gaussian surface and lines of electric field entering and leaving the surface. What can you say about the sign of the electric charge within the surface?

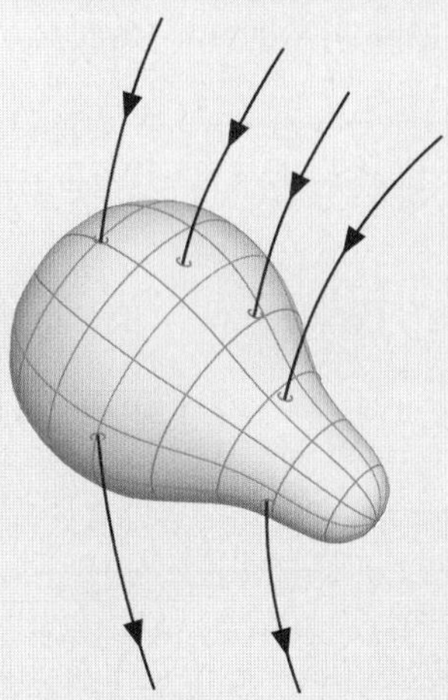

FIGURE 24.25 Gaussian surface and some field lines.

3. Suppose we had adopted some normalization for the number of flux lines per unit charge, say, $1/\epsilon_0$ lines per unit charge. Would this change Gauss' Law?

4. A Gaussian surface contains an electric dipole, and no other charge. What is the electric flux through this surface?

5. A point charge is at the center of a spherical shell. If it is moved to a position just inside the shell surface, does the electric flux through the shell change?

6. Suppose that the electric field of a point charge were not exactly proportional to $1/r^2$, but rather to $1/r^{2+a}$ where a is a small number, $a \ll 1$. Would Gauss' Law still be valid? (Hint: Consider Gauss' Law for the case of a point charge.)

7. Problems soluble by Gauss' Law fall into three categories, according to their symmetry: spherical, cylindrical, and planar. Give some examples in each category.

8. A hemisphere of radius R has a charge Q uniformly distributed over its volume. Can we use Gauss' Law to find the electric field?

9. A spherical rubber balloon of radius R has a charge Q uniformly distributed over its surface. The balloon is placed in a uniform electric field of 120 N/C. What is the net electric field inside the rubber balloon?

10. The electric field at the surface of a conductor in static equilibrium is perpendicular to the surface. Is the gravitational field at the surface of a mass in static equilibrium necessarily perpendicular to the surface? What if the surface is that of a fluid, such as water?

11. Suppose we drop a charged Ping-Pong ball into a cookie tin and quickly close the lid. What happens to the portions of the electric field lines that are outside the cookie tin when we close the lid?

12. When an electric current is flowing through a wire connected between a source and a sink of electric charge—such as the terminals of a battery—there is an electric field inside the wire, even though the wire is a conductor. Why does our conclusion about zero electric field inside a conductor not apply to this case?

13. The free electrons belonging to a metal are uniformly distributed over the entire volume of the metal. Does this contradict the result we derived in Section 24.5, according to which the charges are supposed to reside on the surface of a conductor?

PROBLEMS

24.1 Electric Flux†

1. Consider the thundercloud described in Problem 19 of Chapter 23 and Figure 23.34. What is the total electric flux coming out of the surface of the cloud?

2. A small square surface, 1.0 cm × 1.0 cm, is placed at a distance of 1.0 m from a point charge of 3.0×10^{-9} C. What is the approximate electric flux through this square if it is face on to the electric field (see Fig. 24.26)? If it is tilted by 30°? If it is tilted by 60°?

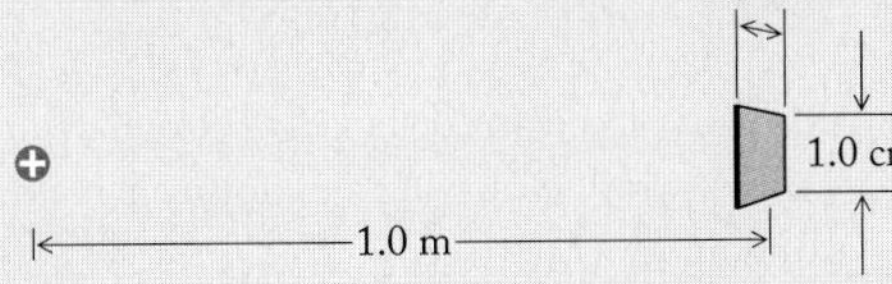

FIGURE 24.26 Point charge and small square surface.

3. The magnitude and the direction of a constant electric field are given by the vector $\mathbf{E} = 2.0\mathbf{i} - 1.0\mathbf{j} + 3.0\mathbf{k}$, where the field is measured in N/C. What is the flux that this electric field produces through the surface shown in Fig. 24.27, which consists of three squares $L \times L = 0.20$ m × 0.20 m each joined along their edges?

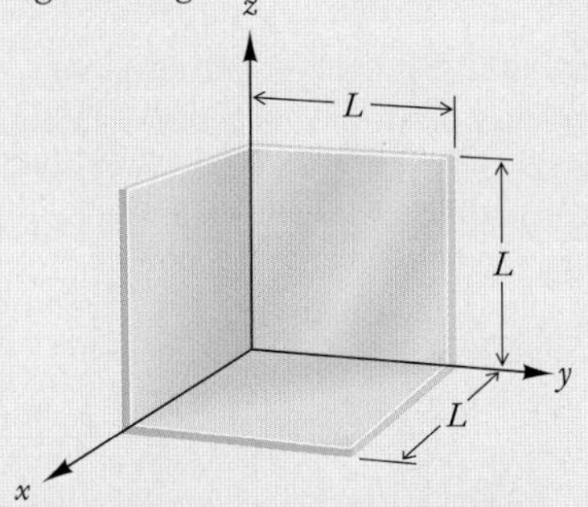

FIGURE 24.27 A surface consisting of three squares joined along their edges.

† For help, see Online Concept Tutorial 28 at www.wwnorton.com/physics

4. A thundercloud produces a vertical electric field of magnitude 2.8×10^4 N/C near ground level. You hold a 22 cm × 28 cm sheet of paper horizontally below the cloud. What is the electric flux through the sheet? What is the flux if you hold the sheet vertically?

5. The four faces of a tetrahedral pyramid are equilateral triangles of side a. Such a pyramid sits with one face flat on an infinite sheet of charge with surface charge density σ. What is the flux through the face of the pyramid that sits on the sheet? What is the flux through each of the other three faces?

6. A uniform electric field is given by $\mathbf{E} = (3.0\ \text{N/C})\mathbf{i} + (2.0\ \text{N/C})\mathbf{j} - (1.0\ \text{N/C})\mathbf{k}$. What is the flux through a flat, 4.0-m^2 area that lies in the y–z plane? What if that same area instead has its surface normal along an octant diagonal, so that the normal unit vector is $\hat{\mathbf{n}} = (1/\sqrt{3})\mathbf{i} + (1/\sqrt{3})\mathbf{j} + (1/\sqrt{3})\mathbf{k}$?

7. The direction of a uniform electric field is in the y–z plane at an angle of 30° from the $+y$ axis, 60° from the $+z$ axis. This uniform field extends throughout the region of a 2.0-m cube similar to the one in Fig. 24.7. What is the electric flux through each of the faces of the cube, labeled 1 through 6 in Fig. 24.7b? What is the net flux?

*8. A point charge of 6.0×10^{-8} C sits at some distance above the x–y plane. What is the electric flux that this charge generates through the (infinite) x–y plane? Does the electric flux depend on the distance?

*9. Consider the (infinite) midplane between two charges of equal magnitudes and opposite signs (see Fig. 23.17). If the charges are $\pm q$, what is the electric flux through this midplane?

*10. A point charge of 6.0×10^{-8} C sits on the wood floor of a room. You hold a small cardboard square measuring 1.0 cm × 1.0 cm face down, and carry it across the room at a height of 2.0 m, passing over the position of the point charge. Roughly plot the flux through the square as a function of position.

*11. A nonuniform electric field points everywhere in the $+x$ direction, so $\mathbf{E} = E_x\mathbf{i}$. The magnitude of the field is independent of the z coordinate and varies with x and y according to $E_x = C(x - 2y)$, where x and y are in meters and C is a constant with value $C = 3.6 \times 10^2$ N/(C·m). For a cube of side 2.0 m at the same position as in Fig. 24.7, what is the flux through the y–z face at $x = 1.0$ m (side 1)?

*12. A very small electric dipole is a distance z from the center of a loop of radius a, with its dipole moment vector $\mathbf{p}$ directed along the axis of the loop. The component of the electric field of the dipole perpendicular to the loop at any point P a distance r from the dipole is given by

$$E_\perp = \frac{p}{4\pi\epsilon_0 r^3}(3\cos^2\theta - 1)$$

where θ is the angle between the axis of the loop and the line from the dipole to P. Show that the electric flux through the loop is given by

$$\Phi_E = \frac{p}{2\epsilon_0}\frac{a^2}{(z^2+a^2)^{3/2}}$$

24.2 Gauss' Law†

13. A point charge of 2.0×10^{-12} C is located at the center of a cubical Gaussian surface. What is the electric flux through *each* of the faces of the cube?

14. A point charge of 1.0×10^{-8} C is placed inside an uncharged metallic can (say, a closed beer can) insulated from the ground. How much flux will emerge from the surface of the can when the point charge is inside?

15. A point charge q is placed at some distance from an infinitely large, flat sheet carrying a uniform distribution of charge σ coulombs per m^2.
    (a) What is the electric flux that this point charge produces through the sheet?
    (b) The electric flux Φ is the integral $\int E_\perp\, dA$; hence the product $\sigma\Phi$ is $\int \sigma E_\perp\, dA$, which equals the electric force that the point charge exerts on the sheet. Using the flux calculated in (a), evaluate this force.
    (c) From the force exerted by the point charge on the sheet, find the force that the sheet exerts on the point charge. From this, find the electric field that the sheet produces at the position of the point charge.

16. A point charge q is placed at a perpendicular distance $d/2$ from the center of a square of size $d \times d$ carrying a uniform charge distribution of σ coulombs per m^2 (see Fig. 24.28).
    (a) What is the electric flux that this point charge produces through the square? (Hint: The square is one of the six faces of a cube surrounding the point charge.)
    (b) By following the steps given in Problem 15, find the electric field that the square produces at the position of the point charge.

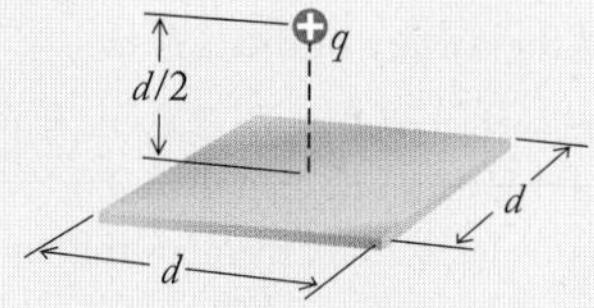

FIGURE 24.28 A point charge q and a charged square.

17. A point charge is placed at a distance $d/2$ from an infinitely long flat strip of width d carrying a uniform charge distribution of σ coulombs per m^2 (see Fig. 24.29).

†For help, see Online Concept Tutorial 29 at www.wwnorton.com/physics

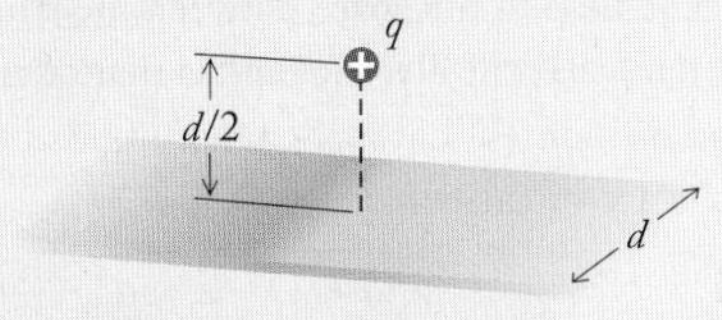

FIGURE 24.29 A point charge q and a charged strip.

(a) What is the electric flux that this point charge produces through the strip? (Hint: The strip is one of the four faces of a rectangular tube, or pipe, surrounding the point charge.)

(b) By following the steps in Problem 15, find the electric field that the strip produces at the position of the point charge.

18. A cube of side 2.0 m, similar to the one shown in Fig. 24.7, is in a region where the electric field is directed outward from both of two opposite cube faces, with uniform magnitude E_0 over each of those two faces. No flux crosses the other four faces. How much charge is inside the cube?

19. A short, thin rod of length 2.5 cm has charge distributed uniformly along its length. The rod is coaxial with and centered inside a much larger, uncharged cylindrical can. The flux through the curved part of the can is $+65\ \text{N}\cdot\text{m}^2/\text{C}$, and the flux through the bottom circular face is $+45\ \text{N}\cdot\text{m}^2/\text{C}$. What is the flux through the top circular face of the can? What is the value of the linear charge density λ of the rod?

20. Consider a cylindrical Gaussian surface of radius a and length $2z$ with a small electric dipole at its center, with its dipole moment $\mathbf{p}$ along the axis of the cylinder. According to the result of Problem 12, the flux though the upper circular face of the cylinder is given by

$$\Phi_E = \frac{p}{2\epsilon_0} \frac{a^2}{(z^2 + a^2)^{3/2}}$$

(a) What is the flux through the lower circular face?

(b) What is the flux through the curved surface of the cylinder?

*21. A spherical Gaussian surface of radius 1.00 m has a small hole of radius 10 cm. A point charge of 2.00×10^{-9} C is placed at the center of this spherical surface (see Fig. 24.30). What is the flux through the surface?

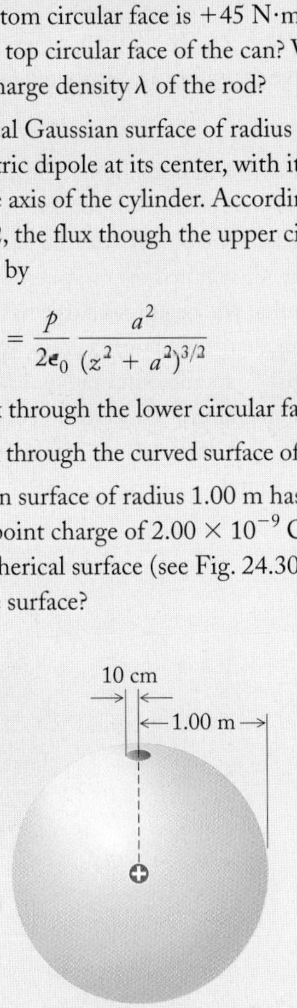

FIGURE 24.30 Spherical Gaussian surface with a small hole.

†For help, see Online Concept Tutorial 29 at www.wwnorton.com/physics

*22. Consider a point charge q located at one vertex of a cubical Gaussian surface, just inside the cube (see Fig. 24.31). What is the electric flux through each face of the cube?

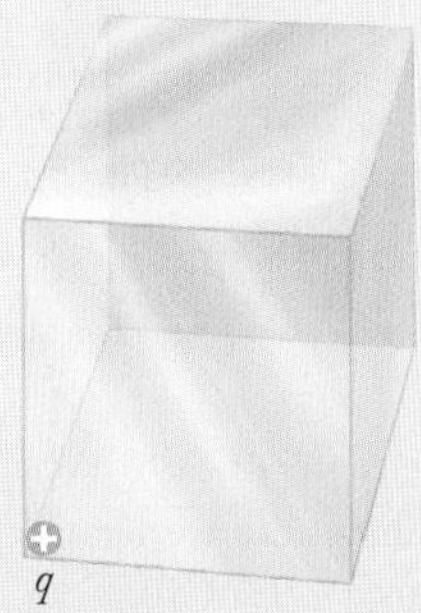

FIGURE 24.31 Point charge located just inside the vertex of a cubical Gaussian surface.

*23. Defining the gravitational field as the gravitational force per unit mass, formulate a Gauss' Law for gravity. Check that your law implies Newton's Law of Universal Gravitation.

*24. The electric field in a region has the following form as a function of x, y, z:

$$E_x = 5.0x \quad E_y = 0 \quad E_x = 0$$

where E is in newtons per coulomb and x is in meters. This represents an electric field in the x direction with a magnitude that increases in direct proportion to x. Show that such an electric field can exist only if the region is filled with some electric charge density. Find the value of the required charge density as a function of x, y, z.

**25. Prove that if the electric field is uniform in some region, then the charge density must be zero in that region. (Hint: Use Gauss' Law.)

†24.3 Applications of Gauss' Law

26. A uranium nucleus is a spherical ball of radius 7.4×10^{-15} m with a charge of $92e$ uniformly distributed over its volume. Find the electric field produced by this charge distribution at $r = 3.0 \times 10^{-15}$ m, 6.0×10^{-15} m, and 9.0×10^{-15} m.

27. Electrons can penetrate the nucleus because the nuclear material exerts no forces on them other than electric forces. Suppose that an electron penetrated a nucleus of lead. What is the electric force on the electron when it is at a distance of 3.0×10^{-15} m from the center of the nucleus? The nucleus of lead is a uniformly charged ball of radius 7.1×10^{-15} m with a total charge of $82e$.

28. In symmetric fission, a uranium nucleus splits into two equal pieces each of which is a palladium nucleus. The palladium nucleus is spherical with a radius of 5.9×10^{-15} m and a charge of $46e$ uniformly distributed over its volume. Suppose that, immediately after fission, the two palladium nuclei are barely touching (Fig. 24.32). What is the value of the total

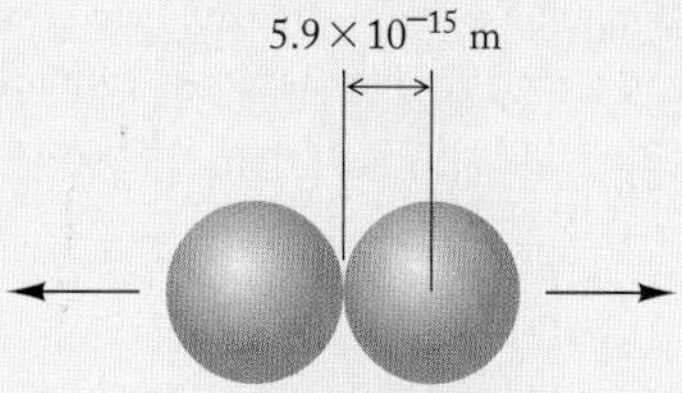

FIGURE 24.32 Two palladium nuclei in contact. Each nucleus is a sphere.

electric field at the center of each? What is the repulsive force between them? What is the acceleration of each? The mass of a palladium nucleus is 1.99×10^{-25} kg.

29. A spherical rubber balloon has a uniform distribution of charge over its surface. Show that the electric field that this charge produces in the (empty) interior of the balloon is exactly zero.

30. The diameter of a proton is about 1.0×10^{-15} m. How many diameters apart are two protons if the Coulomb force between them is 1.0 N?

31. Charge is uniformly distributed over the volume of a very long cylindrical plastic rod of radius R. The amount of charge per meter of length of the rod is λ. Find a formula for the electric field at a distance r from the axis of the rod. Assume $r < R$.

32. What is the maximum amount of electric charge per unit length that can be placed on a long and straight human hair of diameter 8.0×10^{-3} cm if the surrounding air is not to suffer electrical breakdown? The air will suffer breakdown if the electric field exceeds 3.0×10^{6} N/C.

*33. A plastic spherical shell has inner radius a and outer radius b (Fig. 24.33). Electric charge is uniformly distributed over the region $a < r < b$. The amount of charge is ρ coulombs per cubic meter. Find the electric field in the regions $r \leq a$, $a \leq r \leq b$, and $r \geq b$.

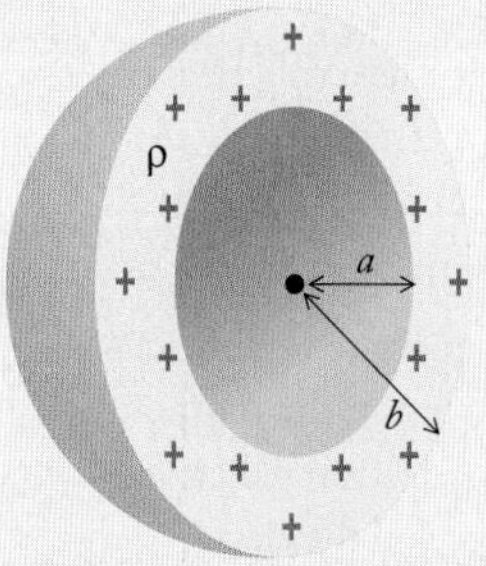

FIGURE 24.33 A spherical shell with uniform volume charge density ρ.

*34. The tube of a Geiger counter consists of a thin conducting wire of radius 1.3×10^{-3} cm stretched along the axis of a conducting shell of radius 1.3 cm (Fig. 24.34). The wire and the cylinder have equal and opposite charges of 7.2×10^{-10} C distributed along their length of 9.0 cm. Find a formula for the electric field in the space between the wire and the cylinder; pretend that the electric field is that of an infinitely long wire and cylinder. What is the magnitude of the electric field at the surface of the wire?

FIGURE 24.34 A thin wire stretched along the axis of a cylindrical conducting shell.

*35. A thick spherical shell of inner radius a and outer radius b has a charge Q uniformly distributed over its volume (Fig. 24.35). Find the electric field in the regions $r \leq a$, $a \leq r \leq b$, and $r \geq b$.

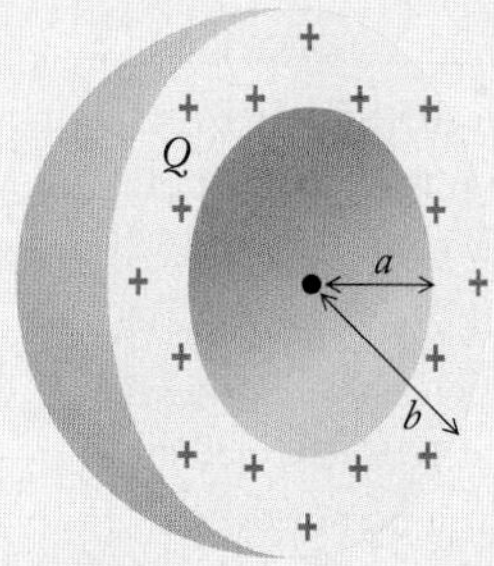

FIGURE 24.35 A thick spherical shell with total charge Q uniformly distributed over its volume.

*36. According to a (crude) model, the neutron consists of an inner core of positive charge surrounded by a shell of negative charge. Suppose that the positive charge has a magnitude $+e$ and is uniformly distributed over a sphere of radius 0.50×10^{-15} m; suppose that the negative charge has a magnitude $-e$ and is uniformly distributed over a concentric shell of inner radius 0.50×10^{-15} m and outer radius 1.0×10^{-15} m (Fig. 24.36). Find the magnitude and direction of the electric field at 1.0×10^{-15}, 0.75×10^{-15}, 0.50×10^{-15}, and 0.25×10^{-15} m from the center.

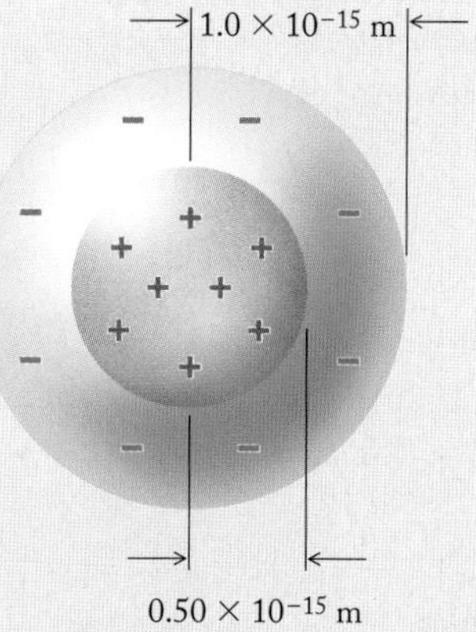

FIGURE 24.36 Hypothetical charge distribution inside a neutron.

*37. According to an old (and erroneous) model due to J. J. Thomson, an atom consists of a cloud of positive charge within which electrons sit like plums in a pudding. The electrons are supposed to emit light when they vibrate about their equilibrium positions in this cloud. Assume that in the case of the hydrogen atom the positive cloud is a sphere of radius R = 0.050 nm with a charge of e uniformly distributed over the volume of this sphere. The (pointlike) electron is held at the center of this charge distribution by the electrostatic attraction.

(a) Show that the restoring force on the electron is $e^2r/(4\pi\epsilon_0R^3)$ when the electron is at a distance r from the center ($r \leq R$).

(b) What is the frequency of small oscillations of the electron moving back and forth along a diameter? Give a *numerical* answer.

*38. According to the Thomson model (see also Problem 37), the atom of helium consists of a uniform spherical cloud of positive charge within which sit two electrons. Assume that the positive cloud is a sphere of radius 5.0×10^{-11} m with a charge of $2e$ uniformly distributed over the volume. The two electrons are symmetrically placed with respect to the center (Fig. 24.37). What is the equilibrium separation of the electrons?

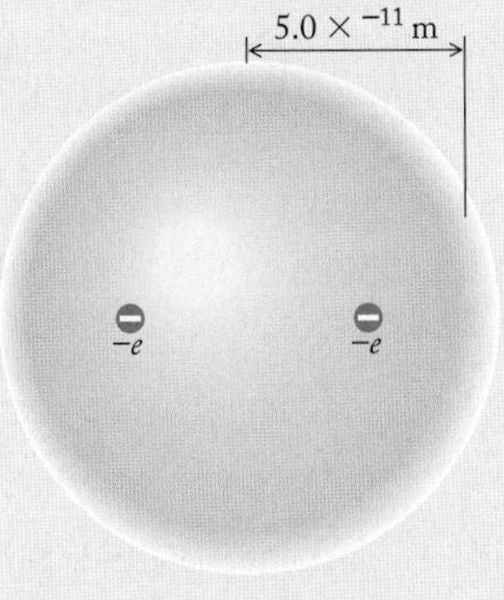

FIGURE 24.37 Thomson model of the helium atom.

*39. If a volume charge distribution varies continuously as a function of position the simple relation $Q = \rho V$ must be replaced by the integral $Q = \int \rho\, dV$. For spherical symmetry, $dV = 4\pi r^2\, dr$ (volume of a spherical shell with thickness dr and area $4\pi r^2$). Consider a sphere of radius R and total charge Q with a **nonuniform volume charge density** $\rho(r) = C/r$, where C is a constant to be determined.

(a) From $Q = \int \rho\, dV$, find the value of the constant C.

(b) Using Gauss' Law, find the electric field for $r \leq R$.

(c) What is the electric field for $r \geq R$?

*40. A thick, insulating spherical shell of inner radius a and outer radius b has a nonuniform volume charge density given by $\rho(r) = C/r^3$, where C is a known constant.

(a) What is the total charge Q in the shell? (Hint: See Problem 39.)

(b) Determine the value of the electric field for $r \leq a$, $a \leq r \leq b$, and $r \geq b$.

*41. A solid insulating sphere has a charge distribution with spherical symmetry, with a nonuniform charge density ρ coulombs per cubic meter described by the formula $\rho = kr^n$, where k is a constant and $n > -3$.

(a) What is the amount of charge $Q(r)$ inside a sphere of radius r? (Hint: See Problem 39.)

(b) What is the magnitude of the electric field as a function of r?

(c) For what value of n is the magnitude of the electric field constant?

(d) Why is it necessary to assume that $n > -3$?

*42. If a charge distribution varies continuously as a function of position, then the simple relation $Q = \rho V$ must be replaced by the integral $Q = \int \rho\, dV$. For cylindrical symmetry, $dV = 2\pi rL\, dr$ (volume of a cylindrical shell with radius r, length L, and thickness dr). Consider a solid cylinder of radius R with a nonuniform volume charge density $\rho(r) = B/r$, where B is a known constant

(a) What is the amount of charge $Q(r)$ inside a cylinder of radius $r \leq R$ and length L?

(b) What is the electric field for $r \leq R$?

*43. Consider a large, insulating slab of thickness d (similar to Example 10), but now with a nonuniform volume charge density $\rho = \rho(x)$, where $x = 0$ is at the center of the slab. Since ρ varies with position, the simple relation $Q = \rho V$ must be replaced by the integral $Q = \int \rho\, dV$. For the volume within the smaller Gaussian surface in Fig. 24.17, $dV = A\, dx$ (volume of a circular disk with area A and thickness dx). If the volume charge distribution is given by $\rho = Cx^2$, where C is a known constant, find the magnitude of the electric field for $|x| \leq d/2$ and for $|x| \geq d/2$.

**44. The electric charge of the proton is not concentrated in a point but, rather, distributed over a volume. According to experimental investigations at the Stanford Linear Accelerator, the charge distribution of the proton can be approximately described by a nonuniform volume charge density that is an exponential function of the radial distance:

$$\rho = \frac{e}{8\pi b^3}\exp\left(-\frac{r}{b}\right)$$

where e is the elementary charge and b is a constant, $b = 0.20 \times 10^{-15}$ m. Find the electric field as a function of the radial distance. What is the magnitude of the electric field at $r = 1.0 \times 10^{-15}$ m? [Hint: See Problem 39. Also, the following integral will be useful: $\int x^2e^{-x}\, dx = -x^2e^{-x} - 2e^{-x}(x + 1)$, where e here refers to the base of the natural logarithms.]

**45. When a point charge q moving at high speed passes by another stationary point charge q', the main effect of the electric forces is to give each charge a transverse impulse. Figure 24.38 shows the charge q moving at (almost) constant velocity v along the x axis in an almost straight line and shows the charge q' sitting at a distance R below the origin. The transverse impulse on q is

$$\int_{-\infty}^{\infty} F_y \, dt = \frac{q}{v} \int_{-\infty}^{\infty} E_y \, dx$$

Evaluate the integral $\int E_y \, dx$ by means of Gauss' Law, and prove that

$$\int_{-\infty}^{\infty} F_y \, dt = \frac{q}{v} \frac{q'}{2\pi\epsilon_0 R}$$

(Hint: Consider $2\pi R \int E_y \, dx$; show that this is the flux that q' produces through the infinite cylindrical surface indicated in Fig. 24.38.)

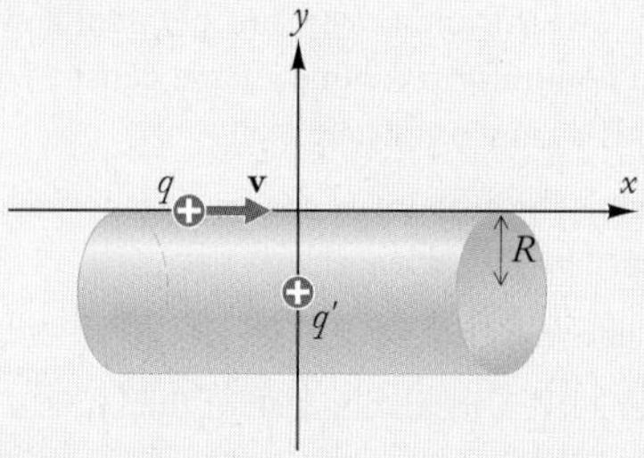

FIGURE 24.38 The charge q moves along the x axis. The charge q' sits at a distance R below the x axis. The cylinder, which extends to infinity, is the Gaussian surface.

*46. The formula derived in Problem 45 gives the transverse momentum that a high-speed charged particle acquires as it passes by a stationary charged particle.
(a) Calculate the transverse momentum that an electron of speed 4.0×10^7 m/s acquires as it passes by a stationary electron at a distance of 0.60×10^{-10} m.
(b) What transverse velocity corresponds to this transverse momentum?
(c) What will be the recoil velocity of the stationary electron (if it is free to move)?

24.4 Superposition of Electric Fields

*47. Positive charge Q is uniformly distributed over the volume of a solid sphere of radius R. Suppose that a spherical cavity of radius $R/2$ is cut out of the solid sphere, the center of the cavity being at a distance of $R/2$ from the center of the original solid sphere (Fig. 24.39); the cut-out material and its charge are discarded. What new electric field does the sphere with the cavity produce at the point P at a distance r from the original center? Assume $r > R$.

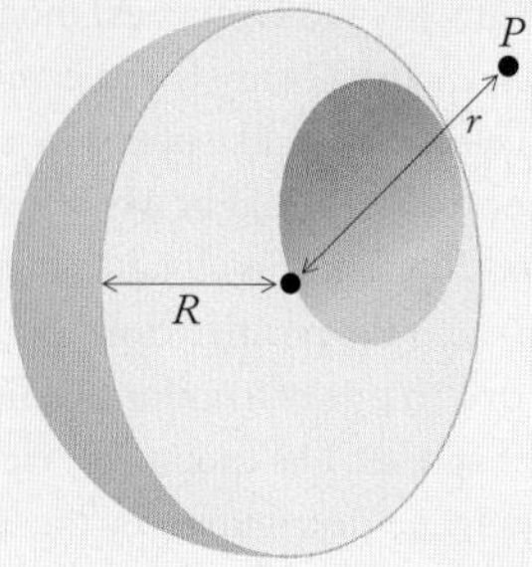

FIGURE 24.39 Sphere with a spherical cavity.

*48. Figure 24.19a shows an infinite cylinder with a positive uniform volume charge density ρ and radius R. The cylinder has a spherical empty cavity of radius $a < R$ centered on its axis. For points a radial distance r from the center of the spherical cavity along a line perpendicular to the cylindrical axis, find the magnitude and direction of the electric field for (a) $r \leq a$; (b) $a \leq r \leq R$; (c) $r \geq R$.

*49. Consider an infinite slab of thickness d parallel to the y–z plane and centered at $x = 0$, similar to the one in Fig. 24.17. This slab, however, has a spherical empty cavity of radius $d/2$ centered at the origin, just touching the slab faces; everywhere other than this cavity, the slab has a positive uniform volume charge density ρ. What are the magnitude and direction of the electric field at $\mathbf{r} = \frac{1}{4}d\mathbf{i}$? At $\mathbf{r} = 2d\mathbf{i}$? At $\mathbf{r} = \frac{1}{2}d\mathbf{i} + d\mathbf{j}$?

*50. A long cylinder of radius R has a uniform charge density ρ distributed over its volume. A cylindrical hole of radius R' has been drilled parallel to the axis of the cylinder along its full length, and the charge on this hole has been discarded. The axis of the hole is at a distance $h > R'$ from the axis of the cylinder (see Fig. 24.40). Find the electric field as a function of the radial distance r on the solid radial line shown in Fig. 24.40. Consider the cases $r \leq h - R'$, $h - R' \leq r \leq h + R'$, $h + R' \leq r \leq R$, and $r \geq R$ separately.

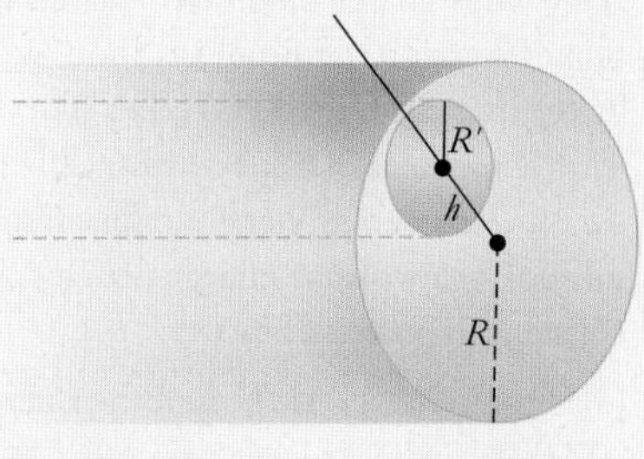

FIGURE 24.40 Cylinder with a cylindrical cavity.

**51. A charge q is uniformly distributed over the volume of a solid sphere of radius R. A spherical cavity is cut out of this solid sphere (Fig. 24.41), and the material and its charge are discarded. Show that the electric field in the cavity will then be uniform, of magnitude $(1/4\pi\epsilon_0)Qd/R^3$, where d is the distance between the centers of the spheres (Fig. 24.41). Make a drawing of the lines of electric field in the cavity.

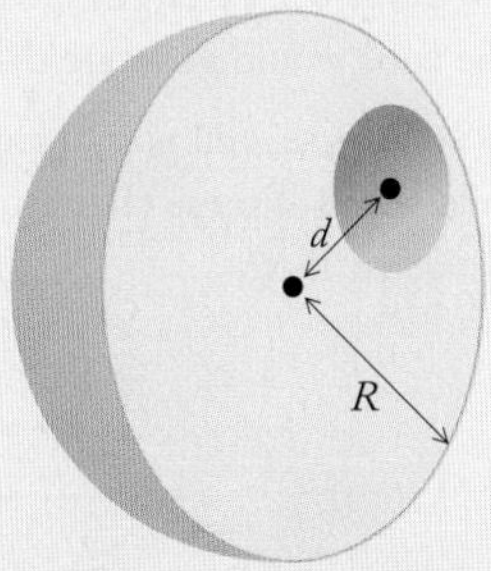

FIGURE 24.41 Sphere with a spherical cavity.

24.5 Conductors and Electric Fields

52. On a clear day, the Earth's atmospheric electric field near the ground has a magnitude of 100 N/C and points vertically down. Inside the ground, the electric field is zero, since the ground is a conductor. Consider a mathematical box of 1.0 m × 1.0 m × 1.0 m, half below the ground and half above. What is the electric flux through the sides of this box? What is the charge enclosed by the box?

53. Suppose we suspend a small ball carrying a charge of 1.0×10^{-6} C in the center of a safe and lock the door. The safe is uncharged and made of solid steel; it has inside dimensions of 0.30 m × 0.30 m × 0.30 m and outside dimensions 0.40 m × 0.40 m × 0.40 m. What is the electric flux through a cubical surface measuring 0.20 m × 0.20 m × 0.20 m centered on the ball? A cubical surface measuring 0.35 m × 0.35 × 0.35 m? A cubical surface measuring 0.50 m × 0.50 m × 0.50 m?

54. Charge is placed on a small metallic sphere which is surrounded by air. If the radius of the sphere is 0.50 cm, how much charge can be placed on the sphere before the air near the sphere suffers electric breakdown? The critical electric field strength that leads to breakdown in air is 3.0×10^6 N/C.

55. The surface of a long cylindrical copper pipe has a charge of λ coulombs per meter (Fig. 24.42). What is the electric field outside the pipe? Inside the pipe?

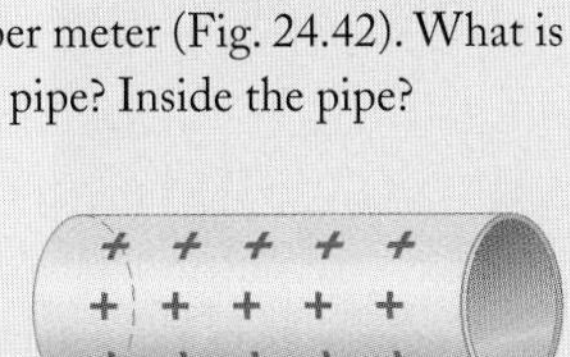

FIGURE 24.42 A long copper pipe with positive charge on its surface.

56. A solid copper sphere of radius 3.0 cm carries a charge of 1.0×10^{-6} C. This sphere is placed concentrically within a spherical, thin copper shell of radius 15 cm carrying a charge of 3.0×10^{-6} C. Find a formula for the electric field in the space between the sphere and the shell. Find a formula for the electric field outside the shell. Plot these electric fields as a function of radius.

57. A thick spherical shell made of solid metal has an inner radius a and an outer radius b and is initially uncharged. A point charge q is placed at the center of the shell. Find the electric field in the regions $r < a$, $a < r < b$, and $r > b$. Find the induced surface charge densities at $r = a$ and $r = b$.

58. Consider a completely isolated, thick, conducting infinite slab. One face of the slab has a uniform surface charge density $+\sigma$.
   (a) What is the electric field inside the slab?
   (b) What is the surface charge density on the other face of the slab?
   (c) What is the electric field outside the slab?

59. On days of fair weather, the atmospheric electric field of the Earth is about 100 N/C; this field points vertically downward (compare Problem 4 of Chapter 23). What is the surface charge density on the ground? Treat the ground as a flat conductor.

60. You wish to generate a uniform electric field of 2.0×10^5 N/C in the space between two flat parallel plates of metal placed face to face. The plates measure 0.30 cm × 0.30 cm. How much electric charge must you put on each plate? Assume that the gap between the plates is small so that the charge distribution and the electric field are approximately uniform, as for infinite plates. Assume the plates carry opposite charges of equal magnitudes.

61. A point charge Q_1 is at the center of two conducting spherical shells. The larger shell has a total charge Q_2 on it; the smaller shell has zero net charge.
   (a) What is the the charge on the inner surface of the smaller shell? On its outer surface?
   (b) What is the electric field at a distance r from the point charge, where r is between the two shells?
   (c) What is the charge on the inner surface of the larger shell? On its outer surface?
   (d) What is the electric field outside both shells, a distance r from the point charge?

*62. A solid dielectric sphere has radius R_1 and uniform volume charge density ρ. It is placed concentrically within a metal shell of inner radius R_2 and outer radius R_3. The shell carries a total charge Q_0. Find the magnitude of the electric field for (a) $r \le R_1$; (b) $R_1 \le r < R_2$; (c) $R_2 < r < R_3$; and (d) $r > R_3$.

REVIEW PROBLEMS

63. A point charge q is placed very close to (almost in contact with) the base of an infinitely long cylindrical Gaussian surface (see Fig. 24.43). What is the flux that the point charge produces through the base? What is the flux that this point charge produces though the curved lateral surface of the cylinder?

FIGURE 24.43 A point charge q inside a cylindrical Gaussian surface.

64. A point charge of 2.0×10^{-8} C is located at the center of a Gaussian surface in the shape of a cube of edge 8.0 cm. What is the average value of $E_{\perp}$ over one face of the cube?

65. Two very large plane surfaces intersect at 90°, forming a floor and a wall (see Fig. 24.44).
   (a) If a point charge q is placed just outside the edge formed by the intersecting surfaces, what is the electric flux through the combined surfaces?
   (b) If the point charge q is placed just inside the edge formed by the intersecting surfaces, what is the electric flux through the combined surfaces?

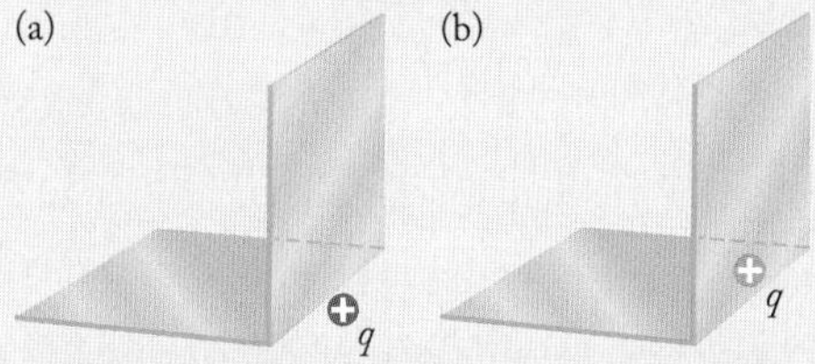

FIGURE 24.44 (a) Point charge located just outside the edge formed by the intersecting surfaces. (b) Point charge located just inside the edge.

66. Consider a cubical Gaussian surface of edge 5.0 cm. You do not know the electric charge or the electric field inside the cube, but you do know the electric field at the surface: at the top of the cube the electric field has a magnitude of 5.0×10^5 N/C and points perpendicularly out of the cube; at the bottom of the cube, the electric field has a magnitude of 2.0×10^5 N/C and points perpendicularly into the cube; on all other faces, the electric field is tangential to the surface of the cube.
   (a) How much electric charge is inside the cube?
   (b) Can you guess what charge distribution inside (and outside) the cube would generate this kind of electric field?

67. Figure 24.45 shows several point charges and several surfaces which enclose some of these charges. For each of the surfaces, what is the electric flux?

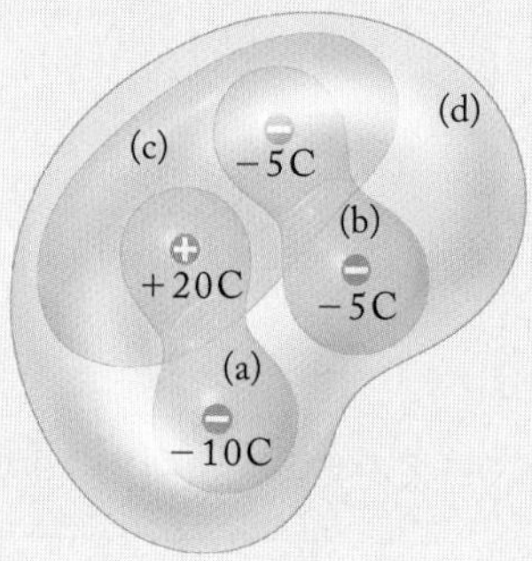

FIGURE 24.45 Several Gaussian surfaces (a–d).

68. In a xerographic copier, during the copying process, electric charge is deposited on the sheet of paper by means of a corona discharge from a thin wire carrying a concentration of charge (in some copiers you can see this corona wire if you open the top of the copier; see Fig. 24.46). The wire has a radius of 2.0×10^{-5} m. To produce the corona discharge, an electric field of 3.0×10^6 N/C is required near the surface of the wire. What amount λ of charge per unit length must be placed on the wire to attain this electric field?

FIGURE 24.46 Corona wire in a xerographic copier.

69. A very long cylinder of radius R has positive charge uniformly distributed over its volume. The amount of charge is λ coulombs per meter of length of the cylinder. A spherical cavity of radius $R' \leq R$, centered on the axis of the cylinder, has been cut out of this cylinder, and the charge in this cavity has been discarded. Find the electric field as a function of distance from the center of the sphere along the axis of the cylinder.

70. A long, thin rod carries a charge of λ coulombs per meter. Consider a square surface of dimensions $d \times d$; the square surface and the rod are in the same plane. The nearest side of the square surface is a distance r from the rod (see Fig. 24.47). What is the electric flux through this square?

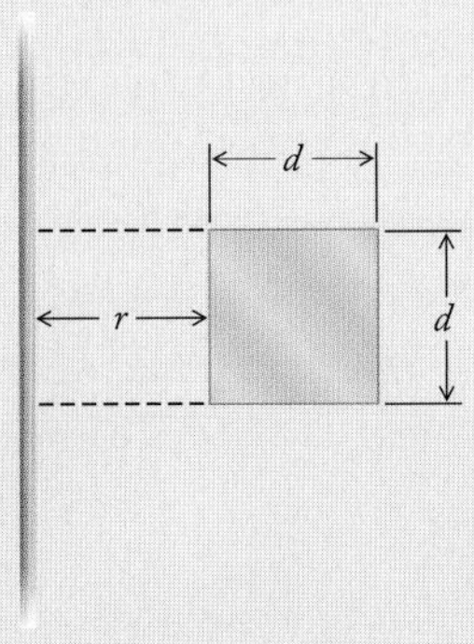

FIGURE 24.47 Charged rod and square surface.

71. A sphere of radius R and uniform volume charge density ρ remains stationary (levitates) when placed above an infinite sheet of paper with a uniform surface charge density σ. What is the mass of the sphere?

72. Two parallel infinite sheets have opposite uniform surface charge densities $\pm\sigma$. What is the electric force that one sheet exerts on a portion of area A of the other sheet?

73. Two parallel rods have the same radius R and opposite linear charge densities $\pm\lambda$. The centers of the rods are separated by a distance $4R$. What is the electric field halfway between the rods? What is the magnitude of the electric field at a point equidistant from the two rods, a distance R from the line parallel to and halfway between the rods?

74. The strength of the atmospheric electric field decreases with altitude. This decrease is due to the presence of positive ions in the atmosphere. Suppose that at ground level the atmospheric electric field has a magnitude of 130 N/C and a downward direction. If the positive ions contribute a volume charge density of 1.0×10^{-14} C/m^3 throughout the air, what is the rate of decrease of the atmospheric electric field with height? What is the magnitude of the atmospheric electric field at a height of 1000 m?

75. A thick cylindrical shell of inner radius a and outer radius b has a charge of λ coulombs per unit length uniformly distributed over its volume. A thin line with a charge of $-\lambda$ coulombs per unit length is stretched along the axis of the shell (see Fig. 24.48). Find the electric field in the regions $r \leq a$, $a \leq r \leq b$, and $r \geq b$.

FIGURE 24.48 A thick cylindrical shell with charge distributed over its volume and a charged line along its axis.

76. Consider two large parallel metallic plates with uniform, opposite charge distributions, as in Fig. 24.49a. Suppose that the magnitude of the charge density on each plate is 2.0×10^{-5} C/m^2. The upper plate is positive and the lower negative.

(a) What is the magnitude of the electric field in the region between the plates?

(b) You now insert a neutral large parallel metallic plate in the space between the two charged plates, as in Fig. 24.49b. Suppose that this plate is 1.0 cm thick. What is the magnitude of the electric field inside this thick plate? What is the magnitude of the electric field in the remaining space above and below this thick plate?

(c) What is the charge density on the upper surface of the thick plate? The lower surface?

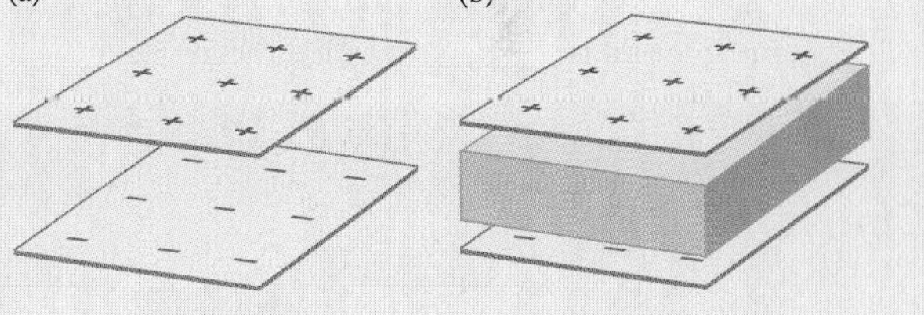

FIGURE 24.49 (a) Parallel plates with opposite charges. (b) Another plate inserted between the first two.

*77. The tau particle carries a negative fundamental charge similar to the electron, but is of much larger mass—its mass is 3.18×10^{-27} kg, about 3490 times the mass of an electron. Nuclear material is transparent to the tau: thus the tau can orbit around inside a nucleus, under the influence of the electric attraction of the nuclear charge. Suppose that a tau is in a circular orbit of radius 2.9×10^{-15} m inside a uranium nucleus. Treat the nucleus as a sphere of radius 7.4×10^{-15} m with a charge $92e$ uniformly distributed over its volume. Find the speed, the kinetic energy, the angular momentum, and the frequency of the orbital motion of the tau.

Answers to Checkups

Checkup 24.1

1. If the surface were horizontal, no field lines would cross it, so the flux would be zero (smaller); if the surface were vertical, more field lines would cross it, so the flux would be larger.
2. No change in either case; the same number of field lines would intersect any closed surface around the charge. As stated in the Comment in Example 2, the net flux depends only on the charge enclosed by the surface. In the next section, we formally explore this relation between the flux through a surface and the enclosed charge, known as *Gauss' Law*.
3. (C) Zero. Since all the field lines that enter the surface also leave it, the net flux through the surface is zero.

Checkup 24.2

1. Gauss' Law states that $\Phi_E = Q_{\text{inside}}/\epsilon_0$, so zero flux means zero total enclosed charge. If the surface has a hole, we can draw no conclusions: we don't know the flux through a closed surface, and "the charge inside" is ill defined for such a surface.
2. Since the two charges are of equal magnitude but opposite sign, the net charge inside the surface is zero, so the net flux through the surface is zero.
3. In Example 3, the net flux through the cube was calculated to be $\Phi_E = 40\ \text{N}\cdot\text{m}^2/\text{C}$, so Gauss' Law tells us that the charge inside that cube is

$$Q_{\text{inside}} = \epsilon_0 \Phi_E = (8.85 \times 10^{-12}\ \text{C}^2/\text{N}\cdot\text{m}^2)(40\ \text{N}\cdot\text{m}^2/\text{C})$$
$$= 0.35\ \text{n}C$$

4. (E) Zero. Since the field is uniform, field lines enter the sphere on one side and leave it on the other; the net flux through the sphere is zero.

Checkup 24.3

1. No. Although Gauss' Law is still valid, there is no surface through which the electric flux may be calculated easily. To calculate the electric field for four point charges, you would have to take the vector sum of the four Coulomb (point-charge) fields.
2. As in Example 7, $E = \lambda/(2\pi\epsilon_0 r) \propto \lambda/r$. So for equal values of λ, the thinner wire will produce a field 10 times larger at its surface.
3. As in Example 5 and Fig. 24.11, the field at the center is zero, and the field is largest at the surface of the sphere [where it takes the value $E = Q/(4\pi\epsilon_0 R^2)$].
4. As in the previous question, the field at the center is zero (zero enclosed charge for $r = 0$). At the surface, only the total charge enclosed matters (not how it is radially distributed), so the field there again takes the value $E = Q/(4\pi\epsilon_0 R^2)$. In this case, however, we do not know where the maximum value of E occurs; depending on the radial distribution, it could be anywhere in the region $0 < r \leq R$.
5. (B) 5.0×10^4 N/C. We saw in Example 7, as in Chapter 23, that the electric field E due to a linear charge distribution varies as $1/r$. Thus at twice the distance, the field is half as large.

Checkup 24.4

1. Yes. Inside the cavity, the field is the same as in Fig. 24.19; outside, there is an additional spherically symmetric contribution from the surface charge density.
2. No. The field due to a uniform cube of charge cannot be readily calculated from one of the high-symmetry (spherical, cylindrical, or planar) applications of Gauss' Law.
3. (B) $\sigma/2\epsilon_0$; $3\sigma/2\epsilon_0$. The field contribution due to the sheet with charge density σ is $\sigma/2\epsilon_0$; that due to the 2σ sheet is σ/ϵ_0. Since both charge densities are positive, each field contribution is directed away from its respective sheet. Thus between the two sheets these contributions oppose to give a net field of magnitude $\sigma/\epsilon_0 - \sigma/2\epsilon_0 = \sigma/2\epsilon_0$, and outside the sheets the two contributions are parallel and give a net field $\sigma/2\epsilon_0 + \sigma/\epsilon_0 = 3\sigma/2\epsilon_0$.

Checkup 24.5

1. No. It is possible in an insulator, but not in a conductor when in electrostatic equilibrium.
2. Yes. Any field lines that intersect the copper surface have to bend to meet the surface at right angles.
3. As we saw above, for a given charge density, the metal plate produces twice the electric field. So for half the charge density on each side of the plate, the electric fields due to the paper sheet and aluminum plate are the same. From far away, the division of the charge density is unimportant; the plate behaves like a single sheet of charge. The division of the charge density into two parts ensures that the electric field inside the metal will be zero.
4. (D) $-Q$; Q. The point charge Q induces an opposite charge $-Q$ on the inner surface of the conductor; this ensures $\mathbf{E} = 0$ inside the conducting shell. Since the shell is overall uncharged, the charge $-Q$ on the inner surface implies there is a charge Q on its outer surface.

CHAPTER

25

Electrostatic Potential and Energy

CONCEPTS IN CONTEXT

Concepts — in — Context

This electrostatic generator at the Boston Science Museum has accumulated a large charge, distributed on the surface of its spherical conductor. When a charged particle is placed somewhere near this charge distribution, it will have a potential energy.

With the concept of electrostatic potential energy introduced in this chapter, we can consider such questions as:

? What is the potential energy of an electron placed near a proton? (Example 2, page 795)

? What is the electrostatic potential energy per unit charge at points outside and inside a charged conducting sphere? (Example 7, page 803)

? How does knowledge of the electrostatic potential energy enable us to calculate the electric field? (Example 9, page 807)

? What is the total work required to build up the charge on a conducting sphere? (Example 12, page 813)

From our study of mechanics we know that to formulate a law of conservation of energy for a particle moving under the influence of some force, we have to construct a potential energy corresponding to this force. In this chapter we will construct the electrostatic potential energy for a charged particle moving in the electric field generated by a static charge distribution. Like the potential energies we examined in Chapter 8, the electrostatic potential energy is helpful in the calculation of the motion of the particle. Furthermore, we will see that knowledge of the electrostatic potential energy at different distances from the charge distribution is equivalent to knowledge of the electric field—the potential energy can be calculated from the electric field and, conversely, the electric field can be calculated from the potential energy. It is therefore possible to calculate the electric field of a charge distribution by first evaluating the electrostatic potential energy that this charge distribution produces when acting on a charged particle placed at different distances. This means that we have available three alternative methods for the calculation of the electric field: via Coulomb's Law, via Gauss' Law, and via the potential energy. If a problem does not yield to the simple and elegant method based on Gauss' Law, it is usually best to use the method based on the potential energy, because the mathematics is likely to be less laborious than with the method based on Coulomb's Law.

In the final parts of this chapter, we will learn how to calculate the electric potential energy of systems of several point charges and of systems of conductors with charges placed on them. Since the charges exert electric forces on each other, it requires a certain amount of work to bring the charges together into their final configuration, starting with an initial configuration of very large (infinite) separation. This work is the electric potential energy of the configuration. It represents energy stored in the configuration during its assembly. We will see that this energy is actually stored in the electric field. The energy is concentrated in those regions of space where the electric field is strong.

25.1 THE ELECTROSTATIC POTENTIAL

Online Concept 30 Tutorial

In Chapter 8 we learned that if a force is conservative, then the work performed by this force on a particle during a displacement can be expressed as a difference between two potential energies, one for the starting point, and one for the end point of the displacement. If we know the potential energy that corresponds to the force, we can immediately construct the conserved mechanical energy, which is simply the sum of the kinetic energy and the potential energy.

The electric force that a static distribution of charges exerts on a point charge is a conservative force. We can easily verify this for the electric force exerted by two uniform distributions of positive and of negative charges placed on two large parallel sheets or plates (see Fig. 25.1). We know from Section 23.2 that such parallel charged plates generate a uniform electric field E_0 in the space between them; the magnitude of this electric field is directly proportional to the amount of charge per unit area on each plate [see Eq. (23.13)]. The force that this electric field exerts on a point charge q is $F = qE_0$, and the work done by this constant force during a displacement from the point y_1 to the point y_2 illustrated in Fig. 25.1 is given by Eq. (7.1),

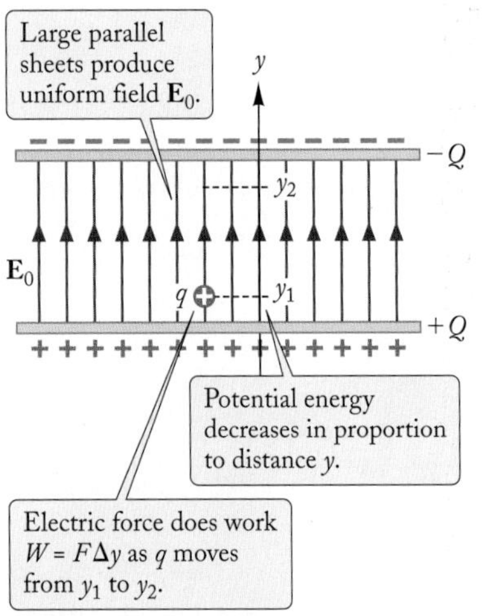

FIGURE 25.1 Uniform electric field in the space between two charged parallel plates.

$$W = F\Delta y = qE_0(y_2 - y_1) \tag{25.1}$$

or

$$W = -qE_0 y_1 + qE_0 y_2 \tag{25.2}$$

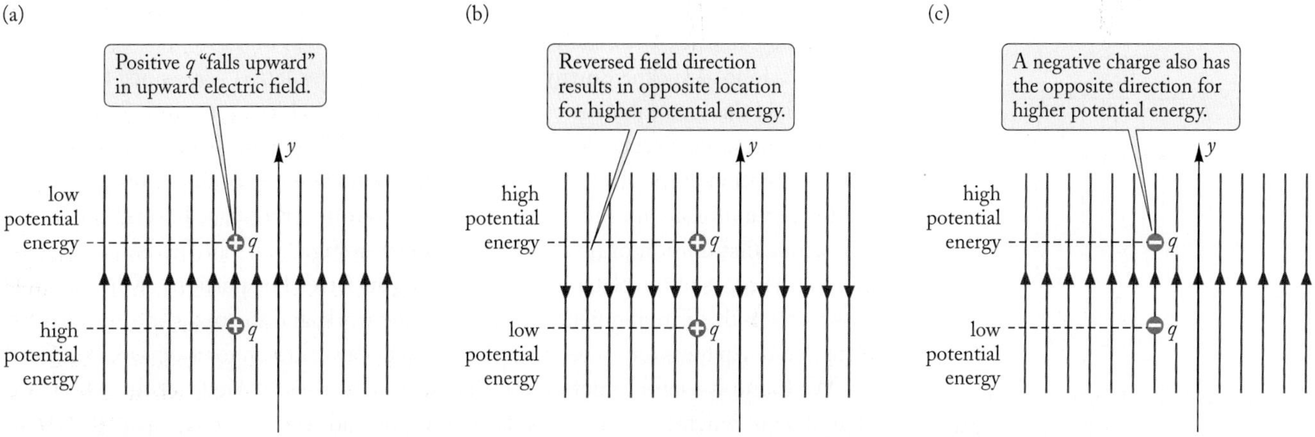

FIGURE 25.2 (a) For a positive charge in an electric field directed vertically upward, the potential energy decreases with height. (b) For a positive charge in an electric field directed vertically downward, the potential energy increases with height. (c) For a negative charge in an electric field directed vertically upward, the potential energy also increases with height.

This shows that if we identify the potential energy U as

$$U = -qE_0y \tag{25.3}$$

then the work is the difference between the potential energies corresponding to the points y_1 and y_2:

$$W = -qE_0y_1 + qE_0y_2 = U_1 - U_2 \tag{25.4}$$

The conserved mechanical energy is the sum of the kinetic energy $K = \frac{1}{2}mv^2$ and the potential energy $U = -qE_0y$:

$$[\text{energy}] = K + U = \tfrac{1}{2}mv^2 - qE_0y = [\text{constant}] \tag{25.5}$$

Note that the potential energy (25.3) is directly proportional to the distance y from the lower plate in Fig. 25.1. This direct proportionality of potential energy and height is reminiscent of the gravitational potential energy mgy. Mathematically, this electric potential energy and the gravitational potential energy are similar because both involve a constant force. But note that the signs of these two potential energies are opposite, because the electric force that the plates in Fig. 25.1 exert on a (positive) charge q is upward, whereas the gravitational force is downward; the signs of these two potential energies would be the same if the electric field in Fig. 25.1 were downward or if the charge q were negative (see Fig. 25.2). Note also the physical implications of changes in potential energy: for instance, in order for a positive charge to move against the electric field (that is, to a position of higher potential energy), the charge either must be pushed by an external agent (which does work) or must lose some kinetic energy.

We recall from Section 8.1 that a general criterion for a conservative force is that the work the force performs on any round trip must be zero. For the electric force exerted by the parallel charged plates in Fig. 25.1, this criterion is satisfied provided the round trip is confined to the region within the plates. For instance, consider the round trip consisting of the four straight segments shown in Fig. 25.3a. The work done by the electric field along the horizontal segments is zero, since the electric field is perpendicular to these segments; the work done along the upward vertical segment (on the right) is positive, and the work done along the downward vertical segment (on the left) is negative, resulting in zero work for the entire round trip.

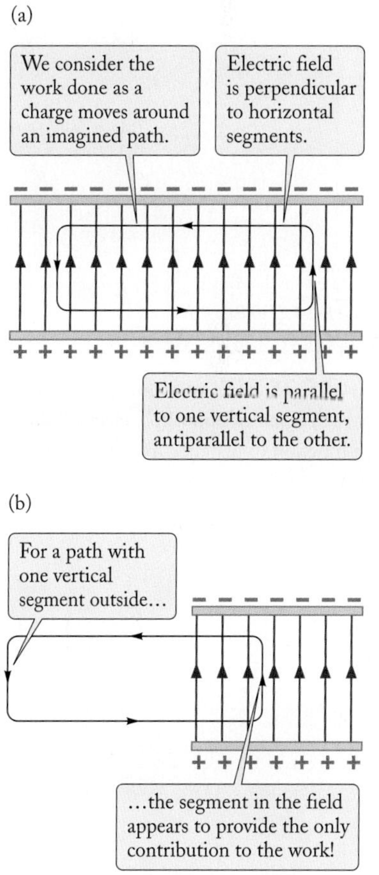

FIGURE 25.3 (a) A closed path between the plates. (b) A closed path with an upward segment between the plates and a downward segment outside the plates.

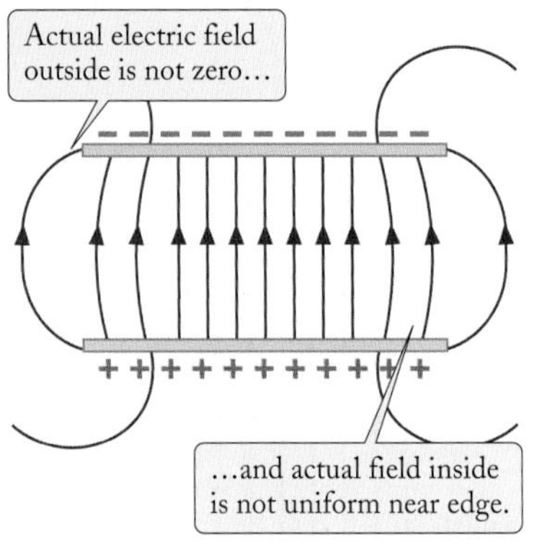

FIGURE 25.4 The fringing field that extends beyond the space between the plates.

If we attempt to apply the criterion of zero work to a different round trip that extends beyond the plates, such as the round trip shown in Fig. 25.3b, we see that now the work done along the downward segment on the far left is zero, since there is no electric field there. In this case it would seem we cannot achieve zero work for the round trip! However, the electric field distribution of Fig. 25.3 is only an approximation; an exact calculation of the electric field shows that near the edges of the plates the electric field is not uniform, and the electric field spills out into the space beyond the plates. The actual distribution of field lines is as shown in Fig. 25.4. The field that spills out beyond the edges of the plates is called the **fringing field**. The portions of the path that pass through this fringing field contribute to the work, and the net work for the complete round trip turns out to be zero, as required by our criterion for a conservative force.

We found it useful to define the electric field as the electric force divided by the charge q on which this force acts. Likewise, we find it useful to define the **electrostatic potential** V as the electric potential energy U divided by the charge q:

electrostatic potential

$$V = \frac{U}{q} \tag{25.6}$$

Thus, the electrostatic potential is the potential energy per unit charge. For example, in the case of a uniform electric field E_0, with a potential energy $U = -qE_0 y$, the electrostatic potential is

$$V = -E_0 y \tag{25.7}$$

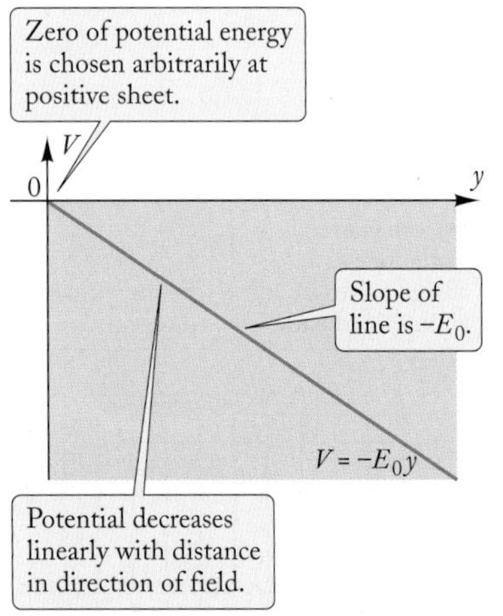

FIGURE 25.5 Electrostatic potential between a pair of oppositely charged sheets as a function of distance from the positive sheet.

Figure 25.5 is a plot of this linear potential as a function of the displacement y in the direction of the uniform field. Recall that it is the difference in potential energy between two points that is meaningful, as in Eq. (25.4). The reference position (the zero of potential energy) can be arbitrarily chosen. The same is true for the electrostatic potential V of Eq. (25.7).

The SI unit of electrostatic potential is the **volt** (V),[1] where

$$1 \text{ volt} = 1 \text{ V} = 1 \text{ joule/coulomb} = 1 \text{ J/C} \tag{25.8}$$

The unit of electric field we have employed in the preceding chapters is the N/C. This unit can be expressed in terms of volts as follows:

$$1\,\frac{\text{N}}{\text{C}} = 1\,\frac{\text{N}\cdot\text{m}}{\text{C}\cdot\text{m}} = 1\,\frac{\text{J}}{\text{C}}\frac{1}{\text{m}} = 1\,\frac{\text{V}}{\text{m}} \tag{25.9}$$

Thus, N/C and V/m are equal units; in practice, volt per meter is the preferred unit for the electric field. Table 25.1 gives some values of electrostatic potentials.

EXAMPLE 1 Suppose that near the ground directly below a thundercloud, the electric field is of a constant magnitude 2.0×10^4 V/m and points upward. What is the potential difference between the ground and a point in the air, 50 m above ground?

SOLUTION: For a constant (uniform) electric field, we use the linear potential of Eq. (25.7),

$$V = -E_0 y = -2.0 \times 10^4 \text{ V/m} \times 50 \text{ m} = -1.0 \times 10^6 \text{ volts}$$

[1] Note that the same letter V is used in physics both as a symbol for potential and as the abbreviation for *volt*. This leads to confusing equations such as $V = 3.0$ V (which is intended to mean that the potential $V = 3.0$ volts). If there is a possibility of confusion, it is best not to abbreviate *volt*.

COMMENTS: This calculation assumes that the ground is flat, without protuberances, such as trees or buildings. It would be wrong to conclude that the potential at the top of a 50-m-high building is -1.0×10^6 volts. Buildings are usually made of conducting materials, and, as we will see, the potential at all points of a conductor is the same. Thus, the building effectively acts as part of the ground, and the potential difference between the top of the building and the ground is zero. The presence of the building or of some other conducting protrusion modifies the electric field, so the assumption of a constant upward electric field is not valid near the building.

We can also verify that the electric force that one point charge exerts on another point charge is a conservative force. The electric force that a point charge q' exerts on another point charge q is $F = (1/4\pi\epsilon_0)qq'/r^2$. To find the potential energy corresponding to this force, we begin by calculating the work done as the charge q moves from, say, position P_1 to position P_2, and we then seek to express this work as a difference of two terms. In Fig. 25.6 the positions P_1 and P_2 are at radial distances r_1 and r_2, respectively, from the fixed charge q'. The work is the integral of the force $F = (1/4\pi\epsilon_0)qq'/r^2$ along the radial path from P_1 to P_2:

$$W = \int F\,dr = \int_{r_1}^{r_2} \frac{1}{4\pi\epsilon_0}\frac{qq'}{r^2}\,dr = \frac{qq'}{4\pi\epsilon_0}\left(-\frac{1}{r}\right)\bigg|_{r_1}^{r_2} = \frac{qq'}{4\pi\epsilon_0}\left(\frac{1}{r_1} - \frac{1}{r_2}\right) \tag{25.10}$$

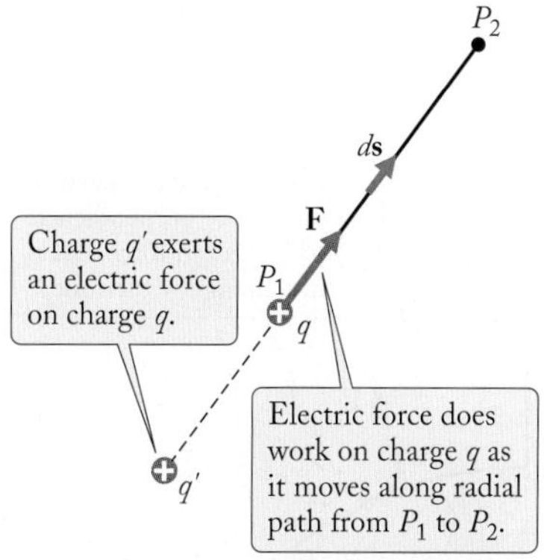

FIGURE 25.6 Two points P_1 and P_2 in the electric field of a charge q'. The path connecting the points P_1 and P_2 is a straight radial line.

TABLE 25.1 SOME POTENTIALS AND POTENTIAL DIFFERENCES

Thundercloud to ground	5×10^7 V
Van de Graaff generator	10^7
High-voltage power line	5×10^5
At nucleus of uranium atom	2×10^5
Power supply for X-ray tube	10^5
Power supply for TV tube	2×10^4
Automobile ignition	10^4
Power supply for neon tube	2×10^3
Household outlet (Europe)	220
Household outlet (USA)	115
At electron orbit in hydrogen atom	27
Automobile battery	12
Dry cell	1.5
Single solar cell	0.6
Resting potential across nerve membrane	0.09
Potential changes on skin (measured by EKG or EEG)	5×10^{-5}
Potential changes due to thermal noise (typical)	10^{-8}

ALESSANDRO, CONTE VOLTA (1745–1827) *Italian physicist. Volta established that the "animal electricity" observed by Luigi Galvani, 1737–1798, in experiments with frog muscle tissue placed in contact with dissimilar metals, was not due to any exceptional property of animal tissues, but was also generated whenever any wet body was sandwiched between dissimilar metals. This led him to develop the first "voltaic pile," or battery, consisting of a large stack of moist disks of cardboard (electrolyte) sandwiched between disks of metal (electrodes).*

As expected, this result shows that the work is the difference between two potential energies. Accordingly, we can identify the electric potential energy as

electric potential energy

$$U = \frac{1}{4\pi\epsilon_0}\frac{qq'}{r} \tag{25.11}$$

In this calculation of the electric potential energy we assumed that the positions P_1 and P_2 lie on the same radial line (see Fig. 25.6). However, this assumption is not really necessary, because, if the points lie on different radial lines, then we can construct a path from one to the other with alternating radial segments and circular arcs (see Fig. 25.7). Recall from Chapter 7 that in general the work done by a force **F** is given by

$$W = \int \mathbf{F} \cdot d\mathbf{s} = \int F \cos\theta \, ds$$

that is, a contribution to the work is the product of the component of the force parallel to the path times the displacement. The total work done along the radial segments is the same as for a single radial line, so this work is as given by Eq. (25.10). The work done along the circular arcs is zero, since the direction of the Coulomb force is perpendicular to these arcs. Hence, the net work, and the change of potential energy, is given by the same expression we found above, when we considered a single radial segment. Note that any arbitrary path between P_1 and P_2 can be approximated by small radial segments and circular arcs, and hence the work done by the electric force is independent of the path connecting P_1 and P_2; it depends only on the initial position P_1 and the final position P_2. This establishes that the electric force exerted by one point charge on another is conservative. More generally, the electric force exerted by any static distribution of electric charges on a point charge is conservative, since any distribution of electric charges consists of point charges, each of which exerts a conservative electric force.

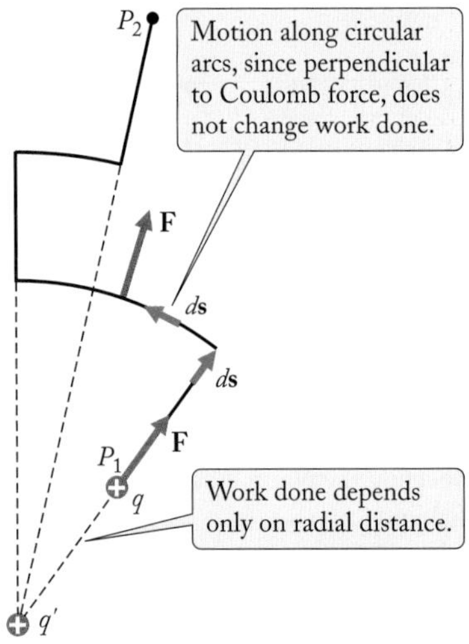

FIGURE 25.7 This path connecting P_1 and P_2 consists of straight radial segments and circular arcs.

For two charges of equal signs, the electric potential energy (25.11) *is positive, and it decreases in inverse proportion to the distance.* A decrease of potential energy with distance is characteristic of a repulsive force. *For two charges of opposite sign, the electric potential energy is negative*, and the magnitude of this negative potential energy decreases also in inverse proportion to distance (the potential energy *increases* from a large negative value to zero). Such an increase of potential energy with distance is characteristic of an attractive force.

The dependence of the electric potential energy of two point charges of opposite signs on distance is the same as the dependence of the gravitational potential energy of two point masses on distance [see Eq. (9.20)]—both are inversely proportional to the distance. This was to be expected, since the electric force is mathematically similar to the gravitational force—both are inversely proportional to the square of the distance.

The electrostatic potential produced by the charge q' is $V = U/q$; that is,

Coulomb potential

$$V = \frac{1}{4\pi\epsilon_0}\frac{q'}{r} \tag{25.12}$$

This electrostatic potential of a point charge is called the **Coulomb potential**.

Figure 25.8a is a plot of the Coulomb potential of a positive point charge q' vs. distance. Figure 25.8b is a plot of the Coulomb potential of a negative point charge q'

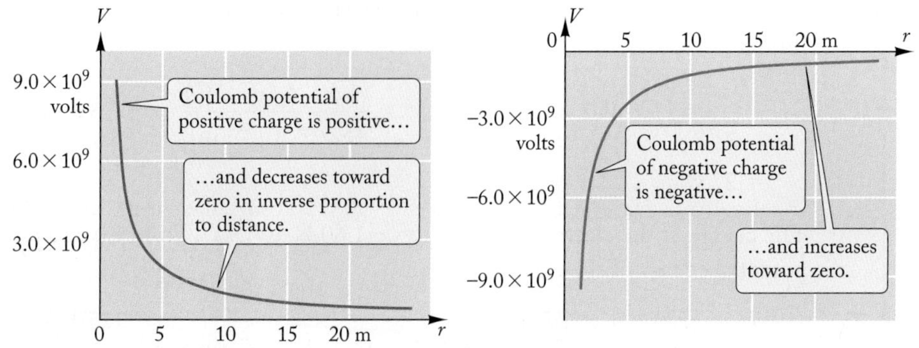

FIGURE 25.8 (a) Electrostatic potential of a positive charge q' vs. distance. For this plot, the magnitude of the charge is taken as $q' = 1$ C. (b) Electrostatic potential for a negative charge of magnitude $q' = -1$ C.

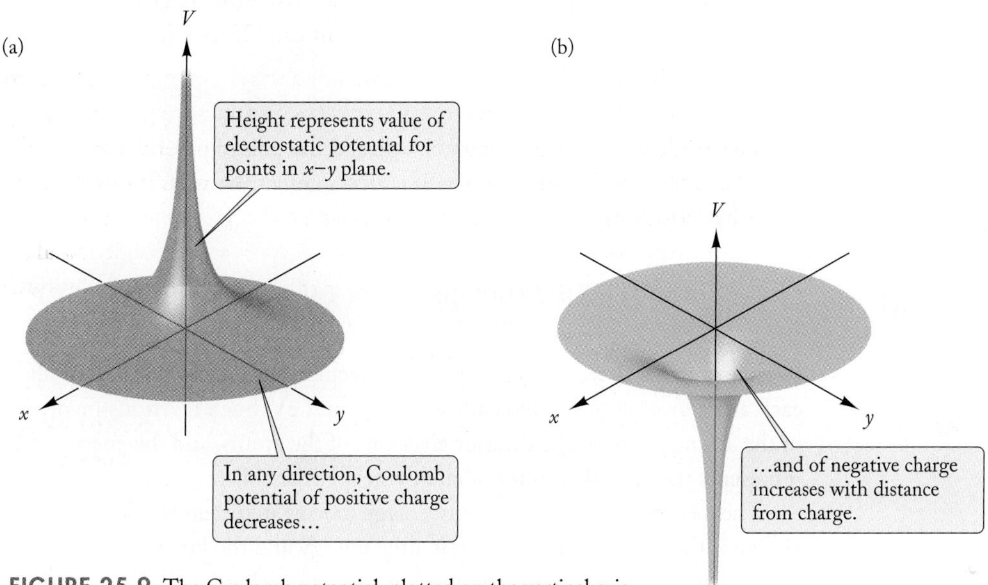

FIGURE 25.9 The Coulomb potential, plotted on the vertical axis, as a function of position in two dimensions. The plot is for positions in a plane containing (a) a positive and (b) a negative point charge.

vs. distance. Figure 25.9 shows plots of the Coulomb potentials of each of these point charges as a function of position in two dimensions.

Equation (25.12) also gives the electrostatic potential outside a solid charged sphere, or a hollow charged spherical shell, or any charge distribution with spherical symmetry. The electric field *outside* such spherical charge distributions is the same as that of a point charge, and therefore the electrostatic potential outside is also the same as for a point charge.

EXAMPLE 2 The electron in a hydrogen atom is at a distance of 5.3×10^{-11} m from the proton (see Fig. 25.10). The proton is a small ball of charge with $q' = e = 1.60 \times 10^{-19}$ C. What is the electrostatic potential generated by the proton at this distance? What is the potential energy of the electron?

Concepts — in — Context

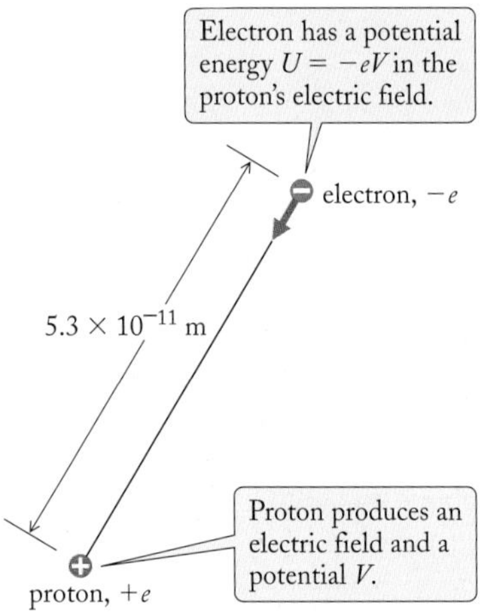

FIGURE 25.10 Proton ($+e$) and electron ($-e$) separated by a distance of 5.3×10^{-11} m.

SOLUTION: The electrostatic potential outside the proton is the same as for a point charge, so according to Eq. (25.12), the electrostatic potential produced by the proton is

$$V = \frac{1}{4\pi\epsilon_0}\frac{q'}{r} = 9.0 \times 10^9 \text{ N·m}^2/\text{C}^2 \times \frac{(1.60 \times 10^{-19}\text{ C})}{(5.3 \times 10^{-11}\text{ m})} \quad (25.13)$$

$$= 27 \text{ volts}$$

The charge of the electron is $q = -e = -1.60 \times 10^{-19}$ C. From Eq. (25.6), the potential energy of the electron is then

$$U = qV = -e \times 27 \text{ volts} \quad (25.14)$$

$$= -1.60 \times 10^{-19}\text{ C} \times 27 \text{ volts} = -4.3 \times 10^{-18}\text{ J} \quad (25.15)$$

For the purposes of atomic physics, the joule is a rather large unit of energy and it is more convenient to leave the answer as in Eq. (27.14), $U = -27e \times$ volt.

The product of the elementary charge and the unit of potential, $e \times$ volt, or eV, is a unit of energy. This unit of energy is called an **electron-volt**. It can be converted to joules by substituting the numerical value for e:

electron-volt (eV)

$$1 \text{ eV} = 1.60 \times 10^{-19}\text{ C} \times 1\text{ V} = 1.60 \times 10^{-19}\text{ J} \quad (25.16)$$

In chemical reactions among atoms or molecules, the energy released or absorbed by each atom or molecule is typically around 1 or 2 eV. Such reactions involve a change in the arrangement of the exterior electrons of the atoms, and the energy of 1 or 2 eV represents the typical amount of energy needed for this rearrangement.

The mechanical energy of a point charge moving in the electric field of another point charge is the sum of the electric potential energy and the kinetic energy. The Law of Conservation of Energy for the motion of a point charge q in the electric field of a fixed point charge q' takes the form

$$[\text{energy}] = K + U = \tfrac{1}{2}mv^2 + \frac{1}{4\pi\epsilon_0}\frac{qq'}{r} = [\text{constant}] \quad (25.17)$$

This total energy remains constant during the motion. As we saw in Chapter 8, examination of the energy reveals some general features of the motion. Obviously, if qq' is negative (opposite charges, attractive Coulomb force), then Eq. (25.17) implies that whenever r increases, v must decrease, and whenever r decreases, v must increase. For qq' positive (like charges, repulsive force), r and v increase together.

EXAMPLE 3 An electron is initially at rest at a very large distance from a proton. Under the influence of the electric attraction, the electron falls toward the proton, which remains (approximately) at rest. What is the speed of the electron when it has fallen to within 5.3×10^{-11} m of the proton? (See Fig. 25.11.)

SOLUTION: The word "initially" suggests using the before-and-after equality of the total energy. Since the electron is initially at rest, the initial kinetic energy is zero (see the discussion of the use of energy conservation in the Problem-Solving Techniques box). Furthermore, "a very large distance" implies the limit $r_0 \to \infty$, where the initial potential energy is also zero. The potential energy of the electron is $U = qV = -eV$. The total energy is the sum of the potential and kinetic energies:[2]

$$K + U = \frac{1}{2}m_e v^2 - eV \tag{25.18}$$

This total energy is conserved. The initial value of the energy is zero; hence the final value of the energy must also be zero:

$$\frac{1}{2}m_e v^2 - eV = 0 \tag{25.19}$$

Solving this for the speed v, we obtain

$$v = \sqrt{\frac{2eV}{m_e}} \tag{25.20}$$

According to Example 2, $V = 27$ volts for $r = 5.3 \times 10^{-11}$ m, so

$$v = \sqrt{\frac{2 \times 1.60 \times 10^{-19}\ \text{C} \times 27\ \text{volts}}{9.11 \times 10^{-31}\ \text{kg}}}$$

$$= 3.1 \times 10^6\ \text{m/s}$$

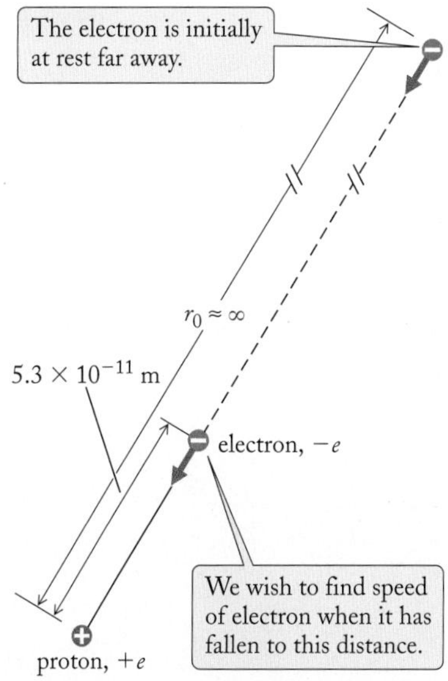

FIGURE 25.11 The approaching electron is instantaneously at a distance of 5.3×10^{-11} m from the proton.

For any nonuniform electric field, we can calculate the electric potential energy of a point charge q by the same method as for the electric field of a point charge: we express the work done by the electric field during a displacement from one position to another as a difference of two terms, one of which depends on the position of the starting point, and the other on the position of the end point. These two terms are the potential energies U_1 and U_2 corresponding to these two points. The work is then

$$W = U_1 - U_2 \tag{25.21}$$

and the conserved mechanical energy is again [see Eq. (25.5)]

$$K + U = \frac{1}{2}mv^2 + U = [\text{constant}]$$

or, with $U = qV$,

$$K + U = \frac{1}{2}mv^2 + qV = [\text{constant}] \tag{25.22}$$

conservation of energy

PROBLEM-SOLVING TECHNIQUES — ENERGY CONSERVATION AND MOTION OF A POINT CHARGE

Note that the use of energy conservation with electric potential energy involves the familiar three steps we encountered with other forms of potential energy:

1. First write an expression for the energy of a charge (kinetic + potential) at one point of the motion [Eq. (25.18)].

2. Then write a similar expression for the energy at another point.

3. Then rely on energy conservation to equate the two expressions [Eq. (25.19)].

This yields one equation, which can be solved for the unknown final speed of the charge (if the final position is known) or the unknown final position (if the final speed is known).

[2] Do not confuse the symbol v (for speed) with the symbol V (for potential).

✔ Checkup 25.1

QUESTION 1: Suppose that the potential of a point charge is 100 V at a distance of 1 m from this charge. What is the potential at a distance of 10 m? 100 m?

QUESTION 2: Given that the electric potential energy for two interacting point charges is positive, what can you conclude about the signs of their charges?

QUESTION 3: According to Fig. 25.8, where is the electric potential of a positive charge q' largest? Smallest? Where is the electric potential of a negative charge q' largest? Smallest?

QUESTION 4: Consider a point charge q moving in the electric field of a stationary point charge q'. If the charges are of the same sign and r increases, the speed of q will

(A) Increase (B) Decrease (C) Remain the same

Online Concept Tutorial 30

25.2 CALCULATION OF THE POTENTIAL FROM THE FIELD

Any arbitrary static distribution of charge can be regarded as consisting of many point charges. Since the electric force generated by a stationary point charge is a conservative force, the electric force generated by such a distribution of charge must also be a conservative force. Thus, when a point charge q moves from a position P_0 to a position P in the electric field of a charge distribution, the work done by the electric force can always be expressed as a difference of two potential-energy terms,

$$W = U_0 - U \tag{25.23}$$

The work is related to the force by the general formula $W = \int \mathbf{F} \cdot d\mathbf{s}$ [Eq. (7.16)], so Eq. (25.23) becomes

$$\int_{P_0}^{P} \mathbf{F} \cdot d\mathbf{s} = U_0 - U$$

Next, we can express the force in terms of the electric field ($\mathbf{F} = q\mathbf{E}$) and the potential energy in terms of the electrostatic potential ($U = qV$):

$$\int_{P_0}^{P} q\mathbf{E} \cdot d\mathbf{s} = qV_0 - qV \tag{25.24}$$

We can then cancel the factor q on both sides of this equation, and solve the equation for the potential V:

$$V = -\int_{P_0}^{P} \mathbf{E} \cdot d\mathbf{s} + V_0 \tag{25.25}$$

For many calculations, we will find it convenient to write Eq. (25.25) in terms of the angle θ between the electric field and the displacement ds:

electrostatic potential from field

$$V = -\int_{P_0}^{P} E \cos\theta \, ds + V_0 \tag{25.26}$$

This formula permits us to calculate the electrostatic potential of any arbitrary charge distribution, if the electric field is known. In such a calculation, we must begin at some position P_0 at which the potential has a given value V_0 and evaluate the integral for some path connecting this position P_0 to the position P. Although the choice of the value of the reference potential V_0 is arbitrary, as discussed above, *it is conventional, whenever possible, to adopt the value* $V_0 = 0$ *at infinity, as in the Coulomb potential.* The following examples illustrate how such calculations are done.

EXAMPLE 4 A sphere of radius R carries a total positive charge Q distributed uniformly throughout its volume (Fig. 25.12). Find the electrostatic potential inside and outside the sphere.

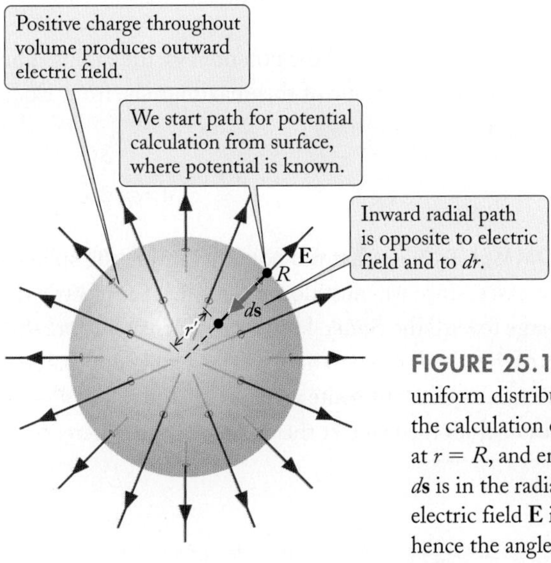

FIGURE 25.12 A sphere of radius R with a uniform distribution of charge. The path for the calculation of the electric potential begins at $r = R$, and ends at $r = r'$. The displacement $d\mathbf{s}$ is in the radially inward direction, and the electric field $\mathbf{E}$ is in the outward direction; hence the angle θ between $\mathbf{E}$ and $d\mathbf{s}$ is 180°, and $\cos\theta = -1$.

SOLUTION: Outside the sphere, the potential is the same as for a point charge,

$$V = \frac{1}{4\pi\epsilon_0}\frac{Q}{r} \tag{25.27}$$

To find the potential at some point r' inside the sphere, we want to evaluate Eq. (25.26) along a radial path that begins at some point where the potential is known. We choose a point on the surface of the sphere, and integrate the electric field from that point (at $r = R$) to a point inside where we want to find the potential (at $r = r'$). We know the potential at $r = R$ from Eq. (25.27),

$$V_0 = \frac{1}{4\pi\epsilon_0}\frac{Q}{R} \tag{25.28}$$

and we know the electric field inside the sphere from Eq. (24.18),

$$E = \frac{Q}{4\pi\epsilon_0}\frac{r}{R^3} \tag{25.29}$$

For our inward radial path, $ds = -dr$, and $\cos\theta = -1$ (see Fig. 25.12). Substituting all of this into Eq. (25.26), we obtain

$$V = -\int_{P_0}^{P} E\cos\theta\, ds + V_0 = -\int_R^{r'} \frac{Q}{4\pi\epsilon_0}\frac{r}{R^3}\,dr + \frac{1}{4\pi\epsilon_0}\frac{Q}{R} \tag{25.30}$$

$$= -\frac{1}{4\pi\epsilon_0}\frac{Q}{R^3}\int_R^{r'} r\,dr + \frac{1}{4\pi\epsilon_0}\frac{Q}{R} = -\frac{1}{4\pi\epsilon_0}\frac{Q}{R^3}\left(\frac{r^2}{2}\right)\Bigg|_R^{r'} + \frac{1}{4\pi\epsilon_0}\frac{Q}{R}$$

$$= -\frac{1}{4\pi\epsilon_0}\frac{Q}{R^3}\left(\frac{r'^2}{2} - \frac{R^2}{2}\right) + \frac{1}{4\pi\epsilon_0}\frac{Q}{R} \tag{25.31}$$

$$= -\frac{1}{4\pi\epsilon_0}\frac{Q}{2R^3}r'^2 + \frac{1}{4\pi\epsilon_0}\frac{3Q}{2R} \tag{25.32}$$

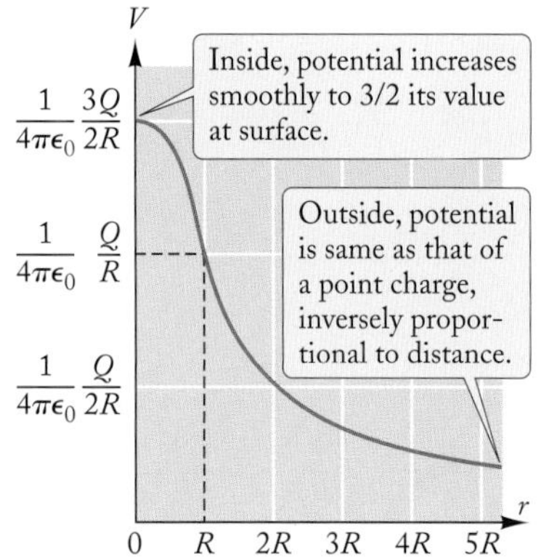

FIGURE 25.13 Electrostatic potential of a uniform spherical distribution of charge.

Figure 25.13 is a plot of the potential vs. the radius. The potential has a maximum at the center. The value of this maximum is, from Eq. (25.32) with $r' = 0$,

$$V = \frac{1}{4\pi\epsilon_0}\frac{3Q}{2R} \tag{25.33}$$

COMMENT: Note that the potential inside the sphere continues to increase as r' decreases, since we must do work against the electric field to move a positive point charge toward the center. However, it increases more slowly than the $\propto Q/r'$ behavior of the Coulomb potential, which would approach infinity as $r' \to 0$. Here, the potential attains the finite value of Eq. (25.33) at $r' = 0$, which is only a factor 3/2 greater than the value at the surface of the sphere.

EXAMPLE 5 A coaxial cable consists of a long, cylindrical conductor of radius a concentric with a thin cylindrical shell of larger radius b (see Fig. 25.14). If the central conductor has a charge per unit length $\lambda = Q/L$ uniformly distributed on its surface, what is the potential difference between the inner and outer conductors? (Assume the space between them is empty.)

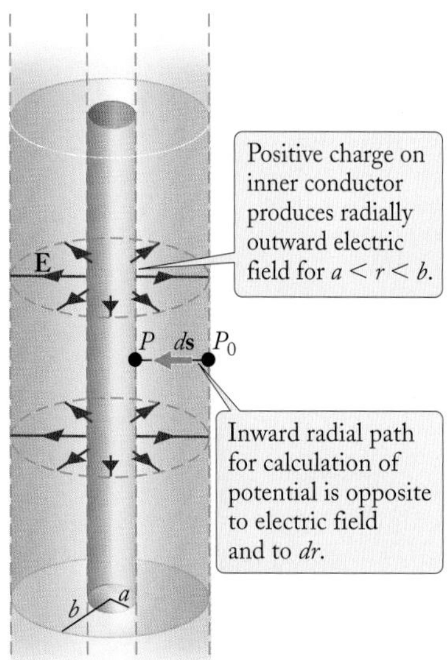

FIGURE 25.14 A coaxial cable.

SOLUTION: We again calculate the potential by integrating the electric field along a path, now from the outer to the inner conductor. Since we are interested only in the potential *difference* between the inner and the outer conductor, we set the reference potential $V_0 = V_b$ in Eq. (25.26). We saw in Chapters 23 and 24 that the electric field outside a cylindrically symmetric charge distribution is given by Eqs. (23.10) and (24.23):

$$E = \frac{1}{2\pi\epsilon_0}\frac{\lambda}{r} \tag{25.34}$$

If we integrate radially inward from the diameter of the outer conductor ($r = b$) to that of the inner conductor ($r = a$), we again have $ds = -dr$, and $\cos\theta = -1$. So we obtain

$$V_a - V_b = -\int_{P_0}^{P} E\cos\theta\, ds = -\int_b^a \frac{1}{2\pi\epsilon_0}\frac{\lambda}{r}\,dr = -\frac{\lambda}{2\pi\epsilon_0}\int_b^a \frac{1}{r}\,dr$$

$$= -\frac{\lambda}{2\pi\epsilon_0}(\ln r)\bigg|_b^a = -\frac{\lambda}{2\pi\epsilon_0}(\ln a - \ln b)$$

$$= \frac{\lambda}{2\pi\epsilon_0}\ln\left(\frac{b}{a}\right) \tag{25.35}$$

where in the last equality we have used $\ln a - \ln b = \ln(a/b) = -\ln(b/a)$.

COMMENT: The outer cylinder played no role in the calculation, since for this symmetric distribution, it affects the electric field only beyond $r = b$. Note that because of the logarithmic dependence of the potential on distance, we could not have calculated the potential of the central conductor alone with respect to infinite distance—this would lead to an infinite result. This is a reflection of the fact that we have assembled an infinite charge distribution. For a coaxial cable in real use, the outer conductor has an opposite charge per unit length $\lambda = -Q/L$, so $E = 0$ outside, and we can set the reference potential $V_0 = 0$ everywhere outside the cable.

In the above examples of the calculation of the electrostatic potential of a charge distribution, we started with the known electric field. But it is also possible to start with the charge distribution and calculate the potential directly, without bothering with the electric field. Sometimes this is a much easier task. For such a direct calculation, we employ Eq. (25.12), which tells us how much a point charge contributes to the potential:

$$V = \frac{1}{4\pi\epsilon_0}\frac{Q}{r}$$

If the charge distribution consists of several point charges, then the electrostatic potential at some point is the sum of the individual Coulomb potentials evaluated at that point,

$$V = \frac{1}{4\pi\epsilon_0}\sum_i \frac{Q_i}{r_i} \tag{25.36}$$

Any other charge distribution can be regarded as consisting of small charge elements dQ, each of which can be treated as a point charge. Each such charge element will contribute an amount dV to the potential at a point,

$$dV = \frac{1}{4\pi\epsilon_0}\frac{dQ}{r} \tag{25.37}$$

The net potential of the charge distribution is then the sum or the integral of all the contributions from all these pointlike charge elements:

$$V = \frac{1}{4\pi\epsilon_0}\int \frac{dQ}{r} \tag{25.38}$$

potential of a continuous charge distribution

Note that since we are considering pointlike charge contributions using the Coulomb potential, the potential calculated will be with respect to $V_0 = 0$ at infinity, the conventional zero of potential.

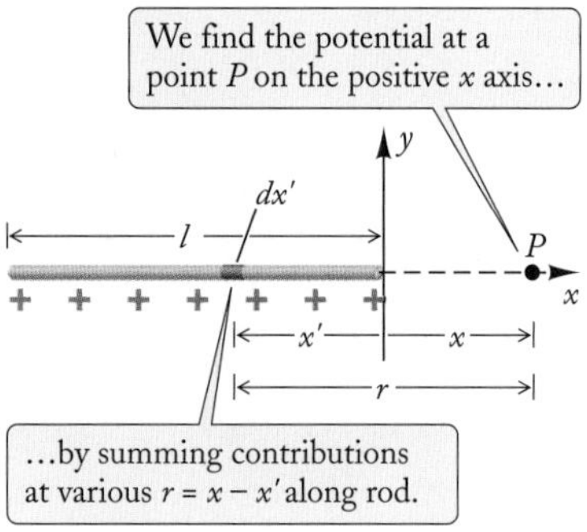

FIGURE 25.15 A rod with a uniform distribution of charge.

EXAMPLE 6 A rod of length l has a charge Q uniformly distributed along its length (see Fig. 25.15). Find the electrostatic potential at a distance x from one end of the rod.

SOLUTION: Consider a small segment dx' of the rod. Since the charge per unit length is Q/l, the charge in this segment is $dQ = (Q/l)\,dx'$. The distance of this segment from the point P is $r = x - x'$ (here, x' is a negative quantity; see Fig. 25.15). According to Eq. (25.37), the segment then makes the following contribution to the potential at the point P:

$$dV = \frac{1}{4\pi\epsilon_0}\frac{(Q/l)dx'}{x - x'}$$

The integral of this over the length of the rod, from $x' = -l$ to $x' = 0$, gives us the total potential at the position P:

$$V(P) = V(x) = \int_{-l}^{0} \frac{1}{4\pi\epsilon_0}\frac{Q/l}{x - x'}dx'$$

$$= \frac{1}{4\pi\epsilon_0}\frac{Q}{l}\left[-\ln(x - x')\right]\Big|_{-l}^{0} = \frac{1}{4\pi\epsilon_0}\frac{Q}{l}\ln\left(\frac{x + l}{x}\right)$$

COMMENT: The distribution of charge along the rod produces a net potential proportional to $\ln[(x + l)/x]$, which, like the Coulomb potential, diverges at $x = 0$. Also, $V(x)$ approaches the Coulomb potential at large distances, where it is difficult to distinguish the rod from a point charge. This can be seen by expanding the logarithm for small values of l/x, since for small z, $\ln(1 + z) \approx z$. Thus

$$V(x) = \frac{1}{4\pi\epsilon_0}\frac{Q}{l}\ln\left(\frac{x + l}{x}\right) = \frac{1}{4\pi\epsilon_0}\frac{Q}{l}\ln\left(1 + \frac{l}{x}\right) \approx \frac{1}{4\pi\epsilon_0}\frac{Q}{l}\frac{l}{x} = \frac{1}{4\pi\epsilon_0}\frac{Q}{x}$$

which is identical to the Coulomb potential.

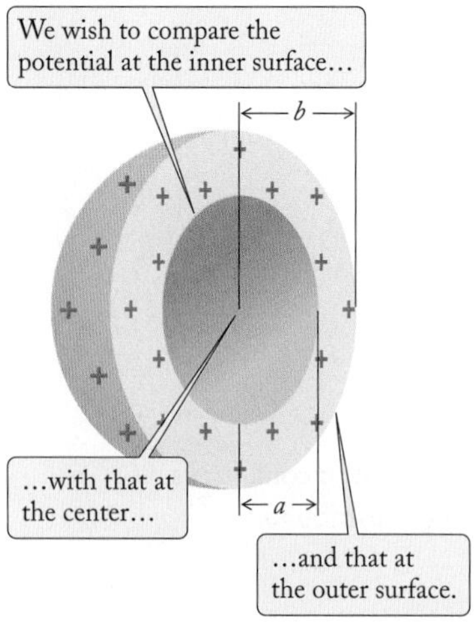

FIGURE 25.16 A charged shell.

Checkup 25.2

QUESTION 1: For a sphere of radius R with a uniform volume distribution of positive charge, where is the electric potential maximum and where is it minimum? For a sphere of radius R with a uniform distribution of negative charge, where is the electric potential maximum and where is it minimum?

QUESTION 2: A hollow spherical charge distribution produces zero electric field in its interior. Does this mean that the electric potential is also zero in its interior?

QUESTION 3: A uniform electric field is directed along the x axis. Does the corresponding electric potential increase or decrease with x?

QUESTION 4: A thin ring of radius R has a charge Q distributed around its circumference. What is the electrostatic potential at the center of the ring? Does the charge need to be distributed uniformly around the ring?

QUESTION 5: A thick, insulating spherical shell with inner radius a and outer radius b has a positive charge Q distributed uniformly throughout its volume; the interior of the shell is empty (see Fig. 25.16). Is the electrostatic potential at $r = a$ larger, smaller, or the same as at $r = b$? Is the electrostatic potential at $r = 0$ larger, smaller, or the same as at $r = a$?

(A) Larger; larger (B) Larger; smaller (C) Larger; the same
(D) Smaller; larger (E) Smaller; the same

25.3 POTENTIAL IN CONDUCTORS

Since the electric field in a conducting body in electrostatic equilibrium is zero, Eq. (25.26) implies that the potential difference between any two points within a conducting body is zero. Thus, *all points within a conducting body are at the same electrostatic potential.* For instance, since the ground is a conductor, all points on the surface of the Earth or within the Earth are at the same electrostatic potential; that is, the surface of the Earth is an equipotential. In experiments with electric circuits, it is usually convenient to adopt the convention that the potential of the surface of the Earth is zero, $V = 0$. The surface of the Earth is said to be the **electric ground**, and any conductors connected to it are said to be grounded. For example, the third terminal (round hole) in an ordinary household electric outlet (see Fig. 25.17) is grounded—it is connected to a plate or a rod buried in the ground outside the house. This grounded terminal is intended as a safety feature—if the insulation in your electric drill or some other appliance fails, the electric current from the "live" (flat) terminals can leak away harmlessly into the ground instead of leaking into your hands.

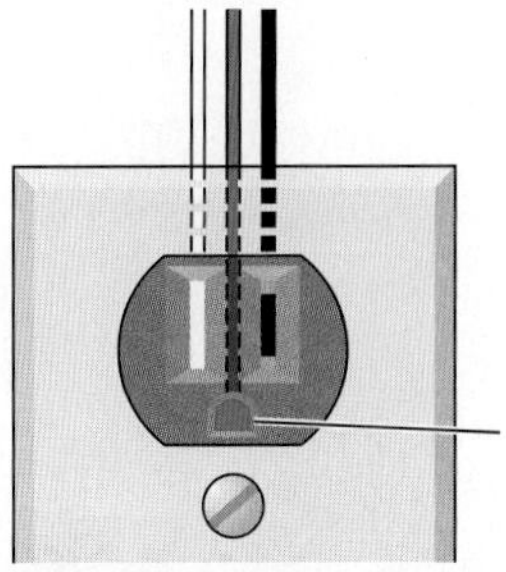

FIGURE 25.17 Grounded terminal of a three-hole outlet.

EXAMPLE 7 A conducting sphere of radius R carries a total positive charge Q uniformly distributed over its surface. What is the electrostatic potential inside and outside the sphere?

Concepts in Context

SOLUTION: As we found in Section 25.1, outside the sphere, the potential is the same as for a point charge,

$$V = \frac{1}{4\pi\epsilon_0}\frac{Q}{r} \quad (r \geq R) \tag{25.39}$$

Inside the conductor, the potential is the same everywhere, and so takes the same value as at the surface of the sphere, where $r = R$,

$$V = \frac{1}{4\pi\epsilon_0}\frac{Q}{R} \quad (r \leq R)$$

A plot of this potential is shown in Fig. 25.18. Note that for such a sphere, the electric field drops discontinuously to zero inside the sphere, but the electrostatic potential is continuous, remaining constant inside the sphere.

COMMENT: We would have obtained the same result if we had considered a hollow conducting shell; in that case, there would be no charge inside the shell, and thus no field and a constant potential.

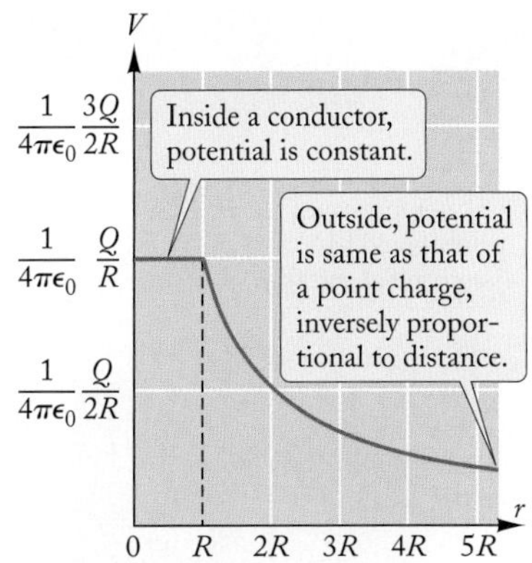

FIGURE 25.18 Electrostatic potential of a solid metal (conducting) sphere.

PHYSICS IN PRACTICE ELECTRIC SHIELDING

The absence of electric fields in closed cavities surrounded by conductors has important practical applications. Delicate electric instruments can be shielded from atmospheric electric fields, and other stray electric fields, by placing them in a box made of sheet metal. Such a box is called a **Faraday cage**. Figure 1 shows a room-sized Faraday cage, also known as a *screen room*. Often, the box is made of fine copper wire mesh rather than sheet metal; although such wire mesh does not have the perfect shielding properties of solid sheet metal, it provides good enough shielding for most purposes. A mesh is convenient, since it is flexible and one can see through it.

The shielding of buildings against lightning relies on the same principle. Figure 2 shows a protective canopy, consisting of an array of wires, for the shielding of a building used for the storage of inflammable or explosive materials. The gaps between the wires permit the penetration of some field lines, and the shielding provided by such a canopy is even less perfect than that of a wire mesh, but the strength of the external atmospheric fields is attenuated to such an extent that lightning flashes are unlikely to reach the building.

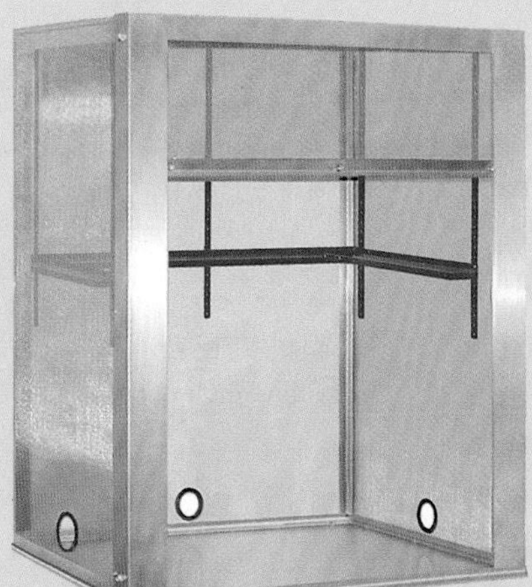

FIGURE 1 A screen room, or room-sized Faraday cage.

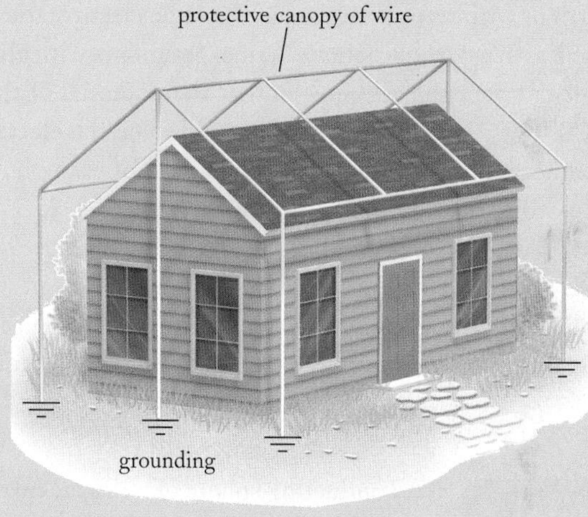

FIGURE 2 A building covering that provides protective electrostatic shielding.

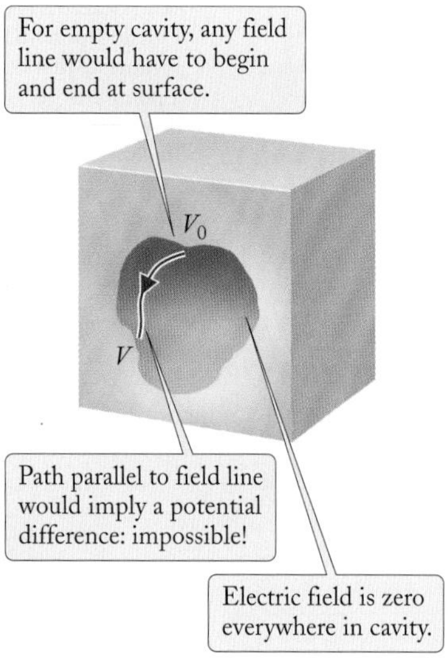

FIGURE 25.19 Empty cavity in a volume of conducting material. The red line is a hypothetical field line in the cavity.

With Eq. (25.26) we can prove an interesting theorem about the electric field in a conducting body with an empty, completely enclosed cavity (see Fig. 25.19): *within a closed, empty cavity inside a homogeneous conductor, the electric field is exactly zero*. The proof of the theorem is by contradiction. If there were an electric field inside this cavity, then there would have to be field lines in the cavity. Consider one of these field lines. Since the cavity is empty (contains no charge), the field line cannot end or begin within the space of the cavity—it must therefore begin and end on the surface of the cavity, as shown in Fig. 25.19 (note that the field line cannot penetrate the conducting material, since the electric field is zero in this material). Now, suppose that at the point where the field line begins, the potential has the value V_0, and evaluate Eq. (25.26) for a path that follows the field line from its beginning to its end. The electric field is everywhere tangent to the field line, so $\theta = 0$ and $\cos\theta = 1$, and therefore $-E\cos\theta\, ds$ is negative. According to Eq. (25.26), this implies that V is smaller than V_0. But such a difference between V and V_0 is impossible, since these potentials are evaluated at the surface of

In the same way, the sheet metal of an automobile provides adequate shielding against lightning. The windows of the automobile leave large gaps, and permit the penetration of some field lines. But the strength of the external atmospheric field is attenuated to such an extent that lightning flashes are unlikely to reach the occupants of the automobile. Figure 3 shows a direct lightning hit on an automobile.

Conductive suits are used to provide shielding for utility workers who are in direct contact with live power lines (see Fig. 4). Many high-voltage power lines are operated at 400 kV, and they are surrounded by fairly strong electric fields. For the routine repair and cleaning of insulators, the worker wears a protective suit of a heavy fabric that incorporates a woven mesh of stainless steel wires, which make it conducting. This suit shields the body from the external electric fields.

FIGURE 3 Lightning striking an automobile.

FIGURE 4 A conductive suit worn by a utility worker performing "bare-handed" repairs on a live power line.

the conductor, and all points on a conductor are necessarily at the same potential. This contradiction establishes the theorem. The fact that the electric field is zero in a conductor's cavity is quite useful, as discussed in Physics in Practice: Electric Shielding.

✔ Checkup 25.3

QUESTION 1: Two conducting spheres are connected by a conducting wire (Fig. 25.20). If one sphere has a radius R_1 and a positive charge Q_1, is this charge uniformly distributed over its surface? Is its potential $V = (1/4\pi\epsilon_0)Q_1/R_1$? Is its potential the same as that of the other sphere?

QUESTION 2: An airplane is flying through a thundercloud, where there are strong electric fields. Do these electric fields affect the occupants of the airplane?

QUESTION 3: An isolated 1.0-cm-diameter steel bearing ball and an isolated 1.0-m-diameter steel demolition ball each carry the same amount of charge. Compared with the small ball, the electrostatic potential of the large steel ball is

(A) Larger (B) Smaller (C) The same

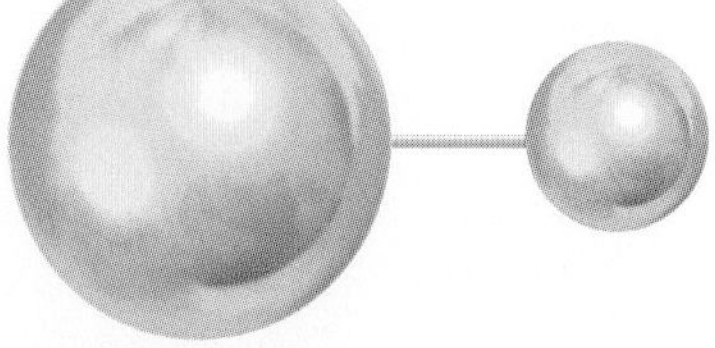

FIGURE 25.20 Two conducting spheres connected by a conducting wire.

Online Concept 30 Tutorial

25.4 CALCULATION OF THE FIELD FROM THE POTENTIAL

In Section 25.2 we saw how to calculate the electrostatic potential from the electric field. Now we will see how to calculate the electric field from the potential. For this purpose, we begin with Eq. (25.26), and we assume for the moment that ds is a small displacement in the direction of the electric field. With $\theta = 0$, Eq. (25.26) can be written as

$$\int E\, ds = -(V - V_0) \tag{25.40}$$

If the integration extends over only a small interval, the electric field is approximately constant, and the change in potential is small, $V - V_0 = dV$. Then Eq. (25.40) reduces to

$$E\, ds = -dV$$

which we can solve for E:

electric field from potential

$$E = -\frac{dV}{ds} \tag{25.41}$$

This says that the electric field equals the negative of the derivative of the potential with respect to the displacement.

EXAMPLE 8 The electrostatic potential generated by a pair of oppositely charged, parallel conducting plates is [Eq. (25.7)]

$$V = -E_0 y \tag{25.42}$$

For this potential, verify that Eq. (25.41) yields the correct value for the electric field.

SOLUTION: According to Eq. (25.41),

$$E = -\frac{dV}{ds} = -\frac{dV}{dy} = -\frac{d}{dy}(-E_0 y) = E_0$$

which is the expected result.

If the small displacement ds is not in the direction of the electric field, then we must retain the factor $\cos\theta$ in Eq. (25.26), and instead of Eq. (25.41) we obtain

$$E\cos\theta = -\frac{dV}{ds}$$

Since $E\cos\theta$ is the component of the electric field in the direction of the displacement, this formula tells us that the component of the electric field in any direction equals the negative of the derivative of the potential with respect to the displacement in that direction. Accordingly, we obtain E_x if we perform a small displacement in the x direction, E_y if we perform a small displacement in the y direction, and E_z if we perform a small displacement in the z direction. In each case, it is assumed that we are holding the other variables constant (e.g., for a displacement in the x direction, y and z are treated as constant). Mathematicians use the *partial derivative* symbol for this operation, for instance, $\partial/\partial x$ instead of d/dx. The components of the electric field are thus given by

$$E_x = -\frac{\partial V}{\partial x} \quad E_y = -\frac{\partial V}{\partial y} \quad E_z = -\frac{\partial V}{\partial z} \tag{25.43}$$

To obtain each component of **E** for a given V, one merely treats the other variables like constants when one takes the derivative with respect to the variable of interest.

EXAMPLE 9 We found in Section 25.1 that the potential outside a charged sphere is the Coulomb potential, $V = (1/4\pi\epsilon_0)Q/r$. The Coulomb potential can be written in rectangular coordinates by using the identity $r = \sqrt{x^2 + y^2 + z^2}$. From that form of the potential, obtain the x, y, and z components of the electric field.

Concepts — in — Context

SOLUTION: The Coulomb potential in rectangular coordinates is simply

$$V = \frac{Q}{4\pi\epsilon_0}\frac{1}{\sqrt{x^2 + y^2 + z^2}}$$

Using Eq. (25.43), we evaluate E_x:

$$E_x = -\frac{\partial}{\partial x}\frac{Q}{4\pi\epsilon_0}\frac{1}{\sqrt{x^2 + y^2 + z^2}} = -\frac{Q}{4\pi\epsilon_0}\frac{\partial}{\partial x}(x^2 + y^2 + z^2)^{-1/2}$$

$$= -\frac{Q}{4\pi\epsilon_0}\left(-\frac{1}{2}\right)(x^2 + y^2 + z^2)^{-3/2} \times (2x) = \frac{Q}{4\pi\epsilon_0}\frac{x}{(x^2 + y^2 + z^2)^{3/2}}$$

Similarly, for E_y and E_z we have

$$E_y = \frac{Q}{4\pi\epsilon_0}\frac{y}{(x^2 + y^2 + z^2)^{3/2}} \quad \text{and} \quad E_z = \frac{Q}{4\pi\epsilon_0}\frac{z}{(x^2 + y^2 + z^2)^{3/2}} \tag{25.44}$$

COMMENT: These can indeed be recognized as the x, y, and z components of the usual radial Coulomb field. For example, the second relation in Eq. (25.44) is

$$E_z = \frac{Q}{4\pi\epsilon_0}\frac{z}{r^3} = \frac{1}{4\pi\epsilon_0}\frac{Q}{r^2}\frac{z}{r} = \frac{1}{4\pi\epsilon_0}\frac{Q}{r^2}\cos\theta$$

where here θ is the angle between the electric field vector and the z axis, as shown in Fig. 25.21.

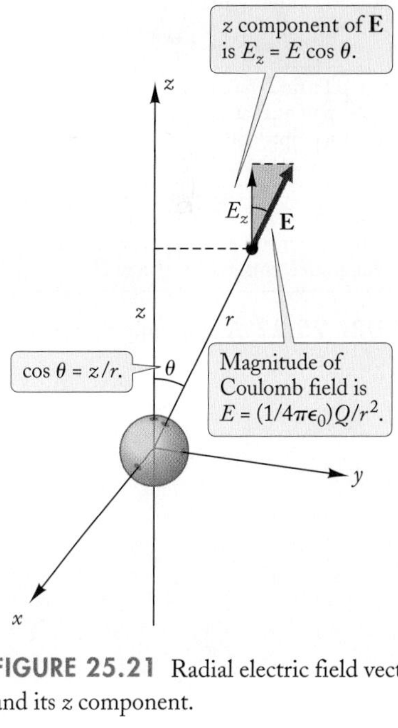

FIGURE 25.21 Radial electric field vector and its z component.

The results obtained in the preceding two examples do not tell us anything new—since we already know the electric field in these examples, the calculation merely verifies the consistency of our formulas. However, if we do not know the electric field, then Eq. (25.41) can be useful in a calculation of the electric field, provided we have some means of finding the potential. As discussed above in Section 25.2, if the charge distribution is specified, we can find the potential by treating the charge distribution as a collection of point charges, each of which contributes a potential of the form (25.37). The net potential is then the sum of all these point-charge-like contributions, given by Eq. (25.38).

The process of calculating the potential from the charge distribution, and then calculating the electric field from the potential, can be be significantly easier than calculating the electric field directly from the charge distribution. This is because the potential is a scalar quantity, so one need not consider vector components when one sums its contributions.

EXAMPLE 10 An amount of charge Q is uniformly distributed along the circumference of a thin ring of radius R (see Fig. 25.22). Find the potential on the axis of the ring, and find the electric field.

SOLUTION: To find the potential, we regard the ring as made up of many small charge elements, each of which can be treated as a point charge (see Fig. 25.22). If one of these elements has a charge dQ, its contribution to the potential at a distance r is

$$dV = \frac{1}{4\pi\epsilon_0}\frac{dQ}{r} = \frac{1}{4\pi\epsilon_0}\frac{dQ}{\sqrt{R^2 + y^2}} \tag{25.45}$$

Since the distance $r = \sqrt{R^2 + y^2}$ does not depend on where along the circumference the charge dQ is located, all the small elements of charge on the ring contribute in the same way, and the net contribution to the potential therefore has the form of Eq. (25.45), with the charge contribution dQ replaced by the total charge $\int dQ = Q$:

$$V = \frac{1}{4\pi\epsilon_0}\frac{Q}{\sqrt{R^2 + y^2}} \tag{25.46}$$

The electric field along the axis is the negative of the derivative of the potential with respect to y:

$$E_y = -\frac{\partial V}{\partial y} = -\frac{1}{4\pi\epsilon_0}\frac{\partial}{\partial y}\frac{Q}{\sqrt{R^2 + y^2}} = \frac{1}{4\pi\epsilon_0}\frac{Qy}{(R^2 + y^2)^{3/2}}$$

We already found this same result in Chapter 23 by the much more difficult technique of integration of the vector electric field, according to Coulomb's Law. Here, the calculation is much easier.

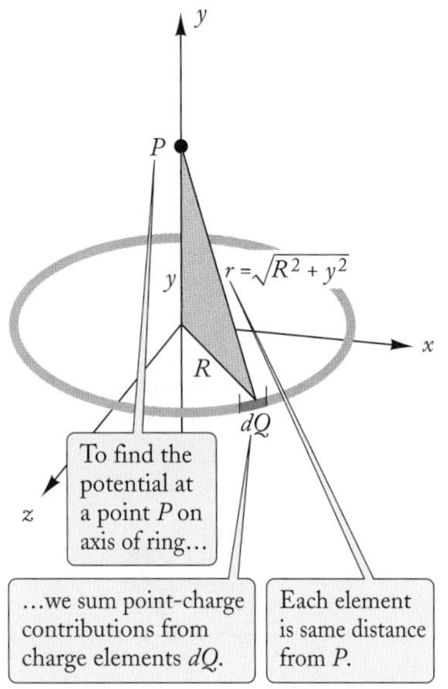

FIGURE 25.22 A uniformly charged ring.

equipotential surface

A mathematical surface on which the electrostatic potential has a fixed, constant value is called an **equipotential surface**. Figure 25.23 shows the equipotential surfaces belonging to the potential of a uniformly charged flat sheet—the equipotential surfaces are parallel planes. Figure 25.24 shows the equipotential surfaces belonging to the potential of a point charge—the equipotential surfaces are concentric spheres. Figure 25.25 shows the equipotential surfaces of a pair of positive and negative charges of equal magnitudes. The equipotential surfaces associated with a given potential provide a graphical representation of the potential.

Note that the electric field is everywhere perpendicular to the equipotentials. This is an immediate consequence of Eq. (25.41): along any equipotential surface, the potential is constant, that is, $dV = 0$; consequently, for a displacement ds along the surface, $E = -dV/ds = 0$, which says that the electric field in the direction tangential to the surface is zero. Thus, the electric field must be entirely perpendicular to the surface (see Fig. 25.26).

Conversely, if the electric field is everywhere perpendicular to a given surface, then this surface must be an equipotential surface. This, also, is a consequence of Eq. (25.41); if the electric field is zero in the direction parallel to the surface, then for a displacement ds parallel to the surface, $dV = 0$, and the potential is constant. Since we already know (from Section 24.5) that along the surface of any conductor in electrostatic equilibrium the electric field is perpendicular to the surface, we can conclude that any conducting surface is an equipotential surface. (This conclusion agrees with the general statement made in Section 25.3—the potential is constant throughout any conductor.)

Keep in mind that equipotential surfaces are surfaces of constant energy for a given charge. Thus it takes no energy to move a charge along a surface of constant potential, since such motion does not act against any electric field (see Fig. 25.26). However, to move perpendicular to such surfaces does require work, since the electric field is perpendicular to the equipotential surface.

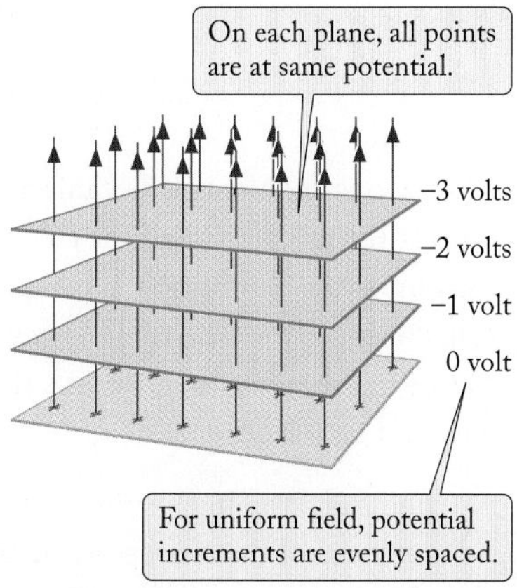

FIGURE 25.23 Equipotential surfaces for a very large flat sheet with a uniform distribution of charge. The equipotentials are flat planes. This indicates that all points at the same height above the sheet are at the same potential.

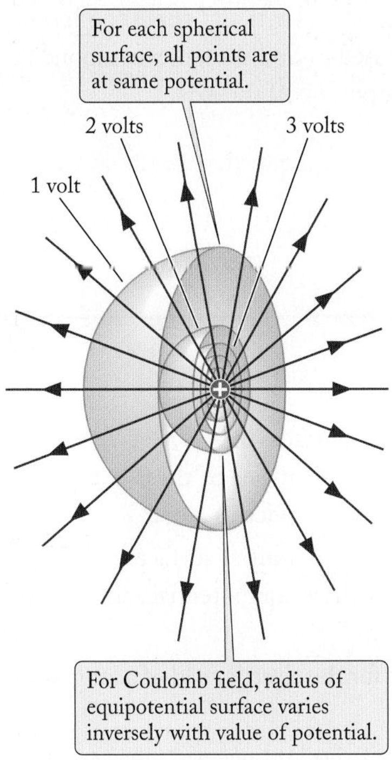

FIGURE 25.24 Equipotential surfaces for a positive point charge. The equipotentials are concentric spheres. This indicates that points at the same distance from the central charge are at the same potential.

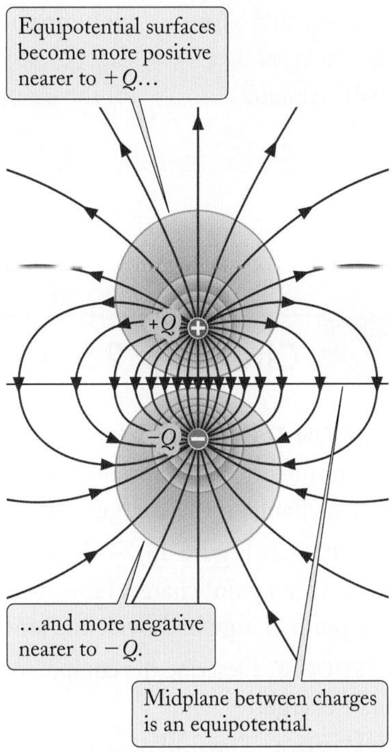

FIGURE 25.25 Equipotential surfaces for a positive and a negative point charge of equal magnitudes. Note that the midplane between the charges is an equipotential.

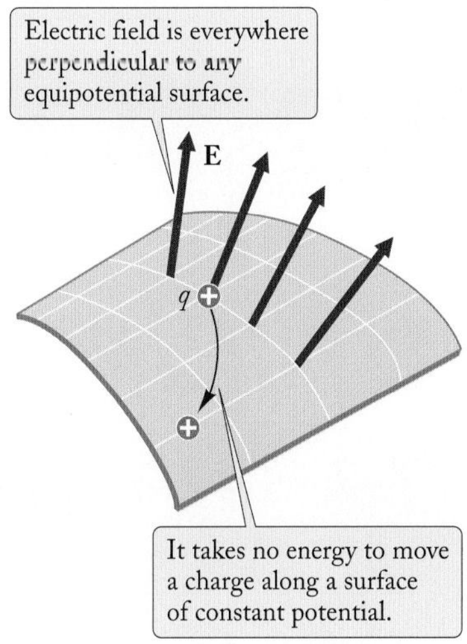

FIGURE 25.26 The electric field is everywhere perpendicular to the equipotential surface. If a charge q moves along this surface, the electric field does no work on it.

PROBLEM-SOLVING TECHNIQUES ELECTROSTATIC POTENTIAL AND FIELD

In previous chapters we became acquainted with two methods for the calculation of the electric field of a charge distribution: from Coulomb's Law, and from Gauss' Law. The calculation of the electric field from the electrostatic potential, as in Example 10, gives us a third method for the calculation of the electric field.

- The first step in this method is to find the electrostatic potential of the charge distribution. For this purpose it is often convenient to introduce x, y, and z coordinates with the origin somewhere near the charge distribution (compare Fig. 25.22). If the charge distribution consists of an arrangement of several point charges, the net electrostatic potential is simply the sum of the individual electrostatic potentials of the point charges. Note that since the potential is a scalar, the sum of the individual potentials is an ordinary sum of ordinary numbers, which is much easier to evaluate than the vector sum required for a calculation of the net electric field via Coulomb's Law. If the charge distribution is continuous (such as the charge distribution along the circumference of the ring in Example 10), you need to regard it as a collection of many small (infinitesimal) charge elements, each of which can be treated as a point charge. The net potential is then the integral over the electrostatic potentials of all these small elements.

- The second step is the calculation of the electric field by differentiation of the potential, according to Eq. (25.41). If the symmetry of the problem permits you to determine the direction of the electric field, then the magnitude of the electric field is the negative derivative of the potential V with respect to the displacement s in that direction,

$$E = -\frac{dV}{ds}$$

- If the direction of the electric field is not obvious, then you can calculate the x, y, and z components of the electric field by differentiating the potential V in the x, y, and z directions:

$$E_x = -\frac{\partial V}{\partial x} \quad E_y = -\frac{\partial V}{\partial y} \quad E_z = -\frac{\partial V}{\partial z}$$

The calculation of all three components requires knowledge of the complete dependence of V on x, y, and z. For instance, Eq. (25.46) tells us the dependence of V on y along the axis of a charged ring, but does not tell us the dependence on x or z at points off the axis; hence Eq. (25.46) cannot be used to find the x or z components of the electric field for points off the axis.

✓ Checkup 25.4

QUESTION 1: Figure 25.23 shows the equipotential surfaces for 0 V, −1 V, −2 V, etc., for a uniformly charged flat sheet. How would the positions of these equipotential surfaces change if the charge density on the sheet were twice as large?

QUESTION 2: Figure 25.24 shows the spherical equipotential surfaces for 1 V, 2 V, 3 V, etc., for a point charge. How would the radii of these equipotential surfaces change if the point charge had twice the amount of charge?

QUESTION 3: Describe the equipotential surfaces for the electric field of a long, straight line of charge.

QUESTION 4: The midplane between a pair of positive and negative charges of equal magnitudes is an equipotential surface (see Fig. 25.25). Is this also true if both charges are of equal signs?

QUESTION 5: Can there be a nonzero electric field at a location where the potential is zero? Can there be a nonzero potential at a location where the electric field is zero?

(A) No; no (B) No; yes (C) Yes; no (D) Yes; yes

25.5 ENERGY OF SYSTEMS OF CHARGES

Arrangements of electric charges are used in physics and engineering to produce electric fields, and also to store electric energy. For instance, the energy for the operation of the flashlamp of a photographic camera is stored in a charge distribution accumulated on a system of parallel plates (called a *capacitor*) in the flash unit. It is therefore often important to calculate the electric energy of an arrangement of charges.

The simplest such arrangement consists of two point charges q and q' separated by a distance r. The electric potential energy of this system is

$$U = \frac{1}{4\pi\epsilon_0}\frac{qq'}{r} \tag{25.47}$$

This potential energy can be regarded as the work required to move q from an infinite distance to within a distance r of q'. It is a *mutual* potential energy that belongs to both q and q'; that is, it is an energy associated with the relative configuration of the pair (q, q').

For arrangements consisting of more than two charges, the net potential energy can be calculated by writing down a term similar to that in Eq. (25.47) for *each pair* of charges. For instance, if we are dealing with three charges q_1, q_2, and q_3 (see Fig. 25.27), we have three possible pairs (q_1, q_2), (q_2, q_3), and (q_1, q_3), so the net potential energy of the system is

potential energy of a system of point charges

$$U = \frac{1}{4\pi\epsilon_0}\frac{q_1q_2}{r_{12}} + \frac{1}{4\pi\epsilon_0}\frac{q_2q_3}{r_{23}} + \frac{1}{4\pi\epsilon_0}\frac{q_1q_3}{r_{13}} \tag{25.48}$$

where r_{12}, r_{23}, and r_{13} are the distances indicated in Fig. 25.27. This potential energy is the work required to assemble the charges in the final configuration shown in Fig. 25.27, starting from an initial condition of infinite separation of all charges.

That this sum of the potential energies of each pair gives the correct total energy can be seen by assembling the charges sequentially. For example, for three charges, it takes no energy to bring a first charge into position (since there are no electric fields to act against). To bring the second charge into position involves moving it in the Coulomb potential of the first charge (thus the q_1q_2 term). Finally, to bring the third charge into position involves working against the Coulomb fields of the first two charges (thus the q_2q_3 and q_1q_3 terms).

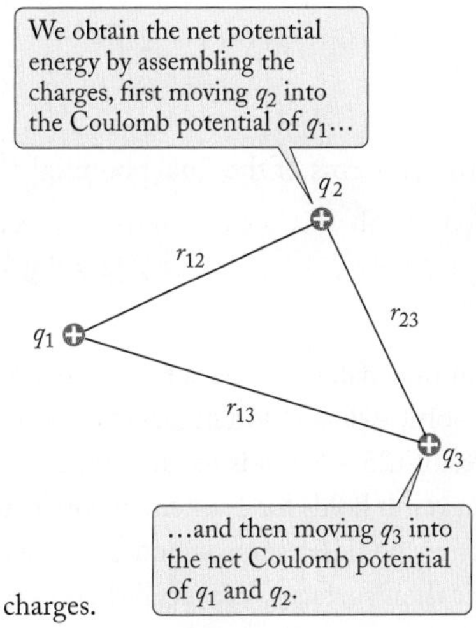

FIGURE 25.27 Three point charges.

EXAMPLE 11 Four charges, $q_1 = q$, $q_2 = 2q$, $q_3 = -q$, and $q_4 = q$, are at the corners of a square of side a, as shown in Fig. 25.28. If $q = 2.0\ \mu\text{C}$ and $a = 7.5$ cm, what was the total energy required to assemble this system of charges?

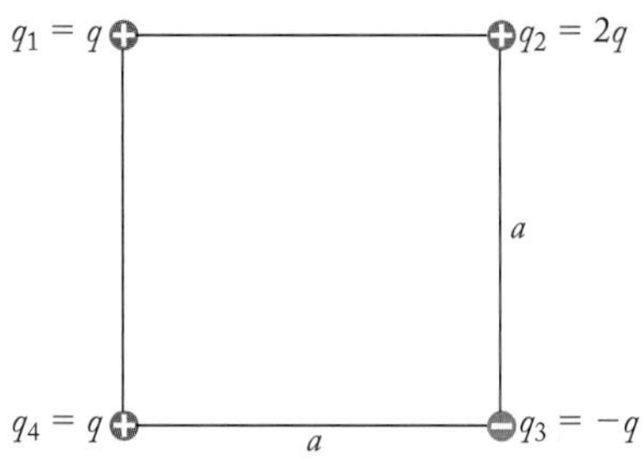

FIGURE 25.28 Four point charges.

SOLUTION: For the total energy, we sum all the potential energies of the individual pairs:

$$U = \frac{1}{4\pi\epsilon_0}\left(\frac{q_1q_2}{r_{12}} + \frac{q_1q_3}{r_{13}} + \frac{q_2q_3}{r_{23}} + \frac{q_1q_4}{r_{14}} + \frac{q_2q_4}{r_{24}} + \frac{q_3q_4}{r_{34}}\right)$$

$$= \frac{1}{4\pi\epsilon_0}\left[\frac{(q)(2q)}{a} + \frac{(q)(-q)}{\sqrt{2}a} + \frac{(2q)(-q)}{a} + \frac{(q)(q)}{a} + \frac{(2q)(q)}{\sqrt{2}a} + \frac{(-q)(q)}{a}\right]$$

$$= \frac{1}{4\pi\epsilon_0}\frac{q^2}{a}\left(2 + \frac{-1}{\sqrt{2}} - 2 + 1 + \frac{2}{\sqrt{2}} - 1\right) = \frac{1}{4\pi\epsilon_0}\frac{q^2}{a}\frac{1}{\sqrt{2}}$$

$$= (9.0 \times 10^9\ \text{N}\cdot\text{m}^2/\text{C}^2)\frac{(2.0 \times 10^{-6}\ \text{C})^2}{(0.075\ \text{m})}\frac{1}{\sqrt{2}} = 0.34\ \text{J}$$

In practical applications, point charges are usually deposited on conductors, each of which carries many point charges. By summing the contributions as we build up such a system, we can calculate the potential energy of a system of charges placed on conductors.

First consider a single conducting sphere of radius R. When it has some charge q, its electrostatic potential $V = (1/4\pi\epsilon_0)(q/R)$. If we imagine changing this amount of charge by a small amount dq, then the potential energy will change by an amount dU given by

$$dU = V\,dq$$

If we build up a total charge Q in small steps from $q = 0$ to $q = Q$, we can sum the contributions to the potential energy during such a process:

$$U = \int dU = \int V\,dq = \int_0^Q \frac{1}{4\pi\epsilon_0}\frac{q}{R}\,dq$$

The radius R of the sphere is not changing while we add charge to it, so this is simply

$$U = \frac{1}{4\pi\epsilon_0}\frac{1}{R}\int_0^Q q\,dq = \frac{1}{4\pi\epsilon_0}\frac{1}{R}\frac{1}{2}Q^2$$

or, writing this in terms of the final potential $V = (1/4\pi\epsilon_0)(Q/R)$,

potential energy of a conductor

$$U = \tfrac{1}{2}QV \tag{25.49}$$

The factor of one-half arises because we are not considering the potential energy of a charge Q in some *external* potential, but instead the energy of interaction of the pieces of Q. The result (25.49) holds for any fixed geometry of conductor.

A similar result holds for a system of conductors. For instance, Fig. 25.29 shows several conducting bodies—such as bodies of metal—with charges Q_1, Q_2, Q_3, . . . distributed over their surfaces. We recall that on each conducting body, the potential has

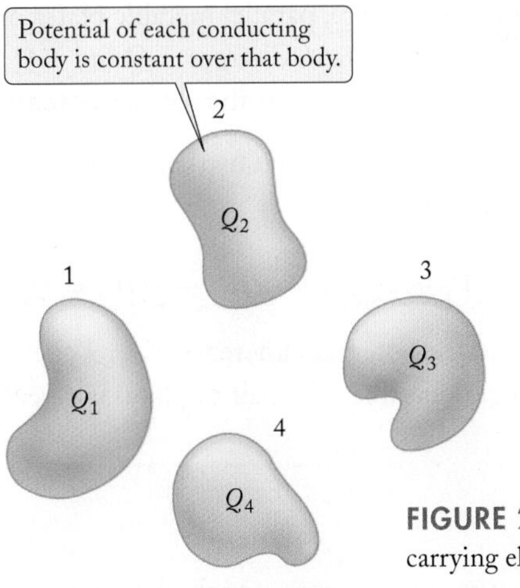

FIGURE 25.29 Four conducting bodies carrying electric charges.

a constant value over the entire volume of the body. Suppose that these electrostatic potentials of the bodies are $V_1, V_2, V_3, \ldots$ It can be shown, although we will not attempt to prove it here, that the net electric potential energy of such a system of conductors is

$$U = \tfrac{1}{2}Q_1V_1 + \tfrac{1}{2}Q_2V_2 + \tfrac{1}{2}Q_3V_3 + \cdots \tag{25.50}$$

potential energy of a system of conductors

EXAMPLE 12 The metallic sphere on the top of a large Van de Graaff generator has a radius of 3.0 m. Suppose that the sphere carries a charge of 5.0×10^{-5} C uniformly distributed over its surface. How much electric energy is stored in this charge distribution?

Concepts — in — Context

SOLUTION: In this problem, there is only one conductor, the metallic sphere. Accordingly, we need only the first term in Eq. (25.50):

$$U = \tfrac{1}{2}Q_1V_1$$

or, with $Q_1 = Q$ and $V_1 = V$, we recover Eq. (25.49):

$$U = \tfrac{1}{2}QV$$

The potential outside a spherically symmetric charge distribution is given by Eq. (25.12). At $r = R$, this potential is

$$V = \frac{1}{4\pi\epsilon_0}\frac{Q}{R}$$

According to Eq. (25.49), the electric energy is then

$$U = \frac{1}{2}QV = \frac{1}{2}Q\frac{1}{4\pi\epsilon_0}\frac{Q}{R} = \frac{1}{8\pi\epsilon_0}\frac{Q^2}{R} \tag{25.51}$$

With $Q = 5.0 \times 10^{-5}$ C and $R = 3.0$ m, this gives

$$U = \frac{1}{8\pi \times 8.85 \times 10^{-12}\ \text{C}^2/\text{N}\cdot\text{m}^2}\frac{(5.0 \times 10^{-5}\ \text{C})^2}{3.0\ \text{m}} = 3.7\ \text{J}$$

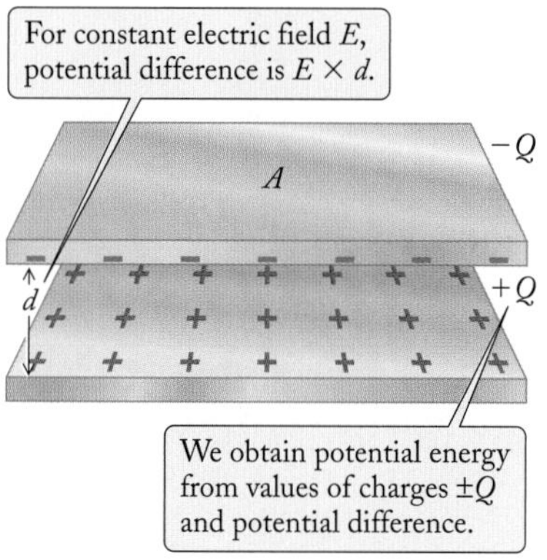

FIGURE 25.30 Two parallel plates, with charges $+Q$ and $-Q$.

EXAMPLE 13 Two large, parallel metallic plates of area A are separated by a distance d. Charges $+Q$ and $-Q$ are placed on the plates, respectively (see Fig. 25.30). What is the electric potential energy?

SOLUTION: Here, we are dealing with two conductors, with $Q_1 = Q$ and $Q_2 = -Q$. Thus, Eq. (25.49) becomes

$$U = \tfrac{1}{2}Q_1V_1 + \tfrac{1}{2}Q_2V_2 = \tfrac{1}{2}QV_1 - \tfrac{1}{2}QV_2 = \tfrac{1}{2}Q(V_1 - V_2) \qquad (25.52)$$

To proceed, we need the potential difference $V_1 - V_2$ between the plates. The electric field in the region between the pair of plates is given by Eq. (24.29),

$$E = \frac{\sigma}{\epsilon_0}$$

where σ is the charge per unit area on each plate. Since the total charge on a plate is Q and the area is A, the charge per unit area is $\sigma = Q/A$, and

$$E = \frac{Q}{\epsilon_0 A} \qquad (25.53)$$

This expression fails near the edges of the plates, where there is an electric fringing field that is not constant. But if the plates are very large, then the edge region is only a very small fraction of the total region between the plates and we can ignore this region without introducing excessive errors in our calculation.

With the constant electric field (25.53) we can calculate the electrostatic potential difference between two plates, using our standard formula for the potential, Eq. (25.26):

$$V_2 = -\int_0^d E\cos\theta\, ds + V_1$$

Since we move parallel to the field going from the positive to the negative plate, $\cos\theta = 1$. For a constant electric field, we then have, similar to Eq. (25.7),

$$V_2 - V_1 = -Ed \qquad (25.54)$$

With $E = Q/\epsilon_0 A$, this is

$$V_1 - V_2 = \frac{Q}{\epsilon_0 A}d$$

Substituting this into Eq. (25.52), we obtain

$$U = \frac{1}{2}Q\frac{Q}{\epsilon_0 A}d = \frac{1}{2}\frac{Q^2 d}{\epsilon_0 A} \qquad (25.55)$$

Equation (25.55) can be rewritten in the following interesting way, in terms of the electric field:

$$U = \frac{1}{2}\frac{Q^2 d}{\epsilon_0 A} = \frac{1}{2}\epsilon_0\left(\frac{Q}{\epsilon_0 A}\right)^2 Ad$$
$$= \tfrac{1}{2}\epsilon_0 E^2 \times Ad \qquad (25.56)$$

Here the product Ad = [area] × [separation] is the volume of the space between the plates, that is, the volume filled with electric field. Thus, the electric potential energy has the form

$$U = \frac{1}{2}\epsilon_0 E^2 \times [\text{volume}] \qquad (25.57)$$

This expression suggests that the energy is distributed over the volume in which there is electric field, with an energy density, or amount of energy per unit volume, $\epsilon_0 E^2/2$. According to this formula, the energy is concentrated where the electric field is strong.

Note that although Eq. (25.57) expresses the energy in terms of the electric field and suggests that the energy is located in the electric field, Eq. (25.49) expresses the energy in terms of the electric charges and suggests that the energy is located on these charges. Thus, these two equations, which are mathematically equivalent, suggest conflicting physical interpretations. To decide which alternative is correct, we need some extra information. The clue is the existence of electric fields that are independent of electric charges. As we will see in Chapter 33, radio waves and light waves consist of electric and magnetic fields traveling through space. Such fields are originally created by electric charges, but they persist even when the charges disappear. For instance, a radio wave or a light beam continues to travel through space long after the radio transmitter has been shut down or the flashlight has been switched off—this indicates that the energy of a radio wave resides in the radio wave itself, in its electric and magnetic fields, and not in the electric charges in the antenna of the radio transmitter. We can then argue that if energy is associated with the traveling electric fields of a radio wave, energy should also be associated with the electric fields of a static charge distribution. A more detailed calculation confirms that the expression we found above for the energy density in the special case of a uniform electric field is valid in general, for uniform or nonuniform electric fields. In any electric field (in vacuum) the energy density u is

$$u = \frac{1}{2}\epsilon_0 E^2 \qquad (25.58)$$

energy density in electric field

EXAMPLE 14 What is the energy density in the electric field of a thundercloud where the electric field $E = 2.0 \times 10^6$ V/m?

SOLUTION: According to Eq. (25.58), the energy density u is

$$u = \frac{1}{2}\epsilon_0 E^2 = \frac{1}{2} \times 8.85 \times 10^{-12}\ \text{C}^2/\text{N}\cdot\text{m}^2 \times (2.0 \times 10^6\ \text{V/m})^2$$

$$= 18\ \text{J/m}^3$$

✔ Checkup 25.5

QUESTION 1: What happens to expression (25.48) if there are only two charges, q_1 and q_2?

QUESTION 2: Is the electric energy U of a system of charges necessarily positive if all the charges q_1, q_2, q_3, etc., are positive? If they are all negative? If some are positive and some are negative?

QUESTION 3: Suppose we have a system of point charges with the electric potential energy given by Eq. (25.48). By what factor will this energy change if we increase the values of all the charges by a factor of 2?

QUESTION 4: Suppose that an energy of 0.50 J is stored in the electric field of a charged spherical balloon coated with a conducting layer. What will the energy in the electric field of this balloon be if we inflate it so its radius becomes twice as large?

QUESTION 5: An electric field has an energy density of 10 J/m^3. What will be the energy density if we increaese the electric field by a factor of 2? A factor of 3?

QUESTION 6: An electric energy of 0.010 J is stored in the electric field between two parallel charged plates. How does this energy change if we increase the plate separation by a factor of 2, while holding the charge constant? How does the energy density change?

QUESTION 7: Suppose that an energy of 1.0 J is stored in the electric field of a charged metallic sphere. What will the energy in the electric field of this sphere be if we allow one-half of the charge to leak away?

(A) 4.0 J (B) 2.0 J (C) 1.0 J (D) 0.50 J (E) 0.25 J

SUMMARY

PROBLEM-SOLVING TECHNIQUES Energy Conservation and Motion of a Point Charge **(page 797)**

PHYSICS IN PRACTICE Electric Shielding **(page 804)**

PROBLEM-SOLVING TECHNIQUES Electrostatic Potential and Field **(page 810)**

ELECTROSTATIC POTENTIAL (POTENTIAL ENERGY PER UNIT CHARGE)

$$V = \frac{U}{q} \quad (25.6)$$

SI UNIT OF POTENTIAL

$$1 \text{ volt} = 1\text{ V} = 1\text{ J/C} \quad (25.8)$$

ELECTROSTATIC POTENTIAL IN A UNIFORM ELECTRIC FIELD E_0 where the distance y is along the field direction.

$$V = -E_0 y \quad (25.7)$$

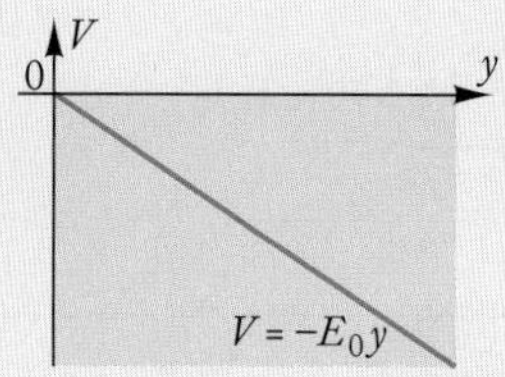

ALTERNATE UNIT OF ENERGY Electron-volt:

$$1\text{ eV} = 1.60 \times 10^{-19}\text{ J} \quad (25.16)$$

CONSERVATION OF ENERGY

$$K + U = \tfrac{1}{2}mv^2 + qV = [\text{constant}] \quad (25.22)$$

POTENTIAL ENERGY OF TWO POINT CHARGES

$$U = \frac{1}{4\pi\epsilon_0}\frac{qq'}{r} \quad (25.11)$$

ELECTROSTATIC POTENTIAL OF POINT CHARGE

$$V = \frac{1}{4\pi\epsilon_0}\frac{q'}{r} \qquad (25.12)$$

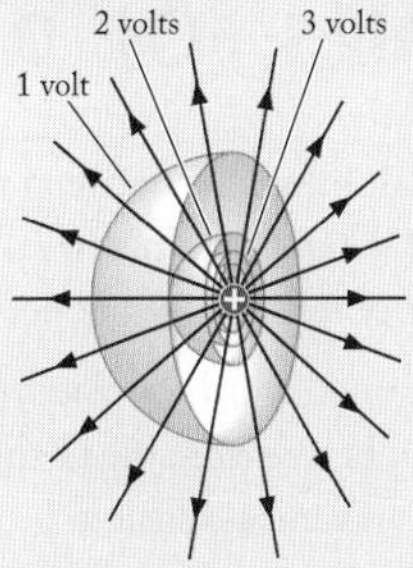

CALCULATION OF POTENTIAL FROM ELECTRIC FIELD If V_0 is the potential at a point P_0, then at a point P,

$$V = -\int_{P_0}^{P} \mathbf{E}\cdot d\mathbf{s} + V_0 \quad \text{or} \quad V = -\int_{P_0}^{P} E\cos\theta\, ds + V_0 \qquad (25.25, 25.26)$$

POTENTIAL OF A CONTINUOUS CHARGE DISTRIBUTION

$$V = \frac{1}{4\pi\epsilon_0}\int \frac{dQ}{r} \qquad (25.38)$$

POTENTIAL AND CONDUCTORS For static charge distributions, the potential throughout a conductor is constant. Within an empty cavity in a conductor, the electric field is exactly zero.

CALCULATION OF THE ELECTRIC FIELD FROM THE POTENTIAL

Along the direction of the field:

$$E = -\frac{dV}{ds} \qquad (25.41)$$

Along x, y, and z directions:

$$E_x = -\frac{\partial V}{\partial x} \quad E_y = -\frac{\partial V}{\partial y} \quad E_z = -\frac{\partial V}{\partial z} \qquad (25.43)$$

EQUIPOTENTIAL SURFACE An imaginary surface on which the electrostatic potential is constant. The electric field is everywhere perpendicular to an equipotential surface.

POTENTIAL ENERGY OF A SYSTEM OF POINT CHARGES

$$U = \frac{1}{4\pi\epsilon_0}\frac{q_1q_2}{r_{12}} + \frac{1}{4\pi\epsilon_0}\frac{q_2q_3}{r_{23}} + \frac{1}{4\pi\epsilon_0}\frac{q_1q_3}{r_{13}} + \cdots \text{[all pairs]} \qquad (25.48)$$

POTENTIAL ENERGY OF A CONDUCTOR

$$U = \tfrac{1}{2}QV \qquad (25.49)$$

POTENTIAL ENERGY OF A SYSTEM OF CONDUCTORS

$$U = \tfrac{1}{2}Q_1V_1 + \tfrac{1}{2}Q_2V_2 + \tfrac{1}{2}Q_3V_3 + \cdots \qquad (25.50)$$

ENERGY DENSITY IN ELECTRIC FIELD

$$u = \tfrac{1}{2}\epsilon_0 E^2 \qquad (25.58)$$

QUESTIONS FOR DISCUSSION

1. The potential difference between the poles of an automobile battery is 12 volts. Explain what this means in terms of the definition of potential as work per unit charge.
2. An old-fashioned word for electrostatic potential is electrostatic *tension*. Is it reasonable to think of the potential as analogous to a mechanical tension?
3. If the electric field is zero in some region, must the potential also be zero? Give an example.
4. A bird sits on a high-voltage power line which is at a potential of 345 000 volts. Does this harm the bird?
5. How would you define the gravitational potential? Are the units of gravitational potential the same as the units of electric potential? According to your definition, what is the gravitational potential difference between the ground and a point 50 m above the ground?
6. Consider an electron moving in the vicinity of a proton. Where is the electrostatic potential produced by the proton highest? Where is the potential energy of the electron highest?
7. Suppose that the electrostatic potential has a minimum at some point. Is this an equilibrium point for a positive charge? For a negative charge?
8. Consider a sphere of radius R with a charge Q uniformly distributed over its volume. Where does the potential have a maximum? Where does the magnitude of the electric field have a maximum?
9. Give an example of a conductor that is not an equipotential. Is this conductor in electrostatic equilibrium?
10. If the potential in a three-dimensional region of space is known to be constant, what can you conclude about the electric field in this region? If the potential on a two-dimensional surface is known to be constant, what can you conclude about the electric field on this surface?
11. In many calculations it is convenient to assign a potential of 0 volt to the ground. If so, what is the potential at the top of the Eiffel Tower? What is the potential at the top of your head? (Hint: Your body is a conductor.)
12. Is it true that the surface of a mass in static equilibrium is a gravitational equipotential surface? What if the surface is that of a fluid, such as water?
13. If a high-voltage power cable falls on top of your automobile, you will probably be safest if you remain inside the automobile? Why?
14. Suppose that several separate metallic bodies have been placed near a charge distribution. Is it necessarily true that all of these bodies will have the same potential?
15. Consider the patterns of field lines displayed in Figs. 23.17 and 23.18. Roughly sketch the equipotentials for these field lines.
16. If we surround some region with a conducting surface, we shield it from external electric fields. Why can we not shield a region from gravitational fields in a similar manner?
17. A cavity is completely surrounded by conducting material. Can you create an electric field in this cavity?
18. Show that different equipotential surfaces cannot intersect.
19. Consider the pattern of field lines shown in Fig. 24.22. Roughly, sketch some of the equipotential surfaces for this case.
20. Consider a metallic sphere carrying a given amount of charge. Explain why the electric energy is large if the radius of the sphere is small. Would you expect a similar inverse proportion between the electric energy and the size of a conductor of arbitrary shape?
21. Equation (25.50) suggests that the electric energy is located at the charges, whereas Eq. (25.57) suggests it is located in the field. How could we perform an experiment to test where the energy is located? (Hint: Energy gravitates.)
22. Figure 25.31 shows a sequence of deformations of a nucleus as it undergos fission. The volume of the nucleus and the electric charge remain constant during these deformations. Which configuration has the highest electric energy? The lowest?

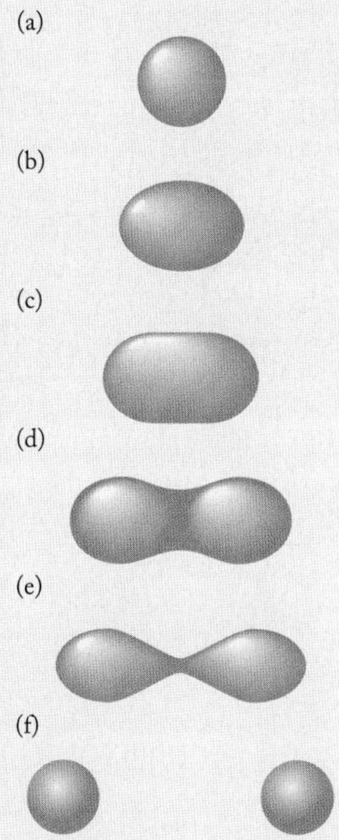

FIGURE 25.31 Fission of a nucleus.

23. Consider a sphere with uniform distribution of charge over its volume. Where is the energy density within the sphere highest? Lowest?
24. Since the electric energy density is never negative, how can the mutual electric potential energy of a pair of opposite charges be negative?

PROBLEMS

†25.1 The Electrostatic Potential

1. The electric potential difference between the positive and negative poles of an automobile battery is 12 volts. In order to charge the battery fully, the charging device must force $+2.0 \times 10^5$ coulombs from the negative terminal of the battery to the positive terminal. How much work must the charging device do during this process?

2. An ordinary flashlight battery has a potential difference of 1.5 V between its positive and negative terminals. How much work must you do to transport an electron from the positive terminal to the negative terminal?

3. On days of fair weather, the atmospheric electric field of the Earth is about 100 V/m; this field points vertically downward (compare Problem 4 of Chapter 23). What is the electric potential difference between the ground and an airplane flying at 600 m? What is the potential difference between the ground and the tip of the Eiffel Tower? Treat the ground as a flat conductor.

4. Consider the arrangement of parallel sheets of charge described in Problem 35 of Chapter 23 (see Fig. 23.37). Find the potential difference between the upper sheet and the lower sheet.

5. A proton is accelerated from rest through a potential of 2.50×10^5 V. What is its final speed?

6. At the Stanford Linear Accelerator (SLAC), electrons are accelerated from an energy of 0 eV to 20×10^9 eV as they travel in a straight evacuated pipe 1600 m in length (Fig. 25.32). The acceleration is due to a strong electric field pushing the electrons along. Assume that the electric field is uniform. What must be its strength?

FIGURE 25.32 Beam pipe at the Stanford Linear Accelerator (SLAC).

7. The potential difference between the two poles of an automobile battery is 12.0 V. Suppose that you place such a battery in empty space and that you release an electron at a point next to the negative pole of the battery. The electron will then be pushed away by the electric force and move off in some direction.

   (a) If the electron strikes the positive pole of the battery, what will be its impact speed?

   (b) If, instead, the electron moves away toward infinity, what will be its ultimate speed?

8. The gap between the electrodes of a spark plug in an automobile is 0.64 mm. In order to produce an electric field of 3.0×10^6 V/m (required to initiate an electric spark), what minimum potential difference must you apply to the spark plug?

9. An electron in a neon tube is accelerated from rest through a 2000-V potential. What velocity does the electron attain?

10. A charge of 2.0×10^{-12} C is placed on a small (pointlike) cork ball. What is the electrostatic potential at a distance of 30 cm from the ball? At a distance of 60 cm?

11. In an electron-beam heater, electrons at rest near a tungsten filament are accelerated toward a metal target at a high electrostatic potential. If the electrons strike the target to be heated with a speed of 1.8×10^7 m/s, what is the potential difference between the target and the filament?

12. An electron in a region of uniform electric field E_0 has an initial velocity $\mathbf{v}_0$ in the direction of the field. How far does the electron travel before stopping?

13. An electron is initially a distance of $r_0 = 4.3 \times 10^{-9}$ m from a proton, traveling directly away from the proton at speed 4.0×10^5 m/s. What is its speed when it is very far from the proton?

14. A conducting spherical shell of radius 12.0 cm is charged to a potential of 50 000 V. What is the value of the electrostatic potential 5.0 cm outside the surface of the sphere?

*15. In a demonstration, a small Styrofoam "worm" of mass 0.20 g sits on a Van de Graaff generator, a conducting spherical shell of radius 15 cm. The shell is at a potential of 75 000 V. When the worm acquires a charge Q, it is repelled by the sphere and travels vertically under the influence of gravity and the electric force. The worm moves upward and achieves equilibrium 0.50 m above the surface of the sphere. What is the charge Q?

†For help, see Online Concept Tutorial 30 at www.wwnorton.com/physics

†25.2 Calculation of the Potential from the Field

16. The nucleus of lead is a uniformly charged sphere with a charge of $82e$ and a radius of 7.1×10^{-15} m. What is the electrostatic potential at the nuclear surface? At the nuclear center?

17. The nucleus of platinum is a uniformly charged sphere with a charge of $78e$ and a radius of 7.0×10^{-15} m. What is the electric potential energy of an incident proton arriving at the nuclear surface? At the nuclear center?

18. An alpha particle of kinetic energy 1.7×10^{-12} J is shot directly toward a platinum nucleus from a very large distance. What will be the distance of closest approach? The electric charge of the alpha particle is $2e$ and that of the platinum nucleus is $78e$. Treat the alpha particle as a point particle, and the nucleus as a spherical charge distribution of radius 5.1×10^{-15} m, and disregard the motion of the nucleus.

19. An alpha particle is initially at a very large distance from a plutonium nucleus. What is the minimum kinetic energy with which the alpha particle must be launched toward the nucleus if it is to make contact with the nuclear surface? The plutonium nucleus is a sphere of radius 7.5×10^{-15} m with a charge of $94e$ uniformly distributed over the volume. For the purpose of this problem, the alpha particle may be regarded as a particle (of negligible radius) with a charge of $2e$.

20. Two positive point charges Q are on the y axis at $y = \pm d/2$. Find the potential for points on the positive x axis.

*21. Eight point charges $+Q$ sit at the corners of a cube of side a. What is the potential at the center of the cube? At the center of a cube face? At the center of a cube edge?

*22. A positive point charge Q is on the y axis at $y = D$; a negative point charge $-2Q$ is at the point $x = D, y = D$. Find the potential for points on the x axis.

*23. An arc of a circle of radius R subtends an angle θ. The arc is a thin rod with a uniform linear charge density λ. What is the electrostatic potential at the center of curvature?

*24. Three large charged sheets are parallel to the x–z plane. The sheets are at $y = 0, y = d$, and $y = 2d$; these sheets have uniform surface charge densities $+\sigma, -2\sigma$, and $+\sigma$, respectively. Assuming the reference potential is zero at $y = 0$, determine the potential as a function of y.

*25. A thorium nucleus emits an alpha particle according to the reaction

$$\text{thorium} \rightarrow \text{radium} + \text{alpha}$$

Assume that the alpha particle is pointlike and that the residual radium nucleus is spherical with a radius of 7.4×10^{-15} m. The charge on the alpha particle is $2e$, and that on the radium nucleus is $88e$.

(a) At the instant the alpha particle emerges from the nuclear surface, what is its electrostatic potential energy?

(b) If the alpha particle has no initial kinetic energy, what will be its final kinetic energy and speed when far away from the nucleus? Assume that the radium nucleus does not move. The mass of the alpha particle is 6.7×10^{-27} kg.

*26. Consider again the arrangement of charges within the thundercloud of Fig. 23.34. Find the electric potential due to these charges at a point which is at a height of 8.0 km and on the vertical line passing through the charges. Find the electric potential at a second point which is at the same height and has a horizontal distance of 5.0 km from the first point.

*27. In a helium atom, at some instant one of the electrons is at a distance of 3.0×10^{-11} m from the nucleus and the other electron is at a distance of 2.0×10^{-11} m, 90° away from the first (Fig. 25.33). Find the electric potential produced jointly by the two electrons and the nucleus at a point P beyond the first electron and at a distance of 6.0×10^{-11} m from the nucleus.

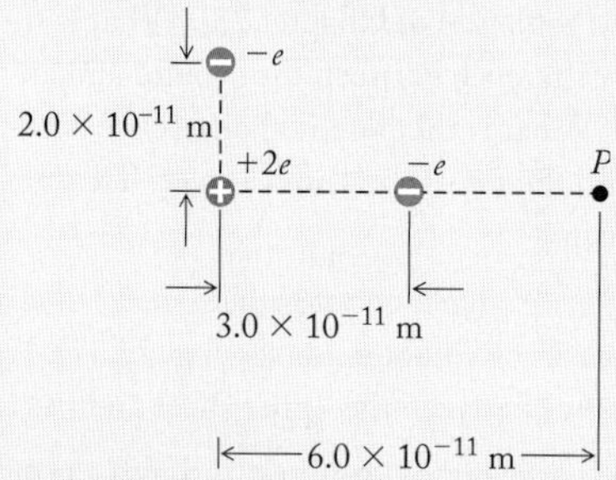

FIGURE 25.33 Instantaneous configuration of the nucleus ($+2e$) and the electrons ($-e$) of an atom of helium.

*28. A total charge Q is distributed uniformly along a straight rod of length l. Find the potential at a point P a distance h from the midpoint of the rod (see Fig. 25.34).

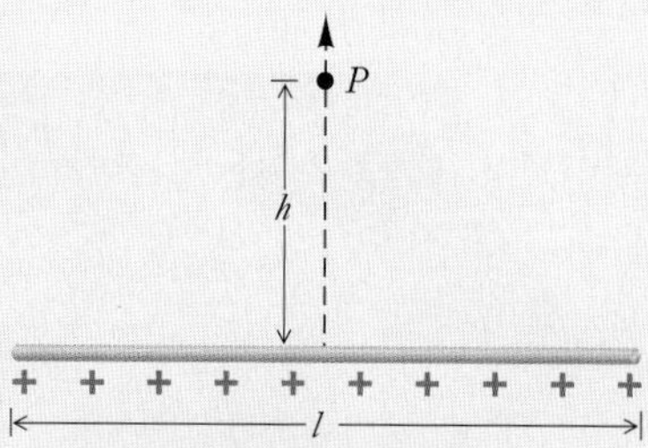

FIGURE 25.34 A uniformly charged rod of length l.

*29. Three thin rods of glass of length l carry charges uniformly distributed along their lengths. The charges on the three rods are $+Q$, $+Q$, and $-Q$, respectively. The rods are arranged along the sides of an equilateral triangle. What is the electrostatic potential at the midpoint of this triangle?

*30. A uniformly charged sphere of radius a is surrounded by a uniformly charged concentric spherical shell of inner radius b and outer radius c (Fig. 25.35). The total charge in the sphere is Q, and that on the outer shell $-Q$. Find the potential at $r = b$, at $r = a$, and at $r = 0$.

†For help, see Online Concept Tutorial 30 at www.wwnorton.com/physics

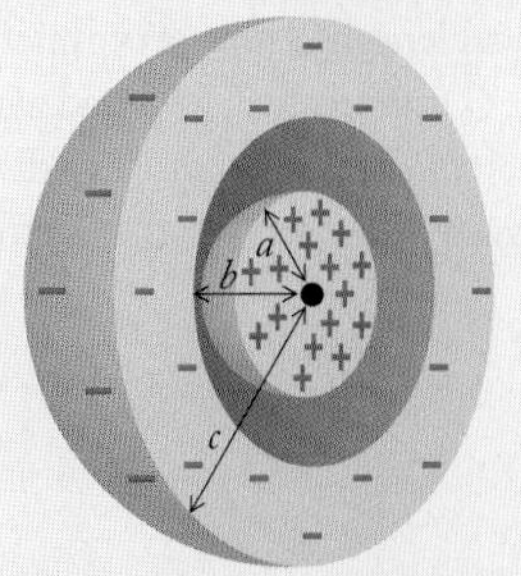

FIGURE 25.35 Charged sphere and concentric spherical shell.

*31. Four rods of length l are arranged along the edges of a square. The rods carry charges $+Q$ uniformly distributed along their lengths (Fig. 35.36). Find the potential at the point P at a distance x from one corner of the square.

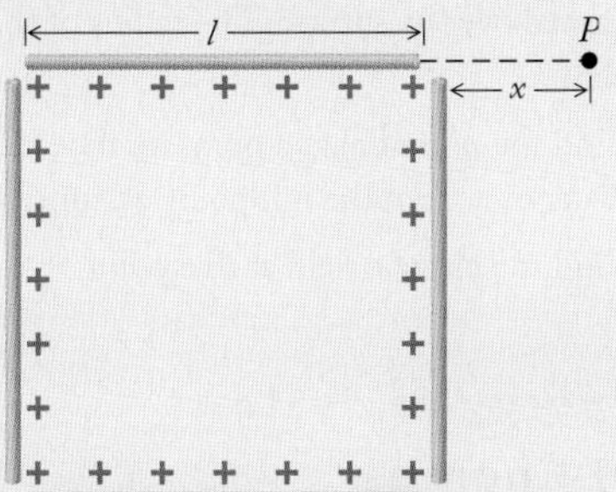

FIGURE 25.36 Four charged rods.

*32. Two semicircular rods and two short, straight rods are joined in the configuration shown in Fig. 25.37. The rods carry a charge of λ coulombs per meter. Calculate the potential at the center of this configuration.

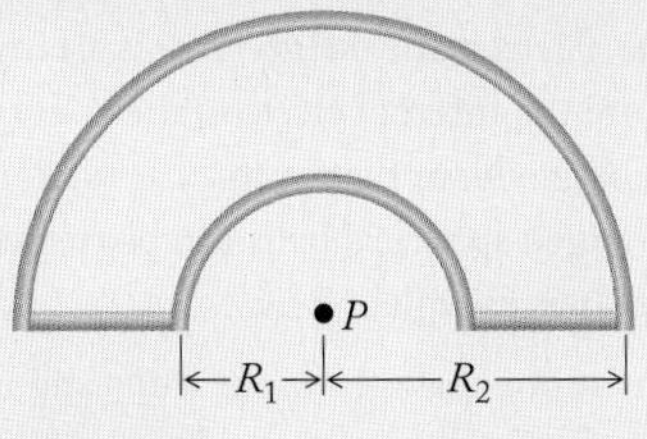

FIGURE 25.37 Two semicircular rods connected by two straight rods.

*33. A long plastic pipe has an inner radius a and an outer radius b. Charge is uniformly distributed over the volume $a \leq r \leq b$. The amount of charge is λ coulombs per meter of length of the tube. Find the potential difference between $r = b$ and $r = 0$. Assume that the plastic has no effect on the electric field.

*34. A flat disk of radius R has a charge Q uniformly distributed over its surface. Find a formula for the potential along the axis of the disk.

*35. The tube of a Geiger counter consists of a thin, straight wire surrounded by a coaxial conducting shell. The diameter of the wire is 0.0025 cm, and that of the shell is 2.5 cm. The length of the tube is 10 cm; however, in your calculation, use the formula for the electric field of an infinitely long line of charge. If the potential difference between the wire and the shell is 1.0×10^3 volts, what is the electric field at the surface of the wire? At the cylinder?

**36. An infinite charge distribution with spherical symmetry has a charge density ρ coulombs per cubic meter given by the formula $\rho = kr^{-5/2}$, where k is a constant. Find the potential as a function of the radius. Assume $V = 0$ at $r = \infty$.

**37. A point charge Q is on the positive z axis at the point $z = h$. A point charge $-Q \times R/h$ (where R is a positive length, $0 < R < h$) is on the z axis at the point $z = R^2/h$. Show that the surface of the sphere of radius R about the origin is an equipotential surface.

25.3 Potential in Conductors

*38. Two large flat parallel sheets have opposite uniform surface charge densities $\pm\sigma$ and are separated a distance d. A large, uncharged conducting slab of thickness $d/3$ is parallel to the charged sheets, centered between them. Find the electrostatic potential as a function of distance y perpendicular to the sheets. Take the reference potential $V_0 = 0$ and the origin $y = 0$ to be at the negative sheet.

*39. A solid conducting sphere of radius R has a charge Q on its surface; the sphere is concentric with a larger, thick conducting spherical shell with inner radius R_1, outer radius R_2, and net charge $-Q$. Find the electrostatic potential for (a) $r \geq R_2$, (b) $R_1 \leq r \leq R_2$, (c) $R \leq r \leq R_1$, and (d) $r \leq R$.

*40. A point charge $-Q$ is at the center of a thick conducting spherical shell of inner radius a and outer radius b. The shell has a net charge of $+3Q$ on it. What is the potential for $r \geq b$? For $a \leq r \leq b$? For $r \leq a$?

*41. A dielectric sphere of radius a has a charge $+Q$ uniformly distributed throughout its volume. The sphere is concentrically placed within a thick conducting shell of inner radius b and outer radius c; the shell has a net charge of $+2Q$. What is the potential outside the shell, for $r \geq c$? In the shell material, for $b \leq r \leq c$? Between the shell and the sphere, for $a \leq r \leq b$? Inside the sphere, for $r \leq a$?

†25.4 Calculation of the Field from the Potential

42. In some region of space, the electrostatic potential is the following function of x and y, but not of z:

$$V = x^2 + 2xy$$

where the potential is measured in volts and the distance in meters. Find the electric field at the point $x = 2, y = 2$.

43. Using the components E_x, E_y, and E_z obtained in Example 9, show that the magnitude $\sqrt{E_x^2 + E_y^2 + E_z^2}$ agrees with the usual expression for the electric field of a point charge.

44. A rod of length l has a charge Q uniformly distributed along its length. In Example 6, we found the potential $V(P)$ at the point P at a distance x from one end of the rod. Find the electric field at this point.

45. Equation (25.32) in Example 4, gives the potential inside a uniformly charged sphere of charge Q and radius R as a function of the distance r from the center:

$$V = -\frac{1}{4\pi\epsilon_0}\frac{Q}{2R^3}r^2 + \frac{1}{4\pi\epsilon_0}\frac{3Q}{2R} \quad (r \leq R)$$

Verify that this potential gives the radial electric field of Eq. (24.18).

46. In a region of space, the electrostatic potential is described by

$$V = x^2y + 3xyz + zy^2$$

Find the electric field in this region.

47. The potential in a region of space is given by $V = \cos(2\pi x/a) \times \cos(2\pi y/b) \times \cos(2\pi z/c)$, where a, b, and c are constants. What are the x, y, and z components of the electric field in this region?

*48. A nucleus of carbon (charge $6e$) and one of helium (charge $2e$) are separated by a distance of 1.2×10^{-13} m and instantaneously at rest. The center of mass of this system is at a distance of 4.0×10^{-14} m from the carbon nucleus. Take this point as origin and take the x axis along the line joining the nuclei, with the carbon nucleus on the negative x axis.

(a) Find the potential V as a function of x, y, and z.

(b) Find E_x and E_y as a function of x, y, and z.

*49. It can be shown that the electrostatic potential of a dipole p placed at the origin and oriented parallel to the z axis is

$$V = \frac{p}{4\pi\epsilon_0}\frac{z}{(x^2 + y^2 + z^2)^{3/2}}$$

By differentiating this potential, find the components E_x, E_y, and E_z of the electric field produced by the dipole. What are the magnitude and the direction of the electric field at points on the z axis? At points on the x axis?

†For help, see Online Concept Tutorial 30 at www.wwnorton.com/physics

*50. Two rods of equal lengths l form a symmetric cross. The rods carry charges $\pm Q$ uniformly distributed along their lengths. Calculate the potential at the point P at a distance x from one end of the cross (Fig. 25.38). Calculate the electric field at this point.

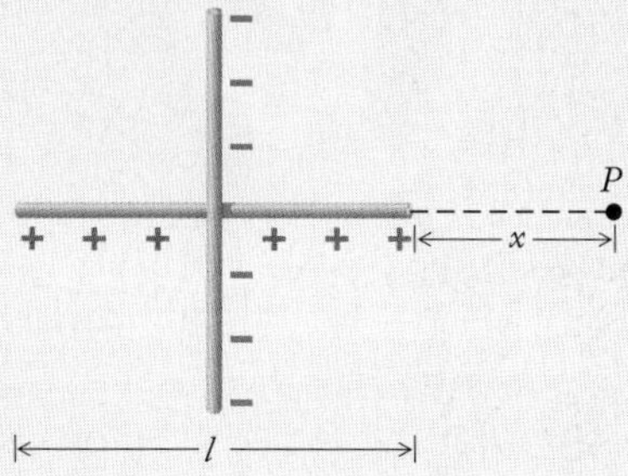

FIGURE 25.38 Two charged rods forming a cross.

**51. A thin cylindrical cardboard tube has a charge Q uniformly distributed over its surface. The radius of the tube is R and the length is l.

(a) Find the potential at a point on the axis of the tube at a distance x from the midpoint. Assume $x > l$.

(b) Find the electric field at this point.

25.5 Energy of Systems of Point Charges

52. Consider once more the distribution of charges within the thundercloud shown in Fig. 23.34. What is the electric potential energy of this charge distribution?

53. In the water molecule, the hydrogen atoms tend to give up their electrons to the oxygen atom. Crudely, the molecule may be regarded as consisting of a uniformly charged ball of charge $-2e$ and two smaller uniformly charged balls of charge $+e$ each. The dimensions of the molecule are given in Fig. 23.27. Calculate the electrostatic energy of this arrangement of three charges; ignore the internal electrostatic energy of the individual balls of charge. Note that in this calculation each of the uniformly charged spherical balls can be treated as though it were a point charge; explain why.

54. Suppose that at one instant the electrons and the nucleus of a helium atom occupy the positions shown in Fig. 25.39; at this instant, the electrons are at a distance of 2.0×10^{-11} m from the nucleus. What is the electric potential energy of this arrangement? Treat the electrons and the nucleus as point charges.

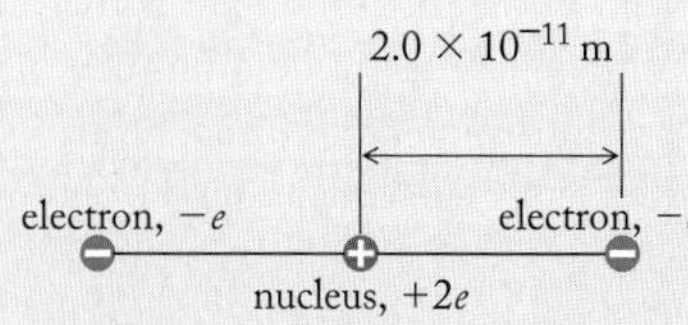

FIGURE 25.39 Instantaneous configuration of the nucleus ($+2e$) and the electrons ($-e$) of an atom of helium.

55. A pair of parallel conducting plates, each measuring 30 cm × 30 cm, are separated by a gap of 1.0 mm. How much work must you do against the electric forces to charge these plates with $+1.0 \times 10^{-6}$ C and -1.0×10^{-6} C, respectively?

56. Two large parallel conducting plates of area 0.20 m^2 are separated by a distance of 0.50 mm. The plates carry opposite charges, and the electric field in the space between them is 5.0×10^5 V/m. What is the electric energy?

57. A penny coin is hung from a silk thread inside a closed tin can placed on the ground. Given that the penny coin carries a charge of 2.0×10^{-6} C and that the potential difference between the tin can and the penny coin is 3.0×10^4 V, find the electric potential energy of this system of two conductors.

58. Near the surface of the nucleus of a lead atom, the electric field has a strength of 3.4×10^{21} V/m. What is the energy density in this field?

59. The atmospheric electric field near the surface of the Earth has a strength of 100 V/m.

(a) What is its energy density?

(b) Assuming that the field has the same magnitude everywhere in the atmosphere up to a height of 10 km, what is the corresponding total energy?

60. Six point charges, alternating in sign $\pm Q$, are at the vertices of a hexagon of side a. What is the electrostatic potential energy of this arrangement?

61. Short, extremely intense laser pulses can be produced with electric fields as high as 2.0×10^{10} V/m. What is the energy density in the region of such an electric field?

*62. Consider an infinite line of ions of alternating charge $\pm e$ separated by a distance a.

(a) Write down the electrostatic potential energy U of one typical ion in the line by summing its potential-energy contributions from increasingly distant ions.

(b) By comparing your expression with the expansion of the logarithmic function $\ln(1+x) = x - \frac{1}{2}x^2 + \frac{1}{3}x^3 - \frac{1}{4}x^4 + \cdots$, show that the potential energy of (a) can be written

$$U = -\frac{e^2 \ln 2}{2\pi\epsilon_0 a}$$

(c) Evaluate the result in (b) for a typical atomic separation, $a = 3.0 \times 10^{-10}$ m. Express this value in eV.

*63. Four equal positive charges of magnitude Q are placed on the four corners of a square of side d. What is the electric energy of this system of charges?

*64. Four positive and four negative point charges of equal magnitudes $\pm Q$ are arranged alternately on the corners of a cube of edge d (see Fig. 25.40). What is the electric energy of this arrangement?

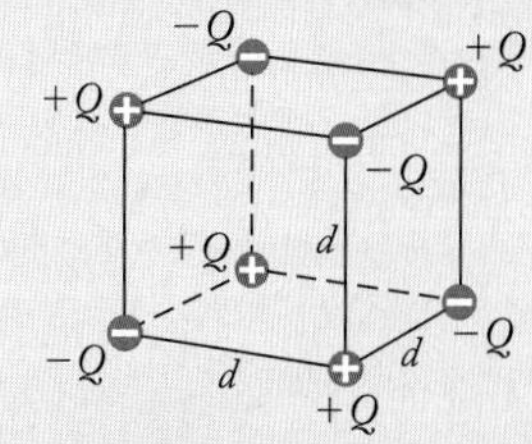

FIGURE 25.40 Point charges at the corners of a cube.

*65. According to the alpha-particle model of the nucleus, some nuclei consist of a regular geometric arrangement of alpha particles. For instance, the nucleus of ^{12}C consists of three alpha particles arranged in an equilateral triangle (Fig. 25.41). Assuming that the distance between pairs of alpha particles is 3.0×10^{-15} m, what is the electric energy (in eV) of this arrangement of alpha particles? Treat the alpha particles as pointlike particles with charge $+2e$.

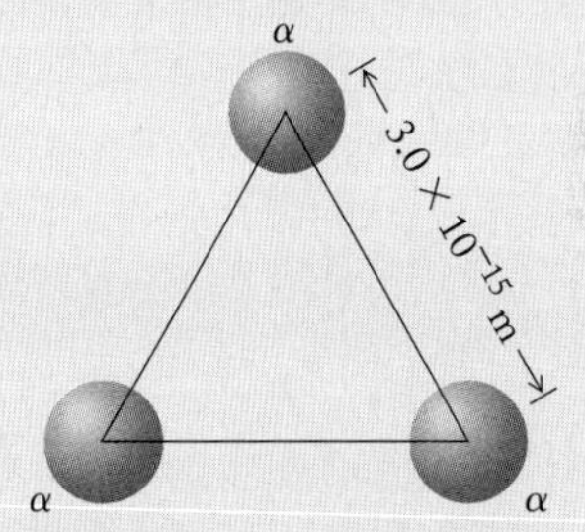

FIGURE 25.41 Three alpha particles.

*66. According to the alpha-particle model (see also the preceding problem), the nucleus of ^{16}O consists of four alpha particles arranged on the vertices of a tetrahedron (Fig. 25.42). If the distance between pairs of alpha particles is 3.0×10^{-15} m, what is the electric energy (in eV) of this configuration of alpha particles? Treat the alpha particles as pointlike particles with charge $+2e$.

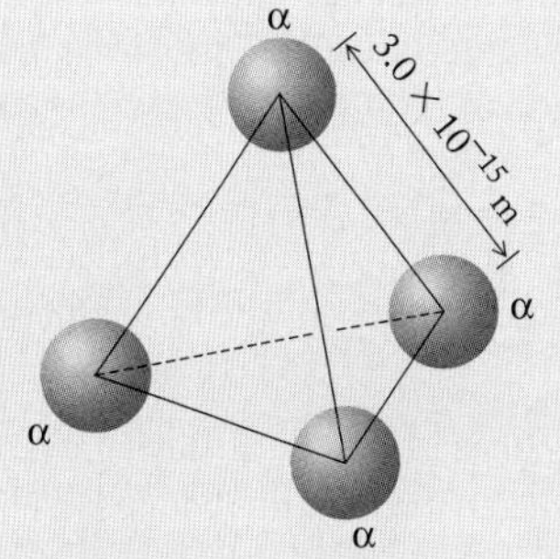

FIGURE 25.42 Four alpha particles.

*67. Problem 37 of Chapter 24 describes the Thomson model of the helium atom. The equilibrium separation of the electrons is 5.0×10^{-11} m. Calculate the electric energy of this configuration. Take into account both the electric energy between each electron and the positive charge, and the electric energy between the two electrons; ignore the energy of the cloud and of the electrons by themselves.

*68. A charge of 7.5×10^{-6} C can be placed on a metallic sphere of radius 15 cm before the surrounding air suffers electrical breakdown. What is the electric energy of the sphere with this charge?

*69. A sphere of radius R has a charge Q uniformly distributed over its volume. A thin conducting shell of radius $2R$ surrounds the sphere concentrically. The shell carries a charge $-Q$ on its interior surface and has no charge on its exterior surface. What is the electric energy of this system?

*70. Pretend that an electron is a conducting sphere of radius R with a charge e distributed uniformly over its surface. In terms of e and the mass m_e of the electron, what must be the radius R if the electric energy is to equal the rest-mass energy $m_e c^2$ of the electron? Numerically, what is the value of R?

*71. A sphere of radius R has a charge Q distributed uniformly over its volume. Show that the electric potential energy of this configuration is $U = (1/4\pi\epsilon_0)\frac{3}{5}Q^2/R$. [Hint: Sum energies of the form (25.49): $U = \int \frac{1}{2} V\, dQ$.]

*72. The nuclei of ^{235}Pu, ^{235}Np, ^{235}U, and ^{235}Pa all have the same radii, about 7.4×10^{-15} m, but their electric charges are $94e$, $93e$, $92e$, and $91e$, respectively. Treating these nuclei as uniformly charged spheres, calculate their electric energies; express your answer in electron-volts. Use the result of Problem 71.

*73. According to a crude model, a proton can be regarded as a uniformly charged sphere of charge e and radius 1.0×10^{-15} m. Find the electric self-energy of the proton. Express your answer in eV. Use the result of Problem 71.

*74. A solid sphere of copper of radius 10 cm with a charge of 1.0×10^{-6} C is placed at the center of a thin, spherical copper shell of radius 20 cm with a charge of -1.0×10^{-6} C. Find a formula for the energy density in the space between the solid sphere and the shell. Find the total electric energy.

*75. In analogy to the electric field **E** (electric force per unit charge), we can define a gravitational feld **g** (gravitational force per unit mass).

(a) The energy density in the electric field is $\epsilon_0 E^2/2$. By analogy, show that the energy density in the gravitational field is $g^2/8\pi G$.

(b) Calculate the gravitational field energy of the Moon due to its own gravity; treat this body as a sphere of uniform density. What is the ratio of the gravitational field energy to the rest-mass energy of the Moon?[3]

[3] Warning: This problem does not take the gravitatioinal *interaction* energy into account. The gravitational interaction energy density is [mass density] × [gravitational potential]. The total gravitational energy is the sum of the field energy and the interaction energy; this total gravitational energy is always negative.

*76. In symmetric fission, the nucleus of uranium (^{238}U) splits into two nuclei of palladium (^{119}Pd). The uranium nucleus is spherical with a radius of 7.4×10^{-15} m. Assume that the two palladium nuclei adopt a spherical shape immediately after fission; at this instant, the configuration is as shown in Fig. 25.43. The size of the nuclei in Fig. 25.43 can be calculated from the size of the uranium nucleus because the nucleus material maintains a constant density (the initial nuclear volume equals the final nuclear volume).

(a) Calculate the electric energy of the uranium nucleus before fission.

(b) Calculate the total electric energy of the palladium nuclei in the configuration shown in Fig. 25.43, immediately after fission. Take into account the mutual electric potential energy of the two nuclei and also the individual electric energies of the two palladium nuclei by themselves.

(c) Calculate the total electric energy a long time after fission when the two palladium nuclei have moved apart by a very large distance.

(d) Ultimately, how much electric energy is released into other forms of energy in the complete fission process (a) through (c)?

(e) If 1.0 kg of uranium undergoes fission, how much electric energy is released?

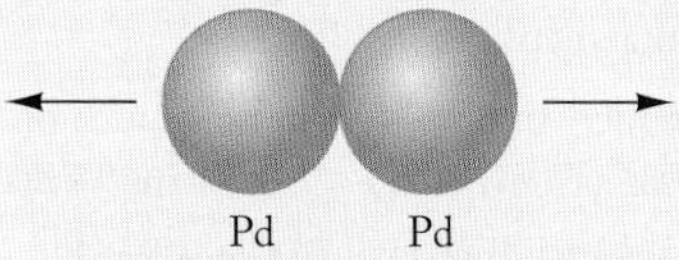

FIGURE 25.43 Two palladium nuclei in contact. The nuclei are spheres.

**77. Consider the Geiger-counter tube described in Problem 35. If the tube is initially uncharged, how much work must be done to bring the tube to its operating voltage of 1.0×10^3 V?

**78. Using the model described in Problem 36 of Chapter 24 for the charge distribution of a neutron, calculate the electric self-energy of a neutron. Express your answer in eV.

**79. A long, thin rod of metal of radius a has a charge of λ coulombs per unit length distributed uniformly over its surface. The rod is surrounded by a concentric cylinder of sheet metal of radius b with a charge of $-\lambda$ coulombs per unit length on its interior surface.

(a) What is the energy density (as a function of radius) in the space between the rod and the cylinder?

(b) What is the total electric energy per unit length?

REVIEW PROBLEMS

80. A positive point charge q, with mass m, is released at a distance d from a fixed positive point charge Q. How fast is the charge q moving when the distance has grown to 3 times the initial value?

81. A proton is decelerated to rest from an initial speed of 6.9×10^6 m/s by a uniform electric field of 2.5×10^6 V/m. How far did the proton travel during this motion? What is the difference in electrostatic potential between the initial and final positions of the proton?

82. A proton sits at the origin of coordinates. How much work must you do against the electric force of the proton to push an electron from the point $x = 1.0 \times 10^{-10}$ m, $y = 0$ in the x–y plane to the point $x = 2.5 \times 10^{-10}$ m, $y = 2.5 \times 10^{-10}$ m?

83. Suppose that, as a function of x, an electric field has an x component

$$E_x = 6x^2y$$

where the electric field is measured in volts per meter and the distances are measured in meters. Find the potential difference between the origin and the point $x = 3$ on the x axis.

*84. The tau particle is similar to an electron and has the same electric charge, but its mass is 3490 times as large as the electron mass. Like the electron, the tau can penetrate nuclear material, experiencing no forces except the electric force. Suppose that a tau is initially at rest a large distance from a lead nucleus. Under the influence of the electric attraction, the tau accelerates toward the nucleus. What is the speed when it crosses the nuclear surface? When it reaches the center of the nucleus? The nucleus of lead is a uniformly charged sphere of radius 7.1×10^{-15} m and charge $82e$.

*85. A long, straight wire of radius 0.80 mm is surrounded by an evacuated concentric conducting shell of radius 1.2 cm. The wire carries a charge of -5.5×10^{-8} coulomb per meter of length. Suppose that you release an electron at the surface of the wire. With what speed will this electron hit the conducting shell?

86. A total charge Q is uniformly distributed along a straight rod of length l.

(a) Find the electrostatic potential at a point P at a distance y from one end of the rod (see Fig. 25.44).

(b) Find the y component of the electric field at this point.

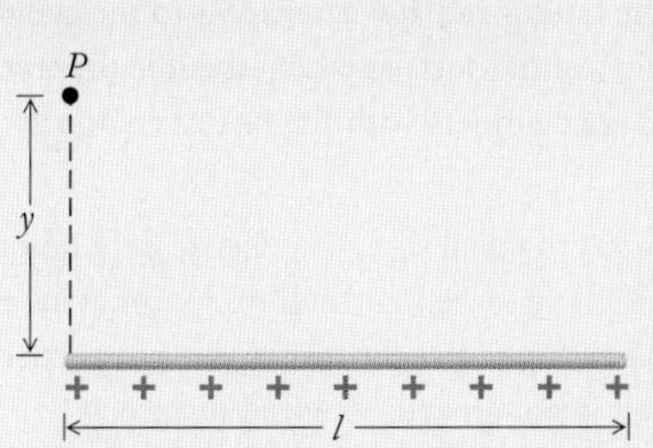

FIGURE 25.44 A charged rod.

*87. An annulus (a disk with a hole) made of paper has an outer radius R and an inner radius $R/2$ (Fig. 25.45). An amount Q of electric charge is uniformly distributed over the paper.

(a) Find the potential as a function of distance on the axis of the annulus.

(b) Find the electric field on the axis of the annulus.

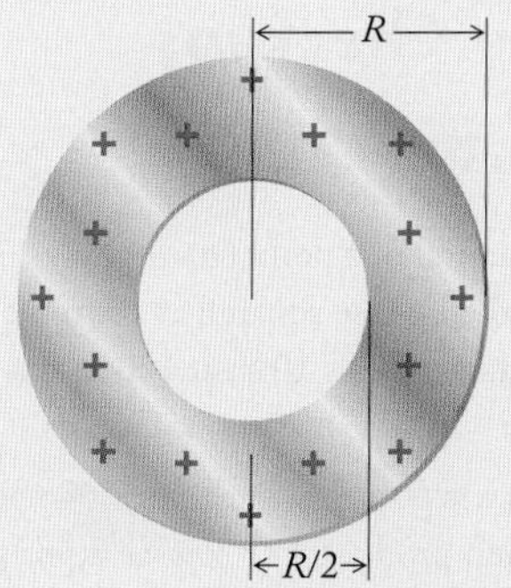

FIGURE 25.45 A charged annulus.

*88. Four equal particles of positive charges q and masses m are initially held at the four corners of a square of side L. If these particles are released simultaneously, what will be their speeds when they have separated by a very large distance?

*89. Two thin rods of length l carry equal charges Q uniformly distributed over their lengths. The rods are aligned, and their nearest ends are separated by a distance x (Fig. 25.46). Calculate the mutual electric potential energy. Ignore the self-energy of each rod.

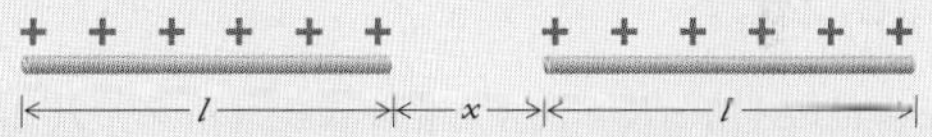

FIGURE 25.46 Two aligned charged rods.

*90. One method for the determination of the radii of nuclei makes use of the known difference of electric energy between two nuclei of the same size but different electric charges. For instance, the nuclei ^{15}O and ^{15}N have the same size, but their charges are $8e$ and $7e$, respectively. Given that the difference in electric energy is 3.7×10^6 eV, what is the nuclear radius?

*91. A spherical shell of inner radius a, outer radius b carries a charge Q uniformly distributed over its volume. What is the electric energy of this charge distribution?

*92. Suppose that a nucleus of charge q, radius R, and electric energy $(1/4\pi\epsilon_0)(3q^2/5R)$ fissions into two equal parts of charge $q/2$ each. The nuclear material in the original nucleus and the final two nuclei has the same density. What is the radius of each of the final two nuclei? How does the sum of the individual electric energies of the two final, separated nuclei compare with the initial electric energy?

*93. In Problem 71 we integrated the potential energy of the charges in a uniformly charged sphere and found that the electric potential is $(1/4\pi\epsilon_0)\frac{3}{5}Q^2/r$. Obtain this result by integrating the energy density $\frac{1}{2}\epsilon_0 E^2$ of the electric field of the sphere. (Hint: The integration needs to be extended over the interior and the exterior of the sphere, since the electric field exists in both the interior and the exterior.)

Answers to Checkups

Checkup 25.1

1. Since the potential of a point charge varies inversely with distance (*not* as the inverse square), at 10 m the potential is reduced a factor of 10 to 10 V; at 100 m it is reduced a factor of 100 to 1 V.

2. Since the electric potential energy for two point charges q and q' involves their product, $U = kqq'/r$, a positive potential energy requires that both charges have the same sign: q and q' are either both positive or both negative.

3. For a positive charge, the potential $V = kq'/r$ is largest for $r = 0$ (where $V = +\infty$) and smallest for $r = \infty$ (where $V = 0$). For a negative charge, the potential is largest, $V = 0$, at $r = \infty$; and the potential is smallest, $V = -\infty$, at $r = 0$.

4. (A) Increase. Since the charges are of the same sign, the potential energy is positive ($U = kqq'/r$). When r increases, such a positive potential energy decreases. For the total energy $K + U$ to remain constant, the kinetic energy $K = \frac{1}{2}mv^2$ must increase. Thus the speed v increases.

Checkup 25.2

1. As calculated in Example 4 and displayed in Fig. 25.13, the potential increases from a minimum of zero at infinite distance to a maximum of $V = (3/2)kQ/R$ at the center of the positively charged sphere, $r = 0$. For the negatively charged sphere, this is inverted, so the minimum is $V = -(3/2)kQ/R$ at $r = 0$, and the maximum is $V = 0$ at $r = \infty$.

2. Zero electric field throughout a region implies that the potential is not *changing* throughout the region. However, the constant value of the potential can be nonzero; indeed, it can take any value, depending on the distribution of charge in the shell.

3. The electric potential decreases in the direction of the field, since a positive charge would lose potential energy as it moved along the field (as the electric force acts on it); it would gain kinetic energy.

4. The potential is obtained by summing, using Eq. (25.38), $V = (1/4\pi\epsilon_0)\int dQ/r$. Since at the center of the ring $r = R$ is constant, and since $\int dQ = Q$, (the total charge), the potential at the center is $V = (1/4\pi\epsilon_0)Q/R$. The result is independent of how the charge is distributed around the ring.

5. (C) Larger; the same. For $b \geq r \geq a$, we have positive charge in a spherically symmetric shell, producing an outward electric field; since we must do work against this electric field to move a positive point charge from $r = b$ to $r = a$, the potential at $r = a$ must be greater than the potential at $r = b$. There is no charge (and thus no field) for $r \leq a$, so no work is required to move in this region, and thus the potential is the same at $r = 0$ as at $r = a$.

Checkup 25.3

1. The charge on the other sphere and the charge on the wire connecting the spheres destroy spherical symmetry, so the charge on the sphere of interest will not be uniformly distributed. Due to contributions from the wire and the other sphere, the potential will not be $V = (1/4\pi\epsilon_0)Q_1/R_1$, but will be larger. Since the spheres are conducting and are connected by a conductor, they are indeed at the same potential.

2. No, not noticeably. The airplane is nearly a closed box of metal, and as such acts as an electric shield; although airplane windows permit some field penetration, external electric fields are severely attenuated.

3. (B) Smaller. The electrostatic potential of a solid metal sphere is simply $V = kQ/R$, so with the same charge, the larger sphere is at a smaller potential.

Checkup 25.4

1. If the charge density doubled, so would the electric field. If the electric field were twice as large, it would take half the distance for the same change in potential (since $V = -E_0 y$).

2. If the charge Q were doubled, the radii for given potentials $V = kQ/r$ would also double.

3. Since the field is radial with respect to the cylindrical axis and the equipotential surfaces are perpendicular to the field, such surfaces are cylinders with the line of charge as their common axis.

4. For charges of equal sign, the electric field points parallel to their midplane (except at their midpoint, where $E = 0$; see Fig. 23.18). Since the electric field is not perpendicular to this surface, it is not an equipotential surface.

5. (D) Yes; yes. The value of the field is given by the negative of the *derivative* of the potential with respect to distance; the particular value of the potential at a point is unimportant. Second, where the electric field is zero, the potential is constant, but not necessarily zero. For example, a charged conducting sphere has zero electric field inside, but the potential is constant there, $V = kQ/R$ (see Fig. 25.18).

Checkup 25.5

1. The expression reduces to a single term, like Eq. (25.47), namely, $U = (1/4\pi\epsilon_0)q_1q_2/r$.
2. If either all are positive or all are negative, then each term in the potential energy (25.48), which contains the product of two charges, is positive, and the total must be positive. However, if some charges are positive and some are negative, then some terms are negative, and the sign of the total depends on the relative sizes of the different terms.
3. Since each term involves the product of two charges, increasing the values of all the charges a factor of 2 will increase the value of each term, and thus the total, by a factor of 4.
4. According to Eq. (25.51), the stored energy varies as $1/R$, so doubling the radius will result in half as much stored electrical energy, or 0.25 J.
5. Since the energy density $u = \frac{1}{2}\epsilon_0 E^2$, doubling the field will quadruple the energy to 40 J/m^3; tripling the electric field will increase the energy density a factor of 9 to 90 J/m^3.
6. Since the charge density on the plates is constant, the electric field does not change. Thus, Eq. (25.56) tells us that when doubling the plate separation d, the total stored energy increases a factor of 2 to 0.020 J. The energy density depends only on the electric field, and so does not change.
7. (E) 0.25 J. The energy of a charged sphere, Eq. (25.51), is proportional to the charge squared. Thus half the charge will have one-fourth of the stored energy, or 0.25 J.

CHAPTER

26 Capacitors and Dielectrics

CONCEPTS IN CONTEXT

Concepts — in — Context

At the National Ignition Facility (NIF) in Livermore, California, intense light pulses from 192 lasers are focused and combined to attain the extremely high energy density required for nuclear fusion. The main amplification for each laser comes from flashlamps powered by the sudden discharge of capacitors, arrangements of conductors that can store charge and energy. The 192 banks of capacitors at the NIF (left photo), each with 20 advanced capacitors (the horizontal cylinders in the right photo), are the highest-energy array ever built.

As we learn about capacitance in this chapter, we can consider such questions as:

? What is the combined capacitance of each 20-capacitor bank? Of all 192 banks? (Example 4, page 835)

? How much charge can be stored on each NIF capacitor? The conductors in each capacitor are separated by an insulator, known as a dielectric. How does the presence of a dielectric modify the properties of a capacitor? (Example 7, page 841)

? How much energy can be stored in each NIF capacitor? In the entire array? (Example 10, page 845)

Any arrangement of conductors that is used to store electric charge is called a **capacitor**. Since work must be done while depositing charge on the conductors, the capacitor also stores electric potential energy whenever it stores electric charge. In our electronic technology, capacitors find widespread application—they are part of the circuitry of radios, CD players, computers, automobile ignition systems, and so on.

The first part of this chapter deals with the properties of capacitors; we will see how the ability of the capacitor to store charge depends on the arrangement and the sizes of the conductors. The second part of the chapter deals with the properties of electric fields in regions of space filled with an electrically insulating material, or **dielectric**. Since capacitors are usually filled with such a dielectric material, the study of the mutual effects between the electric field and the dielectric material is closely linked to the study of capacitors. But the effects of the electric fields and dielectric materials upon one another are also interesting in their own right. For instance, air is a dielectric material, and since many of the electric fields that we encounter in our study of electricity are in air, we ought to inquire how the electric fields in air differ from those in vacuum.

26.1 CAPACITANCE

As a first example of a capacitor, let us consider an isolated metallic sphere of radius R (see Fig. 26.1). Obviously, charge can be stored on this sphere. If the amount of charge placed on the sphere is Q, then the potential of the sphere will be, according to Eq. (25.28),

$$V = \frac{1}{4\pi\epsilon_0}\frac{Q}{R} \qquad (26.1)$$

This says that the amount of charge Q stored on the sphere is directly proportional to the potential V.

This proportionality of Q and V holds in general for any conductor of arbitrary shape. The charge on the conductor produces an electric field whose strength is directly proportional to the amount of charge (twice the charge produces twice the field), and the electric field yields a potential which is directly proportional to the field strength (twice the field strength yields twice the potential); hence, charge and potential are proportional. We write this relationship as

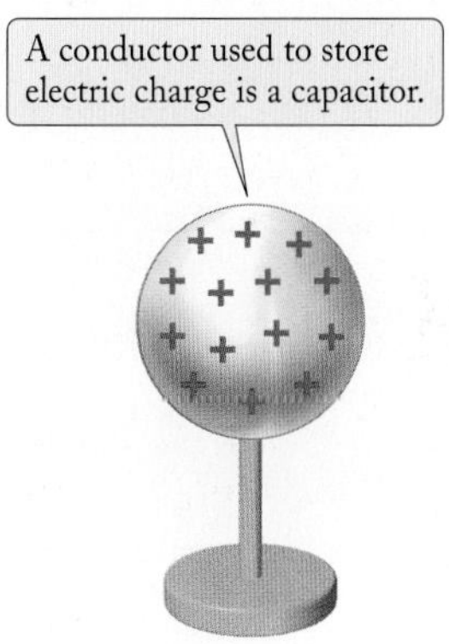

FIGURE 26.1 An isolated metallic sphere.

$$Q = CV \qquad \text{or} \qquad C = \frac{Q}{V} \qquad (26.2)$$

capacitance of a single conductor

where C is the constant of proportionality. This constant is called the **capacitance** of the conductor*. *The capacitance is large if the conductor is capable of storing a large amount of charge at a low potential.* For instance, the capacitance of a spherical conductor is

$$C = \frac{Q}{V} = \frac{Q}{(1/4\pi\epsilon_0)Q/R} = 4\pi\epsilon_0 R \qquad (26.3)$$

*Do not confuse the capacitance C with the abbreviation C for the coulomb, the unit of charge.

Thus, the capacitance of the sphere increases with its radius—a sphere of large radius can store a large amount of charge at a low potential. Note that the value of the capacitance depends only on geometrical properties of the conductor, and not on any particular value of Q or V.

The SI unit of capacitance is the **farad** (F):

farad (F)

$$1 \text{ farad} = 1 \text{ F} = 1 \frac{\text{coulomb}}{\text{volt}} = 1 \frac{\text{C}}{\text{V}} \tag{26.4}$$

This unit of capacitance is rather large; in practice, electrical engineers prefer the **microfarad** and the **picofarad.** A microfarad equals 10^{-6} farad and a picofarad equals 10^{-12} farad:

$$1 \, \mu\text{F} = 10^{-6} \text{ F} \quad \text{and} \quad 1 \text{ pF} = 10^{-12} \text{ F}$$

Note that since $1 \text{ F} = 1 \text{ C/V} = 1 \text{ C}^2/\text{J} = 1 \text{ C}^2/(\text{N}\cdot\text{m})$, the constant ϵ_0 can be written

$$\epsilon_0 = 8.85 \times 10^{-12} \frac{\text{C}^2}{\text{N}\cdot\text{m}^2} = 8.85 \times 10^{-12} \text{ F/m} \tag{26.5}$$

The latter expression is the one usually listed in tables of physical constants.

EXAMPLE 1 What is the capacitance of an isolated metallic sphere of radius 20 cm?

SOLUTION: According to Eq. (26.3),

$$C = 4\pi\epsilon_0 R = 4\pi \times 8.85 \times 10^{-12} \text{ F/m} \times 0.20 \text{ m}$$
$$= 2.2 \times 10^{-11} \text{ F} = 22 \text{ pF}$$

EXAMPLE 2 The ground and the oceans are conductors, and the Earth can therefore be regarded as a conducting sphere. What is its capacitance?

SOLUTION: The radius of the Earth is 6.4×10^6 m; therefore

$$C = 4\pi\epsilon_0 R = 4\pi \times 8.85 \times 10^{-12} \text{ F/m} \times 6.4 \times 10^6 \text{ m}$$
$$= 7.1 \times 10^{-4} \text{ F}$$

COMMENTS: As capacitances go, this is a rather large capacitance. However, note that to alter the potential of the Earth by 1 volt requires a charge of only $Q = CV = 7.1 \times 10^{-4} \text{ F} \times 1 \text{ volt} = 7.1 \times 10^{-4}$ coulomb.

The most common variety of capacitor consists of *two* metallic conductors, insulated from one another and carrying opposite amounts of charge of equal magnitudes, that is, a charge $+Q$ on one conductor and $-Q$ on the other. The capacitance of such a pair of conductors is defined in terms of the *difference* of potential ΔV between the two conductors:

capacitance of pair of conductors

$$Q = C\,\Delta V \quad \text{or} \quad C = \frac{Q}{\Delta V} \tag{26.6}$$

In this expression, both Q and ΔV are taken as positive quantities. Note that the quantity Q is *not* the net charge in the capacitor, but the magnitude of the charge on each plate. The net charge that establishes the potential difference in any two-conductor capacitor is zero.

Figure 26.2 shows such a two-conductor capacitor consisting of two large, parallel metallic plates, each of area A, separated by a distance d. The plates carry charges $+Q$ and $-Q$, respectively, on their facing surfaces. The electric field in the region between the plates is [see Eq. (25.53)]

$$E = \frac{Q}{\epsilon_0 A} \tag{26.7}$$

and the potential difference between the plates is [from Eq. (25.54)]

$$\Delta V = Ed = \frac{Qd}{\epsilon_0 A} \tag{26.8}$$

Hence the capacitance of this configuration is

$$C = \frac{Q}{\Delta V} = \frac{Q}{Qd/\epsilon_0 A} = \frac{\epsilon_0 A}{d}$$

This is the capacitance of a parallel-plate capacitor,

$$C = \frac{\epsilon_0 A}{d} \tag{26.9}$$

capacitance of parallel-plate capacitor

We again see that capacitance depends only on the geometry of the conductors.

From Eq. (26.9) we recognize that to store a large amount of charge at a low potential, we want a large plate area A, but a small plate separation d. Parallel-plate capacitors are often manufactured out of two parallel sheets of aluminum foil, a few centimeters wide, but several meters long. The sheets are placed very close together, but kept from contact by a thin sheet of plastic sandwiched between them (see Fig. 26.3a). For convenience, the entire sandwich is covered with another sheet of plastic and rolled up like a roll of two-ply toilet paper (see Fig. 26.3b). The two sheets of

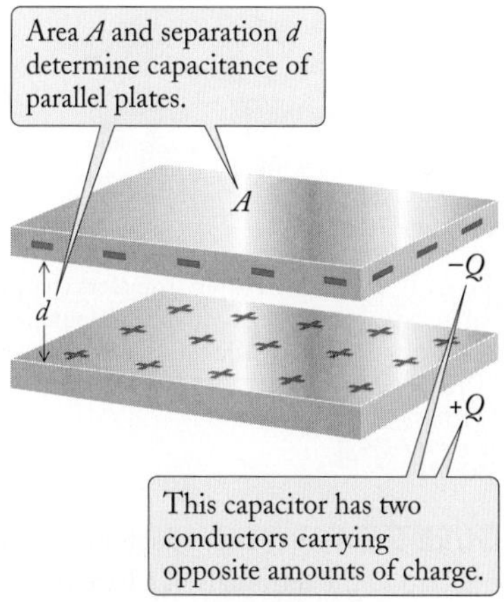

FIGURE 26.2 Two very large parallel plates with opposite electric charges.

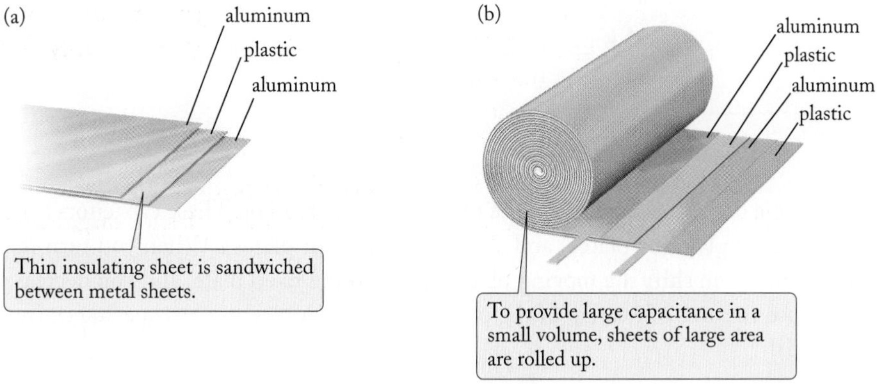

FIGURE 26.3 (a) Sheets of aluminum foil separated by a sheet of plastic. (b) Rolled capacitor.

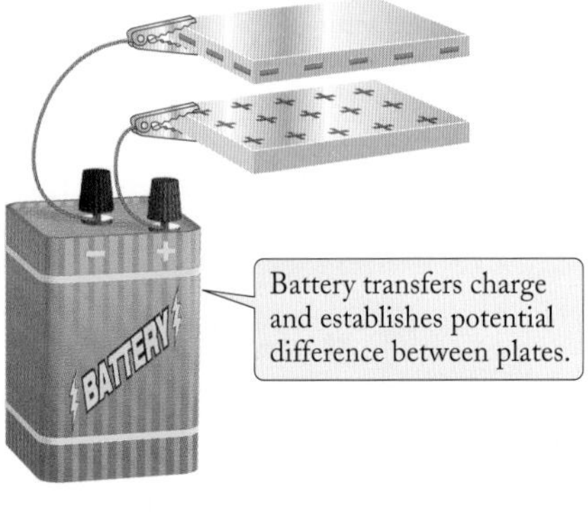

FIGURE 26.4 A two-conductor capacitor connected to the terminals of a battery.

aluminum are connected to the external terminals of the capacitor. To charge such a capacitor, we connect its terminals to the terminals of a battery, which transfers charge from one plate to the other, establishing equal amounts of charge of opposite sign on the capacitor plates, and produces a potential difference between the plates equal to the potential difference (the voltage) of the battery (Fig. 26.4).

Once a capacitor is charged, we can disconnect the battery, and the capacitor will then hold the charge (and the electric potential energy) in storage for a long time. How long depends on how good the insulation is. In some capacitors the insulation between the plates permits some leakage of charge from one plate to another. When the opposite charges of the two plates meet, they neutralize each other, and a capacitor can thereby discharge itself in a few minutes. But some capacitors hold their charge for hours or days. Electronic devices with large capacitors, such as television receivers or computers, often have labels on their casings warning users not to open the casing even when the equipment is disconnected from the power supply, because the capacitors hold their electric charge for a long time and they can give painful and dangerous electric shocks if their terminals accidentally come into contact with the skin of the user.

EXAMPLE 3 A parallel-plate capacitor consists of two strips of aluminum foil, each with an area of $0.20\ \text{m}^2$, separated by a distance of 0.10 mm. The space between the foils is empty. The two strips are connected to the terminals of a battery, which produces a potential difference of 200 volts between them. What is the capacitance of this capacitor? What is the electric charge on each plate? What is the strength of the electric field between the plates?

SOLUTION: According to Eq. (26.9), the capacitance of a parallel-plate capacitor is

$$C = \frac{\epsilon_0 A}{d} = \frac{8.85 \times 10^{-12}\ \text{F/m} \times 0.20\ \text{m}^2}{1.0 \times 10^{-4}\ \text{m}} = 1.8 \times 10^{-8}\ \text{F}$$

$$= 0.018\ \mu\text{F}$$

The charge on each plate is

$$Q = C\,\Delta V = 1.8 \times 10^{-8}\ \text{F} \times 200\ \text{volts} = 3.6 \times 10^{-6}\ \text{coulomb}$$

and the electric field between the plates is, from Eq. (26.8),

$$E = \frac{\Delta V}{d} = \frac{200\ \text{volts}}{1.0 \times 10^{-4}\ \text{m}} = 2.0 \times 10^{6}\ \text{volts/m}$$

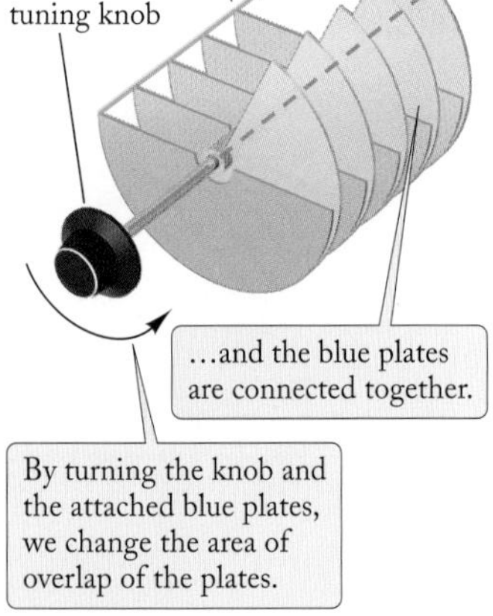

FIGURE 26.5 A variable capacitor.

Variable capacitors are used in the tuning circuits of radios. These capacitors consist of a fixed plate (or plates) and a movable plate (or plates). When you turn the tuning knob, you shift the moving plate parallel to the fixed plate, and you decrease or increase the area of overlap of the plates, and thereby change the capacitance (see Fig. 26.5).

PHYSICS IN PRACTICE CAPACITOR MICROPHONE

A special kind of capacitor is used in the **capacitor microphone**, illustrated in the figure. Another name for a capacitor is a **condenser**; thus, this microphone is also often called a condenser microphone. The flexible diaphragm of this microphone forms one plate of the capacitor, and a rigid disk forms the other plate. When a sound wave impinges on the diaphragm, the periodic fluctuations of the air pressure alternately push and pull the diaphragm toward and away from the rigid plate. The change of distance between the plates produces a change in capacitance, according to Eq. (26.9). Since the plates are connected to a battery, which maintains a steady potential difference between the plates, the change in capacitance results in a change in the amount of electric charge on the plates. The charge that leaves the capacitor plates flows along the wires, forming an electric current. Thus, the capacitor microphone transforms a sound signal into an electric signal, which can be fed into an amplifier and from there into a loudspeaker, a tape recorder, or a digitizer. This kind of microphone has a good sensitivity to a wide range of frequencies, and it is often used in recording studios and telephones.

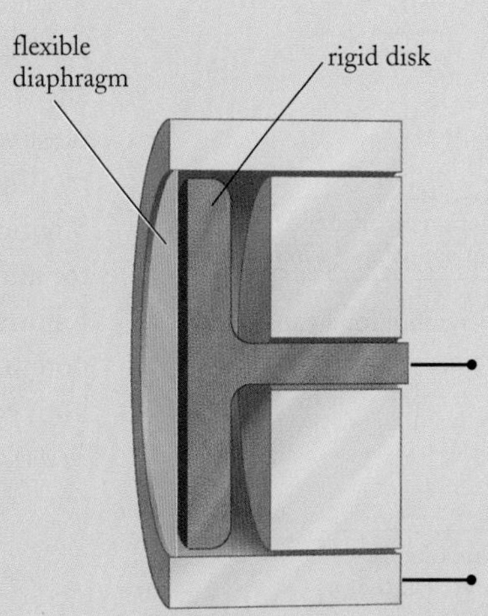

FIGURE 1 A capacitor microphone. When a sound wave pushes on the flexible diaphragm, it alters the distance between the diaphragm and the rigid plate.

Checkup 26.1

QUESTION 1: Is Eq. (26.9) valid for a parallel-plate capacitor with square plates? Rectangular plates? Circular plates?

QUESTION 2: One capacitor has a capacitance of 1 pF; another has a capacitance of 3 pF. If both are charged with 6×10^{-12} C, which has the larger potential difference?

QUESTION 3: Do we increase or decrease the capacitance of a parallel-plate capacitor when we increase the spacing between the plates? When we increase the area of the plates?

QUESTION 4: Suppose that instead of storing the usual equal amounts of charge of opposite signs on the plates of a parallel-plate capacitor, we attempt to store equal charges of the same sign, say, positive, on both plates. Is the capacitance then still given by Eq. (26.9)?

QUESTION 5: If we decrease the separation between the plates of a parallel-plate capacitor by a factor of 2 while holding the electric charge constant, by what factors will we change the electric field, the potential difference, and the capacitance?

QUESTION 6: A parallel-plate capacitor has plates measuring 10 cm × 10 cm and a plate separation of 2.0 mm. If we want to construct a parallel-plate capacitor of the same capacitance but with plates measuring 5.0 cm × 5.0 cm, what plate separation do we need?

(A) 8.0 mm (B) 4.0 mm (C) 2.0 mm (D) 1.0 mm (E) 0.50 mm

26.2 CAPACITORS IN COMBINATION

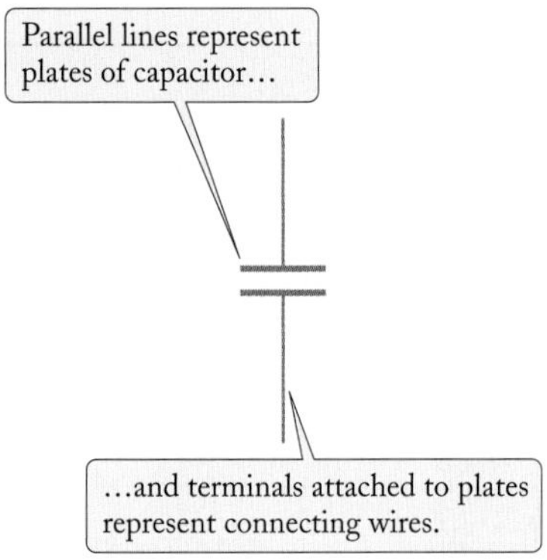

FIGURE 26.6 Symbol for a capacitor in a circuit diagram.

Capacitors used in practical applications in electric circuitry in radios, televisions, computers, etc., commonly are of the two-conductor variety. Schematically, such capacitors are represented in a circuit diagram as two parallel lines with terminals attached to their middles (see Fig. 26.6). The terminals represent wires, each of which we will assume is an ideal conductor. In a circuit, several such capacitors are often wired together and it is then necessary to calculate the net capacitance of the combination. The simplest ways of wiring capacitors together are in **parallel** and in **series**.

Figure 26.7 shows two capacitors connected in *parallel*. If charge is fed into this combination via the two terminals, some of the charge will be stored on the first capacitor and some on the second. The net capacitance of the combination can be found as follows: Since the corresponding plates of the capacitors are joined by conductors, the potentials of corresponding plates must be equal and the potential differences across both capacitors must also be equal. This is true in general: *circuit components connected in parallel have the same voltage across each component*. Thus,

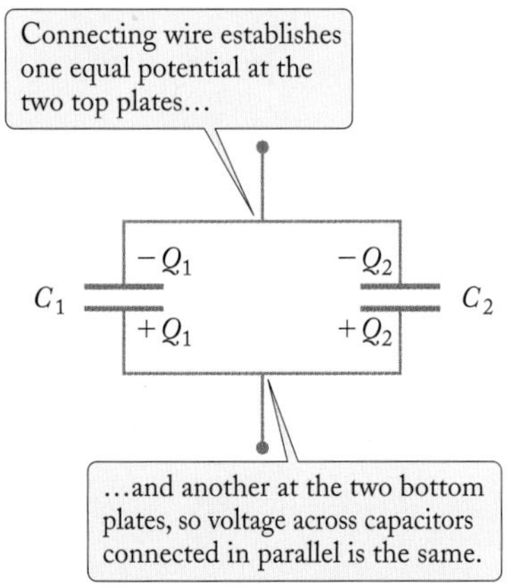

FIGURE 26.7 Two capacitors connected in parallel.

$$\Delta V = \frac{Q_1}{C_1} \quad \text{and} \quad \Delta V = \frac{Q_2}{C_2} \tag{26.10}$$

Therefore the net charge for the capacitor combination can be expressed as

$$Q = Q_1 + Q_2 = C_1\,\Delta V + C_2\,\Delta V$$

that is,

$$Q = (C_1 + C_2)\Delta V \tag{26.11}$$

Comparing this with the definition for capacitance given in Eq. (26.6), we recognize that the combination is equivalent to a single capacitor of capacitance

$$C = C_1 + C_2 \tag{26.12}$$

Thus, *the net capacitance, or equivalent capacitance, of the parallel combination is simply the sum of the individual capacitances.*

It is easy to obtain a similar result for any number of capacitors connected in parallel (see Fig. 26.8). The net capacitance for such a parallel combination is

capacitors in parallel

$$C = C_1 + C_2 + C_3 + \cdots \tag{26.13}$$

Next, we consider the alternative way of connecting capacitors. Figure 26.9 shows two capacitors connected in *series*. Since charge cannot flow across the gap between the plates of the capacitors, any charge fed into the capacitor combination via the two

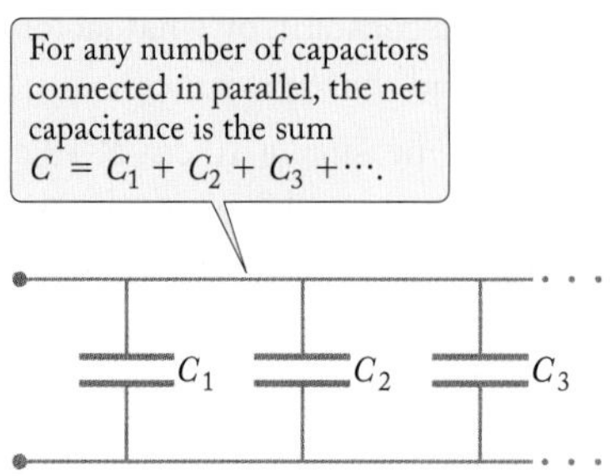

FIGURE 26.8 Several capacitors connected in parallel.

outside terminals will have to remain on the *outside* plates [the lower plate of the bottom capacitor (C_1) and the upper plate of the top one (C_2) in Fig. 26.9]. Thus, the bottom plate will have a charge $+Q$ and the top plate a charge $-Q$. But these charges on the outside plates will induce charges on the inside plates (the upper plate of C_1 and the lower plate of C_2). The charge $+Q$ on the bottom plate of C_1 will attract electrons to the facing plate and a charge $-Q$ will accumulate on this plate. Corresponding to the excess electrons on the upper plate of C_1, there will be a deficit of electrons on the lower plate of C_2, and a charge $+Q$ will accumulate there. In general, *capacitors in series have the same magnitude of charge on each plate.*

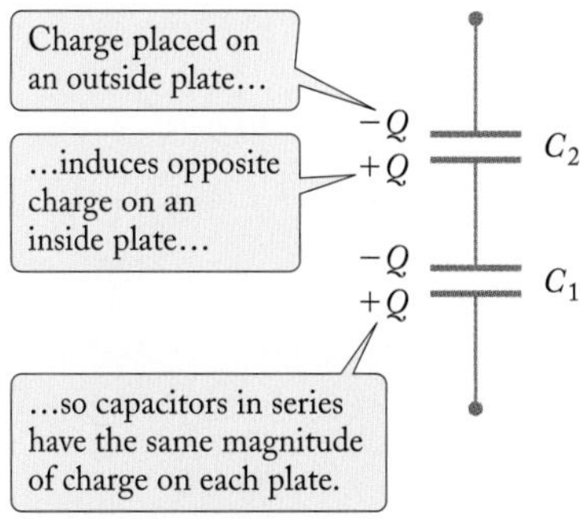

FIGURE 26.9 Two capacitors connected in series.

The capacitance of the combination can then be found as follows. The individual potential differences across the two capacitors are

$$\Delta V_1 = \frac{Q}{C_1} \quad \text{and} \quad \Delta V_2 = \frac{Q}{C_2} \tag{26.14}$$

Since potential is the energy per unit charge to move from one point to another, the net potential difference between the external terminals of two capacitors in series is the sum of the potential differences across the two individual capacitors:

$$\Delta V = \Delta V_1 + \Delta V_2 = \frac{Q}{C_1} + \frac{Q}{C_2} \tag{26.15}$$

from which

$$\Delta V = Q\left(\frac{1}{C_1} + \frac{1}{C_2}\right) \tag{26.16}$$

Comparing this, again, with the definition of capacitance given by Eq. (26.6), we see that the combination is equivalent to a single capacitor with

$$\frac{1}{C} = \frac{1}{C_1} + \frac{1}{C_2} \tag{26.17}$$

or

$$C = \frac{C_1 C_2}{C_1 + C_2}$$

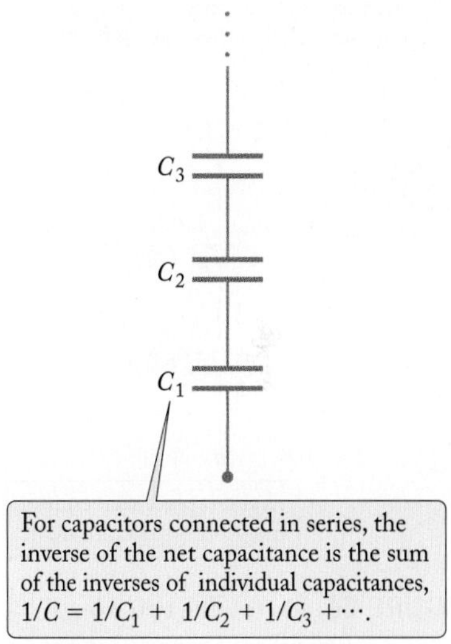

FIGURE 26.10 Several capacitors connected in series.

According to Eq. (26.17), the inverse of the net capacitance of the series combination is obtained by taking the sum of the inverses of the individual capacitances. Note that in series, the net capacitance is always *less* than the individual capacitances; for example, if $C_1 = C_2$, then $C = \frac{1}{2}C_1 = \frac{1}{2}C_2$.

A similar result applies to any number of capacitors connected in series (see Fig. 26.10). The net capacitance C of such a series combination is given by

capacitors in series

$$\frac{1}{C} = \frac{1}{C_1} + \frac{1}{C_2} + \frac{1}{C_3} + \cdots \tag{26.18}$$

EXAMPLE 4 Each of the advanced capacitors at the National Ignition Facility has a capacitance of 300 μF. Each of the 192 laser amplifiers is powered by a bank of 20 such capacitors connected in parallel. What is the net capacitance of each bank? What is the sum of the capacitances of the 192 banks of the entire laser power system?

Concepts in Context

SOLUTION: Since the capacitors in each bank are connected in a simple parallel combination, the net capacitance of a bank is simply the sum of the 20 individual capacitances:

$$C_{\text{bank}} = C_1 + C_2 + C_3 + \cdots = 20C_1$$
$$= 20 \times 300 \times 10^{-6}\,\text{F} = 6.0 \times 10^{-3}\,\text{F}$$

The capacitance of the entire laser power system is

$$C_{\text{total}} = 192 \times C_{\text{bank}} = 192 \times 6.0 \times 10^{-3}\,\text{F}$$
$$= 1.2\,\text{F}$$

This is an immense capacitance, more than a thousand times the capacitance of the Earth, calculated in Example 2. A capacitance value of the order of a farad is hard to achieve.

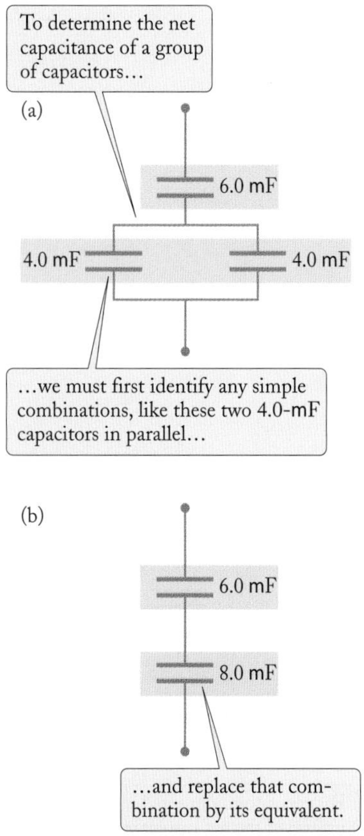

FIGURE 26.11 (a) The two 4.0-μF capacitors (green) are connected in parallel, and the 6.0-μF capacitor (orange) is connected in series with these two. (b) The two 4.0-μF capacitors are equivalent to a single 8.0-μF capacitor, which is connected in series with the 6.0-μF capacitor.

EXAMPLE 5 Two capacitors of 4.0 μF each are connected in parallel and then a third capacitor of 6.0 μF is connected in series with the first two (see Fig. 26.11a). What is the net capacitance of this combination?

SOLUTION: We first look for a group of capacitors that is either a simple series or simple parallel combination. Accordingly, we first calculate the combined capacitance of the two 4.0-μF capacitors. Since they are connected in parallel, Eq. (26.13) tells us that their combined capacitance is the sum 4.0 μF + 4.0 μF = 8.0 μF.

Next, we consider the combined capacitance of this effective 8.0-μF capacitor connected in series with the 6.0-μF capacitor (see Fig. 26.11b). For this series combination, Eq. (26.18) tells us that the net capacitance is given by

$$\frac{1}{C} = \frac{1}{8.0\,\mu\text{F}} + \frac{1}{6.0\,\mu\text{F}} = \frac{6.0\,\mu\text{F} + 8.0\,\mu\text{F}}{8.0\,\mu\text{F} \times 6.0\,\mu\text{F}} = \frac{14}{48\,\mu\text{F}}$$

so

$$C = \frac{48}{14}\,\mu\text{F} = 3.4\,\mu\text{F}$$

This is the net capacitance of the entire combination.

EXAMPLE 6 One way to generate a high voltage is to take a large number of capacitors, charge them while they are connected in parallel, and then reconnect them in series (see Fig. 26.12). Suppose you take 140 capacitors of 0.50 μF and connect them in parallel to a 9.0-volt battery. Once they are fully charged, you reconnect the capacitors in series (without the battery). What is the potential difference across the series combination? How much charge did the capacitors absorb from the battery during the charging? How much charge will they release if you discharge the series combination by connecting the external terminals of the first and last capacitors in the series?

SOLUTION: The potential difference across the series combination is 140 times the potential difference across each capacitor; that is,

$$\Delta V' = 140 \times 9.0\,\text{volts} = 1260\,\text{volts}$$

The capacitance of the original parallel combination is the sum of the capacitances of all the capacitors:

$$C = C_1 + C_2 + C_3 + \cdots = 140 \times 0.50\ \mu\text{F} = 70\ \mu\text{F}$$

Hence the charge absorbed from the battery is

$$Q = C\,\Delta V = 70\ \mu\text{F} \times 9.0 \text{ volts} = 6.3 \times 10^{-4} \text{ coulomb}$$

The capacitance of the series combination is given by

$$\frac{1}{C} = \frac{1}{C_1} + \frac{1}{C_2} + \frac{1}{C_3} + \cdots = 140 \times \frac{1}{0.50\ \mu\text{F}}$$

so

$$C = \frac{0.50\ \mu\text{F}}{140} = 3.57 \times 10^{-3}\ \mu\text{F}$$

Hence the charge of the series combination is

$$Q = C\,\Delta V' = 3.57 \times 10^{-3}\ \mu\text{F} \times 1260 \text{ volts} = 4.5 \times 10^{-6}\ \text{C}$$

Thus, when you discharge the series combination, the charge that flows out of the positive external terminal is $+4.5 \times 10^{-6}$ C, and the charge that flows out of the negative external terminal is -4.5×10^{-6} C.

COMMENTS: In this example, the charge of the series combination is simply the charge of a single capacitor. The charges on all plates except the two connected to the external terminals simply cancel each other and are not counted as charge for the series combination. The initial charge of the parallel combination is 6.3×10^{-4} C, but the charge of the series combination is only 4.5×10^{-6} C, or 140 times smaller.

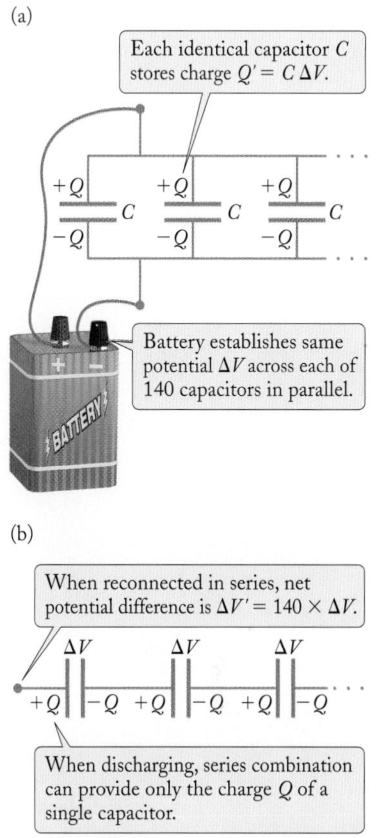

FIGURE 26.12 (a) First the capacitors are connected in parallel and are charged by a battery. (b) Then the capacitors are reconnected in series.

PROBLEM-SOLVING TECHNIQUES COMBINATIONS OF CAPACITORS

Keep in mind that when we speak of the charge Q on a parallel-plate capacitor, we always mean the magnitude of the charge on each plate.

If several such capacitors are connected in *series*:

- They all have the same Q (that is, each capacitor has a charge $+Q$ on the positive plate and $-Q$ on the negative plate), and
- The net potential difference ΔV is the sum $\Delta V_1 + \Delta V_2 + \cdots$ of the potential differences of the individual capacitors.

If several capacitors are connected in *parallel*:

- They all have the same ΔV, and
- The net charge Q is the sum $Q_1 + Q_2 + \cdots$ of the charges of the individual capacitors.

When dealing with a circuit containing several capacitors connected in some complicated manner, proceed in two steps:

1. In the first step, look for groups of capacitors that form simple parallel or series combinations (such as the group of the two 4.0-μF capacitors in Fig. 26.11a). Evaluate the capacitance of each such group.
2. In the second step, see how these effective group capacitances are connected to each other (as in Fig. 26.11b), and evaluate the net capacitance of the combination of the group(s).

In some cases, it may be necessary to repeat these steps until a single equivalent capacitance is obtained.

Checkup 26.2

QUESTION 1: If several capacitors are connected in series, do they all have the same electric field between their plates? What if several capacitors are connected in parallel?

QUESTION 2: Qualitatively, why is the net capacitance of a parallel combination of equal capacitors larger than the individual capacitances? Why is the net capacitance of a series combination of equal capacitors smaller than the individual capacitances?

QUESTION 3: If you connect 10 capacitors of 1.0 μF each in parallel, what is the net capacitance? What if you connect them in series?

(A) 10 μF, 10 μF (B) 10 μF, 0.10 μF
(C) 0.10 μF, 10 μF (D) 0.10 μF, 0.10 μF

26.3 DIELECTRICS

dielectric

So far, in dealing with problems in electrostatics we have assumed that the medium surrounding the electric charges consisted of a vacuum or of air. Vacuum has no effect on the electric field; and air, as we will see, has only a small and often negligible effect on the electric field. However, in dealing with the capacitors used in practical applications, we cannot neglect the effects of the medium that surrounds the electric charges. The space between the plates of such capacitors is usually filled with an electric insulator, or **dielectric**. Such a dielectric drastically changes the electric field from what it would be in a vacuum: *the dielectric reduces the strength of the electric field.*

To understand this, consider a parallel-plate capacitor whose plates carry some amount of charge per unit area. Suppose that a slab of dielectric, such as glass or nylon, fills most of the space between the plates (see Fig. 26.13). This dielectric contains a large number of atomic nuclei and electrons, but, of course, these positive and negative charges balance each other, so the material is electrically neutral. *In an insulator, all the charges are bound*—the electrons are confined within their atoms or molecules and they cannot wander about as in a conductor. Nevertheless, in reponse to the force exerted by the electric field, the charges will move very slightly without leaving their atoms. The electrons move in a direction opposite to that of the electric field; the nuclei move in the direction of the electric field. These opposite displacements separate the positive and negative charges and thereby create electric dipoles within the dielectric. In most dielectrics, the magnitudes of the charge separations and the magnitudes of the induced dipole moments are directly proportional to the strength of the electric field; such dielectrics are said to be **linear**.

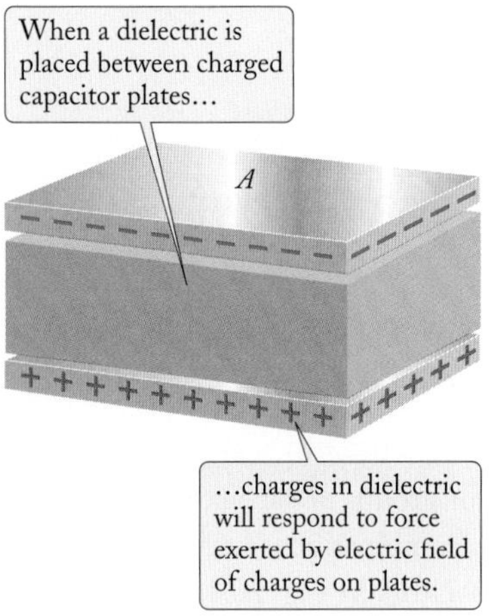

FIGURE 26.13 A slab of dielectric between the plates of a capacitor.

The details of the mechanism of displacement and separation of charge depend on the dielectric. In some dielectrics—such as glass, nylon, and other solids—the creation of dipole moments involves a distortion of the molecules or atoms. By tugging on the electrons and nuclei in opposite directions, the electric field stretches the molecule while producing a charge separation within it (see Fig. 26.14). In other dielectrics—such as distilled water[1] or carbon monoxide—the creation of dipole moments results mainly from a realignment of existing dipoles. In such dielectrics, the molecules have permanent dipole moments which are randomly oriented when the dielectric is left to itself. The randomness of the orientation of the dipoles means that, on the average,

[1] Remember that distilled water is an insulator.

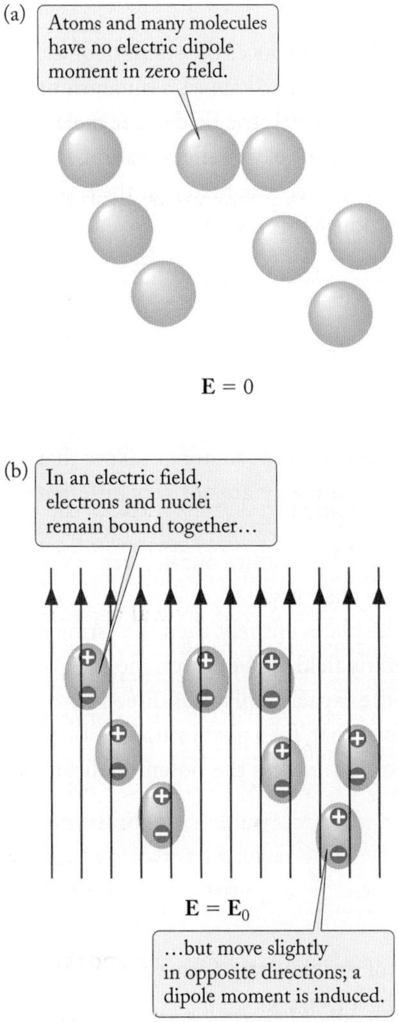

FIGURE 26.14 (a) Undistord molecules. (b) The electric field produces a distortion of the molecules.

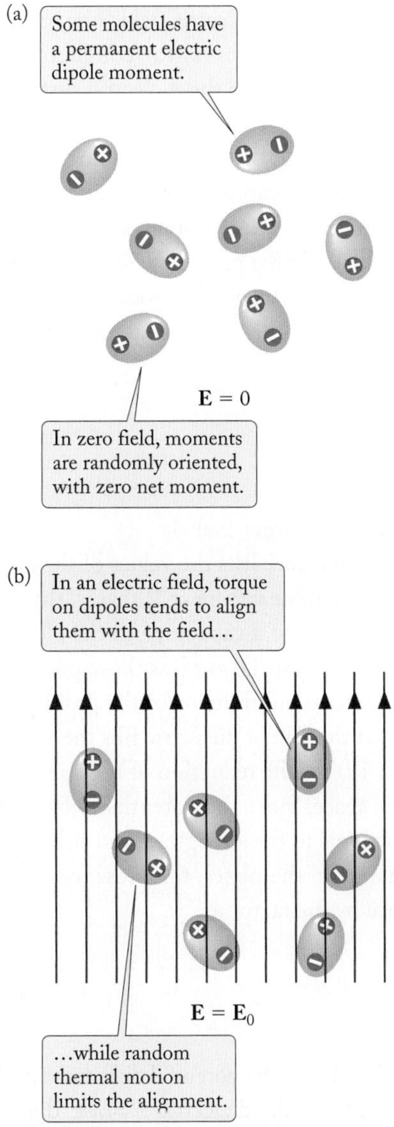

FIGURE 26.15 (a) Unaligned molecules. (b) The electric field produces a (partial) alignment of already distorted molecules.

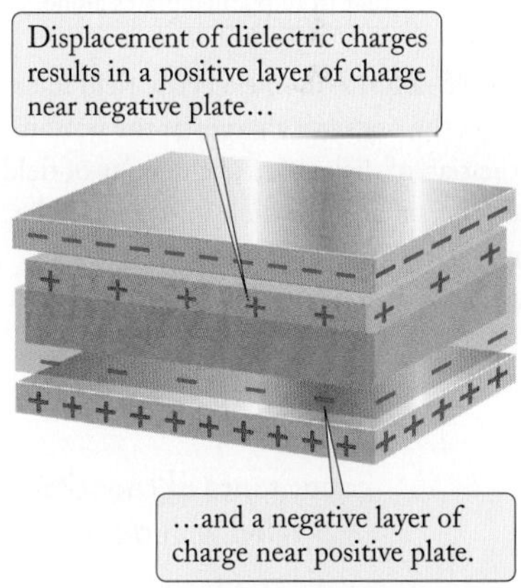

FIGURE 26.16 The distributions of positive charge (brown) and of negative charge (green) of the slab of dielectric fail to overlap precisely. Thus, the electric field has produced a separation of the charges.

there is no charge separation in the dielectric. But as we saw in Section 23.5, when placed in an electric field, permanent dipoles experience a torque that tends to align them with the electric field (see Fig. 26.15). Random thermal motions oppose this alignment, and the molecules achieve an average equilibrium state in which the average amount of alignment is (approximately) proportional to the strength of the electric field. This average alignment is equivalent to an average charge separation.

The displacement of the negative and the positive charges in opposite directions implies that the positive and negative charge distributions of the dielectric cease to overlap precisely (see Fig. 26.16). Consequently, there will be an excess of positive bound charge on one surface of the slab of dielectric and an excess of negative bound charge on the opposite surface. The slab of dielectric is then said to be **polarized**. These surface charges act just like a pair of parallel sheets of positive and negative charge; between the sheets, the charges generate an electric field that is *opposite* to the

polarized dielectric

original applied electric field. The total electric field, consisting of the sum of the field of the free charges on the conducting plates plus the opposing field of the induced bound charges on the dielectric surfaces, is therefore *smaller* than the field of the free charges alone (see Fig. 26.17).

dielectric constant κ

In a linear dielectric, the amount by which the dielectric reduces the strength of the electric field can be characterized by the **dielectric constant κ** (kappa), a dimensionless number. This constant is merely the factor by which the electric field in the dielectric between the parallel plates is reduced; that is, if E_{free} is the electric field that the free charges (on the plates) produced by themselves and E the electric field that these free charges and the bound charges (in the dielectric) produce together, then

electric field in dielectric

$$E = \frac{1}{\kappa} E_{\text{free}} \tag{26.19}$$

where κ is larger than 1.

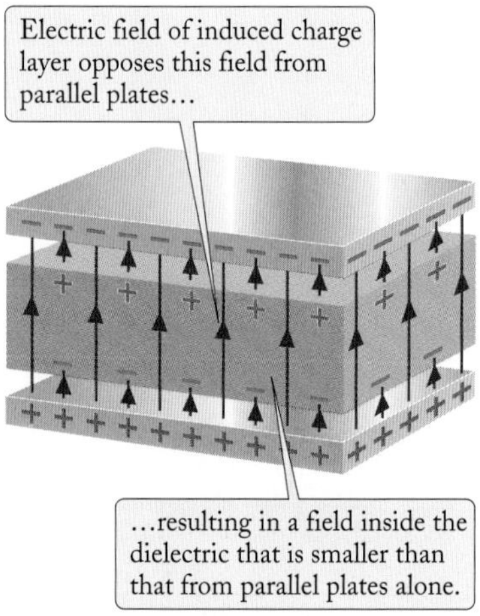

FIGURE 26.17 Some electric field lines stop on the negative charges at the bottom of the slab of dielectric. The density of field lines is smaller in the dielectric than in the empty gaps adjacent to the plates; that is, the electric field is smaller.

Table 26.1 lists the values of the dielectric constants of some materials. Note that air has a value very near 1; that is, the dielectric properties of air are not very different from those of a vacuum, and the electric fields produced by (free) charges placed in air are almost the same as those produced in vacuum. This justifies our neglect of the presence of air in many of the problems in preceding chapters.

If the slab of dielectric fills the space between the plates entirely, then the formula (26.19) for the reduction of the strength of the electric field applies throughout all of this space. Since the potential difference between the capacitor plates is directly proportional to the strength of the electric field, it follows that, for a given amount of free charge on the plates, the presence of the dielectric also reduces the potential difference by the factor κ:

$$\Delta V = \frac{1}{\kappa} \Delta V_0 \tag{26.20}$$

where ΔV_0 is the potential difference in the absence of the dielectric. Consequently, the presence of the dielectric *increases* the capacitance by a factor κ:

$$C = \frac{Q}{\Delta V} = \kappa \frac{Q}{\Delta V_0}$$

or simply

capacitance of capacitor filled with dielectric

$$C = \kappa C_0 \tag{26.21}$$

where $C_0 = Q/\Delta V_0$ is the capacitance in the absence of dielectric. For example, the capacitance of a parallel-plate capacitor filled with Plexiglas (see Table 26.1) is

$$C = \kappa C_0 = \kappa \frac{\epsilon_0 A}{d}$$

$$= 3.4 \times \frac{\epsilon_0 A}{d} \tag{26.22}$$

By filling the space between the capacitor plates with dielectric, we can therefore obtain a substantial gain of capacitance. Furthermore, the dielectric can prevent electric breakdown in the space between the plates. If this space contains air, sparking will

FIGURE 26.18 Electric breakdown in a Plexiglas block in a very strong electric field caused minute perforations in the block and created this beautiful arboreal pattern.

occur between the plates when the electric field reaches a value of about 3×10^6 V/m, and the capacitor will discharge spontaneously. Most dielectrics are better insulators than air and they can tolerate larger electric fields. For instance, Plexiglas can tolerate an electric field of up to 40×10^6 V/m before it suffers electric breakdown (see Fig. 26.18). The maximum electric field that a dielectric can tolerate is called its **dielectric strength**; thus, air has a dielectric strength of 3×10^6 V/m, and Plexiglas a dielectric strength of 40×10^6 V/m.

dielectric strength

EXAMPLE 7 The 300-μF capacitors at the National Ignition Facility are made of two strips of metal foil with an effective parallel-plate area of 123 m^2. The plates are separated by a layer of polypropylene dielectric material 8.0×10^{-6} m thick. What is the dielectric constant? If a potential difference of 24 kV is applied to each of these capacitors what is the magnitude of the free charge on each plate? What is the electric field in the dielectric?

Concepts in Context

SOLUTION: From Eq. (26.21), the dielectric constant is given by the ratio of the actual capacitance to the capacitance in the absence of dielectric:

$$\kappa = \frac{C}{C_0} = \frac{C}{\epsilon_0 A/d} = \frac{300 \times 10^{-6}\ \text{F}}{(8.85 \times 10^{-12}\ \text{F/m} \times 123\ \text{m}^2)/(8.0 \times 10^{-6}\ \text{m})}$$
$$= 2.2$$

The magnitude of the free charge on each plate is

$$Q_{\text{free}} = C\,\Delta V = 300 \times 10^{-6}\ \text{F} \times 24 \times 10^3\ \text{V} = 7.2\ \text{coulombs}$$

The electric field produced by the free charges on the plates in the absence of dielectric would be [see Eq. (26.18)]

$$E_{\text{free}} = \frac{Q}{\epsilon_0 A}$$

and the field in the dielectric is therefore, according to Eq. (26.19),

$$E = \frac{1}{\kappa} E_{\text{free}} = \frac{1}{\kappa}\frac{Q}{\epsilon_0 A} = \frac{1}{2.2} \times \frac{7.2\ \text{coulombs}}{8.85 \times 10^{-12}\ \text{F/m} \times 123\ \text{m}^2}$$
$$= 3.0 \times 10^9\ \text{volts/m}$$

This is a larger electric field, by a factor of about 100, than most common dielectrics can withstand without electric breakdown.

TABLE 26.1

DIELECTRIC CONSTANTS OF SOME MATERIALS[a]

MATERIAL	κ
Vacuum	1
Air	1.000 54
Carbon dioxide	1.000 98
Polyethylene	2.3
Polystyrene	2.5
Rubber, hard	2.8
Transformer oil	≈3
Plexiglas	3.4
Nylon	3.5
Epoxy resin	3.6
Paper	≈4
Glass	≈6
Porcelain	≈7
Water, distilled	80
Strontiun titanate	320

[a] At room temperature (20°C) and 1 atm.

COMMENTS: Alternatively, we can calculate the electric field in the dielectric from the usual relationship between a uniform electric field and its potential,

$$E = \frac{\Delta V}{d} = \frac{24 \times 10^3\ \text{V}}{8.0 \times 10^{-6}\ \text{m}} = 3.0 \times 10^9\ \text{volts/m}$$

This relationship between electric field and potential is not affected by the presence of the dielectric.

The simple formula (26.19) for the reduction of the electric field by a dielectric applies to any configuration in which the dielectric and the distribution of free charges have the same symmetry—for instance, a pair of flat charged plates with a flat slab of dielectric (as in the parallel-plate capacitor) or a spherical distribution of charge surrounded by a concentric spherical shell of dielectric. However, if the symmetries of the charge distribution and the dielectric differ—for instance, a parallel-plate capacitor with a spherical ball of dielectric placed between its plates—then the simple formula (26.19) is not applicable, and the reduction of the electric field in the dielectric becomes rather more difficult to calculate.

EXAMPLE 8 What is the electric field generated by a point charge q surrounded by a large volume of dielectric, for instance, a point charge placed in a large volume of gas?

SOLUTION: If the volume of the dielectric surrounding the point charge is large, then the electric field in the neighborhood of this charge is not significantly influenced by the shape of the (remote) surfaces of the dielectric. Thus, the dielectric can be regarded as providing a spherically symmetric environment for the spherically symmetric point charge. Consequently, Eq. (26.19) is applicable to this problem and

$$E = \frac{1}{\kappa} E_{\text{free}} = \frac{1}{\kappa} \frac{q}{4\pi\epsilon_0 r^2} \tag{26.23}$$

where, as usual, r is the distance from the point charge.

Note that since this inverse-square electric field produced by the point charge in the large volume of dielectric differs from the electric field of a point charge in vacuum only by the factor $1/\kappa$, the arguments that led us to Gauss' Law (in vacuum) in Section 24.2 will now lead us to a modified version of Gauss' Law for a distribution of point charges placed in a large volume of dielectric:

Gauss' Law in dielectrics

$$\oint \kappa E_\perp \, dA = \frac{Q_{\text{free, inside}}}{\epsilon_0} \tag{26.24}$$

This result, which we have reached here by a rather specialized argument involving a large volume of dielectric, is actually valid in general. It holds true for any configuration of free charges and dielectrics, regardless of their symmetry. Equation (26.24) is **Gauss' Law in dielectrics**. We could opt to use this Law, but for the simple problems we will be dealing with, we can always find the electric field from the formula (26.19), by first calculating the electric field of the free charges by themselves.

coaxial cable

EXAMPLE 9 A common type of cable, the **coaxial cable**, consists of a solid cylindrical conductor on the axis of a hollow cylindrical conducting shell; the two conductors are separated by a dielectric (see Fig. 26.19). Calculate the capacitance per centimeter of a coaxial cable with a polystyrene dielectric. The inner conductor of the cable has a diameter of 1.5 mm, and the thin conducting shell has a radius of 3.0 mm.

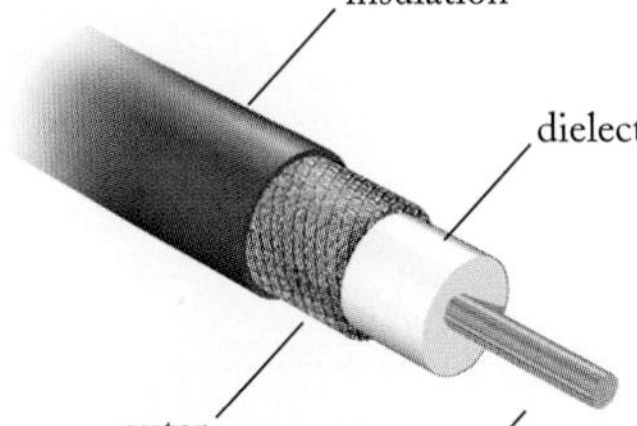

FIGURE 26.19 Coaxial cable.

SOLUTION: The capacitance of a cylindrical capacitor *without* dielectric is defined in the usual way [Eq. (26.6)]:

$$C_0 = \frac{Q}{\Delta V_0}$$

As more length is added, the additional capacitance is in parallel, and so simply adds in proportion to length. So the capacitance per unit length l is

$$\frac{C_0}{l} = \frac{\lambda}{\Delta V_0}$$

where $\lambda = Q/l$ is the linear charge density. In Example 5 of Chapter 25, we calculated the potential difference without dielectric from the field obtained from Gauss' Law; this gave Eq. (25.35),

$$\Delta V_0 = \frac{\lambda}{2\pi\epsilon_0} \ln\left(\frac{b}{a}\right)$$

where b and a are the radii of the outer and inner conductors. Thus we have

capacitance per unit length of cylindrical capacitor

$$\frac{C_0}{l} = 2\pi\epsilon_0 \frac{1}{\ln(b/a)} \tag{26.25}$$

For the capacitor *with* dielectric, we then obtain, with $\kappa = 2.5$ for polystyrene,

$$\frac{C}{l} = \kappa\frac{C_0}{l} = \kappa 2\pi\epsilon_0 \frac{1}{\ln(b/a)}$$

$$= 2.5 \times 2\pi \times 8.85 \times 10^{-12}\ \text{F/m} \times \frac{1}{\ln(3.0\ \text{mm}/0.75\ \text{mm})}$$

$$= 1.0 \times 10^{-10}\ \text{F/m} = 1.0\ \text{pF/cm}$$

A capacitance close to one picofarad per centimeter of length is typical of many coaxial cables. This kind of capacitance plays a role when coaxial cables are used to connect electronic devices, and it is known as **cable capacitance**.

✔ Checkup 26.3

QUESTION 1: Explain why the mechanism of alignment of dipoles described above leads to a dielectric constant $\kappa > 1$.

QUESTION 2: Suppose that the capacitance of a metallic sphere is 3.0×10^{-12} F when this sphere is surrounded by vacuum. What will be the capacitance of this sphere if it is immersed in a large volume of oil with a dielectric constant $\kappa = 3.0$?

(A) 1.0×10^{-12} F (B) 3.0×10^{-12} F (C) 9.0×10^{-12} F

26.4 ENERGY IN CAPACITORS

Capacitors store not only electric charge, but also electric energy. We saw in Section 25.5 that any arrangement of conductors with electric charges has an electric potential energy, which represents the amount of work we must do to bring the electric charges to their positions on the conductors.

Consider a two-conductor capacitor, with charges $\pm Q$ on its plates. If the capacitor contains no dielectric, then we can calculate the electric potential energy directly from Eq. (25.52):

$$U = \tfrac{1}{2}Q(V_1 - V_2)$$

where V_1 and V_2 are the potentials of the plates. Thus, the potential energy can be expressed in terms of the charge and the potential difference:

$$U = \tfrac{1}{2}Q\,\Delta V \tag{26.26}$$

By means of the definition of capacitance, $Q = C\Delta V$, this can be put in the alternative forms

potential energy in capacitor

$$U = \frac{1}{2}C(\Delta V)^2 \quad \text{or} \quad U = \frac{1}{2}\frac{Q^2}{C} \tag{26.27}$$

If the capacitor contains a dielectric, then the calculation of the energy is a bit more involved. The trouble is that the dielectric, with its bound charges, contributes to the electric potential energy. However, in practice we are usually not interested in the total potential energy, but only in that part of the potential energy that changes as we charge (or discharge) the capacitor; that is, we are interested only in the amount of work required to charge (or discharge) the capacitor. It turns out that this amount of work is correctly given by Eqs. (26.26) and (26.27), regardless of whether the capacitor contains a dielectric or not. The quantity Q in these equations is the charge on the plates; i.e., it is the *free* charge. To see this, let us derive Eq. (26.27) from a different starting point, imitating the method we used in section 5 of Chapter 25 for a charged conducting sphere [Eq. (25.49)].

Imagine that we charge the capacitor gradually, starting with an initial charge $q = 0$, adding small amounts of charge step by step, and ending with a final charge $q = Q$. When the plates carry charges $\pm q$, the potential difference between them is q/C and the work that we must perform to increase the charges on the plates by a small amount $\pm dq$ is the product of the transported charge dq and the potential q/C:

$$dU = \frac{q}{C}\,dq \tag{26.28}$$

To find the final potential energy, we add up these small changes in energy by integrating from the initial value of the charge ($q = 0$) to the final value of the charge ($q = Q$):

$$U = \int dU = \frac{1}{C}\int_0^Q q\,dq = \frac{1}{C}\left(\frac{q^2}{2}\right)\Bigg|_0^Q = \frac{1}{C}\left(\frac{Q^2}{2} - 0\right)$$

that is,

$$U = \frac{1}{2}\frac{Q^2}{C} \tag{26.29}$$

This formula is the same as the second relation in Eq. (26.27). This confirms that the relations in Eq. (26.27) remain valid for a capacitor containing a dielectric, provided we use the appropriate value of the capacitance (with the dielectric constant included).

EXAMPLE 10 Consider one of the National Ignition Facility capacitors filled with dielectric, described in Example 7. What is the stored potential energy in this capacitor? What is the total energy stored in all 192 × 20 = 3840 capacitors in the array? What is the energy density in the capacitors?

Concepts in Context

SOLUTION: In Example 7, we were given $\Delta V = 24$ kV and we found the free charge on each plate, $Q = 7.2$ C. By Eq. (26.26), the stored energy in one capacitor is

$$U = \tfrac{1}{2} Q\,\Delta V = \tfrac{1}{2} \times 7.2 \text{ coulombs} \times 24 \times 10^3 \text{ volts} = 8.6 \times 10^4 \text{ J}$$

The total energy stored in the array is the energy stored in each capacitor times the number of capacitors:

$$U_{\text{array}} = 8.6 \times 10^4 \text{ J} \times 3840 = 3.3 \times 10^8 \text{ J}$$

The capacitors can deliver this stored energy, one-third of a billion joules, in a time much less than a nanosecond, corresponding to a peak power greater than billions of gigawatts.

The energy density u is the energy divided by the volume between the plates [Eq. (25.58)]. Using the values from Example 7,

$$u = \frac{U}{Ad} = \frac{8.6 \times 10^4 \text{ J}}{123 \text{ m}^2 \times 8.0 \times 10^{-6} \text{ m}} = 8.7 \times 10^7 \text{ J/m}^3$$

Note that we here calculated the energy density directly from the energy and the volume, instead of using the formula $u = \epsilon_0 E^2/2$ for the energy density from Chapter 25. If we had used this formula, we would have obtained a wrong answer, because the formula from Chapter 25 is not applicable in a dielectric. It is easy to check that in a dielectric material, the correct formula for the energy density is $u = \kappa\epsilon_0 E^2/2$. For example, for a capacitor filled with dielectric, the potential energy is

$$U = \frac{1}{2} C(\Delta V)^2 = \frac{1}{2} \kappa C_0 (\Delta V)^2 = \frac{1}{2} \kappa \frac{\epsilon_0 A}{d} (Ed)^2 = \frac{1}{2} \kappa\epsilon_0 E^2 \cdot Ad$$

and so the energy density is

$$u = \frac{U}{Ad} = \frac{1}{2} \kappa\epsilon_0 E^2 \tag{26.30}$$

energy density in dielectric

EXAMPLE 11 An isolated, charged parallel-plate capacitor with plate separation $d = 0.10$ mm and area $A = 1.0$ m^2 is initially empty (Fig. 26.20a). You slide a Plexiglas dielectric slab with thickness equal to the plate separation but with only half the area of the plates into the region between the plates, as shown. What is the final capacitance? Does the stored energy increase or decrease? Does the capacitor pull the dielectric in or do you have to force the dielectric into the capacitor?

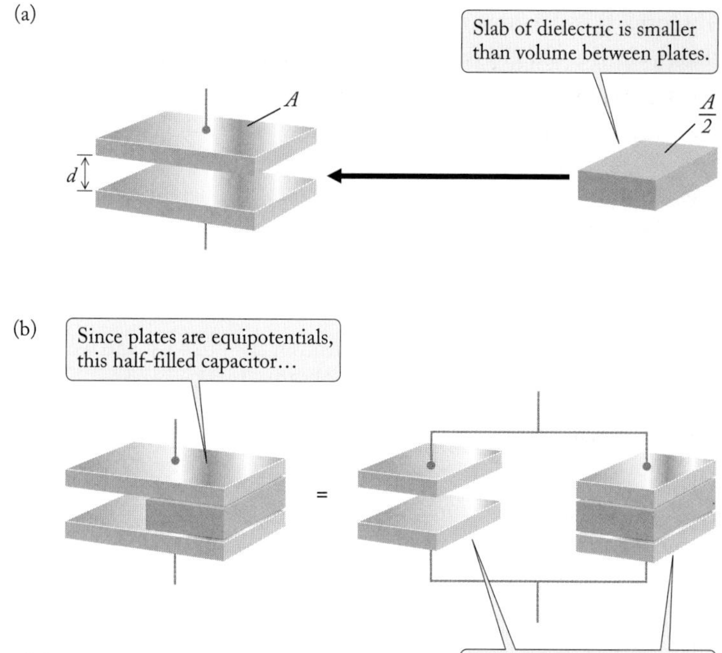

FIGURE 26.20 (a) A parallel-plate capacitor with a dielectric slab that has only half the area of the capacitor. (b) Equivalent capacitor circuit.

SOLUTION: Because the dielectric fills the capacitor only partially, we must consider the right and the left halves of the capacitor separately to determine the equivalent capacitance. Each plate is an equipotential, so we may separate the capacitor into two pieces connected in parallel by wires (which are also equipotentials), as shown in Fig. 26.20b. Each of these two capacitors has half the area of the original, and capacitances in parallel add, so the empty capacitance is

$$C_0 = \frac{\epsilon_0 A}{d} = \frac{\epsilon_0 A/2}{d} + \frac{\epsilon_0 A/2}{d} = \frac{1}{2}C_0 + \frac{1}{2}C_0$$

When one of the two halves is filled, its capacitance increases by a factor of the dielectric constant, so the final capacitance is

$$C = \frac{1}{2}C_0 + \kappa\frac{1}{2}C_0 = \frac{1+\kappa}{2}C_0 = \frac{1+\kappa}{2}\frac{\epsilon_0 A}{d}$$

$$= \frac{1+3.4}{2}\,\frac{8.85\times10^{-12}\ \text{F/m}\times1.0\ \text{m}^2}{10^{-4}\ \text{m}} = 1.9\times10^{-7}\ \text{F}$$

Since the charge is held constant, the energy is most conveniently written

$$U = \frac{1}{2}\frac{Q^2}{C}$$

Because the capacitance increases by a factor of $(1+\kappa)/2$, the stored energy decreases. Since the dielectric attains a position of lower energy, the capacitor is doing work on the dielectric, that is, it produces a force which pulls the dielectric in.

COMMENTS: Note that this example involved an isolated capacitor, that is, the charge on the plates was constant. If the plates had instead been held at a constant potential by keeping them connected to a battery, then the charge would have increased as the dielectric moved into place. Also, for a constant potential difference ΔV between the plates, the stored energy $U = C(\Delta V)^2/2$ would increase by $\Delta U = \Delta C(\Delta V)^2/2$ with increasing C. The increase ΔU in stored energy comes from the battery. The battery actually delivers a larger energy $\Delta Q\,\Delta V = (\Delta C \Delta V)\Delta V = 2\Delta U$; the extra energy is delivered as work on the slab. This work is positive, indicating that the force exerted by the capacitor plates on the slab pulls the slab in. Thus this force has the same direction in the case of constant potential as in the case of constant charge.

✔ Checkup 26.4

QUESTION 1: Two parallel-plate capacitors are identical, except that one has dielectric between its plates and the other does not. If charged to the same voltage, which will have the larger charge on its plates? The larger electric field? The larger energy density? The larger energy?

QUESTION 2: Two parallel-plate capacitors are identical, except that one has dielectric between its plates and the other does not. If equal amounts of charge are placed on their plates, which will have the larger voltage? The larger electric field? The larger energy density? The larger energy?

QUESTION 3: Two identical empty capacitors, each with capacitance C_0, are connected in series. Both are then filled with a material with dielectric constant $\kappa = 2.0$. What is the final net capacitance of the series combination?

(A) $4.0C_0$ (B) $2.0C_0$ (C) C_0 (D) $0.50C_0$

SUMMARY

PHYSICS IN PRACTICE Capacitor Microphone **(page 833)**

PROBLEM-SOLVING TECHNIQUES Combinations of Capacitors **(page 837)**

SI UNIT OF CAPACITANCE 1 F = 1 farad = 1 coulomb/volt **(26.4)**

CAPACITANCE OF A SINGLE CONDUCTOR $C = \frac{Q}{V}$ **(26.2)**

CAPACITANCE OF A PAIR OF CONDUCTORS $C = \frac{Q}{\Delta V}$ **(26.6)**

CAPACITANCE OF AN ISOLATED SPHERE $C = 4\pi\epsilon_0 R$ **(26.3)**

CAPACITANCE OF PARALLEL PLATES $C = \frac{\epsilon_0 A}{d}$ **(26.9)**

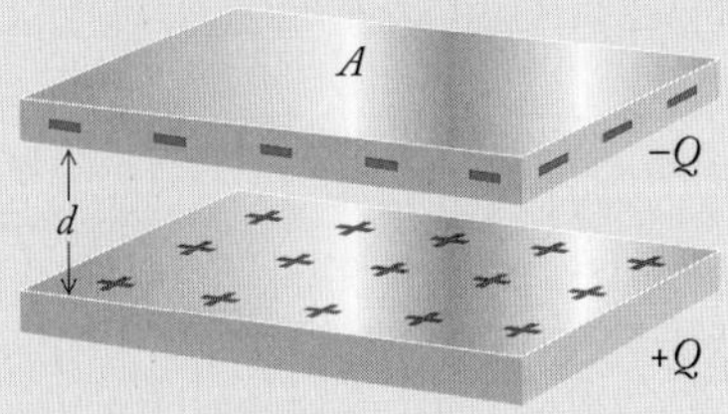

PARALLEL COMBINATION OF CAPACITORS

Capacitors in parallel have the *same potential difference* across each.

$$C = C_1 + C_2 + C_3 + \cdots \quad (26.13)$$

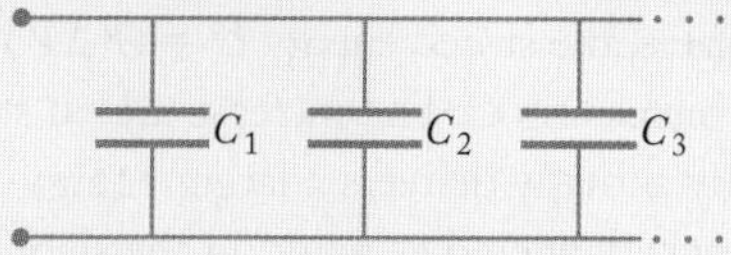

SERIES COMBINATION OF CAPACITORS

Capacitors in series have the *same charge* on each.

$$\frac{1}{C} = \frac{1}{C_1} + \frac{1}{C_2} + \frac{1}{C_3} + \cdots \quad (26.16)$$

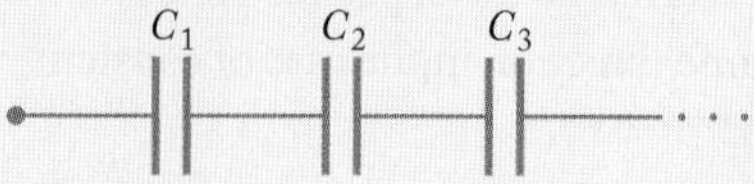

CAPACITANCE PER UNIT LENGTH OF CYLINDRICAL CAPACITOR

$$\frac{C}{l} = \frac{2\pi\epsilon_0}{\ln(b/a)} \quad (26.25)$$

ELECTRIC FIELD IN A DIELECTRIC BETWEEN PARALLEL PLATES

$$E = \frac{1}{\kappa} E_{\text{free}} \quad (26.19)$$

where κ, the dielectric constant, is larger than 1. The relation $E = E_{\text{free}}/\kappa$ also applies in any dielectric with the same symmetry as the distribution of free charges.

CAPACITANCE WITH DIELECTRIC

$$C = \kappa C_0 \quad (26.21)$$

ENERGY STORED IN CAPACITOR

$$U = \frac{1}{2} Q \,\Delta V = \frac{1}{2}\frac{Q^2}{C} = \frac{1}{2} C(\Delta V)^2 \quad (26.26\text{–}27)$$

ENERGY DENSITY IN DIELECTRIC

$$u = \frac{1}{2}\kappa\epsilon_0 E^2 \quad (26.30)$$

QUESTIONS FOR DISCUSSION

1. The large capacitors at the National Ignition Facility have a total capacitance of more than a farad. How is it possible that the capacitance of such a device is larger than the capacitance of the Earth?
2. Suppose we enclose the entire Earth in a conducting shell of radius slightly larger than the Earth's radius. Explain why this would make the capacitance of the Earth much larger than the value calculated in Example 2.
3. Equation (26.9) shows that $C \to \infty$ as $d \to 0$. In practice, why can we not construct a capacitor of arbitrarily large C by making d sufficiently small? (Hint: What happens to E as $d \to 0$ while ΔV is held constant?)
4. If you put more charge on one plate of a parallel-plate capacitor than on the other, what happens to the extra charge?
5. Taking the fringing field into account, would you expect the capacitance of a parallel-plate capacitor to be larger or smaller than the value given by Eq. (26.9)? (Hint: How does the fringing field affect the density of field lines between the plates?)
6. Figure 26.21 shows a capacitor with **guard rings**. These rings fit snugly around the edges of the capacitor plates. In use, the potential on the rings is adjusted to the same value as the potential on the plates. Explain how these rings keep the field of the plates from fringing.

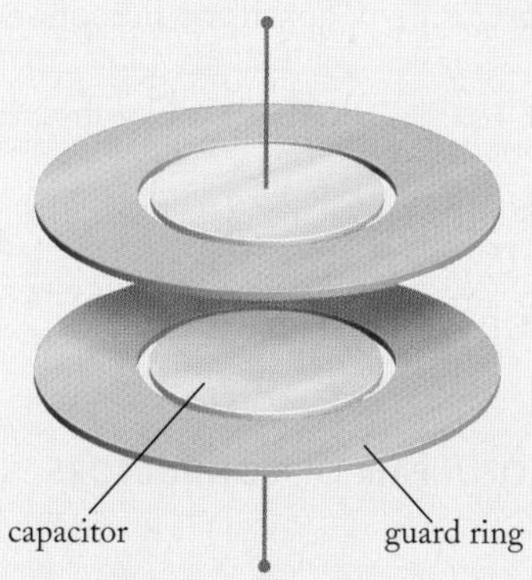

FIGURE 26.21 Capacitor with guard rings.

7. Figure 26.5 shows the design of an adjustable capacitor used in the tuning circuit of a radio. This capacitor can be regarded as several connected capacitors. Are these several capacitors connected in series or in parallel? If we turn the tuning knob

(and the attached blue plates) counterclockwise, does the capacitance increase or decrease?

8. Consider a parallel-plate capacitor. Does the capacitance change if we insert a thin conducting sheet between the two plates, parallel to them?

9. Suppose we insert a thick slab of metal between the plates of a parallel-plate capacitor, parallel to the plates, but not in contact with them. Does the capacitance increase or decrease?

10. Consider a fluid dielectric that consists of molecules with permanent dipole moments. Will the dielectric constant increase or decrease if the temperature increases?

11. Figure 26.22 shows a dielectric slab partially inserted between the plates of a charged, isolated capacitor. Will the electric forces between the slab and the plates pull the slab into the region between the plates or push it out? Explain this in terms of the fringing field.

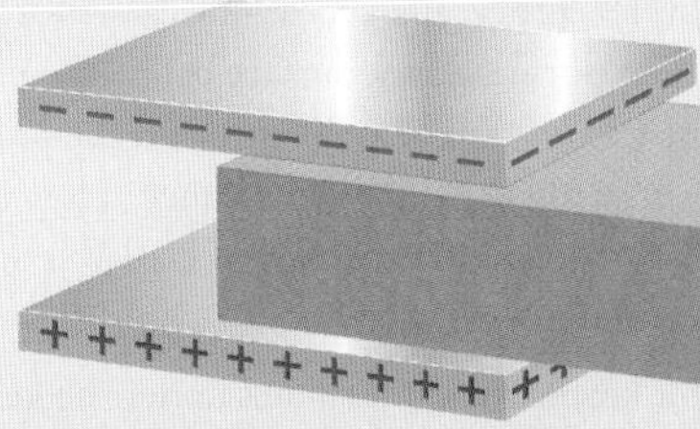

FIGURE 26.22 Dielectric slab partially inserted between the plates of a capacitor.

PROBLEMS

26.1 Capacitance

1. Consider an isolated metallic sphere of radius R and another isolated metallic sphere of radius $3R$. If both spheres are at the same potential, what is the ratio of their charges? If both spheres carry the same charge, what is the ratio of their potentials?

2. The collector of an electrostatic machine is a metal sphere of radius 18 cm.

   (a) What is the capacitance of this sphere?

   (b) How many coulombs of charge must you place on this sphere to raise its potential to 2.0×10^5 V?

3. Your head is (approximately) a conducting sphere of radius 10 cm. What is the capacitance of your head? What will be the charge on your head if, by means of an electrostatic machine, you raise your head (and your body) to a potential of 100 000 V (see Fig. 26.23)?

FIGURE 26.23 A charged head.

4. A capacitor consists of a metal sphere of radius 5.0 cm placed at the center of a thin metal spherical shell of radius 12 cm. The space between is empty. What is the capacitance?

5. A capacitor consists of two parallel conducting disks of radius 20 cm separated by a distance of 1.0 mm. What is the capacitance? How much charge will this capacitor store if connected to a battery of 12 V?

6. What is the electric field in a 3.0-μF capacitor with parallel plates of area 15 m^2 charged to 4.4 volts?

7. A 4.00-μF capacitor has been charged by a 9.00-volt battery. How many electrons must be moved from the negative plate to the positive plate of the capacitor to reverse the electric field inside the capacitor?

8. A capacitor is made from two concentric conducting spherical shells; the inner shell has radius a and the outer shell has radius b. What is the capacitance of this arrangement?

9. A parallel-plate variable capacitor has a fixed plate separation of 0.50 mm, but the plate area can be changed by moving one plate. If the capacitance can be varied from 10.0 pF to 120 pF, what are the corresponding minimum and maximum overlapping plate areas?

10. In digital circuits, "de-spiking" capacitors are often placed near each semiconductor chip to provide a local reservoir of charge. Typically, a 0.10-μF capacitor is placed between the chip's 5.0-volt power-supply connection and ground. How much charge is stored on such a capacitor?

11. Some modern wireless and mobile electronic devices use "supercapacitors" with extremely large capacitance values. How much charge is stored in a supercapacitor with capacitance 50.0 F and potential difference 2.50 V?

12. Spot welders use the sudden discharge of a large capacitor to melt and bond metals. A spot welder uses a 51-mF capacitor at a potential difference of 250 V. How much charge is stored?

13. The local electronic properties of surfaces are sometimes studied using a **scanning capacitance microscope**, where a metal

probe is moved over a surface and acts as one capacitor plate; the other plate is the portion of the surface below the probe. If the effective plate area is 200 nm × 200 nm and the probe is 20 nm from the surface, what is the capacitance? How much charge is on the capacitor when a voltage of 0.10 V is applied between the probe and the surface?

14. In many computer keyboards the switches under the keys consist of small parallel-plate capacitors (see Fig. 26.24). The key is attached to the upper plate, which is movable. When you push the key down, you push the upper plate toward the lower plate, and you alter the plate separation d and the capacitance. The capacitor is connected to an external circuit that maintains a constant potential difference ΔV across the plates. The change of capacitance therefore sends a pulse of charge from the capacitor into the computer circuit. Suppose that the initial plate separation is 5.0 mm and the initial capacitance is 6.0×10^{-13} F. The final plate separation (with the key fully depressed) is 0.20 mm. The constant potential difference is 8.0 V. What is the change in capacitance when you depress the key? What is the amount of electric charge that flows out of the capacitor into the computer circuit?

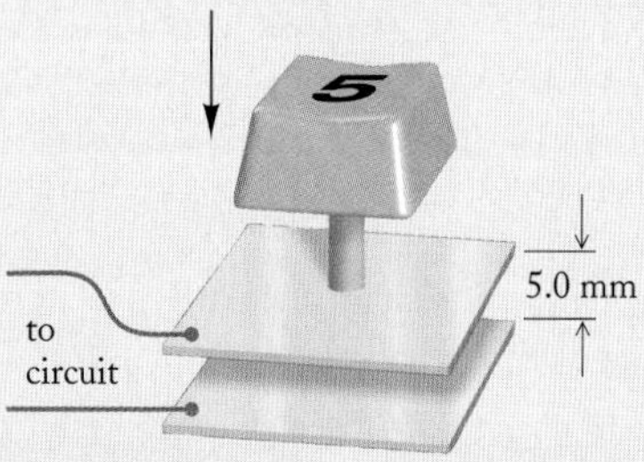

FIGURE 26.24 Capacitive keyboard switch.

*15. What is the capacitance of the Geiger-counter tube described in Problem 35 of Chapter 25? Pretend that the space between the conductors is empty.

26.2 Capacitors in Combination

16. What is the combined capacitance if three capacitors of 3.0, 5.0, and 7.5 μF are connected in parallel? What is the combined capacitance if they are connected in series?

17. Three capacitors with capacitances $C_1 = 5.0\ \mu$F, $C_2 = 3.0\ \mu$F, and $C_3 = 8.0\ \mu$F are connected as shown in Fig. 26.25. Find their combined capacitance.

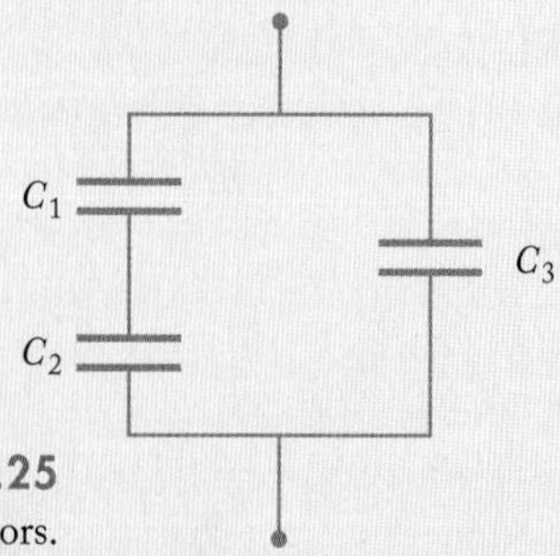

FIGURE 26.25
Three capacitors.

18. Two capacitors of 5.0 μF and 8.0 μF are connected in series to a 24-V battery. What is the potential difference across each capacitor?

19. What is the total charge stored on the three capacitors connected to a 30-V battery as shown in Fig. 26.26?

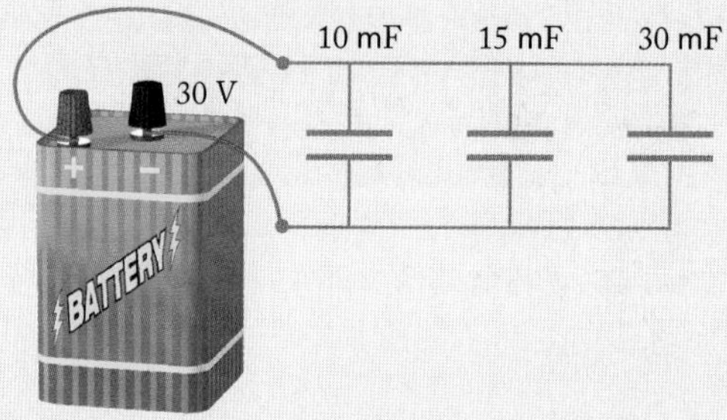

FIGURE 26.26 Three capacitors connected to a battery.

20. Six identical capacitors of capacitance C are connected as shown in Fig. 26.27. What is the net capacitance of the combination?

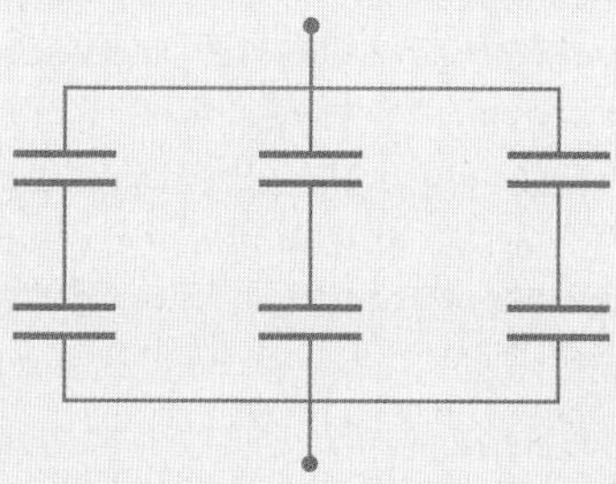
FIGURE 26.27 Six identical capacitors.

21. Six identical capacitors of capacitance C are connected as shown in Fig. 26.28. What is the net capacitance of the combination?

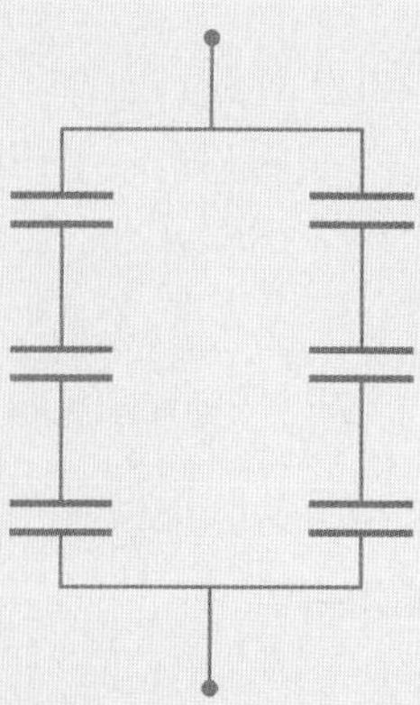
FIGURE 26.28 Six identical capacitors.

22. Seven capacitors are connected as shown in Fig. 26.29. What is the net capacitance of this combination?

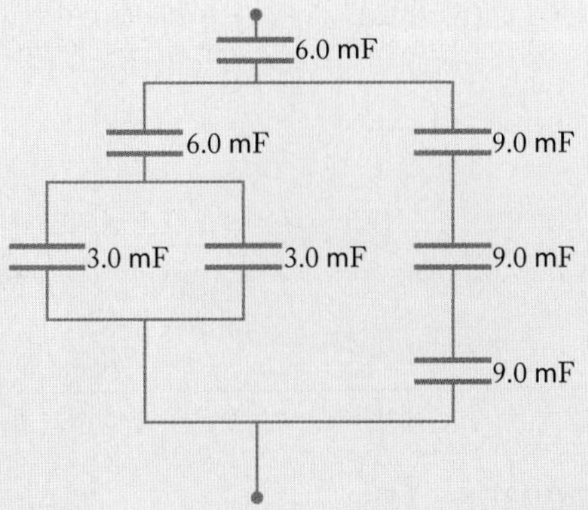

FIGURE 26.29 Seven connected capacitors.

23. When the arrangement of capacitors shown in Fig. 26.11 is connected to a voltage source, each of the 4.0-μF capacitors stores a charge of 6.0 μC. What charge is stored on the 6.0-μF capacitor? What is the voltage of the source?

*24. Two capacitors of 2.0 μF and 5.0 μF are connected in series; the combination is connected to a 1.5-V battery. What is the charge stored on each capacitor? What is the potential difference across each capacitor?

*25. A 2.5-μF capacitor is connected to a 9.0-V battery. The capacitor is then disconnected from the battery and connected to an initially uncharged 5.0-μF capacitor. What is the final charge on each capacitor? What is the final potential difference across the capacitors?

*26. A multiplate capacitor, such as used in radios, consists of four parallel plates arranged one above the other as shown in Fig. 26.30. The area of each plate is A, and the distance between adjacent plates is d. What is the capacitance of this arrangement?

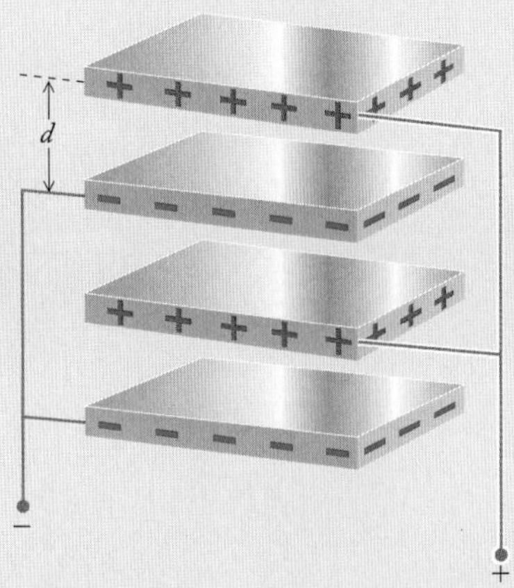

FIGURE 26.30 A multiplate capacitor.

*27. How can you connect four 1.0-μF capacitors so the net capacitance is 1.0 μF?

*28. Two capacitors, of 2.0 and 6.0 μF, respectively, are initially charged to 24 V by connecting each, for a few instants, to a 24-V battery. The battery is then removed and the charged capacitors are connected in a closed series circuit, the positive terminal of each capacitor being connected to the negative terminal of the other (Fig. 26.31). What is the final charge on each capacitor? What is the final potential difference across each?

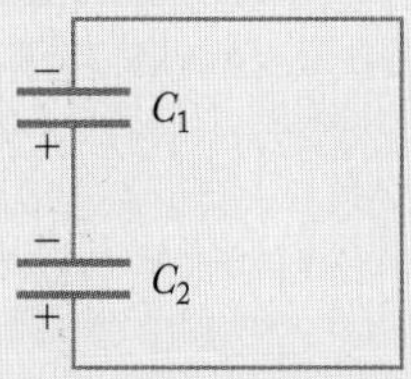

FIGURE 26.31 Two capacitors connected in series after they have been charged.

**29. Three capacitors, of capacitances $C_1 = 2.0\ \mu$F, $C_2 = 5.0\ \mu$F, and $C_3 = 7.0\ \mu$F, are initially charged to 36 V by connecting each, for a few instants, to a 36-V battery. The battery is then removed and the charged capacitors are connected in a closed series circuit, with the positive and negative terminals joined as shown in Fig. 26.32. What is the final charge on each capacitor? What is the voltage across the points PP' in Fig. 26.32?

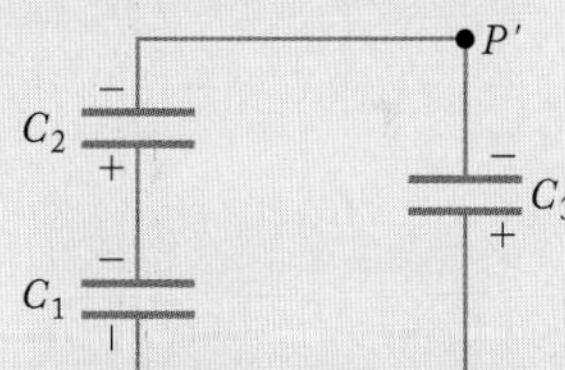

FIGURE 26.32 Three capacitors connected after they have been charged.

26.3 Dielectrics

30. You wish to construct a capacitor out of a sheet of polyethylene of thickness 5.0×10^{-2} mm and $\kappa = 2.3$ sandwiched between two aluminum sheets. If the capacitance is to be 3.0 μF, what must be the area of the sheets?

31. What is the capacitance of a sphere of radius R immersed in a large volume of gas of dielectric constant κ?

32. A parallel-plate capacitor has plates of area 0.050 m^2 separated by a distance of 0.20 mm. The space between the plates is filled with Plexiglas.

(a) What is the capacitance?

(b) What is the maximum potential difference that this capacitor can withstand? The maximum electric field that

Plexiglas can tolerate without electric breakdown is 40×10^6 V/m.

(c) What is the corresponding maximum amount of charge we can place on the plates?

33. High-dielectric-constant thin-film capacitors are attractive for digital memory applications; for example, when barium strontium titanate ($BaSrTi_2O_6$) is used as a 50-nm-thick dielectric material, a capacitance per unit area of 90 μF/cm^2 can be achieved. What is the dielectric constant?

*34. Polystyrene has a dielectric strength of 24×10^6 V/m and nylon has a dielectric strength of 14×10^6 V/m. We wish to make a 1.0-μF parallel-plate capacitor which can withstand 25 V. What is the smallest area for which this can be achieved for each material?

*35. In order to measure the dielectric constant of a dielectric material, a slab of this material 1.5 cm thick is slowly inserted between a pair of parallel conducting plates separated by a distance of 2.0 cm. Before insertion of the dielectric, the potential difference across these capacitor plates is 3.0×10^5 V. During insertion, the charge on the plates remains constant. After insertion, the potential difference is 1.8×10^5 V. What is the value of the dielectric constant?

*36. A parallel-plate capacitor of plate area A and spacing d is filled with two parallel slabs of dielectric of equal thickness with dielectric constants κ_1 and κ_2, respectively (Fig. 26.33). What is the capacitance? (Hint: Check that the configuration of Fig. 26.33 is equivalent to two capacitors in series.)

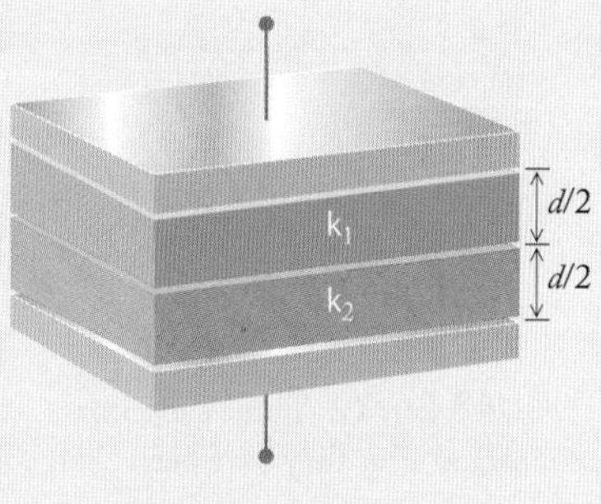

FIGURE 26.33 Parallel-plate capacitor with two slabs of dielectric.

*37. A capacitor with two large parallel plates of area A separated by a distance d is filled with two equal-size slabs of dielectric side by side (Fig. 26.34). The dielectric constants are κ_1 and κ_2. What is the capacitance?

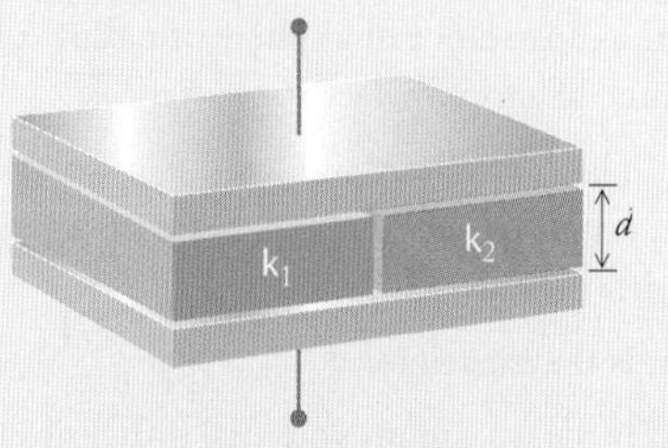

FIGURE 26.34 Parallel-plate capacitor with two slabs of dielectric side by side.

*38. A parallel-plate capacitor of plate area A and separation d contains a slab of dielectric of thickness $d/2$ and constant κ (see Fig. 26.35). What is the capacitance of this capacitor? (Hint: Regard this capacitor as two capacitors in series, one with dielectric, one without.)

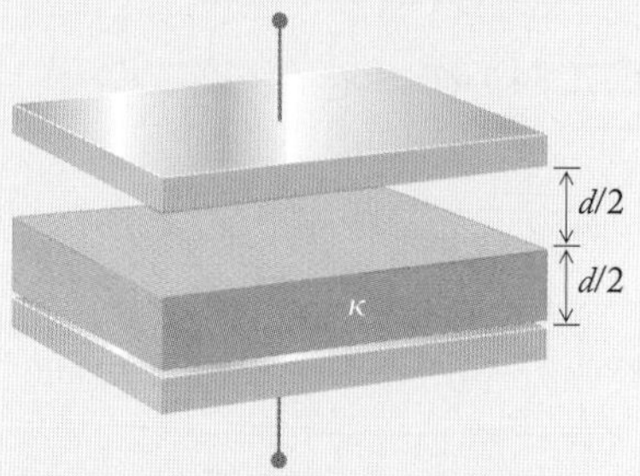

FIGURE 26.35 Parallel-plate capacitor, partially filled with dielectric.

*39. A parallel-plate capacitor of plate area A and separation d contains a slab of dielectric of thickness $d/2$ (Fig. 26.35) and dielectric constant κ. The potential difference between the plates is ΔV.

(a) In terms of the given quantities, find the electric field in the empty region of space between the plates.

(b) Find the electric field inside the dielectric.

(c) Find the density of bound charges on the surface of the dielectric.

*40. Within some limits, the difference between the dielectric constants of air and of vacuum is proportional to the pressure of the air, i.e., $\kappa - 1 \propto p$. Suppose that a parallel-plate capacitor is held at a constant potential difference by means of a battery. What will be the percentage change in the amount of charge on the plates as we increase the air pressure between the plates from 1.0 atm to 3.0 atm?

*41. A sensor for measuring liquid levels is made from a cylindrical capacitor (Fig. 26.36) of length $L = 50$ cm. The inner conductor has radius $a = 1.0$ mm and the outer conducting shell has radius $b = 4.0$ mm. If the sensor is used to detect the level of liquid nitrogen ($\kappa = 1.433$), what is its capacitance when it is (a) empty and (b) full?

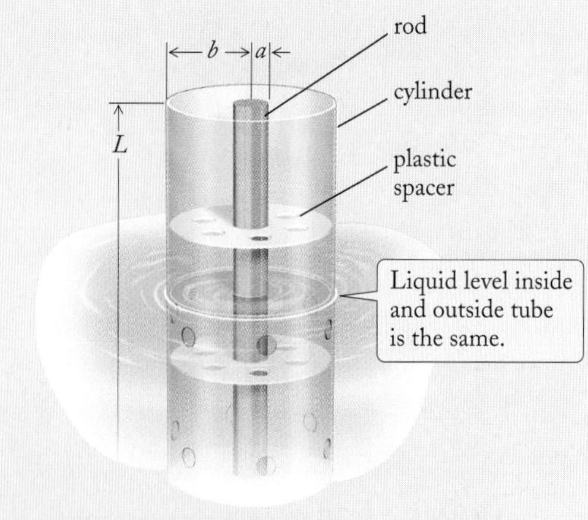

FIGURE 26.36 A capacitance liquid-level sensor.

*42. Two identical capacitors with $C_0 = 2.0\ \mu F$ are in series, initially empty, and connected to a potential difference of $\Delta V = 9.0$ volts (Fig. 26.37).

(a) What is the charge on each capacitor? What is the potential difference across each capacitor?

(b) One capacitor is then filled with a dielectric slab with $\kappa = 2.5$. What are the charge on each capacitor and the potential difference across each capacitor now?

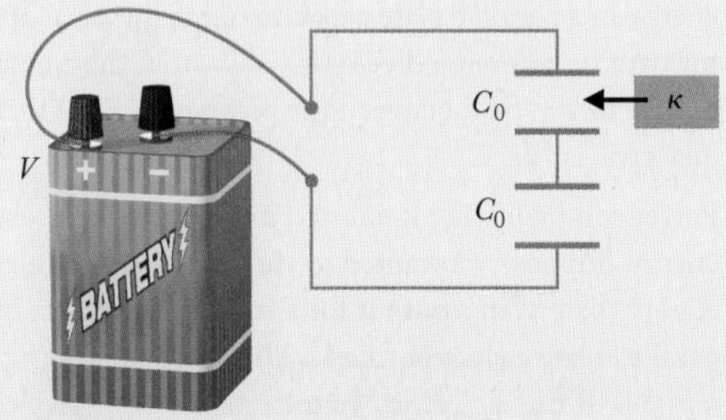

FIGURE 26.37 Two capacitors, connected to a battery.

*43. Two identical capacitors with $C_0 = 2.0\ \mu F$ are in series, initially empty, and are briefly connected to a potential difference of $\Delta V = 9.0$ volts. After charging, the capacitors are disconnected from the potential difference and electrically isolated. One capacitor is then filled with a dielectric slab with $\kappa = 2.5$. What are the charge on each capacitor and the potential difference across each capacitor now?

*44. A spherical capacitor consists of a metallic sphere of radius R_1 surrounded by a concentric thin metallic spherical shell of radius R_2. The space between R_1 and R_2 is filled with a dielectric having a constant κ. Suppose that the free surface-charge density on the metallic sphere at R_1 is $\sigma_{free(1)}$.

(a) What is the free surface-charge density on the metallic shell at R_2?

(b) What is the bound surface-charge density on the dielectric at R_1?

(c) What is the bound surface-charge density on the dielectric at R_2?

*45. A long cylindrical copper wire of radius 0.20 cm is surrounded by a cylindrical sheath of rubber of inner radius 0.20 cm and outer radius 0.30 cm. The rubber has $\kappa = 2.8$. Suppose that the surface of the copper has a free charge density of $4.0 \times 10^{-6}\ C/m^2$.

(a) What will be the bound charge density on the inside surface of the rubber sheath? On the outside surface?

(b) What will be the electric field in the rubber near its inner surface? Near its outer surface?

(c) What will be the electric field just outside the rubber sheath?

*46. A metallic sphere of radius R is surrounded by a concentric dielectric shell of inner radius R, outer radius $3R/2$. This is surrounded by a concentric thin metallic shell of radius $2R$ (Fig. 26.38). The dielectric constant of the dielectric shell is κ. What is the capacitance of this contraption?

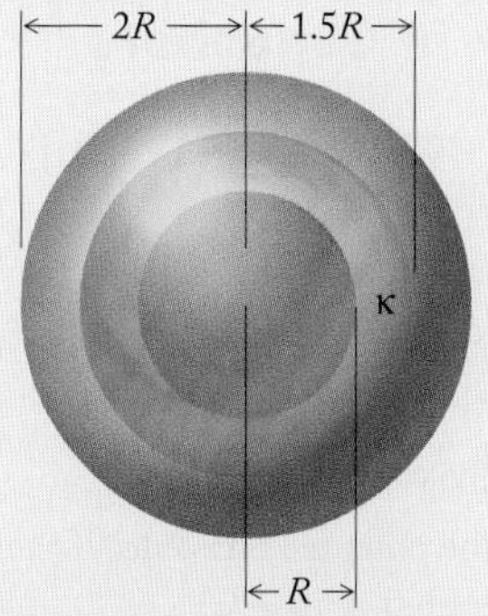

FIGURE 26.38 Spherical capacitor, partially filled with dielectric.

*47. Two small metallic spheres are submerged in a large volume of transformer oil of dielectric constant $\kappa = 3.0$. The spheres carry electric charges of 2.0×10^{-6} C and 3.0×10^{-6} C, respectively, and the distance between them is 0.60 m. What is the force on each?

**48. A hollow sphere of brass floats in a large lake of oil of dielectric constant $\kappa = 3.0$. The sphere is exactly halfway immersed in the oil (Fig. 26.39). The sphere has a net charge of 2.0×10^{-6} C. What fraction of this electric charge will be on the upper hemisphere? On the lower? (Hint: The electric fields in the oil and in the air above the oil are exactly the same.)

FIGURE 26.39 Brass sphere afloat in a lake of oil.

**49. A spherical capacitor consists of two concentric spherical shells of metal of radii R_1 and R_2. The space between these shells is filled with two kinds of dielectric (Fig. 26.40); the dielectric in the upper hemisphere has a constant κ_1, and the dielectric in the lower hemisphere has a constant κ_2. What is the capacitance of this device? (Hint: The electric fields in both dielectrics are exactly the same.)

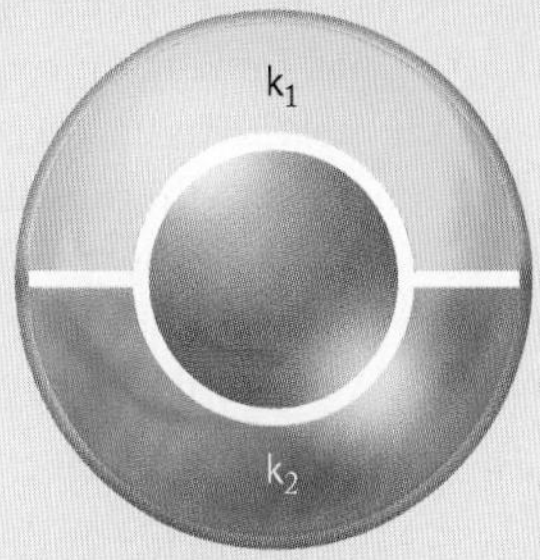

FIGURE 26.40 Spherical capacitor with two kinds of dielectric.

26.4 Energy in Capacitors

50. How much energy is stored in a $3.0 \times 10^3\ \mu F$ capacitor charged to 100 volts?

51. A parallel-plate capacitor has a plate area of 900 cm^2 and a plate separation of 0.50 cm.

    The space between the plates is empty.

    (a) What is the capacitance?

    (b) What is the potential difference if the charges on the plates are $\pm 6.0 \times 10^{-8}$ C?

    (c) What is the electric field between the plates?

    (d) The energy density?

    (e) The total energy?

52. Repeat Problem 51, under the assumption that the space between the plates is filled with Plexiglas.

53. A TV receiver contains a capacitor of 10 μF charged to a potential difference of 2.0×10^4 V. What is the amount of charge stored in this capacitor? The amount of energy?

54. Two parallel conducting plates of area 0.50 m^2 placed in a vacuum have a potential difference of 2.0×10^5 V when charges of $\pm 4.0 \times 10^{-3}$ C are placed on them, respectively.

    (a) What is the capacitance of the pair of plates?

    (b) What is the distance between them?

    (c) What is the electric field between them?

    (d) What is the electric energy?

55. A large capacitor has a capacitance of 20 μF. If you want to store an electric energy of 40 J in this capacitor, what potential difference do you need?

56. An isolated, charged parallel-plate capacitor has plate separation d. If the plates are pulled apart to a separation $3d$, does the stored energy increase or decrease? By what factor?

57. A "supercapacitor" has a huge capacitance of 6.8 F, but can withstand a potential difference of only 2.5 V. A power-supply capacitor has a capacitance of 820 μF and can operate up to 400 V. Which can store more charge? Which can store more energy?

*58. Two identical, empty capacitors are connected in series; the combination is permanently connected to a potential source. When one of the capacitors is filled with a material of dielectric constant κ, does the total stored energy increase or decrease? By what factor?

*59. A parallel-plate capacitor has plates of area 0.040 m^2 separated by a distance of 0.50 mm. Initially, the space between the plates is empty and the capacitor is uncharged. A 12-V battery is available for connection to the capacitor. For each of the following cases, find the resulting capacitance C, potential difference ΔV, and charge Q on the plates; for cases (a) and (b), also find the energy delivered by the battery:

    (a) The battery is connected to the capacitor.

    (b) While the battery remains connected, a slab of dielectric with $\kappa = 3.0$ is inserted between the plates, completely filling the space between them.

    (c) The battery is disconnected, and the dielectric is then pulled out from the capacitor.

    (d) The capacitor is discharged, and the battery is again connected to the empty capacitor for some moments. The battery is then disconnected, and the dielectric is then inserted between the plates of the capacitor.

*60. Two capacitors of 5.0 μF and 8.0 μF are connected in series to a 24-V battery. What is the energy stored in the capacitors?

61. Starting from the general expression $\frac{1}{2}Q\,\Delta V$ for the electric energy in a parallel-plate capacitor (equally valid for a capacitor with or without a dielectric), show that the energy density in the electric field between the plates of a capacitor with a dielectric is $\frac{1}{2}\kappa\epsilon_0 E^2$.

*62. Power companies are interested in the storage of surplus energy. Suppose we wanted to store 10^6 kW·h of electric energy (half a day's output for a large power plant) in a large parallel-plate capacitor filled with a plastic dielectric with $\kappa = 3.0$. If the dielectric can tolerate a maximum electric field of 5.0×10^7 V/m, what is the minimum total volume of dielectric needed to store this energy?

*63. Three capacitors are connected as shown in Fig. 26.41. Their capacitances are $C_1 = 2.0\ \mu F$, $C_2 = 6.0\ \mu F$, and $C_3 = 8.0\ \mu F$. If a voltage of 200 V is applied to the two free terminals, what will be the charge on each capacitor? What will be the energy in each?

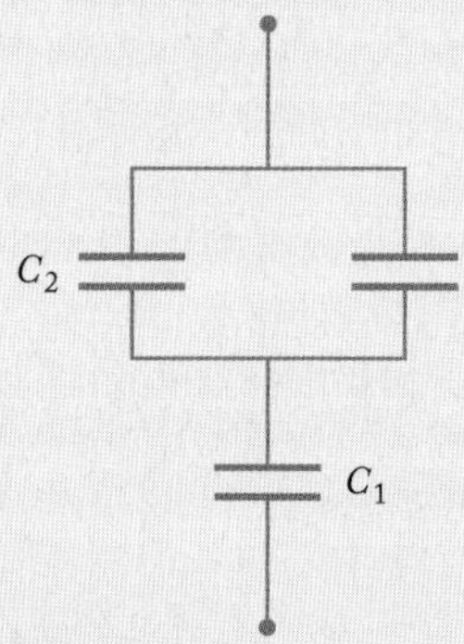

FIGURE 26.41 Three capacitors.

*64. Ten identical 5.0-μF capacitors are connected in parallel to a 240-V battery. The charged capacitors are then disconnected from the battery and reconnected in series, the positive terminal of each capacitor being connected to the negative terminal of the next. What is the potential difference between the negative terminal of the first capacitor and the positive terminal of the last capacitor? If these two terminals are connected via an external circuit, how much charge will flow around this circuit as the series arrangement discharges? How much energy will be released in the discharge? Compare this charge and this energy with the charge and energy stored in the original, parallel arrangement, and explain any discrepancies.

**65. A large dielectric slab partially fills an isolated parallel-plate capacitor with charges $\pm 2.0\ \mu C$. The slab thickness is equal to the plate separation. The parallel plates are squares with 10-cm sides and 1.0 mm separation, and the dielectric constant is 2.5. Find the value of the force on the dielectric when the capacitor is half full.

REVIEW PROBLEMS

66. A parallel-plate capacitor consists of two square conducting plates of area 0.040 m^2 separated by a distance of 0.20 mm. The plates are connected to the terminals of a 12-V battery.
    (a) What is the charge on each plate? What is the electric field between the plates?
    (b) If you move the plates apart, so their separation becomes 0.30 mm, how much charge will flow from each plate to the terminals of the battery? What will be the new electric field?

67. Three capacitors are connected as shown in Fig. 26.42. Their capacitances are $C_1 = 4.0\ \mu F$, $C_2 = 6.0\ \mu F$, and $C_3 = 3.0\ \mu F$. If a voltage of 400 V is applied to the two free terminals, what will be the charge on each capacitor? What will be the potential energy of each?

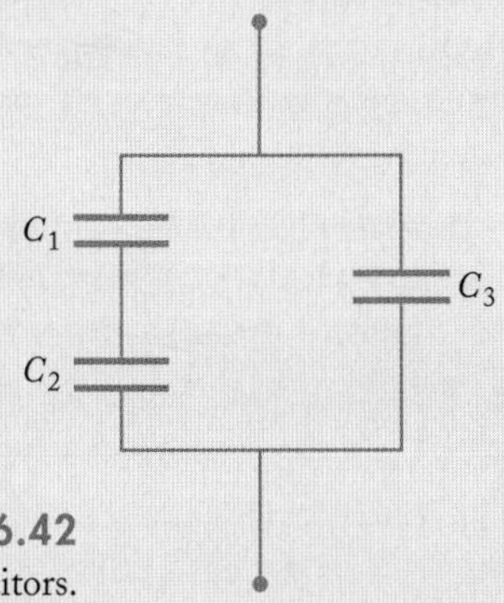

FIGURE 26.42 Three capacitors.

68. You have three capacitors of 1.0 μF, 2.0 μF, and 3.0 μF. If you connect these three capacitors in series, in parallel, or in some combination of series and parallel, what different net capacitances can you construct?

69. Figure 26.43 shows five capacitors of 4.0 μF each connected together.
    (a) Find the net capacitance of this combination between the terminals A and A'.
    (b) Find the net capacitance of this combination between the terminals B and B'.

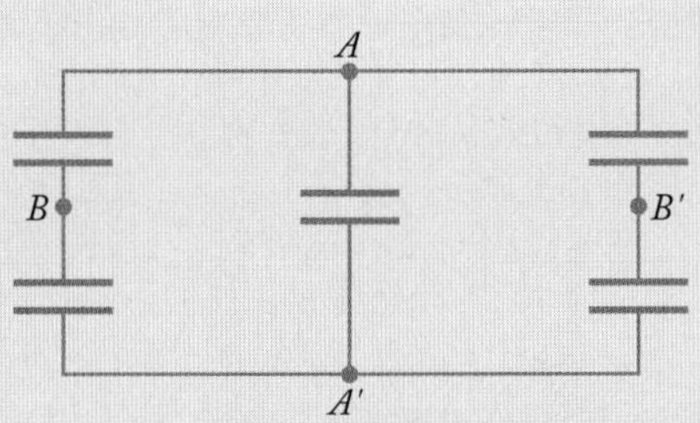

FIGURE 26.43 Five capacitors.

70. A capacitor of 5.0 μF is initially charged by connecting it briefly to a 40-V battery, and a capacitor of 8.0 μF is initially charged by connecting it briefly to a 60-V battery. The batteries are then removed, and the two capacitors are connected in parallel (see Fig. 26.44). What is the final charge on each capacitor? What is the final potential difference across each?

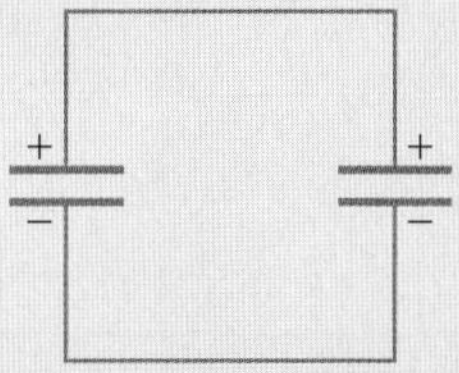

FIGURE 26.44 Two capacitors connected in parallel after they have been charged.

71. The plates of a parallel-plate capacitor are movable and are initially separated by an air gap of width d. A piece of dielectric of dielectric constant κ and a thickness of $3d$ is inserted between the plates. If the ratio of the capacitance before inserting the dielectric to the capacitance after inserting the dielectric is 1.5, what is the value of κ?

72. A parallel-plate capacitor has a capacitance of 25 μF when filled with air and it can withstand a potential difference of 50 V before it suffers electrical breakdown.
    (a) What is the maximum amount of charge we can place on this air-filled capacitor? The dielectric strength of air is 3.0×10^6 V/m.
    (b) If we fill this capacitor with polyethylene, what will be its new capacitance?
    (c) What will be the maximum potential difference that this new capacitor can withstand? What will be the corresponding maximum amount of charge we can place on this capacitor? The dielectric strength of polyethylene is 18×10^6 V/m.

73. A capacitor used to power a strobe light has a capacitance of 200 μF and it is charged to a potential difference of 360 V.
    (a) What is the energy stored in this capacitor?
    (b) Suppose that the dielectric in the capacitor has a dielectric constant of 2.2 and can withstand a maximum electric field of 70×10^6 V/m (the dielectric strength). What is the maximum permitted energy density in the dielectric? What is the minimum volume that the dielectric must have if it is to store the energy calculated in part (a)?

74. A primitive capacitor, originally called a **Leyden jar**, consists of a glass bottle filled with water and wrapped on the outside with metal foil (see Fig. 26.45). The foil plays the role of one plate of the capacitor, and the surface of the water facing the foil plays the role of the other plate; these "plates" are separated by the layer of glass (dielectric). Suppose that, as shown in Fig. 26.45, the wrapped portion of the bottle is

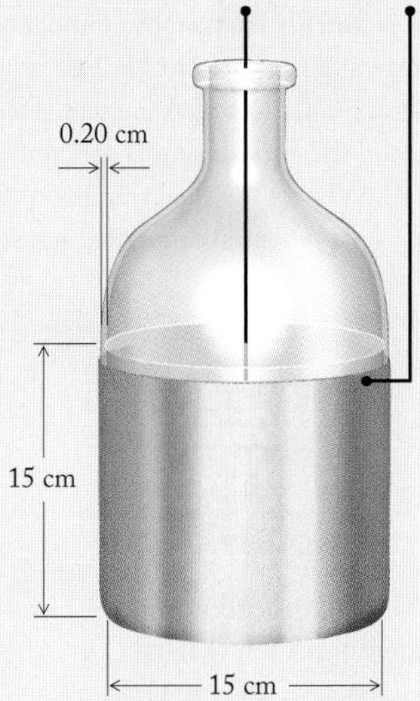

FIGURE 26.45 A Leyden jar.

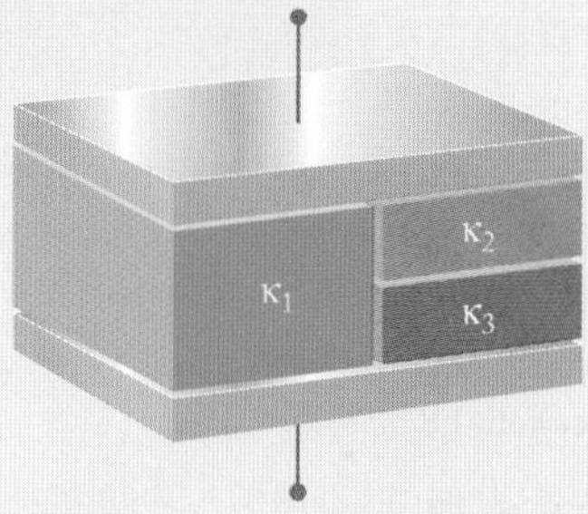

FIGURE 26.46 Parallel-plate capacitor filled with three dielectrics.

cylindrical, of height 15 cm and diameter 15.0 cm (the bottom is not wrapped). The thickness of the glass is 0.20 cm and the dielectric constant is 6.0. Find the capacitance of this contraption (a) by treating it as a cylindrical capacitor and (b) by treating it as a parallel-plate capacitor.

75. Two capacitors are connected in series, and the combination is connected to a 9.0-V battery. One capacitor has a capacitance of 4.0 μF, and the potential across the other is 3.0 V. What is the capacitance of the other capacitor? What is the energy stored in each capacitor?

76. Two identical, empty, isolated capacitors are connected in series; each carries the same charge. If one of the capacitors is then filled with a material of dielectric constant κ, by what factor does the total stored energy decrease?

77. A cylindrical capacitor has an inner conductor of radius a and a coaxial outer conducting shell of radius b. The region between the conductors is filled with two cylindrical shells of dielectric, one of constant κ_1 for $a < r < c$ and another of constant κ_2 for $c < r < b$. What is the capacitance per unit length of this layered cylindrical capacitor?

78. A parallel-plate capacitor has capacitance C_0 when empty. Three slabs, each with area equal to half the plate area, are inserted between the plates as shown in Fig. 26.46: one slab with dielectric constant κ_1 has thickness equal to the plate separation, and the other two slabs with dielectric constants κ_2 and κ_3 have half that thickness. What is the capacitance of the filled capacitor?

*79. A parallel-plate capacitor is filled with carbon dioxide at 1.0 atm pressure. Under these conditions the capacitance is 0.50 μF. We charge the capacitor by means of a 48-V battery and then disconnect the battery so that the electric charge remains constant thereafter. What will be the change in the potential difference if we now pump the carbon dioxide out of the capacitor, leaving it empty?

*80. A spherical capacitor consists of a metallic sphere of radius R_1 surrounded by a concentric metallic shell of radius R_2. The space between R_1 and R_2 is filled with dielectric having a constant κ.

(a) If the free charge on the surface of the inner sphere is Q and that on the outer spherical shell is $-Q$, what is the potential difference between the two conductors?

(b) What is the capacitance of this spherical capacitor?

*81. A parallel-plate capacitor without dielectric has an area A and a charge $\pm Q$ on each plate.

(a) What is the electric force F of attraction between the plates? (Hint: One-half of the electric field between the plates is due to one plate, and one-half is due to the other. Consequently, in the calculation of the electric force on a plate from the product of field times charge, only one-half of the field must be used; the other half gives the electric force of the plate on *itself* and is of no interest.)

(b) How much work must you do against this force in order to increase the plate separation by an amount Δd?

(c) By means of Eq. (26.29), calculate the change ΔU in potential energy during this change.

(d) By comparing (a) and (c), check that $F = -\Delta U/\Delta d$.

Answers to Checkups

Checkup 26.1

1. Yes for all three—as long as the plate separation is small compared with the lateral dimensions, fringing fields will be small, and the particular shape determining the area A in Eq. (26.9) does not matter.
2. A smaller capacitance means it is more difficult to store the same amount of charge, so it is the smaller, 1-pF capacitor that requires the larger potential difference.
3. Increasing the plate separation decreases the capacitance [see Eq. (26.9)]. Increasing the area enables proportionally more charge to be stored, increasing the capacitance [Eq. (26.9)].
4. No. With a single sign of charge, the pair of plates more closely resembles a single isolated conductor, like a charged sphere. We would expect the capacitance to be similar to Eq. (26.3), where the sphere radius R would be replaced by a length on the order of the plate dimension.
5. Holding the charge constant means holding the electric field constant ($E = \sigma/\epsilon_0 = Q/\epsilon_0 A$). The potential decreases a factor of 2 ($\Delta V = Ed$), and the capacitance increases a factor of 2 ($C = Q/\Delta V = \epsilon_0 A/d$).
6. (E) 0.50 mm. Since the area is one-fourth of the original, for a constant capacitance ($C = \epsilon_0 A/d$), the separation needs to decrease by the same factor to 0.50 mm.

Checkup 26.2

1. Not necessarily in either case. For capacitors with the same charge (in series), the electric fields will be the same only if they have the same area ($E = Q/\epsilon_0 A$). For capacitors with the same potential (in parallel), the electric field will be the same only if they have the same plate separation ($E = \Delta V/d$).
2. For a given potential across them, capacitors in parallel have a greater area and so can store a greater amount of charge—thus a larger capacitance. For capacitors in series, a larger total potential difference is required to have a voltage drop across each capacitor sufficient to store a given charge, so the capacitance is smaller.
3. (B) 10 μF; 0.10 μF. Capacitors in parallel add, so ten 1.0-μF capacitors give a net capacitance of 10 μF. Capacitors in series add inverses, so the equivalent capacitance is $1/[10 \times (1/1.0\ \mu\text{F})] = 0.10\ \mu\text{F}$.

Checkup 26.3

1. The electric dipoles in the dielectric tend to align in the electric field generated by the free charges on the conductors, producing a field that partially cancels the original field. A smaller field is what led to our definition (26.19), which requires $\kappa > 1$.
2. (C) 9.0×10^{-12} F. Since $C = \kappa C_0$, where C_0 is the capacitance in the absence of dielectric, we simply have $C = 3.0 \times 3.0 \times 10^{-12}\ \text{F} = 9.0 \times 10^{-12}$ F.

Checkup 26.4

1. Since a dielectric always increases the capacitance according to $C = \kappa C_0$, and since $Q = C\,\Delta V$, at constant potential the capacitor with the dielectric will have the larger stored charge. The electric fields will be the same, since the potential difference and spacing are the same ($\Delta V = Ed$). Since the electric fields are the same, the energy density, $u = \kappa\epsilon_0 E^2/2$, will be greater for the capacitor with dielectric. Also, $U = C(\Delta V)^2/2$, so the one with dielectric has the larger stored energy.
2. Again, the dielectric always increases the capacitance, because $C = \kappa C_0$. So $\Delta V = Q/C$ implies that the empty capacitor will require the larger voltage to store the same charge (larger by a factor of κ). For an identical geometry, $E = \Delta V/d$ requires that the empty capacitor also have the larger electric field (larger by the same factor of κ). For the energy density, $u = \kappa\epsilon_0 E^2/2$; thus the larger electric field in the capacitor without dielectric results in a larger energy density. Finally, $U = Q^2/2C$ tells us that for the same charge, more energy is stored in the capacitor without dielectric.
3. (C) C_0. For capacitors in series, we add inverses to obtain the inverse of the net capacitance [see Eq. (26.17)]. For two identical empty capacitors, the net capacitance is thus $\frac{1}{2}C_0$. When filled, the net capacitance increases by a factor of κ, so $C = 2.0 \times \frac{1}{2}C_0 = C_0$.

CHAPTER

27 Currents and Ohm's Law

CONCEPTS IN CONTEXT

Concepts — in — Context

The electricity we use to power our appliances and electronic devices involves the motion of charge, or current, which flows to our homes along a power-line cable, through the wires and devices in our homes, and back out again. Our appliances, and even ordinary wires and cables, offer an opposition to the flow of charge known as resistance. Special wires and cables called superconductors offer no resistance to the flow of charge.

As we learn about current and resistance in this chapter, we can consider such questions as:

? For a known current, how much total charge flows in a given amount of time? (Examples 1 and 2, pages 860–861)

? How is the resistance of a piece of cable measured? (Example 3, page 867)

? How does the resistance of a piece of wire change with temperature? (Section 27.3, page 877 and Example 6, page 870)

? How do we decide if it is safe to plug several appliances into the same outlet? (Example 10, page 877)

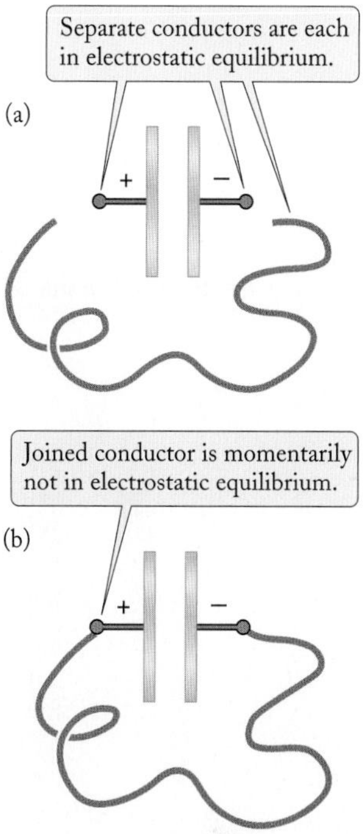

FIGURE 27.1 (a) A charged capacitor and a separate, uncharged piece of copper wire. (b) We suddenly connect the ends of the wire to the capacitor plates. The wire and the plates now form a single conductor, with opposite amounts of charge on its ends.

Under static conditions, no electric field can exist inside a conductor. But suppose that we suddenly deposit opposite amounts of electric charge on the opposite ends of a long metallic conductor, such as a copper wire connected to the opposite plates of a capacitor (see Fig. 27.1). The conductor will then not be in electrostatic equilibrium, and the charges will generate an electric field along and inside the conductor (see Figs. 27.2 and 27.3). This electric field propels the charges toward each other. When the charges meet, they cancel, and the electric field then disappears—the conductor reaches equilibrium.

For a good conductor, such as copper, the approach to equilibrium is fairly rapid: typically, the time required to achieve equilibrium is a very small fraction of a second. However, we can permanently keep a conductor from achieving equilibrium if we continually supply more electric charge to its ends. For example, we can connect the two ends of a copper wire to the terminals of a battery or an electric generator. The terminals of such a device act as a source and sink of electric charge, just as the outlet and the intake of a pump act as source and sink of water. Under these conditions electric charge will continually flow from one terminal to the other. Such a flow of electric charge is called an electric current. In this chapter, we will examine some general properties of electric currents, and in the next chapter we will discuss the inner workings of batteries, generators, and other "pumps" that serve to maintain electric currents.

27.1 ELECTRIC CURRENT

When a wire is connected between the two terminals of a battery or generator, the electric charges are propelled from one end of the wire to the other by the electric field that exists along and within the wire. Most of the field lines originate at the terminals of the battery or generator, but some field lines originate at charges that have accumulated on the wire itself. As Figs. 27.2 and 27.3 show, the field lines tend to concentrate within the conductor, and they tend to follow the conductor. If the conductor has no sharp kinks, the field lines are uniformly distributed over the cross-sectional

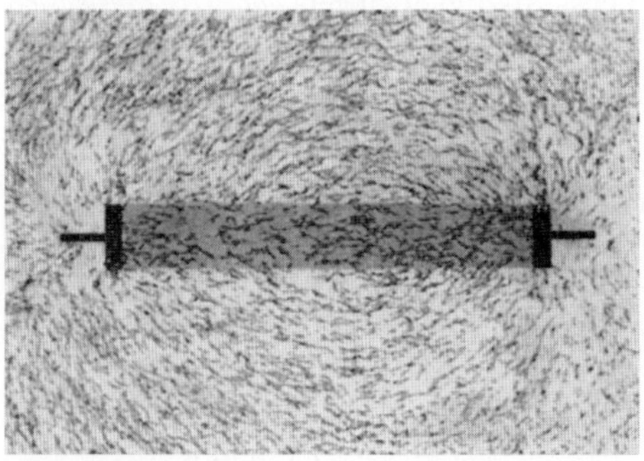

FIGURE 27.2 Electric field lines in and near a straight conductor not in equilibrium. The conductor consists of a strip of metallic paint on a paper surface. The field lines have been made visible by sprinkling grass seeds on the paper.

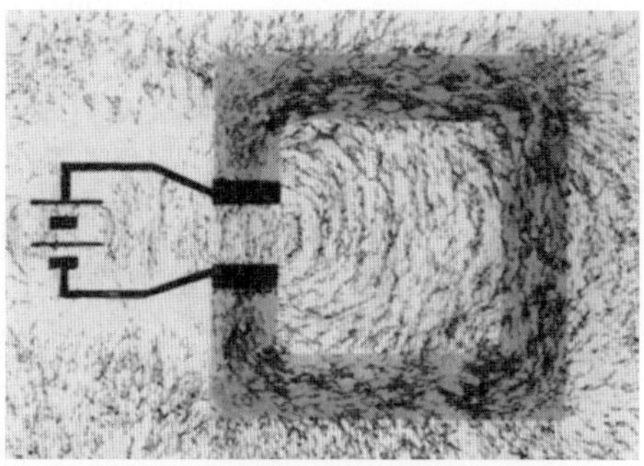

FIGURE 27.3 Electric field lines in and near a rectangular conductor carrying an electric current.

area of the conductor. For instance, if the conductor is a more or less straight wire of constant thickness, then the electric field inside the wire will be of constant magnitude and have a direction parallel to the wire. If the length of the wire is l and if the battery or generator maintains a difference of potential ΔV across its ends, then this constant electric field in the wire has a magnitude [compare Eq. (25.7)]

electric field in the wire

$$E = \frac{\Delta V}{l} \tag{27.1}$$

This electric field causes the flow of charge, or **electric current**, from one end of the wire to the other. Before we can explore the dependence of the current on the electric field, we need a precise definition of the current. Suppose that in a time Δt an amount of charge ΔQ flows past some given place on the wire (for instance, the end of the wire). *The electric current is defined as charge divided by time*:

$$I = \frac{\Delta Q}{\Delta t} \tag{27.2}$$

If the flow is not steady, the current is defined in terms of the small amount of charge dq that flows past a place in a small interval of time dt, that is, it is the instantaneous rate

electric current

$$I = \frac{dq}{dt} \tag{27.3}$$

Note that if the sides of the wire do not leak (good insulation), then the conservation of electric charge requires that *the current is the same everywhere* along the wire; that is, the current is simply the rate at which charge enters the wire at one end or the rate at which charge leaves at the other end.

The SI unit of current is the **ampere** (A); this is a flow of charge of one coulomb per second:

ampere (A)

$$1 \text{ ampere} = 1 \text{ A} = 1 \text{ C/s} \tag{27.4}$$

Table 27.1 (page 862) gives the values of some selected currents.

In metallic conductors, the charge carriers are electrons—a current in a metal is nothing but a flow of electrons. In electrolytes, such as salt water, the charge carriers are positive ions, negative ions, or both—a current in such a conductor is a flow of ions. For the sake of mathematical uniformity, whenever there is need to indicate the direction of the current along a conductor, physicists and engineers *adopt the convention that the current has the direction of a hypothetical flow of positive charge.* This means that we pretend that the moving charges are always positive charges. Of course, in metals the moving charges are actually negative charges (electrons), and hence the above convention assigns to the current a direction opposite to that of the true motion of the charges. However, as regards the transfer of charge, the transport of negative charge in one direction is equivalent to the transfer of positive charge in the opposite direction (see Fig. 27.4). The convention for labeling the direction of the current as the direction of the flow of positive charge takes advantage of this equivalence.

Concepts — in — Context

EXAMPLE 1 In the headlamp of an automobile, a current of 8.0 A flows through the filament of the lightbulb. How much electric charge flows through the filament in a minute? How many electrons?

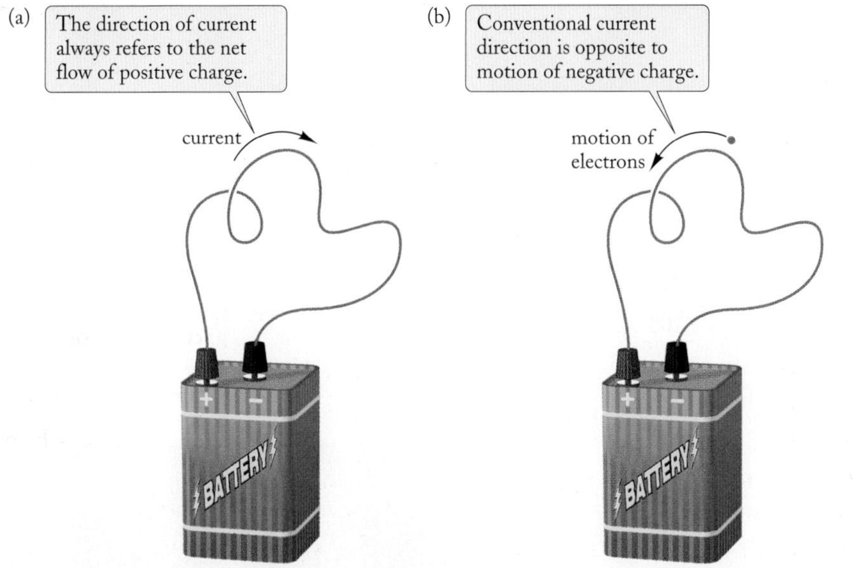

FIGURE 27.4 Flow of charge in a wire connected to a battery. (a) The motion of hypothetical positive charges from the positive terminal to the negative terminal. (b) The actual motion of the electrons in the wire from the negative terminal to the positive. The motions illustrated in (a) and (b) are equivalent in that both transfer positive charge to the negative terminal and negative charge to the positive terminal. According to our convention, the direction of the current is assumed to be from the positive terminal to the negative, even though the actual motion of the electrons is opposite to this.

SOLUTION: The amount of charge that flows through the filament in 1.0 min is, from Eq. (27.2),

$$\Delta Q = I\,\Delta t = 8.0\text{ A} \times 1.0\text{ min} = 8.0\text{ A} \times 60\text{ s} = 480\text{ C}$$

A charge of 480 C moving through the light bulb in one direction is equivalent to an electron charge of −480 C moving in the opposite direction. The number of electrons equals the total charge divided by the charge per electron:

$$[\text{number of electrons}] = \frac{-480\text{ C}}{-e} = \frac{-480\text{ C}}{-1.6 \times 10^{-19}\text{ C}} = 3.0 \times 10^{21}\text{ electrons}$$

EXAMPLE 2 Superconductors can carry large currents without opposition, but to prevent damage, these currents must be changed gradually. For example, a power supply for a superconducting magnet provides a current which starts from zero at time $t = 0$ and steadily increases with time at a gradual rate of 0.080 A/s. How long does it take the current to reach a value of 60 A? At the instant when the current reaches 60 A, how much total charge has flowed through the magnet?

Concepts in Context

SOLUTION: Since the constant rate of increase of current is known, the time to reach 60 A can be obtained by dividing the final value by the rate:

$$\Delta t = \frac{\Delta I}{\Delta I/\Delta t} = \frac{60\text{ A}}{0.080\text{ A/s}} = 750\text{ s}$$

The charge may be obtained by rearranging Eq. (27.3) to solve for dq,

$$dq = I\,dt$$

and then integrating the changing contributions dq to get the total charge q that has flowed:

$$q = \int dq = \int_0^{\Delta t} I\,dt \qquad (27.5)$$

TABLE 27.1 SOME CURRENTS

Lightning stroke (a)	10^4 A
High-tension power line (b)	10^3
Large transformer (c)	10^3
Large electromagnet	200
Starter motor of automobile (d)	100
Alternator of automobile	30
Fuse blows	30
Defibrillation treatment for heart	20
Air conditioner	12
Hair dryer	10
Ordinary lightbulb	1
Flashlight bulb	0.5
Lethal fibrillation of heart	0.1
Barely perceptible by skin	1×10^{-3}
Electronic calculator (e)	1×10^{-4}
Scanning tunneling microscope	1×10^{-12}

(a)

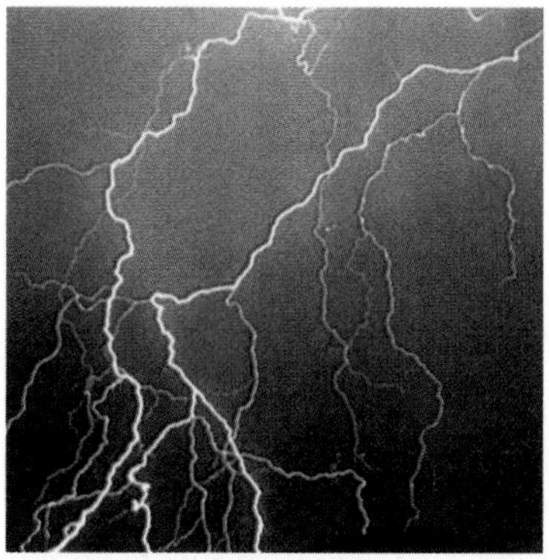

(b)

(c)

(d)

(e)

For a constant rate of change, the current I is simply the rate times the time t, $I = (0.080\ \text{A/s}) \times t$, so

$$q = \int_0^{\Delta t} (0.080\ \text{A/s}) \times t\, dt = (0.080\ \text{A/s})\left(\frac{t^2}{2}\right)\Bigg|_0^{\Delta t}$$

$$= (0.080\ \text{A/s})\left[\frac{(\Delta t)^2}{2} - 0\right] = (0.040\ \text{A/s})(\Delta t)^2$$

Evaluating this expression at $\Delta t = 750$ s, we obtain

$$q = 0.040\ \text{A/s} \times (750\ \text{s})^2 = 22\,500\ \text{C} = 2.3 \times 10^4\ \text{C}$$

Checkup 27.1

QUESTION 1: The flow of an electric current in a wire is analogous to the flow of water in a pipe. However, if we cut the pipe off, the water continues to flow, and it spills out of the pipe. Do electrons continue to flow when we cut a wire carrying a current? Do they spill out? Why or why not?

QUESTION 2: Suppose that a wire has a nonuniform cross section (thicker in some parts than in others). If an electric current of 5 A enters the wire at one end, is the current 5 A everywhere along this wire, even where the wire is thick?

QUESTION 3: In a lightning bolt (see Fig. 22.13) the electric current flows in the upward direction, from the ground to the thundercloud. What is the direction of the electric field in this lightning bolt? What is the direction of the motion of the electrons making up this current?

QUESTION 4: Suppose some currents consist of both positive and negative moving elementary charges, as shown in Fig. 27.5. Which of the currents has the largest magnitude? The smallest magnitude? (Assume all speeds are equal.)

(A) (a), (b) (B) (a), (c) (C) (b), (c) (D) (c), (a) (E) (c), (b)

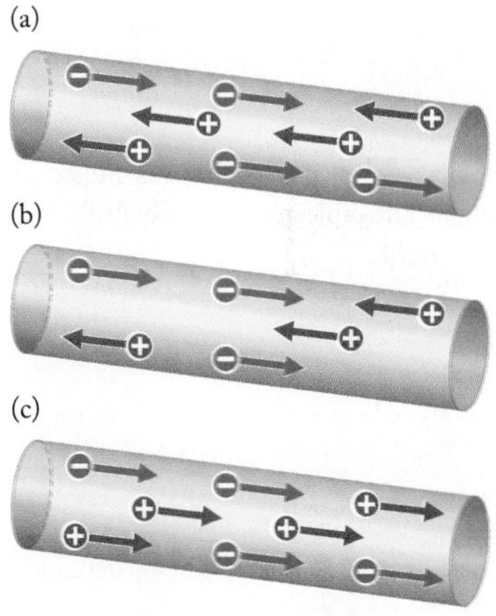

FIGURE 27.5 Several motions of positive and negative charges. All charges have the same magnitude and the same speed.

27.2 RESISTANCE AND OHM'S LAW

We will now examine in detail the behavior of a current in a metallic conductor. Such conductors contain a vast number of free electrons, which are not bound to any atom but are free to roam inside the entire volume of the metal; for example, copper has about 8×10^{22} free electrons per cubic centimeter. These electrons behave like particles of a gas, which are free to roam inside a container. By analogy, we say that the electrons form an **electron gas**, which fills the entire volume of the metal. Of course, in an electrically neutral conductor, the negative charge of the free electrons is exactly balanced by the positive charge of the ions that make up the crystal lattice of the metal. A current in this metallic conductor is simply a flow of the gas of electrons, while the ions remain at rest.

electron gas

The flow of the gas of electrons along a metallic wire is analogous to the flow of water along a canal leading down a gentle slope. In such a canal, the force of gravity acting on the water has a component along the canal; this component pushes the water along. But the water does not accelerate—the friction between the water and the walls of the canal opposes the motion, and the water moves at a constant speed because the friction exactly matches the push of gravity.

Likewise, the electric field in the wire pushes the gas of electrons along. But the gas of electrons does not accelerate—friction between the gas and the crystal lattice of the wire opposes the motion, and the gas moves at a constant speed because such friction exactly matches the push of the electric field.

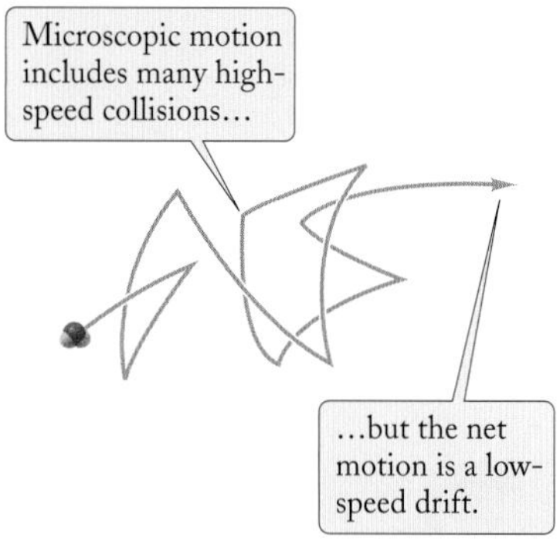

FIGURE 27.6 Path of a water molecule in a canal. The molecule gradually drifts from left to right.

The analogy between the motion of water and the motion of the electron gas extends to the motion of the individual water molecules and individual electrons. Although the water in a canal usually has a fairly low speed, perhaps a few meters per second, the individual molecules within the water have a rather high speed—the typical speed of the random thermal motion of water molecules is about 600 m/s at ordinary temperature. But since this thermal motion consists of rapid random zigzags, which are just as likely to move the molecule backward as forward, this high speed does not contribute to the net downhill motion of water. Figure 27.6 shows the motion of a water molecule in the canal; on a microscopic scale, this motion consists of rapid zigzags on which is superimposed a much slower "drift" along the canal.

Likewise, the electron gas moves along the wire at a rather low speed, perhaps 10^{-4} m/s, but the individual electrons have a much higher speed—the typical speed of the random motion of electrons in a metal is about 10^6 m/s (this very high speed is due to quantum-mechanical effects, which we discuss in Chapter 39). Thus, the net motion of an electron in a metal also consists of rapid zigzags on which is superimposed a much slower drift motion along the wire. Qualitatively, the motion resembles the path of a water molecule shown in Fig. 27.6, but the amount of drift per zigzag is even less than shown in this figure.

The friction between the electron gas and the wire is caused by collisions between the electrons and the ions of the crystal lattice of the wire. An electron moving through a wire of copper will suffer about 10^{14} collisions with ions per second. These collisions change the direction of motion of the electron and produce the zigzags. Because of the disturbing effects of these collisions, the electron never gains much velocity from the electric field that is attempting to accelerate it, and the average drift velocity of the electron along the wire remains low. The collisions dissipate the kinetic energy that the electron acquires from the electric field. This dissipated kinetic energy of the electrons remains in the crystal lattice in the form of random kinetic and potential energy of the ions; that is, it remains in the form of thermal energy or heat. In some instances, the amount of heat produced in a wire is so large as to make the wire glow. The bright glow of a lightbulb is produced in this way, and so is the dull red glow of the coils of an electric range.

We can find the average motion of an electron by examining the losses and gains of momentum of this electron. If the average velocity, or **drift velocity**, of an electron is v_d, then the average momentum is $p = m_e v_d$. We expect that, on the average, a collision will absorb all of this momentum, that is, a collision will destroy the forward drift velocity and leave the electron with only random thermal motion. This means that, in the average time interval τ (tau) between collisions, the electron loses a momentum $m_e v_d$. The averate rate at which the electron loses momentum in collisions is therefore

$$\left(\frac{\Delta p}{\Delta t}\right)_{\text{loss}} = \frac{m_e v_d}{\tau} \tag{27.6}$$

On the other hand, the rate at which the electron gains momentum by the action of the electric force is

$$\left(\frac{\Delta p}{\Delta t}\right)_{\text{gain}} = -eE \tag{27.7}$$

Under steady-state conditions, the rate of loss of momentum must match the rate of gain. By setting the right sides of Eqs. (27.6) and (27.7) equal, we immediately obtain

$$v_d = \frac{-eE\tau}{m_e} \tag{27.8}$$

drift velocity (average velocity)

This is the average velocity with which the electron gas flows along the wire. This relation merely reflects the fact that if the electric field is stronger, the electron gains more velocity between one collision and the next, and therefore attains a proportionally larger average velocity. Similarly, if the time between collisions is longer, a larger average velocity is attained.

The electric current carried by the wire is proportional to the average velocity of the electrons, and it is therefore proportional to the electric field. This is evident if we examine Fig. 27.7, where the charge ΔQ that passes a given area A can be obtained from

$$\Delta Q = [\text{charge per electron}] \times [\text{electrons per unit volume}] \times [\text{volume}] \tag{27.9}$$

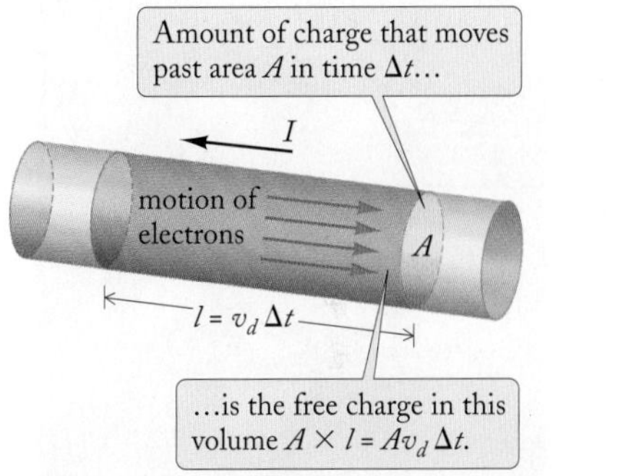

FIGURE 27.7 The current carried by the wire is uniformly distributed over its entire cross section. A wire of twice the cross-sectional area would carry twice the current (with the same electric field).

Let us assume the material is a metal with n electrons per unit volume. From Fig. 27.7, we can see that the volume of moving charge that sweeps past a cross-sectional area of the wire in a time Δt is equal to the area A times the length l, that is, [volume] $= Al = Av_d\ \Delta t$. Thus Eq. (27.9) becomes

$$\Delta Q = (-e) \times (n) \times (A\, v_d\, \Delta t)$$

The current is then given by

$$I = \frac{\Delta Q}{\Delta t} = -enAv_d \tag{27.10}$$

current of free electrons

or, with Eq. (27.8) for v_d,

$$I = \frac{ne^2\tau}{m_e}AE = \frac{1}{\rho}AE \tag{27.11}$$

where ρ is a constant of proportionality that depends on the characteristics of the material of the wire. This constant ρ is called the **resistivity** of the material. It is an intrinsic property of the *type of material,* but does not depend on the shape or size of the particular piece of wire. It is a measure of how strongly a material opposes the flow of charge. Equation (27.11) indicates that the resistivity is determined by the electron's mass, density, charge, and average collision time τ:

resistivity ρ

$$\rho = \frac{m_e}{ne^2\tau} \tag{27.12}$$

resistivity in terms of average collision time

Table 27.2 lists the resistivities of some conducting materials.

With $E = \Delta V/l$ from Eq. (27.1), the current of Eq. (27.11) becomes

$$I = \frac{1}{\rho}A\frac{\Delta V}{l} \tag{27.13}$$

For our wire of area A and length l, it is customary to define the **resistance** R of the wire as

resistance R in terms of resistivity

$$R = \rho \frac{l}{A} \tag{27.14}$$

Thus, the resistance is directly proportional to the length and inversely proportional to the cross-sectional area. Equation (27.13) can then be expressed in the convenient form

Ohm's Law

$$I = \frac{\Delta V}{R} \tag{27.15}$$

GEORG SIMON OHM (1787–1854)
German physicist. Ohm was led to his law by an analogy between the conduction of electricity and the conduction of heat.

This equation is called **Ohm's Law**. It asserts that *the current is proportional to the potential difference between the ends of the conductor.* Ohm's law can be written in three useful forms:

$$I = \frac{\Delta V}{R} \qquad \Delta V = IR \quad \text{and} \quad R = \frac{\Delta V}{I} \tag{27.16}$$

Note that in Eq. (27.15), the resistance R plays the role of a constant of proportionality; the resistance is a quantitative measure of the opposition that the piece of metal presents to the flow of current. For a wire of uniform cross section, the resistance can be calculated from the simple formula (27.14). But Ohm's Law is also valid for conductors of arbitrary shape—such as wires of nonuniform cross section—for which the resistance must be calculated from a more complicated formula tailored to the shape and the size of the conductor (see, for example, Problem 43).

Ohm's Law is valid for metallic conductors and also for many nonmetallic conductors (for example, carbon) in which the current is carried by a flow of electrons. It is also valid for plasmas and for electrolytes, in which the current is carried by a flow

TABLE 27.2 RESISTIVITIES AND TEMPERATURE COEFFICIENTS OF RESISTANCE OF METALS[a]

MATERIAL	ρ	α
Silver	$1.6 \times 10^{-8}\ \Omega\cdot\text{m}$	$3.8 \times 10^{-3}/°\text{C}$
Copper	1.7×10^{-8}	3.9×10^{-3}
Aluminum	2.8×10^{-8}	3.9×10^{-3}
Brass	$\approx 7 \times 10^{-8}$	2×10^{-3}
Nickel	7.8×10^{-8}	6×10^{-3}
Iron	10×10^{-8}	5×10^{-3}
Steel	$\approx 11 \times 10^{-8}$	4×10^{-3}
Constantan	49×10^{-8}	1×10^{-5}
Nichrome	100×10^{-8}	4×10^{-4}

[a] At a temperature of 20°C.

of both electrons and ions. However, we ought to keep in mind that in spite of its wide range of applicability, Ohm's Law is not a general law of nature, such as Gauss' Law. In many materials, Ohm's Law fails when the current is large; and in inhomogeneous materials, such as the layered materials used in semiconductor devices, Ohm's Law fails even when the current is small (in fact, the operation of semiconductor devices usually hinges on the exploitation of this non-ohmic behavior). Thus, Ohm's Law is not of universal applicability; it is merely an assertion about the electrical properties of many conducting materials.

EXAMPLE 3 To measure the resistance of a long piece of cable, a physicist connects this wire between the terminals of a 6.0-volt battery. She finds that this produces a current of 30 A in the wire. What is the resistance of the wire?

Concepts in Context

SOLUTION: According to Ohm's Law,

$$R = \frac{\Delta V}{I}$$

Since the potential difference across the wire is $\Delta V = 6.0$ volts and the current is 30 A, we obtain

$$R = \frac{6.0 \text{ V}}{30 \text{ A}} = 0.20 \text{ V/A}$$

The unit V/A is called the *ohm* (see Section 27.3); thus, the resistance is 0.20 ohm.

EXAMPLE 4 In silver, there is one free electron per atom. Calculate the value of the average collision time τ. The mass density of silver is 10.5×10^3 kg/m^3.

SOLUTION: We can solve for the collision time τ from Eq. (27.12), in terms of the quantities ρ, m_e, e, and n in that relation. From Table 27.2, we see that the resistivity of silver is $\rho = 1.6 \times 10^{-8}$ ohm·m. We know the charge e and the electron mass m_e, so we need only find the electron number density n. The periodic table (Appendix 8) indicates that the atomic mass of silver is 108 g/mole = 0.108 kg/mole. Thus there are $10.5 \times 10^3/0.108$ moles/m^3, and since each mole has 6.02×10^{23} atoms (Avogadro's number), there are $(10.5 \times 10^3/0.108) \times 6.02 \times 10^{23}$ atoms/m^3, each with one free electron. The number density of free electrons is then

$$n = \frac{1 \text{ electron/atom} \times 6.02 \times 10^{23} \text{ atoms/mole} \times 10.5 \times 10^3 \text{ kg/m}^3}{0.108 \text{ kg/mole}}$$

$$= 5.85 \times 10^{28} \text{ electrons/m}^3$$

Solving Eq. (27.12) for the average collision time, we obtain

$$\tau = \frac{m_e}{ne^2 \rho}$$

$$= \frac{9.11 \times 10^{-31} \text{ kg}}{5.85 \times 10^{28} \text{ electrons/m}^3 \times (1.60 \times 10^{-19} \text{ C})^2 \times 1.6 \times 10^{-8} \text{ ohm}\cdot\text{m}}$$

$$= 3.8 \times 10^{-14} \text{ second}$$

Checkup 27.2

QUESTION 1: Suppose that a wire has a nonuniform cross section (thicker in some parts than in others). Is the drift velocity of the electrons the same everywhere along this wire?

QUESTION 2: Two copper wires have the same diameter, but one is 3 times as long as the other. How much larger is its resistance? Two copper wires have the same length, but one has twice the diameter of the other. How much smaller is its resistance?

QUESTION 3: When a thin wire is connected across the terminals of a 6-V battery, the current is 0.1 A. What will be the current if we connect the same wire across the terminals of a 12-V battery? An 18-V battery?

QUESTION 4: When a wire is connected across the terminals of an automobile battery, the current is 0.2 A. What will be the current if we cut this wire to half of its original length and connect such a half wire across the terminals of the battery?

QUESTION 5: If the average collision time gets longer, what happens to the resistivity? What happens to the resistivity when the density of electrons increases?

(A) Increases, increases (B) Increases, decreases
(C) Decreases, increases (D) Decreases, decreases

Online Concept Tutorial 31

27.3 RESISTIVITY OF MATERIALS

As we saw in the preceding section, the resistance R of a uniform wire is directly proportional to the resistivity ρ and the length l, and inversely proportional to the cross-sectional area A:

$$R = \rho \frac{l}{A} \tag{27.17}$$

We can use this formula to calculate the resistance if the resistivity of the material is known, and we can also use it to calculate the resistivity if the resistance has been measured experimentally. The latter calculation is important in the experimental determination of the resistivity of a material, which is done by measuring the potential difference and current in a wire of given length and cross section made of a sample of the material.

As is obvious from Ohm's Law, Eq. (27.15), *the SI unit of resistance is* 1 *volt per ampere; this unit is called the* **ohm**, *abbreviated* Ω:

ohm (Ω)

$$1 \text{ ohm} = 1\ \Omega = 1\ \text{V/A} \tag{27.18}$$

The unit of resistivity is the ohm-meter (Ω·m). Table 27.2 lists the resistivities of diverse conducting materials.

EXAMPLE 5 A wire commonly used for electrical installations in homes is No. 10 copper wire, which has a radius of 0.129 cm. What is the resistance of a piece of this wire 30 m long? What is the potential difference along this wire if it carries a current of 10 A?

SOLUTION: The cross-sectional area of the wire is

$$A = \pi r^2 = \pi \times (0.129 \times 10^{-2}\ \text{m})^2 = 5.2 \times 10^{-6}\ \text{m}^2$$

The resistivity of copper listed in Table 27.2 is $\rho = 1.7 \times 10^{-8}\ \Omega\cdot\text{m}$. By Eq. (27.17), the resistance is

$$R = \rho\frac{l}{A} = 1.7 \times 10^{-8}\ \Omega\cdot\text{m} \times \frac{30\ \text{m}}{5.2 \times 10^{-6}\ \text{m}^2} = 0.098\ \Omega$$

For a current of 10 A, Ohm's Law then gives a potential difference

$$\Delta V = IR = 10\ \text{A} \times 0.098\ \Omega = 0.98\ \text{volt}$$

By rearranging Eq. (27.13) we readily find

$$\frac{I}{A} = \frac{1}{\rho}\frac{\Delta V}{l} \qquad (27.19)$$

The ratio of the current I to the cross-sectional area A is the **current density** j,

$$j = \frac{I}{A} \qquad (27.20)$$

current density j

This is the current per unit area at a point in the conductor. From Eq. (27.1), $\Delta V/l$ is the electric field E in the conductor, so Eq. (27.19) can be written

$$j = \frac{1}{\rho}E \qquad (27.21)$$

Thus, the current density is directly proportional to the electric field. This is an alternative expression for Ohm's Law. This equation relates two local quantities: the current density at one point within the conductor, and the electric field at that point.

The resistivity of materials depends on the temperature. *In ordinary metals, the resistivity increases with the temperature.* This is due to an increase of the rate of collisions between the moving electrons and the atoms of the lattice—at higher temperatures, the atoms jump more violently about their positions in the lattice, and they are then more likely to disturb the motion of the electrons, decreasing the average collision time. The numbers in Table 27.2 give the resistivities at room temperature (20°C). At very low temperatures, the collision rate and resistivity approach constant values, due to the remaining collisions with impurities in the material. Figure 27.8 shows the typical increase of resistivity with increasing temperature in a metal.

Concepts in Context

A drastic increase of the resistance with temperature is observed in ordinary incandescent lightbulbs. The resistance of the filament of a hot lightbulb (at its operating temperature) is about 10 times as large as the resistance of the filament of a cold lightbulb (at room temperature). When you switch on a cold lightbulb, there is a large initial surge of current through the filament, and this current then quickly decreases and levels off while the filament attains its operating temperature. The inital surge, due to the low "cold resistance" of the filament, is the reason filaments often burn out at the instant a light is turned on.

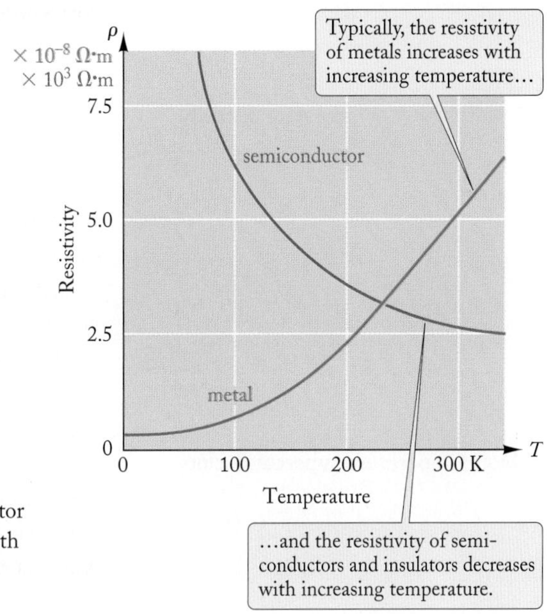

FIGURE 27.8 Resistivity as a function of temperature for a metal and for a semiconductor (silicon). The resistivity of metals increases with increasing temperature, whereas that of semiconductors decreases.

For a reasonably small increment in temperature, the increase in the resistivity and in the resistance of a metal is directly proportional to the increment in temperature. Mathematically, we can express this increase of resistance as

change of resistance with temperature

$$\Delta R = \alpha R_0 \, \Delta T \tag{27.22}$$

temperature coefficient of resistance α

where R_0 is the resistance at the initial temperature. [Note that this equation is reminiscent of the equation for the thermal expansion of the length of a solid; see Eq. (20.6).] The constant of proportionality α is called the **temperature coefficient of resistance**. Table 27.2 (page 866) lists values of this coefficient for some metallic conductors.

Concepts — in — Context

EXAMPLE 6 Suppose that because of a current overload, the temperature of the copper wire in Example 5 increases from 20°C to 50°C. How much does the resistance increase?

SOLUTION: The initial resistance is $R_0 = 0.098\ \Omega$, from Example 5. The temperature increment is $\Delta T = 30°\text{C}$. According to Table 27.2, the temperature coefficient for copper is $\alpha = 3.9 \times 10^{-3}/°\text{C}$. The change of resistance is therefore

$$\Delta R = \alpha R_0 \, \Delta T = 3.9 \times 10^{-3}/°\text{C} \times 0.098\ \Omega \times 30°\text{C}$$
$$= 0.012\ \Omega$$

The new resistance of the wire will then be

$$R = R_0 + \Delta R = 0.098\ \Omega + 0.012\ \Omega = 0.110\ \Omega$$

resistance thermometer

The change of electrical resistance with temperature is exploited in the operation of the **resistance thermometer**. Figure 27.9 shows such a thermometer, made of a coil of fine platinum wire. We can calibrate this thermometer in the same way as, say, a mercury thermometer by first immersing it into an ice–water mixture (0°C) and then into boiling water (100°C). Measurements of the resistance at these two temperatures tell us how many ohms correspond to 100°C, and, by extrapolation, how many ohms correspond to any other temperature.

superconductivity

At very low temperatures, the resistivity of a metal will be substantially less than at room temperature. Some metallic elements, such as lead, tin, zinc, and niobium, as well as many alloys and compounds, exhibit the phenomenon of **superconductivity**: *their resistance vanishes completely at some critical temperature above absolute zero.* For example, Fig. 27.10 displays a plot of resistivity vs. temperature for tin; at an absolute temperature of 3.72 K, the resistivity abruptly vanishes. In one experiment, a current of several hundred amperes was started in a superconducting ring; the current continued on its own with undiminished strength for over a year, without any battery or generator to maintain it.

Concepts — in — Context

high-temperature superconductors

In some copper-oxide–based superconductors, the resistance vanishes at temperatures of up to 138 K (under normal pressure) or even up to 162 K (under high pressure). Such **high-temperature superconductors** were first discovered in 1986, and are now in use in some devices, magnets, and high-current power lines. Although superconducting power cables require the added expense and complications of cryogenic cooling to operate, their benefits include a much higher current capacity than conventional cables, which is especially important in city utility tunnels, where space is at a premium.

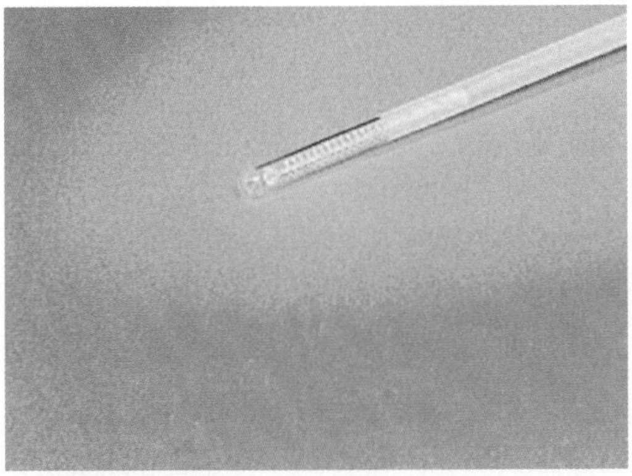

FIGURE 27.9 A coil of fine platinum wire serves as sensor in a resistance thermometer.

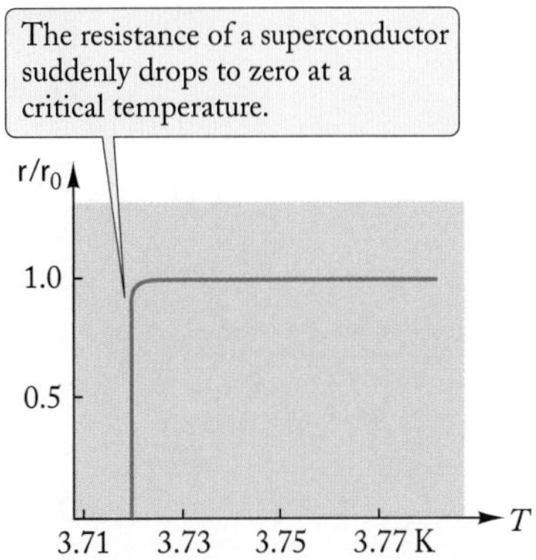

FIGURE 27.10 Resistivity of tin as a function of temperature. Below 3.72 K, the resistivity is zero. For the pupose of this plot, the resistivity has been expressed as a fraction of the resistivity at 4.2 K, the boiling point of helium.

According to the definition we gave in Section 22.5, an ideal insulator is a material that does not permit any motion of electric charge. Real **insulators**, such as porcelain or glass, do permit some very slight motion of charge. What distinguishes them from conductors is their enormously large resistivity. Typically, the resistivities of insulators are more than 10^{20} times as large as those of conductors (compare the values in Table 27.3 with those in Table 27.2). This means that even when we apply a high voltage to a piece of glass, the flow of current will be insignificant (provided, of course, that the material does not suffer electrical breakdown). In fact, on a humid day it is likely that more current will flow along the microscopic film of water that tends to form on the surface of the insulator than through the insulator itself. The resistivities of **semiconductors**, such as carbon and silicon, are between those of conductors and insulators (see Table 27.4).

In both insulators and semiconductors, the temperature coefficient of resistivity α is usually negative, opposite to that of metals. This is because heating the material tends to shake free some of the bound electrons, increasing the density of free electrons [n in Eq. (27.12)], and thus decreasing the resistivity. Values of α for semiconductors at temperatures near room temperature are given in Table 27.4. The temperature dependence of the resistivity of the semiconductor silicon is included in Fig. 27.8. Because of the rapid increase of resistivity with decreasing temperature, semiconductor resistors are often used as thermometers, particularly at low temperatures.

TABLE 27.3 RESISTIVITIES OF INSULATORS

MATERIAL	ρ
Polyethylene	$2 \times 10^{11}\ \Omega\cdot\text{m}$
Glass	$\approx 10^{12}$
Porcelain, unglazed	$\approx 10^{12}$
Rubber, hard	$\approx 10^{13}$
Epoxy	$\approx 10^{15}$

TABLE 27.4 RESISTIVITIES AND TEMPERATURE COEFFICIENTS OF RESISTANCE OF SEMICONDUCTORS[a]

MATERIAL	ρ	α
Carbon (graphite)	$3.5 \times 10^{-5}\ \Omega\cdot\text{m}$	$-5 \times 10^{-4}/°\text{C}$
Silicon	2.6×10^{3}	-8×10^{-2}
Germanium	4.2×10^{-1}	-5×10^{-2}

[a] At a temperature of 20°C.

Checkup 27.3

QUESTION 1: Two wires of iron and of nichrome have the same lengths and same diameters. Which has the larger resistance, and by what factor?

QUESTION 2: Two wires of iron and of silver have equal lengths but different diameters. If these wires are to be of equal resistances, what must be the ratio of their diameters?

QUESTION 3: For a temperature increase of 100°C, what is the percentage increase of resistance of a constantan wire? Of a nickel wire?

QUESTION 4: Equation (27.12), $\rho = m_e/(ne^2\tau)$, relates the resistivity to the density n and average collision time τ of free electrons. Which of these two quantities dominates the temperature dependence of the resistivity for a metal? For a semiconductor?

(A) n, n (B) n, τ (C) τ, τ (D) τ, n

27.4 RESISTANCES IN COMBINATION

The metallic wires of any electric circuit have some resistance. But in electronic devices—CD players, televisions, computers, etc.—the main contribution to the resistance is usually due to devices that are specifically designed to have a high resistance. These devices are called **resistors**, and they are used to control and modify the currents. For instance, the manual volume controls on radios are adjustable resistors. Such a volume control is made of a long, coiled piece of high-resistance wire on which rests a sliding contact (see Fig. 27.11); by moving the sliding contact, you increase or decrease the length of wire that lies between the two external terminals, and you increase or decrease the resistance, and you thereby control the current reaching the loudspeaker. Adjustable, or variable, resistors of this kind are called **rheostats** or **potentiometers** ("pots") when used to limit current or select a voltage, respectively. The name potentiometer is also used in a more technical sense for a device used to measure potential, which incorporates a sliding contact similar to that shown in Fig. 27.11.

resistor

rheostat

potentiometer

Resistors used in the circuits of electronic devices are often made of a piece of pure carbon (graphite) connected between two terminals (Fig. 27.12a). Carbon has a high resistivity, and hence a small piece of carbon can have a higher resistance than a long piece of metallic wire. Such resistors obey Ohm's Law (current proportional to potential difference) over a wide range of values of the current; but if the resistor is overloaded with current, it will heat up, possibly even burn, and Ohm's Law will fail.

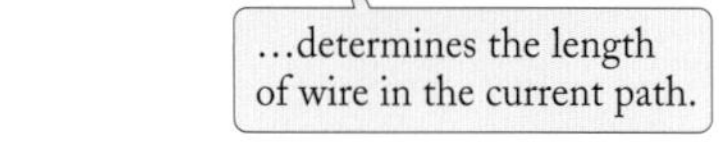

FIGURE 27.11 A rheostat, or potentiometer ("pot"), consisting of a long coiled wire with a sliding contact.

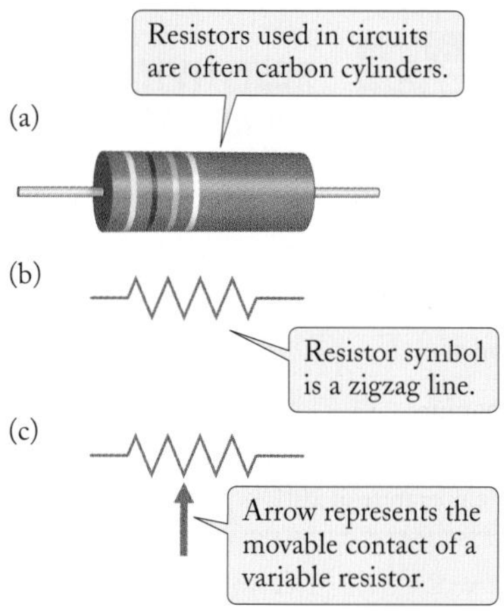

FIGURE 27.12 (a) A resistor consisting of a cylinder of carbon with two terminals attached. (b) Symbol for a resistor in a circuit diagram. (c) Symbol for a variable resistor, also known as a potentiometer.

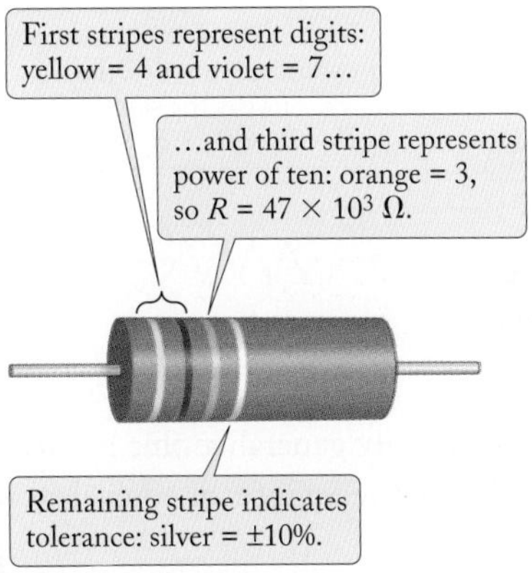

FIGURE 27.13 Resistor color code. The first two stripes represent the first digits of the resitance value; these are multiplied by the power of 10 corresponding to the third stripe. The stripes on this resistor are yellow–violet–orange (4–7–3), corresponding to a resistance value of $47 \times 10^3\ \Omega$, or 47 kΩ. A fourth stripe represents the tolerance, or accuracy, of the specified value (see Table 27.5).

TABLE 27.5 RESISTOR COLOR CODE[a]

COLOR	NUMBER	POWER	TOLERANCE
Black	0	10^0	
Brown	1	10^1	
Red	2	10^2	
Orange	3	10^3	
Yellow	4	10^4	
Green	5	10^5	
Blue	6	10^6	
Violet	7	10^7	
Gray	8	10^8	
White	9	10^9	
Gold		10^{-1}	±5%
Silver		10^{-2}	±10%
None			±20%

[a]Figure 27.13 describes the use of this color code.

In circuit diagrams the symbol for a resistor is a zigzag line, reminiscent of the path of an electron inside a conducting material (see Fig. 27.12b). The potentiometer symbol is the symbol for a resistor with an arrow to represent the sliding contact (Fig. 27.12c). The value of the resistance is sometimes printed on the side of a resistor; more often, the value is encoded in a series of color stripes around the resistor. Table 27.5 gives the numerical values corresponding to the various colors, and Fig. 27.13 shows an example of a color-coded carbon resistor.

Even circuits without resistors, such as the wiring of a house, often have the main contribution to their resistance concentrated, or lumped, in one place. For instance, when an electric toaster or a lightbulb is plugged into an electric outlet, the current flowing through the circuit encounters the most resistance in the relatively short piece of wire within the toaster or the lightbulb. The short, high-resistance piece of wire in the appliance has the same effect on the flow of current as a carbon resistor, and in a circuit diagram it can be treated as though it were a resistor. The external wires of the appliance can often be treated as if they had no resistance; their resistance is negligible compared with that of the device or resistor.

For resistors, as for capacitors, the two simplest ways of connecting several resistors are in series or in parallel. Figure 27.14 shows two resistors connected in **series**. Since each of these resistors offers a resistance to the current entering the wire, our intuition suggests that the net resistance of this combination is the sum of the individual resistances,

$$R = R_1 + R_2 \tag{27.23}$$

The formal derivation of this result begins with the observation that if the potential differences across the individual resistors are ΔV_1 and ΔV_2, then the net potential difference across the combination is

$$\Delta V = \Delta V_1 + \Delta V_2 \tag{27.24}$$

This additivity of the potentials is a direct consequence of the definition of the potential as work per unit charge—the net work done by the electric field on a unit charge that moves through the first resistor and then through the second is simply the sum of the work done in the first and the work done in the second. Furthermore, *the currents in both resistors are exactly the same*, since any charge that flows through the first resistor continues to flow through the second. Hence, using Ohm's Law, $\Delta V = IR$, we find

$$\Delta V = \Delta V_1 + \Delta V_2 = IR_1 + IR_2 = I(R_1 + R_2)$$

From this it is clear that the resistance of the combination is equivalent to a single resistance $R = R_1 + R_2$, as given in Eq. (27.23).

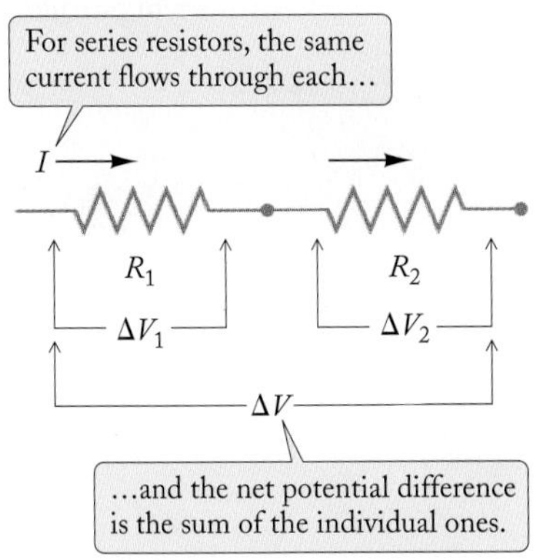

FIGURE 27.14 Two resistors connected in series.

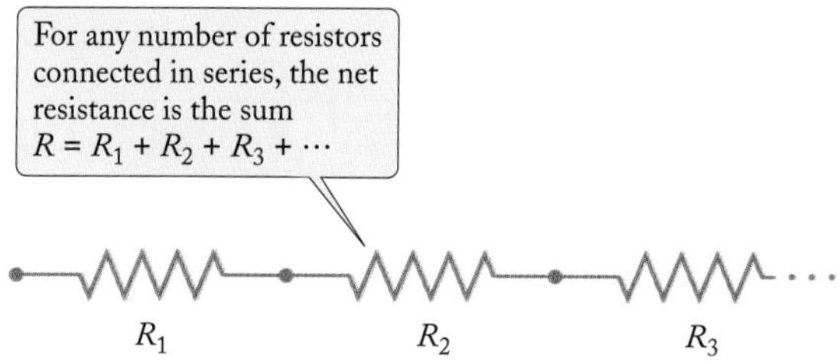

FIGURE 27.15 Several resistors connected in series.

We can easily generalize this result to any number of resistors in series (see Fig. 27.15). The net resistance, or equivalent resistance, of the series combination is

resistors in series

$$R = R_1 + R_2 + R_3 + \cdots \tag{27.25}$$

Figure 27.16 shows two resistors in **parallel**. As with any circuit elements connected *in parallel, the potential difference across each resistor is the same* as the potential difference across the combination. Hence, from Ohm's Law, the currents are

$$I_1 = \frac{\Delta V}{R_1} \quad \text{and} \quad I_2 = \frac{\Delta V}{R_2} \tag{27.26}$$

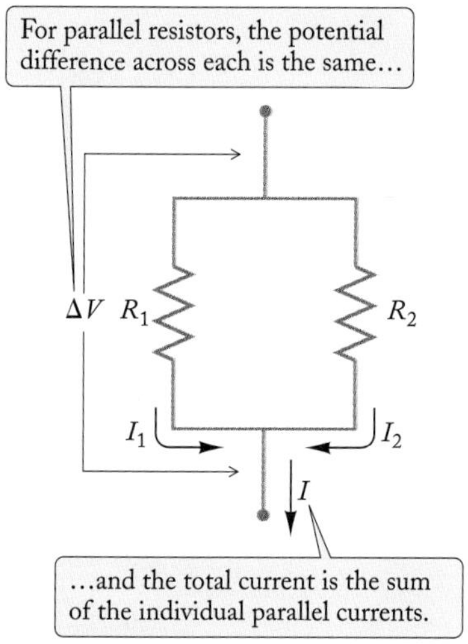

FIGURE 27.16 Two resistors connected in parallel.

Resistors in parallel are analogous to the several parallel lanes of a highway. The total flow of automobiles is the sum of the flows in the individual lanes; likewise, the total flow of charge through a combination of parallel resistors is the sum of the flows in the individual resistors. Thus, the total current through the combination is the sum of the individual parallel currents,

$$I = I_1 + I_2 = \frac{\Delta V}{R_1} + \frac{\Delta V}{R_2} = \left(\frac{1}{R_1} + \frac{1}{R_2}\right)\Delta V \tag{27.27}$$

Comparison of Eq. (27.27) with Ohm's law, $I = \Delta V/R$, shows that the resistance of the combination is therefore equivalent to a single resistance R given by

$$\frac{1}{R} = \frac{1}{R_1} + \frac{1}{R_2} \tag{27.28}$$

Note that the resistance of the parallel combination is less than each of the individual resistances. For example, if $R_1 = R_2 = 1.0\ \Omega$, then $R = 0.5\ \Omega$.

We can also generalize this to any number of resistors in parallel (see Fig. 27.17). The net resistance R of the parallel combination is given by

resistors in parallel

$$\frac{1}{R} = \frac{1}{R_1} + \frac{1}{R_2} + \frac{1}{R_3} + \cdots \tag{27.29}$$

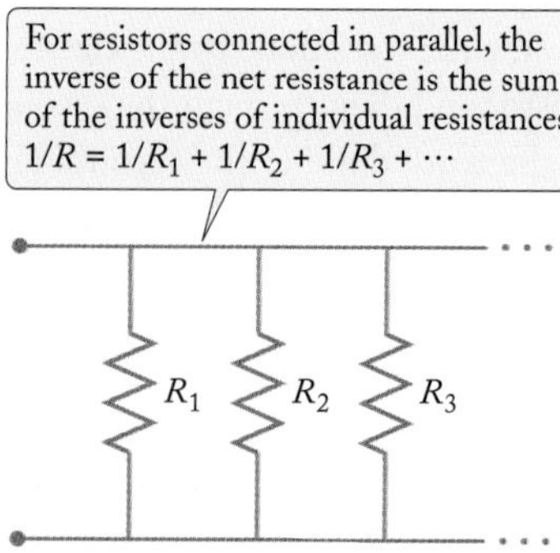

FIGURE 27.17 Several resistors connected in parallel.

EXAMPLE 7 Two resistors, with $R_1 = 10\ \Omega$ and $R_2 = 20\ \Omega$, are connected in series (see Fig. 27.18), and a current of 1.8 A flows through this combination. What is the potential difference across the combination? What is the potential difference across each resistor?

SOLUTION: According to Eq. (27.25), the net resistance is given by

$$R = R_1 + R_2 = 10\ \Omega + 20\ \Omega = 30\ \Omega$$

Ohm's Law applied to this combined system of two resistors tells us that the potential difference is

$$\Delta V = IR = 1.8\ \text{A} \times 30\ \Omega = 54\ \text{volts}$$

Furthermore, Ohm's Law applied to each individual resistor tells us that the individual potential differences are

$$\Delta V_1 = IR_1 = 1.8\ \text{A} \times 10\ \Omega = 18\ \text{volts}$$

$$\Delta V_2 = IR_2 = 1.8\ \text{A} \times 20\ \Omega = 36\ \text{volts}$$

As expected, the sum of these two potential differences is 54 volts, which is the potential difference across the combination. The arrangement of Fig. 27.18 is known as a **voltage divider**, since it serves to divide the total potential difference between the two series-connected resistors in proportion to their resistance.

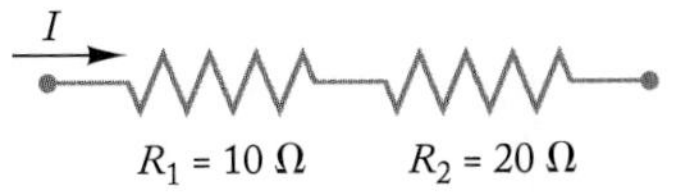

FIGURE 27.18 A current flows through two resistors connected in series.

voltage divider

EXAMPLE 8 Suppose that the same two resistors, with $R_1 = 10\ \Omega$ and $R_2 = 20\ \Omega$, are connected in parallel (see Fig. 27.19) and a net current of 1.8 A flows through this combination. What is the potential difference across the combination? What is the current in each resistor?

SOLUTION: According to Eq. (27.29), the net resistance is given by

$$\frac{1}{R} = \frac{1}{R_1} + \frac{1}{R_2} = \frac{1}{10\ \Omega} + \frac{1}{20\ \Omega} = \frac{30}{200\ \Omega}$$

which yields $R = 200\ \Omega/30 = 6.7\ \Omega$. Hence, Ohm's Law applied to the combination as a whole tells us that the potential difference is

$$\Delta V = IR = 1.8\ \text{A} \times 6.7\ \Omega = 12\ \text{volts}$$

and Ohm's Law applied to the two branches tells us that the individual currents are

$$I_1 = \frac{\Delta V}{R_1} = \frac{12\ \text{volts}}{10\ \Omega} = 1.2\ \text{A}$$

$$I_2 = \frac{\Delta V}{R_2} = \frac{12\ \text{volts}}{20\ \Omega} = 0.60\ \text{A}$$

As expected, the sum of the two currents is 1.8 A, which is the net current through the combination.

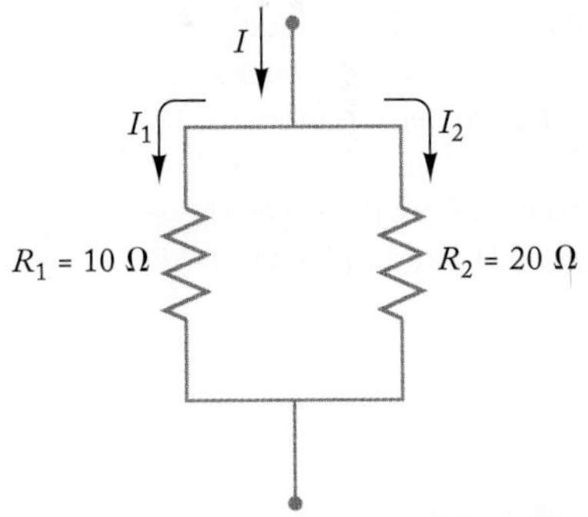

FIGURE 27.19 A current flows through the same two resistors connected in parallel.

In the calculations of the preceding example, we neglected the resistance of the wires connecting the resistors. This is a good approximation if the resistance of these wires is small compared with the resistance of the resistors. If this is not so, then we must take the resistance of the wires into account in our calculation. In circuit diagrams, it

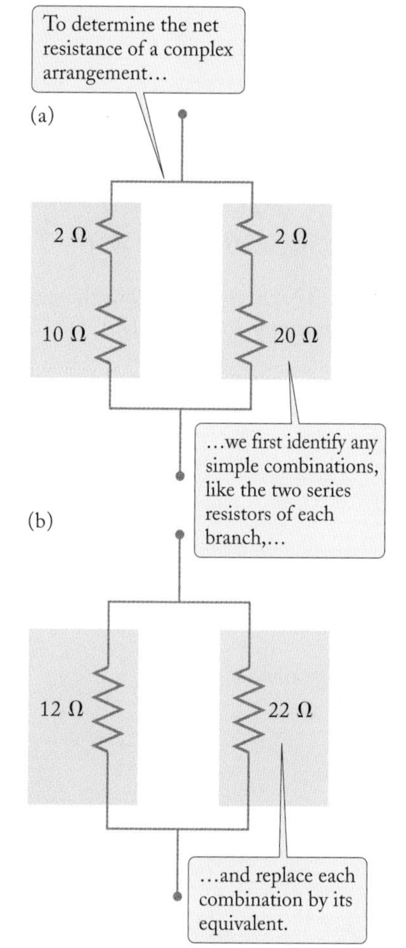

FIGURE 27.20 (a) In this circuit diagram, the resistances of the wires are represented by the two small resistors. (b) The right (green) and the left (orange) branches of the circuit are equivalent to two resistors of 12 Ω and 22 Ω connected in parallel.

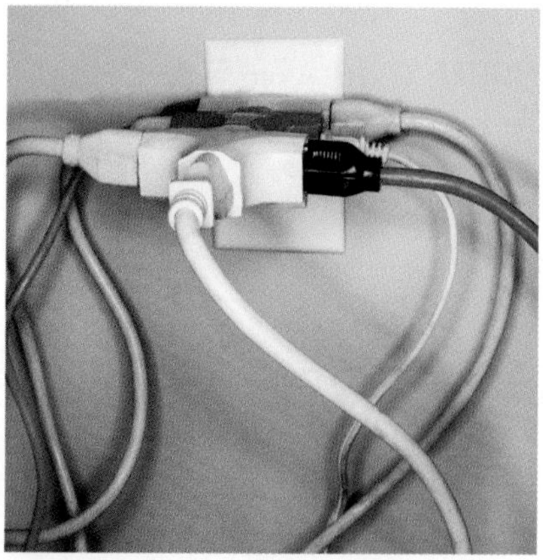

FIGURE 27.21 Several appliances plugged into the power outlets in a house.

is customary to represent the resistance of a wire schematically by an equivalent resistor of suitable magnitude; this means we pretend that all of the resistance of a wire is concentrated, or lumped, in one place. The line segments connecting these equivalent resistors to the rest of the circuit can then be assumed to have zero resistance exactly. For instance, if each of the two wires connecting the resistors in Fig. 27.19 is long and each has a resistance of 2 Ω, then the schematic circuit diagram that includes the resistance of the wires to the connections is as shown in Fig. 27.20. Note that in this diagram, we placed the resistors of 2 Ω that represent the resistances of the wires immediately above each of the resistors R_1 and R_2. We could equally well place these equivalent resistors immediately below each of R_1 and R_2, or one half of each above and one half below—all these arrangements are effectively equivalent, and lead to the same distribution of current among the two branches of the circuit. Also note that we neglected the resistance of the short leads sticking out of the circuit above and below; if these have a significant resistance, then we would also need to take these into account.

EXAMPLE 9 If, as in the preceding example, a current of 1.8 A flows through the combination of resistors and resistive wires shown schematically in Fig. 27.20a, what is the potential difference across the entire combination? What is the current in each resistor?

SOLUTION: To solve this problem we proceed in two steps. First, we find the resistances of the left and the right branches of the circuit. The left branch consists of a resistor R_1 of 10 Ω in series with 2 Ω; the right branch consists of a resistor R_2 of 20 Ω in series with 2 Ω. Hence, the resistance of the left branch of the circuit is $R_1' = R_1 + 2\ \Omega = 12\ \Omega$ and that of the right branch is $R_2' = R_2 + 2\ \Omega = 22\ \Omega$.

Next, we find the net resistance of the circuit by combining R_1' and R_2' in parallel (Fig. 27.20b). This parallel combination of the resistances of the left and the right branches gives

$$\frac{1}{R} = \frac{1}{R_1'} + \frac{1}{R_2'} = \frac{1}{12\ \Omega} + \frac{1}{22\ \Omega} = \frac{34}{264\ \Omega}$$

From this, we find $R = 264\ \Omega/34 = 7.8\ \Omega$ for the net resistance of the circuit. Hence, by Ohm's Law, the potential difference across the combination is

$$\Delta V = IR = 1.8\ \text{A} \times 7.8\ \Omega = 14\ \text{volts}$$

and the individual currents in the branches are

$$I_1 = \frac{\Delta V}{R_1'} = \frac{14\ \text{volts}}{12\ \Omega} = 1.2\ \text{A}$$

$$I_2 = \frac{\Delta V}{R_2'} = \frac{14\ \text{volts}}{22\ \Omega} = 0.64\ \text{A}$$

Ordinary appliances plugged into the power outlets in your home are connected in parallel (see Fig. 27.21). The potential difference across each of these appliances is 115 V, which is the potential difference supplied across the wires that feed electric power into the house.[1] Some of these appliances—such as lightbulbs, electric toasters,

[1] The potential difference supplied across the power wires is actually an oscillating potential difference, which periodically reverses sign. We will deal with the details of such oscillating, or alternating, potentials and currents in Chapter 32.

irons, hot plates, heaters, and blankets—simply consist of a piece of resistive wire in which the dissipation of the electric energy produces heat. More complicated appliances—such as refrigerators, fans, washers, radios, TVs, camcorders, CD players—may contain electric motors and a variety of electronic devices. Such complicated devices usually do not obey Ohm's Law; for instance, if you plug your TV into a 120-V outlet instead of a 115-V outlet, the current will not increase in proportion to the potential. However, for purposes of comparison, it is often useful to assign to a complicated device an effective resistance $R = \Delta V/I$, as though Ohm's Law were valid. Since this effective resistance of a complicated device is *not* constant, it can be used only in calculations at the given, fixed potential difference of, say, 115 V.

EXAMPLE 10 A vacuum cleaner, a hair dryer, and an electric iron are simultaneously plugged into a single outlet (see Fig. 27.22a). Their resistances are 9.0 Ω, 10 Ω, and 12 Ω, respectively. The outlet supplies a potential difference of 115 V. What is the net current flowing through the outlet? If the maximum safe current that the outlet can carry is 30 A, is it advisable to plug all of these appliances into the outlet?

Concepts in Context

SOLUTION: All the appliances are connected in parallel (see Fig. 27.22b); thus the potential difference across each is 115 V. By Ohm's Law, $I = \Delta V/R$, the individual currents are then, respectively,

$$I_1 = 115\ \text{V}/9.0\ \Omega = 12.8\ \text{A}$$

$$I_2 = 115\ \text{V}/10\ \Omega = 11.5\ \text{A}$$

$$I_3 = 115\ \text{V}/12\ \Omega = 9.6\ \text{A}$$

The net current is the sum of the individual parallel currents, 12.8 A + 11.5 A + 9.6 A = 34 A. This is more than the outlet can carry safely. These devices should not be plugged into the same outlet, nor into several outlets supplied by the same wire.

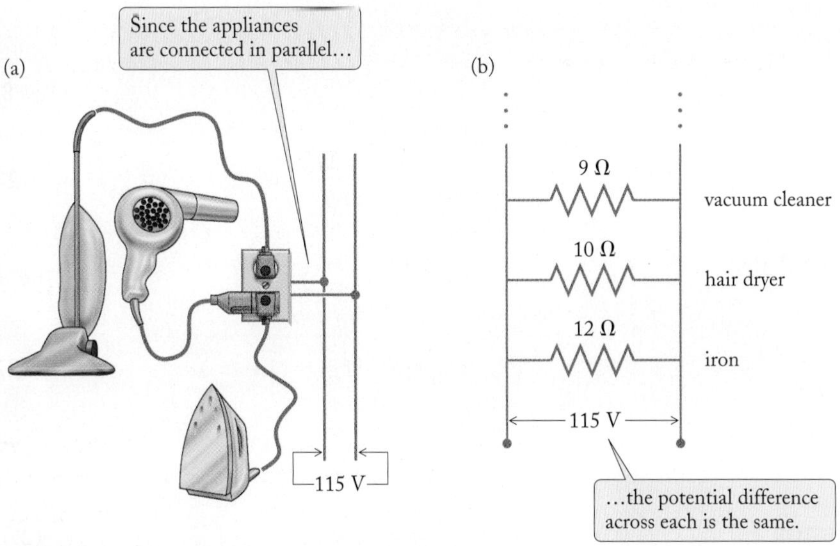

FIGURE 27.22 (a) Three appliances plugged into the same outlet. (b) Schematic diagram showing how the resistances are connected.

PROBLEM-SOLVING TECHNIQUES — COMBINATIONS OF RESISTORS

When seeking the net resistance of a circuit containing several resistors connected in some complicated manner, proceed in steps, as in the case of a circuit with capacitors.

1. In the first step, look for groups of resistors that form simple parallel or series combinations (such as the group of two resistors in the left branch and the group of two resistors in the right branch in Fig. 27.20a). Evaluate the resistance of each such group.
2. In the second step, see how these effective group resistances are connected to each other (as in Fig. 27.20b), and evaluate the net resistance of the combination of the groups.
3. In some cases, you may have to repeat these steps until you obtain a single equivalent resistance.

✔ Checkup 27.4

QUESTION 1: A wire has a resistance of 3 Ω. If we cut this wire in half and connect the two pieces in parallel, what will be the new resistance?

QUESTION 2: Qualitatively, why is the net resistance of a series combination of equal resistors larger than the individual resistances? Why is the net resistance of a parallel combination of equal resistors smaller than the individual resistances?

QUESTION 3: A wire of copper and a wire of silver are connected in parallel to a battery. Both wires have the same lengths and the same diameters. Which carries more current?

QUESTION 4: If you connect 10 resistors of 1 Ω in series, what is the net resistance? What if you connect them in parallel?

(A) 10 Ω, 10 Ω (B) 0.1 Ω, 10 Ω (C) 10 Ω, 0.1 Ω (D) 0.1 Ω, 0.1 Ω

SUMMARY

PROBLEM-SOLVING TECHNIQUES Combinations of Resistors		**(page 878)**
ELECTRIC FIELD IN UNIFORM WIRE OF LENGTH l	$E = \frac{\Delta V}{l}$	**(27.1)**
ELECTRIC CURRENT (FLOW OF CHARGE PER UNIT TIME)	$I = \frac{\Delta Q}{\Delta t}$ or $I = \frac{dq}{dt}$	**(27.2, 3)**
SI UNIT OF ELECTRIC CURRENT	1 ampere = 1 A = 1 C/s	**(27.4)**
DRIFT VELOCITY (AVERAGE VELOCITY) IN TERMS OF THE AVERAGE COLLISION TIME τ	$v_d = \frac{-eE\tau}{m_e}$	**(27.8)**
CURRENT OF FREE ELECTRONS IN A CONDUCTOR where n is the number density of free electrons, v_d is the drift velocity, and A is the cross-sectional area.	$I = -nev_dA$	**(27.10)**
SI UNIT OF RESISTANCE	1 ohm = 1 Ω = 1 V/A	**(27.18)**

RESISTIVITY IN TERMS OF THE AVERAGE COLLISION TIME τ

$$\rho = \frac{m_e}{ne^2\tau} \qquad (27.12)$$

RESISTANCE IN TERMS OF RESISTIVITY

$$R = \rho\frac{l}{A} \qquad (27.14)$$

OHM'S LAW

$$I = \frac{\Delta V}{R} \qquad \Delta V = IR \qquad R = \frac{\Delta V}{I} \qquad (27.16)$$

CURRENT DENSITY

$$j = \frac{I}{A} \qquad (27.20)$$

OHM'S LAW AND CURRENT DENSITY

$$j = \frac{1}{\rho}E \qquad (27.21)$$

CHANGE OF RESISTANCE WITH TEMPERATURE
where α is the temperature coefficient of resistivity (Table 27.2).

$$\Delta R = \alpha R_0\,\Delta T \qquad (27.22)$$

SERIES COMBINATION OF RESISTORS
For resistors connected in series, the same current flows through each, and the net potential difference is the sum of the individual potential differences.

$$R = R_1 + R_2 + R_3 + \cdots \qquad (27.25)$$

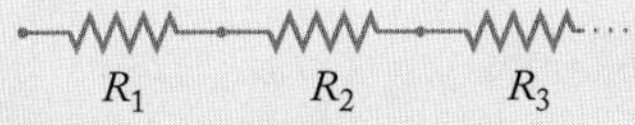

PARALLEL COMBINATION OF RESISTORS
For resistors connected in parallel, the total current is the sum of the individual parallel currents, and the potential difference across each is the same.

$$\frac{1}{R} = \frac{1}{R_1} + \frac{1}{R_2} + \frac{1}{R_3} + \cdots \qquad (27.29)$$

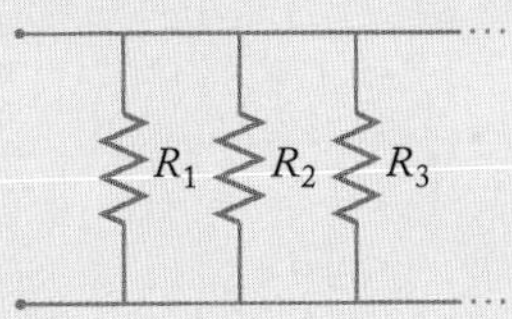

QUESTIONS FOR DISCUSSION

1. A wire is carrying a current of 15 A. Is this wire in electrostatic equilibrium?
2. Can a current flow in a conductor when there is no electric field? Can an electric field exist in a conductor when there is no current?
3. Figure 27.23 shows a putative mechanical analog of an electric conductor with resistance. A marble rolling down the inclined board is stopped every so often by a collision with a pin and therefore maintains a constant average velocity v_d. Is this a good analog, that is, is the average velocity v_d proportional to g? (Hint: Assume that the marble rolls an average distance d between collisions.)

FIGURE 27.23
Mechanical analog of electric conductor with resistance.

4. By what factor must we increase the diameter of a wire to decrease its resistance by a factor of 2?
5. Show that, for a wire of given length made of a given material, the resistance is inversely proportional to the mass of the wire.

6. What deviations from Ohm's Law do you expect if the current is very large?
7. Ohm's Law is an approximate statement about the electrical properties of a conducting body, just as Hooke's Law is an approximate statement about the mechanical properties of an elastic body. Is there an electric analog of the elastic limit?
8. An automobile battery has a potential difference of 12 V between its terminals even when there is no current flowing through the battery. Does this violate Ohm's Law?
9. Figure 27.24 shows two alternative experimental arrangements for determining the resistance of a carbon cylinder. A known potential difference is applied via the copper terminals, the current is measured, and the resistance is calculated from $R = \Delta V/I$. The arrangement in Fig. 27.24a yields a higher resistance than that in Fig. 27.24b. Why?

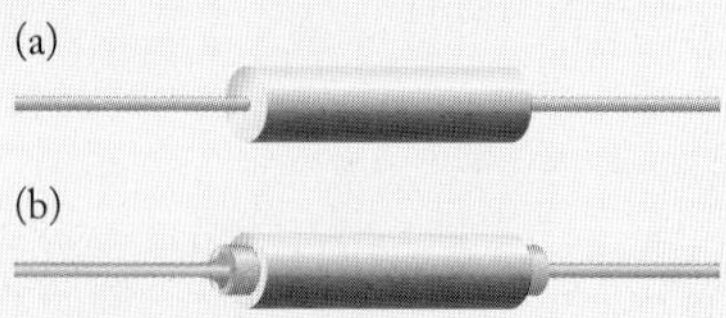

FIGURE 27.24 (a) The wires touch the carbon cylinder directly. (b) The wires end in contact plates which touch the carbon cylinder.

10. Why is it bad practice to operate a high-current appliance off an extension cord?
11. The installation instructions for connecting an outlet to the wiring of a house recommend that the wire be wrapped at least three-quarters of the way around the terminal post (Fig. 27.25). Explain.

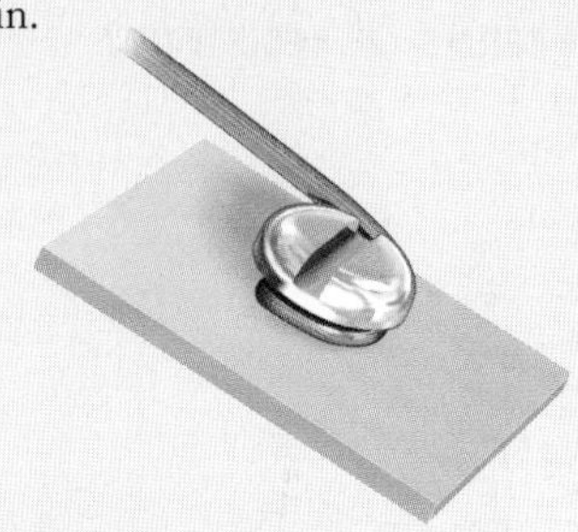

FIGURE 27.25 Terminal post of an outlet.

12. The temperature coefficient of resistivity α is defined as the fractional increase of resistivity per degree Celsius, $\alpha = (1/\rho)d\rho/dT$. According to Table 27.2, what is the value of α for copper?
13. Figure 27.9 shows a platinum resistance thermometer consisting of a coil of fine platinum wire inside a glass tube that can be put in thermal contact with a body. How would you measure the resistance of this wire to determine the temperature of the body?
14. Aluminum wire should never be connected to terminals designed for copper wire. Why not? (Hint: Aluminum has a considerably higher coefficient of thermal expansion than copper.)
15. Two copper wires are connected in parallel. The wires have the same length, but one has twice the diameter of the other. What fraction of the total current flows in each wire?

PROBLEMS

†27.1 Electric Current
27.2 Resistance and Ohm's Law

1. The electric current in the lightbulb of a flashlight is 0.50 A. How much electric charge flows through the lightbulb in one hour? How many electrons pass through the lightbulb?
2. In a typical lightning stroke, the electric current is about 20 000 A and it lasts about 1.0×10^{-4} s. The direction of the current is upward, from the ground to the cloud. What is the charge (magnitude and sign) that this stroke deposits on the ground?
3. A 40-μF capacitor is initially charged with a 9.0-V battery. To reverse the voltage on the capacitor, how long must a constant 3.0-A current flow from the positive to the negative plate of the capacitor?

†For help, see Online Concept Tutorial 31 at www.wwnorton.com/physics

4. What is the capacitance of a capacitor that charges to 1.4 V in 0.50 μs by a constant current of 25 mA?
5. A conducting wire of length 2.0 m is connected between the terminals of a 12-V battery. The resistance of the wire is 3.0 Ω. What is the electric current in the wire? What is the electric field in the wire?
6. When a thin copper wire is connected between the poles of a 1.5-V battery, the current in the wire is 0.50 A. What is the resistance of this wire? What will be the current in the wire if it is connected between the terminals of a 7.5-V battery?
7. A current begins to flow at $t = 0$ and increases with time according to $I(t) = At + Bt^2$, where $A = 0.50\ \text{C/s}^2$ and $B = 0.20\ \text{C/s}^3$. What is the current at $t = 5.0$ s? What total charge has flowed by $t = 5.0$ s?

8. At $t = 0$, a current begins to flow around a simple circuit. The total charge that has flowed clockwise around the circuit is given by $Q(t) = 5.0t - 2.0t^2$, where Q is in coulombs and t is in seconds. What total charge has flowed clockwise around the circuit at $t = 1.0$ s? At $t = 2.0$ s? At $t = 3.0$ s? What current is flowing at each of these times? Comment on the meaning of the algebraic sign of your results at each time.

9. A current begins flowing into an initially uncharged capacitor at $t = 0$ and decreases to zero during $0 \leq t \leq 2.0$ s with a time dependence given by $I = A \times (t - 2.0\text{ s})^2$, where $A = 0.25\text{ C/s}^2$. What is the initial current? The current at 1.0 s? How much charge is on the capacitor when the current stops at $t = 2.0$ s?

10. The collision time for electrons in the metal sodium (atomic symbol Na) is 8.8×10^{-15} s at room temperature. Assuming one free electron per atom, calculate the electrical resistivity of sodium.

11. In silver, the number density of free electrons is $n = 5.8 \times 10^{28}$ per cubic meter.
   (a) Using the value of the resistivity given in Table 27.2, calculate the average time between collisions of one of these electrons.
   (b) Assuming that the electric field in a current-carrying silver wire is 8.0 V/m, calculate the average drift velocity of an electron.

12. The **mean free path** l of an electron is the average distance traveled between collisions, $l = v\tau$. Due to quantum effects, the speed v of an electron in a metal is much greater than the thermal speed of an ideal-gas molecule given by Eq. (19.23); for silver, the actual speed is 12 times greater than the thermal speed. Assuming one free electron per atom, find the mean free path in silver at 20°C.

13. Constantan wire has a high resistivity and is often used as a heater. What is the resistance of a 30-cm length of constantan wire with a diameter of 0.25 mm?

14. Photodiodes supply a current proportional to the power of the light incident on them. The minimum current provided by a low-noise photodiode is 4.0×10^{-14} A. If this current passes though a 1.0-MΩ resistor, what is the voltage difference across that resistor?

15. The resistance of a 150-W, 115-V lightbulb is 88 Ω when the lightbulb is at its operating temperature. What current passes through this lightbulb when in operation? How many electrons per second does this amount to?

16. The resistance of the wire in the windings of an electric starter motor for an automobile is 3.0×10^{-2} Ω. The motor is connected to a 12-V battery. What current will flow through the motor when it is stalled (does not turn)?

17. A long, thin wire of resistance R is cut into eight pieces. Four of these pieces are then placed side by side to form a new wire $\frac{1}{8}$ of the original length. What is the resistance of the new wire?

*18. A circular loop of superconducting material has a radius of 2.0 cm. It carries a current of 4.0 A. What is the orbital angular momentum of the moving electrons in the wire? Take the center of the loop as origin.

*19. An aluminum wire has a resistance of 0.10 Ω. If you draw this wire through a die, making it thinner and twice as long, what will be its new resistance?

*20. A table lamp is connected to an electric outlet by a copper wire of diameter 0.20 cm and length 2.0 m. Assume that the current through the lamp is 1.5 A, and that this current is steady. How long does it take an electron to travel from the outlet to the lamp?

*21. When the starter motor of an automobile is in operation, the cable connecting it to the battery carries a current of 80 A. This cable is made of copper and is 0.50 cm in diameter. What is the electric field in the cable?

*22. A copper cable in a high-voltage transmission line has a diameter of 3.0 cm and carries a current of 750 A. What is the electric field in the wire?

27.3 Resistivity of Materials

23. The electromagnet of a bell is constructed by winding copper wire around a cylindrical core, like thread on a spool. The diameter of the copper wire is 0.45 mm, the number of turns in the winding is 260, and the average radius of a turn is 5.0 mm. What is the resistance of the wire?

24. The following is a list of some types of copper wire manufactured in the United States (Fig. 27.26):

GAUGE NO.	DIAMETER
8	0.3264 cm
9	0.2906
10	0.2588
11	0.2305
12	0.2053

For each type, calculate the resistance for a 100-m segment.

FIGURE 27.26 Copper wire.

25. To measure the resistivity of a metal, an experimenter takes a wire of this metal of diameter 0.500 mm and length 1.10 m and applies a potential difference of 12.0 V to the ends. He finds that the resulting current is 3.75 A. What is the resistivity?

26. A high-voltage transmission line has an aluminum cable of diameter 3.0 cm, 200 km long. What is the resistance of this cable?

27. Consider the aluminum cable described in Problem 26. If the temperature of this cable increases from 20°C to 50°C, how much will its resistance increase?

28. What increase of temperature will increase the resistance of a nickel wire from 0.50 Ω to 0.60 Ω?

29. What is the current density in a 0.259-cm-diameter copper wire carrying a current of 30 A? What is the electric field in the wire?

30. In a scanning tunneling microscope, a small, picoampere current, 1.0×10^{-12} A, flows across an atomic-sized region, roughly 0.30 nm in diameter. What is the current density?

31. Equation (27.21), the alternative form of Ohm's Law, can also be written $j = \sigma E$, where σ is called the **electrical conductivity**. What is the electrical conductivity of copper?

32. Commercial platinum resistance thermometers are often manufactured with a resistance of 100.0 Ω at 0.0°C. The temperature coefficient of resistivity is 3.9×10^{-3}/°C. If the resistance of such a thermometer reads 109.8 Ω, what is the temperature?

33. A piece of silicon has a resistance of 12.0 Ω at a temperature of 20°C. What is the temperature when its resistance is 10.0 Ω?

*34. You want to make a resistor of 1.0 Ω out of carbon rod of diameter 1.0 mm. How long a piece of carbon do you need?

*35. A lightning rod of iron has a diameter of 0.80 cm and a length of 0.50 m. During a lightning stroke, it carries a current of 1.0×10^4 A. What is the potential drop along the rod?

*36. The air conditioner in a home draws current of 12 A.
   (a) Suppose that the pair of wires connecting the air conditioner to the fuse box are No. 10 copper wire with a diameter of 0.259 cm and a length of 25 m each. What is the potential drop along each wire? Suppose that the voltage delivered to the home is exactly 115 V at the fuse box. What is the voltage delivered to the air conditioner?
   (b) Some older homes are wired with No. 12 copper wire with a diameter of 0.205 cm. Repeat the calculation of part (a) for this wire.

*37. Although aluminum has a somewhat higher resistivity than copper, it has the advantage of having a considerably lower density. Find the mass of a 100-m segment of aluminum cable 3.0 cm in diameter. Compare this with that of a copper cable of the same length and the same resistance. The densities of aluminum and of copper are 2.7×10^3 kg/m^3 and 8.9×10^3 kg/m^3, respectively.

*38. According to the National Electrical Code, the maximum permissible current in a No. 12 copper wire (diameter 0.205 cm) with rubber insulation is 25 A.
   (a) What is the potential drop along a 1.0-meter segment of the wire carrying this current?
   (b) What is the electric field in the wire?

*39. An aluminum wire of length 15 m is to carry a current of 25 A with a potential drop of no more than 5.0 V along its length. What is the minimum acceptable diameter of this cable?

*40. According to safety standards set by the American Boat and Yacht Council, the potential drop along a copper wire connecting a 12-V battery to an item of electrical equipment should not exceed 10%, that is, it should not exceed 1.2 V. Suppose that a 9.0-m wire (length measured around the circuit) carries a current of 25 A. What gauge of wire is required for compliance with the above standard? Use the table of wire gauges given in Problem 24. Repeat the calculation for currents of 35 A and 45 A.

*41. A parallel-plate capacitor with a plate area of 8.0×10^{-2} m^2 and a plate separation of 1.0×10^{-4} m is filled with polyethylene. If the potential difference between the plates is 2.0×10^4 V, what will be the current flowing through the polyethylene from one plate to the other?

*42. The windings of high-current electromagnets are often made of copper pipe. The current flows in the walls of the pipe, and cooling water flows in the interior of the pipe. Suppose the copper pipe has an outside diameter of 1.20 cm and an inside diameter of 0.80 cm. What is the resistance of 30 m of this copper pipe? What voltage must be applied to it if the current is to be 600 A?

*43. A 10.0-m length of coaxial cable with a solid inner conductor of diameter 1.00 mm and an outer conducting shell of inner diameter 4.00 mm has a dielectric of polyethylene between the conductors (see Fig. 26.19). What is the resistance of the dielectric cylindrical shell for current flow from one conductor to the other? What current flows when a potential difference of 300 V is applied between the conductors? (Hint: Treat the dielectric as a series combination of thin cylindrical shells connected radially.)

27.4 Resistances in Combination

44. Three resistors of 4.0 Ω, 6.0 Ω, and 8.0 Ω, respectively, are connected in series. What is the resistance of this combination? If the combination is connected to a 12-V battery, what is the current? What is the potential difference across each resistor?

45. Three resistors of 5.0 Ω, 7.0 Ω, and 9.0 Ω, respectively, are connected in parallel. What is the resistance of this combination? If this combination is connected to a 12-V battery, what is the net current? What is the current in each resistor?

46. A brass wire and an iron wire of equal diameters and of equal lengths are connected in parallel. Together they carry a current of 6.0 A. What is the current in each?

47. Consider the brass and iron wires described in Problem 46. What is the current in each if they are at a temperature of 90°C instead of the temperature of 20°C assumed in Table 27.2?

48. Two resistors are connected in parallel; their color codes are brown–red–orange and yellow–violet–red (see Fig. 27.13). What would be the color code of a single, equivalent resistor?

49. You have three resistors: 2.0 Ω, 3.0 Ω, and 4.0 Ω. What values of resistance can be obtained from combinations of some or all of these resistors?

50. When two resistors are connected in series, the net resistance is 80 Ω. When they are connected in parallel, the net resistance is 15 Ω. What are the values of the two resistors?

51. During operation, the effective resistances of a toaster, a microwave oven, and an electric frying pan are 11 Ω, 16 Ω, and 12 Ω, respectively. If the three are all plugged into a 115-volt outlet, what is the total current?

*52. An ordinary lightbulb draws 0.87 A of current when connected to a 115-V outlet. What is the resistance of the bulb? When a series arrangement of two such bulbs is connected to the 115-V outlet, the current through the combination is 0.69 A. What is the resistance of each bulb now? Why did it change?

*53. An electric cable of length 12.0 m consists of a copper wire of diameter 0.30 cm surrounded by a cylindrical layer of rubber insulation of thickness 0.10 cm. A potential difference of 6.0 V is applied to the ends of the cable.

(a) What will be the current in the copper?

(b) Taking into account the finite resistivity of the rubber (see Table 27.3), what will be the current in the rubber?

*54. A copper wire, of length 0.50 m and diameter 0.259 cm, has been accidentally cut by a saw. The region of the cut is 0.40 cm long, and in this region the remaining wire has a cross-sectional area of only one-quarter of the original area. What is the percentage increase of the resistance of the wire caused by this cut?

*55. An underground telephone cable, consisting of a pair of wires, has suffered a short somewhere along its length (Fig. 27.27). The telephone cable is 5.0 km long, and in order to discover where the short is, a technician first measures the resistance across terminals *AB*; then he measures the resistance across terminals *CD*. The first measurement yields 30 Ω; the second, 70 Ω. Where is the short?

FIGURE 27.27 A pair of wires with a short at the point *P* (the wires are touching at *P*).

*56. The air of the atmosphere has a slight conductivity due to the presence of a few free electrons and positive ions.

(a) Near the surface of the Earth, the vertical atmospheric electric field has a strength of about 100 V/m and the atmospheric current density is 4.0×10^{-12} A/m^2. What is the resistivity?

(b) The potential difference between the ionosphere (upper layer of atmosphere) and the surface of the Earth is 4.0×10^5 V. What is the total resistance of the atmosphere? (Hint: For the purposes of this problem assume that the vertical electric field has a constant strength throughout the atomsphere.)

*57. A flexible wire for an extension cord for electric appliances is made of 24 strands of fine copper wire, each of diameter 0.053 cm, tightly twisted together. What is the resistance of a length of 1.0 m of this kind of wire?

*58. Two copper wires of diameters 0.26 cm and 0.21 cm, respectively, are connected in parallel. What is the current in each if the combined current is 18 A?

*59. Commercially manufactured superconducting cables consist of filaments of superconducting wire embedded in a matrix of copper (see Fig. 27.28). As long as the filaments are superconducting, all the current flows in them, and no current flows in the copper. But if superconductivity suddenly fails because of a temperature increase, the current can spill into the copper; this prevents damage to the filaments of the superconductor. Calculate the resistance per meter of length of the copper matrix shown in Fig. 27.28. The diameter of the copper matrix is 0.70 mm, and each of the 896 filaments has a diameter of 0.010 mm.

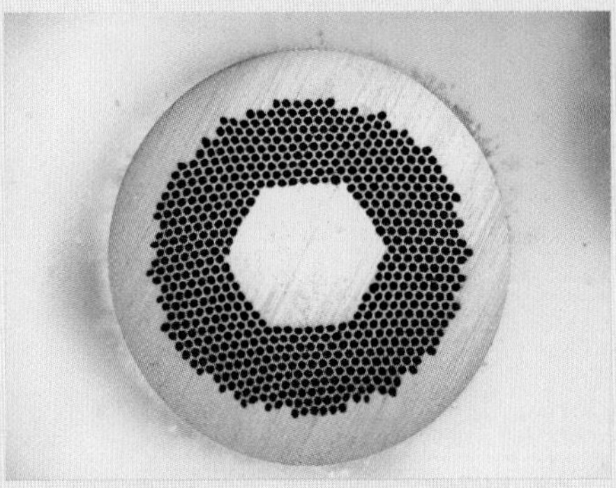

FIGURE 27.28 A cross-section of a strand of a superconducting cable, consisting of 896 filaments of a niobium–titanium superconductor embedded in a copper matrix.

*60. What is the net resistance of the combination of four resistors shown in Fig. 27.29? Each of the resistors has a resistance of 3.0 Ω.

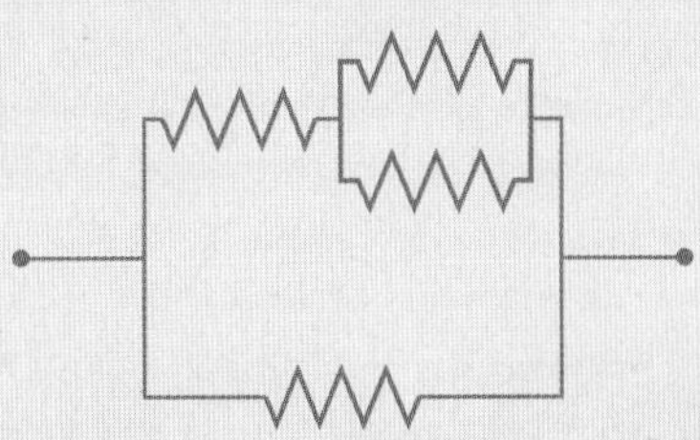

FIGURE 27.29 Four resistors.

*61. Three resistors with $R_1 = 2.0\ \Omega$, $R_2 = 4.0\ \Omega$, and $R_3 = 6.0\ \Omega$ are connected as shown in Fig. 27.30.

(a) Find the net resistance of the combination.

(b) Find the current that passes through the combination if a potential difference of 8.0 V is applied to the terminals.

(c) Find the potential drop and the current for each individual resistor.

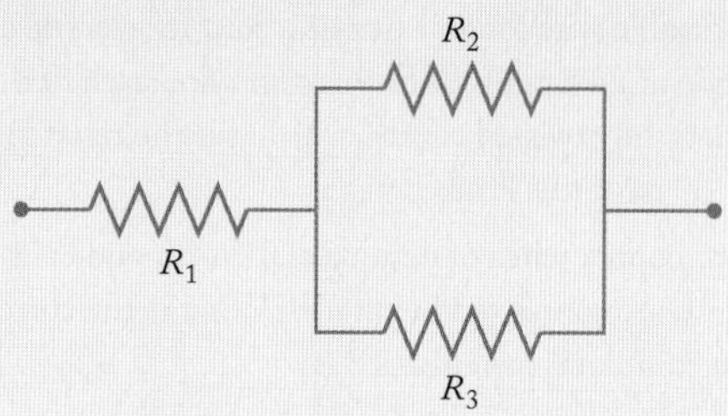

FIGURE 27.30 Three resistors.

*62. Three resistors with $R_1 = 4.0\ \Omega$, $R_2 = 6.0\ \Omega$, and $R_3 = 8.0\ \Omega$ are connected as shown in Fig. 27.31.

(a) Find the net resistance of the combination.
(b) Find the current that passes through the combination if a potential difference of 12.0 V is applied to the terminals.
(c) Find the potential drop and the current for each individual resistor.

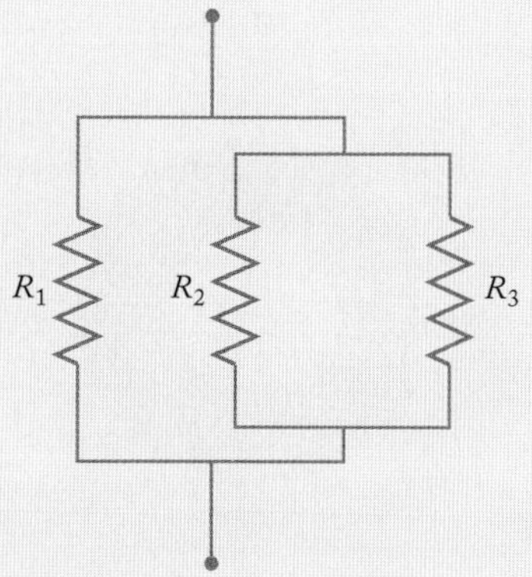

FIGURE 27.31 Three resistors.

*63. Consider the combination of three resistors described in the preceding problem. If we want a current of 6.0 A to flow through resistor R_2, what potential difference must we apply to the external terminals?

**64. Twelve resistors, each of resistance R, are connected along the edges of a cube (Fig. 27.32). What is the resistance between diagonally opposite corners of this cube? (Hint: Calculate $R_{net} = \Delta V/I_{total}$, where ΔV sums the three voltage drops along any path between opposite corners. Use symmetry to determine each current in terms of I_{total}.)

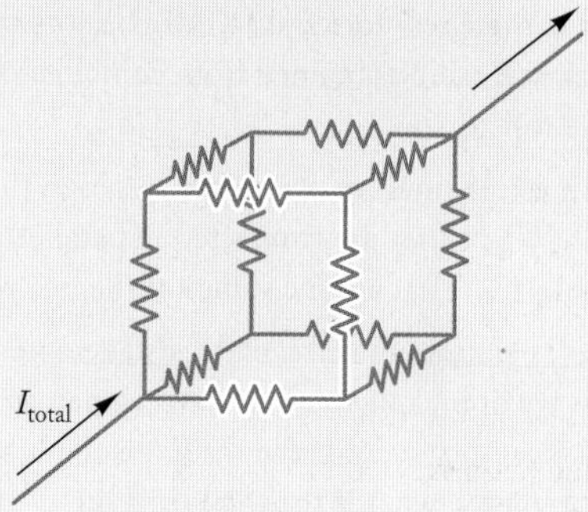

FIGURE 27.32 A cube with a resistor on each edge.

**65. What is the resistance of an infinite ladder of 1.0-Ω resistors connected as shown in Fig. 27.33a? (Hint: The ladder can be regarded as made of two pieces connected in parallel; see Fig. 27.33b.)

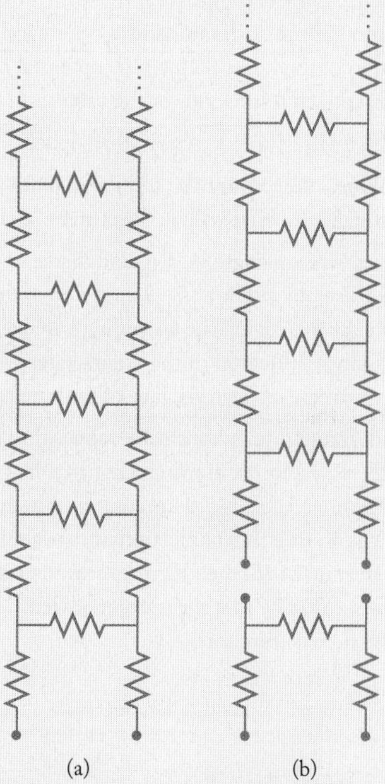

FIGURE 27.33 (a) An infinite "ladder" of resistors. The terminals are marked by the pair of dots. (b) One rung has been cut off from the ladder.

REVIEW PROBLEMS

66. Three resistors, with resistances of 3.0 Ω, 5.0 Ω, and 8.0 Ω, are connected in parallel. If this combination is connected to a 12.0-V battery, what is the current through each resistor? What is the current through the combination?

67. A fully charged automobile battery delivers 40 A for 1.0 h before it runs down. How much electric charge flows through the battery during this time? How many electrons pass through the battery?

68. A high-voltage transmission line consists of a copper cable of diameter 3.0 cm and length 250 km. Assume the cable carries a steady current of 1500 A.
    (a) What is the resistance of the cable?
    (b) What is the electric field inside the cable?

69. The filament of a lightbulb consists of a piece of tungsten wire, 5.0 cm long. If the potential difference across this filament is 115 V, what is the electric field in the wire?

70. Two wires of silver and of copper have the same lengths and the same diameters. If they are to have the same resistance, by how many degrees must the silver wire be warmer than the copper wire?

71. The resistivity of a certain material is found to increase by 4.45% when the material is heated from 20.0°C to 70.0°C. By what percentage will the resistivity of the material increase at 120.0°C compared with 20.0°C?

72. The electromagnet of a bell is wound with 8.2 m of copper wire of diameter 0.45 mm. What is the resistance of the wire? What is the current through the wire if the electromagnet is connected to a 12-V source?

73. The copper cable connecting the positive pole of a 12-V automobile battery to the starter motor is 0.60 m long and 0.50 cm in diameter.
    (a) What is the resistance of this cable?
    (b) When the starter motor is stalled, the current in the cable may be as much as 600 A. What is the potential drop along the cable under these conditions?

74. Ten equal strands of thin wire are held together side by side to form a thick wire of resistance R. Six of these strands are then separated and tied together end to end to form a new wire six times as long as the original wire. What is the resistance of the long, thin wire?

75. How can you connect four 1.0-Ω resistors so the net resistance of the combination is 1.0 Ω?

76. The resistance of the filament of the lightbulb of a flashlight is 8.0 Ω (at its operating temperature). The two batteries of the flashlight, of 1.5 V each, are connected in series, and they therefore supply 3.0 V. What is the current through the filament? What will be the current if you rewire the flashlight so it operates on a single battery? Assume that the resistance of the filament remains the same (actually, with less current, the filament will be cooler, and its resistance will be lower).

77. When connected in parallel across a 12-V battery, two resistors carry currents of 0.50 A and 0.75 A, respectively. What currents will these resistors carry if they are connected in series with the same battery?

78. A water pipe is made of iron with an outside diameter of 2.5 cm and an inside diameter of 2.0 cm. The pipe is used to ground an electric appliance. If a current of 20 A flows from the appliance into the water pipe, what fraction of this current will flow in the iron? What fraction in the water? Assume that water has a resistivity of 0.010 Ω·m.

79. An arc-welder uses No. 2 gauge copper wire (diameter 0.654 cm); during use, the cable carries 110 A. What is the current density? What is the drift velocity of the electrons?

80. Two resistors, 3.0 Ω and 5.0 Ω, are connected in series. The current though the 5.0 Ω resistor is 0.75 A. What is the potential difference across the series combination?

*81. The resistance of a square centimeter of dry human epidermis is about 1.0×10^5 Ω. Suppose that a (foolish) man firmly grasps two wires in his fists. The wires have a radius of 0.13 cm, and the skin of each hand is in full contact with the surface of the wire over a length of 8.0 cm (see Fig. 27.34).
    (a) Calculate the resistance the man offers to a current flowing through his body from one wire to the other. In this calculation you can neglect the resistance of the internal tissues of the human body, because the body fluids are reasonably good conductors and their resistance is small compared with the skin resistance.
    (b) What current will flow through the body of the man if the potential difference between the wires is 12 V? If it is 115 V? If it is 240 kV? What is your prognosis in each case? (A current above 0.001 A can cause injury, and a current above 0.02 A can be fatal; see Section 28.8.)

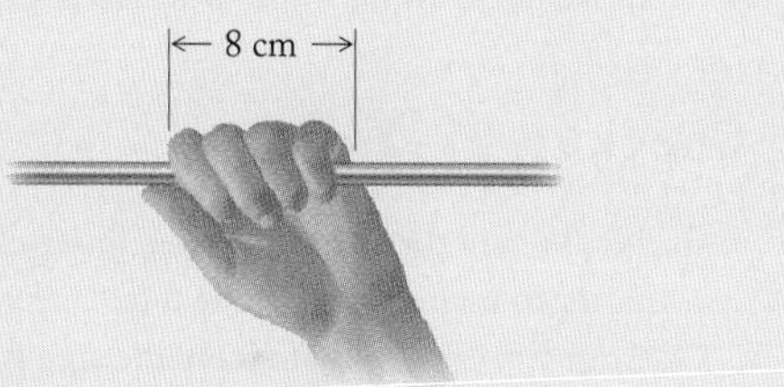

FIGURE 27.34 Contact between hand and wire.

*82. Three resistors, with $R_1 = 4.0$ Ω, $R_2 = 6.0$ Ω, and $R_3 = 2.0$ Ω, are connected as shown in Fig. 27.35. This combination is connected to a 1.5-V battery.
    (a) What is the net resistance of this combination?
    (b) What is the current through the combination?
    (c) What are the potential drop and the current for each resistor?

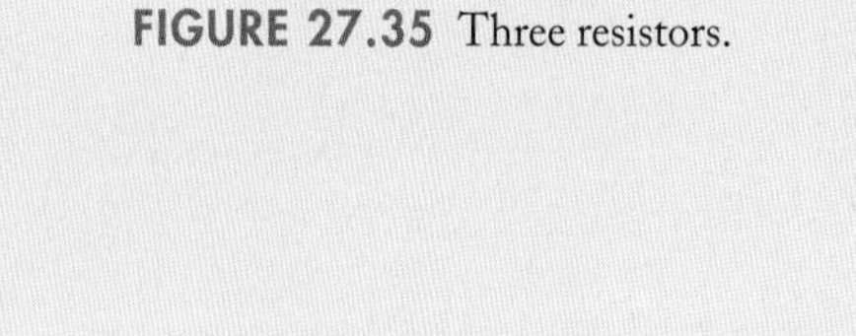

FIGURE 27.35 Three resistors.

Answers to Checkups

Checkup 27.1

1. When we cut a wire, the electrons do not continue to flow, nor do they spill out. Cutting a wire is analogous to closing off a water pipe; since air is not a conductor, the charges remain confined to the wire. Since the current is the same everywhere along a wire, cutting the wire stops the flow everywhere.
2. Yes—current is the flow past a given place; whether the cross-sectional area at that point is small or large does not matter. Of course, where the wire is thicker, that current is spread over a larger cross-sectional area.
3. Since current is defined as the direction of the flow of positive charge, an upwardly flowing current must have been pushed by the force of an upward electric field. Since the current in a lightning bolt is actually a flow of electrons, the motion of the (negative) electrons is downward, which is equivalent to an upward flow of positive charge.
4. (B) (a); (c). In (a), the positive charges move to the left and the negative to the right, so both contribute to a conventional current (the flow of positive charge) to the left. The same is true in (b), but there are fewer charges, so the current in (a) is larger. In (c) equal numbers of positive and negative charges move to the right, so there is zero net current.

Checkup 27.2

1. No. Where the wire is thinner, fewer electrons must move faster to maintain the same total current. One can also see from Eq. (27.11) that where the area is smaller, the electric field must be larger; thus the drift velocity, proportional to the electric field [Eq. (27.8)], is also larger.
2. From $R = \rho l/A$, we see that tripling the length triples the resistance. Also, doubling the diameter increases the area fourfold, and so decreases the resistance by a factor of 4.
3. From Ohm's Law, for a given resistance (for a given piece of wire), the current and voltage are proportional. So the 12-V battery results in 0.2 A of current, and the 18-V battery, 0.3 A.
4. The resistance of the half-length will be half as much as originally, so Ohm's Law tells us that for the same voltage, the current will double to 0.4 A.
5. (D) Decreases; decreases. In both cases (less frequent collisions or more electrons), charge can move more easily, so the resistivity (the opposition to charge flow) decreases.

Checkup 27.3

1. Since the geometrical factors are the same, the resistance will be proportional to the resistivity, by Eq. (27.17). From Table 27.2, the resistivity of nichrome ($100 \times 10^{-8}\ \Omega\cdot$m) exceeds that of iron ($10 \times 10^{-8}\ \Omega\cdot$m) by a factor of 10.
2. For equal lengths and equal resistances, the ratio of areas must be the same as the ratio of resistivities, according to Eq. (27.17). The ratio of the diameter of the iron wire to that of the silver one varies as the square root of the ratio of areas, and so is equal to $(\rho_{\text{iron}}/\rho_{\text{silver}})^{1/2} = (10/1.6)^{1/2} = (100/16)^{1/2} = 10/4 = 2.5$.
3. The percentage increase is 100 times the fractional increase of $\Delta R/R_0 = \alpha\,\Delta T$. For $\Delta T = 100°$C, the values of α in Table 27.2 give 0.1% for constantan and 60% for nickel.
4. (D) τ; n. In metals, the free-electron density n is essentially constant; it is the collision time τ that decreases with increasing temperature due to the increasing thermal motion of lattice atoms. When increasing the temperature of a semiconductor, a similar decrease in τ is overwhelmed by a much larger increase in n, because heating shakes many bound electrons free.

Checkup 27.4

1. Cutting the wire into two equal lengths results in two resistances, each with half the original resistance. Connecting two equal resistors in parallel again halves the resistance, resulting in a new resistance one-quarter of the original, or $\frac{3}{4}\ \Omega$.
2. One easily pictured reason is that resistances in series are analogous to a long wire—the longer the wire, the higher the resistance. Resistances in parallel are analogous to a thicker wire—the thicker the wire, the lower the resistance.
3. Since the potential difference (from the battery), the lengths, and the areas are all the same, the wire with the lower resistivity will have more current (using $I = \Delta V/R$ and $R = \rho l/A$). From Table 27.2, silver has a (slightly) lower resistivity than copper, and so will carry more current.
4. (C) 10 Ω; 0.1 Ω. In series, resistances add, so ten 1-Ω resistors provide a net resistance of 10 Ω. Resistors in parallel add such that the inverse of the net resistance R is the sum of the inverse resistances, so $R = 1/[10 \times 1/(1\ \Omega)] = 0.1\ \Omega$.

CHAPTER

28

Direct Current Circuits

CONCEPTS IN CONTEXT

Concepts in Context

Batteries such as these power our portable tools, appliances, and electronics. We can often represent a device connected to a battery as a resistor.

In this chapter, we will consider circuits with one or more batteries and one or more resistors. We can ask:

? How much energy can a typical battery supply? (Example 1, page 889)

? How does a battery operate? (Section 28.2, page 890)

? In a circuit with several batteries and resistors, how do we determine the current through each resistor? (Example 2, page 894; and Example 5, page 898)

? For such a circuit, what is the power supplied by each battery? What is the power dissipated in each resistor? (Example 6, page 902; and Example 7, page 904)

direct current (DC)

alternating current (AC)

The electric circuits installed in automobiles, homes, and factories carry one or the other of two different kinds of currents: **direct currents (DC)** and **alternating currents (AC)**. The direct current flows steadily along the wires of the circuit; it remains constant, except when it is switched on or off. The alternating current periodically reverses its direction of flow along the wire; like a pendulum swinging back and forth, the alternating current oscillates sinusoidally from one direction (positive) to the opposite direction (negative).

To keep any current flowing in the wires of the circuit, we must connect the ends of the wire to a "pump of electricity," a device that continuously supplies electric charges to one end of the wire and removes them from the other. What kind of current flows in the wire depends on what kind of pump we use. A steady pump—such as an automobile battery—produces a steady current. An alternating pump—such as the generator of the power station to which the circuits in your home are connected—produces an alternating current. In this chapter we will deal with direct currents produced by batteries or by other pumps of electricity that behave like batteries.

28.1 ELECTROMOTIVE FORCE

Online Concept Tutorial 32

Figure 28.1 shows a simple circuit consisting of a single (resistive) wire connected to the terminals of a battery. A steady, time-independent current will then flow around this circuit; in Fig. 28.1, we have indicated the direction of the current according to our convention that it is the direction of flow of (hypothetical) positive charges. The current flowing in this circuit is an example of a direct current—as long as the "strength" of the battery and the resistance of the wire remain constant, the current will also remain constant.

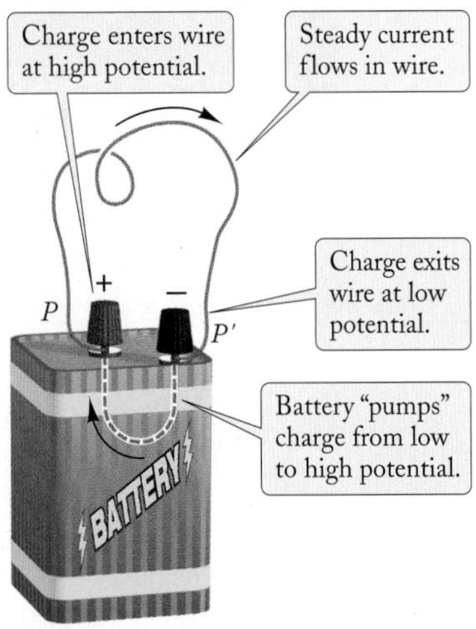

FIGURE 28.1 A resistive wire connected between the terminals of a battery. The potential energy of a (positive) charge is high when it is at the positive terminal P of the battery. The potential energy gradually decreases as the charge moves along the wire toward the negative terminal P'. The potential energy then again increases as the charge passes through the battery (dashed line), from P' to P.

The battery must do work on the charges in order to keep them flowing around the circuit. Suppose that a (hypothetical) positive charge is originally at the point P, at one terminal of the battery. Pushed along by the electric field, the charge moves along the wire. On the average, the kinetic energy that the charge gains from the electric field is dissipated by friction due to collisions within the wire, and the charge reaches the point P', at the other terminal of the battery, with no more kinetic energy than it had originally. Thus, on the average, the kinetic energy does not change. But the potential energy of the charge does change. The electric field is directed along the wire, and it does work on the charge; hence *the electric potential steadily decreases with distance along the wire*, and the charge reaches the point P' with a potential energy lower than its original potential energy. In order to keep the current flowing, the battery must "pump" the charge from the low-potential terminal to the high-potential terminal; that is, the battery must supply electric potential energy to the charge.

Our simple electric circuit is analogous to the hydraulic circuit illustrated in Fig. 28.2, consisting of a channel in which water runs down a hill and a pump that lifts the water from the bottom to the top of the hill. The water then flows around a closed hydraulic circuit, just as charge flows around the closed electric circuit. The wire is analogous to the channel in which the water runs down the hill, losing gravitational potential energy. The battery is analogous to the water pump that lifts water from the bottom to the top of the hill, thereby increasing the gravitational potential energy of the water. The water pump of Fig. 28.2 can be regarded as a source of gravitational potential energy—it produces this energy from an external supply of chemical or mechanical energy. Likewise, the battery, or "pump of electricity," in Fig. 28.1 can be regarded as a source of electric potential energy—it produces this energy from a supply of chemical energy.

To characterize the "strength" of a source of electric potential energy, we introduce the concept of **electromotive force**, or **emf.** The emf of a source of electric potential energy is defined as *the amount of electric energy delivered by the source per coulomb of positive charge as this charge passes through the source from the low-potential terminal to the high-potential terminal.* Since the emf is energy per unit charge, its units are volts. Note that the electromotive "force" is not a force, but an energy per coulomb—the confusing name became attached to it a long time ago, when physicists were not yet making a sharp distinction between force and energy. Because the units of emf are volts, the emf is often simply called the **voltage** of the source.

electromotive force (emf)

If a steady, time-independent current carries one coulomb of charge around the circuit of Fig. 28.1 from P to P' along the wire and from P' to P through the source of emf, then the energy that this charge receives from the source of emf must exactly match the energy it loses within the wire. If so, the charge returns to its starting point with exactly the same energy it had originally, and it can repeat this round trip again and again, in exactly the same manner. We can write this energy balance as

$$\mathcal{E} + \Delta V = 0 \tag{28.1}$$

where $\mathcal{E}$ represents the emf, or the increase of potential energy per coulomb of charge, due to the source and ΔV represents the change of potential energy along the wire (here, $\mathcal{E}$ is positive and ΔV is negative; see Fig. 28.3).

According to Eq. (28.1), the emf $\mathcal{E}$ has the same magnitude as the potential drop in the external circuit connected between the terminals of the source of emf. For example, a battery with an emf of 1.5 V (or 1.5 J/C) connected to an external circuit will do 1.5 J of work on a coulomb of positive charge that passes through the battery in the forward direction (from the − terminal to the + terminal), and the resistors and other devices in the external circuit will do −1.5 J of work on the charge as it flows around this circuit (from the + terminal to the − terminal).

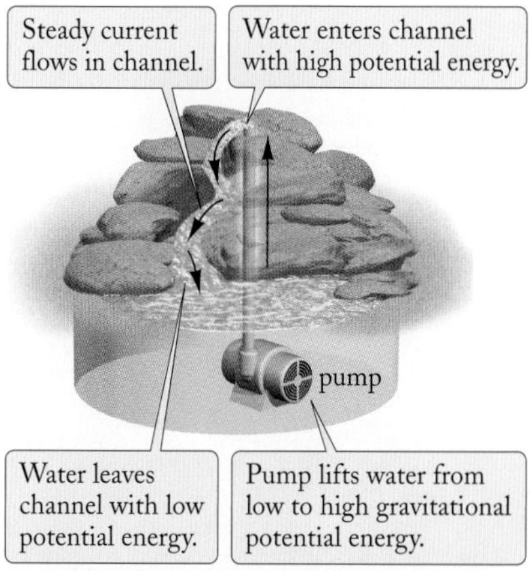

FIGURE 28.2 Mechanical analog of the battery–wire circuit of Fig. 28.1. The potential energy of a parcel of water is high at the top of the hill; it gradually decreases as the water flows down the hill; and it again increases as the pump lifts the water to the top.

EXAMPLE 1 A fresh flashlight battery with a voltage of 1.5 V will deliver a current of 1.0 A for about 1.0 h before running down completely. How much work does the battery do in this time interval?

Concepts — in — Context

SOLUTION: A 1.5-V battery does 1.5 J of work on each coulomb that passes through. If the current is 1.0 A, the charge that passes through in 1.0 h is 1.0 A × 3600 s = 3600 C. The total work is then the work per coulomb multiplied by the number of coulombs, that is, 1.5 J/C × 3600 C = 5400 J.

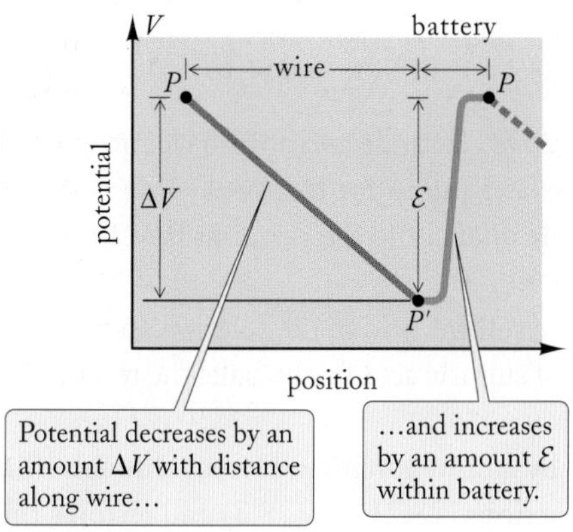

FIGURE 28.3 Plot of the electric potential, or the potential energy per coulomb of charge, as a function of position along the wire for the circuit illustrated in Fig. 28.1. The potential is highest at the point P (the positive pole of the battery), lowest at the point P' (the negative pole).

Note that if one coulomb of positive charge is forced through the battery in the *reverse* direction (from the + terminal to the − terminal), then the charge will deliver electric potential energy to the battery. The charge will then emerge from the battery at a potential that is 1.5 volts lower than the potential with which it entered. The energy delivered by the charge either will be stored within the battery (if it is a reversible, or rechargeable, battery) or else will merely be wasted as heat within the battery (if it is not a rechargeable battery).

Checkup 28.1

QUESTION 1: When a charge moves around a circuit such as shown in Fig. 28.1, the battery delivers energy to the charge. What happens to this energy?

QUESTION 2: The battery delivers positive energy to the current that flows around the circuit as illustrated in Fig. 28.1. Since this current actually consists of a flow of negative electrons, will the battery deliver negative energy to the electrons?

QUESTION 3: Suppose that we change the current through the battery in Example 1 to 2.0 A. As it runs down completely, the battery will then deliver:

(A) The same energy (B) Less energy (C) More energy

Online Concept Tutorial 32

28.2 SOURCES OF ELECTROMOTIVE FORCE

The most important kinds of sources of emf are batteries, electric generators, fuel cells, and solar cells. We will now briefly discuss the physical principles underlying the operation of some of these.

Concepts in Context

Batteries

lead–acid battery

Batteries convert chemical energy into electric energy. A very common type of battery is the **lead–acid battery**, which finds widespread use in automobiles. In its simplest form, this battery consists of two plates of lead—the positive electrode and the negative electrode—immersed in a solution of sulfuric acid (see Fig. 28.4). The positive electrode is covered with a layer of lead dioxide, PbO_2. When the external circuit is closed, the sulfuric acid reacts with the immersed surfaces of the electrodes. As already mentioned in Section 22.4, the reactions that occur at the negative and the positive electrodes are, respectively,

$$Pb + SO_4^{2-} \rightarrow PbSO_4 + 2e^- \quad (28.2)$$

$$PbO_2 + 4H^+ + SO_4^{2-} + 2e^- \rightarrow PbSO_4 + 2H_2O \quad (28.3)$$

These reactions deposit electrons on the negative electrode and absorb electrons from the positive electrode. Thus, the battery acts as a pump for electrons—the negative electrode is the outlet, the positive electrode the intake, and the electrons flow from one to the other via the external circuit.

The reactions (28.2) and (28.3) deplete the sulfuric acid in the solution and deposit lead sulfate on the electrodes. The depletion of sulfuric acid finally halts the reaction—the battery is then "discharged."

The lead–acid battery can be "charged" by simply forcing a current through it in the backward direction. This reverses the reactions (28.2) and (28.3) and restores the

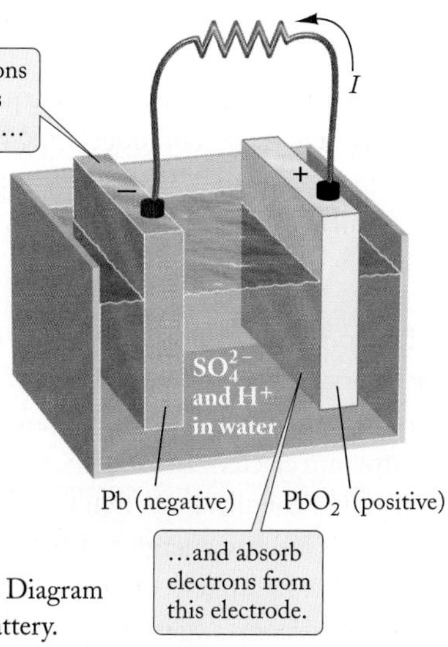

FIGURE 28.4 Diagram of a lead–acid battery.

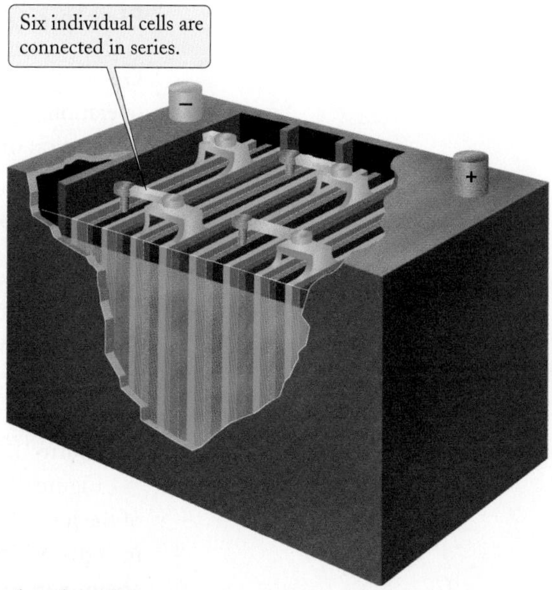

FIGURE 28.5 An automobile battery.

sulfuric acid solution. Note that what is stored in the battery during the charging process is not electric charge, but chemical energy. The number of positive and negative electric charges (protons and electrons) in the battery remains constant; what changes is the concentration of chemical compounds. A charged battery contains chemical compounds (lead, lead dioxide, sulfuric acid) of relatively high internal energy; a discharged battery contains chemical compounds (lead sulfate, water) of lower internal energy. Charging a battery is analogous to winding up the spring of a clock or pumping water into a reservoir on a high hill—in all these cases we are storing energy, which we can subsequently release upon demand.

In the single-cell battery shown in Fig. 28.4, the reactions (28.2) and (28.3) generate an emf of 2.0 V. In an automobile battery (see Fig. 28.5) six such cells are stacked together and connected in series to give an emf of 12.0 V (historically, the word *battery* originated from such stacking of single cells). The energy stored in such a battery is typically about 2×10^6 J.

dry cell

Another familiar type of battery is the **dry cell**, or flashlight battery. In an alkaline dry cell, the positive electrode consists of a carbon rod and the negative electrode of a cylinder of powdered zinc. The electrolyte in which these electrodes are "immersed" is a moist paste of potassium hydroxide and manganese dioxide (see Fig. 28.6). When this battery is connected to an external circuit, the chemical reactions at the electrodes convert chemical energy into electric energy and pump electrons from one electrode to the other via the external circuit. The emf of such a dry cell is 1.5 V. Since there is no liquid to slosh around, these batteries are particularly suitable for portable devices. The energy stored in a flashlight battery is typically of the order of 5000 J.

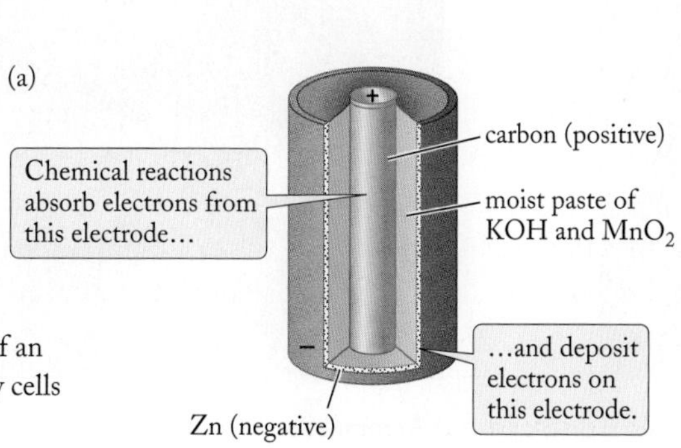

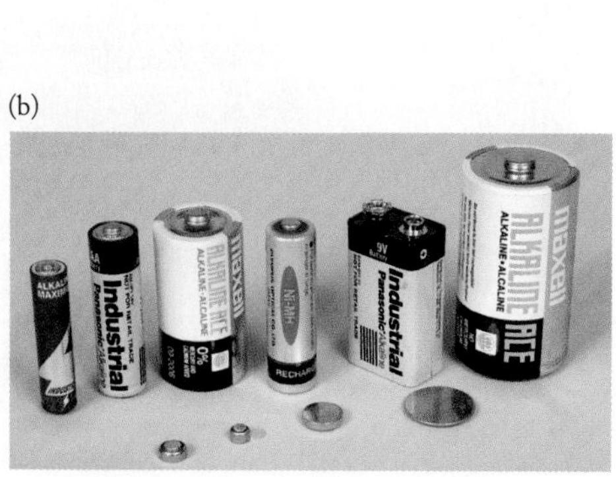

FIGURE 28.6 (a) Diagram of an alkaline dry cell. (b) Several dry cells of different sizes.

Electric Generators

Generators convert mechanical energy (kinetic energy) into electric energy. Their operation involves magnetic fields and the phenomenon of induction. We will leave the description of electric generators for Sections 31.2 and 31.3.

Fuel Cells

Fuel Cells resemble batteries in that they convert chemical energy into electric energy. However, in contrast to a battery, neither the high-energy chemicals nor the low-energy reaction products are stored inside the fuel cell. The former are supplied to the fuel cell from external tanks, and the latter are ejected. Essentially, the fuel cell acts as a combustion chamber in which a controlled chemical reaction takes place. The fuel cell "burns" a high-energy fuel, but produces electric energy rather than heat energy.

Figure 28.7a shows a fuel cell that burns a hydrogen–oxygen fuel. The electrodes of the fuel cell are hollow cylinders of porous carbon; oxygen at high pressure is pumped into the positive electrode and hydrogen into the negative electrode. The electrodes are immersed in a potassium hydroxide electrolyte. The reactions at the negative and positive electrode are, respectively,

$$2H_2 + 4OH^- \rightarrow 4H_2O + 4e^- \tag{28.4}$$

$$O_2 + 2H_2O + 4e^- \rightarrow 4OH^- \tag{28.5}$$

These reactions deposit electrons on the negative electrode and remove electrons from the positive electrode. This pumps electrons from one electrode to the other via the external circuit.

Note that the net result of the sequence of reactions (28.4) and (28.5) is the conversion of oxygen and hydrogen into water. This reaction is the reverse of the electrolysis of water (decomposition of water by an electric current). The excess water is removed from the cell in the form of water vapor.

All fuel cells produce a certain amount of waste heat. The best available fuel cells convert about 45% of the chemical energy of the fuel into electric energy and waste the remainder. Fuel cells have been put to practical use as power sources aboard the Apollo spacecraft and on Skylab (see Fig. 28.7b), and they have also been installed as power sources in experimental models of automobiles (Fig. 28.7c). They are compact and clean; on Skylab, the waste water eliminated from the fuel cell was used both for drinking and for washing.

(a)

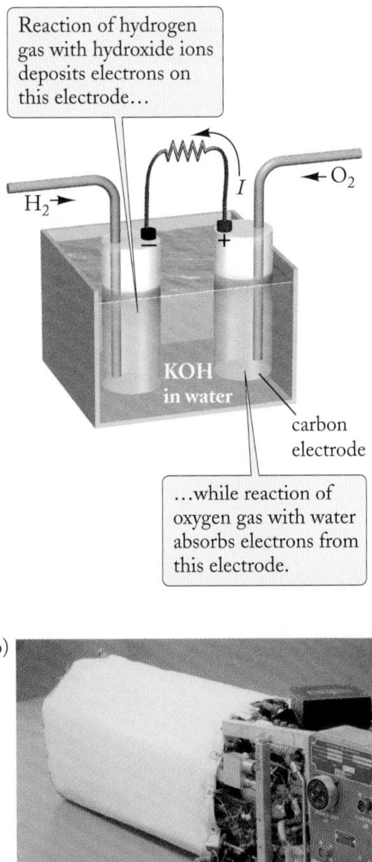

(b)

(c)

FIGURE 28.7 (a) Diagram of a fuel cell. (b) Fuel cell used on Skylab. (c) Automobile using fuel cells.

Solar Cells

Solar cells convert the energy of sunlight directly into electric energy. They are made of thin wafers of a semiconductor, such as silicon. We will discuss semiconductors and their applications in Chapter 39.

Checkup 28.2

QUESTION 1: Does a charged battery have more electric charge than a discharged battery?

QUESTION 2: A battery stores energy. In what form is this energy stored?

QUESTION 3: The fuel of a fuel cell stores energy. In what form is this energy stored?

QUESTION 4: A battery is being used to charge a capacitor, and during this charging the battery delivers energy to the capacitor. What is the form of this energy while it is still in the battery? What is the form of this energy when it is stored in the capacitor?

QUESTION 5: Two identical single-cell batteries are stacked in series; another two are connected in parallel. The two arrangements have a different:

(A) Amount of stored energy (B) Electromotive force (C) Form of stored energy

28.3 SINGLE-LOOP CIRCUITS

Online Concept Tutorial 32

The electric circuits in automobiles and in battery-operated tools or appliances, such as electric drills, flashlights, laptop computers, and portable telephones, contain one or several batteries or other sources of emf connected by wires to lightbulbs, electric motors, display screens, etc. In schematic diagrams of electric circuits, the latter devices can be represented by their resistances. Hence the schematic circuit diagram consists of one or several sources of emf connected to one or several resistors (see Fig. 28.8). In such a diagram, any source with a time-independent emf is represented by a stack of parallel short and long lines suggesting the plates of a lead–acid battery. The high-potential terminal is represented by a long thin line (marked with a plus sign), and the low-potential terminal by a short thick line (marked with a minus sign). If the terminals of such a source are connected to a network with resistances, a steady, direct current,

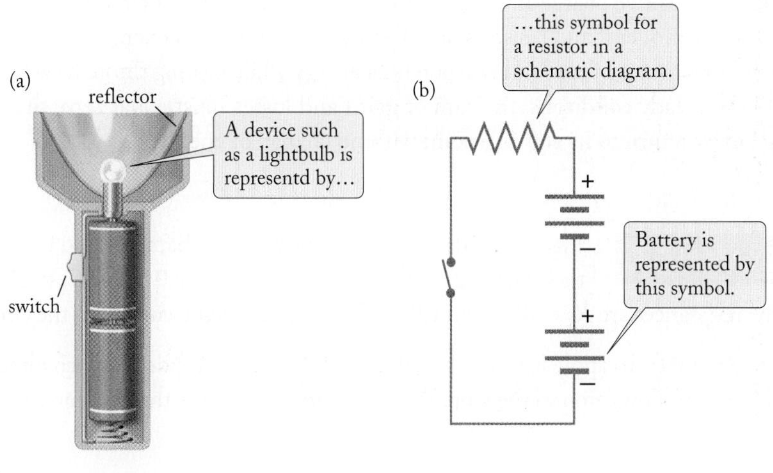

FIGURE 28.8 (a) A flashlight containing a lightbulb connected in series to two batteries. (b) Schematic circuit diagram for the flashlight. Each of the two batteries is represented by a stack of parallel short and long lines with plus and minus signs.

or DC, will flow through the network. In this section and the next we will learn how to calculate the currents that flow through the different parts of a circuit. We begin with simple circuits, consisting of a single closed loop. In a single-loop circuit, there is only one pathway, so *the same current flows past any point on a single loop.*

Figure 28.9 shows a schematic circuit diagram for such a single-loop circuit, consisting of a source of emf, such as a battery, connected to a resistor. The emf of the battery is $\mathcal{E}$ and the resistance of the resistor is R. The wires from the resistor to the battery are assumed to have negligible resistance (if higher accuracy is required, the resistance of the wires must be included in the circuit diagram). The resistance R in Fig. 28.9 could equally well represent a carbon resistor or some other device, such as a lightbulb, endowed with an electrical resistance; the resistance R could even represent the resistance of a wire by itself connected between the terminals of the battery. To find the current that flows through the circuit, we note that, according to Ohm's Law, the potential change across the resistor must be

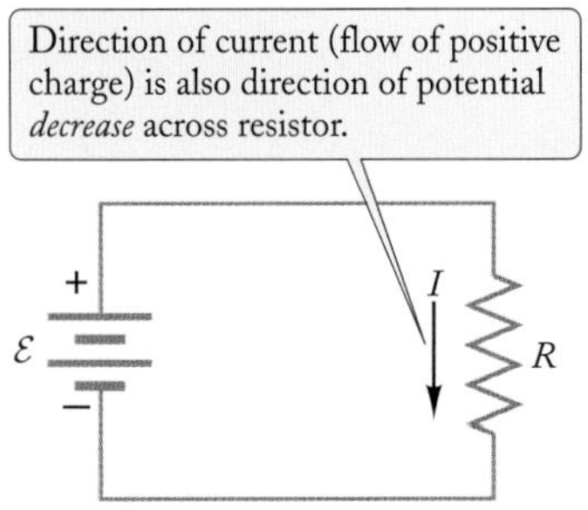

FIGURE 28.9 A simple circuit with a source of emf and a resistor.

$$\Delta V = -IR \tag{28.6}$$

Here the potential change has been reckoned in the direction of the arrow (see Fig. 28.9), from the upper end to the lower end of the resistor. The negative sign on the right side of Eq. (28.6) means that for a positive charge that moves around the circuit in the direction of the arrow, the potential decreases across the resistor. According to Eq. (28.1), the emf plus the potential change must equal zero; hence

$$\mathcal{E} - IR = 0 \tag{28.7}$$

from which

$$I = \frac{\mathcal{E}}{R} \tag{28.8}$$

Kirchhoff's voltage rule

Equation (28.7) is an instance of **Kirchhoff's voltage rule**, which states that when we go *around any closed loop in a circuit, the sum of all the emfs and all the potential changes across resistors and other circuit elements must equal zero.* In this sum, the emf must be considered positive (gain of potential energy) whenever we go through a source of emf in the forward direction, from the − terminal to the + terminal; and negative (loss of potential energy) whenever we go through a source in the backward direction, from the + terminal to the − terminal. Similarly, the potential change across a resistor is reckoned as negative ($-IR$, a voltage drop) when we go through the resistor in the same direction as the current, and as positive ($+IR$, a voltage rise) when we go through the resistor in the direction opposite to the current.

The proof of Kirchhoff's voltage rule is similar to the proof of Eq. (28.1). If one coulomb of positive charge flows once around a closed loop in a circuit with one or several sources of emf and resistors, it will gain or lose potential energy while passing through each source of emf and lose potential energy while passing through each resistor. Under steady conditons, the sum of gains and losses must equal zero, since the charge must return to its starting point with no change of energy.

Concepts — in — Context

EXAMPLE 2 Figure 28.10a shows a circuit with two batteries and two resistors. The emfs of the batteries are $\mathcal{E}_1 = 12.0$ V and $\mathcal{E}_2 = 15.0$ V; the resistances are $R_1 = 4.0\ \Omega$ and $R_2 = 2.0\ \Omega$. What is the current in the circuit?

SOLUTION: To apply Kirchhoff's voltage rule, we must decide in which direction the current flows around the loop. We will arbitrarily assume that the current flows in the clockwise direction (Fig. 28.10a). If this hypothesis is wrong, our calculation of the current, while still correct, will yield a negative value, and this will indicate that the direction of the actual current is opposite to the hypothetical direction.

According to Kirchhoff's voltage rule, the sum of all emfs and all potential changes across resistors is zero. To apply Kirchhoff's voltage rule, we can start at any point on a loop and go around the loop in either direction. If we choose to go around the loop in a clockwise direction, starting and ending at the lower left corner (Fig. 28.10b), we have

$$\mathcal{E}_1 - IR_1 - \mathcal{E}_2 - IR_2 = 0 \tag{28.9}$$

Note that $\mathcal{E}_2$ enters with a negative sign into this equation, since our clockwise path passes through this source of emf in the backward direction. Also, both resistive potential changes are here reckoned as negative, since our clockwise path is in both cases parallel to the assumed current direction. Solving Eq. (28.9) for the current I provides the solution

$$I = \frac{\mathcal{E}_1 - \mathcal{E}_2}{R_1 + R_2} \tag{28.10}$$

or

$$I = \frac{12.0\text{ V} - 15.0\text{ V}}{4.0\ \Omega + 2.0\ \Omega} = -0.50\text{ A} \tag{28.11}$$

Here, the negative sign indicates that the current is *not* clockwise as originally chosen, but counterclockwise instead.

COMMENTS: We could have guessed the direction of the current, since it is obvious that the stronger battery on the right will force the current backward through the weaker battery on the left. But in more complicated circuits the direction of the current will not be so obvious, and we will have to discover the direction by carefully keeping track of the signs in our calculations.

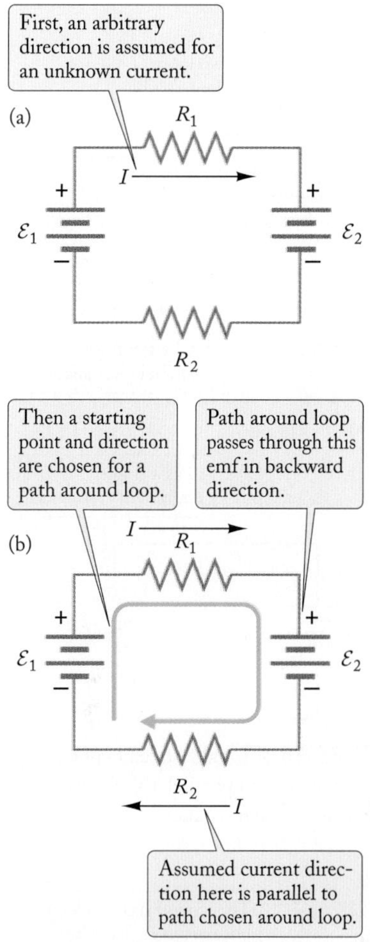

FIGURE 28.10 (a) Two sources of emf and two resistors. (b) Same sources of emf and resistors, with a path chosen for the application of Kirchhoff's voltage rule.

In Example 2 we neglected the **internal resistance** of the batteries. The electrolyte in a battery always has some resistance, and this causes the current to suffer a voltage drop even before it leaves the external terminals of the battery. The nominal emf $\mathcal{E}$ quoted on the labels of batteries refers to the potential difference between the terminals when no current is flowing; this is often called the "open-circuit" voltage. The internal resistance R_i of the battery may be regarded as connected in series with the emf $\mathcal{E}$ (Fig. 28.11). When a current I is flowing, the voltage drops by $\Delta V = -IR_i$ across the internal resistance, and hence the remaining voltage at the external terminals of the battery will be $\mathcal{E} - IR_i$. The internal resistance of a good battery is small, and can often be neglected. But if we need to take it into account, we can do so by simply placing the appropriate internal resistance in series with each battery in the circuit diagram; then we can proceed as usual with the calculation of the currents.

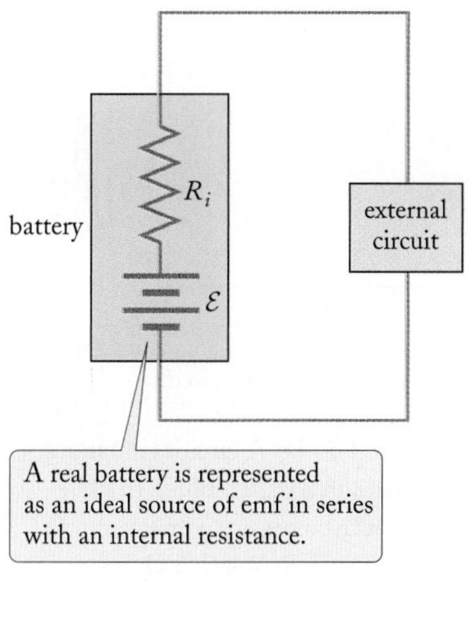

FIGURE 28.11 The internal resistance R_i is in series with the nominal emf $\mathcal{E}$.

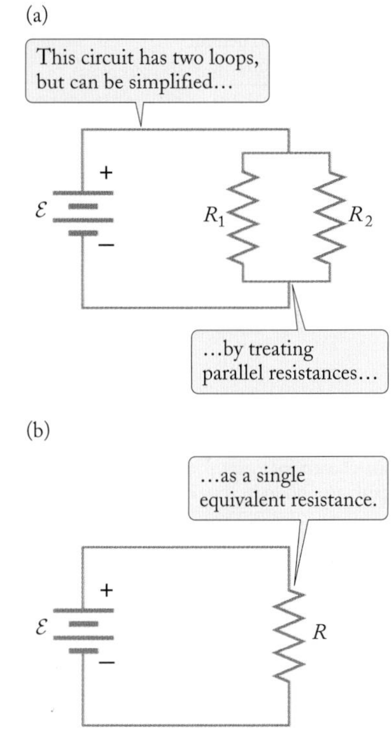

FIGURE 28.12 (a) A circuit containing two resistances in parallel. (b) Equivalent circuit with one resistance.

EXAMPLE 3 An alkaline flashlight battery of nominal emf 1.5 V has an internal resistance of 0.12 Ω. What will be the potential difference across its terminals if the battery is delivering a current of 1 A? What if the battery is delivering a current of 5 A?

SOLUTION: As discussed above, the potential difference available at the terminals will be the emf of the source minus the internal resistive voltage drop. For a current of 1 A, the voltage across the terminals will be

$$\mathcal{E} - IR_i = 1.5\text{ V} - 1\text{ A} \times 0.12\ \Omega = 1.5\text{ V} - 0.1\text{ V} = 1.4\text{ V}$$

For a current of 5 A, the voltage across the terminals will be

$$\mathcal{E} - IR_i = 1.5\text{ V} - 5\text{ A} \times 0.12\ \Omega = 1.5\text{ V} - 0.6\text{ V} = 0.9\text{ V}$$

A circuit that contains two or more resistors in parallel, such as the circuit shown in Fig. 28.12, is not a genuine single-loop circuit, since the parallel resistors and their connecting wires form additional loops. The general method for dealing with multi-loop circuits will be discussed in the next section. However, circuits with parallel resistors, such as the circuit in Fig. 28.12, can be handled by the same method as single-loop circuits because, as we know from the preceding chapter, the parallel resistances are effectively equivalent to a single resistance [see Eq. (27.29)].

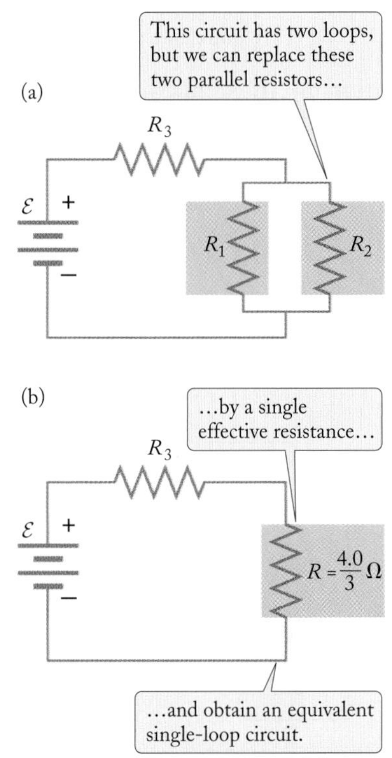

FIGURE 28.13 (a) A circuit containing two resistances in parallel (blue). (b) The two parallel resistances are equivalent to a single resistance of $R = 4.0/3\ \Omega$.

EXAMPLE 4 Suppose that, in the circuit shown in Fig. 28.13a, the emf of the battery is 12 V and the resistances are $R_1 = 4.0\ \Omega$, $R_2 = 2.0\ \Omega$, and $R_3 = 3.0\ \Omega$. What is the current through the battery in this circuit?

SOLUTION: The current through the source flows through R_3 and branches between R_1 and R_2. The net resistance of the parallel combination is given by Eq. (27.29),

$$\frac{1}{R} = \frac{1}{R_1} + \frac{1}{R_2} = \frac{1}{2.0\ \Omega} + \frac{1}{4.0\ \Omega} = \frac{3}{4.0\ \Omega}$$

or

$$R = \frac{4.0}{3}\ \Omega = 1.3\ \Omega$$

The circuit shown in Fig. 28.13a is therefore effectively equivalent to the single-loop circuit shown in Fig. 28.13b. Kirchhoff's rule for this single-loop circuit with resistances of $R = 1.3\ \Omega$ and $R_3 = 3.0\ \Omega$ in series with an emf of 12 V tells us

$$\mathcal{E} - IR - IR_3 = 0$$

We can now solve for the unknown current I:

$$I = \frac{\mathcal{E}}{R + R_3} = \frac{12\text{ V}}{1.3\ \Omega + 3.0\ \Omega} = 2.8\text{ A}$$

COMMENTS: The currents in the individual parallel resistors can be found from this total current in the same way as in Example 8 of Chapter 27. Namely, the potential across each parallel resistor is $\Delta V = IR = 2.8\ \text{A} \times 4/3\ \Omega = 3.7$ volts, so the individual currents are $I_1 = \Delta V/R_1 = 3.7\ \text{volts}/4.0\ \Omega = 0.9$ A and $I_2 = \Delta V/R_2 = 3.7\ \text{volts}/2.0\ \Omega = 1.9$ A.

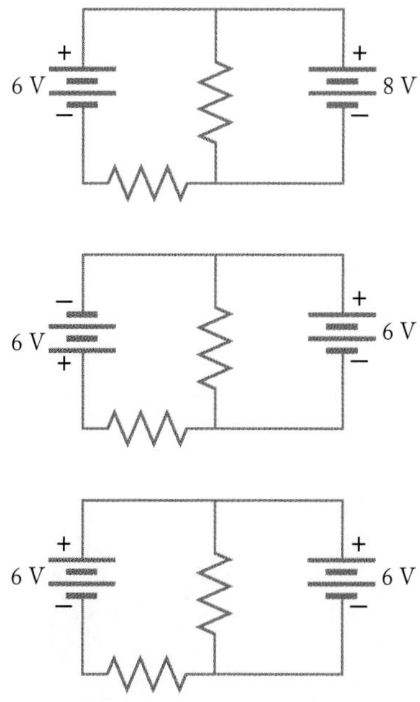

FIGURE 28.14 Several circuits with batteries.

✔ Checkup 28.3

QUESTION 1: A circuit contains several batteries and several resistors. If we replace all the batteries by new batteries of twice the emf, how will the current change? If we next replace all the resistors by new resistors of twice the resistance?

QUESTION 2: Suppose that in a circuit we reverse all the batteries, that is, we turn them end for end. What happens to the current? What if we reverse all the resistors, that is, turn them end for end?

QUESTION 3: Figure 28.14 shows several circuits containing batteries. In each case, what are the directions of the electric currents?

QUESTION 4: Suppose that a battery with emf $\mathcal{E}$ has a somewhat large internal resistance, $R_i \approx 2\ \Omega$. If we connect a thick piece of wire with a resistance $R \approx 0.1\ \Omega$ between the terminals of the battery, the voltage at the terminals of the battery is

(A) Exactly equal to $\mathcal{E}$ (B) Slightly less than $\mathcal{E}$ (C) Much less than $\mathcal{E}$

28.4 MULTI-LOOP CIRCUITS

Online Concept 32 Tutorial

If several sources of emf and several resistors are connected in some complicated circuit with several branches, then the currents will flow along several alternative paths. Kirchhoff's voltage rule can be applied to any loop in the complicated circuit. In doing so, however, we must be able to identify the separate currents in each separate branch. If there is only one branch in some part of a circuit, the current is the same past every point. But when the current encounters a junction, it will split along two or more branches, as illustrated in Fig. 28.15. We can relate the separate currents using **Kirchhoff's current rule**: *the total current flowing into a junction is equal to the total current flowing out of the junction*, or

Kirchhoff's current rule

$$\sum I_{\text{in}} = \sum I_{\text{out}} \tag{28.12}$$

This rule asserts that charge is not accumulating at a junction, nor is it being depleted from the junction; the current is merely flowing through it.

Thus, for the currents flowing into and out of the junction in Fig. 28.15, with the directions as indicated, Kirchhoff's current rule provides the relation

$$I_1 = I_2 + I_3$$

As noted in Example 2, in many applications we might not initially know the directions of the currents. In that case, we choose an arbitrary direction for each separate current. We can then write down the simultaneous set of equations for the unknown currents according to Kirchhoff's rules, assured that when we solve these equations, the algebraic sign of each result will tell us whether that current is in the chosen direction (positive) or opposite to the chosen direction (negative).

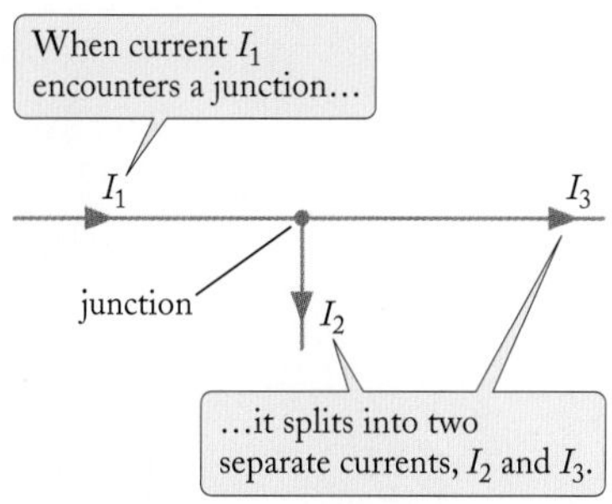

FIGURE 28.15 A junction.

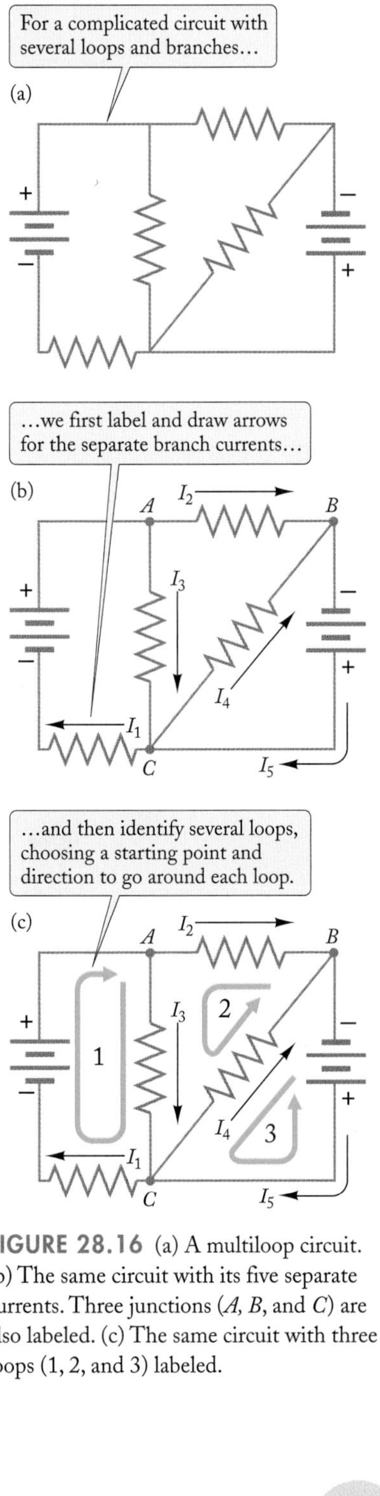

FIGURE 28.16 (a) A multiloop circuit. (b) The same circuit with its five separate currents. Three junctions (*A*, *B*, and *C*) are also labeled. (c) The same circuit with three loops (1, 2, and 3) labeled.

For instance, Fig. 28.16a shows a complicated circuit with several branches. For each branch, we can identify the current and choose a direction for that current, as shown in Fig. 28.16b. Kirchhoff's current rule can be applied at several junctions to relate these currents. Also, the circuit has several loops, and we can apply Kirchhoff's voltage rule to any of these, for example, the rectangular loop (number 1) and two triangular loops (numbered 2 and 3) in Fig. 28.16c. After we choose an arbitrary starting point and direction to go around a particular loop, we sum the potential changes for that path around the loop and set the sum equal to zero. Recall from the previous section that if our path around a loop takes us through a resistor in the same direction as the current, we must write the voltage change as $-IR$ (a voltage drop); if our path takes us through a resistor in the opposite direction to the current, we must write the voltage change as $+IR$ (a voltage rise).

So the procedure for obtaining the necessary equations is as follows:

1. Identify the separate currents through each branch in the circuit, and assign each current a label (I_1, I_2, I_3, . . .) and an arbitrary direction (by drawing an arrow).
2. Write down Kirchhoff's current rule, $\sum I_{in} = \sum I_{out}$, at various junctions. Choose enough junctions so that every current is included in at least one equation, and choose only junctions with at least one current not in any other equation.
3. Regard the circuit as a collection of closed loops. Identify enough loops so that every part of the circuit is included in at least one of the loops. The loops may overlap, but each loop must have at least some circuit element that does not overlap with any other loop.
4. Apply Kirchhoff's voltage rule to each loop: the sum of all the emfs and all the potential changes across resistors must add to zero as we go around a loop. Note that when writing the potential change across a resistor, $\Delta V = \pm IR$, we must take the negative sign if we traverse the resistor in the direction of the current, and the positive sign if we traverse in a direction opposite to the current. Similarly, a source of emf $\mathcal{E}$ must be considered positive, $+\mathcal{E}$, whenever we go through the source in the forward direction, from the $-$ to the $+$ terminals; and negative, $-\mathcal{E}$, whenever we go though the source in the reverse direction, from the $+$ to the $-$ terminal.

This procedure will result in the right number of equations for the unknown currents $I_1, I_2, I_3, \ldots$ We can then solve the equations for these unknowns by the standard mathematical methods for the solution of simultaneous equations for several unknowns (see Appendix A2). If a current turns out to be negative, its direction is opposite to the direction assigned in step 1.

The following example illustrates this method for the two-loop circuit shown in Fig. 28.17.

Concepts in Context

EXAMPLE 5 Suppose that the emfs and the resistors in the circuit shown in Fig. 28.17a are $\mathcal{E}_1 = 12.0$ V, $\mathcal{E}_2 = 8.0$ V, $R_1 = 4.00\ \Omega$, $R_2 = 4.00\ \Omega$, and $R_3 = 2.00\ \Omega$. Find the current in each of the resistors.

SOLUTION: First, we can label and assign directions to the unknown currents at the junction at the top center of the figure, point *A* (see Fig. 28.17b). We apply Kirchhoff's current rule at point *A*:

$$I_1 = I_2 + I_3 \qquad (28.13)$$

Note that for this example, these are the only separate currents in the circuit; there are only three branches in this circuit, so applying Kirchhoff's current rule at a single junction is sufficient. [If we were to apply Kirchhoff's current rule at the junction at the bottom center of the circuit, we would obtain the same information as in Eq. (28.13), since the same three currents encounter that junction.]

Now let us apply Kirchhoff's voltage rule to the left and right loops of the circuit. In this example, we will choose to start at point A and traverse each loop in a clockwise direction, although any starting point and either direction will work for any given loop. For the left loop (number 1) we then obtain

$$-I_2R_2 + \mathcal{E}_1 - I_1R_1 = 0 \tag{28.14}$$

and for the right loop (number 2), we obtain

$$-\mathcal{E}_2 - I_3R_3 + I_2R_2 = 0 \tag{28.15}$$

Note that the signs in Eq. (28.14) and (28.15) have been chosen according to step (4) of the procedure outlined above. For example, the emf in the first equation is written $+\mathcal{E}_1$, since our path went from the $-$ to the $+$ terminal of $\mathcal{E}_1$; whereas the emf in the second equation is $-\mathcal{E}_2$, since that path went through $\mathcal{E}_2$ from the $+$ to the $-$ terminal. Similarly, in the second equation, the voltage change across resistor R_3 is written as $-I_3R_3$, since our path was parallel to the direction chosen for I_3; whereas the voltage change across resistor R_2 is written as $+I_2R_2$, since the path was opposite to the direction chosen for I_2.

Kirchhoff's rules have thus provided the three equations (28.13), (28.14), and (28.15) for the three unknown currents I_1, I_2, and I_3.

Before proceeding to the solution of these equations, it is convenient to substitute the numerical values for the known quantities $\mathcal{E}$ and R. With these numerical values, Eqs. (28.14) and (28.15) become

$$-4I_2 + 12 - 4I_1 = 0 \tag{28.16}$$

$$-8 - 2I_3 + 4I_2 = 0 \tag{28.17}$$

To solve these equations for the three unknowns I_1, I_2, and I_3, we first eliminate I_1 by substituting $I_1 = I_2 + I_3$ from Eq. (28.13) into Eq. (28.16):

$$-4I_2 + 12 - 4(I_2 + I_3) = 0$$

or

$$12 - 8I_2 - 4I_3 = 0 \tag{28.18}$$

Now Eqs. (28.17) and (28.18) contain only the unknowns I_2 and I_3. If we solve Eq. (28.17) for I_2, we find

$$I_2 = \frac{2I_3 + 8}{4} = \frac{1}{2}I_3 + 2 \tag{28.19}$$

We can now eliminate I_2 by substituting Eq. (28.19) into Eq. (28.18):

$$12 - 8(\tfrac{1}{2}I_3 + 2) - 4I_3 = 0$$

or

$$-4 - 8I_3 = 0$$

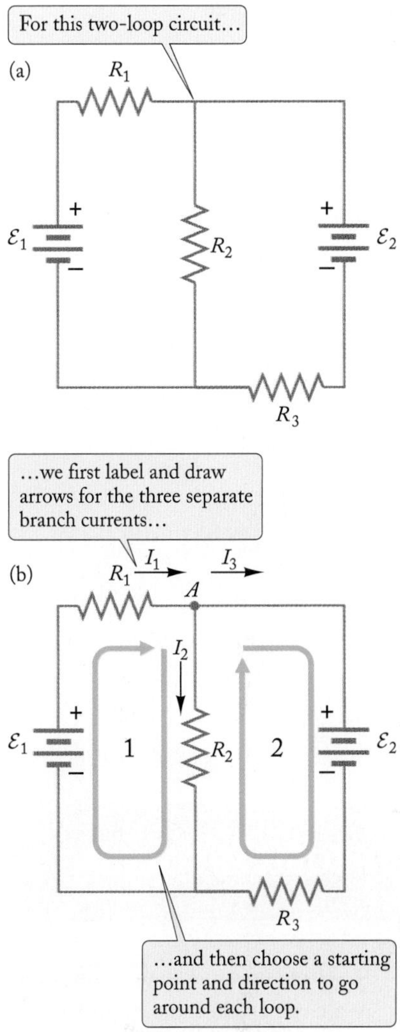

FIGURE 28.17 (a) A two-loop circuit. Any two loops are sufficent to include every part of the circuit. (b) The separate currents I_1, I_2, and I_3 and two loops have been labeled.

This immediately gives

$$I_3 = -\tfrac{1}{2} = -0.50$$

Then Eq. (28.19) gives

$$I_2 = \tfrac{1}{2}I_3 + 2 = -\tfrac{1}{4} + 2 = 1.75$$

and Eq. (28.13) gives

$$I_1 = I_2 + I_3 = 1.75 + (-0.50) = 1.25$$

These currents are, of course, measured in amperes; that is, $I_1 = 1.25$ A, $I_2 = 1.75$ A, and $I_3 = -0.50$ A. The negative sign on I_3 indicates the current in that branch is actually opposite to the direction chosen in Fig. 28.17b.

COMMENT: We could have applied Kirchhoff's rules to a different junction or a different loop in the circuit, but we would have obtained the same results. For example, the larger loop around the outside of the entire circuit could have been substituted for one of the loops used above. However, once loops have been chosen which include every part of the circuit, no new information is obtained by considering other possible loops.

PROBLEM-SOLVING TECHNIQUES — KIRCHHOFF'S RULES AND MULTILOOP CIRCUITS

In setting up the equations for the several junctions and loops in a circuit, you must handle the directions of the currents and the sign conventions for the currents and potentials consistently. To maintain this consistency, follow these steps:

1 First, label the currents in the separate branches and assign a direction for each by drawing an arrow.

2 Choose a junction and apply Kirchhoff's current rule, $\Sigma I_{in} = \Sigma I_{out}$. Continue to do this for other junctions until every separate current is included in at least one equation. Choose only junctions that include at least one current not in any previously chosen junction.

3 Then identify a sufficient number of loops so that all circuit elements of the circuit are included in at least one loop, and each loop has at least some circuit element that is not included in any other loop.

4 Start from any point on each loop and proceed around the loop in either direction, and apply Kirchhoff's voltage rule: the sum of all the potential differences is zero. Note that each emf contributes a positive term if the direction of the chosen path is the forward direction for that emf (from − to + terminal), and it contributes a negative term if the direction is the backward direction. Each resistor contributes a term of the form ∓ [current] × [resistance], where the − sign is used if the direction of the chosen path is the same as the current and the + sign is used if the direction is opposite to the current.

5 Once Kirchhoff's rules have been properly applied, you have enough algebraic equations to find the unknown branch currents. To solve such a system of equations with several unknowns, eliminate unknowns successively, until you obtain one equation with only one unknown [as illustrated by the solution of the system of three equations (28.13), (28.16), and (28.17) in Example 5].

6 If the potential difference between two given points in the circuit is needed, simply sum all the emfs and voltage changes in resistors along a path from one point to another, complying with the same sign conventions as for a closed loop.

Checkup 28.4

QUESTION 1: If we want to use Kirchhoff's rules for the circuit shown in Fig. 28.16, how many closed loops do we need? How many junctions must be evaluated?

QUESTION 2: If we want to use Kirchhoff's rules in Fig. 28.12 (instead of the simple treatment exploiting the parallel resistors used in Example 4), how many loops do we need?

QUESTION 3: You wish to evaluate Kirchhoff's voltage rule for loop number 1 in Fig. 28.16c, using the indicated current directions; the resistances are numbered according to their corresponding currents. You start at point A and proceed clockwise. The correct equation is:

(A) $-I_3R_3 - I_1R_1 + \mathcal{E}_1 = 0$ (B) $-I_3R_3 - I_1R_1 - \mathcal{E}_1 = 0$

(C) $-I_3R_3 + I_1R_1 + \mathcal{E}_1 = 0$ (D) $I_3R_3 - I_1R_1 + \mathcal{E}_1 = 0$

28.5 ENERGY IN CIRCUITS; JOULE HEAT

As we saw in Section 28.1, to keep a current flowing in a circuit, the batteries or other sources of emf must do work. If an amount of charge dq passes through a source of emf, the amount of work the source does will be [compare Eq. (25.6)]

$$dW = \mathcal{E}\, dq \tag{28.20}$$

Hence the rate at which the source does work is

$$\frac{dW}{dt} = \mathcal{E}\frac{dq}{dt} \tag{28.21}$$

The rate of work is the power P [see Eq. (8.34)]; the rate of flow of charge is the current. Equation (28.21) therefore asserts that the electric power P delivered by the source of emf to a current I is

$$P = \mathcal{E}I \tag{28.22}$$

power delivered by emf

Thus, the power delivered by the "pump of electricity" is large if it pumps a large current through a large potential difference. In terms of our analogy with a hydraulic pump, this merely means that the power delivered by a hydraulic pump is large if it pumps water at a fast rate (large water current) and lifts the water to a large height (large gravitational potential difference).

Note that consistency of units in Eq. (28.22) demands that the product of the unit of current (1 ampere) and the unit of emf (1 volt) be equal to the unit of power (1 watt). Indeed,

$$1\text{ A} \times 1\text{ V} = 1\text{ C/s} \times 1\text{ J/C} = 1\text{ J/s} = 1\text{ W}$$

Also note that in Eq. (28.22) we have not yet taken into account the algebraic sign of the power. We will have to attach a positive sign to the power if the current passes through the source in the forward direction and a negative sign if the current passes through in the backward direction. In the former case, the source delivers energy to the current, and in the latter case the source receives energy from the current.

Concepts in Context

EXAMPLE 6 What power do the two batteries described in Example 5 deliver?

SOLUTION: The current through the first battery is 1.25 A, and its emf is 12.0 V. This current passes through the battery in the forward direction; hence the power delivered by the battery is

$$P = 1.25\text{ A} \times 12.0\text{ V} = 15.0\text{ W}$$

Likewise, the 0.50-A current passing through the second battery is also in the forward direction, so the power delivered by this battery is

$$P = 0.50\text{ A} \times 8.0\text{ V} = 4.0\text{ W}$$

The net power delivered by both batteries is 19.0 W.

COMMENT: In this example the power delivered by each battery is positive, so the two simply add to 19.0 W. But in other examples one battery may be delivering power while another receives it. Thus, in the single-loop circuit of Fig. 28.10 (Example 2), with a counterclockwise current of 0.50 A, the 15.0-V battery delivers 7.5 W, while the 12.0-V battery receives 6.0 W. Thus for the circuit of Fig. 28.10, the *net* power delivered by both batteries is 7.5 W − 6.0 W = 1.5 W.

The electric potential energy acquired by the charges is carried along the circuit to the resistors, and it is continually dissipated in the resistors. If within a given resistor the charge dq suffers a potential change ΔV (regarded as a positive quantity), then the loss of potential energy is $dU = \Delta V\, dq$, and the rate at which energy is dissipated is

$$\frac{dU}{dt} = \Delta V \frac{dq}{dt}$$

Hence the power dissipated in the resistor is

power dissipated in resistor

$$P = \Delta V I \tag{28.23}$$

By means of Ohm's Law, $\Delta V = IR$, we can also write the power two more ways,

Joule heating

$$P = I^2 R \quad \text{and} \quad P = \frac{(\Delta V)^2}{R} \tag{28.24}$$

Figure 28.18 provides a summary of the equations relating power, current, potential difference, and resistance. When any two of these quantities are known, the other two can be calculated.

The energy lost by the charges during their passage through a resistor generates thermal energy; that is, it generates disordered, microscopic kinetic and potential energy of the atoms in the resistor. This *conversion of electric energy into thermal energy in a resistor is called* **Joule heating.**

Many simple electric appliances—such as electric toasters, heaters, hot plates, irons, blankets, lightbulbs—rely on Joule heating (see Fig. 28.19). All these devices merely convert electric energy into thermal energy by means of a heating element consisting of a coiled wire of fairly high resistance. For instance, in a typical electric toaster, the ribbon of wire you can see wound around the plates of insulating material inside the bread slots has a resistance of 10 or 15 Ω, which is much larger than the resistance of the wires connecting the toaster to the outlet, or the wires connecting the outlet to the circuit breaker panel of your house. The resistance of the internal wire in the toaster

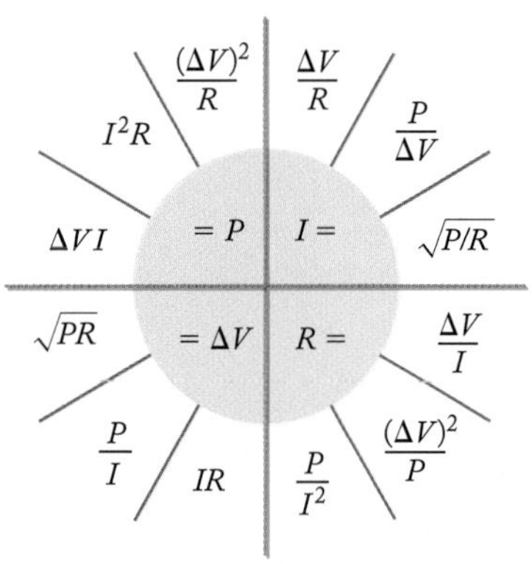

FIGURE 28.18 A summary of relations between current, resistance, potential difference, and power.

must be kept larger than the resistance of the outside wire because otherwise the outside wire would produce more heat than the internal wire, which would be wasteful and hazardous, since overheating of the outside wire could start a fire.

Ordinary incandescent lightbulbs also rely on Joule heating. Within the lightbulb, the current passes through a very fine coiled tungsten filament that acts like a miniature electric heater. The temperature of the tungsten filament reaches 3000°C, and the filament glows, white-hot. The bulb is either partially evacuated or else filled with a chemically inert gas, such as argon. Some of the heat escapes by conduction and convection through the gas in the bulb and through the wires connected to the filament. But a large fraction of the heat escapes by radiation, mostly in the form of infrared light and visible light.

Note that all such appliances consume electric power, but they do not consume electric current or electric charge—any electric current or electric charge that enters the appliance at the high-voltage terminal leaves at the other terminal. However, the electric charge leaves the appliance with less potential energy than it had when it entered, and this decrease represents the electric energy consumed by the appliance.

In any circuit consisting of several sources of emf and several resistors with steady currents, the total power delivered by the sources of emf must equal the total power dissipated in the resistors. This equality can be shown to be mathematically equivalent to Kirchhoff's rules. The net result of the flow of current in such a circuit is therefore a conversion of energy of the sources of emf into an equal amount of energy of heat.

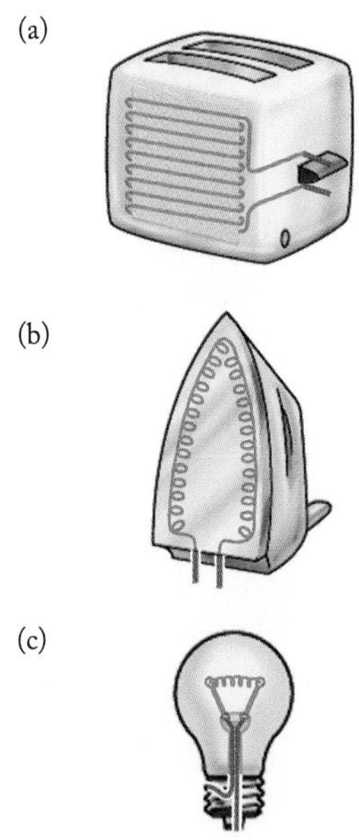

FIGURE 28.19 Some electric appliances that rely on Joule heating. (a) Toaster, (b) iron, (c) incandescent lightbulb.

PHYSICS IN PRACTICE FUSES AND CIRCUIT BREAKERS

Joule heating can also be used for the control of currents by means of fuses and circuit breakers. These safety devices are designed to cut off the current if it exceeds a preset level. In an ordinary fuse (see Fig. 1), the current passes through a thin ribbon of metal of low melting point. The current heats this ribbon, and if the current is excessive, the ribbon melts away and thus cuts off the current. In a circuit breaker, the current sensor is a bimetallic strip (see Fig. 2), which bends when heated by the current. If the bimetallic strip bends far enough upward, a lever permits a contact arm to flip to one side. The contact arm acts as a switch, and when it flips, it opens the contact points and cuts off the current. After the bimetallic strip cools, the contact arm can be pushed back into place, so as to close the contact points and again permit the current to flow.

Maximum currents permitted for wires of different thicknesses are specified in the National Electrical Code. For instance, the maximum current permitted for the typical wire (No. 12 copper) installed in modern houses is 25 A, and the circuit breakers protecting such wire are therefore designed to cut off the current before it exceeds this level.

FIGURE 1 An ordinary fuse.

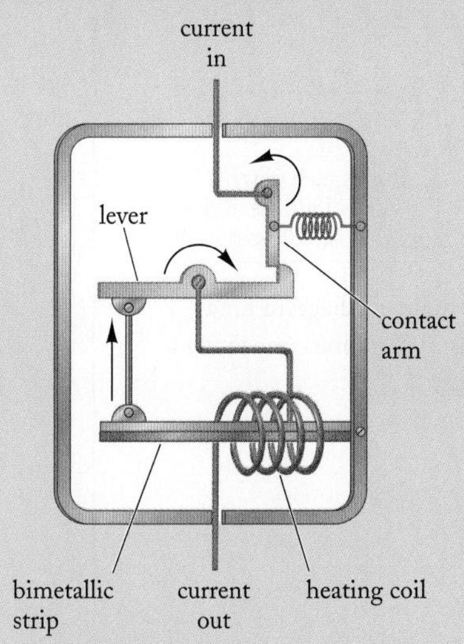

FIGURE 2 A circuit breaker. When heated, the bimetallic strip bends upward and pushes the lever, which releases the spring-loaded contact arm and breaks the circuit.

Concepts in Context

EXAMPLE 7 What is the rate at which Joule heat is produced in the resistors of Example 5?

SOLUTION: The resistances are $R_1 = 4.00\ \Omega$, $R_2 = 4.00\ \Omega$, and $R_3 = 2.00\ \Omega$, the corresponding currents are 1.25 A, 1.75 A, and 0.50 A. Equation (28.24) then gives the power in each resistor:

$$P_1 = I_1^2 R_1 = (1.25\ \text{A})^2 \times 4.00\ \Omega = 6.25\ \text{W}$$

$$P_2 = I_2^2 R_2 = (1.75\ \text{A})^2 \times 4.00\ \Omega = 12.25\ \text{W}$$

$$P_3 = I_3^2 R_3 = (0.50\ \text{A})^2 \times 2.00\ \Omega = 0.50\ \text{W}$$

COMMENTS: Note that the net power dissipated is $P_1 + P_2 + P_3 = 19.0$ W, which agrees with the net power delivered by the batteries (compare with Example 6).

EXAMPLE 8 A high-voltage transmission line that connects a city to a power plant consists of a pair of aluminum cables, each with a resistance of 4.0 Ω. The current flows to the city along one wire and back along the other.

(a) The transmission line delivers to the city 1.7×10^5 kW of power at 2.3×10^5 V. What is the current in the transmission line? How much power is lost as Joule heat in the transmission line? (b) If the transmission line were to deliver the same 1.7×10^5 kW of power at 115 V, how much power would be lost as Joule heat? Is it more efficient to transmit power at high voltage or at low voltage?

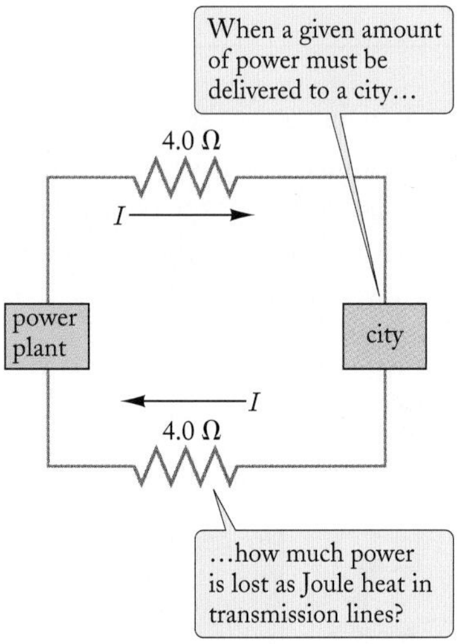

FIGURE 28.20 Circuit diagram for a high-voltage transmission line connecting a city to a power plant.

SOLUTION: (a) Figure 28.20 shows the circuit consisting of power plant, transmission line, and city. In terms of the power and the voltage delivered to the city, the current through the city is, according to Eq. (28.23),

$$I = \frac{P_{\text{delivered}}}{\Delta V_{\text{delivered}}} = \frac{1.7 \times 10^8\ \text{W}}{2.3 \times 10^5\ \text{V}} = 7.4 \times 10^2\ \text{A}$$

The current in both portions of the transmission line must be the same, since any current that flows toward the city must return to the power plant. The combined resistance of both wires is $4.0\ \Omega + 4.0\ \Omega = 8.0\ \Omega$, and hence the power lost in the transmission line is, according to Eq. (28.24),

$$P_{\text{lost}} = I^2 R = (7.4 \times 10^2\ \text{A})^2 \times 8.0\ \Omega = 4.4 \times 10^6\ \text{W}$$

Thus, the power lost is about 3% of the power delivered.

(b) For $\Delta V_{\text{delivered}} = 115$ V, the current is

$$I = \frac{P_{\text{delivered}}}{\Delta V_{\text{delivered}}} = \frac{1.7 \times 10^8\ \text{W}}{115\ \text{V}} = 1.5 \times 10^6\ \text{A}$$

and the power lost is

$$P_{\text{lost}} = I^2 R = (1.5 \times 10^6\ \text{A})^2 \times 8.0\ \Omega = 1.8 \times 10^{13}\ \text{W}$$

Thus, the power lost is much larger than the power delivered! Comparing the results of (a) and (b), we see that transmission at high voltage is much more efficient than transmission at low voltage.

Checkup 28.5

QUESTION 1: A resistor of 10 Ω connected to a battery produces 0.4 W of Joule heat. If we replace this resistor by one of 20 Ω, what will be the rate of production of heat?

QUESTION 2: If we replace both batteries in the circuit described in Example 5 by batteries of twice the emf, by what factor do we increase the rate of production of Joule heat?

QUESTION 3: In Example 8(a), the city receives an electric current of 7.4×10^2 A. Does the city consume this current? Does the city consume electric charge? Electric energy?

QUESTION 4: Figure 28.21a shows three identical resistors connected in parallel to a battery with emf $\mathcal{E}$; Fig. 28.21b shows the same resistors and battery, but now connected in series. What is the ratio of the power dissipated by one of the resistors in Fig. 28.21b to that dissipated by one of the resistors in Fig. 28.21a?

(A) 9 (B) 3 (C) 1 (D) $\frac{1}{3}$ (E) $\frac{1}{9}$

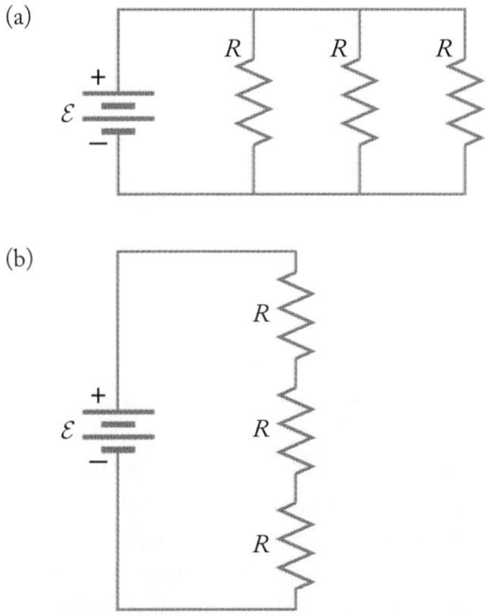

FIGURE 28.21 An emf $\mathcal{E}$ and three identical resistors R connected (a) in parallel and (b) in series.

28.6 ELECTRICAL MEASUREMENTS

Online Concept Tutorial 32

Measurements of currents, potentials, and resistances in electrical circuits require various specialized instruments. Here we will briefly discuss some of the instruments used in DC circuits.

Ammeter and Voltmeter

Most electrical measurements are performed with ammeters and voltmeters (see Fig. 28.22). The ammeter measures the electric current flowing through its terminals, and the voltmeter measures the potential difference applied to its terminals.

The internal mechanisms of the ammeter and the voltmeter are similar. In the old moving-coil instruments, the sensitive element is a small coil of wire, delicately suspended between the poles of a magnet, which experiences a deflection when a current passes through it (see Fig. 30.18 for a view of the internal mechanism of these instruments). Thus, such ammeters and voltmeters both respond to an electric current passing through the instrument. In a modern digital voltmeter, a stable counter measures time while a calibrated current charges a capacitor; the capacitor voltage is compared with the input voltage being measured. When the two are equal, the counter stops; the count is proportional to the input voltage, providing digital output. In a modern ammeter, the current is passed through a known internal resistor, providing a voltage drop that is again measured digitally.

FIGURE 28.22 Two multimeters, which can be used either as ammeters or as voltmeters. The switches are used to select the function and the measurement scale.

internal resistance of meter

The resistance that the meter offers to a current is called the **internal resistance** of the meter. As in the case of a battery, the internal resistance can be regarded as placed in series with the meter. Hence, the current registered by the ammeter equals the current through its internal resistance, and the voltage registered by the voltmeter equals the potential difference across its internal resistance.

Ammeters and voltmeters differ in that *the ammeter has a low internal resistance and permits the passage of whatever current enters its terminals with little hindrance, whereas the voltmeter has a very large internal resistance and draws only an extremely small current,* even when the potential difference applied to its terminals is large.

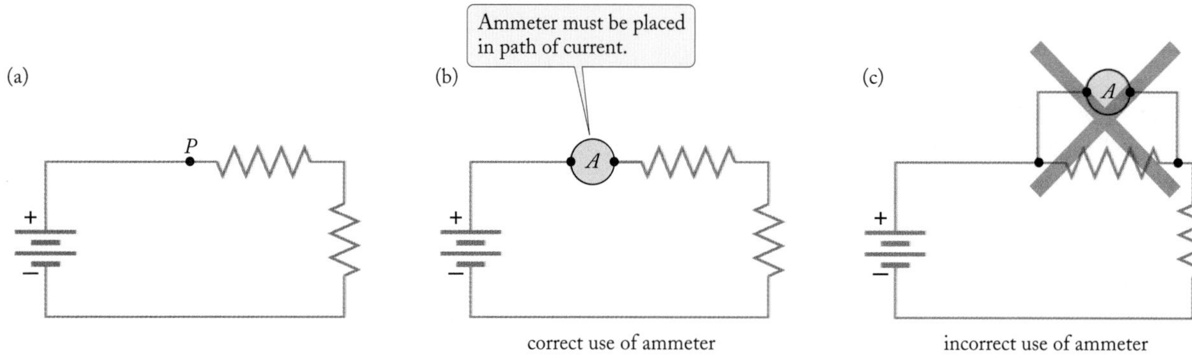

FIGURE 28.23 (a) An electric circuit. (b) Correct connection of the ammeter. The symbol for the ammeter is a circle labeled A, with two terminals. The current to be measured enters through one terminal of the ammeter and leaves through the other. (c) Incorrect connection of the ammeter.

Figure 28.23a shows an example of an electric circuit. To measure the electric current flowing past, say, the point P in this circuit, the experimenter must cut the wire apart and insert the ammeter. Figure 28.23b shows the correct way of connecting the ammeter (for comparison, Fig. 28.23c shows a wrong way). Since the ammeter has a very low internal resistance, its insertion in the circuit usually has an insignificant effect on the current. However, if the resistances in the circuit are very small, then the insertion of an ammeter can have a significant inhibiting effect on the current. For the accurate measurement of the current, the experimenter must select an ammeter whose internal resistance is much lower than the resistances in the circuit.

To measure the potential difference between, say, the points P and P' in a circuit, the experimenter must connect the terminals of the voltmeter to these points. Figure 28.24a shows the correct way of connecting the voltmeter (for comparison, Fig. 28.24b shows a wrong way). Since the voltmeter has a very large internal resistance, it draws only a very small current, and the alteration in the flow of current through the resistance PP', and the consequent alteration of the potential difference between P and P', are usually insignificant. However, if the resistance PP' across which the voltmeter is connected is large, then the voltmeter can draw a significant fraction of the current, with a consequent decrease of potential across the resistance PP'. For an accurate measurement of the potential, the experimenter must select a voltmeter with an internal resistance much larger than the resistance PP'.

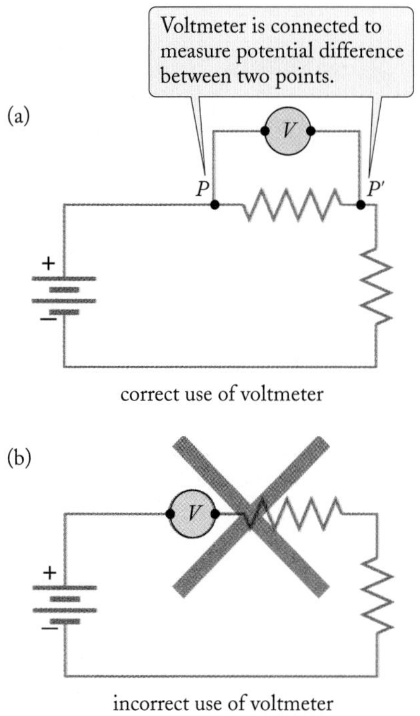

FIGURE 28.24 (a) Correct connection of the voltmeter. The symbol for the voltmeter is a circle labeled V, with two terminals. One terminal is connected to the point P, the other is connected to the point P'. (b) Incorrect connection of the voltmeter.

Wheatstone Bridge

The Wheatstone bridge is used for precise comparisons of an unknown resistance with known reference resistances. Figure 28.25 shows a schematic diagram of a Wheatstone bridge. The resistance R_x is the unknown resistance to be determined. Resistor R_1 is a high-precision variable reference resistor, whereas resistors R_2 and R_3 are fixed reference resistors. In practice, the bridge is "balanced" by varying the resistance R_1 until the current through the ammeter is zero. In that case, a common current flows through R_1 and R_2 in the left leg of the bridge, say, a current I_l. Similarly, a common current I_r flows through R_x and R_3 in the right leg. Also, the potential at points P and Q must be the same, since no current through the ammeter implies no voltage drop across it. Thus we can equate the left and right resistive potential changes for each of the upper and lower parts of the bridge:

$$I_l R_1 = I_r R_x \qquad \text{and} \qquad I_l R_2 = I_r R_3$$

If we divide the two equations to eliminate I_l and I_r, we see that the ratio of R_1 to R_2 must be the same as the ratio of R_x to R_3, and we can solve for the unknown resistance:

$$R_x = \frac{R_3}{R_2} R_1 \qquad (28.25)$$

This equation permits the calculation of the unknown resistance. One advantage of the bridge technique is that the ammeter need not be calibrated; it need only provide a measurement of the zero, or null, value of current, which can be achieved with very high accuracy.

In essence, the operation of the Wheatstone bridge hinges on an adjustment of the ratio of resistances in the circuit. Often, the Wheatstone bridge is set up with $R_3 = R_2$, so that the bridge will balance when $R_1 = R_x$; in this case, the dials on the adjustable, calibrated resistance directly tell us the value of the unknown resistance. Similar bridge circuits can be used to measure the values of other circuit components, such as capacitance.

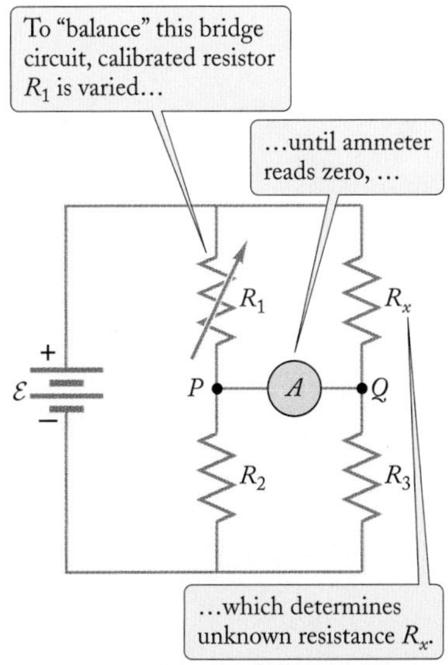

FIGURE 28.25 Schematic diagram for a Wheatstone bridge.

Checkup 28.6

QUESTION 1: What is the main difference between an ammeter and a voltmeter?

QUESTION 2: Exactly what is wrong with the way the ammeter is connected in Fig. 28.23c and the voltmeter in Fig. 28.24b?

QUESTION 3: Consider the incorrect use of the ammeter in Fig. 28.23c. Compared with the original current flowing in the circuit without the ammeter (Fig. 28.23a), the ammeter in Fig. 28.23c will indicate a current that is

(A) Larger (B) The same (C) Smaller

28.7 THE *RC* CIRCUIT

Online Concept 32 Tutorial

Throughout this chapter we have dealt only with steady, time-independent currents. However, *Kirchhoff's rules and the methods for solving circuits we have developed in this chapter apply also to time-dependent currents.* The only restriction is that the emfs and the currents in the circuit must not vary too quickly. (A rough criterion for the applicability of Kirchhoff's rules is that the currents and emfs must not change significantly in an interval equal to the travel time for a light signal around the circuit.)

In Chapter 32 we will deal with a variety of circuits with time-dependent currents. Here we will deal with the simple case of a time-dependent current in a circuit consisting of a resistor R and a capacitor C connected in series and being charged by a battery. Figure 28.26 shows a schematic diagram for such an *RC* circuit. We assume that the capacitor is initially uncharged, and that the battery is suddenly connected at time $t = 0$. Initially, the potential difference across the capacitor is then zero. When the battery is connected, charge flows from the terminals of the battery to the plates of the capacitor. As the charge accumulates on the plates, the potential difference across the plates gradually increases. The flow of charge will come to a halt when the potential difference across the plates matches the emf of the battery. This qualitative discussion of the charging process indicates that the current is initially large, but gradually tapers off, and ultimately approaches zero.

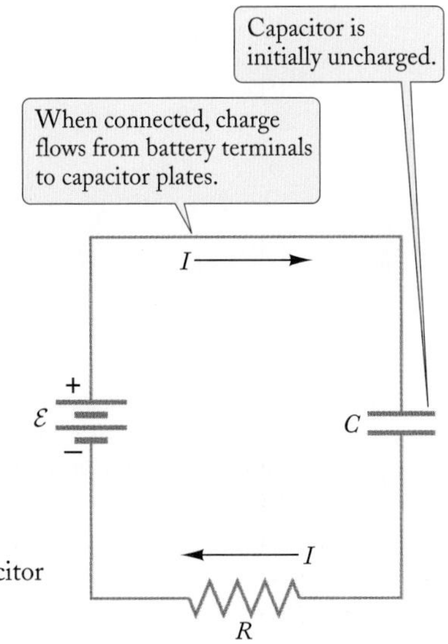

FIGURE 28.26 An *RC* circuit, consisting of a resistor and a capacitor connected in series to a battery.

For a mathematical treatment of the current in the circuit we turn to Kirchhoff's voltage rule: the sum of all the emfs and voltage drops around the circuit must be zero. The emf of the battery is $\mathcal{E}$. If the current at some instant is I, the potential drop across the resistor is $\Delta V_R = -IR$. And if the charge on the capacitor plates at some instant has a magnitude Q, the potential drop across the plates is $\Delta V_C = -Q/C$. Hence

$$\mathcal{E} - IR - \frac{Q}{C} = 0 \tag{28.26}$$

At the initial time, $Q = 0$ and Eq. (28.26) then tells us that the initial current is $I = \mathcal{E}/R$. At the completion of the charging process, $I = 0$ and Eq. (28.26) then tells us that $Q = C\mathcal{E}$. Thus, at the initial time, the entire potential drop occurs in the resistor, and at the completion of the charging process, the entire potential drop occurs in the capacitor. At intermediate times, both the resistor and the capacitor contribute to the potential drop.

To calculate the current and the charge at intermediate times, let us rewrite Eq. (28.26) using $I = dQ/dt$:

$$\mathcal{E} - \frac{dQ}{dt}R - \frac{Q}{C} = 0 \tag{28.27}$$

This equation can be solved by direct integration if we rearrange it to make one side proportional to dQ and the other proportional to dt as follows:

$$\frac{dQ}{Q - C\mathcal{E}} = -\frac{1}{RC}dt$$

We now integrate each side from its initial zero value to the value at some later time t:

$$\int_0^Q \frac{dQ}{Q - C\mathcal{E}} = -\frac{1}{RC}\int_0^t dt' \tag{28.28}$$

The integral on the left is just the natural logarithm $\ln(Q' - C\mathcal{E})$, and the integral on the right is simply t. Evaluation of these at their limits and using $\ln A - \ln B = \ln(A/B)$ gives

$$\ln\left(\frac{Q - C\mathcal{E}}{-C\mathcal{E}}\right) = -\frac{t}{RC} \tag{28.29}$$

If we take the exponential function of each side and recall that $e^{\ln x} = x$, where $e = 2.718\ldots$ is the base of the natural logarithms[1] (see Math Help: The Exponential Function, page 910) we obtain

$$\left(\frac{Q - C\mathcal{E}}{-C\mathcal{E}}\right) = e^{-t/(RC)} \tag{28.30}$$

or, solving for Q, we finally have

charging *RC* circuit

$$Q = C\mathcal{E}\left(1 - e^{-t/(RC)}\right) \tag{28.31}$$

[1] The pure number $e \approx 2.718$ should not be confused with the elementary charge $e \approx 1.60 \times 10^{-19}$ C.

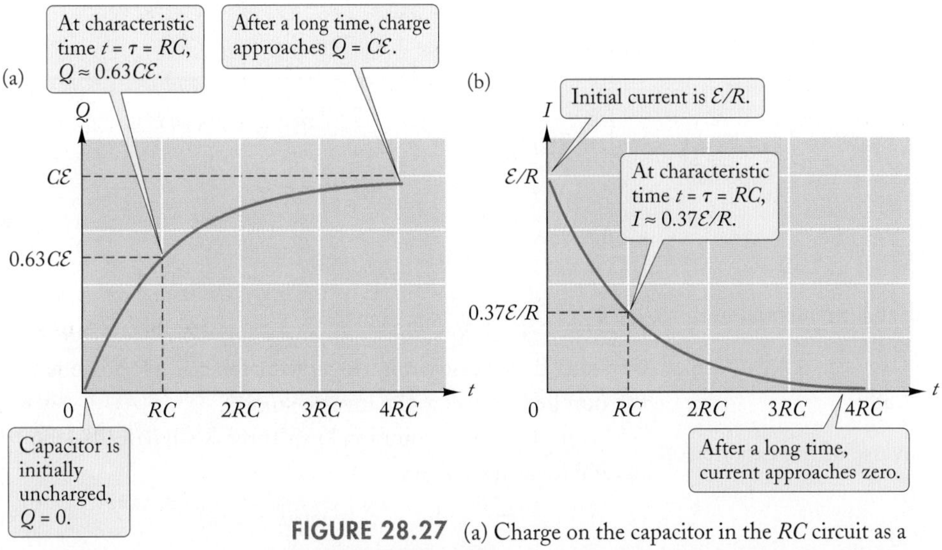

FIGURE 28.27 (a) Charge on the capacitor in the *RC* circuit as a function of time. (b) Current in the *RC* circuit as a function of time.

This behavior of the charge on the capacitor as a function of time is shown in Fig. 28.27a; as expected, it starts at $Q = 0$ (at $t = 0$, we have $e^0 = 1$), and it approaches $Q = C\mathcal{E}$ as the charging process nears completion (for $t = \infty$, we have $e^{-\infty} = 0$).

The current in the circuit is the time derivative of the charge:

$$I = \frac{dQ}{dt} = \frac{d}{dt}\left[C\mathcal{E}\left(1 - e^{-t/(RC)}\right)\right] = C\mathcal{E}\frac{d}{dt}\left(-e^{-t/(RC)}\right) = \frac{C\mathcal{E}}{RC}e^{-t/(RC)} \quad (28.32)$$

or, simply,

$$I = \frac{\mathcal{E}}{R}e^{-t/(RC)} \quad (28.33)$$

current in *RC* circuit

Thus, as shown in Fig. 28.27b, the current is an exponentially decaying function of time.

The product *RC* that occurs in the exponential has the units of time, and it is designated by τ:

$$\tau = RC \quad (28.34)$$

characteristic time (*RC* time constant)

This time is called the **characteristic time** for the *RC* circuit; it is also called the *RC* **time constant**. From Fig. 28.27, we see that most of the charging occurs within the characteristic time. More precisely, at the characteristic time $t = \tau = RC$ the charge reaches the value

$$Q = C\mathcal{E}\left(1 - e^{-1}\right) = C\mathcal{E}\left(1 - \frac{1}{2.718}\right) = C\mathcal{E} \times 0.6321$$

that is, *at the characteristic time, the charge reaches* $\approx$ 63% *of its final value*. In terms of the characteristic time, the charge (28.31) and the current (28.33) can be written

$$Q = C\mathcal{E}\left(1 - e^{-t/\tau}\right) \quad \text{and} \quad I = \frac{\mathcal{E}}{R}e^{-t/\tau} \quad (28.35)$$

***RC* circuit**

MATH HELP THE EXPONENTIAL FUNCTION

The exponential function exp(x), or e^x, is defined as e to the power x, where the number e = 2.7183... is the base of natural logarithms,

$$\exp(x) = e^x = (2.7183...)^x$$

For instance, if $x = 2$, then the value of the exponential function is

$$e^2 = (2.7183...)^2 = 7.3891$$

The natural logarithm ln (x) is the inverse of the exponential function. Thus,

$$\ln(1) = \ln(e^0) = 0$$

$$\ln(e) = 1$$

$$\ln(e^2) = 2$$

$$\ln(e^3) = 3$$

and, in general,

$$\ln(e^x) = x$$

Conversely,

$$e^{\ln x} = x$$

On many calculators, the exponential function is performed by taking the inverse natural logarithm; for example, to obtain the number e, we take the inverse-ln of 1.

As in the case of common logarithms (with base 10), the natural logarithm obeys

$$\ln(y/x) = \ln(y) - \ln(x)$$

For example,

$$\ln(1/2) = \ln(1) - \ln(2) = 0 - \ln 2 = -\ln 2$$

and

$$1/2 = e^{\ln(1/2)} = e^{-\ln 2}$$

EXAMPLE 9 Suppose that in the circuit illustrated in Fig. 28.26, the resistance is $R = 8.0 \times 10^3\ \Omega$, the capacitance is $C = 2.0\ \mu$F, and the emf of the battery is $\mathcal{E} = 1.5$ V. What is the initial value of the current, at the instant after the battery is connected? What is the final value of the charge on the capacitor? What is the charge at $t = \tau$? At $t = 2\tau$? At $t = 5\tau$?

SOLUTION: The initial value of the current is [see the discussion following Eq. (28.26)]

$$I = \frac{\mathcal{E}}{R} = \frac{1.5\ \text{V}}{8.0 \times 10^3\ \Omega} = 1.9 \times 10^{-4}\ \text{A}$$

The final value of the charge is

$$Q = C\mathcal{E} = 2.0 \times 10^{-6}\ \text{F} \times 1.5\ \text{V} = 3.0 \times 10^{-6}\ \text{C}$$

At the time $t = \tau = RC$, the charge is [using Eq. (28.35)]

$$Q = C\mathcal{E}(1 - e^{-1}) = 3.0 \times 10^{-6}\ \text{C} \times \left(1 - \frac{1}{2.718}\right) = 1.9 \times 10^{-6}\ \text{C}$$

at the time $t = 2\tau = 2RC$, the charge is

$$Q = C\mathcal{E}(1 - e^{-2}) = 3.0 \times 10^{-6}\ \text{C} \times \left[1 - \frac{1}{(2.718)^2}\right] = 2.6 \times 10^{-6}\ \text{C}$$

and at the time $t = 5\tau = 5RC$, the charge is

$$Q = C\mathcal{E}(1 - e^{-5}) = 3.0 \times 10^{-6}\ \text{C} \times \left[1 - \frac{1}{(2.718)^5}\right] = 3.0 \times 10^{-6}\ \text{C}$$

After five time constants, the charge has reached within 1% of its final value.

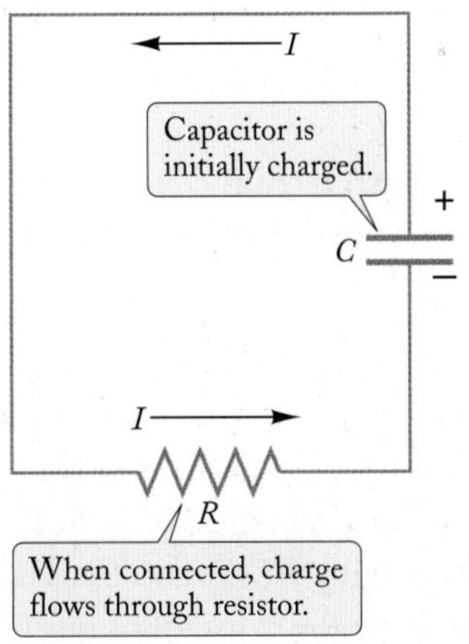

FIGURE 28.28 An *RC* circuit, with the capacitor discharging through the resistor. The arrow indicates the direction of the current.

Once the charging process has been completed and the current in the circuit has stopped, we can disconnect the battery. The charge will then remain on the capacitor plates (except for a slow leakage through the capacitor or into the air). However, suppose we now connect the free terminals of the capacitor and the resistor, as shown in Fig. 28.28. The capacitor will then discharge through the resistor. The current will be large at the initial time, and it will gradually taper off and approach zero as the potential difference across the capacitor decreases. This circuit is simpler than the one we considered above when charging the capacitor. A similar analysis using Kirchhoff's voltage rule and direct integration yields formulas that describe the discharge of the capacitor. The relations involve the same exponential function as in Eqs. (28.31) and (28.33), with the same characteristic time, except now the charge decays to zero with time:

$$Q = C\mathcal{E}e^{-t/(RC)} \tag{28.36}$$

discharging *RC* circuit

and

$$I = -\frac{dQ}{dt} = \frac{\mathcal{E}}{R}e^{-t/(RC)} \tag{28.37}$$

The current now is the negative of the rate of change of the charge because during the discharging process, a decrease of charge, that is, a negative dQ/dt, results in a positive current in the direction shown in Fig. 28.28. Figure 28.29 gives plots of the charge vs. time and the current vs. time for this discharging process.

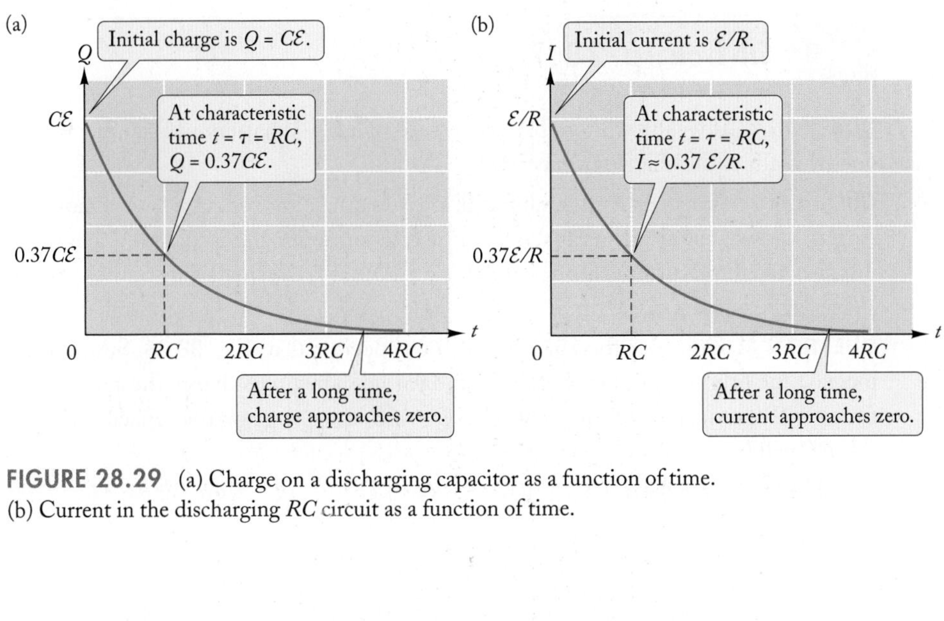

FIGURE 28.29 (a) Charge on a discharging capacitor as a function of time. (b) Current in the discharging *RC* circuit as a function of time.

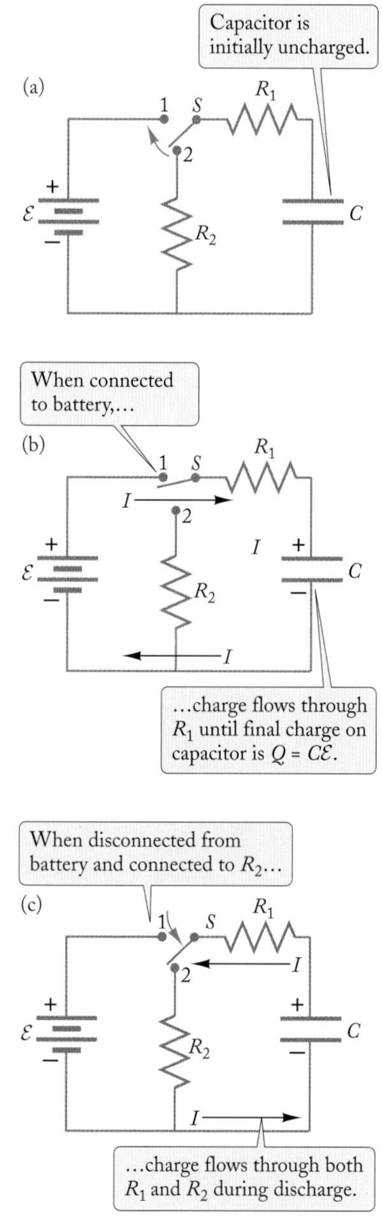

FIGURE 28.30 A switchable RC circuit.

EXAMPLE 10 Consider the circuit of Fig. 28.30a. The switch S has been in position 2 for a long time. (a) At $t = 0$, it is switched to position 1. What is the charge on the capacitor as a function of time t? (b) Much later, at some time $t' = 0$, the switch is returned to position 2. What is the charge on the capacitor as a function of the time t'? What is the voltage across resistor R_2 as a function of the time t'?

SOLUTION: (a) Since the switch has been in position 2 for a long time, the capacitor is uncharged (otherwise it would discharge through the resistors). When the switch is moved to position 1, current flows only in the outer loop (Fig. 28.30b), through R_1, as the capacitor C charges to its final value $Q = C\mathcal{E}$ just as in our previous discussion of a charging RC circuit. Thus the relevant characteristic time is $\tau = R_1C$, and the charge on the capacitor is given by Eq. (28.31) with $R = R_1$:

$$Q = C\mathcal{E}\left(1 - e^{-t/(R_1C)}\right)$$

(b) Much later, at $t >> R_1C$, the switch is moved to position 2. Let this new initial time be $t' = 0$. The capacitor is initially charged with $Q = C\mathcal{E}$. Since the emf is now effectively disconnected, a discharge occurs only in the right loop (Fig. 28.30c), and the current flows counterclockwise through the series combination of resistors, so now the equivalent resistance is $R = R_1 + R_2$. Thus the charge on the capacitor will be given as a function of t' by the discharging function of Eq. (28.36), with $\tau = (R_1 + R_2)C$:

$$Q = C\mathcal{E}e^{-t'/[(R_1 + R_2)C]}$$

The current is counterclockwise in the right loop with magnitude [Eq. (28.37)]

$$I = \frac{\mathcal{E}}{R_1 + R_2}e^{-t'/[(R_1 + R_2)C]}$$

and the voltage across R_2 is given by Ohm's Law:

$$\Delta V = IR_2 = \frac{\mathcal{E}R_2}{R_1 + R_2}e^{-t'/[(R_1 + R_2)C]}$$

Checkup 28.7

QUESTION 1: If we use a 3.0-V battery instead of a 1.5-V battery in Example 9, how does this affect the characteristic time?

QUESTION 2: Suppose that in Example 9, after the capacitor is charged, we remove the battery and we connect the capacitor to the resistor, so it can discharge. What is the initial value of the current through the resistor? What is the final value of the current? What is the characteristic time?

QUESTION 3: Consider the charging of the capacitor C in Fig. 28.26. Suppose we increase the original resistance R to $2R$ and we again begin to charge the capacitor at time $t = 0$. Compared with its final value $Q = C\mathcal{E}$, the charge on the capacitor at time $t = RC$ will be

(A) Less than 63% of $C\mathcal{E}$ (B) $\approx$ 63% of $C\mathcal{E}$ (C) More than 63% of $C\mathcal{E}$

28.8 THE HAZARDS OF ELECTRIC CURRENTS

As a side effect of the widespread use of electric machinery and devices in factories and homes, each year in the United States about 1000 people die by accidental electrocution. A much larger number suffer nonfatal electric shocks. Fortunately, the human skin is a fairly good insulator, providing a protective barrier against injurious electric currents. The resistance of a square centimeter of dry human epidermis in contact with a conductor can be as much as $10^5\ \Omega$. However, the resistance varies in a rather sensitive way with the thickness, moisture, and temperature of the skin, and with the magnitude of the potential difference.[2]

The electric power supplied to factories and homes in the United States is usually in the form of alternating currents, or AC. These are oscillating currents, which periodically reverse direction (the standard period of the alternating current supplied by power companies is $\frac{1}{60}$ second). Since most accidental electric shocks involve alternating currents, the following discussion of the effects of currents on the human body will emphasize alternating currents.

In the typical accidental electric shock, the current enters the body through the hands (in contact with one terminal of the source of emf) and exits through the feet (in contact with the ground, which constitutes the other terminal of the source of emf for most AC circuits). Thus, the body plays the role of a resistor, closing an electric circuit (see Fig. 28.31).

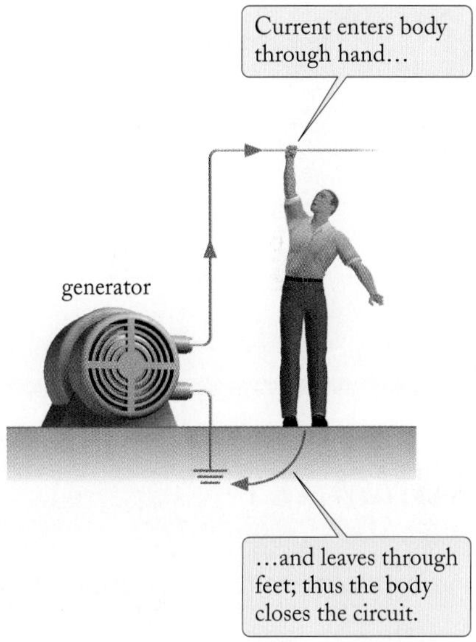

FIGURE 28.31 A human body closing an electric circuit. The current enters through the hand and exits through the feet.

The damage to the body depends on the magnitude of the current passing through it. An alternating current of about 0.001 A produces only a barely detectable tingling sensation. Higher currents produce pain and strong muscular contractions. If the victim has grasped an electric conductor—such as an exposed power cable—with the hand, the muscular contraction may prevent the victim from releasing the hold on the conductor. The magnitude of the "let-go" current, at which the victim can just barely release the hold on the conductor, is about 0.01 A. Higher currents lock the victim's hand to the conductor. Unless the circuit is broken within a few seconds, the skin in contact with the conductor will then suffer burns and blister. Such damage to the skin drastically reduces its resistance and so can lead to a fatal increase of the current.

An alternating current of about 0.02 A flowing through the body from the hands to the feet produces a contraction of the chest muscles that halts breathing; this leads to death by asphyxiation if it lasts for a few minutes. A current of about 0.1 A lasting just a few seconds induces fibrillation of the heart muscles, with cessation of the natural rhythm of the heartbeat and cessation of the pumping of blood. Fibrillation usually continues even when the victim is removed from the electric circuit; the consequences are fatal unless immediate medical assistance is available. The treatment for fibrillation involves the deliberate application of a severe electric shock to the heart by means of electrodes placed against the chest; this arrests the motion of the heart completely. When the shock ends, the heart usually resumes beating with its natural rhythm.

A current of a few amperes produces a seizure of the nervous system and paralysis of the respiratory muscles. Victims of such currents can sometimes be saved by prompt recourse to artificial respiration. At these high values of the current, the effects of AC and DC are not very different. But at lower values, a DC current poses less of a hazard than the comparable AC current, because the former does not trigger the strong muscular contractions triggered by the latter.

[2]The variation of resistance with potential difference implies that skin does not obey Ohm's Law.

In the above we assumed that the path of the current through the body is from the hands to the feet. If the current enters through one leg and exits through the other, no vital organs lie in its path, and the threat to life is lessened. However, an intense current through a limb tends to kill the tissue through which it passes, and may ultimately require the surgical excision of large amounts of dead tissue, and even the amputation of the limb.

Other things being equal, a higher voltage will result in a higher current. The hazard posed by contact with high-voltage sources is therefore obvious. But under exceptional circumstances, even sources of low voltage can be hazardous. Several cases of electrocution by contact with sources of a voltage as low as 12 V have been reported. It seems that in these cases death resulted from an unusally sensitive response of the nervous system; it is also conceivable that an unusually small skin resistance was a contributing factor. Thus, it is advisable to treat even sources of low voltage with respect!

first aid for electric shock

Aid to victims of electric shock should begin with switching off the current. When no switch, plug, or fuse for cutting off the current is accessible, the victim must be pushed or pulled away from the electric conductor by means of a piece of insulating material, such as a piece of dry wood or a rope. The rescuer must be careful to avoid electric contact. If the victim is not breathing, artificial respiration must be started at once. If there is no heartbeat, cardiac massage must be applied by trained personnel until the victim can be treated with a defibrillation apparatus.

SUMMARY

PROBLEM-SOLVING TECHNIQUES Kirchhoff's Rules and Multiloop Circuits **(page 900)**

PHYSICS IN PRACTICE Fuses and Circuit Breakers **(page 903)**

MATH HELP The Exponential Function **(page 910)**

ELECTROMOTIVE FORCE (emf) Energy per unit charge, or voltage, provided by a source.

KIRCHHOFF'S VOLTAGE RULE The sum of emfs and potential changes across resistors around any closed loop in a circuit must equal zero.

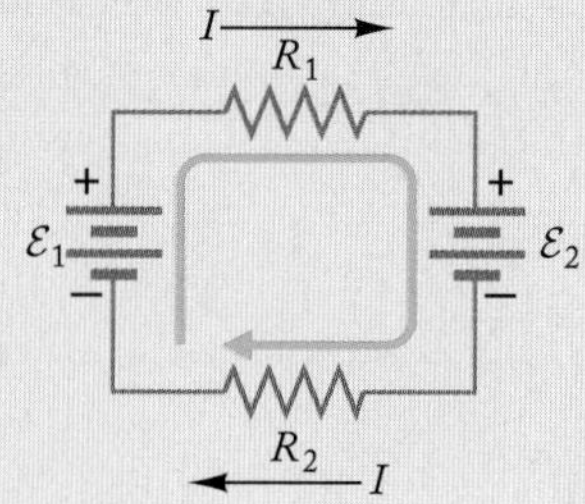

SIGN CONVENTIONS When traversing a source of emf in the forward direction (− to + terminal), the voltage is positive, $(+\mathcal{E})$; in the reverse direction (+ to − terminal), the voltage is negative $(-\mathcal{E})$. Traversing a resistor in the direction of the current provides a voltage drop (negative, $-IR$); doing so in the direction opposite to the current provides a voltage increase (positive, $+IR$).

KIRCHHOFF'S CURRENT RULE The total current flowing into any junction must equal the total current flowing out of the junction:

$$\sum I_{in} = \sum I_{out} \qquad (28.12)$$

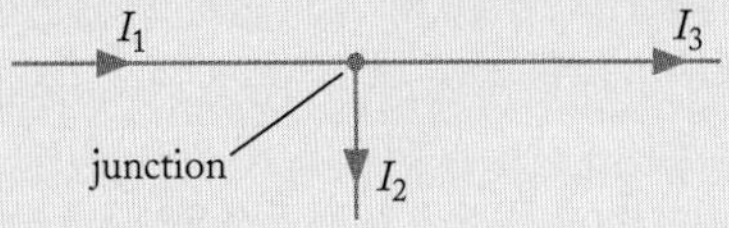

MULTILOOP CIRCUIT ANALYSIS After labeling all separate currents, Kirchhoff's voltage and current rules are applied until enough equations are obtained to solve for the desired unknown quantities.

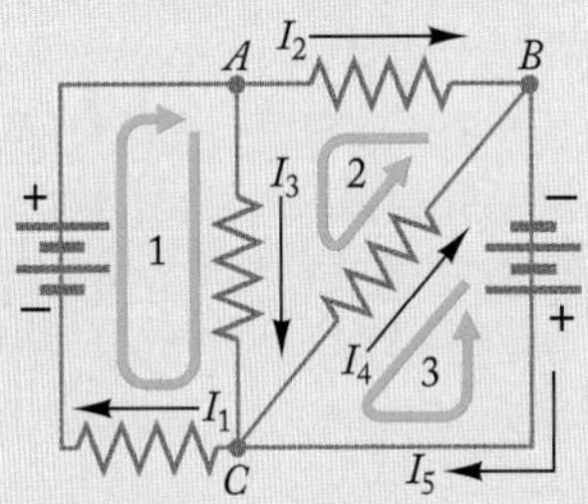

POWER DELIVERED BY A SOURCE OF emf

$$P = \mathcal{E}I \qquad (28.22)$$

POWER DISSIPATED BY A RESISTOR (JOULE HEAT)

$$P = \Delta V I = I^2 R = \frac{(\Delta V)^2}{R} \qquad (28.23, 28.24)$$

(From these relations between current, resistance, potential difference, and power, when any two are known, the other two can be calculated.)

ELECTRICAL MEASUREMENTS Ammeters measure current, have a low internal resistance, and are placed in line; voltmeters measure voltage, have a high internal resistance, and are connected across terminals.

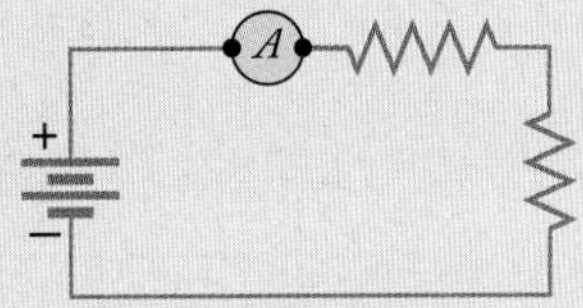

correct use of ammeter

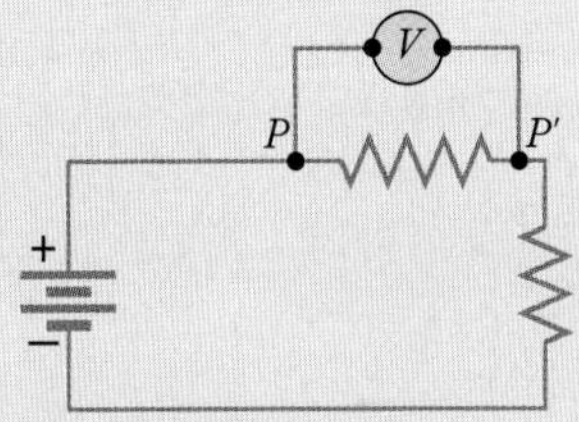

correct use of voltmeter

CHARACTERISTIC TIME FOR *RC* CIRCUIT

$$\tau = RC \qquad (28.34)$$

CHARGE ON CAPACITOR AND CURRENT IN *RC* CIRCUIT

Charging:

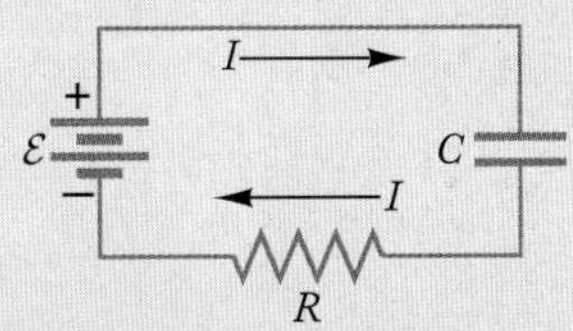

$$Q = C\mathcal{E}\left(1 - e^{-t/\tau}\right) \qquad I = \frac{dQ}{dt} = \frac{\mathcal{E}}{R}e^{-t/\tau} \qquad (28.35)$$

Discharging:

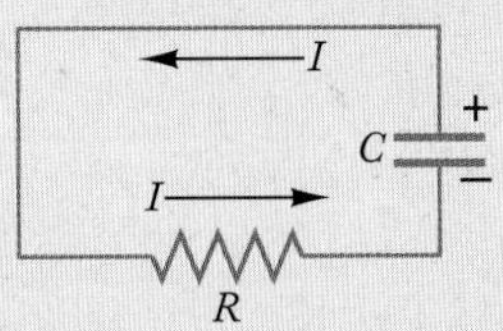

$$Q = C\mathcal{E}e^{-t/\tau} \qquad I = -\frac{dQ}{dt} = \frac{\mathcal{E}}{R}e^{-t/\tau} \qquad (28.36)$$

QUESTIONS FOR DISCUSSION

1. Can we use a capacitor as a pump of electricity in a circuit? In what way would such a pump differ from a battery?
2. Does a fully charged battery have the same emf as a partially charged battery?
3. How would you measure the internal resistance of a battery?
4. Kirchhoff's voltage rule is equivalent to energy conservation in the electric circuit. Explain.
5. What would happen if we were to connect both an ammeter and a voltmeter in series with a source of emf?
6. What would happen if we were to connect the voltmeter incorrectly, as shown in Fig. 28.24b?
7. A mechanic determines the internal resistance of an automobile battery by connecting a rugged, high-current ammeter (of nearly zero resistance) directly across the poles of the battery. The internal resistance of the battery is inversely proportional to the ammeter reading, $R_1 \propto 1/I$. Explain.
8. An **ohmmeter** consists of a reference source of emf (a battery), connected in series with a reference resistance and an ammeter. When the terminals of the ohmmeter are connected to an unknown resistor (Fig. 28.32), the current registered by the ammeter permits the evaluation of the unknown resistance. Explain.

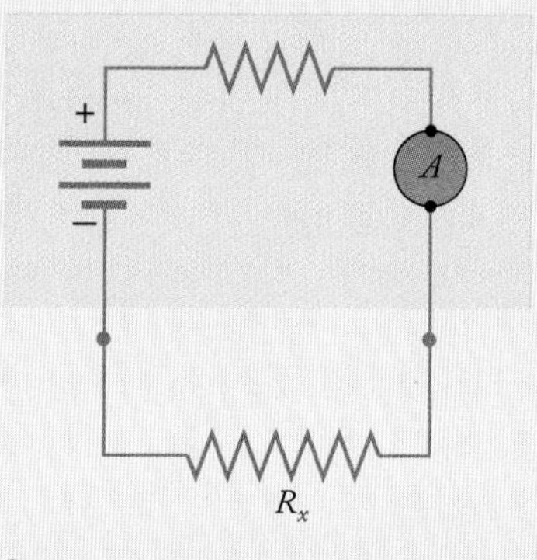

FIGURE 28.32 An ohmmeter, consisting of a battery, a reference resistance, and an ammeter enclosed in a box (color).

9. If a voltmeter and an ammeter are connected to a resistor simultaneously, their readings can be used to evaluate the resistance, according to $R = \Delta V/I$. Figure 28.33 shows two ways of connecting the ammeter and the voltmeter to the resistor. Explain why both of these methods yield a value of R slightly different from the actual value.

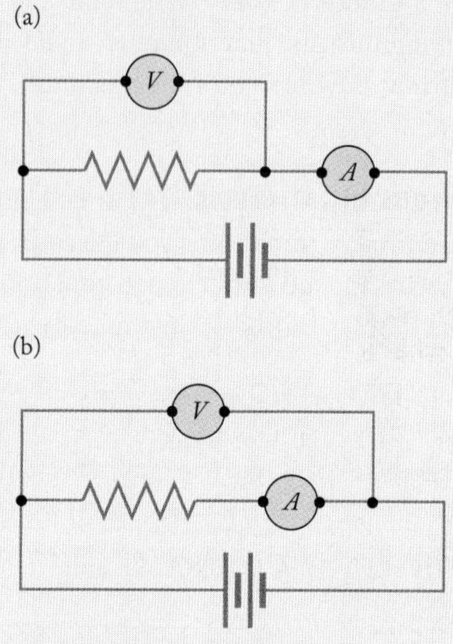

FIGURE 28.33 Simultaneous ammeter and voltmeter connections.

10. A homeowner argues that he should not pay his electric bill, since he is not keeping any of the electrons that the power company delivers to his home—any electron that enters the wiring of his home sooner or later leaves and returns to the power station. How would you answer?
11. The spiral heating elements commonly used in electric ranges *appear* to be made of metal (Fig. 28.34). Why do they not short-circuit when you place an iron pot on them?

FIGURE 28.34 Electric range heating element.

12. What are the advantages and what are the disadvantages of high-voltage power lines?
13. In many European countries, electric power is delivered to homes at 230 V, instead of the 115 V customary in the United States. What are the advantages and what are the disadvantages of 230 V?

PROBLEMS

28.1 Electromotive Force†
28.2 Sources of Electromotive Force†

1. The smallest batteries have a mass of 1.5 grams and store an electric energy of about 5.0×10^{-6} kW·h. The largest batteries (used aboard submarines) have a mass of 270 metric tons and store an electric energy of 5×10^3 kW·h. What is the amount of energy stored per kilogram of battery in each case?
2. A size D flashlight battery will deliver 1.2 A·h at 1.5 V (that is, it will deliver 1.2 A for 1.0 h, or a larger or smaller current for a correspondingly shorter or longer time). An automobile battery will deliver 55 A·h at 12 V. The flashlight battery is cylindrical with a diameter of 3.3 cm, a length of 5.6 cm, and a mass of 0.086 kg. The automobile battery is rectangular with dimensions 30 cm × 17 cm × 23 cm and a mass of 23 kg.
   (a) What is the available electric energy stored in each battery?
   (b) What is the amount of energy stored per cubic centimeter of each battery?
   (c) What is the energy stored per kilogram of each battery?
3. A heavy-duty 12-V battery for a truck is rated at 160 A·h; that is, this battery will deliver 1.0 A for 160 h (or a larger current for a correspondingly shorter time). What is the amount of electric energy that this battery will deliver?
4. Rechargeable 1.2-V nickel–cadmium batteries of sizes AAA, AA, C, and D are rated at 0.22, 0.70, 2.2, and 4.4 A·h, respectively. How much charge in coulombs passes through each battery when fully used as rated? How much energy does each deliver?
5. A 14.4-V metal hydride battery for a laptop computer can deliver a current of 0.65 A for 4.0 h. What is the total energy delivered?

*6. The electric starter motor in an automobile equipped with a 12-V battery draws a current of 80 A when in operation.
   (a) Suppose it takes the starter motor 3.0 s to start the engine. What amount of electric energy has been withdrawn from the battery?
   (b) The automobile is equipped with a generator that delivers 5.0 A to the battery when the engine is running. How long must the engine run so that the generator can restore the energy in the battery to its original level? Assume that all the power delivered to the battery is stored.

28.3 Single-Loop Circuits†

7. In a flashlight, two 1.5-V batteries are connected in series. An 8.0-Ω lightbulb closes the circuit. What is the current in the circuit? What is the electric energy delivered to the lightbulb in 1.0 h?

† For help, see Online Concept Tutorial 32 at www.wwnorton.com/physics

8. Find the currents in the circuit shown in Fig. 28.10 if the battery on the right is reversed.
9. Two small lightbulbs are connected in parallel to a 6.0-V battery. If the resistance of each lightbulb, at its operating temperature, is 2.0 Ω, what is the current in the circuit? What is the current in each lightbulb?
10. An array of solar cells is to be used to charge a 12-V lead–acid battery. If each solar cell generates 0.60 V, how many cells must we connect in series to charge the battery?
11. Consider the circuit shown in Fig. 28.35. When the resistance of R_2 is decreased, do the currents in R_1 and R_3 increase or decrease?

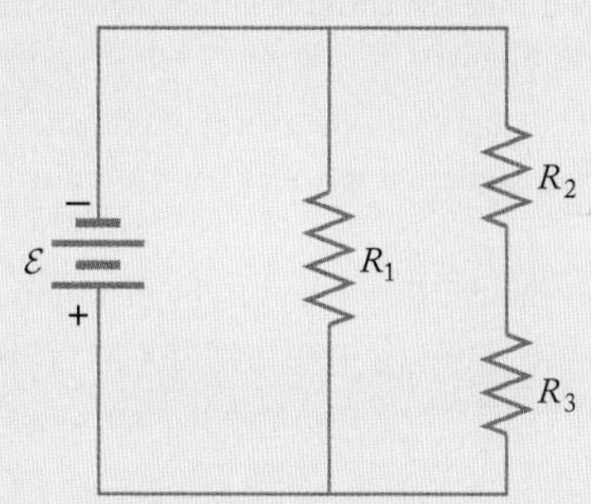

FIGURE 28.35 A source of emf and three resistors.

12. In order to reduce the effective internal resistance of an ammeter, a physicist connects a resistor in parallel across the terminals (see Fig. 28.36). If the resistance of this parallel resistor, or *shunt* resistor, is 1/10 of the resistance of the ammeter, by what factor does the shunt reduce the effective resistance of the modified ammeter? How must the physicist recalibrate the dial of the ammeter?

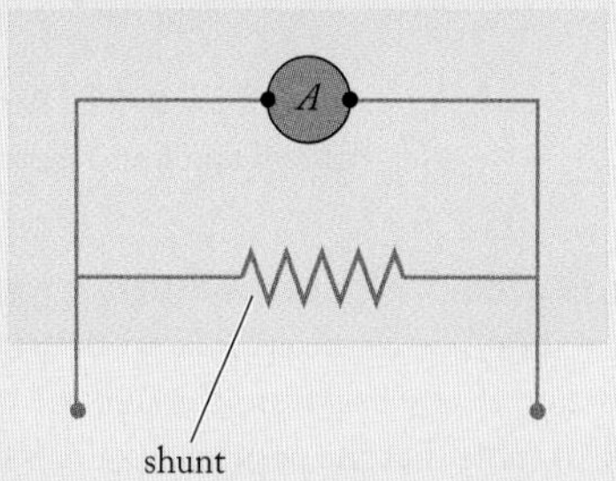

FIGURE 28.36 An ammeter with a parallel resistor, or shunt resistor.

13. A voltmeter of internal resistance 5.0×10^4 Ω is connected across the poles of a 12-V battery of internal resistance 0.020 Ω.
   (a) What is the current flowing through the battery?
   (b) What is the voltage drop across the internal resistance of the battery?
14. A battery has an emf of 9.00 V when no current flows. When the battery is connected to a 47.0-Ω resistor, a current of 0.187 A flows. What is the internal resistance of the battery?

15. In a set of holiday lights, 25 identical miniature lightbulbs are connected in series to a 115-V source; a 0.074-A current flows through the circuit. What are the voltage drop across and the resistance of each bulb? One bulb burns out. What is the potential across that bulb, which then has infinite resistance? For some lights, this high voltage soon causes an automatic short to develop across the leads to the burned filament, subsequently reducing the resistance of that bulb to zero; thus, "when one burns out, the rest stay lit." Now what is the potential across each of the other bulbs?

16. When a battery is connected to a 22.0-Ω resistor, the voltage drop across the resistor is 5.74 V; when connected to a 47.0-Ω resistor, the voltage drop is 6.19 V. What is the emf of the source? What is the internal resistance of the battery?

17. In Fig. 28.37, $\mathcal{E} = 12$ V, $R_1 = 3.0\,\Omega$, $R_2 = 8.0\,\Omega$, $R_3 = 3.0\,\Omega$, and $R_4 = 5.0\,\Omega$. Determine the currents in the circuit with single-loop analysis as follows:

(a) Find the current through both the source of emf and R_1 by first finding the equivalent resistance of the entire resistor network.

(b) Find the currents though R_2, R_3, and R_4 by first subtracting the voltage drop across R_1 from the emf $\mathcal{E}$ to obtain the voltage drop across R_2.

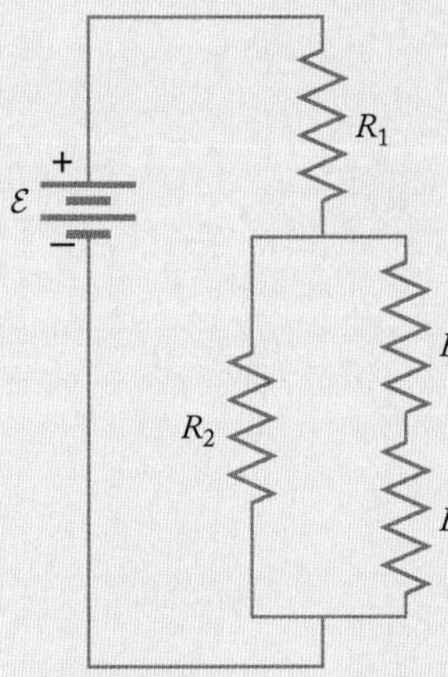

FIGURE 28.37 A source of emf and four resistors.

18. In an effort to recharge a "dead" automobile battery, a strong battery with emf $\mathcal{E}_1 = 12.6$ V is connected to a weak battery with $\mathcal{E}_2 = 11.1$ V, forcing charge backward through the weak battery, as in Fig. 28.10 (note that the positive terminals of the two batteries are connected, as are the negative terminals). Assume that the resistance of the wires and the internal resistance of the batteries total 0.12 Ω, and that the emfs remain constant. How much current flows? How much energy is delivered to the weak batttery in the first 30 seconds? Suppose that, by mistake, the positive terminal of each battery is connected to the negative terminal of the other, a dangerous arrangement. What current flows?

19. To measure the emf and the internal resistance of a battery, an experimenter connects the battery in series to a (resistanceless) ammeter and a variable resistor. She finds that when the variable resistor is set at 1.0 Ω, the current in the circuit is 0.40 A; and when the variable resistor is set at 2.0 Ω, the current is 0.22 A. What emf and what internal resistance for the battery can she deduce from this? (Hint: Draw a circuit diagram including the internal resistance of the battery. What is the total resistance when the variable resistor is set at 1.0 Ω? When it is set at 2.0 Ω?)

20. Four resistors are connected in parallel to a voltage source as shown in Fig. 28.38. The 25-Ω resistor carries a current of 5.0 A. What is the total current supplied by the battery to the resistors?

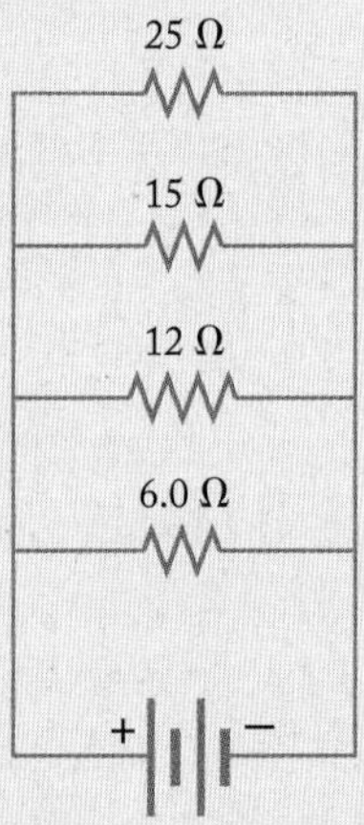

FIGURE 28.38 A source of emf and four resistors.

21. What is the current through the 200-Ω resistor in the circuit shown in Fig. 28.39?

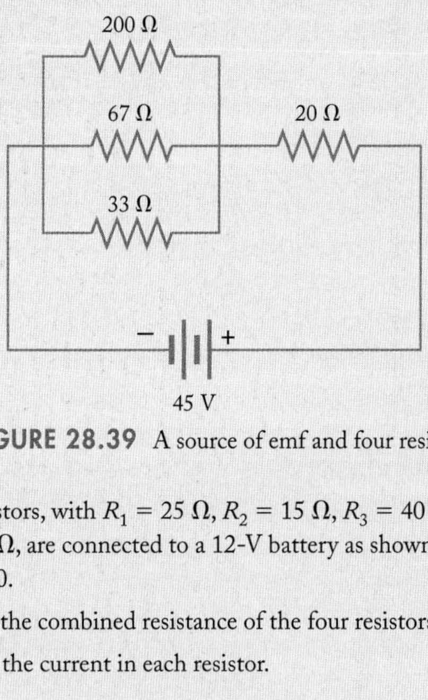

FIGURE 28.39 A source of emf and four resistors.

*22. Four resistors, with $R_1 = 25\ \Omega$, $R_2 = 15\ \Omega$, $R_3 = 40\ \Omega$, and $R_4 = 20\ \Omega$, are connected to a 12-V battery as shown in Fig. 28.40.

(a) Find the combined resistance of the four resistors.

(b) Find the current in each resistor.

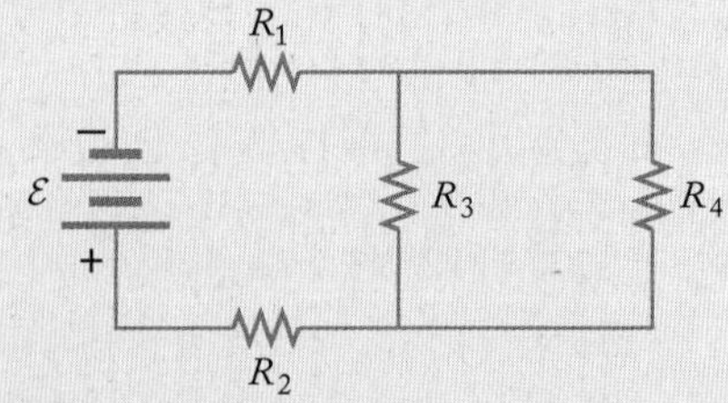

FIGURE 28.40 A source of emf and four resistors.

28.4 Multiloop Circuits†

23. Four wires are connected at a junction. Three of the wires carry currents of 10.0 A, 8.5 A, and 12.5 A into the junction. What are the magnitude and direction of the current in the fourth wire?

24. In the circuit shown in Fig. 28.41, $\mathcal{E}_1 = 6.0$ V, $\mathcal{E}_2 = 4.0$ V, $R_1 = 3.0\ \Omega$, and $R_2 = 5.0\ \Omega$. What is the current in R_1? In R_2? What is the current through the source $\mathcal{E}_1$?

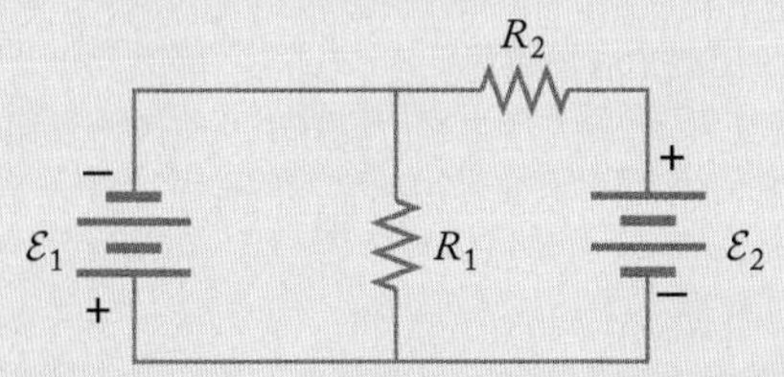

FIGURE 28.41 Two sources of emf and two resistors.

25. Consider the circuit shown in Fig. 28.41. Given that $\mathcal{E}_1 = 6.0$ V and $\mathcal{E}_2 = 10$ V, what must be the value of the resistance R_2 if the current through this resistance is to be 2.0 A?

26. Consider the circuit shown in Fig. 28.42. Given that $\mathcal{E}_1 = \mathcal{E}_2 = 4.0$ V, $R_1 = R_2 = 2.0\ \Omega$, and $R_3 = 1.0\ \Omega$, what is the current in each source? (Hint: Use the symmetry.)

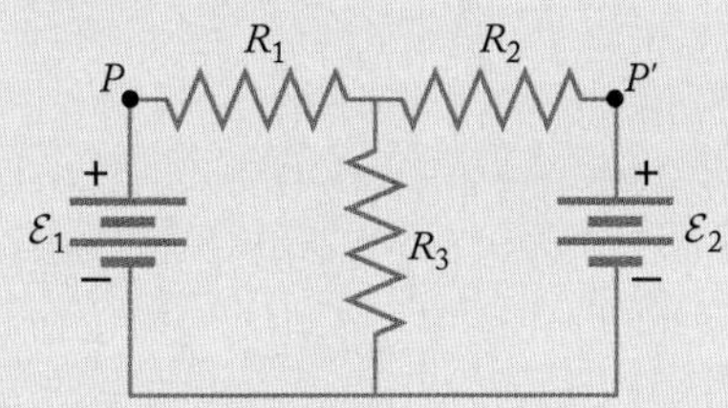

FIGURE 28.42 Two sources of emf and three resistors.

27. In the circuit shown in Fig. 28.43, $R_1 = R_2 = 2.0\ \Omega$, $R_3 = R_4 = 4.0\ \Omega$, and $\mathcal{E}_1 = \mathcal{E}_2 = 10$ V. The value of R_5 is unknown. What is the current through each resistor? (Hint: Use the symmetry.)

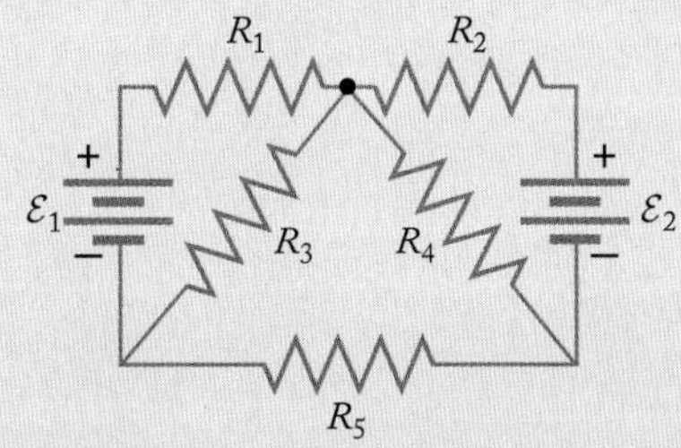

FIGURE 28.43 Two sources of emf and five resistors.

28. Find the currents in the circuit shown in Fig. 28.17a if the battery on the left is reversed.

*29. When using Kirchhoff's voltage rule to obtain the equations for a circuit, we can make several choices for the loops. In Fig. 28.17b, we made one possible choice and obtained Eqs. (28.14) and (28.15). Suppose that instead we make the choice shown in Fig. 28.44. What are the two loop equations in this case? Show that the new loop equations lead to the same result as the old ones.

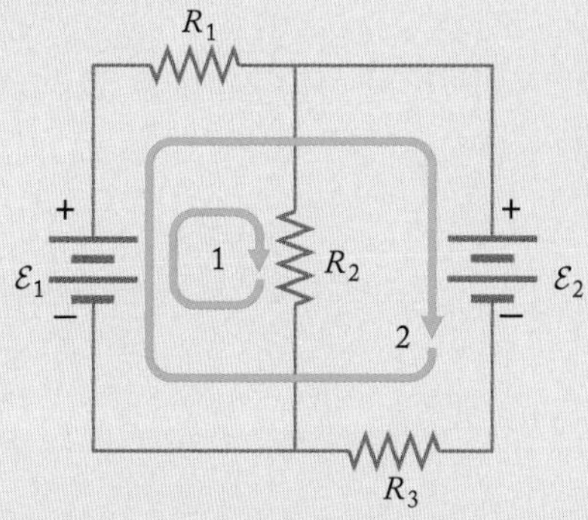

FIGURE 28.44 Alternative choice for voltage loops.

*30. Consider the circuit of Fig. 28.42. Given that $\mathcal{E}_1 = 5.0$ V, $\mathcal{E}_2 = 3.0$ V, $R_1 = 4.0\ \Omega$, $R_2 = 3.0\ \Omega$, and $R_3 = 2.0\ \Omega$, what is the current in each resistor? What is the potential difference between points P and P'?

*31. Consider the circuit shown in Fig. 28.17 with the resistances and emfs given in Example 5. Suppose we replace the emf $\mathcal{E}_1 = 12.0$ V by a larger emf. How large must we make $\mathcal{E}_1$ if the current I_3 is to charge the battery $\mathcal{E}_2$?

*32. Two batteries with internal resistances are connected in the circuit shown in Fig. 28.45. Given that $R_1 = 0.50\ \Omega$, $R_2 = 0.20\ \Omega$, $\mathcal{E} = 12.0$ V, $\mathcal{E}' = 6.0$ V, $R_i = 0.025\ \Omega$, and $R'_i = 0.020\ \Omega$, find the currents in the resistances R_1 and R_2.

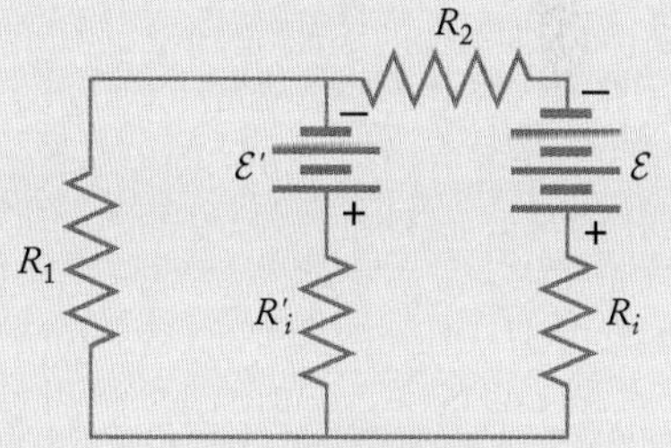

FIGURE 28.45 Batteries with internal resistances R_i and R'_i.

**33. Five resistors, of resistances $R_1 = 2.0\ \Omega$, $R_2 = 4.0\ \Omega$, $R_3 = 6.0\ \Omega$, $R_4 = 2.0\ \Omega$, and $R_5 = 3.0\ \Omega$, are connected to a 12-V battery as shown in Fig. 28.46

(a) What is the current in each resistor?

(b) What is the potential difference between the points P and P'?

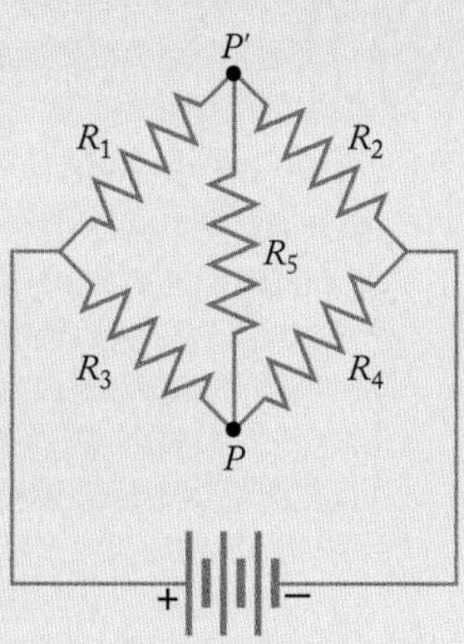

FIGURE 28.46 A source of emf and five resistors.

†For help, see Online Concept Tutorial 32 at www.wwnorton.com/physics

**34. Two batteries, with $\mathcal{E}_1 = 6.0$ V and $\mathcal{E}_2 = 3.0$ V, are connected to three resistors, with $R_1 = 6.0\ \Omega$, $R_2 = 4.0\ \Omega$, and $R_3 = 2.0\ \Omega$, as shown in Fig. 28.47. Find the current in each resistor and the current in each battery.

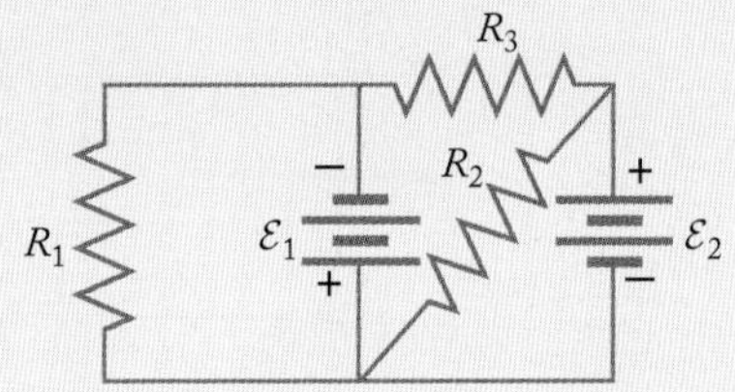

FIGURE 28.47 Two sources of emf and three resistors.

28.5 Energy in Circuits; Joule Heat

35. An electronic calculator operates from a 3.0-V battery. The calculator uses a current of 2.0×10^{-4} A. What power does it use?

36. An electric water heater of resistance 7.5 Ω draws 15 A of current when connected to the voltage supply. What is the cost of operating the water heater for 4.0 h if the electric company charges 15¢ per kilowatt-hour?

37. How much does it cost you to operate a 100-watt lightbulb for 24 hours? The price of electric energy is 15¢ per kilowatt-hour.

38. An air conditioner operating on 115 V uses 1500 W of electric energy. What is the electric current through the air conditioner? What is the resistance of the air conditioner?

39. A cryogenic refrigerator can maintain a low temperature provided that less than 350 microwatts of heat flows into the refrigerator. What maximum current can we permit through a 25-Ω resistor in this refrigerator?

40. Tabletop lasers can produce short pulses of power of 1.0 terawatt (1.0×10^{12} W). If we wish to deliver the same power with a 115-V source of emf, what current must flow? What resistance must be used? The laser pulses are short, approximately 15 femtoseconds (1.5×10^{-14} s). What is the energy per pulse? Suppose that 20 pulses can be produced each second. What is the average power? What current must flow through a 115-V source to produce this average power?

41. An electric heater has two settings, low and high, controlled by a switch. When switched to low, two identical resistors are connected in series and 275 watts are delivered. When switched to high, the two resistors are connected in parallel. What power is delivered then?

42. A photodiode produces a current of 0.50 A per watt of light incident on its surface. In sunlight, 0.10 W of light is incident. The two terminals of the photodiode are connected to a 5.0-Ω resistor. What is the voltage across the resistor? What is the ratio of the power dissipated in the resistor to the power of the incident light? Ignore the internal resistance of the photodiode.

43. A small electric motor operating on 115 V delivers 0.75 hp of mechanical power. Ignoring friction losses within the motor, what current does this motor require?

44. The rate of flow of water over Niagara Falls is 2800 m^3/s; this water falls a vertical distance of 51 m. At night, one-half of the water is diverted to a power plant. If the plant converts all of the gravitational potential energy of this diverted water into electric power, what is the electric power in kilowatts? If this power is fed into a power line at 240 kV, what is the current?

45. A cyclotron accelerator produces a beam of protons of an energy of 700 million eV. The average current of this beam is 1.0×10^{-6} A. What is the number of protons per second delivered by the accelerator? What is the corresponding power delivered by the accelerator?

46. An electric toaster uses 1200 W at 115 V. What is the current through the toaster? What is the resistance of its heating coils?

47. The banks of batteries in a submarine store an electric energy of 5.0×10^3 kW·h. If the submarine has an electric motor developing 1000 hp, how long can it run on these batteries?

48. While cranking the engine, the starter motor of an automobile draws 80 A at 12 V for a time interval of 2.5 s. What is the electric power used by the starter motor? How many horsepower does this amount to? What is the electric energy used up in the given time interval?

49. A 12-V automobile battery stores an energy of 2.5×10^6 J. How long will this battery last if it delivers 1.0 A to a resistor?

50. A house contains appliances with the following power consumptions:

40 lightbulbs, each 75 W	2 air conditioners, 1.2 kW each
1 cooking range, 9.0 kW	2 fans, 0.2 kW each
1 microwave oven, 1.4 kW	1 electric iron, 1.1 kW
1 dishwasher, 1.0 kW	1 hair dryer, 1.4 kW
1 clothes washer, 0.7 kW	1 stereo system, 250 W
1 clothes dryer, 5.0 kW	1 TV, 200 W
1 water pump, 0.7 kW	1 computer, 150 W
1 vacuum cleaner, 1.1 kW	

If all are operating simultaneously, what is the total power required? If the electric power is supplied at 115 V, what is the current?

51. The aluminum cable of a high-voltage transmission line carries a current of 600 A. The cable is 60 km long, and it has a diameter of 2.5 cm. What is the power lost to Joule heating in this cable?

*52. A resistor R, connected to a source of emf $\mathcal{E}$, is dissipating a power P. The voltage is increased to a value 25% larger. How should the resistance be changed to keep the power dissipated by the resistor the same?

*53. A resistor R, connected to a source of emf $\mathcal{E}$, is dissipating a power P. The resistor is then changed to a resistance value 20%

larger. How should the emf be changed to keep the power dissipated by the resistor the same?

*54. To heat 1.00 gram of water (density 1.00 g/cm^3) 1.00°C takes 4.18 joules. How long does it take to heat one liter of water by 50.0°C with an electric heating element of resistance 2.00 Ω carrying a current of 32.0 A?

*55. A hair dryer intended for travelers operates at 115 V and also at 230 V. A switch on the dryer adjusts the dryer for the voltage. At each voltage, the dryer delivers 1000 W of heat. What must be the resistance of the heating coils for each voltage? For such a dryer, design a circuit consisting of two identical heating coils connected to a switch and to the power outlet.

*56. An electric toothbrush draws 7.0 watts. If you use it 4.0 minutes per day and if electric energy costs you 15¢/kW·h, what do you have to pay to use your toothbrush for one year?

*57. In a small electrostatic generator, a rubber belt transports charge from the ground to a spherical collector at 2.0×10^5 V. The rate at which the belt transports charge is 2.5×10^{-6} C/s. What is the rate at which the belt does work against the electrostatic forces?

*58. A solar panel (an assemblage of solar cells) measures 58 cm × 53 cm. When facing the Sun, this panel generates 2.7 A at 14 V. Sunlight delivers an energy of 1.0×10^3 W/m^2 to an area facing it. What is the efficiency of this panel, that is, what fraction of the energy in sunlight is converted into electric energy?

*59. A battery of emf $\mathcal{E}$ and internal resistance R_i is connected to an external circuit of resistance R. In terms of $\mathcal{E}$, R_i, and R, what is the power delivered by the battery to the external circuit? Show that this power is maximum if $R = R_i$.

*60. A 3.0-V battery with an internal resistance of 2.5 Ω is connected to a lightbulb of a resistance of 6.0 Ω. What is the voltage delivered to the lightbulb? How much electric power is delivered to the lightbulb? How much electric power is wasted in the internal resistance?

*61. Suppose that a 12-V battery has an internal resistance of 0.40 Ω.

(a) If this battery delivers a steady current of 1.0 A into an external circuit until it is completely discharged, what fraction of the initial stored energy is wasted in the internal resistance?

(b) What if the battery delivers a steady current of 10.0 A? Is it more efficient to use the battery at low current or at high current?

*62. Two heating coils have resistances 12.0 Ω and 6.0 Ω, respectively.

(a) What is the Joule heat generated in each if they are connected in parallel to a source of emf of 115 V?

(b) What if they are connected in series?

*63. A 40-m cable connecting a lightning rod on a tower to the ground is made of copper and has a diameter of 7.0 mm. Suppose that during a stroke of lightning the cable carries a current of 1.0×10^4 A.

(a) What is the potential drop along the cable?

(b) What is the rate at which Joule heat is produced?

*64. The maximum current recommended for No. 10 copper wire, of diameter 0.259 cm, is 30 A. For such a wire with this current, what is the rate of production of Joule heat per meter of wire? What is the potential drop per meter of wire?

*65. A large electromagnet draws a current of 200 A at 400 V. The coils of the electromagnet are cooled by a flow of water passing over them. The water enters the electromagnet at a temperature of 20°C, absorbs the Joule heat, and leaves at a higher temperature. If the water is to leave with a temperature no higher than 80°C, what must be the minimum rate of flow of water (in liters per minute) through the electromagnet? See Problem 54.

28.6 Electrical Measurements†

66. Most ammeters have different input terminals for measuring large currents and small currents. These terminals have different values of the internal resistance. For one popular digital multimeter, the internal resistance is 11 Ω when measuring 40 mA, but only 0.030 Ω when measuring 3.0 A. What is the power dissipated internally in each case? What would be the power dissipated internally if the 40-mA connections were accidentally used to measure a 3.0-A current?

67. For the Wheatstone bridge of Fig. 28.25, R_2 = 1.0 kΩ and R_3 = 2.0 kΩ. If the ammeter reads zero when R_1 is adjusted to the value 350 Ω, what is the value of the unknown resistance R_x?

*68. A voltmeter of internal resistance 5000 Ω is connected across the poles of a battery of internal resistance 0.20 Ω. The voltmeter reads 1.4993 volts. What is the actual zero-current emf of the battery?

*69. A circuit consists of a resistor of 3.000 Ω connected to a (resistanceless) battery. To measure the current in this circuit, you insert an ammeter of internal resistance 2.0×10^{-3} Ω. This ammeter then reads 3.955 A. What was the current in the circuit before you inserted the ammeter?

*70. A voltmeter reads 11.9 V when connected across the poles of a battery. The internal resistance of the battery is 0.020 Ω. What must be the minimum value of the internal resistance of the voltmeter if the reading of the instrument is to coincide with the emf of the battery to within better than 1.0%?

*71. Experimentalists usually determine resistance with a **four-wire measurement**. Two wires carry a known current to and from the specimen to be measured, and two other wires connect the specimen to a voltmeter, so that the resistance of the wires does not contribute to the measured voltage drop. For a 5.0-Ω specimen, suppose we instead make a two-wire measurement, that is, we include the resistance of the connecting wires. What total resistance is measured if the wires are 0.25-mm-diameter copper with total length 3.0 m? What if the wire was made from constantan? (Constantan wires are often used for thermal isolation.)

†For help, see Online Concept Tutorial 32 at www.wwnorton.com/physics

28.7 The RC Circuit[†]

72. For a laboratory demonstration, you want to construct an *RC* circuit with a characteristic time of 15 s. You have available a capacitor of 20 μF. What resistance do you need?

73. Consider the circuit shown in Fig. 28.48. How long does it take after the switch is closed to charge the capacitor to within 63% of its final voltage?

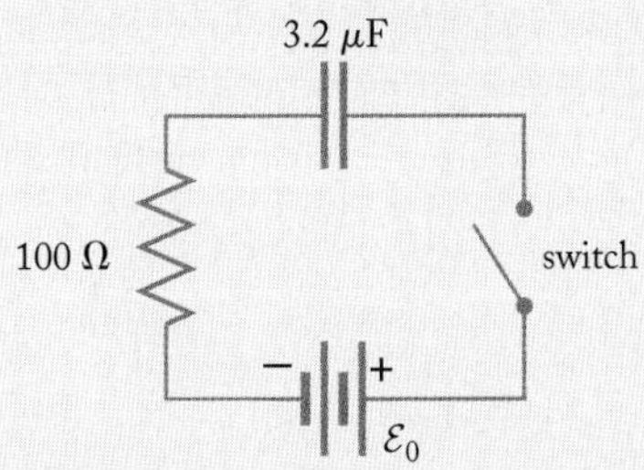

FIGURE 28.48 Charging a capacitor.

74. A capacitor charged through a 75-Ω resistor takes 2.7 ms to come within 63% of its final voltage. What is its capacitance?

75. How long does it take for a fully charged 4.0-μF capacitor to decay to 37% of its initial charge when discharged through a 100-Ω resistor?

76. An *RC* circuit consists of a resistance *R*, a capacitance *C*, and a battery connected in series. At what time is the current 1/10 of its initial (maximum) value? At what time is the current 1/100 of its initial value?

77. Two capacitors of 2.0 μF and 4.0 μF and a resistor of 8.0×10^3 Ω are connected to a battery as shown in Fig. 28.49. What is the characteristic time of this circuit?

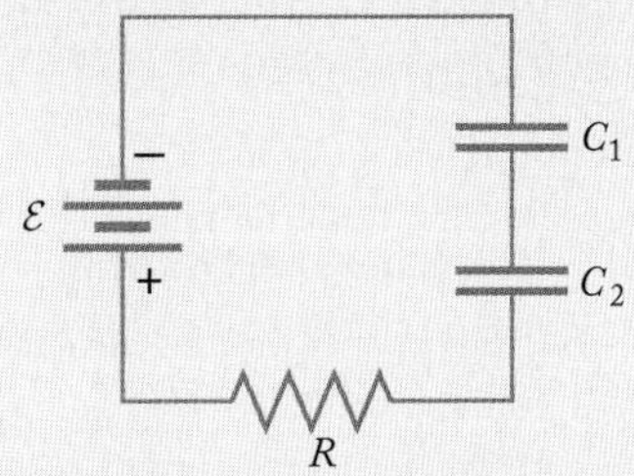

FIGURE 28.49 Charging two capacitors.

78. For the case of a charging capacitor [Eq. (28.31)], what percentage of the final value has the charge reached after two time constants? After three? After five?

79. For the circuit shown in Fig. 28.50, the switch *S* has been in position 2 for a long time. At $t = 0$, it is moved to position 1. (a) Write an expression for the charge *Q* on the capacitor as a function of the time *t*. (b) Determine the voltage across resistor R_1 as a function of the time *t*.

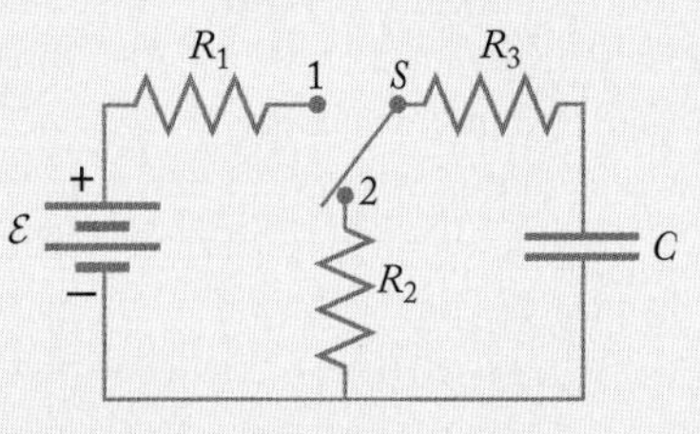

FIGURE 28.50 A switchable *RC* circuit.

[†]For help, see Online Concept Tutorial 32 at www.wwnorton.com/physics

*80. For the circuit of Fig. 28.50, assume that the switch has been in position 1 for a long time, and that at $t = 0$ it is moved to position 2. What is the current in resistor R_2 as a function of time? What is the power dissipated in resistor R_2 as a function of time? What is the total energy dissipated in resistor R_2 for $0 \le t \le \infty$?

*81. In many *RC* circuits, the charge on the capacitor can be written as $Q = A + Be^{-t/\tau}$, where *A*, *B*, and τ can be determined by inspecting the circuit. If we evaluate this expression at $t = 0$, we see that the initial charge equals $A + B$, and if we do so at $t = \infty$, we see that the final charge equals *A*. Consider the circuit shown in Fig. 28.51.

(a) Suppose the switch has been in position 1 for a long time. At $t = 0$, it is moved to position 2. What is *Q* as a function of time, that is, what are *A*, *B*, and τ for this case?

(b) Now suppose that the switch has been in position 2 for a long time. We reset our clock, and at $t = 0$ the switch is moved to position 1. What is *Q* as a function of time for this case?

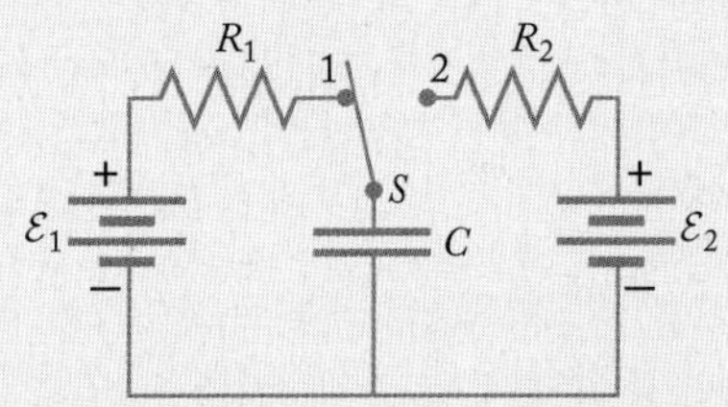

FIGURE 28.51 A switchable *RC* circuit with two sources of emf.

*82. A capacitor with $C = 0.25$ μF is initially charged to a potential of 6.0 V. The capacitor is then connected across a resistor and allowed to discharge. After a time of 5.0×10^{-3} s, the potential across the capacitor has dropped to 1.2 V. What value of the resistance can you deduce from this?

*83. Consider the *RC* circuit described in Example 9. The final (asymptotic) value of the charge in the capacitor is $\mathcal{E}C$. At what time is the charge in the capacitor one-half of this final value? At what time is the electric energy in the capacitor one-half of its final value?

*84. Equation (28.27) is a differential equation for the charge. Substitute the expression (28.31) into this equation and verify explicitly that it is a solution.

**85. Equations (28.31) and (28.33) describe the charge and current in an *RC* circuit with a battery. From these equations, deduce the total energy dissipated as Joule heat in the resistor in the time interval $t = 0$ to $t = \infty$, and deduce the total energy that the battery has to deliver in this time interval.

**86. An *RC* circuit, with $R = 6.0 \times 10^5$ Ω and $C = 9.0$ μF, is connected to a 5.0-V battery until the capacitor is fully charged. Then, the battery is suddenly replaced with a new 3.0-V battery of opposite polarity. At what time after this replacement will the voltage across the capacitor be zero? What will be the current in the resistor at this time? (Hint: see Problem 81.)

REVIEW PROBLEMS

87. Figure 28.52 shows the path of an electric current through a human body when the right hand is in good contact with a high-voltage conductor, such as a power cable, and the right foot is in good contact with the ground. The numbers give the resistances of the body parts in ohms (assume that the current punctures the skin, so the skin resistance is zero).
   (a) What is the net resistance of the body?
   (b) What is the current if the potential at the hand is 600 kV?
   (c) What would be the resistance and the current if instead of contact at the hand and the foot, the body made contact with the cable at the shoulder joint and with the ground at the middle of the thigh?

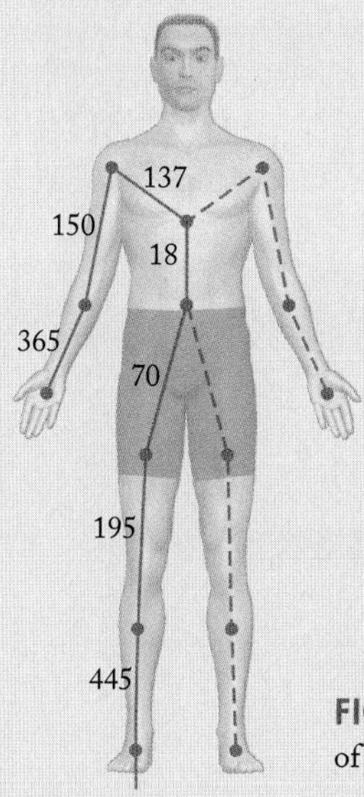

FIGURE 28.52 Resistances of segments of the human body.

88. Two wires of equal lengths and equal diameters are connected in parallel to a source of emf. One wire is made of copper, and the other of aluminum. What is the ratio of the Joule heats produced in these two wires?

89. When connected in parallel across a 115-V source, two lightbulbs deliver 75 W and 130 W, respectively. What powers do these lightbulbs deliver if instead they are connected in series across the same source? Which of the lightbulbs now delivers the larger power? Assume that the resistance of each lightbulb is a constant.

*90. Find the currents in the two resistors in the circuit shown in Fig. 28.53. The resistances are $R_1 = 10\ \Omega$, $R_2 = 8.0\ \Omega$; the emfs are $\mathcal{E}_1 = 10$ V and $\mathcal{E}_2 = 12$ V. Find the power delivered by the 12-V battery.

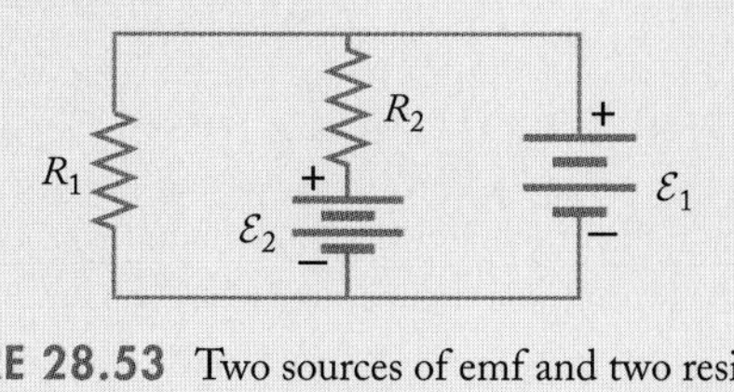

FIGURE 28.53 Two sources of emf and two resistors.

*91. Two batteries of emf $\mathcal{E}$ and of internal resistances R_i and R_i', respectively, are combined in parallel. The combination is connected to an external resistance R. Find the current through each battery.

*92. Two batteries and three resistors are connected as shown in Fig. 28.54. Given that $R_1 = 0.25\ \Omega$, $R_2 = 0.20\ \Omega$, $R_3 = 0.50\ \Omega$, $\mathcal{E}_1 = 6.0$ V, and $\mathcal{E}_2 = 12.0$ V, find the currents in the resistors R_1 and R_3.

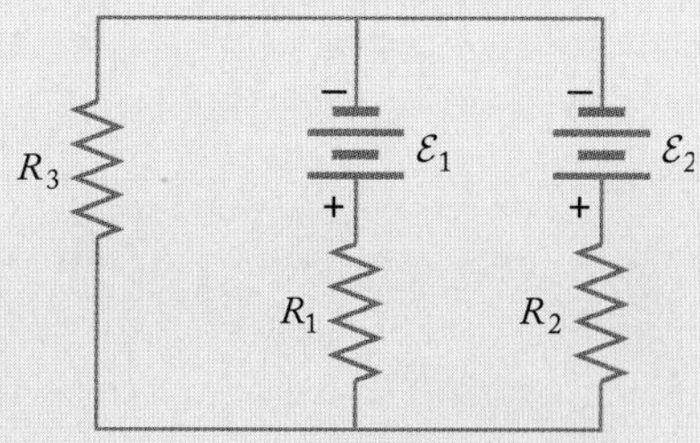

FIGURE 28.54 Two sources of emf and three resistors.

*93. Four resistors, with $R_1 = 10\ \Omega$, $R_2 = 8.0\ \Omega$, $R_3 = 20\ \Omega$, and $R_4 = 10\ \Omega$, are connected to two batteries with $\mathcal{E}_1 = 12$ V and $\mathcal{E}_2 = 16$ V as shown in Fig. 28.55. Find the currents in the batteries, and find the power they deliver.

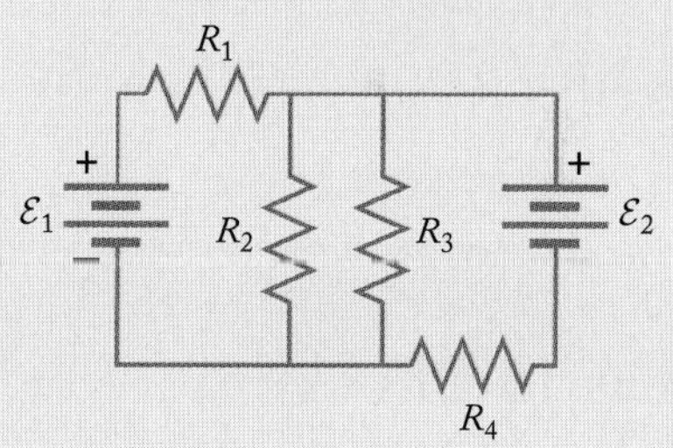

FIGURE 28.55 Two sources of emf and four resistors.

*94. A 12-V battery of internal resistance 0.20 Ω is being charged by an external source of emf delivering 6.0 A.
   (a) What must be the minimum emf of the external source?
   (b) What is the rate at which heat is developed in the internal resistance of the battery?

95. An engineer wants to design a high-voltage transmission line 100 km long. The aluminum cable used in this transmission line is to have a resistance of no more than 6.0 Ω.
   (a) What diameter of the cable is required?
   (b) For a cable of this diameter carrying a current of 750 A, what is the power lost to Joule heating?

96. You want to design a hot-water heater for use in a house. The heater is to operate at 115 V, and it is to heat 150 liters of water from an initial temperature of 10°C to a final temperature of 60°C. If the heater is to take no more than 1.0 h to reach this final temperature, what electric power is required?

What must be the resistance of the electric heating coil in this heater? Assume that no water is taken out of the heater before it reaches its final temperature, and assume that the heater is well insulated and loses no heat to the environment. See Problem 54.

*97. An electric automobile (Fig. 28.56) is equipped with an electric motor supplied by a bank of sixteen 12-V batteries connected in series. When fully charged, each battery stores an energy of 2.2×10^6 J.

(a) What current is required by the motor when it is delivering 12 hp? Ignore friction losses.

(b) With the motor delivering 12 hp, the car has a speed of 65 km/h (on a level road). How far can the automobile travel before its batteries run down?

FIGURE 28.56 An electric automobile operated by batteries.

*98. An electric clothes dryer operates on a voltage of 230 V and draws a current of 20 A. How long does the dryer take to dry a full load of clothes? The clothes weigh 6.0 kg when wet and 3.7 kg when dry. Assume that all the electric energy going into the dryer is used to evaporate the water (the heat of evaporation is 2.26×10^6 J/kg).

*99. The cable connecting the electric starter motor of an automobile with the 12.0-V battery is made of copper and has a diameter of 0.50 cm and a length of 0.60 m. If the starter motor draws 500 A (while stalled), what is the rate at which Joule heat is produced in the cable? What fraction of the power delivered by the battery does this Joule heat represent?

100. Ideal capacitors have an infinite internal resistance between their plates (i.e., the material between the plates is a perfect insulator). However, real capacitors have a finite internal resistance, and consequently the charge will gradually leak from one plate to the opposite, and the capacitor will gradually discharge when left to itself. If a capacitor of 8.0 μF has an internal resistance of $5.0 \times 10^8\ \Omega$, how long does it take for one-half of its initial charge to leak away?

101. A capacitor with $C = 20\ \mu$F and a resistor with $R = 100\ \Omega$ are suddenly connected in series to a battery with $\mathcal{E} = 6.0$ V.

(a) What is the charge on the capacitor at $t = 0$? At $t = 0.0010$ s? At $t = 0.0020$ s?

(b) What is the final value of the charge?

(c) What is the rate of increase of the charge at $t = 0$?

Answers to Checkups

Checkup 28.1

1. The energy is dissipated as heat in the wire.

2. No, the electrons pass through the battery in the direction opposite to the current (from the + terminal to the − terminal), and so they gain potential energy.

3. (A) The same energy. A battery has some given amount of stored chemical energy; in Example 1, the amount is 5400 J. Doubling the current means that the battery will deliver the same energy in half the time.

Checkup 28.2

1. Both charged and uncharged batteries have zero net charge. It is the arrangement of the charges, and thus the stored energy of the system, that is different.

2. A battery stores energy in the arrangement and type of chemical bonds, i.e., chemical energy.

3. Like a battery, the fuel of a fuel cell stores energy in the arrangement and type of chemical bonds, i.e., chemical energy.

4. The energy is chemical in the battery; the capacitor (two oppositely charged conductors) stores electric potential energy in the arrangement of charges.

5. (B) Electromotive force. Both arrangements store the same total energy of the two batteries in the form of chemical energy. For the series-stacked batteries, the potential energy of a charge increases for each battery it passes through (as in the case of the cells of an automobile battery), while for the parallel arrangement, a charge can pass through only one battery or the other. Thus the emf of the series arrangement is twice that of the parallel pair.

Checkup 28.3

1. Doubling the emf of every battery will double each current in the circuit, since current depends linearly on emf (Ohm's Law,

$I = \mathcal{E}/R$). Similarly, if we follow the emf doubling by doubling all of the resistances, the currents will return to their original values, since current is inversely proportional to resistance.

2. Reversing the potentials of the batteries will also reverse the directions of all of the currents in the circuit, but will leave the magnitudes of the currents unchanged. Reversing all or any of the resistors changes nothing, since they still offer the same opposition to current flow; the orientation of a resistor does not matter.

3. The current directions may be determined by considering Kirchhoff's voltage rule for the right and outer loops. From the three right loops, we see that current must flow downward through the center resistor in each circuit. From the outer-loop sources of emf, we see that in the first and second circuits, the currents around the outside must travel counterclockwise. In the third circuit, the balanced potentials result in zero current in the lower resistor; this implies that the downward current in the center resistor must travel back upward through the right source of emf.

4. (C) Much less than $\mathcal{E}$. Since the same current I flows through both the internal resistance and the wire, and since R_i is much greater than R, the terminal voltage IR will be much less than the voltage drop IR_i across the internal resistance.

Checkup 28.4

1. From Fig. 28.16, we see that three closed loops are enough so that every part of the circuit is included in at least one of the loops. However, any such three loops will suffice—it is not required to choose the three shown. We also see that five separate currents can be identified, that is, we have five unknowns. Since we have three voltage equations, we need two more equations; thus, two of the three junctions must be evaluated.

2. For this circuit, any two loops are enough to include every part of the circuit.

3. (A) $-I_3R_3 - I_1R_1 + \mathcal{E}_1 = 0$. Starting at point A and proceeding clockwise, we go through both R_3 and R_1 in the same direction as chosen for each current, so each is a voltage drop and thus includes a minus sign. As we continue around, we go through the source of emf in the forward direction ($-$ to $+$ terminal), so that potential change is an increase, and is written as positive.

Checkup 28.5

1. Since $P = (\Delta V)^2/R$, doubling the resistance will halve the power, so the rate of production of heat will be 0.2 W.

2. Doubling each emf will increase the rate of production of Joule heat by a factor of 4, since the power is $P = (\Delta V)^2/R$.

3. The city does not consume electric charge or current; the charge flows through the city and back to the power plant, and the current is the flow of that charge. The city does consume electric energy; it receives the charge at a high potential energy and returns it at a low potential energy.

4. (E) $\frac{1}{9}$. The power dissipated by one resistor, $P = (\Delta V)^2/R$, is proportional to the square of the potential across the resistor. The full emf $\mathcal{E}$ appears across each of the parallel resistors in Fig. 28.21a, but for the three series resistors of Fig. 28.21b, the potential difference across each is only $\frac{1}{3}\mathcal{E}$. Thus the power per resistor in Fig. 28.21b is $(\frac{1}{3})^2 = \frac{1}{9}$ of that in Fig. 28.21a.

Checkup 28.6

1. An ammeter has a low internal resistance, and measures the current passing through it with (ideally) no appreciable voltage drop across it; a voltmeter has a very high internal resistance, and measures the potential difference between its terminals without (ideally) drawing any appreciable current.

2. In Fig. 28.23c, the ammeter is not in series with the resistance, and so will not measure the current through the resistance (the current to be measured must flow through the ammeter). In Fig. 28.24b, the voltmeter is not connected across the two end terminals of the resistor, and so will not measure the potential difference across the resistor (the potential difference to be measured must be applied across the terminals of the voltmeter).

3. (A) Larger. Since an ammeter has a low internal resistance, placing it in parallel with one of the resistors as in Fig. 28.23c will decrease the overall resistance of the circuit; nearly all of the resulting larger current will flow through the low-resistance ammeter.

Checkup 28.7

1. The characteristic time depends only on the product of the values of R and C, and does not depend on the emf of the voltage source.

2. By Eq. (28.37), the initial current is $I = \mathcal{E}/R = (1.5\text{ V})/(8.0 \times 10^3\ \Omega) = 1.9 \times 10^{-4}$ A and the final current is zero, as in the charging case. The characteristic time is $\tau = RC = 8.0 \times 10^3\ \Omega \times 2.0 \times 10^{-6}\text{ F} = 1.6 \times 10^{-2}$ s.

3. (A) Less than 63% of $C\mathcal{E}$. Doubling the resistance to $2R$ doubles the time constant, so the charge now takes a time $t = 2RC$ to reach $\approx$ 63% of its final value. Thus at half that time, $t = RC$, the charge on the capacitor will be appreciably less than 63% of its final value.

CHAPTER

29

Magnetic Force and Field

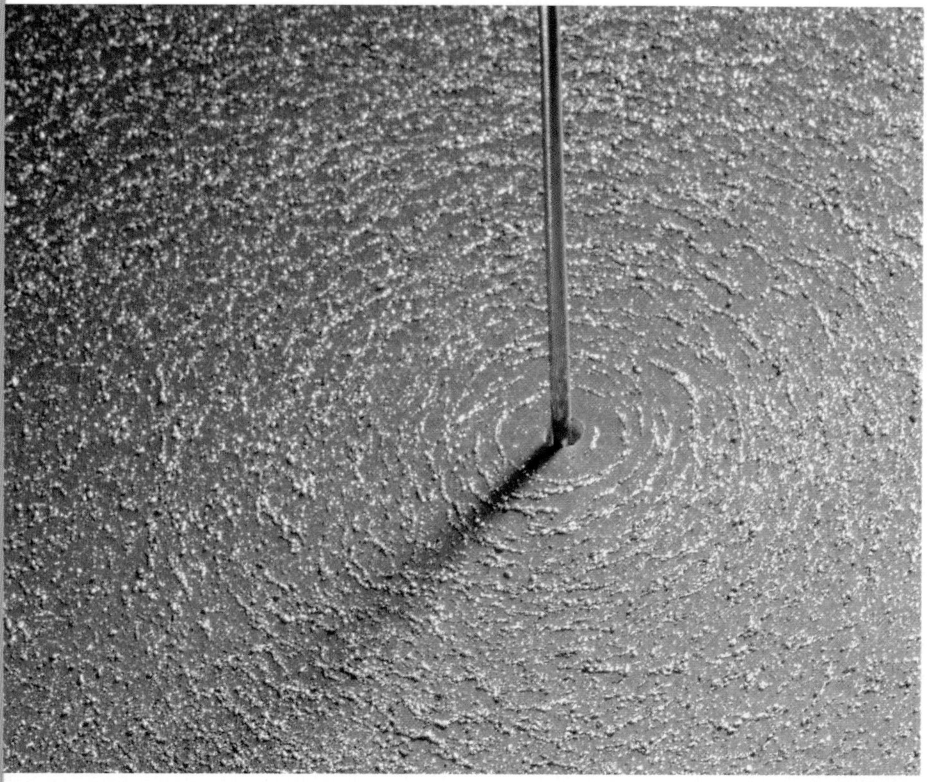

CONCEPTS IN CONTEXT

Concepts — in — Context

The straight wire carries a strong current, and this current forces the small iron particles in the photo into the circular patterns shown. We will see in this chapter that a current exerts a magnetic force on other moving charges or currents, such as the microscopic currents in the iron particles; and we will describe this force in terms of the magnetic field produced by the current.

For a current in a straight wire, we will consider such questions as

? What is the force on a moving charge near a long wire? (Section 29.1, page 928; and Example 1, page 930)

? What is the direction of the magnetic field near the wire? (Section 29.2, page 936; and Example 4, page 940)

? How does the magnetic field outside the wire vary with distance? (Section 29.2, page 939; and Example 4, page 940)? The magnetic field inside the wire? (Example 5, page 940)

? What is the total magnetic field due to two wires? (Example 3, page 938)

? What is the magnetic field near a short, straight wire? (Section 29.5, page 949)

The magnetic forces most familiar from everyday experience are the forces that permanent magnets exert on each other or on pieces of iron or other "magnetic" materials. If you bring two bar magnets together, they will attract or repel. Each magnet has two distinct ends, called its north pole and its south pole. If you bring like poles together, they repel; and if you bring opposite poles together, they attract (see Fig. 29.1). This attraction or repulsion between permanent magnets also accounts for the behavior of a compass needle, which is simply a small magnet free to rotate on a pivot. When we place a compass needle near the, say, south pole of a bar magnet, this south pole will attract the north pole of the compass needle, and the needle will settle into an equilibrium configuration, pointing roughly toward the south pole of the bar magnet (see Fig. 29.2). When we remove the compass needle from the disturbing influence of nearby bar magnets, it will respond to the magnetic force exerted by the Earth, and it will point northward. The core of the Earth acts like a large permanent magnet, whose poles roughly coincide with the geographic poles (see Fig. 29.3; the north pole of the compass needle is attracted by the geographic north pole of the Earth, because this geographic north pole is actually the magnetic south pole of the Earth).

The magnetic forces between permanent magnets were known for many centuries, but only during the nineteenth century did experimenters discover that electric currents also exert magnetic forces on permanent magnets (see Fig. 29.4) and that electric currents exert magnetic forces on one another. Finally, physicists came to understand that *the magnetic force is simply an extra electric force acting between charges in motion.* The ordinary electric force, given by Coulomb's Law, always acts between electric charges, when they are at rest and when they are in motion. The extra magnetic force acts between electric charges only when they are in motion. Whether the charges are moving along a wire, forming currents, or whether the charges are moving through empty space, on their own, makes no essential difference; in both cases *moving charges exert magnetic forces on one another.*

The magnetic forces between permanent magnets involve the same fundamental mechanism as the magnetic forces between moving charges or between currents. The magnetic forces between permanent magnets arise from microscopic currents in the atoms of the magnet. Within these atoms, currents flow around in closed loops.

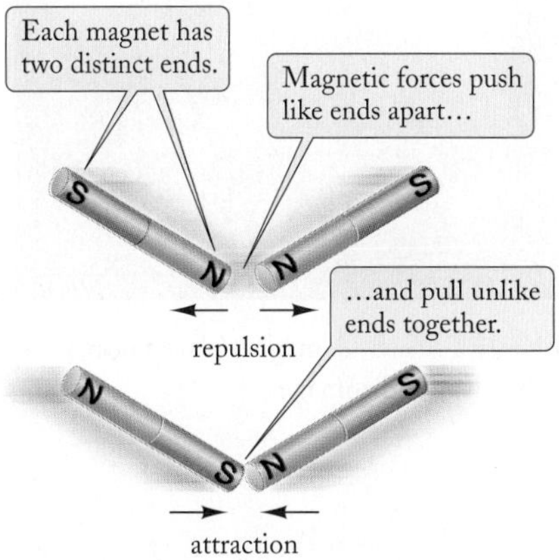

FIGURE 29.1 Equal ends of two bar magnets repel; opposite ends attract.

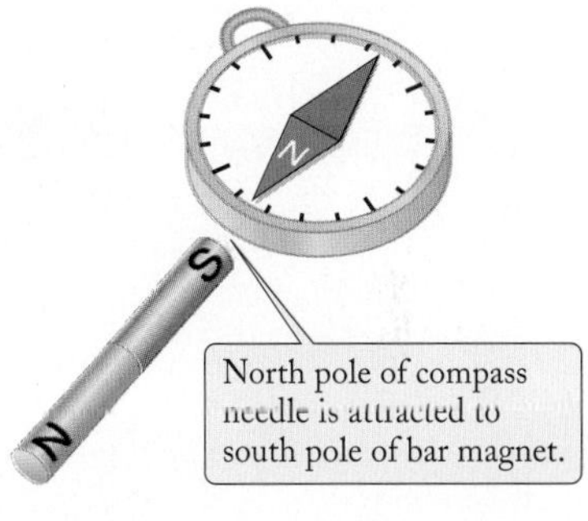

FIGURE 29.2 A compass needle placed near a bar magnet.

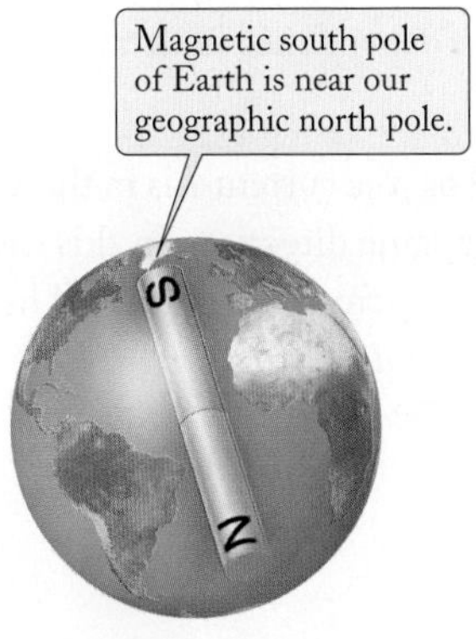

FIGURE 29.3 The Earth behaves like a large permanent magnet with magnetic poles nearly opposite to the geographic poles.

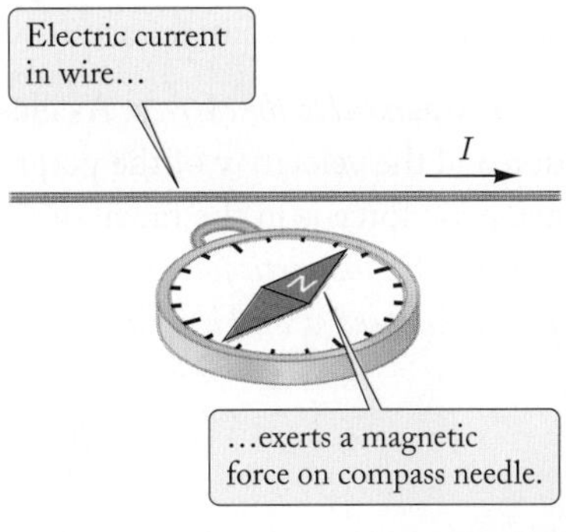

FIGURE 29.4 A compass needle placed near a wire carrying a current. The needle tends to orient at right angles to the wire.

FIGURE 29.5 A magnet placed near a TV tube distorts the image.

Although the individual microscopic rings of current are too weak to produce a noticeable force, the magnet contains many atoms and many rings of current, and the combined effect of all these microscopic rings gives a noticeable macroscopic force on another nearby magnet or on a nearby moving charge or current. You can observe the force exerted by a magnet on a moving charge if you place the magnet near the screen of your TV tube or computer monitor.[1] The magnetic force on the moving electrons in the tube will then deflect the electrons and produce strange distortions of the image on your screen (see Fig. 29.5).

In this chapter we will see that the magnetic force acting on a moving charge can be expressed in terms of a magnetic field, in much the same way that the electric force can be expressed in terms of an electric field. We will examine how magnetic fields are generated by currents, and we will examine several ways to calculate the magnetic fields generated by given distributions of currents.

29.1 THE MAGNETIC FORCE

In principle, it would be desirable to begin the study of magnetic forces with the law for the magnetic force between two moving point charges, just as we began our study of electric forces with the law for the electric force between two point charges at rest (Coulomb's Law). However, the magnitude and direction of the magnetic force that two moving charges exert on each other depends in a complicated manner on the magnitudes and directions of their velocities, and the mathematical formula for the force is rather messy.

Instead of dealing with this complicated case, it is easier to begin with the magnetic force exerted on one moving point charge q by a steady current I flowing on a long, straight wire. We will take this case as the basis for our study of magnetic forces, and we will regard the formula for the magnetic force exerted on a moving point charge by a current flowing on a long, straight wire as a fundamental law of physics, justified by experiments. The formula for this magnetic force is somewhat similar to the formula for the electric force exerted on a point charge by a long, straight charged rod. According to Example 5 of Chapter 23, the electric force that such a charged rod exerts on the point charge is inversely proportional to the distance r between the point charge and the rod. This inverse proportionality to the distance r also holds for the magnetic force $\mathbf{F}$ that a current in a long wire exerts on a moving point charge. However, in contrast to the electric force, the magnetic force also depends on the velocity $\mathbf{v}$ of the moving charge—the magnitude and the direction of the magnetic force $\mathbf{F}$ depend on the magnitude and direction of the velocity $\mathbf{v}$. To spell out this dependence in detail, it is useful to consider three separate cases:

1. $\mathbf{v}$ *parallel to the current.* As illustrated in Fig. 29.6a, the current I is in the x direction and the velocity $\mathbf{v}$ of the point charge q is in the same direction. In this case, the magnetic force is in the radial direction, toward or away from the current. The *magnitude of the magnetic force is directly proportional to the product of the current I, the charge q, and the speed v, and it is inversely proportional to the distance r*:

$$F = -[\text{constant}] \times \frac{qvI}{r} \tag{29.1}$$

[1]You should try this experiment only with an old, unwanted screen or monitor, since you might cause a permanent discoloration of the screen. Some computer monitors possess a "degauss" capability that can reverse such discoloration.

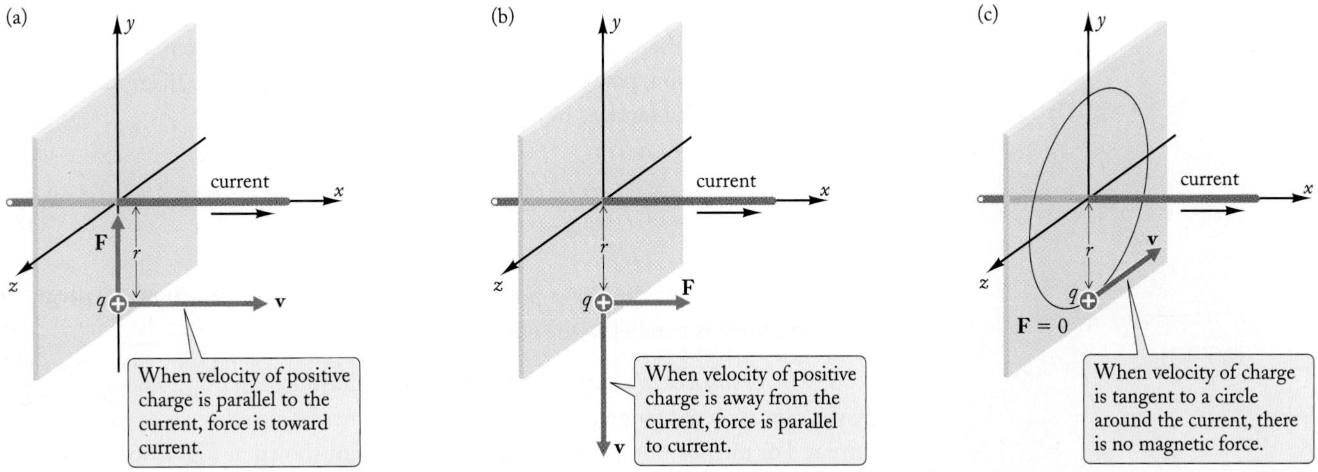

FIGURE 29.6 Direction of the magnetic force for three possible directions of the velocity of the charge q (in these diagrams, q is assumed positive). The current on the wire is in the x direction in each case. (a) The velocity of the charge q is in the x direction. The magnetic force is in the radially inward direction. (b) The velocity of the charge q is perpendicularly away from the current. The magnetic force is in the x direction. (c) The velocity of the charge q is in the tangential direction. The magnetic force is then zero.

The minus sign in this formula indicates that the force is attractive if the charge q is positive and the velocity **v** and the current I are in the same direction. Thus, in contrast to the repulsive electric force, the magnetic force between a positive charge and a current is attractive when the velocity and the current are in the same direction. However, that force is repulsive when the velocity and the current are in opposite directions. More generally, if we reckon v as positive when parallel to the current and negative when antiparallel, then the magnetic force is attractive if the product qvI is positive and repulsive if qvI is negative. Note that the magnetic force is zero unless both the speed v and the current I are different from zero—the charge q and the charges on the wire must *both* be in motion if there is to be a magnetic force between them.

In the SI system of units, the numerical value of the constant of proportionality in Eq. (29.1) is exactly

$$[\text{constant}] = 2 \times 10^{-7}\ \frac{\text{N}\cdot\text{s}^2}{\text{C}^2} \tag{29.2}$$

This constant is conventionally written in the form

$$[\text{constant}] = \frac{\mu_0}{2\pi} \tag{29.3}$$

with

$$\mu_0 = 4\pi \times 10^{-7}\ \frac{\text{N}\cdot\text{s}^2}{\text{C}^2} \approx 1.26 \times 10^{-6}\ \frac{\text{N}\cdot\text{s}^2}{\text{C}^2} \tag{29.4}$$

permeability constant

The quantity μ_0 (pronounced mu-nought) is called the **magnetic constant** or the **permeability constant.** With this constant, our equation for the magnetic force on the point charge q becomes

$$F = -\frac{\mu_0}{2\pi}\frac{qvI}{r} \quad \text{for } \mathbf{v} \text{ parallel to current} \tag{29.5}$$

magnetic force exerted on moving charge by current in long wire

HANS CHRISTIAN OERSTED (1777–1851) *Danish physicist and chemist, professor at Copenhagen. He observed that a compass needle suffers a deflection when placed near a wire carrying an electric current. This discovery gave the first empirical evidence of a connection between electric and magnetic phenomena.*

2. v *in the radial direction.* This case is illustrated in Fig. 29.6b. The velocity **v** is directed perpendicularly away from the direction of the current. The magnetic force is now in the x direction, parallel to the direction of the current. The magnitude of the magnetic force is the same as in the first case:

$$F = +\frac{\mu_0}{2\pi}\frac{qvI}{r} \quad \text{for } \mathbf{v} \text{ in radial direction} \tag{29.6}$$

The + sign indicates that for a positive charge with positive velocity (radially outward), the force is parallel to the current.

3. v *in the tangential direction.* This final case is illustrated in Fig. 29.6c. The velocity **v** is now directed tangentially to the circumference of a circle concentric with the current. For the position of the charge q at the instant shown in Fig. 29.6c, the direction of the velocity is the $-z$ direction. For the tangential case, the magnetic force is *zero* (and has no direction):

$$F = 0 \quad \text{for } \mathbf{v} \text{ in tangential direction} \tag{29.7}$$

If the direction of the velocity **v** is *not* as described in one of the three basic cases 1, 2, or 3 above, then the velocity can be split into three components along the directions described in these three basic cases, and the magnetic force can be obtained by calculating the contribution for each direction separately and then taking the vector sum of the separate magnetic forces. Thus, in principle, we can calculate the magnetic force on a charge q moving in some arbitrary direction by taking a suitable combination of the three basic cases discussed above (in the next section we will learn about a more convenient way of calculating the force on a charge moving in an arbitrary direction, by means of the magnetic field). Note that in all cases, *the magnetic force is perpendicular to the velocity of the point charge*; this perpendicularity of force and velocity is a characteristic feature of the magnetic force.

The formulas (29.5) and (29.6) for the magnetic force exerted by a current on a moving charge are to be regarded as basic laws of physics, rooted in experiments with currents and with moving charges. In our study of magnetism, these laws for the magnetic force play a role analogous to that played in electricity by Coulomb's Law.

Concepts in Context

EXAMPLE 1 A long, straight wire carries a current of 50 A. An electron of speed 2.0×10^6 m/s is (instantaneously) moving parallel to this wire at a distance of 0.030 m. What magnetic force does the current in the wire exert on the electron?

SOLUTION: This is essentially case 1 above, but the charge q is now negative, $q = -e$. If the current I is in the x direction, as in Fig. 29.7, and the velocity of the electron is also in the x direction, then the magnetic force is repulsive, radially outward from the current, and is of magnitude

$$F = -\frac{\mu_0}{2\pi}\frac{qvI}{r} = +\frac{\mu_0}{2\pi}\frac{evI}{r}$$

$$= 2 \times 10^{-7}\frac{\text{N}\cdot\text{s}^2}{\text{C}^2} \times \frac{1.6 \times 10^{-19}\ \text{C} \times 2.0 \times 10^6\ \text{m/s} \times 50\ \text{A}}{0.030\ \text{m}}$$

$$= 1.1 \times 10^{-16}\ \text{N}$$

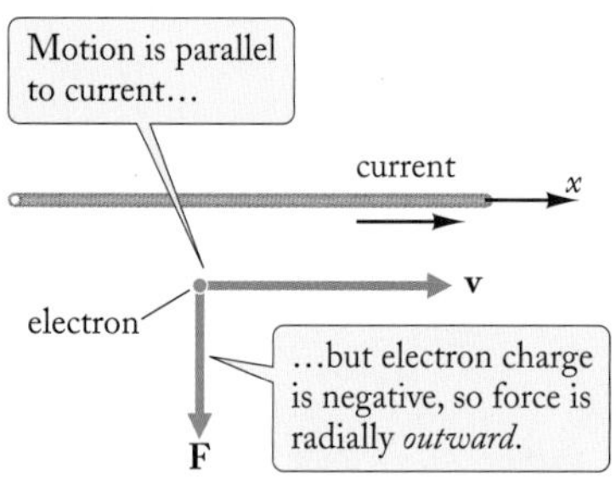

FIGURE 29.7 An electron moves parallel to a wire carrying a current. The magnetic force is directed perpendicularly away from the current.

Another instance of a magnetic force is illustrated in Fig. 29.8, which shows two long, straight, parallel wires carrying currents. The conditions here are similar to those described in case 1. The moving charges of one wire exert magnetic forces on the moving charges of the other wire. If the currents are in the same direction, as in Fig. 29.8a, the magnetic force is attractive. If the currents are in opposite directions (antiparallel), as in Fig. 29.8b, then the magnetic force is repulsive. In a later section we will see how to calculate the net magnitude of the magnetic force between the currents on such wires.

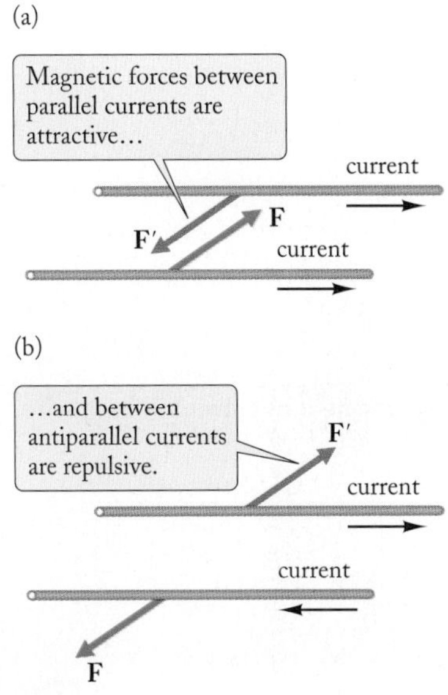

FIGURE 29.8 Two long, straight, parallel wires carrying currents (a) in the same direction and (b) in opposite directions.

✔ Checkup 29.1

QUESTION 1: For each of the following four cases, what is the direction of the magnetic force **F** on the charge q?

(a) In Fig. 29.6a, the positive charge q is moving toward the left (in the $-x$ direction) instead of the right.

(b) In Fig. 29.6b, the positive charge q is moving upward (in the $+y$ direction) instead of downward.

(c) In Fig. 29.6b, a negative charge q is moving downward.

(d) In Fig. 29.6c, the positive charge q is moving in the $+z$ direction, instead of the $-z$ direction.

QUESTION 2: Suppose we reverse the current in Fig. 29.6. How does this change the magnetic force exerted on the charge q?

QUESTION 3: Suppose we increase the distance between the wire and the charge q to twice the distance shown in Fig. 29.7. How does this change the magnetic force exerted on the charge q?

QUESTION 4: Consider a current flowing along the x axis in the $+x$ direction and a charge q a distance r below the origin on the y axis, as in Fig. 29.6. However, now the velocity **v** of the charge is not parallel to an axis. If the speed of the charge is v_0, the force on the charge is largest when

(A) $\mathbf{v} = v_0(\mathbf{i} + \mathbf{j})/\sqrt{2}$ (B) $\mathbf{v} = v_0(\mathbf{i} + \mathbf{k})/\sqrt{2}$ (C) $\mathbf{v} = v_0(\mathbf{j} + \mathbf{k})/\sqrt{2}$

29.2 THE MAGNETIC FIELD

In Section 23.1 we saw that the electric force is communicated from one charge q' to another charge q by an electric field. Likewise, *the magnetic force is communicated from one moving charge q' to another moving charge q through a magnetic field*, which serves as the mediator of the magnetic force according to a scheme similar to that for the electric field:

$$\begin{array}{ccccc} \text{current } I & & \text{magnetic field} & & \text{force} \\ (\text{or moving charge } q') & \Rightarrow & \text{of current } I & \Rightarrow & \text{on moving charge } q \end{array}$$

Starting with our formulas for the magnetic force (29.5) or (29.6) exerted by a current, we define the magnetic field by separating the expression for the force into two factors: one factor comprising quantities associated only with the moving point charge (its charge q, its velocity **v**), and another factor comprising quantities associated with the current (its magnitude I and its distance r from the point charge). Thus, in the case of a point charge moving parallel to the current, we write

$$F = (qv)\left(\frac{\mu_0 I}{2\pi r}\right) \tag{29.8}$$

We identify the second factor (including the constant of proportionality μ_0) as the magnetic field of the current, and we designate this magnetic field by B:

magnetic field of current in long wire

$$B = \frac{\mu_0 I}{2\pi r} \tag{29.9}$$

The expression for the magnitude of the magnetic force (29.5) or (29.6) then becomes

$$F = qvB \tag{29.10}$$

Concepts — in — Context

For a long wire, the direction of the magnetic field of the current is tangent to a circle with the current at its center. Figure 29.9 shows the direction of the magnetic field at several positions around the current. Note that the direction of the magnetic field coincides with that direction of the velocity **v** for which the magnetic force on the moving charge is zero (compare case 3, above). Thus there is no magnetic force when the velocity is parallel to the magnetic field.

right-hand rule for the magnetic field

The direction of the magnetic field of a current on a wire can be determined by a simple **right-hand rule for the magnetic field**: *if the thumb of the right hand is placed along the direction of the current, then the fingers will curl around the wire in the direction of the magnetic field* (see Fig. 29.10). This rule is consistent with the direction of the magnetic field shown in Fig. 29.9.

According to Eq. (29.9), the magnetic field of the current on a long, straight wire is inversely proportional to the distance r. In this regard the magnetic field of the long, straight wire is similar to the electric field of a long straight line of charge [see Eq. (24.23)]. But the directions of these fields differ—the magnetic field is in the tangential direction, whereas the electric field is in the radial direction.

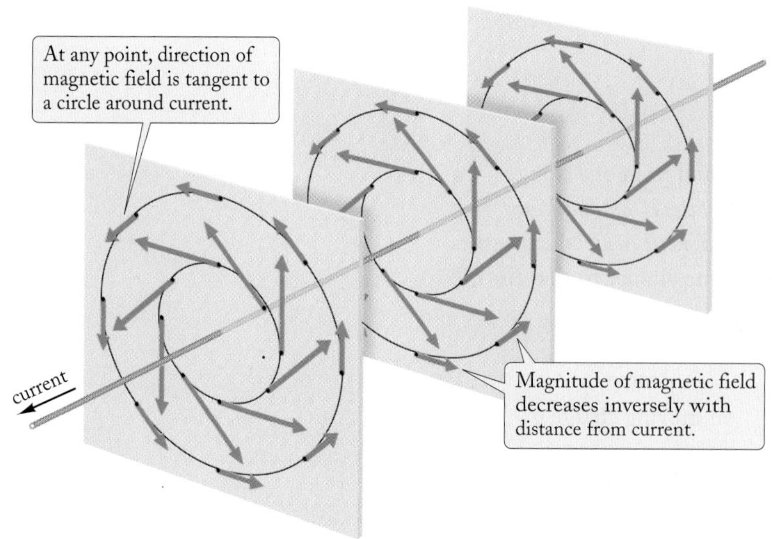

FIGURE 29.9 Magnetic field surrounding a current on a wire.

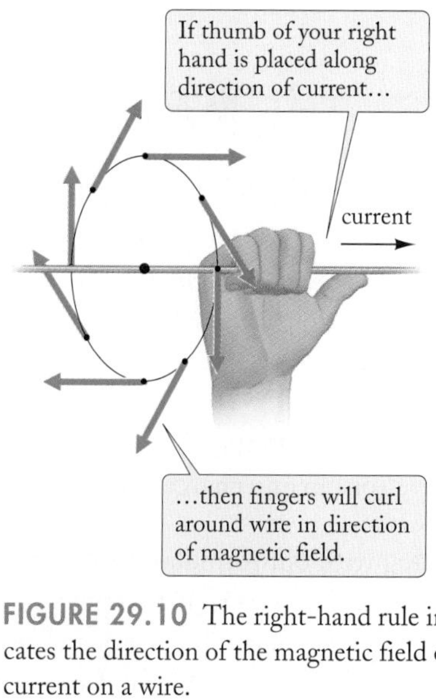

FIGURE 29.10 The right-hand rule indicates the direction of the magnetic field of a current on a wire.

Equation (29.10) is valid for the case of a point charge moving parallel to the current. But we can easily generalize this equation for all the other cases of motion of the point charge by recognizing that in cases 1 and 2 the point charge is moving perpendicularly to the magnetic field, and in case 3 the point charge is moving parallel to the magnetic field (compare Figs. 29.6 and 29.9). The absence of a magnetic force in the case of motion parallel (or antiparallel) to the magnetic field means that if the point charge moves at some general angle α with respect to the magnetic field (see Fig. 29.11), only the component of its velocity perpendicular to the magnetic field generates a magnetic force, whereas the component parallel to the magnetic field does not. But the component of the velocity **v** perpendicular to the magnetic field is $v \sin \alpha$ (see Fig. 29.11). Hence, a general expression for the magnitude of the magnetic force is

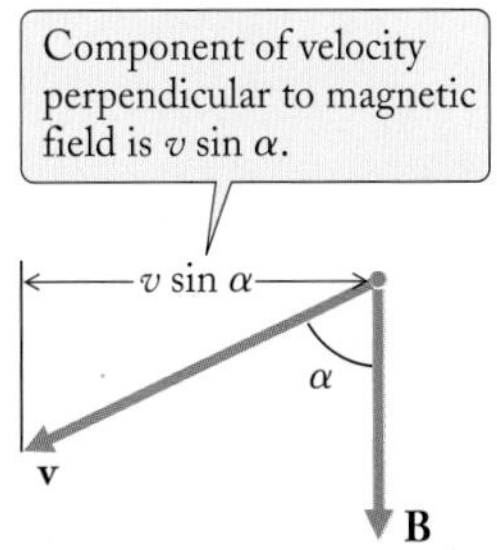

FIGURE 29.11 A point charge q moving at some angle α with respect to the direction of the magnetic field.

$$F = qvB \sin \alpha \tag{29.11}$$

magnitude of magnetic force

This expression automatically gives zero if $\alpha = 0$ (motion parallel to **B**), and it gives $F = qvB$ if $\alpha = 90°$ [motion perpendicular to **B**; see Eq. (29.10)].

By means of the cross product for vectors introduced in Chapter 3, we can write a concise formula that gives both the magnitude and the direction of the force:

$$\mathbf{F} = q\mathbf{v} \times \mathbf{B} \tag{29.12}$$

magnetic force vector

According to the general formula (3.29) for the magnitude of the cross product, the magnitude of $\mathbf{v} \times \mathbf{B}$ is $vB \sin \alpha$, in agreement with Eq. (29.11). Figure 29.12 illustrates how the cross product of the two vectors **v** and **B** yields the direction of the force vector **F**. Note that you must orient your right hand so that the fingers can curl from the first vector **v** to the second vector **B**; the thumb then gives the direction of the cross product $\mathbf{v} \times \mathbf{B}$, and thus, for a positive charge q, the direction of the force **F**.

If both electric and magnetic fields exert forces on a charge q, the net force on the charge is the vector sum of the individual forces. The total force due to both an electric field **E** and a magnetic field **B** is called the **Lorentz force**:

$$\mathbf{F} = q\mathbf{E} + q\mathbf{v} \times \mathbf{B} \tag{29.13}$$

Lorentz force

We will examine situations involving both electric and magnetic fields in the next chapter. For now, we consider magnetic fields only.

Using the definition (29.9) of the magnetic field produced by a current on a long, straight wire as guidance, we can now proceed to define the magnetic field produced by any general distribution of moving charges or currents. To discover the magnetic field at a given position, we use a test charge q and let it move repeatedly through that position,

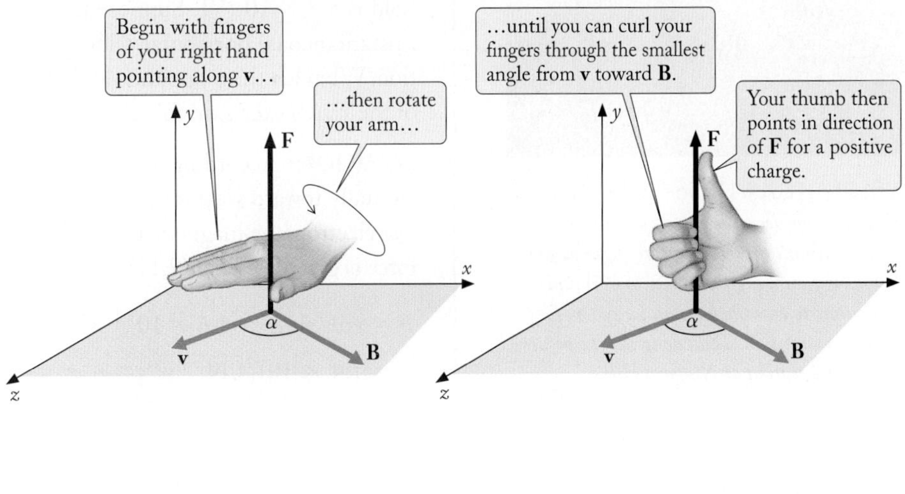

FIGURE 29.12 This right-hand rule for the magnetic force on a moving point charge relates the directions of the magnetic field **B**, the magnetic force **F**, and the velocity **v**. The vector **F** is always perpendicular to the plane defined by the vectors **B** and **v** (horizontal plane in this figure).

with different velocities, and we measure the force on the test charge. The direction and magnitude of the magnetic field **B** are then determined according to the following rules:

- The direction of the magnetic field is parallel or antiparallel to that direction of motion that results in zero force.
- The magnitude of the magnetic field is obtained by dividing the magnitude of the maximum force (the force that acts when the motion of the test charge is perpendicular to the direction of the magnetic field) by the charge and the speed:

$$B = \frac{F}{qv} \quad \text{for } \mathbf{v} \text{ perpendicular to } \mathbf{B} \qquad (29.14)$$

- The remaining ambiguity in the direction of the magnetic field is resolved by the **right-hand rule for the magnetic force**, which is equivalent to the right-hand rule for the cross product of **v** and **B**: if you place the fingers of your right hand along the direction of the velocity **v** and curl the fingers toward the direction of the magnetic field **B** through the smallest angle between **v** and **B**, your thumb will lie along the direction of the force **F** experienced by a positive test charge (see Fig. 29.12). For a negative test charge, the magnetic force is opposite to that for a positive test charge. You can check that the direction of the magnetic force exerted by the current in a wire (see cases 1 and 2 in Section 29.1) agrees with this right-hand rule.

right-hand rule for the magnetic force

With these definitions of the magnitude and the direction of a general magnetic field, the force on a charged particle moving in some arbitrary direction will always be given by the cross-product formula (29.12).

According to Eq. (29.14), the magnetic field is the force per unit charge and unit velocity. The SI unit of magnetic field is N/(C·m/s), the unit of force divided by the unit of charge and of velocity; this unit is called the **tesla** (T):

tesla (T)

gauss (G)

$$1 \text{ tesla} = 1 \text{ T} = 1 \text{ N/(C·m/s)}$$

A non-SI unit of magnetic field in common use is the **gauss** (G):

$$1 \text{ gauss} = 1 \text{ G} = 10^{-4} \text{ T}$$

Table 29.1 lists the values of some typical magnetic fields.

NIKOLA TESLA (1856–1943) *American electrical engineer and inventor. He made many brilliant contributions to high-voltage technology, ranging from new motors and generators to transformers and a system for radio transmission. Tesla designed the power-generating station at Niagara Falls.*

EXAMPLE 2 In Florida, the Earth's magnetic field is in the north–south vertical plane (and toward the north), but directed downward from the horizontal at an angle of 58° (see Fig. 29.13a). The magnitude of this magnetic field is 5.3×10^{-5} T. Suppose that an electron in a TV tube is moving with an (instantaneous) horizontal velocity of 2.0×10^6 m/s in the south to north direction. What are the magnitude and the direction of the force that the magnetic field of the Earth exerts on this electron?

SOLUTION: According to the right-hand rule, the magnetic force would be horizontally toward west for a *positive* charge (see Fig. 29.13b). Hence, for the *negative* electron, the magnetic force is horizontally toward east. The magnitude of this force is given by Eq. (29.11):

$$F = evB \sin\alpha = 1.6 \times 10^{-19} \text{ C} \times 2.0 \times 10^6 \text{ m/s} \times 5.3 \times 10^{-5} \text{ T} \times \sin 58°$$
$$= 1.4 \times 10^{-17} \text{ N}$$

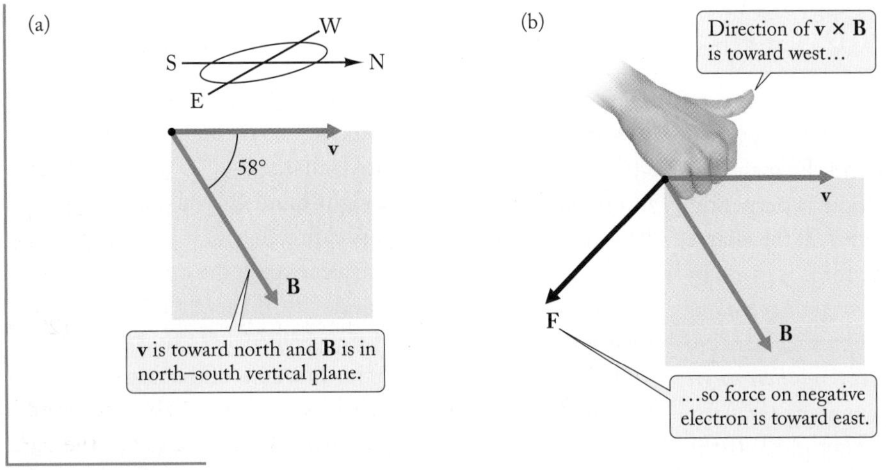

FIGURE 29.13 (a) Direction of the magnetic field of the Earth in Florida. An electron is moving south to north in this magnetic field. (b) Right-hand rule for a *positive* charge. For the *negative* electron, the magnetic force (blue) is opposite to the direction given by the right-hand rule.

TABLE 29.1 SOME MAGNETIC FIELDS

At surface of pulsar	$\approx 10^8$ T
Maximum achieved in laboratory:	
Explosive compression of field lines	1×10^3
Steady	45
In particle accelerator magnet	8
In large bubble-chamber magnet	2
In MRI magnet (a)	1.5
In sunspot (b)	≈ 0.3
Near small ceramic magnet	$\approx 2 \times 10^{-2}$
At surface of Sun	$\approx 10^{-2}$
Near household wiring (c)	$\approx 10^{-4}$
At surface of Earth	$\approx 5 \times 10^{-5}$
In sunlight (rms)	3×10^{-6}
In Crab Nebula (d)	$\approx 10^{-8}$
In radio wave (rms)	$\approx 10^{-9}$
In interstellar galactic space	$\approx 10^{-10}$
Produced by human body	3×10^{-10}
In shielded antimagnetic chamber	2×10^{-14}

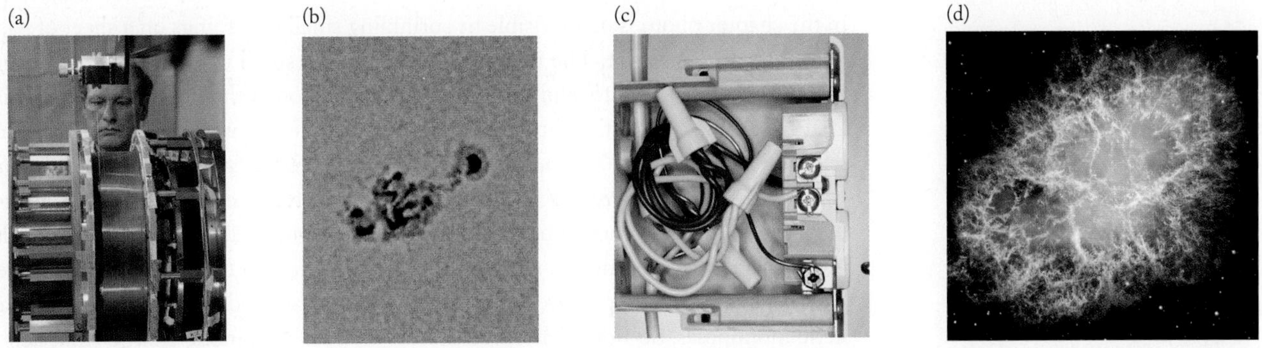

PROBLEM-SOLVING TECHNIQUES DIRECTION OF MAGNETIC FORCE

To determine the direction of a magnetic force $\mathbf{F} = q\mathbf{v} \times \mathbf{B}$ on a charge q, draw the velocity vector **v** and the magnetic field vector **B** tail to tail. The magnetic force is perpendicular to the plane defined by these two vectors. If the charge q is positive, the direction of the magnetic force is given by the right-hand rule: *place the fingers of your right hand along the direction of the velocity* **v** *and curl the fingers toward the direction of the magnetic field* **B** *through the smallest angle between* **v** *and* **B**; *your thumb will then point along the direction of the force* **F** (see Fig. 1 for several examples). If the charge q is negative, the direction of the magnetic force is opposite to that for a positive charge.

Be careful to use your **right hand** (if you use your left hand, you will get the opposite direction).

By trial and error, you can also exploit the right-hand rule to find the direction of the velocity, if the directions of the force and the magnetic field are specified; or to find the direction of the magnetic field, if the directions of the force and the velocity are specified.

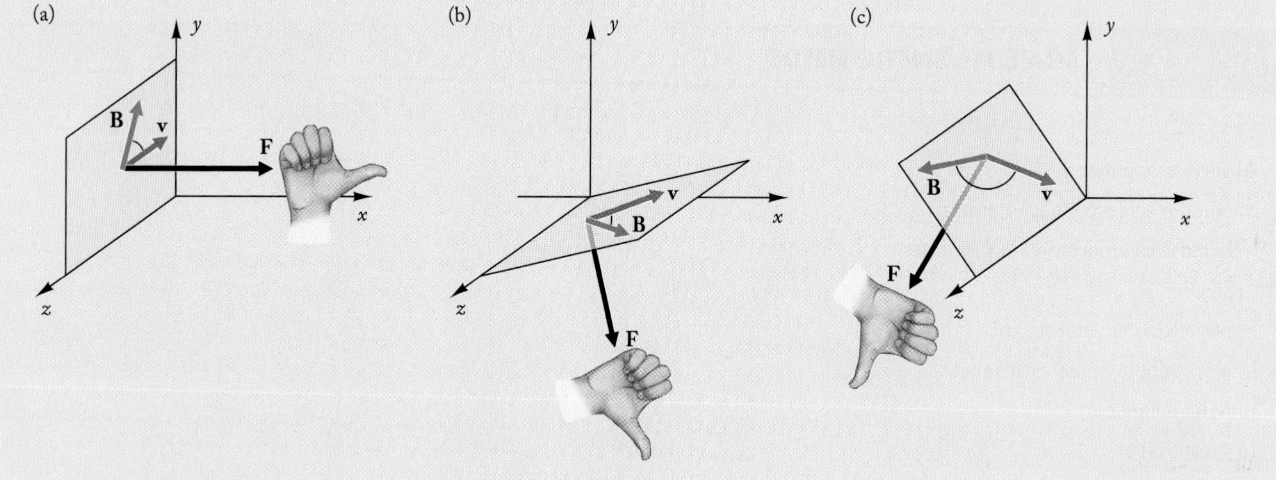

FIGURE 1 Orientations for the right-hand rule.

Concepts — in — Context

The magnetic field can be represented graphically by field lines. As in the case of the electric field, the tangent to the field lines indicates the direction of the field, and the density of field lines indicates the relative strength of the field. Figure 29.14 shows the pattern of magnetic field lines for the magnetic field produced by a current on a long, straight wire. The decrease of the strength of the magnetic field with distance is indicated by the decrease of density of the field lines. The magnetic field of the straight wire in the chapter photo is made visible by sprinkling small iron filings on a sheet of paper placed around the wire. The iron filings behave like small compass needles such as those shown in Fig. 29.15, and they align in the direction of the magnetic field. Figures 29.16 and 29.17 show the pattern of magnetic field lines for the magnetic field produced by a bar magnet. *The end of the magnet from which the field lines emerge is the north pole of the magnet, and the end into which the field lines enter is the south pole.*

Note that the magnetic field lines in Figs. 29.14 and 29.16 form closed loops, that is, the magnetic field lines do not begin or end anywhere in the way that electric field lines begin and end on positive and negative charges. Isolated magnetic poles, or magnetic monopoles, do not exist. Since there is no such "magnetic charge" that acts as source or sink of magnetic field lines, the magnetic field lines of any kind of magnetic

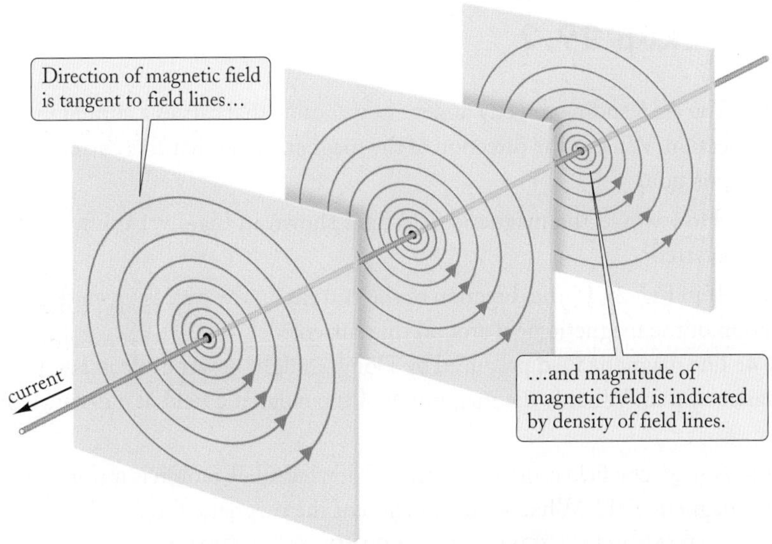

FIGURE 29.14 Magnetic field lines around a current on a long, straight wire.

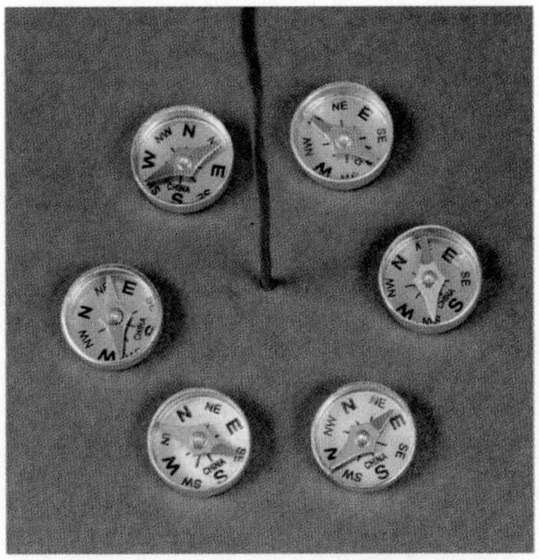

FIGURE 29.15 Circular magnetic field line around a current in a long, straight wire indicated by alignment of compass needles.

field must always form closed loops. Mathematically, we can express this feature of the magnetic field by stating *that the total magnetic flux Φ_B through any closed surface is zero*:

Gauss' Law for the magnetic field

$$\Phi_B = 0 \quad \text{or} \quad \oint \mathbf{B} \cdot d\mathbf{A} = 0 \tag{29.15}$$

Here the magnetic flux is defined in the same way as the electric flux: the magnetic flux through a surface is the integral of the perpendicular component of the magnetic field over the surface. Accordingly, the magnetic flux is proportional to the net number of magnetic field lines intercepted by the surface. Thus, Eq. (29.15) states that the number of magnetic field lines entering any closed surface matches the number leaving the surface (see Fig. 29.18). Equation (29.15) is **Gauss' Law for the magnetic field**.

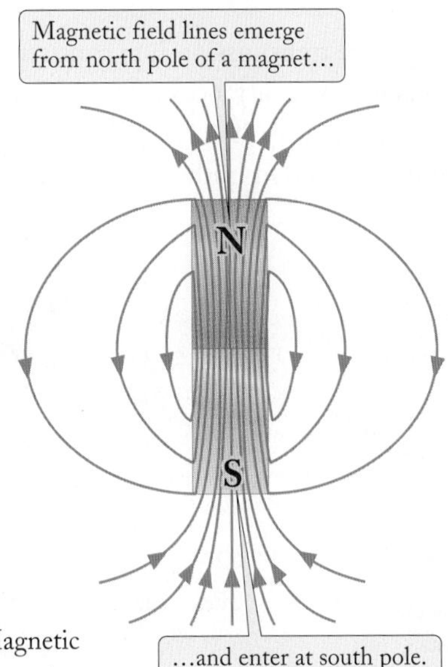

FIGURE 29.16 Magnetic field lines of a bar magnet.

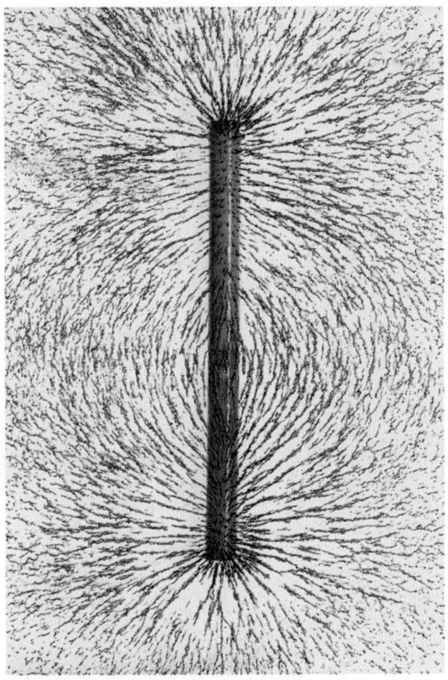

FIGURE 29.17 Magnetic field lines of a bar magnet made visible by iron filings sprinkled on a sheet of paper placed over the magnet.

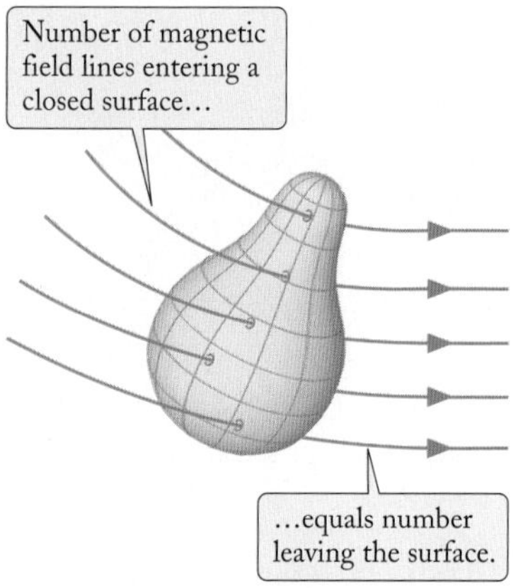

FIGURE 29.18 A closed surface intercepting magnetic field lines.

Checkup 29.2

QUESTION 1: You are standing directly under a power cable that carries a current in the northward direction. What is the direction of the magnetic field that this current produces at your position?

QUESTION 2: How would the magnetic field lines shown in Fig. 29.14 differ if the current were reversed?

QUESTION 3: If in Fig. 29.15 the direction of the current on the wire is upward, what is the direction of the magnetic field around this current?

QUESTION 4: The magnetic field produced by a long, straight power cable is 2×10^{-5} T at a distance of 6 m. What is the magnitude of the magnetic field at a distance of 3 m? 2 m?

QUESTION 5: A magnetic field is directed vertically downward. A proton is moving eastward in this magnetic field. What is the direction of the magnetic force?

(A) Up (B) North (C) South (D) East (E) West

29.3 AMPÈRE'S LAW

The magnetic fields produced by currents flowing in wires are of great practical interest, because many applications of magnetism, such as electromagnets and electric motors, rely on currents in wires to produce magnetic fields. However, the arrangements of wires used in practical applications are usually much more complicated than the single long, straight wire we dealt with in the preceding sections. We therefore need to develop a more general method for the calculation of magnetic fields.

Superposition Principle

The net magnetic field produced by several wires or other current distributions, each of which produces an individual magnetic field, obeys the **Superposition Principle**. This principle states that *the net magnetic field produced by several currents is the (vector) sum of the individual magnetic fields of the individual currents.* The following example illustrates this Superposition Principle.

Concepts in Context

EXAMPLE 3 A high-voltage transmission line consists of two long parallel wires separated by a distance of 2.0 m. The wires carry currents of 800 A in opposite directions (see Fig. 29.19a). What is the net magnetic field that these wires produce jointly at a point midway between them?

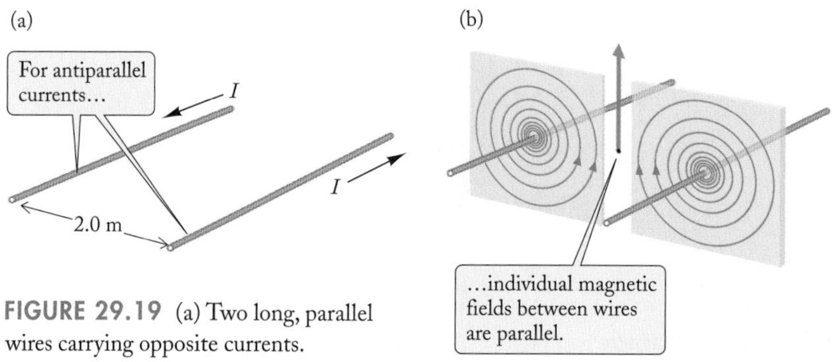

FIGURE 29.19 (a) Two long, parallel wires carrying opposite currents. (b) Magnetic field lines of the wires.

SOLUTION: Figure 29.19b shows the magnetic field lines of each wire. Since the currents are opposite, the field lines curl around the two wires in opposite directions. At the midpoint between the wires, the individual magnetic fields of the two wires are parallel. These individual magnetic fields therefore add, and the net magnetic field is twice as large as the individual magnetic field of each wire. By Eq. (29.9), the individual magnetic field of each wire is $\mu_0 I/2\pi r$, and hence the net magnetic field is

$$B = 2 \times \frac{\mu_0}{2\pi}\frac{I}{r} = 2 \times \frac{4\pi \times 10^{-7}\,\mathrm{N{\cdot}s^2/C^2}}{2\pi} \times \frac{800\ \mathrm{A}}{1.0\ \mathrm{m}}$$
$$= 3.2 \times 10^{-4}\,\mathrm{T}$$

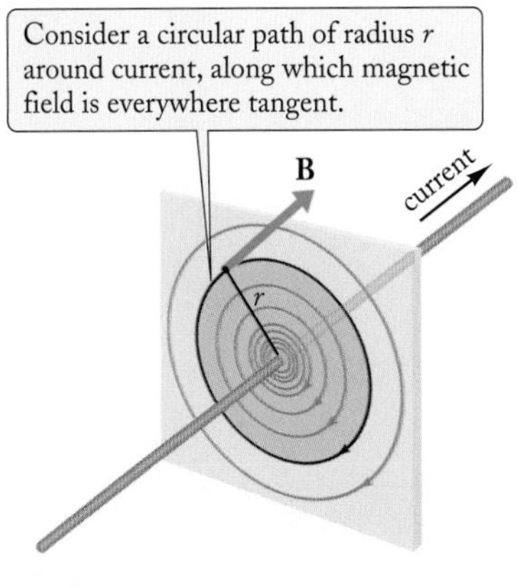

FIGURE 29.20 A circular path of radius r around a long, straight wire carrying a current I. This circular path coincides with a magnetic field line, and the magnetic field is everywhere tangent to this path.

The expression (29.9) for the magnetic field produced by a current on a long, straight wire leads to an interesting relationship between field and current. We can rewrite Eq. (29.9) as

$$2\pi r B = \mu_0 I$$

or as

$$Bs = \mu_0 I \tag{29.16}$$

where $s = 2\pi r$ is the circumference of the circle of radius r (see Fig. 29.20). In words, Eq. (29.16) states that the magnetic field along the circumference of a circle multiplied by the length of this circumference equals μ_0 times the current intercepted by the area within the circle.

In a slightly modified form, this statement is valid for a closed path of arbitrary shape in the magnetic field of an arbitrary distribution of currents. Consider some closed mathematical path (see Fig. 29.21), and designate by $B_\parallel$ the component of the magnetic field along the path, that is, the component parallel, or tangent, to the path. For a path element ds that makes an angle θ with the magnetic field, the product of the component along the path and the length of the path is $B_\parallel\, ds = B\cos\theta\, ds = \mathbf{B}\cdot d\mathbf{s}$. Then **Ampère's Law** states

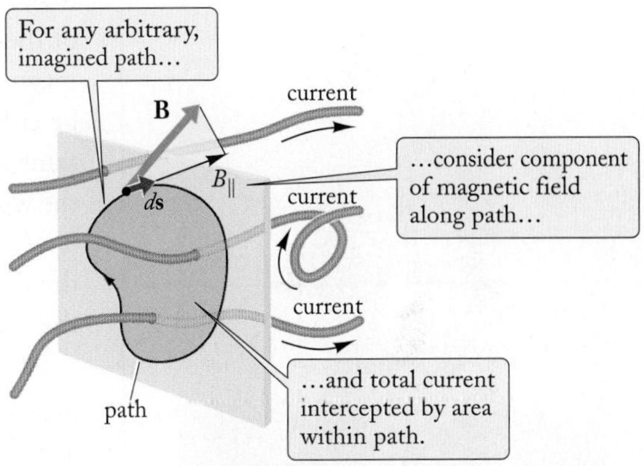

FIGURE 29.21 A closed path in a magnetic field. The area (orange) within the path intercepts some currents flowing on wires or on other conductors.

The integral around a closed path of the component of the magnetic field tangent to the direction of the path equals μ_0 times the current intercepted by the area within the path:

$$\oint \mathbf{B}\cdot d\mathbf{s} = \mu_0 I \tag{29.17}$$

Ampère's Law

or, more simply,

$$\oint B_\parallel\, ds = \mu_0 I \tag{29.18}$$

where the circle on the integral symbol here indicates that we perform the integration of $B_\parallel$ around a *closed* path; that is, we sum [field] × [length] contributions around a closed path.

Ampère's Law is one of the general laws of magnetism. The only restriction on this law is that the currents must be steady; that is, they must be constant in time. Note that the current I in Ampère's Law refers to the current crossing the area *inside* the path; currents outside make no net contribution.

Ampère's Law can be used to calculate the magnetic field of a given distribution of currents, provided that the distribution has a high degree of symmetry. The technique for calculating magnetic fields by means of Ampère's Law is similar to the technique for calculating electric fields by means of Gauss' Law. Using Ampère's Law is somewhat easier, since it involves a length instead of an area. As in Section 24.3, the calculation involves two steps: first determine the *direction* of the magnetic field by appealing to symmetry arguments, and then determine the *magnitude* of the magnetic field by evaluating Ampère's Law along some suitable, cleverly chosen path.

The following example illustrates this procedure in the simple case of a current on a straight wire. For the purposes of this example, we will pretend that we do not know the magnetic field (29.9) of the straight wire, and we will see how this magnetic field can be deduced from Ampère's Law.

Concepts — in — Context

EXAMPLE 4 From Ampère's Law, deduce the magnetic field produced by a current on a very long, straight wire.

SOLUTION: The arrangement of magnetic field lines has to match the symmetry of the current. Since the current has cylindrical symmetry, the arrangement of magnetic field lines must also have cylindrical symmetry. Thus, the field lines must be either concentric circles around the wire, radial lines, or parallel lines in the same direction as the wire. Radial lines would require that the field lines start on the wire, which is impossible, since the field lines must form closed loops. Parallel lines in the direction of the wire are likewise inconsistent with closed loops. Thus, the field lines must necessarily be concentric circles. Furthermore, by symmetry, the magnetic field must have a constant magnitude along each circle.

Taking a path that follows one of these circles, of radius r (see Fig. 29.20), the magnetic field is everywhere parallel to the path, so $B_{\parallel} = B =$ constant, and the left side of Ampère's Law is

$$\oint B_{\parallel}\, ds = B \oint ds = B \times 2\pi r$$

The right side is $\mu_0 I$; hence

$$2\pi r B = \mu_0 I$$

As expected, this yields the magnetic field of Eq. (29.9):

$$B = \frac{\mu_0}{2\pi}\frac{I}{r}$$

ANDRÉ-MARIE AMPÈRE (1775–1836)
French physicist and mathematician. After Oersted's discovery of the generation of magnetic fields by electric currents, Ampère demonstrated experimentally that currents exert magnetic forces on each other. He carefully investigated the relationship between currents and magnetic fields, and he established that a magnet is equivalent to a distribution of currents.

Concepts — in — Context

EXAMPLE 5 A very long, straight conducting wire has a circular cross section of radius R. The wire carries a current I_0 uniformly distributed over this cross-sectional area. (a) What is the magnetic field inside the wire? (b) What is the magnetic field outside the wire?

SOLUTION: The symmetry of this thick wire and its current is the same as the symmetry of the thin wire treated in the preceding example. Hence the magnetic field lines must be concentric circles, both inside the wire and outside the wire.

(a) To calculate the magnetic field inside the wire, we take a path that follows one of the circular field lines, of radius r (see Fig. 29.22). As in the preceding example, $B_{\parallel} = B$ and the left side of Ampère's Law is again

$$\oint B_{\parallel}\, ds = B \times 2\pi r$$

Recall that the current I on the right side of Ampère's Law is only the current *inside* our path. Since the current is uniformly distributed over the cross section of the wire, the amount of current intercepted by the area πr^2 within the path is directly proportional to this area:

$$I = \frac{\pi r^2}{\pi R^2} \times I_0 = \frac{r^2}{R^2} I_0$$

Ampère's Law then becomes

$$2\pi r B = \mu_0 I = \mu_0 \frac{r^2}{R^2} I_0$$

which yields

$$B = \frac{\mu_0}{2\pi} \frac{Ir}{R^2}$$

(b) To calculate the magnetic field outside the wire, we proceed as in the preceding example. Since the magnetic field of a thick wire has the same symmetry as for a thin wire, the calculation presented in the preceding example is equally valid for a thick wire and for a thin wire, and we find the same result as before, Eq. (29.9), except that the net current on the entire wire is now designated by I_0:

$$B = \frac{\mu_0}{2\pi} \frac{I_0}{r}$$

COMMENT: Note that the magnetic field is zero at the center of the wire ($r = 0$); it increases in proportion to the radius r and reaches a maximum value of $\mu_0 I_0/(2\pi R)$ at the surface of the wire. Beyond this point, the magnetic field decreases in proportion to $1/r$.

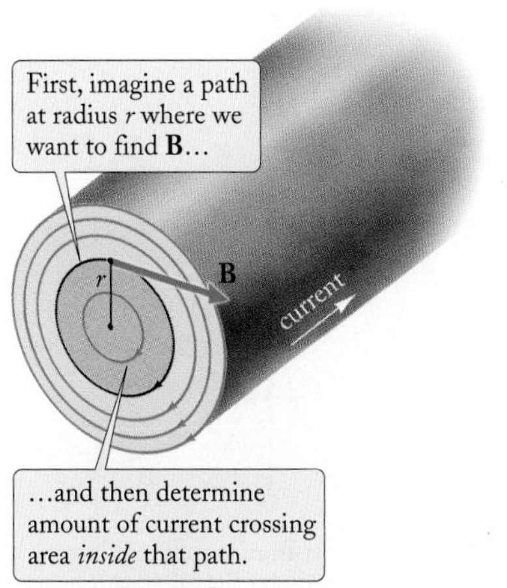

FIGURE 29.22 Circular field lines for the magnetic field inside a long, straight wire of cross-sectional radius R. The path used for Ampère's Law follows one of these field lines, of radius r.

The magnetic field outside a thick cylindrical wire with a uniform distribution of current over its interior is given by the same formula (29.9) as for a thin wire. More generally, this is true of any assembly of concentric cylindrical shells (like those defined by the rings of an ideal, cylindrical tree trunk) when each carries uniformly distributed current. For instance, the magnetic field produced by a power line consisting of a thick aluminum cable, typically a few centimeters across, can be calculated from the formula (29.9).

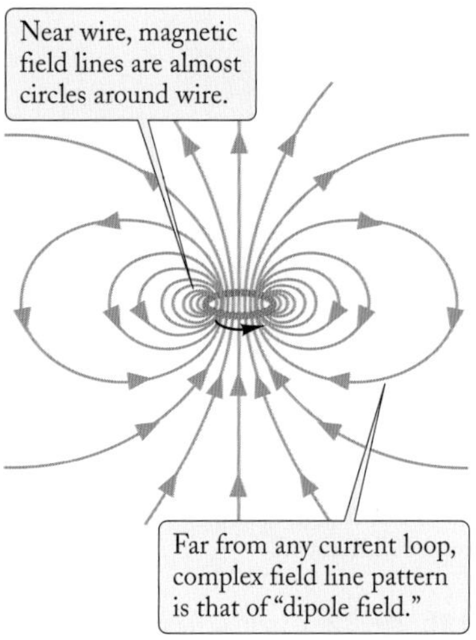

FIGURE 29.23 Magnetic field lines of a ring of current.

Although circular concentric field lines and a magnetic field with the strength given by the formula (29.9) are characteristic of a current on a long, straight wire (thin or thick), this formula also provides an approximation for the magnetic field in the immediate vicinity of any segment of wire, straight or curved, except where the wire has a sharp kink. From points very near the wire, any wire looks almost straight, and the magnetic field at such points is then approximately that of a long straight wire. The magnetic field of a current flowing on a ring illustrates this behavior. The field lines for such a ring are shown in Figs. 29.23 and 29.24; very near the wire, the field lines are almost circles, concentric with the wire. However, far from the wire, the pattern of field lines is much more complex. Similar behavior for two rings is illustrated on the book cover.

The calculation of the complete magnetic field of a ring of current at points near and far from the ring is quite complicated, and we will not attempt to do this calculation. We will show in Section 29.5 that the magnitude of the magnetic field at the center of a ring of radius R carrying a current I is

$$B = \frac{\mu_0 I}{2R} \tag{29.19}$$

right-hand rule for current loop

The direction of the magnetic field at the center is perpendicular to the plane of the ring (see Fig. 29.25), and is given by a **right-hand rule for a current loop**: if the fingers of the right hand curl around the loop in the direction of the current, the thumb gives the direction of the magnetic field at the center of the loop.

Note that the pattern of magnetic field lines at large distance from the ring of current resembles the pattern of electric field lines of an electric dipole (compare Fig. 23.17). This resemblance is also found in the pattern of magnetic field lines at large distance from a closed loop of current of any other shape. At large distance, square loops, rectangular loops, oval loops all produce the same pattern of field lines as a circular ring—the exact shape of the loop has little effect on the distant magnetic field. For example, in the central core of the Earth, currents flow in loops of some kind, and the magnetic field that these currents produce at the surface of the Earth and in the space above is pretty much that of a ring of current (see Fig. 29.26).

FIGURE 29.24 Magnetic field lines of a ring of current made visible with iron filings.

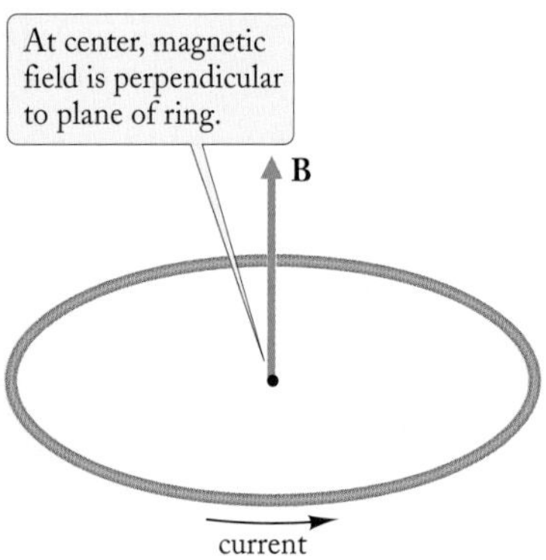

FIGURE 29.25 Direction of the magnetic field at the center of a ring of current.

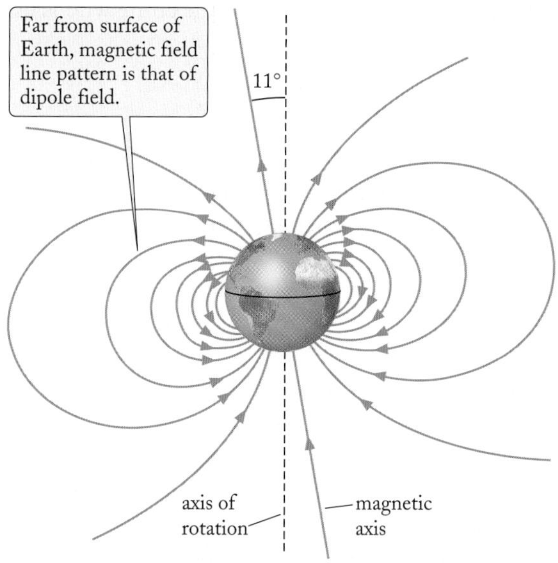

FIGURE 29.26 Magnetic field of the Earth. The axis of the magnetic field makes an angle of about 11° with the axis of rotation of the Earth.

Checkup 29.3

QUESTION 1: Two long parallel wires carry currents of equal magnitudes in the same direction. What is the net magnetic field that these wires produce jointly at a point midway between them?

QUESTION 2: If the direction of the current in the visible segment of the ring in Fig. 29.24 is right to left, what is the direction of the magnetic field along the axis of the ring?

QUESTION 3: If a charged particle is moving vertically upward along the axis of the ring of current shown in Fig. 29.25, what is the magnetic force on this particle?

QUESTION 4: A circular path is located as shown in Fig. 29.27 in the magnetic field of a long straight wire. What is the integral of the tangential magnetic field around this path?

QUESTION 5: A thick wire of radius R carries a current I uniformly distributed over its cross section. What is the magnetic field at the surface of the wire?

(A) $\mu_0 I/(2\pi R)$ (B) $\mu_0 I/(4\pi R)$ (C) $\mu_0 I/R$
(D) $\mu_0 I/(2R)$ (E) $\mu_0 I/(4R)$

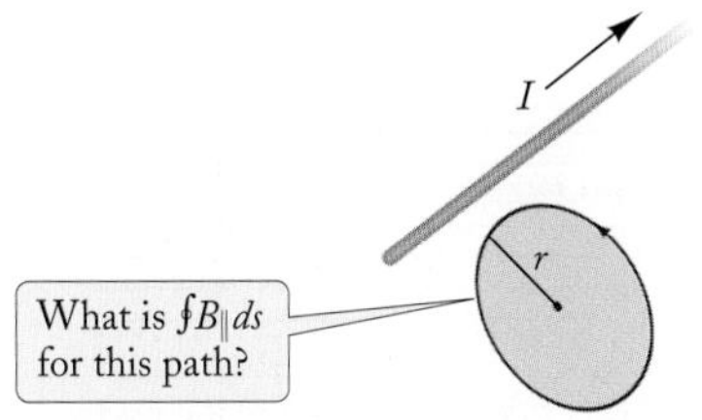

FIGURE 29.27 A path near a current-carrying wire.

29.4 SOLENOIDS AND MAGNETS

A solenoid is a conducting wire wound in a tight helical coil of many turns (see Fig. 29.28). A current in this wire will produce a strong magnetic field within the coil (see Figs. 29.29 and 29.30). Because of the similarity of the current distributions, such a tight coil produces essentially the same magnetic field as a large number of rings stacked next to one another. The calculation of the magnetic field of such a solenoid of finite length is fairly difficult, and we will not attempt it here. Instead, we will calculate the magnetic field of an **ideal solenoid**, that is, a very long (infinitely long) solenoid with very tightly wound coils, so the current distribution on the surface of the solenoid is nearly uniform.

ideal solenoid

To find this magnetic field, we begin with an appeal to symmetry, as in Example 4. The ideal solenoid has translational symmetry (along the axis of the solenoid) and cylindrical symmetry (around the axis). For consistency with these symmetries, the magnetic field lines inside the solenoid will then have to be either concentric circles, or radial lines, or lines parallel to the axis. Concentric circles and radial lines are unacceptable; the former would require the presence of a current along the axis (compare Example 4), and the latter would require that the field lines begin on the

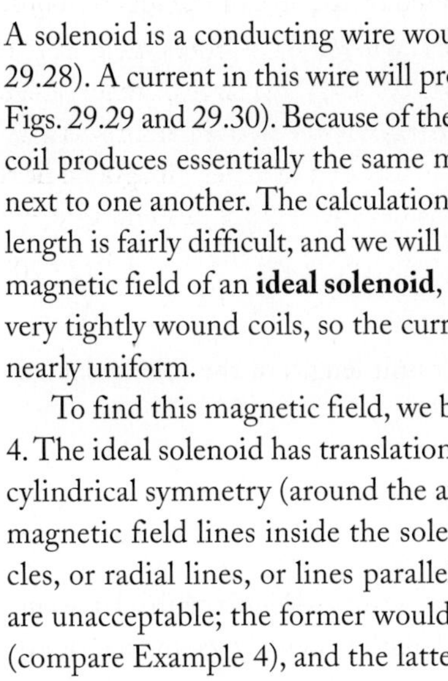

FIGURE 29.28 A solenoid.

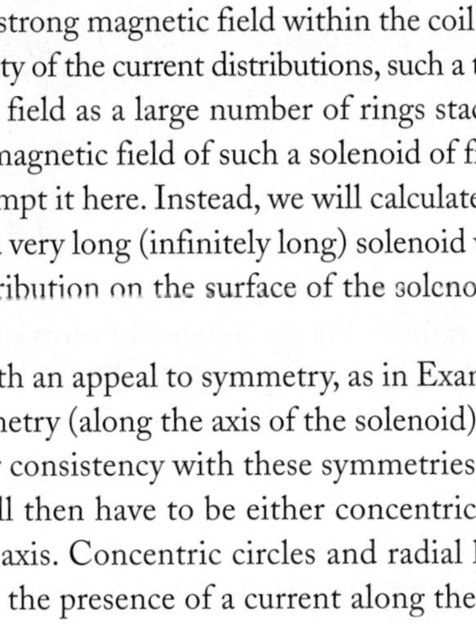

FIGURE 29.29 Magnetic field lines in a segment of a long solenoid.

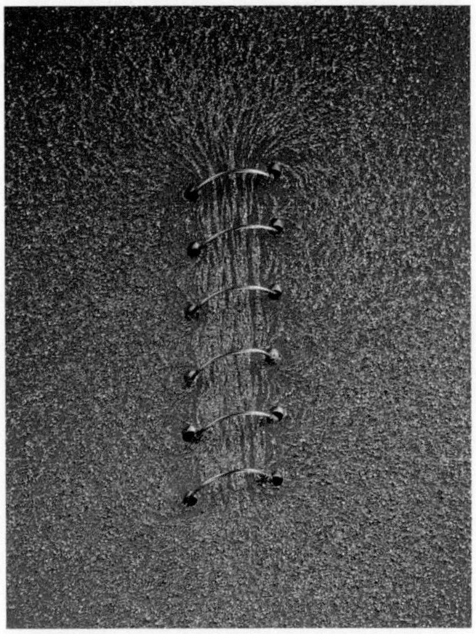

FIGURE 29.30 Iron filings sprinkled on a sheet of paper inserted in a solenoid. Note that inside the solenoid, the distribution of field lines is nearly uniform.

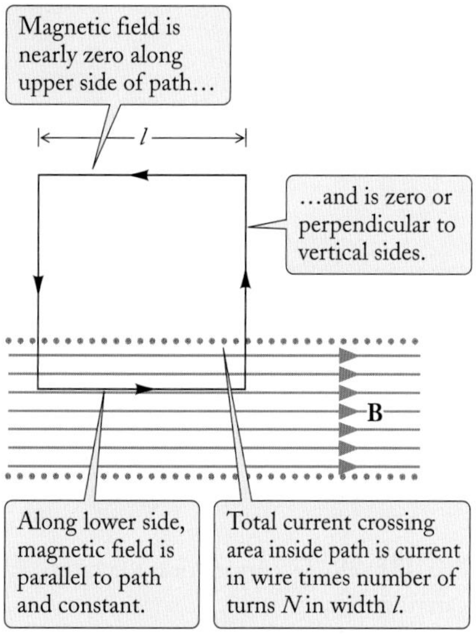

FIGURE 29.31 Magnetic field lines of an ideal, very long solenoid. The path for the application of Ampère's Law is a square of side l.

axis, which is impossible, since magnetic field lines cannot begin or end anywhere. Thus, the field lines inside the solenoid must all be parallel to the axis (see Fig. 29.31). These lines emerge from the (distant) end of the solenoid, and they curve around the exterior of the solenoid and return to the other (distant) end. For an ideal, very long solenoid, these external field lines will then be spread out over a very large region of space; hence, the density of field lines and the magnetic field outside the solenoid are nearly zero.

We can now use Ampère's Law to determine the magnitude of the magnetic field. To evaluate the left and the right sides of Ampère's Law, we must choose a closed path. A convenient choice is the square path shown in Fig. 29.31, which is partly inside the solenoid and partly outside. The magnetic field has a component parallel to the path only along the lower side of the square, within the solenoid; along this side $B_{\parallel} = B$. Note that by symmetry the field cannot vary along the length l of this lower part of the path, and is thus $B =$ constant. Along all other sides, either the magnetic field is zero (outside the solenoid) or the component of the magnetic field tangent to the path is zero (the field is perpendicular to the other parts of the path inside the solenoid). Hence the integral appearing in Ampère's Law involves only the lower side of the square:

$$\oint B_{\parallel}\, ds = 0 + 0 + 0 + \int_0^l B_{\parallel}\, ds = Bl$$

The net current intercepted by the area of the square is $I_0 \times N$, where I_0 is the current in one wire and N is the number of wires intercepted by the area of the square. Thus, Ampère's Law becomes

$$Bl = \mu_0 I_0 N$$

from which

$$B = \mu_0 I_0 \frac{N}{l} \tag{29.20}$$

The ratio N/l is the number of turns of wire per unit length of the solenoid, commonly designated by $n = N/l$. Thus,

magnetic field inside solenoid

$$B = \mu_0 n I_0 \tag{29.21}$$

This shows that to obtain a large magnetic field, we want a solenoid with a large current and a large number of turns of wire per unit length (a densely wound solenoid).

right-hand rule for solenoids

The direction of the magnetic field inside a solenoid is given by a **right-hand rule for solenoids**: if the fingers of the right hand curl around the solenoid in the direction of the current, the thumb gives the direction of the magnetic field inside the solenoid. This rule is consistent with the direction of the magnetic field in Figs. 29.29 and 29.31 for the direction of the current in Fig. 29.28. This rule is similar to the right-hand rule for a current loop given in the previous section, as it must be, since a solenoid is a coil of many current loops.

Note that the result (29.21) is independent of the "depth" to which the square is immersed in the solenoid—we can slide the square deeper into the solenoid (provided its upper side remains outside the solenoid and its lower side remains inside), and we still obtain the same result (29.21) for the magnetic field. Hence the magnetic field has the same magnitude everywhere within the solenoid. This means that the magnetic field within an ideal solenoid is perfectly uniform.

Furthermore, the result (29.21) is also independent of the shape of the cross section of the solenoid. In Fig. 29.28, the solenoid has a circular cross section (circular coils). But it could equally well have a square cross section, or a rectangular cross section, or some other cross section (see Fig. 29.32), since this does not affect the application of Ampère's Law.

A variation on the solenoid is the toroid, shown in Fig. 29.33. This doughnut- or torus-shaped coil resembles a solenoid that has been bent around in a circle so that its two ends meet. The magnetic field lines remain parallel to the axis of the original solenoid and thus now form closed loops in the interior of the toroid. This geometry is useful in electronic devices, since such "contained" field lines will not interfere with nearby circuit components. Although the toroid does not have as high a symmetry as the infinite solenoid, we can still apply Ampère's Law to a circular path of radius r inside the toroid, such as the one through the toroid's midplane shown in Fig. 29.33. For a given value of r, symmetry requires that the field must be constant and tangent to such a path, so the left side of Ampère's Law becomes

$$\oint B_{\parallel}\, ds = B \times 2\pi r$$

For a toroid with N loops of current, each carrying the same current I_0, the right side of Ampère's Law is $\mu_0 N I_0$, so equating the two sides gives

$$2\pi r B = \mu_0 N I_0$$

Solving for the magnetic field, we obtain

$$B = \mu_0 I_0 \frac{N}{2\pi r} \tag{29.22}$$

Note that Eq. (29.22) has the same form as Eq. (29.20), with the solenoid length l replaced by the path length around the torus interior, $2\pi r$. The field varies somewhat with position inside the torus; the field is larger for small r and smaller for large r. Outside, including the region of the doughnut "hole," there are no magnetic field lines, and the field there is zero.

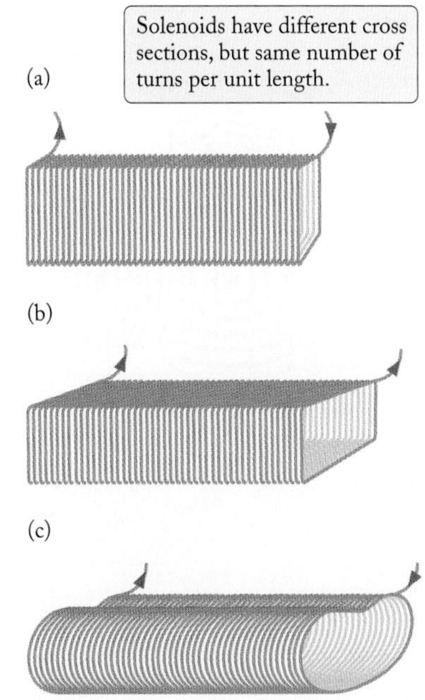

FIGURE 29.32 (a) Square, (b) rectangular, and (c) irregular solenoids. All long solenoids produce equal uniform magnetic fields (for equal currents per unit length).

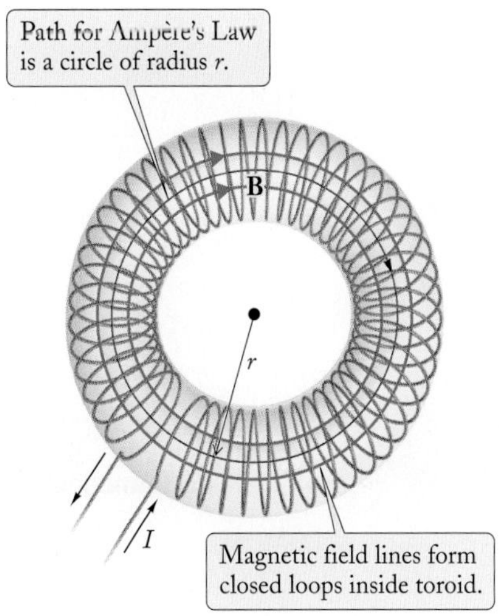

FIGURE 29.33 A toroid.

EXAMPLE 6 A solenoid used for research consists of 180 turns of wire wound on a narrow cardboard tube 19 cm long. The current in the wire is 5.0 A. What is the strength of the magnetic field within the tube?

SOLUTION: The number of turns per unit length is

$$n = 180/0.19\ \text{m} = 9.5 \times 10^2/\text{m}$$

Although this solenoid is not infinitely long, its length is large compared with its width, and therefore the magnetic field at all points within it except those near the ends will be approximately given by Eq. (29.21):

$$B = \mu_0 n I_0 = 1.26 \times 10^{-6}\ \text{N}\cdot\text{m}^2/\text{C}^2 \times 9.5 \times 10^2/\text{m} \times 5.0\ \text{A}$$

$$= 6.0 \times 10^{-3}\ \text{T}$$

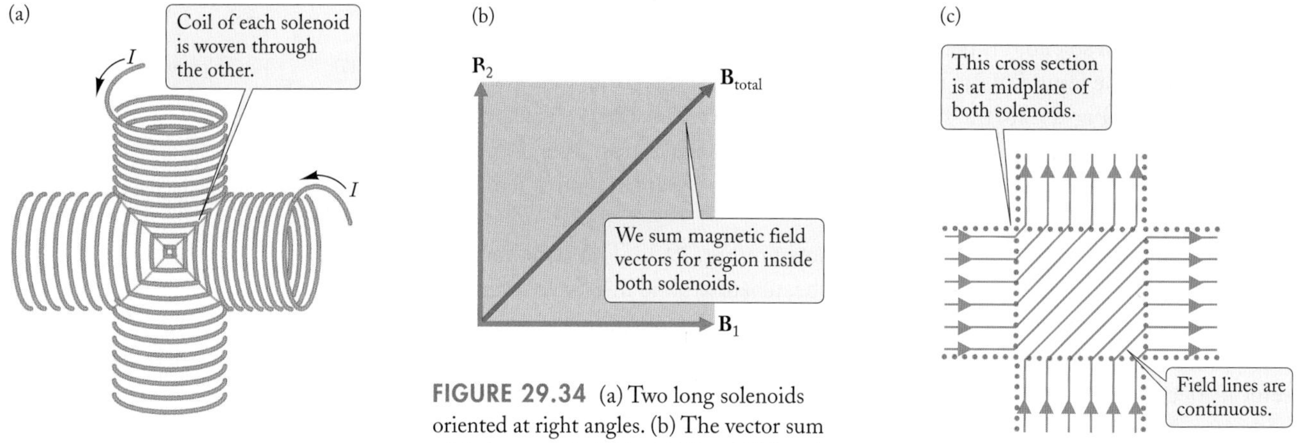

FIGURE 29.34 (a) Two long solenoids oriented at right angles. (b) The vector sum of their magnetic fields. (c) The field lines in the combined solenoids.

EXAMPLE 7 Two long, identical solenoids such as described in Example 6 are oriented at right angles. The coils of one of these solenoids are woven through the coils of the other (see Fig. 29.34a). What are the magnitude and the direction of the magnetic field in the region of intersection?

SOLUTION: The magnetic field of each solenoid has a magnitude $B = \mu_0 n I_0$. The magnetic fields are at right angles to each other, and the magnitude of their vector sum is (see Fig. 29.34b)

$$B_{total} = \sqrt{(B_1)^2 + (B_2)^2} = \sqrt{(\mu_0 n I_0)^2 + (\mu_0 n I_0)^2} = \sqrt{2}\mu_0 n I_0$$

$$= \sqrt{2} \times 6.0 \times 10^{-3}\,\text{T} = 8.5 \times 10^{-3}\,\text{T}$$

This magnetic field makes an angle of 45° with the axis of each solenoid. The field lines are as shown in Fig. 29.34c.

For a more precise calculation of the magnetic field of a short solenoid, we must take the sum of the magnetic fields of the individual rings in this solenoid. But even a solenoid of just a few rings produces a fairly uniform magnetic field in its interior. For instance, Fig. 29.30 shows a photograph of the magnetic field lines, made visible with iron filings, of a short solenoid of just six widely-spaced turns. The magnetic field in the interior is not far from uniform.

electromagnet

An **electromagnet** is essentially a solenoid with a gap, or, what amounts to the same thing, a pair of solenoids with their ends placed close together (see Fig. 29.35a). Magnetic field lines come out of one solenoid and go into the other solenoid (of course, field lines will also have to come out of the solenoids at their other ends, curve around, and close on themselves). The first solenoid is called the north pole of the electromagnet, and the second the south pole, so magnetic field lines emerge from the north pole and enter the south pole. If the gap is small, then the magnetic field in this gap is almost the same as inside the solenoids. The gap makes the magnetic field more accessible; the gap makes it easier to immerse wires, loops, or other equipment in the field.

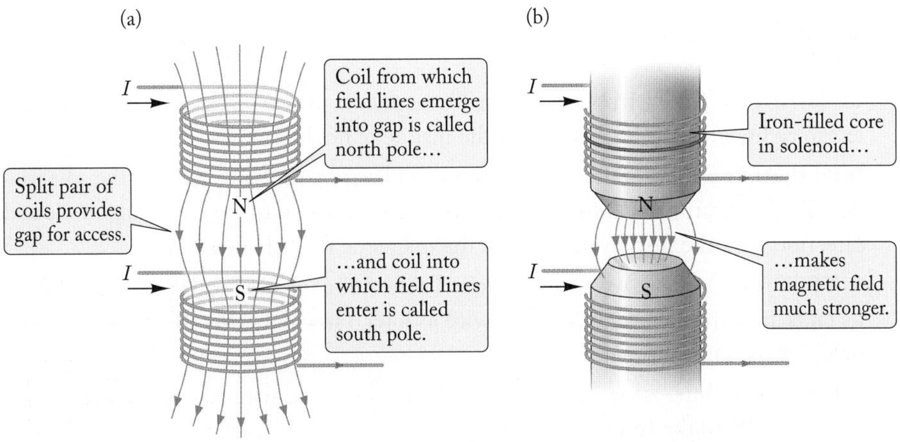

FIGURE 29.35 (a) An electromagnet with two coils. (b) An electromagnet with iron pole pieces.

In most electromagnets, the space inside the solenoids is filled with an iron core (Fig. 29.35b). Iron, and other **ferromagnetic materials**, enhance the magnetic field, making it much stronger than the value given by Eq. (29.21). It is not unusual for the magnetic field to be enhanced by a factor of several thousand. We will discuss this enhancement of magnetic fields in Section 30.4.

ferromagnetic material

PROBLEM-SOLVING TECHNIQUES — AMPÈRE'S LAW FOR CURRENT DISTRIBUTIONS WITH SYMMETRY

The calculation of the magnetic field of a given current distribution by means of Ampère's Law is possible only if the current distribution and the field are highly symmetric. If so, the calculation involves the following steps, which are analogous to those we used in the calculation of the electric field by means of Gauss' Law:

1 Determine the direction of the magnetic field from considerations of symmetry. For example, for current flow along a straight line, the field lines will be loops around the line; for current flow around a cylinder, the field lines will resemble those of an infinite solenoid.

2 Select a closed path such that for some portion of the path the magnetic field is parallel to the path and of constant magnitude B, and for the remaining portion (if any) the magnetic field is perpendicular to the path or is zero. Keep in mind that the path for Ampère's Law is a mathematical path—it does not have to coincide with a wire or with the surface of some physical body. Instead, the path is chosen to coincide with points where the magnetic field B is to be evaluated.

3 Express the integral $\oint B_{\parallel}\, ds$ in terms of the magnitude B of the magnetic field, $\oint B_{\parallel}\, ds = Bl$, where l is the length of that portion of the path where the magnetic field is tangent to the path [the portion (if any) where the magnetic field is perpendicular does not contribute to the integral].

4 Use Ampère's Law to relate Bl to the current enclosed by the path:

$$Bl = \mu_0 I$$

When evaluating I, make sure to include only the portion of the total current that is *enclosed* by the path.

5 Solve the resulting equation for the magnitude of the magnetic field.

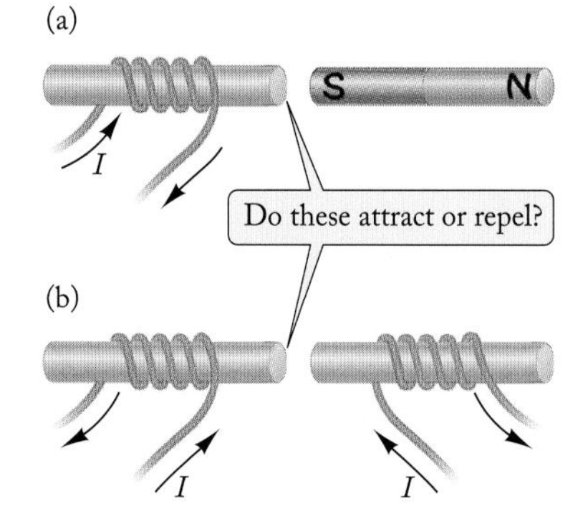

FIGURE 29.36 (a) Electromagnet and permanent magnet. (b) Two electromagnets.

Checkup 29.4

QUESTION 1: If the current in the visible segments of the coil in Fig. 29.30 is from left to right, what is the direction of the magnetic field inside the coil?

QUESTION 2: Suppose that we wind a second layer of 180 turns around the solenoid described in Example 6. If the current in this second layer is of the same magnitude and direction as in the first layer, how will the magnetic field change?

QUESTION 3: Suppose we grasp the coiled wire of the solenoid shown in Fig. 29.28 at its ends and stretch the coil, so it is twice as long. How does this affect the magnetic field, for the same current?

QUESTION 4: Suppose we squeeze the sides of the coiled wire of the solenoid shown in Fig. 29.28, so the coil is flattened uniformly along its full extent and the cross sectional area is reduced to one-half of its original value. How does this affect the magnetic field, for the same current? How does it affect the magnetic flux intercepted by the cross section of the solenoid?

QUESTION 5: In Fig. 29.36a, will the magnets attract or repel each other? In Fig. 29.36b?

(A) Attract, attract (B) Attract, repel (C) Repel, attract (D) Repel, repel

29.5 THE BIOT–SAVART LAW

Although Ampère's Law provides us with a quick way of determining the magnetic field for a current distribution of high symmetry, it is sometimes necessary to calculate the field of an arbitrary current distribution. The prescription for doing this is contained in the **Biot–Savart Law**. Although the implementation of this law usually requires complicated computational methods, there are some simple applications of this technique that readily yield results. Figure 29.37 shows a segment of a current I of length ds and a point P at which we wish to evaluate the magnetic field $\mathbf{B}$. The Biot–Savart Law states,

The contribution $d\mathbf{B}$ to the magnetic field $\mathbf{B}$ from a length ds of a current I is given by

Biot–Savart Law

$$d\mathbf{B} = \frac{\mu_0}{4\pi} \frac{I\, d\mathbf{s} \times \mathbf{r}}{r^3} \tag{29.23}$$

where $\mathbf{r}$ is the displacement vector from the current element to the point P.

The contribution is thus in the direction of $d\mathbf{s} \times \mathbf{r}$; by the right-hand rule, this is perpendicularly into the plane of the paper for the point P shown in Fig. 29.37. If the point P and all contributing current elements are in the same plane, the contributions

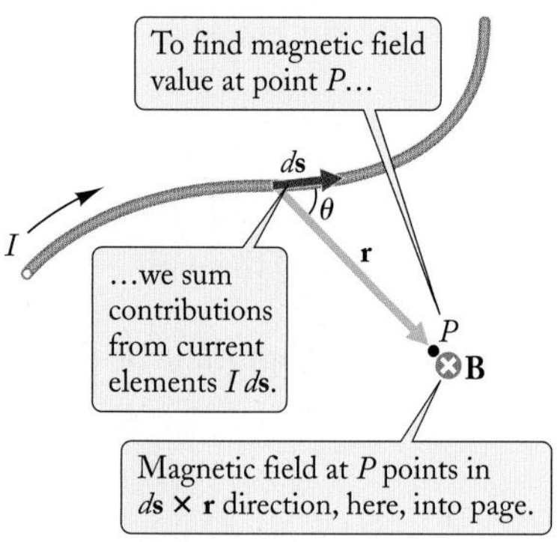

FIGURE 29.37 Evaluation of the magnetic field $\mathbf{B}$ at the point P according to the Biot–Savart Law. The current element $I\,d\mathbf{s}$ contributes to the field, the vector $\mathbf{r}$ points from the current element to the point P, and the angle between $d\mathbf{s}$ and $\mathbf{r}$ is θ.

$d\mathbf{B}$ will all be perpendicular to that plane. For such contributions, we need consider only the magnitude of Eq. (29.23):

$$dB = \frac{\mu_0 I}{4\pi}\frac{ds \sin\theta}{r^2} \tag{29.24}$$

Let us apply the Biot–Savart Law in the form (29.24) to find the magnetic field at the center point P of a circular loop of radius R. As shown in Fig. 29.38, the direction of $\mathbf{B}$ will again be perpendicular to the plane containing the loop and its center. The value of θ for each contribution around the loop is $\theta = 90°$, so $\sin\theta = 1$, and each contribution is equidistant from the center; that is, $r = R =$ constant. So the total field is

$$B = \int dB = \int \frac{\mu_0 I}{4\pi}\frac{ds \sin\theta}{r^2} = \frac{\mu_0 I}{4\pi R^2}\int ds \tag{29.25}$$

where we have brought all quantities that are not varying around the loop outside the integral. There remains only the integral of the length around the loop, $\int ds = 2\pi R$, which yields

$$B = \frac{\mu_0 I}{2R} \tag{29.26}$$

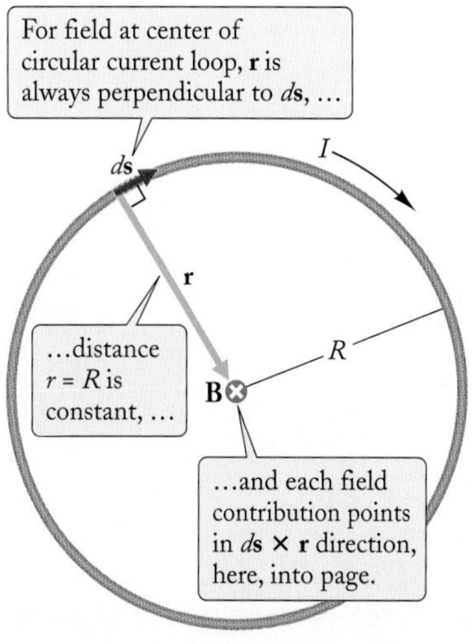

FIGURE 29.38 Geometry used to find the magnetic field at the center of a circular ring of current (also used for the field at the center of curvature of a circular arc).

magnetic field at center of circular current loop

This result was already mentioned in Eq. (29.19).

The result (29.26) can also be extended to find the field at the center of curvature of any circular arc. If an arc subtends an angle $\Delta\theta$, the field at its center of curvature will be reduced by a factor of $\Delta\theta/2\pi$, compared with the result (29.26) for a full loop:

$$B = \frac{\mu_0 I\,\Delta\theta}{4\pi R} \tag{29.27}$$

magnetic field at center of arc

Concepts in Context

A second configuration calculable with the Biot–Savart Law is the case of a finite length of straight wire, as shown in Fig. 29.39. We need to sum the contributions of the form (29.24); now, however, the quantities r, s, and θ all vary along the wire. If we want to sum the contributions, we must put them in terms of a single variable. It so happens that using the angle θ is easiest. From Fig. 29.39, we can relate

$$\tan\theta = R/(-s)$$

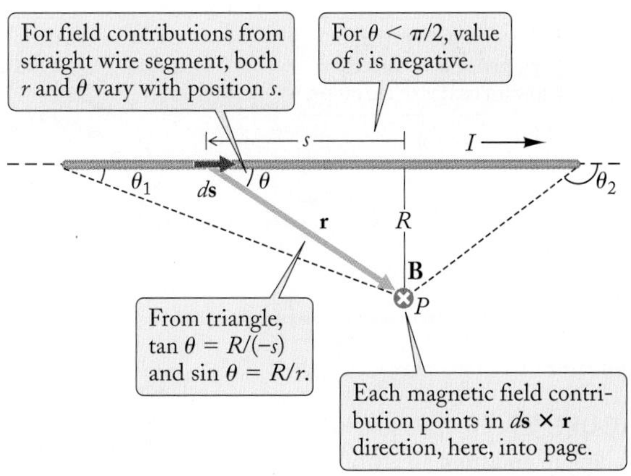

FIGURE 29.39 Geometry used to find the magnetic field of a finite straight wire using the Biot–Savart Law.

and we can take the derivative of $s = -R/\tan\theta$ with respect to θ to obtain

$$ds = \frac{R}{\sin^2\theta}\,d\theta$$

Also from Fig. 29.39, $r = R/\sin\theta$. Substituting these relations into Eq. (29.24), we can write

$$B = \frac{\mu_0 I}{4\pi}\int_{\theta_1}^{\theta_2} \frac{R}{\sin^2\theta}\,\frac{\sin^2\theta}{R^2}\sin\theta\,d\theta = \frac{\mu_0 I}{4\pi R}\int_{\theta_1}^{\theta_2}\sin\theta\,d\theta$$

Using $\int \sin\theta\,d\theta = -\cos\theta$, we obtain

magnetic field of finite wire segment

$$B = \frac{\mu_0 I}{4\pi R}(\cos\theta_1 - \cos\theta_2) \quad (29.28)$$

The result (29.28) can be used repeatedly to sum contibutions from various straight segments of a conductor. Note also that in the case of an infinite wire, we have $\theta_1 = 0$ and $\theta_2 = \pi$, so $\cos\theta_1 - \cos\theta_2 = 1 - (-1) = 2$, and the result (29.28) reduces to the field of an infinite wire, $B = \mu_0 I/2\pi R$, as it must.

EXAMPLE 8 A magnetic probe tip consists of two long wires separated by a distance R and an arc of a circle with the radius also equal to R as shown in Fig. 29.40. If the wire carries a current I, what is the magnetic field at the center of curvature of the arc?

SOLUTION: There are three contributions to the field: one each from the straight segments, and one from the arc. In the top straight segment, the current I is flowing directly toward the point in question, so $\sin\theta = 0$ in Eq. (29.24) and there is no contribution. By the right-hand rule, the other straight segment and the arc each produce a field directed into the plane of the paper, so we may simply sum them. The arc is missing $90° = \pi/2$ with repect to a full loop, and so subtends an angle $3\pi/2$. By Eq. (29.27), the contribution from the arc is

$$B_{\text{arc}} = \frac{\mu_0 I}{4\pi R}\,\frac{3\pi}{2} = \frac{3}{4}\times\frac{\mu_0 I}{2R}$$

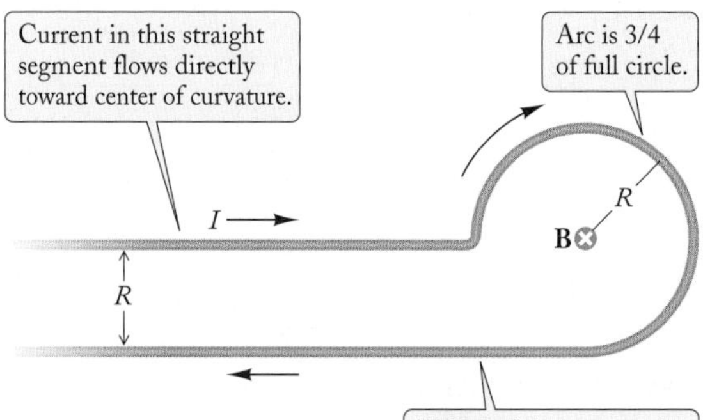

FIGURE 29.40 Two long wires leading to and from an arc of a circle.

JEAN-BAPTISTE BIOT (BEOH) (1774–1862) *French physicist, professor at the Collège de France. His most important work dealt with the refraction and polarization of light, but he was also interested in a broad range of problems in the physical sciences. With Félix Savart, 1791–1841, he confirmed Oersted's discovery of magnetic fields generated by electric currents, and formulated the equation (29.23) for the strength of the magnetic field.*

The bottom straight segment produces a contribution of the form (29.28), with $\theta_1 = 0$ at the far left end and $\theta_2 = \pi/2$ at the right end, directly below the center of curvature. Thus

$$B_{\text{segment}} = \frac{\mu_0 I}{4\pi R}\left[\cos(0) - \cos\left(\frac{\pi}{2}\right)\right] = \frac{\mu_0 I}{4\pi R}$$

which is half the field of the infinite wire. The total field at the center of curvature of the arc is

$$B = B_{\text{arc}} + B_{\text{segment}} = \left(\frac{3}{4} + \frac{1}{2\pi}\right)\frac{\mu_0 I}{2R} \approx 0.91 \times \frac{\mu_0 I}{2R}$$

Checkup 29.5

QUESTION 1: A circular loop of wire carries a current I and produces a magnetic field B_0 at its center. What is the field at the center when the same current flows in a loop of twice the radius?

QUESTION 2: Suppose that two concentric circular loops of wire are in the plane of this page; the larger one carries a clockwise current, the smaller one carries the same current, but counterclockwise. What is the direction of the magnetic field at the center of the loops?

QUESTION 3: Consider a square current loop. The contribution to the magnetic field at the center of the square from a single side of the square is B_1. What is the net magnetic field at the center of the square?

(A) 0 (B) $2B_1$ (C) $2\sqrt{2}B_1$ (D) $4B_1$

SUMMARY

PROBLEM-SOLVING TECHNIQUES Direction of Magnetic Force (right-hand rule) (page 936)

PROBLEM-SOLVING TECHNIQUES Ampère's Law for Current Distributions with Symmetry (page 947)

MAGNETIC FORCE EXERTED ON MOVING CHARGE BY CURRENT IN LONG WIRE where the − sign means **F** is attractive for **v** parallel to I and + means **F** is parallel to I for **v** radially outward. For **v** tangent to circles around the wire, $F = 0$.

$$F = \pm\frac{\mu_0}{2\pi}\frac{qvI}{r} \quad \text{or} \quad F = 0 \qquad (29.5, 29.6)$$

PERMEABILITY CONSTANT

$$\mu_0 = 4\pi \times 10^{-7}\ \text{N·s}^2/\text{C}^2 \approx 1.26 \times 10^{-6}\ \text{N·s}^2/\text{C}^2 \qquad (29.4)$$

FORCE EXERTED BY MAGNETIC FIELD ON MOVING CHARGE

$$\mathbf{F} = q\mathbf{v} \times \mathbf{B} \qquad (29.12)$$

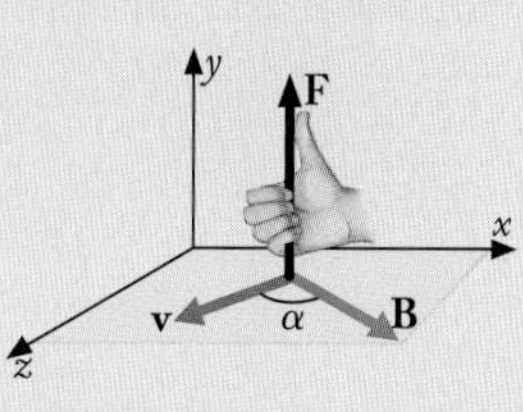

MAGNITUDE OF MAGNETIC FORCE $F = qvB \sin \alpha$ (29.11)

MAGNETIC FIELD OF CURRENT IN LONG WIRE

Right-Hand Rule With thumb along current, fingers curl in direction of field, tangent to circles around current.

$$B = \frac{\mu_0 I}{2\pi r} \quad (29.9)$$

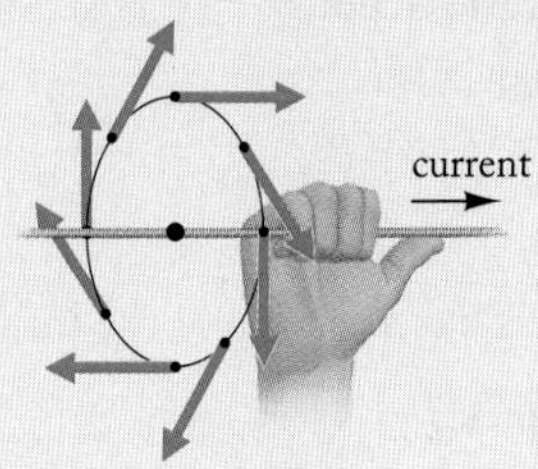

SI UNIT OF MAGNETIC FIELD 1 tesla = 1 T = 1 N/(C·m/s)

LORENTZ FORCE $\mathbf{F} = q\mathbf{E} + q\mathbf{v} \times \mathbf{B}$ (29.13)

GAUSS' LAW FOR MAGNETISM

(for closed surface)

$$\Phi_B = 0 \quad (29.15)$$

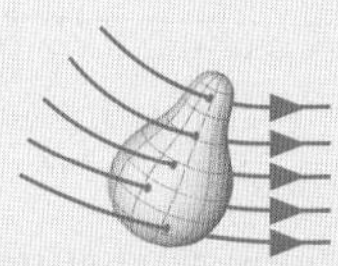

SUPERPOSITION PRINCIPLE Magnetic forces and magnetic fields produced by different currents combine by vector addition.

AMPÈRE'S LAW

$$\oint B_{\parallel}\, ds = \mu_0 I$$

where, for cylindrical symmetry,

$$\oint B_{\parallel}\, ds = \begin{cases} B \times 2\pi r & \text{(current flow along a line)} \\ B \times l & \text{(current flow around a cylinder)} \end{cases} \quad (29.18)$$

MAGNETIC FIELD INSIDE SOLENOID

Right-hand rule: With fingers curled around in direction of current, thumb gives direction of axial field.

$$B = \mu_0 n I \quad \text{where} \quad n = N/l \quad (29.21)$$

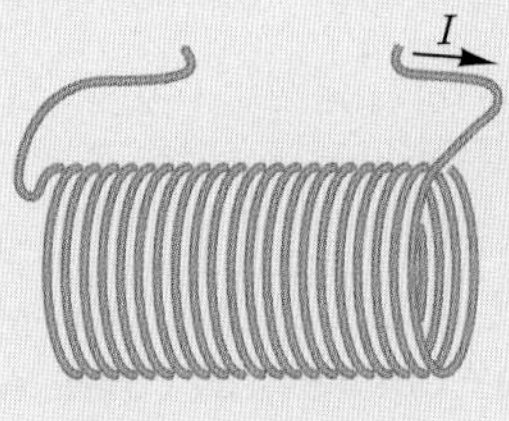

BIOT–SAVART LAW

Contribution to magnetic field: where $\mathbf{r}$ is the vector from the current element $I\,d\mathbf{s}$ to the point P.

$$d\mathbf{B} = \frac{\mu_0}{4\pi}\frac{I\,d\mathbf{s} \times \mathbf{r}}{r^3}$$

If the current and P are in the same plane, then

$$B = \int dB = \int \frac{\mu_0 I}{4\pi}\frac{ds \sin\theta}{r^2} \quad (29.23, 29.24)$$

MAGNETIC FIELD AT CENTER CIRCULAR CURRENT LOOP

Right-hand rule: With fingers curled around in direction of current, thumb gives direction of field at center of loop.

$$B = \frac{\mu_0 I}{2R} \tag{29.26}$$

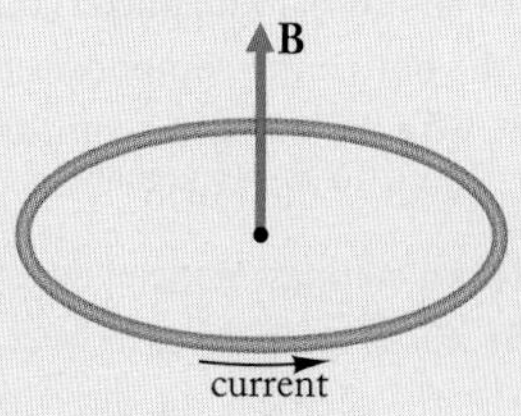

MAGNETIC FIELD AT CENTER OF ARC

$$B = \frac{\mu_0 I \Delta\theta}{4\pi R} \tag{29.27}$$

MAGNETIC FLELD OF FINITE WIRE SEGMENT

$$B = \frac{\mu_0 I}{4\pi R}(\cos\theta_1 - \cos\theta_2) \tag{29.28}$$

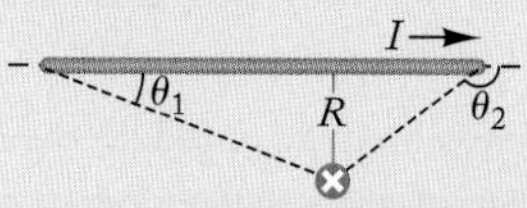

QUESTIONS FOR DISCUSSION

1. Theoretical physicists have proposed the existence of **magnetic monopoles**, which are sources and sinks of magnetic field lines, just as electric charges are sources and sinks of electric field lines. What would the pattern of magnetic field lines of a positive magnetic monopole look like?
2. The Earth's magnetic field at the equator is horizontal, in the northward direction. What is the direction of the magnetic force on an electron moving vertically up?
3. An electron with a vertical velocity passes through a magnetic field without suffering any deflection. What can you conclude about the magnetic field?
4. At an initial time, a charged particle is at some point P in a magnetic field and it has an initial velocity. Under the influence of the magnetic field, the particle moves to a point P'. If you now reverse the velocity of the particle, will it retrace its orbit and return to the point P?
5. An electron moving northward in a magnet is deflected toward the east by the magnetic field. What is the direction of the magnetic field?
6. A Faraday cage made of wire mesh shields electric fields. Does it also shield magnetic fields?
7. A long straight wire carrying a current I is placed in a uniform magnetic field B_0 at right angles to the direction of the field. The net magnetic field is the superposition of the field of the wire and the uniform field; the field lines of this net field are shown in Fig. 29.41. At the point P, the net magnetic field is zero. What is the distance of this point from the wire?

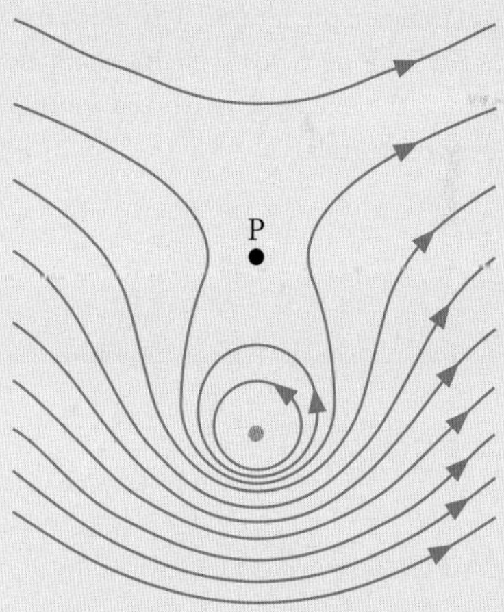

FIGURE 29.41 These magnetic field lines result from the superposition of the field of a long straight wire (directed out of the page and indicated by the colored dot) and a uniform horizontal magnetic field, directed from left to right.

8. Strong electric fields are hazardous—if you place some part of your body in a strong electric field, you are likely to receive an electric shock. Are strong magnetic fields hazardous? Do they produce any effect on your body?
9. Figure 29.26 shows the magnetic field of the Earth. What must be the direction of the currents flowing in the loops inside the Earth to give this magnetic field?

10. The needle of an ordinary magnetic compass indicates the direction of the horizontal component of the Earth's magnetic field. Explain why the magnetic compass is unreliable when used near the poles of the Earth.

11. A **dip needle** is a compass needle that swings about a horizontal axis. If the axis is oriented east–west, then the equilibrium position of the dip needle is the direction of the Earth's magnetic field. The **dip angle** of the dip needle is the angle that it makes with the horizontal. How does the dip angle vary as you transport a dip needle along the surface of the Earth from the South Pole to the North Pole?

12. Suppose we replace the single ring shown in Fig. 29.25 by a coil of N loops. How does this change the formula [Eq. (29.19)] for the magnetic field?

13. In order to eliminate or reduce the magnetic field generated by the pair of wires that connect a piece of electric equipment to an outlet, a physicist twists these wires tightly about each other. How does this help?

14. Consider a circular loop of wire carrying a current. Describe the direction of the magnetic field at different points in the plane of the loop, both inside and outside the loop.

15. An infinite flat conducting sheet lies in the x–y plane. The sheet carries a current in the y direction; this current is uniformly distributed over the entire sheet. What is the direction of the magnetic field above the sheet? Below the sheet?

16. Suppose you evaluate the integral $\int B_{\parallel}\, ds$ for the magnetic field of the Earth along a closed circular path along a meridian passing through the magnetic north and south poles. What is the value of the integral?

17. Consider a long solenoid and a long straight wire along the axis of the solenoid, both carrying some current. Describe the field lines within the solenoid.

18. The drawing of Fig. 29.35 shows the field lines near the gap of an electromagnet. Describe the pattern of field lines beyond the edges of the drawing.

19. A **tangent galvanometer** is an old form of ammeter, consisting of an ordinary magnetic compass mounted at the center of a coil whose axis is horizontal and oriented along the east–west line (Fig. 29.42). If there is no current in the coil, the compass needle points north. Explain how the compass will deviate from north when there is a current in the coil.

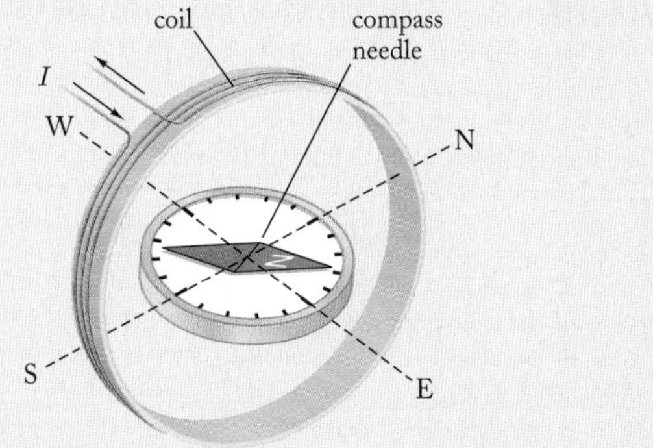

FIGURE 29.42 Tangent galvanometer.

20. A simple indicator of electric current, first used by H. C. Oersted in his early experiments, consists of a compass needle placed below a wire stretched in the northward direction. Explain how the angle of the compass needle indicates the electric current in the wire.

PROBLEMS

29.1 The Magnetic Force

1. Suppose that, instead of moving parallel to the wire, the electron in Example 1 is moving radially away from the wire. In this case, what is the magnitude of the magnetic force on this electron? What is its direction?

2. A proton is instantaneously at a distance of 0.10 m from a long straight wire carrying a current of 30 A. The speed of the proton is 5.0×10^6 m/s, and its direction of motion is radially toward the wire. What is the magnitude of the force on the proton? What is the magnitude of the instantaneous acceleration of the proton? Draw a diagram displaying the direction of the current, and the proton's velocity and acceleration.

3. In a TV tube, electrons of energy 3.0×10^{-15} J are moving on a straight path from the back of the tube to the front. This TV tube is placed near a (single) straight cable carrying a current of 12 A parallel to the path of the electrons, at a radial distance of 0.30 m from the path of the electrons. What is the magnetic force on each electron? What is the corresponding transverse acceleration?

4. A copper atom is 1.5 cm from a long, straight wire and moves parallel to a current of 25 A in the wire with speed 7.0×10^3 m/s. What magnetic force does the current in the wire exert on one electron in the copper atom? On the copper nucleus? On the entire copper atom?

5. A wire lying along the x axis carries a current of 30 A in the $+x$ direction. A proton at $\mathbf{r} = 2.5\mathbf{j}$ has instantaneous velocity $\mathbf{v} = 2.0\mathbf{i} - 3.0\mathbf{j} + 4.0\mathbf{k}$, where $\mathbf{r}$ is in meters and $\mathbf{v}$ is in meters per second. What is the instantaneous magnetic force on this proton?

6. In terms of m, s, and kg, what are the units of $1/\sqrt{\epsilon_0 \mu_0}$?

29.2 The Magnetic Field

7. A charge of q is traveling with a velocity $\mathbf{v}$ at an angle θ with respect to the direction of a magnetic field $\mathbf{B}$ that points along the x axis (see Fig. 29.43). For what angle θ is the magnitude of the magnetic force one-third of the maximum magnetic force?

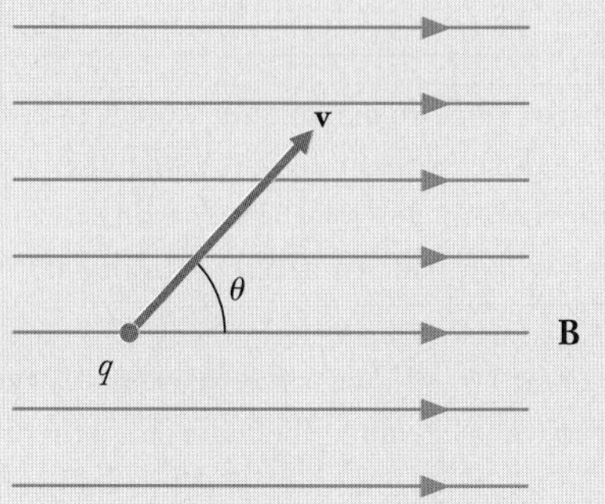

FIGURE 29.43 Velocity vector of a charge q in a uniform magnetic field.

8. An electron is traveling with a velocity of 2.0×10^5 m/s at an angle of 120° to the direction of a 0.33-T magnetic field that points along the x axis (see Fig. 29.44). What are the magnitude and direction of the force on the electron?

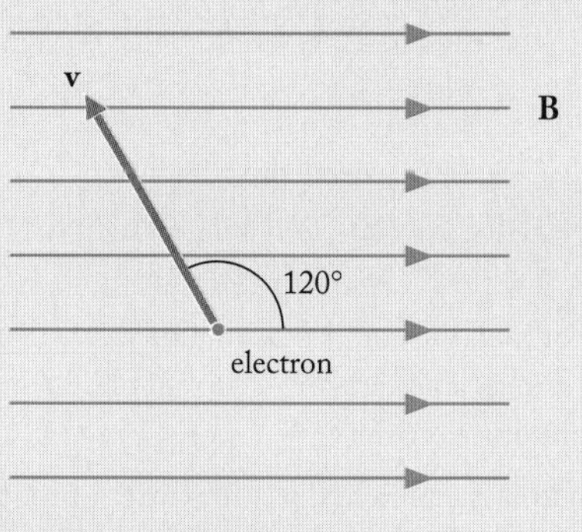

FIGURE 29.44 Velocity vector of an electron in a uniform magnetic field.

9. A proton is moving through a vertical magnetic field. The (instantaneous) velocity of the proton is 8.0×10^5 m/s horizontally in the north direction. The (instantaneous) acceleration produced by the magnetic force is 3.2×10^{14} m/s^2 in the west direction. What is the magnitude of the magnetic field? Is the direction of this field up or down?

10. Suppose you want to momentarily balance the downward gravitational force on a proton by a magnetic force. If the proton is moving horizontally in the east direction with a speed of 6.0×10^4 m/s, what magnetic field do you need (in magnitude and direction)?

11. The current in a lightning bolt may be as much as 2.0×10^4 A. What is the magnetic field at a distance of 1.0 m from a lightning bolt? The bolt can be regarded as a straight line of current.

12. The cable of a high-voltage power line is 25 m above the ground and carries a current of 1.8×10^3 A.
    (a) What magnetic field does this current produce at the ground?
    (b) The strength of the magnetic field of the Earth is 0.60×10^{-4} T at the location of the power line. By what factor do the fields of the power line and of the Earth differ?

13. The magnetic field surrounding the Earth typically has a strength of 5.0×10^{-5} T. Suppose that a cosmic-ray electron of kinetic energy 3.0×10^4 eV is instantaneously moving in a direction perpendicular to the lines of this magnetic field. What is the force on this electron?

14. An alpha particle (charge $+2e$) moves with velocity $\mathbf{v} = 5.0\mathbf{i} - 3.0\mathbf{j}$ in a magnetic field $\mathbf{B} = -4.0\mathbf{i} + 2.5\mathbf{j}$, where $\mathbf{B}$ is in teslas and $\mathbf{v}$ is in meters per second. What is the magnetic force on the alpha particle?

15. In a region where the magnetic field is $\mathbf{B} = 2.5\mathbf{i} + 3.6\mathbf{j} + 1.5\mathbf{k}$, an electron moves with velocity $\mathbf{v} = -3.0\mathbf{i} + 4.0\mathbf{j} - 3.5\mathbf{k}$, where $\mathbf{B}$ is in teslas and $\mathbf{v}$ is in meters per second. What is the magnetic force on this electron?

16. A proton is just above the surface of the Earth on the magnetic equator, where the magnetic field points north and has magnitude $B = 4.2 \times 10^{-5}$ T. In what direction and with what velocity should the proton move in order for the magnetic force to balance the gravitational force?

*17. In New York, the magnetic field of the Earth has a vertical (down) component of 6.0×10^{-5} T and a horizontal (north) component of 1.7×10^{-5} T. What are the magnitude and direction of the magnetic force on an electron of velocity 1.0×10^6 m/s moving (instantaneously) in an east-to-west direction in a television tube?

*18. At the surface of a pulsar, or neutron star, the magnetic field may be as strong as 1.0×10^8 T. Consider the electron in a hydrogen atom on the surface of such a neutron star. The electron is at a distance of 5.3×10^{-11} m from the proton and has a speed of 2.2×10^6 m/s. Compare the electric force that the proton exerts on the electron with the magnetic force that the magnetic field of the neutron star exerts on the electron. Is it reasonable to expect that the hydrogen atom will by strongly deformed by the magnetic field?

*19. The electric field of a long, straight line of charge with λ coulombs per meter is [see Eq. (24.23)]

$$E = \frac{1}{2\pi\epsilon_0}\frac{\lambda}{r}$$

Suppose that we move this line of charge parallel to itself at speed v.
    (a) The moving line of charge constitutes an electric current. What is the magnitude of the current?

(b) What is the magnitude of the magnetic field produced by this current? Show that the magnitude of the magnetic field is proportional to the magnitude of the electric field; that is,

$$B = \mu_0\epsilon_0 vE$$

(However, the directions of the electric and magnetic fields differ. The electric field is radial, whereas the magnetic field is tangential.)

*20. According to the preceding problem, when a line of charge is made to move at speed v parallel to itself, it produces a magnetic field of magnitude proportional to its electric field, $B = \mu_0\epsilon_0 vE$. Use this result to find the magnetic field of a large charged flat sheet of paper with a charge of σ coulombs per square meter moving at a speed v in a direction parallel to itself.

29.3 Ampère's Law

21. A wire of superconducting niobium, 0.20 cm in diameter, can carry a current of up to 1900 A. What is the strength of the magnetic field just outside the wire when it carries this current?

22. A long, straight wire of copper with a radius of 1.0 mm carries a current of 20 A. What are the instantaneous magnetic force and the corresponding acceleration of one of the conduction electrons moving at 1.0×10^6 m/s along the surface of the wire in a direction opposite to that of the current? What is the direction of the acceleration?

23. In a proton accelerator, protons of velocity 3.0×10^8 m/s form a beam of current of 2.0×10^{-3} A. Assume that the beam has a circular cross section of radius 1.0 cm and that the current is uniformly distributed over the cross section. What is the magnetic field that the beam produces at its edge? What is the magnetic force on a proton at the edge of the beam?

24. A ring of superconducting wire carries a current of 2.0 A. The radius of the ring is 1.5 cm. What is the magnitude of the magnetic field at the center of the ring?

25. A circular coil consists of 60 turns of wire wound around the rim of a plywood disk of radius 0.15 m (see Fig. 29.45). If a current of 2.0 A is sent through this coil, what is the magnetic field produced at the center of this disk?

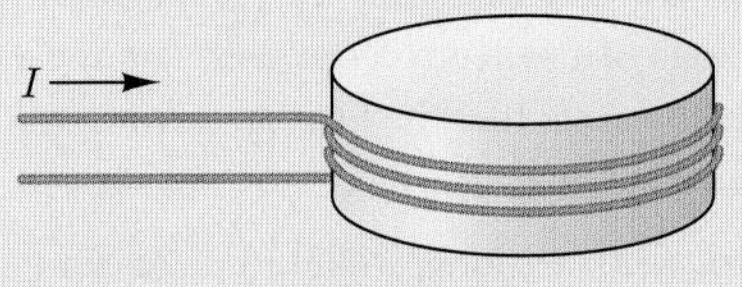

FIGURE 29.45 A circular coil of wire.

26. At the boiling point of liquid helium, a niobium–titanium alloy is superconducting only when the magnetic field is less than 9.5 T. If a wire made of this alloy is 3.0 mm in diameter, what maximum current can it carry before superconductivity at its surface is lost? Assume the current is uniformly distributed over the cross section of the wire.

27. A current I flows in a thin wire bent into a circle of radius R. The axis of the circle coincides with the z axis. What is the value of the integral $\int B_{\|}\, ds$ along the z axis from $z = -\infty$ to $z = +\infty$?

*28. Six parallel aluminum wires of small, but finite, radius lie in the same plane. The wires are separated by equal distances d, and they carry equal currents I in the same direction. Find the magnetic field at the center of the first wire. Assume that the current in each wire is uniformly distributed over its cross section.

*29. A circular ring of wire of diameter 0.60 m carries a current of 35 A. What acceleration will the magnetic force generated by this ring give to an electron that is passing through the center of the ring with a velocity of 1.2×10^6 m/s in the plane of the ring?

*30. Two very long straight wires carry currents I at right angles. One of the wires lies along the x axis; the other lies along the y axis (see Fig. 29.46). Find the magnetic field at a point in the x–y plane in the first quadrant.

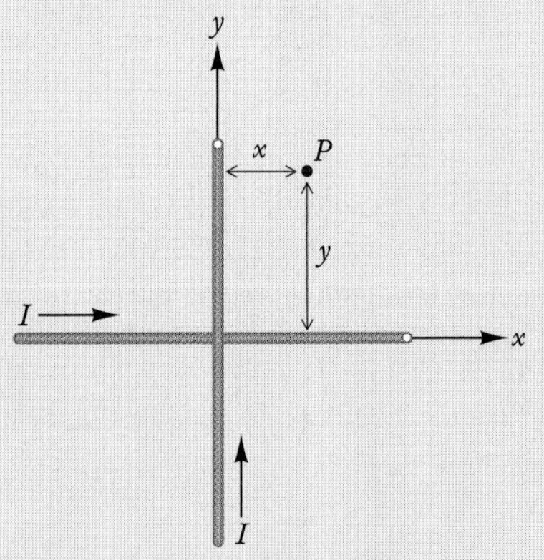

FIGURE 29.46 Two long, straight wires.

*31. In a motorboat, the compass is mounted at a distance of 0.80 m from a cable carrying a current of 20 A from an electric generator to a battery.

(a) What magnetic field does this current produce at the location of the compass? Treat the cable as a long, straight wire.

(b) The horizontal (north) component of the Earth's magnetic field is 1.8×10^{-5} T. Since the compass points in the direction of the net horizontal magnetic field, the current will cause a deviation of the compass. Assume that the magnetic field of the current is horizontal and at right angles to the horizontal component of the Earth's magnetic field. Under these circumstances, by how many degrees will the compass deviate from north?

*32. Two very long, straight, parallel wires separated by distance d carry currents of magnitude I in opposite directions. Find the magnetic field at a point equidistant from the lines, with a distance $2d$ from each line. Draw a diagram showing the direction of the magnetic field.

*33. Three long, parallel wires are spaced as shown in Fig. 29.47. The wires carry equal currents in the same direction. What is the magnetic field at the point P? Draw a diagram giving the direction of the magnetic field.

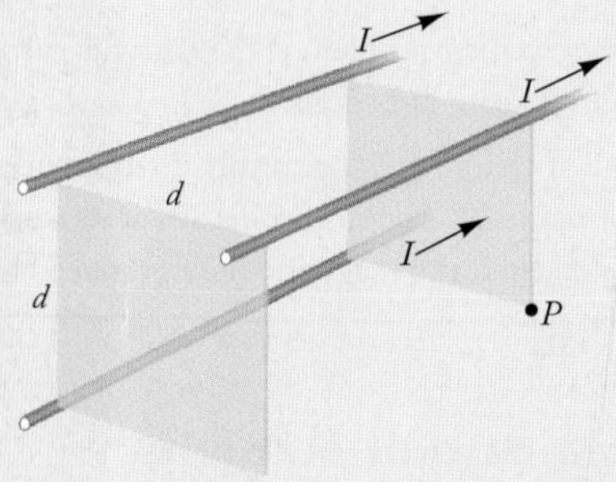

FIGURE 29.47 Three long, parallel wires. The wires pass through three corners of a square.

*34. Two rings of the same radius 0.20 m are placed at right angles (see Fig. 29.48). The rings carry equal currents of 10 A. What is the magnitude of the magnetic field at the center of this arrangement of rings? Draw a diagram showing the direction of this magnetic field.

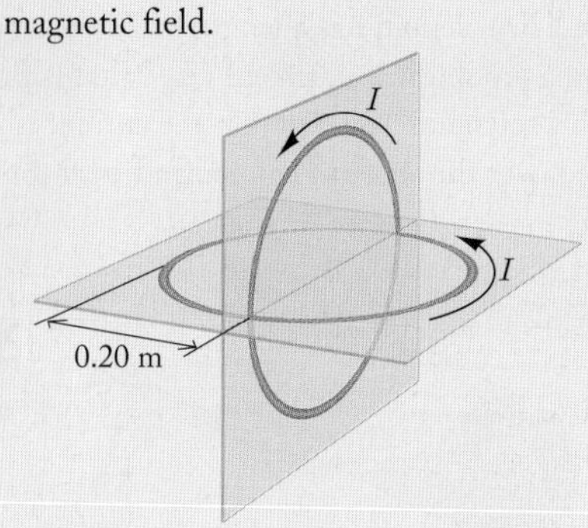

FIGURE 29.48 Two rings at right angles.

*35. A long, straight wire is bent into a circular loop of radius R near its midpoint (see Fig. 29.49). The wire carries a current I. What are the magnitude and direction of the magnetic field at the center of the loop?

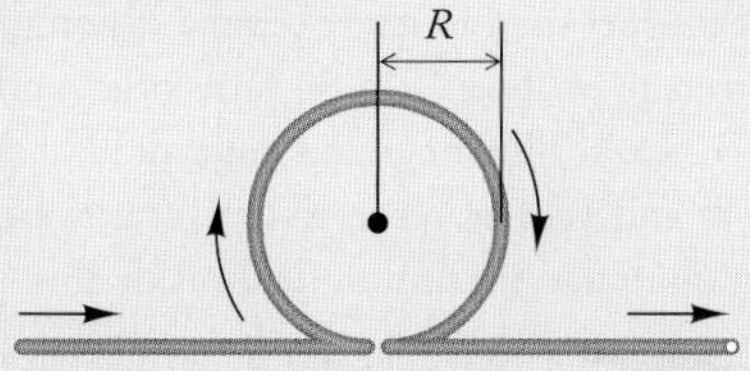

FIGURE 29.49 Long, straight wire deformed into a loop in the same plane.

*36. Two very long, parallel wires separated by a distance of 1.0 cm carry opposite currents of 8.0 A.

(a) Find the magnetic field at the midpoint between the wires.

(b) Find the magnetic field in the plane of the wires, at a distance of 2.0 cm from the midline.

*37. A coaxial cable consists of a long cylindrical copper wire of radius r_1 surrounded by a cylindrical shell of inner radius r_2 and outer radius r_3 (Fig. 29.50). The wire and the shell carry equal and opposite currents I uniformly distributed over their cross sections. Find formulas for the magnetic field in each of the four regions $r < r_1$, $r_1 < r < r_2$, $r_2 < r < r_3$, and $r > r_3$.

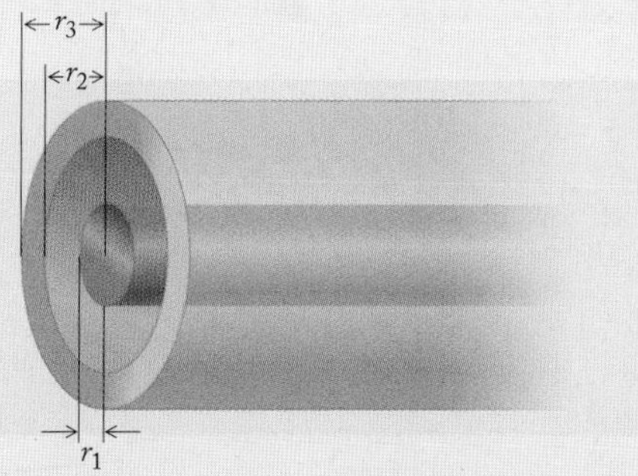

FIGURE 29.50 A cylinder and coaxial cylindrical shell.

*38. A very large, thin conducting plate lies in the x–y plane. The plate carries a current in the y direction. The current is uniformly distributed over the plate with σ amperes flowing across each meter of length perpendicular to the current. Use Ampère's Law to find the magnetic field at some distance from the plate. (Hint: The magnetic field lines are parallel to the plate.)

*39. One very large conducting plate coincides with the x–y plane. Another very large conducting plate coincides with the x–z plane. Each plate carries a uniformly distributed current with σ amperes flowing across each meter of length perpendicular to the current (Fig. 29.51). Find the magnetic field in each of the four quadrants. (Hint: See Problem 38.)

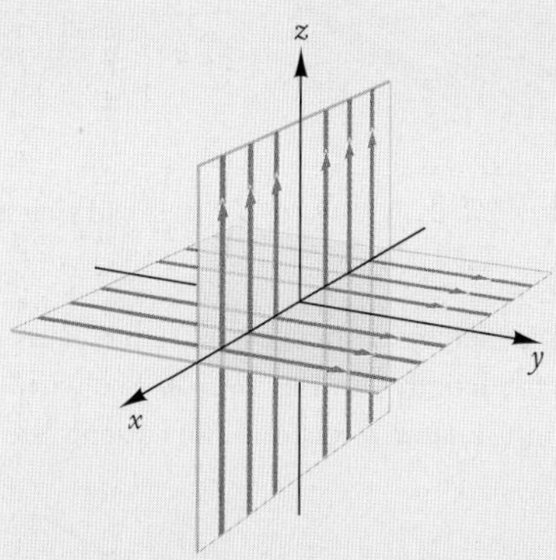

FIGURE 29.51 Two large conducting plates carrying currents.

*40. A cylindrical conductor of radius R_1 has a cylindrical hole of radius $R_2 < R_1/2$ along its length; the hole is adjacent to the surface, as in Fig. 29.52. The conductor carries a current I. Find the magnitude of the magnetic field at points along the vertical dotted line shown (a) below the center, for $r \leq R_1$; (b) below the conductor, for $r \geq R_1$; (c) above the center, for $r \leq (R_1 - 2R_2)$; (d) above the center, for $R_1 \geq r \geq (R_1 - 2R_2)$; and (e) above the conductor, for $r \geq R_1$.

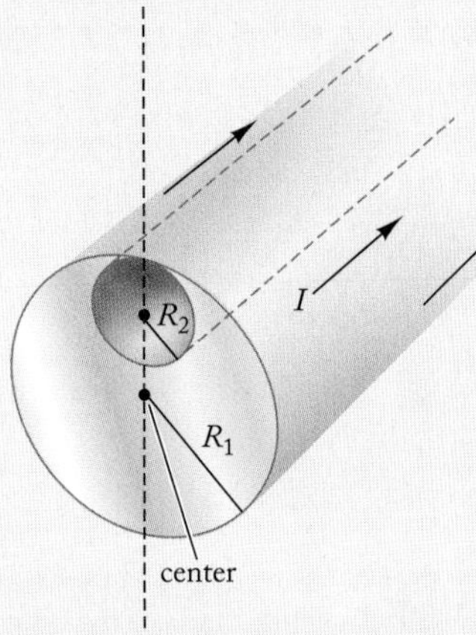

FIGURE 29.52 A cylinder with a cylindrical hole.

**41. A very long strip of copper of width b carries a current I uniformly distributed over the strip. What is the magnetic field at a distance z above the midline of this strip (Fig. 29.53)?

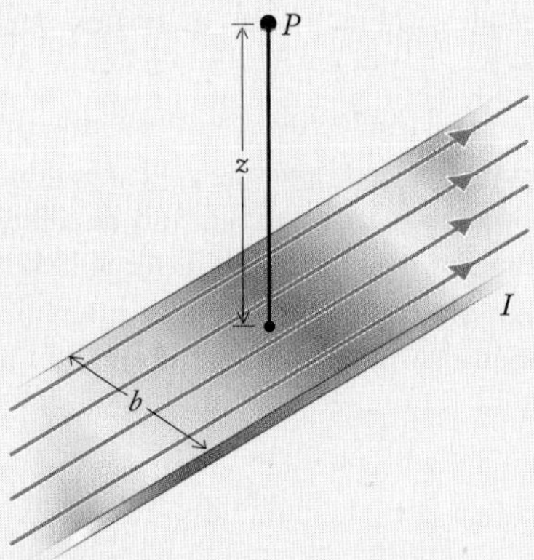

FIGURE 29.53 A long copper strip carrying a current.

**42. Two long, parallel wires of copper of radius R are in contact along their full length (see Fig. 29.54). The wires carry equal currents I, in the same direction. The currents are uniformly distributed over the volumes of the wires. Find the magnetic field in the midplane (z–x plane) of the wires, as a function of the distance z from the point of contact. Where is the magnetic field maximum, and what is the value of the maximum magnetic field?

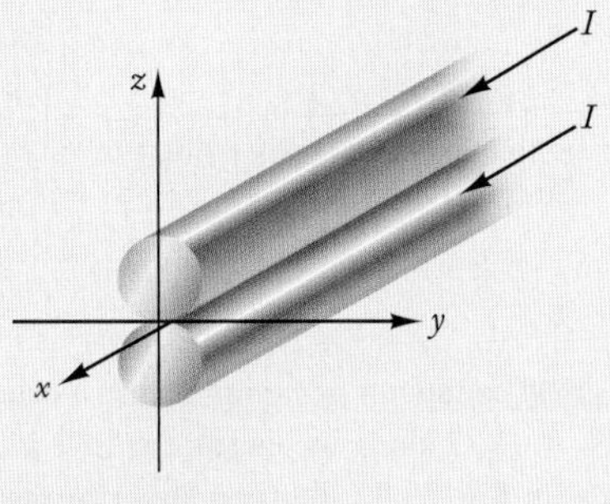

FIGURE 29.54 Two long, straight, thick wires in contact.

**43. Repeat the preceding problem if the currents carried by the wires are in opposite directions.

**44. An amount of charge Q is uniformly distributed over a disk of paper of radius R. The disk spins about its axis with angular velocity ω. Find the magnetic field produced by this disk at its center.

29.4 Solenoids and Magnets

45. A long solenoid has 15 turns per centimeter. What current must we put through its windings if we wish to achieve a magnetic field of 5.0×10^{-2} T in its interior?

46. The electromagnet of a small electric bell is a solenoid with 260 turns in a length of 2.0 cm. What magnetic field will this solenoid produce if the current is 8.0 A?

47. Figure 29.55 shows a "solenoid" made of one turn of a sheet of copper. The solenoid has a length of 20 cm, and the current flowing around it is 2.0×10^3 A. What is the magnetic field in this solenoid? Assume that the current is uniformly distributed over the sheet of copper, and treat the solenoid as very long.

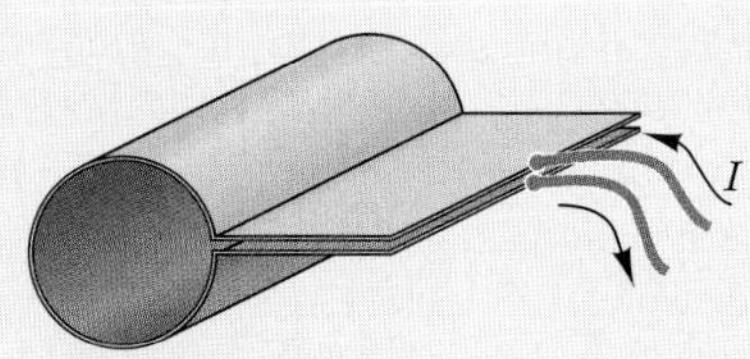

FIGURE 29.55 A sheet of copper shaped into a solenoid.

48. Repeat Example 7, if the current in the second solenoid is twice that in the first. Sketch the field lines for the net magnetic field.

49. A toroidal magnet used to study plasma fusion (a "tokamak") at the University of Texas had 80 turns of a conductor with a current of 125 kA. The inner and outer radii of the torus were 0.60 m and 1.40 m, respectively. What were the values of the minimum and maximum field inside this toroid?

*50. A long solenoid of n turns per unit length carries a current I, and a long, straight wire lying along the axis of this solenoid carries a current I'. Find the net magnetic field within the solenoid, at a distance r from the axis. Describe the shape of the magnetic field lines.

*51. The coil of an electromagnet consists of a large number of layers of very thin wire wound on a cylindrical core. The inner radius of the windings is r_1 and the outer radius is r_2. The number of turns in each layer is n per unit length, and the number of layers is n' per unit length in the radial direction.

The current in the turns of the wire is I. Find formulas for the magnetic field in the region $r < r_1$ and in the region $r_1 < r < r_2$.

*52. A very long rod of radius R has a uniform charge density ρ throughout its volume. The rod rotates about its axis with angular velocity ω. What is the magnetic field a distance $r \leq R$ from the axis?

29.5 The Biot–Savart Law

53. A wire is bent into a square loop of side L and carries a current I. Find an expression for the magnetic field at the center of the square.

54. A wire is bent into a rectangular loop of width $2L$ and height L as shown in Fig. 29.56. If $L = 20$ cm and the wire carries a current $I = 15$ A, find the value of the magnetic field at the center of the rectangle.

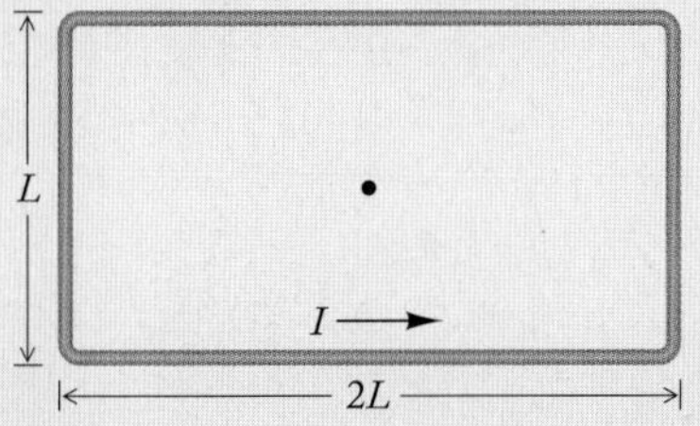

FIGURE 29.56 A rectangular loop of wire.

55. A current $I = 5.0$ A travels along a long, straight wire that is interrupted by a semicircular segment of radius $R = 2.0$ cm as shown in Fig. 29.57. Find the value of the magnetic field at the center of curvature.

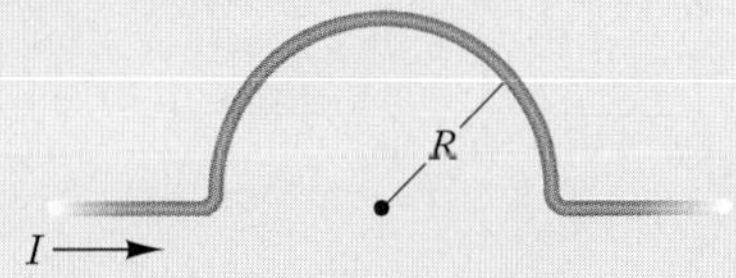

FIGURE 29.57 A straight wire with a semicircular segment.

56. A wire loop is bent into the shape shown in Fig. 29.58, with two semicircles of radii R_1 and R_2 and two straight segments. If a current I flows in the direction shown, what is the direction of the magnetic field at the common center of curvature? Find an expression for the magnetic field there.

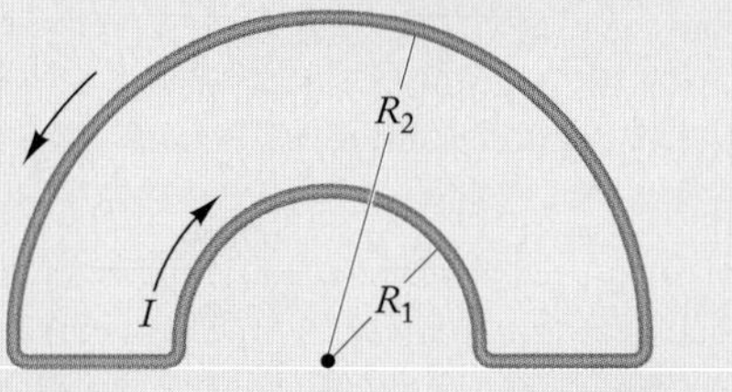

FIGURE 29.58 Two semicircles and two straight segments of wire forming a loop.

57. A long wire carrying a current $I = 10$ A is bent around a corner of a quarter-circle arc of radius $R = 0.50$ cm as shown in Fig. 29.59. Find the value of the magnetic field at the center of curvature.

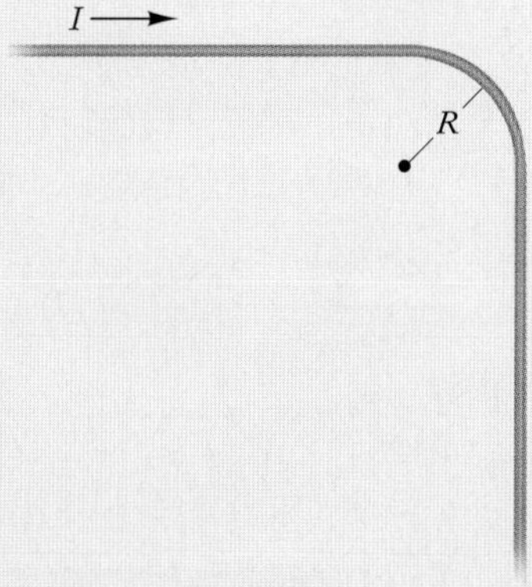

FIGURE 29.59 A long wire with a circular arc.

*58. Assume that in a hydrogen atom, the electron circles the nucleus in a circle of radius $r = 5.29 \times 10^{-11}$ m with an angular momentum $L = mvr = 1.05 \times 10^{-34}$ J·s. Determine the value of the magnetic field at the nucleus due to the current of the electron's orbital motion.

*59. A wire is bent into a square loop of side L and carries a current I. Find an expression for the magnetic field at a point P outside the loop, a perpendicular distance $L/2$ from the midpoint of one edge (see Fig. 29.60).

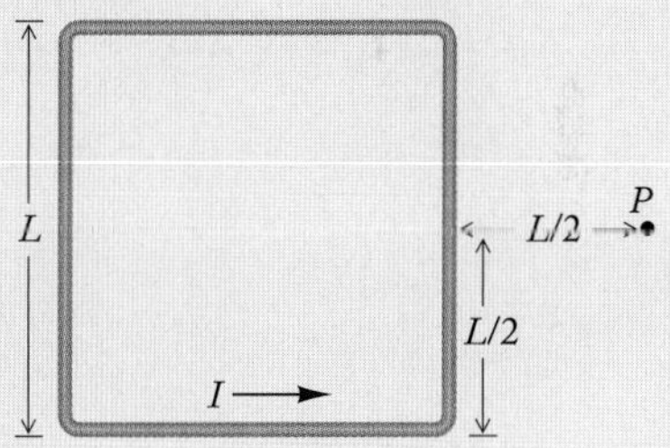

FIGURE 29.60 A square loop of wire.

*60. Consider a point P on the axis of a circular current loop of radius R, at a distance z from the center of the loop, as shown in Fig. 29.61. By symmetry, only the z component of the magnetic field due to a segment $d\mathbf{s}$ of the loop will contribute to the net field. The z component is $dB_z = dB \cos \phi$, where dB is given by Eq. (29.23). From the geometry of the figure, and by summing around the length of the loop, show that the total field at P is

$$B = \frac{\mu_0 I}{2} \frac{R^2}{(z^2 + R^2)^{3/2}}$$

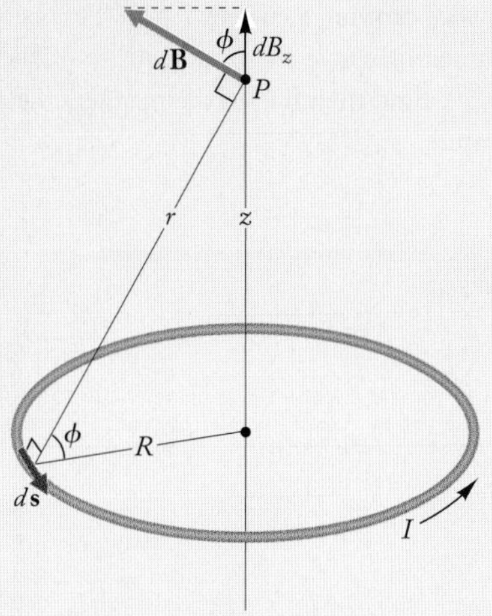

FIGURE 29.61 Geometry to determine the magnetic field on the axis of a circular loop of wire.

*61. A **Helmholtz coil**, useful for producing a region of uniform field, consists of two identical circular coils separated by a distance equal to their radius R, as shown in Fig. 29.62.

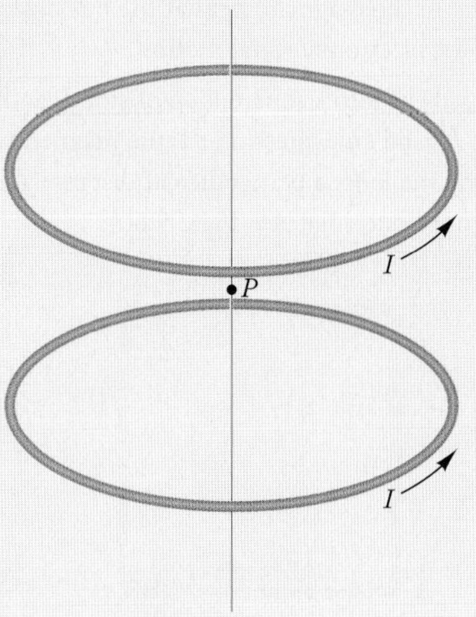

FIGURE 29.62 A Helmholtz coil.

(a) Use the result of Problem 60 to calculate B_z at the midpoint.

(b) Show that dB_z/dz and d^2B_z/dz^2 vanish at the midpoint (d^3B_z/dz^3 also vanishes there, indicating a very homogeneous field at the midpoint).

62. A **finite solenoid of length L and radius R has N turns and carries a current I. Find an expression for the magnetic field at a point on the axis outside the solenoid by treating the solenoid as a stack of current loops in the following way. Sum contributions of the form of the result of Problem 60, replacing B with dB and I with $dI = NI\,dz/L$. Determine the magnetic field at a point a distance z_0 above one end of the stack of loops (see Fig. 29.63) by summing the contributions as a function of z. Obtain the result

$$B = \frac{1}{2}\mu_0\frac{N}{L}I\left[\frac{z_0}{\sqrt{z_0^2 + R^2}} - \frac{z_0 - L}{\sqrt{(z_0 - L)^2 + R^2}}\right]$$

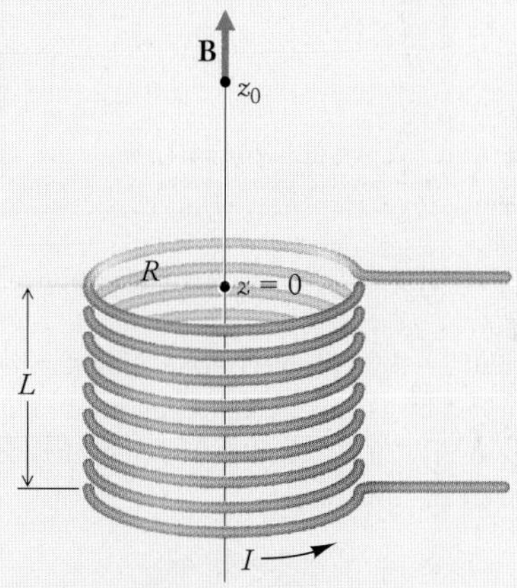

FIGURE 29.63 A finite solenoid.

REVIEW PROBLEMS

63. A proton is instantaneously at a distance of 0.10 m from a long, straight wire carrying a current of 30 A. The speed of the proton is 5.0×10^6 m/s, and its direction of motion is radially toward the wire. What is the magnitude of the magnetic force on the proton? What is the magnitude of the instantaneous acceleration of the proton? Draw a diagram displaying the direction of the current, and the proton's velocity and acceleration.

64. Since the magnetic force is proportional to the speed of a charged particle, it is largest if the particle has the largest speed possible—that is, the speed of light. Suppose that an electron of a speed almost equal to the speed of light is moving radially toward a long straight wire carrying a current of 15 A. What is the magnitude of the magnetic force on this electron when it is at a distance of 0.050 m from the wire?

65. A ring of radius 8.0 cm is placed concentrically around a ring of radius 6.0 cm (see Fig. 29.64). Each ring carries a current of 4.0 A. What is the magnitude of the magnetic field at the center of the rings if the two currents flow in the same direction? If they flow in opposite directions?

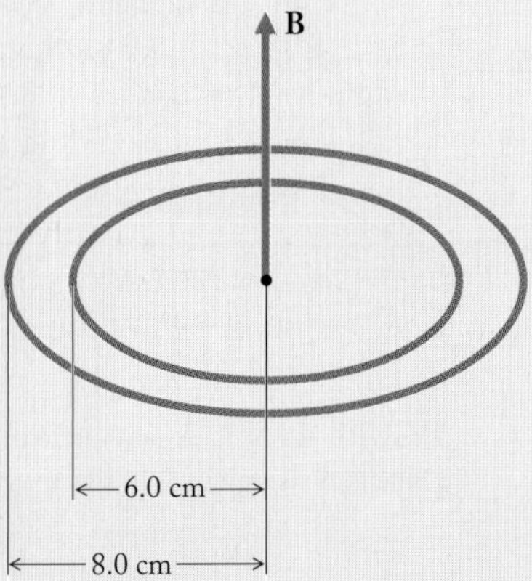

FIGURE 29.64 Two concentric rings.

66. 150 turns of insulated wire are wound around the rim of a plywood disk of radius 20 cm. The resistance of this wire is 20 Ω. What voltage must you supply to the terminals of the wire to generate a magnetic field of 8.0×10^{-4} T at the center of the disk?

67. A long, straight wire is bent into a circular loop of radius R near its midpoint, and this loop is twisted so its plane is perpendicular to the wire (see Fig. 29.65). The wire carries a current I. What are the magnitude and direction of the magnetic field at the center of the loop?

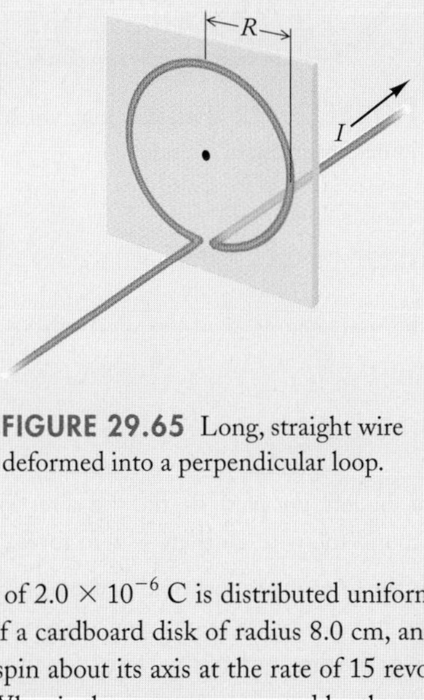

FIGURE 29.65 Long, straight wire deformed into a perpendicular loop.

68. A charge of 2.0×10^{-6} C is distributed uniformly around the rim of a cardboard disk of radius 8.0 cm, and this disk is made to spin about its axis at the rate of 15 revolutions per second. What is the current generated by the motion of the charge? What is the magnetic field that this current produces at the center of the disk?

69. A long solenoid has been placed inside another long solenoid of larger radius (see Fig. 29.66). The solenoids are coaxial, and both have the same number n of turns per unit length and the same current I. What is the formula for the magnetic field in the region within the smaller solenoid? Between the smaller and the larger solenoid?

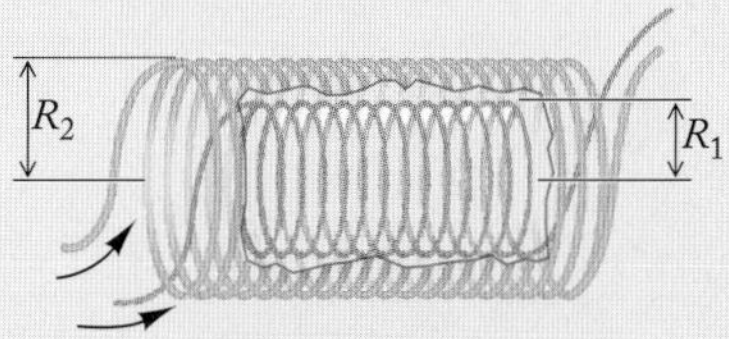

FIGURE 29.66 Two coaxial solenoids.

70. A long solenoid is placed in a constant, uniform magnetic field $\mathbf{B}_0$. The axis of the solenoid is at right angles to this magnetic field (see Fig. 29.67). If the solenoid has n turns per meter and the current in its coils is I, what is the net magnetic field in the solenoid? What is the angle of this magnetic field with the axis of the solenoid? Sketch the field lines of the net magnetic field inside and outside the solenoid.

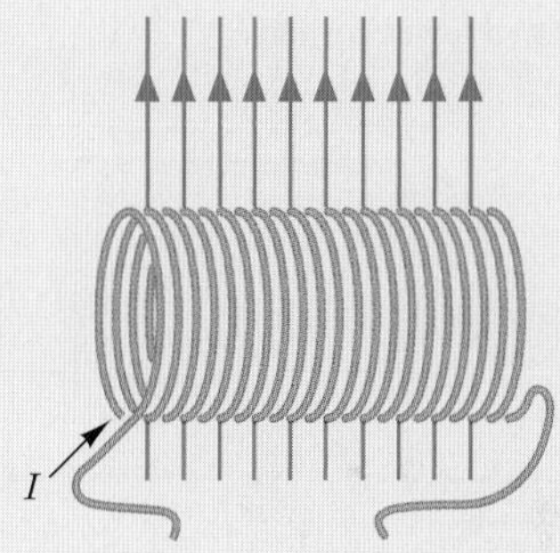

FIGURE 29.67 A solenoid placed in a uniform magnetic field.

71. A large number of thin, long, straight wires are laid parallel to each other on a flat plane (see Fig. 29.68). The number of wires per unit length measured along the plane perpendicularly to the wires is n, and each carries a current I. Use Ampère's Law to find the magnetic field produced by this current distribution. Describe the direction of the magnetic field. Explain why the magnetic field differs in strength from that inside a solenoid [Eq. (29.21)].

FIGURE 29.68 Long, straight, parallel wires.

72. A long copper pipe with thick walls has an inner radius R and an outer radius $2R$ (Fig. 29.69). A current I flows along this pipe, parallel to its axis and uniformly distributed over the cross section of the pipe. Find the magnetic fields at the radius $\frac{3}{2}R$ and at the radius $3R$.

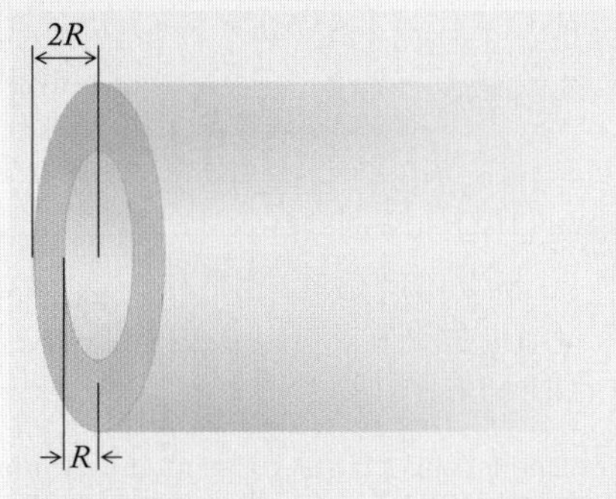

FIGURE 29.69 A thick cylindrical shell.

73. A very long, straight wire with a current I has a square bend near its middle (Fig. 29.70). Each side of the square has length L. What is the magnetic field at the point P, halfway between the two lower corners?

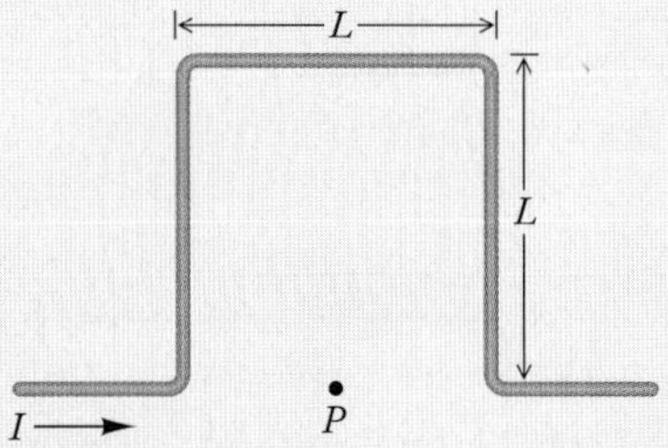

FIGURE 29.70 A very long wire with square bends.

74. A circular loop is folded along a diameter so as to form two semicircles of radius R intersecting at right angles (Fig. 29.71). A current I flows around this loop. What is the magnetic field at the center? Draw a diagram showing the direction of the magnetic field.

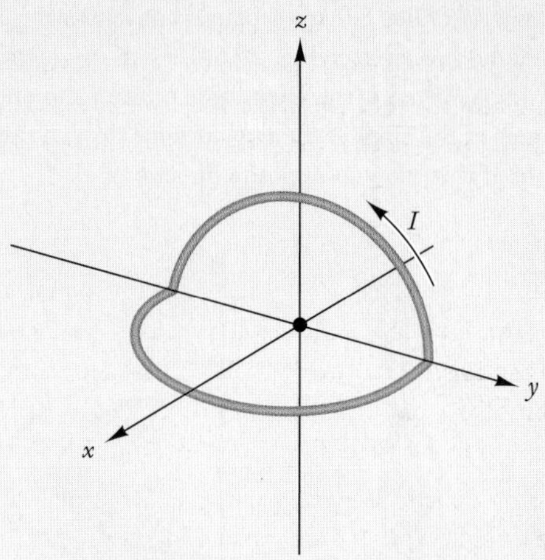

FIGURE 29.71 A semicircle in the z–y plane and a semicircle in the x–y plane.

75. Two semi-infinite wires are in the same plane. The wires make an angle of 45° with each other, and they are joined by an arc of a circle of radius R (Fig. 29.72). The wires carry a current I. Find the magnetic field at the center of curvature of the arc.

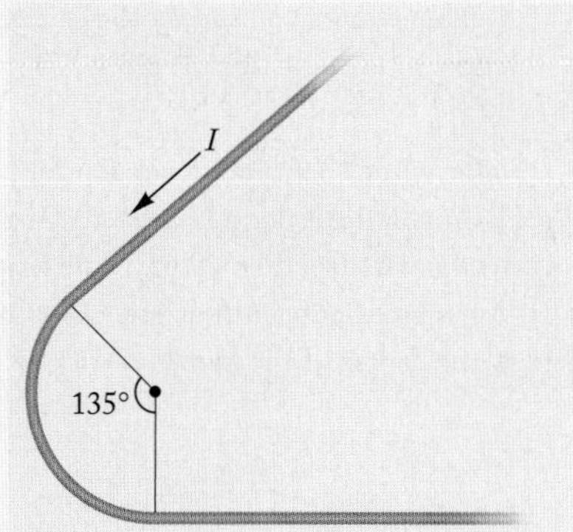

FIGURE 29.72 Two very long wires joined by an arc of a circle.

Answers to Checkups

Checkup 29.1

1. In each case, one can check whether the product qvI in Eq. (29.5) or (29.6) changes sign. For part (a), a positive charge moves to the left, so only v changes sign, and thus the force is then positive (repulsive), i.e., downward. For part (b), the velocity is negative (radially inward), so the force is now in the opposite direction to the current, in Fig. 29.6b, to the left. For part (c), only the sign of the charge changes, so again the force is to the left. For part (d), the $+z$ direction is tangent to a circle around the wire, so there is zero force (no direction).
2. Note that the force is proportional to the product qvI; if the charge and its velocity are unchanged, then reversing the current reverses the direction of the force in each case.
3. Since the force is inversely proportional to the distance r from the wire, doubling the distance will decrease the force by a factor of 2.

4. (A) $\mathbf{v} = v_0(\mathbf{i} + \mathbf{j})/\sqrt{2}$. The contribution to the force is zero for the component of the velocity in the tangential direction; here, that is the **k** direction. Thus the force is largest for the velocity with no **k** component.

Checkup 29.2

1. By the right-hand rule, if we place our thumb in the direction of the current, our fingers curl around the wire in the direction of the magnetic field loops. Thus with our thumb pointing north, the field direction below the wire is westward.

2. Reversing the current reverses the direction of the magnetic field. This is easily seen by the right-hand rule: pointing the thumb in the opposite direction causes the fingers to curl around the wire in the opposite direction.

3. Using the right-hand rule, if the thumb points upward along the wire, the field loops are counterclockwise around the wire (looking from above).

4. Since the magnetic field is inversely proportional to the radial distance from the axis of the wire, the field at half the distance (3 m) will be twice as much, or $B = 4 \times 10^{-5}$ T; and the field at one-third the distance (2 m) will be 3 times as much, or $B = 6 \times 10^{-5}$ T.

5. (B) North. To determine the direction of the force, we first point the fingers of the right hand in the direction of the velocity of the charge (eastward), and then orient the hand so the fingers can curl toward the direction of the field (downward). When the fingers curl from east to down, the thumb points north. Since the proton is a positive charge, the magnetic force is also northward.

Checkup 29.3

1. The magnetic field due to an individual wire forms loops around each wire. Since the currents are in the same direction, the loops have the same sense (unlike the situation in Fig. 29.19), so they exactly oppose at the point halfway between the wires. Thus there is zero net field there.

2. By the right-hand rule for a current loop, we wrap the fingers of the right hand around in the direction of the current, and the thumb gives the direction of the field in the center of the ring; thus, the field direction in the center of the ring will be downward.

3. Since the field is also directed along the axis, the charge will be moving parallel to the magnetic field. For **v** parallel to **B**, there is zero magnetic force.

4. Since no current intercepts the area bordered by the path ($I = 0$), the right side of Ampère's Law is zero, so the left side, which is the integral of the tangential magnetic field, is zero.

5. (A) $\mu_0 I/(2\pi R)$. Since all the current is interior to the surface, the field at the surface will be the same as that for a thin wire, $B = \mu_0 I/(2\pi R)$. Alternatively, one can use either result from Example 5, since at the surface, $r = R$, and the interior field and the exterior field are the same, $B = \mu_0 I/(2\pi R)$.

Checkup 29.4

1. By the right-hand rule given after Eq. (29.21), if we wrap the fingers of our right hand around the coil in the direction of the current, our thumb gives the direction of the field, which is upward.

2. A second layer, also with 180 turns, would double n, the number of turns per meter of the solenoid; since $B = \mu_0 n I_0$, this would double the magnetic field.

3. If we stretch the solenoid coil to twice its length, we are decreasing the number of turns per meter n by a factor of 2; since $B = \mu_0 n I_0$, the field is also half as large.

4. Since in a solenoid the magnetic field is $B = \mu_0 n I_0$, changing the cross-sectional area has no effect on the field inside the solenoid. Since magnetic flux is the product [field] × [area], reducing the cross-sectional area by a factor of 2 makes the magnetic flux half as large.

5. (B) Attract, repel. In Fig. 29.36a, the coil produces a magnetic field that points to the right (using the right-hand rule; see Question 1); thus the right end of the coil is a magnetic north pole, which attracts the south pole of the adjacent permanent magnet. Similarly, in Fig. 29.36b, each coil produces a field that points away from the other coil; thus, two magnetic south poles are adjacent, and these like poles repel.

Checkup 29.5

1. The field at the center of a current loop, $B = \mu_0 I/2R$, varies inversely with the radius of the loop, so the new field is $B_0/2$.

2. By the right-hand rule, the clockwise current contributes a field into the page at the center of the loop; the counterclockwise current produces a field out of the page. Since the field due to a loop is $\mu_0 I/2R$, the smaller loop produces a larger field, and so the net field is out of the page.

3. (D) $4B_1$. By the right-hand rule, the contributions from the four sides are all in the same direction, perpendicular to the loop. Thus the field contributions simply add, and the net field is 4 times the field from one side.

CHAPTER

30

Charges and Currents in Magnetic Fields

CONCEPTS IN CONTEXT

Concepts in Context

This "cyclotron" was one of the first devices used to accelerate protons to very high kinetic energy. The protons are accelerated inside the circular evacuated can placed between the poles of the large electromagnet.

In this chapter we will see that in a uniform magnetic field, such as between the pole faces of the electromagnet shown, a charged particle travels in a circular orbit. We will ask questions such as

? How does a cyclotron accelerate protons? (Section 30.1, page 967)

? The first cyclotron used a magnetic field of 1.4 T. What kinetic energy of the protons could be attained? What was the frequency of the circular motion? (Example 1, page 968)

? An iron-core electromagnet was used to provide the large magnetic field. What is the magnetic field produced solely by the wire coil around the iron core? (Example 6, page 979)

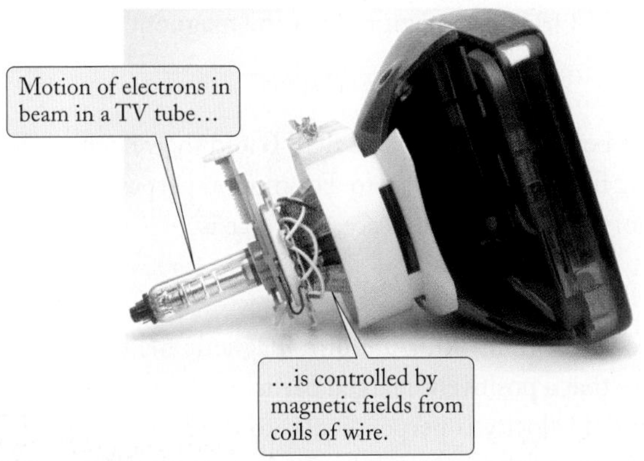

FIGURE 30.1 A TV tube. The beam of electrons from the cathode at the back of the tube strikes the rear side of the screen at the front.

In a TV tube or a cathode-ray tube (CRT) electrons move in a beam from the cathode at the rear end of the tube to the screen at the front end, where their impact causes the screen to glow, forming a visible display (Fig. 30.1). The motion of the electrons is controlled by magnetic fields. These magnetic fields, produced by coils of wire within the tube, deflect the electrons up or down and left or right and thereby determine where they will strike the screen, and what display you will see on the screen.

The magnetic fields in the TV tube are nonuniform, and the motion of the electrons is rather hard to calculate. However, the motion of electrons or other charged particles in uniform magnetic fields is fairly easy to calculate, and in this chapter we will deal with this motion in detail. Such uniform fields are not found in TV tubes, but they are found in magnets used for experiments and applications in science and engineering, such as the magnets in cyclotrons and other accelerators that produce high-energy particles for research in physics, the magnets for materials-science experiments to determine the electrical and magnetic properties of matter, and also the magnets for special kinds of radiation therapy in medicine.

In this chapter we will also examine the magnetic forces acting on wires and on current loops; the enhancement of magnetic fields by the presence of iron or other magnetic materials; and, finally, the potential difference that arises between the sides of a wire carrying a current while placed in a magnetic field (the Hall effect).

Online Concept Tutorial

30.1 CIRCULAR MOTION IN A UNIFORM MAGNETIC FIELD

The direction of the force exerted by a magnetic field on a moving charged particle is always perpendicular to both the magnetic field and the velocity, as specified by the right-hand rule. Since the magnetic force is always perpendicular to the velocity, the acceleration—and the small change of velocity in a small time interval—is perpendicular to the velocity. If the change of velocity is always perpendicular to the velocity, then the velocity can never change in magnitude, but only in direction (see Fig. 30.2). We therefore recognize that *the magnetic force acting on a particle will deflect the particle, keeping the speed constant.* We can also recognize this in another way: the force is always perpendicular to the displacement; therefore it does no work, and therefore the kinetic energy and the speed remain constant.

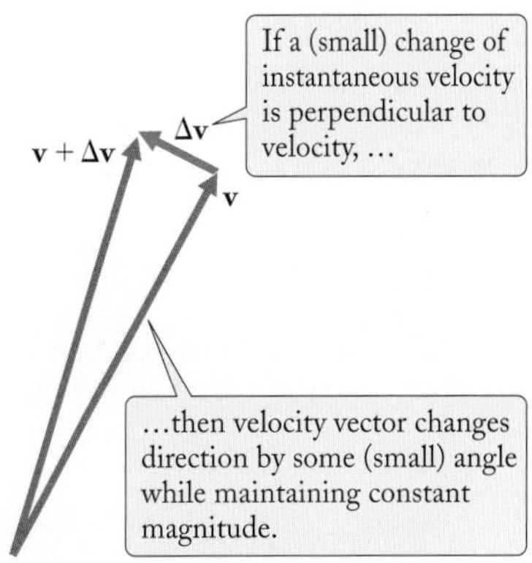

FIGURE 30.2 Instantaneous velocity $\mathbf{v}$, small change in velocity $\Delta\mathbf{v}$ perpendicular to $\mathbf{v}$, and instantaneous velocity $\mathbf{v} + \Delta\mathbf{v}$. The speed is constant.

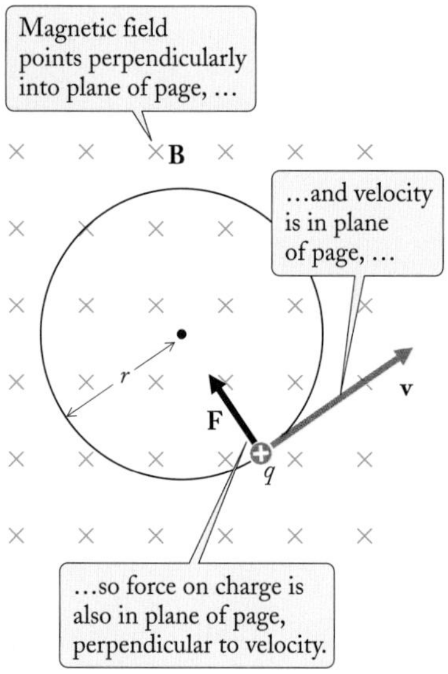

FIGURE 30.3 Positively charged particle moving in a uniform magnetic field. The crosses show the tails of the magnetic field vectors.

According to Eq. (29.11) the magnitude of the magnetic force is

$$F = qvB \sin \alpha \tag{30.1}$$

where α is the angle between the magnetic field **B** and the velocity **v**. For motion parallel to the magnetic field, this force is zero. For motion perpendicular to the magnetic field, the force is not zero. In the latter case, the force is

$$F = qvB \tag{30.2}$$

Figure 30.3 shows a region with a uniform magnetic field, directed into the plane of the page. Suppose that a positively charged particle has an initial velocity in the plane of the page; this initial velocity is perpendicular to the magnetic field. The magnetic force is then in the plane of the page, perpendicular to both the velocity and the magnetic field; its direction is shown in Fig. 30.3. According to Eq. (30.2), the acceleration caused by this force has a constant magnitude

$$a = \frac{F}{m} = \frac{qvB}{m} \tag{30.3}$$

and its direction is also perpendicular to the velocity. Such a constant acceleration perpendicular to the velocity is characteristic of uniform circular motion. Thus, the particle will move in a circle of some radius r, and the acceleration given by Eq. (30.3) will play the role of centripetal acceleration; that is, $a = v^2/r$ [see Eq. (4.49)] and

$$\frac{qvB}{m} = \frac{v^2}{r} \tag{30.4}$$

This leads to the following formula for the radius of the circular orbit:

radius of circular orbit

$$r = \frac{mv}{qB} \tag{30.5}$$

Figure 30.4 is a photograph of a beam of electrons executing such uniform circular motion in a cathode-ray tube placed in a magnetic field.

In terms of the momentum $p = mv$, our equation for the radius of the circular orbit becomes

$$r = \frac{p}{qB} \tag{30.6}$$

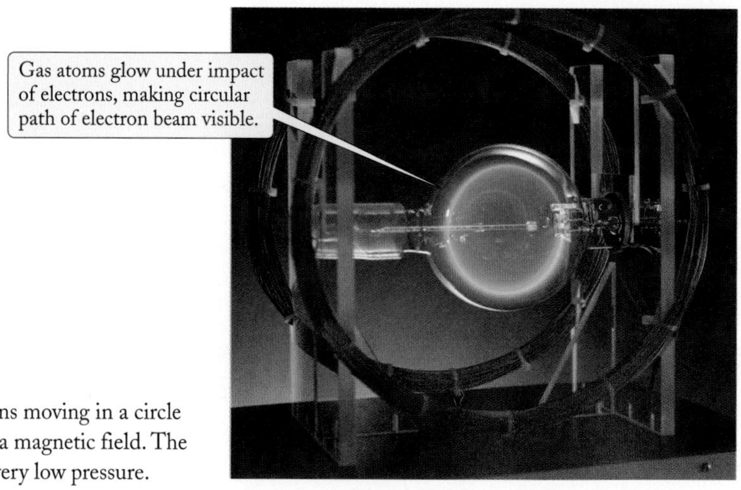

FIGURE 30.4 Electrons moving in a circle in a cathode-ray tube in a magnetic field. The tube contains a gas at a very low pressure.

In this form, the equation is valid even for relativistic particles (that is, particles of a speed close to the speed of light). As we will see in Chapter 36, for such particles, the momentum is not equal to mv, and Eq. (30.5) is not valid; however, Eq. (30.6) remains valid.

The time per revolution is the distance (the circumference) divided by the speed. Thus the frequency f of the circular motion, or the number of revolutions per second, is

$$f = \frac{[\text{speed}]}{[\text{circumference}]} = \frac{v}{2\pi r} = \frac{v}{2\pi (mv/qB)}$$

or

$$f = \frac{qB}{2\pi m} \tag{30.7}$$

cyclotron frequency

This is called the **cyclotron frequency** because the operation of cyclotrons (described below) involves particles moving with this frequency in a magnetic field. The cyclotron frequency is independent of the speed of the circular motion—in a uniform magnetic field, slow particles and fast (but nonrelativistic) particles of a given charge and mass move around circles at the same frequency, but the slow particles move along smaller circles than the fast particles.

Note that this purely circular motion occurs when the velocity of a charged particle is perpendicular to a uniform magnetic field. If the velocity is parallel to the field, the magnetic force is zero, and the particle moves in a straight line at constant speed. If the particle velocity has components both parallel and perpendicular to the uniform magnetic field, then both motions occur: circular motion perpendicular to the field and linear motion parallel to the field. The path of a particle executing such motion traces out a helix. These three possible motions of a charged particle in a uniform magnetic field are illustrated in Fig. 30.5.

cyclotron

Concepts in Context

The **cyclotron** is a device for the acceleration of protons or other ions. It consists of an evacuated container placed between the poles of a large electromagnet; within the container there is a flat metallic can cut into two D-shaped pieces, or *dees* (Fig. 30.6). An oscillating high-voltage generator is connected to the dees; this creates an oscillating electric field in the gap between the dees. The frequency of the voltage generator is adjusted so that it coincides with the cyclotron frequency of Eq. (30.7). An ion source at the center of the cyclotron releases protons or other ions. The electric field in the gap

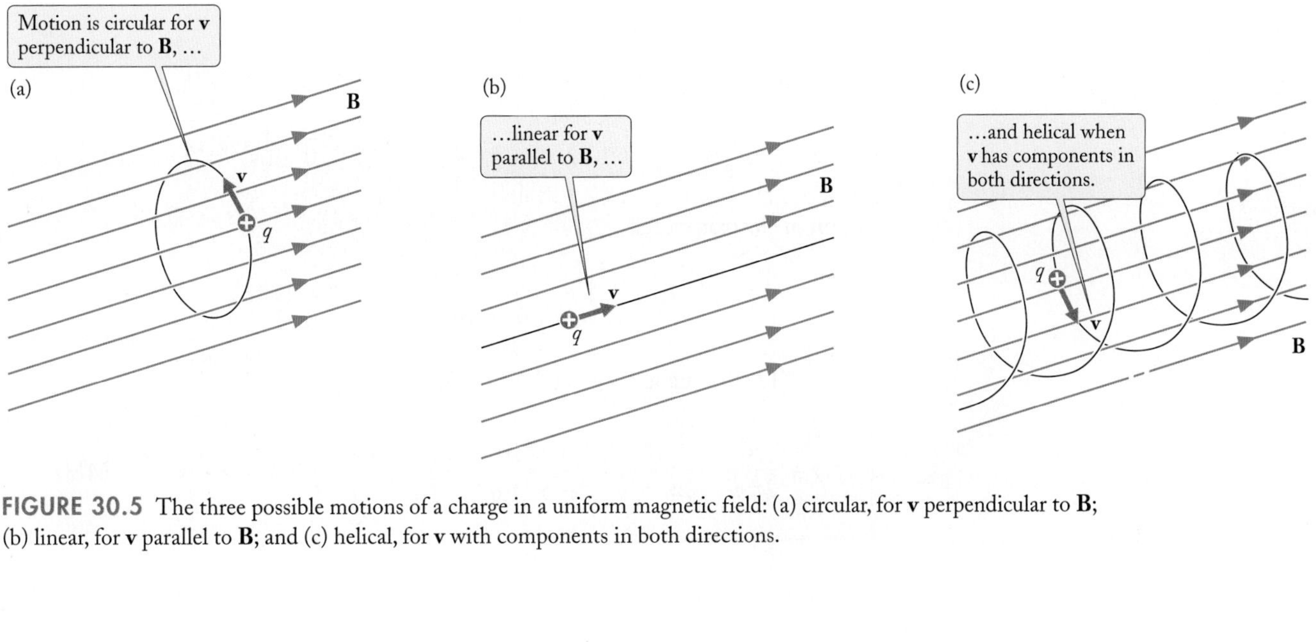

FIGURE 30.5 The three possible motions of a charge in a uniform magnetic field: (a) circular, for **v** perpendicular to **B**; (b) linear, for **v** parallel to **B**; and (c) helical, for **v** with components in both directions.

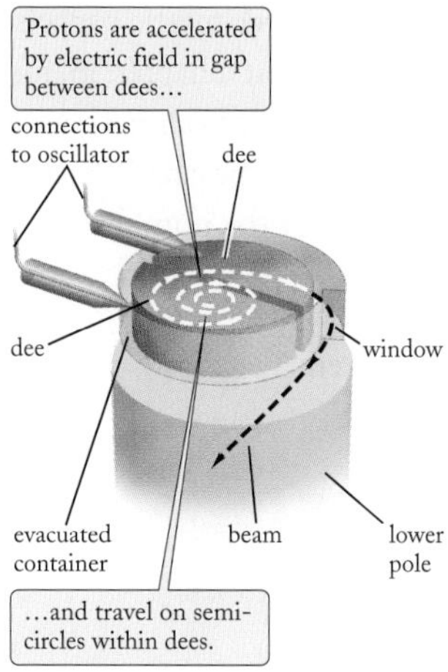

FIGURE 30.6 Trajectory of a particle within the dees of a cyclotron. In this diagram, the upper pole of the electromagnet has been omitted for the sake of clarity.

between the dees gives each of these protons a push, and the uniform magnetic field in the cyclotron then makes the proton travel on a semicircle inside the first dee. When the proton returns to the gap after one half period, the high-voltage generator will have reversed the electric field in the gap; the proton therefore receives an additional push, which sends it into the second dee. There, it travels on a semicircle of slightly larger radius corresponding to its slightly larger energy, and so on. Each time the proton crosses the gap between the dees, it receives an extra push and extra energy. The proton therefore travels along semicircles of stepwise increasing radius. When the protons reach the outer edge of the dees they leave the cyclotron as a high-energy beam.

Concepts — in — Context

EXAMPLE 1 One of the first cyclotrons, built by E. O. Lawrence at Berkeley in 1932, had dees with a diameter of 28 cm, and its magnet was capable of producing a magnetic field of 1.4 T. What was the maximum energy of the protons accelerated by this cyclotron? What was the frequency of the circular motion?

SOLUTION: The energy of the proton is its kinetic energy, $K = \frac{1}{2} m_p v^2$. When the proton reaches its maximum energy, its orbit has a radius of 14 cm. Since the magnetic field is 1.4 T, the speed of such a proton is, according to Eq. (30.5),

$$v = \frac{qBr}{m} = \frac{eBr}{m_p} = \frac{1.6 \times 10^{-19}\ \text{C} \times 1.4\ \text{T} \times 0.14\ \text{m}}{1.67 \times 10^{-27}\ \text{kg}}$$

$$= 1.9 \times 10^7\ \text{m/s}$$

and the energy is

$$K = \tfrac{1}{2} m_p v^2 = \tfrac{1}{2} \times 1.67 \times 10^{-27}\ \text{kg} \times (1.9 \times 10^7\ \text{m/s})^2 = 2.9 \times 10^{-13}\ \text{J}$$

or, in electron-volts,

$$K = 2.9 \times 10^{-13}\ \text{J} \times \frac{1\ \text{eV}}{1.6 \times 10^{-19}\ \text{J}} = 1.8 \times 10^6\ \text{eV} = 1.8\ \text{MeV}$$

The frequency is given by Eq. (30.7),

$$f = \frac{qB}{2\pi m} = \frac{eB}{2\pi m_p} = \frac{1.6 \times 10^{-19}\ \text{C} \times 1.4\ \text{T}}{2\pi \times 1.67 \times 10^{-27}\ \text{kg}} = 2.1 \times 10^7/\text{s} = 21\ \text{MHz}$$

Checkup 30.1

QUESTION 1: Two electrons are moving in a uniform magnetic field. One has an orbital radius of 15 cm. The other has twice the speed of the first. What is its orbital radius?

QUESTION 2: The magnetic field exerts forces on charged particles, but is incapable of doing work on them. Why?

QUESTION 3: A proton and an electron are moving in a uniform magnetic field. Which has the larger cyclotron frequency?

QUESTION 4: Can electrons be accelerated in a cyclotron? In what regard does a cyclotron used for electrons differ from one used for protons, if the magnetic field is the same?

QUESTION 5: Two protons are moving in the same uniform magnetic field. The orbital radius of the first proton is 10 cm, and the orbital radius of the second is 20 cm. Which proton has the larger speed? Which has the larger orbital frequency?

(A) First, first (B) First, second (C) Second, first
(D) Second, second (E) Second, both same

30.2 FORCE ON A WIRE

If a wire carrying a current is placed in a magnetic field, the moving charges within the wire will experience a force. Since the motion of the charges in the wire is constrained by the wire (the charges must move along the wire, regardless of the magnetic force they experience), any force acting on these charges is merely transferred to the wire, and therefore the wire as a whole will experience a force equal to the total force acting on the charges.

Consider a segment of the wire of length dl. To find the amount of moving charge in this segment, suppose that the speed of these moving charges is v; then they take a time $dt = dl/v$ to move out of the segment (see Fig. 30.7). Since the amount of charge is the product of the time and the current, we find that

$$dq = I\,dt = I\,dl/v$$

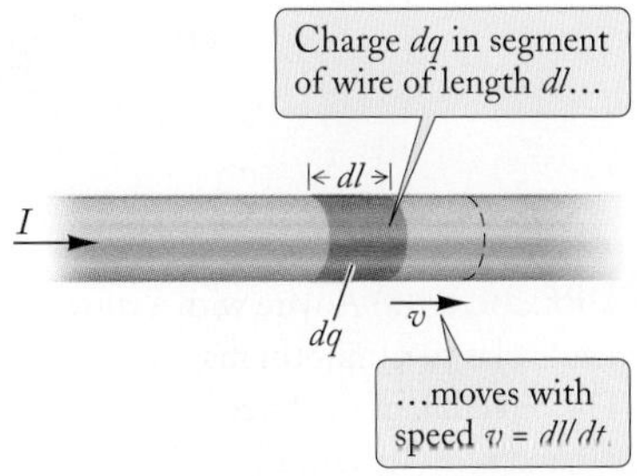

FIGURE 30.7 In a time $dt = dl/v$, the moving charge moves a distance dl toward the right. Hence, in this time, all the moving charge in the segment dl of the wire moves out of this segment toward the right (this charge is replaced by charge entering the segment from the left).

where, as always, we pretend that the moving charge is positive. If the wire segment is oriented at right angles to the magnetic field, then the magnetic force is

$$dF = dq\,vB = I(dl/v)vB = IB\,dl \qquad (30.8)$$

The direction of this force is related to the direction of the current and the direction of the magnetic field by the right-hand rule. For instance, the force on the wire segment oriented perpendicularly to the magnetic field, as illustrated in Fig. 30.8, is perpendicular to the wire and to the magnetic field, out of the plane of the page.

If the wire segment is oriented parallel to the magnetic field, then the force is zero. Other orientations of the wire segment give forces of a magnitude somewhere between zero and the maximum magnitude $IB\,dl$. We can find the dependence of the force on the orientation of the wire segment by noting that, for a charged particle moving at an angle α relative to the magnetic field, the force is reduced by a factor $\sin\alpha$, as in Eq. (30.1). Hence this same factor must appear in the general expression for the force on a wire segment oriented at an angle α to the magnetic field (Fig. 30.9):

force on wire segment

$$dF = IB\,dl\sin\alpha \qquad (30.9)$$

This equation yields Eq. (30.8) as a special case (with $\alpha = 90°$).

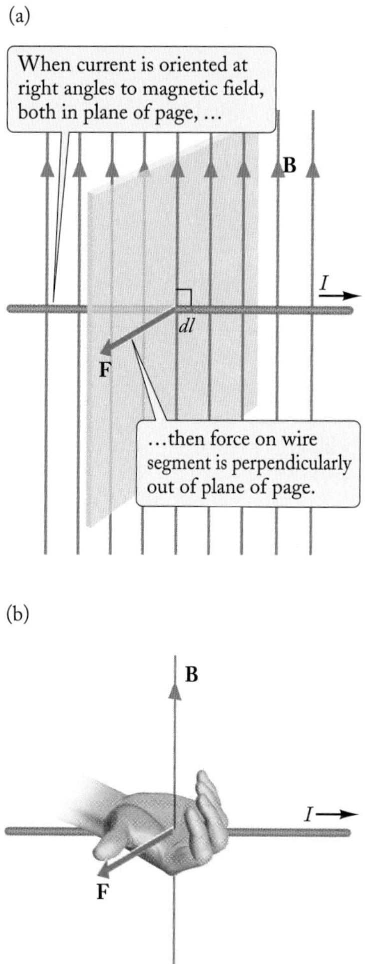

FIGURE 30.8 (a) A wire with a current perpendicular to a uniform magnetic field. (b) The direction of the force is given by the right-hand rule (see also Fig. 29.12).

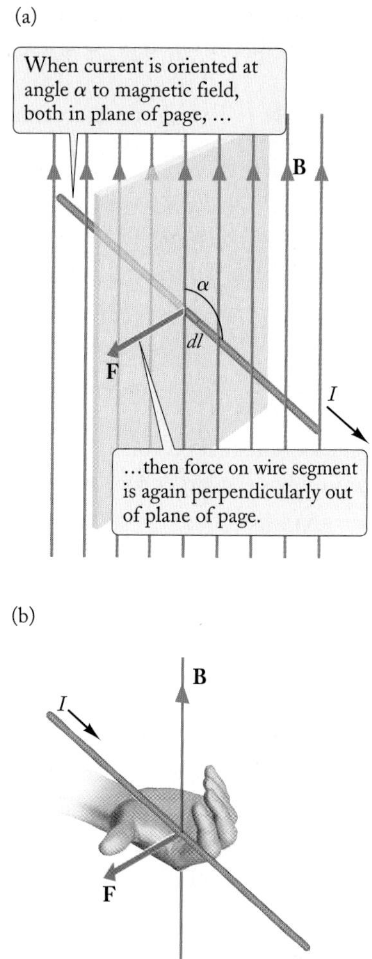

FIGURE 30.9 (a) A wire with a current is oriented at an angle α to a uniform magnetic field. (b) The direction of the force is given by the right-hand rule (see also Fig. 29.12).

To find the force on an entire wire, we must add the contributions (30.9) from all the small segments of this wire vectorially. For a straight wire of length L in a uniform magnetic field, the contributions simply add to give a force of the same form as (30.9),

force on straight wire in uniform field

$$F = ILB \sin\alpha \tag{30.10}$$

If we define a vector $d\mathbf{l}$ of length dl (or $\mathbf{L}$ of length L) in the direction of the current I, then the magnetic force on a wire segment or straight wire can be expressed as, respectively, the cross product

$$d\mathbf{F} = I d\mathbf{l} \times \mathbf{B} \quad \text{or} \quad \mathbf{F} = I\mathbf{L} \times \mathbf{B} \tag{30.11}$$

EXAMPLE 2 Figure 30.10 shows a balance used for the measurement of a magnetic field. A loop of wire carrying a precisely known current is partially immersed in the magnetic field. The force that the magnetic field exerts on the loop can be measured with the balance, and this permits the calculation of the strength of the magnetic field. Suppose that the short side of the loop measures 10.0 cm, the current in the wire is 0.225 A, and the magnetic force is 5.35×10^{-2} N. What is the strength of the magnetic field?

SOLUTION: The upper part of the loop is not in the magnetic field and hence experiences no magnetic force. The vertical segments of the loop carry the same current in opposite directions, and so these force contributions cancel. Thus the net magnetic force is that experienced by the lower, short side of the loop. This segment and the magnetic field are perpendicular, so Eq. (30.10) with $\alpha = 90°$ gives

$$F = ILB$$

The strength of the magnetic field is then

$$B = \frac{F}{IL} = \frac{5.35 \times 10^{-2}\ \text{N}}{0.225\ \text{A} \times 0.100\ \text{m}} = 2.38\ \text{T}$$

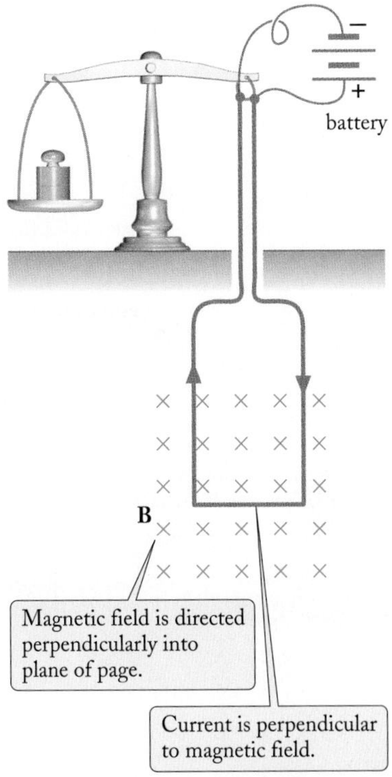

FIGURE 30.10 A current balance.

EXAMPLE 3 Two very long, parallel wires separated by a distance r carry currents I_1 and I_2, respectively. Find the magnetic force that each wire exerts on a segment dl of the other wire.

SOLUTION: Each wire generates a magnetic field, which exerts a force on the other wire. Figure 30.11 shows the magnetic field $\mathbf{B}_1$ that the current I_1 produces in the vicinity of the current I_2. By Eq. (29.9), this magnetic field has a magnitude

$$B_1 = \frac{\mu_0}{2\pi}\frac{I_1}{r}$$

and it is perpendicular to wire 2. By Eq. (30.9), the force on a segment dl of wire 2 has a magnitude

$$dF = I_2 B_1\, dl = \frac{\mu_0}{2\pi}\frac{I_1 I_2}{r}\, dl \qquad (30.12)$$

The force on a segment of wire 1 is of the same magnitude, but opposite direction. The force between the wires is attractive if the currents in the wires are parallel and repulsive if they are antiparallel. Note that for parallel wires, the force dF is a constant, independent of the position of dl. Thus the total force F on a larger segment L is obtained by replacing dl with L in Eq. (30.12), as given in Eq. (30.11).

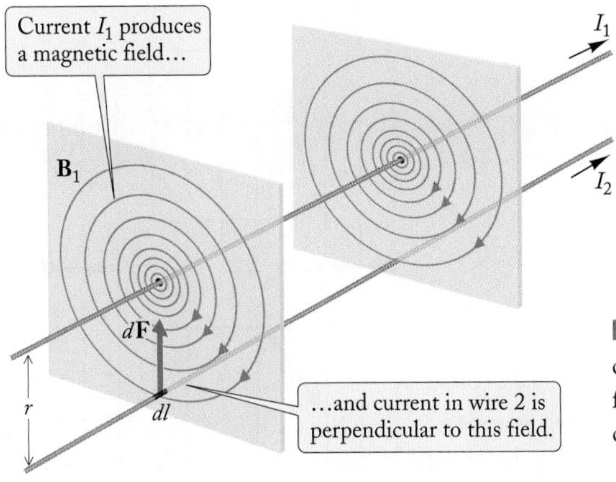

FIGURE 30.11 A long, straight wire carrying a current I_2 in the magnetic field of a long, straight wire carrying a current I_1.

The official definition of the SI **unit of current** is based on the force per meter of length between two long parallel wires. This force can be measured very precisely by holding one wire stationary and suspending the other from a balance; the wires are connected in series so that the currents are exactly equal ($I_1 = I_2$). The force per unit length and the distance r can be measured experimentally, and the value of I_1 (or I_2) calculated from Eq. (30.12) is then the current in amperes. The constant μ_0 appearing in Eq. (30.12) is assigned the value $\mu_0 = 4\pi \times 10^{-7}\ \text{N·s}^2/\text{C}^2$ *by definition.*

coulomb (C)

The official definition of the SI unit of charge is based on the unit of current. *The* **coulomb** *is defined as the amount of charge that a current of one ampere delivers in one second,* 1 coulomb = (1 ampere × 1 second).

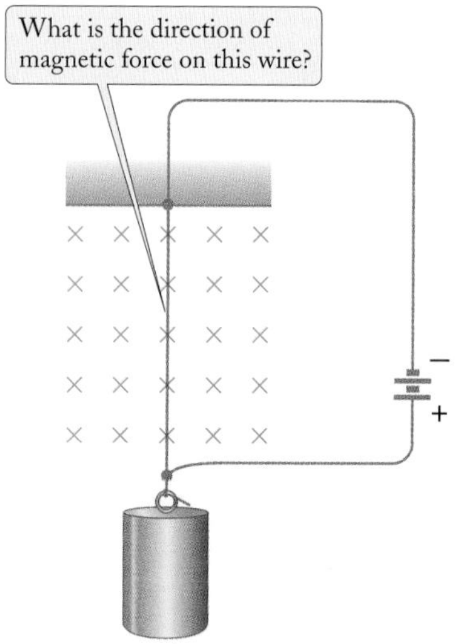

FIGURE 30.12 A wire hanging in a uniform magnetic field.

Checkup 30.2

QUESTION 1: A power cord for an electric appliance consists of two parallel straight wires carrying currents in opposite directions. Is the magnetic force between them attractive or repulsive?

QUESTION 2: A straight wire carrying a current is placed in a uniform horizontal magnetic field. For what orientation of the wire is the magnetic force on the wire zero?

QUESTION 3: A wire hangs in a horizontal magnetic field, held taut by a weight (see Fig. 30.12). We connect this wire to a battery, so a current flows vertically upward. The deflection of the hanging wire will be

(A) To the right (B) To the left (C) Into the page
(D) Out of the page (E) Zero

30.3 TORQUE ON A LOOP

If a loop of wire with a current is placed in a magnetic field, the action of the magnetic field can in general result not only in a net force on the loop, but also in a torque. The force and the torque depend on the shape and the orientation of the loop, and they can be zero for some orientations of the loop. For example, the net force on a current loop in a *uniform* magnetic field is always zero. To see this, let us consider the simple case of a rectangular loop oriented perpendicularly to the magnetic field, illustrated in Fig. 30.13. The forces on the four sides of the loop are given by Eq. (30.10). The forces

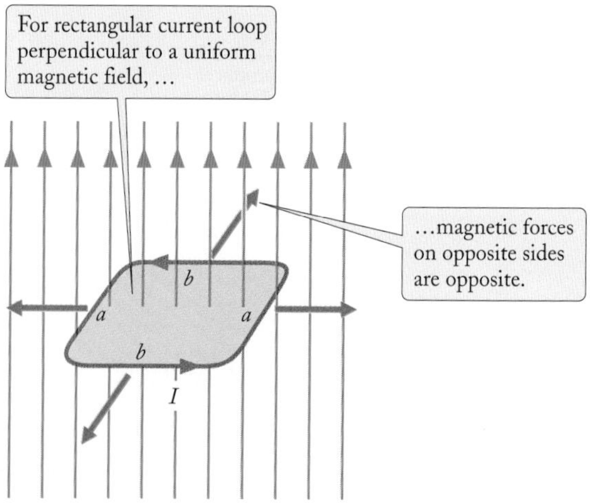

FIGURE 30.13 Forces on a loop of current at right angles to a magnetic field.

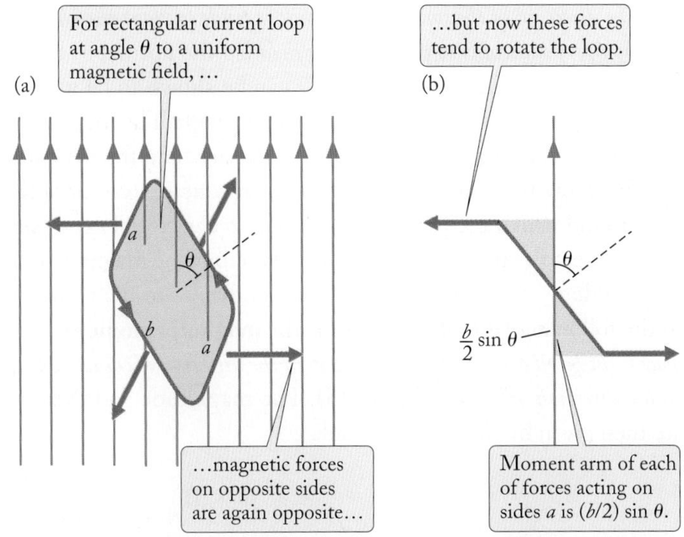

FIGURE 30.14 (a) Forces on a rectangular loop of current oriented at an angle θ with respect to a magnetic field. (b) Side view of the loop showing forces acting on sides a.

on opposite sides are opposite (because the currents are opposite); hence the forces cancel in pairs, and the loop will be in equilibrium. The same is true for a loop of arbitrary shape if we consider all the small contributions to the force: the net force is zero.

Figure 30.14 shows a loop oriented at an angle θ to a uniform magnetic field. In this figure, the angle θ is measured between the magnetic field and the normal to the loop ($\theta = 0$ means the loop is perpendicular to the magnetic field; $\theta = 90°$ means the loop is parallel). Again the forces on opposite sides are opposite, but now the lines of action of the forces pulling on the left and on the right sides do not coincide—these forces exert a torque, which tends to rotate the loop. Since the magnetic field is constant, Eq. (30.10) tells us that the magnitude of the forces on the top and bottom sides is

$$F = IaB \tag{30.13}$$

where a is the length of each of these sides. We recall from Eq. (13.14) that the torque is the product of the force and the moment arm. If, as shown in Fig. 30.14b, the length of each of the other two sides is b, then the moment arm about the center of the loop is $(b/2)\sin\theta$, and the torque due to the pair of forces IaB is therefore

$$\begin{aligned}\tau &= F(b/2)\sin\theta + F(b/2)\sin\theta \\ &= IabB\sin\theta \qquad (30.14)\end{aligned}$$

The product of the area $A = ab$ of the loop and the current flowing around it is called the **magnetic dipole moment** of the loop:

$$\mu = [\text{current}] \times [\text{area}] = IA \tag{30.15}$$

magnetic dipole moment

Note that if the loop consists of several turns of wire, then Eq. (30.15) must be evaluated with the net current flow around the loop. This net current equals the number N of turns of wire times the current I_0 in one turn, so

$$\mu = NI_0A \tag{30.16}$$

In any case, the magnitude of the torque is

$$\tau = \mu B\sin\theta \tag{30.17}$$

torque on current loop

The direction of the torque is such that it tends to twist the plane of the loop into an orientation perpendicular to the magnetic field (so the normal to the loop is parallel to the field, or $\theta = 0$). The formula (30.17) can be shown to be valid not only for the rectangular loop, but also for loops of any other shape. The magnetic moment of any (flat) loop of arbitrary shape is the product of the area and the net current around the loop.

We know from Section 13.4 that the torque is a vector with direction along the axis around which the torque acts. To obtain a vectorial expression for the torque, we need to treat the magnetic moment as a vector. The magnetic moment vector $\boldsymbol{\mu}$ has a magnitude $\mu = IA$ and a direction perpendicular to the plane of the loop, according to the following **right-hand rule for the magnetic moment**: *if you wrap the fingers of your right hand around the current loop in the direction of the current, your thumb will point in the direction of* $\boldsymbol{\mu}$ (see Fig. 30.15). The magnitude and the direction of the torque are then given by the vector formula

right-hand rule for magnetic moment

magnetic torque vector

$$\boldsymbol{\tau} = \boldsymbol{\mu} \times \mathbf{B} \tag{30.18}$$

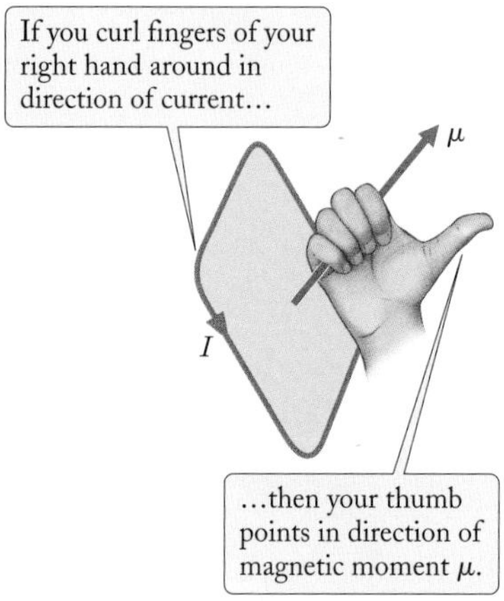

FIGURE 30.15 Right-hand rule for the magnetic moment of a current loop.

A loop of current suitably pivoted on an axis acts as a compass needle; the normal to the loop seeks to align itself with the magnetic field. This similarity is no accident. A compass needle is a small permanent magnet, and as we will see in Section 30.4, a permanent magnet contains a large number of electrons acting as small current loops. The mechanism underlying the alignment of a compass needle with a magnetic field is therefore the same as for a current loop.

The torque on a current loop pivoted on an axis and placed in a strong magnetic field is exploited in electric motors. This torque brings about the rotational motion of the current loop and provides the means of converting electric energy into mechanical energy of rotational motion. In small electric motors, the magnetic field is produced by permanent magnets; but in large motors, the magnetic field is produced by electromagnets. The following example illustrates the operation of a simple electric motor.

EXAMPLE 4 A simple electric motor consists of a rectangular coil of wire that rotates on a longitudinal axle in a magnetic field of 0.50 T (see Fig. 30.16). The coil measures 10 cm × 20 cm; it has 40 turns of wire, and the current in the wire is 8.0 A. (a) In terms of the angle θ betweeen the magnetic field and the normal to the coil, what is the torque that the magnetic field exerts

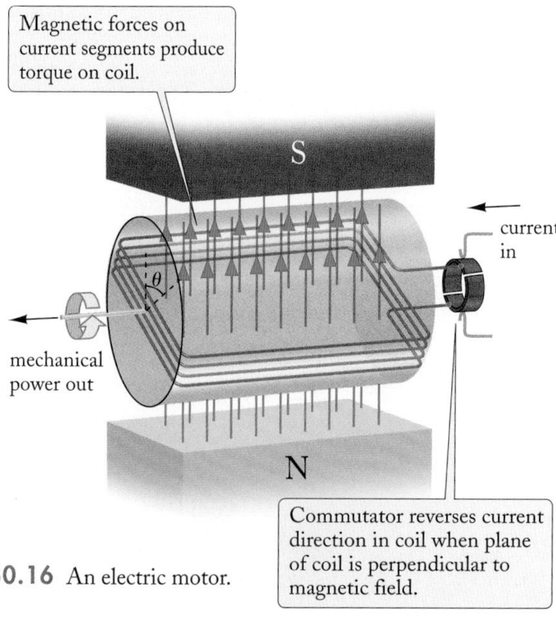

FIGURE 30.16 An electric motor.

on the coil? (b) In order to keep the sign of the torque constant, a switch (commutator) mounted on the axle reverses the current in the coil whenever θ passes through 0° and 180°. Plot this torque vs. the angle θ.

SOLUTION: (a) According to Eq. (30.17), the torque on the coil is

$$\tau = NI_0abB\sin\theta$$
$$= 40 \times 8.0\text{ A} \times 0.10\text{ m} \times 0.20\text{ m} \times 0.50\text{ T} \times \sin\theta$$
$$= 3.2\text{ N}\cdot\text{m}\sin\theta$$

(b) If the torque always has the same sign (say, positive), then the plot of the torque for the interval 180° to 360° merely repeats the plot for the interval 0° to 180° Figure 30.17 shows this plot.

COMMENT: Practical electric motors consist of many such coils, each with its commutator, arranged at regular intervals around the axle. The plot of the net torque of this arrangement is a sum of plots such as shown in Fig. 30.17, but with different starting angles. This averages out the ups and downs of Fig. 30.17 and yields a torque that is nearly constant at all angles.

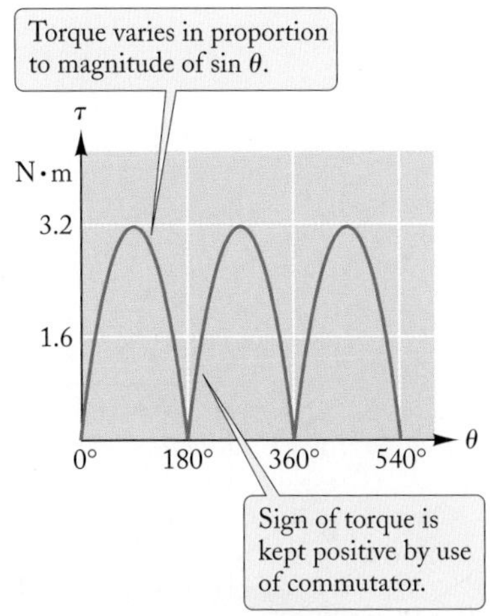

FIGURE 30.17 Torque as a function of angle, according to Example 4.

The torque on a current loop is also exploited in instruments for the measurement of current, generally called **galvanometers**. In such an instrument, the current to be measured is sent into a coil of several turns pivoted on an axis and placed in the magnetic field of a permanent magnet (see Fig. 30.18). A fine spiral spring attached to the moving coil provides a restoring torque that opposes the magnetic torque. Under the influence of these two opposing torques, the moving coil attains an angular deflection that increases with the magnitude of the current. A pointer attached to the moving coil indicates the magnitude of the current on a calibrated scale. Many types of meters operate on this principle.

galvanometer

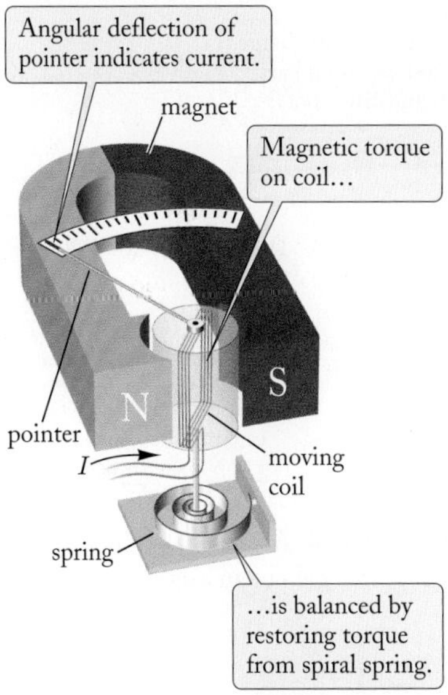

FIGURE 30.18 Mechanism of a galvanometer.

The potential energy of a current loop in a magnetic field can be obtained from the torque as follows. If an external agent pushes against the magnetic torque, turning a loop through an angle $d\theta$, it does work on the loop [see Eq. (13.3)]. The potential energy of the loop will increase:

$$dU = \tau\, d\theta = \mu B \sin\theta\, d\theta \qquad (30.19)$$

If only the angle θ between $\boldsymbol{\mu}$ and $\mathbf{B}$ is changing, the potential energy stored when the loop is turned from some reference angle θ_0 to some final angle θ is

$$U = \int dU = \int \mu B \sin\theta\, d\theta = \mu B \int_{\theta_0}^{\theta} \sin\theta\, d\theta = -\mu B(\cos\theta - \cos\theta_0) \qquad (30.20)$$

If we choose $\theta_0 = 90°$, when $\boldsymbol{\mu}$ is perpendicular to $\mathbf{B}$, as our reference angle, we obtain a simple form for the potential energy:

$$U = -\mu B \cos\theta \qquad (30.21)$$

potential energy of current loop

This can be written in terms of the scalar product (dot product) as

$$U = -\boldsymbol{\mu}\cdot\mathbf{B} \qquad (30.22)$$

This potential energy of a magnetic moment in a magnetic field is smallest (negative) when $\boldsymbol{\mu}$ is parallel to $\mathbf{B}$. The energy difference between parallel and antiparallel alignment of $\boldsymbol{\mu}$ with respect to $\mathbf{B}$ is $\Delta U = 2\mu B$. The magnetic energy (30.21) is an important quantity in the study of atomic physics, magnetic resonance, and magnetic alignment, for example, in magnetic recording materials.

Checkup 30.3

QUESTION 1: The torque given by Eq. (30.17) has a maximum magnitude at $\theta = 90°$ and at $\theta = -90°$. What is the direction of the torque on the loop shown in Fig. 30.14 for $\theta = 90°$? For $\theta = -90°$?

QUESTION 2: The torque given by Eq. (30.17) is zero at $\theta = 0°$ and at $\theta = 180°$, so the loop is in equilibrium. Is the equilibrium stable at 0°? At 180°?

QUESTION 3: What would happen if the motor in Fig. 30.16 had no commutator?

QUESTION 4: For the two loops in Figs. 30.13 and 30.14, what is the direction of the magnetic moment vector?

QUESTION 5: What is the maximum torque delivered by the motor described in Example 4? At what orientation is it maximum? If the motor had 120 turns of wire instead of 40, what would be the maximum torque?

QUESTION 6: A coil in the x-z plane carries a current in the direction shown in Fig. 30.19 and is in a region of uniform magnetic field with direction in the x-y plane as indicated. The direction of the torque is

(A) $-\mathbf{i}$ (B) $\mathbf{j}$ (C) $-\mathbf{j}$
(D) $\mathbf{k}$ (E) $-\mathbf{k}$

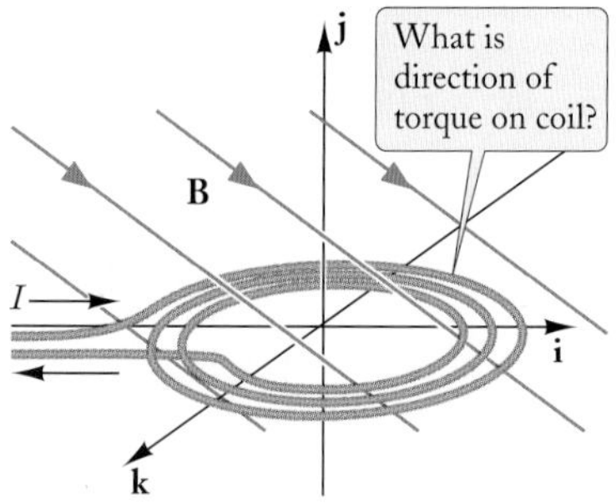

FIGURE 30.19 A coil in a uniform magnetic field.

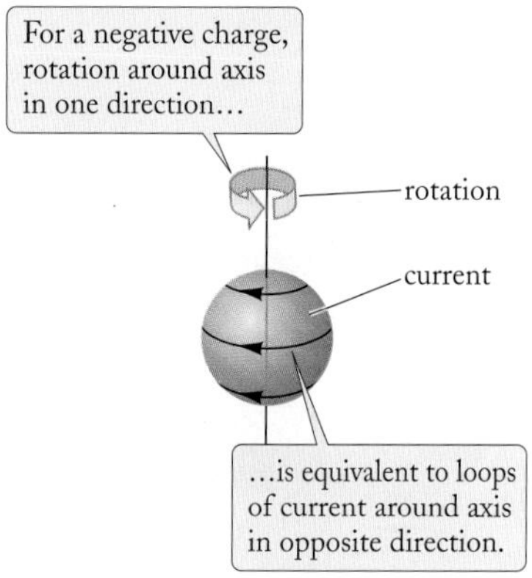

FIGURE 30.20 A small ball of negative charge spinning about an axis.

30.4 MAGNETISM IN MATERIALS

In all the calculations of magnetic fields in Chapter 29 we assumed that the wires carrying the currents were placed in air or in a vacuum. However, in practical applications of magnetism, the current-carrying wires are often placed around cores of solid iron. The presence of iron or other ferromagnetic materials enhances the magnetic field, and often makes it several thousand times stronger than what it would be without the iron. This enhancement arises from an alignment of the axes of spin of the electrons.

Crudely, we can picture the electron as a small ball of negative electric charge spinning about an axis (see Fig. 30.20). The rotational motion of the electron charge is equivalent to currents flowing in loops around the axis of spin, and the magnetic field produced by the electron is therefore that of a small loop of current. The magnetic moment of this loop of current has a fixed value, called a **Bohr magneton**

Bohr magneton

$$\mu_B = \frac{eh}{4\pi m_e} = 9.27 \times 10^{-24}\,\text{J/T} \tag{30.23}$$

Planck's constant

where $h = 6.63 \times 10^{-34}$ J·s is a fundamental constant of nature known as **Planck's constant**, discussed in detail in Chapter 37. In some materials, there are also rings of current due to the orbital motions of electrons around the nuclei in atoms, so in addition to spin magnetic moments, there are also orbital magnetic moments.

The nucleus of an atom can also have a magnetic moment, always much smaller than that of the electron. The proton, the nucleus of the hydrogen atom, has the largest magnetic moment of all nuclei, but even this moment is about 650 times smaller than a Bohr magneton. Nonetheless, the small nuclear magnetic moment can be separately sensed using resonance techniques to provide vivid images of subsurface structures in materials and body tissues; such **magnetic resonance imaging (MRI)** is commonly used in hospitals and in research (see Fig. 30.21).

magnetic resonance imaging (MRI)

The net magnetic moment of an atom is obtained by adding the various spin and orbital magnetic moments, taking into account the directions of these moments. In ferromagnetic materials, the magnetic moment is completely dominated by the contributions from the electron spins.

When a piece of iron is left to itself, the electron current loops are oriented at random, and the magnetic fields of the electrons average to zero. But when the iron is placed in an external magnetic field, such as the field of a solenoid, the axes of the spins of the electrons align with the magnetic field. Although this alignment is triggered by the external magnetic field, which exerts a torque on the small electron current loops (see Section 30.3), the amount of alignment is strongly magnified by a natural tendency for spin alignment in iron and other ferromagnetic materials. This alignment of spins makes all the electron current loops parallel, and *the combined magnetic fields of all these parallel current loops add to the original magnetic field and enhance it drastically.*

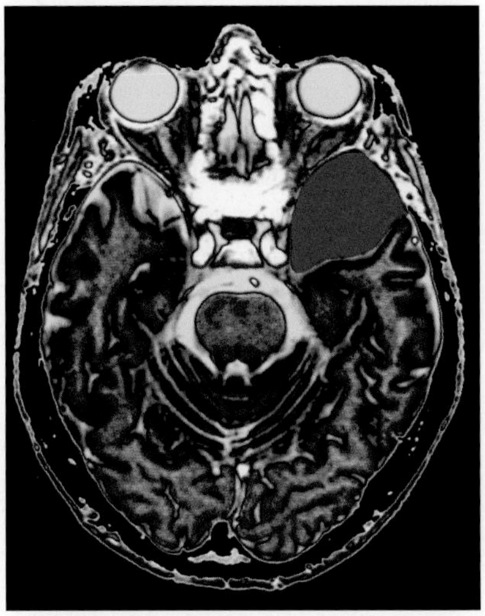

FIGURE 30.21 Nuclear magnetic resonance image (MRI) of a human head.

The enhancement of the magnetic field by this effect is measured by the **magnetic susceptibility** χ (the Greek letter chi). If we denote the magnetic field of the external currents by B_{external} and the field due to the electrons in the material by B_{matter}, then the magnetic susceptibility is defined as the ratio

$$\chi = \frac{B_{\text{matter}}}{B_{\text{external}}} \tag{30.24}$$

magnetic susceptibility

Ferromagnets have very large values of the susceptibility, often on the order of 1000 or more, because of the cooperative alignment of many electron spins.

ferromagnet

For most ordinary "nonmagnetic" materials, χ is very small, with a magnitude around 10^{-5} to 10^{-4}. Such materials with small positive susceptibilities are called **paramagnets.** The external field does not do much to overcome the random orientations of the electron current loops in a paramagnet, except at very low temperatures. Some materials have small negative susceptibilities and are called **diamagnets.**

paramagnet and diamagnet

An alternative measure of magnetism in a material is the **magnetic permeability** μ:[1]

$$\mu = \mu_0(1 + \chi) \tag{30.25}$$

magnetic permeability μ

Thus the magnetic permeabilities of most paramagnetic and diamagnetic materials do not differ appreciably from μ_0 (the permeability constant), because for most materials, $|\chi| \ll 1$. But the magnetic permeability of ferromagnets is much greater than μ_0.

A piece of iron with aligned electron spins is said to be magnetized. If the iron is removed from the solenoid, it will retain some of its magnetization and act as a **permanent magnet**. Some ferromagnetic materials, such as alnico (an alloy mostly of iron, nickel, and cobalt, with some aluminum and copper), retain more magnetization than pure iron, and they make better permanent magnets. Figure 30.22 shows several kinds of permanent magnets, including bar magnets and horseshoe magnets.

Electromagnets and permanent magnets are used in many kinds of electric machinery, such as electric motors (see Example 4 above), electric generators, tape recorders, videocassette recorders, computer disk drives, etc.

FIGURE 30.22 Different kinds of permanent magnets.

[1]Do not confuse the magnetic permeability with the magnetic dipole moment. Both are represented by the same letter μ. But they are different quantities.

PHYSICS IN PRACTICE MAGNETIC RECORDING MEDIA

In the recording head of a magnetic tape recorder, a small electromagnet produces a magnetic field whose strength varies in proportion to the amplitude of the sound signal arriving at the microphone (see Fig. 1). The tape passing over the recording head contains a magnetizable material. Initially, the tape is blank and unmagnetized. When the tape is subjected to the magnetic field of the recording head, it becomes magnetized. The strength of the induced magnetization in the tape is proportional to the strength of the magnetic field of the recording head, which is proportional to the intensity of the sound signal. Thus, as the tape passes over the recording head, it stores the information about the varying sound signal in the form of varying zones of high and low magnetization.

The gap between the poles of the electromagnet in the recording head must be extremely narrow—typically no more than 10^{-5} m—so an instant of the sound signal is recorded on a very narrow, almost pointlike spot or band on the tape. The quality of the recording improves if the speed of the tape over the recording head is high, since this leads to a wider separation between the magnetizations recorded at successive instants. The tape recorders in broadcasting studios typically use a tape speed of 15 in./s.

Digital information is stored on magnetic media by orienting the magnetization of each bit, or unit of information, in either of two directions; since the exact magnitude of the magnetization of each bit is unimportant, digital magnetic

(a)

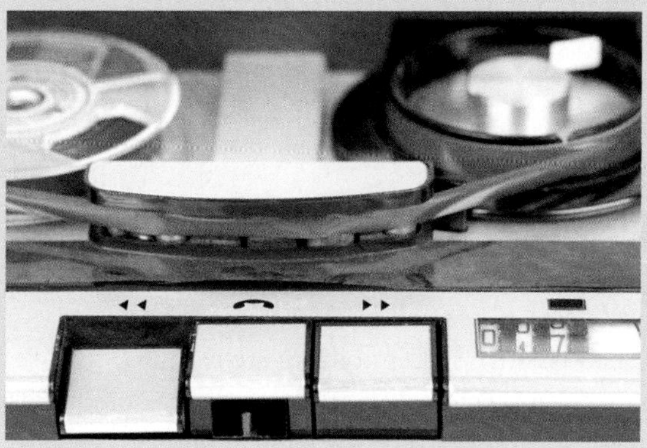

(b)

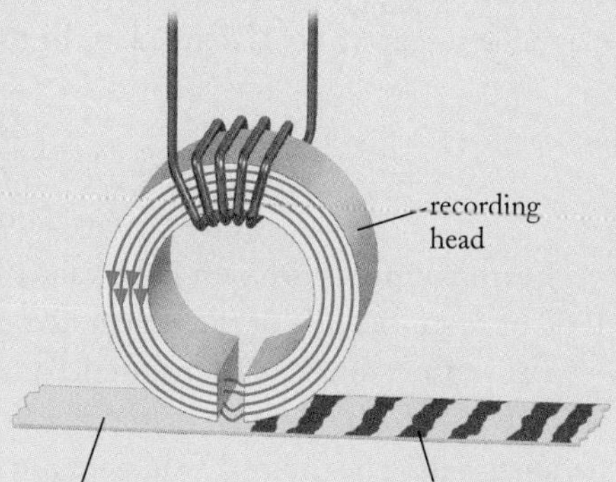

FIGURE 1 Recording head of magnetic tape recorder. (a) Close-up view of reel-to-reel transport, showing the head block. (b) Diagram of interface between recording head and tape.

EXAMPLE 5 Perminvar is an alloy of iron, nickel, and cobalt with a relative permeability of $\mu/\mu_0 = 2200$. Suppose that a long solenoid has 450 turns per meter and carries a current of 1.5 A. If the solenoid is filled with Perminvar, what is the total magnetic field in the solenoid?

SOLUTION: The solenoid provides the external field for the material, given by Eq. (29.21):

$$B_{\text{external}} = \mu_0 nI = 1.26 \times 10^{-6}\ \text{N}\cdot\text{m}^2/\text{C}^2 \times 450/\text{m} \times 1.5\ \text{A} = 8.5 \times 10^{-4}\ \text{T}$$

This external field induces the magnetization in the ferromagnet, and according to Eq. (30.24) produces a field

$$B_{\text{matter}} = \chi B_{\text{external}}$$

The total field inside the solenoid is thus

$$B = B_{\text{external}} + B_{\text{matter}} = (1 + \chi)B_{\text{external}}$$

media are relatively free from random errors, or "noise." So-called floppy disks store more than 10 million bits of information, each in small rectangular regions of magnetization. The magnetic bits on computer hard disks (Fig. 2) can be 100 to 1000 times more dense, with individual bits as small as 10^{-6} m across (see Fig. 3). Such disks often use a cobalt–nickel alloy as their magnetizable material. One of the advantages (and disadvantages!) of magnetic tapes and disks over optical media, such as CDs, is that the magnetic media can be erased and rewritten easily, for an essentially unlimited number of cycles, whereas CDs are harder to rewrite and can tolerate only a limited number of rewriting cycles. Because the data stored on magnetic tapes and disks can be easily erased, it is important to keep such media (including the magnetic stripe on ATM and credit cards) away from strong permanent magnets, which could destroy the recorded information.

FIGURE 2 Computer hard disk.

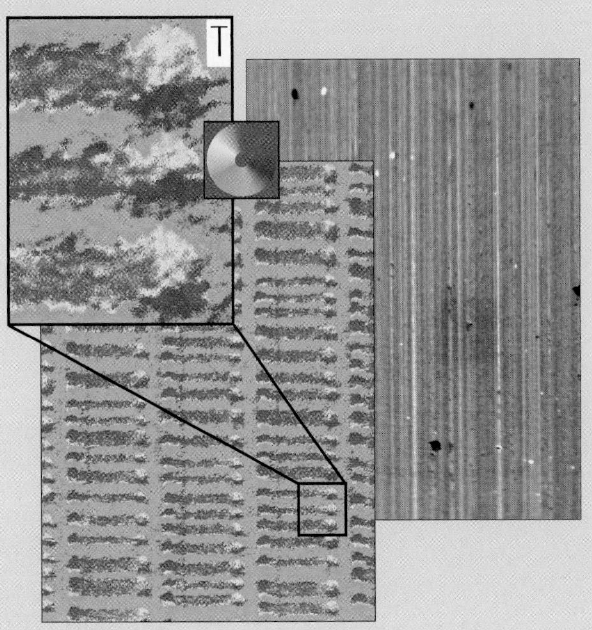

FIGURE 3 Individual bits of a digital hard disk. The magnetic image (color) reveals the bits which are not visible on the corresponding topographic image (gray). The scale bar (top) is 1 μm.

But by Eq. (30.25), $(1 + \chi)$ is just the relative permeability μ/μ_0, so

$$B = \frac{\mu}{\mu_0} B_{\text{external}} = 2200 \times 8.5 \times 10^{-4}\,\text{T}$$

$$= 1.9\,\text{T}$$

EXAMPLE 6 The electromagnet of a cyclotron consists of two iron pole pieces, each surrounded by a large wire coil. Suppose such a coil has 200 turns, and that each turn is a concentric circular loop of radius $R = 30$ cm. To produce a total field of $B = 1.4$ T in the iron at the center of the current loops, what field should the coil alone provide? What should the current in the wire be? The relative permeability of iron is 35 when $B = 1.4$ T.

Concepts in Context

SOLUTION: As in Example 5, the total field B in the iron is related to the externally provided field B_{external} from the current loops by

$$B = \frac{\mu}{\mu_0} B_{\text{external}} \tag{30.26}$$

Solving for B_{external} and using the given permeability gives

$$B_{\text{external}} = \frac{\mu_0}{\mu} B = \frac{1}{35} \times 1.4\ \text{T} = 4.0 \times 10^{-2}\ \text{T}$$

This is the field that must be produced by the wire loops.

For a number N of current loops with radius R, the field produced is N times the field for an individual loop, given by Eq. (29.19):

$$B_{\text{external}} = N \frac{\mu_0 I}{2R}$$

Solving for the current I, and substituting the required value of B_{external}, gives

$$I = \frac{2RB_{\text{external}}}{N\mu_0} = \frac{2 \times 0.30\ \text{m} \times 4.0 \times 10^{-2}\ \text{T}}{200 \times 4\pi \times 10^{-7}\ \text{N}\cdot\text{s}^2/\text{C}^2} = 95\ \text{A}$$

Electromagnets carrying such large currents require water cooling to prevent overheating.

Checkup 30.4

QUESTION 1: Consider the magnetic field of the rotating ball of negative charge shown in Fig. 30.20. What is the direction of the magnetic field that this rotating ball produces along its axis?

QUESTION 2: At high values of the externally applied field, a ferromagnet *saturates*; that is, all the magnetic moments fully align with the field, and no further alignment is possible. In this regime, how does the permeability vary with increasing external field?

(A) Increases (B) Decreases (C) Remains constant

30.5 THE HALL EFFECT

When charges flow along a wire or a conducting rod placed in a uniform magnetic field, a potential difference is produced along opposite sides of the wire. This phenomenon is called the **Hall effect**; it is important in the study of materials and in various practical applications.

Consider a rectangular rod carrying an electric current I. The rod is immersed in a uniform magnetic field **B** perpendicular to the current (see Fig. 30.23). The magnetic force qvB diverts the moving charges within the rod to one side. Since the current is confined to the rod, the diverted charges build up at one side and are depleted from the other side until such charges produce a transverse electric force that exactly balances the magnetic force. If we let $E_\perp$ denote the corresponding electric field, in the direction perpendicular to the current, equilibrium implies

$$qE_\perp = qvB \tag{30.27}$$

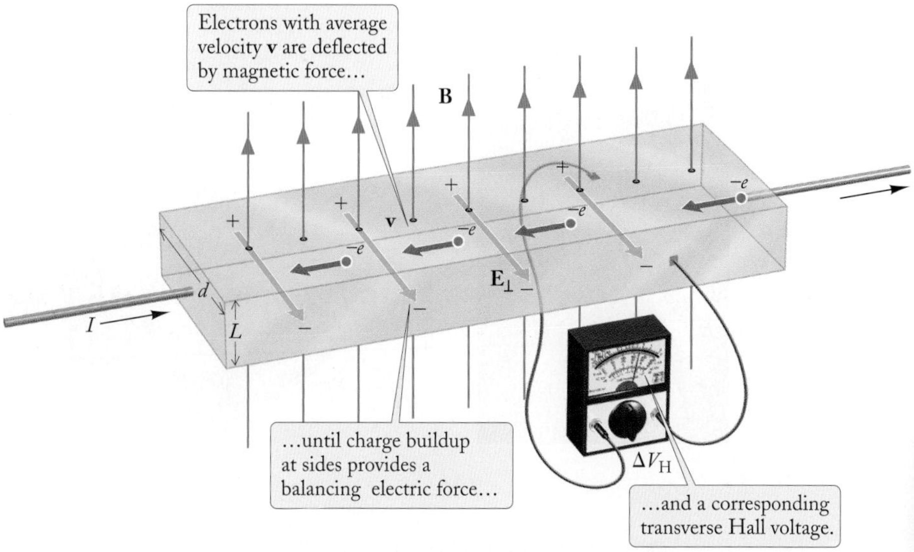

FIGURE 30.23 Geometry used for the Hall effect. A longitudinal current I flows down the length of a conductor in a perpendicular magnetic field **B**; a transverse Hall voltage ΔV_H is generated.

EDWIN HERBERT HALL (1855–1938)
American physicist. While working on his Ph.D. thesis, Hall investigated a question Maxwell had posed: Does the magnetic force act on the conductor or the current? He demonstrated that the force was indeed on the current, and realized that electric charge was pushed to one side of the conductor. His observation of the generation of a transverse electric field in materials carrying current in a magnetic field, or Hall effect, provides the basis for many modern instruments and for our understanding of conduction in solids.

If the width of the rod is d, the transverse electric field produces a potential difference between the sides of the rod known as the **Hall voltage** ΔV_H, where

$$\Delta V_H = E_\perp d$$

or, with Eq. (30.27),

$$\Delta V_H = vBd \tag{30.28}$$

The average velocity v of the charges is related to the current I; for electrons, it has a magnitude given by Eq. (27.10),

$$v = \frac{I}{neA} \tag{30.29}$$

where n is the density of electrons and A is the cross-sectional area of the rod. The Hall voltage is thus

$$\Delta V_H = \frac{IBd}{neA}$$

If the thickness of the rod is L, then the area is $A = Ld$, which gives

Hall voltage

$$\Delta V_H = \frac{IB}{neL} \tag{30.30}$$

Thus the Hall voltage is proportional to the applied current and magnetic field, and is inversely proportional to the density of charge carriers and the thickness of the conductor.

The Hall effect is used to determine the charge carrier density n in a material, as well as the sign of the charge carriers. The Hall effect provides empirical evidence that the charge carriers in most metals are negative charges, that is, electrons. In some

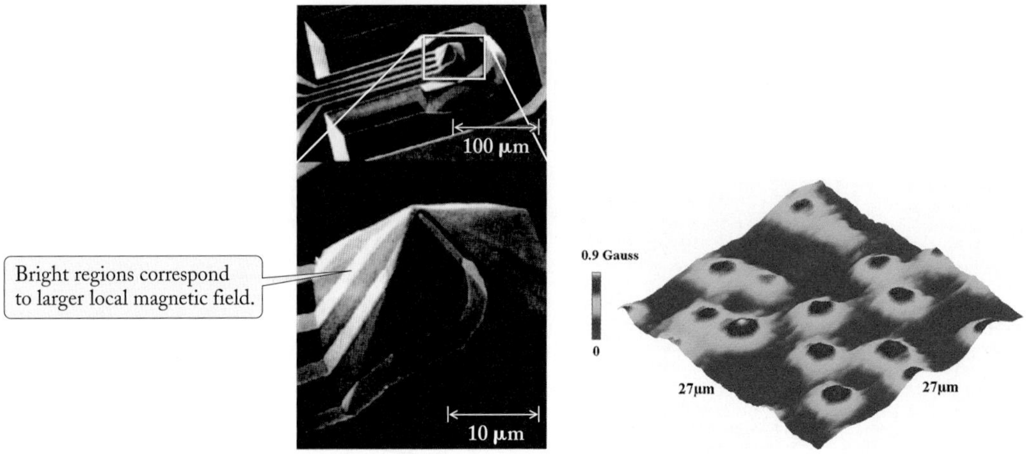

FIGURE 30.24 (a) A microfabricated Hall sensor. (b) A scanning Hall microscope image of the magnetic field as a function of lateral position just above a high-temperature superconductor. The red features represent individual superconducting vortices.

materials with mobile ions, the sign of the Hall voltage identifies positive ions as the charge carriers. Even in some metals, such as aluminum at high magnetic fields, the Hall effect can reveal that the charge carriers are effectively positive. Such "anomalous" behavior is understood in terms of the collective behavior of electrons moving in the electrostatic potential of the crystal lattice of ions (see Section 39.4); the effectively positive charges are known as **holes**.

hole

For metals, with large values of n, Eq. (30.30) tells us that the Hall voltage will be small unless thin wires are used. But for semiconductors, with small values of n, the Hall voltage is large, and thus very sensitive to changes in B. Semiconductors are used as **Hall sensors**, both in simple instruments that measure B, and in devices such as the scanning Hall microscope, where a microfabricated Hall sensor (see Fig. 30.24a) detects small local changes in B. Such a mapping of the local field is shown in Fig. 30.24b, which depicts the magnetic field as a function of position above the surface of a high-temperature superconductor.

Hall sensor

A related phenomenon is observed in thin, nearly two-dimensional samples at very low temperatures. Modern quantum physics reveals that the cyclotron orbits discussed in Section 30.1, like many other quantities, can actually take on only certain discrete sizes. Quantum calculations show that the number of allowed orbits in which electrons may travel, and thus the allowed values of the transverse electric field in the material, depends on the magnetic field B. This gives rise to the **quantum Hall effect**, where precisely spaced steps are observed in the Hall resistance, $\Delta V_H/I$, as a function of the field B, instead of the simple linear relation predicted by Eq. (30.30); both behaviors are shown in Fig. 30.25. These steps occur at exactly inverse-integer multiples of a fundamental Hall resistance given by $h/e^2 = 25\,816.802\ \Omega$, where h is Planck's constant (see Section 30.4 and Chapter 37). The quantum Hall effect permits the precise determination of fundamental constants and also provides an easily reproducible; fundamental standard of resistance.

quantum Hall effect

EXAMPLE 7 A flat metal ribbon of width 2.0 mm and thickness 150 μm is immersed in a perpendicular uniform magnetic field of strength 8.2 T. A current of 1.5 A is passed along the length of the ribbon, and a transverse voltage equal to 375 μV is measured across its width. What is the average velocity of the electrons in the metal? What is the number of conduction electrons per unit volume in this metal?

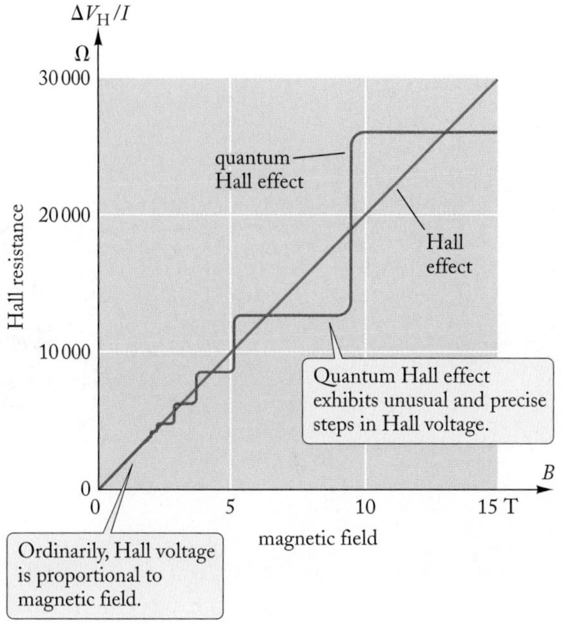

FIGURE 30.25 A plot of the Hall resistance $\Delta V_H/I$ as a function of magnetic field. The straight line is for an ordinary conductor; the stepped curve is for a two-dimensional conductor at low temperatures (the quantum Hall effect).

SOLUTION: The given transverse voltage in a uniform magnetic field is just the Hall voltage, $\Delta V_H = 375\ \mu\text{V}$. The drift velocity can be obtained immediately from Eq. (30.28):

$$v = \frac{\Delta V_H}{Bd} = \frac{375 \times 10^{-6}\ \text{V}}{8.2\ \text{T} \times 2.0 \times 10^{-3}\ \text{m}}$$
$$= 0.023\ \text{m/s}$$

The number of electrons per unit volume, or the charge carrier density, is given by Eq. (30.30).

$$n = \frac{IB}{\Delta V_H e L} = \frac{1.5\text{A} \times 8.2\ \text{T}}{375 \times 10^{-6}\ \text{V} \times 1.6 \times 10^{-19}\ \text{C} \times 150 \times 10^{-6}\ \text{m}}$$
$$= 1.4 \times 10^{27}\ \text{m}^{-3}$$

Checkup 30.5

QUESTION 1: Two wires of the same material are in the same uniform magnetic field, and each carries the same current. If the first wire is 10 times thicker than the second (in the direction of the magnetic field), how do their Hall voltages differ?

QUESTION 2: The Hall effect geometry of Fig. 30.23 shows the case for negative charge carriers. For the current and magnetic field directions shown, consider instead positive charge carriers. Which vectors would reverse direction?

(A) **v** only (B) $\mathbf{E}_\perp$ only (C) Both **v** and $\mathbf{E}_\perp$ (D) Neither **v** nor $\mathbf{E}_\perp$

SUMMARY

PHYSICS IN PRACTICE Magnetic Recording Media **(page 978)**

CIRCULAR ORBIT IN MAGNETIC FIELD

$$r = \frac{mv}{qB} \qquad \textbf{(30.5)}$$

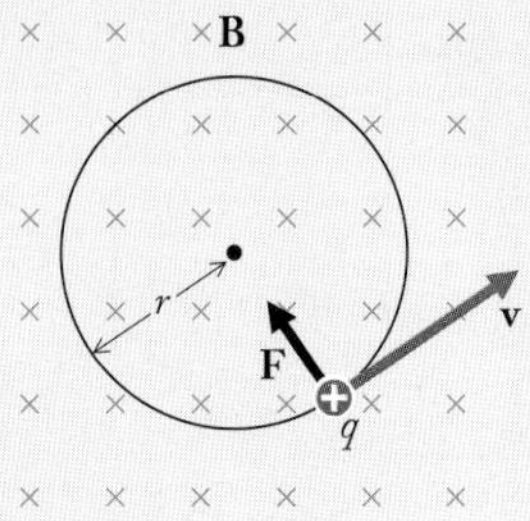

CYCLOTRON FREQUENCY

$$f = \frac{qB}{2\pi m} \qquad \textbf{(30.7)}$$

FORCE ON WIRE SEGMENT

$$dF = I\,dl\,B\sin\alpha$$

or

$$d\mathbf{F} = I\,d\mathbf{l} \times \mathbf{B} \qquad \textbf{(30.9, 30.11)}$$

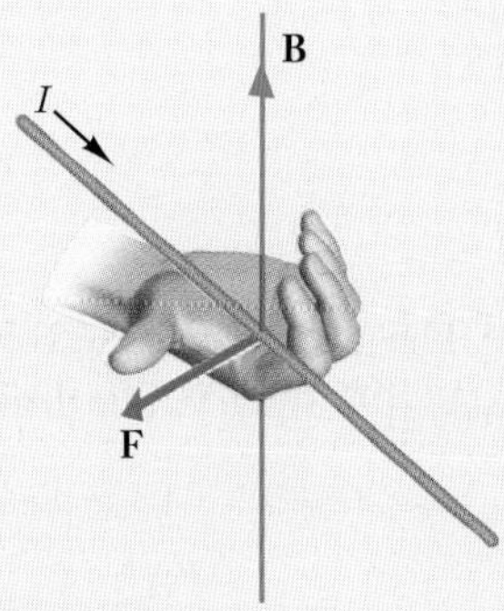

FORCE ON A STRAIGHT WIRE

$$F = ILB\sin\alpha \quad \text{or} \quad \mathbf{F} = I\mathbf{L} \times \mathbf{B} \qquad \textbf{(30.10, 30.11)}$$

FORCE PER UNIT LENGTH BETWEEN PARALLEL WIRES

$$\frac{F}{l} = \frac{\mu_0}{2\pi}\frac{I_1 I_2}{r} \qquad \textbf{(30.12)}$$

MAGNETIC DIPOLE MOMENT OF CURRENT LOOP

$$\mu = [\text{current}] \times [\text{area}] = IA \qquad \textbf{(30.15)}$$

TORQUE ON CURRENT LOOP

$$\boldsymbol{\tau} = \boldsymbol{\mu} \times \mathbf{B} \qquad \textbf{(30.18)}$$

magnitude of torque

$$\tau = \mu B \sin\theta \qquad \textbf{(30.17)}$$

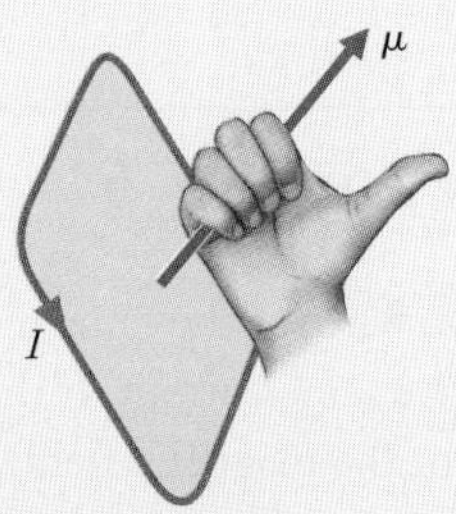

POTENTIAL ENERGY OF CURRENT LOOP

$$U = -\boldsymbol{\mu} \cdot \mathbf{B} = -\mu B\cos\theta \qquad \textbf{(30.21, 30.22)}$$

BOHR MAGNETON

$$\mu_B = \frac{eh}{4\pi m_e} = 9.27 \times 10^{-24}\,\text{J/T} \tag{30.23}$$

MAGNETIC SUSCEPTIBILITY

$$\chi = \frac{B_{\text{matter}}}{B_{\text{external}}} \tag{30.24}$$

χ positive and $\chi \ll 1$: paramagnet

χ positive and $\chi \gg 1$: ferromagnet

χ negative: diamagnet

MAGNETIC PERMEABILITY

$$\mu = \mu_0(1 + \chi) \tag{30.25}$$

HALL VOLTAGE

where n is the number density of charges and d and L are the width and thickness of the conductor.

$$\Delta V_H = vBd = \frac{IB}{neL} \tag{30.28, 30.30}$$

QUESTIONS FOR DISCUSSION

1. Cosmic rays are high-speed charged particles—mostly protons—that crisscross interstellar space and strike the Earth from all directions. Why is it easier for the cosmic rays to penetrate through the magnetic field of the Earth near the poles than anywhere else?
2. Is it possible to define a magnetic potential energy for a charged particle moving in a magnetic field?
3. If we want a proton to orbit all the way around the equator in the Earth's magnetic field, must we send it eastward or westward?
4. Consider the strip of metal placed in a magnetic field, as shown in Fig. 30.23. How does the Hall potential difference between the sides of the strip change if we reverse the current? If we reverse the magnetic field? If we reverse both?
5. If we replace the metallic strip in Fig. 30.23 by a semiconducting strip of p-type semiconductor, that is, a material with positive charge carriers, will the back side remain at the higher potential?
6. The Earth's magnetic field at the equator is horizontal and in the northward direction. Assume that the atmospheric electric field is downward. The electric and magnetic fields are then "crossed" fields. In what direction must we launch an electron if the magnetic and the electric forces are to balance?
7. If a strong current flows through a thick wire, it tends to cause a compression of the wire. Explain.
8. If a wire carrying a current is placed in a magnetic field, the field exerts forces on the moving electrons in the wire. How does this cause a push on the wire?
9. A horizontal wire carries a current in the eastward direction. The wire is located in a magnet producing a uniform magnetic field. What must be the direction of the field if it is to compensate for the weight of the wire?
10. In the SI system, we first define the unit of current (by means of the magnetic force between wires), and we then define the unit of charge in terms of the current. Could we proceed in the opposite manner: first define the unit of electric charge (by means of the electric force between charges), and then define the unit of current as a flow of charge? What would be the advantages and the disadvantages?
11. Suppose that an infinitely long straight wire lies along the axis of a circular loop. Both the wire and the loop carry currents. Explain why the force from the wire on the loop is zero, and explain why the force from the loop on the wire is also zero.
12. Positive charge is uniformly distributed over a sphere. If this sphere spins about a vertical axis in a counterclockwise direction as seen from above, what is the direction of the magnetic moment vector?
13. According to Eq. (30.15), the magnetic moment of a square loop is the same as the magnetic moment of a circular loop of the same area. Why is the shape unimportant? [Hint: The circular loop can be regarded as made of many small (infinitesimal) square loops.]

PROBLEMS

†30.1 Circular Motion in a Uniform Magnetic Field

1. A bubble chamber, used to make the tracks of protons and other charged particles visible, is placed between the poles of a large electromagnet that produces a uniform magnetic field of 2.0 T. A high-energy proton passing through the bubble chamber makes a track that is an arc of a circle of radius 3.5 m. According to Eq. (30.6), what is the momentum of the proton?
2. A proton of kinetic energy 1.0×10^7 eV moves in a circular orbit in the magnetic field near the Earth. The strength of the field is 5.0×10^{-5} T. What is the radius of the orbit?
3. In principle, a proton of the right energy can orbit the Earth in an equatorial orbit under the influence of the Earth's magnetic field. If the orbital radius is to be 6.5×10^3 km and the magnetic field at this radius is 3.3×10^{-5} T, what must be the momentum of the proton? Ignore gravity in this problem.
4. In the Crab Nebula (the remnant of a supernova explosion), electrons of a momentum of up to about 10^{-16} kg·m/s orbit in a magnetic field of about 10^{-8} T. What is the orbital radius of such electrons? Note that it is necessary to use Eq. (30.6) for the calculation of the orbital radius.
5. At the Fermilab accelerator, protons of momentum 5.3×10^{-16} kg·m/s are held in a circular orbit of diameter 2.0 km by a vertical magnetic field. What is the strength of the magnetic field required for this?
6. What is the charge-to-mass ratio of a particle that orbits in a magnetic field of 0.200 T with an angular frequency of 3.51×10^{10} rad/s?
7. In a bubble chamber, a particle is observed to move in a circular orbit of radius 1.2 m. The magnetic field in the bubble chamber has a magnitude of 2.0 T. Assuming the charge of the particle is e, what is the magnitude of its momentum?
8. In a **mass spectrometer**, ions of a precisely selected speed are sent into a uniform magnetic field, where they move along a semicircle and then strike a photographic plate, making a mark (Fig. 30.26). A careful measurement of the distance of the mark from the entrance slit gives us the diameter (and the radius) of the circular motion of the ion in the magnetic field, and from this radius the mass can be calculated by means of Eq. (30.5). Suppose that the magnetic field in the mass spectrometer is 0.050 T. Suppose that an ion of charge e and speed 4.0×10^4 m/s is found to strike the photographic plate at a distance of 0.332 m from the entrance slit. What is the mass of this ion?

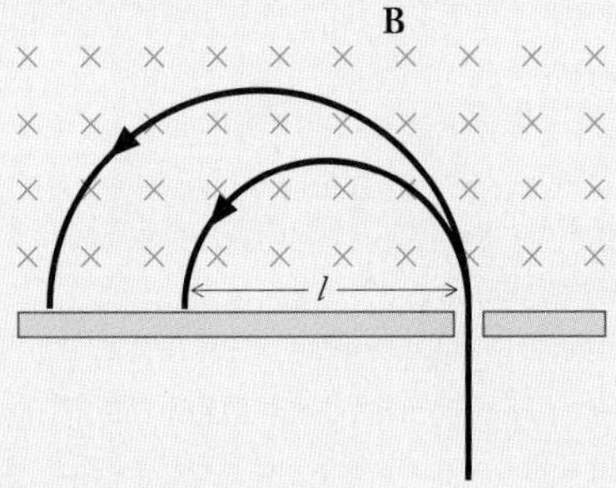

FIGURE 30.26 Paths of ions of different masses in a mass spectrometer. The crosses show the tails of the magnetic field vectors.

9. Astrophysicists believe that the radio waves of 10^9 Hz reaching us from Jupiter are emitted by electrons of fairly low (nonrelativistic) energies orbiting in Jupiter's magnetic field. What must be the strength of this field if the cyclotron frequency is to be 1.0×10^9 Hz?
10. Figure 30.27 shows the tracks of an electron (charge $-e$) and an antielectron (charge $+e$) created in a bubble chamber. When the particles made these tracks, they were under the influence of a magnetic field of magnitude 1.0 T and of a direction perpendicular to and into the plane of the figure. What is the momentum of each particle? Assume that they are moving in the plane of the figure, and that this figure is $\frac{1}{10}$ natural size. Which is the track of the electron, and which is the track of the antielectron?

FIGURE 30.27 Tracks of an electron and an antielectron in a bubble chamber. The tracks spiral because the particles suffer a loss of energy as they pass through the liquid. For the purposes of Problem 10, concentrate on the initial portions of the tracks.

11. A beam containing the nuclei of two isotopes of carbon, ^{12}C (6 protons and 6 neutrons) and ^{14}C (6 protons and 8 neutrons), at the same momentum enters a magnetic field, which bends the trajectories of these two kinds of particles into circles of radii r_{12} and r_{14}, respectively. What is the ratio of the two radii?

†For help, see Online Concept Tutorial 33 at www.wwnorton.com/physics

12. In the Brookhaven AGS accelerator, protons are made to move around a circle of radius 128 m by the magnetic force exerted by a vertical magnetic field. The maximum field that the magnets of this accelerator can generate is 1.3 T.
(a) Calculate the maximum permissible momentum of the protons.
(b) Calculate the orbital frequency of such protons.

*13. You want to confine an electron of energy 3.0×10^4 eV by making it circle inside a solenoid of radius 10 cm under the influence of the force exerted by the magnetic field. The solenoid has 120 turns of wire per centimeter. What minimum current must you put through the wire if the electron is not to hit the wall of the solenoid?

*14. The Earth's magnetic field at the equator has a magnitude of 5.0×10^{-5} T; its direction is horizontal toward the north. Suppose that the Earth's atmospheric electric field is 100 V/m; its direction is vertically down. With what velocity (magnitude and direction) must we launch an electron if the electric force is to cancel the magnetic force?

*15. A high-speed electron is moving in a uniform magnetic field of 5.0×10^{-4} T. The electron has a velocity component of 6.0×10^6 m/s perpendicular to the magnetic field and a velocity component of 4.0×10^6 m/s parallel to the magnetic field.
(a) Describe the motion of the electron qualitatively.
(b) What are the cyclotron frequency and the corresponding period of the transverse motion?
(c) How far does the electron advance parallel to the magnetic field in one such period?

*16. Suppose that a region with a uniform magnetic field **B** also has a uniform electric field **E** perpendicular to the magnetic field, an arrangement called **crossed fields**. Show that for a charged particle moving in such crossed fields in a direction perpendicular to both **E** and **B**, the electric force cancels the magnetic force, provided the particle has a speed

$$v = \frac{E}{B}$$

If the magnetic field is in the vertical upward direction and the electric field is in the northward direction, what must be the direction of motion of the charged particle to produce this cancellation? Crossed fields are used in **velocity selectors**, or velocity filters. In these devices, a beam of charged particles is aimed into a region of crossed fields, and this deflects particles out of the beam, except for those particles that have the critical velocity $v = E/B$.

*17. A velocity selector with crossed **E** and **B** fields (see Problem 16) is to be used to select alpha particles of energy 2.0×10^3 eV from a beam containing alpha particles of several energies. The electric field strength is 1.0×10^6 V/m. What must be the magnetic field strength? Alpha particles have charge $+2e$ and mass 4.0 u.

*18. In a mass spectrometer, a beam of ions is first made to pass through a velocity selector with crossed fields **E** and **B** (see Problem 16). The selected ions are then made to enter a region of uniform magnetic field **B**′, where they move in arcs of circles (Fig. 30.28). The radii of these circles depend on the masses of the ions. Assume that an ion has a single charge e.
(a) Show that in terms of E, B, B', and the impact distance l marked in Fig. 30.28, the mass of the ion is

$$m = \frac{eBB'l}{2E}$$

(b) Assume that in such a spectrometer, ions of the isotope ^{16}O impact at a distance of 29.20 cm and ions of a different isotope of oxygen impact at a distance of 32.86 cm. The mass of the ions of ^{16}O is 16.00 u. What is the mass of the other isotope?

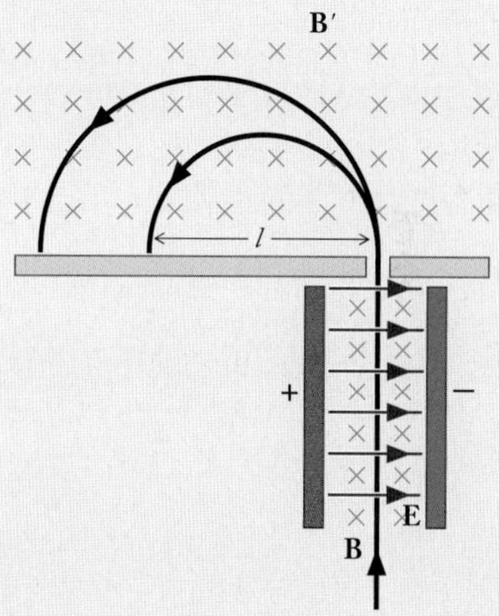

FIGURE 30.28 A mass spectrometer with a velocity selector.

*19. In a uniform magnetic field, an electron and a proton are orbiting in opposite directions along the same circle perpendicular to the magnetic field:
(a) If the speed of the electron is 3.0×10^6 m/s, what must be the speed of the proton? Assume that the only force acting on the magnetic particles is the magnetic force.
(b) If the particles are initially at the same point on the circle (moving in opposite directions), where on the circle will they meet again? Give the answer as an angle, in radians, measured along the circle.

30.2 Force on a Wire

20. A straight wire is placed in a uniform magnetic field; the wire makes an angle of 30° with the magnetic field. The wire carries a current of 6.0 A, and the magnetic field has a strength of 0.40 T. Calculate the force on a 10-cm segment of this wire. Show the direction of the force in a diagram.

21. The electric cable supplying an electric clothes dryer consists of two long straight wires separated by a distance of 1.2 cm. Opposite currents of 20 A flow on these wires. What is the magnetic force experienced by a 1.0-cm segment of wire due to the entire length of the other wire?

22. Two parallel cables of a high-voltage power line carry opposite currents of 1.8×10^3 A. The distance between the cables is 4.0 m. What is the magnetic force pushing on a 50-cm segment of one of these cables? Treat both cables as very long, straight wires.

*23. A closed loop of arbitrary (possibly three-dimensional) shape is placed near a very long, straight wire. Currents I and I', respectively, flow around the loop and along the straight wire. Prove that the force exerted by the straight wire on the loop must be perpendicular to the wire. (Hint: Use Newton's Third Law.)

*24. A straight segment of wire carries a current of 15 A from the point $\mathbf{r}_1 = 0.35\mathbf{i} + 0.50\mathbf{j}$ to the point $\mathbf{r}_2 = 0.50\mathbf{i} - 0.25\mathbf{j} + 0.40\mathbf{k}$, where distances are measured in meters. The wire segment is in a uniform magnetic field of 2.0 T parallel to the positive z axis. What is the magnetic force on the wire?

*25. A long wire carries current of 25 A parallel to the positive x axis, except for three of the four segments that follow the edges of a cube of side 0.25 m, as shown in Fig. 30.29. The wire is in a uniform magnetic field of 2.0 T directed parallel to the positive x axis. What is the net magnetic force on the wire?

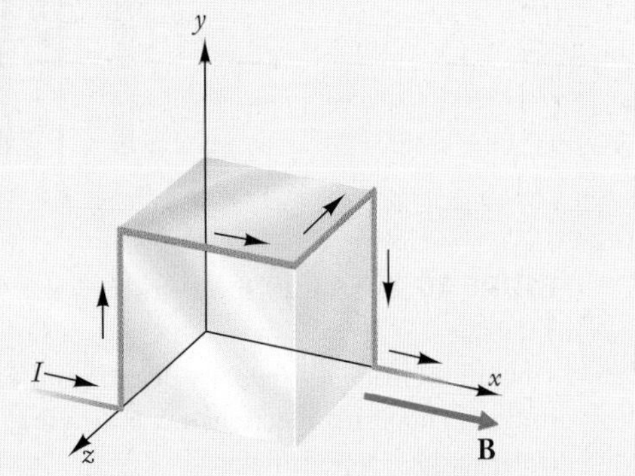

FIGURE 30.29 A long wire bent along four edges of a cube.

*26. Consider the long, straight wire and the U-shaped wire shown in Fig. 30.30. The wires lie in the same plane. The bottom of the U has a length l, and the sides of the U are very long. The wires are separated by a distance d, and they carry currents I and I', respectively. What is the force that the straight wire exerts on the U-shaped wire?

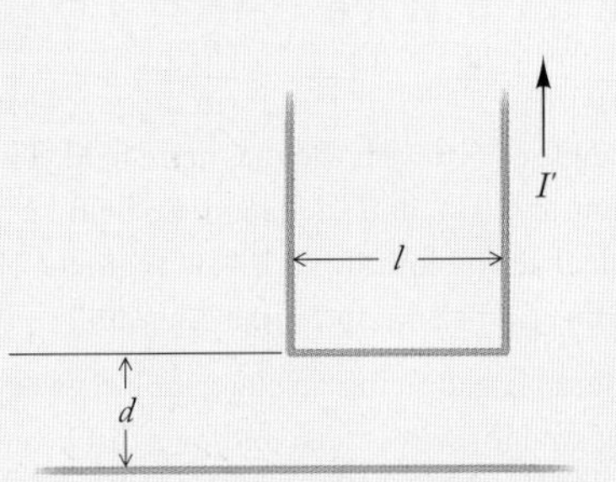

FIGURE 30.30 Straight wire and U-shaped wire.

*27. An electromagnetic launcher, or rail gun, consists of two parallel conducting rails across which is laid a conducting bar, which serves as a projectile. To launch this projectile, the rails are immersed in a magnetic field, and a current is sent through the rails and the bar (see Fig. 30.31). The magnetic force on the current in the bar then accelerates the bar (in actual rail guns of this kind, the magnetic field is itself produced by the current in the rails; but let us ignore this complication). Suppose that the magnetic field has a strength of 0.20 T and the bar has a length of 0.10 m and a mass of 0.20 kg. Ignore friction. What current must you send through the bar to give it an acceleration of 1.0×10^5 m/s^2?

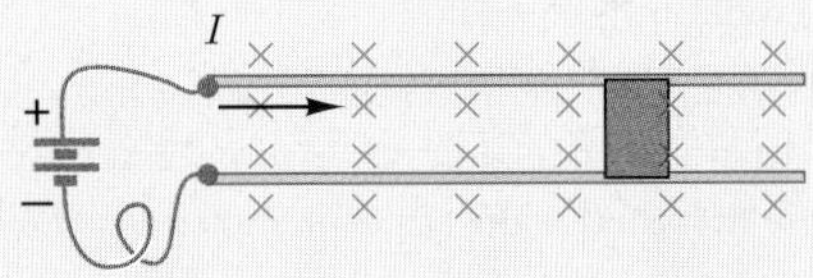

FIGURE 30.31 Electromagnetic launcher.

**28. A thin flexible wire carrying a current I hangs in a uniform magnetic field $\mathbf{B}$ (Fig. 30.32). A weight attached to one end of the wire provides a tension T. Within the magnetic field, the wire will adopt the shape of an arc of a circle.

(a) Show that the radius of this arc of a circle is $r = T/(BI)$.

(b) Show that if we remove the wire and launch a particle of charge $-q$ from the point P with a momentum $p = qT/I$ along the direction of the wire, it will move along the same arc of a circle. (This means that the wire can be used to simulate the orbit of the particle. Experimental physicists sometimes use such wires to check the orbits of particles through systems of magnets.)

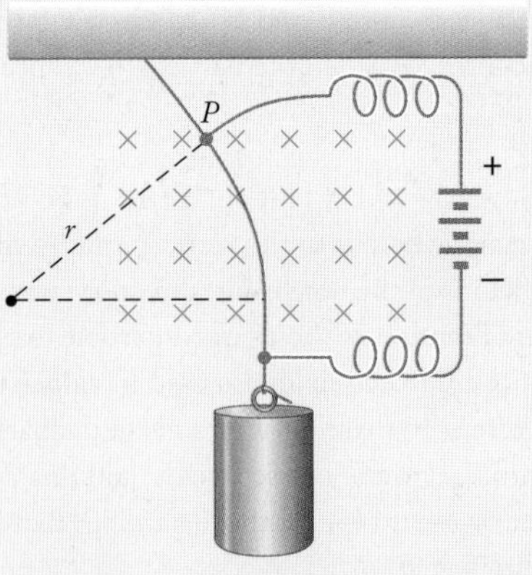

FIGURE 30.32 A wire with a weight hanging in a magnetic field.

30.3 Torque on a Loop

29. The coil in the mechanism of an ammeter is a rectangular loop, measuring 1.0 cm × 2.0 cm, with 120 turns of wire. The coil is immersed in a magnetic field of 0.010 T. What is the torque on this coil when its plane is parallel to the magnetic field and carries a current of 1.0×10^{-3} A?

30. A horizontal circular loop of wire of radius 20 cm carries a current of 25 A. At the location of the loop, the magnetic field of the Earth has a magnitude of 3.9×10^{-5} T and points down at an angle of 16° with the vertical. What is the magnitude of the torque that this magnetic field exerts on the loop?

31. The proton has a magnetic moment of 1.41×10^{-26} A·m².
    (a) If the proton is placed in a uniform magnetic field of 0.80 T, plot the torque as a function of the angle θ.
    (b) If the magnetic moment is initially oriented antiparallel to the field, how much energy will be released when the proton flips into the parallel orientation?

32. Consider the electric motor of Example 4. What is the average value of the torque over a complete rotation? If this motor rotates at the rate of 50 rev/s, what is the average horsepower that it delivers to its axle?

33. A circular coil consists of 60 turns of wire wound around the rim of a plastic disk of radius 2.0 cm. What is the magnetic moment of this coil if the current in the wire is 0.30 A?

*34. A pipe made of a superconducting material has a length of 0.30 m and a radius of 4.0 cm. A current of 4.0×10^3 A flows around the surface of the pipe; the current is uniformly distributed over the surface. What is the magnetic moment of this current distribution? (Hint: Treat the current distribution as a large number of rings stacked one on top of another.)

*35. An amount of charge Q is uniformly distributed over a disk of paper of radius R. The disk spins about its axis with angular velocity ω. Find the magnetic dipole moment of the disk. Sketch the lines of magnetic field and of electric field in the vicinity of the disk.

**36. A rectangular loop of wire of dimension 12 cm × 25 cm faces a long, straight wire. The two long sides of the loop are parallel to the wire, and the two short sides are perpendicular; the midpoint of the loop is 8.0 cm from the wire (Fig. 30.33). Currents of 95 A and 70 A flow in the straight wire and the loop, respectively.
    (a) What translational force does the straight wire exert on the loop?

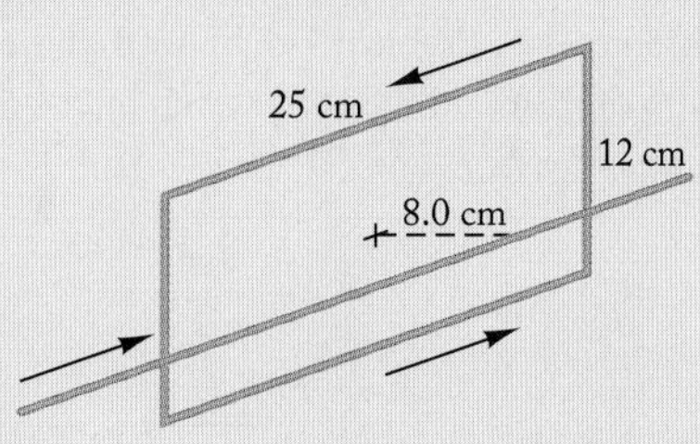

FIGURE 30.33 Long, straight wire and rectangular loop.

    (b) What torque about an axis parallel to the straight wire and through the center of the loop does the straight wire exert on the loop?

**37. A compass needle is attached to an axle that permits it to turn freely in a horizontal plane so that only the horizontal component of the magnetic field affects its motion. The magnetic moment of the needle is 9.0×10^{-3} A·m², its moment of inertia is 2.0×10^{-8} kg·m², and the horizontal component of the Earth's magnetic field is 1.9×10^{-5} T.
    (a) What is the torque on the needle as a function of the angle θ between the needle and the north direction?
    (b) What is the frequency of small oscillations of the needle about the north direction? [Hint: Compare the equation of the rotational motion of the needle with Eq. (15.39) for the motion of a pendulum.]

30.4 Magnetism in Materials

38. Because of its relatively large susceptibility, the paramagnetic metal palladium is used as a calibration standard for magnetic measurements. At room temperature, palladium has a susceptibility of 7.9×10^{-4}. What is the relative permeability of palladium (its permeability in terms of μ_0)?

39. The **magnetization** M of a material is the magnetic field due to the material divided by μ_0; that is, $M = B_{matter}/\mu_0$. The magnetization also defines the magnetic moment per unit volume, or M = [magnetic moment]/[volume]. If iron has a maximum internal field of B_{matter} = 2.19 T, calculate (a) the magnetization of iron, and (b) the magnetic moment per atom of iron (the density of iron atoms is 8.50×10^{28}/m³).

40. A certain long solenoid has 320 turns per meter and carries a current of 5.00 A. Determine the total field inside the solenoid when it is filled with (a) iron, which has a relative permeability of 200 under these conditions, and (b) permalloy, which has a relative permeability of 1500 under these conditions.

41. Some simple elemental superconductors (called type I) are perfect diamagnets, meaning that their internal magnetic field exactly cancels the applied field, or $B_{matter} = -B_{external}$. What are (a) the susceptibility and (b) the permeability of such superconductors?

*42. In so-called type I superconductors, surface currents fully shield the external magnetic field, ensuring zero net field in the interior of the superconductor; such perfect diamagnetism is known as the **Meissner effect**. A 10-cm-long superconducting cylindrical rod of indium metal has its axis parallel to an external field of 2.7×10^{-2} T; the material is a perfect diamagnet. What is the total current flowing around the rod? Such a current is confined to a thin surface layer with a thickness known as the **penetration depth**. For simplicity, assume that the surface current flowing around the rod is uniformly distributed over a penetration depth of 5.0×10^{-8} m. What is the current density in this layer?

30.5 The Hall Effect

43. A copper wire of diameter 3.0 mm is placed at right angles to a magnetic field of 2.0 T. The wire carries a current of 50 A. What is the Hall potential difference between opposite sides of the wire?

44. Repeat the preceding problem, with the assumption that the wire is placed at an angle of 60° with respect to the magnetic field.

45. A Hall sensor made from silicon is 0.30 mm thick and has a charge carrier density of $2.0 \times 10^{19}/\text{m}^3$. When it is placed in an unknown perpendicular magnetic field and carries a current of 5.0 mA along its length, a Hall voltage of 130 mV is measured across its width. What is the value of the magnetic field?

46. A 0.10-mm-thick strip of calcium metal is placed in a perpendicular magnetic field of 8.0 T. When a current of 100 mA is passed along its length, a Hall voltage of 5.4 μV is measured across its width. If the charge carriers in calcium are electrons, how many free electrons per calcium atom are there? (Calcium metal has 2.3×10^{28} atoms/m^3.)

47. A thin strip of rubidium metal, 65 μm thick and 1.0 mm wide, is placed in a field of 6.0 T. A current of 0.30 A is passed along its length. Rubidium has a density of 1.53 g/cm^3, a molar weight of 85.5 g/mole, and one conduction electron per atom.

    (a) Find the drift velocity of these conduction electrons.

    (b) What Hall voltage is measured across this strip?

48. A metal ribbon is 2.0 mm wide and 0.15 mm thick, and carries a current of 1.5 A along its length. In a perpendicular magnetic field of 2.0 T, a Hall voltage of 150 μV is measured across the width of the ribbon.

    (a) What is the number density of charge carriers in this metal?

    (b) What is the drift velocity of these electrons?

49. Related to the Hall voltage is a quantity known as the **Hall coefficient** R_H, which takes the value $R_H = -1/ne$ when the charge carriers are free electrons. What is the value of R_H for potassium? (Potassium has one conduction electron per atom, a density of 0.86 g/cm^3, and a molar weight of 39.1 g/mole.)

REVIEW PROBLEMS

50. An electromagnet produces a magnetic field of 0.60 T in the vertical downward direction. An electron of speed 6.0×10^6 m/s is moving through this magnetic field. What are the magnitude and the direction of the magnetic force on the electron if the electron is moving

    (a) Vertically downward?

    (b) Horizontally toward the east?

    (c) Horizontally toward the south?

    (d) Toward the south and downward at an angle of 30° from horizontal?

51. An electromagnet produces a magnetic field of 0.60 T in the vertical downward direction. A straight wire placed in this magnetic field carries a current of 3.0 A. Calculate the magnitude and direction of the magnetic force on a 0.10-m segment of this wire if the orientation is

    (a) Vertically downward

    (b) Horizontally toward the east

    (c) Horizontally toward the south

    (d) Toward the south and downward at an angle of 30° from horizontal

52. A piece of wire bent to form two perpendicular straight segments carries a current of 5.0 A, first along the x axis from $x = 0.50$ m to the origin, and then along the y axis from the origin to $y = -0.25$ m. The wire segment is in a uniform magnetic field of 1.5 T parallel to the positive z axis. What are the magnitude and direction of the magnetic force on the wire?

53. A proton of speed 4.0×10^5 m/s is seen to move in a circular orbit of radius 0.40 m in a magnetic field. What is the strength of the magnetic field that will give such circular motion? What is the frequency of the motion of the proton?

54. A circular loop of radius $r = 3.0$ cm has a total charge $Q = 5.0\ \mu$C uniformly distributed around its circumference. The loop rotates about its axis with angular velocity $\omega = 6.3 \times 10^4$ rad/s. What is the magnetic moment of the loop?

*55. In 1897, **J. J. Thomson's e/m experiment** determined the charge-to-mass ratio of the electron. First, a beam of electrons of an unknown velocity v was deflected a distance y by being passed through a region of length L of uniform electric field E.

    (a) Show that this deflection distance is (see Section 23.4)

$$y = \frac{eEL^2}{2m_e v^2}$$

    (b) Then a magnetic field B was turned on perpendicular to the electric field and increased until no deflection was observed (see Problem 16). Show that the charge-to-mass ratio of the electron is then given by the experimental values of y, E, L, and B:

$$\frac{e}{m_e} = \frac{2yE}{B^2L^2}$$

*56. A high-speed electron is moving in a uniform magnetic field of 5.0×10^{-4} T and a parallel uniform electric field of 2.0×10^3 V/m. The electron initially has a velocity component of 6.0×10^6 m/s perpendicular to the magnetic and electric fields and a velocity component of 4.0×10^6 m/s parallel to the magnetic and electric fields.

(a) Describe the motion of the electron qualitatively.

(b) What are the cyclotron frequency and the corresponding period of the transverse motion?

(c) How far does the electron advance parallel to the magnetic and electric fields in the first such period? By how much does its parallel velocity increase in any such period?

*57. Two positively charged particles separated by a distance d, each with a charge q and a mass m, are initially moving with the same speed v in opposite directions perpendicular to the line joining them (see Fig. 30.34). A magnetic field applied perpendicularly to the plane of the page will bend the paths of the particles into circles. What strength of magnetic field is necessary to make them collide head-on midway between the two starting points? (Ignore the electric forces between the charges.)

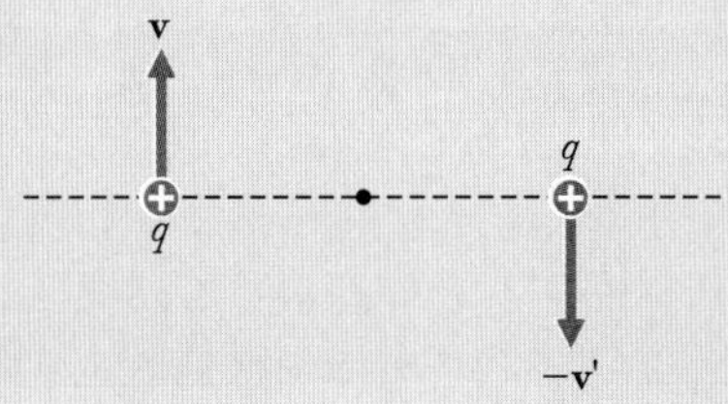

FIGURE 30.34 Two positively charge particles in a magnetic field.

*58. A rectangular loop of wire of dimensions 12 cm × 18 cm is near a long, straight wire. The rectangle and the straight wire lie in the same plane. One of the short sides of the rectangle is parallel to the straight wire and at a distance of 6.0 cm; the long sides are perpendicular to the straight wire (Fig. 30.35). A current of 40 A flows on the straight wire, and a current of 60 A flows around the loop. What are the magnitude and direction of the net magnetic force that the straight wire exerts on the loop?

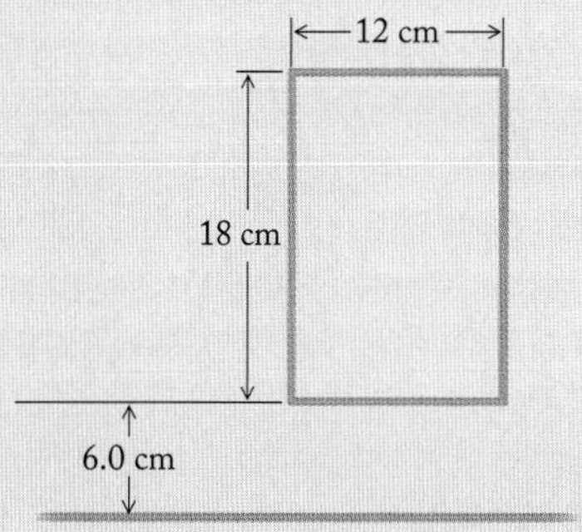

FIGURE 30.35 Long, straight wire and rectangular loop.

*59. A 25-turn circular coil has a mass of 0.050 kg. The coil is immersed in a uniform 0.20-T magnetic field, and is initially held so the plane of the coil is parallel to the magnetic field (see Fig. 30.36). The coil carries 5.0 A of current. When released, what will be the instantaneous angular acceleration of the coil about the horizontal axis?

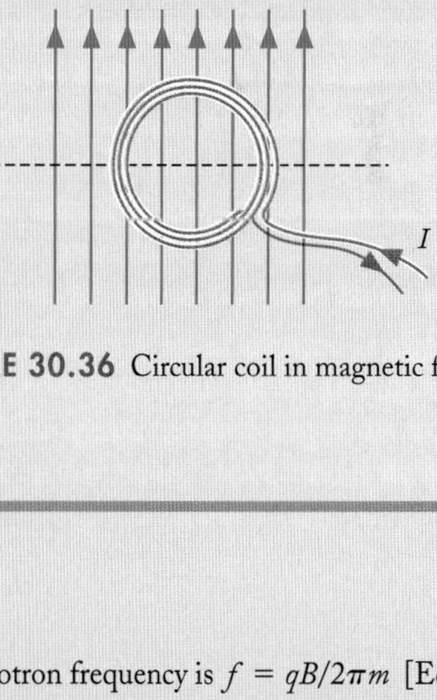

FIGURE 30.36 Circular coil in magnetic field.

Answers to Checkups

Checkup 30.1

1. The radius of circular motion for a charged particle in a magnetic field is given by Eq. (30.5): $r = mv/qB$. Since both electrons are in the same magnetic field, m, q, and B are the same, and the radius varies in proportion to the speed. Thus the second electron has twice the orbital radius of the first, or 30 cm.

2. Since the magnetic force is always perpendicular to the velocity $\mathbf{v} = d\mathbf{s}/dt$, it is always perpendicular to the displacement $d\mathbf{s}$. Thus the incremental amount of work, $dW = \mathbf{F} \cdot d\mathbf{s} = Fds \times \cos\theta$, is always zero, and so the magnetic field can do no work.

3. Since the cyclotron frequency is $f = qB/2\pi m$ [Eq. (30.7)], a smaller mass results in a larger cyclotron frequency for the electron.

4. Yes, any charged particle could be accelerated in a cyclotron. In a given magnetic field, a cyclotron used for electrons would operate at higher frequency (higher by the proton–electron mass ratio). Also, the electrons circulate in the opposite sense to protons (since the magnetic force is in the opposite direction for negative charges).

5. (E) Second, both same. As in Question 1, the larger speed corresponds to the larger radius, since $r = mv/qB$. Both have the same orbital frequency, because the cyclotron frequency $f = qB/2\pi m$ does not depend on radius or speed.

Checkup 30.2

1. The force on them is repulsive. Recall from Eq. (29.5) that the magnetic force on a positive charge moving parallel to a current in a long wire is negative, or radially inward. For currents moving in antiparallel directions, the force is repulsive. This is also easily verified from the vector form $\mathbf{F} = I\mathbf{L} \times \mathbf{B}$.
2. If the wire is straight, the force will be zero when it is parallel to the magnetic field.
3. (B) To the left. From $\mathbf{F} = I\mathbf{L} \times \mathbf{B}$, we see from the right-hand rule that the cross product of an upward current with a field into the page in Fig. 30.12 gives a force that is to the left.

Checkup 30.3

1. For $\theta = 90°$, the direction of the cross product of $\boldsymbol{\mu}$ and $\mathbf{B}$ is out of the page; a torque out of the page implies that the loop will turn counterclockwise (right-hand rule). Similarly, for $\theta = -90°$, $\boldsymbol{\tau}$ is into the page; such a torque implies that the loop will turn clockwise.
2. From $\theta = 0°$, a small displacement results in a configuration with a torque that would return $\boldsymbol{\mu}$ toward $\theta = 0°$, and so $\theta = 0°$ is a stable equilibrium (apply the right-hand rule to Fig. 30.14). However, near $\theta = 180°$, a small displacement results in a torque that would also push $\boldsymbol{\mu}$ toward $\theta = 0°$; thus $\theta = 180°$ is a condition of unstable equilibrium.
3. Without a commutator, the loop would oscillate about its equilibrium orientation (where the coil is at right angles to the field, or the magnetic moment is parallel to the field); with friction, these oscillations would damp out and the magnetic moment would eventually remain parallel to the field.
4. The magnetic moment vector is always perpendicular to the plane of the loop, in the direction given by the right-hand rule: with the fingers wrapped around in the direction of the current, the thumb points upward in Fig. 30.13, and at an angle θ to the right of upward in Fig. 30.14.
5. From the plot in Fig. 30.17, or by setting $\sin\theta = 1$ in the result of part (a) of Example 4, we see that the maximum torque is 3.2 N·m. The torque has this magnitude when the angle $\theta = \pm 90°$, that is, when the plane of the coil is parallel to the magnetic field (magnetic moment perpendicular to the magnetic field). The torque is proportional to the magnetic moment, which is proportional to the number of turns of the coil, so tripling the number of turns to 120 would triple the torque to 9.6 N·m.
6. (D) $\mathbf{k}$. From the right-hand rule preceding Eq. (30.18), the magnetic moment is directed downward, in the $-\mathbf{j}$ direction. By the usual right-hand rule for cross products, the torque $\boldsymbol{\tau} = \boldsymbol{\mu} \times \mathbf{B}$ is then in the $+\mathbf{k}$ direction.

Checkup 30.4

1. Looking from above the ball, the charge rotates counterclockwise. Since this is negative charge, the current rotates in clockwise loops. By the right-hand rule, the field along the axis is thus pointing downward.
2. (B) Decreases. If the alignment of spins cannot increase, then B_{matter} remains constant. With increasing external field, the susceptibility $\chi = B_{\text{matter}}/B_{\text{external}}$ must then decrease, as must the permeability $\mu = \mu_0(1 + \chi)$.

Checkup 30.5

1. By Eq. (30.30), the Hall voltage varies inversely with thickness, so the second wire will have 10 times the Hall voltage of the first.
2. (C) Both $\mathbf{v}$ and $\mathbf{E}_\perp$. For positive charge carriers, the velocity is in the same direction as the current, and so is reversed from the case shown. The magnetic force on the moving positive charges is then in the same direction as previously (both q and $\mathbf{v}$ change sign); consequently, the buildup of charge on the sides of the conductor and the resulting transverse electric field are reversed.

Electromagnetic Induction

CHAPTER

31

CONCEPTS IN CONTEXT

Concepts *in* Context

The coils of this large electric power generator, one of seventeen at Hoover Dam, surround an assembly of electromagnets. When the flow of water provides an external mechanical torque, the magnet assembly rotates, sweeping the magnetic field through the coils.

In this chapter we will see that such a relative motion between magnetic fields and coils converts mechanical work into electrical work by generating a voltage, or an electromotive force (emf), across the coils. We will consider such questions as:

? How does a simple generator produce an emf? (Section 31.2, page 1000)

? How does a rotating coil produce an alternating emf? (Example 4, page 1003)

? How can a rotating coil produce a steady, nonalternating emf? (Section 31.3 and Example 5, page 1009)

? How much magnetic energy is stored in the coils of a large generator such as shown in the photo? (Example 12, page 1015)

In this chapter we will discover that electric fields can be generated not only by charges, but also by changing magnetic fields or by changing magnetic flux. For instance, if we increase the currents in the windings of an electromagnet and thereby increase the strength of the magnetic field, this changing magnetic field generates an electric field, called an *induced electric field*. This kind of electric field exerts the usual electric forces on charges—in this regard the induced electric field does not differ from an ordinary electrostatic electric field.

However, the induced and the electrostatic electric fields differ in that the forces exerted by the latter are conservative, whereas the forces exerted by the former are nonconservative. We recall that, according to the requirement laid down in Chapter 8, what discriminates between conservative and nonconservative forces is the work done during a round trip. For conservative forces, such as the forces exerted by electrostatic electric fields, the work done during a round trip is zero; for nonconservative forces, such as the forces exerted by induced electric fields, the work done during a round trip is nonzero. This means that when a charge moves around a closed circuit, the induced electric fields deliver work to the charge, and they constitute a source of emf. Like the emf of a battery, this **induced emf** is capable of driving a current around the circuit. As we will see, a celebrated law formulated by Michael Faraday asserts that the magnitude of the induced emf is directly proportional to the rate of change of the magnetic flux intercepted by the circuit. Such induced emfs find practical application in electromagnetic generators, widely used for the generation of electric power.

induced emf

We begin with a discussion of the induced emf produced by the motion of a conductor, such as a wire or a rod, in a constant magnetic field. As we will see, such a moving conductor sweeps across magnetic flux and thereby generates an induced emf, called a motional emf.

31.1 MOTIONAL EMF

Suppose that we push a rod of metal with some velocity **v** through a uniform magnetic field, such as the magnetic field of a large electromagnet. The free electrons in the metal will then also have a velocity **v**, and they will experience a magnetic force. For motion perpendicular to the magnetic field, the magnitude of the magnetic force on an electron is evB [see Eq. (29.12)]. If the rod and the velocity **v** of the rod are perpendicular to each other and to the magnetic field, then the direction of this magnetic force on the free electrons is parallel to the rod, as indicated by the right-hand rule (see Fig. 31.1). The electrons will therefore flow along the rod, accumulating negative charge on the upper end and leaving positive charge on the lower end. The flow of charge will stop when the electric repulsion generated by the accumulated charges balances the magnetic force evB. However, if the ends of the rod are in sliding contact with a pair of long wires that provide a stationary return path, then the electrons will flow continuously around the circuit (see Fig. 31.2). Thus, the moving rod acts as a "pump of electricity," or a source of emf, which produces the same effect as a battery connected to the long wires. In Fig. 31.2, the upper end of the rod is the negative terminal of this source, and the lower end is the positive terminal.

The emf associated with the rod is defined as the work done by the driving force per unit positive charge that passes from the negative end of the rod to the positive. Since the magnitude of the force on an electron is evB, the magnitude of the force per unit

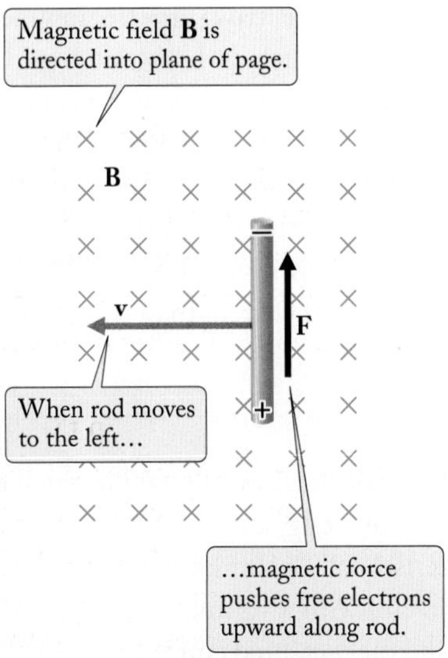

FIGURE 31.1 Conducting rod moving with velocity **v** through a uniform magnetic field. The magnetic field is directed into the plane of the page; the crosses indicate the tails of the magnetic field vectors. According to the right-hand rule, **v** × **B** is in the downward direction. The magnetic force $-e\mathbf{v} \times \mathbf{B}$ on a free electron within this rod is therefore in the upward direction.

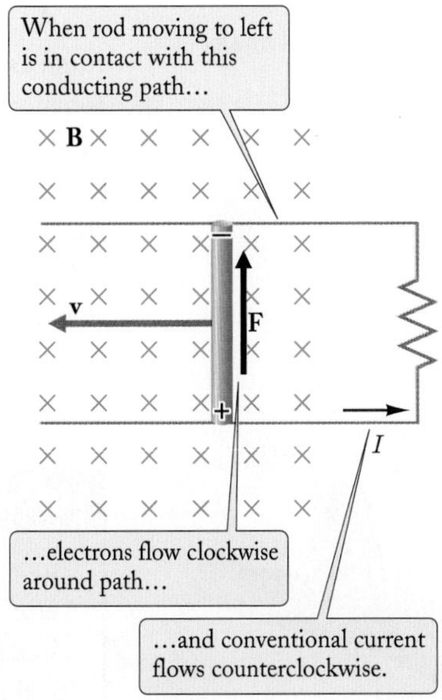

FIGURE 31.2 If the ends of the rod are in sliding contact with a pair of long wires, a current will flow around the circuit. This current can be detected by an ammeter.

charge is $evB/e = vB$. If the length of the rod is l, the work per unit charge equals the distance l times the force per unit charge, and therefore

$$\mathcal{E} = lvB \tag{31.1}$$

motional emf

This is called a **motional emf** because it is generated by the motion of the rod through the magnetic field.

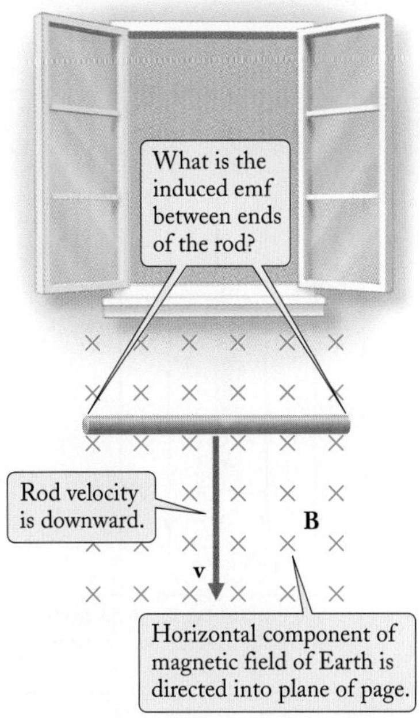

FIGURE 31.3 A rod has been dropped out of a window. The horizontal component of the magnetic field of the Earth is directed into the plane of the page.

EXAMPLE 1 Suppose you drop an aluminum rod of length 1.0 m out of a window at a place where the horizontal magnetic field of the Earth is 2.0×10^{-5} T. The rod is oriented horizontally, at right angles to the magnetic field (see Fig. 31.3). What is the induced emf between the ends of the rod when its instantaneous downward velocity reaches 12 m/s?

SOLUTION: The magnetic field of the Earth has both a horizontal and a vertical component. In our derivation of Eq. (31.1) for the motional emf, we assumed that the magnetic field **B** is perpendicular to the rod and to its velocity. For a vertically falling rod, as in Fig. 31.3, the horizontal component B_x of the Earth's magnetic field is perpendicular to the rod and to its velocity, and hence generates a motional emf. By contrast, the vertical component of the Earth's magnetic field exerts no force on the free electrons in the rod (remember that the magnetic force is zero when the velocity is parallel to the magnetic field), and therefore generates no motional emf.

With $B = B_x = 2.0 \times 10^{-5}$ T, (31.1) gives

$$\mathcal{E} = lvB_x = 1.0 \text{ m} \times 12 \text{ m/s} \times 2.0 \times 10^{-5} \text{ T} = 2.4 \times 10^{-4} \text{ V}$$

electromagnetic flowmeter

A practical application of motional emfs is the **electromagnetic flowmeter** used to measure the speed of flow of a conducting liquid, such as detergent, liquid sodium, tomato pulp, beer, and so on. In this device, the liquid is made to flow in a nonconducting pipe placed at right angles to a magnetic field, and a motional emf of the magnitude given by Eq. (31.1) will then appear across the diameter of the column of liquid, because the diameter acts like a conducting rod. Electrodes inserted into the pipe on opposite sides pick up the emf, which can be measured on a sensitive voltmeter. The velocity of flow can then be calculated from the measured emf, the strength of the magnetic field, and the dimension of the pipe.

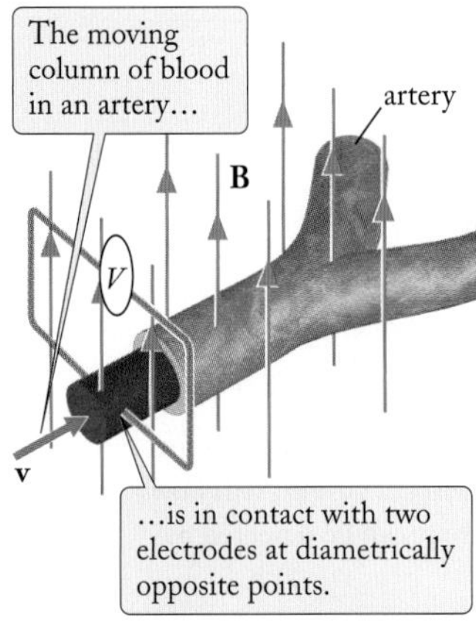

FIGURE 31.4 Technique for measurement of the velocity of the flow of blood.

EXAMPLE 2 To measure the velocity of the flow of blood in the mesenteric artery in the abdomen of a dog, a researcher places the animal in a magnetic field of 3.0×10^{-2} T, inserts small electrodes through the wall of the artery from each side, and measures the emf with a voltmeter (see Fig. 31.4). The inner diameter of the artery is 0.30 cm and the measured emf is 1.8×10^{-6} V. What is the velocity of blood flow?

SOLUTION: According to Eq. (31.1), with $l = 0.30$ cm,

$$v = \frac{\mathcal{E}}{lB} = \frac{1.8 \times 10^{-6}\ \text{V}}{0.30 \times 10^{-2}\ \text{m} \times 3.0 \times 10^{-2}\ \text{T}} = 2.0 \times 10^{-2}\ \text{m/s}$$

It is instructive to reexamine the generation of the induced emf from the point of view of a reference frame in which the rod is at rest. In such a reference frame moving with the rod, the free charges have no velocity (except for perhaps a small drift velocity along the rod), and thus there is *no magnetic force along the rod*. Hence, the free charges must experience some other, nonmagnetic force along the rod that does work on them and supplies an emf. The question is then, what is this new kind of force in the moving reference frame that produces the same effect as the magnetic force in the stationary reference frame? Since the only kinds of force that act on electric charges are the magnetic force and electric force, the "new" kind of force must be electric. It must be due to a "new" kind of electric field, an electric field that exists in the moving reference frame, but not in the stationary reference frame (see Fig. 31.5).

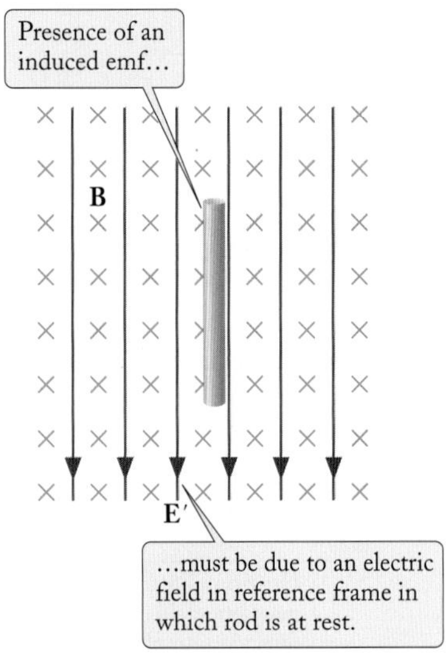

FIGURE 31.5 In the moving reference frame of the rod, there is a magnetic field **B** (blue crosses) and there is also an electric field **E**′ (red lines).

The "new" electric field that exists in the reference frame of the rod moving through a magnetic field is called an **induced electric field**. We can determine the magnitude of the induced electric field from a consistency requirement: the "new" electric force $F' = qE'$ in the moving reference frame must coincide with the magnetic force $F = qvB$ in the stationary reference frame. This tells us that the electric field in the moving reference frame must have a magnitude

$$E' = vB \tag{31.2}$$

This electric field does work on the free charges and therefore represents a source of emf. The work done on a unit positive charge that passes through the rod is

$$\mathcal{E} = E'l \tag{31.3}$$

In view of Eq. (31.2), this value of the emf coincides with the value that we obtained in Eq. (31.1). Thus, the motional emf can be calculated either in a stationary reference frame or else in a moving reference frame; in the former case it arises from a magnetic field and in the latter from an electric field.

✓ Checkup 31.1

QUESTION 1: If the rod shown in Fig. 31.2 moves toward the right instead of the left, what is the direction of the current generated by the motional emf?

QUESTION 2: For the rod in Example 1, which end is the positive terminal for the motional emf? Suppose that instead of dropping the rod downward, we throw it upward. How does this change the emf?

QUESTION 3: When you run through the Earth's magnetic field, a motional emf will be generated from one side of your body to the other. In the Northern Hemisphere, the Earth's magnetic field is mainly downward (the horizontal component is small and can be ignored for the present purposes). If you run eastward, which side of your body will be the positive terminal for the induced emf? If you run westward? If you run northward?

QUESTION 4: What is the direction of the induced electric field in the reference frame of the rod shown in Fig. 31.3?

(A) Up (B) Into page (C) Out of page (D) To left (E) To right

31.2 FARADAY'S LAW

Online Concept Tutorial 34

The quantity lvB appearing on the right side of Eq. (31.1) can be given an interesting interpretation in terms of the magnetic flux. As already mentioned in Section 29.2, the **magnetic flux** is defined in the same way as the electric flux. For any open or closed surface of area A, the *magnetic flux is defined as the integral of the normal component of the magnetic field over the area*:

magnetic flux

$$\Phi_B = \int \mathbf{B} \cdot d\mathbf{A} = \int B_\perp \, dA \qquad (31.4)$$

The magnetic flux is simply proportional to the number of magnetic field lines that are intercepted by the surface, lines crossing the surface in one direction being reckoned as positive, and lines in the opposite direction as negative. Note that for a flat surface in a constant magnetic field, the magnetic flux is simply

$$\Phi_B = \mathbf{B} \cdot \mathbf{A} = B_\perp A = BA \cos\theta \qquad (31.5)$$

where θ is the angle between the magnetic field and the perpendicular erected on the surface (see Fig. 31.6). The unit of magnetic flux is the product of the unit of magnetic field and the unit of area, that is, $\mathrm{T \cdot m^2}$. In the SI system, this unit is called the **weber**:

$$1 \text{ weber} = 1 \text{ Wb} = 1 \, \mathrm{T \cdot m^2}$$

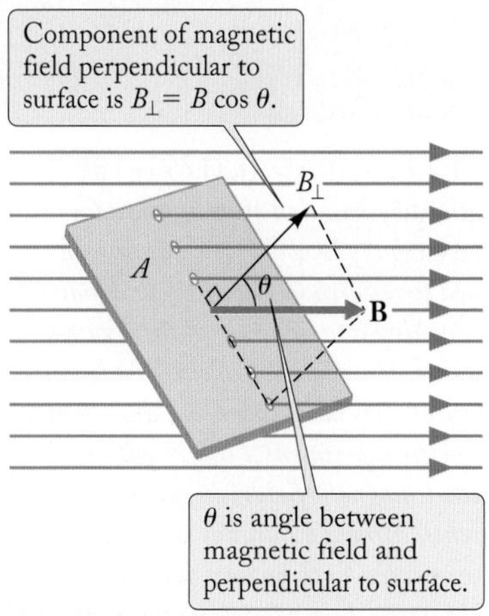

FIGURE 31.6 Flat rectangular surface in a uniform magnetic field.

Now, consider a rod moving through a magnetic field at velocity **v**. In a time interval dt, the rod advances a distance $v\,dt$ and therefore sweeps through an area $l \times v\,dt$, perpendicular to the magnetic field (see Fig. 31.7). The product of this area and the magnetic field is the magnetic flux that the rod sweeps across; that is,

$$lv\,dt\,B = d\Phi_B$$

Thus,

$$lvB = \frac{d\Phi_B}{dt}$$

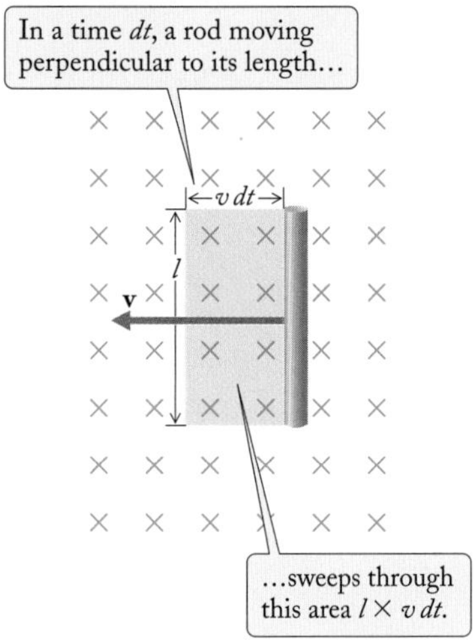

FIGURE 31.7 Rod moving through a magnetic field with speed v.

This tells us that the quantity lvB is the rate at which the rod sweeps across magnetic flux. We can then rewrite Eq. (31.1) as

$$\mathcal{E} = -\frac{d\Phi_B}{dt} \tag{31.6}$$

The minus sign that we have inserted into Eq. (31.6) indicates how the polarity of the induced emf is related to the change of flux; we will discuss the determination of this polarity in the next section.

The advantage of this interpretation of the induced emf in terms of magnetic flux is that Eq. (31.6) is of more general validity than Eq. (31.1)—as we will see, Eq. (31.6) is a general law relating the induced emf to the rate at which the rod sweeps across flux, and it is valid regardless of how the sweeping of flux comes about. For instance, instead of moving the rod relative to the magnet that produces the magnetic field, we can hold the rod fixed and move the magnet. Flux will then sweep across the rod, and Eq. (31.6) suggests that an emf should be induced along the rod. Experiments do indeed confirm that in both cases the induced emf is exactly the same.

But there is another way in which magnetic flux will sweep across the rod: we can hold the rod fixed and increase or decrease the *strength* of the magnetic field. To understand why flux will be swept across the rod under these conditions, we must first take a look at what happens to the field lines of a current when the current changes. Figure 31.8a shows the field lines produced by a current on a long straight wire. If we increase the current, the magnetic field increases, that is, the number of field lines increases. Figure 31.8b shows the field lines of a stronger current. Where do the extra field lines come from? Obviously, the current has to make them. When the current increases, it makes new, small circles of field lines in its vicinity; meanwhile, the circles that already exist in Fig. 31.8a move outward, like ripples on a pond. Thus, the pattern shown in Fig. 31.8a gradually grows into the new pattern shown in Fig. 31.8b. Note that the pattern grows from the inside out.

Thus a change in the strength of the magnetic field involves moving field lines. If a *stationary* rod is located in the vicinity of the changing current, the moving field lines will sweep across the rod, that is, magnetic flux will sweep across the rod. Experiments show that such a sweeping of flux across a stationary rod induces an emf which is given by Eq. (31.6), the same formula as for the case of the sweeping of flux across a moving rod.

From the emf induced in a rod, we can infer the emf induced in a wire of arbitrary shape, since we can regard any such wire as consisting of short straight segments joined

MICHAEL FARADAY (1791–1867)
English physicist and chemist, Faraday's earliest research lay in chemistry, but he soon turned to research in electricity and magnetism, making contributions of the greatest significance. His discovery of electromagnetic induction was no accident, but arose from a systematic experimental investigation of whether magnetic fields can generate electric currents. Although Faraday was essentially an experimenter, with no formal training in mathematics, he made an important theoretical contribution by introducing the concept of field lines and by recognizing that electric and magnetic fields are physical entities.

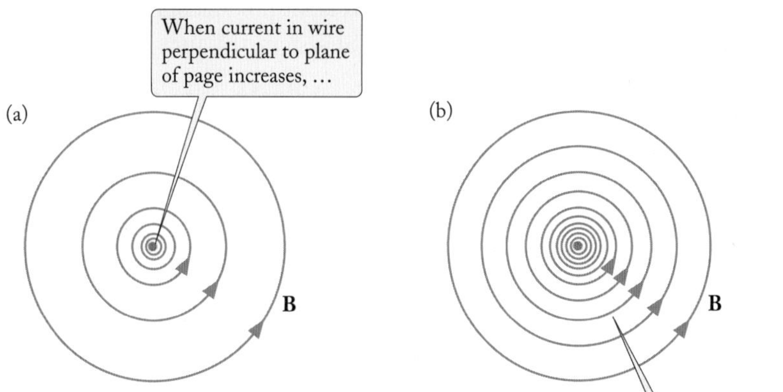

FIGURE 31.8 (a) Magnetic field lines of a current on a very long wire. The wire is perpendicular to the plane of the page. (b) Magnetic field lines of a stronger current.

one to another. The net emf induced between the two ends of the wire is then the sum of the emfs induced in all these short rods, and this net emf equals the net rate at which magnetic flux sweeps across the wire. More generally, we can make an assertion about the emf induced along an arbitrary moving or fixed mathematical path immersed in a constant or changing magnetic field. An emf will be induced between the ends of this path regardless of whether we place a rod or a wire along the path. Whenever a unit positive charge moves along this path, it will gain an amount $\mathcal{E}$ of energy from the induced electric field, regardless of whether the charge moves on a rod or through empty space. The rod or wire merely serves as a convenient conduit for the flow of charge. Of course, for the practical exploitation of the emf, we usually find it convenient to provide a rod, wire, or some other conductor along which the charge can flow, and we will then also have to provide a return path for the charge.

The general statement about the induced emf is known as **Faraday's Law of Induction**,

> ***The induced emf along any moving or fixed path in a constant or changing magnetic field equals the rate at which magnetic flux sweeps across the path:***

$$\mathcal{E} = -\frac{d\Phi_B}{dt} \tag{31.7}$$

Faraday's Law of Induction

For a *closed* path, or a closed circuit, Faraday's Law can be interpreted in terms of the flux intercepted by the surface within the path. For this purpose we imagine that the path is spanned (or closed off) by some smooth mathematical surface, such as shown in Fig. 31.9. Since the magnetic field lines are continuous (without sources or sinks), the exact shape of the surface is of no consequence; we can use any flat or curved surface that has the path as boundary—any such surface intercepts the same flux. If we move the path or if the magnetic field changes, flux will sweep across the path. But any flux that sweeps across the path represents a gain or a loss of flux intercepted by the surface, since any field line that moves across the path either moves into the surface or out of the surface. Therefore, the rate $d\Phi_B/dt$ at which flux sweeps across the closed path equals the rate of change of the flux intercepted by the surface, and we can interpret Eq. (31.7) as stating that the induced emf equals the rate of change of this intercepted flux. This means that for a closed path, Faraday's Law can be stated as follows:

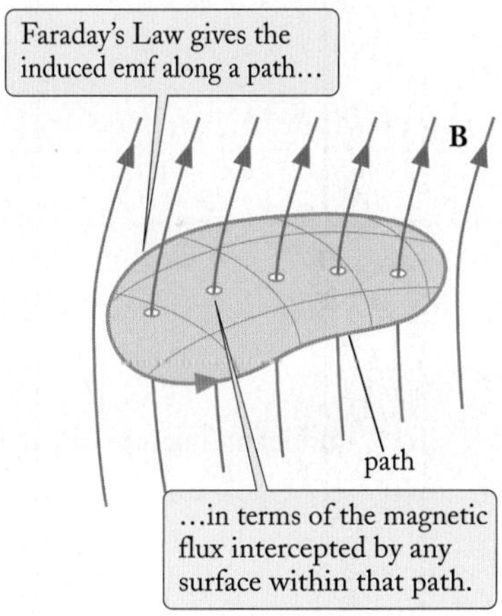

FIGURE 31.9 A closed path spanned by a surface.

> ***The induced emf around a closed path in a magnetic field is equal to the rate of change of the magnetic flux intercepted by the surface within the path.***

Note that the magnetic flux intercepted by the surface within a closed path, say, a wire loop, can be made to change in a variety of ways (see Fig. 31.10). We can move the wire loop in or out of the magnetic field, or deform it so as to increase or decrease its area, or change its orientation, or change the strength of the magnetic field by changing the current in the magnet that produces the field—but in all these cases the emf induced around the loop is related to the change of flux by Eq. (31.7). For example, for a flat loop in a uniform field, Eqs. (31.7) and (31.5) give

$$\mathcal{E} = -\frac{d}{dt}(BA\cos\theta) \tag{31.8}$$

Faraday's Law for flat loop in uniform field

which makes it apparent that changing B, A, or θ will result in an induced emf. Examples of such induced emfs appear in the next section.

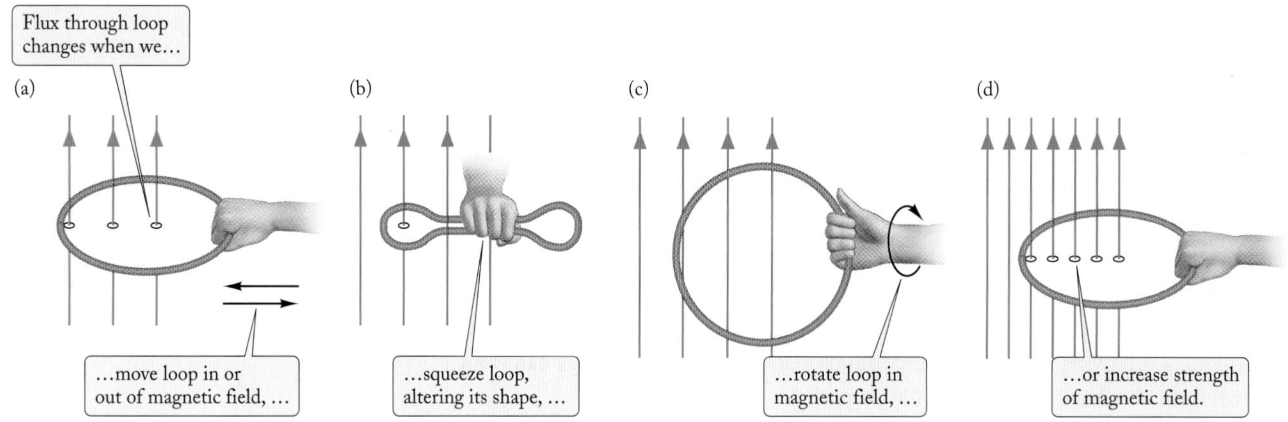

FIGURE 31.10 Several ways the flux through a loop can change.

electromagnetic generator

Concepts in Context

The most important practical application of induction is the **electromagnetic generator**, widely used to generate the electric power that we need in our homes and factories. In its simplest form, the electromagnetic generator consists of a loop of wire that is made to rotate in the magnetic field of a magnet (see Fig. 31.11). When the loop is face on to the magnetic field, the flux intercepted by the surface within the loop is large; when the loop is edge on to the magnetic field, the flux is zero. Thus, the rotation of the loop about its axis leads to a rate of change of flux, which induces an emf and drives a current through the loop and through the external circuit connected to the ends of the loop.

In essence, the generator is an electric motor operating in "reverse." In a motor, we convert electric power into mechanical power—we send a current through a loop in a magnetic field, and we produce rotational motion. In a generator, we convert mechanical power into electric power—we rotate a loop in a magnetic field, and we produce an emf and a current in the loop and in the external circuit. Small or medium-size generators are driven by gasoline or diesel engines; large-size generators at electric power plants are driven by steam turbines or by hydraulic turbines.

induction microphone

There are many familiar, smaller-scale applications of Faraday's Law. For example, the **induction microphone**, or moving-coil microphone, consists of a flexible diaphragm with an attached small coil of wire, which is placed near a bar magnet (see Fig. 31.12). When a sound wave strikes the diaphragm, the pressure fluctuations move the diaphragm

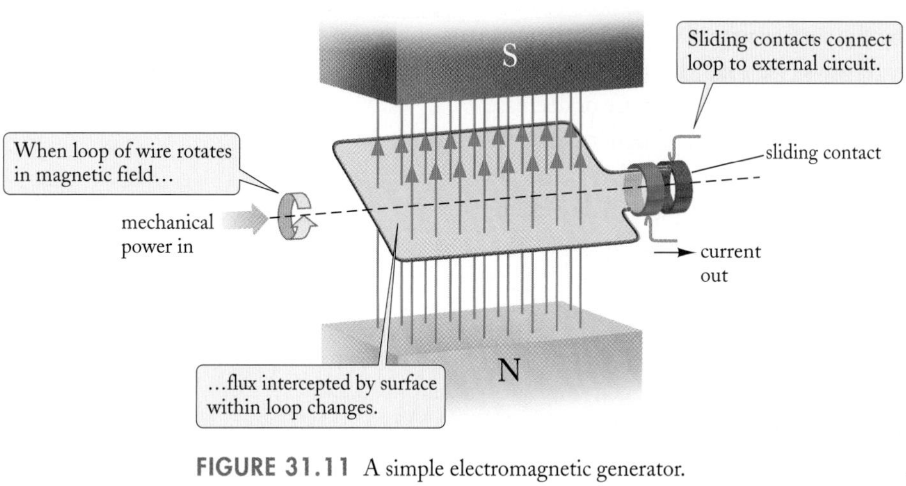

FIGURE 31.11 A simple electromagnetic generator.

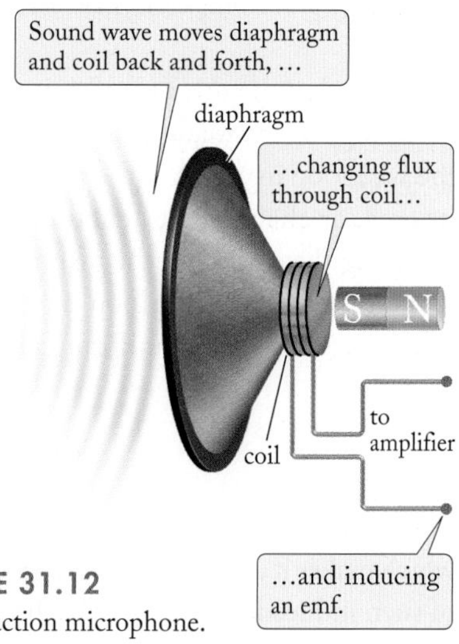

FIGURE 31.12 An induction microphone.

and the coil back and forth, and this motion of the coil produces a changing flux and induces an emf. Thus, the microphone transforms pressure fluctuations into an electric signal, which can be amplified and sent to a loudspeaker. A loudspeaker is simply an induction microphone operating in reverse—a current fed into the coil experiences a force in the magnetic field, which pushes or pulls the coil of the bar magnet and the diaphragm back and forth, producing sound waves. Since the basic mechanism of the loudspeaker is the same as that of the induction microphone, it is actually possible to use a loudspeaker as a microphone—simply hook the loudspeaker into an amplifier and speak into it, and the loudspeaker will transform your sound signal into an electric signal and act as a (somewhat poor) microphone.

✔ Checkup 31.2

QUESTION 1: The magnetic flux through a closed surface is always zero. Why?

QUESTION 2: A closed path is immersed in a magnetic field. If we suddenly reverse the magnetic field (but leave its strength unchanged), will an emf be induced around the path?

QUESTION 3: You shoot a metallic arrow through a magnetic field. Will this generate an induced emf between the tip and the tail of the arrow?

QUESTION 4: Figure 31.13 shows several surfaces in a uniform magnetic field. Which has the largest magnetic flux?

(A) a (B) b (C) c

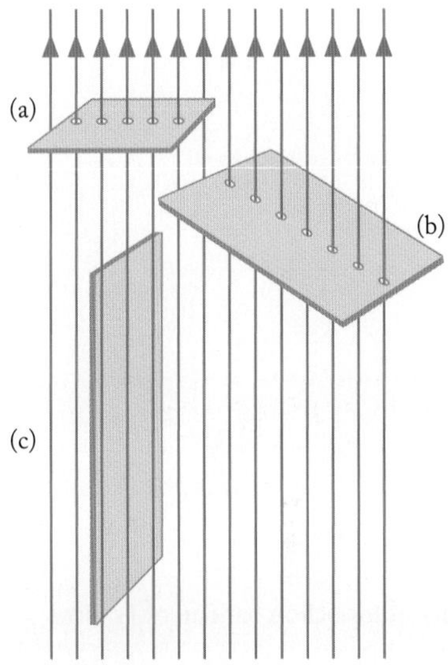

FIGURE 31.13 Several different surfaces in a uniform magnetic field.

31.3 SOME EXAMPLES; LENZ' LAW

Online Concept 34 Tutorial

In this section we will look at some examples of the calculation of induced emfs. We begin by laying down a simple rule for finding the polarity of the induced emf, a rule known as **Lenz' Law:**

> ***The induced current is always such as to oppose the change of flux (or the motion) that generated it.***

For instance, consider the rod moving in a magnetic field illustrated in Fig. 31.14. The rod slides on long parallel wires joined at the right, so as to form a closed circuit. The motion of the rod leads to an increasing magnetic flux due to the increasing area enclosed by the circuit. According to Lenz' Law, the extra flux contributed by the current induced in the rod and the wires must oppose this increase of flux; thus, the magnetic field of this induced current must be opposite to the original magnetic field—within the area enclosed by the circuit, the magnetic field of the current in the wire must be out of the plane of the page in Fig. 31.14, whereas the original magnetic field is into this plane. By the right-hand rule, the current must be counterclockwise around the circuit to produce such a magnetic field, in agreement with the direction indicated in Fig. 31.2.

Lenz' Law

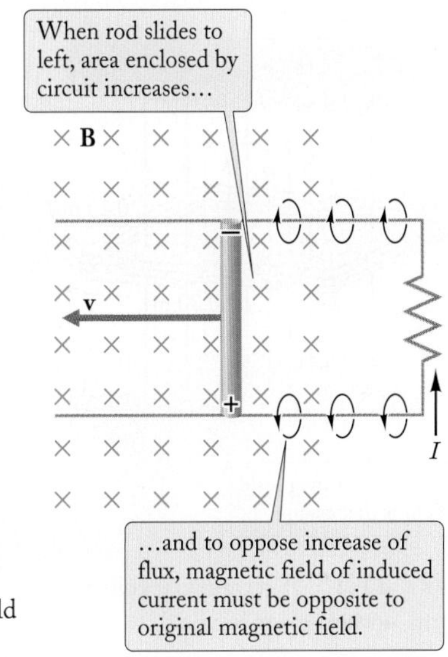

FIGURE 31.14 A rod moving in a uniform magnetic field is in sliding contact with parallel wires. The direction of the uniform magnetic field is perpendicularly into the plane of the page.

(Alternatively, we can deduce the direction of the current by examining the motion of the rod. The rod moves toward the left. According to Lenz' Law, the magnetic force on the current in the rod must oppose this motion, that is, the force must be toward the right. By the right-hand rule, if the magnetic force is to be toward the right, the current in the rod must be downward, which means the current must be counterclockwise around the circuit.)

It is important to understand that the induced current opposes the *change* in the flux, but does not necessarily oppose the flux itself. If the rod in Fig. 31.14 were moving toward the right instead of the left, then this motion would lead to a decreasing magnetic flux within the area enclosed by the circuit. The extra flux contributed by the induced current must oppose this decrease of flux; thus, the magnetic field of the induced current must be in the same direction as the original magnetic field, that is, into the plane of the page in Fig. 31.14. The current must then be clockwise around the circuit.

In practical applications, induced emfs are usually generated in coils consisting of many turns of wire, that is, many loops. In such a coil, each individual loop has a rate of change of flux and generates its own emf. Since the individual loops are connected in series, their emfs add together, and for a coil of N loops, the net emf is N times the emf of one loop:

Law of Induction for coil of N turns

$$\mathcal{E} = -N\frac{d\Phi_B}{dt} \tag{31.9}$$

Alternatively, we can think of the flux through the N loops as crossing an area N times as large as a single loop. From this viewpoint, a single emf is generated, but the total flux is N times as large ($\Phi_{B,\text{total}} = N\Phi_B$), so we obtain the same result (31.9). It is sometimes helpful to be aware of these alternative points of view.

We now look at various examples of the calculation of induced emfs; we will examine changes in flux due to a changing magnetic field, a changing angle between the area and the magnetic field, and a changing area.

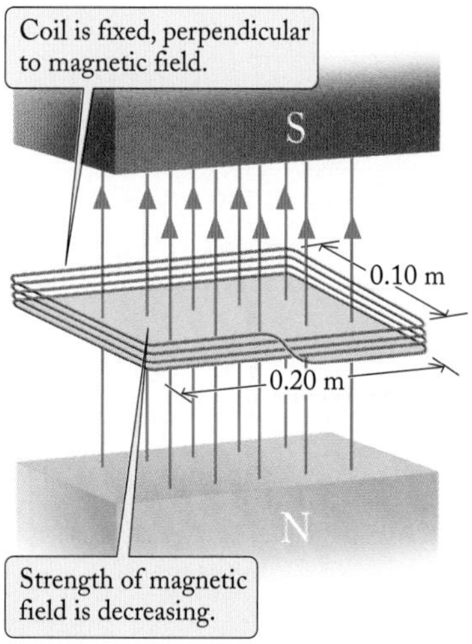

FIGURE 31.15 A rectangular coil in a uniform, upward magnetic field.

EXAMPLE 3 A rectangular coil of 150 loops forming a closed circuit measures 0.20 m × 0.10 m. The resistance of the coil is 5.0 Ω. The coil is placed between the poles of an electromagnet, face on to a magnetic field **B** (see Fig. 31.15). Suppose that when we switch the electromagnet off, the strength of the magnetic field decreases at the rate of 20 teslas per second. What is the induced emf in the coil? What is the direction of the induced current?

SOLUTION: The flux intercepted by each loop in the coil is $\Phi_B = BA$, where B is the magnetic field and A is the area of the loop. If only the magnetic field is changing, the rate of change of flux is

$$\frac{d\Phi_B}{dt} = \frac{dB}{dt}A$$

and Faraday's Law tells us that the total induced emf for the entire coil is

$$\mathcal{E} = -N\frac{d\Phi_B}{dt} = -N\frac{dB}{dt}A$$
$$= -150 \times (-20\ \text{T/s}) \times (0.20\ \text{m} \times 0.10\ \text{m}) = 60\ \text{V}$$

According to Ohm's Law, the magnitude of the induced current is then

$$I = \frac{\mathcal{E}}{R} = \frac{60\ \text{V}}{5.0\ \Omega} = 12\ \text{amperes}$$

To determine the direction of the induced current, we appeal to Lenz' Law. The magnetic field shown in Fig. 31.15 and its flux are decreasing. Hence, the current in the coil must provide a magnetic field that compensates (in part) for this decreasing magnetic flux. According to the right-hand rule, this requires a counterclockwise current around the loop (looking from above), so the magnetic field of this current is upward, in the direction of the original magnetic field.

COMMENTS: Note that for a current to flow, the coil needs to form a closed circuit. If there were a break in the coil, it would not form a closed circuit; an emf would still be generated, but no current could flow, just as in an ordinary open circuit. If instead of a coil, a solid conducting plate had been immersed in the changing field, the induced currents would flow in loops around the conductor. Such currents are called **eddy currents**, because they flow in loops. Eddy currents are sometimes desirable in a device; they remove motional energy and provide damping, for instance, in vibration-isolation devices. When damping is not desired and when eddy currents cause unwanted heating due to I^2R losses, they can be avoided by segmenting the conductor into separate pieces, so that no currents can flow from one piece to the next.

eddy current

EXAMPLE 4 An electromagnetic generator consist of a rectangular coil of N loops of wire that rotates about an axis perpendicular to a constant magnetic field **B**. Sliding contacts connect the coil to an external circuit (see Fig. 31.16). What emf does the coil deliver to the external circuit? The coil has an area A and rotates with an angular frequency ω.

Concepts in Context

SOLUTION: The magnetic flux varies as the coil rotates; it is maximum when the coil is oriented face on to the magnetic field, and mimimum when the coil is oriented parallel to the magnetic field. Instantaneously, the perpendicular to the coil makes an angle $\theta = \omega t$ with the magnetic field (see Fig. 31.16), and the component of the magnetic field perpendicular to the coil is $B_\perp = B\cos\theta = B\cos\omega t$. The instantaneous magnetic flux through the coil is then $\Phi_B = BA\cos\omega t$. To find

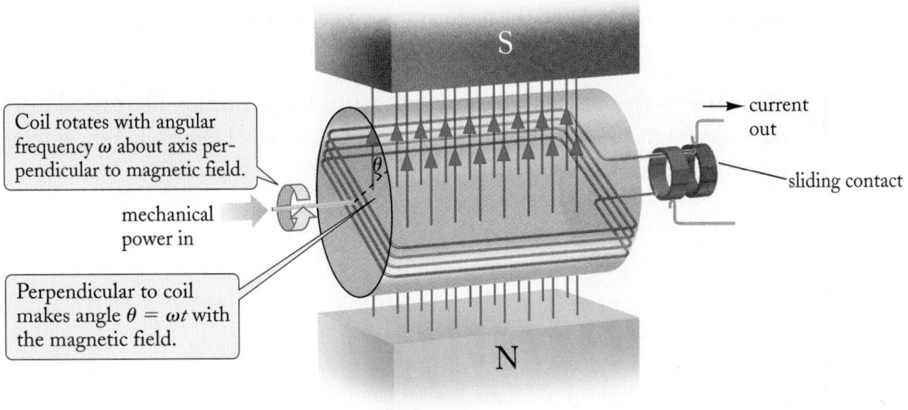

FIGURE 31.16 An electromagnetic generator.

the induced emf around each loop, we need to evaluate the rate of change of this flux, which means we need to evaluate the rate of change of cos ωt,

$$\frac{d}{dt}\cos\omega t = -\omega\sin\omega t$$

Thus, for the magnetic flux $\Phi_B = BA\cos\omega t$, the rate of change is

$$\frac{d\Phi_B}{dt} = -BA\omega\sin\omega t$$

According to Faraday's Law, the total induced emf around all N loops is

$$\mathcal{E} = -N\frac{d\Phi_B}{dt} = NBA\omega\,\sin\omega t \tag{31.10}$$

At the instant shown in Fig. 31.16, the magnetic flux through the coil is increasing; by Lenz' Law, the induced current must therefore flow around the coil in a clockwise direction (viewed from above) so as to contribute an extra magnetic flux that opposes the increase of the original magnetic flux.

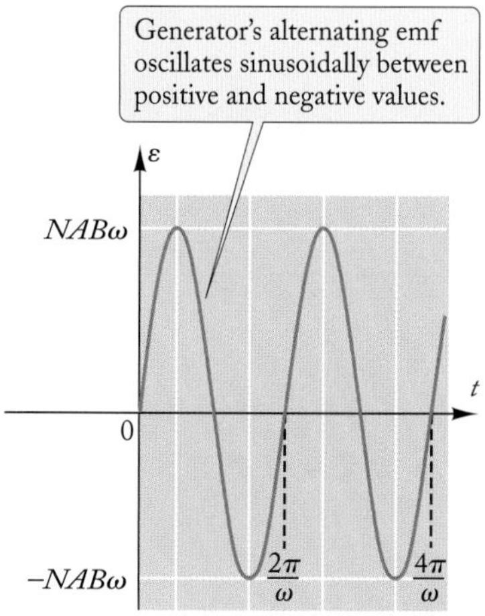

FIGURE 31.17 Emf of the generator; this is an alternating voltage (AC).

alternating emf (AC) and steady emf (DC)

Concepts in Context

The emf given by Eq. (31.10) is an **alternating emf**, or **AC voltage**, that oscillates sinusoidally between positive and negative values. A plot of this emf is shown in Fig. 31.17. Various methods can be used to produce a **steady emf**, or **DC voltage**, that does not alternate between positive and negative values but has a constant value. The most common method relies on a commutator or on an alternator. The **commutator** is simply a mechanical switch that reverses the connection between the generator and the external circuit every half cycle so that the emf delivered to the external circuit always has the same sign. The **alternator** is an electronic switch that uses diodes instead of mechanical devices to accomplish the same thing as a commutator. Another method relies on an entirely different kind of generator, called a **homopolar generator**. Instead of a rotating coil, this generator uses a rotating disk, which is equivalent to a rotating rod, as illustrated by the following example.

EXAMPLE 5 A straight rod of metal is rotating about one end on an axis parallel to a uniform magnetic field B (see Fig. 31.18a). The length of the rod is l, and the angular velocity of rotation is ω. What is the induced emf between the two ends of the rod?

SOLUTION: As the rod turns, it sweeps across magnetic flux. The rod takes a time $\Delta t = 2\pi/\omega$ to sweep out the circular area $\Delta A = \pi l^2$. Hence the rate at which it sweeps out area is

$$\frac{dA}{dt} = \frac{\Delta A}{\Delta t} = \frac{\pi l^2}{2\pi/\omega} = \frac{l^2\omega}{2}$$

The rate at which it sweeps across magnetic flux is

$$\frac{d\Phi_B}{dt} = B\frac{dA}{dt} = \frac{Bl^2\omega}{2}$$

According to Faraday's Law, the induced emf is then

$$\mathcal{E} = -\frac{Bl^2\omega}{2}$$

If the moving end of the rod is in sliding contact with a circular track connected to an external circuit (see Fig. 31.18b), the emf will drive a current around the circuit. By Lenz' Law, the induced current in the rod must be such that the magnetic force on the rod opposes the motion. Thus the magnetic force on the rod must oppose the counterclockwise motion of Fig. 31.18. By the right-hand rule for the magnetic force, this requires a current that flows from the moving end of the rod toward the center. In the external circuit, the current then flows from the center toward the moving end—the center acts as the positive terminal of the source of emf and the moving end as the negative terminal.

COMMENTS: For practical applications, homopolar generators are constructed with a rotating disk, rather than a rotating rod. This does not affect the emf of the generator, but it helps to reduce its internal resistance and mechanical stress. Homopolar generators are used in applications, such as electroplating, requiring a large current but only low voltage.

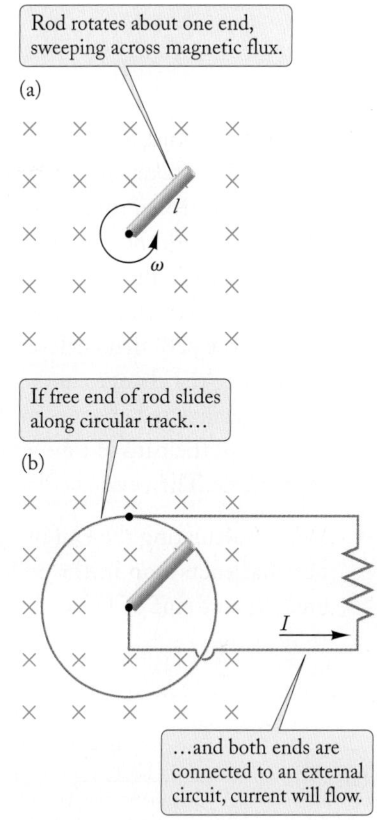

FIGURE 31.18 A homopolar generator. (a) Rod rotating in a uniform magnetic field. (b) Rotating rod connected to an external circuit.

For later use in Chapter 32, we need to reformulate Faraday's Law in terms of the induced electric field. As we saw in Section 31.1 in the example of the rod moving in a magnetic field, the induced emf for a moving straight path sweeping through a constant magnetic field is related to the induced electric field by

$$\mathcal{E} = E'l$$

More generally, the induced emf for an arbitrary moving or fixed path in a constant or changing magnetic field is related to the induced electric field (measured in the reference frame of the path) by

$$\mathcal{E} = \oint \mathbf{E} \cdot d\mathbf{s} = \oint E_{\|}\, ds \tag{31.11}$$

where $E_{\|}$ is the component of the induced electric field tangent to the path and ds is a length element along the path. This permits us to express Faraday's Law as

$$\oint E_{\|}\, ds = -\frac{d\Phi_B}{dt} \tag{31.12}$$

Faraday's Law in terms of induced electric field

This version of Faraday's Law can be used to calculate the induced electric field. Note that this form of Faraday's Law for the electric field due to a changing flux is mathematically similar to Ampère's Law, Eq. (29.18), for the magnetic field due to a current. In most of the problems of this chapter we will be concerned only with the induced emf, and the somewhat simpler version (31.7) of Faraday's Law will suffice. But sometimes we are interested in the induced electric field, as in the following example.

PROBLEM-SOLVING TECHNIQUES FARADAY'S LAW; LENZ' LAW

- Before attempting to calculate an induced emf, ask yourself, where is there a changing magnetic flux? Since the flux in a uniform magnetic field can be written as $\Phi_B = BA\cos\theta$, the change of flux is readily obtained from a change in the magnetic field (Example 3), a change in orientation (Example 4), or a change in the position or shape of the path that encloses the flux (Example 5). Try to visualize the changes in the magnetic field in time or try to visualize the changes in the position, orientation, or shape of the path in time, as though you were watching a movie. Then evaluate the rate of change of the flux.
- When calculating the emf in a coil of N turns, remember that each loop in the coil generates its own emf, and the net emf of the coil is therefore proportional to N.
- The direction of the induced emf and the induced current can be determined by Lenz' Law, which can be applied either to the magnetic flux or to the motion of a conductor: the flux associated with the induced current opposes the original *change* of flux, or the magnetic force on the induced current opposes the original motion. Note that if the device that generates the emf is not actually connected to an external circuit (for instance, the rod in Fig. 31.18a), then it is helpful to imagine connecting it to an external circuit (as in Fig. 31.18b), so it becomes easier to imagine the direction of the induced current.
- The magnitude of the current that flows in an external circuit connected to the induced emf can be calculated in the usual way from the emf $\mathcal{E}$ and the resistance R, according to Ohm's Law, $I = \mathcal{E}/R$.

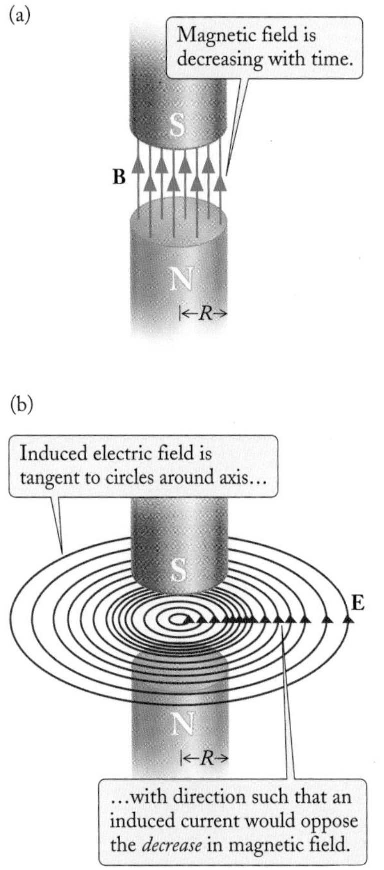

FIGURE 31.19 (a) Magnetic field **B** between the pole faces of an electromagnet. (b) Induced electric field.

EXAMPLE 6 The circular pole faces of the electromagnet shown in Fig. 31.19a have a radius $R = 0.10$ m. The magnetic field is suddenly shut off, decreasing (briefly) at a rate of 2.0×10^4 T/s. What is the induced electric field for $r \leq R$? For $r \geq R$? What is the numerical value of this field at $r = R$? Assume that there is only a magnetic field between the pole faces.

SOLUTION: By analogy with Ampère's Law (see Examples 4 and 5 of Chapter 29), the axial symmetry of the flux dictates that the direction of the induced electric field be tangent to circles around the axis (see Fig. 31.19b). Thus at any value of r, the left side of Faraday's Law in Eq. (31.12) reduces to

$$\oint E_{\parallel}\, ds = 2\pi r E_{\parallel}$$

For the right side of Faraday's Law, we must consider only the flux that crosses the area inside the path of radius r where we want to know the induced electric field. For $r \leq R$, this flux is

$$\Phi_B = BA = B\pi r^2$$

and its rate of change is

$$\frac{d\Phi_B}{dt} = \frac{dB}{dt}\pi r^2$$

Equating the two sides of Eq. (31.12) then gives us

$$2\pi r E_{\parallel} = -\frac{dB}{dt}\pi r^2$$

and we can solve for the electric field:

$$E_{\|} = -\frac{dB}{dt}\frac{r}{2} \qquad \text{for} \quad r \leq R$$

Outside the pole faces, at any value of $r \geq R$ where we want to know the induced electric field, the magnetic flux is just the total flux between the pole faces of area $A = \pi R^2$, so

$$\frac{d\Phi_B}{dt} = \frac{dB}{dt}\pi R^2$$

This gives for Eq. (31.12)

$$2\pi r\, E_{\|} = -\frac{dB}{dt}\pi R^2$$

or

$$E_{\|} = -\frac{dB}{dt}\frac{R^2}{2r} \qquad \text{for} \quad r \geq R$$

At $r = R$, either result gives

$$E_{\|} = -\frac{dB}{dt}\frac{R}{2}$$
$$= -(-2.0 \times 10^4\ \text{T/s}) \times \frac{0.10\ \text{m}}{2} = 1.0 \times 10^3\ \text{V/m}$$

COMMENT: Note that there is an induced electric field even in the region where there is no magnetic field! Such an induced electric field can be large enough to damage delicate electronic equipment near the magnet. In practice, it is advisable to increase and decrease magnetic fields slowly.

Checkup 31.3

QUESTION 1: For each of the cases illustrated in Fig. 31.10, what is the direction of the induced current?

QUESTION 2: If the coil described in Example 3 has 300 loops instead of 150, how does this change the answers?

QUESTION 3: If the magnetic field in Example 3 is increasing rather than decreasing, how does this change the answer?

QUESTION 4: If the current in a long solenoid is increasing, what is the direction of the induced electric field inside and outside the solenoid?

QUESTION 5: A resistor is connected between two railroad tracks; a metal rod across the tracks is pushed toward the resistor (see Fig. 31.20). The Earth's magnetic field is directed into the page. What is the direction of the current, if any, in the resistor?

(A) To the right (B) To the left (C) No direction; current is zero

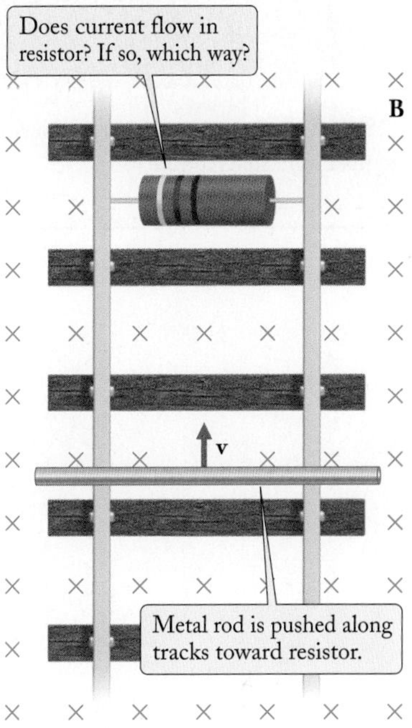

FIGURE 31.20 A metal rod and resistor in contact with railroad tracks.

PHYSICS IN PRACTICE — MAGNETIC LEVITATION

A popular science demonstration features the phenomenon of magnetic levitation, with a permanent magnet floating above a high-temperature superconductor in air (see Fig. 1). Of course, the like poles of two ordinary magnets repel, and so one will float above the other as shown in Fig. 2, but this configuration is not generally stable, and a guide is needed to hold the magnets in place. In the case of the superconductor, stability is readily achieved, even if the magnet is rotating; this behavior is exploited for the construction of high-performance superconducting bearings and gyroscopes.

The origin of superconducting magnetic levitation lies in Faraday's Law; we can understand this with the aid of Fig. 3. As a magnet is brought close to a superconductor, it attempts to increase the flux through the superconductor. But as discussed in Chapter 27, a superconductor has zero electrical resistance and can carry persistent currents. Thus any attempt to generate an emf in the material will result in an induced current sufficient to completely cancel the increase in flux. This simplest behavior, characteristic of type I superconductivity, reflects a property known as perfect diamagnetism: the material responds with an induced current and an internal field which exactly cancels the external field; in terms of the susceptibility discussed in Chapter 30, this means $\chi = -1$ (as mentioned in Section 30.4, ordinary materials can exhibit a much smaller amount of diamagnetism due to induced *atomic* currents). For the superconductor, levitation occurs because the atomic currents in the permanent magnet and the induced

FIGURE 1 A permanent magnet levitating above a high-temperature superconductor.

FIGURE 2 One permanent magnet "floats" above the other, due to their mutual repulsion.

31.4 INDUCTANCE

If a coil carrying a time-dependent current is near some other coil, then the changing magnetic field of the former can induce an emf in the latter. Thus, a time-dependent current in one coil can induce a current in another, nearby coil. For instance, consider the two coils in Fig. 31.21. The first of these coils is connected to some alternating source of emf, such as an AC generator, and it carries a time-dependent current. This coil therefore produces a time-dependent magnetic field $\mathbf{B}_1$. The changing magnetic flux Φ_{B1} through the second coil induces an emf in this coil. According to Faraday's Law, the emf in the second coil is

$$\mathcal{E}_2 = -\frac{d\Phi_{B1}}{dt} \tag{31.13}$$

electric current in the superconductor are oppositely directed, and such currents repel one another.

Commercially useful superconductors do permit some flux to penetrate; these are called type II superconductors. Such penetrating flux can become stuck or "pinned" at defects in the superconductor, aiding in the stability of the supercurrents or the levitation. Even before high-temperature superconductivity, levitation phenomena using conventional type II materials had been exploited. Figure 4 shows a picture of a magnetically levitated train, which uses powerful superconducting magnets made of an alloy of niobium and titanium. Such trains can float at high speed, providing a comfortable and mechanically efficient means of transport. However, the expense of cooling the superconductor has so far been prohibitive, and such trains are not yet in common commercial use.

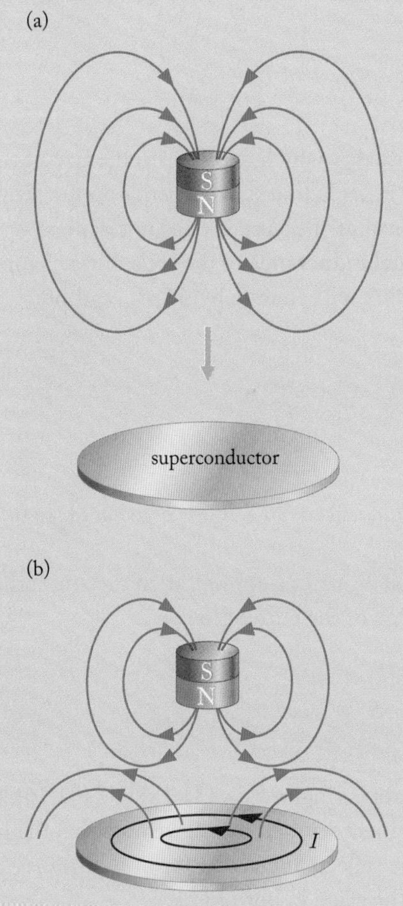

FIGURE 3 (a) A magnet is brought near a superconductor. (b) By Faraday's Law, currents are induced in the superconductor, which oppose the change in flux.

FIGURE 4 A superconductive magnetically levitated (maglev) train.

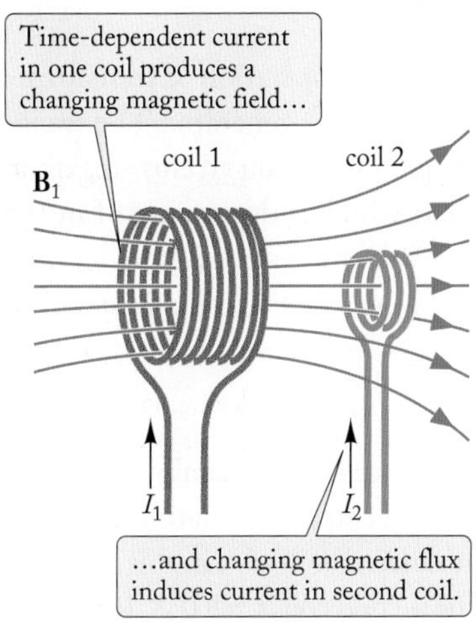

FIGURE 31.21 Coil 1 creates a magnetic field $\mathbf{B}_1$. Some of the field lines pass through coil 2, and create a flux.

The flux Φ_{B1} is proportional to the strength of the magnetic field $\mathbf{B}_1$ in the second coil produced by the current I_1 in the first coil. This field strength is directly proportional to I_1 (an increase of the current by, say, a factor of 2 results in an increase of the magnetic field by the same factor; see Ampère's Law). Hence, the flux Φ_{B1} is also proportional to I_1. We can write the relationship between Φ_{B1} and I_1 as

mutual inductance M

$$\Phi_{B1} = MI_1 \qquad (31.14)$$

where M is a constant of proportionality which depends on the size of the coils, their distance, and the number of turns in each; that is, M depends on the geometry of Fig. 31.21, but M is regarded as a "constant" because it does not depend on the current. The constant M is called the **mutual inductance** of the coils. If we change the current by some small amount dI_1, the flux will change by $M\,dI_1$, and hence Eq. (31.13) becomes

induced emf and inductance

$$\mathcal{E}_2 = -M\frac{dI_1}{dt} \qquad (31.15)$$

This equation states that *the emf induced in coil* 2 *is proportional to the rate of change of the current in coil* 1.

The converse is also true: if coil 2 carries a current, then the emf induced in coil 1 is proportional to the rate of change of the current in coil 2:

$$\mathcal{E}_1 = -M\frac{dI_2}{dt} \qquad (31.16)$$

The constants of proportionality appearing in Eqs. (31.15) and (31.16) are the same. Although we will accept this statement without proof, we note that the result is quite reasonable: the mutual inductance reflects the geometry of the *relative* arrangement of the coils, and that is of course the same in both cases.

The SI unit of inductance is called the **henry** (H):

henry (H)

$$1 \text{ henry} = 1\text{ H} = 1\text{ V}\cdot\text{s/A} \qquad (31.17)$$

Incidentally: The permeability constant μ_0 is commonly expressed in terms of this unit of inductance, so

$$\mu_0 = 4\pi \times 10^{-7}\text{ H/m} \qquad (31.18)$$

transformer

Mutual inductance finds an important application in the operation of **transformers**, used to step up or step down the emf supplied by an AC generator. The transformer consists of two coils—the primary and the secondary—arranged close together. The emf supplied to the primary produces a current which induces an emf in the secondary. This induced emf will be larger or smaller than the original emf, depending on the number of turns in the coils (we will examine transformers in detail in Section 32.6).

EXAMPLE 7 To determine the mutual inductance experimentally, a physicist connects the first coil in Fig. 31.21 to an alternating source of emf and thereby produces a rate of change of the current of 40 A/s in this first coil. He finds that the induced emf measured across the second coil is -8.0×10^{-6} volt. What is the mutual inductance of the two coils?

SOLUTION: Solving Eq. (31.15) for the mutual inductance gives us

$$M = -\frac{\mathcal{E}_2}{dI_1/dt} = -\frac{(-8.0 \times 10^{-6} \text{ volt})}{40 \text{ A/s}} = 2.0 \times 10^{-7} \text{ H}$$

Note that for a given rate of change of current in the first coil, a larger induced voltage in the second coil implies a larger mutual inductance.

EXAMPLE 8 A long solenoid has n turns per unit length. A ring of wire of radius r is placed within the solenoid, perpendicular to the axis (see Fig. 31.22). What is the mutual inductance?

SOLUTION: If the current in the solenoid windings is I_1, the magnetic field is $B_1 = \mu_0 n I_1$ [see Eq. (29.21)], and the flux through the ring is

$$\Phi_{B1} = B_1 \times [\text{area}] = \mu_0 n I_1 \times \pi r^2$$

According to Eq. (31.14), the mutual inductance is

$$M = \frac{\Phi_{B1}}{I_1} = \mu_0 n \pi r^2 \tag{31.19}$$

COMMENTS: Note again that the mutual inductance depends only on the geometry of the two elements, not on any particular values of B_1 or I_1. Note also that if the wire loop instead of the solenoid carried the current, the simple result (31.19) for the mutual inductance also gives us the ratio of the flux through the solenoid to the current I_2, even though that reversed arrangement would be much more difficult to calculate directly.

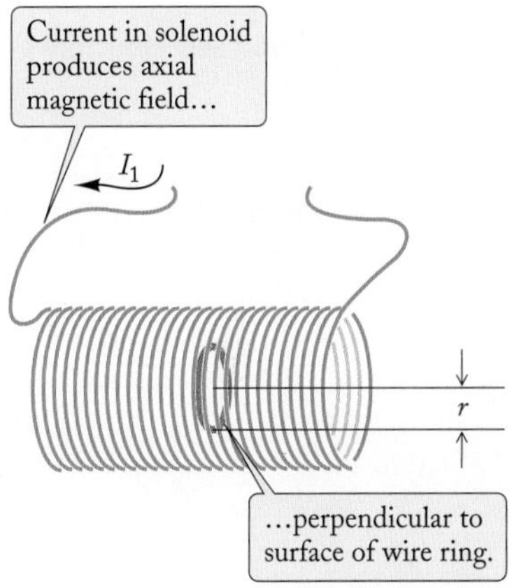

FIGURE 31.22 A ring of wire inside a solenoid.

A conductor by itself has a **self-inductance**. Consider a coil with a time-dependent current. The coil's own magnetic field (see Fig. 31.23) will then produce a time-dependent magnetic flux and, by Faraday's Law, an induced emf. This means that whenever the current is time-dependent, the coil will act back on the current and

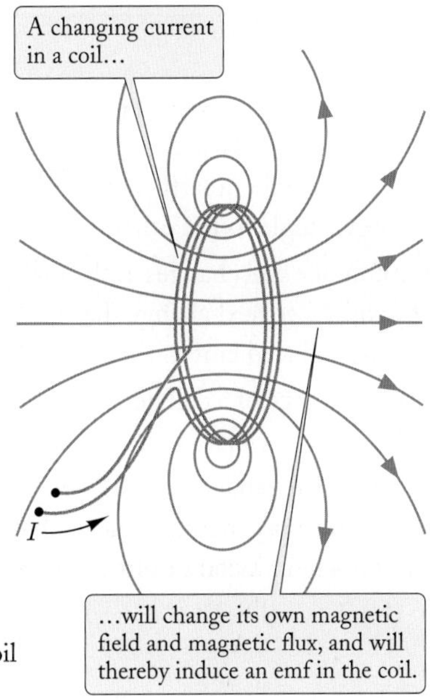

FIGURE 31.23 A coil and its magnetic field.

JOSEPH HENRY (1797–1878)
American experimental physicist. He made important improvements in electromagnets by winding coils of insulated wire around iron pole pieces, and invented an electromagnetic motor and a new, efficient telegraph. He discovered self-induction and investigated how currents in one circuit induce currents in another.

modify it (we will see how to calculate the resultant current in Section 32.3). For this reason the self-induced emf is called a **back emf**. From Lenz' Law, we immediately recognize that *the self-induced emf always acts in such a direction as to oppose the change in the current, that is, it attempts to maintain the current constant.*

The symbol L is used to distinguish the self-inductance from the mutual inductance M. In terms of the flux through the circuit, the definition of the self-inductance is of the same form as Eq. (31.14):

self-inductance L

$$\Phi_B = LI \tag{31.20}$$

and therefore the induced emf is

back emf and self-inductance

$$\mathcal{E} = -L\frac{dI}{dt} \tag{31.21}$$

EXAMPLE 9 A long solenoid has n turns per unit length and a radius R. What is its self-inductance per unit length?

SOLUTION: The magnetic field inside the solenoid is $B = \mu_0 nI$ [see Eq. (29.21)]. The number of loops in a length l is nl; each of these loops has a flux $\pi R^2 B$. Hence the flux through all the loops in a length l is

$$\Phi_B = \pi R^2 B \times [\text{number of turns}] = \pi R^2 Bnl = \pi R^2 \mu_0 n^2 Il$$

and the self-inductance for the entire length l is

$$L = \Phi_B/I = \mu_0 n^2 \pi R^2 l \tag{31.22}$$

The self-inductance per unit length is therefore

self-inductance of solenoid

$$L/l = \mu_0 n^2 \pi R^2 \tag{31.23}$$

✔ Checkup 31.4

QUESTION 1: The two coils in Fig. 31.21 are face to face. How would the mutual inductance change if we were to move them farther apart? Bring them closer together? Turn one coil so it faces at a right angle to the other?

QUESTION 2: When a current in one coil changes at the rate of 10^3 A/s, an emf of 6 V is induced in another, nearby coil. If we next change the current in this other coil at the rate of 10^3 A/s, what will be the induced emf in the first coil?

QUESTION 3: Does the self-inductance of a coil depend on its size? Number of loops? Diameter of the wire?

QUESTION 4: Two solenoids have the same radius and length, but the second one has twice as many turns per unit length as the first one. What is the ratio of the self-inductance of the second to the self-inductance of the first?

(A) 1 (B) $\sqrt{2}$ (C) 2 (D) $2\sqrt{2}$ (E) 4

31.5 MAGNETIC ENERGY

Inductors store magnetic energy, just as capacitors store electric energy. When we connect an external source of emf to an inductor and start a current through the inductor, the back emf will oppose the increase of the current and the external emf must do work in order to overcome this opposition and establish the flow of current. This work is stored in the inductor, and it can be recovered by removing the external source of emf from the circuit and, at the same time, closing the resulting gap in the circuit (this can be done with a suitable switch). The current will then continue to flow for a while, at a gradually decreasing rate, because the inductor supplies a back emf which tends to maintain the current (opposes the decrease). Thus, the inductor delivers energy to the current while the current gradually decreases.

To calculate the amount of stored energy, we note that when the current i in the inductor increases at the rate di/dt, the back emf is

$$\mathcal{E} = -L\frac{di}{dt}$$

The inductor does work on the current at a rate given by the usual formula for the electric power [see Eq. (28.22)]:

$$P = \mathcal{E}i = -L\frac{di}{dt}i \qquad (31.24)$$

Here the negative sign implies that the energy is delivered by the current to the inductor rather than vice versa. In a time dt, the energy U stored in the inductor therefore changes by an amount

$$dU = -P\,dt = -\mathcal{E}i\,dt = Li\,di \qquad (31.25)$$

By integrating this from the initial value of the current ($i = 0$) to the final value ($i = I$), we find the final magnetic energy stored in the inductor:

$$U = \int dU = L\int_0^I i\,di = L\left(\frac{i^2}{2}\right)\Bigg|_0^I = L\left(\frac{I^2}{2} - 0\right)$$

that is,

$$U = \tfrac{1}{2}LI^2 \qquad (31.26)$$

magnetic energy in inductor

Thus, *the total energy stored in the inductor is proportional to the square of the current.*

EXAMPLE 10 A solenoid has a radius of 2.0 cm; its winding has one turn of wire per millimeter. A current of 10 A flows through the winding. What is the amount of energy stored per unit length of the solenoid?

SOLUTION: According to Eq. (31.23), the inductance per unit length is

$$L/l = \mu_0 n^2 \pi R^2$$

and so from Eq. (31.26), the energy per unit length is

$$\frac{U}{l} = \frac{1}{2}\left(\frac{L}{l}\right)I^2 = \frac{1}{2}\mu_0 n^2 \pi R^2 I^2 \tag{31.27}$$

$$= \tfrac{1}{2} \times 1.26 \times 10^{-6}\ \text{H/m} \times (1.0 \times 10^3/\text{m})^2 \times \pi \times (0.020\ \text{m})^2 \times (10\ \text{A})^2$$

$$= 7.9 \times 10^{-2}\ \text{J/m}$$

The energy stored in a solenoid can be expressed in terms of the magnetic field. Consider a length l of the solenoid. From Eq. (31.27), the energy associated with this portion is

$$U = \tfrac{1}{2}\mu_0 n^2 \pi R^2 I^2 l \tag{31.28}$$

Since, for the solenoid, $B = \mu_0 n I$, we can write this as

$$U = \frac{1}{2\mu_0} B^2 \pi R^2 l \tag{31.29}$$

and since $\pi R^2 l$ is the volume filled with magnetic field, the energy is

$$U = \frac{1}{2\mu_0} B^2 \times [\text{volume}] \tag{31.30}$$

According to this equation, the quantity $(1/2\mu_0)B^2$ can be regarded as the magnetic energy per unit volume. Thus, the energy density u in the magnetic field is

energy density in magnetic field

$$[\text{energy density}] = u = \frac{1}{2\mu_0} B^2 \tag{31.31}$$

Although we have derived this equation only for the special case of a long solenoid, it turns out to be generally valid (in vacuum and nonmagnetic materials). The magnetic field, just as the electric field, stores energy. The magnetic energy density is proportional to B^2, just as the electric energy density is proportional to E^2 [compare Eq. (25.58)]. Accordingly, the magnetic energy is concentrated in the regions of space where the magnetic field is strong.

EXAMPLE 11 Near the surface, the Earth's magnetic field typically has a strength of 3.0×10^{-5} T and the Earth's atmospheric electric field typically has a strength of 100 V/m. What is the energy density in each field?

SOLUTION: The magnetic energy density is

$$\frac{1}{2\mu_0} B^2 = \frac{(3.0 \times 10^{-5}\,\text{T})^2}{2 \times 1.26 \times 10^{-6}\ \text{H/m}} = 3.6 \times 10^{-4}\ \text{J/m}^3$$

and the electric energy density is [see Eq. (25.58)]

$$\frac{\epsilon_0}{2} E^2 = \frac{8.85 \times 10^{-12}\ \text{F/m}}{2} \times (100\ \text{V/m})^2 = 4.4 \times 10^{-8}\ \text{J/m}^3$$

Thus, the magnetic energy density in our immediate environment is much larger than the electric energy density.

EXAMPLE 12 The coils of the giant generator shown in the chapter photo have a volume of approximately 45 m^3. These stationary coils surround rotating electromagnets, which produce an average magnetic field within the coils of 0.35 T. What is the total magnetic energy stored in the coil volume?

Concepts — in — Context

SOLUTION: The total magnetic energy stored is the product of the magnetic energy density and the volume:

$$U = u \times [\text{volume}] = \frac{1}{2\mu_0}B^2 \times [\text{volume}]$$

$$= \frac{(0.35\ \text{T})^2}{2 \times 1.26 \times 10^{-6}\ \text{H/m}} \times 45\ \text{m}^3 = 2.2 \times 10^6\ \text{J}$$

Checkup 31.5

QUESTION 1: If we compress a (long) solenoid, making its coils more tight, does the self-inductance change?

QUESTION 2: In the magnetic field of a long, straight wire, where is the magnetic energy density largest?

QUESTION 3: Two inductors, with $L = 10 \times 10^{-6}$ H and $L = 20 \times 10^{-6}$ H, respectively, carry identical electric currents. Which has the larger magnetic energy?

QUESTION 4: Consider the energy stored in the magnetic field of a solenoid. If you increase the current in the solenoid by a factor of 2, by what factor do you increase the magnetic energy?

(A) $\sqrt{2}$ (B) 2 (C) $2\sqrt{2}$ (D) 4

31.6 THE *RL* CIRCUIT

The *RL* circuit consists of a resistor and an inductor connected in series to a battery or some other source of emf. This circuit provides a good illustration of the effects of self-inductance on a current. Figure 31.24 is the schematic diagram for this circuit. In the diagram, the inductor is represented by a coiled line, reminiscent of a coil of wire. The inductance is L, the resistance is R, and the emf of the battery is $\mathcal{E}$. The inductor is supposed to be resistanceless; if the wire in the coils of the inductor has some resistance, then this resistance must be included in R.

We assume that the current in the circuit is initially zero, and that the battery is suddenly connected at the initial time $t = 0$. The current then starts to increase. But the self-inductance will generate an emf across the inductor, which, by Lenz' Law, opposes the increase of the current. Because of this self-induced emf, the current in the circuit cannot increase suddenly; it can increase only gradually. The self-induced emf continues to oppose the current and to restrain its growth until this current attains its final, steady value, $I = \mathcal{E}/R$. This steady, final value of the current is simply $\mathcal{E}/R$, as though the inductor were absent, because when the current becomes steady, the inductor ceases to contribute to the emf in the circuit, that is, it ceases to affect the circuit. Qualitatively, the gradual growth of current in an *RL* circuit is analogous to the increase of charge in an *RC* circuit. As we will see, the plot of the current vs. time

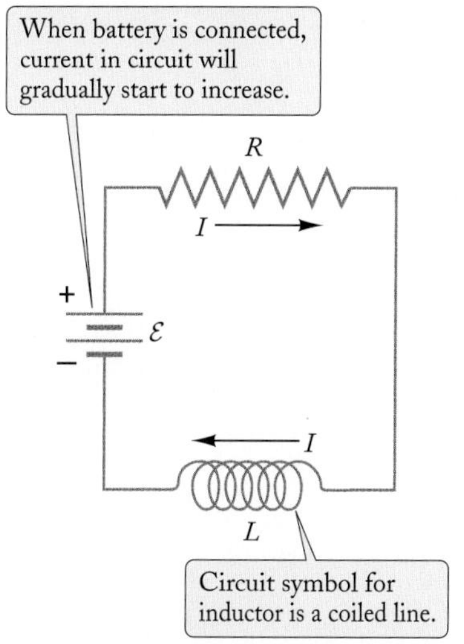

FIGURE 31.24 An *RL* circuit, consisting of a resistor and an inductor connected in series to a battery.

in the RL circuit is mathematically similar to the plot of the charge vs. time in the RC circuit.

For a mathematical treatment of the current in the circuit we turn to Kirchhoff's voltage rule: the sum of all the emfs and voltage drops around the circuit must be zero. The emf of the battery is $\mathcal{E}$. If the current at some instant is I, the voltage drop across the resistor is $-IR$ and the self-induced emf across the inductor is $-L\,(dI/dt)$. Hence

$$\mathcal{E} - IR - L\frac{dI}{dt} = 0 \tag{31.32}$$

At the initial time, $I = 0$ and Eq. (31.32) tells us that the initial potential drop occurs across the inductor, $\mathcal{E} = L(dI/dt)$. After a long time, the current is steady, so $dI/dt = 0$, and the entire potential drop occurs across the resistor, $\mathcal{E} = IR$. At intermediate times, both the resistor and the inductor contribute to the potential drop.

To calculate the current as a function of time, Eq. (31.32) may be solved in a manner identical to that used in Section 28.7 for the RC circuit. Following the technique used in Eqs. (28.28)–(28.33), such direct integration yields for the current

$$I = \frac{\mathcal{E}}{R}\left[1 - e^{-t/(L/R)}\right] \tag{31.33}$$

The quantity L/R that occurs in the exponential has the units of time, and is referred to as the **characteristic time** τ for the process of changing the current in the circuit:

characteristic time τ

$$\tau = \frac{L}{R} \tag{31.34}$$

Thus in terms of the characteristic time, the current is written

increase of current in RL circuit

$$I = \frac{\mathcal{E}}{R}\left(1 - e^{-t/\tau}\right) \tag{31.35}$$

Figure 31.25 is a plot of current vs. time according to this formula, showing the increase in the current from its initial value of zero to its final value of $\mathcal{E}/R$. At the characteristic time $t = \tau = L/R$, the current reaches

$$I = \frac{\mathcal{E}}{R}(1-e^{-1}) = \frac{\mathcal{E}}{R}\left(1 - \frac{1}{2.718}\right) = \frac{\mathcal{E}}{R} \times 0.6321$$

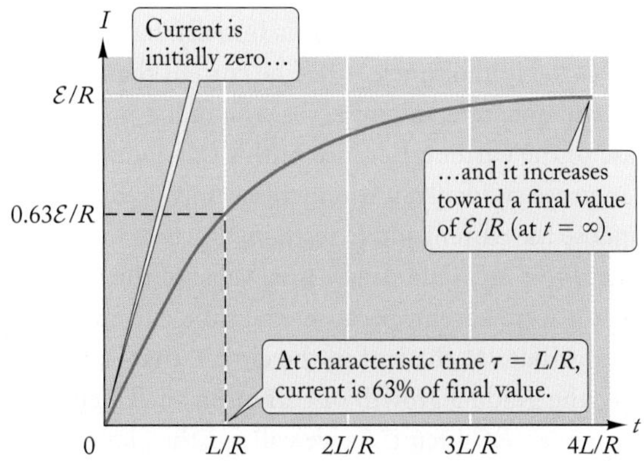

FIGURE 31.25 Growing current in the RL circuit as a function of time.

Thus, at the *characteristic time* $\tau = L/R$, *the current reaches approximately* 63% *of its final value.*

Although in the schematic diagram in Fig. 31.24 the inductance and the resistance are shown separated, this diagram can equally well represent a coil of resistive wire, such as the coil making up the windings of an electromagnet, which has both inductance and resistance. For purposes of calculation, the inductance and the resistance of such a coil may be regarded as placed in series, since each produces its own change of potential in the circuit. From our results for the RL circuit we therefore see that whenever an electromagnet is suddenly switched on, the current and the magnetic field in the magnet take a while to build up to steady values.

EXAMPLE 13 The windings of a large electromagnet have an inductance of 10.0 H and a resistance of 8.00 Ω. This electromagnet is connected to an external emf of 230 V. After the electromagnet is switched on, how long does the current take to build up to 63% of its final value? What is the final, steady value of the current that is attained after a fairly long time?

SOLUTION: The characteristic time for this electromagnet is

$$\tau = \frac{L}{R} = \frac{10.0\ \text{H}}{8.00\ \Omega} = 1.25\ \text{s}$$

This is the time required to reach 63% of the final value of the current.

The final, steady value of the current is

$$I = \frac{\mathcal{E}}{R} = \frac{230\ \text{V}}{8.00\ \Omega} = 28.8\ \text{A}$$

This value of the current is attained after a long time, that is, a time long compared with 1.25 s.

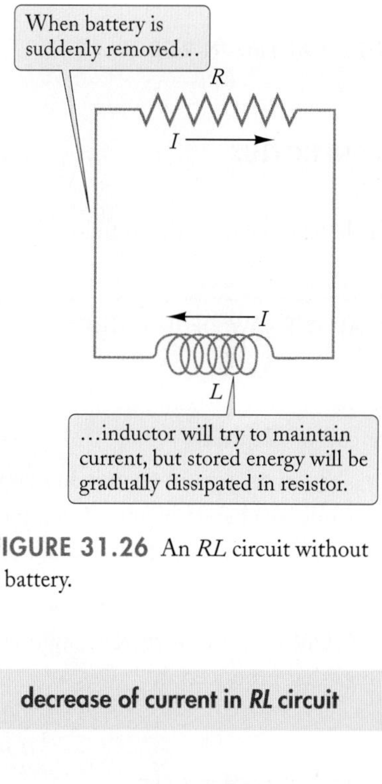

FIGURE 31.26 An RL circuit without a battery.

Once the current has reached its final steady value $\mathcal{E}/R$, it will continue to flow without change as long as the battery continues to provide a steady emf $\mathcal{E}$. But if we remove the battery and, at the same time, close the resulting gap in the circuit (see Fig. 31.26), the inductor will try to maintain the current; however, the stored energy will gradually be dissipated by I^2R loss in the resistor. The result is a gradual decrease of the current from $\mathcal{E}/R$ to 0. The formula that describes the decrease of the current involves the same exponential function as in Eq. (31.33), with the same characteristic time $\tau = L/R$:

$$I = \frac{\mathcal{E}}{R} e^{-t/(L/R)} \qquad (31.36)$$

decrease of current in *RL* circuit

✔ Checkup 31.6

QUESTION 1: Suppose that we switch the electromagnet described in Example 13 off, by removing the emf and, and at the same time, closing the resulting gap with a wire of zero resistance. How long does it take for the current in the windings to fall to 37% of its initial value?

QUESTION 2: An RL circuit with a large inductance has a long relaxation time. Why does this make sense?

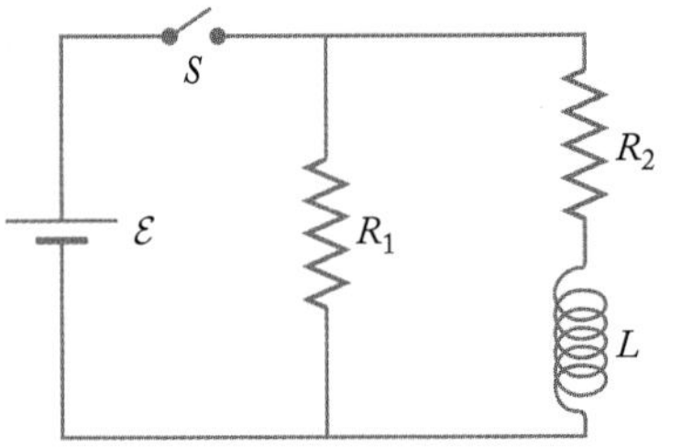

FIGURE 31.27 An RL circuit with two resistors.

QUESTION 3: An RL circuit with a low resistance has a long relaxation time. Why does this make sense?

QUESTION 4: According to Fig. 31.25, at what time does the current in the RL circuit reach one-half of the final value?

QUESTION 5: The switch S in Fig. 31.27 has been open for a long time. At $t = 0$ it is closed. What is the current through the inductor immediately after closing the switch?

(A) 0 (B) $\mathcal{E}/R_1$ (C) $\mathcal{E}/R_2$ (D) $\mathcal{E}/(R_1 + R_2)$ (E) $\mathcal{E}(R_1 + R_2)/R_1R_2$

SUMMARY

PROBLEM-SOLVING TECHNIQUES Faraday's Law; Lenz' Law **(page 1005)**

PHYSICS IN PRACTICE Magnetic Levitation **(page 1008)**

MOTIONAL EMF IN ROD

$$\mathcal{E} = lvB \tag{31.1}$$

MAGNETIC FLUX

$$\Phi_B = \int \mathbf{B} \cdot d\mathbf{A} = \int B_\perp \, dA \tag{31.4}$$

For flat surface in uniform field

$$\Phi_B = B_\perp A = BA\cos\theta \tag{31.5}$$

FARADAY'S LAW OF INDUCTION

$$\mathcal{E} = -\frac{d\Phi_B}{dt} \tag{31.7}$$

For coil of N turns:

$$\mathcal{E} = -N\frac{d\Phi_B}{dt} \tag{31.9}$$

$d\Phi_B/dt$ can arise from change in magnetic field (Example 3), change in area (Example 5), or change in orientation of area (Example 4).

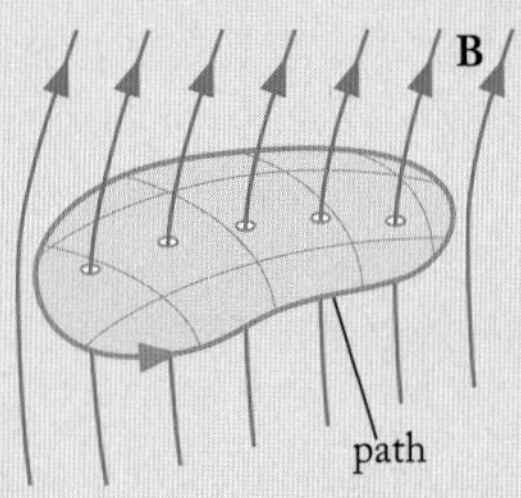

LENZ' LAW Induced current opposes the *change* that produced it.

INDUCED ELECTRIC FIELD

$$\mathcal{E} = \oint \mathbf{E} \cdot d\mathbf{s} = \oint E_\parallel \, ds = -\frac{d\Phi_B}{dt} \tag{31.12}$$

For axial magnetic flux,

$$\oint E_\parallel \, ds = E_\parallel \times 2\pi r$$

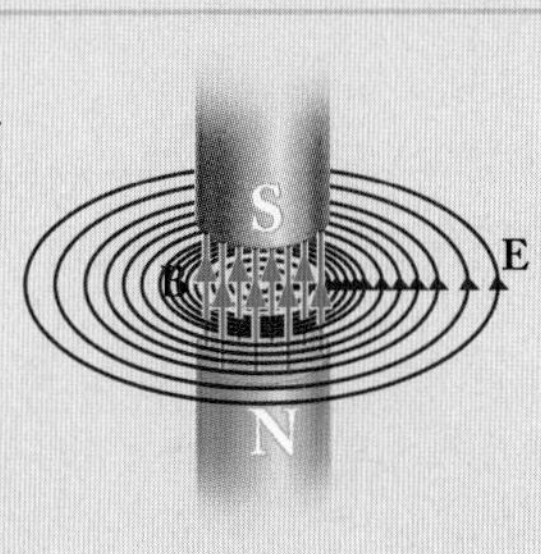

SI UNIT OF INDUCTANCE

$$1\text{ H} = 1\text{ henry} = 1\text{ volt-second/ampere} \tag{31.17}$$

MUTUAL INDUCTANCE M

$$\Phi_{B1} = MI_1 \quad (31.14)$$

$$\mathcal{E}_2 = -M\frac{dI_1}{dt} \quad (31.15)$$

$$\mathcal{E}_1 = -M\frac{dI_2}{dt} \quad (31.16)$$

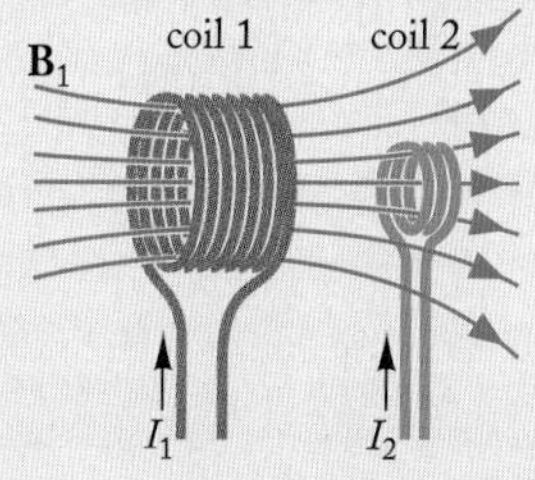

SELF-INDUCTANCE L

$$\Phi_B = LI \quad (31.20)$$

$$\mathcal{E} = -L\frac{dI}{dt} \quad (31.21)$$

SELF-INDUCTANCE OF SOLENOID

$$L = \mu_0 n^2 \pi R^2 l \quad (31.23)$$

MAGNETIC ENERGY IN INDUCTOR

$$U = \tfrac{1}{2}LI^2 \quad (31.26)$$

ENERGY DENSITY IN MAGNETIC FIELD

$$u = \frac{1}{2\mu_0}B^2 \quad (31.31)$$

CHARACTERISTIC TIME OF RL CIRCUIT

$$\tau = L/R \quad (31.34)$$

CURRENT IN RL CIRCUIT

$$I = \frac{\mathcal{E}}{R}(1 - e^{-t/\tau}) \quad \text{(increasing current)} \quad (31.35)$$

$$I = \frac{\mathcal{E}}{R}e^{-t/\tau} \quad \text{(decreasing current)} \quad (31.36)$$

QUESTIONS FOR DISCUSSION

1. At the latitude of the United States, the magnetic field of the Earth has a downward component, larger than the northward component. Suppose that an airplane is flying due west in this magnetic field. Will there be an emf between its wingtips? Which wingtip will be positive? Will there be a flow of current?
2. What is the magnetic flux that the magnetic field of the Earth produces through the surface of the Earth?
3. Is Eq. (31.12) valid for an open path? Is Eq. (31.7) valid for a closed path?
4. A long straight wire carries a steady current. A square conducting loop is in the same plane as the wire. If we push the loop toward the wire, how is the direction of the current induced in the loop related to the direction of the current in the wire?
5. A **flip coil** serves to measure the strength of a magnetic field. It consists of a small coil of many turns connected to a sensitive ammeter. The coil is placed face on in the magnetic field and then suddenly flipped over. How does this indicate the presence of the magnetic field?
6. The **magneto** used in the ignition system of old automobile engines consists of a permanent magnet mounted on the flywheel of the engine. As the flywheel turns, the magnet passes by a stationary coil, which is connected to the spark plug. Explain how this device produces a spark.
7. A long straight wire carries a current that is increasing as a function of time. A rectangular loop is near the wire, in the same plane as the wire. How is the direction of the current induced in the loop related to the direction of the current in the wire?

8. Consider two adjacent rectangular circuits, in the same plane. If the current in one circuit is suddenly switched off, what is the direction of the current induced in the other circuit?

9. Figure 31.28 shows two coils of wire wound around a plastic cylinder. If the current in the left coil is made to increase, what is the direction of the current induced in the coil on the right?

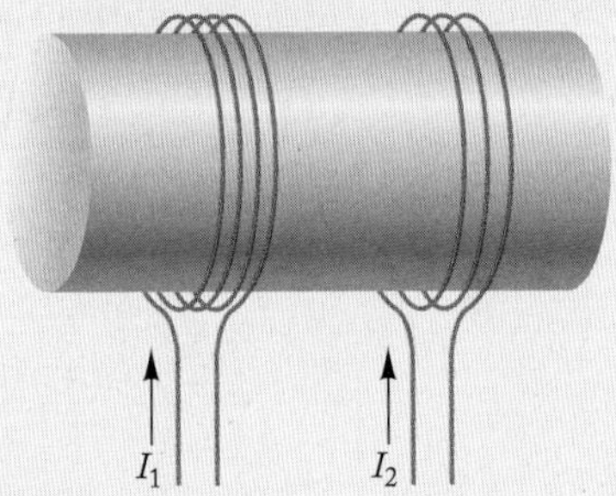

FIGURE 31.28 Two adjacent coils of wire.

10. A circular conducting ring is being pushed toward the north pole of a bar magnet. Describe the direction of the current induced in the ring.

11. Ganot's *Èléments de Physique*, a classic nineteenth-century textbook, states the following rules for the current induced in one loop face to face with another loop:

    I. The distance remaining the same, a continuous and constant current does not induce any current in an adjacent conductor.

    II. A current, at the moment of being closed, produces in an adjacent conductor an inverse current.

    III. A current, at the moment it ceases, produces a direct current.

    IV. A current which is removed, or whose strength diminishes, gives rise to a direct induced current.

    V. A current which is approached, or whose strength increases, gives rise to an inverse induced current.

    Explain these rules on the basis of Lenz' Law.

12. A sheet of aluminum is being pushed between the poles of a horseshoe magnet (Fig. 31.29). Describe the direction of flow of the induced currents, or eddy currents, in the sheet. Explain why there is a strong friction force that opposes the motion of the sheet.

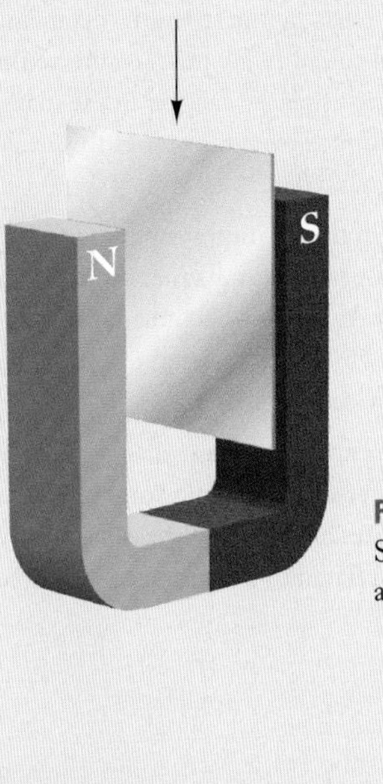

FIGURE 31.29 Sheet of aluminum and horseshoe magnet.

13. Some beam balances use a magnetic damping mechanism to stop excessive swinging of the beam. This mechanism consists of a small conducting plate attached to the beam and a small magnet mounted on a fixed support near this plate. How does this damp the motion of the beam? (Hint: See the preceding question.)

14. A bar magnet, oriented vertically, is dropped toward a flat horizontal copper plate. Explain why there will be a repulsive force between the bar magnet and the copper plate. Is this an elastic force, that is, will the bar magnet bounce if the magnet is very strong?

15. A conducting ring is falling toward a bar magnet (Fig. 31.30). Explain why there will be a repulsive force between the ring and the magnet. Explain why there will be no such force if the ring has a slot cut through it (see dashed lines in Fig. 31.30).

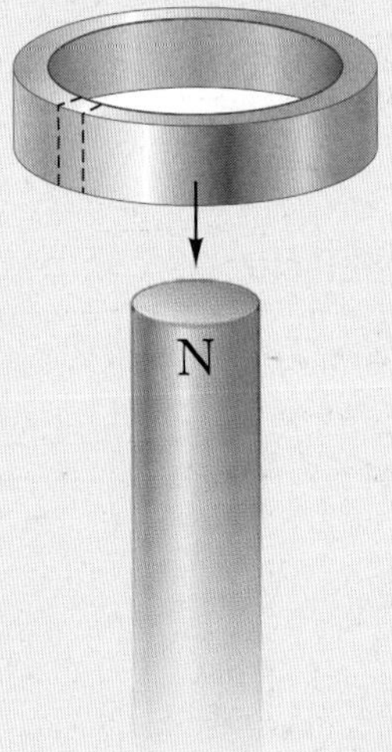

FIGURE 31.30 A conducting ring falling toward a bar magnet. The dashed lines indicate where a slot will be cut through the ring.

16. You have two coils, one of slightly smaller radius than the other. To achieve maximum mutual inductance, should you place these coils face to face or one inside the other?

17. Two circular coils are separated by some distance. Qualitatively, describe how the mutual inductance varies as a function of the orientation of the coils.

18. If a strong current flows in a circuit and you suddenly break the circuit (by opening a switch), a large spark is likely to jump across the switch. Explain.

19. What arguments can you give in favor of the view that the magnetic energy of an inductor is stored in the magnetic field, rather than in the current?

PROBLEMS

31.1 Motional Emf

1. An automobile travels at 88 km/h along a level road. The vertical downward component of the Earth's magnetic field is 5.8×10^{-5} T. What is the induced emf between the right and the left door handles separated by a distance of 1.8 m? Which side is positive and which is negative?

2. The DC-10 jet aircraft has a wingspan of 47 m. If such an aircraft is flying horizontally at 960 km/h at a place where the vertical component of the Earth's magnetic field is 6.0×10^{-5} T, what is the induced emf between its wingtips?

3. In order to detect the movement of water in the ocean, oceanographers sometimes rely on the motional emf generated by this movement of the water through the magnetic field of the Earth. Suppose that, at a place where the vertical magnetic field is 7.0×10^{-5} T, two electrodes are immersed in the water separated by a distance of 200 m measured perpendicularly to the movement of the water. If a sensitive voltmeter connected to the electrodes indicates a potential difference of 7.0×10^{-3} V, what is the speed of the water?

4. A runner runs northward at 9.0 m/s through the Earth's magnetic field, which, at her location, has a downward vertical component of 6.0×10^{-5} T. What is the motional emf between her right shoulder and her left shoulder separated by 50 cm?

5. The rate of flow of a conducting liquid can be measured with an electromagnetic flowmeter that detects the emf induced by the motion of the liquid in a magnetic field. Suppose that a plastic pipe of diameter 10 cm carries beer with a speed of 1.5 m/s. The pipe is in a transverse magnetic field of 1.5×10^{-2} T. What emf will be induced between the opposite sides of the column of liquid?

6. In an X-ray tube, an electron moves parallel to the horizontal component of the Earth's magnetic field with velocity 1.2×10^{7} m/s. The vertical component of the Earth's magnetic field at the tube is 5.5×10^{-5} T. What is the induced electric field in the reference frame of the electron?

*7. Assume that the rod in Fig. 31.2 has a length of 0.86 m, the resistor has value 2.2 Ω, and a magnetic field of 8.0 T is directed into the page. The rod and rails have negligible resistance. The rod is pulled to the left at constant speed. What speed will produce a current of 1.5 A in the resistor? In what direction does the current flow? What pulling force must be applied to maintain a steady current?

*8. An automobile with a vertical radio antenna of length 75 cm travels at 80 km/h eastward. The Earth's magnetic field has a magnitude of 0.70×10^{-4} T and is directed 52° downward with respect to due north. What is the emf generated between the bottom and top of the antenna? Which end is at the higher potential?

†31.2 Faraday's Law

†31.3 Some Examples; Lenz' Law

9. A helicopter has blades of length 4.0 m rotating at 3.0 rev/s in a horizontal plane. If the vertical component of the Earth's magnetic field is 6.5×10^{-5} T, what is the induced emf between the tip of a blade and the hub?

10. The plane of a coil is initially at right angles to the direction of an applied magnetic field. By what angle must the coil be turned to *reverse* the direction of the flux and reduce the magnitude of the flux through the coil to 30% of its initial value?

11. In Idaho, the magnetic field of the Earth points downward at an angle of 69° below the horizontal. The strength of the magnetic field is 5.9×10^{-5} T. What is the magnetic flux through 1.0 m^2 of ground in Idaho?

12. An electromagnetic generator of the kind described in Example 4 has a coil of area 2.0×10^{-4} m^2 with 300 turns of wire. What is the amplitude of the alternating emf delivered by this coil when rotating at a rate of 2000 rev/min in a magnetic field of 0.020 T?

13. A homopolar generator consists of a metal disk rotating about a horizontal axis in a uniform horizontal magnetic field. The external circuit is connected to contact brushes touching the disk at the rim and at the axis. If the radius of the disk is 1.2 m and the strength of the magnetic field is 6.0×10^{-2} T, at what rate (rev/s) must you rotate the disk to obtain an emf of 6.0 V? (Hint: Any radius of the disk can be regarded as a conducting rod; see Example 5.)

14. Pulsars, or neutron stars, rotate at a fairly high speed, and they are surrounded by strong magnetic fields. The material in neutron stars is a good conductor, and hence a motional emf is induced between the center of the neutron star and the rim (this is similar to the emf induced in a rotating metallic rod; see Example 5). Suppose that a neutron star of radius 10 km rotates at the rate of 30 rev/s and that the magnetic field has a strength of 1.0×10^{8} T. What is the emf induced between the center of the star and a point on its equator?

15. Large superconducting magnets are used in hospitals to produce pictures of the interior of the body by magnetic resonance imaging (MRI). For this purpose, the patient is shoved within the coils of the magnet, where the magnetic field is 1.5 T (see Fig. 31.31). Suppose that the patient is shoved into the magnetic field in a time of 10 s. Estimate the emf induced around the patient's trunk, 0.90 m in circumference. Should the patient be shoved into the magnetic field more slowly?

† For help, see Online Concept Tutorial 34 at www.wwnorton.com/physics

FIGURE 31.31 Large magnet used for MRI.

16. The bar magnet shown in Fig. 31.32 is thrust through the coil at a constant speed. Consider the following stages: (a) when the north pole approaches the coil, (b) when the magnet is centered in the coil, and (c) when the south pole leaves the right side of the coil. What is the direction of the current in the resistor at each of these stages?

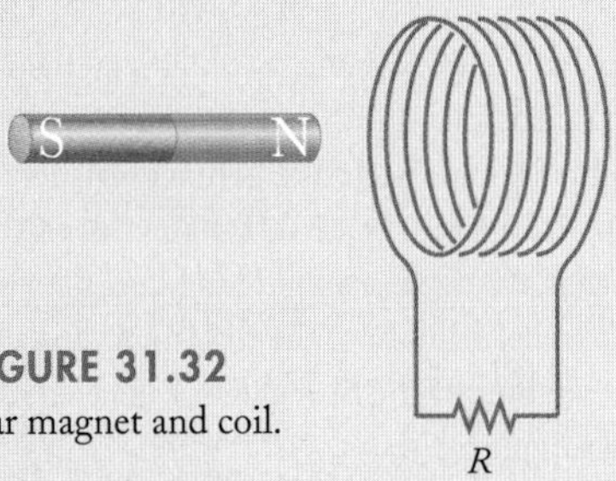

FIGURE 31.32
Bar magnet and coil.

17. A circular coil with 250 turns of 1.0-mm-diameter copper wire has an area of 0.35 m^2. A magnetic field directed at 30° with respect to the plane of the coil increases uniformly from zero to 5.5 T in 35 s. What is the emf induced across the coil during this time? If the ends of the coil are connected, what current flows? How much electrical energy is dissipated during this time?

*18. A circular loop of wire is placed in a magnetic field of 0.30 T while the free ends of the wire are attached to a 15-Ω resistor as shown in Fig. 31.33. When you squeeze the loop, the area of the loop is reduced at a constant rate from 200 to 100 cm^2 in 0.020 s. What are the magnitude and direction of the current in the resistor?

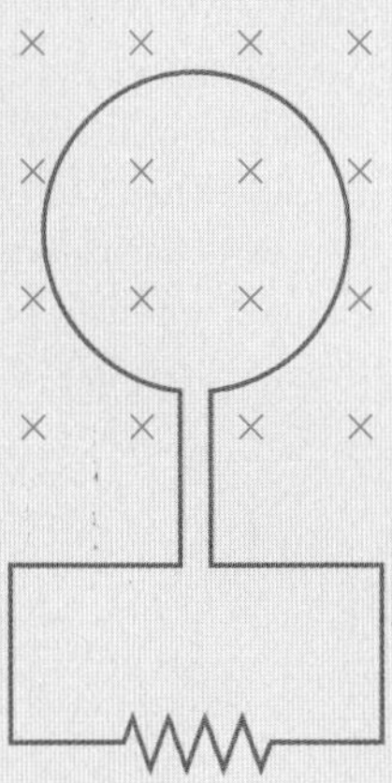

FIGURE 31.33 Circular loop in magnetic field.

*19. A compact disc (CD) is placed in a magnetic field of 1.5 T and rotates at 210 rev/min about an axis parallel to the field. What is the emf generated between a point on its outer track (radius 5.8 cm) and a point on its inner track (radius 2.3 cm)?

*20. A long rectangular conducting loop of width 25 cm is partially in a region of a horizontal magnetic field of 1.8 T perpendicular to the loop, as shown in Fig. 31.34. The mass of the loop is 12 g, and its resistance is 0.17 Ω. If the loop is released, what is its terminal velocity? Assume that the top of the loop stays in the magnetic field. (Hint: The terminal velocity occurs when the magnetic force on the induced current is equal in magnitude to the gravitational force.)

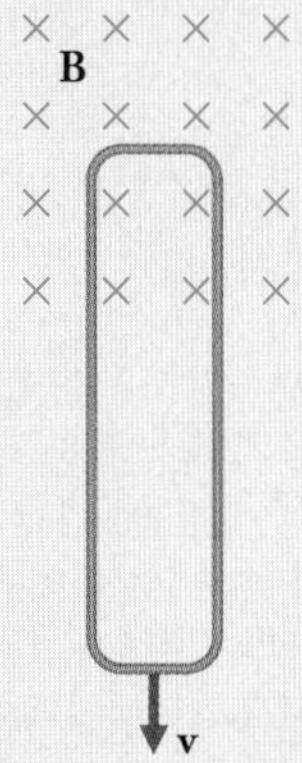

FIGURE 31.34 Falling rectangular loop partially in magnetic field.

*21. A rectangular loop of width $w = 0.30$ m and height $h = 0.50$ m is coplanar with and a distance $d = 0.10$ m from a long wire carrying a current of 2.5 A, as shown in Fig. 31.35. What is the flux through the loop? [Hint: Use Eq. (31.4) to sum the fluxes through vertical strips of area $dA = h\,dr$.]

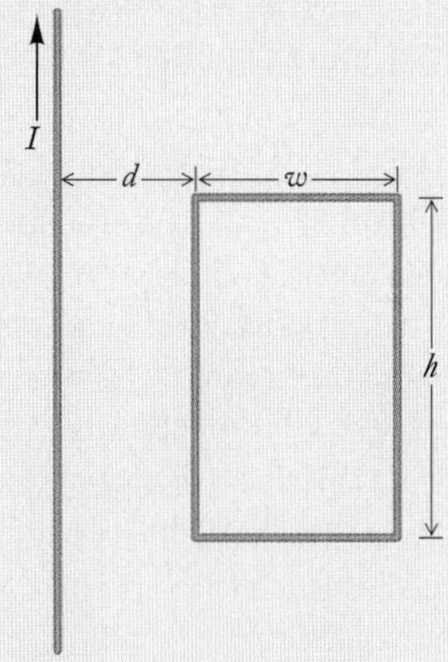

FIGURE 31.35 Long, straight wire and rectangular loop.

*22. A toroid has 150 turns on a rectangular cross section of height 4.0 cm and width 6.5 cm, as shown in Fig. 31.36. The inner radius of the toroid is $R = 7.0$ cm. The current in the wire of the toroid is $I = 2.0$ A. What is the flux through a circular path shown in the figure? [Hint: Use Eq. (31.4) to sum the fluxes through vertical strips of area $dA = h\,dr$.]

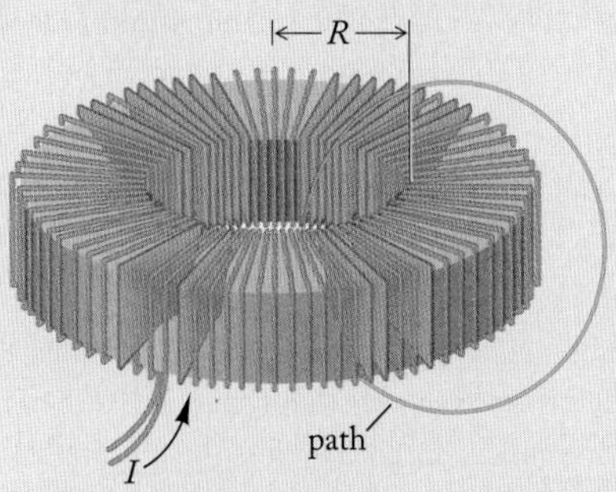

FIGURE 31.36 A toroid with rectangular cross section and a circular path.

*23. The current in a long solenoid of radius $R = 3.0$ cm is increasing uniformly at a rate of 1.5 A/s. The solenoid has 350 turns per meter. What is the induced electric field a distance r from the solenoid axis when (a) $r = 2.0$ cm and (b) $r = 4.0$ cm?

*24. The plane of a circular wire loop of radius 3.5 cm is initially perpendicular to a magnetic field of 8.2 T. The loop is then rotated 90°, so the magnetic field is parallel to the plane of the loop. How much charge flows past any point on the loop during this process? The resistance of the loop is 0.25 Ω.

*25. A metal disk with negligible resistance is 12 cm in diameter. The disk is rotated at 300 rev/s while in a 5.5-T magnetic field perpendicular to the disk. One end of a 33-Ω resistor is in contact with the center of the disk; the other end of the resistor is in contact with the edge of the disk. How much current flows in the resistor? What torque must be supplied to the disk to maintain this current?

*26. A very long train whose metal wheels are separated by a distance of 4 ft 9 in. is traveling at 80 mi/h on a level track. The vertical component of the magnetic field of the Earth is 6.2×10^{-5} T.

(a) What is the induced emf between the right wheels and the left wheels?

(b) The wheels are in contact with the rails, and the rails are connected by metal cross ties which close the circuit and permit a current to flow from one rail to the other. Since the number of cross ties is very large, their combined resistance is nearly zero; most of the resistance of the circuit is within the train. What is the current that flows in each axle of the train? The axles are cylindrical rods of iron of diameter 3 in. and length 4 ft 9 in.

(c) Calculate the power dissipated and the effective friction force on the train.

*27. A square loop of dimension 8.0 cm × 8.0 cm is made of copper wire of radius 1.0 mm. The loop is placed face on in a magnetic field which is increasing at the constant rate of 80 T/s. What induced current will flow around the loop? Draw a diagram showing the direction of the field and the induced current.

*28. A very long solenoid with 20 turns per centimeter of radius 5.0 cm is surrounded by a rectangular loop of copper wire. The rectangular loop measures 10 cm × 30 cm, and its wire has a radius of 0.050 cm. The resistivity of copper is 1.7×10^{-8} Ω·m. Find the induced current in the rectangular loop if the current in the solenoid is increasing at the rate of 5.0×10^4 A/s.

*29. A rectangular loop measuring 20 cm × 80 cm is made of heavy copper wire of radius 0.13 cm. Suppose you shove this loop, short side first, at a speed of 0.40 m/s into a magnetic field of 5.0×10^{-2} T. The rectangle is face on to the magnetic field, and the trailing short side remains outside the magnetic field (see Fig. 31.37). What induced current will flow around the loop?

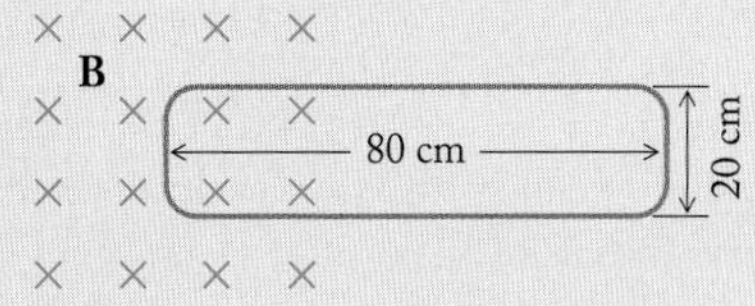

FIGURE 31.37 Rectangular loop partially in magnetic field.

*30. A washer (annulus) of aluminum is lying on top of a vertical solenoid (see Fig. 31.38). When the current in the solenoid is suddenly switched on, the washer flies upward. Carefully explain why the end of the solenoid exerts a repulsive force on the washer under these conditions. (Hint: Take into account that, at the end of the solenoid, the magnetic field lines spread out.)

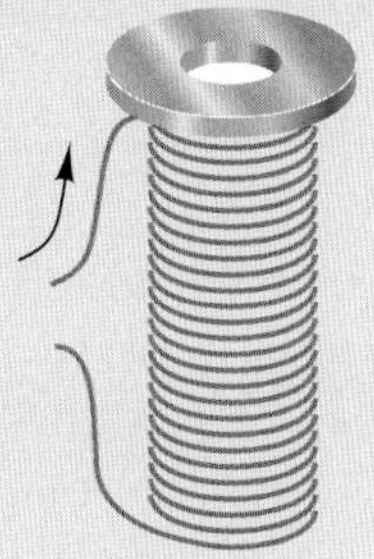

FIGURE 31.38 Washer lying on top of a solenoid.

*31. The magnetic field of a betatron of radius 1.0 m has an amplitude of oscillation of 0.90 T and a frequency of 60 Hz. What is the amplitude of oscillation of the induced electric field at a radius of 0.80 m? At 1.5 m? What is the amplitude of oscillation of the induced emf around circular paths of each of these radii? (Consider the betatron to be a circular region of uniform magnetic field.)

**32. A square loop of dimension $l \times l$ is moving at a speed v toward a straight wire carrying a current I. The wire and the loop are in the same plane, and two of the sides of the loop are parallel to the wire. What is the induced emf of the loop as a function of the distance d between the wire and the nearest side of the loop?

**33. A circular coil of insulated wire has a radius of 9.0 cm and contains 60 turns of wire. The ends of the wire are connected in series with a 15-Ω resistor closing the circuit. The normal to the loop is initially parallel to a constant magnetic field of 5.0×10^{-2} T. If the loop is flipped over, so that the direction of the normal is reversed, a pulse of current will flow through the resistor. What amount of charge will flow through the resistor? Assume that the resistance of the wire is negligible compared with that of the resistor. (Hint: Suppose that the flipping takes a time Δt. What is the average rate of change of the magnetic flux? The average induced emf? The average current?)

31.4 Inductance

34. Two coils are arranged face to face, as in Fig. 31.21. Their mutual inductance is 2.0×10^{-2} H. The current in coil 1 oscillates sinusoidally with a frequency of 60 Hz and an amplitude of 12 A:

$$I_1 = 12 \sin(120\pi t)$$

where the current is measured in amperes and the time in seconds.

(a) What is the magnetic flux that this current generates in coil 2 at time $t = 0$?

(b) What is the induced emf that this current induces in coil 2 at time $t = 0$?

(c) What is the direction of the induced current in coil 2 at time $t = 0$, according to Lenz' Law? Assume that the positive direction for the current I_1 is as shown by the arrow in the figure.

35. A current of 15 A in a coil produces a magnetic flux of 0.10 T·m^2, or 0.10 Wb, through each of the turns of an adjacent coil of 60 turns. What is the mutual inductance?

36. A long solenoid has 400 turns per meter. A coil of radius 1.0 cm with 30 turns of insulated wire is placed inside the solenoid, its axis parallel to the axis of the solenoid. What is the mutual inductance? What emf will be induced around the coil if the current in the solenoid windings changes at the rate of 200 A/s?

37. A loop of wire carrying a current of 100 A generates a magnetic flux of 50 T·m^2, or 50 Wb, through the area bounded by the loop.

(a) What is the self-inductance of the loop?

(b) If the current is decreased at the rate of $dI/dt = 20$ A/s, what is the induced emf?

38. A long solenoid has 2000 turns per meter and a radius of 2.0 cm.

(a) What is the self-inductance for a 1.0-m segment of this solenoid?

(b) What back emf will this segment generate if the current in the solenoid is changing at the rate of 3.0×10^2 A/s?

39. A solenoid of self-inductance 2.2×10^{-3} H in which there is initially no current is suddenly connected in series with the poles of a 24-V battery. What is the instantaneous initial rate of increase of the current in the solenoid?

40. A 7.5-mH inductor carries a time-dependent current given by $I = C_1 t - C_2 t^2$, where t is in seconds, $C_1 = 65$ A/s, and $C_2 = 25$ A/s^2. What is the induced emf across the inductor at $t = 1.0$ s? At $t = 2.0$ s?

41. A superconducting solenoid has an inductance of 25 H. If the current in the solenoid is to increase at a rate of 0.075 A/s, what emf must be supplied across the terminals of the solenoid?

42. In a fast digital circuit, the timing of signals is often limited by the inductance of circuit components. Suppose that a 5.0-V emf is suddenly applied to an effective inductance of 2.5 μH. How long does it take for the current in the inductor to reach 2.0 mA?

*43. A long solenoid of radius R has n turns per unit length. A circular coil of wire of radius R' with 200 turns surrounds the solenoid (Fig. 31.39). What is the mutual inductance? Does the shape of the coil of wire matter?

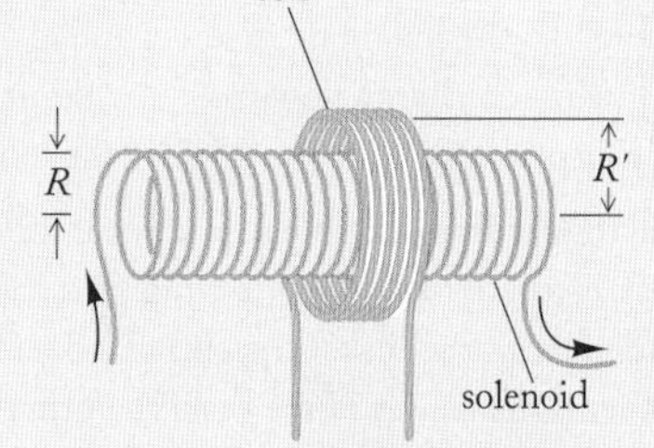

FIGURE 31.39 A long solenoid surrounded by a circular coil.

*44. Two long concentric solenoids of n_1 and n_2 turns per unit length have radii R_1 and R_2, respectively (Fig. 31.40). What is the mutual inductance per unit length of the solenoids? Assume $R_1 < R_2$.

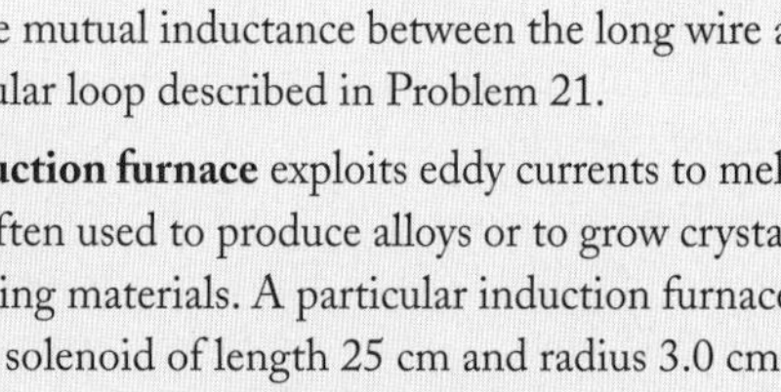

FIGURE 31.40 Two long concentric solenoids.

*45. Find the mutual inductance between the long wire and the rectangular loop described in Problem 21.

*46. An **induction furnace** exploits eddy currents to melt metals and is often used to produce alloys or to grow crystals of conducting materials. A particular induction furnace uses a 15-turn solenoid of length 25 cm and radius 3.0 cm. The current in the solenoid oscillates sinusoidally according to $I = I_{max} \sin \omega t$ where $I_{max} = 2.5$ A and $\omega = 1.2 \times 10^7$ rad/s. What is the maximum voltage induced across the solenoid? Neglect any resistance.

**47. Two inductors of self-inductance L_1 and L_2 are connected in parallel. The inductors are magnetically shielded from one another so that neither produces flux in the other. Show that the self-inductance of the combination is given by $1/L = 1/L_1 + 1/L_2$.

**48. Two inductors of self-inductance L_1 and L_2 are connected in series. The inductors are magnetically shielded from one another so that neither produces flux in the other. Show that the self-inductance of the combination is $L = L_1 + L_2$.

31.5 Magnetic Energy

49. A ring of thick wire has a self-inductance of 4.0×10^{-8} H. How much work must you do to establish a current of 25 A in this ring?

50. In a region of vacuum containing a magnetic field of 1.0 T and an electric field of 1.0 V/m, which field has the larger energy density?

51. What is the magnetic energy density at a point 3.0 mm from a long, thin wire carrying a current of 24 A?

52. The strongest magnetic field achieved in a laboratory is about 1.0×10^3 T. This field can be produced only for a short instant by compressing the magnetic field lines with an explosive device. What is the energy density in this field?

53. For each of the first six entries in Table 29.1, calculate the energy density in the magnetic field.

54. For a crude estimate of the energy in the Earth's magnetic field, pretend that this field has a strength of 5.0×10^{-5} T from the ground up to an altitude of 6.0×10^6 m above the ground. What is the total magnetic energy in this region?

55. A superconducting solenoid carries a current of 55 A, has an inductance of 35 H, and produces a magnetic field of 9.0 T. What energy is stored in the solenoid? What is the volume of the solenoid?

56. Suppose that the magnetic energy stored in an inductor is 2.0×10^{-3} J when the current in this inductor is 30 A. What will be the magnetic energy of the same inductor if the current is 60 A? 90 A?

57. The self-inductance of the aluminum cable of a high-voltage transmission line is 6.6×10^{-4} H per kilometer of length. If the cable carries a current of 800 A, what is the energy in the magnetic field of the cable, per kilometer?

*58. According to one proposal, the surplus energy from a power plant could be temporarily stored in the magnetic field within a very large toroid. If the strength of the magnetic field is 10 T, what volume of the magnetic field would we need to store 1.0×10^5 kW·h of energy? If the toroid has roughly the proportions of a doughnut, roughly what size would it have to be? (Hint: The volume of a toroid equals the cross-sectional area multiplied by 2π times the average radius.)

**59. Two long, straight, concentric tubes made of sheet metal carry equal currents in opposite directions. The inner tube has a radius of 1.5 cm and the outer tube a radius of 3.0 cm. The current on the surface of each is 120 A. What is the magnetic energy in a 1.0-m segment of these tubes?

31.6 The *RL* Circuit

60. An inductor with $L = 2.0$ H and a resistor with $R = 100\ \Omega$ are suddenly connected in series to a battery with $\mathcal{E} = 6.0$ V.
   (a) What is the current at $t = 0$? At $t = 0.010$ s? At $t = 0.020$ s?
   (b) What is the final, steady value of the current?
   (c) What is the rate of increase of the current at $t = 0$?

61. Consider the *RL* circuit shown in Fig. 31.26. At time $t = 0$, the circuit has an initial current $\mathcal{E}/R$. Apply Kirchhoff's rule to this circuit and find a differential equation for dI/dt. Show that Eq. (31.36) is a solution to this equation.

62. Design an *RL* circuit with an arrangement of switches so that the battery can be suddenly switched out of the circuit and the current in the inductor can be suddenly fed into the resistor. If the self-inductance of the inductor is 0.20 H, what resistance do you need to obtain a characteristic time of 10 s in your circuit?

63. A superconducting magnet has an inductance of 45 H and carries a current of 65 A. The terminals supplying current to the magnet are removed immediately after a resistor is connected across them. The current in the magnet drops to zero with a characteristic time of 12 s.
   (a) What is the value of the resistor?
   (b) What is the initial rate of power dissipation in the resistor?
   (c) What is the total energy dissipated in the resistor after a long time?

64. Initially, the switch shown in Fig. 31.41 is open and no currents flow in the circuit. At $t = 0$, the switch is closed.
   (a) What are the currents in the inductor, in resistor R_1, and in resistor R_2 immediately after the switch is closed?
   (b) What are those three currents after the switch has been closed for a long time?

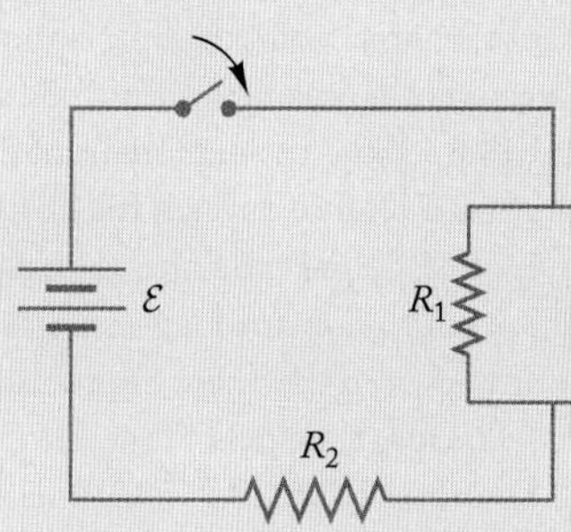

FIGURE 31.41 An *RL* circuit with two resistors.

65. *RL* circuits are sometimes used to generate high-voltage pulses. Consider the circuit shown in Fig. 31.42, where $\mathcal{E} = 12$ V, $R_1 = 1.5$ kΩ, and $R_2 = 6.0\ \Omega$. The switch has been in the upper position, connected to the battery for a long time.

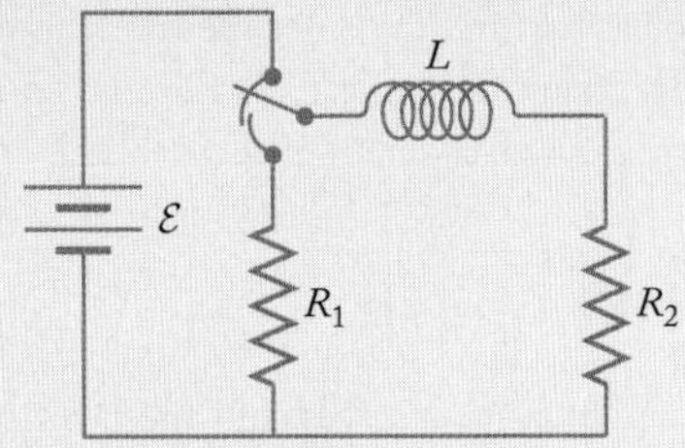

FIGURE 31.42 An RL circuit with two resistors.

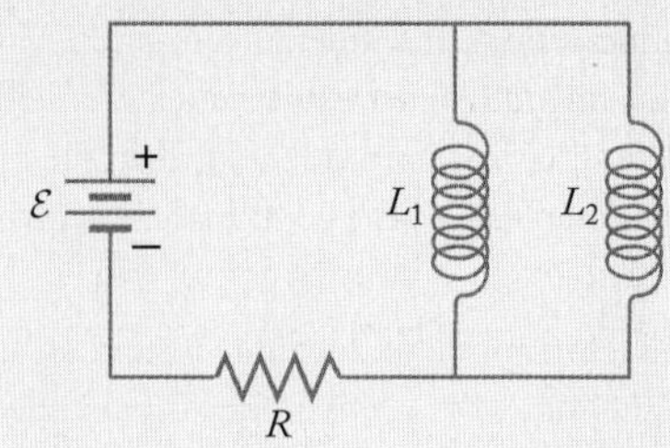

FIGURE 31.43 An RL circuit with two inductors.

(a) What is the current in the inductor?

(b) The switch is suddenly moved to the lower position. What is the voltage across the inductor immediately after the switch is moved to the lower position?

66. Suppose that the switch in Fig. 31.27 has been closed for a long time. At $t = 0$, it is opened. What is the current in the inductor for $t \geq 0$? What is the voltage across R_1 for $t > 0$? Across R_2?

67. Find the time constant of the circuit shown in Fig. 31.42 (a) when the switch is in the upper position and (b) when the switch is in the lower position.

68. An inductor is suddenly connected in series to a resistor with $R = 10\ \Omega$. The initial current in the inductor is 3.4 A. After a time of 6.0×10^{-2} s, the current has dropped to 1.7 A. What is the self-inductance?

*69. An RL circuit consists of two inductors of self-inductance $L_1 =$ 4.0 H and $L_2 = 2.0$ H connected in parallel to each other, and connected in series to a resistor of 6.0 Ω and a battery of 3.0 V (Fig. 31.43). Assume that the inductors have no mutual inductance and no resistance.

(a) When the battery is suddenly connected, what is the initial rate of change of the current in each inductor?

(b) What is the final, steady current in the resistor? What are the final, steady currents in each inductor? (Hint: The currents in the inductors are inversely proportional to their inductances. Why?)

*70. What is the Joule heat dissipated by the current (31.36) in the resistor in the time interval $t = 0$ to $t = \infty$? Compare with the initial magnetic energy in the inductor.

*71. To measure the self-inductance and the internal resistance of an inductor, a physicist first connects the inductor across a 3.0-V battery. Under these conditions, the final, steady current in the inductor is 24 A. The physicist then suddenly short-circuits the inductor with a thick (resistanceless) wire placed across its terminals. The current then decreases from 24 A to 12 A in 0.22 s. What are the self-inductance and the internal resistance of the inductor?

REVIEW PROBLEMS

72. An experiment attempted with the Space Shuttle, but not completed because of mechanical difficulties, was designed to obtain electric power from the motional emf induced by the motion of the Space Shuttle through the Earth's magnetic field. While in orbit around the Earth at an altitude of 296 km at a speed of 7.7 km/s, a 20-km-long wire was to be stretched radially outward between the Space Shuttle and a small "tethered satellite." The magnetic field of the Earth at the altitude of the Space Shuttle is 2.7×10^{-5} T. Calculate the magnitude of the motional emf induced between the ends of such a 20-km wire in this magnetic field. Assume that the motion of the wire is at right angles to the magnetic field.

73. A girl uses a jump rope made of flexible wire. She holds the ends of the rope in her extended hands, so the rope has approximately the shape of a semicircle of radius 0.70 m, and she whirls the rope at the rate of 1.0 revolution per second at a place where the magnetic field of the Earth is nearly vertical, of magnitude 5.0×10^{-5} T. The motion of the wire rope generates an alternating emf between the ends of the rope. What is the amplitude of oscillation of this alternating emf?

74. An electric generator consists of a rectangular coil of wire rotating about its longitudinal axis, which is perpendicular to a magnetic field of 2.00×10^{-2} T. The coil measures 10.0 cm × 20.0 cm and has 120 turns of wire. The ends of the wire are connected to an external circuit. At what speed (in rev/s) must you rotate this coil in order to induce an alternating emf of amplitude 12.0 V between the ends of the wire?

75. A circular current loop has a radius of 1.0 cm and carries a current of 3.0 A. What is the magnetic energy density at a point at the center of the loop?

76. A large toroid built in the early days of plasma research in the former Soviet Union had a major radius of 1.50 m and a minor radius of 0.40 m (Fig. 31.44). If the average magnetic field within such a toroid is 4.0 T, what is the magnetic energy? (Hint: The volume of a toroid equals the cross-sectional area multiplied by 2π times the average radius.)

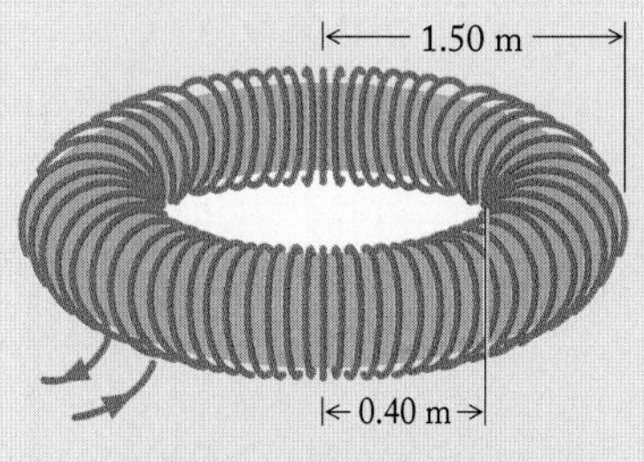

FIGURE 31.44 Giant toroid.

*77. Sharks have delicate sensors on their bodies that permit them to sense small differences in potential (Fig. 31.45). They can sense electrical disturbances created by other fish, and they can also sense the Earth's magnetic field and use this for navigation. Suppose that a shark is swimming horizontally at 25 km/h at a place where the magnetic field has a strength of 4.7×10^{-5} T and points down at an angle of 40° with the vertical. Treat the shark as a cylinder of diameter 30 cm. What is the largest induced emf between diametrically opposite points on the sides of the shark when heading north?

FIGURE 31.45 A shark.

*78. A metal rod of length l and mass m is free to slide, without friction, on two parallel metal tracks. The tracks are connected at one end so that they and the rod form a closed circuit (Fig. 31.46).

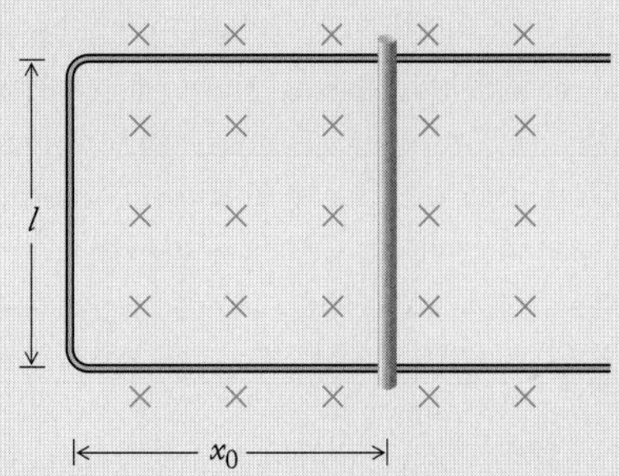

FIGURE 31.46 Rod sliding on parallel tracks. The crosses show the tails of the magnetic field vectors.

The rod has a resistance R, and the tracks have negligible resistance. A uniform magnetic field is perpendicular to the plane of this circuit. The magnetic field is increasing at a constant rate of dB/dt. Initially the magnetic field has strength B_0 and the rod is at rest at a distance x_0 from the connected end of the rails. Express the acceleration of the rod at this instant in terms of the given quantities.

*79. A boxcar of a train is 2.5 m wide, 9.5 m long, and 3.5 m high; it is made of sheet metal and is empty. The boxcar travels at 60 km/h on a level track at a place where the vertical component of the Earth's magnetic field is 6.2×10^{-5} T. Assume that the boxcar is not in electrical contact with the ground.

(a) What is the induced emf between the sides of the boxcar?

(b) Taking into account the electric field contributed by the charges that accumulate on the sides, what is the net electric field inside the boxcar (in the reference frame of the boxcar)?

(c) What is the surface charge density on each side? Treat the sides as two very large parallel plates.

*80. A 25-turn coil of wire has an area of 4.0×10^{-3} m^2 and is oriented perpendicularly to a magnetic field **B**. The coil has a resistance of 15 Ω. At what rate must the magnitude of **B** change for an induced current of 5.0 mA to appear in the coil?

*81. (a) A long solenoid has 300 turns of wire per meter and has a radius of 3.0 cm. If the current in the wire is increasing at the rate of 50 A/s, at what rate does the strength of the magnetic field in the solenoid increase?

(b) The solenoid is surrounded by a coil of wire with 120 turns (Fig. 31.47). The radius of this coil is 6.0 cm. What induced emf will be generated in this coil while the current in the solenoid is increasing?

(c) Suppose we replace the coil of radius 6.0 cm by a new coil of radius 8.0 cm. What will be the induced emf now?

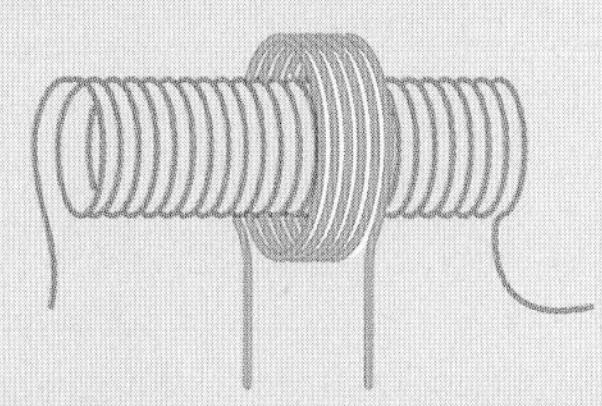

FIGURE 31.47 A long solenoid and a circular coil.

*82. The resistance of the coils of an electric motor is 2.0 Ω when the motor is not in operation (not rotating). When the motor is connected to 115 V and is rotating at full speed, it draws a current of 0.10 A. Deduce the back emf that the coils of the motor produce when rotating.

*83. The current in a 115-volt electric motor jumps from 4.0 A to 36 A when the motor coils are suddenly halted by a brake. What is the back emf when the motor is rotating at full speed?

*84. A current of 5.0 A flows through a cylindrical solenoid of 1500 turns. The solenoid is 40 cm long and has a diameter of 3.0 cm.

(a) Find the magnetic field in the solenoid. Treat the solenoid as very long.

(b) Find the energy density in the magnetic field, and find the magnetic energy stored in the space within the solenoid.

*85. An RL circuit with $L = 0.50$ H and $R = 0.025\ \Omega$ is initially connected to a battery of 1.2 V. When the current reaches its maximum, steady value, the battery is suddenly switched out of the circuit and the current is switched into the resistor.

(a) What is the maximum value of the current? At what time will the current in the inductor drop to 50% of its maximum value?

(b) What is the maximum value of the energy stored in the inductor? What percentage of the energy remains when the current has dropped to 50% of its maximum value?

*86. A long straight wire carries a current that increases at a steady rate dI/dt.

(a) What is the rate of increase of the magnetic field at a radial distance r?

(b) The wire is in the same plane as the rectangular loop shown in Fig. 31.48. What is the induced emf around the loop? Assume that $a \ll r$, that is, the width of the loop is so small that the magnetic field and its rate of change are approximately the same at all points in the loop.

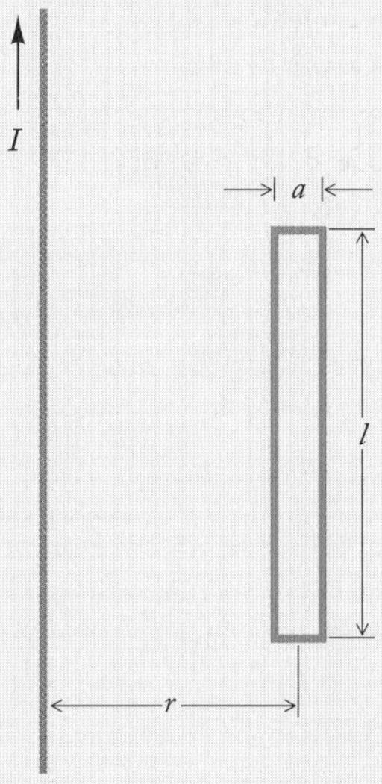

FIGURE 31.48 Long, straight wire and rectangular loop.

Answers to Checkups

Checkup 31.1

1. If the direction of the motion of the rod reverses, the force on the electrons reverses, so the current (the flow of positive charge) also reverses, and flows clockwise around the circuit.

2. Since we are looking in the direction of the horizontal magnetic field (see Fig. 31.3), $\mathbf{v} \times \mathbf{B}$ will point toward the right, so this is the end where positive charge will accumulate. Reversing the velocity (to upward) will reverse the force. making the left end positive.

3. No matter what horizontal direction your velocity is in, for a magnetic field pointing down, $\mathbf{v} \times \mathbf{B}$ will point toward your left side; thus, east, west, and north velocities yield a positive terminal on your north, south, and west side, respectively.

4. (E) To right. The induced electric field in the frame of reference moving with the rod points in the same direction as the magnetic force on a positive charge in the stationary reference frame; in Fig. 31.3, this is toward the right. For a conducting rod, this induced field causes charges to move until the sum of the induced field and the field due to the charges that accumulate at the ends of the rod is zero.

Checkup 31.2

1. Since magnetic field lines do not begin or end anywhere, but instead always form closed loops, any field line that enters a closed surface must also exit the surface, yielding zero net flux.

2. Yes. Reversing the magnetic flux requires it to change direction; an emf will be generated during the time of this sudden change. Since the emf is proportional to the rate of change, for a sudden change, a very large emf is generated during the change.

3. No. For a rod moving in a magnetic field, the induced emf is transverse to the motion, and so would only be generated between the sides of the arrow, not between the tip and the tail.

4. (B) b. The most field lines cross the area shown in Fig. 31.13b, and so this area intercepts the most magnetic flux, even though it is neither the most perpendicular to the magnetic field, nor the largest area.

Checkup 31.3

1. For case (a), if the loop is pulled to the right (out of the field region), the flux through the loop decreases; thus, to oppose

this change, the induced current will be counterclockwise, looking from above (by the right-hand rule). Conversely, if the loop is pushed into the field, the induced current will be clockwise. For case (b), squeezing the loop decreases its area and thus the flux through it, so the induced current will be counterclockwise, looking from above. For case (c), seen from what was the bottom as shown, the induced current will be clockwise to oppose the decrease in flux due to turning the loop. For the increasing field of case (d), the induced current will again be clockwise, seen from above, to oppose this change.

2. The induced emf is proportional to the number of loops, so it will also double to 120 V. However, the resistance will also double for twice as many turns, so the induced current will be the same, 12 A.
3. The induced emf and current will have the same magnitude, but to oppose the increase, they will have the opposite direction.
4. Like the case of the circular electromagnet (Example 6), the electric field will be circles around the axis. Note that for a circle outside the solenoid, there will be an induced electric field, even though there is no magnetic field, because the flux through part of the area bounded by the circle is changing.
5. (A) To the right. As the rod moves, the area of the loop formed by the rod, tracks, and resistor decreases. The induced current will oppose the *decrease* of flux and contribute flux into the page. Thus the current will flow clockwise around the loop. Note that the flux from the induced current opposes the *change* in flux, not the original flux.

Checkup 31.4

1. If we move the coils further apart, the magnetic field due to one will be smaller at the second. Thus the flux will be smaller, and so will the mutual inductance M. If we bring them closer together, the opposite is true: the flux will increase, and so will the mutual inductance. If one coil is at a right angle to the other, there will be no net flux crossing it, and so the mutual inductance will decrease to zero.
2. Since the mutual inductance is the same between a pair of coils regardless of which carries the current, the induced emf will be the same, 6 V.
3. According to Example 9, the self-inductance is inversely proportional to the length of a solenoid and directly proportional to the number of loops squared ($n^2 l = N^2/l$). The diameter of the wire affects how closely a coil can be wound, so its self-inductance can depend on the diameter of the wire; however, for the same number of turns per unit length, the wire diameter does not matter.
4. (E) 4. As in question 3 and Example 9, the self-inductance is proportional to the square of the number of turns, and so is 4 times as large for the second solenoid.

Checkup 31.5

1. From Eq. (31.22), the self-inductance, $L = \Phi_B/I = \mu_0 n^2 \pi R^2 l$, is proportional to $n^2 l = N^2/l$ and so does change (increases) when the solenoid is compressed (when l decreases).
2. The magnetic energy density is largest where the magnetic field is the largest; this occurs at the surface of the wire (see Example 5 of Chapter 29).
3. Since the stored magnetic energy is proportional to the inductance, as in Eq. (31.26), the second inductor has the larger magnetic energy.
4. (D) 4. Since the stored energy is proportional to the square of the current [Eq. (31.26)], doubling the current increases the stored magnetic energy by a factor of 4.

Checkup 31.6

1. The decrease of the current in the electromagnet is governed by the exponential decay of Eq. (31.36); such a decay will reach 37% of its initial value in one characteristic time ($e^{-1} \approx 0.37$). Thus, from Example 13, the time will be $\tau = 1.25$ s.
2. A large inductance implies that the back emf, which opposes the change in current, is large. A large back emf tends to keep the current constant, implying a long relaxation time.
3. A low resistance implies a low "friction," so the current tends to keep going. For example, consider the decaying current of question 1, with some initial current I_0. The initial rate of energy dissipation will be $P = I_0^2 R$, so a lower resistance means that it will take longer for the stored energy in the inductor to dissipate in the resistor.
4. Estimating from the plot in Fig. 31.25, the current reaches one-half of its final value between $t \approx 0.6L/R$ and $0.7L/R$.
5. (A) 0. After the switch has been open for a long time, the current in the inductor must be zero, because there is no applied emf and any current in the right loop will have decayed to zero. Since the current in an inductor must change slowly, the current will still be zero immediately after closing the switch.

CHAPTER

32 Alternating Current Circuits

CONCEPTS IN CONTEXT

Concepts — in — Context

Power companies prefer alternating currents to direct currents because of the ease with which alternating voltages can be stepped up or down by means of transformers, such as the ones shown here. Transformers make it possible to step up the output of a power plant to several hundred thousand volts, transmit the power along a high-voltage line that minimizes Joule heating losses, and finally step down the voltage to 230 volts AC or 115 volts AC just before consumer use in lighting and appliances.

As we learn about AC circuits, we will consider such questions as:

? For the "115-V AC" emf supplied at ordinary outlets in homes, what are the actual maximum and minimum emf? What is the relation between the resistance and the time-average power dissipated in a lightbulb or other resistive device connected to such an outlet? (Section 32.1, pages 1031 and 1033; Example 1, page 1034)

? How does a "dimmer" control the power delivered to a lightbulb? (Example 5, page 1049)

? How does a transformer step up or step down an alternating emf? An alternating current? (Section 32.6, page 1055; and Example 8, page 1056)

The current delivered by power companies to homes and factories is an oscillating, time-dependent current. It periodically flows forward and backward, cycling 60 times per second. This is called **alternating current**, or AC.

alternating current, or AC

All the appliances connected to ordinary outlets in homes therefore involve circuits with oscillating currents. Furthermore, electronic devices—radios, TVs, cellular telephones, computers—involve a variety of circuits with oscillating currents of high frequency. Many of these circuits have a natural frequency of oscillation. Such circuits exhibit the phenomenon of resonance when the natural frequency matches the frequency of a signal applied to the circuit. For instance, in some radios, the tuning relies on an oscillating circuit whose frequency of oscillation is adjusted by means of a variable capacitor (attached to the tuning knob) so that it matches the frequency of the radio signal.

In this chapter, we will discuss how resistors, capacitors, and inductors behave in AC circuits, that is, circuits with oscillating currents. The calculations of currents in AC circuits are similar to the calculations in DC circuits. As we will see, the starting point for these calculations is again Kirchhoff's rule.

32.1 RESISTOR CIRCUIT

Online Concept Tutorial 35

Concepts in Context

In North America, the alternating emf supplied by the power companies to electric outlets in private homes is an alternating emf with an amplitude $\mathcal{E}_{max} = 163$ V and a frequency of 60 Hz. The mathematical formulas describing such an alternating emf are similar to those describing an oscillating particle (see Section 15.1). The angular frequency is $\omega = 2\pi f = 2\pi \times 60$ radians/s, and the time dependence of the emf is given by a cosine function

$$\mathcal{E} = \mathcal{E}_{max} \cos \omega t = 163 \text{ V} \times \cos(2\pi \times 60\, t) \tag{32.1}$$

Recall that the angular frequency $\omega = 2\pi \times 60 \text{ s}^{-1}$ ensures the correct periodicity of the cosine function: each time the time t advances 1/60 second, the argument of the cosine increases by 2π, that is, one cycle. Figure 32.1 shows a plot of the emf as a

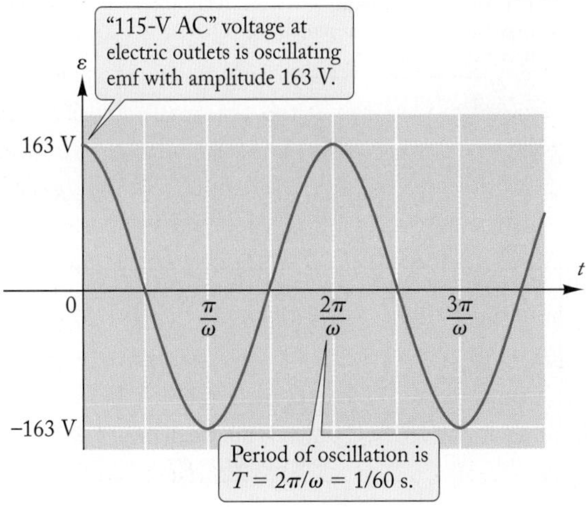

FIGURE 32.1 Emf supplied by an outlet as a function of time.

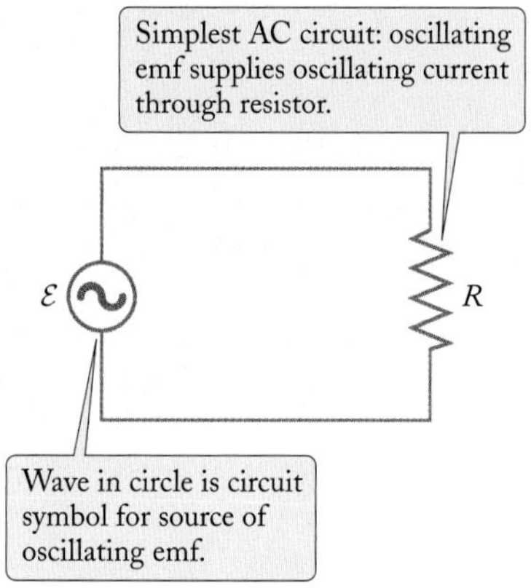

FIGURE 32.2 Resistor connected to a source of alternating emf.

function of time. This kind of voltage is usually called "115 volts AC" for reasons that will become clear shortly. In many European countries, the amplitude of the AC voltage supplied to outlets in homes is twice as large; it is called "230 volts AC." An appliance designed for 115 V AC is likely to overheat and burn out if plugged into 230 V AC. To prevent this, the plugs for 115-V appliances and for 230-V are built in different shapes, so they cannot fit into the wrong kind of outlet.

The simplest possible AC circuit consists of a pure resistor connected to an oscillating source of emf. In the circuit diagram (see Fig. 32.2), the oscillating source of emf is symbolized by a wavy line enclosed in a circle. This circuit might represent an electric heater or an incandescent lamp plugged into an ordinary wall outlet (see Fig. 32.3).

According to Kirchhoff's rule, the sum of emfs and potential changes across resistors around any circuit must be zero. Although we first developed this law for DC circuits, it is equally valid for AC circuits. For the circuit shown in Fig. 32.2, this tells us that, at any instant of time,

$$\mathcal{E} - IR = 0 \tag{32.2}$$

from which we find

current in resistor

$$I = \frac{\mathcal{E}}{R} = \frac{\mathcal{E}_{\text{max}} \cos \omega t}{R} \tag{32.3}$$

FIGURE 32.3 The resistive wire in this toaster becomes orange-hot when connected to a wall outlet, a source of emf.

Thus, *the current oscillates with the same frequency* ω *as the emf,* but whereas the amplitude of the emf is $\mathcal{E}_{\text{max}}$, the amplitude of the current is $\mathcal{E}_{\text{max}}/R$. Figure 32.4 compares plots of the emf and the current. The maxima and the minima of the current and the emf occur simultaneously, and the current is said to be **in phase** with the emf.

The instantaneous electric power dissipated in the resistor is the product of the instantaneous current and the emf, as in the case of a DC circuit [see Eq. (28.22)]:

$$P = \mathcal{E}I = \mathcal{E}_{\text{max}} \cos \omega t \times \frac{\mathcal{E}_{\text{max}} \cos \omega t}{R} = \frac{\mathcal{E}_{\text{max}}^2 \cos^2 \omega t}{R} \tag{32.4}$$

Although the emf and the current both are negative for one-half of each cycle, the power is always positive, because the emf and the current become negative simultaneously (the emf and the current are in phase), and therefore their product is always

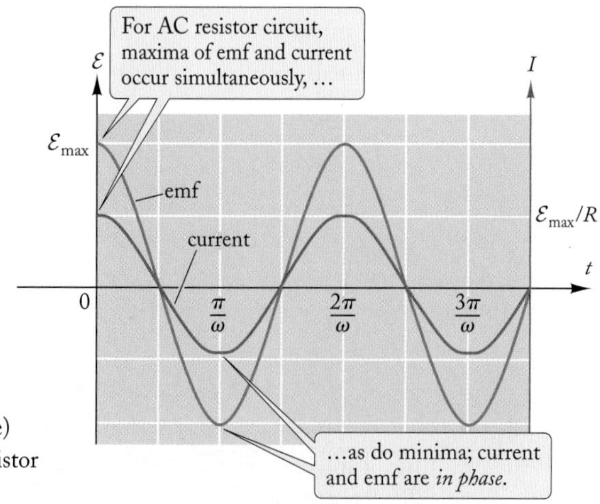

FIGURE 32.4 The emf (blue) and the current (red) in the resistor circuit as a function of time.

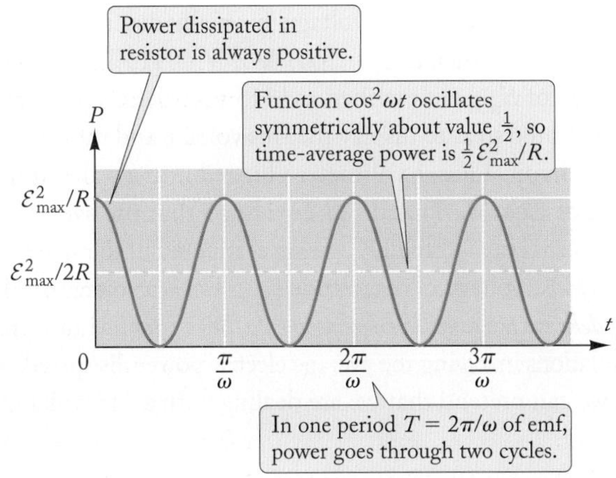

FIGURE 32.5 Instantaneous power dissipated in the resistor as a function of time. Note that the frequency of oscillation of the power is twice that of the emf.

positive. According to Eq. (32.4), the power oscillates between zero and a maximum value $\mathcal{E}_{max}^2/R$. Figure 32.5 is a plot of these oscillations.

For the operation of most appliances involving resistors, such as electric heaters or lightbulbs, we do not care about the oscillations of the power. We care only about the **time-average power** dissipated by the resistor, which tells us the average amount of heat or light produced by the heater or lightbulb. This time-average power can be obtained by averaging the quantity $\cos^2\omega t$ in Eq. (32.4) over one cycle of oscillation. To evaluate this average, note that the average of $\cos^2 \omega t$ is the same as the average of $\sin^2 \omega t$, because both of these quantities have the same number of ups and downs in one cycle. Using overbars to indicate averages, we therefore have

$$\overline{\cos^2 \omega t} = \tfrac{1}{2}(\overline{\cos^2 \omega t} + \overline{\cos^2 \omega t}) = \tfrac{1}{2}(\overline{\cos^2 \omega t} + \overline{\sin^2 \omega t}) \qquad (32.5)$$

But $\cos^2\omega t + \sin^2\omega t = 1$, and hence Eq. (32.5) tells us that $\overline{\cos^2\omega t} = \frac{1}{2}$. This is also evident from Fig. 32.5, where we see that $\cos^2\omega t$ oscillates symmetrically about the value $\frac{1}{2}$. The time-average power is then

$$\overline{P} = \frac{\mathcal{E}_{max}^2}{2R} \qquad (32.6)$$

Concepts in Context

This is usually written in the form

$$\overline{P} = \frac{\mathcal{E}_{rms}^2}{R} \qquad (32.7)$$

power dissipated by resistor

where the quantity $\mathcal{E}_{rms}$, called the **root-mean-square voltage**, or rms voltage, is the maximum voltage divided by $\sqrt{2}$:

$$\mathcal{E}_{rms} = \frac{\mathcal{E}_{max}}{\sqrt{2}} \qquad (32.8)$$

root-mean-square voltage

Note that since $\mathcal{E}^2 = \mathcal{E}_{max}^2 \cos^2 \omega t$, the average value of $\mathcal{E}^2$ is $\frac{1}{2}\mathcal{E}_{max}^2$ and, according to Eq. (32.8), this equals $\mathcal{E}_{rms}^2$. Hence, the square of $\mathcal{E}_{rms}$ is the average, or the mean, of the square of $\mathcal{E}$.

In engineering practice, an AC voltage is usually described in terms of $\mathcal{E}_{rms}$. For example, if $\mathcal{E}_{max} = 163$ V, then $\mathcal{E}_{rms} = 163/\sqrt{2}$ V $= 115$ V. Thus an oscillating voltage with this value of $\mathcal{E}_{max}$ is described as "115 volts AC." Ordinary voltmeters and ammeters are calibrated so as to display the rms voltage and the rms current when connected to an AC circuit. Thus, a voltmeter plugged into a wall outlet will read 115 V.

Comparison of Eqs. (32.7) and (28.24) shows that the average AC power dissipated in the resistor is equal to the DC power dissipated in the same resistor when connected to a steady DC voltage of magnitude $\mathcal{E}_{rms}$. Thus, a voltage of 115 volts AC (*with* $\mathcal{E}_{max} = 163$ V) *delivers the same average power to the resistor as* 115 *volts DC.* This means that in any calculations involving the average electric power dissipated by resistors in such an AC circuit, we can pretend that we are dealing with a 115-volts DC circuit.

Concepts in Context

EXAMPLE 1 A 115-V AC incandescent lightbulb is rated at 150 W. What is the resistance of this lightbulb (when at its operating temperature)?

SOLUTION: We have $\mathcal{E}_{rms} = 115$ V and $\overline{P} = 150$ W. Hence Eq. (32.7) gives us

$$R = \frac{\mathcal{E}_{rms}^2}{\overline{P}} = \frac{(115\text{ V})^2}{150\text{ W}} = 88\ \Omega \tag{32.9}$$

When several appliances are plugged into the same outlet in a house, or into different outlets that are part of the same circuit in the house, all the appliances will be connected in parallel across the supplied emf of 115 V AC. In ordinary wall outlets for three-prong plugs, such as illustrated in Fig. 32.6, the small slot is at a potential of 115 V AC, the larger slot is at a potential of zero, and the round slot is also at a potential of zero. The two flat slots are connected to the two wires coming from the power station, and when an appliance is plugged into the outlet, the current flows into the appliance through one slot and out through the other. The round slot is connected to the ground, usually via a grounding plate or rod buried just outside the house. Under normal circumstances, this slot carries no current. It comes into play only when there is a failure in the electrical insulation of the appliance; then the ground slot permits the leaking currents to flow into the ground (instead of flowing through your body).

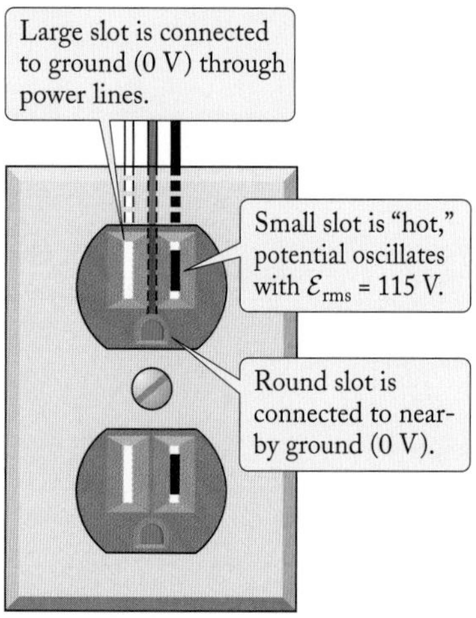

FIGURE 32.6 Standard power outlet for three-prong plug.

Note that the potentials indicated in Fig. 32.6 are open-circuit potentials, that is, they are the potentials before an appliance is plugged into the outlet. When an appliance drawing a heavy current—such as an electric heater or an air conditioner—is plugged into the outlet, the potential at the small flat slot may drop by several volts, and the potential at the large flat slot will rise by an equal amount. These changes in potential are due to the resistance of the wires connecting the outlets to the power station. The wires and the appliance form a series circuit, and the total potential drop of 115 V provided by the power station is distributed among these wires and the appliance in direct proportion to their resistances, as required by Ohm's Law.

✔ Checkup 32.1

QUESTION 1: In Europe, outlets deliver electric power at 230 V AC. What is $\mathcal{E}_{max}$?

QUESTION 2: A lightbulb is connected to an AC outlet operating at 60 Hz. How many minima and how many maxima per second are there in the electric current? In the electric power?

QUESTION 3: Suppose you connect a lightbulb, by means of two wires, to the large slot and the round slot of the ordinary wall outlet illustrated in Fig. 32.6. Will the lightbulb shine?

QUESTION 4: During a "brownout," the oscillating emf supplied by power companies can drop significantly; suppose that the amplitude is $\mathcal{E}_{max} = 141$ V. What average power does this emf supply to a 100-Ω resistor?

(A) 100 W (B) 141 W (C) 200 W (D) 282 W

32.2 CAPACITOR CIRCUIT

Online Concept Tutorial 35

A capacitor connected to an alternating source of emf behaves very differently from a capacitor connected to a constant source of emf, such as a battery. When we connect the capacitor to the battery, there is an initial, brief surge of current that charges the capacitor. The current deposits positive charge on one plate of the capacitor, and negative charge on the other, and thereby increases the voltage across the capacitor. But once the voltage across the charged capacitor matches the emf, no further current flows, and the capacitor thereafter remains in equilibrium.

In contrast, when we connect a capacitor to an alternating source of emf, the current begins to deposit positive charges on one plate and negative charges on the other. But half a cycle later, the emf reverses, and so does the current. Thus, the current now removes the charges it deposited earlier, and begins to deposit reversed charges on the plates. This reversal of the current and of the charges repeats each half cycle. The current flows back and forth along the wires from the emf to the plates, and the signs of the charges on the plates alternate with the same frequency as the emf. Note that the capacitor permits this periodic back-and-forth flow of the current on the wires connected to its plates, even though no current can cross the gap between the plates.

Figure 32.7 shows the circuit diagram for a capacitor connected to our oscillating source of emf. When the instantaneous charge of the capacitor is Q, the instantaneous voltage across the capacitor is Q/C, and therefore Kirchhoff's rule gives

$$\mathcal{E} - \frac{Q}{C} = 0 \qquad (32.10)$$

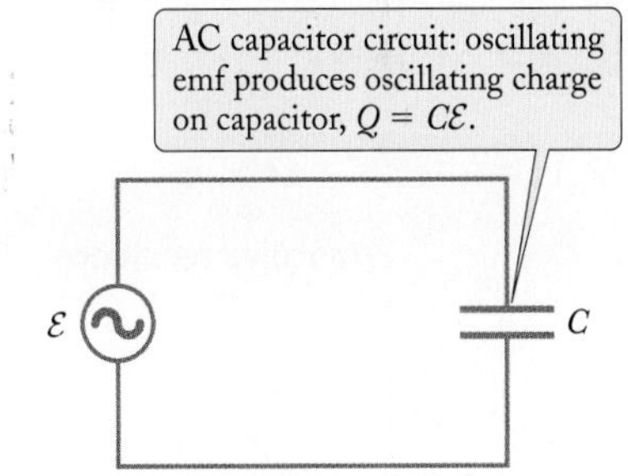

FIGURE 32.7 Capacitor connected to a source of alternating emf.

With the expression (32.1) for our time-dependent $\mathcal{E}$, this yields

$$Q = C\mathcal{E} = C\mathcal{E}_{max}\cos\omega t \qquad (32.11)$$

Note that the charges on the capacitor plates oscillate between a positive value $C\mathcal{E}_{max}$ and a negative value $-C\mathcal{E}_{max}$; that is, the charges on the plates reverse every half cycle, in phase with the oscillating source of emf. The current in the circuit is the rate of change of the charge, $I = dQ/dt$. This means we must evaluate the rate of change of $\cos\omega t$:

$$\frac{d}{dt}\cos\omega t = -\omega\sin\omega t$$

Hence

$$I = \frac{dQ}{dt} = -\omega C\mathcal{E}_{max}\sin\omega t \qquad (32.12)$$

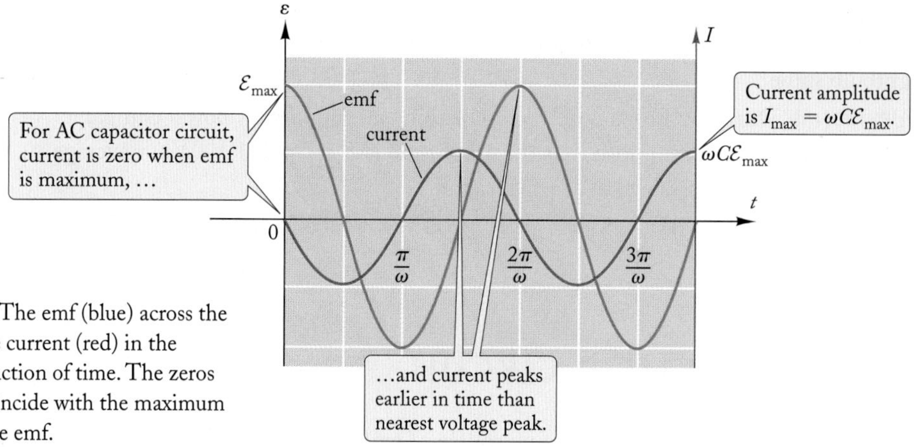

FIGURE 32.8 The emf (blue) across the capacitor and the current (red) in the capacitor as a function of time. The zeros of the current coincide with the maximum magnitudes of the emf.

According to this equation *the current oscillates sinusoidally with a frequency* ω *and with an amplitude* $\omega C\mathcal{E}_{max}$. Figure 32.8 shows plots of the emf and the current. Equation (32.12) tells us that the current has minimum magnitude (0) when the emf has maximum magnitude ($\mathcal{E}_{max}$) and that the current has a maximum magnitude ($\omega C\mathcal{E}_{max}$) when the emf has a minimum magnitude (0). Because of this, the current and the emf are said to be 90° **out of phase**. Because the current in a capacitive circuit peaks earlier than the time of the nearest voltage peak, we say that the *current leads the voltage* by 90°. The mnemonic $IC\mathcal{E}$ (ice) makes this easy to remember: in a capacitive circuit (C), the current I peaks before the voltage $\mathcal{E}$ (see Fig. 32.9). It is customary to write Eq. (32.12) as

current into capacitor

$$I = -\frac{\mathcal{E}_{max}\sin\omega t}{X_C} \tag{32.13}$$

where

capacitive reactance

$$X_C = \frac{1}{\omega C} \tag{32.14}$$

is called the **capacitive reactance**. The quantity X_C plays roughly the same role for a capacitor in an AC circuit as does the resistance R for a resistor [compare Eqs. (32.3) and (32.13)]. Note, however, that the reactance depends not only on the capacitance of the capacitor, but also on the frequency at which we are operating the circuit, whereas the resistance of a resistor does not depend on the frequency. The reactance is large if the frequency is low, and vice versa. The unit of reactance is the ohm, as it is for resistance.

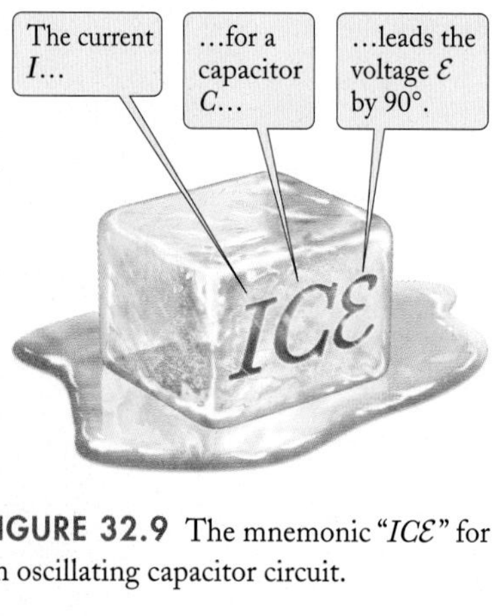

FIGURE 32.9 The mnemonic "$IC\mathcal{E}$" for an oscillating capacitor circuit.

From Eq. (32.12) or (32.13) we see that the current is small if the frequency ω is low. This was to be expected, since it means that the response of the capacitor to a low-frequency emf is nearly the same as its response to a constant emf (DC). As we saw in the discussion at the beginning of this section, for a constant emf, the capacitor blocks the flow of current.

The instantaneous power delivered to the capacitor is, again, the product of the instantaneous current, Eq. (32.12), and the instantaneous emf, Eq. (32.1):

$$P = \mathcal{E}I = -\omega C\mathcal{E}_{max}^2 \cos\omega t \sin\omega t \tag{32.15}$$

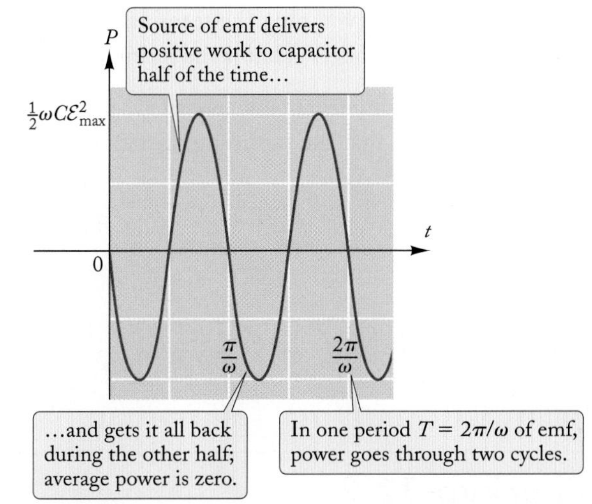

FIGURE 32.10 Instantaneous power delivered to the capacitor as a function of time.

The time dependence of this expression is contained in the factor cos ωt sin ωt. According to a standard trigonometric identity (see Appendix 3), this equals $\frac{1}{2}$ sin $2\omega t$, which shows that the power oscillates at a frequency of 2ω. But the important point is that the average power delivered is zero—as Fig. 32.10 shows, within one cycle, there is as much positive power as negative power. The source of emf does work on the capacitor during parts of the cycle, but the capacitor does work on the source during other parts of the cycle, so **the average power is zero**. The ideal capacitor does not consume power, because it has no means of dissipating electric energy—it can only store charge and return it.

PHYSICS IN PRACTICE FREQUENCY FILTER CIRCUITS

The selective blocking of low-frequency currents by a capacitor is exploited in stereo speaker systems. In order to ensure that sound waves diffract and spread to fill a room, stereo systems route high-frequency signals into small loudspeakers ("tweeters"); meanwhile, to attain adequate amplitudes, low-frequency signals are routed into large loudspeakers ("woofers"). To achieve this, a capacitor is connected in series with the tweeter (see Fig. 1). The capacitor has a large effective resistance (a large reactance) when the frequency is low. Thus, this capacitor acts as a filter that blocks low-frequency currents but permits the passage of high-frequency currents. As we will see in the next section, an inductor has the opposite effect of a capacitor. The inductor has a large effective resistance if the frequency is high. In the stereo speaker system, an inductor is connected in series with the woofer (see Fig. 2); this inductor acts as a filter that blocks high-frequency currents but permits the passage of low-frequency currents. Combinations of capacitors and inductors are used in many other kinds of electronic circuits that filter frequencies.

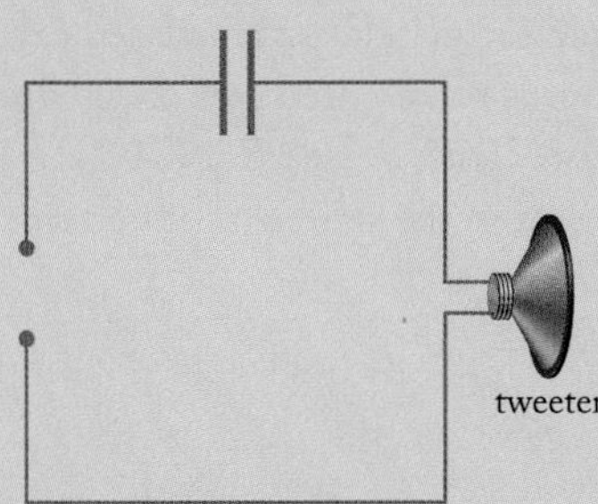

FIGURE 1 Capacitor connected in series with tweeter.

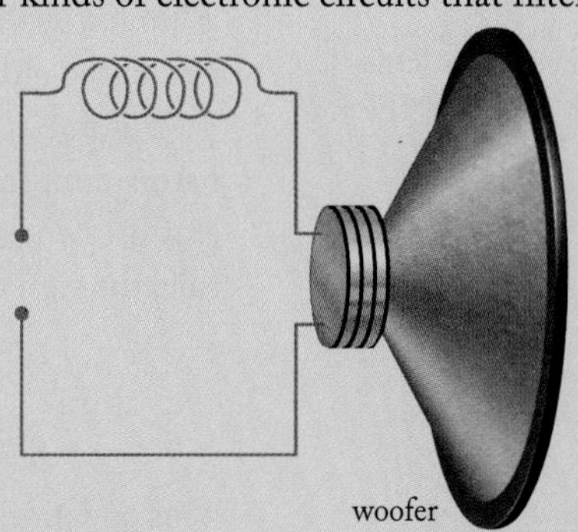

FIGURE 2 Inductor connected in series with woofer.

EXAMPLE 2 Suppose you plug the terminals of a 20-pF capacitor into a 115-V outlet. What is the maximum instantaneous current?

SOLUTION: The maximum instantaneous emf is $\mathcal{E}_{max} = 115\text{ V} \times \sqrt{2} = 163\text{ V}$. From Eq. (32.13) we see that the maximum instantaneous current occurs when $\sin \omega t = -1$, and it has the value

$$I_{max} = \frac{\mathcal{E}_{max}}{X_C} = \omega C \mathcal{E}_{max}$$

$$= (2\pi \times 60\text{ Hz}) \times (20 \times 10^{-12}\text{ F}) \times 163\text{ V} = 1.2 \times 10^{-6}\text{ A}$$

Checkup 32.2

QUESTION 1: Does a capacitor connected to AC obey Ohm's Law?

QUESTION 2: For a capacitor with the emf and the current plotted in Fig. 32.8, make a plot of the electric energy in the capacitor as a function of time. At $t = 0$, is the energy maximum or minimum?

QUESTION 3: A capacitor has a capacitive reactance of $10^9\ \Omega$ at 60 Hz. What is the reactance at 600 Hz? 6000 Hz?

QUESTION 4: According to Eq. (32.12), the current in the capacitor circuit becomes very small if the frequency is low. Does this make sense?

QUESTION 5: Suppose that a capacitor is at first plugged into a 60-Hz, 115-V outlet. If you move the capacitor to a 60-Hz, 230-V outlet, by what factor is the new current amplitude related to the old one? By what factor is the new maximum power related to the old?

(A) 1, 4 (B) 1, 2 (C) 2, 4 (D) 2, 2 (E) 2, 1

32.3 INDUCTOR CIRCUIT

Online Concept Tutorial 35

An ideal inductor is a resistanceless coil of wire. We might expect that when such a coil is connected across a source of emf, a very large current will flow, since there is no resistance to oppose the current. However, for an alternating emf, the current in the inductor is limited by the self-inductance. As we know from Section 31.4, any increase of the current in the inductor generates a back emf, and this opposes the increase in the current.

Figure 32.11 shows the circuit diagram for an inductor connected to a source of alternating emf. In this diagram, the inductor is represented by a coiled line. The induced emf in the inductor, or the back emf, is $-L\,dI/dt$ [see Eq. (31.21)]. By Kirchhoff's rule, the sum of this back emf and the emf of the source must be zero:

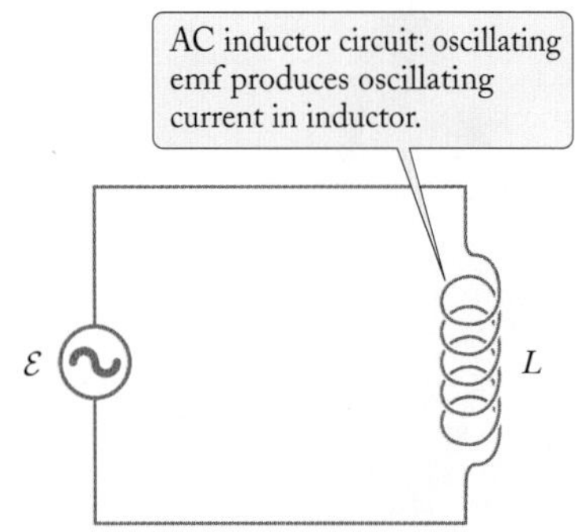

FIGURE 32.11 Inductor connected to a source of alternating emf.

$$\mathcal{E} - L\frac{dI}{dt} = 0 \tag{32.16}$$

This equation gives us

$$\frac{dI}{dt} = \frac{\mathcal{E}}{L} = \frac{\mathcal{E}_{max}\cos \omega t}{L} \tag{32.17}$$

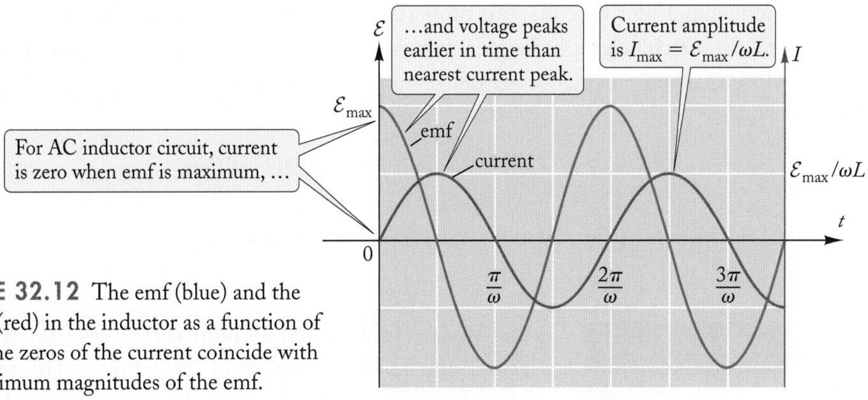

FIGURE 32.12 The emf (blue) and the current (red) in the inductor as a function of time. The zeros of the current coincide with the maximum magnitudes of the emf.

This tells us the rate of change of the current. To discover the current, we note that $\omega \cos \omega t$ is the rate of change of $\sin \omega t$:

$$\frac{d}{dt} \sin \omega t = \omega \cos \omega t$$

Thus, the quantity $\cos \omega t$ appearing in Eq. (32.17) is the rate of change of $(1/\omega) \sin \omega t$. But if the rates of change of two quantities are equal, then the two quantities can differ only by a constant. A constant current is consistent with Eq. (32.16), but it would not persist if there were some (small) resistance in the circuit. Ignoring any constant current, we find that the oscillating current must be

$$I = \frac{\mathcal{E}_{max} \sin \omega t}{\omega L} \tag{32.18}$$

This current oscillates sinusoidally with a frequency ω and an amplitude $\mathcal{E}_{max}/\omega L$. Figure 32.12 gives plots of the current and the emf. Again, comparison of these plots shows that the current is 90° out of phase with the emf—the current has minimum magnitude when the emf has maximum magnitude. Because the voltage in an inductive circuit peaks earlier than the time of the nearest current peak, we say that the *voltage leads the current* by 90°. The mnemonic $\mathcal{E}LI$ (Eli) makes this easy to remember: in an inductive circuit (L), the voltage $\mathcal{E}$ peaks before the current I (Fig. 32.13).

FIGURE 32.13 The mnemonic "$\mathcal{E}LI$ the $IC\mathcal{E}$ man" for oscillating inductor and capacitor circuits.

We can write Eq. (32.18) as

$$I = \frac{\mathcal{E}_{max} \sin \omega t}{X_L} \tag{32.19}$$

current in inductor

where

$$X_L = \omega L \tag{32.20}$$

inductive reactance

is the **inductive reactance**. The inductive reactance depends on both the inductance and the frequency—the reactance is large if the frequency is high, and the reactance is small if the frequency is low. The unit of this reactance is, again, the ohm.

Note that according to Eq. (32.18) or (32.19), the amplitude of the current is small if the frequency is high, and the amplitude of the current is large if the frequency ω is low. This merely means that the response of the inductor to a low-frequency emf is close to its response to a steady emf—for a steady emf (DC), the inductor permits a very large current, since it has no resistance. As already mentioned in Physics in Practice:

Frequency Filter Circuits, this property of an inductor can be exploited in the design of a filter to block high-frequency currents.

The instantaneous power delivered to the inductor is

$$P = \mathcal{E}I = \frac{1}{\omega L}\mathcal{E}_{\text{max}}^2 \cos\omega t \sin\omega t \tag{32.21}$$

As in the case of the capacitor, the average power is zero.

EXAMPLE 3 A resistor, a capacitor, and an inductor are connected in parallel (see Fig. 32.14) to a source of oscillating emf of frequency $\omega = 6.0 \times 10^3$ radians/s and amplitude 1.0×10^{-3} V. The resistance of the resistor is 200 Ω. (a) What is the maximum instantaneous current in the resistor? (b) If we want to make the maximum instantaneous currents in the capacitor and in the inductor equal to that in the resistor, what values of the capacitance and of the inductance must we select?

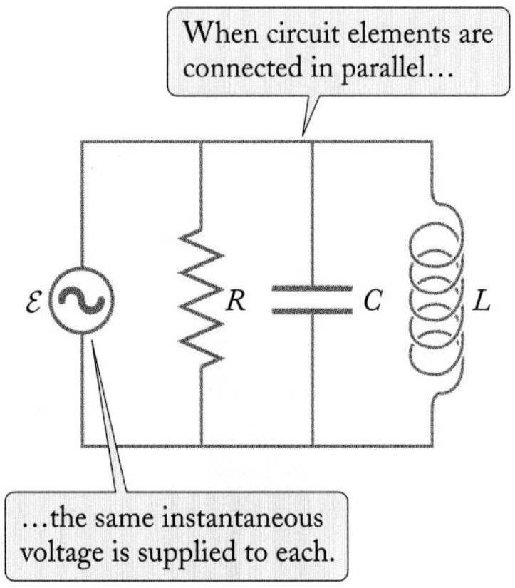

FIGURE 32.14 Resistor, capacitor, and inductor connected in parallel to a source of oscillating emf.

SOLUTION: (a) In a parallel connection, the source of emf supplies the same voltage across each of the circuit elements. Thus, the maximum instantaneous voltage across the resistor is 1.0×10^{-3} V, and the maximum instantaneous current in the resistor is, by Eq. (32.3),

$$I_{\text{max}} = \frac{\mathcal{E}_{\text{max}}}{R} = \frac{1.0 \times 10^{-3}\text{ V}}{200\ \Omega} = 5.0 \times 10^{-6}\text{ A}$$

(b) According to Eqs. (32.13) and (32.19), if the maximum instantaneous currents in the capacitor and the inductor are to match that in the resistor, their reactances must match the resistance:

$$X_C = X_L = 200\ \Omega$$

Thus,

$$\frac{1}{\omega C} = 200\ \Omega \qquad \text{and} \qquad \omega L = 200\ \Omega$$

which gives

$$C = \frac{1}{6.0 \times 10^3\text{ s}^{-1} \times 200\ \Omega} = 8.3 \times 10^{-7}\text{ F}$$

and

$$L = \frac{200\ \Omega}{6.0 \times 10^3\text{ s}^{-1}} = 3.3 \times 10^{-2}\text{ H}$$

COMMENTS: Note that with this choice of C and L, the maximum currents are equal at the frequency of 6.0×10^3 radians/s only. At any other frequency, this choice of C and L will not result in equal maximum currents. Also note that although the currents are equal in amplitude when $X_L = X_C$, they are 180° out of phase (the current in the capacitor leads the emf by 90° and the current in the inductor lags the emf by 90°), so the net instantaneous current in those two branches is zero and the net current is that in the resistor. Related behavior is discussed in more detail in Section 32.5.

Checkup 32.3

QUESTION 1: Does an inductor connected to AC obey Ohm's Law?

QUESTION 2: For an inductor with the emf and the current plotted in Fig. 32.12, make a plot of the magnetic energy in the inductor as a function of time. At $t = 0$ is the energy maximum or minimum?

QUESTION 3: An inductor has an inductive reactance of $10^3\ \Omega$ at 60 Hz. What is the reactance at 600 Hz? 6000 Hz?

QUESTION 4: According to Eq. (32.18), the current in the inductor circuit becomes very large if the frequency is low. Does this make sense?

QUESTION 5: An inductor is connected to a 60-Hz, 115-V emf. Suppose that a piece of iron is inserted into the inductor so that the value of the inductance increases by a factor of 10. By what factor is the new current amplitude related to the old one? By what factor is the new inductive reactance related to the old?

(A) 10, 10 (B) 10, 100 (C) $\frac{1}{10}$, 1 (D) $\frac{1}{10}$, 10 (E) $\frac{1}{10}$, 100

32.4 FREELY OSCILLATING *LC* AND *RLC* CIRCUITS

In this section, we will examine the behavior of circuits when *no external source of alternating emf is present.* We have already examined such so-called free response or natural response for RC and RL circuits (Sections 28.7 and 31.6, respectively); we found that a smooth exponential function, with a characteristic time $\tau = RC$ or L/R, described the response for given initial conditions. In this section, we will consider LC and RLC circuits; we will find that such circuits have a natural tendency to oscillate.

An LC circuit consists of an ideal inductor and an ideal capacitor connected in series (see Fig. 32.15). The circuit has *no source of emf*; nevertheless, a current will flow in this circuit, provided that the capacitor is *initially charged.* The charge on one plate is then initially positive and that on the other plate negative. A current will begin to flow around the circuit from the positive plate to the negative. If the circuit had no inductance, the current would merely neutralize the charges on the plates, that is, the capacitor would discharge, and this would be the end of the current. But the inductor makes a difference: the inductor initially opposes the buildup of the current, but once the current has become established, the inductor will keep it going for some extra time. Hence *more* charge flows from one capacitor plate to the other than required for neutrality, and reversed charges accumulate on the capacitor plates. When the current finally stops, the capacitor will again be fully charged, with reversed charges. And then a reversed current will begin to flow, and so on. Thus, the charge sloshes back and forth around the circuit; because this sloshing is like water in a tank, an LC circuit is sometimes called a "tank circuit." If there is no resistance, these oscillations of the LC circuit continue forever.

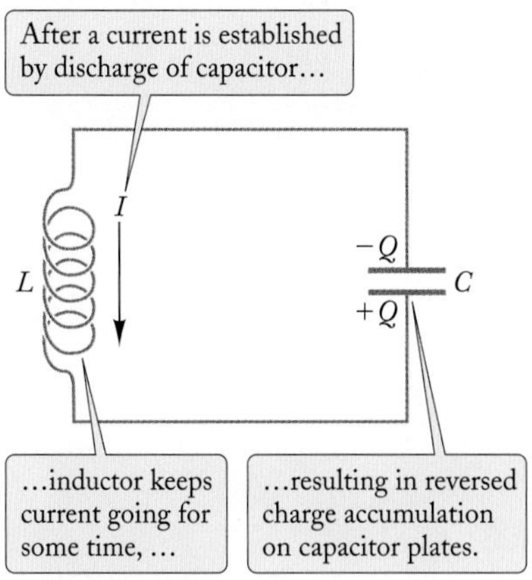

FIGURE 32.15 Inductor and capacitor connected in series.

The LC circuit is analogous to the mass–spring system of Section 15.2. *The inductor is analogous to the mass—it tends to keep the current constant and provides "inertia." The charged capacitor is analogous to the stretched spring—it tends to accelerate the charge and provides a "restoring force."*

The equation of motion of the LC circuit follows from Kirchhoff's rule: the sum of emfs and other voltage changes around the circuit must add to zero. Going around the circuit in the direction of the arrow shown in Fig. 32.15, we find that the induced emf in the inductor (back emf) is

$$-L\frac{dI}{dt}$$

and the voltage across the capacitor is

$$-Q/C$$

Hence

$$-L\frac{dI}{dt} - \frac{Q}{C} = 0 \qquad (32.22)$$

or, equivalently,

$$L\frac{dI}{dt} + \frac{Q}{C} = 0 \qquad (32.23)$$

Note that here Q is reckoned as positive when the charge on the lower plate is positive, and I is reckoned as positive when the charge on the lower plate is increasing.

Equation (32.23) has exactly the same mathematical form as the equation of motion for the simple harmonic oscillator [see Eq. (15.18)],

$$m\frac{dv}{dt} + kx = 0$$

Comparing this with Eq. (32.23), we see that Q plays the role of x, whereas L replaces m and $1/C$ replaces k. The current ($I = dQ/dt$) plays the role of the velocity ($v = dx/dt$). Hence the solution of Eq. (32.23) can be found by recalling the solution for the simple harmonic oscillator [see Eq. (15.22)],

$$x = A\cos\left(\sqrt{\frac{k}{m}}\,t\right)$$

With the above replacements for x, m, and k, we immediately obtain

$$Q = Q_{\text{max}}\cos\left(\frac{1}{\sqrt{LC}}\,t\right) \qquad (32.24)$$

where Q_{max} is the amount of charge on the positive plate at time $t = 0$. Likewise, from the equation for the velocity of the simple harmonic oscillator,

$$v = \frac{dx}{dt} = -\sqrt{\frac{k}{m}}\,A\sin\left(\sqrt{\frac{k}{m}}\,\mathrm{t}\right)$$

we find that the current in the LC circuit must be

$$I = \frac{dQ}{dt} = -\frac{Q_{\text{max}}}{\sqrt{LC}}\sin\left(\frac{1}{\sqrt{LC}}\,t\right) \qquad (32.25)$$

According to Eqs. (32.24) and (32.25), the charge and the current oscillate with a natural frequency, called the **resonant frequency** ω_0:

$$\omega_0 = \frac{1}{\sqrt{LC}} \tag{32.26}$$

resonant frequency ω_0

Equation (32.25) also indicates that the current oscillates with amplitude

$$I_{max} = \frac{Q_{max}}{\sqrt{LC}} = \omega_0 Q_{max}$$

Figure 32.16 is a plot of the charge and the current in an *LC* circuit oscillating according to Eqs. (32.24) and (32.25).

The total stored enery in the *LC* circuit is the sum of the electric and magnetic energies [see Eqs. (26.27) and (31.26)]:

$$U = \frac{1}{2C}Q^2 + \frac{1}{2}LI^2 \tag{32.27}$$

Substituting Q from Eq. (32.24) and I from Eq. (32.25), we can see that the total energy remains constant, equal to the initial stored energy on the capacitor, although the total energy oscillates back and forth between electric and magnetic energy:

$$\begin{aligned} U &= \frac{1}{2C}Q_{max}^2 \cos^2 \omega_0 t + \frac{1}{2}L\frac{Q_{max}^2}{LC}\sin^2 \omega_0 t \\ &= \frac{1}{2C}Q_{max}^2(\cos^2 \omega_0 t + \sin^2 \omega_0 t) \\ &= \frac{1}{2C}Q_{max}^2 \end{aligned} \tag{32.28}$$

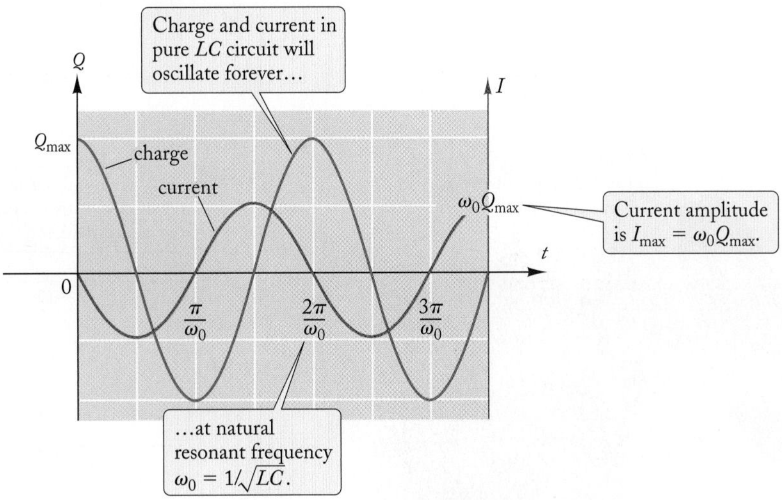

FIGURE 32.16 Charge (blue) on the capacitor and current (red) in the inductor as a function of time.

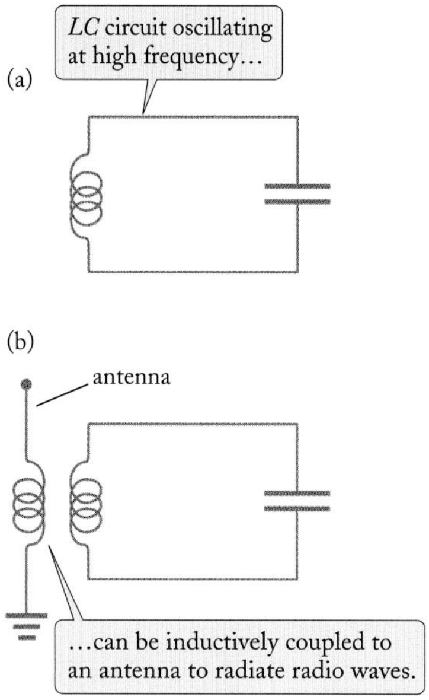

FIGURE 32.17 (a) An LC circuit in a radio. (b) The LC circuit is coupled to the antenna by the mutual inductance of the two inductors.

EXAMPLE 4 A primitive radio transmitter, like those used in the early days of "wireless telegraphy," consists of an LC circuit oscillating at high frequency (Fig. 32.17a). The circuit is inductively coupled to an antenna (Fig. 32.17b), so the oscillating current in the circuit induces an oscillating current on the antenna; the latter current then radiates radio waves (see Chapter 33). Suppose that the inductance in the circuit in Fig. 32.17a is 20 μH. What capacitance do we need if we want to produce oscillations of a frequency of 1.5×10^6 Hz?

SOLUTION: The angular frequency is $\omega_0 = 2\pi f = 2\pi \times 1.5 \times 10^6\ \text{s}^{-1}$. Hence, from Eq. (32.26),

$$C = \frac{1}{\omega_0^2 L} = \frac{1}{(2\pi \times 1.5 \times 10^6\ \text{s}^{-1})^2 \times 20 \times 10^{-6}\ \text{H}}$$

$$= 5.6 \times 10^{-10}\ \text{F} = 560\ \text{pF}$$

A radio receiver employs much the same circuit as shown in Fig. 32.17b to pick up radio signals reaching the antenna. When a radio wave—consisting of oscillating electric and magnetic fields—reaches the antenna, it causes an oscillating current to flow along the antenna, which induces a current in the LC circuit. The current in the circuit will build up to a relatively large value if the frequency of the driving force supplied by the antenna matches the natural frequency of the circuit. To attain this resonance condition, the natural frequency of the circuit must be tuned to the frequency of the radio wave, which is done by adjusting the value of the capacitance. The capacitor in the radio circuit is a variable capacitor, whose capacitance can be controlled with a tuning knob (see Fig. 32.18). Turning this knob makes one of the sets of plates of the capacitor move parallel to the other, thereby changing the amount of overlap between the plates and the effective area of the capacitor.

If we add a resistor to an LC circuit (again without any external alternating emf), we obtain the RLC circuit shown in Fig. 32.19. We can readily analyze this circuit using Kirchhoff's voltage rule; we simply have to add the IR voltage drop across the resistor to our previous result (31.23):

$$L\frac{dI}{dt} + IR + \frac{Q}{C} = 0 \tag{32.29}$$

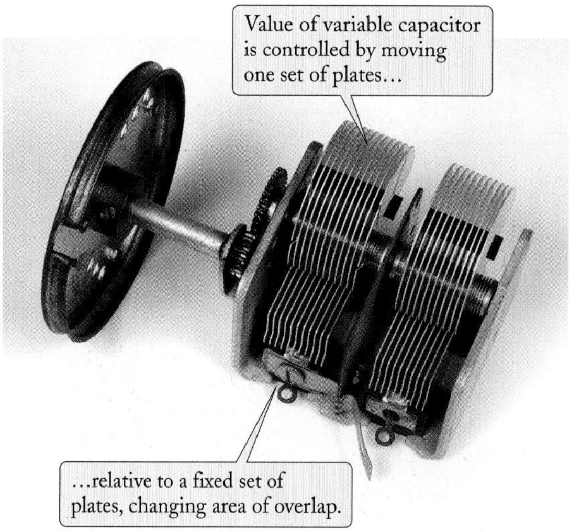

FIGURE 32.18 Adjustable capacitor for a radio. The capacitor consists of two sets of parallel, semicircular plates. The tuning knob controls the amount of overlap of the plates.

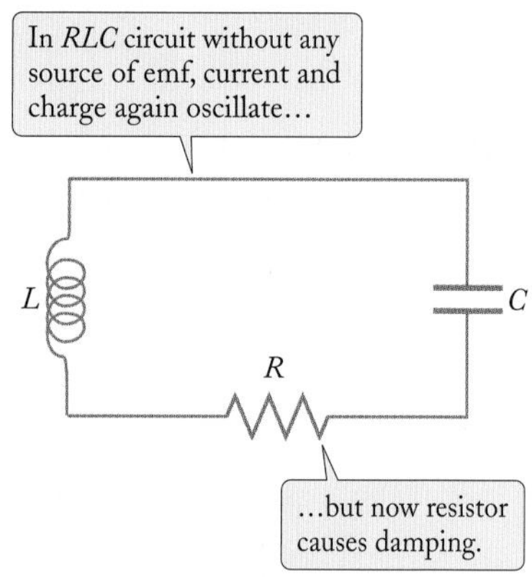

FIGURE 32.19 Resistor, inductor, and capacitor connected in series.

damped harmonic oscillator

Nith $I = dQ/dt$, this equation describes a **damped harmonic oscillator**. In such stem, the charge and current tend to oscillate as they do in the freely oscillating ' circuit, but the amplitude of the oscillations decreases with time, in a manner equiv-ent to the mechanical oscillator with friction discussed in Section 15.5. For the *RLC* ırcuit, the decrease is due to electrical "friction": since the current flows through the esistor, the resulting Joule heating represents some energy loss, or damping, during each cycle of oscillation. If the resistance is small, these losses are small, and the oscillation amplitude decreases slowly, as shown in Fig. 32.20a; the circuit is said to be *underdamped*. If the resistance is too large, the current drops to zero before any oscillations can occur, as in Fig. 32.20b; the circuit is said to be *overdamped* (see Fig. 32.20b). Equation (32.29) can be solved by standard techniques for differential equations, but instead of attempting this here, we will only examine the behavior of the solution. The solution to Eq. (32.29) is an oscillating function multiplied by a decaying exponential function of time:

$$Q = Q_{max} e^{-Rt/2L} \cos \omega_d t \qquad (32.30)$$

where the frequency of damped oscillations ω_d is somewhat smaller than the natural, undamped frequency ω_0, and is given by $\omega_d = \sqrt{\omega_0^2 - R^2/4L^2}$. For sufficiently small values of the resistance R (underdamped case), the oscillation frequency ω_d is almost equal to the natural frequency, $\omega_d \approx \omega_0$. This is the case plotted in Fig. 32.20a.

As mentioned in Section 15.5, the **quality factor**, or Q, is a measure of how freely a system oscillates, and is defined by

$$Q = -2\pi \frac{U}{\Delta U} \qquad (32.31)$$

where U is the stored energy and ΔU is the energy change per cycle. Initially, the stored electrical energy is $Q_{max}^2/2C$. By Eq. (32.30), the electrical energy after a time t has amplitude

$$U = \frac{Q^2}{2C} = \frac{Q_{max}^2}{2C}(e^{-Rt/2L})^2 = \frac{Q_{max}^2}{2C} e^{-Rt/L} \qquad (32.32)$$

For a small change in time Δt, the energy change is the product of the time derivative of U and Δt,

$$\Delta U = \frac{dU}{dt}\Delta t = \frac{Q_{max}^2}{2C} e^{-Rt/L} \times \left(-\frac{R}{L}\right) \times \Delta t = U \times \left(-\frac{R}{L}\right) \times \Delta t$$

Using this expression for ΔU and substituting $\Delta t = T \approx 2\pi/\omega_0$ for one cycle, we find that the quality factor (32.31) is

$$Q = -2\pi \frac{U}{U \times (-R/L) \times (2\pi/\omega_0)}$$

or simply

$$Q = \frac{\omega_0 L}{R} \qquad (32.33)$$

Thus a small resistance results in a large Q.

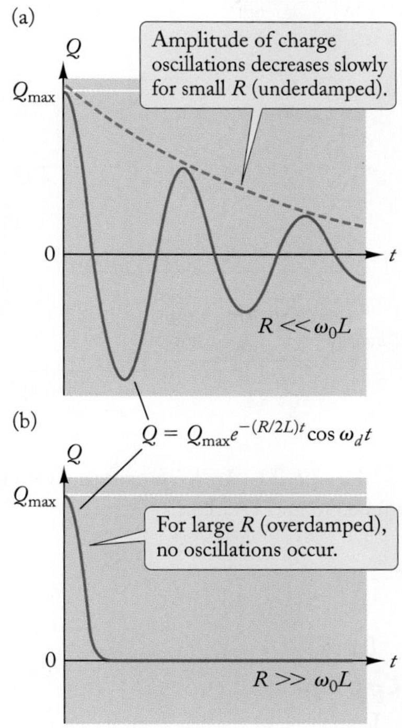

FIGURE 32.20 (a) Decaying oscillation of the charge on the capacitor in an underdamped series *RLC* circuit. (b) Decay of charge on the capacitor in an overdamped *RLC* circuit.

quality factor Q

From Eq. (32.30), the amplitude of the charge oscillations falls to $1/e$ of its initial value when $t = 2L/R$. If $T \approx 2\pi/\omega_0$ is the period of oscillation, the number of oscillations in the time t is

$$N = \frac{t}{T} = \frac{2L/R}{2\pi/\omega_0} = \frac{Q}{\pi} \tag{32.34}$$

Thus, as already mentioned in Section 15.5, the quality factor is a measure of how many oscillations occur before the amplitude of the system decreases appreciably. In simple electrical circuits, a Q of about 100 is often typical; using special techniques (for example, superconducting components), Q's on the order of 10^9 have been achieved.

Checkup 32.4

QUESTION 1: Consider the LC circuit illustrated in Fig. 32.15. If you increase the separation between the plates, what happens to the natural frequency?

QUESTION 2: The circuit illustrated in Fig. 32.17a has a capacitor of 560 pF and a natural frequency of 1.5×10^6 Hz. If we replace this capacitor by two capacitors of 560 pF each connected in parallel, by what factor will the natural frequency change?

QUESTION 3: In an RLC circuit oscillating at resonance, the inductive reactance is 500 Ω and the resistance is 10 Ω. What is the Q of the circuit? After approximately how many oscillations will the amplitude of the charge oscillations fall to $1/e$ of its initial value?

QUESTION 4: In the circuit of Fig. 32.21, the switch has been in position 1 for a long time and the capacitor is initially uncharged. At $t = 0$, the switch is moved to position 2. What is Q_{max}, the maximum charge on the capacitor? What function describes the time dependence of the charge on the capacitor, Q/Q_{max}?

(A) $\dfrac{\mathcal{E}\sqrt{LC}}{R}$, $\cos\omega t$ (B) $\dfrac{\mathcal{E}\sqrt{LC}}{R}$, $\sin\omega t$

(C) $\dfrac{\mathcal{E}}{R\sqrt{LC}}$, $\cos\omega t$ (D) $\dfrac{\mathcal{E}}{R\sqrt{LC}}$, $\sin\omega t$

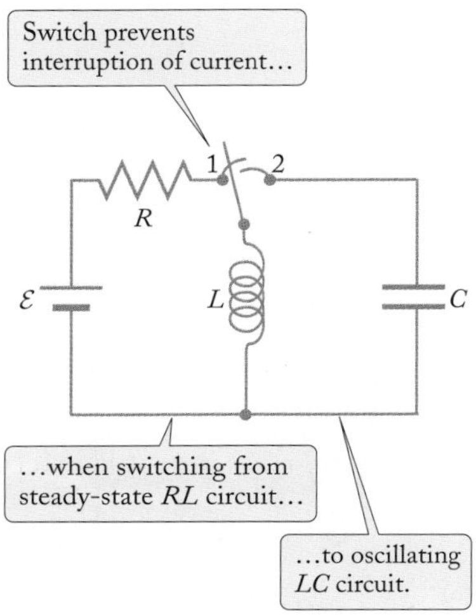

FIGURE 32.21 An RL circuit that can be switched to an LC circuit.

32.5 SERIES CIRCUITS WITH ALTERNATING EMF

Online Concept Tutorial 35

We now consider oscillations in a circuit with more than one R, L, or C component connected in series to an external source of alternating emf, for instance, the RL circuit shown in Fig. 32.22. Such oscillations produced by an applied emf are called driven oscillations or **forced oscillations**. In contrast to the free oscillations, which proceed at the natural resonant frequency [see Eq. (32.26)], the forced oscillations proceed at the frequency of the applied emf. Recall that in a series circuit, the current is the same everywhere. The sum of the instantaneous voltages across the components, by Kirchhoff's voltage rule, must equal the applied emf $\mathcal{E}$:

forced oscillations

$$\mathcal{E} = V_R + V_L \tag{32.35}$$

where V_R and V_L are the instantaneous voltages across the resistor and inductor, respectively. But as we saw above, the voltages across the resistor and inductor have different phases; the resistive voltage is in phase with the current, whereas the inductive voltage leads the current by 90°. Thus we need to know how to add oscillating voltages with different phases. For our circuit, the current may be assigned a reference phase of 0°:

$$I = I_{\max} \cos \omega t \tag{32.36}$$

so the voltages we need to add are

$$V_R = V_{\max,R} \cos \omega t \quad \text{and} \quad V_L = V_{\max,L} \cos(\omega t + 90°) \tag{32.37}$$

These voltages can be summed using trigonometric identities, but that becomes cumbersome for more complicated circuits. A simpler way to sum sinusoidal functions relies on a geometrical construction known as a **phasor**, which is a vector rotating in two dimensions (the two dimensions are fictitious; they have nothing to do with the x and y dimensions of real space). To understand this method, first note from Fig. 32.23 than *an oscillating function*, such as $V_R = V_{\max,R} \cos \omega t$, *can be represented as the x component of a rotating vector*. The rotating vector shown has length equal to the amplitude $V_{\max,R}$ of the oscillating voltage across the resistor, and at any instant makes an angle ωt with respect to the x axis, where ω is the frequency of the applied emf. As is evident in Fig. 32.23, the x component is then indeed the instantaneous value V_R, and will vary from $+V_{\max,R}$ to $-V_{\max,R}$ as the vector rotates (as the oscillations occur).

phasor

Similarly, the oscillating voltage V_L can be represented by the x component of the phasor shown in Fig. 32.24; the only difference is that this vector makes an extra angle of 90° with respect to the x axis. Since V_R and V_L oscillate at the same frequency, they will always have the same relative angle of 90° between them, as was already evident from the functional form of Eq. (32.37).

In order to sum the two voltages, we need to sum the x components of the two voltage phasors shown in Fig. 32.24. This is easy to do if we first perform the vector sum of the two phasors, and *then* take the x component of the sum. Figure 32.25a shows both phasors; for simplicity, the phasors are shown at time $t = 0$. Figure 32.25b shows the vector sum, performed by putting the tail of one phasor on the tip of the

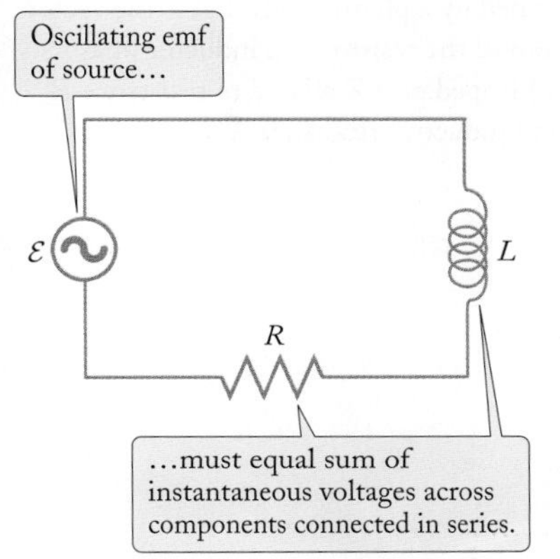

FIGURE 32.22 A series RL circuit with oscillating emf.

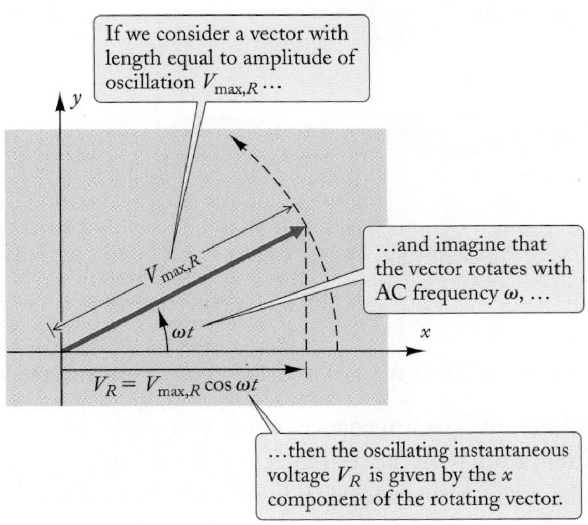

FIGURE 32.23 Phasor representing voltage across a resistor.

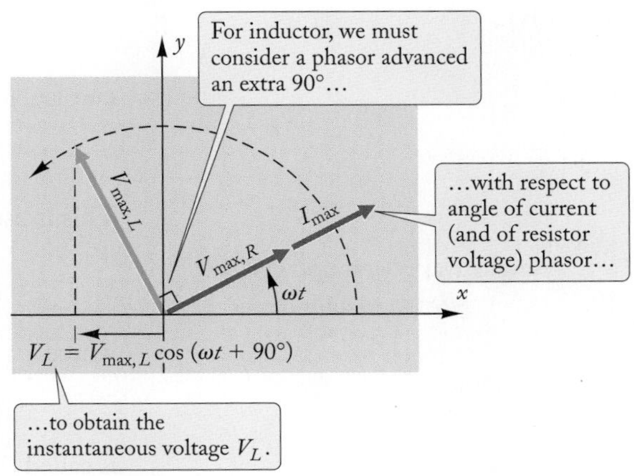

FIGURE 32.24 Three phasors representing series current, voltage across the inductor, and voltage across the resistor. Compared with the current and the resistor voltage, the phase of the inductor voltage leads by 90°.

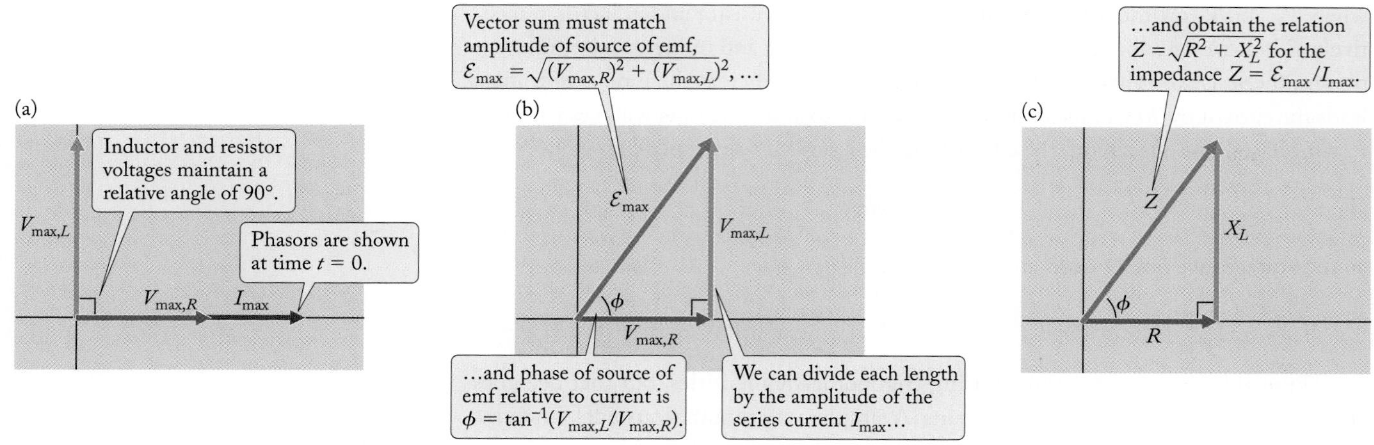

FIGURE 32.25 Addition of oscillating voltages using phasors. (a) Phasors for resistor (blue) and inductor (green) voltages. (b) The total voltage $\mathcal{E}$ can be represented by a phasor (red) that is the vector sum of the resistor and inductor phasors. (c) Impedance Z related to resistance R and inductive reactance X_L.

other. From the right triangle, the sum then has amplitude $\sqrt{(V_{max,R})^2 + (V_{max,L})^2}$, and this must match the amplitude of the applied emf:

$$\mathcal{E}_{max} = \sqrt{(V_{max,R})^2 + (V_{max,L})^2} \tag{32.38}$$

We also see that the applied emf has a phase (with respect to the current I and V_R) of

$$\phi = \tan^{-1}\left(\frac{V_{max,L}}{V_{max,R}}\right) \tag{32.39}$$

The instantaneous source voltage is the x component of the vector sum:

$$\mathcal{E} = \mathcal{E}_{max}\cos(\omega t + \phi) \tag{32.40}$$

In an RL circuit, the source voltage still leads the current, but by an angle less than 90°.

Recall from Sections 32.1–32.3 that the amplitudes of the current and voltage across each individual circuit component are related by

$$V_{max,R} = I_{max}R \qquad \text{and} \qquad V_{max,L} = I_{max}X_L \tag{32.41}$$

Thus Eq. (32.39) may be rewritten as

$$\phi = \tan^{-1}\left(\frac{X_L}{R}\right) \tag{32.42}$$

and Eq. (32.38) as

$$\mathcal{E}_{max} = \sqrt{(I_{max}R)^2 + (I_{max}X_L)^2} = I_{max}\sqrt{R^2 + X_L^2} \tag{32.43}$$

This is usually expressed as

amplitudes of emf and current related to impedance Z

$$\mathcal{E}_{max} = I_{max}Z \tag{32.44}$$

where

$$Z = \sqrt{R^2 + X_L^2} \tag{32.45}$$

is called the **impedance** of the RL circuit. The impedance measures the total oppositon to the flow of alternating current from both the resistance R and the reactance $X_L = \omega L$. The relation (32.45) can be easily reconstructed from Fig. 32.25b by dividing the magnitude of each phasor by the common current amplitude I_{max}. This gives the geometry of Fig. 32.25c.

EXAMPLE 5 A light dimmer is an application of a series *RL* circuit (see Fig. 32.26). Assume for simplicity that the resistance of the lightbulb is constant with, say, $R = 100\ \Omega$. For a dimmer, the bulb is connected in series with a variable inductance L. The circuit is connected to an ordinary emf of 115 V AC oscillating at 60 Hz. What range of L is needed if it is desired to vary the average power from 30 W to 100 W?

Concepts in Context

SOLUTION: The current in the circuit is given by Eq. (32.44),

$$I_{\max} = \frac{\mathcal{E}_{\max}}{Z} = \frac{\mathcal{E}_{\max}}{\sqrt{R^2 + (\omega L)^2}}$$

Thus the average power dissipated in the resistor is

$$\overline{P} = \frac{1}{2} I_{\max}^2 R = \frac{1}{2} \frac{\mathcal{E}_{\max}^2 R}{R^2 + (\omega L)^2}$$

Rearranging to solve for the inductance L, we find

$$L = \frac{1}{\omega}\sqrt{\frac{\mathcal{E}_{\max}^2 R}{2\overline{P}} - R^2}$$

Inserting the values for $\omega = 2\pi f = 377\ \text{s}^{-1}$, $\mathcal{E}_{\max} = \sqrt{2}\mathcal{E}_{\text{rms}} = 163$ V, $R = 100\ \Omega$, and $\overline{P} = 100$ W, we obtain

$$L = \frac{1}{377\ \text{s}^{-1}}\sqrt{\frac{(163\ \text{V})^2 \times 100\ \Omega}{2 \times 100\ \text{W}} - (100\ \Omega)^2} = 0.15\ \text{H}$$

With $\overline{P} = 30$ W, the same calculation yields $L = 0.49$ H.

Thus a variable inductor with a range $L = 0.15–0.49$ H would limit the power delivered to the lightbulb to the desired range.

COMMENTS: A lightbulb could instead be dimmed with a variable resistor, but this would add unnecessary Joule heating. The advantage of using a reactive circuit component to reduce the current is that it does not waste energy; no power is dissipated in an ideal inductor or capacitor. Many common household dimmers work on a much different principle: a semiconductor device is used to turn off the applied emf for a variable fraction of each 60-Hz cycle; such devices typically emit a hefty amount of electromagnetic noise, due to the large induced emf's associated with the sudden changes in current.

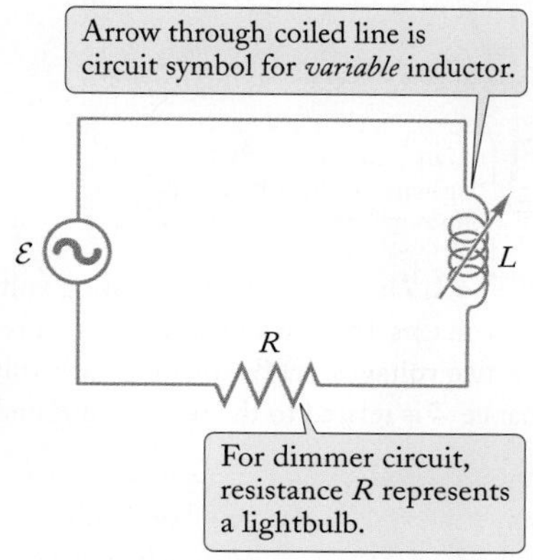

FIGURE 32.26 An *RL* light dimmer circuit with a variable inductance.

An *RC* circuit can be analyzed in the same way as an *RL* circuit; the only difference is that the phase angle is then negative, since the capacitive voltage lags the current. The resulting relations are identical to those found above if we replace X_L with $-X_C$.

More interesting is the behavior of a series *RLC* circuit, as illustrated in Fig. 32.27. The appropriate phasors are shown in Fig. 32.28a; the vector addition is readily performed with the aid of Fig. 32.28b. Since the inductive and capacitive voltages differ in phase by 180°, the corresponding phasors directly oppose one another. The right triangle in Fig. 32.28b immediately provides the relations

$$\mathcal{E}_{\max} = \sqrt{(V_{\max,R})^2 + (V_{\max,L} - V_{\max,C})^2} \tag{32.46}$$

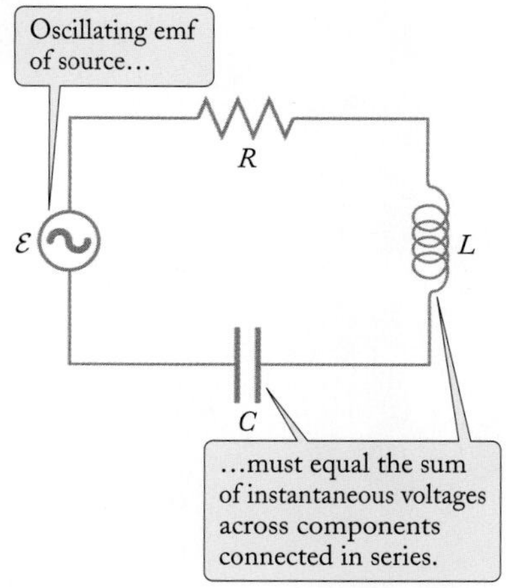

FIGURE 32.27 A series *RLC* circuit with oscillating emf.

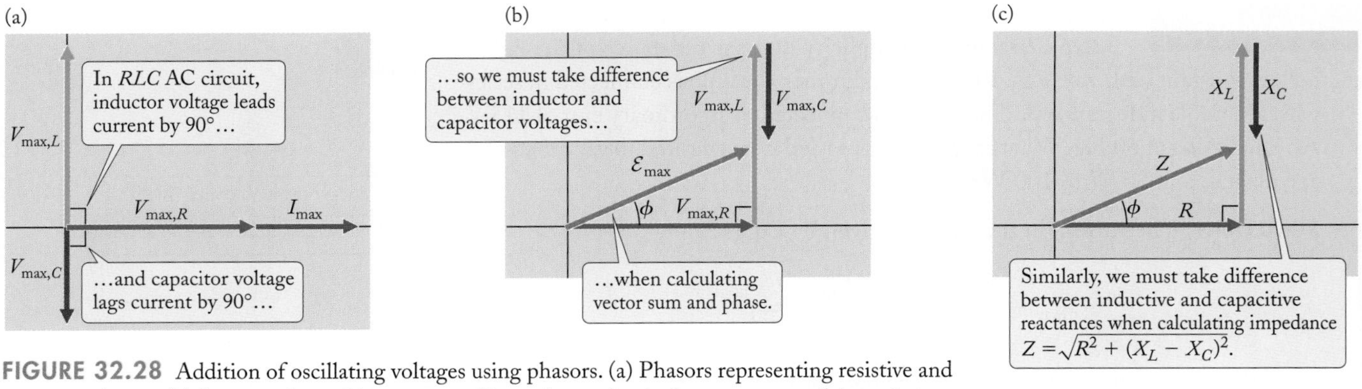

FIGURE 32.28 Addition of oscillating voltages using phasors. (a) Phasors representing resistive and reactive voltages. (b) Source voltage $\mathcal{E}$ is represented by a phasor that is the vector sum of the resistive and reactive voltages. (c) By dividing each voltage amplitude in (b) by the current amplitude I_{max}, the impedance Z is related to the resistance R and the reactances X_L and X_C.

and

$$\phi = \tan^{-1}\left(\frac{V_{max,L} - V_{max,C}}{V_{max,R}}\right) \tag{32.47}$$

As above, if we divide each voltage in Eqs. (32.46) and (32.47) by the current amplitude, we obtain the impedance $Z = \mathcal{E}_{max}/I_{max}$ of the circuit,

impedance Z of RLC circuit

$$Z = \sqrt{R^2 + (X_L - X_C)^2} \tag{32.48}$$

and the phase of the source of emf,

phase ϕ of RLC circuit

$$\phi = \tan^{-1}\left(\frac{X_L - X_C}{R}\right) \tag{32.49}$$

Notice that for the series RLC circuit, the phase will be positive if the inductive reactance dominates (at high frequency, since $X_L = \omega L$), and the phase will be negative if the capacitive reactance dominates [at low frequency, since $X_C = 1/(\omega C)$].

If we substitute $X_L = \omega L$ and $X_C = 1/(\omega C)$ into Eq. (32.48), we can examine the behavior of the impedance Z for different frequencies of the applied emf:

$$Z = \sqrt{R^2 + \left(\omega L - \frac{1}{\omega C}\right)^2} \tag{32.50}$$

For given values of R, L, and C, the impedance will thus vary with the frequency, and will have its minimum value when the term in parentheses is zero, that is, when the two reactances cancel. Thus the minimum value

$$Z_0 = R \tag{32.51}$$

occurs at one particular frequency ω_0, when

$$\omega_0 L - \frac{1}{\omega_0 C} = 0$$

or

$$\omega_0 = \frac{1}{\sqrt{LC}} \tag{32.52}$$

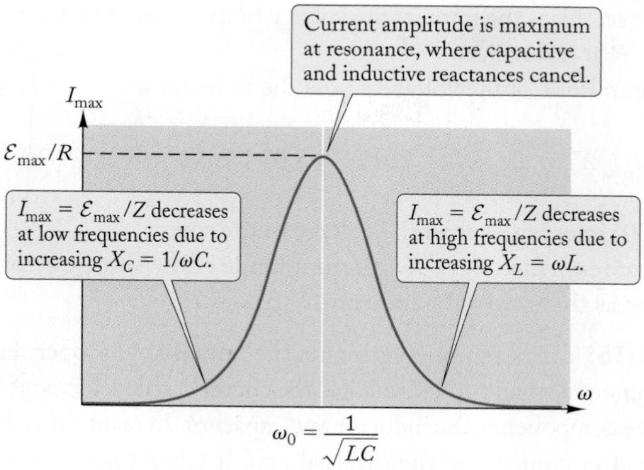

FIGURE 32.29 Current in a series RLC circuit as a function of the frequency of the oscillating emf.

This is called the *resonant frequency* and is the same as the natural resonant frequency of free oscillations of an LC circuit discussed in the previous section. Since the impedance is a minimum at resonance, the current amplitude

$$I_{max} = \frac{\mathcal{E}_{max}}{Z} = \frac{\mathcal{E}_{max}}{\sqrt{R^2 + [\omega L - (1/\omega C)]^2}} \tag{32.53}$$

current amplitude in RLC circuit

and the average power $\overline{P} = \frac{1}{2}I_{max}^2 R$ will both be maximum at resonance. Figure 32.29 is a plot of the current amplitude (32.53) as a function of frequency. Notice the characteristic peak of this resonance curve. The current is large at ω_0 but approaches zero for small and large frequencies. Such behavior is what permits the selection of a particular frequency by the tuning circuit of a radio, as discussed in the previous section.

EXAMPLE 6 A series RLC circuit has an applied AC voltage of amplitude $\mathcal{E}_{max} = 1.0$ V, which oscillates at a frequency equal to the resonant frequency of the circuit. If $R = 0.50\ \Omega$, $L = 20$ mH, and $C = 2.0\ \mu$F, what is the resonant frequency (in hertz)? What is the amplitude of the current? What is the amplitude of the voltage across the resistor? Across the inductor? Across the capacitor?

SOLUTION: The resonant angular frequency is given by Eq. (32.52):

$$\omega_0 = \frac{1}{\sqrt{LC}} = \frac{1}{\sqrt{20 \times 10^{-3}\ \text{H} \times 2.0 \times 10^{-6}\ \text{F}}} = 5.0 \times 10^3\ \text{radians/s}$$

The resonant frequency in hertz is

$$f_0 = \frac{\omega_0}{2\pi} = \frac{5.0 \times 10^3\ \text{radians/s}}{2\pi\ \text{radians/cycle}} = 800\ \text{cycles/s} = 800\ \text{Hz}$$

At resonance, $X_L = X_C$ and $Z = \sqrt{R^2 + (X_L - X_C)^2} = R$, so the amplitude of the current is

$$I_{max} = \frac{\mathcal{E}_{max}}{Z} = \frac{\mathcal{E}_{max}}{R} = \frac{1.0\ \text{V}}{0.50\ \Omega} = 2.0\ \text{A}$$

The amplitude of the voltage across the resistor is

$$V_{max,R} = I_{max}R = 2.0\ \text{A} \times 0.50\ \Omega = 1.0\ \text{V}$$

At resonance, this is the same as the voltage of the source of emf, which it must be, since $Z = R$ at resonance.

The amplitude of the voltage across the inductor is

$$V_{\text{max},L} = I_{\text{max}} X_L = I_{\text{max}} \omega L = 2.0 \text{ A} \times 5.0 \times 10^3 \text{ s}^{-1} \times 20 \times 10^{-3} \text{ H} = 200 \text{ V}$$

This is 200 times larger than the voltage of the source!

Since at resonance $X_L = X_C$, the amplitude of the voltage across the capacitor is the same as that across the inductor, $V_{\text{max},C} = 200$ V.

COMMENTS: This example highlights the remarkable properties of a resonant circuit: even for a small applied voltage, the circuit can develop large voltages across the reactive components, the inductor and capacitor. In reality, this does not happen instantly; after connecting the external emf, it takes a number of cycles for the amplitudes to build up to the values calculated above (the inverse process to the decaying oscillations of the previous section). Thereafter, the amplitudes maintain these steady values.

In Section 32.4, we discussed the quality factor, or Q, of a resonant circuit in terms of freely decaying oscillations. For the forced oscillations examined here, we can employ the same definition of the Q. The energy stored in the circuit is the magnetic energy when the current is a maximum, $\frac{1}{2}LI_{\text{max}}^2$. The energy dissipated per cycle is the average power times the time for one cycle, $\overline{P} \times T = (I_{\text{max}}^2 R/2) \times (2\pi/\omega_0)$. Thus the Q is

$$Q = -2\pi \frac{U}{\Delta U} = -2\pi \frac{\frac{1}{2}LI_{\text{max}}^2}{(-I_{\text{max}}^2 R/2) \times (2\pi/\omega_0)} = \frac{\omega_0 L}{R}$$

as obtained previously in Eq. (32.33). For forced oscillations, we can see that the Q also measures by what factor the reactive voltage amplitude at resonance is increased compared with the applied voltage:

$$Q = \frac{\omega_0 L}{R} = \frac{X_L}{R} = \frac{I_{\text{max}} X_L}{I_{\text{max}} R} = \frac{V_{\text{max},L}}{V_{\text{max},R}} \tag{32.54}$$

or, since $V_{\text{max},R} = \mathcal{E}_{\text{max}}$ at resonance,

$$V_{\text{max},L} = Q\mathcal{E}_{\text{max}}$$

Thus the resonant circuit multiplies the same voltage by a factor of Q.

The Q is also a measure of the sharpness of the resonance peak; it is common to plot the power dissipated as a function of frequency as in Fig. 32.30. If $\Delta\omega$ is the full width of the power peak at half the maximum value, it can be shown (Problem 54) that the above value of the Q is equivalent to

$$Q = \frac{\omega_0}{\Delta\omega} \tag{32.55}$$

Thus a circuit with a large quality factor has a narrow resonance curve (small $\Delta\omega$, see Fig. 32.30) and a circuit with a low quality factor has a broad resonance curve (see also Fig. 15.22).

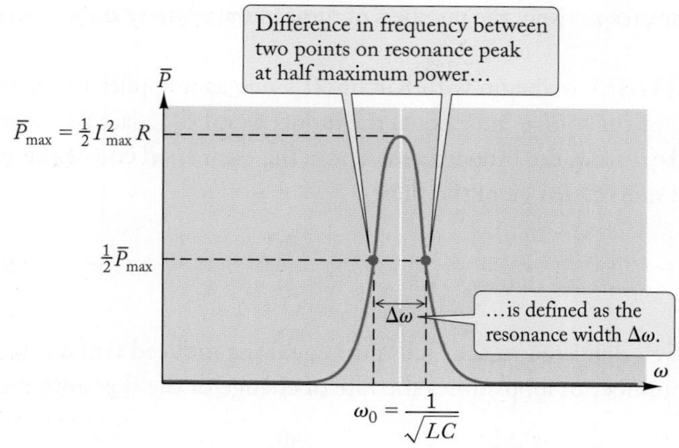

FIGURE 32.30 Average power in a series *RLC* circuit as a function of the frequency of the oscillating emf.

Checkup 32.5

QUESTION 1: An AC circuit consists of a fixed-amplitude emf, a resistor, and one other component. When the frequency increases, the current increases. What is the other component?

QUESTION 2: An AC emf with amplitude 1.0 V is applied to a series *RLC* circuit. The frequency is varied and the current is observed to have a maximum value of 0.20 A when the inductive reactance is $X_L = 50\ \Omega$. What is the value of R? What is the quality factor of the circuit?

QUESTION 3: What is the Q of the *RLC* circuit of Example 6?

QUESTION 4: A series *RLC* circuit is connected to an emf of amplitude 2.0 V. At the resonant frequency, the amplitude of the voltage across the inductor is 50 V. What is the amplitude of the voltage across the resistor?

(A) 102 V (B) 98 V (C) 52 V (D) 48 V (E) 2.0 V

32.6 THE TRANSFORMER

A transformer consists of two coils arranged in such a way that (almost) all the magnetic field lines generated by one of them pass through the other. This can be achieved by winding both coils on a common iron core (see Fig. 32.31). As we saw in Section 30.4, the iron increases the strength of the magnetic field in its interior by a large factor. Since the field is much stronger in the iron than outside, most of the field lines have to stay inside the iron; this means that *the iron tends to keep the field lines together and acts as a conduit for the field lines from one coil to the other.*

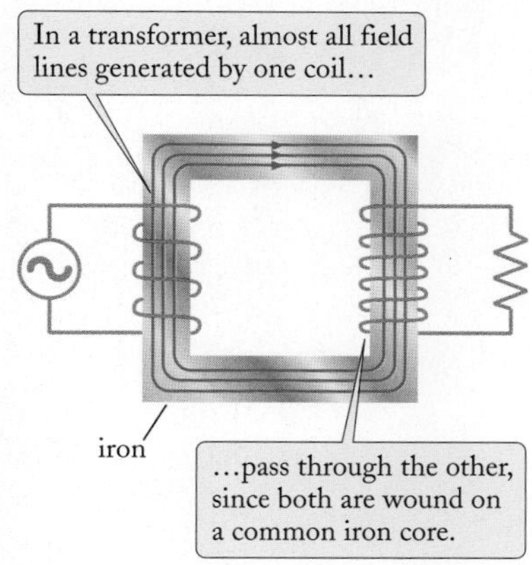

FIGURE 32.31 A transformer.

Each coil is part of a separate electric circuit (see Fig. 32.32). The **primary** circuit has a source of alternating emf, and the **secondary** circuit has a lightbulb, or an electric heater, or some other "load" that consumes electric power. The alternating current in the primary circuit induces an alternating emf in the secondary circuit. We will show that the emf $\mathcal{E}_2$ produced in the secondary circuit is related as follows to the emf $\mathcal{E}_1$ in the primary circuit:

primary circuit and secondary circuit

$$\mathcal{E}_2 = \mathcal{E}_1 \frac{N_2}{N_1} \qquad (32.56)$$

emfs in transformer

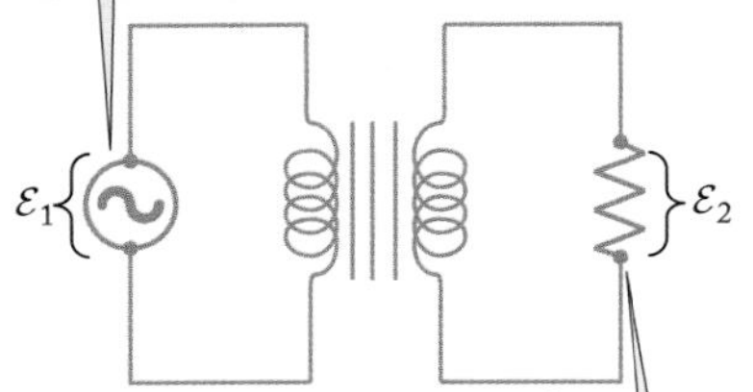

FIGURE 32.32 Circuit diagram for the transformer. The parallel lines represent a solid iron core.

where N_1 and N_2 are, respectively, the numbers of turns in the primary and the secondary coils.

To prove Eq. (32.56), we begin with Kirchhoff's rule as it applies to the primary circuit: the emf $\mathcal{E}_1$ of the source must equal the induced emf $\mathcal{E}_{1,\text{ind}}$ across the primary coil. But by Faraday's Law, the induced emf across the entire coil equals the number of loops times the rate of change of the flux:

$$\mathcal{E}_1 = \mathcal{E}_{1,\text{ind}} = -N_1 \frac{d\Phi_1}{dt} \tag{32.57}$$

Likewise, the emf $\mathcal{E}_2$ delivered to the load must equal the induced emf $\mathcal{E}_{2,\text{ind}}$, which, again, equals the number of loops times the rate of change of the flux in that coil:

$$\mathcal{E}_2 = \mathcal{E}_{2,\text{ind}} = -N_2 \frac{d\Phi_2}{dt} \tag{32.58}$$

Since the same numbers of magnetic field lines pass through both coils, we know that

$$\frac{d\Phi_1}{dt} = \frac{d\Phi_2}{dt} \tag{32.59}$$

Thus, the ratio of Eqs. (32.58) and (32.57) is

$$\frac{\mathcal{E}_2}{\mathcal{E}_1} = \frac{N_2}{N_1} \tag{32.60}$$

which is equivalent to Eq. (32.56).

step-up transformer and step-down transformer

If $N_2 > N_1$, we have a **step-up transformer**, and if $N_2 < N_1$, we have a **step-down transformer**.

EXAMPLE 7 Doorbells and buzzers usually are designed for 12 volts AC and they are powered by small transformers that step down 115 volts AC to 12 volts AC. Suppose that such a transformer has a primary winding with 1500 turns. How many turns are there in the secondary winding?

SOLUTION: Equation (32.56) applies to the instantaneous voltages. It is therefore also valid for the rms voltages. With the appropriate numerical values, Eq. (32.56) gives

$$N_2 = N_1 \frac{\mathcal{E}_2}{\mathcal{E}_1} = 1500 \text{ turns} \times \frac{12 \text{ V}}{115 \text{ V}} = 157 \text{ turns}$$

As long as the secondary circuit is open and carries no current ($I_2 = 0$), *an ideal transformer does not consume electric power*. Under these conditions, the primary circuit consists of nothing but the source of emf and an inductor—it is a pure L circuit. In such a circuit, the power delivered by the source of emf to the inductor averages to zero (see Section 32.3).

If the secondary circuit is closed—that is, if it is connected to some external load—a current will flow ($I_2 \neq 0$). This current contributes to the magnetic flux in the transformer and induces a current in the primary circuit. The current in the latter is then different from that in a pure L circuit, and the power will *not* average to zero over a cycle.

In an ideal transformer, the electric power that the primary circuit takes from the source of emf exactly matches the power that the secondary circuit delivers to the external load. Since the power is the product of the current and the emf, we can express the equality of these instantaneous powers as

$$I_1\mathcal{E}_1 = I_2\mathcal{E}_2 \tag{32.61}$$

where the currents and emfs are also taken to be instantaneous values. Comparison with Eq. (32.60) makes it evident that the currents are in inverse proportion to the corresponding ratio of turns of the coils:

$$\frac{I_1}{I_2} = \frac{N_2}{N_1} \tag{32.62}$$

Concepts in Context

Good transformers approach the ideal condition (32.61) of conservation of electric energy fairly closely: about 99% of the power supplied to the input terminals emerges at the output terminals; the difference is essentially all lost as heat in the iron core and the windings. Such high efficiency is attained by avoiding eddy-current losses (see Section 31.3), by means of an iron core constructed from thin layers separated by insulation. For the immense transformers at power plants (chapter photo), even the loss of a tiny fraction of the power transmitted requires the use of special cooling to carry away the waste heat. Power-plant transformers are typically immersed in oil, and circulation of this liquid carries the heat to large-area metal surfaces; these radiator tubes are air-cooled with arrays of fans (see Fig. 32.33).

Transformers play a large role in our electric technology. As we saw in Example 8 of Chapter 28, transmission lines for electric power operate much more efficiently at high voltage, since this reduces the Joule heating losses caused by the resistance of the wires. To take advantage of this high efficiency, power lines are made to operate

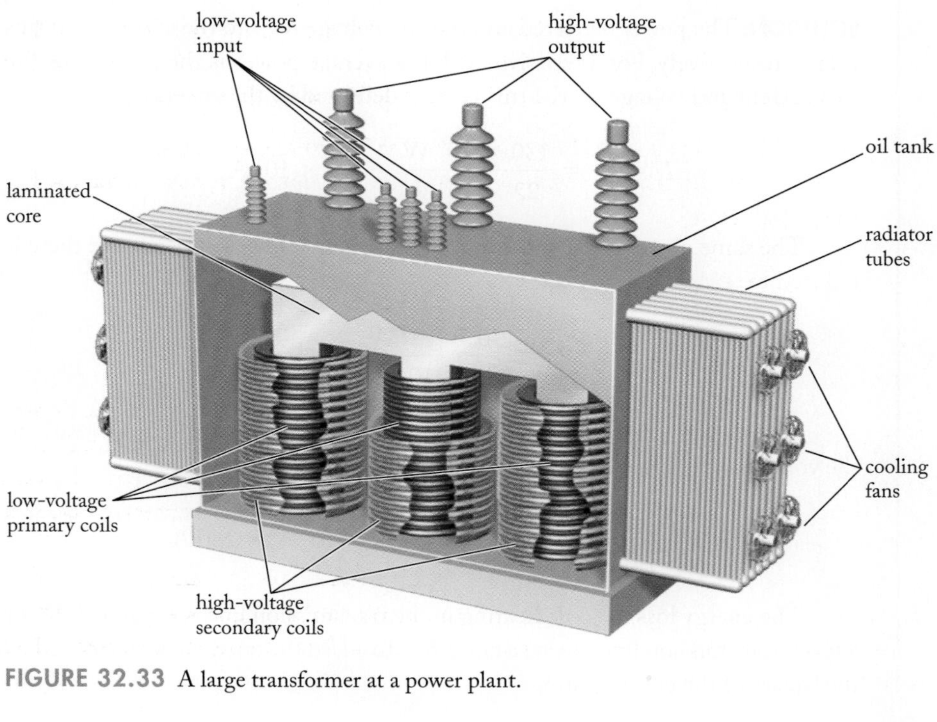

FIGURE 32.33 A large transformer at a power plant.

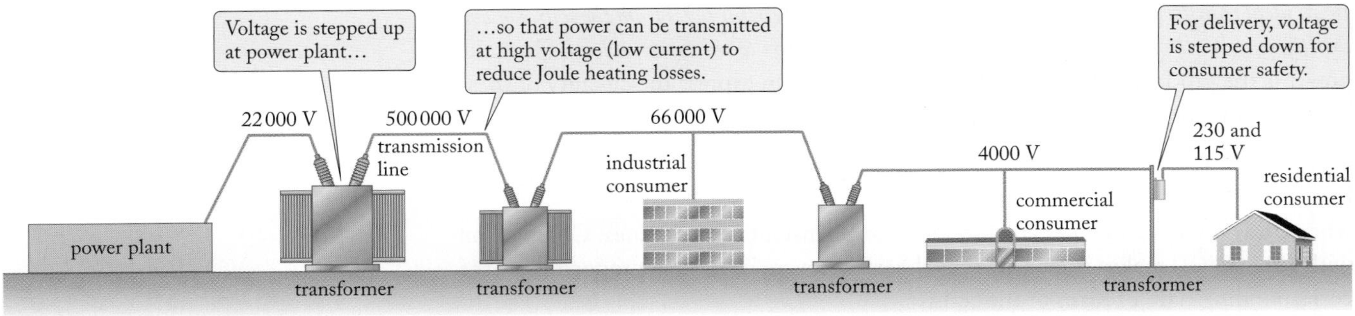

FIGURE 32.34 Typical voltage transformations during the transmission of electric power. The small transformers used to step the power down to 115 V for residential use are often attached to telephone poles.

at several hundred kilovolts. The voltage must be stepped up to this value at the power plant, and for safety's sake, it must be stepped down just before it reaches the consumer (see Fig. 32.34). For these operations, large banks of transformers are needed at both ends. Transformers are also used in many electric and electronic devices, such as TV tubes and X-ray machines, which require high voltages to accelerate beams of electrons.

Concepts in Context

EXAMPLE 8 A single generator of a large power plant delivers an electric power of 130 MW at 22 kilovolts AC. For transmission, this voltage is stepped up to 500 kV by a transformer. What is the rms current delivered by the generator? What is the rms current in the transmission line? By what factor is the energy loss (Joule heating) of the transmission line reduced by using the transformer? Assume that the transformer does not waste any power, and that the loads are purely resistive.

SOLUTION: The power delivered and the AC voltage refer to the average and rms values, respectively. For a resistive load, the average power is the product of the rms current and voltage, so the rms current delivered by the generator is

$$I_1 = \frac{\overline{P}}{\mathcal{E}_1} = \frac{130 \times 10^6\ \text{W}}{22 \times 10^3\ \text{V}} = 5.9 \times 10^3\ \text{A}$$

The same power is delivered to the transmission line, so the current there is reduced to

$$I_2 = \frac{\overline{P}}{\mathcal{E}_2} = \frac{130 \times 10^6\ \text{W}}{500 \times 10^3\ \text{V}} = 260\ \text{A}$$

Alternatively, we can note that the transformer currents are in inverse ratio to the voltages [Eq. (32.61)]:

$$I_2 = \frac{\mathcal{E}_1}{\mathcal{E}_2} I_1 = \frac{22 \times 10^3\ \text{V}}{500 \times 10^3\ \text{V}} \times 5.9 \times 10^3\ \text{A} = 260\ \text{A}$$

The energy loss by Joule heating in the transmission line is equal to I^2R; for a given transmission line, the resistance R is fixed, so the energy loss is reduced by the square of the current ratio,

$$\frac{I_1^2}{I_2^2} = \frac{(5.9 \times 10^3 \text{ A})^2}{(260 \text{ A})^2} = 5.2 \times 10^2$$

Energy losses are reduced by more than a factor of 500! For a 2.0-Ω transmission line, this is the difference between the loss of more than half of the power ($I_1^2 R = (5.9 \times 10^3 \text{ A})^2 \times 2.0\ \Omega = 70$ MW, or 54%) and an almost negligible loss (0.1%). If the power were transmitted at an even lower voltage, as in Example 8 of Chapter 28, the energy loss would become much more extreme.

✔ Checkup 32.6

QUESTION 1: If we want to use a transformer to step up 115 V AC to 230 V AC, by what factor must the number of coils in the secondary differ from that in the primary?

QUESTION 2: Suppose that the transformer described in Example 7 delivers some power to doorbell. Is the current larger in the primary or in the secondary?

QUESTION 3: Suppose that you have two transformers, one with a turns ratio of 2 and one with a turns ratio of 5. Each may be used to step up or to step down a voltage. The two are to be connected one after the other. What are possible values for the ratio of input to output voltages for the pair?

(A) 10, 0.1 (B) 10, 5, 2.5 (C) 25, 4 (D) 10, 5, 2, 0.1 (E) 10, 2.5, 0.4, 0.1

SUMMARY

PHYSICS IN PRACTICE Frequency Filter Circuits **(page 1037)**

CURRENT IN RESISTOR CIRCUIT

$$I = \frac{\mathcal{E}_{max} \cos \omega t}{R} \qquad \textbf{(32.3)}$$

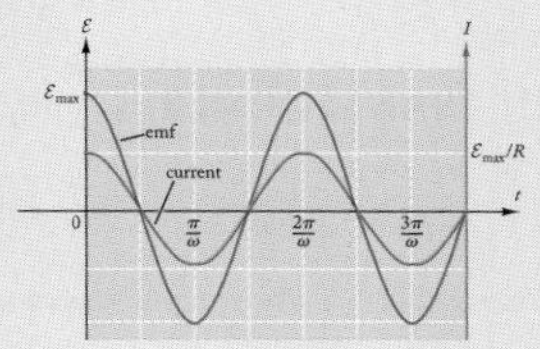

RMS (ROOT-MEAN-SQUARE) VOLTAGE

$$\mathcal{E}_{rms} = \frac{\mathcal{E}_{max}}{\sqrt{2}} \qquad \textbf{(32.8)}$$

AVERAGE AC POWER DISSIPATED BY RESISTOR

$$\overline{P} = \frac{\mathcal{E}_{max}^2}{2R} = \frac{\mathcal{E}_{rms}^2}{R} \qquad \textbf{(32.6, 32.7)}$$

REACTANCE OF CAPACITOR CIRCUIT

$$X_C = \frac{1}{\omega C} \qquad \textbf{(32.14)}$$

CURRENT IN CAPACITOR CIRCUIT

$$I = -\frac{\mathcal{E}_{max} \sin \omega t}{X_C} \qquad \textbf{(32.13)}$$

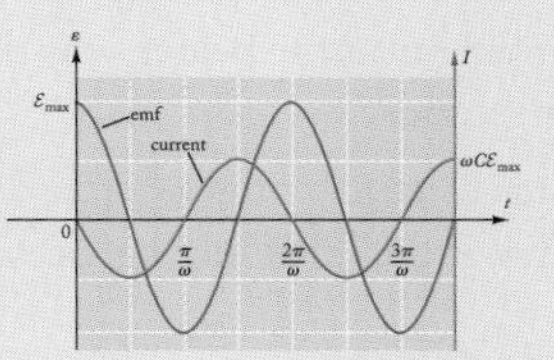

REACTANCE OF INDUCTOR CIRCUIT $$X_L = \omega L \tag{32.20}$$

CURRENT IN INDUCTOR CIRCUIT $$I = \frac{\mathcal{E}_{\text{max}} \sin \omega t}{X_L} \tag{32.19}$$

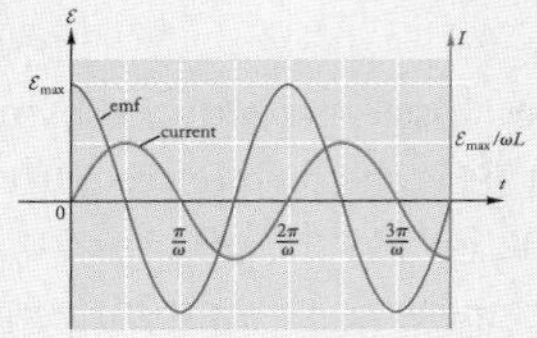

NATURAL FREQUENCY OF *LC* CIRCUIT; RESONANT FREQUENCY $$\omega_0 = \frac{1}{\sqrt{LC}} \tag{32.26}$$

SLOWLY DECAYING OSCILLATIONS IN *RLC* CIRCUIT (damped harmonic oscillator) $$Q = Q_{\text{max}} e^{-Rt/2L} \cos \omega_d t \tag{32.30}$$

***Q* OF *RLC* CIRCUIT** $$Q = \frac{\omega_0 L}{R}$$

IMPEDANCE IN SERIES *RLC* CIRCUIT $$Z = \sqrt{R^2 + (X_L - X_C)^2} \tag{32.48}$$

IMPEDANCE RELATED TO AMPLITUDES OF EMF AND CURRENT $$Z = \frac{\mathcal{E}_{\text{max}}}{I_{\text{max}}} \tag{32.44}$$

PHASE ANGLE OF EMF WITH RESPECT TO CURRENT IN A SERIES *RLC* CIRCUIT $$\phi = \tan^{-1}\left(\frac{X_L - X_C}{R}\right) \tag{32.49}$$

VOLTAGE AMPLITUDES IN A SERIES *RLC* CIRCUIT $$V_{\text{max},R} = I_{\text{max}} R \quad V_{\text{max},L} = I_{\text{max}} X_L \quad V_{\text{max},C} = I_{\text{max}} X_C \tag{32.41}$$

EMFs OF TRANSFORMER $$\mathcal{E}_2 = \mathcal{E}_1 \frac{N_2}{N_1} \tag{32.56}$$

CURRENTS IN TRANSFORMER $$I_2 = I_1 \frac{N_1}{N_2} \tag{32.62}$$

QUESTIONS FOR DISCUSSION

1. In most European countries, the voltage available at outlets in homes is 230 V AC. What is the actual amplitude of oscillation of this voltage?
2. You can perceive the 120-Hz flicker (two peaks of intensity per AC cycle) in a fluorescent light tube (by sweeping your eye quickly across the tube), but you cannot perceive any such flicker in an incandescent lightbulb. Explain.
3. An electric heater operates from 60-Hz, 115-V AC. When standing near the heater, why do we not feel 120 heat pulses per second?
4. Some electric motors operate only on DC, others only on AC. What is the difference between these motors?
5. If you connect a capacitor across a 115-V outlet, does any current flow through the connecting wires? Through the space between the capacitor plates? Does the outlet deliver instaneous electric power? Average electric power?
6. It is sometimes said that a capacitor becomes a short circuit at high frequencies, and that an inductor becomes an open circuit at high frequencies. Explain.
7. Can you blow a fuse by connecting a very large capacitor across an ordinary 115-V outlet?
8. How could you use an *LC* circuit to measure the capacitance of a capacitor?

9. Consider a series *RLC* circuit. Can the voltage across the capacitor ever be larger than $\mathcal{E}_{max}$? Across the inductor? Across the resistor? (Hint: Inspect the phasor diagram.)

10. Roughly plot the phase angle ϕ given by Eq. (32.49) as a function of ω. What is the phase angle at resonance?

11. If you use an AC voltmeter to measure the driving emf and the voltages across the inductor, the capacitor, and the resistor in a series *RLC* circuit, you will find that the emf is greater than or equal to the voltage across the resistor, but smaller than the sum of the voltages across the inductor, capacitor, and resistor. Explain.

12. What is the impedance of a series *RLC* circuit at resonance?

13. If we substitute $R = 0$ in Eq. (32.53), we obtain the equation for the current in a driven *LC* circuit. What must we substitute to obtain the equation for a driven *RC* circuit? A driven *RL* circuit?

14. Show that the equation for the current in a series *RLC* circuit [Eq. (32.53)] includes Eq. (32.3), (32.13), and (32.19) as special cases. What are the individual impedances of a resistor, a capacitor, and an inductor?

15. Why can we not use a transformer to step up the voltage supplied by a battery?

16. Does an electric motor absorb more electric power when working against a mechanical load than when running freely?

PROBLEMS

†32.1 Resistor Circuit

1. An electric heater (Fig. 32.35) plugged into a 115-V AC outlet uses an average electric power of 1200 W.

   (a) What are the rms current and the maximum instantaneous current through the heater?

   (b) What are the maximum instantaneous power and the minimum instantaneous power?

FIGURE 32.35 An electric heater.

2. An immersible heating element used to boil water consumes an (average) electric power of 400 W when connected to a source of 115 volts AC. Suppose that you connect this heating element to a source of 115 volts DC. What power will it consume?

3. A high-voltage power line operates on an rms voltage of 230 000 volts AC and delivers an rms current of 740 A to a resistive load.

   (a) What are the maximum instantaneous voltage and current?

   (b) What are the maximum instantaneous power and the average power delivered?

4. The GG-1 electric locomotive develops 4600 hp; it runs on an rms AC voltage of 1100 V.

   (a) What rms current does this locomotive draw?

   (b) Why is it advantageous to supply the electric power for locomotives at high voltage (and fairly low current)?

5. An electric heater operating with a 115-V AC power supply delivers 1200 W of heat.

   (a) What is the rms current through this heater?

   (b) What is the maximum instantaneous current?

   (c) What is the resistance of this heater?

6. A frequency generator delivers an average power of 2.0 W into a load of 50 Ω. What is the rms voltage? The rms current?

7. In a test of an AM radio station transmitter, an AC voltage amplitude of 2.2 kV is measured across an oil-cooled resistive load of 50 Ω. What is the rms current? What is the average power delivered?

*8. A square loop of nichrome wire with resistance 11 Ω and area 0.15 m^2 is rotated in a magnetic field of 2.0 T. For what frequency of rotation will the average power dissipated in the loop be 1.0 W? 100 W?

†For help, see Online Concept Tutorial 35 at www.wwnorton.com/physics

†32.2 Capacitor Circuit

9. A capacitor of 2.2 μF is connected to a generator supplying 12 V AC. What is the rms current if the angular frequency of the generator is 1.2×10^3 radians/s? What is the rms current if the angular frequency of the generator is 2.4×10^3 radians/s?

10. A capacitor has a capacitance of 0.15 μF. What is the reactance when it is connected to a circuit operating at $\omega = 6.0 \times 10^3$ radians/s? At $\omega = 1.8 \times 10^4$ radians/s?

11. A capacitor has a reactance of 3.0×10^6 Ω at a frequency of 2.0 kHz. What is its reactance at 3.0 kHz? At 4.0 kHz?

12. A capacitor of 600 pF is connected to an audio generator of adjustable frequency. At what frequency is the reactance of this capacitor 2.0×10^5 Ω?

13. A capacitor with $C = 8.0 \times 10^{-7}$ F is connected to an oscillating source of emf. This source provides an emf $\mathcal{E} = \mathcal{E}_{max} \cos \omega t$, with $\mathcal{E}_{max} = 0.20$ V and $\omega = 6.0 \times 10^3$ radians/s.
    (a) What is the reactance of the capacitor?
    (b) What is the maximum current in the circuit?
    (c) What is the current at time $t = 0$? At time $t = \pi/4\omega$?

14. A capacitor of 0.40 μF is connected to an AC source of amplitude 12 V and angular frequency 3.0×10^3 radians/s. What is the instantaneous current in the circuit when the instantaneous voltage is 12 V? What is the instantaneous current one-quarter cycle later? One-half cycle later? Three-quarters cycle later?

15. To measure the capacitance of a capacitor, a physicist connects it to an oscillating source of emf of angular frequency 2.0×10^3 radians/s and of amplitude 4.0×10^{-3} V. She finds that the maximum instantaneous current in the capacitor is 8.0×10^{-9} A. What is the capacitance?

16. We want to generate a maximum current of 5.0×10^{-5} A in a circuit consisting of a capacitor of 2.0 pF connected to a source of oscillating emf of angular frequency 4.0×10^4 radians/s. What amplitude of oscillation of the emf do we need?

17. An emf $\mathcal{E} = \mathcal{E}_{max} \cos \omega t$ with $\omega = 6.0 \times 10^2$ radians/s is applied to a 2.0-μF capacitor. Calculate the instantaneous power delivered to the capacitor at $t = 1.0$ ms, 2.0 ms, 3.0 ms, and 4.0 ms.

18. A capacitor, used as a sensor for measuring the level of liquid nitrogen in a container, has a capacitance of 0.30 nF when the container is empty and 0.43 nF when it is full and the capacitor is immersed in liquid nitrogen. If an emf with an amplitude of 15 V and angular frequency of 6.0×10^4 radians/s is applied to the sensor, what is the amplitude of the current that will flow when the container is empty? When full?

*19. A 47-Ω resistor and a capacitor are to be connected in parallel to a source of emf oscillating with amplitude $\mathcal{E}_{max} = 6.2$ V at a frequency of 60 Hz. For what value of the capacitance C will the amplitude of the current into the capacitor equal the amplitude of the current through the resistor? (Hint: For circuit elements connected in parallel, the instantaneous voltage across each is the same.)

*20. A circuit consists of three capacitors of 0.80 μF each, connected in parallel to the terminals of a generator that delivers 24 V AC at an angular frequency of 1.8×10^3 radians/s (see Fig. 32.36). What is the net rms current in this circuit?

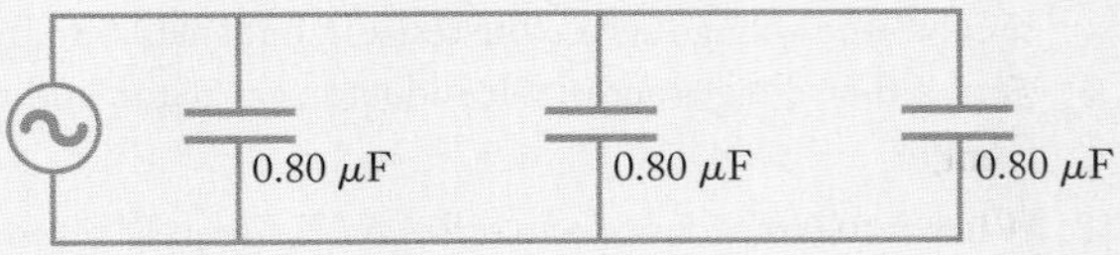

FIGURE 32.36 Three capacitors in parallel connected to a source of emf.

†32.3 Inductor Circuit

21. What is the rate of change of the current in a coil with an inductance of 250 μH that has an instantaneous voltage of 25 V across it?

22. An inductor has an inductance of 0.25 H. What is the reactance when it is connected to a circuit operating at $\omega = 6.0 \times 10^3$ radians/s? At $\omega = 1.8 \times 10^4$ radians/s?

23. An inductor of 0.20 H is connected to an audio generator of adjustable frequency. At what frequency is the reactance of this inductor 2.0×10^3 Ω?

24. A 1.0-km segment of the aluminum cable of a high-voltage transmission line has a resistance of 7.3×10^{-2} Ω, a capacitance of 4.7×10^{-8} F, and a self-inductance of 6.6×10^{-4} H. At the standard frequency of 60 Hz, what are the capacitive and the inductive reactances of this segment? Compare the reactances X_C and X_L and the resistance R. Which of these quantities is largest? Smallest?

25. An inductor of 0.30 H is connected to a generator supplying 12 V AC. What is the rms current if the angular frequency of the generator is 1.2×10^3 radians/s? What is the rms current if the angular frequency of the generator is 2.4×10^3 radians/s?

26. We want to generate a maximum current of 5.0×10^{-5} A in an inductor of 1.0×10^{-2} H connected to a source of oscillating emf of angular frequency 4.0×10^4 radians/s. What amplitude of oscillation of the emf do we need?

27. The primary winding of a transformer has an inductance of 6.2 H. What rms current will flow in this winding if it is connected to an outlet supplying 115 V AC?

28. To measure the inductance of an inductor, an experimenter connects it to an oscillating source of emf of angular frequency 3.0×10^3 radians/s and of amplitude 8.0×10^{-3} V. He finds that the maximum instantaneous current in the inductor is 2.0×10^{-7} A. What is the inductance?

†For help, see Online Concept Tutorial 35 at www.wwnorton.com/physics

†For help, see Online Concept Tutorial 35 at www.wwnorton.com/physics

29. The voltage across a superconducting solenoid of inductance 35 H may not exceed 12 V without malfunction. What is the largest permitted rate of change of current through the solenoid? If an alternating current with amplitude 25 A is to used, what is the maximum frequency at which this current may oscillate?

30. An AM radio antenna has the shape of a coil with inductance 0.90 mH. A radio wave with frequency 1.2 MHz induces a current of amplitude 5.2 nA in the coil. What is the amplitude of the induced voltage across the coil?

31. To prevent radio frequency interference in electronic equipment, inductors are often placed wherever the transmission of only low-frequency signals is desired. If such an **rf choke** has an inductance of 47 μH, what reactance does it present to a 60-Hz power supply current? To a 100-MHz radio signal?

*32. An inductor with $L = 4.0 \times 10^{-2}$ H is connected to a source of alternating emf. This source provides an emf $\mathcal{E} = \mathcal{E}_{max} \times \cos \omega t$, with $\mathcal{E}_{max} = 0.20$ V and $\omega = 6.0 \times 10^3$ radians/s.

(a) What is the reactance of the inductor?

(b) What is the maximum current in the circuit?

(c) What is the current at time $t = 0$? At time $t = \pi/4\omega$?

*33. A 2.0-μF capacitor and a 3.0-mH inductor are connected in parallel with an AC power supply. At what frequency will the inductive reactance be 3 times the capacitive reactance?

*34. Consider the circuit shown in Fig. 32.37. The emf of the source is of the form $\mathcal{E}_{max} \cos \omega t$. In terms of this emf and the capacitance C and the inductance L, find the instantaneous currents through the capacitor and the inductor. Find the instantaneous current and the instantaneous power delivered by the source of emf.

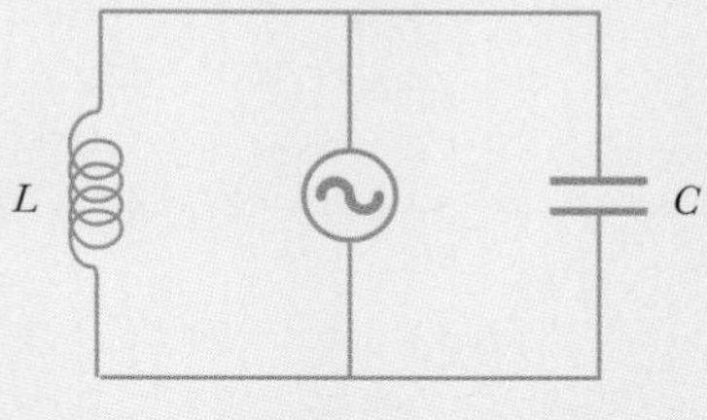

FIGURE 32.37 Inductor and capacitor connected to a source of emf.

*35. In Example 3 we examined the current in a circuit containing a resistor, a capacitor, and an inductor connected in parallel to a source of oscillating emf of an angular frequency 6.0×10^3 radians/s and amplitude 1.0×10^{-3} V. How would the answers to this example change if we reduced the angular frequency to 3.0×10^3 radians/s?

*36. Suppose that in Example 3 the emf is of the form $\mathcal{E} = \mathcal{E}_{max} \times \cos \omega t$. What is the net instantaneous current in the circuit at $t = 0$? At $t = \pi/2\omega$? At $t = \pi/\omega$?

*37. An inductor of 4.5×10^{-2} H and a capacitor of 0.25 μF are connected in parallel to a source of alternating emf of frequency ω. For what value of ω will the rms currents in the inductor and the capacitor be of equal magnitudes?

*38. An inductor is built in the shape of a solenoid of radius 0.20 cm, length 4.0 cm, with 1000 turns. This inductor is connected to a generator supplying an rms voltage of 1.2×10^{-4} V AC at an angular frequency of 9.0×10^3 radians/s.

(a) What is the inductance? Assume that the solenoid can be treated as very long.

(b) What is the rms current flowing through the inductor?

(c) What is the rms energy stored in the inductor?

32.4 Freely Oscillating *LC* and *RLC* Circuits

39. What is the natural frequency for an *LC* circuit consisting of a 2.2×10^{-6} F capacitor and an 8.0×10^{-2} H inductor?

40. You want to construct an *LC* circuit of natural frequency 8.0×10^3 Hz. You have available a capacitor of 0.20 μF. What inductor do you need?

41. What is the natural frequency of oscillation of the circuit shown in Fig. 32.38? The capacitances are 2.4×10^{-5} F each and the inductance is 1.2×10^{-3} H.

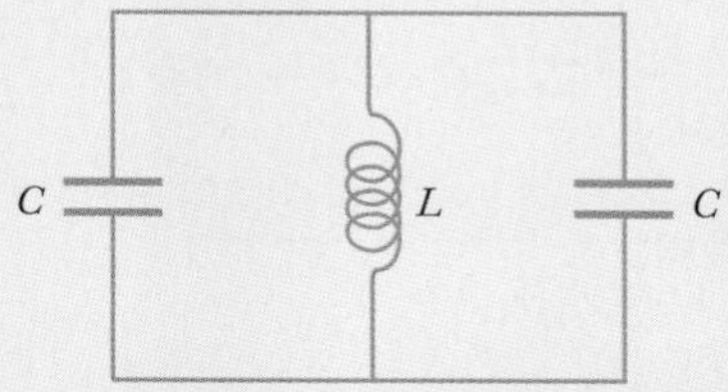

FIGURE 32.38 Two equal capacitors connected to an inductor.

42. A radio receiver contains an *LC* circuit whose natural frequency of oscillation can be adjusted, or tuned, to match the frequency of incoming radio waves. The adjustment is made by means of a variable capacitor. Suppose that the inductance of the circuit is 15 μH. Over what range of capacitances must the capacitor be adjustable if the frequencies of oscillation of the circuit are to span the range from 530 kHz to 1600 kHz?

43. In Fig. 32.39, the emf has value $\mathcal{E} = 12$ V and the capacitance is $C = 22$ μF. After the switch S in the circuit has been in position 1 for a long time, it is moved to position 2. The current in the right loop then oscillates with amplitude 3.5 mA. What is the frequency of oscillation? What is the value of the inductance L?

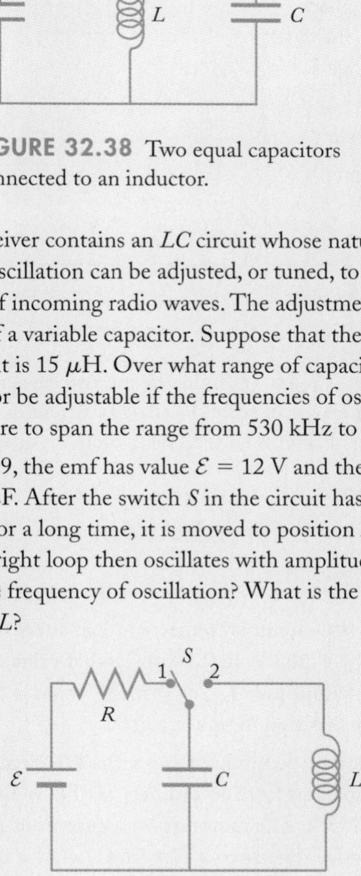

FIGURE 32.39 Switch-selected *RC* and *LC* circuits.

44. Consider an underdamped series *RLC* circuit (Fig. 32.19) with $R = 0.75\ \Omega$, $L = 25\ \mu\text{H}$, and $C = 100$ nF. What is the value of the quality factor Q for this circuit? If the capacitor has an initial charge Q_{max}, how many oscillations will occur before the charge amplitude is reduced to $(1/e) \times Q_{max} \approx 0.37Q_{max}$?

45. An underdamped series *RLC* circuit with $R = 2.2 \times 10^{-3}\ \Omega$ is to have an oscillation frequency of 3.5×10^6 Hz and a quality factor Q of 750. What must be the values of the capacitance C and the inductance L?

*46. The circuit of Fig. 32.40a is oscillating with the switch S closed. The graph of current vs. time is shown in Fig. 32.40b.

(a) At time t_1 the switch S is suddenly opened. Is the frequency of oscillation increased, decreased, or unchanged? In the space on the right in Fig. 32.40b sketch a rough, qualitative graph of current for times after t_1.

(b) At time t_2 the switch S is closed. Sketch the graph of current after this time.

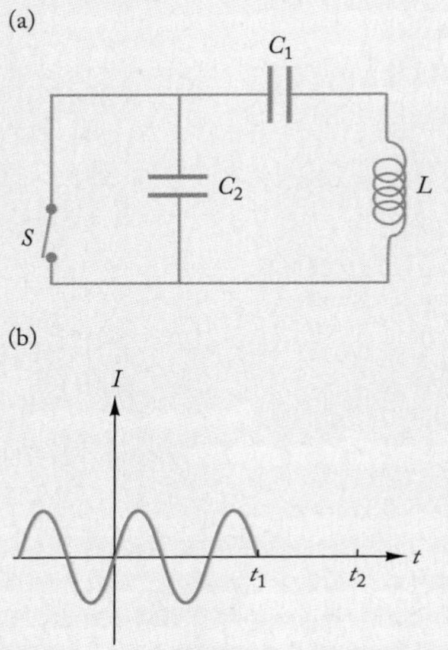

FIGURE 32.40 (a) Inductor and two capacitors in a circuit. The switch S is initially closed. (b) Current in the circuit as a function of time. At $t = t_1$, the switch S is opened.

†32.5 Series Circuits with Alternating Emf

47. A series *RC* circuit is connected to an alternating 60-Hz emf with $\mathcal{E}_{max} = 163$ V. If $R = 50\ \Omega$, what value of C will result in a current amplitude $I_{max} = 0.50$ A? What is the average rate of power dissipation in this circuit?

48. A motor may be modeled as a series *RL* circuit. A particular motor operates from an ordinary 60-Hz wall outlet with $\mathcal{E}_{rms} = 115$ V. The motor draws a current of $I_{rms} = 2.5$ A. If the effective resistance is $R = 16\ \Omega$, what is the inductive reactance of the circuit? What is the value of the inductance?

†For help, see Online Concept Tutorial 35 at www.wwnorton.com/physics

49. In a series AC circuit, average power is dissipated only in the resistor. That power is $\overline{P} = \frac{1}{2}I_{max}^2 R$.

(a) Use Eq. (32.44) to verify that the average power can be written

$$\overline{P} = \tfrac{1}{2}\mathcal{E}_{max} I_{max} \frac{R}{Z}$$

(b) Use the phasor diagram of Fig. 32.28c to rewrite this as

$$\overline{P} = \tfrac{1}{2}\mathcal{E}_{max} I_{max} \cos\phi$$

(c) The quantity $\cos\phi$ is called the **power factor**; it is the factor by which the average power is reduced from that of a purely resistive circuit. A certain series circuit has $\mathcal{E}_{max} = 167$ V, $I_{max} = 14$ A, and dissipates an average power of 750 W. What is the power factor? What is the phase angle? Can we tell whether the circuit contains a capacitor, an inductor, or both?

50. A series *RLC* circuit is connected to a source of alternating emf as shown in Fig. 32.41. The components have values $R = 5.2\ \Omega$, $L = 36$ mH, and $C = 0.41\ \mu\text{F}$. The amplitude of the emf of the source is 100 V. Initially, the frequency of the emf is tuned to resonance.

(a) What is the resonant frequency?

(b) At resonance, what is the amplitude of the voltage across the inductor?

(c) At resonance, what is the amplitude of the voltage between points 1 and 2 of Fig. 32.41?

(d) The frequency is changed to 4.7×10^5 radians per second. Now what is the amplitude of the voltage between points 1 and 2 of Fig. 32.41?

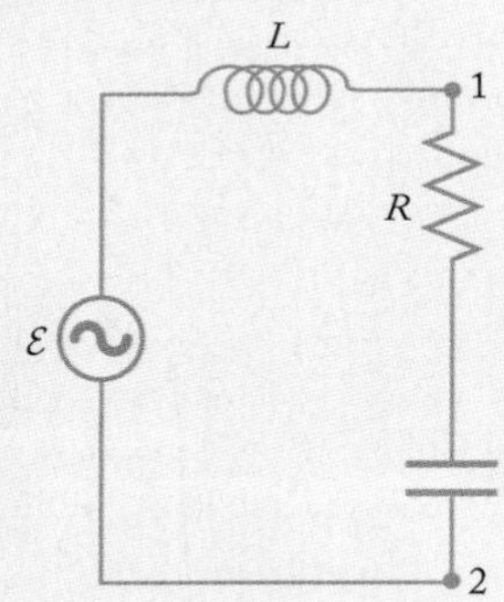

FIGURE 32.41 A series *RLC* circuit.

51. A series *RLC* circuit with $R = 3.0\ \Omega$, $C = 20$ pF, and a resonant frequency of 17 MHz is connected to an alternating emf with $\mathcal{E}_{max} = 12$ V.

(a) What is the amplitude of the voltage across the inductor at resonance?

(b) What is the amplitude of the voltage across the resistor at resonance?

(c) What is the average power dissipated in this circuit?

(d) What is the quality factor Q of this circuit?

52. A series RC circuit is often used as a **high-pass filter**. Consider a resistor R and capacitor C connected in series to an oscillating source of emf with amplitude $\mathcal{E}_{max}$. In terms of $\mathcal{E}_{max}$, what is the amplitude of the voltage across the resistor for a low frequency, when $\omega RC = 1/10$? For a high frequency, when $\omega RC = 10$? At what frequency is the amplitude of the voltage across the resistor equal to $\mathcal{E}_{max}/\sqrt{2}$?

53. A series RL circuit is often used as a **low-pass filter**. Consider a resistor R and inductor L connected in series to an oscillating source of emf with amplitude $\mathcal{E}_{max}$. In terms of $\mathcal{E}_{max}$, what is the amplitude of the voltage across the resistor for a low frequency, when $\omega L/R = 1/10$? For a high frequency, when $\omega L/R = 10$? At what frequency is the amplitude of the voltage across the resistor equal to $\mathcal{E}_{max}/\sqrt{2}$?

*54. The average power dissipated in a series RLC circuit is $\overline{P} = \frac{1}{2}I_{max}^2 R$, where I_{max} is given by Eq. (32.53). We can define the **width of resonance** in terms of the half-power points $\overline{P}_{1/2}$ of the resonance curve; these have half the peak value, or $\overline{P}_{1/2} = \frac{1}{2}(\frac{1}{2}\mathcal{E}_{max}^2/R)$. By setting $\overline{P} = \overline{P}_{1/2}$, obtain an equation for the values of ω at the half-power points. Solve the resulting quadratic equation and show that the two roots are separated by

$$\Delta\omega = \frac{R}{L}$$

Thus we can verify the equivalence of Eqs. (32.33) and (32.55), $Q = \omega_0 L/R = \omega_0/\Delta\omega$.

*55. Consider that an emf is connected to a ***parallel RLC* circuit**. In this case, the voltage V_{max} is the same across the source and each element, but the net current is the sum of the individual currents, so a phasor diagram for the *currents* can be drawn instead. Draw this diagram. By dividing the magnitude of each current phasor by the common value of V_{max}, obtain a phasor diagram that relates impedance, reactances, and resistance. Use the geometry of the diagram to show that

$$Z = \left[\frac{1}{R^2} + \left(\frac{1}{X_L} - \frac{1}{X_C}\right)^2\right]^{-1/2}$$

56. Consider a *parallel RLC* circuit (see Problem 55) with $R = 5.0\ \Omega$, $L = 30\ \mu H$, and $C = 25$ nF. An alternating emf with $\mathcal{E}_{max} = 15$ V is applied. Determine the total current in the circuit when

(a) It operates at the resonant frequency.

(b) It operates at half the resonant frequency.

32.6 The Transformer

57. A transformer used to step up 115 V to 5000 V has a primary coil of 100 turns. What must be the number of turns in the secondary coil?

58. The primary winding of a transformer has 1200 turns and the secondary winding has 80 turns. If the emf supplied to the primary is 115 V AC, what is the emf delivered by the secondary?

59. Consider each of the four transformers illustrated in Fig. 32.34; assume that the last of these delivers 115 V AC. What is the ratio N_1/N_2 of the numbers of turns in the primary and secondary windings of each of these transformers?

60. How much current is flowing in the 1200-V secondary of a transformer if the primary has a voltage of 12 V and a current of 3.0 A?

61. The generators of a large power plant deliver an electric power of 2000 MW at 22 kilovolts AC. For transmission, this rms voltage is stepped up to 400 kV by a transformer. What is the rms current delivered by the generators? What is the rms current in the transmission line? Assume that the transformer does not waste any power.

62. A transformer operating on a primary voltage of 115 volts AC delivers a secondary voltage of 6.0 volts AC to a small electric buzzer (see Fig. 32.42). If the current in the secondary circuit is 3.0 A, what is the rms current in the primary circuit? Assume that no electric power is lost in the transformer.

FIGURE 32.42 Small transformers used for buzzers and bells.

63. In a transformer, the ratio of turns of primary and secondary is 10:1. If the primary voltage is 120 V and the secondary current is 3.0 A, how much power is being absorbed in the primary circuit of the transformer?

64. The largest transformer ever built handles a power of 1.50×10^9 W. This transformer is used to step down 765 kVAC to 345 kVAC. What is the rms current in the primary? What is the current in the secondary? Assume that no electric power is lost by the transformer.

65. The secondary coils of many transformers have *taps*; these are extra connections to an intermediate part of the secondary coil, so that, in addition to the full secondary voltage, a fixed fraction of the secondary voltage is also available. Consider a transformer with a 115-V AC primary and a 12-V AC secondary that has a **center tap**. The 12-V AC secondary voltage is applied to a load that dissipates an average power of 4.0 W, and the center tap (6.0 V AC) to a load of average power 12 W. What is the current in the primary?

66. A **variac** is a variable transformer, usually with a sliding contact that can be moved along the coil to select some integer number of the total turns in the secondary coil. If the 115-V AC primary of a variac has 200 turns and the secondary has 250 turns, what is the largest rms voltage available at the secondary? What is the smallest increment by which the secondary rms voltage may be changed?

REVIEW PROBLEMS

67. A resistor of 100 Ω is connected to an AC source with an amplitude of 163 V. What is the instantaneous current in the resistor when the instantaneous voltage is 163 V? What is the instantaneous current one-quarter cycle later? One-half cycle later? Three-quarters cycle later?

68. A parallel-plate capacitor has plates of area 0.30 m^2 separated by an air gap of 0.0020 cm. This capacitor is connected to a generator supplying an rms voltage of 12 V AC at an angular frequency of 9.0×10^3 radians/s.

(a) What is the capacitance?

(b) What is the rms current flowing into the plates?

(c) What is the rms charge on the capacitor?

(d) What is the rms energy stored in the capacitor?

*69. A circuit consists of a resistor connected in series to a battery; the resistance is 5.0 Ω and the emf of the battery is 12 V. The wires (of negligible resistance) connecting these circuit elements are laid out along a square of 20 cm × 20 cm (Fig. 32.43). The entire circuit is placed face on in an oscillating magnetic field. The instantaneous value of the magnetic field is

$$B = B_0 \sin \omega t$$

with $B_0 = 0.15$ T and $\omega = 360$ radians/s.

(a) Find the instantaneous current in the resistor.

(b) Find the average power dissipated in the resistor.

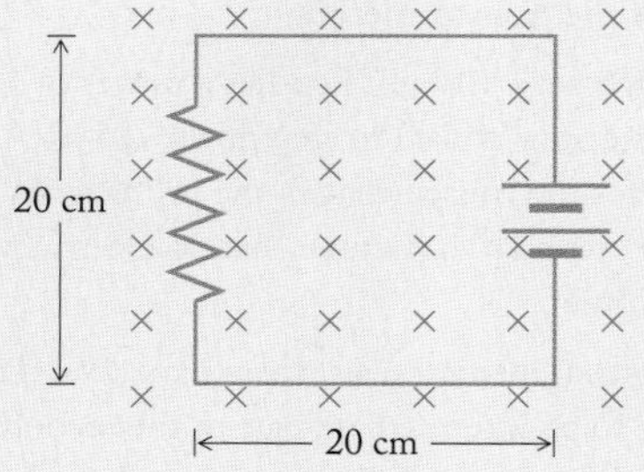

FIGURE 32.43 Circuit placed in magnetic field.

*70. A capacitor with $C = 4.0 \times 10^{-8}$ F is connected to an oscillating source of emf. The source provides an emf $\mathcal{E} = \mathcal{E}_{max} \times \sin \omega t$, with $\mathcal{E}_{max} = 0.80$ V and $\omega = 6.0 \times 10^3$ radians/s.

(a) What is the reactance of the capacitor?

(b) What is the maximum current in the circuit?

(c) What is the current at time $t = 0$? At time $t = \pi/4\omega$?

*71. A circuit consists of two capacitors of 6.0×10^{-8} F and 9.0×10^{-8} F connected in series to an oscillating source of emf (Fig. 32.44). This source delivers a cosinusoidal emf $\mathcal{E} = 1.8 \times \cos(120\,\pi t)$, where $\mathcal{E}$ is in volts and t in seconds.

(a) Find the charge on each capacitor as a function of time.

(b) At what time is the charge on the capacitors maximum? At what time minimum?

(c) What is the maximum energy in the capacitors? What is the time-average energy?

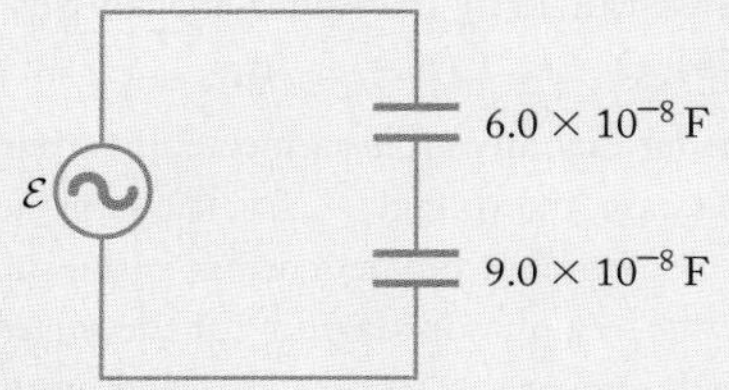

FIGURE 32.44 Two capacitors in series connected to a source of emf.

72. Three capacitors, with $C_1 = 5.0 \times 10^{-8}$ F, $C_2 = 2.0 \times 10^{-8}$ F, and $C_3 = 1.0 \times 10^{-8}$ F, are connected to an oscillating source of emf as shown in Fig. 32.45. The source supplies an emf $\mathcal{E} = 2.0 \cos(120\pi t)$, where $\mathcal{E}$ is in volts and t in seconds.

(a) Find the instantaneous current supplied by the source of emf. What is the maximum value of this current?

(b) Find the instantaneous current that flows through each capacitor, and find the instantaneous potential difference across each capacitor.

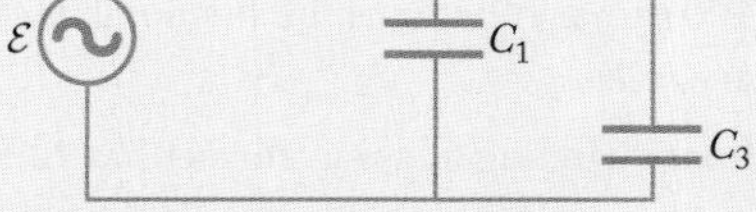

FIGURE 32.45 Three capacitors connected to a source of emf.

73. An inductor of 3.0 mH is connected to an AC source of amplitude 8.0 V and angular frequency 3.0×10^3 radians/s. What is the instantaneous current in the inductor when the instantaneous voltage is 8.0 V? What is the instantaneous current one-quarter cycle later? One-half cycle later? Three-quarters cycle later?

*74. An inductor of 1.6×10^{-3} H is connected to a source of alternating emf. The current in the inductor is $I = I_0 \sin \omega t$, with $I_0 = 180$ A and $\omega = 120\pi$ radians/s.

(a) What is the potential difference across the inductor at time $t = 0$? At time $t = 1/240$ s?

(b) What is the energy in the inductor at time $t = 0$? At time $t = 1/240$ s?

(c) What is the instantaneous power delivered by the source of emf to the inductor at time $t = 0$? At time $t = 1/240$ s?

75. An inductor and a resistor are connected in parallel to a battery and to a source of oscillating emf as shown in Fig. 32.46. The inductance is $L = 5.0 \times 10^{-2}$ H, the resistance is $R = 2000\ \Omega$, the emf of the battery is 3.0 V, and the emf of the oscillating source is $\mathcal{E} = 1.5 \cos(6000\pi t)$, where $\mathcal{E}$ is in volts and t in seconds. At the initial time $t = 0$, the current in the inductor is zero.

(a) From this initial condition, calculate the instantaneous current in the inductor and in the resistor for times after $t = 0$.

(b) Calculate the instantaneous power delivered to the inductor and to the resistor.

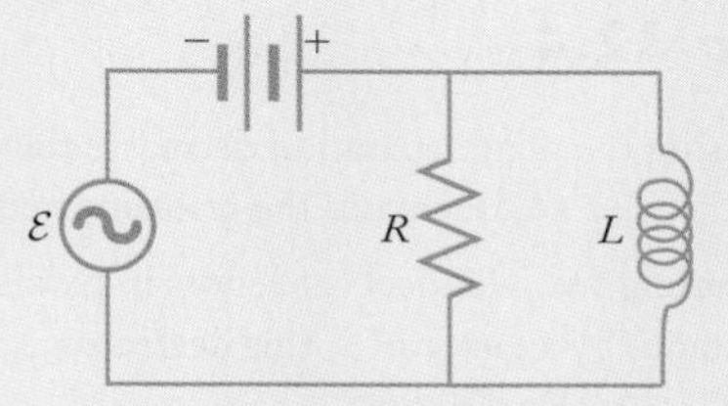

FIGURE 32.46 Inductor and resistor connected to a battery and a source of oscillating emf.

*76. What is the resonant frequency of the circuit shown in Fig. 32.47? The inductance is $L = 1.5$ H, and the capacitances are $C_1 = 20\ \mu$F and $C_2 = 10\ \mu$F.

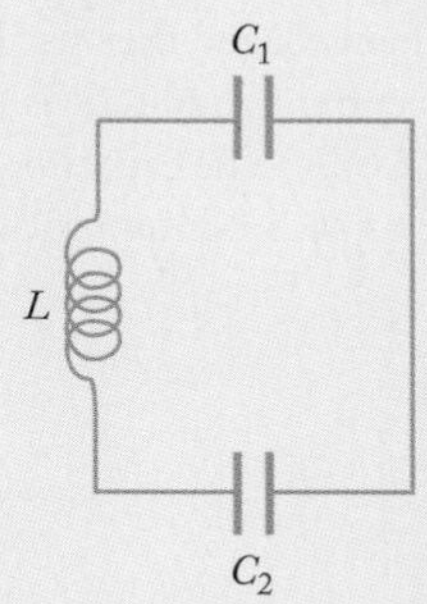

FIGURE 32.47 Two capacitors connected to an inductor.

*77. An LC circuit has an inductance of 5.0×10^{-2} H and a capacitance of 5.0×10^{-6} F. At $t = 0$, the capacitor is fully charged so $Q_{max} = 1.2 \times 10^{-4}$ C. What is the energy in this circuit? At what time, after $t = 0$, will the energy be purely magnetic? At what *later* time will it be purely electric?

78. In a series RL circuit (see Fig. 32.22), a 1.2-kHz oscillating emf has amplitude $\mathcal{E}_{max} = 24$ V and the current amplitude is $I_{max} = 0.45$ A. The ratio L/R is 5.0×10^{-3} s. What is the value of the resistor?

79. A series RLC circuit (see Fig. 32.41) has $\mathcal{E}_{max} = 15$ V, $R = 6.0\ \Omega$, $L = 55$ mH, and $C = 0.25\ \mu$F.

(a) What is the resonant frequency (in Hz)?

(b) Calculate the amplitude of the voltage across the capacitor at resonance.

(c) What is the quality factor Q for this circuit?

80. The primary winding of a transformer has 140 turns and the secondary has 2200 turns. This transformer is connected to a second, identical transformer, so the output of the first transformer serves as input for the second transformer. If the primary of the first transformer is connected to a source of 115 V AC, what is the voltage delivered by the secondary of the second transformer?

81. As illustrated in Fig. 32.34, transformers are used by electric companies to step up or down the voltage from the power plant to your home. If you use a current of 5.0 A in your vacuum cleaner at 115 V AC, how much of a current does this require at the power plant (at 22 000 V)? In the transmission line (at 500 000 V)?

82. A power station feeds 1.0×10^8 W of electric power at 760 kV into a transmission line. Suppose that 10% of this power is lost in Joule heat in the transmission line. What percentage of the power would be lost if the power station were to feed 340 kV into the transmission line instead of 760 kV, other things being equal?

83. A transformer consists of two concentric coils of thin insulated wire of low resistance. One coil has 800 turns and the other has 200 turns. The second coil is wound tightly around the first, so the two have nearly the same radius. If the first coil is connected to a source of oscillating emf supplying a voltage of amplitude 60 V and frequency 1500 Hz, what will be the voltage across the terminals of the second coil?

Answers to Checkups

Checkup 32.1

1. Since this is twice the rms voltage of 115 V AC, the maximum emf will also be doubled to $2 \times 163\ \text{V} = 326\ \text{V}$. Alternatively, one can use Eq. (32.8) to obtain $\mathcal{E}_{max} = \sqrt{2}\mathcal{E}_{rms} = \sqrt{2} \times 230\ \text{V} = 325\ \text{V}$.
2. By the usual definitions, the current is maximum at its most positive value (I_{max}) and is minimum at its most negative value ($-I_{max}$). Thus, for a frequency of 60 Hz, there are 60 minima and 60 maxima per second. The power is always positive for a resistor circuit ($P = I^2R$), and so varies between zero and P_{max} twice per current cycle (once for positive current, once for negative current); thus the power has 120 minima and 120 maxima per second.
3. Both the large slot and the round slot are nearly at a potential of zero volts, so no appreciable current will flow in the lightbulb and it will not shine.
4. (A) 100 W. An emf with a 141-V amplitude has a root-mean-square voltage $\mathcal{E}_{rms} = \mathcal{E}_{max}/\sqrt{2} = 141\ \text{V}/\sqrt{2} = 100\ \text{V}$, and thus an average power $\bar{P} = \mathcal{E}_{rms}^2/R = (100\ \text{V})^2/(100\ \Omega) = 100\ \text{W}$.

Checkup 32.2

1. No—instantaneously, the current is zero when the voltage is maximum, clearly different from Ohm's Law. However, the *amplitudes* of the current and voltage obey a relationship similar to Ohm's Law: $I_{max} = \mathcal{E}_{max}/X_C$, where the capacitive reactance $X_C = 1/(\omega C)$ provides a frequency-dependent opposition to AC.
2. The stored energy varies with the square of the instantaneous emf: $U = \frac{1}{2}C\mathcal{E}^2$ [see Eq. (26.27)]. Since $\mathcal{E} = \mathcal{E}_{max}\cos\omega t$, a plot similar to Fig. 32.5 is obtained, except that the maximum value, occurring at $t = 0,\ \pi/\omega,\ 2\pi/\omega$, etc., is an energy with value $U_{max} = \frac{1}{2}C\mathcal{E}_{max}^2$.
3. The capacitive reactance is inversely proportional to frequency, $X_C = 1/(\omega C)$, and so is 10 times less, or $10^8\ \Omega$, at 600 Hz. Similarly, it is $10^7\ \Omega$ at 6000 Hz.
4. Yes—low frequency means almost static conditions (DC corresponds to zero frequency), where the charge on the capacitor plates does not change at any appreciable rate, so there is essentially no current.
5. (C) 2, 4. For a given capacitor at fixed frequency, the current is proportional to the voltage amplitude, $I_{max} = \mathcal{E}_{max}/X_C$, and so increases a factor of 2. For a capacitor, the maximum power is proportional to the square of the voltage amplitude, and so the new maximum power is related to the old by a factor of 4.

Checkup 32.3

1. No—instantaneously, the current is zero when the voltage is maximum, clearly different from Ohm's Law. However, the *amplitudes* of the current and voltage obey a relationship similar to Ohm's Law: $I_{max} = \mathcal{E}_{max}/X_L$, where the inductive reactance $X_L = \omega L$ provides a frequency-dependent opposition to AC.
2. The stored energy varies with the square of the instantaneous current: $U = LI^2/2$ [see Eq. (31.26)]. Since, from Eq. (32.18), $I = [\mathcal{E}_{max}/(\omega L)]\sin\omega t$, a $\sin^2\omega t$ plot is obtained for the energy (similar to the $\cos^2\omega t$ plot of Fig. 32.5), with a minimum value of zero at $t = 0$ (and at $t = \pi/\omega,\ 2\pi/\omega$, etc.) and maximum values of $U_{max} = \mathcal{E}_{max}^2/(2\omega^2 L)$.
3. The inductive reactance is proportional to the frequency: $X_L = \omega L = 2\pi f L$. Thus at 10 times the frequency (600 Hz), the inductive reactance will be 10 times as large, or $10^4\ \Omega$. Similarly, it is $10^5\ \Omega$ at 6000 Hz.
4. Yes—low frequency means almost static conditions, so there is no back emf to oppose the driving voltage; since there is no resistance, the current will be large.
5. (D) $\frac{1}{10}$; 10. Increasing the inductance by a factor of 10 increases the inductive reactance by the same factor ($X_L = \omega L$), and so *decreases* the current by a factor of 10 ($I_{max} = \mathcal{E}_{max}/X_L$).

Checkup 32.4

1. Increasing the plate separation decreases the capacitance and so, by Eq. (32.26), increases the resonant frequency.
2. Connecting two identical capacitors in parallel increases the capacitance by a factor of 2; this decreases $\omega_0 = 1/\sqrt{LC}$ by a factor of $\sqrt{2}$ [Eq. (32.26)].
3. From Eq. (32.33), $Q = \omega_0 L/R = X_L/R = 500\ \Omega/10\ \Omega = 50$. From Eq. (32.34), one characteristic time for the decaying oscillation amplitude will include $N = Q/\pi = 50/\pi \approx 16$ cycles.
4. (B) $\mathcal{E}\sqrt{LC}/R$; $\sin\omega t$. With the switch in position 1 for a long time, the current in the left loop becomes steady at $I = \mathcal{E}/R$. When switched to position 2, the current in the inductor is initially unchanged, and then oscillates with time. By Eq. (32.25), this current amplitude is $Q_{max}/\sqrt{LC}$ which implies $Q_{max} = \mathcal{E}\sqrt{LC}/R$. The capacitor is initially uncharged, that is, $Q = 0$ at $t = 0$, so $\sin\omega t$ provides the time dependence of the charge.

Checkup 32.5

1. If the current increases, the impedance must have decreased. Only a capacitive reactance decreases with frequency [$X_C = 1/(\omega C)$], so if there is only one other component, it must be a capacitor.
2. The current is maximum at resonance, when $I_{\max} = \mathcal{E}_{\max}/R$, so $R = \mathcal{E}_{\max}/I_{\max} = 1.0\ \text{V}/0.20\ \text{A} = 5.0\ \Omega$. By Eq. (32.54), the quality factor is $Q = X_L/R = 50\ \Omega/5.0\ \Omega = 10$.
3. There are several ways to calculate the Q; for example, Eq. (32.54) may be used: $Q = V_{\max,L}/V_{\max,R} = 200\ \text{V}/1.0\ \text{V} = 200$.
4. (E) 2.0 V. At resonance, the voltages across the inductor and capacitor have equal amplitudes, but are 180° out of phase and sum to zero. Thus, as discussed in Example 6, the amplitude of the voltage across the resistor is the same as the emf; that is, at ω_0, $V_{\max,R} = \mathcal{E}_{\max} = 2.0\ \text{V}$.

Checkup 32.6

1. The ratio of emfs is the same as the ratio of turns in the coils, so the number of turns in the secondary needs to be a factor of 2 greater than the number in the primary.
2. For a step-down transformer, the smaller voltage requires a larger current in the secondary, since the power delivered to the load equals the power drawn from the source [Eq. (32.61)].
3. (E) 10, 2.5, 0.4, 0.1. The transformer with turns ratio 2 will provide a voltage ratio of 2 or $\frac{1}{2}$ depending on whether it is used as a step-up or step-down transformer. Similarly, the other transformer can provide a voltage ratio of 5 or $\frac{1}{5}$. Connected one after the other, we have four possible combinations: $2 \times 5 = 10$, $\frac{1}{2} \times 5 = 2.5$, $\frac{1}{5} \times 2 = 0.4$, and $\frac{1}{5} \times \frac{1}{2} = 0.1$.

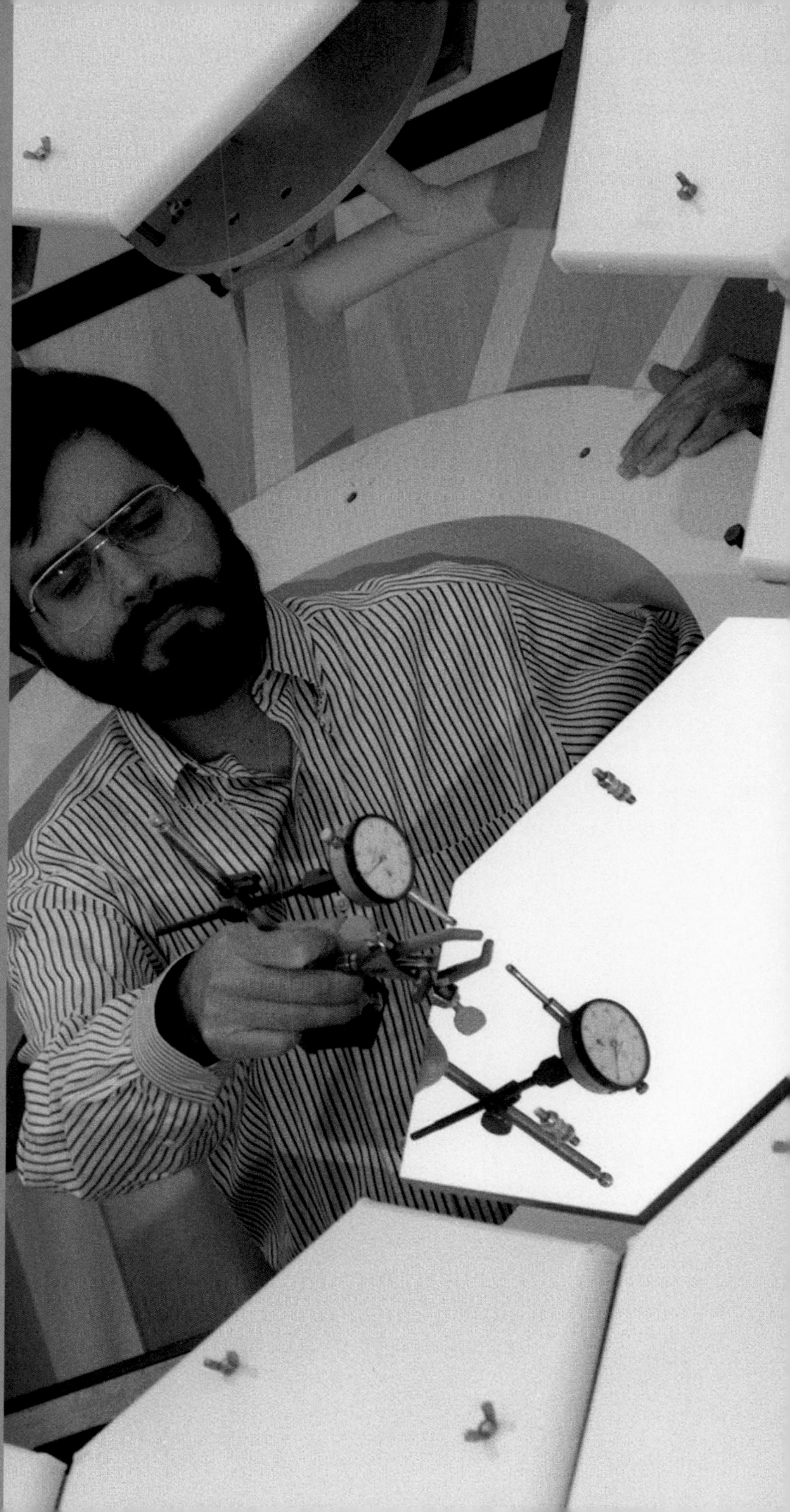

PART

5

Waves and Optics

CONTENTS

Each segment in this 5.5-meter diameter array of hexagonal mirrors has a spherical surface, so the combination of all 144 segments forms an overall approximation of a parabolic surface. A refined optics manufacturing process permits these aluminum segments to attain the smoothness and reflectivity of glass mirrors.

CHAPTER

33 Electromagnetic Waves

CONCEPTS IN CONTEXT

Concepts in Context

Electromagnetic waves are found throughout our environment. The antennas on the tower in the photo emit and receive electromagnetic waves; even the light coming from the photo to your eyes consists of electromagnetic waves.

As we learn about electromagnetic waves, we will consider such questions as:

? How are the electric and magnetic fields of an electromagnetic wave related? (Section 33.3 and Example 3, page 1081; and Example 8, page 1098)

? How does an antenna generate electromagnetic waves? What determines the size of an antenna? (Section 33.4 and Example 5, page 1088)

? In what way is a periodic electromagnetic wave altered for AM and FM radio communications? (Physics in Practice: AM and FM Radio, page 1089)

? How does the strength of an electromagnetic wave decrease with distance from the source? (Example 6 and Section 33.5, page 1094)

We already know that a changing magnetic field induces an electric field, described by Faraday's Law of induction. In this chapter, we will discover that the converse is also true: a changing electric field induces a magnetic field. The law describing this induction effect of electric fields was formulated by James Clerk Maxwell, who thereby achieved a wide-ranging unification of all the laws of electricity and magnetism. These laws became known as Maxwell's equations. The next three chapters of this book are, in essence, nothing but applications of Maxwell's equations.

The mutual induction of electric and magnetic fields gives rise to the phenomenon of self-supporting electromagnetic oscillations in empty space. If, initially, there exists an oscillating electric field, it will induce a magnetic field, and this will induce a new electric field, and so on. Thus, these fields can perpetuate each other. An oscillating charge or current is needed to get the fields started, but after this initiation the fields continue on their own. These self-supporting oscillations are **electromagnetic waves**, either traveling waves or standing waves. Electromagnetic waves are similar to mechanical waves, such as waves on a string or sound waves in air, in that the disturbance at one point causes a disturbance at a neighboring point, and this causes a disturbance at the next neighboring point, etc. But electromagnetic waves differ from mechanical waves in that they propagate through empty space—they propagate without any material medium. Among such electromagnetic waves are radio waves, microwaves, light, and X rays. All of these kinds of waves are qualitatively the same; they consist of oscillating electric and magnetic fields. The only difference between a radio wave and light is in their wavelength and frequency—the radio wave has a much longer wavelength and smaller frequency than light. In this chapter we will examine the properties of electromagnetic waves, and see how they are generated by accelerated electric charges.

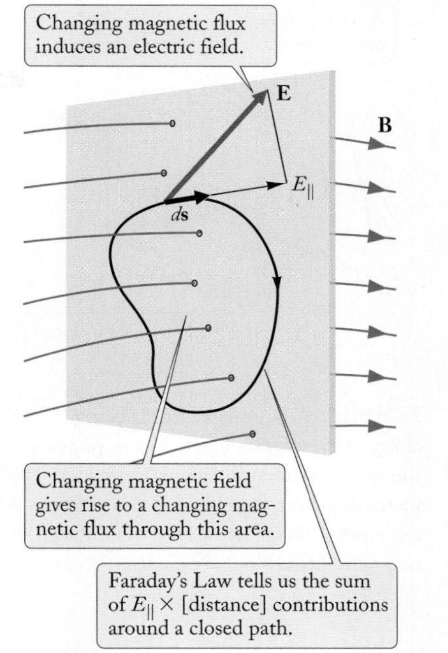

FIGURE 33.1 A closed path placed in a changing magnetic field. $E_{\parallel}$ is the component of the induced electric field tangent to the path.

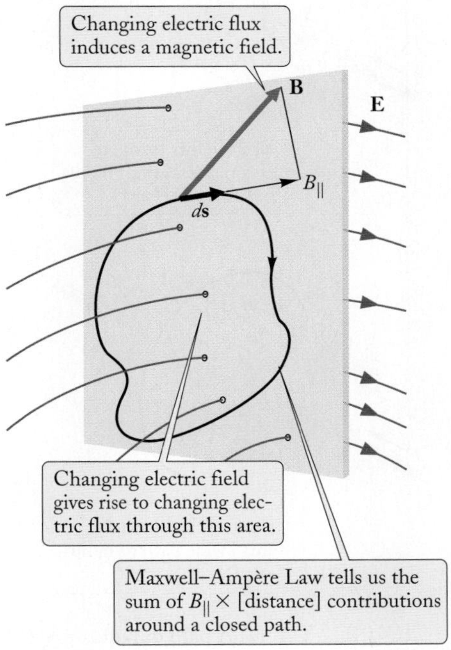

FIGURE 33.2 A closed path placed in a changing electric field. $B_{\parallel}$ is the component of the induced magnetic field tangent to the path.

33.1 INDUCTION OF MAGNETIC FIELDS; MAXWELL'S EQUATIONS

As we saw in Chapter 31 [Eq. (31.12)], Faraday's Law for the electric field induced by a changing magnetic field can be expressed in the form

$$\oint E_{\parallel}\, ds = -\frac{d\Phi_B}{dt} \tag{33.1}$$

where $E_{\parallel}$ is the component of the induced electric field parallel to a closed path (see Fig. 33.1), ds is the magnitude of a displacement along the path, and Φ_B is the magnetic flux intercepted by the surface within the path.

The magnetic field induced by a changing electric field is described by an equation analogous to Faraday's Law. On the left side of the equation that describes this new kind of induction there appears the induced magnetic field (see Fig. 33.2), and on the right side there appears the rate of change of the electric flux intercepted by the surface within the path:

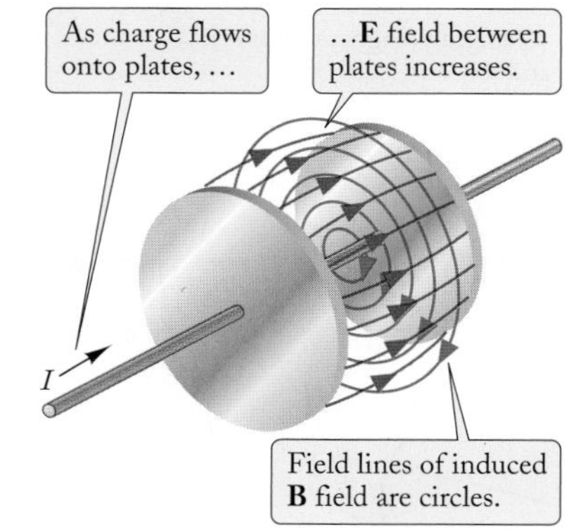

FIGURE 33.3 A capacitor is being charged by wires connected to its plates. In the space between the plates there is an increasing electric field (red) and an increasing electric flux. The lines of the induced magnetic field (blue) are circles.

$$\oint B_{\parallel}\, ds = \mu_0 \epsilon_0 \frac{d\Phi_E}{dt} \tag{33.2}$$

Apart from the minus sign in Eq. (33.1) and the constant $\mu_0\epsilon_0$ in Eq. (33.2), these two equations are "mirror images" of each other. Equation (33.1) states that a changing magnetic flux induces an electric field; Eq. (33.2) states that *a changing electric flux induces a magnetic field.* Thus, these equations can be obtained from each other by exchanging electric and magnetic fields. This indicates a certain symmetry in the effects of electric and magnetic fields on one another. The induction of magnetic fields by a changing electric field was first proposed by Maxwell on purely theoretical considerations; the experimental demonstration of this kind of induction came only later.

For a simple example of a system with a rate of change of electric flux and an induced magnetic field, consider a capacitor that is being charged by currents flowing into its plates (see Fig. 33.3). In the space between the plates there is then an increasing electric field, and there is increasing electric flux. If the plates of the capacitor are circular, as shown in Fig. 33.3, the induced magnetic field lines will have the shape of concentric circles, just like the field lines surrounding a long wire carrying a current. The strength of the magnetic field can be calculated in the manner familiar from the calculations with Ampère's Law in Chapter 29: take a path that follows one of the circular field lines and evaluate each side of Eq. (33.2) for this choice of path. According to Eq. (33.2), we then find that the strength of the induced magnetic field is directly proportional to the rate of change of the electric flux, or to the rate of change of the electric field between the plates.

(a)

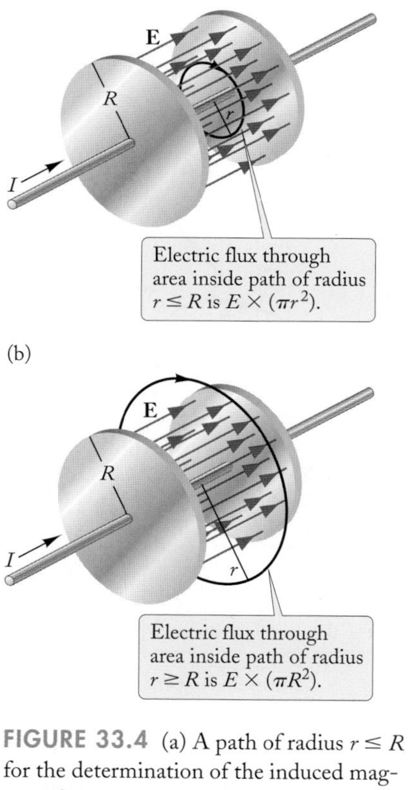

FIGURE 33.4 (a) A path of radius $r \le R$ for the determination of the induced magnetic field between two capacitor plates. (b) A path of radius $r \ge R$ for the determination of the induced magnetic field outside the capacitor plates.

EXAMPLE 1 If the capacitor plates in Fig. 33.3 have radius R and the electric field is increasing at a rate dE/dt, find an expression for the induced magnetic field for $r \le R$. Do the same for $r \ge R$. What is the current in the wire?

SOLUTION: By symmetry, we can see that the magnetic field lines will be tangent to circles around the capacitor axis, as shown in Fig. 33.3. The magnetic field must also have a constant magnitude along a given circular path of some radius r. Thus, similar to what was done with Ampère's Law in Section 29.3, the left side of Eq. (33.2) for any such circular path can be rewritten

$$\oint B_{\parallel}\, ds = B \times 2\pi r$$

For $r \le R$ the electric field is constant and perpendicular to the surface of area $A = \pi r^2$ (see Fig. 33.4a). Thus we can write the rate of change of the electric flux as

$$\frac{d\Phi_E}{dt} = A\frac{d}{dt}E = \pi r^2 \frac{dE}{dt} \qquad (r \le R)$$

Inserting these relations in Eq. (33.2) gives

$$B \times 2\pi r = \mu_0\epsilon_0 \pi r^2 \frac{dE}{dt}$$

or, solving for the magnetic field,

$$B = \frac{\mu_0\epsilon_0}{2} r \frac{dE}{dt} \qquad (r \le R)$$

If $r \geq R$, then the electric field intercepts only part of the surface within a path of radius r. As shown in Fig. 33.4b, this is the part with radius R. Hence

$$\frac{d\Phi_E}{dt} = A\frac{d}{dt}E = \pi R^2 \frac{dE}{dt} \qquad (r \geq R)$$

By equating the two sides of Eq. (33.2) we have

$$B \times 2\pi r = \mu_0 \epsilon_0 \pi R^2 \frac{dE}{dt}$$

and we obtain for the magnetic field

$$B = \frac{\mu_0 \epsilon_0}{2} \frac{R^2}{r} \frac{dE}{dt} \qquad (r \geq R)$$

The current can be determined from the known behavior of a capacitor. The electric field between the plates is $E = \sigma/\epsilon_0$, where $\sigma = Q/(\pi R^2)$ is the charge density on a plate [Eq. (23.13)]. The charge on the plates is $Q = \pi R^2 \sigma = \epsilon_0 \pi R^2 E$, and the current I is

$$I = \frac{dQ}{dt} = \epsilon_0 \pi R^2 \frac{dE}{dt}$$

Notice that if we compare the last two equations, we can conclude

$$B = \frac{\mu_0 I}{2\pi r}$$

which is a result familiar from Ampère's Law [see Eq. (29.9)]. Thus, outside the capacitor, the magnetic field induced by the total changing electric flux is the same as that produced by the current causing that changing flux (but the magnetic field is weaker than this at points inside the capacitor).

Although the magnetic fields of currents (described by Ampère's Law) and the magnetic fields induced by electric fields [described by Eq. (33.2)] can have different sources, the equations describing these fields both involve the tangential component of the field. Hence Ampère's Law and Eq. (33.2) can be combined into one single equation, the **Maxwell–Ampère Law**:

$$\oint \mathbf{B} \cdot d\mathbf{s} = \mu_0 I + \mu_0 \epsilon_0 \frac{d\Phi_E}{dt} \quad \text{or} \quad \oint B_{\|}\, ds = \mu_0 I + \mu_0 \epsilon_0 \frac{d\Phi_E}{dt} \qquad (33.3)$$

Maxwell-Ampère Law

This equation covers all conceivable cases of production of magnetic fields, by currents, by changing electric fields, or by any combination of both. For example, if the space between the capacitor plates in Fig. 33.3 is filled with a slightly conducting material (a leaky capacitor), then there will be both an electric current and an electric flux in this space, and both will contribute to the magnetic field.

The right side of Eq. (33.3) can be written as $\mu_0(I + \epsilon_0\, d\Phi_E/dt)$, which shows that the quantity $\epsilon_0\, d\Phi_E/dt$ has the same effect for the production of magnetic fields as the ordinary electric current I. Accordingly, the quantity $\epsilon_0\, d\Phi_E/dt$ is called the **displacement current** (although there is a good reason for the word *current*, there is no good reason for the word *displacement*).

displacement current

Note that in the case of the capacitor plates illustrated in Fig. 33.3, the displacement current between the plates is large whenever the ordinary current flowing on the external wires connected to the plates is large. As we saw in Example 1, the ordinary current deposits charge on the plates of the capacitor, and whenever the ordinary current is large, the rate of change of the charge is large, and so is the rate of change of the electric field between the plates, and so is the rate of change of the electric flux. Since they both produce the same magnetic field, it is evident from Example 1 that the magnitude of the total displacement current between the plates is equal to the magnitude of the ordinary current flowing on the external wires. *Thus, the displacement current between the capacitor plates can be viewed as a "continuation" of the ordinary current by a corresponding rate of change of electric flux.*

Equation (33.3) is Maxwell's modification of Ampère's Law. The great importance of this equation lies in its general validity—Maxwell boldly proposed that this equation is valid not only for the electric and magnetic fields associated with capacitors, wires, and other such devices, but also for the fields associated with electromagnetic waves.

Equation (33.3) is the last of the fundamental laws that we need for a complete mathematical description of the behavior of electric and magnetic fields. There are four fundamental laws, as follows:

Maxwell's equations

Gauss' Law for electricity [Eq. (24.10)],

$$\oint \mathbf{E} \cdot d\mathbf{A} = \frac{Q_{\text{inside}}}{\epsilon_0} \quad \text{or} \quad \oint E_\perp \, dA = \frac{Q_{\text{inside}}}{\epsilon_0} \tag{33.4}$$

Gauss' Law for magnetism [Eq. (29.15)],

$$\oint \mathbf{B} \cdot d\mathbf{A} = 0 \quad \text{or} \quad \oint B_\perp \, dA = 0 \tag{33.5}$$

Faraday's Law [Eqs. (31.11) and (31.12)],

$$\oint \mathbf{E} \cdot d\mathbf{s} = -\frac{d\Phi_B}{dt} \quad \text{or} \quad \oint E_\parallel \, ds = -\frac{d\Phi_B}{dt} \tag{33.6}$$

and the Maxwell–Ampère Law [Eq. (33.3)],

$$\oint \mathbf{B} \cdot d\mathbf{s} = \mu_0 I + \mu_0 \epsilon_0 \frac{d\Phi_E}{dt} \quad \text{or} \quad \oint B_\parallel \, ds = \mu_0 I + \mu_0 \epsilon_0 \frac{d\Phi_E}{dt} \tag{33.7}$$

The physical basis for each of these four laws may be briefly summarized as follows:

- Gauss' Law is based on Coulomb's Law describing the forces of attraction and repulsion between stationary charges.
- Gauss' Law for magnetism asserts that there are no sources or sinks of magnetic field lines.
- Faraday's Law describes the induction of an electric field by a changing magnetic flux.
- The Maxwell–Ampère Law is based on the law of magnetic force between moving charges and it also contains the induction of a magnetic field by a changing electric flux.

Taken as a whole, the four laws (33.4)–(33.7) are known as **Maxwell's equations**, because Maxwell supplied the missing link between the magnetic and the electric fields [Eq. (33.3)] and thereby placed the capstone on electromagnetic theory. Maxwell recognized that these equations imply a dynamic interplay between electric and magnetic fields, an interplay that couples and unifies electric and magnetic phenomena.

Maxwell's equations provide a complete description of the interactions among charges, currents, electric fields, and magnetic fields. All the properties of the fields can be deduced by mathematical manipulation of these equations. If the distribution of charges and currents is given, then these equations uniquely determine the corresponding fields. Even more important, Maxwell's equations uniquely determine the time evolution of the fields, starting from a given initial condition for these fields—if we know the fields at an initial time, we can calculate them at any later time. Thus, these equations accomplish for the dynamics of electromagnetic fields what Newton's equations of motion accomplish for the dynamics of particles.

Although the experimental foundation on which we based the development of Maxwell's equations was restricted to charges at rest or charges in uniform motion, these equations also govern the fields of accelerated charges and the fields of light and radio waves. In the remaining sections of this chapter we will calculate the fields of electromagnetic waves from our equations and we will see that the results are in agreement with the observed properties of light and radio waves.

✔ Checkup 33.1

QUESTION 1: If a current of 2 A is charging a capacitor of 400 pF, what is the total displacement current between the capacitor plates? What if the capacitance is 800 pF?

QUESTION 2: In some region of space, where there is no electric charge and no current, there are electric and magnetic field lines that form closed loops. Can such electric and magnetic fields be static?

QUESTION 3: Do the field lines of an induced electric field start and stop on electric charges?

QUESTION 4: Figure 33.3 shows a current charging a capacitor. Suppose that at some time the current stops and the capacitor remains with some constant charge. After this time, is there an electric field between the plates? A magnetic field?

QUESTION 5: Consider a constant current I flowing into a capacitor, as in Fig. 33.4. At what radial distance from the center of the capacitor is the magnetic field the largest?

(A) $r = 0$ (B) $r = R/2$ (C) $r = R$ (D) $r = 2R$ (E) $r = \infty$

33.2 THE ELECTROMAGNETIC WAVE PULSE

We started our study of electricity and magnetism with the electric field of a point charge at rest—this is the Coulomb field given by Eq. (23.2). Later, we dealt with the fields of a current, that is, charges in motion with uniform velocity—in addition to the Coulomb field, such moving charges have a magnetic field. Now we will investigate the fields of a charge with *accelerated motion*. We will find that in this case there are extra electric and magnetic fields that spread outward from the position of the charge, like ripples on a pond in which a stone has been dropped, and carry away energy and momentum. These spreading fields are called **radiation fields**, or **wave fields**.

radiation fields

We can gain some insight into the fields spreading out from an accelerated charge by the following argument. Consider a charge that is initially at rest, then is quickly accelerated for some short time interval, and then continues to move with a constant final velocity. The initial electric field lines originate at the initial position of the charge. But the field lines at some later time, after the acceleration ceases, must originate on the new position of the charge. The field lines cannot change from their initial configuration to their final configuration instantaneously; rather a disturbance must travel outward from the position of the charge and gradually change the field lines. The disturbance takes the form of a kink connecting the old and the new field lines.

The disturbance travels at some speed c (we will establish below that the speed of the disturbance is actually the speed of light, and our notation anticipates this result). Figure 33.5 shows the situation at some time after the acceleration has ended. Suppose that the acceleration lasts from time $t = 0$ to time $t = \tau$. Then the disturbance of the electric field lines begins at time $t = 0$ and ends at time $t = \tau$. The leading edge of the disturbance (outer edge of kink) travels outward from the initial position of the charge and in a time t it reaches out to a distances ct. Beyond the sphere of radius ct, the electric field is still the old field with field lines centered on the initial position of the charge.

The disturbance ceases as soon as the acceleration ceases. The field in the vicinity of the uniformly moving charge then settles into the new radial configuration centered on the new position of the charge. The trailing edge of the disturbance (inner edge of kink) marking the cessation of the acceleration travels outward from the position that the charge has at the time $t = \tau$, when the acceleration ceases; and by some later time t it reaches out to a distance $c(t - \tau)$. Within the sphere of radius $c(t - \tau)$, the electric field is the new radial field centered on the new position of the uniformly moving charge.

The disturbance produced by the accelerated charge is confined to the space between the larger and the smaller spheres in Fig. 33.5. The field lines in this zone must connect the lines of the new field of the uniformly moving charge with the lines of the old field of the stationary charge. As shown in Fig. 33.5, each connecting segment of field line makes an angle with the radial line to the position of the charge. Thus, *the electric field in the zone of the kink has both a radial component and a tangential, or trans-*

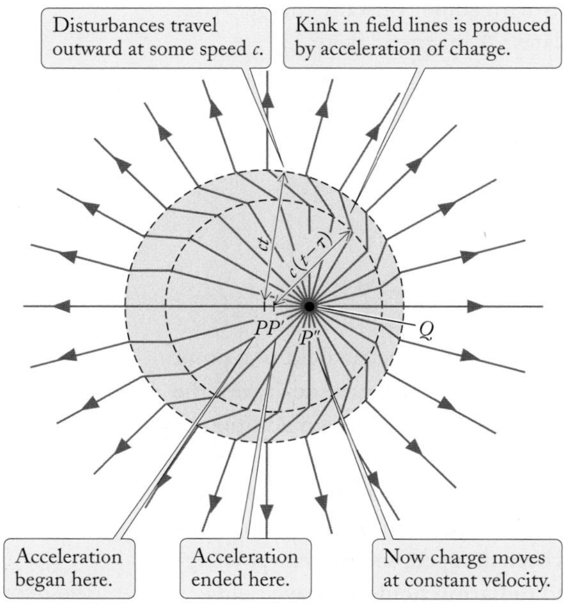

FIGURE 33.5 Electric field lines of a charge that has suffered an acceleration. Here P'' is the present position of the charge, P is the initial position of the charge, and P' is an intermediate position. Between P and P' the charge suffered a constant acceleration. Between P' and P'' the charge moved at constant velocity. The outer dashed sphere (outer edge of the kink) has radius ct and is centered on P; the inner dashed sphere (inner edge of kink) has radius $c(t - \tau)$ and is centered on P'.

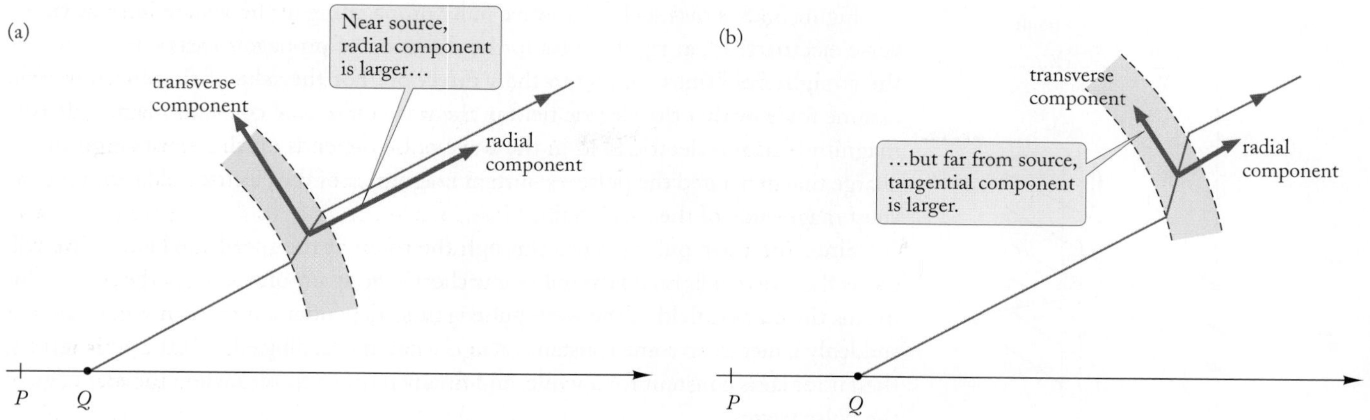

FIGURE 33.6 (a) One of the field lines of Fig. 33.5. In the zone of the kink the electric field has both a radial and a transverse component. (b) The same field line at a later time, when the particle has traveled further to the right and the kink has traveled farther outward. The kink is more pronounced because the radial component of the electric field has decreased more than the transverse component (the radial component decreases in proportion to $1/r^2$; the transverse component decreases in proportion to $1/r$).

verse, component. This transverse component is the radiation field, or the wave field, of the accelerated charge. From Fig. 33.5, we see that this transverse component is maximum for directions perpendicular to the original acceleration, and it is zero along the direction of the original acceleration.

The traveling kink is a single wave pulse, analogous to the single wave pulse we might produce on a stretched string by flicking its end just once. If we want to produce a periodic electromagnetic wave, we have to accelerate the charge back and forth periodically, and thereby generate a succession of kinks in the field lines. This picture of an electromagnetic wave as a succession of kinks in the field lines explains how such waves are able to travel in a vacuum, a puzzle that greatly worried the physicists of the nineteenth century. The electromagnetic wave does not really travel in a vacuum—it is a disturbance in the electric field of the charge, that is, it travels in the electric field of the charge. The space surrounding the charge is not really a vacuum; the space is filled with the electric field of the charge, and this electric field is the medium in which the electromagnetic wave propagates.

As the kink travels outward, the angle between the radial direction and the direction of the electric field in the zone of the kink increases; thus, the kink becomes more pronounced and the electric field becomes more transverse; Fig. 33.6 illustrates this for one field line. Thus, when the kink reaches a large distance from the charge, the electric field in the zone of the kink will be almost entirely transverse. Such an electric field at right angles to the direction of propagation is a characteristic feature of electromagnetic waves.

The transverse kink propagates in the outward direction as a spherical wave pulse, but the strength of this pulse depends on direction. Often we will be interested in the behavior of the wave over only a relatively small interval of distances, an interval small compared with the radius of the spherical wave (see Fig. 33.7). For example, we might be interested in the wave that the moving charges on the antenna of a distant radio transmitter produce in the room in which we are sitting. If the length of the room is, say, 5 m and the distance to the transmitter is 20 km, then the portion of the spherical wave pulse that passes through the room will look flat and uniform, and we can approximate the spherical wave pulse by a plane wave pulse, with plane wave fronts and with straight electric field lines at right angles to the direction of propagation.

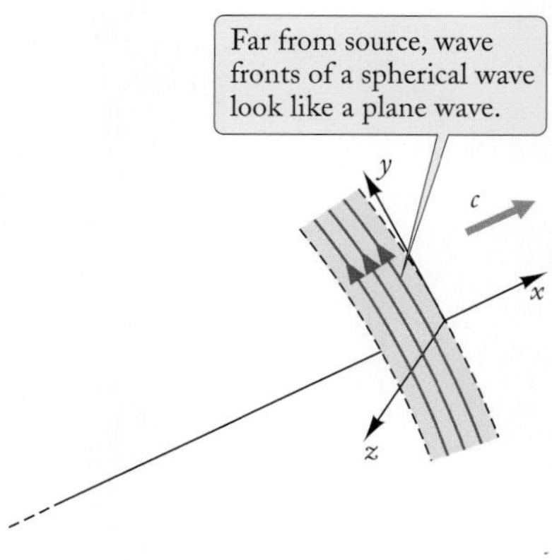

FIGURE 33.7 A small portion of the spherical wave pulse can be regarded as a plane wave pulse. The electric field in this wave pulse is almost perpendicular to the direction of propagation.

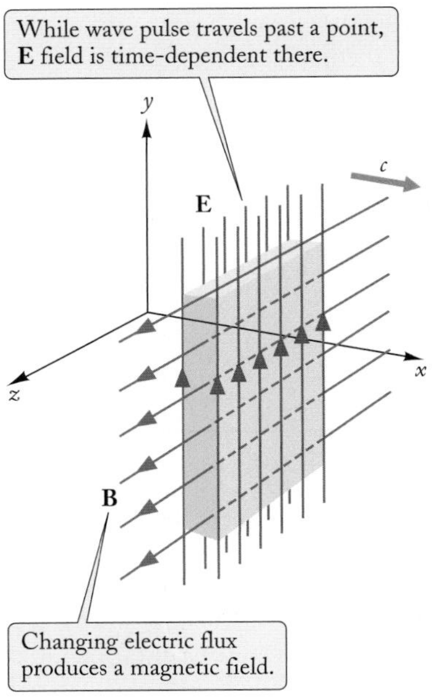

FIGURE 33.8 Electric field lines of a plane wave pulse consisting of a region of uniform electric field **E**. The wave front, or front surface of the pulse, is a flat plane. The magnetic field lines are at right angles to the electric field lines. The entire pattern of field lines travels in the x direction with the wave speed c.

Figure 33.8 shows such a flat wave pulse propagating in the x direction; the transverse electric field, at right angles to the direction of propagation, is represented by the straight field lines parallel to the y direction. For the sake of simplicity, we will assume for now that the electric field in the wave pulse is of constant magnitude (the magnitude of the electric field in the wave pulse depends on the acceleration of the charge that generated the pulse—constant magnitude of the electric field requires constant magnitude of the acceleration).

Since the wave pulse sweeps through the room at the speed c, which, as we will see, is the speed of light, it lasts only some short time at any one point of the room. This means the electric field of the wave pulse is time-dependent: initially it is zero, then it suddenly increases to some constant value E when the leading edge of the pulse arrives, then it remains constant for a while, and finally it drops to zero when the rear edge of the pulse passes.

Besides the transverse electric field, *the wave pulse is also endowed with a transverse magnetic field, at right angles to the electric field and to the direction of propagation* (see Fig. 33.8). This magnetic field arises by induction, from the time-dependent electric field and the time-dependent electric flux. The induced magnetic field can be calculated from the time-dependent electric field by means of the Maxwell–Ampère Law. In the next section, we will relate the magnitude of the induced magnetic field to that of the electric field; here, let us try to understand why its direction is perpendicular to the electric field. For this purpose, consider a small stationary rectangle perpendicular to the electric field. In Fig. 33.9, this rectangle is in the x–z plane. This rectangle registers a change of electric flux when an edge of the wave pulse sweeps across it; thus there is a rate of change of flux, or a displacement current, associated with each edge of the wave pulse. In Fig. 33.9, the displacement current is parallel to the changing vertical electric field lines. Since the effect of such a displacement current is the same as that of a real current, the magnetic field produced by the changing electric flux is the same as that of a vertical plane of real current at the pulse edge. According to the usual right-hand rule for current and magnetic field, such a current produces a magnetic field perpendicular to the current; thus, the magnetic fields contributed by the two edges are in the z direction, and so is the net magnetic field shown in Fig. 33.9. Note that the correct directions of the magnetic field and the electric field are related by a simple **right-hand rule for electromagnetic waves**: if the fingers of the

right-hand rule for electromagnetic waves

FIGURE 33.9 A small stationary rectangle (green) perpendicular to the electric field lines registers a rate of change of electric flux at the instant the front edge of the wave pulse sweeps across the rectangle, and again at the instant the rear edge of the wave pulse sweeps across the rectangle. Hence there is a sheet of displacement current at the front edge of the wave pulse, and another sheet of displacement current at the rear edge. The black lines indicate the flow of the displacement currents. The blue lines indicate the magnetic field produced by these currents.

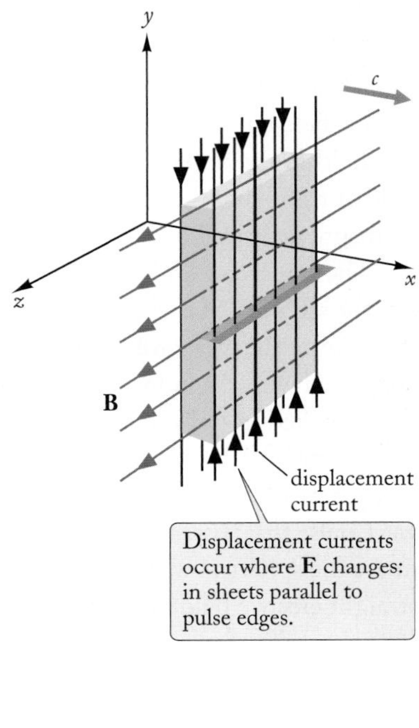

right hand are placed along the direction of the electric field and curl toward the magnetic field, the thumb will point along the direction of propagation of the wave (see Fig. 33.10). In vector notation, the propagation is in the direction of the vector cross product, **E** × **B**.

Like the electric field of the wave pulse, the magnetic field is time-dependent: initially it is zero, then it suddenly increases to some constant value B when the leading edge of the pulse arrives, then it remains constant for a while, and then it drops to zero when the rear edge of the pulse passes. According to Faraday's Law, this time-dependent magnetic field and its time-dependent magnetic flux give rise to an induced electric field. By means of Faraday's Law, it can be verified that this induced electric field coincides with the original electric field E of the wave pulse. Thus, the electric and magnetic fields of the wave pulse mutually induce each other—the electric field induces the magnetic field, and the magnetic field induces the original electric field. These fields thereby become self-supporting, and they continue to exist and to propagate independently of what happens to the charge that initially generated them.

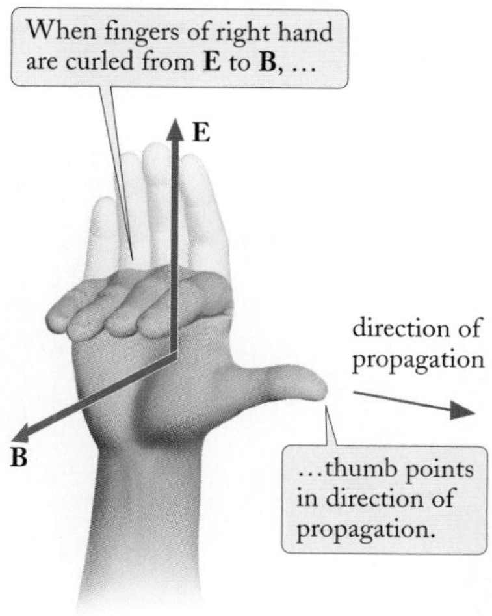

FIGURE 33.10 Right-hand rule for electromagnetic waves. The fingers of the right hand are placed along the direction of the electric field and curl toward the magnetic field; the thumb then points along the direction of propagation of the wave.

✔ Checkup 33.2

QUESTION 1: For each of the following wave pulses, what is the direction of the magnetic field?

(A) Wave pulse is traveling eastward and electric field is southward.

(B) Wave pulse is traveling westward and electric field is southward.

(C) Wave pulse is traveling upward and electric field is southward.

QUESTION 2: For each of the following wave pulses, what is the direction of the electric field?

(A) Wave pulse is traveling eastward and magnetic field is southward.

(B) Wave pulse is traveling westward and magnetic field is southward.

(C) Wave pulse is traveling upward and magnetic field is southward.

QUESTION 3: Which of Maxwell's equations determine how the electric and magnetic fields of the wave pulse induce each other?

QUESTION 4: A wave pulse is traveling southward and the electric field is downward. What is the direction of the magnetic field?

(A) Upward (B) Northward (C) Eastward (D) Westward

33.3 PLANE WAVES; POLARIZATION

Online Concept 36 Tutorial

In this section, we will concentrate on periodic electromagnetic waves, produced by accelerating a charge back and forth repeatedly. *A periodic electromagnetic wave consists of alternating zones of positive and negative electric field and magnetic field.* The positive and negative electric fields define the wave crests and wave troughs of the wave. The wave is characterized by a frequency and a wavelength.

As described in the preceding section, the electric and magnetic fields of the electromagnetic wave mutually induce each other, and the wave thereby becomes self-supporting. The wave is initially generated by some accelerated charges; but once it has been generated, it propagates on its own, independent of the accelerated charges that generated it.

By a detailed theoretical analysis (Section 33.6), based on this mutual induction of the electric and magnetic fields according to the Maxwell–Ampère Law and Faraday's Law, we can establish that the speed c of the electromagnetic wave is given by the expression

speed of electromagnetic wave

$$c = \frac{1}{\sqrt{\mu_0 \epsilon_0}} \tag{33.8}$$

This result was first obtained by Maxwell. Numerically, this theoretical expression for the speed of electromagnetic waves yields

$$\begin{aligned} c &= \frac{1}{\sqrt{\mu_0 \epsilon_0}} = \frac{1}{\sqrt{1.26 \times 10^{-6}\ \text{H/m} \times 8.55 \times 10^{-12}\ \text{F/m}}} \\ &= 3.00 \times 10^{8}\ \text{m/s} \end{aligned} \tag{33.9}$$

This value coincides with the measured value of the speed of light in vacuum, a coincidence that led Maxwell to propose that light waves are electromagnetic waves, consisting of self-supporting, mutually induced electric and magnetic fields.

Maxwell's theory of the propagation of electromagnetic waves and of their generation by accelerated charges received direct experimental confirmation at the hands of Heinrich Hertz, who generated the first artificial radio waves by means of sparks triggered in a gap in a high-frequency *LC* circuit.

wavelength λ and frequency f

The speed c of the electromagnetic wave equals the speed of light. We recall from Chapter 16 that the **wavelength** λ of the wave is the distance between one wave crest and the next, and the **frequency** f is the number of wave crests arriving at some fixed point per second. As for any wave, the product of the wavelength and the frequency of an electromagnetic wave equals the speed of the wave, that is, the speed of light:

wavelength, frequency, and speed of electromagnetic wave

$$\lambda f = c \tag{33.10}$$

HEINRICH RUDOLF HERTZ (1857–1894) *German physicist, and professor at Bonn. He supplied the first experimental evidence for the electromagnetic waves predicted by Maxwell's theory. Hertz generated these waves by means of an electric spark, measured their speed and wavelength, and established their similarity to light waves in the phenomena of reflection, refraction, and polarization.*

EXAMPLE 2 The wavelength of green light is 5.50×10^{-7} m. What is the frequency of this kind of light?

SOLUTION: From Eq. (33.10),

$$f = \frac{c}{\lambda} = \frac{3.00 \times 10^{8}\ \text{m/s}}{5.50 \times 10^{-7}\ \text{m}} = 5.45 \times 10^{14}\ \text{Hz}$$

The most precise modern method for the determination of the speed of light relies on separate measurements of the wavelength and the frequency of the light emitted by a stabilized laser. The speed can then be evaluated as the product of these measured quantities, as in Eq. (33.10). The value of the speed obtained by these means is

$$c = 299\ 792\ 458\ \text{m/s} \tag{33.11}$$

As stated in Chapter 1, this value of the speed of light was adopted as a standard of speed in 1983, and it is now used as the basis for the definition of the meter.[1]

[1]According to modern practice, Eq. (33.9) is then used to calculate the value of ϵ_0 from the value of the speed of light and the value of μ_0.

Concepts in Context

A periodic wave may be regarded as a succession of positive and negative wave pulses. Therefore the arrangement of the electric and magnetic fields in the periodic electromagnetic wave is the same as in the electromagnetic wave pulse, discussed in Section 33.2. *The directions of the electric and magnetic fields are perpendicular to the direction of propagation and they are perpendicular to each other.* The direction of the magnetic field is related to the direction of the electric field by the right-hand rule stated in the preceding section: if you place the fingers of the right hand along the direction of **E** and curl them toward the direction of **B**, the thumb will point in the direction of propagation (see Fig. 33.10).

The theoretical analysis of the mutual induction of the electric and magnetic fields in the wave (Section 33.6) shows that the magnitude of the magnetic field differs from that of the electric field by a factor of c:

magnetic field of wave

$$B = \frac{1}{c}E \tag{33.12}$$

Thus, the magnetic field of the wave is large wherever the electric field is large; that is, the wave crests (and wave troughs) of the magnetic field coincide with the wave crests (and wave troughs) of the electric field.

Figure 33.11a shows the electric and magnetic field lines for such a periodic wave, with the direction of propagation parallel to the x axis, the direction of the electric field parallel to the y axis, and the direction of the magnetic field parallel to the z axis. Figure 33.11b gives plots of the strengths of the electric and magnetic field in a harmonic wave, that is, a wave with a sinusoidal dependence on position. Figure 33.11c displays the wave fronts, or the locations of the wave crests. The maximum strength of the electric field of such a wave, or the strength of the electric field at the wave crests, is called the **amplitude** of the wave, usually designated by E_0. Note that the electric and magnetic fields of the wave are always in phase—the wave crests of the magnetic field are always at the same position as the wave crests of the electric field (but the direction of the magnetic field is perpendicular to the electric field).

amplitude

EXAMPLE 3 The wave reaching a point at some distance from a radio transmitter has an electric field with an amplitude of 2.0×10^{-3} V/m. What is the amplitude of the magnetic field?

Concepts in Context

SOLUTION: By Eq. (33.12), the amplitude of the magnetic field is

$$B_0 = \frac{E_0}{c} = \frac{2.0 \times 10^{-3}\ \text{V/m}}{3.0 \times 10^8\ \text{m/s}} = 6.7 \times 10^{-12}\ \frac{\text{V}\cdot\text{s}}{\text{m}^2}$$

The units V·s/m² on the right side do not look like the unit of the magnetic field, but we can recognize that the units are correct if we perform some substitutions and cancellations:

$$1\,\frac{\text{V}\cdot\text{s}}{\text{m}^2} = 1\,\frac{(\text{J/C})\cdot\text{s}}{\text{m}^2} = 1\,\frac{\text{N}\cdot\text{m}\cdot\text{s}}{\text{C}\cdot\text{m}^2} = 1\,\frac{\text{N}\cdot\text{s}}{\text{C}\cdot\text{m}} = 1\ \text{T}$$

Here the last step used the definition of the tesla in Section 29.2, 1 T = 1 N·s/(C·m).

The wave shown in Fig. 33.11 is a **plane wave**, with uniform electric and magnetic fields over any plane perpendicular to the direction of propagation, such as the planes shown in Fig. 33.11c. Such perpendicular planes, drawn at the location of the wave crests, where the electric field is maximum, are the wave fronts of the wave.

plane wave

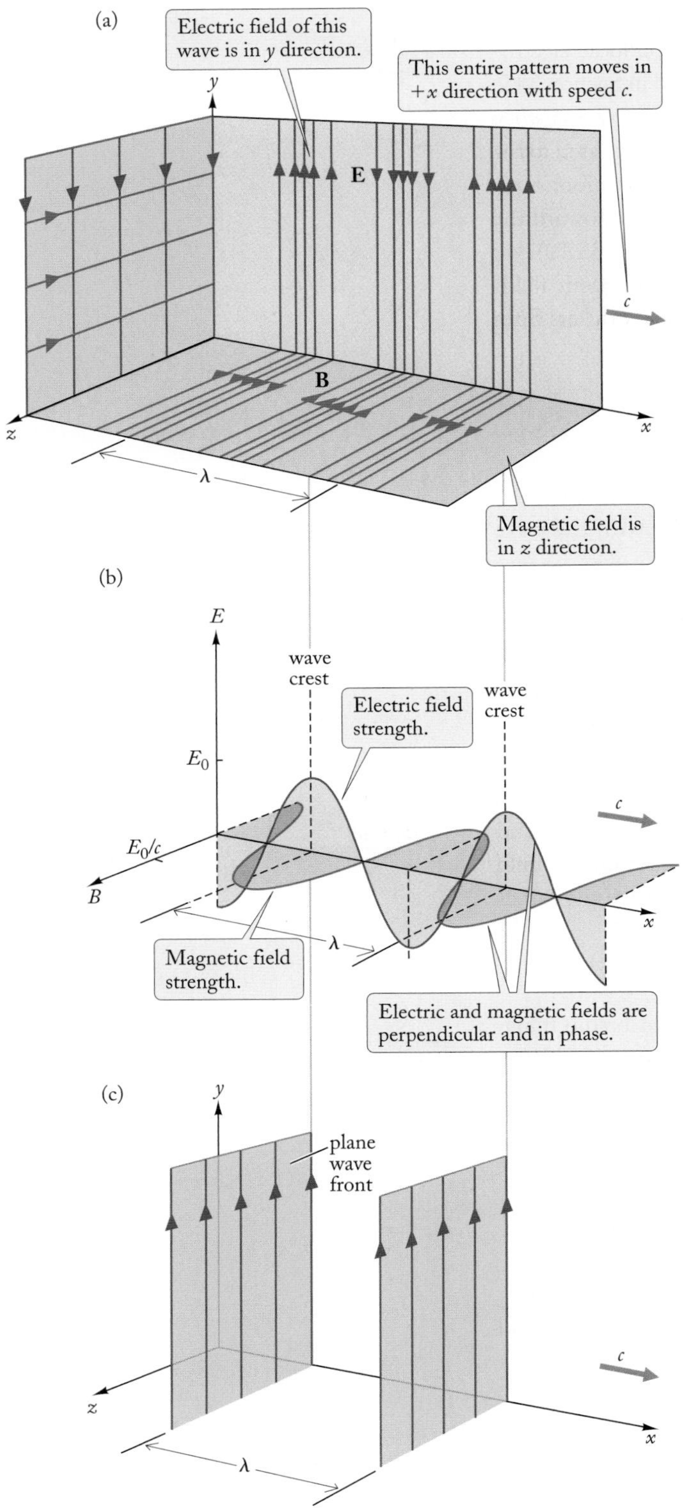

FIGURE 33.11 (a) Electric (red) and magnetic (blue) field lines for a plane wave traveling toward the right, shown at one instant of time. The electric field is vertical, and the magnetic field is horizontal. Only the electric field lines in the x–y plane and the magnetic field lines in the x–z plane are shown. There are many more electric and magnetic field lines parallel to those shown. These lines fill slabs perpendicular to the x axis. (b) Plot of the strengths of the electric and magnetic fields as a function of x. (c) Two of the wave fronts of the plane wave.

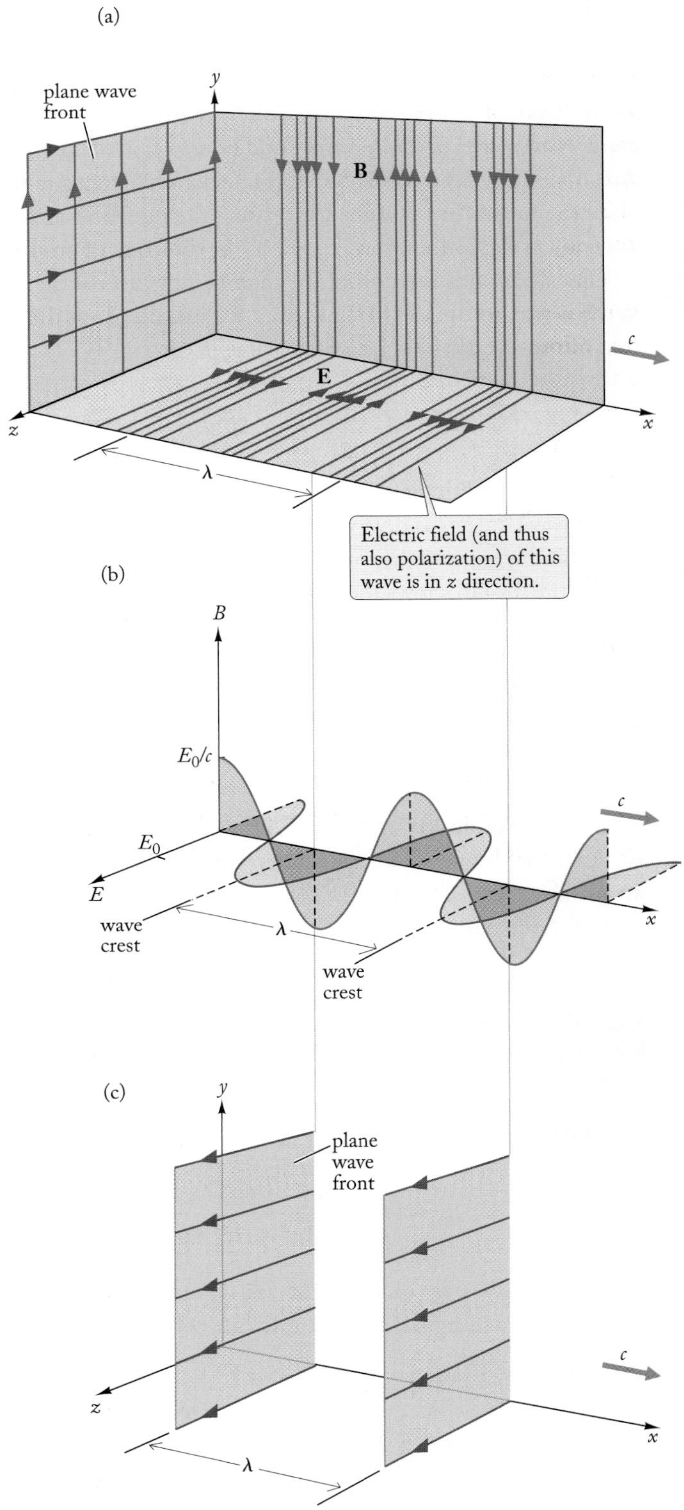

FIGURE 33.12 (a) Electric (red) and magnetic (blue) field lines of another plane wave traveling toward the right, shown at one instant of time. The electric field is horizontal, and the magnetic field is vertical. (b) Plot of the strengths of the electric and magnetic fields as a function of x. (c) Two of the wave fronts of the plane wave.

The direction of the electric field is called the direction of **polarization** of the wave. Thus, the wave shown in Fig. 33.11 is polarized in the y direction (the electric field is in the $\pm y$ directions). By contrast, the wave shown in Fig. 33.12 is polarized in the z direction. To generate a wave polarized in the y direction, we need to accelerate an electric charge back and forth along the y direction; to generate a wave polarized in the z direction, we need to accelerate the charge back and forth along the z direction. It is of course also possible to generate waves polarized in some intermediate direction, say, at 45° to the y and z axes; but such waves are a superposition of those shown in Figs. 33.11 and 33.12, and they therefore are nothing essentially new. Hence, electromagnetic waves (of a given direction of propagation) have only **two independent directions of polarization**. In this regard, electromagnetic waves are analogous to transverse waves on a string; if the string is stretched horizontally, then there are only two independent transverse waves, since we can either shake the string up and down (vertical polarization) or right and left (horizontal polarization). Schematically, the direction of polarization of a light wave is often indicated by a double-headed arrow, which indicates the direction of the positive and negative electric fields of the wave (see Fig. 33.13).

polarization

two independent directions of polarization

Although an individual light wave, like any other kind of electromagnetic wave, is always polarized in some direction or another, the light beams produced by ordinary light sources—the Sun, a lightbulb, a candle—do not exhibit any noticeable polarization. Such "unpolarized" light consists of a superposition of a very large number of plane waves with random directions of polarization (see Fig. 33.14a). Hence, on the average, there is no polarization in the beam. Note that, in Fig. 33.14a, any of the plane waves of an intermediate direction of polarization can be regarded as a superposition of waves of horizontal and of vertical polarizations; hence we can equally well represent an unpolarized light beam as a random mixture of horizontally and vertically polarized waves (see Fig. 33.14b).

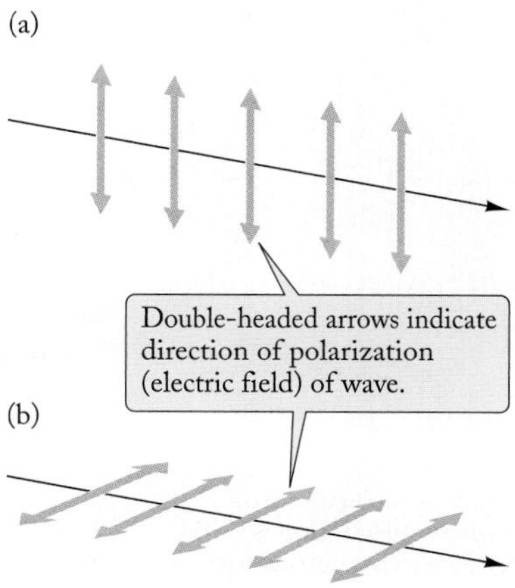

FIGURE 33.13 Schematic representation of the direction of polarization of a light wave. (a) Vertical polarization. (b) Horizontal polarization.

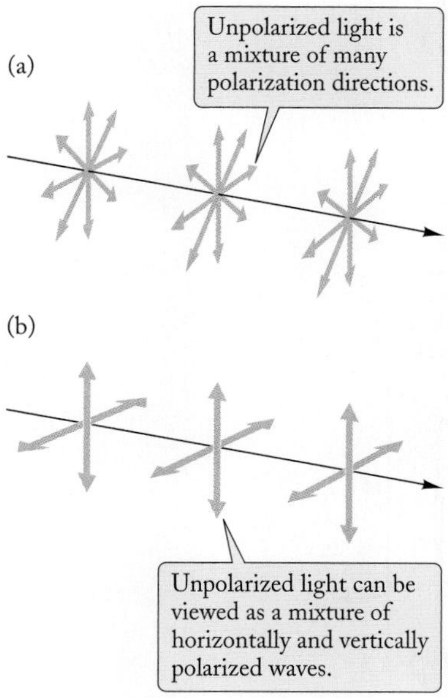

FIGURE 33.14 (a) Unpolarized sunlight consists of a superposition of many plane waves with random directions of polarization. (b) Such light can be represented as a random mixture of horizontally and vertically polarized waves.

polarizer

Unpolarized light can be given a polarization by passing it through a **polarizer** that permits the passage of only the electric field component parallel to the preferential direction in the sheet and absorbs the electric field component perpendicular to the preferential direction. For a wave on a string, the analog of a polarizer is a slot in a plate (see Fig. 33.15). The slot readily permits the passage of a wave with polarization parallel to this slot, but blocks a wave with polarization perpendicular to it. For a light wave, the most commonly employed polarizer is a sheet of Polaroid, which contains long chains of organic molecules arranged parallel to each other. The preferential direction that permits passage of the electric field of a wave is *perpendicular* to the direction of alignment of these molecules. An analogous polarizer for microwaves, or radio waves of short wavelength, can be constructed out of a number of thin conducting rods or wires arranged parallel to each other (see Fig. 33.16). The preferential direction of polarization that permits the passage of the electric field of a wave is then *perpendicular* to the direction of the wires, because the wires have very little effect on a perpendicular electric field. In contrast, an electric field *parallel* to the wires causes strong currents to flow along the wires, which both reflect the wave and dissipate its energy. The preferential

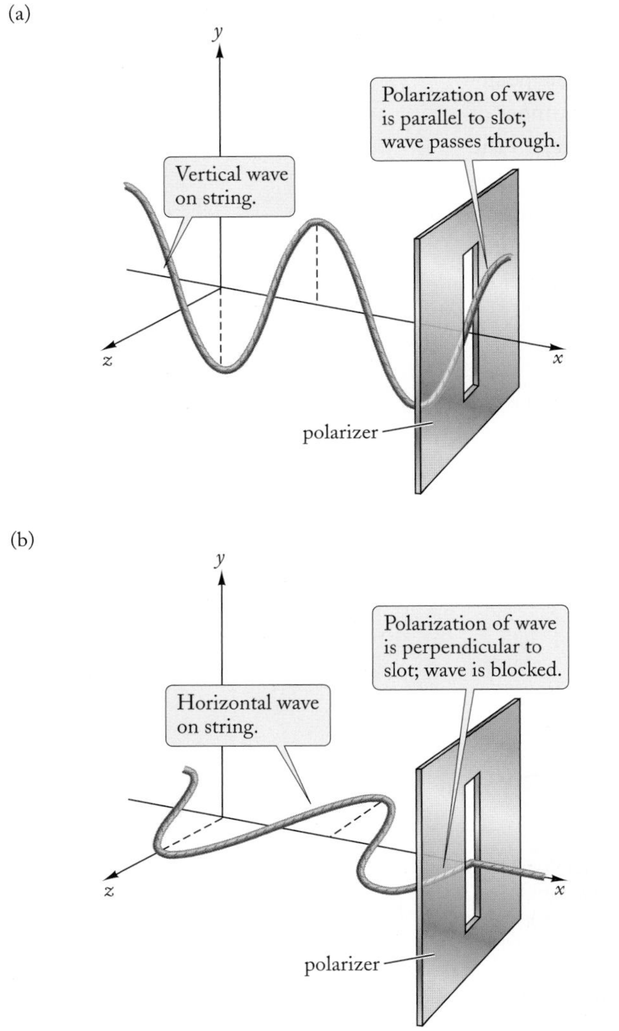

FIGURE 33.15 A wave on a string approaches a plate with a vertical slot. (a) If the direction of polarization of the wave on the string is vertical, the slot permits passage of the wave. (b) If the direction of polarization is horizontal, the slot blocks the wave.

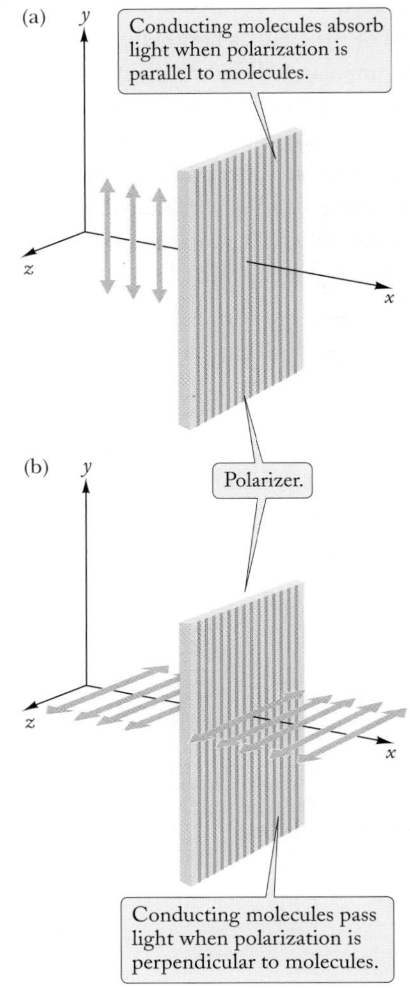

FIGURE 33.16 An array of vertical wires or conducting molecules (a) blocks the passage of a microwave of vertical polarization but (b) permits the passage of a microwave of horizontal polarization.

direction for the passage of waves in Fig. 33.16 is contrary to our intuition—on the basis of the analogy with a wave on a string, we would expect that the slots between the wires in Fig. 33.16a permit the passage of a vertically polarized wave. But in this regard the analogy with the wave on a string misleads us. The wave on a string and the electromagnetic wave respond to the array of wires in a different way.

Polaroid is widely used in sunglasses. On the average, unpolarized sunlight consists of an equal mixture of both directions of polarization, parallel to the preferential direction of the Polaroid and perpendicular. Hence the Polaroid will absorb half the sunlight (actually somewhat more, since the Polaroid in sunglasses is slightly tinted). But the important advantage of Polaroid sunglasses over ordinary sunglasses is that they strongly attenuate the glare of reflected sunlight, such as the sunlight reflected by water or by a road. In distinction to ordinary sunlight, the reflected sunlight is somewhat polarized in the horizontal direction (see Section 34.3); since Polaroid sunglasses have their preferential direction in the vertical plane (unless you tilt your head!), they block the horizontally polarized light.

A variety of interesting experiments can be performed with two or more polarizers arranged in tandem. For instance, Fig. 33.17 shows a simple arrangement of two polarizers. Unpolarized light is incident on the first polarizer, which selects waves of vertical polarization and allows them to pass; the light emerging from this first polarizer is therefore vertically polarized. When this light is incident on the second polarizer, or **analyzer**, its electric field vector makes an angle ϕ with the preferential direction. We can regard a light wave with such an electric field as a superposition of two light waves, whose electric fields are, respectively, parallel and perpendicular to the preferential direction of the analyzer. The analyzer permits the passage of the former wave but blocks the latter. If the amplitude of the wave incident on the analyzer is E_0, then the amplitude of the parallel wave is

analyzer

$$E'_0 = E_0 \cos\phi \tag{33.13}$$

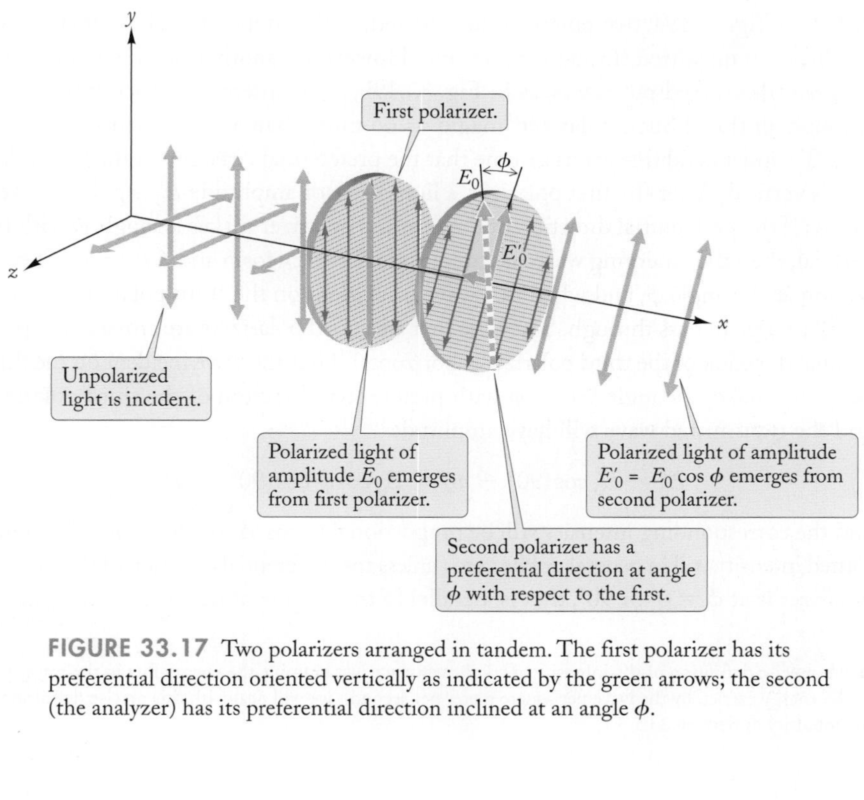

FIGURE 33.17 Two polarizers arranged in tandem. The first polarizer has its preferential direction oriented vertically as indicated by the green arrows; the second (the analyzer) has its preferential direction inclined at an angle ϕ.

Since the intensity of a light wave, like that of any other kind of wave, is proportional to the square of the amplitude,[2] *the intensity I of the transmitted wave emerging from the analyzer is smaller than the intensity I_0 of the incident polarized wave by a factor* $\cos^2\phi$:

$$[\text{transmitted intensity}] = [\text{incident intensity}] \times \cos^2\phi \qquad (33.14)$$

or

Law of Malus

$$I = I_0 \cos^2\phi \qquad (33.15)$$

This relation between the intensity incident on the analyzer and the intensity transmitted by the analyzer is called the **Law of Malus**. Note that if $\phi = 90°$, the transmitted intensity is zero; that is, such "crossed" polarizers block the light completely.

This blocking of light by crossed polarizers can be readily demonstrated by means of two Polaroid sunglasses (see Fig. 33.18). However, such sunglasses cannot be used for a quantitative test of Eq. (33.14), because the reduction of intensity of light by the sunglasses is caused not entirely by polarization but also by the tint of the glass.

FIGURE 33.18 The light that passes through the upper sunglasses becomes polarized in the vertical direction, because this is the preferential direction of the Polaroid in the sunglasses. If the lower sunglasses were oriented parallel to the upper sunglasses ($\phi = 0$), the polarized light could pass and reach us. But if the upper sunglasses are oriented perpendicular to the lower sunglasses ($\phi = 90°$), as shown, the polarized light is blocked.

EXAMPLE 4 Suppose that the preferential direction of the second polarizer makes an angle of 30° with the preferential direction of the first. If unpolarized light is incident on the first polarizer from the left, what fraction of this incident light will pass through both polarizers and emerge on the right?

SOLUTION: In unpolarized light, on the average, one-half of the light waves are polarized in the vertical direction and one-half in the horizontal direction. Hence, one-half of the light will be able to pass through the first polarizer. According to Eq. (33.14), a fraction $\cos^2\phi = \cos^2 30° = 0.75$ of this light will then pass through the second polarizer. Hence the fraction of the light that passes through both polarizers is $0.5 \times 0.75 = 0.375$.

A remarkable demonstration of the behavior of polarizers is illustrated in Fig. 33.19. In Fig. 33.19a, two polarizers are crossed, so that none of the light incident on the first is transmitted through the second. However, if another polarizer is inserted *between* the crossed polarizers, as in Fig. 33.19b, some intensity is then transmitted through all three! Such polarizer "magic" is no more than a consequence of Malus' Law. To understand this, let us assume that the preferential direction of the first polarizer is vertical. After the first polarizer, a light wave of amplitude E_0 is polarized vertically; if the preferential direction of the inserted polarizer makes an angle ϕ with the vertical, then the emerging wave has amplitude $E_0' = E_0 \cos\phi$ and direction of polarization at the angle ϕ, and when this wave is incident on the third polarizer, some of it will be able to pass through. Since the first and third polarizers are crossed, the preferential direction of the third polarizer is horizontal. Thus the wave incident on the third polarizer makes an angle $90° - \phi$ with preferential direction of the third polarizer, and the transmitted wave will have amplitude

$$E_0'' = E_0' \cos(90° - \phi) = E_0 \cos\phi \cos(90° - \phi)$$

and the corresponding intensity will be proportional to $\cos^2\phi \cos^2(90° - \phi)$. The transmitted intensity will in general not be zero (unless the preferential direction of the inserted polarizer is at $\phi = 0°$ or $90°$, that is, parallel to that of one of the crossed polarizers).

ÉTIENNE MALUS (MALÜS) (1775–1812) *French army engineer and physicist, noted for his mathematical and experimental investigations in optics and in double refraction. He discovered the polarization of light by reflection while looking though a calcite crystal at sunlight reflected by the windows of the Luxembourg Palace in Paris.*

[2]For a precise definition of the intensity of a light wave, we just examine the energy flux in the wave (that is, the energy carried by the wave per square meter per second); we will deal with this precise definition of the intensity in Section 33.5.

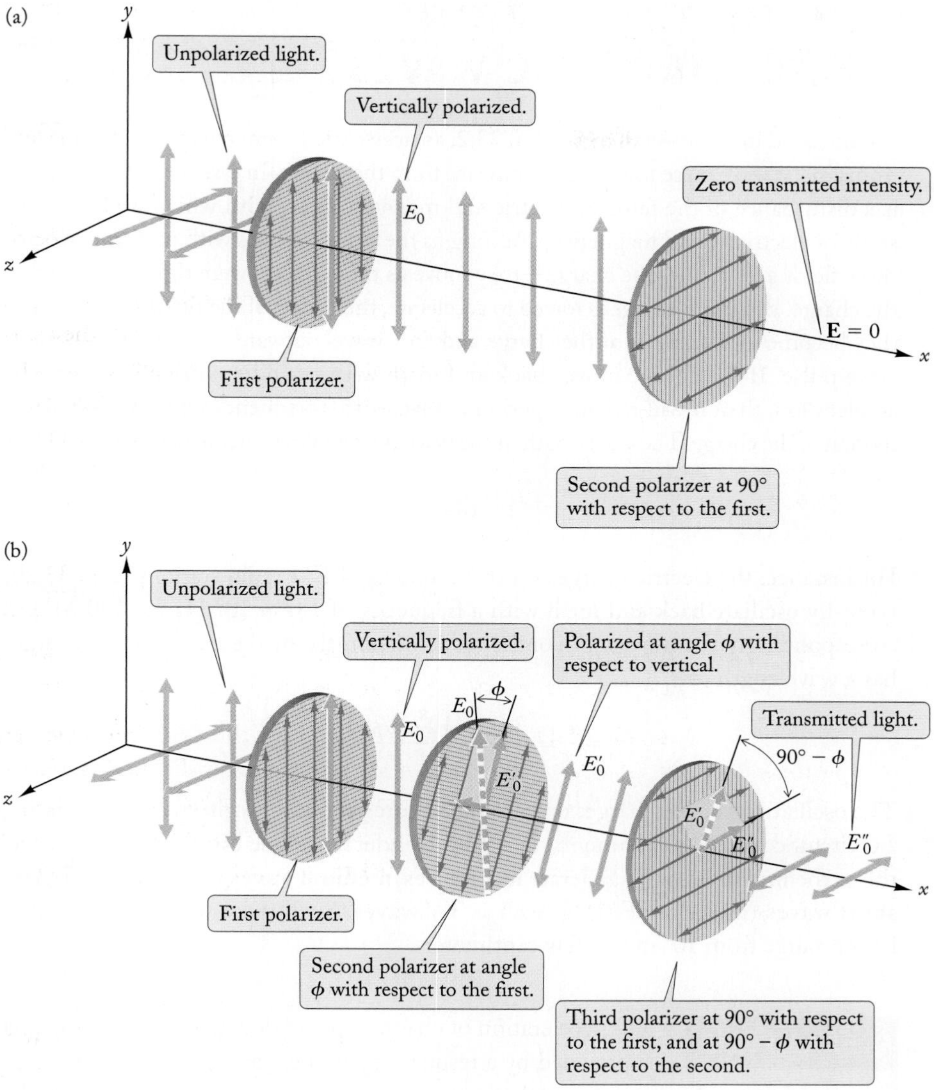

FIGURE 33.19 (a) Two crossed polarizers (perpendicular preferential directions, as indicated by the green arrows); zero intensity is transmitted through such a system. (b) Two crossed polarizers with another polarizer in between. As long as the pass direction of the second polarizer is not parallel to one of the crossed polarizers, some intensity is transmitted through the system.

Checkup 33.3

QUESTION 1: For each of the wave pulses described in Checkup questions 1 and 2 of Section 33.2, what is the direction of polarization?

QUESTION 2: Suppose the y and z axes are vertical and horizontal, respectively, as in Figs. 33.8 and 33.9. An electromagnetic wave traveling in the $+x$ direction is polarized at 45° from the vertical. This wave can be regarded as a superposition of the two waves illustrated in Figs. 33.11 and 33.12. If the amplitude of the wave is E_0, what are the amplitudes of the two waves in the superposition?

QUESTION 3: If the preferential direction of a polarizer is inclined at an angle of 45° to the direction of polarization of an incident wave, by what factor is the transmitted intensity reduced? What if the angle is 60°?

QUESTION 4: Unpolarized light of intensity I_0 is incident upon the first of two polarizers. The second polarizer has its preferential direction at 60° with respect to the first one. What is the final transmitted intensity?

(A) $I_0/8$ (B) $I_0/4$ (C) $3I_0/8$ (D) $\sqrt{3}I_0/4$. (E) $3I_0/4$

33.4 THE GENERATION OF ELECTROMAGNETIC WAVES

Concepts in Context

FIGURE 33.20 Antennas of radio stations atop tall masts.

As discussed in some detail in Section 33.2, an accelerated point charge creates an electromagnetic wave pulse that spreads outward from the charge. In essence, this wave pulse is a disturbance of the familiar electric and magnetic fields with which we began our study of electricity and magnetism. As long as the charge moves with uniform velocity, these fields accompany the charge—they move as though they were rigidly attached to the charge. But if the charge is forced to accelerate, then parts of the fields break away—they become independent of the charge and they travel outward as an electromagnetic wave pulse. If the charge moves back and forth with periodic motion and periodic acceleration, then it will radiate a periodic wave, with a frequency equal to that of the motion of the charge. The wavelength of the wave is related to its frequency by Eq. (33.10),

$$\lambda = \frac{c}{f}$$

For instance, the electric charges on the antennas of FM radio stations (Fig. 33.20) typically oscillate back and forth with a frequency of 1.0×10^8 Hz (or 100 MHz); correspondingly, the wavelength of the radiation emitted by these accelerated charges has a wavelength of

$$\lambda = \frac{c}{f} = \frac{3.0 \times 10^8 \text{ m/s}}{1.0 \times 10^8 \text{ s}^{-1}} = 3.0 \text{ m} \tag{33.16}$$

long waves, medium waves, and short waves

TV waves

The oscillations of the charges on the antenna are produced by means of a resonating *LC* circuit coupled to the antenna by a mutual inductance (see Fig. 32.17). In essence, this is the method used to generate **long waves, medium waves** (including AM), and **short waves** (including FM), as well as **TV waves**. Such radio waves span a wavelength range from 10^5 m to a few centimeters.

Concepts in Context

EXAMPLE 5 The acceleration of charges up and down an efficient antenna is provided by a resonant standing wave, with the travel time up and down the antenna matched to one period of the oscillation. However, the speed v of the wave in the conductor is somewhat slower than the speed of light; for a typical antenna, $v = 0.68c$. How long should such an antenna be for an AM radio station broadcasting at a frequency of 1.0 MHz? For a cellular telephone operating around 1.0 GHz?

SOLUTION: The travel time up and down the antenna is one oscillation period, or T. For an antenna of length l, the distance up and down is $2l$, and this distance must equal the velocity of the wave in the antenna times the travel time:

$$2l = v \times T$$

The period is the inverse of the frequency f, so the length of the antenna is

$$l = \frac{v}{2f}$$

We are told $v = 0.68c$ for a typical antenna. For an antenna transmitting AM radio waves at 1.0 MHz, the length is

$$l = \frac{v}{2f} = \frac{0.68 \times 3.00 \times 10^8 \text{ m/s}}{2 \times 1.0 \times 10^6 \text{ Hz}} = 102 \text{ m}$$

PHYSICS IN PRACTICE AM AND FM RADIO

Concepts in Context

The frequencies of AM or FM radio waves indicated on the dials of radio receivers refer to the frequency of the carrier wave, which is the steady, periodic wave emitted by the radio station during moments of silence, when there is no audio signal (no voice signal and no musical signal). AM radio waves span a frequency range from 550 kHz to 1.7 MHz, and FM radio waves a range from 88 MHz to 108 MHz. AM and FM stations use different methods for imprinting the audio signal on the carrier wave. In the AM (*A*mplitude *M*odulation) method, the *amplitude* of the carrier wave is altered in accordance with the amplitude of the audio signal to be transmitted; thus, the carrier amplitude is increased or decreased when the amplitude of the audio signal increases or decreases (see Fig. 1). In the FM (*F*requency *M*odulation) method, the *frequency* of the carrier wave is altered in accordance with the amplitude of the audio signal; thus, the carrier frequency is increased or decreased when the amplitude of the audio signal increases or decreases (see Fig. 2), but the amplitude of the carrier wave is kept constant. FM attains a higher fidelity than AM because it is quite insensitive to disturbances in the strength of the carrier wave. Any such disturbances in the strength of the FM carrier wave have no effect on its frequency, and hence do not alter the audio signal imprinted on the wave. By contrast, any disturbance in the strength of an AM carrier wave leads to a distortion of the audio signal. FM broadcasts also have higher fidelity because FM stations are permitted to broadcast a wider range of frequencies: In order to avoid overcrowding of the broadcast bands, AM stations are limited by law to imprinting audio signals of no more than 10 kHz on the carrier, which is smaller than the frequency range of human hearing. The modulation may reach up to 20 kHz for FM stations, closer to the full range of human hearing.

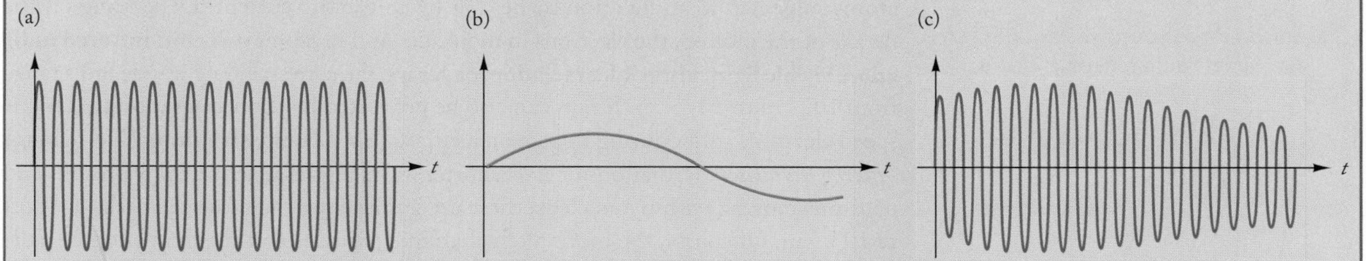

FIGURE 1 (a) Carrier radio wave. (b) Audio, or sound, signal. (c) AM radio wave. The amplitude of the wave is modulated according to the audio signal.

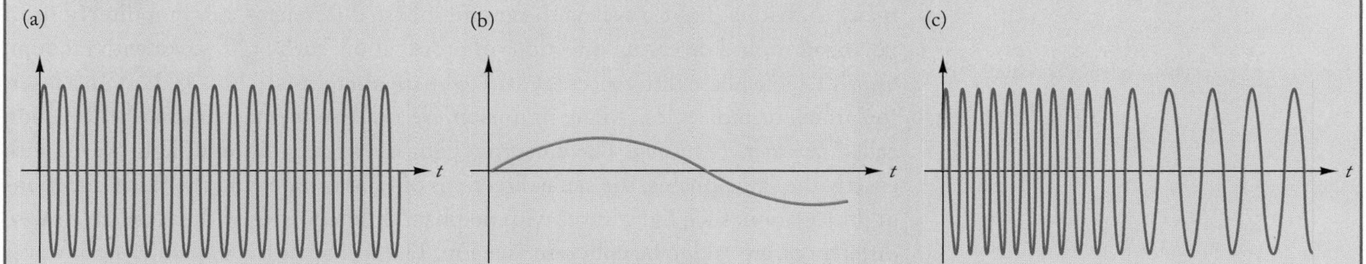

FIGURE 2 (a) Carrier radio wave. (b) Audio signal. (c) FM radio wave. The frequency of the wave is modulated according to the audio signal.

Similarly, for a cellular telephone operating at 1.0 GHz, the length is

$$l = \frac{0.68 \times 3.00 \times 10^8 \text{ m/s}}{2 \times 1.0 \times 10^9 \text{ Hz}} = 0.102 \text{ m} = 10 \text{ cm}$$

This is why an AM radio transmitter antenna is very long, whereas a cellular telephone antenna is quite short. Incidentally: Instead of using a long wire to intercept the electric field, portable AM radio receivers use a small coil of wire to intercept the changing magnetic flux due to the electromagnetic wave.

microwaves

Waves of a wavelength shorter than that of radio waves, called **microwaves**, are best generated by a resonating electromagnetic cavity, consisting of an empty metallic can in which a standing electromagnetic wave is set up by an electron beam passing through, much as a standing sound wave is set up in an organ pipe by a stream of air passing over the blowhole. The antenna that radiates the microwaves is simply a horn attached to the electromagnetic cavity by a metallic pipe, or waveguide, which permits the waves to spill out into space (see Fig. 33.21). This method can be used to generate waves as short as about a millimeter. Shorter wavelengths cannot be generated with currents oscillating in macroscopic laboratory equipment; however, short wavelengths can be easily generated by electrons oscillating within molecules and atoms subjected to stimulation by heat or by an electric current. Depending on the details of the motion, the electrons in molecules and in atoms will emit **infrared radiation**, **visible light**, **ultraviolet radiation**, or **X rays**; the corresponding wavelengths range from 10^{-3} m to 10^{-11} m. X rays can also be generated by the acceleration that high-speed electrons suffer during impact on a target; this is **Bremsstrahlung** (German for *braking radiation*). Radiations of even shorter wavelengths are emitted by protons and neutrons moving within a nucleus; these are **gamma rays**, with wavelengths as short as 10^{-13} m. Of course, the motion of subatomic particles and their emission of radiation cannot be calculated by classical mechanics or classical electricity and magnetism; such calculations require quantum mechanics.

infrared radiation, visible light, ultraviolet radiation, and X rays

Bremsstrahlung

gamma rays

In an ordinary light source, such as an incandescent lightbulb, the individual atoms or molecules radiate independently. The emerging light consists of a superposition of many individual light waves with random phase differences, random directions of polarization, and diverging directions of propagation; such light waves with random, unpredictable phase differences are said to be **incoherent** (see Fig. 33.22a). In a **laser**, the atoms or molecules radiate in unison, by a quantum-mechanical phenomenon called *stimulated emission*. The emerging light is a superposition of light waves with exactly the same phases, the same directions of polarization, and the same directions of propagation; such light waves with no phase differences, or with predictable phase differences, are said to be **coherent** (see Fig. 33.22b). Since the individual light waves in this kind of light combine constructively, the light emerging from the laser is very intense, and it also forms a very narrow, sharp beam.

incoherent and coherent light waves

Another important mechanism for the generation of electromagnetic waves is **cyclotron emission**. This involves high-speed electrons undergoing centripetal acceleration while spiraling in a magnetic field (see Section 30.1). Depending on the speed of the electron and the strength of the magnetic field, the radiation may consist of radio waves, X rays, or anything in between (including visible light). Most of the radio waves reaching us from stars, pulsars, and radio galaxies are generated by this process.

cyclotron emission

Figure 33.23 displays the **electromagnetic spectrum**, the wavelength and the frequency bands of electromagnetic radiation. The bands overlap to some extent, because the names assigned to the different ranges of wavelengths depend not only on the

electromagnetic spectrum

FIGURE 33.21 (a) Horns of microwave antennas. (b) Microwave oven.

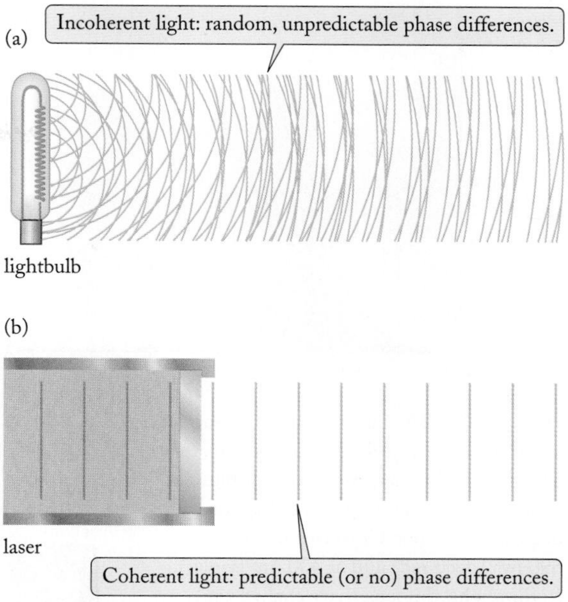

FIGURE 33.22 (a) Wave fronts of incoherent light waves emitted by a lightbulb. (b) Wave fronts of coherent light waves emitted by a laser.

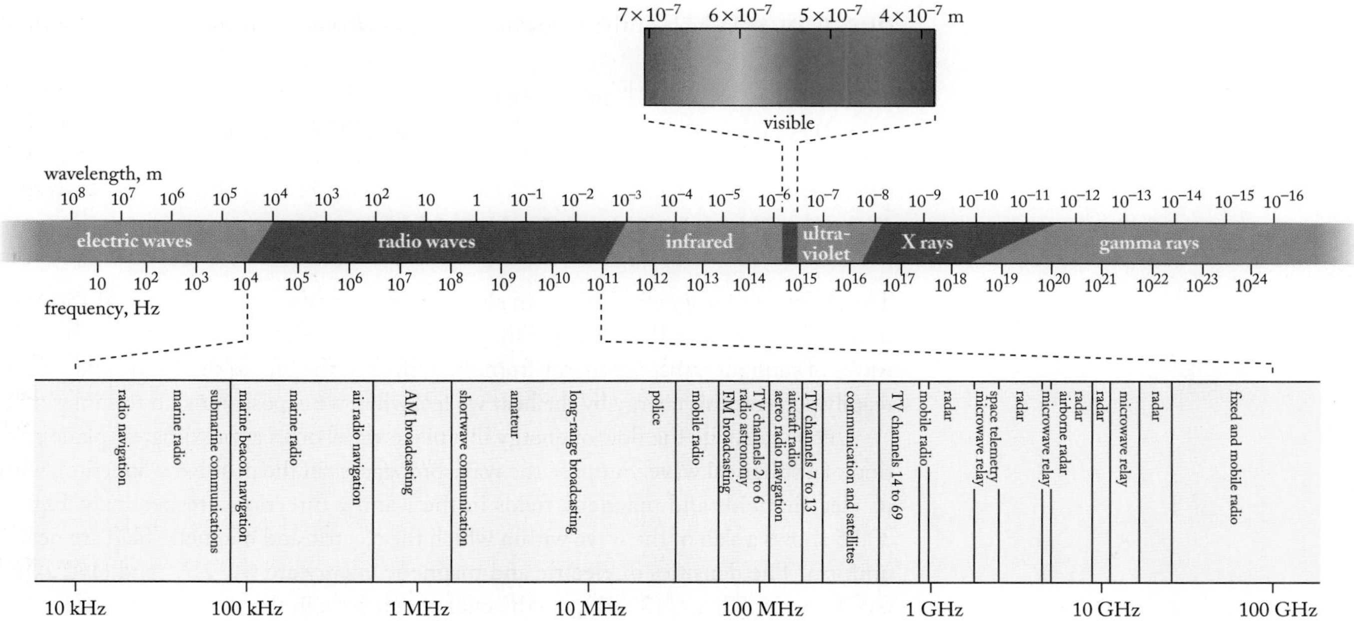

FIGURE 33.23 Wavelength and frequency bands of electromagnetic radiation.

value of the wavelength, but also on the method used to generate or detect the radiation. For example, radiation of a wavelength of a tenth of a millimeter will be called a radio wave (microwave) if detected by a radio receiver, but it will be called infrared radiation if detected by a heat sensor.

The wavelengths of visible light range from about 7×10^{-7} m to 4×10^{-7} m. The wavelength of light is usually expressed in nanometers (1 nm $= 10^{-9}$ m), in terms

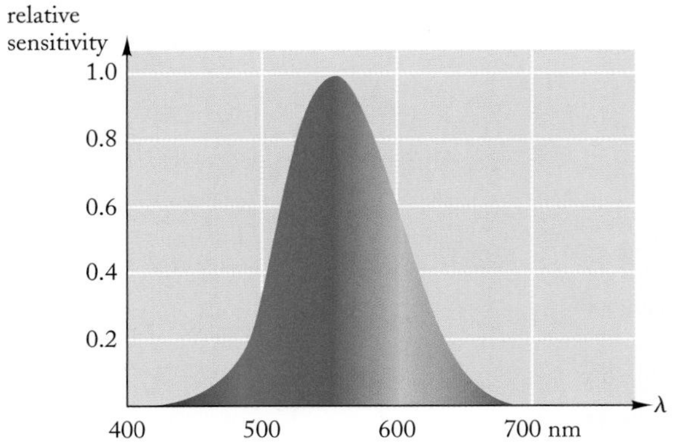

FIGURE 33.24 Sensitivity of the human eye at different wavelengths and colors in bright light. The sensitivity is maximum at about 550 nm (yellow green light) and decreases to about 1% of this maximum at 690 nm (deep red) and at 430 nm (violet). In dim light, the sensitivity changes; it becomes maximum at about 500 nm (blue green).

of which *the range of wavelengths of visible light extends from* 400 nm *to* 700 nm. This is the range of wavelengths over which our eyes are sensitive to light. We perceive different wavelengths within the visible region as having different colors. Figure 33.24 is a plot of the sensitivity of the human eye and shows how colors are correlated with wavelength.

Incidentally: Our eyes are almost completely insensitive to the polarization of light waves. We can detect the polarization only with special equipment, such as Polaroid sunglasses. Radio and TV antennas are, of course, very sensitive to the direction of polarization of radio waves, and they must have the proper orientation to pick up a signal. For instance, a simple antenna in the form of a straight wire or rod is sensitive to a radio wave polarized in the same direction as the antenna, but insensitive to a radio wave polarized at right angles to the antenna. Some TV antennas are equipped with an electric motor so that they can be easily rotated to the optimal orientation for each TV station you want to receive.

✔ Checkup 33.4

QUESTION 1: Why do FM radio waves give higher fidelity than AM?

QUESTION 2: What is the color of light of wavelength 650 nm? 550 nm?

QUESTION 3: What is the wavelength of yellow light?

QUESTION 4: Arrange in order of increasing wavelength: AM, FM, radar, TV.

QUESTION 5: Which of the following objects is closest in size to the wavelength of visible light?

(A) An atom ($\sim 10^{-10}$ m) (B) A virus ($\sim 10^{-6}$ m) (C) A dime (~ 1 cm)
(D) A person (~ 2 m) (E) A football field (~ 100 m)

33.5 ENERGY OF A WAVE

The electric and magnetic fields of an electromagnetic wave contain energy. As the wave moves along, so does this energy—the wave transports energy. For instance, the light waves of sunlight transport energy from the surface of the Sun to the Earth, and we can readily perceive this energy by the heat we feel when we expose our skin to sunlight.

Let us calculate the flow of energy in a plane wave, or an approximately plane portion of a spherical wave. Suppose the wave propagates in the positive x direction, with its electric fields and magnetic fields in the y and z directions, respectively. Figure 33.25 shows a slab of the wave within which the electric and magnetic field are nearly uniform. The densities of electric and magnetic energy are $(\epsilon_0/2)E^2$ and $(1/2\mu_0)B^2$ [see Eqs. (25.58) and (31.31)], so the energy density u in the wave is

energy density in electromagnetic wave

$$u = \frac{\epsilon_0}{2}E^2 + \frac{1}{2\mu_0}B^2 \tag{33.17}$$

If the thickness of the wave slab is dx and the frontal area is A (see Fig. 33.25), then the volume of the slab is $A\,dx$ and the total amount of energy in the slab is the energy density times the volume:

$$dU = \left(\frac{\epsilon_0}{2}E^2 + \frac{1}{2\mu_0}B^2\right) \times A\,dx \tag{33.18}$$

Since $B = E/c$ and $\epsilon_0 = 1/(\mu_0 c^2)$, we can write this energy as

$$dU = \left[\frac{1}{2\mu_0 c^2}E^2 + \frac{1}{2\mu_0}\left(\frac{E}{c}\right)^2\right] \times A\,dx \tag{33.19}$$

$$= \frac{1}{\mu_0 c^2}E^2 A\,dx \tag{33.20}$$

Note that the two terms on the right side of Eq. (33.19) are equal, that is, *the electric and magnetic energy densities in an electromagnetic wave are equal.*

The forward speed of the wave slab is c, and hence the amount of energy dU moves out of the (stationary) volume $A\,dx$ in a time $dt = dx/c$. The rate of flow of energy, or the power, is therefore

$$\frac{dU}{dt} = \frac{dU}{dx/c} = \frac{1}{\mu_0 c}AE^2 \tag{33.21}$$

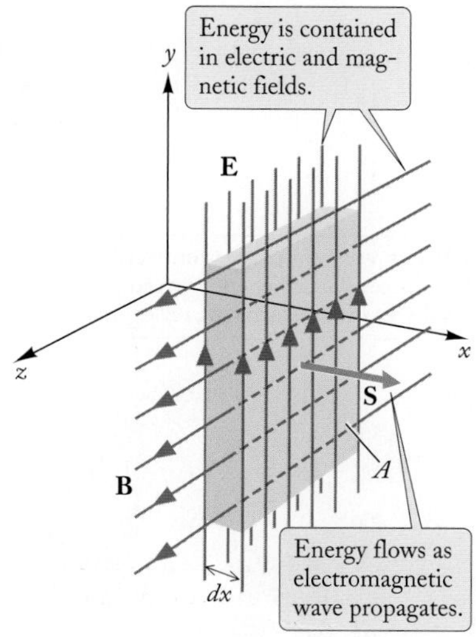

FIGURE 33.25 A slab of electric (red) and magnetic (blue) fields in a plane wave propagating toward the right. The slab of thickness dx and frontal area A moves with the wave.

In Fig. 33.25, the energy flows toward the right, along the direction of propagation of the wave. The **energy flux in the wave** is defined as the rate of energy flow per unit frontal area, that is, $(1/A)dU/dt$. The rate of energy flow is the power; hence the energy flux is the power carried per unit area of wave front. The units for this energy flux are watts per square meter (W/m^2). According to Eq. (33.21), the energy flux for our plane wave is

$$\frac{1}{A}\frac{dU}{dt} = \frac{1}{\mu_0 c}E^2$$

This energy flux is called the **Poynting flux**, usually designated by S:

$$S = \frac{1}{\mu_0 c}E^2 \tag{33.22}$$

energy flux in wave

This energy flux provides us with a precise measure of the instantaneous energy flux of the wave. A wave of high intensity—such as the wave produced by a powerful laser—is a wave of large energy flux.

Since an electromagnetic wave propagates in the $\mathbf{E} \times \mathbf{B}$ direction and $\mathbf{E}$ and $\mathbf{B}$ are perpendicular, the Poynting flux is sometimes expressed as a vector of the form (again using $B = E/c$)

$$\mathbf{S} = \frac{1}{\mu_0}\mathbf{E} \times \mathbf{B} \tag{33.23}$$

Poynting vector S

Thus the **Poynting vector S** has a magnitude equal to the instantaneous energy flux of the wave and points in the direction of propagation.

Since the electric field oscillates in time, so does the energy flux. For a harmonic wave, such as shown in Fig. 33.11b, with a sinusoidal dependence on time, the average of the square of the electric field is one-half of the square of the amplitude E_0 of the electric field, that is, $\overline{E^2} = E_0^2/2$ (compare the calculation of the average of the square of the emf in Section 32.1). Hence the **time-average energy flux** in the wave is

$$\overline{S} = \frac{1}{2\mu_0 c}E_0^2 \tag{33.24}$$

time average energy flux in wave

The time-average energy flux is the **intensity** of the wave.

Concepts in Context

EXAMPLE 6 At a distance of 8.5 km from a radio transmitter, the amplitude of the oscillating electric field in the radio wave is $E_0 = 0.13$ V/m. What is the time-average energy flux there? What is the total power radiated by the radio transmitter? Assume that this transmitter radiates uniformly in all directions.

SOLUTION: From Eq. (33.24), the time-average energy flux is

$$\bar{S} = \frac{1}{2\mu_0 c}E_0^2 = \frac{(0.13\ \text{V/m})^2}{2 \times 1.26 \times 10^{-6}\ \text{H/m} \times 3.00 \times 10^8\ \text{m/s}}$$

$$= 2.2 \times 10^{-5}\ \text{W/m}^2$$

To obtain the total power, we must multiply the calculated power per unit area (the energy flux, or intensity) by the area over which the radio wave has spread. This is the area of a sphere of radius r, namely, an area $A = 4\pi r^2$, with $r = 8.5$ km (see Fig. 33.26). Therefore the total power is

$$\bar{P} = [\text{area}] \times [\text{energy flux}] = 4\pi r^2 \bar{S}$$

$$= 4\pi \times (8.5 \times 10^3\ \text{m})^2 \times 2.2 \times 10^{-5}\ \text{W/m}^2 \tag{33.25}$$

$$= 2.0 \times 10^4\ \text{W} = 20\ \text{kW}$$

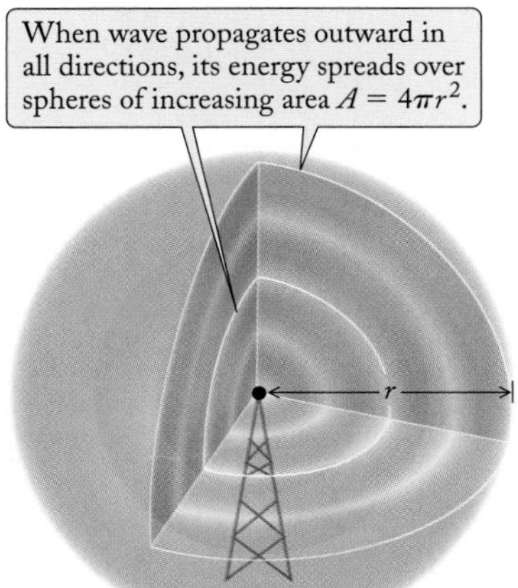

FIGURE 33.26 The energy of the radio wave is spread out over a sphere of radius $r = 8.5$ km.

Concepts in Context

Note that, according to Eq. (33.25), the energy flux $\bar{S}$ for a spherical wave spreading out from a source in all directions is inversely proportional to the square of the distance:

energy flux and power for spherical wave

$$\bar{S} = \frac{\bar{P}}{4\pi r^2} \tag{33.26}$$

We must take this dependence of the flux on distance into account whenever we want to investigate the spreading of the wave over a large range of distances (but we can ignore this dependence over a small range of distances, where the wave can be approximated as a plane wave of constant energy flux).

A wave carries energy, so we might expect that it also carries momentum. Maxwell showed this to be the case. He showed that when a body absorbs an electromagnetic wave with total energy U, the momentum p transferred to the body is given by

$$p = \frac{U}{c} \tag{33.26}$$

From this expression for the momentum, we can deduce that the wave exerts a pressure on the absorbing body. Since the pressure is the force per unit area, or the change in momentum per unit time per unit area, the pressure exerted by the wave on the absorbing body is

$$[\text{pressure}] = \frac{F}{A} = \frac{1}{A}\frac{dp}{dt} = \frac{1}{A} \times \frac{1}{c}\frac{dU}{dt}$$

or, in terms of the energy flux $S = (1/A)\ dU/dt$,

pressure of light

$$[\text{pressure}] = \frac{S}{c} \quad \text{(absorption)} \tag{33.27}$$

If a body reflects rather than absorbs, then the wave is sent back from whence it came; it thus leaves with a momentum opposite to its original value. Thus, for a reflecting body, the momentum transferred and pressure exerted are twice as large as for an absorbing body:

$$[\text{pressure}] = \frac{2S}{c} \quad \text{(reflection)} \tag{33.28}$$

EXAMPLE 7 Near the Earth's orbit, the energy flux from the Sun is approximately 1400 W/m^2 at normal incidence. One proposed form of space propulsion is the "solar sail," driven by sunlight. What is the pressure due to sunlight at normal incidence on a perfectly reflecting sail? What is the force due to this pressure on a sail of area 25 m^2? If such a sail is quite thin, so it has a mass of only 10 grams, what is the Sun's gravitational force on the sail? If released from rest, what will be the speed of the sail after one day? Assume that the sail is released far from the Earth and that the force is essentially constant over one day.

SOLUTION: As mentioned just above, for reflection, the pressure is given by Eq. (33.28):

$$[\text{pressure}] = \frac{2S}{c} = \frac{2 \times 1400\ \text{W/m}^2}{3.00 \times 10^8\ \text{m/s}}$$
$$= 9.3 \times 10^{-6}\ \text{N/m}^2$$

The force due to this pressure on an area of 25 m^2 has magnitude

$$F_{\text{pressure}} = [\text{pressure}] \times [\text{area}] = 9.3 \times 10^{-6}\ \text{N/m}^2 \times 25\ \text{m}^2 = 2.3 \times 10^{-4}\ \text{N}$$

and is directed away from the Sun.

For a mass of 0.010 kg, and using data listed inside the book cover, the gravitational force due to the Sun has magnitude

$$F_{\text{gravitational}} = \frac{GM_S m}{r^2}$$
$$= \frac{6.67 \times 10^{-11}\ \text{N}\cdot\text{m}^2/\text{kg}^2 \times 1.99 \times 10^{30}\ \text{kg} \times 0.010\ \text{kg}}{(1.50 \times 10^{11}\ \text{m})^2}$$
$$= 5.9 \times 10^{-5}\ \text{N}$$

and is directed toward the Sun. This is less than the pressure force, so the net force is directed away from the Sun.

The speed of the sail is the acceleration multiplied by the time, $v = at$. The acceleration is

$$a = \frac{F_{\text{net}}}{m} = \frac{F_{\text{pressure}} - F_{\text{gravitational}}}{m}$$
$$= \frac{2.3 \times 10^{-4}\ \text{N} - 5.9 \times 10^{-5}\ \text{N}}{0.010\ \text{kg}} = 1.7 \times 10^{-2}\ \text{m/s}^2$$

After one day, the velocity will be

$$v = at = 1.7 \times 10^{-2}\ \text{m/s}^2 \times 24 \times 60 \times 60\ \text{s}$$

$$= 1.5 \times 10^3\ \text{m/s} = 1.5\ \text{km/s}$$

Although large speeds could thus be reached for a sail by itself, the speed will be much lower, or the gravitational force will dominate, if a spacecraft of appreciable mass is attached to the sail (unless we manage to "miniaturize" the spacecraft, so its mass is not much more than that of the sail).

Checkup 33.5

QUESTION 1: Given that a light wave and an ultraviolet wave have the same amplitude, which has the higher electric and magnetic energy density?

QUESTION 2: What is the distance between adjacent maxima in the energy density of an electromagnetic wave of wavelength λ?

QUESTION 3: A wave has an amplitude $E_0 = 1.0 \times 10^{-3}$ V/m; another wave has an amplitude $E_0 = 2.0 \times 10^{-3}$ V/m. What is the ratio of the energy densities in these two waves?

QUESTION 4: If the amplitude of the electric field in a radio wave is $E_0 = 0.13$ V/m at a distance of 8.5 km (as in Example 6), what will be the amplitude when this wave spreads to a distance of 17 km? What will be the time-average energy flux?

QUESTION 5: Suppose that a plane wave strikes a polarizer whose preferential direction is at 45° to the direction of polarization of the wave. What fraction of the energy of the wave is transmitted? What happens to the energy that is not transmitted?

QUESTION 6: An electromagnetic wave is absorbed by a body. Another wave with twice the amplitude of the first is reflected by a different body. By what factor do the pressures exerted on the bodies differ?

QUESTION 7: A lightbulb emits 125 W of average power uniformly in all directions. A book is placed 1.0 m away from the bulb. The book has an area of 0.10 m^2 perpendicular to the light and is perfectly absorbing. What is the average energy flux (intensity) reaching the book?

(A) 1.0 W/m^2 (B) 10 W/m^2 (C) 12.5 W/m^2
(D) 125 W/m^2 (E) 1250 W/m^2

33.6 THE WAVE EQUATION

We saw in Section 33.2 how the changing electric field in the wave front of a pulse induces a magnetic field (and vice versa); after that, we examined the behavior of sinusoidal electromagnetic waves. Here, we will see that such sinusoidal waves are solutions of a *wave equation* that we can obtain directly from Faraday's Law and the Maxwell–Ampère Law. We will also see from our analysis that the amplitudes of the fields are indeed related by $B_0 = E_0/c$, as stated previously in Eq. (33.12).

Consider a region of space in which there are perpendicular electric and magnetic fields, but no electric charges or currents. As shown in Fig. 33.27, the electric field **E** is in the y direction and the magnetic field **B** is in the z direction. Suppose these fields vary with position in the x direction. We first evaluate Faraday's Law

$$\oint E_{\parallel}\,ds = -\frac{d\Phi_B}{dt} \qquad (33.29)$$

for the small vertical rectangular path shown in Fig. 33.27, which has a width dx and a height dy. In the integration of $E_{\parallel}$ around this path, only the vertical segments contribute, since **E** is perpendicular to the horizontal segments. The electric field is E along the left segment and $E + dE$ along the right segment, so the path integral is $-E\,dy + (E + dE)\,dy = dE\,dy$. The magnetic flux through the rectangle is $\Phi_B = B\,dx\,dy$, and therefore Eq. (33.29) becomes

$$dE\,dy = -\frac{dB}{dt}\,dx\,dy$$

or

$$\frac{dE}{dx} = -\frac{dB}{dt} \qquad (33.30)$$

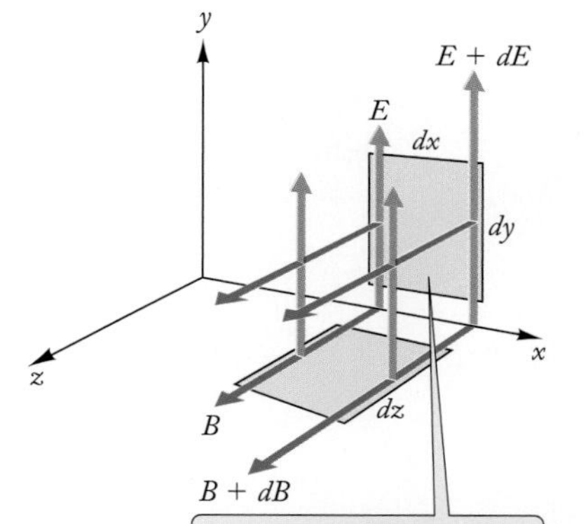

FIGURE 33.27 Electric and magnetic fields in a plane wave; the magnitudes of the field vary as a function of x. Evaluation of Faraday's Law around the vertical tan rectangle (and the Maxwell–Ampère Law around the horizontal one) provide relations between changes in E and B.

Since E and B are functions of x and t, the derivatives are actually partial derivatives (see Appendix 4.2), and the proper form of Eq. (33.30) is

$$\frac{\partial E}{\partial x} = -\frac{\partial B}{\partial t} \qquad (33.31)$$

If we perform the same type of analysis by applying the Maxwell–Ampère Law (with $I = 0$) to the horizontal rectangle in Fig. 33.27, we obtain a similar equation:

$$-\frac{\partial B}{\partial x} = \mu_0\epsilon_0\frac{\partial E}{\partial t} \qquad (33.32)$$

Equations (33.31) and (33.32) can be combined by taking the partial derivative of (33.31) with respect to x. For the left side, we have

$$\frac{\partial}{\partial x}\frac{\partial E}{\partial x} = \frac{\partial^2 E}{\partial x^2}$$

We do the same for the right side of Eq. (33.31), then change the order of derivatives and substitute from Eq. (33.32) as follows:

$$-\frac{\partial}{\partial x}\frac{\partial B}{\partial t} = -\frac{\partial}{\partial t}\frac{\partial B}{\partial x} = +\frac{\partial}{\partial t}\left(\mu_0\epsilon_0\frac{\partial E}{\partial t}\right) = \mu_0\epsilon_0\frac{\partial^2 E}{\partial t^2}$$

Equating these two sides, we obtain

$$\frac{\partial^2 E}{\partial x^2} = \mu_0\epsilon_0\frac{\partial^2 E}{\partial t^2} \qquad (33.33)$$

wave equation

Equation (33.33) is an instance of an important differential equation known as the **wave equation**. Similar wave equations arise in other areas of physics. The study of the solutions of such wave equations plays a large role in microwave and radio communications, radar, acoustics, sonar, and oceanography, as well as in quantum mechanics and high-energy physics.

Among the simplest solutions of Eq. (33.33) are harmonic traveling waves. We can write these in a form familiar from Ch. 16 [see Eq. (16.8)]:

electric field of harmonic traveling wave

$$E = E_0 \cos[k(x - vt)] \tag{33.34}$$

where v is the speed of the wave and k is the wave number, which is related to the wavelength by $k = 2\pi/\lambda$ [see Eq. (16.7)]. If we substitute this tentative solution into Eq. (33.33) and take the space and time derivatives, we obtain

$$-k^2 \cos[k(x - vt)] = -\mu_0 \epsilon_0 k^2 v^2 \cos[k(x - vt)]$$

From this we see that the wave equation is satisfied provided that $\mu_0 \epsilon_0 v^2 = 1$, or

$$v = \frac{1}{\sqrt{\mu_0 \epsilon_0}} \tag{33.35}$$

This is Maxwell's result for the speed c of electromagnetic waves [see Eq. (33.8)].

Finally, we can derive the relation between E and B. Similar to the analysis above, Eqs. (33.31) and (33.32) can be combined to show that the magnetic field B also satisfies the wave equation. These equations also require that the traveling wave solution for the magnetic field be of the same form (33.34) as for the electric field,

magnetic field of harmonic traveling wave

$$B = B_0 \cos[k(x - vt)] \tag{33.36}$$

Substituting the solutions (33.34) and (33.36) into Eq. (33.31), we have

$$\frac{\partial}{\partial x} E_0 \cos[k(x - vt)] = -\frac{\partial}{\partial t} B_0 \cos[k(x - vt)]$$

or

$$-kE_0 \sin[k(x - vt)] = -kvB_0 \sin[k(x - vt)] \tag{33.37}$$

Solving for B_0, we obtain

$$B_0 = \frac{1}{v} E_0 \tag{33.38}$$

Since the speed of electromagnetic waves is $v = c$, we see that Eq. (33.38) gives the same relation between E and B presented earlier [see Eq. (33.12)].

Concepts in Context

EXAMPLE 8 An AM radio signal consists of a plane wave traveling in the $+x$ direction with frequency $f = 1750$ kHz; its oscillating magnetic field is in the $\pm z$ direction and has amplitude 5.0×10^{-10} T. Write down the traveling wave forms (33.34) and (33.36) for the electric and magnetic fields for this wave.

SOLUTION: To express the traveling wave in the forms (33.34) and (33.36), we need the amplitude and wave number. Since B_0 and f are given, we can calculate both the electric field amplitude

$$E_0 = cB_0 = 3.00 \times 10^8 \text{ m/s} \times 5.0 \times 10^{-10} \text{ T} = 0.15 \text{ V/m}$$

and the wave number

$$k = \frac{2\pi}{\lambda} = \frac{2\pi f}{c} = \frac{2\pi \times 1750 \times 10^3\ \text{s}^{-1}}{3.00 \times 10^8\ \text{m/s}} = 3.67 \times 10^{-2}\ \text{m}^{-1}$$

Since the $+x$ direction of propagation is in the $\mathbf{E} \times \mathbf{B}$ direction, for $\mathbf{B}$ in the $+z$ direction, $\mathbf{E}$ must be in the $+y$ direction. Thus the traveling waves may be written in the forms (33.34) and (33.36):

$$E = E_y = (0.15\ \text{V/m}) \times \cos\{(3.67 \times 10^{-2}\ \text{m}^{-1})[x - (3.00 \times 10^8\ \text{m/s})t]\}$$

$$B = B_z = (5.0 \times 10^{-10}\ \text{T}) \times \cos\{(3.67 \times 10^{-2}\ \text{m}^{-1})[x - (3.00 \times 10^8\ \text{m/s})t]\}$$

✔ Checkup 33.6

QUESTION 1: One electromagnetic wave has a wave number k that is twice as large as that of a second electromagnetic wave. How do their wavelengths compare? Their speeds? Their frequencies?

QUESTION 2: The electric field of a traveling electromagnetic wave varies periodically with a wavelength of 300 m. What is the frequency of the corresponding periodic magnetic field?

QUESTION 3: Is the function $E = E_0 x^2/L^2$ a solution of the wave equation? (E_0 and L are constants.)

QUESTION 4: Which of the following functions is *not* a solution of the wave equation? (The quantities k and A and the velocity v are constants.)

(A) $\sin[k(x - vt)]$ (B) $\cos^2[k(x - vt)]$ (C) $A(x + vt)$
(D) $A(x - vt)^2$ (E) $A(x^2 - v^2t^2)$

SUMMARY

PHYSICS IN PRACTICE AM and FM Radio **(page 1089)**

MAXWELL'S EQUATIONS

Gauss' Law $$\oint \mathbf{E} \cdot d\mathbf{A} = \oint E_\perp\, dA = \frac{Q_{\text{inside}}}{\epsilon_0} \quad \textbf{(33.4)}$$

Gauss' Law for magnetism $$\oint \mathbf{B} \cdot d\mathbf{A} = \oint B_\perp\, dA = 0 \quad \textbf{(33.5)}$$

Faraday's Law $$\oint \mathbf{E} \cdot d\mathbf{s} = \oint E_\parallel\, ds = -\frac{d\Phi_B}{dt} \quad \textbf{(33.6)}$$

Maxwell–Ampère Law $$\oint \mathbf{B} \cdot d\mathbf{s} = \oint B_\parallel\, ds = \mu_0 I + \mu_0 \epsilon_0 \frac{d\Phi_E}{dt} \quad \textbf{(33.7)}$$

DISPLACEMENT CURRENT

$$I_{\text{displacement}} = \epsilon_0 \frac{d\Phi_E}{dt}$$

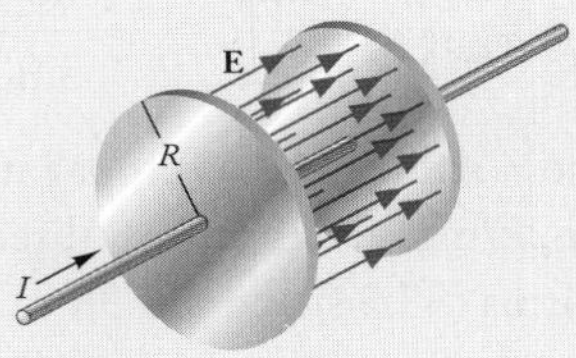

INDUCED MAGNETIC FIELD For axial, cylindrically symmetric electric flux,

$$B \times 2\pi r = \mu_0 \epsilon_0 \frac{d}{dt} E \times [\text{area}]$$ **(Example 1)**

SPEED OF ELECTROMAGNETIC WAVE

$$c = \frac{1}{\sqrt{\mu_0 \epsilon_0}}$$ **(33.8)**

WAVELENGTH λ Distance for one wave oscillation.

FREQUENCY f Number of wave oscillations per second.

WAVELENGTH/FREQUENCY SPEED RELATION

$$\lambda f = c$$ **(33.10)**

MAGNETIC FIELD OF WAVE

$$B = \frac{1}{c} E$$ **(33.12)**

TRANSMITTED INTENSITY THROUGH A POLARIZER

For an unpolarized incident intensity I_0: $I = \frac{1}{2} I_0$ **(Example 4)**

For a polarized incident intensity I_0 (Malus' Law): $I = I_0 \cos^2 \phi$ **(33.15)**

ENERGY DENSITY IN ELECTROMAGNETIC WAVE

$$u = \frac{\epsilon_0}{2} E^2 + \frac{1}{2\mu_0} B^2 = \epsilon_0 E^2$$ **(33.17)**

TIME-AVERAGE ENERGY FLUX (INTENSITY) IN PLANE WAVE

$$\overline{S} = \frac{1}{2\mu_0 c} E_0^2$$ **(33.24)**

POYNTING VECTOR S (propagation direction)

$$\mathbf{S} = \frac{1}{\mu_0} \mathbf{E} \times \mathbf{B}$$ **(33.23)**

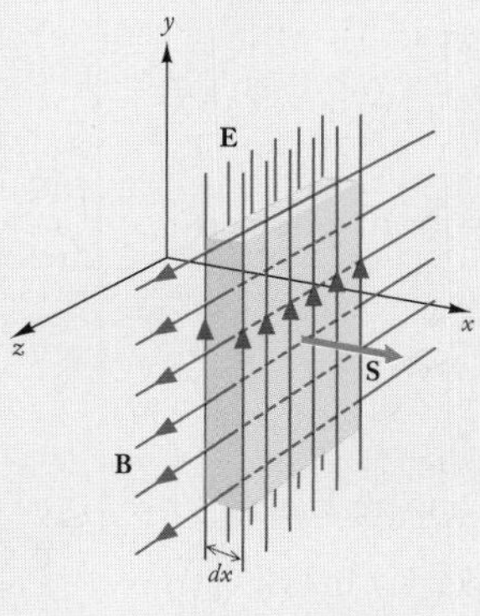

ENERGY FLUX S IN TERMS OF POWER P FOR SPHERICAL WAVE

$$\overline{S} = \frac{\overline{P}}{4\pi r^2} \qquad (33.26)$$

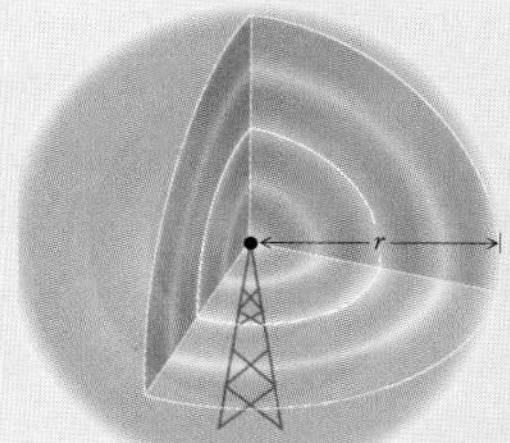

PRESSURE OF ELECTROMAGNETIC WAVE

$$[\text{pressure}] = \begin{cases} \dfrac{S}{c} & \text{(absorption)} \\ \dfrac{2S}{c} & \text{(reflection)} \end{cases} \qquad (33.27, 33.28)$$

WAVE EQUATION

$$\frac{\partial^2 E}{\partial x^2} = \mu_0 \epsilon_0 \frac{\partial^2 E}{\partial t^2} \qquad (33.33)$$

HARMONIC TRAVELING WAVE SOLUTIONS OF THE WAVE EQUATION

$$E = E_0 \cos[k(x - vt)] \qquad (33.34)$$

$$B = B_0 \cos[k(x - vt)] \qquad (33.36)$$

WAVE NUMBER

$$k = \frac{2\pi}{\lambda}$$

QUESTIONS FOR DISCUSSION

1. The displacement current between the plates of a capacitor has the same magnitude as the conduction current in the wires connected to the capacitor and yet the magnetic field produced by the former current near and within the capacitor is much weaker than that produced by the latter current near and within the wire. Explain.
2. Consider the electric field of a single positive electric charge moving at constant velocity. What is the direction of the displacement current intercepted by a circular area perpendicular to the velocity in front of the charge? Behind the charge?
3. Which of Maxwell's equations permits us to deduce the electric Coulomb field of a static charge? Which of Maxwell's equations permits us to deduce the magnetic field of a charge moving with uniform velocity?
4. Suppose that there exist magnetic monopoles, that is, positive and negative magnetic charges that act as sources and sinks of magnetic field lines, analogous to positive and negative electric charges. Which of Maxwell's equations would have to be modified to take into account such monopoles? Qualitatively, what are the required modifications?
5. In practice, the electromagnetic oscillations in a cavity are damped by "frictional" losses. How does this "friction" arise?
6. Efficient waveguides are manufactured out of a very good conductor, such as copper, sometimes with a silver lining. Why is high conductivity essential for high efficiency?
7. Is the transverse electric radiation field E_0 a conservative field?
8. Since the speed of light has now been adopted as the standard of speed and has been assigned the value $2.997\,924\,58 \times 10^8$ m/s *by definition*, why is it still meaningful to compare this number with Maxwell's theoretical prediction for the speed of light?
9. Describe the direction of polarization of the radiation field of an accelerated charge at a few typical points in the space surrounding the charge. Verify that the direction of polarization is never perpendicular to the direction of acceleration.
10. Figure 33.24 is a plot of the sensitivity of the human eye as a function of the wavelength of light. The sensitivity is maximum at about 550 nm and drops to about 1% at 690 nm and at 430 nm. Suppose that the sensitivity of your eye were constant over the entire interval of wavelengths shown in Fig. 33.24. How would this alter your visual perception of some of the things you see in your everyday life?
11. An atom radiates visible light of a wavelength several thousand times longer than the size of the atom. How is this possible?

12. Why does the radio reception fade in the receiver of your automobile when you enter a tunnel?

13. Shortwave radio waves are reflected by the ionosphere of the Earth; this makes them very useful for long-range communications. Explain.

14. In the seventeenth century, the Danish astronomer Ole Roemer noticed that the orbital periods of the moons of Jupiter, as observed from the Earth, exhibit some systematic irregularities; the periods are slightly longer when the Earth is moving away from Jupiter, and slightly shorter when the Earth is moving toward Jupiter. Roemer attributed this apparent irregularity to the finite speed of propagation of light, and exploited it to make the first determination of the speed of light. Explain how one can use the lengthening or the shortening of the observed period to deduce the speed of light. (This shift of period may be regarded as the earliest discovery of a Doppler shift.)

15. It has been proposed that we could eliminate the glare of the headlights of approaching automobiles by covering the windshields and the headlights with sheets of Polaroid. What orientation should we pick for the sheets of Polaroid installed on windshields and on headlights so that the light of every approaching automobile is blocked out, but our own light is not?

16. Suppose you are given a sheet of Polaroid that has no marking identifying its preferential direction. You have available a beam of unpolarized light. How can you determine the preferential direction of the sheet? (Hint: Cut the sheet in two and place one sheet behind the other rotated by 90° around the axis of the beam, so the light is completely blocked. What will happen if you now rotated one sheet about a *transverse axis*, that is, an axis perpendicular to the beam?)

17. The scattered light reaching you from the blue sky is (partially) polarized: if you look straight up, the direction of polarization is perpendicular to the direction of the Sun. How does this polarization arise? [Hint: Consider a beam of (unpolarized) sunlight passing overhead. The electric field of the light waves in this beam accelerates electrons in the molecules of air, and the radiation emitted by these electrons constitutes the scattered light of the sky. Since the direction of acceleration is perpendicular to the incident beam, what can you say about the polarization of the radiation emitted downward, toward you?]

18. A charged particle moving around a circular orbit at uniform speed has a centripetal acceleration and therefore produces a radiation field. However, a uniform current flowing around the circular loop does *not* produce a radiation field. Is this a contradiction? Explain.

PROBLEMS

33.1 Induction of Magnetic Fields; Maxwell's Equations

1. A parallel-plate capacitor of plate area A is being charged by a current I flowing into its plates via external wires. At one instant, the charge on the capacitor plates is Q.
   (a) Assume that the electric field between the plates is uniform. Show that the electric field between the plates at this instant is $E = Q/\epsilon_0 A$.
   (b) Show that the electric flux crossing the mathematical midplane between the plate surfaces is Q/ϵ_0.
   (c) What is the displacement current of this capacitor?
   (d) Show that this displacement current matches the ordinary current I flowing into the plates.

2. Consider the baglike surface shown in Fig. 33.28. If the radius of the mouth of this surface is one-half the radius of the capacitor plates, what is the electric current intercepted by this surface? What is the displacement current intercepted by this surface? What is the sum of the intercepted electric current and displacement current?

FIGURE 33.28 A parallel-plate capacitor with a baglike surface.

3. A parallel-plate capacitor is being charged by a current of 4.0 A.
   (a) What is the displacement current between its plates?
   (b) What is the rate of change of the electric flux intercepted by each plate?

4. A parallel-plate capacitor consists of circular plates of radius 0.30 m separated by a distance of 0.20 cm. The voltage applied to the capacitor is made to increase at a steady rate of 2.0×10^3 V/s. Assume that the electric charge distributes itself uniformly over the plates, and ignore the fringing effects.
   (a) What is the rate of increase of the electric field between the plates?
   (b) What is the displacement current between the plates within a radius of 0.15 m? Within a radius of 0.30 m?
   (c) What is the magnetic field between the plates at a radius of 0.15 m? At 0.30 m?

5. A parallel-plate capacitor has circular plates of radius 20 cm and a uniform electric field between the plates. The capacitor is being charged at a rate of 0.10 A.
   (a) What is the net displacement current between the plates?
   (b) What is the displacement current between the plates within the radial interval of $0 \leq r \leq 5.0$ cm?

6. The electric field in a parallel-plate capacitor with circular plates of radius $R = 5.0$ cm is changing with time t according to $E = Ct^2$, where $C = 5.0 \times 10^4$ V/m·s^2. What is the induced magnetic field at $t = 0.50$ s and $r = 1.0$ cm from the axis? What is the induced magnetic field at $t = 2.0$ s and $r = 6.0$ cm from the axis?

*7. Suppose that the parallel-plate capacitor discussed in Section 33.1 (see Fig. 33.3) is filled with a slab of dielectric with a dielectric constant κ. Assume that, as in the case of the empty capacitor, the displacement current can be viewed as a continuation of the ordinary current. Show that this implies that the displacement current in this filled capacitor must be

$$\kappa\epsilon_0 \frac{d\Phi_E}{dt}$$

and Maxwell's modification of Ampère's Law must be

$$\oint B_{\|}\, ds = \mu_0 I + \kappa\mu_0\epsilon_0 \frac{d\Phi_E}{dt}$$

*8. An emf $\mathcal{E}_{max} \sin \omega t$, with $\mathcal{E}_{max} = 0.50$ V and $\omega = 4.0 \times 10^3$ radians/s, is applied to the terminals of a capacitor with $C = 2.0$ pF. What is the displacement current between the capacitor plates?

*9. The space between the plates of a leaky capacitor is filled with a material of resistance $5.0 \times 10^5\ \Omega$. The capacitor has a capacitance of 2.0×10^{-6} F; its plates are circular, with a radius of 30 cm; and its electric field is uniform. At time $t = 0$, the initial voltage across the capacitor is zero.
   (a) What is the displacement current if we increase the voltage at the steady rate of 1.0×10^3 V/s?
   (b) At what time will the real current leaking through the capacitor equal the displacement current?
   (c) What is the displacement current between the plates within a radius of 20 cm? What is the magnitude of the magnetic field between the plates at radius $r = 20$ cm at $t = 0$? At $t = 1.0$ s? At $t = 2.0$ s?

*10. A parallel-plate capacitor with circular plates of radius 25 cm separated by a distance of 0.15 cm is connected to a source of alternating emf. The voltage across the plates oscillates with an amplitude of 5000 V and a frequency of 60 Hz. What is the amplitude of the magnetic field between the plates at a distance of 20 cm from the axis of the capacitor?

*11. A long, straight wire of radius R and resistivity ρ carries a current which changes direction at $t = 0$ and can be described by the function $I = Ct$, where the rate C is a constant.
   (a) What is the (time-dependent) electric field inside the wire?
   (b) At $t = 0$, what is the induced magnetic field at a distance r from the axis for $r \leq R$?
   (c) What is it for $r \geq R$?

*12. A long solenoid of length l and N turns is wound with wire of cross-sectional area A and resistivity ρ. The wire carries a time-dependent current which changes direction at $t = 0$ and can be described by the function $I = Ct$, where the rate C is a constant. At $t = 0$, what is the induced magnetic field inside the solenoid?

*13. Write Maxwell's equations for a medium of dielectric constant κ.

**14. Suppose that there exist magnetic monopoles, that is, positive and negative magnetic charges that act as sources and sinks of magnetic field lines analogous to positive and negative electric charges. The magnetic field generated by a magnetic charge q_m is an inverse-square field, $B = q_m/4\pi r^2$. Write a new set of Maxwell's equations that take into account the magnetic charge. Be careful with the constants ϵ_0 and μ_0 and with the signs.

33.2 The Electromagnetic Wave Pulse

15. In a collision with an atom, an electron suddenly stops. Describe the directions of the electric and magnetic radiation fields at some distance from the electron at right angles to the acceleration.

16. Draw a diagram analogous to Fig. 33.5, showing the electric field lines of a charge that has suffered an acceleration toward the left (that is, an acceleration opposite to that involved in Fig. 33.5).

17. Suppose that a charge is initially at rest, then is accelerated for a short time interval from $t = 0$ to $t = \tau$, then is decelerated from $t = \tau$ to $t = 2\tau$, and then remains at rest after $t = 2\tau$. Carefully draw a diagram analogous to Fig. 33.5, showing the electric field lines of such a charge, at some time after the deceleration has ended.

18. Figure 33.8 shows the electric and magnetic field lines for a flat wave pulse propagating in the positive x direction. Draw an analogous diagram for a flat wave pulse propagating in the negative x direction.

19. An electromagnetic pulse travels upward from the surface of the Earth. The electric field throughout the pulse points northward. Determine the direction of the magnetic field of the pulse.

20. A sheet of charge in the y–z plane at $x = 0$ is briefly accelerated in the $+y$ direction. As a result, a planar electromagnetic pulse travels outward in the $+x$ and $-x$ directions. Consider a point on the negative x axis. What is the direction of the electric field as the pulse passes that point? The direction of the magnetic field?

*21. Consider the plane wave pulse shown in Figs. 33.8 and 33.9. If the magnitude of the electric field in this pulse is 4.0×10^{-3} V/m, what is the magnitude of the displacement current flowing along the front surface of the pulse per meter of length measured perpendicularly to the current?

†33.3 Plane Waves; Polarization

22. A light-year is defined as the distance that light travels in one year. Calculate how much this is in meters.

23. Laser range finders used by surveyors (see Fig. 33.29) determine the distance traveled to a reflecting target by means of a pulse of laser light, which travels from the range finder to the target and back. The distance is automatically calculated from the travel time of this pulse. If such a range finder is to determine a distance of 100 m to within 1 cm, what is the maximum permitted error in the measurement of the travel time?

FIGURE 33.29 A laser range finder.

24. When the American astronauts on the Moon were in conversation with Mission Control on the Earth, there was a noticeable delay between questions and answers. What is the round-trip travel time for a radio signal from Earth to Moon and back? The distance to the Moon is 3.8×10^{8} m.

25. The mean distance from the Earth to the Sun is 1.50×10^{11} m. How long does it take for light to travel from the Sun to the Earth?

†For help, see Online Concept Tutorial 36 at www.wwnorton.com/physics

26. A transmitter in Los Angeles sends a live broadcast to a listener in Texas; the electromagnetic wave travels 2000 km. How long does it take to travel?

27. A severe limitation on the speed of computation of large electronic computers is imposed by the speed of light, because the electric signals on the connecting wires within the computer are electromagnetic waves ("guided waves"), which travel at a speed roughly equal to the speed of light. If the computer measures about 1.0 m across, what is the minimum travel time required for a typical signal sent from one end of the computer to another? What is the maximum number of signals that can be sent back and forth (sequentially) per second? Is there any way to avoid the limitations imposed by the travel time of signals?

28. (a) One type of antenna for a radio receiver consists of a short piece of straight wire; when the electric field of a radio wave strikes this wire it makes currents flow along it, which are detected and amplified by the receiver. Suppose that the electric field of a radio wave is vertical. What must be the orientation of the wire for maximum sensitivity?
(b) Another type of antenna consists of a circular loop; when the magnetic field of a radio wave strikes this loop it induces a current around it. Suppose that the magnetic field of a radio wave is horizontal. What must be the orientation of the loop for maximum sensitivity?

29. A plane electromagnetic wave travels in the eastward direction. At one instant the electric field at a given point has a magnitude of 0.60 V/m and points down. What are the magnitude and direction of the magnetic field at this instant? Draw a diagram showing the electric field, the magnetic field, and the direction of propagation.

30. Suppose that an unpolarized light beam is incident from the left on the arrangement of two polarizers illustrated in Fig. 33.17. If the intensity of the light emerging on the right is 30% of the incident intensity, what must be the angle between the preferential directions of the polarizers?

31. The preferential directions of two adjacent sheets of Polaroid make an angle of 45°. A beam of polarized light, whose direction of polarization coincides with the preferential direction of the *second* sheet, is incident on the *first* sheet. By what factor is the intensity of the transmitted beam emerging from the second sheet reduced compared with the intensity of the incident beam? Assume that the sheets act as ideal polarizing filters.

32. Two sheets of Polaroid are placed on top of one another. Unpolarized light is perpendicularly incident on the sheets. By what factor is the intensity of the emerging light reduced (relative to the incident light) if the preferential directions of the sheets differ by an angle of 30°? 45°? 60°?

33. Unpolarized light is incident upon the first of two polarizers. The second polarizer has its preferential direction at 80° with respect to the first one. What percentage of the original incident intensity is transmitted?

34. Unpolarized light of intensity I_0 is incident upon the first of three polarizers. The first polarizer has its preferential direc-

tion vertical, the second polarizer has its preferential direction at 60° with respect to vertical, and the third has its preferential direction at 20° with respect to vertical. What is the final transmitted intensity?

35. Polarized light is incident upon the first of two polarizers with its plane of polarization making an angle of 45° with respect to the pass direction of this polarizer. The intensity transmitted through the second polarizer is 30% of the original incident intensity. By what angle do the preferential directions of the two polarizers differ?

*36. (a) Consider two polarizers. The second has its preferential direction at 45° with respect to the first. Polarized light strikes the first polarizer at 45° with respect to its preferential direction. Show that the transmitted intensity is 25% of the original wave. (b) Next, consider three polarizers, each oriented at $90°/3 = 30°$ with respect to the previous one. Polarized light strikes the first polarizer at 30° with respect to its preferential direction. Calculate the percentage of the incident intensity that is transmitted. (c) Finally, consider a large number N of polarizers, each oriented at $90°/N$ with respect to the former. What is the percentage of the incident intensity that is transmitted through all N polarizers for $N = 90$? For $N \rightarrow \infty$?

*37. An electromagnetic wave traveling along the x axis consists of the following superposition of two waves polarized along the y and z directions, respectively:

$$\mathbf{E} = \mathbf{j}E_0 \sin\left(\omega t - \frac{\omega x}{c}\right) + \mathbf{k}E_0 \cos\left(\omega t - \frac{\omega x}{c}\right)$$

where $\omega = kc$. This electromagnetic wave is said to be **circularly polarized**.

(a) Show that the magnitude of the electric field is E_0 at all points of space at all times.

(b) Consider the point $x = y = z = 0$. What is the angle between **E** and the z axis at time $t = 0$? $t = \pi/2\omega$? $t = \pi/\omega$? $t = 3\pi/2\omega$? Draw a diagram showing the y and z axes and the direction of **E** at these times. In a few words, describe the behavior of **E** as a function of time.

*38. If the preferential directions of two adjacent sheets of Polaroid are at right angles, they will completely block a light beam. However, if you insert a third sheet of Polaroid between the other two, then some light will pass through (see Fig. 33.19). A formula for the dependence of the intensity of the transmitted light as a function of the angle that the preferential direction of the inserted sheet makes with that of the first sheet can be obtained from the discussion at the end of Section 33.3. For what orientation of the inserted sheet is the transmitted intensity maximum?

*39. Two sheets of Polaroid are arranged as polarizer and analyzer. Suppose that the preferential direction of the second sheet is rotated by an angle ϕ about the direction of incidence and then rotated by an angle α about the vertical direction (see Fig. 33.30). If unpolarized light of intensity I_0 is incident from the left, what is the intensity of the light emerging on the right?

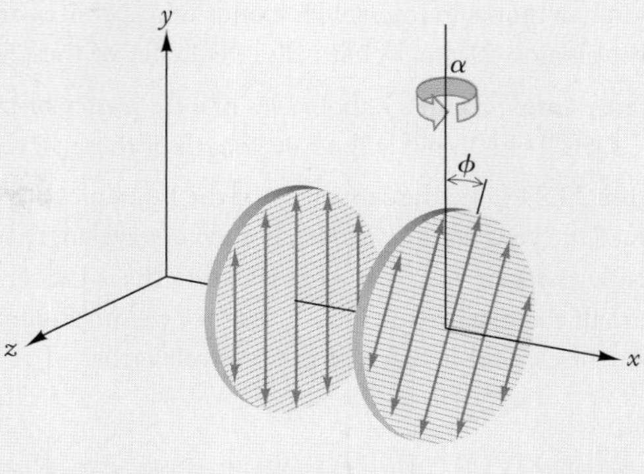

FIGURE 33.30 Two sheets of Polaroid.

33.4 The Generation of Electromagnetic Waves

40. At many coastal locations, radio stations of the National Weather Service transmit continuous weather reports at a frequency of 162.5 MHz. What is the wavelength of these transmissions?

41. The shortest microwaves have a wavelength of about 1.0 mm. What is the frequency of such a wave?

42. Radio station WWV of the National Institute of Standards and Technology, Fort Collins, Colorado, transmits precise time signals at radio frequencies of 2.5, 5.0, 10, 15, and 20 MHz. What are the wavelengths of these transmissions?

43. An ordinary radio receiver, such as found in homes across the country, has an AM dial and an FM dial (Fig. 33.31). The AM dial covers a range from 530 to 1700 kHz and the FM dial a range from 88 to 108 MHz. What is the range of wavelengths for AM? For FM?

FIGURE 33.31 An AM–FM radio.

44. A proton in cyclotron motion moves in a circle in a magnetic field of 8.0 T. Find the wavelength of electromagnetic waves produced at the cyclotron frequency.

45. Find the wavelengths of waves of the following frequencies: (a) 2.0×10^{18} Hz. (b) 3.0×10^{10} Hz. (c) 60 Hz. What part of the electromagnetic spectrum corresponds to each of these frequencies?

46. Hydrogen atoms in interstellar clouds of gas emit radio waves of wavelength 21 cm. What is the frequency of these waves?

47. A radar antenna emits radio waves of a frequency of 1.1×10^{10} Hz. What is the wavelength of these waves?

48. Figure 33.24 gives the sensitivity of the human eye as a function of the wavelength of light. For what wavelength is the sensitivity maximum? For what wavelengths is the sensitivity one-half the maximum? One-quarter of the maximum? What color corresponds to each of these wavelengths?

33.5 Energy of a Wave

49. At one point in an electromagnetic wave, the instantaneous electric field has a magnitude of 80 V/m. What is the energy density?

50. At a distance of several kilometers from a radio transmitter, the electric field of the emitted radio wave has a magnitude of 0.12 V/m at one instant of time. What is the energy density in this electric field? What is the energy density in the magnetic field of the radio wave?

51. A radio wave has an instantaneous magnetic field of 2.0×10^{-10} T. What is the magnitude of the instantaneous Poynting flux?

52. The average energy flux of sunlight incident on the top of the Earth's atmosphere is 1.4×10^3 W/m^2. What are the corresponding amplitudes of oscillation of the electric and magnetic fields?

53. A plane electromagnetic wave travels in the northward direction. At one instant, the electric field at a given point has a magnitude of 0.50 V/m and is in the eastward direction. What are the magnitude and direction of the magnetic field at the given point? What are the magnitude and direction of the Poynting vector?

54. A laser used as a torch, to cut plates of metal (Fig. 33.32), produces a light beam with an average energy flux of 1.0×10^9 W/m^2. What are the magnitudes of the rms electric and magnetic fields in such a light beam?

FIGURE 33.32 Laser used as torch.

55. Starlight arriving at the Earth from the star Capella has an rms energy flux of 1.2×10^{-8} W/m^2. The distance of this star is 4.3×10^{17} m. Calculate the power radiated by this star.

56. The beam of light produced by a small laser is cylindrical, of diameter 2.5 mm. The rms power that the laser feeds into this beam is 1.2 W. Calculate the rms values of the electric and magnetic fields.

57. The Sun emits radiation uniformly in all directions. At the Earth, at a distance of 1.5×10^{11} m, the energy flux of sunlight is 1.4×10^3 W/m^2. Calculate the power radiated by the Sun.

58. The beam of a powerful laser has a diameter of 0.20 cm and carries a power of 6.0 kW. What is the time-average Poynting flux in this beam? What are the rms values of the electric and magnetic fields?

59. A TV transmitter emits a spherical wave, that is, a wave spreading out uniformly in all directions. At a distance of 5.0 km from the transmitter, the amplitude of the wave is 0.22 V/m. What is the magnitude of the time-average Poynting flux at this distance? What is the time-average power emitted by the transmitter?

60. A silicon solar cell (Fig. 33.33) of frontal area 13 cm^2 delivers 0.20 A at 0.45 V when exposed to full sunlight of energy flux 1.0×10^3 W/m^2. What is the efficiency for conversion of light energy into electric energy?

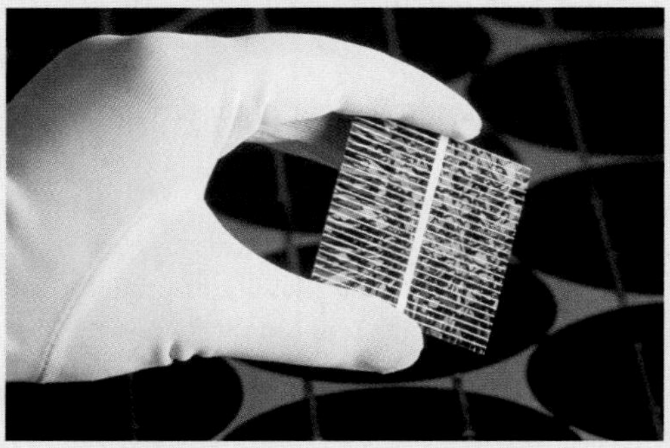

FIGURE 33.33 Solar cells.

61. At a distance of 6.0 km from a radio transmitter, the amplitude of the electric radiation field of the emitted radio wave is $E_0 = 0.13$ V/m. What will be the amplitude of the radio wave when it reaches a distance of 12.0 km? A distance of 18.0 km?

62. A point source, with a power output of 100 W, emits electromagnetic waves uniformly in all directions. At what distance from the source is the amplitude of the electric field 2.0 V/m?

63. A small diode laser pointer emits 5.0 mW of average power in a beam that has a diameter of 1.0 mm. What is the intensity of the beam? What is the amplitude of the electric field in the beam? How much energy is contained in a beam of length 10 m at a given instant?

64. A pulsed laser used for evaporation ("ablation") in materials science has a total energy of 0.50 J in a pulse of ultraviolet light that is 2.0×10^{-8} s long. The beam is 2.0 mm in diameter. What is the power delivered during the pulse? What is the intensity in the pulsed beam? What is the electric field amplitude during the pulse?

65. A point source of light delivers 75 W of average power uniformly in all directions. A circular mirror with a radius of 5.0 cm is placed 2.0 m from the source. The light strikes the mirror (essentially) perpendicularly; the mirror is perfectly reflecting. What is the intensity of light at the mirror? What is the force on the mirror?

66. A radio transmitter broadcasts uniformly in all directions with a power output of 50 kW. What is the time-average Poynting flux 50 km from the transmitter? What is the amplitude of the electric field there?

67. A solar energy "farm" produces 1.0 MW of power for a small town. The solar cells used convert incoming solar energy to electricity with 25% efficiency. Assume that the average solar flux is 500 W/m^2. What total area of solar cells is required?

*68. A magnifying glass of diameter 10 cm focuses sunlight into a spot of diameter 0.50 cm. The energy flux in the sunlight incident on the lens is 0.10 W/cm^2.

(a) What is the energy flux in the focal spot? Assume that all points in the spot receive the same flux.

(b) Will newspaper ignite when placed at the focal spot? Assume that the flux required for ignition is 2.0 W/cm^2.

*69. Binoculars are usually marked with their magnification and lens size (Fig. 33.34). For instance, 7×50 binoculars magnify angles by a factor of 7 and their collecting lenses have an aperture of diameter 50 mm. Your pupil, when dark-adapted, has an aperture of diameter 7.0 mm. When observing a distant pointlike light source at night, by what factor do these binoculars increase the energy flux penetrating your eye? Neglect reflection of light by the lenses.

FIGURE 33.34 Binoculars.

*70. A radio receiver has a sensitivity of 2.0×10^{-4} V/m. At what maximum distance from a radio transmitter emitting a time-average power of 10 kW will this radio receiver still be able to detect a signal? Assume that the transmitter radiates uniformly in all directions.

*71. At normal incidence, the intensity of sunlight at the Earth is 1400 W/m^2. Assume for simplicity that all of the incident energy that strikes the Earth is absorbed. What is the force on the Earth due to radiation pressure from the Sun? Compare this with the gravitational force on the Earth due to the Sun.

*72. A Poynting flux can also be defined for static fields. Consider a long, straight wire of radius r, length l, and resistance R. The wire carries a constant current I uniformly distributed over its cross section. In terms of the given quantities, determine the electric field throughout the wire, the magnetic field at the surface of the wire, and the Poynting flux at a point on the surface of the wire. In what direction is the Poynting vector?

**73. Two plane wave pulses of the kind described in Section 33.3 are traveling in opposite directions. Their polarizations are parallel and the magnitudes of their electric fields are 2.0×10^{-3} V/m.

(a) What are the electric energy density and the magnetic energy density in each pulse?

(b) Suppose that at one instant the two pulses overlap. What are the magnitudes of the electric field and the magnetic field in this superposition?

(c) What are the electric energy density and the magnetic energy density in this superposition?

33.6 The Wave Equation

74. Show that the electric field $E_y = E_0 \sin(kx - \omega t)$ is a solution to the wave equation, Eq. (33.33). In what direction does this wave travel? Write an expression for the magnetic field of this wave.

75. The magnetic field of an electromagnetic wave is described by $B_x = B_0 \cos(kz + \omega t)$. If the frequency of this wave is 3.75 MHz, what is the wavelength? What are the values of k and ω? In what direction does this wave travel? Write an expression for the electric field of this wave.

*76. Show in general that any function f which is a function of the variable $(x - ct)$ is a solution to the wave equation, Eq. (33.33). Then verify this explicitly by checking that the function $E = E_0 \exp[-k^2(x - ct)^2]$ satisfies the wave equation.

*77. Proceeding as in Eqs. (33.29)–(33.31), apply the Maxwell–Ampère law to the horizontal rectangle in Fig. 33.27 to obtain Eq. (33.32).

REVIEW PROBLEMS

78. A capacitor has circular plates, such as illustrated in Fig. 33.3, of radius 0.15 m. Between these plates there is a uniform electric field. Suppose that this electric field is increasing at the rate of 3.8×10^{13} V/m per second. What is the displacement current between the plates? What is the magnitude of the magnetic field at the edge of the capacitor, halfway between the plates?

*79. A parallel-plate capacitor has circular plates of area A separated by a distance d. A thin straight wire of length d lies along the axis of the capacitor and connects the two plates (Fig. 33.35); this wire has a resistance R. The exterior terminals of the plates are connected to a source of alternating emf with a voltage $V = V_0 \sin\omega t$.

(a) What is the current in the thin wire?

(b) What is the displacement current through the capacitor?

(c) What is the current arriving at the outside terminals of the capacitor?

(d) What is the magnetic field between the capacitor plates at a distance r from the axis? Assume that r is less than the radius of the plates.

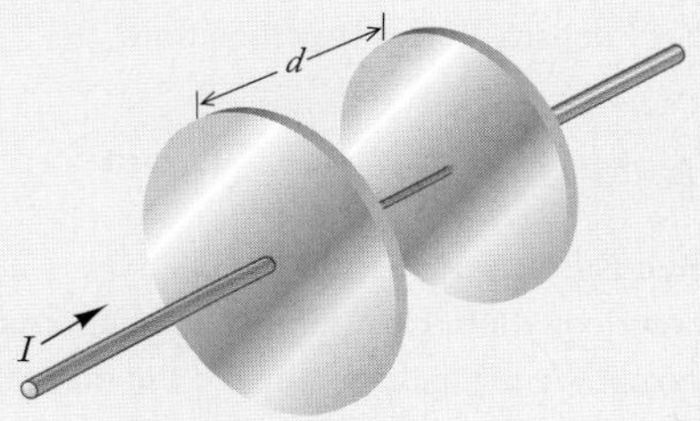

FIGURE 33.35 Parallel-plate capacitor with a thin wire connecting the inside faces of the plates.

80. Sketch the electric field lines for a positive charge that is initially moving at uniform velocity, suddenly stops, and remains stopped.

81. Radio station WWV at Fort Collins, Colorado, continually transmits time signals at several shortwave frequencies. These short waves are reflected by the ionosphere and also by the ground. The waves bounce back and forth between the ionosphere and the ground, and they can therefore roughly follow the curvature of the Earth's surface and travel all the way around the Earth. Suppose that a navigator on a ship in the south of the Indian Ocean, halfway around the Earth, listens to the WWV signals. If he sets his chronometer according to what he hears, roughly how late will he be?

82. Linearly polarized light is incident on a polarizer whose preferential direction is inclined at an angle of 20° relative to the plane of polarization of the incident light. What fraction of the incident light is transmitted? Repeat for an angle of 40°, and repeat for an angle of 60°.

83. A light wave of intensity 0.50 W/m^2 is polarized in the vertical plane. This wave is incident on a polarizer whose preferential direction is inclined at 60° to the vertical, and then on a second polarizer whose preferential direction is vertical. What is the intensity of light that emerges from the second polarizer?

*84. An electromagnetic wave has the form

$$\mathbf{E} = \mathbf{i}E_0 \sin\left(\omega t + \frac{\omega z}{c}\right) + \mathbf{j}2E_0 \sin\left(\omega t + \frac{\omega z}{c}\right)$$

(a) What is the direction of propagation of this wave?

(b) What is the direction of polarization, that is, what angle does the direction of polarization make with the x, y, and z axes?

(c) Write down the formula for the magnetic field of this wave as a function of space and time.

85. An electromagnetic wave is traveling vertically upward. The instantaneous electric field at some point in this wave is eastward, of magnitude 150 V/m. What are the magnitude and the direction of the instantaneous magnetic field?

86. For radio communication with submerged submarines, the U.S. Navy uses ELF (extremely low frequency) radio waves of wavelength 4000 km; such waves can penetrate for some distance below water. What is the frequency of such waves?

87. The beam of light produced by a small laser is cylindrical, of diameter 2.5 mm. The rms power that the laser feeds into this beam is 1.2 W. Calculate the rms values of the electric and magnetic fields.

*88. According to a proposed scheme, solar energy is to be collected by a large power station on a satellite orbiting the Earth. The energy is them to be transmitted down to the surface of the Earth as a beam of microwaves. At the surface of the Earth, the beam is to have a width of about 10 km $\times$ 10 km and is to carry a power of 5.0×10^9 W.

(a) What would be the time-average Poynting flux in this beam?

(b) What would be the amplitude of the electric and magnetic fields in the beam?

89. A radio transmitter emits a time-average power of 5.0 kW in the form of a radio wave with uniform intensity in all directions. What are the amplitudes of the electric and magnetic fields of this radio wave at a distance of 10 km from the transmitter?

*90. A steady current of 12 A flows in a copper wire of radius 0.13 cm.

(a) What is the longitudinal electric field in the wire?

(b) What is the magnetic field at the surface of the wire?

(c) What is the magnitude of the radial Poynting flux at the surface of the wire?

(d) Consider a 1.0-m segment of this wire. According to the Poynting flux, what amount of power flows into this piece of wire from the surrounding space?

(e) Show that the power calculated in part (d) coincides with the power of the Joule heat developed in the 1.0-m segment of wire.

91. The quasar 3C 273 is at a distance of 2.8×10^9 light-years from the Earth. The flux of radio waves reaching the Earth from this quasar is 4.1×10^{-25} W/m^2 in a frequency interval of 1.0 Hz around 1410 Hz (comparable amounts of flux are found at other radio frequencies). What is the power in this 1.0-Hz-wide frequency interval captured by the Arecibo radio telescope, a dish of diameter 300 m? What is the power emitted by the quasar? Assume the quasar radiates uniformly in all directions.

92. In the United States, the accepted standard[3] for the safe maximum level of continuous whole-body exposure to microwave radiation is 10 milliwatts/cm^2.

(a) For this energy flux, what are the corresponding amplitudes of oscillation of the electric and magnetic fields?

(b) Suppose that a man of frontal area 1.0 m^2 completely absorbs microwaves with an intensity of 10 milliwatts/cm^2 incident on this area and that the microwave energy is converted to heat within his body. What is the rate (in calories per second) at which his body develops heat?

*93. At night, the naked, dark-adapted eye can see a star provided the energy reaching the eye is 8.8×10^{-11} W/m^2.

(a) Under these conditions, how many watts of power enter the eye? The diameter of the dark-adapted pupil is 7.0 mm.

(b) Assume that in our neighborhood there are, on the average, 3.5×10^{-3} stars per cubic light-year and that each of these emits the same amount of light as the Sun (3.9×10^{26} W). If so, how far would the faintest visible star be? How many stars could we see in the sky with the naked eye?

[3]It is of interest that in Russia, where many experiments on the effects of microwaves on the human body have been performed, the accepted standard is much lower, 10 microwatts/cm^2.

Answers to Checkups

Checkup 33.1

1. As discussed, the total displacement current $\epsilon_0\, d\Phi_E/dt$ is equivalent to the ordinary current charging the plates, and so is 2 A in either case.

2. No; if the electric field line forms a closed loop, $\oint E_{\|}\, ds \neq 0$ along this line, and by Eq. (33.6), the magnetic flux and the magnetic field must be time-dependent. Similary, if there are no ordinary currents within the closed loop formed by the magnetic field, then Eq. (33.7) implies that such a magnetic field was caused by a time-dependent electric field.

3. No—such field lines form continuous closed loops in space, and thus do not originate on charges.

4. There is an electric field due to the static charges residing on the plates. However, if the amount of charge (and thus the electric field and electric flux) is not changing, then there is no magnetic field between the plates.

5. (C) $r = R$. According to the results of Example 1, for $r \leq R$, the induced magnetic field increases with r; for $r \geq R$, the induced magnetic field decreases with r. The maximum occurs at $r = R$.

Checkup 33.2

1. To determine the direction of the magnetic field, apply the right-hand rule: the direction of **E** × **B** must give the direction of propagation of the wave. Thus for case (A), if the electric field is southward, the magnetic field must be downward for **E** × **B** to be eastward. Similarly, for case (B), for a southward **E** crossed with **B** to give a westward propagation, **B** must be upward. For case (C), **B** must be eastward for **E** × **B** to be upward.

2. Similar to question 1, we need the direction of the E vector which, when crossed with **B**, gives a vector with the given propagation direction. For case (A), **E** must be upward (curling the fingers from upward to southward, the thumb gives the required eastward direction of propagation). Similarly, for case (B), the electric field must be downward. For case (C), a westward **E** crossed with a southward **B** gives an upward direction of propagation.

3. The induction of the electric field by the changing magnetic field is governed by Faraday's Law, $\oint E_{\|}\, ds = -d\Phi_B/dt$. The induced magnetic field is determined by the Maxwell–Ampère Law; for the wave pulse, there is zero real current, so only the displacement-current term contributes, $\oint B_{\|}\, ds = \mu_0\epsilon_0\, d\Phi_E/dt$.

4. (C) Eastward. To determine the direction of the magnetic field, apply the right-hand rule: the direction of **E** × **B** must give the direction of propagation of the wave. Thus if the electric field is downward, the magnetic field must be eastward for **E** × **B** to be southward.

Checkup 33.3

1. The direction of polarization is parallel to the direction of the electric field of the wave or wave pulse; thus, for question 1 of the preceding section, all the cases have southward polariza-

tions. For question 2, the electric fields (and thus the directions of polarization) were in the (A) upward, (B) downward, and (C) westward directions.

2. If the wave of amplitude E_0 is polarized at 45° with respect to the y axis, then the components are equal: $E_y = E_0 \cos 45° = E_0/\sqrt{2}$ and $E_x = E_0 \sin 45° = E_0/\sqrt{2}$.

3. From Malus' Law, the transmitted intensity will be the incident value multiplied by a factor of $\cos^2\phi$. In the first case, this is $\cos^2 45° = \frac{1}{2}$ (intensity reduced by a factor of 2); in the second case we have $\cos^2 60° = \frac{1}{4}$ (intensity reduced by a factor of 4).

4. (A) $I_0/8$. For unpolarized light, a single polarizer reduces the intensity to one-half of the incident value (see Example 4). The second polarizer reduces the polarized light intensity by a factor of $\cos^2\phi = \cos^2(60°) = (1/2)^2 = 1/4$. So the final intensity is $(1/2)I_0 \times (1/4) = I_0/8$.

Checkup 33.4

1. For FM radio waves, the signal is imprinted as variations in frequency, and so is not affected by changes in wave amplitude, whereas AM signals fluctuate with noise and any other amplitude variations (for example, while driving under an overpass); also, regulations permit FM broadcast signals to have a frequency range (20 kHz) twice the AM bandwidth.

2. From Fig. 33.23 or 33.24, we see that light of wavelength 650 nm is red, while light of wavelength 550 nm is green or yellow-green.

3. From Figs. 33.23 and 33.24, we perceive the color yellow over a range of wavelengths near 575–600 nm.

4. By familiarizing ourselves with Fig. 33.23, we see that radar has the shortest wavelength (around 1 cm); most TV stations come next (around 1 m); FM is next (around 3 m); while AM has much longer wavelengths (hundreds of meters).

5. (B) a virus ($\sim 10^{-6}$ m). Visible light has wavelengths in the 400–700 nm range, or about $0.4\text{–}0.7 \times 10^{-6}$ m.

Checkup 33.5

1. The energy densities depend only on the amplitude of the wave, as in Eq. (33.17), so both are equal.

2. The energy density is a maximum whenever the magnitude of the field is a maximum. This happens twice each cycle, when $E = \pm E_0$. Thus the distance between energy density maxima is $\lambda/2$.

3. The energy density is proportional to the square of the amplitude; since the amplitudes have a ratio of 1 to 2, the energy densities have a ratio of 1 to 4.

4. Assuming, as in Example 6, that the wave spreads uniformly in all directions, then for twice the distance the power is spread over an area 4 times as large. Thus the intensity (the time-average energy flux) is one-fourth as large, or $(2.2 \times 10^{-5}\ \text{W/m}^2)/4 = 5.5 \times 10^{-6}\ \text{W/m}^2$. The intensity is proportional to the square of the amplitude, and one-fourth the intensity implies one-half the amplitude, or $(1/2) \times 0.13\ \text{V/m} = 0.065\ \text{V/m}$.

5. By Malus' Law, Eq. 33.15, the fraction of the intensity transmitted is $\cos^2 45° = \frac{1}{2}$. The energy that is not transmitted is absorbed by the polarizer, usually by inducing currents which dissipate the energy as heat.

6. The pressure is proportional to the energy flux, which is proportional to the amplitude squared, so a wave with twice the amplitude exerts 4 times the pressure. However, reflection also produces twice as much pressure as absorption, so overall, the second wave exerts 8 times the pressure of the first.

7. (B) 10 W/m^2. The intensity is the power per unit area at the distance of the book; the 125 W of power is spread over an area $A = 4\pi r^2$ at $r = 1.0$ m, and so the intensity is $I = \overline{P}/A = (125\ \text{W})/[4\pi(1.0\ \text{m})^2] = 10\ \text{W/m}^2$.

Checkup 33.6

1. Since the wave number is proportional to the reciprocal of the wavelength, the first wave will have a wavelength half as large as the second wave. Their speeds will be identical, both equal to c, the speed of light. For equal speeds, frequency is proportional to wave number ($\omega = ck$, or $f = c/\lambda$), so the first wave will have twice the frequency of the second.

2. Both the electric and magnetic fields vary the same way, and both have frequencies given by $f = c/\lambda = (3 \times 10^8\ \text{m/s})/300\ \text{m} = 1 \times 10^6\ \text{s}^{-1} = 1$ MHz.

3. No. This function has no time dependence, and so has no traveling wave character. Mathematically, the second derivative with respect to x is a constant, $2E_0/L^2$, whereas the second derivative with respect to time is zero. Thus this function cannot satisfy Eq. (33.33).

4. (E) $A(x^2 - v^2t^2)$. Each of the other expressions is a function of $(x - vt)$ or $(x + vt)$ which automatically provides a traveling solution; for each function f except (E), direct substitution verifies that $\partial^2 f/\partial x^2 = (1/v^2)\,\partial^2 f/\partial t^2$ is satisfied.

CHAPTER

34

Reflection, Refraction, and Optics

Concepts *in* Context

CONCEPTS IN CONTEXT

The Hubble Space Telescope uses this large concave mirror, with a radius of curvature of 11 m, to image distant galaxies.

With the laws of geometric optics presented in this chapter, we can answer a variety of questions about images formed by the mirror:

? How far from the mirror does the image of a distant galaxy form? (Example 7, page 1129)

? The technician is 4.0 m from the mirror. Where is his image? (Example 10, page 1134)

? The image of the technician is large and upright. How do we calculate such image properties? (Example 13, page 1140)

? How do we measure the magnification of a telescope? (Example 16, page 1151)

So far we have examined the propagation of light waves and other electromagnetic waves only in a vacuum. There, a plane wave will simply propagate in a fixed direction at the constant speed *c*. But if the wave encounters the surface of a region filled with matter—a sheet of metal, a pane of glass, or a layer of water—then the wave will interact with the matter and can suffer a change in speed, direction, intensity, and polarization. One part of the wave will be **reflected** by the surface; that is, it will bounce off. The other part will be **refracted**; that is, it will penetrate into the matter-filled region and continue to propagate, with some change of speed and of direction. You can notice this partial reflection of light by a water surface when you look at the light from streetlamps or other bright sources reflected by the surface of a calm dark pond; the reflected light is not as strong as the direct light. You can notice the refraction of light by looking at your fingers through a drinking glass filled with water; your fingers appear distorted, because the rays of light traveling from your fingers to your eye suffer a change of direction when they enter the drinking glass, and they suffer another change of direction when they leave the drinking glass.

reflection

refraction

Since light is an electromagnetic wave, the changes in speed, direction, intensity, and polarization of the light wave can be calculated from Maxwell's equations, taking into account the motion of the electric charges and the flow of current caused by the action of the wave on the matter. But the complete calculation of all the changes in the wave is rather complicated, and furthermore, Maxwell's equations often tell us more than we want to know. For instance, if a light wave encounters a water surface, we may wish to compute the angle at which it penetrates the water, but we do not always need to know the changes in intensity or polarization.

In this chapter we will see that much can be learned by considering merely one aspect of Maxwell's equations, namely, their implications for the speed of light in matter-filled regions. We will see that in such regions the speed of light is reduced, and that this change of the speed of light leads to the change of direction of propagation during refraction, when the wave penetrates from vacuum into a matter-filled region or when the wave penetrates from one matter-filled region into another.

geometric optics

The laws of reflection and refraction are the basis for **geometric optics**, which we will study in this chapter. Geometric optics relies on the assumption that *light propagates in a fixed direction, along a straight line (rectilinearly), while in a uniform medium; and suffers changes of direction only when it encounters the surface separating two different media.* You can observe this straight-line propagation in the beam of a strong searchlight aimed at the dark sky, or in beams of sunlight piercing through holes in a cloud (Fig. 34.1). But the most dramatic demonstration of straight-line propagation is provided by laser beams, which look like fine straight lines in space (Fig. 34.2).

FIGURE 34.1 Rays of sunlight.

FIGURE 34.2 Beams of laser light.

34.1 HUYGENS' CONSTRUCTION

Online Concept Tutorial 37

The propagation of a light wave or any other electromagnetic wave can be conveniently described by means of the **wave fronts**, or wave crests—that is, the points at which the electric field of the wave has maximum strength at some instant of time. For example, Fig. 34.3 shows the instantaneous wave fronts of the radio wave emitted by a radio station. The interval between one wave front and the next is one wavelength. With the passing of time, each of these wave fronts spreads in the outward direction.

wave front

The rule governing the propagation of wave fronts is **Huygens' Construction**:

Huygens' Construction

To find the change of position of a wave front in a small time interval Δt, draw many small spheres of radius [wave speed] $\times$ Δt with centers on the old wave fronts. The new wave front is the surface that touches the leading edges of these small spheres.

The small spheres employed in this construction are called **wavelets.** Figure 34.4 shows how Huygens' Construction applies to the propagation of the spherical wave fronts in Fig. 34.3. The wave speed in this example is simply c, and hence the radius of the wavelets is $c\,\Delta t$. Erecting wavelets of this radius on the old wave front, we find the new wave front that touches the outer edges of these wavelets; since, in this example, all the wavelets have the same radius, the new wave front is concentric with the old wave front.

wavelet

Figure 34.5 shows a similar construction for the case of the propagation of a plane radio wave or a plane light wave. When we erect the wavelets on the plane wave front, the result is another plane wave front, parallel to the first. This means that the plane continues to propagate in the same direction, without deviating to one side or another. A light beam, such as the light beam from a laser, consists of plane wave fronts, and Huygens' Construction therefore accounts for the straight-line propagation of such a light beam.

Huygens' Construction applies not only to the propagation of light waves in a vacuum, but also to their propagation in any transparent material, such as air, glass, or water. As we will see in the following sections, this construction permits us to derive

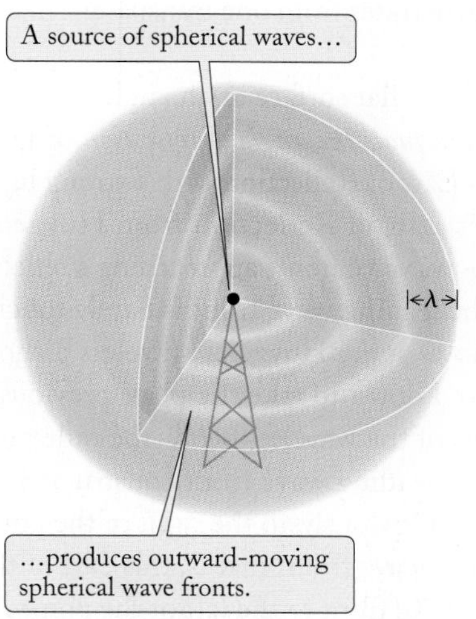

FIGURE 34.3 Spherical wave fronts at one instant of time. At a later time, each of these wave fronts will have moved outward by some distance.

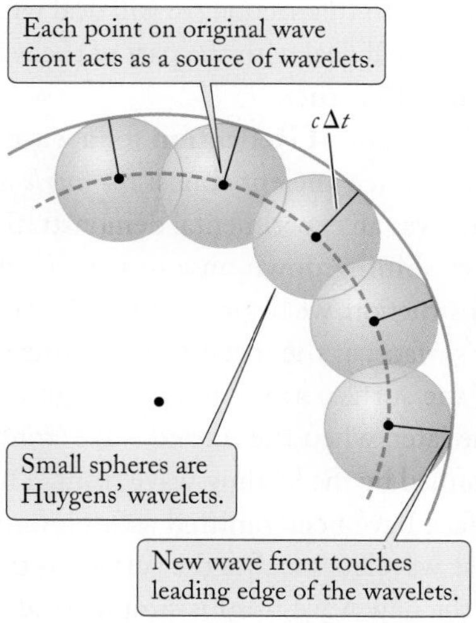

FIGURE 34.4 Huygens' Construction for the propagation of a spherical wave front. The inner (blue dashed) arc shows the wave front at time t; the outer (blue solid) arc shows the propagated wave front at time $t + \Delta t$.

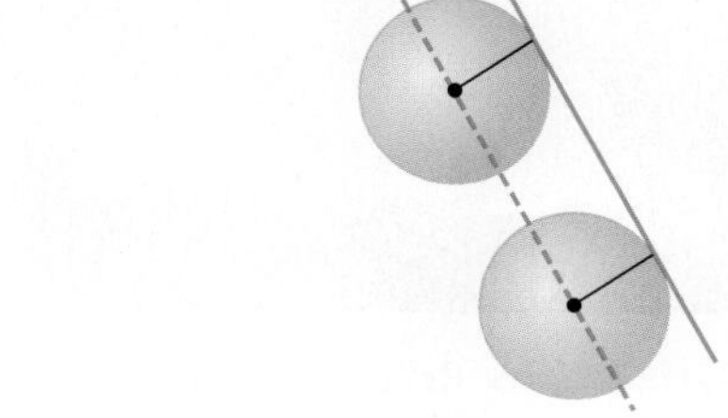

FIGURE 34.5 Huygens' Construction for the propagation of a plane wave front.

the laws of reflection and refraction. Although our emphasis will be on the propagation of light, Huygens' Construction is a general feature of wave propagation—it applies just as well to sound waves and water waves. Apart from differences arising from the different speeds of these waves, the laws of reflection and refraction for all such waves are essentially the same, and many of our results for light waves can be readily generalized to such other waves.

Checkup 34.1

QUESTION 1: Suppose that in Fig. 34.4 the light is propagating *inward*, that is, toward the center of the circle. If the position of the wave front at time t is along the blue dashed arc, what is the propagated wave front at time $t + \Delta t$?

QUESTION 2: Consider the wave front of a water wave on the surface of a pond generated by a brick dropped into the pond with its flat side down. The wave front at the initial instant is then rectangular. According to Huygens' Construction, what is the shape of the wave front a short time Δt later?

QUESTION 3: When a thin stick of length L is dropped onto water (with its length parallel to the water surface), the impact produces a water wave that spreads outward. The wave front soon after the initial instant is long and narrow, like the stick. What is the length-to-width ratio of the wave front when it has traveled a distance L from the stick? When it has traveled a distance much greater than L?

(A) 2, $\frac{3}{2}$ (B) 1, 1 (C) 3, 2 (D) $\frac{3}{2}$, 1 (E) 3, 1

34.2 REFLECTION

Online Concept 37 Tutorial

When a light wave encounters the surface of a transparent material—such as the surface of a pane of glass, or the surface of a pond—part of the wave penetrates the surface and part is reflected. When a light wave encounters the surface of a very smoothly polished metal—such as the silvered surface of a mirror—almost all of the wave is reflected. In this section we will deal with the reflected part of the wave; in the next section we will deal with the part of the wave that penetrates from one transparent material into the other.

The **Law of Reflection** for a wave incident on a flat surface at an angle has been known since ancient times: *the angle of incidence equals the angle of reflection*. Figure 34.6 gives an experimental demonstration of this Law of Reflection with a strong light beam shining down on a mirror. To derive this Law of Reflection from Huygens' Construction, we begin with Fig. 34.7a, which shows wave fronts approaching a reflecting surface; at the instant shown, one edge of the leading wave front is barely touching the surface at the point P. Figure 34.7b shows some Huygens' wavelets a short time later, when the second wave front has moved down to take the place previously occupied by the leading wave front. The portions of the wavelets below the reflecting surface have been omitted as irrelevant. The new leading wave front constructed on these wavelets touches the surface at the point P'. Obviously, to the right of the point P', the new wave front is simply parallel to the old wave front, that is, this part of the wave has not yet been reflected. To find the new wave front to the left of the point P', we draw a straight line that starts at P' and is tangent to the wavelet centered on P. This straight line represents the part of the wave front that has already been reflected. To see that the incident wave front and the reflected wave front make the same angle

FIGURE 34.6 Reflection of a beam of light by a mirror. The angle of incidence equals the angle of reflection.

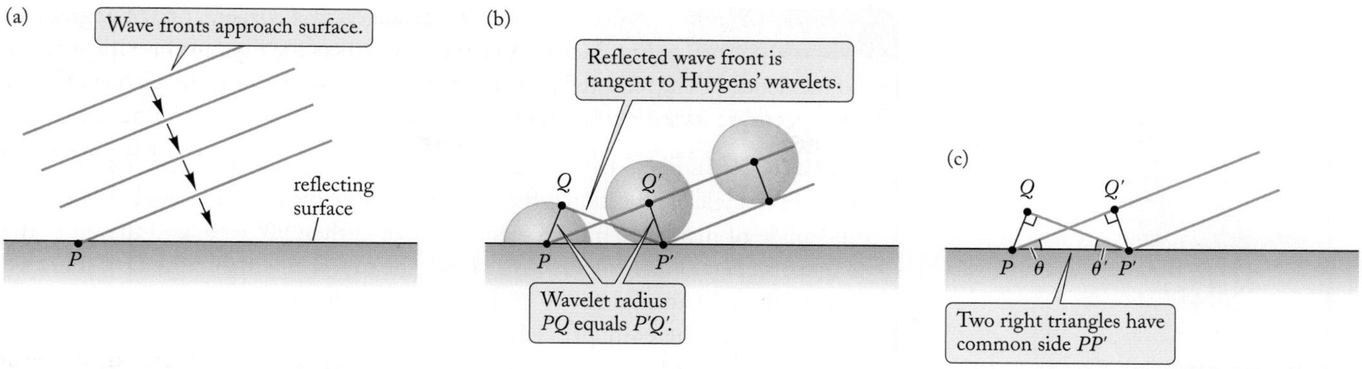

FIGURE 34.7 (a) Wave front approaching a reflecting surface. The leading wave front barely touches the reflecting surface. (b) Huygens' wavelets erected on the leading wave front of part (a). (c) The incident wave front PQ' makes an angle θ with the reflecting surface; the reflected wave front QP' makes an angle θ' with this surface.

with the reflecting surface, we appeal to Fig. 34.7c. The right triangles $PQ'P'$ and $P'QP$ are identical, because they have a common long side (PP') and their short sides (PQ and $P'Q'$) are equal. Hence the angles θ and θ' are equal.

ray

The direction of propagation of a wave is commonly described by the **rays** of the wave. These *rays are lines perpendicular to the wave fronts*. For example, Fig. 34.8 shows the rays associated with the incident and the reflected wave fronts.

angles of incidence and of reflection

The angle θ (or θ') between the wave front and the reflecting surface is equal to the angle between the ray and the perpendicular to the surface. The angles θ and θ' are called the **angles of incidence and of reflection** (see Fig. 34.9). Thus, from the Huygens' Construction, we have deduced that the angle of incidence equals the angle of reflection. This is the Law of Reflection,

Law of Reflection

$$\theta_{\text{incident}} = \theta_{\text{reflected}} \tag{34.1}$$

Note that the incident ray, the reflected ray, and the perpendicular to the surface all lie in the same plane.

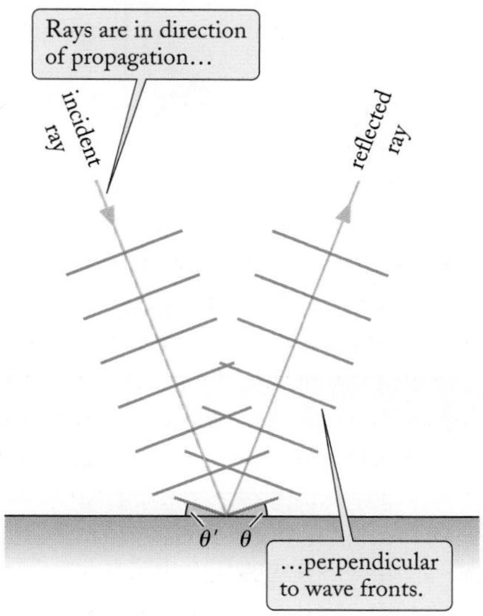

FIGURE 34.8 The rays of the wave are perpendicular to the wave fronts.

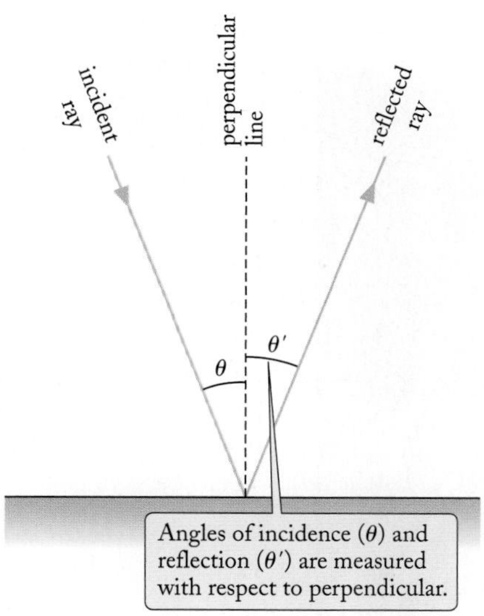

FIGURE 34.9 The angle of incidence θ and the angle of reflection θ'. These angles are the same as in Fig. 34.8.

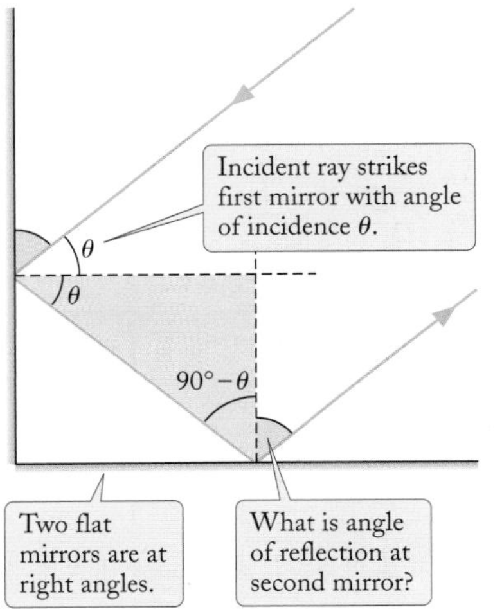

FIGURE 34.10 Two flat mirrors arranged in a corner. The blue angles indicate the directions of the first incident ray and the final reflected ray relative to the surface of the first mirror.

EXAMPLE 1 Two flat mirrors are arranged in a corner, at right angles (see Fig. 34.10). A ray of light strikes the first mirror with an angle of incidence θ. The reflected ray then strikes the second mirror. At what angle does the second reflected ray emerge?

SOLUTION: In Fig. 34.10, the angle of incidence of the ray on the first mirror is θ, and the angle of reflection is the same. Inspecting the colored triangle, we see that the angle of incidence on the second mirror is then $90° - \theta$, and therefore the angle of reflection is also $90° - \theta$. Hence the first incident and the final reflected rays both make an angle of $90° - \theta$ with respect to the surface of the first mirror. These rays are therefore parallel.

COMMENTS: The corner mirrors return the ray in the direction from which it came. This property of reflection by a corner is also valid for a three-dimensional corner, consisting of three mirrors arranged at right angles, called a corner reflector. When a ray is incident on such a corner reflector at some angle, it is returned in the direction from which it came after reflection on the three mirrors. Reflectors on automobiles and bicycles make use of an array of many small corner reflectors (see Fig. 34.11).

When light from some source strikes a flat mirror, the reflection of the light leads to formation of an image of the source. Figure 34.12 shows a point source of light and the rays emerging from it; the figure also shows the reflected rays. If we extrapolate the reflected rays to the far side of the mirror, we find that they *all appear to come from a point source of light placed beyond the mirror*. This apparent point source is the **image**. To an eye looking into the mirror, the image looks like the original source—the eye perceives the mirror image as existing in the space beyond the mirror. But this mirror image is an illusion; the light does not come from beyond the mirror. This kind of illusory image that gives the impression that light rays emerge from where they do not is called a **virtual image**.

image and virtual image

FIGURE 34.11 A reflector on the rear of a bicycle. Each of the small elements is a corner reflector.

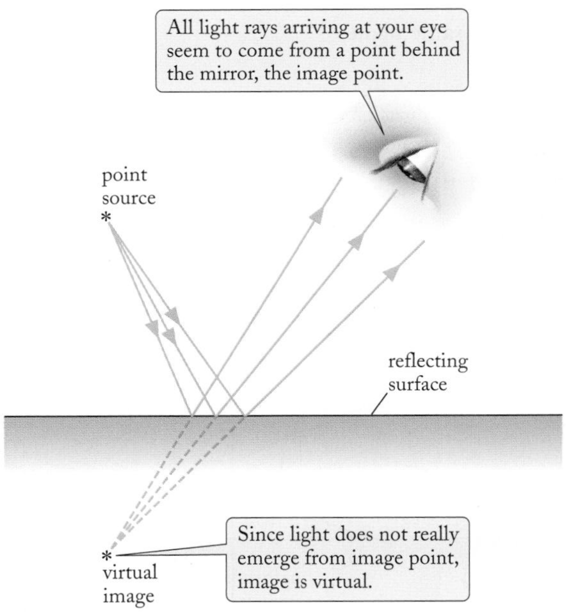

FIGURE 34.12 Rays emerging from a point source are reflected by a mirror. The extrapolated rays (dashed) appear to come from a point source beyond the mirror.

If, instead of a single luminous point, our light source consists of an extended object—such as a book, a hand, or a face—then the mirror image will also be an extended object. If the object is illuminated by sunlight or by a lamp, each point on the surface of the object scatters light and acts as a luminous point. Each such point forms an image in the mirror, and the net result is a mirror image of the entire object. Note that the mirror image of an object is a mirror-reversed object. For instance, Fig. 34.13 shows some written letters and their mirror images. This reversal is commonly referred to as a reversal of left to right. However, it is more accurately described as a reversal of front to back—mirror writing is ordinary writing seen from behind. And the mirror image of, say, a hand facing north is a hand facing south (the reversal is not an ordinary "about face," but involves passing the front of the hand through its back, thereby converting a right hand into a left hand, and vice versa; see Fig. 34.14).

FIGURE 34.13 Some letters and their images in a mirror.

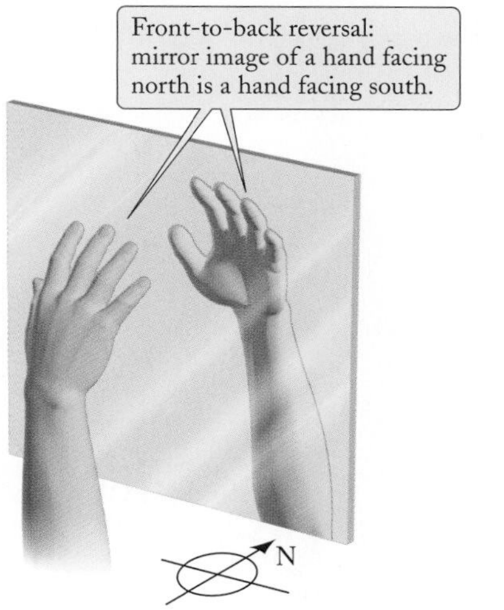

FIGURE 34.14 A hand facing north and its image in a mirror. The north side of the hand (the palm) becomes the south side in the mirror.

✔ Checkup 34.2

QUESTION 1: If a ray makes an angle of 20° with the surface of a mirror, what is the angle of incidence?

QUESTION 2: If the angle of incidence on a flat mirror is 0°, what is the angle of reflection? What happens to the ray?

QUESTION 3: Consider Fig. 34.6. The image of the flashlight is not visible in the photograph. Where is the image?

QUESTION 4: You stand 2 m from a door mirror. What is the distance from you to your image?

QUESTION 5: You run toward a door mirror at 8 m/s. What is the speed of your image relative to you?

QUESTION 6: What is the length of a mirror needed for a person of height h to see his or her reflection from head to toe?

(A) $2h$ (B) $\frac{3}{2}h$ (C) h (D) $\frac{2}{3}h$ (E) $\frac{1}{2}h$

34.3 REFRACTION

Online Concept 37 Tutorial

The speed of light in a transparent material—such as air, water, or glass—differs from the speed of light in vacuum. We can recognize this immediately by recalling the theoretical formula for the speed of light derived from Maxwell's equations, Eq. (33.8):

$$c = \frac{1}{\sqrt{\epsilon_0 \mu_0}} \tag{34.2}$$

We know from Chapter 26 that in a material with given dielectric characteristics, the quantity ϵ_0 in Maxwell's equations must be replaced by $\kappa\epsilon_0$, where κ is the dielectric constant of the material [see Eq. (26.24)].[1] Hence, the formula (34.2) for the speed of light likewise must be replaced by a new formula for the speed,

$$v = \frac{1}{\sqrt{\kappa\epsilon_0 \mu_0}} \tag{34.3}$$

[1]In a material with magnetic properties, μ_0 must also be multiplied by a factor of the relative magnetic permeability (see Section 30.4); but this factor is very near 1 except in ferromagnetic materials. We will therefore ignore any correction factor of μ_0.

This is usually written as

speed of light in transparent material

$$v = \frac{c}{n} \tag{34.4}$$

where $c = 1/\sqrt{\epsilon_0 \mu_0}$ is the standard speed of light in vacuum and

$$n = \sqrt{\kappa} \tag{34.5}$$

index of refraction

The quantity n is called the **index of refraction** of the material. The index of refraction of any material is larger than 1; according to Eq. (34.4), *the speed of light in the material is less than the speed of light in vacuum.* In connection with Eq. (34.5), it is important to keep in mind that the value of the dielectric constant depends on the frequency of the electric field. Hence, the values of the dielectric constants from Table 26.1 cannot be inserted into Eq. (34.5), because the former values apply only to static fields, whereas we are now concerned with the high-frequency fields of a light wave.

Table 34.1 gives the values of the indices of refraction of a few materials. For instance, water has $n = 1.33$, and the speed of light in water is

$$v = \frac{c}{n} = \frac{c}{1.33} = \frac{3.00 \times 10^8 \text{ m/s}}{1.33} = 2.26 \times 10^8 \text{ m/s}$$

The values in the table apply to light waves of medium frequency (yellow-green light). The index of refraction is slightly larger for blue light and slightly smaller for red light; we will deal with this complication later in this section. Note that the index of refraction of air is very close to 1; consequently, in most calculations we can ignore the distinction between air and vacuum.

With the wave speed $v = c/n$, the relation between frequency and wavelength becomes [see Eq. (33.10)]

$$\lambda f = v = \frac{c}{n}$$

or

$$\lambda = \frac{1}{n}\frac{c}{f} \tag{34.6}$$

Since c/f is the wavelength that the wave would have in vacuum [see Eq. (33.10)], we can also write Eq. (34.6) as

wavelength in transparent material

$$\lambda = \frac{\lambda_{\text{vac}}}{n} \tag{34.7}$$

Thus the wavelength of an electromagnetic wave is shorter in a material than in empty space. For example, if a wave penetrates from vacuum or from air into water, where $n =$ 1.33, its speed is reduced by a factor of 1.33, from 3.00×10^8 m/s to 2.26×10^8 m/s, but its frequency remains constant. Consequently, Eq. (34.6) shows that its wavelength will be reduced by a factor of 1.33 (see Fig. 34.15). The fact that the frequency of the wave remains constant can be understood in terms of the atomic mechanism underlying the interaction of the wave with the material. When the wave strikes the water surface, it shakes the electrons of the water molecules; this acceleration of electric charges produces extra waves, which combine with the original wave and result in a slowed-down wave. The frequency of the combined, slowed wave is the same as that of the orig-

TABLE 34.1 INDICES OF REFRACTION OF SOME MATERIALS[a]

MATERIAL	n
Air, 1 atm, 0°C	1.000 29
1 atm, 15°C	1.000 28
1 atm, 30°C	1.000 26
Water	1.33
Ethyl alcohol	1.36
Castor oil	1.48
Quartz, fused	1.46
Glass, crown	1.52
light flint	1.58
heavy flint	1.66

[a]For light of wavelength ≈ 550 nm.

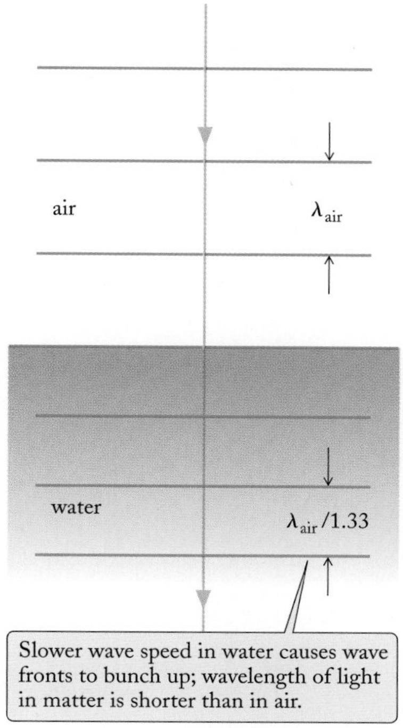

FIGURE 34.15 Change of wavelength of a light wave as it penetrates from air into water. The wavelength in water is shorter by a factor of $n = 1.33$.

inal wave, because the shaking of the electrons proceeds at the original frequency, and the extra waves produced by the electrons are therefore also of the same frequency as the original wave. But the slower speed results in a decreased wavelength, like the decreased spacing of cars when encountering a slowdown in traffic on a highway.

EXAMPLE 2 A light wave of wavelength 550 nm in vacuum enters a plate of glass of index of refraction $n = 1.52$. What is the speed of the light in the glass? What is the wavelength of the light in the glass? What is the frequency of the wave in the glass?

SOLUTION: The speed of light in the glass is

$$v = \frac{c}{n} = \frac{3.00 \times 10^8 \text{ m/s}}{1.52} = 1.97 \times 10^8 \text{ m/s}$$

The wavelength in the glass is

$$\lambda = \frac{\lambda_{\text{vac}}}{n} = \frac{550 \text{ nm}}{1.52} = 362 \text{ nm}$$

The frequency in the glass is the same as the frequency in vacuum,

$$f = \frac{c}{\lambda_{\text{vac}}} = \frac{3.00 \times 10^8 \text{ m/s}}{550 \times 10^{-9} \text{ m}} = 5.45 \times 10^{14} \text{ Hz}$$

Alternatively, we can calculate this frequency from the speed and the wavelength in the glass, with the same result:

$$f = \frac{v}{\lambda} = \frac{1.97 \times 10^8 \text{ m/s}}{362 \times 10^{-9} \text{ m}} = 5.45 \times 10^{14} \text{ Hz}$$

Incidentally: Equation (34.6) does not imply that a light source changes color when immersed in water. The *color* we perceive depends on the *frequency* of the light reaching our eyes, and this frequency is independent of whether the light source, our eyes, or both are immersed in air or water.

When a wave strikes the surface of a transparent material, part of it is reflected and part of it penetrates into the material. In the preceding section we investigated the direction of propagation of the reflected wave; now let us investigate the penetrating wave. Again, we will use Huygens' Construction to find out what the wave does when its strikes the surface of the material. In vacuum the speed of light is c; in the material it is c/n. Figure 34.16a shows the approaching wave fronts at one instant of time. The left edge of one wave front barely touches the surface. Figure 34.16a also shows the Huygens' wavelets that determine the position of this wave front at a later time; only the forward portions of these wavelets are relevant. Above the surface, the wavelet has a radius $c\,\Delta t$; below the surface, in the material, the wavelet has a smaller radius $(c/n)\Delta t$. The reduction of the speed of propagation of one side of the wave front causes the wave front to swing around, changing its direction of advance. This is analogous to the change of the direction of advance of a row of marching soldiers when the soldiers on one edge slow down by taking short steps while the soldiers on the other edge take longer steps.

refraction

This change of direction is called **refraction**. To determine the angle of refraction when light passes from one material into another, we need to examine the triangles in Fig. 34.16b. The right triangles $PP'Q$ and $PP'Q'$ have the long side PP' in common. In terms of the length PP' of this common side, the sines of the angles between the wave fronts and the surface are

$$\sin\theta = \frac{c\,\Delta t}{PP'} \tag{34.8}$$

and

$$\sin\theta' = \frac{(c/n)\Delta t}{PP'} \tag{34.9}$$

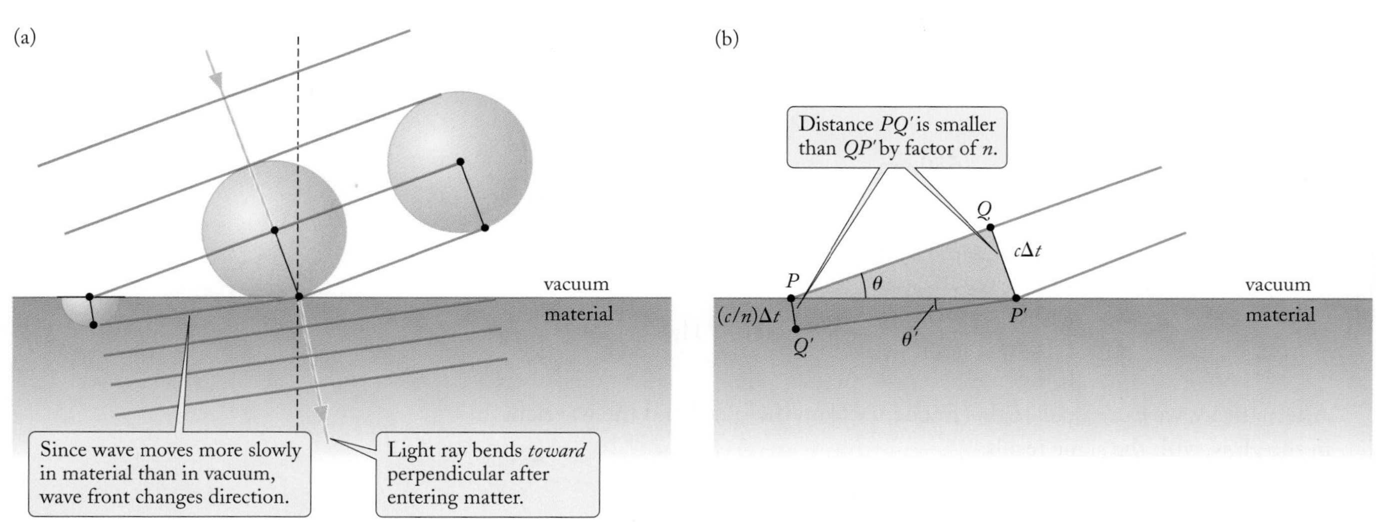

FIGURE 34.16 (a) Huygens' wavelets erected on a wave from whose edge barely touches the dielectric surface. (b) The incident wave front makes an angle θ with the dielectric surface; the refracted wave front makes an angle θ'.

The ratio of the sines is therefore

$$\frac{\sin\theta}{\sin\theta'} = \frac{c}{c/n}$$

from which

$$\sin\theta = n\sin\theta' \qquad (34.10)$$

This equation describes the change of direction of the wave upon penetration into a material. Equation (34.10) is called the **Law of Refraction**, or **Snell's Law**. The angle θ is the angle of incidence, and θ' is the angle of refraction. It is usually more convenient to measure these angles between the two rays and the perpendicular to the surface; all three lie in the same plane. Figure 34.17 shows the incident and refracted rays, and the angles of incidence and refraction. As can be seen in this figure, the ray in the material is bent toward the perpendicular (θ' is smaller than θ).

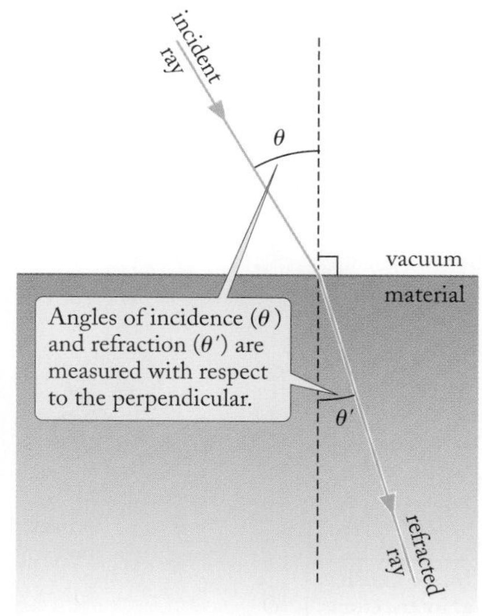

FIGURE 34.17 The angle of incidence θ and the angle of refraction θ'.

Note that we can also use the formula (34.10) for a ray of light that emerges from the material into vacuum. The formula is valid for the ray shown in Fig. 34.17 and also for the ray propagating in the reverse direction, provided we always assume that θ is the angle in vacuum and θ' is the angle in the material, regardless of the direction of propagation.

Our formula (34.10) describing refraction at the interface between a vacuum and a material is a special case of a general formula describing refraction at the interface of two different material media. If the indices of refraction are n_1 and n_2 and the angles between the rays and the perpendicular are θ_1 and θ_2, respectively, then

$$n_1\sin\theta_1 = n_2\sin\theta_2 \qquad (34.11)$$

Law of Refraction (Snell's Law)

EXAMPLE 3 A ray of light enters a thick plate of glass of index of refraction $n = 1.52$ at an angle of incidence of 45° (see Fig. 34.18). (a) What is the angle of refraction of the ray at the upper surface of the glass? (b) When the ray reaches the lower surface of the glass, it is refracted again and it emerges into air. What is the angle at which it emerges?

SOLUTION: (a) With $n = 1.52$ and $\theta = 45°$, the Law of Refraction gives us

$$\sin 45° = 1.52\sin\theta' \qquad (34.12)$$

from which

$$\sin\theta' = \frac{\sin 45°}{1.52} = \frac{0.707}{1.52} = 0.465 \qquad (34.13)$$

With our calculator we then find that the angle of refraction is

$$\theta' = 28°$$

(b) Within the glass, the ray travels at 28° with respect to the perpendicular, and since the two surfaces of a plate of glass are parallel, this will be the angle of incidence at the lower surface. Hence refraction at the lower surface (where the ray proceeds from glass into air) is simply the reverse of the refraction at the upper surface (where the ray had proceeded from air into glass). The angle at which the ray emerges is therefore 45°. Note that the final ray, although shifted, is parallel to the initial ray.

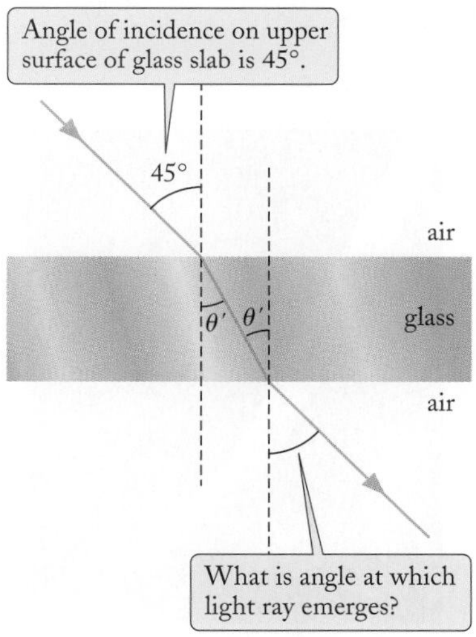

FIGURE 34.18 A ray of light passes through a thick plate of glass.

COMMENTS: This result can also be obtained from the Law of Refraction. We can apply Eq. (34.10) to a ray emerging from glass, provided we assume that θ' in this formula is the angle in the glass and θ is the angle in air; thus

$$\sin\theta = 1.52 \sin 28° \tag{34.14}$$

This again leads to the result $\theta = 45°$.

EXAMPLE 4 A small, shiny fish is in the water 1.0 m below the surface. Where does an angler looking downward into the water see the fish; that is, where is the image of the fish? Assume that the angler's eye is almost directly above the fish.

SOLUTION: Figure 34.19 shows two light rays from the fish to the eye of the angler. The first light ray is perpendicular to the surface and is not bent. The second light ray is bent away from the perpendicular when it emerges from the water. With the assumption that θ' is the angle in water and θ the angle in air, Eq. (34.10) can be applied to this ray emerging from the water. Since the angles are small when the fish is viewed from almost directly above, we can use the familiar approximation that the sine of the angle is approximately equal to the angle expressed in radians that is, $\sin\theta \approx \theta$ and $\sin\theta' \approx \theta'$. Hence Eq. (34.10) becomes approximately

$$\theta \approx n\theta' = 1.33\theta'$$

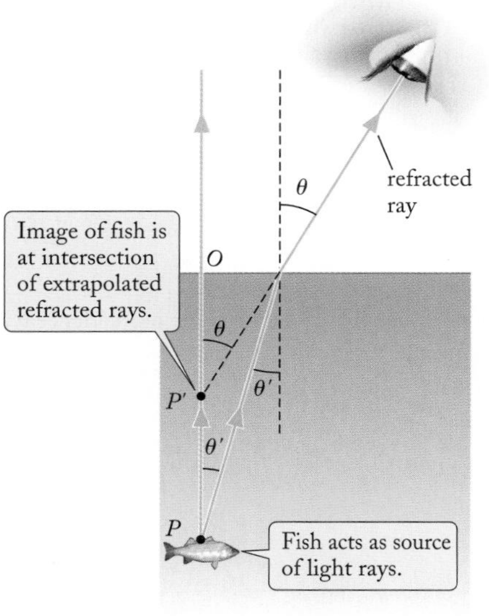

FIGURE 34.19 A small, shiny fish as a source of light. Note that the direction of propagation of the light is opposite to that shown in Fig. 34.17. This does not affect the validity of Eq. (34.10). The extrapolated ray (dashed) appears to come from the point P'. The angles shown are exaggerated for clarity; the location of the image of the fish is independent of angle only for small angles; that is, only when viewed from almost directly above.

where we have used the value $n = 1.33$ for water from Table 34.1. When the eye extrapolates the refracted ray back into the water, it seems to intersect the vertical ray at the point P', above the point P. Hence the image is above the fish. The image distance OP' and the fish distance OP are related as follows (see Fig. 34.19):

$$OP \tan\theta' = OP' \tan\theta$$

For small angles, we can make the approximations $\tan\theta \approx \theta$ and $\tan\theta' \approx \theta'$. Hence

$$\frac{OP'}{OP} \approx \frac{\theta'}{\theta} \approx \frac{1}{1.33}$$

This shows that the image depth is smaller than the fish depth by a factor of 1.33, the index of refraction. For instance, if $OP = 1.0$ m, then $OP' = 0.75$ m. The fish seems to be nearer to the surface than it is (if the angler wants to touch the fish, she must immerse her hand down deeper than where the fish *seems* to be). This smaller **apparent depth** occurs for any flat interface:

$$[\text{apparent depth}] = \frac{[\text{actual depth}]}{n} \tag{34.15}$$

COMMENTS: The apparent bending that you perceive when you look at a straight rod partially immersed in water results from this apparent shrinking of the vertical distance. All the immersed portions of the rod seem to be nearer the water surface than they are, and the rod seems to have a kink at the place where it enters the water (see Fig. 34.20).

When the eye looking into the water is not nearly above the rod, that is, when the angles θ and θ' are not small, there is additional distortion and an exaggeration of the apparent bending.

FIGURE 34.20 A drinking straw partially immersed in water appears bent.

For a ray attempting to leave water, there is a critical angle beyond which refraction is impossible. As the light ray emerges into air, it is bent *away* from the vertical; in the extreme case, it is bent so much that it lies almost along the water surface (see Fig. 34.21); this extreme case corresponds to $\theta = 90°$ in Eq. (34.10):

$$n \sin\theta' = \sin 90° = 1$$

The critical angle for this extreme form of refraction is therefore given by

$$\sin\theta_{crit} = \frac{1}{n} \qquad (34.16)$$

critical angle for total internal reflection

For water, with $n = 1.33$, we then find

$$\theta_{crit} = 49°$$

If a light ray strikes a water surface from below with an angle of incidence larger than this, refraction is impossible. The only alternative is reflection—the water surface behaves as a perfect mirror. This phenomenon is called **total internal reflection**. It can occur whenever the index of refraction of the medium containing the light ray is larger than the index of refraction of the adjacent medium.

Total internal reflection has many important practical applications in optics. For instance, in periscopes the light is reflected down the tube by internal reflection in a prism (see Fig. 34.22); because the reflection is total, without any loss of light, this gives a much better image than reflection in a mirror. Such reflecting prisms are also used in binocular telescopes to reflect the path of the rays of light back and forth, and increase the effective length of the telescope.

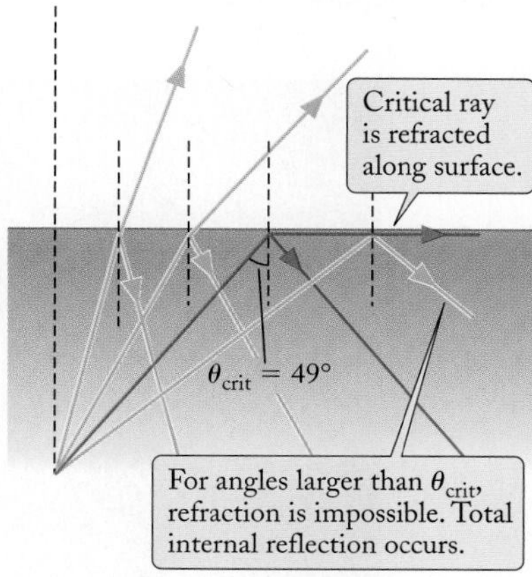

FIGURE 34.21 A ray approaching a water surface from below with an angle of incidence $\theta = 49°$ is refracted along the water surface.

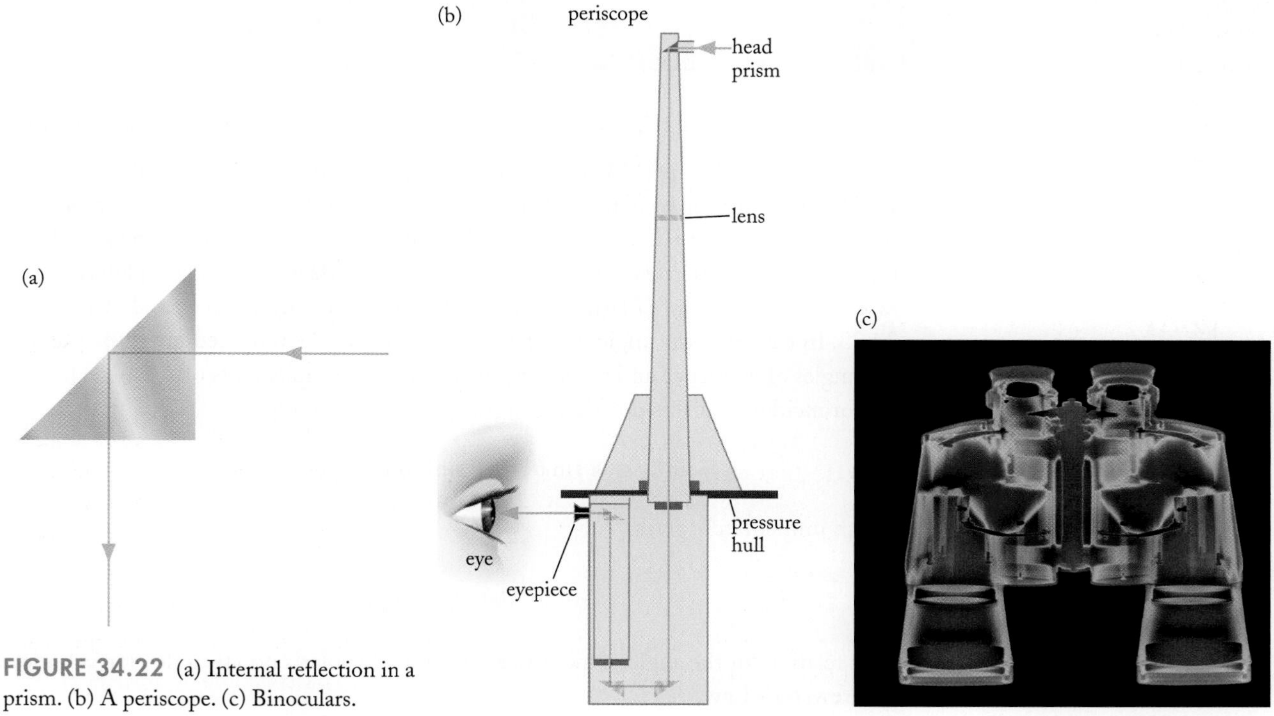

FIGURE 34.22 (a) Internal reflection in a prism. (b) A periscope. (c) Binoculars.

PHYSICS IN PRACTICE — OPTICAL FIBERS

In an optical fiber, light moves along a thin rod made of transparent material; the light zigzags back and forth between the walls of the rod, undergoing a sequence of total internal reflections—the fiber acts as a pipe for light (see Fig. 1). Such optical fibers are being used to replace telephone and other communication cables. The electrical pulses normally carried on a wire cable are converted into pulses of infrared laser light, which can be transmitted in an optical fiber. The efficiency of optical telephone lines is very high because an optical fiber can carry many telephone conversations simultaneously; in modern telephone systems, single optical fibers are being used to carry several hundred telephone conversations simultaneously.

Optical fibers also find application in flexible endoscopes, used by physicians to examine the interior of the intestine and the stomach. This device is a bundle of optical fibers, sometimes over a meter in length, one end of which is inserted into the patients stomach or intestine while the physician looks into the other end. A small lightbulb at the leading end provides illumination (see Fig. 2).

FIGURE 1 Internal reflection in an optical fiber. Light enters the optical fiber at the bottom, travels around several loops of the fiber, and emerges at the top.

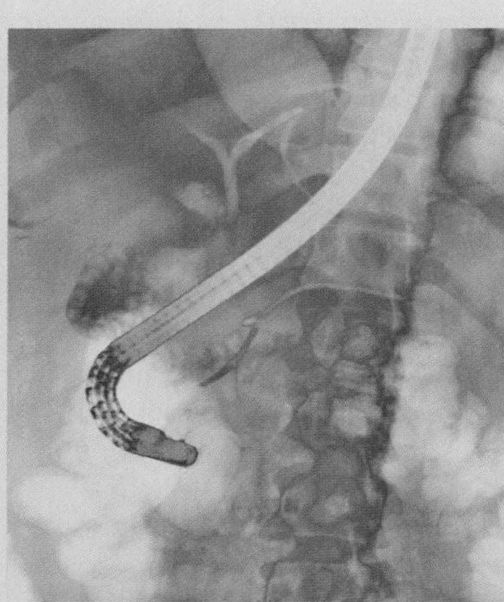

FIGURE 2 Endoscope inside patient.

As we saw in Chapter 33, ordinary light is unpolarized. An experimental observation, which can be derived by detailed analysis of Maxwell's equations, is that *reflection causes partial to complete polarization of light*. In the special case where the reflected and refracted rays are perpendicular to each other, the reflected ray is completely polarized, with the plane of polarization parallel to the reflecting surface, as shown in Fig. 34.23. In this case, the angle between the surface and the refracted ray is the same as the angles of incidence and reflection, since both are complements of the angle labeled ϕ. For incidence at this polarizing angle θ_p, Snell's Law implies

$$\sin\theta_p = n \sin\theta' = n \sin(90^\circ - \theta_p) = n \cos\theta_p$$

which immediately gives

Brewster's Law

$$\tan\theta_p = n \qquad (34.17)$$

This relation for the angle of incidence for complete polarizaton by reflection is known as **Brewster's Law**.

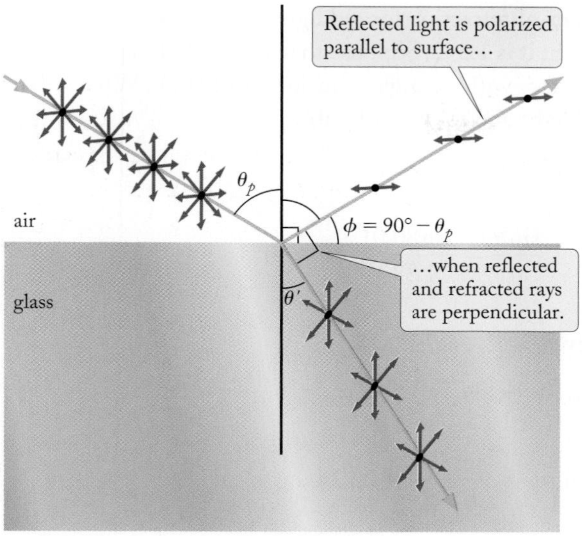

FIGURE 34.23 Polarization by reflection. The incident light can have any polarization, but when incidence is at Brewster's angle, the reflected light is polarized in a plane parallel to the surface.

FIGURE 34.24 A view through polarizing sunglasses (left) reduces reflected glare compared with unpolarized sunglasses (right).

Recall from Section 33.3 that the glare of reflected light, with its partial or complete horizontal polarization, can be attenuated by using polarizing sunglasses, which only pass light polarized in the vertical plane (see Fig. 34.24).

In most materials, the index of refraction depends somewhat on the wavelength of the light. Usually, the index of refraction increases as the wavelength decreases. For instance, Fig. 34.25 displays a plot of the index of refraction of light in water over a range of wavelengths (the wavelengths plotted along the horizontal axis of Fig. 34.25 are measured in air, before the light penetrates the water). When a ray of light containing several wavelengths, or colors, is refracted by a material with an index of refraction that depends on wavelength, the refracted rays of different colors will emerge at somewhat different angles. The separation of a ray by refraction into distinct rays of different colors is called **dispersion**.

dispersion

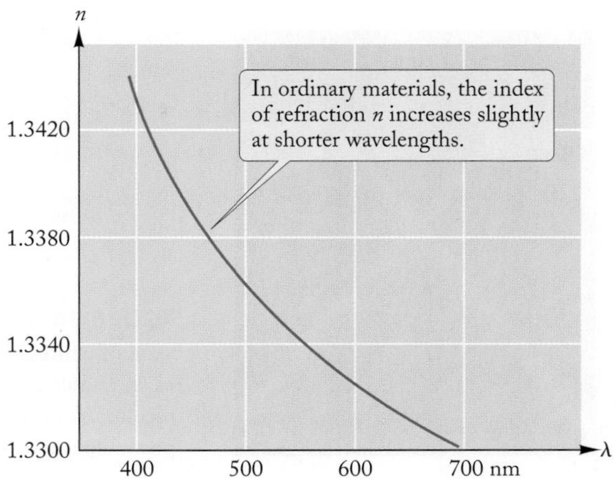

FIGURE 34.25 Index of refraction of light in water as a function of wavelength. The index of refraction varies by about 1% over the range of visible wavelengths.

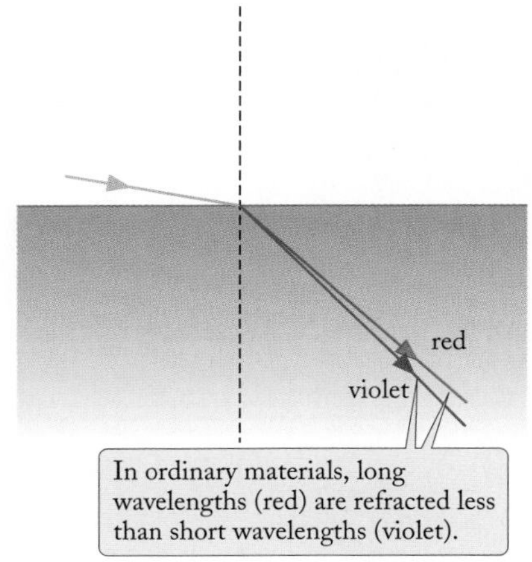

FIGURE 34.26 Refraction of red and of violet light in water. The ray of violet light is refracted more than the ray of red light. The difference between the angles of the refracted rays has been exaggerated for the sake of clarity.

EXAMPLE 5 The index of refraction for red light in water is 1.330, and for violet light it is 1.342. Suppose that a ray of light approaches a water surface with an angle of incidence of 80.0°. What are the angles of refraction for red light and for violet light?

SOLUTION: For $n = 1.330$, Eq. (34.10) yields

$$\sin\theta' = \frac{\sin\theta}{n} = \frac{\sin 80.0^\circ}{1.330} = 0.740$$

for which our calculator gives us $\theta' = 47.7^\circ$.

For $n = 1.342$, Eq. (34.10) yields

$$\sin\theta' = \frac{\sin 80^\circ}{1.342} = 0.734$$

and $\theta' = 47.2^\circ$. The violet light is bent more toward the vertical than the red light (see Fig. 34.26). For $\theta = 80^\circ$, the difference in the angles of refraction 0.5°. Thus, refraction in water separates light rays according to colors. A beautiful demonstration of this effect is found in rainbows, which are produced by the refraction of sunlight in water droplets (sese Problem 47).

prism

A **prism** is the traditional device employed for the separation of light rays into their constituent colors. The basic mechanism is the same as discussed in Example 5: the glass in the prism has slightly different indices of refraction for light of different wavelengths, and hence it bends rays of different colors by different amounts (see Fig. 34.27). In passing through a prism, light is refracted twice, first when it enters the glass, and then when it leaves the glass. Under normal operating conditions, a good prism will introduce a difference of several degrees between the angular directions of the emerging red and and violet rays.

spectrum

The pattern of colors produced by the analysis of light by means of a prism is called the **spectrum** of the light. The white light emitted by the Sun has a continuous spectrum consisting of a blend of all the colors (see Fig. 34.28). The colored light emitted by the atoms of a chemical element in an electric discharge tube, such as a neon tube, has a discrete spectrum consisting of just a few sharp colors. Each of these discrete

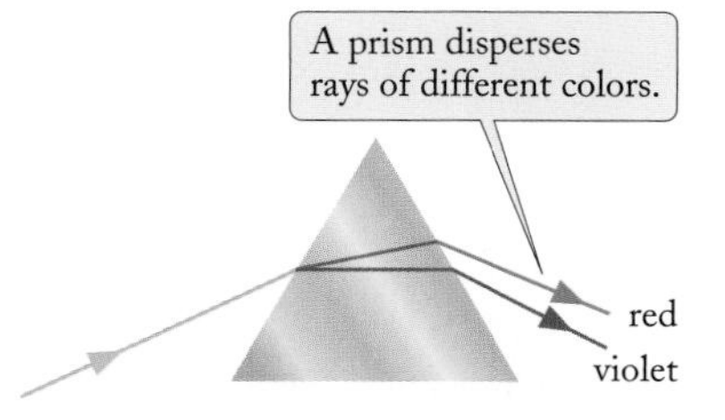

FIGURE 34.27 Refraction of red and of violet light by a prism.

FIGURE 34.28 Analysis of a beam of white light by means of a prism reveals a continuous spectrum of colors, from red to orange, yellow, green, blue, and violet.

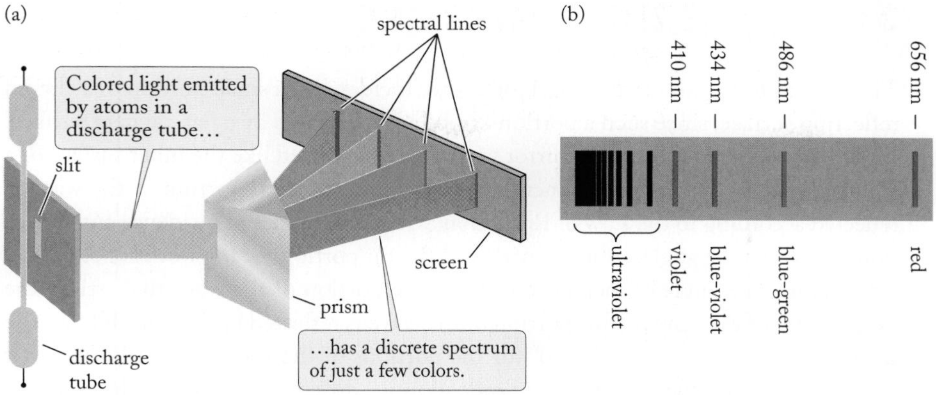

FIGURE 34.29 (a) Arrangement for the analysis of light emitted by the atoms of a chemical element. Each discrete color forms a spectral line on the screen or photographic plate. (b) The spectral lines of hydrogen. (A color print of these and other spectral lines is shown on page 1289.)

colors is essentially pure—it is light of a single wavelength. For example, hydrogen atoms emit the following discrete colors: red, blue green, blue violet, and violet. Such discrete colors are called **spectral lines**; in Chapter 38 we will see how such spectral lines arise from the quantum mechanics of the atom. Figure 34.29 shows the spectral lines of hydrogen as displayed by a prism illuminated with light from a fine slit. Each of the lines in the spectrum is a separate image of the slit made by a separate color after refraction by the prism.

spectral lines

Checkup 34.3

QUESTION 1: A ray of light enters a slab of glass with an angle of incidence of 0°. What is the angle of refraction? Does the answer depend on the index of refraction?

QUESTION 2: Qualitatively, why is the index of refraction of cold air larger than that of hot air?

QUESTION 3: If the wavelength of light in vacuum is 600 nm, what is the wavelength when this light enters water?

QUESTION 4: In each case, state whether a light ray passing from the first medium to the second will be bent closer to the perpendicular or farther away: (a) air to water; (b) water to glass; and (c) quartz to ethyl alcohol.

QUESTION 5: The critical angle for a light ray in water is 49°. What is the final direction of a light ray incident on a water surface from below with an angle barely more than 49°? Barely less than 49°?

QUESTION 6: A ray of light penetrates from a medium of index of refraction 1.3 into a medium with index of refraction 1.6. Can this ray suffer total internal reflection?

QUESTION 7: Light within a material with index of refraction $n = 2.0$ is incident upon an interface with air; the incident ray makes an angle θ with the perpendicular. Which of the following is the full range of angles of incidence for which total internal reflection occurs?

(A) $0 < \theta < 30°$ (B) $0 < \theta < 60°$ (C) $30° < \theta < 60°$
(D) $30° < \theta < 90°$ (E) $60° < \theta < 90°$

Online Concept 38 Tutorial

34.4 SPHERICAL MIRRORS

concave mirror

The Law of Reflection is also valid for a wave incident on a small portion of a curved reflecting surface, since such a portion can be approximated by a flat, tangent surface. Figure 34.30 shows a **concave** mirror with a surface curved like the inner surface of a sphere. When a plane wave is incident on this mirror, each portion of the wave is reflected according to the Law of Reflection, with equal angles of incidence and reflection. However, the angles of incidence for different portions of the wave are different; for instance, the angle of incidence is 0° for the portion of the wave that strikes the exact center of the mirror (this portion of the wave is reflected back on itself), but the angle of incidence is more than 0° for the portions of the wave that strike above or below the center. Since the angles of reflection for different portions of the wave are different, the plane wave incident on the spherical mirror becomes a wave with curved wave fronts.

focal point

As can be seen in Fig. 34.30, the reflected wave fronts converge toward a point, the **focal point** of the mirror. We can describe the direction of propagation of the wave fronts by rays; Fig. 34.31 shows the incident and reflected rays. For incident rays parallel to the axis, the reflected rays intersect at the focal point.

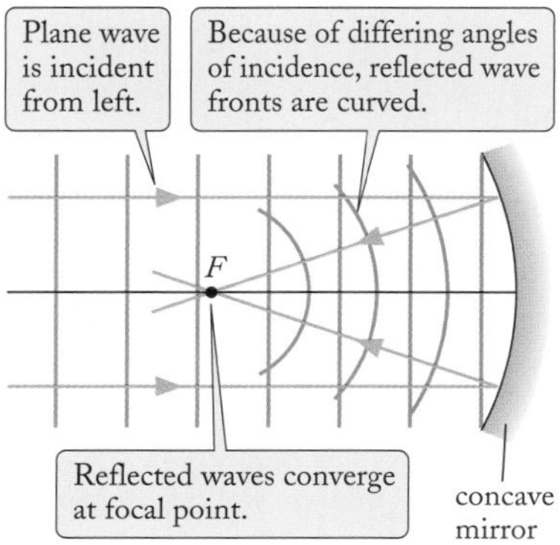

FIGURE 34.30 A concave spherical mirror focuses an incident plane wave to a point.

The focal point of the spherical mirror is halfway between the mirror and the center of curvature of the spherical surface. For a proof, we use Fig. 34.32, which shows the path of a single ray of light. The focal point F is the intersection of the ray with the axial line CA. To find the distance FA from the focal point to the mirror, called the **focal length**, we begin with the observation that in the isosceles triangle CFQ the length CF equals FQ. Under the assumption that angle θ is small (equivalently, that the incident ray is near the axial line), the length FQ is approximately equal to FA. Hence

$$CF = FA \tag{34.18}$$

and thus the point F is halfway between the mirror (A) and the center of curvature (C). This means the focal length is one-half of the radius of curvature of the spherical surface. Designating the focal length by f and the radius of curvature by R, we can write

focal length

$$f = \tfrac{1}{2}R \tag{34.19}$$

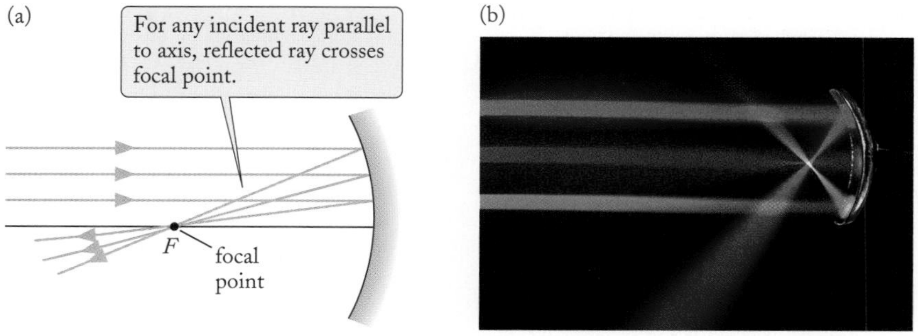

FIGURE 34.31 (a) Reflection of parallel rays by a concave spherical mirror. (b) Rays and concave mirror.

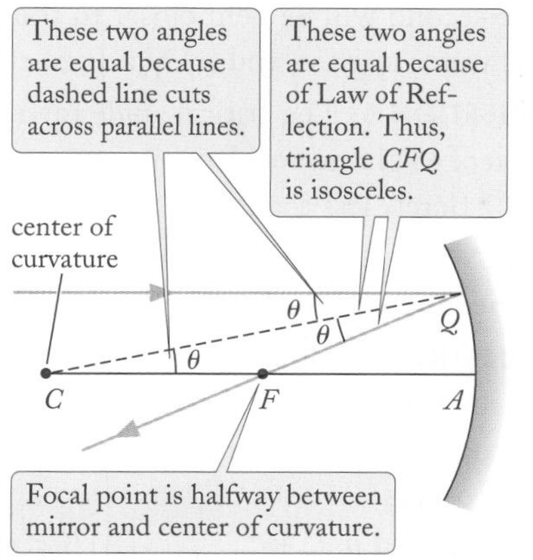

FIGURE 34.32 Reflection of a single ray. The geometry shown implies $f = R/2$.

EXAMPLE 6 A dentist uses a concave mirror to throw light into the patient's mouth (see Fig. 34.33). If a parallel beam of light supplied by a lamp is to be concentrated at a distance of 25 cm from the mirror, what should be the radius of curvature of the mirror?

SOLUTION: For strong concentration of the light, we want the focal point of the mirror to be at the distance of 25 cm from the mirror. According to Eq. (34.19), the radius of curvature of the mirror must be

$$R = 2f = 2 \times 25 \text{ cm} = 50 \text{ cm}$$

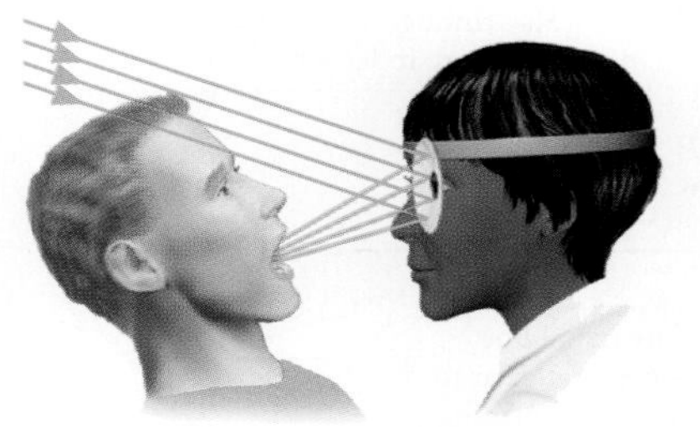

FIGURE 34.33 Mirror used to throw light into the patient's mouth.

EXAMPLE 7 Consider the Hubble Space Telescope mirror described at the beginning of the chapter, with a radius of curvature of 11 m. How far from the mirror does the image of a distant galaxy form?

Concepts in Context

SOLUTION: We know that parallel rays, such as those from a distant source, converge at the focal point. Thus from Eq. (34.19), the image will form at a distance from the mirror equal to half the radius of curvature,

$$f = \frac{R}{2} = \frac{11 \text{ m}}{2} = 5.5 \text{ m}$$

convex mirror

Figure 34.34 shows a **convex** spherical mirror. This kind of mirror is also curved like the surface of a sphere; but, in contrast to the concave mirror, the reflecting surface is the outer surface of the sphere. Parallel rays incident on this mirror diverge upon reflection. If we extrapolate the divergent rays to the far side of the mirror, they all seem to come from a single point, the focal point of the convex mirror. An argument similar to that given above demonstrates that the focal length is again one-half of the radius of the spherical surface:

$$f = -\tfrac{1}{2}R \qquad (34.20)$$

A negative sign has been inserted in Eq. (34.20) to indicate that the focal point is on the far side of the spherical surface for a convex mirror. This negative sign will prove useful in algebraic calculations with mirrors (see below).

Both concave and convex mirrors will form images of light sources placed in front of them. Figure 34.35 shows a point source of light P in front of a concave mirror. To find the position of the image, we must trace some rays of light and find where they intersect. The rays that are easiest to trace are the **principal rays** shown in Fig. 34.35:

1. The first of these rays (PC in Fig. 34.35a) passes through the center of curvature of the mirror. It then strikes the mirror perpendicularly and is therefore reflected back on itself.
2. The second ray (PQ in Fig. 34.35b) begins parallel to the axial line and thus passes through the focal point after being reflected.
3. The third ray (PF in Fig. 34.35c) passes through the focal point. Upon reflection, this ray emerges parallel to the axial line (this behavior of the third ray becomes obvious if we recognize that the third ray is merely the reverse of a ray that arrives at the mirror from a direction parallel to the axial line).
4. The fourth ray (PA in Fig. 34.35d) is reflected symmetrically with respect to the axial line, so the angles θ and θ' are equal.

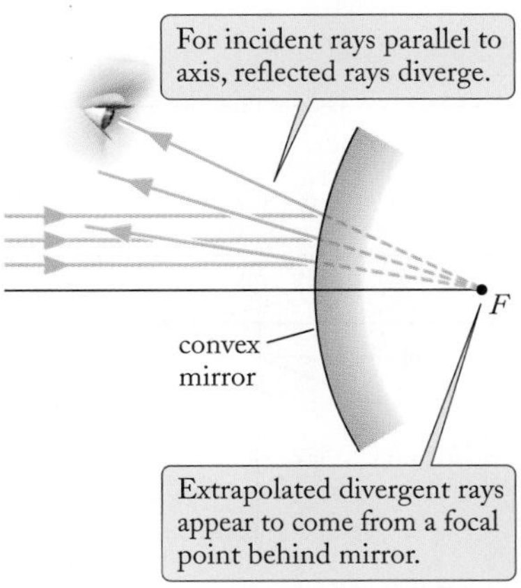

FIGURE 34.34 Reflection of parallel rays by a convex mirror.

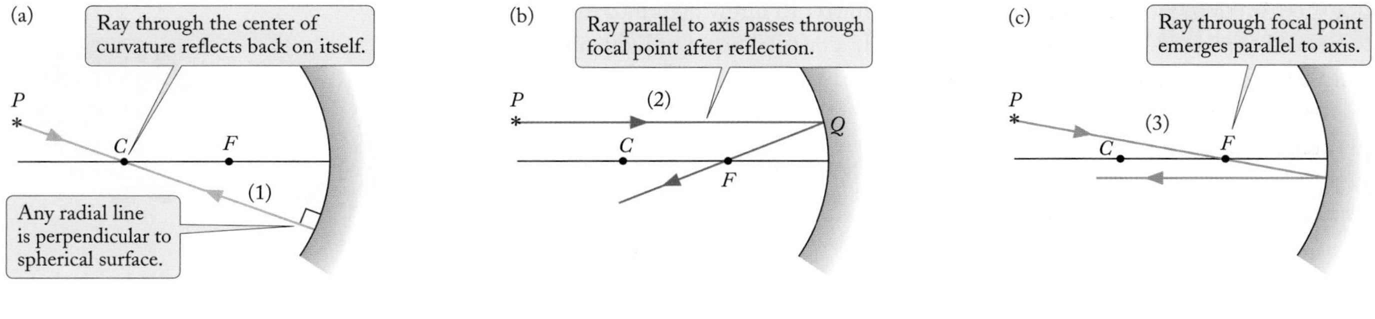

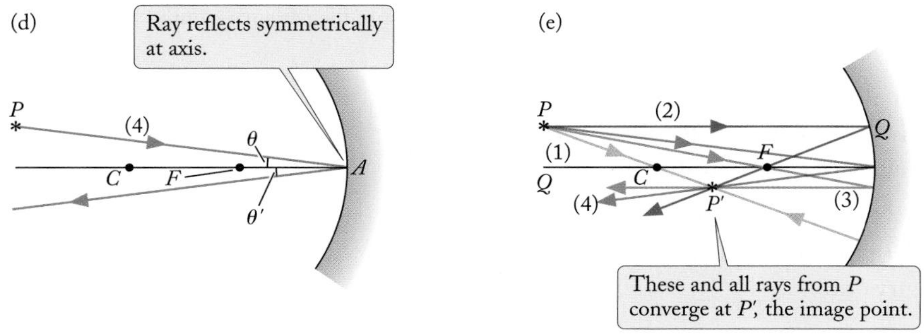

FIGURE 34.35 Concave mirror. (a) Ray from *P* through center of curvature. (b) Ray from *P* parallel to axial line passes through focus. (c) Ray from *P* through focus emerges parallel to axial line. (d) Ray from *P* striking mirror at the axis has angles of incidence and reflection symmetric about the axis. (e) The rays (1), (2), (3), and (4) intersect at the image *P*′

All these rays, and any other rays originating from the point source *P* that also strike the mirror, come together at P' (see Fig. 34.35e). This point P' is the image of the point source *P*. Note that to locate the image, two out of the four rays mentioned above are already sufficient—the other two are redundant but serve as useful checks.

object

The point source, or any source of light in a mirror or lens system, is known as the **object**. If the source of light is an extended object, then we must find the image of each of its points. For instance, a luminous object in the shape of an arrow has an image as shown in Fig. 34.36. We can easily verify this by drawing the rays that emerge from, say, the tip of the arrow, the midpoint of the arrow, the tail of the arrow, and so on. All the rays emerging from the tip of the arrow intersect at the tip of the image; all the rays emerging from the midpoint of the arrow intersect at the midpoint of the image; and so on. However, for finding the position of the image, it often suffices to draw the rays emerging from one or two points of the object. For instance, in Fig. 34.36, we need only the rays emerging from the tip, since we know that the tail of the arrow and the image of the tail both sit on the axial line.

The ray-tracing technique summarized in Fig. 34.36 is a graphical method for finding the image of an object. This method also applies to convex mirrors; an example of this is shown in Fig. 34.37. Notice from Fig. 34.37 that for a convex mirror, a

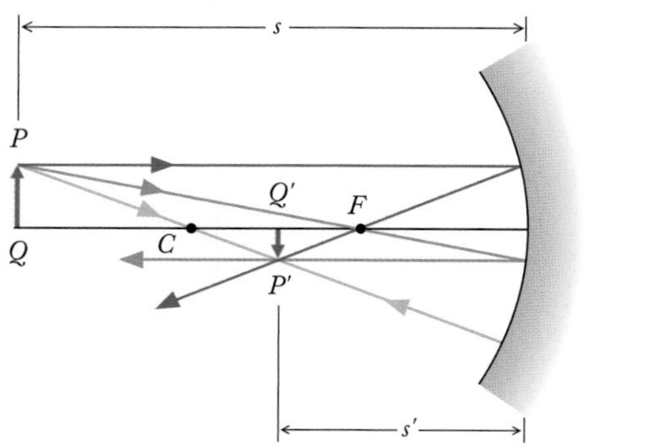

FIGURE 34.36 An object *PQ* in the shape of an arrow and its image $P'Q'$ formed by a concave mirror.

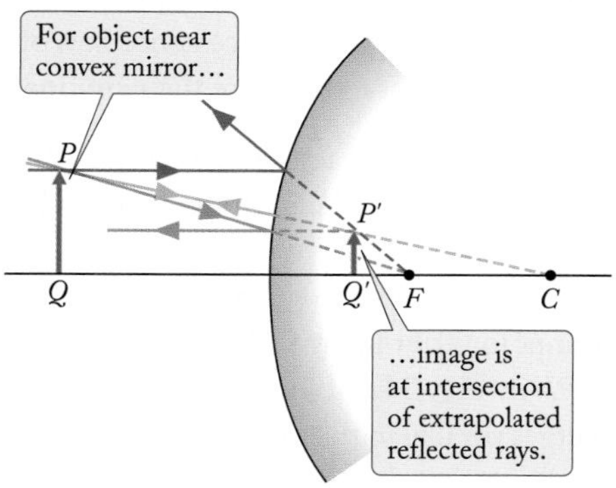

FIGURE 34.37 An object *PQ* and its image $P'Q'$ formed by a convex mirror.

ray parallel to the axis is reflected as if it came from the focal point, a ray heading toward the center of curvature is reflected back on itself; and a ray heading toward the focal point is reflected parallel to the axial line.

Although ray tracing is useful to locate the image roughly and to gain a qualitative understanding of how the positions of object, mirror, and image are related, its precision is limited. For higher precision, we must calculate the position of the image algebraically by means of the **mirror equation**,

$$\frac{1}{s}+\frac{1}{s'}=\frac{1}{f} \tag{34.21}$$

mirror equation

where s is the distance from the object to the mirror and s' is the distance from the image to the mirror, measured along the axial line, or optic axis (see Fig. 34.36). The sign conventions for these distances are as summarized in the box Problem-Solving Techniques: Sign Conventions for Mirrors (see also Fig. 34.38).

For a proof of Eq. (34.21), we make use of Fig. 34.39, which shows an object PQ, its image $P'Q'$, and two rays. The ray PCP' passes through the center of curvature of the spherical surface and is reflected on itself; the ray PAP' strikes the center of the mirror and is reflected symmetrically with respect to the axial line so the angles θ and θ' are equal. The triangles PQA and $P'Q'A$ are similar; that is, they have the same angles. Hence their corresponding sides must be in proportion:

$$\frac{PQ}{P'Q'}=\frac{s}{s'} \tag{34.22}$$

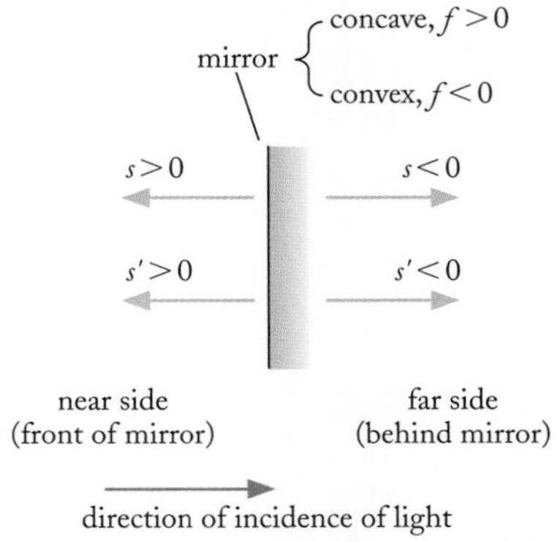

FIGURE 34.38 Summary of sign conventions for spherical mirrors.

The triangles PQC and $P'Q'C$ are also similar; hence

$$\frac{PQ}{P'Q'}=\frac{QC}{Q'C} \tag{34.23}$$

or, since $QC = s - R$ and $Q'C = R - s'$,

$$\frac{PQ}{P'Q'}=\frac{s-R}{R-s'} \tag{34.24}$$

Combining Eqs. (34.22) and (34.24), we find

$$\frac{s}{s'}=\frac{s-R}{R-s'} \tag{34.25}$$

We can rearrange this equation as

$$\frac{R-s'}{s'}=\frac{s-R}{s}$$

or

$$\frac{R}{s'}-1=1-\frac{R}{s}$$

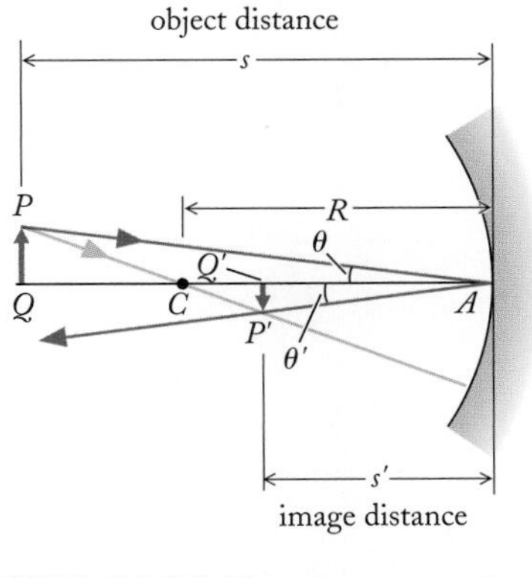

FIGURE 34.39 The geometry used to derive the mirror equation. The angles θ and θ' are equal; hence the right triangles PQA and $P'Q'A$ are similar. The right triangles PQC and $P'Q'C$ are also similar, since they have identical angles at C.

If we divide both sides of this by R and shift terms from one side of the equation to the other, we obtain

$$\frac{1}{s}+\frac{1}{s'}=\frac{2}{R} \tag{34.26}$$

Since $f = R/2$ [see Eq. (34.19)], this is the same as Eq. (34.21).

PROBLEM-SOLVING TECHNIQUES SIGN CONVENTIONS FOR MIRRORS

The object distance s or image distance s' is positive if the object or image is in front of the mirror; the distance s or s' is negative if the object or image is behind the mirror (the object can be behind the mirror if what serves as "object" for the mirror is actually an image produced by another mirror or a lens). As mentioned in the text, f is positive for a concave mirror, negative for a convex mirror. These sign conventions for spherical mirrors are summarized in Fig. 34.38.

EXAMPLE 8 A candle is placed 41.0 cm in front of a convex spherical mirror of radius of curvature 60.0 cm (see Fig. 34.40). Where is the image?

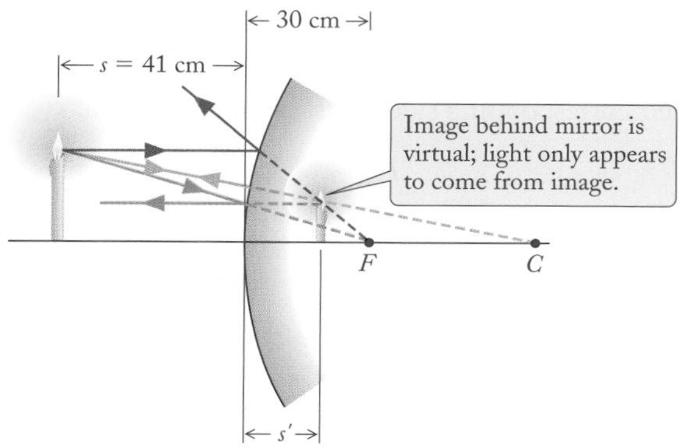

FIGURE 34.40 A candle and its image in a convex mirror. This is a *virtual* image; it lies behind the mirror, and it is upright.

SOLUTION: The focal length is half the radius of curvature, and is negative for a convex mirror [see Eq. (34.20)]. With $s = 41.0$ cm and $f = -R/2 = -30.0$ cm, Eq. (34.21) gives

$$\frac{1}{41.0\ \text{cm}} + \frac{1}{s'} = -\frac{1}{30.0\ \text{cm}}$$

or

$$\frac{1}{s'} = -\frac{1}{30.0\ \text{cm}} - \frac{1}{41.0\ \text{cm}} = \frac{-41.0 - 30.0}{41.0 \times 30.0}\frac{1}{\text{cm}}$$

from which

$$s' = -17.3\ \text{cm}$$

The negative value of s' indicates that the image is behind the mirror.

EXAMPLE 9 Suppose that the same candle is placed 41.0 cm in front of a concave spherical mirror of radius of curvature 60.0 cm (see Fig. 34.41). Where is the image?

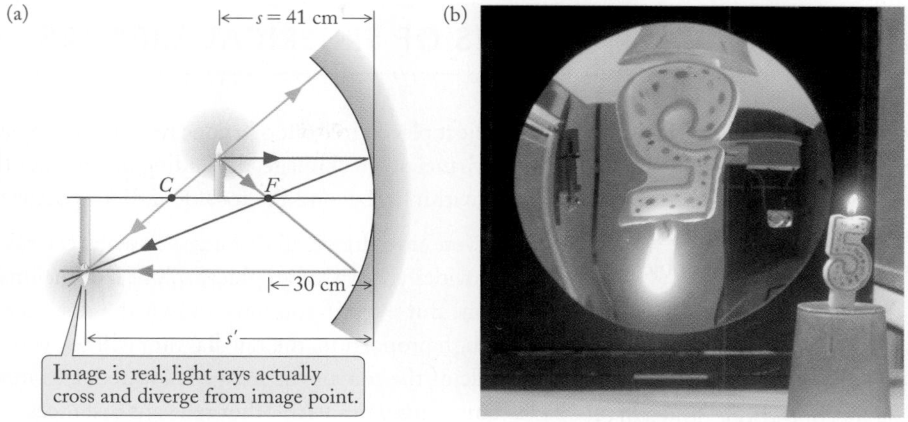

FIGURE 34.41 (a) A candle and its image in a concave mirror. (b) This is a *real* image; it lies in front of the mirror, and it is inverted.

SOLUTION: With $s = 41.0$ cm and now $f = +30.0$ cm, Eq. (34.21) gives

$$\frac{1}{41.0\text{ cm}} + \frac{1}{s'} = \frac{1}{30.0\text{ cm}}$$

or

$$\frac{1}{s'} = \frac{1}{30.0\text{ cm}} - \frac{1}{41.0\text{ cm}} = \frac{41.0 - 30.0}{41.0 \times 30.0}\frac{1}{\text{cm}}$$

from which

$$s' = 112\text{ cm}$$

The positive value of s' indicates that the image is on the same side of the mirror as the object.

real image

The image in Example 9 is a **real image**. The real image is in front of the mirror; if we insert a sheet of paper at its location, the image will be projected on the sheet of paper, and we can see it sharply on this paper. Even without a sheet of paper, we can see the image if we look toward the mirror from the far left of Fig. 34.41; the image then gives the visual impression of a ghostly replica of the object floating in midair. *The light rays that we see when we look toward the mirror not only seem to come from the real image, but they actually do.* As Fig. 34.41 shows, the light rays pass through the real image and diverge from it, just as they diverge from the object. In contrast, the image in Example 8 is a virtual image, similar to the virtual images that we examined in our discussion of flat mirrors. The virtual image is behind the mirror. Light rays do not actually pass through a virtual image; they merely *seem* to come from such an image (see Fig. 34.40).

In both the case of a real image and the case of a virtual image, the geometry of the emerging light rays is as though the image were an object that acts as the source of the rays. This property of images allows us to analyze optical systems consisting of more than one element, for instance, two mirrors, or one mirror and one lens, or two lenses, etc. To find the final image produced by such a compound system, find the image produced by the first element, then treat this image as a new object for the second element, and find the new image that the second element produces of this new object.

PROBLEM-SOLVING TECHNIQUES — IMAGES OF SPHERICAL MIRRORS

1. Begin the solution of a problem involving images in spherical mirrors by drawing a clear, careful diagram showing the mirror, the object, and relevant distances. Draw all the relevant distances to scale, or at least draw all the distances in roughly correct proportions. Represent the object by a heavy arrow, as in several of the diagrams of this section.

2. Trace two or three of the four principal rays that start from the tip of the heavy arrow, and find where they intersect, forming an image. Any two of the four principal rays are sufficient to determine the intersection; but it is sound practice to draw a third ray to serve as a check. Figures 34.35, 34.40, and 34.41 can be used as ray-tracing guides.

3. If the rays do not intersect in front of the mirror after reflection, extrapolate them with dotted lines, and find where the extrapolated rays intersect behind the mirror. If the rays intersect in front of the mirror, the image is real (and inverted); if the rays intersect behind the mirror, the image is virtual (and upright). An object placed in front of a convex mirror always has a virtual image; an object placed in front of a concave mirror can have either a virtual or a real image, depending on whether the object is within or beyond the focal length, respectively.

4. If you have drawn all distances to scale, the ray diagram provides you with a graphical solution of the image problem. But even if you have drawn the distances only in rough proportion, the ray diagram will tell you on which side of the mirror the image is located, and how its distance compares with other relevant distances.

5. Use the mirror equation (34.21) only after you know what to expect on the basis of the ray diagram. Remember to be careful about the sign conventions for s, s', and f; these sign conventions are summarized in Fig. 34.38 and the box on sign conventions on page 1132.

6. To decide whether the image is real or virtual, inspect the ray diagram. Ask yourself, is the image an actual point of intersection of rays (real), or is it only a point of intersection of rays extrapolated backward (virtual)? The sign of the image distance s' can be used to verify your conclusion.

Concepts in Context

EXAMPLE 10 Consider the Hubble Space Telescope mirror described at the beginning of the chapter. The technician is 4.0 m from the mirror, which has a radius of curvature of 11 m. Where is the image of the technician?

SOLUTION: The focal length is $f = R/2 = 5.5$ m, as we found in Example 7. Thus the mirror equation tells us

$$\frac{1}{s'} = \frac{1}{f} - \frac{1}{s} = \frac{1}{5.5\text{ m}} - \frac{1}{4.0\text{ m}} = -0.068\,\frac{1}{\text{m}}$$

from which we obtain

$$s' = -15\text{ m}$$

Thus the image is virtual and 15 m behind the mirror. The appropriate ray diagram is shown in Fig. 34.42.

COMMENT: This arrangement, in comparison with Figs. 34.39 and 34.41, highlights a general property of concave mirrors: when the object is beyond the focal length (as in Figs. 34.39 and 34.41), the image is real and inverted, but when the object is closer to the mirror than the focal point (as in Fig. 34.42 and the chapter photo), the image is virtual and upright.

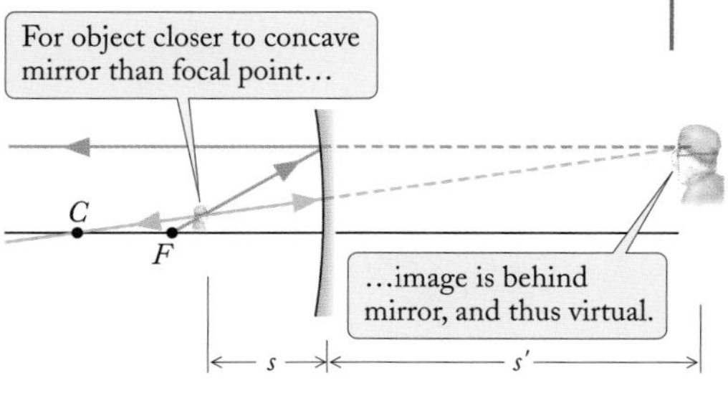

FIGURE 34.42 Ray diagram corresponding to the chapter photo.

Checkup 34.4

QUESTION 1: A fly is at a distance of 60 cm from a concave spherical mirror of radius 30 cm. Is the image behind the mirror or in front? Is it real or virtual? Is it upright or inverted?

QUESTION 2: A fly is at a distance of 10 cm from a convex spherical mirror of radius 30 cm. Is the image behind the mirror or in front? Is it real or virtual? Is it upright or inverted?

QUESTION 3: Can a convex mirror produce a real image?

QUESTION 4: Could the dentist in Example 6 use a convex mirror to concentrate light in the patient's mouth?

QUESTION 5: Consider a concave mirror, with an object closer to the mirror than the focal point. Which of the following is true? The image is:

(A) Virtual and upright, with $s' < 0$ (B) Virtual and inverted, with $s' < 0$
(C) Real and inverted, with $s' > 0$ (D) Real and upright, with $s' < 0$
(E) Virtual and upright, with $s' > 0$

34.5 THIN LENSES

Online Concept Tutorial 38

A lens made of a refracting material with two convex spherical surfaces will focus the parallel rays in a beam of light to a point (see Fig. 34.43). For a thin lens in vacuum or in air, the focal length f is given by the **lens maker's formula**:

lens maker's formula

$$\frac{1}{f} = (n-1)\left(\frac{1}{R_1} + \frac{1}{R_2}\right) \tag{34.27}$$

where n is the index of refraction of the material of the lens and where R_1 and R_2 are the radii of curvature of the two spherical surfaces making up the lens. This equation is based on the assumption that the lens is thin (its thickness is small compared with R_1 and R_2) and that the incident rays are near the axial line.

Equation (34.27) can be verified by tracing rays through the lens, taking into account their refraction at the two curved surfaces. We will not perform this tedious calculation here (see Problems 81–82). We only note that the focusing depends on the

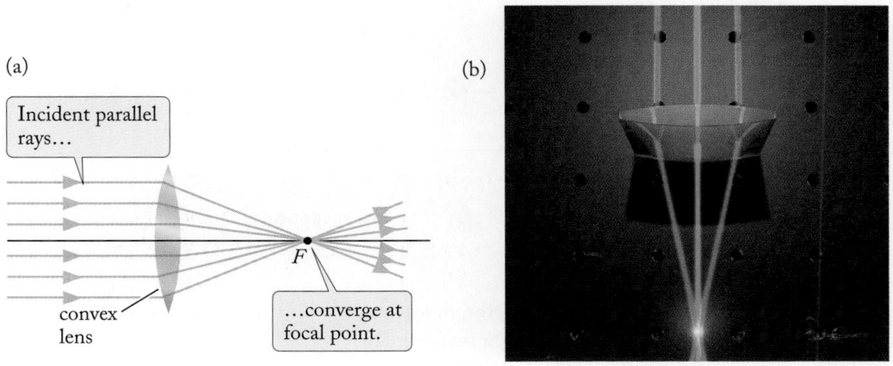

FIGURE 34.43 (a) Refraction of parallel rays by a convex lens. The focal length is positive. (b) Rays passing through a convex lens.

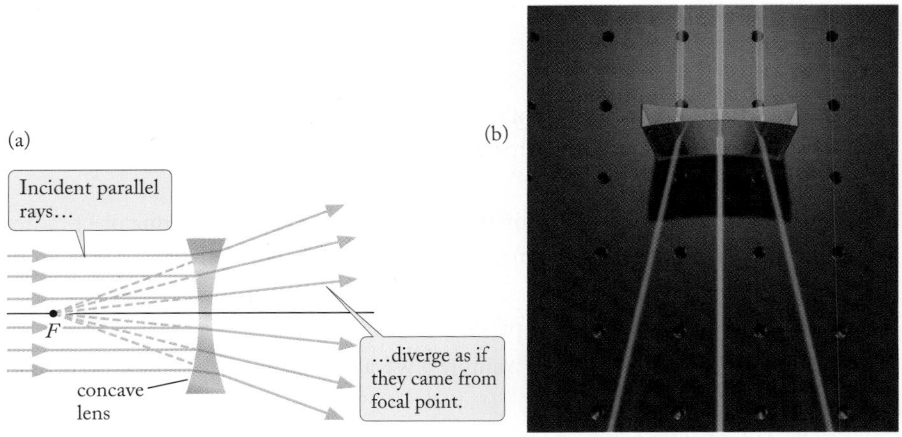

FIGURE 34.44 (a) Refraction of parallel rays by a concave lens. The focal length is negative. (b) Rays passing through a concave lens.

curvature of the lens surface, because of which parallel rays at some distance from the axial line strike the lens surface with larger angles of incidence than rays near the axial line, and consequently they are bent more sharply toward the axial line. This is of course exactly what is required to make these distant rays cross the axial line at the same point (the focus) as the near rays.

Equation (34.27) may also be applied to a lens with two concave surfaces (see Fig. 34.44). In this case, the radii R_1 and R_2 must be reckoned as *negative*, and the focal distance f is then also negative. The meaning of a negative focal length is the same as in the case of mirrors: parallel rays incident on the lens diverge when they emerge from the other side of the lens, and the focal point is where the extrapolated rays appear to intersect (see Fig. 34.44). Furthermore, Eq. (34.27) can be applied to a concave–convex lens, such as shown in Fig. 34.45. Whether such a lens produces net convergence or divergence depends on whether the positive radius (convex) or the negative radius (concave) is smaller; for instance, the lens of Fig. 34.45a produces convergence, because the convex radius of curvature is smaller; and the lens of Fig. 34.45b produces divergence, because the concave radius of curvature is smaller.

Note that *a lens has two focal points at equal distances to the right and the left of the lens*. The point on the right of a converging lens is the focus of a parallel beam coming from the left and conversely, the point on the left is the focus of a parallel beam coming

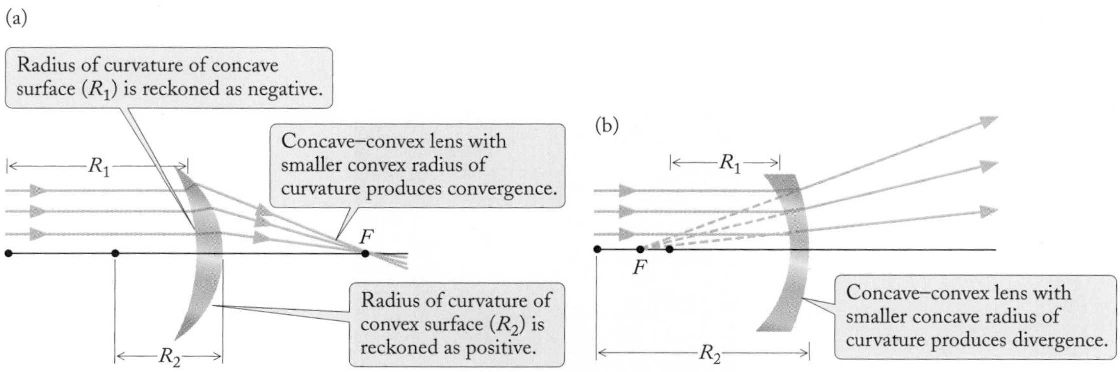

FIGURE 34.45 Refraction of rays by concave–convex lenses. (a) For this lens, the concave surface (left surface) has a larger radius of curvature than the convex surface (right surface). For this lens, the sum $1/R_1 + 1/R_2$ in Eq. (34.27) is positive. (b) For this lens, the concave surface (left surface) has a smaller radius of curvature than the convex surface (right surface). For this lens, the sum $1/R_1 + 1/R_2$ in Eq. 34.27 is negative.

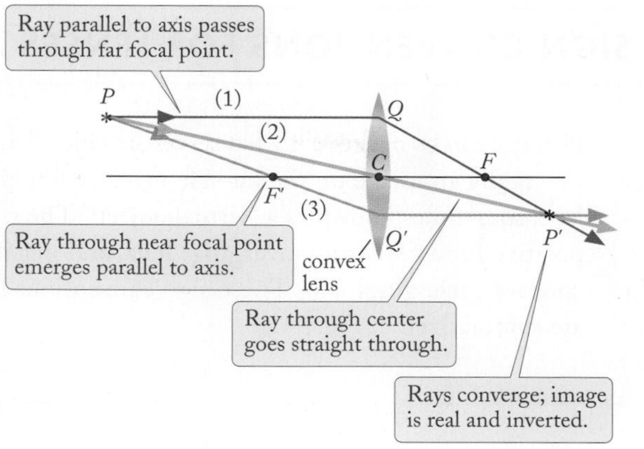

FIGURE 34.46 A point source of light P in front of a convex lens. The rays (1), (2), and (3) intersect at the image P'.

from the right. Moreover, for any thin lens, concave, convex, or concave–convex, the left and right focal distances are equal.

To find the image of an object placed near the lens, we can use a ray-tracing technique similar to that used for mirrors. Figure 34.46 shows the three *principal rays* that are easy to trace:

1. The first ray (PQP' in Fig. 34.46) starts parallel to the axial line. This ray is deflected by the lens so it passes through the far focal point F.
2. The second ray (PCP' in Fig. 34.46) passes through the center of the lens. This ray continues straight through the lens, without deflection (actually, the ray suffers a deflection when it enters the lens and another deflection when it leaves the lens; but for a thin lens, these deflections are opposite and their effects cancel).
3. The third ray ($PQ'P'$ in Fig. 34.46) passes through the near focal point F'. That ray is deflected by the lens so it emerges parallel to the axial line (this becomes obvious if we recognize that the third ray is merely the reverse of a ray that arrives at the lens from the right, parallel to the axial line).

All these rays intersect at the point P', the image point. As in the case of mirrors, two of the above three rays are already sufficient to locate the image. And, of course, the same ray-tracing technique can be applied to find the virtual image formed by a diverging lens. Figure 34.47 gives an example of such ray tracing for a concave lens. Here, the ray parallel to the axis diverges as if it came from the near focal point, and a ray heading for the far focal point emerges parallel to the axis. Note that in Figs. 34.46 and 34.47 (and other figures hereafter) the deflections of rays by a lens are indicated

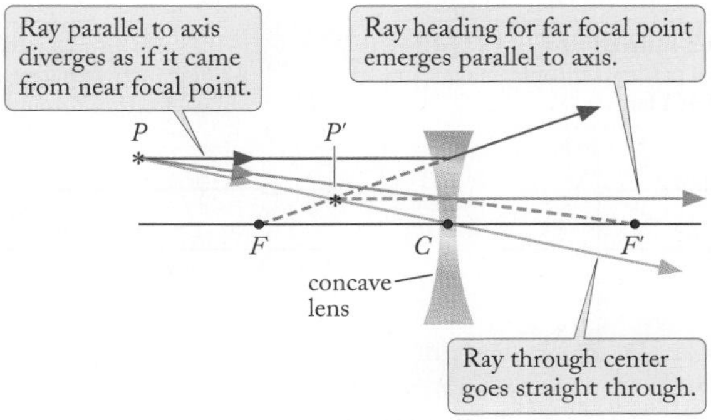

FIGURE 34.47 A point source of light P in front of a concave lens, and the image P'.

PROBLEM-SOLVING TECHNIQUES — SIGN CONVENTIONS FOR LENSES

The object distance s is positive if the object is on the near side of the lens and negative if it is on the far side; the image distance s' is positive if the image is on the far side of the lens and negative if it is on the near side. In this context, the "near" side is the side from which the light rays are incident on the lens (left side in the preceding figures), and the "far" side is the other side (right side in the preceding figures). The object distance can be negative if what serves as "object" for the lens is actually an image on the far side produced by a previous lens, sometimes known as a "virtual object." The sign of f is positive for a convex (converging) lens, and negative for a concave (diverging) lens. These sign conventions for lenses are summarized in Fig. 34.48.

schematically as occurring in the mid plane of the lens, and not at the lens surfaces. This is a convenient simplification for thin lenses.

The equation to be used for the algebraic calculation of the image distances is the same as Eq. (34.21),

thin-lens equation

$$\frac{1}{s}+\frac{1}{s'}=\frac{1}{f} \tag{34.28}$$

also known as the **thin-lens equation**, but the sign conventions are slightly different, as summarized in the box Problem-Solving Techniques: Sign Conventions for Lenses (see also Fig. 34.48).

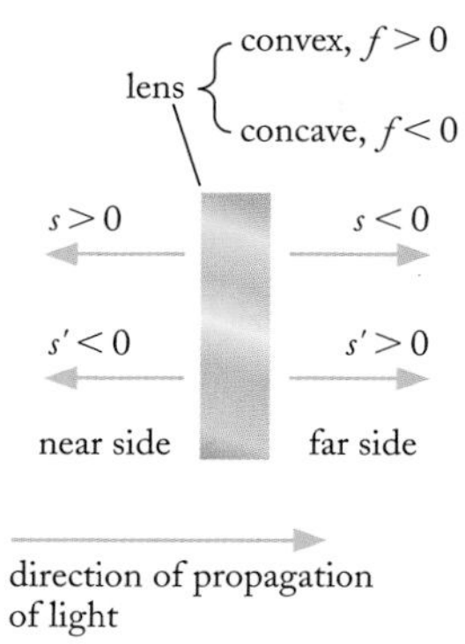

FIGURE 34.48 Summary of sign conventions for lenses.

Although the derivation of the lens equation (34.28) can be based on a geometric argument similar to that used for the mirror formula (34.21), we can bypass this labor by a trick. We begin by noting that a concave mirror is equivalent to one-half of a convex lens placed directly in front of a flat mirror. If the concave mirror and the (entire) convex lens have the same focal length, then the two arrangements shown in Figs. 34.49a and b also have this same focal length. Each ray in Fig. 34.49b has to pass through the half lens twice, once before reflection by the flat mirror and once after; the deflection angle imposed on the ray by the half lens is then the same as the deflection angle imposed by the entire lens, and it is therefore also the same as the deflection angle imposed by the concave mirror (see, for instance, the deflection angles marked in Figs. 34.49a, b, c for the blue ray that passes through the focal point). Thus, the arrangements shown in Figs. 34.49a and b have exactly the same optical properties—in both cases the image distances are the same. If we now remove the flat mirror in Fig. 34.49a and replace the half lens by the entire lens, the image distances will remain the same, but the image will form on the opposite side of the lens (see Fig. 34.49c), that is, the

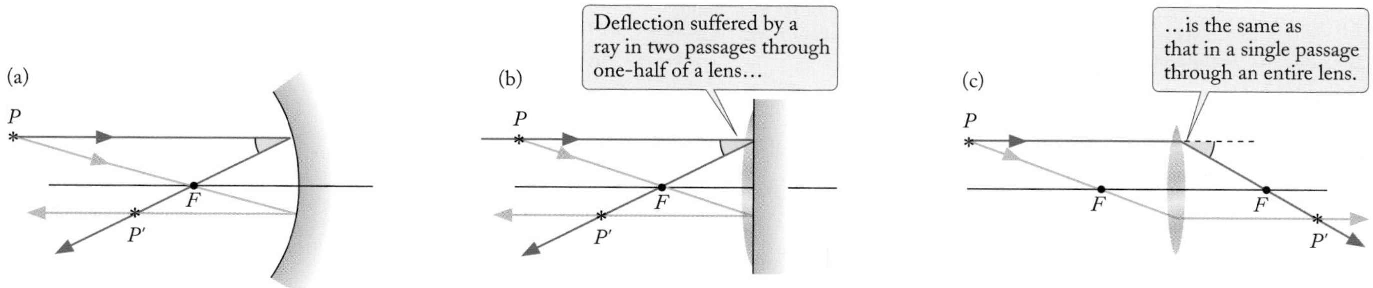

FIGURE 34.49 (a) A point source of light and its image formed by a concave mirror. (b) A similar image is formed by one-half of a convex lens placed in front of a flat mirror. (c) A similar image is also formed by the entire lens, but the image is now on the other side of the lens.

sign of the image distance will be reversed. Consequently, the same equation (34.21) must apply to the concave mirror and to the convex lens; the only difference is that the sign of the image distance is reversed. This reversal of sign has already been taken into account in our description of the sign conventions associated with Eqs. (34.21) and (34.28)—for the mirror, s' is taken as positive if the image is on the *near* side of the mirror, whereas for the lens, s' is taken as positive if it is on the *far* side of the lens. There is, of course, a similar correspondence between a convex mirror and a concave lens.

EXAMPLE 11 A convex lens of focal length 25 cm is placed at a distance of 10 cm from a printed page. What is the image distance? How much larger is the image of the page than the page?

SOLUTION: With $s = 10$ cm and $f = 25$ cm, Eq. (34.28) gives

$$\frac{1}{10\text{ cm}} + \frac{1}{s'} = \frac{1}{25\text{ cm}}$$

from which

$$\frac{1}{s'} = \frac{1}{25\text{ cm}} - \frac{1}{10\text{ cm}} = \frac{10 - 25}{25 \times 10}\frac{1}{\text{cm}}$$

and

$$s' = -17\text{ cm}$$

The negative sign indicates that the image is on the near side of the lens (see Fig. 34.50). The image is virtual (for an observer looking at the lens from the right, light rays *appear* to diverge from the image, but they actually do not).

Since the triangles $P'Q'C$ and PQC are similar, the sizes of image and object are in the ratio

$$\frac{P'Q'}{PQ} = \frac{Q'C}{QC} = -\frac{s'}{s} = \frac{17\text{ cm}}{10\text{ cm}} = 1.7 \tag{34.29}$$

that is, the image is larger than the object by a factor of 1.7. This is the principle involved in the magnifying glass.

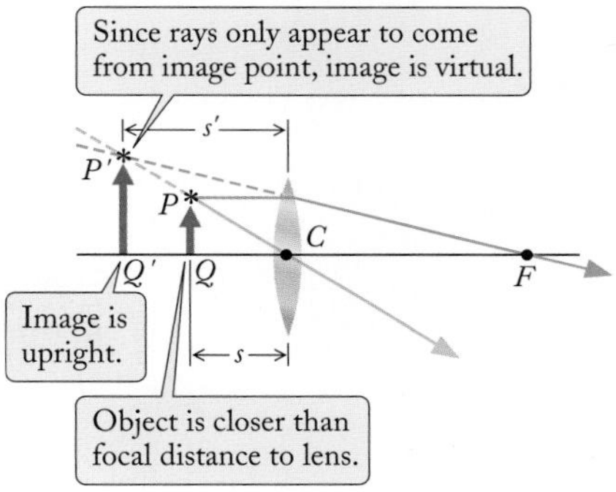

FIGURE 34.50 An object and its image for a convex lens.

EXAMPLE 12 A concave lens of focal length 25 cm is placed at a distance of 10 cm from the same printed page. What is the image distance? How much larger is the image of the page than the page?

SOLUTION: With $s = 10$ cm and, for a concave (diverging) lens, $f = -25$ cm, Eq. (34.28) gives

$$\frac{1}{10\text{ cm}} + \frac{1}{s'} = -\frac{1}{25\text{ cm}}$$

from which

$$\frac{1}{s'} = -\frac{1}{25\text{ cm}} - \frac{1}{10\text{ cm}} = \frac{-10 - 25}{25 \times 10}\frac{1}{\text{cm}}$$

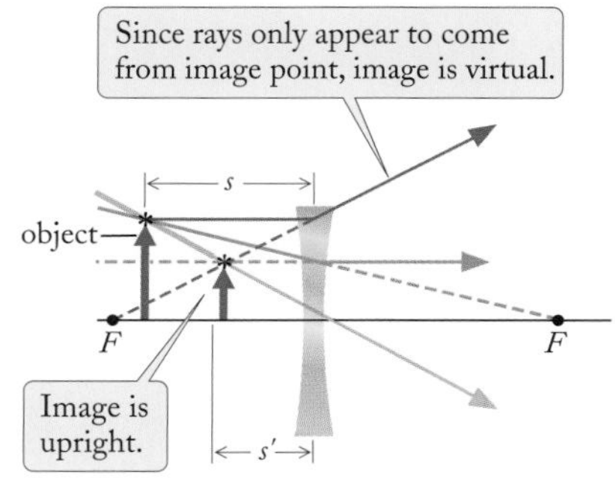

FIGURE 34.51 An object and its image for a concave lens.

and

$$s' = -7.1 \text{ cm}$$

The negative sign again indicates that the image is on the near side of the lens (compared with the object; see Fig. 34.51). The image is virtual.

As in the preceding example, the sizes of image and object are in the ratio

$$-\frac{s'}{s} = \frac{7.1 \text{ cm}}{10 \text{ cm}} = 0.71$$

that is, the image is smaller than the object by a factor of 0.71.

Equation (34.29) is a general result for the magnification produced by a single lens or by a mirror. The size of the image always equals the size of the object multiplied by a factor $-s'/s$. This factor is called the **linear magnification** M:

linear magnification M

$$M = -\frac{s'}{s} \tag{34.30}$$

Thus a magnification $M = 1.7$, as in Example 11, means that the image is 1.7 times the size of the object (and 1.7 times as far from the lens).

The sign of M tells us something about the character of the image. For a single lens, if M is positive, then, as in Examples 11 and 12, the image is upright and virtual (s' is negative). If M is negative, then the image is inverted and real (s' is positive). The sign of M tells us the same information for a mirror.

Concepts in Context

EXAMPLE 13 The image of the technician in the chapter photo is quite large. From the data of Example 10, find the magnification of this image.

SOLUTION: We know the object distance $s = 4.0$ m and the image distance $s' = -15$ m. Thus the magnification is

$$M = -\frac{s'}{s} = -\frac{-15 \text{ m}}{4.0 \text{ m}} = +3.8$$

Thus the image is 3.8 times as large as the object. Also, the positive sign of the magnification for a single mirror tells us that the image is upright and virtual (s' is negative).

For a multiple-lens system, we use the image of the first lens as the object for the second lens, and so on. In this case, the overall magnification is simply the product of the individual lens magnifications; for example, the magnification of a two-lens system is

$$M = M_1 \times M_2 = \left(-\frac{s_1'}{s_1}\right) \times \left(-\frac{s_2'}{s_2}\right) \tag{34.31}$$

For multiple lenses, the overall sign of the magnification still tells us whether the final image is upright with respect to the original object (M positive), or inverted (M negative). However, whether the final image is virtual or real cannot be determined from the sign of the magnification alone.

EXAMPLE 14 A bird at a (large) distance of 100 m (1.00×10^4 cm) is viewed through a two-lens system; the light first travels through a convex lens with $f_1 = 100$ cm, and then through a concave lens, placed 90.0 cm behind the first lens, as shown in Fig. 34.52a. The concave lens has a focal length $f_2 = -8.0$ cm. Locate the final image and determine the overall linear magnification of this system.

SOLUTION: We begin by finding the image produced by the first lens; we will then use this as the object for the second lens. For the first lens, with $f_1 = 100$ cm, the object distance is $s_1 = 100\text{ m} = 1.00 \times 10^4$ cm. We can solve the thin-lens equation (34.28) for the image distance s_1':

$$\frac{1}{s_1'} = \frac{1}{f_1} - \frac{1}{s_1} = \frac{1}{100\text{ cm}} - \frac{1}{1.00 \times 10^4\text{ cm}} = 9.90 \times 10^{-3}\,\frac{1}{\text{cm}}$$

which gives

$$s_1' = 101\text{ cm}$$

Thus, by itself, the first lens would form a real, inverted image, as depicted in the ray diagram of Fig. 34.52b. Now we use this image as the object for the second lens; since the lenses are separated by 90.0 cm, our image (101 cm from the first lens) is 11 cm beyond the second lens. Since it is *beyond* the second lens, the object distance is negative, $s_2 = -11$ cm (see Fig. 34.48 and the box Problem-Solving Techniques: Sign Conventions for Lenses). With $f_2 = -8.0$ cm, we can again apply the thin-lens equation to obtain the image distance from the second lens:

$$\frac{1}{s_2'} = \frac{1}{f_2} - \frac{1}{s_2} = -\frac{1}{8.0\text{ cm}} - \frac{1}{-11\text{ cm}} = -0.034\,\frac{1}{\text{cm}}$$

from which we obtain

$$s_2' = -29\text{ cm}$$

The final image distance is negative, so it is 29 cm before (to the left of) the second lens, and the final image is virtual, as shown in Fig. 34.52c.

As shown in Fig. 34.52c, the final image is upright compared with the original object (inverted with respect to the first image). We can verify the orientation by examining the magnification. The magnification of the first lens is

$$M_1 = -\frac{s_1'}{s_1} = -\frac{101\text{ cm}}{1.00 \times 10^4\text{ cm}} = -0.0101$$

Thus the first image is inverted and about 1/100 the size of the object. The magnification of the second lens is

$$M_2 = -\frac{s_2'}{s_2} = -\frac{-29\text{ cm}}{-11\text{ cm}} = -2.6$$

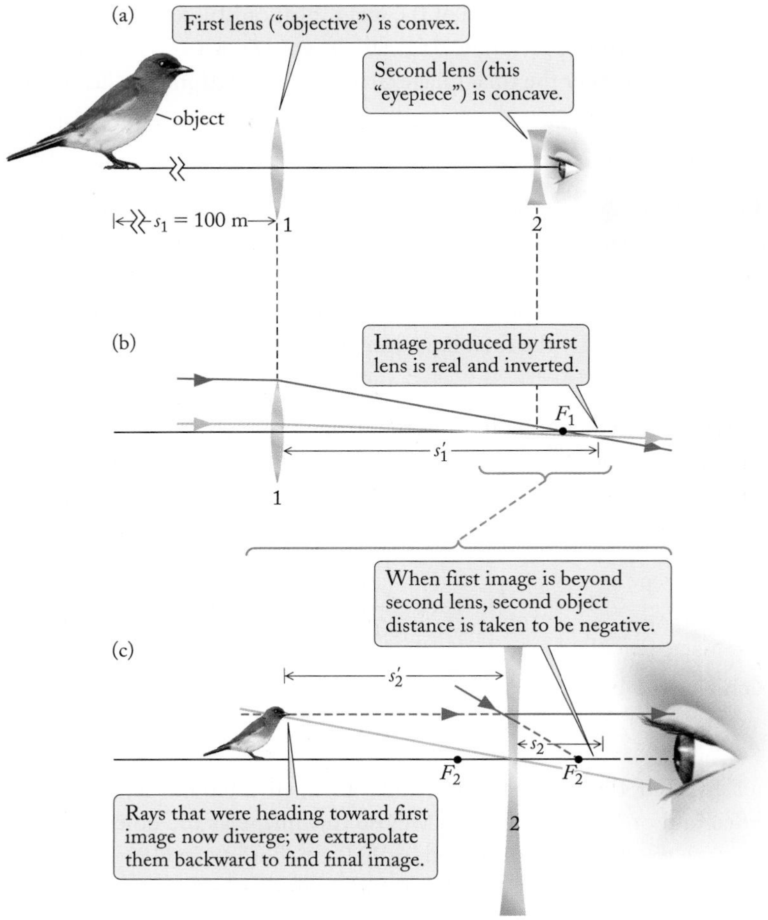

FIGURE 34.52 (a) A two-lens system. (b) The image produced by the first lens is real and inverted. (c) Expanded view with the second lens in place; the final image is virtual and upright.

Thus the second lens inverts the first image and enlarges it 2.6 times. The overall magnification is the product of the two:

$$M = M_1 M_2 = (-0.0101) \times (-2.6) = +0.026$$

The overall magnification is positive, so the final image is indeed upright.

COMMENTS: Although the final image is much smaller than the object ($1/0.026 \approx 40$ times smaller), it is much, much closer to the eye than the object [$(100\text{ m})/(0.29\text{ m}) \approx 300$ times closer], and to the eye the image appears very large, so this lens system acts as a telescope. Thus, for distant objects, the ordinary magnification is not always the best way to characterize a lens system (for example, for a telescope to be useful, an image of the Moon need only appear larger than the Moon appears, but need not be larger than the Moon itself!). In the next section, we will introduce a different measure of magnification, the angular magnification. The lens arrangement of this example is similar to a **Galilean telescope**, often preferred for convenient use because the image remains upright.

Galilean telescope

PROBLEM-SOLVING TECHNIQUES — IMAGES OF LENSES

The procedure for finding the images produced by lenses is similar to the procedure we learned for mirrors:

1 Begin the solution of the problem by drawing a clear, careful diagram showing the lens, the object, and relevant distances. Draw all the relevant distances to scale, or at least draw all the distances in roughly correct proportions. Represent the object by a heavy arrow.

2 Trace the two or three principal rays that start from the tip of the heavy arrow, and find where they intersect, forming an image. Two of the three principal rays are sufficient to determine the intersection; but it is sound practice to draw all three, so the third ray can serve as a check (see Example 12).

3 If the rays do not intersect after passing through the lens, extrapolate them backward with dotted lines, and find where the extrapolated rays intersect on the near side of the lens. If the rays do intersect on the far side of the lens, the image is real (and inverted); if extrapolated rays intersect on the near side of the lens, the image is virtual (and upright). An object placed in front of a concave lens always has a virtual image; an object placed in front of a convex lens can have either a virtual or a real image, depending on whether the object is within or beyond the focal length, respectively.

4 If you have drawn all distances to scale, the ray diagram provides you with a graphical solution of the image problem. But even if you have drawn the distances only in rough proportion, the ray diagram will tell you on which side of the lens the image is located, and how its distance compares with other relevant distances.

5 Use the lens equation (34.28) only after you know what to expect on the basis of the ray diagram. Remember to be careful about the sign conventions for s, s', and f; these sign conventions are summarized in Fig. 34.48.

6 To decide whether the image is real or virtual, inspect the ray diagram. Ask yourself, Is the image an actual point of intersection of rays, or is it only a point of intersection of rays extrapolated backward?

7 For a multiple-lens system, proceed in sequence, beginning from the source of light, using the image distance from the first lens and the separation between lenses to determine the object distance for the second lens. If the image from the first lens is on the far side of the second lens, then the object distance for the second lens is negative (as in Example 14 and in the sign conventions of Fig. 34.48).

8 The relative size of the image and object can be calculated from the magnification of Eq. (34.30), or, for a multiple-lens system, from the product of such magnifications, as in Eq. (34.31).

✔ Checkup 34.5

QUESTION 1: A thin plano–convex lens (one flat side, one convex side) has a focal length of 15 cm when the flat face is aimed toward the light source. If we turn this lens around, so its rounded face is toward the light source, what will be its focal length?

QUESTION 2: A fly is at a distance of 20 cm in front of a convex lens of focal length 30 cm. Is the image on the near side or the far side? Is it real or virtual? Is it upright or inverted?

QUESTION 3. A fly is at a distance of 10 cm in front of a concave lens of focal length 30 cm. Is the image on the near side or on the far side? Is it real or virtual? Is it upright or inverted?

QUESTION 4: Can a concave lens form a real image of a real object?

QUESTION 5: A thin lens, like the one in Fig. 34.45, has one convex surface with radius of curvature 2.0 m and one concave surface with radius of curvature 1.0 m. If the lens is made of glass of index of refraction 1.5, what is its focal length?

(A) 4.0 m (B) −2.0 m (C) 1.3 m (D) 2.0 m (E) −4.0 m

QUESTION 6: Two converging lenses have the same focal length f. An object is a distance $s_1 = 2f$ from the first lens, and the lenses are separated by a distance $5f$. With respect to the upright object, the final image of the second lens is

(A) Upright and real (B) Upright and virtual
(C) Inverted and real (D) Inverted and virtual

34.6 OPTICAL INSTRUMENTS

Optical instruments—cameras, magnifiers, microscopes, and telescopes—employ lenses and mirrors to form images in different ways. Most of the lenses in high-quality optical systems are compound lenses, that is, combinations of several simple lenses chosen in such a way that the optical deficiencies of one lens are canceled by those of another. Such deficiencies include **chromatic aberration**, where rays of different colors are refracted differently by the glass of the lens, and are therefore focused differently. Blurring of images is also caused by **spherical aberration**, where parallel rays that are far from the axis of a lens or mirror miss the focal point. Only parabolic surfaces truly focus parallel rays; spherical surfaces are merely conveniently fabricated approximations of parabolic surfaces. However, in the following discussion of optical instruments, we will ignore these complications, and we will schematically represent the compound lenses in the instrument by single thin lenses of suitable focal lengths.

chromatic aberration

spherical aberration

The Photographic Camera

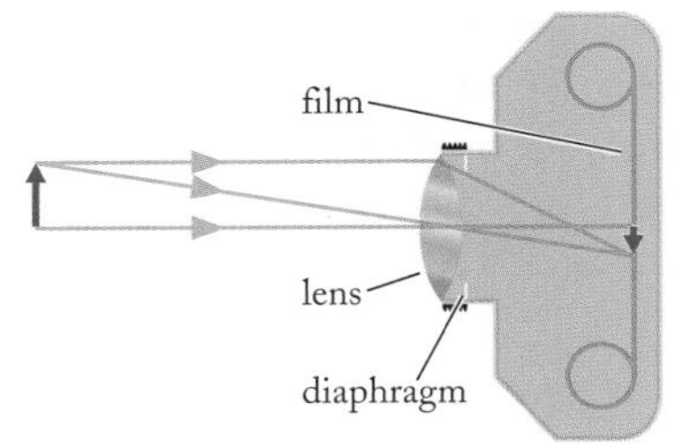

FIGURE 34.53 A photographic camera.

The lens of the camera focuses a real image of the object on the photographic film or digital sensor and thereby imprints the image on the film or sensor (see Fig. 34.53). The distance between the lens and the film is adjusted in focusing, so a properly focused image always falls on the film, regardless of the object distance. A shutter controls the exposure time during which light is admitted to the camera. For a given exposure time, the amount of light entering the camera is proportional to the area of the lens. The size of the camera lens is commonly labeled by the **f number**, which is defined as the ratio of the focal length of the lens to its diameter. For instance, a lens of focal length 55 mm and a diameter of 32 mm has an f number of 55/32, or 1.7. Reducing the diameter of the lens by one-half would double the f number to 3.4. It would also reduce the area of the lens by a factor of 4, and for a given exposure time, this would reduce the amount of light entering the camera by a factor of 4. If we want to compensate for the reduced lens diameter and keep constant the amount of light reaching the film, we would have to increase the exposure time by a factor of 4. A lens of small f number is said to be "fast," because it collects sufficient light for a photograph in a short exposure time.

Good cameras have an adjustable iris diaphragm that can be used to block part of the area of the lens and thereby alter its effective f number. If the diaphragm is closed down so only a small central portion of the lens remains unblocked, the camera will require a long exposure time, but it will have a large **depth of field**—it simultaneously forms sharp images for objects spanning a large range of object distances. This is so because the rays emitted by any object then enter the camera at small angles, and they therefore intersect the image at small angles; thus the rays will stay close together on the photographic film, even if the position of the image does not fall exactly on the

depth of field

film (see Fig. 34.54). Some camera lens systems have a "zoom" lens, permitting the focal length to be increased (other cameras permit substitution of a lens with longer focal length); this feature provides a larger image for a given object distance.

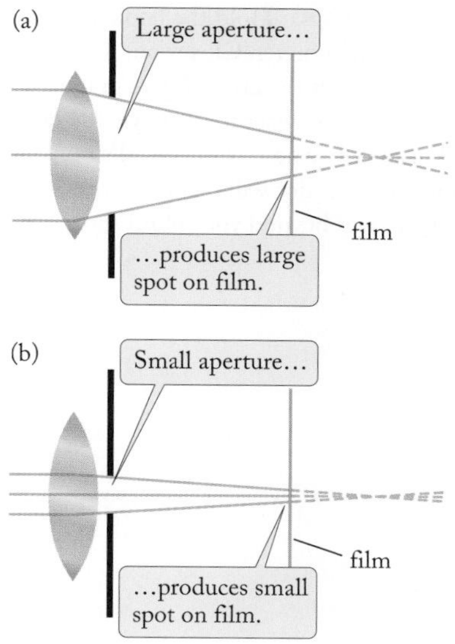

FIGURE 34.54 (a) When the aperture is large, the rays are spread over a large angle. If the focal point is beyond the film, the rays make a large spot on the film. (b) When the aperture is small, the rays are confined to a narrow range of angles. Even if the focal point is beyond the film, the rays make only a small spot on the film.

The Eye

In principle, the eye is similar to a camera. The lens of the eye forms a real image on the retina, a delicate membrane packed with light-receptor cells, which send nerve impulses to the brain. Figure 34.55 shows a horizontal section through the human eye. The diameter of the eyeball is typically 2.3 cm.

The cornea and the aqueous humor act as a lens; they provide most of the refraction for rays entering the eye. The crystalline lens merely provides the fine adjustments of focal length required to make the image of an object fall on the retina, regardless of the object distance. The crystalline lens is flexible and its focal length is adjusted by the ciliary muscles; this adjustment is called accommodation. If the eye is viewing a distant object, the muscles are relaxed and the lens is fairly flat, with a long focal length. If the eye is viewing a nearby object, the muscles are contracted and the lens is more rounded, with a short focal length.

The shortest attainable focal length determines the shortest distance at which an object can be placed from the eye and still be be seen sharply. This shortest distance is called the **near point** of the eye. For a normal young adult, the near point is typically 25 cm (although nearsighted persons can view much closer objects). With advancing age the lens loses its flexibility and the near point recedes; for instance, at an age of 60 years, the near point is typically around 200 cm.

The iris serves the same function as the diaphragm of the camera—it controls the size of the pupil and the amount of light admitted to the eye. The pupil contracts to a diameter of 2 mm in bright light, and expands to 7 mm in faint light, a change of area by a factor of about 12.

The receptor cells lining the retina are of two kinds: about 120 million rods and 6 million cones. The rods are more sensitive to light than the cones, but they do not discriminate among colors. The cones are blind at low light levels, but they give us color perception. See page xlvii of the Prelude for images of the retina and of red cells.

The cones also give us high-acuity vision, because some 2000 of them are densely packed into a small spot, the fovea, located near the center of the retina. However, the density of rod cells is maximum at retinal locations about 20° away from the fovea. For this reason, very faint light is best perceived peripherally, by looking to one side of the light source, instead of staring directly at it.

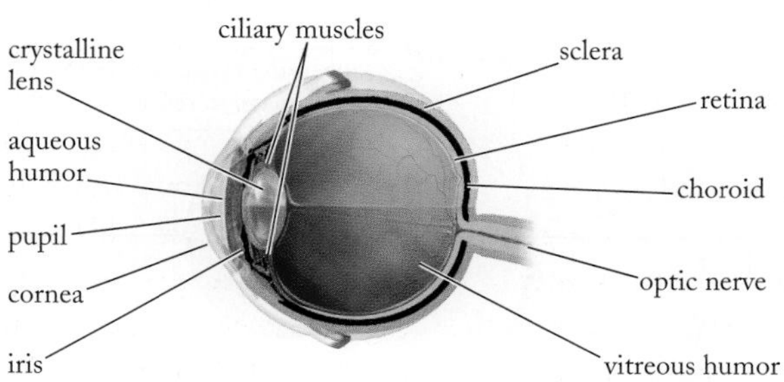

FIGURE 34.55 The human eye (natural size). The space between the cornea and the crystalline lens is filled with a transparent jelly, the aqueous humor. The main body of the eye is also filled with a transparent jelly, the vitreous humor. The index of refraction of the humors is 1.34, nearly the same as for water. The index of refraction of the crystalline lens is 1.44. The sclera is the thick white outer casing of the eye. The choroid is a pigmented black membrane, like the black paint inside cameras.

FIGURE 34.56 Hold the book at a distance of half a meter. Close your left eye and steadily stare at the cross. If you gradually bring the book closer to your eye, the dot will disappear when its image falls on your blind spot; the dot will reappear when the image moves beyond your blind spot.

The bundle of nerve fibers connecting the receptor cells to the brain leaves the eye at a place somewhat to one side of the fovea. This is called the blind spot, because it lacks receptor cells. You are normally not aware of this blind spot, because the brain fills in the visual picture from information acquired during eye movements; but you can demonstrate the existence of the blind spot by means of Fig. 34.56.

The three most common optical defects of the eye are nearsightedness, farsightedness, and astigmatism. *In a nearsighted (myopic) eye, the focal length is excessively short,* even when the ciliary muscles are completely relaxed. Thus, parallel rays from a distant object come to a focus in front of the retina (see Fig. 34.57a) and fail to form a sharp image on the retina—vision of distant objects is blurred. This condition can be corrected by eyeglasses with divergent lenses (see Fig. 34.57b).

In a farsighted (hyperopic) eye, the focal length is excessively long, even when the ciliary muscles are fully contracted (in other words, the near point of the eye is farther away than normal). Hence rays from nearby objects converge toward an image beyond the retina (see Fig. 34.58a) and fail to form a sharp image on the retina—vision of nearby objects is blurred. The condition can be corrected by eyeglasses with convergent lenses (see Fig. 34.58b). In old age, both nearsightedness and farsightedness often occur simultaneously, through loss of flexibility of the crystalline lens and the weakening of the ciliary muscles, with a consequent failure of the eye to accommodate over the full range of distances (presbyopia). The correction then requires bifocal lenses, with a lower convergent portion for near vision and an upper divergent portion for far vision.

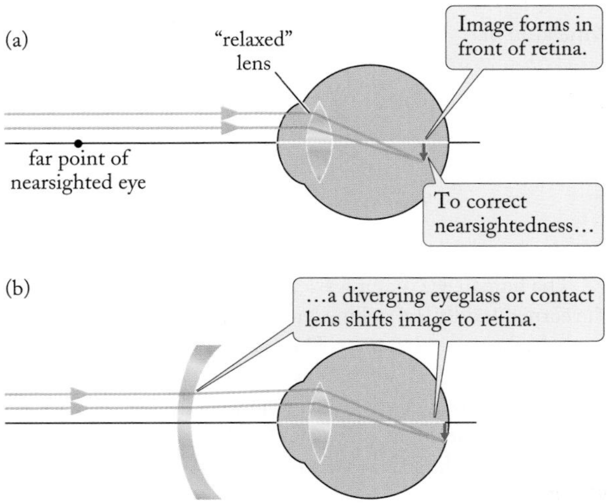

FIGURE 34.57 (a) Nearsighted eye forms an image in front of the retina. (b) Nearsighted eye with corrective lens.

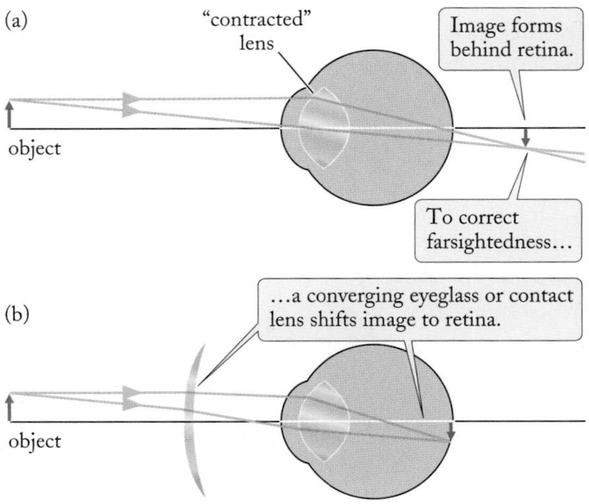

FIGURE 34.58 (a) Farsighted eye forms an image beyond the retina. (b) Farsighted eye with corrective lens.

EXAMPLE 15 A farsighted person has a near point of 200 cm. What kind of glasses does this person need to be able to read a newspaper held at 25 cm from the eyes?

SOLUTION: The object distance is 25 cm (we ignore the small difference between distances measured from the eye and from the glasses). Since the nearest point at which the person can see distinctly is 200 cm away, the lens must form an image of the newspaper at this distance—that is, $s' = -200$ cm. The lens equation then tells us

$$\frac{1}{f} = \frac{1}{s} + \frac{1}{s'} = \frac{1}{25\text{ cm}} + \frac{1}{-200\text{ cm}} = \frac{8-1}{200}\frac{1}{\text{cm}}$$

and

$$f = 29\text{ cm}$$

Incidentally: Eyeglass prescriptions are specified in **diopters**, which is the inverse of the focal length in meters; thus for the farsighted person of this example, reading glasses of strength $1/0.29 = +3.4$ diopters would permit reading at 25 cm. Inexpensive reading glasses of various strengths are available at drug or convenience stores.

diopters

Astigmatism is an inability to focus simultaneously light rays arriving in different planes, for instance, light rays arriving in the vertical plane and light rays arriving in the horizontal plane. This is caused by a slight horizontal or vertical flattening of the cornea—instead of being curved spherically, the cornea is slightly out of round, with more curvature in one direction than in the other. Although astigmatism is quite common, it is often so slight as to be unnoticeable. Figure 34.59 provides a simple test for astigmatism. Correction of this condition requires a lens with a cylindrical surface, which focuses rays of light in the, say, vertical plane but does not deflect rays of light in the horizontal plane.

FIGURE 34.59 Close one eye and focus the other on the central ends of the radial lines. If some of the lines appear less sharp or less black than others, your eye is astigmatic. (To detect mild astigmatism, you may find it helpful to perform this test on each eye and notice the difference, if any.)

The Magnifier

In order to see fine detail with the naked eye, we must bring the object very close to the eye, so the angular size of the object is large and, correspondingly, the image on the retina is large (see Fig. 34.60). This means we want to bring the object to the near point, at a typical distance of 25 cm for the eye of a young adult. To see finer detail, we need a magnifier. This consists of a strong convergent lens placed adjacent to the eye, as shown in Fig. 34.61.[2] Such a lens permits us to bring the object closer to the eye than 25 cm, and thereby increase the size of the image on the retina.

The angular magnification of the magnifier is defined as the ratio of the angular size of the image produced by the magnifier at the largest possible distance (as in Fig. 34.62) to the angular size of the object seen by the naked eye at the standard distance of 25 cm (as in Fig. 34.60). If the magnifier is to form an image at a large distance, we must make the extrapolated rays from the image nearly parallel, which requires that we place the object very near the focal distance of the magnifier. The extrapolated rays (shown dashed in Fig. 34.62) are then nearly parallel, and they intersect far

[2] Note that such a magnifier is not the same thing as a magnifying glass. In common use, the magnifying glass is placed at an appreciable distance from the eye, near the object to be magnified, because this maximizes the magnification. A magnifying glass can be regarded as a magnifier (in the technical sense of the word) only if it is placed near the eye.

from the lens. In the extreme case, we place the object at the focal distance f; the extrapolated rays are then parallel. The intersection of these rays is then at infinite distance, and the image is said to be at infinity.

To evaluate the magnification, we need to compare the angles θ and θ' in Figs. 34.60 and 34.62. Since the angle θ in Fig. 34.60 is small, we can use the familiar formula that this angle (in radians) approximately equals the tangent of the angle, or the size h divided by the distance of 25 cm:

$$\theta = \frac{h}{25 \text{ cm}}$$

Likewise, the angle θ' in Fig. 34.62 is small and approximately equals the size h divided by the focal distance f:

$$\theta' = \frac{h}{f}$$

Taking the ratio of these two angles, we find the angular magnification of the magnifier,

angular magnification of magnifier

$$[\text{angular magnification}] = \frac{\theta'}{\theta} = \frac{25 \text{ cm}}{f} \tag{34.32}$$

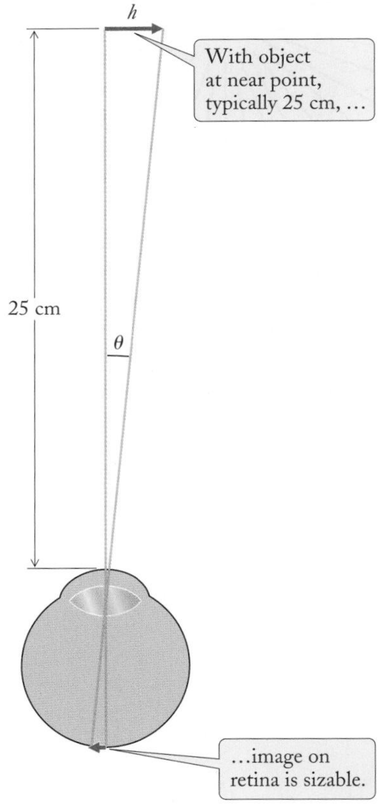

FIGURE 34.60 The angular size of the object determines the size of the image on the retina. Here the object has been placed at a distance of 25 cm from the eye.

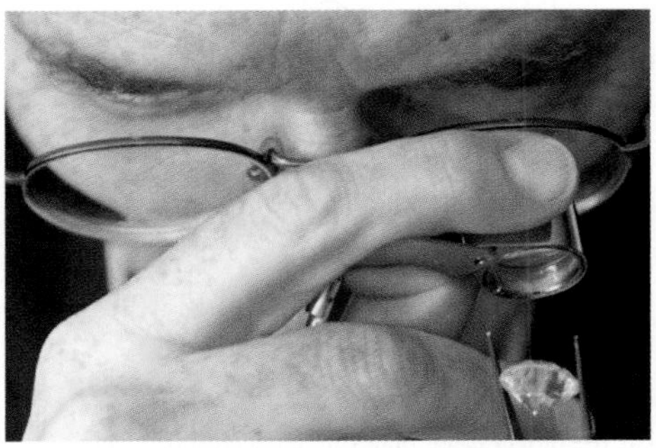

FIGURE 34.61 A magnifier used by a jeweler.

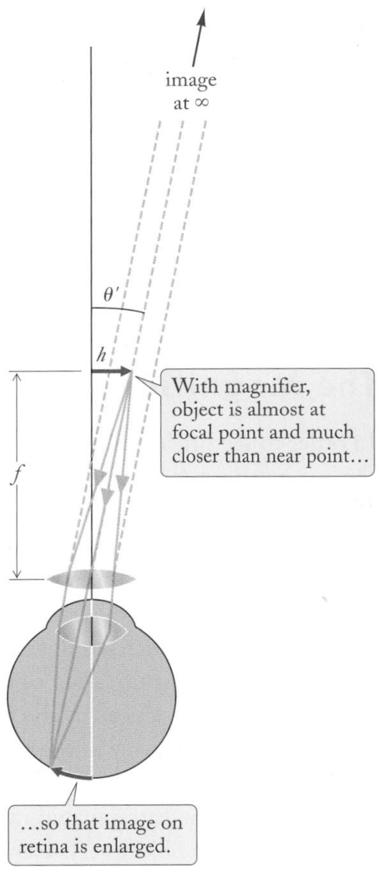

FIGURE 34.62 The magnifier is adjacent to the eye. The object has been placed at a distance slightly shorter than the focal distance, so that the eye sees the image nearly at infinity.

This tells us the magnification relative to that of the typical naked eye; that is, it tells us how much better the magnifier is than the naked eye. For example, a magnifier with $f = 5$ cm has an angular magnification of (25 cm)/(5 cm) = 5. Note that this result is valid only under the assumption that the magnifier is placed adjacent to the eye, and that the object is placed near the focus of the magnifier, so the image is at infinity. The second of these assumptions is not crucial—if the object is placed somewhat closer than the focus, the magnification will be changed only slightly. But the first assumption is crucial—if the magnifier is placed at some appreciable distance from the eye, then the magnification will be quite different.

The Microscope

The microscope consists of two lenses: the objective and the ocular, or eyepiece. Both of these lenses have very short focal lengths. The objective is placed near the object, and it forms a real, magnified image of the object. This image serves as object for the ocular, which acts as a magnifier and forms a virtual image at infinity (see Fig. 34.63). Thus both the objective and the ocular contribute to the magnification of the microscope. The net angular magnification of the microscope is the angular magnification of the ocular multiplied by the magnification of the objective. The angular magnification of the ocular is given by Eq. (34.32) with $f = f_{\text{ocular}}$; and the magnification of the objective is given by Eq. (34.30), where s and s' are, respectively, the object and image distances for the objective (these distances are marked in Fig. 34.63). If we combine these two magnifications, and if we ignore the sign in Eq. (34.30), we find that the net angular magnification of the microscope is

$$[\text{angular magnification}] = \frac{25\text{ cm}}{f_{\text{ocular}}} \times \frac{s'}{s} \qquad (34.33)$$

angular magnification of microscope

Note that, as in the case of the magnifier, this tells us the magnification relative to the (typical) naked eye. Good microscopes operate at magnifications of up to 1400. Although higher magnifications can be achieved, this serves little purpose because diffraction of the light (see the next chapter) at the objective limits the details that can be resolved. To overcome this limitation, we need to use waves of shorter wavelength than light waves, such as the electron waves used in electron microscopes, which will be discussed in Chapter 38.

The Telescope

A simple astronomical telescope consists of an objective of very long focal length and an ocular of short focal length. These two lenses are separated by a distance (nearly) equal to the sum of their individual focal lengths, so their focal points coincide (see Fig. 34.64). The objective forms a real image FP' of a distant object. This image serves as object for the ocular, which forms a magnified virtual image at infinity (Fig. 34.64).

To find the angular magnification produced by this telescope, we begin by noting that the lens equation with object distance $s = \infty$ applied to the objective gives

$$\frac{1}{\infty} + \frac{1}{s'} = \frac{1}{f_{\text{objective}}}$$

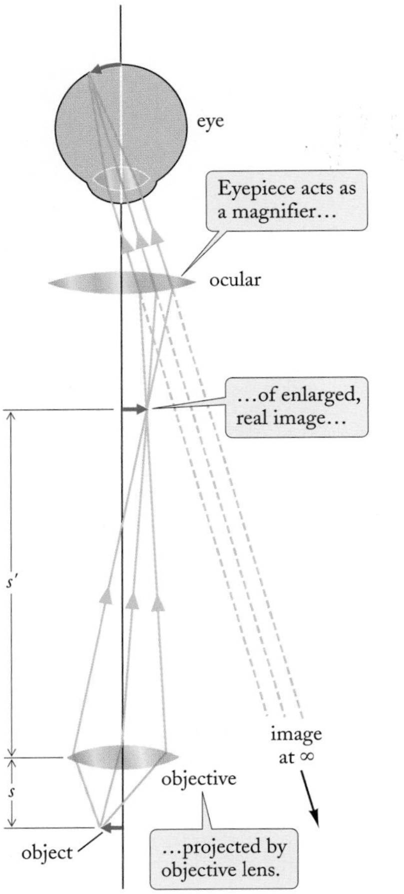

FIGURE 34.63 Arrangement of lenses in a microscope. The object to be magnified is below the objective, and the eye is just above the ocular. The ocular forms a virtual image at infinity, and the lens of the eye focuses on the retina the parallel rays emerging from the ocular.

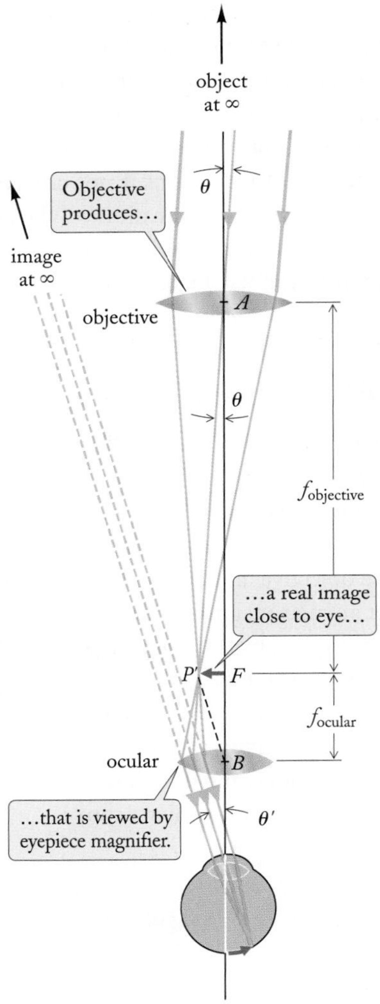

FIGURE 34.64 An astronomical telescope. The object (a star) is at a large distance above the telescope. The observer's eye is below the ocular. The ocular forms a virtual image at infinity, and the lens of the eye focuses on the retina the parallel rays emerging from the ocular.

But $1/\infty = 0$, and hence

$$s' = f_{\text{objective}} \tag{34.34}$$

This means that the first image is at the focal distance of the objective. Accordingly, in Fig. 34.64, the image FP' is shown located at the focal point F. We can now use the geometric relationships contained in this figure. The angular magnification is the ratio of the angles θ and θ' that represent, respectively, the angular sizes of the object and the final image viewed by the eye. Since both these angles are small, their values (in radians) are approximately the transverse distance FP' divided by the distance to the appropriate lens:

$$\frac{\theta'}{\theta} = \frac{FP'/FB}{FP'/FA} = \frac{FA}{FB} = \frac{f_{\text{objective}}}{f_{\text{ocular}}}$$

and so the angular magnification is

angular magnification of telescope

$$[\text{angular magnification}] = \frac{\theta'}{\theta} = \frac{f_{\text{objective}}}{f_{\text{ocular}}} \tag{34.35}$$

For example, an astronomical telescope with $f_{\text{objective}} = 120$ cm and $f_{\text{ocular}} = 2.5$ cm has an angular magnification of $(120 \text{ cm})/(2.5 \text{ cm}) = 48$.

Many astronomical telescopes are reflecting telescopes in which a concave mirror plays the role of the objective. The mirror forms a real image that serves as object for the ocular. Of course, the ocular (or a small mirror diverting light toward the ocular) must be placed in front of the large concave mirror, blocking out some of the light. Figure 34.65 shows an arrangement of concave mirror and ocular. The formula for the angular magnification of such a mirror telescope is, again, Eq. (34.35).

EXAMPLE 16 Suppose that an ocular with focal length $f = 1.5$ cm is used with the Hubble telescope mirror of the chapter photo, discussed in Example 7. What is the system's angular magnification?

Concepts in Context

SOLUTION: The angular magnification is given by Eq. (34.35); thus, using $f_{\text{objective}} = 5.5$ m from Example 7 and $f_{\text{ocular}} = 1.5$ cm,

$$[\text{angular magnification}] = \frac{f_{\text{objective}}}{f_{\text{ocular}}} = \frac{5.5 \text{ m}}{1.5 \times 10^{-2} \text{ m}} = 370$$

COMMENT: In addition to a large focal length and a large magnification, a good astronomical telescope must have other attributes. In order to view faint stars, a telescope must collect as much light as possible, and thus must have a large diameter. A large diameter is also needed to better resolve adjacent distant objects (to be discussed in Section 35.6, Diffraction by a Circular Aperture).

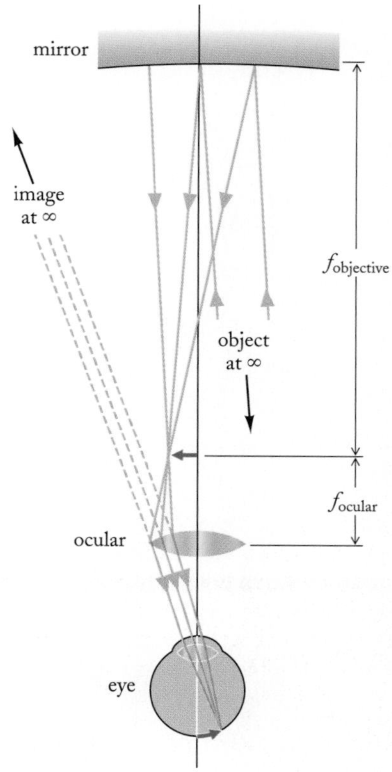

FIGURE 34.65 A reflecting telescope. In this diagram, the object has been placed below, to facilitate comparison with Fig. 34.64.

Because large mirrors of good quality are easier to manufacture than large lenses of good quality, the largest astronomical telescopes all use mirrors. For instance, the telescope on Mount Palomar (see Fig. 34.66a) uses a mirror of diameter 5.1 m and a focal length of 16.8 m. The Hobby–Eberly telescope in Texas and the Keck telescope on the summit of Mauna Kea in Hawaii are the largest reflectors; the former uses an interlocking array of 91 precisely aligned hexagonal mirrors (see Fig. 34.66b), acting as a single mirror of diameter 10.5 m and focal length 13.1 m.

(a)

(b)

FIGURE 34.66 (a) The 200-in. telescope on Mount Palomar in California. The mirror can be seen under the observer cage. (b) The Hobby–Eberly telescope. The large reflector is made up of many hexagonal mirrors.

Checkup 34.6

QUESTION 1: You want to use a camera to make a photograph in dim light. Do you need a camera of small f number or large f number?

QUESTION 2: A farsighted person has difficulty reading in normal light, but can still read well in very bright light. Explain.

QUESTION 3: Consider the microscope illustrated in Fig. 34.63.

(A) Does the objective produce a linear magnification of the object? An angular magnification?
(B) Does the ocular produce a linear magnification? An angular magnification?
(C) Is the image produced by the objective real or virtual? Upright or inverted?
(D) Is the image produced by the ocular real or virtual? Upright or inverted?

QUESTION 4: Consider the telescope illustrated in Fig. 34.64.

(A) Does the objective produce a linear magnification of the object? An angular magnification?
(B) Does the ocular produce a linear magnification? An angular magnification?
(C) Is the image produced by the objective real or virtual? Upright or inverted?
(D) Is the image produced by the ocular real or virtual? Upright or inverted?

QUESTION 5: The arrangement of the lenses in the microscope and the telescope is quite similar. What is the main difference?

QUESTION 6: For which instrument(s) does the angular magnification depend on the (somewhat arbitrary) typical distance of the near point of vision, 25 cm?

(A) Microscope only (B) Magnifier only (C) Telescope only
(D) Microscope and magnifier only (E) Microscope, magnifier, and telescope

SUMMARY

PHYSICS IN PRACTICE Optical Fibers **(page 1124)**

PROBLEM-SOLVING TECHNIQUES Sign Conventions for Mirrors **(page 1132)**

PROBLEM-SOLVING TECHNIQUES Images of Spherical Mirrors **(page 1134)**

PROBLEM-SOLVING TECHNIQUES Sign Conventions for Lenses **(page 1138)**

PROBLEM-SOLVING TECHNIQUES Images of Lenses **(page 1143)**

WAVELET Sphere with radius [wave speed] × Δt centered on old wave front.

HUYGENS' CONSTRUCTION The new wave front is the surface that touches the wavelets erected on the old wave front.

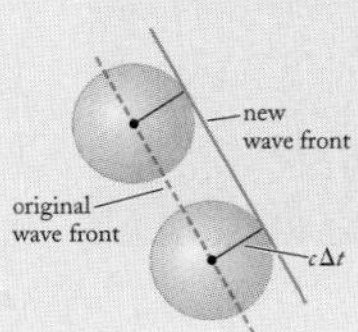

LAW OF REFLECTION

$$\theta_{\text{incident}} = \theta_{\text{reflected}} \qquad (34.1)$$

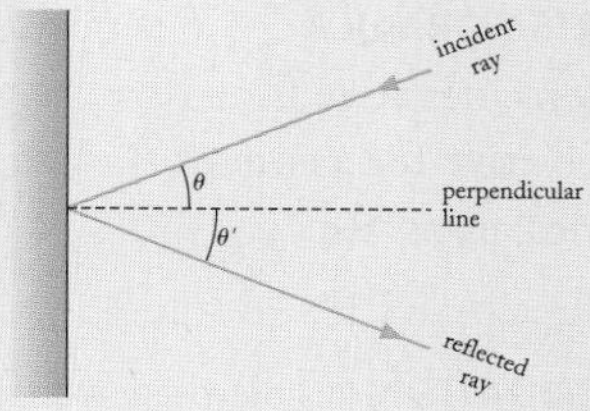

INDEX OF REFRACTION
(v is the speed of light in the material.)

$$n = \frac{c}{v} \qquad (34.4)$$

WAVELENGTH IN MATERIAL

$$\lambda = \frac{\lambda_{\text{vac}}}{n} \qquad (34.7)$$

INDEX OF REFRACTION
(in terms of dielectric constant)

$$n = \sqrt{\kappa} \qquad (34.5)$$

LAW OF REFRACTION (SNELL'S LAW)
where angles are measured with respect to a perpendicular to the surface.

$$n_1 \sin\theta_1 = n_2 \sin\theta_2 \qquad (34.11)$$

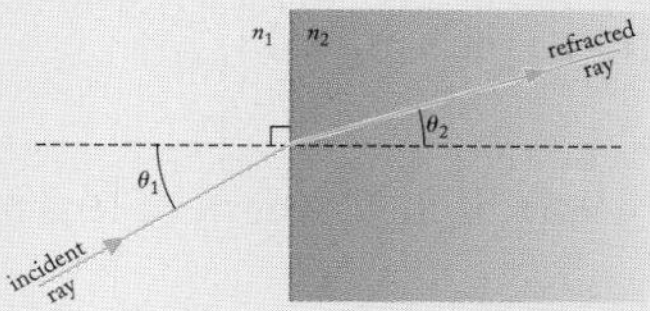

APPARENT DEPTH (for a flat surface, when viewing from almost directly above)

$$[\text{apparent depth}] = \frac{[\text{actual depth}]}{n} \qquad (34.15)$$

CRITICAL ANGLE FOR TOTAL INTERNAL REFLECTION

$$\sin\theta_{\text{crit}} = \frac{1}{n} \qquad (34.16)$$

COMPLETE POLARIZATION BY REFLECTION (BREWSTER'S LAW)

$$\tan\theta_p = n \qquad (34.17)$$

DISPERSION Separation of colors of refracted light due to a wavelength-dependent index of refraction.

FOCAL LENGTH OF SPHERICAL MIRROR (f is positive for concave mirror, negative for convex; R is radius of curvature of mirror surface.)

$$f = \pm\tfrac{1}{2}R \qquad (34.19, 34.20)$$

MIRROR EQUATION (s is the object distance; s' is the image distance. s or s' is positive if object or image is in front of mirror, negative if behind.)

$$\frac{1}{s} + \frac{1}{s'} = \frac{1}{f} \qquad (34.21)$$

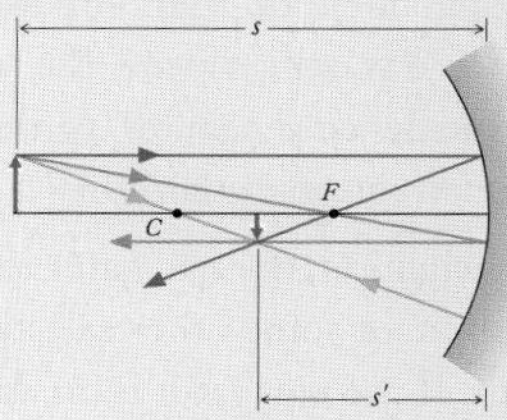

LENS MAKER'S FORMULA (f is positive for a converging lens, negative for a diverging lens; R_1 or R_2 is positive for a convex surface, negative for a concave surface.)

$$\frac{1}{f} = (n-1)\left(\frac{1}{R_1} + \frac{1}{R_2}\right) \quad (34.27)$$

THIN-LENS EQUATION [s is positive if object is on near side of lens, negative if on far side (relative to the source of light); s' is positive if image is on far side of lens, negative if on near side.]

$$\frac{1}{s} + \frac{1}{s'} = \frac{1}{f} \quad (34.28)$$

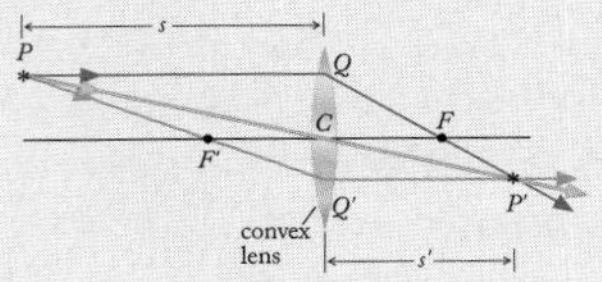

LINEAR MAGNIFICATION (for a lens or a mirror)

$$M = -\frac{s'}{s} \quad (34.30)$$

LINEAR MAGNIFICATION OF A SYSTEM OF LENSES OR MIRRORS (M_1 is the magnification of the first lens or mirror, etc.)

$$M = M_1 \times M_2 \times \cdots \quad (34.37)$$

f NUMBER Ratio of focal length of lens to diameter of aperture.

DEPTH OF FIELD Range of object distances over which a sharp image forms.

NEAR POINT The shortest distance for sharp vision, typically 25 cm.

ANGULAR MAGNIFICATION OF MAGNIFIER

$$\frac{25\text{ cm}}{f} \quad (34.32)$$

ANGULAR MAGNIFICATION OF MICROSCOPE

$$\frac{25\text{ cm}}{f_{\text{ocular}}} \times \frac{s'}{s} \quad (34.33)$$

ANGULAR MAGNIFICATION OF TELESCOPE

$$\frac{f_{\text{objective}}}{f_{\text{ocular}}} \quad (34.35)$$

QUESTIONS FOR DISCUSSION

1. When light is incident on a smooth surface—a glass surface, a painted surface, a water surface—the reflection is strongest if the angle of incidence is near 90° (grazing incidence). Can Huygens' Construction explain this?
2. In celestial navigation, the navigator measures the angle between the Sun, or some other celestial body, and the horizon with a sextant. If the navigator is on dry land, where the horizon is not visible, he can measure instead the angle between the Sun and its reflection in a pan full of water, and divide this angle by 2. Explain.
3. Suppose we release a short flash of light in the space between two parallel mirrors placed face to face. Why does this light flash not travel back and forth between the two mirrors forever?

4. To measure the index of refraction of some small transparent chips of plastic of irregular shape, a physicist places the chips in a glass beaker full of water. She then gradually adds sucrose to the water until the chips suddenly become nearly invisible. Explain why a measurement of the index of refraction of the sucrose solution then yields the index of refraction of the chips of plastic.
5. At sunset, the image of the Sun remains visible for some time after the actual position of the Sun has sunk below the horizon. Explain.
6. After a navigator measures the angle between the Sun and the horizon with a sextant, he must make a correction for the refraction of sunlight by the atmosphere of the Earth. Does this refraction increase or decrease the apparent angle between the Sun and the horizon?
7. Artists are notorious for making mistakes when drawing or painting mirror images. What is wrong with the position and orientation of the image shown in the mirror in Fig. 34.67?

FIGURE 34.67 A detail from a medieval tapestry at the Musée de l'Hôtel de Cluny, Paris. The head of the unicorn is reflected in the mirror. What is wrong with this image?

8. Two parallel mirrors are face to face. Describe what you see if you stand between these mirrors.
9. Figure 34.68 shows spots of sunlight on a wall in the shade of a tree. The spots were made by sunlight that has passed through very small gaps between the leaves of the tree. Explain why all the spots are round and of nearly the same size, even though the gaps are of irregular shape and size. (Hint: The Sun is round.)

FIGURE 34.68 Spots of sunlight on a wall.

10. At amusement parks you find mirrors that make you look very short and fat or very tall and thin. What kinds of mirrors achieve these effects?
11. Two mirrors are arranged at right angles to one another. If an object is near the mirrors, what is the largest number of images of that object that could be visible?
12. Store owners often install convex mirrors at strategic locations in their stores to supervise the customers (Fig. 34.69). What is the advantage of a convex mirror over a flat mirror?

FIGURE 34.69 Convex mirror used to supervise customers in a store.

13. You look toward a lens or a mirror, and you see the image of an object. How can you tell whether this image is real or virtual?
14. Hand mirrors are sometimes concave, but never convex. Why?
15. If you place a book in front of a concave mirror, will it show you mirror writing? Does the answer depend on the distance of the book from the mirror?
16. How could you make a lens that focuses sound waves?
17. If you place a small lightbulb at the focus of a convex lens and look toward the lens from the other side, what will you see?
18. Consider Fig. 34.46. How do we know that the ray $PQ'P'$ emerges parallel to the axis of the lens?
19. Suppose you place a magnifying glass against a flat mirror and look into the glass. What do you see if your face is very near the glass? If it is not very near?
20. Lenses, like mirrors, suffer from aberration. Consider rays incident on a convex lens, parallel to the axis. If the point of incidence is far from the axis, would you expect the ray to pass in front of the focus or behind?

21. Figure 34.70a shows a large **Fresnel lens** used in the lantern of a lighthouse. The lens consists of annular segments, each with a curved surface similar to the curved surface of an ordinary lens (see Fig. 34.70b). Why is this arrangement better than a single curved surface?

(a)

(b)

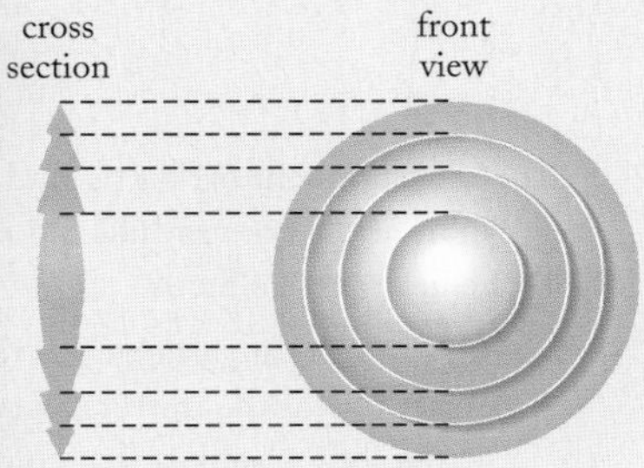

FIGURE 34.70 (a) Fresnel lens of the lighthouse at Point Reyes, California. (b) Two views of a Fresnel lens.

22. A convex lens is made of plastic of index of refraction 1.2. If you immerse this lens in water, will it produce convergence or divergence of incident parallel rays?

23. Binoculars use prisms to reflect the light back and forth (Fig. 34.71). What is the purpose of these prisms, and what is their advantage over mirrors?

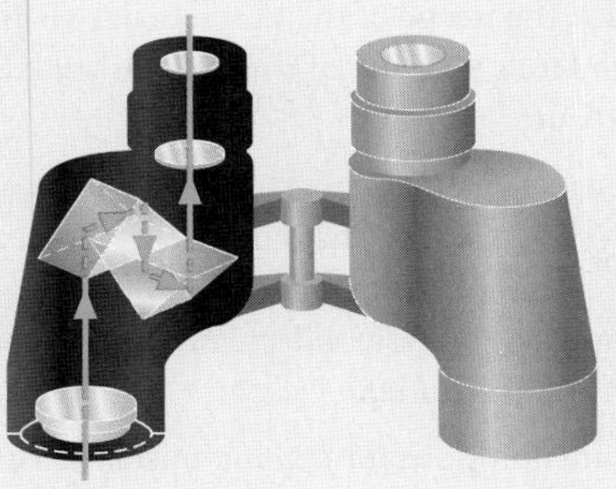

FIGURE 34.71 Arrangement of prisms in a binocular.

PROBLEMS

†34.2 Reflection

1. According to a (questionable) story, Archimedes set fire to the Roman ships besieging Syracuse by focusing the light of the Sun on them with mirrors. Suppose that Archimedes used flat mirrors. How many flat mirrors must simultaneously reflect sunlight at a piece of canvas if it is to catch fire? The energy flux of the sunlight at the surface of the Earth is $0.10\ \text{W/cm}^2$, and the energy flux required for ignition of canvas is $4.0\ \text{W/cm}^2$. Assume that the mirrors reflect the sunlight without loss.

2. Two flat mirrors are joined along their edges so they make an angle of 90° with each other (see Fig. 34.72). A cat sits in the angle formed by the mirrors. Draw a diagram showing the location and orientation of all the images of the cat in this system of mirrors.

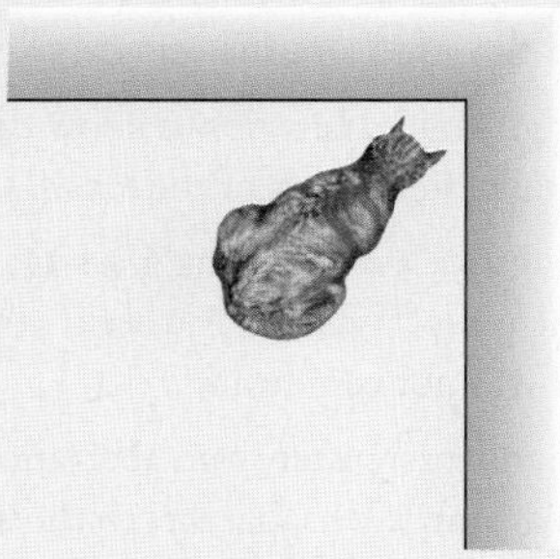

FIGURE 34.72 Cat between two mirrors at an angle of 90°.

†For help, see Online Concept Tutorial 37 at www.wwnorton.com/physics

3. A ray of light strikes a mirror at an angle of θ with respect to the perpendicular. At what angle with respect to the mirror surface does it reflect from the mirror?

4. If you wish to take a picture of yourself in a plane mirror, for what distance should the camera focus be set if you and the camera are 1.2 m from the mirror?

5. A ray of light is incident on a flat mirror at an angle θ. If we then rotate the mirror by 10° (see Fig. 34.73) while keeping the incident ray fixed, by what angle will we rotate the reflected ray?

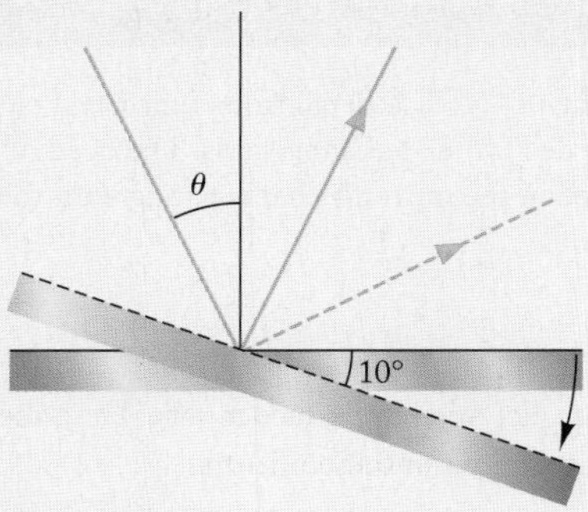

FIGURE 34.73 A flat mirror, before and after rotation.

*6. Two mirrors are arranged so they make an angle of 60° with each other. If a hummingbird hovers near the vertex where the mirrors meet, what is the number of images of itself that the hummingbird can see?

*7. Two mirrors are arranged so they make an angle of 45° with each other. If you stand near the vertex where the two mirrors meet, what is the number of images of yourself that you can see?

*8. A vertical mirror, oriented toward the Sun, throws a rectangular patch of sunlight on the floor in front of the mirror. The size of the mirror is 0.50 m × 0.50 m, and its bottom rests on the floor. If the Sun is 50° above the horizon, what is the size of the patch of sunlight on the floor?

*9. What are the minimum length and minimum width of a mirror hanging on a wall such that you can see your entire body when standing in front of it? Assume that your height is 1.80 m, your width is 0.50 m, and your eyes are separated by 0.06 m.

†34.3 Refraction

10. What is the speed of light in crown glass? Light flint glass? Heavy flint glass?

11. A helium–neon laser, producing the red light commonly used in demonstrations, has a wavelength of 633 nm in air. What is the frequency of the light in air? What is the wavelength of such light in crown glass? What is the frequency in the glass? What is the speed of the light in the glass?

12. The speed of light in diamond is 1.24×10^8 m/s. Calculate the index of refraction of diamond.

13. By what factor does the speed of light change when it goes from castor oil to ethyl alcohol?

14. A ray of sunlight approaches the surface of a smooth pond at an angle of incidence of 40°. What is the angle of refraction?

15. The speed of sound in air is 340 m/s, and in water it is 1500 m/s. If a sound wave in air approaches a water surface with an angle of incidence of 10°, what will be the angle of refraction?

16. Make a plot of the angle of incidence vs. the angle of refraction for light rays incident on a water surface. What is the maximum angle of refraction?

17. A ship's navigator observes the position of the Sun with his sextant and measures that the Sun is 39.000° away from the vertical. Taking into account the refraction of the Sun's light by air, what is the true angular position of the Sun? For the purpose of this problem, assume that the Earth is flat and that the atmosphere can be regarded as a flat, transparent plate of uniform density and index of refraction of 1.0003.

18. A pebble is at the bottom of a 2.0-m-deep swimming pool. What is the apparent depth of the pebble when viewed from outside the water, directly above the pebble?

[3] In all problems, assume the index of refraction of air is exactly 1, unless otherwise stated.

†For help, see Online Concept Tutorial 37 at www.wwnorton.com/physics

19. White light is incident on one of the materials in Table 34.1 at angle of 30.0°. If the blue wavelengths are refracted at an angle of 19.9° and the red wavelengths at an angle of 20.2°, what is the material?

20. A coin is embedded in a plastic holder with index of refraction 1.58. If the coin is at the center of a 10-cm-thick block of plastic, what is its apparent depth when viewed straight on?

21. In a glass (n = 1.62) fish tank filled with water, a ray of light strikes the inner surface of the glass tank at an angle of incidence of 30.0°. It then travels through the glass and out into the air. With what angle with respect to the perpendicular does it exit the glass?

22. Light of wavelength 550 nm in air strikes a surface of a clear material. The light is incident at 45° with respect to the normal; the ray refracts in the material at 28° with respect to the normal. What is the wavelength of light in the material?

23. A scuba diver is underwater; from there, the Sun appears to be at 30° from the vertical. At what actual angle is the Sun located (with respect to directly overhead)?

24. Laser scientists can now produce pulses of light in air almost as short as 1×10^{-15} s = 1 femtosecond. Such pulses are made up of different colors of light. Suppose such a pulse is normally incident on a slab of crown glass that is 1.00 mm thick. Dispersion in the glass can cause appreciable spreading of a short pulse. The index of refraction of blue light is 1.530, and the index of refraction of red light is 1.510. How much sooner does the red part of the pulse get to the other side of the glass compared with the blue part of the pulse?

25. If diamond has an index of refraction of 2.42, what is the critical angle for total internal reflection? (This relatively small angle facilitates total internal reflection, making the diamond more briliant.)

26. One of the angles of a prism is 90°. The prism is made of glass of index of refraction 1.58. If a ray of light enters this prism at an angle of incidence of 70°, as shown in Fig. 34.74, what is the angle at which this ray emerges?

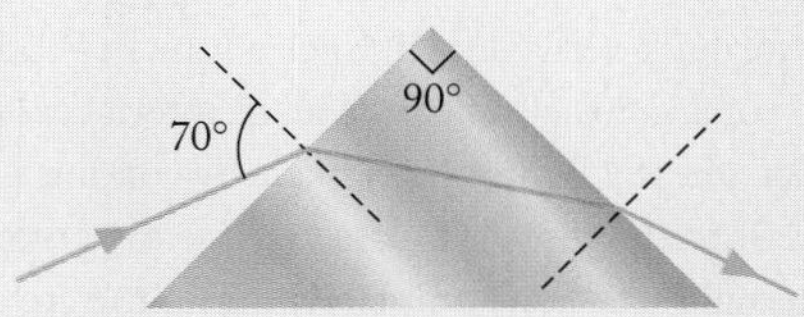

FIGURE 34.74 A prism with a 90° angle.

27. A prism of flint glass has an acute angle of 30.0°. A ray of white light is perpendicularly incident on one face of this prism (see Fig. 34.75). What are the angles at which red, green, and violet light emerge? The indices of refraction for red (660 nm), green (550 nm), and violet (400 nm) light are 1.650, 1.660, and 1.690, respectively.

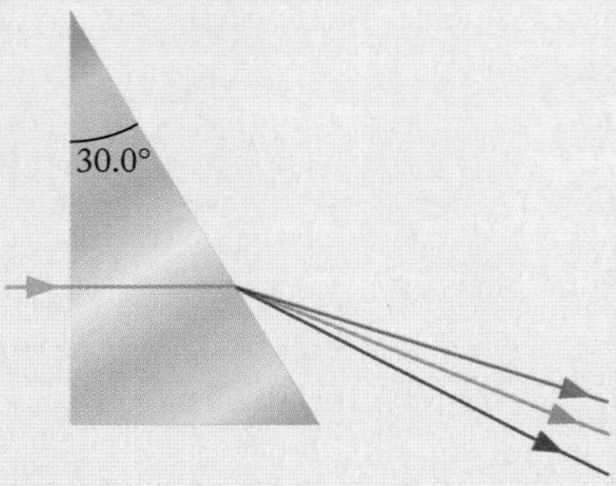

FIGURE 34.75 A 30° prism used to analyze a beam of white light.

28. An optical fiber is made of a thin strand of glass of index of refraction 1.50. If a light ray is to remain trapped within this fiber, what is the largest angle it can make with the surface of the fiber?

29. (a) A transparent medium of index of refraction n_1 adjoins a transparent medium of index of refraction n_2. Assuming $n_1 > n_2$, show that the critical angle for total internal reflection of a ray attempting to leave the first medium is

$$\sin\theta_{\text{crit}} = \frac{n_2}{n_1}$$

(b) A layer of kerosene ($n_2 = 1.20$) floats on a water surface. In this case, what is the critical angle for total internal reflection for a light ray attempting to leave the water?

30. To determine the index of refraction of a liquid, an experimenter sends a ray of light through the liquid from below and measures the critical angle for total internal reflection at the upper surface. She finds that this angle is 56°. What is the index of refraction?

31. What is the critical angle for a ray of light to be internally reflected inside a plate of crown glass? What if this plate is immersed in water? (See Problem 29.)

32. Use the formula derived from Snell's Law in Problem 29 to consider total internal reflection at an interface with a material other than air. If a light pipe is made by coating a cylinder of crown glass with light flint glass, what is the critical angle for total internal reflection completely inside the crown glass? (Such coatings avoid problems with scratches on the outer surface.) Are there angles for which light in the crown glass will refract out into the flint glass but will still be totally internally reflected in the flint glass?

33. A *mirage* is caused by a thin layer of hot air just above the ground, and its appearance is due to total internal reflection. When looking from far away, an observer sees a reflecting surface on the ground (which has the appearance of a water surface), coming from rays with an angle of reflection of 89.4° or more. If the cooler air has an index of refraction $n = 1.000\,26$, what is the index of refraction of the thin layer of hot air? (Hint: Use the result of Problem 29.)

34. When unpolarized light is incident on sapphire at an angle of 60.5°, the reflected light is completely polarized. What is the index of refraction of sapphire?

35. At what angle above the horizon is the Sun when its light, reflected from a calm swimming pool, is linearly polarized?

36. The critical angle for total internal reflection in cubic zirconia (in air) is 27.0°. What is the polarizing angle for reflection from cubic zirconia?

*37. A ray of light strikes a plate of window glass of index of refraction 1.5 and thickness 2.0 mm with an angle of incidence of 50°. Find the transverse displacement between the transmitted ray and the extrapolation of the incident ray.

*38. A point source of light is placed above a thick plate of glass of index of refraction n (Fig. 34.76). The distance from the source to the upper surface of the plate is l, and the thickness of the plate is d. A ray of light from the source may suffer either a single reflection at the upper surface, or a single reflection at the lower surface, or multiple alternating reflections at the lower and upper surfaces. Thus, each ray splits into several rays, giving rise to multiple images. In terms of l and d, find the distance of the first, second, and third images below the upper surface of the plate. Assume that the angle of incidence of the ray is small.

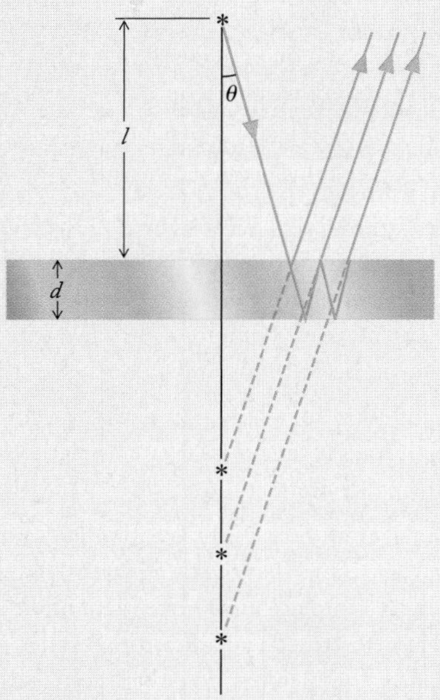

FIGURE 34.76 Ray of light reflected and refracted by a thick plate of glass.

*39. Light polarized in the plane of incidence (see Fig. 34.77) is incident upon water from air. There is no reflected ray. What is the angle of incidence?

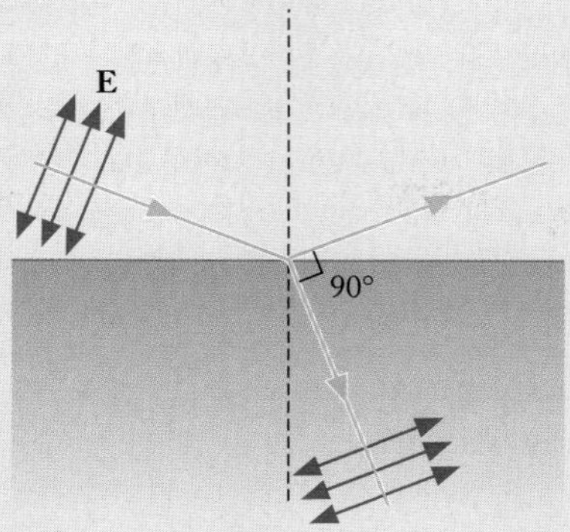

FIGURE 34.77 The light wave is polarized in the plane of the page. The refracted ray is perpendicular to the direction expected for a reflected ray.

*40. Because of refraction in the plate of glass, an object viewed through an ordinary window will seem somewhat nearer than its actual distance.

(a) Consider the rays that strike the glass at a small angle of incidence (nearly normal). Show that if an object is at a distance l from a window, the image is at a distance that is shorter by an amount $\Delta l = (1 - 1/n)d$, where n is the index of refraction and d is the thickness of the glass.

(b) What change in distance does this formula give if the windowpane is ordinary glass with $d = 2.0$ mm and $n = 1.5$?

(c) What change in distance does this formula give if the windowpane is heavy plate glass with $d = 8.0$ mm and $n = 1.5$?

*41. The highly reflective paint used on highway signs contains small glass beads, which reverse the direction of propagation of a light ray, throwing it back toward the source of light (see Fig. 34.78a). The reversal of direction will occur only for those light rays incident on the sphere of glass at a selected distance from the axis, a distance that depends on the index of refraction of the glass. Show that if the reversal of direction of an off-axis ray is to occur at all, the index of refraction of the glass must be in the range $1 < n \leq \sqrt{2}$. (Hint: Consider the light ray shown in Fig. 34.78b; what index of refraction does the reversal of this light ray require?)

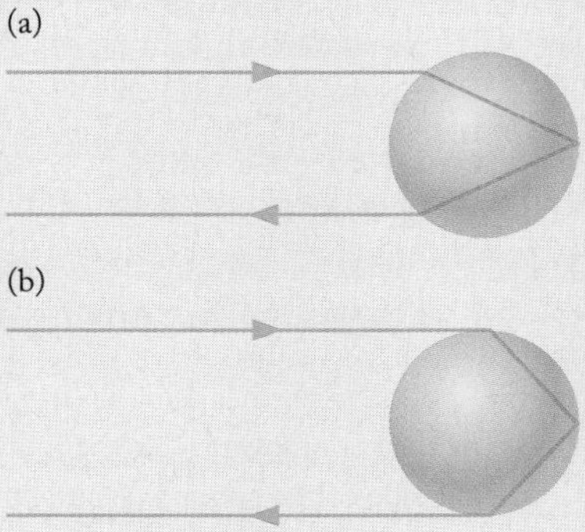

FIGURE 34.78 (a) Path of a light ray that reverses direction in a transparent sphere. (b) Path of a light ray that penetrates at maximum distance from the axis of a sphere.

*42. A signal rocket explodes at a height of 200 m above a ship on the surface of a smooth lake. The explosion sends out sound waves in all directions. Since the speed of sound in water (1500 m/s) is larger than the speed of sound in air (340 m/s), a sound wave can suffer total reflection at a water surface if it strikes at a sufficiently large angle of incidence. At what minimum distance from the ship will a sound wave from the explosion suffer total reflection?

*43. A layer of oil, of index of refraction n', floats on a surface of water. A ray of light coming from below attempts to pass from the water to the oil and from there to the air above. What is the maximum angle of incidence of the ray on the water–oil surface that will permit the ultimate escape of the ray into the air? Does your answer depend on n'? (Hint: Use the formula derived in Problem 29.)

*44. A seagull sits on the (smooth) surface of the sea. A shark swims toward the seagull at a constant depth of 5.0 m. How close (measured horizontally) can the shark approach before the seagull can see it?

*45. To discover the percentage of sucrose (cane sugar) in an aqueous solution, a chemist determines the index of refraction of the solution very precisely and then finds the percentage in a table giving the dependence of index of refraction on sucrose concentration. The chemist determines the index of refraction by immersing a glass prism in the sucrose and measuring the critical angle for total internal reflection of a light ray inside the glass prism.

(a) Suppose that with a prism of index of refraction 1.6640 the critical angle is 57.295°. Use the result of Problem 29 to find the index of refraction of the sucrose solution.

(b) Use the following table, interpolating if necessary, to find the concentration of sucrose to four significant figures:

CONCENTRATION	n
40.00%	1.3997
40.10	1.3999
40.20	1.4001
40.30	1.4003

**46. A prism of glass of index of refraction 1.50 has angles of 45.0°, 45.0°, and 90.0°. A ray of light is incident on one of the short faces at an angle θ (see Fig. 34.79). What is the maximum value of θ for which the ray of light will suffer total internal reflection at the long face?

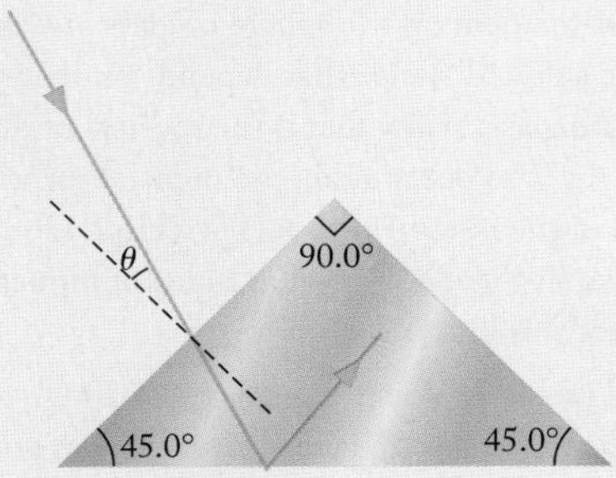

FIGURE 34.79 Total internal reflection in a prism.

**47. Rainbows are produced by the refraction of sunlight by drops of water. Figure 34.80 shows a ray of light entering a spherical drop of water. The ray is refracted at A, reflected at B, and refracted at C. The angle of incidence at A (between the ray and the normal to the surface) is θ, and the angle of refraction is θ'.

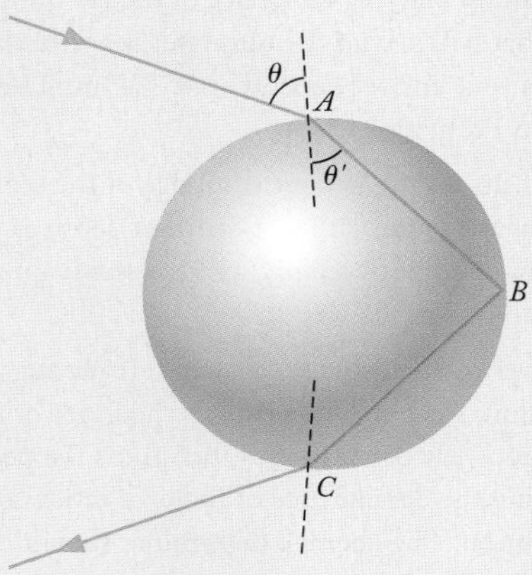

FIGURE 34.80 Path of a light ray in a drop of water.

(a) By geometry, show that the angles of incidence and reflection at B coincide with θ', and that the angles of incidence and refraction at C coincide with θ' and θ, respectively.

(b) Show that the angular deflection of the ray from its path is $\theta - \theta'$ at A, $\pi - 2\theta'$ at B, and $\theta - \theta'$ at C. These angles are all in radians, and they are measured clockwise from the incident path at each point.

(c) The total angular deflection of the ray by the raindrop is $\Delta = 2(\theta - \theta') + \pi - 2\theta'$. A rainbow will form when all the rays within an infinitesimal range $d\theta$ of angles of incidence suffer the same angular deflection, that is, when the derivative $d\Delta/d\theta = 2 - 4d\theta'/d\theta$ is zero. If this condition is satisfied, the rays sent back by the raindrops are concentrated, producing a bright zone in the sky. Show that the critical angle θ_c, at which $d\Delta/d\theta = 0$, is given by

$$\cos^2\theta_c = \tfrac{1}{3}(n^2 - 1)$$

where n is the index of refraction of water.

(d) The index of refraction for red light in water is 1.330. Find θ_c, and find Δ in degrees and minutes of arc. Draw a diagram showing a red ray coming from the Sun, hitting the drop, and reaching the eye of a rainbow watcher.

(e) The index of refraction for violet light in water is 1.342. Find θ_c and find Δ. On top of the preceding diagram, draw a violet ray coming from the Sun, hitting the drop (at a different point), and reaching the eye of the rainbow watcher. Will the watcher see the red color above or below the violet?

†34.4 Spherical Mirrors

48. In the headlamp of an automobile, a lightbulb placed in front of a concave spherical mirror generates a parallel beam of light. If the distance between the lightbulb and the mirror is 4.0 cm, what must the radius of curvature of the mirror be?

49. A lightbulb is placed at a distance of 15.0 cm in front of a convex mirror of radius of curvature 40.0 cm. Where is the image of the lightbulb? Draw a diagram showing the lightbulb, the mirror, and the image. Show three of the principal rays.

50. At what distance from a concave mirror of radius R must you place an object if the image is to be at the same position as the object?

51. A concave mirror has a radius of curvature R. If you want to form a real image, within what range of distances from the mirror must you place the object? If you want to form a virtual image, within what range of distances must you place the object?

52. A candle is placed at a distance of 15.0 cm in front of a concave mirror of radius 40.0 cm. Where is the image of the candle? Draw a diagram showing the candle, the mirror, and the image. Show three of the principal rays.

53. A candy cane is 15.0 cm from a reflecting sphere (a holiday ornament) that is 6.0 cm in diameter. What is the focal length? Where is the image of the candy cane?

54. A concave mirror reflects an image of the Moon 6.0 cm from the mirror. If a coin is placed 8.0 cm from the same mirror, where will its image be?

55. A convex mirror has a focal length of −25.0 cm. What is the object distance if the image is 10.0 cm behind the mirror?

56. A soupspoon is a spherical mirror on either side. When looking at one side of the spoon, a woman sees an image of her eye 7.0 cm from the spoon, on the near side. Without changing positions, she flips the spoon over. Now the image of her eye is 2.0 cm on the far side of the spoon. What is the radius of curvature of the spoon? How far is the woman's eye from the spoon?

†For help, see Online Concept Tutorial 38 at www.wwnorton.com/physics

57. A hand mirror is to show a (virtual) image of your face magnified 1.5 times when held at a distance of 20 cm from your face. What must be the radius of curvature of a spherical mirror that will serve the purpose? Must it be concave or convex? (Hint: Note from the similar triangles in Fig. 34.42 that this magnification is equal to the magnitude of the ratio of the image to object distances.)

58. The surface of a highly polished doorknob of brass has a radius of curvature of 4.5 cm. If you hold this doorknob 15 cm away from your face, where is the image you see? By what factor does the size of the image differ from the size of your face? (Hunt: Note from the similar triangles in Fig. 34.37 that the magnification is equal to the magnitude of the ratio of the image to object distances.)

*59. A concave mirror of radius of curvature 30.0 cm faces a second concave mirror, of radius of curvature 24.0 cm. The distance between the mirrors is 80.0 cm, and their axes coincide. A lightbulb is suspended between the mirrors, at a distance of 20.0 cm from the first mirror.

(a) Where does the first mirror form an image of the lightbulb?

(b) Where does the second mirror form an image of the lightbulb?

*60. A concave mirror of radius of curvature 60.0 cm faces a convex mirror of the same radius of curvature. The distance between the mirrors is 50.0 cm, and their axes coincide. A candle is held between the mirrors, at a distance of 10.0 cm from the convex mirror. Consider rays of light that reflect first off the concave mirror and then off the convex mirror. Where do these rays form an image?

†34.5 Thin Lenses

61. The crystalline lens of a human eye has two convex surfaces with radii of curvature of 10 mm and 6.0 mm. The index of refraction of its material is 1.45. Treating it as a thin lens, calculate its focal length when removed from the eye and placed in air.

62. A thin lens of flint glass with $n = 1.58$ has one concave surface of radius of curvature 15 cm and one flat surface

(a) What is the focal length of this lens?

(b) If you place this lens at a distance of 40 cm from a candle, where will you find the image of the candle?

63. A thin, symmetric convex lens of crown glass with index of refraction $n = 1.52$ is to have a focal length of 20 cm. What are the correct radii of curvature of the spherical surfaces of the lens?

64. A double-convex lens has a focal length of 16 cm, and one of its surfaces has a radius of curvature of 27 cm. The lens glass has an index of refraction of 1.66. What is the radius of curvature of the other surface?

65. A double-concave lens has radii of curvature of 16 cm and 12 cm; the lens glass has an index of refraction of 1.58. What is the focal length of the lens?

†For help, see Online Concept Tutorial 38 at www.wwnorton.com/physics

66. A contact lens has one convex surface of radius of curvature 2.5 cm and one concave surface of radius of curvature 3.0 cm. It is made from a plastic of index of refraction 1.43. What is the focal length of the contact lens?

67. An object is located 15 cm from a concave lens of focal length −20 cm. What is the image distance? Is the image real or virtual? Upright or inverted? Larger or smaller than the object? What is the magnification?

68. The Moon subtends an angle of 0.52° when viewed from Earth. If a lens with a focal length of 150 mm is used to image the Moon, what is the height of the Moon's image?

69. A slide projector has a lens of focal length 13.0 cm. The lens is at a distance of 2.0 m from the screen. What must be the distance from the slide to the lens if a sharp image of the slide is to be seen on the screen? What is the magnification of the slide projector?

70. The convex lens of a magnifying glass has a focal length of 20 cm. At what distance from a postage stamp must you hold this lens to produce a (virtual) stamp twice as large as the stamp?

71. If you place a convex lens of focal length 18 cm at a distance of 30 cm from a small lightbulb, where will you find the image of the lightbulb? Is this a real or a virtual image? Is it upright or inverted?

72. The lens maker's formula, Eq. (34.27), follows from the Law of Refraction at the air–lens surfaces. Argue that if a thin lens of index of refraction n is immersed in a fluid (for example, water) of index of refraction n_0, then the lens maker's formula must be modified as follows:

$$\frac{1}{f} = \left(\frac{n}{n_0} - 1\right)\left(\frac{1}{R_1} + \frac{1}{R_2}\right)$$

(Hint: In the Law of Refraction, only the ratio of the indices of refraction is relevant.)

*73. Show that if two thin lenses of focal lengths f_1 and f_2 are placed next to one another (in contact; see Fig. 34.81), the net focal length f is given by

$$\frac{1}{f} = \frac{1}{f_1} + \frac{1}{f_2}$$

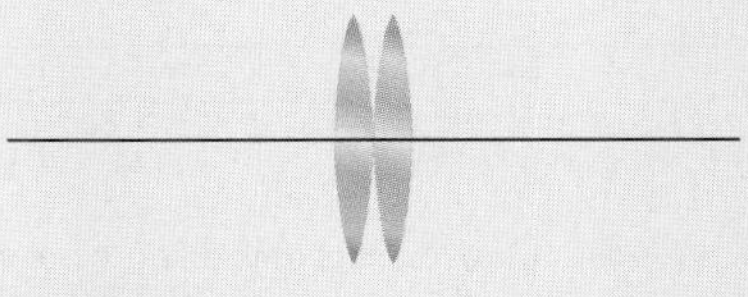

FIGURE 34.81 Two thin lenses in contact.

*74. A convex lens of focal length 25 cm is at a distance of 60 cm from a concave mirror of focal length 20 cm. A lightbulb is 80 cm from the lens (Fig. 34.82).

(a) Where does the lens form an image of the lightbulb?

(b) Where does the mirror form an image of the first image?

(c) Where does the lens form the (final) image of the second image?

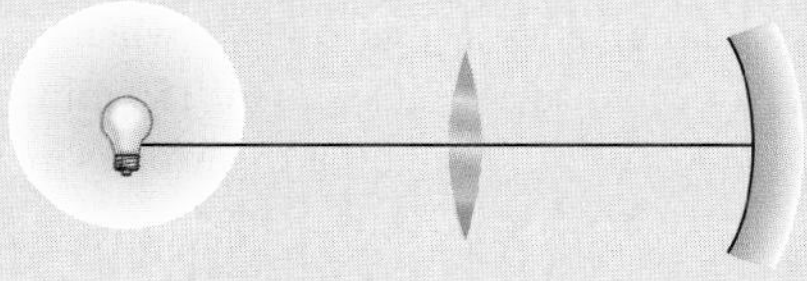

FIGURE 34.82 Lightbulb, convex lens, and concave mirror.

*75. Two lenses, one concave and one convex, have focal lengths of magnitude 30 cm. The lenses are separated by a distance of 10 cm. A candle is 20 cm from the convex lens (Fig. 34.83).

(a) Where does the convex lens form an image?

(a) Where does the concave lens form an image of this image?

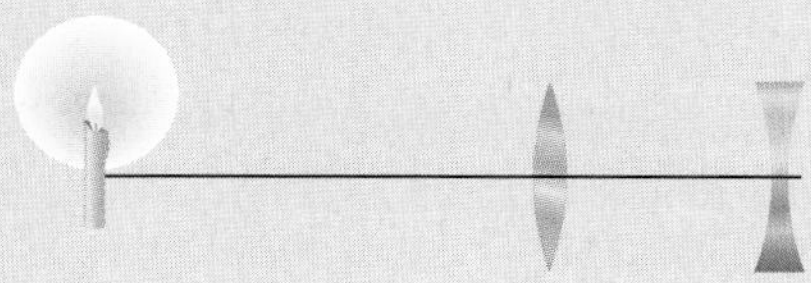

FIGURE 34.83 Candle, convex lens, and concave lens.

*76. An object is a distance $\frac{3}{2}f$ from a convex lens of focal length f. A second, identical lens is a distance $d = f$ beyond the first. Where is the final image? What is the overall magnification of the two-lens system?

*77. Prove that the lens equation can be put in the form $xx' = f^2$, where x and x' are the object and image distances measured from the focal points (see Fig. 34.84). This is called the **Newtonian form** of the lens equation.

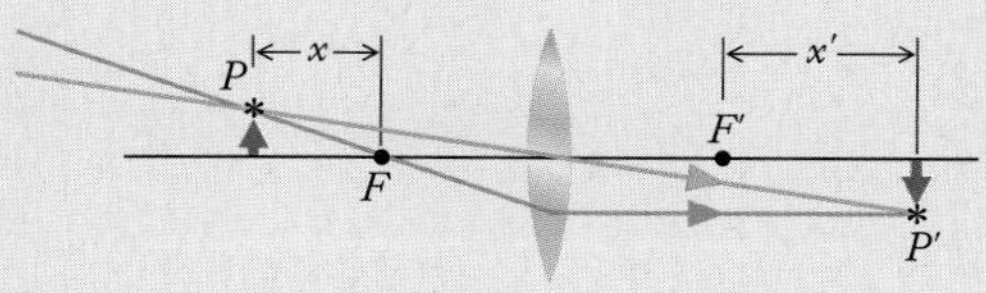

FIGURE 34.84 Convex lens and distances x and x'.

*78. To measure the focal length of a thin concave lens, a physicist places this lens adjacent to a thin convex lens of focal length 12.0 cm. She finds that when the joined lenses face the Sun, the rays of sunlight are focused on a spot 22.0 cm beyond the lenses. What focal length for the concave lens can she deduce from this?

*79. A lightbulb is 15.0 cm in front of a convex mirror of radius of curvature 10.0 cm. A convex lens of focal length 25.0 cm is 5.0 cm beyond the lightbulb (Fig. 34.85). Where do you see the image of the lightbulb if you look through the convex lens at the mirror?

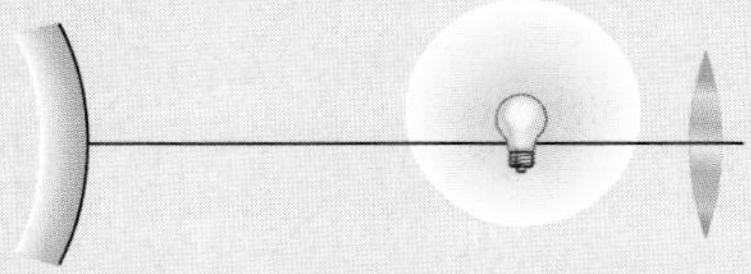

FIGURE 34.85 Lightbulb, convex mirror, and convex lens.

*80. Two convex lenses of focal lengths 40 cm and 60 cm are separated by a distance of 10 cm. Where is the focal point of this system, measured from the center of the lens system, for a parallel beam of light incident from the left? For a beam incident from the right?

*81. Consider **refraction by a spherical surface** between two transparent materials. For instance, Fig. 34.86 shows a sphere of glass of index of refraction n_2 immersed in a medium of index of refraction n_1. For a convex glass surface and $n_2 > n_1$, a real image will form inside the glass if a source is placed sufficiently far outside the glass. For rays making small angles with the axis, show that the equation relating the object distance s and image distance s' shown in Fig. 34.86 is

$$\frac{n_1}{s} + \frac{n_2}{s'} = \frac{n_2 - n_1}{R}$$

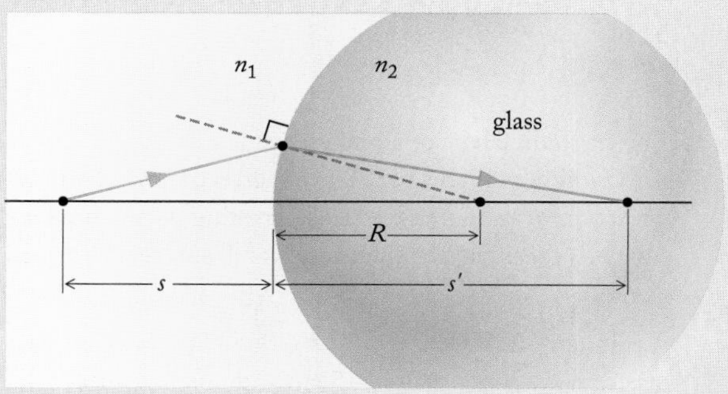

FIGURE 34.86 Spherical refracting surface.

*82. Use the result of Problem 81 to derive the **lens maker's formula** and the **thin-lens equation**, Eqs. (34.27) and (34.28), by considering refraction at two successive convex spherical glass surfaces in air. Treat the image of the first surface as the object for the second surface. For a thin lens, neglect the distance between the surfaces. Recall that the focal length is the image distance for an object at infinity.

**83. The "crystal" ball of a fortune-teller has a diameter of 15.0 cm. It is made of glass of index of refraction 1.60. At what distance from the center will this sphere bring an incident beam of sunlight to a focus? (Hint: Use the result of Problem 81 twice.)

**84. A Fresnel lens consists of a large number of segments each of which is an annular portion of an ordinary lens. Fresnel lenses of diameters more than 1 m are commonly used in the lamps of lighthouses, where ordinary lenses without segments would be much too thick and too heavy (see Fig. 34.70). Consider one of these annular segments in a plano-convex Fresnel lens (Fig. 34.87); for the sake of simplicity, assume that the width of the segment is infinitesimal. Derive a formula for the angle of inclination θ of the surface of this segment in terms of the index of refraction n, the distance r of the segment from the axis of the lens, and the focal length of the lens.

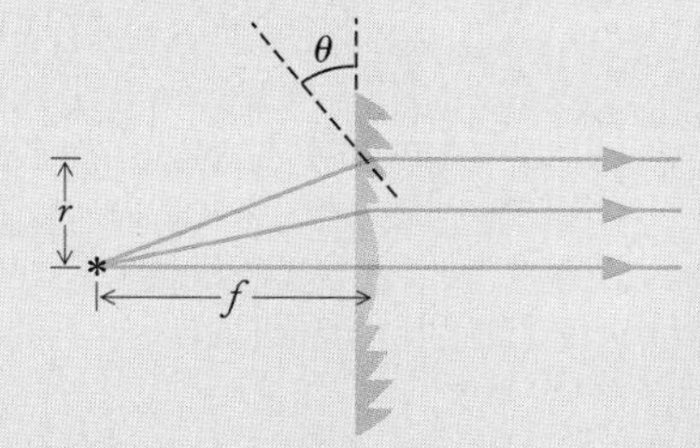

FIGURE 34.87 A ray passing through one of the annular segments of a Fresnel lens.

34.6 Optical Instruments

85. The lens of a camera (Fig. 34.88) has a focal length of 50 mm. The distance of the lens from the film is adjustable over a range from 50 mm to 62 mm. Over what range of object distances (measured from the lens) is this camera capable of producing sharp pictures?

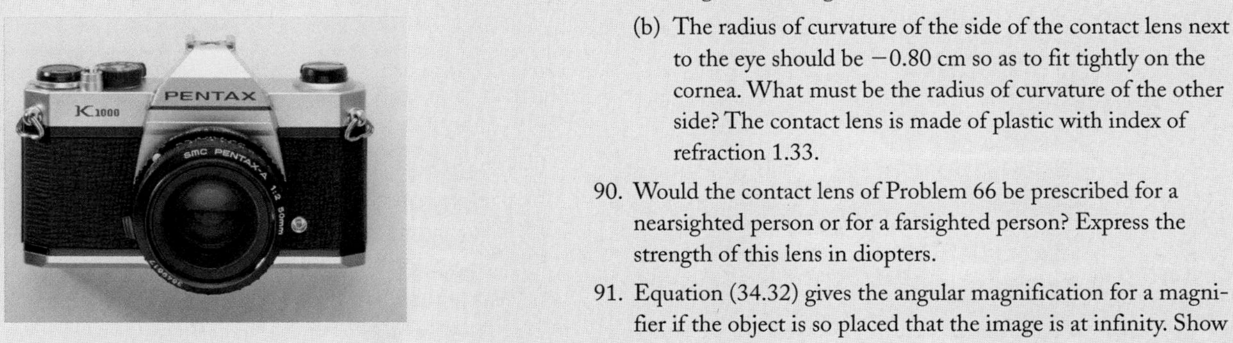

FIGURE 34.88 Lens of a camera.

86. A miniature Minox camera has a lens of focal length 15 mm. This camera can be focused on an object as close as 20 cm, or as far away as infinity. What must be the distance from the lens to the film if the camera is set for 20 cm? What if the camera is set for infinity?

87. The light meter of a 35-mm camera with a lens of f number 1.7 indicates that the correct exposure time for a photograph is $\frac{1}{250}$ s. If the iris diaphragm is closed down so the f number becomes 8, what will be the correct exposure time?

88. Pretend that the cornea, the crystalline lens, and the fluid of the human eye act together as a single thin lens placed at a distance of 2.2 cm from the retina (Fig. 34.89). The lens is deformable; it can change its focal length by changing its shape.
   (a) What must be the focal length if the eye is viewing an object at a very large distance?
   (b) What must be the focal length if the eye is viewing an object at a distance of 25 cm?

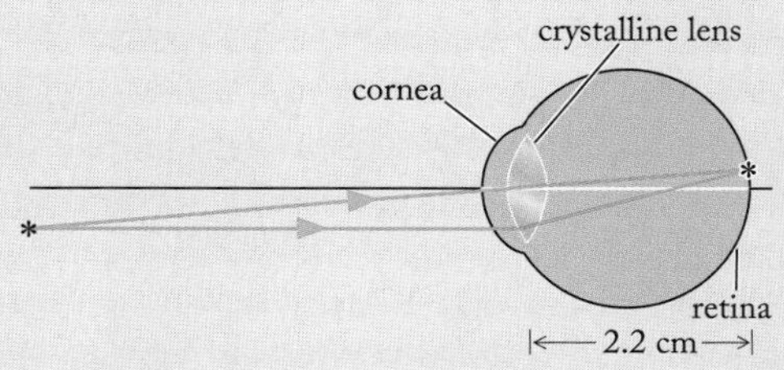

FIGURE 34.89 Lens of human eye forms image on retina.

89. In a nearsighted eye the (relaxed) lens has an abnormally short focal length, and consequently the eye fails to form an image of a distant object on the retina. This defect can be corrected with a contact lens. Pretend that both the lens of the eye and the contact lens are thin lenses so that the formula given in Problem 73 applies.
   (a) Suppose that the focal length of the eye is 2.00 cm. What must be the focal length of the contact lens if it is to increase the net focal length to 2.20 cm? Should it be convergent or divergent?
   (b) The radius of curvature of the side of the contact lens next to the eye should be -0.80 cm so as to fit tightly on the cornea. What must be the radius of curvature of the other side? The contact lens is made of plastic with index of refraction 1.33.

90. Would the contact lens of Problem 66 be prescribed for a nearsighted person or for a farsighted person? Express the strength of this lens in diopters.

91. Equation (34.32) gives the angular magnification for a magnifier if the object is so placed that the image is at infinity. Show that if the object is so placed that the image is at a distance of 25 cm, then the angular magnification is $1 + (25\text{ cm})/f$.

92. A microscope has an objective of focal length 4.0 mm. This lens forms an image at a distance of 224 mm from the lens. If we want to attain a net angular magnification of 550, what choice must we make for the angular magnification of the ocular?

93. A telescope has an objective of focal length 160 cm and an ocular of focal length 2.5 cm. If you look into the *objective* (that is, into the wrong end) of this telescope, you will see distant objects *reduced* in size. By what factor will the angular size of objects be reduced?

94. An amateur astronomer uses a telescope with an objective of focal length 90.0 cm and an ocular of focal length 1.25 cm. What is the angular magnification of this telescope?

95. The large reflecting telescope on Palomar Mountain has a minor of focal length 1680 cm (see Fig. 34.66a). If this telescope is operated with an ocular focal length of 1.25 cm what is the angular magnification?

96. It has been proposed that the six 1.8-m diameter mirrors in the Multiple-Mirror Telescope (Fig. 34.90) be replaced by a single 6.5-m diameter mirror, which would fit into the existing mount. By what factor would this replacement enhance the amount of light collected by the telescope?

97. The telescopes built by Galileo Galilei consisted of a convex objective lens and a concave ocular lens. The lenses are arranged so the focal point of the objective coincides with the focal point of the ocular (similar to Fig. 34.64). Explain how this telescope produces an angular magnification. Show that the eye sees an upright image. (Hint: The concave lens placed *before* the point F in Fig. 34.64 has the same effect as the convex lens placed *beyond* the point F in Fig. 34.64—the lens produces an image at infinity. Is this image upright or inverted?)

FIGURE 34.90 The Multiple-Mirror Telescope on Mt. Hopkins, Arizona.

REVIEW PROBLEMS

98. Two flat mirrors are arranged at an angle of 145° with respect to each other (see Fig. 34.91). A ray of light strikes one of the mirrors at an angle of incidence of 60°, it is reflected, and it then strikes the other mirror. What is the angle of reflection at the second mirror?

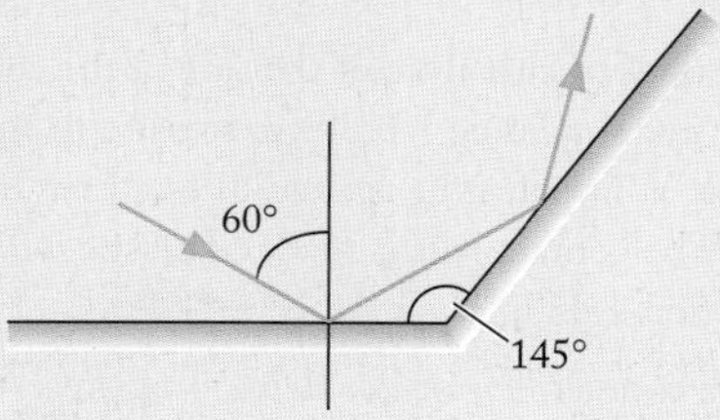

FIGURE 34.91 Two adjacent flat mirrors an angle.

99. A ray of light strikes a plate of window glass of index of refraction 1.5 with an angle of incidence of 50°.

(a) What is the angle of refraction?

(b) At what angle does the ray of light emerge on the other side of the glass?

100. The bottom corner of a swimming pool is 2.0 m below the surface of the water. If a ray of light travels upward from the corner and reaches the water surface at an angle of 40° with the vertical, what will be the angle the ray makes with the vertical when it emerges from the water? If your eye intercepts this ray, how deep will the corner of the swimming pool seem to you?

101. A prism has three equal angles of 60°. The prism is made of flint glass of index of refraction 1.66. At what angle should a ray of light enter this prism so it is refracted symmetrically (see Fig. 34.92)?

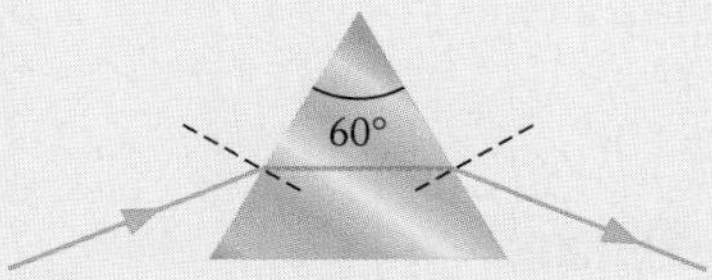

FIGURE 34.92 A symmetric prism.

102. A prism of glass has two angles of 45.0° and one of 90.0°. To determine the index of refraction of the glass in this prism, an experimenter aims a beam of light perpendicularly into a face of this prism and measures the angle at which the ray emerges on the far side (see Fig. 34.93). If this angle is 72.0°, what is the index of refraction?

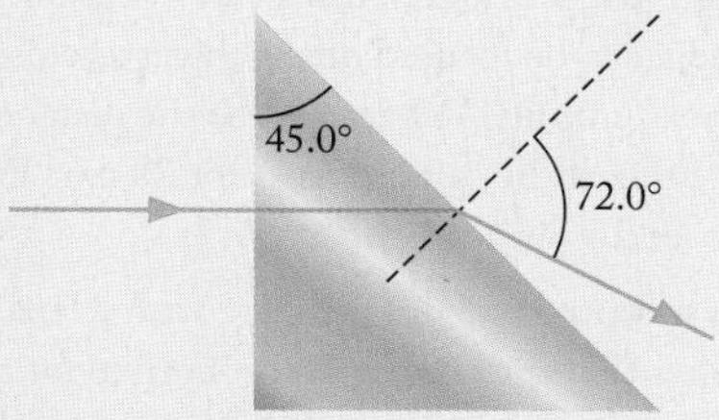

FIGURE 34.93 A 45° prism.

*103. The bottom half of a beaker of depth 20.0 cm is filled with water ($n = 1.33$), and the top half is filled with oil ($n = 1.48$). If you look into this beaker from above, how far below the upper surface of the oil does the bottom of the beaker seem to be?

104. The walls of a (filled) aquarium are made of glass with an index of refraction of 1.5. If a ray strikes the glass from the inside at an angle of incidence of 45°, what is the angle at which it emerges into the air?

105. A prism is to be used for total internal reflection of a ray of light that is perpendicularly incident on one of its faces (see Fig. 34.94). The index of refraction of the glass is 1.60. For what range of vertex angles α of this prism will this total internal reflection occur?

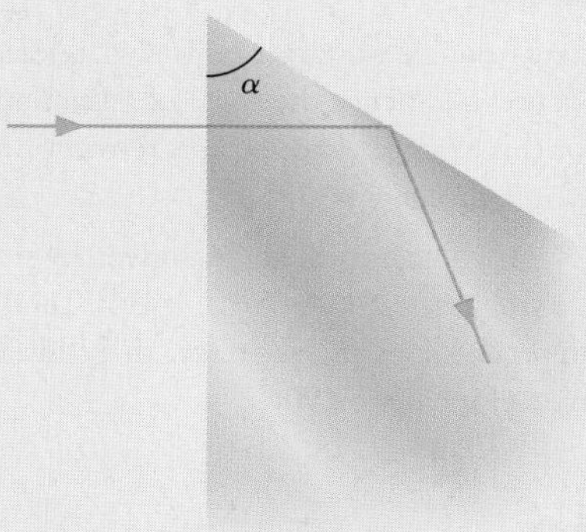

FIGURE 34.94 Total internal reflection in a prism.

106. A swimming pool has a depth of 2.0 m. A waterproof electric light is installed in the bottom of the pool which illuminates the pool at night. When looking down on the pool, you see that this light produces an illuminated circle on the surface, but it leaves the rest of the surface dark. What is the radius of this illuminated circle?

107. You place an object in front of a concave mirror, at a distance smaller than the focal length. Is the image real or virtual? Upright or inverted? Magnified or reduced? What if the distance is greater than the focal length? What if the mirror is convex?

108. The outside rearview mirror on an automobile is convex (Fig. 34.95). When you look into this mirror, you notice that the image of a distant truck or some other distant object is about 1.0 m beyond the mirror. What is the radius of curvature of the mirror?

FIGURE 34.95 Rearview mirror of an automobile.

109. In a grocery store, a convex mirror with $R = 4.0$ m is used for surveillance. Suppose a person of height 1.7 m is 12 m from such a mirror. Where is the image? What is the size of the image?

*110. A concave mirror of radius of curvature 60 cm faces a convex mirror of the same radius of curvature. The distance between the mirrors is 50 cm, and their axes coincide. A candle is held between the mirrors, at a distance of 10 cm from the convex mirror. Consider rays of light that reflect first off the concave mirror and then off the convex mirror. Where do these rays form an image?

111. You place an object in front of a convex lens at a distance smaller than the focal length. Is the image real or virtual? Upright or inverted? Magnified or reduced? What if the distance is greater than the focal length? What if the lens is concave?

112. A candle is placed at a distance of 50 cm in front of a concave lens of focal length 30 cm. Where is the image of the candle? Draw a diagram showing the candle, the lens, and the image. Show three principal rays.

113. A lightbulb is placed at a distance of 15 cm in front of a convex lens of focal length 30 cm. Where is the image of the lightbulb? Draw a diagram showing the lightbulb, the lens, and the image. Show the three principal rays.

*114. Three identical convex lenses of focal length 60 cm each are separated by distances of 12 cm from one to the next. If a small light bulb is placed 20 cm in front of the left lens, where will the final image be? What is the magnification of the final image?

115. A microscope has an objective of focal length 1.9 mm and an ocular of focal length 25 mm. The distance between these lenses is 180 mm.

(a) At what distance from the objective must the object be placed so that the ocular forms an image at infinity, as shown in Fig. 34.63?

(b) What is the net angular magnification of this microscope?

Answers to Checkups

Checkup 34.1

1. If we draw a set of identical spherical wavelets centered on different points on the blue dashed arc (Huygens' Construction), their leading edges (when traveling inward) form a sphere of smaller radius, precisely the reverse of the outward wave.

2. If we draw a set of small, circular wavelets centered on different points around the rectangular edge of the brick, we find a nearly rectangular shape, but with rounded corners, for the propagated wave front.

3. (D) $\frac{3}{2}$; 1. When the wave front moves a distance L out from each tip, it will have a total length $3L$ (including the length L of the stick); out from the sides, its width will be only $2L$ (the stick is thin). Thus, after traveling a distance L, the length-to-width ratio will be $3/2$. Much farther out, the wave front will be nearly circular (for example, after traveling $50L$, it will be $101L$ long and $100L$ wide): the length-to-width ratio approaches 1.

Checkup 34.2

1. Since the angle of incidence is defined as the angle that the ray makes with the *perpendicular to the surface*, a ray making an angle of 20° with the surface has an angle of incidence of 70°.

2. The law of reflection tells us that the angle of reflection is also 0°. Thus both the incident and reflected rays are perpendicular to the surface, and the ray returns back on itself.

3. Similar to the image illustrated schematically in Fig. 34.12, the image of the flashlight in Fig. 34.6 is located below the mirror, at a distance equal to the distance the flashlight is above the mirror.

4. As illustrated in Fig. 34.12, a (virtual) image is the same distance behind a mirror as you are in front: thus, if you are 2 m from the mirror, your image is 4 m away from you.

5. Your image approaches the back of the mirror at the same rate that you approach the front (8 m/s); thus, relative to you, the image travels at 16 m/s.

6. (E) $\frac{1}{2}h$. Since the incident and reflected rays from the foot to the eye make the same angle, and similarly for the rays from the top of the head to the eye, you can see from head to toe in a mirror (appropriately positioned) of only half your height, $\frac{1}{2}h$.

Checkup 34.3

1. The refracted ray is also along the perpendicular, with an angle of refraction of 0° (from Snell's Law). For this one angle, the answer is independent of the index of refraction.

2. Cold air is denser than warm air, and the denser air has a larger dielectric constant, and hence a larger index of refraction [see Eq. (34.5)].

3. The wavelength will be $(600 \text{ nm})/1.33 \approx 450$ nm.

4. The ray will be bent closer to the perpendicular when it enters a medium of greater index of refraction. Thus, from the data of Table 34.1, we can determine that the ray will be bent (a) closer to the perpendicular ($n = 1$ to $n = 1.33$); (b) closer to the perpendicular ($n = 1.33$ to $n = 1.52$–1.66); (c) farther away from the perpendicular ($n = 1.46$ to $n = 1.36$).

5. If the angle is slightly more than the critical angle of 49°, there will be no refracted ray, and only the ray reflected back downward at 49° will occur. If the angle is slightly less than the critical angle of 49°, then the refracted ray will be just above the surface (nearly parallel to it); of course, part of the ray will still be reflected downward with an angle of reflection equal to the angle of incidence.

6. No; for this ray the angle of refraction is always less than the angle of incidence, and hence the angle of refraction can never attain 90°. For this example, the angle of refraction will always satisfy $\sin\theta < 1.3/1.6$, or $\theta < 55°$.

7. (D) $30° < \theta < 90°$. The critical angle is given by $\sin\theta_{crit} = 1/n$. So here, $\sin\theta_{crit} = 1/(2.0)$, or $\theta_{crit} = 30°$. Total internal reflection will occur for all angles greater than this, up to the maximum angle of incidence, 90°.

Checkup 34.4

1. As in Example 9 and Fig. 34.41, an object further away from a concave mirror than the focal point produces a real, inverted image in front of the mirror.

2. For convex mirrors (like flat mirrors), the image is always behind the mirror, and is thus virtual. As in the ray diagram of Fig. 34.40, such images are upright. You should verify that the closer object distance here does not change the nature of the ray diagram.

3. As in Example 8 and Fig. 34.40, and question 2 above, a convex mirror always produces a virtual image. Thus the answer is no.

4. No. The convex mirror does not bring the rays together, but instead always causes them to diverge.

5. (A) Virtual and upright, with $s' < 0$. This situation was addressed in Example 9 and Fig. 34.42.

Checkup 34.5

1. The focal length of any thin lens does not depend on which side the object is on, so its focal length is still 15 cm.

2. This is the "magnifier" situation of Fig. 34.50, where the object is closer to the convex lens than the focal length of the lens. The image is on the near side (the same side as the object); the image is upright and virtual.

3. For a single concave (diverging) lens, the image is always on the same side as the object; it is always upright and virtual (as in Fig. 34.47).

4. No. For a single concave (diverging) lens (a negative f), the thin lens equation implies that the image distance s' will always be negative for a positive object distance s, implying a virtual, upright image.
5. (E) -4.0 m. For the lens maker's equation (34.27), we use $+2.0$ m for the convex radius and -1.0 m for the concave radius, giving $1/f = (n - 1)(1/R_1 + 1/R_2) = (1.5 - 1)[1/(2.0 \text{ m}) + 1/(-1.0 \text{ m})] = -0.25/\text{m}$, so $f = -4.0$ m.
6. (A) Upright and real. The image produced by the first lens is real and inverted; the thin lens equation puts it at a distance $2f$ beyond the first lens. This is the object for the second lens; since the lenses are separated by $5f$, the second object distance is $3f$. Since this is outside the focal length f, a real image forms which is reinverted, or upright with respect to the original object.

Checkup 34.6

1. To get more light into the camera, you need a larger-diameter aperture. Since the f number is inversely proportional to the aperture diameter, you need a small f number.
2. In bright light, the pupil gets smaller. As in a camera, this reduces the amount of light entering the eye, but increases the depth of field: because of the smaller aperture, light rays enter the eye at small angles, and they stay close together on the retina and form a fairly small spot, even when the position of the image does not fall exactly on the retina.
3. (A) The arrangement is such that the object distance is smaller than the image distance, so the objective does produce a linear magnification. Both object and image subtend the same angle, and the object may be at or beyond the typical near point of 25 cm, so no ordinary angular magnification occurs.

   (B) The ocular produces a large linear magnification (in principle, infinite, since the image is usually at infinity). The ocular produces an angular magnification, the same as a magnifier; see Eq. (34.32).

   (C) From Fig. 34.63, the first image, formed by the objective, is real and inverted.

   (D) Again from Fig. 34.63, the final image produced by the ocular is virtual and inverted (with respect to the original object).
4. (A) The objective produces no magnification (or, more accurately, it produces a magnification of zero, since the image distance is finite, and the object is essentially at infinity). As depicted in Fig. 34.64, the object and image subtend the same angle, so there is no angular magnification [see answer 3(A), above].

   (B) As in the case of the microscope of question 2, the ocular does produce a large (in principle, infinite) linear magnification, and does produce an angular magnification, given by Eq. (34.32).

   (C) From Fig. 34.64, the first image, formed by the objective, is real and inverted.

   (D) Again from Fig. 34.64, the final image produced by the ocular is virtual and inverted (with respect to the original object).
5. The microscope uses an objective lens of small focal length to form an enlarged image of a nearby object; the telescope uses an objective lens of large focal length to form an image of a distant object at its focal point.
6. (D) Microscope and magnifier only. As discussed above, it is only for enlarging near objects that the near point of vision plays a role [compare Eqs. (34.32), (34.33), and (34.35)].

CHAPTER

35 Interference and Diffraction

CONCEPTS IN CONTEXT

Concepts — in — Context

Everyday objects, such as these compact discs (CDs), can display spectacular rainbows of color. Each "groove" or track on the CD spreads, or diffracts, a reflected light wave, and waves from different tracks combine to produce the observed patterns.

With the concepts developed in this chapter, we can consider several questions about such interference patterns:

? What do the colors tell us about the source of light? (Section 35.4, page 1185)

? What does the pattern tell us about the spacing between tracks? (Example 5, page 1185)

? How does the pattern depend on the size of a track? (Example 8, page 1195)

Geometric optics relies on the assumption that the sizes of the mirrors or lenses and the separations between them are much larger than the wavelength of light. Under these conditions, we can adequately describe the propagation of light by rays that are rectilinear except when reflected or when refracted by the interface between different media. Thus, in geometric optics the wave properties of light do not show up explicitly (although these wave properties enter into the derivations of the laws of reflection and refraction).

The absence of any explicit wave effects during the propagation of light through a system of mirrors and lenses is related to the small wavelength of light. In comparison with the sizes of typical mirrors and lenses, the wavelength of light is extremely small (about 5×10^{-7} m). We cannot detect this small wavelength and the small-scale wave behavior of light by means of ordinary mirrors and lenses for the same reason that we cannot detect ripples on the surface of the ocean by means of a supertanker—such instruments are too crude. If we want to detect ripples on the surface of the ocean, we need to use small pieces of flotsam, preferably of a size comparable with the wavelength of the ripples. Likewise, to detect the wave behavior of light, we need to let the light interact with a body or an obstacle of a size comparable with a wavelength. In that case, light will display its wave properties by the phenomena of **interference** and **diffraction**. As mentioned in Chapters 16 and 17, *interference is the constructive or destructive superposition of two or more waves meeting at one place; diffraction is the bending and spreading of waves around obstacles.* **Wave optics**, or physical optics, deals with interference and diffraction of light, and other phenomena that directly involve the wave properties of light.

wave optics

In this chapter we will begin by examining the interference of light waves and other electromagnetic waves. Like all electric and magnetic fields, the fields of electromagnetic waves obey the Superposition Principle: if two waves meet at some point, the resultant electric or magnetic field is simply the vector sum of the individual fields. If two waves of equal amplitude meet crest to crest, they combine and produce a wave of double amplitude; if they meet crest to trough, they cancel and give a wave of zero amplitude. The former case is called *constructive interference* and the latter *destructive interference.* We will encounter both cases of interference in the following sections.

35.1 THIN FILMS

The interference between light waves is difficult to observe. For instance, the simplest case of interference is that between two waves of the same frequency traveling in opposite directions, giving a standing wave, with nodes and antinodes. Such a standing wave can be set up by shining light perpendicularly at a mirror, so the incident wave interferes with the reflected wave. However, since the wavelength of light is so small, our eyes cannot perceive the individual nodes and antinodes—we only see the average intensity, without noticeable interference effects.

Spectacular interference effects may become visible when light waves with a range of frequencies and wavelengths are reflected by a thin film, such as a thin film of oil floating on water (see Fig. 35.1) or a thin film of soapy water on the surface of a soap bubble. When the incident waves enter the oil film (see Fig. 35.2), they will set up a multitude of reflected waves due to reflection at the upper surface, reflection at the lower surface, and multiple zigzags between the surfaces. The reflected waves that re-emerge all travel in the same direction, and they can interfere destructively or constructively in the region of space above the film. Note that we are now interested only

FIGURE 35.1 Colored fringes in the light reflected by a thin film of oil floating on water.

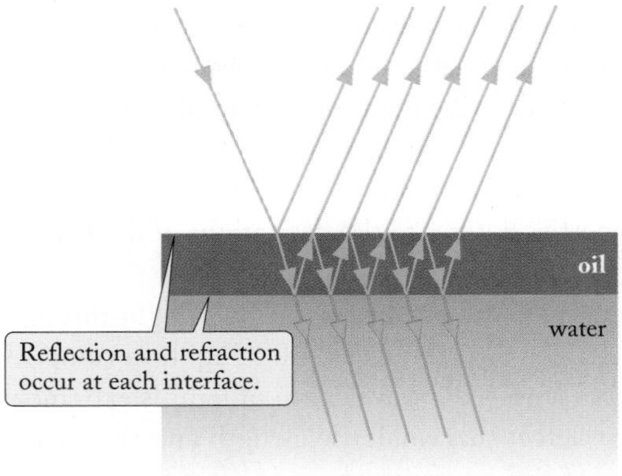

FIGURE 35.2 Incident ray and multiple reflected and refracted rays produced by a thin film.

THOMAS YOUNG (1773–1829)
English physicist, physician, and Egyptologist. Young worked on a wide variety of scientific problems, ranging from the structure of the eye and the mechanism of vision to the decipherment of the Rosetta stone. He revived the wave theory of light and recognised that interference phenomena provide proof of the wave properties of light.

in the interference between the several reflected waves—the incident wave plays no direct role in this. The beautiful shimmering color displays seen in oil slicks and on soap bubbles arise from such interference effects. Different portions of an oil or soap film usually have different thicknesses, and they therefore give constructive interference for different wavelengths. The result is a pattern of bright colored bands, or colored **fringes**.

To find the conditions for constructive and destructive interference between the waves reflected by a thin film, let us make the simplifying assumption that the direction of propagation of the incident wave is nearly perpendicular to the surface of the film and that this wave has a given frequency (and wavelength). The two most intense waves are those that experience only one reflection: the wave that reflects only at the upper surface and the wave that reflects only at the lower surface. Figure 35.3 shows the rays corresponding to these two waves. Let us *first assume that the reflection affects the phases of both waves in the same way*, which is true for the air–oil and oil–water interfaces here. Under what conditions will these waves interfere constructively in the space above the film? Obviously, the wave that is reflected from the lower surface has to travel an extra distance to emerge from the film. If the thickness of the film is d, and if the direction of propagation is nearly perpendicular to the film, then the extra distance the wave has to travel is $2d$, that is, down through the film and then back up.

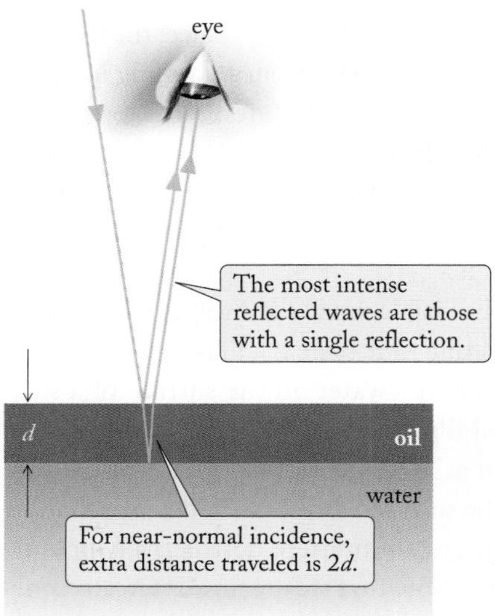

FIGURE 35.3 Incident ray and reflected rays for nearly perpendicular incidence and reflection. The wave reflected at the lower surface has to travel an extra distance of approximately d while propagating down and d while propagating up, that is, an extra distance $2d$.

Provided that this extra distance is equal to one, two, three, etc., wavelengths, the wave reflected from the lower surface will meet crest to crest with the wave reflected from the upper surface (see Fig. 35.4). The condition for constructive interference is therefore

$$2d = \lambda,\ 2\lambda,\ 3\lambda, \ldots \tag{35.1}$$

constructive interference for thin film

Likewise, if the extra distance is equal to one half wavelength, three half wavelengths, five half wavelengths, etc., the wave reflected from the lower surface will meet crest to trough with the wave reflected from the upper surface (see Fig. 35.5). The condition for destructive interference is therefore

$$2d = \tfrac{1}{2}\lambda,\ \tfrac{3}{2}\lambda,\ \tfrac{5}{2}\lambda, \ldots \tag{35.2}$$

destructive interference for thin film

Thus, depending on the thickness of the film and on the wavelength of the light, the reflected wave can be either very strong or very weak. Note that the wavelength λ in Eqs. (35.1) and (35.2) is the wavelength of the light within the film; the wavelength outside the film will differ from this by a factor equal to the film's index of refraction n. If $\lambda_{\text{air}} \approx \lambda_{\text{vacuum}}$ is the wavelength in the air outside the film, recall from Eq. (34.7) that the wavelength λ inside the material of the film is smaller:

$$\lambda = \frac{\lambda_{\text{air}}}{n} \tag{35.3}$$

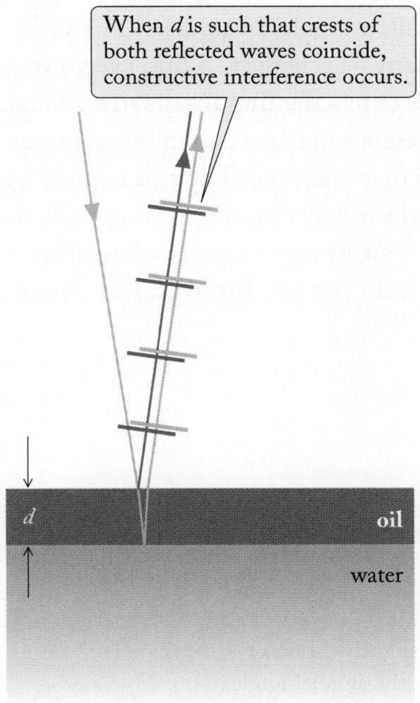

FIGURE 35.4 The wave (orange lines) reflected from the lower surface meets crest to crest with the wave (blue lines) reflected from the upper surface.

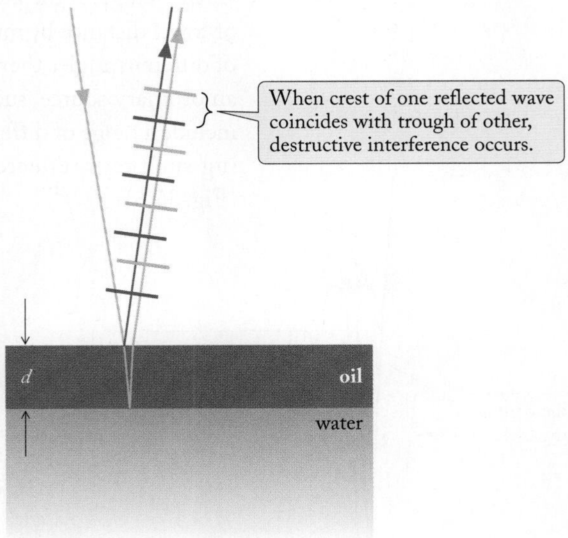

FIGURE 35.5 The wave (orange lines) reflected from the lower surface meets crest to trough with the wave (blue lines) reflected from the upper surface.

EXAMPLE 1 A film of kerosene 450 nm thick floats on water. White light, a mixture of all visible colors, is vertically incident on this film. Which of the wavelengths contained in the white light will give maximum intensity upon reflection? Which will give minimum intensity? The index of refraction of kerosene is 1.20.

SOLUTION: According to Eq. (35.1), for maximum intensity (constructive interference), we need

$$\begin{aligned}\lambda &= 2d, \frac{2d}{2}, \frac{2d}{3}, \ldots \\ &= 900 \text{ nm}, 450 \text{ nm}, 300 \text{ nm}, \ldots\end{aligned}$$

These are the wavelengths in kerosene; to obtain the wavelengths in air, we must multiply by the index of refraction of kerosene, $n = 1.20$. We are interested only in wavelengths in the visible range, approximately 400–700 nm. Of the resulting wavelengths, the only one in the visible region is $1.20 \times 450 \text{ nm} = 540 \text{ nm}$.

For minimum intensity (destructive interference) we need, according to Eq. (35.2),

$$\begin{aligned}\lambda &= 4d, \frac{4}{3}d, \frac{4}{5}d, \ldots \\ &= 1800 \text{ nm}, 600 \text{ nm}, 360 \text{ nm}, \ldots\end{aligned}$$

Upon multiplication by 1.20, we find that the only wavelength in air in the visible region is $1.20 \times 360 \text{ nm} = 432 \text{ nm}$.

COMMENTS: A wavelength of 540 nm corresponds to a yellow green color. The film will therefore be seen to have this color in reflected light.

If the oil film is thick, no interference fringes will be visible. In a thick film, the distance from the top of the film to the bottom is many wavelengths. Thus, even a very small change in the angle at which the ray of light enters the film results in a change of travel distance by many wavelengths. The interference maxima and minima for rays of different angles therefore overlap, and they tend to average out. The rays emitted by an ordinary source, such as the Sun or a lamp, and the rays entering your eye always include a range of different angles; thus, you see only an average intensity when observing such rays reflected by a thick film, and you do not see interference fringes (Fig. 35.6).

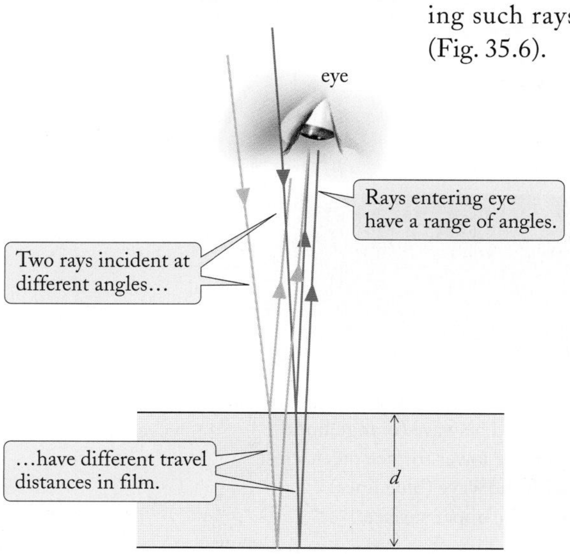

FIGURE 35.6 Two rays arriving at slightly different angles at a thick film. For the ray with the smaller angle of incidence (blue), the travel distance through the film and back is, say, 5000 wavelengths; for the ray with the larger angle of incidence (orange), the travel distance is, say, 5000.5 wavelengths. The former ray gives constructive interference; the latter gives destructive interference. The interference effects tend to average out.

We can also use Eqs. (35.1) and (35.2) for a soap film suspended in air, instead of an oil film floating on water; but we must take into account a minor complication that can be derived from analysis of reflection of electromagnetic waves according to Maxwell's equations. *Whenever an electromagnetic wave propagating in one medium is reflected at the surface of another medium of larger index of refraction, the wave suffers a reversal of its electric field, that is, a change of phase by* 180° (this change of phase upon reflection is similar to that discussed in Section 16.4 for a wave on a string; see Fig. 16.14). In the case of the oil film floating on water, the wave reflected at the upper (or outer) surface of the film suffers such a change of phase, and so does the wave reflected at the lower (or inner) surface; the net result is that the relative phase between the two waves is not altered, and the extra phase changes introduced by the process of reflection can be ignored, as in Example 1. However, in the case of a soap film, or some other film, suspended in air, only the wave reflected at the outer surface suffers the extra change of phase. Thus, besides the phase difference introduced by the thickness of the film, there is an extra phase difference of 180° between the two waves. This means that the waves reflected from the outer and the inner surfaces of the soap film surfaces will be out of phase if the distance $2d$ is equal to one, two, three, etc., wavelengths—and that Eq. (35.1) is now the condition for destructive interference. Conversely, Eq. (35.2) is now the condition for constructive interference.

Similar interference effects can also arise in a narrow gap between two adjacent glass surfaces. Figure 35.7 shows a photograph of the interference fringes produced by the thin film of air between a flat glass plate and a lens of large radius of curvature. The convex surface of the lens is in contact with the plate at the center, but leaves a gradually widening gap for increasing distances from the center (Fig. 35.8). At the bright fringes, the width of the gap is such as to give constructive interference of the reflected light. At the dark fringes, the width is such as to give destructive interference. The rings in Fig. 35.7 are called **Newton's rings**.

Newton's rings

Thin films are of great importance in the manufacture of optical instruments. The lenses of high-quality instruments are often coated with a thin film of dielectric material, so that undesirable reflections of light are eliminated and more light can enter the instrument. With one thin film, we can achieve destructive interference at only one wavelength; but with several layers of thin films, we can achieve destructive interference at several wavelengths.

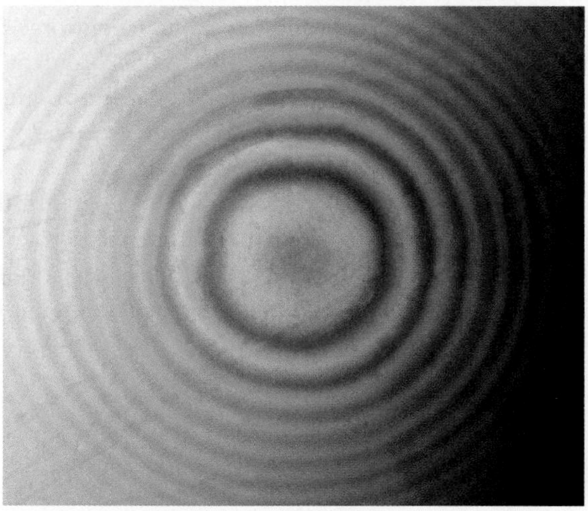

FIGURE 35.7 Fringes of constructive and destructive interference seen in the light reflected in the gap between a flat glass plate and a spherical lens in contact.

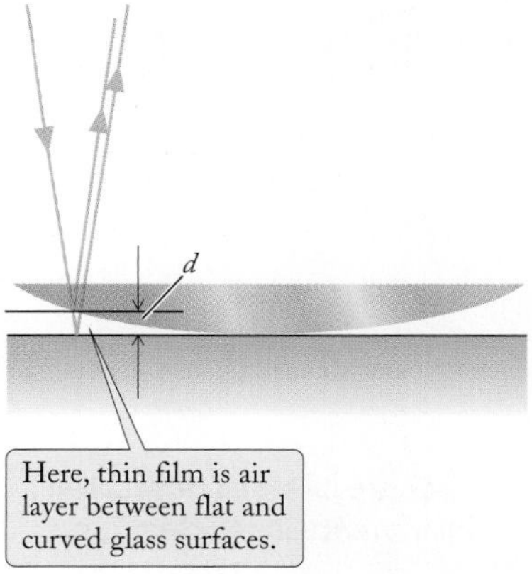

FIGURE 35.8 Rays reflected at the top and the bottom of the air gap between the flat glass plate and the lens.

PROBLEM-SOLVING TECHNIQUES THIN-FILM INTERFERENCE

When examining the interference of light rays reflected in or at thin films, keep in mind that:

- The condition for destructive or constructive interference involves the wave length λ within the film, which is smaller than the wavelength in air by a factor equal to the index of refraction n of the film: $\lambda = \lambda_{air}/n$.
- Whenever the wave reflects on the surface of a medium of higher index of refraction, it suffers an extra phase change of 180°. If neither or both reflections occur at interfaces with a medium of higher index of refraction, there is no net extra effect. However, if only one of the reflections occurs at an interface with a medium of higher index of refraction, then the extra phase change of 180° reverses the conditions for constructive and destructive interference.

✔ Checkup 35.1

QUESTION 1: Why is the central spot in Newton's rings (see Fig. 35.7) dark?

QUESTION 2: If the light in Example 1 is incident on the kerosene film from below (from within the water), are Eqs. (35.1) and (35.2) still valid for constructive and destructive interference? What is different?

QUESTION 3: A "nonreflective" coating is sometimes put on glass lenses to increase the light entering an optical system. For a wavelength of 500 nm in air, what is the smallest coating thickness that could be used to minimize reflection? The index of refraction of the coating is 1.25, and that of the glass is 1.50.

(A) 100 nm (B) 125 nm (C) 156 nm (D) 200 nm (E) 250 nm

35.2 THE MICHELSON INTERFEROMETER

The Michelson interferometer takes advantage of the interference between two light waves to accomplish an extremely precise comparison between two lengths. Figure 35.9 is a schematic diagram of the essential parts of such an interferometer. The apparatus consists of two arms at the ends of which are mounted mirrors M_1 and M_2. Light waves of a given frequency (and wavelength) from a monochromatic source S fall on a semitransparent mirror M (a half-silvered mirror). This mirror splits the light wave into two parts: one part continues straight ahead and reaches mirror M_1, the other part is reflected and reaches mirror M_2. These mirrors reflect the waves back toward the central mirror M, and upon reflection or transmission by this mirror, the waves emerge from the interferometer. When they emerge, they interfere constructively or destructively. Suppose that the lengths MM_1 and MM_2 differ by d. Then one of the waves must travel an extra distance $2d$, and the condition for the constructive interference is

$$2d = 0,\ \lambda,\ 2\lambda, \ldots \tag{35.4}$$

constructive and destructive interference in interferometer

and for destructive interference it is

$$2d = \tfrac{1}{2}\lambda,\ \tfrac{3}{2}\lambda,\ \tfrac{5}{2}\lambda, \ldots \tag{35.5}$$

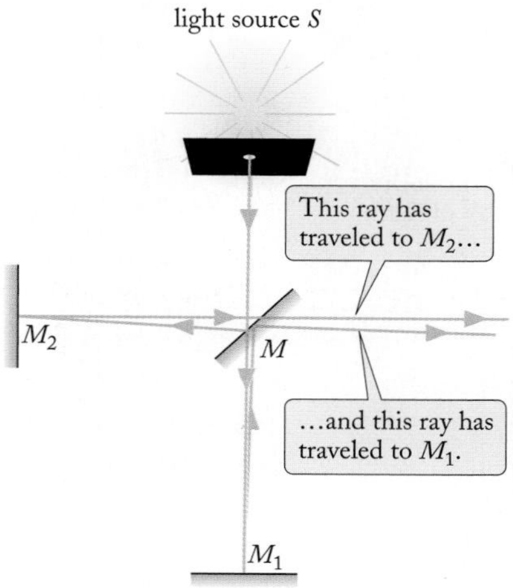

FIGURE 35.9 Paths of rays in a Michelson interferometer. For the sake of clarity, rays are shown reaching the mirrors M_1 and M_2 with a small angle between them; they are actually parallel, and they overlap.

ALBERT ABRAHAM MICHELSON (1852–1931) *American experimental physicist, professor at the Case School of Applied Science and at the University of Chicago. He made precise measurements of the speed of light and used interferometric methods to determine the wavelength of spectral lines in terms of the standard meter. He first performed the "Michelson–Morley" experiment on his own in 1881 and then repeated it several times in collaboration with E. W. Morley, with increasing accuracy. Michelson received the Nobel Prize in 1907.*

To achieve this interference, the mirrors M_1 and M_2 must be very precisely aligned, so that the image of M_1 seen in M is exactly parallel to M_2. The alignment can be achieved by means of adjusting screws on the backs of the mirrors.

The mirror M_1 is usually mounted on a carriage that can be driven along a track by means of an accurately machined screw. If mirror M_1 is slowly moved inward or outward, the interference of the emerging waves will change back and forth between constructive and destructive whenever the mirror is displaced by $\frac{1}{4}$ wavelength—and the intensity of the emerging light will change back and forth between maxima and minima, which are seen as bright and dark fringes. Thus, the displacement of the mirror can be measured very precisely by counting fringes and fractions of a fringe. This measurement expresses the displacement in terms of the wavelength of the light. Modern interferometers used for length measurements, such as that illustrated in Fig. 35.10, are designed to count fringes automatically with a photoelectric sensor. Laboratory interferometers can determine displacements as small as one-millionth of a wavelength. A large-scale instrument, the Laser Interferometer Gravitational Observatory (LIGO), is designed to measure displacements smaller than one-billionth of a wavelength!

Interferometers have played an important role in establishing that the speed of light is independent of the motion of the Earth. This was demonstrated by the famous **Michelson–Morley experiment**, first performed in 1881. The idea behind this experiment is as follows. If light were to propagate in an interplanetary medium in a manner analogous to the propagation of sound in air, then we would expect that the motion of the Earth toward or away from a light wave propagating through the "stationary" interplanetary medium would affect the speed of the light relative to the Earth, just

FIGURE 35.10 Special interferometer used at the International Bureau of Weights and Measures for the comparison of lengths. The interferometer automatically counts the wavelengths when the carriage moves through a given length.

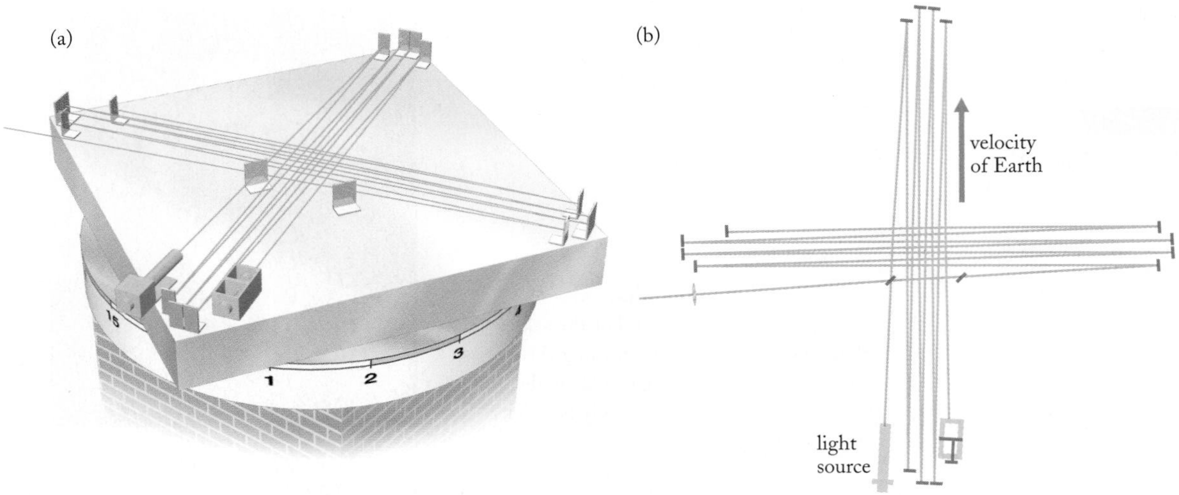

FIGURE 35.11 (a) The interferometer used in the experiment of Michelson and Morley. (b) The many mirrors reflect the light beams back and forth several times, increasing the path length of the light. The vertical arm is parallel to the direction of motion of the Earth and the horizontal arm is perpendicular.

as the motion of a train toward or away from a sound wave propagating through the stationary air affects the speed of the sound relative to the train. Such an alteration of the speed of light could be detected with an interferometer by orienting one of the arms parallel to the direction of motion of the Earth and the other arm perpendicular. A difference in the speeds of light c along the arms would entail a difference in the corresponding wavelengths ($\lambda = c/f$) and alter the conditions (35.4) and (35.5) for bright and dark fringes. The easiest way to detect the speed difference is by rotating the interferometer, so the arm that has been parallel to the motion becomes perpendicular and vice versa. If the speeds were different in the two directions, this rotation could shift the fringes from bright to dark or vice versa.

Michelson and Morley found no observable fringe shift in their experiment. However, a small fringe shift could have been masked by experimental errors. Taking into account possible experimental errors, Michelson and Morley cautiously concluded that the effect of the motion of the Earth on the speed of light was no larger than 5 km/s. This value is substantially less than the speed of the Earth around the Sun ($\approx$30 km/s) and proved beyond all reasonable doubt that the propagation of light through space is *not* analogous to the propagation of sound through air. The demonstration that light did not propagate through some stationary interplanetary medium was fundamental to the development of the theory of Special Relativity, discussed in Chapter 36. Figure 35.11 shows Michelson and Morley's interferometer.

✔ Checkup 35.2

QUESTION 1: Since the light rays in Fig. 35.9 reflect on a medium of larger index of refraction, they suffer phase changes of 180°. Does this alter the conditions (35.4) and (35.5)?

QUESTION 2: Since the light rays in Fig. 35.9 pass through the glass of mirror M, they suffer an extra phase change because the index of refraction of the glass differs from that of air. Does this alter the conditions (35.4) and (35.5)?

QUESTION 3: Using a screw mechanism, one mirror of a Michelson interferometer is moved. If the wavelength of the light used is 500 nm, and if one full turn of the screw results in the observation of 400 bright fringes, through what distance was the mirror moved?

(A) 0.05 mm (B) 0.10 mm (C) 0.16 mm (D) 0.20 mm (E) 0.40 mm

35.3 INTERFERENCE FROM TWO SLITS

Online Concept Tutorial 39

A very clear experimental demonstration of interference effects in light can be performed with two small light sources placed near each other. The light waves spread out from the sources, run into one another, and interfere constructively or destructively, giving rise to a pattern of bright and dark zones (see Fig. 35.12). These interference effects were first discovered by Thomas Young around 1800, and, in conjunction with diffraction effects (see Section 35.5), they settled a long-standing controversy regarding the nature of light. Before Young performed his experiments, some physicists—among them Newton—had argued in favor of a particle nature of light; they had conjectured that light is a stream of particles. But Young's demonstration of interference gave unmistakable evidence in favor of waves. However, the question of what these waves were made of was not settled until much later, when Maxwell formulated his equations and established that light is a wave consisting of electric and magnetic fields.

If the interference pattern produced by two light sources is to remain stationary in space, the two sources must be **coherent**, that is, *they must emit waves of the same frequency* and the same phase (a *constant* phase difference is also acceptable; what is crucial is that the phase difference must not fluctuate in time). Such coherent sources are easily manufactured by aiming a monochromatic wave with plane or spherical wave fronts at an opaque plate with two narrow slits or holes; the waves diverging from the

coherent light

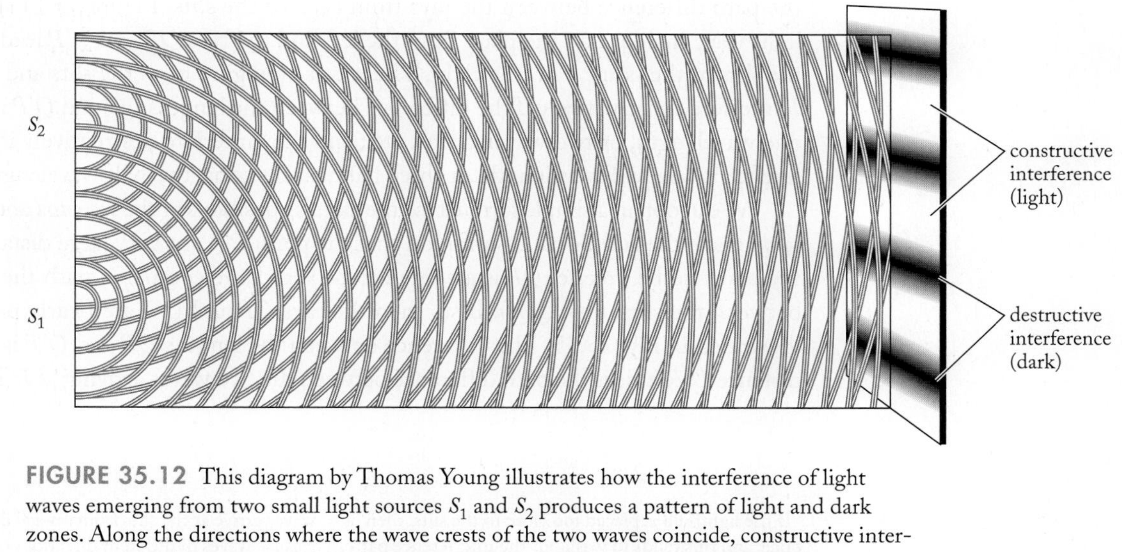

FIGURE 35.12 This diagram by Thomas Young illustrates how the interference of light waves emerging from two small light sources S_1 and S_2 produces a pattern of light and dark zones. Along the directions where the wave crests of the two waves coincide, constructive interference occurs. Along intermediate directions, the wave crests of one wave fall on the wave troughs of the other, giving destructive interference. If viewed from along the edge of the figure, the pattern is more obvious. If the light is intercepted by a screen, these directions of constructive and destructive interference give rise to bright and dark bands, or interference fringes.

two slits are then coherent because they arise from a single original wave (see Fig. 35.13). An extended object, such as an ordinary lightbulb, can be used to illuminate the slits provided it is placed at a *very large* distance. The lightbulb will then effectively act as a point source, illuminating the slits with light waves that consist of a succession of nearly plane wave fronts.[1]

incoherent light

Note that if we use two *separate* lightbulbs as sources, then we will not see any interference pattern. The light from two separate lightbulbs is **incoherent**; it consists of a mixture of a large number of light waves with variable phase differences. We can understand the distinction between coherent and incoherent light sources by means of an analogy with water waves. If we simultaneously drop two large, identical stones into a pond, they act as coherent sources of water waves. The waves from these two sources will interfere constructively at some places, and destructively at others (we can observe the interference effects of the waves by means of bits of cork scattered on the water; on the midline between the sources, the waves interfere constructively, and the bits of cork bob up and down with twice the amplitude they would have had if only one stone had been dropped; at some other places, where the waves interfere destructively, the bits of cork remain at rest). But if, instead of dropping two large stones, we dribble two handfuls of small pebbles into the pond, then the two sources are incoherent. There will then be no distinguishable interference pattern on the surface of the pond—all the bits of cork will bob up and down, more or less at random, although on the average the energy of their bobbing motion will be twice as large as if we had dropped a single handful of pebbles.

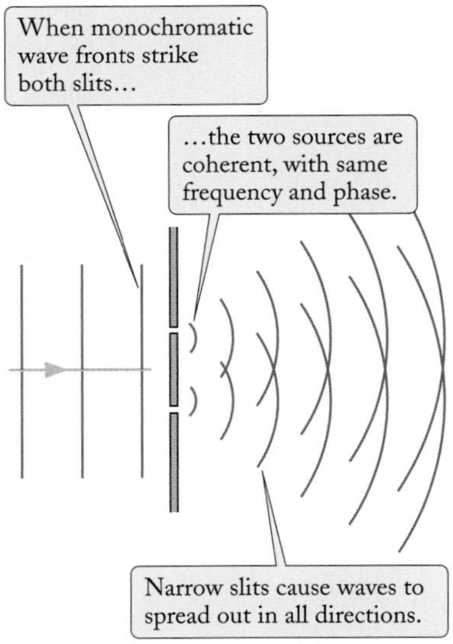

FIGURE 35.13 A plane light wave strikes a plate with two very narrow slits. The slits act as two coherent light sources. Light waves diverge from the two slits and interfere.

In modern practice, a laser is often used to illuminate the two slits, because it provides a very intense plane wave, which makes even faint interference and diffraction effects visible. If the slits in the plate are very narrow, less than one wavelength in width, then the portions of the wave passing through them will diverge from the slits, spreading out in all directions, as illustrated in Fig. 35.13 (the spreading of the wave is a diffraction effect; we will study this in some detail in Section 35.5). This means that the two slits act as two pointlike, coherent sources.

To find the zones of constructive and destructive interference, or the zones of maximum and minimum light intensity, in the space beyond the slits, we need to examine the path difference between the rays from each of the slits. Figure 35.14 shows the light wave incident on the plate from the left and the rays QP and $Q'P$ leading from the slits to the point P on the right. The waves emerging from the slits and reaching P interfere constructively if the difference between the lengths QP and $Q'P$ is zero, or one wavelength, or two wavelengths, etc.; and they interfere destructively if this difference is one half wavelength, or three half wavelengths, five half wavelengths, etc.

We can obtain a simple formula for the angular positions of the maxima and minima if we make the additional assumption that the point P is at a very large distance from the plate (to be precise, QP is assumed to be very large compared with the distance between the slits, $d = QQ'$). If so, then the rays QP and $Q'P$ are nearly parallel, as illustrated in Fig. 35.15. The difference between the lengths QP and $Q'P$ is then the distance $Q'R$; this is the path difference between the two waves arriving at P. This path

[1] If the lightbulb is placed too close to the slits, then light waves arrive at the slits from several directions at once, and this tends to wash out the interference pattern because waves of different directions of incidence will produce overlapping zones of brightness and darkness. For convenience, the lightbulb is sometimes placed fairly near the slits; but it must then be covered with a shield perforated by a single pinhole, so the emerging light comes from just one portion on the surface of the lightbulb; this again makes the light source into a point source.

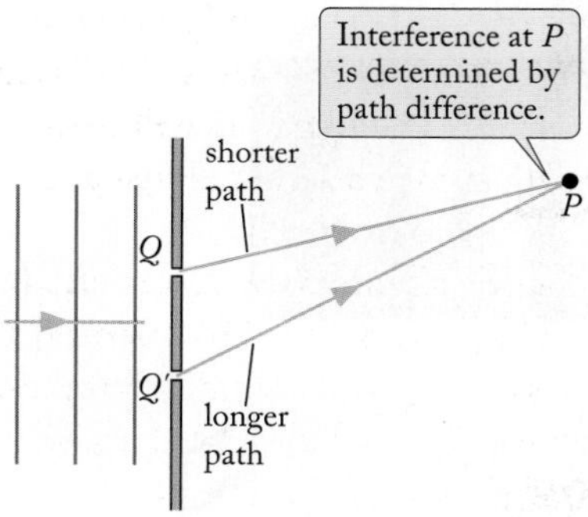

FIGURE 35.14 The waves reaching the point P have different path lengths QP and $Q'P$.

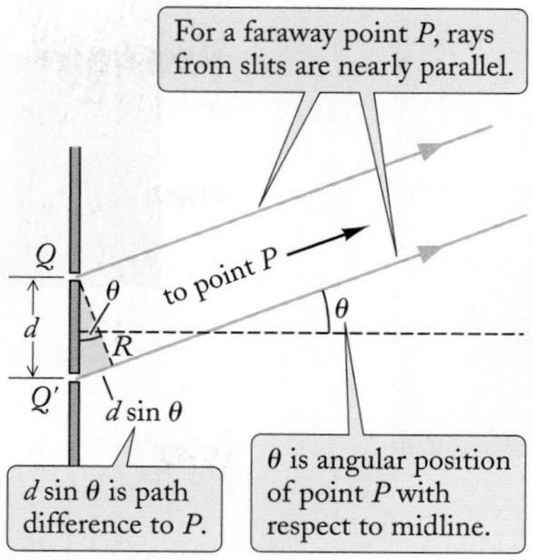

FIGURE 35.15 If P is far away, QP and $Q'P$ are nearly parallel. The lengths QP and $Q'P$ then differ by $d\sin\theta$.

difference $Q'R$ is the short side of the right triangle $QQ'R$. Since the distance d between the slits is the hypotenuse of this triangle, we can express $Q'R$ in terms of d and the angle θ:

$$[\text{path difference}] = Q'R = d\sin\theta \tag{35.6}$$

The angle θ in the triangle equals the angle between the line toward P and the perpendicular midline of the plate (see Fig. 35.15); thus, θ represents the angular position of the point P.

Our condition for constructive interference is then that the path difference $Q'R$ equals zero, one, two, etc., wavelengths:

$$d\sin\theta = 0,\ \lambda,\ 2\lambda, \ldots \tag{35.7}$$

maxima of interference pattern

or

$$d\sin\theta = m\lambda \quad \text{where } m = 0,\ 1,\ 2, \ldots \tag{35.8}$$

Likewise, the condition for destructive interference is

$$d\sin\theta = \tfrac{1}{2}\lambda,\ \tfrac{3}{2}\lambda,\ \tfrac{5}{2}\lambda, \ldots \tag{35.9}$$

minima of interference pattern

For a distant point, these equations give the angular positions of the maxima and the minima of the interference pattern.

The zones of high intensity have the shape of beams fanning out in the space beyond the slits. The central beam (corresponding to $\sin\theta = 0$, or $\theta = 0$) is called the *central maximum*, or the *zero-order maximum*. The beams on each side of this (corresponding to $d\sin\theta = \lambda$, where θ is measured either up or down) are called the *first-order maxima*, etc. Thus, the integer m in Eq. (35.8) is called the *order number*. The beams of high intensity are separated by lines of zero intensity; these are nodal lines, analogous to the nodal points produced by destructive interference in the standing wave on a string (see Section 16.4). The lines of zero intensity between the zero-order maximum and the first-order maxima (corresponding to $d\sin\theta = \frac{1}{2}\lambda$) are called the *first-order minima*, etc.

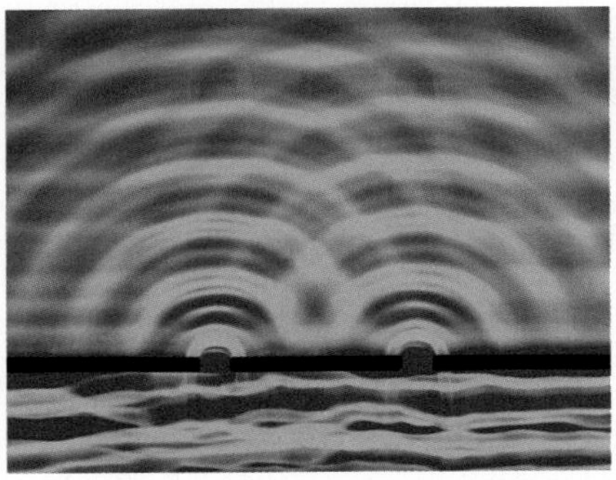

FIGURE 35.16 Interference between water waves spreading out from two coherent pointlike sources in a ripple tank.

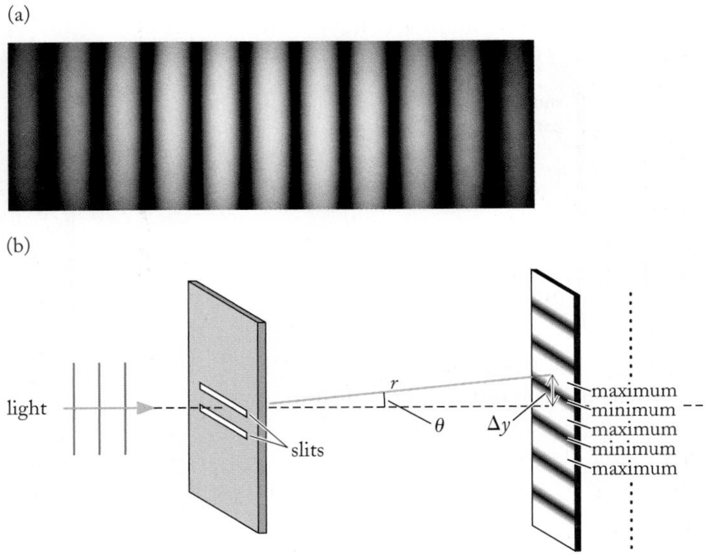

FIGURE 35.17 (a) A photographic film placed beyond two illuminated narrow slits records a regular pattern of bright and dark fringes. (b) Placement of the photographic film beyond the slits. For a small angle θ, the vertical distance Δy is approximately $\Delta y = r\theta$.

Figure 35.16 is a photograph of such a pattern of fanlike beams produced by the interference of water waves spreading out from two pointlike sources [the formulas (35.7) and (35.9) apply to water waves and to any other kinds of waves]. With light waves, it is difficult to photograph the entire pattern of beams at once; instead we must be content with the photograph in Fig. 35.17, which shows the pattern of bright and dark fringes recorded on a photographic film that intercepts the beam at some fixed distance beyond the slits.

EXAMPLE 2 Two narrow slits separated by a distance of 0.12 mm are illuminated with light of wavelength 589 nm from a sodium lamp. What is the angular position of the first-order maximum in the interference pattern? If the light is intercepted by a photographic film placed 2.0 m beyond the slits, what is the distance on the film between this maximum and the central maximum?

SOLUTION: With $d = 1.2 \times 10^{-4}$ m and $\lambda = 5.89 \times 10^{-7}$ m, Eq. (35.7) gives

$$\sin\theta = \frac{\lambda}{d} = \frac{5.89 \times 10^{-7}\,\text{m}}{1.2 \times 10^{-4}\,\text{m}} = 4.9 \times 10^{-3}$$

We recall that, for small angles, the sine (or the tangent) of the angle is approximately equal to the angle in radians. Thus,

$$\theta \approx 4.9 \times 10^{-3} \text{ radian}$$

The distance between the points with $\theta = 0$ and $\theta = 4.9 \times 10^{-3}$ radian on the photographic film is approximately equal to the radial distance r from the slit to the film (see Fig. 35.17b) multiplied by the angle of 4.9×10^{-3} radian:

$$\Delta y \approx r\theta$$

$$= 2.0\text{ m} \times 4.9 \times 10^{-3} = 9.8 \times 10^{-3}\text{ m} = 0.98\text{ mm}$$

Thus, the maxima are separated by nearly 1 cm.

PROBLEM-SOLVING TECHNIQUES — TWO-SLIT INTERFERENCE

- In general, the path difference between the two waves arriving at the observation point P is the difference in the length of the two long sides of the triangle in Fig. 35.14. This path difference can always be calculated by trigonometry. But if the point P is at a very large distance compared with the separation d of the sources, then the path difference takes the simple form $d \sin\theta$.
- Whenever the angular position θ of the observation point P is small, then we can use the approximation $\sin\theta \approx \theta$. Furthermore, the transverse displacement of the observation point from the midline is then approximately $\Delta y \approx r\theta$.

We can calculate the intensity distribution as a function of angle by examining the superposition of the waves contributed by the two slits. The wave propagating outward from each of the slits is of the form [compare Eq. (33.34)]

$$E \propto \cos\left(\omega t - \frac{2\pi r}{\lambda}\right)$$

where ω is the frequency of the wave and r is the distance from the slit to the point P. For the wave emerging from the upper slit, the distance r is

$$r_1 = r_0 - \frac{d}{2}\sin\theta$$

where r_0 is the distance measured from the midpoint between the two slits (see Fig. 35.18). And for the wave emerging from the lower slit, the distance r is

$$r_2 = r_0 + \frac{d}{2}\sin\theta$$

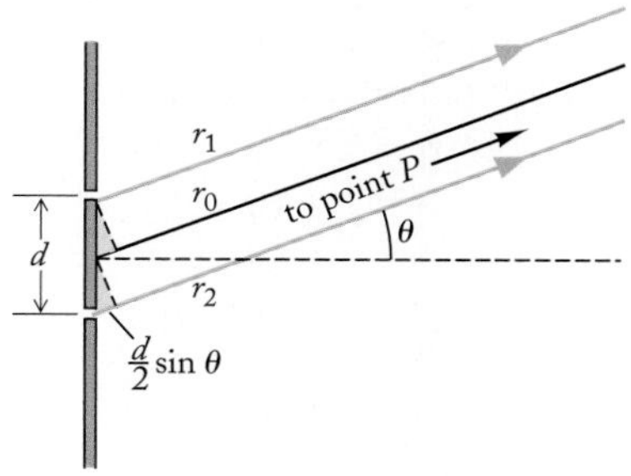

FIGURE 35.18 Rays from the slits to the point P.

At our faraway point P, each wave has essentially the same amplitude E_0. Hence the electric fields contributed by the upper and the lower slits are, respectively,

$$E_1 = E_0 \cos\left(\omega t - \frac{2\pi r_0}{\lambda} + \frac{\pi d}{\lambda}\sin\theta\right)$$

and

$$E_2 = E_0 \cos\left(\omega t - \frac{2\pi r_0}{\lambda} - \frac{\pi d}{\lambda}\sin\theta\right)$$

The superposition of these two electric fields then gives

$$\begin{aligned} E &= E_1 + E_2 \\ &= E_0\left[\cos\left(\omega t - \frac{2\pi r_0}{\lambda} + \frac{\pi d}{\lambda}\sin\theta\right) + \cos\left(\omega t - \frac{2\pi r_0}{\lambda} - \frac{\pi d}{\lambda}\sin\theta\right)\right] \end{aligned}$$

With the trigonometric identity $\cos(\alpha + \beta) + \cos(\alpha - \beta) = 2\cos\alpha\cos\beta$, (see Appendix 14) this becomes

$$E = 2E_0 \cos\left(\omega t - \frac{2\pi r_0}{\lambda}\right)\cos\left(\frac{\pi d}{\lambda}\sin\theta\right) \qquad (35.10)$$

The factor $\cos(\omega t - 2\pi r_0/\lambda)$ is the usual oscillating function of space and time, characteristic of a wave propagating outward. The factor $2E_0 \cos[(\pi d/\lambda) \sin\theta]$ is the amplitude of the wave and depends on the position angle θ.

To determine the interference pattern, we are interested in the intensity as a function of position. The intensity of a wave is proportional to E^2. Thus the maximum intensity of the summed wave (35.10) is 4 times the intensity of the wave arriving from either slit. Since we are interested only in the time-average intensity, we can replace the factor $\cos^2(\omega t - 2\pi r_0/\lambda)$ from Eq. (35.10) by its time-average value $\frac{1}{2}$. The dependence of the intensity I on the position angle is then given by the remaining factor:

intensity for two-slit interference pattern

$$I = I_{max} \cos^2\left(\frac{\pi d}{\lambda} \sin\theta\right) \tag{35.11}$$

where I_{max} is the intensity at the center of the interference pattern, at $\theta = 0$. This formula describes the intensity distribution produced by the two slits. Figure 35.19 is a plot of the intensity as a function of the angle θ. The plot represents the intensity of light found at different angles at some constant distance r_0 from the midpoint of the slits. As expected, the intensity is maximum when $(d/\lambda) \sin\theta = 0, 1, 2, 3$, etc., and it is minimum (zero) when $(d/\lambda) \sin\theta = \frac{1}{2}, \frac{3}{2}, \frac{5}{2}$, etc.

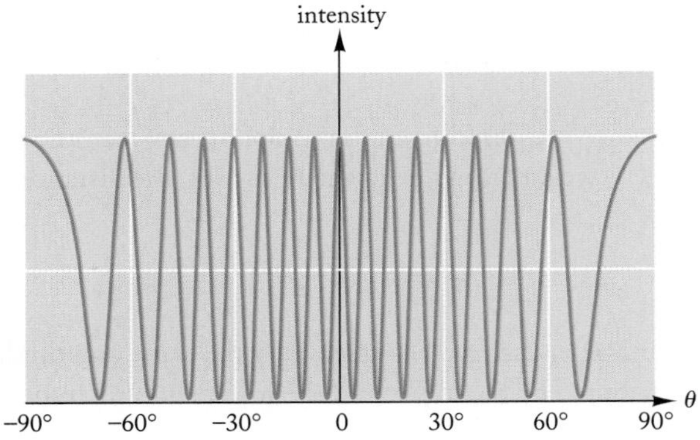

FIGURE 35.19 Intensity as a function of the angle θ for light arriving at a screen placed beyond two slits. The distance between the slits is $d = 8\lambda$.

EXAMPLE 3 Consider the intensity in the two-slit interference pattern on a distant screen at an angle midway between a maximum and a minimum. How does this intensity compare with the intensity at the maximum?

SOLUTION: For the usual experimental arrangement, we can assume, for a distant screen, that the angle θ is small, so the sine of the angle is approximately equal to the angle. The angle midway between the central maximum and the first-order minimum then has $d \times \theta = \frac{1}{4}\lambda$, which is midway between the values $d \times \theta = 0$ (central maximum) and $d \times \theta = \frac{1}{2}\lambda$ (first minimum). According to Eq. (35.11), the intensity at this position is

$$[\text{intensity}] \propto \cos^2\left(\frac{\pi}{\lambda} d \sin\theta\right) \approx \cos^2\left(\frac{\pi}{\lambda} d \times \theta\right) = \cos^2\left(\frac{\pi}{\lambda}\frac{\lambda}{4}\right) = \cos^2\frac{\pi}{4} = \frac{1}{2}$$

Thus, the intensity at this point is one-half the intensity at the central maximum.

Checkup 35.3

QUESTION 1: If, in Example 2, we *decrease* the distance between the slits, does the distance between the first maximum and the central maximum decrease or increase?

QUESTION 2: Suppose that the distance d between the slits equals λ. Describe the intensity distribution for this case.

QUESTION 3: If the first-order maximum is to be at $\theta = 45°$, what wavelength is required for a given slit separation d?

QUESTION 4: Consider a double-slit experiment with $d = 5\lambda$. At what angular position do the third-order minima occur?

(A) 0° (B) 30° (C) 45° (D) 60° (E) 90°

QUESTION 5: In a double-slit experiment, let I_{max} be the intensity at the peak of the central maximum. If the path difference to a point on the screen is $\lambda/6$, what is the intensity at that point?

(A) 0 (B) $\frac{1}{4}I_{max}$ (C) $\frac{1}{2}I_{max}$ (D) $\frac{3}{4}I_{max}$ (E) I_{max}

35.4 INTERFERENCE FROM MULTIPLE SLITS

Online Concept 40 Tutorial

It is easy to see that Eq. (35.8) also applies to the case of a plane wave incident on multiple slits. For instance, Fig. 35.20 shows an opaque plate with three evenly spaced slits. If waves emerging from *adjacent* slits interfere constructively, then the waves emerging from *all three* slits interfere constructively. Hence the condition for maximum intensity is the same as Eq. (35.8),

$$d \sin\theta = m\lambda \qquad \text{where } m = 0, 1, 2, \ldots \tag{35.12}$$

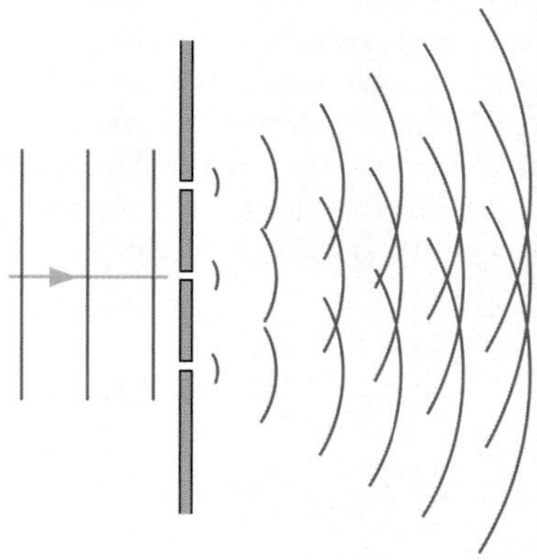

FIGURE 35.20 A plane wave strikes a plate with three very narrow slits.

and this same equation also applies to more than three slits. However, the condition for minimum intensity is *not* Eq. (35.9). For instance, in the case of three slits, destructive interference of waves from adjacent slits will lead to cancellation of the wave originating from one pair of slits, but the wave from the remaining slit will not experience cancellation. Thus, the condition for minima is somewhat more complicated; a technique to determine the detailed behavior between the maxima given by Eq. (35.12) is discussed later in this section. It turns out that between the maxima given by Eq. (35.12), called **principal maxima**, there can be several minima and several **secondary maxima**. Figure 35.21 is a photograph illustrating the strong principal maxima and the weaker secondary maxima of the diffraction pattern of three slits. We will consider secondary maxima in more detail at the end of this section. For now, we will concentrate our attention on the principal maxima, because the secondary maxima are quite weak, especially if the number of slits is large.

principal maxima

secondary maxima

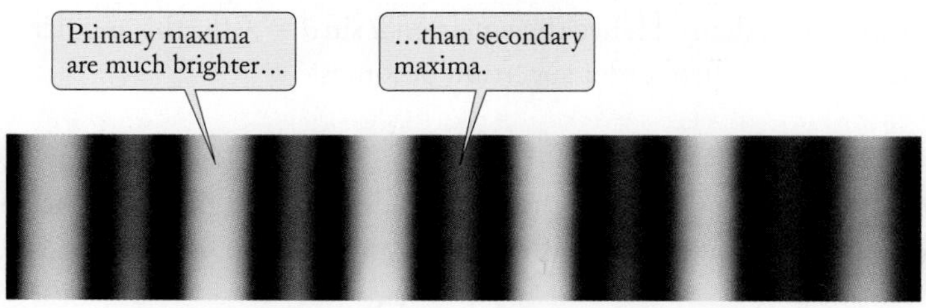

FIGURE 35.21 A photographic film placed beyond the slits shows the strong principal maxima and the weaker secondary maxima between them.

diffraction gratings

Arrangements of large numbers of slits are called **diffraction gratings** or, simply, **gratings**. When the number of slits is large, the principal maxima become very sharply peaked and well separated. Gratings are commonly used in spectroscopy laboratories to analyze light into its colors. If a light beam containing several wavelengths passes though such a grating, the maxima for these different wavelengths will form beams at different angles—according to Eq. (35.12), the angle θ increases with wavelength, and hence the beams of long-wavelength light will be found at larger angles than the beams of short-wavelength light. Thus, the grating separates the light according to color and produces a spectrum in much the same way that a prism does. There is, however, one important difference between the spectra formed by a prism and by a grating: in the prism the long-wavelength light (red) experiences the least deflection; in the grating the long-wavelength light experiences the most deflection.

The grating will produce one complete spectrum for each of the possible alternatives listed for the right side in Eq. (35.12)—that is, each principal maximum, except the central maximum, gets spread out by color. These spectra are called, in turn, the first-order spectrum [$m = 1, d \sin\theta = \lambda$ in Eq. (35.12)], the second-order spectrum ($m = 2, d\sin\theta = 2\lambda$), and so on. Sometimes these spectra overlap; for example, the red end of the second-order spectrum may show up at the same angle as the blue end of the third-order spectrum (see Fig. 35.22).

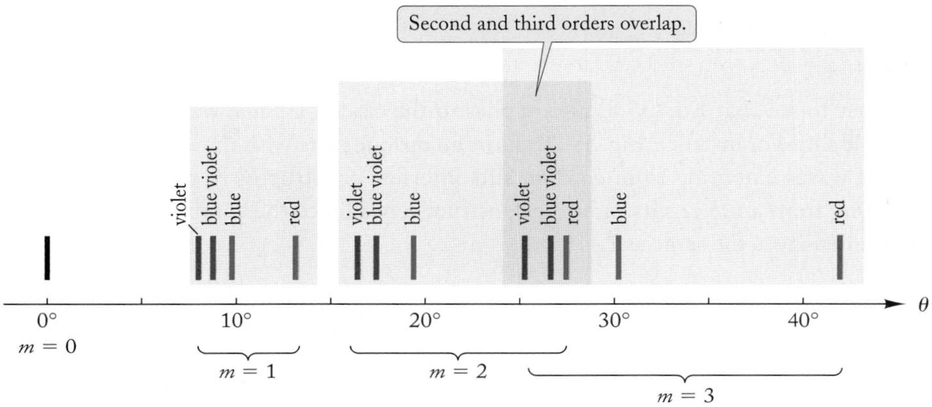

FIGURE 35.22 First-, second-, and third-order spectra ($m = 1$, $m = 2$, and $m = 3$) of hydrogen light produced by a system of N slits. The lines correspond to the principal maxima; the secondary maxima are weak and can be ignored. Each spectrum consists of violet, blue violet, blue, and red spectral lines (compare Fig. 34.29b). The pattern of spectral lines for negative values of θ (negative m) is similar.

EXAMPLE 4 When atoms of iron are heated, they emit numerous spectral lines, among which are spectral lines of wavelengths $\lambda = 500.57\,\text{nm}$ and $\lambda = 500.61\,\text{nm}$. Suppose we analyze the spectrum of iron with a grating with $d = 4.0 \times 10^{-6}$ m. What is the angular separation between the maxima for these wavelengths in the first-order spectrum produced by this grating?

SOLUTION: The first-order maxima occur near

$$\sin\theta = \frac{\lambda}{d} = \frac{500.6 \times 10^{-9}\ \text{m}}{4.0 \times 10^{-6}\ \text{m}} \approx 0.13$$

This is a sufficiently small value that the sine of the angle is approximately equal to the angle in radians. Hence the equation $d\sin\theta = \lambda$ for the angular position of the maxima in the first-order spectrum becomes

$$d \times \theta \approx \lambda$$

Thus, the angle is proportional to the wavelength, and the difference in angle is therefore proportional to the difference in wavelength:

$$d\,\Delta\theta = \Delta\lambda$$

With $\Delta\lambda = 500.61\text{ nm} - 500.57\text{ nm} = 0.04\text{ nm}$, we find

$$\Delta\theta = \frac{\Delta\lambda}{d} = \frac{0.04 \times 10^{-9}\text{ m}}{4.0 \times 10^{-6}\text{ m}} = 1 \times 10^{-5}\text{ radian}$$

The best gratings have a very large number of slits, because this produces sharp, narrow beams and enhances the ability of the grating to discriminate between spectral lines of nearly identical wavelengths. Since it is difficult to cut a large number of slits in an opaque plate, gratings are usually manufactured by cutting fine parallel grooves in a glass or metal surface with a diamond stylus guided by a special ruling machine. When illuminated by a plane light wave, the edges of the grooves act as coherent light sources, in much the same way as slits. High-quality gratings used in spectroscopy have 100 000 or more grooves with distances of about 10^{-6} m between them.

Concepts in Context

reflection gratings

You can verify the effects of a grating on light by a simple experiment with a CD or an LP record; the former is illustrated in the chapter photo. The tracks or grooves on these disks make them behave like crude **reflection gratings**. You can see spectral colors with these crude gratings if you let them reflect the light from some narrow light source, such as sunlight coming through a crack between windows or light from the edge of a bright lamp. Compact discs have tighter grooves than LP records, and they give a more pronounced spread of the spectral colors. When you look across the surface of such a disk, you see light from a range of angles, and thus light of many colors, at the same time. Also, the grooves are curved, and so you see several patterns in different directions. The left CD in the chapter photo shows a continuous spectrum of colors, such as in sunlight or from the hot filament of a lightbulb, whereas the right CD predominantly shows a discrete spectrum, characteristic of the individual atoms of a particular element, in this case hydrogen in a gas discharge tube.

Concepts in Context

EXAMPLE 5 The patterns on the CDs in the photo at the beginning of the chapter result from interference. For the CD on the right side, the light is normally incident on the CD and the third-order blue interference maximum ($\lambda = 486$ nm) is seen when viewing at 60° from the normal. What is the effective slit separation d of the compact-disc grating?

SOLUTION: Since the light is normally incident, it behaves like a wave normally incident on slits. The third-order interference maximum is then given by Eq. (35.12) with $m = 3$,

$$d \sin\theta = 3\lambda$$

so the slit spacing is

$$d = \frac{3\lambda}{\sin\theta} = \frac{3 \times 486\text{ nm}}{\sin 60^\circ} = 1.7 \times 10^{-6}\text{ m} = 1.7\ \mu\text{m}$$

COMMENT: In science and engineering practice, interference phenomena are often used to determine spacings, as in this example. Thus, an X-ray diffractometer or an electron diffraction device each obtain an interference pattern by reflecting a wave from a "grating" of atoms (in a crystal or on a surface, respectively); the measured angles of the interference maxima then determine atomic spacings that are too small to see.

radio telescopes

Radio telescopes, consisting of a regularly spaced array of antennas, act as "gratings" for radio waves. The same formula (35.12) for the angular directions of the maxima applies to these as to ordinary gratings used with light. Of course, there is a difference between the operation of a radio telescope and that of an ordinary grating: in the former the radio waves from a distant source *enter* the antennas and interfere constructively or destructively within the radio receiver, whereas in the latter the light waves *emerge* from the slits and travel to a distant point where they interfere. Nevertheless, the condition for, say, maximum intensity can still be expressed in terms of the direction of the entering or emerging wave by Eq. (35.12), because this condition hinges only on the phase relationships among the waves, and these relationships are independent of the direction of propagation (for instance, in Fig. 35.15, if the waves traveling away from the slits in some direction reach a distant point in phase, then waves traveling in the reverse direction toward the slits from this point will arrive at the slits in phase).

EXAMPLE 6 One branch of the VLA (Very Large Array) radio telescope at Socorro, New Mexico, has nine antennas arranged on a straight line with distances of 2.0 km between one antenna and the next (see Fig. 35.23). These antennas are all connected to a single radio receiver by waveguides of equal length; the receiver then registers a maximum intensity if the radio waves incident on all antennas are in phase. Suppose that radio waves of wavelength 21 cm from a pointlike quasar in the sky strike this radio telescope. When the quasar is at the zenith, the intensity in the radio receiver is maximum. What angular displacement of the quasar from the zenith will also result in maximum intensity?

SOLUTION: Let us assume (and later verify) that the desired angular displacement may be small, so that we can use the approximation that for small angles, the sine of the angle equals the angle in radians. Thus, Eq. (35.12) gives, for the maximum nearest the central maximum,

$$d \times \theta = \lambda \qquad (35.13)$$

Hence

$$\theta = \frac{\lambda}{d} = \frac{0.21\ \text{m}}{2.0 \times 10^3\ \text{m}} = 1.1 \times 10^{-4}\ \text{radian}$$

The angular displacement at the first maximum is indeed small, so our assumption was justified.

FIGURE 35.23 The VLA radio telescope. The distance between the antennas can be adjusted by rolling them along tracks. Here the antennas are shown close together.

For a quantitative analysis of the interference due to multiple slits, we can use the technique of **phasors**, previously introduced in Chapter 32. A phasor is a rotating vector in a fictitious two-dimensional space, useful because any oscillating function can be represented as the x component of a rotating vector. The rotating vector shown in Fig. 35.24 has length equal to the amplitude E_0 of the wave disturbance at some point, and at any instant makes an angle ωt with respect to the x axis. As is evident in Fig. 35.24, the x component then indeed equals the instantaneous value of $E = E_0 \cos \omega t$. The utility of the technique can be seen in Fig. 35.25, where two waves of amplitude E_0 with a phase difference δ between them are summed vectorially to determine the amplitude of the resultant wave. To do this, we merely plot the corresponding phasors tail to tip, with a relative angle equal to the phase difference. From the geometry of Fig. 35.25, we see that the resultant wave has amplitude

phasor

$$E = 2E_0 \cos\left(\frac{\delta}{2}\right) \qquad (35.14)$$

For $\delta = 0$, the resultant is $E = 2E_0$, which corresponds to constructive interference. As the phase difference δ increases, the resultant goes through a series of minima ($\delta = \pi, 3\pi, 5\pi, \ldots$) and maxima ($\delta = 2\pi, 4\pi, 6\pi, \ldots$) as the second phasor alternately cancels or adds to the first phasor. We can apply this to the double-slit experiment. The phase difference between the two waves expressed as a fraction of 2π is the same as the path difference expressed as a fraction of λ:

$$\frac{\delta}{2\pi} = \frac{d \sin\theta}{\lambda} \qquad (35.15)$$

It is easy to verify that with this phase difference δ, Eq. (35.14) gives the same intensity as a function of the position angle θ as obtained previously in Eq. (35.11). Thus either the algebraic result (35.11) or the geometric result (35.14) gives the two-slit intensity pattern with comparable ease.

For multiple, nonidentical, or other complex slit arrangements, the phasor technique can readily provide insight into the locations of the minima and the primary and secondary maxima, as well as the detailed intensity pattern. For example, consider interference from three identical slits, discussed above (see Figs. 35.20 and 35.21). In this case we sum three phasors, as shown in Figs. 35.26a–f. For equally spaced slits, the first and second phasors have the same phase difference δ between them as the second

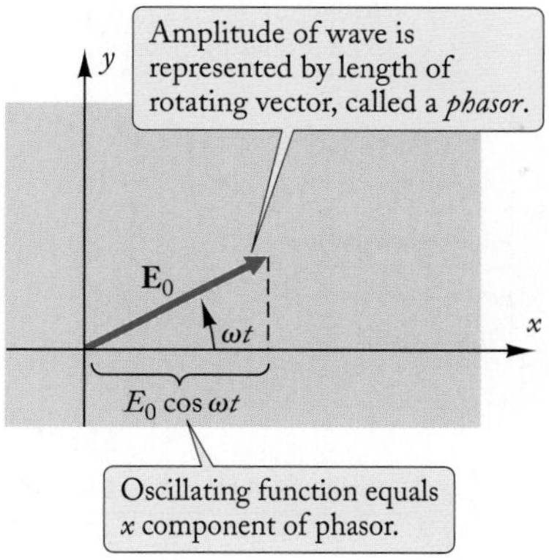

FIGURE 35.24 A rotating vector, or phasor, has magnitude equal to the amplitude of the wave it represents.

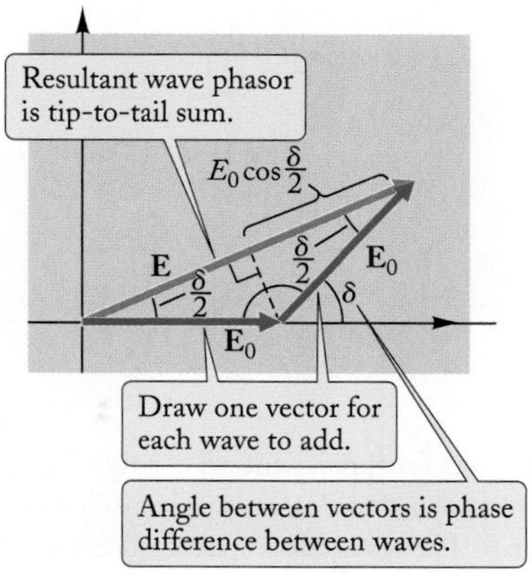

FIGURE 35.25 To add two waves of amplitude E_0, the corresponding phasors are drawn tip to tail and the angle between them is set equal to the phase difference of the waves.

and third phasors [each difference given by Eq. (35.15)]. As we increase our angular position θ, the phase difference δ increases, as usual. As shown in Fig. 35.26, the net amplitude E begins at $3E_0$ (Fig. 35.26a) and decreases smoothly with increasing phase difference until $E = 0$ at $\delta = 2\pi/3$ (Fig. 35.26d), where the phasors form an equilateral triangle. Beyond $\delta = 2\pi/3$, the intensity of the three-slit pattern then increases smoothly until $E = E_0$ at $\delta = \pi$ (Fig. 35.26f), a secondary maximum. Then the sequence of resultant amplitudes continues in reverse, through another minimum at $\delta = 4\pi/3$, until the first-order principal maximum ($E = 3E_0$) is reached at $\delta = 2\pi$; thereafter, the entire sequence repeats for every 2π increase of phase difference between adjacent slits. The resulting intensity pattern (proportional to the square of the amplitude) is shown in Fig. 35.27a (a photograph of this pattern is shown in Fig. 35.21). For the three-slit pattern, there is one secondary maximum between the primary maxima; in general, there will be one more secondary maximum for every additional slit as in Fig. 35.27b.

The same technique of examining the resultant amplitude as the phase difference between adjacent slits increases can be applied to any number of slits, even to unequal slits, which produce waves of different amplitudes. Similarly, for slits that are not evenly spaced, the technique can be applied by maintaining the phase difference between different adjacent pairs of phasors in direct proportion to the spacing between the corresponding pairs of adjacent slits.

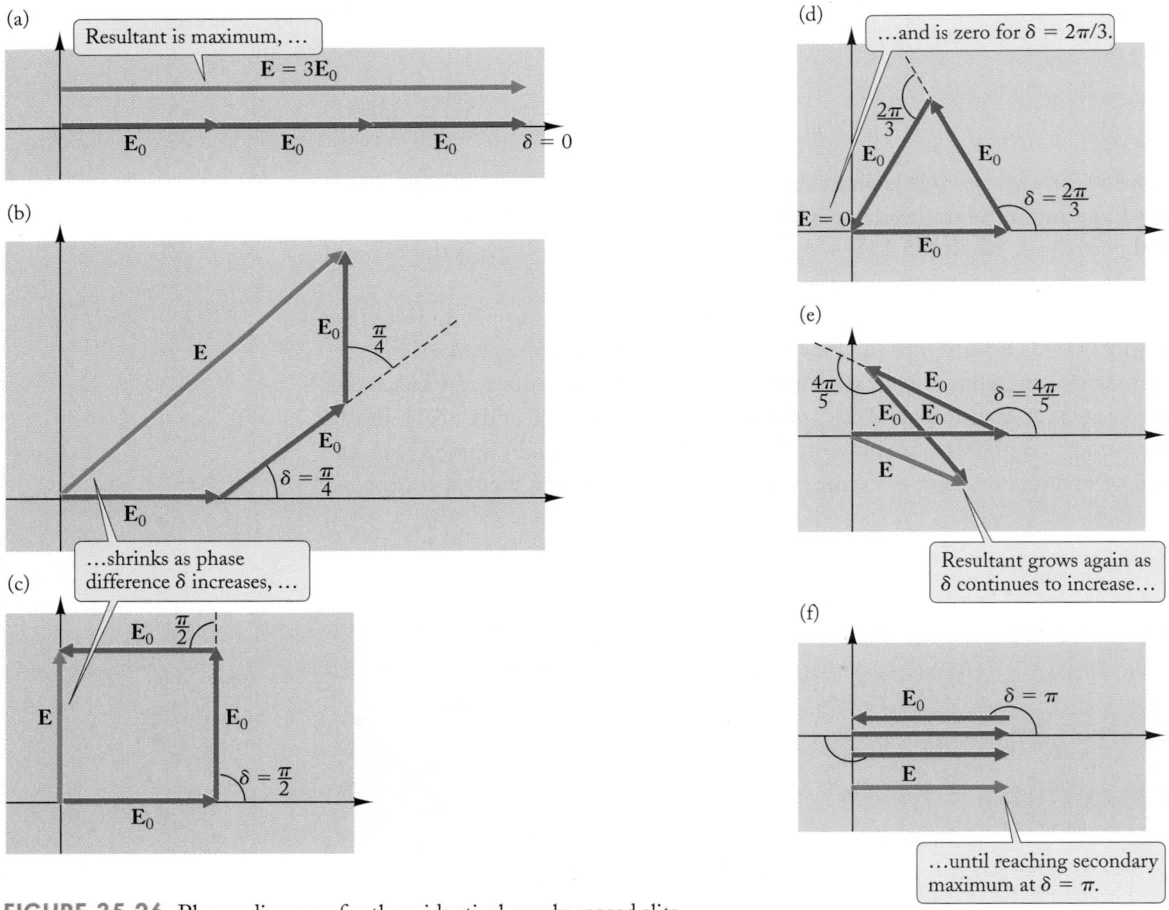

FIGURE 35.26 Phasor diagrams for three identical evenly spaced slits. The phase difference $\delta = (2\pi/\lambda) d \sin\theta$ increases as we consider the resultant for increasing angular position θ. The resultant amplitude decreases from a primary maximum to zero (a–d) and then increases from zero to a secondary maximum (d–f).

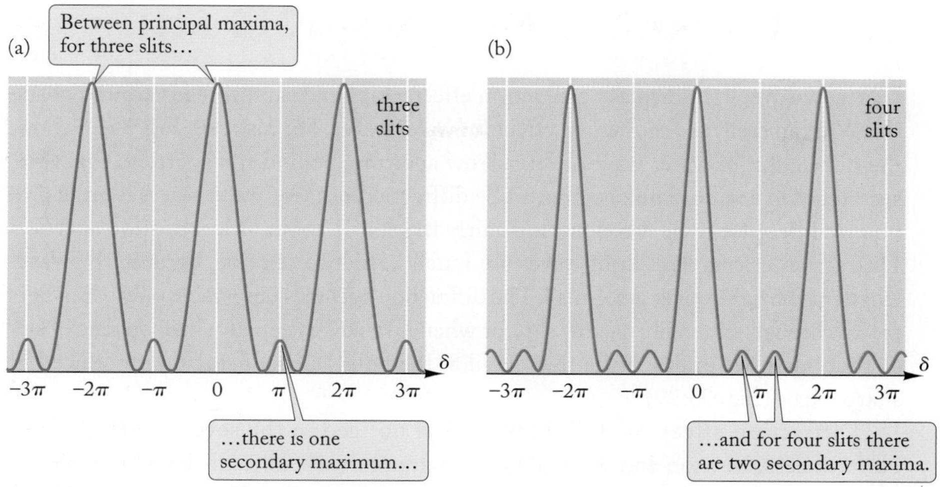

FIGURE 35.27 Intensity as a function of the phase difference δ between adjacent slits for light arriving at a screen placed beyond (a) three slits, and (b) four slits. For three slits, there is one secondary maximum between principal maxima; for four slits, there are two. (Note that these plots can be converted into plots of intensity as a function of the angle θ by using $\delta = (2\pi/\lambda)\sin\theta$.)

For three slits, we saw that the first interference minimum is at $\delta = 2\pi/3$, where the phasors form an equilateral triangle (see Fig. 35.26d). This corresponds to a path difference between adjacent slits of $d\sin\theta = \lambda/3$. Similarly, for any number N of identical, evenly spaced slits, the first minimum occurs when the phasors form an N-sided regular polygon. For N slits, the angle θ of the first minimum is then given by

$$d\sin\theta = \frac{\lambda}{N} \qquad (35.16)$$

first minimum for *N* slits

Thus for large N, such as with a grating, the principal maxima are extremely narrow.

✔ Checkup 35.4

QUESTION 1: In Fig. 35.22, a red spectral line appears between a blue and a blue violet line. How can this be?

QUESTION 2: In Example 4 we calculated the angular separation between two spectral lines in the first-order spectrum produced by a grating. Would the separation between these spectral lines be larger or smaller in the second-order spectrum?

QUESTION 3: Consider the three-slit phasor diagrams of Fig. 35.26. What is the amplitude of the resultant when the phase difference between adjacent slits is $\pi/2$ (90°)?

QUESTION 4: Consider two slits, but now the first is narrow and produces a wave of amplitude E_0 and the second is wider and produces a wave of amplitude $2E_0$. Let I_0 be the intensity due to the first slit alone. Using phasors, answer the following. What is the intensity of each maximum? What is the intensity at each minimum (it is not zero)? What is the intensity half way between a maximum and a minimum?

QUESTION 5: For five identical, evenly spaced slits, the first interference minimum occurs when the phase difference δ between adjacent slits is equal to:

(A) $\pi/10$ (B) $\pi/5$ (C) $2\pi/5$ (D) $\pi/2$ (E) 2π

QUESTION 6: For six identical, evenly spaced slits, how many secondary interference maxima occur between each adjacent pair of primary maxima?

(A) 1 (B) 2 (C) 3 (D) 4 (E) 5

Online Concept 39 Tutorial

35.5 DIFFRACTION BY A SINGLE SLIT

Like any wave, light displays diffraction effects—it bends and spreads around obstacles. We can easily see diffraction effects in water waves. For instance, Fig. 35.28 shows the diffraction of water waves by a narrow aperture; behind the aperture, the waves spread out in a wide, fanlike pattern. The diffraction of these water waves is quite pronounced because their wavelength is fairly large; it is about as large as the aperture. Diffraction effects with light waves are much harder to observe, because the wavelength of the light waves is so small. The diffraction becomes observable only when light passes through extremely narrow slits, or when it strikes extremely small opaque obstacles, or when we examine the fine detail at the boundaries of shadows produced by sharp edges (Fig. 35.29).

Diffraction effects with light were first noticed in the seventeenth century. Nevertheless, Newton and some of his contemporaries held to the belief that light is of a corpuscular nature, and consists of a stream of particles. The importance of diffraction effects in establishing the wave nature of light was not fully appreciated until 1818, when Augustin Fresnel mathematically formulated the theory of diffraction and demonstrated that diffraction is a general and distinctive characteristic of waves. The brilliant success of this theory finally convinced physicists that light is a wave. In this section we will discuss the simple case of diffraction by a single narrow slit.

Figure 35.30 shows a plane light wave approaching a slit in an opaque plate. To find the light distribution in the space beyond the slit, we use the following prescription, called the **Huygens–Fresnel Principle**:

Huygens–Fresnel Principle

Pretend that each point of the wave front reaching the slit can be regarded as a point source of light emitting a spherical wave; the net wave in the region beyond the slit is simply the superposition of all these waves.

This prescription has some obvious similarities to Huygens' Construction (Section 34.1). However, the latter is merely a geometric construction for finding the successive positions of the wave fronts and does not yield any information about the distribution of intensity in different directions, whereas the aim of our new prescription is precisely the calculation of this distribution of intensity. Although the new prescription has a strong intuitive appeal, it it is not all that easy to justify rigorously. Nevertheless, it gives the right answer, or essentially the right answer; this can be

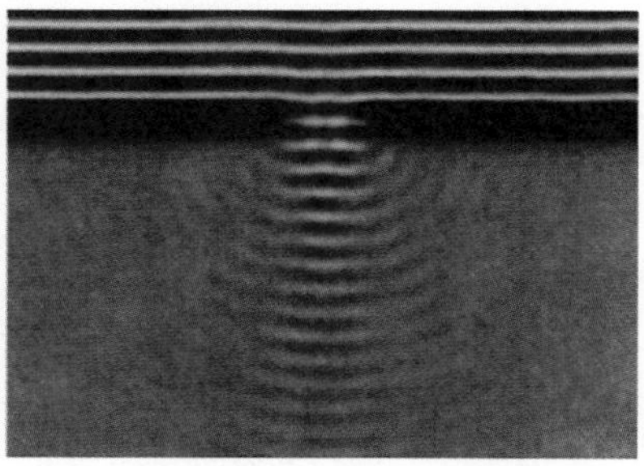

FIGURE 35.28 Diffraction of water waves in a ripple tank. Beyond the narrow aperture, the waves spread out in a fanlike pattern.

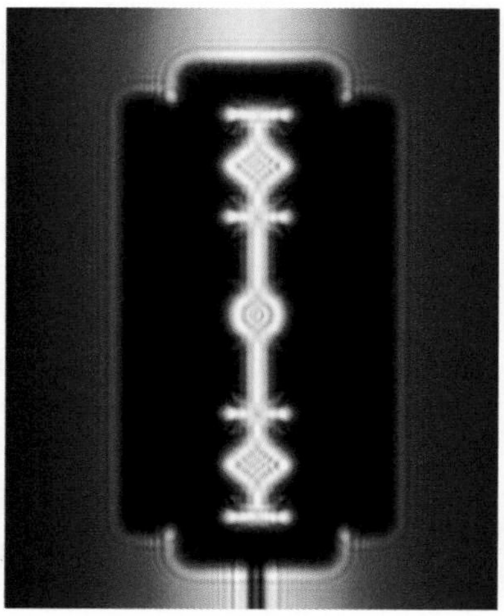

FIGURE 35.29 Diffraction of light around a razor blade. The diffracted light generates a complex pattern of fine fringes at the edges of the shadow. This photograph was prepared by illuminating the razor blade with a distant light source and throwing its shadow on a distant screen.

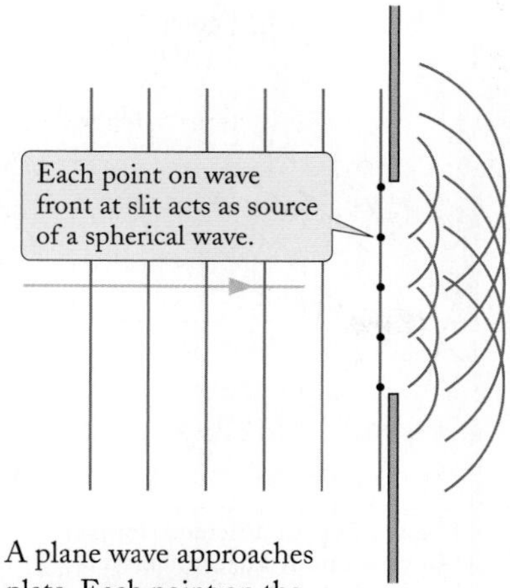

FIGURE 35.30 A plane wave approaches a slit in an opaque plate. Each point on the wave front at the slit gives rise to a spherical wave that spreads out beyond the slit.

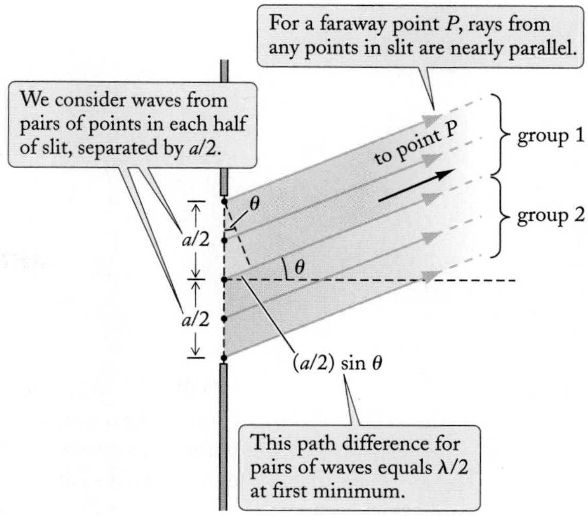

FIGURE 35.31 Rays from points in the slit to the point P. The path difference between the uppermost ray and the middle ray is $(a/2)\sin\theta$.

shown by a somewhat more sophisticated mathematical argument based on Maxwell's equations.

We can use the Huygens–Fresnel Principle to find the positions of the minima in the diffraction pattern of the single slit. Figure 35.31 shows some rays associated with waves that spread out from the points of a wave front at the slit. All the rays lead to a point P in the space beyond the slit. As in our calculation for the case of two slits, we will assume that P is very far away, so all the rays are nearly parallel. We can divide these rays into two equal groups: those that come from the upper part of the slit, and those that come from the lower. Rays from the first group have a shorter distance to travel to the point P than rays from the second group. Consider the uppermost ray from the first group and the uppermost ray from the second group. The path difference between these is $(a/2)\sin\theta$, where a is the width of the slit (see Fig. 35.31). If this path difference equals $\frac{1}{2}\lambda$, the two rays will interfere destructively. Furthermore, pairs of rays that originate at an equal distance below each of the two uppermost rays in the two groups will also interfere destructively. This establishes that all the waves cancel in pairs when

$$\frac{a}{2}\sin\theta = \frac{1}{2}\lambda$$

or

$$a\sin\theta = \lambda \tag{35.17}$$

This is the condition for the first single-slit diffraction minimum.

To find the next minimum, we divide the rays into four equal groups and consider the uppermost rays from the first and second groups (see Fig. 35.32). These rays and other pairs of rays will interfere destructively if their path difference equals $\frac{1}{2}\lambda$:

$$\frac{a}{4}\sin\theta = \frac{1}{2}\lambda \tag{35.18}$$

or

$$a\sin\theta = 2\lambda \tag{35.19}$$

AUGUSTIN FRESNEL (frenel) (1788–1827) *French physicist and engineer. His brilliant experimental and mathematical investigations firmly established the wave theory of light. Fresnel was commissioner of lighthouses, for which he designed large segmented lenses (Fresnel lenses) as a replacement for systems of mirrors.*

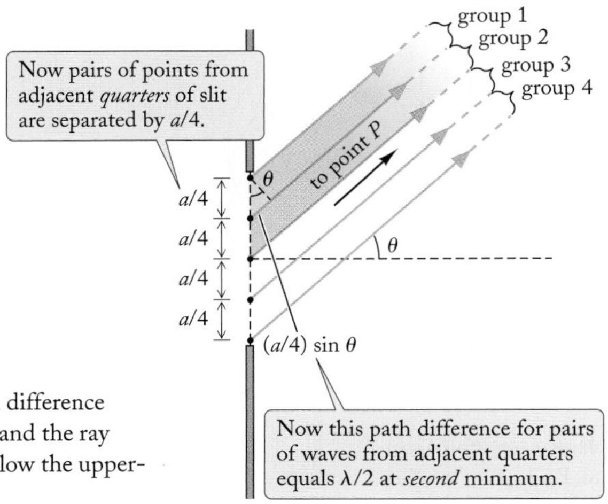

FIGURE 35.32 The path difference between the uppermost ray and the ray starting at a distance $a/4$ below the uppermost ray is $(a/4)\sin\theta$.

By continuing this argument, we find a general condition for single-slit diffraction minima:

minima for diffraction pattern

$$a\sin\theta = \lambda,\ 2\lambda,\ 3\lambda, \ldots \tag{35.20}$$

FIGURE 35.33 A photographic film placed beyond an illuminated single slit records a strong central maximum and successively weaker secondary maxima.

As regards the maxima, there is of course a strong central maximum, at $\theta = 0$, where the path lengths from all of the points in the slit to a distant point on the axis are the same. The secondary maxima cannot be found by any simple argument; we can calculate their locations and heights from the intensity relation derived later in this section. Roughly, their position is halfway between the successive minima. On a photographic film placed at some distance, the maxima and minima show up as a pattern of bright and dark fringes (see Fig. 35.33). The central maximum is much brighter than the first secondary maxima, which are much brighter than the next secondary maxima, and so on.

EXAMPLE 7 Equation (35.20) applies not only to light waves, but also to radio waves and other waves. Suppose that radio waves from a TV transmitter with a wavelength of 0.80 m strike the wall of a large building. In this wall there is a very wide window with a height of 1.4 m (a horizontal slit; see Fig. 35.34). The wall is opaque to radio waves, and the window is transparent. What is the angular width of the central maximum of the diffraction pattern formed by the waves inside the building?

SOLUTION: According to Eq. (35.20), the first minimum is at an angle such that

$$a\sin\theta = \lambda$$

which yields

$$\sin\theta = \frac{\lambda}{a} = \frac{0.80\ \text{m}}{1.4\ \text{m}} = 0.57$$

Our calculator tells us that the angle θ is then 35°. The central maximum extends from −35° to +35°; that is, the width is 70°.

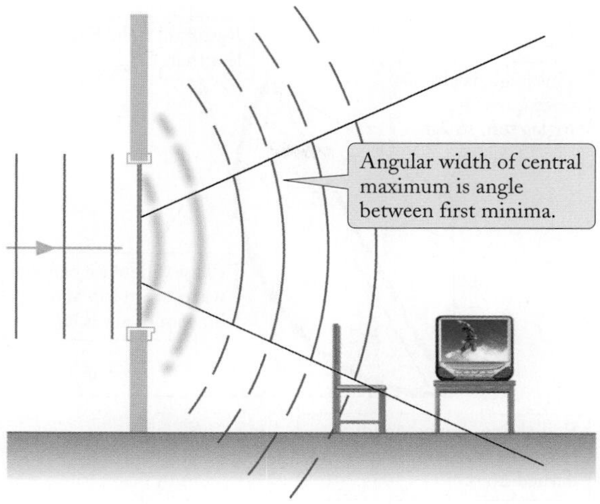

FIGURE 35.34 Diffraction of TV waves at a window.

For a given wavelength, the relation $a \sin\theta = \lambda$ implies that the angle at which the first minimum occurs *increases* when the slit width *a decreases.* Thus a wide slit produces a narrow intensity peak on a distant screen, but for a narrow slit, the pattern spreads out over the screen. This is the fundamental diffraction phenomenon: as the size of the opening or obstruction becomes small compared with a wavelength, the bending and spreading of the waves increases. The angular half-width of the central peak is given by the formula for the first minimum, as in Example 7:

$$\sin\theta = \frac{\lambda}{a} \tag{35.21}$$

The quantity $\sin\theta$ cannot be greater than 1; if $a < \lambda$, then there are no minima, that is, the central peak is so wide that it covers all angles.

To calculate the complete intensity distribution as a function of angle, we must evaluate the superposition of all the waves propagating outward from all the points within the slit (see Fig. 35.35). We divide the slit into a large number k of point sources and sum the electric fields the waves contribute at a distant point P, as previously. Each wave from each point source can be represented by a phasor of length E_0. Figure 35.36a shows the phasor diagram for the central maximum, where all the phasors are in phase and the resultant field is $E_{max} = kE_0$. Figure 35.36b shows the situation away from the central maximum, where the total phase difference between the first and last phasors is ϕ. Because successive points in the slit contribute waves of the same amplitude and equal phase increments, their phasors form a polygonal arc, which we can approximate as an arc of a circle of radius R. The length of the arc is the same total length, E_{max}, as in Fig. 35.36a; the length of the resultant is E. Because the radial lines labeled R are perpendicular to the circular arc, the geometry of Fig. 35.36b shows that the arc subtends the phase angle ϕ. Since angle is defined as arc length over radius, we have

$$\phi = \frac{E_{max}}{R} \tag{35.22}$$

From either of the right triangles formed by the bisector of the angle ϕ we have

$$\sin\left(\frac{\phi}{2}\right) = \frac{E/2}{R} \tag{35.23}$$

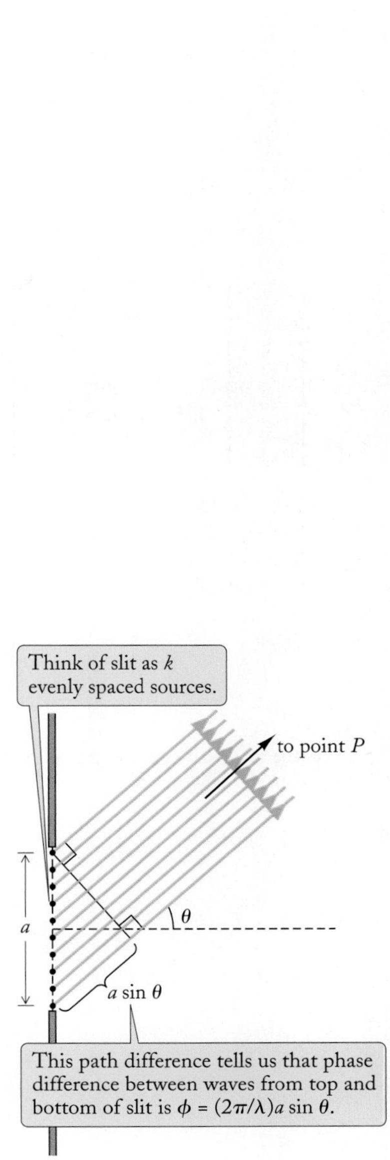

FIGURE 35.35 A single slit considered to be a large number k of point sources.

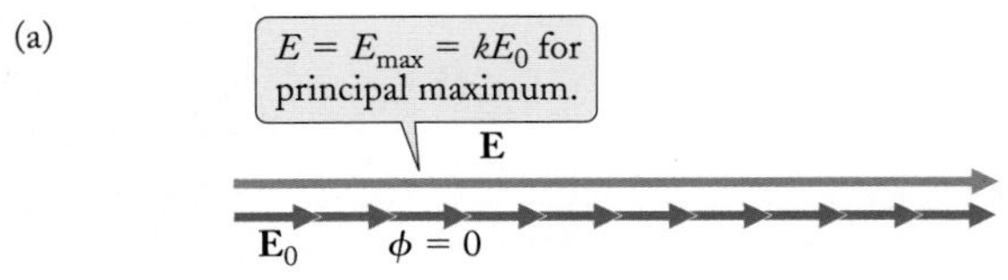

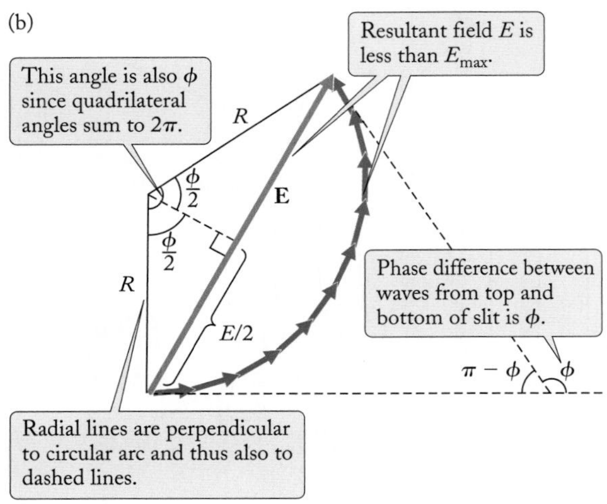

FIGURE 35.36 Phasor diagram for a single slit. (a) At the central maximum, all k phasors are in phase, and the net amplitude is $E_{max} = kE_0$. (b) At other positions, the phasors differ by equal, small phase increments. The k phasors form a polygonal arc of k sides, which we can approximate as an arc of a circle. The central angle of this arc equals the phase difference between the first and last phasors.

Substituting $R = E_{max}/\phi$ from Eq. (35.22) and solving for E gives us

$$E = E_{max}\frac{\sin(\phi/2)}{\phi/2} \tag{35.24}$$

Since intensity is proportional to the square of the amplitude, this yields an intensity

intensity for single-slit diffraction pattern

$$I = I_{max}\left[\frac{\sin(\phi/2)}{\phi/2}\right]^2 \tag{35.25}$$

where I_{max} is the intensity of the central maximum and, as previously [compare Eq. (35.15)], the phase difference ϕ between waves from the top and bottom of the slit is $2\pi/\lambda$ times the path difference $a\sin\theta$ of Fig. (35.35):

$$\phi = \frac{2\pi}{\lambda}a\sin\theta \tag{35.26}$$

Fraunhofer diffraction pattern

Equation (35.25) is the formula for the intensity pattern for a single slit, known as the **Fraunhofer diffraction pattern**. The intensity is plotted as a function of the angle θ toward a distant screen in Fig. 35.37, for the special case of $a = 3\lambda$. As discussed above, the width of the pattern will spread for a narrower slit, and it will shrink for a wider slit.

We previously showed that the intensity goes to zero at the *minima*, where $a\sin\theta = m\lambda$, or where $\phi = 2\pi, 4\pi, 6\pi, \ldots$. The exact location and intensity of the secondary *maxima* may be obtained using differential calculus, by setting the derivative of Eq. (35.25) equal to zero. However, the maxima are fairly close to halfway between the minima. Thus the maxima essentially occur where $\phi = 3\pi, 5\pi, \ldots$. For example, the intensity of the first secondary maximum is approximately

$$I = I_{max}\left[\frac{\sin(3\pi/2)}{3\pi/2}\right]^2 = \frac{4}{9\pi^2}I_{max} \approx 0.045I_{max}$$

or less than 5% of the intensity of the central maximum.

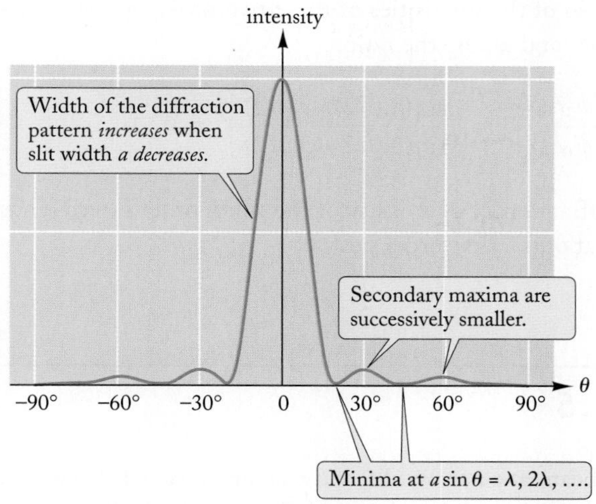

FIGURE 35.37 Intensity as a function of the angle for light arriving at a screen placed behind a narrow slit of width $a = 3\lambda$.

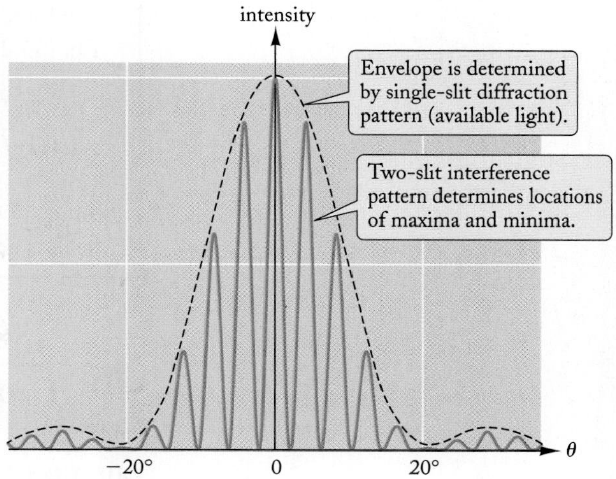

FIGURE 35.38 Intensity as a function of angle for two slits with slit width $a = 3\lambda$ and slit separation $d = 5a$, showing both single-slit diffraction and two-slit interference effects.

Finally, we point out that the single-slit diffraction pattern determines where on a screen light is available to interfere when more than one slit is present. To obtain the full intensity as function of angle for, say, a two-slit interference pattern, we need only to replace the constant I_{max} in the two-slit interference pattern of Eq. (35.11) by the position-dependent single-slit diffraction intensity I of Eq. (35.25). For example, Fig. 35.38 plots the intensity of such a two-slit interference pattern, where the slit spacing d is 5 times the individual slit width a and $a = 3\lambda$. As is evident, there are many two-slit interference maxima within each single-slit diffraction maximum. Also, the fifth-order two-slit interference maxima are missing, since they coincide with the first minima of the single-slit diffraction pattern; there is no *available light* at these points to interfere.

EXAMPLE 8 The compact discs in the chapter photo exhibit interference maxima from several orders. Suppose that the effective slit width for a CD track is $a = 800$ nm. What is the ratio of intensities of the third- and first-order interference maxima for the blue light ($\lambda = 486$ nm)? (See also Example 5.)

Concepts in Context

SOLUTION: In Example 5, we saw that the third-order interference maximum for blue light was at $\theta = 60°$. The phase angle ϕ for the single-slit pattern at the third-order interference maximum is, from Eq. (35.26),

$$\phi_3 = 2\pi \frac{a}{\lambda} \sin\theta = 2\pi \times \frac{800 \text{ nm}}{486 \text{ nm}} \times \sin 60° = 9.0 \text{ radians}$$

The first-order interference maximum occurs at a value of sin θ that is one-third as large, so the single-slit diffraction phase angle at the first-order interference maximum is

$$\phi_1 = \tfrac{1}{3}\phi_3 = \tfrac{1}{3} \times 9.0 \text{ radians} = 3.0 \text{ radians}$$

We then can obtain the ratio of the intensities of these two peaks by evaluating Eq. (35.25) for ϕ_1 and for ϕ_3 and taking the ratio:

$$\frac{I_3}{I_1} = \left[\frac{\sin(\phi_3/2)/(\phi_3/2)}{\sin(\phi_1/2)/(\phi_1/2)}\right]^2 = \left[\frac{\sin(9.0/2)/(9.0/2)}{\sin(3.0/2)/(3.0/2)}\right]^2 = 0.11$$

Thus, due to single-slit diffraction, the intensity of the third-order interference peak is about one-tenth that of the first-order peak.

Checkup 35.5

QUESTION 1: You aim a light beam at a narrow slit, and you aim a particle beam at a similar narrow slit. How do the behaviors of these beams differ if you gradually decrease the width of the slit?

QUESTION 2: Suppose that the width a of a single slit equals $\sqrt{2}\lambda$. Describe the intensity distribution for this case.

QUESTION 3: If $a \gg \lambda$, then the angle of the first minimum [Eq. (35.21)] is very small. Does this mean that light passing through a large aperture will be focused to a narrow spot?

QUESTION 4: Estimate the angle θ (in radians) of the first minimum in a single-slit diffraction experiment. The slit width is $a = 0.1$ mm; the wavelength is $\lambda = 500$ nm.

(A) 5×10^{-3} (B) 1×10^{-3} (C) 5×10^{-4}
(D) 1×10^{-4} (E) 5×10^{-5}

35.6 DIFFRACTION BY A CIRCULAR APERTURE; RAYLEIGH'S CRITERION

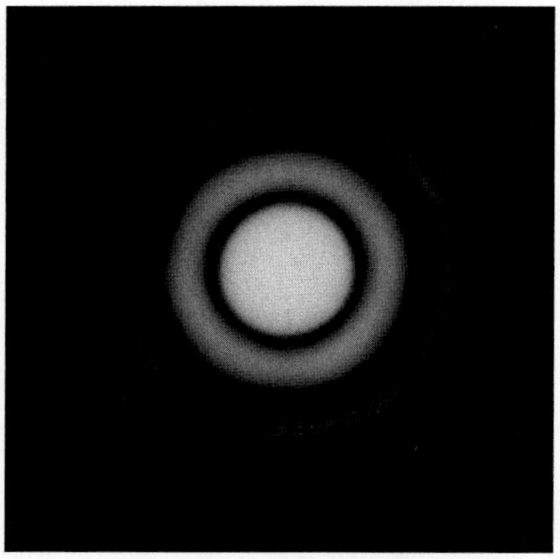

FIGURE 35.39 A photographic film placed beyond a circular aperture records a strong central maximum and annular secondary maxima.

The diffraction of light by a circular aperture is in principle no different from the diffraction by a slit (a very long rectangular aperture). To find the maxima and minima in the diffraction pattern of the light emerging from such a circular aperture, we must sum the waves originating from all points of a wave front at the circular aperture. Because we must sum waves from points throughout a circle and not just across a slit, this calculation is fairly complicated, and we will not attempt it. Figure 35.39 is a photograph that displays the central maximum and the bright and dark fringes generated by diffraction at a circular aperture. The angular position of the first minimum is given by the simple formula

first minimum for circular aperture

$$\sin\theta = 1.22\frac{\lambda}{a} \tag{35.27}$$

where a is the diameter of the circular aperture. Note the similarity with Eq. (35.17).

Many optical instruments—telescopes, microscopes, cameras, etc.—have circular apertures, and these will diffract light. For instance, the objective lens of an astronomical telescope will act like a circular opening in a plate; the parallel wave fronts arriving from some distant star will experience diffraction effects at this aperture and produce an intensity distribution such as that shown in Fig. 35.39. The image of the star as seen through this telescope will then not be a bright point, but a disk surrounded by concentric rings, as in Fig. 35.39. For example, as seen through a telescope with an objective lens 6 cm in diameter, stars look like small disks, about 2×10^{-5} radian (or 4 seconds of arc) across.

This spreading out of the image puts a limit on the detail that can be perceived through the telescope. If two stars are very close together, their images tend to merge, and it may be impossible to tell them apart. Figures 35.40a–d show the images produced by a pair of pointlike light sources upon diffraction by a circular aperture. In the first of these figures, the angular separation of the sources is so small that the two images look like one image. In the second figure, the angular separation is large enough to give a clear indication of the existence of two separate images.

The angular separation of the sources in Fig. 35.40b is such that the central maximum of the diffraction pattern of one source coincides with the minimum of the diffraction pattern of the other source. Since we are now dealing with small angles, we can approximate sin θ by θ, and Eq. (35.26) then tells us that the angular position of the first minimum, or the angular separation between the sources in Fig. 35.40b, is

$$\theta = 1.22\,\frac{\lambda}{a} \tag{35.28}$$

We will regard this as the critical angle that decides whether the two sources are clearly distinguishable: the telescope (or other optical instrument) can resolve the sources if their angular separation is larger than that in Eq. (35.28), and it cannot resolve them if the separation is smaller. This is **Rayleigh's criterion**.

EXAMPLE 9 The star ζ (zeta) Orionis is a binary star; that is, it consists of two stars very close together. The angular separation of the stars is 2.8 seconds of arc. Can the stars be resolved with a telescope having an objective lens 6.0 cm in diameter? Assume that the wavelength of the starlight is 550 nm.

SOLUTION: According to Rayleigh's criterion, a telescope of this aperture can resolve stars as close as

$$\begin{aligned}\theta &= 1.22\,\frac{\lambda}{a} = 1.22 \times \frac{550 \times 10^{-9}\ \text{m}}{0.060\ \text{m}}\\ &= 1.1 \times 10^{-5}\ \text{radian} \times \frac{360\ \text{degrees}}{2\pi\ \text{radians}} \times \frac{3600\ \text{seconds of arc}}{1\ \text{degree}}\\ &= 2.3\ \text{seconds of arc}\end{aligned}$$

Since the angular separation of the stars is 2.8 seconds of arc, the telescope can resolve the double stars.

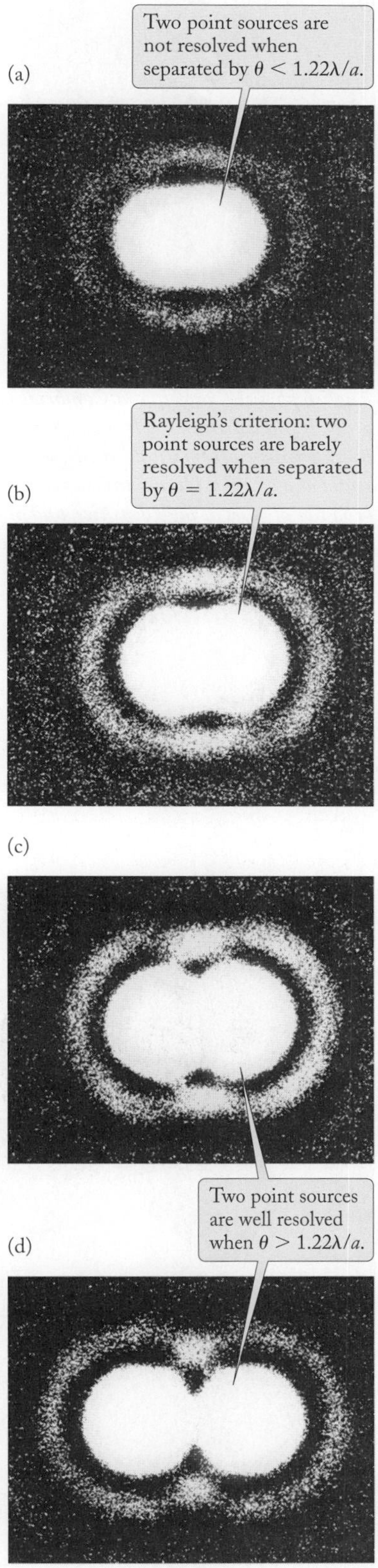

FIGURE 35.40 When the light waves from two pointlike sources arrive at a circular aperture simultaneously, each set of light waves will produce its own diffraction pattern. If the angular separation between the two sources is small, the diffraction patterns overlap. (a) Here the angular separation is small. (b–d) The angular separation is progressively larger.

JOHN WILLIAM STRUTT, 3RD BARON RAYLEIGH (1842–1919)
English physicist, professor at Cambridge and at the Royal Institution. He is best known for his extensive mathematical investigations of sound and of light. He also investigated the behavior of gases at high densities, and he discovered argon; for this he was awarded the Nobel Prize in 1904.

The ability of a telescope to resolve stars or other objects of small angular separation improves with the size of the telescope—a telescope of 30 cm diameter can resolve angular separations as small as 0.5 second of arc. However, beyond 30 cm, the resolution of an Earth-bound telescope does not improve further with size, because fluctuations in the density of the atmosphere introduce irregularities in the path of the light coming down from the sky; this smears out the light and prevents the telescope from reaching its full potential.

The Hubble Space Telescope, launched in 1990, is in orbit above the atmosphere of the Earth, and it is not affected by atmospheric fluctuations. This telescope (see Fig. 35.41) has an aperture of of 2.4 m, and it achieves an angular resolution of about 0.1 second of arc, close to the limit set by Rayleigh's criterion. Thus, the space telescope achieves higher angular resolution than any Earth-bound optical telescope. Furthermore, it is able to observe at ultraviolet and infrared wavelengths, which are blocked by the atmosphere.

For Earth-bound radio telescopes, the atmosphere poses no problem—radio waves do not suffer from the effects of atmospheric density fluctuations. Hence, with increasing size, the resolution of a radio telescope improves indefinitely. Figure 35.42 shows the large radio telescope at Arecibo, Puerto Rico. The concave "mirror" of this telescope has an aperture of 300 m and a radius of curvature which is also 300 m. The shortest wavelength at which this telescope has been operated is 4.0 cm. For this wavelength, Rayleigh's criterion gives a limiting angular resolution

$$\theta = 1.22\frac{\lambda}{a} = 1.22 \times \frac{0.040\text{ m}}{300\text{ m}} = 1.6 \times 10^{-4}\text{ radian}$$

which is about 30 seconds of arc.

FIGURE 35.41 The Hubble Space Telescope.

FIGURE 35.42 The Arecibo radio telescope.

Checkup 35.6

QUESTION 1: If we apply a criterion similar to Rayleigh's to a narrow slit, instead of a circular aperture, what critical angle for the resolution of two images do we obtain?

QUESTION 2: The Arecibo radio telescope has an aperture 125 times as large as the Hubble Space Telescope. However, its resolution is less than that of Hubble. Why?

QUESTION 3: Two dots of red ink on a page are separated by 0.2 mm. Suppose your eye has an aperture of 0.5 cm. Of the following, what is the largest distance at which it is possible to resolve the dots, assuming otherwise perfect eyesight?

(A) 3 cm (B) 10 cm (C) 30 cm (D) 100 cm (E) 300 cm

SUMMARY

PROBLEM-SOLVING TECHNIQUES Thin-Film Interference (page 1174)

PROBLEM-SOLVING TECHNIQUES Two-Slit Interference (page 1181)

CONSTRUCTIVE INTERFERENCE Waves meet crest to crest, making a total wave of larger amplitude.

DESTRUCTIVE INTERFERENCE Waves meet crest to trough, making a total wave of smaller or zero amplitude.

THIN-FILM INTERFERENCE where $\lambda = \dfrac{\lambda_{\text{air}}}{n}$, with film thickness d, for both waves reflecting at an interface with a material of higher index of refraction:
(When only one of the waves reflects at an interface with a material of higher index of refraction, the conditions are reversed.)

Maxima: $2d = \lambda,\ 2\lambda,\ 3\lambda, \ldots$

Minima: $2d = \tfrac{1}{2}\lambda,\ \tfrac{3}{2}\lambda,\ \tfrac{5}{2}\lambda, \ldots$

(35.1, 35.2, 35.3)

TWO-SLIT INTERFERENCE PATTERN with slit separation d, where θ is the angle with respect to the midline. For small angles, the transverse displacement from the midline at distance r is $\Delta y \approx r\theta$.

Maxima: $d \sin\theta = 0,\ \lambda,\ 2\lambda, \ldots$

Minima: $d \sin\theta = \tfrac{1}{2}\lambda,\ \tfrac{3}{2}\lambda,\ \tfrac{5}{2}\lambda, \ldots$

(35.7, 35.9)

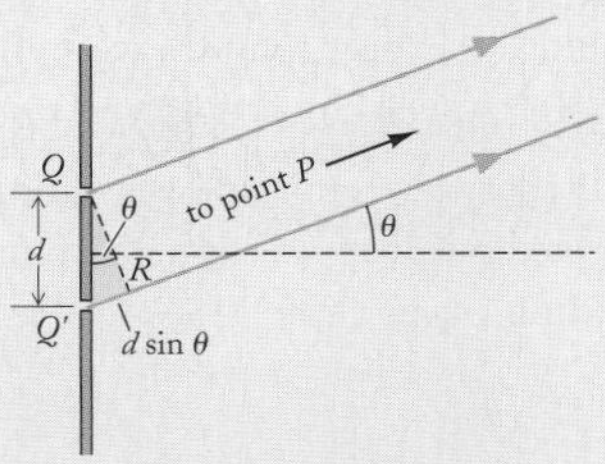

INTENSITY FOR TWO-SLIT INTERFERENCE PATTERN

$$I = I_{max}\cos^2\left(\frac{\pi d}{\lambda}\sin\theta\right) \qquad \textbf{(35.11)}$$

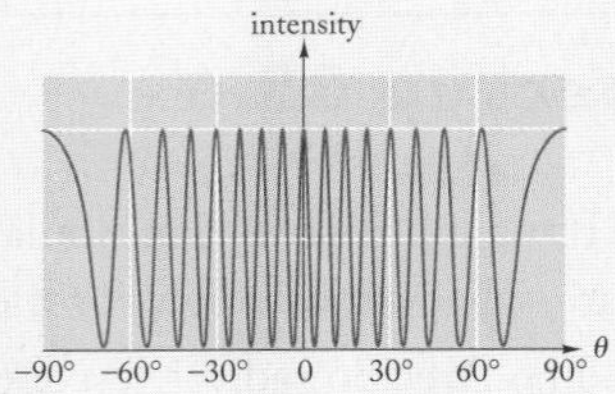

MULTIPLE-SLIT INTERFERENCE PATTERN
(N slits)

Principal maxima: $d\sin\theta = m\lambda$ where $m = 0, 1, 2, \ldots$

First minima: $d\sin\theta = \dfrac{\lambda}{N}$ **(35.12, 35.16)**

PHASOR DIAGRAMS To obtain the resultant wave, add vectors representing the amplitudes and phases of the interfering waves tip to tail.

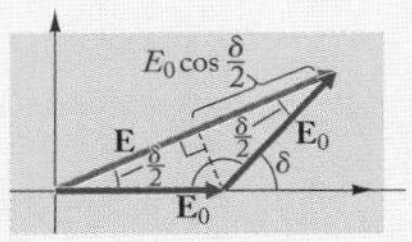

MINIMA FOR SINGLE SLIT DIFFRACTION PATTERN
with slit width a

$$a\sin\theta = \lambda,\ 2\lambda,\ 3\lambda, \ldots \qquad \textbf{(35.20)}$$

INTENSITY FOR SINGLE-SLIT DIFFRACTION PATTERN

where $\phi = \dfrac{2\pi}{\lambda} a\sin\theta$ **(35.25)**

$$I = I_{max}\left[\frac{\sin(\phi/2)}{\phi/2}\right]^2 \qquad \textbf{(35.26)}$$

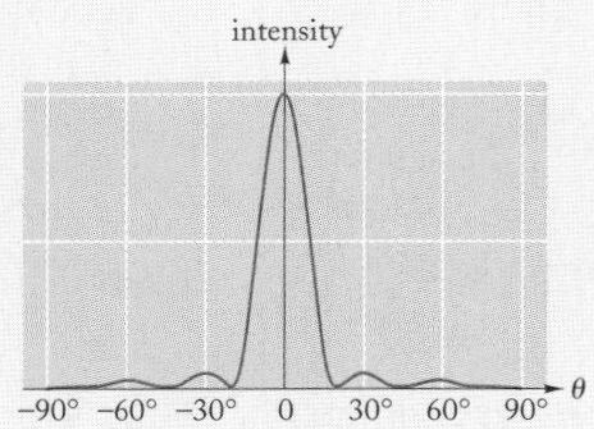

FIRST MINIMUM FOR CIRCULAR APERTURE
with diameter a

$$\sin\theta = 1.22\frac{\lambda}{a} \qquad \textbf{(35.27)}$$

RAYLEIGH'S CRITERION

$$\theta = 1.22\frac{\lambda}{a} \qquad \textbf{(35.28)}$$

QUESTIONS FOR DISCUSSION

1. The light wave inside the tube of a laser is a standing wave, reflected at both ends of the tube. If the distance between the ends is L, what condition must the wavelength of the standing wave satisfy?
2. Would you expect sound waves reflected by a wall to produce a standing wave? How could you test this experimentally?
3. The radar wave emitted by a stationary police radar unit is reflected by an approaching automobile. Does this set up a standing wave?
4. When two waves interfere destructively at one place, what happens to their energy?
5. When two coherent waves of equal intensity interfere constructively at one place, the energy density at that place becomes 4 times as large as the energy density of each individual wave. Is this a violation of the Law of Conservation of Energy?
6. Suppose that N waves of equal intensity meet at one place. If the waves are coherent, the net intensity is N^2 times the intensity of each individual wave. If the waves are incoherent, the

(average) intensity is N times the intensity of each individual wave. Explain, and give an example of each case.

7. When light strikes a windowpane, some rays will be reflected back and forth between the two glass surfaces. Why do these reflected rays not produce visible colored interference fringes?

8. Suppose that a lens is covered with an antireflective coating that eliminates the reflection of perpendicularly incident light of some given color. Will this coating also eliminate the reflection of this light incident at an angle?

9. Is it possible to cover the surface of an aircraft with an antireflective coating so that it does not reflect radar waves of wavelength, say, 5 cm?

10. Explain how Newton's rings may be used as a sensitive test of the rotation symmetry of a lens.

11. Two flat plates of glass are in contact at one edge and separated by a thin spacer at the other edge (Fig. 35.43a). Explain why we see parallel interference fringes in the reflected light if we illuminate these plates from above (Fig. 35.43b).

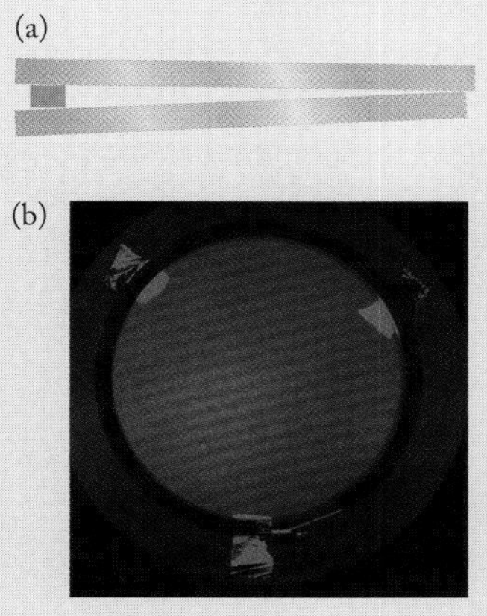

FIGURE 35.43 (a) Two flat plates of glass, separated by a thin wedge of air. (b) Interference fringes.

12. In the experiment that sought to test the dependence of the speed of light on the motion of the Earth, Michelson and Morley used an interferometer with very long arms (about 11 m, obtained by multiple reflections back and forth between sets of mirrors). Why does this make the instrument more sensitive?

13. If you stand next to your radio receiver, your body will sometimes affect the reception. Why?

14. Consider (1) sunlight, (2) sunlight passed through a monochromatic filter selecting one wavelength, (3) light from a neon tube, (4) light from a laser, (5) starlight passed through a monochromatic filter, and (6) radio waves emitted by a radio station. Which of these kinds of light or electromagnetic radiation are sufficiently coherent to give rise to an interference pattern when used to illuminate two slits such as shown in Fig. 35.13?

15. When a pair of stereo loudspeakers are installed, the terminals of the loudspeakers should be connected to the amplifier in the same way, so that the loudspeakers are in phase. What would happen to sound waves of long wavelength if the loudspeakers were out of phase?

16. What happens to the interference pattern plotted in Fig. 35.19 if $d < \lambda$?

17. Suppose we cover one of the slits in Fig. 35.13 with a glass plate. Since the wavelength in the glass is shorter than in air, the path in the glass includes more wavelengths than the path in air, and the phases of the two waves emerging from the slits will be different. If the phase difference between these waves is exactly 180°, how does this change the location of maxima and minima?

18. Consider the interference pattern shown in Fig. 35.19. How would this pattern change if the entire interference apparatus were immersed in water?

19. Suppose that a plane wave is incident at an angle on a plate with two narrow slits (Fig. 35.44). In what direction will we then find the central maximum?

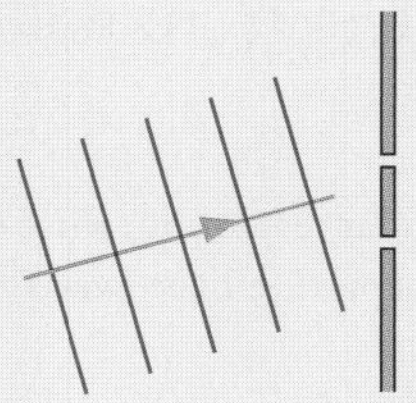

FIGURE 35.44 Plane wave obliquely incident on a plate with two slits.

20. Several radio antennas are arranged at regular intervals along a straight line; the antennas are connected to the same radio transmitter, so they radiate coherently. How can this array be used to concentrate the radio emission in a selected direction?

21. When a crystal is illuminated with X rays, each atom acts as a pointlike source of scattered X rays. The typical separation between adjacent atoms in a crystal is 0.1 nm. Roughly what must be the wavelength of the X rays if they are to exibit distinct interference effects?

22. Gratings used for analyzing light are often called diffraction gratings, but it would be more accurate to call them interference gratings. Why?

23. Consider the diffraction pattern shown in Fig. 35.37. How would this pattern change if the entire apparatus were immersed in water?

24. Figure 17.21 shows the diffraction of water waves at the entrance of a harbor. Could this diffraction be eliminated by making the entrance smaller?

25. In the center of the shadow of a disk or sphere there is a small bright spot, called the **Poisson spot** (Fig. 35.45). This spot is very faint near the disk but becomes more noticeable at large distances. Qualitatively, explain why the diffraction of the light waves around the edges gives rise to this spot.

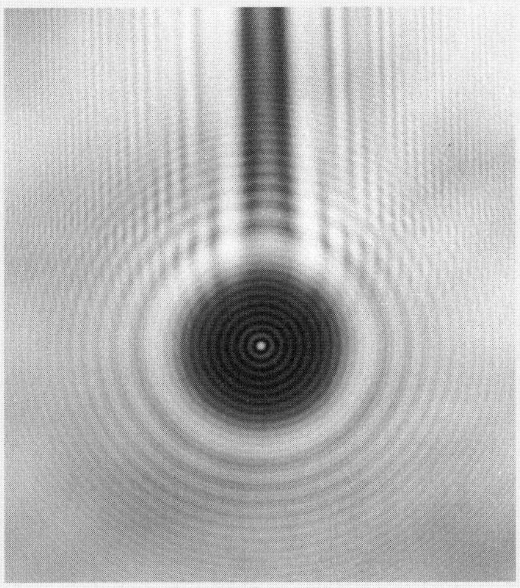

FIGURE 35.45 The Poisson spot.

26. Besides good angular resolution, what other advantage does a telescope of large aperture have over a telescope of small aperture?
27. Spy satellites use cameras with lenses of very large diameter, 30 cm or more. Why are such large diameters necessary?
28. In order to beam a sound wave sharply in one direction with a megaphone, should the horn of the megaphone have a small aperture or a large aperture?
29. The manufacturers of the Questar telescope claim that this telescope can distinguish two stars even if their separation is somewhat smaller than that specified by Rayleigh's criterion. Why is this possible?
30. The maximum useful magnification of an optical microscope is determined by diffraction effects at the objective lens. Explain.
31. Other things being equal, how much resolution can you gain by operating an optical microscope with blue light instead of red light? Why can you not operate an ordinary microscope with ultraviolet light?
32. The picture of atoms in Fig. 35.46 was taken with an electron microscope using electron waves of extremely short wavelength. Roughly, how short must the wavelength be to make individual atoms visible?

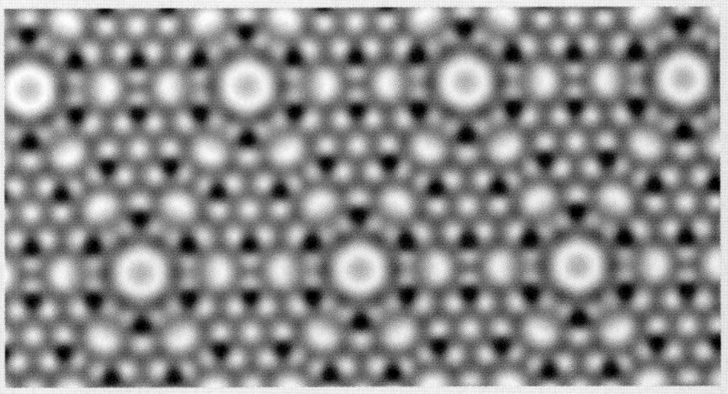

FIGURE 35.46 Atoms in a crystal.

33. The compound eye of insects consists of a large number of small eyes, or ommatidia (Fig. 35.47). Each ommatidium is typically 0.03 mm across; it does not form an image, but merely acts as a sensor of the intensity of light arriving from a narrow cone of directions. What are some of the advantages and disadvantages of such a compound eye as compared with the camera eye of vertebrates?

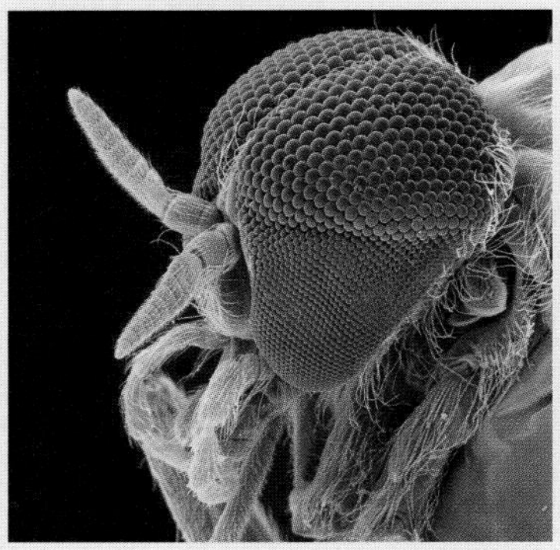

FIGURE 35.47 Compound eyes of a fly.

34. The corona formed by diffraction of sunlight by small droplets of water in a cloud consists of a bright ring surrounding the Sun. What is the color of the outer edge of the corona? The inner edge?
35. Antennas used for the transmission of microwaves in communication links consist of metal dishes. What factors determine the size of these dishes?
36. Very-long-baseline interferometry (VLBI) with two or more radio telescopes placed on different parts of the Earth has been used for precise determination of the rate of continental drift and also the rate of rotation of the Earth (sidereal day). Explain how data collected by radio telescopes monitoring a fixed radio source in the sky can be used for such purposes.

PROBLEMS

35.1 Thin Films

1. In 1887, Hertz gave the first direct experimental demonstration of the existence of the electromagnetic waves that had been predicted by Maxwell's theory. For this experiment, Hertz placed a high-frequency spark generator at some distance in front of a large vertical zinc plate. The wave emitted by the spark reflected off the zinc plate, setting up a standing wave. Hertz found that when the generator was operating at a frequency of about 4.0×10^7 cycles per second, the distance between a node and an antinode in the standing-wave pattern was about 2.0 m. What value of the speed of propagation of the electromagnetic disturbances could he deduce from this?

2. The wall of a soap bubble floating in air has a thickness of 400 nm. If sunlight strikes the wall perpendicularly, what colors seen in the reflected light will be strongly enhanced? The index of refraction of the soap film is 1.35.

3. A thin film with index of refraction $n_{\text{film}} = 1.38$ is used to coat a lens, which has $n_{\text{lens}} = 1.50$. What is the thinnest film that will highly reflect near-ultraviolet light of wavelength 350 nm? What is the thinnest film that will minimize reflection of 700-nm light?

4. Similar to the arrangement in Fig. 35.43, two flat plates of glass are in contact at one edge and a human hair is used as a spacer at the other edge. Green light of wavelength 546 nm is normally incident from above, and 76 dark fringes appear across the plate. What is the thickness of the hair?

5. A Newton's ring arrangement (see Figs. 35.7 and 35.8) exhibits its 25th dark ring out at the edge of the lens when light of wavelength 590 nm is used; that ring has radius 6.0 cm. What is the radius of curvature of the convex surface of the plano–convex lens? (Hint: You may use the formula of Problem 35.13.)

6. In a Newton's ring experiment (see Figs. 35.7 and 35.8), light of wavelength 546 nm is used. When the air space between the plano–convex lens and the flat glass plate is filled with liquid, the radius of the fifth dark ring decreases from 2.0 cm to 1.5 cm. What is the index of refraction of the liquid? (Hint: Use the formula of Problem 35.13.)

7. A coating is used to maximize the light entering a silicon solar cell. If the indices of refraction of silicon and the coating are 3.5 and 1.6 respectively, what minimum coating thickness can be used for 500-nm light?

*8. The tube of a laser has a mirror at each end (Fig. 35.48). A standing electromagnetic wave fills the space between these mirrors. This wave must satisfy the boundary condition that its electric field is zero at all times at the position of the mirrors.

(a) Show that the distance d between the mirrors and the wavelength λ of the wave must be related by

$$d = \left(\frac{m+1}{2}\right)\lambda \qquad m = 0,\ 1,\ 2,\ldots$$

(b) A He–Ne laser with $d = 30.00$ cm operates at $\lambda = 633$ nm. What is the value of m? How many minima and how many maxima does the standing wave have between the two mirrors?

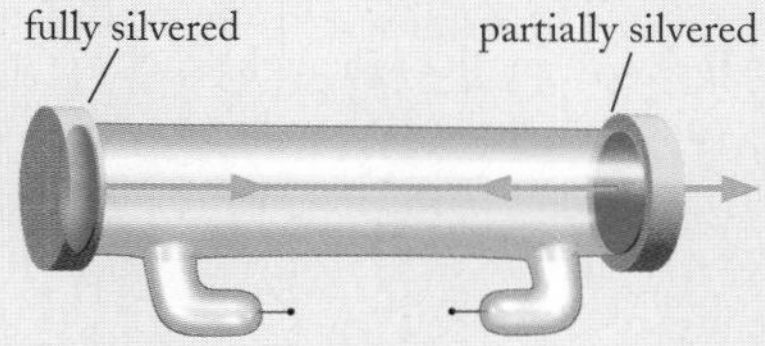

FIGURE 35.48 Tube of a laser with mirrors at its ends.

*9. Consider the light *transmitted* by a thin film. For a light wave with a direction of propagation perpendicular to the surface of the film, what is the condition for constructive interference between the direct wave and the wave that is reflected once by the lower surface and once by the upper surface? Would you expect that this condition for constructive interference in transmission coincides with the condition of destructive interference in reflection?

*10. A thin oil slick, of index of refraction 1.25, floats on water. When a beam of white light strikes this film vertically, the only colors seen enhanced in the reflected beam are red (at a wavelength of 675 nm in air) and violet (at 450 nm). From this information, deduce the thickness of the oil slick.

*11. Two flat parallel plates of glass are separated by thin spacers so as to leave a gap of width d (Fig. 35.49). If light of wavelength λ is normally incident on these plates, what is the condition for constructive interference between the rays reflected by the lower surface of the top plate and the upper surface of the bottom plate?

FIGURE 35.49 Two parallel flat plates of glass.

*12. A layer of oil of thickness 200 nm floats on top of a layer of water of thickness 400 nm resting on a flat metallic mirror. The index of refraction of the oil is 1.24, and that of the water is 1.33. A beam of light is normally incident on these layers. What must be the wavelength of the beam if the light reflected by the top surface of the oil is to interfere destructively with the light reflected by the mirror?

**13. A lens with one flat surface and one convex surface rests on a flat plate of glass (Fig. 35.50). A light ray normally incident on the lens will be partially reflected by the curved surface of the lens and partially reflected by the flat plate of glass. The interference between these reflected rays will be constructive or destructive, depending on the height of the air gap between the lens and the plate. The interference gives rise to the pattern of Newton's rings, shown in Fig. 35.7.

(a) Why is the center of the pattern dark?

(b) Show that the radius of the mth dark ring is

$$r = \sqrt{m\lambda R - m^2\lambda^2/4}$$

where λ is the wavelength of the light and R is the radius of curvature of the convex surface of the lens.

(c) What is the radius of the first dark ring if $\lambda = 500$ nm and $R = 3.0$ m?

FIGURE 35.50 Lens resting on a flat glass plate.

35.2 The Michelson Interferometer

14. In a Michelson interferometer such as shown in Fig. 35.9, one mirror is moved 0.450 mm, and 1422 interference fringes are counted. What wavelength of light was used?

15. In a Michelson interferometer such as shown in Fig. 35.9, light of wavelength 546 nm is used. When one mirror is moved, 930 interference fringes are counted. Through what distance was the mirror moved?

*16. The **Fabry–Perot interferometer** consists of two parallel half-silvered mirrors. A ray of light entering the space between the mirrors may pass straight through, or be reflected once or several times by each mirror (Fig. 35.51). Show that the condition for constructive interference of the emerging light (at large distance from the mirrors) is

$$2d\cos\theta = 0,\ \lambda,\ 2\lambda, \ldots$$

where d is the distance between the mirrors, λ the wavelength of light, and θ the angle of incidence of the light.

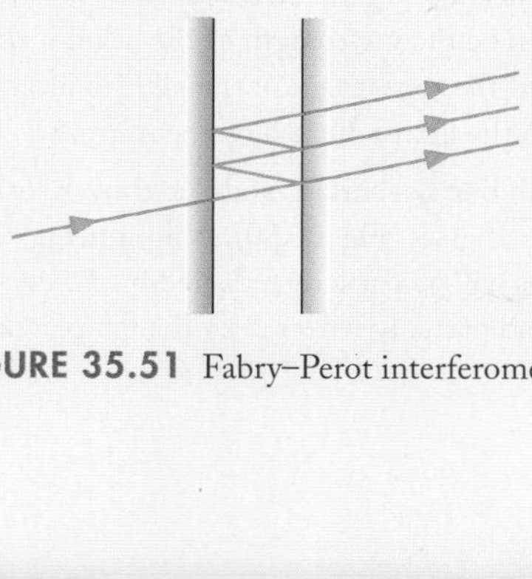

FIGURE 35.51 Fabry–Perot interferometer.

*17. The interferometer of the International Bureau of Weights and Measures can count 19 000 bright fringes (maxima) per second. To achieve this count rate, what must be the speed of motion of the moving mirror (mirror M_1 in Fig. 35.9)? Assume that the interferometer operates with krypton light of wavelength $\lambda = 605.8$ nm.

*18. An **etalon** consists of two mirrors held a fixed distance apart by means of a rigid support (Fig. 35.52a). The distance between the two mirrors can serve as a standard of length. To measure the distance in terms of wavelengths of light, the etalon is installed in a Michelson interferometer, replacing the fixed mirror M_2 (Fig. 35.52b). For a start, the distances MM_1 and MM_3 are made exactly equal; then the mirror M_1 is slowly moved outward, producing a sequence of interference maxima and minima until the distances MM_1 and MM_4 are exactly equal.[2] Suppose that you operate a Michelson interferometer with krypton light of wavelength 605.7802 nm and that you count 36 484.8 interference maxima while moving the mirror M_1 from its initial position to its final position. To within six significant figures, what is the length of the etalon?

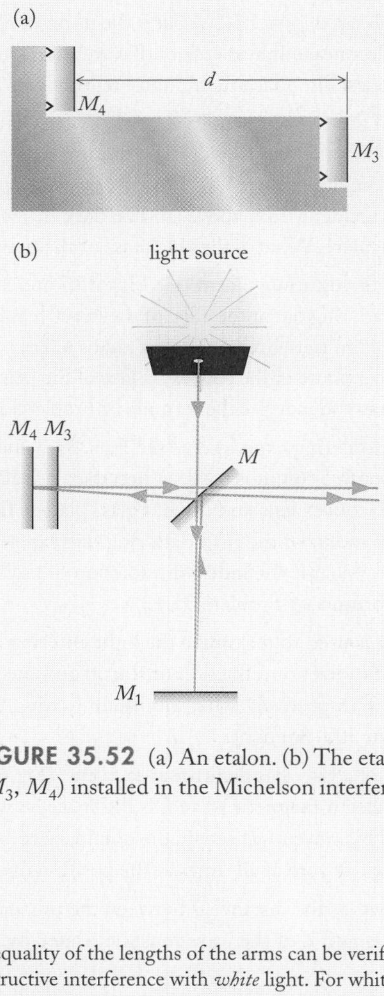

FIGURE 35.52 (a) An etalon. (b) The etalon (M_3, M_4) installed in the Michelson interferometer.

[2] Exact equality of the lengths of the arms can be verified by checking for constructive interference with *white* light. For white light, containing many wavelengths, constructive interference is possible only if the lengths are equal, that is, $d = 0$ in Eq. (35.4).

*19. The index of refraction of air for monochromatic light of some given wavelength can be determined very precisely with a Michelson interferometer one of whose arms (MM_1 in Fig. 35.9) is in air while the other (MM_2) is in a vacuum tank. With the vacuum tank filled with air, the arms are first adjusted so they are exactly equal. Then the vacuum tank is slowly evacuated, while at the same time the length of the arm is gradually increased so as to maintain constructive interference (this increase of length maintains a fixed number of wavelengths in the arm). Show that the index of refraction of air can be expressed in terms of the fractional change of length:

$$n - 1 = \frac{\Delta L}{L}$$

Suppose that for light of wavelength 580 nm, an interferometer with arms of length 40.00 cm requires a readjustment of 0.011 06 cm to maintain constructive interference while the vacuum tank is being evacuated. What is the index of refraction of air at this wavelength?

†35.3 Interference from Two Slits

20. Microwaves of wavelength 2.0 cm from a radio transmitter are aimed at two narrow parallel slits in an aluminum plate. The slits are separated by a distance of 5.0 cm. At what angles at some distance beyond the plate will we find maxima in the interference pattern?

21. Suppose that the two slits, the film, and the light source described in Example 2 are immersed in water. What will now be the distance between the central maximum and the first lateral maximum?

22. Laser light of wavelength 633 nm is incident upon two narrow slits separated by 0.15 mm. On a screen 4.00 m beyond the slits, how far apart are the interference fringes?

23. In a double-slit experiment, green light of wavelength 546 nm is used. On a screen 1.50 m away, what is the distance between the central maximum and the first minimum if the slits are separated by 0.22 mm?

24. Two radio antennas transmit electromagnetic waves of the same frequency and phase. If their separation is $\lambda/2$ what is the angular position of the first interference minimum?

25. Two narrow slits, separated by 0.10 mm, are illuminated with light of wavelength 633 nm. On a screen 1.20 m away, what is the intensity, relative to the maximum, at a point 1.5 cm from the central maximum?

26. A piece of aluminum foil with two narrow slits is being illuminated with red light of wavelength 694.3 nm from a laser. This yields a row of evenly spaced bright bands on a screen placed 3.00 m beyond the slits. The interval between the bright bands is 1.40 cm. What is the distance between the two slits?

27. Consider the water waves shown in Fig. 35.16. With a ruler, measure the wavelength of the waves and the distance between the sources; with a protractor, measure the angular positions of the nodal lines (minima). Do these measured quantities satisfy Eq. (35.9)?

28. For the most detailed mapping of the sky, radio astronomers use radio telescopes placed on different continents, with (straight-line) separation of several thousand kilometers. Suppose that two such radio telescopes are separated by an east–west distance of 5000 km, and that they are connected to a single radio receiver tuned to a wavelength of 21 cm. If a source is symmetrically located above these radio telescopes, the radio waves reaching the receiver will be in phase, and the receiver will register maximum intensity. What westward angular displacement of the source from this location will, again, result in maximum intensity?

29. An interferometric radio telescope consists of two antennas separated by a distance of 1.00 km. The two antennas feed their signals into a common receiver tuned to a frequency of 2300 MHz. The receiver will detect a maximum (constructive interference) if the wave sent out by a radio source in the sky arrives at the two antennas with the same phase. What possible angular positions of the radio source will result in such a maximum? Reckon the angular position from the vertical line erected at the midpoint of the antennas. Treat the antennas as pointlike.

30. Light of wavelength 694.3 nm from a ruby laser is incident on two narrow parallel slits cut in a thin sheet of metal. The slits are separated by a distance of 0.11 mm. A screen is placed 1.5 m beyond the slits. Find the intensity, relative to the central maximum, at a point on the screen 1.2 cm to one side of the central maximum.

31. In a double-slit experiment, the intensity at the peak of the central maximum is I_{max}. If the slit spacing is 12 times the wavelength of light used, what is the intensity at an angular position 1.0° away from the peak of the central maximum? At 2.0°? At 3.0°?

32. In a double-slit experiment, the intensity at the peak of the central maximum is I_{max}. What is the intensity when the path difference to a point on the screen is $\lambda/3$?

*33. In a double-slit experiment such as that in Fig. 35.17, the slit spacing is 0.20 mm, the distance to the screen is 3.00 m, and light of wavelength 633 nm is used. If a thin sheet of polystyrene ($n = 1.49$) is placed over the bottom slit only, the entire interference pattern moves downward on the screen by 4.0 mm. What is the thickness of the polystyrene sheet?

34. A 1.0-kHz sound wave is normally incident on two narrow windows, separated by 2.0 m. If a listener is standing 5.0 m beyond the windows on the midline, how far must she move laterally to be standing at the first interference minimum? Take the speed of sound to be 340 m/s.

†For help, see Online Concept Tutorial 39 at www.wwnorton.com/physics

*35. Light of wavelength λ is obliquely incident on a pair of narrow slits separated by a distance d. The angle of incidence of the light on the slits is ϕ (Fig. 35.53).

(a) Show that the diffracted light emerging at an angle θ interferes constructively if

$$|d\sin\theta - d\sin\phi\,| = 0,\ \lambda,\ 2\lambda, \ldots$$

and destructively if

$$|d\sin\theta - d\sin\phi| = \tfrac{1}{2}\lambda,\ \tfrac{3}{2}\lambda,\ \tfrac{5}{2}\lambda, \ldots$$

(b) Show that if θ is small, the angular separation between the interference maxima and minima is independent of the angle ϕ.

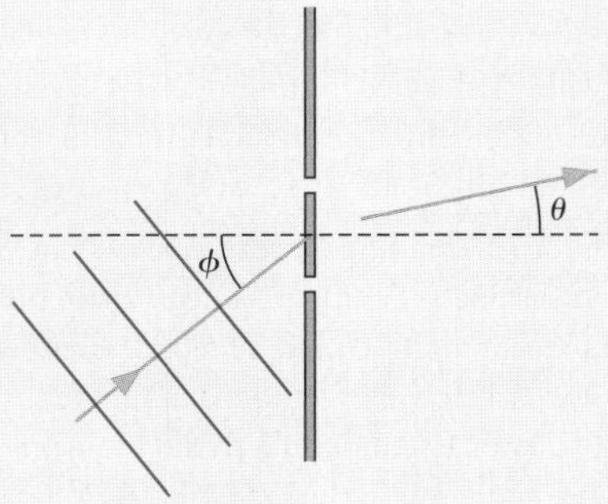

FIGURE 35.53 Plane wave obliquely incident on a pair of slits in a plate.

*36. A device called **Lloyd's mirror** produces interference between a ray reaching a vertical screen directly and a ray reaching the screen after reflection by a horizontal mirror. Show that in terms of the distances z, z_0, and l defined in Fig. 35.54, the condition for constructive interference is

$$\sqrt{l^2 + (z + z_0)^2} - \sqrt{l^2 + (z - z_0)^2} = \tfrac{1}{2}\lambda,\ \tfrac{3}{2}\lambda,\ \tfrac{5}{2}\lambda, \ldots$$

and that for destructive interference is

$$\sqrt{l^2 + (z + z_0)^2} - \sqrt{l^2 + (z - z_0)^2} = 0,\ \lambda,\ 2\lambda, \ldots$$

Note that in this calculation you must take into account the reversal, or change in phase, of the wave during reflection.

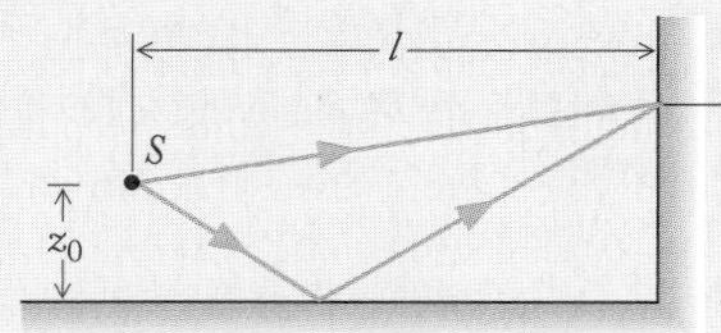

FIGURE 35.54 Lloyd's mirror.

**37. Two radio beacons transmit waves of the same phases and frequencies. The transmitters are on the x axis, at $\pm x_0$. Show that the interference is constructive at those points of the x–y plane satisfying the condition

$$\sqrt{(x + x_0)^2 + y^2} - \sqrt{(x - x_0)^2 + y^2} = m\lambda$$

where $m = 0, 1, 2, \ldots$. Show that for a given nonzero value of m, this is the equation of a hyperbola. (To show this, use either graphic methods to plot the curve or else your knowledge of analytic geometry.)

**38. In the first application of interferometric methods in radio astronomy, Australian astronomers observed the interference between a radio wave arriving at their antenna on a direct path from the Sun and on a path involving the reflection on the surface of the sea (Fig. 35.55). Assuming that the radio waves have a frequency of 6.0×10^7 Hz and that the radio receiver is at a height of 25 m above the level of the sea, what is the least angle of the source above the horizon that will give destructive interference of the waves at the receiver?

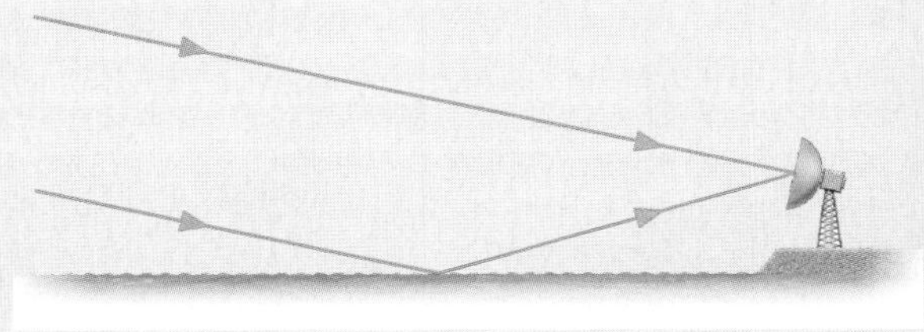

FIGURE 35.55 Direct and reflected radio waves.

†35.4 Interference from Multiple Slits

39. A grating has 5000 lines per centimeter. What are the angular positions of the principal maxima produced by this grating when illuminated with light of wavelength 650 nm?

40. Sodium light with wavelengths 588.99 nm and 589.59 nm is incident on a grating with 5500 lines per centimeter. A screen is placed 3.0 m beyond the grating. What is the distance between the two spectral lines in the first-order spectrum on the screen? In the second-order spectrum?

41. A good grating cut in speculum metal has 5900 lines per centimeter. If this grating is illuminated with white light ranging over wavelengths from 400 nm to 700 nm, it will produce a spectrum ranging over some interval of angles. From what angle to what angle does the first-order spectrum extend? The second-order spectrum? The third-order spectrum? Do these angular intervals overlap? Is the third-order spectrum complete?

42. Six identical, evenly spaced slits are illuminated by light of wavelength 633 nm. What is the angular width of the central maximum if the slit spacing is $d = 0.25$ mm?

†For help, see Online Concept Tutorial 40 at www.wwnorton.com/physics

43. A grating has 40 000 evenly spaced lines over a width of 8.0 cm, illuminated by a beam of monochromatic light. What is the angular width of the central maximum when light of wavelength 550 nm is used?

44. Use the phasor diagrams of Fig. 35.26 to calculate the intensity, relative to a principal maximum, of the three-slit pattern for each of the values of the phase difference δ shown in the figure.

*45. A thin curtain of fine batiste consists of vertical and horizontal threads of cotton forming a net that has a regular array of square holes. While looking through this curtain at the red (670 nm) taillight of an automobile, a physicist notices that the taillight appears as a multiple vertical array of images (an array of principal maxima). The angular separation between adjacent images is 2.0×10^{-3} radian. From this information deduce the vertical spacing between the threads in the batiste curtain.

*46. Consider four evenly spaced identical slits. Use phasor diagrams to calculate the intensity, relative to a principal maximum, when the phase difference between adjacent slits is 0, $\pi/6$, $\pi/4$, $\pi/3$, $\pi/2$, $2\pi/3$, $3\pi/4$, $5\pi/6$, and π. Plot the resulting intensity pattern and compare your plot with Fig. 35.27b.

*47. Three evenly spaced slits are illuminated by monochromatic light. The middle slit is wider than the other two, so the intensity due to the middle slit alone is 4 times the intensity due to either of the top or bottom slits alone. Use phasor diagrams to find the intensity when the phase difference between adjacent slits is 0, $\pi/4$, $\pi/2$, $3\pi/4$, and π. Plot these values of intensity as a function of the phase angle.

*48. Three identical narrow slits are illuminated by monochromatic light. The top and center slits are separated by a distance d; the center and bottom slits are separated by a distance $2d$. Use phasor diagrams to find the intensity on a distant screen when the phase difference between the top and center slits is 0, $\pi/4$, $\pi/3$, $\pi/2$, $2\pi/3$, $3\pi/4$, and π. Make a plot of intensity vs. phase angle.

*49. A **reflection grating** consists of a metal plate in which a large number of closely spaced parallel grooves have been cut.

(a) Show that if light of wavelength λ is obliquely incident on this grating at an angle ϕ (see Fig. 35.56), the light returned by the grating at an angle θ interferes constructively if

$$|d \sin\theta - d \sin\phi| = 0,\ \lambda,\ 2\lambda, \ldots$$

(b) For what angle of incidence ϕ will the direction of the emerging light overlap with the direction of the incident light?

(c) For what angle θ does the reflection grating give constructive interference with white light, consisting of a mixture of many wavelengths?

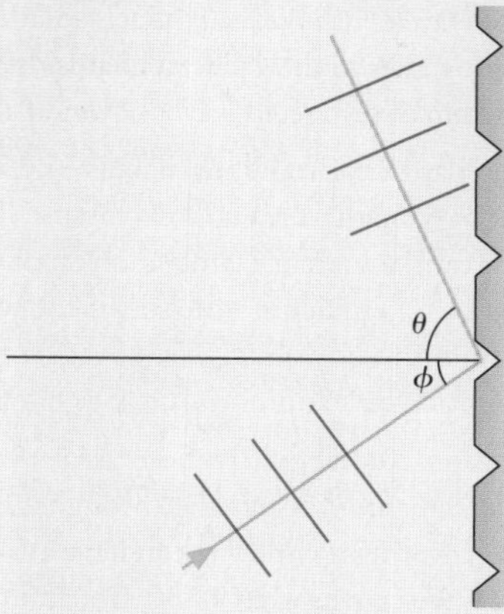

FIGURE 35.56 A reflection grating.

†35.5 Diffraction by a Single Slit

50. Consider the diffraction pattern shown in Fig. 35.33. How would this pattern change if the slit, the light source, and the film were immersed in water?

51. Light of wavelength 632.8 nm from a He–Ne laser illuminates a single slit of width 0.10 mm. What is the width of the central maximum formed on a screen placed 2.0 m beyond the slit?

52. Consider the water waves shown in Fig. 35.57. With a ruler, measure the wavelength of the waves and the length of the gap; with a protractor, measure the angular position of the two nodal lines (minima). Check whether the quantities satisfy Eq. (35.20).

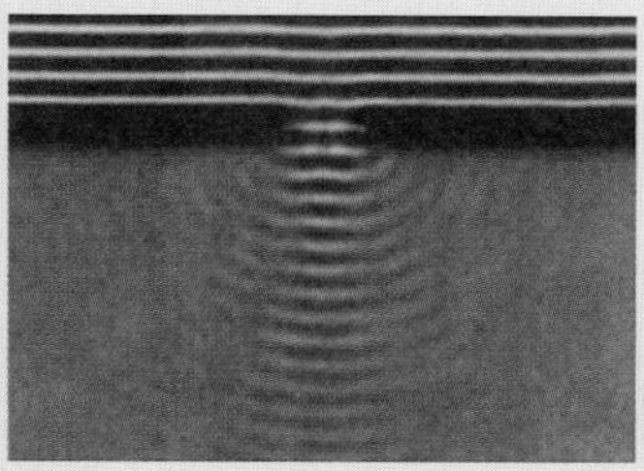

FIGURE 35.57 Diffraction of water waves by an aperture.

53. Light from a 633-nm laser illuminates a single slit. If the full width of the central diffraction maximum on a screen 1.00 m away is 3.0 cm, what is the slit width?

54. A sound wave of frequency 820 Hz passes through a doorway of width 1.0 m. What are the angular directions of the minima of the diffraction pattern? Assume the velocity of sound is 331 m/s.

†For help, see Online Concept Tutorial 39 at www.wwnorton.com/physics

55. Water waves of wavelength 20 m approach a harbor entrance 50 m across at right angles to their path. What is the angular width of the central beam of diffracted waves beyond the entrance?

56. You want to prepare a photograph, such as Fig. 35.33, showing diffraction by a single slit. Suppose you use light of wavelength 577 nm from a mercury lamp and you place your photographic film 2.0 m behind the slit. If the width of the central maximum is to be at least 0.50 cm on your film, what width of the slit do you require?

57. Suppose you want to detect diffraction fringes in the intensity pattern produced by *sunlight* passing through a fine slit cut into the blinds covering a window. You have available a filter that blocks all wavelengths except 550 nm, and you place this filter over the slit.

(a) If the width of the slit is 0.10 mm, will you be able to detect diffraction fringes? Take into account that the Sun's angular diameter is about $\frac{1}{2}°$.

(b) If the width of the slit is 0.010 mm?

(c) If you remove the filter?

58. Relative to the intensity of the central maximum, what is the intensity of the single-slit diffraction pattern at an angular position halfway between the peak of the central maximum and the first minimum? (You may assume small angles.)

59. Relative to the intensity of the central maximum, what is the intensity of the single-slit diffraction pattern at an angular position halfway between the first and the second minimum (that is, near the first secondary maximum)? (You may assume small angles.)

60. Figure 35.58 shows a combined diffraction–interference pattern. How many slits were illuminated? The distance between adjacent slits is 0.12 mm. Deduce the width of each slit. Deduce the wavelength of the light.

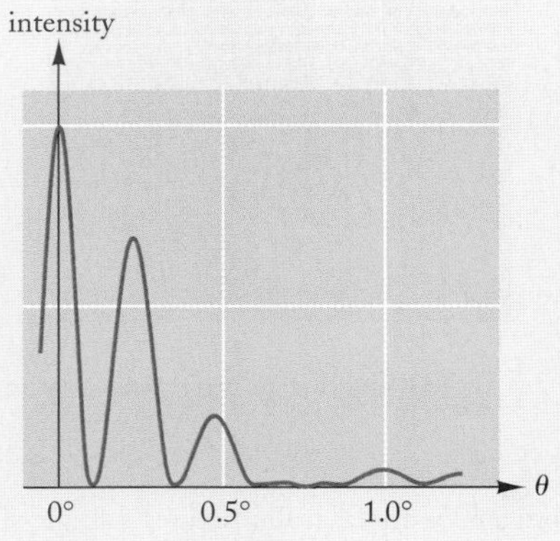

FIGURE 35.58 Combined interference and diffraction pattern.

*61. Two narrow slits of width a are separated by a center-to-center distance d. Suppose that the ratio of d to a is an integer, $d/a = m$.

(a) Show that in the diffraction pattern produced by this arrangement of slits, the mth interference maximum (corresponding to $d \sin\theta = m\lambda$) is suppressed because of coincidence with a diffraction minimum. Show that this is also true for the $2m$th, $3m$th, etc., interference maxima.

(b) How many interference maxima are there between one diffraction minimum on one side and the next on the same side?

*62. Figure 35.59 shows the intensity curve for the interference pattern produced by three very narrow slits separated by distances of 12λ. Suppose the three very thin slits are replaced by three new slits of width 6λ. This will change the intensity curve.

(a) On top of Fig. 35.59 (or on a copy of Fig. 35.59), plot the intensity curve of the diffraction pattern produced by *one* of these new slits.

(b) By suitably combining these two curves, obtain the complete intensity curve for the system of the three new slits.

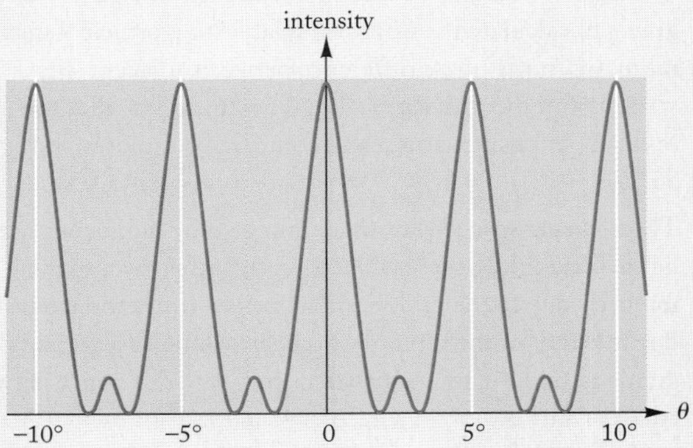

FIGURE 35.59 Interference pattern due to three narrow slits.

*63. Find the actual location of the first secondary maximum in the single-slit diffraction pattern by differentiating Eq. (35.25) and solving for the phase angle to at least three significant figures iteratively (by trial and error). Express your result as a multiple of 3π, the phase angle midway between the first two minima.

*64. Light of wavelength λ is obliquely incident on a slit of width a. The angle of incidence of the light on the slit is ϕ (Fig. 35.60).

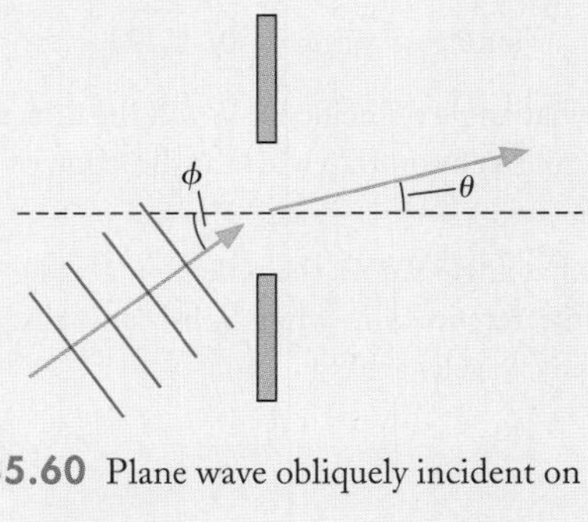

FIGURE 35.60 Plane wave obliquely incident on a single slit.

(a) Show that the diffracted light emerging at an angle θ interferes destructively if

$$a\sin\theta - a\sin\phi = \lambda,\ 2\lambda,\ 3\lambda, \ldots$$

(b) Show that, for small values of θ, the angular separation between the directions of destructive interference is independent of ϕ.

35.6 Diffraction by a Circular Aperture; Rayleigh's Criterion

65. The impressionist painter Georges Seurat used small dots of paint, which form continuous images when viewed from a distance. How small must two adjacent green dots ($\lambda = 550$ nm) be so that a human eye, with a 6.0-mm aperture, can barely resolve them when viewed from 2.0 m (and thus not resolve them from larger distances)?

66. When the eye looks at a star (a point of light), diffraction at the pupil spreads the image of the star on the retina into a small disk.

    (a) When opened to maximum size, the pupil of a human eye has a diameter of 7.0 mm. Assuming the starlight has a wavelength of 550 nm, what is the angular size of the image on the retina?

    (b) The distance from the pupil to the retina is 23 mm. What is the linear size of the image of the star?

    (c) At the midpoint on the retina (fovea), there are 150 000 light-sensitive cells (rods) per mm^2. How many of these cells are illuminated when the eye looks at a star?

67. A sailor uses a speaking trumpet to concentrate his voice into a beam. The opening at the front end of the speaking trumpet has a diameter of 25 cm. If the sailor emits a sound of wavelength 15 cm (this is roughly the wavelength a man emits when yelling "eeeee . . ."), what is the angular width of the central maximum of the beam of sound?

68. The antenna of a small radar transmitter operating at 1.5×10^{10} Hz consists of a circular dish of diameter 1.0 m. What is the angular width of the central maximum of the radar beam? What is the linear width at a distance of 5.0 km from the transmitter?

69. The Hubble Space Telescope placed into orbit above the atomsphere of the Earth has an aperture of 2.4 m. According to Rayleigh's criterion, what angular resolution can this telescope achieve with visible light of wavelength 550 nm? With ultraviolet light of wavelength 120 nm? How much better is this than the angular resolution of 0.5 seconds of arc achieved by telescopes on the surface of the Earth?

70. The radio telescope at Jodrell Bank (England) is a dish with a circular aperture of diameter 76 m. What angular resolution can this radio telescope achieve when operating at a wavelength of 21 cm?

*71. According to newspaper reports, a photographic camera on a Blackbird reconnaissance jet flying at an altitude of 27 km can distinguish detail on the ground as small as the size of a person.

    (a) Roughly, what angular resolution does this require?

    (b) According to Rayleigh's criterion, what minimum diameter must the lens of the camera have?

*72. For an optically perfect lens, the size of the focal spot is limited only by diffraction effects. Suppose that a lens of diameter 10 cm and focal length 18 cm is illuminated with parallel light of wavelength 550 nm. What is the angular width of the central maximum in the diffraction pattern? What is the corresponding linear width at the focal distance?

*73. Kalahari bushmen are said to be able to see the four brightest moons of Jupiter with the naked eye. According to Rayleigh's criterion, what must be the minimum separation between two small light sources placed at the distance of Jupiter if they are to be resolved by the human eye? Compare this with the separations between these four moons of Jupiter. Does the limit on the resolving power of the human eye prevent *you* from seeing the moons? The (average) distance to Jupiter is 6.3×10^8 km; the separations between the moons are typically 4.0×10^5 km. Assume that the diameter of the pupil of the eye is 7.0 mm and that the wavelength of the light is 550 nm.

*74. Mars has a radius of 3400 km; when Mars is at its closest to the Earth, its distance is 7.8×10^7 km. Calculate the angular size of Mars as seen from Earth. Estimate the diameter of the objective lens of the telescope of smallest size that will permit you to tell that Mars has a disk, that is, that the image of Mars is wider than the image of a star.

*75. (a) According to Rayleigh's criterion, what is the angular resolution that the human eye can achieve for light of wavelength 550 nm? The fully distended pupil of the human eye has a diameter of 7.0 mm.

    (b) Even during steady fixation, the eye has a spontaneous tremor that swings it through angles of 20 or 30 seconds of arc. Compare this angular tremor with the angular resolution that you found in part (a). Would the elimination of the tremor greatly improve the acuity of the eye?

*76. At night, on a long stretch of straight road in Nevada, a truck driver sees the distant headlights of another truck. How close must he be to the other truck in order for his eyes to resolve two headlights? Assume that the pupils of the truck driver have a diameter of 5.0 mm, that the headlights are separated by 1.8 m, and that the light has a wavelength of 550 nm.

*77. Some spy satellites carry cameras with lenses 30 cm in diameter and with a focal length of 2.4 m.

    (a) What is the angular resolution of the camera according to Rayleigh's criterion? Assume that the wavelength of light is 550 nm.

(b) If such a satellite looks down on the Earth from a height of 150 km, what is the distance between two points on the ground that the camera can barely resolve?

(c) The lens projects images of the two points on a film at the focal plane of the lens. What is the distance between the two images projected on the film?

*78. "7 × 50" binoculars magnify angles by a factor of 7.0, and their objective lenses have an aperture of 50 mm diameter.

(a) According to Rayleigh's criterion, what is the intrinsic angular resolution of these binoculars? Assume that the light has a wavelength of 550 nm.

(b) At best, the pupil of your eye has an aperture of 7.0 mm diameter. Compare the angular resolution of your eye divided by a factor of 7.0 with the intrinsic angular resolution of the binoculars. Which of the two numbers determines the actual angular resolution you can achieve while looking through the binoculars?

*79. When exposed to strong light, the pupil of the eye of a cat narrows to a fine slit, about 0.30 mm across. Suppose that the cat is looking at two white mice 20 m away and separated by a distance of 5.0 cm. Can the cat distinguish one mouse from the other? Assume that the wavelength of light is 550 nm.

REVIEW PROBLEMS

*80. A lens made of flint glass with an index of refraction of 1.61 is to be coated with a thin layer of magnesium fluoride with an index of refraction of 1.38 in order to reduce reflection.

(a) How thick should the layer be so as to give destructive interference for the perpendicular reflection of light of wavelength 550 nm seen in air?

(b) Does your choice of thickness permit *constructive* interference for the reflection of light of some other wavelength in the visible spectrum? (If it does, you ought to make a better choice.)

*81. When you look at the point on a spherical soap bubble nearest your eye (where the rays of light reaching your eye are perpendicular to the bubble's surface), you see strongly reflected red light, of a wavelength of 650 nm. What can you conclude about the thickness of the wall of the soap bubble? The index of refraction of the soapy water is 1.35.

*82. When you look at a spherical soap bubble, under the condition described in the preceding problem, you see strongly reflected light of a wavelength of 650 nm. Suppose you now inflate the bubble (without adding liquid) until the strongly reflected light becomes blue violet, of a wavelength of 430 nm. By what factor must you inflate the diameter of the bubble to accomplish this? The index of refraction of the soapy water is 1.35.

83. When a beam of monochromatic light is incident on two narrow slits separated by a distance 0.15 mm, the angle between the central beam and the third lateral maximum in the interference pattern is 0.52°. What is the wavelength of the light?

84. The red line in the spectrum of hydrogen has a wavelength of 656.3 nm; the blue violet line in this spectrum has a wavelength of 434.2 nm. If hydrogen light falls on a grating of 6000 slits per centimeter, what will be the angular separation (in degrees) of these two spectral lines as seen in the first-order spectrum?

*85. Two radio beacons are located on an east–west line and separated by a distance of 6.0 km. The radio beacons emit synchronous (in-phase) sinusoidal waves of a frequency of 1.0×10^5 Hz. The navigator of a ship wants to determine his position relative to the radio beacons. His radio receiver indicates zero signal strength at the position of the ship. What are the possible angular bearings of the ship relative to the radio beacons? Assume that the distance between the ship and the radio beacons is much larger than 6 km.

86. Crystal structures are most often determined by the diffraction of X rays from atoms. Consider atomic planes in a crystal as shown in Fig. 35.61. The incident and reflected X rays make an angle θ with the atomic planes. Show that the reflected rays will interfere constructively provided that

$$2d \sin\theta = m\lambda \qquad m = 1, 2, 3, \ldots$$

where d is the distance between planes. This result was first derived by W. L. Bragg and is known as the **Bragg law**; the interference peaks are known as *Bragg reflections.* Note that this relation has solutions only for $\lambda \leq 2d$; thus, for atomic spacings, X rays must be used instead of visible light.

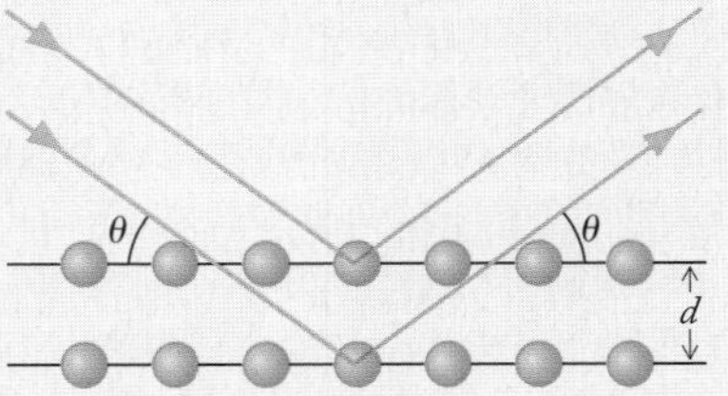

FIGURE 35.61 Reflection of X rays from two planes of atoms in a crystal.

*87. The radio waves from a transmitter to a receiver may follow either a direct path or else an indirect path involving a reflection on the ground (Fig. 35.62). This can lead to destructive interference of the two waves and a consequent fading of the radio signal at certain locations. Suppose that a transmitter and a receiver operating at a wavelength λ are at a height h on tall buildings with bare ground between. The distance from the emitter to the receiver is d.

(a) Show that the condition for destructive interference is

$$\sqrt{d^2 + 4h^2} - d = \lambda,\ 2\lambda,\ 3\lambda, \ldots$$

Note that in this calculation you must take into account the reversal, or change of phase, of the wave during reflection on the ground.

(b) If $h = 60$ m and $d = 2300$ m, what wavelengths will lead to destructive interference? What is the lowest-frequency radio wave for which destructive interference will occur?

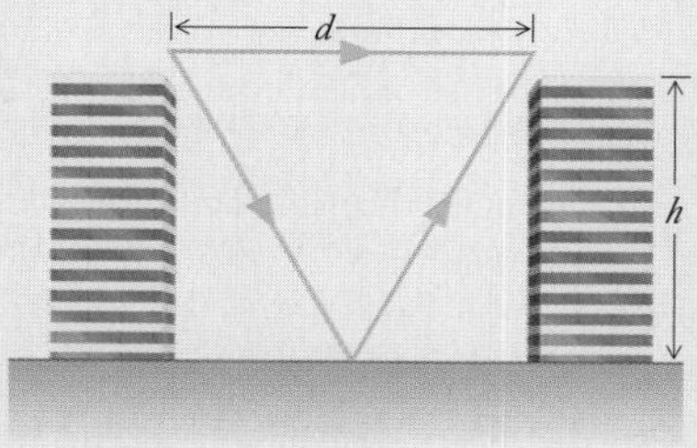

FIGURE 35.62 Antennas on two buildings.

88. Interference effects can be detected with sound waves passing through a picket fence. Suppose that the fence consists of vertical boards or rods separated by a distance of 20 cm. Suppose you stand on one side of the fence and a friend stands on the other side of the fence, at some reasonably large distance at an angle of 30° away from the direction perpendicular to the fence. If you clap your hands, producing white noise, what wavelengths and frequencies will your friend hear strongly?

89. A vehicle approaches at night from a large distance. Only resolving a single source, at first you think it is a motorcycle. But as it gets closer, you can distinguish two headlights. If the headlights are separated by 1.3 m and the pupil of your eye has a diameter of 7.0 mm, how far away was the vehicle when the headlights became distinguishable? Assume $\lambda = 580$ nm and otherwise perfect conditions.

*90. Two radio beacons emit waves of frequency 2.0×10^5 Hz. The beacons are on a north–south line, separated by a distance of 3.0 km. The southern beacon emits waves $\frac{1}{4}$ cycle later than the northern beacon. Find the angular directions for constructive interference. Reckon the angles relative to the east–west line, and assume that the distance between the beacons and the point of observation is much larger than 3 km.

91. The beam of a ruby laser, with light of wavelength 693 nm, is aimed at a slit of width 0.050 mm cut in an aluminum sheet. The diffracted light is intercepted by a screen placed 2.5 m beyond the slit. How far from the centerline in this screen will the first-order minimum be found? The second-order minimum?

*92. A slit of width 0.11 mm cut in a sheet of metal is illuminated with light of wavelength 577 nm from a mercury lamp. A screen is placed 4.0 m beyond the slit.

(a) Find the width of the central maximum in the diffraction pattern on the screen—that is, find the distance between the first minimum on the left and on the right.

(b) Find the width of the secondary maximum—that is, find the distance between the first minimum and the second minimum on the same side.

93. According to a recent proposal, solar energy is to be collected by a large power station on an artificial satellite orbiting the Earth at an altitude of 35 000 km. The power is to be beamed down to the surface of the Earth in the form of microwaves. If the microwaves have a wavelength of 10 cm and if the antenna emitting the microwaves is 1.5 km in diameter, what is the angular width of the central beam emerging from this antenna? What will be the transverse dimension of the beam when it reaches the surface of the Earth?

94. Rumor has it that a photographic camera on a spy satellite can read the license plate of an automobile on the ground.

(a) If the altitude of the satellite is 160 km, roughly what angular resolution does the camera need to read a license plate? Assume that the reading requires a linear resolution of about 5.0 cm.

(b) To attain the angular resolution, what must be the diameter of the aperture of the camera? Assume that the wavelength of light is 550 nm.

*95. A microwave antenna used to relay communication signals has the shape of a circular dish of diameter 1.5 m. The antenna emits waves with $\lambda = 4.0$ cm.

(a) What is the width of the central maximum of the beam of this antenna at a distance of 30 km?

(b) The power emitted by the antenna is 1.5×10^3 W. What is the energy flux directly in front of the antenna? What is the energy flux at a distance of 30 km? Assume that the power is evenly distributed over the width of the central beam.

Answers to Checkups

Checkup 35.1

1. At the center, there is no effective path difference (the flat and curved glass surfaces are essentially touching). The light reflected at the lower (flat) glass surface (a medium of higher index of refraction than air) suffers a 180° phase change. The light reflected from the upper (curved) glass surface (above the air gap) suffers no change in phase; thus these two waves are 180° out of phase and destructively interfere, providing a dark central spot.
2. Equations (35.1) and (35.2) are still valid. When the light was incident from above, the waves reflected from the two interfaces ($n_{air} < n_{oil}$ and $n_{oil} < n_{water}$) each suffer a 180° phase change; when the light is incident from below, neither one does. Thus, in either case, there is no relative phase change due to reflection, and the interference is determined solely by the path difference, as in Eqs. (35.1) and (35.2).
3. (A) 100 nm. For minimal reflection and the smallest thickness, we use the first destructive interference condition of Eq. (35.2) [and the relation $\lambda = \lambda_{air}/n$ of Eq. (35.3)], $d = \lambda/4 = \lambda_{air}/(4n) = (500 \text{ nm})/(4 \times 1.25) = 100$ nm.

Checkup 35.2

1. No; each ray is reflected twice (at an air-to-silver or glass-to-silver interface), and so each suffers a total phase change of 360°. Thus neither has any overall phase change, and there is no relative phase change.
2. Yes—the ray from mirror M_1 passes through the glass of mirror M three times, and the ray from mirror M_2 passes through it once, so there is a relative phase change due to the extra optical path difference. A correction for this extra path difference is provided by a *compensatory,* a slab of glass visible in Fig. 35.11. In practice, the path lengths in the Michelson interferometer need not be known exactly, since it is only the shift in fringes upon changing the interferometer orientation that needs to be determined.
3. (B) 0.10 mm. Because the light travels to and from the mirror, the path difference is twice the mirror motion, so for 400 fringes Eq. (35.4) gives $2d = 400\lambda$ or $d = 200\lambda = 200 \times 500$ nm $= 0.10$ mm.

Checkup 35.3

1. Since the product $d\sin\theta = \lambda$ is constant for the first maximum, decreasing the distance d between the slits results in an increase in the angle θ and thus an increase in the distance between the central and the first maximum.
2. For $d = \lambda$, Eq. (35.11) implies intensity maxima when $\sin\theta =$ 0 or 1, so there is only a central maximum (and a maximum at $\theta = 90°$, which is at the border of the field of view). There will be one minimum in between for $\sin\theta = \frac{1}{2}$, that is, for $\theta = 30°$.
3. To have the first maximum at 45°, Eq. (35.8) or (35.11) requires that $\lambda = d\sin 45° = d/\sqrt{2}$.
4. (B) 30°. As in Eq. (35.9), the minima are at $d\sin\theta = \frac{1}{2}\lambda, \frac{3}{2}\lambda, \frac{5}{2}\lambda, \ldots$. Thus the third-order minima satisfy $d\sin\theta = \frac{5}{2}\lambda$. With the given $d = 5\lambda$, this mean $5\lambda\sin\theta = \frac{5}{2}\lambda$, or $\sin\theta = \frac{1}{2}$. The angle whose sine is $\frac{1}{2}$ is 30°; the third-order minima occur at 30° on either side of the central maximum.
5. (D) $\frac{3}{4}I_{max}$. The intensity is given by Eq. (35.11); for a path difference as small as $\lambda/6$, the point in question is on the central maximum, slightly away from the peak. Since the path difference equals $d\sin\theta$, the intensity is $I = I_{max}\cos^2(\pi d\sin\theta/\lambda) = I_{max}\cos^2(\pi/6) = I_{max}(\sqrt{3}/2)^2 = \frac{3}{4}I_{max}$.

Checkup 35.4

1. The red line belongs to the second-order spectrum ($m = 2$ in $m\lambda = d\sin\theta$); the blue and blue violet lines belong to the third-order spectrum ($m = 3$). Since the wavelength λ is shorter for blue light, the value of 3λ for blue wavelengths occurs at the same θ as the value of 2λ for the longer red wavelengths; the orders overlap.
2. Since the mth-order spectrum is at $d\sin\theta = m\lambda$, each line of the second-order spectrum will be at approximately twice the angle of the first (approximately, since $\sin\theta \approx \theta$ becomes less accurate in the second order); thus, the angular separation will be twice as large in the second-order spectrum as in the first-order spectrum.
3. When the phase difference is $\pi/2$, the three phasors of amplitude E_0 form three sides of a square as in Fig. 35.26c; the resultant is the fourth side, which also has ampliude E_0.
4. At a maximum, the two phasors point the same way, so the amplitudes add to give a total amplitude of $3E_0$; thus, since intensity is proportional to the square of the amplitude, the intensity at the maxima is $9I_0$. At a minimum, the two phasors point in opposite directions, and so sum to give a total amplitude of E_0, corresponding to a minimum intensity of I_0. Halfway between a maximum and a minimum, the phasors point at right angles, so the net amplitude is $\sqrt{2^2 + 1^2}E_0 = \sqrt{5}E_0$, and so the intensity there is $5I_0$.
5. (C) $2\pi/5$. For identical, evenly spaced slits, the first minimum occurs when the five phasors close to form a pentagon; in this case, each is oriented at a phase angle $2\pi/5$ with respect to the previous one. Alternatively, one can see this result by noting that a full 2π of phase difference must be achieved by the five slits.
6. (D) 4. For two slits there are no secondary maxima, for three slits there is one, and so on, so that for six slits there are four secondary maxima. In general, for N slits, there are $N - 2$ secondary maxima.

Checkup 35.5

1. If you decrease the size of the narrow slit (comparable to the wavelength of light), the width of the illuminated zone beyond the slit grows wider for the light beam (a diffraction pattern), but narrower for the particle beam (an ordinary shadow). (We will see in Chapter 37 that particles also have wave properties; however, the corresponding wavelengths are usually much smaller than the wavelength of light and play no role unless the slit is *extremely* narrow.)
2. There will, of course, be a central maximum; the value $a = \sqrt{2}\lambda$ places the first minimum ($a \sin\theta = \lambda$) at $\sin\theta = 1/\sqrt{2}$, or $\theta = 45°$. Beyond that the intensity will increase and be close to the first secondary maximum at $\theta = 90°$, however, there will be no other maxima or minima.
3. No, light passing through a large aperture will form an ordinary shadow; the angle $\theta = \lambda/a$ still gives the direction of the first diffraction minimum at a sufficiently *large* distance from the slit.
4. (A) 5×10^{-3}. The first minimum occurs at $\sin\theta = \lambda/a = (500 \times 10^{-9}\ \text{m})/(1 \times 10^{-4}\ \text{m}) = 5 \times 10^{-3}$. Since this is small, we may use $\sin\theta \approx \theta$ to obtain $\theta \approx 5 \times 10^{-3}$.

Checkup 35.6

1. For a single slit, the first diffraction minimum of one image will coincide with the central diffraction peak of the other when $\theta = \lambda/a$, as in Eq. (35.21), where we assume small angles (as is always the case when resolving distant objects).
2. Although a is larger, the wavelength of radio waves is much larger than that of light ($4\ \text{cm}/500\ \text{nm} = 8 \times 10^4$). Thus, even though the aperture is 125 times larger, the resolution is hundreds of times poorer.
3. (D) 100 cm. The dots subtend a small angle $\theta = y/r$, where $y = 0.2$ mm is the dot separation and r is the distance. Thus, using Rayleigh's criterion, $y/r = \theta = 1.22\lambda/a$, with $\lambda \approx 600$ nm, we can solve for the largest resolving distance $r = ya/(1.22\lambda) = 140$ cm. Thus the dots will be resolved at 100 cm, but not at 300 cm.

PART

6

Relativity, Quanta, and Particles

CONTENTS

The solar cells in the 73-meter long photovoltaic arrays on the International Space Station convert solar energy into electrical power. The arrays are fitted with gimbals that angle the arrays toward the Sun at all times so as to maximize the power supplied to the Space Station.

CHAPTER

36

The Theory of Special Relativity

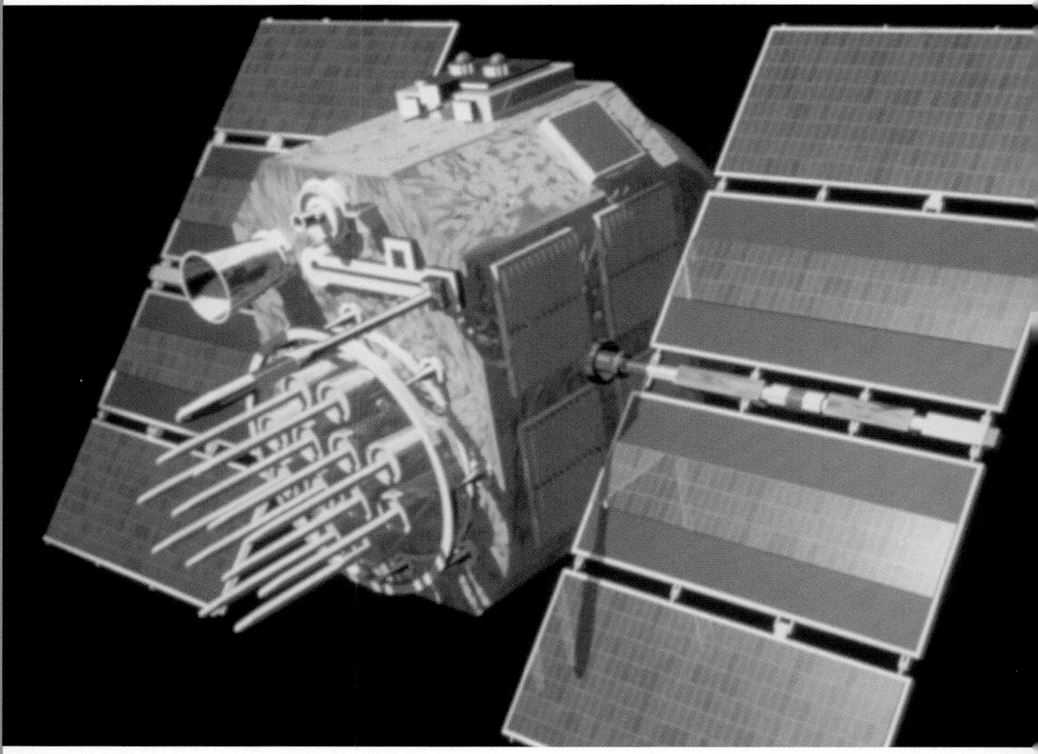

CONCEPTS IN CONTEXT

Concepts in Context

Determination of latitude and longitude by means of radio signals from Global Positioning System (GPS) satellites, such as the one shown here, requires measurement of the travel time from several satellites to the relevant point on the ground. GPS satellites are in relative motion with respect to the Earth's surface, but the speed of their radio signals is not affected by this motion.

In our study of the theory of Special Relativity we will study the propagation of light in different reference frames in relative motion, and we can then ask:

? How are distances calculated from GPS signals, and how do we know that the speed of light is unaffected by the motion of the satellite or by the translational motion of the Earth? (Section 36.1, page 1219)

? Relative to clocks on the surface of the Earth, clocks on GPS satellites are in motion at a somewhat high speed. How does this affect the rate of the clocks? (Example 2, page 1227)

? How does the motion of a GPS satellite relative to the Earth affect the frequency of the radio signal? (Example 3, page 1228)

ALBERT EINSTEIN (1879–1955)
German (and Swiss, and American) theoretical physicist, professor at Zurich, at Berlin, and at the Institute for Advanced Study at Princeton. Einstein was the most celebrated physicist of the twentieth century. He formulated the theory of Special Relativity in 1905 and the theory of General Relativity in 1916. Einstein also made incisive contributions to modern quantum theory, for which he received the Nobel Prize in 1921. Einstein spent the last years of his life in an unsuccessful quest for a unified theory of forces that was supposed to incorporate gravity and electricity in a single set of equations.

As we saw in Chapter 5, Newton's laws of motion are equally valid in every inertial reference frame. All inertial reference frames are unaccelerated, but they can differ in their uniform translational motion. Since Newton's laws make no distinction between different inertial reference frames, no mechanical experiment can detect a uniform translational motion of one inertial reference frame by itself. Such motion can be detected only as a relative motion of one reference frame with respect to another reference frame. This is the Newtonian principle of relativity. For a concrete illustration of this principle, consider the reference frame of a cruise ship moving steadily away from the shore, without acceleration, and consider the reference frame of the shore. Both of these reference frames are inertial, and the behavior of a ball used in a game of tennis on the ship is not different from the behavior of a similar ball on shore—balls on the ship and on the shore accelerate in the same way when subjected to forces, and they obey the same laws of conservation of energy, conservation of momentum, etc.[1] Thus, experiments with such balls aboard the ship will not reveal the uniform motion of the ship relative to the shore. To detect this motion, the crew of the ship must take sightings of points on the shore or use some other navigational technique that fixes the position and velocity of the ship relative to the shore. Hence, in regard to mechanical experiments, uniform translational motion of our inertial reference frame is always *relative motion—it can be detected only as motion of our reference frame with respect to another reference frame.* There is no such thing as absolute motion.

The question naturally arises whether the relativity of motion indicated by mechanical experiments also applies to electric, magnetic, optical, and other experiments. Do any of these experiments permit us to detect an absolute motion or our reference frame? In 1905, Albert Einstein proposed that no experiment of any kind should ever permit us to detect such motion. He postulated a principle of relativity for *all* the laws of physics. This postulate serves as the foundation for Einstein's theory of Special Relativity, widely regarded as one of the greatest achievements of twentieth-century physics. The theory of relativity requires a drastic revision of our concepts of space and time, and it also requires a drastic revision of Newton's laws. *At high speeds—near the speed of light—particles obey new, relativistic laws which are quite different from Newton's laws.* However, at low speeds—small compared with the speed of light—the differences between Einstein's and Newton's theories are usually undetectable. Hence Newton's laws are adequate for describing the behavior of particles and of other bodies at the relatively low speeds we encounter in our everyday experience.

Before we deal with the details of Einstein's theory of relativity, we will briefly describe why nonmechanical experiments—and, especially, experiments with light—might be expected to detect absolute motion, which mechanical experiments cannot detect.

[1] This assumes the ship moves steadily. If the ship lurches forward, or pitches, or rolls, it ceases to be an inertial reference frame, and ball will behave in a manner "inconsistent" with Newton's laws.

Online Concept 41 Tutorial

36.1 THE SPEED OF LIGHT; THE ETHER

Since the laws of mechanics are the same in all inertial reference frames, it might seem quite natural to assume that the laws of electricity, magnetism, and optics are also the same in all inertial reference frames. But this assumption immediately leads to a paradox concerning the speed of light. As we know from Chapter 33, light is an oscillating electric and magnetic disturbance propagating through space, and Maxwell's equations permit us to deduce that the speed of propagation of this disturbance must always be[2] 3.00×10^8 m/s. The trouble with this deduction is that, according to the Galilean addition rule for velocity [Eq. (4.53)], the speed of light ought *not* to be the same in all reference frames. For instance, imagine that an alien spaceship approaching the Earth with a speed of, say, 1.00×10^8 m/s flashes a light signal toward the Earth. If this signal has a speed of 3.00×10^8 m/s in the reference frame of the spaceship, then the Galilean addition rule tells us that it ought to have a speed of 4.00×10^8 m/s in the reference frame of the Earth (see Fig. 36.1).

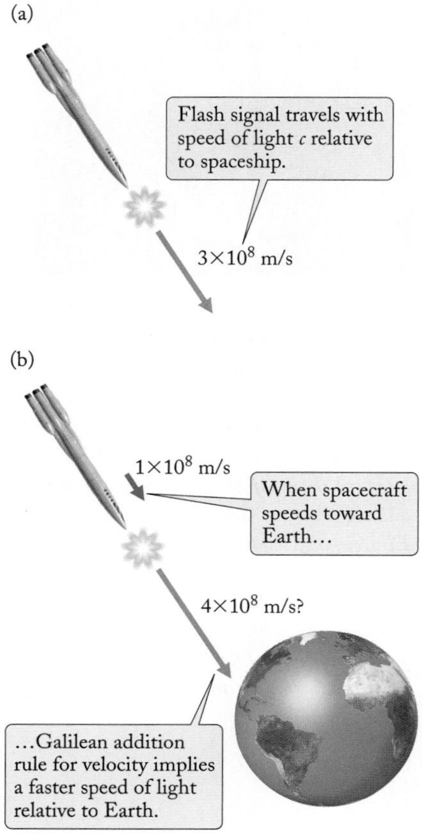

FIGURE 36.1 Velocities according to the Galilean addition rule in (a) reference frame of a spaceship and (b) reference frame of the Earth. A speed of light different from 3×10^8 m/s is at odds with Maxwell's equations.

To resolve this paradox, either we must give up the notion that the laws of electricity and magnetism (and the values of the speed of light) are the same in all inertial reference frames, or else we must give up the Galilean addition rule for velocities. Both alternatives are unpleasant: the former means that we must abandon all hope for a principle of relativity embracing electricity and magnetism, and the latter means that together with the Galilean addition rule we must abandon the transformation rule for position vectors measured in different reference frames [Eq. (4.50)] as well as the intuitively "obvious" notions of absolute time and length from which these rules are derived.

Since the failure of a relativity principle embracing electricity and magnetism might seem to be the lesser of two evils, let us first explore this alternative. Let us assume that there exists a preferred inertial reference frame in which the laws of electricity and magnetism take their simplest form, that is, the form expressed in Maxwell's equations, Eqs. (33.4)–(33.7). In this reference frame, the speed of light has its standard value of $c = 3.00 \times 10^8$ m/s, whereas in any other reference frame it is larger or smaller according to the Galilean addition rule. The propagation of light is then analogous to the propagation of sound. There exists a preferred reference frame in which the equations for the propagation of sound waves in, say, air take their simplest form: the reference frame in which the air is at rest. In this reference frame, sound has its standard speed of 331 m/s. In any other reference frame, the equations for the propagation of sound waves are more complicated, but the speed of propagation can always be obtained directly from the Galilean addition rule. For instance, if a wind of 40 m/s (a hurricane) is blowing over the surface of the Earth, then sound waves have a speed of 331 m/s relative to the air, but their speed relative to the ground depends on direction—downwind the speed is 371 m/s, whereas upwind it is 291 m/s.

This analogy between the propagation of light and of sound suggests that there might exist some pervasive medium whose oscillations bring about the propagation of light, just as the oscillations of air bring about the propagation of sound. Presumably this ghostly medium fills all of space, even the interplanetary and interstellar space, which is normally regarded as a vacuum. The physicists of the nineteenth century called this hypothetical medium the **ether**, and they attempted to describe light waves as oscillations of the ether, analogous to sound waves as oscillations of the air. The preferred reference frame in which light has its standard speed is then the reference frame in which the ether is at rest. The existence of such a preferred reference frame would

ether

[2]The exact value of the speed of light is $2.997\,924\,58 \times 10^8$ m/s, but throughout this chapter we will round this off to 3.00×10^8 m/s.

imply that velocity is absolute—the ether frame would set an absolute standard of rest, and the velocity of any body could always be referred to this frame. For instance, instead of describing the velocity of the Earth relative to some other material body, such as the Sun, we could always describe its velocity relative to the ether.

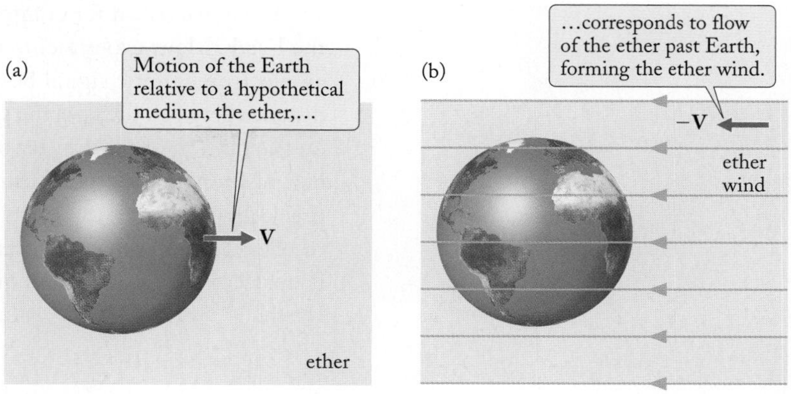

FIGURE 36.2 (a) Motion of Earth in reference frame of ether. (b) Motion of ether in reference frame of Earth.

Presumably the Earth has some nonzero velocity relative to the ether. Even if the Earth were at rest in the ether at one instant, this condition could not last, since the Earth continually changes its motion as it orbits around the Sun. The motion of the ether past the Earth was called the *ether wind* by the nineteenth-century physicists (Fig. 36.2). If the Sun is at rest in the ether, then the ether wind would have a velocity opposite to the velocity of the Earth around the Sun—about 30 km/s; if the Sun is in (steady) motion, then the ether wind would vary with the seasons—it would reach a maximum when the orbital motion of the Earth is parallel to the motion of the Sun, and a minimum when antiparallel.

Experimenters attempted to detect this ether wind by its effects on the propagation of light. A light wave in a laboratory on the Earth would have a greater speed when moving downwind and a smaller speed when moving upwind or across the wind. If the speed of the ether wind "blowing" through the laboratory is V, then the speed of light in this laboratory would be $c + V$ for a light signal with downwind motion, $c - V$ for upwind motion, and $\sqrt{c^2 - V^2}$ for motion perpendicular to the wind (see Fig. 36.3). With a value of 300 000 km/s for c and a value of approximately 30 km/s for V, the increase or decrease of the speed of light amounts to only about 1 part in 10 000, and a very sensitive apparatus is required for the detection of this small change.

Concepts in Context

Michelson-Morley experiment

In a famous experiment first performed in 1881 and often repeated thereafter, A. A. Michelson and E. W. Morley attempted to detect small changes in the speed of light by means of an interferometer. The results of their experiment were negative. As discussed in Section 35.2, the sensitivity of the original experiment of Michelson and Morley was such that an ether wind of 5 km/s would have produced a detectable effect. Since the expected wind is about 30 km/s, the experimental result contradicts the ether theory of the propagation of light. Later, more refined versions of the experiment established that if there were an ether wind, its speed would certainly have to be less than 3 m/s. The experimental evidence therefore establishes conclusively that the motion of the Earth has no effect on the propagation of light. As the Earth moves around the Sun, its velocity is first in one direction, then in another, and another; and the Earth is first in one inertial reference frame, then in another, and another. But all these inertial reference frames appear to be completely equivalent in regard to the propagation of light. There is no preferred reference frame. There is no ether.

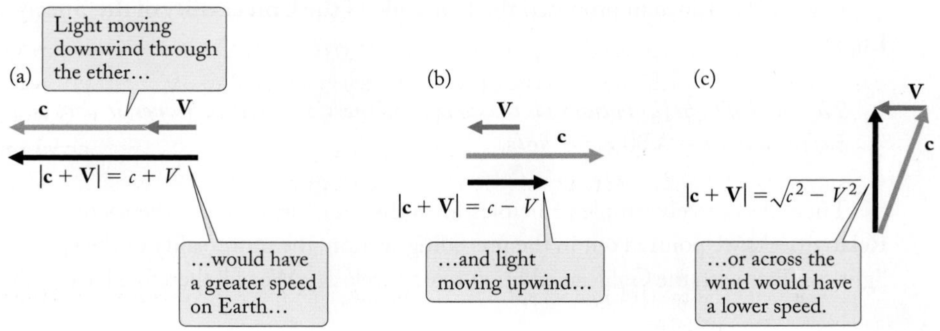

FIGURE 36.3 The velocity of light is **c** relative to the ether, and the velocity of the ether is **V** relative to the laboratory. The velocity of light relative to the laboratory is then the vector sum $\mathbf{c} + \mathbf{V}$. (a) If **c** and **V** are parallel, the magnitude of the vector sum is $c + V$. (b) If **c** and **V** are antiparallel, the magnitude of the vector sum is $c - V$. (c) If **c** and **V** form this right triangle, the magnitude of the vector sum is $\sqrt{c^2 - V^2}$.

Concepts in Context

Today, we take it for granted that the speed of light is unaffected by the motion of the Earth. Many experiments and instruments rely on the fact that the travel time for an electromagnetic signal between an emitter and a receiver depends only on the distance between the two; the speed of light, and thus the travel time, does not depend on the motion of the emitter or of the receiver. For example, accurate determination of a position on the ground using the Global Positioning System (GPS) is achieved by precisely measuring the travel time of signals from several satellites. When a GPS receiver calculates the distances to several satellites, it assumes that the speed of light is unaffected by the motion of the satellites and by the translational motion of the Earth.

Checkup 36.1

QUESTION 1: You are in an open automobile (a convertible) traveling at 80 km/h. Does the speed of sound waves (relative to you) depend on direction?

QUESTION 2: At the equator of the Earth, the rotational speed is 0.46 km/s. Was the original Michelson–Morley experiment capable of detecting the corresponding ether wind?

QUESTION 3: If Michelson and Morley had detected an ether wind of 30 km/s opposite to the motion of the Earth around the Sun, what could they have concluded about the absolute motion of the Earth? The Sun?

QUESTION 4: In 1887, after observations lasting a few days, a null result was obtained in the Michelson–Morley experiment with 5 km/s sensitivity. To be sure that there was no ether, Michelson and Morley had to repeat the experiment

(A) With a different color of light (B) At 1 km/s resolution
(C) At a different time of year

Online Concept Tutorial 41

36.2 EINSTEIN'S PRINCIPLE OF RELATIVITY

Neither the laws of mechanics nor the laws for the propagation of light reveal any intrinsic distinction between different inertial reference frames. This motivated Einstein to take a bold step and to propose a general hypothesis concerning *all* the laws of physics. This hypothesis is the **Principle of Relativity**:

Principle of Relativity

All the laws of physics are the same in all inertial reference frames.

In addition, Einstein proposed the **Principle of the Universality of the Speed of Light**:

Principle of Universality of Speed of Light

The speed of light (in vacuum) is the same in all inertial reference frames; it always has the value $c = 3.00 \times 10^8$ m/s.

These deceptively simple principles form the foundation of the theory of Special Relativity. As we pointed out in the preceding section, the universality of the speed of light conflicts with the Galilean addition rule for velocity. We will therefore have to discard this rule, and we will also have to discard the transformation rule for position vectors or coordinates on which it is based (see Section 4.6).

The universality of the speed of light also requires that we give up some of our intuitive, everyday notions of space and time. Obviously, the fact that a light signal always has a speed of 3.00×10^8 m/s, regardless of how hard we try to move toward it or away from it in a fast aircraft or spaceship, does violence to our intuition. This strange behavior of light is only possible because of a strange behavior of length and time in relativistic physics. As we will see later in this chapter, neither length nor time is absolute—they both depend on the reference frame in which they are measured, and they suffer contraction or dilation when the reference frame changes.

Before we can inquire into the consequences of Einstein's two principles, we must carefully describe the construction of reference frames and the synchronization of their clocks. A reference frame is a coordinate grid erected around some given origin and a set of clocks (see Fig. 36.4), which can be used to determine the space and time coordinates of any **event**. In relativity, as in everyday life, an event is an occurrence that happens at one point of space at one point of time (for example, a climber stepping onto the summit of Mt. McKinley). An event is therefore represented by one point of space and time. The space coordinates of the event are directly given by the coordinate grid intersections at the event. The time coordinate of the event is the time registered by the clock at the event. Different choices of reference frame result in different space coordinates and different time coordinates for the event, and in Section 36.5 we will see how these different coordinates in different reference frames are related.

event

The clocks of any chosen reference frame must be synchronized with each other and with the master clock sitting at its origin of coordinates. Einstein proposed that this synchronization can be accomplished by sending out a flash of light from a point exactly midway between the clock at the origin and the other clock (see Fig. 36.5). The two clocks are synchronized if both show exactly the same time when the light from the midpoint reaches them. Note that this synchronization procedure hinges on the universality of the speed of light. If the speed of light were not a universal constant, but were dependent on the reference frame and on the direction of propagation (say, faster toward the right in Fig. 36.5 and slower toward the left), then we could not achieve synchronization by the simple procedure with a flash of light from the midpoint.

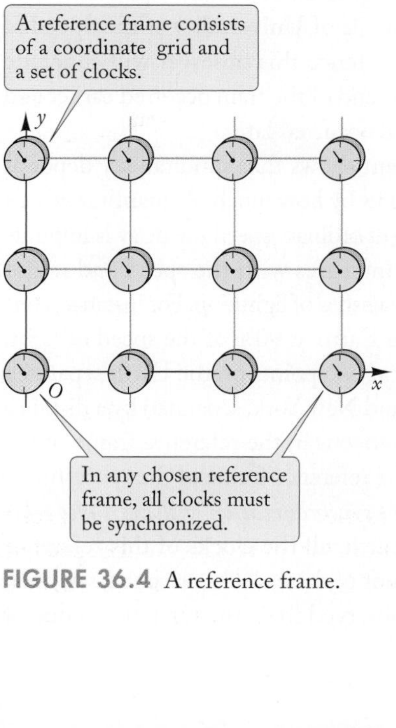

FIGURE 36.4 A reference frame.

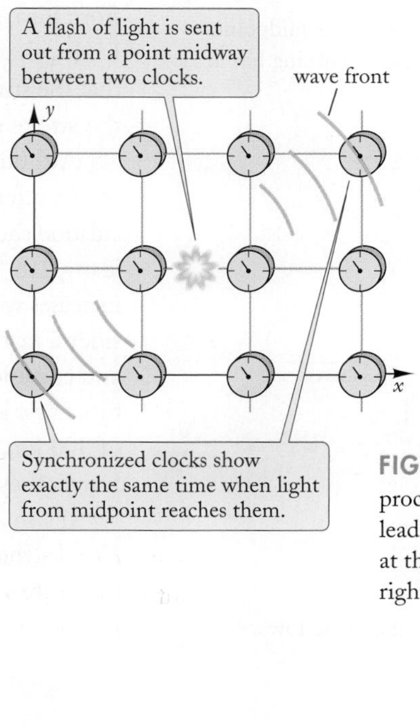

FIGURE 36.5 Synchronization procedure for a pair of clocks. The leading wave front reaches the clocks at the lower left corner and the upper right corner simultaneously.

One immediate consequence of our synchronization procedure is that *simultaneity is relative, that is, the simultaneity of two events depends on the reference frame.* Einstein illustrated this with the following concrete example. Suppose that a train is traveling at high speed along a straight track and two bolts of lightning strike the front end and the rear end of the train, leaving scorch marks on the train and on the ground (the existence of these marks helps us to remember exactly where the lightning struck in each reference frame). Suppose that these two strokes of lightning are exactly simultaneous in the reference frame of the Earth. Then in the reference frame of the train, the two strokes of lightning will not be simultaneous—as judged by the clocks on the train, the stroke at the front end of the train occurs slightly earlier than the stroke at the rear end.

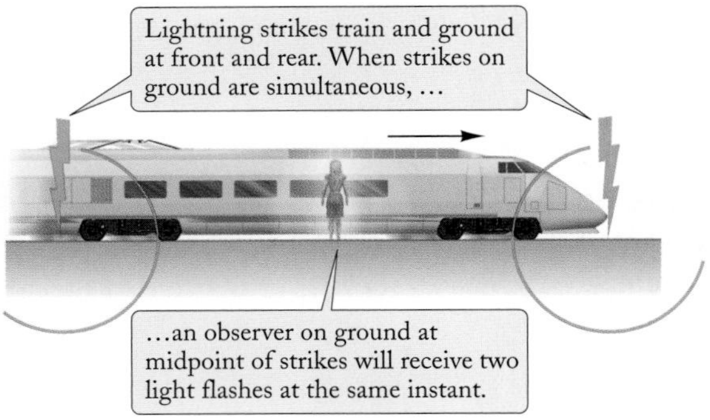

FIGURE 36.6 Observer on the ground at the midpoint between the scorch marks on the track watches for the arrival of flashes of light.

FIGURE 36.7 Observer in the train at the midpoint between the front and rear ends, where lightning has made scorch marks.

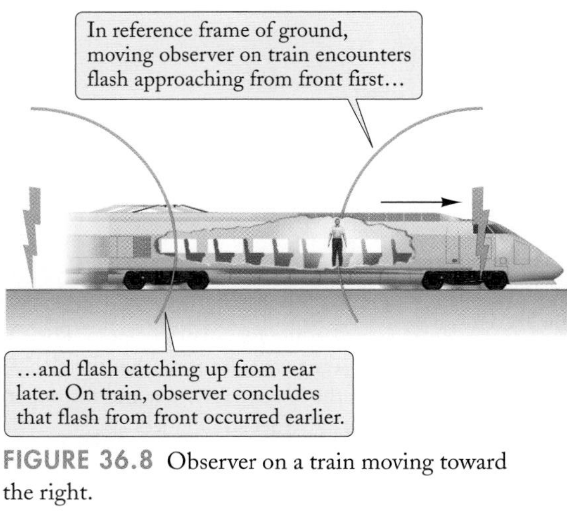

FIGURE 36.8 Observer on a train moving toward the right.

To see how this difference between the two reference frames comes about, let us apply our procedure for testing simultaneity. Suppose that in the reference frame of the Earth, an observer stands near the track at the midpoint between the two scorch marks the lightning made on the ground (see Fig. 36.6); she will then receive flashes of light from the lightning at her left and her right at the same instant. Thus, this observer will confirm that in the reference frame of the Earth, the lightning was simultaneous.

In the reference frame of the train, an observer can likewise test for simultaneity by placing himself exactly at the midpoint between the scorch marks the lightning made at the front and rear ends of the train (see Fig. 36.7) and waiting for the arrival of the flashes of light. Will he receive the flashes of light from the front and the rear of the train at the same instant? We can answer this by examining the motion of this observer and the propagation of the flashes of light in the reference frame of the ground (see Fig. 36.8). This observer is traveling toward the flash of light approaching him from the front end of the train, and he is traveling away from the flash of light trying to catch up with him from the back end of the train. Thus, this observer will encounter the flash of light from the front end before the flash of light from the rear end can catch up. In the reference frame of the ground, this delay between the flashes of light seen by the observer on the train is attributed to his motion toward one flash and away from the other. But in the reference frame of the train, the observer cannot attribute the delay to such a difference in motion—the light flashes from the front and the rear of the train originated at exactly equal distances from him, and, according to the Principle of Universality of the Speed of Light, they traveled at equal speeds. Hence this observer will conclude that the stroke of lightning at the front end of the train occurred earlier and the stroke of lightning at the rear end occurred later.

Although this qualitative argument shows that simultaneity depends on the reference frame, it does not tell us by how much. A quantitative calculation shows that for a train traveling at ordinary speed the delay is insignificant, 10^{-13} s or less. But the delay increases with the speed and it also increases with the distance between the flashes of lightning. For instance, consider a fast spaceship traveling by the Earth at 90% of the speed of light, and consider two flashes of lightning at two points on the Earth separated by a fairly large distance, say, Boston and New York, separated by a distance of 300 km. If these flashes are simultaneous in the reference frame of the Earth, they will differ by 0.001 s in the reference frame of the spaceship.

If simultaneity is relative, then *the synchronization of clocks is also relative.* In the reference frame of the Earth, all the clocks of this reference frame are synchronized, that is, the hands of these clocks reach the, say, noon position simultaneously. But when observed from the reference frame of

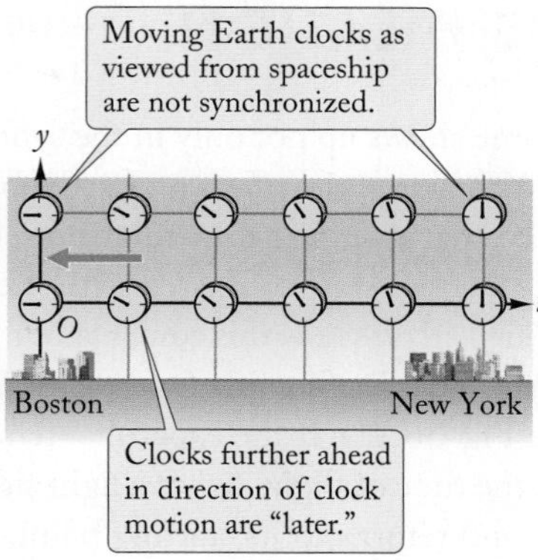

FIGURE 36.9 Clocks of the reference frame of the Earth as observed at one instant of spaceship time.

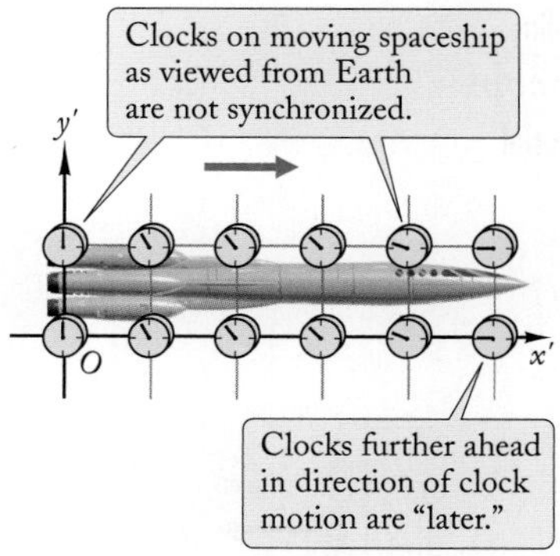

FIGURE 36.10 Clocks of the reference frame of the spaceship as observed at one instant of Earth time.

the train or the spaceship, the clocks in the Earth's reference frame that are further ahead in the direction of clock motion show earlier times. Consequently, they reach the noon position later than those on the right side—they are "late" in the same way as the lightning of the left side is late. This can be seen in Fig. 36.9, which shows the clocks belonging to the reference frame of the Earth as observed at one instant of time from a fast spaceship traveling in the direction from Boston to New York.

The effect is symmetric. In the reference frame of the spaceship, all the clocks onboard are synchronized. But, as observed from the reference frame of the Earth, the clocks on the front part of the spaceship are late. Figure 36.10 shows the clocks belonging to the reference frame of the spaceship as observed at one instant from the Earth. Note that in Fig. 36.9 we are viewing the reference frame of the Earth moving past the spaceship, and in Fig. 36.10 we are viewing the reference frame of the spaceship moving past the Earth. In either case, *the clocks on the leading edge of the reference frame are late.*

The relativity of synchronization is a direct consequence of the universality of the speed of light, since our procedure for testing simultaneity depends crucially on the speed of light. The failure of an absolute synchronization valid for all reference frames implies that there exists no absolute time. Each reference frame has its own way of reckoning time—*time is relative.* Instead of the single absolute time coordinate t we used in Newtonian physics, we now have to use a separate time coordinate for each individual reference frame.

✔ Checkup 36.2

QUESTION 1: A spaceship approaches the Earth at 1.00×10^8 m/s and sends a light signal toward the Earth, as in Fig. 36.1. According to Einstein, what is the speed of this light signal relative to the Earth?

QUESTION 2: Figure 36.10 shows the clocks of the reference frame of a spaceship in motion relative to the Earth. The clocks at the front of the spaceship are late. Does this mean that the crew of the spaceship find that their clocks are out of synchronization?

QUESTION 3: An earthquake occurs in San Francisco, and simultaneously (in Earth time) another earthquake occurs in New York. These earthquakes are not simultaneous as seen in the reference frame of a fast spaceship traveling westward in the direction from New York to San Francisco. Which is late?

QUESTION 4: A satellite, moving away from you at speed v, emits a pulse of radio waves in your direction. The speed of the waves relative to you is

(A) $c + v$ (B) $c - v$ (C) $\sqrt{c^2 - v^2}$ (D) c (E) v

Online Concept Tutorial 42

36.3 TIME DILATION

The relativity of time shows up not only in the synchronization of clocks, but also in the rate of clocks. When experimenters on Earth observe a clock onboard the moving spaceship, they find that it suffers **time dilation** relative to the clocks on Earth: the rate of the moving clock is slow compared with the rate of identically manufactured clocks at rest on Earth. To see how this comes about, imagine that experimenters in the spaceship set up a racetrack of length L perpendicular to the direction of motion of the spaceship (see Fig. 36.11). If the experimenters in the spaceship use one of their clocks to measure the time of flight $\Delta t'$ of a light signal that goes from one end of the track to the other and returns to its starting point, they will then find that the light signal takes a time of $\Delta t' = 2L/c$ to complete the round trip.

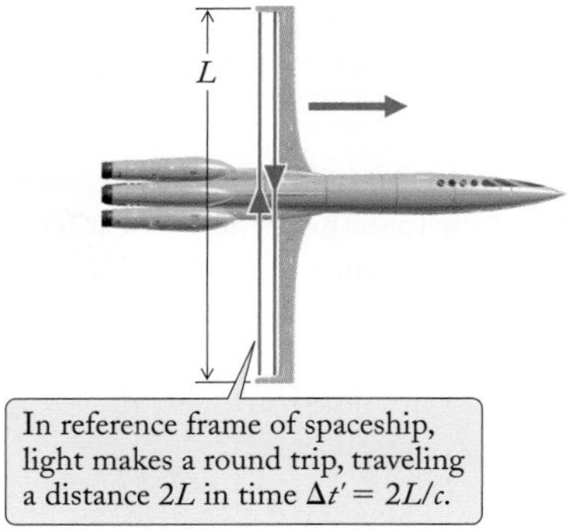

FIGURE 36.11 Spaceship with a "racetrack" for a light pulse.

But experimenters on the Earth see that the light signal has concurrent vertical and horizontal motions (see Fig. 36.12). For the experimenters on Earth, the light signal travels a total distance *larger* than $2L$. Since the speed must still be the standard speed of light c, they will find that according to their clocks the light signal now takes a time Δt *longer* than $2L/c$ to complete the trip. Thus, a given time interval $\Delta t'$ registered by a clock on the spaceship is registered as a longer time interval Δt by the clocks on the Earth. This means that the clock on the spaceship runs slow when judged by the clocks on the Earth. Note that the experiment involves *one* clock on the spaceship (the clock at the starting point of the track), but several clocks on the Earth, because the light signal does not return to the point at which it started on Earth, and the observers on Earth will have to use one (stationary) clock at the starting point and another (stationary) clock at the end point to measure the time of flight.

For a quantitative evaluation of the time dilation, we note that in Fig. 36.12 the upward portion of the path of the light signal is the hypotenuse of a right triangle of sides L and $V\Delta t/2$, where V is the speed of the spaceship relative to the Earth. The total length of the path that the light signal has to cover in the reference frame of the Earth is therefore

$$2 \times \sqrt{L^2 + (V\Delta t/2)^2} \tag{36.1}$$

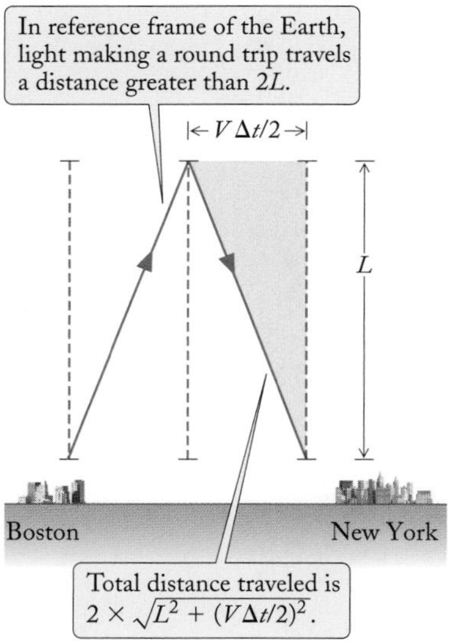

FIGURE 36.12 The trajectory of the light pulse as observed from the Earth. The light pulse has both a vertical motion (up or down) and also a horizontal motion (toward the right).

The time taken to cover this distance is

$$\Delta t = \frac{2 \times \sqrt{L^2 + (V\Delta t/2)^2}}{c} \tag{36.2}$$

If we square both sides of this equation, we obtain

$$(\Delta t)^2 = \frac{4L^2 + V^2(\Delta t)^2}{c^2} \tag{36.3}$$

which we can solve for $(\Delta t)^2$ and then for Δt:

$$(\Delta t)^2 = \frac{4L^2/c^2}{1 - V^2/c^2}$$

and

$$\Delta t = \frac{2L/c}{\sqrt{1 - V^2/c^2}} \tag{36.4}$$

Since, in the reference frame of the spaceship, $2L/c = \Delta t'$, this gives us

$$\Delta t = \frac{\Delta t'}{\sqrt{1 - V^2/c^2}} \quad \begin{cases} \text{clock at rest in spaceship registers } \Delta t' \\ \text{clocks in Earth reference frame measure } \Delta t \end{cases} \tag{36.5}$$

time dilation

This is the time-dilation formula. It shows that the time registered by the clocks on the Earth is longer than the time registered by the clock on the spaceship by a factor of $1/\sqrt{1 - V^2/c^2}$, that is, the clock on the spaceship runs slow when measured with the clocks on the Earth. Figure 36.13 is a plot of the time-dilation factor $1/\sqrt{1 - V^2/c^2}$ for speeds in the range from 0 to c. At low speeds the time-dilation effect is insignificant, but at speeds near c, it becomes quite large.

The slowing down of the rate of lapse of time applies to all physical processes—atomic, nuclear, biological, etc. Thus, the astronauts on the spaceship will perform all their tasks in "slow motion," and they will age slower than normal when measured with the clocks on the Earth. But they themselves will be unaware of this. From their point of view, they are in an inertial reference frame in which the usual laws of physics are valid, and the physical processes in their reference frame proceed at the normal rate, without any indication of anything unusual.

The time-dilation effect is symmetric: as measured by the clocks on the spaceship, a clock on the Earth runs slow by the same factor:

$$\Delta t' = \frac{\Delta t}{\sqrt{1 - V^2/c^2}} \quad \begin{cases} \text{clock at rest on Earth registers } \Delta t \\ \text{clocks in spaceship reference frame measure } \Delta t' \end{cases} \tag{36.6}$$

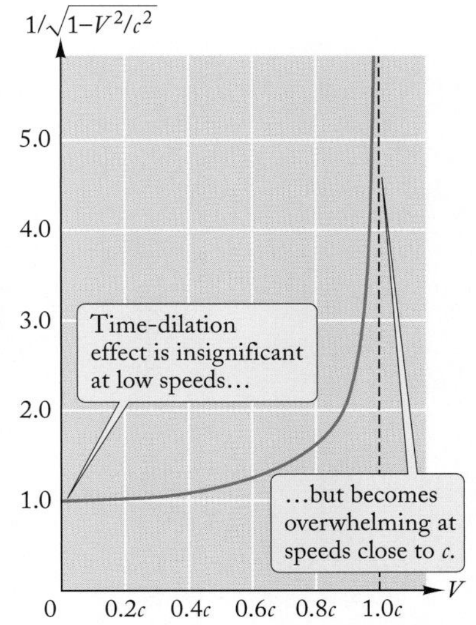

FIGURE 36.13 Time-dilation factor as a function of speed V.

The derivation of Eq. (36.6) can be based on an argument similar to that given above, with a racetrack for light at rest on the Earth. Incidentally: in these arguments we have implicitly assumed that the length of the racetrack is unaffected by the motion of the spaceship or the Earth, that is, we have assumed that the length is absolute. As we will see in the next section, this is true for lengths perpendicular to the direction of motion, although it is not true for lengths along the direction of motion.

EXAMPLE 1 Very drastic time-dilation effects have been observed in the decay of short-lived elementary particles. For instance, a muon particle (see Chapter 41) usually decays in about 2.2×10^{-6} s; but if it is moving at high speed through the laboratory, then the internal processes that produce the decay will slow down as judged by the clocks in our laboratory, and the muon lives a longer time. In accurate experiments performed at the European Organization for Nuclear Research (CERN) accelerator near Geneva, muons with a speed of 99.94% of the speed of light were found to have a lifetime 29 times as long as the lifetime of muons at rest. Is this dilation of the lifetime in agreement with Eq. (36.5)?

SOLUTION: We can regard the muon as a clock at rest in the reference frame of an (imaginary) spaceship with a speed of $V = 0.9994c$. Over the lifetime of the muon, this clock registers a time interval of $\Delta t' = 2.2 \times 10^{-6}$ s. However, the experimenters in the laboratory see the spaceship moving at $V = 0.9994c$, and over the lifetime of the muon they see their own clocks register a larger time Δt than the time $\Delta t'$ registered by the moving clock. According to Eq. (36.5), the times Δt and $\Delta t'$ differ by the time-dilation factor $1/\sqrt{1 - V^2/c^2}$, that is,

$$\frac{1}{\sqrt{1 - V^2/c^2}} = \frac{1}{\sqrt{1 - (0.9994c)^2/c^2}} = \frac{1}{\sqrt{1 - (0.9994)^2}} = \frac{1}{0.035} = 29$$

This time-dilation factor is in agreement with the experimental result.

At everyday speeds, the time-dilation effect is extremely small. For example, consider a clock aboard an airplane traveling at 300 m/s over the ground. The time-dilation factor is then

$$\frac{1}{\sqrt{1 - V^2/c^2}} = \frac{1}{\sqrt{1 - (300)^2/(3.00 \times 10^8)^2}} = \frac{1}{\sqrt{1 - 10^{-12}}} \tag{36.7}$$

Evaluation of this gives 1.000 000 000 000 5, which means that a clock in the airplane will slow down by only 5 parts in 10^{13}! However, detection of such a small change is not beyond the reach of modern atomic clocks. In a notable experiment, scientists from the National Institute of Standards and Technology placed portable atomic clocks aboard a commercial airliner and kept them flying for several days, making a complete round trip around the world. Before and after the trip, the clocks were compared with an identical clock that was kept on the ground. The flying clocks were found to have lost time—in one instance, the total time lost because of the motion of the clock was about 10^{-7} s.

twin paradox

The time-dilation effect leads to the famous **twin paradox**, which we can state as follows: a pair of identical twins, Terra and Stella, celebrate their, say, twentieth birthday on Earth. Then Stella boards a spaceship that carries her at a speed of $V = 0.99c$ to Proxima Centauri, 4 light-years away; the spaceship immediately turns around and brings Stella back to Earth. According to the clocks on Earth, this trip takes about 4 years each way, so Terra's age will be 28 years when the twins meet again. But Stella has benefited from time dilation—relative to the reference frame of the Earth, the spaceship clocks run slow by a factor

$$\frac{1}{\sqrt{1 - V^2/c^2}} = \frac{1}{\sqrt{1 - (0.99)^2}} = \frac{1}{0.14} \tag{36.8}$$

Hence 8 years of travel registered by the Earth clocks amount to only $8 \times 0.14 = 1.1$ years according to the spaceship clocks, and Stella's biological age on return will be only 21 years. Stella will be younger than Terra.

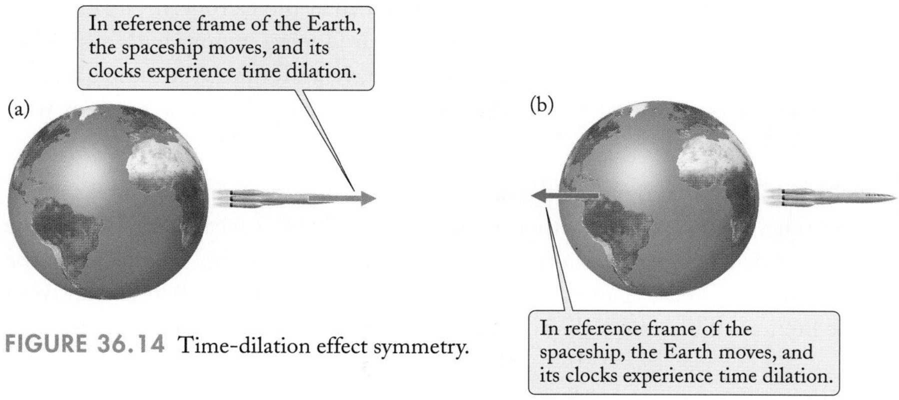

FIGURE 36.14 Time-dilation effect symmetry.

The paradox arises when we examine the elapsed times from the point of view of the reference frame of the spaceship. In this reference frame, the Earth is moving away from the spacecraft (see Fig. 36.14). Hence in this reference frame, the Earth clocks run slow—and Terra should be younger than Stella.

The resolution of this paradox hinges on the fact that our time-dilation formula is valid only if the time of a moving clock is measured from the point of view of an *inertial* reference frame. The reference frame of the Earth is (approximately) inertial, and therefore our calculation of the time dilation of the spaceship clocks is valid. But the reference frame of the spaceship is not inertial—the spaceship must decelerate when it reaches Proxima Centauri, stop, and then accelerate toward the Earth. If the reference frame is not inertial, the Principle of Relativity does not apply. Therefore, we cannot use the simple time-dilation formula to find the time dilation of the Earth clocks from the point of view of the spaceship reference frame. The "paradox" results from the misuse of this formula.

A detailed analysis of the behavior of the Earth clocks from the point of view of the spaceship reference frame establishes that the Earth clocks indeed do also run slow as long as the spaceship is moving with uniform velocity, but that the Earth clocks run *fast* when the spaceship is undergoing its acceleration to turn around at Proxima Centauri. The time that the Earth clocks gain during the accelerated portions of the trip more than compensates for the time they lose during the other portions of the trip. This confirms that Stella will be younger than Terra, even from the point of view of the spaceship reference frame.

EXAMPLE 2 The speed of a GPS satellite relative to a point on the Earth's surface is typically $V = 3.9 \times 10^3$ m/s. Assume that the clock on a GPS satellite is synchronized with the clocks of the Earth's reference frame at one instant. By how much do they differ 1.0 hour later? What is the corresponding distance error for a radio signal? In this calculation, ignore the rotation of the Earth and treat the satellite motion the same as motion with uniform velocity.

Concepts in Context

SOLUTION: A time interval Δt measured on Earth will differ from the interval $\Delta t'$ measured on the GPS satellite by the time-dilation factor:

$$\Delta t = \frac{\Delta t'}{\sqrt{1 - V^2/c^2}}$$

We wish to find the difference $\Delta t - \Delta t'$ when $\Delta t' = 1.0$ hour. Since the speed of the satellite is much less than the speed of light, we can simplify the time-dilation factor using the expansion $(1 + x)^n \approx 1 + nx$ for small x. Here, with $n = -\frac{1}{2}$ and $x = -V^2/c^2$, we have

$$\frac{1}{\sqrt{1 - V^2/c^2}} = \left(1 - \frac{V^2}{c^2}\right)^{-1/2} \approx 1 + \frac{1}{2}\frac{V^2}{c^2} \tag{36.9}$$

Inserting this into the time-dilation expression, we have

$$\Delta t = \Delta t' \times \left(1 + \frac{1}{2}\frac{V^2}{c^2}\right)$$

or

$$\Delta t - \Delta t' = \Delta t' \times \frac{1}{2}\frac{V^2}{c^2} \tag{36.10}$$

With $\Delta t' = 1.0 \text{ h} = 3600 \text{ s}$ and $V = 3.9 \times 10^3$ m/s, we find that the clocks will differ by

$$\Delta t - \Delta t' = 3600 \text{ s} \times \frac{1}{2} \times \left(\frac{3.9 \times 10^3 \text{ m/s}}{3.00 \times 10^8 \text{ m/s}}\right)^2 = 3.0 \times 10^{-7} \text{ s}$$

If such a timing error were uncorrected and the satellite clock were used to determine the distance of a point on the Earth by light travel time, then the error in this distance would be

$$\Delta x = c \times (\Delta t - \Delta t') = 3.00 \times 10^8 \text{ m/s} \times 3.0 \times 10^{-7} \text{ s} = 90 \text{ m} \quad (36.11)$$

Since the accuracy of GPS positioning is required to be much better than that, corrections for the time dilation must be applied, in addition to many more mundane corrections, for example, those due to refraction and the decrease in the speed of radio waves in the atmosphere.

Finally, we point out that the time dilation of relativistic physics affects the **Doppler shift** of the frequency of light waves whenever the emitter is in motion relative to the receiver. Since we want to find the frequency detected by the *receiver*, we consider the reference frame of the receiver, and we pretend that this reference frame serves as the "medium" in which the light propagates. In Newtonian physics the Doppler shift for an emitter moving at speed V through a medium in which the emitted wave has a speed c is simply $f' = f/(1 \pm V/c)$, where f is the frequency radiated by the emitter, f' is the frequency detected by the receiver, and the positive sign applies if the emitter is receding, the negative sign if approaching [see the derivation of Eq. (17.17) for the case of sound waves]. In relativistic physics, the Doppler shift must also include the time dilation of the emitter. The time dilation of the period of the waves corresponds to a *reduction* in the detected frequency by the additional factor $\sqrt{1 - V^2/c^2}$. Including this factor, we have

$$f' = \frac{\sqrt{1 - V^2/c^2}}{1 \pm V/c} f \quad (36.12)$$

where now V is to be interpreted as the speed of the emitter relative to the receiver. Since $1 - V^2/c^2 = (1 - V/c)(1 + V/c)$, Eq. (36.12) can be simplified by appropriate cancellations:

relativistic Doppler shift

$$f' = \sqrt{\frac{1 - V/c}{1 + V/c}}\, f \quad \text{for receding emitter}$$

$$f' = \sqrt{\frac{1 + V/c}{1 - V/c}}\, f \quad \text{for approaching emitter} \quad (36.13)$$

Concepts in Context

EXAMPLE 3 A GPS satellite transmits radio (microwave) signals at a frequency of 1.575 GHz. Assume for simplicity that a GPS satellite is directly approaching your location on the Earth's surface with a relative speed of 3.9×10^3 m/s. By what factor is the frequency that you detect on Earth increased due to the ordinary Doppler shift of Newtonian physics? Due to the time dilation of relativistic physics?

SOLUTION: For an approaching emitter, the Newtonian result for the upward-shifted frequency that you receive is given solely by the denominator in Eq. (36.12) [see also Eq. (17.17)]:

$$f' = \frac{1}{1 - V/c} f$$

Taking the ratio of received to emitted frequencies and inserting the values, we obtain

$$\frac{f'}{f} = \frac{1}{1 - (3.9 \times 10^3 \text{ m/s})/(3.00 \times 10^8 \text{ m/s})} = 1.000\,013$$

This is an upward shift of 13 parts per million, or of about $(1.3 \times 10^{-5}) \times (1.575 \text{ GHz}) = 20$ kHz.

The time dilation factor is

$$\sqrt{1 - \frac{V^2}{c^2}} \approx 1 - \frac{1}{2}\frac{V^2}{c^2}$$

which shifts the received frequency *downward* by a factor that differs from unity by

$$\frac{1}{2}\frac{V^2}{c^2} = \frac{1}{2} \times \frac{(3.9 \times 10^3 \text{ m/s})^2}{(3.00 \times 10^8 \text{ m/s})^2} = 8.4 \times 10^{-11}$$

or less than a tenth of a part per *billion*. Thus for such a "low" speed, the time dilation contribution to the frequency shift is completely negligible compared with the ordinary Doppler effect. We already used this fact when we examined police radar guns in Example 17.7.

✔ Checkup 36.3

QUESTION 1: Distinguish between the relativity of the synchronization of clocks and the relativity of the rates of clocks.

QUESTION 2: Consider the plot of the time-dilation factor given in Fig. 36.13. If we increase the speed of a clock by a factor of 2, do we increase the time-dilation factor by a factor of 2, or by less, or by more?

QUESTION 3: An astronaut aims a beam of light from a green laser pointer at you. If you approach the astronaut at $V = 0.10c$, do you detect blue light or yellow light?

QUESTION 4: A spaceship moves away from Earth at high speed. How do experimenters on the Earth measure a clock in the spaceship to be running? How do those in the spaceship measure a clock on the Earth to be running?

(A) Slow; fast (B) Fast; slow (C) Slow; slow (D) Fast; fast

36.4 LENGTH CONTRACTION

In the preceding sections we have seen that time is relative—both the synchronization of clocks and the rate of clocks depend on the reference frame. Now we will see that length is also relative. A measuring rod, or any other body, onboard the spaceship suffers **length contraction** along the direction of motion. The length of the moving measuring rod will be short when compared with the length of an identically manufactured measuring rod at rest on the Earth. The reason for this is that the length measurement of a moving body depends on simultaneity, and since simultaneity is relative, so is length.

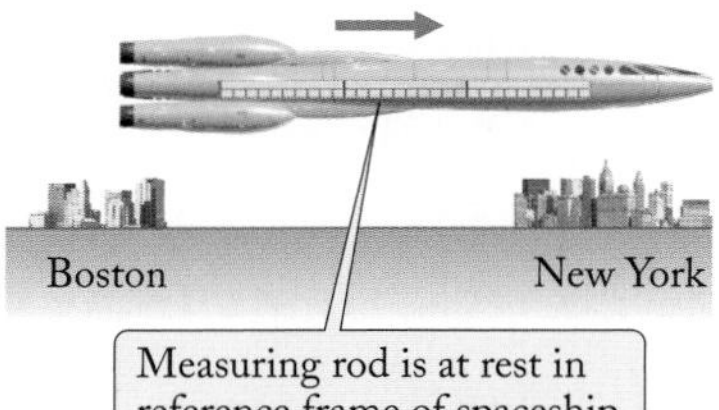

FIGURE 36.15 Spaceship with a measuring rod oriented along the direction of its motion.

Suppose that the spaceship, traveling from Boston toward New York, carries a measuring rod that has a length of, say, 300 km in the reference frame of the spaceship (see Fig. 36.15). To measure the length of this rod in the reference frame of the Earth, we station observers in the vicinity of New York and Boston with instructions to ascertain the positions of the front and the rear ends of the measuring rod at one instant of time, say, at noon.

But when the observers on the Earth do this, the observers on the spaceship will claim that the position measurements were not done simultaneously, and that the observers in the vicinity of Boston measured the position of the rear end at a later time. In the extra time, the rear end moves an extra distance to the right, and hence the distance between the positions measured for the rear and the front ends will be reduced. From the point of view of the observers on the spaceship, it is therefore immediately obvious that the length measured by the observers on the Earth will be *short.* Figures 36.16 and 36.17 show the reference frame of the spaceship moving past the Earth and the reference frame of the Earth moving past the spaceship, respectively. In these figures the length contraction has been included (it was left out in Figs. 36.9 and 36.10). As illustrated in these figures, the length-contraction effect is symmetric: a body at rest in the spaceship will suffer contraction when measured in the reference frame of the Earth, and a body at rest on the Earth will suffer contraction when measured in the reference frame of the spaceship.

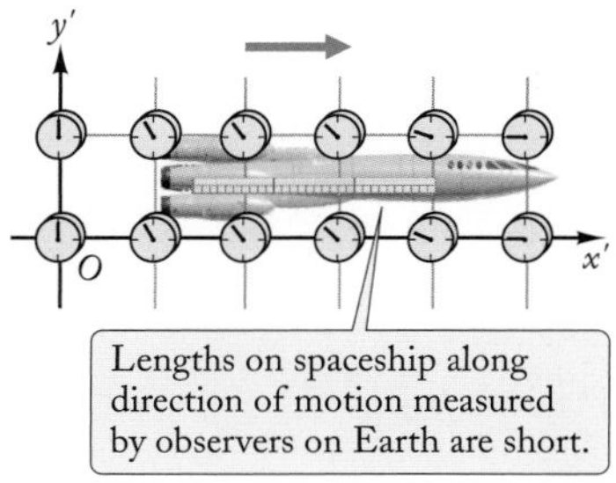

FIGURE 36.16 Reference frame of the spaceship as observed at one instant of Earth time (including length contraction; this length contraction was not included in Fig. 36.10).

We can obtain a formula for the length contraction by exploiting the formula for the time dilation. Consider a rod of length L' at rest in the spaceship. According to the spaceship clocks, an observer on the Earth takes a time $\Delta t' = L'/V$ to travel from one end of the rod to the other. Taking time dilation into account, we then conclude that the observer on the Earth will judge that this takes a shorter time of only

$$\Delta t = \sqrt{1 - \frac{V^2}{c^2}}\,\Delta t' = \sqrt{1 - \frac{V^2}{c^2}}\,\frac{L'}{V} \tag{36.14}$$

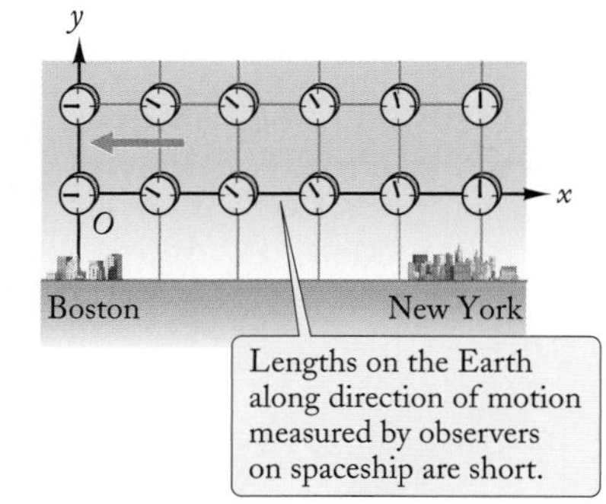

FIGURE 36.17 Reference frame of the Earth as observed at one instant of spaceship time (including length contraction; this length contraction was not included in Fig. 36.9).

However, the observer on the Earth cannot attribute this reduction of the travel time to a slowing of his clock—in his own reference frame his clock runs at a normal rate. Instead, he must attribute the reduction of travel time to a contraction of length of the rod. Measured in the reference frame of the Earth, the rod in the spaceship has some contracted length L, and it moves at speed V. Hence, the rod passes by a point on the Earth in a time L/V. This time must agree with the time Δt calculated in Eq. (36.14), so

$$\frac{L}{V} = \sqrt{1 - \frac{V^2}{c^2}}\,\frac{L'}{V} \tag{36.15}$$

or

$$L = \sqrt{1 - \frac{V^2}{c^2}}\, L' \quad \text{for a rod } L' \text{ at rest in spaceship} \qquad (36.16)$$

length contraction

This is the formula for length contraction. According to this formula, the length of a rod or any body in motion relative to a reference frame is shortened by a factor of $\sqrt{1 - V^2/c^2}$. Figure 36.18 is a plot of the length-contraction factor $\sqrt{1 - V^2/c^2}$ as a function of speed.

As already mentioned, this contraction effect is symetric: if the rod is at rest on the Earth and is measured in the reference frame of the spaceship, the formula for the length contraction is

$$L' = \sqrt{1 - \frac{V^2}{c^2}}\, L \quad \text{for a rod } L \text{ at rest on Earth} \qquad (36.17)$$

The length contraction has not been tested directly by experiment. There is no practical method for a high-precision measurement of the length of a fast-moving body. Our best bet might be high-speed photography, but this is nowhere near accurate enough, since the contraction is extremely small even at the highest speeds that we can impart to a macroscopic body. Note, however, that the experimental evidence for time dilation can be regarded as indirect evidence for length contraction, since, as we saw above, the former implies the latter.

The contraction effect applies only to lengths along the direction of motion of the body. Lengths perpendicular to the direction of motion are unaffected. The proof of this is by contradiction: imagine that we have two identically manufactured pieces of pipe, one at rest on the Earth, one at rest on the spaceship (see Fig. 36.19). If the motion of the spaceship relative to Earth were to bring about a transverse contraction of the spaceship pipe, then, by symmetry, the motion of the Earth relative to the spaceship would bring about a contraction of the Earth pipe. These contraction effects are contradictory, since in one case the spaceship pipe would fit inside the Earth pipe, and in the other case it would fit outside.

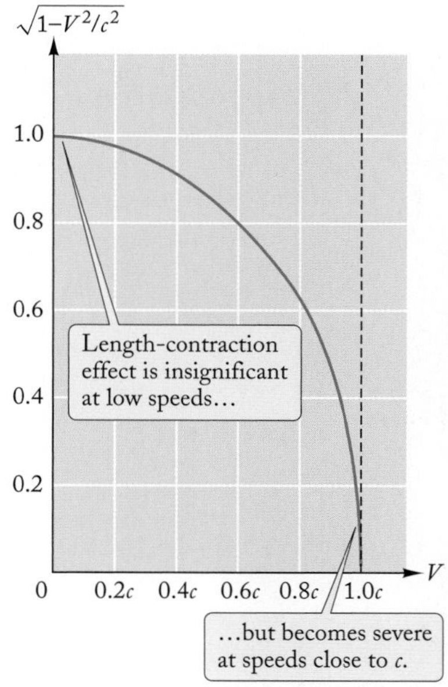

FIGURE 36.18 Length-contraction factor as a function of speed V.

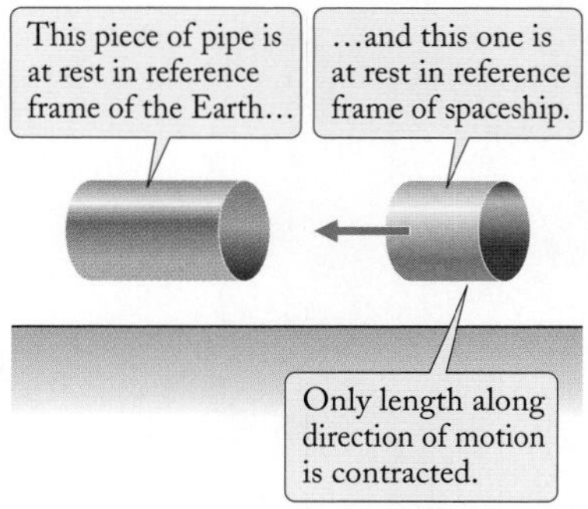

FIGURE 36.19 Two identical pieces of pipe.

EXAMPLE 4 A proton is passing by the Earth at a speed of $0.50c$. In the reference frame of the proton, what is the length of the diameter of the Earth in a direction parallel to that of the motion of the proton? By how much is this shorter than the diameter in the reference frame of the Earth?

SOLUTION: We can regard the reference frame of the proton as the reference frame of a spaceship, and we can regard the diameter of the Earth as a rod at rest in the reference frame of the Earth. Relative to the proton, or the spaceship, the Earth has a speed $V = 0.50c$. In the reference frame of the Earth, the diameter has the familiar value $L = 1.3 \times 10^7$ m. But in the reference frame of the proton, or the spaceship, the Earth has a speed of $V = 0.50c$, and the rod is observed to have a shorter length L'. According to Eq. (36.16),the lengths L and L' differ by the length contraction factor $\sqrt{1 - V^2/c^2}$, that is,

$$\sqrt{1 - V^2/c^2} = \sqrt{1 - (0.50c)^2/c^2} = \sqrt{1 - 0.25} = 0.87$$

Since the diameter of the Earth in its own reference frame is $L = 1.3 \times 10^7$ m, the length L' in the reference frame of the proton is

$$L' = 0.87 \times L = 0.87 \times 1.3 \times 10^7 \text{ m} = 1.1 \times 10^7 \text{ m}$$

This is shorter than 1.3×10^7 m by 2×10^6 m, or 2000 km!

COMMENTS: The dimensions of the Earth perpendicular to the direction of motion do not contract. This implies that, in the reference frame of the proton, the Earth is not a sphere, but an ellipsoid.

From the length contraction of a three-dimensional body we can deduce the volume contraction. The volume of the Earth, which is calculated by taking a product of the dimension parallel to the motion and the two dimensions perpendicular to the motion, will be contracted by just one factor of $\sqrt{1 - V^2/c^2}$, that is, a factor of 0.87 in the case of Example 4.

✔ Checkup 36.4

QUESTION 1: A cannonball is perfectly round in its own reference frame. Describe the shape of the cannonball in a reference frame relative to which it is moving at a high speed, say, $0.5c$.

QUESTION 2: Could we use the argument based on the two identical pieces of pipe (see Fig. 36.19) to prove that lengths *along* the direction of motion are not affected? Why, or Why not?

QUESTION 3: In view of the length contraction, does the density of a material depend on its speed?

QUESTION 4: A track is 100 m long. A particle moves parallel to the track with speed V such that $1 - V^2/c^2 = 0.25$. In the reference frame of the particle, the length of the track is

(A) 25 m (B) 50 m (C) 100 m (D) 200 m (E) 400 m

36.5 THE LORENTZ TRANSFORMATIONS AND THE COMBINATION OF VELOCITIES

Suppose that we measure the space and time coordinates of an event—such as the impact of lightning on a point on the ground—in the reference frame of the Earth and also in the reference frame of a moving spaceship. We will then obtain different values of these coordinates in the Earth and in the spaceship reference frames, but these different values of the coordinates are related by transformation formulas. In Einstein's physics, the transformation formulas for the coordinates are fairly complicated, because they are designed so as to keep the speed of light the same in all reference frames, and they incorporate the length contraction and the time dilation. Before dealing with these complicated formulas, let us examine the much simpler transformation formulas for coordinates in Newton's physics, where there is no length contraction and no time dilation.

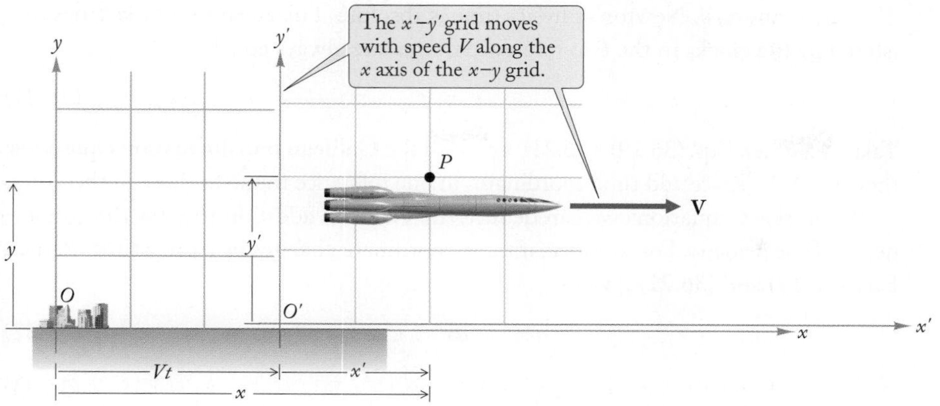

FIGURE 36.20 The coordinate grid x–y belonging to the reference frame of the Earth and the coordinate grid $x'-y'$ belonging to the reference frame of the spaceship. The coordinates of the point P are x, y in the first grid and x', y' in the second grid.

Figure 36.20 shows the coordinate grids x–y and $x'-y'$ of the first (Earth) and the second (spaceship) reference frames. The second reference frame is moving with velocity V along the x axis of the first reference frame. We assume that at time $t = 0$, the origins of the two coordinate grids coincide; at time t, the distance between the origins is then Vt. The coordinates of the point P are x, y in the first reference frame and x', y' in the second reference frame. By inspection of Fig. 36.20, we see that the distance x equals the sum of the distances x' and Vt:

$$x = x' + Vt \tag{36.18}$$

Hence

$$x' = x - Vt \tag{36.19}$$

Furthermore, the distance y equals the distance y'

$$y' = y \tag{36.20}$$

Equations (36.19) and (36.20) are the transformation equations for the x and y coordinates in Newton's physics. These two equations are merely the x and y components of the general vector equation $\mathbf{r}' = \mathbf{r} - \mathbf{V}t$ for the transformation of the position vector we found in Chapter 4 [see Eq. (4.50)]. We could have obtained our equations for the transformation of the x and y coordinates from the general vector equation; but it is just as easy to re-derive these results by inspection of Fig. 36.20. Note that although Fig. 36.20 makes the equations for the transformation of the x and y coordinates seem self-evident, these equations hinge on the absolute character of length in Newton's physics. Absolute length means that the observers in the two reference frames agree on the measurement of any length or any distance between two points. If the observers disagreed on the values of the distances x or x'—for example, if one observer claimed that the distance x was 3.0 m and the other observer claimed that this distance x was contracted to 2.5 m—then Eq. (36.18) would not be valid. The left side of Eq. (36.18) is a distance defined at one instant in the first reference frame, whereas the right side is a sum of a distance x' defined at one instant in the second reference frame and a distance (Vt) defined at one instant in the first reference frame, and such a sum makes no sense unless the observers agree on the values of these distances.

Furthermore, in Newton's physics time is absolute. This means that the times registered by the clocks in the two reference frames are always equal,

$$t' = t \tag{36.21}$$

Galilean transformation equations

Taken together, Eqs. (36.19)–(36.21) are called the **Galilean transformation equations**; they relate the space and time coordinates in one reference frame to those in the other.

From these equations we can deduce the Galilean addition rule for the components of the velocity. For instance, if the x coordinate changes by dx in a time dt, then Eqs. (36.19) and (36.21) give us

$$dx' = dx - V\,dt \tag{36.22}$$

$$dt' = dt \tag{36.23}$$

Dividing these two equations side by side, we obtain

$$\frac{dx'}{dt'} = \frac{dx}{dt} - V \tag{36.24}$$

Here, dx'/dt' is the x velocity of the particle, light signal, or whatever, measured in the second reference frame; and dx/dt is the x velocity measured in the first reference frame. Hence Eq. (36.24) says

$$v'_x = v_x - V \tag{36.25}$$

This, of course, is simply the Galilean addition rule for the x component of the velocity [see Eq. (4.53)].

In Einstein's physics, the Galilean formulas for the transformation of the coordinates and for the addition of velocities must be replaced by more complicated formulas, designed in such a way as to keep the speed of light the same in all reference frames. The transformation equations that accomplish this trick are called the **Lorentz transformations**. If the new reference frame moves, again, with velocity V along the x axis of the first reference frame, and if the origins coincide at time $t = 0$, the Lorentz transformations take the form

Lorentz transformations

$$x' = \frac{x - Vt}{\sqrt{1 - V^2/c^2}} \tag{36.26}$$

$$y' = y \tag{36.27}$$

$$t' = \frac{t - Vx/c^2}{\sqrt{1 - V^2/c^2}} \tag{36.28}$$

These equations cannot be obtained by simple inspection of Fig. 36.20, because the distances displayed in this figure are not absolute in Einstein's physics, and they cannot simply be added as in Newton's physics.

We will not bother with a formal derivation of the Lorentz transformation equations, because we can achieve a clear understanding of the various terms and factors in these equations by comparing them with the Galilean transformation equations. Equation (36.26) differs from the Galilean equation (36.19) only by the factor $1/\sqrt{1 - V^2/c^2}$; this factor represents the length contraction. Equation (36.27) is identical to the Galilean equation, because lengths perpendicular to the direction of motion remain unchanged. And Eq. (36.28) differs from the Galilean equation in two ways: it contains an extra factor $1/\sqrt{1 - V^2/c^2}$ representing the time dilation, and it contains an extra term $-(Vx/c^2)/\sqrt{1 - V^2/c^2}$ representing the relativity of synchronization discussed in Section 36.2. We already gave quantitative treatments of the time dilation and the length contraction, and we therefore do not need to reexamine these here. But we did not yet give a quantitative treatment of the relativity of synchronization, and we will now examine this, to justify the presence of the extra term in Eq. (36.28).

Suppose that observers in the Earth frame send a light signal in the positive x direction from the origin O (with coordinate $x = 0$) to some point P (with coordinate $x > 0$). The signal leaves the origin O at time $t = 0$ and arrives at the point P at time t. According to the observers in the Earth frame, the arrival time is $t = x/c$, since the light signal travels at speed c. We want to know the arrival time t', as seen by the observers in the spaceship frame. For these observers, the light signal moves in the positive x' direction at speed c, while simultaneously the Earth frame and the point P move in the negative x' direction at speed V. Hence the "closing speed" of the light signal and the point P is $c + V$ (this closing speed is larger than c, but that is not objectionable; it merely reflects the fact that if the target and light signal are both moving toward each other, they will meet sooner than if the target is at rest). To calculate the time of arrival of the light signal at the point P, the spaceship observers have to divide the length OP by the closing speed $c + V$. But since the length OP is a moving length, these observers must take the length contraction into account: the length between O and P is not x, but is $x\sqrt{1 - V^2/c^2}$. Accordingly, the observers in the spaceship frame find that the arrival time of the light signal is

$$t' = \frac{x\sqrt{1 - V^2/c^2}}{c + V}$$

Before we compare this with the value of t' given by the Lorentz transformation (36.28), let us multiply the numerator and denominator by $\sqrt{1 - V^2/c^2}$ and rearrange the result:

$$t' = \frac{x\sqrt{1 - V^2/c^2}}{c + V} \times \frac{\sqrt{1 - V^2/c^2}}{\sqrt{1 - V^2/c^2}} = \frac{x(1 - V^2/c^2)}{c(1 + V/c)\sqrt{1 - V^2/c^2}}$$

$$= \frac{x\,(1 + V/c)(1 - V/c)}{c\,(1 + V/c)\sqrt{1 - V^2/c^2}}$$

$$= \frac{x/c - Vx/c^2}{\sqrt{1 - V^2/c^2}} \qquad (36.29)$$

Here, the term x/c in the numerator is simply the time t that the light signal takes to arrive according to observers in the Earth frame. The term $-Vx/c^2$ in the numerator represents the relativity of synchronization. Comparing Eqs. (36.28) and (36.29), (with $x/c = t$), we see they agree exactly—and this agreement provides the justification of the extra term in Eq. (36.28).

HENDRIK ANTOON LORENTZ (1853–1928) *Dutch theoretical physicist, professor at Leiden. He investigated the relationship between electricity, magnetism, and mechanics. In order to explain the observed effect of magnetic fields on emitters of light (Zeeman effect), he postulated the existence of electric charges in the atom, for which he was awarded the Nobel Prize in 1902. He derived the Lorentz transformation equations by examining Maxwell's equations, but he was not aware that this leads to a new concept of space and time.*

EXAMPLE 5 A measuring rod of length $\Delta x'$ is at rest along the x' axis of the spaceship frame, which is moving in the positive x' direction with speed V relative to the Earth frame. What is the length of the measuring rod in the Earth frame according to the Lorentz transformation equations?

SOLUTION: We begin with Eq. (36.26) written as an equation for differences:

$$\Delta x' = \frac{\Delta x - V\Delta t}{\sqrt{1 - V^2/c^2}} \tag{36.30}$$

In the Earth frame, the length of the rod is measured at one instant t of time, so $\Delta t = 0$. Hence Eq. (36.30) reduces to

$$\Delta x' = \frac{\Delta x}{\sqrt{1 - V^2/c^2}}$$

which gives

$$\Delta x = \Delta x' \times \sqrt{1 - \frac{V^2}{c^2}}$$

This is the expected length-contraction formula (36.16).

EXAMPLE 6 A clock is at rest in the Earth frame. If a time Δt elapses as shown by this clock, how much time elapses according to the clocks of a spaceship moving at speed V relative to the Earth frame?

SOLUTION: Again, we begin by writing the Lorentz transformation equation (36.28) in terms of differences:

$$\Delta t' = \frac{\Delta t - V\Delta x/c^2}{\sqrt{1 - V^2/c^2}} \tag{36.31}$$

For the clock at rest in the Earth frame, $\Delta x = 0$, and therefore

$$\Delta t' = \frac{\Delta t}{\sqrt{1 - V^2/c^2}}$$

which is the expected time-dilation formula, Eq. (36.6).

COMMENT: Note that both the Lorentz transformation equations (36.26) and (36.28) have factors of $1/\sqrt{1 - V^2/c^2}$, even though the length contraction has a factor of $\sqrt{1 - V^2/c^2}$, not a factor of $1/\sqrt{1 - V^2/c^2}$. The reason becomes clear from inspection of these two examples: the Lorentz transformation equations (36.26) and (36.28) incorporate the time dilation for a clock at rest on Earth, but Eq. (36.26) provides the length of a rod at rest *in the spaceship frame.* If we want the length contraction for a rod at rest in the Earth frame, we would need to use the **inverse Lorentz transformation equations**, that is, the equations for x, t in terms of x', t'. These can be obtained by solving Eqs. (36.26) and (36.28) for x and t, that is, by pretending x and t are unknowns, to be evaluated by combining these equations to obtain each of them in terms of x' and t' (see Problem 33). The resulting equations have exactly the same form as Eqs. 36.26 and 36.28 with primed and unprimed space and time coordinates exchanged and with V replaced by $-V$.

inverse Lorentz transformations

Note that if the relative velocity between the two reference frames is small compared with the speed of light, then V/c in Eqs. (36.26) and (36.28) is small, and any term involving this quantity can be omitted in the equations. The Lorentz transformations then reduce to

$$x' \approx x - Vt$$

$$y' = y$$

$$t' \approx t$$

Thus, for low speeds, the Lorentz transformations reduce to the Galilean transformations (36.19)–(36.21).

The crucial feature of the Lorentz transformation equations is that they leave the speed of light unchanged. To verify this, we need to find the **relativistic combination rule for velocity**. If the x coordinate changes by dx in a time dt, then the Lorentz transformation equations tell us that

$$dx' = \frac{dx - Vdt}{\sqrt{1 - V^2/c^2}} \tag{36.32}$$

$$dt' = \frac{dt - Vdx/c^2}{\sqrt{1 - V^2/c^2}} \tag{36.33}$$

and dividing these two equations side by side, we find

$$\frac{dx'}{dt'} = \frac{dx - Vdt}{dt - Vdx/c^2} \tag{36.34}$$

On the right side we can divide both the numerator and the denominator by dt, with the result

$$\frac{dx'}{dt'} = \frac{dx/dt - V}{1 - V(dx/dt)/c^2} \tag{36.35}$$

In this expression, dx/dt is the x component of the velocity of a light signal or particle measured in the first reference frame and dx'/dt' is the x component of the velocity measured in the second reference frame. Hence Eq. (36.35) may be written

$$v'_x = \frac{v_x - V}{1 - v_x V/c^2} \tag{36.36}$$

relativistic velocity combination

This is the relativistic combination law for the x components of the velocity (there are somewhat different formulas for the combination of the y and z components of the velocity; see Problem 41).

It is instructive to compare the relativistic combination rule for velocities with the Galilean addition rule

$$v'_x = v_x - V \tag{36.37}$$

It is the denominator in Eq. (36.36) that makes all the difference. For instance, suppose that v_x is the velocity of a light signal propagating along the x axis of the first reference frame. Then $v_x = c$, and Eq. (36.36) yields

$$v'_x = \frac{c - V}{1 - cV/c^2} = \frac{c(1 - V/c)}{1 - V/c} = c \tag{36.38}$$

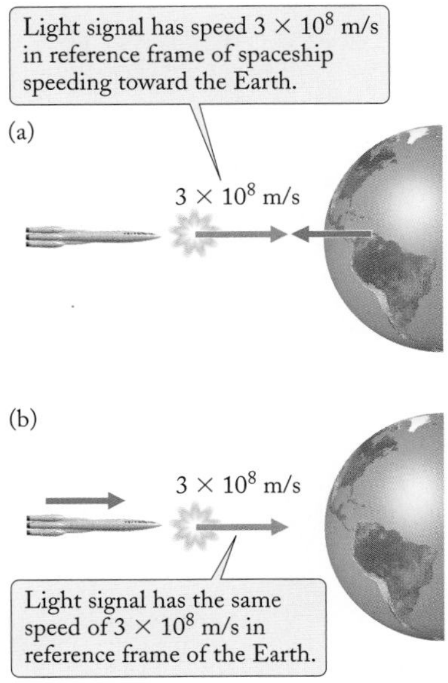

FIGURE 36.21 Addition of velocities according to the relativistic combination rule. A spaceship speeding toward the Earth emits a light signal toward the Earth.
(a) Reference frame of the spaceship.
(b) Reference frame of the Earth.

Thus, as required by the Principle of Universality of the Speed of Light, the velocity of the light signal in the second reference frame has exactly the same magnitude as in the first reference frame (see Fig. 36.21).

The relativistic combination rule for light velocities has been explicitly tested in an experiment at CERN, on the French-Swiss border; involving a beam of very fast pions. These particles decay spontaneously by a reaction that emits a flash of very intense, very energetic light (gamma rays). Hence, such a beam of pions can be regarded as a high-speed light source. In the experiment, the velocity of the pions relative to the laboratory was $V = 0.999\,75c$. The Galilean addition law for velocity would have predicted laboratory velocities of $1.999\,75c$ for light emitted in the forward direction and $0.000\,25c$ for light emitted in the backward direction. But the experiments confirmed the relativistic combination rule—the laboratory velocity of the light had the same magnitude c in all directions.

EXAMPLE 7 An alien spaceship approaching the Earth at a speed of $0.40c$ fires a rocket at the Earth (see Fig. 36.22). If the velocity of the rocket is $0.80c$ in the reference frame of the spaceship, what is its velocity in the reference frame of the Earth?

SOLUTION: The equation for the combination of velocities is easiest to use if v_x is treated as known and v'_x as unknown. Accordingly, we take v_x to be the velocity in the reference frame of the spaceship, and we take v'_x to be the velocity in the reference frame of the Earth. The x axis is directed from the spaceship toward the Earth, and the velocity of the rocket in the reference frame of the spaceship is $v_x = 0.80c$. The velocity V must be taken to be that of the Earth relative to the spaceship; this velocity is negative, $V = -0.40c$. Then Eq. (36.36) gives

$$v'_x = \frac{v_x - V}{1 - v_xV/c^2} = \frac{0.80c - (-0.40c)}{1 - (0.80c)(-0.40c)/c^2} = 0.91c$$

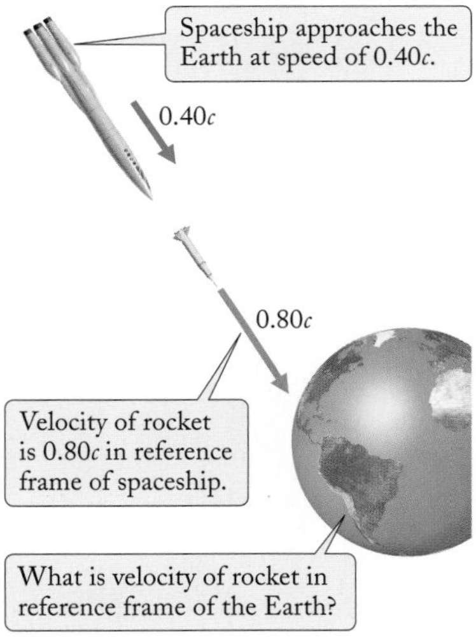

FIGURE 36.22 A spaceship launches a rocket toward Earth.

Checkup 36.5

QUESTION 1: Suppose that the new reference frame moves in the direction of the *negative* x axis of the first reference frame. What are the Lorentz transformation equations in this case?

QUESTION 2: How do we know that the Lorentz transformation equations are consistent with the requirement that the speed of light is left unchanged?

QUESTION 3: If the spaceship in Example 7 is moving away from the Earth instead of approaching the Earth, how does this change the answer for v'_x?

QUESTION 4: A radioactive nucleus approaches Earth at $v = c/2$ and emits an electron toward the Earth at $v = c/2$ relative to the nucleus. What is the electron's speed relative to the Earth?

(A) $c/4$ (B) $4c/5$ (C) $\sqrt{3/4}c$
(D) c (E) $4c/3$

36.6 RELATIVISTIC MOMENTUM AND ENERGY

The drastic revision that the theory of Special Relativity imposes on Newton's concepts of space and time implies a corresponding revision of the concepts of momentum and energy. The formulas for momentum and energy and the equations expressing their conservation are intimately tied to the transformation equations of the space and time coordinates. To see that this is so, we briefly examine the Newtonian (nonrelativistic) case.

In Newton's physics, the momentum of a particle of mass m and velocity $\mathbf{v}$ is

$$\mathbf{p} = m\mathbf{v} \tag{36.39}$$

To find the momentum of this particle in a new reference frame, we note that the Galilean transformation equation for the velocity vector is

$$\mathbf{v}' = \mathbf{v} - \mathbf{V} \tag{36.40}$$

Multiplying this by the mass, we find the transformation equation for the momentum:

$$\mathbf{p}' = m\mathbf{v}' = m\mathbf{v} - m\mathbf{V} \tag{36.41}$$

From this we see that the momentum $\mathbf{p}'$ in the new reference frame differs from the momentum $\mathbf{p}$ in the old reference frame by only the constant quantity $m\mathbf{V}$ (a quantity independent of the velocity $\mathbf{v}$ of the particle). Hence, if the total momentum of a system of colliding particles is conserved in one reference frame, it will also be conserved in the other reference frame—and the Law of Conservation of Momentum obeys the Principle of Relativity. This shows that the nonrelativistic formula for momentum and the nonrelativistic Galilean formula for the addition of velocities match in just the right way.

According to the relativistic physics of Einstein, we must replace the Galilean addition formula for velocities by the relativistic combination rule. If the Law of Conservation of Momentum is to obey the Principle of Relativity, we must then design a new relativistic formula for momentum that matches the new relativistic combination rule for velocities. The required relativistic formula for momentum is

$$\mathbf{p} = \frac{m\mathbf{v}}{\sqrt{1 - v^2/c^2}} \tag{36.42}$$

relativistic momentum

We will not give a proof of this formula.

If the speed of the particle is small compared with the speed of light, then $\sqrt{1 - v^2/c^2} \approx 1$ and Eq. (36.42) becomes approximately

$$\mathbf{p} \approx m\mathbf{v} \tag{36.43}$$

This shows that for low speeds, the relativistic and the Newtonian formulas for the momentum agree. We can therefore regard the Newtonian formula for the momentum as a simple and useful approximation for low speeds. This approximation is quite adequate for the description of all the phenomena we encounter in everyday life and (almost) all the phenomena we encounter in the realm of engineering, such as the phenomena we dealt with in the earlier chapters of this book. But at high speeds, the formulas differ drastically. We must then abandon the Newtonian formula, and rely entirely on the relativistic formula. Note that the relativistic momentum becomes infinite as the speed of the particle approaches the speed of light. Figure 36.23 is a plot of the magnitude of the momentum as a function of the speed.

EXAMPLE 8 An electron in the beam of a TV tube has a speed of 1.0×10^8 m/s. What is the magnitude of the momentum of this electron?

SOLUTION: For this electron, $v/c = (1.0 \times 10^8 \text{ m/s})/(3.0 \times 10^8 \text{ m/s}) = 0.33$.

According to Eq. (36.42), the magnitude of the momentum is then

$$p = \frac{mv}{\sqrt{1 - v^2/c^2}} = \frac{9.11 \times 10^{-31} \text{ kg} \times 1.0 \times 10^8 \text{ m/s}}{\sqrt{1 - (0.33)^2}} = 9.7 \times 10^{-23} \text{ kg·m/s}$$

COMMENTS: Note that if we had calculated the momentum according to the nonrelativistic formula $p = mv$, we would have obtained 9.1×10^{-23} kg·m/s, and we would have been in error by about 6%.

We also need a new relativistic formula for kinetic energy. This formula is

relativistic kinetic energy

$$K = \frac{mc^2}{\sqrt{1 - v^2/c^2}} - mc^2 \qquad (36.44)$$

For low speeds, this relativistic formula for kinetic energy can be shown to agree approximately with the nonrelativistic formula $K = \frac{1}{2}mv^2$.

The relativistic kinetic energy becomes infinite as the speed of the particle approaches the speed of light. This indicates that, for any particle with mass (and for any body), the speed of light is unattainable, since it is impossible to supply a particle with an infinite amount of energy. Figure 36.24 is a plot of the kinetic energy vs. the speed.

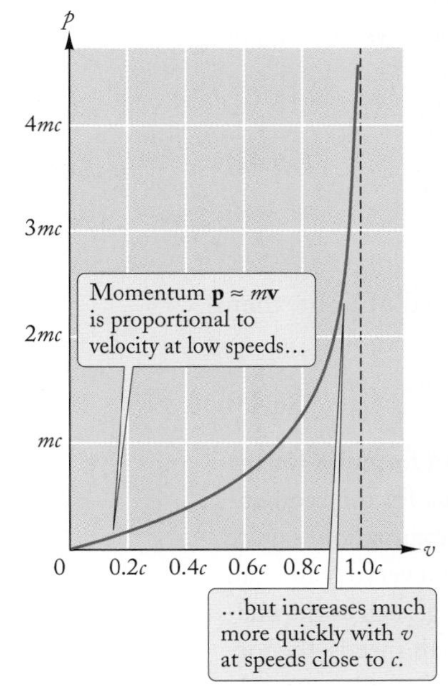

FIGURE 36.23 Momentum of a particle as a function of speed v.

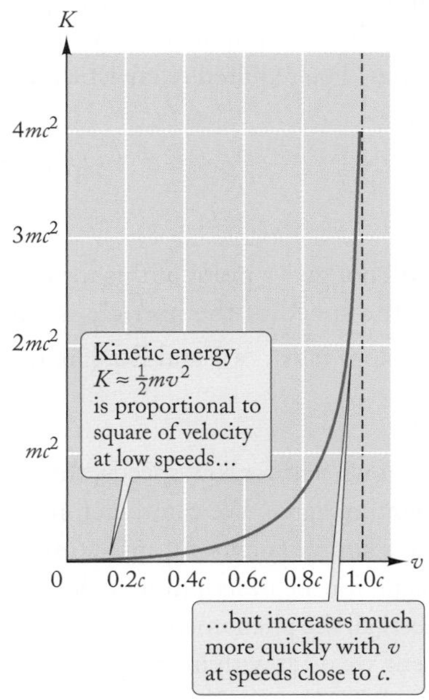

FIGURE 36.24 Kinetic energy of a particle as function of speed v.

EXAMPLE 9 The maximum speed that electrons achieve in the Stanford Linear Accelerator (SLAC) is 0.999 999 999 67c. What is the kinetic energy of an electron moving at this speed?

SOLUTION: The relativistic formula (36.44) contains a factor $\sqrt{1 - v^2/c^2}$. Since v/c for these electrons is extremely close to 1, most calculators are unable to evaluate $\sqrt{1 - v^2/c^2}$. To get around this difficulty, we write

$$\sqrt{1-\frac{V^2}{c^2}} = \sqrt{1+\frac{V}{c}}\sqrt{1-\frac{v}{c}} \approx \sqrt{2}\sqrt{1-\frac{v}{c}} \tag{36.45}$$

and we evaluate $1 - v/c$ "by hand,"

$$1-v/c = 1 - 0.999\,999\,999\,67 = 3.3 \times 10^{-10}$$

The rest of the calculation is within the reach of our calculator:

$$K = mc^2\left(\frac{1}{\sqrt{1-v^2/c^2}} - 1\right) \approx mc^2\left(\frac{1}{\sqrt{2}\sqrt{1-v/c}} - 1\right)$$

$$= 9.11 \times 10^{-31}\ \text{kg} \times (3.00 \times 10^8\ \text{m/s})^2 \times \left(\frac{1}{\sqrt{2}\sqrt{3.3 \times 10^{-10}}} - 1\right)$$

$$= 3.2 \times 10^{-9}\ \text{J}$$

Although the theory of Special Relativity requires a revision of the basic equations of mechanics, it does not require any revision of the basic equations of electricity and magnetism. Maxwell's equations are already relativistic, that is, they match the relativistic behavior of length and of time in just the right way. This concordance between Maxwell's equations and the requirements of relativity is no accident. Maxwell's equations incorporate a universal speed of light—they imply $c = 1/\sqrt{\mu_0\epsilon_0}$ in every reference frame. Einstein's search for a theory of relativity was motivated by his faith in Maxwell's equations and his recognition that if Maxwell's equations were right then the Galilean coordinate transformations had to be wrong.

✔ Checkup 36.6

QUESTION 1: For a given speed, is the relativistic value of the momentum always larger than the Newtonian value? By what factor?

QUESTION 2: Is the relativistic momentum always in the direction of the velocity of the particle?

QUESTION 3: In science fiction stories, spaceships routinely reach speeds equal to or in excess of the speed of light. What is wrong with this?

QUESTION 4: According to the relativistic formula (36.44) for the kinetic energy, what is the kinetic energy of a particle of zero speed?

(A) 0 (B) mc^2 (C) $-mc^2$
(D) $2mc^2$ (E) Infinite

36.7 MASS AND ENERGY

One of the great discoveries that emerged from relativity is that energy can be transformed into mass, and mass can be transformed into energy. Thus, mass is a form of energy. The amount of energy contained in an amount m of mass at rest is given by Einstein's famous formula

rest mass energy

$$E = mc^2 \tag{36.46}$$

The quantity mc^2 is called the **rest-mass energy**.[3] The formula (36.46) can be derived from the theory of relativity, but, as with some other equations in this chapter, we will not give the derivation.

The most spectacular demonstration of Einstein's mass–energy formula is found in the annihilation of matter and antimatter (see Chapter 41). If a proton collides with an antiproton, or an electron with an antielectron, the two colliding particles react violently and they annihilate each other in an explosion that generates an intense flash of very energetic light. According to Eq. (36.46), the annihilation of just 1000 kg of matter and antimatter (500 kg of each) would release an amount of energy

$$E = mc^2 = 1000 \text{ kg} \times (3.00 \times 10^8 \text{ m/s})^2 = 9.00 \times 10^{19} \text{ J} \tag{36.47}$$

This is enough energy to satisfy the needs of the United States for a full year. Unfortunately, antimatter is not readily available in large amounts. On Earth, antiparticles can be obtained only from reactions induced by the impact of beams of high-energy particles on a target. These collisions occasionally result in the creation of a particle–antiparticle pair. Such pair creation is the reverse of pair annihilation. The creation process transforms some of the kinetic energy of the collision into mass, and a subsequent annihilation merely gives back the original energy.

But the relationship between energy and mass in Eq. (36.46) also has another aspect. *Energy has mass.* Whenever the internal energy stored in a body is changed, its rest mass (and weight) is changed. The change in rest mass that accompanies a given change of energy is

mass and energy changes

$$\Delta m = \Delta E/c^2 \tag{36.48}$$

For instance, in the fission of uranium, the nuclear material loses energy, and correspondingly its mass (and weight) decreases. The complete fission of 1.0 kg of uranium releases an energy of 8.2×10^{13} J, and correspondingly the mass of the nuclear material decreases by $\Delta m = (8.2 \times 10^{13} \text{ J})/c^2 = 0.000\,91$ kg, or about 0.1%.

The fact that energy has mass indicates that energy is a form of mass. Conversely, as we have seen above, mass is a form of energy. Hence mass and energy must be regarded as different aspects of essentially the same thing. The laws of conservation of mass and conservation of energy are therefore not two independent laws—each implies the other. For example, consider the fission reaction of uranium inside the reactor vessel of a nuclear power plant (for details, see Chapter 40). The reaction conserves

[3] Throughout this section, *mass* means the mass that a body has when at rest or nearly at rest; to emphasize this, we use the term *rest mass.* The definition and measurement of mass for a body in motion at high (relativistic) speeds is rather tricky, because Newton's equation $m\mathbf{a} = \mathbf{F}$ fails and the direction of the acceleration is not necessarily the direction of the force. The only kind of mass that is unambiguously defined in Einstein's physics is the mass that the body has when at rest, and this is the only kind of mass we will consider.

energy—it merely transforms nuclear energy into heat, light, and kinetic energy but does not change the total amount of energy. The reaction also conserves mass—if the reactor vessel is hermetically sealed and thermally insulated from its environment, then the reaction does not change the mass of the contents of the vessel. However, if the vessel has an opening that lets some of the heat and light escape, then the mass of the residues will not match the mass of the original amount of uranium. The mass of the residues will be about 0.1% smaller than the original mass of the uranium. This mass defect represents the mass carried away by the energy that escapes. Thus, the nuclear fission reaction merely transforms energy into new forms of energy and mass into new forms of mass. In this regard, a nuclear reaction is not fundamentally different from a chemical reaction. The mass of the residues in an exothermic chemical reaction is slightly less than the original mass. The heat released in such a chemical reaction carries away some mass, but, in contrast to a nuclear reaction, this amount of mass is so small as to be quite immeasurable.

The total energy of a free particle in motion is the sum of its rest-mass energy (36.46) and its kinetic energy (36.44):

$$E = mc^2 + K = mc^2 + \frac{mc^2}{\sqrt{1 - v^2/c^2}} - mc^2 \tag{36.49}$$

This leads to a simple formula for the **relativistic total energy** of the particle:

$$E = \frac{mc^2}{\sqrt{1 - v^2/c^2}} \tag{36.50}$$

relativistic total energy

It is easy to verify (see Problem 66) that the relativistic energy can be expressed as follows in terms of the relativistic momentum:

$$E = \sqrt{c^2p^2 + m^2c^4} \tag{36.51}$$

For an ultra-relativistic particle, moving at a speed close to that of light, the first term (c^2p^2) within the square root is much larger than the second term (m^2c^4). Hence, for such a particle we can ignore the second term, and we then obtain the simple result

$$E \approx \sqrt{c^2p^2}$$

or

$$E \approx cp \quad \text{(ultra-relativistic particle)} \tag{36.52}$$

Thus, the momentum and the energy of an ultra-relativistic particle are directly proportional.

EXAMPLE 10 Consider an electron of speed 0.999 999 999 67c, as in Example 9. What is the momentum of such an electron?

SOLUTION: Such as electron is ultra-relativistic. Its kinetic energy is much larger than its rest-mass energy, and the total energy is therefore approximately equal to the kinetic energy, which we have already calculated in Example 9:

$$E = mc^2 + K \approx K = 3.2 \times 10^{-9}\ \text{J}$$

Hence Eq. (36.47) yields

$$p \approx \frac{E}{c} = \frac{3.2 \times 10^{-9}\,\text{J}}{3.0 \times 10^{8}\,\text{m/s}} = 1.1 \times 10^{-17}\,\text{kg·m/s}$$

Checkup 36.7

QUESTION 1: What is the rest-mass energy of a 1.0-kg piece of stone? Why can't we exploit this energy?

QUESTION 2: Does kinetic energy have mass? For instance, does the kinetic energy of the particles of a gas contribute to the overall mass of the gas?

QUESTION 3: For an ultra-relativistic particle (of speed near c), the momentum and the energy are proportional. Is this also true for a particle of lower speed, say, $0.9c$ or lower?

QUESTION 4: We must add energy to a hydrogen atom to ionize it and thus obtain a proton and an electron. A neutron, on the other hand, will spontaneously decay to provide a moving proton and a moving electron. From this information, which has a greater mass, the hydrogen atom or the neutron? Or do they have the same mass?

(A) Hydrogen atom (B) Neutron (C) Both have same mass

SUMMARY

PRINCIPLE OF RELATIVITY All the laws of physics are the same in all inertial reference frames.

PRINCIPLE OF UNIVERSALITY OF SPEED OF LIGHT The speed of light is the same in all inertial reference frames.

TIME DILATION ($\Delta t'$ registered by clock in its own reference frame.)

$$\Delta t = \frac{\Delta t'}{\sqrt{1 - V^2/c^2}} \qquad (36.5)$$

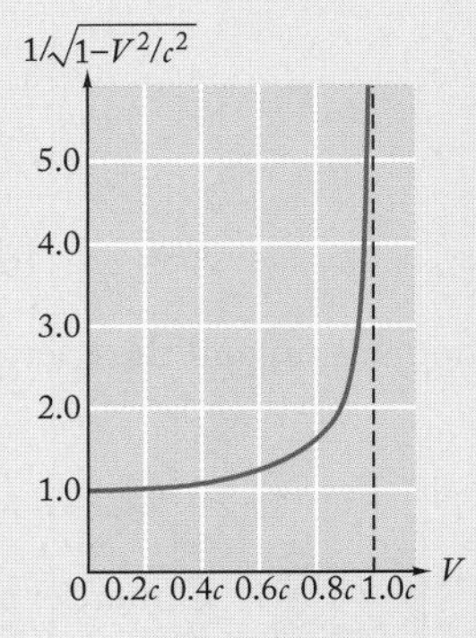

RELATIVISTIC DOPPLER SHIFT

$$f' = \sqrt{\frac{1 - V/c}{1 + V/c}}\, f \quad \text{for receding emitter}$$

$$f' = \sqrt{\frac{1 + V/c}{1 - V/c}}\, f \quad \text{for approaching emitter} \qquad (36.13)$$

LENGTH CONTRACTION (L' is length of body in its own reference frame.)

$$L = \sqrt{1 - V^2/c^2}L' \tag{36.16}$$

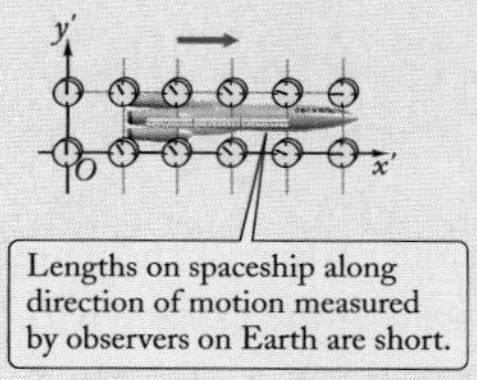

LORENTZ TRANSFORMATIONS

$$x' = \frac{x - Vt}{\sqrt{1 - V^2/c^2}} \tag{36.26}$$

$$y' = y \tag{36.27}$$

$$t' = \frac{t - Vx/c^2}{\sqrt{1 - V^2/c^2}} \tag{36.28}$$

RELATIVISTIC COMBINATION OF VELOCITIES

$$v'_x = \frac{v_x - V}{1 - v_x V/c^2} \tag{36.36}$$

RELATIVISTIC MOMENTUM

$$\mathbf{p} = \frac{m\mathbf{v}}{\sqrt{1 - v^2/c^2}} \tag{36.42}$$

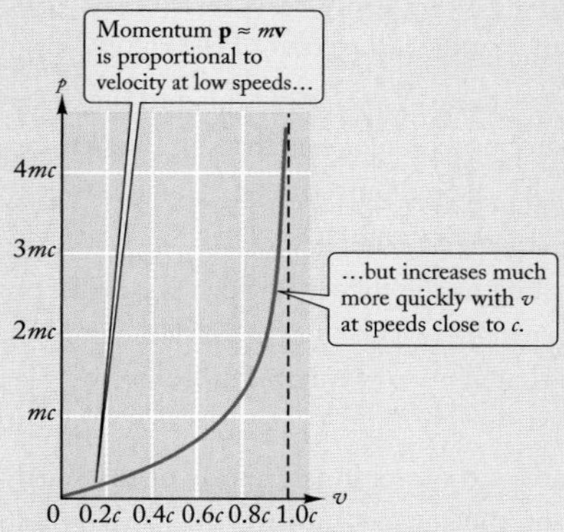

RELATIVISTIC KINETIC ENERGY

$$K = \frac{mc^2}{\sqrt{1 - v^2/c^2}} - mc^2 \tag{36.44}$$

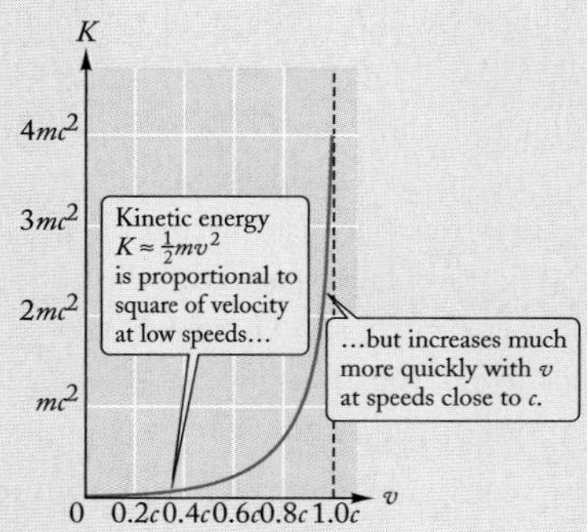

RELATIVISTIC TOTAL ENERGY

$$E = \frac{mc^2}{\sqrt{1 - v^2/c^2}} \tag{36.50}$$

$$E = \sqrt{c^2p^2 + m^2c^4}$$

REST-MASS ENERGY

$$E = mc^2 \tag{36.46}$$

MASS AND ENERGY CHANGES

$$\Delta m = \Delta E/c^2 \tag{36.48}$$

QUESTIONS FOR DISCUSSION

1. An astronaut is inside a closed space capsule coasting through interstellar space. Is there any way the astronaut can measure the speed of the capsule without looking outside?
2. Why did Michelson and Morley use *two* light beams, rather than a single light beam, in their experiment?
3. When Einstein was a boy he wondered about the following question: A runner holds a mirror at arm's length in front of his face. Can he see himself in the mirror if he runs at (almost) the speed of light? Answer this question both according to the ether theory and according to the theory of Special Relativity.
4. Consider the piece of paper on which one page of this book is printed. Which of the following properties of the piece of paper are absolute, that is, which are independent of whether the paper is at rest or in motion relative to you? (a) The thickness of the paper, (b) the mass of the paper, (c) the volume of the paper, (d) the number of atoms in the paper, (e) the chemical composition of the paper, (f) the speed of light reflected by the paper, and (g) the color of the colored print on the paper.
5. Two streetlamps, one in Boston and the other in New York City, are turned on at exactly 6:00 P.M. Eastern Standard Time. Find a reference frame in which the streetlamp in New York was turned on late.
6. According to the theory of Special Relativity, the time order of events can be reversed under certain conditions. Does this mean that a sparrow might fall from the sky before it leaves the nest?
7. Because of the rotational motion of the Earth about its axis, a point on the equator moves with a speed of 460 m/s relative to a point on the North Pole. Does this mean that a clock placed on the equator runs more slowly than a similar clock placed on the pole?
8. According to Jacob Bronowski, author of *The Ascent of Man*, the explanation of time dilation is as follows: If you are moving away from a clock tower at a speed nearly equal to the speed of light, you keep pace with the light that the face of the clock sent out at, say, 11 o'clock. Hence, if you look toward the clock tower, you always see its hands at 11 o'clock. Is this explanation correct? If not, what is wrong with it?
9. Suppose you wanted to travel into the future and see what the twenty-fifth century is like. In principle, how could you do this? Could you ever return to the twenty-first century?
10. According to the arguments of Section 36.3, a light signal traveling along a track placed perpendicular to the direction of motion of the spaceship (see Fig. 36.11) takes a longer time to complete a round trip when measured by the clocks on the Earth than when measured by the clocks on the spaceship. Would the same be true for a light signal traveling along a track placed parallel to the direction of motion? Explain qualitatively.
11. A cannonball is perfectly round in its own reference frame. Describe the shape of this cannonball in a reference frame relative to which it has a speed of $0.95c$. Is the volume of the cannonball the same in both reference frames?
12. A rod at rest in the ground makes an angle of 30° with the x axis in the reference frame of the Earth. Will the angle be larger or smaller in the reference frame of a spaceship moving along the x axis?
13. In the charming tale "City Speed Limit" by George Gamow,[4] the protagonist, Mr. Tompkins, finds himself riding a bicycle in a city where the speed of light is very low, roughly 30 km/h. What weird effects must Mr. Tompkins have noticed under these circumstances?
14. A long spaceship is accelerating away from the Earth. In the reference frame of the Earth, are the instantaneous speeds of the nose and of the tail of the spaceship the same?
15. Suppose that a *very* fast runner holding a long horizontal pole runs through a barn open at both ends. The length of the pole (in its rest frame) is 6 m, and the length of the barn (in *its* rest frame) is 5 m. In the reference frame of the barn, the pole will suffer length contraction and, at one instant of time, all of the pole will be inside the barn. However, in the reference frame of the runner, the barn will suffer length contraction and all of the pole will never be inside the barn at one instant of time. Is this a contradiction?
16. Why can a spaceship not travel as fast as or faster than the speed of light?
17. If the beam from a revolving searchlight is intercepted by a distant cloud, the bright spot will move across the surface of the cloud very quickly, with a speed that can easily exceed the speed of light. Does this conflict with our conclusion of Section 36.6, that the speed of light is unattainable?

[4] George Gamow, *Mr. Tompkins in Wonderland.*

PROBLEMS

†36.1 The Speed of Light; the Ether

1. Consider the case where the Sun moves at a high speed v through the hypothetical ether. What are the minimum and maximum ether-wind speeds on Earth when the Sun moves through the ether at (a) 30 km/s and (b) 60 km/s? Assume the orbital speed of the Earth is 30 km/s.

*2. A Michelson–Morley interferometer determines the shift of two waves traveling in perpendicular directions (see also Section 35.2 and Fig. 35.9).

(a) Assume that one wave travels a distance L along the ether wind with speed $c + V$ to a mirror, and back with speed $c - V$, as in Figs. 36.3a–b. Show that the round-trip time can be written

$$t_{\parallel} = \frac{2L}{c}\left(1 - \frac{V^2}{c^2}\right)^{-1}$$

(b) Assume the other wave travels the same distance perpendicular to the ether wind with speed $\sqrt{c^2 - V^2}$ (see Fig. 36.3*c*). Show that its round-trip time is

$$t_{\perp} = \frac{2L}{c}\left(1 - \frac{V^2}{c^2}\right)^{-1/2}$$

(c) Use the expansion $(1 - x)^n \approx 1 - nx$ for small x to show that the difference in arrival times is

$$\Delta t = t_{\parallel} - t_{\perp} \approx \frac{LV^2}{c^3}$$

(d) What fraction of a full period is this shift for light with $\lambda = 500$ nm? Use the values $L = 11$ m and $V = 30$ km/s.

*3. Ordinarily, the two arms of a Michelson–Morley interferometer cannot be set exactly equal, and instead have two values, L_1 and L_2. Insert these respective values into the results of Problem 2a and b and obtain a new expression for Δt (see Problem 2c). Note that this result alone cannot be used to determine the ether-wind speed, since the difference between L_1 and L_2 is not accurately known. In an actual experiment, the entire apparatus is rotated 90° (thus interchanging L_1 and L_2). Obtain an expression for the net shift by subtracting the differences in arrival times for the two orientations.

†,††36.2 Einstein's Principle of Relativity

4. A spaceship traveling at speed $\frac{1}{2}c$ relative to the Earth ejects a spacepod traveling in the forward direction at speed $\frac{1}{4}c$ relative to the spaceship. The spacepod emits a light signal toward the Earth at speed c relative to the spacepod. What is the speed of the light signal relative to the spaceship? What is the speed of the light signal relative to the Earth? Which observer (on the spaceship or on the spacepod) determines that the light strikes the Earth earlier?

††36.3 Time Dilation

5. If a moving clock is to have a time-dilation factor of 10, what must be its speed?

6. Neutrons have an average lifetime of 15 minutes when at rest in the laboratory. What is the average lifetime of neutrons of a speed of 25% of the speed of light? 50%? 90%?

7. Consider an unstable particle, such as a pion, which has a lifetime of only 2.6×10^{-8} s when at rest in the laboratory. What speed must you give such a particle to make its lifetime twice as long as when at rest in the laboratory?

8. The speed of the Sun around the center of our Galaxy is 200 km/s. Clocks in the Solar System will therefore run slow as compared with clocks at rest in the Galaxy. By what factor are the Solar System clocks slow?

9. The orbital speed of the Earth around the Sun is 30 km/s. In one year, how many seconds do the clocks on the Earth lose with respect to the clocks of an inertial reference frame at rest relative to the Sun? [Hint: If V/c is small, the approximation $\sqrt{1 - (V/c)^2} \approx 1 - \frac{1}{2}(V/c)^2$ is valid.]

10. In 1961, the cosmonaut G. S. Titov circled the Earth for 25 h at a speed of 7.8 km/s. According to Eq. (36.5), what was the time-dilation factor of his body clock relative to the clocks on Earth? By how many seconds did his body clock fall behind during the entire trip? (Hint: Use the approximation given in Problem 9.)

11. At a speed V, the time-dilation factor has some value. Suppose that at speed $2V$, the time-dilation factor has twice the previous value. What is the speed V?

12. An astronaut traveling at $V = 0.80c$ taps her foot 3.0 times per second. What is the frequency of taps determined by an observer on the Earth?

13. An atomic clock aboard a spaceship runs slow compared with an Earth-based atomic clock at a rate of 1.0 second per day. What is the speed of the spaceship?

14. A spaceship equipped with a chronometer is sent on a round-trip to Alpha Centauri, 4.4 light-years away. The spaceship travels at $0.10c$, and returns immediately.

(a) According to clocks on the Earth, how long does this trip take?

(b) According to the chronometer on the spaceship, how long does this trip take?

15. Consider the Doppler-shift formula for a receding source. By what factor does the frequency decrease for $V = 0.50c$? For $V = 0.70c$? For $V = 0.90c$?

† For help, see Online Concept Tutorial 41 at www.wwnorton.com/physics
†† For help, see Online Concept Tutorial 42 at www.wwnorton.com/physics

16. The frequencies of light received from distant galaxies and quasars are shifted due to the Doppler effect. Frequencies 5.0 times smaller than expected for a stationary source have been detected from receding quasars. What is V/c for such a quasar?

*17. In a test of the relativistic time-dilation effect, physicists compared the rates of vibration of nuclei of iron moving at different speeds. One sample of iron nuclei was placed on the rim of a high-speed rotor; another sample of similar nuclei was placed at the center. The radius of the rotor was 10 cm, and it rotated at 35 000 rev/min. Under these conditions, what was the speed of the rim of the rotor relative to the center? According to Eq. (36.5), what was the time-dilation factor of the sample at the rim compared with the sample at the center? (Hint: Use the approximation given in Problem 9.)

*18. If cosmonauts from the Earth wanted to travel to the Andromeda galaxy in a time of no more than 10 years as reckoned by clocks aboard their spaceship, at what (constant) speed would they have to travel? How much time would have elapsed on Earth after 10 years of time on the spaceship? The distance to the Andromeda galaxy is 2.2×10^6 light-years.

*19. Because of the rotation of the Earth, a point on the equator has a speed of 460 m/s relative to a point at the North Pole. According to the time-dilation effect of Special Relativity, by what factor do the rates of two clocks differ if one is located on the equator and the other at the North Pole? After 1.00 year has elapsed, by how many seconds will the clocks differ? Which clock will be ahead? (Although the special-relativistic time dilation slows one clock at the equator, there is an additional gravitational time dilation that slows the other clock. These two time-dilation effects balance, and the two clocks actually run at the same rate.)

**20. The star Alpha Centauri is 4.4 light-years away from us. Suppose that we send a spaceship on an expedition to this star. Relative to the Earth, the spaceship accelerates at a constant rate of $0.10g$ until it reaches the midpoint, 2.2 light-years from Earth. The spaceship then decelerates at a constant rate of $0.10g$ until it reaches Alpha Centauri. The spaceship performs the return trip in the same manner.

(a) What is the time required for the complete trip according to the clocks on the Earth? Ignore the time that the spaceship spends at its destination.

(b) What is the time required for the complete trip according to the clocks on the spaceship? Assume that the *instantaneous* time-dilation factor is still $\sqrt{1 - V^2/c^2}$ even though the speed V is a function of time.

36.4 Length Contraction

21. A meterstick is moving by an observer in a direction parallel to its length. The speed of the meterstick is $0.50c$. What is its measured length in the reference frame of the observer?

22. According to the manufacturer's specifications, a spaceship has a length of 200 m. At what speed (relative to the Earth) will this spaceship have a length of 100 m in the reference frame of the Earth?

23. A cannonball flies through our laboratory at a speed of $0.30c$. Measurement of the transverse diameter of the cannonball gives a result of 0.20 m. What can you predict for the measurement of the length, or the longitudinal diameter, of the cannonball?

24. What is the percent length contraction of an automobile traveling at 96 km/h? (Hint: Use the approximation given in Problem 9.)

25. A hangar for housing spaceships is 100 m long. How fast must a 200-m-long spaceship be traveling to (briefly) fit in the hangar?

26. A right triangle of sheet metal with two 45° angles lies in the x–y plane, with one of its sides along the x axis (see Fig. 36.25). The length of each side is 0.20 m, and the length of the hypotenuse is $\sqrt{2} \times 0.20$ m. Suppose that this triangle is observed from an x', y' reference frame moving at $0.80c$ along the x axis. What are the lengths of the sides and of the hypotenuse in this reference frame? What are the angles?

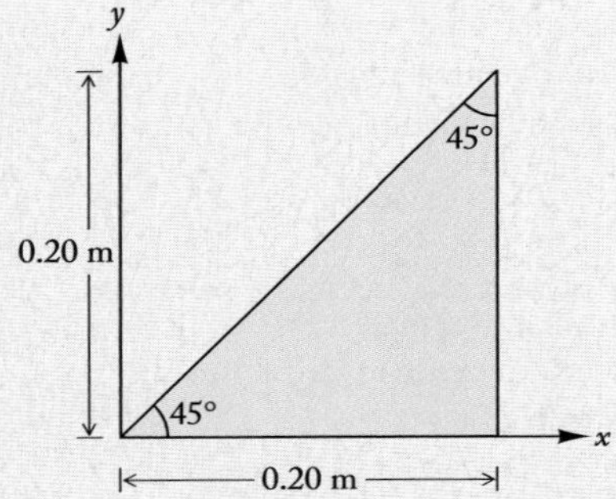

FIGURE 36.25 A triangle.

*27. Two identical spaceships are traveling in the same direction. An observer on Earth measures the first to have speed $0.80c$ and observes the second to be 1.50 times as long as the first one. What is the speed of the second spaceship?

*28. Suppose that a meterstick at rest in the reference frame of the Earth lies in the x–y plane and makes an angle of 30° with the x axis. Suppose that one end of the meterstick is at the origin. At a fixed time t, what are the x and y components of the displacement from this end of the meterstick to the other? At a fixed time t', what are the x' and y' components of the displacement from one end of the meterstick to the other in a new reference frame moving with velocity $V = 0.70c$ in the positive x direction? What is the angle the meterstick makes with the x' axis of this new reference frame?

*29. Electric charge is uniformly distributed throughout a sphere; the charge density is 2.0×10^{-6} C/m^3. If this sphere is put in motion relative to the laboratory at a speed of $0.80c$, what will be the charge density? Keep in mind that the total amount of electric charge is unchanged by the motion of the sphere.

*30. It can be shown that when a point charge moves at uniform velocity of relativistic magnitude, its pattern of electric field lines is contracted by the usual length-contraction factor $\sqrt{1 - V^2/c^2}$ in the longitudinal direction and is unchanged in the transverse direction. Figure 36.26 shows the resulting pattern of field lines for a speed $V = 0.60c$. Draw a similar picture for a speed of $0.80c$.

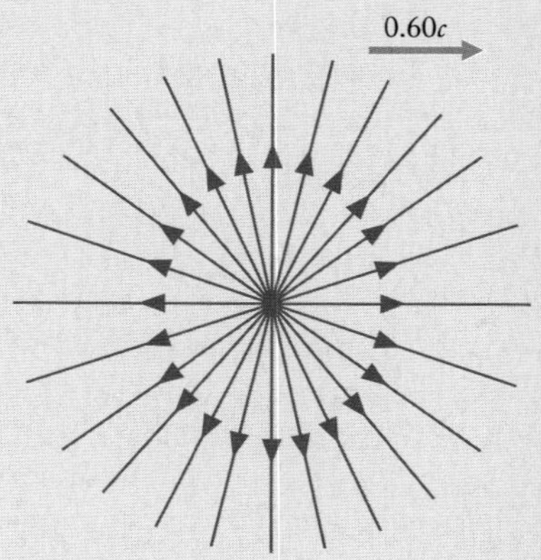

FIGURE. 36.26 Electric field lines of a charge moving at $V = 0.60c$.

*31. A flexible drive belt runs over two flywheels whose axles are mounted on a rigid base (see Fig. 36.27). In the reference frame of the base, the horizontal portions of the belt have a speed v and therefore are subject to length contraction, which tightens the belt around the flywheels. However, in a reference frame moving to the right with the upper portion of the belt, the base is subject to length contraction, which ought to loosen the belt around the flywheels. Resolve this paradox by a qualitative argument. (Hint: Consider the lower portion of the belt as seen in the reference frame of the upper portion.)

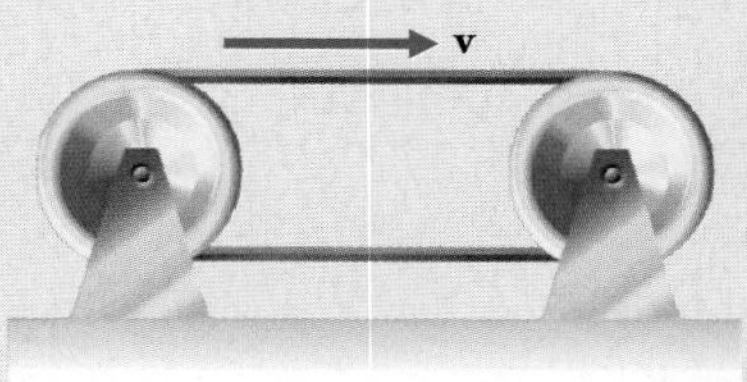

FIGURE 36.27 A drive belt and two flywheels.

36.5 The Lorentz Transformations and the Combination of Velocities

32. In the reference frame of the Earth, a firecracker is observed to explode at $x = 6.0 \times 10^8$ m, at $t = 4.0$ s. According to the Lorentz transformation equations, what are the x' and t' coordinates of this event as observed in the reference frame of a spaceship traveling in the x direction at a speed of $0.50c$? According to the Galilean transformation equations?

33. Obtain the **inverse Lorentz transformation equations** by solving Eqs. (36.26) and (36.28) for x and t, each in terms of x' and t'.

34. A spaceship has a length of 300 m, measured in its own reference frame. It is traveling in the positive x direction at a speed of $0.80c$ relative to the Earth. A strobe light at the nose of the spaceship sends a pulse of light toward the tail of the spaceship.
   (a) As measured in the reference frame of the spaceship, how long does this light pulse take to reach the tail?
   (b) As measured in the reference frame of the Earth, how long does this light pulse take to reach the tail?

35. A spaceship is moving at a speed of $0.60c$ toward the Earth. A second spaceship, following the first one, is moving at a speed of $0.90c$. What is the speed of the second spaceship as observed in the reference frame of the first?

36. Find the inverse of Eq. (36.36); that is, express v_x in terms of v'_x.

37. The captain of a spaceship traveling away from Earth in the x direction at $V = 0.80c$ observes that a nova explosion occurs at a point with spacetime coordinates $t' = -6.0 \times 10^8$ s, $x' = 1.9 \times 10^{17}$ m, $y' = 1.2 \times 10^{17}$ m as measured in the reference frame of the spaceship. He reports this event to the Earth via radio without delay.
   (a) What are the spacetime coordinates of the explosion in the reference frame of the Earth? Assume that the master clock of the spaceship coincides with the master clock of the Earth at the instant $t = t' = 0$ when the midpoint of the spaceship passes by the Earth, and that the origin of the spaceship x', y' coordinates is at the midpoint of the spaceship.
   (b) Will the Earth receive the captain's report before or after astronomers on the Earth see the nova explosion in their telescopes? No calculation is required for this question.

*38. Consider the situation described in Problem 37. Since light takes some time to travel from the nova to the spaceship, the space and time coordinates that the captain reports are not directly measured but, rather, deduced from the time of arrival and the direction of the nova light reaching the spaceship.
   (a) At what time (t' time) did the nova light reach the spaceship?
   (b) If the captain sends a report to Earth via radio as soon as he sees the nova, at what time (t time) does the Earth receive the report?
   (c) At what time do Earth astronomers see the nova?

*39. At $11^h0^m0.0000^s$ A.M. a boiler explodes in the basement of the Museum of Modern Art in New York City. At $11^h0^m0.0003^s$ A.M. a similar boiler explodes in the basement of a soup factory in Camden, New Jersey, at a distance of 150 km from the first explosion. Show that, in the reference frame of a spaceship moving at a speed greater than $V = 0.60c$ from New York toward Camden, the first explosion occurs *after* the second.

*40. A radioactive atom in a beam produced by an accelerator has a speed $0.80c$ relative to the laboratory. The atom decays and ejects an electron of speed $0.50c$ relative to itself. What is the speed of the electron relative to the laboratory, if ejected in the forward direction? If ejected in the backward direction?

*41. In a manner similar to the procedure of Eqs. (36.32)–(36.36), show that relativistic combination formula for the y component of the velocity is

$$v'_y = \frac{v_y\sqrt{1 - V^2/c^2}}{1 - v_x V^2/c^2}$$

*42. Consider two speeds v_x and V, each of which is less than the speed of light. Show that if these speeds are combined by the relativistic combination formula, the result is always less than the speed of light.

*43. The speed of light with respect to a medium is $v'_x = c/n$, where n is the index of refraction. Suppose that the medium, say, flowing water, is moving past a stationary observer in the same direction as the light with speed V. Show that the observer measures the speed of light to be approximately

$$v_x = \frac{c}{n} + \left(1 - \frac{1}{n^2}\right)V$$

This effect was first observed by Fizeau in 1851.

**44. The acceleration of a particle in one reference frame is $a_x = dv_x/dt$, where the particle has instantaneous velocity v_x in that frame. Consider a reference frame moving with speed V parallel to the positive x axis of the first frame. Show that the acceleration in the second frame is given by

$$a'_x = \frac{dv'_x}{dt'} = a_x \frac{(1 - V^2/c^2)^{3/2}}{(1 - v_x V/c^2)^3}$$

36.6 Relativistic Momentum and Energy

45. Consider a particle of mass m moving at a speed of $0.10c$. What is its kinetic energy according to the relativistic formula? What is its kinetic energy according to the Newtonian formula? What is the percent deviation between these two results?

46. Suppose you want to give a rifle bullet of mass 0.010 kg a speed of 1.0% of the speed of light. What kinetic energy must you supply?

47. The yearly energy expenditure of the United Stated is about 8×10^{19} J. Suppose that all of this energy could be converted into kinetic energy of an automobile of mass 1000 kg. What would be the speed of this automobile?

48. The speed of an electron in a hydrogen atom is 2.6×10^6 m/s. For this speed, does your calculator show any difference between the kinetic energies calculated according to the relativistic formula and the Newtonian formula?

49. What is the speed of an electron if its kinetic energy is 1.6×10^{-13} J?

50. What is the momentum and what is the kinetic energy of an electron moving at a speed of one-half the speed of light?

51. Show that the momentum of a particle can be expressed in the concise form

$$\mathbf{p} = \frac{E\mathbf{v}}{c^2}$$

*52. What is the percent difference between the Newtonian and the relativistic values for the momentum of a meteoroid reaching the Earth at a speed of 72 km/s?

*53. Consider three accelerators that produce high-energy particles: the Large Hadron Collider, which will soon produce protons with kinetic energy 7 TeV; the Stanford Linear Accelerator, which produces electrons with kinetic energy 50 GeV; and the Relativistic Heavy Ion Collider, which produces gold nuclei (mass 197 u) with kinetic energy 20 TeV. In each case, calculate the difference $c - v$ between the speed of light and the speed of the particle.

*54. The most energetic cosmic-ray particles have energies of about 50 J. Assume that such a cosmic ray consists of a proton. By how much does the speed of such a proton differ from the speed of light? Express your answer in meters per second. [Hint: Use the approximation given in Eq. (36.45)].

*55. Consider the electrons of a speed of $0.999\,999\,999\,67c$ produced by the Stanford Linear Accelerator. What is the magnitude of the momentum of such an electron?

*56. At the Fermilab accelerator, protons are given kinetic energies of 1.6×10^{-7} J. By how many meters per second does the speed of such a proton differ from the speed of light? What is the magnitude of the momentum of such a proton?

*57. A mass M at rest decays into two particles of masses m_1 and m_2. Use Eq. (36.51) to show that the magnitude of the momentum of each of the two particles is

$$p = \frac{\sqrt{M^2 - (m_1 + m_2)^2}\sqrt{M^2 - (m_1 - m_2)^2}\, c}{2M}$$

*58. A particle of mass m_1 is at rest. A second particle of mass m_2 and kinetic energy K strikes the first particle and sticks to it, a perfectly inelastic collision. Use Eq. (36.51) to show that the mass M of the composite particle is

$$M = \sqrt{(m_1 + m_2)^2 + \frac{2m_1 K}{c^2}}$$

*59. Show that the velocity of a relativistic particle can be expressed as follows:

$$\mathbf{v} = \frac{c\mathbf{p}}{\sqrt{m^2c^2 + p^2}}$$

**60. At the Brookhaven AGS accelerator, protons of kinetic energy 5.3×10^{-9} J are made to collide with protons at rest.

(a) What is the speed of a moving proton in the laboratory reference frame?

(b) What is the speed of a reference frame in which the two colliding protons have the same speed (and are moving in opposite directions)?

(c) What is the total energy of each proton in the latter reference frame?

36.7 Mass and Energy

61. How much energy will be released by the annihilation of one electron and one antielectron (both initially at rest)? Express your answer in electron-volts.

62. The atomic bomb dropped on Hiroshima had an explosive energy equivalent to that of 20 000 tons of TNT, or 8.4×10^{13} J. How many kilograms of rest mass must have been converted into energy in this explosion?

63. The mass of the Sun is 2.0×10^{30} kg. The thermal energy in the Sun is about 2×10^{41} J. How much does the thermal energy contribute to the mass of the Sun? Express your answer in percent.

64. Combustion of one gallon of gasoline releases 1.3×10^8 J of energy. How much mass is converted to energy? Compare this with 2.8 kg, the mass of one gallon of gasoline.

65. The masses of the proton, electron, and neutron are $1.672\,623 \times 10^{-27}$ kg, 9.11×10^{-31} kg, and $1.674\,929 \times 10^{-27}$ kg, respectively. When a neutron decays into a proton and an electron, how much energy is released (other than the energy of the rest mass of the proton and electron)? Compare this extra energy with the energy of the rest mass of the electron.

66. From Eqs. (36.42) and (36.50) show that the relativistic energy and the relativistic momentum are related by

$$E^2 = c^2p^2 + m^2c^4$$

**67. A K^0 particle at rest decays spontaneously into a π^+ particle and a π^- particle. What will be the speed of each of the latter? The mass of the K^0 is 8.87×10^{-28} kg, and the masses of the π^+ and π^- particles are 2.49×10^{-28} kg each.

REVIEW PROBLEMS

*68. Muons are unstable particles which—if at rest in a laboratory—decay after a time of only 2.2×10^{-6} s. Suppose that a muon is created in a collision between a cosmic ray and an oxygen nucleus at the top of the Earth's atmosphere, at an altitude of 20 km above sea level.

(a) If the muon has a downward speed of $v = 0.990c$ relative to the Earth, at what altitude will it decay? Ignore gravity in this calculation.

(b) Without time dilation, at what altitude would the muon have decayed?

*69. Suppose that a special breed of cat (*Felis einsteinii*) lives for exactly 7.0 years according to its own body clock. When such a cat is born, we put it aboard a spaceship and send it off at $V = 0.80c$ toward the star Alpha Centauri. How far from the Earth (reckoned in the reference frame of the Earth) will the cat be when it dies? How long after the departure of the spaceship will a radio signal announcing the death of the cat reach us? The radio signal is sent out from the spaceship at the instant the cat dies.

*70. Suppose that a proton speeds by the Earth at $v = 0.80c$ along a line parallel to the axis of rotation of the Earth.

(a) In the reference frame of the proton, what is the polar diameter of the Earth? The equatorial diameter?

(b) In the reference frame of the proton, how long does the proton take to travel from the point of closest approach to the North Pole to the point of closest approach to the South Pole? In the reference frame of the Earth, how long does this take?

71. A spaceship travels in the positive x direction with speed $0.80c$. A man on Earth makes these observations:

At $t = 0$, $x = 0$, a photon with $\lambda = 400$ nm is emitted at the rear of the ship moving toward the front.

At $t = 1.00\ \mu$s, $x = 960$ m, a photon with $\lambda = 600$ nm is emitted at the front of the spaceship moving toward the rear.

A woman on the spaceship observes the same events.

(a) What time interval does she measure between the two events? Which happens earlier?

(b) What wavelengths does she measure for the two photons?

(c) What is the length of the ship (as determined by the woman on it)?

*72. An observer on Earth sees one spaceship traveling away to the west at speed $0.40c$ and a second spaceship, also traveling away but to the east, at $0.70c$. Each spaceship emits a signal in its own reference frame at 2.00 GHz. What frequency does the Earth observer measure for each signal? What frequency does each spaceship measure for the signal from the other?

*73. Consider a cube measuring 1.0 m × 1.0 m × 1.0 m in its own reference frame. If this cube moves relative to the Earth at a speed of $0.60c$, what are its dimensions in the reference frame of the Earth? What are the areas of its faces? What is its volume? Assume that the cube moves in a direction perpendicular to one of its faces.

*74. A spaceship has a length of 200 m in its own reference frame. It is traveling at $0.95c$ relative to the Earth. Suppose that the tail of the spaceship emits a flash of light.

(a) In the reference frame of the spaceship, how long does the light take to reach the nose?

(b) In the reference frame of the Earth, how long does this take? Calculate the time directly from the motions of the spaceship and the flash of light; then compare it with the results calculated by applying the Lorentz transformations to the result obtained in (a).

75. Suppose that a spaceship is moving at a speed of $V = 0.20c$ relative to the Earth and a meteoroid is moving at a speed of $v_x = 0.10c$ relative to the Earth; in the same direction as the spaceship. What is the speed v'_x of the meteoroid relative to the spaceship according to Eq. (36.36)? What is the percent difference between this relativistic result and the Galilean result?

76. A collision between two gamma rays creates an electron and an antielectron that travel away from the point of creation in opposite directions, each with a speed of $0.95c$ in the laboratory. What is the speed of the antielectron in the reference frame of the electron, and vice versa?

*77. A spaceship traveling at $0.70c$ away from the Earth launches a projectile of muzzle speed $0.90c$ (relative to the spaceship). What is the speed of the projectile relative to the Earth if it is launched in the forward direction? In the backward direction?

*78. Three particles are moving along the positive x axis, in the positive direction. The first particle has a speed of $0.60c$ relative to the second, the second has a speed of $0.80c$ relative to the third, and the third has a speed of $0.50c$ relative to the laboratory. What is the speed of the first particle relative to the laboratory?

79. What is the kinetic energy of a spaceship of rest mass 50 metric tons moving at a speed of $0.50c$? How many metric tons of matter–antimatter mixture would have to be consumed to make this much energy available?

80. A particle has a kinetic energy equal to its rest-mass energy. What is the speed of this particle?

81. Suppose that a spaceship traveling at $0.80c$ through our Solar System suffers a totally inelastic collision with a small meteoroid of mass 2.0 kg.

(a) What is the kinetic energy of the meteoroid in the reference frame of the spaceship?

(b) In the collision all of this kinetic energy suddenly becomes available for inelastic processes that damage the spaceship. The effect on the spaceship is similar to an explosion. How many tons of TNT will release the same explosive energy? One ton of TNT releases 4.2×10^9 J.

*82. At the Brookhaven AGS accelerator, protons of kinetic energy 5.3×10^{-9} J are made to collide with protons at rest.

(a) What is the speed of one of these moving protons in the laboratory reference frame?

(b) What is the magnitude of the momentum?

*83. At the SSC accelerator that was to be built in the United States, protons would have been given kinetic energies of 3.2×10^{-6} J. What is the value of $c - v$ for such a proton, that is, by how many meters per second does the speed differ from the speed of light?

*84. Free neutrons decay spontaneously into a proton, an electron, and an antineutrino:

$$\mathrm{n} \rightarrow \mathrm{p} + \mathrm{e} + \bar{\nu}$$

The neutron has a rest mass of 1.6749×10^{-27} kg; the proton, 1.6726×10^{-27} kg; the electron, 9.11×10^{-31} kg; and the antineutrino nearly zero. Assume that the neutron is at rest. Other than the rest-mass energy of the proton and electron, what is the energy released in this decay?

**85. A K^0 particle moving at a speed of $0.60c$ through the laboratory decays into a muon and an antimuon.

(a) In the reference frame of the K^0, what is the speed of each muon? The mass of the K^0 is 8.87×10^{-28} kg, and the masses of the muon and the antimuon are 1.88×10^{-28} kg each.

(b) Assume that the muon moves in a direction parallel to the original direction of motion of the K^0 and that the antimuon moves in the opposite direction. What are the speeds of the muon and the antimuon with respect to the laboratory?

Answers to Checkups

Checkup 36.1

1. Yes. Sound waves propagate in a medium (air), and thus the speed of sound waves relative to you depends on your speed relative to the air. The speed is largest for sound waves traveling opposite to your motion, from the front to the back of the convertible automobile.

2. An ether wind due only to rotation means that the Earth remains translationally at rest in the ether (the ether moves with the Earth). The original experiment had a sensitivity of about 5 km/s, and thus could not detect an ether wind due only to the rotational speed of 0.46 km/s.

3. Such an observation would have led to the conclusion that the Earth has an absolute motion of 30 km/s relative to the ether. Since this is the same as the velocity of the Earth relative to the Sun, such a result, if observed year round, would have also implied that the Sun is at rest with respect to the ether.

4. (C) At a different time of year. A single null result could have implied that the Earth was (coincidentally) nearly at rest with respect to the ether. Repeating the experiment when the Earth's velocity was in a different direction ensures that this was not the case.

Checkup 36.2

1. The speed of light in vacuum always has the same value, 3.00×10^8 m/s.
2. No; the clocks shown in Fig. 36.10 are as observed from the Earth. For the crew or anyone in the reference frame of the spaceship, the clocks are all synchronized.
3. Relative to a spaceship traveling westward, the Earth reference frame is traveling eastward. Thus, New York is at the leading edge of this reference frame, so the New York clocks, and the New York earthquake, are late.
4. (D) c. The speed of an electromagnetic wave is the speed of light; it does not depend on the speed of the emitter or receiver.

Checkup 36.3

1. Relativity of synchronization means that the times indicated by different clocks in a moving reference frame are different but all these clocks run at the same rate; relativity of rates means that the rate of the clocks in the moving reference frame differs from that of the clocks on the ground.
2. At low speeds, a factor of 2 increase in velocity has a negligible effect on the time-dilation factor, which remains nearly equal to unity. At higher speeds, the time-dilation factor increases more quickly; for example, an increase from $0.45c$ to $0.90c$ increases the time-dilation factor from 1.1 to 2.3, more than a factor of 2, and higher speeds result in larger-factor increases.
3. An approaching receiver is the same as an approaching emitter, since only relative motion matters; and the lower relation in Eq. (36.13) applies. Thus the frequency increases. So the wavelength decreases, and you detect blue light.
4. (C) Slow; slow. The time-dilation effect is symmetric, so observers in each reference frame measure a clock in the other reference frame to be running slow.

Checkup 36.4

1. Like the shape of the Earth moving at high speed relative to the reference frame of the proton in Example 4, the cannonball is an ellipsoid, flattened along the direction of motion, in a reference frame relative to which it is moving at $0.5c$.
2. No; contraction along the direction of motion does not affect which pipe fits inside which. Moreover, any apparent contradiction can be explained in terms of differences in simultaneity at any two different positions along the direction of motion.
3. Yes. Since the volume decreases with increasing speed, the density, or mass per unit volume, increases with increasing speed.
4. (B) 50 m. The 100-m track is contracted by the factor $\sqrt{1 - V^2/c^2} = \sqrt{0.25} = 0.50$, that is, to a length of $0.50 \times 100 \text{ m} = 50 \text{ m}$.

Checkup 36.5

1. The Lorentz transformation equations will be the same as Eqs. (36.26)–(36.28), but now with a negative value of V.
2. We know the Lorentz transformation equations are consistent with a speed of light that is unchanged because they lead directly to the relativistic velocity combination law, Eq. (36.36), which gives $v'_x = c$ when $v_x = c$.
3. This is the same as if the Earth is moving away from the spaceship, that is, V is now positive, and so the relativistic velocity combination rule gives $v'_x = (0.80c - 0.40c)/(1 - 0.80 \times 0.40) = 0.59c$.
4. (B) $4c/5$. As in Example 7, we can obtain the relative speed from the velocity combination rule, Eq. (36.36): $v'_x = [(c/2) - (-c/2)]/[1 - (1/2) \times (-1/2)] = c/(5/4) = (4/5)c$.

Checkup 36.6

1. Yes; other than for speed $v = 0$, the relativistic value of the momentum, given by Eq. (36.42), is always larger than the Newtonian value, by the factor $1/\sqrt{1 - v^2/c^2}$.
2. Yes, the momentum vector $\mathbf{p}$ is proportional to the velocity $\mathbf{v}$, and so $\mathbf{p}$ is always in the same direction as $\mathbf{v}$.
3. For any mass, attaining the speed of light would require infinite kinetic energy, and this is impossible.
4. (A) 0. With $v = 0$ in the first term of Eq. (36.44), the two terms in the realistic kinetic energy cancel.

Checkup 36.7

1. From $E = mc^2$ we have $1.0 \text{ kg} \times (3.00 \times 10^8 \text{ m/s})^2 = 9.0 \times 10^{16}$ J. This energy can't be converted to useful forms of energy, except by annihilation, which would require an equal amount of (unavailable) antimatter.
2. Yes. For example, the mass of a warm container of gas is greater than the mass of a cold container of gas, due to the additional kinetic energy
3. No. This is only true when $v \approx c$. For example, for $v \geq 0.99c$, the energy becomes very nearly proportional to the momentum.
4. (B) Neutron. Since the neutron produces a proton, an electron, and some kinetic energy, its total energy (its mass) must be greater than the hydrogen atom, which requires added energy just to separate the proton and electron.

CHAPTER

37

Quanta of Light

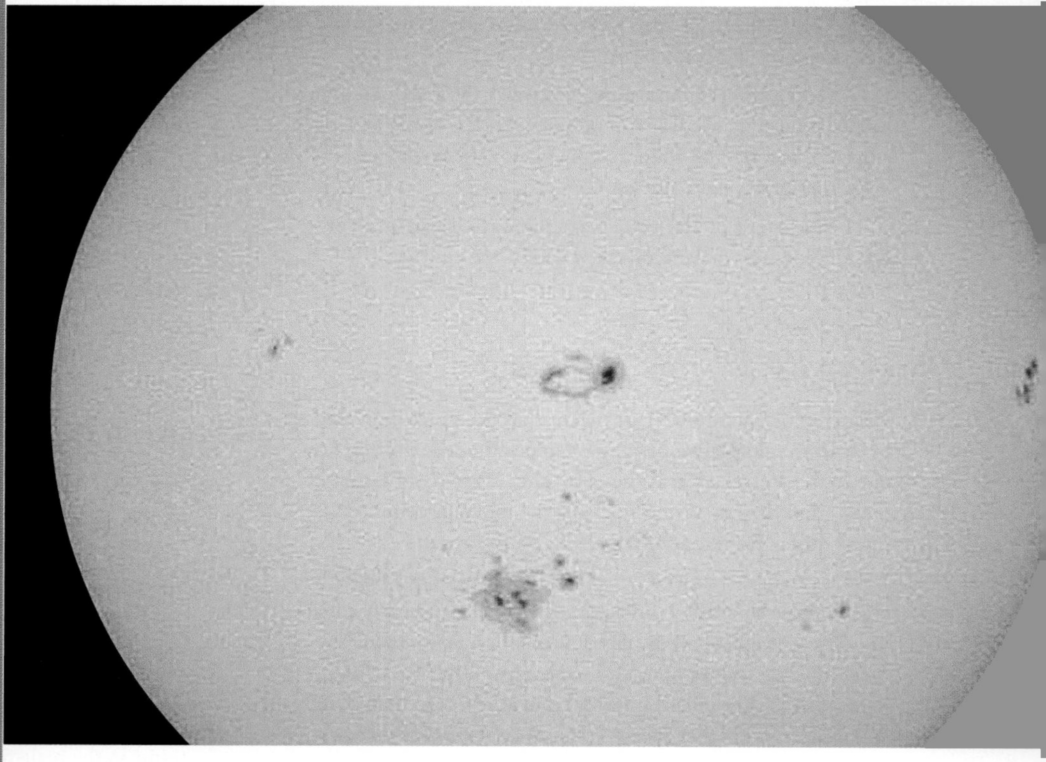

CONCEPTS IN CONTEXT

Concepts *in* Context

The Sun emits thermal radiation consisting of electromagnetic waves of many wavelengths. For the Sun and many hot bodies, there are universal features in the distribution of the emitted wavelengths. The explanation of the wavelength distribution of such thermal radiation requires the revolutionary idea that electromagnetic waves are made up of small, indivisible packets of energy, called quanta of light, or photons.

Our study of thermal radiation will enable us to ask:

? How do we know the temperature of the surface of the Sun? (Section 37.1, page 1257)

? At what wavelength is the Sun's thermal radiation maximum? (Example 1, page 1261)

? How many photons are there in the sunlight that reaches the surface of the Earth? (Example 4, page 1264)

In Chapter 35 we examined the wave properties of light. We saw that light exhibits interference and diffraction, in agreement with Maxwell's theory, according to which light is a wave consisting of oscillating electric and magnetic fields with a smooth, continuous density of energy. In this chapter we will see that light has particle properties. We will discuss experimental evidence that establishes that a light beam consists of a stream of discrete, particle-like energy packets. These energy packets are called *quanta of light* or *photons*.

The discovery of the quantization of light by Max Planck in 1900 initiated the modern era in physics. Physicists quickly came to recognize that *quantization of energy is a pervasive feature of the atomic and subatomic realm*. The energies of the atoms and the energies of the subatomic particles—electrons, protons, and neutrons—are quantized. As we will discuss in the next chapters, such a quantization of energy is in conflict with Newton's laws, and physicists had to find new laws that govern the behavior of atoms and of subatomic particles. The new theory that governs the realm of the atom is called **quantum physics**. In contrast, the old theory of Newton is called **classical physics**.

quantum physics vs. classical physics

The fact that light has the dual attributes of wave and of particle indicates that neither the classical wave nor the classical particle concept we have used in earlier chapters gives an adequate description of light. We have to think of light as a wave–particle object, which sometimes behaves pretty much as a wave, sometimes pretty much as a particle, and sometimes as a bit of both. Furthermore, we will see in the next chapter that electrons, protons, neutrons, and all the other known "particles" also exhibit such dual attributes of wave and of particle.

37.1 BLACKBODY RADIATION

The first hint of a failure of classical physics emerged around 1900 from the study of thermal radiation. When we heat a body to high temperature, it gives off a glow. For instance, when we heat a bar of iron to 1200 or 1300 K, it glows in a bright orange or yellow color. This glow is thermal radiation ("radiant heat"). The color of the thermal radiation depends on the temperature; an extremely hot iron bar glows in a nearly white color ("white-hot"); at an intermediate temperature it glows yellow; and at a lower temperatures it glows orange and then bright red to dull red (Fig. 37.1). You can observe this change of color with temperature by turning on the heating coil of a kitchen range; the coil first glows dull red, and then orange, but it never reaches white-hot. Bodies at room temperature also emit thermal radiation, but the glow is infrared, and not visible to the eye (see the thermogram of Fig. 37.2). *The spectrum of thermal radiation is continuous*—if we analyze the light emitted by a glowing body with a prism, we find that the energy of the light is smoothly distributed over all wavelengths.

FIGURE 37.1 A hot, glowing iron bar. The tip is at the highest temperature ("white-hot"), and the temperature gradually decreases along the bar.

For a quantitative description of the distribution of energy over different wavelengths, we plot the energy flux (or the power per unit area) radiated by the surface of the glowing body vs. the wavelength of the radiation. We can think of such a plot as the intensity distribution seen in the spectrum that an (ideal) prism produces when we use it to analyze thermal radiation. Measurements of the thermal radiation emitted by a glowing body show that the energy flux at very long and at very short wavelengths is quite small, and that the energy flux has a maximum at some intermediate wavelength. The location of this maximum depends on the temperature. For

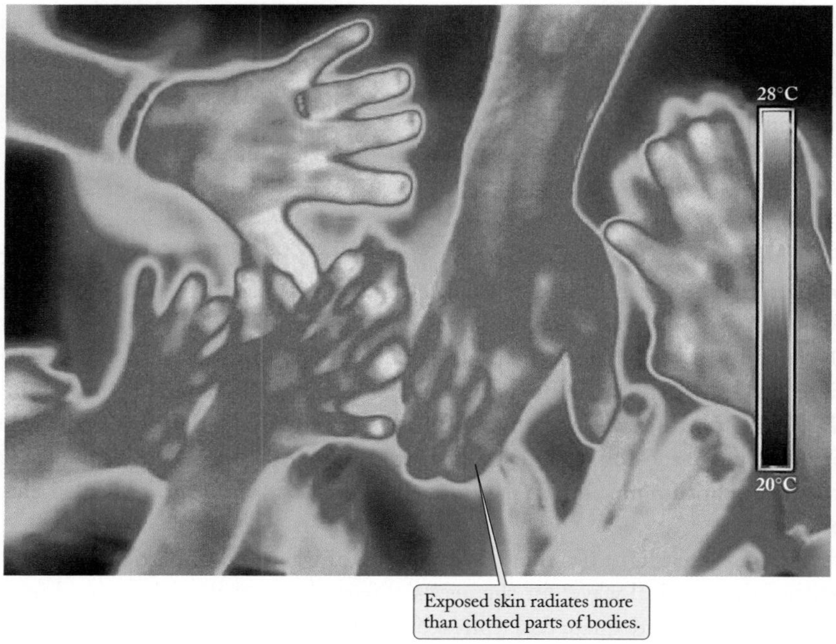

FIGURE 37.2 A thermogram. Each "false" color represents a different wavelength of infrared electromagnetic radiation; unlike the visible spectrum, the brightest colors here are hottest (white, then red, yellow, green, blue, . . .).

example, Fig. 37.3 gives plots of the distribution of energy flux radiated by glowing bodies at 1000 K, 1250 K, and 1450 K. By comparing these plots, we see that an increase of temperature produces more radiation at all wavelengths (the 1450-K curve is everywhere higher than the other curves); and we also see that an increase of temperature shifts the location of the maximum to shorter wavelengths (the peak of the 1450-K curve is located at a shorter wavelength than the peaks of the other curves).

The thermal radiation emerging from the surface of a glowing body is generated within the volume of the body by the random thermal motions of atoms and electrons. Before the radiation reaches the surface and escapes, it is absorbed and re-emitted many times and it attains thermal equilibrium with the atoms and electrons. This equilibration process distributes the radiation continuously over all wavelengths and shapes its spectrum, completely washing out all of the original spectral features that the radiation had when first emitted by the atoms in the body.

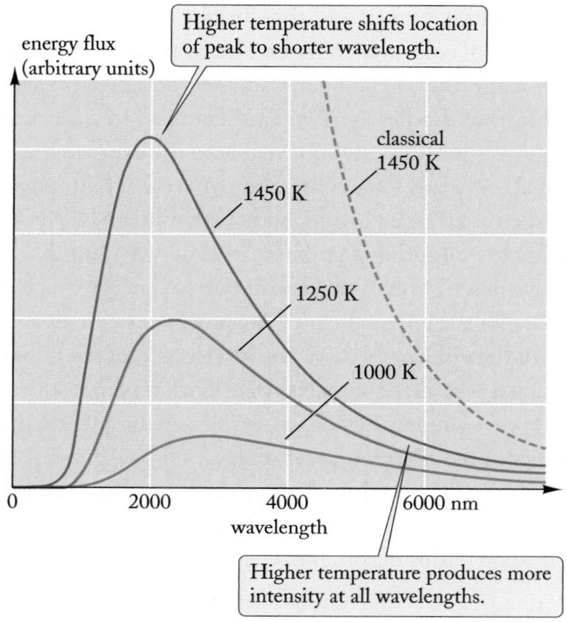

FIGURE 37.3 Distribution of the energy flux in the spectra of thermal radiation emitted by glowing bodies at 1000 K, 1250 K, and 1450 K. The maxima (peaks) of these curves lie in the infrared region. The maximum is at 2900 nm for 1000 K, at 2300 nm for 1250 K, and at 2000 nm for 1450 K. The dashed curve gives the prediction of classical physics according to Rayleigh's calculation.

The flux of thermal radiation emerging from the surface of a glowing body depends to some extent on the characteristics of the surface. The surface usually permits the escape of only a fraction of the flux reaching it from the inside of the body. Correspondingly, if the body is irradiated with an equal flux of thermal radiation from the outside, the surface permits the ingress of only an equal fraction of this flux, reflecting the rest. This equality of the emissive and absorptive characteristics of the surface can be deduced by an argument based on thermodynamics. Thus we are led to a general rule: *a good absorber is a good emitter; and a poor absorber is a poor emitter.* With this rule we can understand how the silvered, mirrorlike glass walls of thermos bottles or dewars provide such excellent thermal insulation. These bottles are constructed with a double glass wall, and the space between these walls is evacuated (Fig. 37.4). Heat cannot flow across the evacuated space by conduction or convection; it can only flow by radiation. To inhibit radiation, the glass walls are silvered and thereby made into mirrorlike reflecting surfaces; these highly reflective surfaces are very poor absorbers and emitters of radiation. This keeps the heat transfer between the walls very small.

blackbody

A body with a perfectly absorbing (and emitting) surface is called a **blackbody**; when such a body is cold and emits no radiation of its own, it looks black because it does not reflect any of the illumination reaching it from the outside. But when a blackbody is hot, its surface emits more thermal radiation than any other hot body at the same temperature. A body covered with black soot is an approximate blackbody. Experimental physicists prefer to achieve the characteristics of an ideal blackbody by a trick: take a body with a cavity, such as a hollow cube, and drill a small hole in one side of the cube (Fig. 37.5). The hole then acts as a blackbody—any radiation incident on the hole from outside will be completely absorbed. Because of this equivalence between a blackbody and a hole in a cavity, the terms **blackbody radiation** and **cavity radiation** are used interchangeably. The curves plotted in Fig. 37.3 are based on measurements of radiation emerging from a small hole in a body with a cavity; thus, these curves represent the spectra of blackbody radiation.

blackbody radiation or cavity radiation

Concepts in Context

The Sun is an almost perfect blackbody radiator. A very small fraction of the Sun's radiation is due to specific chemical features of the Sun; the overwhelming majority of sunlight makes up a spectrum of thermal radiation that precisely follows the shape given in Fig. 37.3. As we have seen, a blackbody spectrum depends on the temperature of the blackbody, so we can tell the temperature of the surface of the Sun from the spectrum of thermal radiation that the Sun sends to the Earth.

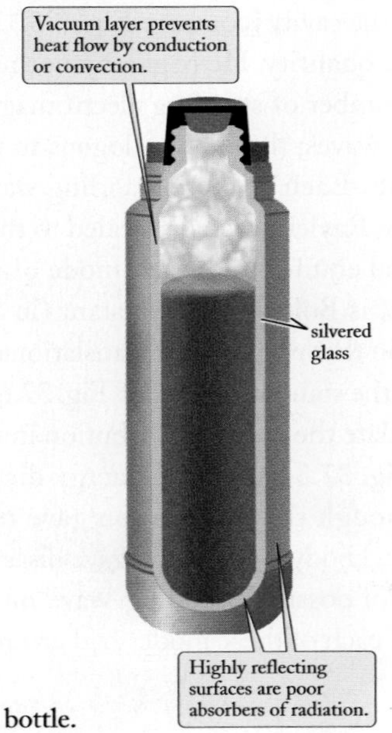

FIGURE 37.4 A thermos bottle.

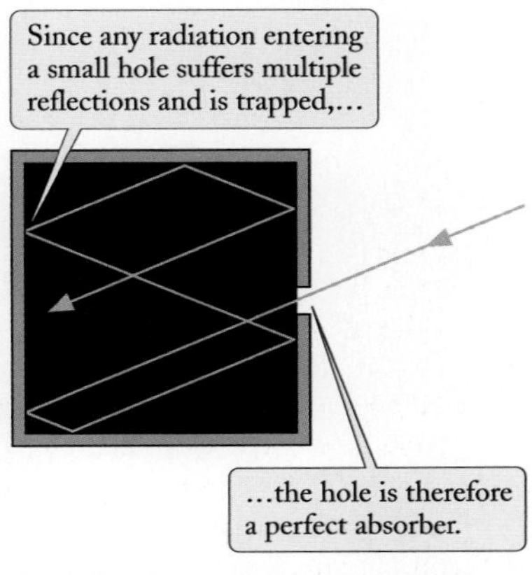

FIGURE 37.5 A cavity with a small hole.

The blackbody plays a special role in the study of thermal radiation because the spectrum of its thermal radiation does not depend on the material of which it is made or on any other characteristics of the body. By an argument based on thermodynamics, it can be established that *the spectrum depends only on the temperature of the blackbody.* The mathematical formula for the distribution of energy in this spectrum is therefore a universal law, and in the last years of the nineteenth century, physicists engaged in an intensive experimental and theoretical effort to find this universal law of blackbody radiation.

✔ Checkup 37.1

QUESTION 1: A wood-burning stove is made of cast iron. How would the performance of the stove change if it were made of polished stainless steel?

QUESTION 2: One house has a roof of dark-colored shingles; another has a roof of light-colored shingles. Which roof absorbs more heat during the day, in sunlight?

QUESTION 3: For space "walks," astronauts wear suits with a shiny, silvery surface layer. What is the purpose of this surface layer?

QUESTION 4: Two ideal cavities emit thermal radiation. The spectrum emitted by the first cavity is most intense at a wavelength of 600 nm, and that of the second at 500 nm. Compared with the second cavity, the radiation emitted by the first cavity is

(A) More intense at all wavelengths
(B) More intense at short wavelengths and less intense at long wavelengths
(C) Less intense at short wavelengths and more intense at long wavelengths
(D) Less intense at all wavelengths

37.2 ENERGY QUANTA

Before 1900, physicists made several attempts at a theoretical explanation of the distribution of energy in the spectrum of blackbody radiation, but they met with disaster. One of the best of these attempts was that of Lord Rayleigh. Since the energy flux of the radiation emerging from the hole in a cavity is directly proportional to the energy density of the radiation inside the cavity [compare Eqs. (33.17) and (33.22)], Rayleigh decided to calculate the latter quantity. He began by noting that the radiation in a cavity is made up of a large number of standing electromagnetic waves. Figure 37.6 shows some of these standing waves; they are analogous to the standing waves in an organ pipe closed at both ends. Each of these standing waves can be regarded as a mode of vibration of the cavity. Rayleigh then appealed to the equipartition theorem, according to which, at thermal equilibrium, each mode of vibration has an average thermal energy of kT, where k is Boltzmann's constant (in Section 19.4 we stated a special case of the equipartition theorem for free translational or rotational motion of a gas molecule). Thus, each of the standing waves of Fig. 37.6 ought to have an energy kT, and from this we can calculate the energy distribution in the spectrum of the radiation (the dashed curve in Fig. 37.3 shows the energy distribution obtained from Rayleigh's calculation). Although this calculation gave reasonable results at the long-wavelength end of the blackbody spectrum, it gave disastrous results at the short-wavelength end: the number of possible standing-wave modes of very short wavelength is infinitely large, and if each of these modes had an energy kT, the total energy

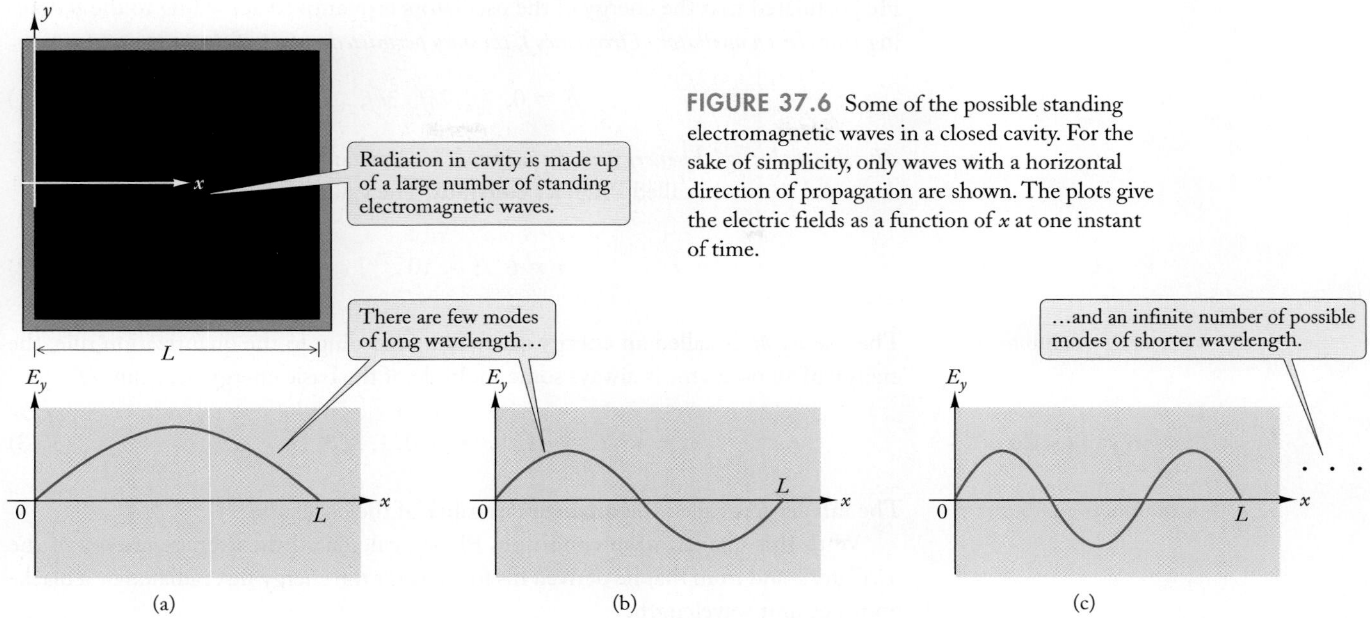

FIGURE 37.6 Some of the possible standing electromagnetic waves in a closed cavity. For the sake of simplicity, only waves with a horizontal direction of propagation are shown. The plots give the electric fields as a function of x at one instant of time.

in the cavity would be infinite! This disastrous failure of classical physics has been called the **ultraviolet catastrophe**.

ultraviolet catastrophe

The correct formula for the distribution of energy in the spectrum of blackbody radiation was finally obtained by Max Planck in 1900. By some inspired guesswork, based on thermodynamics, Planck found a mathematical formula that gave a precise fit to the experimental curves of blackbody radiation, such as those plotted in Fig. 37.3. He then searched for a theoretical explanation for his formula. The search led Planck to a revolutionary discovery: the quantization of energy. This discovery was to bring about the overthrow of classical physics and the birth of quantum physics. For Planck, the postulate of the quantization of energy was "an act of desperation," which he committed because "a theoretical explanation had to be found at any cost, whatever the price."

Planck's derivation of the blackbody radiation formula involves some sophisticated statistical mechanics which we will not reproduce. We will merely give a sketchy outline of this derivation. Planck began by making a theoretical model of the walls of the cavity: he regarded the atoms in the walls as small harmonic oscillators of many different natural frequencies, that is, small masses (with electric charges) attached to springs of many different spring constants. Although this is a rather crude model of the atoms that make up the walls of the cavity, it was adequate for his purposes since, as described in the preceding section, the radiation in a cavity is known to be completely independent of the characteristics of the wall. The random thermal motions of the oscillators result in the emission of electromagnetic radiation. This radiation fills the cavity and acts back on the oscillators. When thermal equilibrium is attained, the average rate of emission of radiation energy by the oscillators matches their rate of absorption of radiation energy. Thus, the oscillators share their energy with the radiation in the cavity, and Planck was able to show that, under equilibrium conditions, the average radiation energy at some frequency f (or at a wavelength $\lambda = c/f$) is directly proportional to the average energy of an oscillator of frequency f.

These steps of Planck's calculation involved nothing but classical mechanics. But in the next step of the calculation, Planck departed radically from classical physics.

He postulated that the energy of the oscillators is quantized according to the following rule: *In an oscillator of frequency f, the only permitted values of the energy are*

$$E = 0,\ hf,\ 2hf,\ 3hf,\ \ldots \tag{37.1}$$

All other values of the energy are forbidden. The constant h in Eq. (37.1) is a new fundamental constant, called **Planck's constant.** The value of this constant is

Planck's constant

$$h = 6.63 \times 10^{-34}\ \text{J·s} \tag{37.2}$$

energy quantum hf

The energy hf is called an **energy quantum**; according to the quantization rule, the energy of an oscillator is always some multiple of the basic energy quantum hf:

energy of oscillator

$$E = nhf \qquad n = 0, 1, 2, 3, \ldots \tag{37.3}$$

quantum number n

The integer n is called the **quantum number** of the oscillator.

With this quantization condition, Planck calculated the average energy of the oscillators; and from that he derived his formula for the energy flux radiated by a blackbody per unit wavelength,

$$S_\lambda = \frac{2\pi hc^2}{\lambda^5 \times (e^{hc/\lambda kT} - 1)} \tag{37.4}$$

The energy flux per unit wavelength means that if $d\lambda$ is a small interval of wavelength centered on a given wavelength λ, then the energy flux dS of electromagnetic waves that have wavelengths in the interval $d\lambda$ is

$$dS = S_\lambda\, d\lambda$$

MAX PLANCK (1858–1947) *German theoretical physicist, professor at Berlin and President of the Kaiser Wilhelm Institute (later renamed the Max Planck Institute). Planck made important contributions to thermodynamics before he became involved with the problem of blackbody radiation. He was deeply troubled by the quantization of energy, because he recognized that this held disastrous consequences for classical mechanics and electromagnetism. He was awarded the Nobel Prize in 1918 for his discovery of energy quanta.*

The distribution function (37.4) agrees with the experimentally measured distributions shown in Fig. 37.3. Although we cannot go into the details of this derivation, we can achieve a rough understanding of how Planck's calculation avoids the ultraviolet catastrophe. The thermal energy of the walls of the cavity is shared at random among all the oscillators in these walls. Some of these oscillators have high frequencies, some have low frequencies. For an oscillator of very high frequency, the energy quantum hf is very large. If this oscillator is initially quiescent ($n = 0$), it cannot begin to move unless it acquires one energy quantum; but since this energy quantum hf is very large, the random thermal disturbances will be insufficient to provide it—the oscillator will remain quiescent. Thus, the quantization of energy tends to inhibit the thermal excitation of the high-frequency oscillators. If the high-frequency oscillators remain quiescent, they will not supply energy to the corresponding high-frequency standing waves in the cavity. These waves will then not have the energy kT predicted by Rayleigh; instead they will have no energy at all. This avoids the ultraviolet catastrophe.

Note that for an oscillator with a frequency of $f \approx 10^{15}$ Hz, which is typical for atomic vibrations, the energy quantum is $hf = 6.6 \times 10^{-34}\ \text{J·s} \times 10^{15}\ \text{Hz} = 6.6 \times 10^{-19}$ J. Since this is a very small amount of energy, quantization does not make itself felt at a macroscopic level. But quantization plays a pervasive role at the atomic level.

Unfortunately, Planck could not offer any basic justification for his postulate of quantization of energy. His postulate gave him a blackbody radiation law which was in complete agreement with the experimentally measured distribution of energy over wavelength (as displayed in Fig. 37.3), but his postulate brought him into conflict with classical physics. Superficially, the quantization of energy is analogous to the quantization of electric charge—we know from Chapter 22 that the electric charge of any particle or body is always some multiple of the fundamental charge e. However, whereas the quantization of charge is consistent with the laws of classical physics, the quantization of energy is inconsistent with these laws. The energy of any oscillator that obeys Newton's laws—such as a mass on a spring—can be changed by a small amount by pushing on the oscillator with a very weak force, and it should therefore be possible to change the energy of the oscillator continuously by any amount we please, not just in discrete steps of one energy quantum. Thus, quantization of energy makes no sense in classical physics. Planck could only justify his postulate by its consequences; but, in theoretical physics, the end does not justify the means. A deeper explanation of the quantization of energy emerged only much later, with the development of quantum mechanics (see Chapter 38).

From Planck's formula (37.4) for the distribution of energy in the spectrum of blackbody radiation one can establish that the energy flux has a maximum at a wavelength

$$\lambda_{max} = \frac{1}{4.965}\frac{hc}{k} \times \frac{1}{T} \tag{37.5}$$

where c is, as always, the speed of light, k is Boltzmann's constant (see Section 19.1), h is Planck's constant, and T is the absolute temperature of the blackbody. Equation (37.5) is called **Wien's displacement Law**. If we insert the numerical values of h, c, and k, Wien's Law takes the simple form

$$\lambda_{max} = \frac{2.90 \times 10^{-3}\ \text{m}\cdot\text{K}}{T} \tag{37.6}$$

Wien's displacement Law

Wien's Law asserts that λ_{max} is inversely proportional to the temperature T. This means that an increase of temperature shifts the location of the maximum to shorter wavelengths, in agreement with the experimental results presented in Fig. 37.3. If the temperature is sufficiently high—6000 K or so—the maximum of the spectrum lies in the visible region. For instance, the Sun, with a surface temperature of 5800 K, emits its largest flux of thermal radiation in the visible region.

EXAMPLE 1 According to Wien's Law, at what wavelength does the thermal radiation emitted by the Sun have its maximum? At what wavelength does the thermal radiation emitted by the tungsten filament in a lightbulb have its maximum? Assume that the Sun and the tungsten filament are approximately blackbodies at temperatures of 5800 K and 3200 K, respectively.[1]

Concepts in Context

SOLUTION: With $T = 5800$ K, Wien's Law gives

$$\lambda_{max} = \frac{2.90 \times 10^{-3}\ \text{m}\cdot\text{K}}{5800\ \text{K}} = 5.0 \times 10^{-7}\ \text{m} = 500\ \text{nm}$$

and with $T = 3200$ K, it gives

$$\lambda_{max} = \frac{2.90 \times 10^{-3}\ \text{m}\cdot\text{K}}{3200\ \text{K}} = 9.1 \times 10^{-7}\ \text{m} = 910\ \text{nm}$$

[1]The blackbody approximation is fairly good for the Sun (except at sunspots; see the chapter photo). But it is not good for an ordinary tungsten filament, because the tungsten surface is not a good absorber. The blackbody approximation becomes better if the filament is tightly coiled, like a solenoid, and if we examine the radiation in the interior of this coil (a cavity).

Thus, the maximum of the thermal radiation from the Sun lies in the visible region (in the green), but the maximum of the thermal radiation from the lightbulb lies in the infrared.

Furthermore, from Planck's formula (37.4) one can calculate the combined energy flux for all wavelengths radiated from the surface of a blackbody by integration of the intensity per unit wavelength, S_λ, over all wavelengths. This gives a total intensity proportional to the fourth power of the temperature,

Stefan–Boltzmann Law

$$S = \sigma T^4 \tag{37.7}$$

This is called the **Stefan–Boltzmann Law**. The constant of proportionality σ in this law can, again, be expressed in terms of h, c, and k. With the appropriate numerical values, the Stefan–Boltzmann constant then has the value

$$\sigma = 5.67 \times 10^{-8}\ \text{W/(m}^2\cdot\text{K}^4) \tag{37.8}$$

Both the Wien and Stefan–Boltzmann laws had been discovered empirically many years before Planck supplied their theoretical foundation. Note that the energy flux S in Eq. (37.7) depends only on the temperature of the blackbody; it does not depend on the material from which the body is made. Since S is the energy flux, or the power per unit area, we can obtain the net power radiated from the body by multiplying S by the surface area; thus, the net power depends on the size of the body.

EXAMPLE 2 On a clear night, the surface of the Earth loses heat by radiation. Suppose that the temperature of the ground is 10°C and that the ground radiates like a blackbody. What is the rate of loss of heat per square meter?

SOLUTION: The temperature of the ground is 10°C, so the absolute temperature of the ground is 283 K. Hence, the Stefan–Boltzmann Law tells us that the radiated flux, or power per unit area, is

$$S = \sigma T^4 = 5.67 \times 10^{-8}\ \text{W/(m}^2\cdot\text{K}^4) \times (283\ \text{K})^4 = 364\ \text{W/m}^2$$

COMMENTS: This large radiative heat loss of the ground explains the sharp drop of temperature experienced during clear nights. The drop of temperature is much less severe if there is an overcast sky. The clouds then reflect most of the radiation back to the ground—they act like a blanket to keep the ground warm.

EXAMPLE 3 In a house, a room is heated by means of a radiator filled with hot water at 82°C. The radiator consists of a large vertical panel. If you place your hand near the panel (see Fig. 37.7a), what is the rate at which thermal radiation is incident on your hand? The area of one side of your hand is 0.016 m^2.

SOLUTION: When your hand is very near the panel, the energy flux reaching your hand is the same as the energy flux, or intensity, emitted by the radiator panel. The power incident on your hand is the incident flux times the area A of your hand,

$$P = SA = \sigma T^4 \times A$$

$$= 5.67 \times 10^{-8}\ \mathrm{W/(m^2 \cdot K^4)} \times (82 + 273)^4\ \mathrm{K^4} \times 0.016\ \mathrm{m^2} = 14\ \mathrm{W}$$

COMMENTS: The distance at which you place your hand from the panel is not crucial. If your hand is somewhat farther away (see Fig. 37.7b), then some of the radiation emitted by the portion of the panel directly facing your hand will miss your hand, but an equal amount of radiation from a more distant portion of the panel will now be able to reach your hand, and these amounts compensate. You can do a simple experiment to verify this: place your hand very near a radiator panel (or near a stove), and gradually move your hand away—you will hardly notice any difference for the first few centimeters of motion. For a radiator panel of finite size, the energy flux incident on your hand will decrease once you move far enough away. But if you are in a room all of whose walls, ceiling, and floor are lined with radiator panels, a compensation similar to that depicted in Fig. 37.7b is valid throughout the room—in such an environment, your hand receives the same flux no matter where it is located or in what direction it faces.

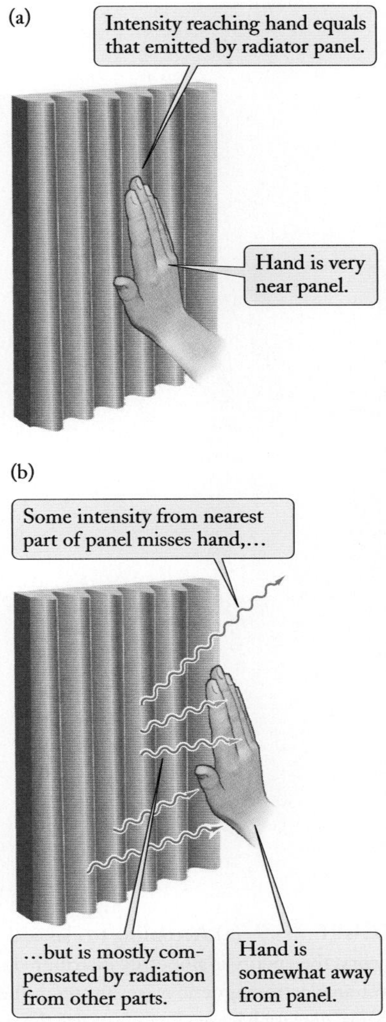

FIGURE 37.7 (a) Hand placed near a flat radiator panel. (b) Hand placed farther away.

Checkup 37.2

QUESTION 1: The caption for Fig. 37.3 gives the wavelengths of the maxima for different temperatures. What is the product of wavelength and temperature in each case? Is this in agreement with Wien's Law?

QUESTION 2: If we increase the temperature of a blackbody from 1000 K to 2000 K, by what factor do we change the wavelength of the maximum of the spectrum? By what factor do we change the total energy radiated from the surface?

QUESTION 3: An oscillator has a quantized energy hf. Is its kinetic energy quantized? Is its potential energy quantized?

QUESTION 4: The pendulum of a clock can be regarded as an oscillator. Is the energy of the pendulum quantized? Why don't we notice the quantization?

QUESTION 5: Suppose that Planck's constant was 6.6×10^{-2} J·s instead of 6.6×10^{-34} J·s. Would we notice the quantization of a pendulum?

QUESTION 6: Figure 37.8 shows photos of three stars. The light emitted by these stars is thermal radiation. Which of these stars is the hottest? The coolest?

(A) Red; yellow (B) Red; blue (C) Yellow; blue
(D) Blue; red (E) Blue; yellow

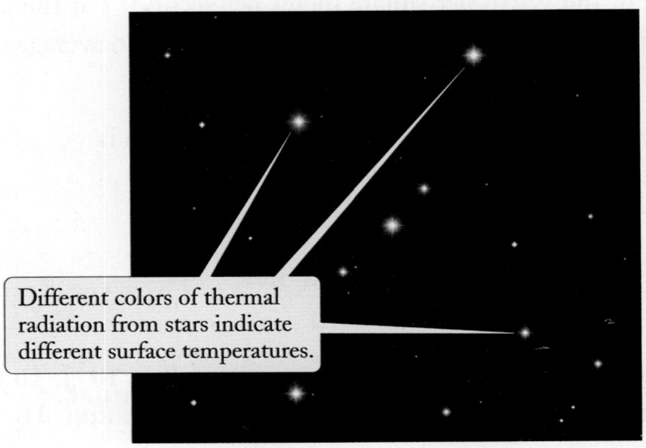

FIGURE 37.8 Betelgeuse, Bellatrix, and Rigel (red, yellow, and blue, respectively) in the constellation Orion.

37.3 PHOTONS AND THE PHOTOELECTRIC EFFECT

In 1905, Einstein showed that Planck's formula could be understood much more simply in terms of a direct quantization of the energy of the radiation. Planck had postulated that the oscillators in the wall of the cavity have discrete quantized energies, but he had treated the electromagnetic radiation as a smooth, continuous density of energy, exactly as it is supposed to be according to classical electromagnetic theory.

In contrast, Einstein proposed that electromagnetic radiation consists of discrete particle-like packets of energy. He regarded a wave of some given frequency f as a stream of more or less localized energy packets, each with one quantum of energy hf (see Fig. 37.9). The particle-like energy packets of magnitude hf are called **photons**. The wave then has an energy hf if it contains only one photon, $2hf$ if it contains two photons, and so on. The thermal radiation in a cavity, with waves traveling randomly in all directions, can then be regarded as a gas of photons. Einstein applied statistical mechanics to calculate the energy spectrum of this gas and he thereby obtained the energy spectrum of the cavity radiation.

photon

The essential difference between Planck's and Einstein's view of the cavity radiation is that Planck quantized only the exchange of radiation with the walls of the cavity, whereas Einstein quantized the radiation itself. Thus, in Einstein's view, electromagnetic radiation is always quantized, regardless of where or how it is produced. Not only is the radiation quantized when it is produced by the oscillators in the walls of a cavity (as in the case of thermal radiation), but also when it is produced outside of a cavity, say, by the acceleration of electric charges on the antenna of a radio transmitter.

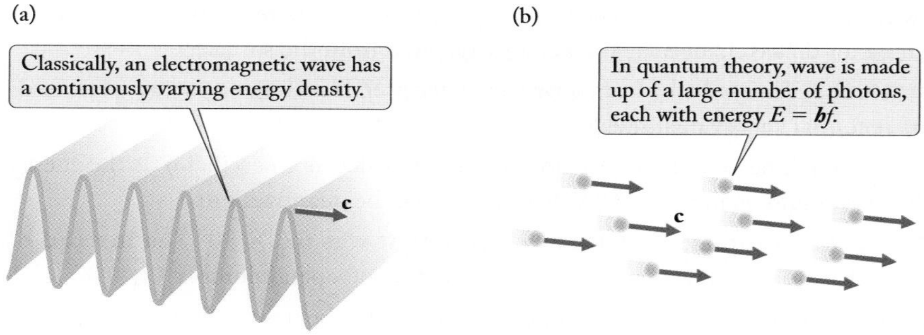

FIGURE 37.9 (a) According to classical theory, the energy is smoothly distributed over an electromagnetic wave, although the energy density has maxima wherever the wave reaches maximum amplitude. (b) According to Einstein, the energy is localized in small energy packets, which move with the wave.

Concepts in Context

EXAMPLE 4 The energy flux of sunlight reaching the surface of the Earth is 1.0×10^3 W/m^2 at normal incidence. How many photons reach the surface of the Earth per square meter per second? For the purposes of this calculation assume that all the photons in sunlight have an average wavelength of 500 nm.

SOLUTION: The energy of a photon of wavelength 500 nm is

$$E = hf = h\frac{c}{\lambda}$$

$$= 6.63 \times 10^{-34}\,\text{J·s} \times \frac{3.00 \times 10^8\,\text{m/s}}{5.0 \times 10^{-7}\,\text{m}} = 4.0 \times 10^{-19}\,\text{J} \qquad (37.9)$$

The energy incident per square meter per second is 1.0×10^3 J. To obtain the number of photons, we must divide this by the energy per photon, 4.0×10^{-19} J. That is,

$$\frac{\text{photons}}{\text{second}} = \frac{\text{energy/second}}{\text{energy/photon}}$$

For a square meter, this gives

$$\frac{1.0 \times 10^3 \text{ J/(s}\cdot\text{m}^2)}{4.0 \times 10^{-19} \text{ J/photon}} = 2.5 \times 10^{21} \text{ photons per square meter per second}$$

COMMENTS: Because the number of photons in sunlight and other common light sources is so large, at the macroscopic level we do not perceive the grainy character of the energy distribution in light.

photoelectric effect

With this concept of light as a stream of photons, Einstein was also able to offer an explanation of the **photoelectric effect**. In some early experiments on the production of radio waves by sparks, Hertz had noticed that light shining on an electrode tended to promote the formation of sparks. Subsequent careful experimental investigations demonstrated that the impact of light on an electrode can eject electrons, which trigger the sparks. The electrons emerge with a kinetic energy which increases directly with the frequency of the light.

Figure 37.10 is a schematic diagram of the apparatus used in the investigation of the photoelectric effect. Light from a lamp illuminates an electrode of metal (C) enclosed in an evacuated tube. Electrons ejected from this electrode travel to the collecting electrode (A), and then flow around the external circuit. A galvanometer (G) detects this flow of electrons. The kinetic energy of the ejected photoelectrons can be determined by applying a potential difference V between the emitting and the collecting electrodes by means of an adjustable source of emf. With the polarity shown in the figure, the collector has a negative potential relative to the emitter, that is, the collector (A) exerts a repulsive force on the photoelectrons. If an electron is to travel from the emitter to the collector, the change in its potential energy $e \times V$ between emitter

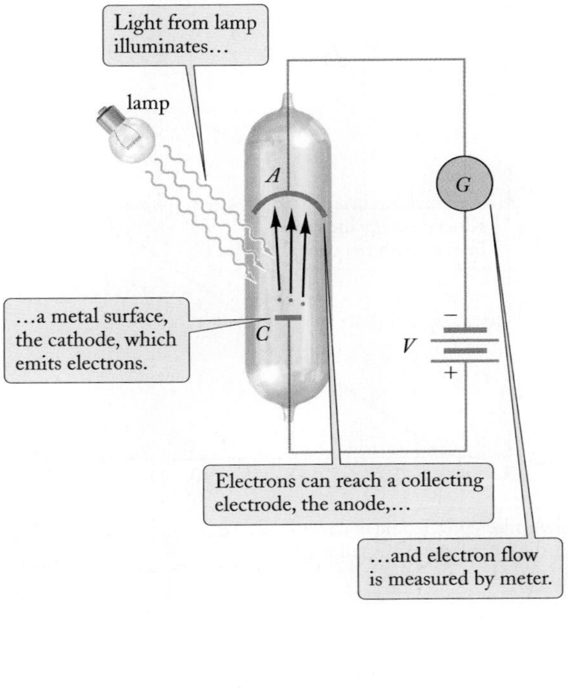

FIGURE 37.10 Schematic diagram of the apparatus for the investigation of the photoelectric effect.

and collector must be less than or equal to its initial kinetic energy. If the change in potential energy exceeds the initial kinetic energy, then the electron will reverse its motion and return to the emitter. The critical potential V_{stop} that stops the flow of electrons from emitter to collector is called the **stopping potential**. The measured value of this stopping potential gives us the initial kinetic energy of the electrons:

stopping potential

$$K = eV_{stop} \tag{37.10}$$

Experimentally, it is found that the kinetic energy determined in this way increases directly with the frequency of the incident light. For example, Fig. 37.11 is a plot of kinetic energy vs. frequency of the light for photoelectrons ejected from sodium. Note that, according to this plot, if the frequency is below 4.4×10^{14} Hz, then the light is incapable of ejecting electrons.

Einstein's quantum theory of light accounts for these experimental observations as follows. The electrons in the illuminated electrode absorb photons from the light, one at a time. When an electron absorbs a photon, it acquires an energy hf. But before this electron can emerge from the electrode, it must overcome the restraining forces that bind it to the metal of the electrode. The energy required for this is called the **work function** of the metal, designated by ϕ. The remaining energy of the electron is then $hf - \phi$, and this must be the kinetic energy of the emerging electron:

work function ϕ

Einstein's photoelectric equation

$$K = hf - \phi \tag{37.11}$$

Some electrons suffer extra energy losses in collisions within the metal before they emerge; thus, $hf - \phi$ actually is the maximum possible kinetic energy with which electrons can emerge.

Equation (37.11) is Einstein's photoelectric equation. It shows that the kinetic energy does indeed increase directly with the frequency, in agreement with the data of Fig. 37.11.

According to Eq. (37.11), a minimum frequency is required to achieve the ejection of an electron. This minimum frequency, called the **threshold frequency**, corresponds to the ejection of an electron of zero kinetic energy; such an electron is just barely ejected. The threshold frequency f_{thresh} is given by

threshold frequency

$$0 = hf_{thresh} - \phi$$

or

$$f_{thresh} = \frac{\phi}{h} \tag{37.12}$$

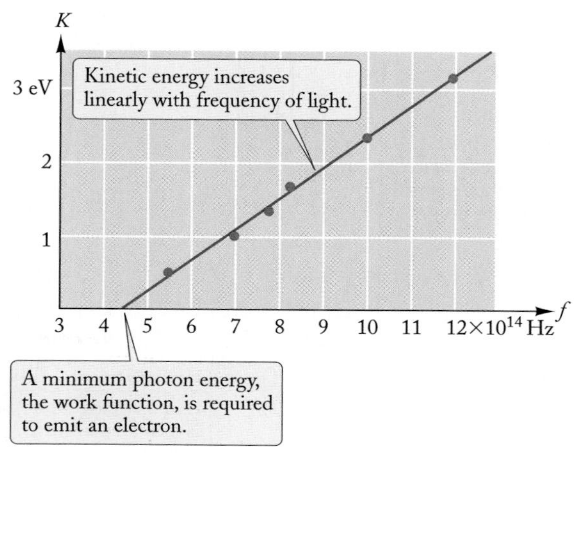

FIGURE 37.11 Kinetic energies (in electron-volts) of photoelectrons ejected from sodium by light of different frequencies.

When an electron absorbs a photon of this frequency, all of the energy of the photon is used to overcome the restraining forces that bind the electron to the metal, and no kinetic energy is left over for the ejected electron.

EXAMPLE 5 According to Fig. 37.11, the threshold frequency for sodium is 4.4×10^{14} Hz (photons of a frequency below this are incapable of ejecting electrons). What is the work function for sodium? Express the answer in eV.

SOLUTION: With $f_{\text{thresh}} = 4.4 \times 10^{14}$ Hz, Eq. (37.12) gives

$$\phi = hf_{\text{thresh}} = 6.63 \times 10^{-34}\ \text{J}\cdot\text{s} \times 4.4 \times 10^{14}\ \text{Hz} = 2.9 \times 10^{-19}\ \text{J}$$

$$= 2.9 \times 10^{-19}\ \text{J} \times \frac{1\ \text{eV}}{1.6 \times 10^{-19}\ \text{J}} = 1.8\ \text{eV}$$

EXAMPLE 6 The work function for platinum is 6.2 eV. If ultraviolet light of frequency 5.0×10^{15} Hz illuminates a platinum electrode, what is the maximum kinetic energy of the ejected electrons? What is the stopping potential?

SOLUTION: With $\phi = 6.2\ \text{eV} = 6.2 \times 1.6 \times 10^{-19}\ \text{J} = 9.9 \times 10^{-19}\ \text{J}$, we find from Eq. (37.11)

$$K = hf - \phi = (6.63 \times 10^{-34}\ \text{J}\cdot\text{s} \times 5.0 \times 10^{15}\ \text{Hz}) - 9.9 \times 10^{-19}\ \text{J}$$

$$= 2.3 \times 10^{-18}\ \text{J}$$

The stopping potential is, from Eq. (37.10),

$$V_{\text{stop}} = \frac{K}{e} = \frac{2.3 \times 10^{-18}\ \text{J}}{1.6 \times 10^{-19}\ \text{C}} = 14\ \text{V} \qquad (37.13)$$

Einstein's photoelectric equation was verified in detail by a long series of meticulous experiments by R. A. Millikan (the data in Fig. 37.11 are due to him). In order to obtain reliable results, Millikan found it necessary to take extreme precautions to avoid contamination of the surface of the photosensitive electrode. Since the surfaces of metals exposed to air quickly accumulate a layer of oxide, he developed a technique for shaving the surfaces of his metals in a vacuum, by means of a magnetically operated knife.

The results of these experiments gave strong support to the quantum theory of light. This success of Einstein's theory was all the more remarkable in view of the failure of the classical wave theory of light to account for the features of the photoelectric effect. According to classical theory, an electromagnetic wave acts on the electron by means of its electric field, which exerts a force on the electron. Therefore the crucial parameter that determines the ejection of a photoelectron should be the intensity of light, since this determines the strength of the electric field in the wave. If an intense wave strikes an electron, it should be able to jolt it loose from the metal, regardless of the frequency of the wave. Furthermore, the kinetic energy of the ejected electron should depend on the intensity of the wave. The observational evidence contradicts these predictions of the classical theory: A wave with a frequency below the threshold frequency never ejects an electron, regardless of its intensity. And, furthermore, the

PHYSICS IN PRACTICE PHOTOMULTIPLIER

The photoelectric effect finds many practical applications in sensitive electronic devices for the detection of light. For instance, in a photomultiplier tube, an incident photon ejects an electron from an electrode at the faceplate of the tube (Fig. 1). To convert this single electron into a measurable pulse of current, electric fields within the tube accelerate this single electron toward a second electrode (called a dynode; see the figure), where its impact ejects several secondary electrons. These, in turn, are accelerated toward a third electrode, where their impact ejects tertiary electrons, and so on. Thus, the single electron from the first electrode generates an avalanche of electrons. In a high-gain photomultiplier tube, a pulse of 10^9 electrons emerges from the last electrode, delivering a pulse of current to an external circuit. In this way, the photomultiplier tube can detect the arrival of individual photons. A similar solid-state device known as an avalanche photodiode is also used for photon counting. Some sensitive video cameras rely on the same multiplier principle to convert the arrival of a photon at a photosensitive faceplate into a pulse of current. This permits these cameras to take pictures in faint light, where there are few photons.

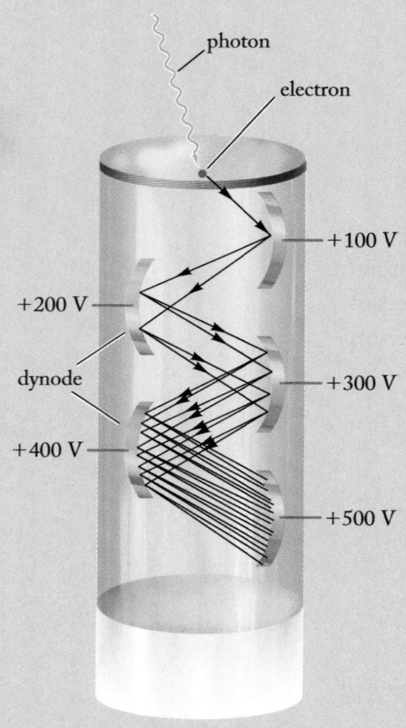

FIGURE 1 Schematic diagram of a photomultiplier tube. The curved electrodes are called dynodes. For the purpose of this diagram, it has been assumed that each electron impact on a dynode releases two electrons. The arrows show an avalanche of electrons.

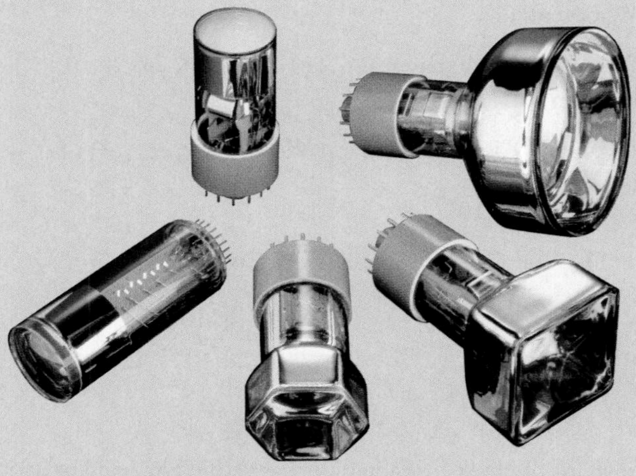

FIGURE 2 Photomultiplier tubes.

kinetic energy depends on the frequency [as specified by Eq. (37.11)] and not on the intensity. High-intensity light ejects more photoelectrons, but does not give the individual electrons more kinetic energy.

Checkup 37.3

QUESTION 1: The colors of light range from red to violet. What color has the most energetic photons?

QUESTION 2: Platinum has a larger work function than sodium. Qualitatively, how does the K vs. f plot for platinum differ from that for sodium (Fig. 37.11)?

QUESTION 3: Figure 37.12 shows a plot of current vs. applied potential for the photoelectric current emitted by the surface of metal illuminated with light of a given wavelength. Qualitatively, why is the current zero if $V < -V_{\text{stop}}$? Why does the current level off for large positive V? Why do the curves differ for different intensities of light?

QUESTION 4: Sodium has a work function of 1.8 eV; for platinum, the value is 6.2 eV. Compared with the plot of the maximum kinetic energy of ejected electrons as a function of the frequency of the incident light for sodium (Fig. 37.11), a similar plot for platinum has

(A) A larger slope (B) A smaller slope (C) The same slope

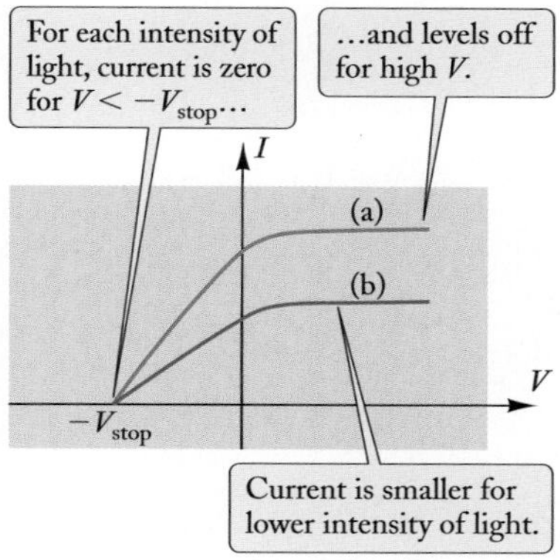

FIGURE 37.12 I vs. V for two different values of the intensity of light: (a) high intensity; (b) low intensity.

37.4 THE COMPTON EFFECT

Very clear experimental evidence for the particle-like behavior of photons was uncovered by Arthur Holly Compton in 1922. Compton had been investigating the scattering, or the deflection, of X rays by a target of graphite (see Fig. 37.13). According to Maxwell's theory, X rays are merely light waves of extremely high frequency. But according to quantum theory, they ought to consist of photons; and since their frequency is much higher than that of ordinary light, the energy of the X-ray photons ought to be much larger than that of photons of ordinary light (X rays will be discussed in more detail in the next section).

When Compton bombarded the graphite with X rays of one selected wavelength, he found that the scattered X rays had wavelengths somewhat larger than that of the incident X rays. Classical theory cannot explain such a change of wavelength. When a classical electromagnetic wave is incident on a graphite target, its oscillating electric fields exert forces on the electrons in the carbon atoms, and this causes the electrons to oscillate at the same frequency as the wave. The oscillating, accelerated electrons then radiate a new electromagnetic wave, which spreads outward in all directions. This radiated wave is the scattered wave. Its frequency is necessarily the same as that of the oscillating electrons, which is the same as that of the incident wave.

Compton soon recognized that the change of wavelength could be understood in terms of collisions of photons with electrons, collisions in which the photons behave like particles. In such a collision the electron of a carbon atom can be regarded as free, because the force binding the electron to the atom is insignificant compared with the force exerted by the incident photon. When the photon bounces off the electron, the electron recoils and thereby picks up some of the photon's energy—the deflected photon

ARTHUR HOLLY COMPTON (1892–1962) *American experimental physicist. For his discovery of the Compton effect, he received the Nobel Prize in 1927.*

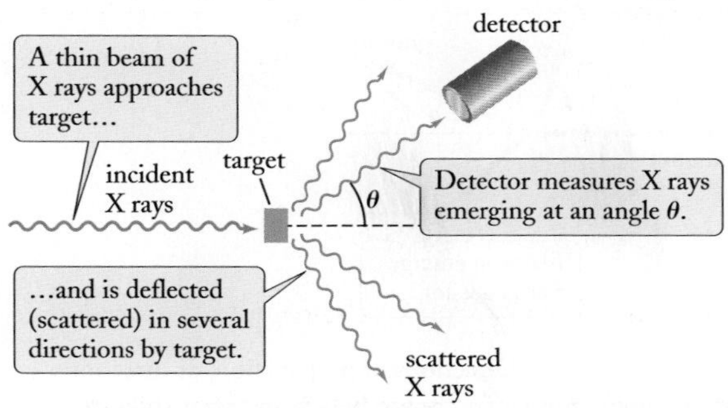

FIGURE 37.13 Scattering of X rays by a target of graphite.

is left with a reduced energy. Since the energy of the photon is $E = hf = hc/\lambda$, a reduction of the energy of the photon implies an increase of wavelength. Qualitatively, we expect that photons that suffer the most violent collisions will be deflected through the largest angles and lose the most energy; and therefore such photons should emerge with the longest wavelength. And this is just what Compton found in his experiments.

For a quantitative discussion of the photon–electron collision, we need an expression for the momentum of the photon. We can derive this expression from the relativistic formula for the momentum of a particle. Since the speed of the photon is the speed of light, the photon must be regarded as an ultra-relativistic particle. For such a particle, Eq. (36.52) tells us that

$$p = \frac{E}{c} \tag{37.14}$$

[Note that according to Eq. (36.50), the energy of a particle with a speed equal to the speed of light can be finite only if the rest mass m is zero; thus, photons must be regarded as particles of zero rest mass.] Since the energy of the photon is $E = hf = hc/\lambda$, the expression for the momentum can be written as

momentum of photon

$$p = \frac{h}{\lambda} \tag{37.15}$$

With the expression (37.15) for the momentum of the photon, Compton calculated the change of energy and the change of wavelength of a photon in an elastic collision with an electron, initially at rest. For a photon that emerges from the collision at an angle θ (the deflection angle; see Fig. 37.14), he calculated from the laws of conservation of energy and momentum that the change of wavelength is

Compton wavelength shift

$$\Delta\lambda = \frac{h}{m_e c}(1 - \cos\theta) \tag{37.16}$$

This change of wavelength is called the **Compton shift**. The change of wavelength is always small; it is always less than 0.005 nm (see the next example). Such changes of wavelength are difficult to detect unless the wavelength of the X rays is itself quite small, so $\Delta\lambda$ is an appreciable fraction of the wavelength.

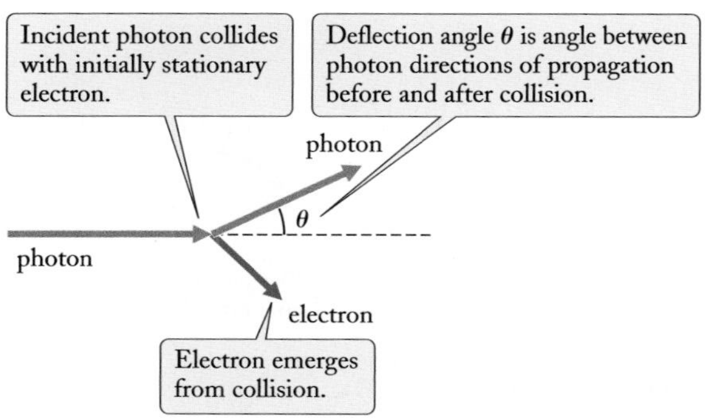

FIGURE 37.14 Directions of propagation, or directions of the momentum, of a photon before and after collision with an electron.

EXAMPLE 7 Evaluate the Compton shift for X rays scattered at 90° and for X rays scattered at 180°.

SOLUTION: If the photon emerges from the collision at an angle of 90°, then $\cos\theta = \cos 90° = 0$, and the wavelength shift of Eq. (37.16) is

$$\Delta\lambda = \frac{h}{m_e c} = \frac{6.63 \times 10^{-34}\ \text{J}\cdot\text{s}}{9.11 \times 10^{-31}\ \text{kg} \times 3.00 \times 10^{8}\ \text{m/s}} = 2.43 \times 10^{-12}\ \text{m}$$

$$= 0.002\,43\ \text{nm}$$

If the photon emerges at 180°, in a direction opposite to that of its initial direction of motion, then $\cos\theta = \cos 180° = -1$, and the wavelength shift is

$$\Delta\lambda = \frac{h}{m_e c} \times [1 - (-1)] = \frac{2h}{m_e c} = 4.85 \times 10^{-12}\ \text{m} = 0.004\,85\ \text{nm}$$

This is the maximum wavelength shift that can be produced by the Compton effect.

EXAMPLE 8 The formula (37.16) for the wavelength shift can be deduced by examining the conservation of energy and momentum in the photon–electron collision. As a simple special case of such a collision, consider a photon that recoils back in the direction it came from [that is, $\theta = 180°$ in Eq. (37.16)] while the electron, initially at rest, acquires a motion in the forward direction. For such a straight-line motion of photon and electron, write down the laws of conservation of energy and momentum, and deduce the wavelength change of the photon.

SOLUTION: We will use the relativistic expressions for momentum and energy, because, in a collision with a high-energy photon, the electron will emerge with an energy large enough to require such a relativistic treatment. In the notation familiar from Chapter 11, we designate the initial momentum of the photon by p_1, the final momentum by p_1'; likewise, we designate the initial momentum of the electron by p_2 (which is zero, since the electron is initially at rest), the final momentum by p_2'. Conservation of momentum then tells us

$$p_1 = p_1' + p_2' \tag{37.17}$$

and conservation of energy tells us

$$E_1 + m_e c^2 = E_1' + E_2' \tag{37.18}$$

where $m_e c^2$ is the initial energy of the electron (the rest-mass energy). To solve this pair of equations for the final photon energy, we need to express the momenta in Eq. (37.17) in terms of the energies. For this purpose, we first rewrite Eq. (37.17) as $p_2' = p_1 - p_1'$ and square both sides:

$$(p_2')^2 = (p_1 - p_1')^2 \tag{37.19}$$

Now we substitute the relativistic relation $(p_2')^2 = (E_2'/c)^2 - m_e^2 c^2$ for the electron [compare Eq. (36.51)] and the relations $p_1 = E_1/c$ and $p_1' = -E_1'/c$ for the photon:

$$\left(\frac{E_2'}{c}\right)^2 - m_e^2 c^2 = \left(\frac{E_1}{c} + \frac{E_1'}{c}\right)^2$$

By means of Eq. (37.18) we can now eliminate the energy of the electron and obtain an equation that relates the initial and final energies of the photon:

$$\left(\frac{E_1}{c} + m_e c - \frac{E_1'}{c}\right)^2 - m_e^2c^2 = \left(\frac{E_1}{c} + \frac{E_1'}{c}\right)^2$$

When we expand the squares in this equation, several terms cancel, leaving us with the expression

$$2m_e c\left(\frac{E_1}{c} - \frac{E_1'}{c}\right) = 4\frac{E_1}{c}\frac{E_1'}{c}$$

Multiplying both sides by $c/(2m_e E_1 E_1')$ gives

$$\frac{1}{E_1'/c} - \frac{1}{E_1/c} = \frac{2}{m_e c} \tag{37.20}$$

But $E_1/c = p_1 = h/\lambda$ from Eq. (37.15), so Eq. (37.20) is equivalent to

$$\frac{\lambda'}{h} - \frac{\lambda}{h} = \frac{2}{m_e c}$$

that is,

$$\Delta\lambda = \frac{2h}{m_e c} \tag{37.21}$$

This result agrees with our general formula (37.16) when we set $\theta = 180°$ and $\cos\theta = -1$.

✔ Checkup 37.4

QUESTION 1: The colors of light range from red to violet. What color has photons of the largest momentum?

QUESTION 2: Does the photon gas in a cavity in a hot body exert a pressure on the surrounding walls?

QUESTION 3: Consider X rays of wavelengths 0.2 nm and 0.6 nm. If these suffer Compton collisions with electrons and emerge at the same scattering angle, which have the larger wavelength shift? Which have the larger percentage change of wavelength?

QUESTION 4: Can the Compton effect occur with visible light? Would it be observable?

QUESTION 5: For Compton scattering, it is impossible that the wavelength shift $\Delta\lambda$ is

(A) Smaller than the wavelength of the incident photon
(B) Larger than the wavelength of the incident photon
(C) Less than 0.001 nm
(D) Greater than 0.005 nm
(E) Larger for larger scattering angles

37.5 X RAYS

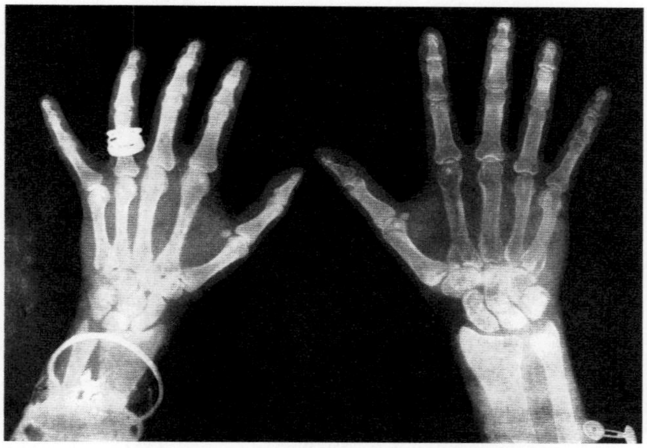

FIGURE 37.15 One of the first X-ray photographs prepared by Röntgen. It shows the hands of his wife; note the sharp image of the ring.

X rays were discovered in 1895 by W. K. Röntgen in experiments with beams of energetic electrons. Röntgen found that when such a beam of electrons moving in an evacuated glass tube struck the wall of the tube, some invisible, mysterious rays were emitted that caused a faint luminosity (fluorescence) on a nearby sheet of paper impregnated with chemicals. He also found that these rays could affect covered and wrapped photographic plates—the rays could penetrate through the wrapping. These rays proved capable of penetrating though thick layers of opaque materials, and Röntgen immediately recognized the possible medical applications of these rays for making images of the tissues inside the human body, especially bones (see Fig. 37.15).

The distance that X rays can penetrate through a material depends on the density of the material. When X rays pass through atoms, they tend to be absorbed by the atomic electrons. Therefore materials of high density, such as lead, with a concomitant high density of electrons, strongly absorb and block X rays. In X-ray photographs of parts of the human body, bones throw sharp shadows because their density is higher than that of the surrounding tissues. Organs made of soft tissues of low density, such as the gastrointestinal tract, do not throw sharp shadows, and to enhance the contrast of the X-ray photograph it is advantageous to fill the organ with a barium solution, a high-density material that blocks the X rays (see Fig. 37.16).

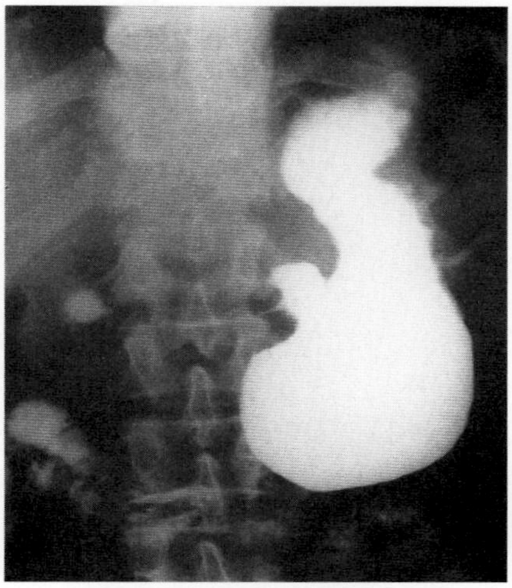

FIGURE 37.16 X-ray photograph of a stomach with barium shadowing.

Röntgen had named his rays "X rays" because their nature was unknown. Although he and his contemporaries suspected that they might be electromagnetic waves of extremely short wavelength, the conclusive experimental proof of this conjecture was not obtained until 1912, when Max von Laue argued that if X rays are waves, they should display interference effects when passing through crystals. The distances between the rows of atoms in a crystal, such as rock salt, are of the same order of magnitude as the wavelengths of X rays, and von Laue proposed that the crystal therefore can play the role of a "grating" for X rays, analogous to a multiple-slit grating used for interference experiments with light. Figure 37.17 shows a photograph of the interference pattern produced by X rays incident on a crystal. In this photograph the interference maxima show up as dark spots. The wavelength of the X rays can be deduced from the angular positions of the interference maxima and the known size of the spacings in the crystal. The wavelengths of X rays range from about 0.001 nm to 10 nm.

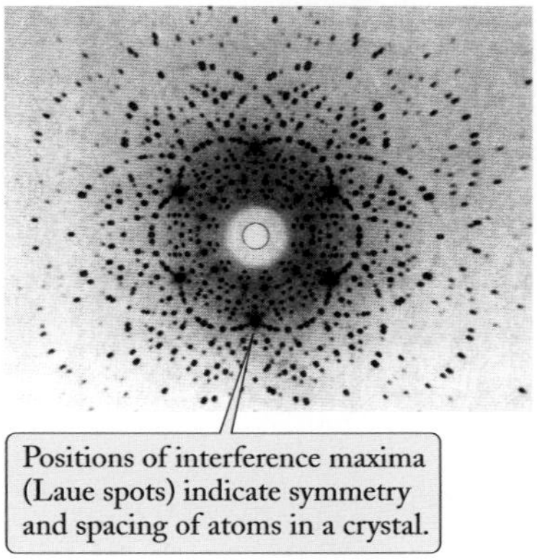

FIGURE 37.17 Interference pattern produced by X rays incident on a crystal.

The spots in such photographs of X-ray interference patterns are called **Laue spots**. The beautiful symmetry of the pattern of Laue spots reflects the symmetry of the arrangement of the atoms in the crystal. In modern crystallography laboratories, the patterns of Laue spots produced by X rays incident on a crystal are often used to investigate the structure of the crystal and that of the molecules in it. For instance, such X-ray interference experiments played a large role in the determination of the structure of DNA.

Since X rays are electromagnetic waves, their generation by the impact of energetic electrons on some sort of obstacle can be understood by the familiar mechanism of the emission of radiation by acceleration of electric charges, which we discussed in Chapter 33. When the electrons in the beam collide with the atoms in the obstacle, they will suffer sudden decelerations and radiate intense electromagnetic waves of short wavelength. This kind of radiation is called **Bremsstrahlung** (German for *braking radiation*). Figure 37.18 shows an X-ray tube in which electrons emerging from a hot filament are accelerated through a potential difference of several kilovolts and then

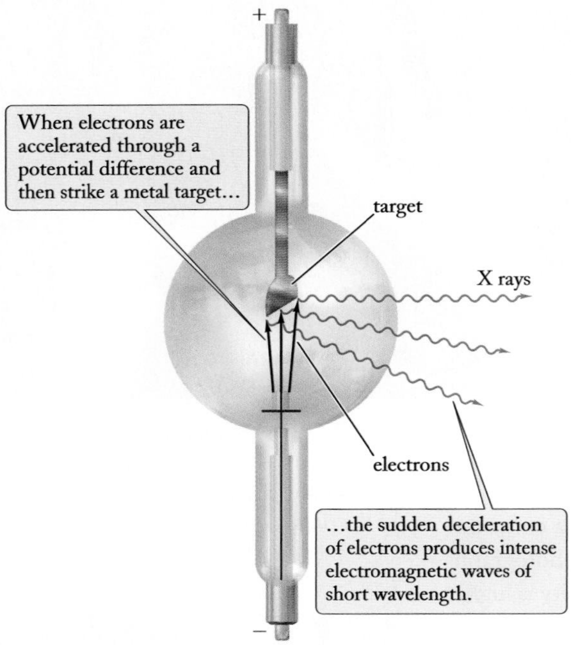

FIGURE 37.18 An X-ray tube.

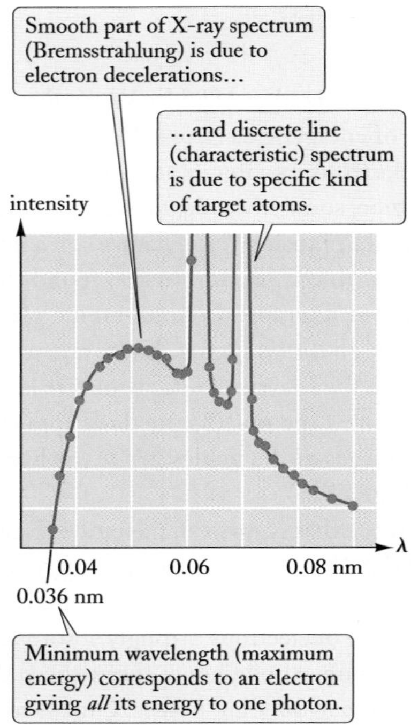

FIGURE 37.19 Distribution of energy in the spectrum of X rays produced by electrons of 35 kilovolts incident on a molybdenum target.

strike an obstacle, or target, made of a heavy metal, such as tungsten or molybdenum. The X rays produced in the decelerations of the electrons escape through the side of the tube. Such X-ray tubes are widely used in medical X-ray machines.

Figure 37.19 is a plot of the distribution of energy of the X rays generated by electrons of 35 kilovolts striking a molybdenum target. Note that the X-ray energy is smoothly distributed over a wide range of wavelengths, but there also are two conspicuous spikes in the energy distribution. The smooth portion of the X-ray spectrum is due to Bremsstrahlung, whereas the discrete spikes are generated in the interior of the atoms of molybdenum, in much the same way as spectral lines of visible light are generated by the atoms (see Chapter 38). In the plot of the energy distribution, the shape and the location of the broad peak of the Bremsstrahlung portion of the spectrum depend on the energy of the incident electrons. But the locations of the spikes do not depend on the electron energy; instead they depend on the material of the target. The spikes are called the **characteristic spectrum**; each kind of target atom has its own distinctive characteristic spectrum, just as each kind of atom has its own distinctive spectrum of visible light (for more on the characteristic spectra of atoms, see the next chapter).

Let us ignore the spikes for now, and concentrate on the smooth Bremsstrahlung spectrum. Note that below a certain wavelength—for instance, below about 0.03 nm in Fig. 37.19—there is no Bremsstrahlung. The minimum wavelength emitted by the electrons is called the **cutoff wavelength**. Experimentally, the cutoff wavelength is found to be inversely proportional to the kinetic energy of the electrons. This cutoff makes no sense from the point of view of classical theory—if an electron suffers a sufficiently violent collision with an atom, it ought to be able to radiate waves of arbitrarily short wavelength (although the intensity of short-wavelength waves is expected to be low). But the cutoff is readily explained by the quantum theory of light, according to which the decelerating electron emits photons. Obviously, the maximum energy that a decelerating electron can give to a photon is *all* its energy. In this case, the electron emits only one single photon, with a maximum energy and a maximum frequency, or **cutoff frequency**, f_{cutoff}. The energy hf_{cutoff} of this single photon of maximum energy then equals the kinetic energy K of the electron:

$$hf_{\text{cutoff}} = K \tag{37.22}$$

from which

$$f_{\text{cutoff}} = \frac{K}{h}$$

The wavelength is related to the frequency by $\lambda = c/f$, so the cutoff wavelength is

$$\lambda_{\text{cutoff}} = \frac{hc}{K} \tag{37.23}$$

cut-off wavelength

EXAMPLE 9 If an X-ray tube is operated at a potential of 35 kilovolts, the electron energy is 35 keV. What is the cutoff wavelength for the X rays generated by electrons of this energy?

SOLUTION: In joules, the electron energy is

$$K = 35 \text{ keV} \times 1.60 \times 10^{-19} \text{ J/eV} = 5.6 \times 10^{-15} \text{ J}$$

Equation (37.23) then gives

$$\lambda_{\text{cutoff}} = \frac{hc}{K} = \frac{6.63 \times 10^{-34} \text{ J}\cdot\text{s} \times 3.00 \times 10^{8} \text{ m/s}}{5.6 \times 10^{-15} \text{ J}}$$

$$= 3.6 \times 10^{-11} \text{ m} = 0.036 \text{ nm}$$

This is in agreement with the cutoff wavelength indicated in Fig. 37.19.

✔ Checkup 37.5

QUESTION 1: A fine slit cut in a sheet of lead produces noticeable diffraction effects for light, but not for X rays. Why not?

QUESTION 2: Figure 37.19 shows the energy distribution for X rays produced by electrons of 35 keV incident on a molybdenum target. If we reduce the electron energy to 17.5 keV, qualitatively, what will change in this figure?

QUESTION 3: Sometimes an X-ray photograph of the intestines reveals bubbles of gas. Keeping in mind that X-ray photographs are usually negatives (they turn dark where the X rays strike, as in Fig. 37.15), would you expect a bubble of gas to look light or dark?

QUESTION 4: When electrons are accelerated through an electrostatic potential V_0 and strike a target, the frequencies f of emitted X rays obey the relation

(A) $f \geq V_0/h$ (B) $f \leq V_0/h$ (C) $f \geq eV_0/h$
(D) $f \leq eV_0/h$ (E) $f \geq eV_0/hc$

37.6 WAVE VS. PARTICLE

From the photoelectric effect and the Compton effect, we learned that photons have particle properties. On the other hand, we know that light displays interference and diffraction phenomena, which prove that photons also have wave properties. Thus photons are neither classical particles nor classical waves. They are some new kind of object, unknown to classical physics, with a subtle combination of both wave and particle properties. Arthur Eddington coined the name **wavicle** for this new kind of object.[2] It is difficult to achieve a clear understanding of the character of a wavicle because these objects are very remote from our everyday experience. We all have an intuitive grasp of the concept of classical particles and classical waves from our experience with, say, billiard balls and water waves, but we have no such experience with wavicles.

wavicle

We can gain some insight into the interplay between the particle behavior and the wave behavior of a photon by a new, detailed examination of a simple diffraction experiment. Figure 37.20 shows a light beam striking a plate with a narrow slit. In the usual diffraction experiments described in Chapter 35, we installed a screen or a photographic film at the far right, and with this we recorded the intensity of the light in the diffraction pattern. Instead, in our new arrangement we will install the faceplate of a very sensitive video camera in place of the customary screen. With this, we can detect the individual photons of the light in the diffraction pattern.

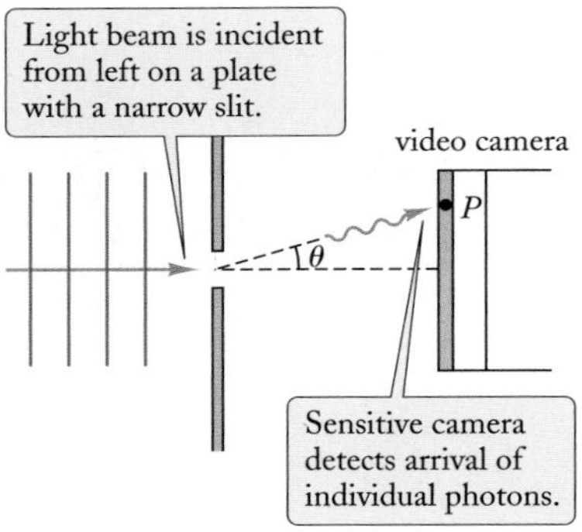

FIGURE 37.20 A diffraction experiment.

If the incident light beam has a very low intensity, so there is only one photon passing through the slit at a time, we can watch the photons arriving one by one at the faceplate of our video camera. Figure 37.21a shows a typical pattern of impacts of 30 photons. The pattern seems quite random. If the photons behaved like classical particles, they would travel along a straight line and they would reach only those points on the faceplate that are within the geometric shadow of the slit. The widely scattered impacts prove that the photons are certainly not traveling along such straight lines. Figure 37.21b shows the pattern of accumulated impacts for 300 photons, and Fig. 37.21c shows it for 3000 photons. In these figures we can recognize a tendency of the photons to cluster in bandlike zones. These zones correspond to the maxima of the diffraction pattern predicted by the wave theory of light. Finally, Fig. 37.21d shows the pattern of accumulated impacts for a very large number of photons; this is simply the familiar intensity pattern of light diffracted by a slit (see also Fig. 35.33).

FIGURE 37.21 Patterns of impacts of photons on the video camera in Fig. 37.20. (a) 30 photons; (b) 300 photons; (c) 3000 photons; (d) a very large number of photons. The first three pictures are computer simulations of accumulated photon impacts; the last picture is a diffraction pattern obtained with a laser beam.

[2]Other names that have been proposed are *quanticle* and *quon*. Most physicists prefer to avoid all such neologisms; if anything, they favor the descriptive phrase *quantum-mechanical particle* or *wave-mechanical particle*.

PROBLEMS

37.1 Blackbody Radiation
37.2 Energy Quanta

1. Consider a seconds pendulum, that is, a pendulum that has a period of 2 seconds. What is the magnitude of one energy quantum for such a pendulum? Would you expect that quantum effects are noticeable in such a pendulum?
2. In molecules, the atoms can vibrate about their equilibrium positions. For instance, in the H_2 molecule, the hydrogen atoms vibrate about their equilibrium positions with a frequency of 1.31×10^{14} Hz. What is the magnitude of the energy quantum for this oscillating system? Note that although there are two masses in this system, they must be regarded as a single oscillator because the vibrational motions of the two masses are always equal.
3. In a solid, each atom is held in its position by elastic forces, which permit the atom to oscillate about its equilibrium point. The frequency of such oscillations of an aluminum atom in solid aluminum is 8.0×10^{11} Hz. The energy of this oscillatory motion is quantized. What is the magnitude of an energy quantum?
4. If we take Planck's model of the walls of a blackbody cavity seriously, we will have to assume that the oscillating masses are electrons. Imagine an electron oscillating with a frequency of 2.0×10^{15} Hz under the influence of a springlike force. What is the amplitude of oscillation of this electron if its energy of oscillation is one energy quantum? Two energy quanta?
5. An oscillator of frequency 3.00×10^{13} Hz consists of a mass of 1.67×10^{-27} kg attached to a spring. What is the amplitude of oscillation of this oscillator if its energy of oscillation is one energy quantum? Two energy quanta?
6. Suppose that two stars have the same size but the temperature of one is twice that of the other. By what factor will the thermal power radiated by the hotter star be larger than that radiated by the cooler star?
7. The theoretical expression for the constant σ in the Stefan–Boltzmann Law is

$$\sigma = \frac{2\pi^5 k^4}{15h^3c^2}$$

Verify that the numerical value implied by this theoretical expression agrees with Eq. (37.8).

8. The flux of thermal radiation from the star Procyon B is observed to have a maximum at a wavelength of 440 nm. Assuming the star radiates like a blackbody, what temperature can you deduce for the surface of this star?
9. Melting iron has a temperature of 1808 K. At what wavelength does the iron radiate a maximum flux? Assume that the iron acts like a blackbody.
10. At what wavelength does your skin radiate a maximum flux of thermal radiation? Assume that the temperature of your skin is 33°C.
11. Nanomechanical oscillators with a frequency of 5.0 GHz and masses as small as 2.0×10^{-20} kg can be fabricated. What is the amplitude of oscillation of such an oscillator if its energy of oscillation is one energy quantum?
12. Thermonuclear reactions near the core of the Sun maintain a temperature of 5.0×10^7 K. Assume that the distribution of this radiation is thermal. At what wavelength does the spectrum of this radiation have a maximum energy flux? What is this part of the electromagnetic spectrum called? What is the energy per second emitted per square meter near the Sun's core due to this radiation?
13. Low temperatures are commonly attained using liquid helium, which has a latent heat such that 1.0 watt will boil approximately 1.0 liter of liquid helium per hour. Consider a region with a surface area of 0.10 m^2. What is the rate of energy transfer into the region due to thermal radiation if it is surrounded by a room-temperature (295 K) environment? If surrounded by a liquid nitrogen (77 K) environment? Which is preferable in order to conserve liquid helium?
14. The nichrome heater wire in a toaster has a radius of 0.20 mm and a total length of 1.0 m. Assume that the wire acts as a blackbody and emits 1200 W of thermal radiation. What is the surface temperature of the wire? At what wavelength is the thermal radiation maximum?
15. A black hole radiates with a thermal spectrum due to quantum effects. A black hole with a radius of 30 km radiates a total of 8.8×10^{-31} W of such thermal radiation, or **Hawking radiation.** What is the temperature of this blackbody?

*16. The tungsten filament of a lightbulb is a wire of diameter 0.080 mm and length 5.0 cm. The filament is at a temperature of 3200 K. Calculate the power radiated by the filament. Assume the filament acts like a blackbody.

*17. The temperature of the surface of the Sun is 5800 K and the Sun's radius is 6.96×10^8 m. If the Sun radiates like a blackbody, what can you predict for the thermal energy flux it emits? What can you predict for the energy flux of sunlight arriving at the Earth, at a distance of 1.5×10^{11} m? Compare with the measured value, 1.35×10^3 W/m^2.

*18. The star Procyon B is at a distance of 11 light-years from Earth. The flux of its starlight reaching us is 1.7×10^{-12} W/m^2, and the surface temperature of the star is 6600 K. Calculate the size of the star.

*19. Consider the integral over wavelengths of Planck's expression, Eq. (37.4), for the distribution of blackbody radiation

$$S = \int dS = \int S_\lambda d\lambda = \int_0^\infty \frac{2\pi hc^2}{\lambda^5 \times (e^{hc/\lambda kT} - 1)} d\lambda$$

Show that the total intensity emitted by a blackbody is proportional to the fourth power of the absolute temperature, as required by the Stefan–Boltzmann Law. (Hint: You need not evaluate the integral; you need only use the substitution $x = hc/\lambda kT$ to make the integral dimensionless.)

*20. At the Earth, the flux of sunlight per unit area facing the Sun is 1.35×10^3 W/m^2. The Earth absorbs heat from the sunlight and reradiates heat as thermal infrared radiation. For equilibrium, the power arriving from the Sun must equal the average power radiated by the surface of the Earth. This permits us to make a rough prediction for the average temperature of the Earth.

(a) Assume that the Earth absorbs all of the sunlight striking it. What is the power of the sunlight absorbed? (Hint: The relevant area is the cross-sectional area πR^2 of the Earth.)

(b) Assume that the surface of the Earth radiates like a blackbody. If the temperature of the surface is T, what is the expression for the power radiated? (Hint: The relevant area is the total surface area $4\pi R^2$.) If this power is to match the power calculated in (a), what must be the value of T?

**21. Deduce the surface temperature of Pluto by the method described in the preceding problem. Pluto is 39 times as far away from the Sun as the Earth, and hence the flux of sunlight at Pluto is $(39)^2$ times smaller than the flux at the Earth.

**22. Maximize Planck's expression for the intensity of blackbody radiation per unit wavelength, Eq. (37.4), and thus obtain Wien's Law, Eq. (37.5). (Hint: Once you obtain the condition for a maximum, consider the dimensionless variable $x = hc/\lambda kT$. Find a solution for x iteratively, for example, using a calculator.)

**23. In Problem 37 of Chapter 24, you will find a description of the Thomson model of the hydrogen atom.

(a) What is the frequency of oscillation of the electron in this atom?

(b) The electron can be regarded as an oscillator. Show that if the energy of the electron is one quantum, the amplitude of oscillation exceeds the radius (5×10^{-11} m) of the atom.

37.3 Photons and the Photoelectric Effect

24. Photons of green light have a wavelength of 550 nm. What is the energy and what is the momentum of one of these photons?

25. You are lying on a beach, tanning in the sun. Roughly how many photons strike your skin in one hour? Assume that the energy flux of sunlight is as described in Example 4.

26. For each of the following kinds of electromagnetic waves, find the energy of a photon: FM radio wave of wavelength 3.0 m, infrared light of 1.0×10^{-5} m, visible light of 5.0×10^{-7} m, ultraviolet light of 1.0×10^{-7} m, X rays of 1.0×10^{-10} m.

27. A radio transmitter radiates 10 kW at a frequency of 8.0×10^5 Hz. How many photons does the transmitter radiate per second?

28. The energy flux in the starlight reaching us from the bright star Capella is 1.2×10^{-8} W/m^2. If you are looking at the star, how many photons per second enter your eyes? The diameter of your pupil is 0.70 cm. Assume that the average wavelength of the light is 500 nm.

29. A laser emits a light beam with a power of 1.0 W. The wavelength of the light is 630 nm. How many photons per second does this laser emit?

30. The energy density in a radio wave of wavelength 3.0 m is 2.0×10^{-13} J/m^3. What is the corresponding density of photons?

31. Calculate the range of energies of photons of visible light, with wavelengths from 700 nm to 400 nm. Express your answers in units of electron-volts.

32. For light-sensing applications that do not require the extreme sensitivity of a photomultiplier (see Physics in Practice: Photomultiplier), photodiodes are often used. A typical photodiode can measure an incident photon flux provided the total power incident is at least 1.0×10^{-14} W; otherwise, the signal is lost in the electrical noise of the photodiode. Laser light of wavelength 633 nm is to be detected. What is the minimum number of photons per second required?

33. Light of wavelength 486 nm is incident on a material with a work function of 2.26 eV. What is the maximum kinetic energy of ejected electrons? When light of wavelength 434 nm is used, what stopping potential is necessary?

34. Show that if we express the energy of a photon in keV and the wavelength in nanometers, then

$$E = 1.24/\lambda$$

35. According to Fig. 37.11, what is the work function of sodium? Express your answer in electron-volts.

36. The work function of potassium is 2.26 eV. What is the threshold frequency for the photoelectric effect in potassium?

37. The work functions of K, Cr, Zn, and W are 2.26, 4.37, 4.24, and 4.49 eV, respectively. Which of these metals will emit photoelectrons when illuminated with red light ($\lambda = 700$ nm)? Blue light ($\lambda = 400$ nm)? Ultraviolet light ($\lambda = 280$ nm)?

38. The photons emitted by a sodium atom have a wavelength of 589 nm when at rest. If this atom is moving away from you at a speed of 2.0×10^6 m/s, what is the energy that you measure for one of these photons?

*39. By inspection of Fig. 37.11, find the slope of the line in eV/Hz. Convert these units into J·s, and verify that the slope is the same as Planck's constant.

*40. The binding energy of an electron in a hydrogen atom is 13.6 eV. Suppose that a photon of wavelength 40 nm strikes the atom and gives up all of its energy to the electron. With what kinetic energy will the electron be ejected from the atom?

37.4 The Compton Effect

41. X rays emitted by molybdenum have a wavelength of 0.072 nm. What are the energy and the momentum of one of the photons of these X rays?

42. In a collision with an initially stationary electron, a photon suffers a wavelength increase of 0.0022 nm. What must have been the deflection angle of the photon?

43. For an experiment on the Compton effect, you want the X rays emerging at 90° from the incident direction to suffer an increase of wavelength by a factor of 2. What wavelength do you need for your incident X rays?

44. A photon of wavelength 0.030 nm collides with a free electron at rest. Calculate the wavelength if the photon emerges from this collision with a deflection of 30°.

45. X rays of wavelength 0.030 nm are incident on a graphite target. Calculate the wavelength of the X rays that emerge from this target with a deflection of 60°. Calculate the wavelength of the X rays that emerge from this target with a deflection of 120°.

46. X-ray photons of wavelength 0.154 nm are produced by a copper source. Suppose that 1.00×10^{18} of these photons are absorbed by a target each second.

(a) What is the total momentum p transferred to the target each second?

(b) What is the total energy E of the photons absorbed by the target each second?

(c) For these values, verify that the force on the target is related to the rate of energy transfer by $dp/dt = (1/c)(dE/dt)$.

47. Calculate the percentage change in wavelength, $(\Delta\lambda/\lambda) \times 100\%$, when an X-ray photon of wavelength $\lambda = 0.0010$ nm is Compton-scattered through an angle of (a) 10°, (b) 30°, and (c) 180°.

48. Suppose that a photon is "Compton-scattered" from a proton instead of an electron. What is the maximum wavelength shift in this case?

*49. In a Compton scattering experiment, X rays of wavelength λ are incident. The maximum energy transferred to an electron is E_e. Find an expression for λ in terms of E_e, h, m_e, and c. What is the value of λ for $E_e = 25$ keV?

*50. Derive the general Compton-shift formula (37.16). [Hint: Repeat the calculation of Example 8, but replace Eq. (37.19) by the relation $p_1^2 - 2p_1p_1' \cos\theta + p_1'^2 = p_2'^2$, obtained from inspection of the diagram representing the vector sum of momentum vectors. Note that here the p's represent magnitudes of momentum vectors and they are positive, whereas in Example 8 the p's are positive or negative depending on the direction of motion.]

*51. What is the maximum energy that a free electron (initially stationary) can acquire in a collision with a photon of energy 4.0×10^3 eV?

*52. A photon of initial wavelength 0.040 nm suffers two successive collisions with two electrons. The deflection in the first collision is 90° and in the second collision it is 60°. What is the final wavelength of the photon?

37.5 X Rays

53. What is the energy of the photons in X rays of a wavelength of 0.050 nm?

54. An X-ray tube is being operated with electrons of energy 25 keV. What is the cutoff wavelength of the emitted X rays?

55. We want to use an X-ray tube to generate X rays of wavelength 1.00 nm. What is the minimum potential difference we must use to accelerate the electrons in the tube?

56. The tube in a medical X-ray machine can be operated at potential differences in the range from 25 kV to 150 kV. What is the cutoff wavelength of the X rays emitted when the tube is operated at 25 kV? At 150 kV?

57. For studies of the structure of materials, a characteristic spectral line of copper, with a wavelength of 0.1542 nm, is often used. For such X rays to be emitted, what is the minimum potential difference that could be used to accelerate electrons toward a copper target?

58. Tungsten targets are often used in X-ray tubes to produce an intense, broad spectrum of Bremsstrahlung. To study the spacing of different planes of atoms in a particular crystal, X-ray wavelengths throughout the range 0.10–40 nm are desired. What is the corresponding minimum accelerating potential for electrons?

37.6 Wave vs. Particle

59. A photon passes through a horizontal slit of width 5.0×10^{-6} m. What uncertainty in the vertical position will this photon have as it emerges from the slit? What uncertainty in the vertical momentum?

60. A photon traveling in the z direction passes through a square hole 2.0×10^{-6} m on a side. What is the uncertainty in each of the x and y components of the momentum? Assuming the uncertainty in the z component of the momentum is negligible, what is the uncertainty in the total momentum?

*61. A particular semiconductor laser emits photons with an uncertainty in position of 1.5 cm along the direction of propagation, known as the longitudinal coherence length. What is the corresponding uncertainty in the momentum of the photons? The photons have an average wavelength of 678 nm. What is the uncertainty in this wavelength?

*62. Consider a radio wave in the form of a pulse lasting 0.0010 s. The pulse then has a length of $0.0010 \text{ s} \times c = 3.0 \times 10^5$ m. Since an individual photon of this radio wave can be anywhere within this pulse, the uncertainty in the position of the photon is $\Delta x = 3.0 \times 10^5$ m along the direction of propagation.

(a) According to Heisenberg's relation, what is the corresponding uncertainty in the momentum of the photon?

(b) What is the uncertainty in the frequency of the photon?

REVIEW PROBLEMS

*63. Consider a mass of 1.0 g attached to a spring of spring constant 3.0×10^{-2} N/m. What is the frequency of oscillation of this system? What is the magnitude of an energy quantum? What is the amplitude of oscillation if the energy of the system equals one quantum?

*64. We know from Chapter 15 that at small amplitudes a pendulum behaves like a simple harmonic oscillator. Suppose that a pendulum consists of a mass of 0.10 kg attached to a (massless) string of length 1.0 m.

(a) Taking into account the quantization of energy, what is the least (nonzero) amount of energy that this pendulum can have?

(b) What is the amplitude of oscillation of the pendulum with this least amount of energy?

*65. Interplanetary and interstellar space is filled with thermal radiation of a temperature of 2.7 K left over from the Big Bang.

(a) At what wavelength is the flux of this radiation maximum? What is this part of the electromagnetic spectrum called?

(b) What is the power incident on the surface of the Earth due to this radiation?

*66. If you stand naked in a room, your skin and the walls of the room will exchange heat by radiation. Suppose the temperature of your skin is 33°C; the total area of your skin is 1.5 m^2. The temperature of the walls is 15°C. Assume your skin and the walls behave like blackbodies.

(a) What is the rate at which your skin radiates heat?

(b) What is the rate at which your skin absorbs heat? What is your net rate of loss of heat?

67. An incandescent bulb radiates 40 W of thermal radiation from a filament of temperature 3200 K. Estimate the number of photons radiated per second; assume that the photons have an average wavelength equal to the λ_{max} given by Wien's Law.

68. In order for a photoelectron to be emitted from a nickel surface, the wavelength of the incident photon must be no longer than 247.5 nm. What is the work function for nickel? When a photon of wavelength 200 nm is used, what stopping potential is required?

69. A photon loses 5.0% of its initial energy when a collision with an electron at rest deflects it by 90°. What is the wavelength of the incident photon? With what speed does the electron emerge from the collision?

70. The energy density of starlight in intergalactic space is 1.0×10^{-15} J/m^3. What is the corresponding density of photons? Assume the average wavelength of the photons is 500 nm.

*71. If you want to make a very faint light beam that delivers 1 photon per square meter per second, what must be the amplitude of the electric field in this light beam? The wavelength of the light is 500 nm.

*72. A photon has an energy of 5.0 eV in the reference frame of the laboratory. What is the energy of this photon in the reference frame of a proton moving through the laboratory at a speed of $\frac{1}{2}c$ in the same direction as the photon? (Hint: Use the Doppler-shift formula given in Section 36.3.)

*73. In a collision with a free electron, a photon of energy 2.0×10^3 eV is deflected by 90°. What energy does the electron acquire in this collision?

*74. A photon of energy 1.6×10^8 eV collides elastically with a *proton* initially at rest. The photon is deflected by 45°. What is its new energy?

*75. Figure 37.19 shows two of the discrete spectral lines of molybdenum. According to this figure, what are the wavelengths of these spectral lines? Would these spectral lines appear if instead of using electrons of 35 keV we were to use electrons of 19 keV in the X-ray tube?

Answers to Checkups

Checkup 37.1

1. If the stove were polished instead of black, it would not be as good an emitter of thermal radiation, since a black surface radiates heat most efficiently, and a shiny surface does not. A good absorber is a good emitter, and a poor absorber is a poor emitter.

2. The roof color provides direct evidence of the light absorbed or emitted: a roof of dark-colored shingles absorbs more sunlight, and a roof of light-colored shingles reflects more sunlight.

3. The shiny surface layer of the suit reflects sunlight, and thus prevents overheating.

4. (D) Less intense at all wavelengths. As discussed in Section 37.1 and shown in Fig. 37.3, a blackbody with a peak at a longer wavelength also produces less radiation at all wavelengths (and also has a lower temperature).

Checkup 37.2

1. For the first peak, $\lambda_{max}T = 1450\text{ K} \times 2000\text{ nm} = 2.9 \times 10^{-3}$ m·K, in agreement with Wien's Law; likewise for the other two peaks.

2. Wien's Law asserts that the wavelength at maximum intensity varies inversely with the absolute temperature, so doubling the temperature changes the wavelength of the maximum of the spectrum by a factor of $\frac{1}{2}$. The Stefan–Boltzmann Law indicates that the total energy radiated varies as the fourth power of temperature, so doubling the temperature increases the total energy radiated from the surface by a factor of 16.
3. No to both—only the total energy of the oscillator is quantized, that is, the sum of kinetic and potential energies is quantized.
4. Yes, the energies are quantized; but for frequencies around 1 Hz, the energy quanta are about $E = hf = 6.63 \times 10^{-34}$ J, and thus individual increments in the motion are much too small to be noticed.
5. For such a large value of Planck's constant, the energy quanta would be nearly one-tenth of one joule, and thus the amplitude of oscillation of a pendulum would jump by a noticeable amount when one quantum of energy is added or removed.
6. (D) Blue; red. By Wien's Law, the hottest star emits more strongly at shorter wavelengths (the blue) and the coolest at longer wavelengths (the red). See Fig. 33.23 for the relation between the colors of the spectrum and wavelength.

Checkup 37.3

1. Violet light has the shortest wavelength, and thus the highest frequency and the most energetic photons. See Fig. 33.23 for the relation between the colors of the spectrum and wavelength.
2. For a larger work function, the threshold frequency is larger, that is, the straight line intercepts the horizontal frequency axis at a larger value of f.
3. For repulsive voltages more negative than $-V_{stop}$, electrons do not have enough kinetic energy to reach the collector. The current levels off for sufficiently high positive voltages because then all ejected electrons have enough energy to reach the collector and, for a given intensity, no more are available. The current decreases for lower intensity since the number of photons, and thus the rate of ejection of electrons, is proportional to intensity.
4. (C) The same slope. By Eq. (37.11), the slope of all such plots is Planck's constant h.

Checkup 37.4

1. Violet light has the shortest wavelength and, since $p = h/\lambda$, the largest momentum. See Fig. 33.23 for the relation between the colors of the spectrum and wavelength.
2. Yes, when a photon bounces off the wall, its momentum reverses, and the wall experiences an impulsive force.
3. At any given scattering angle, the wavelength shift $\Delta\lambda$ is the same for all incident wavelengths. Since the wavelength shift is fixed, the percentage change is largest for the smaller wavelength, 0.2 nm.
4. Yes, visible photons scatter in the same way as X rays and experience a Compton shift. The effect is extremely difficult to observe, since the change is so small, only a few picometers out of hundreds of nanometers.
5. (D) Greater than 0.005 nm. A wavelength shift greater than 0.005 nm is impossible; as in Example 7, the maximum possible wavelength shift is $2h/m_e c = 0.004\ 85$ nm.

Checkup 37.5

1. X-ray wavelengths are much shorter than those of visible light; diffraction effects increase with the ratio of the wavelength to the slit width, and so are much smaller for X rays.
2. Since $\lambda_{cutoff} = hc/K$ [Eq. (37.23)], halving the kinetic energy will double the cutoff wavelength to 2×0.036 nm (compare Example 9). The characteristic spikes will remain at the same wavelengths but disappear for molybdenum since $\lambda_{cutoff} = 0.072$ nm is too large; their locations are determined by the target material.
3. A bubble would allow more X rays to pass than the surrounding tissue, and so would appear dark in a negative.
4. (D) $f \leq eV_0/h$. The maximum energy of the photon is the kinetic energy of the electron, $E = hf = K = eV_0$.

Checkup 37.6

1. We would find a narrow line of the same width as the narrow slit; classically, the slit would merely create an ordinary shadow.
2. Since the probability that a photon arrives at a given point is proportional to the intensity of the wave at that point, and since the intensity is proportional to the square of the electric field amplitude, the point where the amplitude is decreased by a factor of 2 has a photon arrival probability smaller by a factor of 4.
3. No; the uncertainty principle dictates that there must be uncertainty in the momentum p. The electron is thus also not free from uncertainty in the velocity $v = p/m$, or in the kinetic energy $K = p^2/2m$.
4. Yes in both cases. The uncertainty relation applies to the same component of the position and momentum. Additionally, Eq. (37.29) implies that if the uncertainty in, say, p_z is zero, then the uncertainty in z must be infinite.
5. (A) A double-slit interference pattern. Like the case of the single slit, the distribution of photons matches the intensity distribution calculated from the wave theory of light; that is, the probability that a photon arrives at a given point is proportional to the intensity of a classical wave at that point.

CHAPTER

38

Spectral Lines, Bohr's Theory, and Quantum Mechanics

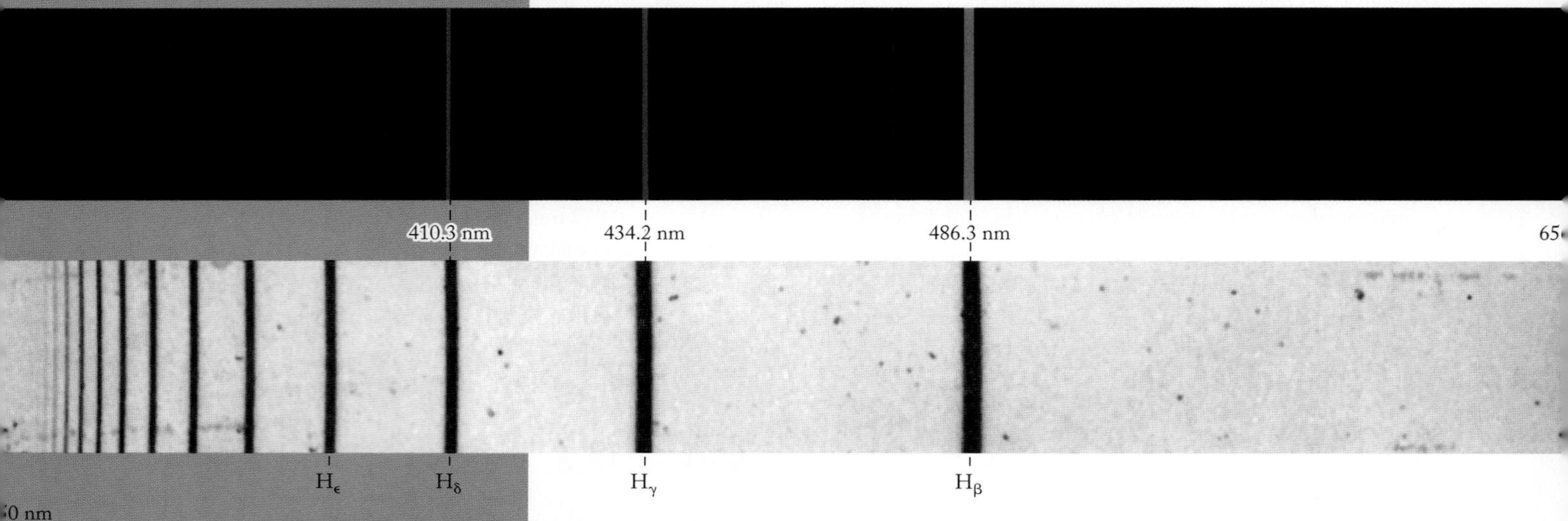

CONCEPTS IN CONTEXT

When a sample of a chemical element is heated, often by means of an electric discharge, it emits characteristic electromagnetic radiaton of discrete wavelengths. The visible light in such a discrete-line spectrum for hydrogen gas is shown in the upper frame of the figure above; the lower frame includes additional lines in the ultraviolet spectrum of hydrogen.

In our study of the spectra of hydrogen, we will be able to ask:

? The spacings between these lines decrease systematically from right to left, from long wavelengths to short wavelengths. This systematic pattern suggests that the wavelengths form a series, described by a simple mathematical formula. What is this formula? (Section 38.2, page 1291)

? Are there other series of spectral lines in the spectrum of hydrogen? What are they, and what is the shortest wavelength in any of these series? (Section 38.2, and Example 1, page 1292)

? How can we calculate the wavelengths of the series from the quantum processes within the hydrogen atom? (Example 2, page 1300)

? What series of wavelengths will hydrogen absorb, when illuminated with light containing a wide range of wavelengths? (Example 3, page 1301)

The photograph shown in Fig. 38.1 gives convincing visual evidence that solids, liquids, and gases are made of atoms, small grains of matter with a diameter of about 10^{-10} m. This photograph was prepared with a powerful electron microscope of a special design. Unfortunately, not even this microscope is sufficiently powerful to reveal the inside of the atom. For the exploration of the internal structure of the atom, we still have to rely on the technique developed by Ernest Rutherford and his associates around 1910: bombard the atom with a beam of particles and use this beam as a probe to "feel" the interior of the atom.

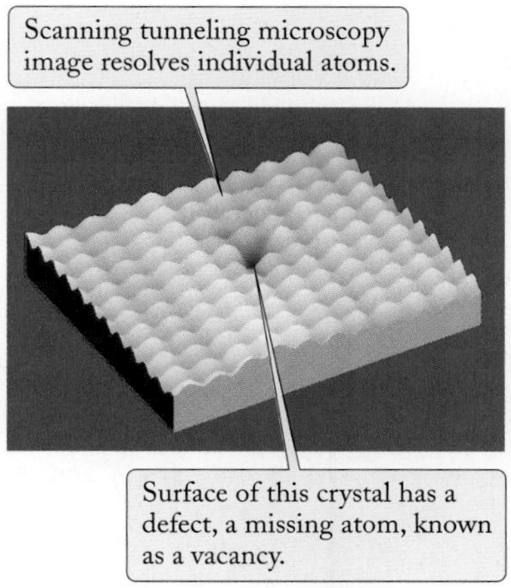

FIGURE 38.1 A scanning tunneling microsope image of the surface of a semiconductor, gallium arsenide (GaAs).

By 1910, most physicists had come to believe that atoms are made of some combination of positive and negative electric charges, and that the attractions and repulsions between these electric charges are the basis for all the chemical and physical phenomena observed in solids, liquids, and gases. Since electrons were known to be present in all of these forms of matter, it seemed reasonable to suppose that each atom consists of a combination of electrons and positive charge. The vibrational motions of the electrons within the atom would then result in the radiation of electromagnetic waves; this was supposed to account for the emission of light by the atom. However, both the arrangement of the electric charges within the atom and the mechanism that accounts for the characteristic colors of the emitted light remained mysteries until Rutherford's discovery of the nucleus and Niels Bohr's discovery of the quantization of atomic states. In this chapter we will look at these two momentous discoveries.

The exploration of the internal structure of the atom led to the inescapable conclusion that in the atomic realm Newton's laws of motion are not valid. Electrons and other subatomic particles obey new equations of motion that are drastically different from the old equations of motion obeyed by planets, billiard balls, or tennis balls. The new equations of motion incorporate the quantization of energy and the wave–particle behaviors we discussed in the preceding chapter, and they extend quantization to the realm of the atom. The new theory of motion that rules the realm of the atom is called **quantum mechanics.** The discovery of an entirely new set of laws of motion was the greatest scientific revolution of the twentieth century.

quantum mechanics

38.1 SPECTRAL LINES

Online Concept 43 Tutorial

The earliest attempts at a theory of atomic structure ended in failure—these early theories were not able to explain the characteristic colors of the light emitted by atoms. These colors show up very distinctly when a small sample of gas is made to emit light by the application of heat or of an electric current. For instance, if we put a few grains of ordinary salt into a flame, the sodium vapor released by the salt will glow with a characteristic yellow color. If we put neon gas into an evacuated glass tube and connect the ends of the tube to a high-voltage generator (Fig. 38.2), the gas will glow with the familiar orange red color of neon signs (Fig. 38.3).

The light emitted by an atom can be precisely analyzed with a prism (Fig. 38.4); this breaks the light up into its component colors. In the arrangement shown in Fig. 38.4, each discrete color generates a bright line, called a **spectral line**. Each kind of atom has its own discrete spectral lines. The color print on page 1289 shows the spectral lines of hydrogen, helium, lithium, mercury, and sodium; the numbers above

spectral line

the visible spectral lines give the wavelengths in nanometers. Hydrogen has four spectral lines in the visible region (already mentioned in Chapter 34; see Fig. 34.29) and many ultraviolet and infrared lines not visible to the human eye. From the color print we see that the set of spectral lines, or **spectrum**, belonging to hydrogen is unmistakably different from the spectra of helium, mercury, lithium, and sodium—the spectrum of an atom can serve as a fingerprint for its identification.

spectrum

Spectroscopy provides us with a useful alternative to the traditional "wet" analysis familiar to students of chemistry. By heating a small sample of an unknown substance—preferably by means of an electric discharge—and inspecting the spectral lines, we can identify the kinds of atoms in the sample. It so happens that *an atom capable of emitting light of a given wavelength is also capable of absorbing light of that wavelength.* In spectroscopy laboratories, scientists often take advantage of this to perform the quantitative analysis of samples of atoms with absorption lines rather than emission lines. When we illuminate a sample of atoms with white light (a mixture containing all colors or wavelengths), the atoms will absorb light of their characteristic wavelength, and upon analyzing the remaining light with a prism, we find dark absorption lines in the continuous background generated by the white light. The last picture in the color print on page 1289 shows such an absorption spectrum for sodium vapor. The dark lines of this absorption spectrum coincide with the bright lines in the emission spectrum (see the next-to-last picture on page 1289).

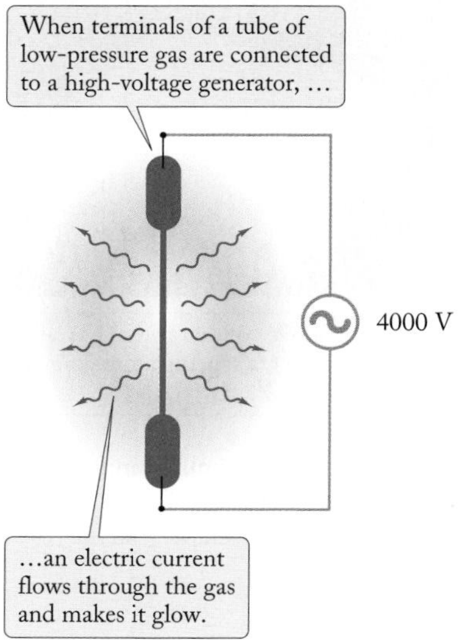

FIGURE 38.2 An electric discharge tube.

One advantage of spectroscopy over "wet" chemistry is that the analysis can be performed even on minuscule amounts of material. What is more, atoms can be identified at a distance. For example, we can identify the atoms on the surface of the Sun by careful analysis of the distribution of colors in sunlight— we do not need to pluck a sample of atoms from the Sun. The power of this technique is best illustrated by the story of the discovery of helium (the "Sun element"—*helios* is Greek for the Sun). In 1868, this gas was yet unknown to chemists when astronomers discovered it on the Sun by means of its spectral lines; 30 years later chemists finally found traces of helium in minerals on the Earth. By spectroscopic techniques, astronomers can identify atoms in remote stars, clouds of interstellar gas, galaxies, and quasars. For example, Fig. 38.5 shows the spectrum of the star Caph in the constellation Cassiopeia; the spectral absorption lines indicate the presence of hydrogen, calcium, iron, manganese, chromium, etc.; these lines are formed in the layers of gas on the surface of the star. Using spectroscopy, astronomers can perform a "chemical analysis" of the material on this star, even though it is many light-years away.

FIGURE 38.3 A neon sign.

Note that apart from these discrete spectral lines the bulk of the starlight is white light, a continuous and more or less uniform mixture of all colors. This is the thermal radiation emitted from the stellar surface. As we know from the preceding chapter, this

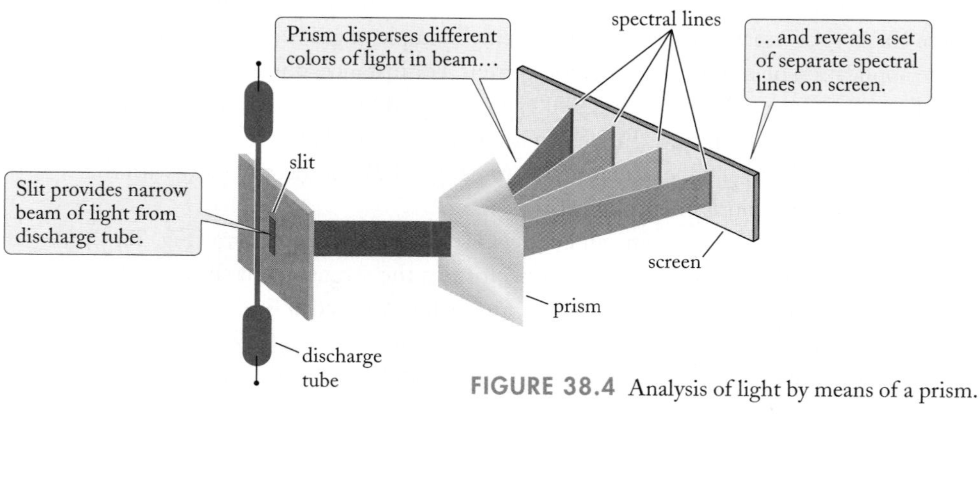

FIGURE 38.4 Analysis of light by means of a prism.

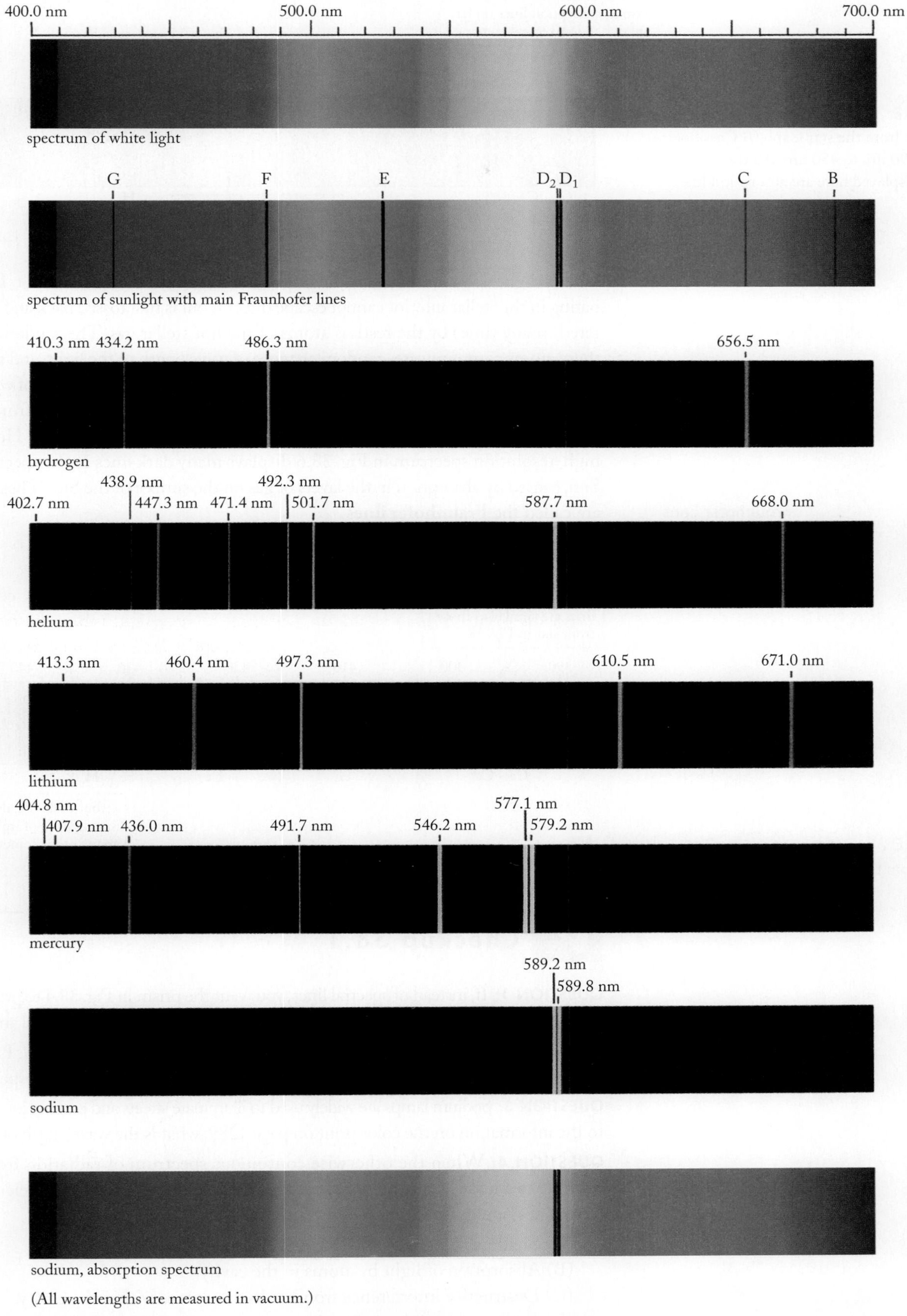

(All wavelengths are measured in vacuum.)

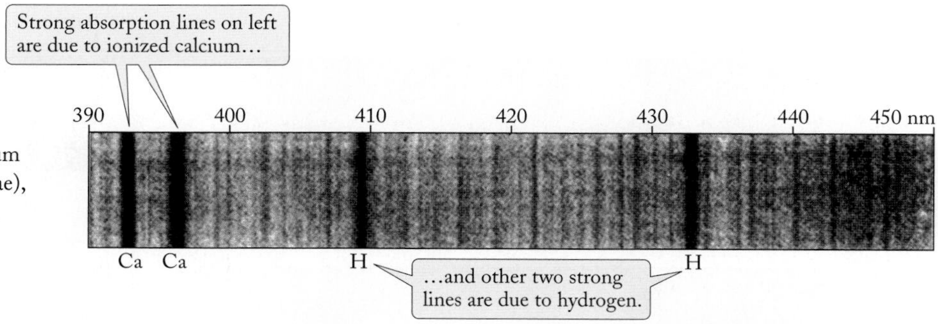

FIGURE 38.5 A portion of the spectrum of light from the star Caph (β Cassiopeiae), from 390 nm to 450 nm. All the spectral lines displayed here are absorption lines.

kind of light does not retain the fingerprint of the atoms that produced it. Light originating in the stellar interior cannot escape directly, but is first tossed back and forth (scattered) many times by the restless atoms of the hot stellar gas. The random motion of these atoms communicates random changes of wavelength to the light, and what finally emerges from the stellar interior is a continuous mixture of a wide range of wavelengths.

Figure 38.6 shows a portion of the spectrum of the white light from our Sun. White light is a continuous and nearly uniform mixture of all colors. However, the high-resolution spectrum in Fig. 38.6 displays many dark lines in the spectrum of the Sun, caused by absorption in the layers of gas on the surface of the Sun. These dark lines are called the **Fraunhofer lines**.

Fraunhofer lines

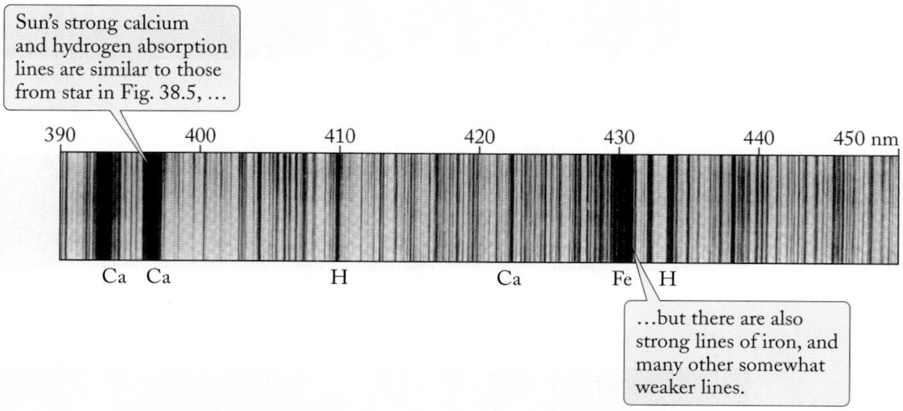

FIGURE 38.6 A portion of the spectrum of light from the Sun, from 390 nm to 450 nm.

✔ Checkup 38.1

QUESTION 1: If, instead of spectral lines, you want the prism in Fig. 38.4 to produce spectral dots (one dot for each color), how do you have to change the experimental arrangement?

QUESTION 2: The flame of the gas burner in a kitchen stove is bluish. Is this color that of the continuum thermal radiation, or the color of one (or several) spectral lines?

QUESTION 3: Sodium lamps are widely used to illuminate streets and parking lots. According to the information on the color print on page 1289, what is the wavelength of the light?

QUESTION 4: When the otherwise continuous spectrum of radiation from a small opening to a hot cavity exhibits reduced intensity at a few particular wavelengths, such "dark lines" are most likely due to

(A) Shadows caused by the shape of the cavity opening.
(B) Absorption of light by atoms in the cavity.
(C) Destructive interference from multiple sources within the cavity.
(D) Diffraction minima related to the size of the cavity opening.

38.2 SPECTRAL SERIES OF HYDROGEN

Online Concept 43 Tutorial

Careful examination of the spectral lines produced by an element reveals certain systematic regularities in the spacings of the lines. These regularities sometimes become especially striking if, instead of examining only the visible region of the spectrum, we examine both the visible region and the adjacent ultraviolet and infrared regions. Figure 38.7 again shows the spectrum of hydrogen in the visible and the near-ultraviolet regions. We can notice immediately that the spacings of the lines decrease systematically as we look at shorter and shorter wavelengths. The hydrogen lines in Fig. 38.7 are said to form a **spectral series**. In the spectra of other elements we find similar series; however, the spectra usually contain several overlapping series, and this makes it somewhat harder to perceive the regularities in the spacings.

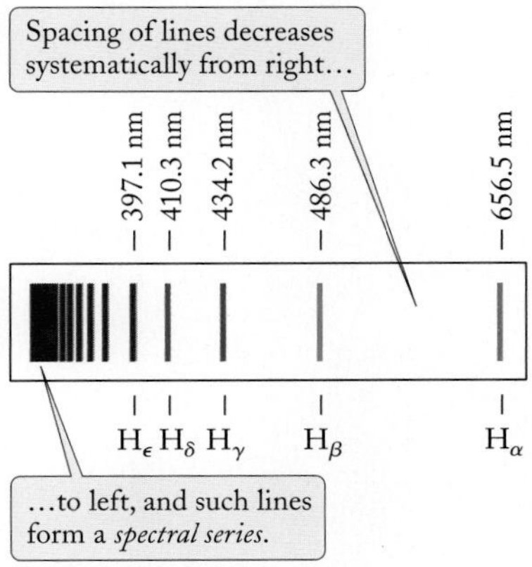

FIGURE 38.7 Spectrum of hydrogen in the visible and near-ultraviolet regions. The spectral lines of this spectral series are consecutively labeled H_α, H_β, H_γ, and so on. The numbers give the wavelengths of the spectral lines.

The systematic pattern in the spacing of the spectral lines of hydrogen shown in Fig. 38.7 suggests that the wavelengths of these lines should be described by some simple mathematical formula. Table 38.1 lists the wavelengths of many of these spectral lines; the spacing between the lines is smaller at shorter wavelengths, where very many spectral lines are crowded together. In 1855, Johann Balmer scrutinized the numbers in such a table and discovered that the wavelengths accurately fit the formula

Concepts in Context

$$\frac{1}{\lambda} = \frac{1}{91.176\text{ nm}}\left(\frac{1}{4} - \frac{1}{n^2}\right) \tag{38.1}$$

with $n = 3, 4, 5$, etc. This infinite series of spectral lines is called the **Balmer series**. Note that if n approaches infinity, the wavelength approaches the ultimate value $\lambda = 4 \times 91.176\text{ nm} = 364.70\text{ nm}$. This value is called the **series limit**, because there are no spectral lines of shorter wavelength beyond this value.

series limit

Balmer's formula is usually written compactly as

$$\frac{1}{\lambda} = R\left(\frac{1}{2^2} - \frac{1}{n^2}\right) \tag{38.2}$$

Balmer series

where R is called the **Rydberg constant**,

$$R = \frac{1}{91.176\text{ nm}} = 1.096\,78 \times 10^7\text{ m}^{-1} \tag{38.3}$$

Rydberg constant

Balmer's formula was purely descriptive, or phenomenological; it did not explain the atomic mechanism responsible for the production of the spectral lines. Nevertheless, it proved very fruitful because it led to more general formulas describing other series

TABLE 38.1 THE BALMER SERIES IN THE HYDROGEN SPECTRUM

WAVELENGTH[a] λ								
656.47 nm	486.27 nm	434.17 nm	410.29 nm	397.12 nm	389.02 nm	383.65 nm	379.90 nm	etc.

[a] Wavelengths are measured in vacuum.

of spectral lines. Balmer proposed that there might be other series in the hydrogen spectrum, with the 2^2 in the parentheses in Eq. (38.2) replaced by 1^2, or 3^2, or 4^2, etc. This yields the series of wavelengths

$$\frac{1}{\lambda} = R\left(\frac{1}{1^2} - \frac{1}{n^2}\right) \qquad n = 2, 3, 4, \ldots \tag{38.4}$$

$$\frac{1}{\lambda} = R\left(\frac{1}{3^2} - \frac{1}{n^2}\right) \qquad n = 4, 5, 6, \ldots \tag{38.5}$$

$$\frac{1}{\lambda} = R\left(\frac{1}{4^2} - \frac{1}{n^2}\right) \qquad n = 5, 6, 7, \ldots \tag{38.6}$$

Lyman, Paschen, and Brackett series

These series of spectral lines were actually discovered some years after Balmer proposed them; they are called, respectively, the **Lyman**, the **Paschen**, and the **Brackett series.** The first of these series lies in the ultraviolet region of the spectrum; the other series lie in the infrared.

Concepts in Context

We can combine all the formulas for all the spectral series of hydrogen into a single general formula

wavelengths of spectral series of hydrogen

$$\frac{1}{\lambda} = R\left(\frac{1}{n_2^2} - \frac{1}{n_1^2}\right) \tag{38.7}$$

where n_1 and n_2 are positive integers and n_1 is larger than n_2.

Concepts in Context

EXAMPLE 1 According to Eq. (38.7), what is the shortest wavelength of light that a hydrogen atom will emit or absorb?

SOLUTION: To find the shortest wavelength, we must chose n_1 and n_2 in Eq. (38.7) so as to obtain the largest possible value for the right side. This means that the positive term must be as large as possible, and the negative term as small as possible, which demands $n_2 = 1$ and $n_1 = \infty$ and gives

$$\frac{1}{\lambda} = R\left(\frac{1}{1^2} - \frac{1}{\infty^2}\right) = R \tag{38.8}$$

and

$$\lambda = 1/R = 91.176 \text{ nm}$$

Since $n_1 = 1$, this spectral line belongs to the Lyman series—it is the series limit for the Lyman series.

✔ Checkup 38.2

QUESTION 1: Figure 38.7 shows that spectral lines of short wavelength in the Balmer series are tightly spaced. Is this also true for the other spectral series of hydrogen?

QUESTION 2: Consider the series limits for the Lyman, Balmer, and Paschen series. Which has the longest wavelength? The shortest?

QUESTION 3: To find the longest wavelength in the Lyman series, what value of n_1 must you select? In the Balmer series? In the Paschen series? In the Brackett series?

QUESTION 4: In Example 1, we found that the shortest wavelength emitted by hydrogen, $\lambda = 91.176$ nm, is in the Lyman series. We previously saw that the Bolmer series limit is $\lambda = 364.70$ nm. One of the spectral lines of hydrogen has $\lambda = 121.57$ nm. In which series is this line?

(A) Lyman (B) Balmer (C) Paschen (D) Brackett

38.3 THE NUCLEAR ATOM

The regularity in the series of spectral lines of the atom must be due to an underlying regularity in the structure of the atom. We may think of an atom as analogous to a musical instrument, such as a flute. The atom can emit only a discrete set of spectral lines, just as the flute can emit only a discrete set of tones which make up a musical scale. The regularity in the spacing of tones in this musical scale is due to an underlying regularity in the structure of the flute—the tube of the instrument has regularly spaced toneholes that determine what kind of standing waves can build up within the tube and what kind of waves will be radiated.

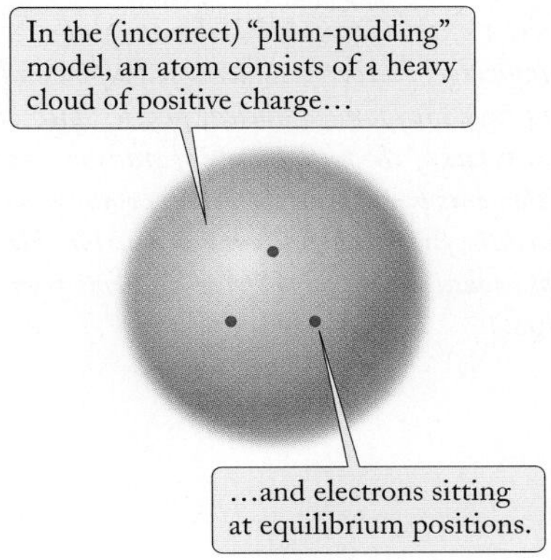

FIGURE 38.8 The lithium atom according to the "plum-pudding" model.

J. J. Thomson, the discoverer of the electron, made one of the first attempts at explaining the emission of light in terms of the structure of the atom. Having established that electrons are a ubiquitous component of matter, Thomson proposed the following picture: An atom consists of a number of electrons, say, Z electrons, embedded in a cloud of positive charge. The cloud is heavy, carrying almost all of the mass of the atom. The positive charge in the cloud is $+Ze$, so it exactly neutralizes the negative charge $-Ze$ of the electrons. In an undisturbed atom, the electrons will sit at their equilibrium positions, where the attraction of the cloud on the electrons balances their mutual repulsion (Fig. 38.8). But if the electrons are disturbed by, say, a collision, then they will vibrate around their equilibrium positions, and this accelerated motion will cause the emission of electromagnetic radiation, that is, the emission of light. This model of the atom, called the **"plum-pudding" model**, does yield frequencies of vibration of the same order of magnitude as the frequency of light, but it does not yield the observed spectral series; for instance, on the basis of this model, hydrogen should have only one single spectral line, in the far ultraviolet. And in 1910, experiments by Ernest Rutherford and his collaborators established conclusively that most of the mass of the atom is *not* spread out over a cloud—instead, the mass is concentrated in a small kernel, or **nucleus**, at the center of the atom.

"plum-pudding" model

nucleus

Rutherford had been studying the emission of alpha particles from radioactive substances. These alpha particles carry a positive charge $+2e$, and they have a mass of 6.64×10^{-27} kg, about 4 times the mass of a proton (alpha particles are nuclei of helium atoms; that is, they are completely ionized helium atoms, with all electrons removed; see Section 40.3). Some radioactive substances, such as radioactive polonium and radioactive bismuth, spontaneously emit alpha particles with kinetic energies of several million electron-volts. These energetic alpha particles readily pass through thin foils of metal, thin sheets of glass, or other materials.

Rutherford was much impressed by the penetrating power of these alpha particles, and it occurred to him that *a beam of these particles can serve as a probe to "feel" the interior of the atom.* When a beam of alpha particles strikes a foil of metal, the alpha particles penetrate the atoms and are deflected by collisions with the subatomic structures; the magnitude of these deflections gives a clue about the subatomic structures. For example, if the interior of the atom had the "plum-pudding"

SIR ERNEST RUTHERFORD (1871–1937) *British experimental physicist and director of the Cavendish Laboratory at Cambridge. Rutherford identified alpha and beta rays. He founded nuclear physics with his discoveries of the nucleus and of transmutation of elements by radioactive decay; he also produced the first artificial nuclear reaction. He was awarded the Nobel Prize in chemistry in 1908.*

structure proposed by J. J. Thomson, then the alpha particles would suffer only very small deflections, since neither the electrons, with their small masses, nor the diffuse cloud of positive charge would be able to disturb the motion of a massive and energetic alpha particle.

The crucial experiments were performed by H. Geiger and E. Marsden working under Rutherford's direction. They used thin foils of gold and of silver as targets and bombarded these with a beam of alpha particles from a radioactive source. After the alpha particles passed through the foil, they were detected on a fluorescent screen, which registers the impact of each particle by a faint scintillation (Fig. 38.9). To Rutherford's amazement, some of the alpha particles were deflected by such a large angle that they came out backward. In Rutherford's own words, "It was quite the most incredible event that has ever happened to me in my life. It was almost as incredible as if you fired a 15-inch shell at a piece of tissue paper and it came back and hit you."

Rutherford immediately recognized that the large deflection must be produced by a close encounter between the alpha particle and a very small but very massive kernel inside the atom. He therefore proposed the following picture: *An atom consists of a small nucleus of charge* $+Ze$ *containing almost all of the mass of the atom; this nucleus is surrounded by a swarm of* Z *electrons, of charge* $-Ze$. Thus, the atom is like a solar system—the nucleus plays the role of Sun and the electrons play the role of planets.

On the basis of this nuclear model of the atom, Rutherford calculated what fraction of the beam of alpha particles should be deflected through what angle. The deflections are caused by the repulsive electric force between the positive alpha particle (charge $+2e$) and the positive nucleus (charge $+Ze$). If an alpha particle passes close to the nucleus, the electric force will be large, and it will be deflected by a large angle; if it passes farther from the nucleus, the electric force will be smaller, and it will be deflected by a smaller angle. Figure 38.10 shows the trajectories of several alpha particles approaching a nucleus; these trajectories are hyperbolas. The perpendicular distance between the nucleus and the original (undeflected) line of motion is called the **impact parameter**. In order to suffer a large deflection, the alpha particle must hit an atom with a very small impact parameter, 10^{-13} m or less; since the alpha particles in the beam strike the foil of metal at random, only very few of them will score such a close hit, and only very few will be deflected by a large angle.

impact parameter

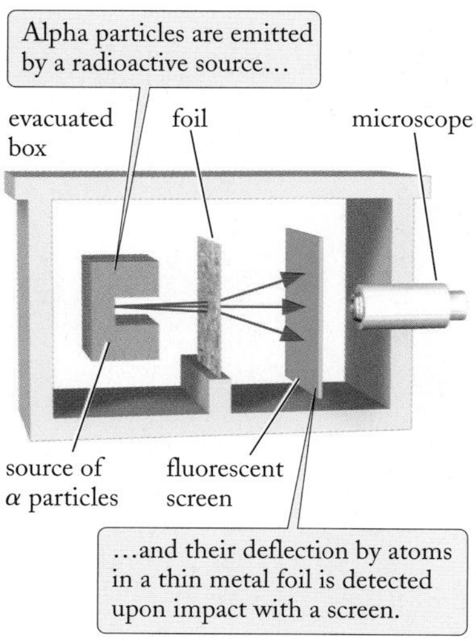

FIGURE 38.9 Rutherford's apparatus.

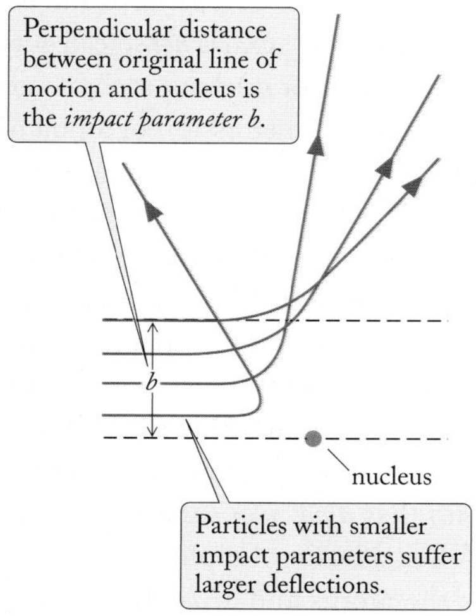

FIGURE 38.10 Four hyperbolic orbits of different impact parameters.

✔ Checkup 38.3

QUESTION 1: The hydrogen atom has one single electron. What does this atom look like according to the "plum-pudding" model? If the cloud of positive electric charge is spherically symmetric, what is the equilibrium position of the electron?

QUESTION 2: If an alpha particle approaches a nucleus with an impact parameter of zero, what will happen to the alpha particle?

QUESTION 3: Is the distance of closest approach between a nucleus and an alpha particle in a hyperbolic orbit (such as shown in Fig. 38.10) larger or smaller than the impact parameter?

QUESTION 4: If the alpha particles in Fig. 38.10 had a negative electric charge $-2e$ instead of their positive charge $+2e$, how would this affect the deflections illustrated in this figure?

QUESTION 5: Rutherford's experiments used foils of silver (atomic number 47) or gold (atomic number 79) as targets. For the same impact parameter, an alpha particle of a given energy incident on a silver target is deflected

(A) Through a smaller angle than when incident on a gold target.
(B) Through the same angle as when incident on a gold target.
(C) Through a larger angle than when incident on a gold target.

38.4 BOHR'S THEORY

Online Concept Tutorial 43

Rutherford's experiments did reveal the gross arrangement of the electrons in the atom, but not the details of their motion. Since the electrons make up the outer layers of an atom, *their arrangement and motion determine the chemical bonds of the atom and the emission of light, that is, they determine all the chemical and spectroscopic properties.* But when physicists tried to calculate the electron motion according to the laws of classical mechanics and electromagnetism, they immediately ran into trouble.

To gain some insight into the source of this trouble, consider the case of the hydrogen atom. Suppose that the single electron of this atom is moving around the nucleus, according to the laws of classical mechanics, in a circular orbit of atomic size, that is, with a radius of about 10^{-10} m. In such an orbit, the electron would have a centripetal acceleration due to the attractive electric force of the nucleus, which would be very large, about 10^{23} m/s^2. Because of this acceleration, the electron would emit high-frequency electromagnetic radiation, that is, it would emit light (see Chapter 33). The energy carried away by the light would have to be supplied by the electron. Hence the emission process would have the same effect on the electron as a friction force—it would remove energy from the electron. This kind of friction would cause the electron to leave its circular orbit and gradually spiral in toward the nucleus, in just the way the residual atmospheric friction on an artificial satellite in a low-altitude orbit around the Earth causes it to spiral down toward the ground. A calculation using the laws of classical mechanics and electricity shows that the rate of emission of light by the orbiting electron in a hydrogen atom would be quite large. Correspondingly, the rate of energy loss of the electron would be large—the electron would spiral inward and collide with the nucleus within a time as short as 10^{-10} s!

Thus, our classical calculation leads us to the incongruous conclusion that hydrogen atoms, and other atoms, ought to be unstable—all the electrons ought to collapse into the nucleus almost instantaneously. Furthermore, the light that the electron emit-

NIELS BOHR (1885–1962) *Danish theoretical physicist. He worked under J. J. Thompson and Rutherford in England and then became director of the Institute of Theoretical Physics in Copenhagen, for the foundation of which he was largely responsible. After formulating the quantum theory of the atom, he played a leading role in the further development of the new quantum mechanics. He received the Nobel Prize in 1922.*

ted during the spiraling motion would have to be a wave of continually increasing amplitude and increasing frequency (in musical terminology, crescendo and glissando); this is so because when the electron came closer to the nucleus, the electric force would become larger, leading to a larger centripetal acceleration and a higher frequency of the orbital motion. Hydrogen atoms do not behave as this classical calculation predicts. Hydrogen atoms are stable, and when they do emit light, they emit discrete frequencies (spectral lines) instead of a continuum of frequencies.

These irreconcilable disagreements between the observed properties of atoms and the calculated properties gave evidence of a serious breakdown of the classical mechanics of Newton and the classical theory of electromagnetism. Although these theories had proved very successful on a macroscopic scale, they were in need of some drastic modifications on an atomic scale.

In 1913, Niels Bohr took a bold step toward resolving these difficulties. He made the radical proposal that, *at the atomic level, the laws of classical mechanics and of classical electromagnetism must be replaced or supplemented by other laws.* Bohr expressed these new laws of atomic mechanics in the form of several postulates:

Bohr's quantum postulates

1. *The orbits and the energies of the electrons in an atom are quantized, that is, only certain discrete orbits and energies are permitted. When an electron is in one of the quantized orbits, it does not emit any electromagnetic radiation; thus, the electron is said to be in a* **stationary state.** *The electron can make a discontinuous transition, or quantum jump, from one stationary state to another. During such a transition from one stationary state to another stationary state of lower energy, the electron does emit radiation.*

stationary state

2. *The laws of classical mechanics apply to the orbital motion of the electrons in a stationary state, but these laws do not apply during the transition from one state to another.*
3. *When an electron makes a transition from one stationary state to another, the excess energy ΔE is released as a single photon of frequency $f = \Delta E/h$.*
4. *The permitted orbits are characterized by quantized values of the orbital angular momentum L. This angular momentum is always an integer multiple of $h/2\pi$:*

quantization of angular momentum

$$L = n\frac{h}{2\pi} \qquad n = 1, 2, 3, \ldots \tag{38.9}$$

The number n is called the angular-momentum quantum number.

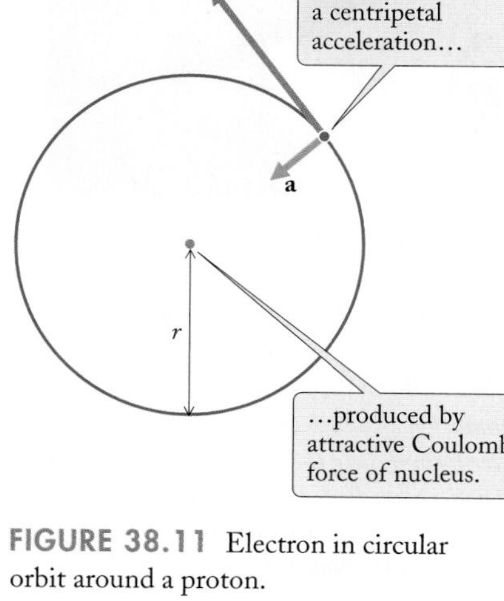

FIGURE 38.11 Electron in circular orbit around a proton.

Let us now see how to calculate the stationary states and the spectrum of the hydrogen atom on the basis of these postulates. For the sake of simplicity, we will assume that the electron moves in a circular orbit around the proton, which remains at rest (Fig. 38.11). The centripetal acceleration of the electron is v^2/r. This centripetal acceleration is produced by the force of attraction between the electron and the proton, that is, the Coulomb force $e^2/(4\pi\epsilon_0 r^2)$. Thus, the equation of motion $m_e a = F$ for the electron is

$$m_e\frac{v^2}{r} = \frac{1}{4\pi\epsilon_0}\frac{e^2}{r^2} \tag{38.10}$$

The orbital angular momentum for a circular orbit is $L = m_e vr$ [see Eq. (13.34)]. According to Bohr's postulate, this orbital angular momentum must be $h/2\pi$ multiplied by an integer n:

$$m_e vr = n\frac{h}{2\pi} \tag{38.11}$$

It is convenient to rewrite this as

$$m_e vr = n\hbar \quad (38.12)$$

where $\hbar$ (pronounced "h-bar") is Planck's constant h divided by 2π:

$$\hbar = \frac{h}{2\pi} = \frac{6.63 \times 10^{-34}\ \text{J}\cdot\text{s}}{2\pi} = 1.05 \times 10^{-34}\ \text{J}\cdot\text{s} \quad (38.13)$$

$\hbar$ (*h*-bar)

From Eq. (38.12), we obtain

$$v = \frac{n\hbar}{m_e r} \quad (38.14)$$

and when we substitute this into Eq. (38.10), we obtain an equation for the radius of the orbit,

$$m_e\left(\frac{n\hbar}{m_e r}\right)^2 \frac{1}{r} = \frac{1}{4\pi\epsilon_0}\frac{e^2}{r^2} \quad (38.15)$$

Solving this for the radius, we find

$$r = \frac{4\pi\epsilon_0 n^2\hbar^2}{m_e e^2} \quad (38.16)$$

From Eq. (38.16) we see that the radii of the permitted orbits are proportional to n^2. The smallest value of n is $n = 1$, and this leads to the radius of the smallest permitted orbit, which is called the **Bohr radius** and designated by a_0:

$$a_0 = \frac{4\pi\epsilon_0\hbar^2}{m_e e^2} = 0.529 \times 10^{-10}\ \text{m} = 0.0529\ \text{nm} \quad (38.17)$$

Bohr radius

The radii of the other permitted orbits are multiples of the Bohr radius:

$$r = \frac{4\pi\epsilon_0 n^2\hbar^2}{m_e e^2} = n^2 a_0 \quad (38.18)$$

For $n = 2, 3, 4, \ldots$, this gives $r = 4a_0, 9a_0, 16a_0, \ldots$, respectively. Figure 38.12 shows the permitted circular orbits, drawn to scale.

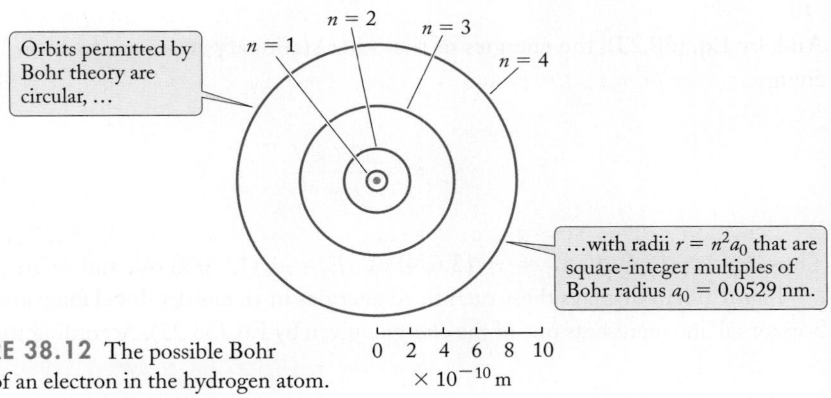

FIGURE 38.12 The possible Bohr orbits of an electron in the hydrogen atom.

The energy of the electron in one of these orbits is a sum of kinetic and potential energies. The kinetic energy is $K = \frac{1}{2}m_e v^2$, and the potential energy is the electrostatic potential energy $U = (-e) \times (1/4\pi\epsilon_0)e/r = -(1/4\pi\epsilon_0)e^2/r$ [see Eq. (25.11)]. Hence the total energy $E = K + U$ is

$$E = \frac{1}{2} m_e v^2 - \frac{1}{4\pi\epsilon_0}\frac{e^2}{r} \tag{38.19}$$

Next, we substitute v^2 from Eq. (38.14) and then r from Eq. (38.16), and we obtain

$$E = \frac{1}{2} m_e \frac{n^2\hbar^2}{m_e^2 r^2} - \frac{1}{4\pi\epsilon_0}\frac{e^2}{r} \tag{38.20}$$

$$= \frac{1}{2} m_e \frac{n^2\hbar^2}{m_e^2}\left(\frac{m_e e^2}{4\pi\epsilon_0 n^2\hbar^2}\right)^2 - \frac{e^2}{4\pi\epsilon_0}\left(\frac{m_e e^2}{4\pi\epsilon_0 n^2\hbar^2}\right)$$

$$= \frac{m_e e^4}{2(4\pi\epsilon_0)^2\hbar^2}\frac{1}{n^2} - \frac{m_e e^4}{(4\pi\epsilon_0)^2\hbar^2}\frac{1}{n^2} \tag{38.21}$$

or

$$E = -\frac{m_e e^4}{2(4\pi\epsilon_0)^2\hbar^2}\frac{1}{n^2} \tag{38.22}$$

This energy depends on the quantum number n, and as a reminder of this dependence, we label this energy with a subscript n:

$$E_n = -\frac{m_e e^4}{2(4\pi\epsilon_0)^2\hbar^2}\frac{1}{n^2} \tag{38.23}$$

According to this equation, the energies of all the stationary states are negative. The stationary state with the least energy, or the most negative energy, is the state with $n = 1$, for which

$$E_1 = -\frac{m_e e^4}{2(4\pi\epsilon_0)^2\hbar^2}$$

$$= -\frac{9.11 \times 10^{-31}\ \text{kg} \times (1.602 \times 10^{-19}\ \text{C})^4}{2 \times (4\pi \times 8.85 \times 10^{-12}\ \text{F/m})^2 \times (1.055 \times 10^{-34}\ \text{J·s})^2}$$

$$= -2.18 \times 10^{-18}\ \text{J} = -13.6\ \text{eV} \tag{38.24}$$

And, by Eq. (28.23), the energies of the other stationary states are fractions of this energy:

energies of stationary states

$$E_n = -\frac{13.6\ \text{eV}}{n^2} \tag{38.25}$$

Thus, $E_1 = -13.6$ eV, $E_2 = -(13.6/4)$ eV, $E_3 = -(13.6/9)$ eV, and so on.

energy-level diagram

Figure 38.13 displays these quantized energies in an **energy-level diagram**. Each horizontal line represents one of the energies given by Eq. (38.25). According to Bohr's

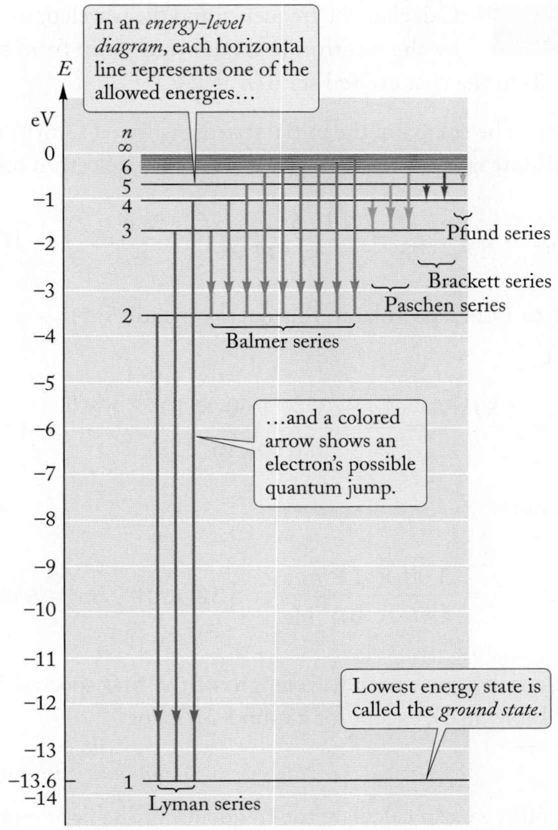

FIGURE 38.13 Energy-level diagram for the hydrogen atom. Jumps (blue) that end in the ground state ($n = 1$) give rise to the Lyman series; jumps (red) that end in the first excited state ($n = 2$) give rise to the Balmer series; etc.

assumptions, the electron emits a photon when it makes a transition, or a quantum jump, from one stationary state to a lower stationary state. Such quantum jumps have been indicated by arrows in Fig. 38.13. The stationary state of lowest energy ($n = 1$) is called the **ground state**, the next state ($n = 2$) is called the **first excited state**, the next state ($n = 3$) is called the **second excited state**, and so on.

ground state and excited states

Ordinarily, the electron of the hydrogen atom is in the ground state, that is, the circular orbit with radius a_0 and energy -13.6 eV. This is the configuration of least energy, and it is the configuration into which the atom tends to settle when it is left undisturbed. As long as the atom remains in the ground state, it does not emit light. To bring about the emission of light, we must first kick the electron into one of the excited states, that is, a circular orbit of larger radius and higher energy. We can do this by heating a sample of atoms or by passing an electric current through the sample. Collisions between the atoms will then disturb the electronic motions and occasionally kick an electron into a larger orbit. From there, the electron will spontaneously jump into a smaller orbit, emitting a photon. Note that the quantum jumps indicated by colored arrows in Fig. 38.13 form several series: one series consists of all those jumps (indicated by blue arrows) that end in the ground state, another series consists of all those jumps (indicated by red arrows) that end in the first excited state, etc. These series of jumps give rise to the series of spectral lines: the Lyman series, the Balmer series, etc.

From our formula for the energies of the states of the hydrogen atom we can calculate the frequency of the light emitted in a quantum jump from some initial state to a final state, as in the following example.

Concepts in Context

EXAMPLE 2 Calculate the frequency and the wavelength of the light emitted by the electron in a quantum jump from the second excited state ($n = 3$) to the first excited state ($n = 2$).

SOLUTION: The energy of the initial state is $E_3 = -(13.6/9)$ eV, and the energy of the final state is $E_2 = -(13.6/4)$ eV. Hence the electron releases an energy

$$\Delta E = E_3 - E_2 = -\frac{13.6}{9}\text{ eV} - \left(-\frac{13.6}{4}\text{ eV}\right) = 1.89\text{ eV}$$

According to Bohr's postulates, this energy is radiated as a single photon of frequency

$$f = \frac{\Delta E}{h} = \frac{1.89\text{ eV}}{h} = \frac{1.89\text{ eV} \times 1.60 \times 10^{-19}\text{ J/eV}}{6.63 \times 10^{-34}\text{ J}\cdot\text{s}} = 4.56 \times 10^{14}\text{ Hz}$$

and wavelength

$$\lambda = \frac{c}{f} = \frac{3.00 \times 10^8\text{ m/s}}{4.56 \times 10^{14}\text{ Hz}} = 6.58 \times 10^{-7}\text{ m} = 658\text{ nm}$$

This wavelength agrees with wavelength of the first spectral line of the Balmer series (see Table 38.1), except for a round-off error.

More generally, we can calculate the frequency of the light emitted in a quantum jump from some initial state of quantum number n_i to some final state of quantum number n_f. The initial energy of the electrons is E_i and the final energy is E_f. Thus, the electron releases an energy $E_i - E_f$,

$$E_i - E_f = \frac{m_e e^4}{2(4\pi\epsilon_0)^2\hbar^2}\left(\frac{1}{n_f^2} - \frac{1}{n_i^2}\right) \tag{38.26}$$

This energy is radiated as a single photon of frequency

frequency of emitted photon

$$f = \frac{\Delta E}{h} = \frac{E_i - E_f}{h} \tag{38.27}$$

From the frequency of the light, we can calculate the wavelength. Since the wavelength is inversely proportional to the frequency, $\lambda = c/f$, Eq. (38.27) implies the following expression for the wavelength of the emitted light:

wavelength of emitted photon

$$\frac{1}{\lambda} = \frac{E_i - E_f}{hc} = \frac{E_i - E_f}{2\pi\hbar c} \tag{38.28}$$

or, with Eq. (38.26),

$$\frac{1}{\lambda} = \frac{m_e e^4}{4\pi(4\pi\epsilon_0)^2\hbar^3 c}\left(\frac{1}{n_f^2} - \frac{1}{n_i^2}\right) \tag{38.29}$$

Comparison of Eqs. (38.7) and (38.29) yields the following theoretical formula for the Rydberg constant:

$$R = \frac{m_e e^4}{4\pi(4\pi\epsilon_0)^2\hbar^3 c} \qquad (38.30)$$

Upon insertion of the accurate values of the fundamental constants given in Appendix 6, we obtain

$$R = \frac{9.109\,53 \times 10^{-31}\ \text{kg} \times (1.602\,189 \times 10^{-19}\ \text{C})^4}{4\pi(4\pi \times 8.854\,178 \times 10^{-12}\ \text{F/m})^2 \times (1.054\,589 \times 10^{-34}\ \text{J}\cdot\text{s})^3 \times 2.997\,925 \times 10^8\ \text{m/s}}$$

$$= 1.097\,37 \times 10^7\ \text{m}^{-1} \qquad (38.31)$$

This theoretical value of R agrees quite well with the experimental value quoted in Eq. (38.3).[1]

EXAMPLE 3 Suppose that the atoms in a sample of hydrogen gas are initially in the ground state. If we illuminate these atoms with light (from some kind of lamp), what frequencies will the atoms absorb?

Concepts in Context

SOLUTION: Absorption of light is the reverse of emission. When an electron in an atom absorbs a photon (supplied by the lamp), it jumps from the initial state to a state of higher energy. The energy of the photon must match the energy difference between the states. Thus, the frequencies of the photons that the electrons can absorb when jumping upward from the ground state are exactly those frequencies that they emit when jumping downward into the ground state, that is, the frequencies of the Lyman series (see Fig. 38.13).

EXAMPLE 4 What is the ionization energy of the hydrogen atom, that is, what energy must we supply to remove the electron from the atom when it is initially in the ground state?

SOLUTION: To remove the electron from the atom, we must lift it into an orbit of infinite radius, that is, an orbit of quantum number $n = \infty$. The energy required to accomplish this transition from $n = 1$ to $n = \infty$ is

$$E_1 - E_\infty = -13.6\ \text{eV} \times \left(\frac{1}{1^2} - \frac{1}{\infty^2}\right) = 13.6\ \text{eV}$$

This is the ionization energy.

With his theory Bohr attained the goal of explaining the regularities in the spectrum of hydrogen in terms of the regularities of the structure of the atom. By showing that this structure is based on a simple numerical sequence, he fulfilled the ancient dream of Pythagoras of a universe based on simple numerical ratios, a dream that arose from an analogy with musical instruments. Bohr's theory tells us how the atom plays its tune.

[1]The small disagreement between the theoretical value of R given in Eq. (38.31) and the experimental value given in Eq. (38.3) is due to the motion of the nucleus of the hydrogen atom, which we have neglected in our calculation. A careful calculation that takes into account the motion of electron and proton about their common center of mass requires the substitution of the so-called reduced mass $m = m_e m_p/(m_e + m_p)$ for the electron mass, and eliminates the disagreement. See also Problem 31.

Checkup 38.4

QUESTION 1: Why is Bohr's postulate of stationary states in direct contradiction with classical mechanics and electromagnetism?

QUESTION 2: If an electron in a hydrogen atom makes a transition from some state to a lower state, does its kinetic energy increase or decrease? Its potential energy? Its orbital angular momentum?

QUESTION 3: Suppose that an electron is initially in the $n = 5$ state. It then makes three quantum jumps in succession, such that the first jump produces a photon in the Paschen series, the second a photon in the Balmer series, and the third a photon in the Lyman series. In the energy-level diagram of Fig. 38.13, find the arrows that indicate these transitions.

QUESTION 4: In the energy-level diagram of Fig. 38.13, what transition would produce the series limit of the Balmer series?

(A) $n = 2$ to $n = 1$ (B) $n = 3$ to $n = 2$ (C) $n = \infty$ to $n = 1$
(D) $n = \infty$ to $n = 2$ (E) $n = 11$ to $n = 2$

38.5 QUANTUM MECHANICS; THE SCHRÖDINGER EQUATION

Bohr's theory is a hybrid. It relies on some basic classical features (orbits) and grafts onto these some quantum features (quantum jumps, quanta of light). In the 1920s the cooperative efforts of several brilliant physicists—L. de Broglie, E. Schrödinger, W. Heisenberg, M. Born, P. Jordan, P. A. M. Dirac—established that the remaining classical features had to be eradicated from the theory of the atom. Bohr's semiclassical theory had to be replaced by a new quantum mechanics with an entirely different equation of motion.

The basis of the new quantum mechanics was laid by the discovery that electrons—as well as protons, neutrons, and all the other "particles" found in nature—have not only particle properties but also wave properties. When a beam of electrons is made to pass through an extremely narrow slit, the electrons exhibit diffraction. This means that electrons are neither classical particles nor classical waves. *Electrons, just like photons, are a new kind of object with a subtle combination of particle and wave properties.* Electrons are wavicles. As in the case of photons [see Eq. (37.15)], *the wavelength associated with an electron or some other wavicle is inversely proportional to its momentum*:

de Broglie wavelength

$$\lambda = \frac{h}{p} \tag{38.32}$$

This formula was postulated by de Broglie, and it is called the **de Broglie wavelength**. This wavelength is quite small, even for electrons of the lowest energies attainable in experiments with beams of electrons.

EXAMPLE 5 What is the de Broglie wavelength of an electron of kinetic energy 1.0 eV, which is about the lowest energy that can be attained in experiments with beams of electrons?

SOLUTION: In joules, a kinetic energy of 1.0 eV is 1.6×10^{-19} J. This kinetic energy is related to the momentum as follows:

$$E = \frac{p^2}{2m_e}$$

Hence

$$p = \sqrt{2m_e E} = \sqrt{2 \times 9.11 \times 10^{-31}\ \text{kg} \times 1.6 \times 10^{-19}\ \text{J}}$$

$$= 5.4 \times 10^{-25}\ \text{kg·m/s}$$

According to the de Broglie equation, the wavelength of the electron is then

$$\lambda = \frac{h}{p} = \frac{6.63 \times 10^{-34}\ \text{J·s}}{5.4 \times 10^{-25}\ \text{kg·m/s}}$$

$$= 1.2 \times 10^{-9}\ \text{m} = 1.2\ \text{nm}$$

LOUIS VICTOR, PRINCE DE BROGLIE (DE BROY) (1885–1962) *French theoretical physicist. He discovered the de Broglie wavelength by reasoning that if waves have particle properties, then maybe particles have wave properties. For his discovery of the wave properties of matter he was awarded the Nobel Prize in 1929, after the existence of these wave properties was confirmed experimentally.*

The wave properties of electrons were confirmed experimentally by C. J. Davisson and L. Germer, who observed interference effects with beams of electrons scattered by a crystal. In these experiments the crystal acts as a grating for electron waves, in much the same way as it acts as a grating for X rays in X-ray interference experiments (see the discussion of X-ray interference and the production of Laue spots in Section 37.5). Davisson and Germer found that constructive interference of the electron waves scattered by the rows of atoms in the crystal produced strong interference maxima in selected directions, and they were able to confirm that these directions agreed with calculations based on the de Broglie formula (38.32).

Schrödinger equation

In the new quantum mechanics, or wave mechanics, the motion of an electron is described by a wave equation, the **Schrödinger equation**. This Schrödinger equation plays the same role for electrons as the wave equation derived from the Maxwell equations plays for photons [see Eq. (33.33)]. As in the case of photons, the behavior of an electron is governed by a probabilistic law. The electron is represented by an electron wavefunction ψ (Greek letter psi) calculated from the Schrödinger equation, and the intensity of this electron wave at some point determines the probability that there is an electron particle at that point [compare Eq. (37.24)]:

probability interpretation of wavefunction

$$[\text{probability for an electron at a point}] \propto |\psi|^2 \tag{38.33}$$

Furthermore, as a consequence of their wave properties, electrons obey the Heisenberg uncertainty relations for position and momentum [see Eq. (37.30)],

$$\Delta y\, \Delta p_y \geq \frac{\hbar}{2} \tag{38.34}$$

These quantum uncertainties are of crucial importance for the behavior of an electron inside an atom. For such an electron, *the uncertainty in the position is very large—about as large as the size of the atom.* This implies that the electron follows no definite orbit. It is therefore not surprising that the Bohr theory should have failed in all attempts at calculating the electron motions in the helium atom and in other atoms with several electrons; what is surprising is that this theory should have succeeded as well as it did in the case of the hydrogen atom.

ERWIN SCHRÖDINGER (1887–1961)
Austrian theoretical physicist. Another of the founders of the new quantum mechanics, he received the Nobel Prize in 1933 for the discovery of his wave equation.

EXAMPLE 6 Consider an electron in the ground state of hydrogen. Show that a well-defined orbit is inconsistent with the Heisenberg uncertainty relations.

SOLUTION: If the electron is to follow a well-defined orbit, the uncertainty in its momentum (in any direction) must be much smaller than the magnitude of the momentum. According to Eq. (38.14), the speed of the electron in the smallest circular orbit ($n = 1$) is $v = \hbar/m_e r$, and the magnitude of the momentum is $p = m_e v = \hbar/r$. For a well-defined orbit, the uncertainty of the momentum must be much smaller than this magnitude of the momentum:

$$\Delta p_y \ll \frac{\hbar}{r} \tag{38.35}$$

Furthermore, for a well-defined orbit, we must require that the uncertainty in the position must be much smaller than the size of the orbit:

$$\Delta y \ll r \tag{38.36}$$

In view of the inequalities (38.35) and (38.36), the product $\Delta y\, \Delta p_y$ must be smaller than $\hbar$:

$$\Delta y\, \Delta p_y \ll \hbar \qquad \text{(for a well-defined orbit)} \tag{38.37}$$

This is inconsistent with the Heisenberg uncertainty relation (38.34).

In wave mechanics, the quantization of the energy in the hydrogen atom and other atoms is an automatic consequence of the wave properties of the electron. The attractive electric force of the nucleus confines the electron wave to some region near the nucleus and causes the wave to reflect back and forth across the region, forming a standing wave. *The different stationary states of the atom correspond to different standing-wave modes.* As in the case of standing waves on a string, the standing electron waves in the atom have a discrete set of eigenfrequencies. For electrons and other particles, just as for photons, the energy is related to the frequency by

energy of stationary state in terms of frequency

$$E = hf \tag{38.38}$$

Thus a discrete set of frequencies for electron standing waves corresponds to a discrete set of energies. The ground state of the atom is analogous to the fundamental mode of the string, the first excited state is analogous to the first overtone, and so on. However, whereas the determination of the eigenfrequencies of standing waves on a string is a quite trivial mathematical exercise, the determination of the eigenfrequencies of the electron waves in an atom is a formidable mathematical problem, which requires an investigation of the solutions of the Schrödinger wave equation.

Although we will not deal here with the mathematical complexities of the Schrödinger wave equation in three dimensions, we can gain some insight into how electron waves determine the discrete energies in the hydrogen atom by means of the following simple calculation. Let us assume that the electron travels around the nucleus along an orbit of radius r, but instead of thinking of the electron as a particle, as in the Bohr theory, let us think of it as a wave. Figure 38.14 shows a "snapshot" of such an electron wave at one instant of time. If the wave is to have a well-defined amplitude at all points, it must repeat whenever we go once around the circumference—if it did not, then the wave amplitude would have two different values at a single point, which makes no sense. Hence, the wave is subject to the condition that some integer number of wavelengths must fit around the circumference:

$$2\pi r = n\lambda \tag{38.39}$$

Although this equation looks like the condition for a standing wave on a string of length $2\pi r$, we are here dealing with a traveling wave, for which the entire wave pattern in Fig. 38.14 rotates rigidly around the center. With the de Broglie relation $\lambda = h/p$, Eq. (38.39) becomes

$$2\pi r = \frac{nh}{p}$$

or

$$rp = \frac{nh}{2\pi} \tag{38.40}$$

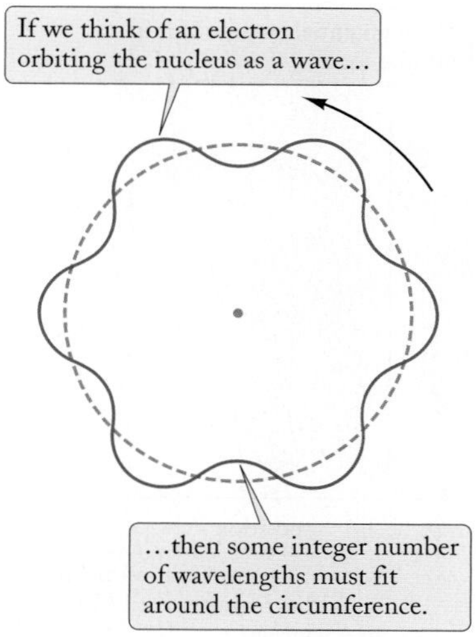

FIGURE 38.14 In this example, six wavelengths of the electron wave fit around the circumference of the orbit.

Since rp is the orbital angular momentum, this equation coincides with Bohr's quantization condition for the angular momentum, Eq. (38.9). Thus, *the wave picture of the electron implies the quantization of the angular momentum and, therefore, the quantization of the energy.* But we must not take this calculation too seriously—its legitimacy is questionable, since it relies in part on the wave picture and in part on the particle picture. Furthermore, analysis of the Schrödinger equation shows that it is not enough to consider the behavior of the wave around the circumference; we must also consider the behavior of the wave outward along each radius. Thus, this simple calculation provides no more than a crude qualitative sketch of the role of the wave properties of electrons in the atom.

There are some applications where a simple one-dimensional calculation gives accurate and meaningful results, such as electron reflection from a barrier, electron transmission through a barrier (known as "tunneling"; see *Physics in Practice: Ultramicroscopes*), and confinement of electrons in one direction in some modern layered semiconductor devices. In one dimension, each allowed stationary state is described by an electron wavefunction $\psi = \psi(x)$ that is a solution to the **time-independent Schrödinger equation**:

$$-\frac{\hbar^2}{2m_e}\frac{d^2}{dx^2}\psi(x) + U(x)\psi(x) = E\,\psi(x) \tag{38.41}$$

time-independent Schrödinger equation

When the potential energy $U(x)$ in this equation is some known function of position, then solutions may be found for the allowed values of the total energy E and the corresponding allowed position-dependent wavefunctions $\psi = \psi(x)$. Since the terms in Eq. (38.41) with U and E involve the potential and total energies, respectively, it is not surprising that the first term represents the kinetic energy $K = p^2/2m_e$. Thus the Schrödinger equation incorporates a wave-mechanical version of the the energy relation $K + U = E$ of classical mechanics.

EXAMPLE 7 Consider an electron in one dimension. The electron is confined to a region of length L, say, to the interval $0 < x < L$ (see Fig. 38.15). Inside this region, the potential energy $U(x) = 0$, and outside $U(x) = \infty$. This is sometimes referred to as a "particle in a box" or the "infinite potential

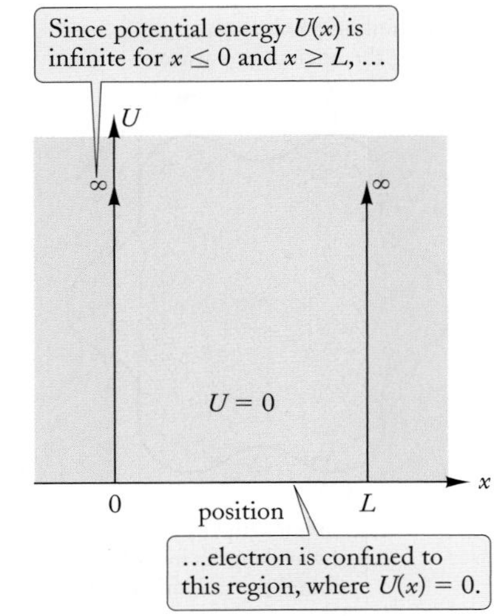

FIGURE 38.15 The one-dimensional infinite potential well. The electron is confined to the region $0 < x < L$.

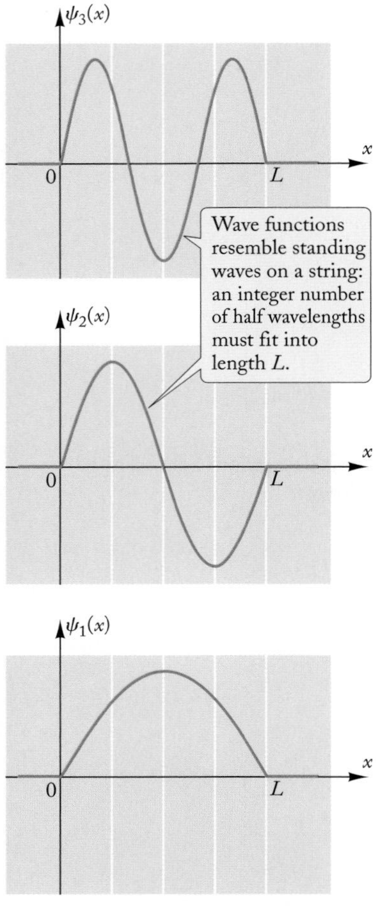

FIGURE 38.16 Wavefunction $\psi(x)$ for the first few states of the infinite potential well.

well." What electron wavefunctions are solutions to the time-independent one-dimensional Schrödinger equation? What are the corresponding allowed energies?

SOLUTION: Since the potential energy $U(x)$ is infinite outside the box, the only possible solution for the Schrödinger equation (38.41) outside the box is $\psi(x) = 0$. We thus know that $\psi(x) = 0$ both at $x = 0$ and at $x = L$. Inside the box, we have $U(x) = 0$, and the Schrödinger equation becomes

$$-\frac{\hbar^2}{2m_e}\frac{d^2}{dx^2}\psi(x) = E\psi(x)$$

We can rearrange this in the more familar form

$$\frac{d^2}{dx^2}\psi(x) = -\frac{2m_eE}{\hbar^2}\psi(x) \tag{38.42}$$

We encountered similar equations, where the second derivative of a function is proportional to the negative of the function, in our study of oscillations in Chapter 15. By analogy with Eq. (15.15), we know solutions to Eq. (38.42) are of the form

$$\psi(x) = A\sin(kx + \delta)$$

where the wave vector $k = 2\pi/\lambda$ of the sinusoidal function is given by

$$k^2 = \frac{2m_e E}{\hbar^2} \tag{38.43}$$

Since we know that $\psi(0) = 0$, we require

$$\psi(0) = A\sin(0 + \delta) = A\sin\delta = 0 \tag{38.44}$$

The amplitude A must be nonzero, since the probability of finding the electron cannot be zero everywhere. Equation (38.44) thus determines that $\delta = 0$. Hence our wavefunctions are of the form

$$\psi(x) = A\sin(kx) \tag{38.45}$$

Similarly, since we know that $\psi(L) = 0$, we require

$$\psi(L) = A\sin(kL) = 0$$

which has solutions when the argument of the sine function is an integer multiple of π. Hence the solutions correspond to discrete values of the wave vector k, which we label with a quantum number n:

$$k_n = n\frac{\pi}{L} \quad \text{for} \quad n = 1, 2, 3, \ldots \tag{38.46}$$

Substituting these values into Eq. (38.45), we see that the corresponding allowed wavefunctions are

$$\psi_n(x) = A\sin\left(n\frac{\pi x}{L}\right) \tag{38.47}$$

The wavefunctions corresponding to $n = 1, 2,$ and 3 are shown in Fig. 38.16; in all cases, an integer number of half wavelengths fits into the length L of the box. This condition on the wavelength is analogous to the condition for standing waves on a string.

From Eq. (38.43) we can express the energy in terms of the wave vector k,

$$E = \frac{\hbar^2 k^2}{2m_e} \tag{38.48}$$

or, inserting the allowed values of k_n, we obtain the allowed energies

$$E_n = \frac{\hbar^2}{2m_e}\left(n\frac{\pi}{L}\right)^2 \tag{38.49}$$

The allowed energies are thus square-integer multiples of the ground-state energy E_1,

$$E_n = n^2 E_1 \qquad \text{for} \qquad n = 1, 2, 3, \ldots \tag{38.50}$$

where the ground-state energy is

$$E_1 = \frac{\hbar^2 \pi^2}{2m_e L^2} \tag{38.51}$$

The energies of the first five wavefunctions are shown in the energy-level diagram of Fig. 38.17.

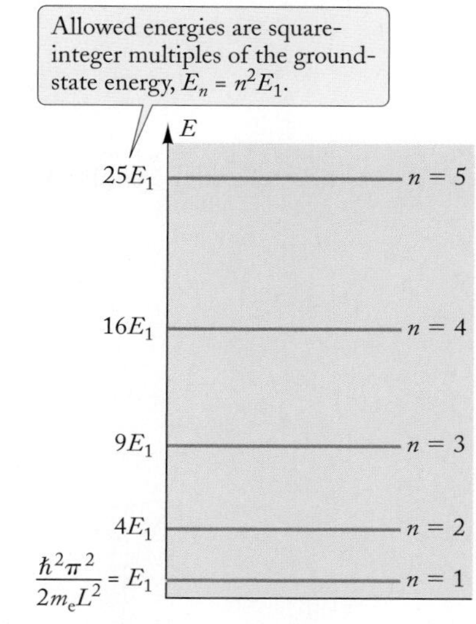

FIGURE 38.17 Energy-level diagram for the one-dimensional infinite potential well.

COMMENTS: Since $U(x) = 0$ inside the box, the energy of an electron confined to a region of length L is all kinetic energy. Note that in the $n = 1$ ground state, this energy is nonzero, and is known as **zero-point energy**; even at absolute zero temperature, the energy of the electron cannot be reduced below this value. This zero-point energy of the infinite potential well is inversely proportional to the square of the length L of the confinement region. Thus confinement costs energy. Note also that the higher kinetic energy of an excited state corresponds to a shorter wavelength. This is also evident from the de Broglie relation, since $K = p^2/2m = h^2/2m\lambda^2$.

zero-point energy

The probabilistic information contained in the wavefunction ψ was mentioned in Eq. (38.33): the probability of finding the electron at a given point in space is proportional to the square of the magnitude of the wavefunction at that point.[2] The probability of finding an electron at a single point is infinitesimal, so the quantity $|\psi|^2$ in Eq. (38.33) represents the probability per unit volume for finding an electron in a small volume around some point. Similarly, in one dimension the quantity $|\psi(x)|^2$ represents a probability per unit length, that is, $|\psi(x)|^2$ is the probability that the electron is in a region of width dx centered at the point x. Thus in practice, the probabilistic relation (38.33) means that the probability P of finding the electron in the region between two points a and b is given by

$$P = \int_a^b |\psi(x)|^2 dx \tag{38.52}$$

probability of finding electron in a region

[2]We use the square of the *magnitude* of ψ to allow for cases where ψ is a complex function. Although in practice the wavefunction is often complex, in this text we will only encounter examples where ψ is a real function.

Since the sum of the probabilities of all possible locations of the electron must be 1, that is, the electron must be somewhere, we must also require that the wavefunction satisfy

normalization condition

$$\int_{-\infty}^{+\infty} |\psi(x)|^2 dx = 1 \tag{38.53}$$

This requirement is known as the **normalization condition**. In addition to satisfying the Schrödinger equation and being continuous and smooth at boundaries, a meaningful wavefunction must satisfy the normalization condition.

EXAMPLE 8 In classical mechanics, an electron with energy E in a one-dimensional box of length L simply bounces back and forth between the walls at constant speed. A classical electron thus has equal probability of being anywhere in the box; the probability per unit length for an electron to be at any point is then $P_{\text{classical}} = 1/L$. For example, classical mechanics predicts that an electron would be found in the middle half of the box with a probability of exactly $\frac{1}{2}$, or 50% of the time. What are the normalized wavefunctions for an electron in the one-dimensional box of Example 7? In the quantum-mechanical ground state, what is the probability per unit length that the electron is at the center of the box? What is the probability of finding the electron in the middle half of the box?

SOLUTION: In order to use the wavefunctions $\psi_n(x) = A\sin(n\pi x/L)$ from Eq. (38.47) to calculate probabilities, we must determine the overall multiplicative constant A from the normalization condition (38.53). Since $\psi(x) = 0$ outside the box, we need only consider the region inside the box. For any quantum number n, inserting the solutions (38.47) into the normalization condition (38.53) gives

$$1 = \int_0^L |\psi_n(x)|^2 dx = \int_0^L A^2 \sin^2\left(n\frac{\pi x}{L}\right) dx = \tfrac{1}{2} LA^2 \tag{38.54}$$

where we have used the fact that for any integer number of half wavelengths, the average value of the square of the sine function is $\frac{1}{2}$, and so the integral over the length L is $\frac{1}{2}L$. Solving for A, we obtain

$$A = \sqrt{\frac{2}{L}} \tag{38.55}$$

independent of the quantum number n. So the normalized wavefunctions are

$$\psi_n(x) = \sqrt{\frac{2}{L}} \sin\left(\frac{n\pi x}{L}\right) \tag{38.56}$$

For the ground state, we use the $n = 1$ wavefunction. The probability per unit length that the electron is at the center of the box is obtained by evaluating $|\psi_1(x)|^2$ at $x = L/2$:

$$\left|\psi_1\left(\frac{L}{2}\right)\right|^2 = \left|\sqrt{\frac{2}{L}} \sin\left(\frac{\pi L/2}{L}\right)\right|^2 = \frac{2}{L}\sin^2\left(\frac{\pi}{2}\right) = \frac{2}{L}$$

Thus, compared with classical mechanics, where $P_{\text{classical}} = 1/L$, quantum mechanics predicts that the electron is twice as likely to be at the center of the box.

The middle half of the box extends from $x = \frac{1}{4}L$ to $x = \frac{3}{4}L$, so the probability of finding the electron between these points is

$$P = \int_{L/4}^{3L/4} |\psi_1(x)|^2 dx = \int_{L/4}^{3L/4} \frac{2}{L} \sin^2\left(\frac{\pi x}{L}\right) dx$$

With a change of variable $u = \pi x/L$, this becomes

$$P = \frac{2}{L}\frac{L}{\pi}\int_{\pi/4}^{3\pi/4} \sin^2 u \, du$$

From our table of integrals (Appendix 4), $\int \sin^2 u\, du = \frac{1}{2}u - \frac{1}{4}\sin 2u$. Hence

$$P = \frac{2}{\pi}\left(\frac{1}{2}u - \frac{1}{4}\sin 2u\right)\Big|_{\pi/4}^{3\pi/4} = \frac{2}{\pi}\left\{\left[\frac{3}{8}\pi - \left(-\frac{1}{4}\right)\right] - \left(\frac{1}{8}\pi - \frac{1}{4}\right)\right\}$$

$$= \frac{1}{2} + \frac{1}{\pi} \approx 0.818$$

Thus quantum mechanics indicates that the electron in the ground state is much more likely to be in the middle half of the box than classical physics would predict: approximately 82% compared with 50%.

COMMENT: The probability distribution depends on the state. For the ground state, the electron is most likely to be found at the center of the box, but for the first excited state, it is least likely to be found there (see Fig. 38.18).

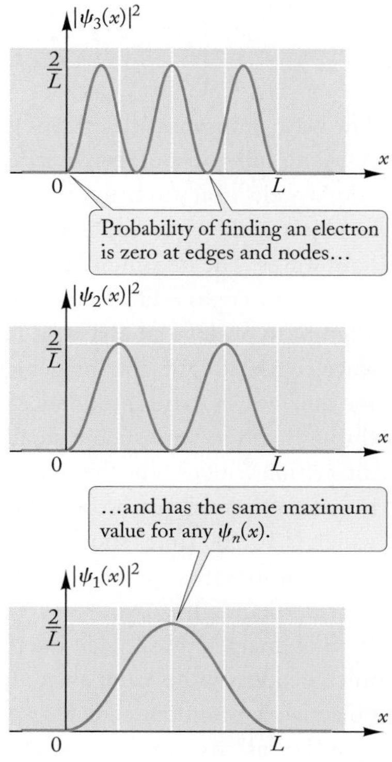

FIGURE 38.18 Probability per unit length, $|\psi(x)|^2$, as a function of position for the first few states of the infinite potential well.

The above examples of an electron in a box demonstrate that we cannot be certain about the position of the electron. Since the electron may be traveling to the right or to the left, its momentum is also uncertain. Within our small confinement region or, similarly, within the atom, electrons always behave very much like waves. But outside the atom, they will sometimes behave pretty much like classical particles. Roughly, we can say that classical mechanics will be a good approximation whenever the quantum uncertainties are small compared with the relevant magnitudes of positions and momenta. For instance, for the electrons in the beam of a TV tube, the quantum uncertainty in the momentum is negligible compared with the magnitude of the momentum. Under these conditions, classical mechanics gives an adequate description of the motion of the electrons.

What we have said about the wave mechanics of electrons also applies to other particles, or wavicles, found in nature—they all have wave properties and they all have quantum uncertainties in their position and momentum. Strictly speaking, even large macroscopic bodies have wave properties. For example, an automobile is a wavicle and it has some quantum uncertainty in its position. However, it turns out that the quantum uncertainties are very small whenever the mass of the body is large compared with atomic masses—the quantum uncertainty in the position of an automobile is typically no more than about 10^{-18} m, a number that can be ignored for all practical purposes. Hence, for automobiles and other macroscopic bodies, quantum effects are completely insignificant and classical mechanics gives an excellent description of the motion of these bodies.

PHYSICS IN PRACTICE ULTRAMICROSCOPES

The wave properties of electrons find practical application in the **transmission electron microscope** (TEM), which employs electron waves to form a highly magnified image of an object in the same way as an ordinary microscope employs light waves. The maximum magnification attainable by ordinary microscopes is limited by the wavelength of light. If we attempt to observe an object as small as or smaller than the wavelength of light, the image becomes indistinct, because the light waves suffer pronounced diffraction when passing through and around such a small object, and the resulting diffraction fringes blur the image. The electron waves used in typical electron microscopes have wavelengths of 0.05 nm, which is 10 000 times shorter than the wavelength of light; the electron waves are therefore much less susceptible to diffraction than light.

The main "optical" elements in a transmission electron microscope are the same as in an ordinary microscope, an objective lens and an ocular, or projector, lens (see Fig. 1). The "lenses" are not made of glass, but of magnetic fields, carefully shaped so as to provide deflections similar to those experienced by light in a glass lens. The electron rays emerging from the projector lens are intercepted by a detector array, which records the image digitally for viewing and analysis.

Another kind of electron microscope is the **scanning electron microscope** (SEM). The principle of operation of this microscope is quite different, and bears no resemblance to the operation of ordinary light microscopes. The scanning electron microscope relies on the particle properties of electrons as well as their wave properties. Instead of forming an image by means of refracted electron waves, the scanning electron microscope uses a diffraction-limited fine beam of electrons to bombard the object. The beam "scans" across the object line by line in a sweep pattern, like the sweep pattern of the electron beam in an ordinary TV tube. A detector picks up the current of electrons that recoil from the object and of secondary electrons knocked out of the object by the incident primary electrons (Fig. 2). This detected current is amplified and fed into a video monitor, which displays a picture. Scanning electron microscopes produce very crisp pictures with an exceptional depth of field and strong shadows that give a vivid three-dimensional impression, but they do not attain the extremely high magnifications of transmission electron microscopes.

(a)

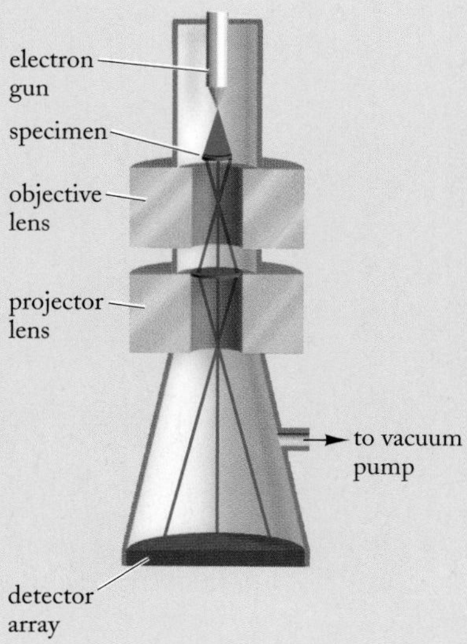

(b)

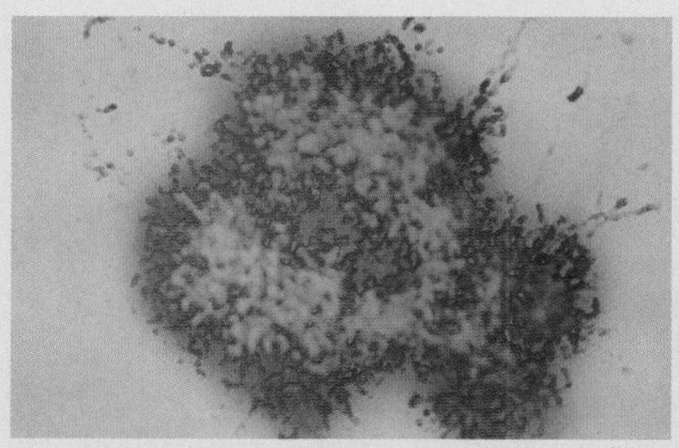

FIGURE 1 (a) Schematic diagram of a transmission electron microscope (TEM). (b) Picture of a human chromosome prepared with a TEM.

(a)

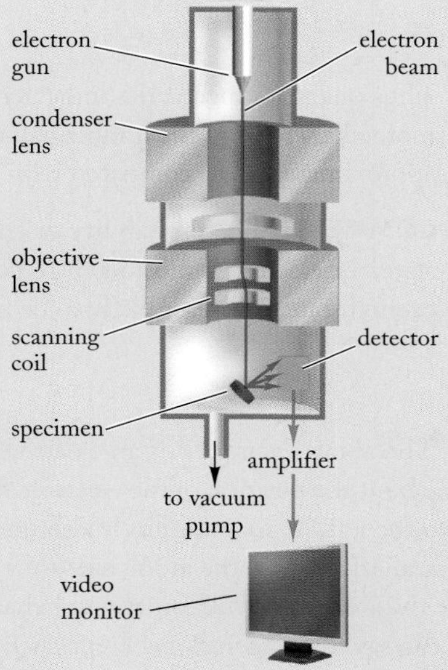

(b)

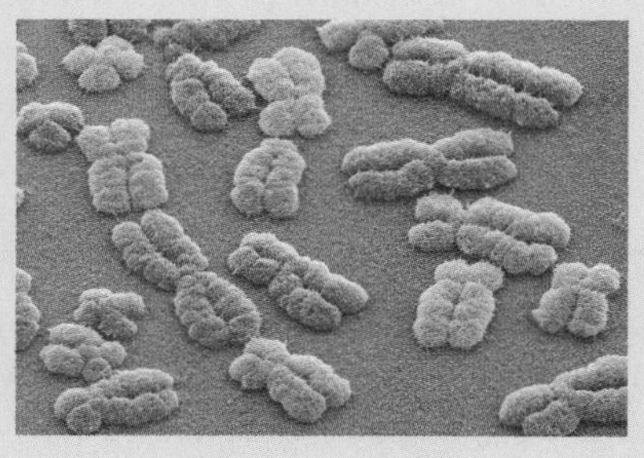

FIGURE 2 (a) Schematic diagram of a scanning electron microscope (SEM). (b) Picture of human chromosomes prepared with an SEM. Figure 1b is a close-up view of one such chromosome.

A newer and more powerful kind of electron microscope is the **scanning tunneling microscope** (STM). This microscope scans the surface of the object with a fine tungsten needle whose motion is precisely controlled by a delicate suspension system (see Fig. 3a). A potential difference is applied between the needle and the surface of the object. The needle is not actually in contact with the surface; it merely sweeps across the surface line by line. The gap between the needle and the surface is effectively an insulator, which tends to block the motion of the electrons from the needle to the surface of the object. However, the wave properties of electrons permit them to spread for some distance into this gap, and leak across to the surface. This kind of leakage of an electron across a gap where its motion is classically forbidden is called **tunneling**. The probability for an electron to succeed in tunneling across the gap depends drastically on the size of the gap. Whenever the needle, during its sweep, comes near any peak in the surface—such as a bulging atom—the tunneling probability starts to increase drastically, and so does the current from the needle to the surface. This incipient increase of the current is detected by the electronic circuit connected to the needle, and an amplified feedback signal is sent to the suspension system to lift the needle, so as to keep it at constant height from the peak. Thus, the needle skims over the peaks and valleys of the surface, maintaining a relatively constant height above the "terrain." The amplified signal used to control the suspension is also sent to a computer, where it is processed and then fed into a video monitor for display. Modern scanning tunneling microscopes attain magnifications of up to 10^8, and they permit us to "see" individual atoms (Fig. 3b).

This technique for constructing a picture of the surface by sweeping a needle across it is also exploited in the **atomic-force microscope** (AFM). But the needle of this kind of microscope is placed in direct contact with the surface—it is pressed against the surface with a force of, say 10^{-8} N, and it is lifted or lowered by the suspension system so as to keep the force constant during the sweep. Thus, the needle actually explores the shape of the surface by directly feeling it, just as you might explore the shape of a surface by feeling it with the tip of your finger. An alternative AFM technique is "tapping mode" (see Example 2 of Chapter 15), where the needle oscillates and taps the surface. While tapping the surface during scanning, the needle is lifted or lowered to keep the oscillation amplitude constant. Magnifications attained with atomic-force microscopes are nearly the same as those attained with scanning tunneling microscopes. The picture displayed in Fig. 4 was prepared with an atomic-force microscope.

(a)

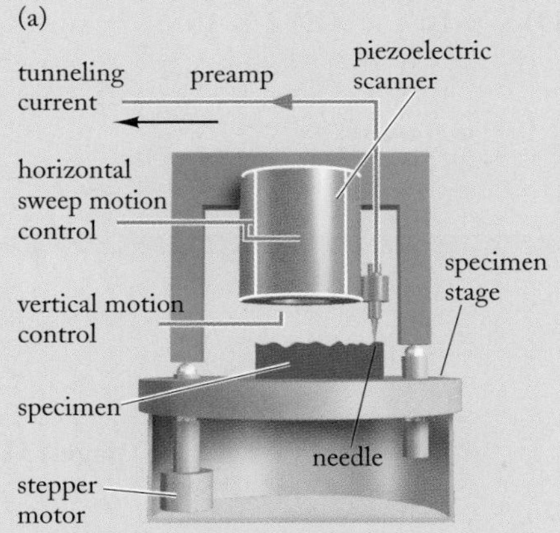

(b)

FIGURE 3 (a) Schematic diagram of a scanning tunneling electron microscope (STM). (b) Picture of iron atoms prepared with an STM. The 48 iron atoms arranged in a circle on the surface of a copper crystal form a confining potential known as a "quantum corral," and the ripples in the image show the wave nature of the surface copper electrons.

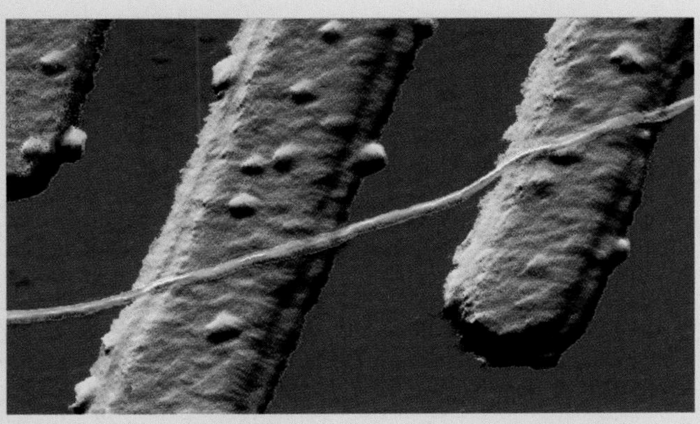

FIGURE 4 Picture of a carbon nanotube lying across ultra thin stripes of gold, prepared with an atomic-force microscope (AFM).

Checkup 38.5

QUESTION 1: One electron has a kinetic energy of 1 eV; another has a kinetic energy of 100 eV. Which has the shorter de Broglie wavelength? By what factor?

QUESTION 2: Suppose that an electron has a de Broglie wavelength of 20 nm. If we increase the speed of this electron by a factor of 4, what will be its new de Broglie wavelength?

QUESTION 3: An electron and a proton have the same kinetic energy. Which has the longer de Broglie wavelength?

QUESTION 4: What is the probability per unit length that an electron in the first excited state ($n = 2$) of a one-dimensional box is at the center of the box? An electron in the second excited state ($n = 3$)? (See Fig. 38.18.)

QUESTION 5: Bohr's theory of the hydrogen atom assumes that the electron remains in a given orbital plane, say, the x–y plane, and that it has no motion in the z direction. What are Δz and Δp_z in this case? Is this consistent with the uncertainty principle?

QUESTION 6: Take the proton-to-electron mass ratio to be $m_p/m_e \approx 2000$. For the ground states of the respective one-dimensional boxes, what is the ratio of the energy of a proton in a region of width 10^{-15} m to that of an electron in a region of width 10^{-10} m?

(A) 0.02 (B) 50 (C) 2000 (D) 5×10^6 (E) 2×10^{13}

SUMMARY

PHYSICS IN PRACTICE Ultramicroscopes **(page 1310)**

SPECTRAL SERIES OF HYDROGEN

$$\frac{1}{\lambda} = R\left(\frac{1}{n_2^2} - \frac{1}{n_1^2}\right) \qquad R = 1.096\ 78 \times 10^7\ \text{m}^{-1} \qquad \textbf{(38.7)}$$

QUANTIZATION OF ANGULAR MOMENTUM

$$L = n\hbar \qquad n = 1, 2, 3, \ldots \qquad \textbf{(38.9)}$$

PLANCK'S CONSTANT ("h-bar")

$$\hbar = \frac{h}{2\pi} = 1.05 \times 10^{-34}\ \text{J}\cdot\text{s} \qquad \textbf{(38.13)}$$

BOHR RADIUS

$$a_0 = \frac{4\pi\epsilon_0\hbar^2}{m_e e^2} = 0.0529\ \text{nm} \qquad \textbf{(38.17)}$$

ENERGY OF STATIONARY STATES OF HYDROGEN

$$E_n = -\frac{m_e e^4}{2(4\pi\epsilon_0)^2\hbar^2}\frac{1}{n^2} = -\frac{13.6\ \text{eV}}{n^2} \qquad \textbf{(38.23, 38.25)}$$

FREQUENCY AND WAVELENGTH OF PHOTON EMITTED IN TRANSITION

$$f = \frac{E_i - E_f}{h} \quad \text{and} \quad \frac{1}{\lambda} = \frac{f}{c} = \frac{E_i - E_f}{hc} \tag{38.27}$$

DE BROGLIE WAVELENGTH OF WAVICLE

$$\lambda = h/p \tag{38.32}$$

ENERGY OF STATIONARY STATE IN TERMS OF FREQUENCY

$$E = hf \tag{38.38}$$

TIME-INDEPENDENT SCHRÖDINGER EQUATION

One-dimension:

$$-\frac{\hbar^2}{2m_e}\frac{d^2}{dx^2}\psi(x) + U(x)\psi(x) = E\psi(x) \tag{38.41}$$

CONDITIONS FOR WAVEFUNCTION

One-dimension:

$\psi(x)$ is a solution to the Schrödinger equation

$\psi(x)$ is everywhere continuous

$\int_{-\infty}^{+\infty} |\psi(x)|^2 dx = 1$ (normalization)

$d\psi(x)/dx$ is everywhere continuous [except if $U(x)$ is infinite]

PROBABILITY FOR FINDING ELECTRON IN A REGION $a < x < b$

$$P = \int_a^b |\psi(x)|^2 dx \tag{38.52}$$

ELECTRON IN A BOX One-dimension, $0 < x < L$:

Wavefunctions: $\psi_n(x) = \sqrt{\frac{2}{L}} \sin\left(\frac{n\pi x}{L}\right)$

Energies: $E_n = n^2 E_1 = n^2 \frac{\hbar^2 \pi^2}{2m_e L^2}$

$n = 1, 2, 3, \ldots$

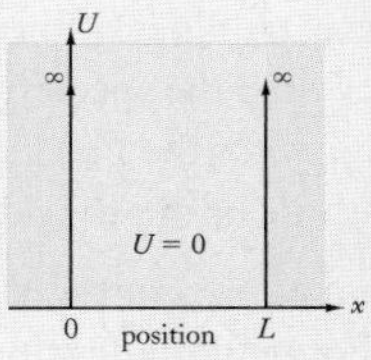

(38.50, 38.51)

QUESTIONS FOR DISCUSSION

1. Do the spectral lines seen in a stellar spectrum (for example, Fig. 38.5) tell us anything about the chemical composition of the stellar interior?
2. The lower spectrum of hydrogen shown in the chapter photo displays all of the spectral lines simultaneously. Since a hydrogen atom emits only one spectral line at a time, how can all the lines be visible simultaneously?
3. The target used in Rutherford's scattering experiment was a very thin foil of metal. What is the advantage of a thin foil over a thick foil in this experiment?
4. How can Rutherford's experiment tell us something about the size of the nucleus?
5. At low temperatures, the absorption spectrum of hydrogen displays only the spectral lines of the Lyman series. At higher temperatures, it also displays other series. Explain.
6. The planets move around the Sun in circular orbits. Is their orbital angular momentum quantized?
7. In **a muonic atom**, a muon orbits around the nucleus. The mass of the muon is 207 times the mass of the electron. What is the Bohr radius for a muonic atom with a hydrogen nucleus?

8. Given that the orbital angular momentum of an atom is quantized, can we conclude that the orbital magnetic moment is also quantized?

9. Bohr's theory of the hydrogen atom can be adapted to the singly ionized helium atom, that is, the helium atom with one missing electron. To what other ionized atoms can Bohr's theory be adapted?

10. According to the **Complementarity Principle**, formulated by Bohr, a wavicle has both wave properties and particle properties, but these properties are never exhibited simultaneously; if the wavicle exhibits wave properties in an experiment, then it will not exhibit particle properties, and conversely. Give some examples of experiments in which wave or particle properties (but not both simultaneously) are exhibited.

11. Show that photons obey the de Broglie relation.

12. If the de Broglie wavelengths of two electrons differ by a factor of 2, by what factor must the kinetic energies differ?

13. According to the de Broglie relation, the wavelength of an electron of very small momentum is very large. Could we take advantage of this to design an experiment that makes the wave properties of the electron obvious?

14. Describe the interference pattern expected for an electron wave incident on a plate with two very narrow parallel slits separated by a small distance.

15. Electron microscopes achieve high resolution because they use electron waves of very short wavelength, usually less than 0.05 nm. Why can we not build a microscope that uses photons of equally short wavelength?

PROBLEMS

†38.2 Spectral Series of Hydrogen

1. Use Eq. (38.4) to calculate the wavelengths of the first four lines of the Lyman series.

2. Show that the spectral lines of the Balmer series all have a higher frequency than the spectral lines of the Paschen series. Do the spectral lines of the Paschen series all have a higher frequency than those of the Brackett series?

3. A hydrogen atom at rest emits a photon. By Eq. (37.15), $p = h/\lambda$, the photon of highest momentum has the shortest wavelength, corresponding to the Lyman series limit. What is the recoil velocity of the hydrogen atom when this photon is emitted?

4. Which of the spectral lines of the Brackett series is closest in wavelength to the first spectral line ($n_2 = 5$; $n_1 = 6$) of the Pfund series? By how much do the wavelengths differ?

*5. When astronomers examine the light of a distant galaxy, they find that all the wavelengths of the spectral lines of the atoms are longer than those of the atoms here on Earth by a common multiplicative factor. This is the *red shift* of light; it is a Doppler shift caused by the motion of recession of the galaxy, away from Earth. In the light of a galaxy beyond the constellation Virgo, astronomers find spectral lines of wavelengths 411.7 nm and 435.7 nm.

(a) Assume that these are two spectral lines of hydrogen, with the wavelengths multiplied by some factor. Identify these lines. What is the factor by which these wavelengths are longer than the normal wavelengths of the two spectral lines? What is the factor by which the frequencies are lower?

(b) If the speed of recession is low compared with the speed of light, the Doppler shift of light obeys a formula similar to that for the Doppler shift of sound [see Eq. (17.17)]. Calculate the speed of recession of the galaxy.

*6. One of the series of spectral lines of the lithium atom is the **principal series**, with the following wavelengths, measured in vacuum: 617.0 nm, 323.4 nm, 274.2 nm, 256.3 nm, 247.6 nm. These wavelengths approximately fit the formula

$$\frac{1}{\lambda} = R\left[\frac{1}{(1+s)^2} - \frac{1}{(n+p)^2}\right] \qquad n = 2, 3, 4, \ldots$$

where $R = 1.097\,29 \times 10^7\ \text{m}^{-1}$ is the Rydberg constant for lithium, and where s and p are constants characteristic of the series. Find the values of these constants. (Hint: Write the equations for $1/\lambda$ for two different wavelengths, eliminate the unknown s by subtracting these formulas, and then solve for p by trial and error.)

*7. Another of the spectral series of the lithium atom is the **diffuse series**, with the following wavelengths, measured in vacuum: 610.5 nm, 460.4 nm, 413.4 nm, 391.6 nm, 379.6 nm. These wavelengths approximately fit the formula

$$\frac{1}{\lambda} = R\left[\frac{1}{(2+p)^2} - \frac{1}{(n+d)^2}\right] \qquad n = 3, 4, 5, \ldots$$

where, as in the preceding problem, $R = 1.097\,29 \times 10^7\ \text{m}^{-1}$ and where p and d are constants.

(a) Find the values of these constants.

(b) The principal series (see Problem 6) and the diffuse series of lithium are analogous to two spectral series of hydrogen. Which two series? (Hint: See the preceding problem.)

†For help, see Online Concept Tutorial 43 at www.wwnorton.com/physics

38.3 The Nuclear Atom

*8. What is the distance of closest approach for a 5.5-MeV alpha particle in a head-on collision with a gold nucleus? With an aluminum nucleus?

*9. The nucleus of platinum has a radius of 6.96×10^{-15} m and an electric charge of $78e$. What must be the minimum energy of an alpha particle in a head-on collision if it is to just barely reach the nuclear surface? Assume the alpha particle is pointlike.

*10. Consider a model of the hydrogen atom with a pointlike nucleus, but assume that the electron charge is uniformly distributed over a sphere of radius $R = 0.053$ nm. The atom is in equilibrium when the nucleus is at the center of the electron charge distribution. Find the frequency of oscillation when the electron sphere is displaced from equilibrium by a distance $x < R$. What is the wavelength of a photon with this frequency?

*11. The probability for a scattering event to occur is equal to the fraction of the area of the target for which that event will occur. The area per nucleus for an event is the **scattering cross section** σ. This area is related to the impact parameter b for that event by $\sigma = \pi b^2$. The impact parameter for a deflection of more than 90° (backscattering) is 1.0×10^{-13} m for alpha particles of energy 5.7 MeV scattering from gold nuclei. What is the scattering cross section for backscattering? Consider an incident beam of alpha particles with a beam radius of 2.0 millimeters and a target with 5.0×10^{17} gold nuclei in the beam area. What is the probability for backscattering?

*12. An alpha particle of energy 5.5 MeV is incident on a silver nucleus with an impact parameter 8.0×10^{-15} m. The distance of closest approach of the particle is 2.7×10^{-14} m. Find the speed at the point of closest approach. (Hint: The angular momentum is conserved.)

**13. A foil of gold, 2.1×10^{-5} cm thick, is being bombarded by alpha particles of energy 7.7 MeV. The particles strike at random over an area of 1.0 cm^2 of the foil of gold.

(a) How many atoms are there within the volume 1.0 $cm^2 \times 2.1 \times 10^{-5}$ cm under bombardment? The density of gold is 19.3 g/cm^3, and the mass of one atom is 3.27×10^{-25} kg.

(b) It can be shown that to suffer a deflection of more than 30°, an alpha particle must strike within 5.5×10^{-14} m of the center of a gold nucleus. What is the probability for this to happen? (Hint: See Problem 11.)

(c) If 1.0×10^{10} alpha particles impact on the foil, how many will suffer deflections of more than 30°?

†38.4 Bohr's Theory

14. If you wanted to give an apple an angular momentum of $\hbar$, at what rate would you have to spin it about its axis? Treat the apple as a uniform sphere of mass 0.20 kg and radius of 4.0 cm.

†For help, see Online Concept Tutorial 43 at www.wwnorton.com/physics

15. What is the angular momentum of a vinyl record (a uniform disk) rotating at $33\frac{1}{3}$ revolutions per minute? The moment of inertia of the record is 1.3×10^{-2} kg·m^2. Express the answer as a multiple of $\hbar$.

16. What are the angular momentum, the kinetic energy, the potential energy, and the net energy of an electron in the smallest ($n = 1$) Bohr orbit in the hydrogen atom? Express your answers in SI units.

17. What is the speed of an electron in the smallest ($n = 1$) Bohr orbit? Express your answer as a fraction of the speed of light. Is it justified to ignore relativistic effects in the Bohr model?

18. What is the centripetal acceleration of an electron in the smallest ($n = 1$) Bohr orbit in the hydrogen atom?

19. Hydrogen atoms in highly excited states with a quantum number as large as $n = 732$ have been detected in interstellar space by radio astronomers. What is the orbital radius of the electron in such an atom? What is the energy of the electron?

20. If a hydrogen atom is initially in the first excited state, what is the *longest* wavelength of light it will absorb? What is the *shortest* wavelength of light it will absorb?

21. If you bombard hydrogen atoms in their ground state with a beam of particles, the collisions will (sometimes) kick atoms into one of their excited states. What must be the minimum kinetic energy of the bombarding particles if they are to achieve such an excitation?

22. Suppose that the electron in a hydrogen atom is initially in the second excited state ($n = 3$). What wavelength will the atom emit if the electron jumps directly to the ground state? What two wavelengths will the atom emit if the electron jumps to the first excited state and then to the ground state?

23. A hydrogen atom emits a photon of wavelength 102.6 nm. From what stationary state to what lower stationary state did the electron jump?

*24. Show that the speed of an electron in a Bohr orbit with quantum number n is given by

$$v = \frac{1}{n}\frac{e^2}{4\pi\epsilon_0\hbar}$$

Also, find an expression for the centripetal acceleration of an electron in the nth Bohr orbit.

*25. The quantity $\hbar/(m_e c)$ is called the **Compton wavelength**. The quantity $e^2/(4\pi\epsilon_0 m_e c^2)$ is called the "**classical electron radius**." Show that the Bohr radius, the Compton wavelength, and the classical electron radius are in the ratios $1{:}\alpha{:}\alpha^2$, where $\alpha = e^2/(4\pi\epsilon_0\hbar c)$. The quantity α is called the **fine-structure constant**. What is the numerical value of this constant?

*26. A hydrogen atom is initially in the ground state. In a collision with an argon atom, the electron of the hydrogen atom absorbs an energy of 15.0 eV. With what speed will the electron be ejected from the hydrogen atom?

*27. Suppose that a sample of hydrogen atoms, initially all in the ground state, is under bombardment by a beam of electrons of kinetic energy 12.2 eV. In an inelastic collision between a hydrogen atom and one of the incident electrons, the hydrogen atom will occasionally absorb all, or almost all, the kinetic energy of the electron and make a transition from the ground state to an excited state. If so, what excited state will the hydrogen atom attain? What are the possible spectral lines the atom can emit subsequently?

*28. The singly ionized helium atom (usually designated HeII or He^+) has one electron in orbit around a nucleus of charge $2e$.

(a) Apply Bohr's theory to this atom and find the energies of the stationary states. What is the value of the ionization energy, that is, the energy that you must supply to remove the electron from the atom when it is in the ground state? Express the answer in electron-volts.

(b) Show that for every spectral line of the hydrogen atom, the ionized helium atom has a spectral line of identical wavelength.

*29. Assume that, as proposed by J. J. Thomson, the hydrogen atom consists of a cloud of positive charge e, uniformly distributed over a sphere of radius R. However, instead of placing the electron in static equilibrium at the center of the sphere, assume that the electron orbits around the center with uniform circular motion under the influence of the electric centripetal force $(e^2/4\pi\epsilon_0)(r/R^3)$. If the angular momentum of this orbiting electron is quantized according to Bohr's theory (so that $L = n\hbar$), what are the radii and the energies of the quantized orbits? What are the frequencies of the photons emitted in transitions from one quantized orbit to another? What must be the value of R if at least two orbits are to fit inside this atom?

**30. In principle, Bohr's theory also applies to the motion of the Earth around the Sun. The Earth plays the role of the electron, the Sun that of the nucleus, and the gravitational force that of the electric force.

(a) Find a formula analogous to Eq. (38.18) for the radii of the permitted circular orbits of the Earth around the Sun.

(b) The actual radius of the Earth's orbit is 1.50×10^{11} m. What value of the quantum number n does this correspond to?

(c) What is the radial distance between the Earth's actual orbit and the next larger orbit?

**31. In our calculation of the energies of the stationary states of hydrogen we pretended that the proton remains at rest. Actually, both the electron and the proton orbit about their common center of mass. Show that the energies of the stationary states, taking into account this motion of the proton, are given by

$$E_n = -\frac{me^4}{2(4\pi\epsilon_0)^2\hbar^2}\frac{1}{n^2}$$

where

$$m = \frac{m_e m_p}{m_e + m_p}$$

(Hint: The electron and the proton move in circles of radii

$$r_e = r\frac{m_p}{m_p + m_e} \quad \text{and} \quad r_p = r\frac{m_e}{m_p + m_e}$$

where r is the distance between the electron and the proton. According to Bohr's theory, the net angular momentum of this system of two particles is quantized, $L = n\hbar$.)

32. The atom of **positronium consists of an electron and a positron (or antielectron) orbiting about their common center of mass. According to Bohr's theory, the net angular momentum of this system is quantized, $L = n\hbar$. What is the radius of the smallest possible circular orbit of this system? What is the wavelength of the photon released in the transition from $n = 2$ to $n = 1$? (Hint: the electron and the positron have the same mass. See also Problem 31.)

38.5 Quantum Mechanics; the Schrödinger Equation

33. What must be the energy of an electron if its wavelength is to equal the wavelength of visible light, about 550 nm?

34. Find the de Broglie wavelength for each of the following electrons with the specified kinetic energy: electron of 20 keV in a TV tube, conduction electron of 5.4 eV in a metal, orbiting electron of 13.6 eV in a hydrogen atom, orbiting electron of 91 keV in a lead atom.

35. What is the de Broglie wavelength of an electron in the ground state of hydrogen? In the first excited state?

36. What is the de Broglie wavelength of a tennis ball of mass 0.060 kg moving at a speed of 1.0 m/s?

37. An electron microscope operates with electrons of kinetic energy 40 keV. What is the wavelength of such electrons? By what factor is this wavelength smaller than that of visible light?

38. A photon and an electron each have an energy of 6.0×10^3 eV. What are their wavelengths?

39. If the de Broglie wavelengths of two electrons differ by a factor of 4, by what factor must their kinetic energies differ?

40. The "thermal" neutrons in a nuclear reactor typically have a kinetic energy of about 0.050 eV. What is the de Broglie wavelength of such a neutron?

41. Suppose that the velocity of an electron has been measured to within an uncertainty of $\pm$ 1.0 cm/s. What minimum uncertainty in the position of the electron does this imply?

42. The nucleus of the aluminum atom has a diameter of 7.2×10^{-15} m. Consider one of the protons in this nucleus. The uncertainty in the position of this proton is necessarily less than 7.2×10^{-15} m. What is the minimum uncertainty in its momentum and velocity?

43. If the position of a parked automobile of mass 2.0×10^3 kg is uncertain by $\pm$ 1.0×10^{-18} m, what is the corresponding uncertainty in its velocity?

44. For an electron confined to a one-dimensional box of length $L = 0.10$ nm, what is the kinetic energy in the ground state? What is the corresponding speed? Is it justified to ignore relativistic effects for such an electron?

45. Suppose that a particle confined to a one-dimensional box of length L is in its first excited state. What is the probability per unit length that the particle is in a small interval near $x = \frac{1}{4}L$? Near $x = \frac{1}{2}L$? Near $x = \frac{3}{4}L$?

46. Consider the possibility that a nucleus might contain an electron. What would be the energy of an electron in the ground state of a one-dimensional box of nuclear size, say 1.0×10^{-14} m? Since nuclear energies are typically only on the order of 1 MeV, do you expect that an electron might be confined within the nucleus?

47. For a particle in the state of quantum number n of a one-dimensional box of length L, at how many points within the box is the probability per unit length of finding the particle zero? At how many points is it maximum? In each case, describe the locations of those points.

48. An electron in a one-dimensional box of length $L = 0.30$ nm makes a transition from the first excited state to the ground state. What is the wavelength of the emitted photon?

49. Use symmetry and the result of Example 8 to determine the probability that an electron in the ground state of a one-dimensional box is in the region $0 \leq x \leq \frac{1}{4}L$.

*50. Neutrinos of energy 10 MeV are emitted by the sun. What is the de Broglie wavelength of such a neutrino? (Hint: Since the neutrino is a particle of very small mass, its momentum at such high kinetic energy must be calculated according to the formula for the momentum of an ultra-relativistic particle.)

*51. What is the de Broglie wavelength of an electron if its kinetic energy is equal to its rest-mass energy? (Hint: For a relativistic particle the momentum must be calculated according to the relativistic formula.)

*52. What is the de Broglie wavelength of the Earth in its motion around the Sun?

*53. Show that the de Broglie wavelength of an electron can be calculated from the formula

$$\lambda = 1.23/\sqrt{K}$$

where the wavelength λ is expressed in nm and the kinetic energy K is expressed in eV.

*54. Consider the solutions to the Schrödinger equation for a particle in a one-dimensional box. For n large, show that the probability per unit length, averaged over a small but suitable interval, yields the classical result $P_{\text{classical}} = 1/L$.

*55. Obtain the energies [Eqs. (38.50–38.51)] of an electron confined to an interval from $x = 0$ to $x = L$ along the x axis by a simple de Broglie wavelength calculation:

(a) What are the de Broglie wavelengths of standing electron waves in this interval? Assume that, as for a standing wave on a string, the amplitude of oscillation is zero at $x = 0$ and $x = L$, so that an integer number of half wavelengths fit into the length L.

(b) What are the momenta corresponding to these wavelengths?

(c) What are the energies of the stationary states?

(d) Evaluate the energies of the ground state and the first three excited states numerically for $L = 0.10$ nm.

56. The **harmonic oscillator potential is $U(x) = \frac{1}{2}m\omega_0^2x^2$; a particle of mass m in this potential oscillates with frequency ω_0. The ground-state wavefunction for a particle in the harmonic oscillator potential has the form

$$\psi(x) = Ae^{-ax^2}$$

(a) By substituting $U(x)$ and $\psi(x)$ into the one-dimensional, time-independent Schrödinger equation [Eq. (38.41)], find expressions for the ground-state energy E and the constant a in terms of m, $\hbar$, and ω_0.

(b) Apply the normalization condition to determine the constant A in terms of m, $\hbar$, and ω_0.

REVIEW PROBLEMS

57. The series limit for the Balmer series is 364.7 nm. What are the series limits for the Lyman, Paschen, and Brackett series?

58. Are the wavelengths of all the spectral lines of the Lyman series shorter than the series limit of the Balmer series? Are the wavelengths of all the spectral lines of the Balmer series shorter than the series limit of the Paschen series? Are the wavelengths of all the spectral lines of the Paschen series shorter than the series limit of the Pfund ($n_2 = 5$) series?

59. If a hydrogen atom is in the ground state, what is the *longest* wavelength it will absorb?

60. What is the frequency of the orbital motion for an electron in the smallest ($n = 1$) Bohr orbit? In the next ($n = 2$) Bohr orbit? Do either of these frequencies coincide with the frequency of the light emitted during the transition from $n = 2$ to $n = 1$?

61. Find the orbital radius, the speed, the angular momentum, and the centripetal acceleration for an electron in the $n = 2$ Bohr orbit of hydrogen.

62. When a hydrogen atom is initially in the ground state, the ionization energy (or the energy for removal of the electron) is

13.6 eV. What is the ionization energy for a hydrogen atom initially in the first excited state?

63. For an electron in the first excited state of a one-dimensional box, what is the probability for finding the electron in the interval $0 \leq x \leq \frac{1}{8}L$? Compare this with the classical result, $P = \frac{1}{8}$.

64. An electron makes a transition from the $n = 3$ state to the $n = 1$ state of a one-dimensional box, emitting a photon of wavelength 450 nm. What is the length of the box?

*65. An alpha particle of energy 5.5 MeV is incident on a copper nucleus with an impact parameter of 5.0×10^{-15} m. Find the distance of closest approach of the particle, and find the speed at the point of closest approach. (Hint: Both the energy and the angular momentum of the alpha particle are conserved.)

*66. The muon (or mu meson) is a particle somewhat similar to an electron; it has a charge $-e$ and a mass 206.8 times as large as the mass of the electron. When this particle orbits around a proton, the two particles form a *muonic hydrogen atom*, similar to an ordinary hydrogen atom, but with the muon playing the role of the electron. Calculate the Bohr radius of this muonic atom and calculate the energies of the stationary states. What is the energy of the photon emitted when the muon makes a transition from the $n = 2$ to the $n = 1$ state?

*67. An electron in a hydrogen atom is initially in the ground state. The electron absorbs a photon from an external light source and thereby makes a transition to the $n = 4$ state. What must have been the energy and the wavelength of the absorbed photon? If the electron now jumps spontaneously to the $n = 3$ state, what are the energy and the wavelength of the emitted photon?

*68. The doubly ionized lithium atom (usually designated LiIII or Li^{2+}) has one electron in orbit around a nucleus of charge $3e$. What is the radius of the smallest Bohr orbit in doubly ionized lithium? What is the energy of this orbit?

*69. Consider a helium atom in interstellar space in a circular orbit around a meteoroid of mass 4.0 kg and radius 10 cm under the influence of the gravitational force. We can apply Bohr's theory to this system; the meteoroid plays the role of the nucleus, the atom that of the electron, and the gravitational force that of the electric force. Find formulas analogous to Eqs. (38.18) and (38.23) for the orbital radii and the energies of the permitted circular orbits. Because of the finite size of the meteoroid, the smallest feasible orbit has a radius of 10 cm. What is the quantum number and what is the energy (in eV) of a helium atom in this orbit?

*70. What is the de Broglie wavelength of a nitrogen molecule (N_2) in air at room temperature (20°C)? Assume that the molecule is moving with the rms speed of molecules at this temperature.

*71. Interferometric methods permit us to measure the position of a macroscopic body to within $\pm 1.0 \times 10^{-12}$ m. Suppose we perform a position measurement of such a precision on a body of mass 0.050 kg. What uncertainty in momentum is implied by the Heisenberg relation? What uncertainty in velocity?

*72. Consider an electron in a circular orbit of quantum number n in a hydrogen atom. The orbit is well defined provided that $\Delta p \ll p$ and $\Delta y \ll r$. Show that if $n \gg 1$, these requirements are *not* in conflict with the Heisenberg uncertainty relations. Thus, *large* orbits in the hydrogen atom are well defined.

Answers to Checkups

Checkup 38.1

1. The original spectra are in the form of lines because of the presence of the slit between the light source and the prism; to produce spectral dots, you would have to replace the slit by a small hole.

2. For the moderately hot temperature of a normal flame, we saw in Chapter 37 that the continuum thermal radiation is peaked in the infrared, not the blue; thus the bluish color is due to spectral lines.

3. According to the color print on page 1289, this yellow light is due to two nearly equal wavelengths (a "doublet") with values 589.2 nm and 589.8 nm.

4. (B) Absorption of light by atoms in the cavity. Before emission from the opening, thermal radiation can strike atoms in the cavity; the atoms will absorb light at their characteristic wavelengths, resulting in dark absorption lines.

Checkup 38.2

1. Yes. In any series, the separation of successive spectral lines is given by the difference of the inverse square of successive integers, which approaches zero as the integers approach infinity.

2. Similar to the result of Example 1, the series limits are given by $1/\lambda = R/n_2^2$. For the series given, the largest wavelength corresponds to $n_2 = 3$ for the Paschen series, and the shortest wavelength to $n_2 = 1$ for the Lyman series.

3. In each case, the longest wavelength is given by the smallest difference in the two terms of Eq. (38.7); this corresponds to choosing $n_1 = n_2 + 1$. Since $n_2 = 1, 2, 3$, and 4 for the Lyman, Balmer, Paschen, and Brackett series, respectively, the corresponding values of n_1 are 2, 3, 4, and 5.
4. (A) Lyman. Similar to Example 1, the shortest wavelength of each series is given by $\lambda = n_2^2/R = n_2^2 \times 91.176$ nm. Thus the series with $n_2 = 2, 3$, and 4 have wavelengths too long to include the given wavelength, so the given wavelength must belong to the $n_2 = 1$, or Lyman series.

Checkup 38.3

1. According to the "plum-pudding" model, the positive charge is spread over a cloud of atomic size, approximately 10^{-10} m, and the electron is a pointlike particle in this cloud. If the cloud of positive charge is spherically symmetric, the electron is in equilibrium when at rest at the center of the cloud, where the electric field is zero.
2. An impact parameter of zero means that the alpha particle approaches the nucleus head-on; thus, the alpha particle bounces back, along its initial line of motion (or if it has sufficient energy, it penetrates into the nucleus).
3. Since the impact parameter is the perpendicular distance between the nucleus and the *undeflected* original line of motion, the distance of closest approach in the *repulsively deflected* hyperbolic orbit is larger than the impact parameter.
4. If the incident particles were negatively charged, the deflections would be attractive; thus, the orbits would go partially around the nucleus, and the alpha particle would emerge in a somewhat downward direction, instead of being deflected upward.
5. (A) Through a smaller angle than when incident on a gold target. For the smaller charge $+Ze$ on the silver nucleus, the repulsive electric force is smaller, and so the deflection is smaller.

Checkup 38.4

1. According to classical mechanics and electromagnetism, an electron in a circular orbit is accelerating and thus must radiate, lose energy, and spiral inward toward the nucleus.
2. The kinetic energy is the first term in Eqs. (38.19)–(38.21), which is positive and inversely proportional to the square of the quantum number n. For a transition to a lower state, n decreases, so the kinetic energy increases. The potential energy, the second term in the same equations, is also inversely proportional to the square of n, but is negative and so decreases as n decreases. According to Bohr's postulate [see Eq. (38.9)], the orbital angular momentum is proportional to n and thus decreases when the hydrogen atom makes a transition to a lower state.
3. Transitions of the Paschen series all end at $n = 3$, so the first transition is from $n = 5$ to $n = 3$. Transitions of the Balmer series all end at $n = 2$, so the second transition is from $n = 3$ to $n = 2$; the final transition must then be from $n = 2$ to $n = 1$. Transitions that end at $n = 1$ are in the Lyman series.
4. (D) $n = \infty$ to $n = 2$. Transitions of the Balmer series all end at $n = 2$; the transition that produces the series limit has the largest energy difference, and so begins at $n = \infty$.

Checkup 38.5

1. Higher energy corresponds to shorter wavelength. Since the kinetic energy is proportional to the square of the momentum (that is, $K = \frac{1}{2}mv^2 = p^2/2m$) and the de Broglie wavelength is $\lambda = h/p$, the wavelength varies inversely with the square root of the energy. So the 100-eV electron has a factor of 10 shorter wavelength than the 1-eV electron.
2. The de Broglie wavelength is inversely proportional to the momentum and, since $p = mv$, also to the speed. A factor of 4 increase in speed thus results in a factor of 4 decrease in wavelength to 5 nm.
3. Since the de Broglie wavelength is $p = h/\lambda$, the kinetic energy is $K = p^2/2m = h^2/2m\lambda^2$. Equality of energies thus implies that the wavelength must be longer for the electron, because of its smaller mass.
4. Since the $n = 2$ state has a node at the center of the box $[\psi(L/2) = 0]$, the probability per unit length that an electron is at the center is zero. The $n = 3$ state has a maximum at the center. In Example 8, we found that all the stationary states of the one-dimensional box have the same wavefunction amplitude, $\sqrt{2/L}$, and thus the probability per unit length at the maximum is $2/L$.
5. If an electron remains within the x–y plane, then it has $\Delta z = 0$. If there is no motion in the z direction, then the electron has $\Delta p_z = 0$. These are inconsistent with the uncertainty relation, which requires $\Delta z \Delta p_z \geq \hbar/2$.
6. (D) 5×10^6. The energy varies inversely with the mass and inversely with the square of the length, so compared with the electron, the proton energy will be both decreased by a factor of 2000 and increased by a factor of 10^{10}, for an overall increase by a factor of 5×10^6. The given lengths correspond to the sizes of nuclear and atomic regions; atomic energies are typically of order eV, and nuclear energies, MeV.

CHAPTER

39

Quantum Structure of Atoms, Molecules, and Solids

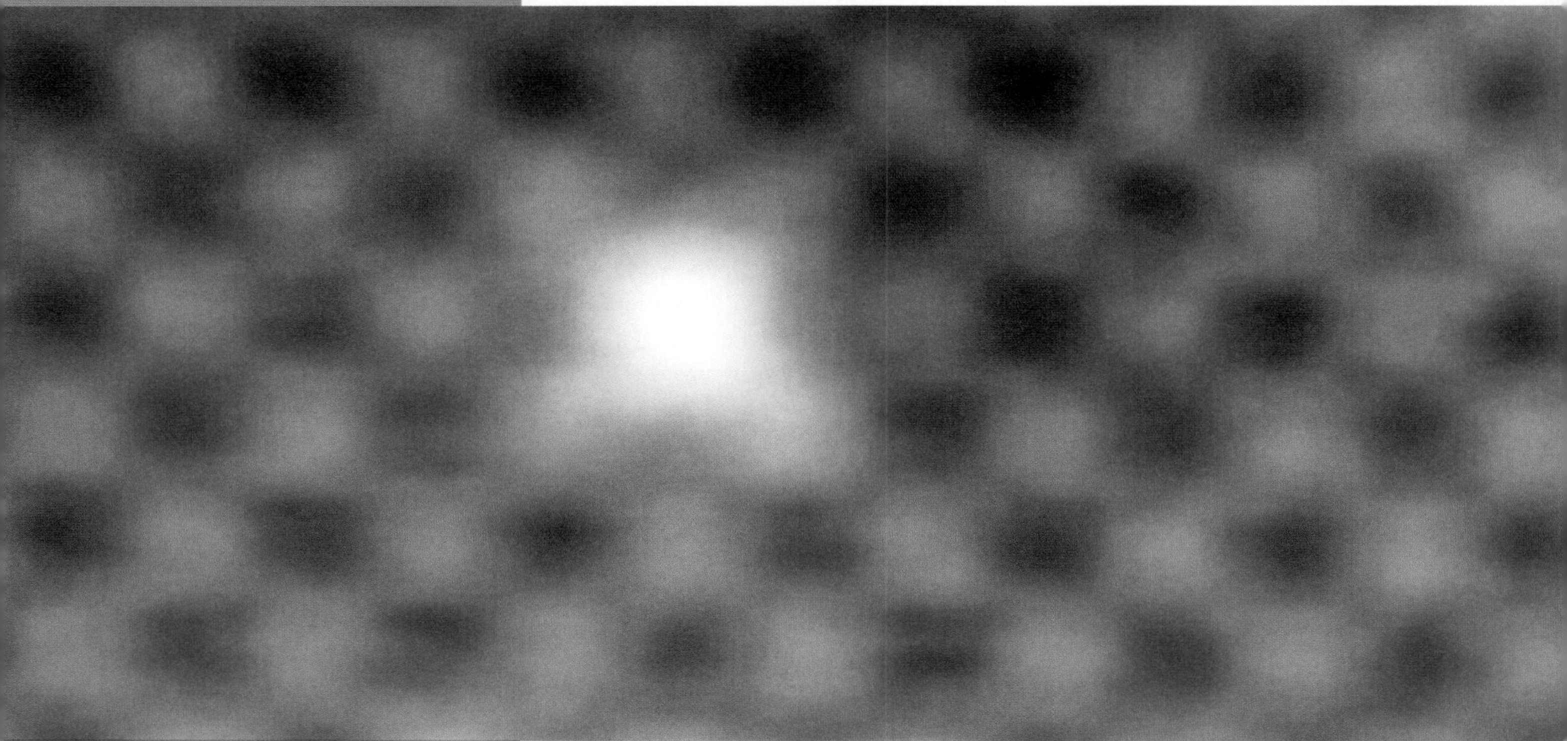

CONCEPTS IN CONTEXT

Concepts in Context

This scanning tunneling microscope (STM) image depicts the electron densities associated with individual atoms on the surface of gallium arsenide, a modern semiconductor preferred in some applications over silicon, the material predominantly used in the fabrication of transistors and integrated circuits. The atoms are arranged in a periodic structure, known as a lattice.

As we learn about the quantum structure of materials, we will consider such questions as:

? The nearly free electrons in a metal interact with the atoms in a lattice, and this restricts the permitted energies of the electrons. How do these restrictions arise? (Section 39.4, page 1336)

? The bright spot in the photo of the crystal lattice shows an indium atom, an impurity. Such impurities are intentionally added to a semiconductor to modify the conductivity. How does the presence of impurity atoms affect the conductivity? (Section 39.4, page 1339)

? How are regions with different impurities combined to make useful devices? (Section 39.5, page 1340)

The physical and chemical properties of atoms and of molecules depend on the quantum behavior of their electrons. These electrons occupy most of the volume of the atom, and their arrangement in different quantum states determines the size and the shape of the atom, the chemical bonds that the atom forms with other atoms, the energy required for ionization, the spectrum of light emitted and absorbed, and so on. Likewise, the physical properties of a solid—such as diamond, silicon, silver, copper—depend on the quantum behavior of the electrons in the solid. The spacing of the crystal lattice of the solid, the electric and thermal conductivities, the magnetic properties, and the mechanical properties (such as elasticity and hardness) all hinge on the arrangement of the electrons. For instance, we will see in Section 39.4 how the arrangement of the electrons determines the ability of the solid to conduct an electric current, and we will see how the differences among conductors, semiconductors, and insulators depend on what stationary quantum states are available in the solid and which of these are occupied by electrons.

The arrangement of the electrons in the stationary states of an atom or a solid is subject to an important restriction: no more than two electrons can occupy the same orbital state. This is called the Pauli Exclusion Principle. In Section 39.1, we will become acquainted with the spin, or the intrinsic angular momentum, of the electron, and in Section 39.2, we will see that the Exclusion Principle is intimately linked to the spin. In later sections we will examine the implications of the Exclusion Principle for the arrangement of the electrons in atoms and in solids.

39.1 PRINCIPAL, ORBITAL, AND MAGNETIC QUANTUM NUMBERS; SPIN

Online 44 Concept Tutorial

In the preceding chapter, we saw that Bohr's theory characterizes the stationary states of the hydrogen atom and their energies by a single quantum number n. However, Bohr's simple theory deals only with circular orbits. We know from the study of planetary orbits (Chapter 9) that the general orbit of a particle moving under the influence of an inverse-square force is an ellipse with one focus at the center of attraction. For an electron in such an elliptical orbit around a nucleus, the quantum number n, now called the **principal quantum number**, characterizes the overall size of the ellipse, that is, its major axis; this quantum number also characterizes the energy of the electron [the formula for the quantized energies of the elliptical orbits is the same as that for the circular orbits, Eq. (38.25)]. But for the complete definition of the ellipse we need two extra quantum numbers l and m that characterize the elongation of the ellipse and the inclination of the ellipse, that is, its orientation in space. These two extra quantum numbers l and m are called, respectively, the **orbital quantum number** and the **magnetic quantum number**.

principal quantum number

orbital quantum number

magnetic quantum number

For a classical orbit, the elongation of the ellipse is closely related to the magnitude of the angular momentum of the electron. Among ellipses of equal sizes (equal major axes), the most elongated has the least angular momentum. Thus, the orbital quantum number l characterizes both the elongation of the ellipse and the magnitude of its angular momentum.

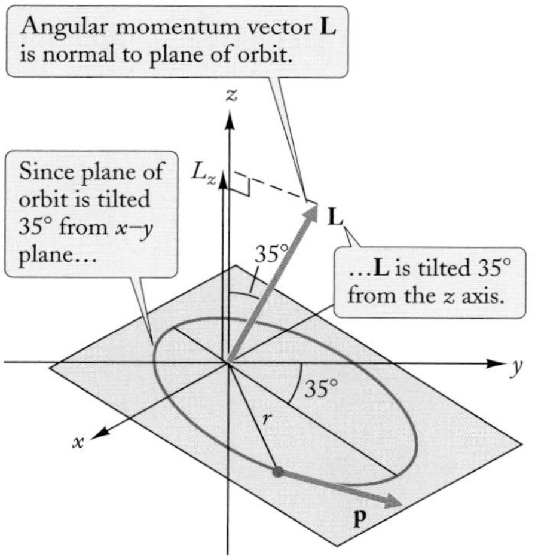

FIGURE 39.1 This elliptical orbit is inclined at an angle of 35° relative to the x–y plane. The angular-momentum vector **L** is perpendicular to the plane of the orbit; this vector is therefore inclined at an angle of 35° relative to the z axis. The z component of the angular-momentum vector is L_z.

The inclination of the ellipse is closely related to the direction of the angular momentum. To understand this relationship, we have to think of the angular momentum as a vector, with a direction perpendicular to the plane of the orbit (see Fig. 39.1). The orientation of the orbit in space is then specified by the direction of the angular-momentum vector; for instance, if the orbit is inclined at an angle of 35° relative to the x–y plane, as in Fig. 39.1, then the angular momentum vector is inclined at an angle of 35° relative to the z axis. As we will see below, the magnetic quantum number m specifies the z component of the angular-momentum vector, and it thereby characterizes both the direction of the angular-momentum vector and the inclination of the orbit. A large value of the magnetic quantum number m corresponds to a large value of this z component and a small inclination of the orbit; a small value of the magnetic quantum number m corresponds to a small value of this z component and a large inclination of the orbit, near 90°. (The quantum number m is called "magnetic" because it acquires a special significance when the atom is placed in a magnetic field; the energy of the orbit then depends not only on the size of the orbit and on the principal quantum number n, but also on the orientation of the orbit and, therefore, on the "magnetic" quantum number m.)

Although the two extra quantum numbers l and m were originally introduced on the basis of semiclassical considerations involving elliptical orbits, a later, more rigorous analysis based on wave mechanics and the Schrödinger equation confirmed that these quantum numbers are, indeed, required for the description of the stationary quantum states. In wave mechanics *a stationary state corresponds to a three-dimensional standing wave, and the quantum numbers n, l, and m characterize the "shape" of this standing wave, and they also characterize the energy, the magnitude of the angular momentum, and the z component of the angular momentum.*

According to wave mechanics, the magnitude of the angular momentum is quantized. However, this quantization differs from that of the simple Bohr theory in that the quantum number for angular momentum is the orbital quantum number l, not the principal quantum number n. Furthermore, the formula for the quantization condition is somewhat more complicated. From the study of the mathematical properties of the Schrödinger equation, it can be demonstrated that *the magnitude L of the angular momentum obeys the quantization condition*

magnitude of angular momentum

$$L = \sqrt{l(l+1)}\hbar \tag{39.1}$$

where the orbital quantum number l is restricted to integer values from 0 to $n - 1$, that is,

$$l = 0, 1, 2, \ldots, n - 1 \tag{39.2}$$

For example, if $n = 3$, then the possible values of the orbital quantum number are $l = 0$, $l = 1$, and $l = 2$; and the corresponding magnitudes of the angular momentum are

$$L = \sqrt{0 \times (0+1)}\hbar = 0$$

$$L = \sqrt{1 \times (1+1)}\hbar = \sqrt{2}\hbar$$

and

$$L = \sqrt{2 \times (2+1)}\hbar = \sqrt{6}\hbar$$

Note that the smallest possible magnitude of the angular momentum is actually zero, in contrast to what was claimed by the Bohr theory, where the smallest magnitude of the angular momentum was $\hbar$.

Since the angular momentum is a vector quantity, it has not only magnitude, but also a direction. However, according to wave mechanics, the direction of the angular-momentum vector is only partially determined. The direction of the angular-momentum vector always has substantial quantum uncertainties, and, correspondingly, the x, y, and z components of this vector have substantial uncertainties. It can be demonstrated that at most *one* of the components is well determined—the other two components are completely uncertain. If we take the well-determined component to be the z component, then the quantization condition for this component is

z component of angular momentum

$$L_z = m\hbar \qquad (39.3)$$

Here, the *magnetic quantum number m is restricted to integer values from* $-l$ *to* $+l$; that is,

$$m = -l,\ -l+1,\ \ldots,\ 0,\ \ldots,\ l+1,\ l \qquad (39.4)$$

Note that, for a given value of l, there are $2l + 1$ possible values of m. Each of these possible values of m corresponds to a possible direction of the angular-momentum vector relative to the z axis. For instance, if $l = 2$, then there are five possible values of m, namely, $m = -2$, $m = -1$, $m = 0$, $m = +1$, and $m = +2$; thus, there are five possible choices for the direction of the angular-momentum vector relative to the z axis. Figure 39.2 shows these possible directions.

From Eq. (39.3) we see that the z component of the angular momentum has a maximum possible value of $l\hbar$; this corresponds to the best attainable alignment of the angular momentum with the z axis. If we compare this maximum possible value of the z component with the magnitude of a nonzero angular momentum, we see that the former is always smaller than the latter, $l\hbar < \sqrt{l(l+1)}\hbar$. This means that the angular-momentum vector is never perfectly aligned with the z axis. The uppermost vector drawn in Fig. 39.2 indicates the best attainable alignment for $l = 2$.

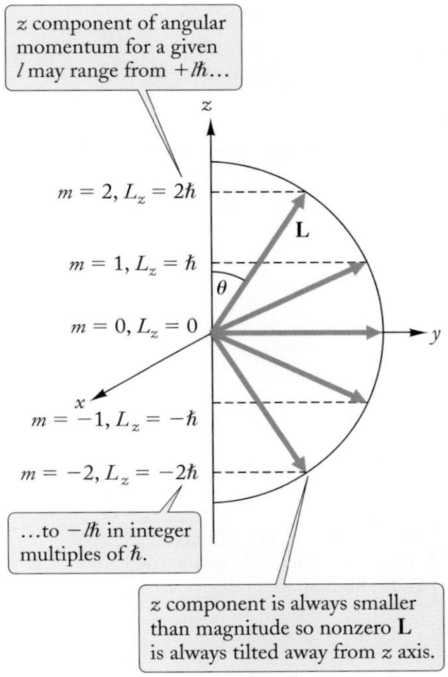

FIGURE 39.2 Possible directions of the angular-momentum vector for the case $l = 2$. In this case, there are five possible values of L_z, characterized by $m = -2, -1, 0, +1$, and $+2$.

EXAMPLE 1 For the angular-momentum vector corresponding to a quantum number $l = 2$, calculate the angle θ between the vector and the z axis for the cases $m = 2$, $m = 1$, and $m = 0$.

SOLUTION: The directions of the angular-momentum vector for these cases are illustrated in Fig. 39.2. The magnitude of the angular-momentum vector is $L = \sqrt{l(l+1)}\hbar = \sqrt{2 \times (2+1)}\hbar = \sqrt{6}\hbar$. In the case $m = 2$, the z component is $L_z = m\hbar = 2\hbar$, and the angle θ is given by

$$\cos\theta = \frac{L_z}{L} = \frac{2\hbar}{\sqrt{6}\hbar} = \frac{2}{\sqrt{6}} = 0.82$$

With a calculator, we find that the angle with this cosine is 35°.

Likewise, in the case $m = 1$, the z component is $L_z = \hbar$, and

$$\cos\theta = \frac{\hbar}{\sqrt{6}\hbar} = 0.41$$

which gives us an angle of 66°.

Finally, in the case $m = 0$, the z component is $L_z = 0$, which gives us an angle of 90°. This angular-momentum vector lies in the x–y plane.

Each possible set of values of n, l, and m corresponds to one kind of three-dimensional standing wave, often called an **orbital**. The simplest of these waves is that for the

orbital

ground state, with $n = 1, l = 0$, and $m = 0$. Figure 39.3a shows a picture of the intensity of this wave, denoted $|\psi|^2$, the square of the amplitude of the wavefunction. In this picture, generated from the theoretical formulas of wave mechanics by a computer, the density of the dots indicates the intensity of the wave; thus, the wave is strongest near the center (near the nucleus of the atom), and gradually fades with increasing distance from the center. The intensity of the wave at any point is proportional to the probability for finding the electron at this point; thus, in the ground state of the hydrogen atom, the most probable position for the electron is right at the nucleus. Because of this probabilistic interpretation, the pictures of wave intensity shown in Fig. 39.3 are often called probability clouds.

Waves with other values of n, l, and m correspond to the excited states; all these waves are more complicated than the wave for the ground state, especially if n, l, and m are large. Figures 39.3b–d give some examples of waves corresponding to some excited states.

spin quantum number m_s

spin, or intrinsic angular momentum

In addition to these quantum numbers n, l, and m, one more quantum number is needed for the complete characterization of the stationary states of an electron in the hydrogen atom. This is **the spin quantum number** m_s that characterizes the **spin**, or **intrinsic angular momentum**, of the electron. The quantum number m_s was originally proposed by Wolfgang Pauli in an attempt to describe the "hyperfine" structure of the spectral lines: when the spectral lines of hydrogen and of other atoms are examined with a spectroscope of high resolving power, they are often found to consist of pairs, or doublets, of very closely spaced lines. This implies that one or both of the atomic energy levels involved in the transition must be a closely spaced pair of energy levels, and it implies that, besides n, l, and m, there must be another quantum number that distinguishes between the two energy levels in the pair. Pauli assigned the values $m_s = +\frac{1}{2}$ and $m_s = -\frac{1}{2}$ to the two energy levels in the pair, but offered no explanation of the physical significance of this new quantum number.

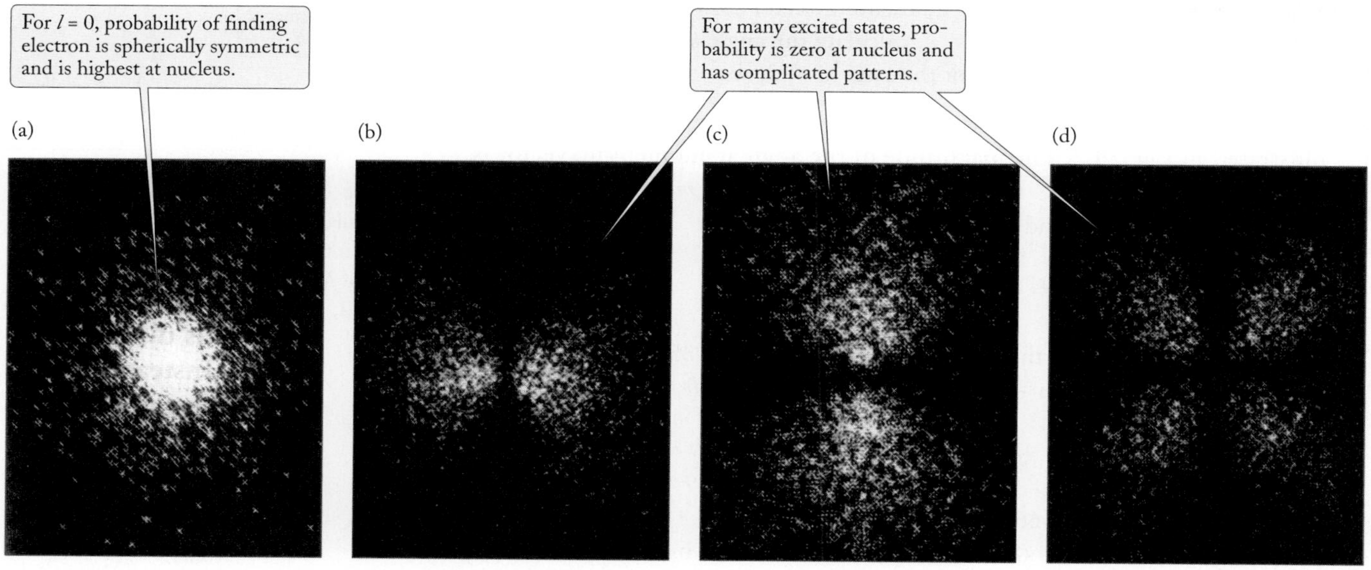

FIGURE 39.3 Intensities of possible standing waves, or orbitals, in the hydrogen atom. The nucleus is at the center of each picture. The density of the dots is proportional to the intensity of the wave, and thus to the probability of finding the electron, $|\psi|^2$.

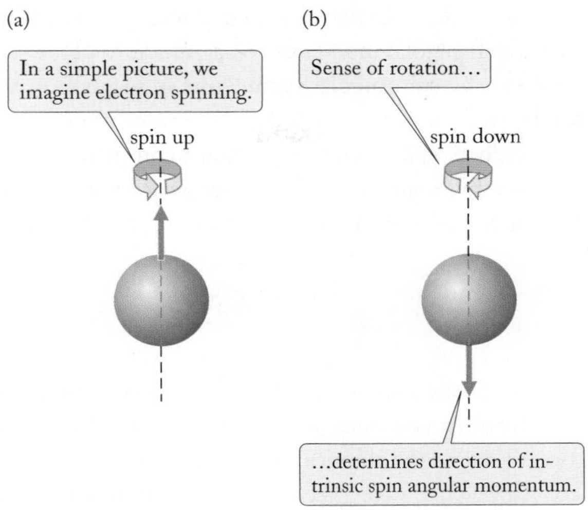

FIGURE 39.4 Simple picture of the electron as a ball of charge spinning about its axis. (a) The electron can spin in the counterclockwise direction, seen from above, so the direction of the angular momentum is up. (b) Alternatively, the electron can spin in the clockwise direction, so the direction of the angular momentum is down.

WOLFGANG PAULI (1900–1958)
Austrian and later Swiss theoretical physicist. For his discovery of the Exclusion Principle, he was awarded the Nobel Prize in 1945. Pauli made an important contribution to the theory of beta decay by proposing that the emission of the beta particle is always accompanied by the emission of a neutrino (see Chapter 40).

Shortly thereafter, S. Goudsmit and G. Uhlenbeck suggested that the electron has an intrinsic spin angular momentum, and that the two values of m_s correspond to different orientations of the axis of spin (see Fig. 39.4). They imagined the electron as a small ball of charge spinning about its axis, like the Earth spinning about its axis, with an angular momentum of magnitude $\sqrt{\frac{1}{2}(\frac{1}{2}+1)}\hbar$, which means that the intrinsic angular momentum quantum number, or the spin quantum number, is $\frac{1}{2}$. If we use the same rule for the possible directions of the spin as for the possible directions of the orbital angular momentum, we find that there are $2 \times \frac{1}{2} + 1 = 2$ possible directions for the spin. *One of these directions is characterized by a magnetic quantum number* $m_s = +\frac{1}{2}$, *and the other by a magnetic quantum number* $m_s = -\frac{1}{2}$; *these two possible spin directions are called spin up and spin down, respectively.*

Goudsmit and Uhlenbeck's simple picture of the spin as due to a rotation of the electron about its axis proved untenable. A consistent, relativistic picture of the spin was later contrived by P. A. M. Dirac. Modern wave mechanics tells us that the spin is an angular momentum generated by a circulating energy flow in the electron wave. The orbital angular momentum is also generated by such a circulating energy flow in the electron wave, but the energy flow that generates the spin is distinguished in that it persists even in the electron wave of an isolated, free electron at rest (that is, an electron outside of the hydrogen atom). Although the simple picture of the electron as a rotating ball of charge is false, it can sometimes serve as a convenient crutch for our imagination. According to this simple picture, we expect that the electron has a magnetic moment, since a piece of charge rotating about an axis amounts to a current loop (see Fig. 39.5). The energy difference between the two energy levels that gives rise to closely-spaced pairs of spectral lines ("doublets", such as seen in the sodium and mercury spectra in the color print on page 1289) is due to this magnetic interaction of the magnetic moment with the magnetic field generated by the motion of the nuclear charge seen in the reference frame of the electron. As already mentioned in Section 30.4, the magnetic moment of the electron also plays an important role in the behavior of ferromagnetic materials.

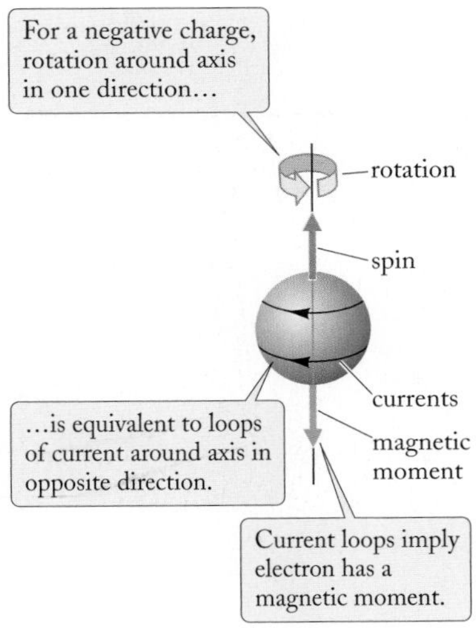

FIGURE 39.5 A ball of negative charge spinning about an axis.

PAUL ADRIEN MAURICE DIRAC (1902–1984) *British theoretical physicist, professor at Cambridge. He formulated the relativistic wave equation for the electron, incorporating spin, and he used this equation to predict the existence of antielectrons. He received the Nobel Prize in 1933, when anti-electrons were confirmed experimentally.*

Many other elementary particles besides electrons have spin. For example, the proton and the neutron have spin angular momenta of magnitude $\sqrt{\frac{1}{2}(\frac{1}{2}+1)}\hbar$ (spin quantum number $\frac{1}{2}$), the photon has a spin angular momentum of magnitude $\sqrt{1(1+1)}\hbar$ (spin quantum number 1), etc.

The quantum numbers n, l, m, and m_s provide a complete characterization of the stationary states of the hydrogen atom. The energies of the stationary state depend mainly on the principal quantum number n. According to Eq. (38.25), the energies are

$$E = -\frac{13.6 \text{ eV}}{n^2} \tag{39.5}$$

Thus, stationary states of the same value of n have (almost) the same energy, regardless of the values of the other quantum numbers. Classically, this means that the energy depends on the overall size of the elliptical orbit, but not on its elongation, or on its orientation in space. However, a wave-mechanics calculation of the energies of the stationary states of the hydrogen atom shows that the energies depend slightly on the orbital quantum number l and on the orientation of the spin relative to the orbital angular momentum. This means that Eq. (39.5) for the energies of the stationary states of the hydrogen atom is not quite accurate. But the deviations from Eq. (39.5) are very small, and we can often ignore them.

Table 39.1 lists the permitted quantum numbers for the stationary states of the hydrogen atom.[1] *These quantum numbers can also be used to characterize the stationary states of atoms other than hydrogen*, but the energies of other atoms are not given by the simple formula (39.5).

TABLE 39.1 QUANTUM NUMBERS OF ELECTRONIC STATES

QUANTUM NUMBER	SYMBOL	VALUES	PHYSICAL QUANTITY
principal	n	$1, 2, 3, \ldots$	energy
orbital	l	$0, 1, 2, \ldots, n-1$	magnitude of orbital angular momentum, $L = \sqrt{l(l+1)}\hbar$
magnetic	m	$-l, -l+1, \ldots, 0, \ldots, l-1, l$	z component of orbital angular momentum, $L_z = m\hbar$
spin	m_s	$+\frac{1}{2}, -\frac{1}{2}$	z component of spin angular momentum, $S_z = \pm\frac{1}{2}\hbar$

EXAMPLE 2 (a) What are the permitted quantum numbers for the case $n = 1$? (b) What are the possible quantum numbers for the case $n = 2$?

SOLUTION: (a) For $n = 1$, Table 39.1 tells us that the only permitted value of l is 0. Furthermore, if $l = 0$, then the only permitted value of m is also 0. The

[1]Strictly, these quantum numbers are appropriate for a hydrogen atom placed in a magnetic field. For a hydrogen atom by itself, the appropriate quantum numbers are n and some intricate combinations of l, m, and m_s. For the sake of simplicity, we will ignore this complication.

permitted values of m_s are $+\frac{1}{2}$ and $-\frac{1}{2}$, regardless of the values of the other quantum numbers. Hence, for $n = 1$, the list of permitted quantum numbers is

$$n = 1; \quad l = 0; \quad m = 0; \quad m_s = -\tfrac{1}{2}$$

$$\text{or} \quad n = 1; \quad l = 0; \quad m = 0; \quad m_s = +\tfrac{1}{2}$$

(b) For $n = 2$, Table 39.1 tells us that the permitted values of l are 0 and 1. If $l = 0$, then the only permitted value of m is 0, and we find the same list as in part (a):

$$n = 2; \quad l = 0; \quad m = 0; \quad m_s = -\tfrac{1}{2}$$

$$\text{or} \quad n = 2; \quad l = 0; \quad m = 0; \quad m_s = +\tfrac{1}{2}$$

However, if $l = 1$, then the permitted values of m are -1, 0, and $+1$, and we find the following list:

$$n = 2; \quad l = 1; \quad m = -1; \quad m_s = -\tfrac{1}{2}$$

$$n = 2; \quad l = 1; \quad m = -1; \quad m_s = +\tfrac{1}{2}$$

$$n = 2; \quad l = 1; \quad m = 0; \quad m_s = -\tfrac{1}{2}$$

$$n = 2; \quad l = 1; \quad m = 0; \quad m_s = +\tfrac{1}{2}$$

$$n = 2; \quad l = 1; \quad m = +1; \quad m_s = -\tfrac{1}{2}$$

$$\text{or} \quad n = 2; \quad l = 1; \quad m = +1; \quad m_s = +\tfrac{1}{2}$$

Each alternative listed here represents a possible electronic state.

✔ Checkup 39.1

QUESTION 1: Consider the ground state of the hydrogen atom. According to wave mechanics, what is the magnitude of the orbital angular momentum? According to Bohr's theory? According to wave mechanics, what is the value of the quantum number n? According to Bohr's theory?

QUESTION 2: If the orbital quantum number is $l = 3$, what is the magnitude of the angular momentum?

QUESTION 3: If the principal quantum number is $n = 3$, what are the possible values of the orbital angular-momentum quantum number?

QUESTION 4: Figures 39.3b, c, and d give the probability clouds for some excited states of the hydrogen atom. According to these figures, what are the probabilities for finding the electron at the center (at the nucleus)?

QUESTION 5: If the orbital angular-momentum quantum number is $l = 1$, what are all the possible values of the magnetic quantum number m?

(A) 0 (B) 0 or 1 (C) 0, 1, or 2 (D) $-\frac{1}{2}$ or $+\frac{1}{2}$ (E) -1, 0, or $+1$

Online Concept Tutorial 44

39.2 THE EXCLUSION PRINCIPLE AND THE STRUCTURE OF ATOMS

atomic number Z

periodic table

The paramount question in atomic structure is the determination of the detailed arrangement of the electrons in their orbital states around the nucleus. An atom of **atomic number** Z has Z electrons in orbitals around a nucleus of positive charge Ze. *The electron arrangement, or configuration, determines all the physical and chemical properties of the atom*—if the electron configuration is known, all the properties of the atom can be deduced by theoretical considerations. For instance, the observed similarities of chemical properties among select groups of elements are due to similarities in their electronic configurations. Chemists list similar elements in columns in the **periodic table** of the elements (see Table 39.2). Thus, the elements helium, neon, argon, krypton, etc., are listed in the last column; these are the noble gases, or inert gases, which are practically unable to react chemically with anything at all. The elements fluorine, chlorine, bromine, iodine, etc., are listed in the column adjacent to the noble gases; these are the halogens, irritating, corrosive substances, all with quite noticeable and distinctive colors (pale yellow, greenish yellow, red, and blue violet, respectively). And the elements lithium, sodium, potassium, rubidium, etc., are listed in the first column; these are the alkalis, silvery white metals, which are extremely reactive. *The pattern of the elements displayed in the periodic table can be explained by a study of the electron configurations.*

For the case of the hydrogen atom, the determination of the electron configuration is trivial: the single electron of this atom is in one or another of the stationary states characterized by the quantum numbers n, l, m, and m_s. If the atom is in the ground state, the values of the quantum numbers of the electron configuration are $n = 1$, $l = 0$, $m = 0$, and $m_s = \pm\frac{1}{2}$; thus, everything is fixed, except the direction of the spin, which can be up or down.

But for atoms with several electrons, the determination of the electron configuration is not so trivial. It might be tempting to suppose that the ground state of the atom (the state of least energy) is attained by placing all the electrons in the lowest stationary state, with $n = 1$, $l = 0$, $m = 0$, $m_s = \pm\frac{1}{2}$, as for the hydrogen atom. But this would imply that all the atoms ought to have a spectral series similar to that of hydrogen, and it would also imply that atoms with a large number of electrons, or with a large atomic number Z, ought to be very small, since the Bohr radius for an atom of nuclear charge Ze is a_0/Z [if the nuclear charge is Ze instead of e, then in the denominator of the formula (38.17) for the Bohr radius a_0 of hydrogen, we must replace *one* of the factors of e by Ze]. These conclusions are in stark conflict with the observed properties of atoms: the spectra of most atoms are quite different, and the sizes of atoms of large Z—such as lead or bismuth—are considerably larger than hydrogen, which is the smallest of all atoms.

The rule that governs the configuration of the electrons in an atom is the **Pauli Exclusion Principle**, which was discovered by Wolfgang Pauli and is often known simply as the Exclusion Principle. For atoms, the Exclusion Principle asserts

Pauli Exclusion Principle

> ***Each stationary state of quantum numbers n, l, m, and m_s can be occupied by no more than one electron.***

Since for each stationary orbital state of quantum numbers n, l, and m there are two possible spin states ($m_s = \pm\frac{1}{2}$), we can also rephrase the Exclusion Principle as follows: each stationary orbital state of quantum numbers n, l, and m can be occupied by no more than two electrons.

TABLE 39.2 THE PERIODIC TABLE OF THE CHEMICAL ELEMENTS[a]

Key (shown with element 4): IIA — Group designation; (2); 4 — Atomic number; Be — Symbol for element; 9.012182 — Atomic mass

Periods	IA (1)	IIA (2)	IIIB (3)	IVB (4)	VB (5)	VIB (6)	VIIB (7)	VIIIB (8)	VIIIB (9)	VIIIB (10)	IB (11)	IIB (12)	IIIA (13)	IVA (14)	VA (15)	VIA (16)	VIIA (17)	VIIIA (18)
1	1 **H** 1.00794																	2 **He** 4.002602
2	3 **Li** 6.941	4 **Be** 9.012182											5 **B** 10.811	6 **C** 12.0107	7 **N** 14.0067	8 **O** 15.9994	9 **F** 18.99840	10 **Ne** 20.1797
3	11 **Na** 22.98977	12 **Mg** 24.3050											13 **Al** 26.98154	14 **Si** 28.0855	15 **P** 30.97376	16 **S** 32.065	17 **Cl** 35.453	18 **Ar** 39.948
4	19 **K** 39.0983	20 **Ca** 40.078	21 **Sc** 44.955910	22 **Ti** 47.867	23 **V** 50.9415	24 **Cr** 51.9961	25 **Mn** 54.938049	26 **Fe** 55.845	27 **Co** 58.93320	28 **Ni** 58.6934	29 **Cu** 63.546	30 **Zn** 65.409	31 **Ga** 69.723	32 **Ge** 72.64	33 **As** 74.92160	34 **Se** 78.96	35 **Br** 79.904	36 **Kr** 83.798
5	37 **Rb** 85.4678	38 **Sr** 87.62	39 **Y** 88.90585	40 **Zr** 91.224	41 **Nb** 92.90638	42 **Mo** 95.94	43 **Tc** 98.9072	44 **Ru** 101.07	45 **Rh** 102.90550	46 **Pd** 106.42	47 **Ag** 107.8682	48 **Cd** 112.411	49 **In** 114.818	50 **Sn** 118.710	51 **Sb** 121.760	52 **Te** 127.60	53 **I** 126.90447	54 **Xe** 131.293
6	55 **Cs** 132.90545	56 **Ba** 137.327	57 ***La** 138.9055	72 **Hf** 178.49	73 **Ta** 180.9479	74 **W** 183.84	75 **Re** 186.207	76 **Os** 190.23	77 **Ir** 192.217	78 **Pt** 195.078	79 **Au** 196.96654	80 **Hg** 200.59	81 **Tl** 204.3833	82 **Pb** 207.2	83 **Bi** 208.98037	84 **Po** 208.9824	85 **At** 209.9871	86 **Rn** 222.0176
7	87 **Fr** 223.0197	88 **Ra** 226.0277	89 **†Ac** 227.0277	104 **Rf** 261.1089	105 **Db** 262.1144	106 **Sg** 263.118	107 **Bh** 262.12	108 **Hs** 265.1306	109 **Mt** (268)	110 **Ds** (271)	111 **Uuu** (272)	112 **Uub** (285)		114 **Uuq** (289)				

*Lanthanides	58 **Ce** 140.116	59 **Pr** 140.90765	60 **Nd** 144.24	61 **Pm** 144.9127	62 **Sm** 150.36	63 **Eu** 151.964	64 **Gd** 157.25	65 **Tb** 158.92534	66 **Dy** 162.50	67 **Ho** 164.93032	68 **Er** 167.26	69 **Tm** 168.93421	70 **Yb** 173.04	71 **Lu** 174.967
†Actinides	90 **Th** 232.0381	91 **Pa** 231.0359	92 **U** 238.0289	93 **Np** 237.0482	94 **Pu** 244.0642	95 **Am** 243.0614	96 **Cm** 247.07003	97 **Bk** 247.0703	98 **Cf** 251.0796	99 **Es** 252.083	100 **Fm** 257.0951	101 **Md** 258.0984	102 **No** 259.1011	103 **Lr** 262.110

[a] In each box, the upper number is the *atomic number.* The lower number is the *atomic mass*, that is, the mass (in grams) of one mole or, equivalently, the mass (in atomic mass units) of one atom. Numbers in parentheses denote the atomic mass of the most stable or best-known isotope of the element; all other numbers represent the average mass of a mixture of several isotopes as found in naturally occuring samples of the element.

Pauli originally proposed this principle as an empirical rule, based on the observed features of atomic spectra. It was later established that the Exclusion Principle is intimately linked to the value of the spin of the electron; the Exclusion Principle can be shown to be a necessary consequence of the quantum theory of particles of half-integer spin. Thus, *protons and neutrons also obey the Exclusion Principle*, a fact of great importance for the configuration of these particles in the interior of the nucleus (see the next chapter). In contrast, *particles of integer spin, such as photons, do not obey the Exclusion Principle.* There is no limit to the number of such particles that can be packed into a given stationary state, for instance, one of the standing-wave states in a cavity filled with blackbody radiation.

For our investigation of the electron configuration of atoms, we will find it convenient to start with a list of all the available states, in order of increasing energy (see Example 2). The states of lowest energy have $n = 1$; there are two such states:

TABLE 39.3 K shell (states with $n = 1$)

$n = 1$	$l = 0$	$m = 0$	$m_s = -\frac{1}{2}$
$n = 1$	$l = 0$	$m = 0$	$m_s = +\frac{1}{2}$

Next, consider $n = 2$; there are eight available states:

TABLE 39.4 L shell (states with $n = 2$)

$n = 2$	$l = 0$	$m = 0$	$m_s = -\frac{1}{2}$
$n = 2$	$l = 0$	$m = 0$	$m_s = +\frac{1}{2}$
$n = 2$	$l = 1$	$m = -1$	$m_s = -\frac{1}{2}$
$n = 2$	$l = 1$	$m = -1$	$m_s = +\frac{1}{2}$
$n = 2$	$l = 1$	$m = 0$	$m_s = -\frac{1}{2}$
$n = 2$	$l = 1$	$m = 0$	$m_s = +\frac{1}{2}$
$n = 2$	$l = 1$	$m = +1$	$m_s = -\frac{1}{2}$
$n = 2$	$l = 1$	$m = +1$	$m_s = +\frac{1}{2}$

Likewise, for $n = 3$, there are 18 available states, and so on.

The groups of states of a given value of n are called **shells**, *and they are conventionally labeled with the letters* K, L, M, etc., from the innermost to the outermost shell. Thus, the two states with $n = 1$ form the K shell, the eight states with $n = 2$ form the L shell, the eighteen states with $n = 3$ form the M shell, etc. In the simple Bohr theory, these groups of states were called shells because the sizes of the orbits in each group are similar; thus, these groups of orbits form layers around the nucleus, like the layers of an onion. But in wave mechanics there is no such simple way to visualize the shells.

According to the Exclusion Principle, *each of the states listed above can accommodate one, and only one, electron.* Thus, if an atom with Z electrons is in its ground state, the electrons will occupy the first Z of the states in the above list. We can therefore build up the configurations for all the atoms in the periodic table of elements by beginning with hydrogen and adding electrons one by one, sequentially filling the states in our list.

The second element in the periodic table is helium, which has two electrons. To obtain its electron configuration, we must add one electron to the hydrogen configuration; since the single electron of hydrogen occupies one of the states of the K shell, we can place the second electron in the other available state in the K shell. Helium therefore has a full K shell.

The third element is lithium, with three electrons. When we add one electron to the helium configuration, we must place this third electron in the L shell, with $n = 2$. Detailed calculations show that in multielectron atoms, the $l = 0$ states are slightly lower in energy than the $l = 1$ states, so this third electron is in one of the two states with quantum numbers $n = 2$, $l = 0$, and $m = 0$.

The next element is beryllium, with four electrons. Thus, we must add one more electron in the L shell, in the other available state with quantum numbers $n = 2$, $l = 0$, and $m = 0$.

We can continue in this way, filling up the states in our list one by one. With the tenth element, neon, we will have filled the L shell. And with the eleventh element, sodium, we must place one electron in the M shell, and so on.

This simple procedure for building up the electron configurations of the atoms provides us with an immediate explanation of the similarities of the elements in a column of the periodic table. For instance, the similarity in the chemical behavior and the similarity in the spectra of helium and neon can be traced to a similarity of their electron configurations: both these atoms have full shells of electrons. The full-shell electron configuration is quite stable, tending neither to capture nor to lose an electron; these and other noble gases are chemically inert. Likewise, the chemical and spectroscopic similarities of hydrogen, lithium, and sodium can be traced to a similarity of their electron configurations: they all have a single electron outside of a full shell of electrons. This single, outer electron tends to come off the atom fairly easily, and in chemical reactions these atoms all tend to lose an electron. In contrast, fluorine and chlorine are one electron short of a full shell, and in chemical reactions they tend to capture an electron to complete their shell.

Thus, the Exclusion Principle in conjunction with a simple counting procedure for the available stationary states is sufficient to explain the broad, qualitative features of the periodic table of elements. Detailed calculations, based on wave mechanics, provide quantitative theoretical results for ionization energies, spectral lines, atomic sizes, etc., in agreement with the observed atomic properties.

For atoms other than hydrogen there is no simple formula for the energies of the stationary states. However, in an atom of fairly large atomic number, say, $Z > 20$, the dominant force on the innermost electrons is the attractive Coulomb force exerted by the positive charge Ze of the nucleus, and the repulsive forces exerted by the other electrons can mostly be neglected. Thus, *these innermost electrons have hydrogenlike orbitals.* The energy of a single electron in such an orbital is given approximately by Eq. (38.23), with one modification: the product $e \times e$ of the electron and the proton charge must be replaced by the product $e \times Ze$ of the electron and the nuclear charge; hence e^4 in Eq. (38.23) must be replaced by e^4Z^2, which leads to the following approximate formula for the energy:

$$E_n \approx -\frac{m_e Z^2 e^4}{2(4\pi\epsilon_0)^2\hbar^2}\frac{1}{n^2} \approx -\frac{Z^2 \times 13.6\ \text{eV}}{n^2} \tag{39.6}$$

From this formula, we can estimate the frequency and the wavelength of light emitted during a quantum jump from some initial state to a final state. However, such a quantum jump between the innermost orbitals of the atom is not possible if the atom has its full complement of electrons—all the orbitals are then occupied by electrons, and the Exclusion Principle forbids quantum jumps into an already occupied orbital. Thus, a jump is possible only if some external disturbance first removes one of the electrons from the atom, leaving a gap into which some other electron can jump. Such a process occurs when the target atoms in an X-ray tube are subjected to the impact of the electron beam. The target atoms are disturbed by this impact, and sometimes an electron in one of the innermost orbitals is ejected, leaving a gap into which another electron can jump. The photon emitted when an electron jumps into the vacant innermost orbital has a very short wavelength; it is an X ray. Thus, the quantum jumps of the innermost electrons of atoms give rise to the characteristic spectrum of X rays mentioned in Section 37.5. These characteristic X rays correspond to the two sharp peaks in Fig. 37.19.

When one of the electrons in the innermost orbital is removed, the one remaining $n = 1$ electron "shields" a part of the nuclear charge from the view of the electron making the quantum jump. For a nearby electron in an $n = 2$ state jumping into a

HENRY MOSELEY (1887–1915) *English physicist, lecturer at Manchester, where he worked under Rutherford. Moseley was a skillful experimenter, and his brilliant investigations of the characteristics of spectral lines of X rays led to firm assignments of atomic numbers for chemical elements. He was killed in action in the futile Gallipoli campaign, at age 28.*

single available $n = 1$ state, the effect of the shielding electron is to reduce the effective nuclear charge by one elementary charge. Thus, we can obtain somewhat better estimates of the energies of these states by replacing the nuclear charge Z in Eq. (39.6) by the shielded charge $Z - 1$:

$$E_n \approx -\frac{m_e(Z-1)^2 e^4}{2(4\pi\epsilon_0)^2\hbar^2}\frac{1}{n^2} \approx -\frac{(Z-1)^2 \times 13.6 \text{ eV}}{n^2} \tag{39.7}$$

Henry Moseley used this predictable behavior of the transitions to the K shell to correct the positions of several heavy metals in the periodic table; these were previously positioned by mass, not by the atomic number Z.

EXAMPLE 3 Suppose that in an atom of copper ($Z = 29$) in the target of an X-ray tube, one of the electrons in the $n = 1$ state is ejected during the impact of the electron beam on the target. Suppose that subsequently one of the other electrons in the atom jumps from the $n = 2$ state into this available empty $n = 1$ state. What are the energy and the wavelength of the photon emitted during this quantum jump?

SOLUTION: The initial energy of the electron is

$$E_2 \approx -\frac{(Z-1)^2 \times 13.6 \text{ eV}}{2^2} = -\frac{(29-1)^2 \times 13.6 \text{ eV}}{4} = -2.67 \times 10^3 \text{ eV}$$

and the final energy is

$$E_1 \approx -\frac{(Z-1)^2 \times 13.6 \text{ eV}}{1^2} = -\frac{(29-1)^2 \times 13.6 \text{ eV}}{1} = -1.066 \times 10^4 \text{ eV}$$

Hence the energy of the photon is

$$E_2 - E_1 \approx 8.00 \times 10^3 \text{ eV} = 1.28 \times 10^{-15} \text{ J}$$

The frequency of the photon is

$$f = \frac{1.28 \times 10^{-15} \text{ J}}{h} = \frac{1.28 \times 10^{-15} \text{ J}}{6.63 \times 10^{-34} \text{ J·s}} = 1.93 \times 10^{18} \text{ Hz}$$

and the wavelength is

$$\lambda = \frac{c}{f} = \frac{3.00 \times 10^8 \text{ m/s}}{1.93 \times 10^{18} \text{ Hz}} = 1.55 \times 10^{-10} \text{ m} = 0.155 \text{ nm}$$

COMMENTS: This agrees well with the experimentally determined wavelength, $\lambda = 0.154\,18$ nm. This characteristic X ray is known as copper K_α radiation, where the α refers to the dominant transition, here to the K shell from the $n = 2$ state. In engineering and materials research, diffractometers often use X rays of this wavelength, somewhat smaller than atomic spacings, to determine crystal structures.

✔ Checkup 39.2

QUESTION 1: Consider the K, L, and M shells. Which of these has the largest number of states? Which the least?

QUESTION 2: When a shell is full, does it always have equal numbers of electrons of spin up and spin down?

QUESTION 3: The ionization energy for the hydrogen atom is 13.6 eV. What is the energy required for the second ionization of a helium atom ($Z = 2$), that is, the energy for the removal of the second electron (if the first electron has already been removed previously)?

(A) 1×13.6 eV (B) 2×13.6 eV (C) 3×13.6 eV
(D) 4×13.6 eV (E) 9×13.6 eV

39.3 ENERGY LEVELS IN MOLECULES

The chemical bonds that bind two or more atoms together in a molecule, such as O_2 or NaCl, arise from a rearrangement of the outer electrons, or valence electrons, of the atoms. In some molecules (e.g., O_2), the outer electrons are shared between the atoms, a sharing that produces an attractive force (covalent bond). In some other molecules (e.g., NaCl), one atom loses an electron to the other atom, and the atom with the missing electron is then electrically attracted by the atom with the extra electron (ionic bond). Hydrogen atoms are especially susceptible to losing electrons to neighboring atoms. When a hydrogen atom is located between two other atoms, it often loses its electron to these neighbors, and the residual proton of the hydrogen atom then electrically attracts the neighboring atoms, holding them together (hydrogen bond).

The chemical bonds are elastic—they behave rather like springs tying the atoms together. The spring holds the atoms at an average equilibrium distance, but permits the atoms to oscillate back and forth about this average distance, with some kinetic and potential energy. This means that the energy of the molecule is the sum of the electronic energy of the atoms and the vibrational energy of the motion of the atoms in relation to each other. We will first focus on the vibrational energy of a molecule and examine its quantization.

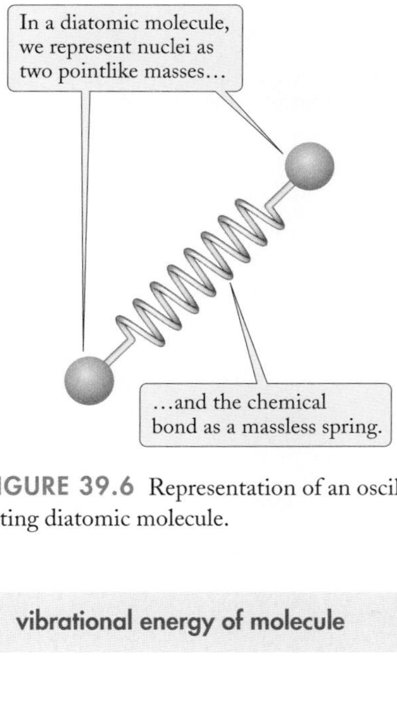

FIGURE 39.6 Representation of an oscillating diatomic molecule.

The mass of the atom is concentrated in its center, in the nucleus, which is much smaller than the interatomic distances in a molecule. We can therefore schematically represent a molecule—for instance, a diatomic molecule—as a system of pointlike masses connected by a massless spring, which represents the bond between the two atoms (see Fig. 39.6). The atoms oscillate in unison relative to the center of mass, which we can regard as fixed. Thus, the system is an oscillator, and the energy of this oscillator is subject to Planck's quantization condition, Eq. (37.3). If the frequency of oscillation is f, *the energy of the vibration is quantized according to*[2]

vibrational energy of molecule

$$E = nhf \qquad n = 0, 1, 2, \ldots \tag{39.8}$$

The corresponding energy-level diagram is shown in Fig. 39.7. The molecule will emit a photon if it makes a transition from an upper level to a lower level. The vibrational transitions in a molecule are restricted by a **selection rule**: *the transition must proceed from one level to the next, that is, transitions spanning two or more levels in one jump are forbidden.*

selection rule

[2]More precisely, the energies are $E = (n + \frac{1}{2})hf$. However, the additional $\frac{1}{2}hf$, the zero-point energy, while important for the calculation of some quantities, does not affect any of the transitions or other properties discussed here.

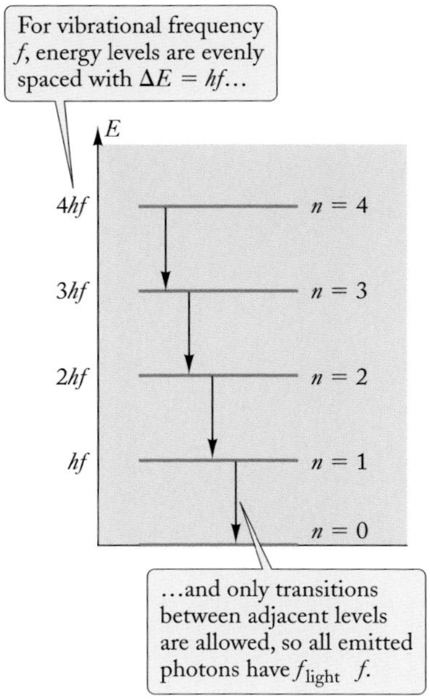

FIGURE 39.7 Energy-level diagram for the oscillating molecule. The arrows indicate the possible transitions.

This selection rule can be deduced from wave mechanics. The arrows in Fig. 39.7 indicate the permitted transitions. The frequencies of the spectral lines emitted during all these transitions are therefore the same,

$$f_{\text{light}} = \frac{\Delta E}{h} = \frac{hf}{h} = f \tag{39.9}$$

Thus, the frequency of the spectral lines equals the frequency of vibration of the molecule. Typically, the frequencies of vibration of molecules are of the order of 10^{13} Hz, and the wavelengths of the emitted spectral lines lie in the infrared.

Besides the vibrational motion, the molecule can also perform rotational motion. For the purposes of this rotational motion, we can regard the molecule as two pointlike masses linked by a massless rod, that is, a dumbbell (see Fig. 39.8). If the moment of inertia of the dumbbell about a perpendicular axis through the center of mass is I, then the kinetic energy of rotation is

$$E = \tfrac{1}{2}I\omega^2 \tag{39.10}$$

where ω is the angular frequency of the rotation. Let us express this in terms of the angular momentum. The magnitude of the angular momentum is $I\omega$, and this is quantized in the usual way [see Eq. (39.1)], so we can write

$$I\omega = \sqrt{J(J+1)}\hbar \qquad J = 0, 1, 2, \ldots \tag{39.11}$$

where J is called the **rotational quantum number**. Solving Eq. (39.11) for ω and substituting this into Eq. (39.10) implies that *the energy of the rotational motion is quantized*, that is,

rotational energy of molecule

$$E = \frac{J(J+1)\hbar^2}{2I} \qquad J = 0, 1, 2, \ldots \tag{39.12}$$

Figure 39.9 displays the energy-level diagram for the rotational states of a molecule. The transitions are, again, subject to the selection rule that they must proceed from one level to the next. Such transitions are indicated by the arrows in Fig. 39.9.

Note that here we considered only rotation about a perpendicular axis through the center of mass. For rotation about the axis through the atoms, the moment of inertia of a diatomic moelcule is many orders of magnitude smaller, since the mass is con-

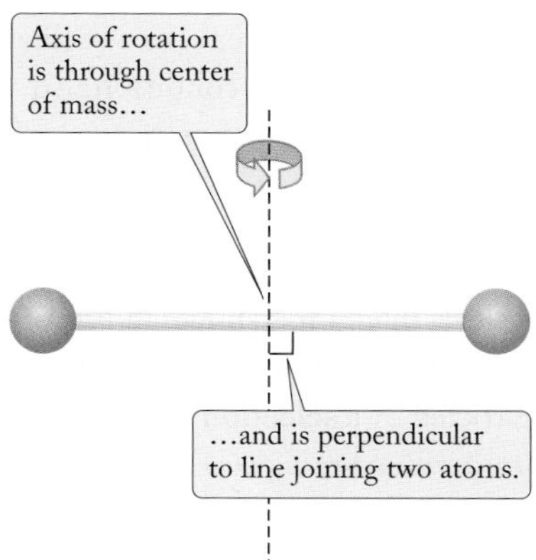

FIGURE 39.8 A rotating diatomic molecule can be regarded as two pointlike masses linked by a massless rigid rod.

centrated in the nuclei. Equation (39.12) tells us that for such small values of I, the energy levels will be very high, too high for practical consequences.

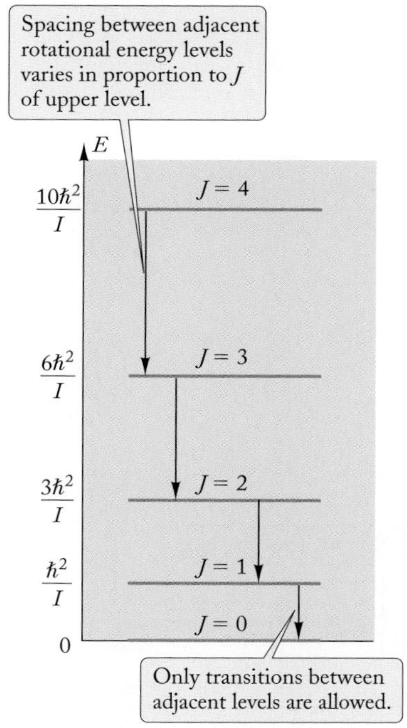

FIGURE 39.9 Energy-level diagram for the rotating molecule. The arrows indicate the possible transitions.

EXAMPLE 4 The moment of inertia of the HCl molecule about its axis of rotation is $2.66 \times 10^{-47}\ \text{kg}\cdot\text{m}^2$. What is the energy of the first excited rotational state of this molecule? The second excited rotational state?

SOLUTION: For the first excited state, $J = 1$ and Eq. (39.12) gives

$$E = \frac{2\hbar^2}{2I} = \frac{(1.05 \times 10^{-34}\ \text{J}\cdot\text{s})^2}{2.66 \times 10^{-47}\ \text{kg}\cdot\text{m}^2} = 4.14 \times 10^{-22}\ \text{J}$$

For the second excited state, $J = 2$ and

$$E = \frac{2 \times 3 \times \hbar^2}{2I} = \frac{3 \times (1.05 \times 10^{-34}\ \text{J}\cdot\text{s})^2}{2.66 \times 10^{-47}\ \text{kg}\cdot\text{m}^2} = 1.24 \times 10^{-21}\ \text{J}$$

The rotational transitions in a molecule involve much smaller energies than the electronic transitions in an atom, such as the electronic transitions in a hydrogen atom. This means that the photons emitted in purely rotational molecular transitions are of rather low energy, and their wavelengths lie in the far infrared region of the spectrum. However, *rotational molecular transitions are often observed in conjunction with a simultaneous electronic transition in one of the atoms of the molecule or with a vibrational transition of the molecule.* This increases the energy of the transition and reduces the wavelength of the spectral line. By inspection of Fig. 39.9 we see that successive rotational transitions have slightly different energies and wavelengths; thus, they give rise to a group, or sequence, of adjacent spectral lines. This is called a **spectral band**. Figure 39.10 is a photograph of several spectral bands in the spectrum of the N_2 (nitrogen) molecule.

spectral band

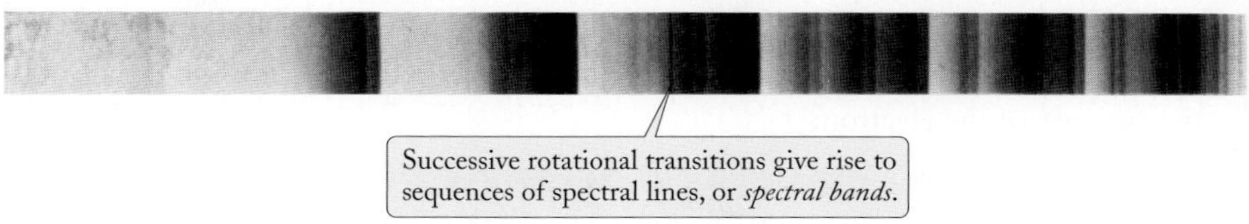

FIGURE 39.10 Several bands of spectral lines emitted by the N_2 molecule.

✔ Checkup 39.3

QUESTION 1: For a vibrational transition in a molecule, the frequency of the radiation equals the frequency of vibration of the molecule. Is it likewise true that for a rotational transition in a molecule, the frequency of the radiation equals the frequency of the rotational motion?

QUESTION 2: Suppose that a molecule is initially in the $J = 4$ state in Fig. 39.9. If it makes the sequence of downward transitions indicated by the arrows in this figure, how many photons does it emit? Which of these photons has the most energy? The least energy? Which has the longest wavelength? The shortest?

QUESTION 3: The moment of inertia of the deuterium molecule (D_2) is twice as large as that of the hydrogen molecule (H_2). Which has a larger energy difference between the ground state and the first excited rotational state? By what factor?

QUESTION 4: The frequencies of the photons emitted during the four transitions shown in Fig. 39.9 are (in units of $\hbar/I$)

(A) 1, 2, 3, 4 (B) 0, 1, 3, 6 (C) 1, 3, 6, 10 (D) 4, 7, 9, 10

39.4 ENERGY BANDS IN SOLIDS

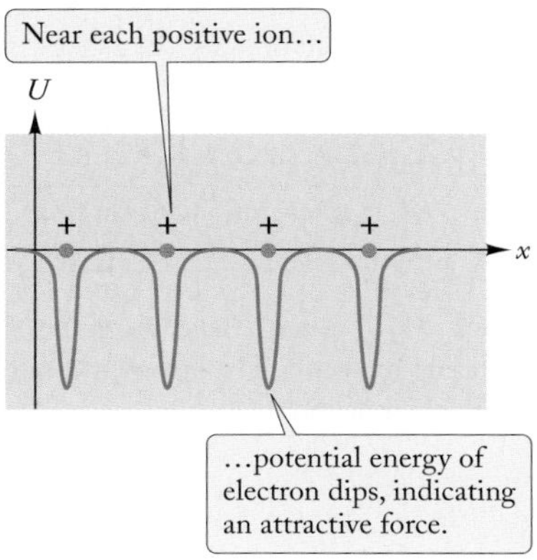

FIGURE 39.11 Potential energy of an electron in a crystal. The plus signs mark the positions of the atoms along a row in the crystal.

As discussed in Section 22.5 in a metal the outermost, or valence, electrons of the atoms are detached from their atoms, and they are free to wander all over the volume of the metal. However, whenever such a "free" electron passes near one of the positively charged atoms of the metal, it experiences an attractive force. Figure 39.11 is a rough sketch of the potential energy for an electron moving along a row of atoms in the crystal lattice of a metal; the dips in the potential energy indicate the attractive force. While the electron moves along this row of atoms, the attractive force acts repetitively, each time the electron passes near an atom. Under suitable conditions, such a repetitive action of a force can lead to a large cumulative effect. According to wave mechanics, we have to think of the electron as a wave, and an encounter with an atom scatters the wave: some fraction of the wave proceeds in its original direction of motion and some fraction is reflected. The repetitive scatterings at the atoms in the row will build up a large reflected wave by constructive interference if the scattering by one atom produces a reflected wave that is in phase with the reflected wave produced by the next atom. This will happen if the extra distance for a round trip from one atom to the next and back is equal to one de Broglie wavelength or an integer multiple of one wavelength. Designating the distance between the atoms by a, we can express the condition for constructive interference of all the reflected waves as

$$2a = \lambda, 2\lambda, 3\lambda, \ldots \tag{39.13}$$

Since the de Broglie wavelength is related to the momentum of the electron by $\lambda = h/p$ [see Eq. (38.32)], we can express the condition for total reflection in terms of the momentum of the electron:

$$2a = \frac{h}{p}, \frac{2h}{p}, \frac{3h}{p}, \ldots \tag{39.14}$$

or

$$p = \frac{h}{2a}, \frac{2h}{2a}, \frac{3h}{2a}, \ldots \tag{39.15}$$

If this condition is satisfied, the reflected wave will gain more and more strength at each reflection at each atom, and finally match the strength of the incident wave. The result is a standing wave, which travels neither right nor left.

Concepts in Context

The energy of such a standing wave is strongly affected by the interaction of the electron with the atoms of the lattice. The energy of the standing wave will be low if the peaks of intensity of the wave coincide with the locations of the atoms (see Fig. 39.12a), because the electron then has a large probability for being found near an atom, where the potential energy is low. The energy of the standing wave will be high if the peaks of intensity of the wave fall between the locations of the atoms (see Fig. 39.12b), because the electron then has a large probability of being found between one atom and the next, where the potential energy is high. Thus, the wavefunction depicted in Fig. 39.12a corresponds to a low-energy state of the electron, and that in Fig. 39.12b corresponds to a high-energy state. The transition from one of these states to the other involves a quantum jump of the electron.

To appreciate the implications of such a quantum jump, suppose we take an electron of low initial momentum and we gradually increase its energy and momentum. At first, the energy of the electron will increase smoothly with the momentum. But at the critical value $p = h/2a$ given by Eq. (39.15), the energy has to increase by a step that represents the energy difference between the low-energy standing wave and the high-energy standing wave. Beyond this critical value, the energy again increases smoothly with the momentum, until we reach the second critical value $p = 2h/2a$, where there is another step in the energy, etc. From this discussion we see that the permitted energies of a free electron in a crystal lattice occur in several permitted intervals, within which the energy increases smoothly with momentum. These permitted intervals are separated by forbidden intervals, where the electron energy increases by a step. Figure 39.13 shows such permitted and forbidden energy intervals on an energy-level diagram. *The permitted intervals, shown with lines in the diagram, are called* **energy bands**; *the forbidden intervals are called* **energy gaps**. The precise widths of the permitted energy bands and the forbidden energy gaps depend on the details of the crystal lattice, but all crystals with "free" electrons have some kind of band pattern in their energy-level diagram.

As in the case of the electron configuration of atoms, we can deduce the electron configuration of crystals by means of the Exclusion Principle. In a crystal in its ground state, the electrons settle in the available states of lowest energy. To discover the electron configuration, we proceed as before: we take all the "free" electrons and pack them, one by one, into the available energy bands. The lowest energy bands will then be completely filled, but the upper energy band will be either filled or partially filled, depending on the number of "free" electrons and the number of available states.

The differences among the electric properties of conductors, semiconductors, and insulators arise from the partial or complete filling of the energy bands. A **conductor**, such as copper or silver, *has a band partially filled with electrons* (see Fig. 39.14). When the electrons in this partially filled band are subjected to an electric field, they absorb energy from the field, and they make transitions to some of the slightly higher, empty states of the band. Thus, the electrons respond to the electric field, and they begin to carry an electric current.

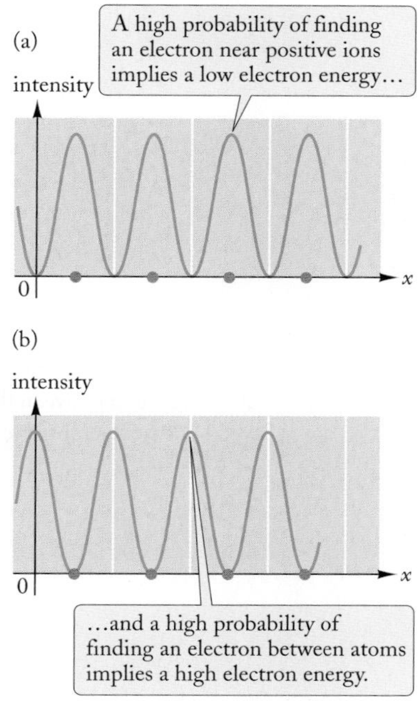

FIGURE 39.12 Standing electron waves in a crystal lattice. The diagrams show the intensity of the wave, which is proportional to the probability of finding the electron, $|\psi|^2$. (a) This standing wave has its peaks of intensity at the locations of the atoms (red dots). (b) This standing wave has its peaks of intensity midway between the locations of the atoms.

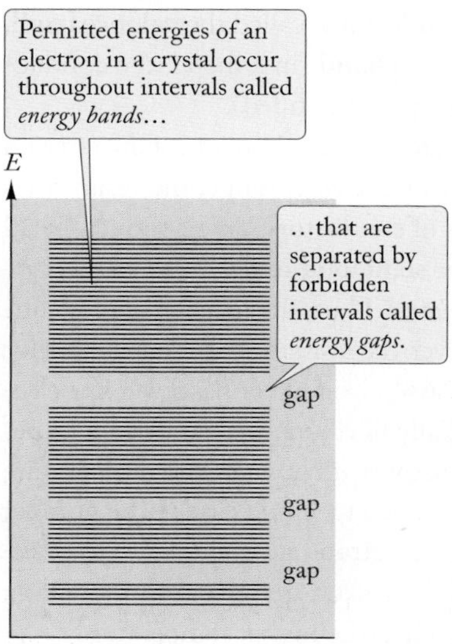

FIGURE 39.13 Energy-level diagram for an electron moving in a crystal.

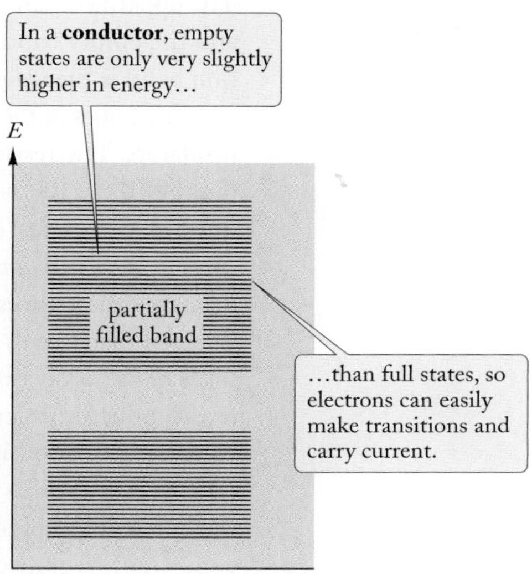

FIGURE 39.14 In a conductor, an upper energy band is only partially filled with electrons. The portion of the band filled with electrons is shown in blue.

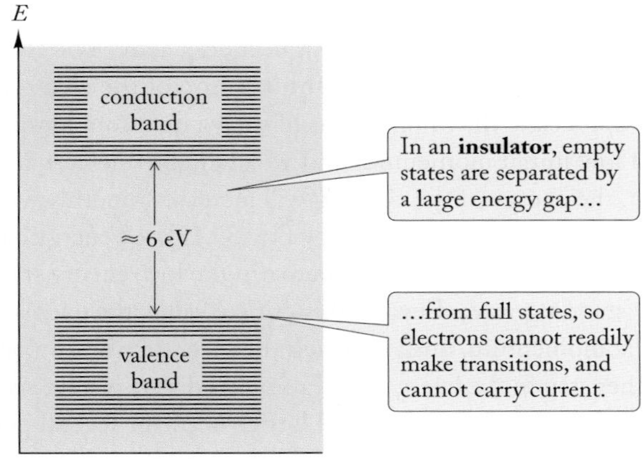

FIGURE 39.15 In an insulator, the lower energy bands are completely filled with electrons. The gap between the uppermost filled band and the next, empty band is fairly large.

In an **insulator**, such as diamond, *the lower bands are completely filled with electrons* (see Fig. 39.15). When the electrons in a full band are subjected to an electric field, they cannot make transitions into other states in the band, because all of these states are already full, and the Exclusion Principle forbids transitions into already full states. This implies that the electrons in the full band cannot respond to the electric field; they therefore cannot accelerate in the direction of the electric force, and they cannot begin to carry a current. The only way the electrons in the full band could respond to the electric field is by making transitions to a higher, empty energy band; but to do this without external help requires that an electron acquire a large amount of energy from random thermal fluctuations, which is difficult, since the typical energy of random thermal fluctuations is small compared with the energy gap at room temperature.

semiconductor

In a **semiconductor**, such as silicon, germanium, or gallium arsenide, the bands are completely filled with electrons, as in an insulator. However, *the energy gap between the last full band and the next, empty band is much smaller than in an insulator* (see Fig. 39.16); typically, the width of the energy gap separating the full band from the next, empty band is less than or approximately equal to 1 eV, whereas in an insulator the gap is often 5 eV or more. In a semiconductor at room temperature, the random thermal fluctuations of the energy will permit many of the electrons at the top of the full band to make a transition to the next, empty band. This means these electrons have nearby empty states; they can respond to the electric field, and they can carry a current. The uppermost full band in an insulator or a semiconductor is called the **valence band**, and the empty band above it is called the **conduction band** (in a metal, the conduction band is the valence band, and this band is only partially filled).

valence band
conduction band

The values of the resistivities of semiconductors are between those of conductors and insulators. The resistivities of semiconductors vary over a wide range; the resistivities may be 10^4 to 10^{15} times as large as the resistivities of conductors. Semiconductors fall into two categories: n type and p type. In an **n-type semiconductor**, *the carriers of current are free electrons that have reached the conduction band.* Thus, the mechanism for conduction is the same as in a metallic conductor. However, the resistance of a semiconductor is higher than that of a metallic conductor, because the semiconductor has fewer free electrons in its conduction band than a metal in its partially filled upper band. Also, a (pure) semiconductor differs from a metal in that the resistivity *decreases* as the temperature increases (see also Fig. 27.8). This curious behavior is due to an increase in the number of free electrons—as the temperature increases, more electrons are excited into the conduction band by random thermal fluctuations, and these extra free electrons more than compensate for the extra friction experienced by each at the higher temperature.

n-type semiconductor

p-type semiconductor

In a **p-type semiconductor**, *the carriers of current are* "**holes**" *of positive charge.* This type of semiconductor has a valence band that is almost, but not quite, filled with elec-

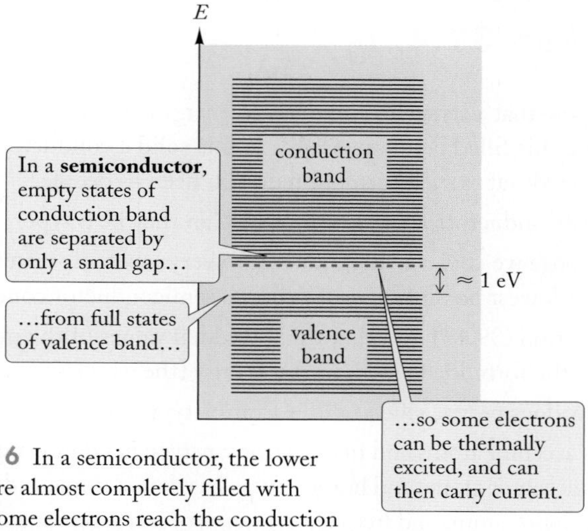

FIGURE 39.16 In a semiconductor, the lower energy bands are almost completely filled with electrons, but some electrons reach the conduction band because the gap between the uppermost filled band and the next, empty band is small.

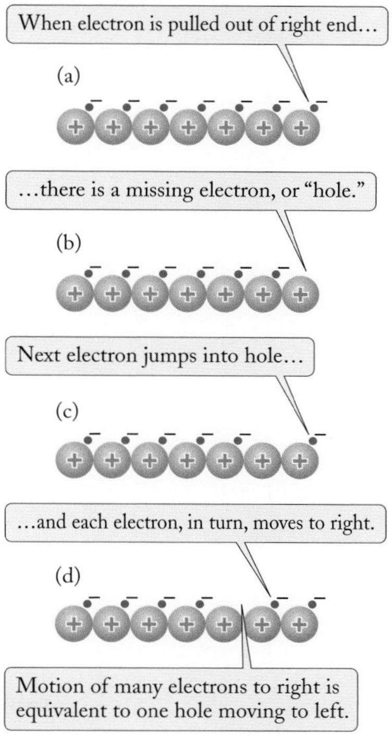

FIGURE 39.17 A row of positive ions (red balls marked +) and electrons (blue dots marked −).

trons. Thus, there are missing electrons, or "holes," in the electron distribution, and if these holes move, they will transport charge. To see how such a transport of charge comes about, consider Fig. 39.17, showing an array of electrons and positive ions. In Fig. 39.17a, these electrons and ions form neutral atoms. Suppose that the right end of this array is connected to the positive pole of a battery (not shown) and the left end to the negative pole. If the battery pulls an electron out of the right end, it will leave the array with a hole, or missing electron, at the position of the last atom (Fig. 39.17b). The electrons will then play a game of musical chairs: the electron from the next-to-last atom will jump into this hole, leaving a hole at the position of the next-to-last atom (Fig. 39.17c); and then the electron from the next atom will jump, etc. The collective motion of the electrons from left to right can be conveniently described as the motion of a hole from right to left. The hole virtually carries positive charge from the right to the left. In essence, this is the mechanism for conduction in a p-type semiconductor. Instead of free electrons, this type of semiconductor has free holes. A flow of current is then a flow of holes, and the direction of the current is the same as the direction of motion of the holes.

Concepts in Context

Semiconductors usually contain both free electrons and free holes. Whether a semiconductor is n type or p type depends on which kind of charge carrier dominates. The concentration of free electrons and of free holes is largely determined by the impurities that are present in the material. **Donor impurities** *consist of atoms that release a valence electron* when placed in the semiconductor, and they increase the number of free electrons. **Acceptor impurities** *consist of atoms that trap electrons* when placed in the semiconductor, and they thereby generate holes. Hence, a semiconductor with donor impurities will be n type and one with acceptor impurities will be p type. For instance, silicon with arsenic impurities, which have one more valence electron than silicon, is an n-type semiconductor, and silicon with boron impurities, which have one less valence electron than silicon, is a p-type semiconductor. Even though the deliberately added impurity atoms may amount to only a few parts per million, they completely change the conductivity because the semiconductor has so few current carriers to start with. For typical devices at room temperature, such "doping" of silicon with impurities increases the density of free charge carriers, and thus the conductivity, by a factor of 10^3 to 10^6 compared with pure silicon.

donor impurities

acceptor impurities

Checkup 39.4

QUESTION 1: Suppose that a crystalline solid has an energy band that is completely filled and all bands above this filled band are empty. Is this solid a conductor, an insulator, or a semiconductor? What extra information do you need to decide?

QUESTION 2: Must conductors always be n type? Can they be p type?

QUESTION 3: Suppose we cool a semiconductor to very low temperature, so all electrons settle into the lowest possible states. Will the semiconductor conduct?

QUESTION 4: Equation (39.14) says that the forbidden values of the momentum are evenly spaced. Are the forbidden values of the energy (the gaps) also evenly spaced?

QUESTION 5: How does p-type silicon differ from n-type silicon?

(A) p-type has acceptor atoms and free electrons; n-type has donor atoms and holes.
(B) p-type has acceptor atoms and holes; n-type has donor atoms and free electrons.
(C) p-type has donor atoms and free electrons; n-type has acceptor atoms and holes.
(D) p-type has donor atoms and holes; n-type has acceptor atoms and free electrons.

Concepts in Context

39.5 SEMICONDUCTOR DEVICES

The manipulation of the resistivity of semiconductor materials by intentional contamination with carefully selected impurities plays a crucial role in the manufacture of semiconductor devices, such as diodes, transistors, and integrated circuits. It is a characteristic feature of semiconductor materials that the addition of impurities to the material has a drastic effect on the resistivity. For instance, the silicon used in electronic devices is usually contaminated, or "doped," with small amounts of arsenic or boron; the addition of just one part per million of arsenic will decrease the resistivity of silicon by a factor of more than 10^5. Pure semiconductor materials are hardly ever used in practical applications. In most cases, the presence of impurities is what gives the semiconductor materials their interesting and useful electric properties.

Rectifier (Diode)

A semiconductor rectifier consists of a piece of n-type and a piece of p-type semiconductor joined together. The n-type semiconductor has free electrons, and the p-type semiconductor has free holes; when they are joined, some of the free electrons will wander from the n region into the p region, and some of the holes will wander from the p region into the n region. Wherever the electrons and the holes meet, they annihilate each other—the electron falls into the hole and fills it, which means that both the electron and the hole disappear. This *recombination* of some electrons and holes leaves residual positive and negative ions near the interface of the two regions, and the electric charges of these ions generate an electric field across the interface (see Fig. 39.18). This electric field opposes any further wandering of holes or electrons from one region into the other. Because of this lack of mobile charge carriers, the interface is known as the *depletion region*.

When such a p–n junction is connected to a battery or some other source of emf, it will permit the flow of current from the p region into the n region, but not in the opposite direction. Figure 39.19 shows the p–n junction connected to the source of emf so the p region is at high potential and the n region at low potential, a configuration

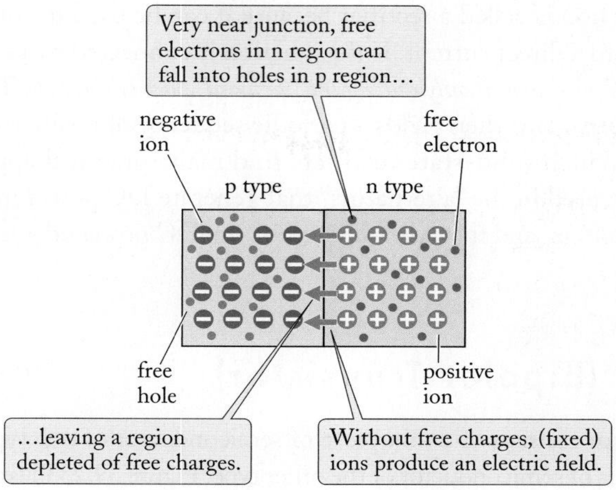

FIGURE 39.18 Pieces of p-type and n-type semiconductor in contact. The plus and minus signs represent the ions of the lattice. The blue dots represent electrons and the red dots holes. The arrows indicate the electric field generated by the ions at the interface.

called "forward bias." The source of emf pumps a steady flow of electrons into the n region and it removes electrons from the p region, which is equivalent to pumping holes into the p region. The electrons and the holes meet at the junction and they recombine. This process can continue indefinitely, and therefore the source of emf can continue to pump current around the circuit indefinitely. Figure 39.20 is a plot of the current vs. the voltage applied to the p–n junction. The current increases steeply with the voltage, because the electric field associated with the applied voltage tends to cancel the internal electric field at the p–n junction, and this makes it easier for the electrons and holes to meet at the center. Evidently, the current is not simply proportional to the voltage, and the p–n junction does not obey Ohm's Law.

Now, consider what happens if the p–n junction is connected to the source of emf so the p region is at low potential and the n region is at high potential, as shown in Fig. 39.21. This configuration is called "reverse bias." Many of the free electrons in the n region then flow away through the wire on the right, and many of the holes in the p region flow away through the wire on the left. Consequently, the region depleted of charge carriers near the interface becomes wider, and the flow of current stops almost immediately—the p–n junction blocks the current.

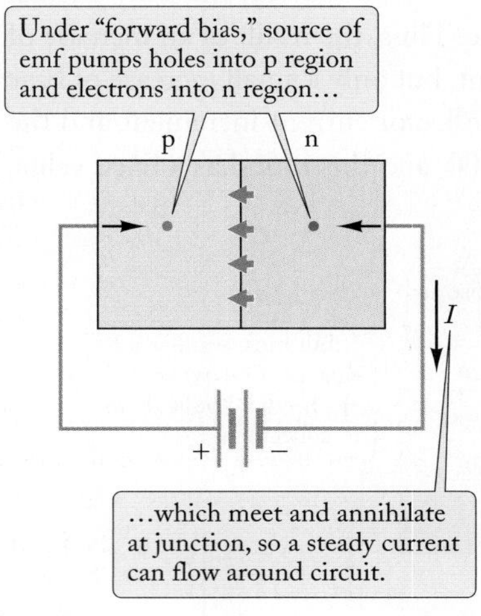

FIGURE 39.19 A p–n junction connected to a source of emf. The n region is at low potential and the p region at high potential (forward bias).

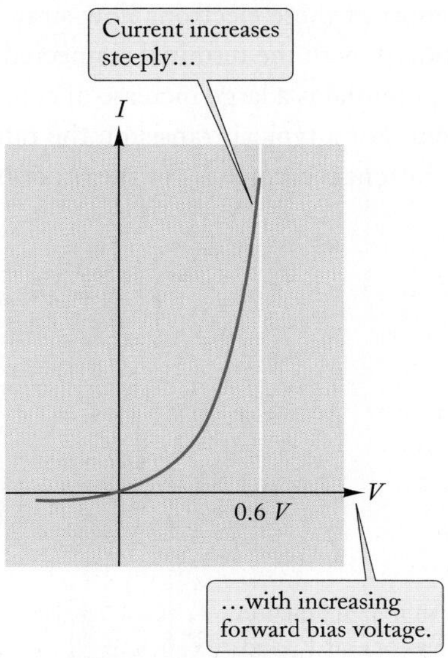

FIGURE 39.20 Plot of current vs. voltage for the p-n junction.

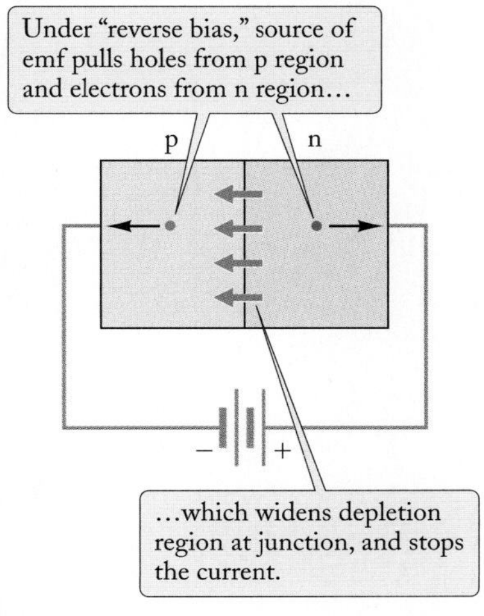

FIGURE 39.21 A p–n junction connected to a source of emf. The n region is at high potential and the p region at low potential (reverse bias).

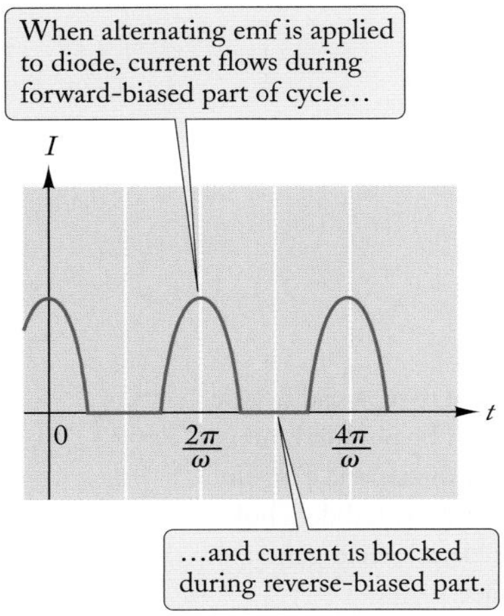

FIGURE 39.22 Current passed by the rectifier vs. time. The negative portions of the alternating current are blocked by the rectifier, and only the positive portions remain.

The p–n junction is called a rectifier because it can be used to convert an alternating current into a direct current. If the junction is connected to a source of alternating emf, *it will pass current only during the "forward" part of the cycle.* The alternating positive and negative emf then yields a periodic sequence of positive current pulses (see Fig. 39.22). Such solid-state rectifiers find many practical applications; for instance, they are used in the "alternators" that generate DC power in the electrical systems of automobiles, and they are found in AC-to-DC power adapters for portable electronics.

Transistor (Bipolar Transistor)

A bipolar transistor consists of a thin piece of semiconductor of one type sandwiched between two pieces of semiconductor of the other type. Figure 39.23 illustrates an n–p–n junction transistor. The thin piece in the middle is called the base, and the pieces at the ends are called the emitter and the collector, respectively. The transistor has three terminals, which are connected to two sources of emf, V_B and V_C, so the emitter-base junction has a forward bias and the base-collector junction has a reverse bias. In this configuration, the emitter-base junction acts as a diode with forward bias, and it permits the flow of electrons from the emitter into the base. However, the electrons that enter the p region fail to recombine with holes, *because this base region is quite thin* and contains only a low density of holes, and the electrons pass through it before they have a chance to meet with a hole. The electrons wander to the base-collector junction, and the electric field across this junction (indicated by the longer red arrows in Fig. 39.23) pulls the electrons into the collector. They leave the collector via the terminal connected to its end, and they continue around the external circuit, forming the external collector current I_C. Of the electrons that enter the base, a small fraction wander to the terminal connected to the base, and they leave via the wire connected there, forming the external base current I_B.

The use of the transistor as an amplifier for currents and voltages hinges on the relationship between the collector and the base currents: a small change in the base current I_B or the base voltage V_B leads to a quite large change in the collector current I_C. As in the case of the diode with forward bias, if we increase the base potential V_B, we cause a drastic increase in the flow of electrons entering the base from the emitter. But most of these electrons flow straight through to the collector, and only a small fraction flow to the terminal connected to the base. Thus, the result of an increase of base potential is a large increase of collector current, but only a small increase of base current. For a typical transistor, the ratio of the collector current increment and the base current increment is of the order of 100 or 200, and this ratio has a fixed value,

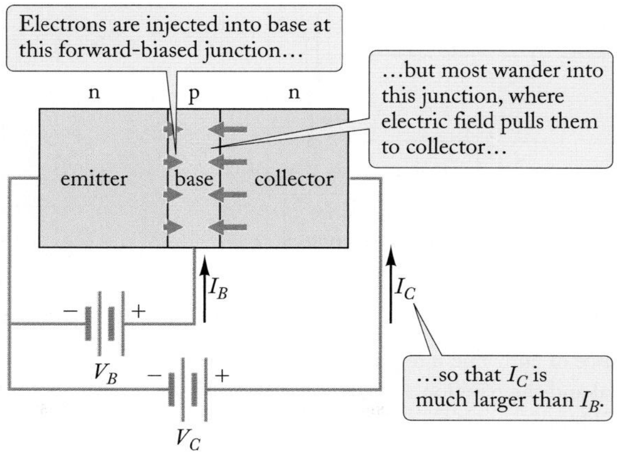

FIGURE 39.23 An n–p–n junction transistor. Two sources of emf V_B and V_C are connected to the base and the collector, respectively.

FIGURE 39.24 Some transistors.

FIGURE 39.25 Integrated circuit with many miniature circuit elements.

over a wide range of currents. The ratio of these current increments is called the current gain factor,[3]

$$[\text{current gain factor}] = \frac{\Delta I_C}{\Delta I_B} \tag{39.16}$$

This means that whenever we change the base current I_B by some amount (by adjusting the voltage V_B), we will change the collector current by an amount a hundred or so times larger. *The transistor amplifies the current—a small input current at the base results in a much larger output current at the collector*. For instance, in the "amplifier" in a radio receiver, the weak current picked up by the radio antenna is amplified by sending it into a transistor, as input current at the base. Further amplification can be achieved by connecting several transistors in tandem, so the output of each serves as input for the next.

Transistors are used in a wide variety of electronic circuits, to amplify and control currents. Figure 39.24 shows some ordinary transistors. In an integrated circuit, such as shown in Fig. 39.25, many transistors and other circuit elements of extremely small size are built up on a single crystal of silicon. The small transistors are not manufactured by sticking together separate pieces of n- and p-type material, but by diffusing suitable concentrations of acceptor and of donor impurities into different adjoining layers of the silicon crystal.

Light-Emitting Diode (LED)

In principle, a light-emitting diode is simply a p–n junction operated with forward bias. At such a junction, electrons arriving from the n region meet holes arriving from the p region, and they recombine, that is, the electrons fall into the holes. But this jump is a transition of the electron from a state of high energy in the conduction band to a state of lower energy in the valence band, a transition that releases energy. In gallium arsenide and some other semiconducting materials, the released energy takes the form of a photon of visible light (see Fig. 39.26). Thus, the p–n junction emits light when an electric current passes through it. Such light-emitting diodes have many practical applications in luminous displays in the dials of measuring instruments, watches, electronic calculators, clocks, automobile speedometers, and so on.

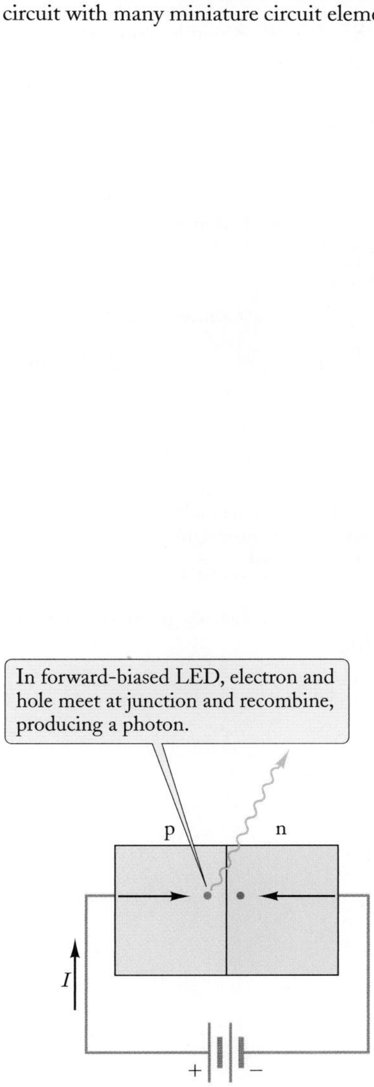

FIGURE 39.26 A p–n junction used as a light-emitting diode.

[3]In electronics catalogs and transistor circuits, the symbol β or h_{fe} is usually used for the current gain factor.

Solar Cell (Photodiode)

A solar cell is simply a light-emitting diode operating in reverse. When sunlight is absorbed at the p–n junction, it excites an electron from the valence band to the conduction band, creating a free electron and a free hole (see Fig. 39.27). The electric field at the junction then pulls the electron toward the n region and the hole toward the p region. This means negative charge flows into the n region and from there into the external wire connected on the right; while positive charge flows into the p region and from there into the external wire connected on the left. Thus, sunlight striking the junction generates an electric current in the external circuit.

Solar cells and other photodiodes are commonly manufactured out of p-type silicon and n-type silicon. The emf of such a silicon solar cell is only about 0.6 V, and the current it delivers is fairly small; most photodiodes produce a few tenths of an ampere of current per watt of incident light. For the solar cell device geometry shown in Fig. 39.28, with an area of about 4 cm^2, the current delivered in full sunlight is about 0.1 A. Many calculators are powered by photodiodes, eliminating the need for batteries.

Solar cells are routinely used to generate electric power on communications satellites and other satellites in orbit around the Earth (see Fig. 39.29). Solar cells have been used to generate power to drive experimental vehicles (see Fig. 39.30). Arrays of solar cells are commercially available (Fig. 39.31). Some attempts have also been made to use them to generate fairly large amounts of electric power for residential and industrial use (see Fig. 39.32). For instance, a solar power station at Pocking, Germany generates about 10 megawatts.

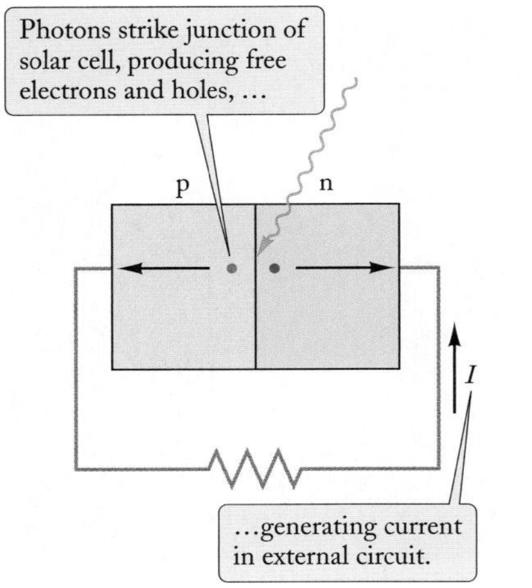

FIGURE 39.27 A p–n junction used as a solar cell.

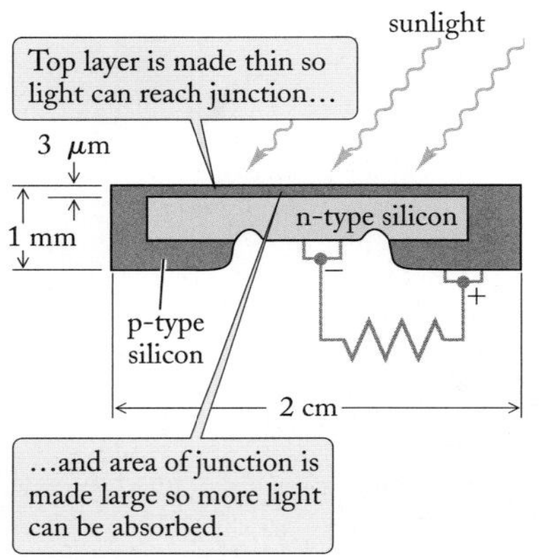

FIGURE 39.28 Device geometry for solar cell (side view).

FIGURE 39.29 Solar panels on the International Space Station.

FIGURE 39.30 A race between experimental vehicles powered by solar cells covering their bodies.

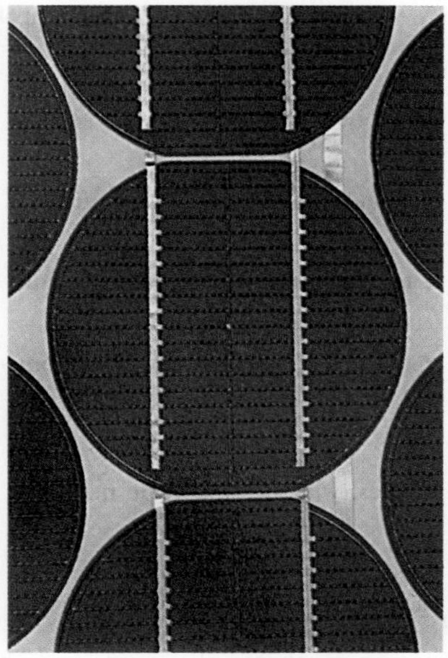

FIGURE 39.31 Panel of solar cells.

FIGURE 39.32 Solar power station at Carrisa Plains in California.

Checkup 39.5

QUESTION 1: If the n region of a p–n junction is at a potential of 3 V and the p region at 5 V, is the bias forward or reverse?

QUESTION 2: In the rectifier, the light-emitting diode, and the solar cell, electrons and holes are created or recombined at the p–n junction. Which devices involve creation, which recombination?

QUESTION 3: You place a piece of n-type silicon in contact with p-type silicon. What is the direction of the electric field at the interface?

(A) Electric field points from n-type region to p-type region.
(B) Electric field points from p-type region to n-type region.
(C) Electric field points parallel to the plane of the interface.
(D) There is no electric field without an external bias.

SUMMARY

QUANTUM NUMBERS OF ATOMIC STATES

$n, l, m,$ and m_s (principal, orbital, magnetic, and spin quantum numbers)

$$n = 1, 2, 3, \ldots$$
$$l = 0, 1, 2, \ldots, n-1$$
$$m = -l, -l+1, \ldots, 0, \ldots, l-1, l$$
$$m_s = -\tfrac{1}{2}, +\tfrac{1}{2}$$

MAGNITUDE OF ANGULAR MOMENTUM

$$L = \sqrt{l(l+1)}\hbar \qquad (39.1)$$

z COMPONENT OF ANGULAR MOMENTUM $$L_z = m\hbar \tag{39.3}$$

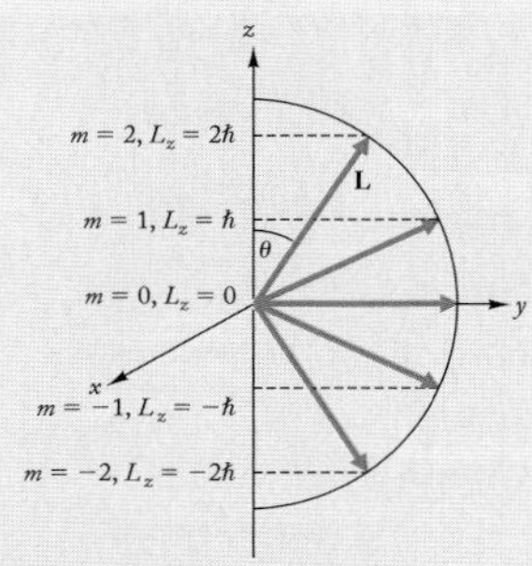

ELECTRON SPIN (INTRINSIC ANGULAR MOMENTUM)

Magnitude: $\sqrt{\frac{1}{2}(\frac{1}{2}+1)}\hbar$

z component: $m_s\hbar$

Spin up: $m_s = +\frac{1}{2}$

Spin down: $m_s = -\frac{1}{2}$

PAULI EXCLUSION PRINCIPLE Each stationary state of quantum numbers n, l, m, and m_s can be occupied by no more than one electron.

VIBRATIONAL ENERGIES OF MOLECULE $$E = nhf \qquad n = 0, 1, 2, \ldots \tag{39.8}$$

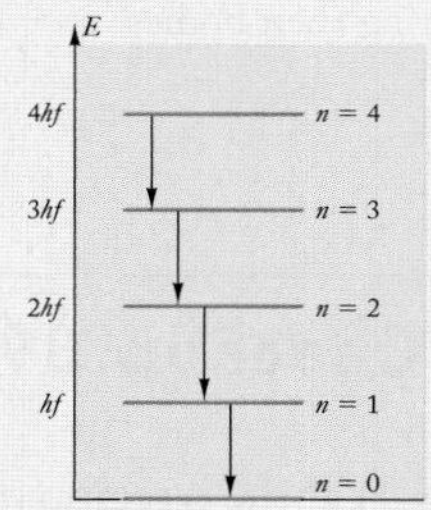

ROTATIONAL ENERGIES OF MOLECULE $$E = \frac{J(J+1)\hbar^2}{2I} \qquad J = 0, 1, 2, \ldots \tag{39.12}$$

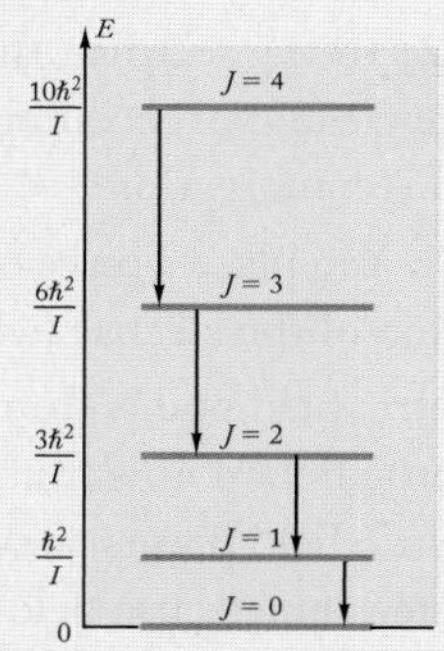

CONDUCTOR Partially filled conduction band.

INSULATOR Full valence band, empty conduction band, large gap between.

SEMICONDUCTOR Full valence band, empty conduction band, small gap between.

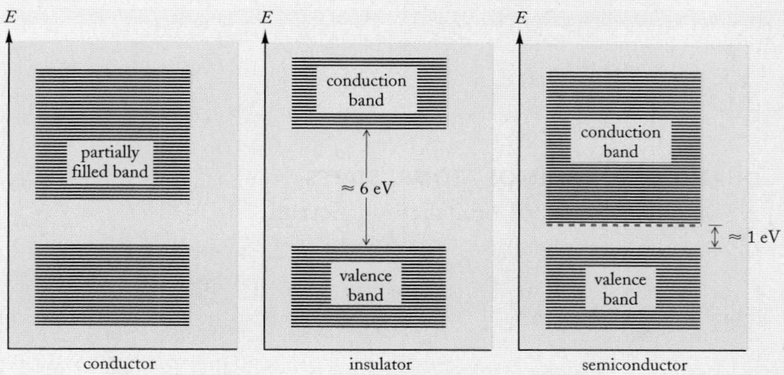

QUESTIONS FOR DISCUSSION

1. According to classical mechanics, what kind of orbit would have zero angular momentum?
2. If there were no Exclusion Principle, what would be the electron configuration of lithium?
3. The bond in the NaCl molecule is similar to the bond in the KBr molecule. Explain this similarity.
4. Rotational transitions in a molecule give a band spectrum, but vibrational transitions do not. Explain.
5. How does n-type silicon differ from p-type?
6. If you dope silicon with phosphorus impurities, will the silicon be n type or p type?
7. What kind of valve in a hydraulic circuit is analogous to a diode rectifier? Draw a picture of a water pipe with such a valve.
8. Do the current I_C and the voltage V_C in a transistor obey Ohm's Law?
9. A spring-loaded butterfly valve in a large water pipe is controlled by a stream of water from a separate pipe (see Fig. 39.33). Is this a reasonable hydraulic analog of a transistor? How could you use such a device to amplify a water current?

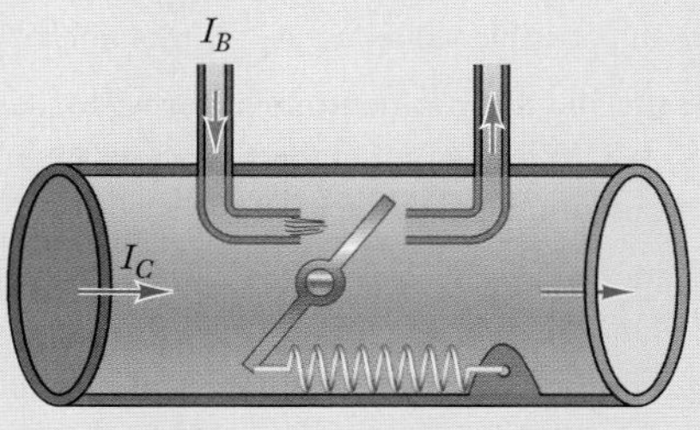

FIGURE 39.33 The butterfly valve in the large pipe is hinged at the center. The stream of water from the small pipe strikes the upper portion of the butterfly valve and pushes it open.

PROBLEMS

†39.1 Principal, Orbital, and Magnetic Quantum Numbers; Spin

1. According to wave mechanics, what are the possible values of the orbital quantum number l if the principal quantum number is $n = 1$? If $n = 2$? If $n = 3$?
2. Suppose that a state in the hydrogen atom has orbital quantum number $l = 5$. What are the possible values of the magnetic quantum number m?
3. Suppose that the magnitude of the orbital angular momentum vector is $\sqrt{20}\hbar$. What are the permitted values of the z component of the angular momentum?
4. Pretend that the electron is a small sphere of uniform density of radius 2.8×10^{-15} m.
   (a) What is the moment of inertia of the electron according to this model?
   (b) What angular velocity of rotation is required to give the sphere an angular momentum of magnitude $\sqrt{\frac{3}{4}}\hbar$? What is the corresponding speed of rotation of a point on the equator of the sphere? Does this model make any sense?
5. An electron in an atom is in a state of quantum numbers $l = 2$, $m = 2$, and $m_s = +\frac{1}{2}$. What is the z component of the total angular momentum of the electron? What is the z component of the total magnetic moment of the electron? [Hint: Eq. (30.23) relates the z component of the spin magnetic moment, $\mu_B = \mu_{z,\text{spin}}$, to the z component of the spin angular momentum, $\frac{1}{2}\hbar$. The proportionality of the orbital magnetic moment to the orbital angular momentum is half as large.]
6. What are the possible values of the orbital angular momentum of the hydrogen atom in its first excited state? Taking into account the spin, what is the maximum possible value of the z component of the total angular momentum? What is the corresponding value of the z component of the magnetic moment of the atom? (Hint: see Problem 5.)
7. Within a shell of an atom, the groups of states of the same l are called a **subshell.** For instance, in the L shell, the two states with $l = 0$ form one subshell, and the six states with $l = 1$ form another subshell.
   (a) How many subshells are there within the M shell? How many states are there in each of these subshells?
   (b) Neon has a complete $l = 0$ subshell and a complete $l = 1$ subshell in the L shell. What atom has similar subshells in the M shell?
8. The circular orbits of orbital angular momentum $n\hbar$ in Bohr's theory roughly correspond to the wave-mechanical states of maximum orbital quantum number, that is, $l = n - 1$. If $n = 2$, compare the magnitude of the angular momentum given by Bohr's theory with the magnitude given by wave mechanics. Repeat for $n = 4$, $n = 10$, and $n = 500$.

†For help, see Online Concept Tutorial 44 at www.wwnorton.com/physics

9. The ρ meson is a particle of spin quantum number 1. What is the magnitude of the spin angular momentum of the ρ meson? What are the possible values of m_s for this particle?

10. In many atoms, the nucleus of the atom has a spin. For example, the nucleus of one of the isotopes of magnesium (^{25}Mg) has spin, with a spin quantum number of 5/2. What is the magnitude of the spin angular momentum of this nucleus? What are the possible values of m_s for this nucleus?

11. Consider the angular-momentum vector with quantum number $l = 3$. What is the smallest possible angle that this angular momentum can make with the z axis?

12. The spin angular momentum of the electron has a magnitude $\sqrt{\frac{1}{2}(\frac{1}{2}+1)}\hbar$ and a magnetic quantum number $m_s = +\frac{1}{2}$ or $m_s = -\frac{1}{2}$. For each of these two values of m_s, calculate the angle between the direction of the spin angular momentum vector and the z axis.

13. The electron orbital angular momentum gives rise to a magnetic moment with z component $\mu_{z,\text{orbital}} = \mu_B m$, where $\mu_B = e\hbar/2m_e$ is the **Bohr magneton** and m is the magnetic quantum number. In Chapter 30, we saw that the electron spin angular momentum gives rise to a magnetic moment with z component $\mu_{z,\text{spin}} = \pm\,\mu_B$. Write an expression that relates $\mu_{z,\text{spin}}$ to the spin quantum number m_s. Is this relation different from that for $\mu_{z,\text{orbital}}$?

14. Consider the angular-momentum quantum numbers for a macroscopic object, such as a toy top with moment of inertia 2.0×10^{-4} kg·m^2 spinning at 25 revolutions per second. The axis of the top makes an angle of 25° with the vertical (with the Earth's gravitational field). What are approximate values of the angular-momentum quantum numbers l and m?

15. Similar to the Bohr magneton [see Problem 13 or Eq. (30.23)], a **nuclear magneton** is defined by $\mu_N = e\hbar/2m_p$, where m_p is the mass of a proton. Unlike the electron, however, the z component of the nuclear spin magnetic moment is not related to the nuclear magneton in a simple way (due to the internal structure of nuclei). For example, the z component of the magnetic moment of a proton is 1.41×10^{-26} J/T, and that of a ^{13}C nucleus (an isotope of carbon) is 3.55×10^{-27} J/T. Express each of these magnetic moments as a mutiple of a nuclear magneton.

*16. As stated in Section 39.1, the energies of the stationary states of the hydrogen atom depend slightly on the orbital angular-momentum quantum number l. An improved formula for the energy of the state of quantum numbers n and l for nonzero l is

$$E_{n,l} = -\frac{m_e e^4}{2(4\pi\epsilon_0)^2\hbar^2 n^2}\left[1 + \frac{e^4}{(4\pi\epsilon_0)^2\hbar^2 c^2 n}\left(\frac{1}{l+\frac{1}{2}\pm\frac{1}{2}} - \frac{3}{4n}\right)\right]$$

where the term $\pm\frac{1}{2}$ corresponds to the spin parallel and antiparallel, respectively, to the orbital angular momentum.

(a) For the case of the first excited state, $n = 2$, $l = 1$, and the spin antiparallel to the orbital angular momentum, find the difference, in eV, between the energy calculated according to Bohr theory and the energy calculated according to the improved formula.

(b) For $n = 2$, $l = 1$, find the difference, in eV, between the energies of the states of spin parallel and antiparallel to the orbital angular momentum calculated according to the improved formula. Which of these states has the lower energy?

*17. For an electron in a state of given orbital (l) and magnetic (m) quantum numbers, the magnitude of the angular momentum and the z component of the angular momentum are well defined, but the x and y components are completely uncertain. Show that nevertheless the sum of the squares of the x and y components is well defined according to the formula

$$(L_x^2 + L_y^2) = l(l+1)\hbar^2 - m^2\hbar^2$$

With the additional assumption that, on the average, L_x^2 and L_y^2 are equal, show that the rms values of the x and y components of the angular momentum are

$$\sqrt{L_x^2} = \sqrt{L_y^2} = \sqrt{\frac{l(l+1) - m^2}{2}}\,\hbar$$

(This is similar to what happens in the case of the velocity components of a molecule of gas in a container. The, say, x component of the velocity is equally likely to be positive or negative, and therefore is completely unpredictable; nevertheless, the rms value $\sqrt{v_x^2}$ is well defined.)

†39.2 The Exclusion Principle and the Structure of Atoms

18. How many possible electron states are there in the M shell ($n = 3$) of a hydrogen atom? Make a list of these states, like the lists in Tables 39.3 and 39.4.

19. List the quantum numbers of all the electrons of a boron atom in its ground state.

20. List the quantum numbers of all the electrons of a carbon atom in its ground state.

21. List the quantum numbers of all the electrons of an Na^+ ion in its ground state.

22. What are the quantum numbers n and l for the outermost electron of the Li atom? The Na atom? The K atom? In what ways are these quantum numbers of these different atoms similar?

23. Suppose that the spin quantum number of the electron were $\frac{3}{2}$ instead of $\frac{1}{2}$. How many possible spin directions would the electrons have in this case? What would be the possible permitted quantum numbers for the case $n = 1$ (K shell)? For the case $n = 2$ (L shell)? How many electrons could be placed in the K shell? The L shell? Compare the resulting periodic table of elements with the familiar periodic table in Section 39.2.

*24. What energy is required to eject one of the electrons from the $n = 1$ state in molybdenum ($Z = 42$) out of the atom? Express your answer in eV.

†For help, see Online Concept Tutorial 44 at www.wwnorton.com/physics

*25. Calculate the energy and the wavelength of the X ray emitted during a quantum jump of an electron from the $n = 3$ to the $n = 1$ state in copper ($Z = 29$).

*26. An X-ray tube has a tungsten target ($Z = 74$). Calculate the energies and the wavelengths of the X rays emitted in the quantum jumps from the $n = 2$ to the $n = 1$ state and from the $n = 3$ to the $n = 1$ state.

*27. For the molybdenum atom ($Z = 42$), the wavelength of the characteristic X ray emitted in the quantum jump from the $n = 2$ to the $n = 1$ state is approximately 0.071 nm. For what atom is the wavelength of the corresponding characteristic X ray twice as large? Half as large?

*28. In addition to X rays from copper and molybdenum targets, commercial X-ray diffractometers often provide X rays of wavelengths 0.229 nm, 0.194 nm, or 0.179 nm. Each of these is due to a transition from the $n = 2$ to the $n = 1$ state of a different element. What are the three elements?

*29. For transitions of electrons to a vacant L-shell state ($n = 2$) from the higher-energy states of a many-electron atom with nuclear charge Z, the nuclear charge is shielded by both K-shell and L-shell electrons. The effective charge seen by an electron making a transition to the L shell is given by $Z_L = Z - 7.4$. What is the wavelength of the X ray emitted when an electron in a gold atom ($Z = 79$) makes a transition from the $n = 3$ state to the $n = 2$ state?

*30. After H. G. J. Moseley measured the energy of characteristic K_α X-ray photons, he plotted the atomic number Z as a function of the inverse of the square root of the photon wavelength, $1/\sqrt{\lambda}$. Such a plot is linear. Find (a) the slope and (b) the y intercept (Z-axis intercept) of such a plot.

*31. A sample of an unknown element is being used as the target in an X-ray tube. It is found that the characteristic X-ray spectrum displays a strong spectral line at $\lambda = 0.0228$ nm. Assume that this spectral line results from the quantum jump of an electron from the $n = 2$ to the $n = 1$ state. Can you identify the unknown element?

39.3 Energy Levels in Molecules

32. The frequency of vibration of the H_2 molecule is 1.31×10^{14} Hz. What are the energies of the vibrational states? What is the frequency of the emitted radiation? The wavelength?

33. The atoms of deuterium (D) and of hydrogen (H) have the same electron configuration (one electron), but the deuterium is a heavier atom than hydrogen, because it has more mass in its nucleus. The mass of the deuterium atom is 2.014 u, whereas the mass of the hydrogen atom is 1.008 u. Given that the frequency of vibration of the H_2 molecule is 1.31×10^{14} Hz, deduce the frequency of vibration of the D_2 molecule.

34. The photon emitted in a vibrational transition of the hydrogen bromide (HBr) molecule has frequency 7.7×10^{13} Hz. What is the spring constant of the chemical bond holding this molecule together? You may assume, for simplicity, that the bromine atom remains essentially at rest, and only the hydrogen atom moves.

*35. In the O_2 molecule, the distance between the two oxygen nuclei is 0.20 nm.

(a) What is the moment of inertia of the molecule for rotation about the perpendicular axis through the center of mass?

(b) What are the energies of the first, second, and third excited rotational states? Express these energies in eV.

*36. The distance between the K and the Br nuclei in the KBr molecule is 0.282 nm, and the center of mass is at a distance of 0.093 nm from the Br nucleus.

(a) What is the moment of inertia of a KBr molecule rotating about its center of mass?

(b) What are the energies of the first, second, and third excited rotational states? Express these energies in eV.

*37. Consider the HD molecule, where one of the atoms of the hydrogen molecule H_2 has been replaced by a deuterium atom. Deuterium and hydrogen have the same electronic structure, but the mass of a deuterium atom is 2.014 u, and that of a hydrogen atom is 1.008 u. Given that the frequency of vibration of the H_2 molecule is 1.31×10^{14} Hz, deduce the frequency of vibration of the HD molecule. [Hint: You may use the result of Problem 41 of Chapter 15, where such an oscillation about the center of mass was shown to have an angular frequency $\omega = \sqrt{k/m}$, where m is the reduced mass $m = m_1m_2/(m_1 + m_2)$.]

*38. In a vibrational transition, a nitric oxide (NO) molecule emits a photon of frequency 5.6×10^{13} Hz. What is the spring constant of the chemical bond holding this molecule together? (Hint: See Problem 37.)

*39. In Section 19.4, we asserted that quantum mechanics enables us to neglect rotations about an axis through the atoms of a diatomic or other linear molecule. Show this explicitly by calculating the energy of the first rotational excited state of H_2 for (a) rotation about the axis through the atoms and (b) rotation about a perpendicular axis. Compare these two energies. For (a), assume the proton is a uniform sphere with radius 1.0×10^{-15} m and the atomic electron is a uniform sphere with radius 5.0×10^{-11} m; for (b), use the interatomic distance of 7.4×10^{-11} m.

*40. What are the ratios of the frequencies emitted in rotational transitions in a molecule from the first excited state to the ground state, from the second excited state to the first, from the third to the second, from the nth to the $(n - 1)$th?

**41. According to spectroscopic measurements, the energy difference between the first and the second excited rotational states of the N_2 molecule is 2.38×10^{-22} J. Deduce the moment of inertia of the molecule. Deduce the center-to-center distance between the N atoms.

39.4 Energy Bands in Solids

42. Consider a crystal with a spacing of 0.10 nm between one atom and the next. What are the de Broglie wavelengths and the energies at which the electron wave cannot propagate through this crystal? Express the energies in electron-volts.

43. In silicon, the energy gap between the valence and the conduction bands is 1.1 eV. If we want to excite an electron from the top of the valence band to the conduction band by means of a photon, what is the maximum permitted wavelength for the photon?

*44. The spacing of atoms in the crystal lattice of a metal, and thus the wavelength for maximum repetitive scattering of electron waves by the atoms, varies with the direction of electron wave propagation in the crystal. Consider a simple cubic arrangement of atoms as in Fig. 39.34, where the length of the cube edge is $a = 2.0 \times 10^{-10}$ m. Find the two longest de Broglie wavelengths and corresponding energies in eV at which the electron wave cannot propagate (a) along a cube edge, (b) along the diagonal of a square face of a cube, and (c) along the body diagonal of a cube.

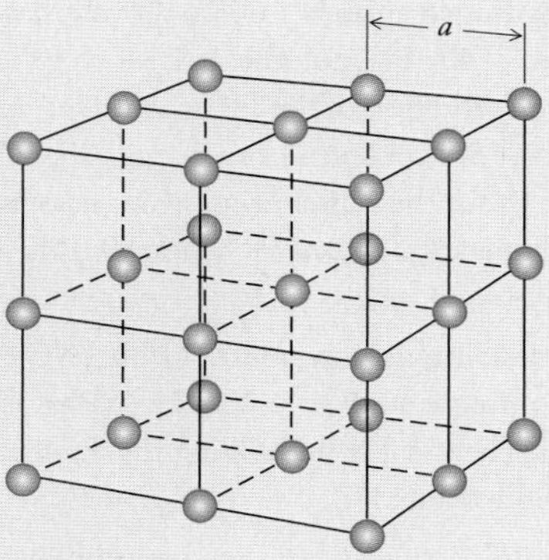

FIGURE 39.34 Structure of a simple cubic crystal.

39.5 Semiconductor Devices

45. Figure 39.23 shows a circuit diagram for an n–p–n transistor. Draw the analogous diagram for the p–n–p transistor, and explain how a small current I_B leads to a large current I_C.

46. When a transistor is connected to a circuit as shown in Fig. 39.35, it serves as an amplifier of voltage. The voltage gain factor is defined as the ratio of the output voltage (measured across the resistor R_C) to the input voltage V_B. Evaluate this ratio if $R_B = 3000\ \Omega$ and $R_C = 6000\ \Omega$. Assume that the current gain factor for this transistor is 100, and that the internal resistance of the emitter–base junction (a diode with forward bias) is negligible.

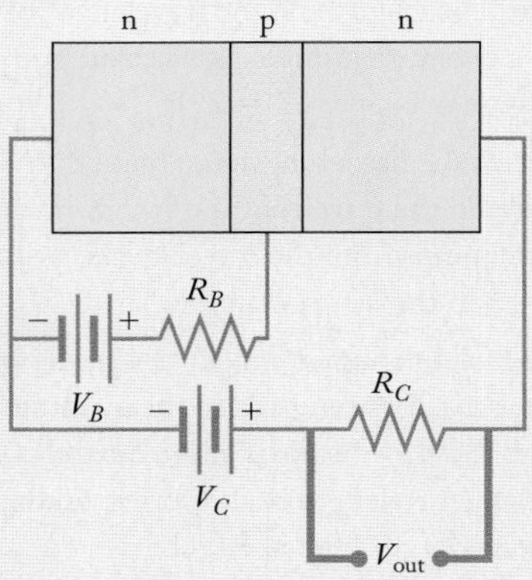

FIGURE 39.35 A transistor acting as a voltage amplifier. The source of emf V_B provides the input signal, and the voltage across the two free terminals constitutes the output signal.

*47. We want to connect two transistors in tandem, so that the net current amplification of the combination is the product of the individual current amplifications of the two transistors. Design a circuit that will accomplish this.

*48. A solar cell delivers 0.10 A at 0.60 V. How many such solar cells do you need, and how must you connect them, to obtain 2.0 A at 6.0 V for charging a battery?

*49. A solar cell of area 5.0 cm^2 facing the Sun delivers 0.10 A at 0.60 V. What is the power delivered by the solar cell? Compare with the incident power of sunlight (1.0 kW/m^2) and deduce the efficiency for the conversion of energy of light into electric energy.

*50. In practical transistor circuits, a single source of emf is used to provide the two emfs V_B and V_C of Fig. 39.23. A typical circuit with a 9.0-V emf is shown in Fig. 39.36. The transistor has a current gain factor of 100 and it is desired to operate with $I_C = 1.0$ mA with the two 1.0-kΩ resistors in place. Assume that the potential across the forward-biased base–emitter junction is 0.60 V. The resistors R_1 and R_2 are known as *bias resistors*. Use Kirchhoff's rules to find the required values of R_2 and R_1.

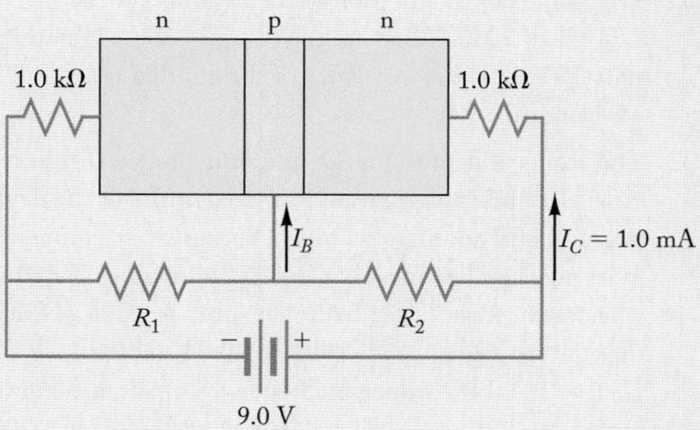

FIGURE 39.36 A transistor circuit with a single source of emf.

REVIEW PROBLEMS

51. (a) How many of the stationary states (counting spin states) of the hydrogen atom have energy -13.6 eV? Energy -3.4 eV?

    (b) How many of the stationary states (counting spin states) of the hydrogen atom have energy -3.4 eV and $l = 0$? Energy -3.4 eV and $l = 1$? Energy -3.4 eV and $l = 3$?

52. Suppose that all you know about a state of the hydrogen atom is that the magnetic quantum number is $m = 3$. What conclusions can you draw about the value of the orbital quantum number? The principal quantum number? The energy of the state? (Hint: Is $l = 0$ compatible with the given value of m? Is $l = 1$? Is $l = 2$?)

*53. Consider the possible directions of the orbital angular-momentum vector for an electron with orbital quantum number $l = 1$. What are the possible values of the magnetic quantum number m? For each value of m, calculate the angle between the angular-momentum vector and the z axis. Draw a diagram showing the possible orientations of the angular-momentum vector for all the different values of m.

*54. Suppose that we regard the proton as a sphere of uniform density with a radius of 1.0×10^{-15} m rotating rigidly about its axis. According to classical mechanics, if the spin angular momentum of this sphere is to have a magnitude of $\sqrt{3/4}\hbar$, what must be the angular velocity of rotation? What must be the speed of a point on the equator?

55. List the quantum numbers of all the electrons of a magnesium atom in its ground state.

56. What are the quantum numbers n and l for the two outermost electrons of a Be atom? The Mg atom? The Ca atom? In what way are these quantum numbers of these different atoms similar?

*57. The conventional range of wavelengths for X rays extends from 10 nm to 0.01 nm. Suppose you want to generate characteristic X rays within this range of wavelengths by means of the transition of an internal electron in an atom, from the first excited state to the ground state. What atomic numbers are suitable? Which atoms do these correspond to?

*58. Calculate the wavelengths of the characteristic X rays emitted by the inner electrons in molybdenum atoms in transitions from the second and the third excited states to the ground state.

*59. When a block of metal serving as target in an X-ray tube is bombarded with a beam of fast electrons, it emits not only Bremsstrahlung but also characteristic X rays of wavelengths 0.167 nm, 0.141 nm, and 0.133 nm. Assuming that these X rays arise in transitions from excited states into the ground state, identify the metal.

60. In the HF molecule, the chemical bonds holding the two atoms together behave like a massless spring of a spring constant $k = 9.7 \times 10^2$ N/m.

    (a) Calculate the frequency of vibration of the molecule. For the sake of simplicity, assume that the fluorine atom remains at rest and only the hydrogen atom moves.

    (b) Calculate the energy of the first excited vibrational state of the molecule.

61. Consider the HCl molecule described in Example 4. Suppose that this molecule is initially in the $J = 3$ rotational state. If the molecule sequentially makes purely rotational transitions to the $J = 2$ state, then to the $J = 1$ state, and finally to the $J = 0$ state, what are the energies and the wavelengths of the photons emitted in each of these transitions?

62. The distance between the two nuclei in the H_2 molecule is 0.074 nm.

    (a) What is the moment of inertia of an H_2 molecule rotating about its center of mass? What are the energies of the first, second, and third excited rotational states? Express these energies in eV.

    (b) Suppose we replace one of the hydrogen atoms by a deuterium atom (D, or ^2H), whose mass is twice that of the hydrogen atom. Where is the new center of mass? What is the moment of inertia of the HD molecule about its center of mass? What are the energies of the first, second, and third excited rotational states? By what factor do these energies differ from those of the H_2 molecule?

63. (a) The binding energy of the NaCl crystal (salt) is 765 kJ/mole; this is the energy required to dissociate the crystal into separate ions. This crystal is cubic, as is illustrated in Fig. 39.37. Pretend that each ion forms a bond with only the six nearest ions. What is the energy per Na^+–Cl^- bond? Express the answer in electron-volts.

    (b) The distance between each ion and the nearest ion is 0.281 nm. What is the Coulomb energy of a pair of ions separated by this distance? Express the answer in electron-volts. Explain why the answers obtained in (a) and (b) are of the same order of magnitude, although not exactly equal.

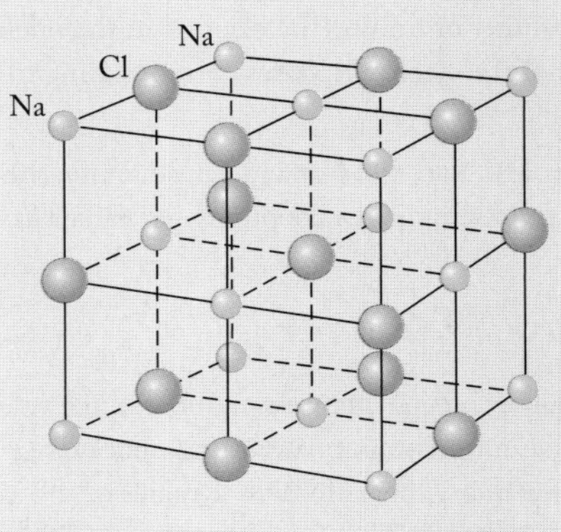

FIGURE 39.37 Structure of NaCl crystal. The yellow balls are sodium ions and the green balls are chlorine ions.

64. Figure 39.38 shows a full-wave rectifier consisting of four diodes connected together. This rectifier not only blocks the negative portion of an entering alternating current, but also reverses this portion, so the current is positive at all times. Describe the flow of current through the four diodes when the entering alternating current is positive and when it is negative.

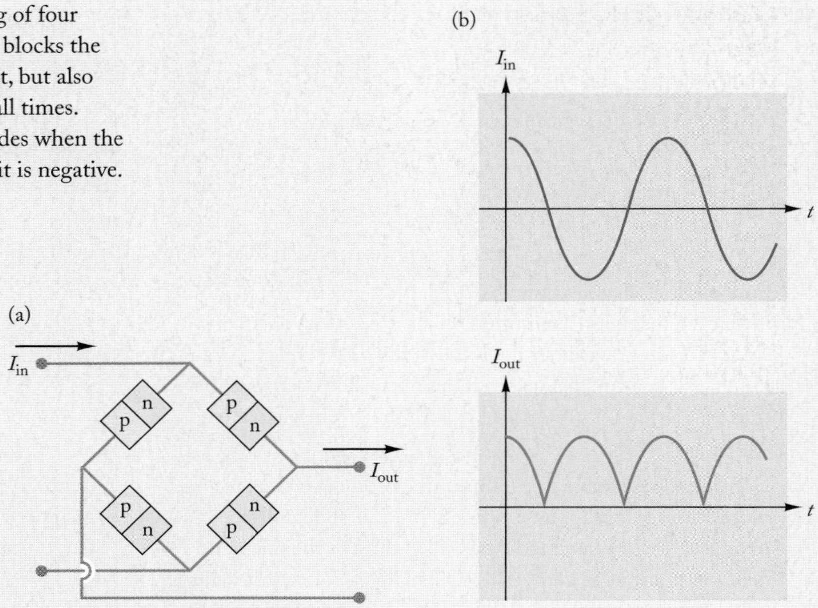

FIGURE 39.38 (a) Four diodes connected to form a full-wave rectifier. (b) The current going into the rectifier, and the current coming out.

Answers to Checkups

Checkup 39.1

1. According to wave mechanics, the ground-state orbital angular momentum is $L = 0$; according to Bohr theory, $L = \hbar$. For both theories, the quantum number $n = 1$ in the ground state.
2. The magnitude of the orbital angular momentum is given by $L = \sqrt{l(l+1)}\hbar$; for $l = 3$, this is $L = \sqrt{3 \times (3+1)}\hbar = 2\sqrt{3}\hbar$.
3. The orbital angular-momentum quantum number may take values from 0 to $n - 1$, so for $n = 3$, the possible values of l are $l = 0$, $l = 1$, and $l = 2$.
4. The intensity drops to zero at the center of each plot, so the probability of finding the electron at the nucleus is zero in each of the electronic states corresponding to Figs. 39.3b, c, and d.
5. (E) -1, 0, and $+1$. For a given value of l, the magnetic quantum number m may take on integer values from $-l$ to $+l$.

Checkup 39.2

1. For any principal quantum number n, the values of the orbital angular momentum quantum number l range from 0 to $n - 1$, and for each l there are $2l + 1$ values of the magnetic quantum number m. There are two values of spin for each state. For the K shell, we have the least number of states, two: $n = 1$, $l = 0$, $m = 0$, and $m_s = \pm\frac{1}{2}$. For the L shell, we have two similar states but with $n = 2$, and six states with $l = 1$ $[2 \times (2l + 1)]$, for a total of eight. For the M shell, we have the largest number: eight states similar to the L-shell states but with $n = 3$, plus ten states with $l = 2$ $[2 \times (2l + 1)]$, for a total of eighteen.
2. Yes, for each orbital state there are a spin up and a spin down state, and if all states are full, for every spin up electron there is a spin down electron.
3. (D) 4×13.6 eV. Neutral helium ($Z = 2$) has two electrons. After one is removed, the single remaining electron is in a purely hydrogenic orbit, but with nuclear charge $Z = 2$. Equation (39.6) gives the energies of such orbits as $E_n \approx Z^2 \times (13.6\text{ eV})/n^2$. To remove the electron, we take it from the $n = 1$ state to $n = \infty$, which requires an energy of removal $\Delta E = Z^2 \times (13.6\text{ eV}) = 4 \times 13.6$ eV.

Checkup 39.3

1. No. From Eq. (39.9), we recall that the energy of rotation is proportional to the square of the frequency of rotation. We know that the frequency of radiation is proportional to the difference of the energies of two states, and so is proportional to the difference of two frequencies squared, and thus is not equal to any single rotational frequency.

2. The molecule emits four photons, corresponding to the four transitions shown. From the spacings in Fig. 39.9, we see that the photon emitted in the first (top) transition has the largest energy, and the photon emitted in the last (bottom) transition has the least energy. Since $E = hf = hc/\lambda$, the first photon has the shortest wavelength, and the last photon the longest.
3. The rotational energy of any state is inversely proportional to the moment of inertia [see Eq. (39.12)], so the lighter H_2 molecule has the larger energy difference, by a factor of 2.
4. (A) 1, 2, 3, 4. We can obtain the energy difference ΔE between two adjacent states directly from the corresponding adjacent values on the vertical axis in Fig. 39.9. The photon frequency is $\omega_{\text{photon}} = \Delta E/\hbar$, so starting with the bottom transition we have $\omega_{\text{photon}} = 1$, 2, 3 and 4 times $\hbar/I$. This sequence continues: for a transition from a state with any J to the next lower ($J - 1$) state, $\Delta E = (\hbar^2/2I) \times [J(J+1) - (J-1)J] = (\hbar^2/I) \times J$. with $J = 1, 2, 3, 4$, etc.

Checkup 39.4

1. A crystalline solid with a full band and only empty bands above it is an insulator or a semiconductor. To decide, we would need to know the magnitude of the energy gap.
2. Most conductors are n type, but a conductor can be p type. A p-type conductor must have holes that behave like moving positive charges (as illustrated in Fig. 39.17); this happens when a band is almost filled. Such behavior occurs in conductors when different bands overlap, so electrons that would otherwise be near the top of one band spill over into the lower-energy states of an overlapping band.
3. No. At sufficiently low temperature, there are, in essence, no electrons with enough thermal energy to cross the energy gap.
4. No. When the lattice potential is not too strong, the evenly spaced values of the momentum and the dominantly kinetic energy ($E = p^2/2m$), imply that the forbidden values of the energy will be spaced roughly in proportion to the square of the corresponding momentum values. (When such behavior is generalized to crystal structures in three dimensions, or when the lattice potential is strong, then more complicated sequences of forbidden values of energy occur.)
5. (B) p-type has acceptor atoms and holes; n-type has donor atoms and free electrons. Acceptor atoms remove electrons from the valence band and leave behind positively charged (p-type) holes; donor atoms contribute electrons (negatively charged, n-type) to the conduction band.

Checkup 39.5

1. Forward. A diode is forward-biased when the p-type region is at a higher potential than the n-type region.
2. The forward-biased rectifier involves recombination, since the external source of emf pumps many electrons into the n region and many holes into the p region, and they recombine at the junction. (A reverse-biased diode involves the creation of an extremely small number of thermally generated electron–hole pairs.) The light-emitting diode is a forward-biased rectifier and so also involves recombination. The solar cell involves creation, since the absorption of photons creates electron–hole pairs.
3. (A) Electric field points from n-type region to p-type region. In the interface region, the mobile carriers are depleted, and since these carriers came from neutral impurity atoms, the impurity ions left behind are opposite in sign to the mobile carriers of each region. The electric field thus points from the positive donor ions of the n-type region to the negative acceptor ions in the p-type region (see Fig. 39.18).

CHAPTER

40 Nuclei

CONCEPTS IN CONTEXT

Concepts in Context

This nuclear reactor, immersed in a pool of water, releases energy by the fission, or splitting, of nuclei of uranium. Such nuclear reactions produce "penetrating radiations," that is, radiations capable of penetrating through cloth, paper, and skin. For protection, nuclear reactors are shielded by thick layers of concrete, or, as in the photo, by a thick layer of water.

With the concepts of this chapter, we can consider such questions as:

? Why does fission release energy, and what is the amount of energy released per fission? (Section 40.5, pages 1377 and 1376)

? What are the penetrating radiations produced in nuclear reactions, and how are they produced? (Section 40.3, page 1365; and Example 5, page 1371)

? What is the energy of the penetrating radiation released in the fission of uranium? (Example 3, page 1367)

? A typical reactor releases a power of 1200 MW. How much uranium does this consume per year? (Example 7, page 1382)

Rutherford's first experiments on the bombardment of atoms with a beam of alpha particles established that the nucleus of the atom is very small, but contains most of the mass of the atom. The nucleus is therefore very dense, and it must be made of massive particles packed very tightly together. In later experiments, Rutherford proceeded to explore the structure of the nucleus, again using a beam of alpha particles as a probe. He found that if the projectiles were energetic enough to penetrate the nucleus, they would often split it into two pieces, two smaller nuclei. The smallest such piece that could be split off was a nucleus of hydrogen, or a **proton**, and Rutherford therefore conjectured that all other nuclei also contain protons. However, all these other nuclei have more mass and less charge than expected if they contained nothing but protons—there must be some neutral particles in the nucleus or, alternatively, some electrically neutral combination of particles of opposite charges. The mystery of the neutral constituent of the nucleus was not solved until 1932, when J. Chadwick discovered the **neutron**, a particle of about the same mass as the proton but of zero electric charge. This discovery led to the modern view of the nucleus as a tightly packed conglomerate of protons and neutrons (see Fig. 40.1).

proton

neutron

Since the average distance between the protons in the nucleus is quite short, the repulsive electric force among the nuclear protons is very large. This force would burst the nucleus apart if there were not an extra, even larger, attractive force holding the protons and the neutrons together. This extra force is the *nuclear force*, or the "*strong*" *force*. Acting on two adjacent protons in a nucleus, this attractive force is about 100 times as large as the repulsive electric force. Thus, the strong force completely overwhelms the electric force. However, in heavy nuclei—such as uranium—with a large number of protons and a large total electric charge, the electric repulsion becomes important. The fission of uranium, as manifested in the explosion of a nuclear bomb, provides a spectacular demonstration of the electric force overpowering the strong force.

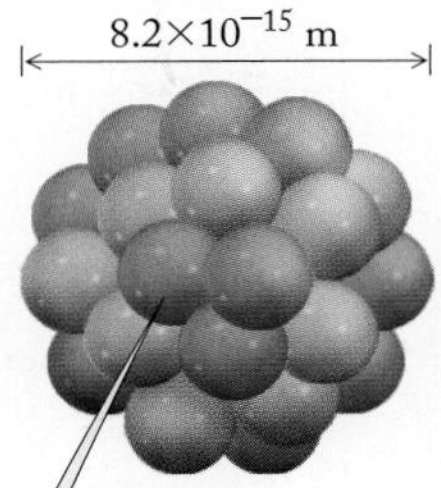

FIGURE 40.1 The nucleus of the argon atom, consisting of 18 protons (red) and 22 neutrons (gray).

40.1 ISOTOPES

Nuclei are made of protons and neutrons. Generically, these two kinds of constituents of the nucleus are called **nucleons**. Table 40.1 lists the main properties of protons and neutrons. The values of the masses listed in this table are expressed in **atomic mass units**, where 1 u $= 1.660\,54 \times 10^{-27}$ kg. The value of $\frac{1}{2}$ listed for the spin is the spin quantum number. According to the usual rule for the magnitude of an angular momentum [see Eq. (39.1)], the magnitude of the spin is actually $\sqrt{\frac{1}{2}(\frac{1}{2}+1)}\hbar$; but physi-

nucleon

atomic mass unit u

TABLE 40.1 THE NUCLEONS

NUCLEON	MASS	SPIN	RADIUS
proton	$m_p = 1.007\,28$ u	$\frac{1}{2}$	1×10^{-15} m
neutron	$m_n = 1.008\,66$ u	$\frac{1}{2}$	1×10^{-15} m

cists often list the values of the spins of particles by giving the spin quantum number, rather than the magnitude of the spin angular momentum.

It is instructive to compare these particles with the electron. Both the proton and the neutron have masses about 1840 times as large as that of the electron.[1] Their spin quantum number is $\frac{1}{2}$, the same as that of the electron. In contrast to the electron, which is a pointlike particle of no discernible size, both *the proton and the neutron are small spheres, of a radius of about* 10^{-15} m.

The number of protons in the nucleus of a (neutral) atom of a given element matches the number of its electrons, that is, it matches the atomic number of the element. For example, the carbon atom has six electrons and it has six protons in its nucleus.

isotopes

All the atoms of a given chemical element, such as carbon, have exactly the same chemical properties, because they all have exactly the same number of electrons and the same electron configuration. However, the atoms of a chemical element can differ in mass, because their nuclei can have different numbers of neutrons. Thus, all carbon atoms have six protons in their nuclei, but some have six neutrons, some have seven, some have eight, and so on. *Atoms with the same number of protons in their nuclei but different numbers of neutrons are called* **isotopes**. Carbon has eleven known isotopes, designated ^{8}C, ^{9}C, ^{10}C, ^{11}C, ^{12}C, ^{13}C, ^{14}C, ^{15}C, ^{16}C, ^{17}C, and ^{18}C, (see Fig. 40.2). The superscript before the chemical symbol (for instance, the superscript "12" in "^{12}C") indicates the sum of the number of protons and the number of neutrons; this sum is called the **mass number**. If we designate the mass number by the symbol A, then

mass number A

$$A = N + Z \tag{40.1}$$

atomic number Z

where N is the number of neutrons and Z is the number of protons, or the **atomic number**. Since the mass of each proton and each neutron is approximately one atomic mass unit, the mass number is approximately equal to the mass of the nucleus in atomic mass units.

Natural samples of atoms of carbon or any other chemical element contain characteristic percentages of different isotopes. Natural carbon, as found in coal, is a mixture of 98.90% of the isotope ^{12}C and 1.10% of the isotope ^{13}C. Carbon dioxide, as found in air, contains not only the isotopes ^{12}C and ^{13}C but also a very small amount (about 2.4×10^{-10}%) of the isotope ^{14}C. The other isotopes of carbon do not occur

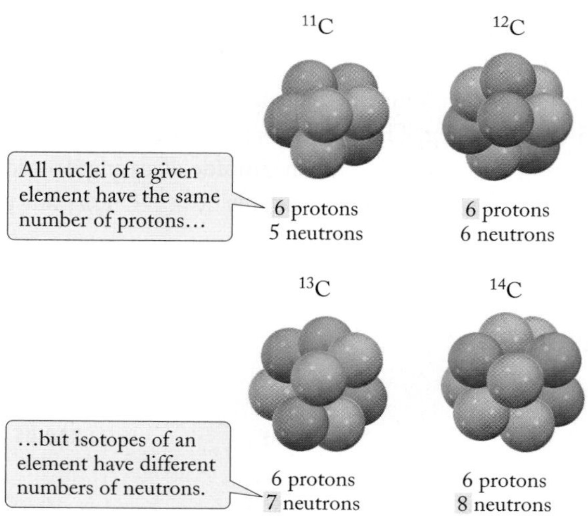

FIGURE 40.2 Some of the isotopes of carbon.

[1]Expressed in atomic mass units, the electron mass is 5.49×10^{-4} u.

naturally; they can be produced only artificially in transmutation of elements in a nuclear reactor or in a particle accelerator.

All chemical elements have several isotopes (see the excerpt from the chart of isotopes in Table 40.2). Hydrogen has three isotopes (^{1}H, or ordinary hydrogen; ^{2}H, or deuterium; ^{3}H, or tritium). Helium has five isotopes, lithium has six, and so on. Some of these isotopes occur in nature, others can be produced only by artificial means.

Most of the isotopes listed in the chart of isotopes are unstable; they decay by a spontaneous nuclear reaction and transmute themselves into another element. The decay is accompanied by the emission of alpha rays, beta rays, or gamma rays. The alpha rays are high-speed alpha particles (that is, helium-4 nuclei), the beta rays are high-speed electrons or antielectrons, and the gamma rays are high-energy photons. We will examine these decay processes in Section 40.3.

TABLE 40.2 EXCERPT FROM THE CHART OF ISOTOPES[a]

Z																			
10							Ne 20.179		17 Ne 0.109 s β^+ 17.0177	18 Ne 1.67 s β^+	19 Ne 1.74 s β^+	20 Ne 90.5% 19.992 439	21 Ne 0.27% 20.993 847	22 Ne 9.22% 21.991 384	23 Ne 37.6 s β^- γ	24 Ne 3.38 min β^- γ	25 Ne 0.61 s β^- γ	26 Ne 26.0005	27 Ne 27.0072
9						F 18.9984		15 F 15.0180	16 F ~10^{-19} s p 16.011	17 F 66.0 s β^+ no γ	18 F 109.8 min β^+ EC	19 F 100% 18.998 403	20 F 11.1 s β^- γ	21 F 4.36 s β^- γ	22 F 4.0 s β^- γ	23 F 2.2 s β^- γ	24 F 24.0093	25 F 25.0138	17
8						O 15.9994	13 O 0.0089 s β^+	14 O 70.5 s β^+ γ	15 O 122 s β^+ no γ 15.003 065	16 O 99.756% 15.994 915	17 O 0.037% 16.999 131	18 O 0.204% 17.999 159	19 O 26.9 s β^- γ	20 O 13.6 s β^- γ	21 O 3.4 s β^- γ	22 O 23.0101	23 O 23.0193	16	
7				N 14.0067		11 N 11.0267	12 N 0.011 s β^+ γ	13 N 9.97 min β^+ no γ 1.007 94	14 N 99.63% 14.003 074	15 N 0.37% 15.000 109	16 N 7.11 s β^- γ 16.006 100	17 N 4.16 s β^- γ	18 N 00.63 s β^- γ	19 N 0.42 s 19.0176	20 N 20.0238	21 N 21.0289	15		
6				C 12.011	9 C 0.127 s β^+	10 C 19.4 s β^+ γ	11 C 20.4 min β^+ EC 11.011 433	12 C 98.89% 12.000 00	13 C 1.11% 13.003 355	14 C 5730 yr β^- no γ 14.003 242	15 C 2.45 s β^- γ	16 C 0.74 s β^-	17 C 17.0226	18 C 18.0267	19 C 19.0370	14			
5		B 10.811			8 B 0.774 s β^+	9 B ~8×10^{-19} s p 9.013 33	10 B 19.8% 10.012 938	11 B 80.2% 11.009 305	12 B 0.020 s β^- γ	13 B 0.017 s β^- γ	14 B β^- γ			17 B 17.0986	13				
4		Be 9.012 18		6 Be $\gtrsim 3 \times 10^{-21}$ s p, α, Li^5 6.019 73	7 Be 53.3 day EC γ 7.016 930	8 Be ~1×10^{-16} s 2α 8.005 305	9 Be 100% 9.012 183	10 Be 1.6×10^6 yr β^- no γ	11 Be 13.8 s β^- γ	12 Be 0.011 s β^-		14 Be 14.0440	11	12					
3	Li 6.941			5 Li ~10^{-21} s p, α 5.0125	6 Li 7.5% 6.015 123	7 Li 92.5% 7.016 005	8 Li 0.85 s β^-	9 Li 0.17 s β^-		11 Li 0.009 s β^-	9	10							
2	He 4.002 60		3 He 0.00013% 3.016 029	4 He ~100% 4.002 603	5 He 2×10^{-21} s n, α 5.0122	6 He 0.802 s β^- no γ		8 He 0.122 s β^- γ	7	8									
1	H 1.0079	1 H 99.985% 1.007 825	2 H 0.015% 2.014 102	3 H 12.33 yr β^- no γ 3.016 049			5	6											
0			1 n 10.6 min β^- 1.008 665		3	4													
		0	1	2					N										

[a]The number Z, increasing vertically along the chart, is the number of protons in the isotope; it coincides with the atomic number. The number N, increasing horizontally, is the number of neutrons. In each box, the number directly below the symbol for the isotope gives the abundance in percent for naturally occurring isotopes, or else the half-life for unstable, artificially produced isotopes (the half-life is the time required for one-half of a sample of unstable isotope to decay). The Greek letters indicate the emissions that accompany the decay: α rays (helium nuclei), β^- rays (electrons), β^+ rays (antielectrons), or γ rays. The bottom number gives the mass of the neutral atom (nucleus plus Z electrons) in atomic mass units. The bottom number in the shaded boxes gives the atomic mass averaged in proportion to the abundance of naturally occuring isotopes.

Experiments on the bombardment of nuclei with alpha particles and other particles indicate that the size of the nucleus is proportional to the cube root of the mass number. Specifically, these experiments indicate that the radius R of a nucleus of mass number A is

radius of nucleus

$$R = (1.2 \times 10^{-15}\ \text{m}) \times A^{1/3} \tag{40.2}$$

For example, the carbon-12 nucleus has a radius

$$R = 1.2 \times 10^{-15}\ \text{m} \times (12)^{1/3} = 2.7 \times 10^{-15}\ \text{m}$$

whereas the uranium-238 nucleus has a radius

$$R = 1.2 \times 10^{-15}\ \text{m} \times (238)^{1/3} = 7.4 \times 10^{-15}\ \text{m}$$

We can gain some feeling for how small these radii are by comparing them with the radius of an atom, typically about 10^{-10} m. The comparison tells us that the radius of the nucleus is less than 1/10000 of the radius of the atom, so the volume of the nucleus is only $1/10^{12}$, or one-trillionth, of the volume of the atom.

The proportionality between R and $A^{1/3}$ implies that *the number of nucleons per unit volume is the same for all nuclei*. Since the volume of a sphere is $\frac{4}{3}\pi R^3$, the number of nucleons per unit volume is

$$\frac{A}{(4\pi/3)R^3} = \frac{A}{(4\pi/3)(1.2 \times 10^{-15}A^{1/3})^3\ \text{m}^3}$$
$$= 1.4 \times 10^{44}\ \text{nucleons/m}^3 \tag{40.3}$$

The mass of each nucleon is about 1.67×10^{-27} kg, and the mass density of the nuclear material is therefore

$$1.67 \times 10^{-27}\ \text{kg} \times \frac{1.4 \times 10^{44}}{\text{m}^3} = 2.3 \times 10^{17}\ \text{kg/m}^3 \tag{40.4}$$

This means that one cubic centimeter, or $10^{-6}\ \text{m}^3$, of nuclear material would have a mass of 230 million tons!

According to Eq. (40.3), the volume per nucleon is $1/(1.4 \times 10^{44})\ \text{m}^3$. We can think of this volume as a cube enclosing the nucleon; the edge of the cube or, equivalently, the distance from one nucleon to its nearest neighbor is therefore the cube root of the volume, $1/(1.4 \times 10^{44})^{1/3}\ \text{m} \approx 2 \times 10^{-15}$ m. By comparing this with the radius of a proton or neutron, about 1×10^{-15} m, we see that inside the nucleus the nucleons are so tightly packed together that they touch or almost touch (see Figs. 40.1 and 40.2).

✔ Checkup 40.1

QUESTION 1: How many protons, how many neutrons, and how many nucleons are there in the nuclei of each of the following carbon isotopes: ^{8}C, ^{9}C, and ^{10}C?

QUESTION 2: How many protons, how many neutrons, and how many nucleons are there in the nuclei of each of the following hydrogen isotopes: ^{1}H, ^{2}H, and ^{3}H?

QUESTION 3: Which of the isotopes in Table 40.2 has the largest number of neutrons? The largest number of protons? The largest mass number?

QUESTION 4: Which of the following isotopes has the largest ratio of the number of neutrons to the number of protons?

(A) ^{2}H (B) ^{3}H (C) ^{4}He (D) ^{6}Li (E) ^{12}C

QUESTION 5: One nucleus has twice the radius of a second nucleus. What is the ratio of the number of nucleons in the first nucleus to the number in the second?

(A) 2 (B) 3 (C) 4 (D) 6 (E) 8

40.2 THE STRONG FORCE AND THE NUCLEAR BINDING ENERGY

Since the protons within a nucleus are at such short distances from one another, they exert very large repulsive electric forces on one another. For two neighboring protons, separated by a center-to-center distance of $r \approx 2.0 \times 10^{-15}$ m, Coulomb's Law gives an electric repulsive force of

$$F = \frac{1}{4\pi\epsilon_0}\frac{e^2}{r^2} = \frac{1}{4\pi\epsilon_0} \times \frac{(1.6 \times 10^{-19}\,\text{C})^2}{(2.0 \times 10^{-15}\,\text{m})^2} \tag{40.5}$$

$$= 58\ \text{N}$$

This is, roughly, four times the weight of this book; acting on a mass of only 10^{-27} kg, the magnitude of this force is colossal.

Obviously, some extra force must be present in the nucleus to prevent it from instantaneously bursting apart under the influence of the mutual electric repulsion of the protons. This extra force is the **strong force**, already mentioned in Section 6.4. *This force acts equally between any two nucleons, regardless of whether they are protons or neutrons* (the force is "charge-independent").

strong force

Figure 40.3 is a plot of the potential energy associated with the strong nucleon–nucleon force, calculated from experimental data on nuclear collisions. From Chapter 8, we know that a potential energy that increases as the nucleons separate corresponds to an attractive force. A decreasing potential energy means that the force does positive work when the nucleons separate; this corresponds to a repulsive force. We therefore see from the plot of the potential energy that the strong force is attractive over a range of internucleon distances from $\approx 0.7 \times 10^{-15}$ m to $\approx 2 \times 10^{-15}$ m. In this range of distances, the strong force is much larger than the electric force, as much as 100 times larger. The strong force is repulsive for internucleon distances less than $\approx 0.7 \times 10^{-15}$ m; this means that the nucleons have a hard core that resists interpenetration. For distances larger than $\approx 2 \times 10^{-15}$ m, the strong force decreases drastically and finally vanishes. Thus, in contrast to the electric force, which fades only gradually and reaches out to large distances, *the strong force cuts off sharply and has only a short range.* In order to feel the strong force, the nucleons must be touching or almost touching; that is, the force acts only between nearest neighbors in the nucleus.

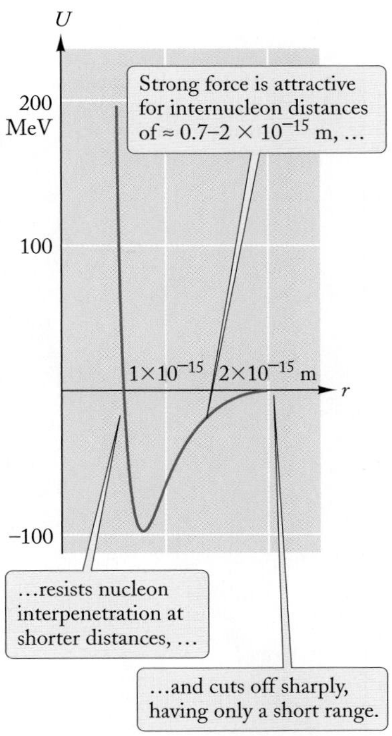

FIGURE 40.3 Potential energy for the strong force acting between two nucleons.

In consequence of the short-range character of the strong force, a nucleon deep inside the nucleus does not experience any net force—the nucleon interacts only with its nearest neighbors, and since these pull it with equal forces in almost all directions, the net force on the nucleon is zero or nearly zero (see Fig. 40.4). However, a nucleon at the nuclear surface has neighbors only on the side that is toward the interior, and hence these will exert a net force pulling the nucleon inward (see Fig. 40.4). Altogether this means that nucleons are more or less free to wander about the interior of the nucleus,

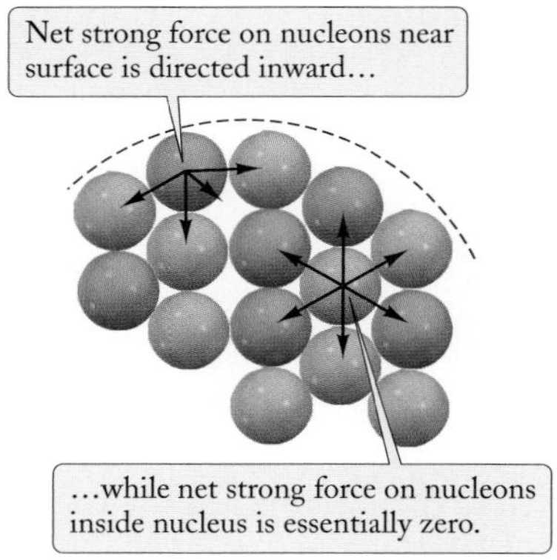

FIGURE 40.4 Strong forces on a nucleon at the nuclear surface, and strong forces on a nucleon in the nuclear interior.

but whenever they approach the nuclear surface, strong forces pull them back and prevent their escape.

This suggests that the nucleons in the nucleus behave somewhat like the molecules in a drop of water; such molecules are free to wander throughout the volume of the drop, but when they approach the water surface, intermolecular forces hold them back. This similarity between nuclei and drops of water rests on a similarity of the laws of force. The intermolecular force has general features rather similar to those displayed in Fig. 40.3; the force is attractive over a short range and then becomes strongly repulsive when the molecules begin to interpenetrate. The hard repulsive core makes the water nearly incompressible, whereas the short-range attraction provides a cohesive force that prevents water droplets from falling apart. The balance of attraction and repulsion encourages water molecules to stay at a particular distance from one another, and this gives water a particular, uniform density.

Because of the similarities between a liquid and the nuclear material, the *nucleus can be crudely regarded as a droplet of incompressible "nuclear fluid" of uniform density*. The fluid is, of course, made of nucleons, but for some purposes we can ignore the individual nucleons, and we can calculate the properties of nuclei in terms of the gross properties of a liquid. For example, the spherical shape adopted by most nuclei can easily be understood as follows: Any nucleon located on the surface of a globule of nuclear fluid experiences an inward force pulling it back into the volume, and consequently the fluid tends to shrink its exposed surface to the smallest value compatible with its (fixed) volume. Since a sphere has the least surface area for a given volume, the globule of fluid will take the shape of a spherical droplet.

In a stable nucleus, the repulsive electric forces among the protons are held in check by the attractive strong forces. To achieve this balance of forces, the presence of neutrons is an advantage: a nucleus with more neutrons will have a larger size and, therefore, a large average distance between pairs of protons—the neutrons in the nucleus dilute the repulsive effect of the electric force. Consequently, all stable nuclei, with the exception of hydrogen and one isotope of helium, contain at least as many neutrons as protons; heavy nuclei, such as uranium, contain substantially more neutrons than protons.

Figure 40.5 is a plot of the number of neutrons vs. the number of protons (N vs. Z). On this plot, blue dots indicate the stable nuclei and red dots indicate unstable nuclei, that is, radioactive isotopes. Note that there is no stable nucleus beyond bismuth ($Z = 83$). However, several elements beyond bismuth have some isotopes with very long lifetimes; these are therefore almost stable, and they occur naturally.

The energy stored in a nucleus is a sum of the potential energies contributed by the electric and the strong forces and the kinetic energies of the nucleons. The potential energy is negative, and its magnitude is larger than that of the kinetic energy. Thus, the "stored" energy is negative, and *energy is released when the nucleus is assembled out of its constituent nucleons*. Conversely, energy must be supplied to take the nucleus apart into its constituent nucleons. The energy that must be supplied to take the nucleus apart, or the energy released during the assembly of the

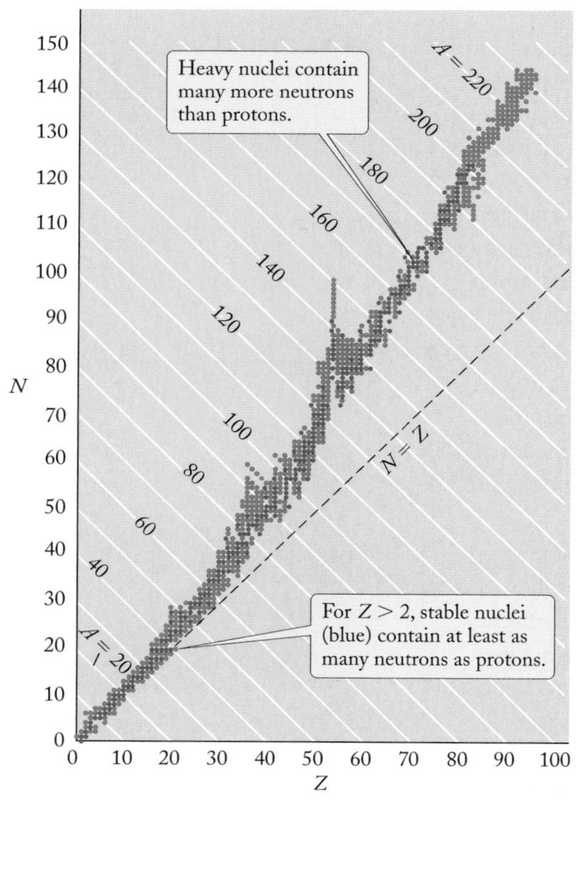

FIGURE 40.5 Number of neutrons (N) vs. number of protons (Z) for stable nuclei (blue dots) and unstable nuclei (red dots).

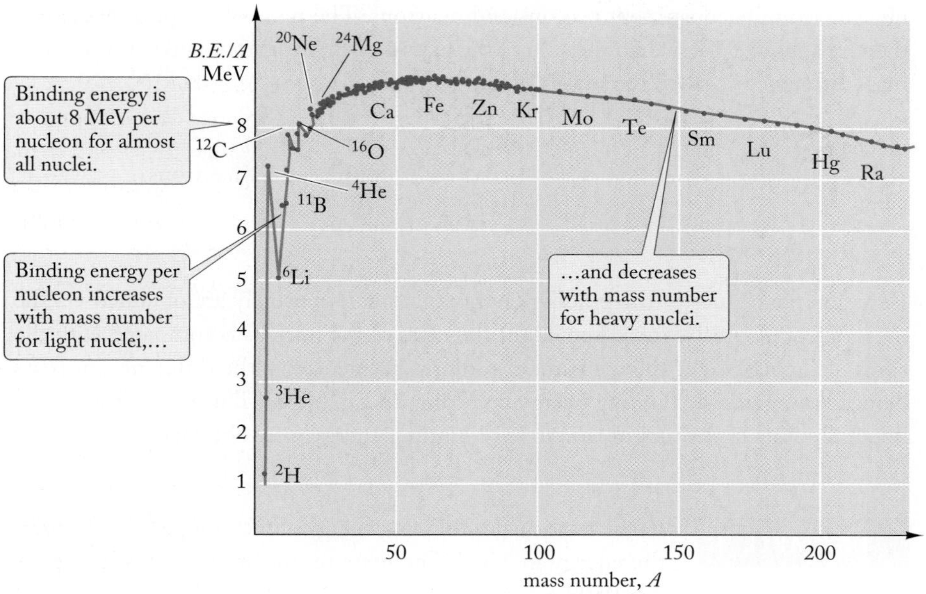

FIGURE 40.6 Average binding energy per nucleon vs. mass number.

nucleus out of its constituent nucleons, is called the **binding energy** (*B.E.*). Figure 40.6 is a plot of *B.E./A*, the binding energy divided by the number of nucleons, or the average binding energy per nucleon. The curve plotted in Fig. 40.6 is called the **curve of binding energy**. The energy unit used in this plot is the MeV, where

curve of binding energy

$$1 \text{ MeV} = 10^6 \text{ eV} = 1.6022 \times 10^{-13} \text{ J} \tag{40.6}$$

This unit is widely used in nuclear physics.

The binding energy of a typical nucleus is a rather large amount of energy. As may be seen from Fig. 40.6, the average binding energy per nucleon is in the vicinity of 8 MeV for almost all nuclei; thus a nucleus with a mass number A typically has a binding energy of about $A \times 8$ MeV. To put this number in perspective, let us compare it with the rest-mass energy of the nucleons. Each nucleon has a mass of about one atomic mass unit. The energy corresponding to one atomic mass unit is

$$1 \text{ u} \times c^2 = 1.6605 \times 10^{-27} \text{ kg} \times (2.9979 \times 10^8 \text{ m/s})^2 = 1.4924 \times 10^{-10} \text{ J}$$

or, in MeV units [see Eq. (40.6)],[2]

$$1 \text{ u} \times c^2 = 1.4924 \times 10^{-10} \text{ J} \times \frac{1 \text{ MeV}}{1.6022 \times 10^{-13} \text{ J}} = 931.5 \text{ MeV} \tag{40.7}$$

energy equivalent of atomic mass unit

Thus, each nucleon has a rest-mass energy of about 930 MeV, and the A nucleons in the nucleus have a rest-mass energy of about $A \times 930$ MeV. The ratio of binding energy to rest-mass energy is then about $8/930 \approx 0.009$, which means the binding energy is nearly 1% of the rest-mass energy!

By Einstein's mass-energy formula, the mass associated with the binding energy is $B.E./c^2$. This mass is carried away by the energy released during the assembly of the

[2]We have retained four significant figures in this result because we will be needing a precise value of the equivalence between energy and mass later on.

nucleus from its constituent protons and neutrons. The mass of a typical nucleus is therefore about 1% less than the sum of the masses of these protons and neutrons. The mass difference is called the **mass defect,**

mass defect

$$\frac{B.E.}{c^2} = [\text{mass defect}] = [\text{mass of } Z \text{ protons and } N \text{ neutrons}] - [\text{mass of nucleus}] \tag{40.8}$$

Experimental values of the binding energy of a nucleus usually are obtained via the mass defect. A precise measurement of the mass of the nucleus is compared with the sums of the masses of the constituent protons and neutrons; the difference, or mass defect, then gives the binding energy according to Eq. (40.8). This is how the experimental points in the plot of the curve of binding energy were obtained.

EXAMPLE 1 What is the nuclear binding energy of the isotope ^{238}U? Express the energy in MeV. The mass of one atom of this isotope is 238.0508 u.

SOLUTION: The atomic number of uranium is $Z = 92$ and uranium-238 has mass number $A = 238$, so this isotope has 92 protons and $N = A - Z = 146$ neutrons. According to Eq. (40.8),

$$\frac{B.E.}{c^2} = [92m_p + 146m_n] - [\text{mass of nucleus}] \tag{40.9}$$

We want to re-express this equation in terms of the mass of the atom, since we know this mass. The mass of the uranium nucleus is the mass of the uranium atom minus the mass of the 92 electrons of this atom. Hence

$$\frac{B.E.}{c^2} = [92m_p + 146m_n] - ([\text{mass of } ^{238}\text{U atom}] - 92m_e)$$

But 92 proton masses plus 92 electron masses equals 92 hydrogen masses, because a hydrogen atom consists of one proton and one electron. We therefore obtain the following convenient formula for the binding energy:

$$\frac{B.E.}{c^2} = 92[\text{mass of } ^1\text{H atom}] + 146m_n - [\text{mass of } ^{238}\text{U atom}]$$

With [mass of ^{1}H atom] = 1.007 825 u (from Table 40.2, page 1357) and $m_n = 1.008\ 66$ u, this yields

$$\frac{B.E.}{c^2} = 92 \times 1.007\ 825 \text{ u} + 146 \times 1.008\ 66 \text{ u} - 238.0508 \text{ u}$$

$$= 1.934 \text{ u}$$

The binding energy is therefore

$$B.E. = 1.934 \text{ u} \times c^2$$

Since $\text{u} \times c^2 = 931.5$ MeV [see Eq. (40.7)], the result is

$$B.E. = 1.934 \times 931.5 \text{ MeV} = 1802 \text{ MeV}$$

Besides its role as nuclear binding force, the strong force also plays a crucial role in nuclear reactions that occur when the nuclei in some target material are bombarded by a beam of protons, neutrons, or alpha particles. When such a projectile strikes the nuclear surface, the strong force pulls the projectile into the nucleus. The projectile then either remains bound within the nucleus, forming a heavier nucleus; or else the projectile disrupts the internal structure of the nucleus to such an extent that the nucleus ejects one or several fragments, that is, one or several protons, neutrons, alpha particles, or even other nuclei. Such nuclear reactions involve **transmutation of elements**: *the original nucleus is changed into a new nucleus of different mass number and atomic number*.

transmutation of elements

The first such transmutation of elements was discovered in 1919 by Rutherford when he bombarded nitrogen with alpha particles. He found that this led to the transmutation of nitrogen into oxygen, according to the reaction[3]

$$\alpha + {}^{14}\text{N} \rightarrow {}^{17}\text{O} + \text{p}$$

As we will see in Section 40.3, the alpha particle is actually the nucleus of a helium-4 atom; hence the reaction can also be written[3]

$${}^{4}\text{He} + {}^{14}\text{N} \rightarrow {}^{17}\text{O} + {}^{1}\text{H} \tag{40.10}$$

FIGURE 40.7 The Cockcroft–Walton accelerator built in 1932.

In his early experiments, Rutherford used a naturally radioactive material as the source of his beam of alpha particles. But in the 1930s, physicists began to build machines for the artificial acceleration of beams of charged particles. The first of these accelerators were electrostatic; they accelerated particles by means of strong static electric fields produced by a large amount of electric charge accumulated on a spherical capacitor (see Fig. 40.7). Many of the later accelerators use a combination of electric and magnetic fields. Thus, in the cyclotron (see Section 30.1), a uniform magnetic field holds protons in a circular orbit, while an electric field acts on them periodically, gradually increasing their energy step by step.

The investigation of nuclear reactions initiated by projectiles from accelerators led to the discovery of a multitude of new isotopes, most of them short-lived and highly radioactive. These investigations also led to a better understanding of the details of the strong force.

The energy of the projectile required to initiate a nuclear reaction or the energy released in a nuclear reaction can be calculated from the rest-mass energies of the isotopes that participate in the reaction. The energy released in the reaction is called its Q value. For a reaction with initial masses $m_1, m_2, \ldots$ and final masses $m'_1, m'_2, \ldots$, the Q value is the difference between the initial and the final sums of rest-mass energies:

$$Q = (m_1c^2 + m_2c^2 + \cdots) - (m'_1c^2 + m'_2c^2 + \cdots) \tag{40.11}$$

Q value of reaction

If the rest-mass energy after the reaction is smaller than that before, the Q value is positive, and energy is released in the reaction. Such a reaction can proceed even if the energy of the projectile is very low (nearly zero).

If the rest-mass energy after the reaction is larger than that before, the Q value is negative, and energy must be supplied to make the reaction proceed; that is, the reaction absorbs energy, instead of releasing energy. The energy required to make the reaction proceed must come from the kinetic energy of the incident projectile; thus, the

[3]These reactions as written include only nucleons, and do not account for the atomic electrons of the atoms or ions involved.

projectile needs a minimum energy to initiate the reaction. The kinetic energy of the projectile is then (partially) converted into rest-mass energy of the reaction products.

EXAMPLE 2 Calculate the Q value for the reaction (40.10). What can you say about the minimum kinetic energy of the alpha particle required to initiate this reaction?

SOLUTION: The Q value is the difference between the total rest-mass energy before the reaction and the total rest-mass energy after the reaction. The total rest-mass energy before the reaction (40.10) is the rest-mass energy of the alpha particle (or ^{4}He nucleus) plus the rest-mass energy of the ^{14}N nucleus. The total rest-mass energy after the reaction is the sum of the rest-mass energies of the ^{17}O nucleus and the proton. Thus,

$$Q = [\text{mass of } ^4\text{He nucleus}]c^2 + [\text{mass of } ^{14}\text{N nucleus}]c^2 - [\text{mass of } ^{17}\text{O nucleus}]c^2 - m_p c^2 \qquad (40.12)$$

Since the chart of isotopes lists the masses of the atoms, rather than the masses of nuclei, we want to express Eq. (40.12) in terms of atomic masses. For this purpose, we add the rest-mass energy of 18 electrons to the first two terms on the right side of Eq. (40.12), and we subtract the rest-mass energy of 18 electrons from the last two terms:

$$Q = [\text{mass of } ^4\text{He nucleus} + 4m_e]c^2 + [\text{mass of } ^{14}\text{N nucleus} + 14m_e]c^2 - [\text{mass of } ^{17}\text{O nucleus} + 17m_e]c^2 - [m_p + m_e]c^2$$

Each of the terms in brackets is now the mass of a complete atom:

$$Q = [\text{mass of } ^4\text{He atom}]c^2 + [\text{mass of } ^{14}\text{N atom}]c^2 - [\text{mass of } ^{17}\text{O atom}]c^2 - [\text{mass of } ^1\text{H atom}]c^2$$

Substituting the values of the masses listed on the chart of isotopes (Table 40.2, page 1357), we then find

$$Q = 4.002\,603 \text{ u} \times c^2 + 14.003\,074 \text{ u} \times c^2 - 16.999\,131 \text{ u} \times c^2 - 1.007\,825 \text{ u} \times c^2 = -0.001\,279 \text{ u} \times c^2$$

Since $\text{u} \times c^2 = 931.5$ MeV, we can express our final result as

$$Q = -0.001\,279 \times 931.5 \text{ MeV} = -1.191 \text{ MeV}$$

This negative Q value indicates that the sum of rest-mass energies after the reaction is larger than before, that is, the reaction absorbs energy. The absorbed energy is 1.191 MeV, and this energy must be supplied by the alpha particle. The minimum kinetic energy of the alpha particle must therefore be at least 1.191 MeV.

COMMENT: Although the minimum energy required for the reaction is 1.191 MeV, the reaction can proceed only with alpha particles of somewhat higher kinetic energy, because momentum conservation dictates that some kinetic energy must be given to the reaction products.

Checkup 40.2

QUESTION 1: By inspection of Fig. 40.3, determine where the strong force has its largest attractive value.

QUESTION 2: In Example 1 we found that the binding energy of ^{238}U is 1802 MeV. Accordingly, what is the value of $B.E./A$ for this isotope?

QUESTION 3: Can a stable nucleus have a negative mass defect, that is, a mass excess?

QUESTION 4: Consider the reverse of the reaction (40.10), that is, bombardment of ^{17}O by protons to form helium and nitrogen, $^{1}H + {}^{17}O \rightarrow {}^{4}He + {}^{14}N$. What is the Q value of this reaction? What can you say about the minimum kinetic energy of the proton required to initiate this reaction? Ignore the recoil of the oxygen nucleus.

QUESTION 5: According to Fig. 40.6, which of the nuclei named in this figure has the largest binding energy per nucleon (largest $B.E./A$)? Which has the largest binding energy (largest $B.E.$)?

(A) ^{2}H, Fe (B) ^{2}H, ^{4}He (C) ^{4}He, Ra (D) Fe, Ra (E) Fe, ^{4}He

40.3 RADIOACTIVITY

Online Concept Tutorial 45

Concepts in Context

Radioactivity was discovered in 1896 by Henri Becquerel. Through an accident, he noticed that samples of uranium minerals emitted invisible rays which could penetrate through sheets of opaque materials and make an imprint on a photographic plate. Subsequent investigations established that uranium, and many other radioactive substances, emit three kinds of rays: **alpha rays** (α), **beta rays** (β), and **gamma rays** (γ). Of these, the alpha rays are the least penetrating; they can be stopped by a thick piece of paper. The beta rays are more penetrating; they can pass through a foil of lead or a plate of aluminum. The gamma rays are the most penetrating; they can pass through a thick wall of concrete. When these three kinds of rays are aimed into a magnetic field, the alpha and beta rays are deflected in opposite directions, whereas the gamma rays proceed without deflection (see Fig. 40.8). This simple experiment demonstrates that alpha rays and beta rays are electrically charged, with charges of opposite signs, whereas gamma rays are neutral.

alpha rays (α), beta rays (β), and gamma rays (γ)

By more detailed experiments. Becquerel identified the beta rays as high-speed electrons. Some years later, Rutherford investigated the nature of the alpha rays and demonstrated that they are identical to nuclei of helium. Thus, electrons and helium nuclei are, somehow, manufactured within the sample of uranium or other radioactive substance, and they are ejected at high speed. Gamma rays are high-energy photons emitted by the radioactive substance; the energy of these gamma-ray photons is typically a thousand times as large as that of X-ray photons.

The "manufacture" and ejection of these high-speed electrons and helium nuclei occurs in the nuclei of the radioactive substance, by nuclear decay reactions. In the following, we will discuss the broad features of these nuclear reactions.

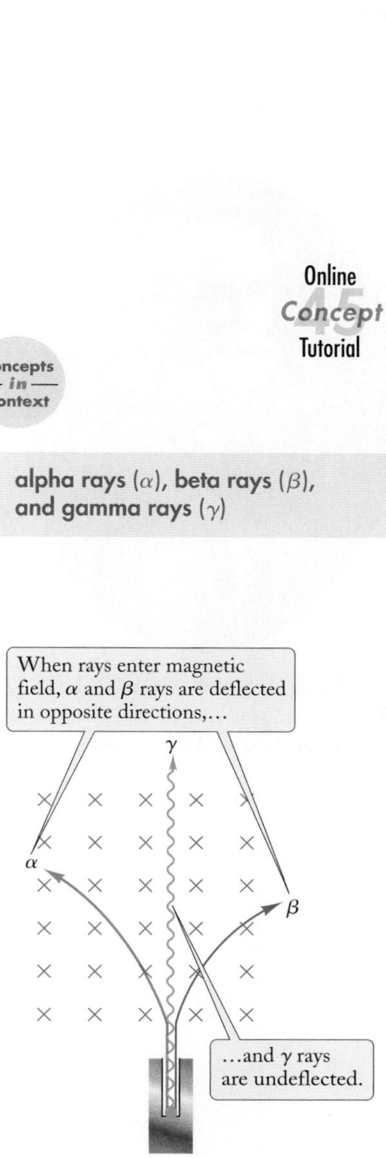

FIGURE 40.8 Alpha (α), beta (β), and gamma (γ) rays emitted by a radioactive source. A magnetic field is perpendicular to the plane of the page.

Alpha Decay

The alpha particle consists of two protons and two neutrons; it has the same structure as the helium-4 nucleus. When a radioactive nucleus ejects an alpha particle consisting of two protons and two neutrons, the radioactive nucleus does not create these protons and neutrons out of nothing—the nucleus merely takes two of its own protons and neutrons and spits them out. Since the nucleus loses two protons and neutrons, both the atomic number and the mass number of the nucleus decrease. The atomic number decreases by 2, and the mass number by 4. Thus, the original isotope becomes a different isotope of a different chemical element. The following examples of alpha decays illustrate such transmutations of chemical elements:

$$^{238}\mathrm{U} \rightarrow {}^{234}\mathrm{Th} + \alpha \tag{40.13}$$

$$^{226}\mathrm{Ra} \rightarrow {}^{222}\mathrm{Rn} + \alpha \tag{40.14}$$

In the first of these alpha-decay reactions, uranium is transmuted into thorium, and in the second, radium into radon. The original nucleus in a decay reaction is called the **parent**, and the resulting nucleus is called the **daughter**.

parent nucleus and daughter nucleus

fission

The alpha-decay reaction can be regarded as a **fission**, *or splitting, of the nucleus into two smaller nuclei.* The, say, uranium nucleus fissions into a thorium nucleus and a helium nucleus. The total number of protons and of neutrons is unchanged in this fission (see Fig. 40.9). The fission occurs spontaneously, because of an instability in the original nucleus. Large nuclei, such as uranium or radium, contain many protons, which exert repulsive electric forces on each other. Although the repulsive electric force is balanced by the attractive strong force, the balance of these forces is rather precarious because the electric force easily reaches from one end of the nucleus to the other, whereas the strong force acts only between adjacent nucleons and therefore, in a large nucleus, cannot reach directly from one end to the other. Thus, any accidental, spontaneous elongation of the nucleus can shift the balance in favor of the electric force—the nucleus then elongates more and more, and ultimately bursts apart into two fragments. In true fission, the two fragments are of approximately equal size. In alpha decay, we are dealing with an extreme case of fission, with two fragments of very unequal size. The ejection of an alpha particle is strongly favored over ejection of, say, a hydrogen nucleus or a lithium nucleus, because the helium nucleus is an exceptionally tightly bound nucleus. In consequence of this large binding energy, the formation of an alpha particle, just before its ejection, makes more energy available for driving the fission reaction.

All the large, heavy nuclei beyond bismuth are unstable; they all are subject to alpha decay or other forms of spontaneous fission. Besides, many isotopes of somewhat smaller nuclei suffer from the same instability.

ANTOINE HENRI BECQUEREL (1852–1908) *French physicist. He was awarded the 1903 Nobel Prize for his discovery of radioactivity.*

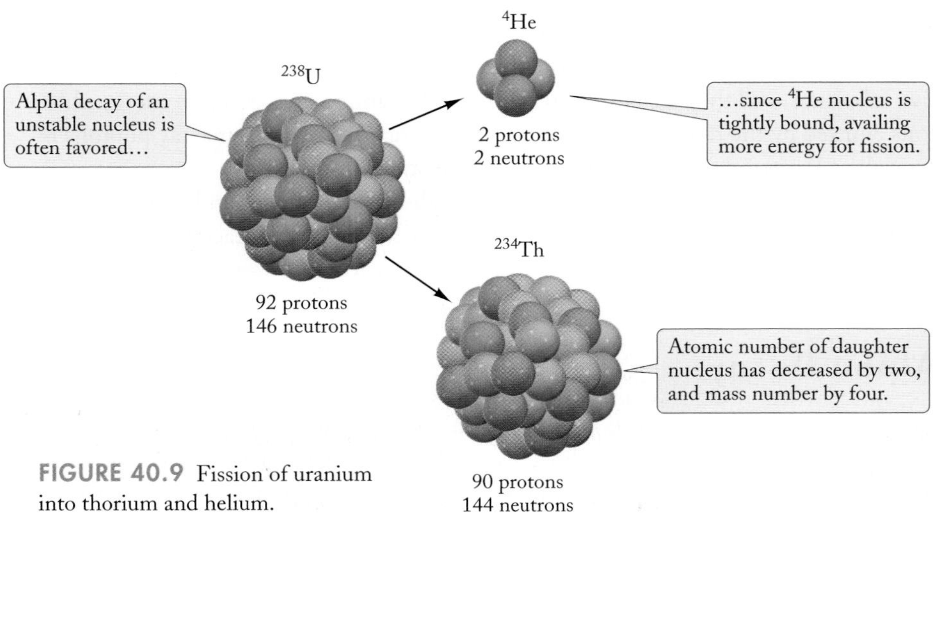

FIGURE 40.9 Fission of uranium into thorium and helium.

The kinetic energy of the alpha particles ejected in reactions such as (40.13) and (40.14) can be calculated from the masses of the participating nuclei and the mass of the alpha particle.

EXAMPLE 3 Calculate the energy released in the alpha decay of ^{238}U, and calculate the kinetic energy of the alpha particle ejected in this reaction. The atomic masses of the isotopes of uranium, thorium, and helium in this reaction are 238.0508 u, 234.0436 u, and 4.0026 u, respectively.

Concepts in Context

SOLUTION: The energy released in the reaction (40.13) is simply the difference between the total rest-mass energy before the reaction and the (smaller) total rest-mass energy after the reaction. The total rest-mass energy before the reaction is the rest-mass energy of the uranium nucleus, [mass of U nucleus]c^2. The total rest-mass energy after the reaction is the sum of the rest-mass energies of the thorium nucleus and the alpha particle, [mass of Th nucleus]c^2 + [mass of He nucleus]c^2. The energy released in the reaction is therefore

$$Q = [\text{mass of } ^{238}\text{U nucleus}]c^2 - [\text{mass of } ^{234}\text{Th nucleus}]c^2 - [\text{mass of } ^{4}\text{He nucleus}]c^2 \quad (40.15)$$

To express this in terms of the masses of the atoms rather than the masses of the nuclei, we add the rest-mass energy of 92 electrons to the first term on the right side of Eq. (40.15) and we subtract the same amount from the other two terms:

$$Q = [\text{mass of } ^{238}\text{U nucleus} + 92m_e]c^2 - [\text{mass of } ^{234}\text{Th nucleus} + 90m_e]c^2 - [\text{mass of } ^{4}\text{He nucleus} + 2m_e]c^2 \quad (40.16)$$

Now each of the terms in square brackets is the mass of a complete atom:

$$Q = [\text{mass of } ^{238}\text{U atom}]c^2 - [\text{mass of } ^{234}\text{Th atom}]c^2 - [\text{mass of } ^{4}\text{He atom}]c^2 \quad (40.17)$$

We know that these masses are 238.0508 u, 234.0436 u, and 4.0026 u, respectively; hence,

$$Q = 238.0508 \text{ u} \times c^2 - 234.0436 \text{ u} \times c^2 - 4.0026 \text{ u} \times c^2 = 0.0046 \text{ u} \times c^2$$

or, with u $\times\, c^2$ = 931.5 MeV,

$$Q = 0.0046 \times 931.5 \text{ MeV} = 4.3 \text{ MeV}$$

Since the thorium nucleus is much heavier than the alpha particle, it suffers almost no recoil, and the alpha particle carries away almost all of the energy released in the reaction. Thus, the kinetic energy of the alpha particle will be 4.3 MeV.

In many cases of alpha decay, the daughter nucleus is also unstable, and decays some time after its formation, either by alpha decay or by beta decay. The daughter of the daughter then decays, and so on. *The sequence of daughters descending from the original parent is called a* **radioactive series**. The series ends when it reaches a stable iso-

radioactive series

tope. For instance, the decay of uranium initiates a radioactive series, which ultimately ends with a stable isotope of lead.

Beta Decay

The simplest beta-decay reaction is the decay of the neutron. The free neutron is unstable, and it decays into a proton, an electron, and an antineutrino:

$$n \to p + e^- + \overline{\nu} \tag{40.18}$$

neutrinos (ν) and antineutrinos ($\overline{\nu}$)

In this equation, the symbol $\overline{\nu}$ (nu-bar) represents the antineutrino. **Neutrinos** (ν) and **antineutrinos** ($\overline{\nu}$) are particles of very small mass and spin quantum number $\frac{1}{2}$, which travel at nearly the speed of light. Thus, they are somewhat similar to photons. However, in contrast to photons, which interact with electric charges, neutrinos and antineutrinos do not interact directly with electric charges—in fact, they hardly interact with anything at all, and they pass through the entire bulk of the Earth with little hindrance. As their names indicate, neutrinos and antineutrinos are *antiparticles* of each other; they can annihilate each other, producing a flash of light (gamma rays).

For a free neutron, the decay reaction (40.18), on the average, takes a time of 15 minutes.[4] For a neutron in a nucleus, the reaction may proceed at a faster rate or at a slower rate, depending on whether the nucleus promotes the reaction by supplying extra energy, or inhibits the reaction by withdrawing energy. When the reaction occurs in a nucleus, the electron and antineutrino are ejected, and the net effect is the conversion of a neutron into a proton, which increases the atomic number of the nucleus by 1, while leaving the mass number unchanged. Thus, the original isotope is transmuted into an isotope of the next chemical element. Two examples of beta decays are the following:

$$^{60}\text{Co} \to {}^{60}\text{Ni} + e^- + \overline{\nu} \tag{40.19}$$

$$^{14}\text{C} \to {}^{14}\text{N} + e^- + \overline{\nu} \tag{40.20}$$

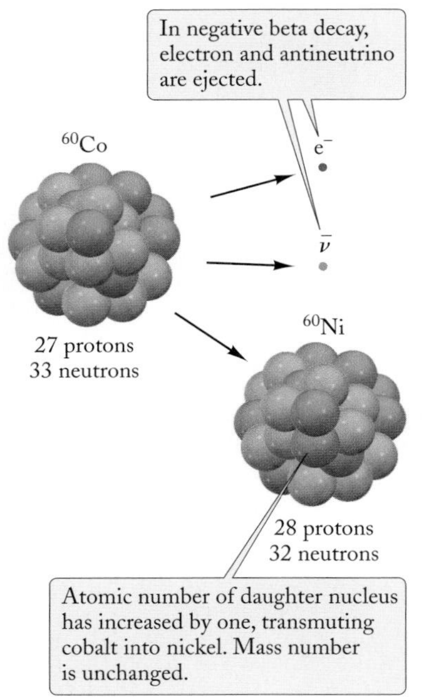

FIGURE 40.10 Beta decay of ^{60}Co into ^{60}Ni. The number of neutrons is 33 before the decay and 32 after; the number of protons is 27 before the decay and 28 after.

In each of these reactions, the number of neutrons (N) decreases by 1, and the number of protons (Z) increases by 1 (see Figs. 40.10 and 40.11).

The antineutrino emitted in these reactions is almost impossible to detect. Its existence, however, can be inferred from the conservation of energy. In the beta decay of, say, radioactive cobalt-60 [Eq. (40.19)], sometimes the electron emerges with one energy, and sometimes with another. This is because the energy of the decay is shared between the electron and the neutrino, and sometimes one of these particles carries off most of the energy, sometimes the other. Hence the energy of the electron is unpredictable—it can be anything from zero up to a maximum amount that corresponds to the electron carrying away *all* the energy of the decay. In fact, it was on the basis of this variability of the energy of the ejected electron that Pauli first proposed the existence of the neutrino, since this was the only way to preserve the Law of Conservation of Energy. When Pauli made this proposal in 1931, there was no direct evidence for the existence of an extra particle; neutrinos and antineutrinos were detected only much later, in experiments with nuclear reactors.

Beta-decay reactions of the type (40.19) and (40.20) involve the conversion of a neutron into a proton and the ejection of an electron and an antineutrino and occur for many nuclei that have *more* neutrons than the stable isotopes of a given element. In nuclei that have *fewer* neutrons than the stable isotopes, there are often beta-decay

[4] The average lifetime of free neutrons is 15.0 min, but the half-life (see next section) is 10.6 min.

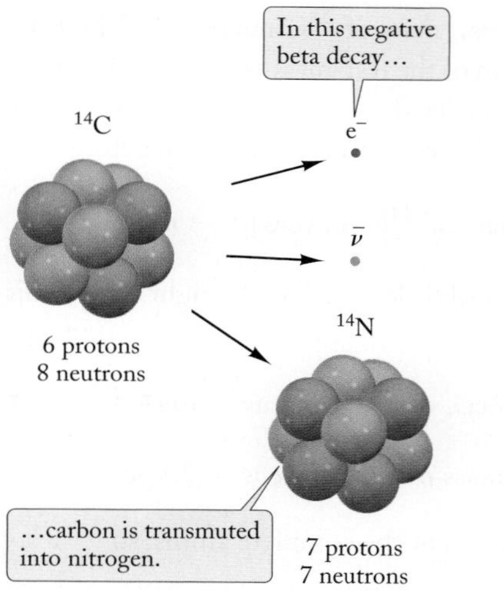

FIGURE 40.11 Beta decay of ^{14}C into ^{14}N with ejection of an electron and an antineutrino. The number of neutrons is 8 before the decay and 7 after; the number of protons is 6 before the decay and 7 after.

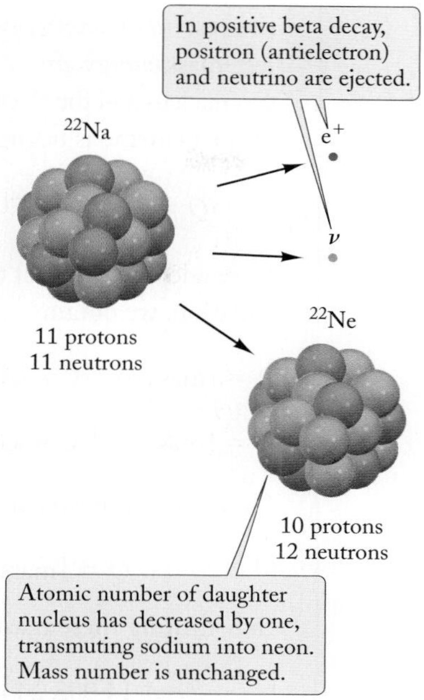

FIGURE 40.12 Beta decay of ^{22}Na into ^{22}Ne, with ejection of an antielectron and a neutrino. The number of neutrons is 11 before the decay and 12 after; the number of protons is 11 before the decay and 10 after.

reactions involving the conversion of a proton into a neutron and the ejection of an antielectron, or **positron**, and a neutrino. The positron is the antiparticle of the electron; it has the same mass and spin as the electron, but the opposite electric charge. An example of a beta-decay reaction with ejection of a positron is

positron (e^+)

$$^{22}\text{Na} \rightarrow {}^{22}\text{Ne} + e^+ + \nu \tag{40.21}$$

Here the symbol e^+ represents the positron. After the positron is ejected in this decay, it sooner or later collides with one of the abundant atomic electrons in the surrounding environment and annihilates with it, producing gamma rays. In the reaction (40.21), the number of protons (Z) decreases by 1, and the number of neutrons (N) increases by 1 (see Fig. 40.12).

"weak" force

All the beta-decay reactions are instigated by a new kind of force, the **"weak" force**. This is one of the four fundamental forces found in matter (see Sections 6.4 and 41.3). Whereas the electromagnetic and the strong forces play an important role in determining the structure of atoms and of nuclei, the weak force plays no role in this. Its only effect is to engender the beta-decay reactions (40.19)–(40.21) and other similar reactions.

EXAMPLE 4 What is the maximum kinetic energy of the beta rays emitted in the beta decay of ^{14}C? The atomic masses of ^{14}C and of ^{14}N are 14.003 24 u and 14.003 07 u, respectively.

SOLUTION: The energy released in the reaction (40.20) is, again, the difference between the total rest-mass energy before the reaction and the (smaller) total rest-mass energy after the reaction. The total rest-mass energy before the reaction is the rest-mass energy of the carbon nucleus, [mass of ^{14}C nucleus]c^2. The total rest-mass energy after the reaction is the sum of the rest-mass energies of the nitrogen nucleus and the electron, [mass of ^{14}N nucleus]$c^2 + m_e c^2$ (the rest-mass energy of the neutrino is negligible). The energy released in the reaction is therefore

$$Q = [\text{mass of } ^{14}\text{C nucleus}]c^2 - [\text{mass of } ^{14}\text{N nucleus}]c^2 - m_e c^2$$

If we add and subtract the rest-mass energy of 6 electrons on the right side of this equation, we obtain

$$Q = [\text{mass of } ^{14}\text{C nucleus} + 6m_e]c^2 - [\text{mass of } ^{14}\text{N nucleus} + 6m_e]c^2 - m_e c^2$$

$$= [\text{mass of } ^{14}\text{C nucleus} + 6m_e]c^2 - [\text{mass of } ^{14}\text{N nucleus} + 7m_e]c^2$$

Here the two terms in brackets are the masses of the complete atoms, so

$$Q = [\text{mass of } ^{14}\text{C atom}]c^2 - [\text{mass of } ^{14}\text{N atom}]c^2$$

Substituting these masses and substituting $\text{u} \times c^2 = 931.5$ MeV, we find

$$Q = 14.003\,24\text{ u} \times c^2 - 14.003\,07\text{ u} \times c^2 = 0.000\,17\text{ u} \times c^2$$

$$= 0.000\,17 \times 931.5\text{ MeV} = 0.16\text{ MeV}$$

The nitrogen nucleus, being much heavier than the electron, remains at rest, or nearly at rest. The decay energy is therefore shared between the electron and the antineutrino. If the antineutrino carries away almost none of the energy, then the electron will acquire nearly all the energy. Thus, the maximum possible electron kinetic energy is 0.16 MeV.

Gamma Emission

Gamma rays are high-energy photons emitted by nuclei when nucleons make transitions from one stationary nuclear state to another. The emission of a gamma ray by a transition of a nucleon is similar to the emission of visible photons or of X rays by a transition of an atomic electron.

When a nucleus suffers alpha decay or beta decay, it is often left in an excited state, and it then eliminates the excitation energy in the form of a gamma ray. Thus, gamma emission is usually one step in a two-step process, with an alpha or beta decay preceding the gamma emission. For example, the beta decay of ^{60}Co leads to subsequent gamma emission according to the sequence of reactions

$$\begin{array}{l} ^{60}\text{Co} \rightarrow {}^{60}\text{Ni} + e^- + \bar{\nu} \\ \qquad\quad \searrow \\ \qquad\qquad {}^{60}\text{Ni} + \gamma \end{array} \qquad (40.22)$$

The isotope ^{60}Co is commonly employed in high-intensity industrial irradiation cells, called cobalt cells. But, as shown in Eq. (40.22), the emitter of the gamma rays is nickel, not cobalt.

As mentioned above, gamma rays are also generated when an electron annihilates with an antielectron. But this is a secondary process, which does not directly involve the nucleus.

PROBLEM-SOLVING TECHNIQUES — NUCLEAR REACTIONS

As illustrated in Examples 2, 3, and 4, the Q value of a nuclear reaction is best calculated by adding electron masses to the nuclear masses, so as to form masses of atoms. The atomic masses can then be copied from charts or tables of isotopes (which always list atomic masses, rather than nuclear masses). However, it is necessary to take some extra care with this technique in the case of beta decays with antielectrons, such as the beta decay in Eq. (40.21). On the left side of Eq. (40.21), the nucleus is sodium, which requires 11 electrons to be a complete sodium atom. On the right side the nucleus is neon, which requires only 10 electrons to become a complete neon atom. Hence, if we add 11 electron masses to the nuclear masses on each side, we will be left with one surplus electron mass on the right side. Furthermore, the right side has an antielectron mass (equal to the electron mass). Thus, the Q value for the reaction (40.21) is

$$Q = [\text{mass of } {}_{22}\text{Na atom}]c^2 - [\text{mass of } {}_{22}\text{Ne atom}]c^2 - 2m_e c^2$$

EXAMPLE 5 Figure 40.13 shows the energy-level diagram for the ^{60}Ni nucleus. When this nucleus is formed in the beta decay of ^{60}Co it is initially in the second excited state (2.50 MeV above the ground state). It makes a transition to the first excited state (1.33 MeV above the ground state), and then a transition to the ground state. What are the energies of the gamma rays emitted in these transitions?

Concepts in Context

SOLUTION: The first transition is from the energy level $E_2 = 2.50$ MeV to the energy level $E_1 = 1.33$ MeV. The energy of the gamma ray emitted in this transition is

$$E = E_2 - E_1 = 2.50 \text{ MeV} - 1.33 \text{ MeV} = 1.17 \text{ MeV}$$

The second transition is from the energy level $E_1 = 1.33$ MeV to the ground state, at $E_0 = 0$. The energy of the gamma ray emitted in this transition is

$$E = E_1 - E_0 = 1.33 \text{ MeV}$$

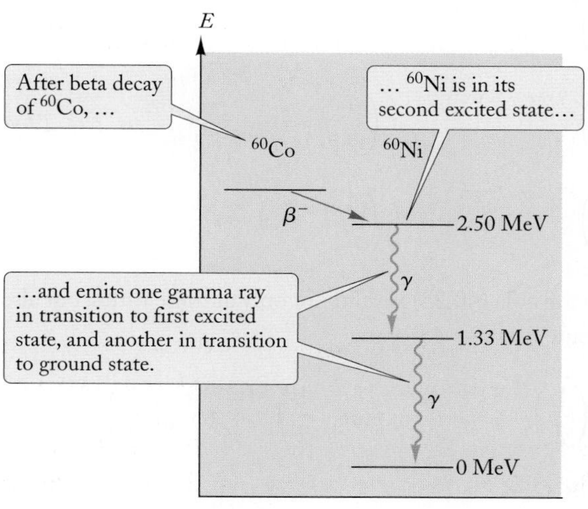

FIGURE 40.13 Energy-level diagram for the ^{60}Ni nucleus formed in the beta decay of ^{60}Co.

Checkup 40.3

QUESTION 1: Why do alpha and beta emissions involve transmutation of elements whereas gamma emission does not?

QUESTION 2: Figure 40.8 shows the deflections of alpha and beta rays in a magnetic field. If instead of consisting of an electron, the beta ray consisted of an antielectron, what would be its direction of deflection?

QUESTION 3: Can a nuclear decay reaction have a negative Q value?

QUESTION 4: An electron is emitted during the beta decay of a cesium-136 nucleus (^{136}Cs). Which of the following nuclei results?

(A) ^{135}Xe (B) ^{136}Xe (C) ^{137}Cs (D) ^{136}Ba (E) ^{137}Ba

Online Concept Tutorial 45

40.4 THE LAW OF RADIOACTIVE DECAY

In radioactive alpha or beta decay, the original isotope, or parent, becomes transmuted into another isotope, or daughter. If we have some given initial amount of parent material, say, 1 gram of radioactive strontium, some nuclei decay at one time, some at another, and the parent material disappears only gradually. In the case of radioactive strontium, one-half of the original amount disappears in 29 years, one-half of the remainder in the next 29 years, one-half of the new remainder in the next 29 years, and so on. Hence the amounts left at times 0, 29, 58, 87 years, etc., are 1, $\frac{1}{2}$, $\frac{1}{4}$, $\frac{1}{8}$ gram, etc. The time interval of 29 years is called the **half-life**, or $t_{1/2}$, of strontium.

half-life $t_{1/2}$

Mathematically, we can represent the amount of strontium at different times by the formula

$$n = n_0\left(\frac{1}{2}\right)^{t/t_{1/2}} \tag{40.23}$$

Here n is the number of strontium nuclei at time t and n_0 is the number at the initial time, $t = 0$. With $t_{1/2} = 29$ years, we can easily verify that at $t = 0$, 29 years, 58 years, etc., the formula (40.23) yields the expected results:

$$n = n_0\left(\frac{1}{2}\right)^0 = n_0 \times 1 = n_0 \qquad \text{at } t = 0$$

$$n = n_0\left(\frac{1}{2}\right)^{29\text{ years}/29\text{ years}} = n_0\left(\frac{1}{2}\right)^1 = n_0 \times \frac{1}{2} \qquad \text{at } t = 29 \text{ years}$$

$$n = n_0\left(\frac{1}{2}\right)^{58\text{ years}/29\text{ years}} = n_0\left(\frac{1}{2}\right)^2 = n_0 \times \frac{1}{4} \qquad \text{at } t = 58 \text{ years}$$

and so on. However, the formula (40.23) is valid not only at these times, but also at intermediate times. For instance, at $t = 45$ years,

$$n = n_0\left(\frac{1}{2}\right)^{45\text{ years}/29\text{ years}} = n_0\left(\frac{1}{2}\right)^{45/29} = n_0\left(\frac{1}{2}\right)^{1.55}$$

$$= n_0 \times 0.34$$

Figure 40.14 is a plot of the number of remaining strontium nuclei vs. time.

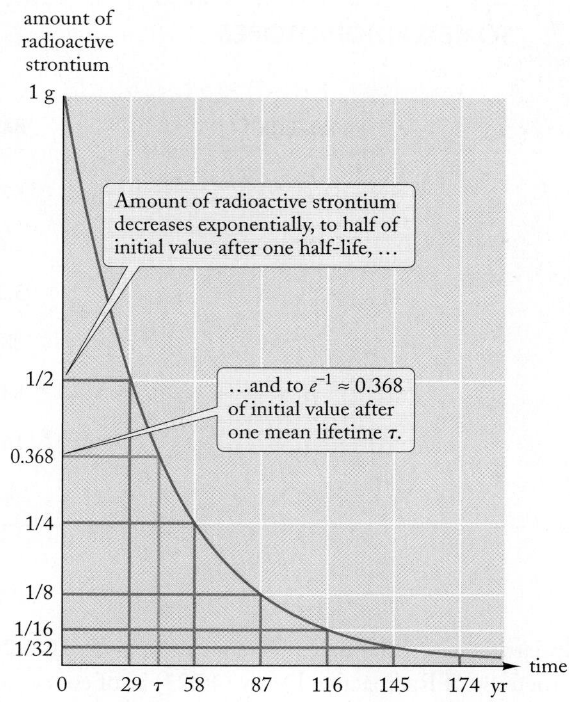

FIGURE 40.14 Amount of remaining radioactive strontium vs. time. In this plot, the amount of strontium is measured in grams. The initial amount is 1 gram, which corresponds to 6.70×10^{21} nuclei. After 29 years the remaining amount is $\frac{1}{2}$ gram, which corresponds to 3.35×10^{21} nuclei.

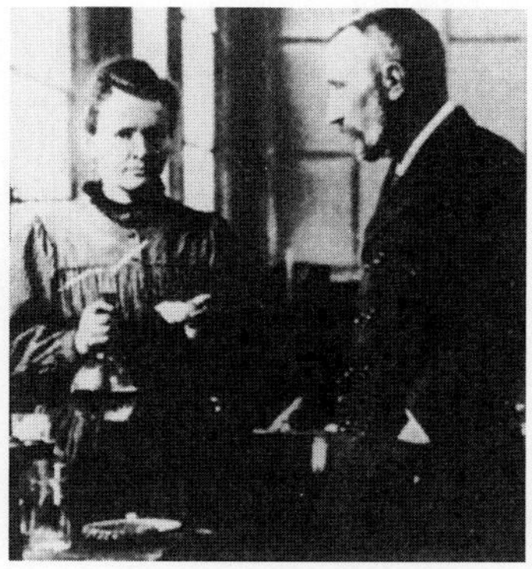

MARIE SKLODOWSKA CURIE (1867–1934) *and* **PIERRE CURIE** (1859–1906) *French physicists and chemists. For their discovery of the radioactive elements radium and polonium, they shared the 1903 Nobel Prize with Henri Becquerel. On the death of Pierre, Marie succeeded him in his professorship at the Sorbonne. She received a second Nobel Prize, in chemistry, for further work on radium. Her daughter Irène Joliot-Curie, shared the 1935 Nobel Prize with Frédéric Joliot-Curie for their work on production of radioactive substances by bombardment with alpha particles.*

We can conveniently express Eq. (40.23) in terms of the exponential function. Since $\frac{1}{2} = e^{\ln 1/2} = e^{-\ln 2}$ (see Math Help: The Exponential Function, page 910 in Chapter 28), we can rewrite Eq. (40.23) as

$$n = n_0(e^{-\ln 2})^{t/t_{1/2}} = n_0 e^{-t(\ln 2)/t_{1/2}}$$

If we adopt the notation

$$\tau = \frac{t_{1/2}}{\ln 2} = \frac{t_{1/2}}{0.693} \qquad (40.24)$$

this becomes

$$n = n_0 e^{-t/\tau} \qquad (40.25)$$

Law of Radioactive Decay

Equation (40.25) is called the **Law of Radioactive Decay**, and τ is called the **mean lifetime**, or the **time constant**. The inverse of τ is often called the **decay constant** $\lambda = 1/\tau$. It can be shown that τ is indeed the average lifetime of all the radioactive nuclei in the initial amount of parent material. Note that the mean lifetime is longer than the half-life; for instance, for strontium, the half-life is 29 years, but the mean lifetime is $\tau = (29 \text{ years})/0.693 = 42$ years. After one time constant, when $t = \tau$, Eq. (40.25) tells us that the number of strontium nuclei has decreased to a factor of $e^{-1} \approx 0.368$ times the original number (see Fig. 40.14).

mean lifetime τ

decay constant $\lambda = 1/\tau$

TABLE 40.3 SOME RADIOISOTOPES

RADIOISOTOPE	RADIOACTIVITY	HALF-LIFE
^{14}C	β	5730 years
^{22}Na	β, γ	2.6 years
^{60}Co	β, γ	5.27 years
^{90}Sr	β, γ	28.8 years
^{131}I	β, γ	8.04 days
^{226}Ra	α, γ	1600 years
^{238}U	α, γ	4.5×10^9 years

With appropriate values of the half-life—and thus, by Eq. (40.24), of the decay time constant—the Law of Radioactive Decay (40.25) is, of course, applicable to the decay of any radioactive isotope. The lifetimes of different radioisotopes vary over a wide range. Some have half-lives of several billions of years; others have half-lives as short as a fraction of a second. Table 40.3 lists the half-lives of some radioisotopes. The chart of isotopes in Table 40.2 lists some more half-lives.

decay rate

Since each decay produces one alpha or beta ray, the rate of emission of rays by the parent material is equal to the **decay rate**, or the number of parent nuclei that decay per second. To find the instantaneous decay rate we differentiate n with respect to time, which gives us the instantaneous rate of change of the number of nuclei:

$$\frac{dn}{dt} = \frac{d}{dt} n_0 e^{-t/\tau} = -\frac{n_0}{\tau} e^{-t/\tau} = -\frac{n}{\tau} \tag{40.26}$$

The negative sign of this rate of change indicates that the number of nuclei is decreasing. The number of nuclei that decay per second is the negative of the rate of change dn/dt:

decay rate in terms of mean lifetime

$$-\frac{dn}{dt} = \frac{n}{\tau} \tag{40.27}$$

Thus, the decay rate is directly proportional to the (instantaneous) number of parent nuclei and inversely proportional to the mean lifetime. This means that the decay rate is large if we have a large sample of radioactive material, and it is large if the mean lifetime is short. Equation (40.27) expresses a characteristic property of many random decay processes: the negative of the rate of change of a quantity is proportional to the quantity. It is straightforward to show by direct integration of (40.27) that for such a proportionality, the exponential decay of Eq. (40.25) follows.

By substituting the definition (40.24) into Eq. (40.27), we can express the decay rate in terms of the half-life:

decay rate in terms of half-life

$$-\frac{dn}{dt} = 0.693\frac{n}{t_{1/2}} \tag{40.28}$$

Since tables of radioactive isotopes list half-lives, the form (40.28) is convenient for calculations.

The radioactive decay rate of a sample, or the number of disintegrations per second, is usually called the **activity** of the sample. The SI unit for the radioactive decay rate, or the activity, is the **becquerel** (Bq).

activity

$$1 \text{ becquerel} = 1 \text{ Bq} = 1 \text{ disintegration/s} \tag{40.29}$$

becquerel (Bq)

In practice, a larger unit is traditionally preferred; this larger unit is the **curie**

$$1 \text{ curie} = 1 \text{ Ci} = 3.7 \times 10^{10} \text{ disintegrations/s} \tag{40.30}$$

curie (Ci)

EXAMPLE 6 What is the decay rate of 1.00 gram of radioactive strontium, ^{90}Sr? The atomic mass of this strontium isotope is 89.9 g.

SOLUTION: Since the atomic mass of this strontium isotope is 89.9 g, the number of moles in 1.00 gram of strontium is 1/89.9 mole, or $(1/89.9) \times 6.02 \times 10^{23}$ atoms $= 6.70 \times 10^{21}$ atoms. With $t_{1/2} = 28.8$ years from Table 40.3, Eq. (40.28) then tells us that the decay rate is

$$-\frac{dn}{dt} = 0.693\frac{n}{t_{1/2}} = 0.693 \times \frac{6.70 \times 10^{21}}{28.8 \text{ years}}$$

$$= 0.693 \times \frac{6.70 \times 10^{21}}{28.8 \times 3.16 \times 10^{7} \text{ s}}$$

$$= 5.10 \times 10^{12} \text{ s}^{-1} = 5.10 \times 10^{12} \text{ Bq}$$

Thus, one gram of strontium-90 emits about 5 trillion beta rays per second!

The effects of penetrating radiation on a material depend on the amount of energy absorbed over time. The amount of absorbed radiation energy per kilogram of material is called the **absorbed dose**. The SI unit of absorbed dose is the **gray** (Gy), where 1 Gy = 1 J/kg. A unit in common use is the **rad**,

absorbed dose

$$1 \text{ rad} = 0.01 \text{ J/kg} = 0.01 \text{ Gy}$$

gray (Gy) **and rad**

The damage that radiation inflicts on biological tissue depends on the kind of radiation; for example, an alpha particle damages many molecules in each cell in its path, whereas gamma rays damage fewer molecules per cell. To obtain a measure of the biological damage, the absorbed dose is multiplied by a correction factor, the **relative biological effectiveness** (RBE). The RBE for alpha particles is in the range 10–20, for beta particles it is 1–2, and for gamma rays and X rays it is 1. The measure of the biologically equivalent absorbed dose is the **rem**[5], where

relative biological effectiveness (RBE)

$$[\text{equivalent dose in rem}] = [\text{dose in rad}] \times \text{RBE}$$

equivalent dose in rem

The SI unit of biological equivalent dose is the **sievert** (Sv), where 1 Sv = 100 rem[5].

sievert (Sv)

[5]An older measure of ionizing radiation is the *roentgen*, defined as the amount of radiation that would ionize a charge of 3.34×10^{-10} C in 1 cm^3 of air at standard temperature and pressure. The *rem* is an acronym for roentgen equivalent in man.

PHYSICS IN PRACTICE RADIOACTIVE DATING

Atmospheric carbon dioxide contains the stable isotopes ^{12}C and ^{13}C, and also the radioactive isotope ^{14}C, with a half-life of 5730 years (the ^{14}C isotope is produced by cosmic rays striking the upper atmosphere). While a plant or animal lives, it absorbs carbon dioxide from the air and therefore maintains the same relative abundance of the isotopes ^{12}C and ^{14}C as in air. When the organism dies, it ceases to absorb carbon, and then the ^{14}C in its body decreases due to beta decay [Eq. (40.25)] according to the Law of Radioactive Decay. Thus, fresh wood or bones have the same abundance of ^{14}C relative to ^{12}C as air, but ancient wood or bones have less ^{14}C.

Archeologists take advantage of this decrease of ^{14}C for the **radioactive dating** of materials found in ancient tombs or other sites. They determine the abundance of carbon (all isotopes) in a sample of material by chemical analysis, and they determine the abundance of ^{14}C by measuring the beta activity of the sample. From this, the age can be calculated. For instance, samples of bone taken from the cadaver of the "iceman" (see Fig. 1) discovered in 1991 in the Alps near the Italian-Austrian border have about half the abundance of ^{14}C relative to ^{12}C as air. We can then conclude that about one half of the "normal" amount of ^{14}C has decayed, and that the age of the sample must be about one half-life, that is, about 5700 years (accurate analysis of the ^{14}C abundance gives 5300 years). This means that the iceman died in the late Stone Age. Radioactive dating by carbon can be employed to establish reliable ages for samples as old as about 60 000 years. In older samples the residue of ^{14}C is too small for accurate measurements of beta activity.

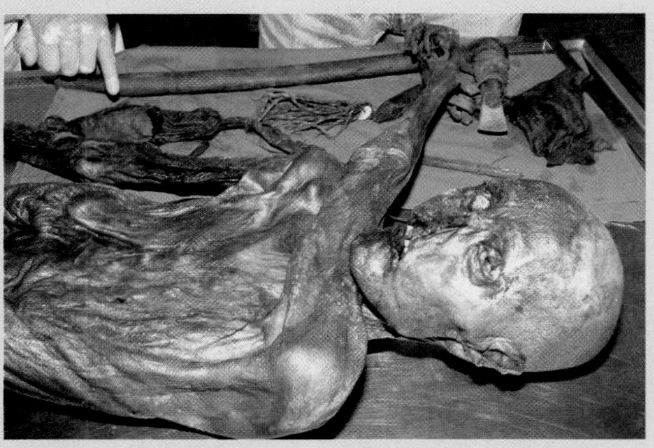

FIGURE 1 This "iceman" was preserved for thousands of years by the ice in which he was buried. An exceptionally warm summer melted the ice and revealed his cadaver.

Checkup 40.4

QUESTION 1: Table 40.3 lists some half-lives of radioactive isotopes. Suppose you have 1.00 μg of each of these isotopes. Which has the highest decay rate? Which the lowest?

QUESTION 2: If you initially have 1 μg of ^{131}I, how long does it take for this to decay to $\frac{1}{2}$ μg? To $\frac{1}{4}$ μg? To $\frac{1}{8}$ μg?

QUESTION 3: According to the Law of Radioactive Decay, at time $t = \tau$, what fraction of the initial amount of parent material remains? For radioactive strontium, $\tau = 42$ years; according to Fig. 40.14, how many grams of strontium remain at this time?

QUESTION 4: The decay rate of a sample of ^{226}Ra (see Table 40.3) is currently 8.0×10^9 disintegrations per second. What will be the number of disintegrations per second after 3200 years?

(A) 4.0×10^9 (B) 2.0×10^9 (C) 1.0×10^9
(D) 5.0×10^8 (E) 2.5×10^5

QUESTION 5: Suppose you have 1.0 μg each of ^{14}C, ^{226}Ra, and ^{238}U. Which has the larger decay rate, that is, which has the larger number of disintegrations per second? (See Table 40.3).

(A) ^{14}C (B) ^{226}Ra (C) ^{238}U

40.5 FISSION

Concepts in Context

In heavy, large nuclei—such as uranium—the balance between the attractive strong force and the repulsive electric force is quite precarious. As already mentioned in our discussion of alpha decay, in such a nucleus the electric force becomes dominant if the nucleus suffers some extra elongation, either spontaneous or caused by an external disturbance, such as the impact of a neutron on the nucleus. The repulsive electric force then overpowers the attractive strong force, and the nucleus splits into two fragments. Such a splitting of a nucleus is called **fission**. Figure 40.15 illustrates the fission of a nucleus after it suffers a disturbance and elongation when hit by a neutron. Fission was first discovered by Otto Hahn, Lise Meitner, and Fritz Strassmann, who bombarded a sample of uranium with a beam of neutrons and found that the uranium fissions into barium and krypton, according to the reaction

fission

$$\text{neutron} + {}^{238}\text{U} \rightarrow {}^{145}\text{Ba} + {}^{94}\text{Kr}$$

More generally, it was found that uranium can fission in several different ways, with the formation of a variety of fission fragments, each with a mass of about one-half of the mass of the original uranium nucleus. The fission fragments are usually radioactive, and they often emit one or several neutrons. Thus, the net reaction is actually

$$\text{neutron} + {}^{238}\text{U} \rightarrow \text{fission fragments} + 2 \text{ or } 3 \text{ neutrons} \tag{40.31}$$

While the fission fragments fly apart, the repulsive electric force does positive work on the fragments, giving them a large kinetic energy. This kinetic energy represents the energy released in the reaction. We can calculate the amount of energy released in the reaction in the usual way, by comparing the mass of the uranium nucleus with the (smaller) sum of the masses of the fission fragments. But even without such a calculation we can see from the curve of binding energy (Fig. 40.6) that the fission of a large nucleus leads to a release of energy. By inspecting this figure we recognize that a large nucleus has a somewhat lower amount of binding energy per nucleon than a medium-sized nucleus (the curve has a maximum at $A = 56$, corresponding to ${}^{56}\text{Fe}$). Hence the fission of a large nucleus into two medium-sized nuclei results in an increase of the net binding energy, that is, it results in a release of energy.

LISE MEITNER (1878–1968) *Austrian physicist. She studied under Boltzmann, and became the first woman to earn the Ph.D. in physics at the University of Vienna. In Berlin she and Otto Hahn collaborated for several decades, and she became the first female full profesor in Germany. She fled Nazi Germany to Sweden in 1938, just before Hahn discovered the splitting of uranium from chemical evidence. With her nephew, Otto Frisch, she coined the term nuclear fission and described its physical mechanism, but only Hahn was awarded the Nobel Prize in Chemistry for nuclear fission in 1944. Hahn, Meitner, and Hahn's assistant Fritz Strassman shared the Enrico Fermi Award for their discovery in 1966. Meitnerium* (Mt), *atomic number 109, was named in her honor.*

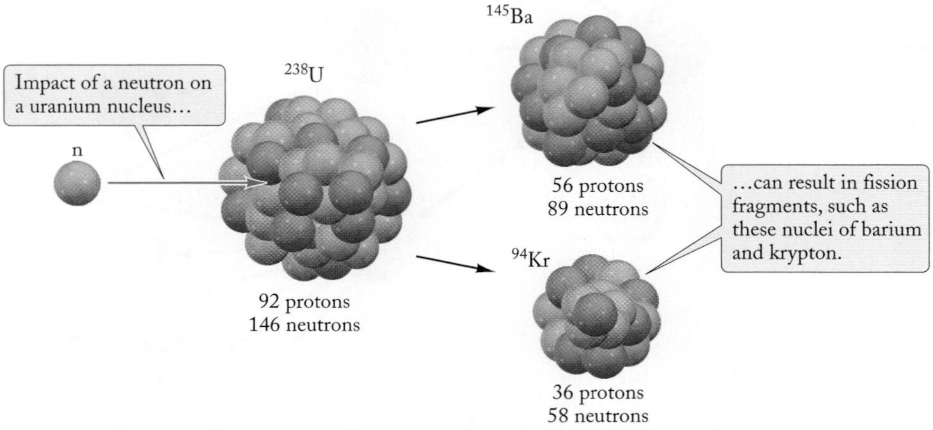

FIGURE 40.15 Fission of uranium triggered by the impact of a neutron.

Concepts in Context

On the average, the energy released in the fission of a uranium nucleus is about 200 MeV. We can gain a better appreciation of the magnitude of this energy by looking at the following numbers. The total energy released by the complete fission of 1 kg of uranium (a lump slightly larger than a golf ball) is 200 MeV times the number of atoms in 1 kg of uranium. Since a mole of uranium is 238 g, 1.00 kg of uranium is $1000/238 = 4.20$ moles, or $6.02 \times 10^{23} \times 4.20$ atoms $= 2.53 \times 10^{24}$ atoms, and the energy released by the fission of the nuclei of these atoms is

$$200 \frac{\text{MeV}}{\text{nucleus}} \times 2.53 \times 10^{24} \text{ nuclei} = 5.1 \times 10^{26} \text{ MeV}$$

$$= 8.1 \times 10^{13} \text{ J} \qquad (40.32)$$

This is equivalent to the energy released in the burning of 2 million liters of gasoline. It is also equivalent to the energy released in the explosion of about 20 000 tons of TNT.

To exploit the energy released by the fission of uranium we rely on the neutrons that emerge in the reaction (40.31). If one initial fission occurs in a lump of uranium, the 2 or 3 neutrons that emerge from this reaction can strike other uranium nuclei and trigger their fission, and the neutrons that emerge from these fissions can trigger further fissions, and so on. *Thus, one initial neutron can initiate an avalanche of neutrons and an avalanche of fission reactions.* Such an avalanche is called a **chain reaction** (see Fig. 40.16). If no neutrons, or very few neutrons, are lost from the chain reaction, the rate of fission and the rate of release of energy grow drastically with time. For instance, if on the average two of the neutrons released in each fission succeed in generating further fission reactions and further neutrons, then the numbers of fission reactions in successive steps of the chain will be 2, 4, 8, 16, ... If this multiplicative growth continues unchecked, the rate of release of energy will become explosive.

chain reaction

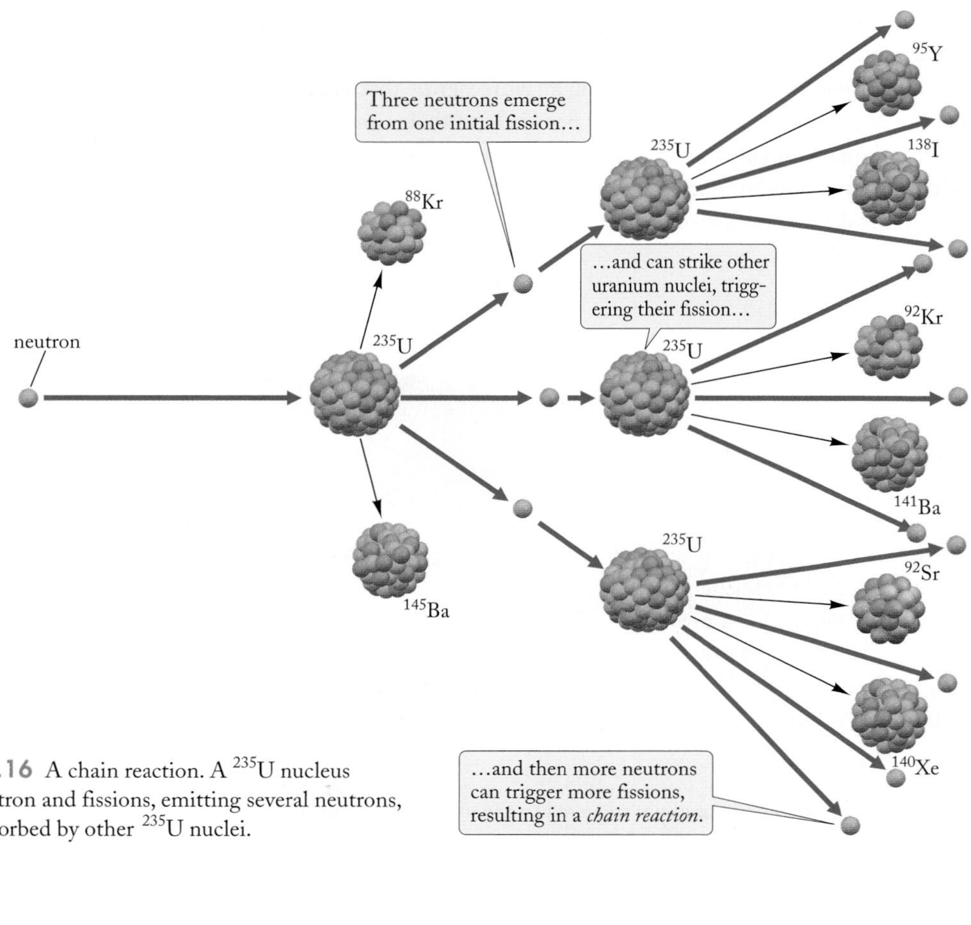

FIGURE 40.16 A chain reaction. A ^{235}U nucleus absorbs a neutron and fissions, emitting several neutrons, which are absorbed by other ^{235}U nuclei.

In the case of ^{238}U, conditions are not favorable for sustaining a chain reaction. Uranium-238 is a fairly stable nucleus, which does not readily fission when a neutron strikes it. Instead, the nucleus often merely absorbs the incident neutron, removing it from the chain reaction. This loss of neutrons blocks the chain reaction. However, naturally occurring uranium also contains a small amount (0.72%) of the isotope ^{235}U, which is unstable and much more susceptible to fission. This isotope is widely used as the nuclear fuel in applications of chain reactions in nuclear bombs and nuclear reactors. Another isotope used in such applications is ^{239}Pu, an isotope of plutonium. But this isotope does not occur naturally; it must be manufactured by nuclear transmutation reactions involving uranium, in a nuclear reactor.

✔ Checkup 40.5

QUESTION 1: Does the energy released in fission of a large, heavy nucleus originate from the initial "strong" potential energy or the electric potential energy?
QUESTION 2: Why are small, light nuclei incapable of fission?
QUESTION 3: Is a chain reaction possible if, on the average, each fission of a nucleus releases less than 1 neutron? Exactly 1 neutron? More than 1 neutron?
QUESTION 4: When a neutron strikes a ^{235}U nucleus resulting in fission, which of the following is true?

(A) The rest mass of the products is greater than the sum of the masses of ^{235}U and the neutron.
(B) The rest mass of the products is less than the sum of the masses of ^{235}U and the neutron.
(C) The products always include Ba (barium) and Kr (krypton).
(D) The total number of nucleons in the products is less than 236.
(E) The total number of nucleons in the products is greater than 236.

40.6 NUCLEAR BOMBS AND NUCLEAR REACTORS

In a given mass of ^{235}U or ^{239}Pu, neutrons produced by spontaneous fission or stray neutrons coming from elsewhere can initiate the first step in the chain reaction. Whether the reaction keeps going depends on how many neutrons are lost from the chain, by absorption without fission (as in ^{238}U) or by escape beyond the boundary of the mass. If the mass is large, few neutrons will reach its boundary before they are intercepted by a nucleus; thus a large mass inhibits escape of the neutrons and favors the chain reaction. The mass is said to be **critical** if the number of neutrons lost from the chain reaction (by escape or by absorption) equals the number of neutrons released by the fissions. In this case the chain reaction merely proceeds at a constant rate—as in a nuclear reactor. The mass is said to be **supercritical** if the number of neutrons lost from the chain is smaller than the number of neutrons released in fission reactions. In this case the chain reaction proceeds at an ever-increasing, runaway rate leading to an explosion—as in a nuclear bomb. For pure ^{235}U arranged in a spherical configuration, the critical mass is about 50 kg.

critical and supercritical mass

The simplest fission bomb, commonly known as an **atomic bomb**, or A-bomb, consists of two pieces of ^{235}U whose separate masses are each less than the critical

atomic bomb

ENRICO FERMI (1901–1954) *Italian and later American physicist. He worked on experimental and theoretical investigations of beta decay, the artificial production of isotopes by neutron bombardment, for which he received the 1938 Nobel Prize, and the fission of uranium. Fermi was one of the leaders of the Manhattan Project, and he provided the first experimental demonstration of a chain reaction.*

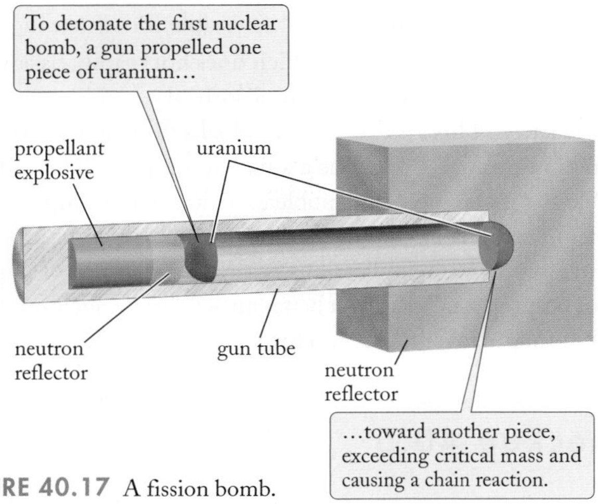

FIGURE 40.17 A fission bomb.

mass but whose combined mass is more than the critical mass. To detonate such a bomb, the two pieces of uranium, initially at a safe distance from each other, must be suddenly brought close together. In the first such bomb (see Fig. 40.17), the device used for the assembly of the two pieces of uranium consisted of a gun which propelled one piece of uranium toward the other at high speed. More sophisticated fission bombs consist of a (barely) subcritical mass of ^{239}Pu; if this mass is suddenly compressed to higher than normal density, it will become supercritical. The sudden compression is achieved by the preliminary explosion of a chemical high explosive arranged in a layer around the mass of plutonium.

The energy released in the explosion of an A-bomb is typically equivalent to that released in the explosion of about 20 kilotons of TNT (see Fig. 40.18).

hydrogen bomb

A much higher explosive yield is achieved by a **hydrogen bomb**, or H-bomb, in which an A-bomb is used to trigger fusion reactions similar to the fusion reactions that power the Sun (see Section 40.7). The energy released in the explosion of an H-bomb is typically equivalent to that in one or several megatons of TNT. Such an explosion would level an entire city, with complete devastation and complete destruction of life by incineration and crushing out to a radius of about 16 km from the center of the explosion.

nuclear reactor

For the peaceful exploitation of nuclear fission in a **nuclear reactor**, we must keep the chain reaction under control, so it releases energy at a steady rate. This means that the configuration of the uranium or other nuclear fuel must be critical rather than super-

FIGURE 40.18 Explosion of a fission bomb.

critical. The most common type of reactor operates with "enriched" uranium consisting of a few percent ^{235}U mixed with more than 95 percent of ^{238}U. Such a uranium mixture cannot by itself maintain a chain reaction—the ^{238}U soaks up too many of the neutrons. However, if the uranium is enmeshed in a substance capable of slowing down the neutrons released in fissions, a chain reaction becomes viable. The substance that slows the neutrons is called the **moderator.** The role of the moderator in a fission reaction is analogous to that of a catalyst in a chemical reaction. The moderator enhances the chain reaction because slow neutrons are more efficient at producing fissions in ^{235}U than fast neutrons, and they are also less likely to be absorbed by ^{238}U.

moderator

Inside the reactor, the uranium is usually placed in long **fuel rods**, and these are immersed in the bulk of the moderator (see Fig. 40.19). Fast neutrons released by fissions travel from the fuel rods into the moderator; there they lose their kinetic energy by collisions with the moderator nuclei; and then they wander back into one or another of the fuel rods and trigger further fissions. The three best moderators are ordinary water (H_2O), heavy water (D_2O), and graphite (pure carbon).

fuel rods

The configuration of the reactor—the size, number, and location of the fuel rods and the shape of the moderator—must be designed so the reactor is nearly critical. Fine adjustments in the number of neutrons and the reaction rate are made by means of **control rods** of boron or cadmium. These substances greedily soak up neutrons, and by pushing the **control rods** in or pulling them out, the reaction rate can be decreased or increased.

control rods

The main application of reactors is for the generation of electric power. In the United States, most of the reactors used for this purpose have water-filled cores. The water acts simultaneously as moderator and as coolant. The water circulates through the core and removes the heat energy released by the fission reactions (see Fig. 40.20).

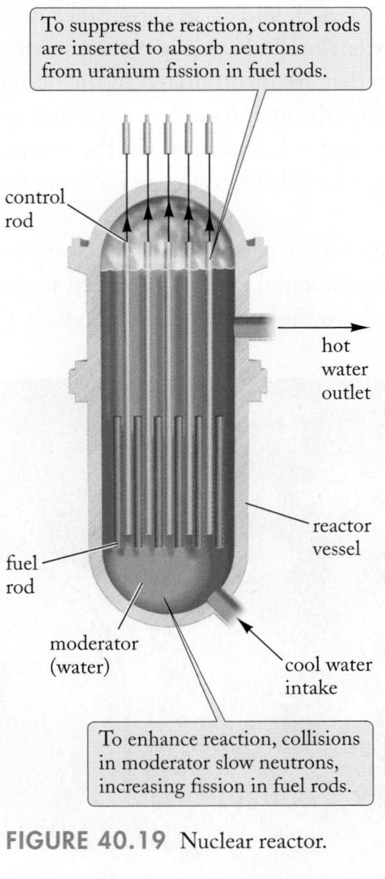

FIGURE 40.19 Nuclear reactor.

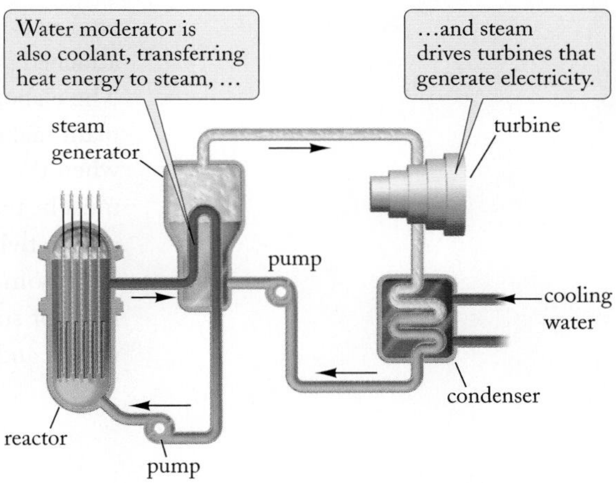

FIGURE 40.20 Schematic diagram of nuclear power plant.

The heat is transferred from the water to steam, which drives a steam turbine connected to an electric generator. Thus, the nuclear reactor plays the role of the furnace of a conventional steam engine and uranium replaces coal or oil as fuel.

Concepts — in — Context

EXAMPLE 7 A nuclear reactor in a power plant produces 1200 MW of heat. How many kilograms of ^{235}U does this reactor consume per year?

SOLUTION: The energy that must be supplied by fission reactions in one year is the product of the power and the time,

$$E = P\Delta t = 1200 \text{ MW} \times 1 \text{ yr}$$

$$= 1200 \times 10^6 \text{ W} \times 3.16 \times 10^7 \text{ s} = 3.8 \times 10^{16} \text{ J}$$

From Eq. (40.32) we know that the energy released in the fission of 1.0 kg of uranium is 8.1×10^{13} J. Hence, the number of kilograms required per year is

$$3.8 \times 10^{16} \text{ J} \times \frac{1.0 \text{ kg}}{8.1 \times 10^{13} \text{ J}} = 470 \text{ kg}$$

Nuclear power plants could meet all our energy requirements for the next several hundred years, or even the next several thousand years, if we exploit low-grade ores containing nuclear fuels. Unfortunately, nuclear fission yields rather dirty energy—the fission reactions generate dangerous radioactive residues. Nuclear power plants must be carefully designed to hold these residues in confinement. The reactor core is enclosed in a massive reactor vessel, and as an extra precaution this vessel and the attached pumps and pipes are enclosed in a strong containment shell (see Fig. 40.21). The elaborate safety features that are incorporated in the design of a nuclear power plant make the construction and the maintenance extremely expensive. Furthermore, when the load of fuel of the reactor has been spent, the residual radioactive wastes must be removed to a safe place to be held in storage for thousands of years until their radioactivity has died away.

If some of the radioactive material contained in a nuclear reactor is released in the form of smoke or dust in an accidental explosion or fire, it can be carried away by winds, and it can then descend to the ground as lethal radioactive fallout. After the

FIGURE 40.21 The containment shells of the two nuclear reactors at the San Onofre nuclear generating station in California.

disastrous nuclear accident that happened at Chernobyl, near Kiev in the former U.S.S.R., in 1986, a plume of fallout spread north and west into Europe and produced hazardous contamination thousands of kilometers away. An area of 30-km radius around the reactor was so heavily contaminated it had to be permanently evacuated. The immediate cause of the Chernobyl disaster was operator error—through a series of almost incredible blunders, the operators lost control of the reactor, and its power level surged to 100 times normal values, which overheated the core, blew it apart, and started a fire in the graphite moderator.

The water-moderated reactors used in the United States are thought to be much safer than the Chernobyl reactor. But some experts question whether they are safe enough. Scenarios for conceivable reactor accidents have been analyzed in great detail. The worst that might happen is a loss-of-coolant accident, that is, the loss of the water that normally circulates through the core of the reactors serving both as moderator and as coolant. Although such a loss of water would shut off the fission chain, the reactor contains a large amount of radioactive residues, and the heat from the radioactivity can by itself overheat the reactor vessel beyond safe limits. As a precaution, reactors are equipped with emergency cooling systems, and they are also encased in a containment shell.

Nuclear power reactors in the United States all use ^{235}U as fuel. Unfortunately, the supply of this nuclear fuel is rather limited, and it is expected to become exhausted in this century. There are, however, several other nuclear fuels in larger supply. One of these is ^{238}U. Although this isotope is incapable of supporting chain reactions, it can be converted into ^{239}Pu, which supports chain reactions. The manufacture of ^{239}Pu is an automatic side effect of the operation of nuclear reactors now in operation; in all of these reactors, the fuel rods contain a mixture of ^{235}U and ^{238}U, and fission neutrons striking the ^{238}U gradually convert it into ^{239}Pu. A reactor fueled with ^{239}Pu not only makes good use of a material that would otherwise go to waste, but, if the reactor is surrounded by a blanket of ^{238}U, it can also manufacture extra ^{239}Pu. The number of neutrons released in the fission of ^{239}Pu is so large that in an efficiently designed reactor neutrons can be diverted to the ^{238}U without hindering the chain reaction. Such a reactor can produce more ^{239}Pu (from ^{238}U) than it consumes (from its original supply). A reactor of this kind is called a **breeder reactor**. Because of worries over the safety of these reactors, they have not been adopted in the United States, but dozens are operating in Europe.

breeder reactor

✔ Checkup 40.6

QUESTION 1: If each fission of a uranium nucleus produces three neutrons, how many of these neutrons can you afford to lose, if you want to maintain a chain reaction?

QUESTION 2: Why is it impossible to maintain a chain reaction in a very small mass of uranium, say, 100 g?

QUESTION 3: Why is the presence of ^{238}U detrimental to the maintenance of a chain reaction in a mass of ^{235}U?

QUESTION 4: A breeder reactor produces more nuclear fuel than it consumes. Is this in conflict with conservation of energy?

QUESTION 5: In a nuclear reactor, the moderator

(A) Coordinates the activities of the technicians operating the reactor.
(B) Controls the steam that drives the turbines of the electrical generator.
(C) Absorbs neutrons before they decay into protons and electrons.
(D) Slows neutrons so that they can be absorbed by ^{235}U in fuel rods later.
(E) Slows the reaction by permitting more neutrons to be absorbed by ^{238}U.

40.7 FUSION

fusion

Fusion is the merging of two small nuclei to form a larger nucleus. The fusion of two small nuclei, such as the nuclei of hydrogen, deuterium, or tritium, releases energy. For instance, the heat of the Sun is generated by a fusion reaction called hydrogen burning, in which hydrogen nuclei fuse together to make helium nuclei. By inspection of the curve of binding energy, we can immediately recognize that the fusion of small nuclei leads to a release of energy. Figure 40.6 shows us that the smallest nuclei have exceptionally low binding energies per nucleon. Hence, the fusion of two such nuclei into a larger nucleus of higher binding energy results in release of energy. Note that *the process of fusion is the reverse of fission: small nuclei (such as hydrogen) release energy when they fuse; large nuclei (such as uranium) release energy when they split.*

HANS ALBRECHT BETHE (1906–2005)
German and later American Physicist. During World War II he worked on the Manhattan Project as director of the Theoretical Physics Division. He received the Nobel Prize in 1967 for his investigations of nuclear reactions in stars.

The hydrogen burning in the Sun proceeds in three steps; first hydrogen nuclei fuse to make deuterium (^{2}H), then deuterium fuses with hydrogen to make helium-3 (^{3}He), and finally helium-3 fuses with helium-3 to make helium-4 (^{4}He):

$$^1\text{H} + {}^1\text{H} \rightarrow {}^2\text{H} + e^+ + \nu \tag{40.33}$$

$$^1\text{H} + {}^2\text{H} \rightarrow {}^3\text{He} + \gamma \tag{40.34}$$

$$^3\text{He} + {}^3\text{He} \rightarrow {}^4\text{He} + {}^1\text{H} + {}^1\text{H} \tag{40.35}$$

Each of the first two reactions must occur twice for the last reaction to occur once. Each of these reactions releases energy; when the first two reactions occur twice and the last occurs once (with the formation of one nucleus of ^{4}He), the net energy released is 24.7 MeV. Four nuclei of ^{1}H are consumed in this process (6 are consumed when the first two reactions occur twice, but 2 are regenerated when the last reaction occurs); hence the amount of energy released per nucleon of "fuel" is 24.7 MeV per 4 nucleons, or 6.2 MeV per nucleon. This number is to be compared with the energy released in the fission of uranium, about 200 MeV per 235 nucleons of "fuel," or 0.85 MeV per nucleon. Thus, fusion of a given mass of ^{1}H in the Sun releases about 7 times as much energy as the fission of an equal mass of ^{235}U.

thermonuclear reaction

The fusion reactions (40.33)–(40.35) are called **thermonuclear**, because they can proceed only at extremely high temperatures and pressures, such as the temperature of about 1.5×10^7 K in the core of the Sun. The high temperature is needed to overcome the electric repulsion that the hydrogen nuclei experience whenever they come close together. At high temperatures, the hydrogen nuclei have high speeds, and their collisions are sufficiently violent to overcome the electric repulsion and bring the nuclei into the intimate contact required for fusion.

The thermonuclear reaction based on hydrogen fuel cannot be duplicated on Earth, because we cannot attain the pressure found in the core of the Sun. However, there are some fusion reactions based on deuterium (^{2}H) and tritium (^{3}H) that proceed at lower pressures, for example,

$$^2\text{H} + {}^3\text{H} \rightarrow {}^4\text{He} + \text{neutron} \tag{40.36}$$

Although the reaction (40.36) can proceed at attainable pressures, it requires a temperature of about 10^8 K, even higher than in the center of the Sun. Such temperatures have been attained in the explosions of H-bombs, where the preliminary explosion of a fission bomb heats and compresses a mixture of deuterium and tritium and thereby initiates fusion.

Experimental attempts at a peaceful exploitation of fusion power also rely on a deuterium–tritium fuel mixture, which is heated to the required temperature of about 10^8 K either by an intense electric current, or by a particle beam, or by a laser beam. At such a temperature, the deuterium and tritium will be in the form of a plasma, that is, a totally ionized gas, consisting of a mixture of independent nuclei and electrons. The plasma cannot be contained by the walls of a conventional reactor vessel, since any contact with the wall of the vessel would cool the plasma, and would also melt the wall. Instead, the plasma must be held suspended in the middle of the vessel, away from contact with the walls.

One scheme for controlled fusion attempts to suspend the plasma by means of magnetic fields, that is, by **magnetic confinement**. Figure 40.22 shows the Princeton Tokamak Test Reactor, in which the plasma is confined inside a large toroidal solenoid (a solenoid shaped like a donut). The plasma is heated by a combination of electric currents induced in the plasma and particle beams aimed into it. A larger test reactor, a 500 MW fusion power generator known as ITER (International Thermonuclear Experimental Reactor), is under construction at Cadarache, France.

magnetic confinement

Another scheme for fusion attempts to extract energy by exploding small pellets of a deuterium–tritium mixture in a combustion chamber by hitting them with intense laser beams. This scheme is called **inertial confinement** because the inertia of the pellet holds it together long enough for the reaction to occur. The beams heat the pellet so suddenly that it has no time to disperse before fusion begins. The pellet then explodes like a miniature hydrogen bomb. After the thermal energy has been extracted from the combustion chamber, the next pellet is placed in the chamber, and so on. Figure 40.23 shows the combustion chamber of the National Ignition Facility (NIF) at Lawrence Livermore National Laboratory, where 192 high-power laser beams converge to heat a deuterium–tritium pellet (see also Chapter 26). Although both magnetic and inertial confinement schemes have been successful in initiating fusion, the amounts of energy released in fusion have remained far below the amount of energy that had to be fed into the reactor chambers to heat the plasma.

inertial confinement

Nuclear fusion is an attractive source of energy because it bypasses many of the safety problems associated with nuclear fission, especially the production of heavy, long-lived radioactive nuclei. However, tritium is radioactive and thus raises safety concerns, both when used as a fuel and when produced in the liquid lithium coolant of proposed fusion reactors by absorption of fast neutrons. Since the half-life of tritium is fairly short (12.3 years) and extraneous tritium could be recycled for fuel,

FIGURE 40.22 The Princeton Tokamak Test Reactor.

FIGURE 40.23 Combustion chamber at the National Ignition Facility in Livermore, California.

fusion is considered a potential source of much cleaner energy than fission. Besides, we have available an enormous supply of deuterium. There is enough heavy water mixed with the ordinary water in the oceans of the Earth to satisfy our energy needs for millions of years.

✔ Checkup 40.7

QUESTION 1: Why are large, heavy nuclei incapable of fusion?

QUESTION 2: What is the net result of the sequence of fusion reactions (40.33)–(40.35), that is, what nuclei are consumed, and what nuclei are produced?

QUESTION 3: Fusion requires high temperature. Is this also true for fission?

QUESTION 4: Do the fusion reactions (40.33)–(40.35) produce any radioactive nuclei?

QUESTION 5: Tritium is produced in the reaction

$$^{2}\text{H} + {}^{2}\text{H} \rightarrow {}^{3}\text{H} + {}^{1}\text{H}$$

What is the net result of this reaction followed by reaction (40.36)?

(A) $^{1}\text{H} + {}^{2}\text{H} + {}^{3}\text{H} \rightarrow {}^{4}\text{He} + {}^{1}\text{H}$ + neutron.
(B) $^{1}\text{H} + {}^{2}\text{H} + {}^{2}\text{H} \rightarrow {}^{3}\text{He} + {}^{1}\text{H}$ + neutron.
(C) $^{2}\text{H} + {}^{2}\text{H} + {}^{3}\text{H} \rightarrow {}^{4}\text{He} + {}^{1}\text{H}$ + 2 neutron.
(D) $^{2}\text{H} + {}^{2}\text{H} + {}^{2}\text{H} \rightarrow {}^{4}\text{He} + {}^{1}\text{H}$ + neutron.
(E) $^{2}\text{H} + {}^{2}\text{H} + {}^{2}\text{H} \rightarrow {}^{3}\text{He} + {}^{1}\text{H}$ + 2 neutron.

SUMMARY

PROBLEM-SOLVING TECHNIQUES Nuclear Reactions **(page 1371)**

PHYSICS IN PRACTICE Radioactive Dating **(page 1376)**

ISOTOPES Atoms with the same number of protons in their nuclei, but different numbers of neutrons.

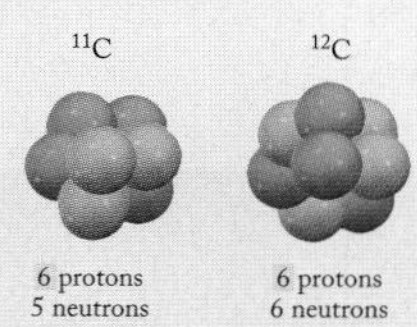

RADIUS OF NUCLEUS (A is mass number.) $R = (1.2 \times 10^{-15}\ \text{m}) \times A^{1/3}$ **(40.2)**

ATOMIC MASS UNIT u $1\ \text{u} = 1.660\,54 \times 10^{-27}\ \text{kg}$

ENERGY EQUIVALENT OF ATOMIC MASS UNIT $1\ \text{u} \times c^2 = 931.5\ \text{MeV}$ **(40.7)**

MASS DEFECT AND BINDING ENERGY ($B.E.$) [mass of N neutrons and Z protons] − [mass of nucleus] = $B.E./c^2$ **(40.8)**

Q VALUE OF REACTION Initial masses $m_1, m_2, \ldots$ and final masses $m_1', m_2', \ldots$ $Q = (m_1c^2 + m_2c^2 + \cdots) - (m_1'c^2 + m_2'c^2 + \cdots)$ **(40.11)**

KINDS OF RADIOACTIVITY

alpha decay: α particle (^{4}He nucleus)
beta decay: β particle (electron or antielectron)
gamma decay: γ ray (high-energy photon)

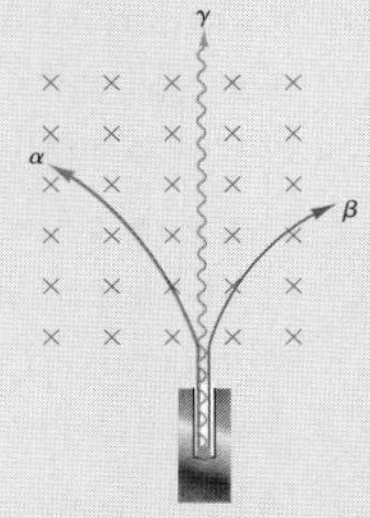

LAW OF RADIOACTIVE DECAY

$$n = n_0 e^{-t/\tau} \quad \text{or} \quad n = n_0 \left(\frac{1}{2}\right)^{t/t_{1/2}} \qquad (40.23)$$

where $\tau = \dfrac{t_{1/2}}{\ln 2} = \dfrac{t_{1/2}}{0.693}$, $t_{1/2}$ is the half-life, and τ is the mean lifetime, or time constant, of the decay. **(40.25)**

DECAY RATE OR ACTIVITY

$$-\frac{dn}{dt} = \frac{n}{\tau} = 0.693\frac{n}{t_{1/2}} \qquad (40.28)$$

BECQUEREL

$$1 \text{ Bq} = 1 \text{ disintegration/s} \qquad (40.29)$$

CURIE

$$1 \text{ Ci} = 3.7 \times 10^{10} \text{ disintegrations/s} \qquad (40.30)$$

SI UNIT OF ABSORBED DOSE

1 gray = 1 Gy = 1 J/kg = 100 rad

RELATIVE BIOLOGICAL EFFECTIVENESS (RBE)

10–20 for alpha
1–2 for beta
1 for gamma

EQUIVALENT ABSORBED DOSE

[equivalent dose in rem] = [dose in rad] × RBE

SI UNIT OF EQUIVALENT ABSORBED DOSE

1 sievert = 1 Sv = 100 rem

FISSION The splitting of a nucleus into two fragments, usually with the emission of one or more neutrons.

FUSION The merging of two small nuclei to form a larger nucleus.

QUESTIONS FOR DISCUSSION

1. What is the average overall density of a typical atom, such as iron? Why is this much smaller than the density of a nucleus, is given by Eq. (40.4)?
2. Naturally occurring magnesium has an atomic mass of 24.305. On the basis of this information, can you conclude that natural magnesium contains a mixture of several isotopes? Can you guess which isotopes?
3. According to Fig. 40.6, which isotope has the largest binding energy?
4. Why do alpha and beta emissions involve transmutation of elements whereas gamma emission does not?
5. If you irradiate a sample of material with alpha, beta, or gamma rays, is the sample likely to become radioactive? What if you irradiate it with neutrons?

6. Radiation is sometimes used to sterilize food or surgical supplies. What are the advantages or disadvantages of this procedure over sterilization by heating?
7. Can radioactive dating with ^{14}C be used to determine the age of a book printed during the last century? The age of deposits of mineral oil? The age of ancient Egyptian gold jewelry? Ancient Egyptian furniture?
8. Why is an absorbed dose of alpha radiation more likely to kill a biological cell than the same absorbed dose of gamma radiation?
9. Tritium (^3H) is a radioactive isotope of hydrogen which occurs naturally in small concentrations in the ordinary water found in the environment. The half-life of this isotope is 12 years. Describe how you could take advantage of this isotope to determine the age of a bottle full of wine that your wine merchant claims is 25 years old.
10. Alpha particles are more massive than beta particles, yet for equal kinetic energies, the alpha particles are stopped by a thinner layer of material than the beta particles. Can you explain why an alpha particle loses its energy more quickly? (Hint: Think of the electric forces that act on the particles in a collision with an atom.)
11. All the best moderators for reactors consist of fairly lightweight nuclei. Why is a material with heavy nuclei, such as lead, not a good moderator?
12. Why is it difficult to separate the isotopes of ^{238}U and ^{235}U?
13. Some artificial satellites carry small nuclear reactors as power supplies. What danger does this pose for people on Earth?
14. Nuclear fission bombs produce a long-lived radioactive isotope of strontium (^{90}Sr), an element that is chemically similar to calcium. Explain why strontium poses a severe hazard to humans and other vertebrates.
15. Hydrogen bombs operate with tritium, a radioactive isotope of hydrogen with a half-life of 12.3 years. When a hydrogen bomb is held in storage, its tritium gradually decays, and must be replenished with fresh tritium every few years. According to a recent proposal, arms control could be achieved by halting production of tritium. Every few years, the superpowers would then have to take some of their bombs out of service, extract the remaining tritium, and use it to replenish their other bombs. Suppose the United States and Russia each have 2000 hydrogen bombs now. Without fresh tritium, how many bombs will each superpower have in 25 years? In 50 years?
16. Why is it more difficult to achieve controlled fusion in a reactor than controlled fission?
17. A fusion reactor would produce a large amount of tritium (^3H), a radioactive isotope of hydrogen. If the tritium were accidentally released into the environment, it would be likely to contaminate the water. Explain.

PROBLEMS

40.1 Isotopes

1. How many protons and neutrons are there in the nucleus of the isotope ^{16}O? ^{56}Fe? ^{238}U?
2. What isotope has 82 protons and 122 neutrons in its nucleus?
3. What are the number of protons and the number of neutrons in each of the following isotopes: ^{24}Na, ^{27}Al, ^{52}Cr, ^{52}Mn, ^{63}Cu, ^{63}Zn, ^{124}Xe, ^{138}La?
4. What isotope has 17 protons and 18 neutrons in its nucleus? What isotope has 18 protons and 17 neutrons? Such isotopes in which the numbers of protons and neutrons are exchanged are called **mirror nuclei**. Find another example of mirror nuclei in the chart of isotopes in Table 40.2.
5. Use the chart of isotopes in Table 40.2 to make a list of all the isotopes of oxygen. What are the number of protons and the number of neutrons in the nucleus of each of these isotopes?
6. By inspection of Fig. 40.5, estimate the ratio of neutrons to protons in stable light nuclei ($Z \approx 20$) and in stable heavy nuclei ($Z \approx 80$).
7. In any nuclear reaction, the nuclear electric charge must be conserved and the mass number must be conserved. Are the following reactions in accord with these conservation laws? Explain.

$$^{2}H + {}^{12}C \rightarrow {}^{4}He + {}^{9}B$$

$$^{4}He + {}^{10}B \rightarrow {}^{13}C + n$$

$$n + {}^{238}U \rightarrow {}^{121}Ag + {}^{118}Pd$$

8. Naturally occurring boron is a mixture of 80.2% ^{11}B and 19.8% ^{10}B. From the masses of these isotopes (listed in the chart of isotopes in Table 40.2) calculate the atomic mass of naturally occurring boron.
9. According to Eq. (40.2), what is the nuclear radius of the smallest of the isotopes of carbon? The largest of these isotopes?
10. The largest known nucleus is that of the isotope 266 of element 109, which was recently named meitnerium (Mt). What is the radius of the ^{266}Mt nucleus?

11. In some reference books, isotopes are listed with their atomic number as a left subscript; for example, deuterium (^{2}H) would be listed as ${}^2_1$H, and carbon-14 would be ${}^{14}_{6}$C. Determine which of the following are incorrectly written: ${}^{31}_{15}$P, ${}^{44}_{22}$Sc, ${}^{118}_{50}$Sn, ${}^{207}_{83}$Pb. Since the subscript is unique for each element, we could consider using a notation without the atomic symbol of the element. What isotope would be meant by ${}^{89}_{39}$? By ${}^{197}_{79}$?

12. Many stable nuclei have an even mass number A, and all but a few stable nuclei have an even number N of neutrons or an even atomic number Z. Certain even numbers for N and Z correspond to very stable nuclei; these "magic nuclei" have (Z or N) = 2, 8, 14, 20, 28, 50, 82, or 126. These suggest a nuclear shell structure analogous to atomic shells, with independent shells for protons and neutrons. In the chart of isotopes in Table 40.2 for what value of N are there the most stable isotopes (that is, for what value of N are there the most naturally occurring isotopes, as indicated by any listed values for the percent abundance)? For what value of Z? What do the stable isotopes ^{36}S, ^{37}Cl, ^{38}Ar, ^{39}K, and ^{40}Ca have in common? The most abundant stable isotopes of barium, lanthanum, cerium, praseodymium, and neodymium are ^{138}Ba, ^{139}La, ^{140}Ce, ^{141}Pr and ^{142}Nd. What do these have in common?

13. What stable isotope has one-half the radius of a ^{104}Ru nucleus? One-half the radius of a ^{128}Xe nucleus? See the chart of isotopes in Table 40.2.

14. Chlorine has two stable isotopes, ^{35}Cl and ^{37}Cl, with atomic masses 34.968 85 u and 36.965 90 u, respectively. In the periodic table, the average atomic mass of chlorine is 35.453 u. From these data, calculate the percent abundance of the two stable chlorine isotopes.

15. Nuclei with the same mass number but different atomic numbers are called **isobars**. Find the number of isobars for each mass number from 1 to 17. Which nuclei have no isobars? Use the chart of isotopes in Table 40.2.

*16. Neutron stars are made (almost) entirely of neutrons, and they have approximately the same density as that of a nucleus. What is the radius of a neutron star of mass 0.50 times the mass of the Sun?

*17. What fraction of the volume of your body is filled with nuclear material? The average density of your body is about 1000 kg/m^3.

*18. Suppose that an alpha particle of energy 4.4 MeV collides head-on with a stationary gold nucleus. What is the distance of closest approach? Does the alpha particle make contact with the surface of the nucleus? Since the gold nucleus is so heavy, assume for simplicity that it gains no energy in the collision.

40.2 The Strong Force and the Nuclear Binding Energy

19. The binding energy of the electron in the ground state of the hydrogen atom is 13.6 eV. Calculate the corresponding mass defect of the hydrogen atom; express this in atomic mass units. (Since the mass defects associated with atomic binding energies are small, they are usually ignored in nuclear physics.)

20. According to Fig. 40.6, what is the binding energy of ^{56}Fe? From this binding energy, calculate the mass of the isotope (in atomic mass units).

21. What are the mass defect (in atomic mass units) and the nuclear binding energy (in MeV) of the isotope ^{4}He? The mass of ^{4}He is given in the chart of isotopes in Table 40.2.

22. What are the mass defect (in atomic mass units) and the nuclear binding energy (in MeV) of the isotope ^{12}C?

23. Nuclei with 2, 8, 14, 20, 28, 50, 82, or 126 protons or neutrons are called magic nuclei because they have exceptionally large binding energies and they are exceptionally stable (see also Problem 12). Compare the binding energy of the magic nucleus ^{16}O with that of the nonmagic nucleus ^{16}F. The masses of these isotopes are given in Table 40.2.

24. How much energy (in MeV) is required to remove one neutron from the nucleus of the isotope ^{14}N? (Hint: Compare the binding energies of ^{14}N and ^{13}N; the masses of these isotopes are given in the chart of isotopes in Table 40.2.

25. Nuclei with "magic numbers" of protons or neutrons are exceptionally stable (see also Problems 12 and 23), analogous to a closed shell in atomic physics. When a nucleus has only one or a few nucleons outside a closed shell, such nucleons are often relatively weakly bound. Consider cerium-140, which has a magic number of 82 neutrons. Compare ^{140}Ce and ^{142}Ce to find the average binding energy for each of the two extra neutrons in ^{142}Ce. Compare your result with the average binding energy per nucleon in ^{140}Ce. The atomic masses of ^{140}Ce, ^{142}Ce, and the neutron are 139.9054 u, 141.9092 u, and 1.0087 u.

26. Two nuclei with the same mass number but different atomic numbers are called isobars. Calculate the binding energy per nucleon for the stable isobars germanium-74 and selenium-74. Which nucleus is more tightly bound? The atomic masses of ^{74}Ge and ^{74}Se are 73.9218 u and 73.9225 u, respectively.

27. Verify that in the reaction (40.10), both sides have the same number of protons and neutrons.

*28. When boron is bombarded with alpha particles, the following reaction is observed:

$$^4\text{He} + {}^{10}\text{B} \rightarrow {}^{13}\text{C} + {}^1\text{H}$$

How much energy is released in this reaction?

*29. When ^{7}Li is bombarded with protons, the following reaction occurs:

$$^1\text{H} + {}^7\text{Li} \rightarrow {}^7\text{Be} + \text{n}$$

What is the minimum kinetic energy required for the proton to initiate this reaction? Neglect the recoil energy of the ^{7}Be nucleus.

*30. Consider the reaction

$$^1\text{H} + {}^3\text{H} \rightarrow {}^3\text{He} + \text{n}$$

(a) What is the energy absorbed in this reaction?

(b) We can initiate this reaction either by bombarding a tritium target with protons, or by bombarding a hydrogen target with tritium ions. To initiate this reaction, the kinetic energy of the bombarding particles must be somewhat larger than the energy absorbed in the reaction, since the recoil of the reaction products retains some of the energy supplied by the bombarding particle. If tritium ions are used as bombarding particles, the required kinetic energy is larger than if the protons are used as bombarding particles. Explain the difference.

*31. Consider the spontaneous fission of uranium into barium and krypton, according to the reaction

$$^{238}\text{U} \rightarrow {}^{145}\text{Ba} + {}^{93}\text{Kr}$$

The masses of the isotopes ^{238}U, ^{145}Ba, and ^{93}Kr are 238.050 79 u, 144.926 87 u, and 92.931 12 u, respectively.

(a) Find the Q value for this reaction.

(b) The uranium nucleus is initially at rest. Find the final speeds of the barium and krypton nuclei after the reaction, and find their individual energies.

*32. For large nuclei, the total electrostatic potential energy can be approximated by that of a sphere of uniform positive charge density ρ. This energy can be calculated by building up spherical shells of charge.

(a) Show that the potential at the surface of a small sphere of charge q and radius r is

$$V = \frac{1}{4\pi\epsilon_0}\frac{q}{r} = \frac{\rho r^2}{3\epsilon_0}$$

(b) Adding a shell of charge $dq = \rho\, d\,(\text{volume}) = \rho\, 4\pi r^2 dr$ requires an energy

$$dU = V\,dq$$

Integrate such contributions from $r = 0$ to $r = R$ to obtain the total electrostatic energy of a sphere of radius R and total charge Q.

(c) Use Eq. (40.2) and the result from (b) to estimate the average electrostatic energy per proton for ^{238}U ($Z = 92$).

†40.3 Radioactivity

33. What isotope is formed by the alpha decay of ^{216}Po? ^{239}Pu?

34. What isotope is formed by the negative beta decay of ^{85}Kr? ^{63}Ni?

35. What isotope is formed by the positive beta decay, or beta plus decay, of ^{22}Na? ^{64}Cu?

36. The alpha decay of uranium [see Eq. (40.13)] is the first step in a radioactive series of decays. The next four steps are a negative beta decay, another negative beta decay, another alpha decay, and another alpha decay. What are the daughter nuclei produced in these four steps?

37. For some radioactive nuclei, the inverse of beta decay is favorable, a process known as **electron capture;** typically, the nucleus absorbs one of its own atomic electrons. A neutrino is created in this process. What is the energy of the emitted neutrino when radioactive ^{37}Ar transmutes via electron capture into the stable nucleus ^{37}Cl? The atomic masses of ^{37}Ar and ^{37}Cl are 36.966 78 u and 36.965 90 u, respectively.

38. When ^{129}I beta-decays into ^{129}Xe, what is the maximum possible kinetic energy of the emitted electron? The maximum possible energy of the emitted neutrino? What are the minimum energies? The atomic masses of ^{129}I and ^{129}Xe are 128.904 98 u and 128.904 78 u, respectively.

39. The nucleus ^{57}Co is often used as a a source of gamma rays for experiments in magnetism, since these gamma rays are readily absorbed by iron nuclei. Suppose an atom of ^{57}Co at rest emits a 0.707-MeV gamma ray. What is the recoil velocity of the cobalt atom?

*40. The alpha decay of ^{210}Po results in ^{206}Pb. Calculate the energy of the emitted alpha particle. The masses of these two isotopes are 209.9829 u and 205.9745 u, respectively.

*41. What is the maximum kinetic energy of the beta ray emitted in the beta decay of a neutron, according to the reaction (40.18)? Ignore the recoil of the proton.

*42. If neutrons had a somewhat smaller mass, then the (slow) electrons in atoms could combine with protons in the nucleus according to the reaction

$$e^- + p \rightarrow n + \nu$$

How much smaller would the mass of the neutron have to be to make this reaction viable? What consequences would this reaction have for the existence of atoms and the existence of life?

*43. What isotope is formed in the negative beta decay of ^{14}C? Calculate the maximum energy of the beta rays emitted in this decay. The masses of the relevant isotopes are given in the chart of isotopes in Table 40.2.

†40.4 The Law of Radioactive Decay

44. According to Fig. 40.14, at what time will the remaining amount of radioactive strontium have fallen to 1/10 of the initial amount?

45. According to the best available data, the half-life of ^{14}C is believed to be 5730 years. However, according to previous data, the half-life was thought to be 5570 years, and age determinations based on this value of the half-life were in error. What percentage error in age determination does the error in the half-life introduce? For a sample from the year 3000 B.C., what is the error in years?

46. The isotope ^{238}U found on the Earth was originally synthesized in nuclear reactions in supernovas that exploded in our Galaxy about 6.8×10^9 years ago and scattered debris through the Galaxy, some of which became trapped in the Earth during its formation, about 4.6×10^9 years ago. What fraction of the original amount of uranium synthesized in the

†For help, see Online Concept Tutorial 45 at www.wwnorton.com/physics

Galaxy is still in existence? What fraction of the amount of uranium trapped in the Earth during its formation is still in existence? The half-life of ^{238}U is 4.5×10^9 years.

47. The activity of an industrial ^{60}Co irradiation cell is 1.0×10^{16} Bq when the cell is new. What will be the activity after the cell has been in use for 2.0 years? The half-life of ^{60}Co is 5.3 years.

48. What is the activity of 1.00 gram of pure radium, ^{226}Ra? The half-life of this isotope is listed in Table 40.3.

49. What is the activity of 1.0 microgram of ^{22}Na? The half-life of this isotope is listed in Table 40.3.

50. Some modern theories of elementary particles propose that the proton might be unstable with a very long half-life. Suppose the half-life for proton decay is 1.0×10^{33} years. What would be the decay rate for the protons in an experiment using 5.0×10^4 liters of water? What is the inverse of this decay rate, that is, how long on average would you have to wait for one disintegration?

51. The activity of a sample of ^{57}Co is initially 12.2 mCi (millicuries). After 2.00 years, the activity has dropped to 1.87 mCi. From these data, what is the half-life of ^{57}Co? If the initial sample was pure ^{57}Co (atomic mass 56.9), what was the initial mass of this sample?

52. 1.00 mole of naturally occurring rubidium emits beta rays at a decay rate of 7.75×10^4 Bq due to the radioactive isotope ^{87}Rb, which has a natural isotopic abundance of 27.8%. From these data, calculate the half-life of ^{87}Rb.

53. A 75-kg worker receives an accidental whole-body dose of 15 rads of beta radiation with an RBE of 1.7. What energy is absorbed? What is the absorbed dose in grays? What is the equivalent absorbed dose in rems? In sieverts?

54. What absorbed dose in rads of alpha radiation with an RBE of 20 causes the same amount of biological damage as 10 rads of gamma radiation?

55. During a radiation treatment of a 75-g tumor, a patient absorbs 5.0 J of gamma radiation. The tumor absorbs 30% of the radiation; by rotating the gamma-ray beam, the remaining energy is distributed throughout 1.5 kg of surrounding tissue. What is the radiation dose absorbed by the tumor? By the surrounding tissue?

40.5 Fission

40.6 Nuclear Bombs and Nuclear Reactors

56. According to the curve of binding energy in Fig. 40.6, what is the binding energy of a nucleus with mass number $A = 200$? What is the binding energy of a nucleus with mass number $A = 100$? What is the energy released in the fission of the $A = 200$ nucleus into two $A = 100$ nuclei?

57. The bomb exploded at Hiroshima had an explosive yield of about 20 kilotons of TNT, or 8.4×10^{13} J. How many kilograms of uranium actually underwent fission to release this amount of energy?

58. Use the curve of binding energy to find the nucleus of smallest mass number A for which fission into two nuclei of mass number $A/2$ is energetically viable. (Hint: For fission to be viable, the height of the curve at $A/2$ must be above the height at A. Why?)

59. Naturally occurring uranium contains 99.28% of the isotope ^{238}U and 0.72% of the isotope ^{235}U. How many kilograms of natural uranium must we process to extract the 50 kg of ^{235}U required for a bomb? Pretend that all of the ^{235}U can be extracted.

60. Some artificial space probes that travel far from the Sun are powered by nuclear energy, typically the alpha decay of plutonium-238. Suppose such a satellite uses 2.0 kg of ^{238}Pu. At what rate is energy generated from alpha decay? The atomic masses of ^{238}Pu and ^{234}U are 238.0496 u and 234.0409 u, respectively, and the half-life of ^{238}Pu is 88 years.

61. How much energy is released in the following fission reaction?

$$n + {}^{235}\text{U} \rightarrow {}^{87}\text{Rb} + {}^{137}\text{Cs} + 12\,n$$

The atomic masses of rubidium-87 and cesium-137 are 86.9092 u and 136.9071 u, respectively.

*62. Consider the fission reaction

$$n + {}^{235}\text{U} \rightarrow {}^{144}\text{Ba} + {}^{92}\text{Kr}$$

Given that the masses of the isotopes ^{235}U, ^{144}Ba, and ^{92}Kr are 235.043 94 u, 143.922 85 u, and 91.926 27 u, respectively, calculate the energy released in this reaction.

40.7 Fusion

63. Some designs for fusion reactors include a liquid lithium blanket for the absorption of fast neutrons and energy. The lithium also generates more tritium for fuel according to the reaction

$$n + {}^{6}\text{Li} \rightarrow {}^{3}\text{H} + {}^{4}\text{He}$$

How much energy is generated in this reaction? Use the chart of isotopes in Table 40.2.

64. To achieve fusion, a sufficient number density n of nuclei must be confined for a long enough time τ to provide a sustained reaction. For deuterium–tritium fusion the condition is $n\tau \geq 1.0 \times 10^{20}$ s/m^3, known as **Lawson's criterion**.

(a) In magnetic confinement test reactors, confinement times of about 0.50 s have been achieved. What is the corresponding required density of nuclei? How does this compare with the density of an ideal gas at standard temperature and pressure?

(b) For inertial confinement, suppose the fuel is in a liquid pellet with the same number density as that of protons in water. The laser pulse lasts only 1.0×10^{-10} s; take this to be the confinement time. By what factor must the pulse compress the volume of the liquid to achieve Lawson's criterion?

*65. For deuterium and tritium to fuse, they must get close enough to touch.

(a) Use Eq. (40.2) to estimate the radii of deuterium and tritium and thus obtain their center-to-center distance when barely touching.

(b) Calculate the electrostatic potential energy of the two nuclei when touching.

(c) If the nuclei approach head-on, each with kinetic energy equal to about half the amount obtained in (b), they can fuse. At a temperature T, there will be a broad distribution of thermal velocities. Suppose that at a temperature T, enough nuclei with kinetic energy equal to 10 times the average thermal kinetic energy $\frac{3}{2}kT$ are present for fusion. From these considerations, estimate the temperature needed for fusion.

*66. Consider the three fusion reactions (40.33)–(40.35) that occur in the core of the Sun. From the masses of the nuclei that participate in these reactions (see the chart of isotopes in Table 40.2), calculate the energy released in each reaction. What is the net energy if the first two reactions occur twice and the last reaction occurs once?

*67. (a) The rate at which the Sun radiates heat and light is 3.9×10^{26} W. If the energy for all of this radiation comes from the fusion reactions (40.33)–(40.35), what is the rate at which the Sun consumes hydrogen?

(b) The amount of hydrogen in the Sun is about 1.5×10^{30} kg. How long will this fuel last if the Sun goes on radiating at a steady rate?

*68. Whenever the first two of the fusion reactions (40.33)–(40.35) proceed twice and the last proceeds once, 24.7 MeV is released, and so are two neutrinos. Hence, the number of neutrinos released is 1 per 12.3 MeV. Assume that the fusion reactions (40.33)–(40.35) account for all of the 3.9×10^{26} W of heat and light radiated by the Sun. Calculate the rate at which the Sun releases neutrinos.

**69. A sequence of reactions that can be used in controlled fusion is the burning of deuterium "fuel," as follows:

$$^{2}H + {}^{2}H \rightarrow {}^{3}He + n$$

$$^{2}H + {}^{2}H \rightarrow {}^{3}H + {}^{1}H$$

$$^{2}H + {}^{3}H \rightarrow {}^{4}He + n$$

$$^{2}H + {}^{3}He \rightarrow {}^{4}He + {}^{1}H$$

The net result of this sequence of reactions is the transmutation of six nuclei of deuterium (^{2}H) into two nuclei of ^{4}He, two protons (^{1}H), and two neutrons. Calculate the energy released in each of these reactions. What is the net energy released per nucleon of fuel?

REVIEW PROBLEMS

70. How many protons and neutrons are there in the nucleus of the isotope ^{18}Na? ^{39}K? ^{231}Pu?

*71. Suppose you bombard a target of magnesium with alpha particles. If, in a head-on collision with a stationary magnesium nucleus, an alpha particle is to reach the nuclear surface just barely before being halted by the repulsive electric force, what must be the energy of the alpha particle? Express your answer in MeV.

72. A nucleus of radius 7.4×10^{-15} m fissions into two equal spherical pieces. What is the radius of each piece?

73. Chadwick discovered the neutron when he bombarded boron with alpha particles, which resulted in the reaction

$$^{4}He + {}^{11}B \rightarrow {}^{14}N + n$$

Calculate the energy released in this reaction.

74. Consider the following nuclear reaction:

$$^{2}H + {}^{12}C \rightarrow {}^{4}He + {}^{10}B$$

What is the energy absorbed in this reaction? Can this reaction be initiated by bombarding carbon with a beam of deuterons (nuclei of deuterium) of 1.2 MeV? Neglect the recoil of the carbon nucleus.

75. Bombardment of ^{13}C with protons produces an isotope of nitrogen, according to the reaction

$$^{1}H + {}^{13}C \rightarrow {}^{13}N + n$$

What is the minimum energy that the proton must have to initiate this reaction? Neglect the recoil of the carbon nucleus.

*76. Nuclear physicists sometimes deduce the mass of an isotope from the energy of the beta rays emitted by the isotope. For instance, the maximum energy of the beta rays emitted in the decay of ^{27}Mg into ^{27}Al is 2.610 MeV. Given that the mass of ^{27}Al is 26.9815 41 u, what mass can you deduce for ^{27}Mg?

*77. Find the maximum kinetic energy of the beta rays emitted in the decay

$$^{16}N \rightarrow {}^{16}O + e^{-} + \bar{\nu}$$

78. One of the dangerous radioactive isotopes in the radioactive waste produced by nuclear reactors is ^{85}Kr, with a half-life of 10.8 yr. How long must we hold the radioactive waste in stor-

age to reduce the activity of the ^{85}Kr material to 0.10% of its initial value?

79. Bananas contain potassium, in the form of ^{39}K (93.258%), ^{40}K (0.0117%), and ^{41}K (6.7302%). The isotope ^{40}K is radioactive, with a half-life of 1.28×10^9 yr. The total amount of potassium in a banana is 0.50 g. Calculate the activity of a banana.

80. For diagnostic medical imaging, the isotope iodine-123 is absorbed by healthy thyroid tissue and emits 1.4-MeV gamma rays. Similarly, technetium-99 is absorbed by abnormal brain tissue and emits 0.14-MeV gamma rays. Thus, the absence or presence of gamma emission from a region of the thyroid or brain, respectively, is indicative of abnormal tissue. Assume that a compound containing 0.10 μg of each isotope is injected into the bloodstream of a 25-kg child, and that 10% of the radiation is absorbed throughout the body. What is the absorbed dose in each case?

*81. The partial eradication of the thyroid in patients suffering from hyperthyroidism can be accomplished by injecting a compound containing radioactive iodine ^{131}I into the body; the iodine then concentrates in the thyroid and kills its cells. If the thyroid is to be subjected to an activity of 0.10 Ci, how many grams of ^{131}I should be injected? The half-life of ^{131}I is 8.04 days.

82. In a **mass spectrometer**, ions are accelerated through a potential and enter a region of uniform magnetic field, where they travel in a semicircular arc before reaching a detector (see Fig. 30.26). Atoms from an unknown substance are singly ionized and analyzed using a magnetic field of 0.425 T. The detector is fixed for an arc radius of 6.50 cm. When the accelerating voltage is varied, a large number of ions are detected at 5249 V, and a small number at 6123 V. From the chart of isotopes in Table 40.2, what is the unknown substance?

83. When a neutron combines with a proton to form a deuteron (^{2}H), a gamma ray with energy 2.223 MeV is released. The atomic masses of ^{1}H and ^{2}H are 1.007 825 u and 2.014 102 u, respectively. From these data, calculate the mass of the neutron.

84. Consider the fission of a ^{238}U nucleus into two ^{119}Pd nuclei. In the first instant after this fission, the two ^{119}Pd nuclei are just barely in contact, and they then accelerate away from each other, impelled by their mutual electrostatic repulsion. If the center-to-center distance between the ^{119}Pd nuclei is twice their radius (see Fig. 40.24). what is the electric potential energy? What is the force? What is the acceleration of each fragment?

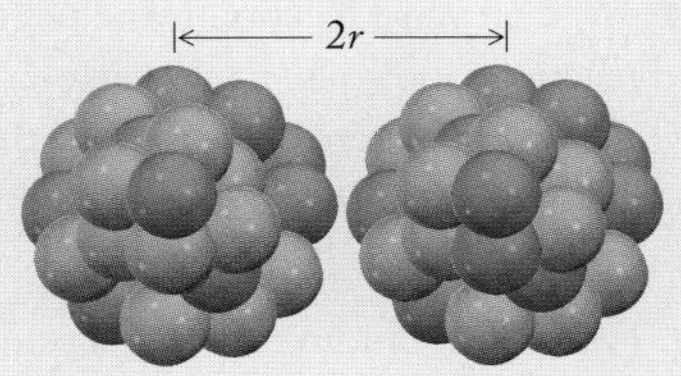

FIGURE 40.24 Two palladium nuclei.

85. The total worldwide supply of ^{235}U in high-grade ores is estimated at 2.6×10^8 kg. The total worldwide demand for energy is about 5.0×10^{20} J/yr. Suppose that the entire energy demand were supplied by 1000-MW nuclear reactors fueled with ^{235}U.

(a) How many reactors would we need?

(b) How long would the high-grade fuel last?

Answers to Checkups

Checkup 40.1

1. The number of protons is 6 for all three isotopes; all isotopes of a given chemical element have the same atomic number. The number of nucleons, or mass number, is given by the superscript before the element symbol, and so is 8, 9, and 10 for the respective isotopes. The number of neutrons is the number of nucleons minus the number of protons, and so is 2, 3, and 4, respectively.

2. Each hydrogen isotope has one proton (all isotopes of a given element have the same atomic number), and the three isotopes have 0, 1, and 2 neutrons, and thus 1, 2, and 3 nucleons, respectively.

3. The rightmost column, containing only ^{27}Ne, corresponds to the largest number of neutrons ($N = 17$). The top row contains isotopes of Ne; these have the most protons ($Z = 10$). The isotope ^{27}Ne has the largest mass number in the chart.

4. (B) ^{3}H. Tritium, or ^{3}H, has two neutrons and one proton; each of the other isotope choices has an equal number of neutrons and protons.

5. (E) 8. Since Eq. (40.2) asserts that the radius is proportional to the cube root of the atomic number, the atomic number is thus proportional to the cube of the radius, so the ratio is $2^3 = 8$.

Checkup 40.2

1. Since the force is the negative of the gradient of the potential, the largest attractive (negative) force occurs where the slope of the potential vs. distance is most positive. From the plot of Fig. 40.3, this is at about $0.9-1.0 \times 10^{-15}$ m.
2. The binding energy per nucleon for ^{238}U is 1802 MeV/(238 nucleons) $\approx$ 7.6 MeV/nucleon, at the heavy-nuclei (low) end of the $\approx$ 8 MeV values common to all but the lightest nuclei.
3. No; this would mean the nucleus has more energy than the separate protons and neutrons, and the nucleus could and would fly apart.
4. The Q value of the reverse reaction is the negative of that of the forward reaction, and so is +1.191 MeV. The energy required would be nearly zero, although for the reaction to occur, some energy is required to overcome the electric repulsion of the nucleus.
5. (D) Fe; Ra. The largest binding energy per nucleon occurs at the peak of the curve in Fig. 40.6, or, of the nuclei named, for Fe (iron). Since the curve per nucleon is almost constant, decreasing very slowly for large mass numbers, the largest binding energy occurs at the far right end of the curve, or for the named nucleus with the largest mass number, Ra (radium).

Checkup 40.3

1. Because alpha and beta decays involve the emission of charged particles, they result in a change of the atomic number of the nucleus; gamma decay involves the emission of a photon, which carries no charge, and so does not result in transmutation.
2. The direction of deflection for a positron would be to the left, in the same direction as the α ray, since the antielectron carries an elementary positive charge.
3. No. To decay, that is, to spontaneously overcome the binding energy and emit particles, the initial energy of the nucleus must be greater than the final energy of the system, and so the Q value (initial minus final rest-mass energy) must be positive.
4. (D) ^{136}Ba. The negative beta decay results in a unit *increase* in atomic number Z, transmuting cesium into barium. Beta decay does not change the mass number A = 136, so the decay product is ^{136}Ba.

Checkup 40.4

1. The decay rate is proportional to the number of nuclei and inversely proportional to the half-life [Eq. (40.28)]. Since the atomic masses of the isotopes in Table 40.3 differ by less than a factor of 20, so do the number of nuclei in 1.0 μg. The extreme values of half-life in the table differ from the nearest values by much more than a factor of 20, so the smallest and largest half-life determine the highest and lowest decay rate, respectively. Thus, ^{131}I has the highest decay rate and ^{238}U, the lowest.
2. As discussed before and after Eq. (40.23), the number of radioactive nuclei is $\frac{1}{2}$ the initial number after one half-life, or 8.04 days for ^{131}I (see Table 40.3); the number is $\frac{1}{4}$ the initial number after two half-lives, or 16.08 days, and is $\frac{1}{8}$ the initial number after three half-lives, or 24.12 days.
3. The fraction of radioactive nuclei remaining at $t = \tau$ is, according to Eq. (40.25), $e^{-1} \approx 0.368$. According to Fig. 40.14, the initial amount is 1.00 g, and so the amount remaining after τ = 42 years is 0.37 g.
4. (B) 2.0×10^9. From Table 40.3, 3200 years is two half-lives, so the number of radioactive nuclei and the decay rate will decrease by a factor of $(\frac{1}{2})^2 = \frac{1}{4}$ to the value 2.0×10^9.
5. (A) ^{14}C. ^{226}Ra has a half-life 1600/5730 $\approx$ 0.3 times that of ^{14}C. However, 1.0 μg of ^{14}C has more nuclei by a factor of 226/14 $\approx$ 16 than 1.0 μg of ^{226}Ra, and thus has a higher decay rate by a factor of about $0.3 \times 16 \approx 5$.

Checkup 40.5

1. It is the stored electric potential energy of protons held close together that is released when the nuclear fragments separate.
2. From the curve of binding energy, Fig. 40.6, we can see that the fission of a light nucleus would result in a smaller binding energy per nucleon; that is, it would require a net increase of energy.
3. A release of less than one neutron per fission would result in less than one additional fission, and so would not provide a chain reaction; similarly, exactly one neutron could in principle provide a steady reaction, but since some neutrons would be lost, it would not provide a true chain reaction. A release of even slightly more than one neutron per fission could indeed result in a chain reaction, since one or more additional fissions could be caused by each fission.
4. (B) The rest mass of the products is less than the sum of the rest masses of ^{235}U and the neutron. Since energy is released in the reaction, the rest mass of the system decreases. The other choices are not valid, because a variety of products are possible, and because the number of nucleons, 236, remains the same.

Checkup 40.6

1. To ensure a chain reaction, you could lose up to two; that is, to continue a steady reaction, each fission must produce one neutron that is later absorbed.
2. A mass of 100 g is too small in size to ensure that enough neutrons are reabsorbed to support a chain reaction; that is, too many neutrons escape from the mass.
3. The ^{238}U nuclei absorb too many neutrons in nonfission events, preventing those neutrons from causing the fission of ^{235}U.
4. No, the new nuclear fuel is produced from extra uranium placed around the reactor, not from fuel being consumed inside the reactor.

5. (D) Slows neutrons so that they are more likely to be absorbed by ^{235}U in fuel rods later. Thus, the moderator enhances the reaction.

Checkup 40.7

1. Heavy nuclei contain so many protons that electrostatic repulsive forces push the nuclei apart, so that fusion is not energetically favorable.
2. The reaction (40.35) requires that each of the reactions (40.33) and (40.34) occur twice. The net result is that four ^{1}H nuclei are consumed and one ^{4}He nuclei is produced (along with two positrons, two neutrinos, and two gamma rays).
3. No; fission can proceed spontaneously or with the absorption of slow (low-temperature) neutrons.
4. No; all the nuclear products of reactions (40.33)–(40.35) are stable nuclei. However, penetrating radiation is produced, including the gamma ray produced in reaction (40.34) and the gamma rays produced when the positron of reaction (40.33) annihilates an electron.
5. (D) $^{2}H + {}^{2}H + {}^{2}H \rightarrow {}^{4}He + {}^{1}H$ + neutron. While all the net reactions listed are in principle possible, only (D) represents the sum of the given reaction (production of tritium from two deuterium nuclei) and the reaction (40.36) (production of ^{4}He from the tritium product of the given reaction and a third deuterium nucleus).

CHAPTER

41 Elementary Particles and Cosmology

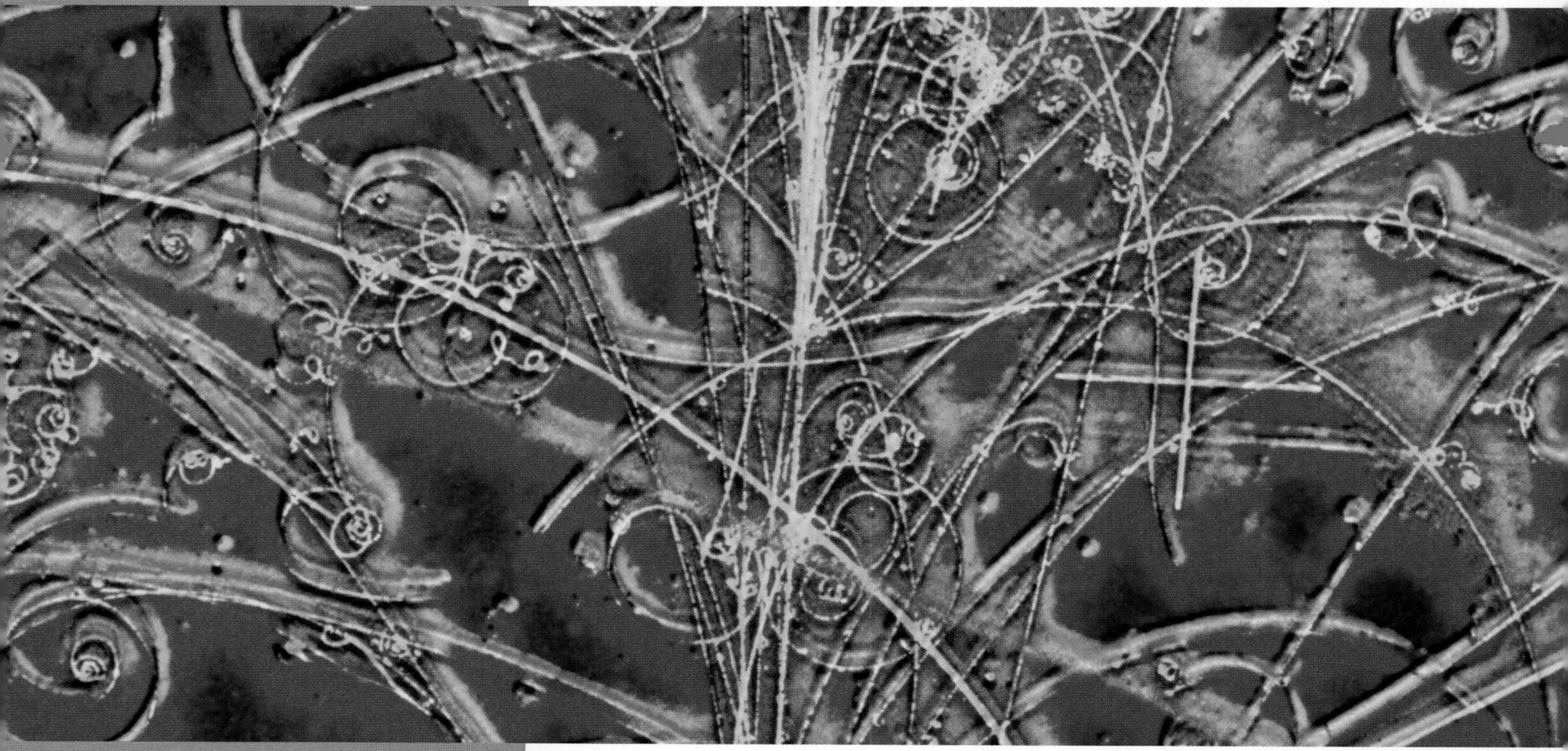

CONCEPTS IN CONTEXT

Concepts *in* Context

Trails of small bubbles reveal the passage of high-energy particles in a bubble chamber filled with liquid hydrogen. A high-energy proton entered the chamber and struck a proton in a hydrogen nucleus, and the collision produced a spray of several kinds of new particles.

While learning about elementary particles in this chapter, we will consider such questions as:

? How do charged particles make trails of bubbles in a bubble chamber? (Section 41.1, page 1399)

? The several tracks emerging from the collision shown in Fig. 41.5 reveal the creation of new particles in the collision. What reaction created these particles? (Section 41.1, page 1400)

? How can we calculate the momentum of a particle from the radius of curvature of the track? (Example 2, page 1401)

? Besides energy and momentum, what quantities are conserved in a reaction that creates particles, such as the reaction that created the new particles in the photos? (Section 41.3, page 1406)

In this final chapter we briefly explore two fascinating questions: What is the structure of the physical world on the smallest scale? And what is the structure of the physical world on the largest scale? To answer the first of these questions we must turn to particle physics and the search for the elementary, indivisible building blocks of matter. And to answer the second question we must turn to cosmology and the investigation of the dynamics of the Universe on a large scale. We will see that there are profound links between the physics of elementary particles and the evolution of the Universe. The study of these links relies on a wide range of tools—from high-energy accelerators used to re-create the kinds of particles that populated the Universe immediately after the Big Bang, to large optical and radio telescopes needed to observe distant regions of the Universe.

Early in the last century physicists discovered that the atom is not an elementary, indivisible unit—each atom consists of electrons orbiting around a nucleus. The electron is a truly elementary particle, the first elementary particle to be discovered. But the nucleus is not an elementary, indivisible unit—each nucleus consists of protons and neutrons packed tightly together. In the 1930s, physicists began to build accelerating machines producing beams of energetic protons or electrons that could serve as projectiles; with these atom smashers physicists could split the nucleus. In the 1950s, they built much larger and more powerful accelerating machines; with these new machines they attempted to split the proton and neutron. But the result of these attempts was chaos: when struck by very energetic projectiles, the proton and neutron do not split into any simple subprotonic pieces. Instead, such violent collisions generate a multitude of new, exotic particles by the conversion of kinetic energy into mass. For want of a better name, these new particles were called "elementary particles." However, most of these new particles are more massive and more complicated than protons and neutrons—they do not appear to be truly elementary, indivisible units. After much effort, theoretical physicists imposed some order on this chaos. They found convincing evidence that protons, neutrons, and many other "elementary particles" are made of very small, compact subunits. The subprotonic building blocks are called **quarks**.

41.1 THE TOOLS OF HIGH-ENERGY PHYSICS

Protons and neutrons are much "harder" than atoms or nuclei. A projectile of a bombarding energy of a few eV can shatter an atom, and a projectile of a bombarding energy of a few MeV can shatter a nucleus. But to make a dent in a proton, we need a bombarding energy of a few hundred or thousand MeV. Elementary-particle physicists like to measure the energies of their projectiles in billion electron-volts (GeV) or in trillion electron-volts (TeV). Expressed in joules, these energy units are

$$1\ \text{GeV} = 10^9\ \text{eV} = 1.60 \times 10^{-10}\ \text{J}$$

$$1\ \text{TeV} = 10^{12}\ \text{eV} = 1.60 \times 10^{-7}\ \text{J}$$

The acceleration of a projectile to such a high energy requires large, sophisticated accelerator machines. High energy—one GeV to thousands of GeV—characterizes the realm of particle physics. Collisions between particles at these high energies are capable of creating a large variety of new particles by the conversion of kinetic energy into mass.

accelerator

The two largest **accelerators** are the proton synchrotrons at the Fermi National Accelerator Laboratory (Fermilab, near Chicago) and at the *Conseil Européen pour la Recherche Nucléaire* (CERN, on the Swiss–French border near Geneva). Figure 41.1 shows an overall view of Fermilab; the accelerator is buried underground in a circular tunnel 6 km in circumference. The CERN accelerator tunnel is even larger, about 27 km in circumference. The large hadron collider (LHC) under construction at CERN will provide proton collisions with an energy of 14 TeV. The Fermilab Tevatron accelerator provides the largest energy now available, producing a beam of protons with an energy of 1 TeV and a speed of 99.9995% of the speed of light. The protons travel in an evacuated circular beam pipe (Fig. 41.2). Large magnets placed along this pipe exert forces on the protons, preventing their escape—the protons move as if on a circular racetrack. At regular intervals the pipe is joined to cavities connected to high-voltage oscillators; in each of these cavities, oscillating electric fields impel the protons to higher energy. After several hundred thousand circuits around the racetrack, the protons reach their final energy of 1 TeV.

Before the protons are allowed to enter the main accelerator ring, they must pass through several smaller preliminary accelerators. The protons generated by a proton gun first pass through an electrostatic generator, then through a linear accelerator, and then through a "small" circular accelerator. Only after the protons have passed through these preliminary stages do they enter the main ring. Once the particles have been given their maximum energy, they are guided out of the accelerator by steering magnets and made to crash against a target consisting of a block of metal or a tankful of liquid. Within the material of the target, the high-energy particles collide violently with the protons or neutrons of the nuclei. The reactions that take place in the collisions create new particles by conversion of energy into mass. Unfortunately, not all of the kinetic energy of the incident particles can participate in these reactions. As we saw in Section 11.3, the velocity of the center of mass remains constant during the collision, and therefore the particles must retain some kinetic energy. A relativistic calculation indicates that, in the collision of a high-energy proton and a stationary proton, the parti-

FIGURE 41.1 Panoramic view of Fermilab.

FIGURE 41.2 View of the underground tunnel housing the Tevatron accelerator at Fermilab. The long row of magnets encases the beam pipe.

cles retain most of the kinetic energy, and only a small fraction becomes available for reactions. Such collisions are quite inefficient.

The efficiency improves drastically if two high-energy protons are made to collide head-on. The available energy is then the sum of the energies of the two protons. At Fermilab, a second beam pipe has been installed in the tunnel housing the original beam pipe. A beam of antiprotons travels in this beam pipe, in a direction opposite to that of the protons in the original beam pipe. These two opposite beams cross at two intersection points (Fig. 41.3), where the violent head-on collisions create a wide variety of new particles.

FIGURE 41.3 The Collider Detector at Fermilab (CDF).

In order to observe the particles that emerge from the scene of these collisions, physicists have used several kinds of particle detectors. Some of these—scintillation counters and Cerenkov counters—signal the passage of each electrically charged particle by giving off brief (and weak) flashes of light; these flashes of light are detected by sensitive photomultiplier tubes (see Chapter 37, *Physics in Practice: Photomultiplier*, page 1268). Other detectors—bubble chambers and multiwire chambers—render the tracks of electrically charged particles visible, either on a photographic record or on a computer-generated picture.

bubble chamber

Concepts in Context

A **bubble chamber** is a tank filled with superheated liquid, usually liquid hydrogen, whose temperature is slightly above the boiling point. Such a liquid is unstable—it is about to start boiling but it will usually not start until some disturbance triggers the formation of the first few bubbles. A charged particle zipping through the chamber provides just the kind of disturbance the liquid is waiting for—a fine trail of small bubbles forms in the wake of the particle's passage. High-speed cameras can take a picture of these bubble tracks before they disperse and disappear in the turmoil of subsequent widespread bubbling and boiling of the liquid.

Figure 41.4 shows the BEBC bubble chamber that was used at CERN for many years in a wide variety of experiments. The bubble chamber is surrounded by a large electromagnet, which aids in the identification of the particles passing through the bubble chamber. The magnetic field generated by this magnet pushes the particles into curved orbits as they pass through the chamber. The direction of the curvature depends on the sign of the electric charge, and the radius of the curved orbit is proportional to the momentum

(a)

(b)

FIGURE 41.4 (a) The Big European Bubble Chamber (BEBC) at CERN, surrounded by the magnet, which almost completely hides it from view. (b) The BEBC before its installation into the magnet.

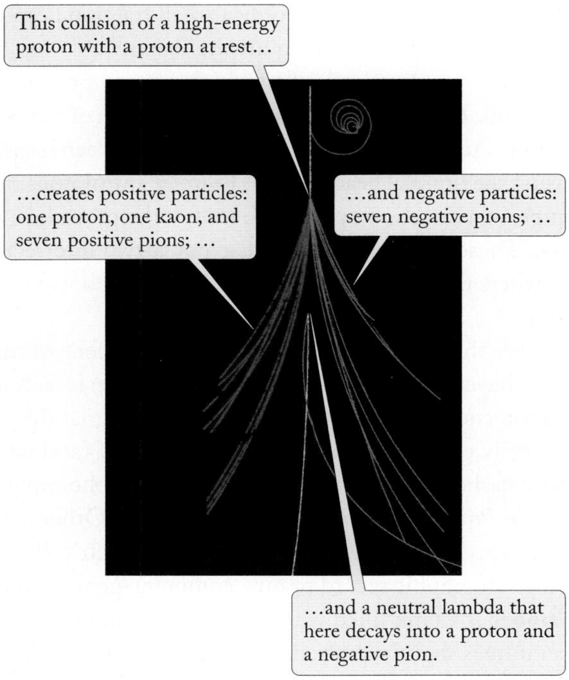

FIGURE 41.5 Tracks of particles in a bubble chamber (the tracks have been colored to distinguish different kinds of particles; in the original bubble-chamber photograph the tracks are all white). A proton enters the bubble chamber (yellow track from top of photograph) and collides with one of the protons at rest in the nucleus of one of the hydrogen atoms in the liquid filling the chamber. The tracks of positively charged particles (red) and negatively charged particles (blue) created in this collision can be seen emerging from the scene of the accident. Besides the charged particles, a neutral lambda was created in the collision. Such a neutral particle leaves no track in the bubble chamber, but it reveals itself when it decays into a proton (yellow track at bottom) and a negative pion (purple track)

of the particles (Section 30.1). Thus, measurement of the radius of a bubble track tells us the sign of the electric charge of the particle and the momentum of the particle.

Concepts — in — Context

The photograph of Fig. 41.5 was taken with a bubble chamber. It shows the track of a high-energy proton that entered the bubble chamber and collided with a proton at rest in the core of one of the hydrogen atoms in the liquid filling the chamber. The collision destroyed one of the two protons, but produced a handful of new particles: seven positive pions, seven negative pions, a positive kaon, and a neutral lambda particle. Using the respective symbols for these particles (see Tables 41.2 and 41.3), the reaction can be summarized as follows:

$$\mathrm{p} + \mathrm{p} \rightarrow \mathrm{p} + 7\pi^+ + 7\pi^- + \mathrm{K}^+ + \Lambda \tag{41.1}$$

The lambda particle subsequently decayed into a proton and a negative pion:

$$\Lambda \rightarrow \mathrm{p} + \pi^- \tag{41.2}$$

The sum of the rest masses of the particles after the reaction (41.1) is much larger than the sum of the rest masses of two protons. The excess mass comes from the conversion of energy into mass—some of the kinetic energy of the incident proton has been converted into mass.

This conversion of kinetic energy into mass plays a crucial role in the discovery of new particles. Almost all the new particles discovered during the past 30 years are considerably heavier than protons and neutrons. Physicists need powerful accelerators to produce the large kinetic energies that must be supplied for the manufacture of these new heavy particles.

EXAMPLE 1 Elementary-particle physicists prefer to measure the masses of particles in units of MeV/c^2, using the conversion 1 u = 931.5 MeV/c^2. Thus a proton has a mass of 938 MeV/c^2, a pion has 139.6 MeV/c^2, a kaon has 494 MeV/c^2, and lambda has 1115 MeV/c^2 (see Tables 41.2 and 41.3).

Given these masses, calculate the minimum kinetic energy required for each proton if we want to initiate the reaction (41.1) in a head-on collison between two protons of equal kinetic energies.

SOLUTION: In a head-on collison between two protons of equal energies, all of the kinetic energy can be converted into rest-mass energy (this would *not* be true for the collision of a high-energy proton with a stationary proton; that is, it would not be true for the collision as photographed in Fig. 41.5). If the kinetic energy of each proton is K, the kinetic energy of both is $2K$, and this must equal the difference of rest-mass energy before and after the reaction:

$$\begin{aligned} 2K &= m_p c^2 + 14[\text{mass of pion}]c^2 + [\text{mass of kaon}]c^2 \\ &\quad + [\text{mass of lambda}]c^2 - 2m_p c^2 \\ &= 14 \times 139.6\ \text{MeV} + 494\ \text{MeV} + 1115\ \text{MeV} - 938\ \text{MeV} \\ &= 2625\ \text{MeV} \end{aligned} \tag{41.3}$$

Thus, the kinetic energy of each proton must be $2625\ \text{MeV}/2 = 1313\ \text{MeV}$.

EXAMPLE 2 In a detector region with a uniform magnetic field of 4.38 T, a pion leaves a track with a radius of curvature of 1.46 m. What is the momentum of the pion? The magnetic field is applied in the downward direction, and the motion of the pion is horizontal, clockwise when looking from above. What is the sign of the elementary charge on the pion?

Concepts — in — Context

SOLUTION: From Eq. (30.6), the magnitude of the momentum of a pion in circular motion in a uniform magnetic field is

$$p = qBr = 1.60 \times 10^{-19}\ \text{C} \times 4.38\ \text{T} \times 1.46\ \text{m} = 1.02 \times 10^{-18}\ \text{kg·m/s}$$

As mentioned in Section 30.1, this expression for the momentum is relativistically correct. In relativistic calculations, the momentum is often expressed in MeV/c, or GeV/c:

$$\begin{aligned} p &= 1.02 \times 10^{-18}\ \text{kg·m/s} \times \frac{1.00\ \text{MeV}}{1.60 \times 10^{-13}\ \text{J}} \times \frac{3.00 \times 10^8\ \text{m/s}}{c} \\ &= 1.92 \times 10^3\ \text{MeV}/c = 1.92\ \text{GeV}/c \end{aligned}$$

From this form it is apparent that p^2c^2 is much greater than the square of the pion rest-mass energy, 140 MeV (Example 1). Recalling that $E^2 = p^2c^2 + (mc^2)^2$, we see that $E^2 \gg (mc^2)^2$, which means that this pion is extremely relativistic.

With the magnetic field downward, the magnetic force $\mathbf{F} = q\mathbf{v} \times \mathbf{B}$ implies counterclockwise motion for a positive particle. The clockwise motion implies that the motion is that of a negative pion.

Although bubble chambers yield excellent pictures of particle tracks, they are very complex, very large, and very expensive machines, and they cannot take picures fast enough to satisfy the needs of experimenters. Bubble chambers have now been replaced by various types of electronic **tracking chambers**, for example, **multiwire chambers**. These chambers are strung with thousands of fine wires, in an array that resembles a

tracking chamber

multiwire chamber

FIGURE 41.6 Wires strung in a cylindrical array for a multiwire chamber. The array is shown during construction, before it was placed inside the vacuum chamber.

coordinate grid (Fig. 41.6). The wires are connected to voltage supplies and to current sensors. When a charged particle passes through the gas in the chamber, it ionizes the gas along its track, and the electrons released in this ionization are attracted to the wires near the track, where they form miniature electric discharges, which are detected as current pulses by the sensors individually connected to the wires. These current pulses reveal which wires are near the track, and a computer can then immediately reconstruct the location of the track and draw a picture of the track. An important advantage of multiwire chambers over bubble chambers is that the raw data are in the form of current pulses, which can be directly processed and stored in digital form by a computer. In contrast, the raw data from bubble chambers were in the form of photographs, and the tracks recorded on these photographs had to be measured and converted into numbers before they could be processed and analyzed by a computer; the conversion of photographic data into digital data is very time-consuming.

The UA1 detector at CERN incorporates several multiwire chambers. Figure 41.7 shows a computer-generated picture of tracks of particles in the multiwire chamber of this detector.

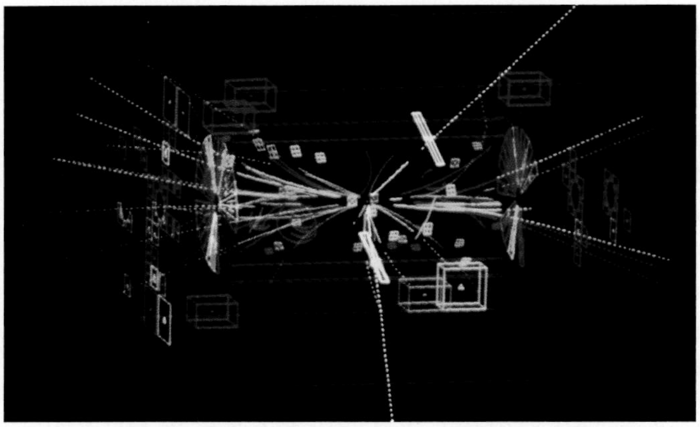

FIGURE 41.7 Computer-generated picture of particle tracks in the multiwire chamber of the UA1 detector at CERN. These are the tracks of particles produced in the collision of a proton and an antiproton that entered the chamber horizontally from opposite directions.

✔ Checkup 41.1

QUESTION 1: Do neutrons make tracks in bubble chambers? In multiwire chambers?

QUESTION 2: For what purpose are particle detectors placed in magnetic fields?

QUESTION 3: Figure 41.5 shows the tracks of positive and of negative particles in a bubble chamber immersed in a magnetic field. The magnetic field is perpendicular to the plane of the page. Is it directed into the page or out of the page?

QUESTION 4: Is the Q value for the reaction (41.1) positive or negative? For the reaction (41.2)?

QUESTION 5: What is the advantage of head-on collisions between protons of opposite motion over collisions with a stationary target?

(A) The head-on collision is easier to achieve than hitting the stationary target.
(B) In the head-on collision, the full kinetic energy is available for reaction.
(C) In the head-on collision, the backward recoil makes detection easier.
(D) In the head-on collision, both protons and antiprotons can be used.
(E) In the head-on collision, exactly twice as much energy is available.

41.2 THE MULTITUDE OF PARTICLES

As physicists built accelerators of higher and higher energy, they discovered more and more new particles. By now, several hundred particles are known. Whenever a new, more powerful accelerator makes available more energy for collisions, more particles and heavier particles are produced—there seems to be no end in sight.

The known particles fall into three groups: leptons, baryons, and mesons. The leptons are distinguished from baryons and mesons by the interactions in which they participate. Leptons interact via the electromagnetic force and the weak force (we already became acquainted with the weak force in Chapter 40, where we found that the weak force produces some radioactive decay reactions involving neutrinos). In contrast, baryons and mesons interact via the electromagnetic force, the weak force, *and* the strong force. Because the strong force is stronger than the other forces, it predominates in reactions among baryons and mesons; accordingly, baryons and mesons are called strongly interacting particles, or **hadrons** (from the Greek *hadros*, thick). The distinction between baryons and mesons is based on their spin: baryons have half-integer spin ($\frac{1}{2}\hbar, \frac{3}{2}\hbar, \frac{5}{2}\hbar, \ldots$) and mesons have integer spin ($0, \hbar\ 2\hbar, \ldots$). Particles with half-integer spin are called **fermions**, and particles with integer spin are called **bosons**. Thus, baryons are fermions, and mesons are bosons.

hadrons

bosons and fermions

leptons

There are six different **leptons** (see Table 41.1). Among these, the electron (e) is the most familiar. The muon (μ) is very similar to the electron; it has the same electric charge as an electron but its mass is about 200 times as large. The tau (τ) is also similar to the electron, but its mass is even larger than that of the muon. The neutrinos (ν_e, ν_μ, and ν_τ) are particles of zero electric charge and very small mass. Until recently, neutrinos were believed to have zero mass, like photons. But now there is clear evidence that neutrino masses are not zero. In Table 41.1, the masses of the particles are expressed in units of MeV/c^2 and the electric charge and the spin are expressed in multiples of the elementary charge e and in multiples of Planck's constant $\hbar$. The choice of MeV/c^2 for the unit of mass is very convenient in calculations involving the rest-mass energies; for instance, from the electron mass of 0.511 MeV/c^2 listed in the table, we immediately see that the electron rest-mass energy is 0.511 MeV. All the leptons have spin $\frac{1}{2}$.

Besides the six leptons of Table 41.1, there are six antileptons: the antielectron (or positron), the antimuon, the antitau, and the three antineutrinos. These antiparticles have the opposite electric charge and some other opposite properties, but they have

TABLE 41.1 THE LEPTONS

PARTICLE	SYMBOL	MASS[a]	SPIN	CHARGE	COMMENTS
electron	e	0.511 MeV/c^2	$\frac{1}{2}$	−1	constituent of atoms
muon	μ	105.6 MeV/c^2	$\frac{1}{2}$	−1	produced in decay of pion; abundant in cosmic rays
tau	τ	1784 MeV/c^2	$\frac{1}{2}$	−1	
electron neutrino	ν_e	< 3 eV/c^2	$\frac{1}{2}$	0	produced in beta decay; abundantly emitted by Sun
muon neutrino	ν_μ	< 0.19 MeV/c^2	$\frac{1}{2}$	0	
tau neutrino	ν_τ	< 18.2 MeV/c^2	$\frac{1}{2}$	0	

[a] Neutrinos are now known to undergo quantum-mechanical oscillations between different mass states. Many experiments are underway to clarify such processes.

TABLE 41.2 SOME BARYONS

PARTICLE	SYMBOL	MASS	SPIN	CHARGE	QUARK CONTENT	COMMENTS
proton	p	938.3 MeV/c^2	$\frac{1}{2}$	+1	uud	constituent of nucleus
neutron	n	939.6	$\frac{1}{2}$	0	ddu	constituent of nucleus
lambda	Λ	1115	$\frac{1}{2}$	0	uds	has strangeness −1
sigma-plus	Σ^+	1189	$\frac{1}{2}$	+1	uus	has strangeness −1
sigma-minus	Σ^-	1197	$\frac{1}{2}$	−1	dds	has strangeness −1
sigma-zero	Σ^0	1192	$\frac{1}{2}$	0	uds	has strangeness −1
xi-minus	Ξ^-	1321	$\frac{1}{2}$	−1	dss	has strangeness −2
xi-zero	Ξ^0	1315	$\frac{1}{2}$	0	uss	has strangeness −2
omega-minus	Ω^-	1672	$\frac{3}{2}$	−1	sss	has strangeness −3
charmed lambda	Λ_c	2280	$\frac{1}{2}$	+1	udc	contains charmed quark

exactly the same mass and spin as the corresponding particles. (The notation for an antiparticle is a bar over the letter or, alternatively, a superscript indicating the electric charge; thus, the notation for the antielectron is $\bar{e}$ or e^+.)

baryons

The **baryons** are the most numerous group of particles. The most familiar baryons are the proton and the neutron; these are the baryons of the least mass. Table 41.2 lists some of the other baryons. For every baryon in Table 41.2, there exists an antibaryon. As in the case of leptons, these antiparticles have an opposite charge but the same mass and spin as the corresponding particles.

mesons

Finally, the **mesons** are another numerous group of particles. Some of them are listed in Table 41.3. The most familiar mesons are the three pions (with electric charges 0, $+e$, $-e$); these were the first mesons to be discovered.

For every meson there exists an antimeson; however, these antiparticles have already been included in Table 41.3. For example, the antiparticle to the π^- is the π^+, and vice versa. The antiparticle to the π^0 is the π^0; this means that when two π^0 mesons meet, they can annihilate each other [see Section 40.3 and the discussion before Eq. (36.47)]. Also, the π^0 usually decays into two photons, in essence annihilating itself.

Most of the particles are unstable; they decay, spontaneously falling apart into several other particles. The only absolutely stable particles are the electron, the proton, and the neutrinos. The lifetimes of the other particles range from a fairly long 15 minutes for the neutron to about 10^{-23} s for many of the exotic particles.

A particle that lives only 10^{-23} s is incapable of making a measurable track in a tracking chamber or other detector. Thus, such a short-lived particle cannot be detected directly, but its existence can be inferred from a careful study of the rates of reactions of longer-lived particles engaged in collisions. The short-lived particle participates in these reactions as an intermediate, ephemeral state, which causes a characteristic increase of the reaction rate whenever the energy of the colliding particles coincides with the energy required for the production of the short-lived particle.

resonances

Because of their enhancing effects on reaction rates, the short-lived particles are often called **resonances**.

TABLE 41.3 SOME MESONS

PARTICLE	SYMBOL	MASS	SPIN	CHARGE	QUARK CONTENT	COMMENTS
pion-zero	π^0	135.0 MeV/c^2	0	0	$u\bar{u}$ and $d\bar{d}$	abundant in cosmic rays; carrier of nuclear force
pion-plus	π^+	139.6	0	+1	$u\bar{d}$	abundant in cosmic rays; carrier of nuclear force
pion-minus	π^-	139.6	0	−1	$d\bar{u}$	abundant in cosmic rays; carrier of nuclear force
kaon-zero	K^0	498	0	0	$d\bar{s}$	has strangeness +1
kaon-plus	K^+	494	0	+1	$u\bar{s}$	has strangeness +1
kaon-minus	K^-	494	0	−1	$s\bar{u}$	has strangeness −1
J/psi	J/ψ	3097	1	0	$c\bar{c}$	contains charmed quarks
D-zero	D^0	1865	0	0	$c\bar{u}$	contains charmed quark
D-plus	D^+	1869	0	+1	$c\bar{d}$	contains charmed quark
upsilon	Υ	9460	1	0	$b\bar{b}$	contains bottom quarks

Checkup 41.2

QUESTION 1: Do all leptons have spin $\frac{1}{2}$? All baryons?

QUESTION 2: What are the mass, spin, and electric charge of an antitau? An electron antineutrino?

QUESTION 3: Which is the heaviest particle listed in Tables 41.1–41.3?

QUESTION 4: Particles not listed in Tables 41.1–41.3 include the neutral eta particle (η^0), which has spin $\frac{1}{2}$; and the neutral B particle (B^0), a massive particle with spin 0. To what group of particles does the η^0 belong? The B^0?

(A) Lepton, baryon (B) Lepton, meson (C) Baryon, lepton
(D) Baryon, meson (E) Meson, baryon

41.3 INTERACTIONS AND CONSERVATION LAWS

The reactions that occur among the particles in a high-energy collision are governed by the four fundamental forces: the "strong" force, the electromagnetic force, the "weak" force, and the gravitational force. However, at the microscopic level, particles are subject to quantum uncertainties in position and momentum, and the force acting on them is not well defined. Hence, physicists usually prefer to speak of four fundamental types of **interactions**, instead of forces.

strong force

The **strong force** acts on baryons and mesons, but not on leptons. As we will see in Section 41.5, the strong force is a special case of an even stronger force, called the "color" force.

electromagnetic force

The **electromagnetic force** acts primarily on charged particles, but it also acts on neutral particles—such as the neutron—that contain an internal distribution of electric charge and are endowed with magnetic moments.

weak force

The **weak force** acts on leptons, baryons, and mesons. However, its effect on baryons and mesons is often hidden behind the much larger effects produced by the strong or the electromagnetic force. To see the purest manifestation of the weak force, we have to examine reactions involving leptons. The weak force is deeply involved in many reactions that bring about the decay of unstable particles. For example, the weak force is responsible for the beta decay of the neutron. The weak force is also responsible for other beta decays.

gravitational force

The **gravitational force** is of no direct interest in particle physics. Although all particles and all forms of energy interact gravitationally, the gravitational effects produced by individual particles are too feeble to be of any significance at the energies available in laboratories on Earth. However the gravitational force is of great interest to theorists who study the ultrahigh energies of the early Universe.

Table 41.4 lists the strength of each of the fundamental forces and also the range, or the maximum distance over which this force can reach from one particle to another. In the table, the strengths of the forces are expressed relative to that of the strong force, to which a strength of 1 has been assigned arbitrarily.[1]

Concepts in Context

All the forces, and all the reactions that they produce, obey the usual conservation laws for energy, momentum, angular momentum, and electric charge. Besides these conservation laws for familiar quantities, experiments with high-energy particles have led to the discovery of new conservation laws involving several esoteric quantities, such as baryon number, lepton number, isospin, strangeness, and parity. All reactions obey a conservation law for **baryon number**, which is a generalization of the conservation law for mass number from nuclear physics. Each baryon has a baryon number of $+1$, each antibaryon -1, and all other particles have baryon number 0. The conservation law for baryon number then simply states that the net baryon number remains unchanged in any reaction. This is a mathematical expression of the requirement that any baryons that disappear in a reaction must be replaced by an equal number of other baryons, with antibaryons counting as negative baryons. For example, consider the reaction (41.1). The baryon numbers for p, $\pi^{\pm}$, K^{+}, and Λ are 1, 0, 0, and 1, respectively; hence, in the reaction (41.1), the net baryon number is $1 + 1$ before and $1 + 0 + 0 + 0 + 1$ after; that is, the net baryon number remains unchanged.

baryon number

TABLE 41.4 THE FOUR FUNDAMENTAL FORCES

FORCE	ACTS ON	RELATIVE STRENGTH	RANGE
strong/color	baryons and mesons (hadrons)	1	$\approx 10^{-15}$ m
electromagnetic	particles with charge or magnetic moment	10^{-2}	infinite
weak	leptons, baryons, and mesons	10^{-6}	$\approx 10^{-18}$ m
gravitational	all forms of matter	10^{-38}	infinite

[1]The strengths of the forces depend on the energies of the particles. The values in the table are appropriate for low energies.

lepton number

Lepton number is for leptons what baryon number is for baryons. Each lepton has lepton number +1, each antilepton −1, and all other particles have lepton number 0. The net lepton number remains unchanged in any reaction. See also Problem 18.

isospin

Isospin is somewhat more complicated because it is a quantity with several components; that is, isospin is like a vector quantity. The conservation law for isospin states that the net sum of the isospin "vectors" of all the particles involved in a reaction remains unchanged.

strangeness

Strangeness is similar to the baryon and lepton numbers. Each hadron has a strangeness number: the proton has strangeness 0, the kaon-plus has +1, the lambda has −1, the pion has 0, etc. (values of the strangeness are given in the "Comments" column in Tables 41.2 and 41.3). The strangeness of an antiparticle is the negative of the strangeness of the corresponding particle. The conservation law for strangeness states that the net strangeness number remains unchanged in a reaction. For example, in the reaction (41.1), the net strangeness is $0 + 0$ before and $0 + 0 + 0 + 1 - 1$ after; that is, there is no change.

parity

Parity characterizes the behavior of a quantum-mechanical wave under a reversal of the x, y, and z coordinates. Such a reversal is physically equivalent to forming a mirror image of the wave. It can be shown that the mirror image of the quantum-mechanical wave for a stationary state is either equal to the original wave (parity $= +1$) or else equal to the negative of the original wave (parity $= -1$). Conservation of parity means that the net parity (the product of all the individual parities) of all the particles participating in a reaction is unchanged.

The conservation laws for energy, momentum, angular momentum, electric charge, baryon number, and lepton number are *absolute*—no violation of any of them has ever been discovered. In contrast, the conservation laws for isospin, strangeness, and parity are *approximate*—they are valid for some reactions, but fail in some others. This, of course, raises the question of what possible meaning can be attached to a "law" that works sometimes and fails sometimes. The answer is that whether a conservation law is obeyed or not depends on the kind of interaction, or the kind of force, that drives the reaction. The reactions caused by the strong force obey all the conservation laws, but reactions caused by the electromagnetic or the weak force do not. It is usually easy to tell what force is involved in a reaction: reactions involving the strong force tend to be fast; reactions involving the other forces tend to be (relatively) slow. For example, reaction (41.1) is a fast reaction brought about by the strong force, whereas reaction (41.2) is a slow reaction brought about by the weak force. The difference is evident in Fig. 41.5, where reaction (41.1) appears to occur at a single point whereas reaction (41.2) occurs only after the lambda particle has traveled a measurable distance.

Table 41.5 lists the conservation laws obeyed by the strong (color), electromagnetic, and weak forces.

TABLE 41.5 FORCES AND CONSERVED QUANTITIES

FORCE	ENERGY, MOMENTUM, AND ANGULAR MOMENTUM	CHARGE	BARYON NUMBER	LEPTON NUMBER	STRANGENESS	PARITY	ISOSPIN
strong/color	✓	✓	✓	✓	✓	✓	✓
electromagnetic	✓	✓	✓	✓	✓	✓	
weak	✓	✓	✓	✓			

EXAMPLE 3 Do the following reactions conserve electric charge? Do they conserve baryon number? Do they conserve strangeness? Are the reactions possible?

$$\Lambda \rightarrow n + \pi^0 \tag{41.4}$$

$$\Xi^- \rightarrow K^- + \pi^0 \tag{41.5}$$

SOLUTION: In the first reaction, all particles are electrically neutral, and in the second, there is one negative elementary charge before and after, so both reactions conserve electric charge. Reaction (41.4) conserves baryon number, since the baryon number is 1 for the Λ and $1 + 0$ for $n + \pi^0$. But Eq. (41.5) does not conserve baryon number, since the baryon number is 1 for the Ξ^- and $0 + 0$ for $K^- + \pi^0$. This violation of an absolute conservation law implies that the reaction (41.5) is not a possible reaction.

Reaction (41.4) does not conserve strangeness, since the strangeness is -1 for the Λ and $0 + 0$ for $n + \pi^0$. Reaction (41.4) is possible, but since strangeness is not conserved, it is not mediated by the strong force. Reaction (41.5) does not conserve strangeness, since the strangeness is -2 for the Ξ^- and $-1 + 0$ for $K^- + \pi^0$.

✔ Checkup 41.3

QUESTION 1: What is a hadron? Is the proton a hadron? The neutron? The electron? The photon? The tau neutrino?

QUESTION 2: What are the mass, spin, electric charge, and strangeness of an anti-xi-zero ($\overline{\Xi}^0$)? An anti-xi-minus, or xi-plus ($\overline{\Xi}^-$, or Ξ^+)?

QUESTION 3: What are the quantities conserved in all interactions? Conserved in electromagnetic interactions, but not in weak interactions? Conserved in strong interactions, but not in electromagnetic?

QUESTION 4: Consider the wavefunction $\psi = \sin kx$. What is the parity of this wavefunction, that is, what happens to it when you change x to $-x$?

QUESTION 5: The reaction $\Lambda \rightarrow n + \pi^0$ does not conserve strangeness. Which of the strong, electromagnetic, or weak interactions can cause this reaction?

(A) Strong only
(B) Weak only
(C) Electromagnetic only
(D) Electromagnetic or weak only
(E) Strong, electromagnetic, or weak

QUESTION 6: In which of the four interactions does the electron participate? In which does the neutron participate?

(A) All, all
(B) All, all except electromagnetic
(C) All except weak, all except strong
(D) All except strong, all
(E) All except weak, all except electromagnetic

41.4 FIELDS AND QUANTA

We know from our study of electricity and magnetism that forces are mediated by fields. Distant particles do not act on one another directly; rather, each particle generates a field of force, and this field acts on the other particles. As we saw in Section 23.1, the existence of fields is required by conservation of energy and momentum. Fields play the role of storehouses of energy and momentum; the energy and momentum stored in the fields balance any excess or deficit in the energy and momentum of the interacting particles engaged in (nonuniform) motion. The most spectacular example of the conversion of particle energy into field energy occurs in the annihilation of matter with antimatter: if an electron collides with an antielectron, the two particles annihilate one another, giving off a burst of very energetic light, or gamma rays. According to classical physics, such a burst of light consists of electric and magnetic fields; thus, the annihilation converts all of the energy of the particles—including their rest-mass energy—into field energy. The reverse reaction is also possible: if a gamma ray collides with a charged particle, it can create an electron–antielectron pair. In such **pair creation**, the field energy of the gamma ray is converted into the energy of the pair of particles (Fig. 41.8)

Each of the four fundamental forces is mediated by fields of its own. Hence there are gravitational fields, electromagnetic fields, strong fields, and weak fields. According to quantum physics, the energy stored in fields is not smoothly distributed; rather the energy is found in **quanta**, that is, small packets or lumps of energy that can be regarded as particles.

In Chapter 37, we became familiar with the quanta of the electromagnetic field; these quanta are the photons. Each of the other fundamental fields has quanta of its own. Table 41.6 lists the quanta of all four fundamental fields. Like photons, the quanta of the gravitational field, or **gravitons**, and the quanta of the strong field, or **gluons**, are both massless. In contrast, the quanta of the weak field, or $\mathbf{W^+}$, $\mathbf{W^-}$, and $\mathbf{Z^0}$ **particles**, are endowed with mass.

At the quantum level, we can picture the field of force generated by a particle as a swarm of quanta buzzing around the particle. For example, we can picture the electric field surrounding an electron, or any other charged particle, as a swarm of photons. The swarm is in a state of everlasting activity—the charged particle continually emits and reabsorbs the photons of the swarm. Emission is creation of a photon; absorption is annihilation of a photon. Hence we can say that the electric field arises from the continual interplay of three fundamental processes: a photon is emitted by one particle, propagates through

FIGURE 41.8 The two spiraling tracks in this bubble-chamber photograph were made by an electron (green) and an antielectron (red). These particles were created by a high-energy gamma ray in a collision with the electron of a hydrogen atom in the bubble chamber.

gravitons
gluons
W^+ particles
W^- particles
Z^0 particles

TABLE 41.6 FIELDS AND THEIR QUANTA

FIELD	QUANTA	MASS
gravitational	gravitons	0 MeV/c^2
weak	W particles	80.4
	Z particles	91.2
electromagnetic	photons	0
strong/color	gluons	0

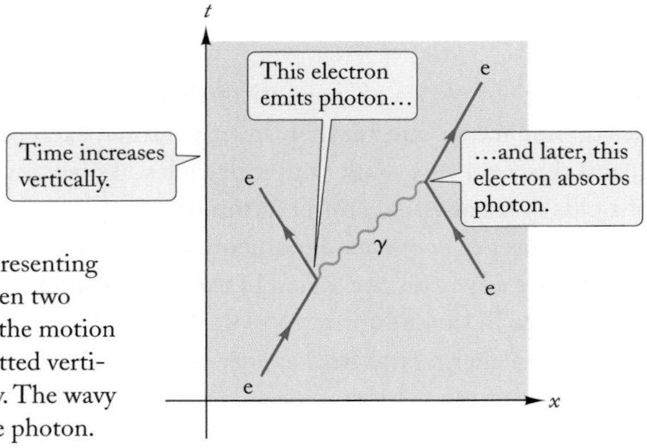

FIGURE 41.9 Feynman diagram representing the exchange of a virtual photon between two electrons. The solid blue lines indicate the motion of the two electrons, with the t axis plotted vertically and the x axis plotted horizontally. The wavy colored line indicates the motion of the photon.

Feynman diagram

virtual photon

carrier

the intervening distance, and is absorbed by the other particle. This exchange process can be represented graphically by a **Feynman diagram** (Fig. 41.9), invented by Richard Feynman. The photon exchanged between the two electrons is called a **virtual photon**, because it lasts only a very short time and, being reabsorbed by an electron, is undetectable by any direct experiment. *The steady attractive or repulsive force between two charged particles is generated by continual repetition of the photon exchange process.* The photon is the **carrier** of the electromagnetic forces. This is action-by-contact with a vengeance—at a fundamental level, all the electromagnetic forces reduce to local acts of creation and destruction involving particles and photons in direct contact.

In terms of a simple analogy, we can easily understand how the exchange of particles brings about forces. Imagine two boys tossing a ball back and forth between them (Fig. 41.10); it is obvious from momentum conservation that this produces a net repulsive force between the boys due to the recoil they suffer when throwing or catching the ball. Our intuition suggests that no such exchange process can ever produce attraction. However, imagine two Australian boys tossing a boomerang back and forth between them in the manner shown in Fig. 41.11; it is then obvious that this produces an attractive force between the boys. Whether a photon exchanged between two charges behaves like a ball or like a boomerang depends on the signs of the charges. Quantum calculations, which take into account the wave nature of all the particles involved, show that the net force is attractive for unlike charges and repulsive for like charges, as it should be.

RICHARD PHILLIPS FEYNMAN (1918–1988) *American physicist. His invention of the Feynman diagram revolutionized the computations of relativistic quantum processes. He shared the 1965 Nobel Prize with* **Julian Schwinger** (1918–1994), *American physicist, and* **Sin-Itiro Tomonaga** (1906–1979), *Japanese physicist, for work on the quantum theory of electrons.*

FIGURE 41.10 Two boys throwing a ball back and forth.

FIGURE 41.11 Two boys throwing a boomerang back and forth.

The gravitational, weak, and strong forces are also generated by the exchange of virtual particles: gravitons, W and Z particles, and gluons. It is a general rule of quantum theory that the range of the force is inversely related to the mass of the particle that serves as the carrier of the force. Thus, the photons and the gravitons that are the carriers of the electromagnetic and the gravitational force have zero mass—the ranges of these forces are infinite. The W and Z particles that are the carriers of the weak force have a very large mass—the range of this force is very short. The gluons that are carriers of the strong force have zero mass, and accordingly the range of the force ought to be infinite. But because this force is very strong, its action over any appreciable distance triggers the spontaneous creation of particle-antiparticle pairs, and this effectively obstructs the force and limits it to short distances.

Although the four fundamental forces seem drastically different, theoretical physicists have sought to formulate a unified theory that treats several or all of these forces as aspects of a single, more fundamental force. Electromagnetism is the most familiar example of a **unified field theory**; that is, it is a theory that treats electric and magnetic forces as two aspects of a single, underlying force. To appreciate fully the unification of electricity and magnetism we would have to examine what relativity says about electric and magnetic fields; we could then see that electric and magnetic forces are merely two aspects of a single force called the electromagnetic force. But even without adopting a relativistic point of view, we can see from Maxwell's equations that electricity and magnetism are intimately connected.

unified field theory

If we seek to unify the electromagnetic and weak forces, we must regard the carriers of these forces—the quanta whose exchange generates the force—as closely related. This would seem to contradict the large mass difference between these particles: the photon is massless, but the W and Z particles are the heaviest particles now known. The unified theory of the **electroweak force** formulated by S. Weinberg, A. Salam, and S. Glashow attributes this mass difference to an imperfect symmetry (a "broken

electroweak force

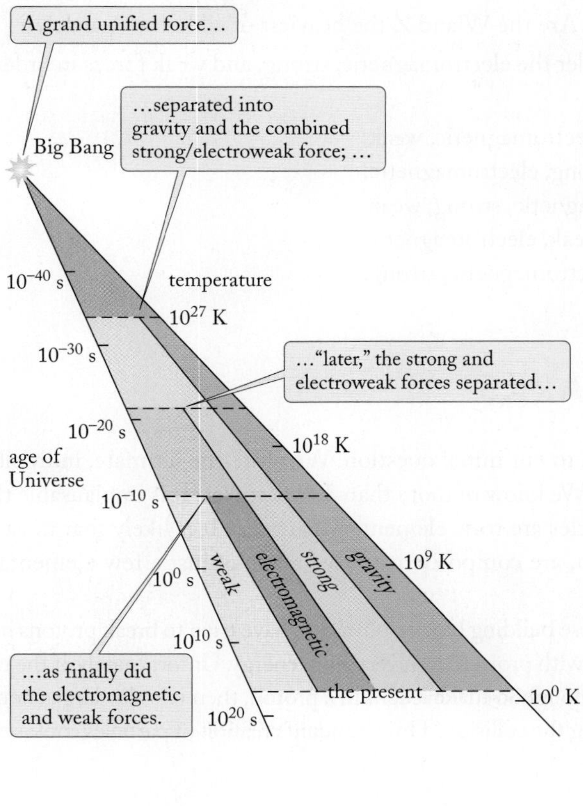

FIGURE 41.12 The evolution of the Universe after the Big Bang.

symmetry") between the photon and the W and Z particles. The theory asserts that perfect symmetry between these particles can be restored by giving them very high energies, in excess of 100 GeV; at such high energies, the photon and the W and Z particles should become essentially identical. Such high energies are difficult to achieve in our laboratories, but they were readily available during the early stages of the Big Bang (see Section 41.6), when the Universe was younger than 10^{-10} s and had a temperature in excess of 10^{15} K. It is believed that at these early times, there was no difference between the photon and the W and Z particles, and there was no difference between the electromagnetic and weak forces. Furthermore, many theoretical physicists are now striving to formulate a **grand unified theory** (GUT) which presumes that the gravitational, electroweak, and strong forces were all originally a single, unified force. It is believed that at about 10^{-34} s after the beginning of the Big Bang, gravity separated from the still-unified strong and electroweak forces, and the strong force separated from the electroweak force slightly later (see Fig. 41.12).

grand unified theory

The most impressive success of the unified electroweak theory was its prediction of the masses of the W and Z particles. The W and Z particles were detected in 1982 in experiments at the proton–antiproton collider at CERN (and at the electron–position collider at Stanford). The experiments at CERN involved the observation of about a billion head-on collisions between protons and antiprotons of the same energy, 270 GeV. A few dozen W and Z particles were produced in these collisions (Fig. 41.13). The measured masses of the W and Z particles are, respectively, 80 GeV/c^2 and 91 GeV/c^2. These measured values are within a fraction of a GeV/c^2 of the theoretically predicted values. This excellent agreement constitutes a brilliant confirmation of the unified theory of weak and electromagnetic interactions.

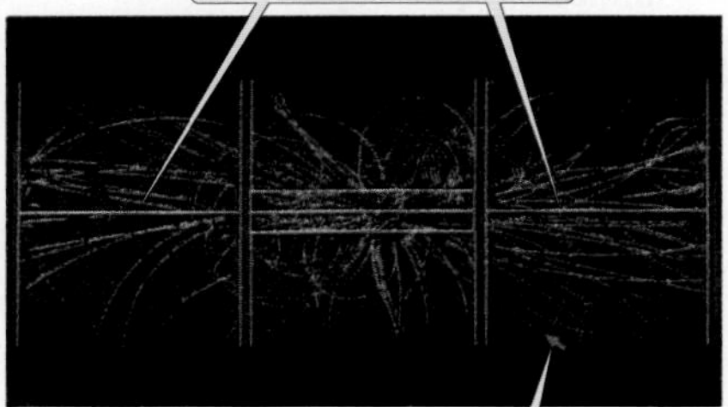

FIGURE 41.13 Tracks of particles produced in a very energetic head-on collision between a proton of energy 270 GeV that entered from the right and an antiproton of energy 270 GeV that entered from the left. In this collision a W^- particle was created. It immediately decayed into an electron and a neutrino. The track of the neutrino is not visible.

Checkup 41.4

QUESTION 1: Compare the masses of the W and Z particles with the masses listed in Tables 41.1–41.3. Are the W and Z the heaviest of all known particles?

QUESTION 2: Order the electromagnetic, strong, and weak forces in order of increasing range.

(A) Strong, electromagnetic, weak.
(B) Weak, strong, electromagnetic.
(C) Electromagnetic, strong, weak.
(D) Strong, weak, electromagnetic.
(E) Weak, electromagnetic, strong.

41.5 QUARKS

Let us now return to our initial question. What are the ultimate, indivisible building blocks of matter? We know of more than 300 particles. It is implausable that all these hundreds of particles are truly elementary particles. It is likely that most of them, or maybe all of them, are composite particles made of just a few elementary building blocks.

To discover these building blocks, physicists have tried to break protons into pieces by bombarding them with projectiles of very high energy. Unfortunately, if the energy of the projectile is large enough to make a dent in a proton, then it is also large enough to create new particles during the collision. This abundant creation of particles confuses the issue—

we can never be quite sure which of the pieces that come flying out of the scene of the collision are newly created particles and which are fragments of the original proton. In fact, none of the pieces ever found in such collision experiments seems a likely candidate for an elementary building block. Typically, the particles that emerge from a collision between a high-energy projectile and a proton are pions, kaons, lambdas, deltas, and so on, all of which seem to be less elementary than the proton.

Although brute-force collision experiments have failed to fragment protons into elementary building blocks, somewhat more subtle experiments have provided us with some evidence that distinct building blocks do indeed exist inside protons. At the Stanford Linear Accelerator Center (SLAC), very high-energy electrons were shot at protons; these electrons served as probes to "feel" the interior of the protons. The experiments showed that occasionally the bombarding electrons were deflected through large angles, bouncing off sharply from the interior of a proton. These deflections indicate the presence of some lumps or hard kernels in the interior of the proton, just as, in Rutherford's experiments, the large deflections of alpha particles by atoms indicated the presence of a hard kernel (nucleus) in the interior of the atom. Protons and all the other baryons and mesons seem to be composite bodies made of several distinct pieces. In contrast, electrons and the other leptons seem to be indivisible bodies with no internal structure. In some experiments the electron was probed with beams of extremely energetic particles to within 10^{-18} m of its center. Even at these extremely short distances, no substructures of any kind were found. Thus, the electron seems to be a pointlike particle with no size at all, a truly elementary particle.

Even before the experimental evidence for lumps inside protons became available, theoretical physicists had noticed that particles could be classified into groups or families of similar particles on the basis of their quantum numbers and their behavior in reactions. To explain these similarities, they had proposed theories in which all the known baryons and mesons are regarded as constructed out of a few fundamental building blocks. According to these theories, the similarity between particles within a given family reflects the similarity of their internal construction, just as the similarities between atoms in a group of the periodic table reflect the similarity of their internal construction.

The most successful of these theories was the quark model proposed by M. Gell-Mann and by G. Zweig. In this model, *all particles were constructed of three kinds of fundamental building blocks called* **quarks**. (Gell-Mann took the word quark from *Finnegan's Wake*, a book by James Joyce.) The three quarks are labeled **up**, **down**, and **strange**, or simply u, d, and s. They all have spin $\frac{1}{2}$ and they have electric charges of $\frac{2}{3}$, $-\frac{1}{3}$, and $-\frac{1}{3}$ elementary charge, respectively (see Table 41.7). Each quark—like any other particle—has an antiparticle, of opposite electric charge.

quarks
up quark (u)
down quark (d)
strange quark (s)

TABLE 41.7 THE FIRST THREE QUARKS

QUARK	MASS[a]	SPIN	ELECTRIC CHARGE	STRANGENESS
u	5 MeV/c^2	$\frac{1}{2}$	$\frac{2}{3}$	0
d	10	$\frac{1}{2}$	$-\frac{1}{3}$	0
s	200	$\frac{1}{2}$	$-\frac{1}{3}$	−1

[a]Based on theoretical estimates.

To make the ordinary particles out of the quarks, the latter must be glued together in diverse ways. For example, a proton is made of two u quarks and one d quark (Fig. 41.14). A neutron is made of two d quarks and one u quark (Fig. 41.15). A positive pion is made of one u quark and one d antiquark (Fig. 41.16), and so on (Tables 41.2 and 41.3 list the quark composition of baryons and mesons). By gluing quarks together, we can build up all the known baryons and mesons and explain their quantum numbers and their similarities. Besides, we can predict some properties, such as magnetic moments and reaction rates, of the composite particles from the assumed properties of the quarks.

There is only one snag: all experimental searches for free quarks have been unsuccessful. Physicists now believe that quarks are permanently confined inside the ordinary particles so there is no way to break the quark out of, say, a proton. It seems that the quarks are held in place by an exceptionally strong force, which prevents their escape. This new force is the "**color**" **force**.

color force

The concept of "color" was first introduced to remedy an unacceptable violation of the Exclusion Principle (see Chapter 39). Quarks, like leptons and other particles of half-integer spin, ought to obey the Exclusion Principle. But investigations of the quantum states of quarks within protons, neutrons, sigmas, etc., disclosed that several apparently identical quarks were often found in the *same* quantum state. To avoid this apparent violation of the Exclusion Principle, physicists postulated that each of the quarks exists in three varieties, and that whenever two apparently identical quarks are found in the same quantum state, they actually are of different varieties. The varieties of quarks are characterized by a new property called **color**. Of course this "color" has nothing to do with real color; it is merely a (somewhat unimaginative) name for a new property of matter. The different quark colors are **red**, **green**, and **blue**. Thus there is a *red* u quark, a *green* u quark, and a *blue* u quark, and so on. The antiquarks have anticolors; the different antiquark colors are anti*red*, anti*green*, and anti*blue.* Color is a very subtle property of matter; it usually remains hidden inside the ordinary particles. All the normal particles are "colorless"—they consist of several quarks with an equal mixture of all three colors. For instance, one of the three quarks inside the proton is *red*, one is *green*, and one is *blue.* Nevertheless, color plays a crucial role in the theory of the forces that confine the quarks inside the ordinary particles.

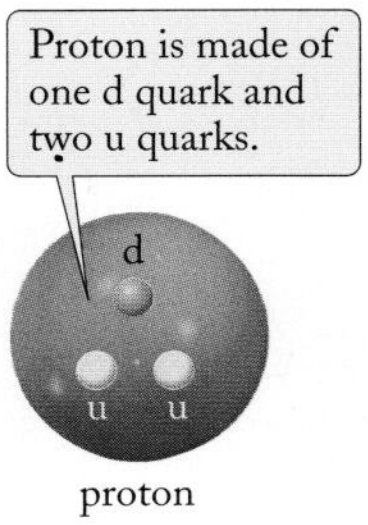

FIGURE 41.14 Structure of the proton. The sizes of the quarks are not drawn to scale. Their sizes are probably much smaller than the size of the proton.

The quarks are confined by extremely strong mutually attractive forces. These forces between quarks are called color forces because the source of these forces is the color just as the source of the electric force is the electric charge. Each of the three varieties of color is analogous to a kind of electric charge. A body is color-neutral, or "colorless," if it contains equal amounts of all three colors or if it contains equal amounts of color and anticolor, just as a body is electrically neutral if it contains equal amounts of positive and negative electric charge.

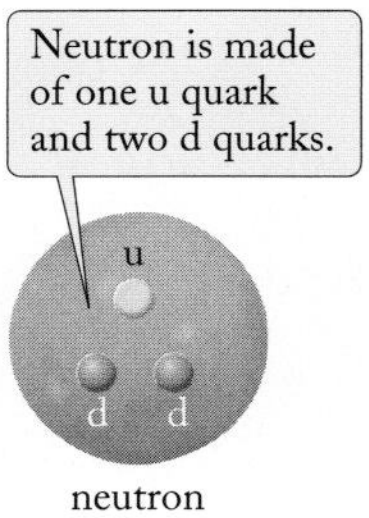

FIGURE 41.15 Structure of the neutron.

The color force is a fundamental force that is included in our table of fundamental forces together with the strong force (Table 41.4). The color force is closely related to the strong force—the latter is actually a special instance of the former. The relationship between the color force and the strong force is analogous to the relationship between the electric force and the intermolecular force. As we saw in Section 22.1 the force between two electrically neutral atoms or molecules is a residual electric force resulting from an imperfect cancellation among the attractions and repulsions of the charges in the two atoms or molecules. Likewise, the strong force between, say, two "colorless" protons is a residual color force resulting from an imperfect cancellation among the attractions and repulsions of the quarks in the two protons. Thus the "strong" force between two protons is no more than a pale reflection of the much stronger color forces acting within each proton.

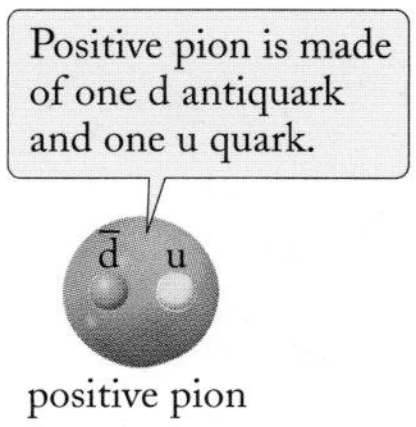

FIGURE 41.16 Structure of a positive pion (π^+).

At a fundamental level, the color force between two quarks is due to an exchange of virtual particles between the quarks. The particle that acts as the carrier of the color force is the gluon, the same as the carrier of the strong force. Figure 41.17 shows a Feynman diagram representing the exchange of a gluon between two quarks. Such an exchange of a gluon between two quarks is analogous to the exchange of a photon between two charged particles (see Fig. 41.9).

Another modification of the simple quark model with three quarks arose from the theory of the unification of the electromagnetic and weak forces. In order to make this theory fit some experimental data, physicists had to postulate the existence of a fourth quark, different from the u, d, and s quarks. This new hypothetical quark was labeled **charmed**, or simply c.

charmed quark (c)
bottom quark (b)
top quark (t)

The hypothesis of the charmed quark soon received firm experimental support. In 1974, teams of experimenters at the Brookhaven accelerator and at the Stanford accelerator discovered the J/ψ meson and some other related mesons (see Table 41.3), all of which have exceptionally long lifetimes. These particles contain charmed quarks with an electric charge of $\frac{2}{3}$ and a mass of 60 MeV/c^2.

The proliferation of quarks did not stop with four quarks. In 1977, a team of experimenters at Fermilab discovered the Υ (Greek *Upsilon*) mesons. These are by far the most massive mesons known (see Table 41.3). Each of these contains a new quark; this fifth kind of quark has been labeled **bottom**, or b. Theoretical considerations then suggested that there should exist a sixth quark, labeled **top**, or t. In 1995, another team of experimenters at Fermilab confirmed the existence of the top quark.

Figure 41.18 summarizes all the constituents of particles and the carriers of forces; these form the foundation of the **Standard Model** of particle physics. This model describes all the known particles in terms of the elementary particles listed in Fig. 41.18, and it describes all interactions in terms of four kinds of carriers (however, the graviton is not fully incorporated into the model, because there is as yet no complete and coherent theory of quantum gravity). The four kinds of carriers are the gluons (which come in various color combinations), the photon, the W^+, W^-, Z, and the graviton. The six leptons fall into three families: the electron family, the muon family, and the tau family. Correspondingly, the six quarks also fall into three families: the up–down family, the strange–charm family, and the bottom–top family. Each of the six leptons has an antiparticle. Each of the six quarks (up, down, strange, charmed, top, bottom) comes in three colors (*red, green, blue*); furthermore, each quark has an antiquark, which comes in three varieties of anticolor (anti*red*, anti*green*, anti*blue*). Altogether, this amounts to 36 quarks, 12 leptons, and 24 carriers (counting all the color combinations of gluons and their

Standard Model

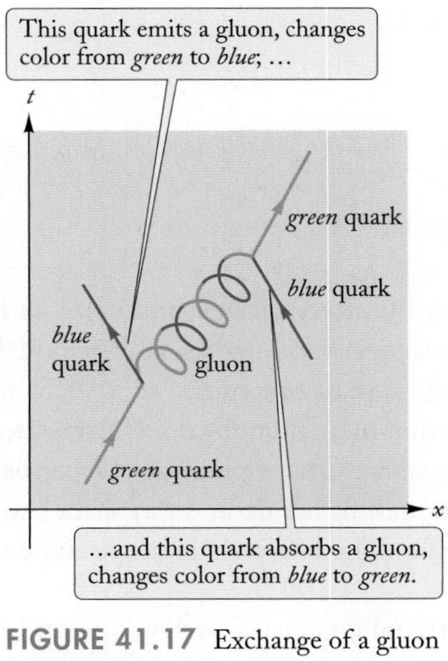

FIGURE 41.17 Exchange of a gluon between two quarks.

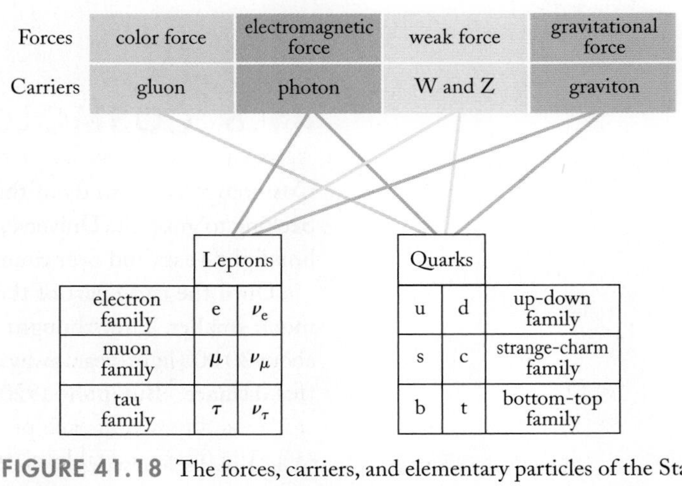

FIGURE 41.18 The forces, carriers, and elementary particles of the Standard Model.

antiparticles). This proliferation of constituents raises the question whether matter really has such a large number of elementary building blocks. Pushing forward our search for the ultimate building blocks, we have uncovered layers of structures within layers of structures—electrons and nuclei within atoms, protons and neutrons within nuclei, quarks within protons and neutrons. Is there another, more elementary layer within quarks?

string theories

Seeking to answer this question, some theorists have been exploring **string theories**, according to which all particles consist of tiny, vibrating strings, only about 10^{-33} cm in length. The different masses, spins, electric charges, and other properties of the particles are supposed to arise from different vibrational patterns of the strings. For this scheme to work, the strings may have to exist in a 9-dimensional space, that is, a space with 6 more dimensions than the 3-dimensional space of our everyday experience. The extra 6 dimensions are thought to be imperceptible because they are tightly curled up, so they extend over distances of only about 10^{-33} cm. It is the ambition of string theorists to deduce the masses and all the other properties of particles from the characteristics of the strings. But in spite of prodigious effort, they have not yet succeeded in this, and they have not demonstrated any clear experimental connection between string theories and our physical world.

✔ Checkup 41.5

QUESTION 1: Do the charges of two u quarks and one d quark add up to the charge of the proton? Do the charges of two d quarks and one u quark add up to the charge of a neutron?

QUESTION 2: Each quark has spin $\frac{1}{2}$. How can the net spin of three quarks add to form the spin $\frac{1}{2}$ of the proton?

QUESTION 3: What is a gluon?

QUESTION 4: What is the electric charge of the u antiquark? The d antiquark? The s antiquark?

(A) $+\frac{2}{3}e, -\frac{1}{3}e, -\frac{1}{3}e$
(B) $-\frac{2}{3}e, -\frac{1}{3}e, -\frac{1}{3}e$
(C) $+\frac{2}{3}e, +\frac{1}{3}e, +\frac{1}{3}e$
(D) $-\frac{2}{3}e, -\frac{2}{3}e, -\frac{1}{3}e$
(E) $-\frac{2}{3}e, +\frac{1}{3}e, +\frac{1}{3}e$

41.6 COSMOLOGY

Cosmology is the study of the Universe at large, its size, its shape, and its evolution. Seeking to grasp the Universe, our minds must wander over distances as great as 10 billion light-years and over times as long as 10 billion years or longer.

Until the early part of the last century, astronomers thought the Universe to be much smaller. They thought that the farthest stars at the edge of our Galaxy were about 30 000 light-years away and that there was nothing but dark, empty space beyond that distance.[2] But in the 1920s, Edwin Hubble used the 100-inch telescope on Mount

[2]A light-year is the distance traveled by light in 1 year, 9.5×10^{15} m.

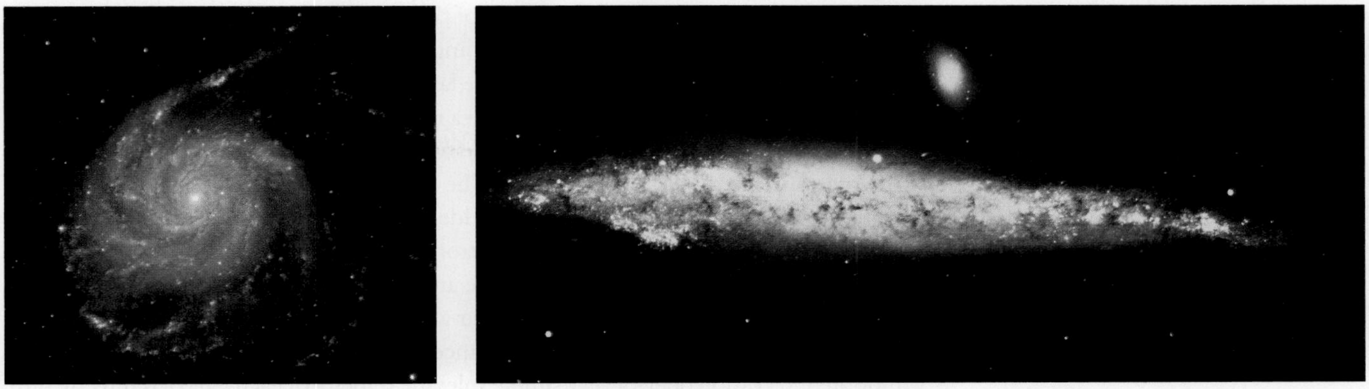

FIGURE 41.19 These pictures taken with the Hubble Space Telescope give an idea what our Milky Way Galaxy looks like when viewed (a) face on and (b) edge on. The pictures are actually photographs of two distant galaxies (NGC 5457 and NGC 4631) similar to our Galaxy.

Wilson to establish that the faint, wispy "nebulae" found in all parts of the sky were actually gigantic conglomerations of stars, similar to our own Galaxy but at very great distance from us.

Our **Milky Way Galaxy** contains about 10^{11} stars arranged in an irregular, disklike region some 10^5 light-years in diameter. The disk has a central bulge, and it has spiral arms along which stars are concentrated (Fig. 41.19). There are many external galaxies beyond our Galaxy. With our large telescopes we can see altogether about 10^{11} galaxies! There are supergiant galaxies with 10^{13} stars each, and there are dwarf galaxies with "only" 10^6 stars. There are spherical galaxies and elliptical galaxies, like luminous globes and eggs; there are spiral galaxies and barrel-shaped galaxies, like whirling pinwheels; and there are irregular galaxies with the weirdest shapes (see Figs. 41.20a–c).

FIGURE 41.20 (a) Spiral galaxy (NGC 3031) in Ursa Major. The plane of this galaxy is inclined to our line of sight; face on, this galaxy would look circular. (b) Unusual galaxy (NGC 5128) in Centaurus. Note the thick lane of dust surrounding this galaxy. (c) Spiral galaxy (NGC 4565) in Coma Berenices. Note how thin this galaxy is.

All the galaxies are in motion. Many of them congregate in clusters, orbiting about each other, sometimes accidentally colliding. But let us ignore the fine details of the motion of galaxies and concentrate on the large-scale features of the motion. We then find that on a large scale, the galaxies have a motion of recession—all the distant galaxies are moving away from us. The recessional velocities of galaxies are determined by the **red-shift** method (see Fig. 41.21). The light from the receding stars is Doppler-shifted to lower frequencies, or longer (redder) wavelengths [see Eq. (36.13) in Section 36.3]. To find the recessional velocity, astronomers need only measure how much the wavelength of the light received from the atoms in some distant galaxy is shifted relative to the wavelength emitted by similar atoms in our laboratories on Earth.

red-shift method

By measuring the velocities and distances of galaxies, Hubble discovered that the motion of recession obeys a very simple rule: the velocity of each galaxy is directly proportional to its distance. This means that from the point of view of our Galaxy, nearby galaxies move slowly and distant galaxies move fast (see Fig. 41.22). This proportionality between velocity v and distance r is called **Hubble's Law** and can be expressed as

Hubble's Law

$$v = H_0 r \tag{41.6}$$

where H_0 is the Hubble constant. If r is expressed in billions of light-years, the numerical value of the Hubble constant is

Hubble constant

$$H_0 = 2.1 \times 10^7 \text{ (m/s)/(billion light-years)} \tag{41.7}$$

Astronomers determine the enormous cosmic distances from us to other galaxies by the brightness or "headlight" method. In essence, this method relies on the following. Stars in faraway galaxies look faint to us, and stars in nearby galaxies look bright—just as, on a dark road, the headlights of a faraway automobile look faint and the headlights of a nearby automobile look bright. If all stars generated precisely the same quantity of light, then differences in their apparent brightness as observed by our telescopes would be entirely due to differences in their distances—there would then be a simple mathematical relationship between apparent brightness and distance.

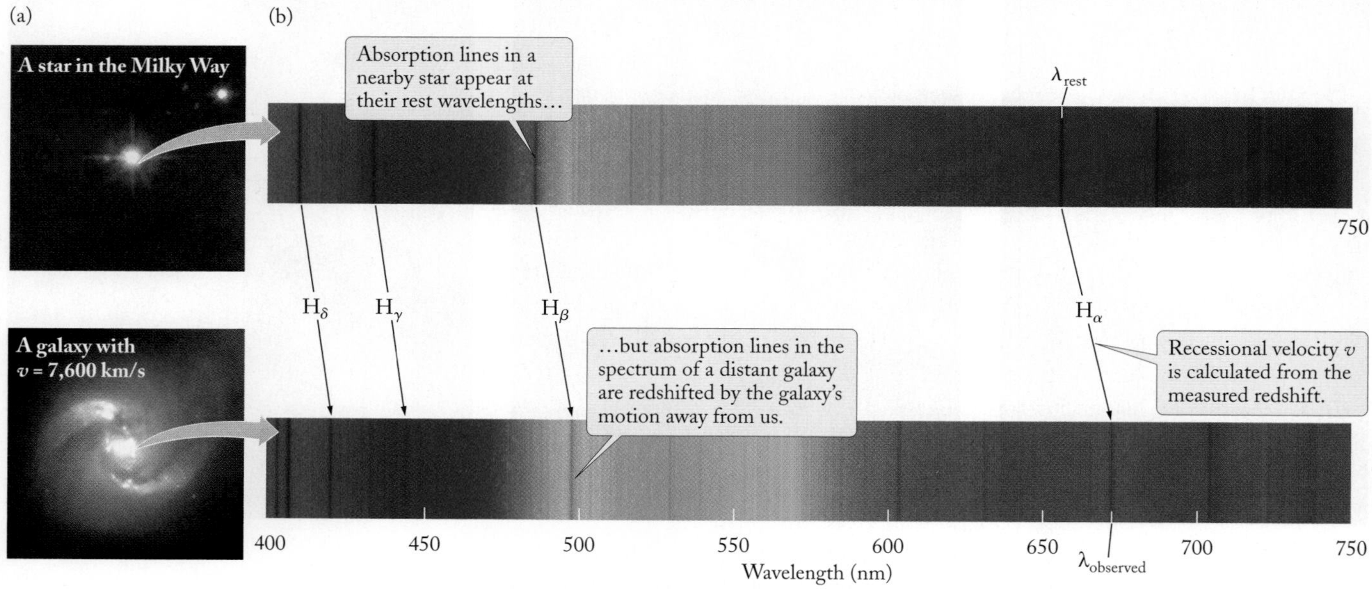

FIGURE 41.21 (a) A distant galaxy seen together with a star in our galaxy. (b) The spectrum of each, shown on the same scale. Note that lines in the galaxy spectrum are red-shifted to longer wavelengths. The H_α, H_β, H_γ and H_δ absorption lines are the first four lines of Bolmer Series of hydrogen (see Fig. 38.7).

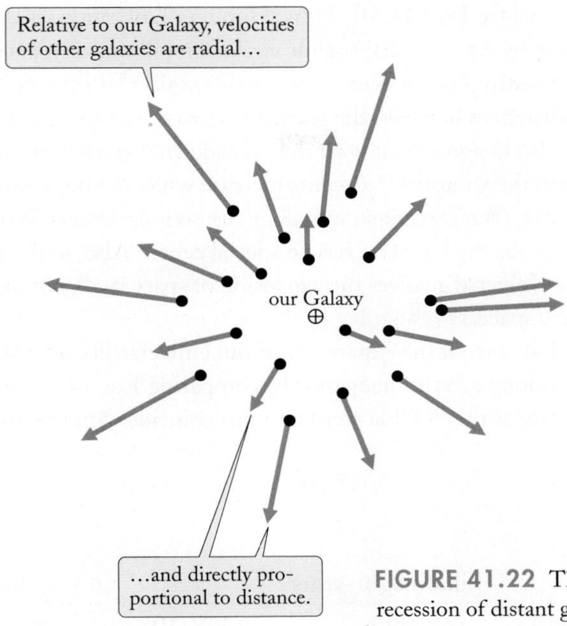

FIGURE 41.22 The motion of recession of distant galaxies.

However, in practice, there can be some complications; two stars can be at the same distance yet differ in apparent brightness because one generates more light than the other. Such intrinsic differences between stars must be taken into account when using the brightness method. Astronomers have developed clever techniques for selecting stars of standard brightness, but some observational errors remain in the distance determinations. Numerous analyses support the direct proportionality expressed by Eq. (41.6), but there are substantial uncertainties in the numerical value of Eq. (41.7)—the Hubble constant is uncertain by at least $\pm$ 10%.

Although Fig. 41.22 gives the misleading impression that our Galaxy is at the center of the Universe and that all other galaxies are fleeing away from us, our Galaxy does not occupy any special spot in the Universe. The other galaxies are not just fleeing away from us; they are fleeing away from each other. The Universe is expanding. An extraterrestrial astronomer sitting on a distant galaxy would see our Galaxy and all other galaxies fleeing away from her. Hence our spot in the Universe is pretty much the same as every other spot. Cosmologists believe that this overall uniformity holds not only in regard to the expansion, but also in regard to all other general features of the Universe. For instance, the numbers and types of galaxies that the extraterrestrial astronomer finds in her neighborhood will, on the average, be the same as we find in our neighborhood—the Universe is pretty much the same everywhere. This assertion of the large-scale uniformity of the Universe is called the **Cosmological Principle**.

Cosmological Principle

The motion of recession of galaxies can be described by a very crude analogy. When a grenade explodes in midair, the fragments of shrapnel spurt out in all directions. Different fragments may have different velocities, and in a given amount of time, they will reach different distances. After a time t, the position of a fragment having velocity v will be

$$r = vt \qquad (41.8)$$

If we rewrite this as

$$v = r/t \qquad (41.9)$$

we see that at any given time the fragments that are at the greatest distances are those with the highest velocities. This proportionality of velocity and distance has the same form as Hubble's Law [see Eq. (41.6)]. Thus, Hubble's Law suggests that the galaxies were set in motion by a primordial cosmic explosion billions of years ago and have been more or less coasting along ever since. Incidentally: In the expansion of the Universe, only the distances between the galaxies increase; the galaxies themselves do not expand. This is also in agreement with the grenade analogy, where, of course, only the distances between the shrapnel fragments increase while the fragments themselves remain of constant size. (But the grenade analogy has serious defects: It posits a center for the explosion, whereas the Universe has no special center. Also, in the modern view, the expansion of the Universe involves the expansion of space itself, not just the motion of matter in ordinary space.)

Big Bang

The explosion that started the expansion of the Universe is called the **Big Bang**. We can calculate how long ago this happened by comparing Eqs. (41.6) and (41.9). We thus see that the inverse of the Hubble constant must coincide with the expansion time:

$$t = 1/H_0 \tag{41.10}$$

or

$$t = \frac{1}{2.1 \times 10^7} \times \frac{\text{billion light-years}}{\text{m/s}} = \frac{1}{2.1 \times 10^7} \times \frac{9.5 \times 10^{24}\ \text{m}}{\text{m/s}}$$

$$= 4.5 \times 10^{17}\ \text{s} = 1.4 \times 10^{10}\ \text{years} \tag{41.11}$$

However, in this calculation of the age of the Universe, we have ignored the possibility that the velocity of galaxies may change with time. For instance, since gravity pulls the galaxies toward each other, we might expect that gravity tends to inhibit the motion of recession and tends to slow down the expansion of the Universe. This would imply that the velocities of all galaxies were somewhat larger in the past and, consequently, the true age of the Universe ought to be somewhat smaller than the 14 billion years indicated by our naive calculation. But in any case, this number gives us a rough estimate of the age.

Since the Universe started some finite time ago, only light from those parts of it that are sufficiently near can have reached us. The speed of light is $c = 3.00 \times 10^8$ m/s = 1 light-year/year, and in the time t light travels a distance

$$ct = (1\ \text{light-year/year}) \times (1.4 \times 10^{10}\ \text{year})$$

$$= 1.4 \times 10^{10}\ \text{light-years}$$

observable universe

This distance is the radius of the **observable universe**[3]. Everything within this radius we can see (given sufficiently powerful telescopes); anything beyond we cannot see because the light has not yet had time to reach us. Note that as time increases, the radius ct increases, that is, the observable universe includes more and more of the total Universe.

The idea of the Big Bang can be put to several direct tests. First, nothing we observe can be older than about 14 billion years. Our Sun has an age of only 5 billion years. The oldest objects that we can reliably date are the globular clusters of stars found near our Galaxy (see Fig. 41.23). Old stars enter a red-giant stage, turning a reddish color and swelling to several hundred times normal size. A young cluster contains few red giants, and an old cluster contains many. Careful calculations of stellar evolution indicate that the oldest globular clusters have ages of about 10 billion to 13 billion years, in good agreement with the expansion time of Eq. (41.11).

[3] This value of the radius of the observable universe is only an approximation. For an exact calculation, we need to use the theory of General Relativity, taking into account that space and time are curved.

Another method by which we can date the birth of stars is by the age of chemical elements. All of the elements, with the exception of hydrogen and helium, were synthesized in very massive stars soon after our Galaxy came into being. These stars survived only a short time and then exploded as supernovas, scattering these elements into the cloud of gas and dust that was destined to become our Solar System. Radioactive dating tells us that these elements are somewhere between 7 and 15 billion years old, in agreement with the expansion time to within experimental uncertainties.

FIGURE 41.23 The globular cluster (NGC 5272) in Canes Venatici.

More evidence for the Big Bang emerges from studies of the abundances of hydrogen and helium. These elements were formed in the hot, primordial fireball of the Big Bang. When the Universe was about 3 minutes old, the temperature was 10^9 K, subjecting hydrogen to nuclear fusion. Theoretical calculations show such fusion led to abundances of about 75% hydrogen and 25% helium, in remarkable agreement with the observed abundances of 74% and 24% (plus traces of other elements).[4] This confirms our picture of a hot Big Bang.

But the most decisive item of evidence for the Big Bang is that some of the radiant heat given off by the primordial explosion can still be found in the sky today. Originally, the radiant heat emitted by the dense, primordial fireball was in the form of very penetrating gamma rays and X rays. But as the Universe expanded, the electromagnetic radiation expanded with it, and at present the wavelengths of the fireball radiation are much longer. Theoretical calculations indicate that the radiant heat must have a blackbody spectrum (see Chapter 37) with a peak around a wavelength of 1 mm, the wavelength of microwaves.

This kind of radiation was discovered in 1964 by Arno A. Penzias and Robert W. Wilson, two scientists working at Bell Laboratories with very sensitive microwave communication equipment (Fig. 41.24). They found that the entire sky is noisy; there is radiation coming at the Earth from all directions. Physicists at Princeton immediately recognized the cosmological significance of this discovery, and identified the radiation as **cosmic background radiation**, a relic of the Big Bang. This radiation is that of a blackbody with temperature of about 3 K (more precise data give 2.73 K). What has happened here is that the extremely hot radiant heat from the primordial fireball has gradually cooled down as the Universe expanded, and by now its temperature has come pretty close to absolute zero. The cosmic background radiation is residual radiant heat left in the sky after the Big Bang. It is direct material evidence for the Big Bang.

cosmic background radiation

FIGURE 41.24 Horn antenna at Bell Laboratories, Holmdel, New Jersey. This antenna was designed for microwave communication experiments with the Echo and Telstar satellites.

[4] Fusion within stars continues to produce helium, but the total amount of helium produced by all the stars since the Big Bang is only a few percent.

The big question in cosmology is this: Will the Universe expand forever? Or will the gravitational attraction between galaxies eventually stop the expansion and cause the Universe to contract, ultimately leading to to a terminal cosmic implosion? To help find the answer, we must calculate the deceleration of the motion of recession caused by gravity.

Consider a spherical region of our Universe, and assume that galaxies are uniformly distributed throughout this spherical region and the entire Universe (see Fig. 41.25). Gauss' Law applies to the inverse-square gravitational force as it does to the inverse-square electric force. Accordingly, we know that the motion of the galaxies in the spherical region is unaffected by the rest of the Universe. Consider now one of the galaxies at the surface of this region, presently at a radial distance r_0 from us. That galaxy will continue to recede from us forever provided its present velocity v_0 is greater than the escape velocity, Eq. (9.27),

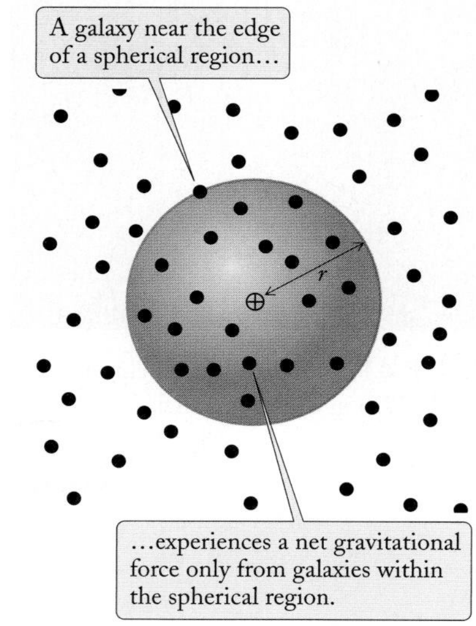

FIGURE 41.25 A spherical region of our Universe.

$$v_0 > \sqrt{2GM/r_0} \tag{41.12}$$

where M is the mass inside the spherical region. This mass can be expressed in terms of the average density ρ_0 of mass in the Universe,

$$M = \rho_0 \frac{4\pi}{3} r_0^3 \tag{41.13}$$

and the velocity is given by Eq. (41.6), $v_0 = H_0 r_0$. Thus the condition for a permanently expanding Universe becomes

$$H_0 r_0 > \sqrt{\frac{8\pi}{3} G r_0^2 \rho_0} \tag{41.14}$$

or

$$\rho_0 < \frac{3}{8\pi G} H_0^2 \tag{41.15}$$

If we insert the numerical value $H_0 = 1/(4.5 \times 10^{17}\text{ s})$ [see Eq. (41.11)] and the numerical value for the gravitational constant, we obtain the conditions

$$\rho_0 < 9 \times 10^{-27}\text{ kg/m}^3 \qquad \text{(permanent expansion)}$$

$$\rho_0 > 9 \times 10^{-27}\text{ kg/m}^3 \qquad \text{(ultimate contraction)} \tag{41.16}$$

In principle, this makes it very simple to predict the future evolution of the Universe; but in practice, we are severely handicapped by the uncertainty in the mass density.

When reckoning the mass of the Universe, we must take into account the mass belonging to galaxies and also whatever mass is to be found in the intergalactic space between galaxies. Table 41.8 lists all the known contributions to the mas density of the Universe, as a percentage of the critical mass density 9×10^{-27} kg/m^3. The visible, luminous matter contained in stars makes only a small contribution, less than $\frac{1}{2}$% of the critical mass. Thus, the bulk of the mass of the Universe is in nonluminous, invisible forms, which can be detected only by indirected means.

The mass of baryonic matter (protons and neutrons) listed in the table is inferred from studies of the abundance of deuterium formed in the fireball of the Big Bang. Apparently, most of this baryonic matter did not get captured into luminous stars during the later evolution of the Universe, and it must still be lurking in and around galaxies, in the form of small, nonluminous stars ("brown dwarfs" and cold stellar remnants), perhaps interstellar planets, and large clouds of low-density gas surrounding the galaxies.

The mass of the nonbaryonic **dark matter** is inferred from studies of galactic dynamics. Astronomers can detect the presence of dark matter by its gravitational effects on the rotational motion of galaxies and on the orbital motion of galaxies around each other. From such studies, we know that galaxies are surrounded by large, extragalactic clouds, or halos, of some kind of dark matter. This is sometimes called the "missing mass" because we cannot see it, and we do not know what it consists of. In the absence of any information about this dark matter, astronomers and physicists have felt free to speculate about various possibilities. One such speculation is that this dark matter might consist of clouds of neutrinos. We know that neutrinos have a small, nonzero mass, but the magnitude of this mass has so far proved too small to be measured. If it is about 4 eV/c^2, then the clouds of neutrinos left over from the Big Bang in the form of thermal neutrino blackbody radiation might be sufficient to account for the dark matter. Another speculation is that the dark mass might consist of some new, exotic particles invented by theorists working on unified theories of interactions. But no such particles have so far been confirmed experimentally.

dark matter

TABLE 41.8 CONTRIBUTIONS TO THE MASS DENSITY OF THE UNIVERSE

FORM OF MATTER	PERCENTAGE OF CRITICAL DENSITY
luminous matter (stars)	0.4%
baryonic matter	4%
dark matter (nonbaryonic)	25%
dark energy	71%

The final contribution to the mass of the Universe listed in Table 41.8 is the most mysterious. Beginning in 1998, astronomers combined luminosity and red-shift data for a class of very bright supernovas to determine their recessional velocities and distances. These data indicate that instead of the presumed state of slowing expansion, the Universe is actually experiencing an *accelerating expansion.* To account for an accelerating Universe, cosmologists now assume there exists some sort of **dark energy** that permeates the space between galaxies. This dark energy apparently makes up about 70% of the critical density and provides an effective repulsive gravitational force that accelerates the expansion. We do not yet have any firm grasp on exactly what this dark energy might be.

dark energy

Will the Universe continue to expand or will it ultimately contract? A forever-expanding and accelerating Universe gives the best fit to all the facts as we currently know them. But there are enough uncertainties in our measurements and enough loopholes in our arguments that the possibility of an ultimately contracting Universe cannot be dismissed. It will be a while before we know the ultimate fate of the Universe.

✔ Checkup 41.6

QUESTION 1: Why do astronomers not rely on triangulation to measure the distances of galaxies?

QUESTION 2: If the expansion of the Universe is accelerating or is slowing down, can the Hubble constant be truly a constant?

QUESTION 3: Imagine a galaxy very close to the edge of the observable Universe. This galaxy is moving away from us at a speed very close to the speed of light. Does this mean that this galaxy will reach the edge of the observable universe and disappear from sight?

QUESTION 4: Where were the atoms of most of the elements inside your body made?

(A) In the core of the Earth soon after the formation of the Sun
(B) In supernovas soon after our Galaxy was born
(C) In the Sun, later transmitted to Earth by cosmic radiation
(D) In nuclear fission of heavier elements such as uranium
(E) In the first 10^{-34} s of the Big Bang

SUMMARY

PARTICLES

- fermions: leptons, baryons
- hadrons: baryons, mesons
- bosons: mesons, carriers

FORCES

	Strength	Carrier
strong/color	1	gluon
electromagnetic	10^{-2}	photon
weak	10^{-6}	W and Z
gravitational	10^{-38}	graviton

CONSERVED QUANTITIES

Absolute	Approximate
energy	isospin
momentum	strangeness
angular momentum	parity
electric charge	
baryon number	
lepton number	

LEPTONS electron (e), muon (μ), tau (τ), and neutrinos (ν_e, ν_μ, and ν_τ).

QUARKS up (u), down (d), strange (s), charmed (c), bottom (b), and top (t).

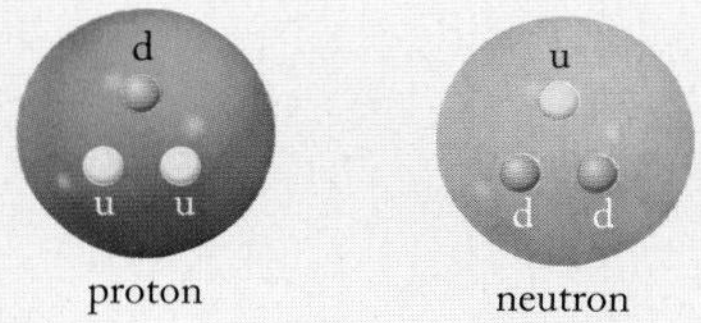

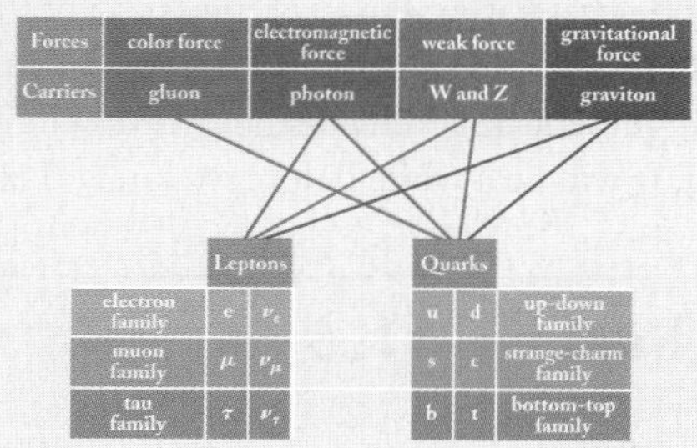

HUBBLE'S LAW (v is the recessional velocity of a galaxy and r is the distance to the galaxy.)

$$v = H_0 r \quad (41.6)$$

$$H_0 = 2.1 \times 10^7 \text{ (m/s)/(billion light-years)} \quad (41.7)$$

COSMOLOGICAL PRINCIPLE On a large scale, the Universe is uniform.

QUESTIONS FOR DISCUSSION

1. Why are high-energy accelerators necessary for the production and discovery of new, massive particles?
2. Physicists were planning to construct the Superconducting Super Collider (SSC), a 20-TeV accelerator, which would have cost some $4 billion. Can such an expenditure be justified?
3. The names *baryon, meson*, and *lepton* come from the Greek *barys* (heavy), *mesos* (middle), and *leptos* (thin, slender). These names were originally intended to indicate the masses of the particles. According to the lists of particles and masses given in this chapter, is it true that the baryons have the largest masses and the leptons the smallest?

4. How does the antineutron differ from the neutron?
5. In experiments at CERN, an antielectron and an antiproton have been put together to form an antihydrogen atom. In principle, two of these atoms can form an antihydrogen molecule. Could we confine a sample of antihydrogen gas in an ordinary steel bottle? Can you think of any way of confining it?
6. How would you refute the proposition that the Sun is made of antimatter?
7. Why does a particle that lives only 10^{-23} s not make a track in a bubble chamber? (Hint: Suppose the particle moves at the maximum conceivable speed; how far will it travel in 10^{-23} s?)
8. The strengths of the fundamental forces depend on the energies of the particles. In the case of the gravitational force, the strength increases with the energy. Why would you expect this to be true?
9. The boomerang analogy described in Fig. 41.11 is defective in that the boomerang requires the presence of air. What would be the motion of a boomerang in vacuum?
10. In Fig. 41.7, a large number of particles emerge in the longitudinal direction (toward the right and the left). Why is this expected, whereas the emergence of particles in the transverse direction (upward and downward) is surprising? (Hint: Consider a head-on collision between two rockets; which way do you expect most fragments to spurt out?)
11. Neglecting dark energy, describe the difference between the final states of the Universe for $\rho_0 < 3H_0^2/(8\pi G)$ and $\rho_0 > 3H_0^2/(8\pi G)$.
12. How will life ultimately end if the Universe continues to expand forever? If it contracts?

PROBLEMS

41.1 The Tools of High-Energy Physics
41.2 The Multitude of Particles

1. According to Eq. (30.6), the radius of the orbit of a charged particle of momentum p in a magnetic field is

$$r = \frac{p}{qB}$$

Expressed in this way, in terms of the momentum, this formula remains valid even if the particle is relativistic (although $p \neq mv$ for a relativistic particle).

(a) Show that for an ultra-relativistic particle the formula becomes

$$r = \frac{E}{qcB}$$

(b) At a CERN accelerator, protons of energy 450 GeV travel in a circular orbit of radius 1.1 km. Calculate the strength of the magnetic field required to achieve this orbital radius.

2. The Large Hadron Collider (LHC) accelerator at CERN will have a radius of 4.2 km and produce protons of momentum 7.0 TeV/c, or 3.7×10^{-13} kg·m/s.

(a) What magnetic field is required to hold the protons in a circular orbit of this radius? (Hint: Use the formula given in Problem 1.)

(b) What is the period of the orbital motion? (Hint: The speed of the proton is nearly equal to the speed of light.)

3. The relativistic expression $\sqrt{2mc^2(2mc^2 + K)}$ gives the energy available for inelastic reactions when a particle of mass m and kinetic energy K is incident on a stationary particle of the same mass. Use this formula to calculate the available energy (in GeV) for an antiproton incident on a stationary proton in the following cases:

(a) The kinetic energy of the incident antiproton is 10 MeV.

(b) The kinetic energy of the incident antiproton is 1.0 TeV (as in the Tevatron).

(c) The kinetic energy of the incident antiproton is 7.0 TeV (as in the Large Hadron Collider).

4. Which is the most massive particle listed in the tables of Section 41.2? Express the mass of this particle in atomic mass units and compare the mass with that of the helium atom and that of the lithium atom.

5. Count the number of particles (including antiparticles) that are leptons.

*6. A particle detector is in a uniform magnetic field of 1.20 T. For motion perpendicular to the field, what are the radii of curvature for the following particles and kinetic energies: (a) a 5.0-GeV proton; (b) a 1.0-MeV electron; and (c) a 300-MeV positive pion.

*7. Show that for a particle of (total) energy E, the time-dilation factor can be expressed as E/mc^2. What is the time-dilation factor for a muon of energy 950 MeV? The lifetime of this muon is 2.2×10^{-6} s in its own reference frame. What is its lifetime in the laboratory reference frame?

*8. Suppose we want to produce a Σ^+ particle by the head-on collision of two protons of equal energies according to the reaction

$$\mathrm{p} + \mathrm{p} \rightarrow \mathrm{p} + \mathrm{K}^0 + \Sigma^+$$

What is the minimum kinetic energy required for each proton to initiate this reaction? The masses of the particles are given in Tables 41.2 and 41.3.

*9. In 1908, a meteoroid struck near Tunguska, in Siberia. The destructive effects of this impact have been estimated as equivalent to the explosion of 12.5 megatons of TNT, or 5.3×10^{16} J. One possible explanation for this explosion is that the meteoroid was made of antimatter, which annihilated with an equal amount of matter when it came in contact with the Earth's surface. If so, how much antimatter would account for the explosion?

*10. The Ω^- particle decays in two alternative ways:

$$\Omega^- \rightarrow \Lambda + K^-$$

and

$$\Omega^- \rightarrow \Xi^0 + \pi^-$$

Which of these reactions releases the most energy and gives the decay products the largest kinetic energy?

*11. A Σ^- particle at rest decays into a pion and a neutron:

$$\Sigma^- \rightarrow \pi^- + n$$

What is the net kinetic energy of the decay products?

*12. Consider the annihilation of an electron and an antielectron, resulting in two gamma rays,

$$e^- + e^+ \rightarrow \gamma + \gamma$$

Suppose that the electron and antielectron are initially at rest. What are the energies of the resulting gamma rays? What are their wavelengths?

*13. The π^0 meson decays into two gamma rays:

$$\pi^0 \rightarrow \gamma + \gamma$$

If the pion is initially at rest, what are the energies of the two gamma rays? What are their wavelengths?

*14. A K^0 particle at rest decays into two pions:

$$K^0 \rightarrow \pi^0 + \pi^0$$

What is the kinetic energy of each of these pions? What is the momentum of each? (Hint: Use the relativistic relation between energy and momentum.)

41.3 Interactions and Conservation Laws

15. Verify that the reaction (41.2) conserves baryon number. Does the reaction conserve strangeness?

16. Which of the following reactions are forbidden by an absolute conservation law?

$$\pi^+ + p \rightarrow \Lambda + K^0$$

$$K^- + p \rightarrow K^- + p + \pi^0$$

$$\pi^- + n \rightarrow \pi^- + \pi^0 + \pi^0$$

$$K^- + n \rightarrow \Sigma^- + \pi^0$$

$$e^- + \nu_e \rightarrow \pi^- + \pi^0$$

17. Is strangeness conserved in the following reactions?

$$\pi^+ + n \rightarrow K^+ + \Lambda$$

$$\Lambda \rightarrow p + \pi^-$$

$$K^0 + n \rightarrow \pi^+ + \pi^-$$

18. Conservation of lepton number is actually three separate conservation laws; experiments indicate that **electron lepton number, muon lepton number**, and **tau lepton number** are each individually conserved. For example, the electron lepton number is +1 for e^- and ν_e, it is −1 for e^+ and $\bar{\nu}_e$, and it is zero for all other particles. For the following reactions, determine what missing particles are needed to conserve all lepton numbers:

$$\mu^- \rightarrow e^- + \_\_\_ + \_\_\_$$

$$p \rightarrow n + e^+ + \_\_\_$$

$$\tau^+ \rightarrow \nu_\mu + \_\_\_ + \_\_\_$$

19. Show that the emission of a photon by a free electron is impossible because it conflicts with energy conservation. (Hint: Consider the emission process in the reference frame in which the electron is initially at rest.)

20. Show that the annihilation of an electron and an antielectron into a single photon ($e^- + e^+ \rightarrow \gamma$) is impossible, because it conflicts with conservation of momentum. (Hint: Consider the reaction from the reference frame in which the two electrons have opposite velocities of equal magnitudes.)

21. Consider the following reactions produced by a beam of K^- particles in a bubble chamber filled with liquid hydrogen:

$$K^- + p \rightarrow \Sigma^+ + \pi^-$$

$$K^- + p \rightarrow \Sigma^0 + \pi^0$$

$$K^- + p \rightarrow \Lambda + \pi^0$$

$$K^- + p \rightarrow \Lambda + \pi^+ + \pi^-$$

Verify that all of these reactions conserve baryon number and strangeness.

22. The Ξ^- particle decays according to the reaction

$$\Xi^- \rightarrow \Lambda + \pi^-$$

Does this reaction conserve baryon number? Strangeness?

41.4 Fields and Quanta

23. The W particle can have either a positive charge (W^+) or a negative charge (W^-). Figure 41.26 shows the Feynman diagram for the decay of the neutron (n) via exchange of a W^-; the end products are a proton (p), an electron (e), and an electron antineutrino ($\bar{\nu}_e$). Can you guess the Feynman diagram for the decay of the antineutron?

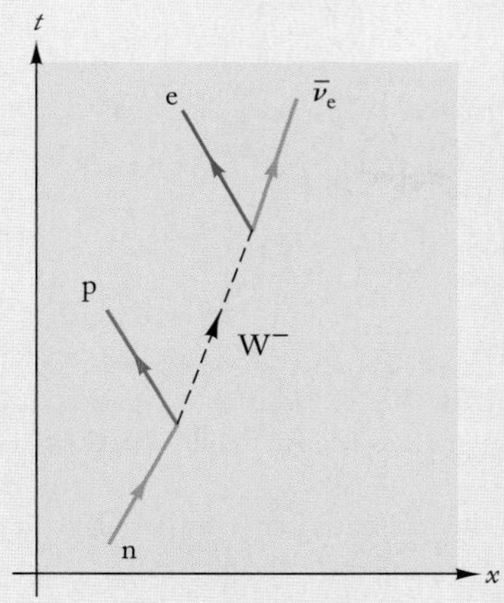

FIGURE 41.26 Feynman diagram for the decay of a neutron.

24. The strong force can also be crudely interpreted as pion exchange. Figure 41.27 shows the Feynman diagram for the exchange of a π^+ between a proton and a neutron; note that the proton changes into a neutron, and vice versa, so electric charge is conserved at each vertex. Draw corresponding diagrams for the exchange of a π^- and of a π^0.

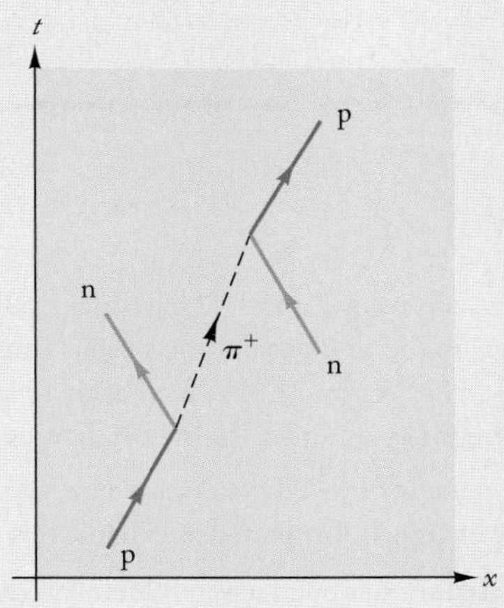

FIGURE 41.27 Feynman diagram for the exchange of a π^+ between a proton and a neutron.

25. A virtual particle may exist for a time Δt provided that its energy is less than the uncertainty in energy ΔE. The two are related by an uncertainty principle, $\Delta E\,\Delta t \approx \hbar$. To mediate the strong nuclear force, how long may a virtual pion of rest-mass energy 140 MeV exist? To mediate the weak nuclear force, how long may a virtual Z^0 of rest-mass energy 91 GeV exist? How far does each particle travel in this time? (For simplicity, use $v = c$ to estimate the range of the force.)

*26. According to a speculative theory of the strong interactions, the proton should be unstable and decay with a lifetime of about 10^{33} years. Consider the protons in a mass of 1.0×10^6 kg of water. How many of these protons would decay in one year?

*27. In the hot, early Universe, at a temperature of 10^{15} K, what was the average kinetic energy of the random thermal motion of gas particles? Compare this energy with the energy of about 100 GeV required to achieve symmetry between W particles and photons.

41.5 Quarks

28. Table 41.2 lists the quark composition of baryons. For all the Σ and Ξ particles, verify that the listed quark composition gives the correct values of electric charge, baryon number, and strangeness.

29. The antiproton $\bar{p}$ is made of three quarks. What kind of quarks are these?

30. How many quarks are there in a hydrogen atom? In a water molecule? (The oxygen nucleus contains eight protons and eight neutrons.)

31. How many quarks are created in the reaction (41.1)?

32. A particle is made of one d quark and one u antiquark. What is the electric charge of this particle? What is this particle?

33. Antiparticles contain antiquarks; for example, the proton quark content is uud and the antiproton quark content is $\bar{u}\,\bar{u}\,\bar{d}$. Are any of the baryons in Table 41.2 their own antiparticle? Are any of the mesons in Table 41.3 their own antiparticle? Which ones? What can you say about the quark content of a particle that is its own antiparticle?

34. According to a theoretical prediction based on the quark model, the masses of the nucleon (N = proton or neutron), the Ξ, the Λ, and the Σ should be related by

$$(m_N + m_\Xi) = \tfrac{1}{2}(3m_\Lambda + m_\Sigma)$$

Since this is intended as an approximate relation, the mass differences between the positive, negative, and neutral kinds of nucleon, Ξ, or Σ are to be neglected. Check this relation against the experimental values of the masses.

*35. As described in Section 41.5, high-energy electrons produced by the Stanford Linear Accelerator were used to probe the internal structure of protons. To detect small lumps in the proton, the wavelength of the electrons must be smaller than the lumps. The electrons had an energy of 20 GeV (which is an ultra relativistic energy). What is the de Broglie wavelength, $\lambda = h/p$, of these electrons?

41.6 Cosmology

36. Figure 41.28 shows the positions of four galaxies, with our own Galaxy at the center. Draw a figure showing the positions of these five galaxies at a later time, when the Universe is twice as old.

FIGURE 41.28 Four galaxies (●) at various distances from our Galaxy (⊕).

37. Consider the galaxies shown in Fig. 41.29; the arrows are the velocity vectors of those galaxies in the reference frame of the Earth. According to the Galilean addition law for velocities, find the velocity vectors of these galaxies (including our Galaxy) in the reference frame of the galaxy P; do this graphically, by subtracting the velocity vectors.

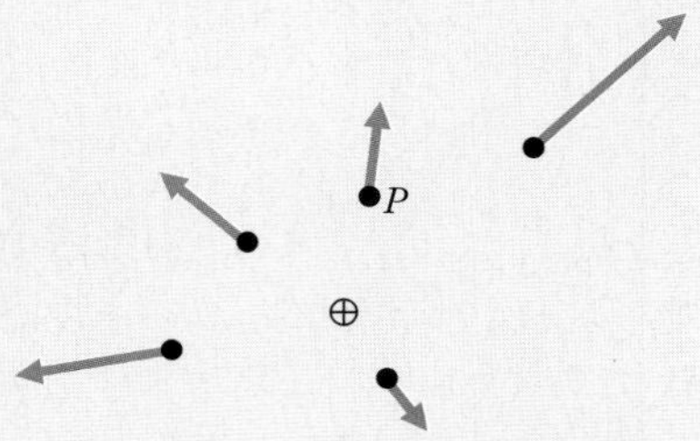

FIGURE 41.29 Motion of a galaxy P and other galaxies relative to our Galaxy (⊕).

38. If the velocity of recession of the galaxies were not proportional to distance ($v = H_0 r$) but, rather, proportional to the distance squared ($v = H'_0 r^2$) or proportional to some other power of the distance, then our Galaxy would occupy a preferred, central spot in the Universe. Explain.

39. The value of the Hubble constant that Hubble had deduced from the available data in 1936 was 1.0×10^8 (m/s)/(billion light-years). The corresponding expansion time [see Eq. (41.11)] is 1.9×10^9 years. How does this compare with the age of the Earth and the age of globular clusters? With what problem was Hubble faced?

REVIEW PROBLEMS

*40. A particle at rest decays into two pions, one positive and one negative. In a magnetic field of 1.25 T, each pion leaves a curved track perpendicular to the field with a radius of 54.9 cm. What is the mass of the particle that decayed? What is this particle?

*41. The neutron decays to a proton, an electron, and an electron antineutrino. Assume that for one particular decay, the neutron is at rest and the neutrino carries away negligible energy and momentum. What is the net kinetic energy of the proton and electron?

*42. In Dan Brown's novel *Angels and Demons*, 0.25 gram of antimatter is released about 3.0 km above ground level. Assuming that the subsequent annihilation results in isotropic radiation, what is the total energy incident on a square meter of ground directly below the annihilation? If the annihilation occurs over 10 seconds, compare the average energy flux at ground level during the annihilation with that of direct sunlight, approximately 1.0 kW/m^2.

43. In each of the following forbidden reactions, what conservation law is violated?

$$\mathrm{p} + \bar{\mathrm{p}} \rightarrow \mathrm{p} + \pi^-$$

$$\mathrm{p} + \bar{\mathrm{n}} \rightarrow \bar{\mathrm{p}} + \mathrm{n}$$

$$\gamma + \mu^+ + \mathrm{n} \rightarrow \pi^0 + \mathrm{p}$$

44. Two of the baryons in Table 41.2 have identical quark content. Which two are these? Consider the various conserved quantities discussed in Section 41.3. Which of these quantities do you think might be different for the two baryons?

45. A particle is made of two d quarks and one s quark. What is the electric charge of this particle? What is this particle?

46. The Δ^{++} particle has quark content uuu and decays into two particles in a reaction where only one quark–antiquark pair is created. If one particle is a proton, what is the other particle in the decay?

47. Sketch a Feynman diagram in which two masses are initially moving apart along the x axis, then come to rest after exchanging a graviton, and then approach after exchanging another graviton.

48. The recessional velocity of a galaxy in the constellation Ursa Major is 1.5×10^7 m/s, and the distance to this galaxy is approximately 1.0×10^9 light-years. Deduce a value of the Hubble constant from these data alone. How does your value compare with the average accepted value in Eq. (41.7)? Obtain an estimate of the age of the Universe in years from your Hubble constant.

Answers to Checkups

Checkup 41.1

1. No to both. Charged particles are required to boil the liquid in a bubble chamber to create a bubble track. Similarly, charged particles are required to ionize the gas in a multiwire chamber and produce the electrons that create current pulses.

2. The magnetic field in a particle detector diverts a charged particle into a circular orbit that reveals both the sign of the charge and the momentum of the particle.

3. The positively charged particles (red tracks) are diverted to the left as they travel downward. With the initial velocity downward, the magnetic field **B** must be directed out of the page in order to produce a magnetic force $\mathbf{F} = q\mathbf{v} \times \mathbf{B}$ to the left.

4. In reaction (41.1), the masses of the products greatly exceed the masses of the two initial protons, so the Q value is negative. We saw that reaction (41.2) proceeds spontaneously, so its Q value must be positive.

5. (B) In the head-on collision between protons of opposite motion, the full kinetic energy is available for reaction. For a stationary target, the nonzero momentum of the projectile must be conserved, requiring much of the final energy to remain kinetic.

Checkup 41.2

1. All leptons have spin $\frac{1}{2}$, but some baryons have other half-integer spins; for example, the Ω^- has spin $\frac{3}{2}$.

2. An antiparticle has the same mass and spin but the opposite charge as the corresponding particle; thus, from Table 41.1, the antitau has mass 1784 MeV/c^2, spin $\frac{1}{2}$, and electric charge $+e$, and the electron antineutrino has a mass less than 3 eV/c^2, spin $\frac{1}{2}$, and zero electric charge.

3. The heaviest particle listed is the upsilon (Υ) particle, one of the mesons of Table 41.3.

4. (D) Baryon; meson. All known leptons are listed in Table 41.1; there are no others. Other particles with half-integer spin, such as the η^0, are baryons; particles with mass and with integer spin, such as the B^0, are mesons.

Checkup 41.3

1. A hadron is a particle that participates in the strong interaction, that is, a baryon or a meson. Thus the proton and neutron are hadrons, but the electron and the tau neutrino, both leptons, are not. Similarly, the photon, the quantum of electromagnetism, is not a hadron.

2. An antiparticle has the same mass and spin as the particle, but the opposite charge and strangeness. From the xi-zero data of Table 41.4, the anti-xi-zero ($\overline{\Xi}^0$) then has mass 1315 MeV/c^2, spin $\frac{1}{2}$, zero charge, and strangeness +2; similarly, the ant-xi-minus, or xi-plus, has mass 1321 MeV/c^2, spin $\frac{1}{2}$, charge +1, and strangeness +2.

3. All interactions conserve energy, momentum, angular momentum, charge, baryon number, and lepton number. Strangeness and parity are conserved in electromagnetic (and strong) interactions, but not in weak interactions. Isospin is conserved in strong interactions, but not in electromagnetic (nor in weak) interactions.

4. The parity of sin kx is -1, since the function becomes the negative of itself when you reverse the x coordinate, sin $(-k_y) = \sin kx$.

5. (B) Weak only. Only the weak interaction does not conserve strangeness.

6. (D) All except strong; all. As a lepton, the electron does not participate in strong interactions. The neutron participates in all four—even though the neutron is net charge-neutral, it has an internal charge distribution and a magnetic moment, and participates in the electromagnetic interaction.

Checkup 41.4

1. Yes. Although heavier particles continue to be sought, the W and Z are the heaviest particles now known.

2. (B) Weak, strong, electromagnetic. The range of a force is inversely related to the mass of the particle that mediates the force. The W and Z particles (weak force) have the largest masses, photons (electromagnetic force) and gluons (strong force) are massless, but the range of the strong force is limited by pair production.

Checkup 41.5

1. Yes to both. Since the charge of the u quark is $+\frac{2}{3}e$ and the charge of the d quark is $-\frac{1}{3}e$, the charge of 2u + d is $+e$ and the charge of 2d + u is zero.

2. Two of the spins are up, one is down, resulting in a net spin of $\frac{1}{2}$.

3. The gluon is the particle that acts as carrier of the color force and the strong force.

4. (E) $-\frac{2}{3}e$, $+\frac{1}{3}e$, $+\frac{1}{3}e$. The charge of an antiquark is the negative of the charge of the corresponding quark.

Checkup 41.6

1. Other galaxies are so distant that the angles required for triangulation cannot be measured accurately. Even with the diameter of the Earth's orbit as a baseline, no difference in the angle to a distant galaxy from either end of such a baseline can be detected.
2. No. The Hubble "constant" reflects the present rate of expansion, and its value would increase if the expansion accelerates or decrease if the expansion slows; it would even change sign if contraction ultimately occurs.
3. No. The edge of the observable universe is expanding at the speed of light, and so a galaxy near the edge, moving slower than the speed of light, becomes further from the edge, and will not disappear.
4. (B) In supernovas soon after our Galaxy was born. These early supernovas spewed out dust and gas with the heavy elements that later coalesced to become our Solar System.

Appendix 1: Greek Alphabet

Α	α	alpha	Ν	ν	nu
Β	β	beta	Ξ	ξ	xi
Γ	γ	gamma	Ο	o	omicron
Δ	δ	delta	Π	π	pi
Ε	ϵ	epsilon	Ρ	ρ	rho
Ζ	ζ	zeta	Σ	σ	sigma
Η	η	eta	Τ	τ	tau
Θ	θ	theta	Υ	υ	upsilon
Ι	ι	iota	Φ	ϕ	phi
Κ	κ	kappa	Χ	χ	chi
Λ	λ	lambda	Ψ	ψ	psi
Μ	μ	mu	Ω	ω	omega

Appendix 2: Mathematics Review

A 2.1 Symbols

$a = b$ means a equals b
$a \neq b$ means a is not equal to b
$a > b$ means a is greater than b
$a < b$ means a is less than b
$a \geq b$ means a is not less than b
$a \leq b$ means a is not greater than b
$a \propto b$ means a is proportional to b
$a \approx b$ means a is approximately equal to b
$a \gg b$ means a is much greater than b
$a \ll b$ means a is much less than b
$\pi = 3.141\,59\ldots$
$e = 2.718\,28\ldots$

A 2.2 Powers and Roots

For any number a, the nth *power* of the number is the number multiplied by itself n times. This is written as a^n, and n is called the **exponent**. Thus,

$$a^1 = a \quad a^2 = a \cdot a \quad a^3 = a \cdot a \cdot a \quad a^4 = a \cdot a \cdot a \cdot a \quad \text{etc.}$$

For instance,

$$3^2 = 3 \times 3 = 9 \quad 3^3 = 3 \times 3 \times 3 = 27 \quad 3^4 = 3 \times 3 \times 3 \times 3 = 81 \quad \text{etc.}$$

A negative exponent indicates that the number is to be divided n times into 1; thus

$$a^{-1} = \frac{1}{a} \quad a^{-2} = \frac{1}{a^2} \quad a^{-3} = \frac{1}{a^3} \quad \text{etc.}$$

A zero exponent yields 1, regardless of the value of a:

$$a^0 = 1$$

The rules for the combination of exponents in products, in ratios, and in powers of powers are

$$a^n \cdot a^m = a^{m+n}$$

$$\frac{a^n}{a^m} = a^{n-m}$$

$$(a^n)^m = a^{nm}$$

For instance, it is easy to verify that

$$3^2 \times 3^3 = 3^5$$

$$\frac{3^2}{3^3} = 3^{-1} = \frac{1}{3}$$

$$(3^2)^3 = 3^{2\times 3} = 3^6$$

Note that for any two numbers a and b

$$(a \cdot b)^n = a^n \cdot b^n$$

For instance,

$$(2 \times 3)^3 = 2^3 \times 3^3$$

The nth root of a is a number such that its nth power equals a. The nth root is written $a^{1/n}$. The second root $a^{1/2}$ is usually called the square root, and designated by $\sqrt{a}$:

$$a^{1/2} = \sqrt{a}$$

As suggested by the notation $a^{1/n}$, roots are fractional powers, and they obey the usual rules for the combination of exponents:

$$(a^{1/n})^n = a^{n/n} = a$$

$$(a^{1/n})^m = a^{m/n}$$

A 2.3 Arithmetic in Scientific Notation

The scientific notation for numbers (see the first page of the Prelude) is quite handy for the multiplication and the division of very large or very small numbers, because we can deal with the decimal parts and the power-of-10 parts in the numbers separately. For example, to multiply 4×10^{10} by 5×10^{12}, we multiply 4 by 5 and 10^{10} by 10^{12}, as follows:

$$(4 \times 10^{10}) \times (5 \times 10^{12}) = (4 \times 5) \times (10^{10} \times 10^{12})$$

$$= 20 \times 10^{10+12} = 20 \times 10^{22} = 2 \times 10^{23}$$

To divide these numbers, we proceed likewise:

$$\frac{4 \times 10^{10}}{5 \times 10^{12}} = \frac{4}{5} \times \frac{10^{10}}{10^{12}} = 0.8 \times 10^{10-12} = 0.8 \times 10^{-2} = 8 \times 10^{-3}$$

When performing additions or subtractions of numbers in scientific notation, we must be careful to begin by expressing the numbers with the same power of 10. For example, the sum of 1.5×10^9 and 3×10^8 is

$$1.5 \times 10^9 + 3 \times 10^8 = 1.5 \times 10^9 + 0.3 \times 10^9 = 1.8 \times 10^9$$

A 2.4 Algebra

An equation is a mathematical statement that tells us that one quantity or a combination of quantities is equal to another quantity or combination. We often have to solve for one of the quantities in the equation in terms of the other quantities. For instance, we may have to solve the equation

$$x + a = b$$

for x in terms of a and b. Here a and b are numerical constants or mathematical expressions which are regarded as known, and x is regarded as unknown.

The rules of algebra instruct us how to manipulate equations and accomplish their solution. The three most important rules are:

1. Any equation remains valid if equal terms are added or subtracted from its left side and its right side.

This rule is useful for solving the equation $x + a = b$. We simply subtract a from both sides of this equation and find

$$x + a - a = b - a$$

that is,

$$x = b - a$$

To see how this works in a concrete numerical example, consider the equation

$$x + 7 = 5$$

Subtracting 7 from both sides, we obtain

$$x = 5 - 7$$

or

$$x = -2$$

Note that given an equation of the form $x + a = b$, we may want to solve for a in terms of x and b, if x is already known from some other information but a is a mathematical quantity that is not yet known. If so, we must subtract x from both sides of the equation, and we obtain

$$a = b - x$$

Most equations in physics contain several mathematical quantities which sometimes play the role of known quantities, sometimes the role of unknown quantities, depending on circumstances. Correspondingly, we will sometimes want to solve the equation for one quantity (such as x), sometimes for another (such as a).

2. Any equation remains valid if the left and the right sides are multiplied or divided by the same factor.

This rule is useful for solving

$$ax = b$$

We simply divide both sides by a, which yields

$$\frac{ax}{a} = \frac{b}{a}$$

or

$$x = \frac{b}{a}$$

Often it will be necessary to combine both of the above rules. For instance, to solve the equation

$$2x + 10 = 16$$

we begin by subtracting 10 from both sides, obtaining

$$2x = 16 - 10$$

or

$$2x = 6$$

and then we divide both sides by 2, with the result

$$x = \frac{6}{2}$$

or

$$x = 3$$

3. Any equation remains valid if both sides are raised to the same power.

This rule permits us to solve the equation

$$x^3 = b$$

Raising both sides to the power $\frac{1}{3}$, we find

$$(x^3)^{1/3} = b^{1/3}$$

or

$$x = b^{1/3}$$

As a final example, let us consider the equation

$$x = -\tfrac{1}{2}gt^2 + x_0$$

(as established in Chapter 2, this equation describes the vertical position of a particle that starts at a height x_0 and falls for a time t; but the meaning of the equation need not concern us here). Suppose that we want to solve for t in terms of the other quantities in the equation. This will require the use of all our rules of algebra. First, subtract x from both sides and then add $\frac{1}{2}gt^2$ to both sides. This leads to

$$0 = -\tfrac{1}{2}gt^2 + x_0 - x$$

and then to

$$\tfrac{1}{2}gt^2 = x_0 - x$$

Next, multiply both sides by 2 and divide both sides by g; this yields

$$t^2 = \frac{2}{g}(x_0 - x)$$

Finally, raise both sides to the power $\frac{1}{2}$, or, equivalently, extract the square root of both sides. This gives us the final result

$$t = \sqrt{\frac{2}{g}(x_0 - x)}$$

A 2.5 Equations with Two Unknowns

If we seek to solve for two unknowns simultaneously, then we need two independent equations containing these two unknowns. The solution of such simultaneous equations can be carried out by the method of elimination: begin by using one equation to solve for the first unknown in terms of the second, then use this result to eliminate the first unknown from the other equation. An example will help to make this clear. Consider the following two simultaneous equations with two unknowns x and y:

$$4x + 2y = 8$$

$$2x - y = -2$$

To solve the first equation for x in terms of y, subtract $2y$ from both sides and then divide both sides by 4:

$$x = \frac{8 - 2y}{4}$$

Next, substitute this expression for x into the second equation:

$$2 \times \frac{8 - 2y}{4} - y = -2$$

To simplify this equation, multiply both sides by 4:

$$2 \times (8 - 2y) - 4y = -8$$

and combine the two terms containing y:

$$16 - 8y = -8$$

This is an ordinary equation for the single unknown y, and it can be solved by the methods we discussed in the preceding section, with the result

$$y = 3$$

It then follows from the above expression for x that

$$x = \frac{8 - 2y}{4} = \frac{8 - 2 \times 3}{4} = \frac{2}{4} = \frac{1}{2}$$

A 2.6 The Quadratic Formula

The quadratic equation $ax^2 + bx + c = 0$ has two solutions:

$$x = \frac{-b \pm \sqrt{b^2 - 4ac}}{2a}$$

A 2.7 Logarithms and the Exponential Function

The **base-10 logarithm** of a (positive) number is the power to which 10 must be raised to obtain this number. Thus, from $10 = 10^1$ and $100 = 10^2$ and $1000 = 10^3$ and $10\,000 = 10^4$ we immediately deduce that

$$\log 10 = 1$$
$$\log 100 = 2$$
$$\log 1000 = 3$$
$$\log 10\,000 = 4, \text{ etc.}$$

Likewise

$$\log 1 = 0$$
$$\log 0.1 = -1$$
$$\log 0.01 = -2$$
$$\log 0.001 = -3, \text{etc.}$$

Thus, the logarithm of a number between 1 and 10 is somewhere between 0 and 1, but to find the logarithm of such a number, we need the help of a computer program (many calculators have built-in computer programs that yield the value of the logarithm at the touch of a button). For some calculations, it is convenient to remember that $\log 2 = 0.301 \approx 0.3$ and $\log 5 = 0.699 \approx 0.7$.

The logarithm of the product of two numbers is the sum of the individual logarithms, and the logarithm of the ratio of two numbers is the difference of the individual logarithms. This rule makes it easy to find the logarithm of a number expressed in scientific notation. For example, the logarithm of 2×10^6 is

$$\log(2 \times 10^6) = \log 2 + \log 10^6 = 0.301 + 6 = 6.301$$

Note that the logarithm of any (positive) number smaller than 1 is negative. For example,

$$\log(5 \times 10^{-3}) = \log 5 + \log 10^{-3} = 0.699 - 3 = -2.301$$

The **exponential function** $\exp(x)$ is defined by the following infinite series:

$$\exp(x) = 1 + x + \frac{x^2}{2} + \frac{x^3}{3 \cdot 2} + \frac{x^4}{4 \cdot 3 \cdot 2} + \cdots$$

This function is equivalent to raising the constant $e = 2.718\,28\ldots$ to the power x:

$$\exp(x) = e^x$$

The **natural logarithm** $\ln x$ is the inverse of the exponential function, so

$$x = e^{\ln x}$$

and

$$x = \ln(e^x)$$

Natural logarithms obey the usual rules for logarithms,

$$\ln(x \cdot y) = \ln x + \ln y$$

$$\ln\left(\frac{x}{y}\right) = \ln x - \ln y$$

$$\ln(x^a) = a \ln x$$

Note that

$$\ln e = 1$$

and

$$\ln 10 = 2.3026$$

If we designate the base-10 logarithm, or **common logarithm**, by $\log x$, then the relationship between the two kinds of logarithm is as follows:

$$\ln x = \ln(10^{\log x}) = (\log x)(\ln 10) = 2.3026 \log x$$

Appendix 3: Geometry and Trigonometry Review

A3.1 Perimeters, Areas, and Volumes

[perimeter of a circle of radius r] $= 2\pi r$
[area of a circle of radius r] $= \pi r^2$
[area of a triangle of base b, altitude h] $= hb/2$
[surface area of a sphere of radius r] $= 4\pi r^2$
[volume of a sphere of radius r] $= 4\pi r^3/3$
[area of curved surface of a cylinder of radius r, height h] $= 2\pi rh$
[volume of a cylinder of radius r, height h] $= \pi r^2 h$

A3.2 Angles

The angle between two intersecting straight lines is defined as the fraction of a complete circle included between these lines (Fig. A3.1). To express the angle in **degrees**, we assign an angular magnitude of 360° to the complete circle; any arbitrary angle is then an appropriate fraction of 360°. To express the angle in **radians**, we assign an angular magnitude of 2π radians to the complete circle; any arbitrary angle is then an appropriate fraction of 2π. For example, the angle shown in Fig. A3.1 is $\frac{1}{12}$ of a complete circle, that is, 30°, or $\pi/6$ radian. In view of the definition of angle, the length of arc included between the two intersecting straight lines is proportional to the angle θ between these lines; if the angle is expressed in radians, then the constant of proportionality is simply the radius:

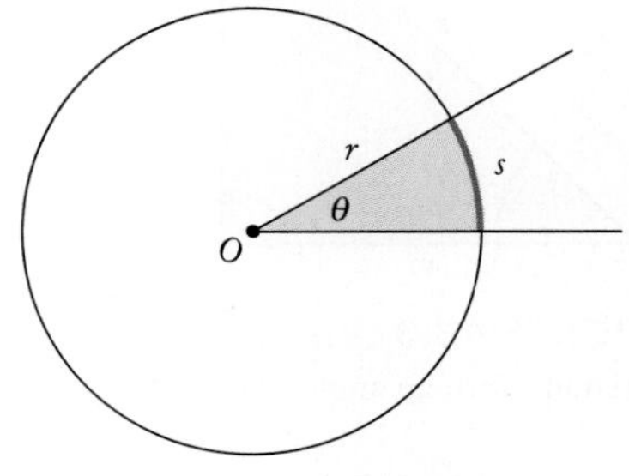

FIGURE A3.1 The angle θ in this diagram is $\theta = 30°$, or $\pi/6$ radian.

$$s = r\theta \tag{1}$$

Since 2π radians $= 360°$, it follows that

$$1 \text{ radian} = \frac{360°}{2\pi} = \frac{360°}{2 \times 3.141\,59} = 57.2958° \tag{2}$$

Each degree is divided into 60 minutes of arc (arcminutes), and each of these into 60 seconds of arc (arcseconds). In degrees, minutes of arc, and seconds of arc, the radian is

$$1 \text{ radian} = 57°\,17'\,44.8'' \tag{3}$$

A3.3 The Trigonometric Functions

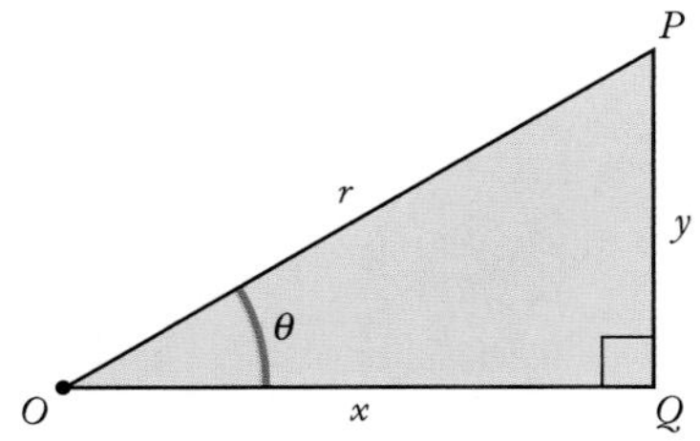

FIGURE A3.2 A right triangle.

The trigonometric functions of an angle are defined as ratios of the lengths of the sides of a right triangle erected on this angle. Figure A3.2 shows an acute angle θ and a right triangle, one of whose angles coincides with θ. The adjacent side OQ has a length x, the opposite side QP a length y, and the hypotenuse OP a length r. The **sine, cosine, tangent, cotangent, secant,** and **cosecant** of the angle θ are then defined as follows:

sine	$\sin\theta = y/r$	(4)
cosine	$\cos\theta = x/r$	(5)
tangent	$\tan\theta = y/x$	(6)
cotangent	$\cot\theta = x/y$	(7)
secant	$\sec\theta = r/x$	(8)
cosecant	$\csc\theta = r/y$	(9)

EXAMPLE 1 Find the sine, cosine, and tangent for angles of 0°, 90°, and 45°.

SOLUTION: For an angle of 0°, the opposite side is zero ($y = 0$), and the adjacent side coincides with the hypotenuse ($x = r$). Hence

$$\sin 0° = 0 \qquad \cos 0° = 1 \qquad \tan 0° = 0 \tag{10}$$

For an angle of 90°, the adjacent side is zero ($x = 0$), and the opposite side coincides with the hypotenuse ($y = r$). Hence

$$\sin 90° = 1 \quad \cos 90° = 0 \quad \tan 90° = \infty \tag{11}$$

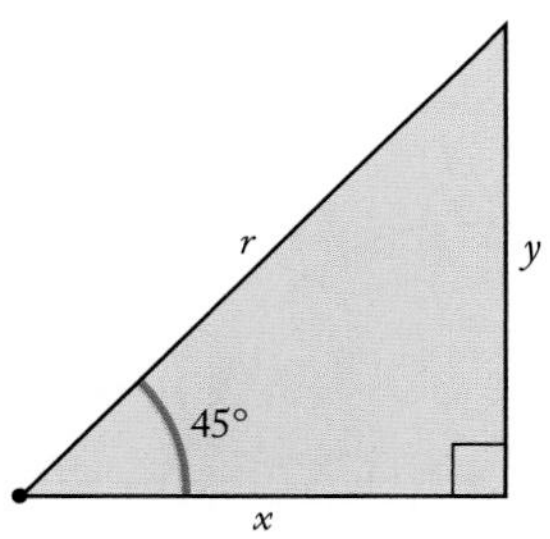

FIGURE A3.3 A right triangle with an angle of 45°.

Finally, for an angle of 45° (Fig. A3.3), the adjacent and the opposite sides have the same length ($x = y$) and the hypotenuse has a length of $\sqrt{2}$ times the length of either side ($r = \sqrt{2}x = \sqrt{2}y$). Hence

$$\sin 45° = \frac{1}{\sqrt{2}} \qquad \cos 45° = \frac{1}{\sqrt{2}} \qquad \tan 45° = 1 \tag{12}$$

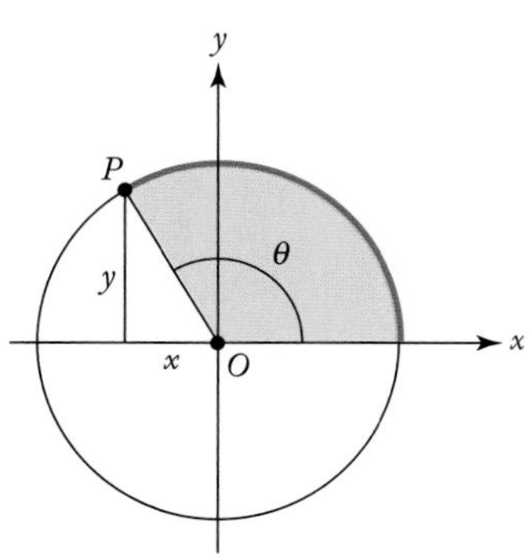

FIGURE A3.4 The angle θ in this diagram is larger than 90°.

The definitions (4)–(9) are also valid for angles greater than 90°, such as the angle shown in Fig. A3.4. In the general case, the quantities x and y must be interpreted as the rectangular coordinates of the point P. For any angle larger than 90°, one or both of the coordinates x and y are negative. Hence some of the trigonometric functions will also be negative. For instance,

$$\sin 135° = \frac{1}{\sqrt{2}} \qquad \cos 135° = -\frac{1}{\sqrt{2}} \qquad \tan 135° = -1 \tag{13}$$

Figure A3.5 shows plots of the sine, cosine, and tangent vs. θ.

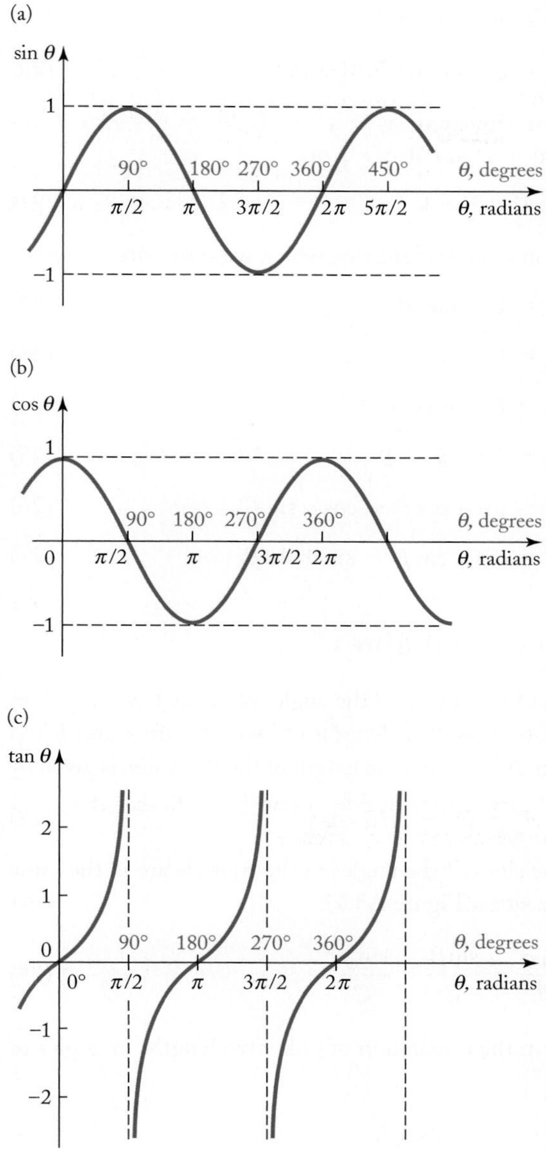

FIGURE A3.5 Plots of the sine, cosine, and tangent functions.

A3.4 Trigonometric Identities

From the definitions (4)–(9) we immediately find the following identities:

$$\tan\theta = \sin\theta/\cos\theta \tag{14}$$

$$\cot\theta = 1/\tan\theta \tag{15}$$

$$\sec\theta = 1/\cos\theta \tag{16}$$

$$\csc\theta = 1/\sin\theta \tag{17}$$

Figure A3.6 shows a right triangle with angles θ and $90° - \theta$. Since the adjacent side for the angle θ is the opposite side for the angle $90° - \theta$ and vice versa, we see that the trigonometric functions also obey the following identities:

$$\sin(90° - \theta) = \cos\theta \tag{18}$$

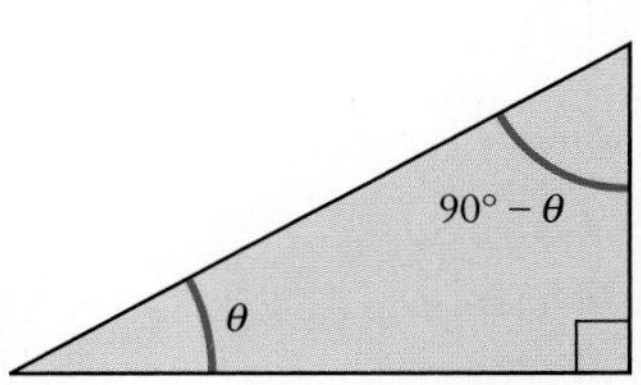

FIGURE A3.6 A right triangle with angles θ and $90° - \theta$.

$$\cos(90° - \theta) = \sin\theta \tag{19}$$

$$\tan(90° - \theta) = \cot\theta = 1/\tan\theta \tag{20}$$

According to the **Pythagorean theorem**, $x^2 + y^2 = r^2$. With $x = r\cos\theta$ and $y = r\sin\theta$, this becomes $r^2\cos^2\theta + r^2\sin^2\theta = r^2$, or

$$\cos^2\theta + \sin^2\theta = 1 \tag{21}$$

The following are a few other trigonometric identities, which we state without proof:

$$\sec^2\theta = 1 + \tan^2\theta \tag{22}$$

$$\csc^2\theta = 1 + \cot^2\theta \tag{23}$$

$$\sin 2\theta = 2\sin\theta\cos\theta \tag{24}$$

$$\cos 2\theta = 2\cos^2\theta - 1 \tag{25}$$

$$\sin(\alpha + \beta) = \sin\alpha\cos\beta + \cos\alpha\sin\beta \tag{26}$$

$$\cos(\alpha + \beta) = \cos\alpha\cos\beta - \sin\alpha\sin\beta \tag{27}$$

A3.5 The Laws of Cosines and Sines

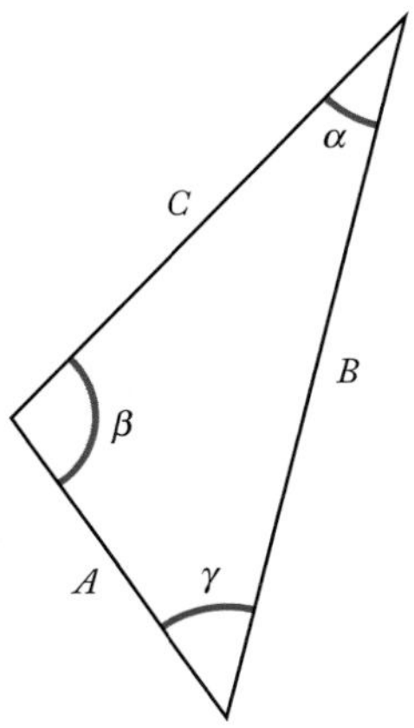

FIGURE A3.7 An arbitrary triangle.

In an arbitrary triangle the lengths of the sides and the angles obey the laws of cosines and of sines. The **law of cosines** states that if the lengths of two sides are A and B and the angle between them is γ (Figure A3.7), then the length of the third side is given by

$$C^2 = A^2 + B^2 - 2AB\cos\gamma \tag{28}$$

The **law of sines** states that the sines of the angles of the triangle are in the same ratio as the lengths of the opposite sides (Figure A3.7):

$$\frac{\sin\alpha}{A} = \frac{\sin\beta}{B} = \frac{\sin\gamma}{C} \tag{29}$$

Both of these laws are very useful in the calculation of unknown lengths or angles of a triangle.

Appendix 4: Calculus Review

A4.1 Derivatives

We saw in Section 2.3 that if the position of a particle is some function of time, say, $x = x(t)$, then the instantaneous velocity of the particle is the derivative of x with respect to t:

$$v = \frac{dx}{dt} \tag{1}$$

This derivative is defined by first looking at a small increment Δx that results from a small increment Δt, and then evaluating the ratio $\Delta x/\Delta t$, in the limit when both Δx and Δt tend toward zero. Thus

$$\frac{dx}{dt} = \lim_{\Delta t \to 0} \frac{\Delta x}{\Delta t} \tag{2}$$

Graphically, in a plot of position vs. time, the derivative dx/dt is the slope of the straight line tangent to the curved line at the time t (see Figure A4.1).

In general, if $f = f(u)$ is some given function of a variable u, the **derivative** of f with respect to u is defined by

$$\frac{df}{du} = \lim_{\Delta u \to 0} \frac{\Delta f}{\Delta u} \tag{3}$$

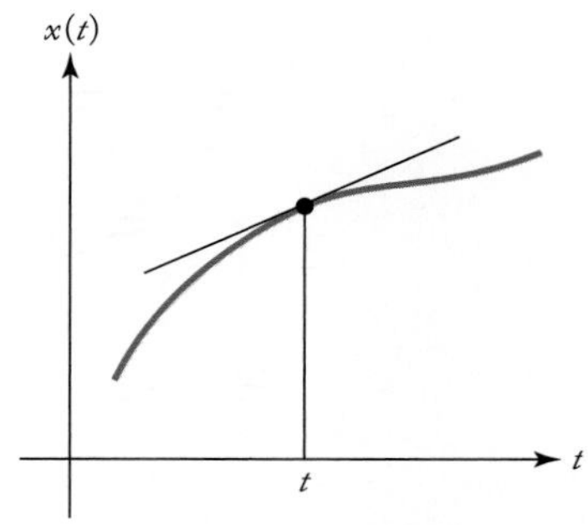

FIGURE A4.1 The derivative of $x(t)$ at t is the slope of the straight line tangent to the curve at t.

In a plot of f vs. u, this derivative is the slope of the straight line tangent to the curve representing $f(u)$.

Starting with the definition (3) we can find the derivative of any function (provided the function is sufficiently smooth so the derivative exists!). For example, consider the function $f(u) = u^2$. If we increase u to $u + \Delta u$, the function $f(u)$ increases to

$$f + \Delta f = (u + \Delta u)^2 \tag{4}$$

and therefore

$$\begin{aligned} \Delta f &= (u + \Delta u)^2 - f = (u + \Delta u)^2 - u^2 \\ &= 2u\,\Delta u + (\Delta u)^2 \end{aligned} \tag{5}$$

The derivative df/du is then

$$\frac{df}{du} = \lim_{\Delta u \to 0} \frac{\Delta f}{\Delta u} = \lim_{\Delta u \to 0} \frac{2u\,\Delta u + (\Delta u)^2}{\Delta u} \tag{6}$$

$$= \lim_{\Delta u \to 0} (2u) + \lim_{\Delta u \to 0} (\Delta u) \tag{7}$$

The second term on the right side vanishes in the limit $\Delta u \to 0$; the first term is simply $2u$. Hence

$$\frac{df}{du} = 2u \tag{8}$$

or

$$\frac{d}{du}(u^2) = 2u \tag{9}$$

This is one instance of the general rule for the differentiation of u^n:

$$\frac{d}{du}(u^n) = nu^{n-1} \tag{10}$$

This general rule is valid for any positive or negative number n, including zero. The proof of this rule can be constructed by an argument similar to that above. Table A4.1 lists the derivatives of the most common functions.

TABLE A4.1

SOME DERIVATIVES

$\frac{d}{du} u^n = nu^{n-1}$

$\frac{d}{du} \ln u = \frac{1}{u}$

$\frac{d}{du} e^u = e^u$

(In all the following formulas, u is in radian:)

$\frac{d}{du} \sin u = \cos u$

$\frac{d}{du} \cos u = -\sin u$

$\frac{d}{du} \tan u = \sec^2 u$

$\frac{d}{du} \cot u = -\csc^2 u$

$\frac{d}{du} \sec u = \tan u \sec u$

$\frac{d}{du} \csc u = -\cot u \csc u$

$\frac{d}{du} \sin^{-1} u = 1/\sqrt{1 - u^2}$

$\frac{d}{du} \cos^{-1} u = -1/\sqrt{1 - u^2}$

$\frac{d}{du} \tan^{-1} u = \frac{1}{1 + u^2}$

A4.2 Important Rules for Differentiation

1. Derivative of a constant times a function:

$$\frac{d}{du}(cf) = c\frac{df}{du} \tag{11}$$

For instance,

$$\frac{d}{du}(6u^2) = 6\frac{d}{du}(u^2) = 6 \times 2u = 12u$$

2. Derivative of the sum of two functions:

$$\frac{d}{du}(f + g) = \frac{df}{du} + \frac{dg}{du} \tag{12}$$

For instance,

$$\frac{d}{du}(6u^2 + u) = \frac{d}{du}(6u^2) + \frac{d}{du}(u) = 12u + 1$$

3. Derivative of the product of two functions:

$$\frac{d}{du}(f \times g) = g\frac{df}{du} + f\frac{dg}{du} \tag{13}$$

For instance,

$$\begin{aligned}\frac{d}{du}(u^2 \sin u) &= \sin u \frac{d}{du}u^2 + u^2\frac{d}{du}\sin u \\ &= \sin u \times 2u + u^2 \times \cos u\end{aligned}$$

4. Chain rule for derivatives: If f is a function of g and g is a function of u, then

$$\frac{d}{du}f(g) = \frac{df}{dg}\frac{dg}{du} \tag{14}$$

For instance, if $g = 2u$ and $f(g) = \sin g$, then

$$\begin{aligned}\frac{d}{du}\sin(2u) &= \frac{d\sin(2u)}{d(2u)}\frac{d(2u)}{du} \\ &= \cos(2u) \times 2\end{aligned}$$

5. Partial derivatives: If f is a function of more than one variable, then the *partial derivative* of f with respect to one of the variables, say x, is denoted $\partial f/\partial x$, and is obtained by treating all the *other* variables as constants when differentiating.
For instance, if $f = x^2y + y^2z$, then

$$\frac{\partial f}{\partial x} = 2xy, \quad \frac{\partial f}{\partial y} = x^2 + 2yz, \quad \text{and} \quad \frac{\partial f}{\partial z} = y^2$$

A4.3 Integrals

We have learned that if the position of a particle is known as a function of time, then we can find the instantaneous velocity by differentiation. What about the converse problem: if the instantaneous velocity is known as a function of time, how can we find the position? In Section 2.5 we learned how to deal with this problem in the special case of motion with constant acceleration. The velocity is then a fairly simple function of time [see Eq. (2.17)]

$$v = v_0 + at \tag{15}$$

and the position deduced from this velocity is [see Eq. (2.22)]

$$x = x_0 + v_0 t + \tfrac{1}{2}at^2 \tag{16}$$

where x_0 and v_0 are the initial position and velocity at the initial time $t_0 = 0$. Now we want to deal with the general case of a velocity that is an arbitrary function of time,

$$v = v(t) \tag{17}$$

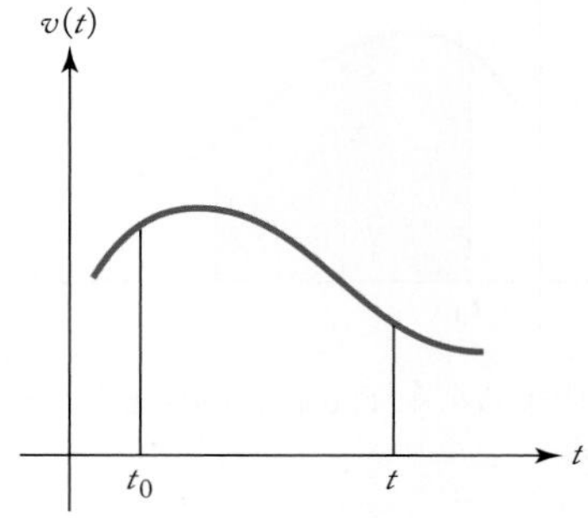

FIGURE A4.2 Plot of a function $v(t)$.

Figure A4.2 shows what a plot of v vs. t might look like. At the initial time t_0, the particle has an initial position x_0 (for the sake of generality we now assume that $t_0 \neq 0$). We want to find the position at some later time t. For this purpose, let us divide the time interval $t - t_0$ into a large number of small time intervals, each of the duration Δt. The total number of intervals is N, so $t - t_0 = N\,\Delta t$. The first of these intervals lasts from t_0 to $t_0 + \Delta t$; the second from $t_0 + \Delta t$ to $t_0 + 2\Delta t$; etc.

In Figure A4.3 the beginnings and the ends of these intervals have been marked t_0, t_1, t_2, etc., with $t_1 = t_0 + \Delta t$, $t_2 = t_0 + 2\Delta t$, etc. If Δt is sufficiently small, then during the first time interval the velocity is approximately $v(t_0)$; during the second, $v(t_1)$; etc. This amounts to replacing the smooth function $v(t)$ by a series of steps (see Fig. A4.3). Thus, during the first time interval, the displacement of the particle is approximately $v(t_0)\,\Delta t$; during the second interval, $v(t_1)\,\Delta t$; etc. The net displacement of the particle during the entire interval $t - t_0$ is the sum of all these small displacements:

$$x(t) - x_0 \approx v(t_0)\,\Delta t + v(t_1)\,\Delta t + v(t_2)\,\Delta t + \cdots \tag{18}$$

Using the standard mathematical notation for summation, we can write this as

$$x(t) - x_0 \approx \sum_{i=0}^{N-1} v(t_i)\,\Delta t \tag{19}$$

We can give this sum the following graphical interpretation: since $v(t_i)\,\Delta t$ is the area of the rectangle of height $v(t_i)$ and width Δt, the sum is the net area of all the rectangles shown in Figure A4.3, i.e., it is approximately the area under the velocity curve. Note that if the velocity is negative, the area must be reckoned as negative!

Of course, Eq. (19) is only an approximation. To find the exact displacement of the particle we must let the step size Δt tend to zero (while the number of steps N tends to infinity). In this limit, the steplike horizontal and vertical line segments in Fig. A4.3 approach the smooth curve. Thus,

$$x(t) - x_0 = \lim_{\substack{\Delta t \to 0 \\ N \to \infty}} \sum_{i=0}^{N-1} v(t_i)\,\Delta t \tag{20}$$

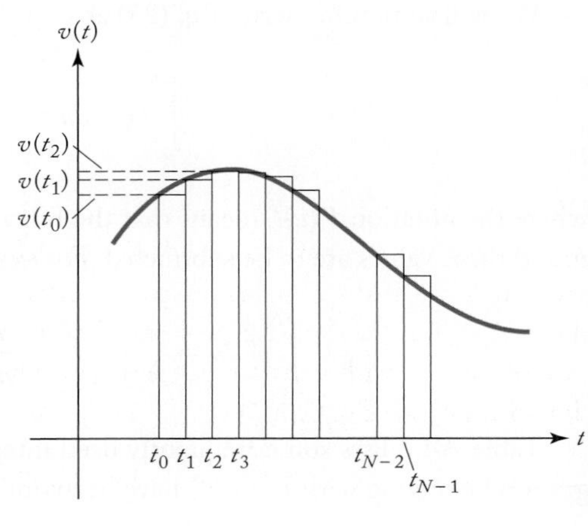

FIGURE A4.3 The interval $t - t_0$ has been divided into N equal intervals of duration Δt, so $t_1 = t_0 + \Delta t$, etc.

In the notation of calculus, the right side of Eq. (20) is usually written in the following fashion:

$$x(t) - x_0 = \int_{t_0}^{t} v(t')\, dt' \tag{21}$$

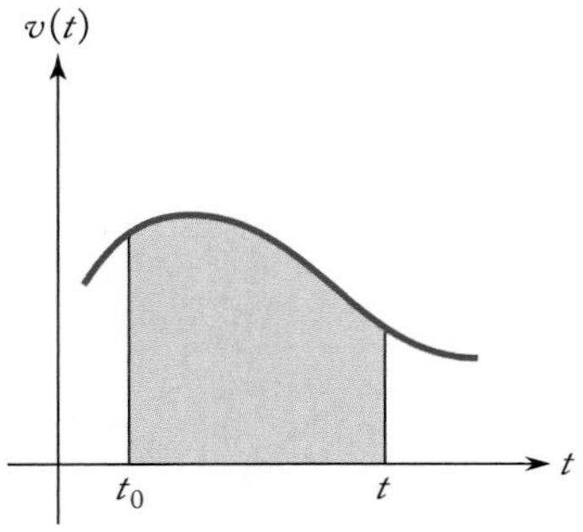

FIGURE A4.4 The area under the velocity curve.

The right side is called the **integral** of the function $v(t)$. The subscript and the superscript on the integration symbol $\int$ are called, respectively, the lower and the upper limit of integration; and t' is called the variable of integration (the prime on the variable of integration t' merely serves to distinguish that variable from the limit of integration t). Graphically, the integral is the exact area under the velocity curve between the limits t_0 and t in a plot of v vs. t (see Fig. A4.4). Areas below the t axis must be reckoned as negative.

In general, if $f(u)$ is a function of u, then the integral of this function is defined by a limiting procedure similar to that described above for the special case of the function $v(t)$. The integral over an interval from $u = a$ to $u = b$ is

$$\int_a^b f(u)\, du = \lim_{\substack{\Delta u \to 0 \\ N \to \infty}} \sum_{i=0}^{N-1} f(u_i)\, \Delta u \tag{22}$$

where $u_i = a + i\,\Delta u$. As in the case of the integral of $v(t)$, this integral can again be interpreted as an area: it is the area under the curve between the limits a and b in a plot of f vs. u.

For the explicit evaluation of integrals we can take advantage of the connection between integrals and antiderivatives. An **antiderivative** of a function $f(u)$ is simply a function $F(u)$ such that $dF/du = f$. For example, if $f(u) = u^n$ and $n \neq -1$, then an antiderivative of $f(u)$ is $F(u) = u^{n+1}/(n+1)$. The fundamental theorem of calculus states that the integral of any function $f(u)$ can be expressed in terms of antiderivatives:

$$\int_a^b f(u)\, du = F(b) - F(a) \tag{23}$$

In essence, this means that integration is the inverse of differentiation. We will not prove this theorem here, but we remark that such an inverse relationship between integration and differentiation should not come as a surprise. We have already run across an obvious instance of such a relationship: we know that velocity is the derivative of the position, and we have seen above that the position is the integral of the velocity.

We will sometimes write Eq. (23) as

$$\int_a^b f(u)\, du = F(u)\Big|_a^b \tag{24}$$

where the notation $F(u)|_a^b$ means that the function $F(u)$ is to be evaluated at a and at b, and these values are to be subtracted. For example, if $n \neq -1$,

$$\int_a^b u^n\, du = \frac{u^{n+1}}{n+1}\Big|_a^b = \frac{b^{n+1}}{n+1} - \frac{a^{n+1}}{n+1} \tag{25}$$

Table A4.2 lists some frequently used integrals. In this table, the limits of integration belonging with Eq. (24) have been omitted for the sake of brevity.

TABLE A4.2 SOME INTEGRALS

$$\int u^n du = \frac{u^{n+1}}{n+1} \quad \text{for } n \neq -1$$

$$\int \frac{1}{u} du = \ln u \quad \text{for } u > 0$$

$$\int e^{ku}\, du = \frac{e^{ku}}{k}$$

$$\int \ln u\, du = u \ln u - u$$

$$\int \sin(ku)\, du = -\frac{1}{k}\cos(ku) \quad \text{(where } ku \text{ is in radians)}$$

$$\int \cos(ku)\, du = \frac{1}{k}\sin(ku) \quad \text{(where } ku \text{ is in radians)}$$

$$\int \frac{du}{1+ku} = \frac{1}{k}\ln(1+ku)$$

$$\int \frac{du}{\sqrt{k^2-u^2}} = \sin^{-1}\left(\frac{u}{k}\right)$$

$$\int \frac{du}{\sqrt{u^2 \pm k^2}} = \ln\left(u + \sqrt{u^2 \pm k^2}\right)$$

$$\int \sqrt{k^2-u^2}\, du = \frac{1}{2}\left[u\sqrt{k^2-u^2} + k^2\sin^{-1}\left(\frac{u}{k}\right)\right]$$

$$\int \frac{du}{k^2+u^2} = \frac{1}{k}\tan^{-1}\left(\frac{u}{k}\right)$$

$$\int \frac{du}{u\sqrt{k^2 \pm u^2}} = -\frac{1}{k}\ln\left(\frac{k+\sqrt{k^2 \pm u^2}}{u}\right)$$

$$\int \frac{du}{(u^2+k^2)^{3/2}} = \frac{u}{k^2\sqrt{u^2+k^2}}$$

A4.4 Important Rules for Integration

1. Integral of a constant times a function:

$$\int_a^b cf(u)\, du = c\int_a^b f(u)\, du \tag{26}$$

For instance,

$$\int_a^b 5u^2\, du = 5\int_a^b u^2\, du = 5\left(\frac{b^3}{3} - \frac{a^3}{3}\right) \tag{27}$$

2. Integral of a sum of two functions:

$$\int_a^b [f(u) + g(u)]\, du = \int_a^b f(u)\, du + \int_a^b g(u)\, du \tag{28}$$

For instance,

$$\int_a^b (5u^2 + u)\, du = \int_a^b 5u^2\, du + \int_a^b u\, du = 5\left(\frac{b^3}{3} - \frac{a^3}{3}\right) + \left(\frac{b^2}{2} - \frac{a^2}{2}\right) \quad (29)$$

3. Change of limits of integration:

$$\int_a^b f(u)\, du = \int_a^c f(u)\, du + \int_c^b f(u)\, du \quad (30)$$

$$\int_a^b f(u)\, du = -\int_b^a f(u)\, du \quad (31)$$

4. Change of variable of integration: If u is a function of v, then

$$\int_a^b f(u)\, du = \int_{v(a)}^{v(b)} f(u) \frac{du}{dv}\, dv \quad (32)$$

For instance, with $u = v^2$,

$$\int_a^b u^3\, du = \int_a^b v^6\, du = \int_{\sqrt{a}}^{\sqrt{b}} v^6(2v)\, dv \quad (33)$$

Finally, let us apply these general results to some specific examples of integration of the velocity.

EXAMPLE 1 A particle with constant acceleration has the following velocity as a function of time [compare Eq. (15)]:

$$v(t) = v_0 + at$$

where v_0 is the velocity at $t = 0$.

By integration, find the position as a function of time.

SOLUTION: According to Eq. (21), with $t_0 = 0$,

$$x(t) - x_0 = \int_0^t v(t')\, dt' = \int_0^t (v_0 + at')\, dt'$$

Using rule 2 and rule 1, we find that this equals

$$x(t) - x_0 = \int_0^t v_0\, dt' + \int_0^t at'\, dt' = v_0 \int_0^t dt' + a\int_0^t t'\, dt' \quad (34)$$

The first entry listed in Table A4.2 gives $\int dt' = t'$ (for $n = 0$) and $\int t'\, dt' = t'^2/2$ (for $n = 1$). Thus,

$$x(t) - x_0 = v_0\, t' \Big|_0^t + \tfrac{1}{2}\, at'^2 \Big|_0^t$$

$$= v_0 t + \tfrac{1}{2}\, at^2 \quad (35)$$

This, of course, agrees with Eq. (16).

EXAMPLE 2 The instantaneous velocity of a projectile traveling through air is the following function of time:

$$v(t) = 655.9 - 61.14t + 3.26t^2$$

where $v(t)$ is measured in meters per second and t is measured in seconds. Assuming that $x = 0$ at $t = 0$, what is the position as a function of time? What is the position at $t = 3.0$ s?

SOLUTION: With $x_0 = 0$ and $t_0 = 0$, Eq. (21) becomes

$$\begin{aligned} x(t) &= \int_0^t (655.9 - 61.14t' + 3.26t'^2)\,dt' \\ &= 655.9\int_0^t dt' - 61.14\int_0^t t'\,dt' + 3.26\int_0^t t'^2\,dt' \\ &= 655.9(t')\Big|_0^t - 61.14(t'^2/2)\Big|_0^t + 3.26(t'^3/3)\Big|_0^t \\ &= 655.9t - 61.14t^2/2 + 3.26t^3/3 \end{aligned}$$

When evaluated at $t = 3.0$ s, this yields

$$\begin{aligned} x(3.0) &= 655.9 \times 3.0 - 61.14 \times (3.0)^2/2 + 3.26 \times (3.0)^3/3 \\ &= 1722 \text{ m} \end{aligned}$$

EXAMPLE 3 The acceleration of a mass pushed back and forth by an elastic spring is

$$a(t) = B\cos\omega t \tag{36}$$

where B and ω are constants. Find the position as a function of time. Assume $v = 0$ and $x = 0$ at $t = 0$.

SOLUTION: The calculation involves two steps: first we must integrate the acceleration to find the velocity, then we must integrate the velocity to find the position. For the first step we use an equation analogous to Eq. (21),

$$v(t) - v_0 = \int_{t_0}^t a(t')\,dt' \tag{37}$$

This equation becomes obvious if we remember that the relationship between acceleration and velocity is analogous to that between velocity and position. With $v_0 = 0$ and $t_0 = 0$, we obtain from Eq. (29)

$$\begin{aligned} v(t) &= \int_0^t B\cos\omega t'\,dt' = B\frac{1}{\omega}\sin\omega t'\Big|_0^t \\ &= \frac{B}{\omega}\sin\omega t \end{aligned} \tag{38}$$

Next,

$$\begin{aligned} x(t) &= \int_0^t v(t')\,dt' = \int_0^t \frac{B}{\omega}\sin\omega t'\,dt' = \frac{B}{\omega}\left(-\frac{1}{\omega}\cos\omega t'\right)\Big|_0^t \\ &= -\frac{B}{\omega^2}\cos\omega t + \frac{B}{\omega^2} \end{aligned} \tag{39}$$

A4.5 The Taylor Series

Suppose that $f(u)$ is a smooth function of u in some neighborhood of a given point $u = a$, so the function has continuous derivatives of all orders. Then the value of the function at an arbitrary point near a can be expressed in terms of the following infinite series, where all the derivatives are evaluated at the point a:

$$f(u) = f(a) + \frac{df}{du}(u - a) + \frac{1}{2}\frac{d^2f}{du^2}(u - a)^2 + \frac{1}{3 \cdot 2}\frac{d^3f}{du^3}(u - a)^3 + \cdots \quad (40)$$

This is called the **Taylor series** for the function $f(u)$ about the point a. The series converges, and is valid, provided u is sufficiently close to a. How close is "sufficiently close" depends on the function f and on the point a. Some functions, such as $\sin u$, $\cos u$, and e^u, are extremely well behaved, and their Taylor series converge for any choice of u and of a. The Taylor series gives us a convenient method for the approximate evaluation of a function.

EXAMPLE 4 Find the Taylor series for $\sin u$ about the point $u = 0$.

SOLUTION: The derivatives of $\sin u$ evaluated at $u = 0$ are

$$\frac{d}{du}\sin u = \cos u = 1$$

$$\frac{d^2}{du^2}\sin u = \frac{d}{du}\cos u = -\sin u = 0$$

$$\frac{d^3}{du^3}\sin u = \frac{d}{du}(-\sin u) = -\cos u = -1$$

$$\frac{d^4}{du^4}\sin u = \frac{d}{du}(-\cos u) = \sin u = 0, \quad \text{etc.}$$

Hence Eq. (32) gives

$$\sin u = 0 + 1 \times (u - 0) + \frac{1}{2} \times 0 \times (u - 0)^2 + \frac{1}{3 \cdot 2} \times (-1) \times (u - 0)^3$$
$$+ \frac{1}{4 \cdot 3 \cdot 2} \times 0 \times (u - 0)^4 + \cdots$$
$$= u - \frac{1}{6}u^3 + \cdots$$

Note that for very small values of u, we can neglect all higher powers of u, so $\sin u \approx u$, which is an approximation often used in this book.

A4.6 Some Approximations

By constructing Taylor series, we can obtain the following useful approximations, all of which are valid for small values of u. It is often sufficient to keep just the first one or two terms on the right side.

$$\sqrt{1 + u} = 1 + \frac{1}{2}u - \frac{1}{8}u^2 + \frac{1}{16}u^3 + \cdots \quad (41)$$

$$\frac{1}{1+u} = 1 - u + u^2 - u^3 + \cdots \tag{42}$$

$$\frac{1}{\sqrt{1+u}} = 1 - \frac{1}{2}u + \frac{3}{8}u^2 - \frac{5}{16}u^3 + \cdots \tag{43}$$

$$\frac{1}{(1+u)^n} = 1 - nu + \frac{n(n+1)}{2}u^2 - \frac{n(n+1)(n+2)}{2\cdot 3}u^3 + \cdots \tag{44}$$

$$e^u = 1 + u + \frac{1}{2}u^2 + \frac{1}{2\cdot 3}u^3 + \cdots \tag{45}$$

$$\ln(1+u) = u - \frac{1}{2}u^2 + \frac{1}{3}u^3 + \cdots \tag{46}$$

In all the following formulas, u is in radians:

$$\sin u = u - \frac{1}{6}u^3 + \frac{1}{120}u^5 + \cdots \tag{47}$$

$$\cos u = 1 - \frac{1}{2}u^2 + \frac{1}{24}u^4 - \frac{1}{720}u^6 + \cdots \tag{48}$$

$$\tan u = u + \frac{1}{3}u^3 + \frac{2}{15}u^5 + \cdots \tag{49}$$

$$\sin^{-1} u = u + \frac{1}{6}u^3 + \frac{3}{40}u^5 + \cdots \tag{50}$$

$$\tan^{-1} u = u - \frac{1}{3}u^3 + \frac{1}{5}u^5 + \cdots \tag{51}$$

Appendix 5: Propagating Uncertainties

Experimentalists carefully work to measure physical quantities and to determine the uncertainty in each quantity. We must often calculate a new result from a measured quantity or from several quantities; we must therefore understand the propagation of uncertainties through functions and formulas.

To keep things simple, we will make the assumption that the uncertainties in each quantity are symmetrically distributed about its measured value and that the various measured quantities are independent of each other. This is not always true. But by ignoring correlations and assuming symmetry, we can reduce all the necessary propagation of uncertainties to some simple formulas.

Suppose we have a measured quantity and its uncertainty, $x \pm \Delta x$, where Δx is a positive quantity and has the same units as x, and is also known as the *absolute uncertainty* in x. What, then, is the uncertainty of some function, $f(x)$, of this data? Under the assumption that the uncertainty is small, we can obtain the uncertainty from the first terms of the Taylor series expansion of f: $f(x + \Delta x) = f(x) + (df(x)/dx)\Delta x + \cdots$ From this we find the uncertainty $\Delta f = |f(x + \Delta x) - f(x)|$ in the function value $f(x)$ is

$$\Delta f = \left| \frac{df}{dx} \Delta x \right| \tag{1}$$

with the derivative evaluated at the point x. We can generalize this result to functions of several variables as follows: given the data $x \pm \Delta x$, $y \pm \Delta y$, . . ., the function $f(x, y, \ldots)$ has the associated uncertainty

$$\Delta f = \left|\frac{\partial f}{\partial x}\Delta x\right| + \left|\frac{\partial f}{\partial y}\Delta y\right| + \cdots \tag{2}$$

where all the partial derivatives (see App. 4.2) are evaluated at the point $x, y, \ldots$. If we recall that we defined absolute uncertainties to be positive, we can write this as

$$\Delta f = \left|\frac{\partial f}{\partial x}\right|\Delta x + \left|\frac{\partial f}{\partial y}\right|\Delta y + \cdots \tag{3}$$

From this relationship, we can derive several simple results for uncertainty propagation.

EXAMPLE 1 **Addition and Subtraction.**

Given $f(x, y) = 3x + y - z + 5$, find Δf:

$$\begin{aligned}\Delta f &= \left|\frac{\partial f}{\partial x}\right|\Delta x + \left|\frac{\partial f}{\partial y}\right|\Delta y + \left|\frac{\partial f}{\partial z}\right|\Delta z \\ &= |3|\,\Delta x + |1|\,\Delta y + |-1|\,\Delta z \\ &= 3\Delta x + \Delta y + \Delta z\end{aligned}$$

Thus in addition or subtraction, the uncertainties add, and in multiplication by a constant, the uncertainty is multiplied by the same constant.

EXAMPLE 2 **Multiplication, Division, and Exponentiation.**

Given $f(x, y) = x^2y/(5z)$, find Δf:

$$\Delta f = \left|\frac{\partial f}{\partial x}\right|\Delta x + \left|\frac{\partial f}{\partial y}\right|\Delta y + \left|\frac{\partial f}{\partial z}\right|\Delta z$$

$$= |2xy/(5z)|\,\Delta x + |x^2/(5z)|\,\Delta y + |-x^2y/(5z^2)|\,\Delta z$$

Equivalently, for multiplication and division, we add *relative uncertainties* (e.g., $\Delta x/x$), and for exponentiation, we multiply the relative uncertainty by the magnitude of the exponent, to get the relative uncertainty of the product, quotient, or power.

EXAMPLE 3 **Numerical Application to Ohm's Law, $V = IR$.**

Given $V = 1.5 \pm 0.1$ Volt and $I = 0.50 \pm 0.02$ A, find R and ΔR:

Rearranging we find $R = V/I = (1.5\text{ Volt})/(0.50\text{ A}) = 3.0\ \Omega$, and

$$\Delta R = \left|\frac{\partial R}{\partial V}\right| \Delta V + \left|\frac{\partial R}{\partial I}\right| \Delta I$$

$$= \left|\frac{1}{I}\right| \Delta V + \left|\frac{-V}{I^2}\right| \Delta I$$

$$= \left|\frac{1}{0.50\ \text{A}}\right| (0.1\ \text{Volt}) + \left|\frac{-1.5\ \text{Volt}}{(0.50\ \text{A})^2}\right| (0.02\text{A})$$

$$= 0.2\ \Omega + 0.12\ \Omega = 0.4\ \Omega$$

Note in the last step that unlike an ordinary calculation, we have rounded this final result up; uncertainties should always be rounded up, never down.

Appendix 6: The International System of Units (SI)

A6.1 Base Units

The SI system of units is the modern version of the metric system. The SI system recognizes seven fundamental, or base, units for length, mass, time, electric current, thermodynamic temperature, amount of substance, and luminous intensity.[b] The following definitions of the base units were adopted by the Conférence Générale des Poids et Mesures in the years indicated:

meter (m) "The metre is the length of the path travelled by light in vacuum during a time interval of 1/299 792 458 of a second." (Adopted in 1983.)

kilogram (kg) "The kilogram is . . . the mass of the international prototype of the kilogram." (Adopted in 1889 and in 1901.)

second (s) "The second is the duration of 9 192 631 770 periods of the radiation corresponding to the transition between the two hyperfine levels of the ground state of the cesium-133 atom." (Adopted in 1967.)

ampere (A) "The ampere is that constant current which, if maintained in two straight parallel conductors of infinite length, of negligible circular cross section, and placed one meter apart in vacuum, would produce between these conductors a force equal to 2×10^{-7} newton per meter of length." (Adopted in 1948.)

kelvin (K) "The kelvin . . . is the fraction 1/273.16 of the thermodynamic temperature of the triple point of water." (Adopted in 1967.)

[b] At least two of the seven base units of the SI system are redundant. The mole is merely a certain number of atoms or molecules, in the same sense that a dozen is a number; there is no need to designate this number as a unit. The candela is equivalent to $\frac{1}{683}$ watt per steradian; it serves no purpose that is not served equally well by watt per steradian. Two other base units could be made redundant by adopting new definitions of the unit of temperature and of the unit of electric charge. Temperature could be measured in energy units because, according to the equipartition theorem, temperature is proportional to the energy per degree of freedom. Hence the kelvin could be defined as a derived unit, with $1\ \text{K} = \frac{1}{2} \times 1.38 \times 10^{-23}$ joule per degree of freedom. Electric charge could also be defined as a derived unit, to be measured with a suitable combination of the units of force and distance, as is done in the cgs system.

Furthermore, the definitions of the supplementary units—radian and steradian—are gratuitous. These definitions properly belong in the province of mathematics and there is no need to include them in a system of physical units.

TABLE A6.1 NAMES OF DERIVED UNITS

QUANTITY	DERIVED UNIT	NAME	SYMBOL
frequency	1/s	hertz	Hz
force	$kg\cdot m/s^2$	newton	N
pressure	N/m^2	pascal	Pa
energy	N·m	joule	J
power	J/s	watt	W
electric charge	A·s	coulomb	C
electric potential	J/C	volt	V
electric capacitance	C/V	farad	F
electric resistance	V/A	ohm	Ω
conductance	A/V	siemen	S
magnetic flux	V·s	weber	Wb
magnetic field	$V\cdot s/m^2$	tesla	T
inductance	V·s/A	henry	H
temperature	K	degree Celsius	°C
luminous flux	cd·sr	lumen	lm
illuminance	$cd\cdot sr/m^2$	lux	lx
radioactivity	1/s	becquerel	Bq
absorbed dose	J/kg	gray	Gy
dose equivalent	J/kg	sievert	Sv

TABLE A6.2 PREFIXES FOR UNITS

FACTOR	PREFIX	SYMBOL
10^{24}	yotta	Y
10^{21}	zetta	Z
10^{18}	exa	E
10^{15}	peta	P
10^{12}	tera	T
10^{9}	giga	G
10^{6}	mega	M
10^{3}	kilo	k
10^{2}	hecto	h
10	deka	da
10^{-1}	deci	d
10^{-2}	centi	c
10^{-3}	milli	m
10^{-6}	micro	μ
10^{-9}	nano	n
10^{-12}	pico	p
10^{-15}	femto	f
10^{-18}	atto	a
10^{-21}	zepto	z
10^{-24}	yocto	y

mole "The mole is the amount of substance of a system which contains as many elementary entities as there are atoms in 0.012 kilogram of carbon-12." (Adopted in 1967.)

candela (cd) "The candela is the luminous intensity, in a given direction, of a source that emits monochromatic radiation of frequency 540×10^{12} Hz and that has a radiant intensity in that direction of $\frac{1}{683}$ watt per steradian." (Adopted in 1979.)

Besides these seven base units, the SI system also recognizes two supplementary units of angle and solid angle:

radian (rad) "The radian is the plane angle between two radii of a circle which cut off on the circumference an arc equal in length to the radius."

steradian (sr) "The steradian is the solid angle which, having its vertex in the center of a sphere, cuts off an area equal to that of a [flat] square with sides of length equal to the radius of the sphere."

A6.2 Derived Units

The derived units are formed out of products and ratios of the base units. Table A6.1 lists those derived units that have been glorified with special names. (Other derived units are listed in the tables of conversion factors in Appendix 8.)

A6.3 Prefixes

Multiples and submultiples of SI units are indicated by prefixes, such as the familiar *kilo, centi*, and *milli* used in *kilometer, centimeter*, and *millimeter*, etc. Table A6.2 lists all the accepted prefixes. Some enjoy more popularity than others; it is best to avoid the use of uncommon prefixes, such as *atto* and *exa*, since hardly anybody will recognize those.

Appendix 7: Best Values of Fundamental Constants

The values in the following table are the "2002 CODATA Recommended Values" by P. J. Mohr and B. N. Taylor Listed at the website physics.nist.gov/constants of the National Institute of Standards and Technology. The digits in parentheses are the one–standard deviation uncertainty in the last digits of the given value.

TABLE A7.1 BEST VALUES OF FUNDAMENTAL CONSTANTS

QUANTITY	SYMBOL	VALUE	UNITS	RELATIVE UNCERTAINTY (PARTS PER MILLION)
UNIVERSAL CONSTANTS				
speed of light in vacuum	c	299 792 458	$m{\cdot}s^{-1}$	(exact)
magnetic constant	μ_0	$4\pi \times 10^{-7}$	$N{\cdot}A^{-2}$	
		$= 12.566\ 370\ 614 ... \times 10^{-7}$	$N{\cdot}A^{-2}$	(exact)
electric constant $1/\mu_0 c^2$	ϵ_0	$8.854\ 187\ 817 ... \times 10^{-12}$	$F{\cdot}m^{-1}$	(exact)
gravitational constant	G	$6.6742(10) \times 10^{-11}$	$m^3{\cdot}kg^{-1}{\cdot}s^{-2}$	1.5×10^{-4}
Planck constant	h	$6.626\ 0693(11) \times 10^{-34}$	J·s	1.7×10^{-7}
in eV·s	$\hbar$	$4.135\ 667\ 43(35) \times 10^{-15}$	eV·s	8.5×10^{-8}
$h/2\pi$		$1.054\ 571\ 68(18) \times 10^{-34}$	J·s	1.7×10^{-7}
in eV·s		$6.582\ 119\ 15(56) \times 10^{-16}$	eV·s	8.5×10^{-8}
ELECTROMAGNETIC CONSTANTS				
elementary charge	e	$1.602\ 176\ 53(14) \times 10^{-19}$	C	8.5×10^{-8}
magnetic flux quantum $h/2e$	Φ_0	$2.067\ 833\ 72(18) \times 10^{-15}$	Wb	8.5×10^{-8}
quantum	$2e^2/h$	$7.748\ 091\ 733(26) \times 10^{-5}$	S	3.3×10^{-9}
Josephson constant	$2e/h$	$483\ 597.879(41) \times 10^9$	$Hz{\cdot}V^{-1}$	8.5×10^{-8}
Bohr magneton $e\hbar/2m_e$	μ_B	$927.400\ 949(80) \times 10^{-26}$	$J{\cdot}T^{-1}$	8.6×10^{-8}
in $eV{\cdot}T^{-1}$		$5.788\ 381\ 804(39) \times 10^{-5}$	$eV{\cdot}T^{-1}$	6.7×10^{-9}
nuclear magneton $e\hbar/2m_p$	μ_N	$5.050\ 783\ 43(43) \times 10^{-27}$	$J{\cdot}T^{-1}$	8.6×10^{-8}
in $eV{\cdot}T^{-1}$		$3.152\ 451\ 259(21) \times 10^{-8}$	$eV{\cdot}T^{-1}$	6.7×10^{-9}

(continued)

QUANTITY	SYMBOL	VALUE	UNITS	RELATIVE UNCERTAINTY (PARTS PER MILLION)
ATOMIC AND NUCLEAR CONSTANTS				
General				
fine-structure constant $e^2/4\pi\epsilon_0\hbar c$	α	$7.297\ 352\ 568(24) \times 10^{-3}$		3.3×10^{-9}
inverse fine-structure constant	α^{-1}	$137.035\ 999\ 11(46)$		3.3×10^{-9}
Rydberg constant $\alpha^2 m_e c/2h$	R_∞	$10\ 973\ 731.568\ 525(73)$	m^{-1}	6.6×10^{-12}
Bohr radius $4\pi\epsilon_0\hbar^2/m_e e^2$	a_0	$0.529\ 177\ 2108(18) \times 10^{-10}$	m	3.3×10^{-9}
Electron				
electron mass	m_e	$9.109\ 3826(16) \times 10^{-31}$	kg	1.7×10^{-7}
in u		$5.485\ 799\ 0945(24) \times 10^{-4}$	u	4.4×10^{-10}
energy equivalent in Me V	$m_e c^2$	$0.510\ 998\ 918(44)$	MeV	8.6×10^{-8}
electron-proton mass ratio	m_e/m_p	$5.446\ 170\ 2173(25) \times 10^{-4}$		4.6×10^{-10}
electron charge to mass quotient	$-e/m_e$	$-1.758\ 820\ 12(15) \times 10^{11}$	$C{\cdot}kg^{-1}$	8.6×10^{-8}
Compton wavelength $h/m_e c$	λ_C	$2.426\ 310\ 238(16) \times 10^{-12}$	m	6.7×10^{-9}
classical electron radius $\alpha^2 a_0$	r_e	$2.817\ 940\ 325(28) \times 10^{-15}$	m	1.0×10^{-8}
Thomson cross section $(8\pi/3)r_e^2$	σ_e	$0.665\ 245\ 837(13) \times 10^{-28}$	m^2	2.0×10^{-8}
electron magnetic moment	μ_e	$-928.476\ 412(80) \times 10^{-26}$	$J{\cdot}T^{-1}$	8.6×10^{-8}
to Bohr magneton ratio	μ_e/μ_B	$-1.001\ 159\ 652\ 1859(38)$		3.8×10^{-12}
to nuclear magneton ratio	μ_e/μ_N	$-1838.281\ 971\ 07(85)$		4.6×10^{-10}
electron magnetic moment anomaly $\lvert\mu_e\rvert/\mu_B - 1$	a_e	$1.159\ 652\ 1859(38) \times 10^{-3}$		3.2×10^{-9}
electron g-factor $-2(1 + a_e)$	g_e	$-2.002\ 319\ 304\ 3718(75)$		3.8×10^{-12}
Muon				
muon mass	m_μ	$1.883\ 531\ 40(33) \times 10^{-28}$	kg	1.7×10^{-7}
in u				
energy equivalent in MeV	$m_\mu c^2$	$0.113\ 428\ 9264(30)$	u	2.6×10^{-8}
		$105.658\ 3692(94)$	MeV	8.9×10^{-8}
muon-electron mass ratio	m_μ/m_e	$206.768\ 2838(54)$		2.6×10^{-8}
muon Compton wavelength $h/m_\mu c$	$\lambda_{C,\mu}$	$11.734\ 441\ 05(30) \times 10^{-15}$	m	2.5×10^{-8}
muon magnetic moment	μ_μ	$-4.490\ 447\ 99(40) \times 10^{-26}$	$J{\cdot}T^{-1}$	8.9×10^{-8}
to Bohr magneton ratio	μ_μ/μ_B	$-4.841\ 970\ 45(13) \times 10^{-3}$		2.6×10^{-8}
muon magnetic moment anomaly $\lvert\mu_\mu\rvert/(e\hbar/2m_\mu) - 1$	a_μ	$1.165\ 919\ 81(62) \times 10^{-3}$		5.3×10^{-7}
muon g-factor $-2(1 + a_\mu)$	g_μ	$-2.002\ 331\ 8396(12)$		6.2×10^{-10}
Proton				
proton mass	m_p	$1.672\ 621\ 71(29) \times 10^{-27}$	kg	1.7×10^{-7}
in u		$1.007\ 276\ 466\ 88(13)$	u	1.3×10^{-10}
energy equivalent in MeV	$m_p c^2$	$938.272\ 029(80)$	MeV	8.6×10^{-8}
proton-electron mass ratio	m_p/m_e	$1836.152\ 672\ 61(85)$		4.6×10^{-10}
proton-neutron mass ratio	m_p/m_n	$0.998\ 623\ 478\ 72(58)$		5.8×10^{-10}
proton charge to mass quotient	e/m_p	$9.578\ 833\ 76(82) \times 10^{7}$	$C{\cdot}kg^{-1}$	8.6×10^{-8}
proton Compton wavelength $h/m_p c$	$\lambda_{C,p}$	$1.321\ 409\ 8555(88) \times 10^{-15}$	m	6.7×10^{-9}
proton magnetic moment	μ_p	$1.410\ 606\ 71(12) \times 10^{-26}$	$J{\cdot}T^{-1}$	8.7×10^{-8}
to Bohr magneton ratio	μ_p/μ_B	$1.521\ 032\ 206(15) \times 10^{-3}$		1.0×10^{-8}
to nuclear magneton ratio	μ_p/μ_N	$2.792\ 847\ 351(28)$		1.0×10^{-8}
Neutron				
neutron mass	m_n	$1.674\ 927\ 28(29) \times 10^{-27}$	kg	1.7×10^{-7}
in u		$1.008\ 664\ 915\ 60(55)$	u	5.5×10^{-10}
energy equivalent in MeV	$m_n c^2$	$939.565\ 360(81)$	MeV	8.6×10^{-8}
neutron-electron mass ratio	m_n/m_e	$1838.683\ 6598(13)$		7.0×10^{-10}
neutron-proton mass ratio	m_n/m_p	$1.001\ 378\ 418\ 70(58)$		5.8×10^{-10}

(*continued*)

QUANTITY	SYMBOL	VALUE	UNITS	RELATIVE UNCERTAINTY (PARTS PER MILLION)
neutron Compton wavelength $h/m_n c$	$\lambda_{C,n}$	$1.319\ 590\ 9067(88) \times 10^{-15}$	m	6.7×10^{-9}
neutron magnetic moment	μ_n	$-0.966\ 236\ 45(24) \times 10^{-26}$	$J{\cdot}T^{-1}$	2.5×10^{-7}
to Bohr magneton ratio	μ_n/μ_B	$-1.041\ 875\ 63(25) \times 10^{-3}$		2.4×10^{-7}
to nuclear magneton ratio	μ_n/μ_N	$-1.913\ 042\ 73(45)$		2.4×10^{-7}
Deuteron				
deuteron mass	m_d	$3.343\ 583\ 35(57) \times 10^{-27}$	kg	1.7×10^{-7}
in u		$2.013\ 553\ 212\ 70(35)$	u	1.7×10^{-10}
energy equivalent in MeV	$m_d c^2$	$1875.612\ 82(16)$	MeV	8.6×10^{-8}
deuteron-electron mass ratio	m_d/m_e	$3670.482\ 9652(18)$		4.8×10^{-10}
deuteron-proton mass ratio	m_d/m_p	$1.999\ 007\ 500\ 82(41)$		2.0×10^{-10}
deuteron magnetic moment	μ_d	$0.433\ 073\ 482(38) \times 10^{-26}$	$J{\cdot}T^{-1}$	8.7×10^{-8}
to Bohr magneton ratio	μ_d/μ_B	$0.466\ 975\ 4567(50) \times 10^{-3}$		1.1×10^{-8}
to nuclear magneton ratio	μ_d/μ_N	$0.857\ 438\ 2329(92)$		1.1×10^{-8}
Alpha Particle				
alpha particle mass	m_α	$6.644\ 6565(11) \times 10^{-27}$	kg	1.7×10^{-7}
in u		$4.001\ 506\ 179\ 149(56)$	u	1.4×10^{-11}
energy equivalent in MeV	$m_\alpha c^2$	$3727.379\ 17(32)$	MeV	8.6×10^{-8}
alpha particle to electron mass ratio	m_α/m_e	$7294.299\ 5363(32)$		4.4×10^{-10}
alpha particle to proton mass ratio	m_α/m_p	$3.972\ 599\ 689\ 07(52)$		1.3×10^{-10}
PHYSICO-CHEMICAL CONSTANTS				
Avogadro constant	N_A	$6.022\ 1415(10) \times 10^{23}$	$mole^{-1}$	1.7×10^{-7}
atomic mass constant $m_u = \frac{1}{12} m(^{12}C) = 1\,u$	m_u	$1.660\ 538\ 86(28) \times 10^{-27}$	kg	1.7×10^{-7}
energy equivalent in MeV	$m_u c^2$	$931.494\ 043(80)$	MeV	8.6×10^{-8}
Faraday constant $N_A e$	F	$96\ 485.3383(83)$	$C{\cdot}mole^{-1}$	8.6×10^{-8}
molar gas constant	R	$8.314\ 472\ (15)$	$J{\cdot}mole^{-1}{\cdot}K^{-1}$	1.7×10^{-6}
Boltzmann constant R/N_A	k	$1.380\ 6505(24) \times 10^{-23}$	$J{\cdot}K^{-1}$	1.8×10^{-6}
in $eV{\cdot}K^{-1}$		$8.617\ 343(15) \times 10^{-5}$	$eV{\cdot}K^{-1}$	1.8×10^{-6}
molar volume of ideal gas RT/p $T = 273.15$ K, $p = 101.325$ kPa	V_m	$22.413\ 996(39) \times 10^{-3}$	$m^3{\cdot}mole^{-1}$	1.7×10^{-6}
Loschmidt constant N_A/V_m	n_0	$2.686\ 7773(47) \times 10^{25}$	m^{-3}	1.8×10^{-6}
Stefan-Boltzmann constant $(\pi^2/60)k^4/\hbar^3 c^2$	σ	$5.670\ 400(40) \times 10^{-8}$	$W{\cdot}m^{-2}{\cdot}K^{-4}$	7.0×10^{-6}
Wien displacement law constant $b = \lambda_{max} T$	b	$2.897\ 7685(51) \times 10^{-3}$	m·K	1.7×10^{-6}

Appendix 8: Conversion Factors

The units for each quantity are listed alphabetically, except that the SI unit is always listed first. The numbers are based on "American National Standard; Metric Practice" published by the Institute of Electrical and Electronics Engineers, 1982.

Angle

1 radian $= 57.30° = 3.438 \times 10^3{}' = (1/2\pi)$ rev $= 2.063 \times 10^5{}''$

1 degree (°) $= 1.745 \times 10^{-2}$ radian $= 60' = 3600'' = \frac{1}{360}$ rev

1 minute of arc (′) $= 2.909 \times 10^{-4}$ radian $= \frac{1}{60}{}° = 4.630 \times 10^{-5}$ rev $= 60''$

1 revolution (rev) $= 2\pi$ radians $= 360° = 2.160 \times 10^4{}' = 1.296 \times 10^6{}''$

1 second of arc (″) $= 4.848 \times 10^{-6}$ radian $= \frac{1}{3600}{}° = \frac{1}{60}{}' = 7.716 \times 10^{-7}$ rev

Length

1 meter (m) $= 1 \times 10^{-9}$ nm $= 1 \times 10^{10}$ Å $= 6.685 \times 10^{-12}$ AU $= 100$ cm $= 1 \times 10^{15}$ fm $= 3.281$ ft $= 39.37$ in. $= 1 \times 10^{-3}$ km $= 1.057 \times 10^{-16}$ light-year $= 1 \times 10^6$ μm $= 5.400 \times 10^{-4}$ nmi $= 6.214 \times 10^{-4}$ mi $= 3.241 \times 10^{-17}$ pc $= 1.094$ yd

1 angstrom (Å) $= 1 \times 10^{-10}$ m $= 1 \times 10^{-8}$ cm $= 1 \times 10$ nm $= 1 \times 10^{-5}$ fm $= 3.281 \times 10^{-10}$ ft $= 1 \times 10^{-4}$ μm

1 astronomical unit (AU) $= 1.496 \times 10^{11}$ m $= 1.496 \times 10^{13}$ cm $= 1.496 \times 10^8$ km $= 1.581 \times 10^{-5}$ light-year $= 4.848 \times 10^{-6}$ pc

1 centimeter (cm) $= 0.01$ m $= 1 \times 10^8$ Å $= 1 \times 10^{13}$ fm $= 3.281 \times 10^{-2}$ ft $= 0.3937$ in. $= 1 \times 10^{-5}$ km $= 1.057 \times 10^{-18}$ light-year $= 1 \times 10^4$ μm

1 fermi, or **femtometer** (fm) $= 1 \times 10^{-15}$ m $= 1 \times 10^{-13}$ cm $= 1 \times 10^5$ Å

1 foot (ft) $= 0.3048$ m $= 30.48$ cm $= 12$ in. $= 3.048 \times 10^5$ μm $= 1.894 \times 10^{-4}$ mi $= \frac{1}{3}$ yd

1 inch (in.) $= 2.540 \times 10^{-2}$ m $= 2.54$ cm $= \frac{1}{12}$ ft $= 2.54 \times 10^4$ μm $= \frac{1}{36}$ yd

1 kilometer (km) $= 1 \times 10^3$ m $= 1 \times 10^5$ cm $= 3.281 \times 10^3$ ft $= 0.5400$ nmi $= 0.6214$ mi $= 1.094 \times 10^3$ yd

1 light-year $= 9.461 \times 10^{15}$ m $= 6.324 \times 10^4$ AU $= 9.461 \times 10^{17}$ cm $= 9.461 \times 10^{12}$ km $= 5.879 \times 10^{12}$ mi $= 0.3066$ pc

1 micron, or **micrometer** (μm) $= 1 \times 10^{-6}$ m $= 1 \times 10^4$ Å $= 1 \times 10^{-4}$ cm $= 3.281 \times 10^{-6}$ ft $= 3.937 \times 10^{-5}$ in.

1 nautical mile (nmi) $= 1.852 \times 10^3$ m $= 1.852 \times 10^5$ cm $= 6.076 \times 10^3$ ft $= 1.852$ km $= 1.151$ mi

1 statute mile (mi) $= 1.609 \times 10^3$ m $= 1.609 \times 10^5$ cm $= 5280$ ft $= 1.609$ km $= 0.8690$ nmi $= 1760$ yd

1 parsec (pc) $= 3.086 \times 10^{16}$ m $= 2.063 \times 10^5$ AU $= 3.086 \times 10^{18}$ cm $= 3.086 \times 10^{13}$ km $= 3.262$ light-years

1 yard (yd) $= 0.9144$ m $= 91.44$ cm $= 3$ ft $= 36$ in. $= \frac{1}{1760}$ mi

Time

1 second (s) $= 1.157 \times 10^{-5}$ day $= \frac{1}{3600}$ h $= \frac{1}{60}$ min $= 1.161 \times 10^{-5}$ sidereal day $= 3.169 \times 10^{-8}$ yr

1 day $= 8.640 \times 10^4$ s $= 24$ h $= 1440$ min $= 1.003$ sidereal days $= 2.738 \times 10^{-3}$ yr

1 hour (h) $= 3600$ s $= \frac{1}{24}$ day $= 60$ min $= 1.141 \times 10^{-4}$ yr

1 minute (min) $= 60$ s $= 6.944 \times 10^{-4}$ day $= \frac{1}{60}$ h $= 1.901 \times 10^{-6}$ yr
1 sidereal day $= 8.616 \times 10^4$ s $= 0.9973$ day $= 23.93$ h $= 1.436 \times 10^3$ min $= 2.730 \times 10^{-3}$ yr
1 year (yr) $= 3.156 \times 10^7$ s $= 365.24$ days $= 8.766 \times 10^3$ h $= 5.259 \times 10^5$ min $= 366.24$ sidereal days

Mass

1 kilogram (kg) $= 6.024 \times 10^{26}$ u $= 5000$ carats $= 1.543 \times 10^4$ grains $= 1000$ g $= 1 \times 10^{-3}$ t $= 35.27$ oz $= 2.205$ lb $= 1.102 \times 10^{-3}$ short ton $= 6.852 \times 10^{-2}$ slug
1 atomic mass unit (u) $= 1.6605 \times 10^{-27}$ kg $= 1.6605 \times 10^{-24}$ g
1 carat $= 2 \times 10^{-4}$ kg $= 0.2$ g $= 7.055 \times 10^{-3}$ oz $= 4.409 \times 10^{-4}$ lb
1 grain $= 6.480 \times 10^{-5}$ kg $= 6.480 \times 10^{-2}$ g $= 2.286 \times 10^{-3}$ oz $= \frac{1}{7000}$ lb
1 gram (g) $= 1 \times 10^{-3}$ kg $= 6.024 \times 10^{23}$ u $= 5$ carats $= 15.43$ grains $= 1 \times 10^{-6}$ t $= 3.527 \times 10^{-2}$ oz $= 2.205 \times 10^{-3}$ lb $= 1.102 \times 10^{-6}$ short ton $= 6.852 \times 10^{-5}$ slug
1 metric ton, or **tonne** (t) $= 1 \times 10^3$ kg $= 1 \times 10^6$ g $= 2.205 \times 10^3$ lb $= 1.102$ short tons $= 68.52$ slugs
1 ounce (oz) $= 2.835 \times 10^{-2}$ kg $= 141.7$ carats $= 437.5$ grains $= 28.35$ g $= \frac{1}{16}$ lb
1 pound (lb)[c] $= 0.4536$ kg $= 453.6$ g $= 4.536 \times 10^{-4}$ t $= 16$ oz $= \frac{1}{2000}$ short ton $= 3.108 \times 10^{-2}$ slug
1 short ton $= 907.2$ kg $= 9.072 \times 10^5$ g $= 0.9072$ t $= 2000$ lb
1 slug $= 14.59$ kg $= 1.459 \times 10^4$ g $= 32.17$ lb

Area

1 square meter (m^2) $= 1 \times 10^4$ cm^2 $= 10.76$ ft^2 $= 1.550 \times 10^3$ $in.^2$ $= 1 \times 10^{-6}$ km^2 $= 3.861 \times 10^{-7}$ mi^2 $= 1.196$ yd^2
1 barn $= 1 \times 10^{-28}$ m^2 $= 1 \times 10^{-24}$ cm^2
1 square centimeter (cm^2) $= 1 \times 10^{-4}$ m^2 $= 1.076 \times 10^{-3}$ ft^2 $= 0.1550$ $in.^2$ $= 1 \times 10^{-10}$ km^2 $= 3.861 \times 10^{-11}$ mi^2
1 square foot (ft^2) $= 9.290 \times 10^{-2}$ m^2 $= 929.0$ cm^2 $= 144$ $in.^2$ $= 3.587 \times 10^{-8}$ mi^2 $= \frac{1}{9}$ yd^2
1 square inch ($in.^2$) $= 6.452 \times 10^{-4}$ m^2 $= 6.452$ cm^2 $= \frac{1}{144}$ ft^2
1 square kilometer (km^2) $= 1 \times 10^6$ m^2 $= 1 \times 10^{10}$ cm^2 $= 1.076 \times 10^7$ ft^2 $= 0.3861$ mi^2
1 square statute mile (mi^2) $= 2.590 \times 10^6$ m^2 $= 2.590 \times 10^{10}$ cm^2 $= 2.788 \times 10^7$ ft^2 $= 2.590$ km^2
1 square yard (yd^2) $= 0.8361$ m^2 $= 8.361 \times 10^3$ cm^2 $= 9$ ft^2 $= 1296$ $in.^2$

Volume

1 cubic meter (m^3) $= 1 \times 10^6$ cm^3 $= 35.31$ ft^3 $= 264.2$ gal $= 6.102 \times 10^4$ $in.^3$ $= 1 \times 10^3$ liters $= 1.308$ yd^3
1 cubic centimeter (cm^3) $= 1 \times 10^{-6}$ m^3 $= 3.531 \times 10^{-5}$ ft^3 $= 2.642 \times 10^{-4}$ gal $= 6.102 \times 10^{-2}$ $in.^3$ $= 1 \times 10^{-3}$ liter
1 cubic foot (ft^3) $= 2.832 \times 10^{-2}$ m^3 $= 2.832 \times 10^4$ cm^3 $= 7.481$ gal $= 1728$ $in.^3$ $= 28.32$ liters $= \frac{1}{27}$ yd^3

[c] This is the "avoirdupois" pound. The "troy" or "apothecary" pound is 0.3732 kg, or 0.8229 lb avoirdupois.

1 gallon (gal)[d] $= 3.785 \times 10^{-3}$ m^3 = 0.1337 ft^3
1 cubic inch (in.^{3}) $= 1.639 \times 10^{-5}$ m^3 = 16.39 cm^3 $= 5.787 \times 10^{-4}$ ft^3
1 liter (l) $= 1 \times 10^{-3}$ m^3 = 1000 cm^3 $= 3.531 \times 10^{-2}$ ft^3 = 0.2642 gal
1 cubic yard (yd^3) = 0.7646 m^3 $= 7.646 \times 10^5$ cm^3 = 27 ft^3 = 202.0 gal

Density

1 kilogram per cubic meter (kg/m^3) $= 1 \times 10^{-3}$ g/cm^3 = 6.243×10^{-2} lb/ft^3 = 8.345×10^{-3} lb/gal = 3.613×10^{-5} lb/in.^{3} = 8.428×10^{-4} short ton/yd^3 = 1.940×10^{-3} slug/ft^3
1 gram per cubic centimeter (g/cm^3) $= 1 \times 10^3$ kg/m^3 = 62.43 lb/ft^3 = 8.345 lb/gal = 3.613×10^{-2} lb/in.^{3} = 0.8428 short ton/yd^3 = 1.940 slugs/ft^3
1 lb per cubic foot (lb/ft^3) = 16.02 kg/m^3 = 1.602×10^{-2} g/cm^3 = 0.1337 lb/gal = 1.350×10^{-2} short ton/yd^3 = 3.108×10^{-2} slug/ft^3
1 pound-per gallon (1 lb/gal) = 119.8 kg/m^3 = 7.481 lb/ft^3 = 0.2325 slug/ft^3
1 short ton per cubic yard (short ton/yd^3) = 1.187×10^3 kg/m^3 = 74.07 lb/ft^3
1 slug per cubic foot (slug/ft^3) = 515.4 kg/m^3 = 0.5154 g/cm^3 = 32.17 lb/ft^3 = 4.301 lb/gal

Speed

1 meter per second (m/s) = 100 cm/s = 3.281 ft/s = 3.600 km/h = 1.944 knots = 2.237 mi/h
1 centimeter per second (cm/s) = 0.01 m/s = 3.281×10^{-2} ft/s = 3.600×10^{-2} km/h = 1.944×10^{-2} knot = 2.237×10^{-2} mi/h
1 foot per second (ft/s) = 0.3048 m/s = 30.48 cm/s = 1.097 km/h = 0.5925 knot = 0.6818 mi/h
1 kilometer per hour (km/h) = 0.2778 m/s = 27.78 cm/s = 0.9113 ft/s = 0.5400 knot = 0.6214 mi/h
1 knot, or **nautical mile per hour** = 0.5144 m/s = 51.44 cm/s = 1.688 ft/s = 1.852 km/h = 1.151 mi/h
1 mile per hour (mi/h) = 0.4470 m/s = 44.70 cm/s = 1.467 ft/s = 1.609 km/h = 0.8690 knot

Acceleration

1 meter per second squared (m/s^2) = 100 cm/s^2 = 3.281 ft/s^2 = 0.1020 g
1 centimeter per second squared (cm/s^2) = 0.01 m/s^2 = 3.281×10^{-2} ft/s^2 = 1.020×10^{-3} g
1 foot per second squared (ft/s^2) = 0.3048 m/s^2 = 30.48 cm/s^2 = 3.108×10^{-2} g
1 g = 9.807 m/s^2 = 980.7 cm/s^2 = 32.17 ft/s^2

Force

1 newton (N) $= 1 \times 10^5$ dynes = 0.2248 lb-f = 1.124×10^{-4} short ton-force
1 dyne $= 1 \times 10^{-5}$ N = 2.248×10^{-6} lb-f = 1.124×10^{-9} short ton-force
1 pound-force (lb-f) = 4.448 N = 4.448×10^5 dynes = $\frac{1}{2000}$ = short ton-force
1 short ton-force = 8.896×10^3 N = 8.896×10^8 dynes = 2000 lb-f

[d] This is the U.S. gallon; the U.K. and the Canadian gallon are 4.546×10^{-3} m^3, or 1.201 U.S. gallons.

Energy

1 joule (J) = 9.478×10^{-4} Btu = 0.2388 cal = 1×10^{7} ergs = 6.242×10^{18} eV = 0.7376 ft·lb-f = 2.778×10^{-7} kW·h

1 British thermal unit (Btu)[e] = 1.055×10^{3} J = 252.0 cal = 1.055×10^{10} ergs = 778.2 ft·lb-f = 2.931×10^{-4} kW·h

1 calorie (cal)[f] = 4.187 J = 3.968×10^{-3} Btu = 4.187×10^{7} ergs = 3.088 ft·lb-f = 1×10^{-3} kcal = 1.163×10^{-6} kW·h

1 erg = 1×10^{-7} J = 9.478×10^{-7} Btu = 2.388×10^{-8} cal = 6.242×10^{11} eV = 7.376×10^{-8} ft·lb-f = 2.778×10^{-14} kW·h

1 electron-volt (eV) = 1.602×10^{-19} J = 1.602×10^{-12} erg = 1.182×10^{-19} ft·lb-f

1 foot-pound-force (ft·lb-f) = 1.356 J = 1.285×10^{-3} Btu = 0.3239 cal = 1.356×10^{7} ergs = 8.464×10^{18} eV = 3.766×10^{-7} kW·h

1 kilocalorie (kcal), or **large calorie** (Cal) = 4.187×10^{3} J = 1×10^{3} cal

1 kilowatt-hour (kW·h) = 3.600×10^{6} J = 3412 Btu = 8.598×10^{5} cal = 3.6×10^{13} ergs = 2.655×10^{6} ft·lb-f

Power

1 watt (W) = 3.412 Btu/h = 0.2388 cal/s = 1×10^{7} ergs/s = 0.7376 ft·lb-f/s = 1.341×10^{-3} hp

1 British thermal unit per hour (Btu/h) = 0.2931 W = 7.000×10^{-2} cal/s = 0.2162 ft·lb-f/s = 3.930×10^{-4} hp

1 calorie per second (cal/s) = 4.187 W = 14.29 Btu/h = 4.187×10^{7} ergs/s = 3.088 ft·lb-f/s = 5.615×10^{-3} hp

1 erg per second (erg/s) = 1×10^{-7} W = 2.388×10^{-8} cal/s = 7.376×10^{-8} ft·lb-f/s = 1.341×10^{-10} hp

1 foot-pound-force per second (ft·lb-f/s) = 1.356 W = 0.3238 cal/s = 4.626 Btu/h = 1.356×10^{7} ergs/s = 1.818×10^{-3} hp

1 horsepower (hp)[g] = 745.7 W = 2.544×10^{3} Btu/h = 178.1 cal/s = 550 ft·lb-f/s

1 kilowatt (kW) = 1×10^{3} W = 3.412×10^{3} Btu/h = 238.8 cal/s = 737.6 ft·lb-f/s = 1.341 hp

Pressure

1 newton per square meter (N/m^2), or **pascal** (Pa) = 9.869×10^{-6} atm = 1×10^{-5} bar = 7.501×10^{-3} mm-Hg = 10 dynes/cm^2 = 2.953×10^{-4} in.-Hg = 2.089×10^{-2} lb-f/ft^2 = 1.450×10^{-4} lb-f/$in.^2$ = 7.501×10^{-3} torr

1 atmosphere (atm) = 1.013×10^{5} N/m^2 = 760.0 mm-Hg = 1.013×10^{6} dynes/cm^2 = 29.92 in.-Hg = 2.116×10^{3} lb-f/ft^2 = 14.70 lb-f/$in.^2$

1 bar = 1×10^{5} N/m^2 = 0.9869 atm = 750.1 mm-Hg

1 dyne per square centimeter (dyne/cm^2) = 0.1 N/m^2 = 9.869×10^{-7} atm = 7.501×10^{-4} mm-Hg = 2.089×10^{-3} lb-f/ft^2 = 1.450×10^{-5} lb-f/$in.^2$

[e] This is the "International Table" Btu; there are several other Btus.

[f] This is the "International Table" calorie, which equals exactly 4.1868 J. There are several other calories; for instance, the thermochemical calorie, which equals 4.184 J.

[g] There are several other horsepowers; for instance, the metric horsepower, which equals 735.5 W.

1 inch of mercury (in.-Hg) $= 3.386 \times 10^3$ N/m^2 $= 3.342 \times 10^{-2}$ atm $=$ 25.40 mm-Hg $= 0.4912$ lb-f/in.^{2}
1 pound-force per square inch (lb-f/in.^{2}, or psi) $= 6.895 \times 10^3$ N/m^2 $=$ 6.805×10^{-2} atm $= 6.895 \times 10^4$ dynes/cm^2 $= 2.036$ in.-Hg $=$ 7.031×10^{-2} kp/cm^2
1 torr, or **millimeter of mercury** (mm-Hg) $= 1.333 \times 10^2$ N/m^2 $= 1/760$ atm $=$ 1.333×10^{-3} bar $= 1.333 \times 10^3$ dynes/cm^2 $= 0.03937$ in.-Hg $= 0.01934$ lb-f/in.^{2}

Electric Charge[h]

1 coulomb (C) $\Leftrightarrow 2.998 \times 10^9$ statcoulombs, or esu of charge $\Leftrightarrow$ 0.1 abcoulomb, or emu of charge

Electric Current

1 ampere (A) $\Leftrightarrow 2.998 \times 10^9$ statamperes, or esu of current $\Leftrightarrow$ 0.1 abampere, or emu of current

Electric Potential

1 volt (V) $\Leftrightarrow 3.336 \times 10^{-3}$ statvolt, or esu of potential $\Leftrightarrow 1 \times 10^8$ abvolts, or emu of potential

Electric Field

1 volt per meter (V/m) $\Leftrightarrow 3.336 \times 10^{-5}$ statvolt/cm $\Leftrightarrow 1 \times 10^6$ abvolts/cm

Magnetic Field

1 tesla (T), or **weber per square meter** (Wb/m^2) $= 1 \times 10^4$ gauss

Electric Resistance

1 ohm (Ω) $\Leftrightarrow 1.113 \times 10^{-12}$ statohm, or esu of resistance $\Leftrightarrow 1 \times 10^9$ abohms, or emu of resistance

Electric Resistivity

1 ohm-meter ($\Omega \cdot$m) $\Leftrightarrow 1.113 \times 10^{-10}$ statohm-cm $\Leftrightarrow 1 \times 10^{11}$ abohm-cm

Capacitance

1 farad (F) $\Leftrightarrow 8.988 \times 10^{11}$ statfarads, or esu of capacitance $\Leftrightarrow 1 \times 10^{-9}$ abfarad, or emu of capacitance

Inductance

1 henry (H) $\Leftrightarrow 1.113 \times 10^{-12}$ stathenry, or esu of inductance $\Leftrightarrow 1 \times 10^9$ abhenrys, or emu of inductance

[h] The dimensions of the electric quantities in SI units, electrostatic units (esu), and electromagnetic units (emu) are usually different; hence the relationships among most of these units are correspondences ($\Leftrightarrow$) rather than equalities (=).

Appendix 9: The Periodic Table and Chemical Elements

TABLE A9.1 THE PERIODIC TABLE

IIA (2) — Group designation
4 — Atomic number
Be — Symbol for element
9.012182 — Atomic mass

Periods	IA (1)	IIA (2)	IIIB (3)	IVB (4)	VB (5)	VIB (6)	VIIB (7)	VIIIB (8)	VIIIB (9)	VIIIB (10)	IB (11)	IIB (12)	IIIA (13)	IVA (14)	VA (15)	VIA (16)	VIIA (17)	VIIIA (18)
1	1 **H** 1.00794																	2 **He** 4.002602
2	3 **Li** 6.941	4 **Be** 9.012182											5 **B** 10.811	6 **C** 12.0107	7 **N** 14.0067	8 **O** 15.9994	9 **F** 18.99840	10 **Ne** 20.1797
3	11 **Na** 22.98977	12 **Mg** 24.3050											13 **Al** 26.98154	14 **Si** 28.0855	15 **P** 30.97376	16 **S** 32.065	17 **Cl** 35.453	18 **Ar** 39.948
4	19 **K** 39.0983	20 **Ca** 40.078	21 **Sc** 44.955910	22 **Ti** 47.867	23 **V** 50.9415	24 **Cr** 51.9961	25 **Mn** 54.938049	26 **Fe** 55.845	27 **Co** 58.93320	28 **Ni** 58.6934	29 **Cu** 63.546	30 **Zn** 65.409	31 **Ga** 69.723	32 **Ge** 72.64	33 **As** 74.92160	34 **Se** 78.96	35 **Br** 79.904	36 **Kr** 83.798
5	37 **Rb** 85.4678	38 **Sr** 87.62	39 **Y** 88.90585	40 **Zr** 91.224	41 **Nb** 92.90638	42 **Mo** 95.94	43 **Tc** 98.9072	44 **Ru** 101.07	45 **Rh** 102.90550	46 **Pd** 106.42	47 **Ag** 107.8682	48 **Cd** 112.411	49 **In** 114.818	50 **Sn** 118.710	51 **Sb** 121.760	52 **Te** 127.60	53 **I** 126.90447	54 **Xe** 131.293
6	55 **Cs** 132.90545	56 **Ba** 137.327	57 ***La** 138.9055	72 **Hf** 178.49	73 **Ta** 180.9479	74 **W** 183.84	75 **Re** 186.207	76 **Os** 190.23	77 **Ir** 192.217	78 **Pt** 195.078	79 **Au** 196.96654	80 **Hg** 200.59	81 **Tl** 204.3833	82 **Pb** 207.2	83 **Bi** 208.98037	84 **Po** 208.9824	85 **At** 209.9871	86 **Rn** 222.0176
7	87 **Fr** 223.0197	88 **Ra** 226.0277	89 **†Ac** 227.0277	104 **Rf** 261.1089	105 **Db** 262.1144	106 **Sg** 263.118	107 **Bh** 262.12	108 **Hs** 265.1306	109 **Mt** (268)	110 **Ds** (271)	111 **Uuu** (272)	112 **Uub** (285)		114 **Uuq** (289)				

*Lanthanides	58 **Ce** 140.116	59 **Pr** 140.90765	60 **Nd** 144.24	61 **Pm** 144.9127	62 **Sm** 150.36	63 **Eu** 151.964	64 **Gd** 157.25	65 **Tb** 158.92534	66 **Dy** 162.50	67 **Ho** 164.93032	68 **Er** 167.26	69 **Tm** 168.93421	70 **Yb** 173.04	71 **Lu** 174.967
†Actinides	90 **Th** 232.0381	91 **Pa** 231.0359	92 **U** 238.0289	93 **Np** 237.0482	94 **Pu** 244.0642	95 **Am** 243.0614	96 **Cm** 247.07003	97 **Bk** 247.0703	98 **Cf** 251.0796	99 **Es** 252.083	100 **Fm** 257.0951	101 **Md** 258.0984	102 **No** 259.1011	103 **Lr** 262.110

TABLE A9.2 ATOMIC MASSES AND ATOMIC NUMBERS OF CHEMICAL ELEMENTS

Data were obtained from the National Institute for Standards and Technology; values are for the elements as they exist naturally on Earth or for the most stable isotope, with carbon-12 (the reference standard) having a mass of exactly 12 u. The estimated uncertainties in values between ± and ± 9 units in the last digit of an atomic mass are in parentheses after the atomic mass.
(Source: http://physics.nist.gov/PhysRefData/Compositions/index.html)

ELEMENT	SYMBOL	ATOMIC NUMBER	ATOMIC MASS (u)
Actinium	Ac	89	227.027 7
Aluminum	Al	13	26.981 538 (2)
Americium	Am	95	243.061 4
Antimony	Sb	51	121.760 (1)
Argon	Ar	18	39.948 (1)
Arsenic	As	33	74.921 60 (2)
Astatine	At	85	209.987 1
Barium	Ba	56	137.327 (7)
Berkelium	Bk	97	247.070 3
Beryllium	Be	4	9.012 182 (3)
Bismuth	Bi	83	208.980 38 (2)
Bohrium	Bh	107	264.12
Boron	B	5	10.811 (7)
Bromine	Br	35	79.904 (1)
Cadmium	Cd	48	112.411 (8)
Calcium	Ca	20	40.078 (4)
Californium	Cf	98	251.079 6
Carbon	C	6	12.010 7 (8)
Cerium	Ce	58	140.116 (1)
Cesium	Cs	55	132.905 45 (2)
Chlorine	Cl	17	35.453 (9)
Chromium	Cr	24	51.996 1 (6)
Cobalt	Co	27	58.933 200 (9)
Copper	Cu	29	63.546 (3)
Curium	Cm	96	247.070 3
Darmstadtium	Ds	110	271
Dubnium	Db	105	262.114 4
Dysprosium	Dy	66	162.500 (1)
Einsteinium	Es	99	252.083
Erbium	Er	68	167.259 (3)
Europium	Eu	63	151.964 (1)
Fermium	Fm	100	257.095 1
Fluorine	F	9	18.998 403 2 (5)
Francium	Fr	87	223.019 7
Gadolinium	Gd	64	157.25 (3)
Gallium	Ga	31	69.723 (1)
Germanium	Ge	32	72.64 (1)
Gold	Au	79	196.966 55 (2)
Hafnium	Hf	72	178.49 (2)
Hassium	Hs	108	265.130 6
Helium	He	2	4.002 602 (2)
Holmium	Ho	67	164.930 32 (2)
Hydrogen	H	1	1.007 94 (7)
Indium	In	49	114.818 (3)
Iodine	I	53	126.904 47 (3)
Iridium	Ir	77	192.217 (3)
Iron	Fe	26	55.845 (2)
Krypton	Kr	36	83.798 (2)
Lanthanum	La	57	138.905 5 (2)
Lawrencium	Lr	103	262.110
Lead	Pb	82	207.2 (1)
Lithium	Li	3	6.941 (2)
Lutetium	Lu	71	174.967 (1)
Magnesium	Mg	12	24.305 0 (6)
Manganese	Mn	25	54.938 049 (9)
Meitnerium	Mt	109	268
Mendelevium	Md	101	258.098 4
Mercury	Hg	80	200.59 (2)
Molybdenum	Mo	42	95.94 (1)
Neodymium	Nd	60	144.24 (3)
Neon	Ne	10	20.179 7 (6)
Neptunium	Np	93	237.048 2
Nickel	Ni	28	58.693 4 (2)
Niobium	Nb	41	92.906 38 (2)
Nitrogen	N	7	14.006 7 (2)
Nobelium	No	102	259.101 1
Osmium	Os	76	190.23 (3)
Oxygen	O	8	15.999 4 (3)
Palladium	Pd	46	106.42 (1)
Phosphorus	P	15	30.973 761 (2)
Platinum	Pt	78	195.078 (2)
Plutonium	Pu	94	244.064 2
Polonium	Po	84	208.982 4
Potassium	K	19	39.098 3 (1)
Praseodymium	Pr	59	140.907 65 (2)
Promethium	Pm	61	144.912 7
Protactinium	Pa	91	231.035 88 (2)
Radium	Ra	88	226.025 4
Radon	Rn	86	222.017 6
Rhenium	Re	75	186.207 (1)
Rhodium	Rh	45	102.905 50 (2)
Rubidium	Rb	37	85.467 8 (3)
Ruthenium	Ru	44	101.07 (2)
Rutherfordium	Rf	104	261.108 9
Samarium	Sm	62	150.36 (3)
Scandium	Sc	21	44.955 910 (8)
Seaborgium	Sg	106	263.118 6
Selenium	Se	34	78.96 (3)
Silicon	Si	14	28.085 5 (3)
Silver	Ag	47	107.868 2 (2)
Sodium	Na	11	22.989 770 (2)
Strontium	Sr	38	87.62 (1)
Sulfur	S	16	32.065 (6)
Tantalum	Ta	73	180.947 9 (1)
Technetium	Tc	43	98.907 2
Tellurium	Te	52	127.60 (3)
Terbium	Tb	65	158.925 34 (2)
Thallium	Tl	81	204.383 3 (2)
Thorium	Th	90	232.038 1 (1)
Thulium	Tm	69	168.934 21 (2)
Tin	Sn	50	118.710 (7)
Titanium	Ti	22	47.867 (1)
Tungsten	W	74	183.84 (1)
Ununbium	Uub	112	285
Unununium	Uuu	111	272
Ununquadium	Uuq	114	289
Uranium	U	92	238.028 9 (1)
Vanadium	V	23	50.941 5 (1)
Xenon	Xe	54	131.293 (2)
Ytterbium	Yb	70	173.04 (3)
Yttrium	Y	39	88.905 85 (2)
Zinc	Zn	30	65.409 (4)
Zirconium	Zr	40	91.224 (2)

Appendix 10: Formula Sheets

Chapters 1–21

$v = dx/dt$

$a = dv/dt = d^2x/dt^2$

$x = x_0 + v_0t + \frac{1}{2}at^2$

$a(x - x_0) = \frac{1}{2}(v^2 - v_0^2)$

$A_x = A\cos\theta$

$A = \sqrt{A_x^2 + A_y^2 + A_z^2}$

$\mathbf{A} \cdot \mathbf{B} = AB\cos\phi$

$= A_xB_x + A_yB_y + A_zB_z$

$|\mathbf{A} \times \mathbf{B}| = AB\sin\phi$

$a = v^2/r$

$\mathbf{v}' = \mathbf{v} - \mathbf{V}_O$

$m\mathbf{a} = \mathbf{F}_{net}$

$w = mg$

$f_k = \mu_k N$

$f_s \leq \mu_s N$

$F = -kx$

$W = F_x\,\Delta x$

$W = \mathbf{F} \cdot \mathbf{s}$

$W = \int \mathbf{F} \cdot d\mathbf{s}$

$K = \frac{1}{2}mv^2$

$U = mgy$

$E = K + U = [\text{constant}]$

$U(x) = -\int_{x_0}^{x} F_x(x')\,dx'$

$F_x = -\frac{dU}{dx}$

$U = \frac{1}{2}kx^2$

$E = mc^2$

$P = dW/dt$

$P = \mathbf{F} \cdot \mathbf{v}$

$F = GMm/r^2$

$v^2 = GM_S/r$

$g = GM_E/R_E^2$

$U = -GMm/r$

$\mathbf{p} = m\mathbf{v}$

$\mathbf{r}_{CM} = \frac{1}{M}\int \mathbf{r}\,dm$

$\mathbf{I} = \int_0^{\Delta t} \mathbf{F}\,dt$

$v_1' = \frac{m_1 - m_2}{m_1 + m_2}v_1;\ v_2' = \frac{2m_1}{m_1 + m_2}v_1$

$\omega = d\phi/dt$

$\alpha = d\omega/dt = d^2\phi/dt^2$

$v = R\omega$

$K = \frac{1}{2}I\omega^2$

$I = \int R^2\,dm$

$I_{CM} = MR^2$ (hoop); $\frac{1}{2}MR^2$ (disk);

$\frac{2}{5}MR^2$ (sphere); $\frac{1}{12}ML^2$ (rod).

$I = I_{CM} + Md^2$

$\tau = FR\sin\theta$

$I\alpha = \tau$

$P = \tau\omega$

$L = I\omega$

$\mathbf{L} = \mathbf{r} \times \mathbf{p}$

$\frac{d\mathbf{L}}{dt} = \mathbf{r} \times \mathbf{F}$

$x = A\cos(\omega t + \delta)$

$T = 2\pi/\omega;\ f = 1/T = \omega/2\pi$

$m\,d^2x/dt^2 = -kx$

$\omega = \sqrt{k/m}$

$\omega = \sqrt{g/l};\ T = 2\pi\sqrt{l/g}$

$\omega = \sqrt{mgd/I}$

$y = A\cos k(x \pm vt) = A\cos(kx \pm \omega t)$

$\lambda = 2\pi/k;\ f = v/\lambda;\ \omega = 2\pi f$

$v = \sqrt{F/(M/L)}$

$f_{beat} = f_1 - f_2$

$f' = f(1 \pm V_R/v)$

$f' = f/(1 \mp V_E/v)$

$\sin\theta = v/V_E$

$p - p_0 = -\rho gy$

$\frac{1}{2}\rho v^2 + \rho gy + p = [\text{constant}]$

$pV = NkT$

$T_C = T - 273.15$

$v_{rms} = \sqrt{3kT/m}$

$TV^{\gamma-1} = [\text{constant}];$

$pV^{\gamma} = [\text{constant}];\ \gamma = C_p/C_V$

$\Delta E = Q - W$

$e = 1 - T_2/T_1$

$\Delta S = \int_A^B dQ/T$

$g = 9.81\ \text{m/s}^2$

$G = 6.67 \times 10^{-11}\ \text{N·m}^2/\text{kg}^2$

$M_E = 5.98 \times 10^{24}\ \text{kg}$

$R_E = 6.37 \times 10^6\ \text{m}$

$m_e = 9.11 \times 10^{-31}\ \text{kg}$

$m_p = 1.67 \times 10^{-27}\ \text{kg}$

$c = 3.00 \times 10^8\ \text{m/s}$

$N_A = 6.02 \times 10^{23}/\text{mole}$

$k = 1.38 \times 10^{-23}\ \text{J/K}$

$1\ \text{cal} = 4.19\ \text{J}$

$$F = \frac{1}{4\pi\epsilon_0}\frac{qq'}{r^2}$$

$$E = \frac{1}{4\pi\epsilon_0}\frac{q'}{r^2}$$

$$E = \sigma/2\epsilon_0$$

$$p = lQ$$

$$\tau = \mathbf{p} \times \mathbf{E}$$

$$U = -\mathbf{p} \cdot \mathbf{E}$$

$$\Phi_E = \int \mathbf{E} \cdot d\mathbf{A}$$

$$\oint \mathbf{E} \cdot d\mathbf{A} = \oint E_\perp \, dA = \frac{Q_{\text{inside}}}{\epsilon_0}$$

$$V = \frac{1}{4\pi\epsilon_0}\frac{q'}{r}$$

$$E_x = -\frac{\partial V}{\partial x}, \quad E_y = -\frac{\partial V}{\partial y}, \quad E_z = -\frac{\partial V}{\partial z}$$

$$U = \tfrac{1}{2}Q_1V_1 + \tfrac{1}{2}Q_2V_2 + \tfrac{1}{2}Q_3V_3 + \cdots$$

$$u = \tfrac{1}{2}\epsilon_0 E^2$$

$$C = Q/\Delta V$$

$$C = \epsilon_0 A/d$$

$$E = E_{\text{free}}/\kappa$$

$$\oint \kappa E_\perp \, dA = \frac{Q_{\text{free, inside}}}{\epsilon_0}$$

$$\Delta Y = -\int_{P_0}^{P} F \cdot \mathrm{ds}$$

$$u = \tfrac{1}{2}\kappa\epsilon_0 E^2$$

$$I = \Delta V/R$$

$$R = \rho l/A$$

$$P = I\mathcal{E}\,; \; P = I\,\Delta V$$

$$F = \frac{\mu_0}{2\pi}\frac{qvI}{r}$$

$$\mathbf{F} = q\mathbf{v} \times \mathbf{B}$$

$$d\mathbf{B} = \frac{\mu_0}{4\pi}\frac{I\,d\mathbf{s} \times \mathbf{r}}{r^3}$$

$$\oint \mathbf{B} \cdot d\mathbf{s} = \oint B_\parallel \, ds = \mu_0 I$$

$$B = \mu_0 n I$$

$$r = \frac{p}{qB}$$

$$d\mathbf{F} = I\,d\mathbf{l} \times \mathbf{B}$$

$$\mu = I \times [\text{area of loop}]$$

$$\boldsymbol{\tau} = \boldsymbol{\mu} \times \mathbf{B}$$

$$U = -\boldsymbol{\mu} \cdot \mathbf{B}$$

$$\mathcal{E} = vBl$$

$$\mathcal{E} = -\frac{d\Phi_B}{dt}$$

$$\Phi_B = \int \mathbf{B} \cdot d\mathbf{A}$$

$$\oint \mathbf{E} \cdot d\mathbf{s} = \oint E_\parallel \, ds = -\frac{d\Phi_B}{dt}$$

$$\Phi_B = LI$$

$$\mathcal{E} = -L\frac{dI}{dt}$$

$$U = \tfrac{1}{2}LI^2$$

$$u = \frac{1}{2\mu_0}B^2$$

$$\omega_0 = 1/\sqrt{LC}$$

$$Z = \sqrt{R^2 + \left(\omega L - \frac{1}{\omega C}\right)^2}$$

$$\mathcal{E}_2 = \mathcal{E}_1\frac{N_2}{N_1}$$

$$\oint \mathbf{B} \cdot d\mathbf{A} = \oint B_1 dA = 0$$

$$\oint \mathbf{B} \cdot d\mathbf{s} = \oint B_\parallel \, ds = \mu_0 I + \mu_0\epsilon_0\frac{d\Phi_E}{dt}$$

$$B = E/c$$

$$\mathbf{S} = \frac{1}{\mu_0}\mathbf{E} \times \mathbf{B}$$

$$[\text{pressure}] = S/c$$

$$\frac{\partial^2 E}{\partial x^2} = \mu_0\epsilon_0\frac{\partial^2 E}{\partial t^2}$$

$$c = 1/\sqrt{\mu_0\epsilon_0}$$

$$v = c/n$$

$$n_1 \sin\theta_1 = n_2 \sin\theta_2$$

$$f = \pm\tfrac{1}{2}R$$

$$\frac{1}{s} + \frac{1}{s'} = \frac{1}{f}$$

Interference minima:

$$d \sin\theta = \tfrac{1}{2}\lambda, \tfrac{3}{2}\lambda, \tfrac{5}{2}\lambda, \ldots$$

Interference maxima:

$$d \sin\theta = 0, \lambda, 2\lambda, \ldots$$

Diffraction minima:

$$a \sin\theta = \lambda, 2\lambda, 3\lambda, \ldots$$

$$a \sin\theta = 1.22\lambda$$

$$f' = \sqrt{\frac{1 \mp v/c}{1 \pm v/c}}\,f$$

$$x' = \frac{x - Vt}{\sqrt{1 - V^2/c^2}}$$

$$y' = y$$

$$t' = \frac{t - Vx/c^2}{\sqrt{1 - V^2/c^2}}$$

$$\Delta t = \frac{\Delta t'}{\sqrt{1 - V^2/c^2}}$$

$$L = \sqrt{1 - V^2/c^2}\,L'$$

$$v'_x = \frac{v_x - V}{1 - v_x V/c^2}$$

$$\mathbf{p} = \frac{m\mathbf{v}}{\sqrt{1 - v^2/c^2}}; \quad E = \frac{mc^2}{\sqrt{1 - v^2/c^2}}$$

$$E^2 = p^2c^2 + m^2c^4$$

$$E = hf$$

$$p = hf/c$$

$$\Delta y \, \Delta p_y \geq h/4\pi$$

$$L = n\hbar$$

$$E_n = -\frac{m_e e^4}{2(4\pi\epsilon_0)^2\hbar^2}\frac{1}{n^2} = -\frac{13.6 \text{ eV}}{n^2}$$

$$\lambda = h/p$$

$$\mu_{\text{spin}} = \frac{e\hbar}{2m_e}$$

$$E = \frac{J(J+1)\hbar^2}{2I}$$

$$R = (1.2 \times 10^{-15} \text{ m}) \times A^{1/3}$$

$$n = n_0 e^{-t/\tau}; \quad \tau = t_{1/2}/0.693$$

$e = 1.60 \times 10^{-19}$ C

$\epsilon_0 = 8.85 \times 10^{-12}$ F/m

$\mu_0 = 1.26 \times 10^{-6}$ H/m

$c = 3.00 \times 10^8$ m/s

$h = 2\pi\hbar = 6.63 \times 10^{-34}$ J·s

$m_e = 9.11 \times 10^{-31}$ kg

$m_p = 1.67 \times 10^{-27}$ kg

Appendix 11: Answers to Odd-Numbered Problems and Review Problems

Chapter 1

1. 5.87 ft; 1.78 m (Assuming a height of 5 ft 10 in)
3. 48.7 m
5. 66 picas long and 51 picas wide
7. 12.7 mm; 6.35 mm; 3.18 mm; 1.59 mm; 0.794 mm; 0.397 mm
9. a) 1 mm; 3×10^6 m (Assuming grapefruit diameter = 0.1 m); b) 7 mm; 0.5 km (Assuming head diameter = 0.2 m)
11. 1 mm
13. 4.41 μm; 6.94 μm
15. 6.3×10^6 m
17. 1.4×10^{17} s
19. 7761 s
21. 23 h 56 min
23. 12 days
25. 3.7×10^7 beats/year
27. 0.25 min of arc; 0.463 km
29. 0.134 % in planets; 99.9% in sun
31. 0.021 % electrons; 99.98 % nucleus
33. 373.24 g
35. a) 8.4×10^{24} molecules; b) 4.3×10^{46} molecules; c) 1680 molecules
37. 28.95 g/mol
39. 6.9×10^8 m
41. 2.1×10^{22} m
43. a) 1 pc = 2.06×10^5 AU; b) 1 pc = 3.08×10^{16} m; c) 1 pc = 3.25 ly
45. 35.31 ft^3
47. 2.72 m
49. 8.9×10^3 kg/m^3; 5.6×10^2 lb/ft^3; 0.32 lb/in^3
51. 8.0 m^3/day
53. 10^8; 10^{13}
55. a) 7.4×10^2; b) 1.855×10^2; c) 8.47×10^{-3}
57. 6.0×10^7 metric tons/cm^3
59. 5.00×10^{-3} m^3/s; 5.00 kg/s
61. 7.1×10^{-15} m; 3.0×10^{-15} m
63. 354 m^2
65. 11°; 5.7°; 570 atoms
67. 359.76°; 1440.0°
69. 8.9 m; 9295 tons
71. 3.902×10^{-25} kg; 235.0 u
73. 2.8×10^{19} molecules
75. 0.125 mm
77. 88.5 km/h; 80.7 ft/s; 24.6 m/s
79. 3.81×10^9 s
81. yes, because the distance traveled while gliding = 18.7 km
83. a) 3840 km; b) 296 km; c) 0.315 or 1:3.2

Chapter 2

1. 0.3 s
3. 6.3×10^{-7} m/s; 5.4 cm/day
5. 32.5 km/h
7. 600 km/h
9. 14 km/h
11. 2.5×10^4 yr; 2.5×10^7 yr
13. 12.8 m/s; 46 km/h
15. 5.87 h; 150 h
17. 0.06 m
19. a) 14 s; 380 m; b) 72 m
21. 4.83 m/s
23. 2.0 m/s
25. a)

PLANET	ORBIT CIRC (km)	PERIOD (s)	SPEED	LOG SPEED	LOG CIRC
Mercury	3.64×10^8	7.61×10^6	47.8	1.68	8.56
Venus	6.79×10^8	1.94×10^7	35.0	1.54	8.83
Earth	9.42×10^8	3.16×10^7	29.8	1.47	8.97
Mars	1.43×10^9	5.93×10^7	24.1	1.38	9.16
Jupiter	4.89×10^9	3.76×10^8	13.0	1.11	9.69
Saturn	8.98×10^9	9.31×10^8	9.65	0.985	9.95
Uranus	1.80×10^{10}	2.65×10^9	6.79	0.832	10.26
Neptune	2.83×10^{10}	5.21×10^9	5.43	0.735	10.45
Pluto	3.71×10^{10}	7.83×10^9	4.74	0.676	10.57

b)

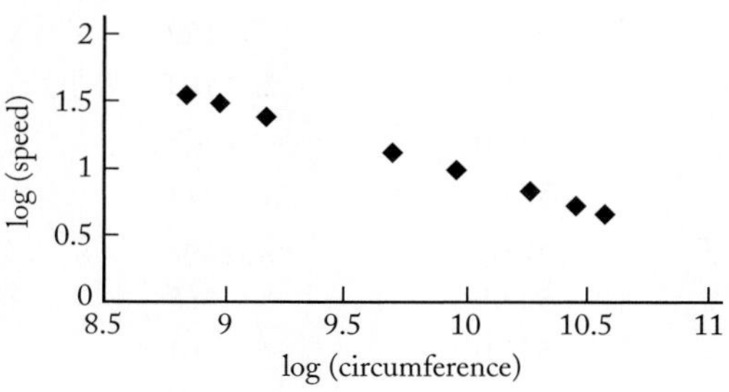

slope = −2.01

27. 20 m/s; 16.3 m/s

29. 1.2 m/s; 0.5 m/s

31. 12 m/s; 0 m/s

33. 0.67 m/s; 0.53 m/s

35. 32.4 m/s

37. 3.4×10^3 m/s^2

39. a)

t(s)	$\bar{a}$(m/s^2)	$\bar{a}$(in g)
0	6.1	0.62
10	1.4	0.14
20	0.83	0.085
30	0.56	0.057
40	0.49	0.050

Method:
i) draw tangent to curve
ii) get slope of line by counting squares
iii) to find Δv and Δt convert from km/h to m/s

b)

t(s)	$\bar{a}$(m/s^2)	$\bar{a}$(in g)
0	−0.74	−0.075
10	−0.44	−0.045
20	−0.44	−0.045
30	−0.31	−0.032
40	−0.22	−0.022

41. 0 s; 1 s; $x(0) = 0$ m; $x(1) = 1.2$ m

43. 1 m/s^2; 0.9 m/s^2; 1.3 m/s^2

45. at $t = 0$, $a = 0$; at $t = 2$ s, a $= -2.5$ m/s^2; at $t \to \infty$, $a \to 0$

47. a)

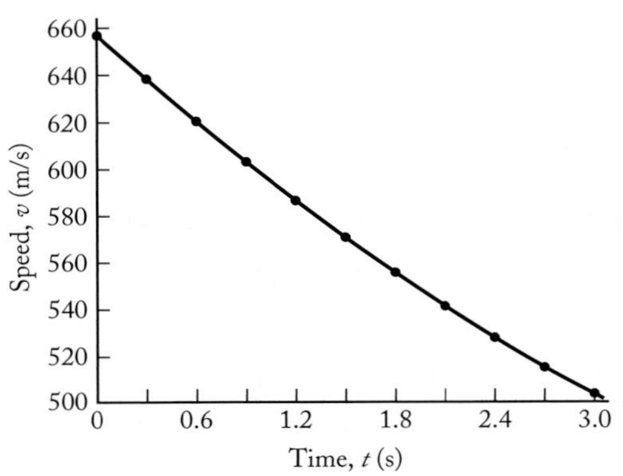

b)

TIME INTERVAL (s)	AVG SPEED (m/s)	DISTANCE TRAVELED (m)
0–0.3	647.5	194
0.3–0.6	628.5	189
0.6–0.9	611.5	183
0.9–1.2	596.0	179
1.2–1.5	579.5	174
1.5–1.8	564.0	169
1.8–2.1	549.5	165
2.1–2.4	535.0	161
2.4–2.7	521.0	156
2.7–3.0	508.0	152

c) 1722 ± 2 m

49. a)

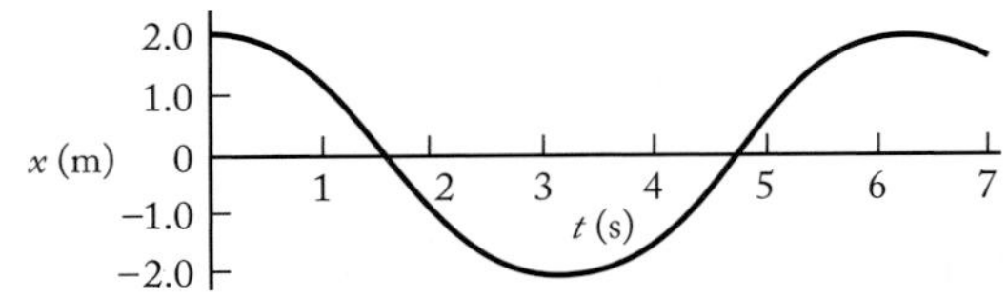

b) 1.6 s; 4.7 s; c) 0 s; 3.1 s; 6.3 s; $v(0) = v(3.1) = v(6.3) =$ 0 m/s; $a(0) = a(6.3) = -2$ m/s^2; $a(3.1) = 2$ m/s^2

51. 2.4 m/s^2

53. 6.36×10^7 s; 6.2×10^8 m/s

55. -350 m/s^2; will probably survive

57. -7.1 m/s^2; 3.8 s

59. 30 m/s; 300 m

61. 16 s

63.

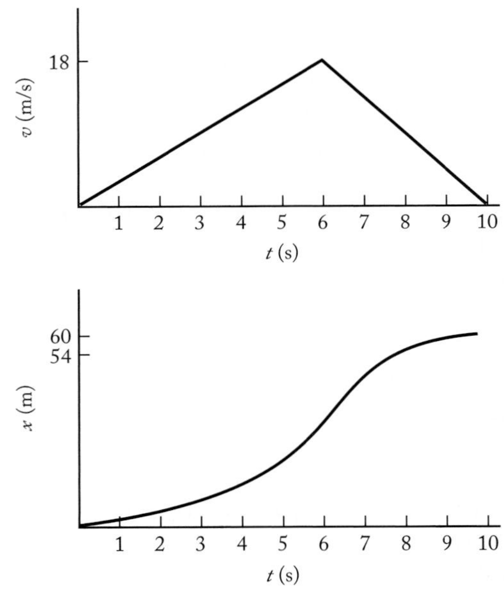

65. 32.9 m/s; 40.4 m/s

67. 0.875 m/s^2; 4.4 m/s

69.

v_0(km/h)	v_0(m/s)	$v_0\,\Delta t$(m)	$-\frac{v_0^2}{2a}$ (m)	TOTAL STOPPING DISTANCE (m)
15	4.17	8.3	1.1	9.4
30	8.33	16.7	4.3	21.0
45	12.5	25.0	10	35.0
60	26.7	33.3	18	51.3
75	20.8	41.7	27	68.7
90	25.0	50.0	39	89.0

71. 15.5 m/s
73. $a = -ge^{-t/\tau}$
75. 66 m
77. 6.1 m/s
79. 44 m
81. 7.96 m or 3 floors; 22.1 m or 8 floors; 43.5 m or 15 floors
83. 2.8 s; 14 m/s up
85. 3.7 m above launching point; 8.6 m/s
87. 1.1 m/s; 5.5 m/s
89. 0.22%
91. 1.6×10^4 m/s^2
93. 1.9×10^3 m/s; 2.6×10^2 s
95. 80^2 m/s; 1.9 s
97. 14.9 m/s; 5.1 m/s
99. a) $n\sqrt{2h/g}$; b) $(3/4)h$ above the ground; c) $(2/3)h$
101. 18.3 m/s; 26.7 m/s; 33.3 m
105. 13.7 m
107. average speed = 1.3 m/s; average velocity = 0 m/s
109. 0.95 s; 28.8 m
111. 2.9 s
113. a) 4.3 m; b) 3.0 m/s; c) -6.0 m/s^2
115. 21.1 m/s
117. 33.1 m/s; 2.21×10^3 m/s^2
119. a) 8.10 m above ground; 11.1 m above ground; b) 9.8 m/s down; c) 0 m/s^2

Chapter 3

1. 11.8 km, 30° N of E
3. 11.2 km, 27.7° S of E
5. 612 m, 11.3° W of S
7. 436 km, 7.4° W of N
9. $\mathbf{B} = (-1.26\text{ m})\mathbf{i} + (3.2\text{ m})\mathbf{j}$
11. 13.6 nmi, 88° E of N
13. 1.88×10^4 km, 1.98×10^4 km
15. 6.07 mi, 78.3° W of S
17. 9.19 km N, 7.71 km W
19. 1.7 m
21. $(2\mathbf{i} + 5\mathbf{j})$ cm
23. $A_z = \pm 4.2$ units
25. a) $-3\mathbf{i} - 2\mathbf{j} - 2\mathbf{k}$ b) $-7\mathbf{i} - 4\mathbf{j} + 4\mathbf{k}$ c) $-16\mathbf{i} - 9\mathbf{j} + 11\mathbf{k}$
27. $x = -9.9$ m, $y = 9.9$ m
29. $(1/3)\,\mathbf{i} + (2/3)\,\mathbf{j} + (2/3)\,\mathbf{k}$
31. $c_1 = -8/7$, $c_2 = 9/7$
33. 4940 km
35. $(6/7)\,\mathbf{i} - (12/7)\,\mathbf{j} + (4/7)\,\mathbf{k}$
37. 9
39. -8, 112°
41. 56.1°
43. 45°
45. $(-3.9 \times 10^6\text{ m})\mathbf{k}$
47. Because the vectors are nonzero, a zero result for the dot product means they must be perpendicular.
49. $B_x = -6.83$, $B_z = -4.5$, $C_z = 1.34$
51. $0.44\mathbf{i} - 0.22\mathbf{j} - 0.87\mathbf{k}$
53. $0.49\mathbf{i} + 0.81\mathbf{j} + 0.32\mathbf{k}$
55. 24
59. $-12\mathbf{i} - 14\mathbf{j} - 9\mathbf{k}$
61. $0.45\mathbf{i} - 0.59\mathbf{j} - 0.67\mathbf{k}$
65. Coordinate system rotated at $-26.6°$
67. 415 m, 29.8° W of N
69. $x = 1.0$, $y = 1.7$
71. $\mathbf{A} + \mathbf{B} = 5.4\mathbf{i} - 12.7\mathbf{j}$; $\mathbf{A} - \mathbf{B} = 5.4\mathbf{i} + 6.5\mathbf{j}$
73. 4.58
75. -304 m^2
77. 4.0, 5.0

Chapter 4

1. a) 7 km, 5 degrees E of N; b) 5.6 km/h, 5 degrees E of N; c) 8.24 km/h
3. 3.93 m
5. a) $2\,\mathbf{i} + (5 + 8t)\,\mathbf{j} - (2 + 6t)\,\mathbf{k}$; b) $8\,\mathbf{j} - 6\,\mathbf{k}$, magnitude = 10 m/s^2, direction = 37° below the y-axis in the y-z plane
7. 19.6 m at 90° below the direction of travel of the airplane at 2 s; 24.7 m at 83° below the direction of travel of the plane at 3 s
9. -13.3 km/h $\mathbf{i}$ $-$ 123 km/h $\mathbf{j}$
11. velocity = $(90\,\mathbf{i} - 15\,\mathbf{j})$ m/s, speed = 91 m/s; direction = 9.5° below the x-axis
13. a) $\mathbf{v} = (3t\,\mathbf{i} + 2t\,\mathbf{j})$ m/s; b) $\mathbf{r} = [(3t^2/2)]\,\mathbf{i} + t^2\,\mathbf{j}$ m
15. 2.4 m/s
17. 38 m/s
19. 65.8 m/s, 93.4 m/s
21. a) 7.25° b) 13 m
23. 1.74 sec, 14.9 m, 59.5 m
25. 3.13×10^3 m/s, 2.5×10^5 m/s, 452 sec
27. 64.8 m, 3.04 sec
29. 76°
31. 12 m/s, $\mathbf{r} = -21$ m $\mathbf{i}$ + 55 m $\mathbf{j}$
33. 21 m/s
35. The lake surface is 34.3 m below the release point and the horizontal distance from release point is 68.8 m
37. Yes, puck passes 2.2 m above the goal, 0.391 sec
41. 63.4°
43. 9.29° and 80.5°
45. 5.19°, when angle off 0.03° in vertical direction arrow still hits bull's-eye (arrow hits 4.6 cm off center, which is still within 12 cm diameter), when angle off 0.03° in horizontal

direction arrow still hits bull's-eye (arrow hits 4.7 cm off center, which is still within 12 cm diameter)

47. 7.49°

49. No, projectiles will never collide

51. 45.8 km

53. $\theta = \dfrac{1}{2}(d + \pi/2)$

55. $h = R(1 + \sin\theta) + \dfrac{u^2\cos^2\theta}{2g}$; the maximum possible height is 4.4 m

57. $\theta = 9.94°$, 17 km

59. 2.1 rev/min

61. 9.4 m/s^2

63. 3.95 m/s^2, $4 \times 10^5\ g$

65. 8.99×10^{13} m/s^2, $9.16 \times 10^{12}\ g$

67. 5.9×10^{-13} m/s^2

69. $a_{eq} = 3.39 \times 10^{-2}$ m/s^2, $a_{45} = 2.4 \times 10^{-2}$ m/s^2

71. $a_M = 0.0395$ m/s^2, $a_V = 0.0113$ m/s^2, $a_E = 0.00595$ m/s^2

73. 5.5 m/s, 2.5 m/s

75. 633 m/s $\mathbf{i}$ + 226 m/s $\mathbf{j}$

77. $V = 12$ m/s, $\theta = 83°$

79. 4.60 m/s

81. 60 cm/s at 34° above the horizontal

83. speed = 27 km/h, $\theta = 33°$

85. a) 50×10^3 m; b) 33×10^3 m, 67×10^3 m

87. 15.1 km/h at 15° E of N

89. $v_{rel} = \dfrac{2v_0}{(h^2/4v_0^2t^2 + 1)^{1/2}}$

91. 528 km/h at 8.5° N of W

93. a) vertical component = 62.1 km/h down, horizontal component = 232 km/h in direction of plane's travel; (b) 1.9 min

95. $v_E = 3.9$ km/h, $v_N = -1.4$ km/h (1.4 km/h south)

97. a) 13 m/s; b) 56.3°

99. a) 25 km; b) 50 km; c) No

101. 26 m/s

103. 8.9 m/s^2

105. 10.8 sec

Chapter 5

1. 442 kg

3. 2.69×10^{-26} kg; 3.7×10^{25} atoms

5. 3.8 m/s^2; 6.2×10^3 N

7. 6.6×10^3 N; 12 times the weight

9. -1.2×10^3 N

11. -1.8×10^3 m/s^2; -1.3×10^5 N

13. 35 N

15. -4.2 m/s^2; -2.4 m/s^2; -1.0×10^3 N; -5.7×10^2 N

17. 0.063 m/s^2

19. -2.90×10^3 N; -1.82×10^3 N

21. $v = bx_0 \sin(bt)$; $a = b^2x_0\cos(bt)$; $F = mb^2x_0\cos(bt)$; $F = -mb^2(x - x_0)$

23. No, since the tension in the rope = 150 000 N > breaking tension

25. 36° south of east; 260 N

27. 3.7 m/s^2; 23.4° east of north

29. 4.7×10^{20} N; 25° clockwise from the Moon-Sun direction

31. 770 N in the positive x-direction

33. 2.6 kg; 34 N

35. 285 N on Mars; 1900 N on Jupiter

37. a) 9.9904×10^{-4}; b) no

39. 128 N in the upper cord; 29.4 N in the lower cord

41. $F = T_1 = (m_1 + m_2 + m_3)g$; $T_2 = (m_2 + m_3)g$; $T_3 = m_3g$

43. Mg; $Mg/2$

45. $T = \dfrac{\pi d^2 l\rho g}{4}$ at the upper end; $T = \dfrac{\pi d^2 l\rho g}{8}$ at the midpoint

47. 1.2 m/s^2; 36 N to the right; 36 N to the left

49. 600 N

51. 5.2×10^3 N

53. 165 N; 19.5° clockwise from positive x

57. 1.8×10^3 N upward

59. F in the first cable; $F\left(\dfrac{m_2 + m_3}{m_1 + m_2 + m_3}\right)$ in the second cable; $F\left(\dfrac{m_3}{m_1 + m_2 + m_3}\right)$ in the third cable; $a = \dfrac{F - (m_1 + m_2 + m_3)g}{m_1 + m_2 + m_3}$

61. 1.14×10^3 N or 265 lb; 820 N or 184 lb

63. 0.51 N

65. 1.9 m/s^2; 14 m/s

67. 64 m; 5.1 s

69. $\dfrac{mgR}{\sqrt{l(l + 2R)}}$

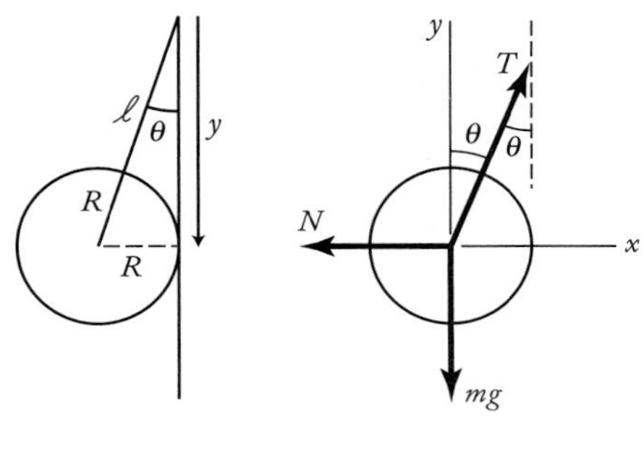

71. N = 680 N, F = 340 N

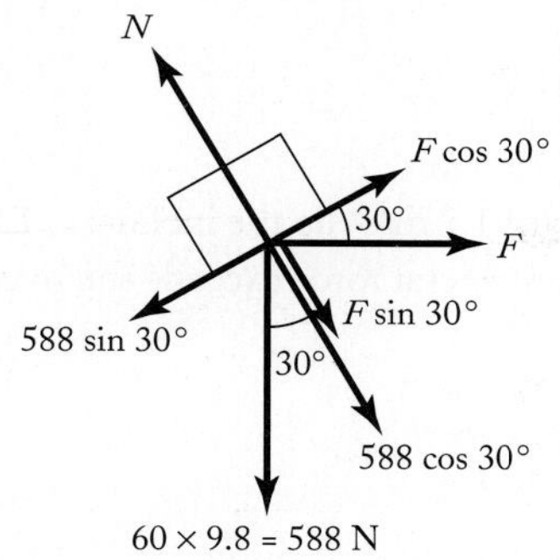

73. a) Incline forward; b) 22.3 m/s^2

75. $\pi/4$; $2\sqrt{\dfrac{\Delta x}{g}}$

77. $a = \dfrac{2g}{l}x$; $x(t) = x_0 e^{\sqrt{2g/l}\,t}$

79. 7.9×10^{-5} m/s^2; 0.14 m

81. a) $\mathbf{F}_{\text{net}} = -2\mathbf{i} + 3\mathbf{j} + 4\mathbf{k}$ N; b) $\mathbf{a} = -0.33\mathbf{i} + 0.50\mathbf{j} + 0.67\mathbf{k}$ m/s^2, 0.9 m/s^2

83. a)

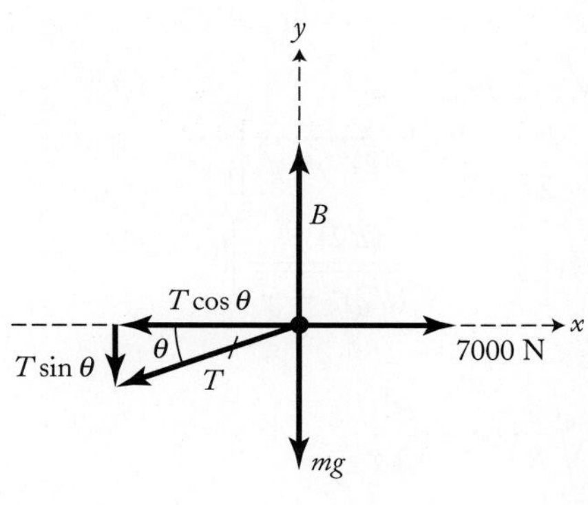

b) 7.1×10^3 N; c) 2.6×10^4 N

85. a)

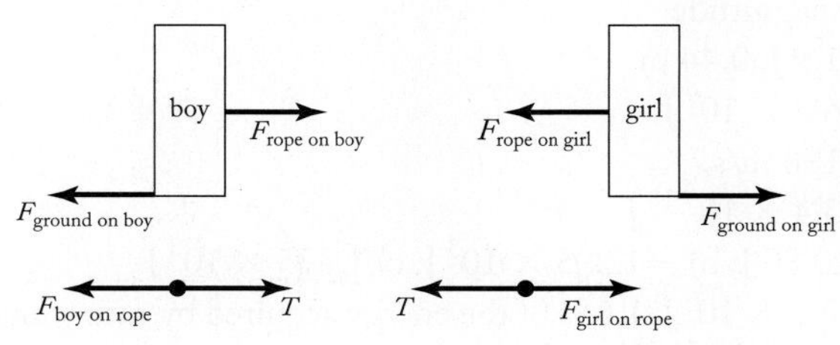

b) 250 N; c) 250 N; d) 250 N

87. a)

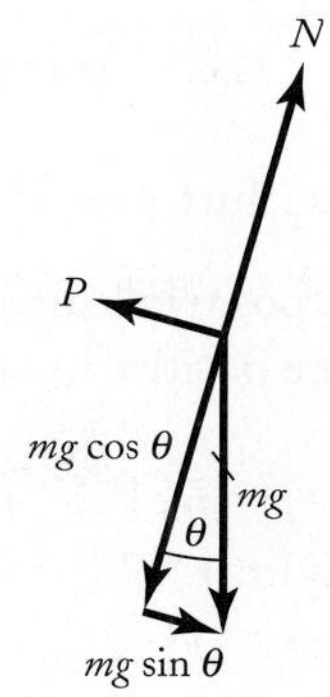

b) 1.8×10^4 N; c) same as b)

89. a) 590 N; b) 700 N; c) 590 N; d) 0 N

91. a) -0.98 m/s^2; b) 99 m; c) 50 km/h, same as speed when first decoupled

93. a) $a = \dfrac{m_2}{m_1 + m_2}g$; b) $T = \dfrac{m_1 m_2}{m_1 + m_2}g$

Chapter 6

1. 5.7×10^3 people
3. 0.83
5. 1.6×10^2 m
7. 53 m
9. 3.4 m, so he will reach the plate; 1.9 m/s
11. 0.48
13. 2.1×10^3 N
15. 0.27
17. 2.0×10^2 N
19. 2.8 m/s
21. 1.9°
23. 0.78
25. 1.4×10^{-3} m/s (1.4 mm/s)
27. 39.5 m
29. a)

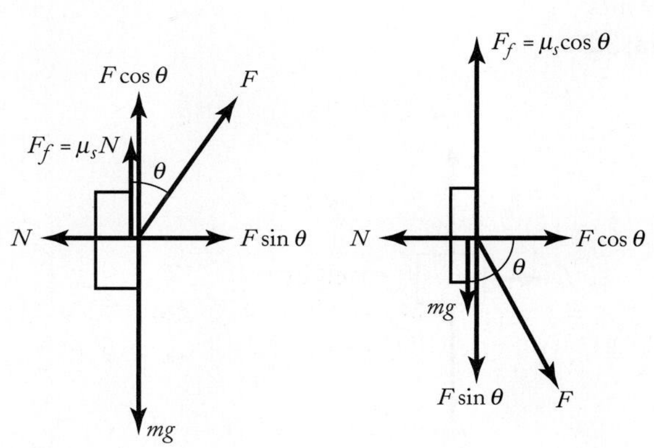

b) $F = \dfrac{mg}{\cos\theta + \mu_s \sin\theta}$; c) $\theta = \tan^{-1}\mu_s$; $F = \dfrac{mg}{\sqrt{(1+\mu_s^2)}}$; d) $\theta = \tan^{-1}\mu_s$, but now θ is the angle between the force **F** and the positive x direction

31. The quantity $\cos\theta - \mu_k \sin\theta$ must be positive in order to find a solution for P
This gives the condition $\tan\theta > \dfrac{1}{\mu_k}$

33. 3.6 m/s^2

35. $T = \dfrac{\mu_k m_1 m_2 g \cos\theta}{(m_1 + m_2)\cos\phi - (m_1 + 2m_2)\mu_k \sin\phi}$

37. 23 N; 11 N

39. No, since k is not constant

41. 50 N/m

43. 8.8×10^2 N/m

45. 1.1×10^{-3} m

49. 4.4×10^2 N

51. 7.8×10^2 N at top; 8.1×10^2 N at bottom

53. 6.3°

55. 0.13 m

57. 0.224 N

59. 6.9 m/s

61. 22 m/s

63. 68°

65. $\sqrt{gl \sin\theta \tan\theta}$

67. 1.40×10^3 m

69. The equilibrium conditions when the balls are at maximum angular displacement is $\cos\theta_2/\cos\theta_1 = m_1/m_2$, and the condition when they are both vertical is $(m_1 - m_2)g = \dfrac{m_2 v_2^2 - m_1 v_1^2}{l}$. These conditions cannot both be satisfied, so the motion described is impossible

71. $T = \dfrac{mv^2}{2\pi r}$

73. $\phi = \tan^{-1}\left(\dfrac{\tan\theta}{1 - 4\pi^2 R/t^2 g}\right) - \theta$; at $\theta = 45°$, $\phi = 0.099°$

75. 0.89 m/s

77. 40 m; 3.2 s

79. a)

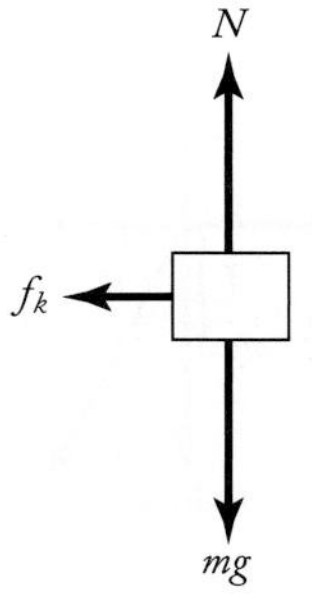

b) $3.9 \times 10^2\,\mathbf{j}$ N; c) $-3.1 \times 10^2\,\mathbf{i}$ N.; d) $-3.9 \times 10^2\,\mathbf{j}$ N; $-3.1 \times 10^2\,\mathbf{i}$ N; e) a = $-7.85\mathbf{i}$ m/s^2; 31 m

81. 0.15 m

83. $a_1 = \dfrac{m_2 - \mu_k m_1}{m_1 + m_2} g$

85. 4.1 cm; 2.5 cm, 1.6 cm

87. a) 1.2×10^{-2}m; b) 1.7 m/s^2 up the incline; c) 1.0×10^{-2} m

89. Yes, since the centripetal force exceeds the maximum frictional force

Chapter 7

1. 1.5×10^3 J

3. 252 J

5. 3.7×10^3 J

7. 2.35×10^5 J; 357 J/s

9. 2.2×10^7 J by first tugboat; 1.0×10^7 J by second tugboat; 3.2×10^7 J total

11. 2.6×10^3 J

13. 7×10^4 J by gravity; -7×10^4 J by friction

15. 54°

17. a) 1.3×10^4 J; b) 290 N; 1.3×10^4 J

19. a) 1.4×10^4 N; 8.3×10^3 N; b) 4.3×10^3 J; c) 2.2×10^4 J

21. a) 7.1×10^3 N; b) 2.2×10^5 J; 8.1×10^3 N

23. 6 J

25. -26 J

27. $3W_0$; $(2N+1)W_0$

29. a) $k\left[y^2 + \dfrac{l^2}{2} - \dfrac{l}{2}\sqrt{(l/2)^2 + y^2}\right]$;
b) $P = 2ky\left[1 - \dfrac{(l/2)}{\sqrt{(l/2)^2 + y^2}}\right]$

31. 17 J

33. a) $x_{eq} = \sqrt[6]{\dfrac{A}{B}}$; b) $-\dfrac{B^2}{12A}$

35. 2.7×10^{33} J

37. 1.3×10^5 J; 5.8×10^3 J; 22

39. a) 4.0×10^5 J; b) 2.5×10^4 J; c) 1.2×10^6 J

41. 4.1×10^6 J

43. $K_{ball} = 46$ J; $K_{person} = 38$ J; they are of the same order of magnitude

45. 1.9 J; 0.44 m

47. 6.2×10^9 J

49. 196 m/s

51. 3.4×10^{-18} J

53. a) 80 J; b) -1.295×10^3 J; c) 1.375×10^3 J

55. 7.4 $\times 10^3$ J; 1.6 % of the energy acquired by eating an apple

57. 8.2×10^6 m^3

59. 5.1 m
61. 99 m/s; 9.8×10^{10} J; 23 tons
63. 0.16
65. 79%
67. 7.7 m/s
69. 1.1×10^4 J
71. 53 N
73. 100 m/s
77. a) 24 m/s; b) 7.1 m; c) 26 m
79. 48.2°
81. 2.1×10^3 J
83. 1.69×10^5 J; 2.06×10^5 J
85. a) -8.8 m/s^2; 35.4 m; b) 1.06×10^4 N; 3.75×10^5 J
87. a) $U = 2.35 \times 10^7$ J; $K = 2.89 \times 10^6$ J; b) $U = 17.7 \times 10^6$ J; $K = 8.7 \times 10^6$ J; 120 m/s (or 430 km/h)
89. a) 150 J; b) 150 J; c) 122 m/s; d) 1520 m; e) 122 m/s
91. 4.1×10^4 J
93. a) 25.8 m/s; b) 10.3 m/s

Chapter 8

1. 0.076 m
5. $U(x) = \dfrac{Ax^4}{4}$, assuming that $x_0 = 0$; 5.6 m/s
7. $U(x) = x^2 + \dfrac{x^4}{4}$, assuming that $x_0 = 0$; 1.3 J; 8 J; 29.3 J
9. $F = -4x - 4x^3$
11. 64.5 m/s
13. a) bxy; b) $-bxy$
15. 2.61×10^6 J
17. $U(x) = \dfrac{A}{12x^{12}} - \dfrac{B}{6x^6}$
19. a) 13 kN; b) 13 kN. The force is independent of the rope length
21. $\mathbf{F} = a\dfrac{x\mathbf{i} + y\mathbf{j}}{(x^2 + y^2)^{3/2}}$
23. 1.89×10^5 N
25. a) -4.58 m; 14.9 m/s; b) -14.7 m; 21.7 m/s^2
27. a)

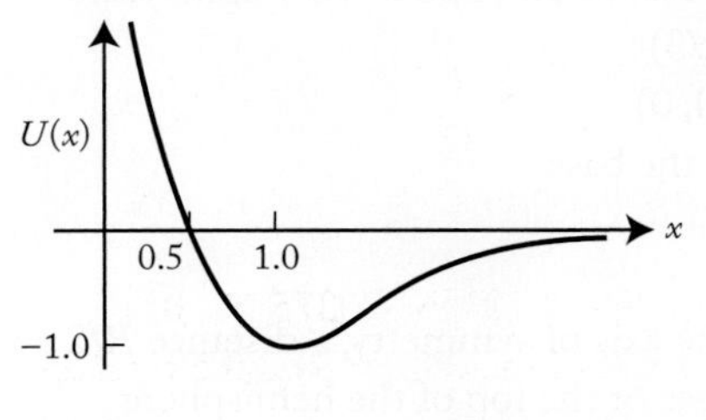

b) $x_{1,2} = (2 \pm \sqrt{2})(b/c)$; c) $x = (\sqrt{6} - 2)(b/c)$

29. $x = \pm \dfrac{mv^2}{2A}$
31. $x = \pm 1.0$ m; unbound for $E > 0$ J
33. a) 0.382 nm; b) -1.67×10^{-19} J; c) ± 0.34 nm, ± 0.89 nm
35. 6.3 eV/molecule
37.

VEHICLE	ENERGY PER MILE (J/mi)	ENERGY PER PASSENGER PER mi (J/ passenger-mi)
Motorcycle	1/60 gal/mi $\times 1.3 \times 10^8$ J/gal $= 2.2 \times 10^6$ J/mi	2.2×10^6
Snowmobile	$1/12 \times 1.3 \times 10^8 = 1.1 \times 10^7$	1.1×10^7
Automobile	$1/12 \times 1.3 \times 10^8 = 1.1 \times 10^7$	$1.1 \times 10^7/4 = 2.7 \times 10^6$
Bus	$1/5 \times 1.3 \times 10^8 = 2.6 \times 10^7$	$2.6 \times 10^7/45 = 5.8 \times 10^5$
Jetliner	$1/0.12 \times 1.3 \times 10^8 = 1.1 \times 10^9$	$1.1 \times 10^9/110 = 9.8 \times 10^6$
Concorde	$1/0.1 \times 1.3 \times 10^8 = 1.3 \times 10^9$	$1.3 \times 10^9/360 = 3.6 \times 10^6$

Most efficient is the bus, least efficient is the snowmobile.

39. 183 m, assuming a mass of 70 kg for the climber
41. 1.05×10^4 kJ
43. Walking 1.7 kcal/kg; Slow running plus standing 2.8 kcal/kg; Fast running plus standing 2.8 kcal/kg
45. 1.88×10^9 eV
47. Thermal energy is 0.0001 % of mass energy
49. 511 keV; 939 MeV
51. 1.4×10^{-9} kg; .00000005% of the mass of the gasoline
53. 9.40×10^8 eV
55. 542 kcal
57. 18 kWh
59. 2.88×10^8 J
61. 1.5×10^{-3} hp; 23 kcal
63. a) 769 gal; b) 5.8 kW
65. 526 kWh/year; $79
67. 1100 W; 0 W
69. 0.61 hp
71. 746 J
73. 50°
75. -2.0 W
77. 4.24×10^5 W
79. 2500 km^2
81. a) 1.7×10^{10} J; b) 17 min; c) 3.73 km
83. 1.2×10^{-4} W
85. 195 m diameter
87. a) 487 hp; b) 2593 hp
89. 52 hp
91. 37%
93. a) 3.2×10^4 W; b) -784 W; c) -3.1×10^4 W
95. 2.3 W
97. 2.1×10^7 kW
99. 3.4 kW

101. a) 4.3×10^{-12} J; b) 6.39×10^{14} J; c) 6.1×10^{11} kg/s; d) 7.8×10^{10} years
103. a) 5 J; b) −4 J; no
105. a) 7.2×10^{6} N; b) 0.414 m/s
107. 4.2×10^{9} W
109. a) $P = -1.82 \times 10^{6} + 3.63 \times 10^{5}\, t - 2.71 \times 10^{4}\, t^{2} + 964\, t^{3}$; b) 9.757×10^{6} J; 5.714×10^{6} J; c) -1.35×10^{6} W
111. 14 min
113. a) 1.6×10^{7} kWh; b) 3.8×10^{2} m^3/s

Chapter 9

1. 8.2 N
3. 3.46×10^{8} m
5. $F_{Sun} = 0.41$ N, $F_{Moon} = 2.3 \times 10^{-3}$ N
7. $F_{Alpha} = 1.5 \times 10^{17}$ N, $F_{Earth} = 3.5 \times 10^{22}$ N
9. $a_J = 24.9$ m/s^2, $a_S = 10.5$ m/s^2, $a_U = 8.99$ m/s^2
11. 1×10^{9} N
13. 2.54×10^{-10} N at 52°
15. 2.76×10^{-4} g
17. $\Delta a_{Earth\text{-}Moon} = 2.21 \times 10^{-6}$ m/s^2, $\Delta a_{Earth\text{-}Moon}/g = 2.25 \times 10^{-7}$, $\Delta a_{\text{Jupiter - Io}} = 0.0123$ m/s^2, $\Delta a_{\text{Jupiter - Io}}/g = 0.00687$
19. 101 m/s
21. 3.08×10^{3} m/s
23. 5.8×10^{15} sec $= 1.8 \times 10^{8}$ years, 3.1×10^{5} m/s
25. $T_{Io} = 1.77$ days, $T_{\text{Europa}} = 3.55$ days, $T_{\text{Ganymede}} = 7.15$ days
27. 0.927 days
29. About 10 times
31. Same latitude 22.6° West, around Lincoln, Nebraska
33. $m_1/m_2 = 1.6$
35. 3.0×10^{10} m
37. a) 7.50×10^{3} m/s, 8.32×10^{3} m/s; b) 3.94×10^{8} J, 4.85×10^{8} J
39. 8.2×10^{3} m/s
41. $T_S = 96.5$ min, $T_E = 115$ min
43. 5.33×10^{10} km, about 10 times Pluto's mean orbital distance from sun
45. 7.8×10^{3} m/s, -1.4×10^{11} J
47. $\sqrt{2}$
49. $U = -1.04 \times 10^{6}$ J, $K = 5.2 \times 10^{5}$ J, $E = -5.2 \times 10^{5}$ J
51. 8.86 mm
53. 0.253
55. a) 1.11×10^{4} m/s; b) 1.23×10^{11} J = 29 tons of TNT; c) 1.23×10^{5} m/s^2
57. 2270 m/s, 1.11×10^{4} m/s
59. elliptical
61. a) speed = 1680 m/s, time = 6510 sec = 109 min; b) this will give an elliptical orbit; c) this orbit will not be closed because the launch speed is greater than the escape speed
63. a) No, speed is less than that needed for circular orbit; b) 1.22×10^{4} m/s
65. $E_S = -4.4 \times 10^{9}$ J, $E_E = -6.06 \times 10^{8}$ J
67. $v_{perigee} = 6.96 \times 10^{4}$ m/s, $v_{apogee} = 5.75 \times 10^{3}$ m/s
71. $h = 4.26 \times 10^{6}$ m, $v = 817$ m/s
73. a) 2.6×10^{3} m/s; b) 2.8×10^{3} m/s; c) 0.401 years; d) Venus moves 234.7°, Earth moves 144°
75. 1 rev/min
77. 1.6×10^{-9} N
79. 4.9 years
81. 1.90×10^{27} kg
83. a) $v_{rel} = 1.53 \times 10^{4}$ m/s; b) 3.91 g
85. height above earth $= 9.89 \times 10^{6}$ m
87. a) -3.30×10^{9} J; b) 3.30×10^{9} J

Chapter 10

1. 9.0 kg·m/s; 3.2 kg·m/s.
3. 1.8×10^{29} kg·m/s; 4.3×10^{7} kg·m/s; 3.8×10^{4} kg·m/s; 95 kg·m/s; 2.0×10^{-24} kg·m/s
5. $(1.6 \times 10^{-25}\,\mathbf{i} + 7.7 \times 10^{-26}\,\mathbf{j})$ kg·m/s
7. a) $(9.7\mathbf{i} + 5.6\mathbf{j})$ kg·m/s; b) $(9.7\mathbf{i} + 0\mathbf{j})$ kg·m/s; c) $(9.7\mathbf{i} - 5.6\mathbf{j})$ kg·m/s
9. 9.81 kg·m/s down; 98.1 kg·m/s down
11. $-9.0\,\mathbf{i}$ m/s
13. -2.2×10^{5} j
15. $-2.0^{-2} \times 10^{-5}\,\mathbf{i}$ m/s
17. 8.26×10^{3} m/s; 1.29×10^{-17} J
19. $-(1.3$ m/s$)\mathbf{i} + 0\mathbf{j}$
21. 66 N; 1.3×10^{6} J
23. 150 N·s
25. $-(4.10 \times 10^{3}\mathbf{i} + 929\mathbf{j})$ m/s
27. 5.2 N
29. 5×10^{-11} kg/s; 5×10^{-5} N
31. $u \dfrac{m}{M} \sum_{k=1}^{n} \dfrac{1}{1 - km/M}$
33. 1.9 m from woman
35. 7.42×10^{5} m; 0.107% of the sun's radius
37. $h/3$ along the height, away from the unequal side
39. 0.027 cm directly away from the 40 g piece
41. 0.23 nm from the hydrogen atom
43. $(950 \times 10^{3}, 180 \times 10^{3}, 820 \times 10^{3})$ light-years
45. $(L/3, L/3, L/3)$
47. $(-0.061L, 0, 0)$
49. 950 m from the base
51. 31.5 J; 63 J
53. 9.0×10^{7} J
55. CM is on the axis of symmetry, a distance $R/2$ away from either the base or the top of the hemisphere
57. 6.9×10^{6} m/s in the direction of motion of the proton
59. 953 kg

61. 1.05 m
63. 1.6 m/s in the direction of motion of the bullet
65. $4D$ from the launch point
67. $(-17.4$ km/s, -17.4 km/s$)$
69. $K_{CM} = 3.8 \times 10^4$ J; $K_{TOT} = 6.38 \times 10^5$ J
71. 4.76×10^4 J; 3.6×10^5 J; 4.76×10^4 J
73. 3.9×10^3 J; 4.0×10^3 J
75. 0.955×10^{-3}
77. 1.1 m/s
79. $a_{boy} = 5.0$ m/s^2 toward girl; $a_{girl} = 6.7$ m/s^2 toward boy; 1.7 m from boy
81. 8 km/h
83. 0.0927 nm
85. 59 cm
87. If the stack of two books is at the top of the triangle, the CM is at a point halfway between the other two books and 0.43 m above the line connecting them
89. Halfway along the line joining the centers of the plates

Chapter 11

1. 1.13×10^3 kg·m/s, 2.3×10^4 N
3. a) 12 m/s, 7m/s, 3 m/s, 1 m/s, -1 m/s, 250 m/s^2, 200 m/s^2, 100 m/s^2, 100 m/s^2; b) 4.2×10^5 N, 3.4×10^5 N, 1.7×10^5 N, 1.7×10^5 N; c) 1.1×10^5 N·s
5. 12.6 kg·m/s, 4200 N
7. -1400 N
9. -1.8 kg·m/s, -1.35×10^4 N
11. 8.1 kg·m/s, 0.045 s
13. 18 kg·m/s
15. 7.5×10^{-2} s, 2.8×10^4 N
17. a) $v_{proj}' = -0.27$ m/s, $v_{targ}' = 0.53$ m/s; b) $K_{proj} = 1.9 \times 10^{-2}$ J, $K_{targ} = 0$ J $K_{proj}' = 2 \times 10^{-3}$ J, $K_{targ}' = 1.7 \times 10^{-2}$ J
19. 39 m/s
21. 0.57 J
23. $0.22v$
25. $\dfrac{M}{7}$
27. $v_1' = 15$ m/s, $v_2' = 17$ m/s
29. Last ball has velocity $= v$ and other two balls have velocity $= 0$
31. a) mass m rises to $\dfrac{h}{9}$, mass $2m$ rises to $\dfrac{4h}{9}$; b) mass m rises to h, mass $2m$ stops and does not rise
33. 13.5 m/s
35. a) The 1400 kg mass has a velocity $= 1.3$ m/s and the 800 kg mass has a velocity $= 6.1$ m/s; b) t $= 0.98$ s $x = 1.7$ m
37. Yes
39. a) -7.5 m/s; b) -15 m/s; c) -15 m/s
41. -0.17 m/s, 0.18 m/s, 0.41 m/s, -0.34 m/s
43. 9.3 m/s
45. a) 9.8 m/s; b) 4.8×10^5 J; c) -130 m/s^2, 850 m/s^2
47. 4.0×10^{-13} J
49. a) 3.9×10^5 J, 3.9×10^5 J; b) 7.8×10^5 J, 3.9×10^5 J
51. 210 m/s
53. a) 440 m/s; b) -1200 J; c) 9.6 J; d) missing kinetic energy is energy that shows up as heat in bullet and block, compression/deformation, and noise
55. 620 m/s
57. 860 m/s
59. (a) 3 m/s $\hat{\mathbf{i}}$ (b) 79 m/s^2
61. 21 m/s
63. a) $\pi - 2\sin^{-1}\left(\dfrac{b}{2R}\right)$; b) $2mv\sin^{-1}\left(1 - \dfrac{b^2}{4R^2}\right)$
65. $v_1' = \dfrac{v}{2}(1 + \cos\theta)\,\mathbf{i} + \dfrac{v}{9}\sin\theta\,\mathbf{j}$; $v_2' = \dfrac{v}{2}(1 + \cos\theta)\,\mathbf{i} - \dfrac{v}{2}\sin\theta\,\mathbf{j}$ $v_1' \cdot v_2' = \dfrac{v^2}{4}(1 - (\sin^2\theta + \cos^2\theta)) = 0$ so $v_1' \perp v_2'$
67. 3.2 kg·m/s, 3200 N
69. $v_1' = 2.6$ km/h, $v_2' = 13$ km/h
71. 1.2×10^{-12} J
73. a) $v_1' = 0$ and $v_2' = 20$ m/s; b) The ball that had an initial velocity lands on ground next to fence and the ball with no initial velocity lands 11 meters away from the fence
75. 0.964, 0.036
77. a) 3.3×10^{-12} m/s; b) 2.4×10^7 tons of TNT; c) 8.1×10^{14} N
79. a) Ball height $h = \ell(1 - \cos\theta)$ where θ is the angle with the vertical; b) height $h = L/4\,(1 - \cos\theta)$ where $\theta = \cos^{-1}\left(\frac{3}{4} - \frac{1}{4}\cos\theta\right)$
80. a) 21 m/s and 11° west of north; b) 1.4×10^5 J

Chapter 12

1. 1.7×10^{-3} rad/s; 3.5×10^{-4} m/s
3. 81.5 rad/s; 13 rev/s
5. a) 0.52 m/s; 0.17 m/s; b) 1.8 m/s^2; 0.61 m/s^2
7. 22.0 rad/s; 1.28 m/s; 55.4 rad/s; 8.83 rev/s
9. 9.4 rad/s; 0.94 m/s; 1.9 rad/s^2
11. 88 rev/s for aluminum; 2.1 rev/s for steel
13. At $t = 0$, $\phi = 0$; $\omega = 0$; $\alpha = 40$ rad/s^2; at $t = 1.0$ s, $\phi = 15$ rad; $\omega = 25$ rad/s; $\alpha = 10$ rad/s^2; at $t = 2.0$ s, $\phi = 40$ rad; $\omega = 20$ rad/s; $\alpha = -20$ rad/s^2
15. 5.5×10^{-3} cm/s; 51 cm/s; 22 cm/s
17. 611 rad/s^2; 70 revolutions
19. 1.5×10^2 rad/s^2
21. -4.36×10^{-2} rad/s^2; 0.89 revolutions

23. $-23\ \text{rad/s}^2$; 1.7 s
25. $-9.5 \times 10^{-22}\ \text{rad/s}^2$
27. 10 rad/s; 8 rad
31. 3.8×10^4 J
33. 1.21×10^{-10} m
35. $6.50 \times 10^{-46}\ \text{kg·m}^2$
37. $0.44\ \text{kg·m}^2$
39. $0.46\ \text{kg·m}^2$
41. 2.13×10^{29} J
43. $0.96\ \text{kg·m}^2$; 1.5×10^7 J
45. $-1.10 \times 10^{22}\ \text{kg·m}^2$
47. $0.379 M_E R^2{}_E$
49. $\dfrac{ML^2 \sin^2\theta}{12}$
51. $I = \dfrac{Ml^2}{3}$
53. $0.426\ MR^2$
55. $\left(\dfrac{1}{2} - \dfrac{1}{10 + 30h/4R}\right)MR^2$
57. $\dfrac{2}{5}\dfrac{M(R_2^5 - R_1^5)}{R_2^3 - R_1^3}$
59. $I = \dfrac{3}{5}M_I^2$
61. a) $225\ \text{kg·m}^2$; b) 4.4×10^3 J
63. 3.49×10^5 J; 3.9%
65. 2×10^{25} J/s; 1.05×10^{15} s or 3.3×10^7 yr
67. $\dfrac{MR^2}{4}$
69. $\dfrac{Ml^2}{6}$
71. $\dfrac{MR^2}{4}$
73. $\dfrac{3}{10}mR^2$
75. (160 km/h) **i** at top; 0 km/h at bottom; (80 km/h) **i** − (80 km/h) **j** at front
77. $12\ \text{rad/s}^2$; 74.1 rad/s; 35 turns
79. 0.012 J; 0.37 m/s
81. $I_1 = 2 \times \dfrac{2}{5}MR^2 = \dfrac{4}{5}MR^2$;
$I_2 = 2\left(\dfrac{2}{5}MR^2 + MR^2\right) = \dfrac{14}{5}MR^2$
83. $MR^2\left(\dfrac{5}{8} + \dfrac{1}{27\pi}\right)$
85. 4.4 m/s

Chapter 13

1. (4610 N·m)R, 613 kg, 940 kg
3. 310 N
5. 59 N·m
7. 130 hp, 176 N·m
9. 2900 N·m
11. 5.4 m/s, 7.7 m/s
13. 230 W
15. 1.9×10^6 J
17. 19 J, 75 J
19. 5.6×10^4 W, 1.1×10^4 N·m
21. 4.6 rad/s
23. a) 1140 N·m; b) $2.1\ \text{m/s}^2$
25. 820 N
27. 2.7×10^4 N·m
29. $9.7\ \text{m/s}^2$
31. Proof required.
33. $9.6\ \text{m/s}^2$
35. rolls to the right, $f = 2F/3$
37. 17 rev
39. 0.024 N·m, 0.16 N
41. Proof required.
43. $2.83\ \text{m/s}^2$, $2.94\ \text{m/s}^2$, 3.89 %
45. 1.6 m
47. $2.8 \times 10^{34}\ \text{kg·m}^2/\text{s}$
49. $1.6\ \text{kg·m}^2/\text{s}$
51. $1.05 \times 10^{-34}\ \text{kg·m}^2/\text{s}$, $2.11 \times 10^{-34}\ \text{kg·m}^2/\text{s}$, $3.15 \times 10^{-34}\ \text{kg·m}^2/\text{s}$
53. a) $1.8 \times 10^9\ \text{kg·m}^2/\text{s}$, upward; b) zero $\text{kg·m}^2/\text{s}$
55. 7.9×10^{22} rad/s, 7.9×10^7 m/s, 2.1×10^{-12} J, 1.5×10^{-10} J
57. $0.051\ \text{kg·m}^2/\text{s}^2$
59. $5.6 \times 10^{41}\ \text{kg·m}^2/\text{s}$, $3.14 \times 10^{43}\ \text{kg·m}^2/\text{s}$, 1.8 %
61. 0.57 rad/s, 5.5 rad/s
63. 14 rad/s
65. 1.5×10^{-19} rev/day
67. $v_1 = \sqrt{\dfrac{GM_E}{r_1}}$ and $v_2 = \sqrt{\dfrac{GM_E}{r_2}}$. The satellite closer to earth has the greater speed. $L_1 = m\sqrt{r_1 GM_E}$ and $L_2 = m\sqrt{r_2 GM_E}$. The satellite closer to earth has the smaller angular momentum.
69. $-4.3 \times 10^{-22}\ \text{rad/s}^2$, 3.5×10^{16} N·m, 2.6×10^{12} W
71. $-7°$
73. a) 5.0 m/s; b) −0.009 rad/s; c) −6860° or 19 rev
75. $\dfrac{m^2 v^2}{2g(m + \frac{1}{3}M)(m + \frac{1}{2}M)}$
77. $\dfrac{mv_0^3}{g\sqrt{32}}$, $\dfrac{mv_0^3}{g\sqrt{2}}$
79. 0.37 rad/s

81. The instantaneous change in angular momentum opposes the original direction of the angular momentum and makes the tilt worse
83. a) 110 kg·m^2/s; b) 34 N·m; c) 34 N·m
85. 1.6×10^8 kg·m^2/s, east or west, 1.2×10^4 N·m, 1.0×10^4 N
87. 3100 N·m
89. 3100 J
91. a) 76.2°; b) 290 N, −250 N
93. a) 4.10×10^{-5} kg·m^2; b) 0.12 rad/s^2; c) 4.8×10^{-6} N·m; d) 0.0023 J
95. 420 N
97. $8.50\mu_s$
99. a) 2.0×10^{-10}; b) 4.9×10^9; c) 4.9×10^9 rev/month or 1860 rev/s
101. 1.3×10^{14} kg·m^2/s, north

Chapter 14

1. 590 N
3. 5.5×10^6 N; 5.1×10^6 N
5. 5200 N
7. 8 cm; 18 cm
9. 5.88 kg
11. 3500 N; 6800 N
13. 51 N; 29 N
15. 1420 N; 2500 N
17. 30°
19. 7.65 m
21. Proof required.
23. $F = Mg\dfrac{\sqrt{R^2 - (R-h)^2}}{(R-h)}$; $F = Mg\dfrac{R}{\sqrt{R^2 - (R-h)^2}}$
25. a) 9×10^3 N; b) 2.6×10^3 N
27. 0.408 Mg
29. $T = Mg\left(\dfrac{L+R}{\sqrt{(L+R)^2 - R^2}}\right)$; $N = Mg\dfrac{R}{\sqrt{(L+R)^2 - R^2}}$
31. 0.62 mg
33. a); b) 26.6°; c) 56 J

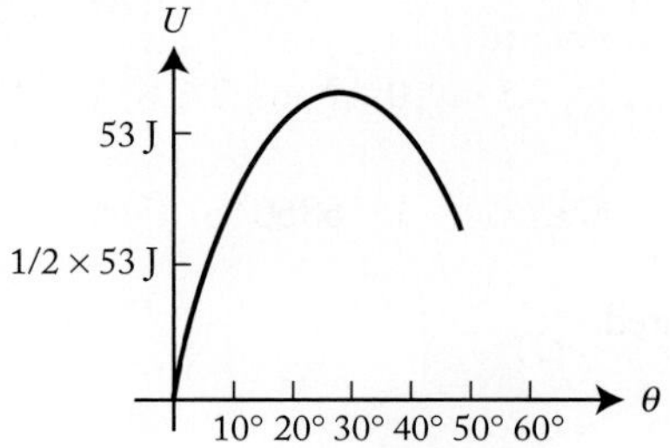

35. 1.04 L
37. 17.5 m/s
39. 7.2 m/s^2
41. a) 8.8 m/s^2; b) 5.4 m/s^2; c) 3.8 m/s^2
43. $2mg$; $\dfrac{mg}{\sqrt{3}}$; $\dfrac{mg}{\sqrt{3}}$
45. $\tan^{-1}(2\mu_s)$
47. $T_1 = \dfrac{\tau}{R}\left(\dfrac{1}{e^{\mu_s\pi} - 1}\right)$ and $T_2 = \dfrac{\tau}{R}\left(\dfrac{1}{1 - e^{-\mu_s\pi}}\right)$
49. 1580 N; 1340 N
51. 400 N
53. 240 N
55. 0.010 micrograms resolution; 0.20000 milligrams max load
57. 736 N
59. a) $\frac{1}{2}mg(R_1 - R_2)$; b) $F = \frac{1}{2}mg\dfrac{(R_1 - R_2)}{l}$; c) $\dfrac{F'}{F} = \dfrac{2l}{R_1 - R_2}$
61. 8.9
63. 393
65. 36
67. 2×10^{-3} m
69. 3.5×10^4 m
71. a) 4×10^{-3} m; b) 3.5×10^3 N
73. 0.057 m
75. 0.033%
77. 4.3×10^{-6} m
79. 0.52 cm
81. 1.5°
83. 3.96×10^8 N/m^2
85. 624 rad/s
87. 425 rad/s
89. 360 N
91. 1.23×10^4 N; 2.13×10^4 N
93. 2400 N·m
95. 0.577 Mg; 0.289 Mg
97. 490 N
99. 1200 N
101. 0.71 mm
103. 5.0×10^3 m
105. 4.85×10^{-5} N

Chapter 15

1. a) 3.0 m; 0.318 Hz; 2.0 rad/s; 3.14 s; b) $t_{midpoint} = 0.785$ s; $t_{turn.point} = 1.57$ s
3. a) 0.83 Hz; 5.2 rad/s; b) 0.20 m; c) 0.30 s; 0.40 s; d) 1.05 m/s; 0.91 m/s
5. a) 251 m/s; b) 251 m/s
7. 211 N; 4.2 m/s

9. $A = \sqrt{x_0^2 + \frac{v_0^2}{\omega^2}}$; $\delta = -\tan^{-1}\left(\frac{v_0}{\omega x_0}\right)$
11. 3.98×10^{-6} m
13. $\delta = \frac{3\pi}{2}$
15. a) $x = 0.292 \cos(6\pi t - 0.815)$; b) 0.043 s; -103 m/s^2
17. 2.8×10^6 N/m; 3.16 Hz
19. 1.9×10^4 N/m
21. 1.13×10^{14} Hz
23. 0.20 m; 7.3 rad/s; 0 rad; 1.46 m/s; 10.7 m/s^2
25. 5.51×10^3 Hz; 1.0×10^{-4} g; 6.7×10^{-6} cm
27. $x = 0.27 \cos 6t$, with the axis chosen so the initial position of the unstretched spring is at $x = 0.27$ m.
29. $\frac{mg \sin\theta}{k}$; $\frac{1}{2\pi}\sqrt{\frac{k}{m}}$
31. $f = \frac{1}{2\pi}\sqrt{\frac{k}{(m + 6M)}}$
33. 3.6 J
35. 2.12 Hz; E same; A same; 2.0 m/s; 26.7 m/s^2
37. 0.34 m
39. $Fd - \frac{kd^2}{2}$
43. 10.4 s
45. 34.8 s
47. 0.188 Hz
49. a) 0.73 min/day; b) 1.0 mm
51. 24.8 m
53. 3.0 s
55. a) $\sqrt{\frac{2\kappa}{MR^2}}$; b) $\theta_0\sqrt{\frac{2\kappa}{MR^2}}$; c) $\theta_0 = 1$ radian
57. $2.09\sqrt{\frac{L}{g}}$
59. a) 1.26×10^{-3} J; b) 0.145 m/s
61. 1.64 s
63. 9.8×10^{-3} m/s^2
65. 1.6 s
67. $2\pi\sqrt{\frac{L}{g}\left(\frac{m_1 + m_2}{m_1 - m_2}\right)}$
69. 3.6 rad/s; 3.3 m/s
71. $m = \frac{\kappa}{g}$
73. $2\pi\sqrt{\frac{4L}{5g}}$
75. a) $mg\left[1 - \frac{A^2}{2} + \frac{3A^2}{2}\sin^2\left(\sqrt{\frac{g}{l}}\,t\right)\right]$;
b) $\frac{\pi}{2}\sqrt{\frac{l}{g}}$; $mg\,(1 + A^2)$
77. 1.0×10^3
79. 92; 0.32 W
81. 30
83. 395
85. 21
87. 1.5 cm; 66.7 Hz
89. a) midpoint at 2 s; 6 s, 10 . . .; turning point at 0 s, 4 s, 8 s, . . .; b) midpoint at 0 s, 4 s, 8 s, . . .; turning point at 2 s, 6 s, 10 s, . . .
91. 26.7 m/s; 1.68×10^4 m/s^2; 2.0×10^4 N
93. 2.12 Hz
95. 1.25 Hz
97. 18.75 J; 3.54 m/s
99. 0.375 m
103. 0.35 Hz

Chapter 16

1. 4.3×10^{14} Hz to 7.5×10^{14} Hz (violet)
3. 2.08 cycles/hour, 356 km
5. a) 4.4 m/s; b) 205 m/s
7. a) 10.8 hour; b) 2.5 /hour
9. a) $v_{max} = 0.27$ m/s as it passes through equilibrium between crests; b) $|a_{max}| = 6.2$ m/s^2 at the wave crests.
11. a) 0.02 m; b) 1.4 Hz; c) 10 m
13. a) 0.2 sec, 5.0 Hz, 31 rad/sec, 5.2 m^{-1}; b) $y = 0.020 \cos(5.2\,x - 31.4\,t)$
15. wavelength decreases by 20 cm
17. $|a_{max}| = 2.41$ m/s^2; wavelength $= 156$ m
19. 0.45 m/s
21. 0.97 sec
23. 2.0 kg/m
25. 250 m/s
27. 1280 N
29. 184 km
31. $v = v\sqrt{\frac{\mu}{\mu}}$
33. 0.017 sec
35. $2\sqrt{l/g}$
37. $\frac{mV^2}{2\pi R}$; Proof required.
41. a) $\frac{\pi}{2}$; b) 0.20 m
43. 6.0 m, .6 m, 0.80 Hz; -3.2 m
45. 1.14 m, 6.28 m

47. a) 0.0194 m; b) −0.0179 m

49. a) $y = 0.0060 \cos(400\pi t) + 0.0040 \cos(1200\pi t)$; b)

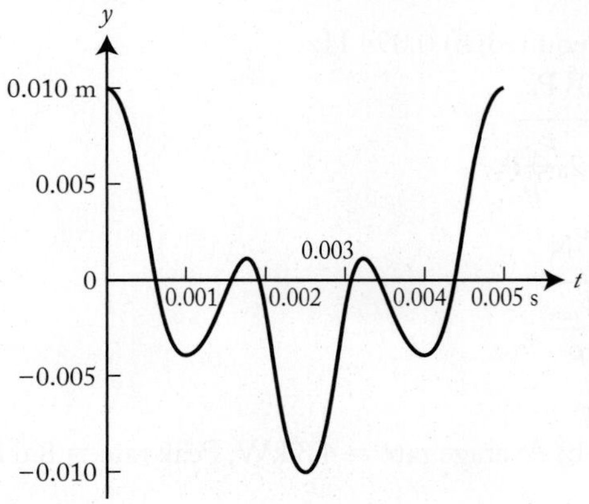

51. 0.028 is the fractional increase or decrease. We cannot tell which from the given information.

53. 392 Hz, 588 Hz, 784 Hz, 980 Hz

55. 1.58 Hz, 3.16 Hz

57. 28 Hz

59. 7.07 m, 66.6 m/s, 628 m/s^2

61. 9.3 ms, down, 54 Hz

63. 8.16 Hz, 16.3 Hz, 24.5 Hz, 32.7 Hz, 40.8 Hz

65. 71 N

67. 1.3×10^{-7} m, 2.5×10^{-7} m

69. 3.7 m/s, 6.8×10^3 m/s^2

71. a) 4.82×10^3 N; b) 8.2 Hz

73. $y = (0.20 \text{ mm}) \sin(2\pi x + 880\pi t) + (0.20 \text{ mm}) \sin(2\pi x - 880\pi t)$, $A = 0.20$ mm, $v = 440$ m/s

75. a) $v(x) = \sqrt{\dfrac{F}{A + Bx}}$; b) $\lambda(x) = \dfrac{1}{f}\sqrt{\dfrac{F}{A + Bx}}$;

c) $f = \dfrac{n}{2L}\sqrt{\dfrac{F}{A + BL}}$

77. Large amplitude is at $x = \dfrac{n\pi}{k}$. Smallest amplitude is at $x = \dfrac{(2n + 1)\pi}{k}$ where n is an integer.

79. a) 0.030 m; b) 5.2 m; c) crests: 0 m, 5.2 m, 10.4 m, . . .; troughs: 2.6 m, 7.9 m, 13.5 m, . . .

81. 2.1 m/s^2, 4.9 m/s

83. 13 m/s, 7.9×10^3 m/s^2

85. 0.0731 kg/m, 261 m/s

87. a) $T_1 = \dfrac{\sqrt{2}}{1 + \sqrt{3}} T$, $T_2 = \dfrac{2}{1 + \sqrt{3}} T$;

b) $v_1 = 7.2$ m/s, $v_2 = 8.6$ m/s

89. 12 Hz

91. a) $v(x) = \sqrt{g(L - x)}$, 14 m/s, 9.9 m/s, 0 m/s; b) 2.9 sec

93. $\lambda = 4L, \dfrac{4L}{3}, \dfrac{4L}{5}, \dfrac{4L}{7}, \ldots$, where $4L$ is the longest possible wavelength

Chapter 17

1. 17 m (20 Hz) to 1.7 cm (20 kHz)

3. 765 m; 166 m

5. about 9 cm

7. 1.9 mm and 0.10 mm

9. 6.8 m/s

11. D–D# 1.9 cm, D#–E 1.8 cm, E–F 1.7 cm, F–F# 1.6 cm, F#–G 1.5 cm, G–G# 1.5 cm, G#–A 1.3 cm, A–A# 1.3 cm, A#–B 1.2 cm, B–C 1.1 cm, C–C# 1.1 cm, C#–D 1.0 cm

13. 1.0×10^3 W/m^2

15. D#, 6 octaves above the one listed in Table 17.1

17. −3.0 dB

19. 83 dB

21. 0.11 W

23. 130 times (intensity measured in W/m^2), 21 dB

25. 9.1 sec

27. 249 m

29. a) $f = 3.0 \times 10^{-5}$ Hz; $T = 3.3 \times 10^4$ s (about 9 h); b) It's possible because the period of the first overtone is close to 1/4 of the tidal period.

31. a) 3.0×10^{-5} s; b) glass

33. 272 Hz, 3.9%

35. about 4000 Hz

37. 2.0 km

39. 3.3 km in sea water

41. 2.8 km

43. 92.4 m

45. 337 m/s

49. a) 0.632 m; b) C–C# 3.5 cm, C#–D 3.4 cm, D–D# 3.2 cm, D#–E 3.0 cm, E–F 2.8 cm, F–F# 2.7 cm, F#–G 2.5 cm, G–G# 2.4 cm, G#–A 2.2 cm, A–A# 2.1 cm, A#–B 2.0 cm, B–C 1.9 cm

51. $\lambda_n = \dfrac{2L}{n}$

$f_n = \dfrac{nv}{2L}$, $n = 1, 2, 3, \ldots$

53. 619 Hz

55. 21.5 m/s, 0.215 Hz

57. 381 m/s

59. 2.63 m/s

61. 30°

63. 405 Hz

65. 594 Hz, 595 Hz

67. 476 Hz

69. 481 Hz

71. 29.4°

73. a) Proof required; b) 165 m/s
75. 0.15 mm
77. 1.5 mm, 0.33 mm
79. 5.0 m/s, 3.2×10^7g
81. 3 women
83. 3.0 dB
85. a) 3.0* 10^{-3} sec; b) The bat will think distances are 0.77 times the real distances.
87. a) 33.5°; b) 30.2 sec
89. a) 660 Hz; b) 691 Hz; c) 723 Hz

Chapter 18

1. In the hose: 1.39 m/s; 2.8 m/s; 4.2 m/s; In the nozzle: 22.3 m/s; 25.1 m/s; 23.9 m/s
3. 7.23×10^5 W
5. 12 m/s
7. 8.84 cm/s; 8.84 m/s
9. a) 11.5 m; b) 8.1 cm; 11.7 cm
11. 84 m
13. 1370 lbf = 6090 N
15. 132 cm^2
17. 2.34×10^5 Pa
19. 5.08×10^4 N
21. 48.6 cm^2
23. 2.0×10^4 Pa; 7.5×10^6 Pa
25. a) 360 N; b) 330 N
27. 3.56×10^5 Pa; 3.60×10^5 Pa; 4×10^3 N
29. 0.85 m
31. 10.3 m
33. 3.3×10^4 Pa
35. 2.94×10^5 Pa
37. 2.1%
39. 3.1×10^8 N
41. a) Proof required; b) 5.0×10^9 Pa
43. a) $F_b = \rho g h \pi R_2^2$; $F_\perp = \dfrac{\pi \rho g h (R_2 - R_1)(R_1 + 2R_2)}{3}$;

$$F_{total} = -\frac{\pi \rho g h\left(R_1^2 + R_1R_2 + R_2^2\right)}{3};$$

b) $F_b = \rho g h \pi R_1^2$; $F_\perp = -\dfrac{\pi \rho g h (R_2 - R_1)(2R_1 + R_2)}{3}$;

$$F_{total} = -\frac{\pi \rho g h\left(R_1^2 + R_1R_2 + R_2^2\right)}{3}$$

45. a) 4.7×10^7 m^3; b) 4.3×10^{10} kg
47. Yes
49. 0.32 m
51. 61 m/s^2
53. 113 kg
55. 31 m
57. 12.6 cm
59. a) Proof required; b) $-2g$; c) ∞
61. 4.4 cm
63. a) Proof required; b) 0.094 Hz
65. 1.13×10^5 Pa
67. $\sqrt{2gh + 2\dfrac{p_{atm}}{\rho}}$
69. 1.9×10^4 N
71. $v = \sqrt{\dfrac{2p_{tank}}{\rho}}$
73. 0.013 m/s
75. a) 332 N; b) Average rate = 4.3 kW; Peak rate = 8.6 kW
77. 12.4 hp
79. 2.7 m/s; 190 liters/s
81. 7.3×10^3 N
83. 1.12×10^{12} Pa
87. 8.06 mm-Hg
89. 1.21×10^5 Pa; 1.01×10^5 Pa; 2×10^3 N
91. 0.73 cm
93. 2.5×10^5 kg
95. 5.3×10^{-3} m^3
97. 1.07×10^3 kg/m^3
99. 2.0×10^{-3} m^3/s
101. The pressure inside increases by 209 Pa; Smaller

Chapter 19

1. 32°F, − 380°F, − 423.4°F, − 452.2°F, − 454°F, − 459.67°F
3. 5.3×10^{21} atoms
5. 78% N_2, 22% O_2
7. The frequency decreases by 14 Hz
9. 1.9 atm
11. a) $p_{O2} = 7.5 \times 10^4$ Pa, $p_{N2} = 8.6 \times 10^4$ Pa; b) 1.6×10^5 Pa
13. 3.4 atm
15. $\rho_i/\rho_0 = 1.05$
17. 1.4×10^{-9} Pa
19. 12 kg/m^3
21. 4.3 atm
23. 500 kg/m^3
25. 100 kg/m^3
27. 1.3 cm
29. 4.5×10^7 Pa
31. 96.3 g
33. 3150 kg, 2.8×10^3 m^3
35. a) Water rises 1.2 m; b) 6.8 kg at 2.6×10^5 Pa
37. 29 g/mol
39. Differentiating $p - p_0 = -\rho g y$ yields $dp = -\rho g y$. Using the Ideal-Gas Law and $n = \dfrac{m}{M}$, $\rho = \dfrac{m}{V} = \dfrac{pM}{RT}$.

Substituting ρ into the differential equation: $dp = -\frac{pMg}{RT}dy$

or, $\frac{dp}{p} = -\frac{Mg}{RT}dy.$

41. 615 m/s
43. 1200 K
45. 4100 m/s, 1.4×10^{-20} J
47. 0.12 m/s
49. 5.65×10^{-21} J
51. For O_2, $v_{rms} = 428$ m/s, For O_3, $v_{rms} = 349$ m/s, For O_2, (translational) $K = 4.87 \times 10^{-21}$ J, for O_3, (translational) $K = 4.85 \times 10^{-21}$ J
53. 0.43%
55. 9.7×10^6 K
57. 0.47 m/s
59. Using the hint, the volume swept out per molecule with an effective radius $2R_0$ going a distance l is cylindrically shaped with volume $V/N = \pi (2R_0)^2 l$. Solving for l yields the desired result.
61. a) for 1 atm: − 0.091%; b) for 1000 atm: − 91%
63. 1.9×10^5 J, translational 0.6, rotational 0.4
65. A 7% increase in kinetic energy by changing temperature, no change in the kinetic energy by changing the pressure.
67. 1.0×10^5 J
69. 291 J
71. 1.88×10^{25} molecules
73. 1.3×10^3 N
75. 1.1×10^{23} nitrogen molecules, 2.9×10^{22} oxygen molecules, 1.4×10^{23} total
77. a) 1.8×10^{32} particles/m^3 3.7×10^{16} Pa; b) 9.0×10^{31} particles/m^3, 1.9×10^{16} Pa; c) 4.5×10^{31} particles/m^3, 9.3×10^{15} Pa
79. From the Ideal-Gas Law at constant temperature $pV = p'V'$. So,

$$\frac{\Delta Z}{V} = \frac{V - V'}{V} = 1 - \frac{V'}{V} = 1 - \frac{p}{p'} = \frac{p' - p}{p'} = \frac{\Delta p}{p'}.$$

This can be rearranged to $V = \Delta V\left(\frac{p'}{\Delta p}\right)$, where ΔV is the *decrease* in volume and Δp is the corresponding *increase* in pressure.

Furthermore, $\frac{p'}{\Delta p} = \frac{p + \Delta p}{\Delta p} = 1 + \frac{p}{\Delta p}$ and

$$V = \Delta V\left(1 + \frac{p}{\Delta p}\right) \approx \Delta V\left(\frac{p}{\Delta p}\right) \text{ for } \Delta p << p.$$

Using the specifics of the problem, $V_c = V_s + V$ or $V_s =$

$$V_c - V = V_c - \Delta V\left(\frac{p}{\Delta p}\right)$$

81. 615 m/s
83. a) $\sqrt{2}$; b) 1; c) 1; d) 0.5
85. 375 K

Chapter 20

1. 540 s
3. 0.28°C
5. 1.6 km
7. 750 s
9. 8500 steps
11. 0.17°C
13. 1.7°C
15. 1.7×10^{-4} °C/km
17. The heat produced from electric power 2.6% of the incident solar heat. This is enough to slightly increase the local temperature.
19. 1.1×10^{-3} m^3/s
21. a) 1.7×10^3 N·m; b) 2.3×10^3 W; c) 4.0×10^{-4} °C
23. 27°C
25. 136°C
27. 0.67°C
29. 38°C
31. 0.18 liter
33. 0.028 J of work done by iron, 2.7×10^7 J of heat absorbed by iron, amount of work is 1.0×10^{-6} times the heat absorbed
35. 4.9×10^{-4} m, 17 N
37. a) 3.8×10^{-4}, 1.9×10^{-4}; b) 16 s
39. Proof required.
41. 100.28°C
43. 23000 W, the rate through window is 13 times greater than the rate through the wall
45. a) 2.4 m^2·s·°C/J; b) 13.6 ft^2·h·°F/BTU
47. 4.2×10^3 W
49. The solution is a proof.
51. 11 W, 79°C
53. The solution is a proof.
55. 0.51 cm/h
57. a) 4.26×10^{14} J; b) 5.3 bombs
59. 270 g
61. 1.1×10^9 J, 1.2×10^4 W
63. 3.9 kg
65. a) 2.0×10^{11} kg; b) 1.1×10^{17} cal; c) 2.9×10^{15} J = 7.0×10^{14} cal; d) 1.0×10^{13} J = 2.4×10^{12} cal. The kinetic energy is smaller than the potential energy due to frictional losses with the air.
67. 0.092 kg
69. 4.3 km^3/h
71. 41°C

73.

GAS	C_V (J/(kG·K)
He	3.12×10^3
Ar	3.13×10^2
N_2	7.42×10^2
O_2	6.50×10^2
CO	7.39×10^2
NH_3	1.60×10^3
CH_4	1.69×10^3

The gas with the highest specific heat per kilogram is helium; and that with the lowest is argon.

75. 971 m/s

77. $\Delta V = 3.7 \times 10^{-2}$ m^3, $W = 3.7 \times 10^3$ J

79. $C_p = 26.3$ J/(mol·K), $C_V = 18.0$ J/(mol·K)

81. 110 kcal/h

83. 1 K

85. a) 0.072 m^3; b) 0.42 m^3, 145 K

87. 214 K

89. 3.6°C

91. 160 liters/h

93. 0.33 m, 0.050 m, 46 m^2

95. 2.3×10^{-5} kg/s

97. 0.52 kg

99. 880 J, 1500 J

101. a) 1700 J; b) 1200 J; c) 0.029 m^3, 1.7×10^4 N/m^2; d) 5.0×10^2 J, -2.4×10^3 J

Chapter 21

1. $Q = 1.9 \times 10^3$ J, $\Delta E = 1.1 \times 10^3$ J

3. a) $W = 0$, $Q = \Delta E = 610$ J; b) $W = 810$ J, $Q = 2.0 \times 10^3$ J, $\Delta E = 1.2 \times 10^3$ J

5. -470 J

7. a) 4.29 moles; b) $W = 1010$ J, $\Delta E = 2490$ J; c) 5/2, diatomic

9. a) $W = -9.19$ J, $\Delta E = 3.34 \times 10^5$ J; b) $-$ 18.4 J, the heat of vaporization remains unchanged

11. $W = -37.9$ J, $\Delta E = -37.1$ J

13. 4.87×10^4 J, 2.56 J

15. 4.3×10^4 J

17. 0.014 m^3, 7.2×10^5 Pa

19. 2.43×10^6 J/kg

21. 35%

23. 14%

25. 60%, 1.2×10^7 J

27. 44.5%

29. 5.5%, $W = m_{\text{you}}$ (9.81 m/s^2) (3.0 m),

$$Q = \frac{m_{\text{you}}(9.81\ \text{m/s}^2)\ (3.0\ \text{m})}{0.055}\left(\frac{1\ \text{kcal}}{4187\ \text{J}}\right)$$

31. 8.2×10^6 J/s, 0.19 kg/s

33. 0.999 999 997

35. 1.4×10^5 J

37. 44%, 1100 W

39. 75 W

41. 19.5

43. 8.5×10^3 J, 3.4

45. a) -1.3×10^3 J; b) Heat is absorbed during step 2 and rejected in step 3; c) heat is rejected by the system in step 1, d) 0.39

47. a) 0.067; b) 1.39×10^7 W; c) 180 kg/s

49. a) 48 W; b) 20 times

51. a) $e_{\text{turbine}} = 0.34$, $e_{\text{engine}} = 0.42$; b) 0.62, the two efficiencies are the same

53. 9.5×10^3 J/K

55. 3.0 W/K

57. 12 400 W/K

59. 3×10^3 J/K

61. $\Delta S_{\text{Al}} = 430$ J/K, $\Delta S_{\text{Fe}} = 150$ J/K, $\Delta S_{\text{Ag}} = 80$ J/K, $\Delta S_{\text{Hg}} =$ 47 J/K. The change in entropy seems to decrease with increasing atomic number. Largest is aluminum; smallest mercury.

63. 41 W/K

65. 9.5×10^6 W/K

67. 0.94 W/K

69. 5.8 J/K

71. a) Proof required; b $\Delta S = 780$ J/K

73. 37 J/K

75. Proof required.

77. 120 K

79. a) 4.16×10^5 Pa, 2.27×10^5 Pa; b) $W = \Delta Q = -3.0 \times 10^2$ J

81. 24%, 4.8×10^4 W

83. 4.0×10^2 W

85. a) Beginning with the point at the upper left, the gas undergoes an isobaric expansion in step 1 as the volume increases, followed by a isovolumetric reduction of pressure in step 2 as the temperature is reduced. The gas is then compressed isobarically in step 3 by reducing the volume, before an isovolumetric increase in pressure in step 4 by increasing the temperature.
b) $W_1 = 2100$ J, $W_2 = 0$ J, $W_3 = -700$ J, $W_4 = 0$ J; c) $Q_1 = 5260$ J, $Q_2 = -3160$ J, $Q_3 = -1750$ J, $Q_4 = 1050$ J; d) 44%

87. 48%, 1×10^7 W

89. $\Delta S_{\text{N}} = 2600$ J/K, $\Delta S_{\text{O}} = 2300$ J/K, $\Delta S_{\text{H}} = 22\,500$ J/K. Hydrogen is largest and oxygen smallest.

91. a) $\Delta S = 1.1$ J/K; b) $Q_2 = 340$ J, $\Delta S = 1.1$ J/K; c) $\Delta S =$ 0 J/K

Chapter 22

1. 5.8×10^5 N
3. 58 N, 3.5×10^{28} m/s^2
5. 51 N
7. 1.6×10^{20} electrons
9. 2.39×10^{-7} N
11. $F_g = 6.7 \times 10^{-13}$ N, $F_e = 2.3 \times 10^{-8}$ N
13. 9.63×10^4 C
15. 4.7×10^{13} electrons
17. 3.2×10^{19} N
19. 1.3
21. 99.9 %
23. 2.9×10^{-9} N/m
25. $(-2.3 \times 10^{-5}\text{ N})\mathbf{i} + (-3.5 \times 10^{-5}\text{ N})\mathbf{j}$, $(2.3 \times 10^{-5}\text{ N})\mathbf{i} + (3.5 \times 10^{-5}\text{ N})\mathbf{j}$
27. 6.81×10^{32} electrons on Earth and 1.9×10^{32} electrons on Moon
29. 1.0×10^{-9} at 1 m, 1.0×10^{-5} at 1×10^4 m
31. 3.8×10^{-39} C, ratio $= 2.8 \times 10^6$ (attractive)
33. $\dfrac{2kqxQ}{[\frac{d^2}{4} + x^2]^{3/2}}$, $+x$ direction
35. $-(1.9 \times 10^{-7}\text{ N})\,\mathbf{i} - (1.7 \times 10^{-7}\text{ N})\,\mathbf{j}$
37. 1.0×10^{-7} C
39. $-1.35k\dfrac{Q^2}{L^2}\mathbf{i} - 1.35k\dfrac{Q^2}{L^2}\mathbf{j} = -1.35k\dfrac{Q^2}{L^2}(\mathbf{i}-\mathbf{j})$
41. $-(3.1 \times 10^{-15}\text{ N})\,\mathbf{i} + (6.9 \times 10^{-16}\text{ N})\,\mathbf{j}$
43. 1.2×10^{-4} kg or 0.12 g
45. $2\sqrt{2}\left(\frac{\sqrt{3}}{2} + 1\right)^2 = 9.85$
47. $kQq\left[\dfrac{1}{x^2} - \dfrac{4}{(d + 2x)^2}\right]$
49. $k\dfrac{Q^2}{a^2}(1.116\,\mathbf{i} - 1.75\,\mathbf{j} - 0.5\mathbf{k})$
51. $0.35d$
53. $p + p \rightarrow n + n + \pi^+$, $p + p \rightarrow n + p + \pi^0$, $p + p \rightarrow n + p + \pi^0 + \pi^-$
55. 1
57. 5.6×10^{21} electrons
59. 1.9×10^{-9} kg
61. any negative charge
63. 3.6×10^{-8} N, 6.9×10^{-2} m/s^2
65. 1.0×10^{-6} C, 6.5×10^{12} electrons
67. 0 C, $+e$, $+e$, 0 C

Chapter 23

1. $F = 5.4 \times 10^{-14}$ N, $a = 6.0 \times 10^{16}$ m/s^2
3. $\theta = -20°$
5. $\mathbf{E} = -2.1 \times 10^5$ N/C $\mathbf{j}$
7. $F_e = 0.6\,F_g$
9. 6.3×10^{-7} m
11. $E = 5.1 \times 10^{11}$ N/C
13. $E = 5.1 \times 10^{12}$ N/C
15. 28 N/C
17. $\mathbf{E}_A = \left(1.15 \times 10^{10}\dfrac{\text{N}\cdot\text{m}^2}{\text{C}^2}\right)\dfrac{Q}{L^2}\mathbf{i}$,

$\mathbf{E}_B = \left(1.15 \times 10^{10}\dfrac{\text{N}\cdot\text{m}^2}{\text{C}^2}\right)\dfrac{Q}{L^2}\mathbf{j}$,

$\mathbf{E}_C = -\left(1.15 \times 10^{10}\dfrac{\text{N}\cdot\text{m}^2}{\text{C}^2}\right)\dfrac{Q}{L^2}\mathbf{i}$,

$\mathbf{E}_D = -\left(1.15 \times 10^{10}\dfrac{\text{N}\cdot\text{m}^2}{\text{C}^2}\right)\dfrac{Q}{L^2}\mathbf{j}$

19. $E_P = 9.5 \times 10^3$ N/C $\mathbf{i} - 2.8 \times 10^4$ N/C $\mathbf{j}$
21. $$E(x) = 2 \times \left(8.99 \times 10^9\dfrac{\text{N}\cdot\text{m}^2}{\text{C}^2}\right)\left[\dfrac{(-40\text{ C})(10000\text{ m})}{[x^2 + (10000\text{ m})^2]^{3/2}} + \dfrac{(30\text{ C})(4000\text{ m})}{[x^2 + (4000\text{ m})^2]^{3/2}}\right]$$
23. $E \approx \dfrac{Q}{2\pi\epsilon_0 x^2}$
25. a) E_{max} is at $y = \pm R\frac{\sqrt{2}}{2}$
b) The field distribution for the ring is the same as that for two positive charges rotated around the y axis.
27. $E_{0.5} = 7.2 \times 10^{-4}$ N/m, $E_{1.0} = 3.6 \times 10^{-4}$ N/m, $E_{1.5} = 2.4 \times 10^{-4}$ N/m
29. $\mathbf{E} = \left(2.6 \times 10^{10}\dfrac{\text{N}\cdot\text{m}^2}{\text{C}^2}\right)\dfrac{\lambda}{L}(\mathbf{i}+\mathbf{j})$
31. $\mathbf{E} = \dfrac{1}{\pi\epsilon_0}\dfrac{\lambda}{2d}(\cos 30°)\,\mathbf{j}$
33. $E_x = \dfrac{1}{4\pi\epsilon_0}\left[\dfrac{-\lambda}{\sqrt{x^2 + y^2}}\right]$, $E_y = \dfrac{\lambda}{4\pi\epsilon_0 y}\left[1 + \dfrac{x}{\sqrt{x^2 + y^2}}\right]$
35. $\mathbf{E}_A = 1.13 \times 10^5$ N/C $\mathbf{j}$, $\mathbf{E}_B = -1.13 \times 10^5$ N/C $\mathbf{j}$, $\mathbf{E}_C = -3.39 \times 10^5$ N/C $\mathbf{j}$, $\mathbf{E}_D = -1.13 \times 10^5$ N/C $\mathbf{j}$
37. $E = 2.4 \times 10^7$ N/C directed at an angle of 45° with respect to each sheet
39. $E_z = -\dfrac{z\sigma}{2\epsilon_0\sqrt{z^2 + R^2}}$

41. a) $\mathbf{E}_P = \frac{1}{4\pi\epsilon_0}\frac{Q}{l}\left(\frac{1}{x} - \frac{1}{x+l}\right)\mathbf{i}$,

b) $\mathbf{E}_{P'} = \frac{1}{2\pi\varepsilon_0}\frac{Q}{y}\frac{1}{\sqrt{l^2+4y^2}}\mathbf{j}$

43. a) $\lambda_0 = \frac{2Q}{l}$,

b) $\mathbf{E}_p = \frac{1}{2\pi\epsilon_0}\frac{-Q}{l^2}\left[\frac{1}{2}\ln\left(\frac{x+l}{x}\right)^2 + \frac{x}{x+l} - 1\right]\mathbf{i}$,

c) When $x >> l$, the electric field resembles that of a point charge and goes to zero as x goes to infinity.

45. $\mathbf{E} = \frac{\lambda\sqrt{2}}{\pi\epsilon_0 l}\mathbf{i} - \frac{\lambda\sqrt{2}}{\pi\epsilon_0 l}\mathbf{j}$

47. $\mathbf{E} = -\frac{\lambda}{4\pi\epsilon_0}\left\{\left[\frac{1}{x}\left(1+\frac{y}{(x^2+y^2)^{1/2}}\right) + \frac{1}{(x^2+y^2)^{1/2}}\right]\mathbf{i} + \left[\frac{1}{y}\left(1+\frac{x}{(x^2+y^2)^{1/2}}\right) + \frac{1}{(x^2+y^2)^{1/2}}\right]\mathbf{j}\right\}$

49. $E = 0.0609\frac{\lambda}{\epsilon_0 R}$

51. The solution is a sketch of the electric field.

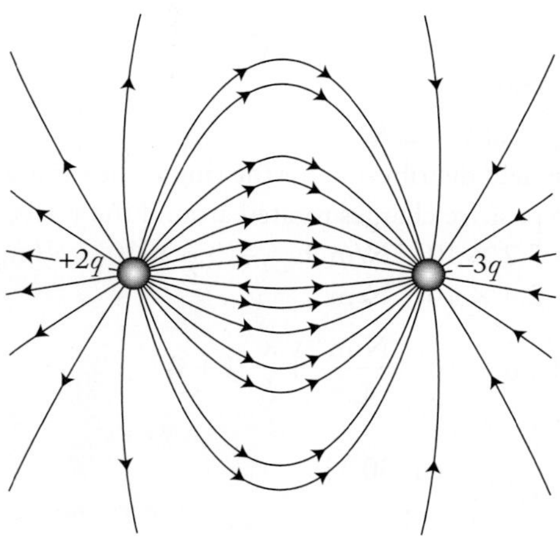

53. The solution is a sketch of the electric field.

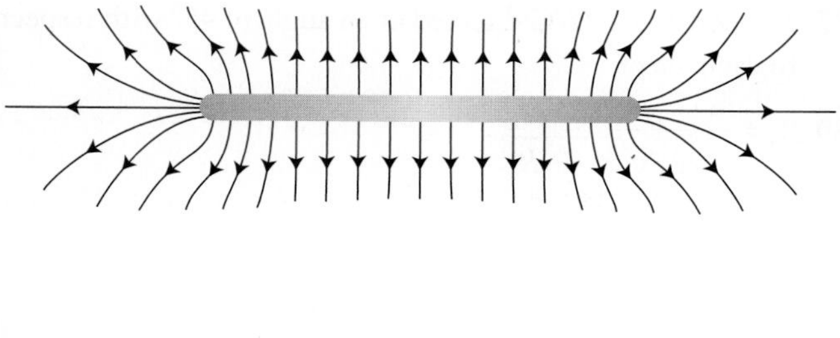

55. The solution is a sketch of the electric field.

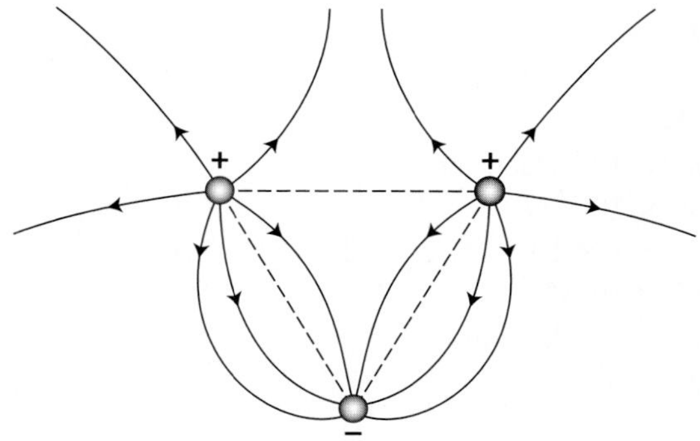

57. $E = 3.1 \times 10^5$ N/C
59. $v_0 = 1.4 \times 10^5$ m/s
61. $v_0 = 110$ m/s
63. 1085 electrons
65. $p = 2.0 \times 10^5$ C·m
67. a) $p = 1.6 \times 10^{-29}$ C·m, b) The dipole moment is reduced due to the motion of electrons.
69. $E = 520$ N/C
71. $E = 1.1 \times 10^7$ N/C, $\theta = 3°$ with respect to the y-axis
73. $E = 1.21 \times 10^4$ N/C, $\theta = 74°$ with respect to the y-axis

75. (a) $E = \frac{1}{4\pi\epsilon_0}\left(\frac{\lambda}{d}\right)$, (b) $F = \frac{q}{4\pi\epsilon_0}\left(\frac{\lambda}{d}\right)$,

(c) The magnitude of the force is the same as in (b), but its direction is opposite that of the rod on the charge.

77. $\mathbf{E} = \frac{Q}{2\pi\epsilon_0}\left(\frac{1}{y^2} - \frac{1}{(y^2+d^2)^{\frac{3}{2}}}\right)\mathbf{j}$

79. $E = \frac{1}{4\pi\epsilon_0}\frac{Q}{y^2}$

Chapter 24

1. 1.1×10^{12} N·m²/C
3. 0.16 N·m²/C
5. $-\frac{\sqrt{3}}{8}\frac{\sigma}{\epsilon_0}a^2, \frac{\sqrt{3}}{24}\frac{\sigma}{\epsilon_0}a^2$
7. $\phi_{E,1} = \phi_{E,6} = 0, \phi_{E,2} = -\phi_{E,4} = 2.0$ N·m²/C, $\phi_{E,3} = -\phi_{E,5} = -3.5$ N·m²/C, $\phi_{total} = 0$
9. $\phi = \frac{q}{\epsilon_0}$
11. 1.4×10^3 N·m²/C
13. 0.038 N·m²/C
15. (a) $\frac{q}{2\epsilon_0}$ (b) $\frac{\sigma q}{2\epsilon_0}$ (c) $F = \frac{\sigma q}{2\epsilon_0}, E = \frac{\sigma}{2\epsilon_0}$

17. (a) $\frac{q}{4\epsilon_0}$ (b) $\frac{\sigma}{4\epsilon_0}$

19. $+45\ \text{N}\cdot\text{m}^2/\text{C}$, 5.5×10^{-8} C/m

21. $2.3 \times 10^2\ \text{N}\cdot\text{m}^2/\text{C}$

23. $\phi_G = 4\pi G m_{inside}$

25. Imagine a small cube of volume dV. If the cube itself contains no charge and it's located in a uniform electric field, the net flux through the six faces of the cube must be zero, no matter how the cube is oriented in the field. If the cube contains charge, then the flux through it cannot be zero and the field cannot be uniform. If the field is uniform, Gauss' law tells us that the charge density inside the cube, which is the charge inside divided by the volume dV, must be zero.

27. 160 N

29. $EA = \frac{q_{enc}}{\epsilon_0} \Rightarrow q_{enc} = 0 \therefore EA = 0$

since $A \neq 0, E = 0$

31. $\frac{\lambda r}{2\pi R^2\, \epsilon_0}$ in direction perpendicular to axis

33. $r \leq a\ E = 0, a \leq r \leq b\ E = \frac{\rho}{3\epsilon_0}\frac{(r^3 - a^3)}{r^2}$,

$r \geq b\, E = \frac{\rho}{3\epsilon_0}\frac{(b^3 - a^3)}{r^2}$

35. $r \leq a\, E = 0, a \leq r \leq b\ E = \frac{Q}{4\pi\epsilon_0 r^2}\frac{(r^3 - a^3)}{(b^3 - a^3)}$,

$r \geq b\ E = \frac{Q}{4\pi\epsilon_0 r^2}$

37. (a) $F = qE$ where $q = -e$. $E = \frac{Q}{4\pi\epsilon_0 r^2}\, r$ where

$Q = +e\left(\frac{r^3}{R^3}\right)$. Substituting yields $\vec{F} = -\frac{e^2 r}{4\pi\epsilon_0 R^3}\,\hat{r}$

where the minus sign indicates this is a restoring force and

the magnitude of that force is $F = \frac{e^2 r}{4\pi\epsilon_0 R^3}$. (b) 7.2×10^{15} Hz

39. (a) $C = \frac{Q}{2\pi R^2}$ (b) $E = \frac{Q}{4\pi\epsilon_0 R^2}$ (c) $E = \frac{Q}{4\pi\epsilon_0 r^2}$

41. (a) $Q = \frac{4\pi k}{n + 3}\, r^{n+3}$ (b) $E = \frac{k r^{n+1}}{\epsilon_0(n + 3)}$ (c) $n = -1$,

$E = \text{k}/2$ (d) if $n = -3, Q = \infty$

43. For $|x| \leq \frac{d}{2}, E = \frac{Cx^3}{3\epsilon_0}$. For $|x| \geq \frac{d}{2}\ E = \frac{Cd^{\,3}}{24\epsilon_0}$.

45. $\phi = \int_{-\infty}^{\infty} E_y 2\pi R dx = 2\pi R \int_{-\infty}^{\infty} E_y dx = \frac{q'}{\epsilon_0}$

$$\int_{-\infty}^{\infty} E_y dx = \frac{q'}{2\pi R\epsilon_0} \quad \therefore \frac{q}{v}\int_{-\infty}^{\infty} E_y dx = \frac{q}{v}\frac{q'}{2\pi\epsilon_0 R}$$

$$\therefore \int_{-\infty}^{\infty} F_y dt = \frac{q}{v}\frac{q'}{2\pi\epsilon_0 R}$$

47. $E = \frac{Q}{4\pi\epsilon_0}\left(\frac{1}{r^2} - \frac{1}{8\left(r - \frac{R}{2}\right)^2}\right) r$

49. $\mathbf{r} = \frac{1}{4}d\hat{i}\ \vec{E} = \frac{\rho d}{6\epsilon_0}\hat{i}$, $\mathbf{r} = 2d\hat{i}\ \mathbf{E} = \frac{47\rho d}{96\epsilon_0}\hat{i}$,

$$\vec{r} = \frac{1}{2}d\hat{i} + d\hat{j}\ \vec{E} = \frac{\rho d}{\epsilon_0}\left[\left(\frac{1}{2} - \frac{1}{30\sqrt{5}}\right)\hat{i} - \frac{1}{15\sqrt{5}}\hat{j}\right]$$

51. $\mathbf{E}_{bigsphere} = \frac{q\mathbf{r}}{4\pi\epsilon_0 R^3}$ and $\mathbf{E}_{smallsphere} = \frac{q\mathbf{r}'}{4\pi\epsilon_0 R^3}$

where $\mathbf{r}' = \mathbf{d} - \mathbf{r}\ \mathbf{E}_{total} = \mathbf{E}_{bigsphere} + \mathbf{E}_{smallsphere}$

$\therefore \mathbf{E}_{total} = \frac{q\mathbf{d}}{4\pi\epsilon_0 R^3} \therefore E = \frac{qd}{4\pi\epsilon_0 R^3}$

53. 1.1×10^5 C, 0 C, 1.1×10^5 C

55. $E = \frac{\lambda}{2\pi\epsilon_0 x}$, $E = 0$

57. $r < aE = \frac{q}{4\pi\epsilon_0 r^2}$, $a < r < b\ E = 0, r > b$

$E = \frac{q}{4\pi\epsilon_0 r^2}$, $\sigma_a = -\frac{q}{4\pi a^2}$, $\sigma_b = \frac{q}{4\pi b^2}$

59. $8.85 \times 10^{-10}\ \text{C/m}^2$

61. (a) $Q_{small,inner} = -Q_1$, $Q_{small,outer} = +Q_1$

(b) $E = \frac{Q_1}{4\pi\epsilon_0 r^2}$ (c) $Q_{large,inner} = -Q_1$,

$Q_{large,outer} = Q_1 + Q_2$ (d) $E = \frac{Q_1 + Q_2}{4\pi\epsilon_0 r^2}$

63. $\phi_{base} = \frac{q}{2\epsilon_0}$, $\phi_{curved} = \frac{q}{2\epsilon_0}$

65. (a) $\phi = \frac{q}{4\epsilon_0}$ (b) $\phi = \frac{3q}{4\epsilon_0}$

67. (a) $1.1 \times 10^{12}\ \text{N}\cdot\text{m}^2/\text{C}$ (b) $-1.1 \times 10^{12}\ \text{N}\cdot\text{m}^2/\text{C}$ (c) $1.7 \times 10^{12}\ \text{N}\cdot\text{m}^2/\text{C}$ (d) 0

69. $r \leq R'\ \ E = -\frac{r\lambda}{3\pi R^2\epsilon_0}$ where the minus sign indicates the field points towards the center of the sphere. $r \geq R'$

$E = -\dfrac{R'^3\lambda}{3\pi\epsilon_0 R^2 r^2}$ where again, the minus sign indicates the field points towards the center of the sphere.

71. $m = \dfrac{4\pi R^3 \rho\sigma}{6g\epsilon_0}$

73. $E = \dfrac{\lambda}{2\pi\epsilon_0 R}$, $\vec{E}_{total} = \dfrac{2\lambda}{5\pi\epsilon_o R}\vec{i}$

75. $r \le a$ $E = -\dfrac{\lambda}{2\pi\epsilon_0 r}$, $a \le r \le b$ $E = -\dfrac{\lambda}{2\pi\epsilon_0 r}\dfrac{b^2 - r^2}{b^2 - a^2}$, $r \ge b$ $E = 0$

77. 1.2×10^7 m/s, 2.3×10^{-13} J, 1.1×10^{-34} kg·m²/s, 6.6×10^{20} Hz

Chapter 25

1. 2.4×10^6 J

3. 60,000 V

5. 6.9×10^6 m/s

7. a) 2.1×10^6 m/s; b) 1.45×10^6 m/s

9. 2.7×10^7 m/s

11. 9.2×10^2 V

13. 2.05×10^5 m/s

15. 7.36×10^{-8} C

17. 3.86×10^{-12} J

19. 5.8×10^{-12} J

21. $\dfrac{Q}{a\pi\epsilon_0}\left(1 + \dfrac{2}{\sqrt{5}} + \dfrac{1}{3}\right)$

23. $\dfrac{\lambda\theta}{4\pi\epsilon_0}$

25. a) 5.5×10^{-12} J; b) 4.0×10^7 m/s

27. -23 V

29. $\dfrac{Q}{l}\dfrac{1}{4\pi\epsilon_0}\ln\left(\dfrac{2+\sqrt{3}}{2-\sqrt{3}}\right)$

31. $\dfrac{Q}{l}\dfrac{1}{4\pi\epsilon_0}\left[\ln\left(\dfrac{l+x}{x}\right) + \ln\left(\dfrac{l + \sqrt{l^2 + (x+l)^2}}{l+x}\right) + \ln\left(\dfrac{l + \sqrt{x^2 + l^2}}{x}\right) + \ln\left(\dfrac{x + l + \sqrt{l^2 + (x+l)^2}}{x + \sqrt{x^2 + l^2}}\right)\right]$

33. $\dfrac{\lambda}{2\pi\epsilon_0}\left[\dfrac{a^2\ln(b/a)}{b^2 - a^2} - \dfrac{1}{2}\right]$

35. 1.16×10^4 V/m for the wire and the same for the cylinder.

37. $V(r) = \dfrac{4k}{\epsilon_0\sqrt{r}}$ Volt

Take any point (x,y,z) on the surface of the sphere. Then $x^2 + y^2 + z^2 = R^2$

$$V = \frac{1}{4\pi\epsilon_0}\left[\frac{Q}{r_1} + \frac{-Q(R/h)}{2}\right]$$

r_1 and r_2 as shown

$$r_1 = \sqrt{x^2 + y^2 + (h - z)^2}$$

$$r_2 = \sqrt{x^2 + y^2 + \left(\frac{R^2}{h} - z\right)^2}$$

Therefore,

$$V(x,y,z) = \frac{Q}{4\pi\epsilon_0}\left[\frac{1}{\sqrt{x^2 + y^2(h-z)^2}} - \frac{R}{h}\frac{1}{\sqrt{x + y + \left(\frac{R^2}{h} - z\right)^2}}\right]$$

$$= \frac{Q}{4\pi\epsilon_0}\left[\frac{1}{(x^2 + y^2 + z^2 - 2zh + h^2)^{1/2}} = \frac{R}{h}\frac{1}{(x^2 + y^2 + z^2 - 2R^2z/h + R^4/h^2)^{1/2}}\right]$$

$$= \frac{Q}{4\pi\epsilon_0}\left[\frac{1}{(R^2 - 2zh + h^2)^{1/2}} - \frac{1}{h/r(R^2 - 2R^2z/h + R^4/h^2)^{1/2}}\right]$$

$$= \frac{Q}{4\pi\epsilon_0}\left[\frac{1}{(R^2 - 2zh + h^2)^{1/2}} - \frac{1}{(h^2 - 2zh + R^2)^{1/2}}\right] = 0$$

Therefore, potential constant is at 0 on entire surface.

39. (a) $r > R_2$, $V_a = 0$

(b) $R_1 < r < R_2$, $V_b = 0$

(c) $R < r < R_1$, $V_c = \dfrac{Q}{4\pi\epsilon_0 R_1} - \dfrac{Q}{4\pi\epsilon_0 r}$

(d) $R_0 < r < R$, $V_d = \dfrac{Q}{4\pi\epsilon_0 R_1} - \dfrac{Q}{4\pi\epsilon_0 r}$

41. $\dfrac{3Q}{4\pi\epsilon_0 r}$; $\dfrac{3Q}{4\pi\epsilon_0 c}$; $\dfrac{3Q}{4\pi\epsilon_0 c} + \dfrac{Q}{4\pi\epsilon_0 r} - \dfrac{Q}{4\pi\epsilon_0 b}$; $\dfrac{Q}{4\pi\epsilon_0}\left(\dfrac{3}{c} + \dfrac{1}{a} - \dfrac{1}{b} - \dfrac{r^2}{a^3} + \dfrac{1}{2a}\right)$

47. $E_x = -\dfrac{2\pi}{a}\sin\dfrac{2\pi x}{a}\cos\dfrac{2\pi y}{b}\cos\dfrac{2\pi z}{c}$;

$E_y = -\dfrac{2\pi}{b}\cos\dfrac{2\pi x}{a}\sin\dfrac{2\pi y}{b}\cos\dfrac{2\pi z}{c}$;

$$E_z = -\frac{2\pi}{c}\cos\frac{2\pi x}{a}\cos\frac{2\pi y}{b}\sin\frac{2\pi z}{c}$$

49. $E_x = \frac{p}{4\pi\epsilon_0}\frac{3xz}{(x^2+y^2+z^2)^{5/2}}$;

$E_y = \frac{p}{4\pi\epsilon_0}\frac{3yz}{(x^2+y^2+z^2)^{5/2}}$;

$E_z = \frac{p}{4\pi\epsilon_0}\left[\frac{3z^2}{(x^2+y^2+z^2)^{5/2}} - \frac{1}{(x^2+y^2+z^2)^{3/2}}\right]$;

On the z axis: $E_x = E_y = 0$, $E_z = \frac{p}{2\pi\epsilon_0 z^3}$; On the x axis:

$E_x = E_y = 0$, $E_z = -\frac{p}{2\pi\epsilon_0 x^3}$

51. a) $\frac{Q}{l}\frac{1}{4\pi\epsilon_0}\left[\ln\left(\frac{x + l/2 + \sqrt{R^2 + (x+l/2)^2}}{x - l/2 + \sqrt{(x+l/2)^2 + R^2}}\right)\right]$;

b) $-\frac{Q}{4\pi\epsilon_0}$

$$\frac{4l}{\sqrt{l^2 + 4lx + 4(R^2+x^2)}\left(-l + 2x + \sqrt{l^2 + 4lx + 4(R^2+x^2)}\right)}$$

53. -8.1×10^{-18} J

55. 6.3×10^{-4} J

57. 3×10^{-2} J

59. a) 4.4×10^{-8} J/m^3; b) 2.3×10^{11} J

61. 1.8×10^9 J/m^3

63. $\frac{Q^2}{4\pi\epsilon_0 d}(4 + \sqrt{2})$

65. 5.8×10^6 eV

67. -50 eV

69. $\frac{Q^2}{8\pi\epsilon_0}\left(\frac{7}{10R}\right)$

73. 8.6×10^5 eV

75. b) $U_{\text{grav}} = \frac{3}{5}\frac{\text{GM}^2}{\text{R}} = 1.24 \times 10^{29}\,J$

$\frac{U_{\text{grav}}}{U_{\text{rest mass}}} = \frac{(3/5)(GM^2/R)}{Mc^2} = \frac{3}{5}\frac{GM}{Rc^2} = 1.9 \times 10^{11}$

77. 4.1×10^{-7} J.

79. a) $\frac{\lambda^2}{8\pi^2\epsilon_0 r^2}$; b) $\frac{\lambda^2}{4\pi\epsilon_0}\left(\frac{1}{4} + \ln\frac{b}{a}\right)$

81. 9.96×10^{-2} m; 2.5×10^5 V

83. 0 V

85. 3.1×10^7 m/s

87. a) $\frac{2Q}{3\pi\epsilon_0 R^2}\left(\sqrt{R^2+x^2} - \sqrt{\frac{R^2}{4} + x^2}\right)$;

b) $-\frac{2Q}{3\pi\epsilon_0 R^2}\left(\frac{x}{\sqrt{R^2+x^2}} - \frac{x}{\sqrt{\frac{R^2}{4}+x^2}}\right)\mathbf{i}$

89. $V = \frac{Q^2}{4\pi\epsilon_0 l^2}\left[x\ln\frac{(x+2l)x}{(x+l)^2} + 2l\ln\frac{(x+2l)}{(x+l)}\right]$

91. $U_{\text{total}} = \frac{Q^2}{8\pi\epsilon_0 b(b^3 - a^3)^2}\left(\frac{6}{5}b^6 - 3a^3b^3 + \frac{9}{5}a^5 b\right)$

Chapter 26

1. $Q_2 = 3Q_1$, $V_1 = 3V_2$

3. $C = 1.1 \times 10^{-11}$ F, $Q = 1.1 \times 10^{-6}$ C

5. $C = 1.1 \times 10^{-9}$ F, $Q = 1.3 \times 10^{-8}$ C

7. $n = 4.5 \times 10^{14}$ electrons

9. $A_{min} = 5.6 \times 10^{-4}$ m^2, $A_{max} = 6.8 \times 10^{-3}$ m^2

11. $Q = 125$ C

13. $C = 1.8 \times 10^{-17}$ F, $Q = 1.8 \times 10^{-18}$ C

15. $C = 8.0 \times 10^{-12}$ F/m

17. $C = 9.9\ \mu$F

19. $Q = 1.7 \times 10^{-3}$ C

21. $C_{net} = \frac{2}{3}$ C

23. $Q = 12.0\ \mu$C, $\Delta V = 3.5$ V

25. $Q = 1.1 \times 10^{-5}$ C, $\Delta V_{2.5\mu F} = 4.5$ V, $\Delta V_{5.0\mu F} = 2.3$ V

27. The only arrangement of capacitors to give the same net capacitance is two pairs of two capacitors in series, connected in parallel.

29. $Q_1 = 3.0 \times 10^{-4}$ C, $Q_2 = 4.1 \times 10^{-4}$ C, $Q_3 = 4.8 \times 10^{-4}$ C, $\Delta V_{PP'} = 68$ V

31. $C = 4\pi\epsilon_0\kappa R$

33. $k = 5100$

35. $k = 1.7$

37. $C = (\kappa_1 + \kappa_2)\left(\frac{\varepsilon_0 A}{2d}\right)$

39. a) $E_{\text{free}} = \frac{2\Delta V}{d}\left(\frac{\kappa}{1+\kappa}\right)$, b) $E = \frac{2\Delta V}{d}\left(\frac{1}{1+\kappa}\right)$,

c) $\sigma_{\text{bound}} = \frac{2\epsilon_0\Delta V}{d}\left(\frac{1}{1+\kappa}\right)$

41. a) $C_0 = 2.0 \times 10^{-11}$ F, b) $C = 2.9 \times 10^{-11}$ F

43. $Q = 9.0 \times 10^{-6}$ C, $\Delta V_{filled} = 1.8$V, $\Delta V_{empty} = 4.5$ V

45. a) $\sigma_{bound,inside} = -2.6 \times 10^{-6}$ C/m^2,

$\sigma_{bound,outside} = 1.7 \times 10^{-6}$ C/m^2,

b) $E_{inner} = 1.6 \times 10^5$ V/m, $E_{outer} = 1.1 \times 10^5$ V/m,
c) $E_{free} = 3.0 \times 10^5$ V/m

47. $F = 0.050$ N

49. $C = 2\pi\epsilon_0\left(\frac{1}{R_1} - \frac{1}{R_2}\right)^{-1}(\kappa_{\text{top}} + \kappa_{\text{bottom}})$

51. a) $C = 1.6 \times 10^{-10}$ F, b) $\Delta V = 380$ V, c) $E = 7.5 \times 10^4$ V/m, d) $u = 0.025$ J/m^3, e) $U = 1.1 \times 10^{-5}$ J

53. $Q = 0.2$ C, $U = 2000$ J

55. $\Delta V = 2000$ V

57. $Q_{super} = 17$ C, $Q_{supply} = 0.33$ C (52 times more charge in the supercapacitor), $U_{super} = 21$ J, $U_{supply} = 66$ J (3 times more energy in the supply capacitor)

59. a) $C = 7.1 \times 10^{-10}$ F, $Q = 8.5 \times 10^{-9}$ C, $\Delta V = 12$V, $U = 5.1 \times 10^{-8}$ J,
b) $C = 2.1 \times 10^{-9}$ F, $Q = 2.5 \times 10^{-8}$ C, $\Delta V = 12$ V, $\Delta U = 1.0 \times 10^{-7}$ J,
c) $C = 7.1 \times 10^{-10}$ F, $Q = 2.5 \times 10^{-8}$ C, $\Delta V = 36$ V,
d) $C = 2.1 \times 10^{-9}$ F, $Q = 8.5 \times 10^{-9}$ C, $\Delta V = 4.0$ V

61. The solution is a proof.

63. $Q_1 = 3.5 \times 10^{-4}$ C, $Q_2 = 1.5 \times 10^{-4}$ C, $Q_3 = 2.0 \times 10^{-4}$ C, $U_1 = 3.1 \times 10^{-2}$ J, $U_2 = 1.9 \times 10^{-3}$ J, $U_3 = 2.5 \times 10^{-3}$ J

65. $F = 11$ N

67. $Q_1 = Q_2 = 9.6 \times 10^{-4}$ C, $Q_3 = 1.2 \times 10^{-3}$ C, $U_1 = 0.115$ J, $U_2 = 0.077$ J, $U_3 = 0.24$ J

69. a) $C_{net} = 8.0 \times 10^{-6}$ F, b) $C_{\text{net}} = 8.0 \times 10^{-7}$ F

71. $k = 2.0$

73. a) $U = 13$ J, b) $u = 4.8 \times 10^4$ J/m^3, $Volume = 2.7 \times 10^{-4}$ m^3

75. $C_2 = 8.0\ \mu$F, $U_1 = 7.2 \times 10^{-5}$ J, $U_2 = 3.6 \times 10^{-5}$ J

77. $\frac{C}{l} = 2\pi\epsilon_0\left(\frac{\kappa_1\kappa_2}{\kappa_2\ln\left(\frac{c}{a}\right) + \kappa_1\ln\left(\frac{b}{c}\right)}\right)$

79. $\Delta V = 4.7 \times 10^{-2}$ V

81. a) $F = \frac{Q^2}{2A\epsilon_0}$, b) $W = -\frac{Q^2}{2A\epsilon_0}\Delta d$, c) $\Delta U = \frac{Q^2}{2\epsilon_0 A}\Delta d$,
d) Compare the answers from parts a) and c).

Chapter 27

1. $Q = 1800$ C/h, $n = 1.1 \times 10^{22}$ electrons/h

3. $t = 1.2 \times 10^{-4}$ s

5. $I = 4.0$ A, $E = 6.0$ V/m

7. $I = 7.5$ A, $Q = 15$ C

9. $I_{(0.0s)} = 1.0$ A, $I_{(1.0s)} = 0.25$ A, $Q_{(2.0s)} = 0.67$ C

11. $\tau = 3.8 \times 10^{-14}$ s, $v_d = 0.054$ m/s

13. $R = 3.0\ \Omega$

15. $I = 1.3$ A, $n = 8.2 \times 10^{18}$ electrons/s

17. $R_{net} = R/32$

19. $R = 0.40\ \Omega$

21. $E = 0.069$ V/m

23. $R = 0.87\ \Omega$

25. $\rho = 5.7 \times 10^{-7}\ \Omega\cdot$m

27. $\Delta R = 0.92\ \Omega$

29. $j = 5.7 \times 10^6$ A/m^2, $E = 0.097$ V/m

31. $\sigma = 5.9 \times 10^7(\Omega\cdot\text{m})^{-1}$

33. $T = 22°$C

35. $\Delta V = 9.9$ V

37. $m_{Cu} = 380$ kg, $m_{Al} = 190$ kg

39. $d = 0.16$ cm

41. $I = 8.0 \times 10^{-5}$ A

43. $R = 4.4 \times 10^{10}\ \Omega$, $I = 6.8 \times 10^{-9}$ A

45. $R_{eq} = 2.2\ \Omega$, $I = 5.4$ A, $I_1 = 2.4$ A, $I_2 = 1.7$ A, $I_3 = 1.3$ A

47. $I_{iron} = 1.7$ A, $I_{brass} = 4.3$ A

49. In series: $R_{eq} = 9\ \Omega$, in parallel: $R_{eq} = 0.92\ \Omega$, in combinations of series and parallel: $R_{eq} = 3.7\ \Omega$, $4.3\ \Omega$, $5.2\ \Omega$, $1.6\ \Omega$, $2.0\ \Omega$, $2.2\ \Omega$

51. $I = 27$ A

53. a) $I_{Cu} = 210$ A, b) $I_{rubber} = 2.7 \times 10^{-19}$ A

55. $d = 1.5$ km from point A

57. $R = 3.2 \times 10^{-3}\ \Omega$

59. $R = 5.4 \times 10^{-2}\ \Omega$/m

61. a) $R_{net} = 4.4\ \Omega$, b) $I = 1.8$ A, c) $I_1 = 1.8$ A, $I_2 = 1.1$ A, $I_3 = 0.7$ A, $\Delta V_1 = 3.6$ V, $\Delta V_2 = \Delta V_3 = 4.4$ V

63. $\Delta V = 36$ V

65. $R = 2.73\ \Omega$

67. $Q = 1.4 \times 10^5$ C, $n = 9.0 \times 10^{23}$ electrons

69. $E = 2300$ V/m

71. $\frac{\Delta R}{R_0} = 0.089$ or 8.9%

73. $R = 5.2 \times 10^{-4}\ \Omega$, $\Delta V = 0.31$ V

75. Connecting two sets of two resistors in parallel gives $R_{net} = 1.0\ \Omega$

77. $I = 0.30$ A

79. $j = 3.3 \times 10^6$ A/m^2, $v_d = 2.4 \times 10^{-4}$ m/s

81. a) $R_{eq} = 1.3 \times 10^6\ \Omega$, b) $I_1 = 9.2\ \mu$A, harmless, $I_2 = 88\ \mu$A, harmless, $I_3 = 0.18$ A, fatal

Chapter 28

1. smallest battery: $3.3 \times 10^{-3}\ \frac{\text{kW}\cdot\text{h}}{\text{kg}} = 1.2 \times 10^4\ \frac{\text{J}}{\text{kg}}$;
largest battery: $1.9 \times 10^{-2}\ \frac{\text{kW}\cdot\text{h}}{\text{kg}} = 6.7 \times 10^4\ \frac{\text{J}}{\text{kg}}$

3. 6.9×10^6 J

5. 1.3×10^5 J
7. 0.38 A, 4.0×10^3 J
9. 6.0 A, $I_2 = I_1 = 3.0$ A
11. current through R_1 does not change, but it increases through the other two resistors
13. (a) 2.4×10^{-4} A; (b) 4.8×10^{-6} V
15. 62 Ω, 115 V, 4.8 V
17. (a) 1.7 A, (b) 0.86 A through resistors 2, 3, and 4
19. 0.49 V
21. 0.11 A
23. 31.0 A, out of the junction
25. 8.0 Ω
27. $I_1 = -15$ A, $I_2 = 25$ A, $I_3 = -5$ A, $I_4 = 15$ A, $I_5 = -20$ A, $R_5 = 4.0\ \Omega$
29. $I_1 = 1.25$ A, $I_2 = 1.75$ A, $I_3 = -0.50$ A
31. 16 V
33. The current through R_1 is 2.58 A, through R_2 is 1.71 A, through R_3 is 1.28 A, through R_4 is 2.15 A, through R_5 is 0.85 A. $V = 2.6$ V
35. 6.0×10^{-4} W
37. \$0.36
39. 3.7×10^{-3} A
41. 1100 W
43. 4.9 A
45. 6.3×10^{12} protons/s, 7.0×10^2 W
47. 6.7 h
49. 58 h
51. 1.2×10^6 W
53. 10%
55. $R_{115} = 13.23\ \Omega$, $R_{230} = 52.90\ \Omega$
57. 0.50 W
59. The solution is a proof.
61. (a) 3.3%, (b) 33%, lower current is more efficient
63. (a) 180 V, (b) 1.8×10^6 W
65. 19 liters/min,
67. $7.0 \times 10^2\ \Omega$
69. 3.958 A
71. $R_{Cu} = 5.0\ \Omega$, $R_{constantan} = 5.6\ \Omega$
73. 3.2×10^{-4} s
75. 4.0×10^{-4} s
77. 0.011 s
79. $Q = C\varepsilon\left(1 - e^{-t/(R_1+R_2)C}\right)$,

$$\Delta V = \frac{ER_1}{(R_1 + R_2)}\left(e^{-t/(R_1+R_2)C}\right)$$

81. (a) $A = E_2C$, $B = C(E_1 - E_2)$, $\tau = R_2C$, $Q(t) = E_2C + (E_1 - E_2)Ce^{-t/R_2C}$
(b) $A = E_1C$, $B = C(E_2 - E_1)$, $\tau = R_1CI$, $Q(t) = E_1C + (E_2 - E_1)Ce^{-t/R_1C}$
83. 0.020 s
85. 5.3 s, 5.0×10^{-6} A
87. (a) 1380 Ω, (b) 430 A, (c) 2670 A
89. 3.0×10^1 W, 1.7×10^1 W
91. $I_1 = \dfrac{ER_i'}{R_iR' + RR' - RR_i'}$, $I_2 = \dfrac{E}{R_i' + R\left(\dfrac{R'}{R_i} - 1\right)}$,
93. $I_1 = 0.45$ A, $I_2 = 1.31$ A, $I_3 = 0.85$ A, $P_1 = 5.4$ W, $P_2 = 14$ W
95. (a) 0.024 m, (b) 3.4×10^6 W
97. (a) 47 A, 7.1×10^3 m
99. 2.2%
101. (a) 7.6×10^{-5} C, (b) 1.2×10^{-4} C, (c) 6.0×10^{-2} A

Chapter 29

1. $|F| = 1.07 \times 10^{-16}$ N, opposite the direction of the current
3. $|F| = 1.04 \times 10^{-16}$ N, $a = 1.14 \times 10^{14}$ m/s^2
5. $|F| = 1.38 \times 10^{-24}$ N
7. $\theta = 19.5°$
9. $B = 4.18$ T, pointing downward
11. $B = 1.44 \times 10^{-5}$ T, $B_{earth} = 4.2\ B$
13. $F = 8.2 \times 10^{-16}$ N
15. $\mathbf{F} = (-2.98 \times 10^{-18}\,\mathbf{i} + 6.8 \times 10^{-19}\,\mathbf{j} + 3.33 \times 10^{-18}\,\mathbf{k})$ N
17. $F = 9.98 \times 10^{-18}$ N and points in direction 16° above the horizontal in North direction.
19. a) $I = \lambda v$,
b) $B = \dfrac{\mu_0 I}{2\pi r} = \dfrac{\mu_0\lambda v}{2\pi r} = \dfrac{\mu_0 v}{2\pi r}2\pi\epsilon_0 Er = \epsilon_0\mu_0 vE$
21. $B = 0.19$ T
23. $B = 4 \times 10^{-8}$ T, $F = 1.92 \times 10^{-18}$ N
25. $B = 5.03 \times 10^{-3}$ T
27. $\displaystyle\int B\,ds = \frac{\mu_0 I}{R}$
29. $a = 1.5 \times 10^{13}$ m/s^2
31. a) $B = 5 \times 10^{-6}$ T, b) $\theta = 16°$
33. $B = \dfrac{\mu_0}{2\pi}\dfrac{I}{d}\dfrac{3}{2}\sqrt{2}$
35. $B = \dfrac{\mu_0 I}{2R} - \dfrac{\mu_0 I}{2\pi R}$, pointing into the page.
37. $r < r_1 : B = \dfrac{\mu_0 Ir}{2\pi r_1^2}$, $r_1 < r < r_2 : B = \dfrac{\mu_0 I}{2\pi r}$, $r_2 < r < r_3$:
$B = \dfrac{\mu_0 I}{2\pi r}\dfrac{r_3^2 - r^2}{r_3^2 - r_2^2}$, $r < r_3 : B = 0$.
39. For $y > 0$, $z > 0$, $-\infty < x < +\infty : \mathbf{B} = 0$,
for $y < 0$, $z > 0$, $-\infty < x < +\infty : \mathbf{B} = \mu_0\sigma\,\mathbf{i}$,
for $y < 0$, $z < 0$, $-\infty < x < +\infty : \mathbf{B} = 0$,
for $y > 0$, $z < 0$, $-\infty < x < +\infty : \mathbf{B} = -\mu_0\sigma\,\mathbf{i}$

41. $B = \frac{\mu_0}{\pi}\frac{I}{b}\tan^{-1}\left(\frac{b}{2z}\right)$

43. For $z > 2R$, $B_1 = \frac{\mu_0 IR}{\pi(z^2 - R^2)}$, for $R < z < 2R$,

$$B_2 = \frac{\mu_0 I}{2\pi}\left(\frac{z-R}{R^2} - \frac{1}{z+R}\right), \text{ for } 0 < z < R,$$

$$B_3 = -\frac{\mu_0 I}{2\pi}\left(\frac{R-z}{R^2} + \frac{1}{z+R}\right), \text{ for } -R < z < 0,$$

$$B_4 = -\frac{\mu_0 I}{2\pi}\left(\frac{1}{R-z} + \frac{R+z}{R^2}\right), \text{ for } -2R < z < -R,$$

$$B_5 = \frac{\mu_0 I}{2\pi}\left(\frac{1}{R-z} - \frac{z+R}{R^2}\right), \text{ for } z < -2R,$$

$$B_6 = \frac{\mu_0 IR}{\pi(z^2 - R^2)},\ B_{max} = B_3(z=0) = \frac{\mu_0 I}{\pi R}$$

45. $I = 26.5$ A

47. $B = 1.26 \times 10^{-2}$ T

49. $B_{max} = 3.34$ T, $B_{min} = 1.43$ T

51. $B = \mu_0 nn'(r_2 - r_1)I$ *for* $r < r_1$;
$B = \mu_0 nn'(r_2 - r)I$ *for* $r_1 < r < r_2$

53. $B = \frac{4\mu_0 I}{\sqrt{2}\pi L}$

55. $B = 7.9 \times 10^{-5}$ T, directed into the paper

57. $B = \frac{\mu_0 I}{2\pi R} + \frac{\mu_0 I}{8R}$, directed into the paper

59. $B = 0.11\frac{\mu_0 I}{L}$, directed into the paper

61. a) $B_P = \frac{8\mu_0 I}{\sqrt{125}R}$,

b) At any point z, the fields produced by the coils are

$$B_{bottom} = \frac{\mu_0 I}{2}\frac{R^2}{(z^2 + R^2)^{3/2}}$$

$$B_{top} = \frac{\mu_0 I}{2}\frac{R^2}{[(R-z)^2 + R^2]^{3/2}}$$

Their first derivatives are

$$\frac{dB_{bottom}}{dz} = -\frac{3\mu_0 I}{2}\frac{zR^2}{(z^2 + R^2)^{5/2}}$$

$$\frac{dB_{top}}{dz} = \frac{3\mu_0 I}{2}\frac{(R-z)R^2}{[(R-z)^2 + R^2]^{5/2}}$$

Both of these derivatives cancel each other at $z = R/2$.
The second derivatives are

$$\frac{d^2B_{bottom}}{dz^2} = -\frac{3\mu_0 IR^2(R^2 - 4z^2)}{2(z^2 + R^2)^{7/2}}$$

$$\frac{d^2B_{top}}{dz^2} = \frac{3\mu_0 IR^2\left[R^2 - 4(R-z)^2\right]}{2\left[(R-z)^2 + R^2\right]^{7/2}}$$

Both of these derivatives cancel each other at $z = R/2$.

63. $F = 4.8 \times 10^{-17}$ N, $a = 2.87 \times 10^{10}$ m/s^2, both directed opposite the current.

65. For currents in the same direction, $B = 7.33 \times 10^{-5}$ T, for currents in the oposite direction, $B = 1.05 \times 10^{-5}$ T.

67. $|\mathbf{B}| = 1.05\frac{\mu_0 I}{R}$, the angle between **B** and the straight wire is $\theta = 17.7°$

69. For $r < R_1$, $B = 2\mu_0 nI$; for $R_1 < r < R_2$, $B = \mu_0 NI$; for $r > R_2$, $B = 0$.

71. $B = \frac{\mu_0 nI}{2\pi}$, directed in a plane parallel to the plane of the wires and perpendicular to the current, with the exception along the edges.

73. $B = \frac{\mu_0 I\sqrt{5}}{2\pi L}$, directed into the paper.

75. $B = \frac{\mu_0 I}{2R}\left(\frac{1}{\pi} + \frac{3}{8}\right)$, directed out of the paper.

Chapter 30

1. $p = 1.1 \times 10^{-17}$ kg·m/s

3. $p = 3.4 \times 10^{-17}$ kg·m/s

5. $B = 3.3$ T

7. $p = 3.8 \times 10^{-19}$ kg·m/s

9. $B = 0.036$ T

11. $\frac{r_{12}}{r_{14}} = 1$

13. $I = 0.39$ A

15. (a) The electron will follow a circular path that spirals in the direction of the magnetic field;
(b) $f = 1.4 \times 10^7$ Hz, $T = 7.2 \times 10^{-8}$ s;
(c) $x = 0.29$ m.

17. $B = 3.2$ T

19. (a) $v_p = 1.6 \times 10^3$ m/s; (b) $\theta_p = 3.3 \times 10^{-3}$ rad

21. $F = 6.7 \times 10^{-5}$ N

23. Consider an element **dl** of the long, straight wire. As the current I flows through the loop, the magnetic field **B** produced by the loop exerts a magnetic force **F** on the element d**l** given by d**F** $= I\,d$**l** ? **B**. This force is perpendicular to element d**l**. Applying Newton's third law of motion, the element d**l** exerts a force that is directed oppositely to d**F**, which is also perpendicular to element d**l**.

25. **F** $= -13$ N **j**

27. $I = 1.0 \times 10^6$ A

29. $\tau = 2.4 \times 10^{-7}$ N·m

31. (a) $\tau = (1.1 \times 10^{-26} \sin\theta)$ N·m, sinusoidal behavior;
(b) $W = 2.25 \times 10^{-26}$ J

33. $\mu = 0.023\ \text{A}\cdot\text{m}^2$

35. $\mu = \frac{1}{4}Q\omega R^2$ The magnetic and electric fields surrounding the spinning paper disk are illustrated as follows:

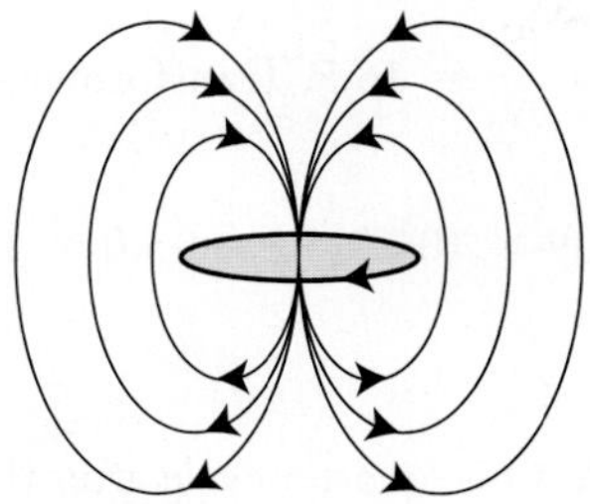

37. (a) $\tau = (1.71 \times 10^{-7} \sin\theta)\ \text{N}\cdot\text{m}$; (b) $f = 0.47$ Hz

39. (a) $M = 1.7 \times 10^6$ A/m; (b) $\frac{\mu}{V} = 2.0 \times 10^{-23}\ \text{A}\cdot\text{m}^2$

41. (a) $\chi = -1$; (b) $\mu = 0$

43. $\Delta V_H = 3.1\ \mu\text{V}$

45. $B = 2.5 \times 10^{-2}$ T

47. (a) $v_d = 2.7 \times 10^{-3}$ m/s; (b) $\Delta V_H = 1.6 \times 10^{-5}$ V

49. $R_H = -4.7 \times 10^{-10}\ \text{m}^3/\text{C}$

51. (a) $F = 0$ N; (b) $F = 0.18$ N due north; (c) $F = 0.18$ N due east; (b) $F = 0.16$ N due east.

53. $B = 0.010$ T, $f = 1.6 \times 10^5$ Hz

55. (a) $y = \frac{1}{2}a_y t^2 = \frac{1}{2}\left(\frac{F}{m}\right)\left(\frac{L}{v_{0x}}\right)^2 = \frac{eEL^2}{2mv^2}$,

(b) $y = \frac{eEL^2}{2mv}$, therefore $\frac{e}{m} = \frac{2yE}{B^2L^2}$

57. $B = \frac{4mv}{qd}$

59. $a = 3.1 \times 10^3\ \text{rad/s}^2$

Chapter 31

1. 3.0×10^{-3} V; left side is positive and right side is negative

3. 0.5 m/s

5. 2.2×10^{-3} V

7. $v = 0.48$ m/s, current flows counterclockwise, $F = 10$ N to the left

9. 9.8×10^{-3} V

11. $-5.5 \times 10^{-5}\ \text{T}\cdot\text{m}^2$

13. 22 rev/sec

15. 9.7×10^{-3} V; patient does not need to be pushed more slowly

17. $|E| = 6.88$ V, $I = 0.606$ A, $W = 146$ J

19. 0.047 V

21. $3.5 \times 10^{-7}\ \text{T}\cdot\text{m}^2$

23. (a) 6.6×10^{-6} V/m; (b) 7.4×10^{-6} V/m

25. $I = 0.565$ A, $\tau = 5.59 \times 10^{-3}\ \text{N}\cdot\text{m}$

27. 3.0×10^2 A

29. 0.63 A

31. At $r = 0.80$ m: $E_0 = 140$ V/m $\varepsilon_0 = 704$ V; at $r = 1.5$ m: $E_0 = 117$ V/m $\varepsilon_0 = 1100$ V

33. 0.010 C

35. 0.40 H

37. (a) 0.5 H; (b) 10 V

39. 1.1×10^4 A/s

41. -1.9 V

43. $M = 200\mu_0 n\pi R^2$; the shape of the coil wire does not matter

45. 1.1×10^{-7} H

49. 1.2×10^{-5} J

51. $1.0\ \text{J/m}^3$

53.

B-field (T)	u (J/m^3)
10^8	4.0×10^{21}
10^3	4.0×10^{11}
45	8.1×10^8
8	2.5×10^7
2	1.6×10^6
1.5	9.0×10^5

55. $U = 5.29 \times 10^4$ J; $V = 1.64 \times 10^{-3}\ \text{m}^3$

57. 211 J/km

59. 1.0×10^{-3} J

61. $IR + L\frac{dI}{dt} = 0$; $I = \frac{E}{R}e^{-t/\tau}$ satisfies the differential equation if $= L/R$

63. (a) $3.75\ \Omega$; (b) 1.58×10^4 W; (c) 9.51×10^4 J

65. (a) 2.0 A; (b) -3.0×10^3 V

67. (a) $\frac{L}{R_2}$; (b) $\frac{L}{R_1 + R_2}$

69. (a) $\frac{dI_1}{dt} = 0.75$ A/s, $\frac{dI_2}{dt} = 1.5$ A/s; (b) $I_{Resistor} = 0.50$ A, $I_{L_1} = 0.17$ A, $I_{L_2} = 0.33$ A

71. $R = 0.125\ \Omega$, $L = 0.0396$ H

73. 2.4×10^{-4} V

75. $0.014\ \text{J/m}^3$

77. 7.5×10^{-5} V

79. (a) 2.6×10^{-3} V; (b) 0; (c) $9.2 \times 10^{-15}\ \text{C/m}^2$

81. (a) 0.0188 T/s; (b) 6.4×10^{-3} V; (c) 6.4×10^{-3} V

83. 102 V

85. (a) $I_{max} = 48$ A, $t = 14$ s; (b) $U_{max} = 576$ J, percentage of energy that remains = 25%

Chapter 32

1. (a) $I_{rms} = 10.4$ A, $I_{max} = 14.8$ A; (b) $P_{max} = 2400$ W, $P_{min} = 0$ W

3. (a) $\mathcal{E}_{max} = 3.25 \times 10^5$ V, $I_{max} = 1.05 \times 10^3$ A;
(b) $P_{max} = 3.4 \times 10^8$ W, $P_{ave} = 1.7 \times 10^8$ W

5. $I_{rms} = 10.4$ A, $I_{max} = 14.8$ A, $R = 11.1\ \Omega$

7. $I_{rms} = 31.1$ A, $P_{ave} = 48.5$ kW

9. 0.032 A, 0.064 A

11. $2.0 \times 10^6\ \Omega$, $1.5 \times 10^6\ \Omega$

13. (a) 208 Ω; (b) 9.6×10^{-4} A;
(c) $I(t = 0) = 0$ A, $I(t = 4\pi/\omega) = -6.8 \times 10^{-4}$ A

15. 1.0×10^{-9} F

17. For an amplitude of 1.00 V:

t (s)	$0.0012\cos^2(600t)$ (W)
0.001	8.17E-04
0.002	1.58E-04
0.003	6.19E-05
0.004	6.52E-04

19. 5.6×10^{-5} F

21. 1.0×10^5 A/s

23. $f = 1.6 \times 10^3$ Hz

25. 0.033 A, 0.017 A

27. 0.049 A

29. 0.343 A/s, 0.00218 Hz

31. 0.018 Ω at 60 Hz, $3.0 \times 10^4\ \Omega$ at 100 MHz

33. 3.6×10^3 Hz

35. (a) $I_{max} = 5.0 \times 10^{-6}$ A,
(b) $C = 1.66 \times 10^{-6}$ F, $L = 6.6 \times 10^{-2}$ H

37. 9.4×10^3 rad/s

39. 379 Hz

41. 663 Hz

43. 2.11 Hz, 259 H

45. 2.76×10^{-8} F, 7.5×10^{-8} H

47. 8.23×10^{-6} F, 6.25 W

49. (a) $\overline{P} = \frac{1}{2} I_{max}\mathcal{E}_{max}\left(\frac{R}{Z}\right)$

(b) From the impedance triangle, $\cos\phi = \frac{R}{Z}$

(c) $\cos\phi = 0.642$, $\phi = \pm 50°$. No.

51. (a) 1.9×10^3 V; (b) 12 V; (c) 24 W; (d) 1.6×10^2

53. $0.995\mathcal{E}_{max}$; $\mathcal{E}_{max}/10$; $\omega = \frac{R}{L}$

55. I_C

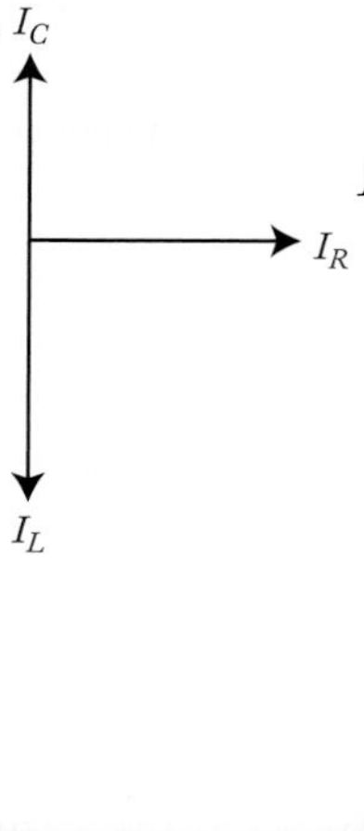

$$I_{max} = \sqrt{I_R^2 + (I_C - I_L)^2} = \sqrt{\left(\frac{\mathcal{E}_{max}}{R}\right)^2 + \left(\frac{\mathcal{E}_{max}}{X_C} - \frac{\mathcal{E}_{max}}{X_L}\right)^2}$$

57. 4348 turns

59. Transformer 1: $\frac{N_1}{N_2} = 0.044$, Transformer 2: $\frac{N_1}{N_2} = 7.6$,
Transformer 3: $\frac{N_1}{N_2} = 16.5$, Transformer 4: $\frac{N_1}{N_2} = 34.8$

61. $I_{generator} = 9.09 \times 10^4$ A, $I_{line} = 5.0 \times 10^3$ A

63. 36 W

65. 0.139 A

67. $I_{max} = 1.63$ A. One-quarter cycle after the maximum $I = 0$. One-half cycle after maximum $I = -1.63$ A, three-quarters cycle after maximum $I = 0$.

69. (a) $I = (2.4 + 0.43\cos 360t)$A; (b) 29.3 W

71. (a) $Q = 6.5 \times 10^{-8} \sin(120\,\pi t)$ F; (b) First max. at $t = 1/240$ sec, first min. at $t = 0$; (c) 5.9×10^{-8} J, 2.95×10^{-8} J

73. $I_{max} = 0.89$ A. One quarter cycle later, $I = 0$. One half cycle later, $I = -0.89$ A. Three-quarters of a cycle later, $I = 0$.

75. (a) $I_R(t) = 1.5 \times 10^{-3}\ \text{A} + (7.5 \times 10^{-4}\ \text{A})\cos(6000\pi t)$,
$I_L(t) = 60t + (1.6 \times 10^{-3})\sin(6000\pi t)$ A.
(b) $P_R = 4.5 \times 10^{-3} + 1.1 \times 10^{-3}\cos^2(6000\pi t)$ W,
$P_L = 180t + 90t\cos(6000\pi t) + (4.8 \times 10^{-3})\sin(6000\pi t)$
$+ (2.4 \times 10^{-3}\ \sin(6000\pi t\ \cos(6000\pi t)$ W

77. 1.4×10^{-3} J; 7.9×10^{-4} s for fully magnetic energy; 1.6×10^{-3} s for fully electric energy.

79. (a) 1.36×10^3 Hz, (b) 1.2×10^3 V, (c) 78

81. 0.026 A at the power plant, 0.0012 A in the transmission line

83. $\mathcal{E}_2 = 15\sin(3000\pi t)$ V

Chapter 33

1. (c) $I_{disp} = \varepsilon_0 \frac{d\Phi_E}{dt} = \frac{dQ}{dt}$

3. (a) 4.0 A; (b) 4.5×10^{11} V·m/s

5. (a) 0.10 A; (b) 6.3×10^{-3} A

9. (a) 2.0×10^{-3} A; (b) 1.0 sec; (c) 8.9×10^{-4} A,
8.9×10^{-9} T, 1.8×10^{-8} T, 2.7×10^{-8} T

11. (a) $E = \frac{\rho C t}{\pi R^2}$; (b) $B = \frac{\mu_0\epsilon_0\rho C r}{2\pi R^2}$; (c) $B = \frac{\mu_0\epsilon_0\rho C}{2\pi r}$

13. $\oint \mathbf{E}\cdot d\mathbf{A} = \frac{Q}{\kappa\epsilon_0}$, $\oint \mathbf{B}\cdot d\mathbf{A} = 0$, $-\oint \mathbf{E}\cdot d\mathbf{s} = -\frac{d\Phi_B}{dt}$,
$\oint \mathbf{B}\cdot d\mathbf{s} = \mu_0\left(I + \kappa\varepsilon_0\frac{d\Phi_E}{dt}\right)$

15. **E** is $\parallel$ to acceleration. **B** is $\perp$ to acceleration and propagation.

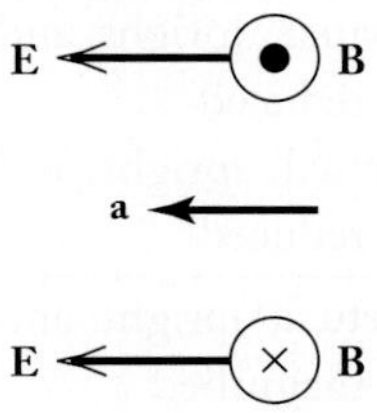

17.

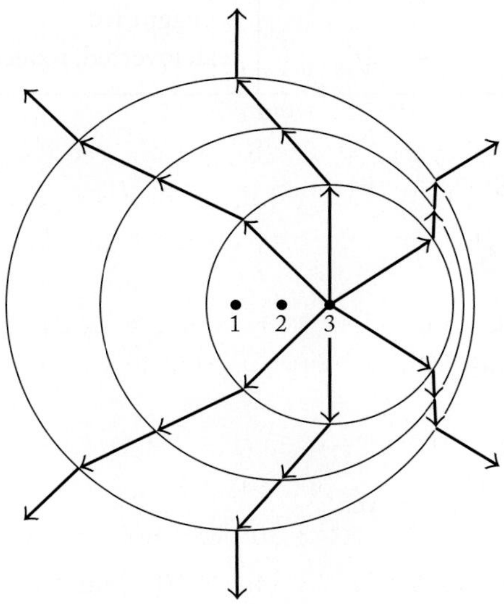

19. west

21. 1.1×10^{-5} A/m

23. 3.3×10^{-11} s

25. 5.00×10^{3} s (8.33 min)

27. 3.3×10^{-9} s, 0.30 GHz, no

29. North, 2.0×10^{-9} T

31. 1/4

33. 1.5%

35. 39°

37. (b) $\theta = 0°$, $\theta = 90°$, $\theta = 180°$, $\theta = 270°$; The field rotates clockwise with a period of $2\pi/\omega$

39. $$\frac{I_0 \cos^2\phi}{2(\sin^2\phi \cos^2\alpha + \cos^2\phi)}$$

41. 3.0×10^{11} Hz

43. AM: 566 m to 187 m; FM: 3.41 m to 2.78 m

45. (a) 1.5×10^{-10} m (X-ray); (b) 1.0 cm (microwave radio); (c) 5.0×10^{6} m ("electric" wave)

47. 0.027 m

49. 2.8×10^{-8} J/m^3

51. 9.5×10^{-6} W/m^2

53. 1.7×10^{-9} T down; 6.6×10^{-4} W/m^2 north

55. 2.8×10^{28} W

57. 4.0×10^{26} W

59. 6.42×10^{-5} W/m^2, 2.02×10^{4} W

61. 0.065 V/m, 0.043 V/m

63. 6.37×10^{3} W/m^2, 2.19×10^{3} V/m, 1.67×10^{-10} J

65. 1.49 W/m^2, 7.81×10^{-11} N

67. 8.0×10^{3} m^2

69. 51

71. 6.0×10^{8} N, 3.5×10^{22} N

73. (a) 1.77×10^{-17} J/m^3 each; (b) 4.0×10^{-3} V/m, and 0 T, or 0 V/m and 1.33×10^{-11} T; (c) one is 7.08×10^{-17} J/m^3, the other is zero.

75. 80.0 m, 0.0785 m^{-1}, 2.36×10^{7} s^{-1}, minus z direction, $E_y = cB_0 \cos(kx + \omega t)$

79. $V_0 \sin\omega t/R$, $\varepsilon_0 \frac{A}{d}\omega V_0 \cos\omega t$, $\frac{V_0 \sin\omega t}{R} + \varepsilon_0 \frac{A}{d} V_0\, \omega\cos\omega t$, $\frac{\mu_0}{2\pi}\left(\frac{V_0 \sin\omega t}{rR} + \frac{\varepsilon_0 \pi r}{d} V_0\, \omega\cos\omega t\right)$

81. 0.067 sec

83. 0.31 W/m^2

85. 5.00×10^{-7} T north

87. 9.60×10^{3} V/m, 3.2×10^{-5} T

89. 5.5×10^{-2} V/m, 1.8×10^{-10} T

91. 2.9×10^{-20} W, 3.6×10^{27} W

93. (a) 3.4×10^{-15} W; (b) 5.9×10^{17} m = 63 ly, 3600 stars

Chapter 34

1. 40

3. $90° - \theta$

5. 20°

7. 7

9. $H = 0.90$ m; $W = 0.31$ m

11. 4.74×10^{14} Hz, 416 nm, 1.97×10^{8} m/s

13. 1.09

15. 50°

17. 39.014°

19. Fused Quartz

21. 41.7°

23. 42°

25. 24.4°

27. 55.6°, 56.1°, 57.7°

29. 64.5°

31. 41.1°, 61.0°

33. 1.000 21

35. 36.9°

37. 0.77 mm

39. 53.1°

41. $1 \leq n \leq \sqrt{2}$

43. 48.8°, no dependence on n'

45. 1.4002, 40.25%.

47. 137°29′, 139°14′

49. −10.9 cm

51. $s > \frac{1}{2}R$, $s < \frac{1}{2}R$

53. −2.5 cm

55. 16.7 cm

57. 120.0 cm, concave

59. 60.0 cm, 30.0 cm

61. 8.3 mm

63. 21 cm

65. 12 cm

67. −8.6 cm, virtual, upright, smaller, 0.57

69. 14 cm, 14

71. 45 cm, inverted, enlarged

73. The solution is a proof.

75. −60 cm, −21 cm

77. The solution is a proof.

79. 475 cm behind the lens (also behind the mirror)

81. The solution is a proof.

83. 3.9 cm to the right of the center of the ball

85. 25.8 cm to infinity

87. 8.9×10^{-2} s

89. −22 cm, diverging lens, −0.72 cm

91. The solution is a proof.

93. 1/64

95. 1340

97. The solution is a proof.

99. 31°, 50°

101. 56°

103. 7.5 cm

105. $\alpha \geq 38.7°$

107.

mirror	object distance	image characteristics
concave	$s < f$	virtual, upright, and magnified
	$f < s < 2f$	real, inverted, and magnified
	$s > 2f$	real, inverted, and reduced
convex	all	virtual, upright, and reduced

109. −1.7 m, 0.24 m

111.

lens	object distance	image characteristics
concave	$s < f$	virtual, upright, and reduced
	$s > f$	virtual, upright, and reduced
convex	$s < f$	virtual, upright, and magnified
	$f < s < 2f$	real, inverted, magnified
	$s > 2f$	real, inverted, reduced

113. −30 cm

115. 1.9 mm, 108

Chapter 35

1. 3.2×10^8 m/s

3. 127 nm for both

5. 244 m

7. 78 nm

9. $2d = \frac{\lambda}{n_2}, \frac{2\lambda}{n_2}, \frac{3\lambda}{n_2}, \ldots$; yes

11. $2d = \frac{1}{2}\lambda, \frac{3}{2}\lambda, \frac{5}{2}\lambda, \ldots$

13. (a) Only one reflected ray suffers a phase reversal; (c) 1.22 mm

15. 0.257 mm

17. 5.76 mm/s

19. 1.000277

21. 7.4 mm

23. 1.9 mm

25. 0.994

27. measured angles: 5.5°, 20.5°, 35.5°, 51.5°; predicted angles: 6.1°, 18.6°, 32.2°, 48.2°

29. 0, ± 0.0074°, ± 0.0149°, ± 0.0223°, . . .

31.

θ (degrees)	θ (radians)	I/I_{max}
1.0	0.0175	0.626
2.0	0.0349	0.064
3.0	0.0524	0.154

33. 5.4×10^{-7} m

39. 19.0°, 40.5°, 77.2°

41. First order: $\theta_{400} = 13.7°$, $\theta_{700} = 24.4°$;
Second order: $\theta_{400} = 28.2°$, $\theta_{700} = 55.7°$
Third order: $\theta_{400} = 45.1°$; no third order max for 700 nm
Second and third orders overlap

43. 1.38×10^{-5} rad

45. 0.34 mm

47.

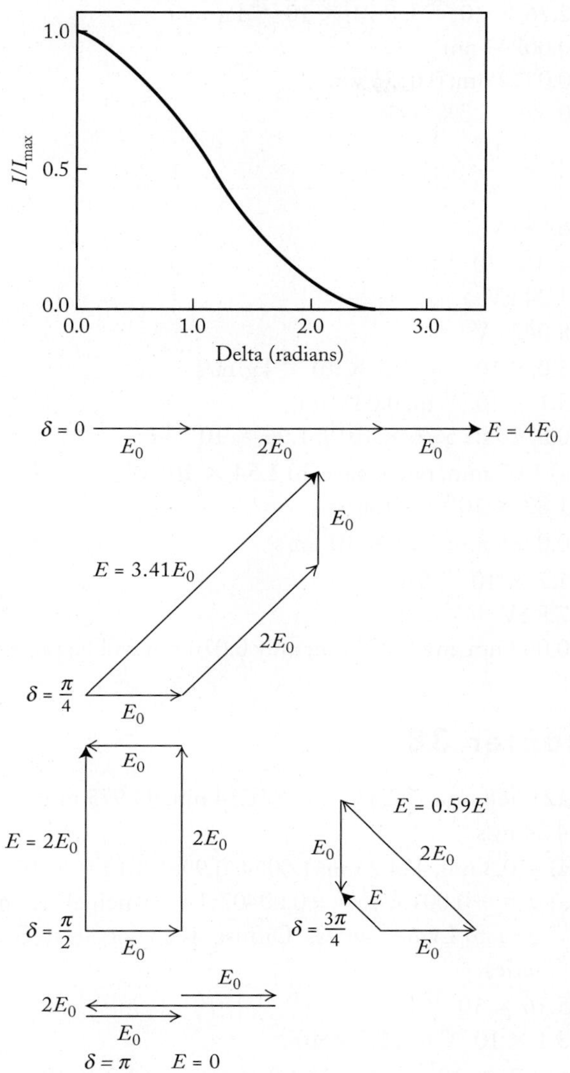

49. (b) $\phi = 0°$; (c) $\theta = \phi$
51. 2.5 cm
53. 0.042 mm
55. 47°
57. (a) not resolved; (b) resolved; (c) colors smeared, not resolved
59. $0.045I_{max}$
61. (b) m interference maxima between adjacent diffraction minima
63. $0.952(3\pi)$
65. about 0.22 mm in diameter
67. 94°
69. about 0.06 arcsec, about eight times better than from Earth
71. (a) 6.7×10^{-5} rad; (b) about 1 cm in diameter
73. minimum separation $\approx 6 \times 10^7$ m; it is possible to resolve the moons but they can only be seen as points of light
75. (a) 9.6×10^{-5} radian; (b) tremor is 9.7×10^{-5} radian to 1.5×10^{-4} radian, eliminating tremor would improve resolution somewhat
77. (a) 2.24×10^{-6} radian; (b) 0.34 m; (c) 0.0054 mm
79. cat barely distinguishes the mice
81. film is at least 120 nm thick
83. 454 nm
85. ±14.5, ±48.6°
87. (b) 95.9 MHz
89. 13 km
91. 4.2 cm, 2.1 cm
93. 1.63×10^{-3} radian; 5.7 km
95. (a) 1.95 km; (b) 5.02×10^{-4} W/m²

Chapter 36

1. (a) 0, 60 km/s; (b) 30 km/s, 90 km/s
3. $$\Delta t_1 = \frac{2}{c}\left[(L_1 - L_2) + \left(L_1 - \frac{L_2}{2}\right)\left(\frac{V}{c}\right)^2\right];$$
$$\Delta t_2 = \frac{2}{c}\left[(L_2 - L_1) + \left(L_2 - \frac{L_1}{2}\right)\left(\frac{V}{c}\right)^2\right];$$
$$\Delta t_1 - \Delta t_2 = \frac{2(L_1 - L_2)}{c}\left[2 - \frac{3}{2}\left(\frac{V}{c}\right)^2\right]$$
5. $0.995c$
7. $0.866c$ (2.60×10^8 m/s)
9. 0.16 s.
11. $0.447c$
13. $4.8 \times 10^{-3}c$ (1.4×10^6 m/s)
15. $\frac{f'}{f} = 0.58; 0.42; 0.23$
17. 366 m/s; time dilation factor $= 1 + 7.4 \times 10^{-13}$
19. time dilation factor $= 1 + 1.17 \times 10^{-12}$; the clock at the North Pole will be ahead by 3.7×10^{-5} s in one year.
21. 0.866 m
23. 0.19 m
25. $0.866c$
27. $0.44c$
29. 3.3×10^{-6} C/m³
31. Relative to the upper part of the belt, the lower belt segment is moving with a speed greater than V and appears shortened even more than the base. Therefore the belt is tightened in both reference frames.
35. $0.65c$
37. (a) $x = 7.67 \times 10^{16}$ m, $y = 1.2 \times 10^{17}$ m, $t = -1.56 \times 10^8$ s; (b) The light from the nova will arrive at earth before the radio message.

45. $K_{rel} = 5.04 \times 10^{-3}mc^2$, $K_N = 5.00 \times 10^{-3}mc^2$, deviation = 0.8%
47. $0.85c$
49. $0.94c$
53. 2.7 m/s; 0.016 m/s; 1.3×10^4 m/s.
55. 1.06×10^{-17} kg·m/s
61. 1.02×10^6 eV
63. $\dfrac{m_{thermal}}{m_{sun}} = 1.1 \times 10^{-4}\%$
65. 1.26×10^{-13} J, 1.5 times the rest mass energy of the electron.
67. $0.828c$
69. 9.33 light years from earth, message arrives 21 years after departure
71. (a) To the ship observer, the 600 nm pulse is emitted 2.60 μs before the 400 nm pulse; (b) 133 nm instead of 400 nm, 1800 nm instead of 600 nm; (c) 1.2 km
73. Dimensions will be 1.0 m × 1.0 m × 0.8 m. The area of the two faces perpendicular to the direction of motion will be 1.0 m^2. The four faces whose planes are parallel to the direction of motion will have area = 0.80 m^2. The volume of the cube will be 0.80 m^3.
75. The deviation is 2%.
77. $0.98c$ if forward, $0.54c$ if backward
79. 6.96×10^{20} J; 7.74 metric tons
81. 1.2×10^{17} J; 2.9×10^7 tons of TNT
83. 0.33 m/s
85. (a) $0.906c$; (b) $|v_{muon}| = 0.98c$, $|v_{antimuon}| = 0.67c$

Chapter 37

1. 3.3×10^{-34} J; no
3. 5.3×10^{-22} J
5. 0.0259 nm; 0.0366 nm
9. 1.60×10^{-6} m
11. 5.8×10^{-13} m
13. 43 W; 0.2 W; liquid nitrogen
15. 6.1×10^{-9} K
17. 1.38×10^3 W/m^2
21. 44 K
23. a) 7.16×10^{15} Hz; b) 0.07 nm
25. 7×10^{24} photons
27. 1.9×10^{31} photons/s
29. 3.2×10^{18} photons/s
31. 1.77 eV; 3.10 eV
33. 0.29 eV; 0.60 V
35. 1.86 eV
37. Red light (700 nm): None; Blue light (400 nm): K; UV light (280 nm): K, Cr, Zn
39. 6.8×10^{-34} J·s
41. 2.76×10^{-15} J; 9.21×10^{-24} kg·m/s
43. 0.00243 nm
45. 0.0312 nm; 0.0336 nm
47. 0.37%; 3.3%; 49%
49. $E_c = \dfrac{2h^2}{m_c^2}$ 0.016 nm
51. 62.5 eV
53. 4.0×10^{-15} J
55. 1.24 kV
57. 8.042 kV
59. 5.0×10^{-6} m; 1.3×10^{-28} kg·m/s
61. 3.1×10^{-11} m; 0.031 nm
63. 0.872 Hz; 5.76×10^{34}j; 1.96×10^{-16} m
65. a) 1.07 mm, microwave; b) 1.54×10^9 W
67. 1.82×10^{20} photons/s
69. 0.0486 nm; 2.12×10^7 m/s
71. 1.7×10^{-8} V/m
73. 7.8 eV
75. 0.062 nm and 0.070 nm; only 0.070 nm will be present

Chapter 38

1. 121.568 nm, 102.573 nm, 97.254 nm, 94.975 nm
3. 4.34 m/s
5. a) 410.3 nm, 434.2 nm; 1.0034; 0.9966; b) 1.0×10^6 m/s
7. a) $d = -0.0013$; $p = -0.00407$; b) "Principal" is analogous to Lyman series, "Diffuse" is analogous to Balmer series.
9. 5.16×10^{-12} J
11. 3.1×10^{-26} m^2; 1.3×10^{-3}
13. a) 1.2×10^{18} atoms ; b) 1.2×10^{-4}; c) 1.2×10^6
15. $4.3 \times 10^{32}\ \hbar$
17. 2.2×10^6 m/s $= 0.0073c$
19. 0.0283 mm; -2.5×10^{-5} eV
21. 10.2 eV
23. From the $n = 3$ to the $n = 1$ state.
25. 7.2×10^{-3}
27. $n = 3$, the first two lines of the Lyman series and the first line of the Balmer series.
29. $E = \dfrac{n\hbar e}{\sqrt{4\pi\varepsilon_0 m_e R^3}} - \dfrac{e^2}{4\pi\varepsilon_0}\left(\dfrac{3}{2R}\right)$;

$$f = \frac{(n_2 - n_1)}{2\pi}\left(\frac{e}{\sqrt{4\pi\varepsilon_0 m_e R^3}}\right);\ R = 4a_0$$

33. 4.98×10^{-6} eV
35. $2\pi a_0$; $4\pi a_0$
37. 0.00614 nm; 1.1×10^{-5}

39. If $\lambda_2/\lambda_1 = 4$, then $E_2/E_1 = 1/16$
41. 5.8 mm
43. 2.7×10^{-20} m/s
45. $\frac{2}{L}, 0, \frac{2}{L}$
47. $n + 1$ locations where the probability per unit length = 0, at $x = \frac{mL}{n}$, $m = 0,1,2,\dots n$; n locations where the probability per unit length = 1, at $x = \frac{(2m-1)L}{2n}$, $m = 1,2,\dots n$
49. 0.091
51. 0.0014 nm
55. a) $\frac{2L}{n}$, $n = 1,2,3,\dots$; b) $\frac{nh}{2L}$; c) $\frac{n^2h^2}{8mL^2}$; d) 37.7 eV, 151 eV, 339 eV, 603 eV
57. 91.176 nm; 820.58 nm; 1458.8 nm
59. 122 nm
61. 0.212 nm; 1.10×10^6 m/s; 2.12×10^{-34} J·s; 5.71×10^{21} m/s^2
63. 0.0454
65. 1.67×10^{-14} m; 4.87×10^6 m/s
67. 12.8 eV; 97.3 nm; 0.661 eV; 1.88×10^3 nm
69. $r_n = \frac{n^2\hbar^2}{Gm^2M}$; $E_n = -\frac{(GM)^2m^3}{2n^2\hbar^2}$; $n = 323$; -5.9×10^{-17}eV
71. 5.3×10^{-23} kg·m/s; 1.1×10^{-21} m/s

Chapter 39

1. $n = 1, l = 0$ only, for $n = 2, l = 0$ or 1, and for $n = 3, l =$ 0, 1, or 2
3. $-4\hbar, -3\hbar, -2\hbar - \hbar, 0, \hbar, 2\hbar, 3\hbar$, or $4\hbar$
5. $\frac{5}{2}\hbar, \frac{3e\hbar}{2m_e}$
7. (a) 18 states, (b) argon
9. $\hbar$; $m_s = -1, 0, +1$
11. 30°
13. $\pm m_s g\mu_B$, where g is approximately equal to 2 for the spin moment and equal to 1 for the orbital moment.
15. $\mu_p = 2.79\mu_N$, $\mu_C = 0.703\ \mu_N$
17. $L_x^2 + L_y^2 = l(l+1)\hbar^2 - m^2\hbar^2$ and

$$L_x = L_y = \hbar\sqrt{\frac{l(l+1) - m^2}{2}}$$

19. In its ground state, two electrons occupy the $n = 1$ level with $l = 0$, $m = 0$, and $m_s = \pm 1/2$, two electrons occupy the $n =$ 2 level with $l = 0$, $m = 0$, and $m_s = \pm 1/2$, and the remaining electron has $n = 2, l = 1, m = 0$, and $m_s = -1/2$.
21.

n	l	m	m_s	Number of States
1	0	0	±1/2	2
2	0	0	±1/2	2
	1	−1	±1/2	2
	1	0	±1/2	2
	1	+1	±1/2	2

23. K shell: $n = 1, l = 0, m = 0, m_s = -3/2, +3/2, -1/2$, and $+1/2$ (4 states)
L shell: $n = 2, l = 0, m = 0, m_s = -3/2, +3/2, -1/2$, and $+1/2$ (4 states) and $l = 1, m = 0$ and ±1, $m_s = -3/2, +3/2, -1/2$, and $+1/2$ (12 states)
In comparing the hypothetical periodic table with the known periodic table, we see that the two additional spin states cause significant changes. For example, the completed shells would no longer be the present noble gases, but would be Be, Mg, Ca, Sr, … Instead of two elements with $n = 1$ ground states, there would be four: H, He, Li, Be. Instead of Li through Ne (8 elements) having $n = 2$, there are now 16 elements, C through Ca.
25. 9.48×10^3 eV, 1.31×10^{-10} m
27. 30 (zinc), 60 (neodymium)
29. 0.128 nm
31. 74 (tungsten)
33. 9.27×10^{13} Hz
35. (a) 5.43×10^{-46} kg·m^2, (b) 1.3×10^{-4} eV, 3.9×10^{-4} eV, 7.8×10^{-4} eV
37. 1.13×10^{14} Hz
39. (a) 37.9 eV, (b) 0.0150 eV
41. 9.26×10^{-47} kg·m^2, 0.089 nm
43. 1.1×10^{-6} m
45. Because of the steep slope of the curve, a small change in voltage across the emitter and base produces a large increase in voltage crossing the emitter-base junction. This increased current appears in the current I_C leaving the collector.

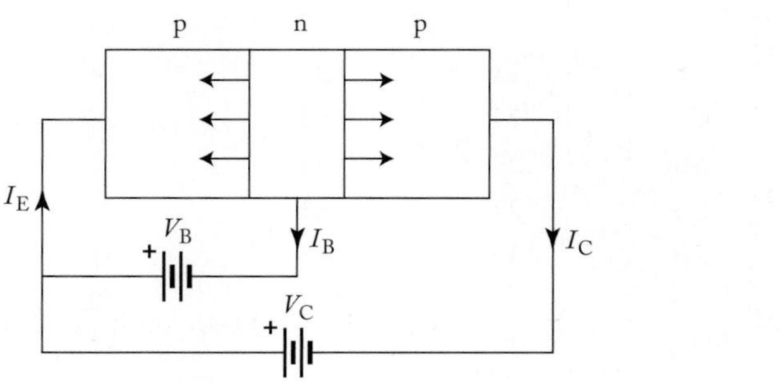

47.

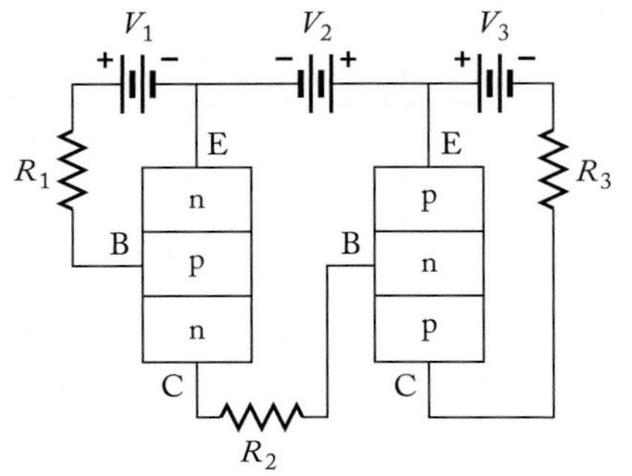

49. 0.060 W, 12%
51. (a) for $n = 1$, two electrons and for $n = 2$, eight electrons, (b) two, six, and zero
53. $m = -1, 0$, or $+1$, $-45°$, $90°$ and $45°$

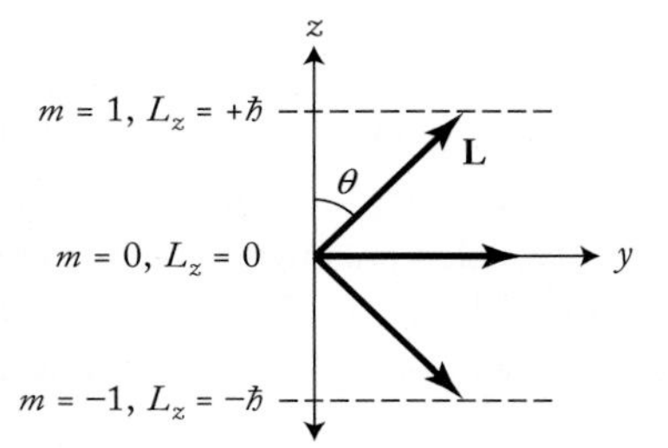

55. two electrons occupy the $n = 1$ level with $l = 0$, $m = 0$, and $m_s = \pm 1/2$, two electrons occupy the $n = 2$ level with $l = 0$, $m = 0$, and $m_s = \pm 1/2$, and six electrons have $n = 2$, $l = 1$, $m = 0$ and ± 1, and $m_s = \pm 1/2$. The remaining two electrons have $n = 3$, $l = 0$, $m = 0$, and $m_s = \pm 1/2$.
57. 5 (boron), 111 (transuranic elements)
59. 28 (nickel)
61. 4.14×10^{-22} J, 4.80×10^{-4} m, 8.27×10^{-22} J, 2.40×10^{-4} m, 1.24×10^{-21} J, 1.60×10^{-4} m
63. (a) 1.32 eV/bond, (b) 5.12 eV

Chapter 40

1. ^{16}O: $Z = 8, A = 16, N = 8$; ^{56}Fe: $Z = 26, A = 56, N = 30$; ^{238}U: $Z = 92, A = 236, N = 146$
3.

ISOTOPE	Z	A	N = A - Z
^{24}Na	11	24	13
^{27}Al	13	27	14
^{52}Cr	24	52	28
^{52}Mn	25	52	27
^{63}Cu	29	63	34
^{63}Zn	30	63	33
^{124}Xe	54	124	70
^{138}La	57	138	81

5.

Z = 8	^{13}O	^{14}O	^{15}O	^{16}O	^{17}O	^{18}O	^{19}O	^{20}O	^{21}O	^{22}O	^{23}O
N = A - 8	5	6	7	8	9	10	11	12	13	14	15

7. First reaction is not possible (Z is OK, A is wrong); Second reaction is not possible (Z is wrong, A is OK); Third reaction is not possible (Z is wrong, A is OK)
9. 2.5×10^{-15} m; 3.2×10^{-15} m
11. $^{31}_{15}P$ is correct; $^{44}_{22}Sc$ is not correct; $^{118}_{50}Sn$ is correct; $^{207}_{83}Pb$ is not correct; $^{89}_{39}Y$; $^{197}_{79}Au$
13. ^{13}C; ^{16}O
15.

A	1	2	3	4	5	6	7	8	9	10	11	12	13	14	15	16	17
No. of isobars	1	1	2	1	2	3	2	4	4	3	5	4	4	5	4	4	6

^1H, ^2H, ^4He have no isobars
17. 4.3×10^{-15}
19. 1.46×10^{-8} u
21. 28.3 MeV
23. 128 MeV for ^{16}O, 112 MeV for ^{16}F
25. 6.3 MeV/neutron vs 8.38 MeV/nucleon in ^{140}Ce
27. $Z_{TOT} = 9$ on both sides, $N_{TOT} = 9$ on both sides
29. 1.64 MeV
31. (a)180 MeV; (b) 1.51×10^7 m/s for Kr; 9.69×10^6 m/s for Bar; 110 MeV for Kr; 70.1 MeV for Bar
33. ^{212}Pb; ^{235}U
35. ^{22}Ne; ^{64}Ni
37. 0.820 MeV
39. 4.0 km/s
41. 0.782 MeV
43. ^{14}N; 0.156 MeV
45. 2.8; 140 yr
47. 7.7×10^{15} Bq
49. 2.2×10^8 Bq
51. 271 days; 1.44×10^{-6} g
53. 11 J; 0.15 Gy; 0.26 Sv (26 rem)
55. 20 Gy; 2.3 Gy
57. 1.0 kg
59. 6.94×10^3 kg
61. 123 MeV
63. 4.78 MeV
65. (a) 3.24×10^{-15} m; (b) 7.10×10^{-14} J; (c) 1.7×10^8 K
67. (a) 6.6×10^{11} kg/s; (b) 7.2×10^{10} yr
69. 3.27 MeV; 4.03 MeV; 17.6 MeV; 18.4 MeV; 7.2 MeV/nucleon
71. 6.4 MeV
73. 0.157 MeV
75. 3.00 MeV
77. 9.91 MeV
79. 1.5 Bq
81. 8.1×10^{-7} g

83. 1.0087 u

85. (a) 16,000 reactors; (b) 42 yr.

Chapter 41

1. (b) 1.36 T

3. (a) 1.88 GeV, (b) 43.4 GeV, (c) 115 GeV

5. 12 leptons altogether

7. Time dilation factor = 9.0; $t = 1.98 \times 10^{-5}$ s

9. 0.294 kg

11. 118 MeV

13. 67.5 MeV, 1.84×10^{-14} m for each of the γ-rays

15. Baryon number is conserved; strangeness is not conserved.

17. Strangeness is conserved in the first reaction; strangeness is not conserved in the second and third reactions.

19. In the rest frame of the electron, the energy before the emission equals the rest mass energy of the electron. After the emission, the total energy equals the rest mass energy of the electron plus the energy of the photon. This violates the conservation of energy law.

21.

Reaction	left side baryon #	right side baryon #	left side strangeness	right side strangeness
$K^- + p \rightarrow \Sigma^+ + \pi^-$	0+1=1	1+0=1	−1+0=−1	−1+0=−1
$K^- + p \rightarrow \Sigma^0 + \pi^0$	0+1=1	1+0=1	−1+0=−1	−1+0=−1
$K^- + p \rightarrow \Lambda + \pi^0$	0+1=1	1+0=1	−1+0=−1	−1+0=−1
$K^- + p \rightarrow \Lambda + \pi^+ + \pi^-$	0+1=1	1+0+0=1	−1+0=−1	−1+0+0=−1

23.

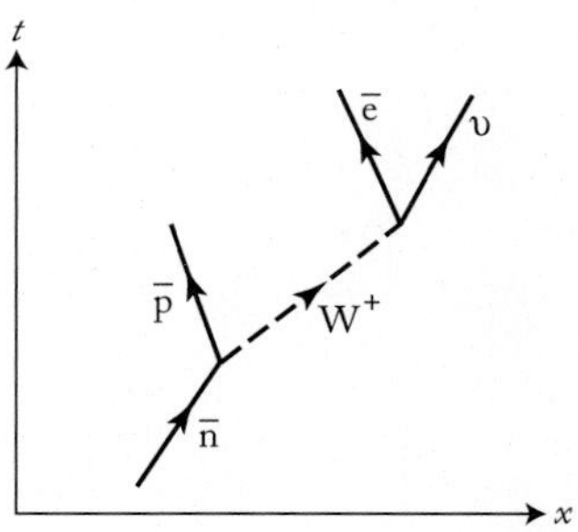

25. $\Delta t_\pi = 4.70 \times 10^{-24}$ s, $d_\pi = 1.41 \times 10^{-15}$ m; $\Delta t_Z = 7.23 \times 10^{-27}$ s, $d_Z = 2.17 \times 10^{-18}$ m

27. 129 GeV, greater than the energy required to achieve symmetry between W and photons.

29. $\bar{u}\,\bar{u}\,\bar{d}$

31. 30 quarks are created.

33. In Table 41.2, no particle is its own antiparticle. In Table 41.3, π^0 is its own antiparticle, J/ψ is its own antiparticle, Υ is its own antiparticle. The latter three consist of a quark and the corresponding antiquark.

35. 9.87×10^{-9} nm

37.

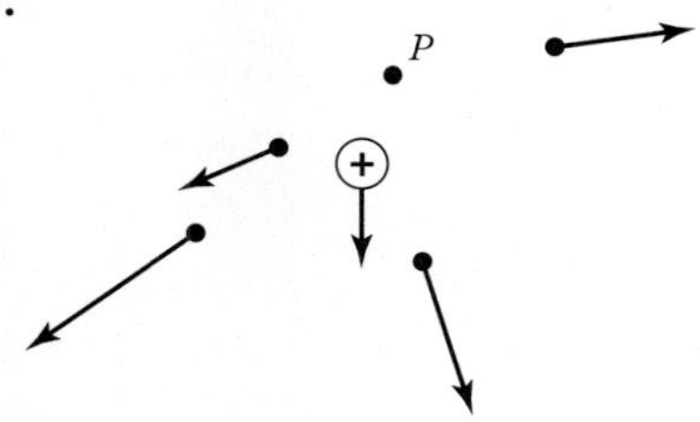

39. The Earth and the oldest globular clusters were older than the estimated total age of the universe according to Hubble's Law.

41. 0.789 MeV

43. Baryon number is not conserved in the first reaction, electric charge is not conserved in the second reaction, energy is not conserved in the third reaction.

45. Electric charge is −1, the particle is Σ^-.

47.

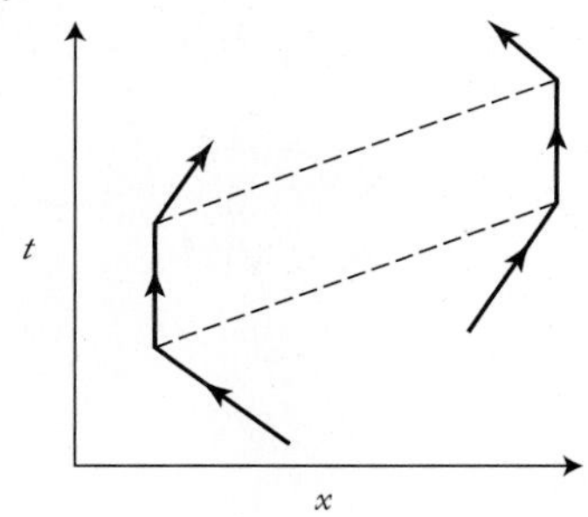

Photo credits

Owners Manual: **p. xxv**: Michael Kim/Corbis; **p. xxvi**: James L. Amos/Corbis; **p. xxix**: NASA/Johnson Space Center; **p. xxix**: Mark Bolton/Corbis; **p. xxix**: Robert Laberge/Allsport/Getty Images; **p. xxix**: Robert Laberge/Allsport/Getty Images; **p. xxx**: Duomo/Corbis; **p. xxxi**: George Hall/Corbis.

Prelude:
$\mathbf{10^0}$: John Markert/Junenoire Photography; $\mathbf{10^1}$: John Markert/Junenoire Photography; $\mathbf{10^2}$: John Markert/Junenoire Photography; $\mathbf{10^3}$: Jim Mairs; $\mathbf{10^4}$: Aerial Imagery Courtesy of GlobeXplorer.com; $\mathbf{10^5}$: Aerial Imagery Courtesy of GlobeXplorer.com; $\mathbf{10^6}$: Aerial Imagery Courtesy of GlobeXplorer.com; $\mathbf{10^7}$: Jacques Descloitres, MODIS Rapid Response Team, NASA/GSFC; $\mathbf{10^8}$: NASA/Corbis; p.5: $\mathbf{10^9}$: NASA/ JSC; $\mathbf{10^{21}}$: Dr. Fred Espenak/Photo Researchers, Inc.; $\mathbf{10^{22}}$: David Malin; $\mathbf{10^{23}}$: The Hubble Heritage Team AURA/STScI/NASA; $\mathbf{10^{24}}$: NASA/ESA/R. Thompson (University of Arizona); $\mathbf{10^{26}}$: Max Tegmark/SDSS Collaboration; $\mathbf{10^0}$: John Markert/Junenoire Photography; $\mathbf{10^{-1}}$: John Markert/Junenoire Photography; $\mathbf{10^{-2}}$: Professor Pietro M. Motta/Photo Researchers, Inc.; $\mathbf{10^{-3}}$: OMIKRON/Photo Researchers, Inc.; $\mathbf{10^{-4}}$: SPL/Photo Researchers, Inc.; $\mathbf{10^{-5}}$: G. Murti/Photo Researchers, Inc.; $\mathbf{10^{-6}}$: DOE/Science Source.; $\mathbf{10^{-7}}$: Kenneth Eward/Photo Researchers, Inc..; $\mathbf{10^{-8}}$: Eurelios/Phototake.

Part Openers: **Part Opener 1**: NASA/GRIN; **Part Opener 2**: L.Weinstein, NASA/Photo Researchers, Inc.; **Part Opener 3**: Alfred Pasieka/Photo Researchers, Inc.

Chapter Opener 1: Roger Ressmeyer/Corbis; **fig. 1.5**: Photo by H. Mark Helfer/NIST; **fig. 1.7**: Photo by Barry Gardner; **fig. 1.8**: ©2004 Bruce Erik Steffine; **table 1.1a**: NASA/JSC; **table 1.1b**: Charles O Rear/Corbis; **table 1.1c**: U.S. Mint Handout/Reuters/Corbis; **table 1.1d**: Courtesy of Robert G. Milne, Plant Virus Institute, National Research Council, Turin, Italy; **fig. 1.9**: NIST; **fig. 1.10**: Robert Rathe/NIST; **fig. 1.11**: BIPM (International Bureau of Weights and Measures/Bureau International des Poids et Mesures), www.bipm.org; **table 1.7a**: NASA/JSC; **table 1.7b**: Gene Blevins/LA Daily News/Corbis; **table 1.7c**: Royalty-Free/Corbis; **table 1.7d**: Clouds Hill Imaging Ltd./Corbis; **fig. 1.13**: Reuters/Corbis; **fig. 1.15**: National Maritime Museum, UK; **fig. 1.19**: John Brecher/Corbis.

Chapter Opener 2: Reuters/Corbis; **table 2.1a**: Corbis; **table 2.1b**: Randy Wells/Corbis; **table 2.1c**: David Muench/Corbis; **fig. 2.19**: James Sugar/Black Star; **p.51**: The Granger Collection,

Every effort has been made to contact the copyright holders of the material used in **Physics for Engineers and Scientists 3rd Edition.** Please contact us with any updated information.

New York; **fig. 2.22**: Tom Sanders/Photri/Microstock; **fig. 2.23**: Wally McNamee/Corbis; **fig. 2.26**: Taxi/Getty Images; **fig. 2.31**: NASA/JPL.

Chapter Opener 3: NOAA/Hurricane Research Division; **fig. 3.3**: Ron Watts/Corbis; **p.71**: (Pip) Galen Rowell/Corbis.

Chapter Opener 4: Dr. J. Alean (Stromboli); **p.99**: (Pip fig. 2) Photo Courtesy of Mark Wernet/NASA/Glenn Research Center, Cleveland Ohio; **fig. 4.7**: Krafft-Explorer/Photo Researchers, Inc.; **fig. 4.9**: Richard Megna/Fundamental Photographs; **fig. 4.19**: Aero Graphics/Corbis; **fig. 4.15**: Richard Megna/Fundamental Photographs; **fig. 4.16**: Wally McNamee/Corbis; **fig. 4.22**: NASA/JSC; **fig. 4.27**: Roger Ressmeyer/Corbis; **fig. 4.34**: Fermilab; **fig. 4.35**: Robert Harrington, www.BobQat.com.

Chapter Opener 5: Bob Krist/Corbis; **fig. 5.1**: Courtesy of John Markert; **p.131**: The Granger Collection, New York; **fig. 5.5**: Courtesy DYNCorp/NASA/JSC; **table 5.1a**: NASA/JSC; **table 5.1b**: Mark Bolton/Corbis; **fig. 5.6**: Bettmann/Corbis; **fig. 5.8**: TEK Image/Photo Researchers, Inc.; **fig. 5.14**: Tim Kiusalaas/Corbis; **fig. 5.15**: Jim Sugar/Corbis; **fig. 5.16**: Roger Ressmeyer/Corbis; **fig. 5.44**: Science Museum, London.

Chapter Opener 6: Michael Kim/Corbis; **p.174**: James L. Amos/Corbis; **fig. 6.17**: Kai Pfaffenbach/Reuters/Corbis; **p.189**: (Pip fig. 2) Photodisc Green/Getty Images; **p.189**: (Pip fig. 3) Courtesy Beckman Coulter, Inc.; **fig. 6.25**: Duomo/Corbis; **fig. 6.26**: George Hall/Corbis; **fig. 6.27**: October 2001 Physics Today (Volume 54, Number 10, p.39), Courtesy of John Yasaitis, Analog Devices, Inc.; **fig. 6.28**: Brian Erler/Corbis; **fig. 6.39**: Duomo/Corbis.

Chapter Opener 7: Lester Lefkowitz/Corbis; **p.207**: Hulton Deustch/Corbis; **fig. 7.17**: Jim Cummins/Corbis; **fig. 7.19**: Courtesy Klockit, Inc.; **p.221**: Bettmann/Corbis; **fig. 7.23**: David Cumming/Eye Ubiquitous/Corbis; **fig. 7.32**: Jim Cummins/Corbis; **fig. 7.34**: Lester Lefkowitz/Corbis; **fig. 7.37**: Image Bank/Getty Images.

Chapter Opener 8: Courtesy Blenheim-Gilboa Pumped Storage Power Project/New York Power Authority; **p.236**: The Granger Collection, New York; **p.242**: (Pip) Courtesy New York Power Authority; **fig. 8.11**: Paul A. Souder/Corbis; **p.248**: Corbis; **table 8.1a**: Courtesy International Dark Sky Association, Defense Meteorological Satellite Program (DMSP).**table 8.1b**: Royalty-Free/Corbis; **table 8.1c**: Mika/Zefa/Corbis; **p.254**: North Wind Picture Archives; **table 8.2a**: Earth Observatory/NASA; **table 8.2b**: Stone/Getty Images; **table 8.2c**: Taxi/Getty Images; **fig. 8.17**: Courtesy New York Power Authority; **fig. 8.18**: Corbis; **fig. 8.21**: Paul A. Souders/Corbis; **fig. 8.27**: National Archives.

Chapter Opener 9: Bettmann/Corbis; **fig. 9.5**: Science Museum and Society Picture Library; **p.277**: The Granger Collection, New York; **p.279**: The Granger Collection, New York; **fig. 9.8**: NASA/Corbis; **p.281**: (Pip) NASA; **table 9.1a**: JPL/NASA; **p.285**: The Granger Collection, New York; **fig. 9.15**: Bettmann/Corbis; **fig. 9.28**: NASA; **fig. 9.30**: ©2002 Calvin J. Hamilton; **fig. 9.36**: National Astronomy Observatiries/AP Images; **fig. 9.37**: Dennis di Cicco/Corbis; **fig. 9.40**: Photofest; **fig. 9.41**: JPL/NASA; **fig. 9.42**: Photographer: Mark Avino ©Copyright 1995 by Smithsonian Institution.

Chapter Opener 10: Reuters/Corbis; **fig. 10.6**: Richard Megna 1990 Fundamental Photographs; **fig. 10.13**: ©2005 Estate of Alexander Calder/Artist Rights Society (ARS), New York; **fig. 10.15**: Roger Ressmeyer/Corbis; **p.320**: (Pip fig. 1) Macduff Everton/Corbis; **p.320**: (Pip fig. 2) Right George D. Lepp/Corbis; **fig. 10.18**: Jason Reed/Reuters/Corbis; **fig. 10.22**: NASA/Roger Ressmeyer/Corbis; **fig. 10.25**: Philip James Corwin/Corbis.

Chapter Opener 11: Tom Wright/Corbis; **fig. 11.1**: Daimlerchrysler; **p.343**: (Pip: top and bottom) Robert Laberge/Allsport/Getty Images; **fig. 11.6**: Smithsonian Institute; **fig. 11.11**: Charles D. Winters/Photo Researchers, Inc; **fig. 11.17**: Reuters/Corbis; **fig. 11.20**: Oxford University Press, UK; **fig. 11.21**: Charles & Josette Lenars/Corbis.

Chapter Opener 12: Courtesy Sandia National Laboratories; **fig. 12.1**: Courtesy Palm Press, Inc.; **fig. 12.3c**: Royalty-Free/Corbis; **fig. 12.3b**: Picture Arts/Corbis; **fig. 12.3d**: Hans C. Ohanian; **fig. 12.3d**: Richard T. Nowitz/Corbis; **fig. 12.3a**: Hans C. Ohanian; **fig. 12.20**: Corbis; **fig. 12.32**: Neil Rabinowitz/Corbis.

Chapter Opener 13: Don Harlan/Gravity Probe B; **fig. 13.6**: Courtesy of Gravity Probe B Photo Archive, Stanford University; **fig. 13.11**: Wally McNamee/Corbis; **fig. 13.14**: Nancy Ney/Corbis; **p.414**: (Pip fig. 2) Ron Keller, N.M. Museum of Space; **p.414**: (Pip fig. 1) Science & Society Picture Library; **fig. 13.25**: Lawrence Lucier/Getty Images; **fig. 13.34**: Reuters/Corbis; **fig. 13.37**: NASA/JSC.

Chapter Opener 14: Kroll Cranes A/S - Denmark; **fig. 14.5**: Corbis; **fig. 14.29**: Tom Pantages; **fig. 14.50**: Tim De Waele/Isosports/Corbis.

Chapter Opener 15: NASA/JSC; **fig. 15.6**: Courtesy of David Hammond; **fig. 15.8**: Courtesy of John Markert; **fig. 15.15**: Loren Winters/Visuals Unlimited; **fig. 15.19**: National Maritime Museum; **fig. 15.23**: Bibliotheque Nationale de France; **p.493**: (Pip fig. 3) Courtesy John Markert; **fig. 15.26**: P.B. Umbanhowar, F. Melo, and H. L. Swinney, "Localized excitations in a vertically vibrated granular layer," Nature 382, 793-796 (1996); **fig. 15.27**: Jim Craigmyle/Corbis; **fig. 15.36**: Courtesy of John Markert; **fig. 15.37**: NASA/JSC.

Chapter Opener 16: Vladimir Smolyakov/Stolichnaya Vechernyaya Gazeta /AP Images; **fig. 16.1**: Richard Megna Fundamental Photo; **fig. 16.17**: Richard Megna Fundamental Photo; **fig. 16.18**: Keystone/Getty Images; **fig. 16.19**: Royalty-Free/Getty Images; **fig. 16.20**: David Nock of British Car Specialists; **fig. 16.21**: Royalty-Free/Corbis; **fig. 16.27**: Francois Gohier/Photo Researchers, Inc.

Chapter Opener 17: Stone/Getty Images; **fig. 17.1**: Aaron Horowitz/Corbis; **fig. 17.2** *Engineering Applications of Lasers and Holography* by Winston E.Kock, Plenum Publishing Co., New York 1975; **fig. 17.5**: William B. Joyce; **fig. 17.7**: Courtesy of C.F. Quate and L. Lam Hansen Laboratory; **p.546**: (Pip fig. 1) www.777life.com/free photo stock; **p.546**: (Pip fig. 3) John Ross Buschert, Goshen College, IN; **p.549**: AIP Emilio Segre Visual Archives; **fig. 17.19**: Gary S. Settles/Photo Researchers, Inc.; **p.552**: The Granger Collection, New York; **fig. 17.20**: Museum of Flight/Corbis; **fig. 17.21**: John Shelton Photography; **fig. 17.22**: PSSC Physics 2nd edition 1965 DC Health & Company and Educational Development Center, Inc., Newtown, MA; **fig.**

17.23: PSSC Physics 2nd edition 1965 DC Health & Company and Educational Development Center, Inc., Newtown, MA; **fig. 17.27**: Tim Bird/Corbis.

Chapter Opener 18: Woods Hole Oceanographic Institution **fig. 18.3**: Hans C. Ohanian; **fig. 18.8**: AD Moore from Introduction to Electromagnetic; **fig. 18.9**: D.C. Hazen and R.F. Lehnert, Subsonic Aerodynamics Laboratory, Princeton; **fig. 18.10**: D.C. Hazen and R.F. Lehnert, Subsonic Aerodynamics Laboratory, Princeton; **fig. 18.11**: Royalty-Free/Corbis; **fig. 18.13**: Image Design by Nature; **fig. 18.14**: Paul Edmondson/Corbis; **fig. 18.17**: Tom Kleindinst, Woods Hole Oceanographic Institution; **p.574**: Stefano Bianchetti/Corbis; **fig. 18.20**: AC Hydraulic; **p.579**: (Pip fig. 1) Anonia Reeve/Photo Researchers, Inc.; **fig. 18.26**: DW Stock Picture Library/S. Drossinos; **fig. 18.28**: Stan White Photography; **p.581**: Archivo Iconografico, S.A./Corbis; **p.585**: Bettmann/Corbis; **fig. 18.34**: Columbia University Physics Department; **fig. 18.37**: Principals of Physics, 1994; **fig. 18.38**: Peter Finger/Corbis; **fig. 18.40**: Roger Ressmeyer/Corbis; **fig. 18.41**: Bettmann/Corbis; **fig. 18.48**: Royalty-Free/Corbis; **fig. 18.49**: Roger Ressmeyer/Corbis; **fig. 18.50**: AFP/Hyundai Heavy Industries.

Chapter Opener 19: Vince Streano/Corbis; **p.604**: AIP Emilio Segre Visual Archives; **p.606**: Chemical Heritage Foundation Collection; **p.608**: Bettmann/Corbis; **fig. 19.7**: Royalty-Free/Corbis; **fig. 19.8**: Tom Pantages; **fig. 19.9**: Crown copyright 1999, Reproduced permission; **fig. 19.10**: Courtesy of Cole-Parmer Instrument Company; **fig. 19.12**: Liquid Crystal Resources, Glenview, IL; **table 19.1a**: Julian Baum/Photo Researchers, Inc.; **table 19.1b**: Visuals Unlimited; **table 19.1c**: Dr. Arthur Tucher/Photo Researchers, Inc.; **fig. 19.17**: Courtesy of Worthington Cylinders; **fig. 19.18**: Dr. Kimberly Strong, University of Toronto; **fig. 19.20**: National Institute of Standards and Technology.

Chapter Opener 20: Peter Arnold, Inc./Alamy; **p.629**: (bio) Burnstein Collection/Corbis; **fig. 20.1**: Bettmann/Corbis; **fig. 20.6**: Norbert Wu; **fig. 20.9**: Visuals Unimited; **fig. 20.10**: AP Images; **table 20.4a**: David Taylor/Corbis; **table 20.4b**: David Pollack/Corbis; **table 20.4c**: D.Winters/Photo Researchers, Inc.; **fig. 20.19**: Lowell Georgia/Corbis; **fig. 20.20**: Alfred Pasieka/Photo Researchers, Inc.; **fig. 20.22**; Private Collection; **fig. 20.24**: Stanford University.

Chapter Opener 21: The Image Bank/Getty Images; **fig. 21.5**: Science & Society Picture Library; **fig. 21.6**: Courtesy of BMW World; **fig. 21.8**: Jim Cummins/Corbis. **p.667**: Bridgeman Art Library; **fig. 21.17**: Tom Pantages; **fig. 21.19**: Tom Pantages; **p.678**: Bettmann/Corbis; **p.681**: Bettmann/Corbis; **fig. 21.23**: ©2005 The M.C. Escher Company - Holland. All rights reserved; **fig. 21.27**: Picture Arts/Corbis; **fig. 21.32**: Inga Spence/Visuals Unlimited.

Part Opener 4: Andrew Syred/Photo Researchers, Inc.; **Part Opener 5**: NASA/Marshall Space Flight Center.

Chapter Opener 22: Canon Inc.; **fig. 22.3**: Departamento de Física Faculdade de Ciências e Tecnologia Universidade de Coimbra; **p. 698**: Art Resource, NY; **fig. 22.10a**: Corbis; **fig. 22.10c**: Department of Physics, University of Oslo, Norway; **fig. 2210d**: IBMRL/Visuals Unlimited; **fig. 22.9a**: Jun Yang, Ting-Jie Wang, Hong He, Fei Wei, Yong Jin. *Particle size distribution and morphology of in situ suspension polymerized toner*. Industrial & Engineering Chemistry Research. 2003, 42 (22): 5568-5575; **fig. 22.10b**: Wolfson Nanometrology Laboratory at the University of Strathclyde by Gregor Welsh; **fig. 22.13**: Chuck Doswell/Visuals Unlimited; **fig. 22.14**:

H. David Seawell/Corbis; **p. 710**: North Carolina Museum of Art/Corbis; **fig. 22.15**: Lightingmaster, Streamer Delaying Air Terminal; **fig. 22.17**: Larry Stepanowicz/Visuals Unlimited.

Chapter Opener 23: Kent Wood/Photo Researchers, Inc.; **p. 735**: (Pip fig. 2): PPC Industries; **fig. 23.22**: Courtesy of Ionoptika Ltd., England.

Chapter Opener 24: Trustees of Princeton University; **p. 763**: American Institute of Physics; **fig. 24.22**: Trustees of Princeton University; **fig. 24.46**: Tom Pantages.

Chapter Opener 25: Hank Morgan/Photo Researchers, Inc.; **p. 793**: Bettmann/Corbis; **p. 805**: (Pip fig. 3) IT Stock Int'l/indexphoto.com; **p. 805**: (Pip fig. 4) Euclid Garmet/Georgia Power; **fig. 25.32**: Stanford University.

Chapter Opener 26: (left and right) National Ignition Facility; **fig. 26.18**: Edward Kinsman/Photo Researchers, Inc.; **fig. 26.23**: Edward Kinsman/Photo Researchers, Inc.

Chapter Opener 27: William Taufic/Corbis; **fig. 27.2**: American Journal of Physics 30, 19, 1963; **fig. 27.3**: American Journal of Physics 30, 19, 1963; **Table 27.1a**: National Center for Atmospheric Research; **Table 27.1b**: Patrick Bennett/Corbis; **Table 27.1c**: Brownie Harris/Corbis; **Table 27d**: Courtesy of Bosch; **Table 27e**: Corbis; **p. 866**: Bettmann/Corbis; **fig. 27.9**: © Crown copyright 1999. Reproduced by permission of the Controller of HMSO and the Queen's Printer for Scotland; **fig. 27.21**: John Wilkes Studio/Corbis; **fig. 27.26**: *Principals of Physics* by Hans Ohanian, 1994; **fig. 27.28**: Courtesy of Superpower, Inc.

Chapter Opener 28: Martyn F. Chillmaid/Photo Researchers, Inc.; **fig 28.6b**: Loren Winters/Visuals Unlimited; **fig. 28.7b**: NASA/Johnson Space Center; **fig. 28.7c**: AP Photos; **fig. 28.22**: Loren Winters/Visuals Unlimited; **fig. 28.34**: Mark C. Burnett/Photo Researchers, Inc.; **fig. 28.56**: Kim Kulish/Corbis.

Chapter Opener 29: © 1990 Richard Megna, Fundamental Photographs, NYC; **fig. 29.5**: Tom Pantages; **p. 930**: American Institute of Physics, Emilio Segrè Visual Archives.; **p. 934**: Science Museum/Science and Society Picture Library; **Table 29.1a**: Maximilian Stock Ltd./Photo Researchers, Inc.; **Table 29.1b**: John Chumack/Photo Researchers, Inc.; **Table 29.1c**: Adrianna Williams/zefa/Corbis; **Table 29.1d**: NASA/HST/ASU/J. Hester et al; **fig. 29.15**: Loren Winters/Visuals Unlimited; **fig. 29.17**: Loren Winters/Visuals Unlimited; **p. 940**: Science Museum/Science and Society Picture Library; **fig. 29.24**: © 1990 Richard Megna, Fundamental Photographs, NYC; **p. 943**: © 1990 Richard Megna, Fundamental Photographs, NYC; **p. 950**: Science Museum/Science and Society Picture Library.

Chapter Opener 30: Lawrence Berkley Laboratory; **fig. 30.1**: Science Museum/Science and Society Picture Library; **fig. 30.4**: © 2002 Richard Megna, Fundamental Photographs, NYC; **fig. 30.21**: Neil Borden/Photo Researchers, Inc.; **fig. 30.22**: © 1986 Richard Megna, Fundamental Photographs, NYC; **p. 978** (Pip fig. 1): Up The Resolution (uptheres)/Alamy; **p. 979**: (Pip fig. 2) Tek Image/Photo Researchers, Inc.; **p. 979**: (Pip fig. 3) Courtesy of John Markert; **p. 981**: AIP Emilio Segrè Visual Archives; **fig. 30.24a**: Appl. Phys. Lett., Vol. 61, No. 16, 19 October 1992; **fig. 30.24b**: Appl. Phys. Lett., Vol. 61, No. 16, 19 October 1992; **fig. 30.27**: LBNL/Photo Researchers, Inc.

Chapter Opener 31: Porterfield-Chickering/Photo Researchers, Inc.; **p. 998**: Hulton-Deutsch Collection/Corbis; **Table 31.1**: Phototake Inc./Alamy; **Table 31.2**: Image courtesy of Indigo (R) Instruments, www. indigo.com; **Table 31.4**: AllOver Photography/Alamy; **p. 1011**: Bettmann/Corbis; **fig. 31.31**: Maximilian Stock Ltd./Photo Researchers, Inc.; **fig. 31.45**: Marty Snyderman/Visuals Unlimited.

Chapter Opener 32: (left) Lester Lefkowitz/Getty Images; (right) Chris Knapton/Photo Researchers, Inc.; **fig. 32.3**: David Michael Zimmerman/Corbis; **fig. 32.18**: Loren Winters/Visuals Unlimited; **fig. 32.35**: Steve Callahan/Visuals Unlimited; **fig. 32.42**: Tom Pantages.

Chapter Opener 33: The Imagebank/Corbis; **p. 1080**: American Institute of Physics **p. 1086**: Library of the Academy of Science, Paris. Courtesy of AIP, Niels Bohr Library; **fig. 33 18**: © 1992 Diane Hirsch, Fundamental Photographs, NYC; **fig. 33 20**: Philip Bailey/Corbis; **fig. 33.21a**: Corbis; **fig. 33.21b**: Food pix/Getty Images; **fig. 33.29**: Jim Sugar/Corbis; **fig. 33.31**: Courtesy of Hans Ohanian; **fig. 33.32**: Maximilian Stock Ltd./Photo Researchers, Inc.; **fig. 33.33**: Steve Percival/Photo Researchers, Inc.; **fig. 33.34**: Courtesy of Hans Ohanian.

Chapter Opener 34: NASA/Science and Society Picture Library; **fig. 34.1**: Marli Miller/Visuals Unlimited; **fig. 34.2**: Digital Vision/Getty Images; **fig. 34.6**: © 1997 Richard Megna, Fundamental Photographs, NYC; **fig. 34.11**: Courtesy of Hans Ohanian; **fig. 34.20**: © 1990 Richard Megna, Fundamental Photographs, NYC; **fig. 34.22** : Hugh Turvey/Photo Researchers, Inc.; **p. 1124**: (Pip fig. 1) Hank Morgan/Photo Researchers, Inc.; **p. 1124**: (Pip fig. 2): PHT/Photo Researchers, Inc.; **fig. 34.24**: Junenoire Photography; **fig. 34.28**: Will/Demi McIntyre/Photo Researchers, Inc.; **fig. 34.31**: © 1987 Ken Kay, Fundamental Photographs, NYC; **fig. 34.41b**: Courtesy of John Markert; **fig. 34.43**: David Parker/Photo Researchers, Inc.; **fig. 34.44**: David Parker/Photo Researchers, Inc.; **fig. 34.61**: Reuters/Corbis; **fig. 34.66a**: Roger Ressmeyer/Corbis; **fig. 34.66b**: McDonald Observatory; **fig. 34.67**: The Lady and the Unicorn: 'Sight' (tapestry), French School, (15th century)/Musée National du Moyen Age et des Thermes de Cluny, Paris/The Bridgeman Art Library; **fig. 34.68**: Courtesy of Hans Ohanian; **fig. 34.69**: © 1990 Paul Silverman, Fundamental Photographs, NYC; **fig. 34.70**: Point Reyes National Seashore; **fig. 34.88**: DK Limited/Corbis; **fig. 34.90**: Roger Ressmeyer/Corbis; **fig. 34.95**: Bo Zaunders/Corbis.

Chapter Opener 35 (left): Steve Percival/Photo Researchers, Inc.; (right): Peter Steiner/Alamy; **fig. 35.1**: Peter Aprahamian/Photo Researchers, Inc.; **p. 1170**: The Granger Collection, NY; **fig. 35.7**: Sciencephotos/Alamy; **fig. 35.10**: Bureau of International des Poids et Mesures, Sevres, France; **p. 1175**: The Granger Collection, NY; **fig. 35.16**: Erich Schrempp/Photo Researchers, Inc.; **fig. 35.17**: Courtesy of Chris C. Jones; **fig. 35.21**: Courtesy of Chris C. Jones; **fig. 35.23**: Roger Ressmeyer/Corbis; **fig. 35.28**: *Principals of Physics* by Hans Ohanian, 1994; **fig. 35.29**: © 1987 Ken Kay, Fundamental Photographs, NYC; **p. 1191**: Stefano Bianchetti/Corbis; **fig. 35.33**: Courtesy of Chris C. Jones; **fig. 35.38**: Courtesy of Chris C. Jones; **fig. 35.39**: Courtesy of Chris C. Jones; **fig. 35.40**: Courtesy of Chris C. Jones; **fig. 35.41**: Denis Scott/Corbis; **fig. 35.42**: Stephanie Maze/Corbis; **p. 1198**: Mary Evans Picture Library/Alamy; **fig. 35.43**: Courtesy of Michael Lockwood, University of Illinois Urbana-Champaign.; **fig. 35.45**: Courtesy of Chris C. Jones; **fig. 35.46**: Northwestern University/Photo Researchers, Inc.; **fig. 35.47**: Eye of Science/Photo Researchers, Inc.; **fig. 35.57**: *Principals of Physics* by Hans Ohanian, 1994.

Part Opener 6: NASA/Marshall Space Flight Center.

Chapter Opener 36: ESA/CE/Eurocontrol/Photo Researchers, Inc.; **p. 1217**: Hulton Archive/Getty Images; **p. 1235**: Science Photo Library/Photo Researchers, Inc.

Chapter Opener 37: Larry Landolfi/Photo Researchers, Inc.; **fig. 37.1**: Simon Lewis/Photo Researchers, Inc.; **fig. 37.2**: Edward Kinsman/Photo Researchers, Inc.; **p. 1260**: AIP Emilio Segrè Visual Archives; **fig. 37.8**: Photo Researchers, Inc.; **p. 1268**: (Pip fig. 2): Courtesy of Burle Industries; **p. 1269**: AIP Emilio Segrè Visual Archives; **fig. 37.15**: Science Museum/Science and Society Picture Library; **fig. 37.16**: © Mediscan/Visuals Unlimited; **fig. 37.17**: © Science VU/Visuals Unlimited; **fig. 37.21a-c**: Courtesy of E.R. Huggins; **fig. 37.21d**: Courtesy of Chris C. Jones; **p. 1277**: AIP Emilio Segrè Visual Archives; **p. 1278**: AIP Niels Bohr Library.

Chapter Opener 38: (top) Courtesy of Eastman Kodak, Inc. (bottom) Courtesy of Hans C. Ohanian; **fig. 38.1**: Prof. C.K. Shih, Department of Physics, The University of Texas at Austin; **fig. 38.3**: Adam Jones/Visuals Unlimited; **p. 1289**: Courtesy of Eastman Kodak; **fig. 38.5**: Courtesy of Hans C. Ohanian; **fig. 38.6**: Courtesy of Hans C. Ohanian; **p. 1294**: Cavendish Laboratory, Cambridge; **p. 1295**: Princeton University/American Institute of Physics/Science Photo Library; **p. 1301**: AIP Emilio Segrè Visual Archives; **p. 1304**: AIP Emilio Segrè Visual Archives; **p. 1310**: (Pip fig. 1b): Richard J. Green/Photo Researchers, Inc.; **p. 1310**: (Pip fig. 2b): Andrew Syred/SPL/ Photo Researchers, Inc.; **p. 1311**: (Pip fig. 3b): © IBM Research; **p. 1311**: (Pip fig. 4): Delft University of Technology/Photo Researchers, Inc.

Chapter Opener 39: Professor C.K. Shih, Department of Physics, The University of Texas at Austin; **fig. 39.3a-d**: A.F. Burr and A. Fisher, New Mexico State University; **p. 1325**: AIP Emilio Segrè Visual Archives; **p. 1326**: Granger Collection, NY; **p. 1332**: Courtesy of The University of Oxford, Museum of the History of Science; **fig. 39.10**: Courtesy of John Markert; **fig. 39.24**: Ton Kinsbergen/Photo Researchers, Inc.; **fig. 39.25**: Astrid & Hanns-Frieder Michler/Photo Researchers, Inc.; **fig. 39.29**: NASA/Corbis; **fig. 39.30**: Reuters/Corbis; **fig. 39.31**: Corbis; **fig. 39.32**: Roger Ressmeyer/Corbis.

Chapter Opener 40: U.S. Department of Energy/Photo Researchers, Inc.; fig. 40.7: John Cockcroft, Cambridge, UK; **p. 1366**: Jean-Loup Charmet/Science Photo Library; **p. 1373**: Physics Today Collection/AIP/Science Photo Library; **p. 1376**: (Pip fig. 1): Reuters/Corbis; **p. 1377**: Ullstein bild/The Granger Collection, NY; **p. 1380**: AIP Emilio Segrè Visual Archives; **fig. 40.18**: Corbis; **fig. 40.21**: Corbis; **p. 1384**: Corbis; **fig. 40.22**: Roger Ressmeyer/Corbis; **fig. 40.23**: National Ignition Facility (NIF).

Chapter Opener 41: CERN, P. Lopez/Photo Researchers, Inc.; **fig. 41.1**: Fermilab/Photo Researchers, Inc.; **fig. 41.2**: Fermilab/Photo Researchers, Inc.; **fig. 41.3**: David Parker/Photo Researchers, Inc.; **fig. 41.4a**: CERN/SPL/ Photo Researcher, Inc.; **fig. 41.4b**: CERN/SPL/ Photo Researcher, Inc.; **fig. 41.6**: Stanford Linear Accelerator Center/SPL/Photo Researchers, Inc.; **fig. 41.7**: CERN/Photo Researchers, Inc.; **fig. 41.8**: LBNL/Photo Researchers, Inc.; **p. 1410**: Harvey of Pasadena/American Institute of Physics/Science Photo Library; **fig. 41.19a**: Jean-Charles Cuillandre/CFHT/Photo Researchers, Inc.; **fig. 41.19b**: NOAO/Photo Researchers, Inc.; **fig. 41.20a**: Dr. Jean Lorre/Photo Researchers, Inc.; **fig. 41.20b**: Celestial Image Co./Photo Researchers, Inc.; **fig. 41.20c**: C. Butler/Photo Researchers, Inc.; **fig. 41.21a**: NASA; **fig. 41.23**: Courtesy of Jeff Hester; **fig. 41.22**: NASA; **fig. 41.23**: NOAO/AURA/NSF/Photo Researchers, Inc.; **fig. 41.24**: Roger Ressmeyer/Corbis.

Index

Page numbers in *italics* refer to biographies. Page numbers in **boldface** refer to figures. Page numbers followed by "*t*" refer to tables.